The Geology of England and Wales

Geological Society books refereeing procedures

The Society makes every effort to ensure that the scientific and production quality of its books matches that of its journals. Since 1997, all book proposals have been refereed by specialist reviewers as well as by the Society's Books Editorial Committee. If the referees identify weaknesses in the proposal, these must be addressed before the proposal is accepted.

Once the book is accepted, the Society Book Editors ensure that the volume editors follow strict guidelines on refereeing and quality control. We insist that individual papers can only be accepted after satisfactory review by two independent referees. The questions on the review forms are similar to those for *Journal of the Geological Society*. The referees' forms and comments must be available to the Society's Book Editors on request.

Although many of the books result from meetings, the editors are expected to commission papers that were not presented at the meeting to ensure that the book provides a balanced coverage of the subject. Being accepted for presentation at the meeting does not guarantee inclusion in the book.

More information about submitting a proposal and producing a book for the Society can be found on its web site: www.geolsoc.org.uk.

Brenchley, P.J. & Rawson, P.F. (eds) 2006. *The Geology of England and Wales*. The Geological Society, London.

Woodcock, N.H. & Soper, N.J. 2006. The Acadian Orogeny: The mid-Devonian phase of deformation that formed slate belts in England and Wales. *In*: Brenchley, P.J. & Rawson, P.F. (eds) 2006. *The Geology of England and Wales*. The Geological Society, London, 131–146.

The Geology of England and Wales

SECOND EDITION

EDITED BY

P. J. BRENCHLEY

University of Liverpool, UK

and

P. F. RAWSON

University College London, UK

2006

Published by

The Geological Society

London

THE GEOLOGICAL SOCIETY

The Geological Society of London (GSL) was founded in 1807. It is the oldest national geological society in the world and the largest in Europe. It was incorporated under Royal Charter in 1825 and is Registered Charity 210161.

The Society is the UK national learned and professional society for geology with a worldwide Fellowship (FGS) of over 9000. The Society has the power to confer Chartered status on suitably qualified Fellows, and about 2000 of the Fellowship carry the title (CGeol). Chartered Geologists may also obtain the equivalent European title, European Geologist (EurGeol). One fifth of the Society's fellowship resides outside the UK. To find out more about the Society, log on to www.geolsoc.org.uk.

The Geological Society Publishing House (Bath, UK) produces the Society's international journals and books, and acts as European distributor for selected publications of the American Association of Petroleum Geologists (AAPG), the Indonesian Petroleum Association (IPA), the Geological Society of America (GSA), the Society for Sedimentary Geology (SEPM) and the Geologists' Association (GA). Joint marketing agreements ensure that GSL Fellows may purchase these societies' publications at a discount. The Society's online bookshop (accessible from www.geolsoc.org.uk) offers secure book purchasing with your credit or debit card.

To find out about joining the Society and benefiting from substantial discounts on publications of GSL and other societies worldwide, consult www.geolsoc.org.uk, or contact the Fellowship Department at: The Geological Society, Burlington House, Piccadilly, London W1J 0BG: Tel. +44 (0)20 7434 9944; Fax +44 (0)20 7439 8975; E-mail: enquiries@geolsoc.org.uk.

For information about the Society's meetings, consult *Events* on www.geolsoc.org.uk. To find out more about the Society's Corporate Affiliates Scheme, write to enquiries@geolsoc.org.uk

Published by The Geological Society from: The Geological Society Publishing House, Unit 7, Brassmill Enterprise Centre, Brassmill Lane, Bath BA1 3JN, UK

(*Orders*: Tel. +44 (0)1225 445046, Fax +44 (0)1225 442836)
Online bookshop: www.geolsoc.org.uk/bookshop

British Library Cataloguing in Publication Data
A catalogue record for this book is available from the British Library.

ISBN-10 1-86239-199-8 (hardback)
ISBN-13 978-1-86239-199-4 (hardback)
ISBN-10 1-86239-200-5 (softback)
ISBN-13 978-1-86239-200-7 (softback)

Typeset by Charlesworth, UK
Printed by Cambrian printers, Aberystwyth, UK

Distributors

North America
For trade and institutional orders: The Geological Society, c/o AIDC, 82 Winter Sport Lane, Williston, VT 05495, USA
Orders: Tel +1 800-972-9892
Fax +1 802-864-7626
Email gsl.orders@aidevt.com

For individual and corporate orders:
AAPG Bookstore, PO Box 979, Tulsa, OK 74101-0979, USA
Orders: Tel +1 918-584-2555
Fax +1 918-560-2652
Email bookstore@aapg.org
Website http://bookstore.aapg.org

India
Affiliated East-West Press Private Ltd, Marketing Division, G-1/16 Ansari Road, Darya Ganj, New Delhi 110 002, India

Contents

Acknowledgements

Bringing this volume together has relied on the help and commitment of a large team of people, foremost of whom are the authors of the individual chapters. They in turn gratefully acknowledge the assistance of the following for their help in critically reading draft manuscripts, drawing figures or providing information, including photographs:

Chapter 1. Professor P. Henderson, Dr J. H. Powell.
Chapter 2. Professor M. D. Brasier, Mr M. Garton, Dr G. J. Potts, Dr W. Gibbons.
Chapter 3. Dr D. E. B. Bates, Professor M. G. Bassett, Dr L. R. M. Cocks, Dr J. R. Davies, Professor R. A. Fortey, Dr M. Lewis, Dr P. Kokelaar, Dr S. G. Molyneux, Dr A. Rees, Dr N. J. P. Smith, Dr R. A. Waters, Dr R. B. Wilson, Dr J. A. Zalasiewicz.
Chapter 4. Professor M. G. Bassett, Dr S. G. Molyneux, Dr D. I. Schofield, Dr D. Wilson.
Chapter 5. Dr M. J. Branney, Dr. P. Kokelaar, Dr D. Millward, Dr R. B. Rickards, Professor N. J. Soper.
Chapter 6. Dr R. P. Barnes, Dr R. P. Merriman, Dr R. J. Stone, Dr R. A. Strachan.
Chapter 7. Professor R. Macdonald, Dr T. C. Pharaoh, Professor N. J. Soper, Mr B. Young.
Chapter 8. Dr P. F. Friend, Mr S. Howells, Mr L. B. Morrisey, Dr S. B. Marriott. Shell UK are thanked for providing the funding for the colour figures 8.5 and 8.6.
Chapter 9. Mr W. J. Barclay, Mr J. I. Chisholm, Dr P. J. Cossey, Dr J. R. Davies, Mr M. T. Dean, Mr R. Demaine, Dr E. W. Johnson, Mr P. Lappage, Mr M. McCormac, Mr A. McMillan, Dr N. J. Riley, Dr J. H. Powell, Mr G. J. Tuggey, Dr R. A. Waters, Dr D. Wilson, Mr B. Young. In addition Fig 9.7 was drafted by JS Publications, Newmarket, Suffolk.
Chapter 10. Dr M. T. Holder, the late Mr A. J. J. Goode, Dr R. K. Shail.
Chapter 11. Dr L. M. Hollick, Mr. M. J. Howe, Professor D. J. Manning, Dr B. E. Leveridge.
Chapter 12. Miss M. Pringle, Dr N. E. Worley.
Chapter 13. Mr S. Chew, Dr M. Geluk, Dr A. S. Howard, Dr. D. W. Holliday, Dr R. Ireland, Mr R. Porter.
Chapter 14. Professor J. H. Callomon, Dr B. M. Cox, Dr S. P. Hesselbo, Mrs L. C. Norton, Dr M. G. Sumbler, Professor P. B. Wignall.
Chapter 15. Dr S. P. Hesselbo, Dr A. Ruffell, Mr C. J. Wood.
Chapter 16. Professor M. E. Collinson, Mr R. Ellison, Dr J. J. Hooker, Dr J. Skipper, Mr D. J. Wood.
Chapter 17. Ms E. McBay.
Chapter 18. Environment Agency, Halcrow, Dr L. Naylor.

Contributions to this volume by Dr R. P. Barnes, Dr J. R. Davies, Dr D. W. Holliday, Dr B. E. Leveridge, Dr R. C. Scrivener, Dr P. Stone, Dr C. N. Waters and Dr. R. A. Waters are published with the permission of the Executive Director of the British Geological Survey (NERC).

The authors have made every reasonable effort to identify and acknowledge the copyright owners of material reproduced in this book. They are grateful to the Executive Director of the British Geological Survey for permission to reproduce numerous photographs and maps (NERC all rights preserved). These are acknowledged individually in the figure and plate captions. Prof L. R. M. Cocks and Dr A. W. A. Rushton publish by permission of The Natural History Museum. They also thank the following organizations for permission to reproduce figures or access archives:

Blackwell Science
Yorkshire Geological Society
Cambridge University Press

The editors are grateful to the following for refereeing individual chapters: Professor R. J. Aldridge, Dr M. P. Atherton, Professor J. A. Catt, Dr A. Currant, Dr J. I. Chisholm, Dr P. J. Cossey, Professor R. A. Fortey, Dr P. F. Friend, Dr W. Gibbons, Dr K.W. Glennie, Dr S. P. Hesselbo, Dr D. W. Holliday, Dr A. S. Howard, Dr R. W. O'B. Knox, Dr S. B. Marriott, Professor R. J. Merriman, Dr D. Millward, Professor R. N. Mortimore, Professor K. T. Pickering, Professor J. Rose, Dr D. I. Schofield, Professor N. J. Soper, Dr E. Steurbaut, Dr R. A. Strachan, Professor M. E. Tucker, Professor P. B. Wignall and Dr J. A. Zalasiewicz.

We would also like to thank Angharad Hills, Helen Floyd-Walker and their colleagues at the Geological Society's Publishing House for all their help during the long gestation of this book. Finally, we thank our wives, Barbara Brenchley and Sue Rawson, for their patience, especially when the editing lasted well into our 'retirement'.

Contributing authors

Robert P. Barnes
British Geological Survey, Edinburgh: rpb@bgs.ac.uk
Michael J. Branney
Department of Geology, University of Leicester: mjb26@leicester.ac.uk
Patrick J. Brenchley
Department of Earth and Ocean Sciences, University ofLiverpool: pat@bbrenchley.fsnet.co.uk
John A. Catt
Department of Geography, University College London: j.catt@ucl.ac.uk
Richard Cave
Institute of Geography and Earth Sciences, University of Wales, Aberystwyth
Lesley Cherns
Department of Earth Sciences, Ocean and Planetary, Cardiff University: Cherns@cardiff.ac.uk
L. Robin M. Cocks
Department of Palaeontology, The Natural History Museum, London: r.cocks@nhm.ac.uk
John C. W. Cope
National Museum of Wales, Cardiff: john.cope@nmgw.ac.uk
Jeremy R. Davies
British Geological Survey, Keyworth: jrdav@bgs.ac.uk
Sarah J. Davies
Department of Geology, University of Leicester: sjd27@le.ac.uk
Philip L. Gibbard
Cambridge Quaternary Department of Geography, University of Cambridge: plg1@cam.ac.uk
Adrian J. Hartley
Department of Geology and Petroleum Geology, Aberdeen University: a.hartley@abdn.ac.uk
Robert D. Hillier
Department of Geology, National Museum of Wales, Cardiff: robhillier@aol.com
Douglas W. Holliday
British Geological Survey, Keyworth: dwho@bgs.ac.uk
Jana Horák
Department of Geology, National Museum of Wales, Cardiff: jana.horak@nmgw.ac.uk
Mark W. Hounslow
Geography Department, Lancaster Environmental Centre, Lancaster University: m.hounslow@lancaster.ac.uk
M. F. Howells
22 Davies Avenue, Leeds, LS8 1JY: fletcher@howells 22.force 9.co.uk
Chris King
16A Park Road, Bridport, Dorset: chrking@globalnet.co.uk
Brian E. Leveridge
British Geological Survey, Exeter: bel@bgs.ac.uk
John J. Lowe
Department of Geography, Royal Holloway, University of London: j.lowe@rhul.ac.uk
Danny McCarroll
Department of Geography, University College, Swansea: D.McCarroll@swansea.ac.uk
Duncan McIlroy
Canada Research Chair in Petroleum Geosciences, Department of Earth Sciences, Memorial University, Newfoundland, Canada: dmc@esc.mun.ca
David Millward
British Geological Survey, Edinburgh: dmill@bgs.ac.uk
Jane E. Rawson
The Environment Agency, Peterborough: jane.rawson@environment-agency.gov.uk
Peter F. Rawson
Department of Earth Sciences, University College London: p.rawson@ucl.ac.uk
Alastair H. Ruffell
School of Geography, Queen's University, Belfast: a.ruffell@qub.ac.uk
Adrian W. A. Rushton
Department of Palaeontology, The Natural History Museum, London: a.rushton@nhm.ac.uk
Richard C. Scrivener
British Geological Survey, Exeter: rcs@bgs.ac.uk
James D. Scourse
School of Ocean Sciences, University of Wales (Bangor): j.scourse@bangor.ac.uk
Alan G. Smith
Department of Earth Sciences, University of Cambridge: ags1@esc.cam.ac.uk
Denys B. Smith
79 Kenton Road, Newcastle-upon-Tyne
N. Jack Soper
Gams Bank, Threshfield, Skipton, BD23 SNP
Philip Stone
British Geological Survey, Edinburgh: psto@bgs.ac.uk
Mike J. C. Walker
Department of Archaeology and Anthropology, University of Wales (Lampeter): walker@lamp.ac.uk
Colin N. Waters
British Geological Survey, Keyworth: cnw@bgs.ac.uk
Richard A. Waters
Llwynonn, Penglais Road, Aberystwyth, SY23 2E: *dick.waters@ukgateway.net*
Brian P. J. Williams
Department of Geology and Petroleum Geology, University of Aberdeen: brianpjwilliams@hotmail.com
Mark Williams
Department of Geology, Leicester University, Leicester, LE1 7RH: mri@leicester.ac.uk
Nigel H. Woodcock
Department of Earth Sciences, University of Cambridge: nhw1@esc.cam.ac.uk
John J. Wymer
17 Duke Street, Bildeston, Ipswich

Preface to the second edition

This new edition of *The Geology of England and Wales* follows the recent publication of *The Geology of Scotland* (Trewin ed. 2002) to complete a duo of books offering an up-to-date account of the geology of Great Britain.

Although it is a little more than a decade since the First Edition of this book was published (1992) there have been significant developments in our understanding of major geological processes and a huge increase in the available database on British geology. Of particular importance has been the growth in our knowledge of the timing and paths of plate movements and the distribution of plates on the Earth's surface. In the light of this knowledge it is possible to infer how the plate movements have driven deformation of the Earth's crust. Thus we can view the sequence of geological events in a dynamic context, unfolding through time. When planning this new edition the editors and the Publications Committee of the Geological Society agreed that a dynamic view of the evolution of our region should be reflected in the volume while it would retain a sound stratigraphical basis – following as far as possible the schemes produced in the Geological Society's Special Report Series, which has a separate report on the stratigraphy and correlation of each system in Britain. Thus the book is organized primarily as an account of the evolution of the area, with chapters arranged on a temporal basis.

For each interval of time, authors were asked to place the geology of England and Wales into its global and regional context; summarize the basic lithostratigraphy and chronostratigraphy; and discuss the regional development basin-by-basin. As part of the plan to integrate as many aspects of geology as possible, offshore geology is no longer dealt with in a separate chapter and chapters specifically on 'Structure' and 'Igneous Intrusions' have not been repeated. Instead the main orogenic episodes, the Acadian and Variscan orogenies and their resulting fold belts, are described in separate chapters, as are the phases of major intrusion, the Early Palaeozoic plutons and the Cornubian granites of SW England. Following the decline in metalliferous mining and coal mining there is no specific chapter on economic geology, but significant economic deposits, such as coal, hydrocarbons and building resources, are mentioned wherever appropriate.

The diversity of expertise that is necessary to do justice to the complexity of modern geology and the multiplicity of approaches to its understanding has entailed a substantial increase in the number of authors of this volume, to 40. Few chapters have a single author, which is likely to be the pattern in the future. Inevitably in such a multi-author compilation there has been some variation in the interpretation of the editorial guidelines, and some authors perceived a need to expand considerably the discussion of some of the broader aspects, such as palaeogeography, where there are strongly conflicting interpretations in the previous literature. The editors have taken a liberal approach to this variation in the hope that it will lead to a greater understanding of the fascinating geology of England and Wales, highlight some of the remaining problems, and provide a valuable account for both geoscientists in the British Isles and visitors from overseas.

Patrick J. Brenchley & Peter F. Rawson

1 England and Wales through geological time

P. J. BRENCHLEY & P. F. RAWSON

The present-day landmass of England and Wales offers a wealth of geological variety. Remarkably, all the geological systems are represented within this relatively small area (Fig. 1.1), which encompasses strata ranging in age from Neoproterozoic (late 'Precambrian') to Holocene. This partly explains why so many early advances in geology were made here, both in mapping rocks and in developing a geological timescale. In the late 18th and early 19th centuries, William Smith recorded and mapped the sequence of Upper Palaeozoic and Mesozoic rocks in England, and made the first detailed geological map of England and Wales and part of Scotland, published in 1815 (Fig. 1.2). This pioneering effort was aimed primarily at practical people such as mine owners and surveyors, and was an early and fundamental demonstration of the applicability of geology to the broader needs of mankind. But the work of Smith 1799 and others also indicated the need for a geological timescale against which the local sequences could be calibrated.

It was within England and Wales that some of the first major subdivisions of the geological record were made. Conybeare & Phillips (1822) combined three established rock units, 'The Mountain Limestone', the 'Millstone Grit' and the 'Coal Measures' to form a new division – 'The Carboniferous' – the same year that d'Omalius d'Halloy proposed the Cretaceous System in France. In the following two decades, work on the complex geology of Wales and the Welsh Borderland led to the establishment of the Cambrian and Silurian systems by Adam Sedgwick (1835) and Roderick Murchison (1839), respectively (Chapter 3), while the same two authors jointly named the Devonian System in 1839, based on the folded sequence in Devon and Cornwall (Sedgwick & Murchison 1839). Much later, Charles Lapworth proposed the Ordovician System for rocks that had previously been assigned to either the Cambrian or the Silurian (Lapworth 1879), so resolving a long-standing conflict. Hence four of the six Palaeozoic systems were based on Welsh or English successions.

The establishment of this relative timescale during the 19th century was a fundamental step forward in understanding Earth's history. The need then arose to put numbers to that scale, a process pioneered in the first half of the 20th century by another English geologist, Arthur Holmes. Then came another fundamental development in the latter half of the 20th century, the theory of plate tectonics, which has enabled us to bring together so many disparate observations and theories to provide a more dynamic view of Earth's history. Although England and Wales form only a tiny part of Earth's surface their geological evolution is a fair reflection of what was happening over our globe. Hence, in this introductory chapter, we firstly summarize the broader tectonic, sea-level and climatic changes that have influenced the evolution of our area, then outline the main events that can be recognized in the evolution of England and Wales. These are expanded upon in the following chapters, to which we give appropriate cross-reference. A final chapter (Chapter 18) looks at what may happen in the future.

The global setting

Tectonic events

In the first edition of this book, Derek Ager pointed to the fortuitous coincidence that the boundary between Scotland and England closely follows a fundamental structural lineament, the Solway line. This line, almost invisible at the surface, forms part of the Iapetus Suture, which marks the final closure of the Early Palaeozoic ocean, Iapetus. At its maximum, this ancient ocean may have been several thousand kilometres wide – with present-day Scotland on one side and England and Wales on the other. Thus, the closure brought together two highly contrasting geological areas. To the north of the suture lies the highly complex set of terranes that form Scotland, while to the south is a somewhat simpler area occupied principally by part of the ancient Avalonian microplate. That microplate in turn was torn apart by the opening North Atlantic Ocean during the Cretaceous, so that now the western half lies in Maritime Canada while eastern Avalonia forms most of the small area of crust that now comprises England and Wales.

The Early Palaeozoic evolution of our area is largely influenced by the changing manner in which plate tectonics affected Avalonia, which had been formed by accretion of mainly volcanic fragments offshore from the margin of Gondwana and then temporarily accreted with Gondwana in the latest Proterozoic. It later separated from Gondwana during the early Ordovician. Its subsequent Early Palaeozoic history was influenced by its voyage towards Laurentia, with which it and Baltica collided obliquely along the line of the Iapetus Suture during the Silurian (Fig. 1.3). The collision resulted in the Caledonian Orogeny. However, its oblique component gave a complex history involving both transtension and transpression, culminating in a compressional event, the Acadian Orogeny, of mid-Devonian age (Chapter 6), which additionally may have involved part of the Rheic Ocean to the south of Avalonia. Finally, the Variscan Orogeny (Devonian–Carboniferous) was caused by the oblique closure of the Rheic Ocean, with both extensional and compressional effects (Chapter 10).

These orogenic phases brought together the previously separate landmasses of the globe to form one supercontinent, Pangaea, by the end of the Carboniferous. The subsequent break-up of this supercontinent, which started during the Jurassic, was marked by a long-lasting, dominantly tensional regime that led to the opening of the Atlantic. Subsequently, some important compressional phases occurred, including the 'Alpine' Orogeny (Chapter 16). This marked the convergence of the European and African plates, which peaked in a collision during the Miocene. Although the collision zone lay far from the British Isles the orogeny did affect successions in southern England.

Sea-level changes

Changes in sea level play an important part in the palaeogeographic and climatic evolution of an area. With sea-level rise,

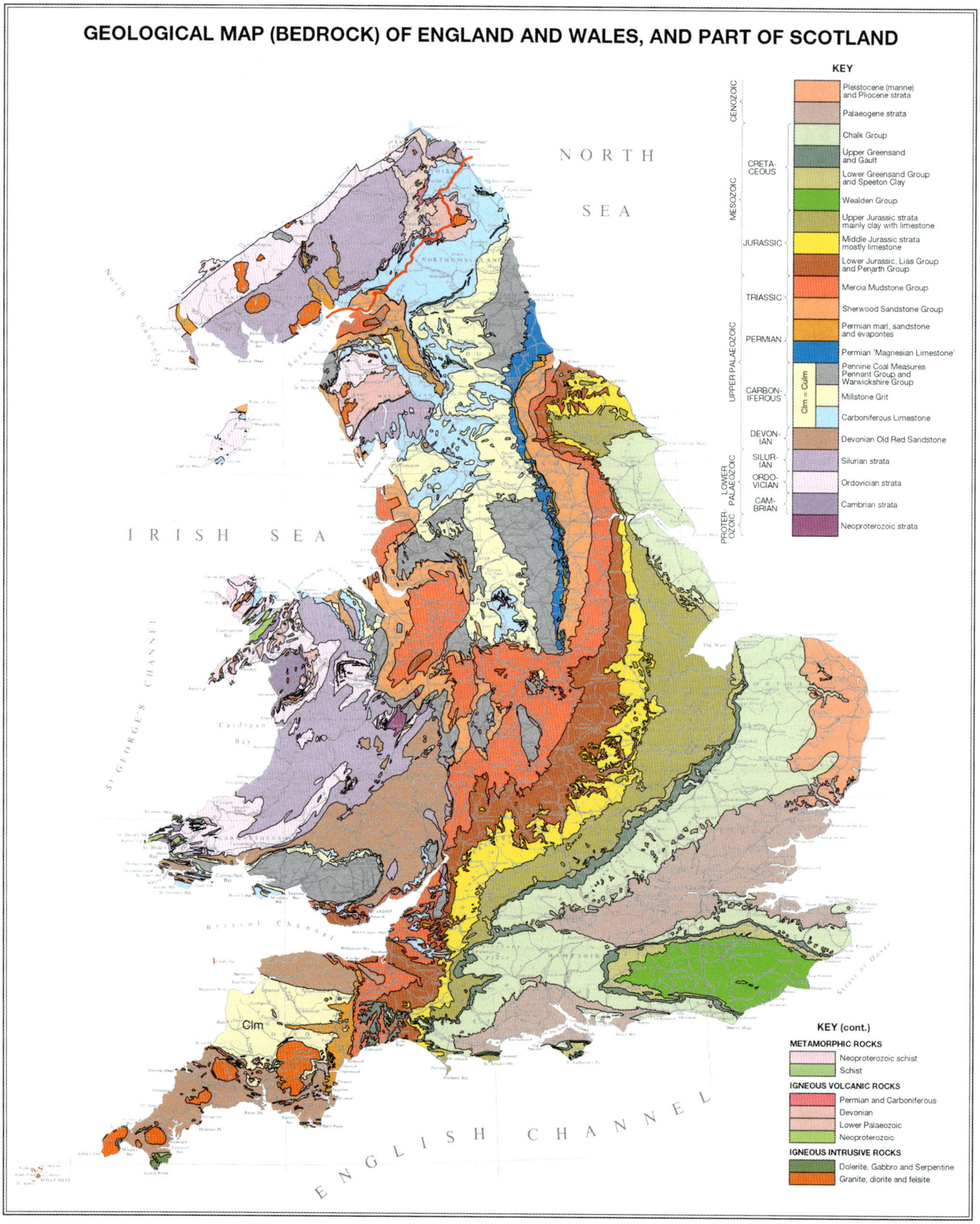

Fig. 1.1. Geological map of England and Wales. Reproduced by permission of the British Geological Survey. (C) NERC. All rights reserved.

land area is reduced, which may result in a more maritime (humid) climate. With sea-level fall, land areas increase and a more continental (arid) climate may develop. The timescale of global changes in sea level varies greatly, from a few thousand to many millions of years. Slow and progressive changes are thought to arise from changes in the spreading rate and overall volume of mid-ocean ridges, which may occur over tens of millions of years (Fig. 1.4). Such changes are well documented temporally, although they are difficult to quantify in terms of metres above present-day sea level. In the Phanerozoic, sea

Fig. 1.2. The 1815 geological map of William Smith. Reproduced by permission of the British Geological Survey. (C) NERC. All rights reserved.

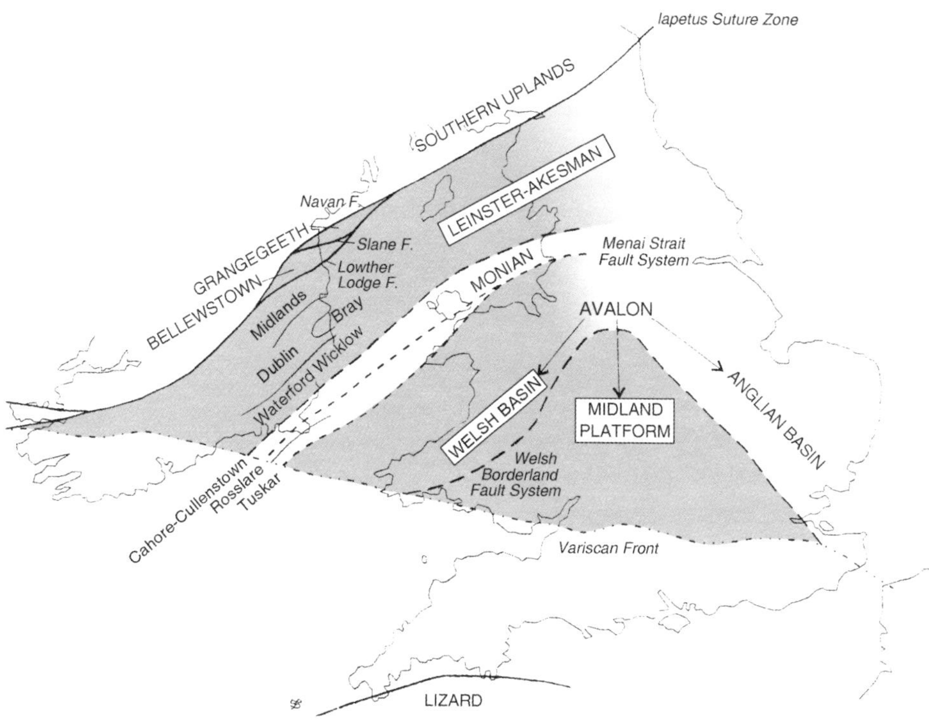

Fig. 1.3. Geological terrane map of England and Wales, showing the three main terranes and major structural lineaments in England and Wales referred to in Chapter 3 (modified from Woodcock 2000*d*).

levels were generally high during the Early Palaeozoic, with peaks at about the Cambrian–Ordovician boundary and in the Late Silurian. They fell during the later Palaeozoic, reflecting the consolidation of Pangaea, to reach a low in Triassic times. As Pangaea began to break up during the Jurassic so sea levels rose again to reach a possible Phanerozoic maximum during the Late Cretaceous. They began to fall again in latest Cretaceous times towards the present-day low.

In contrast to these slow changes, rapid global (eustatic) changes in sea level may occur, particularly in response to the growth and decay of major ice caps. The effects of both the Carboniferous–Permian (Gondwanan) and Pleistocene glacially induced sea-level changes are seen in our successions. Rapid changes in relative sea level may occur also on a regional scale as a result of tectonic factors, particularly in the change of the rate of subsidence relative to the rate of sediment accumulation – which often produces cyclic sequences. The understanding of the effects of both eustatic and tectonically induced sea level has advanced greatly with the application of sequence stratigraphy, which has, at its core, models of the effects of relative sea-level change on the architecture of sedimentary sequences.

Climate

Through the Phanerozoic there have been well-documented shifts in global climate from 'greenhouse' to 'icehouse' conditions (Fig. 1.4), though this may be an oversimplification in that small icecaps may have existed during some of the 'greenhouse' phases (Chapter 15). In general, the 'greenhouse' phases have coincided closely with intervals of higher global sea level. The climate of England and Wales has changed considerably through time in response to these global changes. But it has also evolved as a result of plate movements that, over a long period of time, have gradually carried our area northwards through different latitudes (e.g. Chapter 3, Fig. 3.1), whose climate is commonly reflected in the marine biota and particularly clearly in distinctive sedimentary facies in continental regions. Throughout the Early Palaeozoic, the Avalonian terrane lay in mid latitudes in the southern hemisphere and generally experienced temperate climates. By late Silurian times the closure of the Iapetus Ocean had joined Avalonia and Laurentia to form all but the southernmost part of the present-day British Isles, while the remainder (the Lizard Terrane) was added in the Carboniferous. During the Devonian and Carboniferous these terranes moved from the southern hemisphere arid belt to humid equatorial latitudes. After the Variscan uplift of the region and with further northward drift the climate became more generally continental and arid through the Permo-Triassic. Continued Mesozoic and Cenozoic drift moved England and Wales to their present-day northern temperate-climate latitudes.

Key events in the geological evolution of England and Wales

The Neoproterozoic basement crust

The eastern Avalonian crust is mainly Neoproterozoic in age (700–500 Ma); it is thinner and younger compared with the Precambrian crust of the adjacent Laurentian plate (*c.* 3000 Ma) which lies below much of Scotland. The Avalonian crust originally formed near the margin of Gondwana, mainly from calc-alkaline igneous complexes, which together with associated sedimentary basins accreted to form the eastern part of Avalonia. Fossils belonging to the Neoproterozoic Ediacara fauna have been found in one of the volcano-sedimentary sequences in Charnwood Forest, Leicestershire (Chapter 2).

Early Palaeozoic events

During the movement of Avalonia towards Laurentia, England and Wales experienced periods of extension leading to the formation of the Welsh, Anglian and Lakesman basins (Chapter 3). Substantial thicknesses of mudstones and turbidites accumulated in the Welsh Basin through much of the Cambrian. The development of a volcanic arc transformed the basin into a

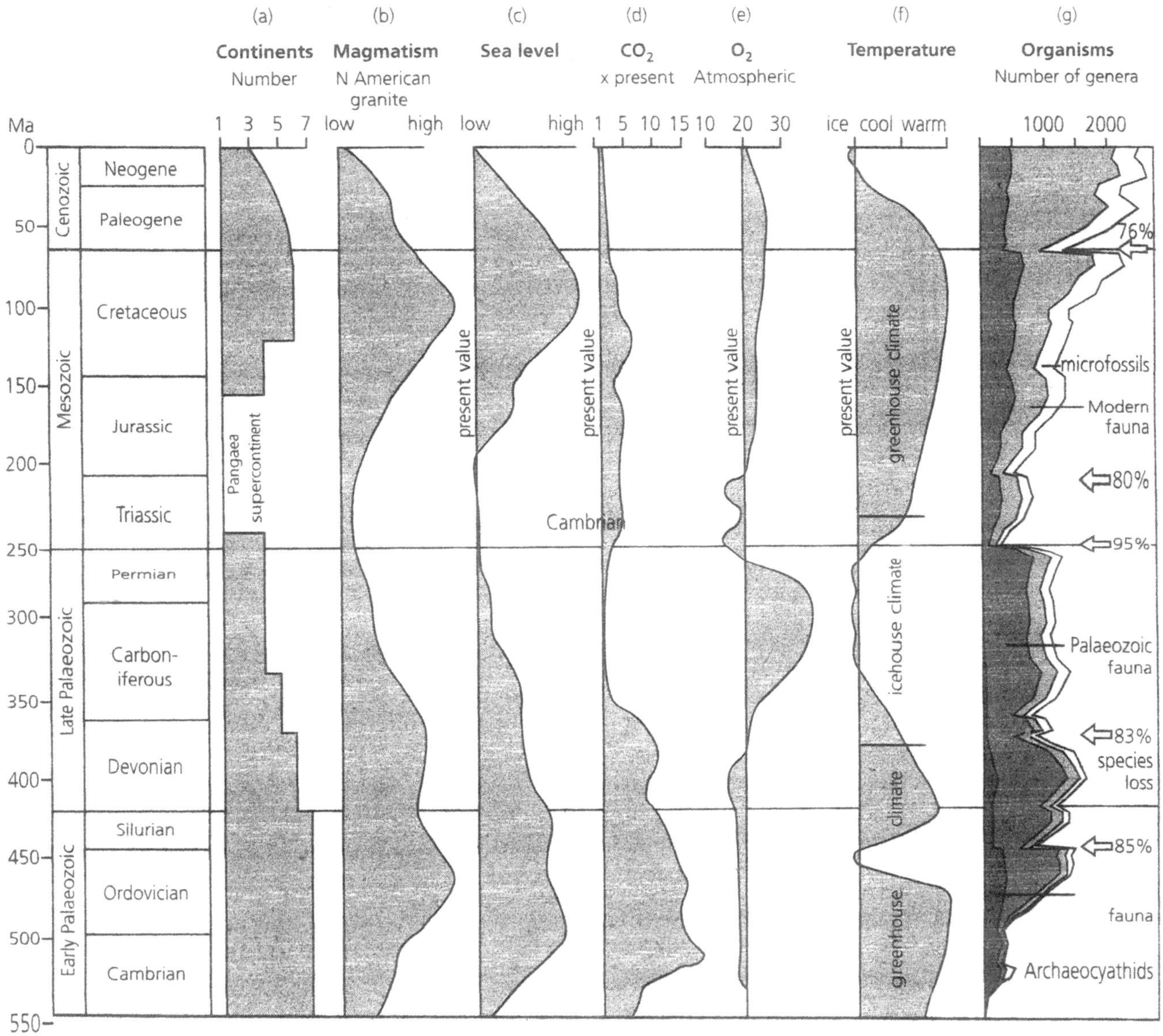

Fig. 1.4. Aspects of global change through Phanerozoic time. Reproduced from Woodcock & Strachan (2000) by permission of Blackwell Science.

back-arc basin in the early Ordovician, which was subsequently characterized by volcanic activity until the mid Caradoc. It then evolved into an active extensional, mainly non-volcanic, basin until the end of the Silurian. The Lakesman Basin lay oceanwards of the Welsh Basin and differed in that there was a particularly thick accumulation of mudstones and turbidites in the early Ordovician. Major volcanic activity did not start there until the Late Ordovician. The Lakesman Basin was transformed into a foreland basin in the Silurian as the lithosphere was loaded by the late thrust sheets of southern Scotland during closure of the Iapetus Ocean. Within central England, the Midland Platform formed a triangular stable area surrounded by extensional basins of Early Palaeozoic age (Fig. 1.3). It generally experienced low subsidence rates, but periodically sustained basins of modest size, such as the narrow rifts of Tremadoc age at the site of the Worcester Graben and in the central part of the platform, where scant borehole evidence suggests their presence.

Volcanism was associated with major developments in the evolution of both the Welsh and Lakesman basins. Small volcanic arcs developed during the Tremadoc in west and north Wales, where they were associated with local folding of basin strata and emergence of some areas. A major development in the nature of the basin occurred in the early Arenig when there was a change in both the location and chemistry of the volcanism, which records the transformation of the Welsh Basin into a back-arc basin. There followed an early Arenig transgression that spread shallow-marine sediments across a widespread unconformity surface. Volcanism continued in the Welsh Basin until the late Caradoc (Late Ordovician), while in the Lakesman Basin a large volcanic arc, with associated intrusions (Chapter 5), developed in the area of the Lake District in the early Caradoc.

Following the end of volcanism in both the Welsh and Lakesman basins there was an interval with modest thermal subsidence in both basins and local tilting and folding in the Welsh Basin and Welsh Borderland Fault System (the Shelveian event). The interval finished with a large and rapid glacio-eustatic fall and rise of sea level reflecting the growth and decay of the large southern ice cap on Gondwana. The sea-level rise, which started in latest Ordovician times and continued into the early Silurian, is marked in the Welsh Basin by black euxinic mudstones. Sea level continued to rise through the early Silurian, so that in areas around the Welsh Basin that had been affected by late Ordovician folding, transgressive sediments lie above a marked unconformity. Through most of the Silurian, extension in the Welsh Basin produced internal half grabens in which substantial thicknesses of sediments accumulated (Chapter 4).

The sparsity of large carbonate accumulations before the Silurian reflects the mid latitudinal position of the Avalonian terrane.

The Caledonian Orogeny

The term Caledonian Orogeny is now used to include all Cambrian–Devonian events associated with the development and closure of the Iapetus Ocean (McKerrow *et al.* 2000*a*). It comprises a number of phases often distinguished as separate orogenies. In England and Wales the first effect of final Iapetus closure is the early Silurian development of a foreland basin on the Lakesman Terrane due to the loading of Avalonian lithosphere by the advancing Laurentian margin. The volume of Silurian turbidites supplied to this basin (Windermere Supergroup, Chapter 5) eventually exceeded flexural subsidence rates and caused a late Ludlow to Přídolí shallowing into non-marine, 'Old Red Sandstone' facies, a transition also seen in the Welsh Basin.

Old Red Sandstone sedimentation continued over substantial areas of England and Wales through Early Devonian time (Chapter 8), until terminated by the Acadian Orogeny in the late Early Devonian.

The Acadian Orogeny

The Acadian Orogeny has generally been seen as the culminating event of Iapetus closure and therefore of the Caledonian Orogeny. However, recent views (Chapter 6) have suggested that the event is separated from the Iapetus closure by an important Early Devonian transtensional event, which also triggered extensive 'late Caledonian' magmatism. Renewed Acadian shortening may have been due to collision of a continental fragment or thick oceanic lithosphere south of Avalonia and is therefore not strictly Caledonian at all.

This mid Devonian orogeny produced the main folding and cleavage in the Cambrian–Devonian rocks of the basin rimming the northern margin of Avalonia. Earlier Devonian sequences were locally uplifted, and the deeper Lower Palaeozoic basinal sediments were deformed to form the slate belts exposed in Wales and the Lake District (Chapter 6). Approximately coeval with these events, large granite intrusions, the Shap, Weardale and Skiddaw plutons, were emplaced in northern England (Chapter 7).

The Variscan Orogeny

Contemporaneously with deposition of the Devonian Old Red Sandstone, marine basins developed within the Variscan Orogenic Province in SW England. Basins within the orogenic belt initially resulted from continental rupture and the formation of ocean crust in the south of the belt and rifting of the Avalonian continental margin on its northern side. Infill of the Gramscatho Basin in the south of the orogenic province was related to uplift resulting from the convergence that affected the southern margin of the province during the mid to late Devonian. At approximately the same time the basins to the north were undergoing extension (Chapter 10). Deformation within the orogenic belt continued until the end of the Carboniferous and resulted in a complex fold belt in Devon and Cornwall. In the final stages of the orogeny the 'Cornubian' granite plutons of SW England, were emplaced (Chapter 11).

North of the Variscan Front, extensional basins of Devonian and Carboniferous age formed within the Midland Platform. The Devonian and Carboniferous successions reflect the interaction between the tropical to subtropical climate and sea-level changes. The Devonian was an essentially arid interval and sea level was low, so that so that red-beds accumulated in terrestrial areas. Early Carboniferous transgression resulted in the widespread development of marine carbonates. Sea-level fall later in the Carboniferous led to the development of extensive deltas on which tropical forests flourished, to form the Coal Measures (Plate 15). The southern hemisphere glaciation, which fluctuated in area as the ice advanced and retreated, caused cyclic episodes of eustatic sea-level rise and fall that are reflected in the cyclic nature of many of our Carboniferous sequences (Chapter 9).

Many of the Devonian and Carboniferous basins were uplifted in a final, late Carboniferous phase of the orogeny. This terminal Variscan deformation within central England, Wales and adjacent offshore areas reactivated existing faults and created new ones, resulting in both the inversion of basins and the formation of new ones. This created a new palaeogeography of basins and highs in the early Permian (Chapter 12).

Post-Variscan events

The Variscan was the last orogeny to affect our area directly: after that the British Isles became part of a stable continent and remain so today. However, events further afield made their effects felt, especially the opening of the North Atlantic and the Alpine Orogeny. The net result was that over most of the time from the Permian onward, tensional stresses have dominated, though there have been some significant compressional events too.

Our understanding of the Permian–Recent geology of onshore England and Wales has changed dramatically over the last 40 years as the adjacent offshore areas have been explored so thoroughly for hydrocarbons. It is now clear that through most of Permian, Mesozoic and Cenozoic time, the main areas of subsidence lay in the present-day offshore areas – the North Sea, Channel and Celtic Sea basinal systems. The magnificent coastal Mesozoic sections in Dorset and Yorkshire represent marginal areas of the North Sea and Channel basins, respectively, which were often linked by shallower-water areas extending across the relatively stable Midland Platform.

Permo-Triassic extension

During the Permo-Triassic interval there were important phases of rifting, and the Central Graben, Western Approaches and Celtic Sea basins were all well defined, as were the onshore Worcester Graben and Cheshire basins. The Permo-Triassic climate was mainly arid to semi-arid, resulting in the deposition of red beds containing a range of aeolian and fluviatile facies (Chapters 12 and 13). The extensive shallow-marine areas of the Zechstein Sea are characterized by evaporites, which form a cap to the hydrocarbon reservoirs in the Rotliegende sandstone reservoirs below.

Jurassic–Cretaceous events

A major eustatic rise in sea level that commenced in latest Triassic times led to a rapid Early Jurassic marine transgression across former land areas in England and Wales and the surrounding offshore region. Concurrently the climate became more humid. Continuing extension was accompanied by thermal subsidence, creating broader basins and extensive shallow-marine platforms (Chapter 14). Regional uplift in the central North Sea in the mid Jurassic created new sources of sediment. This was followed by renewed extension during the mid–late Jurassic and earliest Cretaceous which reactivated many of the rifts.

The main focus of movement moved to the west of Britain during Early Cretaceous times, associated with the opening of the North Atlantic Ocean (Chapter 15). This led to regional uplift in the west and reactivation of E–W faults of Variscan origin. The faults bounded the Palaeozoic massifs that supplied sediments to the intervening E–W-trending basins such as the

Weald Basin. By mid Cretaceous times rifting had effectively ceased in the North Sea Basin, which became a region of more uniform subsidence. At about the same time a global rise in sea level commenced, to radically alter the geography by submerging most of the region. This resulted in widespread deposition of a calcareous nannofossil ooze, forming a blanket of chalk, which at maximum sea level is believed to have covered most of England and Wales and the surrounding offshore area. Temperatures were generally high throughout the Cretaceous, rising towards the end of the period to the peak of the Mesozoic greenhouse climatic regime (Chapter 15).

Cenozoic history

Sea level fell again at the end of the Cretaceous and much of the onshore area of England and Wales became land. However, marine sediments continued to accumulated in the onshore areas and across parts of eastern and southern England (Chapter 16). Most of the Mesozoic and younger rocks have experienced only gentle tectonic deformation into broad open folds or gently tilted strata, the latter giving rise to a characteristic scarp and dip-slope present-day topography. The main exception to this is in the Howardian–Flamborough fault belt in Yorkshire and in the Isle of Wight and adjacent parts of the mainland, where latest Cretaceous–Miocene 'Alpine' movements reactivated basement faults to deform the overlying Mesozoic and early Tertiary strata. The same movements led to basin inversion in both onshore (Cleveland and Weald) and offshore basins, while in the southern North Sea Basin the movement of Zechstein salt from early Mesozoic times onward has created some spectacular halokinetic features which have produced traps in which major reserves of natural gas.have accumulated.

In the Palaeocene a major 'hot spot', above a mantle plume, developed beneath Greenland and caused thermal uplift and widespread volcanism to the west of northern Britain (Chapter 16). In the late Palaeocene–early Miocene a continental rupture occurred across the thermal bulge, initiating sea-floor spreading. Consequently, NW Europe, including the North Sea Basin, became part of a thermally subsiding passive continental margin. Alternatively, it has been suggested that thermal uplift, possibly resulting from magmatic underplating, occurred on the western side of Britain. This would have had a major influence on the development of the present-day topography of England and Wales, creating the mountainous areas of the Lake District and North Wales and an overall slope towards the east into the subsiding region of the North Sea. It was, however, the Pleistocene glaciations that eroded the mountain regions into the deeply dissected topography that makes the striking landscapes we know today, and extensively reshaped the lowland areas of central and eastern England (Chapter 17).

2 Neoproterozoic: the late Precambrian terranes that formed Eastern Avalonia

D. MCILROY & J. M. HORÁK

The late Proterozoic rocks of England and Wales comprise part of eastern Avalonia. Characterization of the basement rocks in southern Britain allows the recognition of five distinct terranes in this part of eastern Avalonia known as the Monian Composite Terrane, the Cymru Terrane, the Wrekin Terrane, the Charnwood Terrane and the Fenland Terrane (Gibbons & Horák 1996; Pharaoh & Carney 2000; Fig. 2.1). During the Neoproterozoic, eastern Avalonia was situated on the NE margin of Gondwana on the southern margin of the Ran Sea (Nance & Murphy 1996; Hartz & Torvik 2002; Fig. 2.2a–d). The Ran Sea itself was formed as a result of rifting of the older Rodinian continental landmass (Fig. 2.2). Avalonian tectonics during this late Proterozoic period were driven by subduction on the NE margin of Gondwana, resulting in associated magmatism and arc basin development. With progressive obliquity of subduction, arc magmatism was replaced by a regime dominated by large-scale transform faulting that progressively dissected and dispersed the arc. The switch from arc magmatism to intra-continental wrench-related volcanism and magmatism was diachronous, and is first seen in western Avalonia (Murphy *et al.* 2000). Neoproterozoic sediments of the Avalon Terrane are almost exclusively siliciclastic or volcaniclastic and were deposited within numerous geographically restricted strike-slip basins (Pharaoh *et al.* 1987*a*; Nance *et al.* 1991; McIlroy *et al.* 1998; Hartz & Torvik 2002; Fig. 2.3).

The Neoproterozoic strata of England and Wales typically have only limited exposure but, through many years of intensive study, the stratigraphic and structural history of most units has become better understood in the light of plate tectonic theory (e.g. Baker 1969; Wright 1969; Thorpe 1972; Wood 1974; Barber & Max 1979). Of particular importance was the realization that a major shear zone with demonstrable strike-slip movement separate the Monian rocks of Anglesey from mainland southern Britain. Recognition of a major shear zone separating Monian rocks from the Proterozoic rocks of the rest of southern Britain led to the application of the suspect terrane concept (Gibbons 1987, 1990; Pharaoh & Carney 2000). This led to the recognition of three terranes on Anglesey that are grouped into the Monian Composite Terrane and, together with improved dating of Neoproterozoic igneous bodies, has led to a re-evaluation of the remaining Neoproterozoic terranes of southern Britain (Pharaoh *et al.* 1987*a*, *b*; Tucker & Pharaoh 1991; Horák *et al.* 1996; Compston *et al.* 2002).

Examples of proven pre-Neoproterozoic (*c.* 2100 Ma) basement are exposed in the Icartian Terrane of the Channel Islands, NW Brittany and Normandy (Auvrey *et al.* 1980; Samson & D'Lemos 1998). Such basement is, however, unexposed in southern Britain. Indirect evidence for the nature of the pre-Neoproterozoic basement must therefore be gained from isotopic data. The strongest evidence is provided by inherited zircons in igneous rocks. Recent data from the Malverns Complex of the Wrekin Terrane (Tucker & Pharaoh 1991) indicate an inherited component of 1600 Ma, suggesting that older crust underlies the region. In the adjacent Charnwood Terrane, zircon data also suggest the presence of crust of similar, or older, age (Compston *et al.* 2002). Sm–Nd depleted mantle model ages for the Sarn Complex (Cymru Terrane) and Fenland Terrane record involvement of crust with a minimum age of 1460 and 1400 Ma, respectively, and support the previous interpretation of crust of 1600 Ma in the Wrekin Terrane (Davies *et al.* 1985; Horák 1993; Noble *et al.* 1993).

Stratigraphy and correlation

In the absence of abundant biostratigraphic markers, which is the norm for rocks of this age, direct dating of igneous rocks – in conjunction with lithostratigraphic and geochemical characterization along with tectonostratigraphic approaches – has gradually allowed correlation between the different terranes to be established. In recent years, the base of the Cambrian has been defined at a point in rock at the global stratotype section for the Proterozoic–Cambrian boundary (Brasier *et al.* 1994). This point is at the base of the *Phycodes pedum* ichnozone, and gives a high degree of confidence to recognizing rocks of Cambrian age on ichnological grounds (e.g. McIlroy *et al.* 1998 and the section on Anglesey below). In addition, a tectonostratigraphic approach to correlation has been taken, combining both geochronological and palaeontological evidence to recognize and correlate region-wide events (e.g. Nance *et al.* 1991; McIlroy *et al.* 1998; Fig. 2.3).

Fenland Terrane

This unexposed terrane is the most easterly expression on the British Avalonian and is characterized by limited data from acid ash-flow tuffs from the Glinton, Oxendon Hall and Orton boreholes (Fig. 2.1). This material records crustally-derived, or mature crust-contaminated, volcanism at around 616–612 Ma (Noble *et al.* 1993). The tuffs are, however, geochemically distinct from volcanic rocks of the adjacent Charnwood Terrane. The boundary with the adjacent Charnwood Terrane is unexposed but is taken at the postulated Northeast Charnwood Boundary Fault (Carney *et al.* 2000; Fig. 2.1).

Charnwood Terrane, Central England

The Charnwood Terrane lies to the southwest of the Northeast Charnwood Boundary Fault (Carney *et al.* 2000) and is

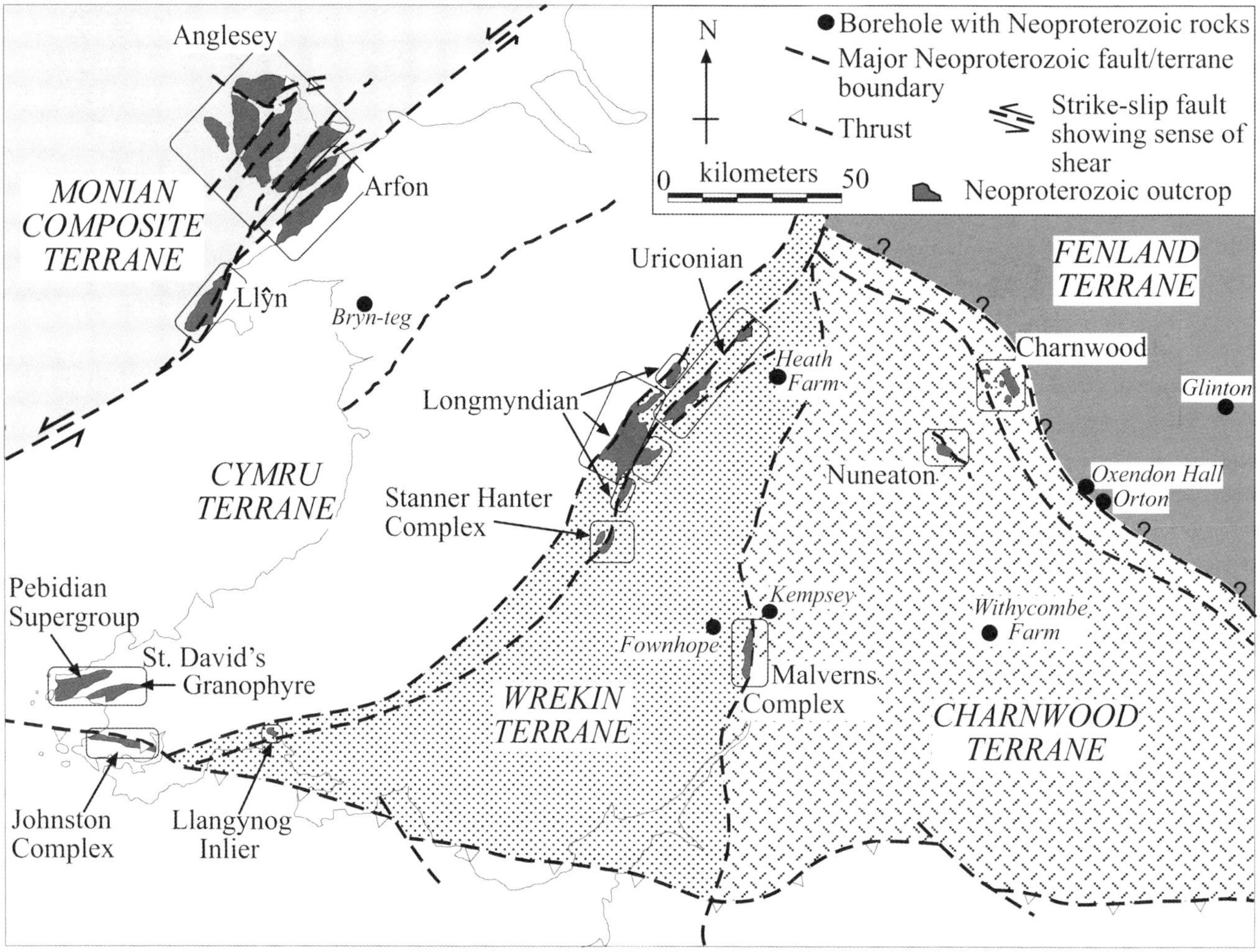

Fig. 2.1. The distribution of Neoproterozoic rocks in England and Wales with respect to Neoproterozoic terrane boundaries (redrawn from Carney *et al.* 2000).

separated from the Wrekin Terrane by the Malvern Lineament (Fig. 2.1). It includes two major outcrops of volcano-sedimentary sequences and plutonic intrusions, at Charnwood Forest and Nuneaton. In addition, core from the Kempsey and Withycombe Farm boreholes demonstrate its southerly extent. The thick Proterozoic succession at Charnwood Forest comprises one of the classic areas of British geology. The volcano-sedimentary succession comprises six formations and includes strata from the mid-Neoproterozoic (NP2) through to the recently recognized lower Cambrian (Bland & Goldring 1995; McIlroy *et al.* 1998) deposited in a marine basin. The completeness of this succession provides a stratigraphic standard for comparisons with strata elsewhere in the Charnwood Terrane. The numerous small exposures describe a broad anticline (Fig. 2.4) that was deeply eroded during the Triassic (Watts 1903; Bosworth 1912). This Triassic palaeotopography, with Neoproterozoic and Cambrian rocks emerging through a Triassic cover, is reflected in the patchy outcrops of the present day.

Volcano-sedimentary succession

The lowest known strata of the Charnwood Terrane are feldspar-phyric dacites of the Morley Lane Volcanic Formation, recorded from the Morley Quarry Borehole No. 1, which were interpreted as lava flows by Pharaoh & Evans (1987). These lavas are overlain by the fine-grained Ives Head Formation, which contains Ediacaran body fossils in the upper part (Boynton & Ford 1995; see discussion below). The volcaniclastic Ives Head Formation mostly comprises parallel-laminated fine-grained sandstones and mudstones, in repetitive upward-fining packages. This is taken to indicate deposition from turbidity currents, representing Bouma d–e divisions, probably representing deposition on the basin floor by distal turbidity currents. The upper member of the Ives Head Formation is the South Quarry Breccia Member, which contains pebble-sized clasts of volcanic material and reworked sedimentary clasts. This coarse-grained interval separates the Ives Head Formation from the sedimentologically similar Blackbrook Reservoir Formation and is not inferred to be a unit of great stratigraphic significance. The overlying Beacon Hill Formation comprises dacitic tuffs, containing abundant evidence for wave reworking and possible *Planolites* burrows, that pass laterally into the pyroclastic deposits of the Charnwood Lodge Volcanic Formation (Carney *et al.* 2000). The base of the Bradgate Formation is marked by the distinctive Sliding Stone Slump Breccia, which contains contorted clasts of mudstone in a volcaniclastic sandstone matrix. The transition into the overlying volcaniclastic Hallgate Member is a gradual upward-fining succession that contains some of the most important fossiliferous horizons in the area (described below). The bulk of the Bradgate Formation contains turbidites with normally graded beds and slump horizons. The uppermost unit of the Neoproterozoic part of the sedimentary succession is the Hanging Rocks Formation, which has an erosional base and

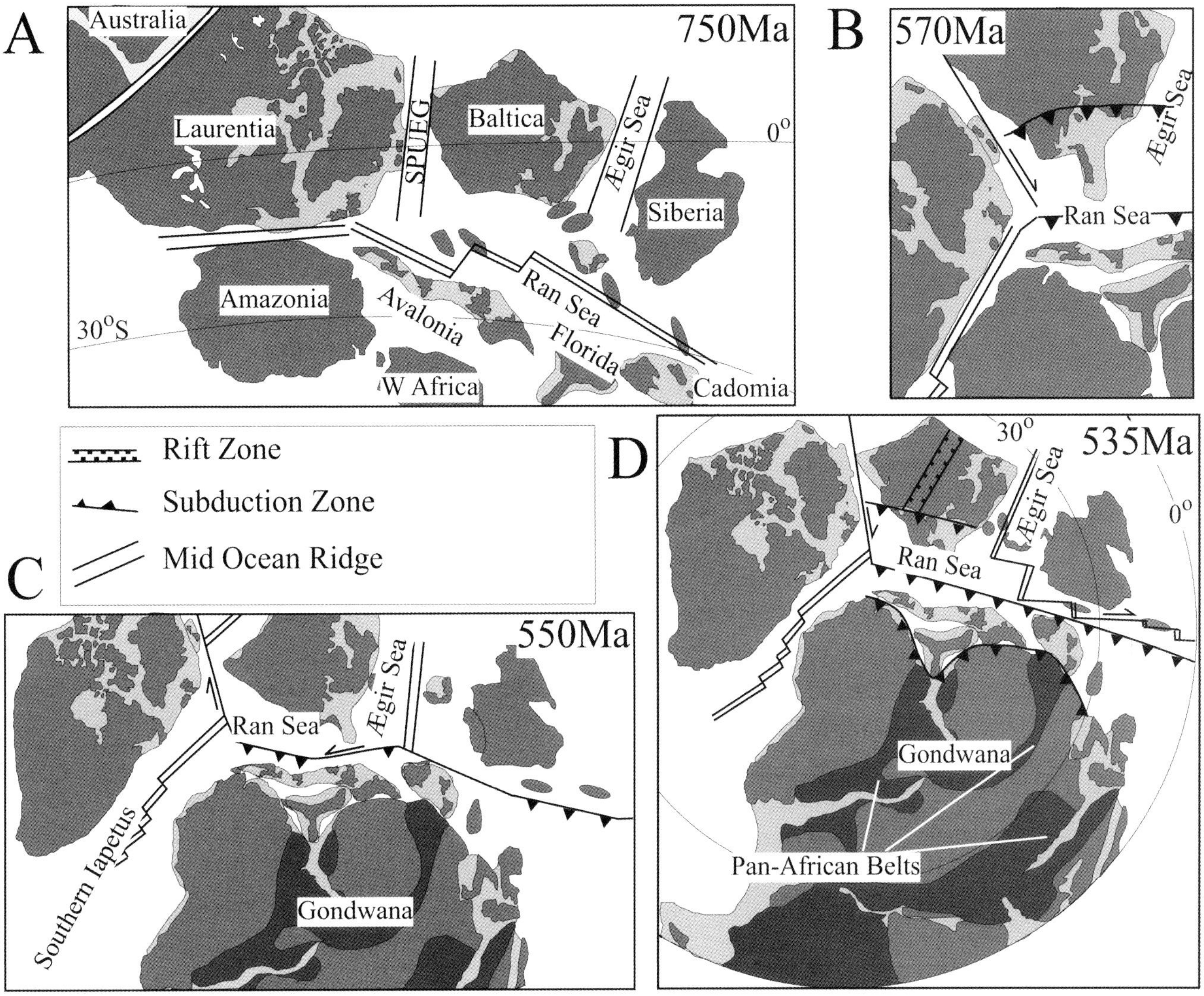

Fig. 2.2. Neoproterozoic palaeogeography of Avalonia and related areas showing the location of Avalonia relative to major tectonic boundaries from the Neoproterozoic 700 Ma to lower Cambrian 530 Ma (redrawn from Hartz & Torvik 2002).

fines upwards into purple pelites with 2–20 mm thick rippled sandstone horizons (McIlroy *et al.* 1998). The rest of the succession at Charnwood Forest has recently been reinterpreted as being lower Cambrian in age on palaeontological and provenance data (Bland & Goldring 1995; McIlroy *et al.* 1998), with the basal Cambrian unconformity identified at the base of the Stable Pit Member of the Brand Hills Formation, Brand Group (McIlroy *et al.* 1998).

Intrusive bodies
Subsequent to the deposition of the thick volcano-sedimentary succession, two suites of intrusive bodies were emplaced in the Charnwood Forest area. These are geochemically and texturally distinct from one another. The earlier North Charnwood Diorites are geochemically similar to the upper part of the Charnian volcanic succession, being of basic composition, whereas the later South Charnwood Diorites 'Markfieldite' are characterized by their granophyric texture and are geochemically similar to high-K calc-alkaline suites. This geochemical change has been interpreted by Pharaoh *et al.* (1987*b*) to reflect a thickening of the crust below the Charnwood Terrane during the latest Proterozoic.

The Neoproterozoic succession at Charnwood Forest has long been compared with that 20 km away at Nuneaton. The volcanic succession at Nuneaton is approximately 130 m thick and is lithologically and geochemically most similar to the Bradgate Formation (Carney & Pharaoh 1993; Carney *et al.* 2000). The sequence comprises dacitic volcanics including crystal-lithic tuffs deposited as subaqueous pyroclastic flows and crystal lapilli tuffs interbedded with tuffaceous siltstones (Bridge *et al.* 1998). Like Charnwood Forest, the late Neoproterozoic volcano-sedimentary succession at Nuneaton is cut by granophyric diorites. These diorites have yielded a U–Pb age of 603 ± 2 Ma (Tucker & Pharaoh 1991) and are linked to the South Charnwood Diorites and Withycombe Volcanics by their similar $\varepsilon_{Nd}(t)$ isotopic values that suggesting derivation from a similar melt at more or less the same time (Thorogood 1990; McIlroy *et al.* 1998). This information is not easily reconcilable with data presented by Compston *et al.* (2002), which demonstrate that the South Charnwood Diorites are younger than the Beacon Hill Formation (dated at 559.3 ± 2 Ma). This issue can only be resolved by direct dating of the South Charnwood Diorites, which would also provide an upper age limit for the Ediacara fauna of the Charnwood region.

Palaeontology
The Charnwood terrane has yielded fossils from numerous isolated outcrops within the Charnwood inlier, but most prolifically from the Hallgate Member of the Bradgate Formation.

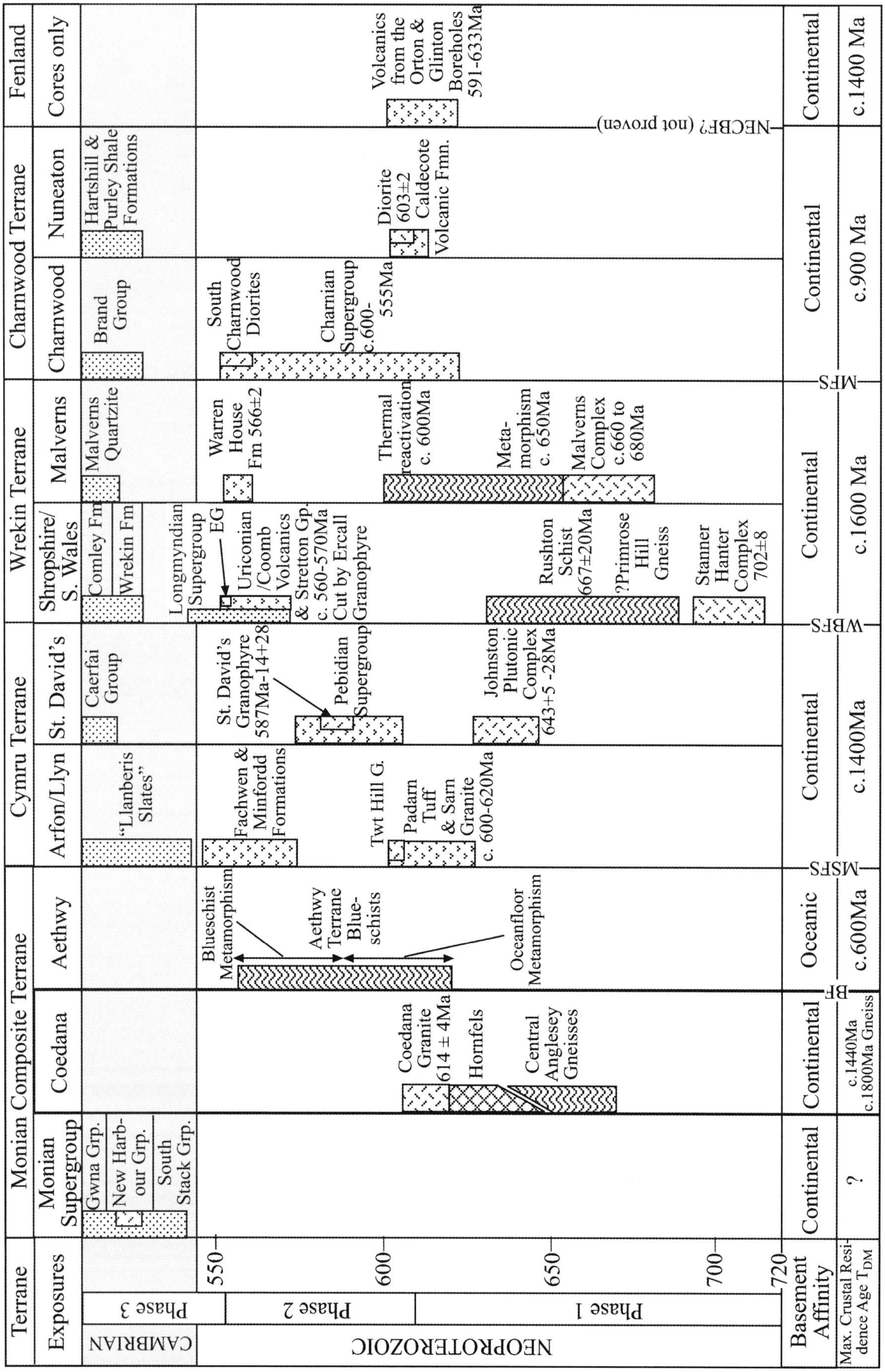

Fig. 2.3. Correlation diagram relating the main outcrops of Neoproterozoic rocks in England and Wales. Geochronological data is taken from sources mentioned in the text and averaged where appropriate. MSFS, Menai Strait Fault System; BF, Berw Fault; WBFS, Welsh Borderlands Fault System; MFS, Malverns Fault System; NECBF, North East Charnwood Boundary Fault (hypothetical).

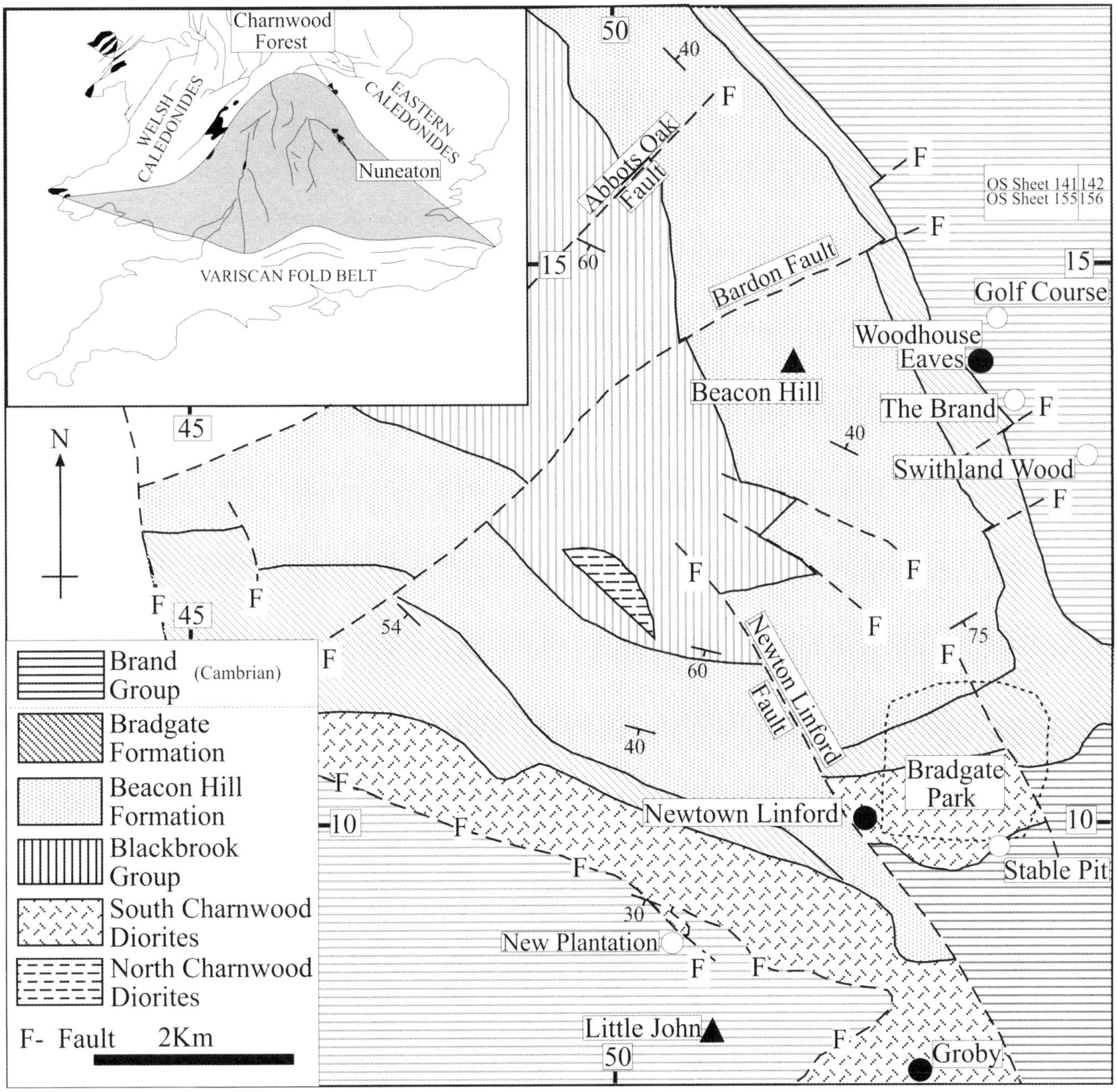

Fig. 2.4. Geological map showing the distribution of Proterozoic–Cambrian strata of the Charnwood Forest area, Leicestershire (based on Worssam & Old 1988).

Since the description of the frond-like Ediacaran fossil *Charnia masoni* (Fig. 2.5) by Ford (1958), several classic elements of the Ediacara fauna have been described including the giant *C. grandis* (in excess of 60 cm long) along with disc-like forms related to *Cyclomedusa* (Boynton & Ford 1995) and the simple tubular trace fossil *Planolites* isp. The rich diversity of the fauna invites comparison with the classic Ediacara faunas of Newfoundland, Canada and the White Sea region of Russia (e.g. Clapham *et al.* 2003; Fedonkin 1981). Uncertainties about the age of the Charnwood Forest Ediacara fauna highlighted by McIlroy *et al.* (1998) have recently been resolved by the U–Pb zircon dating of the underlying Beacon Hill Formation, which demonstrates that the Ediacara fauna in Charnwood Forest is younger than 559.3 ± 2 Ma (Compston *et al.* 2002).

The oldest palaeontological remains in the Charnwood Terrane are documented from the Ives Head Formation of the Blackbrook Group. The forms were originally reported by Boynton (1978) as dubiofossils, but reinterpreted as discs and fronds by Boynton & Ford (1995). Wrinkling of sedimentary laminations is a common feature of Proterozoic sediments (cf. McIlroy & Walter 1997; Hagadorn & Bottjer 1999) and, given that irregular margins are characteristic of the Ives Head occurrences, this material should be carefully re-evaluated. This is especially important since the material significantly predates the beds dated by Compston *et al.* (2002), and is probably in the order of around 580 Ma, which would make it one of the earliest Ediacara faunas in the world.

Wrekin Terrane – Welsh Borderland, Llangynog Inlier and Johnston Complex

Most outcrops of Proterozoic rocks in the Wrekin Terrane are associated with either the Malvern Lineament or the Welsh Borderlands Fault System, that bound the Wrekin Terrane to

Fig. 2.5. The frond-like *Charnia masoni* Ford from the Bradgate Formation, Charnwood Forest.

the east and NW, respectively. Exposures can be found in the Malvern Hills, the Welsh Borders and in the Llangynog Inlier of South Wales (see Fig. 2.1). In all cases, the oldest igneous rocks are calc-alkaline plutonics (700–600 Ma) associated with younger volcanic-dominated successions (570–560 Ma) that show a variable within-plate geochemical signature indicative of arc-rifting. The bedded volcanics of the Coomb Volcanic Formation and the Uriconian Group are both cut by intrusions. Small, typically poorly exposed and fault-bounded occurrences of metamorphic rock have been interpreted as representing 'basement' to the Wrekin Terrane. These include metasedimentary schists and quartzites south of the Malvern Hills and the Primrose Hill Gneiss, along the Welsh Borderland Fault System. An alternative origin for the latter, which contain mixed plutonic and volcanic protoliths, may be the shearing of Malvern Complex and Uriconian Group lithologies. The Rushton Schists, occurring a few kilometres to the NW of the Primrose Hill Gneiss, are greenschist facies quartz–feldspar–chlorite–epidote phyllitic schists that have been retrogressed from amphibolite facies by shearing. Whereas the meaning of a Rb–Sr whole-rock age of 667 ± 20 Ma (Thorpe *et al.* 1984) and a Rb–Sr biotite age of 536 ± 8 Ma remain unclear, the high $^{87}Sr/^{86}Sr$ initial ratio and maximum model Sm–Nd $T_{(DM)}$ ages of 1800 Ma indicate a significant protolith crustal history.

Intrusive bodies

The Wrekin Terrane contains some of the best exposures of basement rocks in England and Wales (Fig. 2.1); the oldest of which are to be found in the Stanner–Hanter Complex. The Stanner–Hanter Complex itself has been interpreted as small magma chamber that probably formed a feeder zone to an unpreserved volcanic complex (Carney *et al.* 2000). Four intrusive stages have been identified: cross-cutting relationships and chilled margins indicate early dolerite, followed by quartz-bearing and quartz-free gabbros, both of which were cross-cut by and locally hybridized with minor granite and granophyre. The presence of hybridization and local mingling textures between acid and basic components confirms their coeval nature. Patchett *et al.* (1980) obtained a whole-rock Rb–Sr date of 702 ± 8 Ma, from the granophyre at Stanner Hill that is interpreted as the magmatic age of the intrusion. Outcrops of the Stanner–Hanter complex are located on three hills immediately adjacent to the southerly extension of the terrane-bounding Church Stretton Fault System. Both fault movement and post-magmatic hydrothermal activity have resulted in pervasive, brittle fracturing and extensive retrogression of the primary magmatic mineral assemblage (Holgate & Hallowes 1941; Woodcock 1988). To date, this retrogression has precluded meaningful interpretation of the original geochemical signature of the complex.

The Johnston Complex is only exposed as an allochthonous slice within a Variscan thrust zone (Fig. 2.1). The pluton has a calc-alkaline geochemistry and is composed of diorite and granodiorite-granite with evidence of localized hybridization (Thorpe 1972). These features support comparison with the Malverns Complex and form the basis for its inclusion within the Wrekin Terrane (Carney *et al.* 2000).

On the opposite margin of the Wrekin Terrane, the Malvern Complex consists of a plutonic suite dominated by diorite and tonalite but also including more minor components of gabbro, ultramafic granite and hybridized lithologies (Dearnley 1990). Although Proterozoic metamorphism and shearing have modified the original igneous textures, the geochemistry still demonstrates a clear, calc-alkaline, magmatic arc signature (Thorpe 1974). A considerable amount of isotopic dating has been undertaken on this suite of rocks, but a 677 ± 2 Ma U–Pb zircon age is considered to most precisely date the age of emplacement (Tucker & Pharaoh 1991). The age of the subsequent deformation is bracketed by upper greenschist–lower amphibolite metamorphic mineral cooling ages of approximately 650 Ma and the presence of cross-cutting, undeformed, 610 Ma pegmatite veins (Strachan *et al.* 1996). Isotopic evidence indicates that significant crustal recycling was involved in the formation of the complex with Sm–Nd $T_{(DM)}$ model ages in the range 1043–1147 Ma (Murphy *et al.* 2000) and an inherited zircon component of $1598 \pm 30/32$ Ma (Tucker & Pharaoh 1991) provides some constraint on the antiquity of this crust.

The latest phase of intrusion is found in the Wrekin Range, where dolerite dykes and the later Ercall Granophyre cut the volcanic rocks of the Uriconian Group. The stratigraphic relationship of these intrusives were resolved by Bonney (1879) and dated at 560 ± 1 Ma by Tucker & Pharaoh (1991). These late Neoproterozoic intrusives are similar in age to the South Charnwood Diorites (Fig. 2.3).

Volcano-sedimentary successions

The early magmatism and metamorphism, represented in the Stanner–Hanter/Malvernian complexes, was followed by a sedimentary phase. This is largely volcaniclastic in the Malvern, Wrekin and Llangynog areas and dominantly siliciclastic in the Longmynd area. Much debate has surrounded the relationship between the Uriconian and Longmyndian sequences. The junction has classically been interpreted as an unconformity (e.g. Blake 1890), although more recent work has reinterpreted the contact as a fault and proposed that the units are coeval (Greig *et al.* 1968; Pauley 1990, 1991; Thorogood 1990). This interpretation is supported by U–Pb zircon data from Longmyndian Supergroup bentonites (Compston *et al.* 2002).

The Uriconian Group lies to the SE of, and as fault-bounded slices within, the Church Stretton Fault System (Fig. 2.6). The

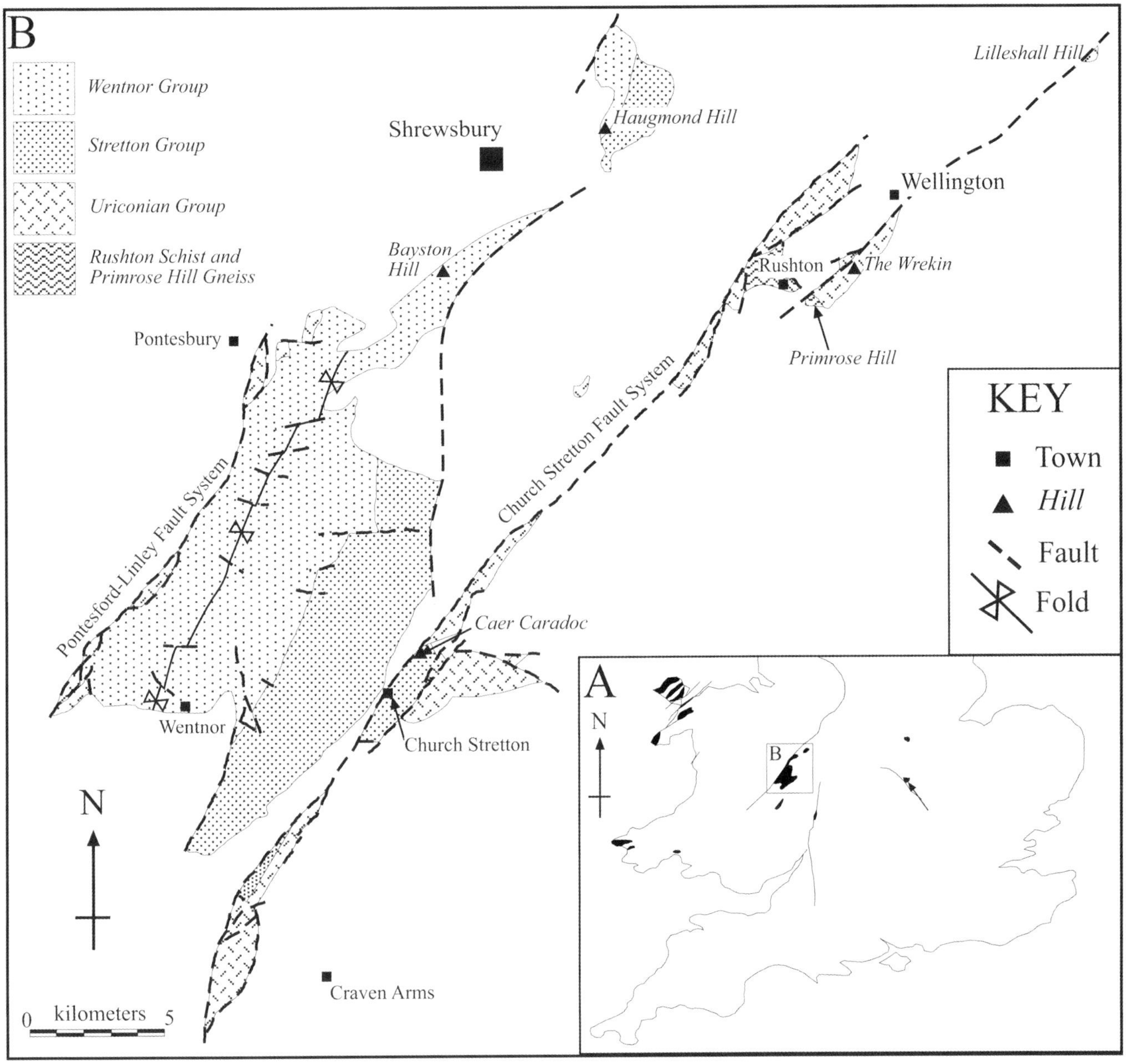

Fig. 2.6. Geological map showing the distribution of Proterozoic strata in the Shropshire area (redrawn from Pauley 1991).

Uriconian succession is a dominantly bimodal volcanic suite with basic and intermediate to acid components. These have a within-plate geochemical signature, reflecting only a small influence of subduction zone components (Pharaoh *et al.* 1987*b*). This has been interpreted to reflect marginal basin volcanicity behind the main volcanic arc (Pharaoh & Gibbons 1994). This could be more specifically interpreted as magmatism linked to transtension in the arc caused by oblique subduction.

U–Pb zircon data indicate that the Uriconian Group was produced by relatively rapid deposition. The age of volcanicity is pinned at around 566 ± 2 Ma with a minimum age of 560 ± 1 Ma provided by the cross cutting Ercall Granophyre (Tucker & Pharaoh 1991). The predominance of rhyolitic, pyroclastic material, including welded tuffs and coarse volcanic breccias, suggests proximity to a highly explosive volcanic centre (Carney *et al.* 2000). In contrast to the Charnwood Terrane, volcanic deposits in this sequence show textural evidence of deposition in subaerial environments.

The Coomb Volcanic Formation defines the most SW extension of the Wrekin Terrane, in the Llangynog Inlier, SW Wales, and is the most recently identified suite of Neoproterozoic rocks in southern Britain (Cope & Bevins 1993; Carney *et al.* 2000). Although exposed over a relatively small area, of just 3 km^2, it consists of over 1000 m of subaqueously deposited volcanic rocks. These are interpreted as having been formed on ephemeral, rapidly reworked, islands (Cope & Bevins 1993). The initial phase of volcanism records explosively generated ash-flow tuffs and more passive, rhyolitic flows or domes. This is followed by a bimodal suite of lavas, ash-flow tuffs, minor intrusions and volcaniclastic rocks. Some of the tuffs contain Ediacaran-type body fossils and trace fossils (Cope 1977, 1983). The geochemistry of this formation indicates that the lavas and tuffs are linked by fractional crystallization and show both within-plate and subduction-zone signatures (Bevins *et al.* 1995*a*). It is this distinct geochemistry, similar to that of the Uriconian volcanics, that identifies the Coomb Volcanic Formation as part of the Wrekin Terrane.

The late Neoproterozoic volcanic successions of the Warren House Formation in the Malvern Hills are in thrust contact with the Malvern Complex. The Warren House Formation includes altered and metasomatized pillow basalts, intermediate lavas and tuffs, altered rhyolites and both welded and non-welded acid tuffs (Carney *et al.* 2000). A U–Pb zircon age from the acid tuff dates the ages of eruption at 566 ± 2 Ma (Tucker & Pharaoh 1991). The geochemistry of the formation shows that all components are linked by fractional crystallization. In contrast with other volcanic suites described from the Wrekin Terrane, the geochemical signature is most similar to primitive volcanic arcs founded on oceanic or thin, immature, continental crust (Pharaoh *et al.* 1987*b*).

The last area of Neoproterozoic sedimentation to be considered is the Longmyndian Supergroup, which comprises a sedimentary succession around 6000 m in thickness, encompassing a range of sedimentary facies from basin-floor shales and distal turbidites through deltaic and fluviatile successions in a broad-scale regressive sequence (Pauley 1990; Fig. 2.7). The succession is subdivided into two groups, the basal Stretton Group and the Wentnor Group and lies between the Pontesford–Linley Fault System and the Church Stretton Fault System (Woodcock 1984*b*; Woodcock & Gibbons 1988; Fig. 2.6). It has been folded in response to strike-slip fault movement along these major structures, most probably during the late Neoproterozoic (Pauley 1990), to form a tight NNE–SSW-trending synform that plunges gently southwards. Minor volcanic rocks are associated with this dominantly siliciclastic succession, including intermediate felsic tuffs of the Batch Volcanics (Synalds Formation, Stretton Group) and several thin lapilli tuffs. Compston *et al.* (2002) obtained a U–Pb zircon age of 566.6 ± 2.9 Ma, from a thin bentonite at the base Stretton Group (Stretton Shale Formation) and 555.9 ± 3.5 from a lapilli tuff towards the top (Lightspout Formation) (see Fig. 2.7).

Palaeontology

The Wrekin Terrane contains abundant evidence of the Ediacara fauna in the Longmynd, Shropshire and also the Llangynog Inlier of South Wales. Macroscopic organic remains were first recorded from the Longmyndian by Salter (1856, 1857) (who erroneously considered the numerous circular structures to be the tops of U-tubes of *Arenicolites*) and described subsequently by Cobbold (1900), Watts (1925) and Bland (1984). While some authors have considered these structures to be inorganic (James 1952; Greig *et al.* 1968) re-evaluation and comparison with material from other sections worldwide confirm an organic origin (McIlroy *et al.* 2005). The taxa include the body fossil taxa *Beltanelliformis brunsae* and *B. minutae*, and the trace fossils *Intrites* and *Medusinites* (McIlroy *et al.* 2005; Fig. 2.8). The purported Ediacaran body fossil Arumberia is also recorded (see Bland 1984 for review), but this has been re-interpreted as a pseudo-fossil formed by current modification of microbially bound sediment surfaces (McIlroy & Walter 1997; McIlroy *et al.* 2005). The documentation of filamentous microfossils from the Arumberia-bearing Lightspout Formation documented by Peat (1984) may lend support to this argument.

The Ediacara fauna of the Llangynog inlier is likewise composed exclusively of discoidal forms in great abundance (Cope 1977, 1983), but lacks the typical fronds and rarer elements of the Ediacara fauna and still awaits full taxonomic description. The Llangynog section is noteable for containing the trace fossil *Palaeopascichnus delicatus* Palij, which is a component of the late Neoproterozoic *Harlaniella podolica* ichnozone and is not known from Cambrian or younger rocks.

Cymru Terrane – North Wales and St David's Peninsula

The Cymru Terrane comprises an extensive area, constrained to the NW by the Menai Strait Fault System and to the SE by Pontesford Lineament–Myddfai Steep Belt (Fig. 2.9). Proterozoic rocks of the Cymru Terrane in North Wales are found in ridges south and SW of Bangor (Arfon Group) and small exposures on the Llŷn (Sarn Complex and Twt Hill Granite). In South Wales, exposures of Neoproterozoic rocks are found around the coastline of St David's Peninsula (Pebidian Supergroup, St David's Granophyre) (Fig. 2.1). Presumed Proterozoic volcanics are also known from the Bryn-Teg Borehole, near Trawsfynydd, Gwynedd (Allen & Jackson 1978) and possibly as isolated slivers within the Menai Strait Fault System (Gibbons & Horák 1996). The contemporeneity of the Padarn Tuff of Gwynedd and the Sarn Complex was established by Horák *et al.* (1996), and the unconformity between Padarn Tuffs and Fachwen Formation has been constrained by dates of 604.7 ± 1.6 Ma from the Padarn Tuffs and 572.5 ± 1.2 Ma from the Fachwen Formation (Compston *et al.* 2002). Direct evidence for ancient basement to this region is entirely lacking, with the only inherited zircons dating from 617 to 638 Ma (Compston *et al.* 2002). Exposures of highly retrogressed granitoid gneiss and garnet amphibolite (Parwyd Gneiss) occur within the Llŷn Shear Zone, the SW extension of the Menai Strait Fault System (Fig. 2.1). These may be interpreted as either an exotic sliver within the Llŷn Shear Zone or a metamorphic slice sheared from the Monian Composite Terrane. Sm–Nd data for the granitoid gneisses show maximum T_{DM} model ages of 1350 Ma, similar to that of the Sarn Granite, thereby precluding their interpretation as ancient basement.

Intrusive bodies

The St David's Granophyre, is a small, high-level intrusion that cuts the Pebidian Supergroup volcanic sequence. Although the intrusion has been classified as an 'alaskite' or trondhjemite, on account of the low-K feldspar content, geochemical data show that this a product of alteration and its true character is a calc-alkaline granite, with a volcanic arc signature (Bloxham & Dirk 1988). As such, it has more affinity with the Arfon Group of North Wales than the nearby Coomb Volcanic Formation (Bevins *et al.* 1995*a*). A U–Pb age of 587 $+25/-14$ Ma, although confirming a Neoproterozoic age, has poor precision and an alternative age of 625 ± 25 Ma, produced by excluding metamict zircon analyses has also been generated (Patchett & Jocelyn 1979). The granophyre played a lead role in one of the most famous controversies in British geology, between Sir Archibald Geikie (the then director of the Geological Survey) and Henry Hicks. This centred around Geikie's interpretations of the Pebidian volcanics as part of the Cambrian sequence, in direct opposition to Hicks (1877, 1878) who considered the cross-cutting St David's Granophyre to be Precambrian in age. Green (1908) finally resolved this famous debate by means of a small excavation in which he clearly demonstrated an unconformity, marked by a basal conglomerate, between the granophyre and the Cambrian sediments. The details of this heated debate are well documented by Pearson & Nicolas (1992).

The Sarn Complex, although the largest plutonic unit within the Cymru Terrane, is very poorly exposed, being covered by thick glacial drift. It is sheared by the Llŷn Shear Zone, the SW extension of the Menai Straits Fault System, to the west and is overlain by Arenig sediments to the east. It is essentially a bimodal suite of gabbro–diorite, altered to a greenschist-facies mineralogy, biotite monzogranite (Sarn Granite) and heterogeneous granodiorite. The leucocratic and relatively homogeneous Sarn Granite outcrops over an area of approximately 6 km^2, whereas the gabbro and diorite occur as small, scattered, exposures. A Neoproterozoic intrusive age for this calc-alkaline suite has been confirmed by a U–Pb ziron age of 614 ± 2 Ma, from the dioritic component (Horák 1993), which also constrains the timing of deformation in the Llŷn Shear Zone.

Stratigraphy			Sedimentology	Palaeontology & Geochronology
Longmyndian Supergroup	Wentnor Group	Bridges Formation	*The basal contact of this formation is gradational but marked. The massive and cross bedded sandstones of the Bayston-Oakswood Formation become interbedded with heterolithic purple siltstones and thin ripple cross-laminated sandstones. Thick cross bedded sandstones are also recorded with sharp erosional bases and are interpreted as fluvial channels associated with mud-rich alluvial plain deposits.*	
Longmyndian Supergroup	Wentnor Group	Bayston-Oakswood Formation	*This formation comprises mainly fine to medium grained cross-stratified sandstones with mudstone rip-up clasts and sub-rounded lithic clasts. These are interbedded with cross-laminated finer grained sandstones in subtle upward-fining successions.* *The formation also contains locally developed conglomeratic members which are variably clast- to matrix-supported, containing subrounded lithic clasts and subangular clasts of sedimentary origin. These conglomerates are interpreted as braided fluvial deposits.* *No fossils are recorded from this unit.*	
Longmyndian Supergroup	Stretton Group	Portway Formation	*Red mudstones and siltstones interbedded with fine grained sandstone. Sandier and coarser than the underlying Synalds and Lightspout formationsbut similarly interpreted as alluvial plain and fluvial deposits. Possibly with marine incursions associated with fossil bearing horizons. The base is marked by a minor conglomerate (Huckster Conglomerate) deposited in a braided fluvial palaeoenvironment.*	*B. minutae* Arumberia
Longmyndian Supergroup	Stretton Group	Lightspout Formation	*Upward coarsening sucession of laminated mudstones and upward fining sandstones interpreted as alluvial plain, fluvial and ?deltaic deposits.*	*B. minutae* Arumberia 555.9 ± 3.5
Longmyndian Supergroup	Stretton Group	Synalds Formation	*Mudstones with inter-bedded laminated and cross stratified sandstones interpreted as alluvial plain, fluvial and ?deltaic deposits. Mudstones green at the base and red at the top with rare matgrounds, mudcracks and possibly marine incursions associated with fossiliferous horizons.*	*B. brunsae* *B. minutae* Arumberia "Elephant-skin texture"
Longmyndian Supergroup	Stretton Group	Burway Formation	*The base of the Burway Formation is taken at the base of the rhyolitic tuffs of the Buxton Rock Member. The Burway Formation shows a coarsening upward trend from thin to thick bedded turbidite lobe facies followed by shallow marine mudstones and deltaic sandstones. The Cardingmill Grit at the top is interpreted as a fluvial deposit.*	*B. brunsae* *B. minutae* *Intrites punctatus* *M.* aff. *asteroides* Arumberia
Longmyndian Supergroup	Stretton Group	Stretton Shale Formation	*The Stretton Shale is grey/green with normally graded siltstone beds and is interpreted as distal turbidite and basin floor facies. The Helmeth Grit marks the base of the formation and comprises medium-grained sandstone similar in composition to the underlying Ragleth Tuff Formation*	566 ± 2.9
Longmyndian Supergroup	Stretton Group	Ragleth Tuff Formation	*Coarse epiclastic sediments that may pass laterally and stratigraphically upward into the Helmeth Grit. Contacts are faulted and stratigraphic position tentative.*	

Fig. 2.7. Stratigraphic column showing the broadly progradational nature of the Longmyndian sedimentary succession (based on Pauley 1991, with geochronological data from Compston *et al.* 2002).

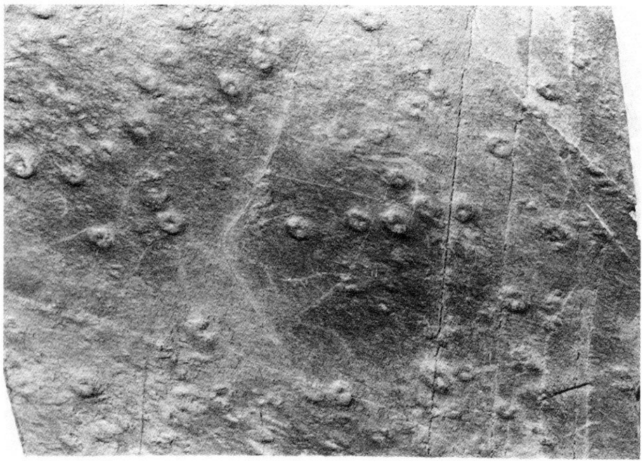

Fig. 2.8. The Neoproterozoic trace fossil *Intrites punctatus* from the Longmyndian Supergroup of the Wrekin Terrane preserved as impressions on the under-surface of a sandstone bed.

Volcano-sedimentary succession

Thick Neoproterozoic volcanic successions are present in both North and South Wales, in the form of the Arfon Group and Pebidian Supergroup, respectively. Although it is likely that these are broadly coeval they are geochemically distinct, the Arfon Group being dominantly acidic in composition and the Pebidian dominantly basic. Both are cut by minor granitic intrusions.

The Pebidian Supergroup comprises a thick volcanic pile of basic lava and tuff interbedded with acid pyroclastic and volcaniclastic deposits, cut by the St David's Granophyre and minor basic sheets. Evidence of both subaerial and subaqueous deposition reflects periodic emergence of the volcanic pile above sea level. Like many other Neoproterozoic igneous rocks in Wales, the primary mineralogy and textures of these rocks are modified by greenschist or subgreenschist metamorphism. Identification of this alteration has enabled lithologies identified previously as andesite to be re-identified as silicified basalt, essentially making this a bimodal suite (Bevins *et al.* 1995*a*).

Initial basaltic magmatism (of the lower part of the Pebidian Group) is represented by a sequence of columnar jointed

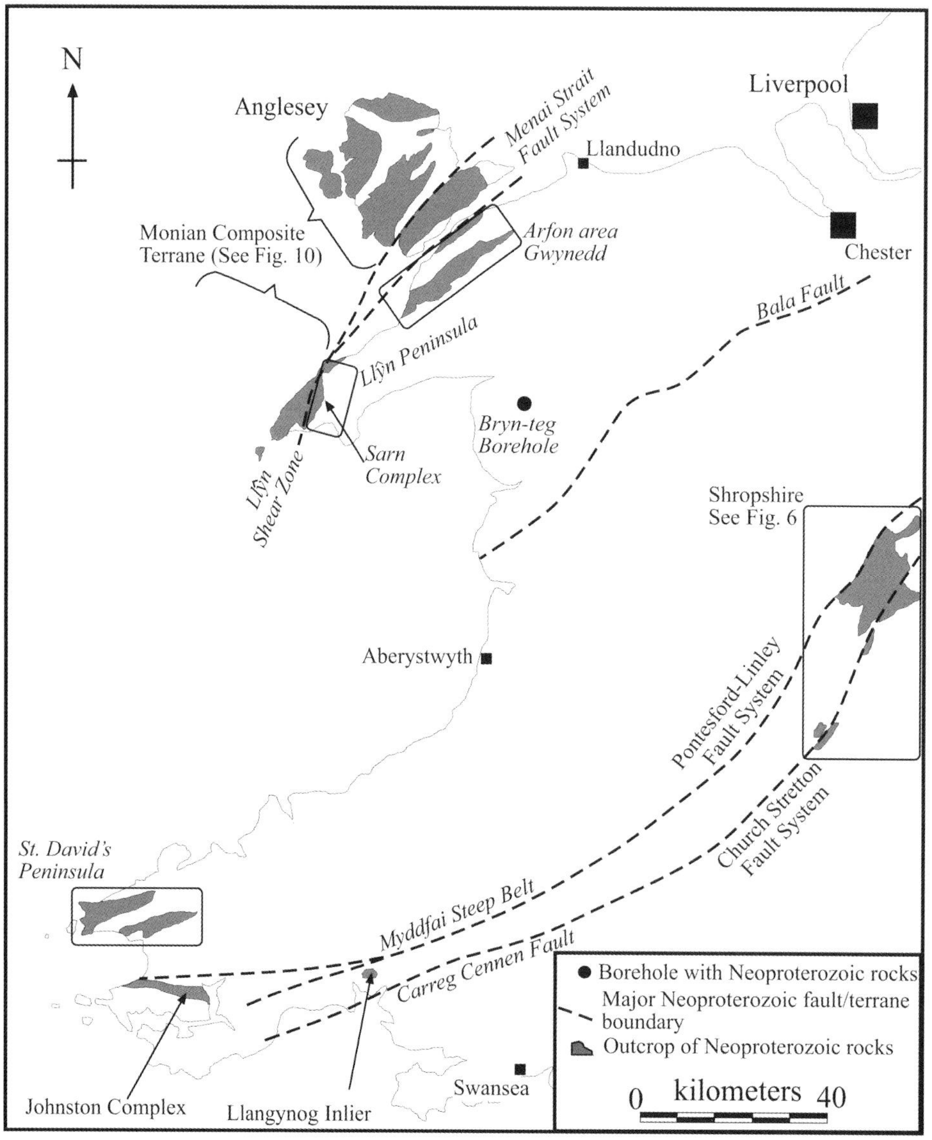

Fig. 2.9. Map showing the distribution of Proterozoic sections in the Cymru Terrane.

and autobrecciated lavas, scoriaceous tuffs and fine-grained tuffs with intercalations of fine-grained tuffaceous sediments. Interbedded with this are minor silicic tuffs that become volumetrically more important towards the top of the group and in the overlying Caerbwdy and Ramsey Sound Groups. The latter units, which may be laterally contiguous, contain silicic crystal and lithic tuffs, locally preserving welding and flattening fabrics characteristic of welded ash-flow tuffs (Carney *et al.* 2000). Towards the top of the Pebidian succession there is a return to basaltic magmatism. This is preserved in the Rhosson Group as autobrecciated lava, tuffs and scoria deposits and in the overlying Ogofgolchfa Group by finer grained, now chloritic, tuff, interbedded with purple slates. The Ogofgolchfa Group is cut by basic intrusive sheets and overlain unconformably by the lower Cambrian Caerfi Group (Green 1908).

The Arfon Group in North Wales comprises over 4000 m of sediments with evidence of syn-depositional faulting and explosive volcanicity. In the past, some or all of these rocks were considered to be Cambrian in age (Howells *et al.* 1985); however, detailed mapping and U–Pb data confirm that the lower units and the whole succession described here is Neoproterozoic in age. The only detailed, modern account of the Arfon Group is provided by Reedman *et al.* (1984). The lowest unit (Pardarn Tuff Formation) is exposed along two parallel ridges, one between Bangor and Caernarfon and a second to the SE around Llyn Padarn. The SE ridge comprises a thick, typically homogeneous, sequence of acid ash-flow tuff. This is locally welded and associated with thin air-fall tuffs and rhyolitic flows, but individual flows are not distinguishable within the sequence. Variations in thickness from east to west of 800–2000 m indicate rapid deposition in a half-graben (or graben) bounded to the NW by the Dinorwic Fault. This depression is estimated to have been 15 km wide and up to 60 km long. Although previously considered Cambrian in age (e.g. Howells *et al.* 1985) a U–Pb zircon date of 614 ± 2 Ma from the lower part of the succession (Tucker & Pharoah 1991) and 604.7 ± 1.6 Ma towards the top of it (Compston *et al.* 2002) confirms its Neoproterozoic age.

The strata unconformably overlying the Padarn Tuff show continued fault-controlled differential subsidence and uplift. Most notably this produces an angular discordance between the Minfordd and Bangor Formations and a marked difference in the thickness of strata either side of the Aber Dinlle Fault. Between the Aber Dinlle Fault and the Dinorwic Fault to the NW, the sandstone-dominated epiclastic and tuffaceous sequence of the Minfordd Formation records a broad fining-upward trend with evidence of instability marked by grain flow deposits. Sporadic volcanism throughout this sequence is evinced by the presence of thick welded and non-welded ash flow tuffs. A similar sequence is present in the thinner, and more restricted, Bangor Formation (Reedman *et al.* 1984; Howells *et al.* 1985). The basal conglomerate of the Minfordd Formation contains clasts of granite, vein quartz, quartzitic sandstone and quartzose schists, and was previously considered to represent the base of the Cambrian in North Wales. Decreasing fault influence is seen towards the top of this formation with deposition of the marine lower Cambrian, Llanberis Slates whose age is constrained by a combination of biostratigraphy, ichnology and event stratigraphy (Rushton 1974; Landing 1996; McIlroy *et al.* 1998; McIlroy & Logan 1999).

To the east of the Aber Dinlle Fault the Fachwen Formation is considered the lateral equivalent of the Minfordd and Bangor Formations. The succession of conglomerates, sandstones and interbedded acid tuff thickens rapidly towards the fault (40–500 m) and is interpreted as alluvial fan and related fluvial deposits. A similar suite of conglomeratic clasts (quartzite, basalt, mica schist, granite and jasper) to that in the Bangor Formation suggests increased uplift and exposure of sources other than the Padarn Tuff.

Although dominated by acid volcanism, the Arfon Group contains minor basic tuffs in the Minfordd and Fachwen formations that preserve delicate scoria textures indicative of a proximal source. In addition, thin basaltic intrusions cut the entire Arfon Group succession in the Llyn Padarn region (Reedman in Carney *et al.* 2000). Further intrusive activity is recorded by the Twt Hill Granite, which is a small fine-grained, leucocratic intrusion that cuts the Padarn Tuff at the western end of the Bangor to Caernarfon ridge.

An outlier of eutaxitic welded acid tuff on Anglesey (Bwlch Gwyn Tuff) occurs as a small sliver along the Berw Fault. Similarities with Arfon Group ash-flow tuff have led to it assuming a significance inversely proportional to its outcrop size as it was used as evidence to constrain a pre-Arfon Group age for docking of the Monian Composite Terrane (Tucker & Pharaoh 1991; Compston *et al.* 2002). However, this evidence is invalidated by recognition of the fault bounded nature of the Bwlch Gwyn outcrop, which can be interpreted as resulting from post-Berw Shear Zone and pre-Arenig faulting.

A further insight into the Neoproterozoic rocks of the Welsh Basin, was provided by the Bryn-teg borehole in the centre of the Harlech Dome, near Trawsfynedd (Fig. 2.9). The Neoproterozoic succession is in excess of 140 m thick and is in sharp erosional contact with the lower Cambrian Dolwen Formation. A detailed description of the Bryn-teg Volcanic Formation is provided by Allen & Jackson (1978) who describe a sequence of basic and intermediate lavas and tuffs. These were deposited along with volcaniclastic mudstone, siltstone and sandstone in a subaqueous environment and cut by thin basic dykes. Geochemical analyses suggest a calc-alkaline affinity for this suite. No direct correlation with the Arfon Group is possible, although the two suites are widely considered to be contemporaneous (Carney *et al.* 2000; McIlroy *et al.* 2001).

Palaeontology

To date, the Cymru Terrane has not yielded an Ediacara fauna despite numerous searches of the well-exposed Pebidian Volcanics that are lithologically similar to the Ediacara-bearing beds of Charnwood Forest, the Longmynd and the Llangynog Inlier (see above). This may be due to the slightly greater antiquity of the Pebidian (Fig. 2.3). The age of the Bryn-teg Volcanic Formation is constrained by the presence of the lower Cambrian foraminiferan *Platysolenites cooperi* in the overlying beds, which is otherwise only known from the Placentian of SE Newfoundland (Rushton in Allen & Jackson 1978; McIlroy *et al.* 2001).

The Monian Composite Terrane of Anglesey and the Llyfn Peninsula

The Monian Composite Terrane, originally termed the Mona Complex, was mapped and studied in great detail by Greenly (1919). It represents the largest area of Neoproterozoic rocks exposed in southern Britain and is also the most geologically complex (Fig. 2.10). This results from the many contacts between units that are either unexposed or are tectonic in nature, with brittle and ductile faults obfuscating original relationships. This difficulty is exacerbated by the lack of precise radiometric ages caused by thermal resetting and/or an absence of datable lithologies. Application of the suspect terrane concept has invoked terrane analysis to clarify some of this complexity, by identifying fault-bounded units with contrasting geological histories (Gibbons 1983*b*; Gibbons & Horák 1990). As a consequence, the Monian rocks can no longer be considered as forming one 'Mona Complex', within which the component rocks were produced without significant displacement between them (Gibbons & Horák 1990; Horák 2003). The key to further interpretation of this area of complex geology lies in firmly establishing the age of units and the timing of linkage between the terranes.

Of the three constituent Monian terranes, the igneous and low-grade metamorphic rocks of the Coedana Complex are of proven Neoproterozoic age (Moorbath & Shackleton 1966; Fitch *et al.* 1969; Tucker & Pharaoh 1991). Similarly, blueschist-facies rocks from the Aethwy Terrane (also know as the Eastern Schist Belt) show late Neoproterozoic $^{40}Ar/^{39}Ar$

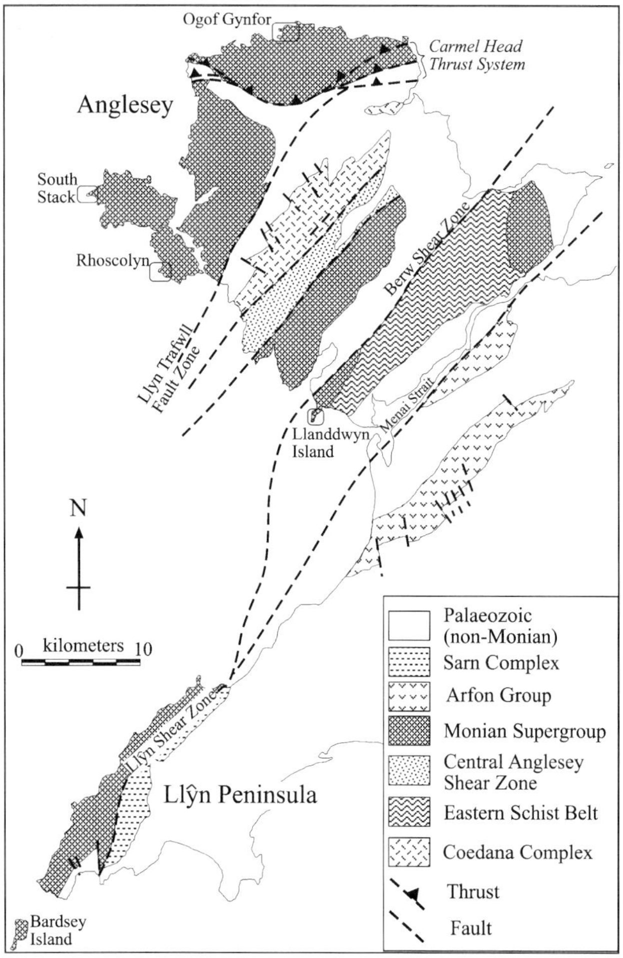

Fig. 2.10. Geological map showing the distribution of Proterozoic rocks (and the Monian Supergroup) of the Monian Composite Terrane in relation to major faults and the Proterozoic section of the northern part of the Cymru Terrane (redrawn from Horák & Gibbons in Carney *et al.* 2000).

cooling ages of 560–550 Ma (Dallmeyer & Gibbons 1987). In contrast, the age of the extensive Monian Supergroup metasediments, traditionally considered as Neoproterozoic, is now questioned by trace fossil evidence (see below).

The Coedana Terrane

This terrane is only exposed in Anglesey, lying to the south of the Llyn Trafwll Fault Zone. The Coedana Granite forms the central, elongated, and most geographically extensive, unit within the terrane. To the NW, the granite has a faulted or unexposed contact with the Coedana Complex Gneisses. It also can be seen to intrude, and is partially rimmed to the SE and NE by, low-grade hornfels. Fault-bounded exposures of gneiss (Gader and Nebo Inliers) and granite (Deri Inlier) within the Acadian Carmel Head Thrust System suggest that the Coedana Complex continues at depth below the Monian Supergroup as far as the northern coast of Anglesey. To the south, slivers of mylonite, sheared amphibolite and metasediment along the Berw Shear Zone, the Holland Arms Gneisses (Greenly 1919; Beckinsale & Thorpe 1979) have protoliths resembling the granite and gneisses of the Coedana Complex. If this comparison is correct, it provides evidence of linkage between the Coedana Complex and the blueschist-bearing Berw Shear Zone (Horák 1993).

The Coedana Granite has an evolved monzogranitic composition and a calc-alakaline signature. Four facies (porphyritic, non-porphyritic, muscovite-bearing and aplitic facies) can be identified, all of which have been affected by an initial phase of steep ductile deformation converted the intrusion into a proto-mylonite. This ductile deformation was affected by later brittle deformation and retrogression that (Horák 1993). A U–Pb zircon age of 614 ± 4 Ma (Tucker & Pharaoh 1991) dates the age of granite intrusion. In addition, both the inherited zircon component (1443 ± 34 Ma) and Sm–Nd data (1350–1430 Ma, Davies *et al.* 1985; Horák 1993) reflect the presence of older crust that was incorporated during magmatism. The age and geochemical signature of the Coedana Granite are consistent with generation in the Avalonian arc. The hornfels associated with the granite are typically fine-grained and grey-green with a mineralogy dominated by quartz, plagioclase, muscovite and chlorite. Such a mineralogical suite is generally taken to be indicative of fine-grained sediments metamorphosed under low-grade conditions.

The Coedana Complex gneisses (formerly the Central Anglesey Gneisses) comprise a heterogeneous sequence of high-grade metamorphosed migmatitic pelites, semipelites, minor (meta)sandstones and impure carbonates. These are associated with a series of minor intrusions, or lavas, of basaltic composition. Low-grade retrogression hampers precise calculation of the peak metamorphic temperature and pressure. However, the presence of sillimanite in migmatized biotite-garnet gneisses and plagioclase–hornblende geothermometry, suggest upper amphibolite conditions in the region of 4–5 kb at 650–750 °C. Attempts to constrain the age of metamorphism have, to date, failed. Published Palaeozoic ages appear to record retrogression or isotopic resetting (Horák 1993). Sm–Nd $T_{(DM)}$ ages from the amphibolite provide a maximum age of formation and metamorphism of the complex of 1000 Ma. These data correlate well with $T_{(DM)}$ values of 1100 Ma for amphibolite from the Rosslare Complex in SE Ireland, which is considered to be an extension of the Monian Composite Terrane. A similar correlation can been established for the paragneiss and granite gneiss in these two units; both of which have crustal residency ages in the range 1770–1830 Ma (Davies *et al.* 1985; Horák 1993). Although Tucker & Pharaoh (1991) consider the gneisses to be intruded by the Coedana Granite, this is not substantiated by field evidence, leaving the relationship between these two units of the Coedana Complex unproven. The possibility exists, therefore, that the gneisses are not linked to this terrane and, as such, may be a separate suspect terrane. Such an interpretation has been proposed for the Rosslare Complex (Murphy 1990).

Blueschist Belt – Aethwy Terrane

The smallest of the Monian Composite Terranes lies within the Menai Strait Fault System and forms a poorly exposed belt of blueschist facies rocks (Blueschist Belt) approximately 5 km wide, and is also known as the Eastern Schist Belt (Carney *et al.* 2000). Although preservation of Neoproterozoic blueschist is not common, this is not the sole significance of this unit. The high-pressure, low-temperature subduction-zone metamorphism, recorded by the blueschist mineralogy, contrasts strongly with the arc-related magmatic and sedimentary environments identified in all other Neoproterozoic terranes in southern Britain.

The blueschists consist of a metamorphosed sequence of basalts and sediments. The MORB geochemical signature of the metabasites evinces their origin as oceanic crustal rocks that underwent an early static, oceanic-floor metamorphism (Gibbons & Gyopari 1986; Thorpe 1972). This has been dated at 580–590 Ma and provides the minimum age for the generation of this oceanic crust (Dallmeyer & Gibbons 1987). Subsequent blueschist-facies deformation and metamorphism produced an intense foliation and isoclinal folding, interleaving the sediments – now converted to phengite-mica schists and lawsonite schist – with the oceanic crust, now forming crossite–glaucophane metabasites (Gibbons & Mann 1983; Gibbons & Gyopari 1986). $^{40}Ar/^{39}Ar$ dating of this mineralogy records cooling and uplift of the blueschists by oblique (NE), strike-slips faulting within in the Menai Strait Fault System at 550–560 Ma (Dallmeyer & Gibbons 1987).

The Monian Supergroup Terrane

The Monian Supergroup, comprises a thick (>7000 m) sequence of low-grade metasedimentary and meta-igneous rocks divided into three groups; in ascending order the South Stack, New Harbour and Gwna groups. It contrasts with the other Monian Terranes in being exposed both in Anglesey and on NW Llŷn (Fig. 2.10). The original description of these strata by Greenly (1919) was reassessed by Shackleton (1954, 1969), following the discovery of way-up structures that required inversion of the original sequence, and was further refined by Gibbons & Ball (1991) into the tripartite stratigraphy outlined above. As neither the base of the sequence nor its contact with other Monian rocks are exposed, its status within the Monian Composite Terrane remains unproven. The Monian Supergroup shows a polyphase deformation and metamorphic history with anchizone grade dominating, rising to lower greenschist-grade-facies conditions at the base of the sequence (Phillips 1991a). The structure has been studied most intensely on Holy Island (e.g. Shackleton 1969; Cosgrove 1980; Phillips 1991*a*), with most recent interpretations suggesting that early upright folds, such as the Rhoscolyn Anticline, were tightened and overturned to the SE by a subsequent deformational phase (Treagus *et al.* 2003). Comparable deformation is observed in the Gwna Group on Llŷn (Gibbons & McCarroll 1993). The precise age of deformation is unconstrained, although a pre-Caradoc age for at least the early deformation is provided by boulders of foliated New Harbour Group in the sediments of NE and NW Anglesey (Gibbons *et al.* 1994).

Although previously considered Proterozoic, there is growing palaeontological and stratigraphic evidence that the Monian Supergroup is probably of Palaeozoic in age. The minimum age for the Monian Supergroup must predate deformation along the Llŷn Shear Zone and the Central Anglesey Shear Zone, which both involve Gwna Group sediments (Gibbons 1983*a*; Mann 1986) (Fig. 2.10). At present this is only constrained by the pre-Arenig overstep sequence (Matley 1899, 1900; Greenly 1919; Bates 1974). New U–Pb detrital zircon data provide a maximum age of deposition for the base of the Monian Supergroup sequence (South Stack Formation) of 522 ± 6 Ma (Collins & Buchan 2004), thus confirming the Palaeozoic age of this suite of rocks.

The oldest part of the sedimentary succession, the South Stack Group, is best exposed on Holy Island, where a lower mud-rich unit (South Stack Formation) is overlain by progressively more sand-rich facies (Holyhead and Rhoscolyn formations, including the Holyhead Quartzite). The overlying New Harbour Group shows a tectonized sedimentary contact with the South Stack Group (Greenly 1919; Phillips 1991*b*) and lateral facies variation between Holy Island and northern Anglesey. In the latter area, the basal pelites (Bodelwyn Formation) grade upwards into massive, metamorphosed volcaniclastic sandstones (Lynas Formation and Skerries Formation). The Skerries Formation lithologies (Church Bay Tuffs and Skerries Grits), considered a separate unit by earlier workers (e.g. Greenly 1919; Shackleton 1954, 1969), were incorporated into the New Harbour Group by Gibbons & Ball (1991). This revision is supported by petrographic data (Phillips 1991*b*). On Holy Island, the New Harbour Group is of similar composition to the Bodelwyn Formation, being dominated by chlorite–mica schist. In addition, it contains disrupted horizons of metabasaltic lavas and tuffs at the top and base of the group and an intervening zone of ultramafic and metagabbroic rocks. The coarser volcaniclastic component seen in the Lynas and Skerries Formations is not represented in this area, and is interpreted as having wedged out southwards into pelitic facies (Phillips 1991*b*).

Considerable debate has centred on whether these basic and ultrabasic rocks are primary intrusions, thrust slices or were emplaced by gravity slides (for full discussion see Phillips 1989). The association of metabasic and ultrabasic rocks within the New Harbour Group has led to their interpretation as constituents of an ancient ophiolite (e.g. Wood 1974). However, uncertainty regarding the mode of emplacement of the intrusive rocks and the contrasting affinities of metabasalts (arc) and serpentinites/metagabbros (mid-ocean-ridge) in the New Harbour Group means that this issue remains unresolved (Thorpe 1972; Thorpe *et al.* 1984; Phillips 1989).

The South Stack and New Harbour Groups represent sedimentation in a NE–SW-trending basin (Wood 1974; Phillips 1991*b*). Sediments in the South Stack Group have a continental provenance and were derived from the SE. Flute marks, tool marks and water-escape structures in the lowest (South Stack Formation) have been taken to indicate deposition in a sand-rich turbidite fan system (Phillips 1991*b*). However, this evidence needs to be reconciled with the presence of: (a) rare vertical burrows of *Skolithos* isp., which is unknown in Proterozoic and lowermost Palaeozoic basin floor turbidites; and (b) the presence of hummocky and swaley cross bedding in the overlying Holyhead Formation. These features are all indicative of a shallow-water environment of deposition for the South Stack and Holyhead Formations. A more parsimonious interpretation of the graded beds with sole marks might be shelfal tempestites.

Unequivocal turbidite sedimentation is, however, recorded by the uppermost formation of the South Stack Group and the whole of the New Harbour Group. The New Harbour Group turbidites are derived from the north or northwest and contain immature volcanic arc detritus reflecting high rates of sedimentation during the unroofing and dissection of an arc system (Phillips 1991*b*).

The coherent stratigraphy of the New Harbour Group is overlain by the Gwna Group, a thick (>3000 m), chaotic unit with a highly disrupted stratigraphy for which Greenly (1919) introduced the term mélange to the geological literature. The mélange includes clasts on all scales within a muddy or silty matrix, with megablocks up to several kilometres in length (Greenly 1919; Gibbons & McCarroll 1993). Some of the most spectacular exposures of mélange are megaclasts of quartzite, described in detail by Gibbons & McCarroll (1993), exposed along the southwestern tip of the Llŷn coast (Plate1). Lithologies can be grouped broadly into those with an oceanic or deeper water affinity (basalt pillows and tubes, micritic limestone, jaspery chert) and those from a shallow water or continental setting (orthoquartzite microbial limestone, feldspathic and tuffaceous sandstone). Plutonic rocks are uncommon, represented by just two exposures of metagabbro on Llŷn and thin sheets of granite on Bardsey Island and in northern Anglesey (Gibbons & McCarroll 1993; Horák *et al.* 1996). Neither the top nor the base of the mélange is exposed in Llŷn, but in northern Anglesey a disrupted zone at the New Harbour Group–Gwna Group junction is considered to have formed by tectonism of semi-lithified New Harbour Group sediments during generation of the mélange (Gibbons & Ball 1991). The Gwna Group represents the remnants of a regional-scale deposit that Shackleton (1954) interpreted to have originated by large-scale, olisthostromal, slumping. This was evinced by the presence of a coherent, undisrupted, stratigraphy overlying the mélange (Fydlyn Beds and Gwyddel Beds). However, re-mapping has shown these beds to be megaclasts within the melange and consequently casts doubt on this interpretation (Gibbons & McCarroll 1993). Gibbons (1983*b*) considered that, although a subduction-generated accretionary prism model was possible for the mélange, a collisional event initiating olistostromal formation could better explain the mixture of oceanic and continental lithologies.

Palaeontology

Palaeontological studies of the sedimentary Monian Supergroup have focused on description of stromatolites in the

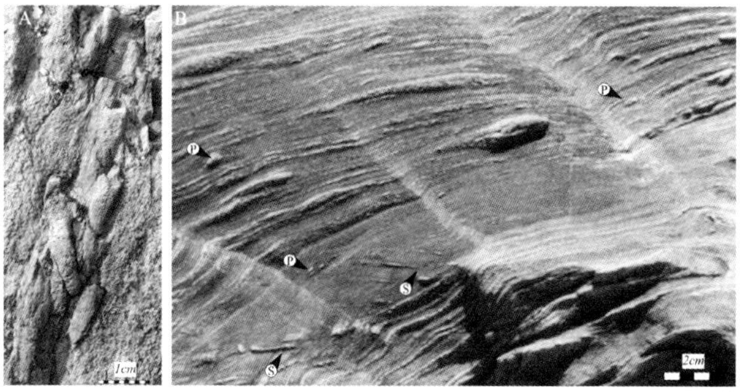

Fig. 2.11. Palaeozoic-type trace fossils from the Monian Supergroup: (**a**) horizontal Teptichnus pedum from the Holyhead Formation; (**b**) vertical burrows of Skolithos (S) and cross section of horizontal Planolites (P) from the South Stack Formation.

Gwna Group mélange and the search for microfossils. The enormous limestone blocks within the mélange contain a simple unbranched stromatolite attributed to *Conophyton* by Wood & Nicholls (1973). A stromatolite with a more complex irregular branching pattern was noted by Muir *et al.* (1979), placed within the genus *Georginia*, but not figured. Both of these types of stromatolite have representatives in the latest Neoproterozoic and lower Cambrian and are thus of little utility in determining if the Monian Supergroup is Proterozoic or Cambrian in age. In contrast however, the bedded jaspers of the Gwna Group on Llanddwyn Island preserve a contended microflora which purportedly includes microbial filaments and a poorly preserved acritarch assemblage of the disphaeromorph acritarch *Granomarginata squamacea* (lower Cambrian first occurrence datum) and cf. *Archaeosacculina* sp. (Vendian–lower Cambrian), along with the tentatively identified ?*Micrhystridium lanatum* (lower Cambrian first occurrence datum), ?*Deunffia flagellata* (lower Cambrian) and unidentified sphaeroidal microfossils (Muir *et al.* 1979; although questioned by Peat 1984). The poor quality of this palynological material and the lack of associated organic matter suggest that until this material is carefully reassessed this material does not comprise compelling biostratigraphic data.

Compelling evidence that the Monian Supergroup is Cambrian in age can be found in the ichnology of the South Stack Group. Vertical tubular forms of *Skolithos* isp. have been described by Greenly (1919) and Barber & Max (1979) from the heterolithic South Stack Formation (note that the trace fossil is not spelt 'Skolithus' as in many papers on the Monian!; Fig. 2.11b). The suggestions of Muir *et al.* (1979), that the purported trace fossils are not tubular does not stand close scrutiny of the material, and the absence of branching and backfill is not a reasonable argument because many trace fossils (*Skolithos* included) neither branch nor are backfilled. Muir *et al.* (1979) prefer to interpret the tubes as dewatering structures; however, the small size (5–10 mm in diameter) and the high density (comprising up to 80% of the sediment fabric locally) is not reconcilable with their interpretation. Ichnological study of the upper part of the Holyhead Formation demonstrates an abundance of horizontal tubes in the quartzites, these are unlined and unbranched, and are best described as *Planolites* isp. (Fig. 2.11b). In addition, a heterolithic portion of the succession revealed a large serially branching form with alternate branches similar to the form originally called *Phycodes pedum* (Fig. 2.11a), although *P. pedum*-like forms are currently under taxonomic review. The importance of this observation is that it allows direct comparison with the Global Stratotype Section for the Proterozoic–Cambrian boundary in SE Newfoundland. The branching behaviour of the trace figured herein is a component of the *Phycodes pedum* ichnozone and is therefore by definition a Phanerozoic trace fossil (Brasier *et al.* 1994). In addition, the size of trace fossils and the intensity of bioturbation of such coarse-grained high-energy facies is a normal feature of Palaeozoic bioturbation but unknown in the Proterozoic (McIlroy & Logan 1999). A minimum age for the Monian Supergroup is provided by an unconformity that is overlain by rocks of Arenig age at several sites (Matley 1899, 1900; Greenly 1919; Shackleton 1969; Bates 1974).

The Monian Supergroup can thus be seen to contain a succession of Cambrian deposits that predate the widespread Ordovician (Arenig) unconformity. It therefore appears highly likely that the Monian Supergroup is younger than both the Coedana Complex Terrane and blueschist metamorphism of the Aethwy Terrane. Since the contact relationship with these terranes is unproven, the possibility also exists that the Monian Supergroup is exotic with respect to them both. The Monian Supergroup, as with the blueschists, may have been generated some distance from its present location prior to being tectonically juxtaposed against the other units of the Monian Composite Terrane. To verify this model, clarification of the timing of deformation of Monian Supergroup Sediments within the Aethwy Terrane and Central Anglesey Shear Zone is required.

Summary

The Neoproterozoic strata of England and Wales constitute a complex mosaic of fault-bounded terranes that formed the eastern part of Avalonia between 700 and approximately 530 Ma. A tectonostratigraphic approach has facilitated identification of two distinct suites of rocks separated by the Menai Strait Fault System, which is recognized as a major terrane boundary (Fig. 2.1). The Neoproterozoic terranes SE of the Menai Strait Fault System, although characterized by differing igneous geochemical signatures, can all be related to subduction-zone activity as is seen widely across the Avalonian arc (Fig. 2.3). The main phase of Avalonian arc magmatism lasted from around 620 to 585 Ma. In Maritime Canada (Western Avalonia) basin development at this time is linked to strike-slip faulting and a similar interpretation of fault-bounded, arc basins is applicable to the sedimentary sequences of southern Britain. From 590 Ma onwards a diachronous change from arc magmatism to extensional magmatism occurred, witnessed initially in New England (Nance & Murphy 1996) and represented in southern Britain by the within-plate signatures of the 550–560 Ma volcanic suites of the Wrekin Terrane. Although major faults bounding the terranes to the SE of the Menai Strait Fault System show evidence of Neoproterozoic activity (e.g. Malvern Lineament, Thorpe 1987; Welsh Borderland Fault System, Woodcock & Gibbons 1988), proof of significant lateral displacement has not been demonstrated. In contrast to the terranes from within the magmatic arc, the Monian Composite Terrane preserves more outboard elements of the subduction system (blueschist and mélange) than are seen elsewhere in the Avalonian Superterrane. Part of the complexity seen in these units north west of the Menai Strait Fault lies in the modification of the original structural relationship between constituent terranes by later (Palaeozoic) tectonics. The blueschist belt represents the only demonstrably exotic component within the Avalonian system of Southern Britain, but the width of the sinistral shear zone between the Monian Composite Terranes and the remaining terranes of southern Britain also implies a displaced status for the Coedana Granite. This shows many similarities with granitoid magmatism elsewhere within the Avalonian arc, but must have originated in the arc to the northeast of its current location. An increase in the obliquity of subduction caused cessation of subduction-dominated magmatism, as described above, and induced transcurrent faulting. Slivers of arc, such as that represented by the Coedana Granite, were shaved off during SW-directed lateral displacement and were juxtaposed

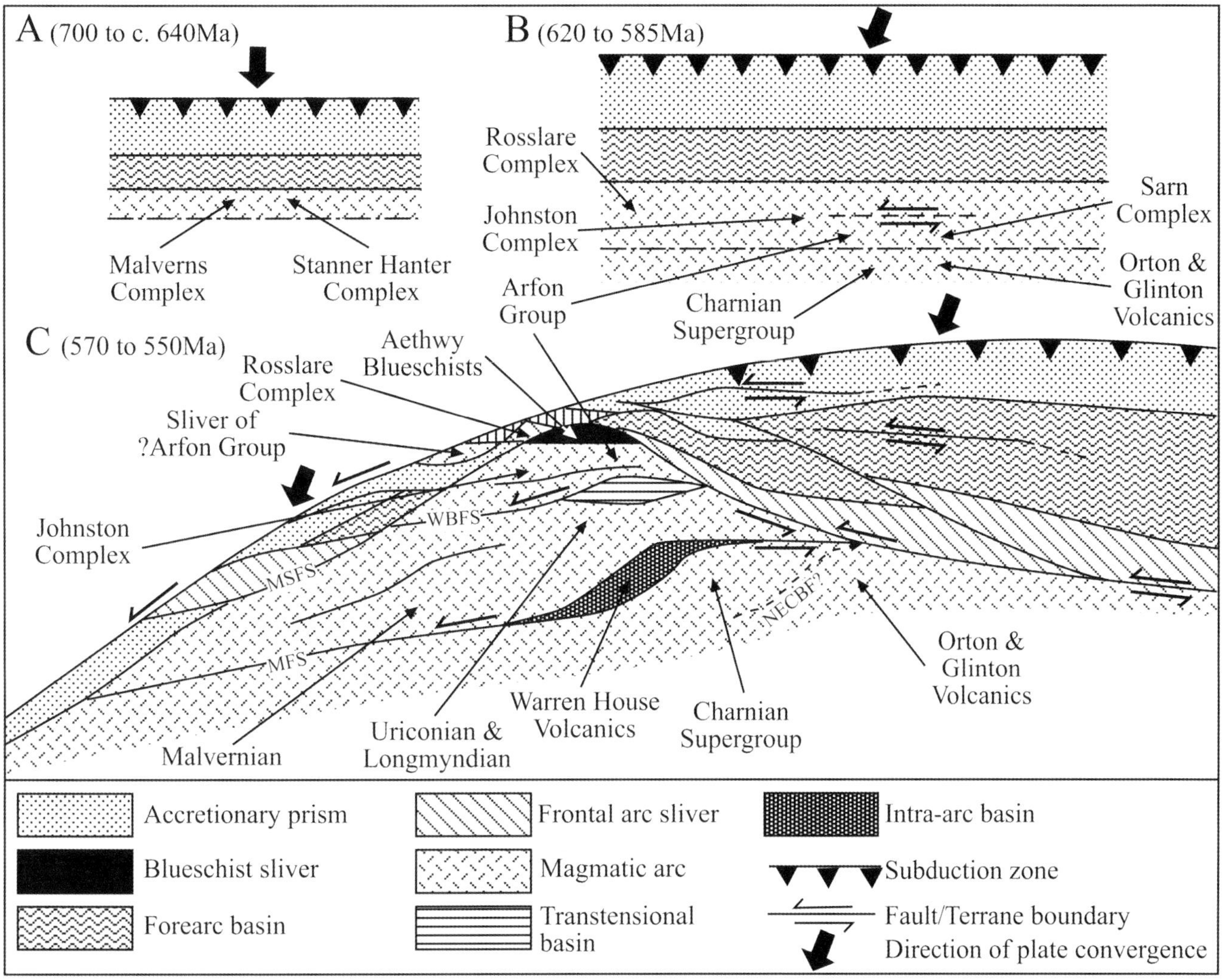

Fig. 2.12. Model showing the Proterozoic evolution of subduction in the Eastern Avalon region (modified from Gibbons & Hórak 1996).

against other outboard components of the subduction complex. The model outlined in Fig. 2.12, while by no means comprehensive, gives a reasonable summary of the inter-relationships between the numerous elements of the Neoproterozoic of England and Wales. An additional component of this model is the SW extension of the Coedana Complex, the Rosslare Complex of SE Ireland, a unit previously identified as having a suspect status (Tietzsch-Tyler & Phillips 1989). Correlation between part of the assumed sedimentary cover to the Rosslare Complex, the Cullenstown Formation, and the lower part of the Monian Supergroup has been made on a lithostratigraphic basis (Tietzsch-Tyler & Phillips 1989) and, if valid, suggests a Palaeozoic age; a conclusion that is supported independently by palaeontological evidence. Such an interpretation clearly has implications for the timing of amalgamation and docking of the Monian Composite Terrane.

3 Cambrian and Ordovician: the early Palaeozoic tectonostratigraphic evolution of the Welsh Basin, Midland and Monian Terranes of Eastern Avalonia

P. J. BRENCHLEY, A. W. A. RUSHTON, M. HOWELLS & R. CAVE

The Early Palaeozoic history of England and Wales was substantially influenced by the separation of Avalonia from Gondwana and its subsequent migration towards Laurentia. At the start of the Early Palaeozoic, the vast palaeocontinent of Gondwana straddled the South Pole and extended northwards into low latitudes. On the margin that hosted North Africa and North and South America there were areas of crust that were later to become detached terranes. The largest of these was Avalonia, the remnants of which now extend from NE USA, through the Atlantic Provinces of Canada and through England and Wales to Belgium and North Germany (Cocks 2000). Elsewhere, at lower latitudes on the Gondwana margin, there were crustal segments that were later to become the terranes of Armorica (Britanny, Normandy and the Massif Central regions of France), Perunica (much of central Europe, but mainly preserved in the Bohemian part of the Czech Republic) and Iberia (Spain). In addition there are some smaller continental fragments whose history is difficult to establish.

The area of England and Wales lay within Eastern Avalonia, which consisted of an initial crustal fragment that separated from Gondwana (Fig. 3.1a) and then accreted smaller terranes as it moved towards Laurentia. The core Avalon Terrane was probably assembled by accretion of crustal fragments on the Gondwana margin in the Late Precambrian or early Cambrian. At about the same time, this terrane accreted both the basement of the Welsh Basin (Woodcock & Gibbons 1988) and an amalgamation of subterranes forming the Monian Terrane. The latter joined with the Welsh Basin Terrane along the line of the Menai Strait Fault System (Fig. 3.2) (Woodcock & Gibbons 1988). The juxtaposition of the Monian and Leinster–Lakesman Terranes may have been established as early as the beginning of the Palaeozoic and, at latest, by the Caradoc (Woodcock 2000*a*). Throughout Cambrian times Eastern Avalonia was attached to Gondwana, but the development of the Welsh Basin and the enhanced subsidence of the Midland Platform suggest that the region was already being extended. Following the onset of arc volcanism in the Tremadoc, Avalonia began to migrate outwards from Gondwana, opening up the Rheic Ocean behind it (Fig. 3.1a, b). By Caradoc time small but distinct differences had developed between the benthic faunas on opposite sides of this ocean, though some genera were virtually cosmopolitan, indicating some faunal communication between terranes and relatively uniform global temperatures (Cocks & Torsvik 2002). The northward migration of Avalonia and the narrowing of the Iapetus Ocean continued as a result of marginal subduction, which is reflected in the volcanism in the Leinster–Lakesman and Welsh Basins until the Caradoc. By the mid Caradoc Avalonia was approaching Baltica and the Tornquist Ocean, which had separated the two terranes, was closing (Fig. 3.1c) allowing faunal interchange between Avalonia and Baltica. At the end of the Ordovician the two terranes amalgamated and then moved obliquely towards Laurentia. Subsequent collision with Laurentia ranged from mid-Silurian to mid-Devonian times, initially causing the Caledonian and later the Acadian orogenic events (Fig. 1d).

The Welsh Basin, Midland and Monian Terranes

Early Palaeozoic lineaments, basins and platforms

In Eastern Avalonia the junctions between terranes are inferred to be major fractures, whose later normal and strike-slip movements largely determined the margins of basins and platforms (Fig. 3.2) (Woodcock 2000*d*). The centre of England is occupied by the Midland Platform, a relatively stable area throughout the Lower Palaeozoic, except during the Tremadoc when several narrow basins developed. On the eastern side of the platform was the Anglian Basin, which is now a deformed region, including some Palaeozoic rocks, known very incompletely

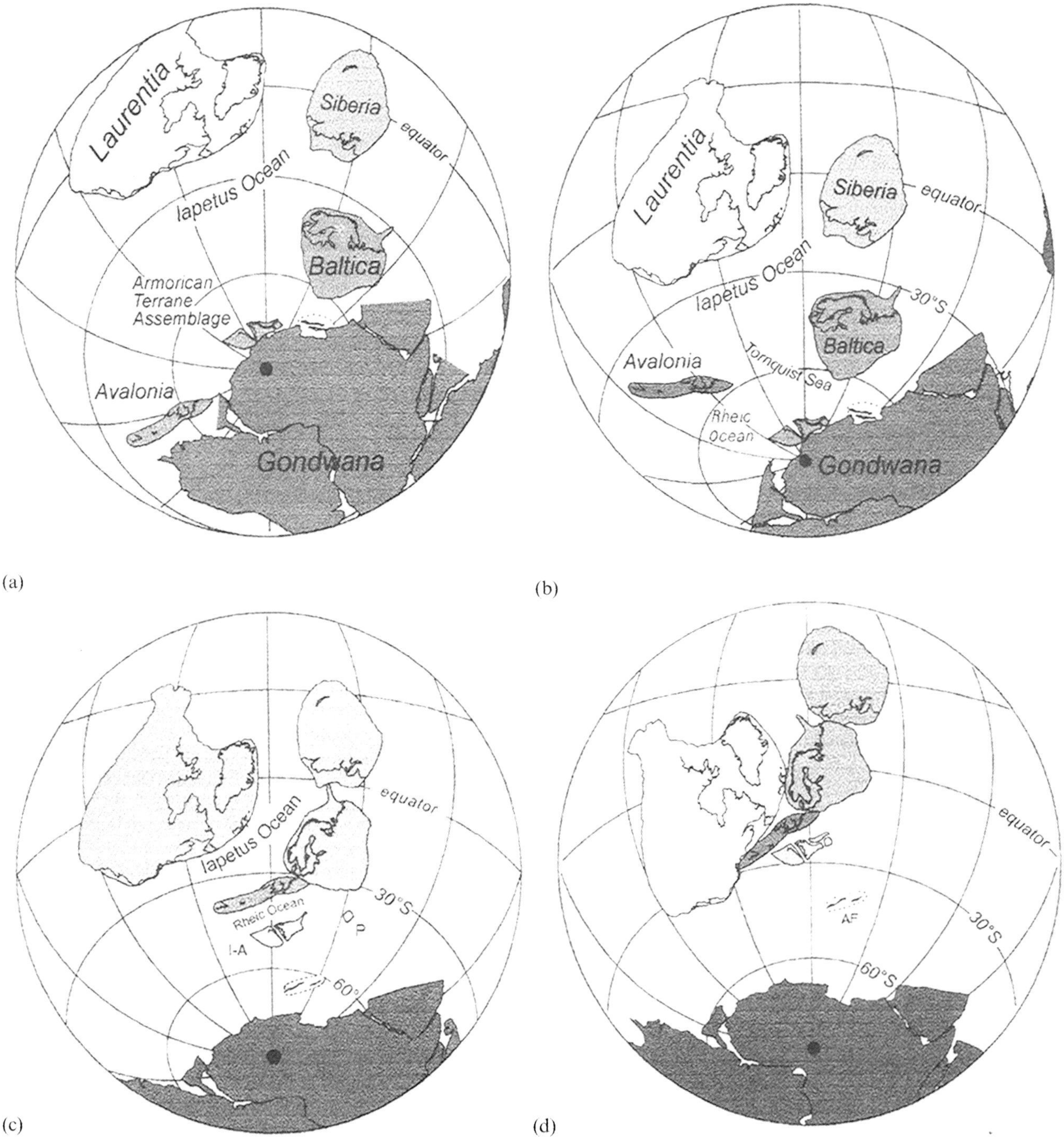

Fig. 3.1. The changing position of terranes in the Lower Palaeozoic: (**a**) Arenig, showing Avalonia having separated from Gondwana in its later Gondwanan position near South America: (**b**) Llanvirn, with Avalonia drifting north across the Iapetus Ocean: (**c**) late Ashgill, with Avalonia colliding with Baltica: (**d**) Přídolí, with the amalgamation of Laurentia, Baltica, Avalonia and Laurussia largely completed. From Cocks (2000).

from boreholes at Wyboston, Great Paxton, Huntingdon and further north at Eyam in Derbyshire (Rushton in Fortey *et al.* 2000). The NW side of the platform is bounded by the Welsh Borderland Fault Complex, which forms the SE margin to the Welsh Basin (Fig. 3.2). The opposite side of the Welsh Basin was bounded mainly by the Menai Strait Fault System, beyond which was the Monian Terrane (the Irish Sea Horst Complex of some authors), a complex assemblage of gneisses, blueschists, granites and segments of the Monian Supergroup. The Wexford Boundary Lineament, between the Cahore–Cullenstown and Wexford–Wicklow terranes, formed the contact between the northwest margin of the 'Irish Sea Horst Complex' and the Leinster–Lakesman Basin (Fig. 3.2). Strike-slip movements may possibly have later displaced the Leinster–Lakesman Basin relative to the Welsh Basin. The NW Ordovician margin to the Leinster–Lakesman Basin is not preserved, so it is not possible to determine whether the basin was bounded there by uplifted crust or whether the basin lay on thinned continental crust adjacent to the Iapetus Ocean. However, by mid to late Silurian times the Iapetus Ocean was closing and the Leinster–Lakesman Basin was progressively influenced by southward-directed thrust sheets from the Laurentian margin.

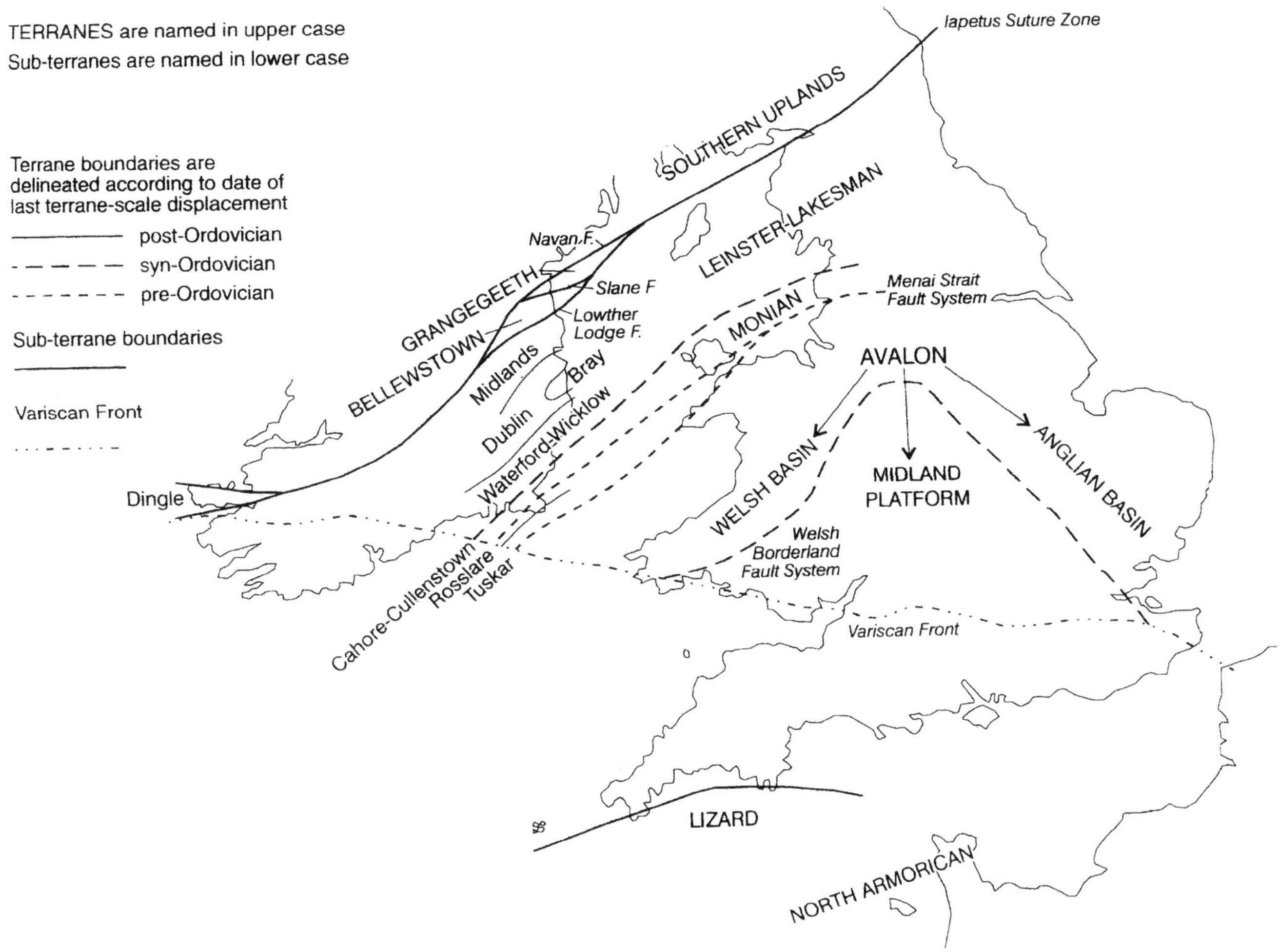

Fig. 3.2. The main Lower Palaeozoic lineaments in England and Wales (modified from Woodcock 2000*a*).

Early Palaeozoic tectonostratigraphic phases

The separation of Avalonia from Gondwana and its subsequent migration towards Laurentia created a succession of changing stress fields within Eastern Avalonia that influenced the evolution of the Welsh and Leinster–Lakesman basins and adjacent areas. Periods of extension were associated with accelerated rates of subsidence within basins (Prigmore *et al.* 1997) and phases of compression resulted in deformation that produced regional unconformities. The unconformities within Lower Palaeozoic sequences in the Welsh Basin have been used to identify three major unconformity-bounded sequences, the Dyfed, Gwynedd and Powys supergroups. The megasequence boundaries are attributed primarily to tectonic and volcano-tectonic events, rather than than purely eustatic changes (Woodcock 1990*a*) (Fig. 3.3). In this account, the early Palaeozoic fill of the Welsh Basin is described in tectonostratigraphic terms similar to those of Woodcock (2000), but with more emphasis on the role of volcanism and major changes in sea-level. Four sequences reflecting four major phases of basin evolution in the Early Palaeozoic are identified, with their limits partly coinciding with the unconformity-bounded supergroups of Woodcock (1990*a*) (Fig. 3.3). Substantial changes in sea level played an important role in shaping the stratigraphic record of the Early Palaeozoic. Sea level in Britain shows three substantial fluctuations in the Cambrian, five in the Ordovician and at least three in the Silurian. The major eustatic transgression, that occurred in the Early Cambrian, deposited sediment unconformably over folded Precambrian rocks, marking the start of the first phase in the evolution of the Welsh Basin and surrounding areas. In the Ordovician a minor regressive interval at the top of the Tremadoc was followed by an early Arenig transgression across regions of folded rocks, initiating the second phase.

Towards the end of Arenig time a general shallowing has been identified at several locations globally and an early Llanvirn transgression is widely recognized, followed by a deep-water interval ending with a modest late Llanvirn regression. The early Caradoc '*gracilis* transgression' caused the largest Ordovician rise in sea level, accompanied by extensive coastal onlap. High sea level prevailed throughout Caradoc and Ashgill times until the large and rapid Hirnantian glacio-eustatic fall and subsequent sea-level rise (Brenchley & Newall 1980) that continued into the early Silurian, coinciding with the start of the fourth tectonostratigraphic phase. Apart from the Hirnantian event, the Ordovician sea-level curve in North America (Ross & Ross 1992) is somewhat different from that in Europe, which suggests that although some of the sea-level changes may be eustatic, others may have been due to local or regional tectonics.

The timing of events in the Leinster–Lakesman Basin is different in some respects from that in the Welsh Basin. Most particularly, arc volcanism occurs in the Welsh basin in the late Tremadoc but not until the early Caradoc in the Leinster–Lakesman Basin. This disparity might be the result of steepening of the subduction zone beween the Tremadoc and Caradoc, or fore-arc widening (Woodcock 2000*b*), but strike-slip displacement of the Lakesman arc relative to that in the Welsh Basin is also possible (Kokelaar 1988).

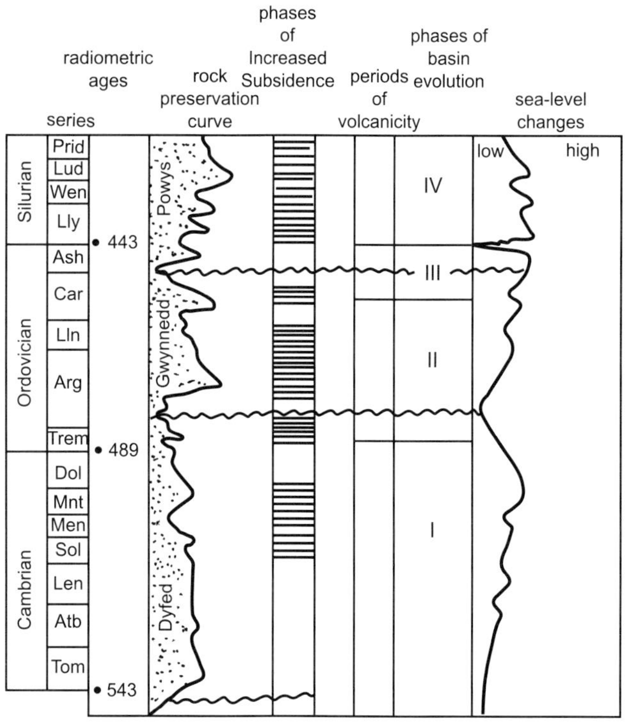

Fig. 3.3. Tectonostratigraphic divisions of the Lower Palaeozoic. Rock preservation curve from Woodcock 1990*a*, phases of increased subsidence (horizontal shading) from Prigmore *et al.* 1997, and bathymetric curve modified from Leggett *et al.* 1981 and Fortey 1984).

The initiation of the Welsh extensional basin

The early Cambrian transgression flooded some areas with deformed Precambrian rocks and other areas where sedimentation continued uninterruptedly into the Cambrian. Subsequently transtensional extension caused rapid subsidence and the initiation of the Welsh and possibly the Leinster–Lakesman basins, which were filled with mainly deep-water or offshore marine sediments, although changes in subsidence rates and sea level in the Welsh Basin resulted in shallower-water intervals, whilst the Leinster-Lakesman Basin remained continuously deep.

Arenig–mid-Caradoc subduction-related volcanic basins
This phase was initiated by arc-related volcanism in the Welsh Basin during the late Tremadoc. The volcanism was preceded by a phase of deformation and followed by the early Arenig transgression that created the unconformity marking the base of the Gwynedd supergroup (Woodcock 1990*a*). The subsequent back-arc-basin volcanism started in the late Arenig and peaked in the Llanvirn and Caradoc, although the division between the two peaks is locally obscure. The Llanvirn activity is most clearly expressed in the coastal cliffs of North Pembrokeshire, near Builth Wells, in the Shelve area of the Welsh Borderlands and around the Harlech Dome in North Wales. The Caradoc activity is mainly confined to North Wales where it dominates the sequence through central and north Snowdonia and the Lleyn Peninsula. The arc volcanism in the Leinster–Lakesman Basin did not commence until the early Caradoc, when thick volcanic sequences accumulated in the Lake District. The volcanicity was related to southwards directed subduction as Avalonia moved northwards. In the Welsh Basin extrusive volcanic rocks are interbedded with marine sediments, although locally there were some emergent volcanic islands. In the region of the former Leinster–Lakesman Basin the Borrowdale Volcanic Group represents the residual part of an emergent volcanic arc of early Caradoc age.

Mid Caradoc to end-Ashgill; the basins in their post-volcanic phase
This phase followed the cessation of major volcanic activity, and, by inference, of southward subduction, possibly determined by Avalonia overriding the Iapetus spreading axis (Woodcock 2000*d*). Subsequently Avalonia continued to move northwards through continued subduction at the Laurentian margin. In Wales the volcanic accumulations cooled and as thermal subsidence progressed they became submerged to form a surface on which sedimentation rates were low. In contrast the higher relief of the arc volcanics of the Lake District was initially progressively transgressed over a substantially longer period, although later sedimentation rates accelerated, possibly in response to flexure in the developing foreland basin.

Silurian extensional and foreland basins
This phase was initiated with the widespread accumulation of anaerobic and dysaerobic muds during the latest Ordovician and earliest Silurian glacio-eustatic rise in sea level. Subsequent tectonism caused rapid basin subsidence and local marginal uplift that stimulated turbiditic sedimentation. Half grabens within the Welsh Basin suggest that it was an extensional basin. Similar extension is not recognized in the Leinster–Lakesman Basin during Llandovery and Wenlock times. There, Silurian rates of sedimentation increased progressively as a foreland basin developed due to crustal loading by marginal parts of Laurentia or its associated accretionary prisms. The main subsidence effects appear to have been diachronous, from early Ludlow in the Lakesman basin to mid-Ludlow to Přídolí in the Welsh Basin (King 1994*a*).

The origins of Lower Palaeozoic stratigraphy

The major divisions of the Lower Palaeozoic had their origins in of the 19th century, when the science of geology was still young. The three systems of the Lower Palaeozoic all derive their names from Wales and the Welsh Borderland. The Cambrian takes its name from Cambria, a Latin variant of the Celtic name Cumbria, the old name for Wales. The Silurian and Ordovician are both named after Celtic and British hill tribes. The majority of the series within the three systems (Fig. 3.4) also take their names from locations in Wales and the Welsh Borderland.

The initial subdivisions of the Lower Palaeozoic sequence, known as the 'Transition rocks' or 'Transition series', arose largely from the geological mapping and traverses across Wales and the Welsh Borderland by Adam Sedgwick and Roderick Murchison in the 1830s. Following earlier visits to North Wales, Sedgwick met Murchison and they agreed to divide up the country and produce systematic accounts. Sedgwick's first traverse in this programme in 1831 was westwards from a base in Shrewsbury into the Berwyns, and on, into Snowdonia to the Caernarvon region, where he reached an understanding of both the complex structure and the stratigraphy. During this first traverse he was accompanied for several weeks by Darwin who was learning the skills of geological mapping (Bassett 1991).

Initially the sequence was only divided into Lower and Upper Cambrian, but later a lithological sequence from the Lingula Flags to the Bala Group was established (Sedgwick 1852).

In the same year (1831) that Sedgwick was working in NW Wales, Murchison undertook his first field season with a traverse starting in the Llandovery area and moving north-eastwards to Builth and then into the Welsh Borderland, where he identified abundant fossils in beds below the Old Red sandstone. These he was able to follow along strike into the Ludlow area so giving him a key to the stratigraphy and structure of the region (Bassett 1985); by September 1831 he was able to present

System		Series
SILURIAN*		Přídolí
		Ludlow*
		Wenlock*
		Llandovery*
ORDOVICIAN*	U	Ashgill
		Caradoc
	M	Llanvirn
	M / Lr	Arenig
	Lr	Tremadoc*
CAMBRIAN*	U	Merioneth
	M	St David's
	Lr	Branch
		Placentia

Fig. 3.4. The systems and series of the Lower Palaeozoic. Divisions marked with an asterisk are internationally recognized.

coloured maps of the 'Transition Rocks' to an inaugural meeting of the British Association for the Advancement of Science in York. However, it was not until 1835 that Murchison and Sedgwick presented their combined research. The first use of the name 'Cambrian' for the sequence established from Sedgwick's traverses in Wales came in the title of a joint paper with Murchison presented to the British Association (Sedgwick & Murchison 1835). By then Murchison had already proposed a Silurian System, consisting of four divisions, equivalent to the Llandeilo, Caradoc, Wenlock and Ludlow (Bassett 1991). The succession below the Llandeilo he regarded as within the province of Sedgwick. A full account of the Silurian System was published within four years (Murchison 1839). Subsequent detailed work by the Geological Survey in central Wales during the middle part of the century showed that the Upper Cambrian (Bala Group) of Sedgwick was equivalent to the Llandeilo and Caradoc of Murchison. Consequently, the level at which the boundary between the Cambrian and Silurian should be drawn became a matter of controversy between Sedgwick and Murchison and their respective followers that lasted for several decades. Finally, the issue was largely resolved by the gradual acceptance of a tripartite division of the Lower Palaeozoic, justified by the creation of the Ordovician System by Lapworth (1879). The new system spanned the interval from the unconformity at the base of the Arenig to the unconformity at the base of the Llandovery. As such it embraced the upper part of Sedgwick's Cambrian and the lower part of Muchison's Silurian. Subsequently, many of the boundaries of the three Lower Palaeozoic systems and their component series have been moved from those established within the 19th century and have been more rigorously defined. A substantial change has been made to the base of the Cambrian, which has been defined in Newfoundland at a lower level to take account of fossiliferous sequences that are now recognized to be older than the base of the Cambrian as set in Britain. Two new series, the Branch and Placentia Series, now occupy the downward extension of the Lower Cambrian. (Fig. 3.4).

The Ordovician System has been extended downwards, following the recommendation that the lower boundary be moved to the base of the Tremadoc (Norford 1988), thus removing the series from the Cambrian. The subdivision of the Ordovician system into five series (Fig. 3.4) arose somewhat haphazardly. The series are not based on continuous sections and have their type areas in different regions, three in Wales, one in the Welsh Borderland and one in the Lake District. Nevertheless, the divisions have proved moderately useful and the definition of their basal boundaries and the amalgamation of the Llanvirn and Llandeilo into a single Llanvirn Series have enhanced their value as regional series (Fortey *et al.* 1995). Only the Tremadoc is regarded as an international series. Attempts to agree on three internationally recognized series that would replace the loosely defined divisions of the Ordovician into Lower, Middle and Upper have been continuing for several years. The base of the Upper Ordovician is now acceped as being coincident with the base of the *gracilis* Zone and to correlate with the base of the Caradoc. The base of the Middle Ordovician is now set at a level that correlates approximately with a level in the mid part of the Arenig. Otherwise, for the time being, the British regional series work well in Britain and are commonly referred to elsewhere in Europe.

Following nearly a decade of appraisal, the level of the upper and lower boundaries of the Silurian were ratified and the limits of the four series were set (Bassett 1985; Holland 1985; Holland & Bassett 1989). The main change was in the level of the Devonian–Silurian boundary that was traditionally set in the Welsh Borderlands at the boundary between the old Red Sandstone and the marine 'Silurian' below.The top of the Silurian is now defined by the base of the Devonian System at a higher stratigraphic level, within a marine sequence in the Czech Republic. The revised Silurian System includes a fourth, Přídolí Series (Fig. 3.4), which correlates with some of the lower levels of the Old Red Sandstone, that are now Silurian rather than Devonian.

The Welsh Basin, Midland Platform and Monian Terranes

The Welsh Basin, which occupies most of the Welsh Basin Terrane, is defined by the Menai Strait Fault System in the NW and the Welsh Borderland Fault System to the SE (Figs 3.2 & 3.5). Movements on the component faults varied from strike-slip to normal during their history. Some of the largest of the faults, such as the Central Wales lineament, were orientated NE–SW, although N–S fractures were also influential (Fitches & Campbell 1987; Kokelaar 1988). The entire basin and its margins generally subsided throughout most of their history, but the location and rates of subsidence varied both within the basin and within sub-basins in the platforms close to the basin margin, as in Anglesey and South Shropshire (Fig. 3.6) (Prigmore *et al.* 1997). The variation in subsidence was commonly related to intra-basinal faults which influenced the thickness and facies variations of the basinal sediments. Where subsidence was greater than the rate of sediment accumulation, basins became topographic depressions in which deep-marine sediments accumulated. Sub-basins both within the central parts of the basin and close to its margin were commonly the location of volcanic activity, as in the Shelve Sub-basin close to the eastern margin of the Welsh Basin (Woodcock 1984*a*; Gibbons 1987).

To the east of the Welsh Basin the Midland Platform was generally stable, apart from a period during the Tremadoc

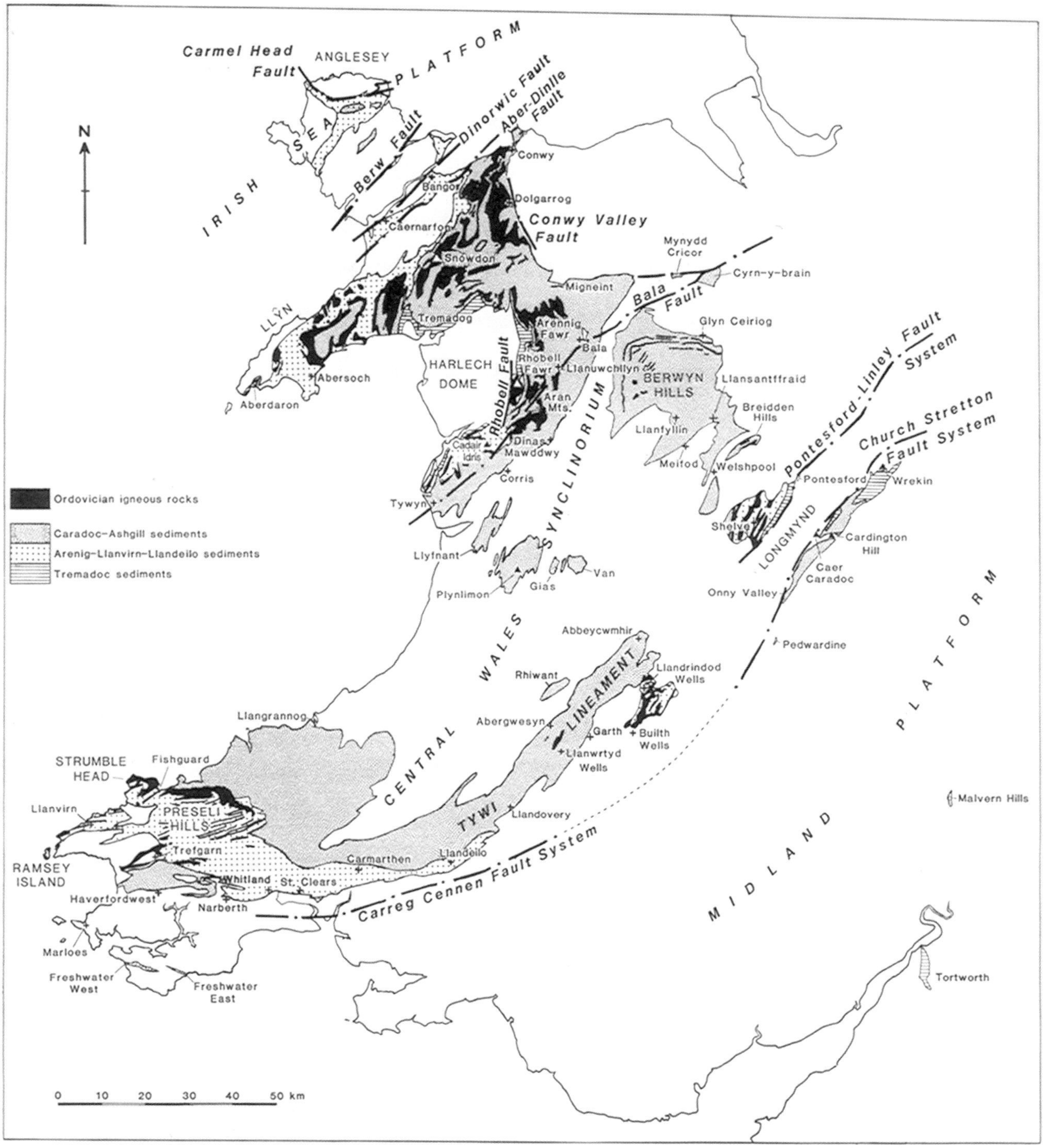

Fig. 3.5. Geography of Wales and the Welsh Borders, with the outcrop of Ordovician rocks and the main structures (from Bassett *et al.* 1992*b*).

when thick sequences of deep-marine sediments accumulated in narrow rift basins. Post-Tremadoc rocks of Ordovician age are mainly absent on the platform and the region may have been largely emergent throughout most of the Early Palaeozoic. The nature of the Irish Sea Horst Complex, at the northwest margin of the basin, is difficult to interpret. Palaeocurrents indicate that the platform was intermittently a source of sediment, but the Lower Palaeozoic rocks preserved on Anglesey are mainly basinal sediments and emergent areas were apparently relatively small and transient. However, it is possible that most of the platform areas have been translated elsewhere by later strike-slip displacements and only the basins have been more generally preserved.

The Cambrian and Tremadoc: the pre-volcanic basin and Midland Terrane

Definition and subdivision

The Cambrian System was originally conceived to encompass an assemblage of ancient rocks in North Wales, parts of which are now considered to be of Precambrian and Ordovician age. Subsequently, when the Cambrian had been characterized palaeontologically, the concept underwent refinement and its use was extended worldwide. Further criteria for its recognition, such as radiometric ages and palaeomagnetic signature, have since been applied and the concept of the Cambrian System is

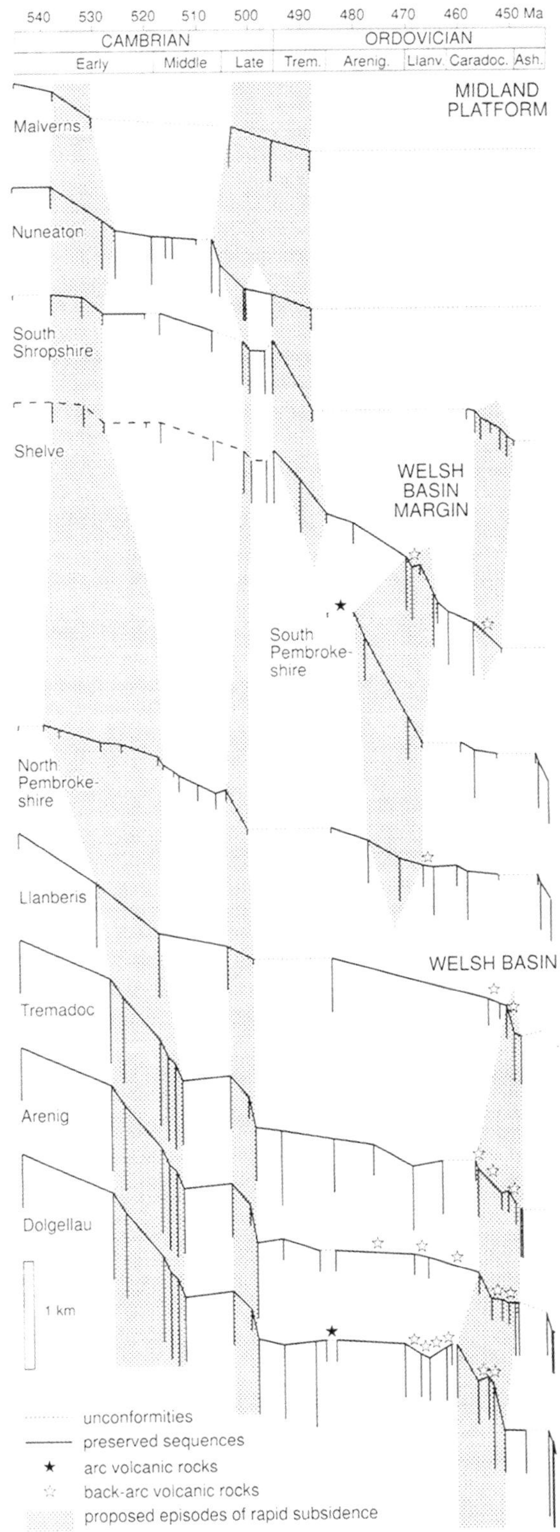

Fig. 3.6. Lower Palaeozoic subsidence history of the Welsh basin (from Prigmore *et al.* 1997). Cambrian subsidence at Shelve is assumed to be similar to South Shropshire. Vertical lines show uncertainties in water depth. Main episodes of enhanced subsidence are patterned.

now stabilized by the formal definition of its base and by the definition of the base of the overlying Ordovician System.

The basal stratotype of the Cambrian is at the base of the *Phycodes pedum* trace-fossil zone at Fortune Head in SE Newfoundland (Brasier *et al.* 1994); the age of this horizon is thought to be about 543 Ma (Grotzinger *et al.* 1995). It is at a lower level than was formerly customary for the base of the Cambrian in Britain, and consequently some strata previously considered of Precambrian age are now regarded as Early Cambrian (Carney *et al.* 2000; and see Chapter 2). The base of the Ordovician, which marks the top of the Cambrian, is taken at Green Point in NW Newfoundland, at the level at which the conodont *Iapetognathus fluctivagus* appears. This is close to the level at which Rushton (1982) proposed for the base of the Tremadoc Series in Wales and whose age is 489 ± 0.6 Ma (Landing *et al.* 2000).

Formal subdivisions of the Cambrian (series, stages) have not been agreed internationally, but the terms 'Lower', 'Middle' and 'Upper' are widely used. The levels of the boundaries between these divisions has long differed somewhat in different areas (Shergold in Whittington *et al.* 1997, pp. 306–311) and Cowie *et al.* (1972) therefore proposed the regional series names 'Comley', 'St David's' and 'Merioneth' to encompass the Lower, Middle and Upper Cambrian as then understood in Britain. Subsequent work in SE Newfoundland indicates that the term 'Comley' should be replaced by two more comprehensive divisions, the Placentia Series (Landing *et al.* 1989) for the earlier Early Cambrian and Branch Series (Landing 1992) for the later, trilobite-bearing, Early Cambrian. Davidek *et al.* (1998) discussed estimates of the isotopic ages for the base of the Branch Series (about 520 Ma), the base of the Middle Cambrian (about 510 Ma) and base of the Upper Cambrian (about 500 Ma).

There is no agreed system of local or regional stages for use in Britain. The biostratigraphical divisions in current use mainly exploit trilobite faunas, as shown in Figure 3.7. These are based on Scandinavian faunal, essentially trilobitic, successions, apart from the Early Cambrian for which the Avalon area of Newfoundland supplies a more applicable standard.

Rushton (1974) gave an account of research into the Cambrian rocks of Wales and England, and recorded the lithology and palaeontology of each division, with an extensive bibliography to earlier work. Cowie *et al.* (1972) gave correlation charts of British Cambrian strata (including Tremadoc) as then recognized. Brasier *et al.* (1992) reviewed the palaeogeography, whilst Woodcock & Strachan (2000, pp. 141–152) surveyed the sedimentary processes and the tectonic setting, together with a discussion of some of the problems associated with the interpretation of the Cambrian in Britain. For an overview of the British Cambrian and Tremadoc rocks, with detailed accounts of numerous representative localities, see Rushton *et al.* (2000).

Overview of Cambrian deposition

Outcrops of Cambrian rocks in Wales and England occur in disconnected inliers, mostly of small extent, in many of which the Cambrian is succeeded conformably or paraconformably by Tremadoc rocks of early Ordovician age (Fig. 3.8); together these constitute the Dyfed Supergroup of Woodcock (1990*a*) which is equivalent to the first of the Early Palaeozoic tectonostratigraphic phases identified in this volume. Cambrian, and especially Tremadoc, rocks have a wide subsurface distribution in central England (Fig. 3.8). In Wales the Cambrian generally consists of thick basinal deposits, and these contrast markedly with the thinner inner- to outer-shelf deposits of England. Because the outcrops are scattered and relatively small (Fig. 3.8), it is not possible to establish the size and shape of the Welsh Basin in the Cambrian with any certainty. The NW margin of the basin lay to the SE of the Irish Sea Horst Complex and the SE margin must have lain somewhere to the NE of the Welsh Borderland fault system; however, there is some evidence that during the early and mid-Cambrian the Welsh Basin was considerably less extensive than during the Silurian (Fig. 3.9). The extension of the basin to the NE, to connect with the 'eastern England Caledonides', is conjectural. The stable Midlands Microcraton that lay to the east of the Welsh Basin was the

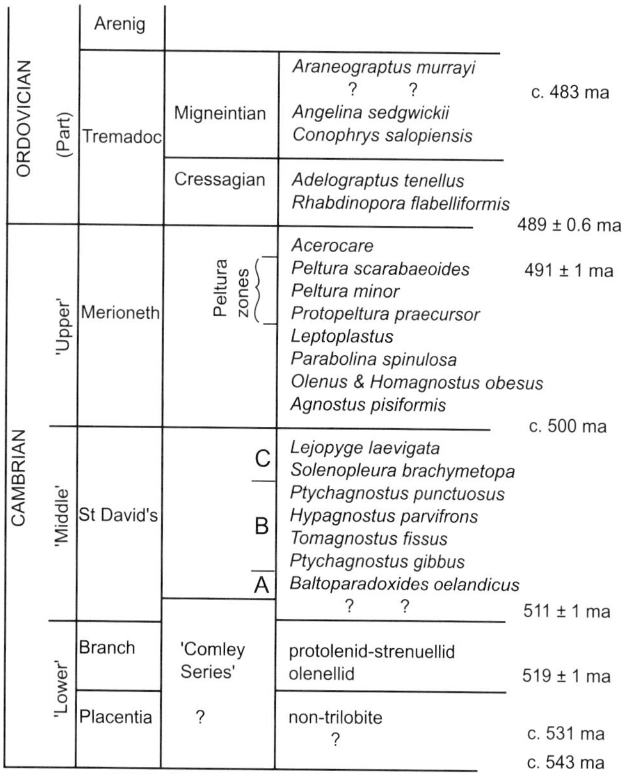

Fig. 3.7. Cambrian and Tremadoc chronostratigraphy and biostratigraphy used in England and Wales. Some isotopic ages are shown.

main area where shelf deposits accumulated (Fig. 3.8). Between the Welsh Basin and the main part of the Midland Platform, in the small outcrops of the Welsh Borderland and Malvern fault systems, the Cambrian is represented by shallow-water deposits that were laid down in local fault-bound basins.

An early Cambrian marine transgression, induced by a eustatic rise in sea level, deposited shallow-water marine sediments across an eroded terrain of variegated Precambrian rocks, largely of volcanic origin. Furthermore, movements associated with the faulted margins of the Irish Sea Horst Complex (Anglesey–Wexford) led to the formation of local basins. The basal Cambrian rocks are arenaceous but differ from area to area, and, despite the difficulty of establishing their precise correlation, it is evident that the earliest Cambrian rocks in different areas are diachronous. Prigmore *et al.* (1997) identified an Early Cambrian episode of rapid regional subsidence that led to the formation of the varied Lower Cambrian sequences.

The Middle and Upper Cambrian sequences are more uniform and are punctuated by episodes of low clastic input, during which dark grey mudstones were widely deposited. At about the beginning of Tremadoc times, however, renewed movements, probably associated with the movement of Avalonia in relation to Gondwana (Prigmore *et al.* 1997), created rifts in the Midlands Microcraton in which thick Tremadoc strata were deposited. By the end of the Tremadoc deposition had practically ceased and erosion of older strata began, such that almost everywhere rocks of Arenig and later age overstep the entire Cambrian succession.

Cambrian–Tremadoc rocks in North Wales

Harlech Dome

The Harlech Dome succession (Fig. 3.10) is the most complete representation of the Cambrian System in Wales. Two principal cycles of sedimentation have been identified, corresponding to the Harlech Grits Group and the overlying Mawddach Group (Fig. 3.10). The deposits accumulated in an extensional basin, possibly formed in response to transtensional movements along the margin of Gondwana (Murphy & Nance 1989). Where known, the formations reach their greatest thicknesses near the centre or towards the east of the Harlech Dome, and thin northwestwards towards the Irish Sea Horst Complex (Fig. 3.10). Gravity profiles suggest that rocks of a similar density to the Harlech Grits are thinner south of the Bala Fault (Dr N. J. P. Smith pers. comm.), and this, together with the proximal character of the coarse, easterly-derived greywackes of the earlier part of the Cambrian, indicates that the Harlech basin was, at that time, much less extensive than the better-documented Welsh Basin during the Silurian.

The Harlech Grits Group, consisting mainly of greywacke sandstones and silty mudstones that form a sequence from the Dolwen Formation to the top of the Gamlan Formation, together more than 2 km in thickness (Fig. 3.10). This records the initial stages in the evolution of the Welsh Basin, at first with shallow-marine deposits, followed by proximal and distal turbidites. Allen & Jackson (1985) described and defined each formation. Crimes (1970) gave a review of the sedimentary structures, mostly flute and groove casts and cross-bedding, and the trace-fossils, most of which are nondescript burrows. Griffiths (1987) gave a full account of the sedimentology of the succession. Body fossils are very rare, but the *Platysolenites* from the Dolwen Formation (Rushton in Allen & Jackson 1978) is thought to represent the lower Lower Cambrian (Placentia Series). The top of the Rhinog Formation is correlated with beds yielding trilobites of the highest lower Cambrian (upper Branch Series) and the upper part of the Gamlan Formation has yielded paradoxidids and other trilobites of the Middle Cambrian (Fig. 3.10).

The Dolwen Formation, the basal conglomerate of which rests on Bryn-teg volcanic rocks, is a series of cross-bedded deltaic sandstones and the overlying Llanbedr Formation consists of prodeltaic silty mudstones (Allen & Jackson 1985, p.5), corresponding to the base of a transgressive sequence. The succeeding thick greywackes of the Rhinog and Barmouth formations are typical proximal turbidites whilst the intervening Hafotty and Gamlan formations are mainly siltstones with lesser sandstone. The turbiditic divisions in the Rhinog, Barmouth and Gamlan formations show abundant sedimentary structures. In general, in the Rhinog Formation, bottom-structures indicate currents flowing from the north but cross-bedding indicates flow from the east (Crimes 1970; Griffiths 1987); in the Barmouth Formation, however, sole-marks indicate flow from the south whilst cross-bedding indicates flow from the east or SE. The source area has not been identified, but may have lain to the SE, on the upthrow side of the Bala Fault. In the later Cambrian formations sedimentary structures consistently indicate flow from the south (Crimes 1970). Certain horizons are manganiferous. The manganese ore-bed near the base of the Hafotty Formation is a diagenetically modified chemical precipitate, whilst the upper part of the Gamlan contains coticules (Bennett 1987). There have been various interpretations of the water-depth and environment of deposition of the manganiferous beds at the base of the Hafotty Formation; it is now accepted that they were part of a basinal sequence (Allen & Jackson 1985; Bennett 1987) rather than shallow-water deposits.

A sequence similar to that of the Harlech Dome is exposed in the cliffs of St Tudwal's Peninsula and was revised by Young *et al.* (1994), using local statigraphical names (Fig. 3.10). They noted certain differences from the Harlech succession: their Cilan Formation is homotaxial with the Barmouth Formation, but differs in having sedimentary structures showing flow from the NE, rather than the SE. Although the Ceiriad Formation is homotaxial with the Gamlan Formation, it is finer-grained and Young *et al.* (1994) excluded it from the Harlech Grits Group. A key horizon at the top of the Hell's Mouth Formation yielded late Early Cambrian trilobites (Bassett *et al.* 1976) and

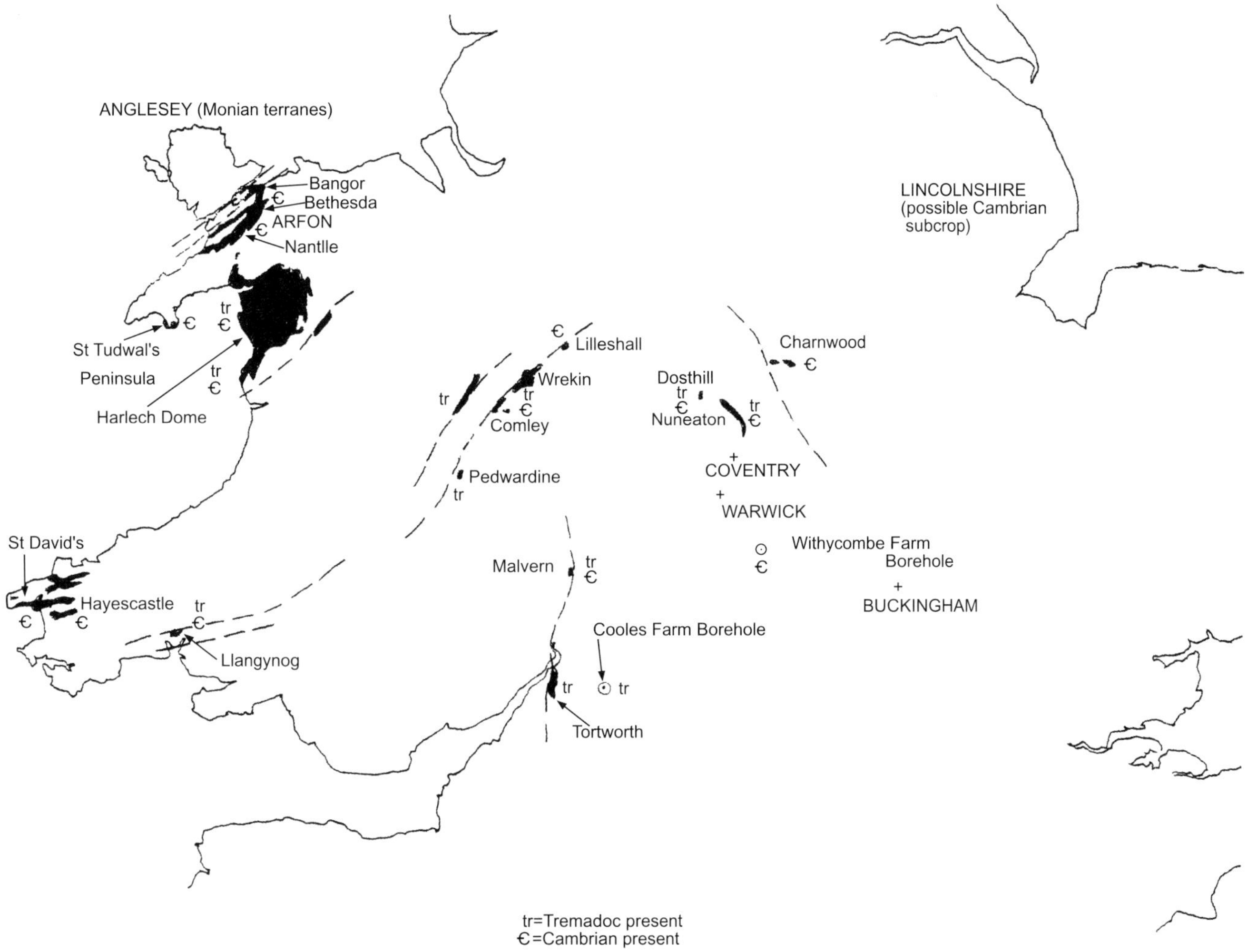

Fig. 3.8. Distribution of Cambrian and Tremadoc (early Ordovician) rocks in England and Wales, with key localities. Outcrops are shown in black. They represent Cambrian (€) or Tremadoc (tr) or both (see also Fig. 3.7). The principal faults and lineaments are named in Figures 3.5, 3.10–3.12.

the manganiferous beds at the base of the overlying Trwyn y Fulfran Formation allows correlation with the Hafotty Formation.

The Mawddach Group. Allen *et al.* (1981) defined the Mawddach Group by the appearance of strata containing layers of black pyritic mudstone and by the presence of mature quartzose sandstone beds. At the base they recognized the Clogau Formation, which, with the largely correlative Nant-y-big Formation of the St Tudwal's Peninsula (Fig. 3.10), consists of dark laminated mudstones deposited during the later mid-Cambrian, and is marked by the appearance of black pyritous layers, as described by Pratt *et al.* (1995, p.6). It includes phosphate nodules and interbeds of bentonite (Roberts & Merriman 1990), and contains a relatively diverse benthic fauna of trilobites (including agnostoids and paradoxidoids), sponges, brachiopods and other shelly fossils that can be correlated more widely than the meagre faunal elements known from the underlying formations. The litho- and biofacies that characterise the Clogau Formation are found at the same horizon in South Wales and the English Midlands. A marked regression near the end of the middle Cambrian is reflected in the biostratigraphical and sedimentological break at the top of the Clogau and Nant-y-big formations, and likewise at the top of the Abbey Shales in Central England. At St Tudwal's, a composite calcareous bed at the junction between the Nant-y-big Formation and the overlying beds typical of the Maentwrog Formation contains reworked fossils of late Middle Cambrian age and is suggestive of erosion during the regressive interval (Young *et al.* 1994). This hiatus is identified throughout Avalonia and was attributed to epeirogenesis by Landing (1996) and Taylor & Rushton (1972), but Nielsen (1996) has shown that a corresponding break in Scandinavia was probably caused by eustasy.

The main thickness of the Mawddach Group is made up of alternations of fine sandstone and silty mudstone, comprising the Maentwrog and Ffestiniog formations (Fig. 3.10). The Maentwrog includes turbidite beds (Allen *et al.* 1981), although much of the sand-grade material was reworked by persistent bottom currents. Directional sedimentary structures show that throughout the Upper Cambrian the prevailing flow was from the south, presumably from the putative positive area, Pretannia, that lay in the Bristol Channel area (Cope *et al.* 1992). If so, it is likely that, by Late Cambrian times, the more proximal and easterly area that had been the source of the earlier Cambrian formations was submerged. The Maentwrog Formation was deposited in a basinal setting (Allen & Jackson 1985) that Crimes (1970) considered deep. However, at St Tudwal's Peninsula, correlative strata show hummocky cross-stratification and are interpreted as having been deposited under the influence of storms in a shallower sea (Bose 1983), presumably reflecting proximity to the Irish Sea Horst Complex. The upper parts of the Maentwrog Formation tend to be more argillaceous. Certain of the finer beds have yielded a succession of early Late Cambrian olenid trilobite faunas, though the basal beds of the formation are thought to be of late Mid-Cambrian age (Allen *et al.* 1981, p. 307).

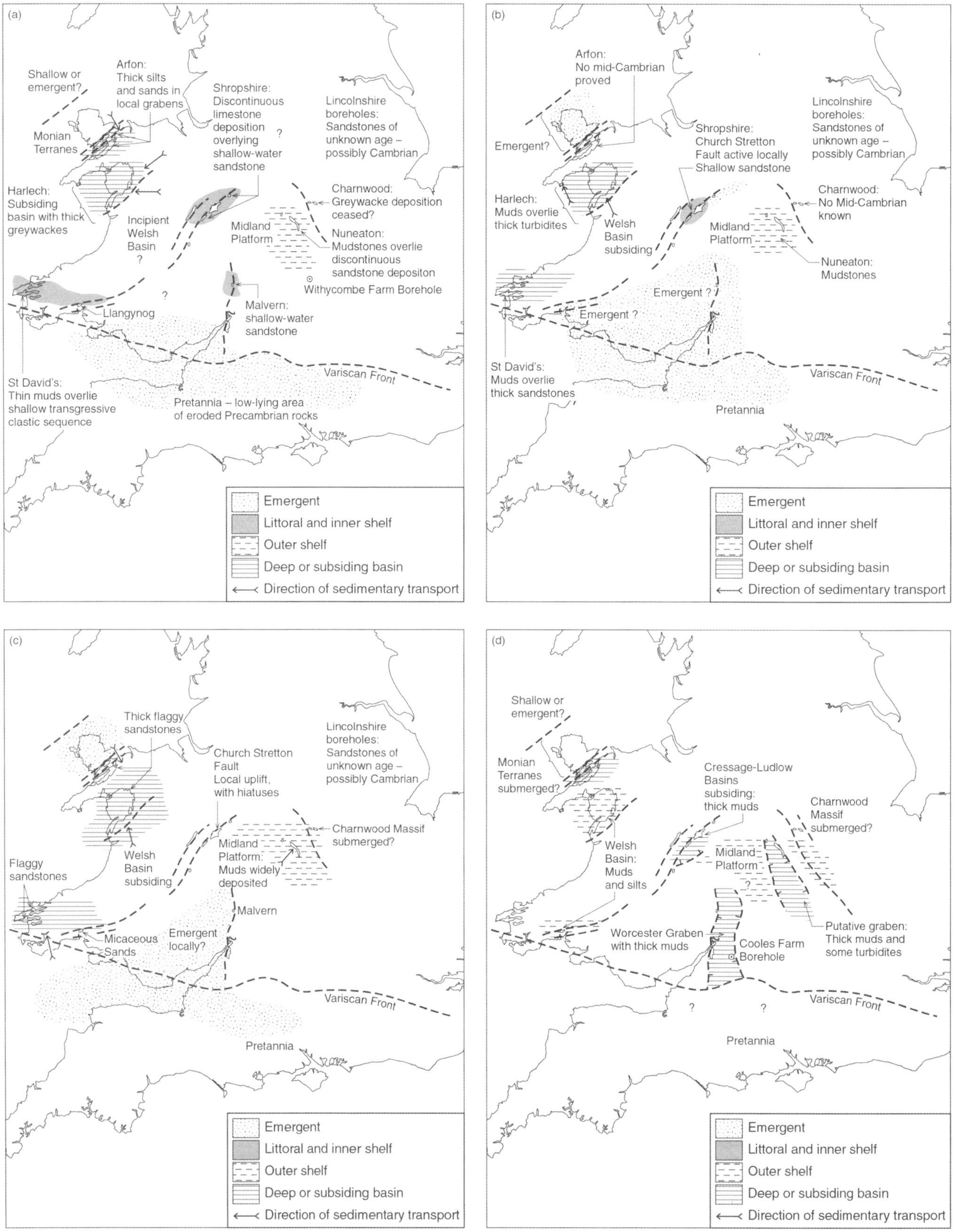

Fig.3.9. Cambrian palaeogeographical maps. (**a**) Early Cambrian: about the base of Branch Series (*Callavia*-bearing interval; olenellid Zone). (**b**) Middle Cambrian: mid-St David's Series (about *Tomagnostus fissus* Zone). (**c**)Late Cambrian: mid-Merioneth Series (upper part of *Olenus* Zone). (**d**) Early Ordovician: low Tremadoc Series (Cressagian Stage).

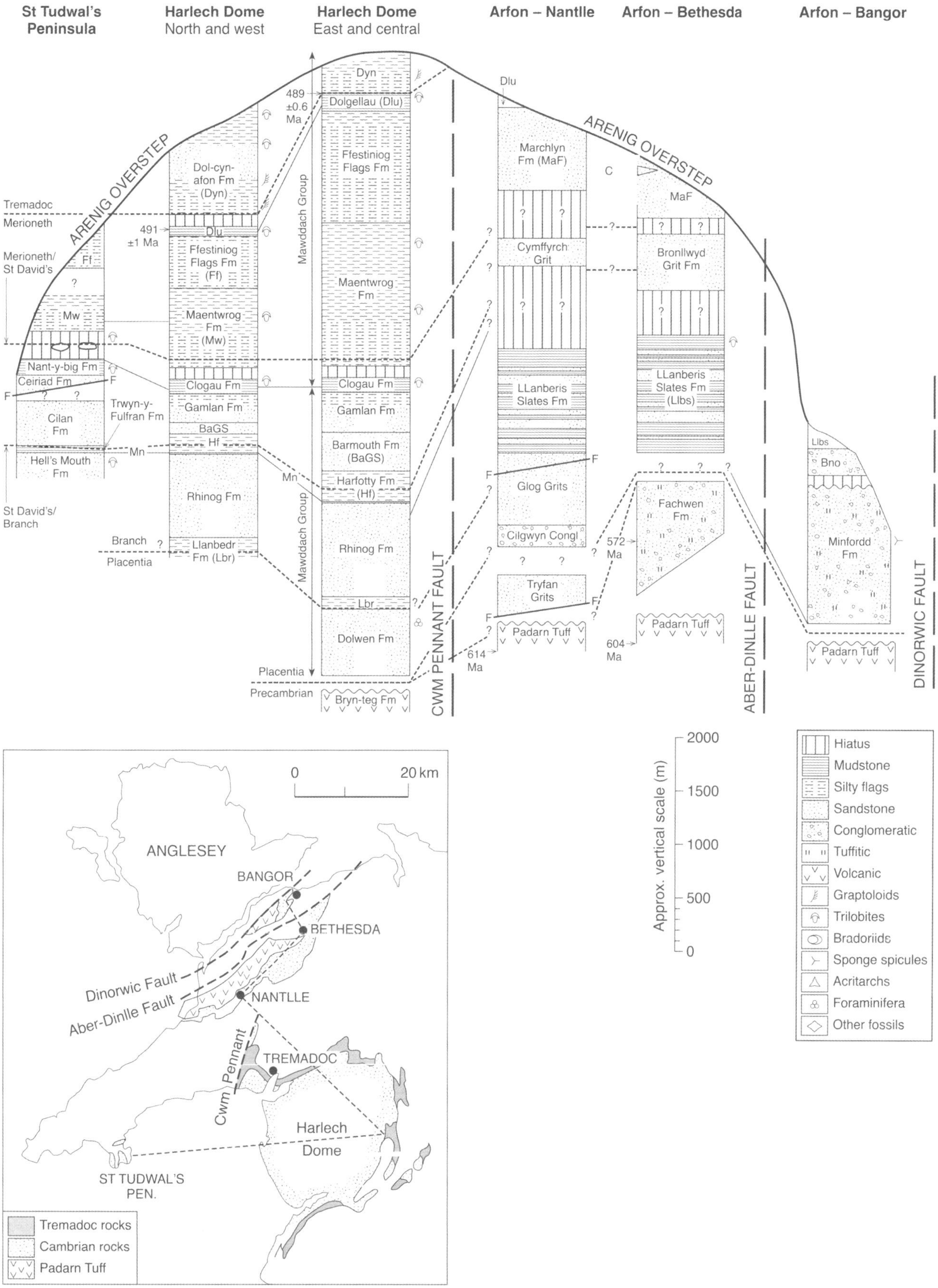

Fig. 3.10. Generalized vertical sections through the Cambrian and Tremadoc rocks of North Wales, with their locations shown in the inset map. Some fossiliferous levels and inferred correlations are shown, and also the horizons at which some isotopic ages were recently obtained. Abbreviations for stratigraphical divisions, mostly evident on the figure, follow British Geological Survey usage (see Rushton & Howells 1999), and include **Bno**, Bangor Formation and **CGr**, Carnedd y Filiast Grit. Other abbreviations: F–F, fault; Fm, Formation; Ma, million years; Mn, managanese-rich level.

The conformably succeeding *Ffestiniog Flags Formation*, about 1000 m thick in the SE of the Harlech Dome and 650 m to the NW, is composed predominantly of flaggy sandstone alternating with silty mudstone. It contains an abundance of the brachiopod *Lingulella davisii* and varied trace fossils (*Skolithos*, *Cruziana*, *Rusophycus*), and rare trilobites (Allen *et al.* 1981, p. 311), which indicate that the formation was deposited rapidly, approximately within the compass of one trilobite subzone. The sandstone beds show hummocky cross-stratification, ripples and scour-and-fill structures; mud-cracks are reported. Indications are that the formation was deposited in shallow water, above wave-base, but that subsidence, kept pace with deposition and was relatively rapid (Prigmore *et al.* 1997); it passes up gradationally into the darker, finer-grained Dolgellau Formation.

The Dolgellau Formation consists of black pyritic and thinly laminated carbonaceous mudstone with small phosphate nodules and occasional calcareous concretions. The sediments are hemipelagic deposits and low-concentration turbidites deposited in quiet conditions in poorly oxygenated (although not necessarily deep) waters. Such rocks, characteristic of the 'olenid biofacies', are widespread, extending across the Welsh Basin and the Midlands Shelf to other regions of Avalonia and Baltica. The formation is about 150 m thick in the SE and thins towards the NW to about 45 m. It is condensed and contains a succession of olenid trilobite faunas, representing at least ten zonal units (Howells & Smith 1997, p. 14). Near the base are beds crowded with the brachiopod *Orusia lenticularis*, thought to be an opportunistic species that colonized the olenid sea during a eustatic fall of sea level that has been identified on the Baltic platform (Nielsen 1996). Towards the top of the Dolgellau Formation there are beds of tuffitic material that have yielded radiometric ages of 491 ± 1 Ma for the *scarabaeoides* Zone (Davidek *et al.* 1998) and 489 ± 0.6 Ma for the top of the *Acerocare* Zone (Landing *et al.* 2000).

At about the start of Tremadoc times there was a eustatic fall and then a rise (Fortey 1984; but see Landing 1993). At about the same time fault-bounded basins formed in parts of southern Britain, rates of sedimentation increased and the sea floor became better oxygenated; a distinctive shelly fauna, accompanied by the planktonic graptoloid *Rhabdinopora*, became established (Howells & Smith 1997, p. 19). The passage from the black Dolgellau mudstones to the grey Dol-cyn-afon Formation ('Tremadoc Slates') is seen in the east of the Harlech Dome at Bryn-llin-fawr and Dol-cyn-afon (Rushton *et al.* 2000, pp. 101, 105) and, in a more condensed form with possible minor non-sequence, in the west at Ogof Ddû (Rushton *et al.* 2000, p. 38). The silty mudstones of the Dol-cyn-afon Formation (Pratt *et al.* 1995; Howells & Smith 1997) were deposited over the whole region, with sandstone units developed locally, especially in the north. The mudstone units show fine lamination, commonly disrupted by bioturbation, and the sandstone units may be graded and show ripples, cross-bedding and channelling.

During later Tremadoc times, uplift and erosion in the eastern part of the Harlech Dome preceded the eruption of the subduction-related Rhobell Volcanic Group of late Tremadoc age (Allen & Jackson 1985). Approximately contemporaneously, in the Upper Sandstone Member of the Dol-cyn-afon Formation of the northern part of the Harlech Dome, Howells & Smith (1997) identified rapid changes of thickness across active faults, suggesting synsedimentary movements that gave rise to the formation of a local N–S trough. Across the whole of North Wales there is a tectonically driven unconformity between the latest Tremadoc rocks, namely the silty slates with the trilobite *Angelina sedgwickii*, and the earliest local representatives of the Arenig Series.

Arfon

The Cambrian of Arfon (Carnarvonshire) was deposited in local fault-controlled basins, principally the Arfon Basin, that developed along the SE margin of the Irish Sea Horst; they probably developed in response to strike-slip movement within the Menai Strait Fault Zone. The sequences (Fig. 3.10) are less continuous than those of the Harlech Dome, but some of the individual formations, which bear only a limited resemblance to those of Harlech, are very thick. They rest with unconformity or paraconformity on a basement of Padarn Tuff (Reedman *et al.* 1984), which has yielded a radiometric ages of 614 ± 2 Ma (Tucker & Pharaoh 1991) and 604.7 ± 1.6 Ma (Compston *et al.* 2002).

In the NW, in a half-graben developed between the Dinorwic and Aber-Dinlle faults (Fig. 3.10), Reedman *et al.* (1984) recognized the Minffordd Formation, up to 1600 m thick and composed of sandstone and conglomerate with beds of acid tuffite. Basal beds are composed of material derived from the Padarn Tuff, but higher in the formation the sandstones are more heterogeneous, with lithic and basaltic detritus. Reedman *et al.* (1984) noted marked overlap of the higher parts of the Minffordd Formation from west to east. The age of the formation is uncertain, but the presence of sponge spicules indicates a Cambrian rather than Precambrian age. The unconformably overlying Bangor Formation (*c.* 240 m thick) is lithologically similar to the Minffordd Formation, and likewise includes tuffitic beds. It is followed conformably by the Llanberis Slate Formation, discussed below. The Careg-onen Beds of eastern Anglesey occupy outliers unconformably overlying rocks of the Mona Supergroup. Like the Minffordd Formation they contain sponge spicules, and are regarded as of Cambrian age (Rushton 1974, p. 79).

To the SE of the Aber-Dinlle Fault the Padarn Tuff is overlain unconformably or paraconformably by the Fachwen Formation, hitherto regarded as of early Cambrian age (e.g. Rushton *et al.* 2000, p. 42). This consists of conglomerates and sandstones that are composed very largely of fragments from the Padarn Tuff and includes beds of welded ash-flow tuff, from one of which Compston *et al.* (2002) recorded an age of 572.5 ± 1.2 Ma, thereby indicating that the Fachwen Formation is substantially older than the base of the Cambrian. North of Llyn Padarn the thickness of the Fachwen Formation varies rapidly from 40 to 450 m. To the SW, near Nantlle, it appears to pass laterally into some of the divisions mapped by Morris & Fearnsides (1926) as the Tryfan Group, Cilgwyn Conglomerate and Glog Group, which together exceed 1000 m in thickness (Fig. 3.10).

The Llanberis Slate Formation (estimated to be nearly 1000 m thick) is mapped as succeeding the Fachwen and Bangor formations conformably, and is composed of well cleaved laminated silty mudstones with turbiditic sandstone interbeds. The stratigraphy is fairly uniform along strike (Morris & Fearnsides 1926). Early Cambrian trilobites found in the uppermost unit of the slates are of mid-Branch age (<520 Ma) and are older than those from the top of the Hell's Mouth Formation at St Tudwal's Peninsula (Plate 2) (Bassett *et al.* 1976). If the Llanberis Slate Formation is accepted as following the Fachwen Formation and its lateral equivalents essentially conformably, it would indicate that the deposition of this succession extended over a period approaching 40 Ma – such a possibility is implicit in charts by Compston *et al.* (2002, fig. 18) and Landing (1996, fig. 7). However, the Arfon area has long been the subject of controversy, in particular on the nature and significance of several lithostratigraphical contacts (e.g. Rushton 1974, p. 74), and the finding by Compston *et al.* (2002) that the Fachwen Formation is Precambrian in age should prompt a review of the supposed continuity of the Fachwen–Llanberis succession. The overlying Bronllwyd Grit Formation consists of graded greywackes with cleaved slaty beds, some of which are manganiferous, together 500 m thick. The nature of both the basal and upper contact is disputed, each having been claimed to represent an erosional gap or to be conformable.

In the absence of palaeontological evidence the age and correlation of the Bronllwyd Grit remains uncertain (Rushton 1974, p. 81). The British Geological Survey (1988) favoured conformity with the overlying Marchllyn Formation, thus implying a considerable stratigraphical break at the base of the Bronllwyd Grits.

The Marchllyn Formation (240–430 m thick), consisting of silty mudstones and laminated sandstone, is a coarser-grained equivalent of the Maentwrog and Ffestiniog formations of the Harlech Dome. Towards the top, the Carnedd y Filiast Grit Member is made up of coarse quartzose sandstones displaying sedimentary structures (Plate 3) and distinctive trace fossils (*Cruziana, Skolithos*) indicative of shallow, perhaps intertidal, conditions (Crimes 1970). Palaeocurrents indicate flow from the northwest, from the direction of the Irish Sea Horst Complex, with a SE-facing palaeoslope. This contrasts with the contemporaneous Ffestiniog Formation which was derived from the south. In Cwm y Ffynnon a local outlier of fossiliferous lower Dolgellau Formation overlies the Marchllyn Formation (Howells & Smith 1997, p. 13).

South Wales

In South Wales the Cambrian is found in two areas. One, in NW Pembrokeshire, is the long-known Cambrian of the St David's region, which, together with the area immediately to the east, represents an area of deposition towards the southern margin of the Welsh Basin. The other, near Llangynog, SW of Carmarthen, was discovered much more recently through the initiative of J. C. W. Cope; it lies close to the Tywi lineament, which marks the break between shelf and basin through much of the Lower Palaeozoic time, and shows a rather different succession.

St David's area

The succession (Fig. 3.11) is well exposed on the coast, but much less so inland and to the east (Rushton 1974, p. 85). There are two major cycles of basin-filling (although Crimes 1970 tentatively identified three cycles), the first being represented by the Caerfai, Solva and Menevian groups, and the second by the 'Lingula Flags'. Sedimentological and palaeontological information was synthesized by Prigmore & Rushton (in Rushton *et al.* 2000, pp. 53–67).

The Caerfai Group ('Caerfai Beds' of the Institute of Geological Sciences 1973) represents an Early Cambrian transgressive sequence, of Branch and possibly late Placentia age, accumulating at the southern edge of the newly formed Welsh Basin; it is much less thick than the early Cambrian rocks of North Wales, and deposition in South Wales may have started significantly later. The lithologies and thicknesses of the Caerfai Group are, apart from the basal conglomerate, fairly constant across the area. The basal conglomerate (0–50 m thick) is made up mainly of quartzose pebbles, mostly differing from the known underlying Precambrian. The overlying St Non's Sandstone (140 m), composed of green feldspathic sandstones deposited in shallow water, shows hummocky cross-stratification and appears to have been transported from the SSW (A. J. Rees written. comm., 2002). The succeeding Caerfai Bay Shales (15 m thick) are cleaved bioturbated red mudstones deposited in deeper, oxygenated water, within the influence of storm waves. There are thin interbeds of water-laid crystal tuffs that gave a radiometric age of 519 ± 1 Ma (Landing *et al.* 1998), and the presence in the red mudstones of *Indiana lentiformis* (Cobbold) allowed Siveter & Williams (1995) to correlate the division with the *Callavia* Sandstone (Lower Branch Series) of Shropshire. Landing *et al.* (1998) proposed a correlation of the base of the Caerfai Bay Shales with an unconformity in the succession in SE Newfoundland, but so far no angular unconformity has been identified between the St Non's and Caerfai Bay formations. The Caerbwdy Sandstone (150 m thick) is a succession of dull purple sandstones probably deposited in fairly shallow water. The sandstones are thoroughly bioturbated and occur in thick beds. No fossils are known and the age of the beds is unresolved. There is difficulty in forming a regional interpretation of the succession from the Caerfai into the Solva Group: in the coastal sections the Caerbwdy Sandstone appears to pass up through coarse pebbly beds into the Solva, but to the east, in inland exposures around Hayscastle and Tremaenhir, the purple sandstones are succeeded by coarser green or yellow sandstones; it is not known whether these are equivalent to the upper Caerbwdy Sandstone or part of the Solva Group. Biostratigraphically, the interval suggested between the *Indiana lentiformis* in the Caerfai Bay Shales and the paradoxidid fauna at the base of the Solva Group seems to represent the whole Moroccan Tissafinian Stage with three trilobite zones–a remarkably long interval to be represented by 150 m of sandstone. If the biostratigraphical correlation is accepted, condensation and/or a cryptic hiatus may be suspected.

The Solva Group. There are three main divisions of the Solva Group, known informally as the lower, middle and upper Solva (summary in Rushton 1974). The lower division consists of about 50 m of coarse greenish-grey sandstones, tidally re-worked and deposited rapidly in a shallow-water delta fan by currents flowing from a southeasterly quarter. The fauna of trilobites (*Condylopyge, Plutonides, Acadoparadoxides*?), hyoliths and sponges indicate an early, although not basal, Middle Cambrian age. This succession is followed by an alluvial incursion (A. J. Rees pers. comm.). The succeeding middle Solva consists of medium-grained, partly turbiditic sandstones in massive beds with local conglomerate horizons. It is reported to attain a thickness of 400 m around Solva. Sedimentary structures and fossils are rare. The upper Solva, 50 m thick, is made up of grey fine-grained flaggy sandstones, including coarse sandy and conglomeratic horizons near the base, showing burrows and ripple cross-lamination; they alternate with mudstones and were probably deposited below wave-base (Prigmore in Rushton *et al.* 2000, p. 63), possibly on a delta-slope (A. J. Rees written comm., 2002). The upper Solva beds pass up into the Menevian Group. Body fossils are scarce but include mid-St David's trilobites, about equivalent to the *gibbus* Zone of Scandinavia.

The Menevian Group. This is divided into three informal units (summary in Rushton *et al.* 2000, p. 60). The lower beds, about 100 m thick, are laminated dark and light grey mudstones with thin sandstone layers and beds of metabentonite (Roberts & Merriman 1990). There are slump-structures in the basal beds and the whole division is interpreted as lower-delta slope and pro-delta deposits. The trilobite fauna is typified by *Tomagnostus fissus* and *Plutonides hicksii*. The middle Menevian beds are similar, but darker and more pyritous, with fewer sandy beds. They contain phosphatic nodules and successive trilobite faunas representing the *Hypagnostus parvifrons* and *Ptychagnostus punctuosus* zones, the latter including *Paradoxides davidis*. These units were probably deposited below wave-base from low-concentration turbidity currents, in conditions of low but variable oxygenation; they resemble the contemporaneous Clogau Formation in North Wales. The upper Menevian beds are marked by the abrupt appearance of massive sandstone beds with black mudstone interbeds typical of the underlying middle Menevian, together about 30 m thick. The sandstones yield the brachiopod *Billingsella*? *hicksii*, and Hicks (1892) also recorded trilobites. Despite this, the boundary between the upper Menevian beds and the base of the succeeding 'Lingula Flags', of late Cambrian (Merioneth) age, is not easily drawn in the field.

'Lingula Flags'. This name, originally used for the Maentwrog and Ffestiniog formations of North Wales, has been used for a thick comparable sequence of alternating sandstone and silty mudstone beds in South Wales. Locally the names Treffgarne

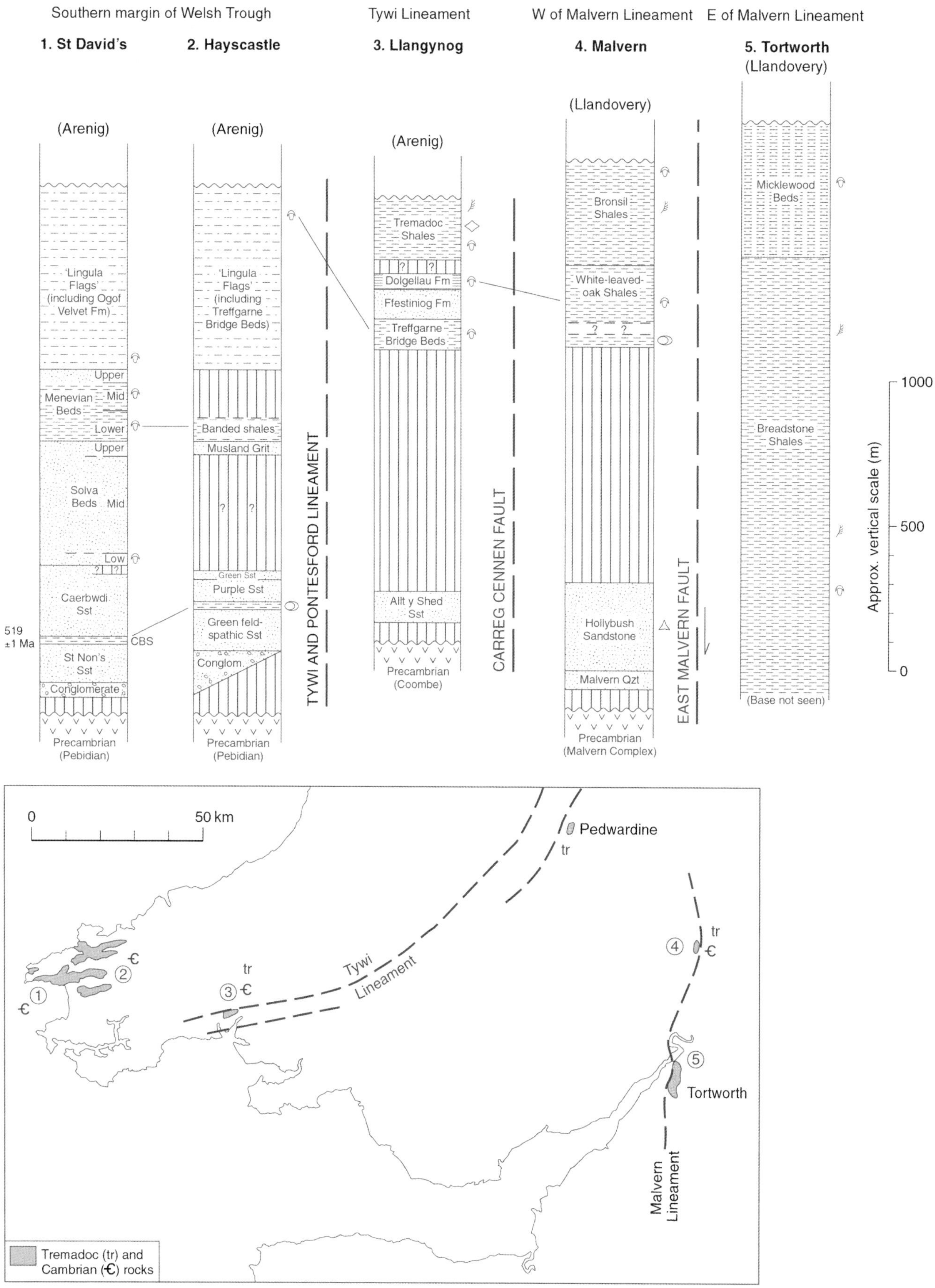

Fig. 3.11. Generalized vertical sections through the Cambrian and Tremadoc rocks of South Wales and western England, with their locations shown in the inset map. Some fossiliferous levels and inferred correlations are shown, and also the horizon at which an isotopic age was obtained. Key: see Figure 3.10. Abbreviations: Fm, formation; Sst, sandstone; Qzt, Quartzite. In columns 1 and 2, **CBS**, Caerfai Bay Shales; Ma, million years.

Bridge and Ogof Velvet have been applied to parts of the succession, but these have yet to be fully characterized and their distributions documented. The Lingula Flags occupy the greater part of the Cambrian outcrop in South Wales and are estimated to be 600 m thick. The sandstones show hummocky cross-stratification and ripples suggestive of deposition in shallow water under the influence of currents flowing from the SW or south, in contrast to the prevailing SE derivation of the Solva and Menevian groups. Deposition more or less kept pace with local subsidence, in a manner similar to the contemporaneous beds in North Wales. The sandy detritus was presumably derived from the southerly landmass of Pretannia inferred by Cope & Bassett (1987) and is reworked and mature. Over most of the outcrop the only fossil recorded is the brachiopod *Lingulella davisii,* but in the lower beds near Porth-y-rhaw Dr M. Lewis (pers. comm.) collected the earliest Merioneth trilobite *Agnostus pisiformis*. In a more argillaceous development to the southeast, distinguished as the Treffgarne Bridge Beds, are the Merioneth trilobites *Homagnostus obesus* and *Olenus cataractes*, that represent the suceeding *Olenus* Zone. Arenig strata overstep the whole formation, and no younger Merioneth strata are known in the St David's area.

Llangynog area

The Cambrian inlier at Llangynog is small and the rocks are poorly exposed, but by exploiting temporary and artificial exposures, the following succession (Fig. 3.11) has been assembled (Cope & Rushton 1992). The Allt y Shed Sandstones rest unconformably on a Precambrian basement and may be about 100 m thick. They are mainly green and olive siltstones and sandstones, but include a limestone unit that yields agglutinating foraminiferans (Cope & McIlroy 1998). The age is uncertain, but an Early Cambrian age is inferred. Middle Cambrian rocks are unknown, the Allt y Shed being overlain, presumably unconformably, by beds of conglomerate, sandstone, shales and siltstones (100 m thick). The finer beds resemble those of the Treffgarne Bridge Beds and have yielded a few trilobites of comparable age (Cope & Rushton 1992). The succeeding micaceous sandstones and siltstones are a typical, although attenuated, representative of the Ffestiniog Flags Formation of North Wales, and are followed by black mudstones of the Dolgellau Formation. The latter succession appears incomplete at its base and top, but contains olenid trilobites indicative of several horizons in the *Peltura* zones (Cope & Rushton 1992).

The contact of the overlying Tremadoc Series with the Dolgellau Formation is not seen but may be disconformable. The Tremadoc rocks are more than 200 m thick; they are grey-green mudstones with local sandstone beds that generally resemble those of the Shineton Shales of the Welsh Borderland in lithology, described in the next section. The faunas of trilobites, brachiopods and graptoloids indicate the lower zones of the Tremadoc Series (Owens *et al.* 1982).

England

In England the Cambrian is exposed in scattered outcrops that reveal two contrasting facies types. Over parts of the Midland Platform the Cambrian is represented by basal sandstones overlain by a continuous succession of mudstones deposited on a cratonic shelf, whereas at the northwestern edge of the platform, along the active Welsh Borderland fault systems, the Cambrian is largely of shallow-water sandstones, the successions being marked by several breaks in deposition. In both settings the Cambrian extends up into Tremadoc (basal Ordovician) mudstone sequences that are locally very thick.

Midlands Platform

The platform succession is best known in the Nuneaton inlier (Fig. 3.12), other outcrops being very incompletely exposed, but borehole records indicate that there is a wide subsurface distribution. The basal sandstones are best represented by the Hartshill Formation, up to 270 m thick, the base of which includes coarse immature deposits derived from fault-scarps when they were drowned by the Early Cambrian transgression (Bridge *et al.* 1998). These are overlain by shallow-water shoreface and storm-influenced sands of the Park Hill and Tuttle Hill members. As the sea deepened the latter was succeeded by condensed arenaceous divisions, including the carbonate and phosphate-rich Home Farm Member, with fossils of Placentia age (Brasier 1984), that was deposited in deeper water. Brasier (1989, pp. 85–90) summarized the succession of faunas and facies and noted hiatuses in deposition. To the east, at the top of the Charnian Supergroup, which is mainly of Neoproterozoic volcanic rocks, is the Brand Group, part of which, the Swithland Greywacke Formation, is assigned to an early part of the Early Cambrian (Placentia Series) (Carney *et al.* 2000, p. 201; Bland & Goldring 1995; see also Chapter 2).

Overlying the Hartshill Formation is the Stockingford Shale Group, comprising 660 m of Cambrian strata and passing up into an unknown thickness of Tremadoc shales. The whole group was laid down on a stable but gently subsiding cratonic area and is divided into seven formations, based on differences in the colour and fissility of the mudstones and on the presence of sandy beds. The formations alternate between mudstone units, commonly bioturbated, which were deposited in oxygenated conditions and include sandstone beds that show ripples and cross-lamination, with dark pyritous mudstones indicative of low levels of oxygenation, as reviewed in Bridge *et al.* (1998, p. 30). The Purley Shale Formation was formed in oxygenated conditions and is sparingly fossiliferous, but contains high Branch and low St David's trilobite faunas (Rushton 1966). The lowest of the pyritous divisions is the variegated Abbey Shale Formation, which resembles the Menevian and Clogau formations of Wales in lithology (including the presence of phosphate nodules and layers of bentonite), and also in the diverse benthic fauna of trilobites, brachiopods, sponges and hyoliths. A mild unconformity at the top of the Abbey Shale Formation corresponds to a late Middle Cambrian regression during the *brachymetopa* Zone (Rushton 1978). The thin basal bed of the succeeding Mancetter Shale Formation is conglomeratic and resembles a shoreface deposit. A few fossils in the overlying grey mudstones indicate the highest Middle Cambrian *Lejopyge laevigata* Zone (Rushton 1978). The Outwoods Shale Formation is mainly of grey mudstones, but includes many black pyritous layers that yield trilobites typical of the dysaerobic olenid biofacies and indicate the lowest Merioneth zones of *Agnostus pisiformis* and *Olenus* (Rushton 1983). The Moor Wood Sandstone Formation is thin (15 m) and only locally developed (Bridge *et al.* 1998, p. 35). It consists of flaggy units of well-sorted quartz and micaceous detritus, commonly showing convolute bedding with signs of having slumped down a north-easterly facing palaeoslope (Taylor & Rushton 1972). The Monks Park Shale Formation consists largely of black pyritous mudstones starved of arenaceous matter and deposited in dysaerobic conditions. Olenid trilobites occur through much of the succession and indicate the presence of five biozones, divided into 12 subzones (Taylor & Rushton 1972). The formation passes up into the Merevale Shale Formation. Although this is mainly of Tremadoc age, the Merioneth–Tremadoc transition has not yielded distinctive faunas. The Monks Park and Merevale formations are partly exposed in the small inlier at Dosthill (Taylor & Rushton 1972). Typical Tremadoc mudstones, as discussed in the next section, are widely distributed.

Welsh Borderland

In the Welsh Borderland the Cambrian is exposed in small inliers that lie along the sporadically active fault complexes of the Church Stretton (Fig. 3.12) and the Malvern (Fig. 3.11) lineaments. The Lower Cambrian successions, pieced together from faulted outcrops, are largely made up of sandstones with

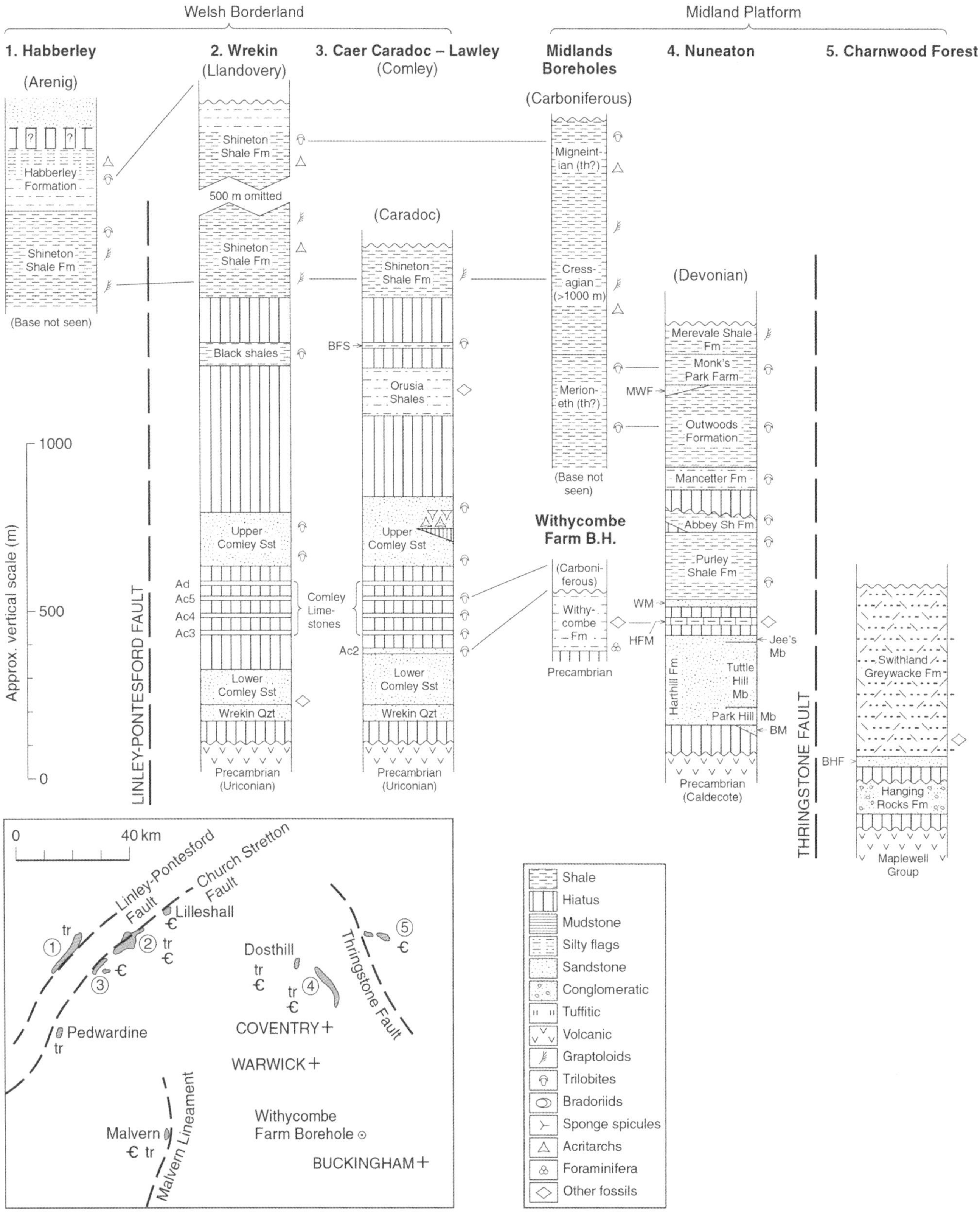

Fig. 3.12. Generalized vertical sections through the Cambrian and Tremadoc rocks of the Welsh Borderlands and Central England. The locations of the columns are shown in the inset map. Some fossiliferous levels and inferred correlations are shown. Abbreviations for stratigraphical divisions: In columns 2 and 3 **Ac2–Ac5** and **Ad** refer to Cobbold's (1921) divisions of the Comley Limestones (totalling *c.* 2 m in thickness). In column 3 **BFS**, Bentleyford Shales (4 m thick). In column 4 **MWF**, Moor Wood Formation; **WM**, Woodlands Member; **HFM**, Home Farm Member (*c.* 2 m thick); **BM**, Boon's Member. In column 5 **BHF**, Brand Hills Formation. Fm, Formation; Mb, Member; Qzt, quartzite; Sst, sandstone. In map € = Cambrian, tr = Tremadoc.

greater proportions of siltstones and shales higher in the succession. In the Wrekin and Comley areas (Fig. 3.12) the basal beds are hard quartzose shoreface sands (Wrekin Quartzite), up to 50 m thick, succeeded by the Lower Comley Sandstone, which is commonly green, micaceous and glauconitic and up to about 150 m thick; the faunas of 'small shelly fossils' in the lower beds may be of Placentia age (Brasier 1989), and the lowest trilobite (representing the Branch Series) is recorded at about the middle of the formation. A comparable sandstone succession is known from Malvern (Worssam *et al.* 1989) and a small part of the upper levels are known in the Croft Borehole at Lilleshall (Rushton *et al.* 1988).

The lower Comley Sandstone is followed by the thin Comley Limestones (2 m thick), a succession of five beds, progressively more calcareous and phosphatic upwards, each separated from the bed below by a surface of discontinuity (summary in Rushton 1974), collectively representing substantial hiatuses (Landing 1996). The limestones each contain a distinct fauna, and they range in age from mid- to latest Branch. The topmost bed, the *Lapworthella* Limestone, contains calcareous cyanophyte(?) algae (Danielli in Rushton *et al.* 1988) and is of uncertain early to middle Cambrian age; it locally oversteps the other limestone units and is itself overstepped by the Upper Comley Sandstone which is known to be of mid-Cambrian (St David's) age. The Comley Limestones were recognized in the Wrekin area, and partly at Lilleshall, but no equivalent beds are known at Malvern. Clasts of fossiliferous rocks akin to the Comley Limestones were found in a borehole core of the Clent Breccia (Permian) at Nechells, Birmingham, suggesting that comparable strata may have been deposited on a structural high to the west of the Knowle Basin (Powell *et al.* 2000*b*).

The Upper Comley Sandstone consists of sandstones and mudstones up to 200 m thick (where fully developed) and yields a discontinuous succession of faunas of St David's age (Greig *et al.* 1968). At both Comley and the Wrekin (Fig. 3.12) there is evidence for the development of growth faults, on the upthrow sides of which there are significant breaks in deposition, whereas successions on the downthrow sides are more continuous (Smith & Rushton 1993). There are other breaks in the succession, especially about the St David's–Merioneth boundary interval. The St David's Series is absent at Malvern (Fig. 3.11), Llangynog and Lilleshall, and these areas are regarded as tectonic highs.The overlying Merioneth Series, where present, is represented largely by shales and mudstones. The lowest Merioneth beds are not proved and the succeeding beds of the *Olenus* Biozone have been recognized only at one place in the Malvern area (Williams & Siveter 1998) and is absent near Comley and at Lilleshall. The *Parabolina spinulosa* Zone is well represented at Comley by the Orusia Shales (150 m thick), by about 15 m of black shale with thin sandstones at Lilleshall (Rushton *et al.* 1988), and appears to be present in the Malvern area. Higher Merioneth beds are represented everywhere by black mudstones with olenid trilobites that, like the Dolgellau Formation of Wales, represent several zones. These beds are thin near Comley (Stubblefield 1930) and the Wrekin, but rather thicker (with sandstone beds) at Lilleshall (>25 m) and much thicker at Malvern, if the reported thickness of 150 m or more (Worssam *et al.* 1989) is accepted. Thomas *et al.* (1984, fig. 6) tabulated the zones recognized in each area, but there is no record of the highest Cambrian *Acerocare* Zone underlying the Tremadoc Series.

The Tremadoc

The Tremadoc rocks of England are mainly grey or green shales with sandstone interbeds at some levels. They occur overlying the various Cambrian outcrops (apart from those at Charnwood and Lilleshall) and also occur at Habberley (Fig. 3.12), Pedwardine and Tortworth (Fig. 3.11), where their base is not seen. Boreholes at Shrewton and Cooles Farm, and several around Coventry (Bridge *et al.* 1998), Warwick and Banbury (Old *et al.* 1987), Birmingham (Powell *et al.* 2000*b*) and Buckingham (Molyneux 1991) have proved an extensive subcrop. The Tremadoc rocks of Tortworth, coupled with those in the Cooles Farm Borehole and with seismic evidence, indicate that a great thickness (*c.* 2 km) of beds was deposited in a forerunner Worcester Graben of Tremadoc age; and similarly more than 1 km of Shineton Shales were deposited in a half-graben, the 'Cressage–Cardington sub-basin'. These grabens are interpreted as trans-tensional basins that formed in response to the opening of the Rheic Ocean (Smith & Rushton 1993), apparently in the early Tremadoc, since the local late Merioneth rocks, where known, are thin and condensed black shales.The fullest Tremadoc succession known is in the Shineton Shale Formation of the Wrekin district (Fig. 3.12) (Stubblefield & Bulman 1927). The lower division, the Cressagian stage, is mainly of mudstone with graptolitic horizons but includes sandy beds in its upper part. The upper division, the Migneintian Stage, includes trilobitic mudstones and turbidite sandstones of the *Conophrys salopiensis* Biozone (Fortey & Owens 1991) and, in the Habberley outcrop, extends up into younger shallower-water arenites of the Habberley Formation (Fortey & Owens 1992); the latter is in contact with the Stiperstones Quartzite (lower Arenig), but possibly with non-sequence. In other areas only the Cressagian is exposed (Fortey *et al.* 2000, fig.16) or present at depth, e.g. around Warwick (Old *et al.* 1987); in the Tattenhoe Borehole it includes fine-grained graded turbidites that are steeply dipping and uncleaved (Bulman & Rushton 1973). Erosion prior to the Llandovery transgression seems to have removed much of the Tremadoc, for despite the great thickness and wide subcrop, the Migneintian has seldom been proved (Fortey *et al.* 2000, p. 30).

Ordovician volcanic basins and their margins

Ordovician chronostratigraphy and biostratigraphy

Ordovician chronostratigraphy, biostratigraphy and radiometric age determinations are shown in Figure 3.13. The regional series names are as recently revised (see Fortey *et al.* 1995) and are used in this account because several of the series boundaries in Laurentia, which have been proposed for international recognition (Fig. 3.13), cannot be recognized in Avalonia or in some instances in Gondwana. The main change to the regional series in recent years is the abandonment of the Llandeilo Series so that its lower part is incorporated in the Llanvirn and its upper part in the Caradoc. Stages have now been defined for the entire Ordovician sequence (Fortey *et al.* 1995). Three stages in the Arenig, the Moridunian, Whitlandian and Fennian, were established in South Wales (Fortey & Owens 1987) and four new stages have replaced the many stages of the Shropshire Caradoc, which have been redesignated as substages (Fig. 3.13). Graptolite biozones are effective for regional correlation for most of the Ordovician, but only the late Llanvirn *teretiusculus* Biozone, the earliest Caradoc *gracilis* Biozone, the mid Caradoc *clingani* Biozone and late Ashgill *pacificus, extraordinarius* and *persculptus* biozones are useful in inter-regional correlation.Where graptolites are relatively rare, useful biozones have been established, based on trilobite faunas, as in the Arenig of South Wales (Fortey & Owens 1987).The trilobite and graptolite biozones through the upper Llanvirn and lower Caradoc interval have been revised and correlated using graphic correlation (Bettley *et al.* 2001) and the base of the *Nemagraptus gracilis* Biozone has been shown to correlate with the base of the Caradoc, which level has now been chosen as the international base of the Upper Ordovician (at Fågelsång, southern Sweden). In the more arenaceous facies of the Caradoc, particularly in the shallow-marine sequence of South Shropshire, the substages are determined on the basis of their trinucleid faunas and in some instances the associated brachiopods (Dean 1964; Bettley *et al.* 2001; Bowdler-Hicks *et al.* 2002). The Ashgillian stages,

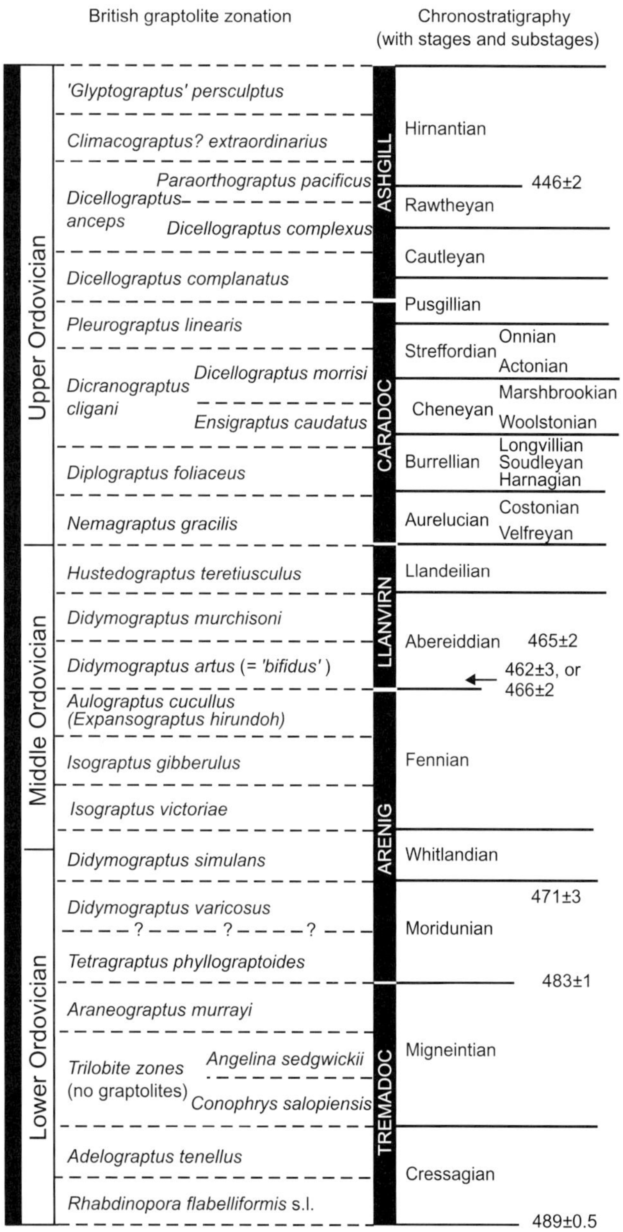

Fig. 3.13. Ordovician stratigraphy (modified from Fortey *et al.* 2000). See text (below) for a modification of the correlation of the graptolite zones with the stages of the Ashgill.

Pusgillian, Cautleyan and Rawtheyan, were established in the Cautley inliers of northern England, mainly on the basis of their trilobite faunas (Ingham 1966) and correlated with the *linearis*, *complanatus* and *anceps* graptolite biozones, on the evidence from sections in southern Scotland where graptolites and shelly faunas were found together. Recent discoveries of varied graptolite faunas in the Cautley inliers (Rickards 2002) suggests that the *linearis* Biozone extends from the Pusgillian to the lower part of the Rawtheyan, which would confine the *complanatus* and *anceps* zones to a relatively thin interval in the upper part of the Rawtheyan (Fig. 3.13). A full account of Ordovician sequences in the British Isles and their correlation is to be found in Fortey *et al.* (2000).

Arenig–mid-Caradoc: subduction-related basin evolution

Throughout the Ordovician, the Welsh Basin and its margins were dissected by faults that delineated sub-basins and intervening highs whose position changed with time in response to normal faulting or strike-slip displacements. The sub-basins showed high rates of subsidence (Fig. 3.6) and particularly thick successions (Fig. 3.14) and hosted the most prominent volcanic centres. The intervening highs were characterized by lower subsidence rates and thin shallow marine successions, commonly with stratigraphic breaks (Fig. 3.14).

Volcanic activity

Volcanic activity in the Welsh Basin started during late Tremadoc times, and had virtually ceased by mid Caradoc times. The volcanism occurred in two distinctly different phases, the first was a phase of arc volcanism, the second phase was a period of back-arc volcanism. In the first phase, Late Tremadoc subduction-generated NE–SW and N–S fractures, which became conduits for arc volcanism in the vicinity of Rhobell Fawr on the SE edge of the Harlech Dome and near Treffgarne in Pembrokeshire (Fig. 3.5) (Kokelaar 1988). The volcanic products in both these areas comprise basaltic and intermediate volcanic rocks with a calc-alkaline geochemical signature which reflect an island-arc episode of volcanism on continental crust, resulting from the southward subduction of Iapetus Ocean crust beneath Eastern Avalonia. A marked change in the composition of volcanic rocks occurred at the beginning of the Arenig, recorded by the appearance of mainly bimodal, basaltic and rhyolitic volcanic rocks, which continued to accumulate through to the Caradoc. These rocks form the spectacular profiles through the Snowdonia National Park, the coastline near Fishguard in north Pembrokeshire (Fig. 3.5) and are also recognized along the SE margin of the basin in the Welsh Borderland. The tholeiitic character of these rocks reflect a back-arc, extensional basin setting. The relatively minor volumes of intermediate andesite–rhyodacite magmas were the result of low-pressure fractional crystallization of the tholeiitic basalts. However, their small volumes, together with the lack of geophysical evidence for large mafic intrusions, have been used to suggest that the large volumes of acid volcanics resulted from partial melting of the crust. Most recent studies have concluded that the whole range of compositions was derived by crystal fractionation from three basaltic parents and not from a single zoned magma chamber. The volcanic rocks range from effusive lavas to highly fragmented pyroclastic deposits resulting from highly explosive eruptions. Many of the thicker accumulations, whether basic or acid, show evidence for a range from high level intrusion through to extrusion. Abundant evidence indicates that the environment of accumulation was dominantly marine. Any sub aerial surfaces or volcanic islands were quickly eroded and variously reworked into the adjacent sedimentary systems.

Dolerite intrusions, typically transgressive sills, are common throughout the Ordovician in the Welsh Basin. For many years it was considered that these intrusions were end-Silurian and not an integral part of the Ordovician volcanism. However, the intrusions are restricted to the volcanic sequences, such as the Caradoc outcrop in Snowdonia, and recently they have been incorporated into interpretations of the overall geochemical evolution of the volcanism.

Arc volcanism

The only arc-volcanism in the Welsh Basin occurred in a short interval in the late Tremadoc and is confined to the area of Rhobell Fawr in North Wales and Treffgarne in South Wales (Fig. 3.15). At Rhobell Fawr, a succession of commonly brecciated basaltic lavas, *c.* 1000 m thick, is associated with some sedimentary breccias and alluvial deposits (Rhobell Volcanic Group), which unconformably overlie Tremadoc and Cambrian strata. The basalts were erupted subaerially from N–S-trending fissures and successive basalt flows overstep eastwards across a highly irregular surface. The fissures, intruded by a suite of subvolcanic dolerite and diorite intrusions, were associated with the Rhobell Fracture which was responsible for the uplift of the volcanic complex prior to its erosion and the

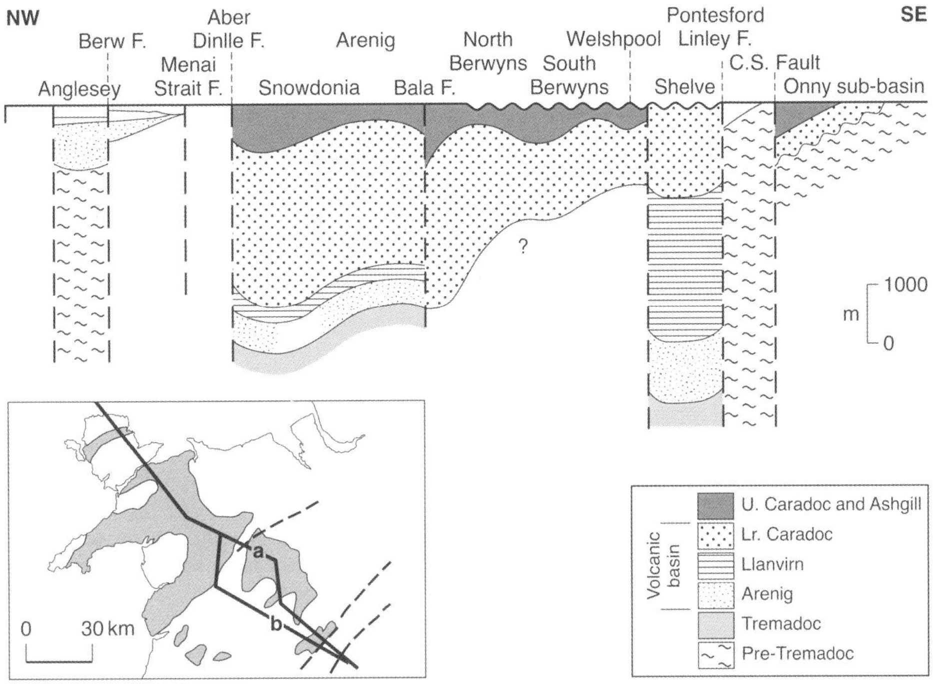

Fig. 3.14. NW–SE stratigraphic profile of the Ordovician across the Welsh basin (line of section a), showing the sub-basins associated with strike-slip movements in Anglesey, Shelve and South Shropshire (Onny Sub-basin, and the sub-basin in Snowdonia associated with fault-related volcanicity). The stippled area on the inset shows the Ordovician outcrop.

succeeding deposition of the basal Arenig sedimentary rocks. At Treffgarne, in north Pembrokeshire, basaltic andesite lavas with some intercalated volcaniclastic sediments (Treffgarne Volcanic Group), lie unconformably below early Arenig sandstone turbidites. The lavas are geochemically similar to the Rhobell Fawr rocks and are possibly their equivalents.

The sub-Arenig unconformity
A widespread unconformity of variable magnitude separates the Arenig from the underlying rocks within the Welsh Basin and on the adjacent Irish Sea Horst Complex, which had been subjected to variable degrees of deformation, possibly as a result of compression, arising from oblique convergence of the Iapetus oceanic plate. The compression caused sinistral strike slip movements within and adjacent to the Mona Terrane which culminated in transpressive deformation of the Mona Group (Fig. 3.15) and variable degrees of deformation within the basin, which is partly reflected in variations in the magnitude of the resulting unconformity. In the central and east Harlech Dome Arenig lies on upper Cambrian strata and oversteps northwards onto late Precambrian and early Cambrian rocks in Arfon (Fig. 3.10).There is a similar overstep from the central Harlech Dome to St Tudwal's in the Llŷn Peninsula (Fig. 3.10) At Rhobell Fawr Arenig overlies the arc volcanics, which in turn lie unconformably on the Tremadoc. In the southern part of the basin, near St David's and on Ramsey Island, Arenig rocks rest with substantial unconformity on upper Cambrian rocks, but to the SE of Carmarthen they appear to be conformable above the Tremadoc. Near the eastern margin of the basin, in the Shelve area, there is also little or no break between the Tremadoc and Arenig. On the Midland Platform the locally thick Tremadoc sequences were gently folded and uplifted, but little or no Arenig cover is preserved.

Arenig sediments, biofacies and volcanic rocks: an introduction

Following end-Tremadoc deformation, basin subsidence was resumed and marine sediments accumulated as sea level began to rise. East of the Welsh Borderland Fault System no rocks of unequivocal Arenig age are recorded and the area may have

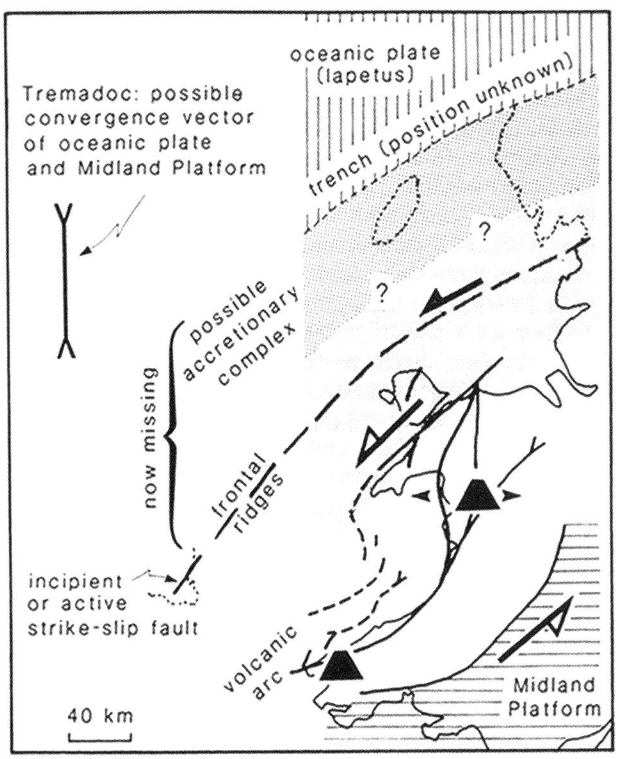

Fig. 3.15. A plate-tectonic model to explain late Tremadoc tectonism in Wales. Half arrows indicate possible shear sense, small arrows the local extension and topped triangles the location of arc volcanoes. The coastline of the English Lake District, Isle of Man and SE Ireland are represented by dotted lines (from Kokelaar 1988).

been emergent following late Tremadoc deformation. However, the Lickey Quartzite on the Midland Platform, near Birmingham, is probably of Ordovician age (Molyneux in Old *et al.*

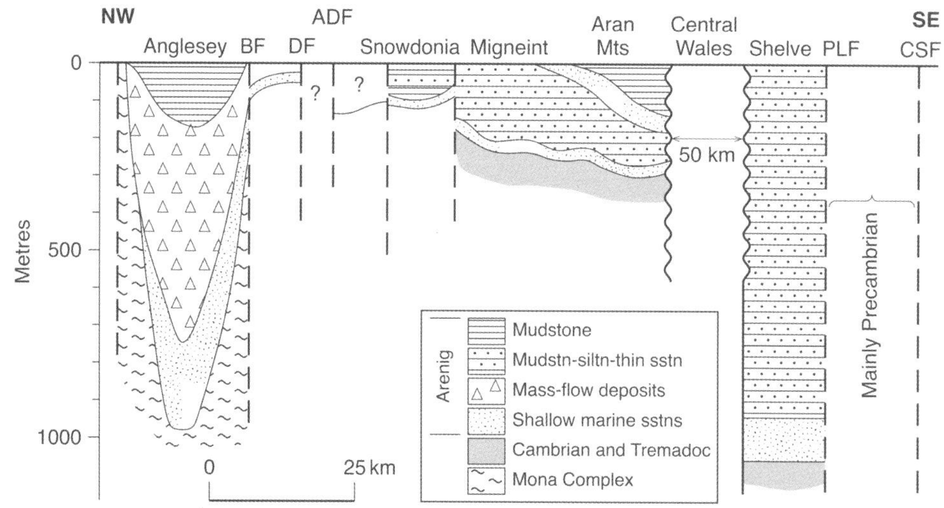

Fig. 3.16. A schematic NW–SE profile of the Arenig across the Welsh Basin and its margins, from Anglesey to the Welsh Borderland Fault Complex (line of section b on inset in Figure 3.14). Anglesey, Snowdon and Migneint are located in Figure 3.5.

1991) and possibly of Arenig age, which would suggest that Arenig sediments might have been more widely deposited and subsequently removed. On the opposite margin, the Irish Sea Horst Complex, Anglesey, was not submerged until the Fennian so might have been a source of sediment for the Welsh Basin in the earlier part of the Arenig.

The extension associated with the development of the back arc basin resulted in only moderately low rates of subsidence, which suggests that the thermal effects of the volcanism may have counteracted any tendency for basement subsidence (Prigmore *et al.* 1997). However, thick sequences did accumulate close to the margins of the basin, in the Shelve area, immediately west of the Pontesford–Linley Fault and in fault-bounded basins in Anglesey and Llŷn (Fig. 3.16). Within the axial part of the Welsh Basin, Arenig sequences are of a deeper marine facies and generally thicken southwards, where subsidence was particularly rapid (Fig. 3.6). More locally, depositional patterns were influenced by both NE–SW and E–W faults.

The laterally variable facies within the Arenig commonly results in problems of correlation. The thick sequences in South Wales have been divided into a series of trilobite biozones (Fortey & Owens 1987), which can be approximately correlated with the British graptolite zonation (Fig. 3.17). In North Wales species of the trilobite *Neseuretus* and the graptolite *Azygograptus* were used by Beckly (1987) to distinguish the Moridunian, Whitlandian and Fennian stages.

The Arenig transgression

The early Arenig transgression was probably global, and in many regions from Saudi Arabia to Portugal the lower part of the Arenig is represented by one of the most extensive blankets of quartzites in Early Palaeozoic history. In Britain this is represented only by the Stiperstones Quartzite in the Shelve area and possibly by the Lickey Quartzite in the Midlands. Elsewhere in the Welsh Basin and on the Irish Sea Horst Complex the basal transgressive deposits are generally coarse, immature and of local origin.

The Arenig transgression was diachronous, ranging in age from Moridunian to Fennian. The facies and thickness of Arenig sequences were influenced by topographic highs and lows, related to late Tremadoc faulting (Beckly 1987; Kokelaar *et al.* 1984). However, regardless of the age of the transgression, the unconformity is overlain by coarse sediments followed by an upward-fining sequence, incorporating a wide variety of laterally changing facies reflecting local environments (Traynor 1990). Back-arc volcanicity commenced only locally in the Arenig, mainly being confined to an early stage of the Aran Volcanic Group in North Wales, and to volcaniclastic sediments and 'distal' acidic tuffs in middle and late Arenig sequences on Ramsey Island (Kokelaar *et al.* 1985).

STAGES	TRILOBITE BIOZONES	GRAPTOLITE BIOZONES
FENNIAN	Dionide levigena	Expansograptus hirundo
	Bergamia rushtoni	Isograptus gibberulus
	Stapeleyella abyfrons	
WHITLANDIAN	Gymnistomix gibbsi	Didymograptus simulans
	Furcalithus radix	
MORIDUNIAN	Merlinia rhyacos	Didymograptus varicosus
	Merlinia selwynii	
	?= earlier biozone in Lake District	Tetragraptus phyllograptoides

Fig. 3.17. Arenig biostratigraphy.

Basins in the Irish Sea Platform-Anglesey and western Llŷn

During the Moridunian and Whitlandian the area of Anglesey was emergent, but it was transgressed in the Fennian and then locally subsided rapidly to accommodate more than 700 m of basinal deposits, nearly as much as any succession within the Welsh Basin.

Arenig rocks crop out between Rhosneigr, Carmel Head and Dulas Bay and in small outliers, which occur discontinuously on the north Anglesey coast, north of the Carmel Head thrust (Fig. 3.18). Less-well exposed outcrops occur near Llangwyllog and in the Berw fault complex (Fig. 3.18). The Arenig succession appears to have accumulated in a single narrow steep-sided basin, assuming there has been no displacement between the separate outcrops. The main outcrop in the Principal Area (Fig. 3.18) represents the axial part of the basin, where shallow-marine sandstones and subordinate conglomerates rest on the Mona Complex. There are marked vertical and lateral facies changes between conglomerates, pebbly sandstones and sandstones, some with cross-stratification or hummocky cross-stratification. Thickness variations probably mainly reflect the topographic relief across the unconformity, but a more substantial tectonically-induced change occurs across the Ceido fault, where different shallow-marine sequences represented by the Carmel and Foel formations occur on opposite sides. Within the Berw Fault complex the unconformity is not seen, but

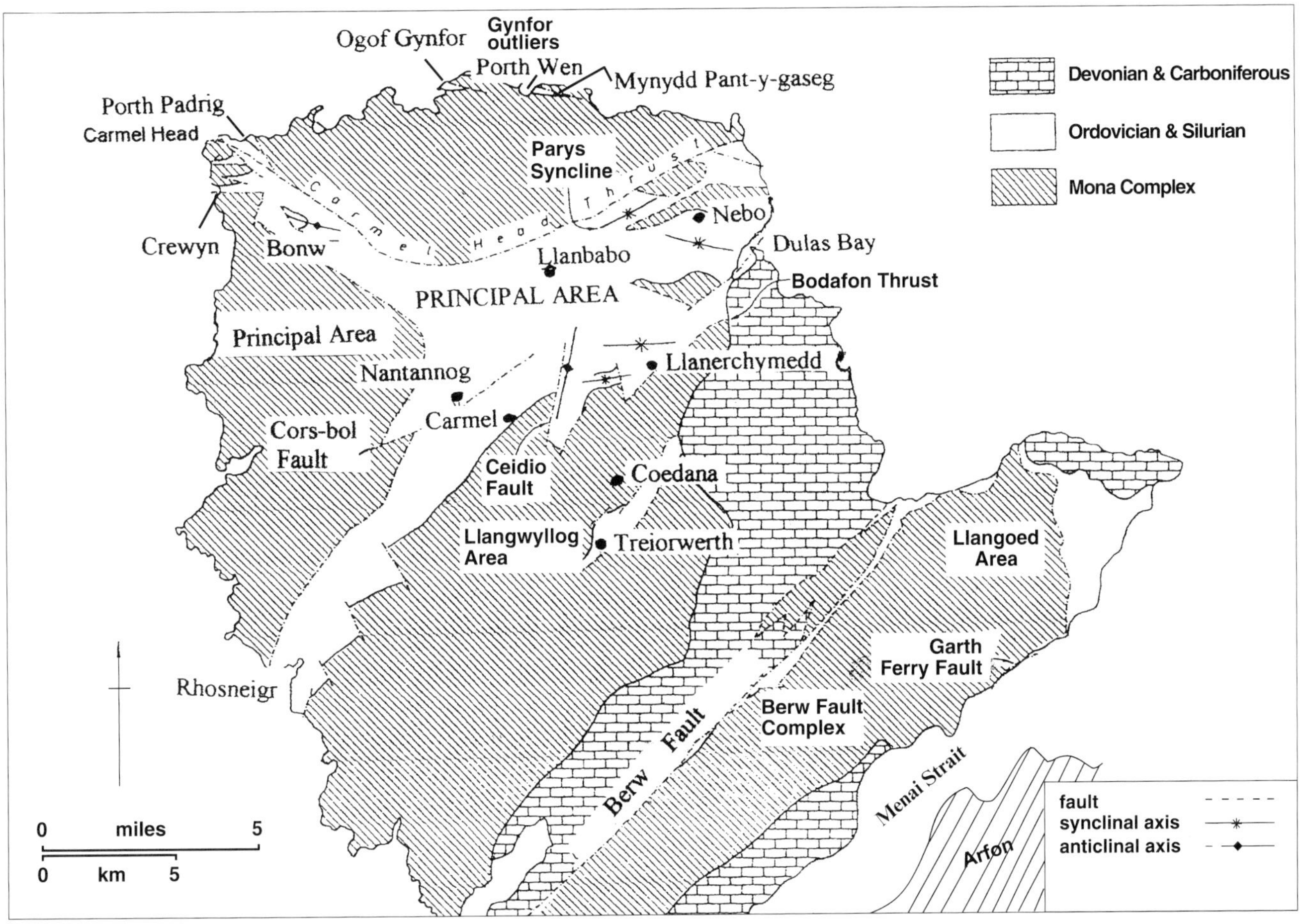

Fig. 3.18. Distribution of Arenig outcrop in Anglesey and selected localities (adapted from fig.1, sketch map of Anglesey, with main tectonic features in Bates 1972. Reproduced by permission: John Wiley & Sons Ltd).

~20–30 m of coarse grained sandstones (the Berw-uchaf Grits) are discontinuously exposed and locally contain a rich brachiopod fauna (Bates 1972). In the Llangoed area the Garth Ferry Grit, a coarse–granule and pebbly sandstone, rests on an irregular unconformity, exposed on the shore of the Menai Strait.

Above the basal formations, coarse-grained gravity-flow deposits (Treiworth Formation) (Fig. 3.19), up to 700 m thick in the Treiworth area, record a transitional change into coarse sandstones and granule and pebble conglomerates with siltstone and shale partings. In the lower parts of the transitional beds there are lenticular shell accumulations with as many as 18 species of brachiopods, indicating that shallow-marine environments still prevailed (Neuman & Bates 1978). Higher in the formation, conglomerates and breccias predominate, some containing angular blocks derived from the New Harbour Group of the Mona Complex. Most of the beds appear to have been deposited in deep-marine environments from debris flows or from high-concentration turbidity currents. Laterally and vertically the Treiworth Formation passes into the Nantannog Formation which consists predominantly of silty mudstones with a variable admixture of sand- to pebble-sized clasts. A large proportion of the sequence appears to have been deposited from relatively dilute muddy debris flows, but interbedded there are sporadic beds of granule–pebble conglomerate, and at Nantannog there is a 20 m thick pebble–cobble conglomerate. The formation appears to pass laterally eastwards and vertically upwards into mudstones and siltstones. The large volume of coarse sediment within the Treiworth Formation reflects a large, narrowly-constricted, breccia and sand fan similar to those described by Surlyk (1978) from the Cretaceous of East Greenland. The more muddy gravity flows in the Nantannog Formation apparently formed thinner and wider fans. The massive influx of coarse sediments suggests that the development of the basin was the result of rapid subsidence, accompanied by uplift along the margins.

In the marginal successions of the Gynfor outliers, massive pebble–cobble conglomerates (Porth Cynfor Conglomerate Formation) fill the lower part of deep incisions into the underlying Mona Group (Fig. 3.20). The clasts are subangular–rounded and the deposits are generally clast supported and moderately sorted. They are possibly a beach accumulation, but in view of their massive nature they are more likely to be gravity-flow deposits derived from a nearshore location, The succeeding Torllwyn Formation consists of a variety of thick-bedded coarse sandstones and poorly sorted breccias, with quartzite blocks up to 1.7 m in diameter that came to rest unconformably on the Gwna Group (Plate 4). In some sections interbedded grey and cream shales and thinner-bedded sandstones have yielded varied brachiopod faunas (Bates 1972). The poorly-sorted breccias were most probably deposited from debris flows or slides and the graded sandstones are likely to be turbidites. The associated brachiopod faunas in the finer sandstones are apparently not transported and record a relatively shallow-marine environment, albeit at the base of a slope. The Torllwyn Formation overlaps the Porth Cynfor Conglomerate and onlaps the slopes of the depressions, until they are filled, so that the upper part of the formation may have covered the whole belt of the Gynfor outliers (Fig. 3.20).

On the Llŷn Peninsula, an Arenig sequence (Wig Bach Formation) ranges in age from Moridunian to Fennian and is approximately 550 m thick (Gibbons & McCarroll 1993). A

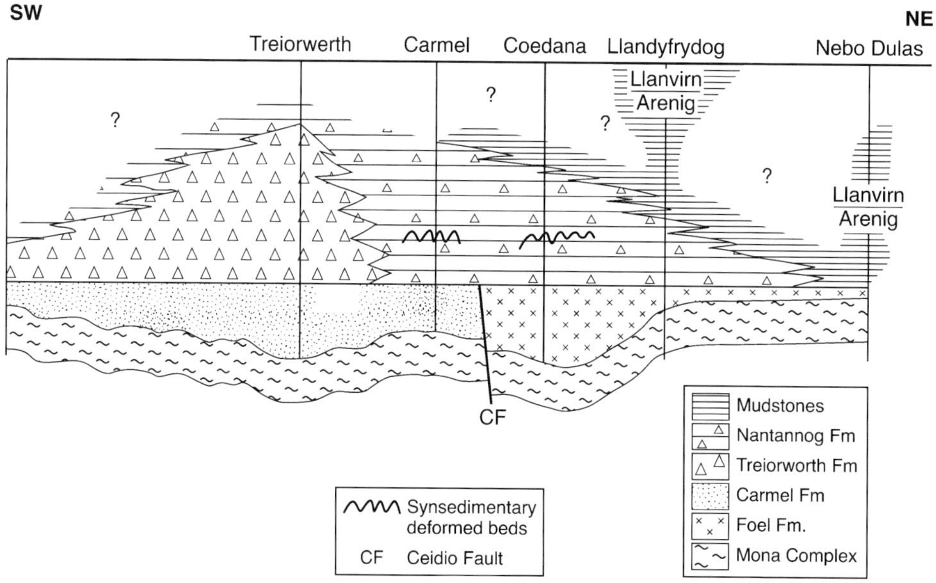

Fig. 3.19. Approximately SW–NE profile of Arenig facies in the 'Principal area' of Anglesey from Treiorworth to Dulas Bay (data from Bates 1972).

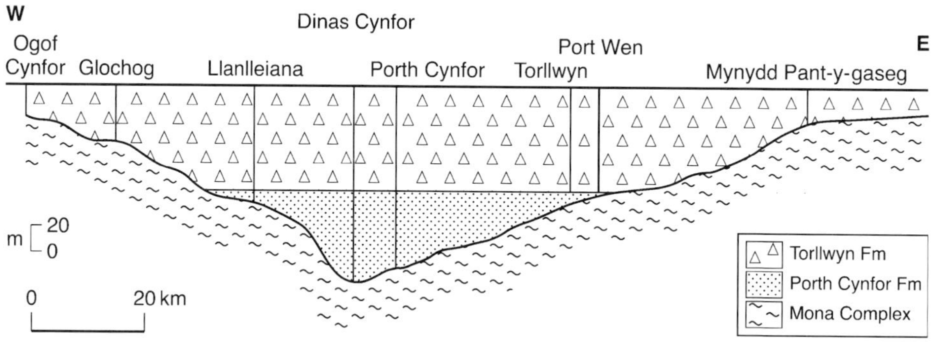

Fig. 3.20. Approximately W–E profile of Arenig facies in the Gynfor Outliers along the north coast of Anglesey (adapted from fig. 6, Vertical sections through the rocks of the principal area and the Gynfor outliers along the north coast of Anglesey; from Bates 1972. Reproduced by permission: John Wiley & Sons Ltd).

basal conglomerate is overlain by a conglomeratic sandstone with ferruginous ('chamositic'), phosphatized ooliths and is interpreted as a condensed sequence on a drowned surface (Gibbons & McCarroll *et al*, 1993), formed as the basin subsided and sea-level rose in the early Arenig. The succeeding strata are composed of dark grey mudstones, laminated siltstones and bioturbated sandstones. The two main sandstones contain a shallow-marine ichnofauna including *Skolithos, Phycodes* and *Teichichnus,* but the mudstones and siltstones generally reflect deeper water environments (Crimes 1970b). On the St Tudwal's Peninsula, a very different Arenig sequence consists of the St Tudwal's Sandstone Formation, of Fennian age, resting unconformably on folded Cambrian rocks. A thin basal conglomerate is overlain by about 150 m of predominantly thinly interbedded fine sandstones and shales with intervals containing a 'fodichnia' ichnofauna, including *Teichichnus* and *Phycodes* and other intervals with rare trace fossils. The changes in ichnofauna reflect changes in environment from above to below wave-base (Crimes 1970). Three sandstone units containing beds with *Skolithos* record the shallowest environments, while beds with *Cruziana* indicate deeper water environments.

The NW part of the Welsh Basin, North Wales

The Arenig in North Wales is predominantly a shallow-marine, upward-fining sequence, recording the early Arenig transgression and the renewed onset of basin subsidence associated with the early stages of a new back-arc basin. The main outcrop is around the Harlech Dome (Fig. 3.21), where the sequence (Allt Lŵyd Formation) is of Fennian age. However, a different sequence is present to the east in the Arenig district where the lower part of the Carnedd Iago Formation is of Moridunian age, but Whitlandian and lower Fennian rocks appear to be missing (Fig. 3.21) (Beckly 1987; Fortey *et al.* 2000). Throughout the north and east flanks of the Harlech Dome the basal Garth Grit Member consists of shallow-marine coarse sandstones belonging to the *Neseuretus* biofacies (Fortey & Rushton 1992), deposited in a variety of deltaic and nearshore tidal environments (Traynor 1990). These beds commonly contain winnowed oncolites, previously referred to as '*Bolopora undosa*'. In the NW, the Garth Grit includes conglomerates deposited from non-cohesive gravity flows and interbedded shallow-marine cross-stratified sandstones that were interpreted as being deposits within fan deltas (Fig. 3.21) (Traynor 1990). Further east the fan deltas pass into fluviodeltaic deposits, with channels filled with stacked cross-stratified coarse and pebbly sandstones, and intervening bioturbated sandstone and shale beds containing *Cruziana, Skolithos* and *Planolites.* Laterally, around the NE of the Harlech Dome, the Garth Grit is composed of coarse to- very coarse sandstones with cross-stratified sets commonly 0.5–1 m thick and occasionally up to 3 m. Palaeocurrents are bipolar and reactivation surfaces are common indicating a dominantly tidal environment with sandwaves and intervening channels (Fig. 3.21) (Traynor 1990).

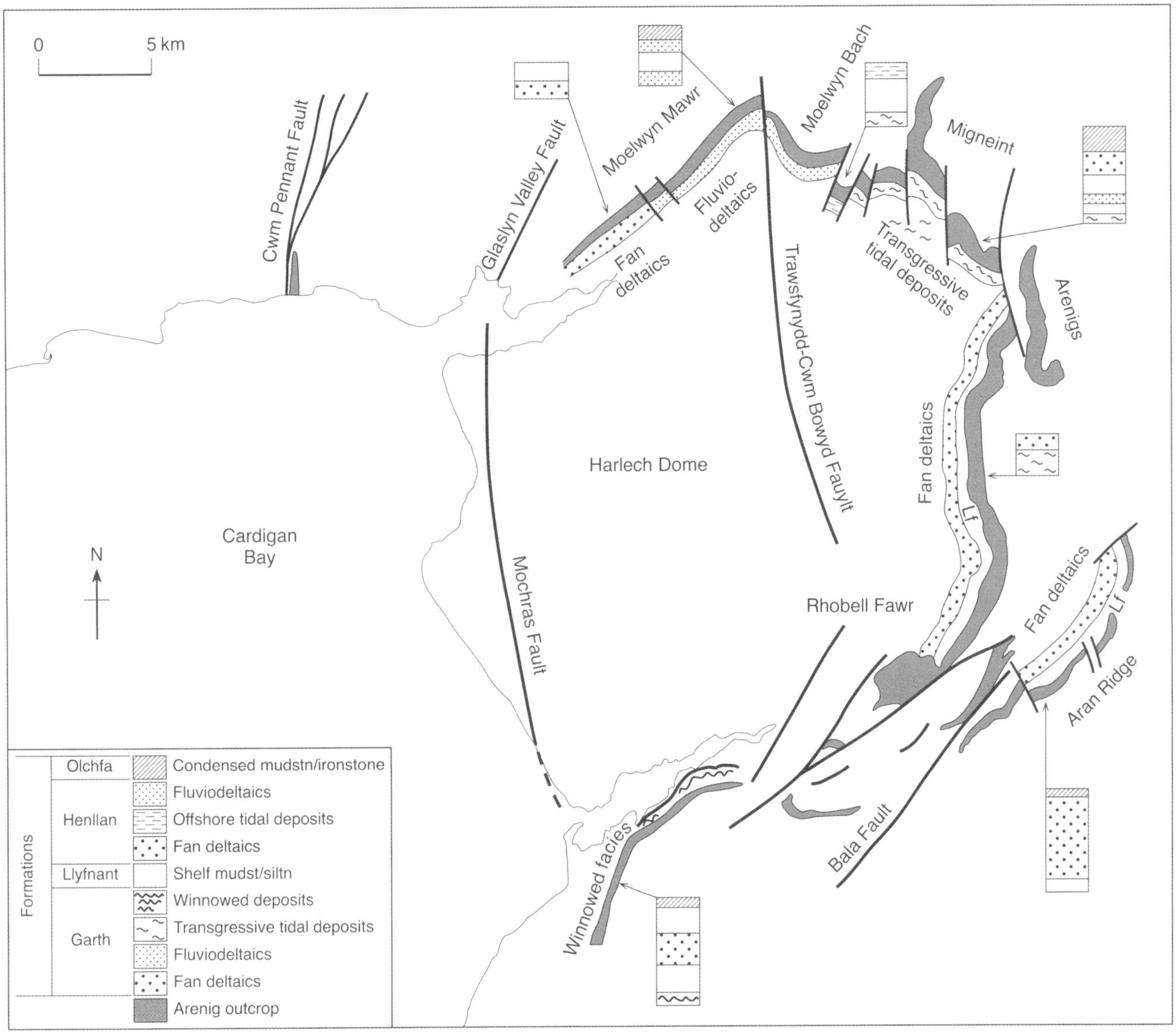

Fig. 3.21. Facies variations in the Allt Llwyd Formation around the Harlech Dome, with representative sedimentary logs. The lateral changes in depositional environments of the Garth Grit Member around the north and south flanks are shown, and also the distribution of fan deltaics of the Henllan Member on the east flank (modified from Traynor 1990).

To the south, the Garth Member thins and is present as a winnowed basal lag or is absent (Fig. 3.21). The Llyfnant Member succeeds the Garth Member, or overlaps it to rest directly on the unconformity, reflecting the continuing transgression. The Llyfnant Member comprises flat-laminated mudstones and siltstones with thin-bedded fine sandstones, resembling turbidites, but of storm origin. The typical ichnofauna includes *Teichichnus, Phycodes, Chondrites* and *Planolites,* which in the absence of wave ripples suggests a depositional environment a little below wave-base. The Llyfnant member grades up into the coarse-grained Henllan Member, which from the Aran ridge to the Migneint (Fig. 3.21) represents a fan delta with its source to the south. The proximal sequence is composed mainly of gravity-flow deposits of sandstone, conglomerate and breccia interbedded with marine deposits. Within this sequence is the spectacular Aran Boulder Bed (Ridgeway 1975), a composite unit of clast-supported, pebble–boulder conglomerates and rare breccias in lenses 60–300 m thick, 400 m wide and 2500 m long. They are interpreted as debris flows derived from a shallow intrusion or extrusive dome and deposited as alluvial fans. Away from the Aran Ridge, facies change progressively northwards to more distal sandstones with a rich shelly fauna (Zalasiewicz 1984) and an ichnofauna including *Skolithos, Teichichnus* and *Phycodes.* Laterally, on the north flank of the Harlech Dome, the facies change to volcaniclastic sandstones, interpreted as offshore, tidally-affected sand shoals; to the south of the Harlech Dome volcaniclastic sands were locally deposited in a fluviodeltaic environment (Traynor 1990). The lateral variability of the facies has been interpreted as being related to fault bounded, ENE–WSW-orientated basins, closed at both ends by N–S faults. Sediments were fed across the bounding faults to form fans and deltas and in some places were redistributed along the margins of the basins by tidal currents. Additionally some material was supplied from the incipient Harlech Dome in the centre of the basin.

A substantial transgression, during which the Olchfa Member was deposited, followed the deposition of the Henllan Member. To the west of Arenig the Olchfa Member consists of black fissile mudstones. South of the Arenig Mountains the member is a typical condensed trangressive unit, occurring as black silty pyritous mudstones containing phosphate nodules with ooids and pisoliths of phosphate replacing chamosite; associated are sponge spicules, brachiopods, gastropods and trilobite fragments.

Central Wales, South Wales and the Welsh Borderlands

Southeast of the Harlech dome, throughout central Wales, there are no outcrops of Arenig rocks. However, at the eastern margin of the basin, adjacent to the Pontesford–Linley Fault Complex, the Arenig is approximately 1000 m thick in the Shelve area (Fig. 3.16). The basal Stiperstones Quartzite rests with minor disconformity on Tremadoc rocks and records the early Arenig transgression. It contains *Neseuretus ramseyensis* of Moridunian age and a varied shallow-marine ichnofauna with common *Skolithos,* and *Chondrites*. The sandstones are of cross-stratified tidal- and storm-influenced facies, overlain by a transition into the Mytton Flags consisting of muddy siltstones and siltstones, with subordinate sandstones. The succession is pervasively bioturbated and sedimentary structures are rarely preserved, but the presence of *Neseuretus* sporadically through the sequence (Whittard 1979) indicates a relatively shallow-marine environment. The accumulation of about 900 m of nearly uniform facies suggests a delicate balance between subsidence and sedimentation. The relationship of the Shelve region to the Welsh Basin is not clear, but it is likely that it was a shallow-marine sub-basin (Cave & Hains 2001) bounded to the east by the Pontesford Fault Complex and to the west by the Severn Valley Fault.

In the SW part of the Welsh Basin the Arenig is apparently up to approximately 1000 m in the Whitland area (Fig. 3.22). The sequence is thinner to the east near the Carreg Cennen Fault complex in the Carmarthen area and westwards near St David's (Figs 3.22 and 3.23). The late Tremadoc volcanics in the Treffgarne area were uplifted along pre-existing volcano-tectonic faults into a topographic high, creating two sub-basins (Traynor 1988) (Fig. 3.22). Within the lower Arenig succession of north Pembrokeshire there is widespread evidence of sporadic volcanic activity, sometimes associated with tectonic movements. The lower part of the Ogof Hen Formation (Arenig) is distinctively contaminated by volcanic debris and some thick tuffs of intermediate composition (informally referred to as the Llandeloy Ashes) have been recognized. Some associated mass flow deposits of volcaniclastic sandstones reflect tectonically induced disruption of unconsolidated debris. In the vicinity, fine-grained devitrified rhyolites and acid tuffs (Roch Rhyolite Group) have been considered to reflect contemporaneous volcanic activity although the field relationships are much obscured. Further east, around Carmarthen, Arenig mudstones include volcaniclastic turbidites with much acidic volcanic debris. On Ramsey Island to the west, thin beds of acidic tuffs in the Abermawr Formation indicate 'distal' volcanism and major syndepositional slides below the Road Uchaf Formation indicate the continued tectonic instability.

The basal Arenig sediments in SW Wales (Fig. 3.23), as represented by the Ogof Hen Formation, are generally shallow-marine sandstones of the *Neseuretus* biofacies. A thick sequence of fluviodeltaic sandstones occurs near Carmarthen, while nearby sandstones at Llangynog have yielded an unusually rich shallow-marine fauna, including a variety of bivalves (Cope 1996*a*), *Ramseyocrinus cambrensis* one of the earliest crinoids (Donovan & Cope 1989) an early parablastoid, *Blastoidocrinus antecedens* (Paul & Cope 1982) and an asteroid *Petraster ramseyensis*. The Ogof Hen Formation is not seen at Whitland, where the lowest part of the Arenig is unexposed. At St Davids the Ogof Hen Formation has a basal conglomerate with oncolites above the sub-Arenig unconformity, overlain by mudstones and storm-influenced sandstones which grade into more generally planar laminated siltstones and mudstones, formed as the water depth increased. The succeeding sequence, throughout the SW part of the basin, is predominantly composed of mudstones, some with interbedded turbidites, but shales and mudstones predominate towards the top of the Arenig sequence (Figs 3.22 and 3.23). In the Carmarthen Formation on the

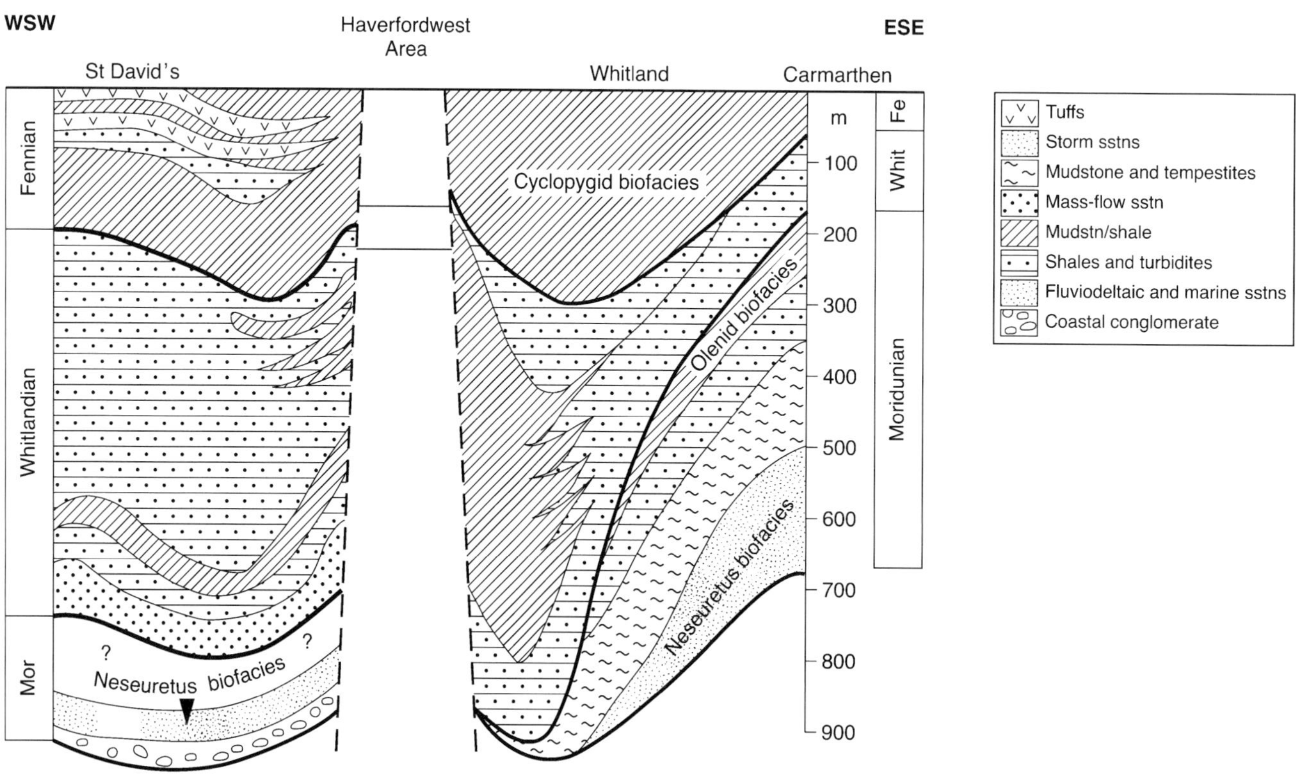

Fig. 3.22. Schematic profile of the basin in SW Wales during the Arenig. Thicknesses of units, and facies-interpretations are from Traynor (1990). Biofacies are from Fortey & Owens (1987). The positive area west of Whitland is inferred, but is consistent with the kind of sub-basins envisaged by Traynor (1988).

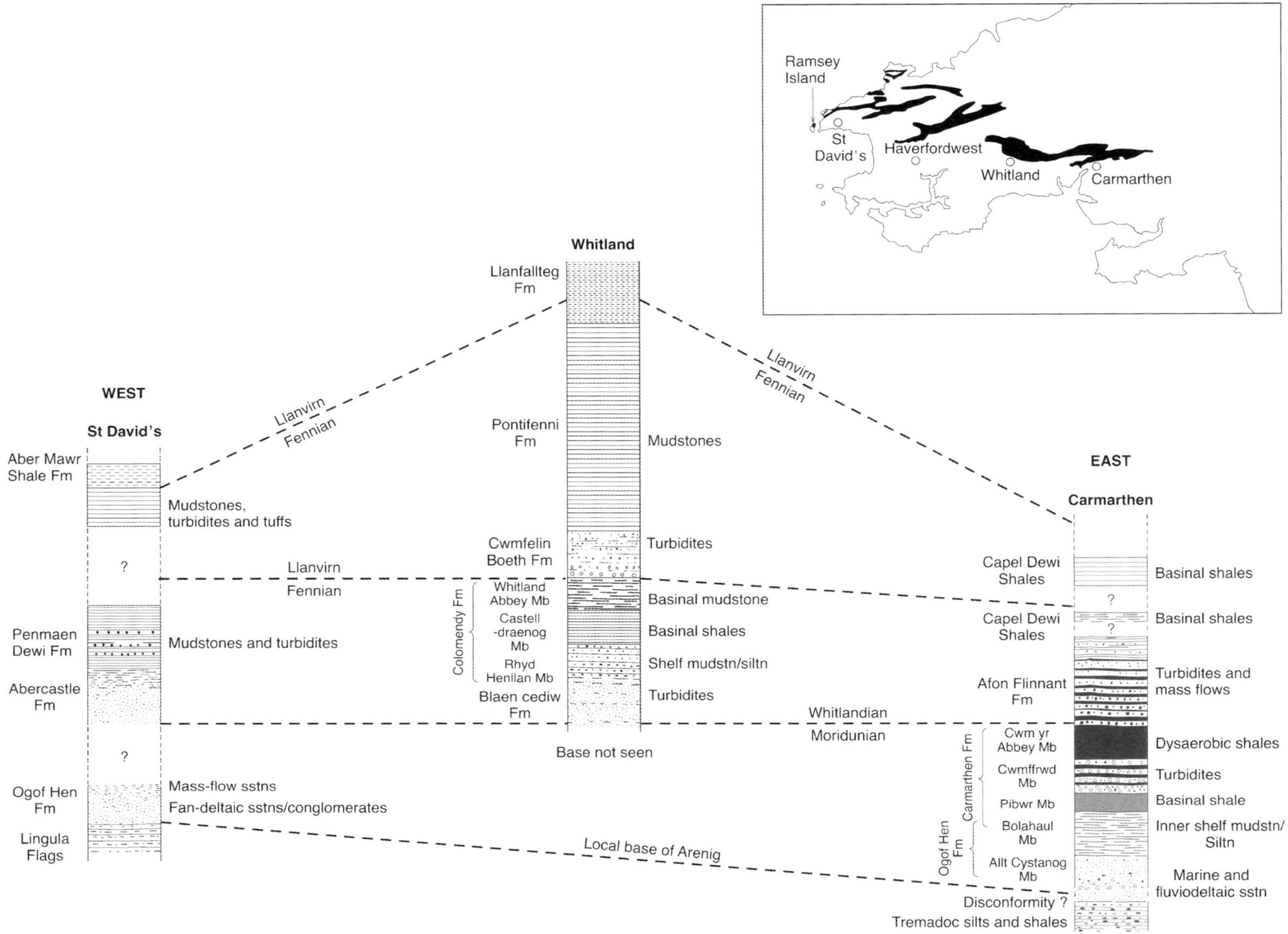

Fig. 3.23. Stratigraphy of Arenig sequences in SW Wales with environmental interpretations of the facies (stratigraphy from Fortey & Owens (1987) and environmental interpretations after Traynor 1988).

SE side of the basin, dark shales of the olenid biofacies are interbedded with turbidites in the lower part of the deep-water sequence. A depression, bounded to the northwest by a topographic high (Fig. 3.22) has been invoked to produce an environment with restricted circulation in which dysaerobic sediments might have accumulated (Fortey & Rushton 1992). Within the shales in the upper part of the Arenig, the Pontyfenni Formation (Fig. 3.23) records a major addition to the trilobite fauna with about 20 new species, many of which are cyclopygids, this is the first known occurrence of the cyclopygid biofacies. Large eyed pelagic forms are associated with the blind or nearly blind (atheloptic) species. 'Atheloptic assemblages' have been interpreted as living in oceanic sites where the sedimentary surface was below 300 m (Fortey & Owens 1987). This would suggest that in the late Arenig the southern part of the Welsh Basin was deeper or more open to the Iapetus Ocean than previously thought.

An eustatic regression occurred at the end of the Arenig, marked globally by shallower facies or widespread disconformities (Fortey 1984), but not recorded in many British outcrops. However, the event is well defined at the boundary between the Pontyfenni and Llanfallteg formations in South Wales, where deep-water dark shales, with a cyclopygid biofacies, give way to light grey shales with a large-eyed trilobite biofacies (Fortey & Owens 1987).

Llanvirn sedimentary and volcanic rocks

The late Arenig regression was followed by a substantial early Llanvirn eustatic transgression (Fortey 1984), which is widely recognized in Avalonia and Gondwana, where Llanvirn shales commonly succeed Arenig quartzites. There is, however, no record of Llanvirn on the Midland Platform, but Llanvirn rocks, deposited within the Anglian Basin, have been recorded from boreholes at Great Paxton, Huntingdon and Eyam in Derbyshire. Throughout the Welsh Basin and the basins within the Irish Sea Horst Complex, deep-marine shales and mudstones prevail in the early Llanvirn sequences. Subsequently, continued basinal subsidence maintained muddy basinal environments in many areas throughout Llanvirn times. However, thick volcanic sequences in the Shelve area, in the Builth inlier and near Fishguard and Ramsey Island in SW Wales accumulated in sub-basins (Figs 3.24 and 3.25). The large volcanic centre at Cadair Idris was also active through Llanvirn times and there was a lesser volcanic centre near Manod in Snowdonia (Fig. 3.25). The volcanics were mainly submarine, but some emerged as volcanic islands and became local sources of coarse volcaniclastic sediments.

The Irish Sea Horst Complex (Anglesey and western Llŷn)
In central Anglesey basinal shales persisted into the Llanvirn, but some local uplift of the basin margins is indicated by the

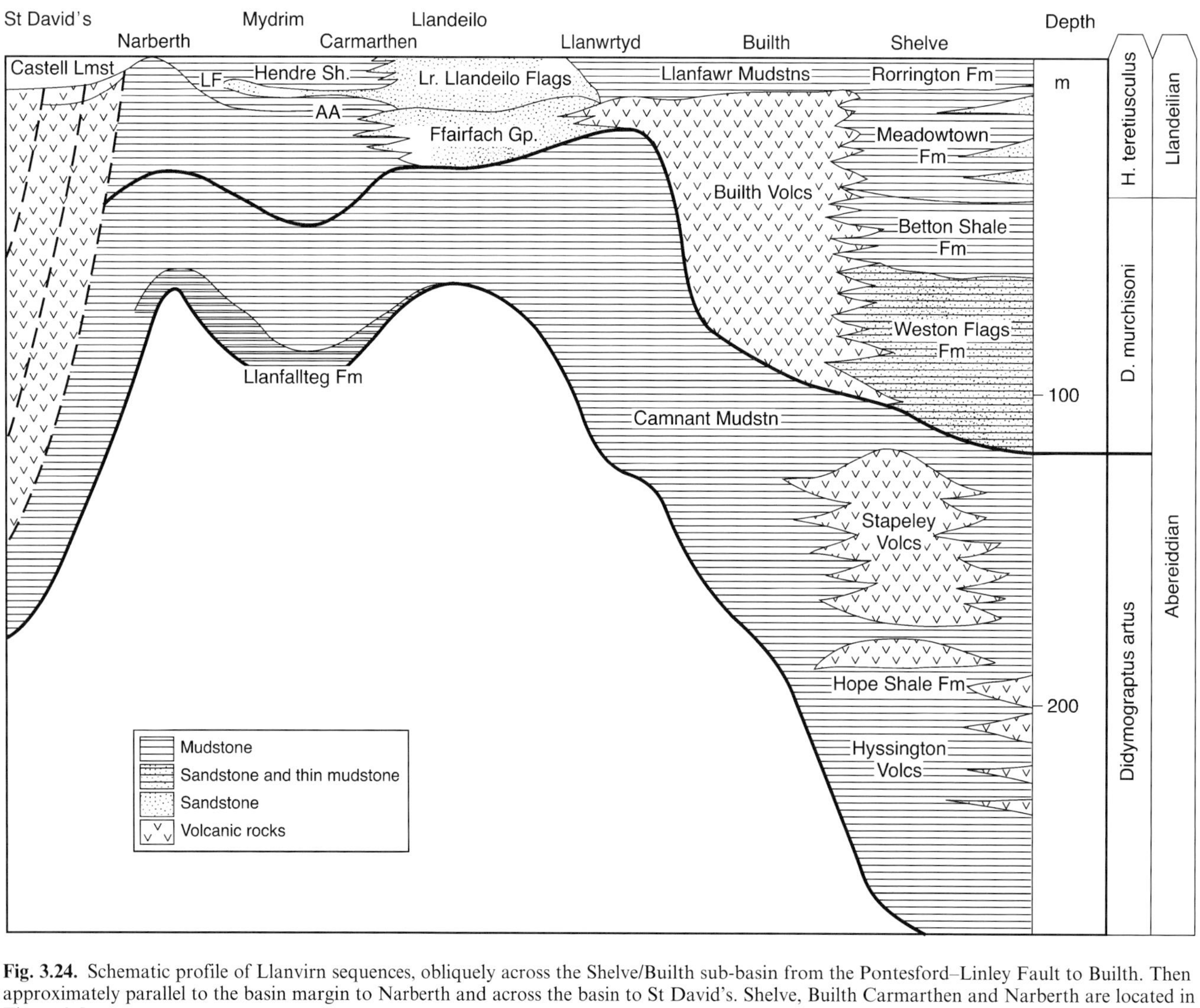

Fig. 3.24. Schematic profile of Llanvirn sequences, obliquely across the Shelve/Builth sub-basin from the Pontesford–Linley Fault to Builth. Then approximately parallel to the basin margin to Narberth and across the basin to St David's. Shelve, Builth Carmarthen and Narberth are located in Figure 3.5.

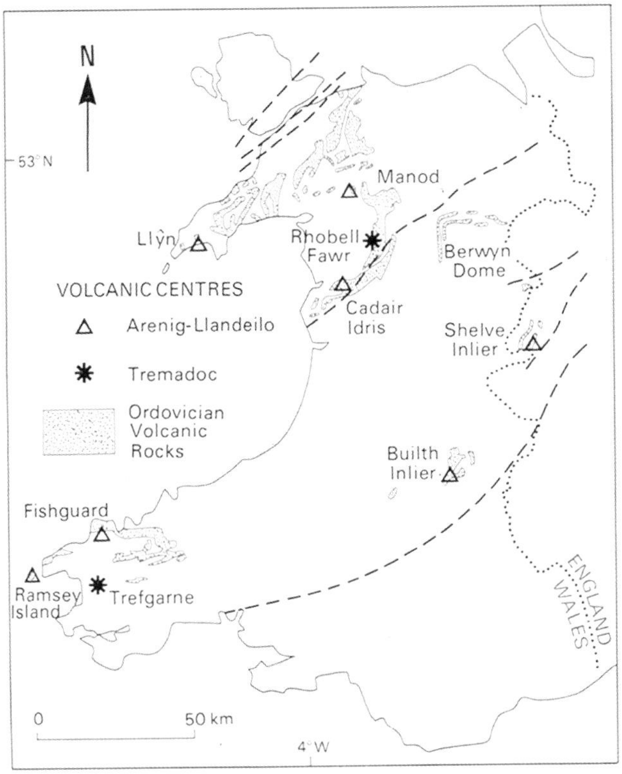

Fig. 3.25. Location of Llanvirn volcanic activity in the Welsh basin (from fig. 4 in Howells *et al.* 1991). (Reproduced by permission of the British Geological Survey@NERC. All rights reserved. IPR/60-06C.)

thick, lenticular, cobble-conglomerate debris flows forming the Bod Deiniol Formation (180 m thick) in the Llanbabo area. Towards the northern basin-margin, in the Gynfor outliers, the Llanvirn is missing where the Lower Caradoc oversteps on to the Arenig. In western Llŷn the base of the Llanvirn is a thin (< 1 m) pebbly sandstone at the top of the Porth Meudwy Formation, overlain by 2–3 m of bioturbated siltstones and fine sandstones. Above are approximately 3 m of tuffaceous beds, succeeded by a silty pisolitic ironstone 1.5 m thick, containing graptolites and abundant, mainly fragmented chitinozoans. Ooids are most abundant in the upper part of the bed, where some are phosphatized and in places associated with a phosphatic cement. At some other locations, the pisolitic horizon is manganiferous (Gibbons & McCarroll *et al*, 1993). The pebbly sandstones and bioturbated sandstones probably reflect the late Arenig regression and the ironstone a flooding surface at the onset of the Llanvirn transgression. Nevertheless, the succeeding tuffaceous sediments and bioturbated sandstones were deposited in relatively shallow water.

North and South Wales

In North Wales the late Arenig–Llanvirn transgression is reflected in the relatively deep marine siltstones and mudstones that constitute the Nant Ffrancon Subgroup, which crops out between the Llŷn and Snowdonia (Fig. 3.5). The biota in the mudstones is restricted to a sparse graptolite fauna associated with acritarchs and chitinozoans (Howells & Smith 1997). Near Ffestiniog, in the southern part of of Snowdonia, rhyolite and rhyodacite lavas, high-level intrusive domes and acidic ash-flow tuffs (Rhiw Bach Volcanic Formation) are interbedded with silty mudstones of Llanvirn age. To the west, towards Tremadoc, and to the east, towards Migneint (located on Fig. 3.5), the sequence is truncated by a sub-Caradoc unconformity of variable magnitude (Fig. 3.26). In the area between Moel Hebog, Nantmor and Rhyd in SW Snowdonia the unconformity is overlain by a mélange with blocks and rafts of Arenig and upper Cambrian age. The allochthonous rafts are up to 300 m in length and 70 m wide, within a matrix of disrupted mudstones and siltstones. The mélange is the product of gravity sliding and mass flow, related to a slide-plane keyed to a fault scarp (Howells & Smith 1987). Elsewhere in North Wales, the Aran Volcanic Group dominates the massif that extends through Cadair Idris into Aran Fawddwy, from where it thins northwards along the eastern flank of the Harlech Dome. The thick sequence of acid and basic lavas, tuffs, volcaniclastic sedimentary rocks and associated intrusions is mainly of Llanvirn age but extends up into the early Caradoc. On Cadair Idris, welded and non-welded acid tuffs (Offrwm Volcanic Formation), at the base of the Llanvirn sequence, are the first major expression of volcanism. Thin interbedded mudstones and fine-grained devitrified dust tuffs are common and the lack of current structures suggests that the tuffs were emplaced in relatively deep water. The formation thins eastwards and, south of Dolgellau, it oversteps onto Upper Cambrian strata. The subaqueous setting is further corroborated in the succeeding Cregennen Formation, which comprises dark grey mudstones with both acid and basic tuffs, tuffites and rare pillowed basalt. The volcaniclastic deposits reflect substantial and complex resedimentation of both acid and basic volcanic debris to form upward-fining, coarse-grained tuffites in broad migrating channels and thinner, fine-grained tuffites from low-concentration turbidity currents. Large tabular and contorted clasts of bedded tuffs in debris-flow deposits suggest incorporation from the erosion and collapse of channel margins. Northeast of Dolgellau, a wedge of basic tuffs and tuffite and pillowed and brecciated basalts with intercalated siltstones form the Melau Formation, which overlies the Offrwm Volcanic Formation. Adjacent to the Bala Fault, at Creigiau Brithion, a rhyolite dome is interpreted as the source of locally restricted welded acidic ash-flow tuffs (Brithion Formation). The tuffs are sharply truncated at their top, which together with associated conglomerates, suggest that the dome surfaced temporarily and was eroded in shallow water. The uppermost Llyn y Gafr Volcanic Formation is dominantly of basalt lavas and bedded basic tuffs with a few acid tuffs and intercalated siltstones. There is marked lateral variation, but precise correlation is obscured by basaltic intrusions. Local accumulations of pyroclastic breccias, with blocks of highly vesicular basalt in a sparse matrix of comminuted basaltic debris probably accumulated close to the vent and the massive and pillowed lavas reflect the first major effusive episode in southern Snowdonia (Howells & Smith 1997).

Within the discontinuous outcrop of Llanvirn rocks along the southeast margin of the basin, there is a record of substantial volcanic activity in the Shelve area and near Builth Wells (Fig. 3.24). At Shelve, the sub-basin adjacent to the Pontesford–Linley lineament (Fig. 3.5) accumulated an estimated 1500 m of shales, mudstones and volcanic rocks (Hope Shale Formation). The shales contain graptolites and a diverse trilobite fauna (Whittard 1955, 1967, 1979), mostly blind members of the atheloptic community, but also including some large-eyed members of the raphiophorid trilobite community (Cave & Hains 2001). The fauna reflects a dysaerobic, deep-neritic environment that prevailed throughout the deposition of the formation (Cave & Hains 2001). Two volcanic accumulations, the Hyssington Volcanic and Stapeley Volcanic Members, were deposited in the Early Llanvirn (Fig. 3.24). The lower, Hyssington Member, consists of acid tuffs and volcaniclastic sandstones interbedded with black mudstones, while the Stapeley Volcanic Member is dominated by basic volcaniclastic debris, mainly crystals and scoriaceous fragments, with no clear evidence of primary volcanic rocks. The presence of debris-flow deposits, some with well-rounded boulders up to 0.8 m across, have been interpreted as evidence of an emergent volcanic island as a source area, although generally the mudstones and the accumulations of volcanic detritus in the Hope Shale Formation were deposited in a relatively deep-water environment.

(a)

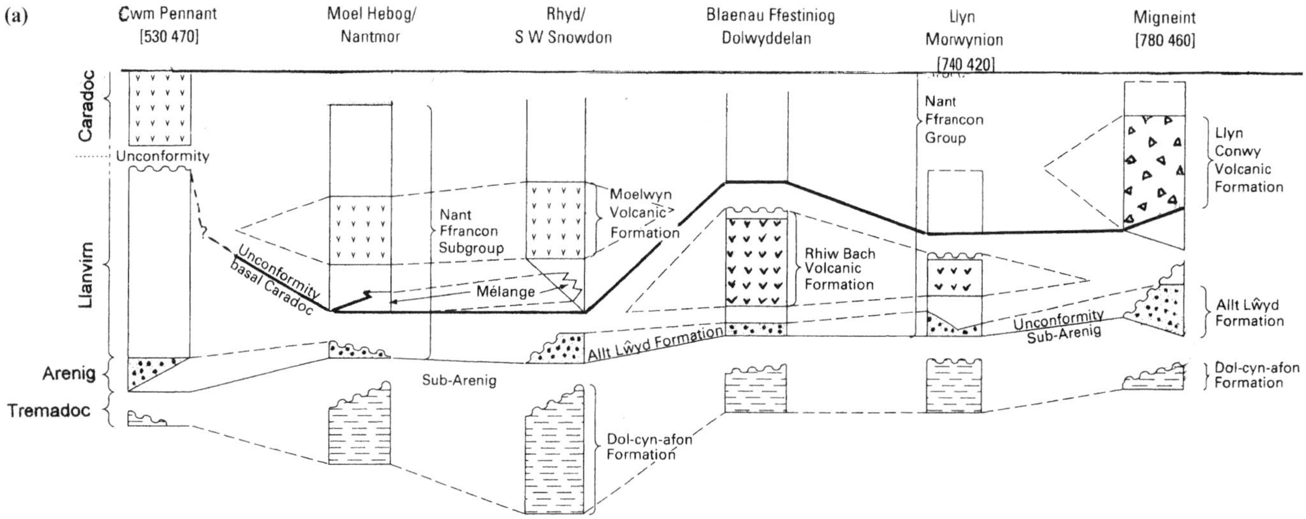

(b)

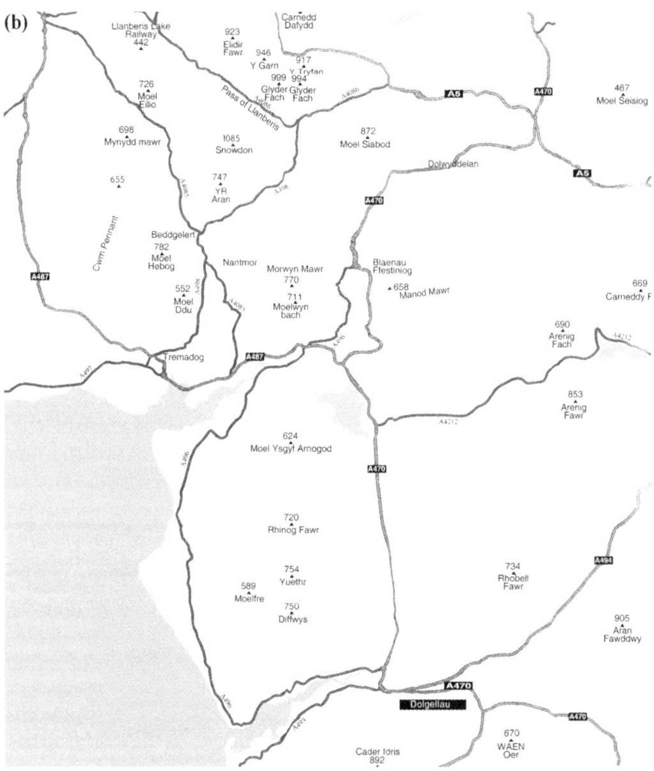

Fig. 3.26. (**a**) Stratigraphic profile along southern Snowdonia, showing the Llanvirn volcanics (the Rhiw Bach Volcanic Formation) and its relationship to the basal Caradoc unconformity (thick line) and the mélange (from fig. 6 of Howells & Smith 1997). (**b**) Location within the Snowdon Volcanic Group referred to in the text (reproduced by permission of the British Geological Survey@NERC.

About 50 km to the SW is the Builth volcanic sequence, possibly sited at the closure of a Shelve–Builth sub-basin, where the bounding faults are convergent. The Builth Volcanic Formation, up to 700 m thick, comprises mainly acid tuffs, agglomerates, and tuffaceous sandstones with interbedded graptolitic mudstones. In the south of the inlier this sequence is overlain by spilitic, pillowed, porphyritic andesitic basalt with, in places, high level basalt intrusions, locally overlain by boulder beds and coarse felspathic sandstones. These relationships were interpreted (Jones & Pugh 1942) to represent sedimentation around sea stacks on the margin of a contemporary volcanic island, identifiable 'down to the last boulder on the beach'. The overlying finely laminated mudstones of the Llanfawr Mudstone Formation contain a trilobite association with abundant benthic taxa, dominated by asaphids, raphiophorids, nileids and trinucleids together with rare cyclopygids. The association suggests a water depth of 150–200 m, below storm wave base, but within the range of benthic taxa and too shallow for many mesopelagic cyclopygids (Owens 2002). West of the Builth area, volcanic rocks at Llanwrtyd lie on what was possibly the basinal side of an intervening structural high. There, volcanic activity continued into the Caradoc, with basalts that interacted with seawater to form peperitic margins: these are overlain by dysaerobic dark mudstones and thin siltstone and sandstone turbidites. The gravity-flow deposits may have been partly derived from the volcanic centre, but most were transported from the top of a NW-facing slope (Cave & Rushton 1996).

Southwest of Builth Wells, along the southeastern margin of the Tywi Lineament, (Fig. 3.5) Lower Llanvirn graptolitic shales persist laterally from Llandeilo to Carmarthen, but contain minor thin tuff layers in the Llandeilo area. These tuffs increase in abundance upwards and grade into shallow marine volcaniclastic sandstones, tuffs and conglomerates of the Ffairfâch Group. The group is divided into five formations that embrace three regressive cycles separated by sharp contacts

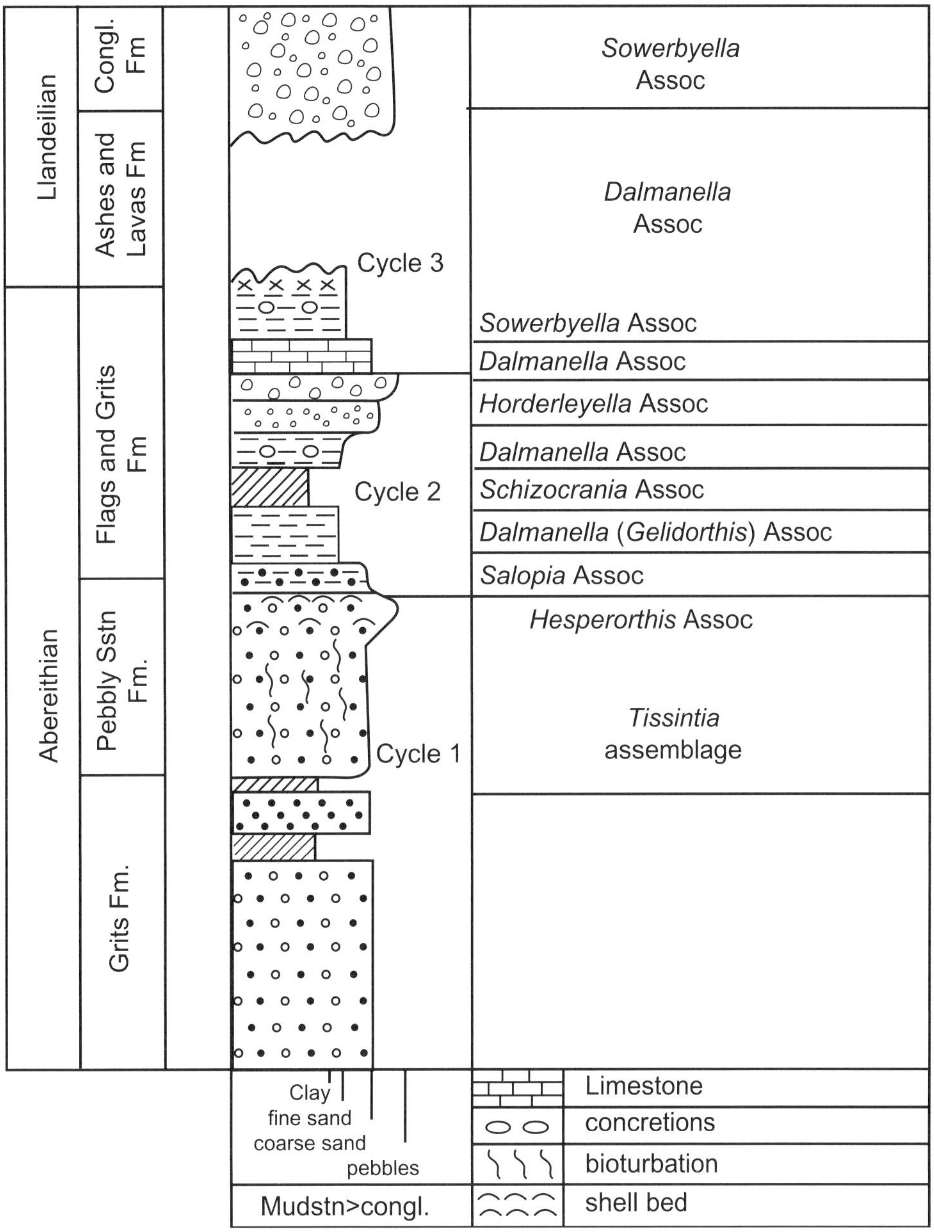

Fig.3.27. Sedimentary log of the Ffairfach Group, showing three regressive cycles and associated depth related brachiopod associations (adapted from Williams *et al.* 1981).

between pebbly sandstones at the top of a cycle and finer-grained facies marking transgressive events (Fig. 3.27). The bathymetric changes are reflected in the succession of depth-related brachiopod associations and particularly by the presence of the shallow-marine *Hesperorthis* and *Horderleyella* associations at the top of cycles 1 and 2, respectively (Fig. 3.27) (Williams *et al.* 1981). The overlying Lower Llandeilo Flags mark the initiation of a major transgressive cycle that followed the end of volcanicity in this region of the basin (Williams *et al.* 1981).

In the SW part of the Welsh Basin there was intense volcanic activity during Early Llanvirn times (Fig. 3.25). The activity is most clearly displayed in the cliff sections on Ramsey Island and along the north Pembrokeshire coast. On Ramsey Island there is considerable evidence for local uplift and emergence of a volcanic island during the deposition of a sequence embracing the black mudstones and coarse debris flows of the Porth Llauog Formation, the rhyolitic tuffs and lavas of the Carn Llundain Formation and coeval high-level intrusions of rhyolite and microtonalite. To the west of the volcano-tectonic Ramsey Fault, conglomerates represent contemporaneous littoral conditions in the vicinity of the island. The Cader Rhwydog Tuff, which occurs on both sides of the fault, is interpreted as a proximal, subaqueously emplaced, welded ash-flow tuff overlain by graded ash deposits that settled out from ash suspended in the water column (Kokelaar *et al.* 1985). Within most of the district, the volcanic activity occurred with background sedimentation of black graptolitic muds.

On the mainland the volcanic activity is most clearly displayed in the cliff sections between Solva and Fishguard. Here, the thickening of the volcanic sequence northwards across the Fishguard–Cardigan Fault Belt (Fig. 3.39), marks the southern edge of an ENE-orientated graben, where fractures provided the principal conduit for magma to propogate to the surface (Davies *et al.* 2002). The volcanic activity, of early Upper Llanvirn age, is dominantly bimodal, rhyolitic and basaltic, with lavas, intrusions and fragmental rocks of both compositions forming the Fishguard Volcanic Group. The primary volcanic rocks are associated with much reworked volcaniclastic debris in debris flow deposits, turbidites and tuffaceous sediments contaminated by the silty mudstone background sediment. The earliest activity was the emplacement of rhyolite and rhyodacite lavas, which, in places, are pillowed (Porth Maen Melyn Formation) and are overlain, on the north side of Porth Maen Melyn, by basalts with minor basic tuffs, hyaloclastites and few silty mudstone turbidite intercalations (Strumble Head Volcanic Formation up to 1000 m thick). The fluid basic magmas were emplaced mainly as interdigitating pillowed and sheet flows with the minor tuffs and hyaloclastites indicating periodic explosive activity. The thick accumulation is thought to have been restricted to a subsiding graben. The volcanic episode terminated with the emplacement of welded and non-welded acidic ash-flow tuffs, thick rhyolite lavas with flow-banded and perlitic cores and autobrecciated carapaces (Goodwick Volcanic Formation). The sequence includes much reworked volcanic debris in debris-flows and turbidite deposits interbedded with the silty mudstone background sediments of Upper Llanvirn age. The sequence, with increasingly distal volcaniclastic turbidites, can be traced eastwards into Mynydd Preseli and into feather edges near Narbeth. To the SW, in Abereiddi Bay and the historical type area of the Llanvirn, rhyolitic tuffs and coarse- and fine-grained volcaniclastic deposits (Llanrian Volcanic Formation) may be equivalent to part of the Fishguard Volcanic Group. At its top, the Abereiddi Tuff, previously termed the 'Murchisoni Ash', constitutes mainly basaltic fragments in two fining upwards sequences, grading both laterally and vertically from massive coarse lapilli tuffs to well-bedded fine-grained tuffs. These were mainly emplaced by slumping from the unstable flanks of a scoriae cone and represent the youngest volcanic episode preserved in SW Wales.

Throughout this Llanvirn volcanic sequence in western Pembrokeshire there is a complex of gabbro, dolerite and diorite intrusions which deformed the wet sediments indicating that they were largely coeval with the volcanism. Included in this complex are the layered gabbros of Carn Llidi and St David's Head that were considered to be separate intrusions but are probably parts of the same intrusion, on the limbs of a NE-trending syncline. The intrusion complex extends westwards, beyond the outcrop of the volcanic rocks and is particularly prominent on Mynydd Preseli.

At the end of volcanic activity the area was covered by a mantle of black pelagic mudstones, constituting the Caerhys Formation of Late Llanvirn age. The mudstones are tuffaceous in their lower part, but have thin silty distal turbidites above and contain a fauna including brachiopods, trilobites and bivalves near their contact with the overlying Castell Limestone. The micritic and silty limestones are notable for yielding abundant trinucleid trilobites.

Early Caradoc sedimentation and volcanism

During the early part of the Caradoc the lineaments bounding the Welsh Basin were tectonically active. Renewed uplift along faults within the Irish Sea Horst Complex resulted in an influx of coarse debris, deposited from gravity flows into the basin within Anglesey. At the eastern margin of the Welsh Basin, new sub-basins were initiated by strike-slip movements on faults within the Welsh Borderland Lineament. In addition to these tectonic events, basin evolution was profoundly influenced by both the large '*gracilis* transgression' and the shift in the focus of volcanic activity from South and Mid Wales to North Wales. Subsidence was greatest in areas of volcanic activity, such as Snowdonia (Fig. 3.14), but rapid subsidence also occurred in the small, non-volcanic sub-basins associated with the Welsh Borderland Fault System (Fig. 3.6). Volcanic activity was most widespread during the Soudleyan and Longvillian Substages, but persisted until the end of the Woolstonian. Most of the activity was associated with shallow marine environments in which ephemeral volcanic islands were constructed. Close to the eastern margin of the basin, in the Breidden Hills and Shelve areas, volcanoes were situated in deeper marine settings, but nevertheless some became emergent.

The '*gracilis* transgression' was a prolonged eustatic event, probably starting in the late Llanvirn and persisting into its peak in late Caradoc times. Within the Welsh Basin it caused a diachronous change to prevalently argillaceous facies in the Llandeilian–Aurelucian, with the '*Dicranograptus*' Shales in the St David's area, the Hendre Shales near Meidrim and Carmarthen, the *gracilis* Shales at Llandeilo and the Rorrington Shales in the Shelve area (Fig. 3.28). In North Wales, in the argillaceous Nant Ffrancon Subgroup, the start of the transgression is not discernible and is similarly obscure in the argillaceous sequences in South Wales.

The sub-basins in South Shropshire

At the eastern margin of the basin the sea transgressed across Precambrian basement and folded Cambrian and Tremadoc rocks into three small sub-basins, the Cressage–Cardington, the Onny and the Ludlow sub-basins, bounded on their west side by the Church Stretton Fault (Fig. 3.29) (Smith & Rushton 1993). The transgression did not apparently extend east of the sub-basins, as upper Ordovician rocks are not recorded in that direction. The transgression started in the Aurelician, but continued in the Onny Sub-basin until the early Burrelian, when Harnagian sediments lapped on to the irregular surface of the Uriconian basement at Hope Bowdler and filled neptunean dykes in the Uriconian at Hazler Hill (Strachan *et al.* 1948) on the Ragleth High (Fig. 3.29). The sequence of Costonian–Actonian age, with mainly shallow-marine deposits of sandstone and deeper-marine mudstones above, is broadly similar throughout south Shropshire. However, there are some differences between the successions recorded in the Cressage–Cardington and the Onny Sub-basins (Fig. 3.30), The most striking of the facies differences is the prevalence of storm-affected sediments with hummocky cross-stratification in the Horderley Sandstone and Cheney Longville Formations of the Onny Sub-basin (Brenchley & Newall 1982) and their rarity in coeval formations in the Cressage–Cardington Sub-basin.

Anglesey

On the opposite side of the Welsh Basin, within the Irish Sea Platform, on Anglesey, early Caradoc deposits overlie a disconformity or locally an unconformity. The basal shallow marine sediments above the unconformity, are sandstones in the Crewyn Formation and quartz conglomerates in the Gader Formation. Elsewhere, basal deposits of the Garn Formation are massive, poorly sorted breccias interbedded with shales, deposited in deeper water as mass-flow deposits. Locally the unconformity is overlain by a breccia with Monian supergroup clasts and a few angular blocks of Ordovician limestones up to 2 m across. One limestone block yielded a diverse brachiopod and trilobite fauna with associated bryozoans, crinoid ossicles and inarticulate brachiopods; similar limestone is not recognized *in situ* in Anglesey, but might possibly have been derived from the north. At Ogof Gynfor (Fig. 3.18) the unconformity is covered by a thin breccia of quartzite clasts in a black shale matrix, probably deposited at the same depth as that of the overlying dark grey silty shales of the Gynfor Formation. The presence of shallow-marine sediments above the unconformity at some locations, but bouldery debris-flow deposits elsewhere, suggests that the *gracilis* trangression across Anglesey was coeval with rapid basin subsidence and synchronous footwall uplift that created shallow-marine areas above fault scarps from which the debris flows were sourced.

Volcanic activity in the Welsh Basin

Within the Welsh Basin, deposition of shales and mudstones prevailed throughout the early Caradoc (Velfreyan–Harnagian)

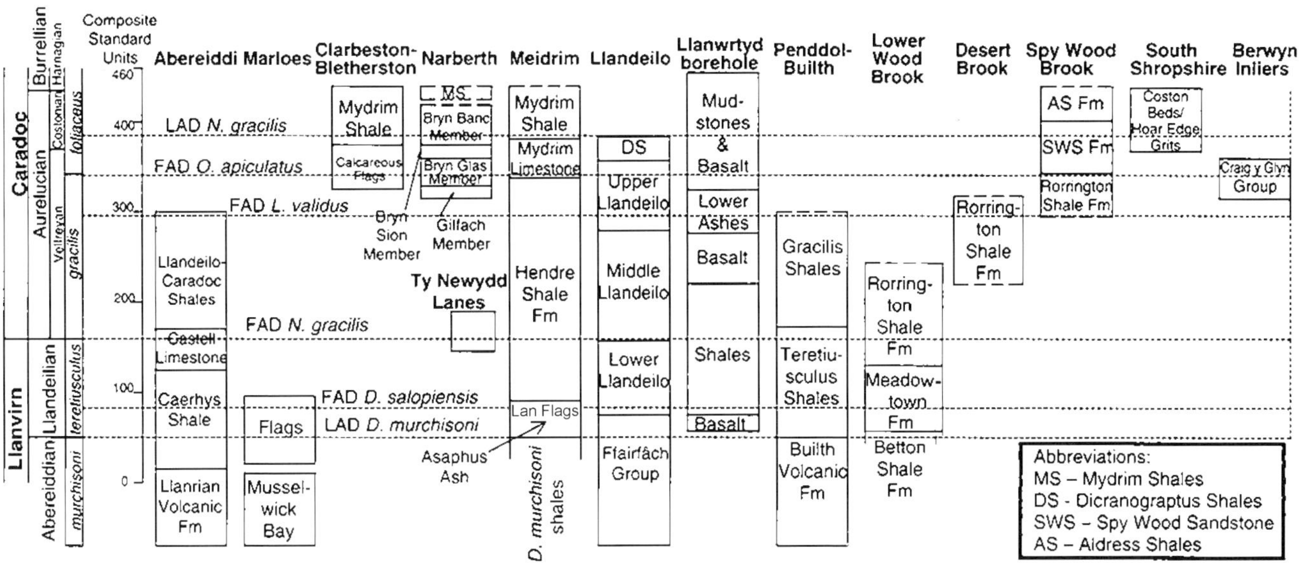

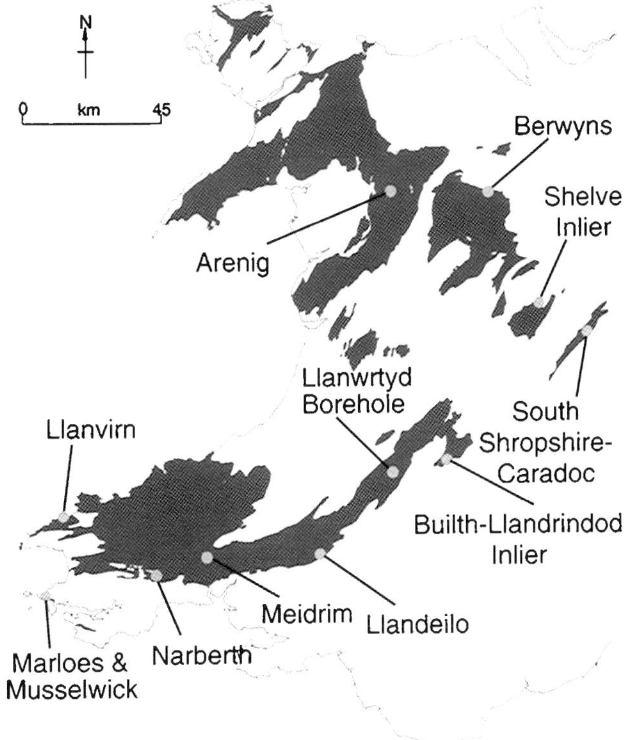

Fig. 3.28. Correlation chart of upper Llanvirn and lower Caradoc lithological units with reference to a composite standard based on graptolites. First appearance datum (FAD) and last appearance datum (LAD) are shown for selected species. (from Bettley *et al.* 2001).

with volcanic activity in some places. Across Cadair Idris Llanvirn volcanic rocks of the Aran Volcanic Group are overlain by the early Caradoc Ty'r Gawen Formation, with the Fron Newydd Member at its base. This member is a phosphatic mudstone with oolitic ironstone, which probably reflects a flooding surface associated with the *gracilis* transgression. West of the Llanegryn Fault, the mudstone lies unconformably on the Llanvirn volcanic sequence but east of the fault it is conformable. To the east the Ty'r Gawen mudstones interdigitate with the Pen y Gadair Volcanic Formation, which is predominantly composed of pillowed and autobrecciated basalt lavas. However, in the west, on Tyrau Mawr, these interdigitate with two wedges of acidic ash-flow tuffs and crystal-tuffs, that probably developed from a local centre, and with silty mudstones. Further east, the Benglog Formation, at approximately the same stratigraphical level, comprises crystal-rich and blocky tuffs with intercalated mudstones. The coarse tuffs are massive, locally reverse-graded at the base and normally-graded at the top. The variable admixture of mudstone matrix suggests either incorporation during transport or eruption through unlithified sediments. The uppermost Craig Cau Formation, comprises thick, massive, volcaniclastic debris-flow deposits, with a significant mudstone matrix, overlain by acidic ash-flow tuffs with interbedded siltstones. The lithological heterogeneity of the pyroclastic and epiclastic components, and the volumes involved suggest that they were formed by explosive eruption, of acid pyroclastic debris at a number of centres, through an unlithified basinal marine sequence. The formation persists through Aran Fawddwy to the north of the Bala Fault, and throughout its outcrop it is overlain by mainly mudstones

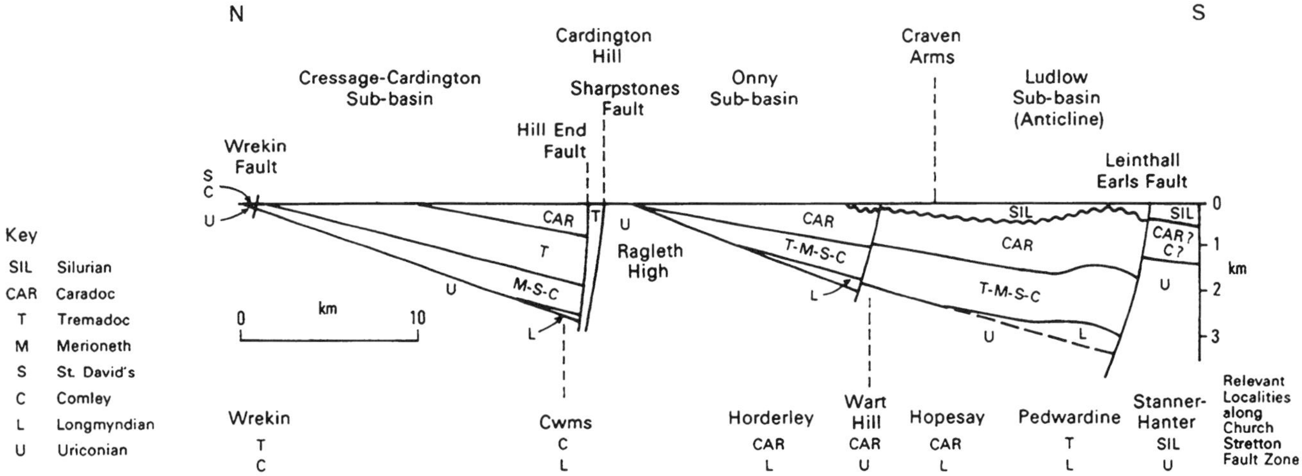

Fig. 3.29. Stratigraphic profiles of the Lower Palaeozoic fill of sub-basins associated with the Welsh Borderland Fault System in South Shropshire. (adapted from Smith & Rushton 1993).

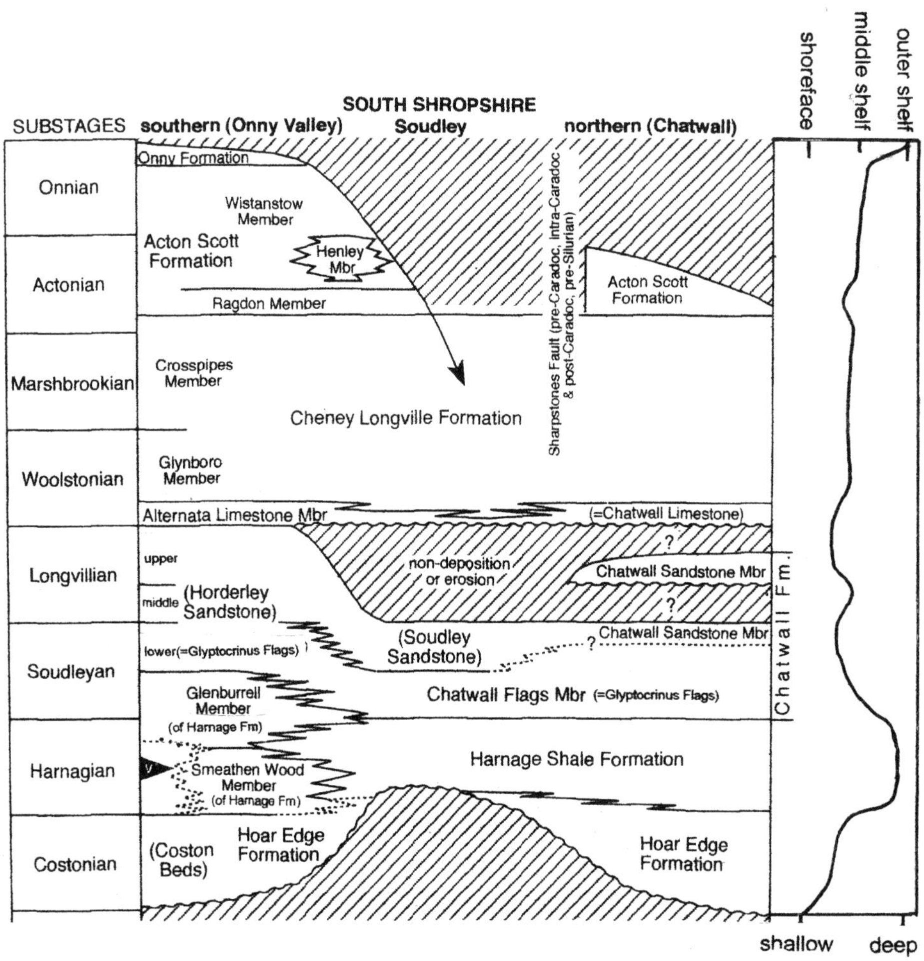

Fig. 3.30. Lower Caradoc stratigraphy in the the Onny valley (Onny sub-basin) and near Chatwall (Cressage–Cardington sub-basin) (modified from Bowdler-Hicks *et al.* 2002 with permission of the Palaeontological Association).

and siltstones (Ceiswyn Formation) of possible Costonian–Soudleyan age. The upward increase of laminated black mudstone and decrease in the turbiditic component probably reflects the gradual drowning of the shelf by eustatic sea-level rise.

The earliest expressions of Caradoc volcanism in central Snowdonia are of Costonian age. In the scarp of the Moelwyns Hills (Fig. 3.26b) is the Moelwyn Volcanic Formation, lying close to the margins of the earlier Aran Volcanic Group. The formation comprises volcaniclastic debris-flow deposits, some with rounded rhyolite boulders, which are spatially and lithologically related to concordant rhyolite bodies, suggesting high-level intrusion and extrusion into a shallow-marine, littoral zone. The debris flows resulted from the local uplift of the rhyolites to within wave-base, where they were reworked and collapsed through tectonic instability. To the east, across the Migneint, the Llyn Conwy Formation comprises acid ash-flow tuffs, tuffites and tuffaceous mudstones and siltstones closely associated with rhyolite and rhyodacite bodies. Further SE, in the Arenigs, similar acidic crystal tuffs, tuffites and debris-flow deposits are associated with high-level penecontemporaneous dacite sheets (the Lower and Upper Ashes of Arenig). On the east side of Arenig Fawr the overlying Derfel Limestone of the *multidens* Biozone, marks the end of volcanism in the area. In central Wales, near Builth Wells, the Trelowgoed Volcanic Formation of probable Caradoc age is separated by less than 500 m of Llanfawr Mudstones from the underlying Builth Volcanic Group. To the west of Builth, at Llanwrtyd, agglomeratic tuffs, rhyolitic breccias and basalts are interbedded with mudstones, thin turbidites and mass-flow deposits of Caradoc age (Cave & Rushton 1996).

The peak of Caradoc volcanic activity throughout North Wales was in the Soudleyan–Longvillian times, although some localized activity persisted until the end of the Woolstonian. In central and northern Snowdonia the upper part of the Nant Ffrancon Formation contains beds of pisolitic ironstone with trilobites and brachiopods, reflecting periods of shoaling preceding the transgression in the the late Llanvirn–early Caradoc. A little higher in the sequence, intraformational disturbed beds probably caused by contemporaneous seismic activity heralded the onset of intense volcanic activity. This formed two eruptive cycles expressed as a lower, Llewelyn Volcanic Group and an upper, Snowdon Volcanic Group, separated by the dominantly sedimentary Cwm Eigiau Formation (Fig. 3.31) (Howells *et al.* 1991). This volcanic sequence is confined to the Soudleyan, Longvillian and Woolstonian Sub-stages, but even so dominates the entire Ordovician sequence through central and northern Snowdonia. The early stages of the First Eruptive Cycle developed from four separate, partly contemporaneous centres, across the Carneddau in northern Snowdonia, which are reflected in four formations with a number of subvolcanic intrusions, which are spatially and genetically associated. Locally great thicknesses of extrusive volcanic rocks accumulated and for the most part lack evidence of extensive erosion or reworking. The inference is that most of the extrusive rocks were ponded in subsiding, probably fault bounded sectors of the sea floor. This differential subsidence is clearly indicated by lateral variations in the thickness of strata betweeen successive extensive ash-flow tuffs (Fig. 3.32). In the north, the Conwy Rhyolite Formation comprises intrusive and extrusive rhyolites and acidic tuffs, which probably developed close to a local centre, although some of the rhyolite lavas indicate some considerable mobility in traversing many kilometres. To the south, these acidic rocks abut and interdigitate with trachyandesite lavas and tuffs of the Foel Fras Volcanic Complex. Further south, basalt lavas (Foel Grach Basalt Formation) and rhyolite lavas and welded acid ash-flow tuffs (Braich tu du Volcanic Formation) are the broadly lateral equivalents. The sedimentary background of this sequence is consistently of offshore, marine silty mudstones. The lack of significant reworking in the volcanic accumulations and the dearth of volcaniclastic debris in the associated sediments suggest that emplacement was dominantly below wave base. However, to the south, the sequence wedges out in a succession of volcaniclastic sediments between the Nant Ffrancon and Llanberis passes.

Following this early phase of volcanic activity there was a prolonged period of sedimentation, which, in the area of the Carneddau, coincided with progressive shallowing of the marine environment. Conglomeratic alluvial fan and braid plain facies, in the north, pass laterally into coarsening-upwards siltstone–sandstone sequences, representing prograding fan deltas in the south (fig. 3.33) (Orton1988). It was upon the surface of these sediments that the first major expression of acidic ash-flow tuff volcanism (Capel Curig Volcanic Formation) in Snowdonia was emplaced at the end of the First Eruptive Cycle. The tuffs were erupted from three centres, two in a subaerial setting in the north and a third, in the south, that was, to a large extent, subaqueous. The ash flows, which were erupted subaerially, transgressed southwards across a shoreline into a subaqueous environment. In the final phases the activity was centred in the Glyders, to the south of the shoreline, and it comprises

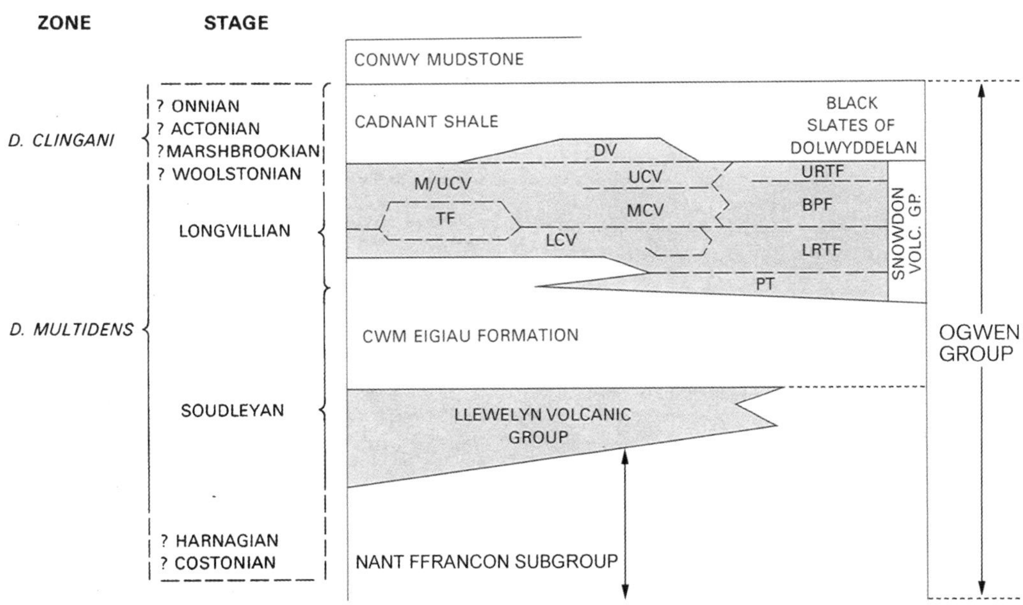

Fig. 3.31. Stratigraphy of the Caradoc sequences in Snowdonia containing the Llewelyn and Snowdon Volcanic Groups: PT, Pitts Head Tuff Formation: LRT, Lower Rhyolite Tuff Formation: URTF, Upper Rhyolite Tuff Formation: LCV, Lower Crafnant Volcanic Formation: M/UCV, Middle/Upper Crafnant Volcanic formation: TF: Tal y Fan Volcanic Formation, DV: Dolgarrog Volcanic Formation (from fig. 9 in Howells *et al.* 1991. Reproduced by permission of the British Geological Survey@NERC. All rights reserved IPR/60-06C). (Rushton & Howells 1999).

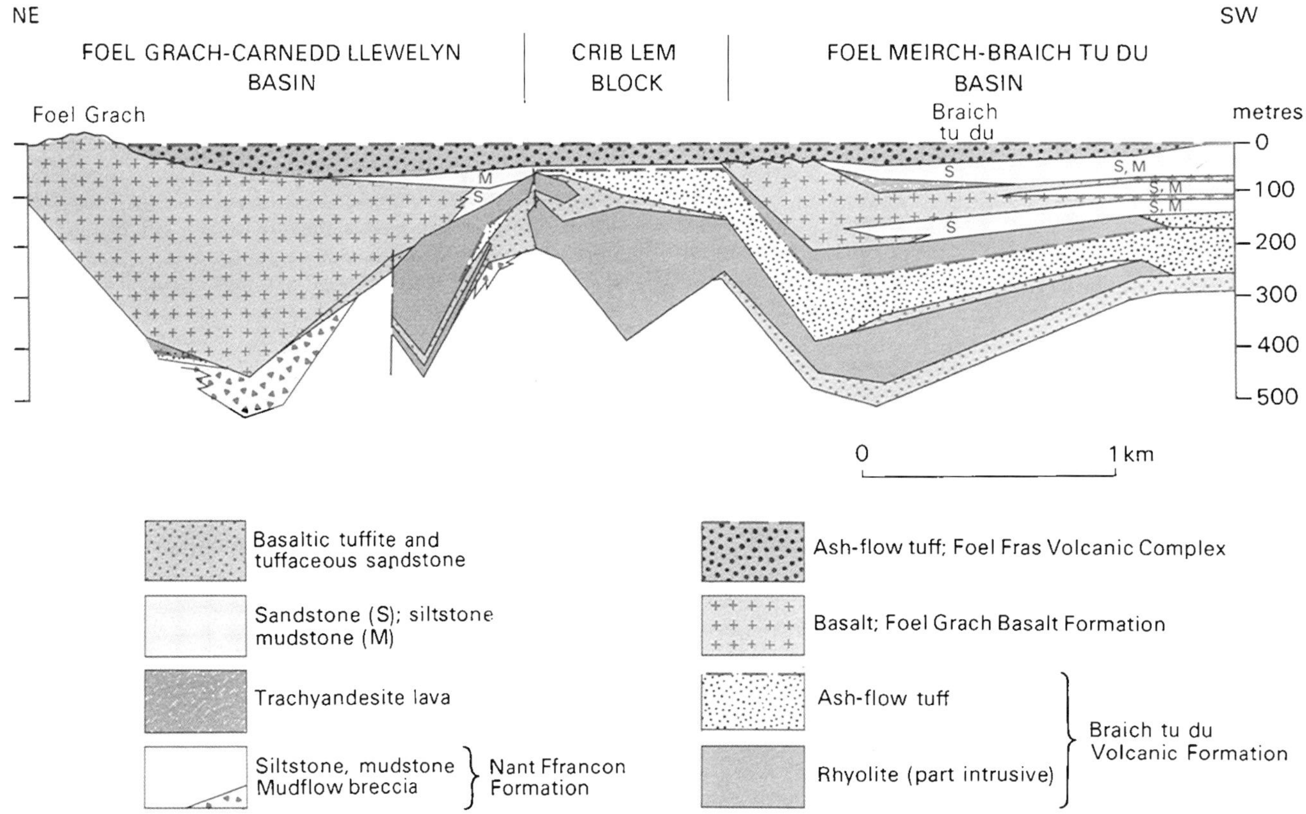

Fig. 3.32. Block and basin topographic control on accumulation of the lower formations of the Llewelyn Volcanic Group at the southern end of its outcrop (from fig. 17 in Howells *et al.* 1991. Reproduced by permission of the British Geological Survey@NERC. All rights reserved. IPR/60-06C).

block and ash flow tuffs, slumped and reworked tuffs. Common accretionary lapilli tuffs indicate that at least part of the volcanism was subaerial. Following the emplacement of the Capel Curig Volcanic Formation there was a period of prolonged sedimentation (Cwm Eigiau Formation) in late Soudleyan–early Longvillian times. In the north the volcanic surface is overlain by fluvio-deltaic, coarse-grained sandstones that pass upwards into shallow-marine siltstones and sandstones. To the south, the sequence thickens to approximately 1 km in the vicinity of Capel Curig and is dominated by offshore, silty mudstones. However, to the SW, between Moel Hebog and Snowdon, a sandstone-dominated sequence records deltaic and alluvial fan progradation (Orton 1988) (Fig. 3.34) and it was on to this surface that the initial ash-flow tuff (Pitts Head Tuff Formation), was emplaced, heralding the Second Eruptive Cycle. The tuff was erupted subaerially from a vent and formed an ash flow that travelled overland until the flow crossed the shore zone further west in the vicinity of Llwyd Mawr. It then continued, for at least 10 km, underwater to the vicinity of Llyn Ogwen (Howells *et al.* 1991). Wet sea-floor sediments were heated by the ash flow and locally fluidized by steam, resulting in blocks of tuff detaching and sliding downslope. Extensional breaks in the tuff sheet caused the locally fluidized sediment to form rootless vents (Kokelaar & Königer 2000).

The activity of the Second Eruptive Cycle developed from a number of centres across central and NE Snowdonia, but, most importantly, it was dominated by the development of a major caldera centred about Snowdon. The siting of the caldera and its subsequent development was controlled by pre-existing, deep-seated, NE–SW fault lines that controlled the formation of the Snowdon Graben (Fig. 3.35). The Snowdon caldera evolved in three phases of rhyolite intrusion–extrusion, the first immediately preceding the main caldera-forming eruptions, the second during resurgence and reworking of the Lower Rhyolitic Tuff Formation, and the third during the emplacement of the Upper Rhyolitic Tuff Formation. The distribution patterns of the rhyolites were controlled by the linear and arcuate volcano-tectonic structures involved in caldera formation. At the SE edge of the graben, early acidic ash-flows, pyroclastic breccias and rhyolite domes were erupted about a vent along a major fracture in the vicinity of Yr Arddu. Elsewhere, about Llyn Idwal and Llanberis Pass, early activity is represented by small submarine accumulations of pillowed basalts, breccias and hyaloclastites around local centres. However, the dominant event was the eruption and emplacement of the acidic tuffs of the Lower Rhyolitic Tuff Formation in association with the collapse of the Snowdon Caldera (Fig. 3.36). In the early stages, welded ash-flow tuffs were emplaced close to central fractures in the Gwynant Valley, but the climactic eruptions, dominated by remarkably homogeneous, non-welded tuffs, were largely contained within the subsiding and largely submerged caldera. The base of the thickest accumulation, approximately 600 m, on the south side of Llanberis Pass, is marked by a coarse pyroclastic breccia and the top was reworked in shallow-marine conditions, particularly where it had been uplifted by the emplacement (resurgence) of high-level rhyolite domes. To the north, a single outflow tuff, which crossed the caldera margin, can be traced, within the marine sedimentary sequence, into NE Snowdonia. In the vicinity of the caldera margin the outflow tuff is overlain by reworked tuff turbidites which sloughed off the margin and became progressively more contaminated with epiclastic debris.

The resurgence and shallow-marine reworking across the top of the caldera was followed by an episode of basaltic volcanic activity forming the Bedded Pyroclastic Formation, which is preserved in the core of the syncline between Llyn Idwal and

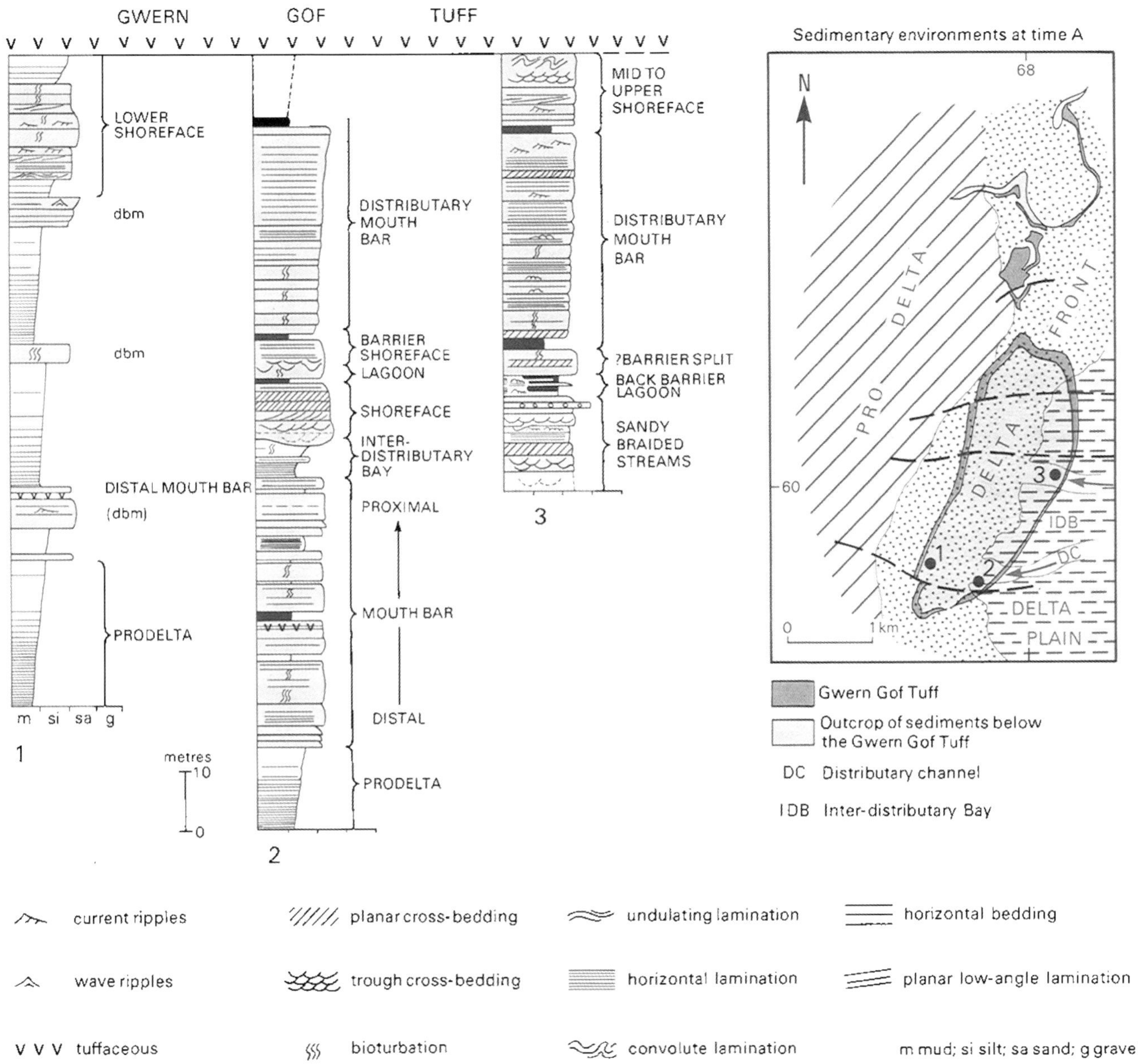

Fig. 3.33. Sedimentological and environmental interpretation of sequences below the Gwern Gof Tuff in the core of the Tryfan anticline (modified from Orton 1988). Reproduced from Howells *et al.* 1991. Reproduced by permission of the British Geological Survey@NERC. All rights reserved. IPR/60-06C.

Moel Hebog, and in the Dolwyddelan Syncline. The formation, up to 450 m thick, is dominated by basic volcaniclastic sediments, high-level intrusive and extrusive basalts and hyaloclastites (Kokelaar 1992). The sequence developed from a number of small centres, of which, only a few are preserved. In the most extensive outcrop, between Cwm Glas and the NE face of Snowdon (Fig. 3.37), nine stages of development have been determined. The stages involve subsidence, subaqeous basaltic eruptions, subaerial agglomeratic filled vents, the development of Strombolian scoria and ash cones, and small islands, uplift, local emergence of rhyolite and littoral erosion and reworking (Fig. 3.38). In central Snowdonia the deposits of the basaltic activity are overlain by a sequence of acidic ash-flow tuffs, bedded tuffs and tuffites with a few tuffaceous siltstones (Upper Rhyolitic Tuff Formation), which crop out along Crib y ddysgl ridge on Snowdon and in the core of the Dolwyddelan Syncline. The sequence marked reactivation of the Snowdon Centre and the final expression was the intrusion of rhyolites of peralkaline composition. During the evolution of the Snowdon caldera three phases of rhyolite intrusion–extrusion have been recognized: immediately preceding the main caldera-forming eruptions, during resurgence and reworking of the Lower Rhyolitic Tuff Formation and during the emplacement of the Upper Rhyolitic Tuff Formation. The distribution patterns of the rhyolites was controlled by the linear and arcuate volcanotectonic structures involved in caldera formation.

In addition to the rhyolites, both acid and basic intrusions were an integral part of the Caradoc volcanism throughout Snowdonia and Llŷn although most of the isotopic ages were reset by hydration reactions during low grade regional metamorphism. The acid intrusions include microgranites and feldspar porphyries and basic intrusions, dolerites and basalts. The only intermediate intrusions are the Llŷn Teyrn andesite, in the core of the Snowdon Horseshoe, the stock-like Manod quartz latite, east of Blaenau Ffestiniog and the suite of pre-Areig microdiorite sills about the Harlech Dome. Along the SE side of

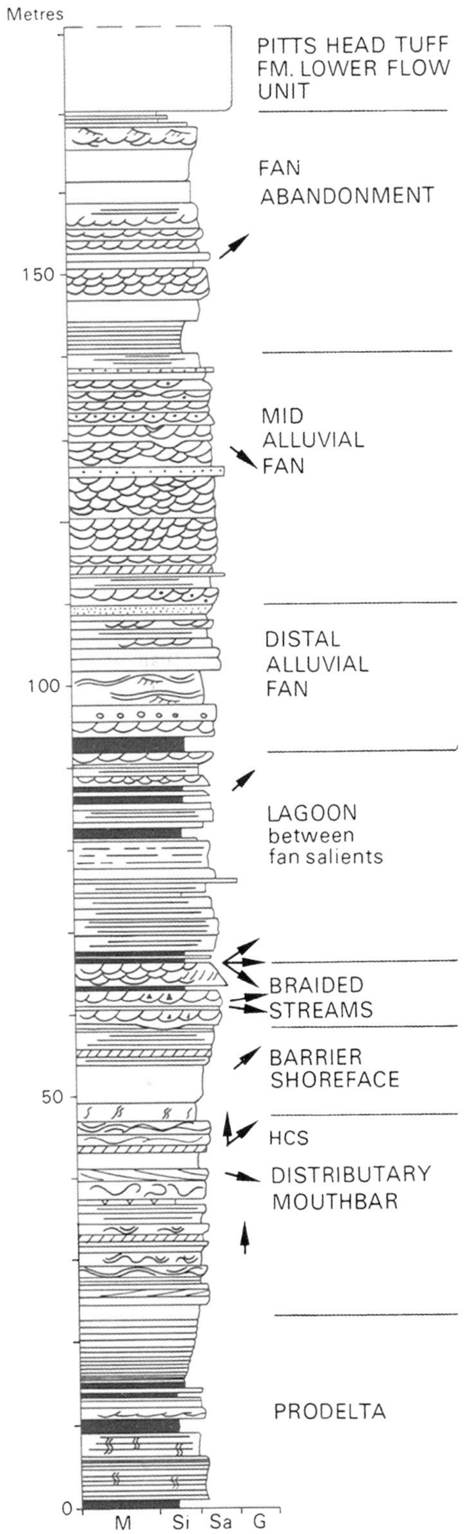

Fig. 3.34. Sedimentary log and environmental interpretation of the sequence below the Pitts Head tuff Formation at Moel Hebog (after Orton 1988) from Howells *et al.* 1991. Reproduced by permission of the British Geological Survey@NERC. All rights reserved. IPR/60-06C.

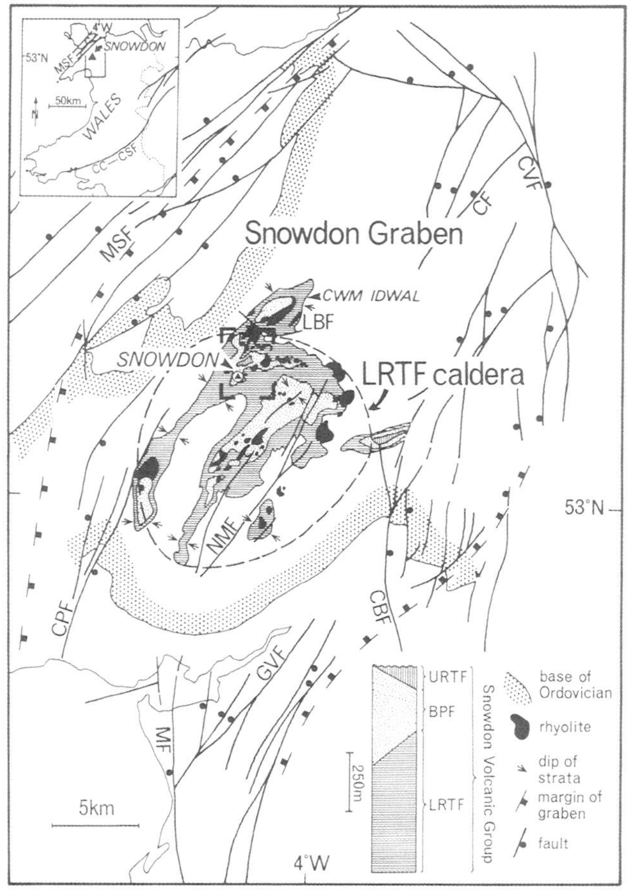

Fig. 3.35. Structures associated with the Snowdon Graben, influencing the location and development of the Snowdon Caldera. (based on Campbell *et al.* 1987; Howells *et al.* 1986; Kokelaar 1988). LRTF, Lower Rhyolite Tuff Formation; BPF, Bedded Pyroclastic Formation; URTF, Upper Rhyolite Tuff Formation. Inset: CC–CSF, Carreg Cennen and Church Stretton fault system; MSF, Menai Straight fault system. Main figure: CBF, Cwm Bowyd fault; CF, Cowlyd fault; CPF, Cwm Pennant fault system: CVF, Conway Valley Fault; LBF, Llanberis Pass fault; MF, Mochras fault; NMF, Nant Mor fault. Reproduced from Kokelaar 1992, with the permission of the Geological Society of America.

the Gwynant Valley sills, up to 100 m thick, represent multiple intrusions with precursor basic margins and cores of either feldspar porphyry or rhyolite which grade, in the thicker sills, into microgranite. In central Snowdonia there are three acid plutons, at Tan y Grisiau, Mynydd Mawr and Ogwen. The Tan y Grisiau granite crops out over an area of 4 km^2, but the field and geophysical evidence indicates a much larger body at depth. Its lack of a significant feature at surface is similar to the Ogwen microgranite. In contrast the Mynydd Mawr microgranite forms an upstanding boss-like feature and erosion has exposed a coarser, more porphyritic facies, intruding the fine-grained margins. The Gurn Ddu, Bwlch Mawr and Garn Boduan microgranites, and Saergribin granophyre, form the upstanding features along the north edge of the Llŷn, which are presented to the southern edge of Anglesey. Similarly, along the southern edge of the Ordovician outcrop, the microgranite transgressive sills, up to 500 m thick, across Cregennan and Cadair Idris contribute an important element to the topographic profile.

Following the cessation of volcanic activity the primary character of the rocks was modified by devitrification of primary glass, hydrothermal alteration and regional metamorphism. In both welded acidic ash-flow tuffs and rhyolites, spherulitic and snowflake devitrification fabrics are common and some of the numerous large siliceous nodules may also have developed as spherulites. Mineralogical changes were caused by the interaction between lavas and hot pyroclastic flows with water and by magma intruded into incompletely dewatered sediments. Hydrothermal activity was common in the vicinity of volcanic

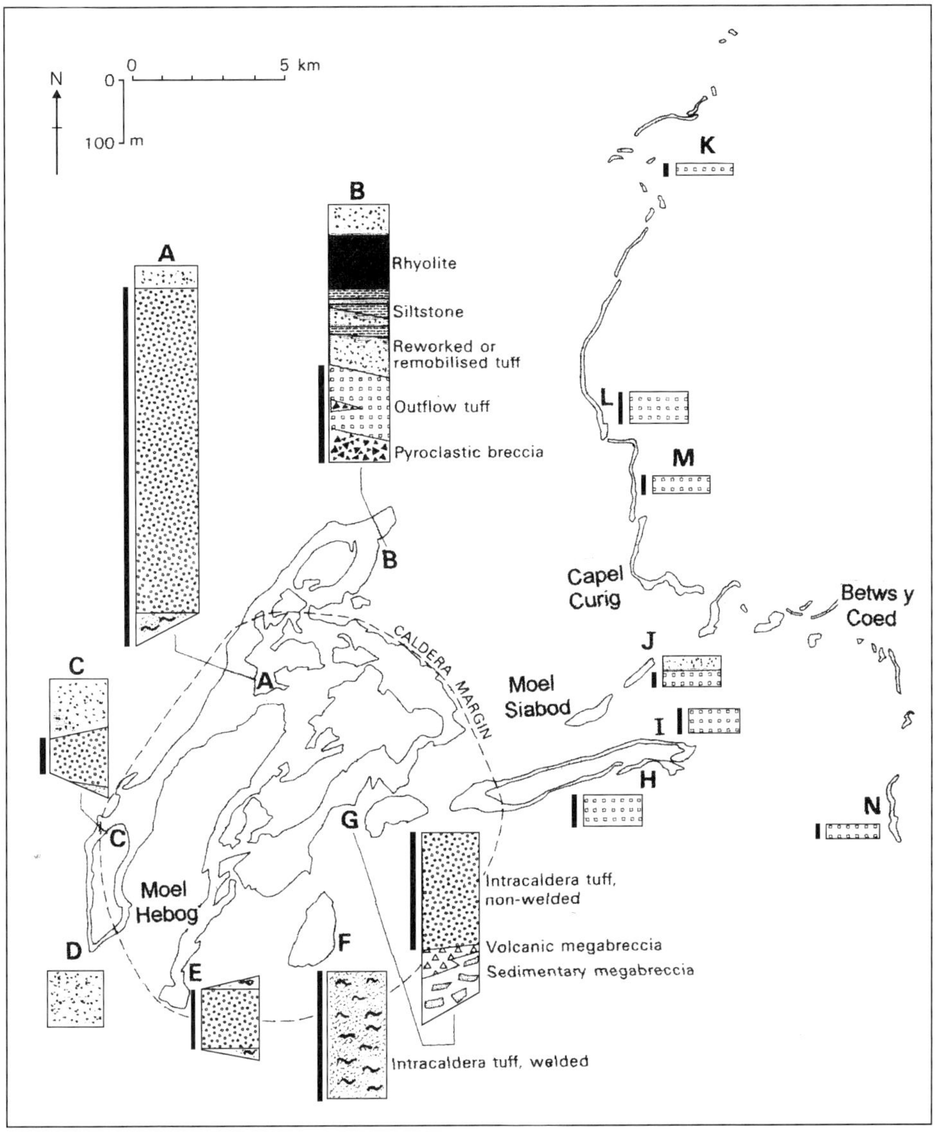

Fig. 3.36. Generalized vertical sections of the Lower Rhyolite Tuff Formation related to its outcrop and conjectural position of the caldera margin. Vertical bars to the left of the sections indicate the thickness of the primary ash-flow tuff (from Howells *et al.* 1986).

centres and caused a range of mineralogical, chemical and textural changes resulting from interaction with hot water. At the Snowdon caldera, Cu, Pb and Zn sulphides, in thin veins and disseminations, were intimately associated with hydrothermal alteration and these deposits were extensively mined in the 19th century.

During the volcanic development of the Snowdon centre the marine environment deepened towards the east and northeast and in this vicinity a contemporaneous eruptive centre developed. The activity was dominantly acidic with primary ash-flow tuffs, block and ash-flow tuffs, disrupted and slumped beds and sequences of admixed pyroclastic and epiclastic debris (Crafnant Volcanic Group), but a local basaltic centre (Tal y Fan Volcanic Formation) with pillowed basalts, breccias and hyaloclastites, was active at the same time. The sedimentary associations were entirely of silts and muds and the mud component increased progressively with time. The absence of coarse terrigenes and shallow-marine reworking reflects the deep subaqueous context of the volcanism. Soon after the end of the acidic volcanism, massive and pillowed basalts with hyaloclastites (Dolgarrog Volcanic Formation) accumulated in the vicinity of the Crafnant centre and marked the cessation of volcanism in NW Wales. There is no evidence of shallow-marine reworking of either the basic or acid volcanic rocks and both are conformably overlain by black, pyritic graptolitic mudstones (Cadnant Shales). Similar mudstones overlie the Upper Rhyolitic Tuff Formation in the core of the Dolwyddelan Syncline and elsewhere in North Wales, reflecting a new, more quiescent phase of basin evolution.

North Wales, west of Snowdonia

To the west of Snowdonia, Caradoc volcanic activity has been recognized in the vicinity of Pwllheli and Llanbedrog on the Llŷn Peninsula. From the current outcrops it seems as though the activity developed from a number of local centres and its distribution was restricted, although it was for the most part contemporaneous with the activity in central Snowdonia. In the vicinity of Pwllheli, acidic ash-flow tuffs of the Pitts Head Tuff Formation lie at the base of the Upper Lodge Volcanic Group. Further west, near Boduan, the group comprised a basaltic trachyandesite lava, overlain by an acidic ash-flow tuff, approximately coincident with the Soudleyan–Longvillian boundary. Higher in the sequence the Llanbedrog Volcanic Group displays a similar pattern of accumulation with the initial emplacement of basaltic trachyandesite lavas and tuffs, locally reworked, succeeded by rhyodacite lavas and tuffs (Foel Ddu Rhyodacite

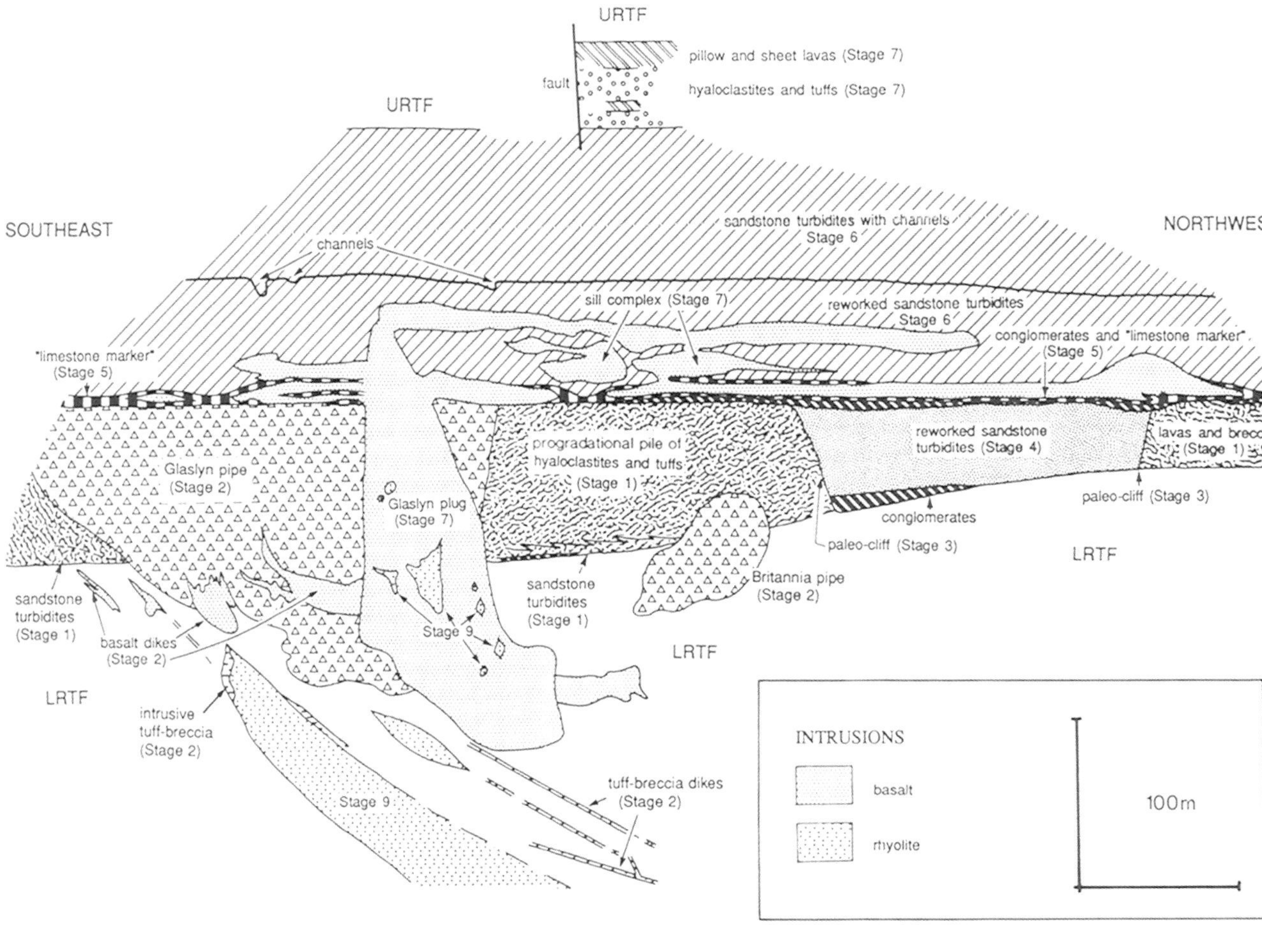

Fig. 3.37. A restored cross section of the Bedded Pyroclastic Formation and related intrusions in the northeast face of Snowdon (from Kokelaar 1992, with permission from the Geological Society of America). URTF = Upper Rhyolite Tuff Formation, LRTF = Lower Rhyolitic Tuff Formation.

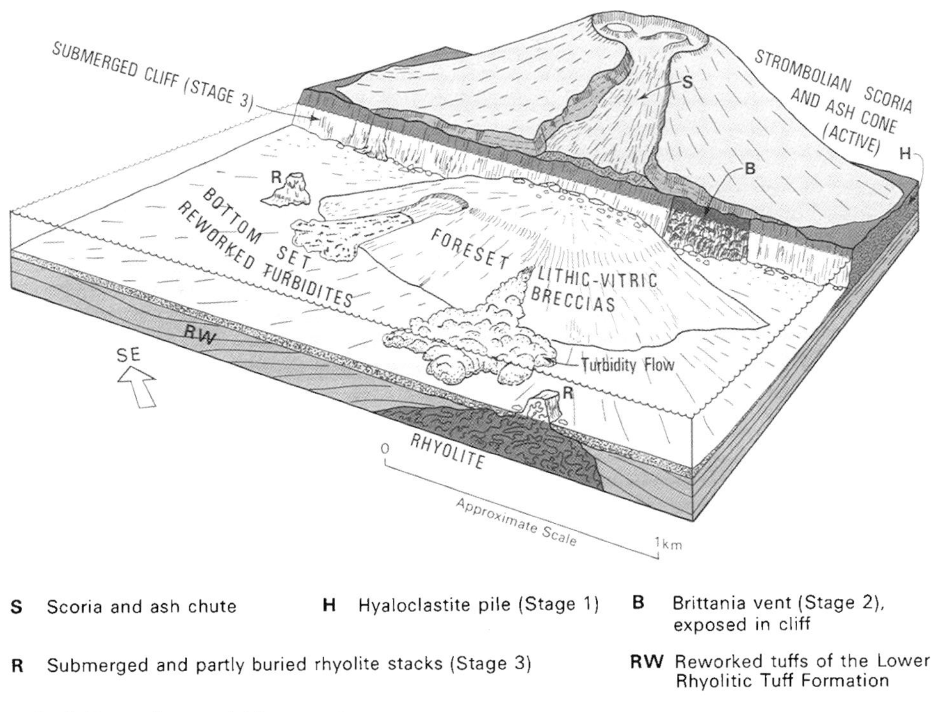

Fig. 3.38. Bedded Pyroclastic Formation: model for evolution of Stage 4 deposits. S-scoria and ash chute (stage 1), H-hyaloclastic pile, B-Britannia vent exposed in cliffs (stage 2), R-submerged and partly buried rhyolite stacks (stage 3), RW- reworked tuffs of the Lower Rhyolite Tuff Formation (from Howells *et al.* 1991). Reproduced with permission of the British Geological Survey@NERC. All rights reserved. IPR/60-06C.

Formation), and, finally, by two acidic ash-flow tuffs (Carneddol Rhyolitic Tuff Formation). These volcanic rocks overlie silty mudstones with a shelly fauna of Woolstonian age and are overlain by an apron of volcaniclastic siltstones, sandstones and conglomerates (York House Formation) derived from the shallow marine reworking of the primary volcanic rocks. The sequence is overlain by blue-black, silty, graptolitic mudstones (Cadnant Shales) typical of the widespread mid- to late Caradocian facies.

North Wales, east of Snowdonia

To the east of the Snowdon graben, in the core of the Derwen Anticline, to the north of Bala, three thin tuffs (Cefn Gwyn Tuff, Frondderw Tuff, Pont Ceunant Ash/Tuff) form the basis for subdivision of the mudstones of the Ceiswyn Formation. The crystal and vitric tuffs contain variable admixtures of epiclastic components from small rounded clasts to mud and locally grade into tuffaceous sediments. The sedimentary structures reflect a setting influenced both by tidal currents and wave action and a volcanic source to the northeast has been suggested. Further east, in the Berwyn Hills (Fig. 3.5), the equivalent sequence has similarly been subdivided on the basis of included tuff horizons, in upward order the Cwm Clwyd, Swch Gorge, Pandy and Pen y Graig tuffs. The lowermost, the Cwm Clwyd Tuff Formation, is the thickest with flaggy, bedded, crystal and crystal-pumice tuffs with low-angle cross-stratification suggestive of surge deposits. These tuffs, together with beds of accretionary lapilli and channels filled with coarse, graded, lithic and blocky tuffs, indicate a subaerial environment (Brenchley 1972). The volcanic centre probably lay to the east and volcaniclastic material was progressively reworked across a subaerial cone into the marine environment to the SW. The Swch Gorge and Pandy tuffs are both acidic ash-flow tuffs emplaced into a shallow-marine environment, but appear to have become briefly emergent (Brenchley in Stevenson *et al.* 1999). The Swch Gorge Tuff in the NW Berwyns consists of a sequence of welded ash-flows, but eastwards the sequence passes laterally into shallow-marine volcaniclastic sandstones with a brachiopod and trilobite fauna and desiccation cracks at one horizon, seen in the Glyn Ceriog valley (Brenchley in Stevenson *et al.* 1999). The Pandy Tuff consists of a laterally extensive sheet of non-welded ash-flow tuff with angular clasts of less than approximately 7 cm diameter. This sheet was incised by steep sided gullies that were filled with vitric welded tuff. The formations between the tuffs consist mainly of silty mudstones and sandstones with common hummocky cross-stratification in the thicker beds and wave ripples in the thinner ones. The associated fauna consists of varied brachiopod and trilobite assemblages, consistent with shallow subtidal environments (Pickerill & Brenchley 1979). Both of the sequences with volcanic units, to the east of the Snowdonia graben, are at a similar stratigraphic level to the volcanic activity of the First Eruptive Cycle and it is possible that some of the thinner tuffs in the sequence represent distal deposits of this activity. However, the thicker volcaniclastic deposits and primary ash-flow tuffs were most probably sourced locally.

The Shelve area

In the Shelve area, the early Caradoc shales of the Rorrington Formation have intact delicate laminae and a fauna of graptolites and sparse trilobites that reflect hemipelagic sedimentation in relatively deep-water environments that prevailed at the start of the *gracilis* transgression. The overlying thin-bedded sandstones of the Spy Wood Sandstone mark an influx of shallow marine sediment into this deeper water environment (Cave & Hains 2001). The overlying sequence is mainly composed of mudstones, but includes two intercalated volcanic formations. At the base of the sequence is the Aldress Shale Formation consisting of soft shaly mudstones with thin and widely spaced sandstone beds and a few thicker sandstone The overlying Hagley Volcanic Formation comprises volcaniclastic sandstones, conglomerates and breccias, andesitic and rhyolitic tuffs, and rhyolitic andesitic lavas. The massive poorly sorted volcaniclastic sediments, with flute and groove casts at the base of some beds, suggests deposition from gravity flows (Cave & Hains 2001). The succeeding mudstones of the Hagley Shale Formation contain a sparse fauna of graptolites and trilobites, and the sandstone beds have abundant bedding-parallel trails and burrows and a brachiopod fauna of the *Nicolella* Association, which is commonly an indicator of a mid- to outer-shelf environment. However, the position of the fauna above mass-flow deposits suggests a somewhat deeper environment at the base of a slope. The components of the succeeding Whittery Volcanic Formation are similar to those of the Hagley Volcanic Formation, including volcaniclastic sandstones, coarse pyroclastic andesitic and dacitic breccias, and acidic lavas. In addition there is large outcrop at Rockabank showing massive breccias with clasts up to a metre in size. The blocks of lava are preserved in various stages of mechanical brecciation suggesting that they are associated with a volcanic vent (Cave & Hains 2001). The overlying Whittery Shale Formation is very similar to the Hagley Shale and is the uppermost Ordovician formation present in the Shelve area. Approximately coeval shales and sandstones (Forden Mudstone Formation), west of Shelve form a sequence estimated to be as much as 1000 m thick (Cave & Hains 2001). Some of the thicker sandstones are cross-bedded and some have flute or groove-casts at their base and are interpreted as either turbidites or tempestites (Cave & Hains 2001).

During the Soudleyan there were volcanic centres, in addition to those of the Hagley Volcanic Formation, in nearby areas. Most of the volcanic accumulations were submarine, with marine fossils; but cobble conglomerates with well-rounded clasts are present in the Hagley Volcanic Formation and in the Bulthy Formation in the Breidden Hills and at Castle rock, Montgomery, suggest that, in places, the volcanic centres formed small volcanic islands. Volcaniclastic conglomerates in the Forden Mudstone Formation around Castle Hill, Montgomery and near Forden Castle contain andesite clasts up to 1 m across within a felspathic sandstone matrix. The deposits were probably formed from debris flows that infilled channels (Cave & Hains 2001). Similarly conglomerates in parts of the Bulthy Formation in the Breidden Hills are interbedded with turbiditic volcaniclastic sediments with a varied brachiopod and trilobite fauna (Dixon 1988) and are interpreted as a turbiditic slope-apron sequence. Beneath the Bulthy Formation is the Middletown Quarry Member of the Stone House Shale Formation. In it are the Lower and Upper Rhyolite Tuff beds, separated by graptolitic shales. The tuff beds are composed largely of lapilli tuffs, and turbidites with accumulations of pumice at their tops. Laterally this sequence is intruded by the Moel-y-Golfa Andesite with marginal pillows and tubes that locally enclose some of the host conglomerate. It is not clear whether this intrusion reached the surface, but the abundance of andesitic cobbles in the Bulthy Formation suggests that it did. Here are also several dolerite sills and a thick gabbroic-dolerite laccolith exposed on the north face of Breidden Hill in the Criggion Quarry, which is a major source of roadstone.

Coeval with the volcanic deposits of North Wales, mudstones (the Dicranograptus Shales, Hendre Shales and the lower part of the Mydrim Shales), were prevalent throughout most of the southern part of the Welsh Basin, except for a short interval in the late Velfreyan to early Costonian when limestones and sandstones were widespread (Fig. 3.28). These facies formed the Bryn Glas Member at Narberth, the nodular biomicrites of the Mydrim Limestone at Meidrim and the calcareous sandstones of the 'Upper Llandeilo', all of which suggest an interval of shallower marine sedimentation. In the north Pembrokeshire–Cardigan area, there was a new phase of tectonic activity at

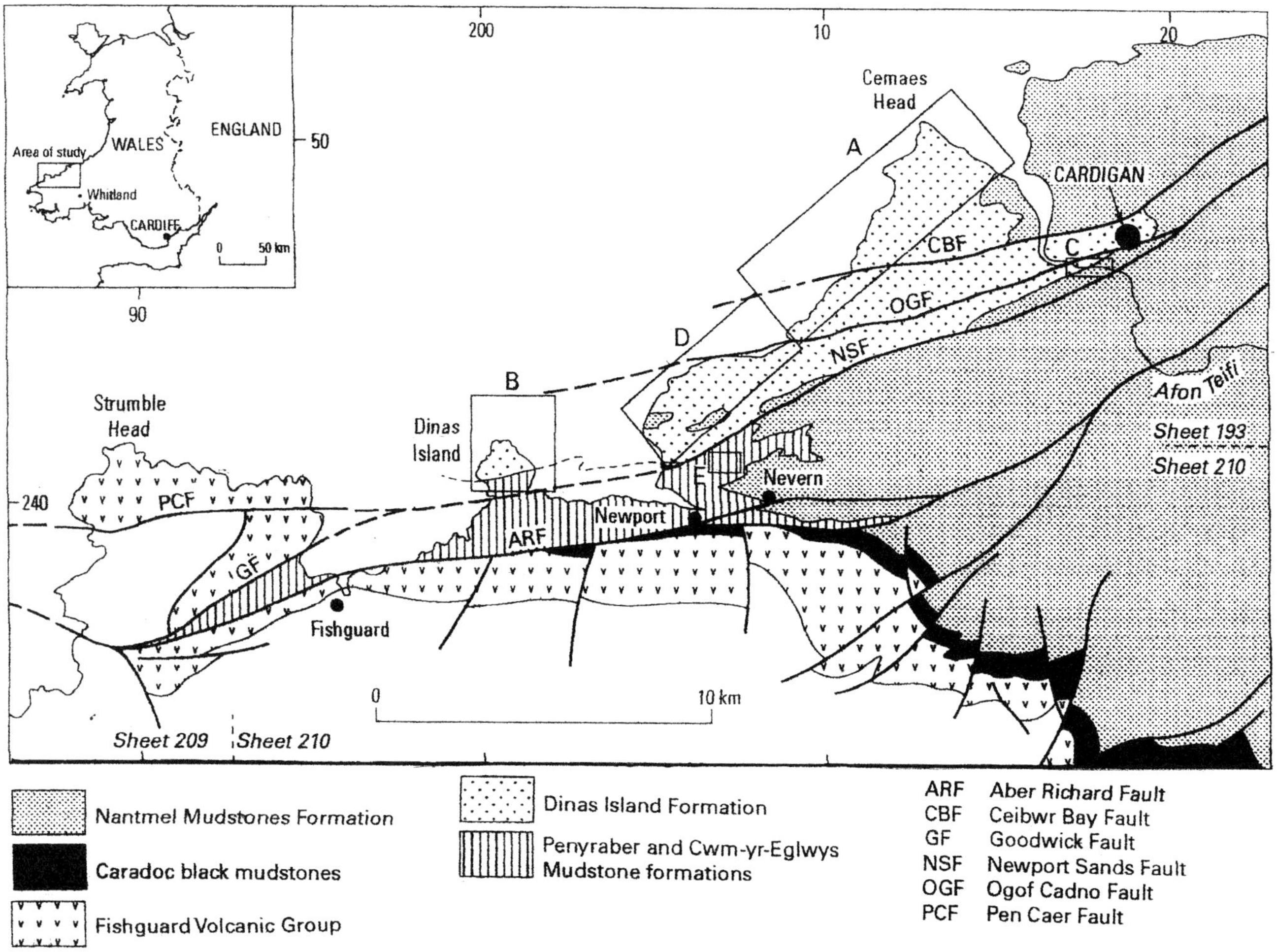

Fig. 3.39. Geological map of the Fishguard–Cardigan fault belt (from Davies *et al* 2002). Reproduced with permission of the British Geological Survey@NERC. All rights reserved. IPR/60-06C.

approximately the same time, with uplift along the Fishguard–Cardigan Fault Belt (Fig. 3.39). Rocks of the Fishguard Volcanic Formation were exposed and eroded along the southern edge of the graben and subsidence began in the trough that had previously been the site of volcanic active in the Llanvirn (Figs. 3.24 & 3.40). During the following transgression shallow marine shelly and bioturbated sandstones (Castle Point Member), that form the lower part of the Penyraber Mudstone Formation, were deposited unconformably on the Fishguard Volcanic Group (Davies *et al.* 2002). The overlying silty mudstones (Saddle Point Member) contain angular sandstone and volcanic blocks, some in excess of 1 m and rafts of black mudstone several metres long, with slump folds indicating mass movement of sediments (Fig. 3.40). The overlying 500 m of the Penyraber Formation is equivalent to part of the former Parrog Shales (McCann 1992) and consists of grey turbiditic siltstones interbedded with some sandstones containing fine shell debris and with thin beds of laminated hemipelagic mudstones including phosphate nodules, which accumulated in a moderately deep dysaerobic environment. The formation was confined to the southern edge and slope of the graben and does not extend north of the Castle Point Fault, but coeval black graptolitic muds were deposited outside the basin to the south and east (Fig. 3.40). Overlying the Penyraber Formation, close to the southern margin of the graben, are dark, anoxic, hemipelagic mudstones and thin, fine-grained turbidites (Cwm-yr-Eglwys Mudstone Formation) ranging in age from the *clingani* to the *linearis* Zone. North of the Newport Sands Fault the formation interfingers laterally with the coarser turbidites of the Dinas Island Formation, deposited in the axial part of the graben (Fig. 3.40). The base of the Cwm-yr-Eglwys Formation is approximately equivalent to the middle of the *caudatus* Subzone, which correlates with the top of the Woolstonian Substage when most volcanicity in the Welsh Basin ceased.

Late Caradoc and Ashgill – the post-volcanic Ordovician Welsh Basin

Three major events directed the further evolution of the Ordovician Welsh Basin The initial event occurred at the close of the Woolstonian; it was a plate-tectonic event possibly the result of Avalonia overiding the spreading axis (Woodcock 2000*c*, p. 164) or an arc–continent collision during the closure of the Iapetus Ocean (Campbell 1984), which caused southward subduction of oceanic crust beneath Eastern Avalonia to cease, thus ending the mid-Caradoc back-arc volcanicity in North Wales. In southern Wales the event had little visible effect on sedimentation. Euxinic muds had been collecting there throughout the earlier Caradoc (the Mydrim or Dicranograptus Shales) and their depositon continued. In contrast the effect in North Wales was drastic. The mainly shallow-marine volcanic pile and associated abundance of coarse clastic sediment were subjected to rapid post-volcanic subsidence and the province became relatively starved of clastic sediment. The euxinic depositional conditions of the rest of the basin then extended to North Wales so producing the black shales that constitute the Nod Glas Formation. The absence from the Nod Glas of all but the finest sediment attests to the generation of the underlying, coarse,

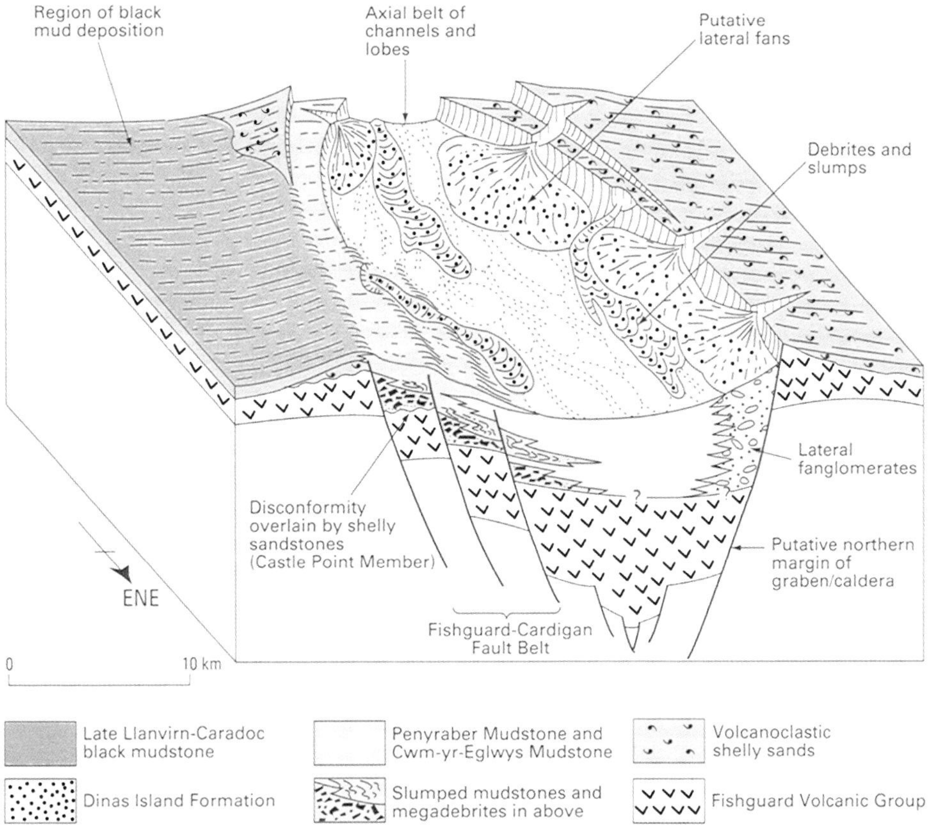

Fig. 3.40. Reconstruction of the Fishguard–Cardigan trough (from Davies *et al.* 2002). Reproduced with permission of the British Geological Survey@NERC. All rights reserved. IPR/60-06C.

commonly volcaniclastic, sediments as having been purely indigenous to North Wales.

The second evolutionary event occurred at, or just after, the start of the Ashgill when there was an abrupt change from black shale to pale grey, *Chondrites*-mottled, and commonly calcareous, mudstone so characteristic of low Ashgill sequences in Wales. The colour change is striking and reflects a rapid switch from euxinic to oxygenated deposition, caused probably by a global fall in sea level. This change was synchronous across the basin. More fundamentally, however, the first two events signalled the initiation of a new phase of the fault-controlled extensional Welsh Sedimentary Basin, which evolved incrementally to its demise towards the end of the Silurian.

The third event that significantly affected the Welsh Basin was the late Ashgill (Hirnantian) glaciation centred upon Gondwana. The concomitant initial fall in sea level (by *c.* 100 m) brought the essentially low-energy mud deposition to an end. Coarsely clastic sediment was introduced into the basin, sedimentation rates increased and slope instabilities were created. At the eventual collapse of the ice sheet, sea levels rose introducing a major marine transgression that greatly reduced the marginal land areas, thus dampening detrital input to the basin. Mud deposition became dominant once more. The rapid fall and then the rise in sea levels virtually define the start and end of the Hirnantian Stage.

Late Caradoc: the Nod Glas and correlative rocks

In North Wales the topmost formation of the Caradoc Series is the Nod Glas. The name for the formation was adopted by Pugh (1923) from a slate quarrymen's term meaning blue mark, which referred to a soot-black, pyritous mudrock of marine origin. This name was adopted by Rushton & Howells (1999) for all the correlative black shale formations in North Wales, and rocks of similar nature and age occur in eastern and southern Wales (Fig. 3.41) under other names and are almost certainly present beneath the younger strata of mid-Wales. Within this ubiquitous deposit of black mud are significantly different facies: basal calcareous phosphates around the Berwyn Dome: turbidites in a fault controlled trough near Cardigan, and even volcanics in the Dolgarrog area. Together with areas of non-deposition, or condensed deposition, these are distributed as shown in Figure 3.41.

In its type area of North Wales the Nod Glas occupies approximately the interval between the end of the Woolstonian (Upper Longvillian of older literature) Substage and the early Pusgillian Stage of Ashgill age. Thus defined it was deposited within the interregnum between the first two evolutionary events, described earlier.

North Wales

This description traces the formation from its fullest and simplest development in the type area, where the sea was deep, northeastward up the basin slope in the direction of increasing sediment starvation, across the area south of Bala to the 'Berwyn high', where condensed 'precipitates' occur at the base. From there it tracks westward across the Bala Fault, over an area where the formation was removed during the early Ashgill, to Snowdonia, the area of maximum post-volcanic subsidence.

In the area around Corris and Aberllefenni, (Fig. 3.41) (Pugh 1923) and Towyn (Jehu 1926) the Nod Glas is some 25 m thick. At 'Nant yr Nod' [SH 8164 1382] between Aberllefenni and Aberangell, Pugh (1928) recorded what he regarded as the best section. Lockley (1980, p. 36) formally designated this as the type locality. Northeastwards from Corris the Nod Glas is progressively thinner and beyond Bwlch y Groes (9, Fig. 3.41), it is absent. This absence, due mainly to early Ashgill uplift and erosion, persists over the western Berwyn Hills and a large area to the NW of Bala (Fig. 3.41), but some 10 km east of Bwlch y Groes the Nod Glas is present again. In Nant Achlas [SJ 013

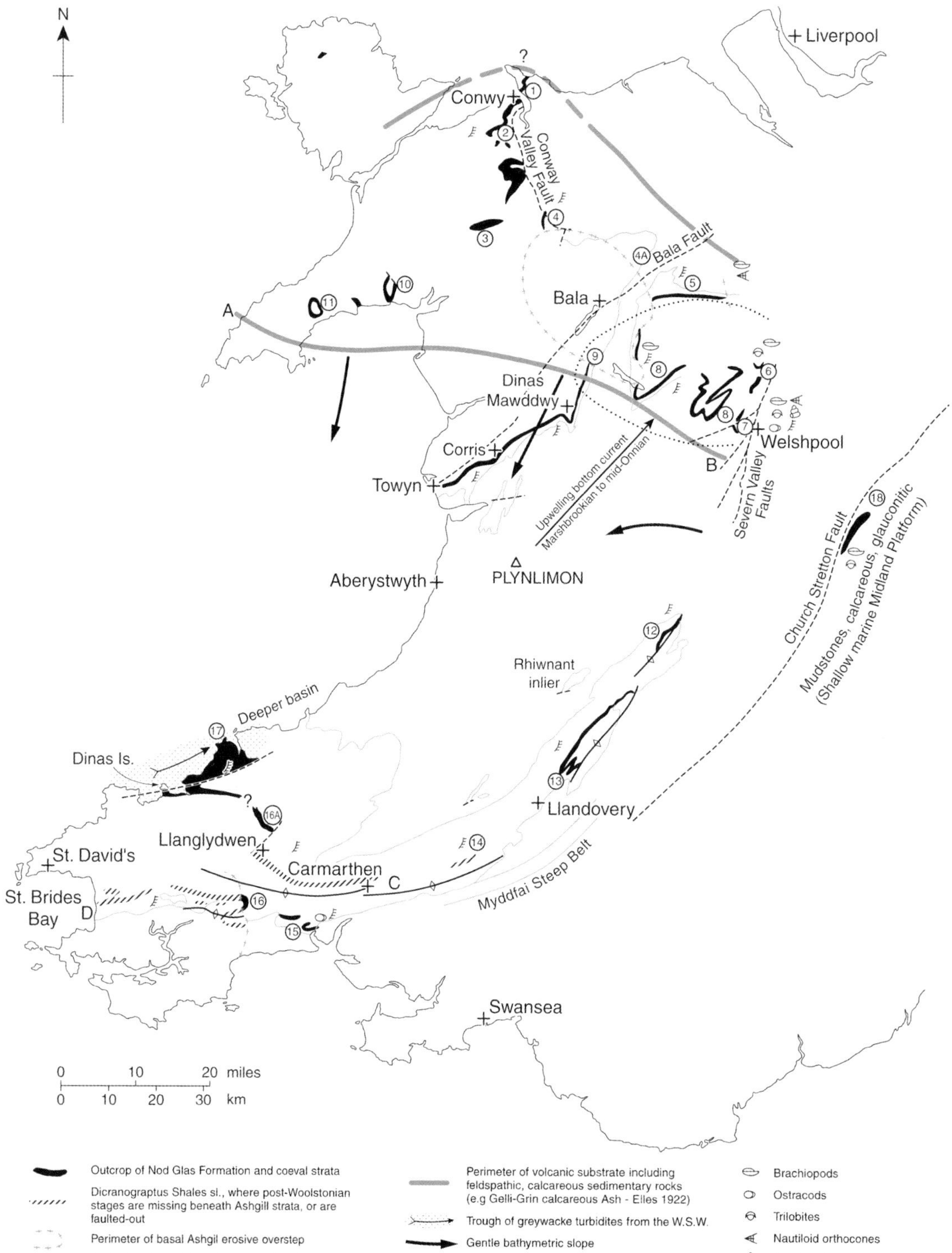

Fig. 3.41. The outcrop of the Nod Glas and related formations in Wales and the Welsh Borderland. The base of the Silurian is also outlined. (1) Cadnant Shales, Conwy (Elles 1909; Wood & Harper 1962). (2) Llanrhychwyn Slates, Trefriw (Davies 1936; Stevenson 1971; Howells *et al.* 1991). (3) Black Slates of Dolwyddelan (Williams & Bulman 1931; Howells & Smith 1997). (4) & (4A) Tyn y glyn Mudstone Formation (Alder 1976). (5) Blaen-y-cwm Beds/Formation, Glyn Ceiriog (Wedd *et al.* 1927; Brenchley 1978). (6) Pen-y-garnedd Shales, Llansantffraid (Whittington 1938). (7) Gwern-y-Brain Shales, Welshpool (Wade 1911), Nod Glas (Cave 1965). (8) Pen-y-garnedd Shales, Meifod (King 1928) and Pen-y-garnedd (Wedd *et al.* 1929). (9) Nod Glas, Llanymawddwy (Pugh 1929). (10) 'Black graptolitic shales', Llanystwmdwy (Harper 1956). (11) Penarwel Mudstones, Pwllheli and Madryn (Matley 1938). (12) Carmel Group (*pars*) (Roberts 1929) = St Cynllo's Church Formation (*pars*), Davies *et al.* 1997). (13) Dicranograptus Shales,Llanwrtyd Wells (Roberts 1929; Jones 1949) = St Cynllo's Church Formation (*pars*) Davies *et al.* (1997). (14) Dicranograptus Shales (pre-*gracilis* Biozone only exposed), Llandeilo (Williams 1953). (15) Mydrim Shales, Mylet (Evans 1906; Strahan *et al.* 1909). (16) Mydrim Shales, (Dicranograptus Shales (*pars*), Whitland (Strahan *et al.* 1914; Zalasiewicz *et al.* 1995); (16A) Mydrim Shales, Crymych; = Crymmech (Evans 1945). (17) Poppit Formation, Cardigan (Latter 1925; Challinor 1927; James 1975; McCann 1992). Also Dinas Island and Cwm-yr-Eglwys Mudstone formations (Davies *et al.* 2002). (18) Onny (Shales) Formation (Bancroft 1929, 1933, 1945; Dean 1958; Hurst 1979). The numbers in brackets are used also to identify Ashgill localities mentioned in the text.

266] it is 19.5 m thick, but thins northwards and southwards terminates abruptly (Brenchley 1978, p. 155).

Although there are gaps in the mapping, east of Nant Achlas the outcrop probably forms a continuous ribbon, mainly as NNE–SSW strips in the flanks of major folds. However, the soft nature of the shales has resulted in exposure being sporadic and rarely extensive. On approach to the N–S fault system at Welshpool (7, Fig. 3.41) the outcrop fails again, but here the failure is the result of overstep by Llandovery strata. In this southeastern adjunct of the Berwyn Hills the Nod Glas shales are less silty than they are in the SW, towards Towyn, and the smaller thicknesses, in the range 14–16 m, may in part be a reflection of this. On the north side of the Berwyn Hills, west of Glyn Ceiriog (5, Fig. 3.41), the formation occurs in an E–W outcrop some 10 km long. From a maximum approaching 50 m, near the western end (Brenchley 1978), the thickness of the shales diminishes sharply westwards and more gradually eastward. In this area too the shales are more silty than to the SE.

Mention has been made of a large area around Bala where the Nod Glas is totally absent (Fig. 3.41), but beyond this to the NW, in eastern and northern Snowdonia, it reappears under several local names (1–4, Fig. 3.41). Here the soot-black shales occur at the top of a thick pile of Caradoc volcanic and associated sedimentary rocks and, as in the southwestern outcrop, they are strongly cleaved. Indeed, in the tight Dolwyddelan syncline (3, Fig. 3.41) at Chwarel Ddu, where they rest upon the Upper Rhyolitic Tuff Formation, they have been quarried for slate, although normally the pyritous nature of the Nod Glas renders it unfit for this purpose. As the top of the formation is not preserved in the syncline a thickness there cannot be ascertained.

Farther north the black shales crop out as the Llanrhychwyn Slates and the Cadnant Shales (2 and 1, respectively, Fig. 3.41). The former are 60 m thick at Betws y Coed (IGS 1976), 65 m at Trefriw (Davies 1936) and variously 83–400 m proximal to the very localized basaltic Dolgarrog Volcanic Formation (IGS 1981). This volcanic formation rests above the Upper Crafnant Volcanic Formation and is interbedded with the black shales suggesting that here, was a very late burst of volcanic activity of Nod Glas age. However, there is no biostratigraphical proof of this age, thus leaving the possibility that the Cadnant Shales and Llanrhychwyn Slates extend to lower horizons than do the Black Slates of Dolwyddelan.

Black 'Dicranograptus Shales', with horizons probably comparable with those of the Cadnant Shales, crop out at Llanbabo Church, Anglesey (Greenly 1919), but there is no evidence that they rest upon a volcanic substrate, comparable with the rest of North Wales, or even that they don't pass down to incorporate lower horizons as such shales do in mid-Wales and South Wales.

The remaining outcrops of Nod Glas in this counter-clockwise review of North Wales occur in the Llŷn Peninsula (10 and 11, Fig. 3.41). Here the black shales do rest on a formation rich in volcanic detritus, comparable with the sequences in and around the Berwyn Hills. Indeed, the 'black graptolitic shales' of Llanystwmdwy have a comparable thickness (c.16 m). The stratigraphy and thicknesses of the various outcrops of the nearby 'Penarwel shales' to the west of Pwllheli (11, Fig. 3.41) are less resolved, although *D. clingani* is present and thus probably the *caudatus* Subzone (Rushton & Zalasiewicz in Young *et al.* 2002, p. 55). This suggests that the Penarwel Mudstones extend down into low horizons as at Tywyn, but no lower than Woolstonian in view of the well documented Longvillian and Woolstonian shelly faunas (also of the *caudatus* Subzone) of the subjacent strata of the Llŷn (Matley 1938; Harper 1956, p. 387–389).

Southern Wales

For the southern half of Wales the use of the term Nod Glas is inappropriate. In many places graptolite biostratigraphy proves that the higher parts of the black shales here are indeed equivalent to the Nod Glas, but the facies descends uninterruptedly to much lower levels. This extended range of the facies constitutes the Dicranograptus Shales of older literature. Thus, for southern Wales, Figure 3.41 depicts the outcrop of the Dicranograptus Shales as the basis for the identification of the portion at the top which is equivalent in age to the Nod Glas.

Near Abbeycwmhir (12, Fig. 3.41) Davies *et al.* (1997) renamed the Dicranograptus Shales as the St Cynllo's Church Formation. They noted a tripart division in which the top third, 50 m thick, is darker in colour than the rest and is devoid of the sandstones that characterize the mid-part. The same shales, 40 m thick, crop out around and to the south of Llanwrtyd (13, Fig. 3.41) where they have the name Sugar Loaf Formation (James 1990).

At Llandeilo (14, Fig. 3.41) at least the lower parts of the Dicranograptus Shales crop out and while the upper parts were almost certainly deposited, their apparent absence now may be due to overthrust faulting. The same reason may account for the similar absence along the outcrop from just east of Carmarthen to Llanglydwen in the west. In this the Dicranograptus beds of Evans (1906) and the Geological Survey map Sheet 229 (IGS 1967) (15, Fig. 3.41) dip northwards and are strike-faulted against Ashgill mudstones to the north.

In SW Wales the Dicranograptus Shales are generally divided into three parts, the mid-part being the Mydrim Limestone and the top part being the Mydrim Shales (Strahan *et al.* 1909). Both of these are of Caradoc age, but it is the upper part of the Mydrim Shales which correlates with the Nod Glas, for instance at Mylet (15, Fig. 3.41) where there is passage upwards into basal Ashgill limestone (Strahan *et al.* 1909, p. 52). A similar passage is exposed nearby, on the road approximately 600 m west of Llanddowror, but the best exposure is farther west in a road cutting on the A40 approximately 3 km west of Whitland (16, Fig. 3.41). Here Zalaseiwicz *et al.* (1995) placed the mid-*clingani* Zone (the junction between the *caudatus* and *morrisi* subzones) at 32 or 33 m below the base of the Pusgillian Stage [SN 1697 1699]. This, therefore, is the thickness of that part of the Mydrim Shales which equates with the Nod Glas. West of here, to St Bride's Bay (Fig. 3.41), the top of the Mydrim Shales is missing beneath unconformable Ashgill rocks. How much of the Mydrim Shales here, if anything, equates with the Nod Glas is not clear.

The remaining outcrops of late Caradoc Welsh rocks are the Mydrim Shales of Crymmech Arms (Crymych) (16A, Fig. 3.41) and the thick, fault controlled trough-sequence of black shales and turbidites in the Cardigan area (17, Fig. 3.41). The thickness of the latter may be as much as 1.3 km. but probably only the upper parts belong to the substages Marshbrookian–Onnian. Whether, or how, these two outcrops join across the intervening ground is not known yet.

Associated facies. The pyritous black graptolitic mudstone facies is so characteric of the Nod Glas throughout the basin that any variant is noteworthy and there are three.

(1) *The Pen y Garnedd Phosphorite.* Around the Berwyn Hills, especially in the south and southeast, the basal metre or so carries calcareous phosphorites (Cave 1965). At the very base there may be a dark grey, crystalline phosphatic limestone, as at Welshpool (7, Fig. 3.41) and Pen y garnedd (8), where it is overlain by several thinner beds of rubbly shelly mudstone, commonly containing round, hard, phosphatic nodules, interbedded with the black shale. To the SW, this phosphatic base extends almost to Dinas Mawddwy (Pugh 1928) (Fig. 3.41) and here Lockley gave the facies the status of Dyfi Member (Lockley 1980, p. 38). Small plectambonitid brachiopods and ostracods characterize all of these basal phosphatic and calcareous beds. From

Figure 3.41 it can be seen that there is a striking coincidence between this area where phosphorites are developed at the base and the area where the Nod Glas is depositionally thinnest (as distinct from the effects of sub-Ashgill erosion).

(2) *Shelly facies.* There is one important exception to the singularly graptolitic nature of the basinal black shale and that is in Gwern-y-Brain, Welshpool (Cave 1965; 7, Fig. 3.41). Here brachiopods and trilobites thrived despite the euxinic nature of the environment of their eventual preservation. In other respects these black shales are not different from those elsewhere.

(3) *Turbiditic facies.* The third variant of the Nod Glas is the sequence forming the Dinas Island Formation, comprising sandstone with mudstone turbidites, and subordinate mass-flow conglomerates (debrites) and pyritic, laminated hemipelagic mudstones; the sequence is punctuated by several units each up to 100 m thick, of interbedded turbiditic and hemipelagic mudstone. Underlying this sequence are the mudstones of the Cwm-yr-Eglwys Formation (Davies *et al.* 2002). The coarse deposits of the Dinas Island Formation, up to 1.3 km thick, accumulated preferentially on the northern, downthrow, sides of several contemporaneously active WSW–ENE faults across which they stepped progressively southward. Thus the base of the Dinas Island Formation youngs southward by passage, across the faulting, of the lower parts of the Dinas Island Formation into the upper parts of the Cwm-yr-Eglwys Mudstone Formation. The coarse detritus was fed into this fault-bounded trough from the WSW where it had been scoured, or eroded, from the Fishguard Volcanic Group of Llanvirn age.

Both of the formations described above lie within the *clingani* Biozone possibly extending into the *linearis* Biozone. Thus they embrace deposits coeval with the Nod Glas. However, at their bases they may include sub-Nod Glas horizons and thus the outcrop illustrated in Figure 3.41 might be exaggerated.

Shelf sequence. A Caradoc type-sequence which includes the substages Woolstonian–Onnian occurs near Craven Arms (18, Fig. 3.41). It lies, east of the Welsh Basin, on the western edge of the Midland Platform within and immediately east of the Church Stretton Fault complex.

These Woolstonian–Onnian strata consist of approximately 200 m of an upward-fining sequence of sandstones, siltstones and mudstones that follow above Costonian–Longvillian strata without a break. From their sedimentary facies, together with the species diversity and associations within their shelly faunas, Hurst (1979*a–c*) revealed that they were deposited in a deepening marine environment which evolved from a proximal to a distal shelf position, i.e. from the Alternata Limestone Formation (basal Woolstonian) with proximal storm layers, to the dysaerobic deposits of the Onny Shales Formation (Onnian). There is no sign here of the stratigraphical break so evident below the Nod Glas barely 30 km to the NW at Welshpool.

Biostratigraphy. Historically, the type Caradoc Series was based upon the shelly sequence in South Shropshire (18, Fig. 3.41) and from it Bancroft (1933) erected nine stages (now substages). The Nod Glas, however, is almost exclusively graptolitc and the graptolitic biozonation is based in the Lower Hartfell Shale Formation of the Southern Uplands of Scotland and this has a fauna only partially duplicated on the southern side of Iapetus. Thus, while there is an obvious problem in relating the Nod Glas to the type shelly sequence, there is a measure of incompatability with the type graptolite sequence too.

Five fossil taxa from the Nod Glas near Welshpool and one from the beds immediately below its basal limestone are crucial to the biostratigraphic understanding of the formation. From below the limestone the trilobite *Estoniops alifrons* (M'Coy) [*Pterygometopus jukesi* of older literature], is a reliable indicator of the Woolstonian in Wales; from above the limestone, the trilobites *Onnia gracilis* (Bancroft) and *Flexicalymene onniensis* (Shirley), the brachiopod *Onniella broeggeri* Bancroft and the graptolite *Dicellograptus morrisi* Hopkinson are all Onnian, the last being the index for the upper subzone of the *clingani* Biozone. *Dicranograptus clingani* Carruthers itself has not been found near Welshpool, but the sequence there demonstrates that the Nod Glas above the basal limestone is of Onnian age, while the beds beneath the limestone are Woolstonian; only the limestone remains to fill the unidentified Marshbrookian–low Onnian substages. This leaves three possibilities: the limestone is low Onnian with a large non-sequence below; or it forms a condensed sequence spanning the Marshbrookian–Onnian; or it represents just a part of that interval with a smaller non-sequence below. Furthermore, it is logical to believe that wherever the phosphorite is present these same conditions obtain, because over most of North Wales (11–7, Fig. 3.41) the substrate can be dated as Woolstonian by the presence of *E. alifrons.* However, SW of Llanymawddwy the basal phosphorite and underlying shelly fauna are absent so that there the black shale facies probably progressively occupies lower horizons in the Actonian and Marshbrookian. Lockley (1980) suggests that the Nod Glas may extend down to the Woolstonian and Longvillian and this is supported by the record of *D. clingani* from the thick Corris to Towyn outcrop (Pugh 1923; Jehu 1926). *D. clingani* tends to be restricted to the lower half of that biozone, which is equivalent to the upper part of the Longvillian and the Woolstonian, and it is present in the downward extended black shale sequence of mid and South Wales. *D. morrisi*, or *'Dicellograptus forchhammeri'* (Geinitz) is significant of the upper part of the *clingani* Biozone and it is recorded widely in the Nod Glas of North Wales as well as in the Mydrim Shales of South Wales. It is also in the turbiditic sequence of Cardigan (Challinor 1927; Williams *et al.* 2003).

In general the presence within these sequences of the topmost Caradoc biozone, that of *Pleurograptus linearis*, has been accepted only with reservation, perhaps because *P. linearis* itself has never been found in Wales. However, Harper (1956) was in no doubt that the zone is present at Llanystumdwy (10, Fig. 3.41) and Zalaseiwicz *et al.* (1995) accepted its presence at Whitland (16, Fig. 3.41) with reserve; indeed, many accounts have been published recording *Climacograptus styloideus* Elles & Wood (e.g. Strachan *et al.* 1914; Wedd *et al.* 1929) and this fossil taxon, although not easily recognized, is restricted to the *linearis* Biozone in Scotland (Zalaseiwicz *et al.* 1995).

Environment. It is generally accepted that basinal euxinic deposition commonly accompanies rising sea level with consequential transgression of the adjacent platform (Leggett 1978) and it is known that just such an event characterized the Caradoc epoch. It is probably the reason behind the dark grey–black mudstones of southern Wales. In North Wales volcanism dominated the Longvillian and there this displaced the euxinic conditions. Then, when the volcanism ceased, in or just after the Woolstonian, post-volcanic subsidence set in which effectively linked northern and southern areas into a single euxinic 'basin'. The progressive inundation of eastern platform areas additionally explains the paucity and fineness of terrigenous detritus reaching this basin from that quarter (Fig. 3.41) (Cave 1965).

The overall thickening and coarsening of the North Wales Nod Glas, southwestwards towards Towyn, may be in part the distal effect of the input of turbidites to the deepest region of the basin from the far SW (17, Fig. 3.41) as well as the replacement of the basal phosphorite by normal deposits. Missing late Caradoc sequences, below the Ashgill of western Pembrokeshire, give credence to the belief that this region was being subjected to erosion and was delivering detritus northwards.

Phosphorites are uncommon rocks. In the Nod Glas they are precipitates of phosphatic minerals, requiring very specfic and unusual conditions for their formation, some of which they

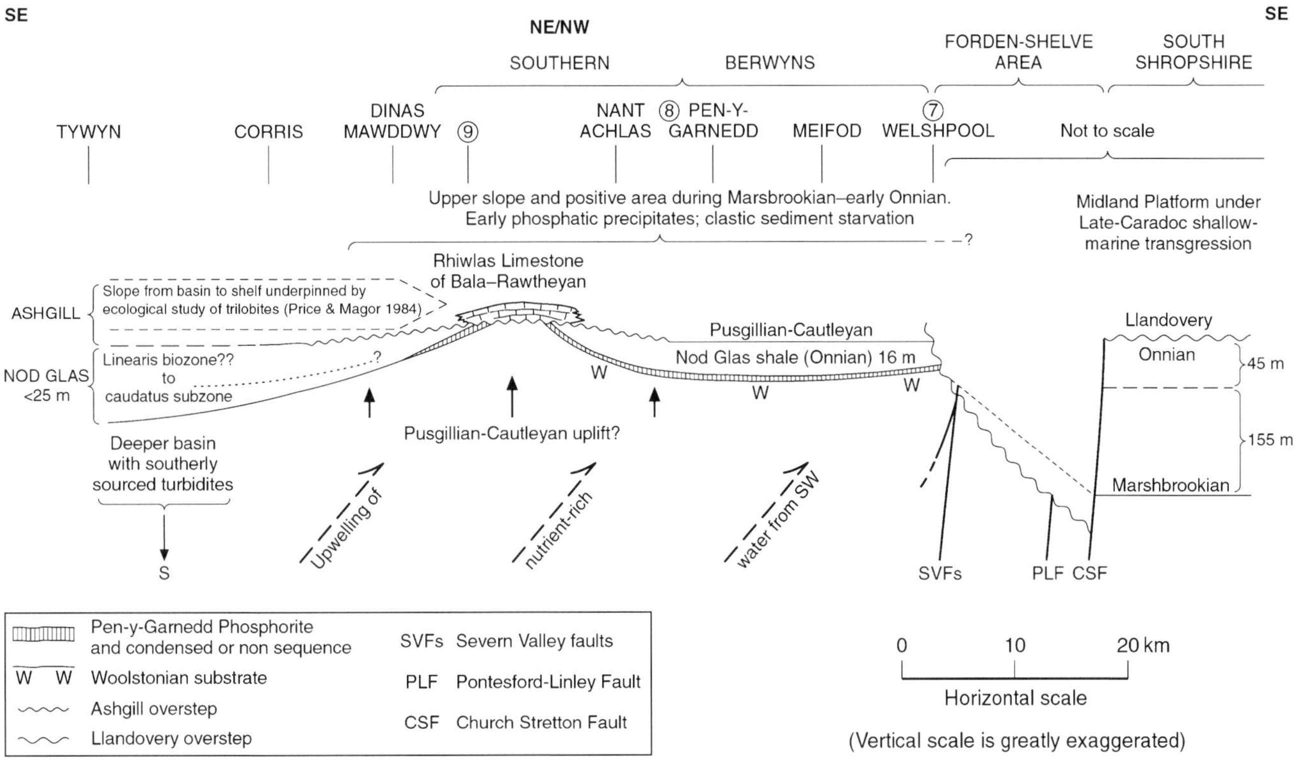

Fig. 3.42. Section of late Caradoc and early Ashgill strata across southern North Wales into Shropshire.

share with black mudstones; hence, the two rock types are often associates. More specifically phosphorites tend to form on and around the flanks of bathymetric rises, or positive areas, starved of terrigenous sediments. Thus they commonly accompany stratigraphical hiatuses, or condensed sequences, where they are often multilayer, suggesting repeated resurgences of the optimum conditions (Notholt 1980, p. 794). In a model suggested by Cave (1965, pp. 293–294) the Pen y Garnedd Phosphorite reflects all these conditions. Situated on a bathymetric high in the Berwyns (Figs 3.41 & 3.42) the phosphorite was nourished by the upwelling of basinal water from the SW, but starved of detritus from the east. Leggett (1980) has suggested a comparison with the model of Heckel (1977) in which a shallow, but strong, thermocline can develop over shelf areas and this will trap anoxic nutrient-rich water beneath it. Leggett considered that the Caradoc marine transgression of the shelf might have produced such a situation.

The bathymetric high around Bala and the Berwyns has not been explained. It might be a tectonic upwarp; equally it could have arisen as a result of variations in the rate of the post-volcanic subsidence of North Wales. The most active subsidence could be expected to have affected the most active volcanic areas. These were in Snowdonia. Bala and the Berwyns were distal from the volcanism and the recipients mainly of its reworked detritus. Subsidence would have been less active there, leaving the areas upstanding.

Ashgill stratigraphy and environments

The locus of Welsh Ashgillian tectonism lay mainly along eastern mid-Wales extending into SW Wales. The responses to the tectonism were a North Wales which remained comparatively shallow marine (mid-deep neritic), a mid-Wales which experienced a basin-floor collapse creating a sink for a large thickness of rapidly accumulated sediment and a South Wales which supported a shelf environment passing northwards into deep neritic and possibly unstable, basin-slope conditions. The tectonic movements are reflected in the unconformity below Ashgill (Cautleyan and Rawtheyan) rocks between Nant Achlas and Pentre Foelas (Figs 3.41 & 3.42). It has been suggested that movement on a precursor to the Bala Fault could be responsible (Bassett *et al.* 1966). An unconformity also occurs at the base of Ashgill rocks in western South Wales and pre-Hirnantian faulting on the line of the Tywi axis substantially affected sedimentation. In addition, an important episode of folding and faulting of the Shelve and Welshpool Ordovician rocks was of post-Rawtheyan, but pre-Telychian age. At the base of the Ashgill there is an abrupt change in many places from the anoxic sediments of the Nod Glas to the the typical deposits in and around the basin, which were oxygenated mud and silt but with important argillaceous limestones in North Wales and South Wales.

The estimated 100 m glacio-eustatic fall in the final Hirnantian Stage (Brenchley & Newall 1984) drained the peripheral platform seas, exposing them to subaerial erosion. This stripped off much of the recently deposited littoral and sublittoral sediment of the Midland Platform which was then removed westward downslope, to be redeposited in deeper basinal areas. New littoral and sublittoral areas formed temporarily in a narrow fringe around the mid-Wales deeper basin and in parts of the newly exposed platform edge; karstic surfaces were eroded into recently cemented earlier (Rawtheyan) sediments. When the glaciation collapsed towards the end of the Hirnantian, the karstic surfaces became preserved beneath the sands of the readvancing shores and these sands were followed above by shelly argillaceous sediments (8 and 12) (Brenchley *et al.* 2006).

Pusgillian, Cautleyan and Rawtheyan stages

North Wales

The basal Ashgill sequences in North Wales consist typically of bioturbated mudstones with argillaceous limestones, but there is substantial diachroneity (Fig. 3.43) and the Pusgillian and Cautleyan are only sparsely represented. The sequences

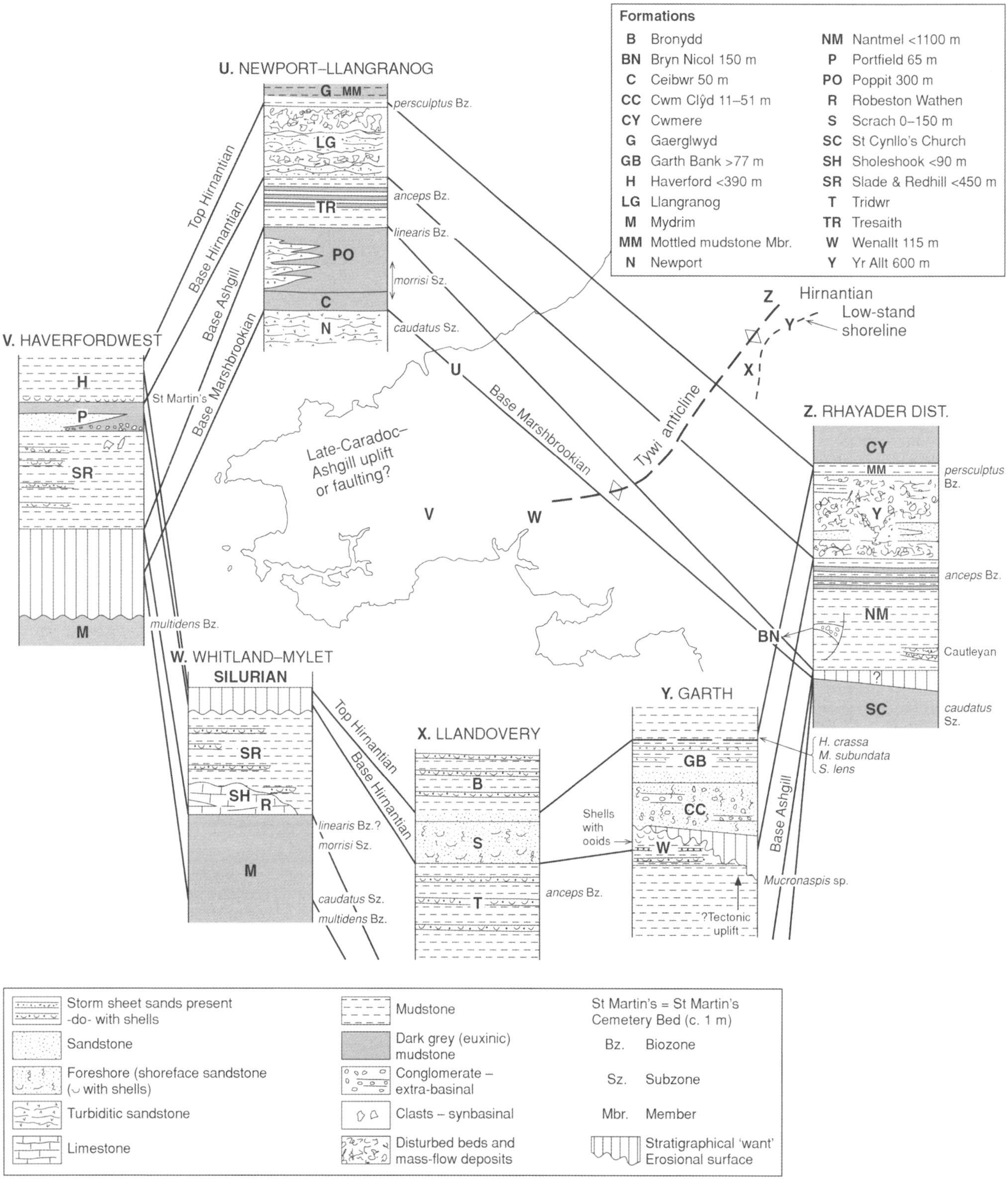

Fig. 3.43. Ashgill and late Caradoc sequences, South Wales.

possess diverse brachiopod-trilobite faunas of small individuals which, in places, are prolific e.g. in the lowest parts of such formations as the Crugan Mudstones (10 and 11) (Rawtheyan on Nod Glas), the Maerdy Mudstone (4 and 4A), the Moelfryn Mudstones at Bala (Rawtheyan), the Ty'n-y-twmpath Member i.e. basal Dolhir Formation (5) (Pusgillian on Nod Glas), the Lower Tre-wylan Formation (6) (Pusgillian on Nod Glas), the Trawscoed Mudstone Formation (7) (Cautleyan on Nod Glas) and equivalents at Meifod and Pen-y-garnedd (8) (Pusgillian on Nod Glas). They are interpreted as having been deep-neritic sediments of a low-energy input. The argillaceous limestones occur as local facies overlying Caradoc rocks. The best known is the Rhiwlas Limestone Member (Rawtheyan on Woolstonian) of the Moelfryn Mudstones Formation around Bala (Bassett *et al.* 1966) and sporadically as far west as Capel Garmon (4A), where it is known as the Hafod Ifan Limestone. A second limestone is the Dolhir Limestone Member (Cautleyan) of the Dolhir Formation at Glyn Ceiriog (5) (Wedd *et al.* 1927; & Hiller 1981). These two limestones might be the product of slightly shallower 'shoals' and significantly both occur where the Ashgill is unconformable (Wills & Smith 1922, p. 189; Bancroft 1928; Campbell 1993) and where a late Caradoc bathymetric high might have survived (Fig. 3.42).

Above the calcareous beds the bulk of each of these formations, including the Bodeidda Mudstones and Deganwy Mudstones (1), continues upwards rather monotonously, more silty in places and less fossiliferous. Beds of siltstone are present, for instance in the Moelfryn Mudstones at Bala, where they may be up to 8 cm thick, each probably representing a single storm event. Thus, the depositional environment overall was non-energetic and fairly uniform. Bioturbation, commonly as *Chondrites*-mottling, persisted and so did the marine oxicity, except in the late Rawtheyan; the Deganwy Mudstones (1), for instance, record that a degree of oxygen starvation may have returned, as graptolites of the *anceps* Biozone are preserved. In the upper limits of these mudstones and in the Foel y Ddinas Mudstones of Bala, *Mucronaspis mucronata* (Brongniart) is common and is said to be indicative of the Hirnantian (Brenchley & Cullen 1984, p. 124).

Thicknesses of these largely Rawtheyan formations differ across the region. They are thinnest in the west, e.g. 120 m (1) 300 m (10 and 11), thickest around Bala (up to 700 m) and to the north (up to 1500 m) (4A), thinning again eastward from 600 m to 300 m (5), but there is no obvious linear pattern to this. It has been suggested (Bassett *et al.* 1966, p. 255), at least for areas close to the Bala Fault, that the differences are erosional rather than depositional, caused by up-throw on the NW side of a precursor of the fault.

Biostratigraphy. Biostratigraphic evidence suggests that in many places there is an unbroken passage across the basal Ashgill boundary. Price (1984), for instance, using tretaspid trinucleids, indicated that at several places on the eastern side of the Berwyns (5, 6 and 8), the Pusgillian Stage, at least in part, and the Cautleyan Stage are indeed represented in basal Ashgill mudstones. Here there could be stratigraphical continuity up from the Nod Glas. However, elsewhere there is a stratigraphic break of variable size. At Welshpool (4), for example, a Cautleyan fauna occurs but a few metres from the top of the Nod Glas leaving little room for the Pusgillian. Elsewhere the lowest rocks of the North Wales Ashgill sequences are characterised by the trilobites *Arthrorhachis tarda* (Barrande) and *Philipsinella parabola* (Barrande). In the type sequences at Cautley *P. parabola* does not make an appearance lower than Cautleyan Zone 3 and *A. tarda* comes in only above Rawtheyan Zone 5 (Ingham 1966) and thus they support the general belief that there is a fairly widespread non-sequence under the Ashgill rocks that spans the Pusgillian and Cautleyan. A larger break occurs below the Rhiwlas Limestone at Bala, which is of Rawtheyan age, resting unconformably on mid-Caradoc strata.

Mid-Wales (south of line A–B and north of the line 14–St.David's–Fig. 3.41).
The Ashgill rocks of mid-Wales were deposited in two bathymetrically definable areas. The first is the basin in the west, with a relatively gentle slope rising northwards and northeastwards to the shallower environments of North Wales described above. Secondly, a steeper, tectonically active, slope rose in the east and SE to a shelf margin which lay along the line of the impending Tywi Anticline, or Lineament (Fig. 3.41). The deposits of these areas are, as in North Wales, separable into two sequences: the lower (earlier) part of the Ashgill consists mainly of uniform mudstones, pale grey, rhythmically thin-bedded and burrow mottled; the upper part is far more varied and coarser, with large and laterally impersistent bodies of bedded siltstone and mudstone, sandstone and conglomerate commonly with extra-basinal clasts. The lower sequence probably falls within the Pusgillian to Rawtheyan stages the upper is mainly Hirnantian, but in part Rawtheyan in the SE.

The presence of the Pusgillian stage is unproven and, as in parts of North Wales, pale grey mottled (*Chrondites* burrows) mudstones succeed black shale sharply and with conformity (Pugh 1928, 1929; Cave & Hains 1986; Anketell 1987; Leng in Pratt *et al.* 1995; and Davies *et al.* 1997). Thus, the possibility remains that the Pusgillian (and Cautleyan) are present somewhere at the base of those sequences, but undetected and perhaps partly in the top of the underlying Nod Glas. The facies is considered to be one of mud turbidites, parts possessing the thin basal fine sandstones of Bouma Td and Tc intervals. Black phosphatic concretions are common at the bases of the units of pale grey burrowed mudstone, which, in part, is formed of oxygenated hemipelagite, and which caps the turbiditic intervals (Cave & Hains 1986; Davies *et al.* 1997). These turbidites were deposited as a slope-apron extending into the basin. The change in the sequence from dark grey shale below into bioturbated pale mudstone resulted from the same change in depositional oxicity as was seen in North Wales, caused by the same marine regression. The pale grey facies is represented by the Tresaith Formation in the deep basin north of Cardigan (U, Fig. 3.43) and by the Nant-y-Moch Formation west of Plynlimon. The facies is also present as the Broad Vein–Abercwmeiddau Group in the Towyn to Dinas Mawddwy outcrop and as the Nantmel Formation (12–13) of the eastern slope. Towards the top of all these formations, and mirroring the same event(s) recorded by graptolitic beds, for example, in the Deganwy Mudstones (1, Fig. 3.41) of North Wales, there are interbedded packets of dark grey laminated hemipelagite. All have yielded graptolites of the *anceps* Biozone, including the zone fossil, and thus their age can be accepted as Rawtheyan. In places they occur as two or three packets which, at Plynlimon, alternate with packets of thin, sand-dominated, mudstone–sandstone turbiditic couplets which are bioturbated. It is possible that these alternations in lithology represent late Rawtheyan pulses of marine transgression and regression.

The outcrop of the Broad Vein–Abercwmeiddau Group from Towyn to Dinas Mawddwy traverses obliquely from the base of the slope to upper slope with deposits exemplifying this low-energy distributive regime. Price & Magor (1984) revealed that along this traverse there are three trilobite associations, each related to depth. They are replaced one by another from deep to shallow, as the outcrop traverses northward. Eventually the traverse crosses a diffuse slope-to-shelf break into the shelly Moelfryn Mudstones of Bala with its basal Rhiwlas Limestone. The E–SE slope-to-shelf break, on the other hand, is much less diffuse and the facies changes reveal that it is aligned closely with the Tywi Anticline (Fig. 3.43), a fundamental curvilinear structure associated with major faults.

At the northern extremity of the anticline around Abbey Cwmhir, near Rhayader (12, Fig. 3.41), Pusgillian to Rawtheyan deposits (Nantmel Formation) are largely the same as those to the north and west of mid-Wales, namely bioturbated rhythmic slope-apron mud turbidites.The proximity of the Midland Platform, is suggested by minor beds of shelf-derived shelly sandstones near the base, and by at least two major units (Cautleyan or Rawtheyan) of sandy conglomeratic debrites, within the anticline and farther to the SE (Davies *et al.* 1997). These major units appear to have been trapped in a contemporary fault-generated trough on the upper slope. Southwestward, along the anticline towards Llandovery, the depositional regime was different. The switch from anoxic Caradoc mudstones to oxic basal Ashgill mudstones (Craig Las Formation) is as elsewhere, but higher in the succession the sandy conglomeratic, debrite/turbidite facies within the Nantmel Formation has become coarser and more prevalent and here it forms the Bryn Nicol Formation, of Mackie & Smallwood (1987). Evidence is emerging (D. Wilson (B.G.S.) Pers. comm.) that these debrites and turbitites are also bounded upslope by strike-faults; these all indicate that down-to-basin faults were actively generating sediment traps on the mid- to upper-slope where the fault tips failed to break surface of a rapidly accumulating slope-apron (Smallwood 1986). It is here, near Llandovery, that the Tywi anticlinal structures (Fig. 3.43) bend sharply towards Pembrokeshire and also approach another fault structure, the Myddfai Steep Belt (Fig. 3.41). Tectonic stresses were likely to have been focused here.

Southwestward, approaching South Wales, the slope deposits continue to be present on the NW side of the Tywi Anticline. Sandstones are minor elements and the input of sediment was less dynamic. However one conglomerate, with an exotic Rawtheyan fauna of shallow water shells, does occur here in slope-apron muds which are continuous with the Bryn Nicol and Nantmel formations. One intriguing aspect of the Rawtheyan at Llandeilo is the record of a bed 'a few feet thick' of andesitic ash (Williams 1953). There is no other such record of Ashgill volcanism in Wales.

Although the shelf facies of the SE side of the Tywi Anticline is poorly preserved, it is exposed near Garth, some 20 km NE of Llandovery, where mid-late Rawtheyan (Zones 6–7) sandy siltstones yield a deep-shelf, shelly fauna of brachiopods with trilobites and at one locality *Mucronaspis* sp. was recorded by Williams & Wright (1981, p. 7).

The deep neritic mudstones of the Tridwr Formation exposed in the anticlinal 'window' 14 km SSW of Llandovery (Cocks *et al.* 1984) (X, Fig. 3.43) are another expression of the Rawtheyan outer shelf. Traction currents are indicated by the presence of thin, distal storm-sheet sandstones with a shelly fauna of Rawtheyan age. Immediately above lies the Scrach Formation with a Hirnantian fauna. The Tridwr Formation must, therefore, pass southwestwards into the similar shelly Slade and Redhill Formation of South Wales.

South Wales (Fig. 3.41)

South Wales lay on the southern margin of the Welsh Basin (Fig. 3.41), as part of a probably broad marine shelf. As in North Wales at the end of the Caradoc, this rapidly became oxic in a shallowing, regressive environment. Thus there is a thin upward passage from pyritous black anoxic Mydrim Shales into bioturbated and silty mudstones, or limestones. A line C–D which extends westward from the Tywi Anticline at Carmarthen to the centre of St. Bride's Bay (Fig. 3.41) continues the Tywi facies divide between shallow shelf and deep shelf/break of slope, but much of the detail of the facies pattern is obscured in a plexus of strike faults which disrupts the outcrop.

South of the line C–D (Fig. 3.41)

Limestones. South of the line C–D the dominant facies overlying the Mydrim Shales is argillaceous limestone rising into sandy limestone. In the east, at Llandeilo, (14, Fig. 3.41) this is known as the Crug Limestone and the Birdshill Limestone, while west of Carmarthen it is the Robeston Wathen Limestone and the Sholeshook Limestone. Well exposed sequences up from the Mydrim Shales are sparse, the only two documented being at Mylet (15, Fig. 3.41) (Strahan *et al.* 1909, p. 56) and in a road cutting on the A40 just west of Whitland (16, Fig. 3.41) (Zalaseiwicz *et al.* 1995). At these two places the passage seems to be unbroken, although in both a bed of phosphatic nodules marks the base of the limestone and might indicate a minor diastem. To the east of these, the outcrops are faulted, so that stratigraphical relationships there are uncertain. Westward of Whitland, however, the limestones rest unconformably upon horizons of the Mydrim Shales as low as the low *multidens* Biozone, near Trewern [SN173 177] for instance (Strahan *et al.* 1914).

The nature and thickness of the limestones (Rushton *et al.* 2000) differ from place to place, possibly reflecting a deposition in different depth zones of the marine shelf; Price (1973, p. 125) suggested that the Birdshill, a pale grey, sparry bioclastic limestone, formed under shallow water, whilst the Sholeshook, much more argillaceous and less shelly, was the product of deeper less energetic water. Their respective trilobite faunas differ accordingly (Price 1980*a*, p. 883). He pointed out that there are much closer genetic similarities between the Sholeshook and the Rhiwlas of North Wales than between the various South Wales limestones.

It has been customary to consider the Robeston Wathen as a separate limestone, darker, purer and coralliferous, lying below the Sholeshook and separated from it by a few metres of calcareous mudstone. However limestones so defined are not present everywhere and it seems likely that such a separation is merely a local phenomenon and that the Robeston Wathen Limestone is really the basal part of the Sholeshook (Price 1973, p. 244); Zalaseiwicz *et al* (1995) did not even seek to identify a separate Robeston Wathen Limestone in the Whitland section. The thickest development of limestone is to the SW of Carmarthen (W, Fig. 3.43) where it attains approximately 90 m. Westward it is thinner and beyond Haverfordwest it is absent. The base, where exposed around Llanddowror, has yielded *T. moeldenensis moeldenensis*, while at Whitland there is a comparable species (Zalasiewicz *et al.* 1995, p. 15). Both taxa are indicative of high Pusgillian to low Cautleyan. The Birdshill Limestone has yielded what is probably the same taxon, while it and the Crug Limestone have yielded conodonts suggestive of a low Cautleyan age (Bergström 1964, p. 53; Orchard 1980, p. 13; Price 1984, p. 103; Savage & Bassett 1985). The top of the limestone formation has been established to be within Rawtheyan Zone 5 at several places between Carmarthen and Haverfordwest (Price 1977, p. 779, 1980*b*). Thus, across that distance it appears not to be diachronous, but to the NW and further to the west the abrupt absence of the limestone from the sequence, leaves the Slade and Redhill Formation resting unconformably upon Mydrim Shales.

The exact nature of the lateral replacement of the Sholeshook Limestone by the Slade and Redhill Formation is not clear, nor is the genesis of the unconformity beneath. The latter is clearly angular, presumably due to tectonic uplift in SW Carmarthenshire and southern Pembrokeshire and because the unconformity lies at the base of the Sholeshook Limestone, the tectonic event appears to have been of Pusgillian age. However, if that were the only factor then the Sholeshook Limestone here would have been the product either of a transgressive, deepening sea, whereas elsewhere the Ashgill was introduced by regression, or tectonic subsidence.

Mudstones. As discussed above, the top of the Sholeshook Limestone lies in the basal Rawtheyan and, except in the west where the Sholeshook Limestone is absent, the Slade and Redhill Formation follows conformably above. This formation records an abrupt change to mud deposition with a rich brachiopod–trilobite fauna and its basal parts, like the top of the Sholeshook Limestone, are also of Zone 5 age. The formation attains a thickness of up to 450 m, but upward constraints on age are poor; Cocks & Price (1975, p. 709) consider that, at Haverfordwest, even the highest beds are of pre-Hirnantian age, probably within Rawtheyan Zone 6 leaving the overlying black, Scotchwell Shale Member (46 m) of the Portfield Formation still within the Rawtheyan. This member has yielded only the conjoined valves of a brachiopod perhaps referrable to *Onniella,* but the brachiopod and black shale suggest a change to anoxic, low-energy marine conditions of deposition and, if the generic identificatoin is correct, they were rather deep. Such a deepening, and the anoxicity, would accord with the late Rawtheyan (*anceps* Biozone) anoxic deposits of the mid-Wales Basin. The same member is noted also for impersistent beds of coarse feldspathic sandstone and conglomerate, largely of extrabasinal clasts of 'felsite' and vein quartz up to 13 cm long.

The presence of these initially suggested this member represented the base of the Silurian as part of the so called 'Basement Beds'. But later the St Martin's Cemetery bed higher in the sequence was shown to be of Hirnantian age (Ingham & Wright 1970; Cocks & Price 1975). Nevertheless, the beds are of considerable significance and reflect again Ashgill tectonic instability of the basin margin nearby. The rest of the Portfield Formation, i.e. the Cethings Sandstones and the Cuckoo Shales members, is unfossiliferous and could represent the lower parts of the Hirnantian.

North of the line C–D (Fig. 3.41)

The shallow-water shelly shelf limestones south of the anticlinal line C–D are not present north of it. In fact their absence from the sequence occurs even to the south of that line in the west, where the anticlinal structure is spread across several folds and is attached on the north to the structural high of the Prescelly Hills. North of Carmarthen and the anticlinal core of older rocks, the Ashgill reappears in the northern limb of the Tywi Anticline dipping towards the mid-Wales Basin. Here it is a single formation of mudstones, an expanded Slade-and-Redhill Formation over 800 m thick. Geological maps depict its contact with the older, core rocks of the Tywi Anticline as a very long strike fault, and thus lower parts of the Ashgill are concealed; there is no documented evidence of the Pusgillian, Cautleyan or Hirnantian and little is known of this northerly outcrop of the Slade and Redhill Formation. However, available evidence suggests that it was deposited in a deep neritic marine environment of the outer shelf to top or slope, not unlike the Tridwr Formation of Llandovery, but its age is ill defined. It is believed to include the Rawtheyan on faunal similarities with the formation in the south, but the shelly fauna is sparse. In the west around Llawhaden [075 175], below the Cethings Sandstone and probably of topmost Rawtheyan age, there are interbedded sandstones and conglomerates, the latter including clasts of Mydrim Shales up to 1 m long within which graptolites of a low *multidens*, or *gracilis*, Biozone occur. It is unlikely that traction currents transported clasts of this size so that gravity-flow seems more likely. That implies a relatively steep slope at the top of which was a scarifiable surface of low Caradoc rocks. Such cannot have lain to the SE where the Caradoc–Ashgill sequence is entire, so that the source area must have been in the western quadrants, where much of the late Caradoc sequence is absent. Rawtheyan, as well as late Caradoc, tectonic uplift of western and northern Pembrokeshire must be considered as a probability.

Hirnantian. The third event affecting the basin occurred towards the end of the Ashgill and is recorded in the Hirnantian rocks. Rapidly changing sedimentary patterns resulted from the major Hirnantian glacio-eustatic fall and rise in sea level. The sediments are characteristically coarser than the preceeding Ashgill rocks, so that the base of the Hirnantian is generally marked by a facies change toward coarsening and high energy (gravity-flow) in regressive basinal sequences and, on the shelf fringe, the presence of sandier facies with a the *Hirnantia* fauna. The effects of the regression in shelf areas was to produce widespread erosion and some channelling, with little associated deposition. Detritus, predominantly mud, was carried on to the slopes and into the basin, leading to accretion of unstable muddy slopes and mixed mud and sand lobes on the basin floor. The transition from Rawtheyan mud formations into coarser Hirnantian rocks is abrupt, for instance the Deganwy Mudstones into the overlying Conway Castle Grit (1, Fig. 3.41), suggesting that the glaciation was established very rapidly. During the early stages of the subsequent transgression across the shelves, channels were filled with cross-stratified sandstones, which also spread across the interchannel areas. In places there is a conglomeratic base to the sand body. At the shelf edge the transgression is locally reflected in an upward fining sequence. In the basin, with a reduced supply of detritus from the shelf, it is more generally recorded by a change to well bedded muddy sediments. Thus this, the upper part of the Hirnantian (the *persculptus* Biozone) is represented by shales deposited initially largely under oxygenated conditions, but later, as the marine transgression expanded, euxinic conditions prevailed and continued into the Silurian *acuminatus* Biozone.

North Wales. The sequence of events on the shelf is well illustrated at Meifod (8), where a broad channel was cut into Rawtheyan Mudstones. The lower part of the channel is filled with bioclastic limestone containing intraclasts and phosphatic nodules derived from the underlying Rawtheyan. The limestones were subsequently karstified during further regression, but a coating of encrusting brozoans, crinoid holdfasts and attached cystoids record the succeeding transgression during which the karst was buried below cross-stratified sandstones (the Graig-wen Sandstone) with a *Hirnantia* fauna (Brenchley *et al.* 2006). Similar sandstones are to be found around the Berwyns where they range from about 50 m to less than 10 m thick. Other similarly shallow-marine Hirnantian facies include: the Calettwr Quartzite north of Bala, the oolitic Hirnant Limestone also near Bala, siliceous and conglomeratic sandstones of the Corwen and Glyn 'grits' (4 and 5) and the calcareous conglomeratic sandstones of the Conway Castle Grit (1), with synbasinal mud clasts and numerous corals, brachiopods and even trilobites, from nearby shallow water.

An unusual situation arose north of Bala where the Calettwr Quartzite occupies a position near the top of the Foel-y-Ddinas Mudstones Formation and overlies a large non-sequence. Much of the earlier part of this Hirnantian formation (Bassett *et al.* 1966, p. 262) seems to have been removed during the marine low-stand, suggesting that a bathymetric high here suffered erosion and possibly a local emergence, before the arrival of shallow water sands which now form the quartzite. This might be another record of a transgressive late Hirnantian shoreface with a *Hirnantia* fauna.

In a study of the *Hirnantia* fauna (s.1), Brenchley & Cullen (1984) divided it into three brachiopod associations which are essentially depth communities. A shallow shelf assemblage dominated by *Hindella* was prevalent in the eastern Berwyns, and an *Eostropheodonta* and *Hirnantia* Association occupied a deeper shelf position around Conwy and north of Bala. The third, with a reduced presence of *Hindella*, is the *Dalmanella* Association. This is positioned between the two extremes in the western Berwyns.

Mid-Wales. The Hirnantian rocks of mid-Wales are essentially a record of the effects of the sea-level changes upon the Welsh Basin and its E–SE slopes. The shoreline against the Midland Platform receded rapidly westward, almost to the line marking the position of the developing Tywi Anticline. Thus was created, very suddenly, a period of scavenging erosion which recycled a considerable thickness of the preceding Ordovician shelf deposits, including most of those of Ashgill age. In a short time these deposits were swept up by the fluvial systems of a vulnerable, plant-free, landscape, transported to the shelf-slope break and there dumped as rapid accumulations of mud with sand and some gravel. The latter two are deposits characteristic of the eastern and SE margins and were accumulated in channels and depressions on the slope. Much of it was then to suffer destabilization and transportation farther downslope as slumps and in gravity-flows. It was a fast-moving sedimentary system, given additional impetus by tectonic activity along the Tywi Anticline where there is evidence of fault-controlled sedimentary thickness differences. The thicknesses of lower-slope and basinal rocks are of between 600 m and 1200 m. In western mid-Wales the lower-slope and basinal formations are the Garnedd-wen, Drosgol + Bryn-glas and Llangranog Formations. Along the traverse between Towyn and Dinas Mawddwy (Fig. 3.41) the thickness of the Garnedd-wen Formation reduces steadily from 1200 m to 100 m, concomitant with a gradual loss of sandstones in an up-slope direction. This is a thickness trend similar to that of the underlying Broad Vein (Rawtheyan), but between the Broad Vein and Garnedd-wen there is a thin (13–15 m) formation of dark grey laminated mudstone, the Narrow Vein. This is a slate generally of good quality and evenly bedded like the Broad Vein below. In this it contrasts sharply with the Hirnantian Garnedd-wen above, suggesting a pre-Hirnantian age. When traced northeastward, up-slope, Pugh (1929, p. 263) was in no doubt that it became linked with the lower part of the Garnedd-wen and passed into the lower part of the Foel y Ddinas Mudstones of Bala, a formation claimed by Bassett *et al.* (1966, p. 262) to be of Hirnantian age on the presence of *E. hirnantensis* and *H. sagittifera* (Bassett *et al.* 1966, p. 253). However, in a counter claim by Pratt *et al.* (1995 p. 51) it is stated that the Narrow Vein passes northeastward into the upper parts of the Moelfryn Mudstones (Rawtheyan). Thus while its lithology suggests Rawtheyan, palaeontologically it seems to be Hirnantian. The traverse does not provide any evidence such as channels or canyons that the large volumes of sediment which were delivered to the mid-Wales formations travelled via this slope; rather must they have gone via slopes farther to the SE.

The base of the western mid-Wales formations is another illustration of the rapidity of the onset of sea-level change in response to the glaciation. In both the Drosgol and the Garnedd-wen, the base marks an abrupt change from low energy mud-turbidites to 10 m of mudstones with thick beds of coarse sand-turbidites, channelised by mass-flow pebbly mudstones, before being overwhelmed by thick structureless muds, showing some interdigitation with bedded sand turbidites. The latter sand-turbidites feature most strongly in the extreme SW around Llangranog. The top of the formation is sharp everywhere in mid-Wales and is overlain by the Mottled Mudstone Member of the Cwmere Formation. Its surface probably marks the end of the marine regression and thus the marine low-stand, i.e. when the karstic surface of Meifod formed in North Wales. If this is so, then it implies that the initial deposits of the marine transgression caused by the glacial collapse, i.e. the graptolitic Mottled Mudstone which is of *persculptus* Biozone age, are coeval with the shelly Graig-wen Sandstone of Meifod.

Traversing eastward up-slope from the deepest part of the basin in the west, through the Rhiwnant inlier, (Fig. 3.41) to the NW flank of the Tywi Anticline near Rhayader (Z, Fig. 3.43), the deposits differ only in detail and they are at least 600 m thick, forming Yr Allt Formation (Davies *et al.* 1997). The main difference is the much greater proportion of well-bedded strata, in units between disturbed mudstones. This suggests that the area lay near the upper limits of slope destabilization where, initially, the mass-flows bypassed parts of the upper-slope, but coalesced downslope before they differentiated again into graded turbidites in the Llangranog area.

Along the flank of the Tywi Anticline to the SW the formation possesses an increasing proportion of sand in the form of turbiditic lobes and channel-fills so that, near Llanwrtyd, it becomes the Cynnant Fach and Cefn Ystradffin formations. No doubt the tectonic activity of Rawtheyan times here continued into the Hirnantian, contributing to destabilization and creating more fault-bounded sediment traps. Further SW towards Llandeilo these slope-sandstones diminish in importance again, possibly because the shelf was broader and thus shore-face sources were more distant. Representatives of the shallow-marine shelf environment, on the SE side of the Tywi Anticline, are largely concealed under, or were deeply eroded prior to, the Silurian overstep. Over the closure of the anticline near Abbeycwmhir the topmost Ordovician strata are conglomerates and sandstones succeeding the regressive Yr Allt Formation. Their bedforms are characteristic of a subtidal shallow environment reflecting the effects that occasional storms had along the shoreface just to the east. A brachiopod fauna with *E. hirnantensis* indicates the presence of the *Hirnantia* fauna.

To the SW, in the isolated outcrops near Garth, (Y, Fig. 3.43) Rawtheyan stata grade upwards into oolitic siltstones of the Wenallt Formation (< 115 m remaining) with a prolific Hirnantian brachiopod fauna. These were differentially eroded, very deeply in places, presumably during the marine low-stand, before renewed deposition covered the resultant uneven surface with a variably coarse, clastic shoreline sediment of the Cwm Clyd Formation (11–51 m). This has yielded one specimen of *E. hirnantensis* from near the base, with *Hindella crassa* present in overlying strata, and must reflect the same, very early, phase of the end-glacial marine transgression which is recorded at Meifod by the Graig-wen sandstone.

A late Hirnantian shoreline facies, the Scrach Formation (0–150 m), recurs in an outcrop NE and SW of Llandovery. Its age is constrained only by graptolites of a high *anceps* Biozone in Rawtheyan strata below and a fauna indicative of the *persculptus* Biozone above. The shoreline itself is thought to have been controlled here by faults of the Myddfai Steep Belt (Woodcock & Smallwood 1987).

South Wales. In Pembrokeshire, south of the line C–D, the Portfield Formation of Rawtheyan age at the base, but very possibly of Hirnantian age in the unfossiliferous higher parts, is overlain on the south side of its outcrop by the St Martin's Cemetery bed of Haverfordwest (1 m) forming the base of the Haverford Mudstone Formation (Cocks & Price 1975). The paler colour and obvious bioturbation of the bed contrast with the mudstone below and are suggestive of regression. This is supported by the fauna which includes *Mucronaspis mucronata* together with more definative elements of the *Hirnantia* fauna. The rest of the Haverford Formation (= Cartlett beds below and Gasworks Mudstone above, of Strahan *et al.* 1914) consists of about 350–390 m of mudstone, almost lacking fossils throughout its lower 250 m. It has yielded further *M. mucronata* and other taxa likely to be of Hirnantian age, but near the top the fauna is suggestive of the Rhuddanian. The remaining 140 m of the formation has a very rich Rhuddanian shelly fauna and thus there appears to be an unbroken succession from Rawtheyan into the Silurian; so that it is possible that, when compared with Meifod in North Wales, this area of Hirnantian shelf lay a little further basinward and accordingly there was no emergence here at the nadir of the regression and the succeeding transgression is not marked by a major facies change, as is common elsewhere.

4 Silurian: the influence of extensional tectonics and sea-level changes on sedimentation in the Welsh Basin and on the Midland Platform

L. CHERNS, L. R. M. COCKS, J. R. DAVIES, R. D. HILLIER, R. A. WATERS & M. WILLIAMS

The evolution of Silurian geology in Britain was strongly influenced by the final stages of the closure of the Iapetus Ocean (Fig. 4.1). Avalonia united with Baltica at around the time of the Ordovician-Silurian boundary (443ma) and impacted with the Laurentian margin (Fig. 4.1), between 440 Ma and 420 Ma, probably during the Llandovery or Wenlock (Soper & Woodcock 1990; Cocks & Torsvik 2002). The Avalonian margin apparently remained tectonically active throughout the final stages of convergence as indicated by sporadic volcanicity and evidence of early to mid Silurian crustal extension within the Welsh Basin (Fig. 4.2). The initial effects of the collision of Laurentia and Avalonia particularly affected the development of the Lakesman Basin (see section on Lakesman Basin), but later, as Laurentia and Avalonia were driven more tightly together, crustal shortening ended the regime of extensional subsidence in the Welsh Basin. An influx of sediment from the uplifted land to the north and from the land area (Pretannia) to the south caused the basin to become progressively filled in the Late Silurian (Woodcock *in* Aldridge *et al.* 2000) as the basin underwent thermal subsidence prior to the onset of compression. However, this was punctuated by an interval of south-easterly migrating crustal flexure and increased subsidence, as Laurentia or the advancing accretionary prism overrode Avalonia from the north (King 1994).

The climate of Avalonia during the Silurian was probably tropical as suggested by the palaeogeographic reconstructions of plate distribution. This is supported by palaeomagnetic data from the late Llandovery lavas at Tortworth, in Gloucestershire, which indicate a palaeolatitude of 16°S (Torsvik *et al.* 1994). The presence of reefs in the Much Wenlock Limestone Formation is also consistent with a tropical ocean. The evolution of Silurian sequences was influenced by a series of events caused by rises and falls in relative sea-level, as well as changes in climate and related ocean state. The former were driven by eustatic or tectonic movements, and in some cases by both (Fig. 4.1). Changes in climate and related ocean state are reflected in 'Primo' (P) and 'Secundo'(S) episodes, with intervening 'events', identified by Jeppson (1990) and Aldridge *et al.* (1993). Primo episodes were characterized by cool high-latitude climates, cold ocean bottom waters and high nutrient supply, which supported abundant and diverse plankton communities. They are thought to result in sea-level lowstands of variable magnitude. Secundo episodes were characterized by warmer high-latitude climates, salinity-dense ocean bottom water, low-diversity planktonic communities and carbonate formation in shallow water. They were thought to result in minor sea-level highstands. However, a comparison of the global sea-level curve with P and S episodes identified in Scandanavia (Fig. 4.1) reveals no simple correlation with the major eustatic events, whereas there is a distinct correlation between Gondwanan glaciations and sea-level movements.

Silurian stratigraphy

The Silurian System was named by Roderick Murchison in 1835. It was based initially on rock successions in the Welsh Borderland, but during the later 1830s was extended into much of Wales. Three of the four series have their type sections in Wales and the Welsh Borderland, whilst the fourth is in the Czech Republic. Graptolites offer the most highly resolved biostratigraphy for the Silurian, with at least 34 biozones differentiated in Britain (Fig. 4.3). Some of these biozones may represent as little as 500 000 years of time. The worldwide distribution of certain graptolite species enables rock successions of similar age to be correlated both regionally and globally, and the bases of the Llandovery, Wenlock, Ludlow and Přídolí series are all correlated with graptolite biozones, though in the British succession the uppermost Ludlow and Přídolí are non-graptolitic. Other macrofossil groups used for correlation include brachiopods, trilobites and corals. These, particularly brachiopods, are useful in the shallower marine

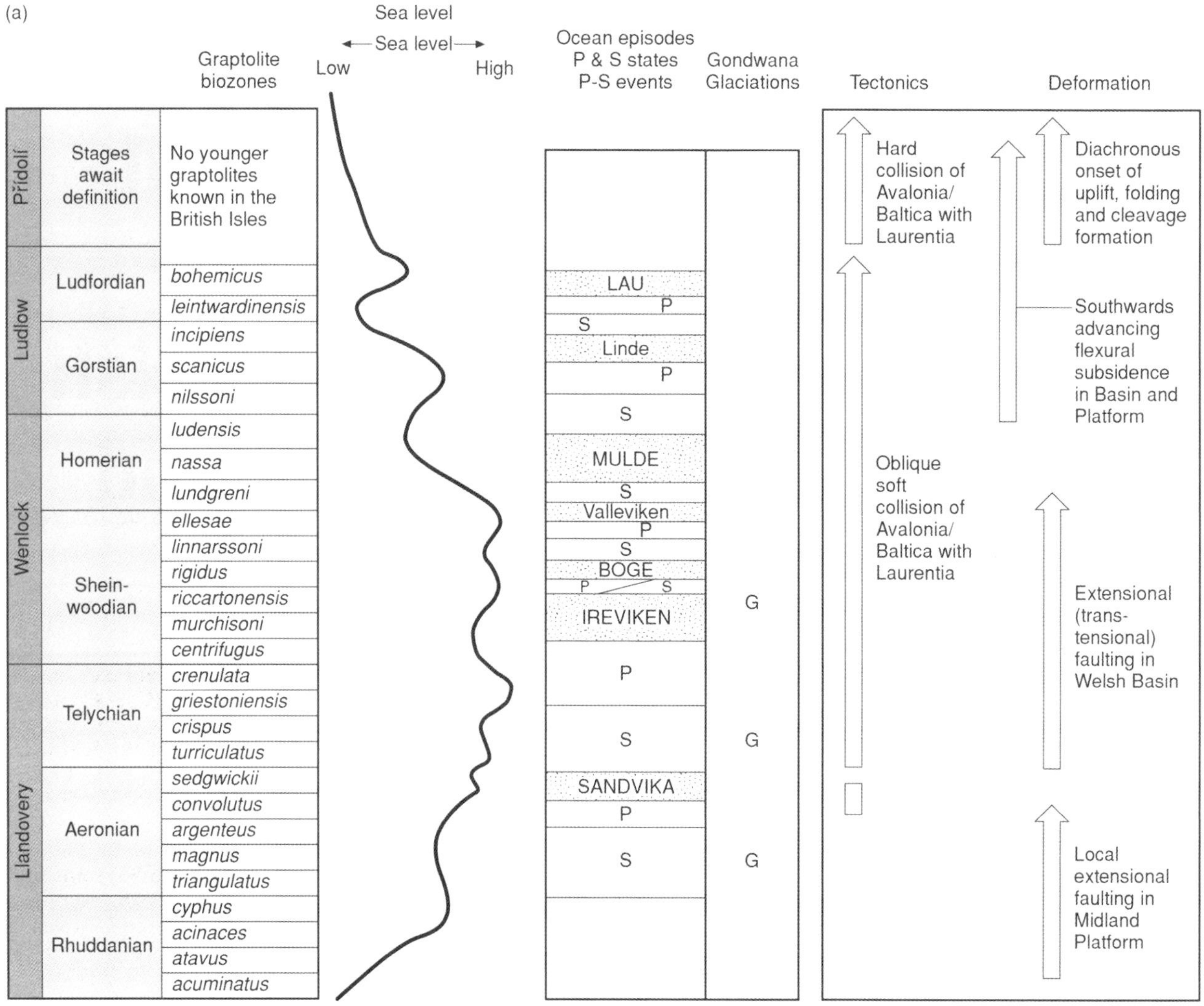

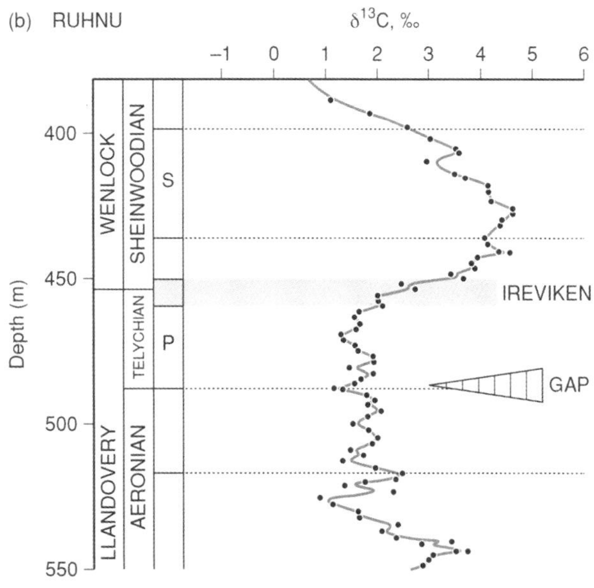

Fig. 4.1. Silurian events. (**a**) Sea-level movements, ocean episodes and tectonics. The global sea-level curve is based on Woodcock in Aldridge *et al.* (2000), after Johnson *et al.* (1991). Ocean episodes are recorded as P and S states, and six named P–S ocean events are shown with ornament (modified from Calner *et al.* 2004). The cycles were mainly defined in Scandinavia (Aldridge *et al.* 1993; Jeppson *et al.* 1995) and have not been determined in detail in Britain. Three glacial episodes are marked **G**. The age of the Silurian Gondwana glaciations was established by field evidence (Caputo 1998 and references therein). (**b**) The stable carbon isotope record from the Ruhnu core (Estonia) is one of three similar Silurian carbon isotope records described from three Estonian cores by Kaljo *et al.* (2003); it shows the major positive excursion in the Wenlock, with the start of the excursion coinciding with the Ireviken event and the Gondwana glaciation.

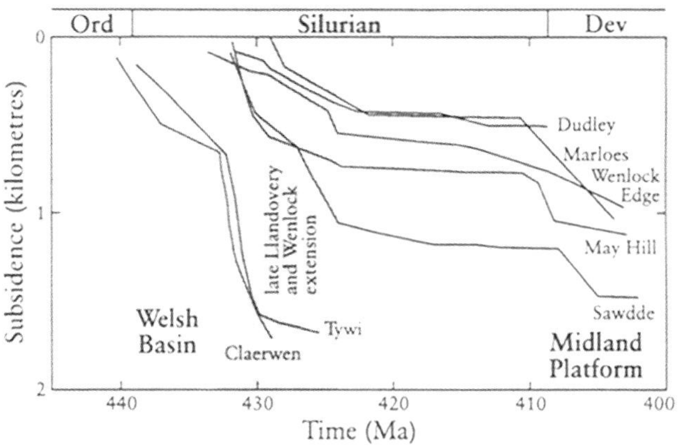

Fig. 4.2. Backstripped, water-loaded subsidence curves for the Silurian of the Midland Platform and central Welsh Basin (from Woodcock *et al.* 1996 with permission of the Joint Nature Conservancy Committee).

System	Series	Stages	Graptolite biozones		Chitinozoan biozones
Silurian System	Přídolí (419 m.a. / 410.7 m.a.)	No stages designated	No younger graptolite biozones recognised in Britain		No younger chitinozoan biozones recognised in Britain
	Ludlow (423 m.a. / 424 m.a.)	Ludfordian		Bohemograptus proliferation	philipi
				leintwardinensis	
		Gorstian		tumescens/incipiens	elongata
				scanicus	3 biozones in the Ludlow area (Sutherland 1994)
				nilssoni	
	Wenlock (428 m.a. / 430.4 m.a.)	Homerian		ludensis	?
				nassa	
				lundgreni	cribrosa; subcyatha
		Sheinwoodian		ellesae	pachycephala
				flexilis/linnarssoni	
				rigidus	cingulata
				riccartonensis	mamilla-tuba
				firmus	bella
				murchisoni	bouniensis
	Llandovery (443 m.a. / 439 m.a.)	Telychian		centrifugus	margaritana
			crenulata s. l	insectus	
				lapworthi	banwyensis; acuminata; longicollis
				spiralis	
				crenulata s.s.	
				griestoniensis	
				sartorius	
				crispus	dolioliformis
			turriculatus s. l.	turriculatus s.s.	
				guerichi	
		Aeronian	sedgwickii s.l.	halli	
				sedgwickii s.s.	
				convolutus	alargada
			gregarius	leptotheca/argenteus	
				magnus	
				triangulatus	maenili
		Rhuddanian		cyphus	electa
			vesiculosus	acinaces	postrobusta
				atavus	
				acuminatus	fragilis

Fig. 4.3. Silurian chronostratigraphy and recent graptolite and chitinozoan biozonation of the British Isles. In the column for 'Series' the radiometric dates on the right-hand side are from Harland *et al.* (1989). Graptolite data are summarized from Rickards (1976) and Loydell (1998). Chitinozoan data are from Sutherland (1994), Verniers (1999), Mullins & Loydell (2001, 2002) and De Parmentier & Verniers (2002).

rock successions of the Midland Platform where graptolites are rare. Biostratigraphical schemes for organic walled microfossils such as plant spores, acritarchs and chitinozoans are all well developed (Fig. 4.3). These, together with other microfossils such as conodonts and ostracods, have become increasingly important for correlation, particularly in rock successions where macrofossils are absent. Organic walled microfossils such as acritarchs occur across a spectrum of shallow and deeper marine deposited rock successions, enabling correlation from the shallow marine Welsh Borderland successions into the deep-marine basin successions of central and north Wales. In some areas, such as the Llandovery area of south Wales, the intermingling of deeper marine graptolite-bearing rocks and shallower marine shell-bearing rocks enables the various graptolite-based, shell-based and microfossil-based biostratigraphies to be correlated.

Recent estimates (Tucker & McKerrow 1995; Gradstein *et al.* 2004) suggest that the Silurian lasted from 443 to 416 Ma, some 27 Ma, with the majority of time (443–428 Ma), some 15 Ma, belonging to the Llandovery. The Wenlock (428–423 Ma) and Ludlow (423–419 Ma) are 4–5 Ma each, whilst the Přídolí was probably less than 3 Ma in duration.

The three lowermost stages of the Llandovery, the Rhuddanian, Aeronian and Telychian, are distinguished mainly by graptolite biostratigraphy (Fig. 4.3). The global boundary stratotype for the Ordovician–Silurian boundary, and for the base of the Llandovery Series and Rhuddanian Stage, is located at Dob's Linn in the Southern Uplands of Scotland. It is placed within the Lower Birkhill Shales Formation at the first occurrence of the graptolites *Akidograptus ascensus* and *Parakidograptus acuminatus*. These graptolites are indicative of the lowermost Silurian *acuminatus* Biozone (Williams & Ingham 1989).

Although some graptolites of zonal significance have been found both in the type Llandovery area (Cocks *et al.* 1984) and also in Shropshire (Cocks & Rickards 1969), not all the graptolite zones are represented inshore of the Welsh Basin, and thus most of the dating of the many Welsh Borderland sequences have been achieved by the recognition and use of evolving sequences of brachiopods, particularly the pentameride *Stricklandia* and the rhynchonellide *Eocoelia*. *Stricklandia* and its lineal descendant *Costistricklandia*, were present and evolved quickly during the whole Llandovery and the early Wenlock. The earlier Rhuddanian and Aeronian species and subspecies were described in a classic paper by Williams (1951), who mostly used material from the type Llandovery area: the later forms of that lineage, in the Telychian and early Wenlock, were elucidated by a series of authors, using material from the United States, Canada, Norway, Estonia and Siberia as well as Britain.

Eocoelia is first known only from middle Aeronian times, but, as mentioned above, it was a widespread genus in the shallower-water associations. It too evolved rapidly, and Ziegler (1966) was able to establish four species in a well-documented stratigraphical sequence. Later work has identified a fifth species which extended on into the early Wenlock. It is the brachiopod biostratigraphy that has established the progressive stages of the Llandovery transgression.

Graptolite-bearing strata are rare in the uppermost Ordovician and lowermost Silurian strata of the type Llandovery area, and correlation of the base of the Rhuddanian remains problematic, but it certainly lies above *Hirnantia*-bearing beds of the Upper Ordovician. Nevertheless, graptolites indicative of the *acuminatus, acinaces* and *cyphus* graptolite biozones of the Rhuddanian have been reported in the Llandovery area, as have the Rhuddanian brachiopods *Stricklandia lens lens* and *Stricklandia lens prima* (Cocks *et al.* 1984). Elsewhere, *Parakidograptus acuminatus* and the zonal index species of the three other graptolite biozones of the Rhuddanian are widespread in the deeper marine deposited mudstones of mid and north Wales, and NW England (Fig. 4.4) and the boundary between the Ordovician and Silurian can be more easily correlated. In the Welsh Borderland and south Pembrokeshire successions, brachiopod and graptolite evidence indicates that the Rhuddanian is largely absent, except in the Haverfordwest area of Pembrokeshire (see Cocks *et al.* 1992).

The Aeronian is the second stage of the Llandovery. Its base is correlated with the base of the *triangulatus* graptolite

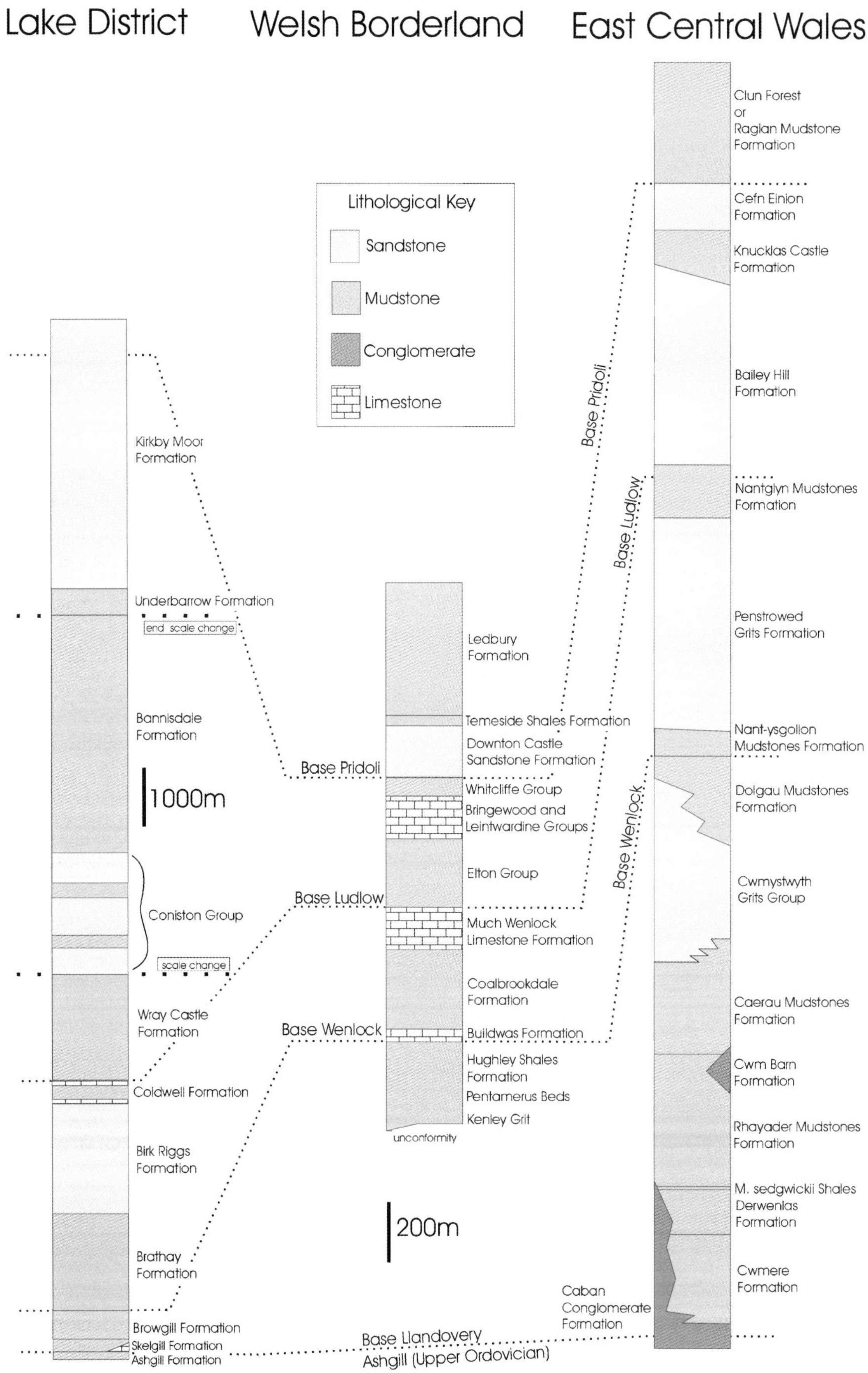

Fig. 4.4. Representative lithostratigraphical successions from the two main palaeogeographic areas of Silurian in England and Wales. The columns are based on Woodcock (2000) with additional information from Kneller *et al.* (1994), Davies *et al.* (1997) and Cave & Hains (2001). The scales and relative thicknesses are only approximate; note scale change from the Coniston Group and Bannisdale Formation in column 1.

Biozone. Near the stratotype boundary in the type Llandovery area the fauna is dominated by shelly fossils, but graptolites are also present, the *triangulatus* Biozone being indicated by *Monograptus austerus sequens*. In the neighbouring area the same stratigraphical level yields a more prolific *triangulatus* Biozone graptolite fauna including *Monograptus triangulatus* (Cocks *et al.* 1984). Up to six graptolite biozones are recognized for the Aeronian (Fig. 4.3), enabling precise correlation of rock units across much of mid Wales and, in the upper Aeronian, with the Welsh Borderland, where the diagnostic graptolites occur intermittently. Brachiopods such as *Stricklandia lens progressa, Eocoelia hemisphaerica* and *Eoecelia intermedia* provide a degree of stratigraphical resolution approaching that of the graptolites at higher levels in the Aeronian (Bassett 1989*a*).

Acritarchs and chitinozoans are represented by up to six biozones in the Rhuddanian and Aeronian interval, with the *microcladum* and *estillis* acritarch biozones of the upper Aeronian offering a degree of stratigraphical resolution approaching that of the graptolites. In contrast, most of the other microfossil groups provide limited help in correlating rock units.

The uppermost stage of the Llandovery is the Telychian. The base of the stage approximates to the base of the *guerichi* graptolite Biozone (= the lower part of the *turriculatus* Biozone *s.l.* as used by some authors). Although there are no graptolites at the boundary stratotype in the type Llandovery area, graptolites from elsewhere in the district support the correlation and this datum is used to identify the base of the Telychian across Wales and in NW England. Just below the boundary in the Llandovery area is a horizon yielding 25 different species of macrofauna, including the highest records of the brachiopods *Eocoelia intermedia* and *Stricklandia lens progressa*. The stratotype base of the Telychian is defined immediately above the last occurrence of *Eocoelia intermedia*, with the brachiopods *Stricklandia laevis* and *Eocoelia curtisi* diagnostic for the overlying rock sequence (Cocks *et al.* 1984). The base of the *monospinosa* acritarch Biozone may also correlate approximately with the base of the Telychian Stage. The nine graptolite biozones of the Telychian, many subdivided into subzones, offer a degree of biostratigraphical precision which is probably unsurpassed at other levels in the Silurian. These enable precise correlation of rock units across mid Wales (Loydell 1991; Davies *et al.* 1997).

Graptolite biostratigraphy is of paramount importance for subdivision of the Wenlock and for correlation of Wenlock-age strata, and as many as 10 graptolite biozones are distinguished (see Zalasiewicz & Williams 1999). Informal recognition of lower and upper Wenlock assemblages of brachiopods, trilobites and ostracods allows broad recognition of the two stages regionally, and biozonal schemes for acritarchs, chitinozoans and spores provide additional tools for correlation (Fig. 4.3).

The base of the Wenlock Series and of the Sheinwoodian Stage has been traditionally equated to the base of the *centrifugus* graptolite Biozone, but evidence from chitinozoans in the stratotype boundary section at Hughley Brook suggests that the exact level lies in the upper part of that biozone, or the lower part of the succeeding *murchisoni* graptolite Biozone (Mullins & Aldridge 2004). The global stratotype for the Llandovery–Wenlock boundary is in the Welsh Borderland near the town of Much Wenlock, Shropshire (Bassett *et al.* 1975). The sequence immediately above and below the boundary is non-graptolitic, but is rich in many microfossil groups, particularly conodonts, ostracods, acritarchs and chitinozoans (Siveter 2000*a*). Fossils indicative of the *amorphognathoides* conodont Biozone, *chulus–nanus* spore Biozone and *brevispinosa* acritarch Biozone all span the boundary. Many shelly fossils also cross the boundary, including the typical late Telychian brachiopods *Costistricklandia lirata s.l.* and *Eocoelia sulcata*. Nevertheless, age-diagnostic Wenlock species such as the brachiopods *Dicoelosia biloba, Eoplectodonta duvalii, Resserella sabrinae* and *Isorthis elegantulina* replace con-generic species from the underlying Llandovery. Although the Wenlock graptolite faunas of the Welsh Borderland are generally poorer than those of north and mid Wales, the upper *centrifugus*–lower *murchisoni* Biozone interval allows approximate correlation of the series boundary from the Welsh Borderland into Wales and NW England (Fig. 4.4; see also Zalasiewicz & Williams 1999).

The upper stage of the Wenlock is the Homerian. The base of this stage is correlated with the first occurrence of the graptolite *Cyrtograptus lundgreni* in graptolitic mudstones at the boundary stratotype section in Shropshire (Bassett *et al.* 1975). *Cyrtograptus lundgreni* is a widespread graptolite in mid and north Wales and in NW England (Zalasiewicz & Williams 1999), and provides a widespread datum for correlating the Sheinwoodian–Homerian boundary. The upper two graptolite biozones of the Homerian, the *nassa* and *ludensis* biozones, are sometimes less clearly defined, even within the Welsh Basin. Nevertheless, *Colonograptus ludensis* can be used to infer a late Wenlock age for part of the Much Wenlock Limestone in the Welsh Borderland (Bassett *et al.* 1975).

Locally, some microfossil groups such as chitinozoans, provide a refined tool for correlating the Wenlock which approaches the resolution of the graptolites (Fig. 4.3). Thus, in central Wales the base of the *subcyatha* chitinozoan Biozone approximates to the base of the *lundgreni* Biozone, and therefore to the base of the Homerian, whilst the upper part of the *lundgreni* Biozone is equivalent to the *cribrosa* chitinozoan Biozone (Verniers 1999).

The third series of the Silurian, and the lowermost chronostratigraphical division of the upper Silurian, is the Ludlow, named after the Shropshire town of the same name. The Ludlow district was identified by Murchison (1839) as the reference area for his 'Ludlow Rocks'. The Ludlow Series is divided into two stages, the Gorstian and Ludfordian. In Britain, all but the upper part of the Ludfordian is graptolite-bearing and the boundaries of the stages are equated approximately with the bases of the *nilssoni* and *leintwardinensis* graptolite biozones, respectively (Fig. 4.3).

Many Wenlock taxa appear to cross into the Ludlow, and microfossils such as ostracods, acritarchs or conodonts cannot be used as precise indicators of the series boundary. Thus, conodonts from the boundary sequence comprise only the long-ranging species *Ozarkodina excavata* and *Panderodus equicostatus*. Similarly, much of the lower Ludlow macrofauna includes Wenlock relicts. However, graptolites from just above the stratotype boundary at Pitch Coppice near Ludlow include possible *Neodiversograptus nilssoni* and *Saetograptus varians*. These, and other occurrences of graptolites in the neighbouring area, are consistent with the notion that the base of the Ludlow equates approximately to the base of the *nilssoni* graptolite Biozone, providing a widespread datum for regional correlation of the base of the Ludlow Series and Gorstian Stage from the Welsh Borderland, across Wales and into NW England.

The stratotype base for the upper Ludlow Ludfordian Stage, also located near Ludlow (see Siveter 2000*b*), is marked by the disappearance of many shelly elements of the lower Ludlow fauna. However, the boundary is defined most clearly by the appearance of the graptolite *Saetograptus leintwardinensis*, which is widespread at this level in Wales. The base of the Ludfordian therefore equates approximately to the base of the *leintwardinensis* graptolite Biozone. Brachiopods such as *Protochonetes ludloviensis* and *Shaleria ornatella* are useful for regional recognition of the Ludfordian in shelly facies, as is the ostracod *Neobeyrichia lauensis*.

The uppermost series of the Silurian is the Přídolí. This is the only Silurian series not to be subdivided into stages. The global boundary stratotype for the base of the series is placed in fully marine graptolite-bearing rocks in the Czech Republic. In Wales and the Welsh Borderland the Přídolí is represented by lithostratigraphical units such as the Downton Group (e.g. Downton Castle Sandstone Formation of Fig. 4.4), embracing mainly continental Old Red Sandstone type successions in which graptolites are absent. The base of the Downton Group is traditionally demarcated by the thelodont-rich Ludlow Bone Bed. Through a chain of correlation via eastern Europe, involving microfossil groups such as ostracods, the base of the Downton Group can be shown to lie stratigraphically close to the base of the *Monograptus parultimus* graptolite Biozone, and therefore close to the base of the Přídolí Series as defined in the Czech Republic. Conodonts such as *Ozarkodina* cf. *crispa* just below the base of the Downton Group in the Ludlow area support this correlation (see White & Lawson 1989, p. 138), as does the record of the chitinozoan *Eisenackitina barrandei* from within 0.26 m below the top of the Ludlow Series near Downton (Richardson & Rasul 1990).

Significant changes in the fauna and flora occur at the base of the Downton Group in Wales and the Welsh Borderland (White & Lawson 1989; Miller 1995; Miller *et al.* 1997). For example, the lowermost Přídolí is characterized by the ostracods *Nodibeyrichia verrucosa* and *Frostiella groenvalliana*. The latter ostracod is particularly useful for correlation and also occurs in the upper Kirkby Moor Formation ('Scout Hill Flags') of NW England (see Fig. 4.4). Also important for regional correlation within the Downton Group are vertebrates such as thelodonts and acanthodians (Turner 2000). Spores of the *tripapillatus-spicula* Biozone and other informally recognized spore biozones are useful for correlation in Wales and the Welsh Borderland (Richardson & Edwards 1989).

Silurian successions of the Welsh Basin and Midland Platform

The Silurian outcrop of Wales and the bordering regions of western England is traversed by a variety of tectonic structures (Fig. 4.5), dominated by SW–NE-trending faulted synforms and antiforms, and including most notably the curvilinear Welsh Borderland Fault System (WBFS). The latter comprises three deep-seated long-lived fault and fold belts, namely the Tywi and Pontesford lineaments and the Church Stretton Fault Zone. During the Silurian, this structural belt separated a region of enhanced subsidence – the Welsh Basin – from a more stable region to the SE termed the Midland Microcraton (or Platform). The latter was possibly contiguous with the landmass of Pretannia, cited as the source for detritus periodically supplied to the Welsh Basin from the south, across its Pembrokshire margin. The WBFS was also influential during the subsequent late, early Devonian, Acadian deformation, broadly separating cleaved Silurian rocks to the west from virtually undeformed rocks to the east. A similar structural belt – the Menai Straits Fault Zone – defined the northern margin of the Welsh Basin, separating it from the putative Irish Sea Horst (the Anglesey Geanticline of earlier literature) sited to the north. A further belt of fractures – the Bala Lineament – and the associated Derwen Anticline (or Horst) mark the location of an intra-basinal ridge, which served to separate a North Wales sub-basin from a larger sub-basin to the south, which occupied much of central Wales. Major deep-seated fractures in the southern basin periodically controlled sedimentation and comprise the Bronnant Fault (Glandyfi Lineament) in the west and the Central Wales Lineament in the centre. Within the Midland Platform, the N–S-trending structure which brings up Precambrian rocks in the Malvern Hills is known as the Malvern Line. Together with the NE–SW Vale of Neath Disturbance, it delimits two intra-shelf sub-basins, the Woolhope and Usk, developed on the western side of the Midland Platform. The Skomer Sub-basin in SW Pembrokeshire, defined by the Ritec and Benton faults, is a further, probably related intra-shelf sub-basin.

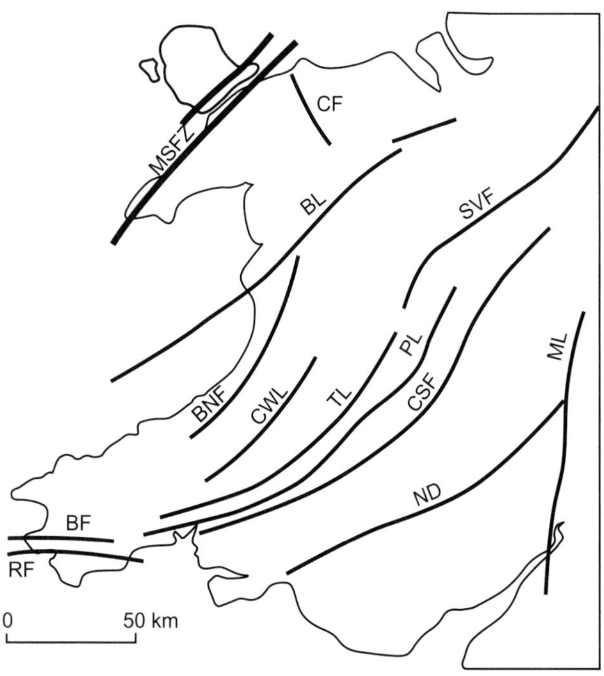

Fig. 4.5. Tectonic structures inferred to have been active during the Silurian. BF, Benton Fault; BNF, Bronnant Fault/Glandyfi Lineament; BL, Bala Lineament; CF, Conwy Fault; CWL, Central Wales Lineament; CSF, Church Stretton Fault; ML, Malvern Line; MSFZ, Menai Straits Fault Zone; ND, Neath Disturbance; PL, Pontesford Lineament; RF, Ritec Fault; SVF, Severn Valley Faults; TL, Tywi Lineament.

The sediments of the Midland Platform are shelfal with shallow-water sandstones and carbonates dominant in nearshore settings and mudstones in offshore settings. Unconformities are locally present. Basinal sequences are characterized by a variety of small- and large-scale sandy and muddy turbidite systems with subordinate hemipelagites. The sedimentary sequence that accumulated in the Welsh Basin, and across its margins, is recognized as the late Ordovician–early Devonian Powys Supergroup, the fourth of the Lower Palaeozoic tectonostratigraphic units of Woodcock (1990*a*) (Fig. 4.4). It is characterized by having either major facies changes, or an unconformity at its base, the latter particularly pronounced in the Welsh Borderland, and a passage at its top into the Devonian portion of the Old Red Sandstone.

The Midland Platform, which stretched from Nottinghamshire, in the north, to Belgium in the south, and from the eastern margin of the Welsh Basin in the west to the central North Sea in the east, was relatively stable throughout the Silurian and was probably emergent as land before the end of the Ordovician. This emergence was caused by the relatively mild intra-Ashgill tectonism related to the oblique soft docking of Avalonia with the much larger terrane of Baltica. The effects of this tectonism are well seen along the Welsh Borderland Fault System notably in the Shelve and Builth Wells areas and along the Church Stretton Fault Zone and it has been labelled the Shelvian Orogeny. The centre of the Midland Platform probably remained as land until early Devonian times, but transgression on both its eastern and western margins proceeded in pulses with intervening regressions during the Silurian time.

The tectonic influence on the sedimentary accumulations in the Welsh Basin was reflected in the development of half-grabens in which accumulated thick developments of hemipelagic sediments and turbidites. The latter were supplied from uplifted areas around the margin of the basin (Woodcock *et al.* 1996) and from a larger region of land (Pretannia) to the SW (Cope & Bassett 1987). Subsidence curves (Woodcock *et al.* 1996) show particularly high rates of subsidence in the basin during the late Llandovery–mid Wenlock, but more moderate subsidence rates on the Midland Platform (Fig. 4.2). Subsidence rates were very low on the platform throughout the Ludlow but high in the basin in North Wales (King 1994). They increased again on the platform in the Přídolí and early Devonian as uplifted land to the north supplied increasing amounts of sediment to the area.

The Silurian of eastern England is entirely concealed below Mesozoic rocks, but many boreholes have revealed rocks which are Llandovery to Přídolí in age under East Anglia (Cocks *et al.* 1992). Woodcock & Pharaoh (1993) have mapped the facies there and demonstrated shallower-water rocks in the west of that area and deeper-water to the east, which define the Anglian Basin; the Ware borehole in Hertfordshire, for example, includes late Wenlock rocks with calcareous mudstones and limestones yielding shallow-water brachiopods and other fossils: rocks and fauna very similar to those of the Much Wenlock Limestone of Shropshire on the western side of the microcraton. In contrast, the contiguous Silurian rocks in Belgium are entirely within the deeper-water facies from which graptolites and microplankton such as acritarchs are the only fossils known.

The Llandovery

Introduction

The chief outcrop areas of the Llandovery in Britain, south of the Lake District, include those associated with the Welsh Basin and its eastern and southern margins, which stretch in an arc from Conwy in North Wales through the Welsh Borderland and then westwards to the Pembrokeshire coast. The basin was the site of exclusively deep water, predominately turbiditic deposition, whereas its margins are characterized by sediments of shelfal aspect punctuated by unconformities. Though it is this subdivision into successions of shelfal or deep-water facies which provides the descriptive structure for this account, emphasis has also been placed on the principal events (oceanographic and/or tectonic) which appear to have been a controlling influence on sedimentation and biota in both types of setting.

Woodcock *et al.* (1996) and Davies *et al.* (1997) have reviewed the sequence stratigraphy of the Central Welsh Basin and the surrounding areas, and differentiate two separate depositional episodes: (a) an early–mid Llandovery phase during which eustasy appears to have been the dominant control; and (b) a late Llandovery–Wenlock phase driven more emphatically by tectonic activity. The latter was probably a reflection of the partial oblique collision of Avalonia–Baltica with Laurentia.

By the late Ordovician, intra-Ashgill tectonism (Shelvian Orogeny) along the Welsh Borderland Fault System had folded, faulted and uplifted the Ordovician sequence as seen today in the Shelve, Welshpool and Builth Wells areas. Subsequent erosion locally formed a deeply dissected topography. The melting of the ice of the southern hemisphere Gondwana glaciation, caused a major sea-level rise which started in the late Hirnantian and continued into the early Llandovery. The effect of the rise of the sea level was to cause anoxia, preserving black graptolitic shales (hemipelagites) in the basin. On the shelf, in a narrow zone along the margins of the Midland Platform, transgressive shoreface sandstones and conglomerates, ranging from the latest Hirnantian to late Rhuddanian in age, unconformably overlie Ordovician rocks (Fig. 4.6).

Temporary falls in global sea level, during the mid and late Aeronian preceded further transgressive pulses during the *sedgwickii s.l.* and *turriculatus s.l.* biozones. With each deepening event, Llandovery shelf deposition encroached further eastwards (Fig. 4.7a, b). The *sedgwickii* transgression submerged the Builth Wells and Shelve areas; the *turriculatus* inundation extended as far east as Coventry on the Midland Platform.

Between the Shelvian Orogeny and the late Aeronian, there is little evidence for widespread contemporaneous tectonism in the basin (Davies *et al.* 1997; but see Smith 2004). Localized extension, however, led to the opening of small rift sub-basins within the Midland Platform at Usk and Woolhope (Butler *et al.* 1997). A similar basin, the Skomer Sub-basin, in SW Pembrokeshire, was the site of Aeronian volcanism (Skomer Volcanic Group) (Powell 1987, 1989; Hillier 2000). In contrast, the Telychian and early Wenlock saw tectonism as the dominant influence on sedimentation within the basin as the onset of oblique plate collision caused the reactivation of underlying basement fractures. Non-sequences and condensed successions record contemporary uplift along the margins of the basin associated with the rejuvenation of sediment source areas.

Shelfal Llandovery successions

The transgressive events which led to an eastward and southward extension of Llandovery deposition onto the Midland Platform (Figs 4.6–4.8) are recognized in small, commonly isolated, outcrops in South Wales and the Welsh Borderland, where facies of near-shore aspect predominate. In contrast, off-shore sediments dominate the successions along the basin margin to the east of the Tywi Lineament in mid Wales, at Llandovery and Garth, and in outcrops located between the Severn Valley and the shelf margin, for example at Meifod, Welshpool and Berriew.

Ziegler *et al.* (1968*a*) identified five repetitive brachiopod-dominated associations in the Upper Llandovery, which reflect the original depth-related benthic communities. The associations are (from shallowest to deepest) the *Lingula, Eocoelia, Pentamerus, Stricklandia* and *Clorinda* associations and commonly record the progressive rise in sea level (Fig. 4.8). *Eocoelia* and *Pentamerus*, and many of their associated fossils, are not present in the Rhuddanian and early Aeronian rocks, and their benthic ecological niche was occupied by the *Cryptothyrella* Association (Cocks & McKerrow 1978). In the early Rhuddanian, shelf margin successions, such as those of the type Llandovery area and nearby Garth, and further north at Meifod, the basal Llandovery (Bronydd Formation at Llandovery and Garth) carries *Cryptothyrella* assemblages, indicating that deposits representing the shallowest-water *Lingula* Association have not been preserved there. However, in the middle Llandovery Aeronian Stage, sandstones further east (the Kenley Grit) bearing a *Lingula* Association, are widespread in Shropshire both east and west of the Church Stretton Fault Zone.

The Welsh Borderland

This area of the Midland Platform was first drowned by the late Aeronian transgression, quickly followed by a further pulse in the early Telychian (Figs 4.7 & 4.8). The sub-Llandovery landscape over which the sea transgressed was by no means smooth, Bridges (1975) documenting that the Longmynd and Shelve areas of Shropshire were local highs with a ria topography over which the transgressions progressed unevenly. Comparable sequences with uneven topography underlying the transgressive Llandovery rocks are seen elsewhere; for example, around the Builth Ordovician Inlier where discontinuous sandstone lenses of the Trecoed Formation bearing late Aeronian mid-shelf brachiopods including *Pentamerus* and *Stricklandia* are unconformable upon the underlying Ordovician (Ziegler *et al.* 1968*b*). The shelf sediments were locally-derived sandstones, siltstones and shales (e.g. the Bog Quartzite, Venusbank Formation and Minsterley Formation) which in general display a set of overall fining-up sequences, although there are many small local exceptions to that. There is some evidence from subsidence analysis (King 1994) and sedimentary architecture (Woodcock 1984; Cocks 1990) that at least some of the elements of the Welsh Borderland Fault System were periodically active during the early Silurian and thus influenced both the shape of the local depocentres and also the sediment supply to them.

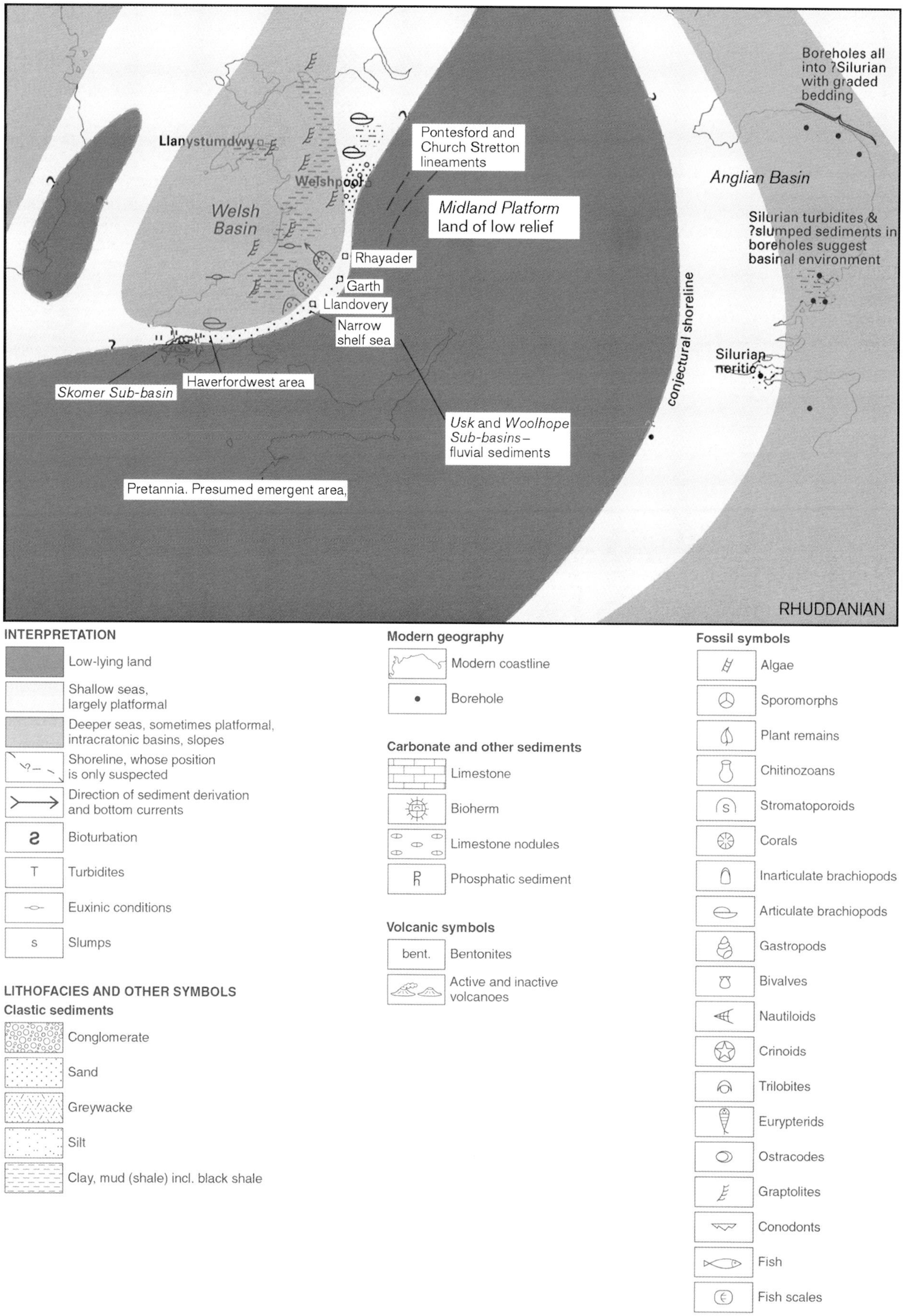

Fig. 4.6. Llandovery facies map for the early Llandovery (Rhuddanian) (from Bassett *et al.* 1992*a*).

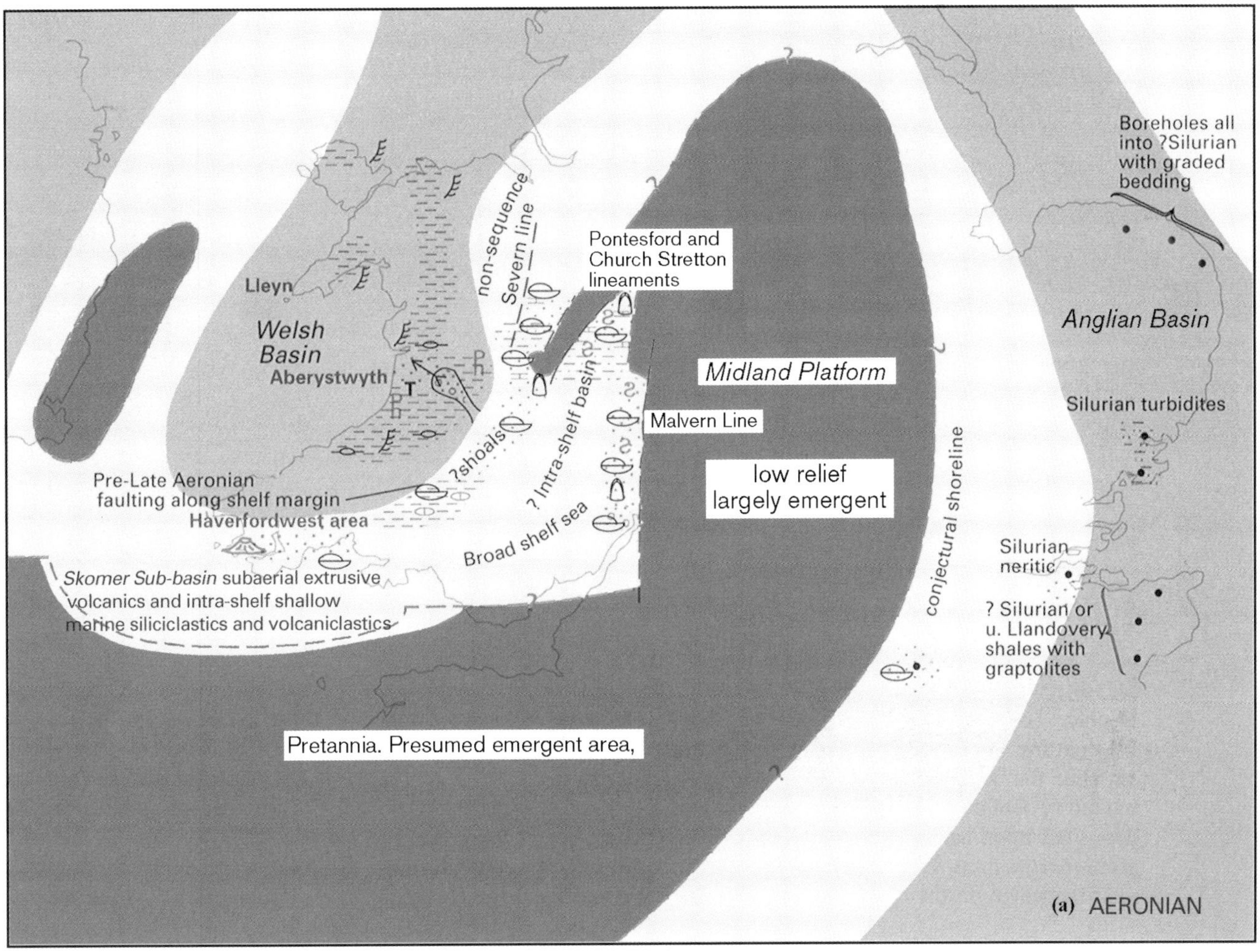

Fig. 4.7. Llandovery facies map for the (**a**) Mid Llandovery (Aeronian); (**b**) Late Llandovery (Telychian) (from Basset *et al.* 1992*a*). Key as for Fig. 4.6.

South Wales

The southern part of the platform was dominated by intra-shelfal extensional tectonics and associated sedimentation during the Early and Mid-Llandovery (King 1994; Butler *et al.* 1997; Hillier 2000, 2002). In close association was the development of volcanic centres, the best exposed being the Aeronian Skomer Volcanic Group (SVG) in SW Pembrokeshire. The SVG is confined to half-grabens comprising the Skomer Sub-basin which is delineated by a series of inverted normal faults, namely the Wenall, Ritec and Benton faults, and their corresponding footwall shortcuts (the Musselwick and Johnston Thrusts; Powell 1987, 1989) (Fig. 4.9). The Group comprises approximately 760 m of basic and rhyolitic lavas, pyroclastic and associated siliciclastic sediments (Thorpe *et al.* 1989). Overall, the SVG demonstrates the classic sedimentological signature of a continental rift-suite in that interdigitated fluvial/lacustrine sediments and volcanics at the base of the Group (rift onset) pass upward into cyclic arrangements of lavas/pyroclastics and shallow marine sediments (syn-rift), the latter transgressions facilitated by volcanic-related subsidence (Ziegler *et al.* 1969; Bridges 1976). An increase in marine-influenced sedimentation is seen in an easterly direction within the SVG, and it is probable that the half-grabens defined cuvettes that connected to the Welsh Basin proper in this direction. The topmost SVG comprises 45 m of wave-influenced, distal-shelf mudstones with good open marine faunas of late Aeronian age (Ziegler *et al.* 1969). This transgression represented a phase of probable thermal subsidence where the sedimentation rate was outpaced by the development of accommodation space. Eustatic controls on sea level, cannot be ignored, and it is probable that this transgression was accentuated by the *sedgwickii s.l.* Biozone global highstand of sea level.

The late Aeronian transgression was punctuated by an episode of tectonic uplift in southern Pembrokeshire resulting in subaerial exposure of the shelf for most of the Telychian and a significant basinward shift in sedimentary environments (Fig. 4.9) (Hillier 2002). This is manifested by a well-defined angular unconformity in the Marloes Peninsula, which clearly terminates extension faults within the topmost SVG, demonstrating that extensional strain was still being accommodated within the cover sequence. During the Telychian, shelf-incision and sediment bypass ensued with spatially limited, coarse-grained, low-sinuosity fluvial channels of the basal Coralliferous Group transporting sediment to the basinal sandy turbidite systems (Aberystwyth and Cwmystwyth Grits Groups) (Fig. 4.9a). Palaeocurrents from this unit indicate a northward transport direction, being antecedent to the local structure and topographic trend. Renewed transgression during the late Telychian (*crenulata s.l.* Biozone) saw lowstand gravels being ravined and reworked within high energy, wave-dominated shoreface/inner shelf environments. Thin, coarsening-upward sequences are preserved, and the shoreline probably had significant topographic irregularities (embayments and rias) (Fig 4.9b).

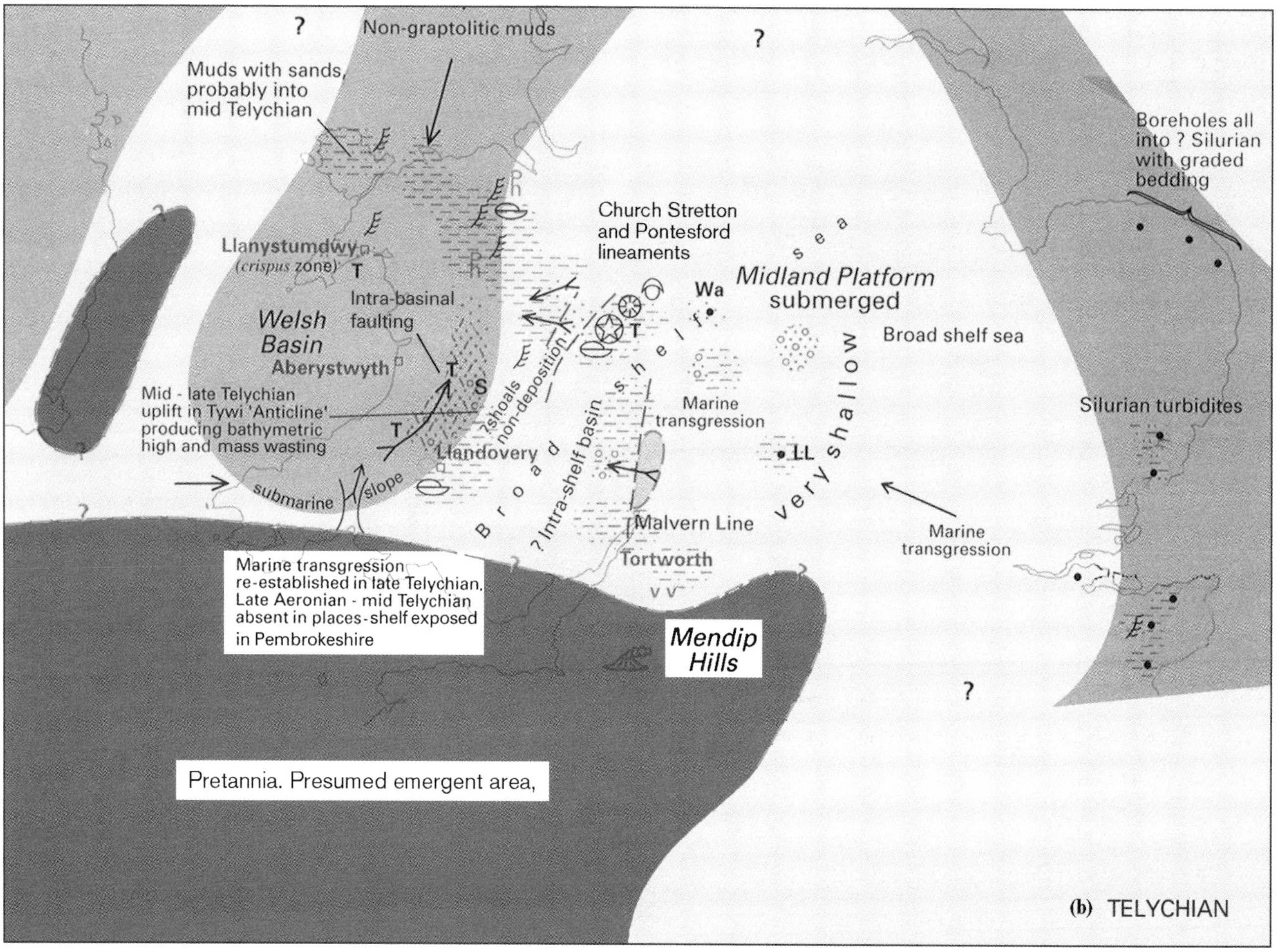

Fig. 4.7. Continued.

Renewed subsidence led to further coastal onlap leading to a reduction in significant coarse-grained sediment supply. The remaining 100 m of Coralliferous Group is dominated by wave-influenced silts, mud-shales and thin sandstones with good open marine faunas (Walmsley & Bassett 1976) (Fig. 4.9c). Corals, brachiopods and conodonts confirm a late Telychian–Wenlock age. Further north, open-marine, mud-prone conditions are indicated by the Rosemarket Formation, resting unconformably on Precambrian rocks in the hangingwall of the Benton Fault. *Pentamerus oblongus* is common, indicating a late Aeronian or early Telychian age (Aldridge *et al.* 2002).

At Haverfordwest, open marine conditions prevailed during Llandovery times recording a continuum of sedimentation across the Ordovician–Silurian boundary (Cocks & Price 1975; Aldridge 2000). The Rhuddanian Haverford Mudstone Formation (300–390 m thick; Cocks & Price 1975; Cocks *et al.* 1992) reflects the postglacial rise in sea level and contains a rich, diverse marine fauna, with brachiopods being the most common. The sediments and faunas indicate deposition within a mudstone-prone offshore shelf environment (Aldridge 2000). The mid Aeronian regression is probably reflected by the overlying Rhuddanian to Aeronian Gasworks Sandstone Formation. This contains dm-thick sandstone beds that have sharp erosive bases, low-angle cross-stratification and fine-upward into mudstones containing the brachiopod *Cyclocrinites* and other benthic faunas of the *Clorinda* community (Cocks & Price 1975; Aldridge 2000). Sanzen-Baker (1972) ascribed deposition by turbidity currents, but the presence of a shallow-water fauna, and the burrow *Palaeophycus* suggests deposition within a shoreface/shelf environment, with sandstones being storm-emplaced. The overlying Millin Mudstone Formation is probably entirely of Telychian age, although the lowest beds are unfossiliferous (Aldridge 2000). It reflects the early Telychian transgression and has a macrofauna indicative of low-energy, mid–outer shelf environments.

Within the Usk and Woolhope sub-basins, Rhuddanian low-sinuosity fluvial deposits drained a local volcanic source region (Butler *et al.* 1997). The Woolhope–Usk sub-basins were the sink for sediment transported from the Midland Platform to the east, especially during falls of relative sea level (James 2003). This area was transgressed during the late Aeronian sea level rise, with sedimentation dominated by marginal/shallow-marine environments, an analagous scenario to that in the Skomer Sub-basin.

Llandovery volcanics are encountered in boreholes in south Wales and across the Midland Platform, and late Telychian lavas crop out in the Tortworth Inlier, unconformably overlying sediments of Tremadoc age (Benton & Thackeray 1995). Here, the Lower and Upper Trap lavas comprise altered basaltic flows, the latter extruded within a submarine environment (Curtis 1972; Cave 1977; Kellaway & Welch 1993). The intervening Damery Formation contains fine-grained sandstones with interbedded silt- and mudstones deposited in a mid-shelf position. Cave (1977) identified probable storm-induced sediment input within an overall transgressive environment. The

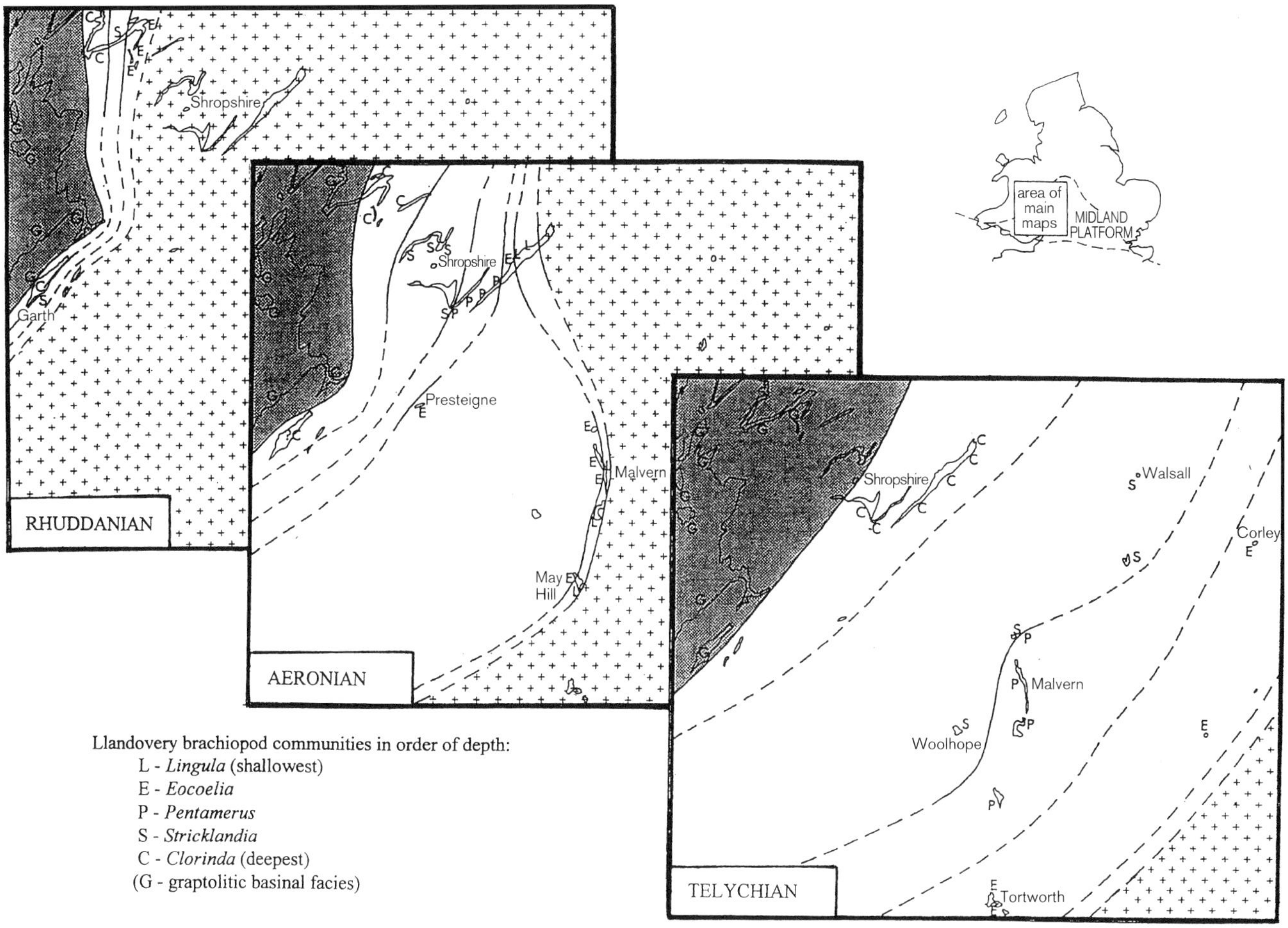

Fig. 4.8. Three maps (**a**) Rhuddanian, (**b**) Aeronian and (**c**) Telychian, showing the progressive Llandovery transgression across Wales and the Welsh Borderland (from Ziegler *et al.* 1968*b* with permission of the Palaeontological Association).

presence of *Eocoelia curtisi* confirms a late Telychian age. Many sandstone beds show tool marks resulting from chaotic transport of skeletal debris, with an ichnofauna dominated by *Palaeophycus* burrows (Benton & Hiscock 1996). Similar environments are recorded by the latest Telychian Tortworth Formation that rests conformably on the Upper Trap lava.

Volcanic activity continued into Wenlock times in the East Mendips Inlier of Somerset (Hancock 1982), and numerous bentonitic ash-fall tuffs are recorded across the margins and the eastern part of the basin (Teale & Spears 1986; Cave & Loydell 1998; Hillier 2000). The volcanic centres delineate a WNW-trending active zone on the southern margin of Avalonia. Similar trending growth faults have been identified across the southern margin of the Welsh Basin. Indeed, this trend continues through the Upper Palaeozoic Mendips and into the Mesozoic Wessex Basin, and it seems likely that Mid–Late Llandovery extension faults within these basins were re-utilized during later basin development. It is unclear whether the Mid–Late Llandovery rifting on the platform was related to the development of the Rheic Ocean to the south (Pharaoh *et al.* 1991), or to transtension associated with the collision of Eastern Avalonia and Laurentia (Soper & Woodcock 1990).

Mid Wales

Evidence for the eastern margin of the Welsh Basin is clearly seen in the type Llandovery and Garth (Fig. 4.8) areas of mid Wales where marked changes in lithofacies and biofacies occur to the east of the Tywi Lineament across fractures such as the Llanwrtyd and Garth faults (strands of the Tywi and Pontesford lineaments). To the west of these structures, the lowstand mudstones (Yr Allt Formation) deposited during the late Ordovician glaciation are conformably succeeded by the late Hirnantian to Aeronian, proximal slope-apron facies comprising turbidite mudstones and hemipelagites (Tycwtta Mudstone Formation) (Figs 4.10 & 4.11). The formation has yielded rare graptolites, but the only shelly faunas are an assemblage of reworked Rhuddanian brachiopods obtained from the turbiditic Allt-y-clych Conglomerate Member (Davies *et al.* 1997).

In the Garth area (Williams &Wright 1981; British Geological Survey 2005), east of the Garth Fault, late Ashgill, more proximal sandy and muddy siltstones and sandstones, deposited during the glacioeustatic lowstand, are overstepped by transgressive, latest Hirnantian shoreface deposits (Cwm Clyd Sandstone Formation). The latter is overlain by deep water unfossiliferous mudstones and lenticular sandstones (Garth House Formation). This post-glacial deepening event equates with the onset of anoxic conditions in the basin marked by the Cwmere Formation. Overlying the Garth House Formation, progradational sequences of shallow-marine burrowed sandy mudstones and muddy sandstones contain a shelly fauna of Rhuddanian–Aeronian age. A widespread disconformity above these strata formed during the mid Aeronian eustatic fall in sea level; an event which imposed oxic conditions across the basin floor to the west. The ensuing late Aeronian sea level rise, an event which re-introduction anoxic facies (M. sedgwickii Shales Member) in the basin, is marked in the Garth area by the

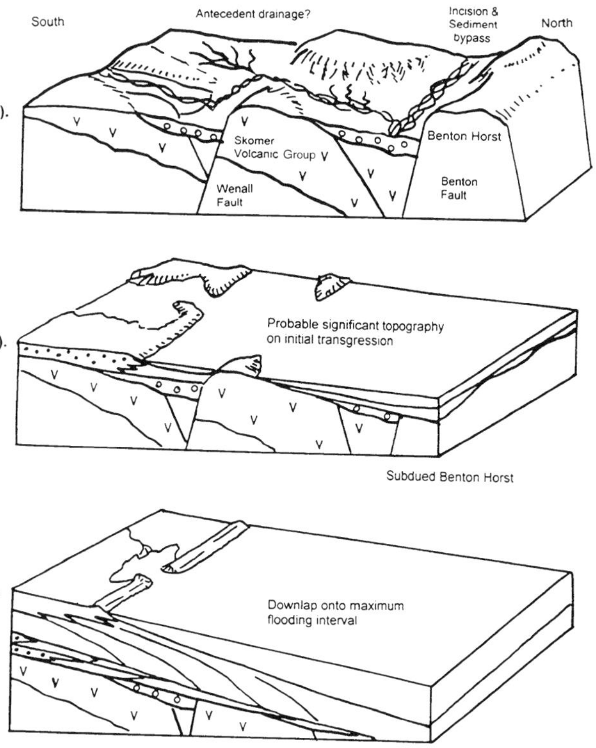

Fig. 4.9. Conceptual block-diagrams illustrating the latest Aeronian–Telychian evolution of southern Pembrokeshire (the Coralliferous Group). (**a**) Fluvial incision of the southern shelf and sediment by pass to the northerly Welsh Basin; (**b**) deposition of coarse-grained transgressive deposits associated with ravinement and reworking of gravels within a wave-dominated shoreface environment; (**c**) during the highstand of relative sea-level, deposition was dominated by fine-grained shelf-siliciclastics (from Hillier 2002).

transgressive, pentamerid-bearing Tyncoed Sandstone Member. The latest Aeronian regression is manifested by shoreface conglomeratic sandstones of the Ty-mawr Formation. The following Telychian transgression saw the deposition of green, burrow-mottled mid to distal shelf mudstones of the Cerig Formation. They pass upwards gradationally into anoxic mudstones of the lowermost Wenlock, the transition being marked by a sequence, exceptionally up to 130 m thick, of alternating burrow-mottled and laminated mudstones, locally recorded as the Dolfawr Mudstones Formation

In the Llandovery area, detailed formal subdivision of the Llandovery sequence was made by Jones (1925, 1949), and updated by Cocks *et al.* (1984). The latter defined the global boundary stratotypes for the base of the Aeronian and Telychian Stages of the Llandovery Series in the area (Cocks 1989). Cocks *et al.* (1984) recognized a conformable succession from the Ordovician, with sedimentation being at or close to the shelf-edge throughout the Llandovery. The base of the lowest Llandovery formation at Llandovery, the Bronydd Formation, rests apparently conformably on the Scrach Formation, which carries the latest Ordovician *Hirnantia* brachiopod fauna and represents the shallow-water deposits of the latest Ordovician glacioeustatic lowstand (Cocks *et al.* 1984). The Rhuddanian Bronydd and Crychan Formations manifest the post-glacial sea-level rise and comprise an upward-coarsening sequence that represents a progradational pro-delta lobe sourced from the SE (Cocks *et al.* 1984; Aldridge *et al.* 2000). A diverse brachiopod fauna of the *Stricklandia* benthic community is present. The mid Aeronian regression is represented by sandstones in the upper part of the overlying Trefawr Formation. The late Aeronian transgression saw mud deposition over a wide area as the Rhyddings Formation. To the SW the Coldbrook Formation, being the time equivalent of the Bronydd, Crychan and Trefawr Formations, is believed to have been deposited further from the source of siliciclastic material, yielding a *magnus* Biozone graptolite fauna. To the SW of Llandovery, the Goleugoed Formation is the lateral equivalent of the Coldbrook Formation, being mudstone-prone and presumably deposited in an offshore pro-delta environment (Cocks *et al.* 1984).

The overlying latest Aeronian–earliest Telychian Wormwood Formation comprises well-sorted sandstones and calcareous mudstones, a manifestation of the late Aeronian regression. Brachiopod communities indicate a shallowing (*Clorinda*, through *Pentamerus*) followed by a deepening of sea level through the formation. This deepening, indicating the early Telychian transgression, continued into the overlying Cerig Formation, a blanket of offshore mudstones over the prodelta.

In the area stretching from Meifod to Welshpool progressively younger parts of the Llandovery are represented by sediments unconformable on older rocks (Ziegler *et al.* 1968*b*). At Meifod, Rhuddanian and later formations overlie the Graig-wen Sandstone, which yields a latest Ashgill *Hirnantia* Fauna of brachiopods and other shelly forms, whilst at Welshpool the Aeronian Powis Castle Conglomerate and the succeeding Laundry Mudstone Formation overlie late Ordovician rocks with an angular unconformity (Ziegler *et al.* 1968*b*; Cave & Dixon 1993).

Deep-water Llandovery successions

Mid Wales

The Llandovery rocks of mid Wales were deposited in the southern portion of the Welsh Basin. They occupy a broad outcrop centred on the Cambrian Mountains and extending from the coast of Cardigan Bay to the Builth Wells area, and are only linked with the North Wales successions through the core of the narrow Llanderfel Syncline. The deep-water basinal succession, which includes the thickest Llandovery sequence (over 4 km) in southern Britain, is largely constructed from resedimented lithologies including slumps and debrites, but dominated by a spectrum of different forms of turbidite. Thin hemipelagic mudstones preserved between the resedimented 'event beds' record the slow accumulation of fine-grained sediment from suspension. Although it is a minor component of the basin-fill in terms of volume, this background sediment provides an important record of the state of oxicity of the basin waters at different times. Delicately laminated, dark-grey, hemipelagic mudstones, which preserve abundant graptolites, testify to deposition beneath stagnant, anaerobic bottom waters during periods which are described as anoxic. In contrast, paler, burrow-mottled, hemipelagic mudstones formed during intervening oxic intervals when the higher oxygen content in the water allowed colonization of the sea floor by a bioturbating infauna (Cave 1979; Smith 1987*b*).

The deep-water Llandovery rocks of the Welsh Basin were the subject of some of the early, seminal studies of turbidite facies including those by Wood & Smith (1959) and by Kelling and his co-workers (e.g. Kelling & Woollands 1969). More recently, the British Geological Survey and its collaborators (e.g. Cave & Hains 1986; Smith 1987*a*, *c*; Woodcock *et al.* 1996; Davies *et al.* 1997) have advanced sophisticated architectural and evolutionary models, which distinguish a series of deep-water sedimentary systems within the basin fill. They have also assessed the relative roles of eustasy, intrabasinal extensional tectonism and external sediment supply as controls on system development. Essential to this improved understanding has been the enhanced resolution of graptolite biozonal schemes used to date the basinal sequence achieved by Loydell (1991, 1992) and Zalasiewicz (1990).

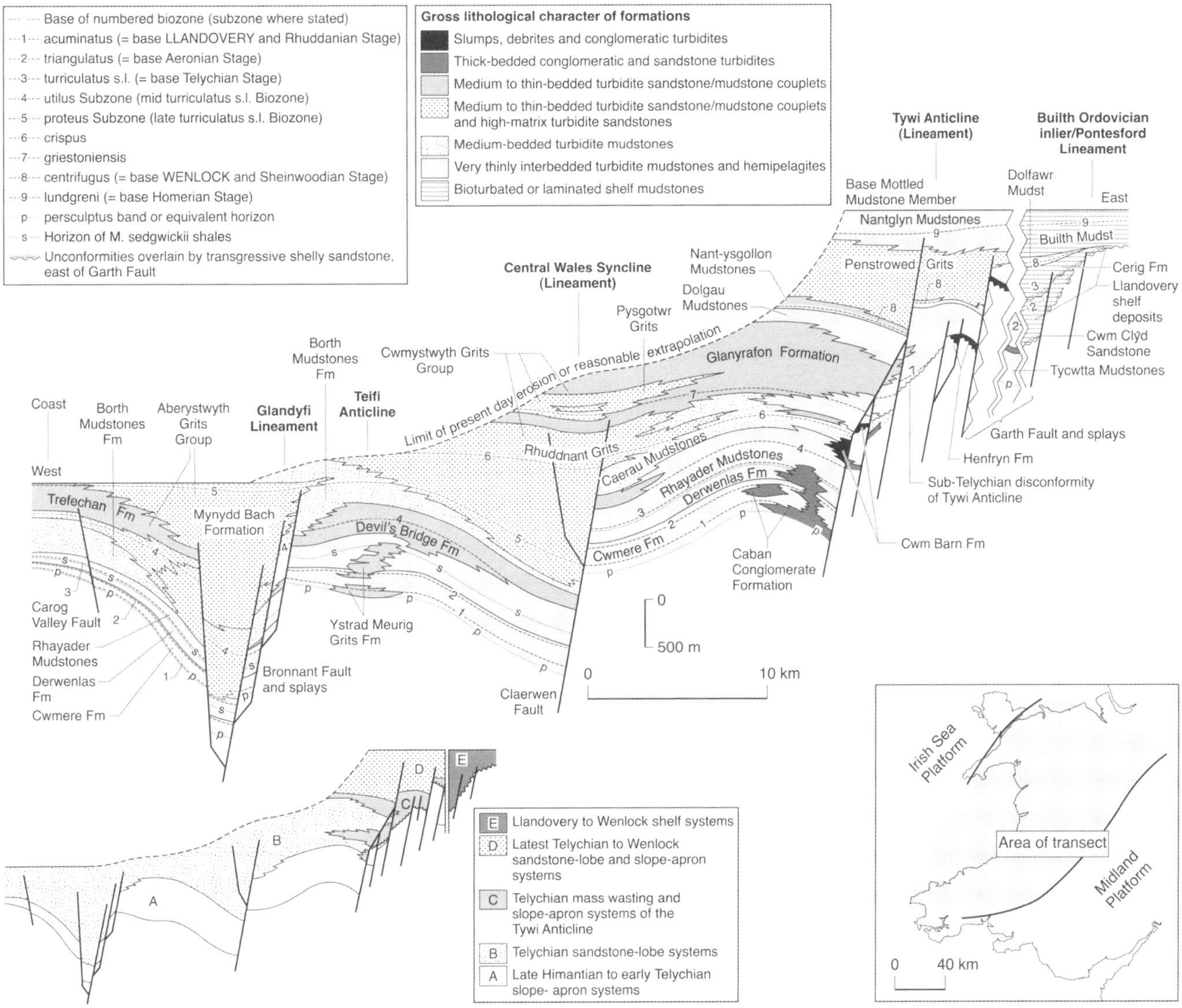

Fig. 4.10. Schematic pre-Acadian lithostratigraphical cross-section of the latest Hirnantian (Ashgill) to Homerian (Wenlock) succession of Central Wales (after Davies *et al.* 1997). Bottom left inset shows simplified architectural elements.

Turbidite systems. Davies *et al.* (1997) recognize two principal types of turbidite system within the late Ashgill–Wenlock basin fill: slope-apron systems and sandstone-lobe systems (Fig. 4.10). These systems are distinguishable not only by differences in facies, but also in sediment provenance as revealed by sandstone heavy mineral analysis (Morton *et al.* 1992) and mudstone geochemistry (Ball *et al.* 1992), as well as by palaeocurrent indicators.

The 'slope-apron systems' comprise wedge-shaped accumulations of turbiditic and hemipelagic mudstone that thin westwards from the eastern basin margin of the basin. Additionally they may locally envelope lenticular bodies or coarse-grained turbidite facies of particular interest as hydrocarbon reservoir analogues. Both the fine- and coarse-grained turbidites, and the associated hemipelgic sediment were supplied from the eastern margin of the basin. Each system comprises a lower anoxic division and an upper predominantly oxic unit. Slope-apron systems formed during periods of tectonic quiescence and, hence, the alternation of anoxic and oxic facies appears closely to correspond to reported eustatic events. Periods of anoxic sedimentation correlate with periods of marine transgression (Fig. 4.11a), and are thought to be linked to high rates of organic production and the formation of a strongly stratified water column. In contrast, marine regressions (Fig. 4.11b) promoted the greater mixing of surface and bottom waters marked by periods of oxic facies deposition.

Although slope-apron deposition was active over much of the basin throughout the early and mid Llandovery, during the Telychian a series of southerly-supplied 'sandstone-lobe systems' invaded the basin (Fig. 4.10). The arrival of these sandy turbidites coincided with a period of intrabasinal growth faulting. Thick sequences, locally in excess of 2 km, reflect high rates of sedimentation, whilst marked lateral variations in turbidite facies reveal the influence of fault induced sea-floor topography. The distribution of oxic and anoxic facies within these lobe systems appears unrelated to contemporary eustasy, but instead record local, tectonically driven changes in marine base level and/or basinal circulation patterns. The marked increase in grade and volume of sediment being supplied to the basin during the Telychian and early Wenlock has been linked to plate collision and the uplift of sediment source areas.

During the same period, faulting and uplift along the Tywi Lineament resulted in submarine mass-wasting of the earlier Ashgill and Llandovery slope-apron sequences. Slumps,

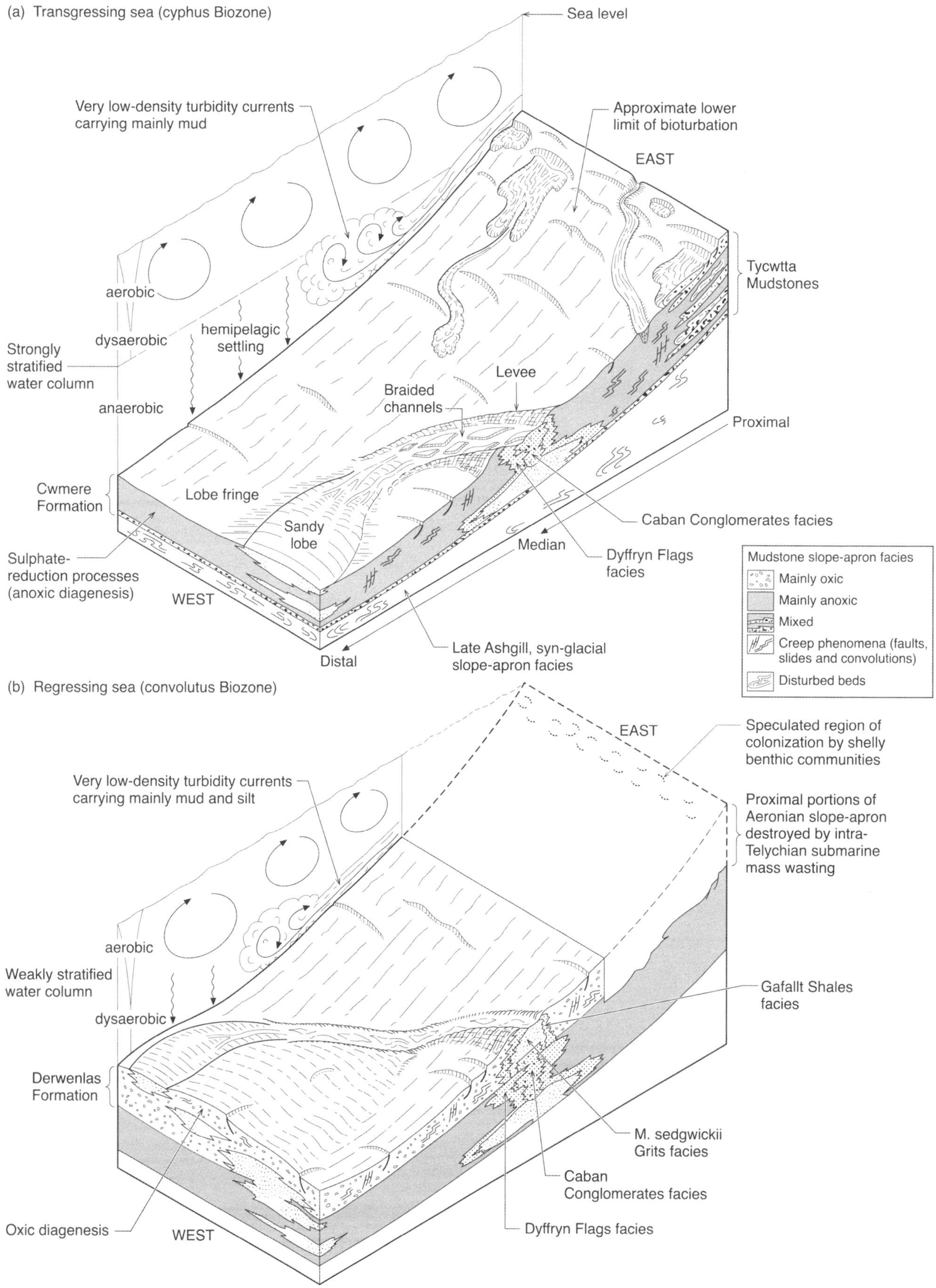

Fig. 4.11. Depositional model for latest Ashgill to Llandovery slope apron, nested channels and lobes in central Wales between the Pontesford Lineament and Ystradmeurig. (**a**) during a transgressive sea (*cyphus* Biozone); and (**b**) during a regressive sea (*convolutus* Biozone) (modified after Davies *et al.* 1997).

debrites and conglomeratic turbidites were shed westwards onto the former slope-apron surface. Slope-apron facies deposition resumed in the late Telychian burying both the deposits and the disconformity, which date from this mass-wasting episode. East of the Tywi Lineament, basin-margin facies pass into coeval Llandovery shelf deposits exposed in the Llandovery, Garth, Builth Wells, Shelve and Welshpool areas. Bioturbation levels in mudstone facies, transgressive and regressive sand-bodies, subaerial disconformities and the succession of shelly benthic communities (Ziegler *et al.* 1968*a*) all reflect the direct influence of contemporary sea-level movements on a shallow shelf, many of which can be correlated with the changes in oxygen levels observed in the basinal succession.

In synthesizing the distribution and interplay of these various basinal system types Davies *et al.* (1997) recognize four architectural elements within the Llandovery–Wenlock succession of Central Wales:

(1) latest Hirnantian–early Telychian, easterly-supplied mudstone slope-apron systems with associated nested channel fills and sandy lobes;
(2) Telychian, southerly-supplied sandstone-lobe systems;
(3) mass-wasting and slope-apron systems of the Tywi Lineament;
(4) Latest Telychian–Wenlock sandstone-lobe and slope-apron systems.

The last of these, comprising rocks predominantly of Wenlock age, is described with the other strata of that age.

Latest Hirnantian–early Telychian easterly supplied mudstone slope-apron systems. The post-glacial latest Hirnantian to early Llandovery transgression (Figs 4.10 & 4.11) in mid Wales is marked by an abrupt change from the largely slumped silty mudstones, sandstones and conglomerates of the late Ordovician glacial lowstand sequence to the smooth, anoxic slope-apron mudstones of the Cwmere Formation (Rhuddanian–early Aeronian). This formation comprises up to 500 m of thinly interbedded turbidite mudstones and thinly laminated, graptolitic, hemipelagic mudstones. At the base of the formation is a widely recognised unit of oxic grey mudstone called the Mottled Mudstone Member. In places along the faulted eastern side of the Tywi Anticline, close to the eastern shelf margin of the Welsh Basin, the remnants of the proximal part of the Rhuddanian to Aeronian slope-apron sequence are preserved as the Tycwtta Mudstones Formation. This mixed oxic-anoxic formation is 750 m thick and differs from its correlative, the deeper water Cwmere Formation, in its greater proportion of burrowed oxic mudstone (about 50%) and the presence of slumped units. It also includes the Allt-y-clych Conglomerate Member, a distinctive 20 m unit of massive debrites, turbiditic pebble-cobble conglomerates and sandstones.

The Cwmere Formation is conformably succeeded by the Claerwen Group (Fig. 4.10), an up to 650 m-thick sequence of predominantly oxic mudstone slope-apron facies. This Aeronian to Telychian succession is subdivided into the Derwenlas Formation and succeeding Rhayader Mudstones Formation, and consists mainly of thinly interbedded, variably silt-laminated greenish-grey turbidite mudstones and burrowed hemipelagites with thin horizons of anoxic mudstone. The widespread change from anoxic to oxic facies deposition at the base of the Derwenlas Formation equates with the onset, during the *magnus* Biozone of the mid-Aeronian eustatic marine regression; an event which reached its acme during *convolutus* Biozone times. The M. sedgwickii Shales, a thin horizon of black, laminated hemipelagic mudstones found throughout mid Wales, forms the basal unit of the overlying, up to 450 m-thick, Rhayader Mudstones Formation. It records a second, late Aeronian, pulse of Llandovery transgression, which briefly reintroduced anoxic bottom conditions into the basin (Davies *et al.* 1997). The succeeding part of the formation records largely uninterrupted oxic mudstone deposition which persisted through early Telychian times.

The late Hirnantian–Telychian slope-apron mudstones of mid Wales also envelope sequences of coarse-grained turbidite facies. Thus, along the western margin of the Tywi Anticline, south of the mid Wales town of Rhayader, and around the adjacent Rhiwnant Anticline, the mudstones of the Cwmere Formation and Claerwen Group pass locally into facies of the Caban Conglomerate Formation (Kelling & Woollands 1969; Davies & Waters 1995) (Fig. 4.10). There the thick units of turbidite conglomerates contained within the eastern slope-apron succession were deposited in a nested array of narrow submarine channels flanked by broad levees (Fig. 4.11). These channels supplied sand grade sediment westwards to a series of small, laterally migrating sandstone lobes sited on the distal slope-apron surface. These lobe-like sandstone bodies now collectively comprise the Ystrad Meurig Grits Formation (Davies & Waters 1995). The Caban Conglomerate and Ystrad Meurig Grits formations represent a long-lived pathway via which easterly sourced coarse-grade sediment was routed across the slope apron (Figs 4.10 & 4.11). However, the putative canyon which supplied this system has not been identified, possibly due to structural displacement along the Tywi Lineament (Davies *et al.* 1997).

A silt- and sand-rich facies in the upper part of the Rhayader Mudstones Formation may herald the first introduction of tectonically generated detritus during the Telychian. It is associated with the expansion of turbidite fan systems, which introduced the Devil's Bridge Formation (Fig. 4.10) into the basin. These are represented by extensive sequences of thinly interbedded, mainly oxic, turbidite sandstones and mudstones. The Devil's Bridge Formation (Cave & Hains 1986), some 400–500 m thick, occupies large areas of the Aberystwyth district west of the Central Wales Lineament and extends northwards into the Corris area. It was deposited contemporaneously with the Rhayader Mudstones Formation east of the Central Wales Lineament. The Devil's Bridge Formation resembles parts of the southerly derived turbidite systems (see below). However, palaeocurrents and heavy-mineral data (Morton *et al.* 1992) suggest, at least in part, an easterly source for the sediment, which may have been similar to that of the Rhayader Mudstones Formation. Deposition of the Devil's Bridge Formation in part coincides with the influx of tectonically generated detritus west of the Bronnant Fault/Glandyfi Lineament (the Aberystwyth Grits Group), and may reflect the same episode of source area rejuvenation.

Telychian southerly supplied sandstone-lobe systems. During the late Llandovery, the oblique collision of Avalonia with Laurentia was responsible for a major phase of extensional faulting in the basin and its margins. This resulted in major uplift of areas supplying sediment to the basin and a major intra-Telychian unconformity on the shelf in SW Pembrokeshire. It was accompanied by major subsidence in the basin, which provided space for the introduction of sandstone turbidites which progressively replaced the mud-dominated slope-apron facies. Large-scale faults were important in controlling the disposition of turbidite systems such as the Aberystwyth Grits and Cwmystwyth Grits groups.

Patterns of sediment influx during the Telychian show successive initiation of southerly derived turbidite deposition from west to east, and imply movement on the faults associated with the Bronnant Fault/Glandyfi Lineament and Central Wales Lineament (Fig. 4.10). Movements on each fault were intermittent, periods of activity being indicated by confinement of turbidite systems to the west of each fault, and periods of quiescence by eastward overstep across them. These systems mainly consist of medium- to thin-bedded turbidite sandstone-mudstone couplets with both laminated and burrowed hemipelagites. Over wide areas, these typical Bouma turbidites are intercalated with packets of thick-bedded, high-matrix turbidite sandstone, relatively mud-depleted sandstone, and subordinate pebbly mudstone (Clayton 1994; Smith 1987*a*).

The Aberystwyth Grits Group (Wilson *et al.* 1992) is the most westerly of these systems, lying west of the Bronnant Fault and occupying the coastal tract between New Quay and Clarach. It comprises a division with abundant thick-bedded, high-matrix sandstones, the Mynydd Bach Formation, up to (2 km thick), which interdigitates westwards with the coeval Trefechan Formation made up exclusively of Bouma turbidites (Fig. 4.12). The group was deposited during early Telychian times, in part contemporaneously with the Devil's Bridge and Rhayader Mudstones formations further east (Fig. 4.10). A limited amount of eastward overstep across the Bronnant Fault commenced in early *turriculatus* graptolite Biozone *s.l.* times. Although turbidites continued to be deposited west of the Bronnant Fault as late as the upper *turriculatus* graptolite Biozone *s.l.* of the Telychian, the focus of the main turbidite deposition shifted eastwards, depositing the Cwmystwyth Grits Group. The latter outcrops in the core of the Central Wales Syncline and likewise comprises divisions with abundant high-matrix sandstones, the Rhuddnant Grits and Pysgotwr Grits, which interdigitate with the Glanyrafon Formation of thinly bedded Bouma-type turbidites (Fig. 4.10). The oldest, sandstone-rich, member of this Group, the Llyn Teifi Member of the Rhuddnant Grits Formation, is confined to the west of the Claerwen Fault (part of the Central Wales Lineament) (Fig. 4.10). It is probably of mid to late *turriculatus* Biozone *s.l.* age; younger, late *turriculatus* Biozone *s.l.* graptolites have been recorded 200 m above the base of the Rhuddnant Grits Formation (Davies *et al.* 1997). Sequences of thin- to medium-bedded, grey turbiditic mudstone with both burrowed and laminated hemipelagite interdigitate with, and grade into, the turbidite sandstone facies. They represent the distal and laterally fringing mudstone facies of the lobes now represented by several formations (e.g. the Borth Mudstones) (Fig. 4.10).

Initial deposition of the sandstone-lobe turbidite facies of the Cwmystwyth Grits west of the Claerwen Fault was contemporaneous with the later stages of deposition of the Rhayader Mudstone Formation and succeeding southerly derived turbidite mudstones of the Caerau Mudstone Formation in the east (Fig. 4.10). Eastward overlap of the sandstone-lobe facies of the Rhuddnant Grits Formation across the Claerwen Fault began in late *turriculatus* Biozone *s.l.* times (*carnicus* Subzone), but the greatest lateral extent of the Rhuddnant Grits Formation was during the mid Telychian (early *crispus* Biozone). Thereafter, widespread abandonment of its sandstone-lobe facies occurred in late *crispus* Biozone times.

Sandstone-lobe development resumed in latest *crispus* to *griestoniensis* Biozone times with deposition of the youngest unit of the Cwmystwyth Grits Group, the Pysgotwr Grits Formation. The thickest and oldest part of this formation suggests that the Claerwen Fault may have been the focus of deposition. The eastern limits of both the Rhuddnant and Pysgotwr Grits formations suggest that sediments were lapping against a W-facing confining slope (Fig. 4.10).

The easternmost units of the southerly-derived turbidite facies are the Caerau Mudstones and Glanyrafon Formation. Between the Central Wales and Tywi lineaments these interfinger with oxic, mud-dominated, easterly derived, slope-apron deposits of the Dolgau Mudstones Formation (Fig. 4.10). Eastwards, lateral failure of these lobe-fringe facies during *crispus* and *griestoniensis* Biozone times appears to reflect the confining effect of west-facing slopes.

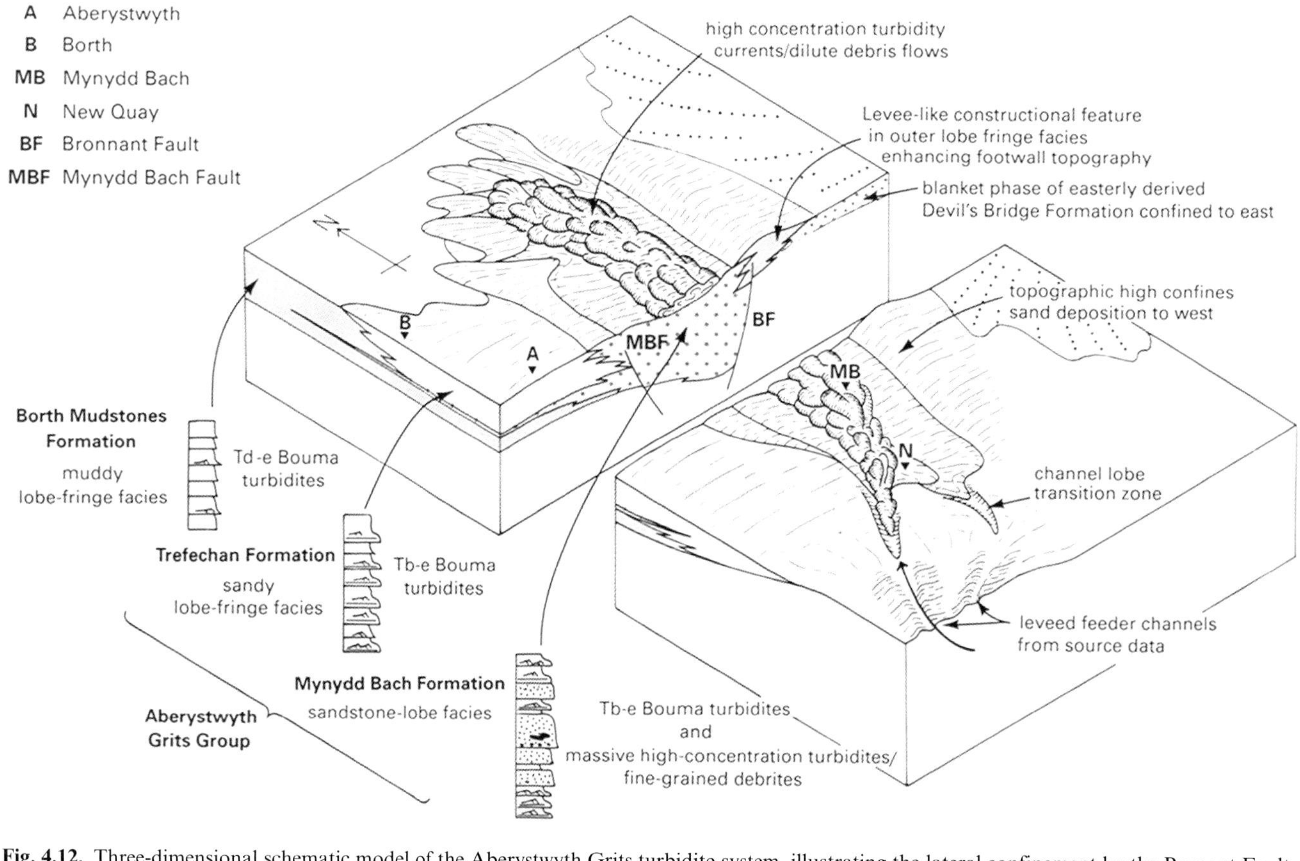

Fig. 4.12. Three-dimensional schematic model of the Aberystwyth Grits turbidite system, illustrating the lateral confinement by the Bronant Fault and resulting facies distribution. Palaeocurrent flow is to the north (from Davies *et al.* 1997).

Telychian mass-wasting and slope-apron systems of the Tywi Lineament. Late Llandovery extensional tectonism also resulted in uplift and unconformity along the marginal areas of the basin and the adjacent Midland Platform, the effects being most noticeable across the major reactivated structures of the Welsh Borderland Fault System (Fig. 4.10). Over the Tywi Anticline (Lineament), much of the previously deposited Llandovery succession may have been removed by submarine mass-wasting. The Dolgau Mudstones Formation in this region rests on sequences of thinly interbedded shelly, phosphatic conglomerates, sandstones and pebbly mudstones, represented by the Cwm Barn Formation and Henfryn Formation. The pattern of overstep, and the nature of the facies below and above the unconformity, suggest contemporaneous faulting. The Dolgau Mudstones Formation is overlain by the predominantly anoxic, latest Telychian–Wenlock slope-apron deposits of the Nant-ysgollon Mudstones of the Tywi Anticline. The Dolgau Mudstones Formation gives way to a distal shelf to slope facies comprising oxic, burrow-mottled mudstones and siltstones of the Cerig Formation; the latter is well represented in the more complete shelf sequences of the Llandovery area.

North Wales

The basinal sequence in North Wales is largely mudstone-prone, lacking large-scale sandy turbidite systems. There is also little evidence for widespread contemporaneous tectonism as seen in the southern part of the basin, especially in the Telychian, but this may reflect the fact that the area has not been studied in detail for a considerable time. A full sequence is observed only in the Conwy Valley; comparable, but incomplete sequences are preserved on Parys Mountain, on Anglesey, and at Llanystumdwy, on Llyn (Cocks *et al.* 1992). In the Conwy Valley, the Bryn Dowsi Mudstone Formation of latest Hirnantian–Aeronian age rests conformably on late Ashgill mudstones (Warren *et al.* 1984). The lower quarter of the 400 m thick formation is dominated by anoxic graptolitic (hemipelagic) and turbiditic mudstones reflecting the latest Hirnantian–Rhuddanian post-glacial rise in sea level. Above the *atavus* Biozone the formation comprises a mixed sequence of pale grey oxic burrow mottled and dark grey graptolitic hemipelagic mudstones interbedded with turbidite mudstones. There is insufficient stratigraphical and sedimentological information to recognize the signatures of the sea-level movements seen in the southern part of the basin between the mid Aeronian and early Telychian. The succeeding Pale Slates are up to 118 m thick and span the Telychian Stage (Warren *et al.* 1984). Basal passage beds, still with graptolitic levels, are overlain by latest Telychian (*crenulata* Biozone *s.l.*) pale, locally burrow-mottled, grey-green mudstones of turbiditic and hemipelagic origin. These mark the period of tectonic quiescence during which oxic bottom conditions were sustained following the early Telychian reconfiguration of the basin. These latter effects are reflected in the Llandovery successions of the Derwen Horst/Anticline. Here, the middle Llandovery Bron y graig Formation, resting unconformably on Ashgill strata, comprises a succession of sandstones and siltstones. It is overstepped by the local representative of the Pale Slates (British Geological Survey 1993). To the east, in fuller Llandovery successions, latest Ashgill–Aeronian silty and locally graptolitic mudstones are termed the Fron Frys Formation, and precede the pale coloured, Telychian mudstones, locally known as the Ty-draw Slates.

The Wenlock

Introduction

Wenlock rocks crop in the eastern part of the Welsh Basin and over extensive areas of the Midland Platform, notably in the Welsh Borderland. The general palaeogeographic position in central and southern England and Wales did not change much from late Llandovery to Wenlock times. The basin remained the site of deep-water, predominantly turbiditic, deposition, while the platform was mainly the site of continuous shelf sedimentation, apart from localized areas where disconformities are developed (Fig. 4.13). Several oceanographic and tectonic events can be recognized in the Wenlock sequence (Fig. 4.1).

The global sea-level curve includes a lowstand beginning in the latest Llandovery and reaching its acme in the early Sheinwoodian. On the Midland Platform this is manifested in the Welsh Borders and the English Midlands by the local development of shallow water carbonates. In South Wales, the Wenlock records an overall regressive signature, punctuated by high order changes in relative sea level. This regressive episode documents a combination of increased sediment supply related to oblique plate collision, a global trend of falling sea level, and intra-shelf tectonism. In contrast, the latest Llandovery in the basin and its immediate margins saw the onset of anaerobic bottom conditions leading to the deposition of graptolitic anoxic hemipelagites that persisted as the background sediment for most of the Wenlock. The fact that the onset of this anoxic interlude correlates with a global lowstand rather than an anticipated highstand, suggests that a eustatic model for this interval is unlikely and a major climatic or oceanographic change, possibly resulting from the contemporary narrowing of the Iapetus seaway, is more likely. The subsequent mid Sheinwoodian transgression and Homerian regression exerted strong influences on contemporary shelf facies of the Midland Platform.

In the early Wenlock, ongoing oblique docking of Avalonia/Baltica with Laurentia in the Caledonide orogeny led to further phases of extensional fault movement within and along the margins of the Welsh Basin. As in the Llandovery, such structures influenced sediment source areas and controlled the distribution and thicknesses of sedimentary facies.

Associated early Wenlock volcanism is known from the east Mendips of Somerset (Hancock 1982) and from the shelf marginal area of west Shropshire (Sanderson & Cave 1980). Subsequently, in the late Wenlock, this phase of transtensional tectonism gave way to a period of more widespread but less dramatic subsidence, thought to reflect the thermal contraction of the attenuated lithosphere (King 1994).

Shelfal Wenlock successions

The Welsh Borderland

With the main transgression over Wales and the Welsh Borderland complete by the late Llandovery, the Midland Platform appears to have been less structurally mobile during the succeeding Wenlock and Ludlow epochs. However, the platform continued to subside slowly, enabling widespread deposition of 'Wenlock Shale', although by the end of the Wenlock subsidence had slowed, enabling widespread deposition of carbonates, including bioherms, most notably on Wenlock Edge and in the West Midlands.

Bassett (1974) has reviewed the stratigraphy of the various Wenlock outcrops in South Wales and the Welsh Borderland. The calcareous developments at the base of the Wenlock are best seen in the 60 m thick Nash Scar Limestone, which includes several bioherms at Presteigne (Old Radnor). In Radnorshire, bioherms are also present in the Woolhope Limestone, which has its maximum development in the Woolhope (Squirrel & Tucker 1960) and May Hill (Lawson 1955) inliers of Herefordshire. To the north, in Shropshire, there are no discrete limestone horizons at equivalent levels, but calcareous nodules up to 1 m across distinguish the Buildwas Formation, which overlies the Upper Llandovery Hughley (or Purple) Shales conformably there (Bassett *et al.* 1975). However, the brachiopod-dominated benthic associations present in Shropshire indicate deeper water than at Woolhope and May Hill. The main bulk of the Wenlockian sequence is taken up by the

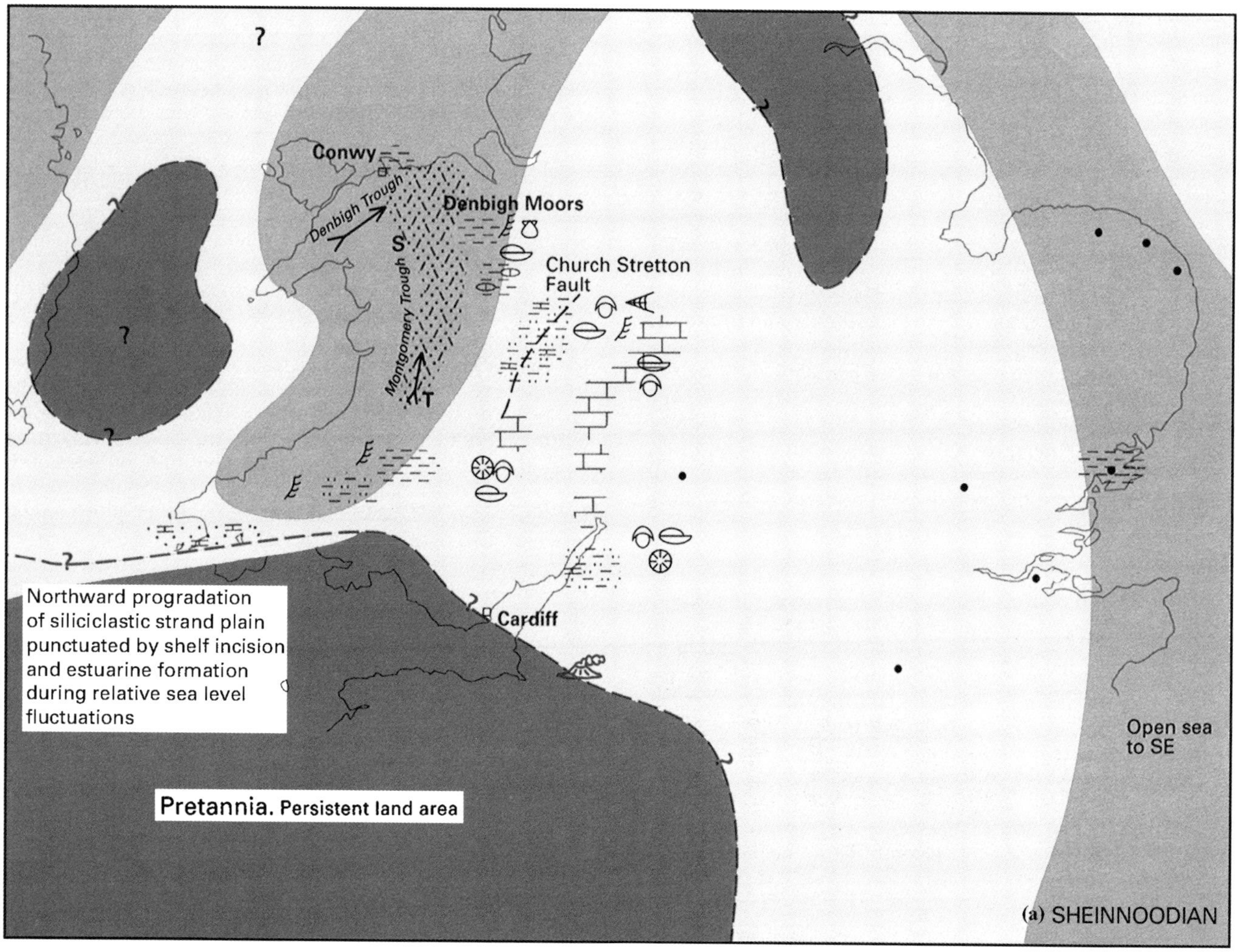

Fig. 4.13. Wenlock facies map for: (**a**) early Wenlock (Sheinwoodian); and (**b**) late Wenlock (Homerian) (from Bassett *et al.* 1992*a*). Key as for Fig. 4.6.

Wenlock Shale, more properly termed the Coalbrookdale Formation in the Welsh Borderland. Hurst *et al.* (1978) have described the stratigraphy and palaeogeography of the whole Welsh Borderland Wenlock, using the benthic depth-related brachiopod-dominated associations in clastic deposits identified by Calef & Hancock (1974) in a similar way to the Llandovery associations mentioned above. Following a general deepening after the deposition of the Woolhope and Buildwas formations, during the early Wenlock (Sheinwoodian), there was a mid- to late-Wenlock reversal (Fig. 4.1); during the succeeding Homerian Stage, the rocks were deposited under progressively shallower water in the centre of the Welsh Borderland and eastwards round the Midland Platform. The Coalbrookdale Formation is up to 255 m thick in the type Wenlock area (Bassett *et al.* 1975), but the top 10 m contains enough interbedded platey carbonates to be distinguished as a separate Farley Member. Above this there is the famous Much Wenlock Limestone Formation. Because it caps the 40 km long scarp of Wenlock Edge and its southern extension, that formation seems more substantial than its actual thickness there of up to 29 m. However, the Much Wenlock Limestone Formation extends for a considerable distance laterally, from May Hill in the south to where it disappears under the Upper Palaeozoic near Shrewsbury in the north, and from the Church Stretton Fault Zone in the west to at least as far as the Dudley area near Birmingham in the east. There at Dudley it is 62 m thick, but the thickest Much Wenlock Limestone Formation measured is within the N–S-trending Malvern Line, where 152 m is measured at the Abberley Hills, north of the Malverns. Details of the Wenlock successions in the Malvern and Abberley areas are to be found in Phipps & Reeve (1967).

The sedimentology of the Much Wenlock Limestone Formation is well known as consisting of a series of large bioherms, which have a basic framework of stromatoporoids and algae, as well as subsidiary corals. The bioherms can be up to 20 m across and 10 m thick: They are surounded by a variety of carbonate facies, many of which also have a substantial clastic component, usually of fine muds. From these peri-biohermal limestones four marine level-bottom benthic brachiopod-dominated associations have been documented by Hurst (1975): from shallow to deeper shelf, these are the *Sphaerirhynchia wilsoni, Isorthis clivosa, Eoplectodonta duvalii* and *Visbyella trewerna* associations. The distribution of these associations once again reflects a progressive increase in water depth rather than a correlation with the major limestone facies types.

In the Tortworth area, Gloucestershire, in the south of the Welsh Borderland, Benton & Thackeray (1995) have demonstrated that the Brinkmarsh Formation, which consists of 244 m of siltstones, mudstones and calcareous sandstones originally described in his review of that area by Curtis (1972), apparently spans most of Wenlock time. The calcareous sandstones were probably storm-induced deposits, and represent rare events in a relatively protected embayment within the shallow shelf.

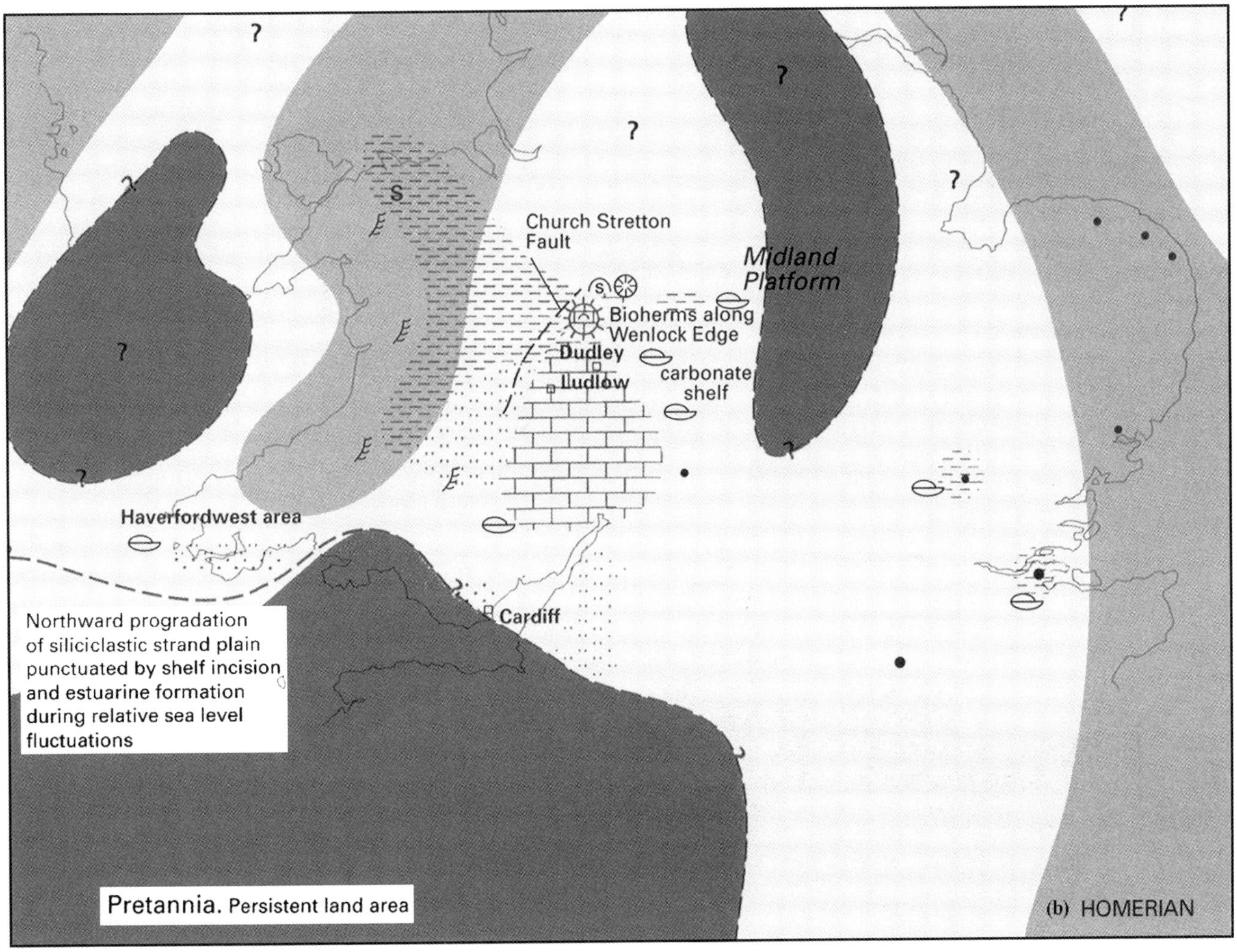

Fig. 4.13. Continued.

South Wales

The late Llandovery transgression led to a significant area of South Wales being dominated by shallow-marine environments during early Wenlock times (Hurst *et al.* 1978; Bassett *et al.* 1992*a*). In the west, proximity to a significant source of siliciclastic detritus inhibited the formation of carbonate shoals and reefs that were well established in the Welsh Borderland. In southern Pembrokeshire, the basal Sheinwoodian is dominated by mudstone-prone open marine siliciclastics of the Coralliferous Group. By the mid to late Sheinwoodian, deposition is represented by the Gray Sandstone Group which describes a northerly progradation of a thick (500 m maximum thickness in the Marloes Peninsula), predominantly shallow-marine wedge across the Skomer Sub-basin (Fig. 4.14). The Gray Sandstone Group is in part diachronous, with Wenlock distal-shelf mudstones of the Coralliferous Group being age equivalent within the Winsle Inlier to the north. To the south, at Freshwater East and West, the section is seen to thin dramatically, resting unconformably upon Llanvirn age rocks. It is probable that the thickening of the section in the Marloes peninsula is due to increased accommodation space developed in association with growth across the Benton and Wenall faults.

Changes in relative sea-level had profound effects on the depositional environment and architecture within the Gray Sandstone Group, with several regressive–transgressive sequences being identified within the Marloes Peninsula.

Falls in relative sea level resulted in subaerial exposure and valley incision of the shelf, with sediment by-pass to the northerly, deeper parts of the Welsh Basin to supply the sandy deep-water turbidite systems such as the Penstrowed Grits. During the subsequent rise in sea-level the valleys were filled with estuarine sediments, deposited in high-sinuosity estuarine channels and tidal flats, that drained the southerly hinterland. These estuarine sediments represent the distal equivalents of the embryonic Old Red Sandstone (Hillier 2000; Hillier & Williams 2004) (Fig. 4.14). They comprise strongly heterolithic sandstone/mudstone bedsets with tidal flat sediments exhibiting desiccation cracks. Such emergence indicators predominate within the youngest (Homerian) estuarine sequence, heralding the onset of true continental deposition and the conformable transition into the overlying Old Red Sandstone Red Cliff Formation (of Late Ludfordian age) within the Marloes Peninsula (Hillier 2000; Hillier & Williams 2004).

Wave/storm-dominated shallow-marine units comprise stacked upward-coarsening and thickening cycles 10–20 m thick. At the base is an erosion surface (ravinement) cut into the underlying valley fill and marked by a lags of bioturbated coarse-grained sandstone, and abundant phosphate nodule development. Sandstones are predominantly well sorted, fine-grained with wave ripples and hummocky cross-stratification. Mudstone units are bioturbated with abundant *Planolites*, *Palaeophycus* and *Chondrites* reflecting the open-marine conditions on the shelf. The coarsening-upward cycles demonstrate northward strand plain progradation of a probable wave/storm-dominated delta which was fed by a ready supply of

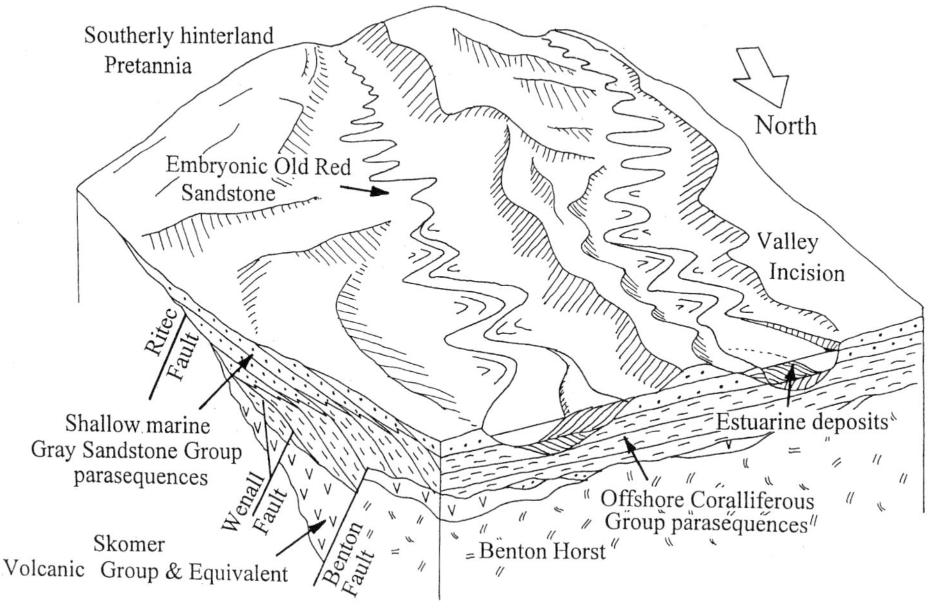

Fig. 4.14. Schematic sequence development and palaeogeography during deposition of the Gray Sandstone Group. Note Skomer Sub-basin delineated to the north by the Benton Horst and the northwards progradation of the Gray Sandstone Group siliciclastic wedge. Falls in relative sea level led to shelf incision and estuarine deposition – the distal equivalent of the early Old Red Sandstone continental deposits (from Hillier 2000).

eroded material from the emergent southerly hinterland Pretannia, or from a westerly source region associated with Avalonian–Laurentian collision.

It is uncertain whether valley incision is related to changing base-levels associated with Avalon–Laurentian collision, fault-block movement associated with crustal extension or eustatic sea-level changes.

To the east, in the Sawdde Gorge, a similar regressive shallow-marine section is preserved (Aldridge *et al.* 2000). The early Sheinwoodian Sawdde Siltstone Formation represents offshore marine mudstones analogous to the Coralliferous Group mudstones. The Sawdde Sandstone and Homerian Ffinant Sandstone formations comprise wave-influenced shallow-marine sandstones similar to those of the Gray Sandstone Group, and it seems likely that a wave-influenced delta prograded across the shelf in the Llandovery area.

In the Cardiff district, the Pen-y-lan Mudstone was deposited in a mid-shelf setting below normal wave-base. It is highly fossiliferous, with common brachiopods, rugose corals, bivalves, cephalopods and trilobites (Bassett 1969; Waters & Lawrence 1987). The overlying Homerian Cae-Castell Formation contains the Rhymney Grit, a high-energy subtidal sandbar with a low diversity fauna, presumably deposited close to the shoreline of Pretannia (Siveter 2000*a*). The remainder of the Cae Castell Formation is sandstone-dominated, being deposited above storm wave base in an inner shelf/shoreface environment (Waters & Lawrence 1987).

Elsewhere on the shelf in the Tortworth Inlier, thin storm-induced sandstones of the Brinkmarsh Formation were deposited possibly within a protected embayment below fair-weather wave-base (Curtis 1972; Cave 1977; Kellaway & Welch 1993; Benton & Thackeray 1995). The background sedimentation includes carbonate mudshales, reflecting the areas distance from the significant siliciclastic source to the west, and proximity to the carbonate-dominated Midland Platform to the north. In the East Mendips Inlier, rare Wenlock-age andesites, tuffs and vent agglomerates comprising the Moon's Hill Volcanics, represent some of the last volcanic activity within the Welsh Basin (Hancock 1982). Interbedded sediments at the base of the succession contain brachiopod faunas, but the lavas and pyroclastics were presumably subaerially deposited.

Mid Wales

Within the broad zone between the Pontesford and Church Stretton lineaments, basinal Wenlock facies pass laterally into sequences transitional with the shelf facies of the Welsh Borderland. Thus, the turbiditic and laminated hemipelagic mudstones of the Nant-ysgollon and Nantglyn mudstones formations give way to shelly, bioturbated silty and calcareous mudstones with local thin impure limestones of the Bromleysmill Shales and Aston Mudstone Formation south of the Shelve area in Shropshire. These last-named formations are lateral equivalents of the Buildwas and Coalbrookdale formations of the Midland Platform. The Builth Mudstones Formation, which outcrops around the south and west flanks of the Builth–Llandrindod Ordovician Inlier, represents a somewhat deeper-water facies of variably calcareous, finely laminated graptolitic mudstones with locally abundant shell detritus and slumped beds (Zalasiewicz & Williams 1999).

In early Wenlock times, muddy and variably calcareous sediments, collectively referred to as the 'Wenlock Shales', were deposited across much of the shelf in east and south Wales (see above). The carbonate element diminishes towards the Welshpool and Corwen areas, closer to the basin edge. These mudstones contain a rich shelly fauna of brachiopods, trilobites and corals, and a sparse graptolite fauna which is more clearly determined at the shelf edge.

Deep-water Wenlock successions

Mid Wales

Deep-water Wenlock facies in the southern Welsh Basin crop out in mid Wales along the Llanderfel Syncline, in a broad tract south of the Berwyn Hills, and between Newtown and Builth Wells. The pattern of deposition during the Wenlock inherited the features of the underlying Llandovery with both slope-apron mudstone facies and southerly-sourced sandstone lobe turbidite systems being recognized. The sequence comprises the fourth of the basinal architectural elements recognized by Davies *et al.* (1997) (see above).

The predominantly oxic, late Telychian, Dolgau Mudstones (= Tarannon Shale Formation) are succeeded by the predominantly anoxic early Sheinwoodian Nant-ysgollon Mudstones, comprising thinly bedded graptolitic hemipelagites and turbidite mudstones (Fig. 4.10). They are viewed in their lower part as a slope-apron facies and in their upper part as a mudstone fringing facies of the succeeding Sheinwoodian, Penstrowed Grits Formation sandstone lobe system (Dimberline & Woodcock 1987). The latter, a local equivalent of the Denbigh Grits of North Wales, comprises an up to 1.1 km-thick sequence dominated by both high matrix and Bouma turbidite sandstones, interbedded with turbiditic and laminated hemipelagic mudstones. In contrast to the earlier, Telychian, southerly-derived turbidite systems of the Welsh Basin, the sandstones are characterized by abundant large mica flakes, leached bioclastic detritus and carbonaceous remains (Davies *et al.* 1997). Intra-basinal and basin-bounding faults of the Tywi and Pontesford lineaments influenced the growth of the system (Fig. 4.10). The regional pattern of westerly, basinwards onlap demonstrates that the initial focus for sand deposition lay along a downwarp situated along a strand of the Pontesford Lineament. The westerly younging top and westerly thickening of the formation suggests that the deposional focus subsequently moved west in the late Sheinwoodian to earliest Homerian. Both the uplifted southerly source areas in south west Pembrokeshire and beyond, and the fault control of the system itself reflect the continuing oblique collision of Avalonia with Laurentia. The lowstand valleys incised into the coeval shelfal sucession (Gray Sandstone Group) in Pembrokeshire probably fed the deltas that sourced the high concentration turbidites of the Penstrowed Grits (Hillier 2000).

As the sandstone lobe facies migrated west with time, anoxic slope apron facies deposition resumed in the east as the Nantglyn Mudstones Formation (Fig. 4.10). Following abandonment of the sandstone lobe system by the early Homerian, Nantglyn Mudstones facies spread across the basin and into North Wales. Although the presence of some slumped units reflects some intrabasinal faulting, the tectonic control of southerly source areas and system architecture had largely ceased.

In the late Homerian, the anoxic bottom conditions of the slope apron were interrupted by the appearance of the Mottled Mudstone, comprising up to two thin units of burrow mottled oxic mudstones with limestone nodules and a fauna of brachiopods, trilobites and bivalves. This event demonstrates that the basin had shallowed to within the colonisation depth range of shelly benthos and represents the acme of the late Homerian eustatic regression.

North Wales

Unlike the underlying Telychian, the early to mid Wenlock succession of North Wales exhibits evidence of the phase of extensional tectonics associated with the oblique collision of Avalonia and Laurentia. Wenlock strata crop out in the northern Welsh Basin principally to the west and south of the Denbigh Moors and in the Clwyd Hills. So called 'ribbon-banded' mudstones (Warren *et al.* 1984), in which structureless turbidite mudstones alternate with laminated hemipelagite mudstones on a 1–2 cm scale, enter in the upper part of the Telychian Pale Slates and, in the east, south of the Bala Fault, make-up the entire, 600 m thick, Wenlock sequence. To the west, north of the Bala Fault, the entrance of the westward and northward thickening Sheinwoodian-age Denbigh Grits Formation, divides this ribbon-banded sequence into the basal Nant-ysgollon Mudstones and the younger Nantglyn Mudstones. The Denbigh Grits were deposited in an east–west sub-basin, the Denbigh Trough, confined to the west by the Conwy Valley Fault and to the SE by the Derwen Horst. Within the formation, packets of thick to thin-bedded, commonly coarse-grained, turbidite sandstones and 'disturbed beds' (slumps and debrites) are subordinate to silt-laminated turbidite mudstones in a sequence ranging up to 1.1 km in thick. The geometry of the group, consistent with palaeocurrent and palaeoslope vectors, suggests accumulation was against an active N–S-orientated Conwy Valley Fault scarp. Movement on the fault triggered frequent westerly sourced turbidity currents and slumps in nearshore sands and muds forming on a narrow shelf to the west and southwest. Clasts of metaquartzite, pelitic schist and acid and intermediate volcanics, associated with heavy mineral assemblages rich in chlorite, zircon, tourmaline, apatite and rutile, suggest that sand and pebble-grade detritus was supplied from emergent tracts of local Precambrian and Ordovician strata (Warren *et al.* 1984). On the south side of the Denbigh Trough, cross-cutting relationships at the base of the formation, associated with a pronounced local thickening and the presence of coarse conglomerates, may mark the site and fill of canyon feature or major slide scar (Cave *et al.* 1992).

Ribbon-banded facies, coeval with the Denbigh Grits, record anoxic mud deposition in parts of the basin beyond the reach of sand laden turbidity currents. As the volume and grade of detritus supplied to the Denbigh Grits apron declined and eventually ceased in the late Sheinwoodian, perhaps in response to declining activity on the Conwy Valley Fault, these finer anoxic facies encroached westwards as the Nantglyn Mudstones. Accumulation of these facies persisted well into the succeeding Ludlow Series, but was interrupted, near the close of Wenlock times, by two closely spaced units of shelly and bioturbated mudstones representing the Mottled Mudstone (Warren *et al.* 1984), a manifestation of the widely recognized Homerian eustatic regression.

The Ludlow and Přídolí

Introduction

The main outcrop of Ludlow rocks extends from the Welsh Borderland into east-central Wales, with inliers farther to the east, south and west. There is another outcrop area in North Wales where the upper part of the succession is missing because of erosion (Warren *et al.* 1984; Figure 4.4, 4.15 & 4.17). The Přídolí outcrop area (Fig. 4.15) flanks Ludlow rocks from the Welsh Borderland into east central Wales; there are also inliers in the West Midlands, and outcrop in SW Wales.

Ludlow shelf facies covered the Midland Platform across the Welsh Borderland and English Midlands into SE Wales. In South Wales the shelf extended across to SW Wales flanking an uplifted southern land source, Pretannia (Fig. 4.16) (Bassett *et al.* 1992*a*). Shelf successions (Fig. 4.17) are continuous but relatively thin in the main outcrop of the Ludlow type area, indicating limited subsidence on the platform (Woodcock *et al.* 1996); in inshore shelf inliers farther east and south, successions are thinner and include localized disconformities.

Global sea-level curves (Johnson *et al.* 1991) show late Wenlock (*ludensis* Biozone) lowstand followed by transgression into the early Ludlow (*nilssoni–scanicus* biozones; Fig. 4.1). In shelf areas of the Midland Platform, rising sea level brought increasing clastic input of mud and silt that killed off carbonate sedimentation across the late Wenlock reef belt. Maximum flooding across the eastern shelf in early Ludlow times was followed by slow regression, recorded by benthic assemblage sequences in the brachiopod dominated shelly faunas (Boucot 1975), through to the onset of non-marine sedimentation (Cherns & Bassett 1999). Within this sequence, a mid-Ludlow shelf carbonate interval ('Aymestry Limestone') was followed by a short-lived episode of non-deposition across inshore shelf areas, shelf edge instability and valley incision in the *leintwardinensis* Biozone (Whitaker 1962, 1994; Cherns 1980, 1988).

West of the fault controlled shelf margin (Church Stretton Fault–Pontesford Lineament), in the basinal areas of east central Wales, early Ludlow transgression saw graptolitic, hemipelagic muds replace late Wenlock lowstand deposits

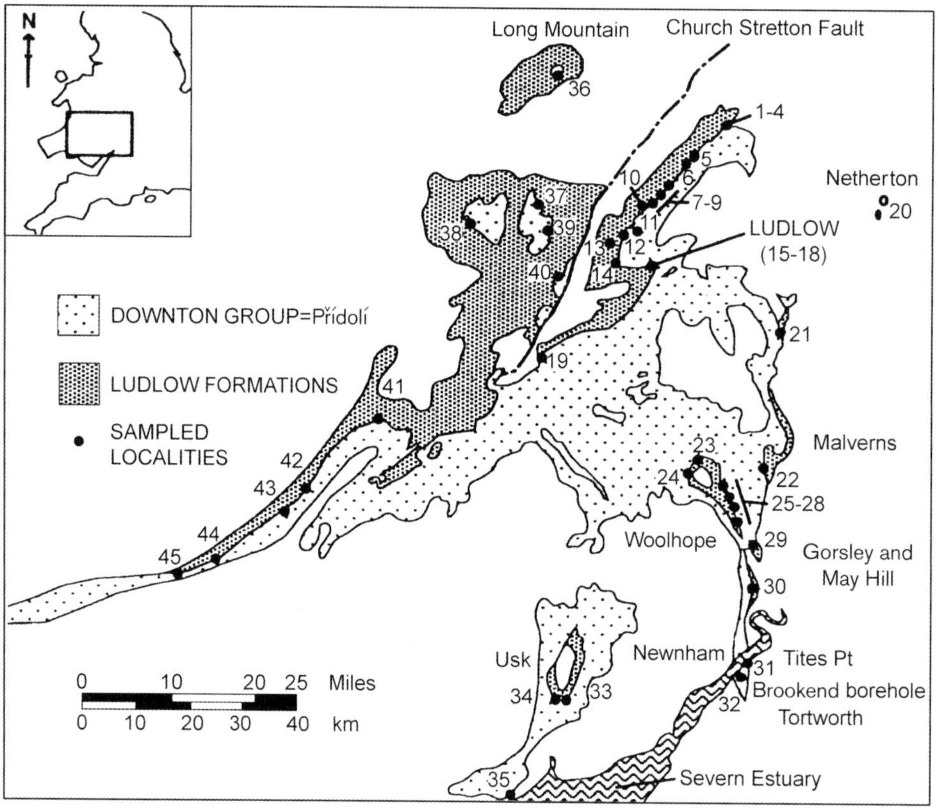

Fig. 4.15. Outcrop map of Ludlow and Přídolí with key localities (after Bassett *et al.* 1982).

(a)

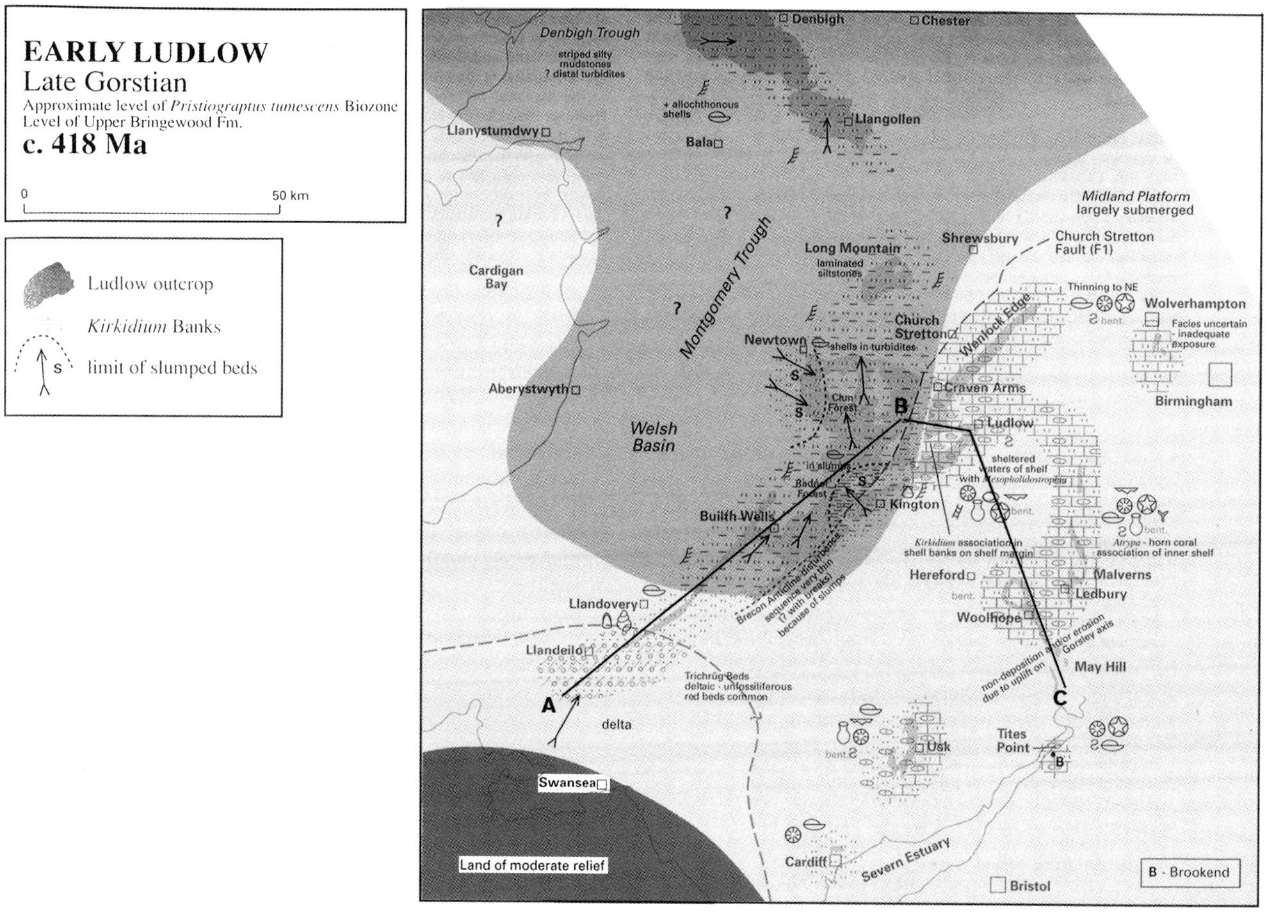

Fig. 4.16. Facies maps for: (**a**) the Early Ludlow (late Gorstian); (**b**) the late Ludlow (early Ludfordian); and (**c**) the earliest Přídolí. Key as for Fig. 4.6 (from Bassett *et al.* 1992*a*).

(b)

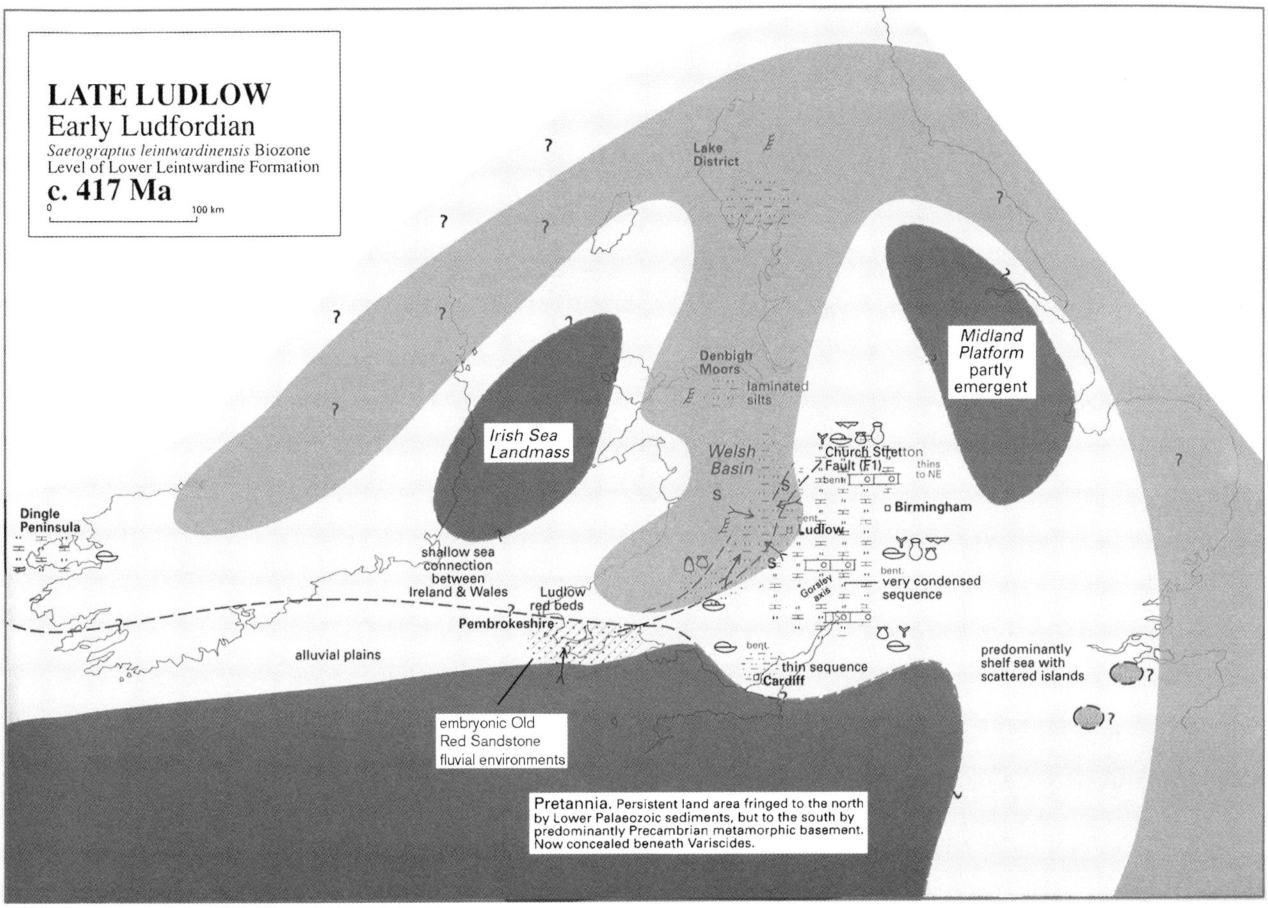

Fig. 4.16. Continued.

represented by the bioturbated and shelly Mottled Mudstone Member of the Nantglyn Mudstones Formation. High rates of subsidence through early Ludlow times are indicated along the eastern part of the basin in eastern mid Wales, locally known as the N–S-orientated Montgomery Trough, by thick successions of southerly derived turbidites, with slump sequences off the shelf margins (Fig. 4.16a). In North Wales (Denbigh–Conwy area), a thick lower Ludlow succession is developed in anoxic mudstones and mainly distal turbidites. High subsidence rates in the E–W-orientated Denbigh Trough have been related to lithospheric flexure and deposition in a foreland basin ahead of the Laurentia–Avalonia collision zone (King 1994). In the Montgomery Trough, after initial Ludlow transgression, there was prolonged regressive shallowing and upward coarsening, from basinal up into proximal shelf facies through to the marginal marine Přídolí. Decreasing accommodation space in the basin as subsidence slowed was followed by tectonic inversion and regional uplift as collision proceeded, and increasing sediment was shed from newly formed uplands into the basin.

The onset of non-marine sedimentation is diachronous, occurring first in southwest Wales by late Wenlock times (Hillier 2000; Hillier & Williams 2004), but more generally in early Přídolí times across central Wales and the Welsh Borderland. This transition in the Welsh Borderland is marked regionally by a bone bed (Ludlow Bone Bed Member). Late Ludlow–Přídolí global sea-level fall (Johnson *et al.* 1991) accompanied the change from marine to non-marine sedimentation (except in SW Wales: Hillier 2000; Hillier & Williams 2004), but those depositional changes took place in an increasingly enclosed depocentre controlled by tectonic inversion, regional uplift and increasing sediment input.

Přídolí rocks extend from shelf areas of the West Midlands, Welsh Borderland and SW Wales across into former basinal areas of central Wales (Knighton, Builth and the Long Mountain). Přídolí facies grade through marine shelf, marginal, alluvial plain and up into fluviatile red beds as the remnant Welsh Basin rapidly silted up. Steepening subsidence curves after the Ludlow, indicated by thicker Přídolí successions, may represent renewed crustal extension in late Přídolí–Devonian times (Woodcock *et al.* 1996). The subcrop across to eastern England indicates southward retreat of the marine basin (Bassett *et al.* 1992*a*).

Shelfal Ludlow and Přídolí successions

Welsh Borderland

Wenlock–Ludlow transgression (Johnson 1996; Loydell 1998) replaced platform carbonate environments shelf-wide by fine siliciclastic facies (Cherns & Bassett 1999). Ludlow shelf faunas are mostly brachiopod-dominated, and include the full spectrum of onshore-offshore benthic assemblages (BA 1–6; Boucot 1975; Fig. 4.17). The full bathymetric range for these associations was interpreted as less than 200 m (Brett *et al.* 1993), which suggests that shelf biofacies changes may represent relatively minor oscillations in sea level.

The Ludlow type area at Ludlow lay in mid-shelf silt environments, which graded westwards into muddier facies in shelf edge regions around Leintwardine and Aymestry (Figs 4.16 & 4.17). The main outcrop belt continues NE into the Wenlock district, and there are inshore shelf inliers from the West Midlands across the Welsh Borderland south to Newnham and to Usk. Successions thicken westwards across the shelf towards

(c)

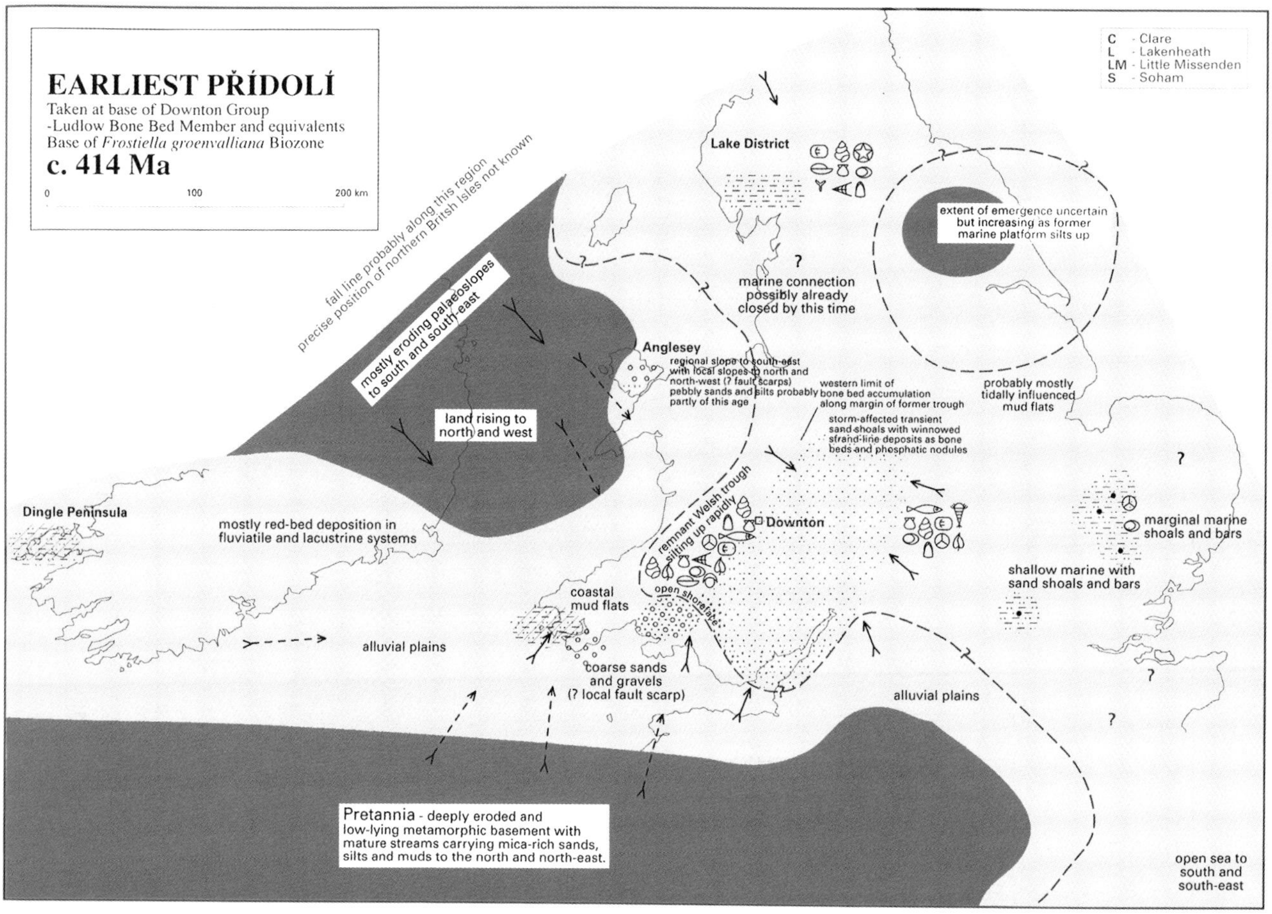

Fig. 4.16. Continued.

the shelf edge (Fig. 4.18; *c.* 345 m at Ludlow–*c.* 575 m at Leintwardine; Whitaker 1962; Holland & Lawson 1963), and then increase rapidly farther west across the fault-controlled shelf margin into the basin (Fig. 4.18). Between inliers at north Woolhope and Usk (both <395 m; Walmsley 1959; Squirrell & Tucker 1960), the succession thins rapidly (south Woolhope *c.* 60 m, Squirrell & Tucker 1960; May Hill 75 m, Lawson 1955) and is interrupted by stratigraphical breaks towards the structural high of the Gorsley axis, where the total Ludlow is reduced to <4 m (Lawson 1954).

The Přídolí rocks of the Downton Group overlie the Ludlow along the main outcrop and in inliers (Fig. 4.15). Marginal marine conditions developed across a wide coastal belt of sandy shoals before transition up into terrestrial coastal plain environments of the Old Red Sandstone facies that accompanied basin inversion. The Přídolí succession is *c.* 300–600 m on the Midland Platform (Clee Hills, Downton Group 420–515 m; Forest of Dean, Long Quarry–Raglan Marl Group 305–610 m). More rapid subsidence continuing in remnant basinal areas is indicated by the thickness of 675 m in central Wales (Allen 1979; Lawson & White 1989).

Ludlow and Přídolí rocks known from the subsurface in eastern and southern England (Cocks *et al.* 1992; Woodcock & Pharaoh 1993) indicate marine environments continuing towards the south and SE (Bassett *et al.* 1992*a*).

Ludlow. The Wenlock–Ludlow boundary in the type Ludlow area (Pitch Coppice; Lawson & White 1989; Siveter 2000*b*) shows a rapid gradation at the top of the Much Wenlock Limestone Formation through to nodular shales and siltstones of the Lower Elton Formation. A mid-shelf, large brachiopod–coral assemblage becomes replaced by a more offshore, small brachiopod assemblage (Cherns & Bassett 1999). The end of carbonate platform sedimentation corresponds to early Ludlow transgression and increasing clastic sediment input. The boundary level correlates with the base of the *nilssoni* Biozone (Lawson & White 1989), and the facies change appears coeval across to the Wenlock district (Bassett 1989*b*) and to shelf inliers (e.g. May Hill; Lawson 1955; Siveter 2000*b*).

Continuing early Gorstian deepening is indicated by mudstone facies of quiet, offshore environments, with graptolites and only scattered shelly faunas (nautiloid cephalopods, trilobites, small brachiopods). Laminated sediments indicating bottom environments inimical to benthic faunas become increasingly common ('laminated shale facies' of Watkins 1979). In the shelf-edge region, typically laminated sediments have graptolites with much sparser shelly faunas. Maximum deepening corresponds to Middle Elton Formation, *scanicus* Biozone times (Cherns & Bassett 1999). Higher up the shelf, in the Wenlock district (Shergold & Shirley 1968) and in shelf inliers (e.g. Malverns, May Hill), mudstone sequences are thinner, more calcareous and shelly, and graptolites are scarce.

Replacement of the mudstones by pervasively bioturbated, shelly calcareous siltstones of the Lower Bringewood Formation, with strophomenid-rich brachiopod assemblages, indicates somewhat shallower, subtidal environments (Lawson 1999). These silty facies are increasingly nodular and calcareous up into limestones of the Upper Bringewood Formation that form a Ludlow shelf carbonate platform ('Aymestry Limestone' facies; Fig. 4.17). The limestone outcrop belt forms a prominent scarp feature to the east of the Much Wenlock Limestone scarp of Wenlock Edge, and from the western Ludlow anticline across towards the shelf-edge at Aymestrey where it thins out rapidly westwards into muddy siltstone facies (Lawson 1973). Massive,

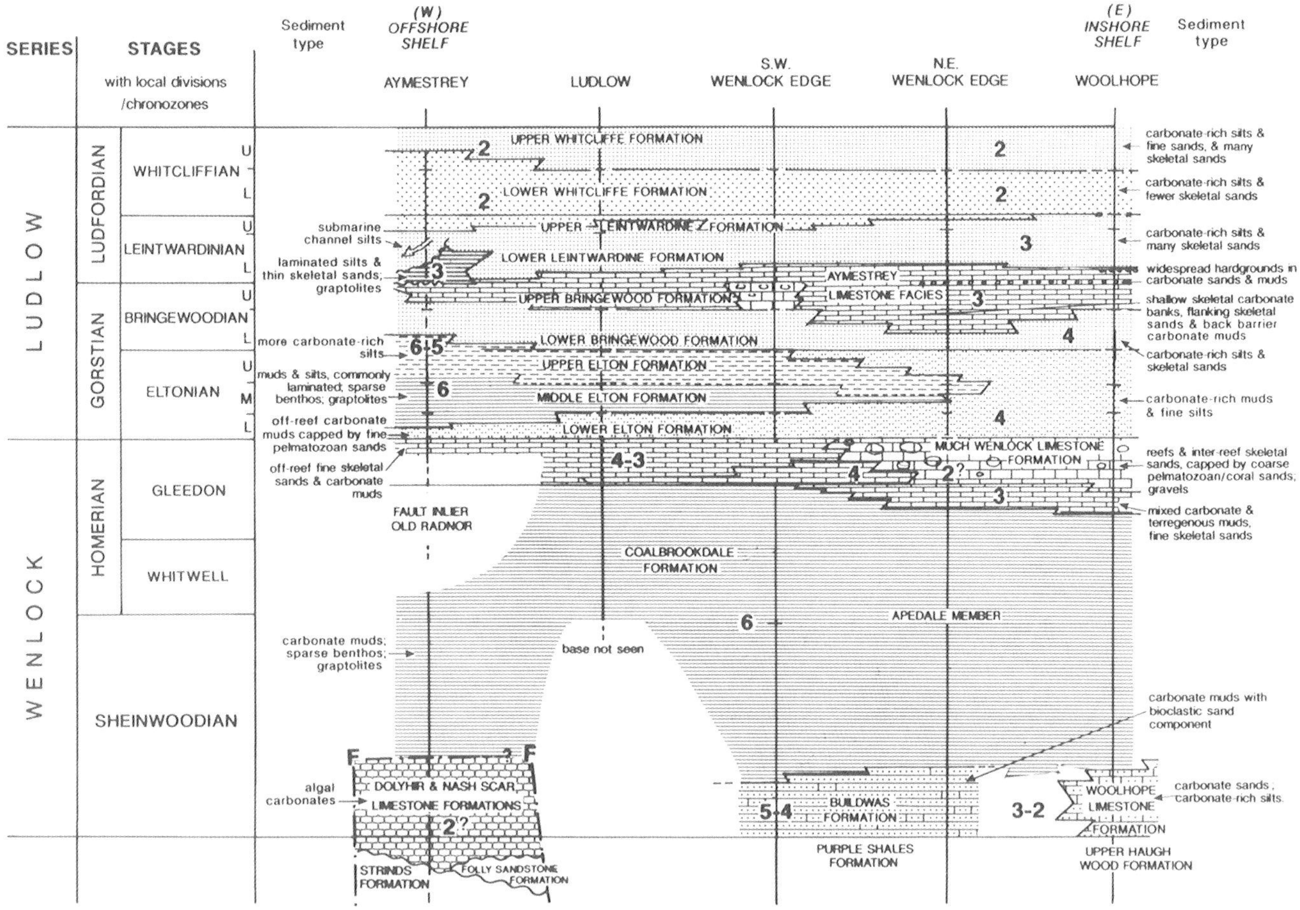

Fig. 4.17. Ludlow shelf communities and facies. Text along each margin denotes original sediment type and fauna. Large bold numbers refer to benthic assemblages: 1 – 2 3 4 5 6 (from Fig. 5.2 of Cherns & Bassett 1999, reproduced with permission of Cambridge University Press, Cambridge).

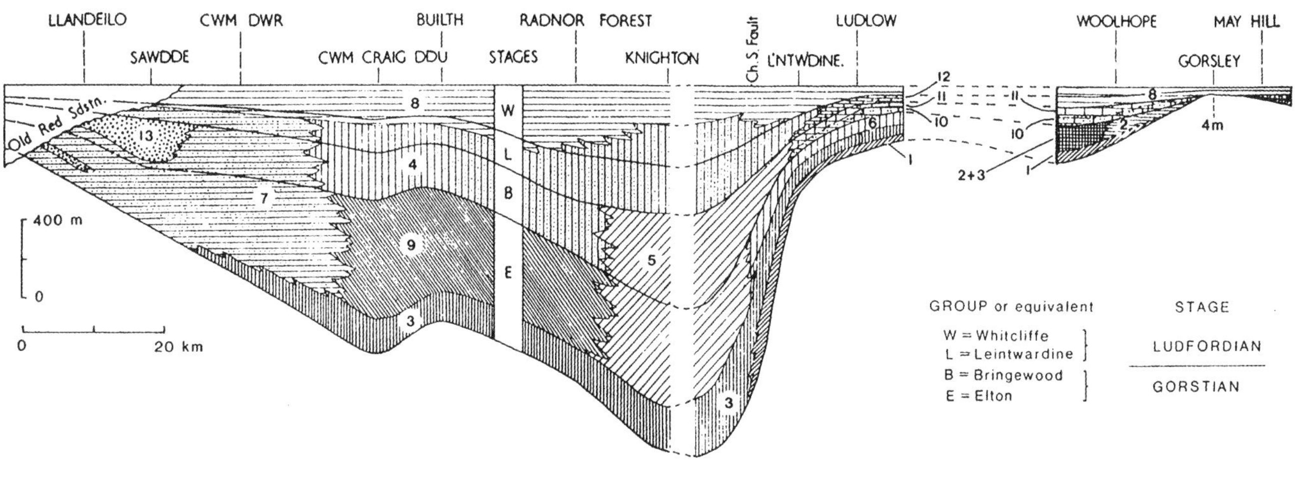

Fig. 4.18. Approximate E–W profile across Wales and the Welsh Borderland, showing thickness changes in the Ludlow succession across the shelf-basin margin. For line of profile (A–B–C) see Fig. 4.16a (from Holland & Lawson 1963, reproduced in Bassett *et al.* 1992*a*, fig. 4).

tabular and cross-bedded limestones include high energy, pentamerid brachiopod (*Kirkidium knightii*) biostromes that contributed to a narrow, shelf-edge wave-resistant carbonate barrier (e.g. View Edge, Bow Bridge; Siveter *et al.* 1989, loc. 3.10; Siveter 2000*b*). This restricted a lower energy, muddy outer-shelf carbonate belt where thick-bedded, nodular, argillaceous and silty limestones have a characteristic *Atrypa*-coral shelly fauna (Watkins 1979; Watkins & Aithie 1980). The basal transition up into carbonate-rich facies appears diachronous, with the Lower Bringewood Formation already in very calcareous siltstones/limestones in NE Wenlock Edge (Shergold & Shirley 1968; Figure 4.17), the Malverns ('calcisiltite facies' of Watkins 1979; see also Phipps & Reeve 1967) and locally at Usk (Walmsley 1959). The Upper Bringewood Formation is cut out by stratigraphical breaks at south Woolhope and May Hill (Bassett *et al.* 1992*a*), while in the Gorsley inlier Ludfordian rocks lie directly on the Wenlock Gorsley Limestone (Lawson 1954; Cocks *et al.* 1992; Siveter 2000*b*).

The top of the Aymestry Limestone facies is also diachronous, extending into the Ludfordian at the Ludlow stratotype (Sunnyhill Quarry; Cherns 1988; Siveter 2000*b*), and also in north Woolhope (Cherns 1988, 1999). The Lower Leintwardine Formation, which correlates with the *leintwardinensis* Biozone in basinal successions, is recognized by changes in the shelly macrofauna. In southern inliers (e.g. May Hill, Usk, Tites Point, Brookend borehole), limestone conglomerates at this level include hardgrounds that represent episodes of sediment starvation and erosion (Cave & White 1968, 1971; Cherns 1980, 1988). A shelf-wide facies transition into calcareous siltstone facies indicates increased clastic sediment input, and accompanies a change to rich brachiopod-dominated assemblages (*Dayia navicula, Isorthis orbicularis* and *Sphaerirhynchia wilsoni*; Cherns 1999) (Fig. 4.17). Shelf circulation increased with the loss of the mid-Ludlow carbonate barrier, but mid-shelf brachiopod assemblages indicate no more than minor transgression across the shelf (Watkins 1979; Cherns & Bassett 1999; cf. Phipps & Reeve 1967).

However, at the shelf edge at Aymestrey, there is a sharp, locally erosional contact at the top of the Upper Bringewood Formation (Aymestry Limestone), overlain by shaly laminated siltstones with *leintwardinensis* Biozone graptolites. The lower siltstones include small brachiopod (*D. navicula*-rich) assemblages in lenses and filling linear scours (Cherns 1988, 1999; Siveter 2000*b*). The abrupt facies change up into fine clastic sediments of distal storm-influenced subtidal environments indicates early Ludfordian deepening in this hinge zone. NE–SW-trending submarine canyon heads (e.g. Mocktree Quarries, Church Hill; Whitaker 1962; Siveter *et al.* 1989, loc. 3.8; Siveter 2000*b*) (Fig. 4.16) incised through the Ludlow succession in the Leintwardine area. The channels were both cut and filled in early Ludfordian times, the extent of downcutting increasing westwards. The shaly and locally slumped siltstones of channel fills include exotic faunas (e.g. eurypterids, phyllocarids, fish, starfish; Whitaker 1962; Siveter 2000*b*). Early Ludfordian tectonic instability in the shelf edge region is confirmed by mass sliding and slumping downslope in the Wigmore Rolls area (Whitaker 1994).

Through the Ludfordian across the shelf, storm influence on typically subtidal sedimentation is evident from shelly coquinas and 'sheet-laminated' beds interspersed among background, well bioturbated sediments that have dispersed shell material (Fig. 4.16). Faunas are derived locally in individual storm sheets, which are very thin (<10 cm) in the inshore shelf inlier sequences but somewhat thicker (10–30 cm) in the mid-shelf at Ludlow (Cherns 1988). Late Ludfordian (Upper Leintwardine–Upper Whitcliffe formations) facies transition in shelf edge areas from shaly laminated siltstones up into thicker bedded, flaggy calcareous and sandy siltstones indicates rapid progradation offshore of shelf facies in response to regression. The Lower and Upper Whitcliffe formations have a characteristic, common but low diversity brachiopod assemblage (*Salopina lunata* Community, Bradfield 1999; *Protochonetes ludloviensis* Association, Watkins 1979) shelf wide, which also spread with the calcareous siltstone facies into east central Wales as the basin shallowed. At Ludlow, an increasing proportion of laminated storm sheets among the bioturbated siltstones accompanied shallowing into more proximal subtidal zones (Watkins 1979; Bassett *et al.* 1992*a*). In several of the southern shelf inliers (e.g. Gorsley, May Hill, Tites Point; Lawson 1954, 1955; Cave & White 1971), and at Usk (Squirrel & Downing 1969), reduced sedimentation, with periodic shallowing into marginal marine conditions, is evident from late Ludfordian phosphatized pebble and bone beds that contain fish scales and denticles.

Přídolí. The Ludlow Bone Bed Member at Ludlow defines the base of the Downton Group (Downton Castle Sandstone Formation) and of the Přídolí of Wales and the borders (Ludford Corner; Bassett *et al.* 1982; Cocks *et al.* 1992; Lane 2000). It marks the transition from uppermost Ludlow marine faunas into generally limited, reduced diversity fossil assemblages of marginal marine environments, and represents a brief transgressive episode within the overall regression when lag concentrates of vertebrate material formed on tidal strands (Bassett *et al.* 1992*a*) (Fig. 4.16c). Thin lenticular bone beds (0.2 m) that represent intertidal skeletal lags comprise mostly fish scales and denticles, but also include eurypterids, the earliest terrestrial arthropods and early land plants (Lane 2000). The overlying 2 m of mudstones (Platyschisma Shale Member) up into approximately 15 m of dominantly yellowish, fine sandstones (Sandstone Member) were laid down in fluctuating shoreface–shallow-subtidal environments, with storm reworking (hummocky cross-stratification) and shallow channelling (Bassett *et al.* 1982; Smith & Ainsworth 1989; White & Lawson 1989). Thin bone beds and phosphatized pebble beds occur widely across the shelf inliers at the base of the Přídolí, e.g. Netherton, Turner's Hill, Abberleys, Gorsley, May Hill (Bassett *et al.* 1982; Lane 2000; Siveter 2000*b*). Overlying siltstone/sandstone successions are thin (e.g. Perton, Siveter 2000*b*) and contain limited shelly faunas (lingulides, molluscs, ostracodes) which, as at Ludlow, indicate periodic marine influence through the lower beds, together with freshwater fish debris, early land plants and spores (e.g. *Cooksonia*; Edwards *et al.* 1979).

The Ludlow Bone Bed in the Leintwardine area is overlain by a mudstone up into yellowish, fine sandstone succession (14 m) similar to Ludlow, but with a local gastropod-fish horizon at the top of the mudstones ('Downton Bone Bed', Whitaker 1962). The Ludlow Bone Bed extends west of the shelf edge fault zone into east central Wales around Kington (Holland & Williams 1985) and the Long Mountain (Lower Wallop Quarry; Miller 1995; Lane 2000). In the latter area, it occurs within grey, micaceous sandstones and siltstones (Causemountain Formation) that have palynofacies and limited skeletal faunas indicating very shallow, marine to restricted environments (Richardson & Rasul 1990; Lane 2000). Further into the remnant marine Welsh Basin, shallowing across the Ludlow–Přídolí boundary without a bone bed corresponds to faunal change to limited, mollusc–lingulid–ostracode faunas of intertidal–shallow-subtidal environments (Bassett *et al.* 1982; Bradfield 1999). The lower Přídolí succession represents a wide shoreface area of sandy shoals across a broad shelf affected by both terrestrial and marine influences, into which sandy sediment was brought from several source areas. Those include local highs (e.g., Gornal Sandstone in the West Midlands derived from the east) and more extensive land areas of the Irish Sea Platform and Pretannia, which lay to the west and south (Allen 1985) (Fig. 4.16c).

In the Downton area, the higher Přídolí comprises the Temeside Shale and Ledbury formations (Lawson & White 1989). Green-brown mudstones and siltstones with pedogenic horizons (Temeside Shales Formation, <27 m) contain few fossils, but a local vertebrate lag (Temeside Bone Bed, Antia 1981) and occasional skeletal horizons indicate periodic marine influence in intertidal–exposed mud-flat environments. The Ledbury Formation (330–460 m) comprises dominantly thick red mudstone sequences with pedogenic horizons, subordinate micaceous red and green sandstones and occasional intraformational conglomerates. Bone beds are absent through the upper Přídolí, but rare skeletal horizons (lingulide brachiopods, molluscs) indicate temporary marine incursions continuing to flood across wide areas of coastal mudflats. Drifted plant debris and vertebrates become more common. Rivers brought in sandy sediment from newly uplifted, distant uplands, while ephemeral streams drained more locally (Allen 1979, 1985). Widespread pedogenic horizons (calcretes) formed stable surfaces starved of sediment input for long periods, in conditions of low seasonal rainfall and hot climates (Allen 1985; Bassett *et al.* 1992*a*).

Air fall tuffs of the Townsend Tuff Bed (Allen & Williams 1981) form a useful mid-Přídolí marker horizon, enabling correlation across the Welsh Borderland and southern shelf. The Přídolí–Devonian boundary is not easy to define in non-marine Old Red Sandstone facies, but vertebrate and spore correlations suggest it should be placed in the upper Ledbury Formation, approximately at the Downton–Ditton boundary (Lawson & White 1989; Bluck *et al.* 1992). At this level, red sandstones with calcretes replace the red mudstones, and the prominent Bishops Frome Limestone calcrete forms a useful marker horizon in the Welsh Borderland.

The subsurface Ludlow and Přídolí successions across eastern and southern England provide limited evidence of marine environments continuing south. Below East Anglia, bioturbated mudstones represent outer-shelf, possibly distal turbidite, environments (Woodcock & Pharaoh 1993). Westward shallowing during the Přídolí is inferred from more westerly boreholes with storm-influenced sandy shelf facies that shallow up into shoal environments with subtidal sand bars. The early–mid-Přídolí marginal sandy facies passes southwards into more open-marine facies with a brachiopod–mollusc fauna (Little Missenden borehole; Straw 1933; Bassett *et al.* 1992*a*).

South Wales

Ludlow. Across the southern part of the shelf, a diachronous transition from marine to continental deposition began during the Ludlow (Allen 1985). This shoaling was in response to a reduction in accommodation space associated with Avalonian–Laurentian collision, and an oversupply of sediment from the south and west. In southern Pembrokeshire, there is a conformable transition from estuarine deposits of the Gray Sandstone Group into the overlying Ludlow Red Cliff Formation, the oldest Old Red Sandstone facies preserved within Wales and the borders (Fig. 4.14). The Red Cliff Formation comprises a mudstone-rich interval with subordinate fluvial fine-grained, sheet-like sandstones (Allen & Williams 1978; Hillier 2000; Hillier & Williams 2004). At the base of the unit are thin, laterally-accreted bedsets that possibly share a genetic origin with the underlying estuarine channels of the Gray Sandstone Group. It is uncertain as to whether the Red Cliff Formation was deposited within a valley fill, or on a low-relief plain formed after the infill of initial valley topography. Renewed extension-related tectonism in southern Pembrokeshire led to the deposition of petrographically distinctive, multi-storey pebbly sandstones of the Albion Sands Formation, deposited within the hanging-wall valley of the active E–W-trending Wenall Fault (Allen & Williams 1978; Hillier 2000; Hillier & Williams 2004). Debris-flow-dominated fans shed from both the hanging wall and footwall of the Wenall Fault deposited the Lindsway Bay Formation, an exotic-clast conglomerate dominated unit sourced predominantly from the south and west.

To the east in Carmarthenshire, in the Sawdde Gorge, a basal Ludlow transgression is preserved on the southern flank of the Tywi Anticline with siltstones of the Tresglen Formation that yield a *Dicoelosia* Community, after which an upward shoaling, northward prograding delta deposited sandstones and conglomerates of the Black Cock Formation across a muddy shelf (Potter & Price 1965, Siveter 2000*b*). Fluvial red beds are common within the Late Gorstian Trichrûg Formation, although inarticulate brachiopods and gastropods indicate marine influences (Siveter *et al.* 1989). Links between this section and those in Pembrokeshire are uncertain. The Trichrûg Formation delta was itself transgressed during the Ludfordian by thin siltstones and flaggy sandstones of the shallow-marine Upper Cwm Clŷd and Lower Roman Camp Formations prior to the establishment of continental conditions during the Přídolí.

To the SE, marine conditions persisted at Cardiff, where shelly siltstones predominate in the Cardiff Group (Waters & Lawrence 1987). Condensation is evident with the Ty Mawr Ironstone, an oolitic ferruginous limestone 9 m above the base of the Hill Gardens Formation. The base of the latter records a transgression with a rich marine fauna similar to that of the underlying Cae Castell Formation (Siveter 2000*b*). The Llanederyn Formation records a gradual regression heralding the onset of continental conditions during the Přídolían Raglan Mudstone Formation (Waters & Lawrence 1987). Ludlow age sediments are exposed in the northern part of the Tortworth Inlier at Tites Point (Cave & White 1971; Siveter 2000*b*). These are dominated by shelfal mudstones and silty limestones with minor, thin sandstone beds. Faunas include brachiopods, bivalves, nautiloids, conodonts and fish denticles. A fish horizon within channelled sandstones of the basal Downton Group is taken to represent the Ludlow Bone Bed. This succession is succeeded in an obscured but assumed conformable passage into the Přídolían Thornbury Beds of Old Red Sandstone facies (Cocks *et al.* 1992).

Přídolí. In Pembrokeshire, Old Red Sandstone deposition is predominantly mudrock-dominated, comprising pedified mudflats, reworked soil peds and thin fine-grained sheet sandstones of the Sandy Haven Formation. Sediment was being sourced from a distant Laurentian provenance by Ludlow times (Sherlock *et al.* 2002). Important regionally correlatable air-fall tuffs are preserved within this interval.

In the Llandovery area, marine influences persisted during the Přídolí. The Long Quarry Sandstone Formation (formerly Tilestones) unconformably overlies Ludlow-aged sediments in the area, presumably reflecting movement along the Tywi Lineament. The Long Quarry Sandstone Formation comprises micaceous, fine-grained sandstones with hummocky cross-stratification and decalcified brachiopods, bivalves and gastropods (Potter & Price 1965; Almond *et al.* 1993). Also preserved as coalified impressions are important early vascular land plants, including *Cooksonia* sp. and *Steganotheca striata* (Edwards 1970). The topmost Long Quarry Sandstone Formation is dominated by current- and wave-rippled heterolithics believed to have been deposited as subtidal–intertidal shoals, prior to the conformable transition into continental green and red mudrocks of the Gwynfe Formation. The latter have vertisol pedification. Thin sandbodies within this unit are interpreted as tidally-influenced, low-sinuosity channels deposited in the lower reaches of a coastal plain.

Mid Wales (basin–shelf transition – the Montgomery Trough)

Ludlow and Přídolí. Developments within the basin that began during late Llandovery times continued through the Wenlock and into the early Ludlow, with a progressive eastward shift of deep-water sedimentation. The eastern margin of the Welsh Basin was relatively well defined, being located along the Church Stretton Lineament. After a period of rising sea level during the early Ludlow a prolonged period of regression set in, culminating in the shallow marine and fluviatile red bed facies of the Přídolí. The thick sequences of these rocks which accumulated in east central Wales and the Welsh Borders imply high rates of subsidence and have been related to the development of an Acadian foreland basin. The latter formed as a consequence of Laurentian and Avalonian plate collision, southward nappe emplacement and downloading of the Avalonian crust (King 1994).

Lowermost Ludlow rocks are generally referred to as the 'Lower Ludlow Graptolitic Shales' (Cocks *et al.* 1992), but include thinly interbedded, ribbon-banded turbidite mudstones and laminated hemipelagic mudstones representative of the Nantglyn Mudstones Formation and, further east in the Welsh Borderland area, the graptolitic, shaly hemipelagic mudstones of the Oakeley Mynd Shale Formation (Cocks *et al.* 1992). The bulk of the overlying Ludlow succession in the basin is represented by the Bailey Hill Formation (Tyler & Woodcock 1987), a sequence variously interpreted as turbiditic (Cummins

1959; Bailey 1969; Woodcock 1976) or storm-generated distal shelf deposits (Tyler & Woodcock 1987). It comprises up to 1.3 km of thickly bedded, fine-grained, shelly sandstones and silty mudstones with subordinate, silty, laminated hemipelagic mudstones. Large parts of the formation are slumped, the slumped units commonly being associated with bentonites. The Bailey Hill Formation is gradationally overlain by the Knucklas Castle Formation (Holland 1959), a sequence of up to 500 m of thinly laminated, bioturbated siltstones and subordinate thin sandstones, in part, of early Ludfordian *leintwardinensis* Biozone age. It differs from the underlying formation in its greater degree of bioturbation, which in turn reflects increased levels of oxicity within the basin at the time. It passes upwards gradually into the Cefn Einion Formation (Cave & Hains 2001), equivalent to the former Wern Quarry Beds and Llan-wen Hill Beds of the Knighton area. The Cefn Einion Formation is 200–300 m thick and comprises thin to thick-bedded, hummocky cross-stratified sandstones and bioturbated silty mudstones, deposited in a distal shelf setting influenced by storm activity. The overlying Clun Forest Formation (Cave & Hains 2001) is a lithologically heterogeneous deposit of interbedded grey, green and reddish brown silty mudstones and massive to laminated micaceous sandstones with variable amounts of carbonaceous material (previously the Platyschisma Beds through to the Red Downton succession). It represents transitional facies from the fully marine sequences of the Cefn Einion Formation to brackish water and alluvial plain deposits that characterise the Přídolí of the Welsh Borderland.

Deep water Ludlow successions

North Wales

Ludlow rocks occupying the northern Welsh Basin (Denbigh Trough) underlie much of the Denbigh Moors and the Clwyd Hills and occupy the core of the Llangollen Syncline (Warren *et al.* 1984; Davies *et al.* 2005). Throughout these areas the base of the series falls in the upper part of the Nantglyn Mudstones. The base of the succeeding Gorstian Elwy Formation marked a major resurgence in the supply of westerly-derived silt-laminated turbidite mudstones and disturbed beds to the basin (Fig. 4.16). A sequence in excess of 1.75 km in thickness accumulated in the west, but a little over 600 m is present in the Llangollen Syncline and just 450 m in the Clwyd Hills, a lateral variation which suggest that the Conwy Valley Fault was again influential. Packets of turbidite sandstones are less common, and the individual turbidites generally fine-grained and of more distal aspect than in the otherwise comparable Wenlock Denbigh Grits (Warren *et al.* 1984). In the Llangollen area the Elwy Formation is overlain by the more mudstone-prone Dinas Bran Beds, characterized by a relatively shallow-water fauna. The unit is thought to range into the late Ludfordian (Siveter 2000*b*). No rocks higher than this are preserved in North Wales.

The thick mid Ludlow sequence in the Denbigh Trough is thought to reflect deposition in a south easterly migrating foreland basin, generated by the depression of the lithosphere by the overriding Laurentian continent or the accretionary prism at its leading edge (King 1994).

5 The Lakesman Terrane: the Lower Palaeozoic record of the deep marine Lakesman Basin, a volcanic arc and foreland basin

R. P. BARNES, M. J. BRANNEY, P. STONE & N. H. WOODCOCK

The Lake District, Isle of Man and Leinster basins lie within the Leinster–Lakesman Terrane, which is situated between the Monian Terrane and the Iapetus Suture (Fig. 5.2). It is sometimes convenient to refer to the Isle of Man and Lake District pre-Caradoc basins together as the Lakesman Basin because they have a similar early Ordovician history characterized by a thick accumulation of mudrocks and turbidites, although the basins may have been separate entities.

Following the early phase of basin evolution there was an extensive development of volcanism in the Lake District. The mildly alkali tholeitic composition of the volcanic rocks of the Eycott Group might reflect an immature volcanic arc built on a thin outer edge of the continent. The Borrowdale Volcanic Group records the climactic phase of arc volcanism.

After the end of most, but not all, volcanic activity the Lake District Basin subsided, accommodating a thick sequence of clastic sediments in what is interpreted as a foreland basin, which filled and shallowed in its final stages. The Lakesman terrane was subsequently deformed during the Acadian Orogeny (Chapter 4).

The Tremadoc to Llanvirn successions of the Lakesman Terrane

The Manx (Isle of Man), Skiddaw (English Lake District) and Ingleton (Craven district, NW England) groups crop out immediately to the south of the inferred position of the early Palaeozoic Iapetus Ocean (Fig. 5.1). Although composed of broadly similar, mud-rich turbidite facies, they are different in detail, suggesting deposition in localized sedimentary systems. The Manx and Skiddaw groups (Fig. 5.2) were deposited contemporaneously during the early to mid-Ordovician in relatively close proximity on the northern margin of Avalonia. The Ingleton Group has a very different sandstone composition and may be older than the other two groups, to which its relationship is unclear.

During the early Ordovician, the Avalonian margin of the Iapetus Ocean lay at a latitude of around 60°S (Scotese & McKerrow 1990, 1991) within the circum-polar 'Atlantic' faunal province characterized by an abundant, but low-diversity, graptolite fauna (Skevington 1974; Cooper *et al.* 1995). The limited occurrences of early Ordovician trilobites in the Skiddaw Group are also indicative of a cool, peri-Gondwanan marine environment (Fortey *et al.* 1989). The depositional basins lay close to the triple junction of the Iapetus Ocean (between Laurentia and Gondwana) with the Tornquist Sea (between Baltica and Gondwana) (Fig. 5.1). Various dynamic configurations of the three continental plates and their intervening oceans have been suggested (e.g. Soper & Hutton 1984; Hutton 1987; Soper *et al.* 1987; Pickering & Smith 1995). Most of these palaeogeographic and tectonic reconstructions have emphasized the role of the Iapetus Ocean and subduction at its margins (McKerrow *et al.* 1991 and references therein; Soper *et al.* 1992*a*, *b*). However, some aspects of Lake District geology seem more comparable with the concealed Caledonides of eastern England and the Brabant Massif, which relate more to the evolution of the Tornquist Sea (Pharaoh *et al.* 1991). The likelihood of a marine, and possibly tectonic, continuity between the two regions during the early Palaeozoic rather reduces the significance of the distinction.

Manx Group

Lower Palaeozoic metasedimentary rocks crop out over much of the onshore and nearshore area of the Isle of Man (Fig. 5.3). The succession is essentially steeply dipping to vertical and NNE- to ENE-striking, such that outcrops of the different formations lie along the axis of the island (Fig. 5.3). First described in considerable detail by Lamplugh (1903), the 'Manx Slate' was, until recently, thought to be entirely of Ordovician (Arenig) age (e.g. Cooper *et al.* 1995). Howe (1999) showed that part of the succession is Silurian (probably Wenlock) in age and hence it was separated as the 'Dalby Group' by Morris *et al.* (1999). The 'Manx Group' was correspondingly redefined by Woodcock *et al.* (1999*b*).

The Manx Group is dominated by laminated mudstone and siltstone but varies from mudstone to units composed largely of thin- to medium-bedded, typically fine-grained sandstone. Debrite (pebbly mudstone) composed of intrabasinal material

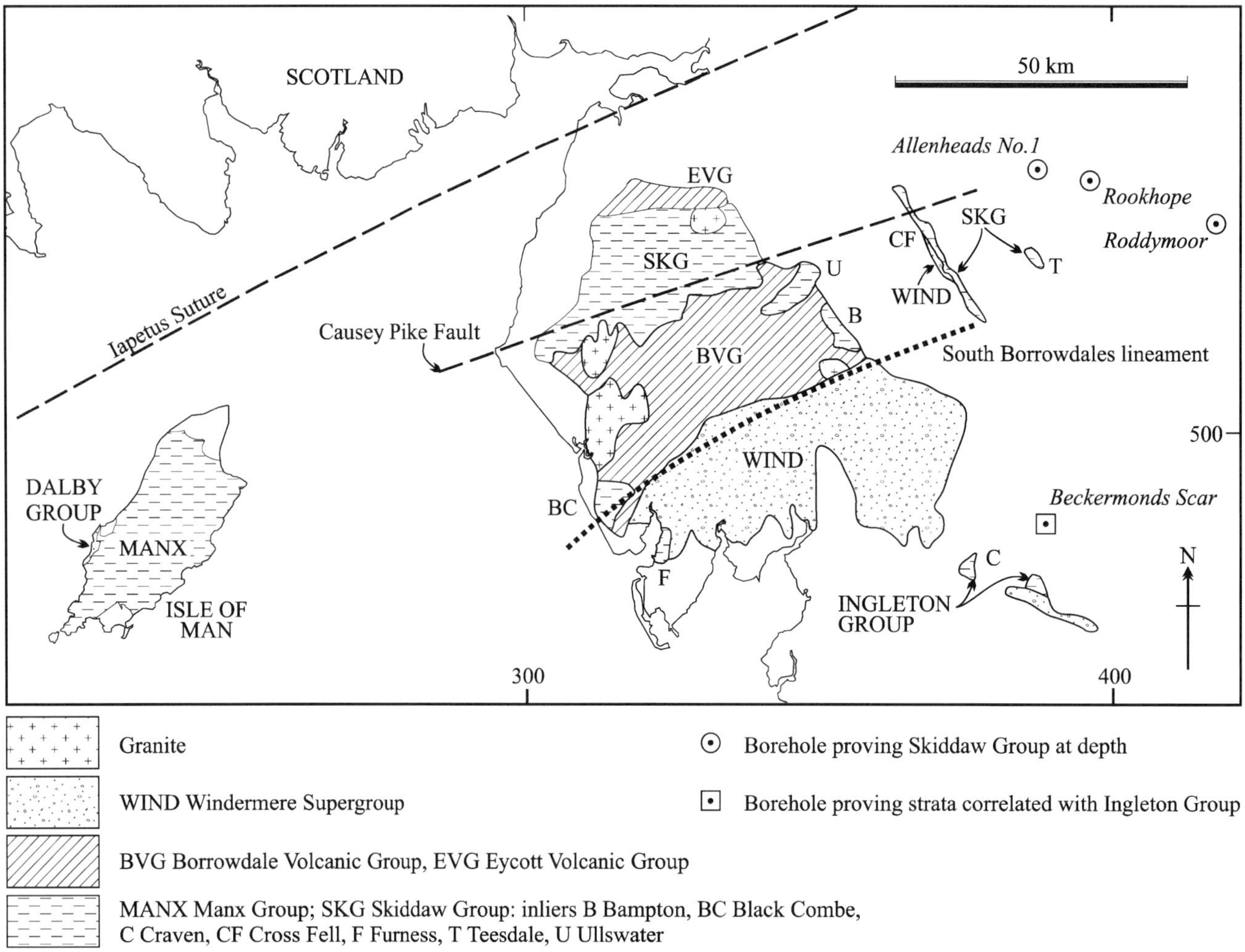

Fig. 5.1. The distribution and outline geology of the Lower Palaeozoic outcrops of northern England and the Isle of Man.

occurs in parts of the succession and small amounts of volcanic rocks occur at two localities. Minor intrusions are abundant, many pre-dating or contemporaneous with deformation of the succession in late Silurian or early Devonian. Three larger granitic intrusions range from syn- to post-tectonic and may represent a much larger granitic body at depth (Kimbell in Chadwick *et al.* 2001).

The stratigraphy and order of succession of the Manx Group has long been a matter of debate. From recent work (Woodcock *et al.* 1999*b*; Barnes *et al.* in Chadwick *et al.* 2001; Burnett & Quirk 2001) the rocks are considered to crop out in NW-younging succession across most of the island (Fig. 5.3), although they are folded and disrupted by strike-parallel faults which locally repeat parts of the succession. The oldest and youngest parts of the sequence are structurally repeated in the north of the outcrop. It is difficult to estimate the thickness of the individual formations, but most are likely to be of the order of several hundreds of metres to 1 km. The Lonan Formation is thicker, probably in excess of 2.5 km. The total thickness of the exposed succession is likely to be in excess of 5 km.

Late Tremadoc–early Arenig turbidite facies

The oldest rocks preserved are the Glen Dhoo Formation in the north, and the Lonan Formation, which crops out over much of the southern and eastern part of the Isle of Man (Figs 5.2 & 5.3). Both have produced several late Tremadoc–early Arenig acritarch faunas (Molyneux 2001), suggesting that they are of equivalent age. They are dominated by mudstone with a variable proportion of siltstone to very fine-grained sandstone in laminae to very thin beds, the latter commonly being planar- or cross-laminated. Locally, fine-grained immature sandstone in thin to very thick (up to 2 m) beds is dominant within the Glen Dhoo Formation which also includes minor occurrences of andesitic volcanic breccia. Sandstone-dominant members are better exposed in the Lonan Formation where they make up much of the SE coast of the Isle of Man and include two distinct sandstone types.

The Keristal Member (Lonan Formation) is only a few metres thick and lenticular on a scale of 100 m, but is laterally persistent over at least 15 km. It comprises medium to very thick beds of distinctive pale, fine-grained quartz arenite. The beds are ungraded or weakly graded, commonly massive or with parallel-lamination, but in some cases with ripple cross-lamination in thinner beds. They may be organized in either thinning-up or thickening-up sequences. In thinning-up sequences, the basal bed is commonly strongly erosional into underlying thin-bedded turbidites.

The Santon Member (>600 m thick) and the along-strike, but slightly younger, Ny Garvain Member (Lonan Formation) are characterized by thin- to medium-bedded, locally thick-bedded, light grey or greenish grey fine-grained wacke intercalated with a laminated or thin-bedded facies more typical of the Lonan Formation generally. Sandstone beds grade into the intervening mudstone. Parallel- and ripple cross-lamination is common and some bed bases preserve flute marks and horizontal burrows. Locally, isolated beds or packets of medium to thick beds of quartz arenite punctuate the less mature wacke successions.

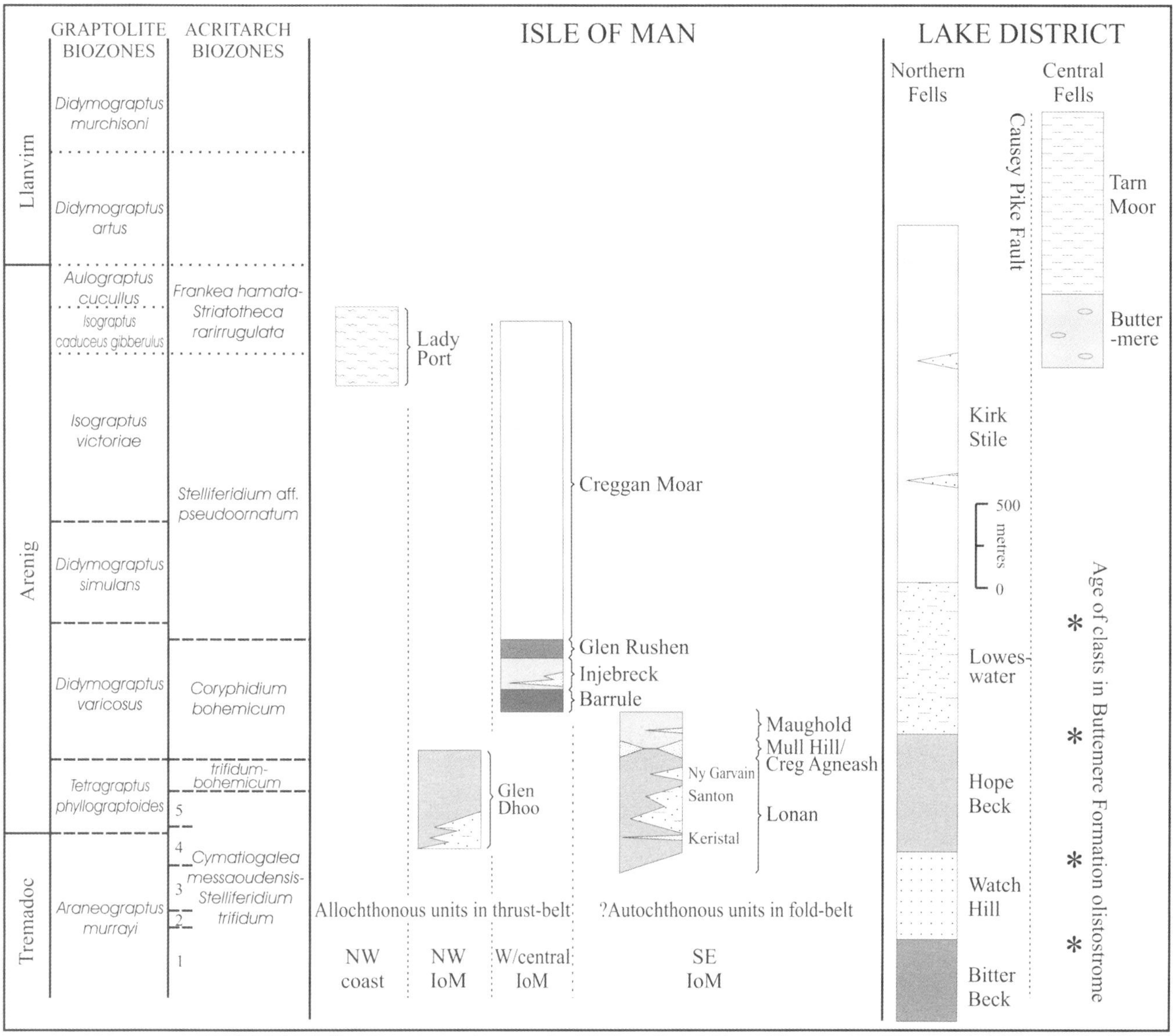

Fig. 5.2. Biostratigraphical correlation of the component formations of the Manx and Skiddaw groups. Ornamentation and shading relate the formations to their outcrop on the geological maps shown in Figs 5.3 and 5. 4.

The laminated to very thinly bedded facies are interpreted by Woodcock & Barnes (1999) as deposits from low-concentration turbidity flows in an oxygenated marine environment, with periodic phases of high- to medium-concentration turbidity flows forming the wacke-dominated units. Palaeoflow directions from sole marks are consistently W-directed. A 50 m-thick, unusually thick-bedded, relatively coarse-grained sandstone succession in the Santon Member at Purt Veg [SC 3255 7037] may represent a trunk distributary channel. The Keristal Member represents a short-lived system of sand-rich gravity flows which appears to have tapped a source of relatively clean quartz sand (Woodcock & Barnes 1999), also apparent in the sporadic quartz arenite beds in the Santon Formation.

The base of the Lonan Formation does not occur at outcrop, but at least 2.5 km of strata are exposed. The top of the Lonan Formation is defined by the depositional transition into the Creg Agneash Formation in the NE, and the Mull Hill Formation in the south (Fig. 5.3). It is more subtle in the centre of the island where the Lonan Formation passes directly up into the lithologically similar Maughold Formation.

The high- and medium-concentration quartz arenite turbidity flows represented only locally in the Lonan Formation become dominant in the overlying Creg Agneash and Mull Hill formations. Both are characterized by white–light grey quartz arenite, ranging from thin- to very thick-bedded with a variable amount of interbedded mudstone. The sandstone beds are massive or weakly graded from medium or fine sand to very fine sand or silt. A few beds have coarse- to very-coarse-grained sandstone bases with mudstone rip-up clasts. Internal structure is typically a weakly defined, upward-thinning parallel lamination, in some cases with a thin ripple cross-laminated division at the top. Slump folding on a scale of tens of centimetres to metres is commonly developed, with close to tight disharmonic folds of widely varying orientation. The top of the Mull Hill Formation is not seen due to faulting. In the Creg Agneash Formation the background of low-concentration turbidite facies becomes more mudstone-dominated upwards, passing into the Maughold Formation where the sandstone beds die out.

The age of the Lonan Formation beneath the Santon Member is constrained by acritarch floras from several localities (Molyneux 1999, 2001). These include *Stelliferidium trifidum?*, indicative of the late Tremadoc–early Arenig interval. Acritarchs from the Santon Member itself established correlation with the earliest Arenig *Tetragraptus phyllograptoides*

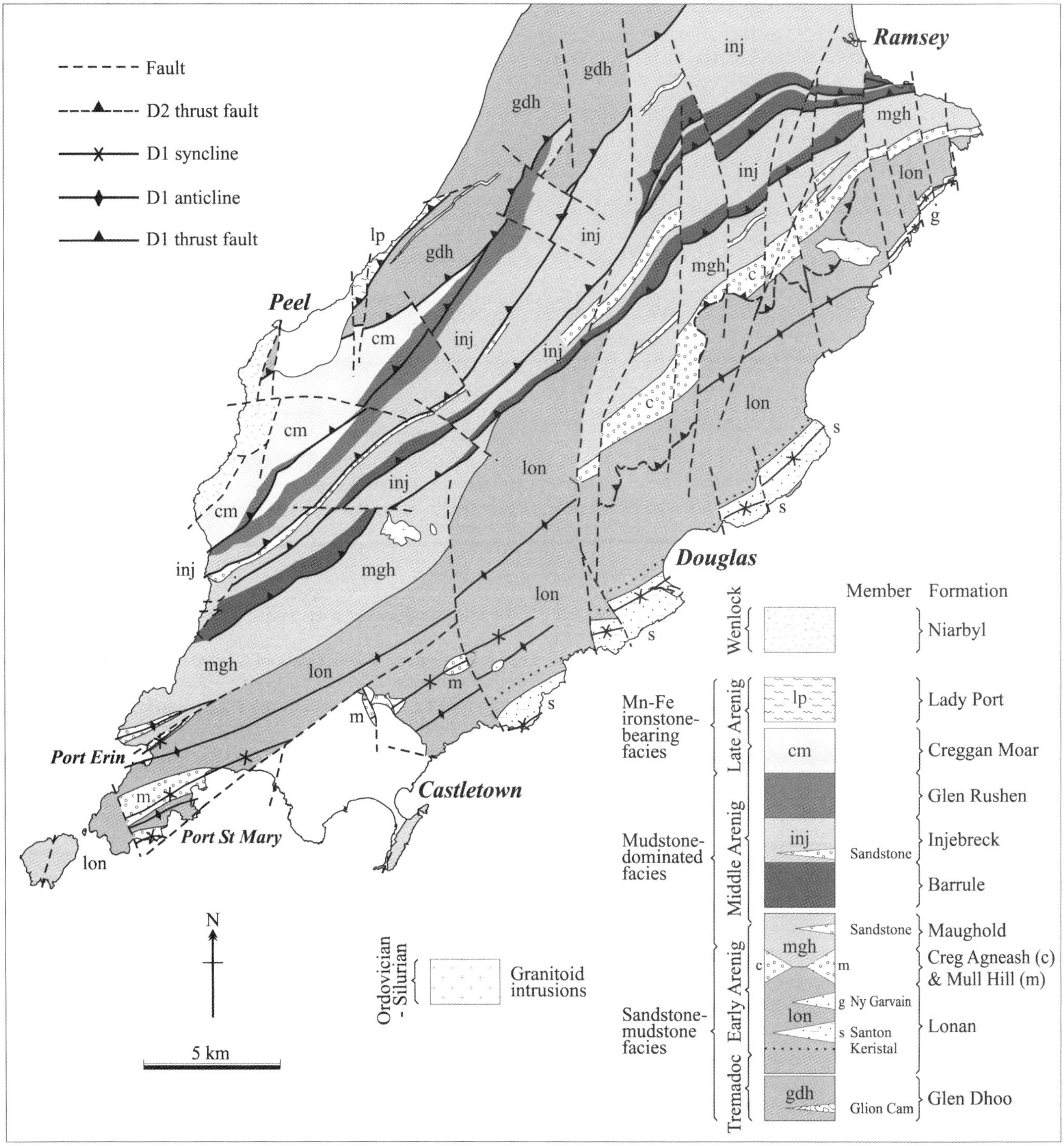

Fig. 5.3. The distribution of the Manx Group's component formations and members across the southern part of the Isle of Man.

Biozone. Rushton (1993) reported graptolites from the Santon Member but their poor preservation allowed only a general Arenig age to be allocated.

Early to mid-Arenig mudstone-dominated facies

The upland spine of the Isle of Man is composed of mudstone-dominated formations (Fig. 5.3). The Maughold Formation is composed mainly of dark grey mudstone in bedded or disrupted facies, but locally includes significant amounts of quartz arenite. The laminated mudstone, typically with a millimetre-scale 'pinstripe' lamination, includes a variable proportion of pale grey siltstone in laminae to very thin beds and sporadic, thin to medium beds of fine-grained quartz arenite. Bioturbation is common as spots or as discordant silt-filled or mud-filled burrows. Pebbly mudstone is absent in the NE part of the outcrop, but it becomes important to the south. Matrix-supported, the clasts are generally up to a few centimetres and have a maximum size of 30 cm. They are mostly intraformational but include rare fragments of a fine-grained igneous rock. Medium to thickly bedded pale or very pale greyish-white, fine- to coarse-grained quartz arenite occurs in isolated beds and is locally dominant in packets up to about 25 m thick. The sandstone beds have sharp top and basal contacts. They are not discernibly graded and commonly contain tabular cross-sets through the thickness of each bed.

The contact of the Maughold Formation with the Barrule Formation is interpreted as faulted (Fig. 5.3), but an originally conformable stratigraphical contact between the two formations seems likely. The outcrop of the Barrule Formation, duplicated by thrusting, consists of homogeneous or faintly laminated grey–black mudstone interpreted as hemipelagic sediment deposited in an anoxic seabed environment. The formation may be several hundred metres to 1 km thick, depending on internal structure, and passes gradationally into the Injebreck Formation with the appearance of distinctive silty laminae.

The Injebreck Formation is in many respects similar to the Maughold Formation. It is dominated by dark mudstone with pale grey silty laminae and a variable proportion of thinly interbedded siltstone and fine-grained sandstone. Sandstone beds typically have sharp tops and bases, and are parallel- and ripple cross-laminated. Sporadic thin–medium beds and packets of quartz arenite occur throughout the formation. One unit a few tens of metres to 100 m thick, dominated by thin- to medium-bedded quartzose arenite, persists 200–300 m above the base of the formation. Pebbly mudstone occurs locally in the SW of the outcrop, but becomes dominant through much of the lower part of the formation in the NE. Clasts range up to 10 cm, occasionally to 35 cm and are matrix-supported. They are dominated by intrabasinal material but include fine-grained igneous rock.

The Injebreck Formation is interpreted by Barnes *et al.* (in Chadwick *et al.* 2001) to pass stratigraphically up into the Glen Rushen Formation, although the evidence is equivocal (cf. Woodcock *et al.* 1999*b*). The Glen Rushen Formation is composed predominantly of laminated, medium–dark grey or bluish-grey mudstone with a variable proportion of pale siltstone laminae generally less than 1 mm thick. Bioturbation is absent and the laminae are laterally persistent. Thin quartzose sandstone beds and pebbly mudstone occur rarely. The Glen Rushen Formation probably passes stratigraphically into the Creggan Moar Formation with a gradual increase in the proportion of siltstone laminae and the appearance of very thin manganiferous ironstone beds.

The mudstone facies are interpreted to result from hemipelagic fall-out and low-concentration turbidity flows into a marine, periodically oxygenated basin. Sandstone-dominated facies record periods of medium- to high-concentration turbidity flows, some of which were highly quartzose. Periodic basin instability resulted in disruption and resedimentation of the bedded facies as pebbly mudstone facies.

The age of the Maughold, Barrule and Injebreck formations is poorly constrained. Their interpreted stratigraphical position, above the Tremadoc–early Arenig Lonan Formation and below the middle Arenig Creggan Moar Formation, implies an early to mid-Arenig age. Samples from all of the formations have yielded sparse, low-diversity assemblages of sphaeromorph and acanthomorph acritarchs (Molyneux 2001) consistent with the lower Arenig (*Didymograptus varicosus* Biozone) in the Lake District (Cooper *et al.* 2004). Most samples from the Glen Rushen Formation have also yielded sparse assemblages but two samples close to the transition with the Creggan Moar Formation have yielded acritarchs diagnostic of the mid-Arenig *S.* aff. *pseudoornatum* Biozone in the Lake District (Molyneux 2001).

Mid- to late Arenig manganiferous ironstone–bearing facies

Two units that crop out in the NW of the Isle of Man are biostratigraphically constrained as the youngest parts of the Manx Group, ranging from mid- to late Arenig in age. The Creggan Moar Formation (Kennan & Morris 1999) crops out in sequence above the Glen Rushen Formation but is structurally truncated along its western margin (Fig. 5.3). The Lady Port Formation (Woodcock & Morris 1999) forms a tectonically isolated sliver on the NW coast of the Isle of Man.

The Creggan Moar Formation was formed mainly from low-concentration turbidity flows, punctuated by chemical deposits, into oxygenated bottom waters. It is dominated by very-thin- or thin-bedded, dark grey mudstone, siltstone and very fine-grained pale sandstone. The siltstone and sandstone beds are typically parallel-laminated, rarely cross-laminated, grading through bioturbated tops into the mudstone. Very thin beds of distinctive reddish brown or dark brown–black weathering manganiferous ironstone occur intermittently or in clusters. Parallel-laminated siltstone is locally dominant, forming thick monotonous successions with rare mudstone or manganiferous beds. Quartz arenite occurs sporadically as pale grey, faintly laminated, thin–medium beds.

The age of the Creggan Moar Formation was only recently constrained as mid- to late Arenig (Molyneux 2001). Acritarchs, sparse near the transition from the Glen Rushen Formation but becoming more abundant and more varied elsewhere, commonly include the zonal fossil *Stelliferidium* aff. *pseudoornatum*. However, one locality (Creggan Moar [SC 2152 7711]) in the middle of the outcrop, yielded a variety of forms that suggest the late Arenig *hamata–rarirrugulata* Biozone. The top of the Creggan Moar Formation is not seen; in the north the Glen Dhoo Formation has been thrust above it and in the south it is faulted into juxtaposition with Silurian rocks of the Dalby Group.

The Lady Port Formation is heterogeneous and extensively disrupted, recording major basin instability. The dominant feature is matrix-supported pebbly mudstone containing a varying proportion of sandstone and siltstone clasts commonly up to 20 cm in size and rafts of bedded sediment up to at least tens of metres in size. Fragments of intact lithostratigraphy include grey to black mudstone with siltstone laminae (locally resembling the Glen Rushen Formation) and thin- to medium-bedded, light grey–greenish grey wacke. Two sections include millimetre-scale buff–black manganiferous ironstone in association with siltstone and fine-grained quartzose sandstone in a facies similar to that of the Creggan Moar Formation (Kennan & Morris 1999). The Lady Port Formation has yielded poorly preserved acritarchs considered to be late Arenig by Molyneux (1979, 1999), including evidence of the *hamata–rarirrugulata* Biozone. However, its complex nature and the possibility that the larger stratified sections may be rafts derived from underlying formations means that this may only be a maximum age.

Skiddaw Group

The Skiddaw Group comprises a thick (*c*. 5 km) succession consisting largely of turbiditic sandstone and mudstone. It contains both a graptolite fauna and an acritarch flora from which can be established a Tremadoc, Arenig and Llanvirn biostratigraphy (Cooper *et al.* 1995, 2004). The principal outcrop is in the main Skiddaw inlier of the northern Lake District, with the smaller Ullswater and Bampton inliers nearby to the SE, and the Black Combe and Furness inliers futher away to the south (Fig. 5.1). The group also appears further east in inliers at Cross Fell (Cooper & Molyneux 1990 and references therein) and Teesdale (Johnson 1961), and has been proved by boreholes to underlie the Carboniferous rocks of the Alston Block (Woolacot 1923; Burgess 1971). The only evidence for pre-Ordovician rocks within the Skiddaw Group is provided by a probable Cambrian acritarch microflora from the vicinity of Eycott Hill, in the NE of the Skiddaw inlier (Millward & Molyneux 1992).

In the largest of the inliers (480 km^2 including the eponymous mountain of Skiddaw), two distinct stratigraphical belts are present within the Skiddaw Group, namely the Northern Fells Belt and the Central Fells Belt (Figs. 5.4 & 5.5). These belts

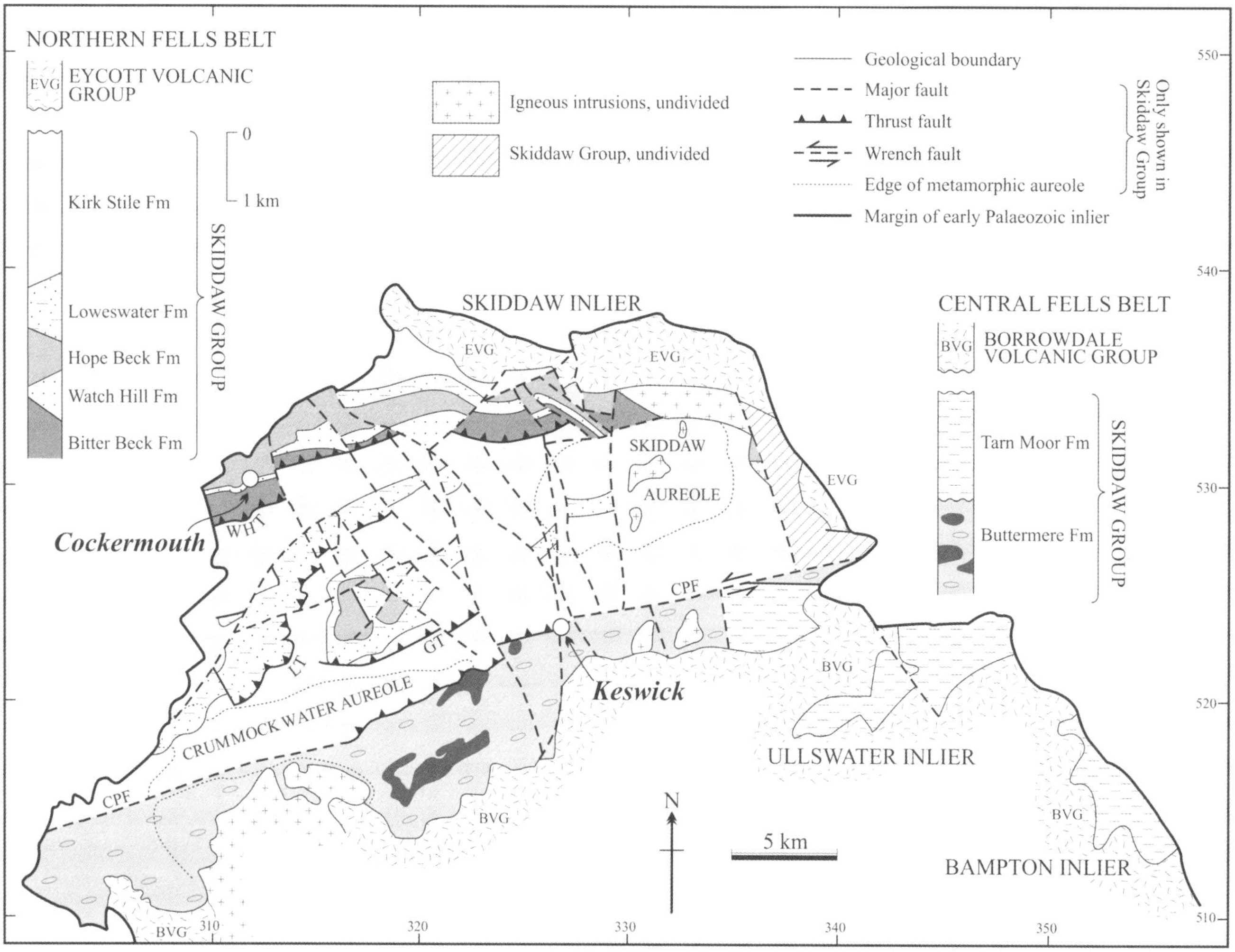

Fig. 5.4. The principal stratigraphical and structural features of the Skiddaw Group outcrop in the northern Lake District. CPF, Causey Pike Fault; GT, Gasgale Thrust; LT, Loweswater Thrust; WHT, Watch Hill Thrust.

are separated by a major wrench and thrust fault system, the Causey Pike Fault, which can be traced across the Lake District and eastwards to Cross Fell (Fig. 5.1; Cooper & Molyneux 1990). The fault coincides with a major geophysical lineament, interpreted as a basement structure along which wrench movement has occurred and into which a concealed, *c*. 400 Ma, elongate granitic body has been intruded (Cooper *et al*. 1988; Hughes *et al*. 1993).

To the north of the Causey Pike Fault, in the Northern Fells Belt of Cooper *et al*., are preserved some 5 km of mainly mudstone turbidites that were deposited between the Tremadoc and the early Llanvirn, although there is some evidence that sedimentation may have commenced in the Cambrian (Millward & Molyneux 1992). Wacke-type sandstone beds occur sporadically throughout the succession but become dominant at two levels, the Loweswater and Watch Hill formations (Figs 5.2 & 5.4). The provenance was deduced by Cooper *et al*. (1995) to be largely an old, inactive, continental volcanic arc lying to the SE. South of the Causey Pike Fault, in the Central Fells Belt of Cooper *et al*., is a major olistostrome, the Buttermere Formation, that was emplaced from the south during the late Arenig (Webb & Cooper 1988). It is overlain by 1 km or more of late Arenig–Llanvirn siltstone and mudstone, with sporadic interbedded bentonite, forming the Tarn Moor Formation. Correlative strata to those forming the Skiddaw inlier can be identified in the smaller inliers to the south and east (Figs 5.1 & 5.4).

Tremadoc–early Llanvirn turbidite facies of the Northern Fells Belt

In the Northern Fells Belt the Skiddaw Group comprises five formations overlain by the Eycott Volcanic Group (Fig. 5.4). In ascending order these formations are: Bitter Beck, Watch Hill, Hope Beck, Loweswater and Kirk Stile. Boundaries between the formations are transitional.

The Bitter Beck Formation comprises thinly laminated, dark grey mudstone, silty mudstone and siltstone with minor amounts of fine grained, wacke-type sandstone. The latter is concentrated in the lower part of the formation where it forms up to 20% of the succession in thin–medium beds (3–30 cm). The graded siltstone beds, with homogeneous mudstone laminae, represent distal turbidites, typically of Bouma-type T_{de}. The sandstone component is pale grey, fine-grained, and forms generally thinly parallel to wavy laminated beds with irregular bases and tops. In places there is abundant slump folding with dislocations sub parallel to bedding. The formation is at least 500 m thick and lies within the upper Tremadoc *murrayi* graptolite Biozone (fig. 2 of Cooper *et al*. 2004).

The Watch Hill Formation comprises interbedded lithic sandstone (locally coarsening to microconglomerate) and mudstone. The sandstone is typically fine grained with abundant siltstone and mudstone intraclasts; bed thickness ranges up to about 1 m. Normal grading, parallel lamination, wavy lamination and rare ripple cross-lamination are present and locally

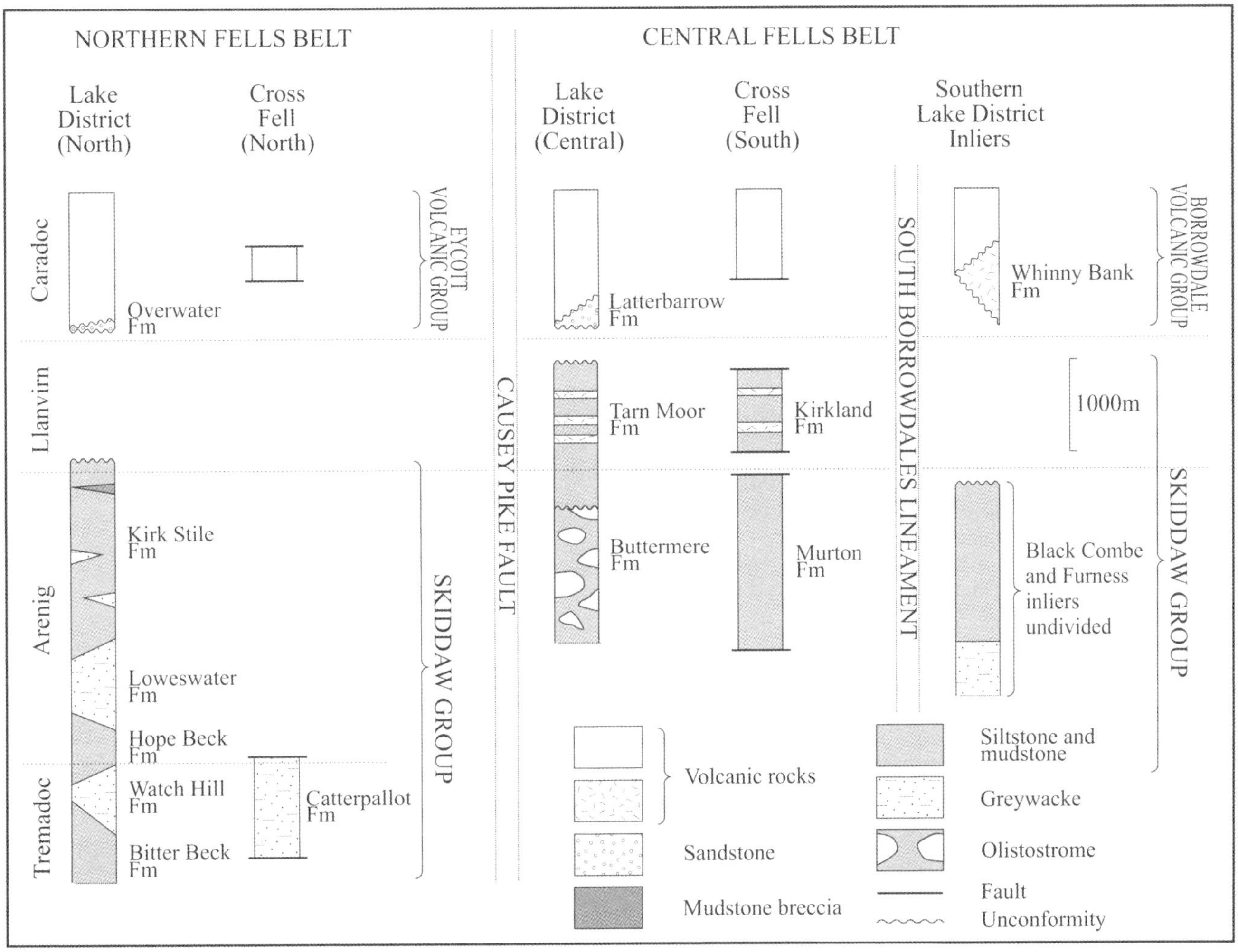

Fig. 5.5. Stratigraphical contrasts and correlations within and between the principal Skiddaw Group inliers of the Lake District and Cross Fell.

occur together in Bouma-type T_{abc} turbidite units. Sandstone–mudstone proportions change vertically and laterally throughout the succession, but are typically around 60–70% sandstone and 30–40% mudstone. The maximum thickness of the formation is somewhere between 550 m and 800 m, but it thins both westwards to about 100 m, and eastwards to 40 m, suggesting deposition in a system of channels. Basal flute casts in the type area suggest palaeocurrent flow from an easterly direction (Moore 1992). The age of the formation is within the latest Tremadoc *murrayi* graptolite Biozone (Cooper *et al.* 2004). Equivalent strata further east in the Cross Fell inlier (Fig. 5.1) there form the Catterpallot Formation (Fig. 5.4) as described by Cooper & Molyneux (1990).

The Hope Beck Formation comprises laminated and thinly bedded mudstone with sporadic, thin sandstone and pebbly mudstone beds. Bioturbation is common, mainly as burrows subhorizontal to bedding, some of which are filled with faecal pellets. The sandstone is quartz-rich lithic wacke, generally medium- to coarse grained, but including numerous beds with granules and pebbles. The clast assemblage is dominated by monocrystalline plutonic quartz and polycrystalline metamorphic quartz. Sandstone beds are T_{abc}, T_{bc} and T_c Bouma units; the mudstone forms both turbiditic T_d units and hemipelagic T_e units. Pebbly mudstones are matrix-supported (*c.* 50% matrix) and probably represent debris flows. The thickness of the formation is between 600 and 800 m with a biostratigraphical range from latest Tremadoc to early Arenig (Fig. 5.2), spanning the *phyllograptoides* graptolite Biozone (Cooper *et al.* 2004), equivalent to the *approximatus* Biozone of Fortey *et al.* (1991).

The Loweswater Formation is dominantly formed of quartz-rich wacke; grains are predominantly of strained monocrystalline quartz with some large unstrained monocrystalline clasts. The wacke is fine to medium grained, parallel-laminated and ripple cross laminated sandstone corresponding to Bouma-type T_{bc} turbidite units. Bed thickness increases gradually upwards to a maximum of 1 m near the middle of the formation; concurrently the maximum grain size increases to very coarse grained, although most beds are still fine- to medium grained. The upper part of the formation is a mirror image of the lower part, bed thickness and grain size decreasing gradually while the percentage of siltstone and mudstone increases. Overall, the sandstone beds are slightly lenticular. Most have planar bases, but channels, groove casts and flute casts occur locally. Palaeocurrent flow was mainly from the SE to SW quadrant (Moore 1992; Cooper *et al.* 2004). Small scale slump folds, up to a few metres across are widespread (Webb & Cooper 1988). The Loweswater Formation has a maximum thickness of approximately 900 m, decreasing northwards to an estimated 450 m. Biostratigraphically it occupies a position in the middle part of the Arenig, spanning the *varicosus* and *simulans* graptolite biozones (Fig. 5.2).

The Kirk Stile Formation is between 1500 and 2500 m thick, but some of this variation may be due to stacking of slumped masses. The principal lithology is thinly laminated mudstone deposited from low density turbidite currents. Interbedded with the mudstone are local lenticular units dominated by lithic wacke; these sandstone-rich units are are typically thinly bedded with parallel and ripple cross-lamination (Bouma-type

T_{bc} turbidite units). Strained quartz is the most abundant grain type, feldspar is rare, but various lithic grains, mostly volcaniclastic siltstone and altered mafic volcanic rock, are relatively common. In the upper part of the formation, sporadic units of sedimentary breccia and slumped strata, from 2 m to more than 40 m thick, include intrabasinal clasts up to about 0.5 m in diameter contained in a silty mudstone matrix. Biostratigraphically, the formation ranges from the mid-Arenig to the early Llanvirn, spanning the *simulans*, *gibberulus*, *hirundo* and *artus* graptolite biozones (Fig. 5.2); coeval strata occupy the Furness and Black Combe inliers in the south of the Lake District (Cooper *et al.* 2004; Rushton & Molyneux 1989).

A late Arenig olistostrome and the Llanvirn turbidite facies of the Central Fells Belt

South of the Causey Pike Fault, the Skiddaw Group comprises two divisions: the massive olistostrome of the Buttermere Formation, unconformably overlain by the mudstone-dominated Tarn Moor Formation, which also makes up all of the exposed succession in the small inliers of Ullswater and Bampton. The Tarn Moor Formation is equivalent to the Murton Formation further east in the Cross Fell inlier, where the olistostrome is represented by the Kirkland Formation (Figs 5.1 & 5.5) (Cooper & Molyneux 1990).

The Buttermere Formation is an olistostrome deposit, at least 1500 m thick, comprising mudstone, siltstone and sandstone turbidite olistoliths set in an argillaceous matrix (Webb & Cooper 1988). The olistoliths range in size from granules up to 1 km or more across with most in the 5–10 m range; the largest are formed by wacke-type sandstone. Both clasts and matrix are intensely deformed by minor folds and shears, many of which were generated during emplacement of the olistostrome. Fossils from the Buttermere Formation (trilobites – Rushton 1988 – as well as graptolites and acritarchs) indicate a range of ages from early Tremadoc to late Arenig, whilst the overlying Tarn Moor Formation is of latest Arenig and Llanvirn age (Cooper *et al.* 1995). Hence the age of olistostrome emplacement is inferred to be in the late Arenig, possibly at about the boundary between the *gibberulus* and *cucullus* graptolite biozones. From the geometry of the major slump folds, Webb & Cooper (1988) inferred that the olistostrome was emplaced by down-slope movement towards the NNW.

The Tarn Moor Formation (Cooper *et al.* 1995) comprises the late Arenig–early Llanvirn mudstone and siltstone, with minor volcaniclastic turbidite and bentonite beds, that occurs in the SE part of the main Skiddaw inlier and forms all of the Ullswater and Bampton inliers (Fig. 5.1). The formation rests unconformably on the highly disrupted Buttermere Formation, although the time-break involved would appear to be very small. The top of the Tarn Moor Formation is cut off unconformably by the Borrowdale Volcanic Group (Wadge 1972) and the youngest strata seen are those in the Tarn Moor Tunnel described by Wadge *et al.* (1969). The thickness of the formation is probably around 1000–1500 m. In its lower part, laminated mudstone is similar in overall character to the Kirk Stile Formation of the Northern Fells Belt, with which it partly overlaps in age, spanning the *cucullus* graptolite Biozone of the upper Arenig and the *artus* Biozone of the lower Llanvirn. The upper part of the formation is characterized by mudstone with up to 5% of bentonite and volcaniclastic interbeds. The bentonite ranges up from very thin laminae to discrete beds a few centimetres thick. Graptolites indicate the *artus* Biozone (Cooper *et al.* 1995; cf. Wadge 1972) in the Skiddaw and Bampton inliers but range up into the *murchisoni* Biozone in the Ullswater Inlier (Wadge *et al.* 1969, 1972).

The depositional basin

The thickness, duration and geographical extent of the Skiddaw Group successions suggest deposition in a large basin with a long history of subsidence. The provenance was deduced by Cooper *et al.* (1995) to be largely an old, inactive, continental volcanic arc lying to the SE; Nd isotope results reported by Stone & Evans (1997, 2002) confirm the absence of a large juvenile component since, across the whole group, sandstones have an εNd range of −4.1 to −9.3 whilst mudstone εNd ranges down to −10.6. Manx Group sandstone data are comparable, ranging from −4.7 down to −7.2. An extensional passive margin therefore seems a more likely site for deposition than an inter-arc or back-arc zone. In this apparent absence of coeval volcanism until late in its depositional history, the Skiddaw Group contrasts with the Manx Group, which contains the interbedded Peel volcanic assemblage described by Simpson (1963) as andesitic lava, tuff and agglomerate. The Manx Group siltstone adjacent to the volcanic rocks contains an Arenig microflora (Molyneux 1979; Cooper *et al.* 1995). There is also a major contrast with contemporaneous developments in the Welsh Basin, where there was considerable subduction-related volcanicity during the Tremadoc and Arenig.

Some idea of the possible basin configuration is given by Webb & Cooper (1988) from an examination of the widespread slump folds. Within the Northern Fells Belt these are predominantly orientated towards the SE and, although the origin of the larger folds has been debated (Hughes *et al.* 1993), there remains evidence of a likely SE-facing palaeoslope, at least during the Tremadoc and early Arenig. Conversely, south of the Causey Pike Fault in the Central Fells Belt, the Buttermere Formation is a large olistostrome emplaced towards the northwest; only relatively thin debris flow beds are seen in the Northern Fells Belt at an equivalent stratigraphical level. From the contrasting scale and style and orientation of the slump movements Webb & Cooper (1988) deduced a steeper, faulted southerly margin to the depositional basin.

An extensive analysis of palaeocurrent evidence led Moore (1992) to propose two periods of submarine fan development (essentially the Watch Hill and Loweswater formations) each with a different basin configuration. During deposition of the first (Watch Hill) axial flow was inferred along a trough orientated approximately E–W. Subsequent deposition (Loweswater) was spatially more complex, with greater influence by sea-floor topography. The latter was thought to arise from syn-depositional extensional faults trending NW–SE within the depositional basin, with the intervening fault blocks tilted to the NE. The overall picture that emerges is of a N-facing extensional half-graben system with the major boundary fault (or faults) on the SE side. The observed stratigraphical contrast across the Causey Pike Fault would certainly have been made more conspicuous if the basin was originally composite, and so perhaps that structure was initiated as a NW-downthrowing normal fault partitioning the Skiddaw Group depositional basin. Additional support for this is derived from the provenance implications of whole rock Nd isotope data (Stone & Evans 2002), which show a marked difference across the Causey Pike Fault.

Further evidence in support of an extensional basin model is provided by the illite crystallinity and clay mineral assemblages of the Skiddaw Group mudstones. Fortey *et al.* (1993) and Merriman & Frey (1999) suggest that early burial metamorphism, characterized by late diagenetic to low anchizonal grades, occurred under a higher geothermal gradient (>35 °C km^{-1}) than would normally be expected in an ensialic basin. They related this to high heat flow arising from extension and crustal thinning. Similar results have been reported from the Manx Group (Roberts *et al.* 1990). However, there must remain some uncertainty as to the relationship of burial metamorphism in the depositional basin during its putative extension (Tremadoc–Llanvirn), and that during subsequent supra-subduction zone basin uplift and volcanicity (Llanvirn–Caradoc; discussed below) when a high geothermal gradient would also be expected.

Basin uplift and volcanism

The first indications of volcanic activity during Skiddaw Group deposition are seen in the early Llanvirn (Cooper *et al.* 1995), with sporadic interbeds of volcaniclastic turbidite sandstone and bentonite ash in the Tarn Moor Formation of the Central Fells Belt (Fig. 5.4). The abundance and thickness of volcanic interbeds in the dominantly mudstone succession each appears to increase upwards, but Nd isotope data for the mudstones (Stone *et al.* 1999; Stone & Evans 2002) show that the juvenile component must be restricted to the discrete volcaniclastic beds since the background provenance remained relatively ancient. The values of εNd for the mudstones (calculated at a depositional age of 480 Ma) range from −8 to −10.6, equivalent to depleted mantle model ages of 1.52–2.01 Ga and compatible with derivation from a Proterozoic source. It must be stressed that these isotope data give an average value for all the components of the rock and do not relate to a single unique source. Nevertheless, they rule out any significant juvenile contribution to the Tarn Moor Formation mudstones. The Llanvirn initiation of volcanism therefore seems to have been either at some distance from the depositional basin, or to have involved relatively small-scale and localized eruptions.

The onset of volcanicity in the early Llanvirn followed the late Arenig emplacement of the major Buttermere Formation olistostrome (Hughes & Kokelaar 1993; Cooper *et al.* 1995) and seismic activity, as a precursor to the volcanic episode, may have triggered the mass-flow movements. However, the olistostrome was clearly emplaced by down-slope movement into a still-extant basin and so it may equally have been instigated by normal, extensional movement on the basin boundary fault(s). Whatever the trigger mechanism, the extensive slumping throughout much of the Skiddaw Group succession caused considerable stratigraphical disruption. Further disruption was inevitable during subduction-related uplift of the continental margin and inversion of the Skiddaw Group basin. Uplift was most likely caused by the generation and rise of andesitic melts above the subducted oceanic crust (Branney & Soper 1988; Hughes *et al.* 1993). It resulted in the deep marine basinal strata being converted into the subaerial basement to the ensuing, mainly Caradoc, Borrowdale and Eycott volcanic groups.

The magnitude of the pre-volcanic stratigraphical disruption, caused by the combination of gravity-driven, mass slump movement and the subsequent basin inversion, may be gauged by the wide range of biostratigraphical zones determined immediately subjacent to the overstepping unconformity. South of the Causey Pike Fault, below the Borrowdale Volcanic Group, various Arenig–Llanvirn biostratigraphical levels occur close to the unconformity cut across the Buttermere Formation olistostrome and the overlying Tarn Moor Formation (Cooper & Hughes 1993). North of the Causey Pike Fault the Skiddaw Group strata immediately subjacent to the Eycott Volcanic Group ranges in age from possibly Cambrian to Llanvirn (Millward & Molyneux 1992). In this northern area, much of the variation seems likely to have been caused by fault-block rotation prior to volcanicity since there is stratigraphical coherence over wide areas of the outcrop. Rotation could have been either an extensional or compressional effect and was probably superimposed both on disrupted zones caused by the earlier slump movements and on areas that had escaped such disruption.

It should be stressed that there is no evidence for a compressive tectonic event producing penetrative deformation prior to the eruption of the volcanic rocks. This has been a long-running controversy in Lake District geology and is reviewed by Hughes *et al.* (1993). The earliest regional cleavage cuts both the Skiddaw Group and overlying volcanic and sedimentary formations that range up to Přídolí in age. It is a product of the early Devonian, Acadian Orogeny, which is discussed in more detail below. The only pre-volcanic but post-depositional fabric present in the Skiddaw Group is a widespread bedding-parallel compaction cleavage. This is developed particularly well in the most argillaceous lithologies and probably results from burial accentuation of the original bedding lamination either prior to, or during, basin uplift. The high heat flow thought to have been prevalent at the time (Merriman & Frey 1999, discussed above) may have assisted the formation of this fabric by accelerating recrystallization of the clay minerals (Merriman & Peacor 1999). There is no indication that any penetrative tectonic fabric was imposed during pre-volcanic uplift and stratal disorganization. This situation has given rise to dispute in the past with Simpson (1967) and Helm (1970) proposing mid-Ordovician orogenesis before Soper (1970) and Soper & Roberts (1971) established an early Devonian age for formation of the first tectonic cleavage.

It is now thought that the volcanic rocks that occur on either side of the main Skiddaw Group inlier, the Eycott Volcanic Group to the north and the Borrowdale Volcanic Group to the south (Figs 5.4 & 5.5) are penecontemporaneous and largely Caradoc in age (Millward & Molyneux 1992). Previously, the Eycott Volcanic Group was thought to be significantly older than the Borrowdale Volcanic Group on the basis of an acritarch flora described from sedimentary interbeds at the base of the former by Downie & Soper (1972). Both of the volcanic groups had a continental margin, supra-subduction zone origin which requires an arc-trench gap, probably exceeding 100 km of forearc, extending to the north; a situation supportive of an ensialic setting for the Skiddaw Group basin. A network of volcanotectonic, caldera-related faults disrupts the Borrowdale Volcanic Group (Branney & Soper 1988) with proved displacement commonly in excess of 400 m (Branney & Kokelaar 1994). It is highly probable that similar structures would have affected those parts of the Skiddaw Group currently exposed, with reactivation of the existing fault framework. Major structures such as the Causey Pike Fault would have been a likely focus for such movement.

Ingleton Group

To the SE of the Lake District, strata commonly correlated with the Skiddaw Group crop out in the Craven inliers of Chapel le Dale and Ribblesdale (Fig. 5.1). These form the Ingleton Group, 600 m or more of grey-green turbiditic sandstone and siltstone with sporadic intraformational conglomerate (Arthurton *et al.* 1988, pp. 4–5). There is much evidence for soft-sediment deformation and slumping (Leedal & Walker 1950). Faunal evidence of age was obtained from the nearby Beckermonds Scar borehole [SD 8635 8016] that penetrated the possible lateral equivalents of the Ingleton Group beneath 260 m of Carboniferous cover (Wilson & Cornwall 1982). A single sample from a depth of 396.2 m contained a probable Arenig acritarch microflora.

In overall lithofacies the Ingleton Group appears broadly comparable to the Skiddaw Group. The one area of significant difference is the more juvenile character of the Ingleton Group clastic detritus as deduced from neodymium isotope ratios (Stone & Evans 1997). These data, εNd values between −1.6 and −2.1 (Davies *et al.* 1985; Miller & O'Nions 1984), show a significant juvenile volcaniclastic input and so preclude the Skiddaw and Ingleton groups from having a common sedimentary provenance. The most recent assessment (Soper & Dunning 2005) regards the Ingleton Group as considerably older than the Skiddaw Group.

Post-volcanic convergence of Avalonia and Laurentia

The relative brevity but great intensity of the Borrowdale–Eycott, Caradoc volcanic episode may have been the result of ridge subduction at the Avalonian margin of the Iapetus Ocean

(Pickering & Smith 1995). These authors point out that such a situation could have two important outcomes. First, it would allow creation of a window in the subducting slab which would explain the abrupt cessation of volcanism at the Avalonian margin. Secondly, it would effectively transfer Avalonia onto a N-moving plate that was being subducted beneath the Laurentian margin of the ocean.

Continued subduction at the northern, Laurentian margin of the Iapetus Ocean, effected continental convergence by the late Llandovery or Wenlock (e.g. Soper *et al.* 1992*a*, *b*). The accretionary complex that had developed at the leading edge of Laurentia over-rode the Avalonian margin and continued southwards as a foreland fold and thrust belt preceded by a foreland basin. The initiation of the foreland basin in southern Scotland has been discussed by Stone *et al.* (1987); its progression onto Avalonia and across the Lake District is detailed by Kneller (1991) and Kneller *et al.* (1993). If the Skiddaw Group was caught up in this process, in which S-directed thrusts would have been the dominant structure, the principal such structures within the main outcrop may have been initiated at this stage, i.e. the Watch Hill, Loweswater and Gasgale thrusts and thrusts within the Causey Pike Fault System (Figs 5.2 & 5.4) (Hughes *et al.* 1993). The timing of any thrust propagation through the Skiddaw Group could be indicated by the widespread resetting, to about 420–430 Ma, of the Rb–Sr isotopic systems in adjacent igneous rocks (Hughes *et al.* 1996; Stone *et al.* 1999). This would have been a hydration event arising from the increased permeability, through fracturing and uplift, associated with the onset of thrusting.

A Rb–Sr isochron of 427 ± 34 Ma from late Arenig Skiddaw Group mudstone (Kirk Stile Formation) at Dodd Wood [NY 2363 2785] is also relevant to the resetting debate (Stone *et al.* 1999). Despite the large error there is no overlap with the depositional age of about 480 Ma. In mudstones such resetting is a dehydration reaction and has been linked to re-equilibration of the Rb–Sr system during the illite to smectite transition (Evans *et al.* 1995). Movement of fluids out of the mudstone during a dehydration reaction is thus coincident with fluid movement into the igneous rocks during hydration, both phenomena re-setting the Rb–Sr systems to the 420–430 Ma interval.

Early Devonian, Acadian deformation

Despite indications of late Ordovician and Silurian disruption in the Skiddaw–Manx group's depositional basin, penetrative tectonic deformation did not occur until the early Devonian, Acadian Orogeny (see Chapter 4). The structure of both groups is highly complex, and its apparently polyphase character led initially to proposals for a protracted deformation history (e.g. Simpson 1963, 1967; Helm 1970). The recognition of the earliest deformations as soft-sediment, slump-related phenomena (e.g. Webb & Cooper 1988) and confirmation that the first, regional tectonic cleavage was indeed Acadian (e.g. Soper & Roberts 1971; Soper *et al.* 1987; Merriman *et al.* 1995) have largely resolved the regional interpretational controversies. At a more local scale, the structure of the Manx and Skiddaw groups remains open to different interpretations, whilst the detailed structure of the Ingleton Group is not well established beyond the recognition of some soft-sediment deformation and upright, tight to isoclinal, tectonic folds.

Manx Group structure

The Manx Group succession on the Isle of Man has a relatively simple outcrop pattern due to predominantly steep dip and consistent NE-strike, although rotating to E–W in the NE of the outcrop. At the exposure scale, however, dip may vary widely due to two main phases of deformation: the first is a pervasive slaty cleavage associated with gently to moderately plunging folds on a range of scales; the second is a widely developed, gently dipping crenulation cleavage associated with small folds that verge in the direction of dip of bedding. These affect both the Ordovician Manx Group and the late Silurian Dalby Group. The early deformation affects many of the minor intrusions and the Dhoon granite (Power & Barnes 1999) and possibly metamorphic phases in the aureole of the early Devonian Foxdale granite, although not the granite itself (Simpson 1965). The deformation is therefore considered to be late Silurian or early Devonian in age, equivalent to the Acadian deformation of NW England (Soper *et al.* 1987). D_1 and D_2 are associated with low-grade regional metamorphism, although the pattern is strongly influenced by the original composition of the various units (Power & Barnes 1999) and possibly granite batholiths at depth (Roberts *et al.* 1990). A later phase of localized deformation seen as a broadly N-trending, steeply dipping crenulation cleavage associated with upright folds may be Carboniferous in age (Fitches *et al.* 1999).

Simpson (1963) proposed a number of kilometric scale folds in the outcrop of the Lonan Formation (Fig. 5.3) that have largely been corroborated by recent work (e.g. Fitches *et al.* 1999). Determination of the overall structure of the Isle of Man has, however, been limited by the absence of a well-established stratigraphical framework. Early interpretations (e.g. Lamplugh 1903; Simpson 1963) suggested fold models based on lithological correlations that have since been shown to be wrong. Fitches *et al.* (1999) considered for the first time the possibility that major strike-parallel faults may separate tracts of reasonably coherent stratigraphy. Developing from this, Barnes *et al.* (in Chadwick *et al.* 2001) have suggested a broad zone of SE-directed thrusting with the upper part of the sequence imbricated onto the folded 'foreland'.

Several ductile shear zones are exposed on the west coast of the Isle of Man and are generally marked by narrow zones of intense fabric development with indications of predominantly sininstral shear. One, at Lag ny Keeilley [SC 215 745], corresponds with the the fault that imbricates the top of the Barrule Formation and the base of the Injebreck Formation. A relatively wide (150 m) zone of ductile deformation at The Niarbyl [SC 215 776] is more enigmatic (e.g. Fitches *et al.* 1999). It is situated near the boundary between the Dalby Group and the Creggan Moar Formation but the sinistral phyllonitic mylonite fabrics are almost perpendicular to the junction which is obscured by a later brittle fault.

Geophysical data in the vicinity of the Isle of Man (e.g. Kimbell & Quirk 1999; Kimbell in Chadwick *et al.* 2001) suggest a zone of complex structure in the mid-crust, underlying and trending parallel to the zone of faulting in the NW part of the Isle of Man. This, together with the apparent northwestwards increase in the structural complexity of the Manx Group, may be related to the proximity of the Isle of Man to the Iapetus Suture.

Skiddaw Group structure

The Acadian slaty cleavage forms a regional arc in the Skiddaw Group, with a west to east variation in trend between NE–SW and E–W (Soper *et al.* 1987). The cleavage is axial planar to gently plunging, steeply-inclined, open to isoclinal folds with amplitudes of hundreds of metres. Its fabric is of penetrative, pressure-solution type in some areas but of spaced, fracture type in others. Where the bedding-parallel compaction fabric is well developed the regional Acadian cleavage may have the appearance of a crenulation fabric.

In the Skiddaw Group of the northern Lake District the regional Acadian cleavage is commonly crenulated by fabrics that are axial planar to open, gently plunging minor folds with gently inclined axial planes. The attitude of these folds is variable and more than one generation is apparent with a widespread, gently dipping crenulation plane itself crenulated by later, more variable fabrics. Some of the crenulation fabrics are associated with minor folds related to a set of S-directed thrusts (Roberts 1992). Further south, in the Black Combe inlier (Rushton & Molyneux 1989), the Skiddaw Group has been severely deformed within a major, S-directed thrust zone. On the north side of this zone the rocks are intensely cleaved, sheared and metasomatized with much quartz–tourmaline veining (Johnson 1992); south of the thrust zone the combination of slaty and crenulation cleavages is very similar to that seen in the main, northern inlier. However, only a short distance south from Black Combe, in the Furness inlier, only the regional slaty cleavage is present (Soper 1970). The overall impression is of highly domainal crenulation fabrics imposed on an earlier, regional slaty cleavage during the later phases of the Acadian Orogeny. Any pre-existing thrust system would have been reactivated at this time to form the principal domainal boundaries (Hughes *et al.* 1993). Acadian, thrust re-activation of the Causey Pike Fault can be established from the offset of the metamorphic aureole to a concealed granite that was intruded into the fault zone at 401 ± 3 Ma (Cooper *et al.* 1988).

The Borrowdale Volcanic Group and Ordovician continental arc volcanism in northern England

Introduction

In the late Ordovician, northern England was part of a volcanic arc on the northern margin of eastern Avalonia, related to subduction of the Iapetus Ocean and possibly also the Tornquist Ocean. Remnants of the arc crop out in northern England, SE Ireland (Max *et al.* 1990) and Wales, and lie concealed under eastern England (Pharaoh *et al.* 1993).

In northern England the most instructive volcanic rocks lie in the central part of the Lake District (Fig. 5.6), where an inlier of Lower Palaeozoic rocks forms the distinctive landscape of the Lake District National Park. The central fells of the park are formed of Borrowdale Volcanic Group rocks, which lie between older marine sedimentary rocks to the north (the Skiddaw Group) and younger marine sedimentary rocks to the south (the Windermere Supergroup). The volcanic rocks overlie a subvolcanic batholith and are little deformed. They are superbly exposed. Deep incision by glaciers, following Tertiary uplift, reveals an array of radial sections through large caldera volcanoes and their substructures (Plate 10). At modern volcanoes, such as Yellowstone or Toba, caldera structures and the deposits that fill them are generally concealed under water or younger volcanic deposits, and the Borrowdale Volcanic Group significantly informs our understanding of how large caldera volcanoes evolve though time and space.

A less well-exposed subaerial volcanic succession in the north of the Lake District National Park (Fig. 5.6) is similar to the lower part of the Borrowdale Volcanic Group and was given a separate name, the Eycott Volcanic Group, largely because it was thought to be older (Llanvirn) (Downie & Soper 1972), although no age difference can now be proved (Millward *et al.* 2000*c*). Smaller, less-instructive outcrops of similar volcanic rocks occur in northern England at Cross Fell, Teasdale and Furness (see review by Millward 2002), and younger, minor volcanic intercalations occur in the Dent Group of the main inlier and smaller inliers to the south and east (see next section).

Arc volcanism associated with subduction beneath Avalonia lasted at least from late Tremadoc to Ashgill times. Subduction was underway in Tremadoc times, as recorded in Wales by the Rhobell and Trefgarn volcanic groups (Kokelaar 1988), and in Arenig times as recorded in SE Ireland (e.g. McConnell & Morris 1997) and the Isle of Man (Woodcock *et al.* 1999*a*). On the English mainland, early arc volcanism is recorded by volcaniclastic detritus within the Ingleton Group (Arenig; Aitkenhead *et al.* 2002) and by marine volcaniclastic turbidites and bentonites within the Llanvirn Tarn Moor and Kirkland formations of Cumbria; the latter successions lack proximal volcanic facies and probably record impersistent explosive eruptions and sedimentary reworking from elsewhere in the arc. Andesite and felsite sheets within the Skiddaw Group were once taken to indicate that volcanism in northern England dated back to Arenig times, but they have since been re-interpreted as intrusive and are probably Caradoc (in the Watch Hill and Kirkland formations; Hughes & Kokelaar 1993).

By early Caradoc times northern England had developed into a site of major volcanism, recorded by proximal volcanic facies of the Borrowdale and Eycott volcanic groups. The onset of volcanism in this area was preceded by substantial uplift, registered as a prominent unconformity at which deep marine Tremadoc–Llanvirn sedimentary rocks of the Skiddaw Group are overstepped by the predominantly subaerial volcanic successions (Fig. 5.6). The marked overstep probably reflects removal by sliding and erosion of the uplifted sediments. Localized shallow-marine and alluvial sedimentary units containing little volcanic material (respectively the Overwater and Latterbarrow Sandstone formations of Allen & Cooper 1986) have been assigned to the base of the volcanic groups, primarily because they are thought also to rest with angular unconformity upon the uplifted marine Skiddaw Group (Millward & Molyneux 1992). There is no evidence for pre-volcanic, cleavage-forming Ordovician orogenesis in Northern England (Soper 1970; Hughes *et al.* 1993) as was once thought. The regional uplift probably reflects a significance change in tectonism heralded by major slumping in the upper Skiddaw Group, and the effects of increased heat flow associated with the generation and rise of arc magmas (Branney & Soper 1988). It is noteworthy that the emergence occurred at a time of major eustatic sea-level rise (Woodcock 2000). Uplift also preceded volcanism elsewhere in the arc, for example in SE Ireland (Gardiner 1974; Max *et al.* 1990), although the unconformities at bases of volcanic units in Wales appear to record smaller, more localized uplift events, some associated with sliding of unconsolidated sediments (Kokelaar 1988; Howells & Smith 1997).

The volcanism in the Lake District was predominantly subaerial, and the area probably underwent somewhat lower rates of intra-arc extension than parts of the arc represented in Snowdonia and Llŷn where bimodal (basalt–rhyolite) compositions are common and where shallow marine conditions were widespread (Kokelaar 1988). Intra-arc sinistral transtension in Wales occurred across S- and SW-trending crust-penetrating faults in Tremadoc–Caradoc times (Kokelaar 1988) and it is likely that other prominent SW-trending crust-penetrating faults further north, in Anglesey, Ireland and inferred beneath the Lake District (Figs. 1.5 and 5.6) were active at least until the cessation of arc volcanism. During the Ordovician overall, the volcanic arc moved from 60°S towards lower latitudes (*c.* 30°S; Cocks & Torsvik 2002) although there is palaeomagnetic evidence for a short-lived southerly excursion from *c.* 22°S to *c.* 43°S during the volcanism (Piper *et al.* 1997).

SE-directed subduction under the Caradoc arc is commonly assumed, but the subduction vectors and the evolving tectonic configuration of the arc are poorly constrained, despite attempts to derive some systematic relationships from the temporal and spatial variations exhibited (see Fitton *et al.* 1982; Woodcock 2000). This is because: (a) to this day, the relative positions of English, Welsh and Irish volcanics in Caradoc times are not known (e.g. Kokelaar 1988; Murphy *et al.* 1991); (b) the geochemical characteristics of these areas are now known to be more variable than was evident in early simplistic characterisations; and (c) it is now better appreciated that analogous active continental arcs, such as at Japan, the Philippines and New Zealand, show rather varied spatial and temporal variations, such as in magma chemistry, volcanic facies, and also in rates of tectonic extension and subsidence. In such arcs, for example, large andesite stratovolcanoes stand immediately adjacent to, and along strike of, caldera complexes with bimodal (basalt-rhyolite) composition.

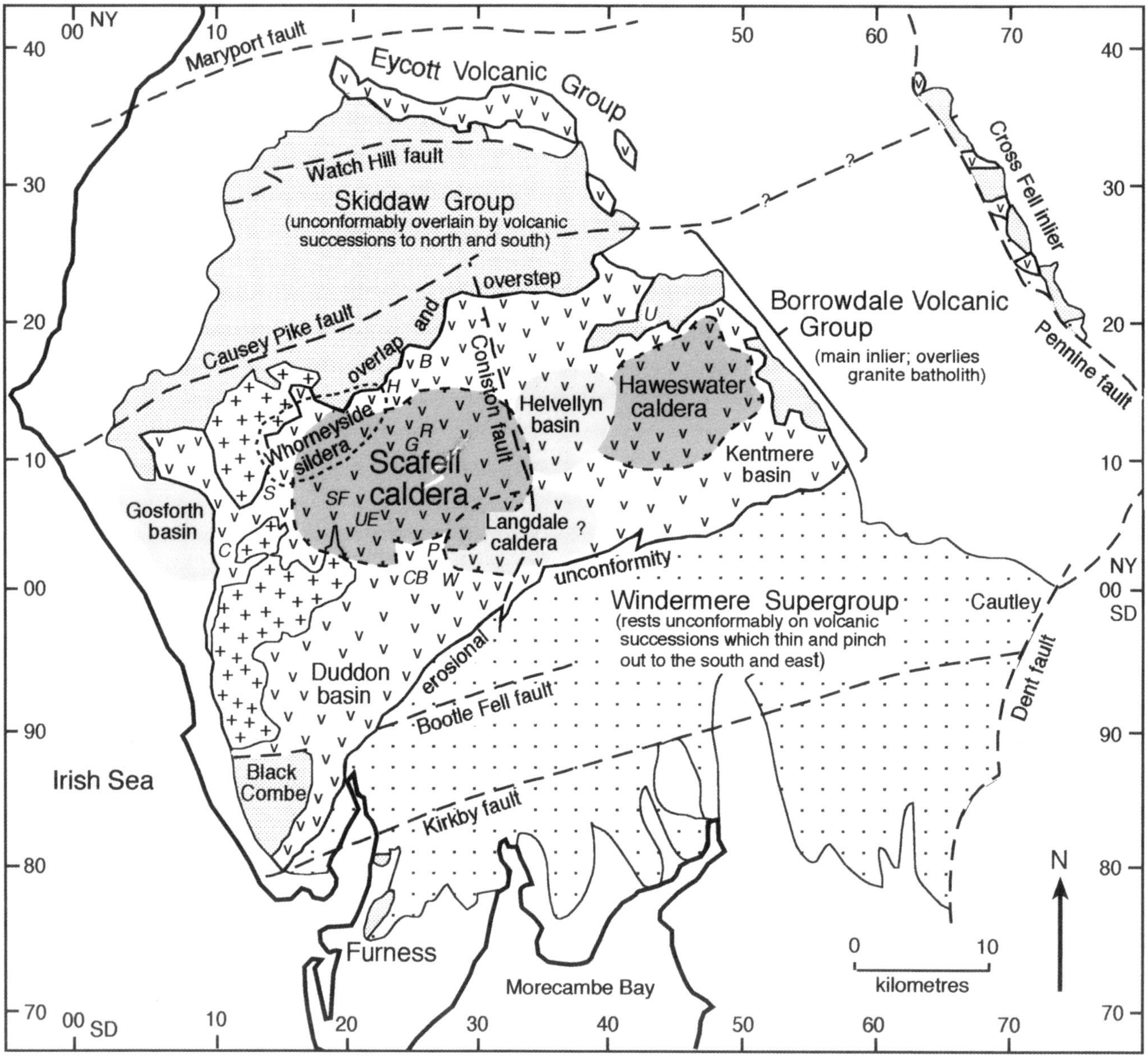

Fig. 5.6. Map of the English Lake District, NW England, showing the location of the volcanic inliers (v ornament), the main caldera volcanoes and basinal depocentres, and place names mentioned in the text: *B*, Borrowdale; *C*, Craghouse centre; *CB*, Cockley Beck; *E*, Upper Eskdale; *G*, Glaramara; *H*, Honister; *P*, Pike o' Blisco; *R*, Rosthwaite Fell; *S*, Seatallan; *SF*, Scafell; *U*, Ullswater; *W*, Wrynose Pass. Subvolcanic granitic intrusions (crosses) and inferred basement faults (dashed lines). Modified from BGS maps and Millward (2002).

Volcanic environments in northern England

Several overlapping major eruptive centres, lava piles, and basins infilled with volcaniclastic detritus are represented in northern England. The larger volcanoes were polygenetic, involving multiple eruptions of lava or ash. Some underwent repeated caldera collapse during successive large-volume explosive eruptions. Lava flows and domes were extruded, and the volcanoes were intruded by numerous dykes, sills and some larger tabular-shaped subvolcanic magma bodies of similar compositions to the extrusives. The volcanic products were subjected to rapid subaerial erosion and aqueous resedimentation. As at modern caldera volcanoes, hydrothermal alteration was locally pervasive, with chemical alteration and the growth of minerals such as epidote, chlorite and sericite.

Evidence for the subaerial conditions include preserved tuff rings with cross-stratified accretionary-lapilli-bearing tuffs; lake deposits with desiccation cracks and small wave ripples; alluvial and laharic deposits; v-shaped erosion rills draped by ashfall layers; steep-sided box-canyons cut into unlithified ignimbrites and sediments; and a general absence of marine fossils and bioturbation (Branney 1988). There are also numerous cross-stratified tuff layers, which record fully dilute pyroclastic density currents with densities too low to have entered standing water.

The emerging picture for northern England is of a vigorously active volcanic area characterised by construction of low-profile lava shields and ignimbrite volcanoes with frequent volcanotectonic subsidence, flooding and infill. Marked subsidence is indicated by the survival from erosion of over 6 km thickness of predominantly subaerial volcanic strata; by the widespread intercalation of proximal volcanic facies with lacustrine volcaniclastic sediments; by close-spaced large-displacement volcanotectonic growth-faults, some of which bound contrasting local successions that are overstepped by higher units in the volcanic succession; and by numerous fine-grained phreatomagmatic pyroclastic layers (Fig. 5.7B) which suggest that large volumes of water periodically gained rapid

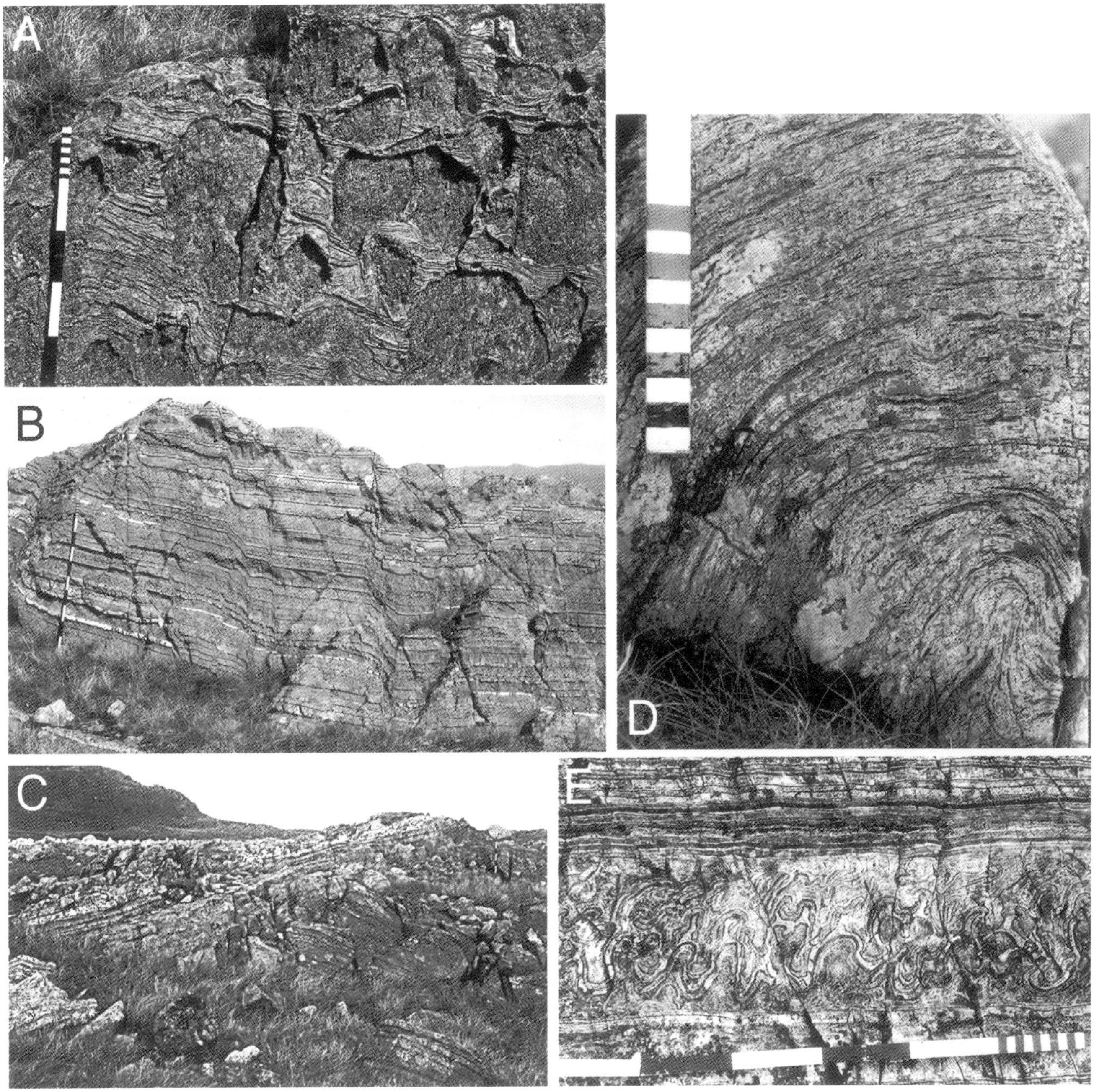

Fig. 5.7. Representative lithofacies from the continental arc volcanics of the Borrowdale Volcanic Group (Photos: M. Branney). (**A**) Laminated sandstone and siltstone infills and buries the upper autobreccia of a pre-caldera subaerial andesite block-lava at Wrynose Pass, Cumbria (Fig. 5.6). (**B**) Part of the 35 m-thick subaerial ashfall deposit from the vast Whorneyside phreatoplinian eruption. The fine tuff layers contain small ash pellets. The abundant soft-state faults formed prior to lithification, during caldera collapse. (**C**) Subaerial unconformities in the volcanic successions are common. Here, silicic ashfall and thin ignimbrites (Stonesty Tuff, Fig. 5.3; pale, dipping left) unconformably drape the andesitic Whorneyside phreatoplinian tuff (darker, dipping right), Crinkle Crags. No time-gap or uplift is inferred – the slumped Whorneyside ash was immediately buried before it could be eroded away. (**D**) Rheomorphic fold in welded silicic caldera-fill ignimbrite (Crinkle Tuffs; Fig. 5.3). The lenticular fiamme represent collapsed pumice lapilli. (**E**) Caldera lake siltstones with soft-sediment convolutions caused by rapid accumulation and post-climactic seismic shock at Scafell caldera. Seathwaite Fell Formation (Fig. 5.8), Scafell Pike, Cumbria.

access to erupting magmas. The area may have resembled low-lying parts of the present-day Philippine and Japanese arcs, where volcanic and sedimentary units accumulate in active basins associated with large andesite–rhyodacite caldera volcanoes.

Environmental reconstructions must consider the likely preservation bias in favour of lavas, sills, welded rocks and lacustrine deposits because, in the absence of vegetation, layers of non-welded ash and pumice would have been particularly vulnerable to subaerial erosion unless rapidly buried (Branney 1988). To underscore this, most of the 200 m-thick ignimbrite from the 5–8 km^3 1991 explosive eruption of Mount Pinatubo in the Philippines was stripped away by erosion in just 10 years.

Basal relationships of the volcanic rocks

Relations between the volcanic successions and older marine sedimentary rocks vary in a complex way across the Lake District. The contact is locally faulted and, in addition to the overstep described above, there is marked overlap of successive volcanic units within the lower part of the volcanic succession. For example, some volcanic units at the contact along the northern margin of the main volcanic outcrop, e.g. near Borrowdale and Ullswater (Fig. 5.6), are not the oldest in the Borrowdale Volcanic Group; away from the contact, they rest upon older units within the volcanic succession, as though ponded against irregular SE-facing scarp(s) developed in the uplifted sedimentary rocks. When the strata (e.g. as mapped by Moseley 1964) are restored to horizontal, the contacts with the Skiddaw Group take the form of irregular southward-facing palaeotopographic slopes. Where subsidence was greatest, such as beneath Scafell caldera, the base of the volcanic succession is not seen. Growth structures with progressive subsidence and infill by successive volcanic units, rather than a single tall scarp, are indicated by a general scarcity of Skiddaw Group clasts in Borrowdale Volcanic Group sedimentary rocks above the lowermost exposed strata.

Timing and duration of the volcanism and related intrusions

Radiometric dates from the Borrowdale Volcanic Group include 457 ± 4 Ma (Sm–Nd on garnet–whole-rock pairs; Thirlwall & Fitton 1983), and 451.6 ± 1.4 and 452.8 ± 0.7 Ma (U–Pb zircon ages; Millward & Evans 2003). These dates are within error of those of closely related subvolcanic granitic intrusions (Threlkeld, Carrock, Ennerdale and Eskdale; Soper *et al.* 1987; Hughes *et al.* 1996; Millward & Evans 2003).

It has been proposed that some of the larger intrusions post-dated the volcanism causing post-volcanic uplift and erosion (Millward 2004). However, large magma chambers must have been already in place at the time the of climactic caldera-forming eruptions. These will have continued to evolve and intrude long after caldera collapse, so the present intrusions do not represent the magmas at the time of collapse. Nevertheless, a geographical association between the locations of the intrusions and the thickest preserved volcanic successions, suggests that the intrusions are predominantly associated with subsidence rather than with post-volcanic uplift.

The Borrowdale and Eycott volcanic groups are enclosed by fossil-bearing strata that constrain the main volcanic episode in northern England as having occurred within approximately 15 Ma, from late Llanvirn to mid-Caradoc times. Onset of volcanism in late Llanvirn (Llandeilian) times is possible in as much as the pre-volcanic sedimentary strata in the Lake District and Cross Fell inliers extend up into the *D. murchisoni* Biozone of the Llanvirn (Tarn Moor and Kirkland formations; see previous chapter), and no unequivocal uppermost Llanvirn rocks have been reported from northern England (Fortey *et al.* 2000). In the north of the Lake District and at Cross Fell to the east, the volcanic successions are succeeded by Longvillian marine sedimentary strata. In the south of the Lake District the volcanic rocks are unconformably overlain by Ashgillian strata (Windermere Supergroup; see next section) and the gap in the record suggests that the arc rocks may have remained subaerial in this area until Ashgill times.

It is not certain whether the thick volcanic succession represents a large or trivial part of the approximately 15 Ma time interval represented by the stratigraphic envelope: volcanic successions can accumulate very rapidly, but there are numerous internal erosional unconformities and caldera fields elsewhere remain active for periods ranging from approximately 1.6 Ma (Houghton *et al.* 1995) to ≥ 15 Ma (e.g. Lipman 1984). Palaeomagnetic evidence indicates that the Borrowdale–Eycott volcanic episode may have lasted less than 5 Ma, within a single palaeomagnetic polarity chron that extended from late Llanvirn to early Caradoc times (Piper *et al.* 1997). This is consistent with available radiometric data (see Millward & Evans 2003) but remains unconfirmed because radiometric dates have not been obtained from the oldest and youngest volcanic units.

Magma compositions

In northern England the Ordovician arc rocks range from basalt to rhyolite, with calc-alkaline andesite and rhyodacite being most abundant (Beddoe-Stephens *et al.* 1995; Kanaris-Sotiriou *et al.* 2002). Many are porphyritic with plagioclase phenocrysts, and some contain one or more of olivine, orthopyroxene, clinopyroxene, alkali feldspars, biotite, iron-titanium oxides and almandine-pyrope garnet in more silicic units. There is evidence of mingling between silicic, intermediate and mafic magmas (Walker 1904; Allen *et al.* 1987). The mafic magmas all appear to have been derived from a similar subcontinental mantle source. Incorporation of various additions from the subducting slab has given rise to several magmatic series with no obvious overall spatial or temporal trend (Mathieson 1986; Macdonald *et al.* 1988; Beddoe-Stephens *et al.* 1995; Millward *et al.* 2000*c*). Radiogenic isotopes indicate the addition of crust to some of the magmas (Thirlwall 1989; McConnell *et al.* 2002). Subordinate suites with a tholeiitic iron-enrichment trend occur within both the Borrowdale and Eycott groups (e.g. Macdonald *et al.* 1988).

Internal stratigraphy: problems and methods

The internal stratigraphy of the Borrowdale and Eycott volcanic groups is far from 'layer-cake': vertical sections at different locations are dissimilar and this reflects the occurrence of different local successions arranged adjacent to each other. Resolving the stratigraphic relations has proved challenging. The only fossils, other than trace fossils, found in the groups are acritarchs thought to have been reworked from subjacent Skiddaw Group rocks (Holehouse Gill and Whinny Bank formations; Molyneux, in Millward 2002), so resolution of the volcanic stratigraphy relies entirely upon lithostratigraphic correlation. This presents a non-trivial problem because the successions are intensely faulted, including repetitions by S- to SE-verging thrusts, and because there are numerous irregular subaerial erosion surfaces, impersistent units, and abrupt lateral facies and thickness changes. This intricacy resulted from irregular volcanic aggradation and infill of a fast-changing subaerial topography that was subjected to rapid and variable volcanotectonic subsidence and, probably, localized uplift. The localized rather than regional nature of the subsidence is revealed by adding together the maximum thicknesses of each localized stratal succession. This gives a cumulative thickness for the Borrowdale Volcanic Group of approximately 13 km, more than twice the true thickness of the group at any one location. It is thus not surprising that there are a bewildering number of local formations.

Early workers attempted to place successions of similar rocks types, such as andesites, into a local formation and then distinguish this from a nearby similar grouping of, say, clastic units, designated as another formation. Clastic units were typically referred to as 'tuffs' even though many are sedimentary rather than pyroclastic in origin. The limited success of this approach became evident when workers were unable to correlate between the locally established successions. Also problematic was that many of the andesite sheets regarded of as lavas proved to be sills.

Andesite sills occur at many levels within the volcanic successions. Most do not exhibit baked margins and many resemble lavas with blocky autobreccias (Branney & Suthren 1988). They are vesicular, autobrecciated sills intruded at shallow levels into wet host deposits, so that film-boiling of pore water fluidized the sedimentary host around the intruding magma (Kokelaar 1982). Their intrusive origin is revealed locally at upper contacts by peperite textures formed by mingling of the magma with wet sediment. There also is evidence for hydroclastic fragmentation of the magma and for disaggregation and vesiculation of the clastic host rocks (Branney & Suthren 1988). Peperitic sills have subsequently been identified elsewhere in both the Borrowdale and Eycott volcanic groups (Branney 1991; Millward &

Molyneux 1992; Millward *et al.* 2000*b*; Kanaris-Sotiriou *et al.* 2002). Andesite and basalt lavas also occur abundantly within the successions (Fig. 5.7A); their extrusive origin is locally revealed by the presence of clasts reworked from the upper autobreccias into overlying and adjacent sediments (Branney & Suthren 1988). However, many of the andesite sheets do not expose diagnostic features, and their extrusive v. intrusive status remains equivocal; this renders them, as well as the known sills, redundant for lithostratigraphic purposes.

A breakthrough in resolving the volcanic stratigraphy was the application of process-based tephrostratigraphy, in which widespread pyroclastic layers deposited by fallout and from long run-out pyroclastic density currents are treated as near-isochronous stratigraphic markers (Fig. 5.10B, C). Although some of these layers have trivial thicknesses (< 1 m), many have distinctive, internal, mm to cm-scale micro-stratigraphies that are laterally persistent. This allows them to be traced with confidence through the otherwise highly variable succession of impersistent volcanic and sedimentary units. Some pyroclastic layers show systematic regional variations, such as in the maximum size of lithic clasts, and these may be used to infer source locations (Branney 1991; Brown 2001). Ashfall layers just a few centimetres thick have been traced for tens of kilometres, in the same way as is typically undertaken at modern explosive volcanoes (e.g. Pyle 1989). Tracing these layers has led to: (1) the discovery that the intervening volcanic units exhibit abrupt lateral changes in thickness and facies; (2) the discovery of numerous large-displacement (> 1 km) dip-slip faults; (3) the designation to different stratigraphic levels of adjacent successions of andesite sheets that were formerly considered to be correlatives; and (4) the discovery of successions repeated by thrusts (e.g. at Great Langdale and Wrynose). However, detailed tephrostratigraphy is time-consuming and has yet to be carried out across much of the volcanic outcrop, where stratigraphic uncertainties remain despite detailed geological mapping. For example, the relative ages of some of the inferred volcanic centres (see below) are not known.

Interpreting the volcanic successions

The andesite volcanoes

As first noted by Suthren (1977), there is no evidence for the tall, steep-sided stratovolcanoes that are so prominent along modern destructive plate margins, such as Popocatépetl and Cotopaxi in central Mexico and the Andes. Such steep edifices have poor preservation potential as they are prone to rapid erosion and periodic sector collapse. Also, given their typical wide spacing along continental arcs, it is possible that the locations of tall stratovolcanoes in the arc simply did not coincide with current outcrops. Rather, the andesite volcanoes in the Lake District probably had the form of low-profile shields and lava fields that aggraded within volcanic and tectonic-subsidence structures. The abundant presence of sedimentary layers between the andesite sheets, and the great thickness of individual lava successions relative to the limited topography evident on their upper surfaces suggests that the andesite volcanoes were inverse structures that grew predominantly by filling accommodation spaces created by subsidence. For example, one ignimbrite (the Whorneyside ignimbrite; Fig. 5.8), which is no more than 120 m thick, overlies several different adjacent lava successions, each of which is several hundreds of metres thick (Fig. 5.9). This indicates that the stratovolcanoes had a very subdued topography (*c.* 70 m, after allowing for ignimbrite compaction) and were easily buried. Partially ponded lava successions of this type are common within modern calderas worldwide, and can reach significant thicknesses without developing steep topographic flanks. For example, successive calderas at Santorini volcano, Greece, have been progressively filled by shield-like effusive centres (Druitt *et al.* 1999). The 'inverse stratovolcanoes' of the Lake District are lava-dominated in parts, but the lower parts of some are dominated by sill (e.g. at Honister Pass; Fig. 5.6).

Lava fields occur around Wrynose, upper Eskdale, and Seatallan (Branney & Suthren 1988; Petterson *et al.* 1992) (Fig. 5.6). These include successions of thin aa basalt lavas and thick dacite block lavas, but most lavas are andesitic, typically 5–50 m thick, with upper, lower and marginal blocky autobreccias (Fig. 5.7 A) surrounding a massive or flow-banded central part (e.g. in the Lingcove Formation of Branney & Soper 1988; the Birker Fell Formation, Petterson *et al.* 1992; and, further north, in the lower part of the Eycott Volcanic Group, Millward *et al.* 2000*b*). In several locations they are associated with shallow-level sills.

It is difficult to estimate the durations of the effusive eruptions, and of the intervening repose periods, because no soils or regoliths are identified. Thicknesses (or absences) of volcaniclastic layers between the lavas give little indication of the length of time gaps between eruptions because volcaniclastic sedimentation is typically convulsive and ephemeral, and thicknesses of sedimentary layers on subaerial volcanoes primarily reflect topographic context, accommodation, and preservation factors rather than time.

Explosive eruptions

Some of the oldest known units in the Borrowdale Volcanic Group record phreatic and phreatomagmatic explosive eruptions and sedimentary reworking. These include the laterally impersistent tuffs in the west (the basaltic Devoke Water Member of Millward *et al.* 2000*b*) and in the SW (the Whinny Bank Formation of Mathieson 1986; and Johnson *et al.* 2001) that represent remnant tuff rings. They contain clasts derived from the Skiddaw Group, and some layers contain ash pellets composed of Skiddaw Group mud and silt. This indicates that the early eruptions were phreatomagmatic, with explosive interaction between rising hot magma and groundwater within the uplifted Skiddaw Group. The pellet layers contains acritarchs that had once been taken to indicate that the volcanic rocks were Lower Ordovician, but the acritarchs are now thought to be derived from the Skiddaw Group (Rushton & Molyneux 1989). It seems that the phreatomagmatic eruptions ejected Skiddaw Group sediments along with the contained acritarchs high into the atmosphere, to fallout as ash pellets. Silicic phreatomagmatic tuff rings are recorded at higher stratigraphic levels, e.g. at Glaramara (Glaramara Tuff of Brown 2001; Fig. 5.8) and around Kentmere (Woundale Tuff Formation of Millward 2004).

The lower part of the Borrowdale Volcanic Group is grouped together on BGS maps as the Birker Fell Formation, and is commonly reported as being dominated by lavas (Millward 2004). The nearly 3 km-thick succession, however, is a varied succession of basaltic, andesitic and dacitic lavas, sills, pyroclastic layers and volcaniclastic sedimentary deposits. Andesitic volcanism typically involves explosive activity, and the record of explosivity is probably under-represented because loose subaerial pyroclastic deposits tend to erode readily, and become reworked by lahars and rivers. Despite this, significant early explosive episodes are indicated by numerous pyroclastic layers and their reworked derivatives, including the following. (1) Andesitic volcaniclastic lake sediments intruded by andesite sills occur in contact with the Skiddaw Group at Honister Pass (Fig. 5.6) and may be a lacustrine record of explosive eruptions preserved in an early volcanic subsidence structure; they resemble caldera lake successions elsewhere. (2) A Plinian fall layer overlain by an andesitic–dacitic, compositionally zoned ignimbrite succession, more than 1000 m thick, may record a major caldera eruption in the vicinity of Craghouse (Fig. 5.6; Petterson *et al.* 1992). (3) Dacitic breccias are thought to record lava-derived block-and-ash flows around Ennerdale, and east of Haweswater (Millward 2004). (4) Garnet-bearing tuffs and lapilli-tuffs record explosive eruptions associated with a

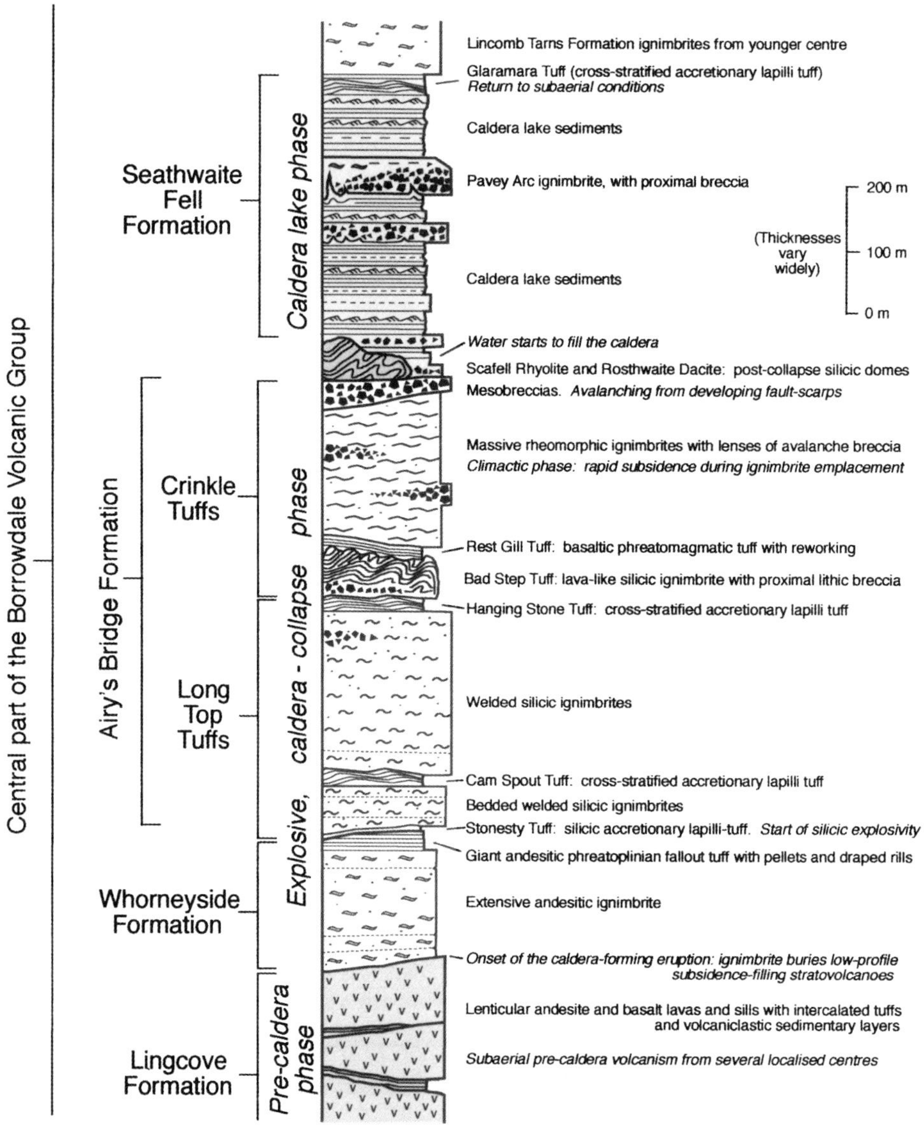

Fig. 5.8. Simplified stratigraphy of Scafell caldera volcano, just one of several nested volcanic centres in the English Lake District (see Fig. 5.6). The localized succession of subaerial tuffs record a single, protracted caldera collapse eruption and are succeeded by a caldera lake succession.

dacite dome near Cockly Beck in the Duddon valley (Fig. 5.6) (Millward 2004). (5) A diverse succession, 120 m thick, of ignimbrites (both welded and non-welded), block- and ash-flow deposits, Plinian pumice fall and ashfall layers occur interbedded with fluvial and lahar deposits, some containing metre-sized blocks, and record a protracted explosive eruption-episode (Blisco Formation of Branney & Soper 1988). This succession is thickest at Pike 'o Blisco (Fig. 5.6), where it abuts steep andesite palaeotopography, and can be traced southwards for 16 km along the Duddon valley. It includes a silicic ignimbrite (Little Stand Tuff) and an extensive dacite blocky-lava (Great Whinscale Dacite), which is nearly aphyric and thought to have been erupted at high temperatures (Kanaris-Sotiriou *et al.* 1991). (6) Accretionary lapilli-bearing stratified basaltic tuffs occur intercalated with aa basalt lavas, such as at Throstle Garth in upper Eskdale, and record explosive basaltic eruptions.

Explosive caldera volcanoes and ignimbrites

Several explosive eruptions produced devastating pyroclastic density currents, which deposited large-volume ignimbrites. These range from andesitic to rhyolitic. Many of the ignimbrites were emplaced hot and developed columnar cooling joints. They vary from non-welded to intensely welded, as revealed by microscopic vitroclastic textures and the presence of relic spherulites, which indicate the matrix was once a vitrophyre. Some can be traced to their source volcanoes, for example where coarse lithic breccias indicate proximity to source vents (e.g. Branney *et al.* 1992). Source locations of others are unknown, and cannot easily be derived from thickness variations because ignimbrites tend to thicken into topographic depressions even away from source. Estimation of eruption volumes from the caldera volcanoes such as Scafell caldera (see below) is hindered by the lack of correlated outflow sheets. However, the dimensions of the calderas suggests that some of the eruptions were super-eruptions of very large-magnitude, probably VEI 7 on the volcanic explosivity index of Newhall & Self (1982). Ignimbrites known locally as the Lincomb Tarns Formation (Oliver 1961) have been correlated widely across the Lake District and may represent an ignimbrite shield, approximately 40 km in diameter and as much as 800 m thick, east of the Coniston fault (Millward 2004) (Fig. 5.6). In west Cumbria a localized succession of thick welded ignimbrites has been discovered in boreholes under Gosforth (Fig. 5.6). The ignimbrites

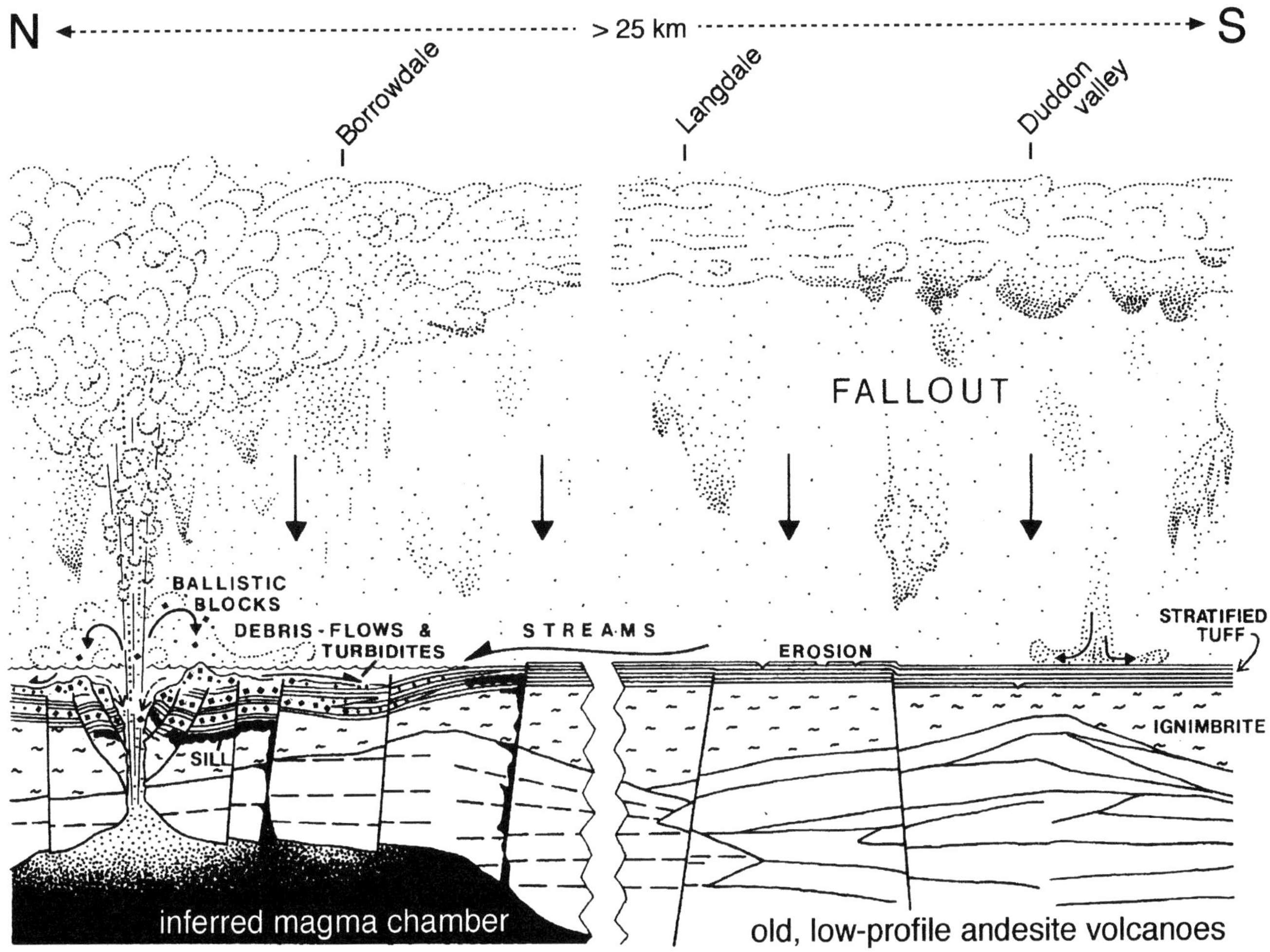

Fig. 5.9. The catastrophic phreatoplinian phase of the explosive Whorneyside eruption in the English Lake District (modified after Branney 1991). The initial ignimbrite eruption caused proximal subsidence (left), which allowed ingress of water to the vent area, transforming the eruption to violent phreatoplinian. Rainfall accompanied ash fallout, causing minor rill erosion of the subaerial tephra and reworking into the rapidly subsiding subaqueous depocentre, where subsiding proximal deposits were being intruded by peperitic andesite sills (black). The eruption may have lasted from weeks to months.

display rapid thickness variations, and the area may represent a concealed caldera volcano (Akhurst *et al.* 1997).

An interesting facet of the volcanic succession is the paucity of obvious Plinian pumice fall layers in comparison with many modern subaerial continental arc volcanoes, where they are typically abundant. It may be an artefact of poor pumice-preservation or it might indicate that the large-scale explosivity was predominantly phreatomagmatic rather than magmatic (Plinian); this is consistent with the abundance throughout the succession of fine ashfall layers with ash pellets (see below).

Scafell caldera volcano

Scafell caldera volcano in the central fells of the Lake District is the best known volcano in northern England. It is a large andesite–rhyolite volcano with a caldera at least 17 km from east to west and 14 km from north to south (Fig. 5.6), filled with an alternation of welded ignimbrites and phreatomagmatic tuffs (Fig. 5.8) (Branney & Kokelaar 1994). The alternation resulted from rapid changes in the style of the explosive eruptions as faults moved, vents migrated and the explosivity fluctuated between magmatic and phreatomagmatic as lake- or groundwater gained intermittent access to the erupting magma. The resultant layering of the caldera-fill usefully both picks out the faults and allows the eruption and subsidence history to be reconstructed with a degree of detail that is not possible at many caldera volcanoes, which apparently have non-layered fills (e.g. Lipman 1984). Outflow correlatives of the intra-caldera ignimbrites have not yet been identified.

Caldera collapse started with the eruptions of a large-volume andesitic ignimbrite (Whorneyside ignimbrite; Fig. 5.10A) and caused rapid ingress of water to the vent region, northwest of Borrowdale. This transformed the explosive eruption style to phreatoplinian explosivity, in which steam expansion enhanced what was already a vigorously explosive eruption (Fig. 5.9). The resultant fine-grained, parallel-laminated ashfall deposit (Whorneyside phreatomagmatic tuff of Branney 1991, fig. 369B) contains abundant ash pellets that fell from a vast, moist umbrella cloud (Fig. 5.9). It is remarkably thick (35 m), and has been traced for over 25 km from source with no discernible decrease in thickness; it probably represents the largest-known eruption of its type worldwide. Such giant phreatomagmatic eruptions create rainfall, and during this eruption the run-off caused minor rill erosion and reworking of the subaerial tephra into a proximal lake. Degassed andesite leaked up into the watery deposits of the subsiding proximal lacustrine depocentre (Fig. 5.9), to form a stack of peperitic sills that exceeds 400 m thick. The resultant sill-filled volcanic subsidence structure, or 'sildera', in Borrowdale is approximately 10 km across (Fig. 5.6), and had a subdued topographic expression with the internal form of an inverse stratovolcano.

Subsidence-induced faulting started to tap rhyolitic magma even before the andesitic Whorneyside eruption had ceased,

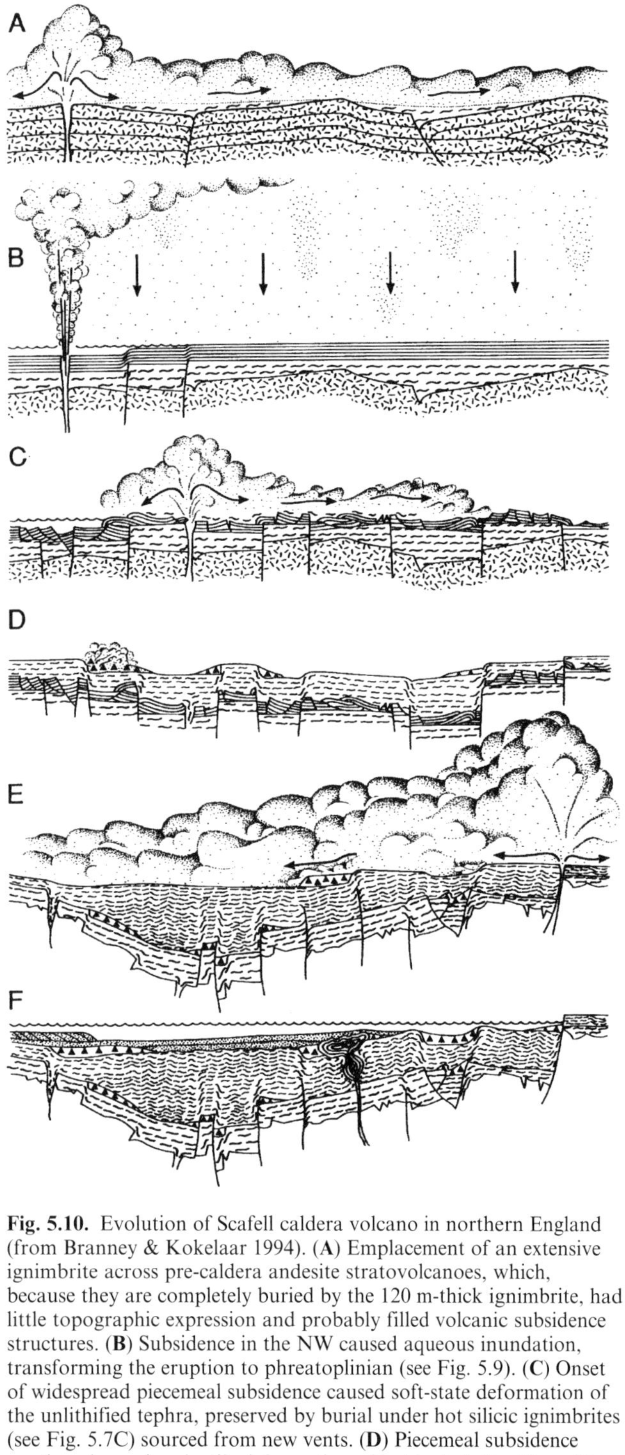

Fig. 5.10. Evolution of Scafell caldera volcano in northern England (from Branney & Kokelaar 1994). (**A**) Emplacement of an extensive ignimbrite across pre-caldera andesite stratovolcanoes, which, because they are completely buried by the 120 m-thick ignimbrite, had little topographic expression and probably filled volcanic subsidence structures. (**B**) Subsidence in the NW caused aqueous inundation, transforming the eruption to phreatoplinian (see Fig. 5.9). (**C**) Onset of widespread piecemeal subsidence caused soft-state deformation of the unlithified tephra, preserved by burial under hot silicic ignimbrites (see Fig. 5.7C) sourced from new vents. (**D**) Piecemeal subsidence continued, causing ductile deformation of hot ignimbrites (Long Top Tuffs) and fault-scarp collapse. (**E**) The eruption climaxed with emplacement of very hot rheomorphic ignimbrites (Crinkle Tuffs; Fig. 5.7D), ponded where subsidence was greatest. Fault blocks tilted and tuff-filled fissures formed. (**F**) A caldera lake occupied the central fells of the Lake District. Volcaniclastic lake sediments (Fig. 5.7E) bury the caldera-filling ignimbrites and post-collapse silicic domes.

and the site of explosivity shifted southwards (Fig. 5.10C), with the rapid emplacement of a succession of welded silicic ignimbrites intercalated with thin, cross-stratified accretionary-lapilli-bearing deposits, collectively known as the Long Top Tuffs (Fig. 5.8) (Branney & Soper 1988). The extent of caldera subsidence migrated and widened to the SE across the Langdale region. The phreatomagmatic ash layers are affected by steep zones less than 10 m wide, of particularly intense soft-state deformation that relate closely to caldera floor faults, and show that the faults moved tens of metres prior to lithification. Steep ductile shear zones in welded ignimbrites along the faults (Fig. 5.10D) show that the caldera-floor faults moved before the hot caldera-fill ignimbrites cooled (Branney & Kokelaar 1994).

The explosive eruption episode at Scafell caldera climaxed with the emplacement of thick rhyolitic ignimbrites, known as the Crinkle Tuffs (Fig. 5.8). These were so hot on deposition that they continued to spread and flow downslope along the ground rather like a lava flow, a process known as rheomorphism. They exhibit spectacular flow-folds and, although most of them preserve fiamme (Fig. 5.7 D) and welded shards, some have developed flow-banding and look almost identical to lava (the Bad Step Tuff; Branney *et al.* 1992). Such eruptions are particularly awesome, and have occurred in more recent times; for example, in the Snake-River–Yellowstone volcanic province of NW USA about 12–8 Ma ago (Bonnichsen 1982) but are not well understood.

Scafell caldera is probably the best example of a piecemeal caldera; that is, a caldera that, rather than subsiding wholesale like a piston, subsided in a piecemeal fashion along numerous close-spaced caldera floor faults (Fig. 5.10). The timing and changing pattern of the successive increments of subsidence, block tilting and fault reactivation is revealed by marked lateral thickness variations in successive ignimbrites (Branney & Kokelaar 1994). In Langdale, for example, the Bad Step Tuff (Fig. 5.8) changes from 40 m to ≥200 m thick across the Oxendale fault, and from 300 to 500 m across the Grave Gill Fault, indicating that downthrow displacements of hundreds of metres occurred, probably in just minutes. Such rapid displacements are at least an order of magnitude larger than is reported for tectonic faults. Tall fault scarps formed and rapidly collapsed, shedding debris into adjacent graben, where it became intercalated with the ignimbrites (Fig. 5.10E); bases of some of these avalanche breccias show load-structures where blocks fell and sank into extremely hot and still plastic ignimbrite. As subsidence progressed, avalanching increased and breccia lenses towards the top of the caldera-fill locally reach 70 m in thickness (Fig. 5.8). Palaeomagnetic evidence confirms that downsag (inward tilting) accompanied caldera subsidence (Fig. 5.10E) prior to post-volcanic tectonism (Channell & McCabe 1992; Piper *et al.* 1997).

After the climactic phase of the Scafell eruption, relatively degassed silicic magma extruded as viscous lava domes and coulees in the vicinity of Scafell and Rosthwaite Fell (Fig. 5.6). These have flow-folded flow banding, autobreccia carapaces and aprons of blocky talus. Current exposures reveal cross-sections though the eruption conduits. The subaerial caldera filled with water (Fig. 5.10F), and the Seathwaite Fell Formation (Fig. 5.7E) is one of the best-exposed caldera-lake sequences in the world. Rapid sedimentation in the lake occurred from turbidity currents and debris flows, locally forming prograding delta fans, with input from further explosive eruptions, and with local current reworking. Wave reworking is restricted to the top and bottom of the sequence, where shallow water depths are recorded. High sediment accumulation rates coupled with continuing seismic instability and de-watering caused spectacular soft-state deformation and convolution (Fig. 5.7E). One large explosive eruption produced a pyroclastic density current, part of which flowed into the lake, producing a lacustrine ignimbrite with a spatter-bearing proximal breccia, some 120 m thick (Pavey Ark ignimbrite; Kokelaar *et al.* 2006) (Fig. 5.8). Similar, spatter-bearing proximal breccias formed during caldera-collapse eruptions are found on the shores of modern flooded calderas such as Santorini (Greece), Taal (Philippines) and Phlegrean Fields (Italy); they may record explosive disruption of hydrothermal systems formed under the standing water (Mellors & Sparks 1991; Rosi *et al.* 1996). The lacustrine succession is complex and widely exceeds 500 m in thickness. Upper parts record emergence, with a return to subaerial conditions recorded by cross-stratified accretionary-lapilli-bearing tuffs that record passage of fully dilute pyroclastic density currents (e.g. the Glaramara Tuff, Brown 2001) (Fig. 5.8).

Overall, the location of subsidence at Scafell caldera migrated ESE during its lifetime, from early subsidence in the NW (the Whorneyside 'sildera'; Fig. 5.6) followed by subsidence associated with the Long Top Tuffs west of Langdale (Branney & Kokelaar 1994), that associated with the Bad Step Tuff in Langdale (Branney *et al.* 1992) and subsidence accompanying emplacement of the Pavey Arc ignimbrite predominantly to the east of Langdale (Kokelaar *et al.* 2006, fig.1).

A caldera at Haweswater

A succession remarkably similar to that of Scafell caldera occurs in the east of the Lake District National Park (Fig. 5.6) and has been interpreted as another caldera volcano (Millward 2004). The lowermost pyroclastic units are andesitic to dacitic tuffs (Scalehow Formation) and are overlain by a succession, up to 550 m thick (Whelter Knotts Formation) of bedded silicic ignimbrites intercalated with thin phreatomagmatic cross-stratified and ashfall layers (similar to those of the Long Top Tuffs of Scafell caldera) overlain by more massive and more intensely welded, and rheomorphic layers (similar to the Crinkle Tuffs of Scafell caldera). Abrupt and marked thickness changes of ignimbrites and localised zones of soft-state deformation of tephra suggest that caldera subsidence was piecemeal (Millward 2004). The succession occupies a broadly similar overall stratigraphic position to Scafell caldera. Its similarity to the Scafell caldera succession suggests a possible direct correlation, but E–W correlation through Borrowdale volcanic succession is not sufficiently well established to determine whether or not one is older than the other.

Other depocentres

Distinct thick successions that include sills, ignimbrites, fallout tuffs, lavas, and sedimentary rocks occur at Duddon, Kentmere, Helvellyn, and Langdale within the upper part of the Borrowdale Volcanic Group. These define separate and partly fault-bound basins (Fig. 5.6); none have been correlated more widely (see Millward 2004). Some may represent eruptive centres, others tectonically controlled basins (see Johnson *et al.* 2001; Millward 2002; Millward *et al.* 2000*b* and references therein), and their volcanic and structural development merit further investigation. In the vicinity of the Duddon, a 2 km-thick succession of ignimbrites is overlain by 1 km-thick succession of andesitic volcaniclastic rocks intruded by andesite sills and may represent a caldera fill (Millward 2004). Some of the ignimbrites are densely welded and contain coarse lithic breccias. The Kentmere succession, approximately 3 km thick, comprises several successions of andesite sheets with fluvio-lacustrine volcaniclastic sediments, tuffs, and a silicic ignimbrite over 500 m thick. Some of the andesites resemble lavas and sills in the lower part of the Borrowdale Volcanic Group and may represent low-profile lava shields within the depocentre. Around Helvellyn (Fig. 5.6), a localized basinal succession, *c.* approximately 1600 m thick, occurs dominated by volcaniclastic sedimentary rocks with subordinate tuffs and andesites (Millward 2004).

In Langdale (Fig. 5.6), an explosive caldera volcano is indicated where the Side Pike ignimbrite, which post-dates Scafell caldera, thickens dramatically eastwards from 20 more than to 500 m thick across close-spaced volcanotectonic faults. The overlying andesitic lacustrine volcaniclastic succession thickens from 30 m to *c.* 1 km in the same direction, and includes some spectacular turbidites, exposed in quarries around Langdale (Fig. 5.11) and intruded by andesite sills. Between Side Pike and Wrynose Pass, a chaotic megabreccia with giant local and exotic megablocks up to 1 km in size, may represent the

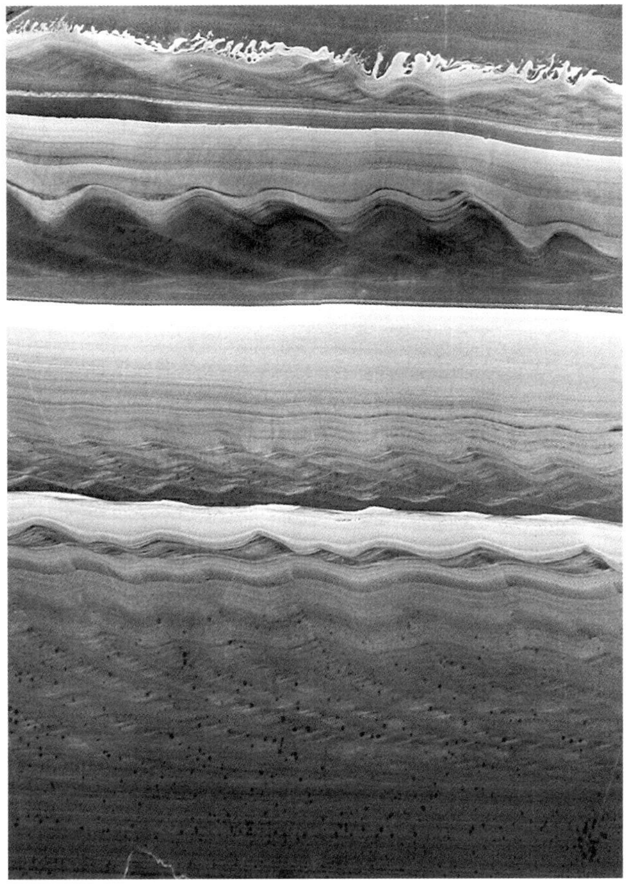

Fig. 5.11. Lacustrine turbidites from Langdale caldera lake (Fig. 5.6). Parallel stratified and climbing ripple-laminated sandstone divisions pass up into parallel-laminated siltstone and mudstone drapes (pale) and record rapid deposition from waning turbidity currents.
Note how the pattern of ripple-crest migration steepens during aggradation. Slab is 1.5 m high.

collapsed western margin of this caldera (Fig. 5.6) (Branney 1988; Brown 2001). This caldera probably probably relates to the Ambleside depocentre of Millward (2002) and would reward further investigation.

Closing stages of arc volcanism

Volcanism in the Lake District ceased in Caradoc–Ashgill times. This suggests that either subduction ceased or the location of arc magmatism shifted elsewhere. It has been suggested that this was linked to collision of Avalonia with Baltica, or a soft collision with Laurentia. An unconformity at the top of most Ordovician volcanic successions in northern England reflects marine inundation as a result of tectonic and/or thermal subsidence of the eroded subaerial arc as magmatism waned. Regional subsidence must have been substantial during the Ashgill marine transgression across the subaerial eroded volcanoes, because the transgression occurred at a time of global sea-level fall (Woodcock 2000). There is no evidence that tectonic shortening was involved in the formation of the unconformity at the base of the Windermere Supergroup; what were once thought to be folds unconformably overstepped by the base of the Windermere Group are reinterpreted as strata faulted and tilted during arc volcano-tectonic activity and subsidence prior the marine transgression (Branney & Soper 1988). Volcanism continued impersistently, until Ashgill times, with silicic explosive eruptions recorded in several erosional remnants within the Dent Group (Appletreeworth, High Haume and Yarlside formations, Millward 2004; and the Cautley and Dam House Bridge members – Fig. 5.16 – in the succeeding Windermere

Supergroup; Arthurton *et al.* 1988). The limited thickness of these remnants gives little impression of the erupted volumes, and more likely reflects limited accommodation space at a time when subsidence was less rapid than it had been earlier in Caradoc times. The Yarlside Volcanic Formation comprises pyroclastic deposits and an enigmatic 185 m-thick lava-like rhyolite, the Yarlside (formerly Stockdale) Rhyolite, which may be an extensive silicic lava or a rheomorphic ignimbrite (Stephenson *et al.* 1999). No proximal volcanic facies are preserved higher in the succession, although K-bentonite beds record more distant explosive silicic eruptions (Fortey *et al.* 1996).

There is still much to learn about the volcanoes of northern England. The main interest lies, perhaps, not just in reconstructing ancient regional geotectonics from a fragmentary record, but in taking advantage of the extraordinary exposure and deep dissection in the high fells of the Lake District to learn about how large explosive caldera volcanoes evolve with time. Such research is complimentary to studies of active caldera volcanoes, where exposure is limited to surficial levels, and where the caldera floors, faults and eruption conduits are generally concealed. The Lake District rocks have revealed spectacular volcanic phenomena that have not been gleaned from modern volcanoes, such as subsiding inverse volcanoes (silderas), and the piecemeal style of caldera collapse in which centres of subsidence and vent patterns shift and change during a single, protracted eruption. They have revealed phreatoplinian eruptions far larger than are known from modern volcanoes, and the area offers some of the world's best-exposed caldera lake successions (Figs 5.7D & 5.11), with abundant evidence for what happens when lavas and pyroclastic currents encounter water. The area would richly reward further research into, for example, what precursory activity typically heralds the start of explosive super-eruptions, and how vent locations and fault patterns evolve during caldera collapse.

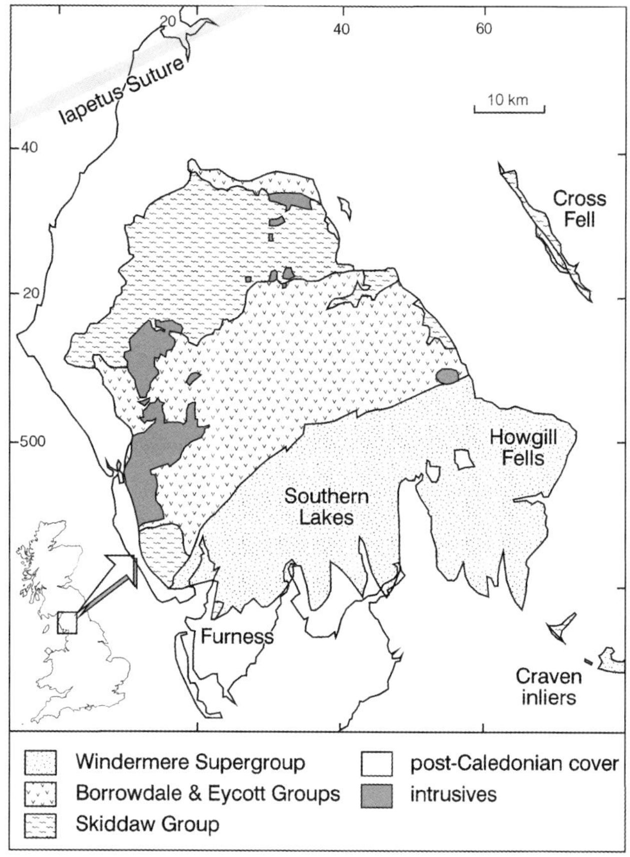

Fig. 5.12. Geological map of the Lake District, showing the outcrop of the volcanic arc succession (Borrowdale & Eycott Groups) and the post-volcanic succession (Windermere Supergroup) in relation to the surface trace of the Iapetus Suture. Named regions are referred to on Fig. 5.3.

The Late Ordovician and Silurian Windermere Supergroup: a post-volcanic foreland basin

Basin setting and subdivision

The eruptive activity of the Borrowdale Volcanic Group continued at least into early Caradoc (early Burrellian) time, based on a marine microflora, possibly reworked, near the top of the group (Molyneux 1988). By late Burrellian time, the predominantly non-volcanic and marine sediments of the Windermere Supergroup were being deposited unconformably on the eroded remnants of the volcanic arc. The time interval between volcanic shutdown and the first marine flooding of the dead arc cannot have been more than about 5 Ma, and the rapid erosion potential of pyroclastic sediments would make a significantly shorter interval geologically plausible.

The Windermere Supergroup is assumed to have been deposited over the whole of the area of northwest England (Fig. 5.12), although the unit is now preserved in one extensive outcrop in the southern Lake District and in smaller inliers at Cross Fell and Craven (Fig. 5.12). Although the Windermere Supergroup is variable in detail across this area, its stratigraphy can readily be generalized at the group level (Fig. 5.13). Each group has a consistent first-order character, sedimentary environment and deposition rate.

The Dent Group (middle Caradoc–Ashgill) is a slowly-deposited, mostly shallow-marine succession of more or less calcareous oxic mudstone and limestone with some volcanic intercalations (Fig. 5.13). A trend of transgressive deepening is interrupted by a brief (<1 Ma) shallowing in the late Ashgill, recording the Hirnantian glacio-eustatic sea-level fall. The overlying Stockdale Group (mostly Llandovery) apparently continued the deepening trend with slowly deposited mudstone. The basin was by now sediment-starved and, except during late Llandovery time, predominantly anoxic. Basin anoxia continued during deposition of much of the succeeding Tranearth Group (Wenlock–lowest Ludlow). Sediment input, mostly of silt or very fine sand, steadily increased and some more substantial deep-water turbidite sands occur. Calcareous mudstone near the top of the group marks short-lived oxygenation of the basin rather than a return to shallow-marine conditions.

The upper part of the Windermere Supergroup accounts for little more than 10% of its time span, but about 80% of its thickness. The Coniston Group (lower Ludlow) comprises thick packets of rapidly deposited turbidite sandstone interspersed with more slowly deposited anoxic siltstones. This signature continued into the overlying Kendal Group (upper Ludlow–Přídolí), but beds thinned and basin anoxia waned as water depths decreased. Shallow-marine sandstone successions dominate the upper part of the Kendal Group. These units may have passed upwards again into thick non-marine successions, later eroded during the uplift associated with the Acadian Orogeny (Soper & Woodcock 2003).

The progressive increase in deposition rate up through the Windermere Supergroup (Fig. 5.13) requires a similar increase in sediment accommodation space and corresponding subsidence of the Avalonian basement. Increasing subsidence rates through time are characteristic of flexural foreland basins, and a strong case has been made (Kneller 1991; Kneller *et al.* 1993) that the Silurian Lake District Basin was formed by flexure ahead of the southward-advancing load of the Laurentian continent over Avalonia. The surface trace of the Iapetus

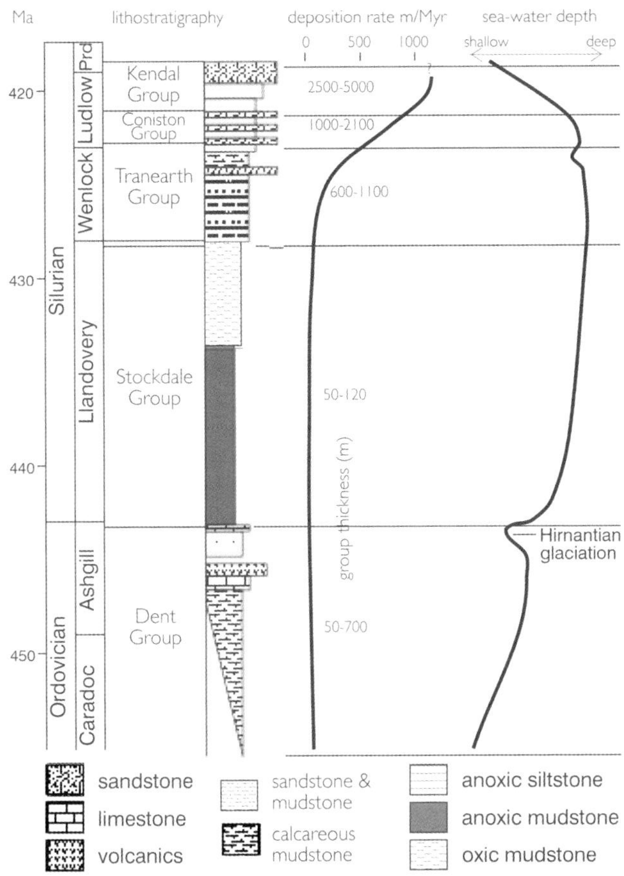

Fig. 5.13. A generalized lithological log of the Windermere Supergroup plotted against a linear chronometric scale in Ma (Rickards & Woodcock 2005). The range of thickness (m) within each lithostratigraphic group is used to derive the curve of deposition rate (m Ma^{-1}). The schematic first-order water-depth curve is deduced from local facies only. The time scale is from Gradstein & Ogg (1996).

Suture now lies about 50 km north of the preserved outcrop of the basin fill (Fig. 5.12). With this regional context in mind, the internal architecture of the components of the Windermere Supergroup can now be detailed.

Dent Group: mid-Caradoc–Ashgill

The Dent Group was deposited over the deeply incised topography of the formerly subaerial Borrowdale Volcanic Group. The Borrowdale succession in the southwest Lake District (Fig. 5.12) is at least 4 km thinner than in the Central Fells (Fortey *et al.* 2000), although mostly due to differential syndepositional subsidence rather than to intra-Caradoc erosion. In Furness and the Craven Inliers (Fig. 5.12) the Windermere Supergroup sits directly on pre-Borrowdale rocks, the Skiddaw and Ingleton Groups respectively. At a time of globally falling sea level (Fortey & Cocks 1986), the accommodation space for the Dent Group was probably created by thermal contraction of the subvolcanic lithosphere as magmatism waned.

A chronostratigraphic correlation chart (Fig. 5.14) shows the wide variation in the age of the base of the Dent Group, recording the strongly diachronous nature of the marine transgression over the extinct Borrowdale volcanic field. Attempts to map the progressive shoreline to this transgression (Ingham & Rickards 1974; McNamara 1979) are refined in Figure 5.15 using the shape of the Lake District batholith as a proxy for the post-volcanic topography. On this basis, the area had a steeper northern slope and a more gentle southeastern slope, forming the flank of a gulf with an opposing slope through the Craven area.

Accumulation of the Dent Group was by no means continuous (Fig. 5.14). Frequent unconformities suggest shallow-marine water depths, such that any relative sea-level fall could subaerially expose much of the Dent Group basin. A thickness plot across the basin (Fig. 5.16) shows the extent of these unconformities more clearly. The plot also shows the correlation between the total thickness of the Dent Group and its depositional site with respect to the Borrowdale volcano. Sites close to the volcano and transgressed late (e.g. sites 2–4, Fig. 5.16), accumulated thin successions whereas more distant sites such as Cross Fell and Cautley have thicker Dent Group sections. Despite the regional thermal subsidence, the volcano clearly remained a positive, perhaps partly subaerial, feature throughout late Ordovician time.

The Dent Group has a basal unit of sandstone or conglomerate with debris derived directly from the local land surface. This unit is typically less than 15 m thick, but thickens to 150 m in the Cross Fell Inlier (Fig. 5.16). It records a variety of marginal marine, mainly shoreface environments, together with possible subaerial debris-flow, lacustrine and fluviatile deposits (Millward *et al.* 2000*a*). Above this basal clastic unit, the predominant lithologies in the lower part of the Dent Group – in the Kirkley Bank, Cautley Mudstone and Dufton Shale formations (Fig. 5.15) for instance – are calcareous mudstone and nodular limestone, with more sporadic calcareous sandstone. A marine fauna, abundant in places, includes trilobites, corals, bryozoans, brachiopods, gastropods, ostracodes and echinoderms. The sandstones and some clastic limestones preserve sedimentary structures consistent with shallow marine deposition on a storm-influenced shelf. Sandstone units, such as the Billy's Beck and Wilsey Beck members, may have been supplied from relict islands of the Borrowdale volcano persisting in the carbonate-dominated shelf sea. In the finer grained limestone and mudstone units, bioturbation has typically destroyed most of the primary sedimentary structure. Such limestones are often concretionary and diagenetic in origin. One example is the Broughton Moor Formation (Figs 5.15 & 5.16), which apparently forms an unconformity-bounded sequence in the main Lakes outcrop.

The deposition of the Dent Group was interrupted by two episodes of acidic volcanism. Most voluminous of these episodes is the Yarlside Volcanic Formation (Cautleyan, Fig. 5.16 1), reaching up to 180 m thick. This unit comprises flow-banded rhyolite, interpreted as an ignimbrite (Millward & Lawrence 1985), and associated bedded tuffs and volcaniclastic sediments. Thinner, but more laterally extensive, units of primary or reworked rhyolitic ash occur in the Rawtheyan – the Appletreeworth Formation and the Dam House Bridge Tuff and Cautley Volcanic members (Figs 5.15 & 5.16). These volcanic episodes occurred some millions of years after shutdown of the Borrowdale arc, and have no obvious spatial or genetic relationship to it.

The Ashgill Formation forms the uppermost depositional sequence of the Dent Group, lying with significant disconformity on underlying units (Figs 5.15 & 5.16). The basal few metres comprise a nodular calcareous siltstone, the Cystoid Limestone Member, but the typical lithology of the Ashgill Formation is poorly calcareous bioturbated mudstone with a low-diversity shelly fauna, the *Hirnantia* Association. This restricted fauna is interpreted as a response to global cooling leading to the latest Ordovician glacial event (e.g. Brenchley & Newall 1984). In some sections, the climax of this event is marked by an increase in clastic input, giving laminated rather than bioturbated siltstones and fine sandstones, or even by coarser sandstones and conglomerates – the Rebecca Member and the Wharfe Conglomerate Member (Fig. 5.16). These units are assumed to record the glacio-eustatic sea-level fall and the resulting proximity of an eroding landmass (Brenchley 1988). Across the

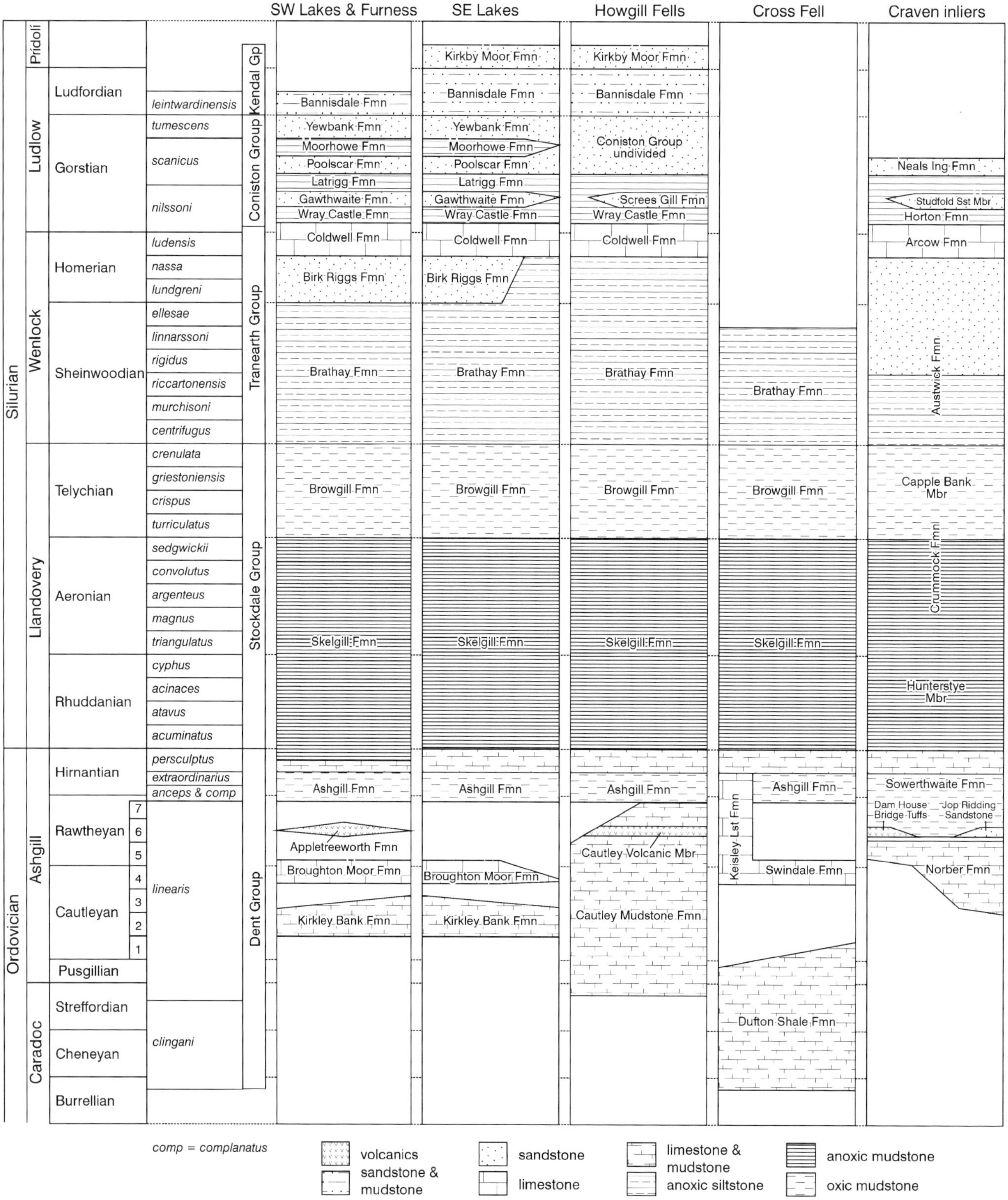

Fig. 5.14. Correlation chart for the Windermere Supergroup in five areas of NW England (see Fig. 5.1). The lithostratigraphy is mainly shown at formation level, but includes some correlative members. Only the predominant lithology in each formation is shown. Mainly after Kneller *et al.* (1994) and Fortey *et al.* (2000), based on shelly faunas. Rickards (2002) showed that the previous graptolite zonation for the Ashgill is in error, and that the *linearis* Biozone extends as high as Rawtheyan Zone 7.

outcrop area of the Ashgill Formation, however, deposition was continuous through this substantial (> 100 m?) fall in global sea level. This continuity is notable, given than lower-amplitude falls earlier in Dent Group deposition had resulted in long unconformities. The continuity suggests that accommodation space for the Ashgill Formation was created by renewed basement subsidence, perhaps the first hint on the Avalonian margin of the flexural subsidence that was to maintain the Lake District Basin through Silurian time.

Stockdale Group: Llandovery

The base of the Stockdale Group coincides with the rapid rise in global sea level that marked the end of the Ordovician

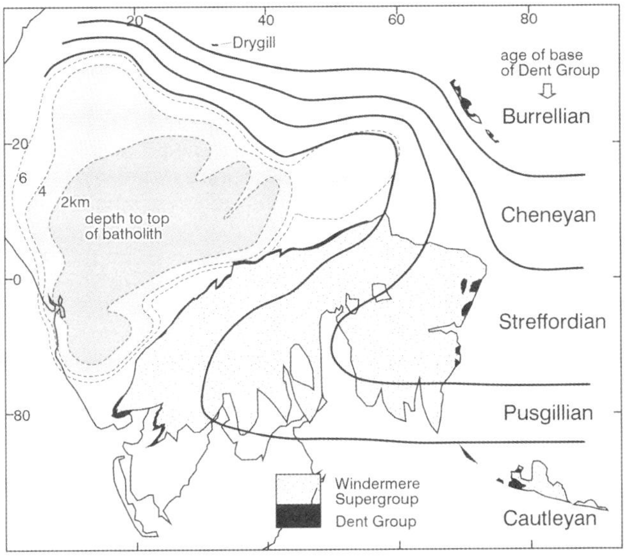

Fig. 5.15. Age of the unconformable base of the Dent Group. Onlap contours modified after McNamara (1979) to conform to the shape of the underlying Lake District batholith. Depth-to-batholith contours modified from Millward *et al.* (2000*a*).

glaciation. The distinctive lithology of the group's lower formation – the Skelgill Formation – is a carbonaceous, pyritic black shale, containing abundant graptolites of the *acuminatus* Biozone onwards (Fig. 5.15). However, underlying the black shale unit in many sections is a thin (typically less than 1 m) unit of calcareous mudstone with abundant brachiopods, trilobites, crinoids, bryozoans, corals and other shelly fossils (Fig. 5.15). This Spengill Member, in its more calcareous developments, contains no graptolites. However, a weakly calcareous correlative in the southwestern Lake District contains both a shelly fauna and graptolites of the *persculptus* Biozone (Rickards 1989). This critical evidence suggests that, as elsewhere in the world, the rapid post-glacial transgression began within the *persculptus* Biozone and therefore within Ordovician time (Brenchley 1988).

The high biodiversity of the Spengill Member compared with the underlying Ashgill Formation can be attributed to warmer, deeper and more nutrient-rich water flooding over the post-glacial shelf. The relict sea-floor topography that had influenced Dent Group deposition (Fig. 5.15) may have similarly influenced facies variation in the early stages of the transgression (Rickards 1989). The latest Ordovician bottom waters remained oxygenated, but by the earliest Silurian the high organic productivity due to the transgression (Leggett 1980) had induced anaerobic conditions. Lower–Middle Llandovery successions are dominated by the resulting black shales, presumably deposited by pelagic fallout or very dilute density flows. Only 20–40 m of sediment were deposited in this 10 Ma time interval, representing slow deposition distant from any sediment source. Water depths are uncertain, but could be progressively increasing in phase with flexural subsidence.

In the upper (Aeronian) part of the Skelgill Formation, blue-grey mudstones are progressively intercalated with the graptolitic shales. Their bioturbation and a dwarf shelly fauna suggest dysaerobic conditions, which became predominant upwards into the overlying Browgill Formation (Fig. 5.15). This unit is dominated by pale green-grey siltstone, almost barren of fossils but presumed to record dysaerobic conditions, with subordinate interbeds of anaerobic black graptolitic mudstone.

Two distinctive units near at the top of the Browgill Formation continue the trend of increasing bottom-water oxygenation. The extensive Far House Member comprises massive grey siltstones. The more localized Hebblethwaite Member

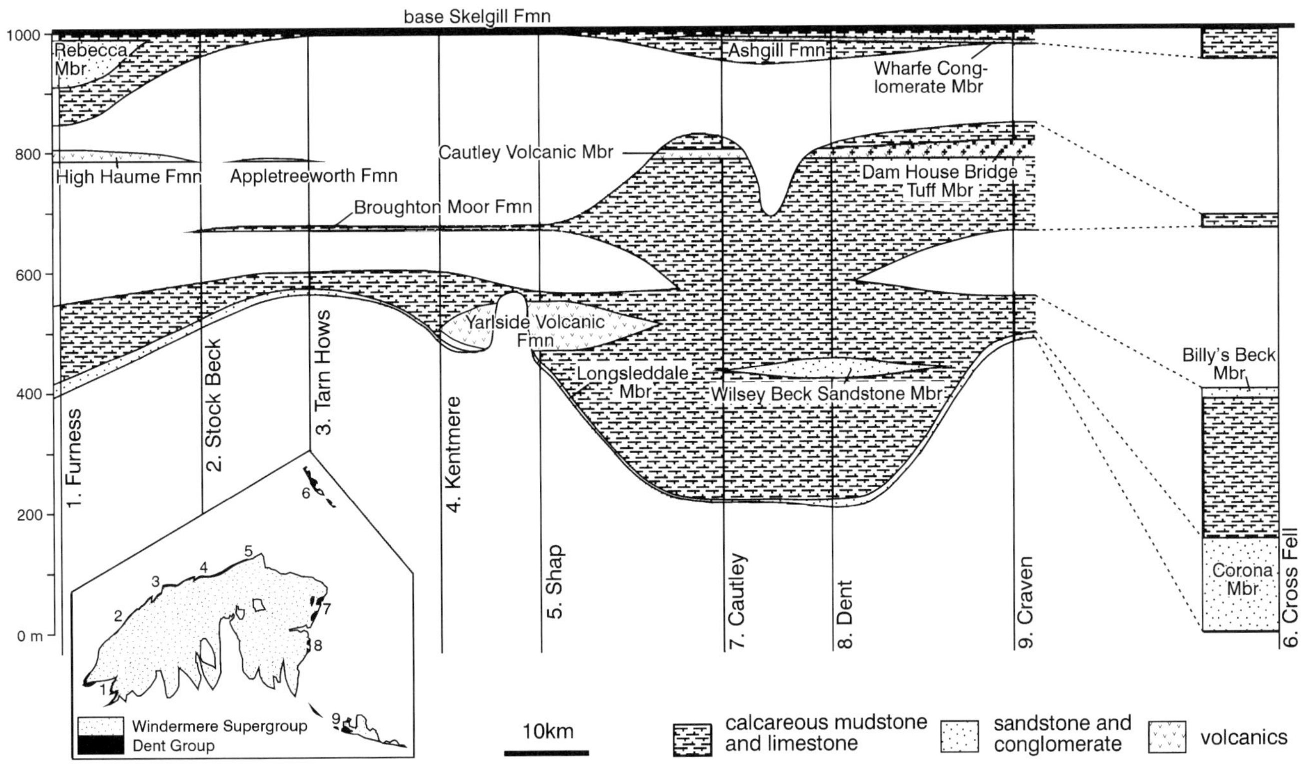

Fig. 5.16. Diagrammatic plot of the thickness variation of the Dent Group across NW England. Control sections are numbered on inset map. Only selected distinctive units are named. Based on Ingham & McNamara (1978) and McNamara (1979).

is composed of distinctive red-brown siltstones containing a marine fauna (Rickards 1973, 1989). The member varies in thickness laterally, perhaps being thicker in submarine hollows than on highs (Rickards 1978). Ziegler & McKerrow (1975) suggested that the widespread marine red beds in the uppermost Llandovery were due to an abundant supply of oxidized sediment washed off the arid Silurian land surface during the post-glacial transgressions.

Tranearth Group: Wenlock

The tendency towards better-oxygenated bottom waters that had developed during the later 8 Ma or so of Llandovery time was rapidly reversed at the end of the epoch. The 5 Ma years of Wenlock time were prone to bottom-water anoxia throughout the world's oceans, and anoxia continued to imprint the Lake District sequence for the succeeding 4 Ma of the Ludlow.

The distinctive lithology of the Lower Wenlock successions is the very thinly laminated siltstone of the Brathay Formation. The lamination is defined by alternating layers of quartz silt and organic carbon, each silt–carbon couplet having a spacing of about 0.2 mm in the compacted rock. In the basal 10–30 m of the Brathay Formation this laminated siltstone is interbedded with blue-grey burrow-mottled mudstone, marking the upward transition from the aerobic sediments of the Browgill Formation. The anaerobic environment of the laminated facies is suggested by its lack of bioturbation and its preservation only of pelagic graptolites and nektonic shelly faunas (e.g. Rickards 1964). The origin of the lamination has been widely discussed (e.g. Rickards 1964; Dimberline *et al.* 1990; Kemp 1991; King 1992). At issue is whether the lamination represents a periodic, possibly annual, fluctuation in silt and organic supply and whether the sediment was supplied entirely by hemipelagic fallout or partly by dilute turbidity flows. In any case, the Brathay Formation represents deposition distant from a supply of coarse sediment, but at a rate (50 m Ma^{-1}) at least an order of magnitude higher than during Llandovery time.

Although the hemipelagic environment of the Brathay Formation persisted into the late Wenlock in the Howgill Fells (Fig. 5.14), it was overwhelmed in other areas by sand turbidites, first in the Craven Inliers (Austwick Formation) and later in the main Lake District outcrop (Birk Riggs Formation) (Figs 5.14 & 5.17). Judging by erosional sole marks, the Wenlock turbidite systems were supplied from differing local directions (Fig. 5.18). For instance, the Austwick flows came from the southeast (McCabe 1972) and the Birk Riggs flows from the NE or SW (Millward *et al.* 2000*a*). Current directions deduced from ripples are more consistently from the NW or north, and may be at a high angle or even opposed to those deduced from sole structures (Fig. 5.18). This mismatch of erosional and depositional palaeoflows persists into the Coniston Group, and will be discussed under that head.

The renewed influx of sand into the Lake District basin has been taken by Soper & Woodcock (1990) to indicate newly uplifted source areas within the developing collision zone between Laurentia and Avalonia. Early provenance work suggested that this sand was derived directly from Laurentia (Furness 1965). However, McCaffrey & Kneller (1996) have used geochemical criteria to demonstrate derivation from the Baltica–Laurentia collision zone of the Scandian Orogen further NE, a conclusion supported by studies on both palynology (McCaffrey *et al.* 1992) and neodymium isotopes (McCaffrey 1994). Recycling of this sediment through the Laurentian accretionary prism is not precluded.

Two units of bioturbated, calcareous siltstone in the uppermost part of the Tranearth Group mark times when the persistent basin anoxia broke down. These aerobic siltstones have a shelly fauna, and relict grading that suggests deposition as event beds, possibly from storms (Kneller *et al.* 1994). The siltstones, together with intervening laminated hemipelagite, comprise the Coldwell Formation, and mark widespread, probably eustatic, falls in sea level (Fig. 5.13).

A time equivalent of the Tranearth Group has been identified in the Isle of Man, about 100 km west of the Windermere Supergroup outcrop (Morris *et al.* 1999). The Dalby Group (Niarbyl Formation) comprises sandstone–mudstone turbidites and laminated hemipelagites. The hemipelagites are identical to those in the Tranearth Group. The turbidites are thin to thick-bedded, and interpreted as part of a largely non-channelized mid–lower fan. Primary palaeoflows were from the NW. Although general comparisons can be made between the Niarbyl Formation and the sand turbidites of the Birk Riggs Formation (Morris *et al.* 1999), comparisons with parts of the Riccarton Group of the Scottish Southern Uplands are equally valid. These detailed correlations may be immaterial in that, by Wenlock time, all these areas were in depositional continuity with each other in the advancing foreland basin.

Coniston Group: lower Ludlow

Despite the increasing sedimentation rates in the Tranearth Group (Fig. 5.13), its thickness only exceeds 500 metres where it includes the sandstone turbidite systems of the Birk Riggs and Austwick Formations (Fig. 5.17). By contrast, the overlying Coniston Group comprises between 1000 and 2500 m of sediment, accumulated at a rate (>500 m Ma^{-1}) about an order of magnitude higher than that of the laminated hemipelagites of the Tranearth. No evidence for shallow-water deposition is seen within the Coniston Group, implying increasing basin subsidence of the same order as sediment accumulation (King 1994*a*), compatible with the continuing flexural load of the Laurentian content (Kneller *et al.* 1993).

The anoxic hemipelagite lithology persisted as a background sediment throughout the Coniston Group. However, increasing input of silt during the early Ludlow progressively expanded the average spacing of the carbon laminae, from 0.3 mm at the base of the group to 0.8 mm at the top (King 1992). Graded thin-bedded silt-mud turbidites progressively diluted the hemipelagic background, even within the finer grained formations of the group, the Wray Castle, Latrigg, Moorhowe and Horton Formations (Fig. 5.17). However, the bulk of the thickness of the Coniston Group is accounted for by the sandstone–turbidite formations (Figs 5.15 & 5.17), each comprising numerous thin to very thick-bedded, fine-grained sandstone beds grading rapidly up to mudstone. Individual beds are typically laterally impersistent on outcrop scales of hundreds of metres, though packets of beds are more persistent. Channelling is rare and little vertical cyclicity in sand bed thickness can be discerned (King 1992). The Coniston Group sandstone bodies were probably deposited in poorly organized sand–turbidite lobes a few kilometres in extent. These lobes switched and overlapped through time to produce the lobate systems on a scale of tens of kilometres that are now mapped as formations (Fig. 5.17).

Broadly, the Coniston Group is thickest in the west (sections 1–4, Fig. 5.17) and thins southeastward through the Howgill Fells into the Craven Inliers. This pattern may partly reflect the orogenic load of the Laurentian continent in the NW, as well as along-basin variation. In any case, local thinning predominates across three major fault belts, those of Coniston–Brathay, Whinfell and Dent (Fig. 5.17). This thinning suggests that these belts were active during early Ludlow time, perhaps allowing differential flexure of the downgoing Avalonian plate. The fault belts also partitioned the basin into sub-basins with contrasting depositional architecture (Fig. 5.17). Along the main northwestern crop (sections 1–5), three laterally persistent turbidite systems – represented by the Gawthwaite, Poolscar and Yewbank formations – punctuate the hemipelagic background. This organization breaks down east of the Whinfell Faults. Here (sections 6, 7) the Screes Gill Formation is probably the earliest of the Coniston Group sand systems. It is overlain by a

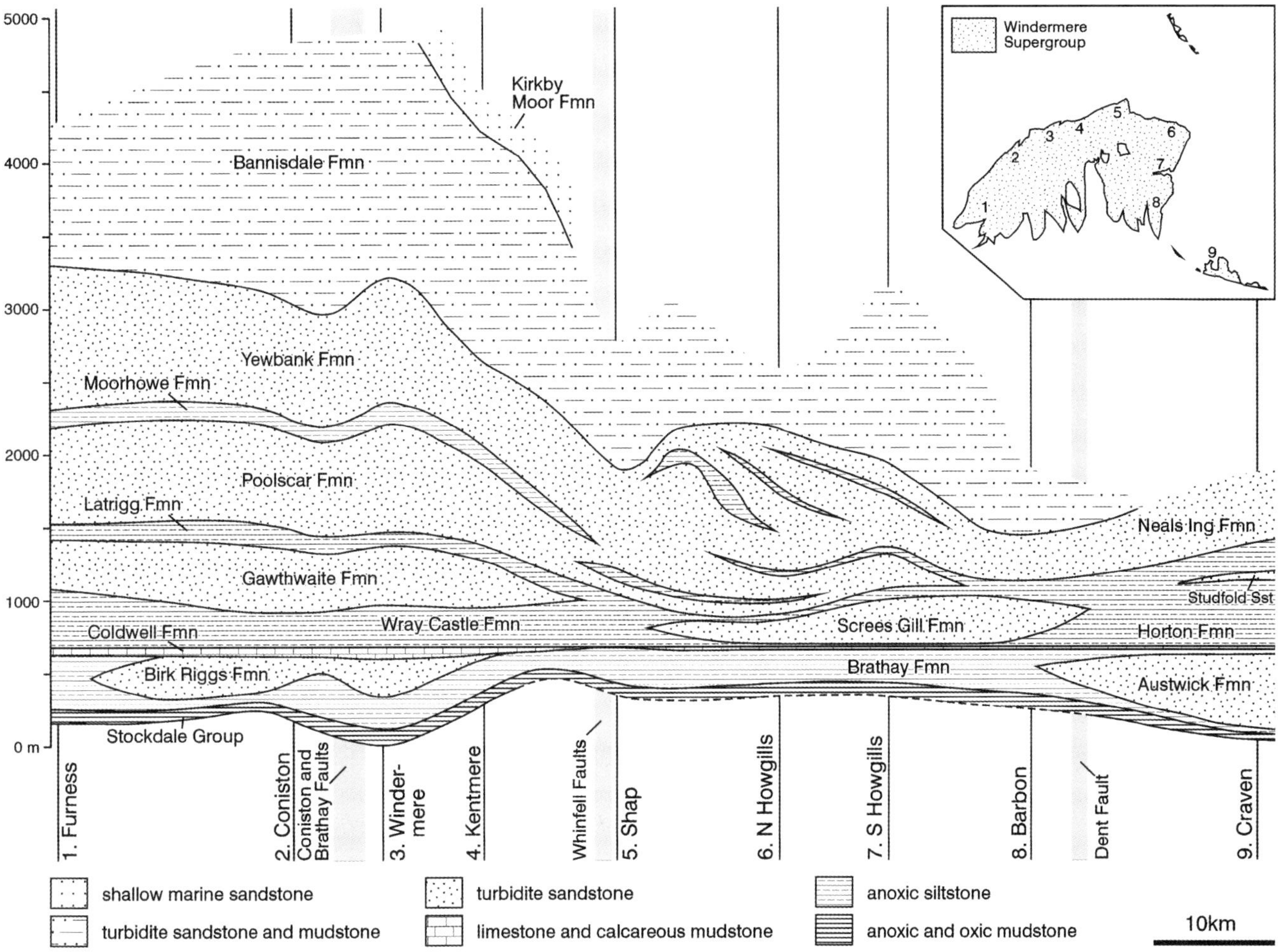

Fig. 5.17. Diagrammatic plot of the thickness variation of Silurian rocks across NW England (Rickards & Woodcock 2006), based on control transects numbered on inset map . Correlations between transects 8 and 9 are particularly uncertain. Datum is the top of the Coldwell Formation. Lake District data mainly from Millward *et al.* (2000*a*). Shap and Howgill Fells data from Soper (1999) and Woodcock & Rickards (1999), and Craven data from Arthurton *et al.* (1988).

system with a complex stratal geometry (Soper 1999), possibly indicating southeastward progradation. Southeast of the Dent Fault, the Screes Gill Formation seems to pinch out into the thick lower Ludlow hemipelagites of the Horton Formation. In the Craven area two thin sandstone systems – the Studfold Sandstone Member and the Neals Ing Formation – have an uncertain correlation with systems further NW.

Palaeoflow indicators from the Coniston Group and the overlying Bannisdale Formation (Fig. 5.18) show, as in the Wenlock, a strong mismatch between erosional and depositional indicators. Current ripple cross-lamination shows a fairly consistent mean flow from the northwest or north. Erosional structures are much more variable and, in the eastern half of the outcrop, show marked divergence between flutes and grooves (King 1992). Kneller *et al.* (1991) took the erosional sole structures in the west (Fig. 5.18) to indicate a dominantly southwestward primary axial flow along a topographically confined trough. They saw the transverse depositional structures as due to reflection of solitary waves within the flow off steep bounding slopes to the trough. King (1994*b*) interpreted the predominantly southeastward primary flows in the east as following the other arm of an arcuate basin margin. However, the divergence between flutes and grooves here, and the presence of further southeastward primary flows in the far west of the Windermere Group outcrop (Fig. 5.18), suggests an even more complex local sea-bed topography. Perhaps the flutes and grooves were cut by the high concentration, bottom-hugging parts of the flow, and tend to follow local topography, partly draped above the major fault belts (Fig. 5.18). The current ripples were then formed by the thicker low-concentration upper parts of the flow, controlled by the regional slope of the sediment surface. Similar contrasts between different parts of turbidity flows have been proposed elsewhere in the Lower Palaeozoic, for instance in the Manx Group (Woodcock & Barnes 1999).

Kendal Group: upper Ludlow

The top of the Coniston Group is marked by a decrease in the proportion and thickness of sandstone turbidite beds. The overlying Bannisdale Formation (Figs 5.15 & 5.17) is characterized by very thin or thin beds of siltstone or fine sandstone grading up into mudstone, interpreted as the deposits of dilute turbidity flows. Medium to thick sandstone beds form mappable packets in the lower part of the Bannisdales, making for a transitional boundary with the underlying Coniston Group. A minor but persistent component of the lower Bannisdale Formation is laminated hemipelagite. The lamination is defined, as in the Tranearth and Coniston groups by alternating laminae of silt and organic carbon, but the lamination spacing is now greater, between 0.8 and 1.2 mm (King 1992).

Traced up-sequence, the laminated hemipelagite becomes less abundant and the mudstone tops of the graded siltstones and sandstones become bioturbated. Convolute lamination becomes more common and small calcareous and phosphatic

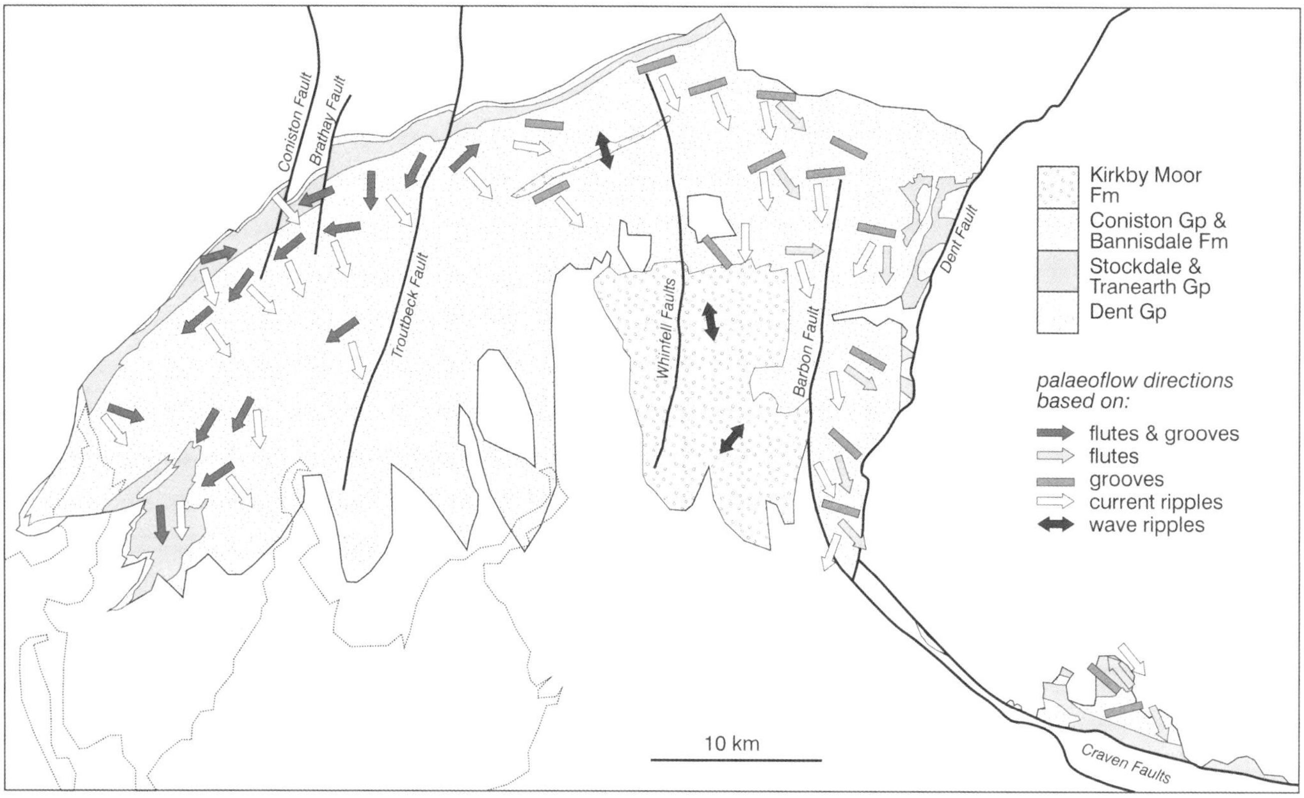

Fig. 5.18. Palaeoflow directions, from flutes, grooves and ripple cross-lamination, for the turbidite systems of the Tranearth and Coniston groups and Bannisdale Formation and, from wave ripples, from the shallow-water systems of the Kirkby Moor Formation. Data for the eastern region from King (1992) and for the western region from Kneller (1990). Flutes and grooves are not separately distinguished in the west.

nodules are present. These changes are interpreted in terms of increasing bottom-water oxygenation and a transition into the shallow-marine environments of the overlying Kirkby Moor Formation (King 1994*b*). The transitional facies has previously been distinguished as the Underbarrow Formation (Shaw 1971; Lawrence *et al.* 1986). However, subsequent work (Soper 1999) has established that this formation cannot be mapped as a coherent unit.

The Kirkby Moor Formation is dominated by medium to thick fine-grained sandstone beds, containing planar and low-angle hummocky lamination, symmetrical ripple cross lamination and convolute lamination. Shell material forms lags at the base of sandstone beds, and their mudstone tops are commonly bioturbated. Thin-bedded units are subordinate. The formation was deposited from storm-induced waning flows, modified by fairweather waves (King 1994*b*). This diagnosis implies steadily shallowing water depths through the underlying Bannisdale Formation. An intercalation of reddened wave-rippled heterolithic siltstones, lacking storm beds, near the top of the Kirkby

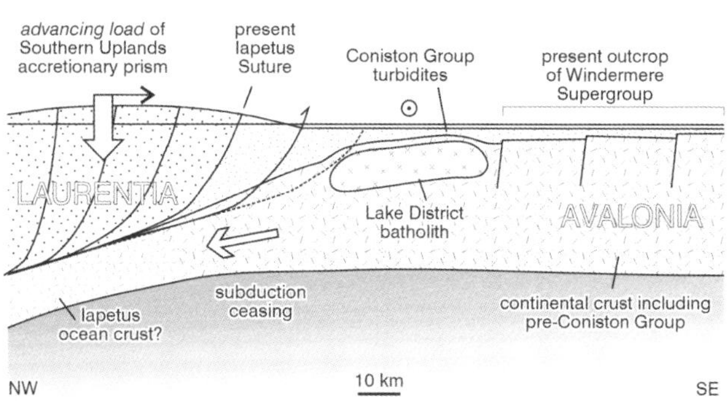

Fig. 5.19. Cartoon showing the tectonic setting late in the deposition of the Windermere Supergroup, with the Laurentian accretionary prism down-flexing the Avalonian crust (Rickards & Woodcock 2006).

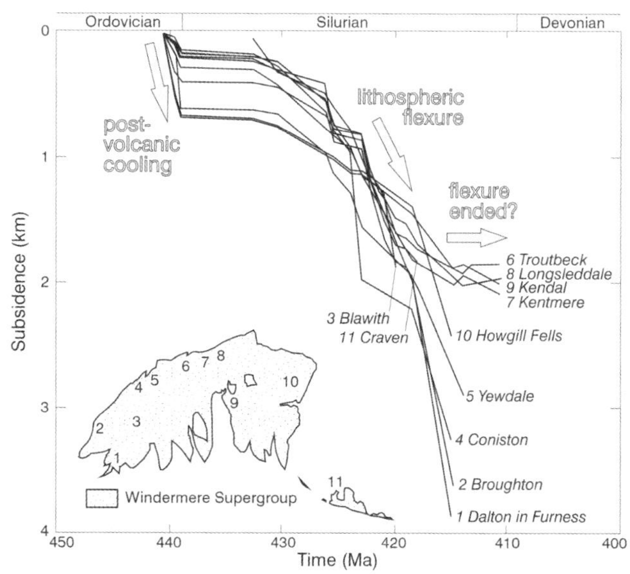

Fig. 5.20. Backstripped subsidence curves for selected sections through the Windermere Supergroup (inset). Curves show a phase of post-volcanic thermal subsidence followed by flexural subsidence due to lithospheric loading. Data from King (1994*a*). The time scale is from Harland (1990) and therefore differs in detail from that in Figure 5.2.

Moor Formation may represent lagoonal or at least more sheltered conditions (King 1994*b*).

Palaeoflow data within the Bannisdale Formation mimics closely the flow patterns of the underlying Coniston Group (King 1994*b*). Wave ripples in the Kirkby Moor Formation imply current between NNW and NE (Fig. 5.18). There are too few data to test King's hypothesis that this pattern reflects an arcuate shoreline to the north. King (1994*b*) envisaged turbidites still being deposited to the south of this storm-influenced shelf, as part of a general southward facies shift in the migrating foreland basin. However, the lithological and biostratigraphic mapping of Shaw (1971), on which this diachronous relationship is based, has not been verified by more recent mapping (Soper 1999).

The thickness of the Kendal Group, particularly of the Bannisdale Formation, is poorly constrained, due to the complexity of its deformation and the difficulty of mapping marker units through it. The schematic diagram (Fig. 5.17) displays conservative thickness estimates, and values of over 4 km have been proposed along the Coniston and Windermere transects (Millward *et al.* 2000*a*). However, in the Shap Fells, where capped by the Kirkby Moor Formation, the Bannisdale thickness is little more that 1.5 km (Soper 1999). It is tempting to see the Bannisdale Formation as thinning southeastward, as does the underlying Coniston Group. However, the formation is still about 1.5 km thick in the Howgill Fells, and apparent thinning towards the Craven Inliers is most probably due to pre-Carboniferous erosion rather than non-deposition. Metamorphic grade within the Windermere Supergroup suggests that this missing sedimentary cover was at least 3.5 km thick (King 1994*b*; Soper & Woodcock 2003). The upward-shallowing signature seen in the Kendal Group suggests that a significant fraction of this cover was a non-marine 'Old Red Sandstone' facies.

Basin evolution

Seen as a whole (Fig. 5.13) the Windermere Supergroup represents a natural cycle of basin evolution. So, basal transgressive deposits pass up into a thick deep-water succession and then into a culminating regressive sequence, albeit truncated by basin uplift and erosion. The progressive increase in sediment accumulation rates through this cycle (Fig. 5.13) suggested to Kneller (Kneller 1991; Kneller *et al.* 1993) a basin origin by flexural loading rather than extension. As the last remnants of the Iapetus Ocean were destroyed at the subduction zone beneath Laurentia (Fig. 5.19), the Avalonian continental margin was bent down into the trench. The load formed by the leading edge of the Laurentian continent advanced southwards relative to the margin of Avalonia, driving the increasing subsidence rates through time. The cartoon (Fig 5.19) is drawn in late Silurian time, when convergence was waning or ended, and clastic deposition almost filled a continuous basin between the Laurentian and Avalonian margins. Steep faults on Avalonia – oriented parallel as well as perpendicular to the section line – are being reactivated to control the detailed architecture of the basin fill. Southward-verging thrust faults have sliced into the sub-Windermere succession on Avalonia.

Derivation of more detailed subsidence curves (Fig 5.20) from the basin stratigraphy confirm the diagnostic convex-up signature of a flexural basin (King 1994*a*). The flexure was preceded by the last phase of the post-volcanic thermal subsidence. Marked flexural subsidence is first apparent in mid-Llandovery time. However, water-depth constraints are poor in this part of the succession and earlier (late Ashgill?) onset of flexural subsidence has been suggested above. After the post-glacial transgression, no shallow-marine environments are apparent until the late Ludlow. Through Llandovery, Wenlock and most of Ludlow time, therefore, the basin was underfilled (cf. Crampton & Allen 1995), with sediment supply unable to keep pace with steadily increasing basement subsidence rates. Shallowing environments in latest Ludlow and Přídolí time – immediately preceding the time represented in the cartoon (Fig. 5.19) – mark the transition to an overfilled basin. This was partly due to increasing sediment supply rates from the uplifting collisional mountain belt, but also to decreasing basin subsidence rates as flexural loading ended (Fig. 5.20). Some kilometres of non-marine 'Old Red Sandstone' cover to the basin was probably deposited (Chapter 4) before being eroded again during the mid-Devonian uplift associated with the Acadian Orogeny. Soper & Woodcock (2003) proposed an extensional basin setting for this Lower Devonian sedimentation, an event that would have separated the Iapetan flexural basin formation from the Acadian Orogeny.

6 The Acadian Orogeny: the mid-Devonian phase of deformation that formed slate belts in England and Wales

N. H. WOODCOCK & N. J. SOPER

Deformation of the Early Palaeozoic rocks of England and Wales has traditionally been ascribed to a late phase of the Caledonian Orogeny that occurred in end-Silurian time. More recently, it has been recognized that this deformation took place towards the end of the Early Devonian and forms part of the Acadian orogenic belt, which extends from the NE Appalachians through western Europe to Poland (McKerrow 1988) (Fig. 6.1).

The Caledonian mountains, as originally defined by Suess in the early 20th century, extended between Scotland and Norway and were thought to result from the deformation of an Early Palaeozoic geosyncline. It was soon recognized that the Caledonian orogen continues into the Appalachians and East Greenland, flanked to the west by a faunally distinctive foreland sequence of Cambro-Ordovician platform deposits, which includes the Durness succession of NW Scotland. In a seminal pre-plate tectonic interpretation Wilson (1966) saw the Caledonides as the result of closure of a 'Proto-Atlantic' ocean that existed in late Precambrian and Early Palaeozoic time, subsequently named the Iapetus Ocean by Harland & Gayer (1972).

In the first plate tectonic model to interpret Caledonian orogenesis in terms of lithospheric convergence driven by ocean spreading and subduction, Dewey (1969*a*) envisaged closure between two major continental plates, North America–Greenland and Europe, which produced the N–S-striking North Atlantic Caledonides and NE–SW Appalachians. The British Caledonides, located at the intersection of the two margins, were divided into two zones. An orthotectonic zone comprised the high grade, poly-deformed and mainly Proterozoic rocks of the Scottish Highlands. The paratectonic Caledonides (or 'slate belts' Soper 1986b) were characterized by the weaker deformation of the Early Palaeozoic strata of the Scottish Southern Uplands, England and Wales. Developments of this 'two plate' model (e.g. Phillips *et al.* 1976; Dewey 1982; McKerrow 1982) envisaged roughly E–W convergence (in the present geographical frame) during the terminal collision phase, inducing dextral transpression in the NE–SW-trending British–Irish Caledonides.

However, the presence of late Caledonian deformation belts in western Europe, particularly the North German–Polish Caledonides (Ziegler 1982) suggested a more complex plate tectonic scenario. The provinciality of early Ordovician shallow-water faunas indicated wide oceanic separation at that time between Laurentia, Baltica and a third element of Gondwanan origin that included southern Britain (Cocks & Fortey 1982). Soper & Hutton (1984) introduced a three-plate model in which this microcontinent ('Cadomia') was accreted northwards onto Laurentia–Baltica in the Silurian, inducing sinistral transpression in the British slate belts and strike-slip displacements in the orthotectonic zone (Watson 1984).

This basic three-plate model has been developed and modified in the light of new data and a recognition of along-strike links between the British sector of the Caledonides and the northern Appalachians (Fig. 6.2a). Britain and Ireland south of the Iapetus Suture and adjacent parts of continental Europe are now seen as part of a terrane separate from Armorica, named Eastern Avalonia (Soper 1986*b*) from its similarities with the Avalon terrane of eastern Newfoundland (Piqué 1982; Williams & Hatcher 1982). Deformation of the slate belts of southern Britain and the Ardennes was shown to be of Devonian age and thus an extension of the Acadian orogen of maritime Canada (Soper *et al.* 1987); as subsequently demonstrated, this was a short-lived event of Emsian age (McKerrow 1988; McKerrow *et al.* 2000). The model envisaged sinistrally oblique collision between Baltica and Laurentia in the Silurian (Scandian Orogeny) and northward accretion of Avalonia onto the Appalachian margin in the Siluro-Devonian, culminating in the Acadian Orogeny.

Discussion has subsequently centred on the timing of the accretion of Eastern Avalonia onto Laurentia and Baltica and on the relationship, if any, of these convergence regimes to the Acadian deformation. Divergent views for the timing of these collisions, ranging from mid-Ordovician to Early Devonian, have arisen because of differing emphasis placed on the various lines of evidence – palaeomagnetic, faunal, magmatic, stratigraphic and tectonic – which relate to different stages of the closure process and have their own precision levels.

It is now agreed that Avalonia accreted to Laurentia during the mid-Silurian. In Ireland a unified depositional system was established across the Iapetus Suture in the Wenlock (Murphy & Hutton 1986); 'soft docking' is indicated by tranquil hemipelagic sedimentation in the Southern Uplands and NW

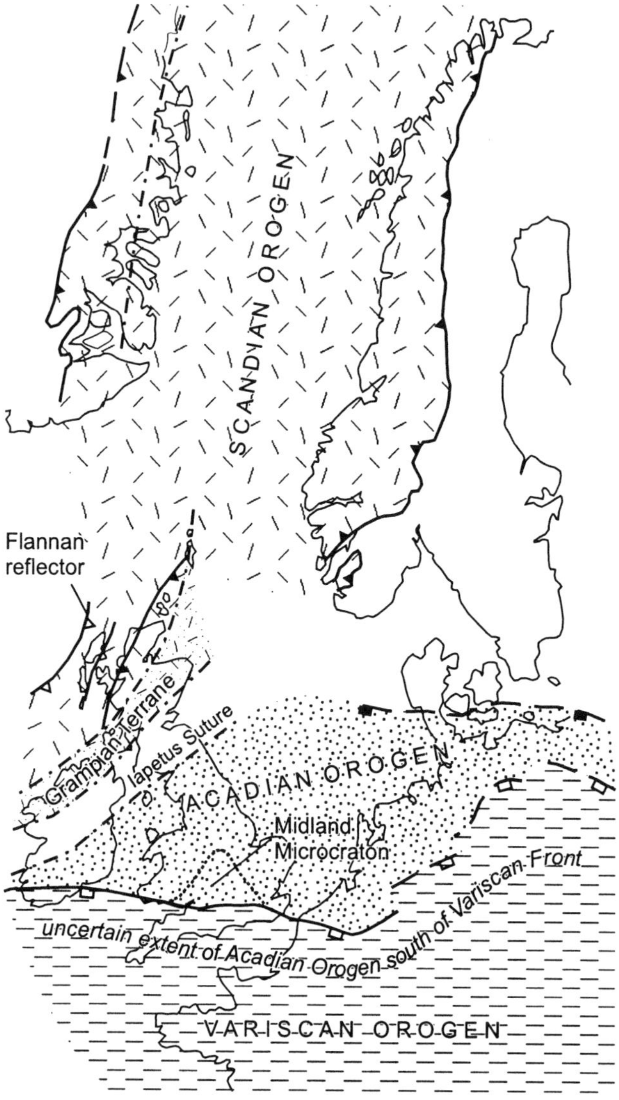

Fig. 6.1. Outline map of the North Atlantic Caledonides on the pre-Atlantic reconstruction of Mosar *et al.* (2002) showing tectonic context of the Acadian slate belts of England and Wales. Note the mismatch between the Scottish and East Greenland Caledonides due to post-Caledonian/pre-Atlantic deformation, and uncertainty at the North Sea 'triple junction'.

England at that time (Kemp 1987; Soper & Woodcock 1990; Kneller 1991).

The timing of Baltica–Avalonia collision across the Tornquist suture is more controversial. Palaeomagnetic evidence suggests a late Ordovician age (Trench & Torsvik 1992) coinciding with the amalgamation of benthic faunas (Cocks & Fortey 1982). The cessation of subduction-related volcanicity in England and Wales during the late Ordovician (van Staal *et al.* 1998) and a weak Ashgill deformation episode in the Welsh borders (McKerrow *et al.* 2000) have been taken as supporting evidence. However, the palaeomagnetic and faunal evidence require only proximity of the continental margins and a shallow-marine connection between them, and the burst of volcanicity in the Caradoc followed by shut-down more likely reflect the subduction of a Iapetan ridge-transform system, not ocean closure (Woodcock 1990*a*).

Given that Avalonia was in contact with the Laurentian margin by mid-Silurian time (e.g. Stone & Evans 1995) if rigidly attached to Baltica, as has been suggested, it would have experienced Scandian scale orogenic deformation during Baltica–Laurentia convergence, not the mild flexural response observed. An alternative model for Iapetus closure involves independent sinistrally transpressive convergence of Baltica and Avalonia with Laurentia during the Silurian (Fig. 6.2a). In this model, convergence ceased some 20 Ma before the Acadian deformation episode created the slate belts of England, Wales and Belgium in the Early Devonian (Soper *et al.* 1992; Trench & Torsvik 1992; Soper & Woodcock 2003). The Acadian Orogeny in England and Wales was therefore not related to Iapetus closure but to a subsequent convergence across the mid-European Rheic Ocean (Fig. 6.2b).

Acadian structure and metamorphism

Both the style and trend of Acadian structures in England and Wales reflect the geometry of the Early Palaeozoic basins that flank the Midland Platform or Microcraton (Fig. 6.3). On the platform, a thin cover of dominantly shallow marine Lower Palaeozoic strata is little affected by Acadian deformation and metamorphism. The thick sequences of dominantly turbiditic sediments and volcanic rocks in the flanking basins show a typical slate belt style of deformation, with folding and cleavage developed at anchizonal and epizonal grades. The Acadian structural trend appears to wrap around the platform (Fig. 6.3); the origin of this regional arcuation is discussed at the end of this chapter.

Northwest England

Lower Palaeozoic rocks crop out over about 2500 km^2 in the Lake District and Howgill Fells, while the smaller inliers of Cross Fell, Craven and Teesdale, together with borehole evidence, show that similar rocks underlie much of the Upper Palaeozoic cover of the northern Pennines (Plate 12). Their Acadian structure has been elucidated during the BGS resurvey of the Lake District that is nearing completion (Millward *et al.* 2000*b*; Johnson *et al.* 2002) and is the basis of the synoptic tectonic map shown in Fig. 6.4. The deformation affects strata as young as Přídolí and predates the unconformable Mell Fell conglomerate (MFC, Fig. 6.4), a clastic wedge of ORS (Old Red Sandstone) facies that is probably of Upper Devonian age.

The regional-scale structure of the main Lake District inlier may be described as anticlinal, in that older rocks are flanked at outcrop by younger. The Skiddaw Group, dominantly muddy and silty turbidite deposits of Tremadoc–Llanvirn age, forms a core that occupies the northern Lake District and is flanked to the north and south by unconformable Caradoc volcanic rocks of the Eycott and Borrowdale groups, respectively (Chapter 3). The Borrowdale Group is overstepped to the south by another dominantly turbiditic sequence of Ashgill–Přídolí age, the Windermere Supergroup, and similar strata are thought to underlie the Solway Basin to the north of the Lake District, contiguous with the Silurian rocks of the Southern Uplands. The structure of the Windermere Supergroup is described first as it has the least complex tectonic history.

Windermere Supergroup

Figure 6.4 shows that the structure of the WSG is dominated by a single fold, the Bannisdale syncline. This, in fact, forms the lower hinge line of the major SE-facing Westmorland monocline that affects both the WSG and southern part of the Borrowdale Volcanic Group and has a maximum amplitude of some ten kilometres (Kneller & Bell 1993). In the SW Lake

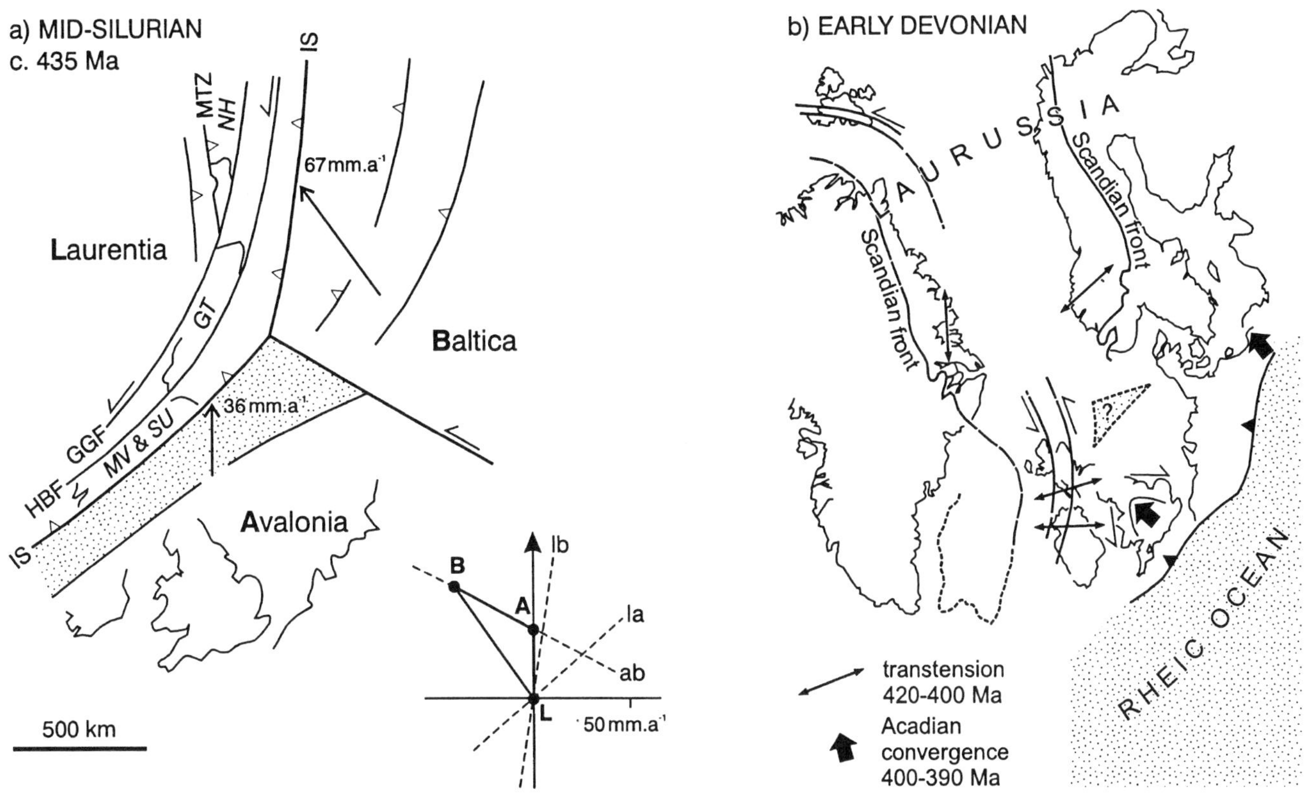

Fig. 6.2. Plate tectonic setting of England and Wales during the Siluro-Devonian. (**a**) Two dimensional representation of Iapetan convergence geometry in mid-Silurian time. Principal sources: Soper *et al.* (1992), basic three-plate geometry; Cocks & Fortey (1982); Trench & Torsvik (1992), late Ordovician juxtaposition of Baltica and Avalonia; Strachan *et al.* (2002, fig. 4.36), Tornquist continental transform; Dewey & Strachan (2003, fig. 5a), Baltica–Laurentia convergence vector. GGF, Great Glen Fault; GT, Grampian terrane; HBF, Highland Boundary Fault; MTZ, Moine Thrust Zone; MV, SU, Midland Valley and Southern Uplands terrane; NH, Northern Highlands terrane; TT, Tornquist transform boundary. Velocity diagram drawn relative to a fixed Laurentia. (**b**) Cartoon showing post-Iapetan tectonic setting of Britain in Early Devonian time. Modified from Soper *et al.* (1992, fig. 6c) to show an open Rheic ocean and inferred Andean convergent margin south of Britain (Torsvik & Cocks 2004, fig. 5). Transtensional incremental stretch directions from Dewey & Strachan (2003, fig. 4) and Soper & Woodcock (2003, fig. 10).

District the syncline trends 055° with a 30° plunge to the ENE, about half of which is due to post-Carboniferous block tilting. Traced northeastwards the trace swings to about 065° and the axis becomes subhorizontal. The regional arcuation is best displayed by medium-scale folds to the south of the main syncline; these trend 060° in the SW, E–W in the Howgill Fells and 110° in the Ribblesdale inlier (Fig. 6.3).

A single Acadian cleavage is developed in the WSG. It typically combines spaced stress-dissolution seams, dominant in sandy lithologies, with a more penetrative phyllosilicate fabric dominant in mudrocks. The cleavage is steeply inclined and shows a weaker arcuation than the folds, from about 065° in the SW Lake District where it strikes about 10° clockwise to the fold traces, through 090° around Kendal to about 105° in Ribblesdale where it transects the folds at a small anticlockwise angle. The origin of this arcuation and transection pattern is discussed below.

Thrusts are not a dominant feature of the WSG. The Stockdale and Park Gill thrusts (Fig. 6.4) are associated with particular formations and excise instead of repeating strata; they appear to be rootless 'out-of-syncline' accommodation structures that cut down-section in the displacement direction, which is to the NW. North-vergent reverse faults in the SW of the district and many smaller examples in the Coniston and Bannisdale formations replace synclines and tip-out laterally, affecting a limited volume of rock. There is no evidence that these faults are linked to a basal detachment, or that the Acadian deformation of the Windermere Supergroup was of thin-skinned fold-and-thrust type.

Figure 6.4 illustrates the pattern of northerly trending faults that dissect the main crop of the WSG. In some areas the pattern resolves into distinct NNW and NNE sets, and individual faults often tip-out laterally. As first demonstrated by Moseley (1968) many of these faults separate tracts with different fold and cleavage patterns, accommodating differing flexural and ductile strains and thus with variable displacements along their length. The well-known Coniston, Brathay and Troutbeck faults are of this type. They appear to be reactivated synvolcanic faults that give rise to kilometre-scale map offsets of the sub-Windermere unconformity but lose displacement southwards in the younger formations. They are interpreted as pre-existing basement fractures that propagated upwards both during Early Devonian regional extension (Soper & Woodcock 2003) and during the Acadian deformation, when they acted to partition strain in the Silurian rocks. It is evident from Figure 6.4 that many have also had a post-Acadian history of reactivation, particularly during Permo-Triassic extension, as they define the boundaries of Carboniferous tilt-blocks in the southern Lake District.

Borrowdale and Eycott volcanic groups

As described in Chapter 3 the BVG was affected by synvolcanic faulting associated with piecemeal caldera collapse. The most important of these faults are shown in Fig. 6.4 together with the volcanic sag basins that they help to define (Millward 2002). The most important Acadian structure superimposed on the previously block-faulted volcanic tract was the Westmorland Monocline, which coincides with the southern flank of the

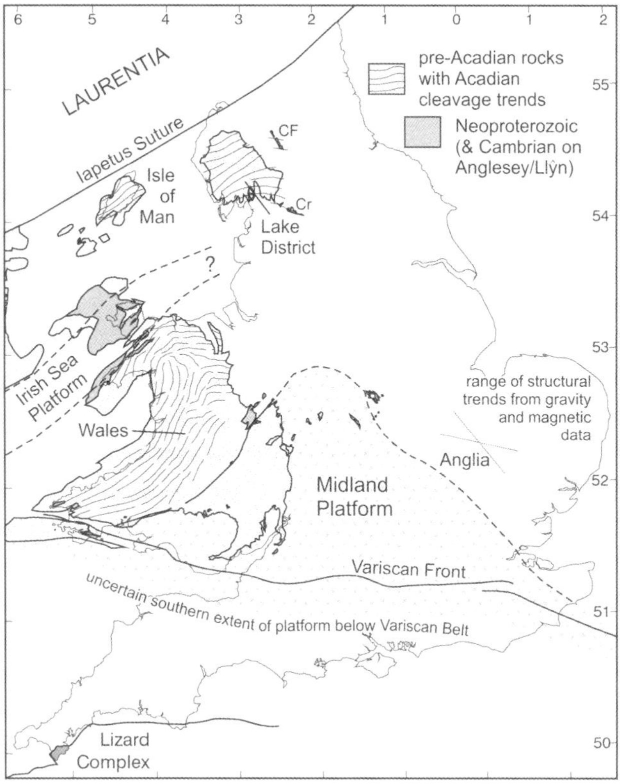

Fig. 6.3. Map of Acadian cleavage trends in England and Wales. CF Cross Fell Inlier; Cr Craven Inliers. Simplified after Soper & Woodcock (2003).

Ordovician subvolcanic batholith (Fig. 6.4). The monocline is less continuous in the volcanic rocks than in the WSG where it constitutes the NW limb of the Bannisdale Syncline. It is disrupted by the Troutbeck fault system, which crosses a 'col' in the batholith, and is terminated to the SW by the Duddon Basin. Few other folds of purely Acadian origin have been recognized in the BVG. Cleavage is strong in the monocline, forming the Tilberthwaite slate belt in which the green slate industry is now concentrated, but weak over the central part of the batholith, strengthening again in the Honister slate belt that coincides with its northern flank.

The Eycott Volcanic Group occupies the northern part of the Lake District inlier. It crops out over a smaller area than the BVG but a substantial extension beneath the Upper Palaeozoic cover may be inferred from its associated aeromagnetic anomaly (Fig. 6.4). Steep dips in the volcanic rocks define a N-facing monocline of about 8 km amplitude, the Cumberland Monocline. The regional Lake District 'anticline' thus resolves into a pair of outward facing monoclines, asymmetrically related to the subjacent batholith (Fig. 6.5).

Skiddaw Group

In common with many mudrock-dominated sequences that have been multiply deformed at low metamorphic grade, the Skiddaw Group shows complex patterns of small-scale structures. The following structural sequence has been established in the main inlier (Hughes *et al.* 1993):

Soft-sediment structures occur on all scales, from an olistostrome with kilometre-size blocks that comprises the Buttermere Formation (Chapter 3), to trains of slump folds with curvilinear axes and occasionally an associated cleavage, intraformational breccias and 'fluxoturbidites'. A bedding-parallel fabric is widely developed in the finer lithologies and is thought to result from compaction enhanced by mimetic recrystallization.

Main phase structures. The dominant cleavage is generally weaker than in equivalent Silurian lithologies, suggesting that the Skiddaw Group experienced a degree of thermal metamorphism before the Acadian deformation. This conclusion is supported by clay mineral assemblages and bedding-parallel microfabrics that indicate burial metamorphism (Merriman 2002). The cleavage is usually of spaced, stress-dissolution type, but in fine-grained lithologies with a pronounced bedding-parallel fissility it takes the form of a crenulation fabric. It is of broadly 'Caledonoid' trend, but with much local variation due to refolding, and is axial planar to folds on scales up to several hundred metres in amplitude.

Later structures. In many areas, folds with gently inclined axial planes and an associated crenulation cleavage are superimposed on the main cleavage, presumably a result of vertical shortening under gravitational stress, after thickening of the sedimentary pile during the main phase of deformation. Later still are trains of brittle minor structures, kink-bands, box folds and the like, in many different orientations, that have given rise to extended 'D-number' sequences in some areas.

Thrusts and faults. Frequently, these late structures are associated with post-main cleavage displacement on thrust faults, of which the principal examples in the main inlier – Causey Pike, Gasgale, Loweswater and Watch Hill – are shown in Figure 6.4. These appear to be of composite origin. For example, the Causey Pike structure is mapped as a subvertical fault with inferred strike-slip displacement (*cpf* on Fig. 6.4), except for its central section which is a reversed dip-slip fault (*cpt*). Webb & Cooper (1988) viewed the Loweswater and Gasgale thrusts as detachments associated with large scale slump folding. P. Stone (pers. comm.) has suggested a similar early origin for the Watch Hill Thrust, which is pinned by the late Ordovician Embleton dioritic intrusions (Fig. 6.4). It is likely that the Acadian deformation involved reactivation of favourably oriented segments of these early dislocations, as well as the initiation of new thrusts.

Steep NNW-trending faults dominate the northern Lake District, in contrast to NNE trends in the south (Fig. 6.4). Again, a protracted reactivation history is inferred, involving control of depositional patterns in the Skiddaw Group (Moore 1992) and extensional displacements during the Caradoc volcanicity, as well as elements of the Siluro-Devonian history of fault reactivation described above for the Windermere Supergroup.

Interpretive problems

The structure of the Skiddaw Group has proved difficult to elucidate and continues to present problems that bear on an understanding of the structure of NW England and Acadian tectonics generally. Structural complexity of the Skiddaw mudrocks compared to the overlying Borrowdale Volcanic Group led to the suggestion of an intra-Ordovician orogeny (Simpson 1967; Helm 1970). This was disproved by the demonstration of fabric continuity between the two groups (Soper 1970), a post-Silurian age for the main cleavage (Soper & Roberts 1971) and the importance of soft sediment deformation in the Skiddaw Group (Webb & Cooper 1988). However, stratigraphic conformity that had been inferred between the Skiddaw Group and overlying volcanic rocks (Soper 1970) was also shown to be incorrect (Millward & Molyneux 1992; Cooper & Hughes 1993); there was substantial pre-volcanic uplift and erosion, attributed to the production and rise of andesitic melts during subduction (Branney & Soper 1988).

The structural interpretation of the Skiddaw Group developed during the BGS resurvey (Hughes *et al.* 1993; Stone *et al.* 1999) envisaged a fold-and-thrust system that prograded

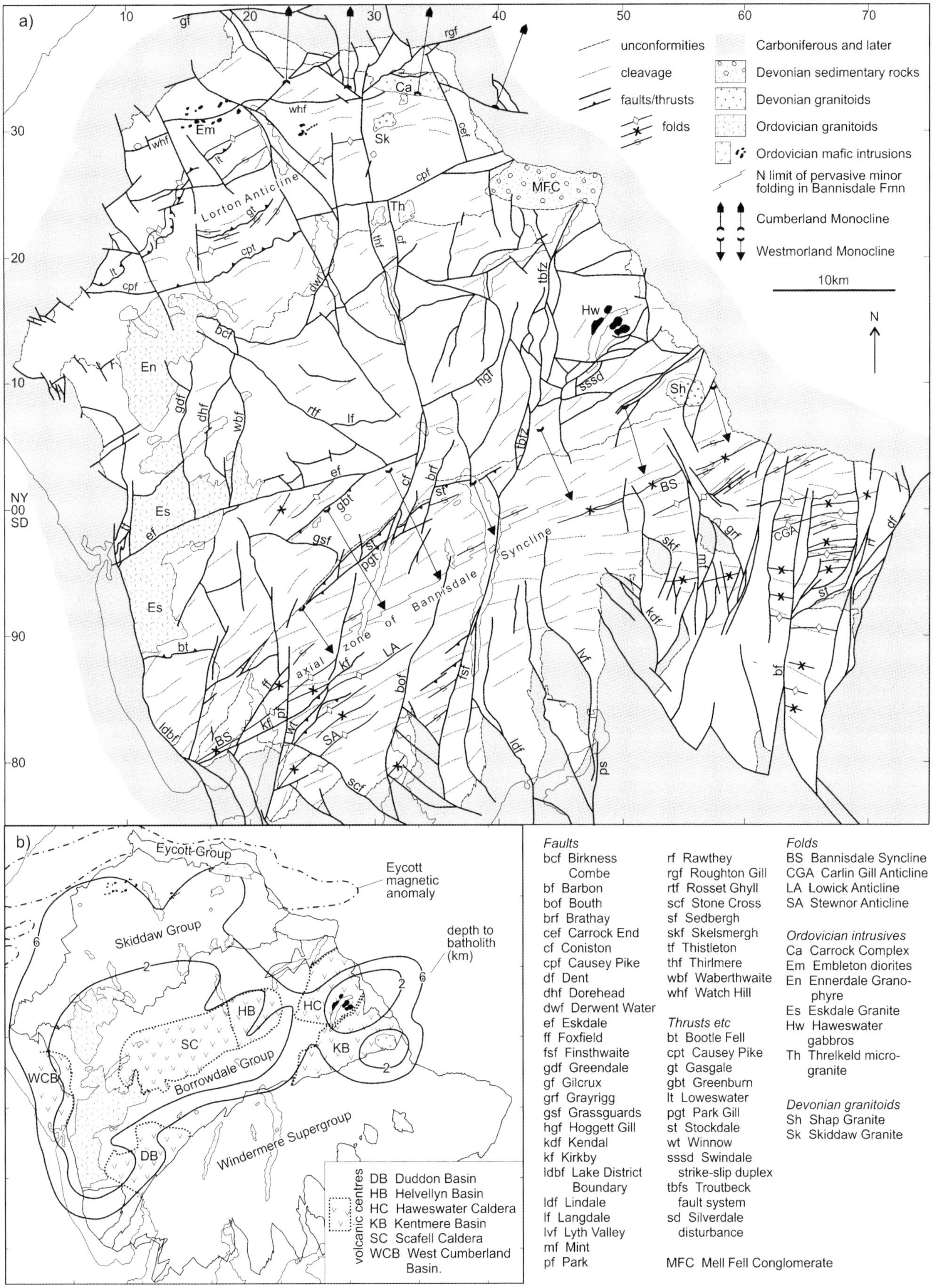

Fig. 6.4. Synoptic structural map of the Lake District compiled by NJS from published BGS maps, with additional information provided by D. Millward and NHW. Depth contours to top of the Lake District batholith and aeromagnetic anomaly associated with the Eycott Volcanic Group from Lee (1989).

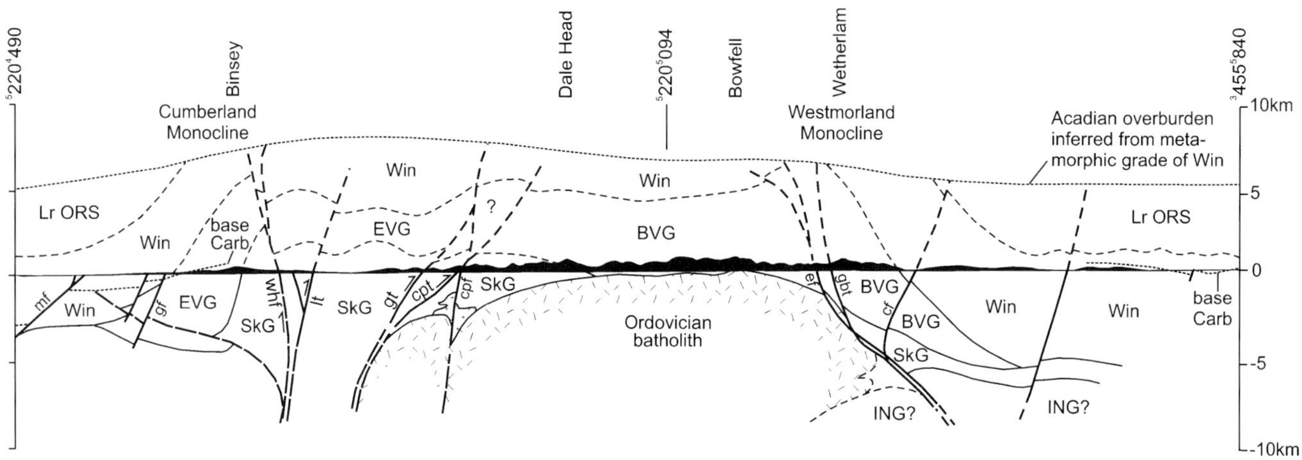

Fig. 6.5. Interpretive thick-skinned section across the Lake District at the time of peak Acadian metamorphism, partly after Soper & Woodcock (2003, fig. 4) and incorporating information provided by Dr P. Stone. Most of the structures depicted are thought to be of composite origin. The paired monoclines result from Acadian shortening superimposed on earlier extensional block tilting; fault displacements within the Skiddaw Group may comprise early Ordovician (soft-sediment), late Ordovician (volcanotectonic) and Acadian (compressive) components. BVG, Borrowdale Volcanic Group; EVG, Eycott Volcanic Group; mf, Maryport Fault; Win, Windermere Supergroup; other abbreviations as Fig. 6.4.

southwards during the Silurian. This model linked two well-established tectonic regimes: the southward-propagating deformation of the Southern Uplands associated with Iapetan convergence (Barnes *et al.* 1989); and the development of the Windermere Supergroup foreland basin on the under-riding Avalonian margin during the Silurian (Kneller 1991). It was proposed that thrust sheets were emplaced over the BVG in the late Silurian, inversion of the Windermere basin commenced in the Přídolí and culminated in deformation and low grade metamorphism of the Silurian strata in the Early Devonian (Kneller *et al.* 1993). The model thus attributes the Acadian deformation to continued Avalonia–Laurentia convergence, and sees the Westmorland monocline as a mountain front structure of earliest Devonian age (Kneller & Bell 1993).

Several objections have been raised to this interpretation. The main deformation of the Skiddaw Group is not Silurian: it overlapped the time of emplacement of the Crummock and Skiddaw granites, which are of Devonian age (Fig. 6.13). The deformation style is not of fold-and-thrust type: the thrusts post-date the main cleavage, cut across major folds and disrupt the cleavage-related metamorphic pattern (Fortey *et al.* 1993); further, they are steeply inclined at outcrop and appear to be reverse faults, unrelated to shallow detachments.

The stratigraphical shortcomings of the model are crucial. The envisaged pile of emergent thrust sheets above the BVG in Ludlow time would have been a source of proximal fans in the Bannisdale Formation, which are not seen. Inversion of the Windermere foreland basin could not have begun at the end of the Silurian because sedimentation continued well into Devonian time. A substantial thickness of Early Devonian strata must have been deposited on top of the Silurian to provide sufficient cover to generate the anchizone–low epizone burial metamorphism associated with Acadian cleavage development (Merriman *et al.* 1995; Soper & Woodcock 2003).

It may be inferred that the Acadian deformation in NW England was not a result of continuing convergence but that it took place some 20 Ma after Iapetus closure under a separate tectonic regime. This theme is developed in its Avalonian context at the end of this chapter. Figure 6.5 is an interpretive cross-section of the Lake District showing the lost cover of Lower Old Red Sandstone at peak Acadian metamorphism, and a thick-skinned interpretation of the paired-monocline structure which apparently involved buttressing of Acadian strain against the subvolcanic batholith.

Isle of Man

The Isle of Man is the Acadian terrain exposed closest to the trace of the Iapetus Suture (Fig. 6.3). In common with NW England, the main structures show some evidence of influence by buried intrusions, but also a strong linkage with NW-dipping shear zones that may root in the suture itself. Although the main structures are assumed to be Acadian in age, the possibility that some accompany rather than post-date Iapetus closure cannot be ruled out.

The structure of the Isle of Man is detailed in Chapter 3. In outline, the NE–SW structural grain of the island (Fig. 6.6) is defined by a series of main phase (D_1) thrusts, with associated folds and cleavage. Southeast of the central spine of the island, the facing on the D_1 folds and cleavage is upwards to the NW whereas, over and to the NW of the spine, it is to the SE (Fig. 6.7). In both areas a D_2 deformation produces a gently dipping crenulation cleavage and associated small folds (Simpson 1965). There is a later N-striking crenulation cleavage (Fitches *et al.* 1999), often localized close to N–S faults that offset earlier structures (Fig. 6.6). The two main phases of deformation are regarded as Acadian, in that they affect both the Ordovician Manx Group – correlated with the Skiddaw Group of northwest England – and the Silurian (Wenlock) Dalby Group – correlated with the Windermere Supergroup. This Acadian age is questioned, however, by the absence of the main deformation in the Peel Sandstone Group, regarded as Lower Devonian on palaeomagnetic grounds (Piper & Crowley 1999). Resolution of this problem is hampered by the late steep N–S faults that now obscure the original stratigraphic relationship of the Peel Group to the Manx and Dalby groups.

The NE–SW spine of the Isle of Man hosts a number of granitic sheets and the two granitic plutons of Foxdale and Dhoon (Fig. 6.7). Gravity modelling (Cornwell 1972) shows that buried parts of these plutons extend to 10 km depth and for some 10–15 km along strike, although they may not join to form a major batholith as in the Lake District (Chadwick *et al.* 2001). Nevertheless, these intrusions approximately coincide with the vergence confrontation in the D_1 structures and it is tempting to propose that they formed a buttress to top-to-SE overthrusting (Fig. 6.7). A problem in ascribing this strain contrast to buttressing by an Ordovician granite – as in NW England – is the uncertain intrusion ages: the Dhoon pluton was emplaced before or during the supposed Acadian deformation, but the Foxdale pluton may be syn- or even post-deformation (Power & Barnes 1999).

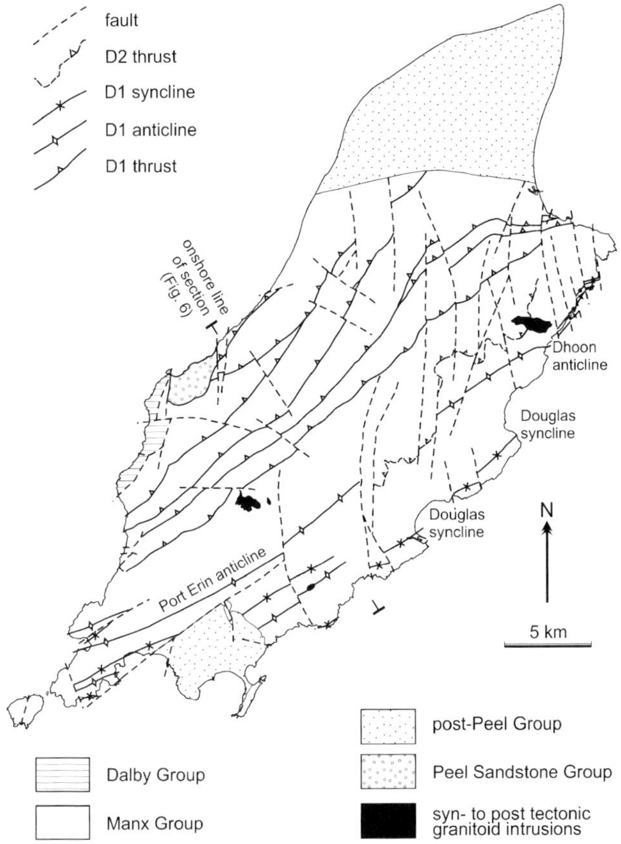

Fig. 6.6. Structural map of the Isle of Man (after British Geological Survey 2001). Line of section on Fig. 6.7 is shown.

The northwestern part of the island hosts a number of ductile shear zones generally marked by narrow zones of intense fabric development, often polyphase, but with indications of sinistral shear. One such onshore zone, Barrule Thrust Zone (Figs 4.6 & 4.7), can plausibly be linked with dipping reflectors imaged offshore and down-dip to the WNW (Quirk *et al.* 1999; Chadwick *et al.* 2001). The sole thrust of this zone can be traced with a dip of about 20° to about 8 km depth. There it could either steepen to join a ramp at the northwestern edge of the magnetic Avalonian crust (Kimbell & Quirk 1999) or more likely shallow to link with the lower crustal interface interpreted as the Iapetus Suture (Hall *et al.* 1984; Chadwick *et al.* 2001). Seismic evidence north of the Lake District (Chadwick & Holliday 1991) confirms that the thrust influence on Acadian deformation may be a general feature close to the Iapetus convergence zone. To what extent the thrusting relates to Iapetus closure rather that to the later Acadian shortening is an unresolved question.

Wales

Boundaries of the Welsh Acadides

Pre-Acadian rocks comprise most of Wales, together with a contiguous area of England in the southern Welsh Borderland (Fig. 6.8). However the southeastern part of this outcrop overlaps the western end of the Midlands Microcraton, relatively stable crust that contains no Acadian cleavage and few Acadian folds. This Acadian Foreland is crossed by fault belts, such as the Neath Disturbance and the Malvern Lineament (Fig. 6.8), typically dominated by Variscan displacements but sometimes with inferred Acadian or pre-Acadian slip (e.g. Butler *et al.* 1997).

The southeastern front of strong Acadian deformation is broadly coincident with the Welsh Borderland Fault System (Woodcock & Gibbons 1988). More specifically, the Tywi Lineament approximates to the strong cleavage front and the Pontesford Lineament to the limit of tight folding (Fig. 6.8). During Early Palaeozoic time, these two lineaments also marked the southeastern edge of the deep marine Welsh Basin (e.g. Woodcock *et al.* 1996). It therefore seems that strong Acadian deformation in Wales was concentrated in the basinal areas where Avalonian continental crust had previously been thinned by Cambrian–Silurian extension and transtension.

Traced into west Wales, the Acadian structures swing from SW- to WSW-trending, and appear progressively to be cut out by the WNW-trending Variscan Front, which forms the northern boundary of this Late Carboniferous deformation belt (Chapter 10). However, nowhere is the Variscan cleavage seen to be superimposed on rocks already bearing the Acadian cleavage. Either the Variscan and Acadian fronts become essentially coincident here or, more likely, the rocks in the Variscides that bear superposed structures were shifted westwards by dextral displacement on Variscan faults. The Welsh Acadides must then continue southwestward below the Mesozoic basins of the Celtic Sea.

To the west, north and east of Wales, Acadian structures are either overstepped by Carboniferous or Permian deposits or truncated by the bounding faults of post-Variscan basins. Only in NW Wales are original first-order structures exposed near the limits of the Welsh Acadides. Here the Menai Strait Fault System (Fig. 6.8) has been interpreted as a major late Proterozoic–Early Cambrian terrane boundary, later reactivated as the local northwest margin to the Welsh Basin (Gibbons 1987, 1989). Anglesey and NW Llŷn were part of an

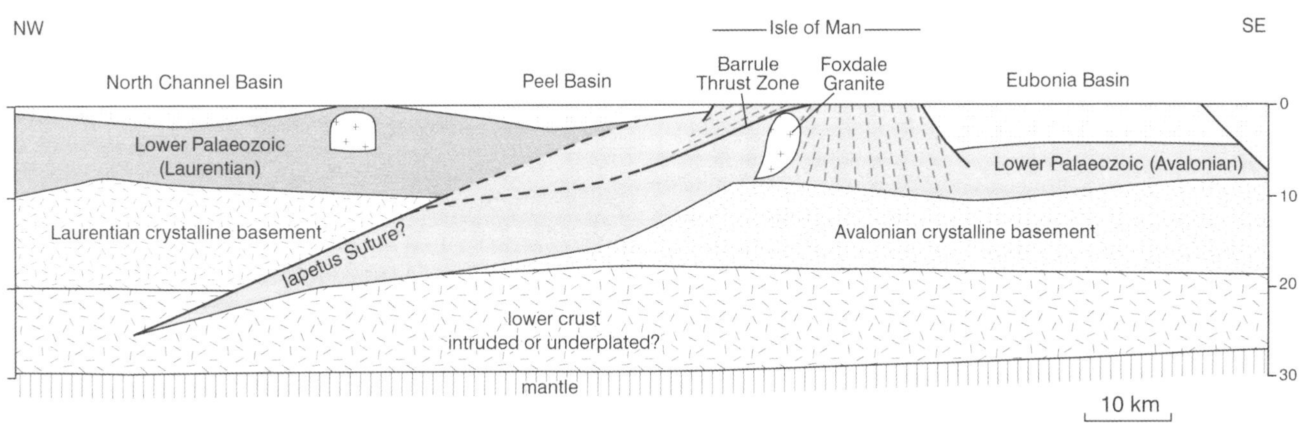

Fig. 6.7. Structural cross sections across Iapetus Suture and the Isle of Man (after British Geological Survey 2001). Section line is shown on Fig. 6.6.

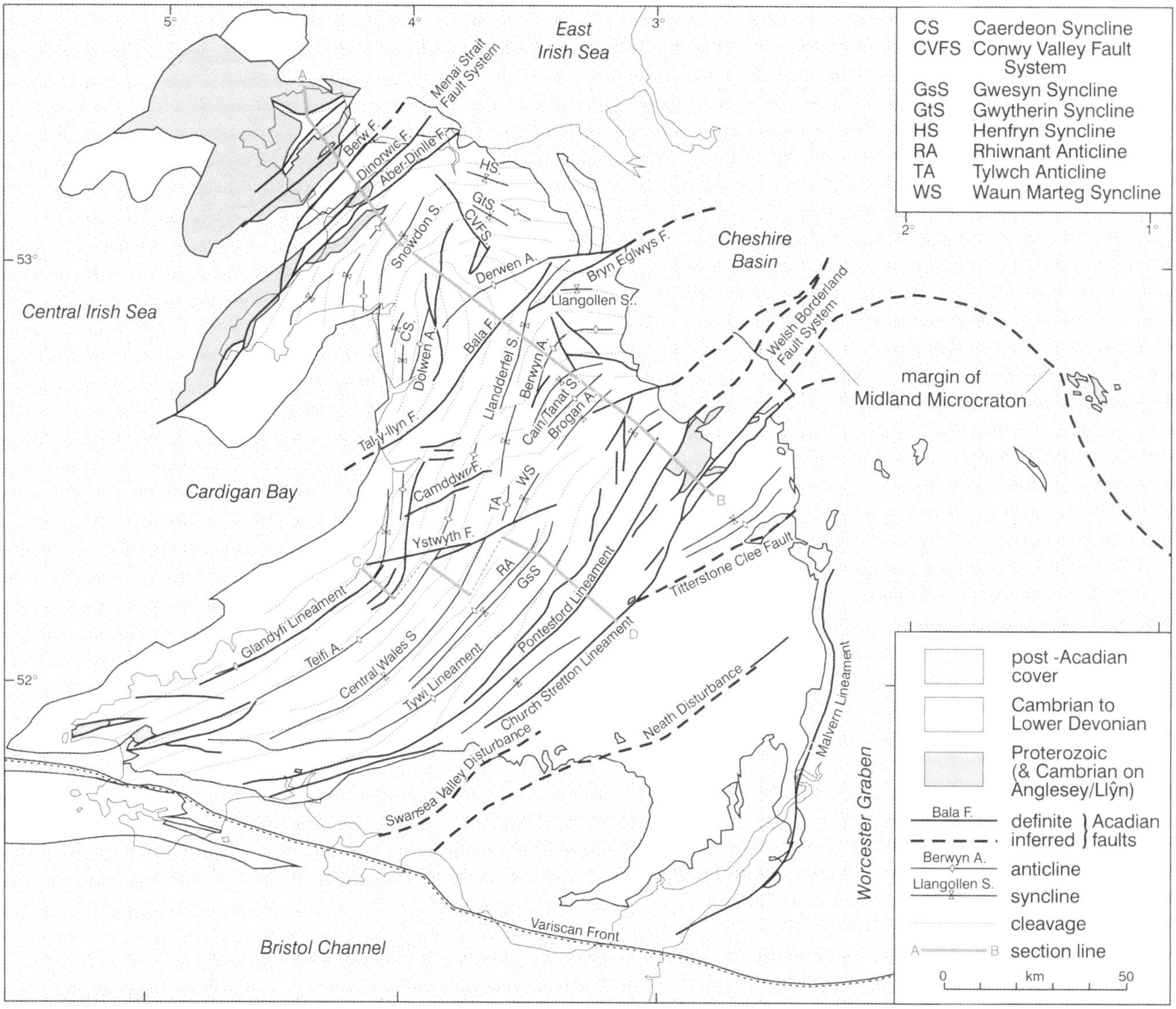

Fig. 6.8. Structural map of Wales. Modified after Woodcock (1984*a*) and the Tectonic Map of Britain and Ireland (British Geological Survey 1996). Cleavage traces modified after Johnson (1992). A–B and C–D are structural cross-section lines in Figure 6.9.

Irish Sea Platform NW of the basin. Although the platform hosted a thinner and more discontinuous cover succession than the basin, this cover and its basement were still strongly affected by the Acadian deformation (Bates 1974; Barber & Max 1979). The Menai Strait Fault System is not therefore a northwestern 'Acadian Front'. Indeed, structures on the Isle of Man, described above, suggest that Acadian deformation affected the Avalonian margin as far outboard as the Iapetus Suture.

Folds

Welsh Acadian structures form a sigmoidal pattern with N–S trends in central Wales turning clockwise into more E–W trends in South and North Wales (Fig. 6.8). More specifically, this sigmoid is mainly defined by folds and cleavage rather than by faults.

The summary structural map (Fig. 6.8) is able to portray only large-scale folds, those with half-wavelengths of about 5 km or more. In many cases these major folds are composite both in map view and in cross-section (Fig. 6.9), comprising kilometric-scale anticlines and synclines. Examples are the Teifi Anticline (or 'anticlinorium') and the Central Wales Syncline. Some of the major folds are also periclinal, with plunge culminations and depressions along their hinge lines that produce boat-shaped outcrop patterns; the Dolwen and Rhiwnant anticlines and the Snowdon Syncline, for instance.

The sigmoidal pattern of Acadian folds (Fig. 6.8) is least pronounced in southwest and central Wales, where it is defined by the Teifi, Central Wales, Rhiwnant, Gwesyn and Tywi folds. None of these structures continues intact north of the Ystwyth Fault, but the same northerly trends are continued by the Tylwch, Waun Marteg and Tarannon folds. Further north again, however, the Llandderfel, Berwyn, Cain/Tanat and Brogan folds all start to swing clockwise around the arc of the 'Berwyn Dome', although none of these folds is as continuous around the arc as has sometimes been portrayed. A yet larger and more pronounced arcuation is cumulatively defined by the folds north of the Bala Fault – the Dolwen, Caerdeon, Derwen, Snowdon, Henfryn and Gwytherin structures.

In general, the larger Welsh Acadian folds – with wavelengths upwards of 10 m – are open to close, horizontal or gently plunging, upright to steeply inclined, and face upwards on their axial planes. Minor folds are broadly coaxial with major folds, though significant departures from this rule have been described (e.g. Craig 1987). However, the axial planes of minor folds tend to fan around major structures, so show a greater range of dip.

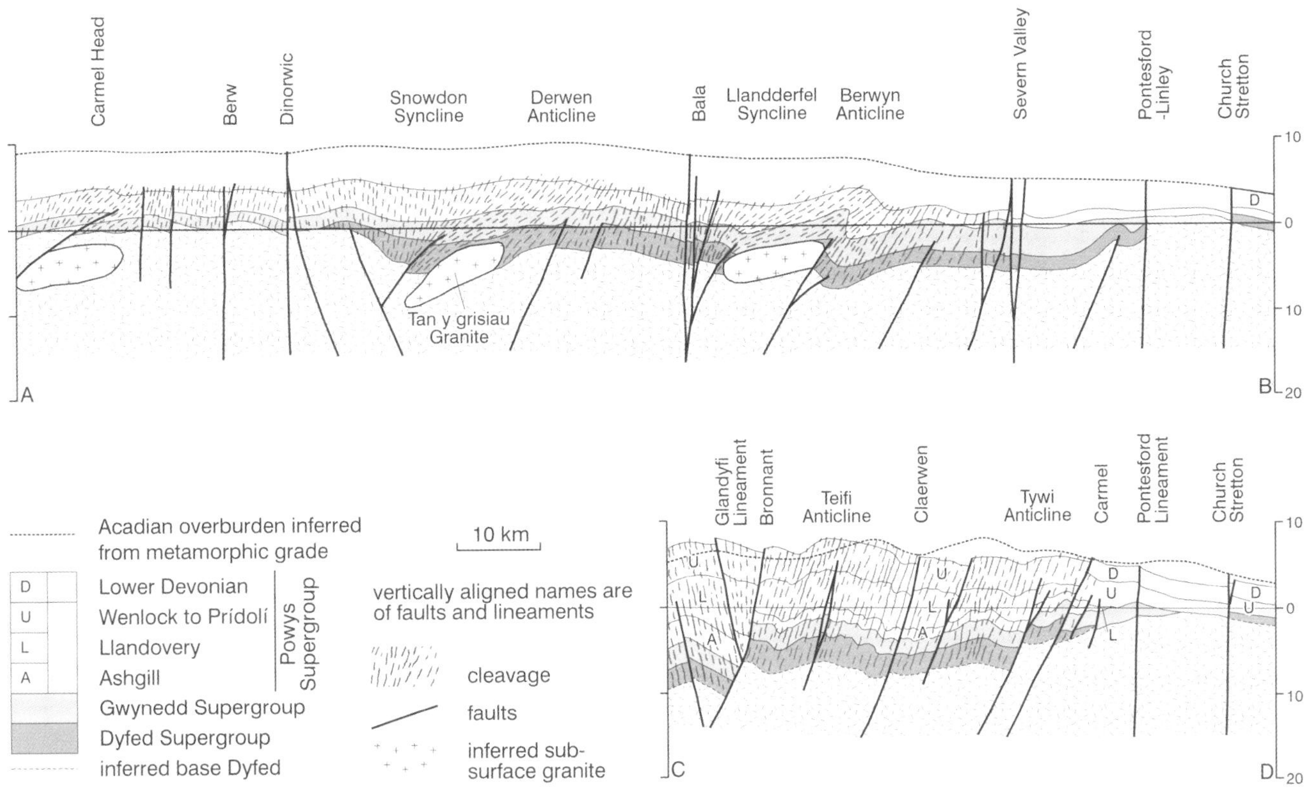

Fig. 6.9. Structural cross-sections across North Wales (A–B) and Mid-Wales (C–D). A–B based on Shackleton (1954), C–D based on Davies *et al.* (1997).

Minor folds have an asymmetry determined by their position on the host major folds. However the first-order structures such as those depicted on Fig. 6.8, show a predominant vergence to the SE or south (Fig. 6.9). Two important zones show west or NW vergence (Figs 6.9 & 6.10): in NW Wales, close to the former margin of the Welsh Basin, and in central Wales west of the Glandyfi Lineament.

The size of sub-kilometric Acadian folds is probably controlled mainly by the bed-thickness characteristics of the host succession. Hence the Silurian turbidites of mid-Wales tend to be thinner bedded than the Ordovician volcanic rocks of North Wales, and promote correspondingly smaller folds with wavelengths of the order of metres to tens of metres. Most Welsh successions have formation-scale variations that promote folds with wavelengths of hundreds of metres or more. By contrast, the positions of the largest Acadian folds – with apparent wavelengths of kilometres to tens of kilometres – are commonly determined by pre-existing fault zones that had controlled Early Palaeozoic sedimentation or volcanism. Examples are the Central Wales Syncline (Smith 1987*b*) and the Tywi Anticline (James & James 1969; Davies *et al.* 1997), which ponded turbidite systems in central Wales, and the Snowdon Syncline, one of the folds nucleated against volcanotectonic faults in north Wales (Wilkinson & Smith 1988; Howells & Smith 1997).

The majority of the folding in the Welsh Acadides seems to date from one protracted phase in Early to mid- Devonian time. On this interpretation, the periclinal nature of many Acadian folds is primary, rather than due to later 'cross-folding' (see discussions in Bassett 1969; Dewey 1969*b*), and the non-coaxiality of some minor folds with their host major fold is due to rotational strains during transpression (e.g. Craig 1987; Johnson 1992). Nevertheless, significant local enclaves of early folds do exist. Examples are: (a) the N-trending possibly volcano-tectonic folds associated with the Rhobell Volcanic Group in North Wales (Allen & Jackson 1985; Kokelaar 1988); (b) the tilting of about the same (Tremadoc) age evidenced by a sub-Arenig angular unconformity on the Llŷn Peninsula (Nicholas 1915); and (c) the NE–SW folding in the Builth and Shelve inliers assigned to a pre-Early Silurian Shelvian phase (Woodcock & Gibbons 1988; Toghill 1992). Sporadic refolding relations within the north Wales arcs between more northerly and more easterly folds are inconsistent in their age relations (e.g. Bassett 1969; Dewey 1969*b*; Baird & McCaffrey 1999) and probably represent Acadian complications rather than organized overprinting.

Cleavage

The main cleavage forms a similar sigmoidal pattern through Wales to the Acadian folds (Fig. 6.8). However, in places, such as around the northern end of the Harlech Dome, this arcuation is even more extreme. Mostly, the cleavage dips westward or northwestward, in conformity with the general SE vergence of the major folds (Fig. 6.10). In the two main zones of west or NW fold vergence (Fig. 6.10); however, the cleavage comes through the vertical to dip east or SE.

Although some previous structural maps of Wales (e.g. Dewey 1969*b*) have shown a number of different phases of cleavage, more recent opinion (e.g. Woodcock *et al.* 1988; Johnson 1992) has reverted to the view of Shackleton (1954) that the main cleavage essentially results from one heterogeneous Acadian event. Crenulation cleavages and kink bands undoubtedly occur superimposed on more penetrative fabrics (e.g. Fitches & Johnson 1978; Smith 1987*b*; Pratt 1992; Baird & McCaffrey 1999), but these are typically of local extent and sometimes of possible Variscan age. Shortening estimates across the main cleavage have been made from deformed objects. High shortening values of around 70% obtained by Wood (1971) from reduction spots in North Wales have been shown to be erroneous, the reduction spots probably post-dating cleavage formation (Nakamura & Borradaile 2001).

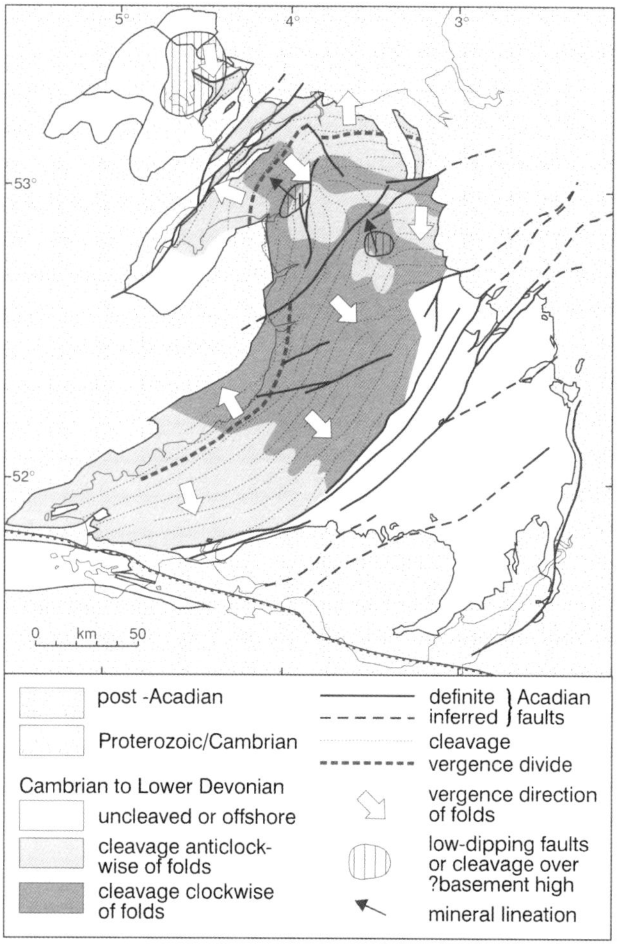

Fig. 6.10. Structural map of Wales showing the sense of transection of folds by cleavage, the vergence direction of major folds, and the areas where low-dipping structures overlie presumed granites or basement highs. Transection data from Johnson (1992).

Shortening values of around 50% obtained by Wilkinson (1987) from clasts in North Welsh volcanics are probably more realistic. Johnson (1992) used line-length balancing across folds to obtain average shortening values of about 50% for North Wales and 40% for Mid-Wales.

The strong local cleavage arcs in several areas of North Wales have been shown to relate spatially to buried granites or structural horsts. The mostly concealed Tan y grisiau microgranite (Fig. 6.10) forms the focus of the arc around the north end of the Harlech Dome (Fig. 6.10). Campbell *et al.* (1985) showed how the Acadian cleavage not only wraps around but also shallows over the top of this Ordovician pluton. Top-to-SE tectonic transport is implied by a NW-plunging mineral lineation in Snowdonia and by strain estimates from deformed objects (Wilkinson 1987). A similar geometry, with a wrapped, shallowed and lineated cleavage, occurs further SE around the Berwyn Dome (Awan & Woodcock 1993). Here the basement feature is only inferred from gravity data, and the overthrust direction is more SSE. Nevertheless, the attribution of local cleavage arcuation to moulding around basement features strongly supports the view of Shackleton (1954) and Coward & Siddans (1979) that the sigmoidal pattern of the Welsh Acadides is primary rather than due to later refolding (e.g. Helm *et al.* 1963).

The main Acadian cleavage is approximately axial planar to folds of the same age. However, systematic deviations from axial planarity occur in two ways; refraction in profile view and transection in axial (plan) view. Refraction is most pronounced where lithologies with strongly contrasting rheology are interbedded. In the turbidites of central Wales, for instance, cleavage is near axial-planar in mudstones but refracts towards bed-normal in sandstones (Plate 13).

Axial transection, by contrast, is more subtly developed, mostly with angles less than 20° and typically less than 10°. Despite this, numerous examples have now been described from throughout central Wales (e.g. James 1986; Smallwood 1986; Craig 1987; Smith 1987*b*; Woodcock 1990*b*; Johnson 1992; Davies *et al.* 1997) and north Wales (e.g. Wilkinson & Smith 1988; Awan & Woodcock 1993; Pratt & Fitches 1993). Cleavage tends to be clockwise of the N- and NNE-trending folds in central Wales, but anticlockwise of the more E-trending folds in North and South Wales (Fig. 6.10) (Woodcock *et al.* 1988*a*; Johnson 1992). On the basis that transection develops by non-coaxial strain, typically during transpressive tectonics (Soper 1986*a*), these transection senses have been taken to imply sinistral transpression in central Wales and dextral transpression in North and South Wales (Fig. 6.10). The overall pattern can been interpreted in terms of NNW–SSE Acadian shortening direction across Wales (Woodcock *et al.* 1988*a*; Johnson 1992). This shortening direction approximately parallels the mineral lineation developed in North Wales (Fig. 6.10).

Faults and lineaments

The most important faults in Wales might seem to be the various low-dipping thrusts identified in earlier literature. However, current opinion is that such thrusting is not a widespread component of Acadian deformation. For instance, the Carmel Head Thrust in Anglesey was assigned by Greenly (1919) a displacement of at least 30 km, an estimate shown by Bates (1974) to be based on an inaccurate reading of stratigraphic relations. The Tremadoc Thrust Zone in southern Snowdonia (Fearnsides 1910) has been shown by Smith (1987*a*) to be a zone of intrastratal slumping. The thrusts along the Welsh Borderland Fault System proposed by Davies (1981), have not been substantiated by later mapping (e.g. Smallwood 1986; Woodcock 1987; Davies *et al.* 1997), although the existence of poorly exposed thrusts in South Wales (Cope 1979) remains plausible. However, most Acadian thrusts are local structures associated with the steep limbs of large Acadian folds.

Although evidence of important low-dip Acadian faulting is lacking, Wales is pervaded by numerous steep faults (Figs 6.8 & 6.9), both parallel and transverse to the strike of folds and cleavage. Only major faults can be shown on Figure 6.8, and the lateral continuity even of these is there overestimated. Some of these structures are denoted as fault or fracture zones, recognizing that they comprise a plexus of separate fault strands. Other structures are distinguished as lineaments, particularly where some of their length is represented by tight folds or cleavage rather than faults, or where they have been reactivated from some earlier structure.

Strike faults are most common at the old margins of the Welsh Basin. In the NW, the Berw, Dinorwic and Aber-Dinlle faults comprise the Menai Strait Fault System (Gibbons 1987). This formed a major late Proterozoic or earliest Cambrian terrane boundary, but was reactivated in dip-slip both during Early Palaeozoic sedimentation and Acadian shortening. On the SE basin margin, the Welsh Borderland Fault System includes fault strands such as the Pontesford–Linley and Church Stretton, as well as many smaller faults. Any pre-Ordovician history of these faults is poorly constrained, but late Ordovician dextral strike-slip on the Pontesford Lineament and late Ordovician–early Silurian down-to-basin dip-slip along the faults of the Tywi Lineament are well documented (e.g. James 1983; Woodcock 1984*b*; Woodcock & Gibbons 1988; Davies *et al.* 1997). Acadian displacements include reverse slip along faults of the Tywi Lineament, passing up-section into folded cover (Davies *et al.* 1997), and sinistral strike-slip on the

Pontesford Lineament (Woodcock 1984*b*). Sinistral strike-slip on the Church Stretton Fault is probably Acadian, though Variscan components cannot be ruled out (Woodcock 1988).

Within the old Welsh Basin, many strike faults show evidence of Acadian reactivation of older structures. The Glandyfi Lineament (Fig. 6.9), for instance, includes the Bronnant Fault which bounded one of the Llandovery turbidite sub-basins (Wilson *et al.* 1992; Davies *et al.* 1997), a relationship paralleled along the Central Wales Lineament (Smith 1987*b*; Woodcock 1990*b*). The pattern of Ordovician volcanism was influenced by N–S faults such as the Trawsfynydd Fault (Kokelaar 1988), and NNE and NE faults in Snowdonia (Howells *et al.* 1991; Howells & Smith 1997), all of which suffered Acadian reactivation. The Bala Lineament is even more complex, with evidence for both pre- and post-Acadian movements as well as a record of Acadian sinistral transpression (Fitches & Campbell 1987; Pratt 1992; Pratt & Fitches 1993)

Some sectors of the Bala Lineament, notably the Bryn Eglwys Fault, cross rather than follow the strike of Acadian cleavage and folds, a relationship shown by a number of other prominent faults (Fig. 6.8). Most such 'transverse' faults demonstrably post-date the Acadian cleavage, and therefore assigning them an Acadian rather than Variscan age is often problematic. However, two of the larger ENE-striking faults, the Camddwr and Ystwyth (Fig. 6.8), host Early Devonian mineralization pointing to at least a component of Acadian displacement (Davies *et al.* 1997). Most probably these NNE faults and the NW– striking faults that comprise the other prominent set (Fig. 6.8) had both Acadian and Variscan components, a hypothesis argued explicitly for the Conwy Valley faults (Warren *et al.* 1984). In common with strike faults, some of the transverse faults show evidence for pre-Acadian influence on sedimentation (Woodcock *et al.* 1988).

Metamorphism

The metamorphic pattern in Wales was first charted using mineral assemblages in metabasic rocks (Roberts 1981; Bevins & Rowbotham 1983; Harris 1985), but later crystallinity studies in mudstones and slates allowed more detailed metamorphic maps to be made (Merriman & Roberts 1985; Roberts & Merriman 1985; Robinson & Bevins 1986). Broadly, such maps (Fig. 6.11) show grade increasing from the margin to the centre of the Welsh Basin, but with a tendency for higher grades in the NW than in the SE.

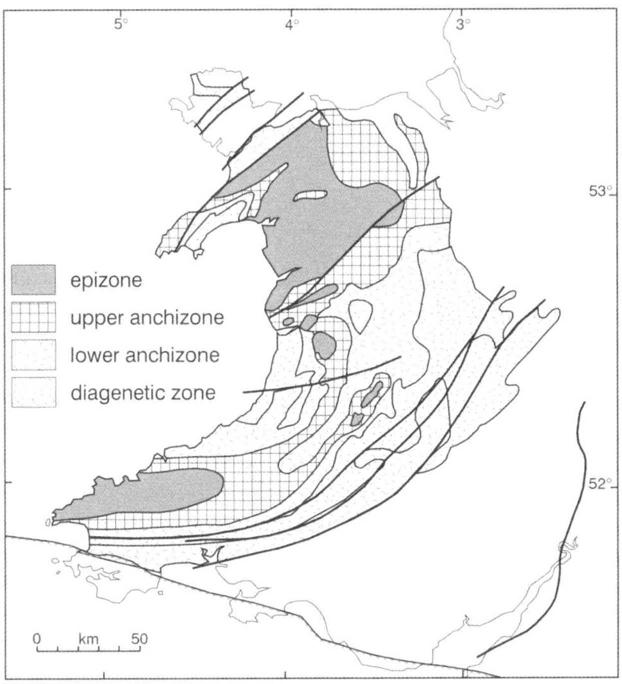

Fig. 6.11. Map of metamorphic grade in Wales, determined from illite crystallinity data (modified from Merriman 2002).

The significance of this grade pattern has been much debated. Was it attained during basin extension, at temperatures that closely reflect the thickness of the basin fill (Bevins & Robinson 1988, 1989)? Alternatively, did the metamorphic peak accompany basin shortening, folding and cleavage formation (Roberts *et al.* 1989, 1991)? Restoration of metamorphic isocrysts through Mid-Wales suggests that they were imposed on rocks that were already folded and reverse faulted (Roberts *et al.* 1996), whereas thermal modelling supports the diastathermal extension model (Robinson *et al.* 1999). A correlation of increasing grade with stratigraphic depth has been verified in a number of areas (e.g. Roberts *et al.* 1991; Davies *et al.* 1997), and is acknowledged to reflect overburden thickness, whether during extension or contraction.

The overburden thicknesses implied by the metamorphic grade are marked on the cross sections (Fig. 6.9) assuming a thermal gradient of 36 °C km^{-1} (see discussion in Roberts *et al.* 1996). For the cleaved rocks of Mid-Wales (section C–D), the overburden might have been made up only of Silurian rocks, but in North Wales (section A–B) 5 or 6 km of Lower Devonian cover are implied by the epizone grades. This implied thickness of 'lost' Old Red Sandstone may represent a distinct phase of pre-Acadian basin subsidence (Soper & Woodcock 2003). Followed SE towards the Midland Platform, both Welsh sections require some Lower Devonian cover, but of the same order of magnitude as that currently preserved on the platform (Fig. 6.9).

Early radiometric age determinations had a Devonian cluster around 400 Ma, assumed to record the formation of the Acadian cleavage (Evans 1989*a*, *b*, 1991), and therefore proving at least some component of syncontraction metamorphism. However, later studies have shown a more complex age pattern, particularly from Mid-Wales. Late Silurian–earliest Devonian ages (430–415 Ma) record recrystallization before any textural or stratigraphic evidence for the onset of basin shortening (Evans 1996; Evans & Zalasiewicz 1996; Dong *et al.* 1997). Mineral recrystallization and resetting of isotopic systems therefore occurred over a protracted period in Welsh Basin history, spanning both extensional and contractional tectonic episodes. For example, U–Pb and Pb–Pb data for authigenic monazite nodules date diagenesis during compactional dewatering (Evans & Zalasiewicz 1996) and Rb–Sr whole-rock ages from mudrocks may record the pro-grade transition from smectite to illite (Evans 1996) (Fig. 6.13). Both these events were probably promoted by increased mid-Silurian subsidence and sedimentation rates. By contrast, K–Ar dates from illite and Rb–Sr whole-rock dates form igneous rocks cluster around 400 Ma, record retrograde thermal closure, interpreted as occurring during the Acadian contractional climax.

In summary, it is probable that the metamorphic grade based on illite crystallinity (Fig. 6.11) approximately correlates with overburden thickness during the Acadian shortening. Sufficient overburden can only be provided if it includes some kilometres of Lower Devonian rocks over North Wales for which there is only indirect evidence (Allen & Crowley 1983; Soper & Woodcock 2003).

Kinematic summary

The accumulated evidence leads to a kinematic model of Acadian deformation in Wales having the following components:

(1) Strong, cleavage-forming, Acadian deformation was mainly confined to the former Welsh Basin. Here the crust had been heterogeneously thinned by Early Palaeozoic extension and transtension, which probably continued

into Early Devonian time, and had been overlain by a thick cover of deformable sediments and volcanics.

(2) During Acadian shortening, this cover was not significantly detached from its crystalline basement. The deformation was essentially thick skinned and basement-involved.

(3) Granites and basement horsts acted as buttresses that the cover wrapped around and over in localized cleavage/fold arcs. The large-scale sigmoidal arcuation pattern of Acadian structures through Wales is probably also predetermined by basement structure.

(4) Large folds commonly formed above reactivated basement faults, and not necessarily perpendicular to regional shortening.

(5) The folds are typically transected by the Acadian cleavage. The transection pattern is compatible with a broadly southeastward shortening vector, tending to produce clockwise transection in Mid-Wales and anticlockwise transection in south and northernmost Wales.

(6) The predominant southeast structural vergence implies top-to-the-SE kinematics, perhaps induced by the synorogenic topographic slope implied by metamorphic grade estimates.

(7) The grade implies a thick (5–6 km) cover of Lower Devonian sediments in North Wales. Devonian transtension has been favoured over lithospheric flexure by an over-riding Laurentian plate as the main way of accommodating such a thick succession (Soper & Woodcock 2003).

East Anglia

An Acadian deformation belt below eastern England, essentially proposed by Turner (1949), is now well documented (e.g. Pharaoh *et al.* 1987, 1995), even though most of it is concealed below post-Acadian cover (Fig. 6.12). An Acadian age for deformation is implied by cleaved Přídolí rocks in boreholes at Soham, Lakenheath and Clare (Pharaoh *et al.* 1987; Woodcock 1991). The revised stratigraphy of the exposed Charnian Supergroup, with its cleaved Cambrian as well as Proterozoic units, allows deformation there to be Acadian, an interpretation supported by $^{40}Ar/^{39}Ar$ dating (Carney 2002).

The SW boundary of the Anglian Acadides with the Midlands Microcraton can be drawn, by analogy with Wales, at the limit of cleavage and strong folding (Pharaoh *et al.* 1987; Woodcock 1991) (Fig. 6.12). As in Wales, such a boundary approximately coincides with the transition from deep marine Silurian facies in the Anglian Basin to shallow-marine facies on the Midland Platform (Woodcock & Pharaoh 1993). Ordovician igneous rocks and raised metamorphic grades are also confined to the NE of the craton margin (Fig. 6.12). The NE boundary of the Acadian belt is unclear. A SW-dipping mantle reflector in the region of the Mesozoic Dowsing faults (Fig. 6.12a) has been attributed to Ordovician subduction (England 1995), and would be expected to have a structural role in Acadian deformation.

Gravity and magnetic lineaments in eastern England (Lee *et al.* 1990, 1991), are the best guide to possible structural trends in the Acadian basement, even though they may reflect some

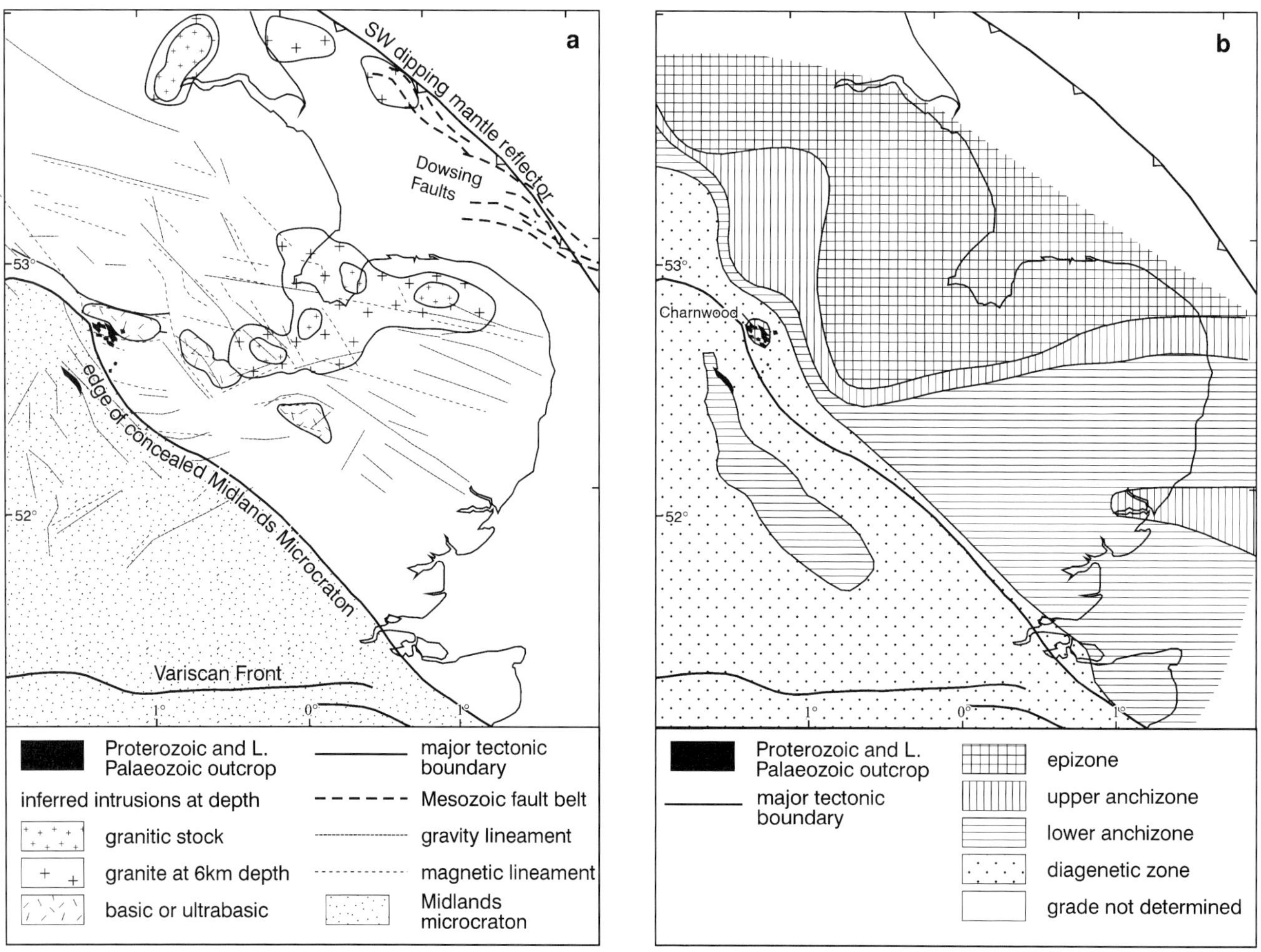

Fig. 6.12. Maps of eastern England showing (a) major tectonic boundaries and igneous bodies (England 1995; British Geological Survey 1996) and gravity and aeromagnetic lineaments (Lee *et al.* 1990, 1991), and (b) estimates of metamorphic grade (Merriman *et al.* 1993).

degree of Carboniferous and later reactivation. Distinct SE and ESE trends are observed (Fig. 6.12b). However, there is little evidence as to the age of these trends, either or both of which could be Proterozoic or late Ordovician as well as Acadian (Lee *et al.* 1991). In general, however, the Anglian Acadides continue the arcuate swing in trends seen through northern England around the north edge of the Midlands Microcraton.

Southwest England

In contrast to the more northerly areas described above, where the Acadian deformation is marked by a basement–cover unconformity at which much of the Devonian is missing, Devon and Cornwall saw continuous sedimentation through Devonian time. Alluvial sediments of ORS facies prograded south into marine deposits of the Cornubian basins, major regressive cycles marking global rises in sea level (Allen 1979) (Chapter 8). The marine deposits contain much evidence of contemporaneous tectonic and volcanic activity. Moreover, the presence of mafic and ultramafic rocks at the Lizard, interpreted as an ophiolite, suggested the closure of a small ocean basin due to subduction during the Devonian (Holder & Leveridge 1986), predating the Variscan Orogeny of Late Devonian–Carboniferous time.

Recent research has led to reinterpretation of the Lizard peridotite as a rift-related mantle diapir (Cook *et al.* 2000). U–Pb zircon dating (Sandeman *et al.* 2000; Nutman *et al.* 2001) has demonstrated that thrust stacking of the peridotite, metamorphism and crustal anatexis to produce the associated Kennack gneiss took place within the time span 390–377 Ma (Fig. 6.13). A possible igneous date for the complex is the 397±2 age of Clark *et al.* (1998) This dating links the pre-Variscan deformation of the Lizard complex and Cornubian basins to the Acadian slate belt deformation further north. The Cornubian basins appear to be an example of continuing marine deposition during orogenic shortening, a situation which is likely to occur when the crust/lithosphere thickness ratio is less than a critical value (Dewey 1982).

Age of the Acadian Orogeny in England and Wales

Compared to its major development in the northern Appalachians and Canadian maritime states (e.g. Murphy & Keppie 2005), where Acadian tectonometamorphism spanned much of Devonian time, deformation of the slate belts in southern Britain was limited to a brief period in Early to mid-Devonian time (McKerrow *et al.* 2000). To their stratigraphical compilation can be added the well-constrained relations on the Dingle Peninsula, western Ireland. A high-angle unconformity between the Dingle and Smerwick groups – of late Silurian–Pragian or early Emsian age – and the overlying ORS succession that starts with the Eifelian Pointagare Group (Richmond & Williams 2000) is taken to record the Acadian deformation.

Geochronological evidence from Eastern Avalonia (Fig. 6.13). also points to an Acadian deformation climax within the later Emsian to Eifelian, specifically the period 390–400 Ma (Fig. 6.13). Direct K–Ar dates on illites from cleaved anchizonal mudrocks have yielded mean ages of 397±7 Ma (Merriman *et al.* 1995), 399±3 Ma (Evans 1996) and, most precisely, 396.1±1.4 (Sherlock *et al.* 2003). Textural relationships of dated intrusions in the Lake District to their country rocks point to comparable deformation ages. The Skiddaw granite was emplaced into Skiddaw Group mudrocks late in the formation of the main cleavage: chiastolite in the aureole overgrows the S1 fabric but is weakly wrapped by it, and is crenulated by S2 (Soper & Roberts 1971). The aureole of the Shap granite overprints cleavage in the surrounding Borrowdale and Windermere rocks but some cogenetic dykes are cleaved at their margins (Soper & Kneller 1990). The metasomatic aureole of the concealed Crummock intrusion overprints cleavage in the Skiddaw Group but predates movement on the Causey Pike Thrust (Cooper *et al.* 1988). The mean age of the five age determinations on these late tectonic intrusions (Fig. 6.13) is 397±2 Ma. Isotopic closure is thought to have taken place within a few million years of granitoid emplacement, as erosional unloading and cooling appear to have been rapid. The date may therefore closely approximate the time of peak Acadian metamorphism.

Predating the inferred Acadian climax by 10–20 Ma is a cluster of age determinations of uncertain tectonic significance, already partially discussed in a metamorphic context in section on Wales. The Rb–Sr whole rock and Ar/Ar ages from mudrocks in both Wales and the Lake District probably all date prograde mineral reactions, but whether during tectonic cleavage formation or pretectonic subsidence is unknown (e.g. Evans 1996; Dong *et al.* 1997). The Wenlock (average 426±2 Ma) resetting of Rb–Sr systems in Lake District granitoids and volcanics has been ascribed to thrust-induced crustal uplift (Stone *et al.* 1999), although the Causey Pike Thrust, at least, moved during or after the *c.* 400 Ma event. The few earliest Devonian (average 413±7 Ma) dates from lamprophyre dykes are consistent with the precleavage age apparently implied by their foliated margins (Nixon *et al.* 1984).

Overlapping and post-dating the Acadian climax are the formation and deformation ages from the meta-igneous rocks of the Lizard complex. The possible primary igneous age, presumably reflecting at least local extension or transtension, is synchronous with the window of Acadian shortening. However, subsequent tectonic emplacement, deformation and metamorphism occurred within the protracted period 390–377 Ma, significantly later than the Acadian climax further north. Nutman *et al.* (2001) are of the view that these events may well have occupied a brief period, perhaps 2 Ma, within that span.

Three conclusions may be drawn. Within error, the Acadian tectonometamorphism occurred synchronously in England and Wales, in the Early to mid-Devonian. It appears to have been a short-lived episode, although the higher grade metamorphic rocks of the Lizard may have seen a more protracted cooling history than the low grade slate belts further north, and were soon to be involved in Variscan deformation. Secondly, it is unlikely that the late Silurian–earliest Devonian ages record the onset of Acadian shortening. Thirdly, the spatial pattern of stratigraphic and radiometric dates lend no support to the idea that the Acadian deformation front prograded southwards through England and Wales in the Early Devonian, as proposed by King (1994*a*).

Pre-Acadian ORS sedimentation

The Old Red Sandstone of Britain and Ireland has been seen as a classic example of post-orogenic molasse, derived from erosion of the Scottish Caledonides and prograding southwards through late Silurian and Devonian time (Allen 1979). Interpretive problems arise in applying this model to the late Caledonian (Acadian) deformation of southern Britain, in relation to which the Lower ORS is pre-tectonic. Friend *et al.* (2000) attempted to link sedimentation south of the Iapetus suture to southward progradation of a foreland basin ahead of the Acadian deformation front, following King (1994*a*) and Woodcock (2000). Accommodation space for ORS sediment was supposedly produced by lithospheric flexure ahead of an advancing mountain front, with deformation occurring at depth as sedimentation continued at the surface. This model accounts for the general southward recycling of sediment, but requires a more protracted history of cleavage formation that the radiometric data

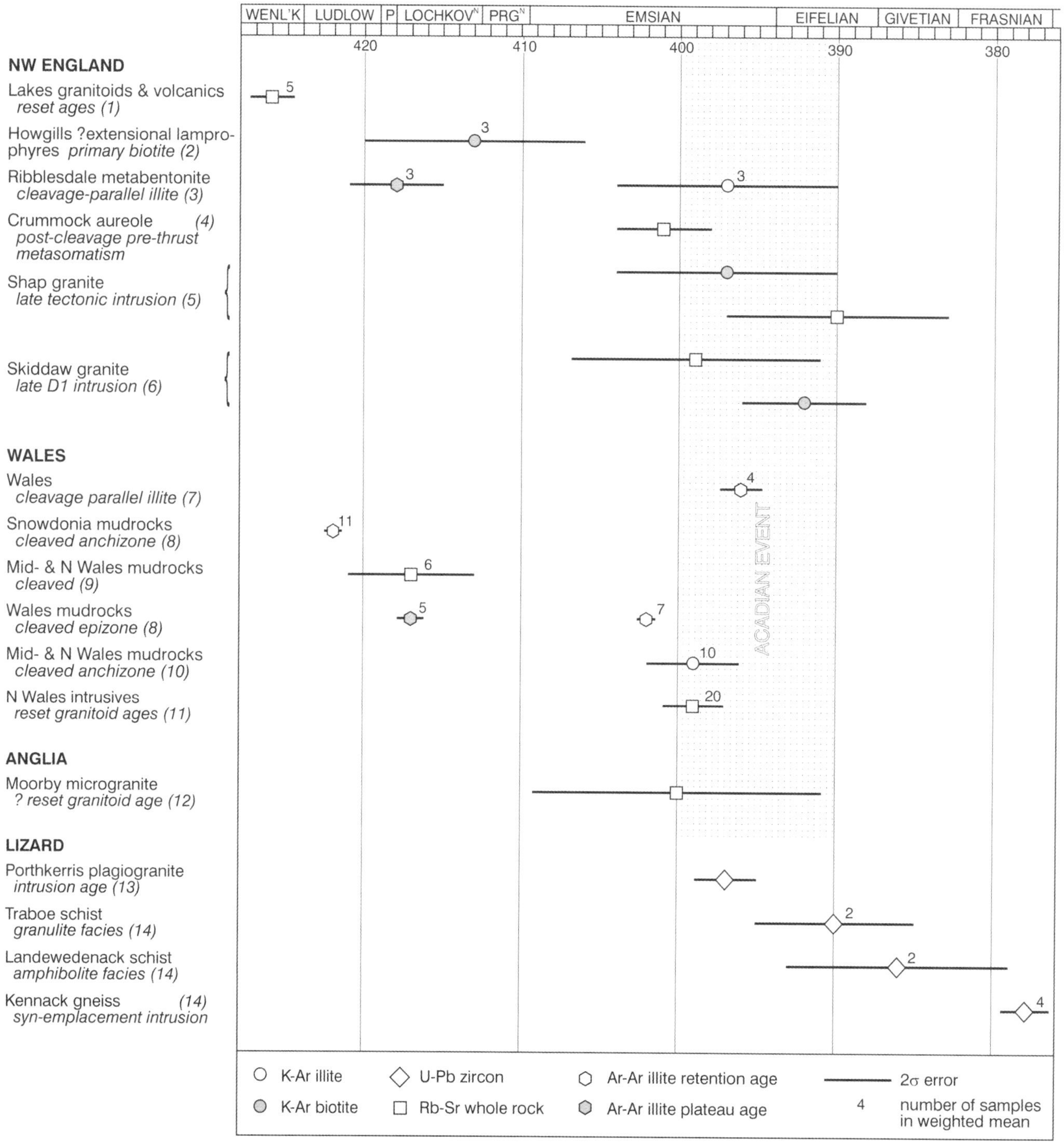

Fig. 6.13. Selected radiometric age determinations relevant to the timing of the Acadian Orogeny. The timescale is that of Tucker *et al.* (1998), as modified by Williams *et al.* (2000). Data sources are: (1) Rundle (1979; 1987), Gale *et al.* (1979); (2) Nixon *et al.* (1984); (3) Merriman *et al.* (1995); (4) Cooper *et al.* (1988); (5) Rundle (1992), Pidgeon & Aftalion (1978); (6) Rundle (1981), Shepherd *et al.* (1976); (7) Sherlock *et al.* (2003); (8) Dong *et al.* (1997); (9) Evans (1989*a*; 1996); (10) Evans & Zalasiewicz (1996); (11) Evans (1991); (12) Pharaoh *et al.* (1997); (13) Clark *et al.* (1998); (14) Nutman *et al.* (2001).

suggest. Neither is it clear that a sufficient overburden of ORS could be deposited to generate cleavage at depth, particularly in northern England, before it was eroded and recycled southward. As shown above, the Acadian deformation and metamorphism apparently did not prograde in any particular direction.

Friend *et al.* (2000) and Dewey & Strachan (2003) inferred transtensional or extensional collapse settings for some ORS basins. In particular lithospheric extension and sinistral strike-slip played a part in the development of Lower ORS basins close to the Iapetus Suture, for example the Midland Valley (Smith 1995) and the west of Ireland (Richmond & Williams 2000; Todd 2000). Soper & Woodcock (2003) have proposed that regional transtension also affected southern Britain between the end of Iapetus convergence in the late Silurian and the onset of

Acadian transpression in the Emsian. They suggest that N–S normal faults – later reactivated as Acadian structures (Fig. 6.4) – allowed basin subsidence to accommodate the lost Lower ORS sediment. More vigorous rifting clearly took place towards the southern edge of Avalonia, with marine sedimentation and diapiric emplacement of the Lizard peridotite (Nutman *et al.* 2001), although problematically the only radiometric date coincides with, rather than precedes, the Acadian climax (Fig. 6.13).

In conclusion, the evidence on Acadian timing is now inconsistent with this deformation being simply the Devonian continuation of deformation associated with the Silurian closure of the Iapetus Ocean. This Iapetan deformation is recorded in the southward advancing Silurian thrust stack comprising the Scottish Southern Uplands (Barnes *et al.* 1989), and in the flexural foreland basin developed on Avalonian crust to the south (Fig. 6.14) (Kneller 1991). However, regional transtension then dominated from latest Silurian through to Early Devonian time (420–400 Ma, Fig. 6.14). As well as accommodating thick continental sediments, the extension triggered magmatism across the now sealed suture; first lamprophyres and then more voluminous granites, especially north of the suture (Soper & Woodcock 2003). The Acadian shortening was a discrete and relatively short-lived event spanning late Emsian–Eifelian time (400–390 Ma, Fig. 6.14).

Acadian geotectonics

Regionally, the map pattern of Acadian structures in Britain and Ireland south of the Solway Line describes a major arc (Fig. 6.3). Caledonoid (NW–SE) trends dominate to the west in the Leinster massif, central Wales, the Isle of Man and Lake District; E–W trends in west and NW Wales and the Howgill Fells; and ESE to SE trends in the Ribblesdale inliers and subsurface in East Anglia. This convex-north arc appears to wrap around the Midlands microcraton which, as proposed by Soper

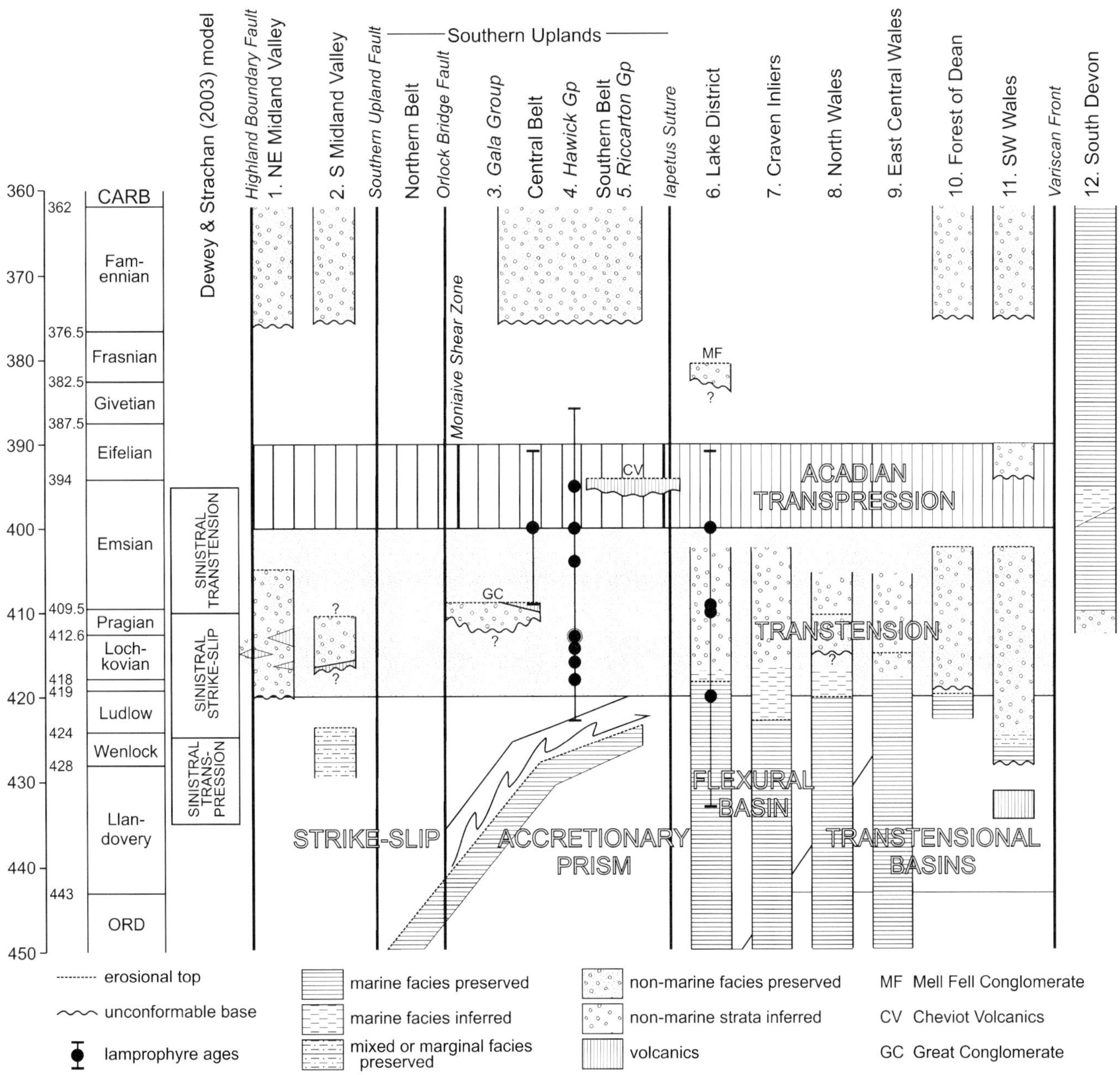

Fig. 6.14. Stratigraphic columns along a N–S transect across Britain, south of the Highland Boundary. From Soper & Woodcock (2003). Lamprophyre dates from Rock *et al.* (1986), Rundle (1979) and Nixon *et al.* (1984).

et al. (1987), acted as a rigid indenter against which the flanking turbidite-dominated basins were moulded during the Acadian deformation. This idea is based on an interpretation of regional variations in the sense of axial transection, that is whether the mean cleavage orientation strikes clockwise or anticlockwise of the fold axial trend. In previously undeformed sequences that undergo folding and cleavage formation by the flattened buckle mechanism (Ramsay 1962) in a transpressive strain field, sinistral transpression leads to clockwise transection and vice versa (Soper 1986*a*). In an arcuate structure where the transection sense changes from clockwise on one flank to anticlockwise on the other, the regional displacement or shortening direction is inferred to lie normal to the structural trend where the transection angle passes through zero.

This is the situation in the Silurian rocks of NW England (Fig. 6.3) where the change from clockwise transection in Caledonoid trending structures of the southern Lake District to weakly anticlockwise transection in the Ribblesdale inliers occurs in the SE Lake District and Howgill Fells where the structures trend a little south of east and statistically the cleavage is axial planar. This gives a shortening direction about 010° across NW England (Soper *et al.* 1987). As previously noted, the sigmoidal cleavage trends in Wales (Fig. 6.10) yield a NNW Acadian shortening direction (Woodcock *et al.* 1988; Johnson 1992).

In view of these two results, it might be anticipated that the shortening direction in the concealed SE-striking Anglian Acadides lies in the NE quadrant. The overall pattern indicates a radial component of contact strain around the northern flanks of the Midlands microcraton, consequent on regional N–S Acadian compression of the Eastern Avalonian lithosphere (Fig. 6.3). Sinistral transpression to the west of the microcraton in Wales and dextral transpression to the east (Fig. 6.2b) involve vertical axis rotations, anticlockwise and clockwise, respectively, which are difficult to quantify but must be taken into account when restoring palaeomagnetic directional data. Piper (1997) has noted a westerly shift of Ordovician magnetizations from Wales with respect to the northern Lake District, whose anticlockwise sense is consistent with the transpression model. His explanation involving rigid rotation and a dextral deformation zone between southern Britain and the Lake District is unsupported by independent evidence.

The Acadian deformation is not now thought to represent the final stages of docking of Eastern Avalonia with Laurentia. More likely it was associated with Rheic convergence: a 'push from the south'. Comparisons with the Acadian of Western Avalonia (Murphy *et al.* 1999) suggest that this push was due either to shallowing of a Rheic subduction zone beneath Avalonia or to collision with another Gondwana-derived fragment. If the Early Devonian reconstruction of Torsvik & Cocks (2004, fig. 5) is correct in showing an open Rheic Ocean and Andean margin to the south of Britain (Fig. 6.2b) then the former origin for the British slate belts is favoured.

If one accepts the traditional definition of the Caledonian Orogeny as comprising the various tectonic events associated with the closure of Iapetus (McKerrow *et al.* 2000), then the Acadian 'slate belt' deformation of England and Wales, seen as the earliest of a series of compressional events in the later Palaeozoic consequent on northward movements of Gondwana, should be regarded as 'proto-Variscan' rather than 'late Caledonian'.

7 Caledonian intrusive rocks of northern England and the Midlands

D. MILLWARD

Intrusive igneous rocks contributed very significantly to crustal growth in northern England and the English Midlands during Ordovician and Devonian times, and the buoyancy effect of these mainly granitic rocks beneath the Lake District, Cheviot Hills and the Askrigg and Alston areas of the north Pennines was to have a profound effect on the later development of Carboniferous and Permo-Triassic extensional basins in the region. During late Ordovician times voluminous subduction-related intrusive rocks were emplaced at the margin of Eastern Avalonia in a zone extending at least from Ireland, through the Lake District and the Midlands, to Belgium as a result of closure of the Iapetus Ocean and Tornquist Sea. By contrast, Early Devonian granitic plutons in northern England and the Southern Uplands belong to suites that were emplaced across the Iapetus suture zone, before, during and after the Acadian Orogeny. A similar sequence may be present in the Isle of Man.

In the Lake District, the main components of a large, subvolcanic, granitic batholith (Fig. 7.1), along with a wide variety of minor intrusions, were emplaced contemporaneously with the Caradoc Eycott and Borrowdale volcanic groups (Millward 2002, 2004 and references therein). Some of the exposed intrusions are linked geochemically to the volcanic rocks, whereas others have distinctive geochemical signatures (e.g. O'Brien *et al.* 1985).

To the SE, within the Askrigg Block and in central England, late Ordovician calc-alkaline intrusive and volcanic rocks comprise a relatively small number of exposures within Lower Palaeozoic and Precambrian inliers, and provings from deep boreholes beneath Mesozoic cover rocks (Fig. 7.1). Many of the localities and borehole sites in this region lie on the NE margin of the Midlands Microcraton, though some occur within it. However, together with gravity and magnetic data there is convincing evidence for a substantial, mainly concealed, NW–SE belt of Caledonian subduction-related magmatic rocks (Pharaoh *et al.* 1993).

The Early Devonian Shap, Skiddaw and Weardale plutons in northern England are high heat-production granites. Farther north, forming the easternmost member of the Galloway Suite, is the Cheviot Granite Pluton, which is comagmatic and coeval with the Cheviot Volcanic Formation (Thirlwall 1988). Swarms of microgranite dykes emanate from these centres. Sporadic lamprophyre dykes occur throughout northern England and southern Scotland and, although apparently small in volume, they are a significant part of this episode.

The sequence of magmatic events summarized above is established mainly from published isotopic ages of the plutonic and minor intrusive suites, and associated volcanic rocks, in the Caledonides of northern and central England (Fig. 7.2). Recently acquired, very precise, U–Pb zircon dates, particularly from the Lake District, enable the very short-lived Caradoc and Ashgill events there to be separated, and a sharp distinction to be made between these and emplacement of the Early Devonian plutons (Hughes *et al.* 1996; Millward & Evans 2003). Early Devonian magmatism appears to be absent in central England.

Lake District batholith

The principal intrusions exposed in the Lake District include the Ennerdale, Eskdale, Broad Oak, Threlkeld, Skiddaw and Shap masses (Fig. 7.3). Their outcrops were mapped in detail by the latter part of the 19th century, but it was Bott's (1967, 1974) pioneering interpretations of the Bouguer gravity anomalies that linked these surface exposures to large concealed granitic masses underlying northern England. The subsurface extent of the Lake District batholith is more than 1500 km^2, and detailed interpretation of gravity and magnetic data permits recognition of a substantial number of components (Lee 1989).

The form of the concealed intrusions was defined further from seismic reflection profiles across the western margin of the batholith. These profiles exhibit zones of intense, subhorizontal reflections intercalated with non-reflective ones (Fig. 7.4); these are interpreted, respectively, as Skiddaw Group country rock and a set of stacked lenticular granitic sheets (Evans *et al.* 1994). The saw-tooth-like western margin of the tabular sheet complex is broadly coincident with the Lake District Boundary Fault. This interpretation is consistent with best-fit modelling of the gravity data, which showed the Ennerdale Microgranite and Broad Oak Granodiorite as laccoliths 1–2 km thick (Lee 1989). Country-rock xenoliths are very sparse within the exposed granites, favouring a model based on the subhorizontal injection of tabular sheets approximately along the Skiddaw–Borrowdale Volcanic Group contact, rather than stoping during forceful pluton emplacement. In view of the Caradoc ages of the exposed tabular sheets and the fact that they broadly underlie the Scafell Caldera, it is likely that the postulated Buttermere, Dunmail and Rydal intrusions are of the same age (Fig. 7.3) (Millward 2002).

The batholith is divided by a narrow 'neck' that coincides with the trace of the Troutbeck Fault Zone (Fig. 7.3). To the east, there are two bodies, the Haweswater and Shap granites (Lee 1986). The former is almost non-magnetic compared with the relatively high magnetic susceptibility of the latter. The Haweswater Granite is possibly Ordovician, on grounds that it underlies a caldera-related pyroclastic succession around Haweswater (Millward 2002, 2004) and that the oldest Dent Group strata in the Shap area are Cautleyan in age, as in the SW Lake District (see Fig. 5.15 in Chapter 5).

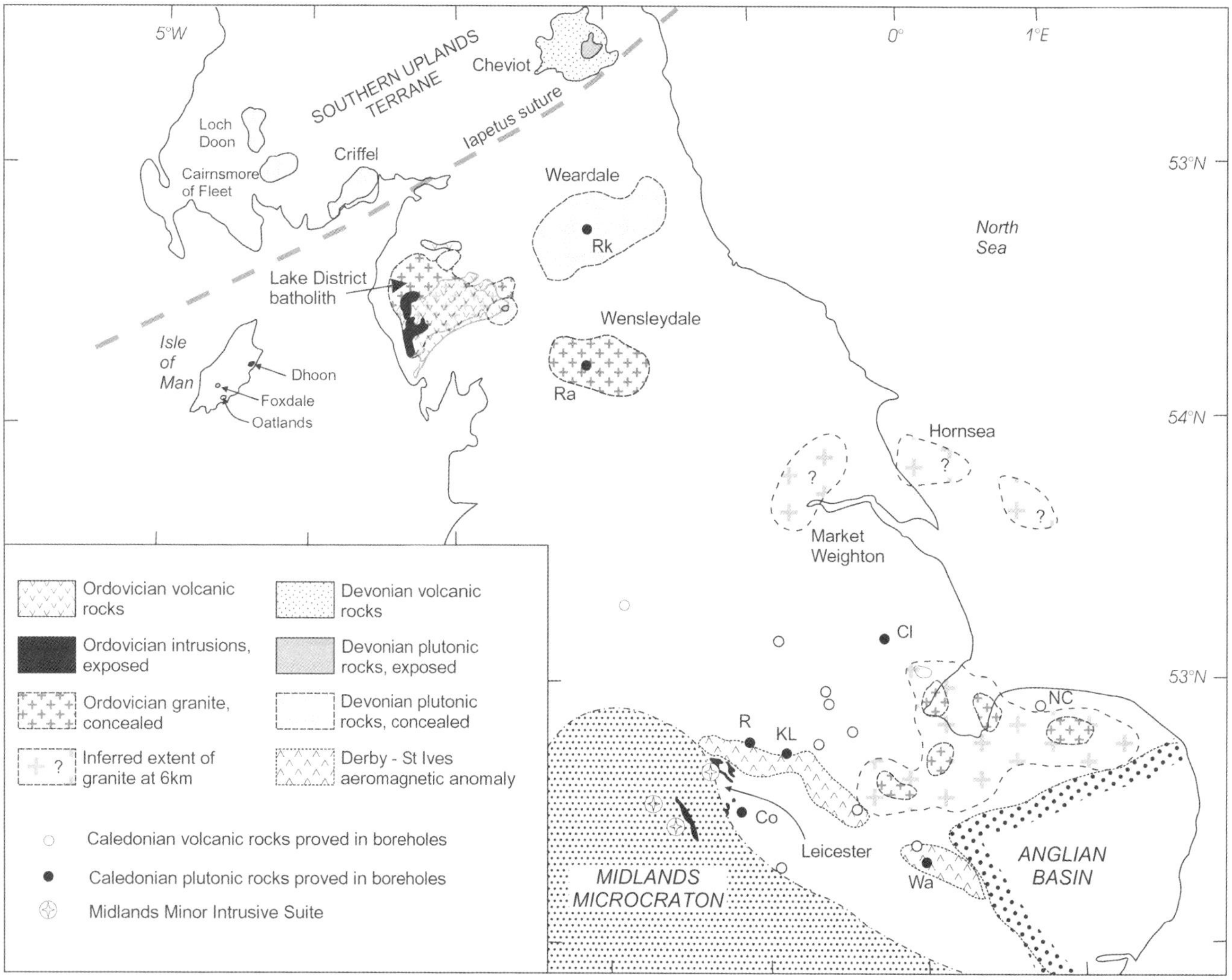

Fig. 7.1. Outline map showing the occurrences of Caledonian igneous rocks in northern England and the Midlands. Information from various Geological Survey maps and from Pharaoh *et al.* (1993). Boreholes: Cl, Claxby; Co, Countesthorpe; KL, Kirby Lane; NC, North Creake; R, Rempstone; Ra, Raydale; Rk, Rookhope; Wa, Warboys.

The steep-sided, Early Devonian Shap and Skiddaw granite plutons are located at the margin of the batholith. By contrast, the concealed Crummock Water Granite, just to the north of the Causey Pike Fault in the western Lake District, is narrowly elongate. The form of this granitic body was inferred from the zone of intensely bleached Skiddaw Group mudstones and from interpretation of gravity data (Cooper *et al.* 1988). A similar zone of bleached Skiddaw Group rocks, possibly of the same age, also occurs in the northern part of the Black Combe inlier (Johnson *et al.* 2001) and may be related to the concealed Ulpha 'Granite' (Fig. 7.3). Like the Shap and Skiddaw plutons, the Ulpha 'Granite' is marginal to the batholith and could be the source of the swarm of Early Devonian microgranite dykes, and of the tungsten, tin and bismuth mineralization that occur in adjacent areas.

A close geographical and structural relationship has long been advocated between the batholith and the abundant, and formerly economic, copper, lead, zinc, iron and other mineral veins that cut the Skiddaw Group and igneous rocks (Firman 1978). The Coniston copper veins and the tungsten-bearing veins of Grainsgill, for example, are thought to be genetically related to the late Ordovician or Early Devonian magmatism, respectively (Shepherd *et al.* 1976; Millward *et al.* 1999), but other veins dominated by lead and zinc ores are clearly later (Stanley & Vaughan 1982).

Ordovician intrusions in northern England

Granitic masses

The *Ennerdale Granite Intrusion* (Fig. 7.3), commonly known as the Ennerdale Granophyre, crops out NW of Wast Water in the western Lake District. This intrusion is mostly in contact with the Skiddaw and Borrowdale Volcanic groups, but at its southern extent it is possibly faulted against the Eskdale Granite Pluton. The Ennerdale Granite is a relatively thin, tabular body with an outcrop area of approximately 53 km^2 (Evans *et al.* 1994). Porphyritic granophyric granite dominates, but dolerite, dioritic, and hybridized dioritic, granodioritic and melanocratic granitic rocks occur locally, adjacent to the margin of the intrusion. The dolerite possibly represents sidewall cumulates from crystallization of an early dioritic magma, but emplacement of the more voluminous granitic magma whilst the former was still hot and partly crystallized led locally to the formation of hybrid rocks (Millward *et al.* 2000).

The *Eskdale Granite Pluton* (Fig. 7.3) is mainly a semi-concordant sheet within the lowest part of the Borrowdale Volcanic Group, rising in the northern part of its outcrop to within a few hundred metres of the base of the upper part of the volcanic succession. The intrusion, which has an outcrop area of about 53 km^2, consists of medium-grained muscovite

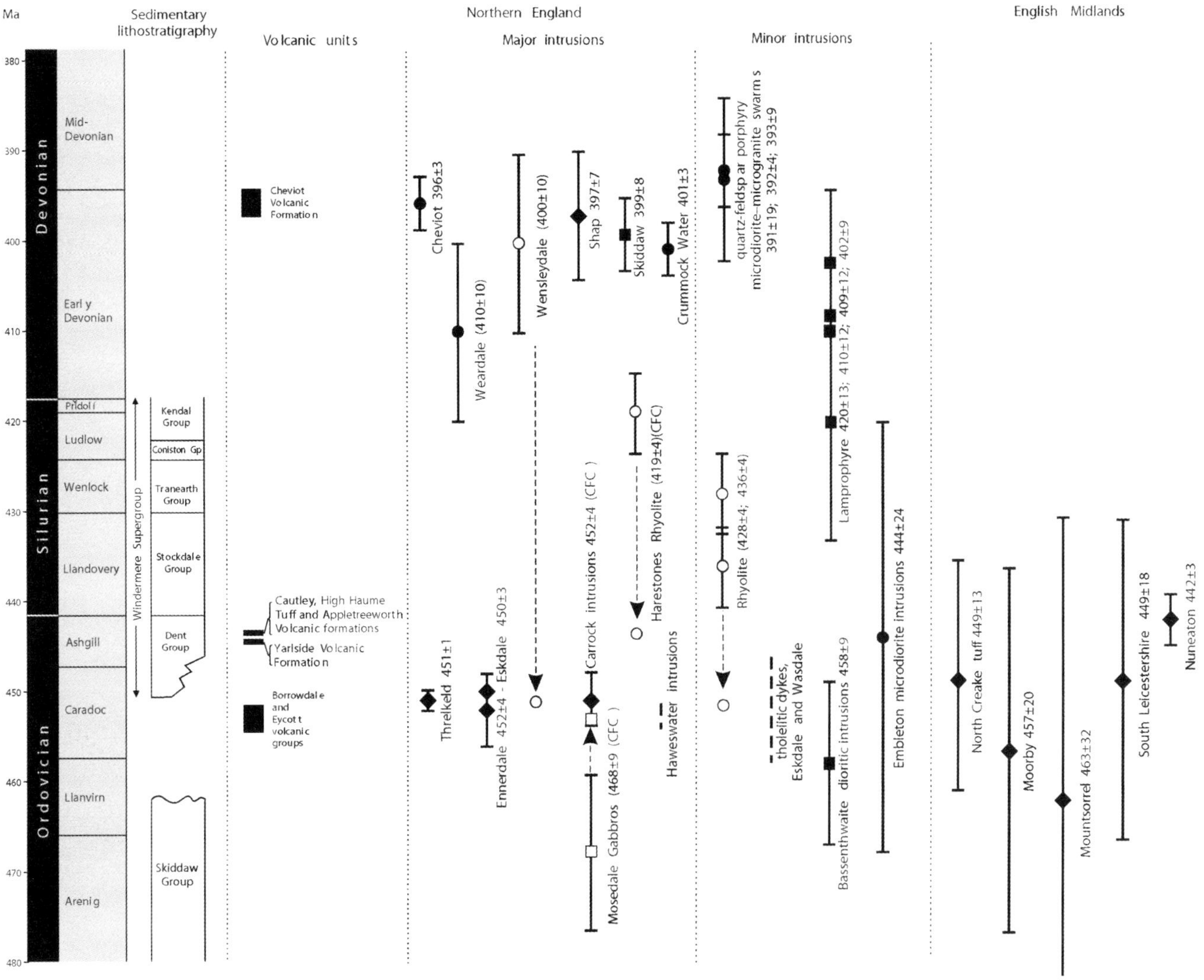

Fig. 7.2. Summary chart showing the stratigraphical distribution and radiometric ages of Caledonian igneous rocks in northern and central England. Radiometric dates are in Ma with errors quoted as 2 standard deviations: diamonds U–Pb; circles, Rb–Sr; squares, K–Ar. Open symbols represent radiometric data for igneous bodies, the ages of which are considered to have been reset; the inferred emplacement age is shown without errors. No radiometric ages have been determined for the Haweswater Intrusions and the tholeiitic dykes. See text for explanation of the inferred ages of Wensleydale Intrusion, Mosedale Gabbros and Harestones Rhyolite. Age of the Weardale Intrusion has been recalculated using more recent constants; the rhyolite minor intrusions are geochemically similar to the Ennerdale Granite Intrusion (Millward *et al.* 2000). CFC, Carrock Fell Intrusion-complex. Modified from Millward (2002) with additional data from Nixon *et al.* (1984), Noble *et al.* (1993) and Millward & Evans (2003).

granite, aphyric and megacrystic microgranite, and coarse–very coarse granite. Microgranite is most common in the northern part where the low-dipping marginal contacts and inliers of hornfelsed volcanic rocks suggest that the roof zone of the intrusion is exposed. Metasomatic recrystallization to quartz–white mica and quartz–topaz greisens occurs locally within all facies except the coarse granite (Young *et al.* 1988). A distinctive quartz–andalusite rock is associated with topaz–greisen adjacent to the contact with the Skiddaw Group near Devoke Water, and scattered, small pockets of arsenopyrite, bismuthinite, native bismuth and molybdenite in this greisen may be genetically related to it.

The *Broad Oak Granodiorite Pluton* (Fig. 7.3) was previously referred to, rather confusingly, as the Eskdale Granodiorite. It is a discordant mass, about 23 km^2 in outcrop area, intruded into rocks of the lower and upper parts of the Borrowdale Volcanic Group west of the synvolcanic Duddon Basin. The main facies is typically medium grained, with hornblende and biotite, but muscovite is absent. In the NW part of the pluton is a biotite-rich facies that also contains almandine garnet. A marginal microgranodiorite is developed along the contact. In the south, the granodiorite displays strong argillic alteration. This intrusion is considered to be Ordovician and its Wenlock Rb–Sr age of 428 ± 71 Ma (Rundle 1979) to be reset; such a disturbance in Rb–Sr systems is a common feature of the Ordovician intrusions and volcanic rocks of the Lake District (Stone *et al.* 1999).

The garnet-bearing *Threlkeld Microgranite* (Fig. 7.3) was intruded into the Skiddaw Group and basal part of the Borrowdale Volcanic Group. Interpretation of gravity anomalies east of Keswick has shown that the separate microgranite outcrops there are connected to a largely concealed laccolith, 500–1000 m thick, approximately 12 km^2 in area, that lies above the main part of the batholith (Lee 1989). The microgranite contains microphenocrysts of albite and orthoclase, and phenocrysts of corroded quartz and sporadic almandine garnet. The rock geochemistry and style of its sericite, carbonate and chlorite alteration are very similar to that seen in silicic pyroclastic rocks in the Borrowdale Volcanic Group, and this association has been confirmed by a Caradoc U–Pb zircon age (Fig. 7.2) (Millward & Evans 2003).

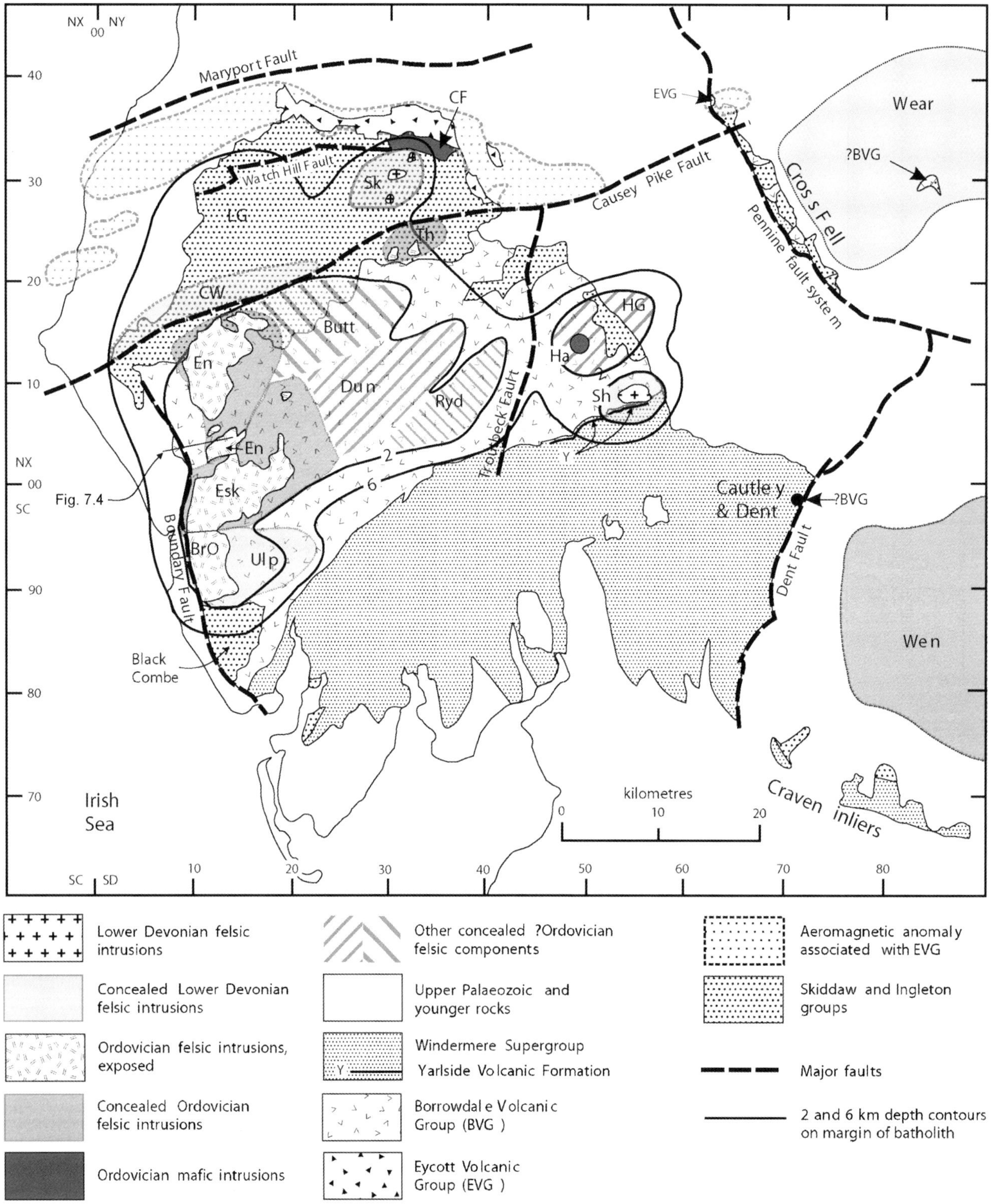

Fig. 7.3. Caledonian igneous rocks of the Lake District and adjacent areas. BrO, Broad Oak Granodiorite Pluton; Butt, Buttermere 'Granite'; CF, Carrock Fell Intrusion-complex; CW, Crummock Water Granite; Dun, Dunmail 'Granite'; En, Ennerdale Granite Intrusion; Esk, Eskdale Granite Pluton; Ha, Haweswater Intrusions; HG, Haweswater 'Granite'; LG, Loweswater 'Granite'; Ryd, Rydal 'Granite'; Sh, Shap Granite Pluton; Sk, Skiddaw Granite Pluton; Th, Threlkeld Microgranite Intrusion; Ulp, Ulpha 'Granite'; Wear, Weardale Intrusion; Wen, Wensleydale Intrusion. After Millward (2002), incorporating data from Lee (1989). Approximate line of section in Fig. 7.4 is shown.

The presence of the granitic *Wensleydale Intrusion* within the Askrigg Block, first suspected from the negative gravity anomaly over the area, was confirmed by sinking of the Raydale Borehole (Fig. 7.1). Core recovered from this comprises a pink, medium-grained granite (Kirby *et al.* 2000), and, although it has an Early Devonian Rb–Sr isochron age (Dunham & Wilson 1985), the presence of cleavage, its heat production value and geochemistry suggest that it belongs to

ERRATUM

***The Geology of England and Wales*, 2nd edn.**

Cope, J. C .W. 2006. Chapter 14. Jurassic: The returning seas.

The caption for Figure 14.20 (on p. 354) should be transposed with the caption under Figure 14.21 (on p.355). Thus the correct captions should read:

Fig. 14.20. Correlation of the Corallian Group along the outcrop from north Dorset to Oxfordshire. Modified after Wright & Cox (2001) with permission of JNCC.

Fig. 14.21. Development of the Corallian Group across north Dorset. Note how thicknesses vary across the syndepositional faults. Based on Bristow *et al.* (1995, fig. 38) by permission of the British Geological Survey.

Catt, J. A. et al. 2006. Chapter 17. Quaternary: ice sheets and their legacy.

Figure 17.7 is reproduced as Figure 17.5 in error. The correct Figure 17.5 is shown below:

Fig. 17.5. Red Crag Formation exposed in 1978 at Butley (TL 371510), Suffolk, showing large scale cross-stratification interpreted as sandwave deposition in a shallow marine environment. The laminae dip towards the SW indicating a dominant current direction similar to that of the present-day southern North Sea. Photograph by Philip Gibbard.

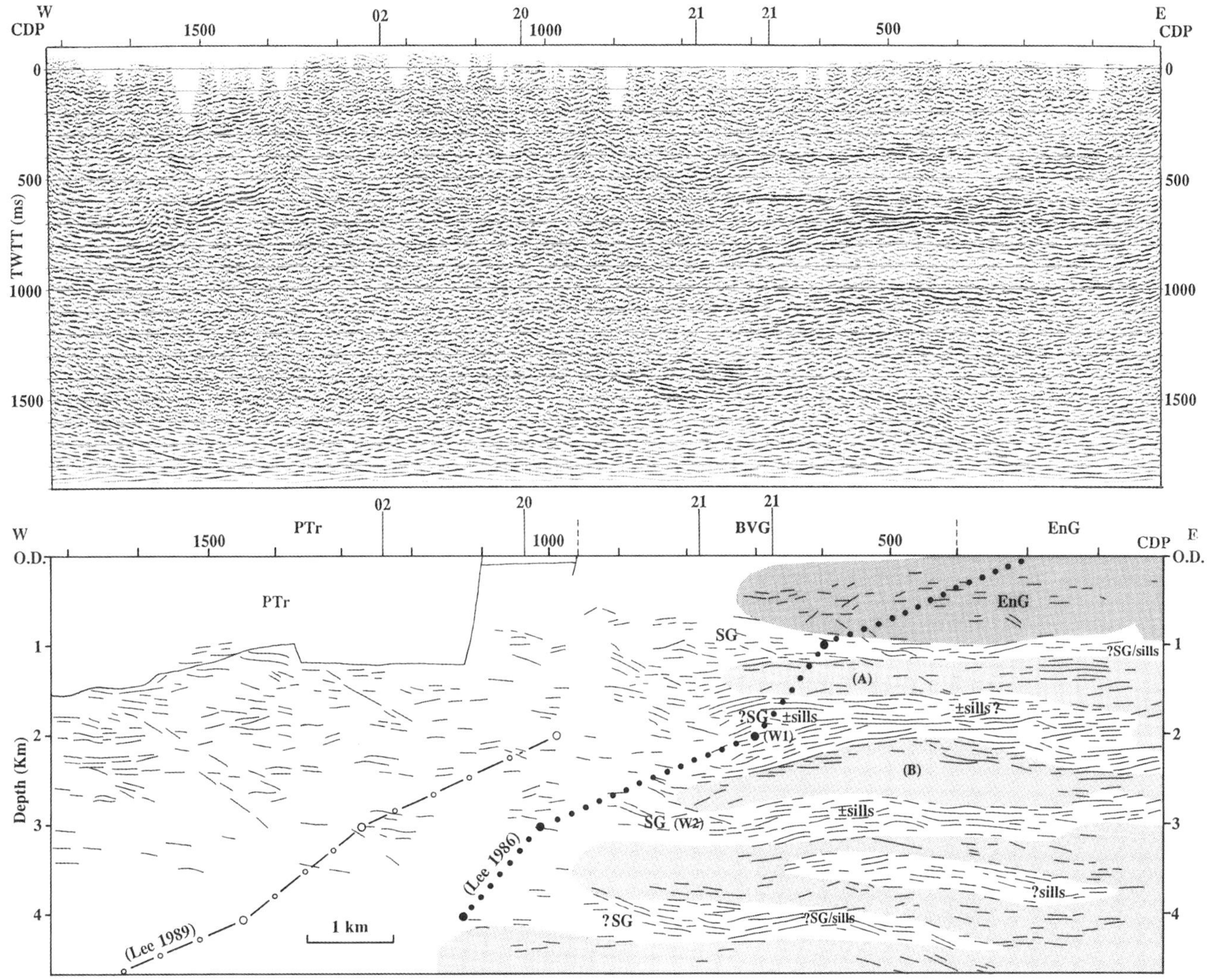

Fig. 7.4. Seismic reflection line and depth-converted line-drawing interpretation of GDGG-90-09. Light shading illustrates areas of what are interpreted to be less reflective granite, and darker shading represents the area interpreted as Ennerdale Intrusion. The depths to the top of the batholith (Lee 1986, 1989) are also plotted. BVG, Borrowdale Volcanic Group; EnG, Ennerdale Intrusion; PTr, Permo-Triassic strata; SG, Skiddaw Group. From Evans *et al.* (1994).

the Caradoc granite suite of the Lake District (Pharaoh *et al.* 1997). A zoned orefield, dominated by veins containing lead and zinc sulphides, is associated with Carboniferous rocks overlying the granite. Heat flow from it may have contributed to emplacement of these deposits.

Mafic intrusions

In northern England, there are only two Lower Palaeozoic mafic intrusive masses of any volume, the gabbro–microgranite intrusion-complex at Carrock Fell and the Haweswater intrusions (Fig. 7.2). There is no geophysical evidence to suggest that there are further concealed mafic plutons.

The dyke-like *Carrock Fell Intrusion-complex* was emplaced at the junction between the Skiddaw and Eycott Volcanic groups and comprises mafic, mafic–felsic and felsic intrusions with distinctive geochemical affinities (Fig. 7.3; see summary in Stephenson *et al.* 1999). The earliest intrusions, the *Mosedale Gabbros*, form the southern part of the complex and were intruded as a subhorizontal sill at the base of the volcanic group (Harris & Dagger 1987). Although the layered cumulate rocks are considered to be cogenetic with the volcanic rocks (Fitton 1971), the K–Ar biotite-age of 468 ± 10 Ma for the gabbros (Rundle 1979) does not accord with the probable Caradoc age for the volcanic rocks, inferred from palaeomagnetic data (Piper *et al.* 1997).

Cutting the Mosedale Gabbros with near-vertical, intrusive contacts are the *Carrock Intrusions*. These comprise a main mass of micrographic microgranite, commonly known as the Carrock Granophyre, intruded by a comagmatic, dyke-like body of microgabbro; there is a narrow zone of apatite-bearing iron-rich microdioritic rocks along the southern margin of the microgranite. Typically, the microgranite comprises zoned albite–oligoclase phenocrysts within micrographic intergrowths of quartz and alkali feldspar, and interstitial spherulitic intergrowths. There are small amounts of Ca-rich pyroxene, amphibole and titanomagnetite, and accessory apatite and zircon. The Carrock Intrusions have a Caradoc U–Pb zircon age (Fig. 7.2) (Millward & Evans 2003) and are interpreted to represent the subsided roof zone of a near-vertical magma chamber in which crystal fractionation of low-Mg tholeiitic basaltic magma had occurred (Hunter 1980).

Later, lenticular intrusions of micrographic microgranite were emplaced along the Roughton Gill Fault, at the western extent of the complex. The latest intrusion at Carrock Fell is the quartz–phyric Harestones Rhyolite. This was emplaced into, and metamorphoses, the Longvillian Drygill Shales and crops out within a fault-slice along the northern margin of the complex. The Harestones Rhyolite has a late Silurian Rb–Sr isochron age (Rundle 1979), but it is suspected that this has been reset and that the rhyolite belongs to the Ashgill phase of magmatism, along with the Yarlside Volcanic Formation (Millward & Evans 2003).

Small plugs and dykes of hornblende-bearing dolerite and gabbro, with subordinate microdiorite, were emplaced contemporaneously within the Borrowdale Volcanic Group around Haweswater in the eastern Lake District. The intrusions occur over 19 km^2 but have small outcrops that together total 2.6 km^2. The rocks are weakly cleaved locally and are geochemically similar to the volcanic rocks. There is no evidence to indicate that these intrusions are linked to a major basic mass at depth; to the contrary, the gravity low over the region indicates a concealed, shallow, granitic body (Lee 1986).

Minor intrusions

Minor intrusions of Ordovician age are spatially and genetically associated with the major intrusions, and with the Eycott and Borrowdale volcanic groups. The intrusions are diverse in composition and some suites have a significantly greater range of compositions than is represented by the larger bodies. Sporadically distributed dykes of basalt, andesite and rhyolite represent intrusive members of the two volcanic groups, but more abundant are high-level sills, mainly of andesite typically with peperitic breccias at their margins (Branney & Suthren 1988; see Chapter 5, section on Ordovician continental-arc volcanism). A cluster of magma-mingled, composite basalt–andesite dykes and minor intrusions were intruded into the upper part of the Borrowdale Volcanic Group in the area between Scafell and Bowfell (McConnell *et al.* 2002). At least some of these invaded sediments that were not fully lithified at the time. Of more silicic composition are the garnet-bearing dacite dykes in Eskdale and Wasdale which are probably coeval with the Scafell Dacite, a lava dome erupted during the final phase of activity at the Scafell Caldera (Beddoe-Stephens & Mason 1991).

Some andesite bodies within the Skiddaw Group, previously thought to be lavas, are now interpreted as sills related to the volcanic groups (Hughes & Kokelaar 1993). However, the Skiddaw Group is cut by many other dykes, sheets and small plutons belonging to the Bassenthwaite Intrusions. These comprise calc-alkaline, augite–phyric meladiorite, microdiorite, dolerite and olivine–augite hornblendite, with geochemical characteristics similar to the Borrowdale Volcanic Group (Fortey *et al.* 1994), an association supported by their K–Ar age (Fig. 7.2) (Rundle 1979). These masses are distinct from the Embleton Intrusions, a group of sills and small plugs of aphyric basalt, andesite and microdiorite adjacent to the Watch Hill Fault (Fig. 7.3), east of Cockermouth. These are geochemically similar to the Eycott Volcanic Group and Mosedale Gabbros, and one of them has an Ordovician Rb–Sr isochron age (Fig. 7.2) (Rundle 1979).

Cutting the Borrowdale Volcanic Group in the western Lake District are aphyric basalt and dolerite dykes, comprising calc-alkaline and high Fe–Ti tholeiitic varieties (Macdonald *et al.* 1988). The former may be associated with basalt lavas within the volcanic group, but the other is unique in the region. They were emplaced both before and after the Eskdale Granite, but no radiometric ages have been determined. An altered dolerite intrusion at Seathwaite in Borrowdale, possibly Ordovician in age (Millward 2004), is associated with a series of pipe-like deposits containing remarkably pure nodules of graphite, a paragenesis that has few parallels worldwide.

Pink, aphyric and sparsely microporphyritic rhyolite dykes are locally abundant adjacent to the Eskdale and Ennerdale granites. Geochemical similarities with the Ennerdale mass suggest that the Rb–Sr isochron ages for the dykes of 436–428 Ma have been reset (Rundle 1992; Al Jawadi 1987).

Ordovician intrusions in the English Midlands

Three groups of calc-alkaline plutonic rocks are recognized in this area. The first of these, referred to as the South Leicestershire Diorite Complex, crops out on the margin of the Midlands Microcraton, SW of Leicester (Fig. 7.1). The second group is associated with a belt of strongly positive aeromagnetic anomalies up to 10 km wide and extending about 125 km from Hathern, near Derby to St Ives in Huntingdonshire (Allsop 1987); examples of these plutonic rocks include the exposed Mountsorrel Complex and occurrences in the Rempstone, Kirby Lane and Warboys 1 boreholes (Fig. 7.1). The xenolithic granodiorite from Rempstone is petrographically similar to Mountsorrel. Further NE, in south Lincolnshire, the Moorby Microgranite encountered in the Claxby borehole is part of a third group of silicic plutons interpreted from geophysical anomalies in the area around The Wash (Busby *et al.* 1993). The granophyric microgranite is cleaved and has a Caradoc age (Noble *et al.* 1993). Geochemically, it is similar to contemporaneous intrusions in the Lake District (Pharaoh *et al.* 1997).

The Caradoc *South Leicestershire Diorite Complex* comprises quartz-diorite, diorite, tonalite and microtonalite. The diorite contains hornblende and sparse augite. The numerous, widely scattered occurrences of diorite, along with its presence in the Countesthorpe Borehole (Fig. 7.1), have been inferred to indicate a composite pluton about 14 km wide, intruding the Cambrian Stockingford Shale Group (Le Bas 1972).

The *Mountsorrel Complex*, about 10 km north of Leicester, comprises gabbro, diorite and granodiorite. The gabbro is composed of labradorite and ophitic brown hornblende enclosing relict augite; pseudomorphs after olivine may be present. Although similar to the South Leicestershire Complex, the diorite contains more biotite and the most extensive rock type is a biotite granodiorite. Tungsten, tin and molybdenum vein and joint mineralization occurs at Mountsorrel.

The Midlands Minor Intrusive Suite

Lamprophyre and diorite dykes and sills were intruded into the Cambrian and Tremadoc rocks within, and at the margin of, the Midlands Microcraton, probably during Ashgill times (Fig. 7.1) (Noble *et al.* 1993; Thorpe *et al.* 1993). Exposures of these rocks are in quarries close to Nuneaton, near the Wrekin in Shropshire and in the Malvern Hills. The suite is typified by the 50 m-thick composite sill of spessartite, hornblende diorite and hornblende meladiorite at Griff Hollow, near Nuneaton. The presence of olivine and absence of quartz distinguishes this group from the other Caledonian igneous rocks of central England.

Aphyric microdiorite intrusions may also have been contemporaneous with volcaniclastic rocks of the Tremadoc Barnt Green Volcanic Formation at the southern end of the Lickey Hills, south of Birmingham (Old *et al.* 1991). Geochemical similarities with the Midlands Minor Intrusive Suite may imply that this magma type was available over a substantial time span.

Early Devonian intrusions of northern England

The roof zone of the broadly cylindrical *Skiddaw Granite Pluton* is exposed in valleys within the Skiddaw Massif (Fig. 7.3). It is a coarse, biotite granite that is locally porphyritic, except in Grainsgill in the northern part of the pluton, where granite passes into quartz–white mica greisen, and is associated with the formerly economic tungsten vein mineralization that cuts both the granite and the adjacent Carrock Fell Intrusion-complex (Shepherd *et al.* 1976). The pluton, at least 4.5 km in diameter, was emplaced into the Skiddaw Group and is surrounded by a concentric thermal aureole that grades outwards through garnet, biotite–cordierite, chiastolite and chloritoid isograds (Burdon 1999).

In the eastern Lake District the *Shap Granite Pluton*, approximately 8 km^2 in outcrop area, cuts steeply dipping strata of the Borrowdale Volcanic Group and adjacent Windermere Supergroup. The surrounding thermal aureole and interpretation of gravity and aeromagnetic data indicate a steeply conical form at subcrop (Lee 1986). This pink and grey biotite monzogranite contains alkali feldspar megacrysts that have been interpreted either as phenocrysts or as porphyroblasts resulting from late-stage potassium metasomatism (see Stephenson *et al.* 1999 for summary). Microdioritic enclaves and xenoliths of country rock are abundant locally. Hydrothermal mineralization and fissure metasomatism are conspicuous features of the granite and its aureole respectively, but no economic quantities have been found. A wide variety of minerals in the former include molybdenite, bismuthinite and minor scheelite (Firman 1978), whereas the latter is dominated by garnet–epidote–hornblende veins (Firman 1957).

To the north of the Ennerdale Granite, bleached and recrystallized Skiddaw Group rocks, accompanied by locally abundant tourmaline veins, occur in an elongate, ENE-trending zone, 24 km long and up to 3 km wide adjacent to the Causey Pike Fault (Fig. 7.3; Cooper *et al.* 1988). The metasomatic event has been dated at approximately 400 Ma and this overprints a thermal metamorphic event probably associated with the concealed Crummock Water Granite. The bulk loss of metals from the bleached rocks compared with unbleached Skiddaw Group rocks suggests that these sedimentary rocks may have been a source of ore metals in the Lake District.

The northern Pennines are underpinned by the granitic *Weardale Intrusion*, 60 × 25 km in extent, the existence of which was detected by gravity surveys and confirmed by the Rookhope Borehole (Fig. 7.1) (Bott 1967; Dunham *et al.* 1965). The core recovered consists of an aphyric two-mica granite, with pegmatite and aplitic facies, and a shallow-dipping, gneissose-like foliation (Dunham 1990, plate 2). It has a high heat-flow value, is peraluminous in composition and geochemically similar to the Skiddaw granite, a correlation that is supported by its Rb–Sr isochron age (Holland & Lambert 1970). The pink *Dufton Microgranite* of the Cross Fell Inlier is one of several cupolas or separate intrusions present beneath the Alston Block. High heat flow from the granite is believed to have had an important role in driving the convection cells responsible for the later formation of the zoned North Pennine Orefield (Dunham 1990). The granite may have been the source of elements such as Sn, F and Bi within the mineralization.

The *Cheviot Granite Pluton* forms the core of the Cheviot Hills in north Northumberland (Fig. 7.1). Its outcrop covers an area of about 62 km^2, and at a depth of 4 km it has a diameter of approximately 20 km. An outer zone of grey, quartz monzonite–quartz monzodiorite was emplaced first, followed by an inner zone of medium to coarse monzogranite; the final phase of intrusion is represented by granophyric granite (Jhingran 1942). The contemporaneous Cheviot Volcanic Formation rests unconformably on deformed Silurian sedimentary rocks of the Southern Uplands terrane. Thus, these igneous rocks were the last of this episode in northern England, emplaced after the Acadian Orogeny.

Minor intrusion suites

Sporadic calc-alkaline lamprophyre dykes cut all Lower Palaeozoic lithostratigraphical units of the Lake District, Cross Fell, Ingleton, Cautley, Dent and Teesdale inliers. Macdonald *et al.* (1985) recognized two compositional groups, clinopyroxene–phlogopite–Ca-amphibole-phyric kersantite, and strongly biotite-phyric kersantite and minette. Some of the dykes are cleaved, whilst others appear not to be, but their late Silurian–Early Devonian K–Ar dates suggest intrusion ahead of the granite plutons (Fig. 7.2) (Rundle 1979; Nixon *et al.* 1984).

Swarms of Early Devonian feldspar and quartz–feldspar-phyric microgranite, felsite and subordinate microdiorite dykes and minor intrusions occur throughout the Lake District and Cross Fell inliers. In the former area these are particularly concentrated around Scafell, in the Duddon valley and radiating away from the Shap Pluton. These are geochemically similar to the Shap granite (Al Jawadi 1987). A suite of felsic dykes emanates in a broadly radial pattern south from the Cheviot Granite.

Isle of Man

Intrusive rocks are abundant within the Manx and Dalby groups in the Isle of Man. These igneous rocks are small in extend, mostly dykes and minor intrusions, although granite may underlie much of the island at depth (Chadwick *et al.* 2001). Three main masses present are: the Dhoon Granodiorite and Foxdale Granite plutons which have outcrop areas of approximately 2 km, and the smaller Oatlands complex (Fig. 7.1).

The *Dhoon Granodiorite Pluton* is biotite-bearing and locally porphyritic, but it is strikingly altered with plagioclase replaced by zoisite and muscovite, interpreted to be the products of greenschist-facies metamorphism (Chadwick *et al.* 2001). There is no radiometric age determination on the granodiorite, but relationships with the deformation sequence affecting the Manx Group rocks indicate that the pluton was emplaced early in the sequence of events. Together with the volcanic-arc affinity suggested by its trace element geochemistry, the Dhoon Pluton may have been emplaced contemporaneously with the Ordovician arc magmatism in the Lake District, although equally a later age is possible. The granitic through to gabbroic *Oatlands complex* is considered to be more in common with the Dhoon Granodiorite than with the Foxdale Granite.

The *Foxdale Granite Pluton* is muscovite-bearing and the accessory mineral assemblage includes garnet; porphyritic microgranite and pegmatitic varieties occur locally (Chadwick *et al.* 2001). The relationship to the deformation sequence, the relatively high content of radioactive elements and the radiometric age of around 400 Ma indicate an association with the Early Devonian granites of northern England.

Many of the dykes and minor intrusions, which range in composition from mafic to felsic, are deformed and metamorphosed, and thus may have been emplaced early in the tectonic history. Altered basalt and basaltic andesite sills at Poortown have a volcanic arc signature and, though the age of these sills is uncertain, they may be equivalent to the Caradoc volcanic rocks in the Lake District. A few mafic dykes cutting Silurian strata confirm that at least some dykes were emplaced later. The suite of microgranite sheets occurring along the central spine of the island has been linked to the Dhoon Pluton.

Petrogenesis

The mineralogical and geochemical changes exhibited by these Caledonian igneous rocks as a consequence of hydrothermal alteration and low-grade metamorphism inhibit precise modelling of petrogenetic processes involved in their derivation. Despite this, our understanding of the likely types of process involved has evolved considerably from early models that showed a trend in the region throughout this time from early tholeiitic magmatism to increasingly evolved, calc-alkaline granites (e.g. O'Brien *et al.* 1985). The recent high-precision U–Pb dates enable the inferred subduction-related magmatic events during late Ordovician times in the Lake District and the

Midlands to be clearly separated from the lamprophyric and high heat-flow granitic magmas of the Early Devonian transtensional and orogenic regimes seen in northern Britain.

The Ordovician mafic magmas, including those of the Eycott and Borrowdale volcanic groups, were derived by partial melting of a common lithospheric mantle source (Macdonald *et al.* 1988), enriched to different levels in incompatible elements through subduction-derived aqueous fluids. Subsequent geochemical divergence into suites with tholeiitic, calc-alkaline and transitional affinities resulted from varied amounts of partial melting, different fractionation assemblages and from the assimilation of crustal material (Beddoe-Stephens *et al.* 1995). The geochemistry thus does not provide robust evidence for relative geotectonic positions within the continental-arc system as originally proposed by Fitton & Hughes (1970).

The Threlkeld and Ennerdale granites, the Broad Oak Granodiorite and associated minor intrusions exhibit continental-margin, volcanic-arc geochemical signatures, and thus show many similarities with the silicic volcanic rocks which they underpin (Millward *et al.* 2000). By contrast, many of the distinctive features of the Eskdale Granite are more typical of granites with sedimentary protoliths ('S-type'), including a restricted high-SiO_2 content, peraluminous composition, the presence of muscovite and garnet in the mineral assemblage, and high Sr and O isotope compositions. Given that these granites were all emplaced during a very short period during the Caradoc, the compositional differences must relate to processes such as the amount of crustal material assimilated and differing crystal fractionation histories.

The Early Devonian lamprophyric magmas were emplaced during the transtensional tectonic regime that resulted in the accumulation of a molasse pile (see Chapter 6) and the granites during the latter part of, and after, the ensuing Acadian deformation episode, which caused crustal thickening and rapid uplift of the orogen. The lamprophyres are interpreted as high melt fractions of depleted oceanic lithosphere that had been metasomatized by aqueous fluids derived from the subduction dehydration of Iapetus oceanic crust and by a CO_2-rich phase from a deeper mantle source (Macdonald *et al.* 1985). Variable but small degrees of fractional crystallization were involved before final emplacement.

Despite the prevailing geotectonic environment of the Acadian Orogeny, a sedimentary protolith is considered unlikely as the source of the Early Devonian granites (O'Brien *et al.* 1985). Rather, the parental magmas were probably derived from a mafic source that contained residual garnet. These calc-alkaline granites are compositionally evolved rocks, yet show little evidence of an extended history of crystal fractionation. However, as with the Ordovician magmas before them, assimilation of Skiddaw Group material is implicated, particularly by the Sr, Pb and O isotope compositions (Thirlwall 1989).

8 The alluvial Old Red Sandstone: fluvial basins

R. D. HILLIER & B. P. J. Williams

Classically, the Old Red Sandstone (ORS) embraces the continental, predominantly siliciclastic deposits of Devonian age, being in part the terrestrial correlatives of the marine Devonian of SW England. Subsequent stratigraphical revision (e.g. House 1977) demonstrated that the base of the ORS is actually of Silurian age in many places. The ORS has long been of interest due to the presence of early vascular plants and vertebrate faunas. Studies of the ORS have spawned significant sedimentological advances, for example the now classic analysis of high-sinuosity fluvial channels (Allen 1965a, 1970). Today, the ORS is a term applied to tectono-stratigraphic units of Upper Silurian–Carboniferous age bordering the North Atlantic Ocean (Friend 1969, Friend *et al.* 2000). It has long been seen as representing the syn- to post-orogenic depositional response (molasse) to the Caledonian Orogeny, being modified by synchronous tectonism and volcanicity. The influence of Variscan tectonics on basin formation and subsequent deformation has recently being highlighted (Friend *et al.* 2000). Lithologically, it embraces a wide range of textural grades from mudrocks to conglomerates. Fluvial, lacustrine, aeolian, pedogenic and marginal marine deposits have been recognized. In England and Wales, the ORS crops out in four main areas (Fig. 8.1): the Anglo-Welsh Basin, Anglesey, Edenside (Cumbria) and North Devon.

Palaeocontinental framework

Basins containing fills of ORS sediments in the North Atlantic region developed on the continent of Laurussia (Fig. 8.2). This 'Old Red Sandstone Continent' formed by the amalgamation of Laurentia with Baltica and the microcontinent Avalonia in Silurian–Devonian times (Friend *et al.* 2000; Woodcock & Strachan 2000). The Iapetus Ocean, and its eastern arm, the Tornquist's Sea, separated the northern Laurentian continent from the southern landmass of Gondwana. Closure of the Tornquist's Sea in the late Ordovician was associated with the collision of Avalonia with Baltica. The main Caledonian event in the British Isles was the result of the oblique collision of the Eastern Avalonia microplate with Laurentia by mid-Llandoverian times (Soper *et al.* 1992*b*). The Scottish Highlands overthrust their NW foreland during this time as a consequence of the Scandian collision of Laurentia and Baltica. By late-Llandovery times major orogen-parallel, sinistral strike-slip displacements occurred along major NE–SW-trending transcurrent faults as a direct consequence of oblique collision (Dewey & Strachan 2003). By Ludlovian times, the Iapetus Ocean closed along the line of former subduction beneath the Southern Uplands. Basins on the northern edge of Eastern Avalonia deepened as a consequence of crustal flexure south of the suture line. These basins then rapidly shoaled due to tectonic uplift and oversupply of sediment from the advancing mountain front. Marine conditions were replaced by terrestrial environments in a diachronous manner across Eastern Avalonia (Allen 1985; King 1994*a*), the Welsh Basin shoaling by mid-Přídolían times (Bassett *et al.* 1992*a*).

The preserved fills of ORS basins record differing depositional architectures reflecting variations in subsidence and uplift histories, provenance and marine influence. The UK basins can be classified in a broad sense in relation to their position relative to the Caledonian Mountain belt (Woodcock & Strachan 2000). The internal or intramontane basins (e.g. the Orcadian and Midland Valley basins) are those found within the mountain belt and not having direct connectivity to the southerly sea for most of their history. Basins to the south of the mountain front, the *external* basins, had fluvial networks draining to contemporaneous coastlines to the south and east during their evolution. The most important spatially are the Anglo-Welsh Basin, and the Dingle/Munster basins in southern Ireland.

The large-scale fluvial networks that drained the region have prompted varying interpretations. Orogen-parallel drainage from the region of Scandian uplift to the NE has been proposed as the provenance for the ORS of the Midland Valley of Scotland (Bluck 2000). It is possible that this drainage system was deflected southward through the present Irish Sea region (Simon & Bluck 1982), the Anglo-Welsh Basin, terminating at the southerly coastline arguably as early as the late Ludfordian (Sherlock *et al.* 2002). Orogen-parallel drainage was disputed by Allen & Crowley (1983) who advocated that the rivers terminated internally within the Midland Valley; it is also possible they continued flowing to the SW into Northern Ireland (Friend *et al.* 2000). The Anglo-Welsh Basin may have received much of its infill from more proximal, mainly Lower Palaeozoic sources stretching from southern Ireland into southern Scotland (Allen & Crowley 1983).

Lower Palaeozoic and Lower Devonian successions across Eastern Avalonia were not significantly deformed until Emsian/Eifelian times during the Acadian Orogeny (Chapter 6; McKerrow 1988; Dewey & Strachan 2003; Soper & Woodcock 2003). This event generated SSE-directed shortening and sinistrally-transpressive cleavages that cut across associated folds in a clockwise sense. Cleavage patterns are locally moulded around pre-existing Ordovician plutons in both the Lake District and North Wales (Woodcock & Strachan 2000). Intrusion of the 'Newer Granites' occurred during the late stage of folding and cleavage development as far south as the Lake District. Isostatic uplift associated with this intrusive phase contributing to the generation of the late Early–Mid Devonian unconformity across much of England and Wales. Other causes of the Acadian event are uncertain, being possibly the 'final' docking stage of the Eastern Avalonian terrane (Friend *et al.* 2000), or the Rheic convergence of Armorica with Eastern Avalonia (Chapter 6; Woodcock & Strachan 2000; Dewey & Strachan 2003). The latter generated dextral transpression leading to the formation of extensional deformation along the southern margin of Avalonia, that is, the external ORS basins and their marine equivalents marginal to the Rheic Ocean (Chapter 10).

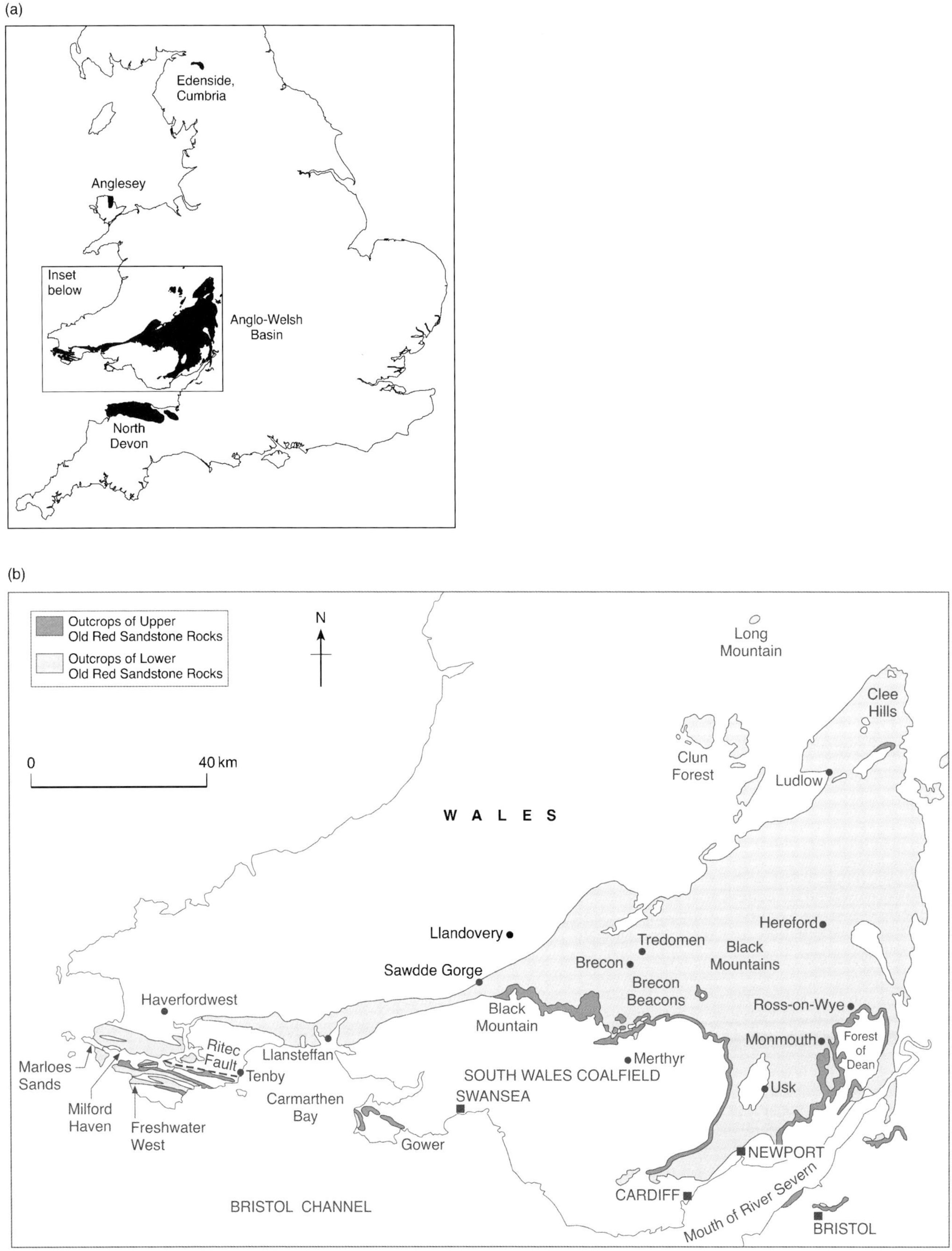

Fig. 8.1. (**a**) Location of ORS exposures within England and Wales. (**b**). Detail of the Anglo-Welsh Basin.

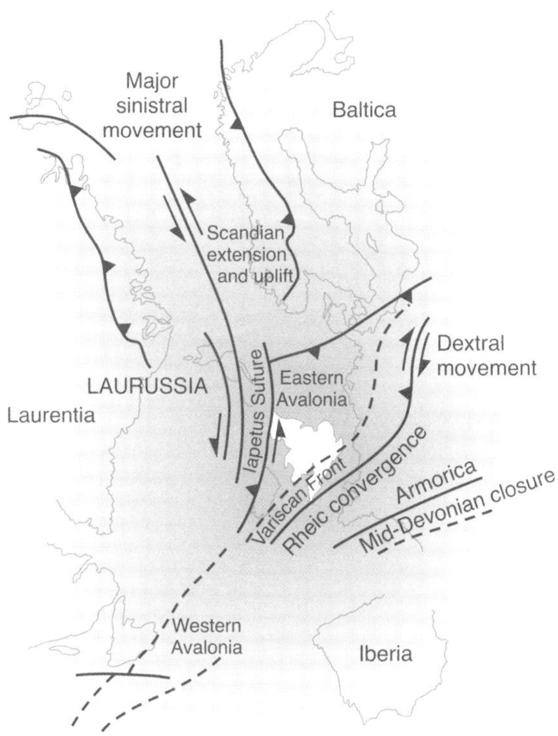

Fig. 8.2. Palaeozoic 'Old Red Sandstone' continent of Laurussia, formed by the amalgamation of various terranes during Siluro-Devonian times. After Woodcock & Strachan 2000.

Oceanic crust was generated within the Gramscatho Basin in the Early Devonian (Clark *et al.* 1998*a*), and it is possible that related extensional faulting was important in SW. Pembrokeshire possibly as early as Ludlovian times (Hillier 2000).

Geodetically, England and Wales lay in subtropical latitudes, with palaeomagnetic data from the Lower ORS of South Wales suggesting latitudes of 17±5°S (Channel *et al.* 1992). Palaeoclimate interpretations have mainly derived from studies of palaeosols preserved within the ORS. Analysis of calcic vertisols in the Anglo-Welsh Basin led Allen (1974*a*, 1985) to suggest that the Lower ORS accumulated under warm to hot, semi-arid conditions with a seasonal rainfall. The absence of a dense vegetative cover other than that close to perennial water stores would have led to extremes of eroded sediment yield following precipitation, causing widespread flash-flooding.

ORS stratigraphy

Recognition and correlation of stage boundaries within the ORS are hampered by the poor preservation of flora and fauna within the chiefly alluvial stratigraphy. The defined Global Stratigraphic Section and Points (GSSPs) for the Devonian system were established within marine, pelagic facies in Bohemia, Belgium, France, Morocco and Germany (House 2000). These were calibrated largely by ammonoid, conodont and graptolite assemblages, and, not surprisingly, correlation of these assemblages with mainly continental ORS flora/fauna remains a significant problem. Also, much of the Lower ORS is of Přídolían age, the formal stages of which have yet to be defined. As such, local stratigraphies have been erected using lithostratigraphic correlation criteria. Loosely defined local subgroups have been erected for the Anglo-Welsh Basin (Downtonian, Dittonian, Breconian and Farlovian, Fig. 8.3). The three subdivisions of the ORS, first recognized by Murchison namely Lower, Middle and Upper, are still utilized as informal lithostratigraphic terms. Fish and spores are the chief biostratigraphic tools to aid correlation, but they rarely occur together, although they may be found in close proximity. The best preservation of fish is seen in the Welsh Borderland (Dineley 1999*a*, *b*), and analysis of micro-invertebrate assemblages (e.g. acanthodian, chondrychian and thelodont scales) are contributing to the development of zonation schemes with potential for correlation with marine sequences (Turner 1973, 1984; Vergoossen 1999, 2000). Vertebrate communities and their preservation are discussed by Dineley & Loeffler (1999). In areas where fish remains are scarce, a greater reliance has been placed on spore assemblages of which there are two zonation schemes; Richardson & McGregor (1986) utilizing Anglo-Welsh Basin data, and Streel *et al.* (1987) using assemblages from the Ardenne–Rhenish regions.

During the last decade, a chronometric time scale has been established almost exclusively from the radiometric dating of air-fall tuffs that occur within ORS basin-fills, although well-exposed tuffs within the Anglo-Welsh Basin have yet to be age-dated (Figure 8.3) (Williams *et al.* 2000). Furthermore, the position of the Silurian–Devonian boundary has yet to identified with any certainty. The aerially extensive Townsend Tuff Bed in the Anglo-Welsh Basin was proposed as the Silurian–Devonian boundary marker horizon in the region (Allen & Williams 1981). Subsequent biostratigraphic analysis in the Forest of Dean places the boundary 30 m below the Bishop's Frome Limestone Member (previously termed the *Psammosteus* Limestone, and its correlatives the Chapel Point Calcrete; Richardson *et al.* 2000), a basin-wide lithostratigraphic marker approximately 100 m above the Townsend Tuff Bed that reflects a period of sustained geomorphic stability across the basin.

The Anglo-Welsh Basin

The ORS in the Anglo-Welsh Basin crops out within the Variscan Foldbelt of SW Pembrokeshire, extending eastward through the Carmarthen Fans and the Brecon Beacons, skirting the north and eastern extent of the South Wales Coalfield, and extends into the Welsh Borderland (Figure 8.1). Borehole intersections in SE England demonstrate continuity across the Midland Platform where marine influences become more prevalent. The basin preserves a relatively uniform fill of up to 2.4 km in central-eastern areas, thickening to 3.4–4.4 km within the hanging wall of the Benton Fault in SW Pembrokeshire (Williams *et al.* 1982; Friend *et al.* 2000). The present-day northern limit to the basin is delineated by important structural lineaments comprising the Benton–Llandefaelog–Welsh Borderland fault systems, which also defines the southern limit of significant Acadian deformation (Chapter 6). It is probable that the Anglo-Welsh Basin had a greater spatial extent, with Lower ORS sedimentation being significant across the former Welsh Basin, connecting with the Anglesey, Clun Forest and Long Mountain outliers. Subsequent Acadian inversion concentrated within the confines of the Welsh Basin led to the erosion of most of the ORS cover, recycling sediment into the North Devon ORS depocentre during the Middle Devonian (Tunbridge 1986).

Using backstripped subsidence analysis for South Wales, James (1987) advocated Arenig stretching, followed by thermal subsidence up to Přídolían times. Load-induced flexural subsidence within a foreland basin was inferred during the Early Devonian. King (1994*a*) proposed extensional subsidence

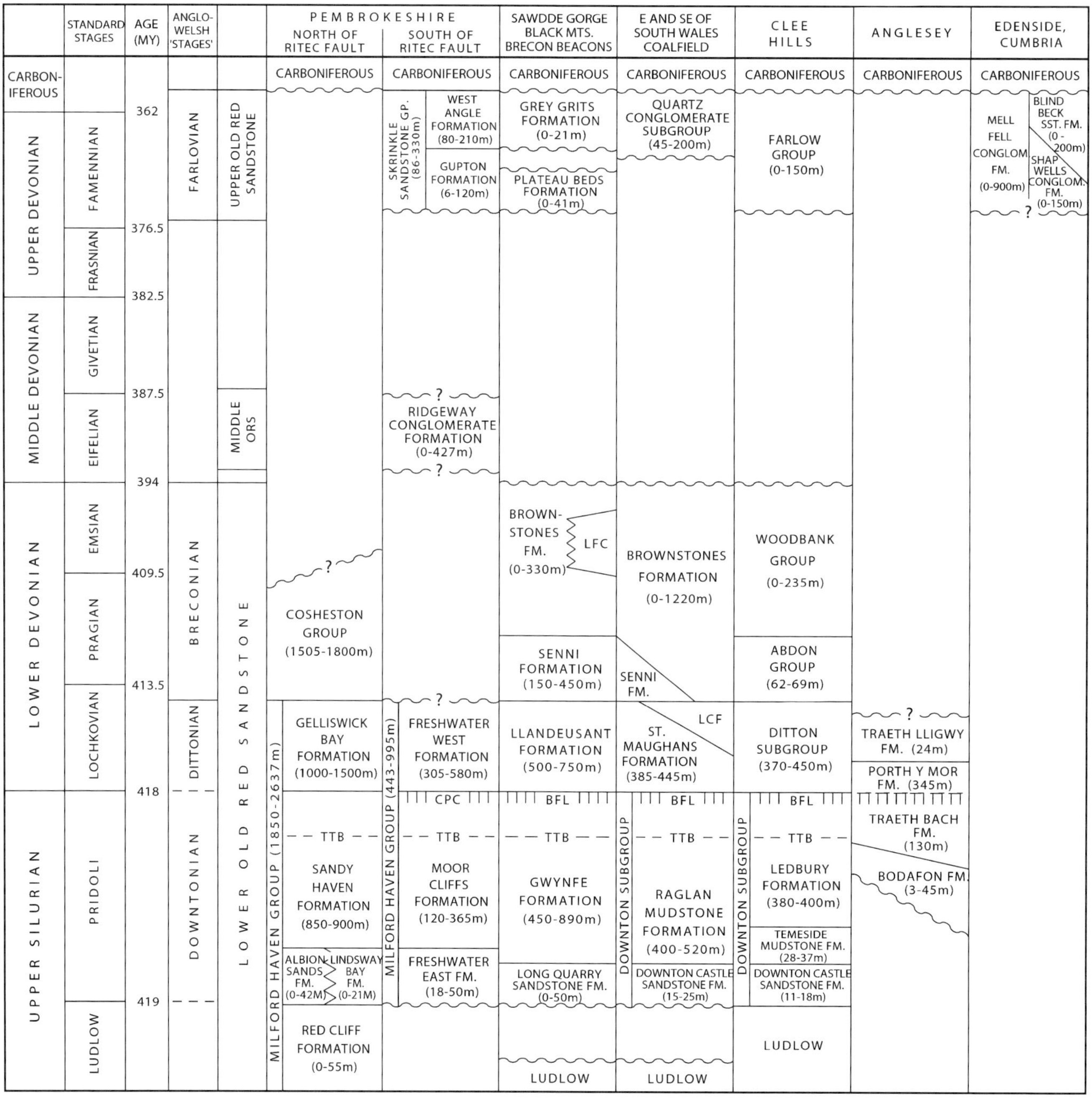

Fig. 8.3. Stratigraphic correlation of selected Anglo-Welsh Basin exposures. BFL, Bishop's Frome Limestone Member; CPC, Chapel Point Calcrete; LFC, Llyn-y-fan Conglomerate; LCF, Llanishen Conglomerate Formation; TTB, Townsend Tuff Bed. Data compiled from Allen (1965*a*, 1974*b*), House *et al.* (1977), Allen & Williams (1981), Allen *et al.* (1982), Williams *et al.* (1982, 2000), Tunbridge (1986), Almond *et al.* (1993), Wellman *et al.* (1998) and Hillier & Williams (2004).

during the Silurian (probably transtensional), followed by flexural loading as early as mid-Ludlovian times in North Wales, and by early Přídolían–Lower Devonian times in South Wales reflecting the southward propagation of a foreland basin ahead of the advancing mountain front. Problems with this interpretation arise as Acadian deformation and metamorphism appear not to have propagated in any particular direction (Chapter 6, Soper & Woodcock 2003). South of the Variscan Front, ENE- trending extension faults are anomalous to the model of sinistral transpression associated with Acadian deformation, as faults of this orientation should have a reverse throw.

In SW Pembrokeshire extensional faulting probably controlled ORS sedimentation as early as Ludlow times (Hillier 2000, Hillier & Williams 2004), being important throughout deposition of the Lower and Upper ORS (Marshall 2000*a*,*b*; Williams & Hillier 2004). Similar extension is recognised in southern Ireland associated with the development of the Mid to Late Devonian Munster Basin (Williams 2000). This extensional regime may be linked to rifting across the southern margin of Eastern Avalonia leading to the formation of oceanic crust within the Gramscatho Basin during the Early Devonian (Chapter 8). Extensional basins were possibly developed as a consequence of dextral transpression associated with the Rheic convergence of Avalonia with Armorica (Chapter 6).

The Emsian/Eifelian Acadian uplift across the basin led to unconformable contacts between the Lower and Upper ORS, with tilting producing an enhanced N–S palaeoslope. Lower

ORS sediments were eroded off the uplifted areas of the former Welsh Basin, being transported to the marine basins in North Devon and Cornwall. A local high developed in the Bristol Channel shedding sediment both northward and southward during the Middle Devonian (Tunbridge 1986).

Lower ORS Basin infill

Ludlow

ORS deposition commenced within the basin in SW Pembrokeshire with deposition of the Ludlovian Red Cliff Formation, the basal formation of the Milford Haven Group (Allen & Williams 1978; Allen *et al.* 1982). At Marloes Sands (Fig. 8.1) estuarine channels of the Wenlock Gray Sandstone Group are interdigitated with red pedified mudstones of the Red Cliff Formation strongly indicating that the transition between the marine Silurian and the overlying ORS is locally conformable, as advocated by Cantrill *et al.* (1916); Sanzen-Baker (1972); Walmsley & Bassett (1976); Hurst *et al.* (1978) and Hillier (2000). Walmsley & Bassett (1976) and Hurst *et al.* (1978) date the transition as late Wenlock–early Ludlow. This was disputed by Allen & Williams (1978) who argued on lithological and sedimentological grounds for a significant hiatus between the marine-Silurian and overlying ORS section. Subsequent recovery of palynomorphs from the basal Red Cliff Formation has confirmed a Ludfordian age (Hillier & Williams 2004), suggesting the Lower Ludlow is thin or absent. The junction between the Red Cliff Formation and the underlying Gray Sandstone Group at Red Cliff on the Marloes Peninsula was marked by a chiselled arrow-head by Cantrill *et al.* (1916), signifying what they thought was the Silurian–Devonian boundary. The Red Cliff Formation is dominated by red mudrocks with immature vertic palaeosols. It is possible that some of the mud was transported as silt-grade pedogenic aggregates a process initially proposed for the Mid Devonian Ridgeway Conglomerate (Ekes 1993) or possibly, in part, as loess. Sandstone units are sheet-like in nature, with sedimentary structures indicative of ephemeral flood events (Allen & Williams 1978; Hillier &Williams 2004). Analysis of detrital white mica from within the Red Cliff Formation indicates a Laurentian (Scottish Highlands) provenance (Sherlock *et al.* 2002). Such a scenario is at odds with local palaeocurrent evidence for the Red Cliff Formation (Fig. 8.4) which indicates derivation from the south and west (Allen & Williams 1978; Allen *et al.* 1981; Allen 1985; Hillier & Williams 2004) from a persistent land area in the Lower Palaeozoic called Pretannia (Cope & Bassett 1987). Long-distance sediment transport from the Scottish Highlands during Ludlow times would have necessitated an extremely tortuous drainage network, as a marine basin with presumed significant bathymetry existed immediately to the north (Bassett *et al.* 1992*a*).

To the east of Llandeilo, deltaic conditions were established by Late Ludlovian times, and red conglomerates and sandstones of the Trichrûg Beds presumably indicate periodic subaerial exposure along the southern shelf of the Welsh Basin (Bassett *et al.* 1992*a*).

Early Přídolí

In Pembrokeshire, at the top of the Red Cliff Formation, an abrupt change in sediment grade, provenance and facies association is observed with the deposition of the presumed Přídolían Albion Sands and, in part, coeval Lindsway Bay formations (Fig. 8.8) (Allen & Williams 1978; Hillier 2000). The Albion Sands Formation is a pebbly sandstone with exotic clasts deposited as multistorey, low-sinuosity sheets (Fig. 8.5a). The Lindsway Bay Formation comprises extraformational conglomerates with clasts of acidic extrusive and intrusive igneous rocks, vein quartz, sandstones and mudstones (Allen & Williams 1978). Deposition took place on a debris-flow-dominated alluvial fan (Hillier & Williams 2004).

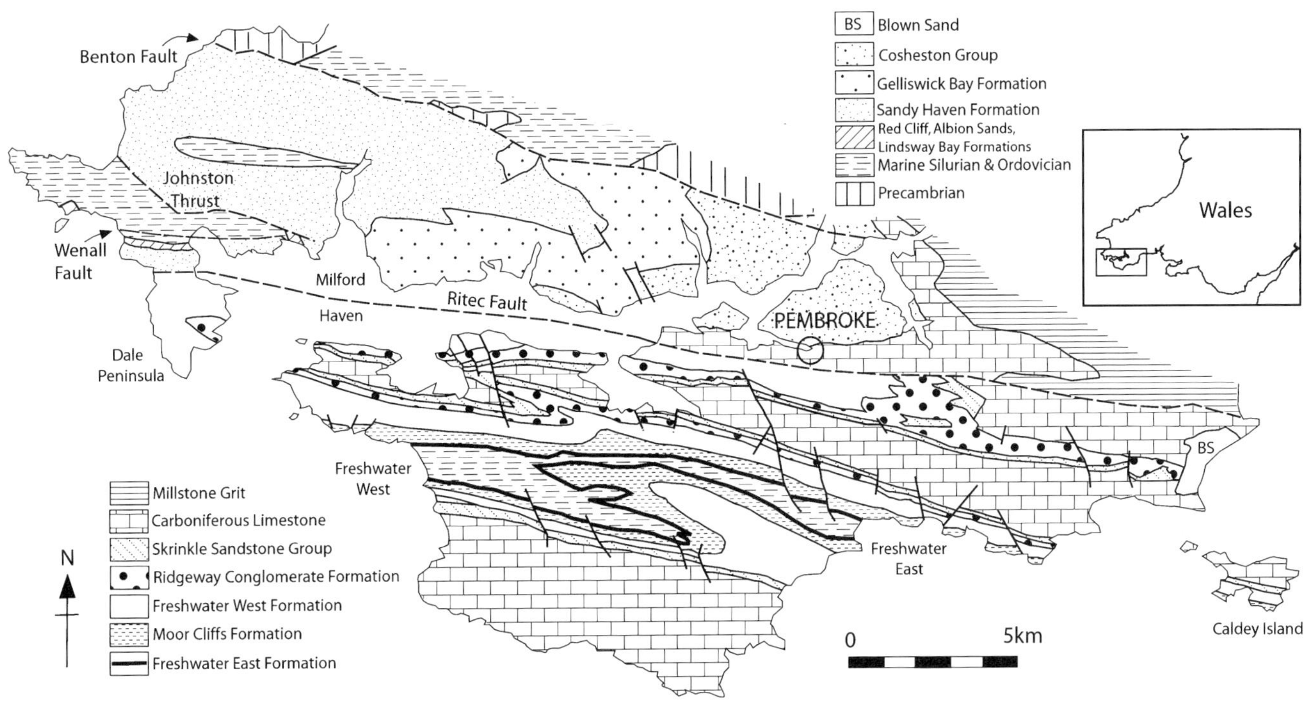

Fig. 8.4. Geological map detailing ORS formations in southern Pembrokeshire. From Williams & Hillier (2004).

Fig. 8.5. (**a**) Multistorey, pebbly sandstone low-sinuosity channel of the Albion Sands Formation, Albion Sands, Pembrokeshire (SM 772077). Scale bar 5 cm. (**b**) Basal conglomerate of the Freshwater East Formation, Freshwater West (SS 023982). The hammer is 30 cm long. (**c**) Moor Cliffs Formation mudstone with Vertisol, Priest's Nose (SS 059972). Note vertic calcrete rods and pseudoanticlinal slip planes. Scale bar 5 cm. (**d**) Wave-rippled Townsend Tuff Bed, Old Castle Head (SS 077967). (**e**) Platform comprising Vertisol calcrete profiles of the Chapel Point Calcrete, Chapel Point, Caldey Island (SS 142958) Way up to right, thickness of calcrete approximately 10 m. (**f**) Aerial view of Priest's Nose, Manorbier Bay looking east (SS 0597). Identified within the Moor Cliffs Formation are: RCT, Rooks Cave Tuff; TT, Townsend Tuff; PBT, Pickard Bay Tuff; CPC, Chapel Point Calcrete; c, prominent calcrete layer. (**g**) Aerial view looking ENE of Moor Cliffs Formation illustrating layer-cake stratigraphy; Black Cave, Pembroke Peninsula (SM 8700). (Photographs f. and g courtesy of Sid Howells). (**h**). Inclined heterolithic stratification within the Conigar Pit Sandstone Member, Swanlake Bay (SS 044980). Note: A, channel base; B, inclined heterolithic units; C- calcretized mudrock. Scale bar 1 m.

Although lowstands of sea level have been identified during the Late Silurian (Johnson *et al.* 1998) it is probable that the transition from the Red Cliff to Albion Sands Formation was tectonically enhanced, as synsedimentary extension faults point to extensional strain within the cover sequence during deposition of the Albion Sands Formation. Semi-regionally, isopach and facies-assemblage variations indicate that the Wenall Fault was a controlling structure on sedimentation certainly post-Red Cliff Formation deposition. Along a traverse W–E parallel to the fault, the Lower ORS formations thin dramatically (Fig. 8.8) (Hillier & Williams 2004) due to decreasing accommodation space generated within the hanging wall of the fault. Palaeocurrents within the Albion Sands Formation show sediment transport predominantly from the west, that is, parallel to the fault trace. As such, the Albion Sands Formation represents fluvial deposition within an axial valley with the greatest depositional thickness being preserved at the position of fault-generated maximum subsidence adjacent to the fault. The abrupt change in sediment transport, grade and facies association presumably reflects movement along the Benton, Wenall and Ritec Faults (Fig. 8.8). The cause of this activity may be transtension associated with the Avalonia–Laurentian collision, or extension related to rifting associated with opening of the Rheic Ocean to the south. The presence of numerous air-fall tuffs within the Albion Sands Formation probably reflects this tectonic activity. The ultimate depocentre for these sediments is a moot point, but the remnants of the rapidly infilling Welsh Basin to the north and east are a likely sink (Allen 1985; Bassett *et al.* 1992*a*).

The Lindsway Bay Formation was derived from the south and west. It is possible that the fan was developed at its maximum extent along the relay ramp adjacent to the tip point of the Wenall Fault. The interdigitation of Lindsway Bay Formation with Albion Sands Formation suggests that the fan advanced into the valley axis, and as such was reworked by fluvial processes within the valley proper (Fig. 8.8). Smaller-scale fans are envisaged to have been generated from the footwall of the Wenall Fault, supplying sediment southward and to the NE where a thin succession is preserved (Allen & Williams 1978). The exact provenance of this unit is unknown. Lavas and pyroclastics are similar to those seen within the Llandoverian Skomer Volcanic Group, and it is possible that the margins of the Skomer Rift Basin were uplifted and exposed during this phase of fault movement. It is also possible that the unit is derived from reworked Neoproterozoic volcanics that are believed to underlie much of SW Wales (Bevins *et al.* 1995*b*). Soil profiles within this Lower ORS fill are thin and immature, indicating that sediments were in the zone of active pedogenesis for brief to moderate residence-time periods reflecting rapid subsidence within the hanging wall of the Wenall Fault.

The remainder of the early Přídolí in Pembrokeshire is represented by a mudrock-dominated interval with locally important air-fall tuffs encompassing the lower–mid Milford Haven Group. These mudrocks were deposited on a low-relief alluvial plain bordering the sea to the south. Marine influence is restricted to the lower Sandy Haven Formation to the north of Milford Haven (Allen & Williams 1978; Allen *et al.* 1982) and the Freshwater East Formation to the south (Fig. 8.5b) (Williams *et al.* 1982). In both areas, thin sheet-like sandstones and conglomerates contain onchids, *Pacytheca* and lingulids within presumed tidally-influenced channels that dissected the coastal plain. The Freshwater East Formation has yielded a rich flora comprising *Cooksonia*, *Steganotheca*, *Tortilicaulis* and *Psilophytites* (Edwards 1979).

Further east, in the Sawdde Gorge (Fig. 8.1), marine shelf conditions of the Ludlovian were replaced by marginal marine conditions during the early Přídolí (Almond *et al.* 1993). Basal units of the Gwynfe Formation comprise a heterolithic association of quartzitic sandstones, silt and mudstones interpreted as a lagoonal sequence with washover sandstone sheets. This passes upward into upward-fining heterolithic tidal channel fills with a nearshore marine fauna (Almond *et al.* 1993). A gradual transition into red mudstones with incipient calcrete nodules indicates prolonged subaerial exposure heralding onset of ORS deposition. The presence of modiolopsids and lingulids in life position associated with flaser and lenticular-bedded sandstones indicates a marine influence.

Across the Welsh Borderland–Midland Platform area, marine, wave-dominated conditions persisted within the Downton Subgroup (the Downton Castle Sandstone Formation and its correlatives, Allen 1985); the remnant Welsh Basin was now rapidly shoaling (Fig. 8.8). Marine connections probably existed northward into the Lake District and to the SE of England where borehole intersections indicate shallow-marine sand shoals (Bassett *et al.* 1992*a*). The presence of arthropod cuticles including at least two centipedes, a trigonatarbid arachnid, scorpions, a possible spider and arthropleurids from a horizon less than 1 m above the basal Přídolían Ludlow Bone Bed provides evidence that predatory faunas and detritus feeders had colonized the terrestrial ecosystems to the north and west (Jeram *et al.* 1990; Shear & Selden 2001), possibly within a rhyniophytoid salt-marsh environment.

Mid–late Přídolí

In SW Pembrokeshire, the mid–late Přídolían Milford Haven Group comprises mainly siltstones/mudstones displaying varying stages of pedogenesis (Fig. 8.7) (Allen 1974*a*; Love &Williams 2000; Marriott & Wright 1993, 1996, 2004; Wright & Marriott 1996*a*). The mudrocks have traditionally been regarded as overbank deposits (Allen & Williams 1979), but a diverse range of mechanisms have recently been proposed for mud emplacement. Commonly observed within the Milford Haven Group are low-angle accretion surfaces several metres thick composed of interbedded ripple-drift laminated fine-grade sandstone and muddy siltstones overlying intraformational calcrete and mud-chip conglomerates. These have been interpreted as ephemeral, sinuous channels that reworked the floodplain during seasonal flooding. Modern analogies for mudstone deposition within the ORS exist within the Channel Country of central Australia, in which reworking of the low-gradient peneplain takes place following flooding by a system of anas tomosing channels, with the fine-grained sediment formed by pedogenic processes being transported as cohesive, silt and sand-grade aggregates (Ekes 1993; Marriott & Wright 2004; Marriott *et al.* 2005; Williams & Hillier 2004). Sinuous trunk streams were envisaged to concentrate low-stage ephemeral flow, with increasing discharge activating a peripheral zone of anastomosing mud-aggregate channels. Only during infrequent, high-discharge, events would flooding reach and modify distal parts of the floodplain (Marriott *et al.* 2005). The rare occurrence of ephemeral lakes is indicated by thin bedsets of laminated, fine-gained sandstone and mudstone couplets.

Soil profiles are typically between 1 and 10 m thick, being characterized by desiccation cracks, carbonate nodules with various morphologies and pseudoanticlinal fracture surfaces with common drab haloes (Figs 8.5c, 8.7). The profiles are interpreted as vertisols formed in a seasonal, subhumid climate (Love &Williams 2000). Differing patterns of soil profiles reflect varying stages of soil maturity, a reflection of soil residence time influenced by sedimentation rate (Marriott & Wright 1993, 1996), and variations in topographic relief and substrate (Love & Williams 2000). Maturity is seen to increase above locally important volcanic ash horizons that periodically blanketed the topography, maturity being enhanced by the rapid weathering of ash that would release compounds from which soils develop rapidly (Love & Williams 2000). Soil profiles are commonly terminated by erosion surfaces overlain with intraformational conglomerates of reworked calcrete nodules

Fig. 8.6. (**a**) Desiccated mudrocks and thin sheet sandstones of the Conigar Pit sandstone Member, Freshwater West (SR 885993). Pencil 14 cm long. (**b**) Bedding-plane view of *Beaconites baretti* burrows, Red Marl Group (Conigar Pit Sandstone Member equivalent), Llansteffan (SN 345098). (**c**) Multistorey sandstone channel, younging to right. Conigar Pit Sandstone Member, Freshwater West (SR 886990). (**d**) Bedding-plane view of siltstone with *Diplichnites gouldi* trackways, St Maughans Formation, Tredomen Quarry (SO 116304). Scale bar 3 cm; photograph courtesy of L.B. Morrissey. (**e**) Channelized Senni Formation sandbodies and associated heterolithics, Craig Ddu (SN 321100). (**f**) Dune forms within Ridgeway Conglomerate Formation younging to upper-left corner. Freshwater East (SR 887987). Scale bar 5 cm. (**g**) *Retusotriletis triangulates*, Lochkovian, St Maughans Formation, Gardner's Bank. Spore Width 59 μm. (**h**) *Cymbohilates* sp. A; spore width 44 μm, St Maughans Formation, Bryn Glas tunnels borehole (see Wellman *et al.* 2000). Photographs g and h courtesy of C. H. Wellman. (**i**) Lacustrine Heterolithic Association of the Gupton Formation, Freshwater West (SR 887987). See Marshall (2000*b*). Scale bar 2 cm.

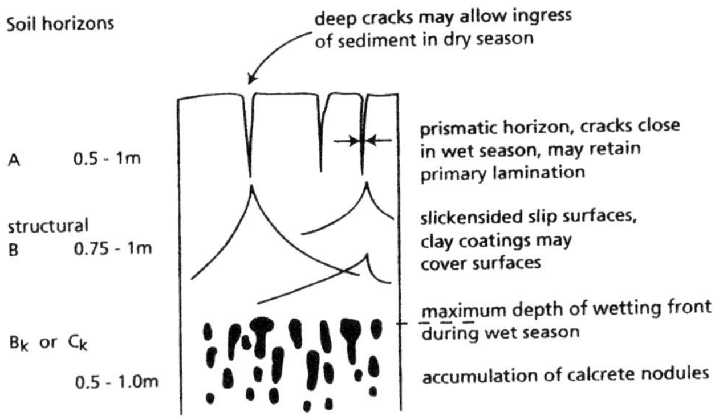

Fig. 8.7. Main features and horizons from palaeosols from the Lower ORS (from Marriott & Wright 2004). Copyright John Wiley & Sons. Reproduced with permission.

and mud-chips. Reworking was common within interfluvial areas that drained into the trunk streams (Allen & Williams 1979). Seasonal drying of the plain would have promoted mud-grade material being reworked by aeolian processes (Marriott & Wright 2004).

Sandbodies within the Přídolían Milford Haven Group can be traced along strike for almost 2 km (Fig. 8.5g), their bases being mildly erosive into underlying mudrocks. Texturally, the sandstones are fine to coarse grained, with sporadic development of intraformational mudstone rip-up clasts and extraformational vein quartz pebbles/granules and exotic clasts. The sandbodies comprise dm-thick bedsets of trough cross-stratification and planar lamination. Decimetre-scale convolute lamination is also present, being terminated at well-defined erosion surfaces. Locally, bedsets are separated by thin mudrocks. Although the sandbodies are sheet-like in external geometry, having width to depth ratios in excess of 500:1; internally they represent the amalgamation of bedsets having both planar and lenticular geometries, the latter infilling broad (30–50 m), shallow (dm) scours. These fluvial sandbodies clearly represent significant influxes of coarse-grained, exotic material into the basin, presumably by composite flow events within relatively unconfined channels. The absence of significant vegetation would have impeded the development of levées, thus promoting broad, shallow channels across the low-relief plain. Palaeocurrents indicate a northerly provenance with a metamorphic-dominated source (Allen 1985). Incised sandbodies up to 4 m thick, infilling concave-up scours with up to 10 m of relief also exist (Allen & Williams 1982). The bedsets thin and fine-upward. Channel bases are often strewn with intraformational mudstone chips, and reworked calcrete nodules. Overlying sandstones are fine to medium grained, displaying trough and planar lamination, with thinner bedsets predominantly current-rippled. Palaeocurrents within these channels display bimodality, commonly with a north and south component (Allen & Williams 1982). This possibly indicates a tidal influence, although palaeontological and sedimentary structures diagnostic of marine influence are lacking.

Strong vertical cyclicity of soil maturities associated with the proximity to sandbodies has been observed within the Moor Cliffs Formation (Love & Williams 2000). The most mature palaeosols consistently occur in positions furthest from the sandstones in both vertical and lateral directions. This is consistent with the 'pedofacies relationship' (Bown & Kraus 1987) where soil maturity decreases with proximity to active fluvial channels. Lower ORS sections on the Pembroke Peninsula are seen to thicken significantly northwards across Milford Haven (Williams & Hillier 2004). It seems likely that the basin-bounding structure was the Benton Fault and its footwall, the Benton Horst. Clearly, significant accommodation space was developed within the hanging wall immediate to the Benton Fault to generate the isopach variations. The section is also seen to thin eastwards, indicating a decrease of accommodation space associated with variable throw along the Benton Fault. The exact role of the Wenall and Ritec faults during this time interval is unclear.

Important air-fall tuffs (e.g. the Townsend, Pickard Bay and Rook's Cave Tuffs, Fig. 8.5f) occur within the Moor Cliffs Formation, these aiding correlation both locally and regionally (Allen & Williams 1981, 1982; Marriott & Wright 1993; Love & Williams 2000). The Townsend Tuff Bed comprises three distinctive, normally-graded air-fall tuffs typically 2–4 m thick, that blanketed the low-relief plain. Internal bedding surfaces preserve wave ripples, (Fig. 8.5d), delicate trackways, faecal pellets and burrow entrances. The tuff has been identified across the Anglo-Welsh Basin and its position close to the top of the Přídolí led Allen & Williams (1981) to propose it as the regional Silurian–Devonian boundary. Clearly the need to age-date the Townsend Tuff is paramount. Also of stratigraphic use is the development of a thick sequence of vertisols approximately 100 m above the Townsend Tuff Bed, the Chapel Point Calcrete (Fig. 8.5e, and its correlative the Bishop's Frome Limestone Member; Allen 1974*a*; Brandon 1989; Barclay & Smith 2002; Hawley & Owen 2006). This interval crops out over much of the Anglo-Welsh Basin, forming a prominent escarpment in the Welsh Borderland. The calcrete reflects a period of significant geomorphic stability and tectonic quiescence across the region that coincides with a major change in fauna, depositional environment and sediment provenance (Allen 1985). The thick development of calcrete possibly indicates a period of extended aridity.

In the Sawdde Gorge, the Townsend and Pickard Bay Tuffs are exposed within the Middle–Upper Divisions of the Gwynfe Formation (Almond *et al.* 1993). Here, the red mudrock-dominated succession is up to 846 m thick, the thickness probably reflecting accommodation space developed along the Carreg Cennen Disturbance. The Upper Division coarsens-upward, a reflection of increasing sandbody content. The sandbodies comprise upward-fining bedsets of medium- to fine-grained, micaceous litharenites. Planar and trough-bedded sets thin upward into cross-laminated sandstones and muds. The units are interpreted as low to moderately sinuous channels, possibly exhibiting tidal influences (Almond *et al.* 1993). The equivalent of the Chapel Point Calcrete is not developed in this area, although well-developed calcretes are observed within the lower and upper parts of the Upper Division. The apparent absence of significant palaeosols probably reflects low soil residence times dictated by high sedimentation rates associated with movement along the Carreg Cennen Disturbance.

Further east, Jenkins (1998) identified a brackish-marine incursion within the Upper Downtonian Subgroup at Lydney using a combination of sedimentological (bipolar palaeocurrents and wave-influenced siltstones), geochemical (marine hydromorphism of palaeosols) and palaeontological evidence (modiolopsid bivalves). The marine incursion must have been restricted, possibly a seasonal wetland of low gradient being influenced by marine storm surges, processes advocated by Wright & Marriott (1996*b*) for the Lochkovian St Maughans Formation of the Welsh Borderland. In the Welsh Borderland, the mudstone-dominated Temeside Mudstone Formation and its correlatives are intertidal in origin, yielding vascular plants and a restricted marine invertebrate fauna (Allen 1985). The later Downton Subgroup red mudrocks in the area appear to have been deposited on extensive mudflats with fluvial input from the north and west (Allen 1985). Bone-beds are absent in the Raglan Mudstone Formation, and drifted plant material and articulated vertebrates become increasingly common upwards through the section. The latter are dominated by disarticulated headshields and body scales of *Traquairaspis*

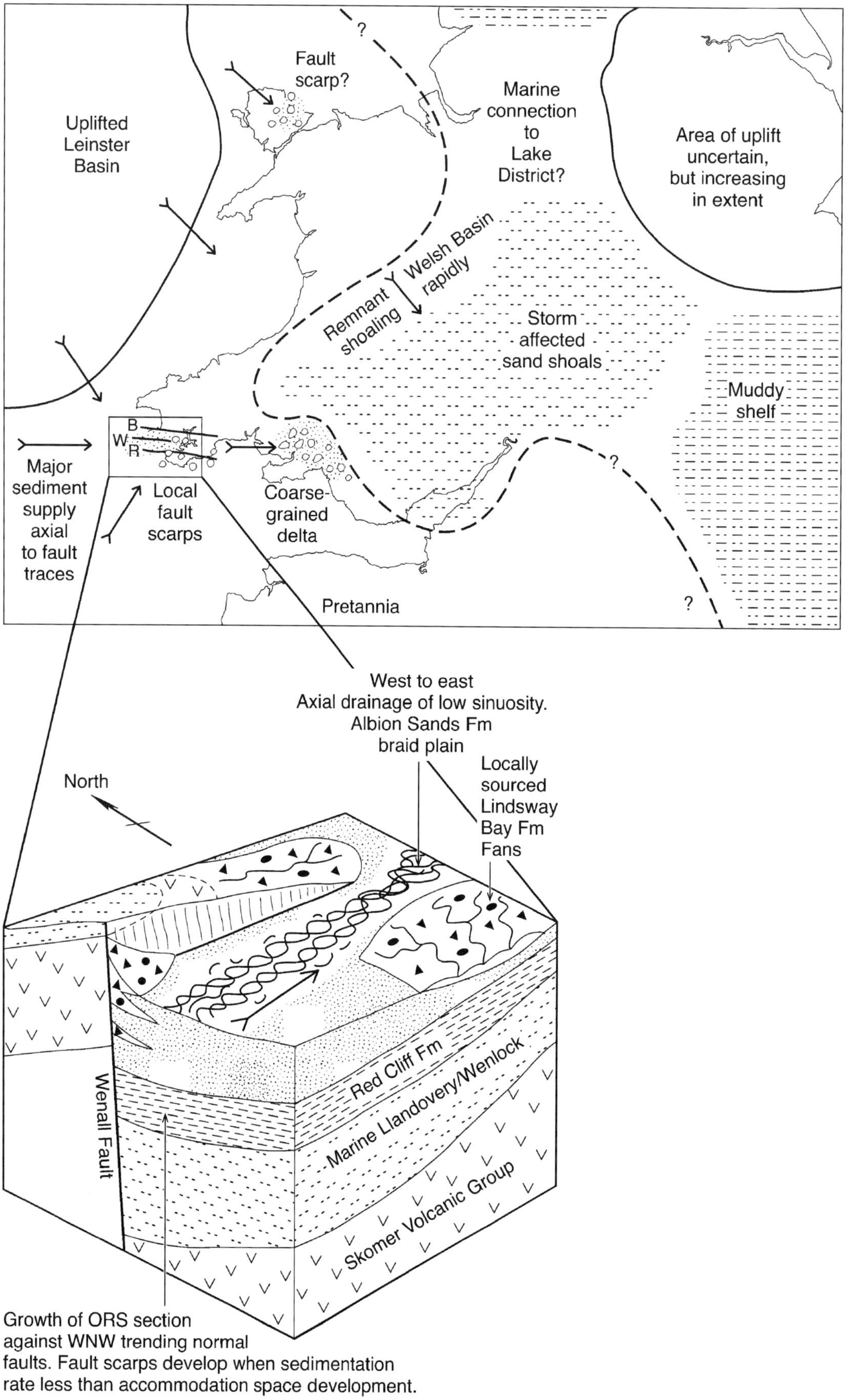

Fig. 8.8. Early Přídolían depositional environments, with inset block diagram illustrating tectonic controls on sedimentation in SW Pembrokeshire. Modified from Bassett *et al.* (1992*a*) and Hillier & Williams (2004). W, Wenall Fault; B, Benton Fault; R, Ritec Fault.

(Philiaspis) symondsi, *Tesseraspis tesselata*, osteostracans, cyathaspids and acanthodians (Dineley 1999*a*), although few Přídolían fish species extend above the overlying Bishop's Frome Limestone Member possibly reflecting extreme aridity associated with its formation. It is still uncertain whether the Přídolían fish faunas were solely fresh-water dwellers, or occupied brackish water associated with tidal-influenced rivers (see discussion in Dineley 1999*a*). Clearly salt-water intrusion at the base of water columns within low-gradient rivers could extend several tens of kilometres inland. Alternatively, the fish may have migrated seasonally from marine into fresh-water ecosystems to spawn.

Early Devonian

Lochkovian–Emsian

Following the geomorphic stability demonstrated by the development of the Chapel Point Calcrete/Bishop's Frome Limestone across the Anglo-Welsh Basin, changes in depositional style, textural grade and provenance, mark the onset of Devonian sedimentation. A decline in the importance of garnet, mica and metamorphic grains and a change to Lower Palaeozoic-sourced sedimentary and igneous grains with an increase in zircon, tourmaline and rutile infers either progressive unroofing of a new provenance, or significant modification of drainage basins (Allen & Crowley 1983; Allen 1985). It seems possible that parts of the former Lower Palaeozoic Welsh Basin and Irish Sea areas were being uplifted at this time, with palaeocurrents from fluvial channels flowing consistently from the north and northwest (Fig. 8.9) (Allen 1985). Recovery of mid-Lochkovian cryptospores and trilete spores constrains the depositional age of this influx of exotic detritus (Higgs 2004).

In Pembrokeshire, the Conigar Pit Sandstone Member of the Freshwater West Formation (Williams *et al.* 1982) comprises four distinctive fluvial architectural styles (Williams & Hillier 2004). The basal association that sits erosively on the Chapel Point Calcrete comprises an upward-fining and -thinning, coarse-grained, pebbly sandstone up to 5 m thick, worked into dm-thick sets of trough cross-stratification. These thin upwards into finer-grained sandstones and siltstones that are predominantly current-rippled. Thin interbedded mudstones exhibit well-formed desiccation cracks. Bedsets above the basal conglomerate are predominantly low angle in nature, indicating deposition on laterally-accreted bars. This basal association is well exposed along a 50 km traverse through southern Pembrokeshire and Carmarthenshire. However, at Llansteffan, the conglomerates comprise bedsets of reworked intraformational calcrete nodules and mud-flakes, deposited in a region clearly devoid of the extrabasinal detritus common in southern Pembrokeshire. Important volumetrically within the lower Conigar Pit Sandstone Member are fine-grained heterolithic associations of laterally-accreted mudrocks and fine-grained sandstones similar to those described by Marriott & Wright (2004) and Marriott *et al.* (2005) as ephemeral, sinuous channels (Fig. 8.5h). The Conigar Pit Sandstone Member coarsens upwards, as channel and sheet-like sandstones become more important (Figs 8.6a, c). The former comprise impressive amalgamations of grey/green, fine- to medium-grained sandstone up to 10 m thick. Although sheet-like in geometry, it is likely that these sandbodies were constrained within low amplitude

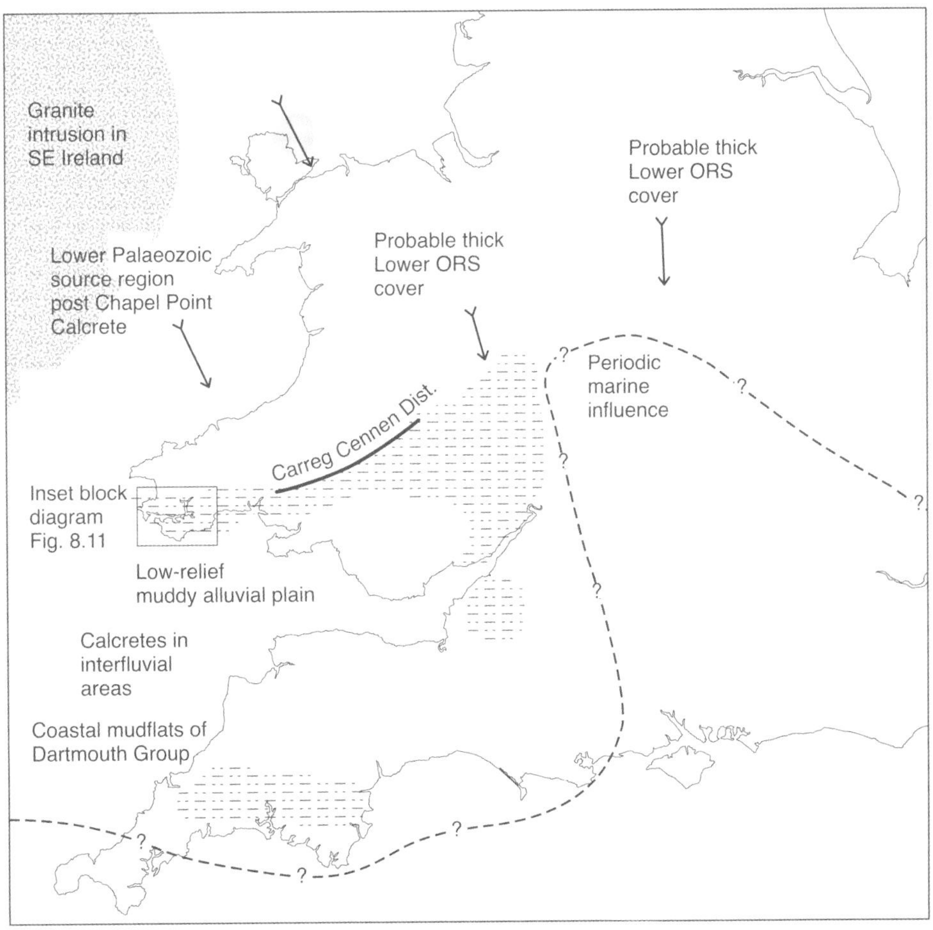

Fig. 8.9. Schematic palaeogeography during Early Devonian times (modified after Bluck *et al.* 1992).

depressions defining primary channel axes, with sandbodies thinning into dm-thick sheets on interfluves recording superflood deposition. The multistorey units clearly represent significant influxes of arenaceous material into the basin, recording periods of pulsed discharge. Variations in sandbody architecture within the Conigar Pit Sandstone Member are difficult to explain. Owen & Hawley (2000) have advocated tectonic uplift and a change in depositional slope to explain changes from mudstone-prone channel belts to low-sinuosity multistorey sandbodies within the Senni Formation of central South Wales. Tectonic pulses would generate the exposure of a sand-prone provenance either within the inverted Welsh Basin or from local fault blocks. The textural maturity of the multistorey sandbodies implies significant progressive sorting and therefore probable lengthy drainage networks. The low-sinuosity channels would have aggraded within areas of increasing accommodation space, possibly explaining the multistorey nature of many observed sandbodies within the hanging wall of the Benton Fault (Williams & Hillier 2004). Although fine-grained fluvial systems appear to develop in low-gradient drainage basins (Orton & Reading 1993), experimental studies by Ouchi (1985) indicate that high-sinuosity channels increase their sinuosity after a gradient increase, and it is possible that the mudstone-prone laterally-accreted bedsets indicate such tectonic activity.

The Conigar Pit Sandstone Member is seen to thicken northward significantly into the correlative Gelliswick Bay Formation, demonstrating growth along the Benton Fault (Fig. 8.11). The absence of the Chapel Point Calcrete north of Milford Haven may be explained by subsequent erosion during the onset of deposition of the Conigar Pit Sandstone Member/Gelliswick Bay Formation. Alternatively, its absence from within the basin proper may be explained by its elevated position on the hanging wall, with increased residence times promoted by its position far removed from the maximum development of structural accommodation space (Williams & Hillier 2004). South of Milford Haven, the Conigar Pit Sandstone Member is succeeded conformably by the Rat Island Mudstone Member (Williams *et al.* 1982). This is a mudrock-dominated succession comprising upward-fining and -thinning units interpreted as sinuous ephemeral channels (Marriott *et al.* 2005), and thin sheet sandstones. The Member also thickens northward into the Ritec Fault, but is absent north of Milford Haven. This possibly indicates the onset of Ritec Fault movement, a lineament that became dominant in the area during the Middle and Late Devonian. The Freshwater West Formation and its correlatives contain a rich, diverse ichnofauna dominated by *Beaconites barretti* (Morrissey & Braddy 2004) (Fig. 8.6b). Trackways indicate a gregarious arthropod epifauna of arachnids (*Paleohelcura*), and myriapods (*Diplichnites* and *Diplopodichnus*, Fig. 8.6d). Additional ichnotaxa comprise arthropod foraging and resting traces, worm burrows and faecal pellets characteristic of the *Scoyenia* ichnofacies (Morrissey & Braddy 2004). The occurrence of coprolites elsewhere in the Anglo-Welsh Basin, each comprising an amalgamated mass of spores, provides evidence of detritivores, probably myriapods (millipedes and enarthropleurids; Edwards *et al.* 1995).

In the Sawdde Gorge, marine influence was lost by mid Lochkovian times with an extensive alluvial plain developing through to Emsian times (Almond *et al.* 1993). An upward-coarsening sequence up to 1700 m thick is developed, from the Llandesuant Formation, through the Senni Formation and into the Brownstones Formation (see below). This upward-coarsening profile is interpreted as an episodic, progradational response to source area uplift. The lower, mudstone-prone Llandeusant Formation (and its correlatives the Raglan Mudstone Formation and Ledbury Formation) demonstrate deposition similar to that in the Late Přídolí, that is a low-gradient alluvial plain traversed by low-sinuosity rivers and unconfined sheetfloods (Barclay 1989). To the east, brackish-water faunas have been identified within the Lochkovian St Maughans Formation in the Welsh Borders at Ammons Hill by Barclay *et al.* 1994, and at Gardners Bank by Jenkins (1998, figure 8.9). Morrisey *et al.* (2004) document the oldest known trails of swimming fish within the St Maughans Formation (*Undichnia unisulca*, *U.* cf *simplicitas* and *U. trisulcata* isp. nov., Fig. 8.10), as well as body fossils of heterostracans and osteostracans, their inferred producers. The exceptional preservation of the Trigonatarbid arachnid, *Arianrhoda bennetti* attests to a well-developed predatory insect fauna (Dunlop & Selden 2004) (Plate 14) The discovery of marine-shark micro-remains (*Polymelepsis whitei*) 132 m above the Bishop's Frome Limestone Member in the Welsh Borderland clearly alludes to periodic marine incursions (Vergoossen 1999). The exact nature of marine influence, however, is difficult to ascertain, probably reflecting tidal influence or storm surges across the low relief plain (Wright & Marriott 1996*b*). The St Maughans Formation contains an abundance of fish-bearing intraformational conglomerates, with common cephalaspids and pteraspids (Barclay 1989).

Floras recovered from the St Maughans Formation provide evidence of differing assemblages between upland and lowland areas (Figs 8.6g, h). Lowland floodplains with seasonal, possibly permanently damp environments reflecting high water tables, are believed to have been dominated by 'Lilliputian forests' of small plants, whereas larger and longer-lived plants of the 'zosterophyll communities' are envisaged to have colonized more stable, but dryer upland areas (Wellman *et al.* 2000). It is possible that interfluves provided refugia/habitats for cryptospore-producing floras, being able to colonize areas with variable, intermittent water supply, reproducing during periods of high humidity with rapid completion of life cycles followed by dormacy as spores (Edwards & Richardson 2004). The presence of coalified plant mesofossils strongly indicates the presence of wildfires within such ecosystems.

In the Bristol Channel area, a presumed structural high shed quartzite and phyllite clasts northward into the Cardiff area during Pragian times, being deposited as the laterally restricted Llanishen Conglomerate Formation (Fig. 8.13) (Allen 1975; Waters & Lawrence 1987). This structural high became important as a source of sediment during the Middle Devonian (Tunbridge 1986), possibly reflecting the periodic reactivation of the Pretannian block, important during Lower Palaeozoic sedimentation (Williams 1971; Cope & Bassett 1987).

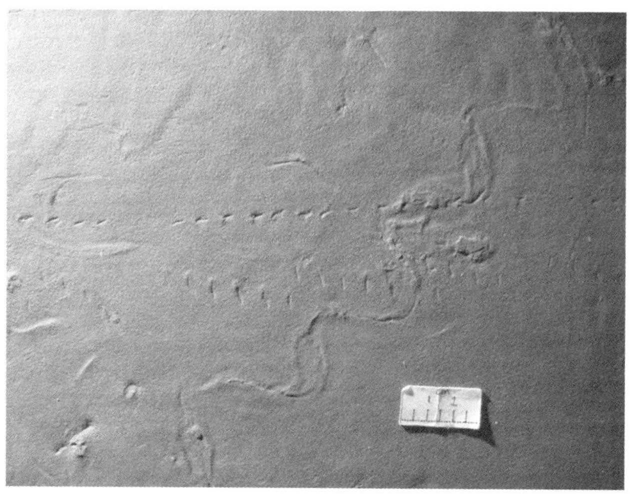

Fig. 8.10. Sinusoidal fish-fin trail *Undichna*, the oldest recorded trails of swimming fish (from Morrissey *et al.* 2004). Copyright John Wiley & Sons. Reproduced with permission). *Undichna* is cut and overprinted by *Diplichnites* trackway. St Maughans Formation, Tredomen Quarry (SO 116304). Scale bar 3 cm.

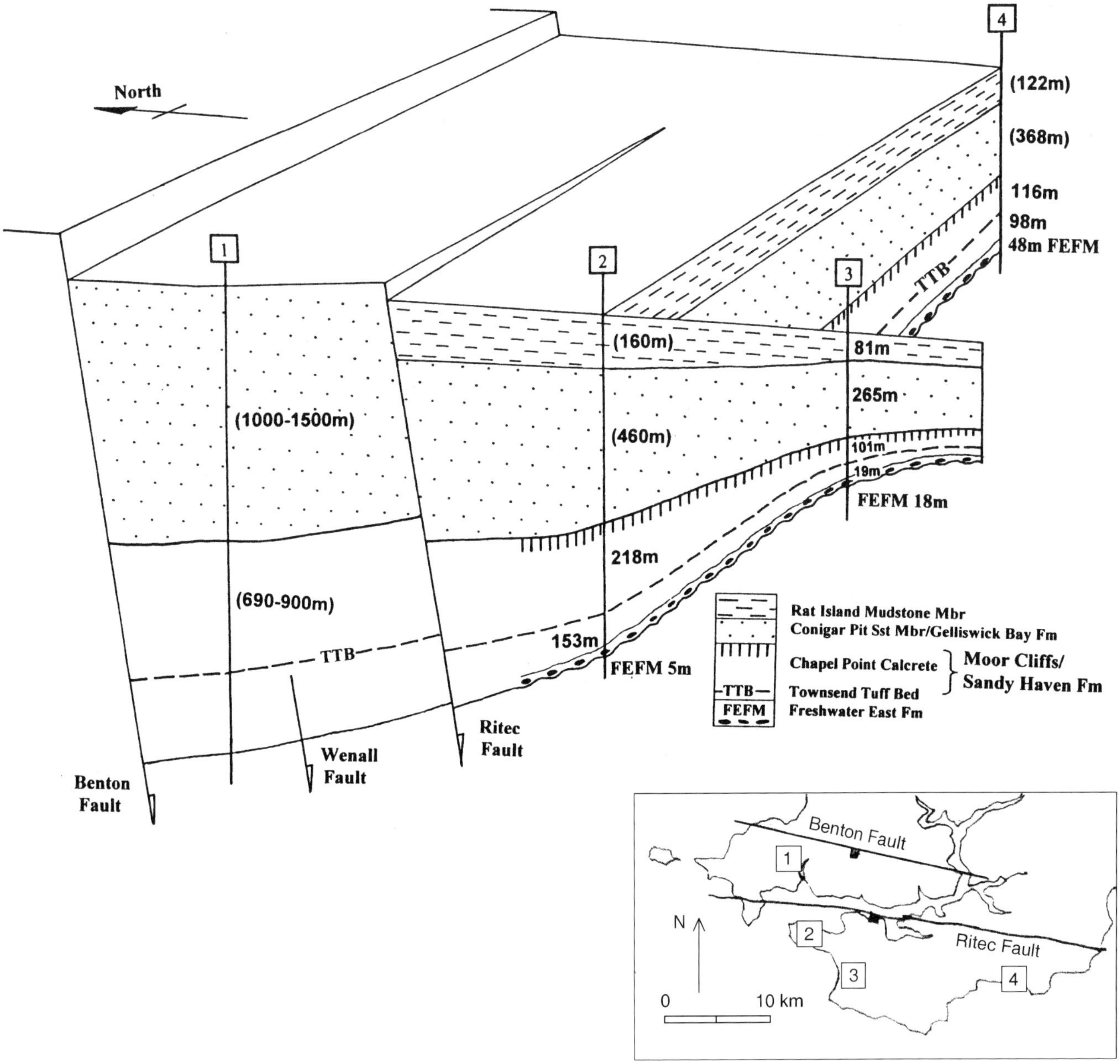

Fig. 8.11. Fault control on Lower ORS basin infill, SW Pembrokeshire. Note thickening of Lower ORS interval within hangingwall of Benton Fault, and presumed topographic control on Chapel Point Calcrete development. Thicknesses in brackets are estimated values. Modified from Williams & Hillier (2004). Copyright John Wiley & Sons. Reproduced with permission.

To the north of Milford Haven, the Gelliswick Bay Formation is overlain by the Lochkovian–Emsian Cosheston Group (Thomas 1978; Allen *et al.* 1982; Wellman *et al.* 1998). The latter has exotic clasts that appear in significant amounts toward the top of the Lower ORS across the Welsh Basin (Allen & Crowley 1983). Conglomerates within the Cosheston Group contain extraformational clasts of vein-quartz, rhyolitic and felsic lavas and tuffs, and a range of lithic sandstones. Also identified are minor components of various metamorphic rocks, cherts and jasper (Allen *et al.* 1982). The conglomerates are interpreted as braided stream deposits, but palaeocurrents and relationships to tectonism are unclear. Also present are heterolithic, upward-fining units interpreted as moderate sinuosity channels with associated calcretes. Sandstones commonly exhibit large-scale soft-sediment deformation, and contain heavy mineral assemblages dominated by garnet (Allen *et al.* 1982). The diverse exotic clast assemblage possibly heralds the onset of Mid Devonian Acadian uplift across the Anglo-Welsh Basin (see below). Plant assemblages within the Cosheston Group are less diverse than the age-equivalent Senni Formation of central South Wales, but comprise *Deheubarthia splendens*, *Gosslingia breconensis*, *Zosterophyllum* and *Sennicaulis* (Wellman *et al.* 1998). A prolific spore assemblage has also been described (Thomas 1978).

The Senni Formation is dominated by grey-green sandstones with subordinate siltstones with immature calcretes. In the Carmarthen area, the Senni Formation comprises ephemeral, low sinuosity channels up to 6.5 m deep that flowed to the SE (Fig. 8.6e) (Owen 1995). Architectural elements indicate episodic floodplain incision, followed by aggradation of channel fills that fine upwards from intraformational conglomerates into cross-bedded sandstones. Channel abandonment is recorded by mudrock fills with subsequent calcretisation. In central South Wales, sandbodies exhibit variable architectures demonstrating a variety of depositional processes. Small (10 m wide) braided gravel-dominated channels indicate pulsed, though permanent, discharge, being found in close association with larger-scale sandy low sinuosity channels with compound cross-bedded bars (Owen & Hawley 2000). It is possible that the climate was wetter at this time, but again with periods of dry interludes during which interbedded vertisols developed (Owen 1995). Heterolithic, mudrock-dominated intervals have been interpreted as high-sinuosity systems, the overall controlling influence on sedimentation is believed to be movement on the Carreg Cennen Disturbance. Footwall uplift is believed to have encouraged incision to the north of the Disturbance, with increased gradients and bedload concentrations promoting low-sinuosity channel development south of the fault line (Owen & Hawley 2000). As such, the recurring facies styles are believed to be tectonically-driven. Heterolithic units contain abundant *Diplichnites* trackways indicating the presence of eoarthropleurids and kampecarid arthropods (Smith *et al.* 2003). The Senni Formation has yielded a rich and diverse macroflora including *Gosslingia*, *Cooksonia*, *Drepanophycus*, *Sawdonia*, *Zosterophyllum* and *Thrinkophyton* (Fig. 8.12) (Edwards & Richardson 1978; Kenrick & Edwards 1988; Edwards *et al.* 1989), that presumably flourished close to fluvial

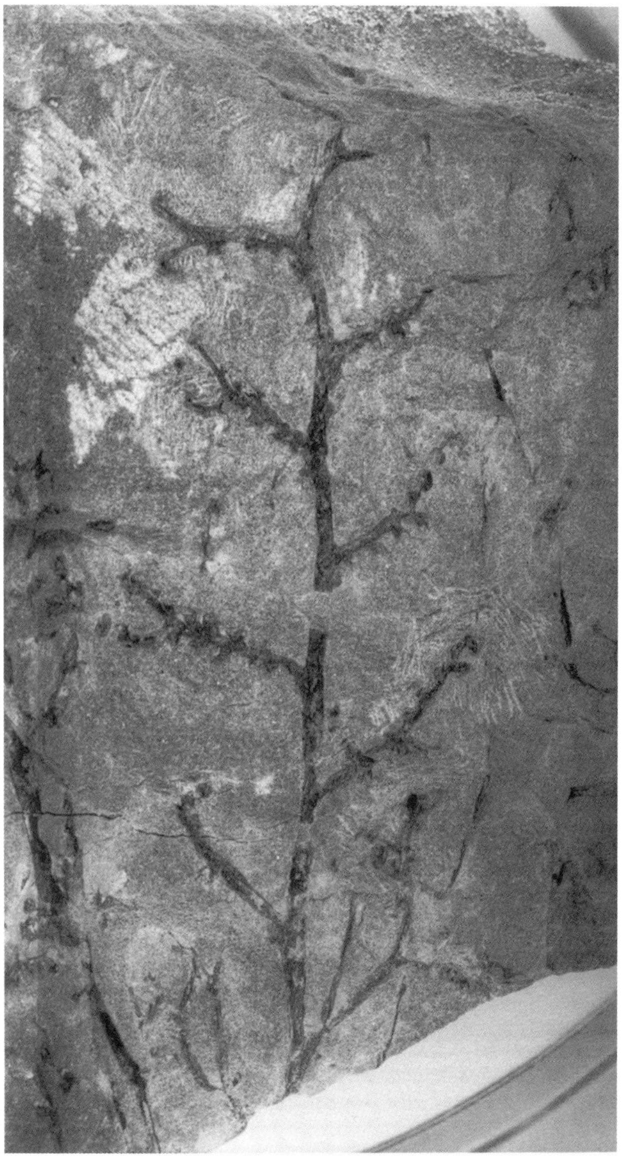

Fig. 8.12. *Thrinkophyton formosum*: fertile axis, Senni Formation, Craig Ddu, Carmarthenshire (Kenrick & Edwards 1988; photograph courtesy of D. Edwards). Photograph length 10 cm.

channels and temporal lake shores. These assemblages provide evidence for a significant radiation of vascular plants, particularly zosterophylls, at the end of the Lochkovian (Wellman *et al.* 1998). Also present are an assemblage of dispersed miospores and the pteraspidid *Althaspis senniensis* (Loeffler & Thomas 1980).

The highest Lower ORS formation in South Wales comprises the Brownstones Formation, cropping out in a belt extending from Carmarthen Bay to the Forest of Dean, being magnificently exposed in the scarp faces of the Brecon Beacons and Black Mountains (Tunbridge 1981*b*; Hawley & Owen 2005). Here, the interval comprises a 425 m thickness of laterally extensive sheet sandstones and interbedded mudrocks, thickening significantly eastward where approximately 800 m crop out to the north of the Forest of Dean (Allen & Dineley 1976). The Brownstones Formation has yet to yield stratigraphically useful plants or miospores (Waters & Lawrence 1987), although rootlets and plant debris are locally common. Disarticulated fragments of the ostracoderm *Althaspis leachi* (White) and the thelodont *Turinia pagei* (Powrie) from Wilderness Quarry in the Forest of Dean are, to date, the only recorded vertebrates discovered within the Brownstones Formation, indicating a Dittonian (Lochkovian–Pragian?) age (Allen *et al.* 1968). Overall, the interval coarsens upwards, marked by a lower succession of mudstones with thin beds of parallel-laminated sheet sandstones interpreted as unconfined sheetfloods deposited over a low-gradient plain. Sandbodies increase in importance upward, comprising fine- to medium-grained lithic arenites arranged in upward-fining, multistorey units up to 11 m thick (Tunbridge 1981). Individual sandbodies extend for 2000 m along scarp-face exposures of the Brecon Beacons, and are believed to represent low-sinuosity channels. At Ross-on-Wye, Allen (1983) described the internal architecture of the mid to upper Brownstones Formation. Bedsets of cross- or plane-bedded medium- to coarse-grained sandstones are consistent with deposition in strongly three-dimensional dunes. These bedsets are arranged into laterally extensive, conglomerate-floored sheets, with internal geometries indicating both forward and lateral accretion of low-sinuosity barforms. Similar lithofacies crop out within the Brownstones Formation of the Gower, the Welsh Borderland and Clee Hills (the Woodbank Group; Allen 1974*b*), and at Portishead on the southern edge of the Bristol Channel (Pick 1964). Palaeocurrents within the Brownstones Formation across the Anglo-Welsh Basin document a persistent N–S palaeoslope (Fig. 8.13, Tunbridge 1981*b*). The upward-coarsening signature is possibly linked to a source-area uplift, heralding Mid Devonian Acadian deformation. In central South Wales, the Emsian? Llyn-y-fan Conglomerate Formation was possibly shed off the Towy Anticline, reflecting such uplift (Tunbridge 1986).

Mid Devonian

Eifelian–Givetian

Across the Anglo-Welsh Basin, the Lower ORS terminates in a well-defined unconformity representing uplift and deformation related to the Acadian Orogeny. Significant deformation occurred within the former Welsh Basin, with the SE front of deformation being defined by the Welsh Borderland Fault System. In west Wales, deformation is observed to the north of the Benton Fault. Geochronological evidence points toward a short-lived deformation episode of Emsian/Eifelian age (Chapter 6; Soper & Woodcock 2003).

Apart from southern Pembrokeshire, Middle Devonian rocks are absent across the Anglo-Welsh Basin reflecting widespread uplift and recycling of Lower ORS sediment during this period. Lithofacies similarities between the ORS of Anglesey and those of the Anglo-Welsh Basin (Allen 1965*a*) imply connectivity

Plate 1. A quartzite olistolith, about 3 m long, in the Gwna Melange (Gwna Group), an extensive Cambrian olistostrome, until recently regarded as late Precambrian in age. Coastal outcrops near the SW tip of the Llŷn Peninsula, North Wales. Photograph: P. J. Brenchley. (Ch. 2).

Plate 2. Thick bedded turbites in the Hell's Mouth Grits (Lower Cambrian) at Trwyn carreg-y-tir, St Tudwal's, off the west coast of the Lleyn Peninsula, looking south. Photograph: A. W. A. Rushton. (Ch.3).

Plate 3. Large-scale wave ripples (< 15 cm high) in granule sandstone of the Ffestiniog Stage (Upper Cambrian). On the northern face of Cwn Grainog, by Carnedd-Filiast, on the west-side of Nant Francon, North Wales. Photograph: P. J. Brenchley. (Ch. 2).

Plate 4. Unconformity between Arenig sandstone and blocks forming marine debris slides (above 10 cm scale) on to the irregular surface of quartzite olistoliths in the Cambrian Gwna Melange. At the base of the cliffs at Ogof Gynfor about 2 km west of Camaes Bay on the north coast of Anglesey. Photograph: P. J. Brenchley. (Ch. 3).

Plate 5. Heterolithic wavy and lenticular-bedded sandstones and mudstones, modified by soft sediment deformation (Silurian, Wenlock) Gray sandstone Group at Marloes Sands, Pembrokeshire. Photograph: R. Hillier (Ch. 4).

Plate 6. Bioturbation, with sandstone-lined *Palaeophycus* burrows in thin bedded storm-emplaced sandstones and mudstones, deposited within a lower shoreface environment on the southern shelf of the Welsh Basin. Marloes Sands, Pembrokeshire. Photograph: R. Hillier (Ch. 4).

Plate 7. Laminated- to thinly-bedded sandstones of the Lonan Formation (Manx Group, Lower Ordovician). Lenan, Maughold Head, Isle of Man. BGS photograph. (Ch. 5).

Plate 8. Thin- to medium-bedded sandstone of the Niarbyl Formation (Manx Group, Lower Ordovician) folded into a D.1 syncline. Coastal exposure south of Glen Maye, Isle of Man. BGS photograph. (Ch. 5).

Plate 9. Slump folds in hornfelsed Shiddaw Group (Lower Ordovician) siltstones within the Crummock water aureole. The folds are commonly disharmonic with sheared limbs and obvious plastic deformation of the bedding, which weathers out better than in unaltered siltstones. Lad Hows, east of Crummock Water [NY 1729 1925]. Photograph: T. S. Bain. BGS photograph. (Ch. 5).

Plate 10. Borrowdale Vocanic group seen in the Langdale Pikes in the heart of the English Lake District. The group forms part of a volcanic arc and the glaciated valleys (centre) expose levels deep into the pyroclastic fill of the Scafell caldera. Peaks along the skyline expose the caldera-lake succession. Pavy Arc, the dark crag on the right is a 200 m thick proximal ignimbrite emplaced into the lake. Foreground slopes, including the Sike Pike (centre, right) are a chaotic megabreccia developed along the western margin of the younger Langdale caldera. Photograph: M. Branney (Ch. 5).

Plate 11. Glaciated trap topography in St-Johns-in-the-Vale, near Keswick, (Cumbria, Northern England). Formed of andesite sheets (the crags) separated by thin layers of stratified andesitic volcaniclastic material (the benches). The sequence is within the lower part of the Borrowdale Volcanic Group. Photograph: P. J. Brenchley (Ch. 5).

Plate 12. The Acadian unconformity with Carboniferous limestone over folded and cleaved Silurian. Arcow Wood Quarry, Ribblesdale, NW England. Photograph: N. H. Woodcock (Ch. 6).

Plate 13. Graded Llandovery turbidites showing spaced pressure solution cleavage in sandstones and penetrative slaty cleavage in mudstones. Cwn Berwyn Central Wales. Photograph: N. H. Woodcock (Ch. 6).

Plate 14. Trigonotarbid arachnid (*Ariarhoda bennetti*) from the Lower Devonian, St Maughans Formation (Old Red Sandstone), Powys (Dunlop & Selden 2004). Scale in mm. Photograph courtesy of R. M. Owens, reproduced with permission of the National Museums and Galleries of Wales. (Ch. 8).

Plate 15. North and west faces of the Sudeley opencast site, near Nuneaton, Warwickshire. The lower part of the face exposes strata of the Pennine Middle Coal Measures and the upper part is within sandstones of the Etruria Formation (Warwickshire Group). The Nine Feet coal seam is at the level of the foreground excavators, and the Ell and Two yard seams (by the upper excavators) represent splits of the Warwickshire Thick Coal. Above, erosively-based minor channel sandstones occur at the base of the Etruria Formation. BGS photograph A 14576 (Ch. 9).

Plate 16. Mam Tor in the Derbyshire Peak District [SK131835]. The black shales of the Bowland Shale Formation (former Edale Shales) are exposed in the lower left part of the photograph. The thin-bedded turbiditic sandstones of the Mam Tor Sandstone, (Millstone Grit Group), form the main c. 75 m high scarp slope. BGS photograph: Number 1213 July, 1969) (Ch. 9).

Plate 17. Variscan folds. Southward facing structures in the inverted limb of the major inversion overfold of Crackington Formation rocks at Millook Haven, Cornwall [SS 184 002]. Photograph: A. J. J. Goode (Ch. 10).

Plate 18. The Land's End Granite. A coarse-grained, megacrystic pluton at Land's End, Cornwall (Ch. 10). The image is from 'End of the Land' by Andrew Nadolski. Acknowledged with thanks.

Plate 19. The Whin Sill at Bamburgh Castle, Northumberland. The sill is dated at 295 Ma and here intrudes Carboniferous sandstones. Associated latest Carboniferous–earliest Permian igneous rocks are found throughout the North Sea and comprise tangible evidence for Coward's (1995) 'hotspot'. (Ch. 12).

Plate 20. Typical outcrop of the Zechstein of the NE coast of England: the seastack at Marsden Bay known as 'Lot's Wife', comprising Concretionary Limestone (ZS3, equivalent of the Edlington Formation – highstand systems tract). Photograph: M. Tucker. (Ch. 12).

Plate 21. The Penrith Sandstone in the banks of the River Eden at Appleby (NY 687201). Interepreted as dominantly aeolian sandstones by virtue of the dune-form cross-stratification shown. Horizons with rip-up clasts and massive (homgenous) sandstones remain enigmatic. (Ch. 12).

Plate 22. Plane-bedded sandstones interbedded with red and green coloured mudstones and siltstones, with rippled and desiccation cracked surfaces. Arden Sandstone Formation, Newnham on Severn, Gloucestershire. 30 cm long hammer for scale. Photograph: R. Porter. (Ch. 13).

Plate 23. Aeolian dune-bedded Kinnerton Sandstone Formation overlain by the fluvial Chester Pebble Beds Formation, Burton Point, the Wirral, Cheshire. Photograph: R. Porter. (Ch. 13).

Plate 24. Fluvial channel sandstones with interbedded mudstones in the lower part of the St Bees Sandstone, St Bees Head, Cumbria. Photograph R. Porter. (Ch. 13)

Plate 25. The transition between the top of the Branscombe Mudstone Formation and the base of the Blue Anchor Formation at St Audries Bay, Somerset. The strong apparent cyclicity is due to alternation between red, green, grey and pale carbonate rich beds. The upper-most grey unit of dolomitic mudstones and dolostones is the 'grey marls' of Richardson (1911). Photograph: R. Porter. (Ch. 13).

Plate 26. Saltwick Nab and (behind) East Cliff, Whitby, North Yorkshire. The scars in the foreground are formed by the 'Jet Rock, the lower part of the Mulgrave Shale Member. East Cliff exposes the Alum Shale Member overlain by the thin Dogger Formation followed by the Saltwick Formation. The lower part of the cliff on the left hand side of the photographgraph is the remains of an alum working. Photograph: P. Rawson. (Ch. 14).

Plate 27. Upper part of the Kimmeridge Clay, Portland Sand and lower part of Portland Stone formations. In the foreground the waterfall is at Freshwater Steps and Hounstout Cliff is the principal cliff in the picture, Isle of Purbeck, Dorset. This is the most complete section through this interval in northwest Europe. Photograph: J. Cope. (Ch. 14).

Plate 28. Seaford Head and Cuckmere Haven, Sussex, from the air: the cliffs expose much of the White Chalk Subgroup, from the upper part of the Lewes Nodular Chalk Formation to the basal part of the Culver Chalk Formation. Photograph: R. Mortimore (Ch. 15).

Plate 29. The middle C Beds of the Speeton Clay, Middle Cliff, Speeton, East Yorkshire. The thin yellow line zig-zagging along the middle of the cliff is a 5 mm pyretic layer of probable volcanic origin in bed C7E; the brown-weathering bed just beneath is C7F. Photograph: P. Rawson. (Ch. 15).

Plate 30. The Welton Chalk Formation at Little Thornwick Bay, Flamborough, East Yorkshire. Photograph: P. Rawson. (Ch. 15).

Plate 31. The Sub-Paleogene surface, Alum Bay, Isle of Wight (Hampshire 'Basin'). A generally planar surface of Campanian chalk, tilted by mid Cenozoic Alpine tectonics, exhumed by recent erosion. Here it is overlain by the Early Eocene Reading Formation, mostly stripped off in this area. Some deep karstic solution hollows in the sub-Paleogene surface are still partly filled by Eocene sands. (Photograph: C. King. (Ch. 16).

Plate 32. Lower Eocene, Alum Bay, Isle of Wight. Yellow and white fluvial and alluvial sands and kaolinitic clays of the Poole Formation (to the right), underlying thinly bedded heterolithic tidally-influenced marginal-marine sediments of the Wittering Formation. The erosional contact (immediately to the right of the highest point) marks a sequence boundary. The sediments were folded and uplifted by Neogene reactivation of basement structures in response to compressive Alpine tectonism. Length of section: c. 30 m. Photograph: C. King. (Ch. 16).

Plate 33. Upper Eocene Solent Group of the Hampshire Basin. The type-section of the Bembridge Limestone Formation, a dominantly freshwater limestone, deposited during an episode of clastic starvation during the early stages of sea-level rise. Seven lacustrine cycles are identified here (Daley & Balson 1999), probably parasequences reflecting minor sea-level fluctuations. A transgressive ('ravinement') surface overlies the Bembridge Limestone, followed by a brief marine incursion (Bemridge Oyster Bed). This forms the base of the Bouldnor Formation, poorly exposed at the top of the cliff, predominantly marginal-marine and coastal plain clays. Whitecliff Bay, Isle of Wight. Photograph: C. King. (Ch. 16).

Plate 34. Hummocky moraine, Jervaux Park (SE 1785), Wensleydale, N. Yorkshire. Photograph: J. Catt. (Ch. 17).

Plate 35. Cretaceous cliffs at Hunstanton, Norfolk, consisting of Chalk cap, Hunstanton Formation, and Carstone Formation at the base. Cliffs such as these are subject to infrequent failures and their strength tends to make them less susceptible to the increased wave action likely to result from climate change. Photograph: M. Page. (Ch. 18).

Plate 36. Blakeney Point, Norfolk. This shingle spit is over 15 km in length; an actively moving but geologically relict feature which provides defence to villages such as Cley and Blakeney. The spit has extended westwards as a result of longshore drift, and during storms the shingle is thrown landward across it, gradually overwhelming the marsh as it moves inland. Features such as these are less resilient to the dynamic forces of nature acting at the coast. Photograph: M. Page. (Ch. 18).

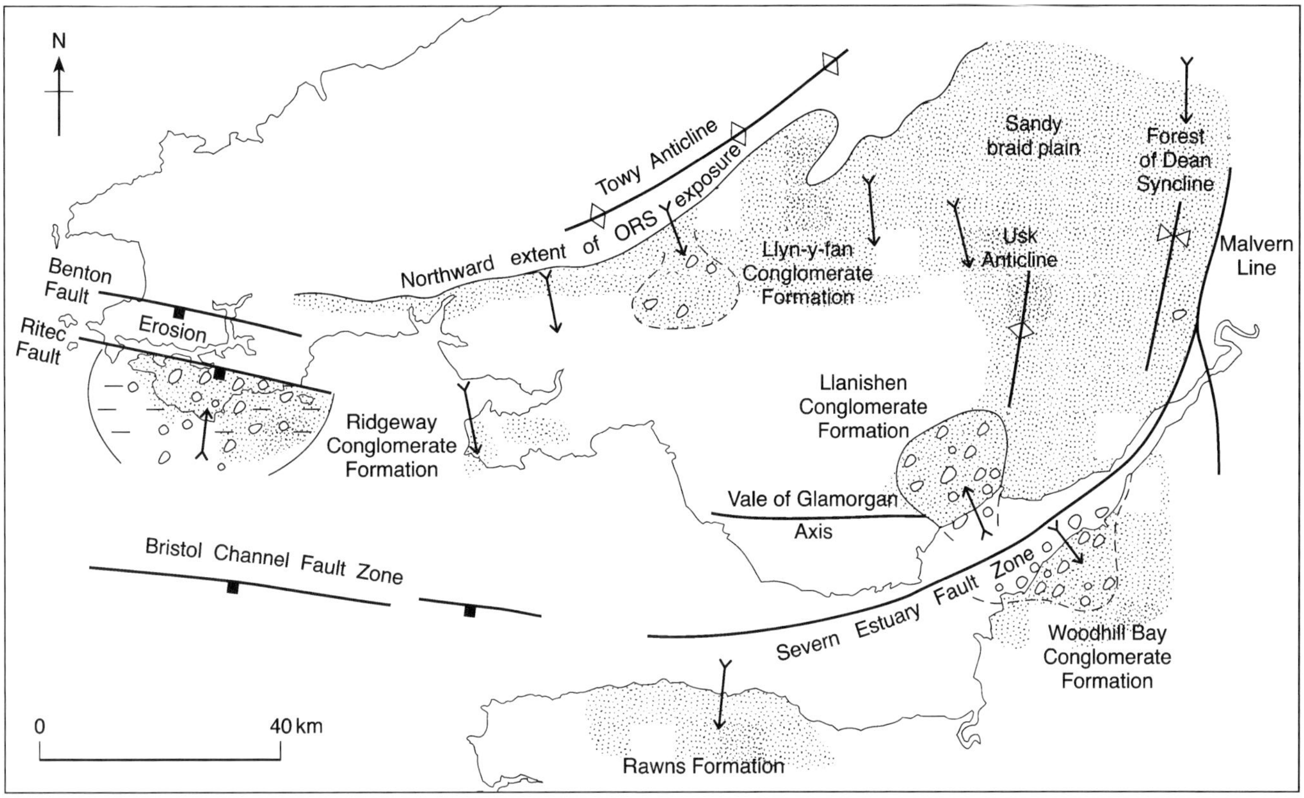

Fig. 8.13. Latest Lower ORS palaeogeography, indicating presumed tectonic controls and sources of Lower–Mid Devonian extraformational conglomerates. Data sourced from Tunbridge (1981, 1986) and Wilson *et al.* (1988).

between these areas, as does provenance and palaeocurrent evidence demonstrating that the Lower ORS must have been transported across the former Welsh Basin from a distal Laurentian source (Allen 1974*b*; Allen & Crowley 1983). Significant Lower ORS cover must have also been in place to generate the overburden necessary for metamorphism within the Welsh Basin (Chapter 6), with at least 3.5 km cover over North Wales (Soper & Woodcock 2003). It is likely that the Long Mountain and Clee Hills outliers are the erosional remnants of this cover.

From petrological and palaeocurrent evidence, Tunbridge (1984) claimed that the Eifelian Trentishoe Formation of North Devon was sourced from the recycling of Lower ORS sediments from the Anglo-Welsh Basin, arguably the Brownstones. The overlying Rawns Formation, however, differs in petrological character, comprising coarse pebbly sandstones with exotic clasts of porphyry, tuffs, quartzites and lithic arenites. Palaeocurrent evidence demonstrates a northern source to this low-sinuosity system, probably an uplifted structural high within the Bristol Channel (Tunbridge 1986). In the Bristol area, the Woodhill Bay Conglomerate (Eifelian age?, Tunbridge 1986) was also possibly shed from this structural feature (Fig. 8.13) (Cope & Bassett 1987).

In southern Pembrokeshire, the unconformity-bounded Ridgeway Conglomerate Formation is of probable Mid Devonian age (Fig. 8.6f) (Allen *et al.* 1981; Williams *et al.* 1982). This is an upward-coarsening sequence of petromict, exotic-clast conglomerates, lithic sandstones and mudrocks with vertic palaeosols. The conglomerates exhibit crude lamination, with imbrication and clast-supported nature implying a stream-flood origin. Clasts comprise protoquartzites, phyllites, lithic sandstones and felsites. Sandbodies are rich in phyllarenites, arranged in sheets and local channels of flat or cross-laminated sets, probably deposited by sheetfloods. The Ridgeway Conglomerate Formation was deposited on the medial fringe of an alluvial fan that prograded northward into a playa flat complex. A Precambrian–Lower Palaeozoic provenance within the Bristol Channel has been proposed (Williams 1971; Allen *et al.* 1981; Cope & Bassett 1987), probably an uplifted horst block (Tunbridge 1986). The sequence thickens markedly into the hanging wall of the Ritec Fault (Powell 1989), suggesting movement and development of accommodation space or the infilling of a remnant topography. Tunbridge (1986) suggested that fault movement within the Bristol Channel area was related to strike-slip tectonics, probably dextral transtension linked to the convergence of Avalonia and Armorica (Wilson *et al.* 1988).

Late Devonian (Upper ORS)

Frasnian–Famennian

In stark contrast to the Lower ORS, the Upper ORS constitutes a thin veneer of sediment, locally preserved within southern Pembrokeshire, areas skirting the South Wales Coalfield, the Forest of Dean and the Clee Hills. In Pembrokeshire, the Skrinkle Sandstones (Frasnian–Tournaisian) accumulated to the south of the Ritec Fault (Fig. 8.14, Marshall 2000*a*, *b*). The succession thickens from 100 m close to the fault to 330 m to the south at Freshwater West, implying differential subsidence. The southerly bounding fault presumably exists within the Bristol Channel subsurface. The onset of deposition within the Gupton Formation is recorded by a siliciclastic lacustrine interval dominated by red mudstones and ripple cross-laminated sandstones (Fig. 8.6i). Upward-coarsening profiles suggest a progradational, deltaic influence with sediment transport to the SE, that is into the structurally defined bathymetric low. The basin shoaled, being infilled by a high-energy sandy braidplain that flowed to the SE. The overlying West Angle Formation is marked by a change in texture and mineralogy, with lithic sandstones and conglomerates rich in sandstone, igneous and phyllitic clasts (Marshall 2000*a*). The basal Conglomerate

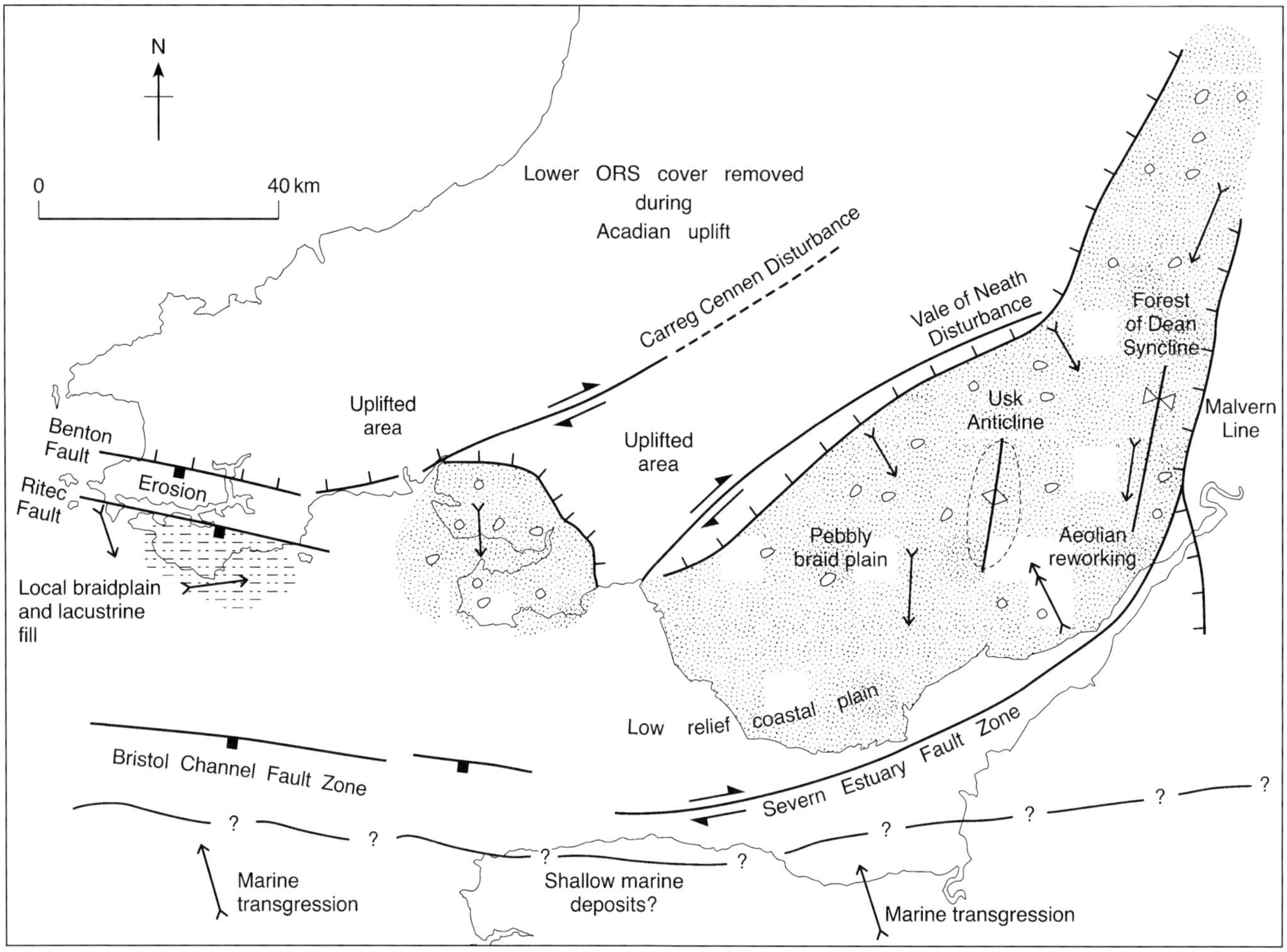

Fig. 8.14. Upper ORS palaeogeography of South Wales, and intra-basinal tectonism (after Lovell 1978*a*; Wilson *et al.* 1988; Marshall 2000*a*, *b*).

Member is dominated by conglomerate-filled channels and sheets being presumably eroded from the Precambrian or Lower Palaeozoic sources of north Pembrokeshire, or by reworking clasts from the Cosheston Group to the north of the Ritec Fault. Upward-fining heterolithic units are also present, indicating low-sinuosity channels were also in operation. The upper half of the West Angle Formation, the Red-Grey Member, records evidence of increasing marine influence, heralding the onset of Lower Carboniferous transgression. The Red-Grey Member is a heterolithic association of sand and mudstones with marine faunas interpreted as a barrier-island–lagoon complex. Palaeocurrents within the barrier and washover sandbodies indicate a northward direction to the marine transgression which overstepped the Ritec Fault by Courceyan times.

In areas skirting the north and east of the South Wales Coalfield, the 'Plateau Beds Formation' comprises a tripartite division (Lovell 1978*a*). The base of the sequence is defined by a quartz-granule rich mudstone interpreted as a mudflow deposit (Lovell 1978*a*, *b*). The basal unit contains occasionally pebbly, trough cross-bedded, locally channelized sandstones and associated red mudstones deposited on a low-sinuosity braidplain that flowed southwards. The central unit contains fine- to medium-grained, cross-bedded sandstones in m-scale bedsets. These are interpreted by Lovell (1978*a*) as aeolian deposits with interbedded fluvial sediments, wind-blown units being transported to the NW. Allen (1965*b*) and Taylor & Thomas (1974, 1975), however, consider this unit as being solely fluvial in origin. The upper unit is heterolithic in nature with channelized, cross-bedded sandstones and mudstones suggesting a tide-influenced environment. Lingulids, spiriferids and fish fragments (*Bothriolepsis* and *Holoptychius*) confirm a Frasnian or early Fammenian age (Lovell 1978*b*; Barclay 1989), the marine signature possibly reflecting eustatic highstand of sea level (Johnson *et al.* 1985).

Isopach values of the 'Plateau Beds Formation' are reduced by the erosive overstep of the overlying Grey Grits Formation and its correlative the Quartz Conglomerate Subgroup along the northern and eastern crop of the South Wales Coalfield. The latter has three constituent formations, in ascending order the Wern Watkin, Craig-y-cwm and Garn-gofen Formations (Lovell 1978*a*; Barclay 1989). The contact between the Plateau Beds Formation and underlying units is locally angular in nature, with thickening of the sequence eastward reflecting development of accommodation space on the flanks of the Usk Anticline. The latter is arguably the hanging-wall anticline to the reverse Usk Fault Line that developed by E–W shortening associated with dextral movement on the Neath and Severn Estuary Fault Zones (Butler *et al.* 1997); dextral displacement became dominant during the Upper Palaeozoic along NE-trending faults with corresponding folding along N–S-trending structures (principally the Usk Anticline and Forest of Dean Syncline, Wilson *et al.* 1988) (Fig. 8.13). The Wern Watkin Formation comprises fine-grained quartzitic sandstones, locally with pebbles of vein quartz. Decimetre-thick tabular sets are flat laminated and trough-cross bedded, being deposited by low-sinuosity rivers that drained to the south and SE. On the Gower the Quartz Conglomerate Subgroup is a coarser-grained correlative unit of low-sinuosity channels again with a southerly dispersal. It is possible that the latter were derived from the uplifted Towy Anticline (Lovell 1978*a*). Similar facies associations are developed within the Craig-y-cwm Formation, with the addition of laterally extensive calcretized mudrocks. The overlying Garn-gofen Formation is predominantly sandstone-prone, having a high mica content and immature detrital mineralogy suggesting limited transportation and sorting by sinuous, bed-load dominated rivers that flowed to the SE on the Northern Crop, and to the SW on the Eastern Crop (Lovell 1978*a*). It is possible that the Eastern Crop deposits, rich in garnet and feldspar, were derived from an exposed metamorphic Malvern Axis source, presumably uplifted during Acadian deformation (Butler *et al.* 1997). At the top of the Formation, marine reworking is observed, heralding marine transgression of the Lower Limestone Shales. It is likely that the Garn-gofen Formation is the lateral equivalent of the Tintern Sandstone Formation of the Forest of Dean (Barclay 1989).

In the Clee Hills the Upper ORS is represented by the Farlow Group (Allen 1974*b*). Approximately 150 m of pebble/cobble conglomerates, cross-bedded pebbly sandstones and calcretized mudrocks are present. Palaeocurrents were consistently to the south, in rivers of moderate sinuosity. Allen (1974*b*) suggested that the Farlow Group was derived locally from exposed Lower Palaeozoic and Lower ORS source rocks.

South of the Bristol Channel, the Portishead Formation correlates with the Quartz Conglomerate Subgroup of South Wales (Pick 1964). The formation is mainly fluvial in origin (Williams & Hancock 1977; Kellaway & Welch 1993), with minor aeolian and lacustrine deposits (Dodd 1986). It is of considerable biostratigraphic importance due to its unique fish fauna (Dineley 1999*b*). The presence of *Groenlandaspis*, the only British record of this widely dispersed genus, implying connectivity with other parts of Laurasia.

Other alluvial basins

Anglesey

The ORS of Anglesey is important as it documents the former existence of Lower ORS strata across the Welsh Basin prior to Acadian uplift and erosion. Allen (1965*a*) subdivided the section on lithostratigraphic criteria into four units. The basal, but partly diachronous, Bodafon Formation comprises pebbly sandstones and conglomerates locally derived from the Precambrian and Lower Palaeozoic subcrop. Beds are predominantly massive, and are interpreted as alluvial-fan deposits shed off the basin margin to the NW. The Traeth Bach Formation is dominated by red siltstones with calcretes and dolocretes, locally interbedded with conglomerates of the Bodafon Formation. The Traeth Bach Formation is interpreted as a non-saline playa complex distal to the basin-margin fans. The succeeding Porth y mor Formation comprises heterolithic units of upward-fining and -thinning deposits. Epsilon cross-beds are well developed, and these high-sinuosity fluvial channels were the first to be described in the rock record. The channels flowed consistently to the SE. The youngest Traeth Lligwy Formation comprises thinly bedded sandstones and mudstones, the latter being almost devoid of calcretes. Sandstones are extensively bioturbated, and it is possible that deposition took place within a semi-permanent lake. Allen (1965*a*) argued on lithostratigraphic criteria that the Anglesey ORS could be correlated with the Lower ORS of the Anglo-Welsh Basin, being part of the same depositional tract. It seems likely that the Traeth Bach Formation correlates with the Milford Haven Group and its equivalents, the thick pedogenic horizons at the top of the unit being the correlative of the Chapel Point Calcrete (Fig. 8.3). The overlying Porth y mor Formation is presumably the equivalent to the Lochkovian succession of the Anglo-Welsh Basin. Heavy mineral assemblages correlate well with those seen in Pembrokeshire, but not with those of the Welsh Borderland.

Edenside, Cumbria

In NW England exposures of predominantly coarse-grained ORS deposits crop out on the NE flank of the Lake District. The deposits occur within two down-faulted grabens, the Shap Wells and Mell Fell Troughs (McCormac 2001). Accurate age-dates have not yet been ascertained, but field relationships suggest a Mid–Late Devonian age. The absence of a cleavage indicates post-Acadian deposition, whereas intrusion by bosses and dykes with a geochemical signature similar to Courceyan-age basalt lavas in the Cockermouth area (Mitchell *et al.* 1978) constrain upper-age limits.

Shap Wells Trough

Two lithostratigraphic formations are present. The lower Shap Wells Conglomerate Formation comprises up to 150 m of poorly sorted, lithic cobble-grade conglomerate interbedded with mudstones. The formation thickens into the southern and western margins of the graben, and is believed to have been deposited by coalescent alluvial debris fans contemporaneous to a muddy playa lake environment. The, in part, coeval Blind Beck Sandstone Formation consists of up to 200 m of sandstone and mudstone deposited by low-sinuosity rivers. The presence of well-rounded sand grains with haematite coatings possibly indicate an aeolian origin to some of the sandstones (Capewell 1954; McCormac 2001).

Mell Fell Trough

A thickness of approximately 1000 m of ORS deposits accumulated within this small (15 km^2) fault-bounded graben (McCormac 2001). The Mell Fell Conglomerate Formation is subdivided into three units (Capewell 1954), the Basal Breccia and Lower Group contains poorly sorted, cobble to boulder grade clasts of greywackes, with reworked Ludlovian faunas derived from the Windermere Supergroup. Minor constituents are lavas and pyroclastics of Borrowdale Volcanic Group affinity. Large clasts are often well rounded suggesting reworking from a pre-existing conglomerate source (McCormac 2001). The Middle and Upper Groups comprise finer-grained, red-brown lithic sandstones with lenses of cobble-grade conglomerate. The latter comprise clasts of acid intrusive and intermediate volcanics and greywackes. The conglomerates are believed to have been sourced locally, being deposited from flash-floods draining the uplifted Lake District hinterland to the south and SW. A subsidiary occurrence of conglomerate similar to the Mell Fell Conglomerate Formation at Heltondale Beck alludes to the possibility of a third graben, the Heltondale Trough (McCormac 2001).

9 Carboniferous: extensional basins, advancing deltas and coal swamps

C. N. WATERS & S. J. DAVIES

The structural evolution of England and Wales during the Carboniferous was primarily a consequence of an oblique (dextral) collision between Gondwana and Laurussia (Warr 2000). Several phases can be recognized. The Rhenohercynian Ocean opened during Early–Mid Devonian regional back-arc transtension between Avalonia and Armorica (Fig. 9.1), possibly associated with northward-directed subduction along the southern margin of Armorica. A narrow seaway floored by oceanic crust developed, extending across southwest England, northern France and Germany. Cessation of the subduction, associated with the Ligerian orogenic phase of central Europe, resulted from the collision of the Iberian and Armorican microplates (Fig. 9.1). During the Late Devonian, transpressive closure of this restricted ocean, associated with the Bretonian orogenic phase, may have occurred in response to short-lived southward-directed subduction of the Rhenohercynian oceanic plate beneath Armorica.

A return to northward-directed subduction of the Theic oceanic plate along the southern margin of Iberia/Armorica (Fig. 9.1) resulted in a Late Devonian–Early Carboniferous phase of back-arc extension within the Avalonian part of the Laurussian plate (Warr 2000). The resultant N–S rifting affected all of central and northern England and North Wales, initiating development of a series of graben and half-grabens, separated by platforms and tilt-block highs (Leeder 1982, 1988). The orientations of the bounding faults, and hence of the basins, was inherited from structures within the pre-Carboniferous basement (Fraser *et al.* 1990; Fraser & Gawthorpe 2003). Principal amongst the basement areas is the Midlands Microcraton, a triangular terrane of Neoproterozoic rocks overlain by relatively undeformed early Palaeozoic strata.

Northern England, north of the Craven Fault System (Fig. 9.2a) comprises a 'Block and Basin' province dominated by the Northumberland Trough-Solway Basin and Stainmore Trough, and by the Alston and Askrigg blocks (Figs 9.2 & 9.3), persistent fault-bound highs associated with buoyant late Caledonian granites. To the west, the volcanic remnants of the Caledonian Orogeny persisted as the Lake District High during the Dinantian. The Southern Uplands High and its offshore continuation into the Mid North Sea High, broadly separated the Northumberland Trough–Solway Basin from the Midland Valley of Scotland. The barrier was breached by a series of narrow NNW-trending basins. At various times the basins may have been connected during sea-level highstands or possibly via through-drainage during periods of low sea level.

To the south of the Craven Fault System, the generally high subsidence rates created a province dominated by basinal facies. A series of half-grabens, including the Gainsborough Trough and Widmerpool Gulf of the East Midlands (Fig. 9.4), the Craven Basin of Lancashire and Harrogate Basin of Yorkshire, formed connected narrow embayments. The intervening platforms include the linked East Midlands Shelf and Derbyshire High, the Holme High, Hathern Shelf and the North Wales Shelf (Fig. 9.2). These platform areas represent the northern margin of the Wales–Brabant High, which formed a persistent topographical feature throughout the Carboniferous.

Along the southern margin of the Wales–Brabant High there is a gradual change southward from a shelf in the South Wales and Bristol region into a deep marine back-arc seaway of the Culm Basin of Cornwall and Devon (Fig. 9.2), that developed from the Late Devonian to Silesian times (see Chapter 10).

By late Dinantian time the magnitude of regional N–S extension had greatly reduced. The positive thermal anomaly generated in the lithosphere during the main phase of extension gradually decayed with time and thermal relaxation subsidence became the dominant structural control on basin evolution in England and Wales during the late Carboniferous (Leeder 1988). The Pennine Basin evolved through Namurian and Westphalian times as a broad region of subsidence between the Southern Uplands and Wales–Brabant High, with a depocentre from south Lancashire to north Staffordshire. The thermal subsidence co-existed with small-scale, pulsatory phases of localized extension and compression, which reactivated basement lineaments (Waters *et al.* 1994).

Closure of the Theic Ocean and initial collision of Gondwana and Iberia/Armorica during the Late Carboniferous (Fig. 9.1) resulted in the Sudetian orogenic phase of crustal thickening, with foreland basin development along the northern margin of the orogenic belt, including SW England and South Wales (Warr 2000). The end Carboniferous late collisional stages, the Asturian orogenic phase, is associated with the northward advance of the Variscan thrust-fold belt into the foreland basin.

The Wales–Brabant High appears to have formed a buttress that prevented further propagation of the NNW-directed thrusting that affected southern England and South Wales (see Chapter 10). During the late Carboniferous Variscan crustal compression culminated in the regional uplift of the Pennine Basin, and erosion of the basin fill formed a major angular unconformity beneath Permo-Triassic deposits. The Variscan compression appears to have reactivated the major basement structures, resulting in the wide range of Variscan structural orientations (Fig. 9.5) and Dinantian basin-bounding normal faults became partially inverted as reverse or obliquely reverse faults. Typically, the faults maintain net normal displacements at depth, whereas, at higher structural levels, local reverse faults, anticlines and monoclines developed. In northern England a deformation phase, prior to Stephanian intrusion

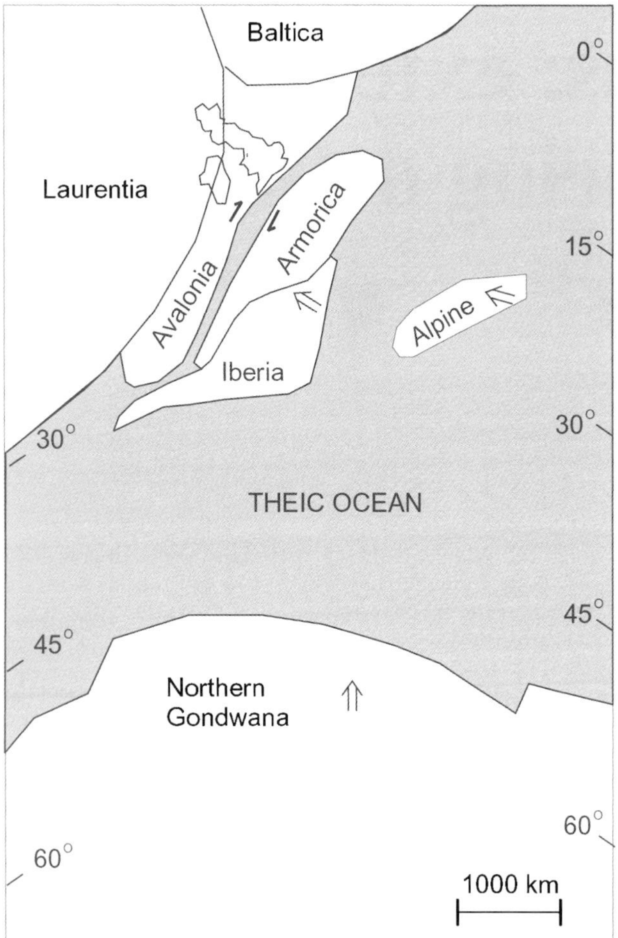

Fig. 9.1. Plate tectonic model for the Mid to Late Devonian (Warr 2000).

of the Whin dykes, resulted in N-trending anticlines and ENE-trending dextral wrench faults in the Northumberland Trough and thrusts along the western margin of the Alston block (Critchley 1984). Intrusion of the Whin Sill is associated with a phase of uplift of the Alston Block and development of conjugate ENE and NNW normal faults, with the former fault set associated with the principal ore veins in the region. Following intrusion of the Whin Sill, a subsequent phase of east–west early Permian compression resulted in the formation of ESE-trending mineral veins associated with sinistral strike-slip movements (Critchley 1984). North-trending structures include the reverse Dent Fault, forming the western margin of the Askrigg Block, and the Pennine Line, a west-verging asymmetrical anticline (Fig. 9.5). Inversion of the Dinantian Craven Basin resulted in the development of ENE-trending anticlines of the Ribblesdale Fold Belt and reactivation of the bounding Pendle Fault to form the Pendle Monocline (Fig. 9.5). The Edale Basin, Widmerpool Gulf, Gainsborough Trough and East Midlands Shelf are associated with NW-trending anticlines and monoclines, which developed in response to reverse reactivation of smaller intra-basinal faults (Aitkenhead *et al.* 2002). In the West Midlands Variscan deformation is associated with east–west compression, reactivation of basement lineaments and development of local unconformities ranging from Bolsovian to Permian age (Waters *et al.* 1994).

Climate

Palaeomagnetic evidence suggests that Britain was situated in near equatorial latitudes in Dinantian times (Scotese & McKerrow 1990). Tournaisian terrestrial strata, laid down in small isolated basins across northern and central England and North Wales, contain calcareous and dolomitic pedogenic horizons ('cornstones'), suggesting a semi-arid climate (Wright 1990). In the Northumberland Trough interbedded mudstone, sandstone and dolostone ('cementstone') developed on marginal marine flats that were subject to periodic desiccation and fluctuating salinity. The early Viséan climate was probably semi-arid with seasonal, possibly monsoonal wetting (Fig. 9.6). Evidence includes growth rings in fossil wood (Falcon-Lang 1999) and the development of arid phase mature calcrete palaeosols, with intermittent development of humid phase palaeokarsts and leached vertisols (Wright 1990). Worldwide facies distributions indicate that during the late Viséan Britain occupied an arid climatic zone present to the south of a humid equatorial belt (Witzke 1990). The regular superposition of calcretes and palaeokarsts can be interpreted as a probable cyclical fluctuation from semi-arid to humid conditions during Asbian and Brigantian times (Wright & Vanstone 2001).

During the Silesian, the British part of Laurentia lay within humid equatorial latitudes (Scotese & McKerrow 1990; Witzke 1990). By late Dinantian times the pole-to-equator temperature gradient began to increase (Raymond *et al.* 1989) and large ice sheets extended across Gondwana in the later Carboniferous. Intense glaciations may have brought about short-term seasonally drier climates, whereas ice melting resulted in sea-level rises and a potentially wetter equatorial climate (Fig. 9.6). During the late Westphalian and Stephanian increasingly arid conditions developed across Laurussia. Initially, Bolsovian primary red-bed facies developed on the flanks of the Wales–Brabant High in response to a lowering of water tables due to tectonic uplift or sediment aggradation, rather than as a consequence of increased aridity (Besly 1988*b*). Subsequent Westphalian D and Stephanian red-beds contain calcretes, evidence for a semi-arid climate. The increasing aridity may reflect a rain-shadow developed to the NW of the evolving Variscan mountain belt (Leeder 1988).

Global eustasy and sediment cyclicity

The Carboniferous was a period of glacial eustasy and therefore sea-level fluctuations were likely to have played a significant role in the development of sedimentary successions. Geographically widespread marine bands have long been accepted as the products of glacio-eustatic transgressions (Ramsbottom 1973).

Significant land-ice probably began to develop during the Dinantian. The Tournaisian–Viséan boundary is marked by a major hiatus in the Bristol area (Fig. 9.6) inferred to represent a large and rapid sea-level fall (Holdsworth & Collinson 1988; Martinsen 1990). Evidence from stable isotope analyses of brachiopods suggests a late Tournaisian phase of glacial activity (Bruckschen & Veizer 1997). Sea-level fall in late Chadian–Arundian times (Fig. 9.6) is suggested by the development of marked non-sequences within the platform carbonates of northern England, Derbyshire High, North Wales and South Wales. This was followed by a phase of late Arundian–Holkerian marine transgression (Ramsbottom 1973).

The onset of the main phase of glaciation began in the early Namurian and peaked in the late Westphalian and Stephanian (Gonzalez-Bonorino & Eyles 1995). At the peak, an ice sheet of approximately the same dimensions as the Pleistocene ice sheet is thought to have developed across large areas of the Gondwanan supercontinent (Caputo & Crowell 1985; Veevers & Powell 1987; Crowley & Baum 1991).

As in the Quaternary (Morton & Suter 1996), Carboniferous ice sheets are believed to have accumulated slowly, but wasted rapidly, producing an asymmetric sea-level curve (Church & Gawthorpe 1994; Brettle 2001). High-magnitude, high-frequency glacio-eustatic sea-level change is considered to be an important characteristic of the Carboniferous (Hampson *et al.* 1997). Wright & Vanstone (2001) use Asbian–early

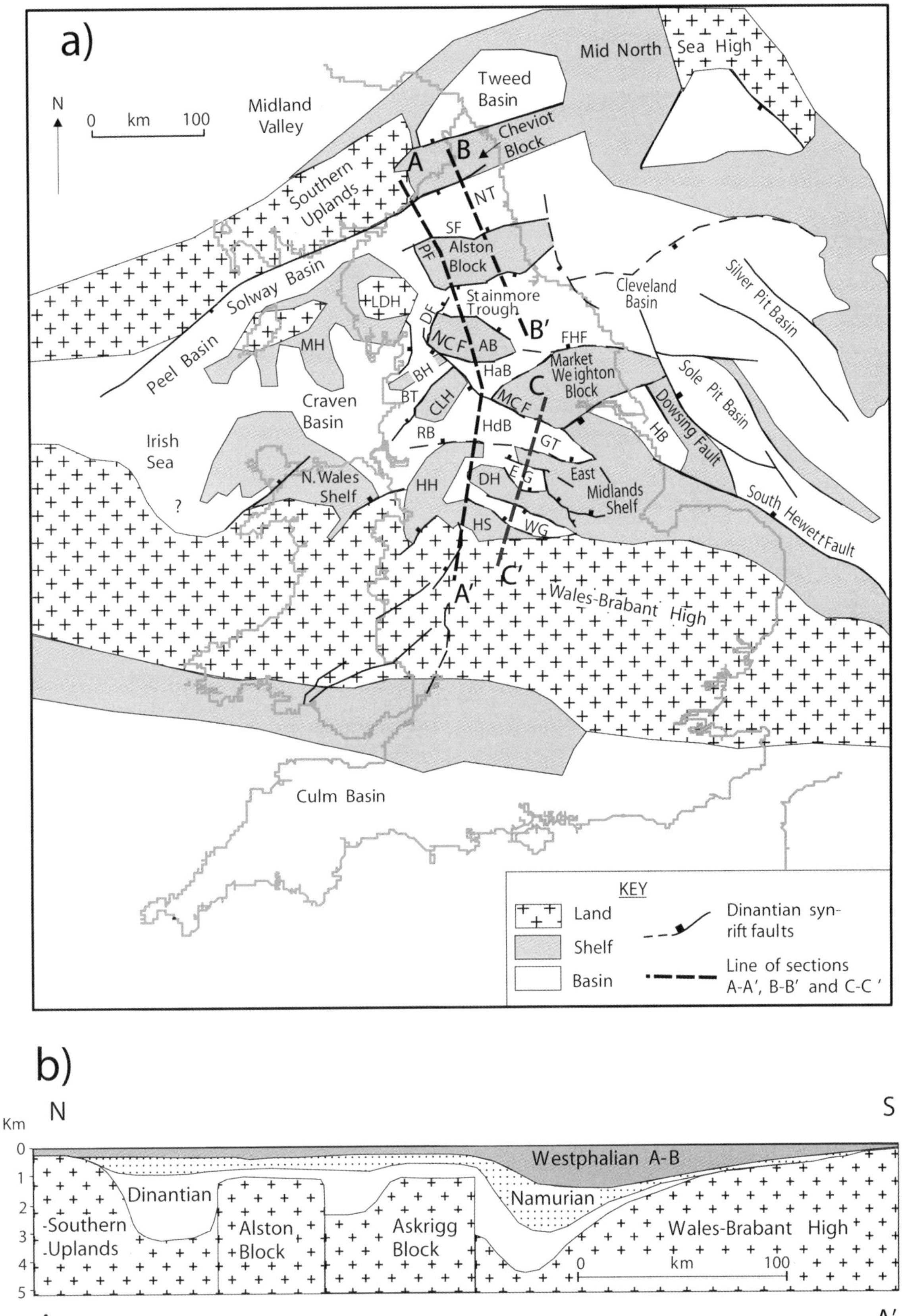

Fig. 9.2. (**a**) Dinantian palaeogeography of England and Wales showing the main tectonic features, including extensional basins and platforms (Waters *et al.* In press). Location of cross-sections: A–A' (Fig. 9.2b), B–B' (Fig. 9.3) and C–C' (Fig. 9.4). AB, Askrigg Block; BH, Bowland High; BT, Bowland Trough; CLH, Central Lancashire High; DF, Dent Fault; DH, Derbyshire High; EG, Edale Gulf; FHF, Flamborough Head Fault; GT, Gainsborough Trough; HaB, Harrogate Basin; HdB, Huddersfield Basin; HB, Humber Basin; HH, Holme High and Heywood High; HS, Hathern Shelf; LDH, Lake District High; MCF, Morley–Campsall Fault; MH, Manx High; NCF, North Craven Fault; NT, Northumberland Trough; PF, Pennine Fault; RB, Rossendale Basin; SF, Stublick Fault; WG, Widmerpool Gulf. (**b**) Schematic cross-section of the Pennine Province showing the change from several Dinantian 'rift' basins to a single Westphalian 'sag' basin (after Leeder 1982).

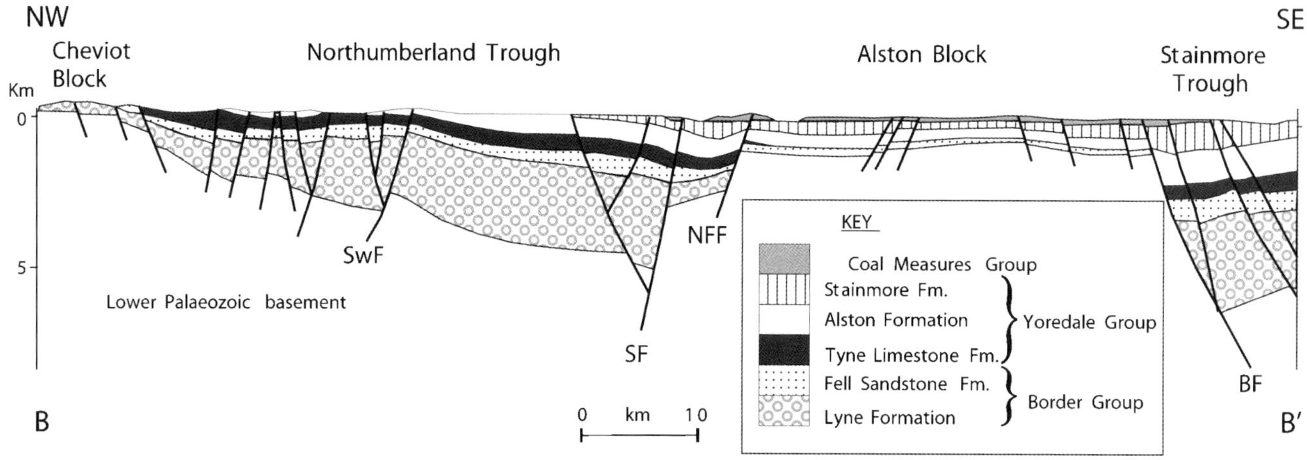

Fig. 9.3. Cross-section showing the Carboniferous infill of the Northumberland and Stainmore troughs, separated by the Alston Block (after Chadwick *et al.* 1995). Line of section is shown in Fig. 9.2a. BF, Butterknowle Fault; NFF, Ninety Fathom Fault; SF, Stublick Fault; SwF, Sweethope Fault.

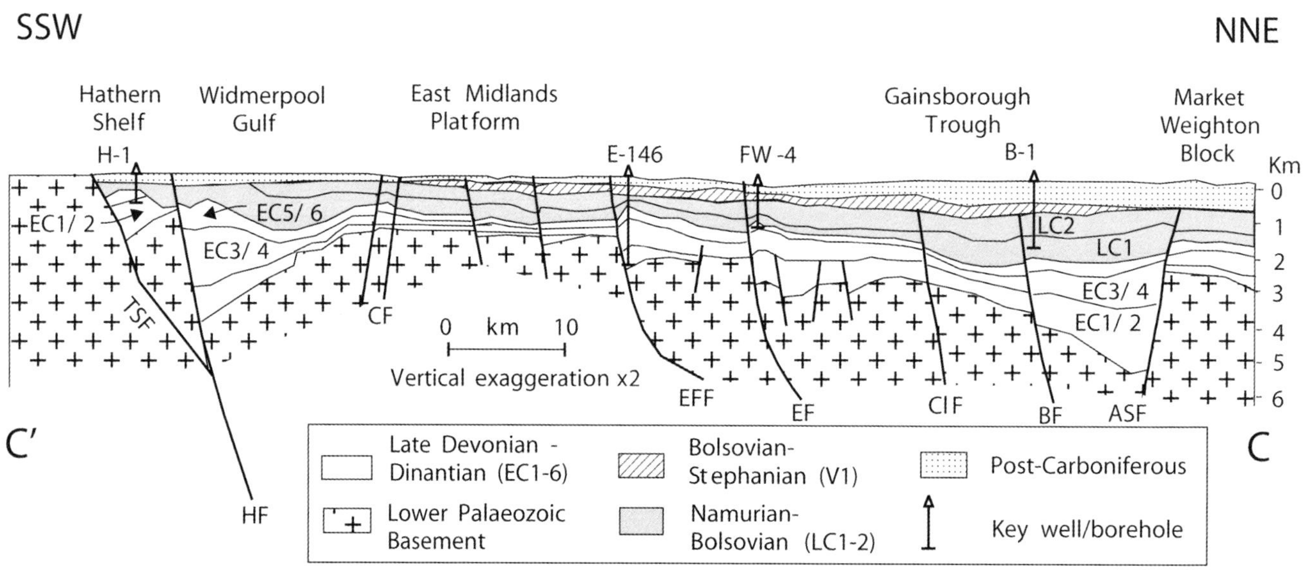

Fig. 9.4. Cross-section of the East Midlands showing the increased thickness of Carboniferous strata within the half-graben of the Widmerpool Gulf and Gainborough Trough (after Fraser *et al.* 1990). Line of section is shown in Figure 9.2a. Seismic sequences from Fraser *et al.* (1990) are summarized in Figure 9.10. ASF, Askern–Spittal Fault; BF, Beckingham Fault; CF, Cinderhill Fault; ClF, Clarborough Fault; EF, Egmanton Fault; EFF, Eakring–Foston Fault; HF, Hoton Fault; TSF, Thringstone–Sileby Fault. Boreholes and wells: B-1, Beckingham No.1; E-146, Eakring No. 146; FW-4, Farleys Wood No. 4; H-1, Hathern No. 1.

Brigantian cycles to estimate sea-level fluctuations in the order of 10–50 m, reaching up to 95 m by late Brigantian times. The magnitude of Silesian sea-level fluctuations has been estimated as 60 ± 15 m (Church & Gawthorpe 1994). Power spectral analysis estimates of marine band periodicities for Kinderscoutian–Duckmantian strata of 120 ka (Maynard & Leeder 1992) are based on a time duration of 8 ± 3.2 Ma. With 60 marine bands recognized from Namurian strata (Holdsworth & Collinson 1988; Martinsen 1990) and the average duration of the Namurian of 6 Ma (Claoué-Long *et al.* 1995), marine band periodicity may have been as short as 65 ka (Riley *et al.* 1995).

Igneous activity

Carboniferous and Permian igneous rocks, north of the Variscan Front, are described fully by Stephenson *et al.* (2003). Carboniferous intrusive and extrusive igneous rocks crop out in a number of relatively small and isolated centres in the Northumberland Trough, Solway Basin and Tweed Basin in northern England, the Derbyshire Peak District, the Black Country of the West Midlands, the Welsh Borderlands and the Bristol/Gloucester areas (Fig. 9.7). Boreholes for oil and coal exploration in the East Midlands (see Fig. 9.4), Oxfordshire and Berkshire have proved additional Carboniferous igneous rocks at depth. All the igneous rocks are basic and interpreted as within-plate activity on the Laurussian continent, produced in response to lithospheric thinning during back-arc extension (Leeder 1982). The nature of the igneous activity evolved in response to changes in tectonic processes and can be broadly subdivided into events of Dinantian and Silesian age (Fig. 9.7). Tournaisian and Viséan lavas in the Solway and Tweed basins and Northumberland Trough and the late Viséan lavas and sills of Derbyshire are tholeiitic. Tournaisian and Viséan lavas in southern England are typically alkaline.

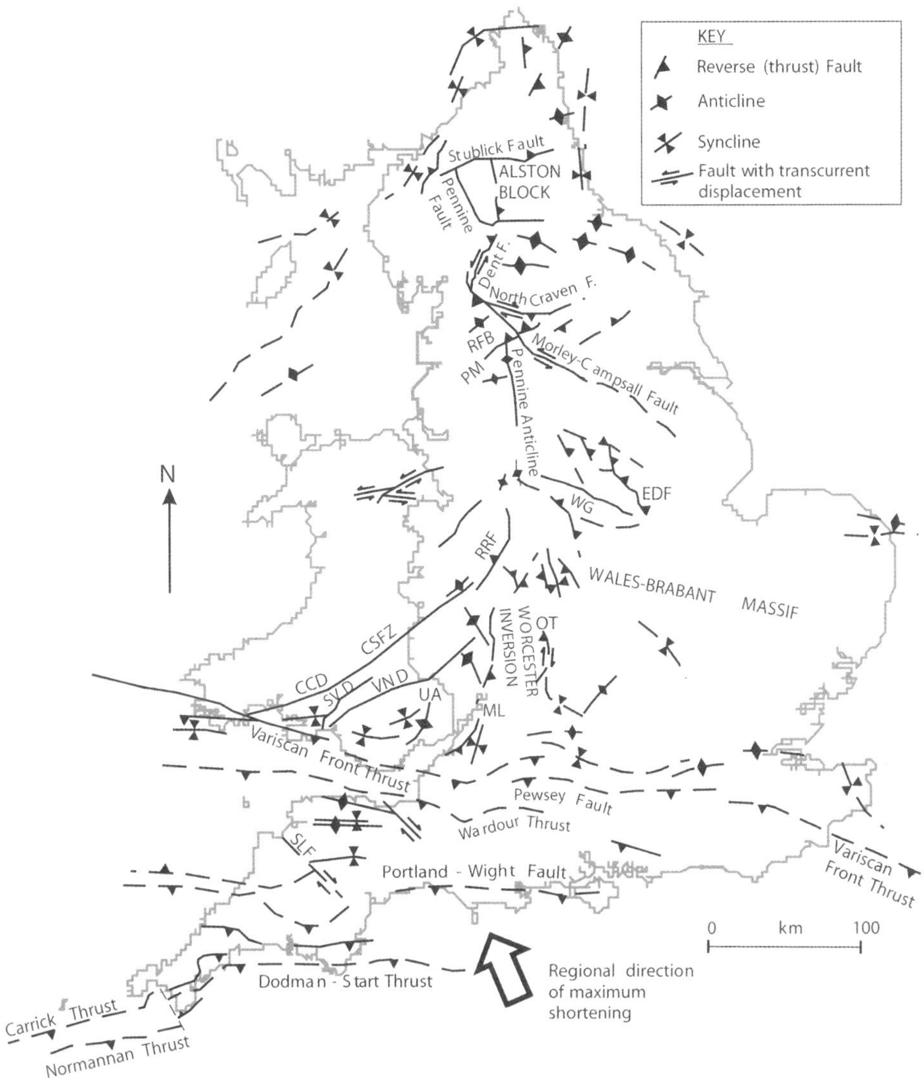

Fig. 9.5. Map showing the main late Carboniferous inversion structures associated with Variscan compression. CCD, Carreg Cennen Disturbance; CSFZ, Church Stretton Fault Zone; EDF, Eakring-Denton Fault; ML, Malvern Lineament; OT, Oxfordshire Thrust; PM, Pendle Monocline; RFB, Ribblesdale Fold Belt; RRF, Red Rock Fault; SLF, Sticklepath–Lustleigh Fault; SVD, Swansea Valley Disturbance; UA, Usk Anticline; VND, Vale of Neath Disturbance; WG, Widmerpool Gulf.

The Namurian appears to be a time of relatively little igneous activity in England and Wales, limited to thin bentonite horizons interpreted as altered vitric tuff deposits sourced from volcanic activity within the Variscan destructive plate margin (Spears *et al.* 1999). A resurgence in activity during the Westphalian and Stephanian is mainly associated with alkaline sills and dykes in central and southern England, although some tholeiitic lavas also occur in the Westphalian of the East Midlands. In NE England, widespread intrusion of tholeiitic magmas as dykes and sills, such as the Whin Sill (Plate 19) occurred during the Stephanian. The products of all these events show a typical lack of differentiation in comparison with their Scottish equivalents, probably because only small volumes of magma were produced and the eruptive activity was short-lived (Francis 1970). Thin tonsteins, kaolinite mudstones formed from the alteration of volcanic ash, are common within the Westphalian strata of England and Wales. Two types of tonstein have been recognized, one formed from basaltic ash of local origin and the other from rhyolite or rhyodacite ash of distant origin, possibly from the Variscan destructive plate margin (Spears & Lyons 1995).

Carboniferous and Permian igneous rocks, south of the Variscan Front, are described fully by Floyd *et al.* (1993*a*). These include Dinantian–Permian alkali-basalt lavas and pyroclastic rocks and the calc-alkaline granite batholith of SW England, which was intruded during late Carboniferous–early Permian times (Chapter 11). These represent magmatism generated in response to the northward subduction of the Rheic oceanic plate (Upton 1982).

Mineral resources

The thick limestones of the Carboniferous Limestone Supergroup are of great importance in the construction industry, being used as building stones and in cement and crushed rock aggregate manufacture. In the iron and steel industry limestone is used as a flux and dolostone as a refractory during smelting. Limestone has a wide range of uses in the chemical industry, glass manufacture and as an agricultural fertilizer. The use of limestone in power station flue gas control has become increasingly important. The main limestones of the Yoredale Group are typically thinner, and their quality is somewhat

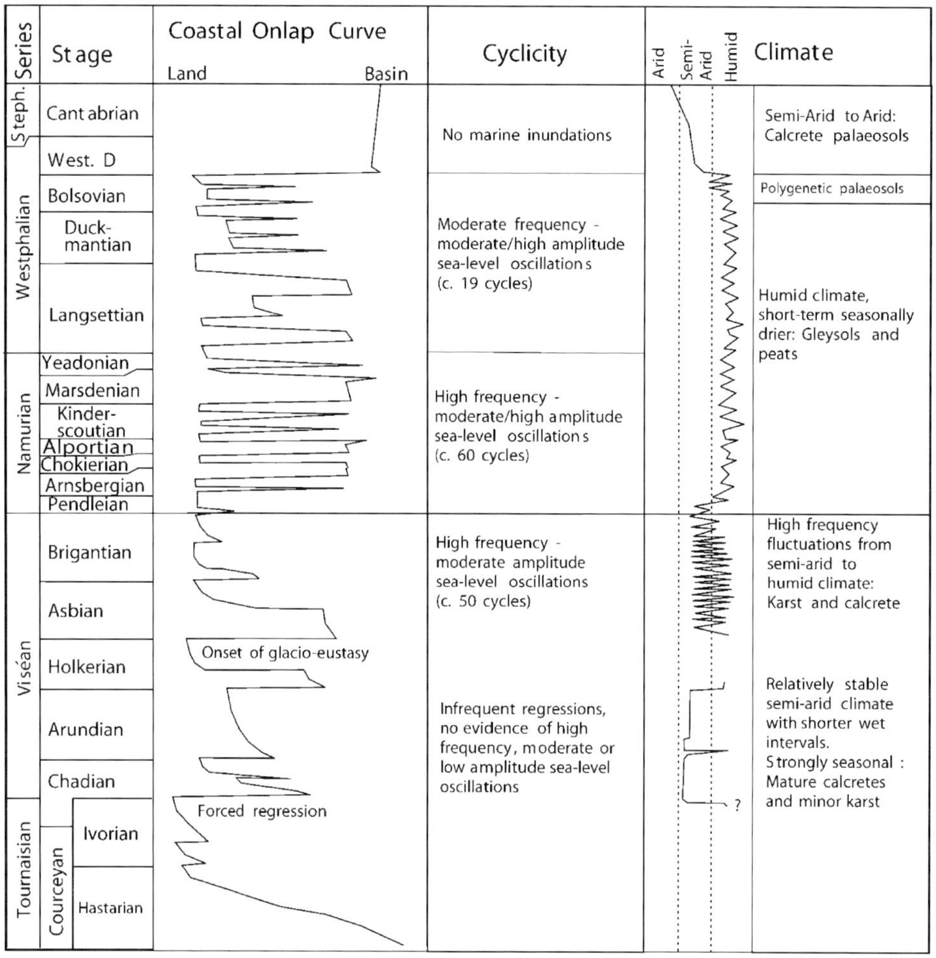

Fig. 9.6. Carboniferous sea-level and climatic record, for the Dinantian, after Wright & Vanstone (2001) and Silesian, after Ramsbottom (1977) and Ramsbottom *et al.* (1978).

reduced by the presence of clay and chert. Within this group the Great Limestone is quarried at numerous sites in County Durham and Northumberland for aggregates and cement.

The main Dinantian platform carbonates also play host to metalliferous veins, particularly lead and zinc, in the Peak District of Derbyshire, and the Askrigg and Alston Blocks of the northern Pennines. Calcite, fluorspar and barytes are commonly associated with this mineralization. Vein calcite and fluorspar are still worked at several sites within the Peak District. In the northern Pennines, barytes was extensively worked in the past and a small number of mines continue extraction in County Durham, Cumbria and Derbyshire. The metalliferous mineralization also occurs at several relatively impermeable intervals within the Yoredale Group and also in the lower part of the Millstone Grit Group of the Craven Basin. There has been a long history of mining these minerals, although current extraction is very limited in scale. In west and south Cumbria hematite is present as replacement mineral flats or cones ('sops') within the Great Scar Limestone Group. Formerly, these deposits sourced a significant iron and steel industry, but extraction is now reduced to a single mine.

Mudstones and shales of the Coal Measures and to a lesser extent of the Millstone Grit Group, are worked for bricks, tiles and ceramic building materials, although extraction has decreased over the past century. The Etruria Formation of the Warwickshire Group continues to be worked at numerous clay pits within the West Midlands and North Staffordshire for facing and engineering bricks and roofing and floor tiles.

Sandstones from the Coal Measures and Millstone Grit Group have historically been worked for building stone, although current demand is for specialized items such as monumental stone, paving flags and roof tiles. Within the Elland Flags of West Yorkshire the value of certain sandstone beds was sufficient to justify underground working, although such activities have now ceased. The working of sandstone for crushed aggregate and sand has become increasingly important and now represents the main bulk extraction industry in parts of the Pennines. In South Wales and Gloucestershire, sandstones of the Pennant Sandstone Formation continue to be extensively worked for building stone, roadstone and high specification aggregates.

The Coal Measures are rich in economic minerals and its presence was of utmost importance in the initiation of the industrial revolution in England and Wales. Many large industrial conurbations in northern and central England and South Wales owe their location to the close proximity of these strata. Coal seams have been extensively worked at depth in mines and at the surface as opencasts (Plate 15). Extraction continues, though at a much-reduced scale and relying increasingly on opencast operations. The thinner coal seams present within the Millstone Grit Group and Yoredale Group have also been worked, but on a much smaller scale. Fireclays are kaolinite-rich palaeosols, or seatearths, which were often extracted along with their overlying coal seams. Fireclays are used for production of refractory firebricks, pipes and sanitary ware, but modern extraction is on a much-reduced scale, with workings in Lower Coal Measures of Yorkshire, Leicestershire and Shropshire. Ganisters are

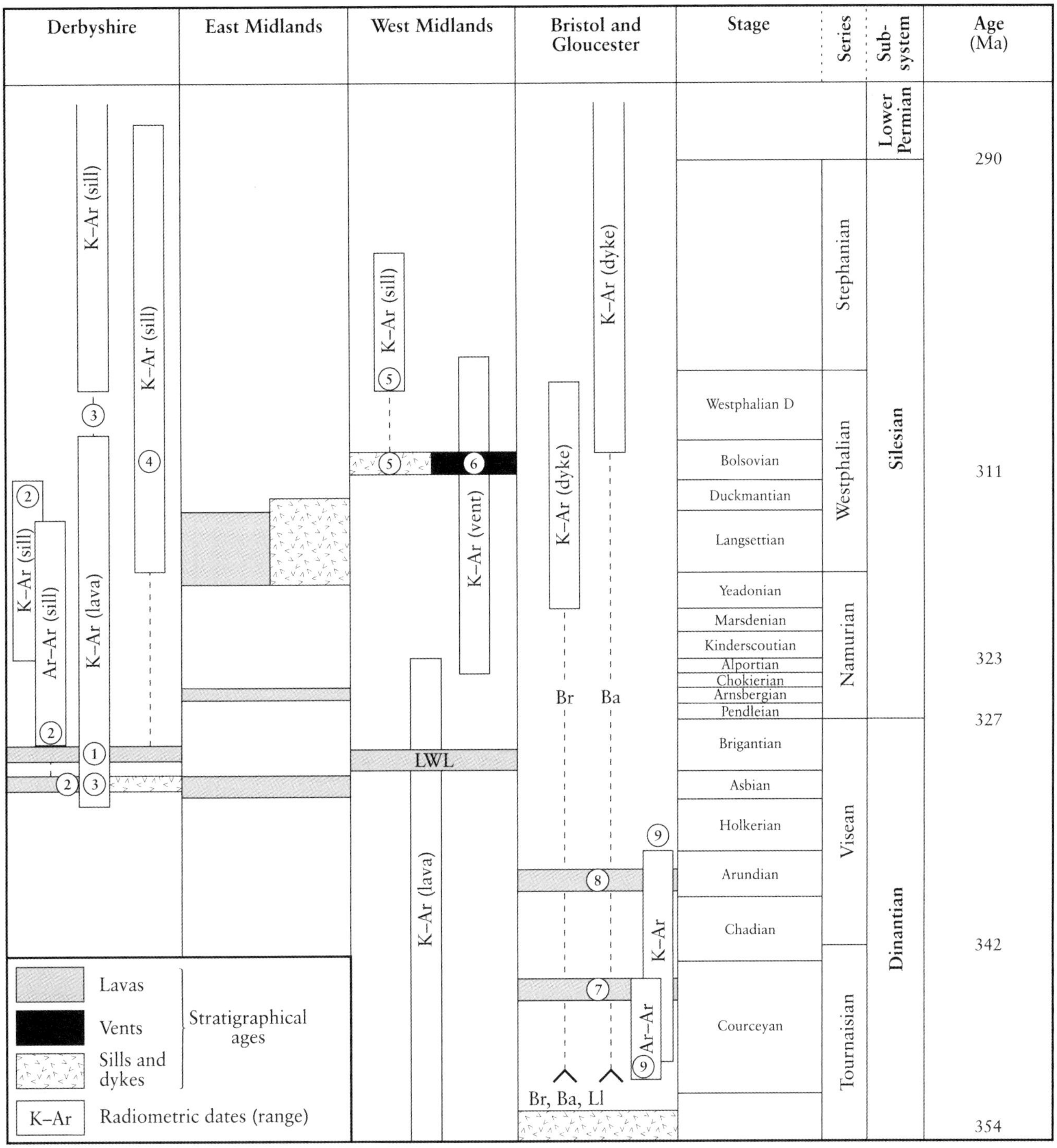

Fig. 9.7. Age and distribution of intrusive and extrusive igneous activity within central and southern England and Wales (excluding Cornwall and Devon) (from Waters 2003*a*, with permission from JNCC 2004). GCR sites: 1 – Litton Mill, 2 – Water Swallows, 3 – Tideswell Dale, 4 – Calton Hill, 5 – Clee Hills, 6 – Barrow Hill, 7 – Middle Hope, 8 – Spring Cove, 9 – Golden Hill. Ba, Bartestree Dyke; Br, Brockhill Dyke; Ll, Llanllywel Monchiquite Dyke; LWL, Little Wenlock Lavas. Timescale is that of Gradstein & Ogg (1996).

leached siliceous palaeosols, principally worked for silica bricks and furnace linings for the iron and steel industry. The main worked ganisters have historically been extracted in the Pennines from the lower part of the Lower Coal Measures, also often in association with fireclays. Ironstones were locally worked and sourced early iron and steel industries, notably in South Wales and Yorkshire. However, Coal Measures ironstones are of no importance today due to their thin and laterally impersistent nature and relatively poor quality, with the ironstones typically containing only 30% metallic iron.

Hydrocarbon production from the Carboniferous of onshore northern England commenced in the 1920s and continues in a number of small fields. Natural gas is extracted in Yorkshire and Lincolnshire from thick Coal Measures and Millstone Grit sandstone reservoirs, including the Oaks Rock, Rough Rock and Chatsworth Grit. Oil is primarily extracted in Lincolnshire and Nottinghamshire, mainly from the Mexborough Rock, Crawshaw Sandstone, Rough Rock and Chatsworth Grit. The main hydrocarbon source is the late Dinantian and early Namurian mudstone of the Bowland Shale Formation,

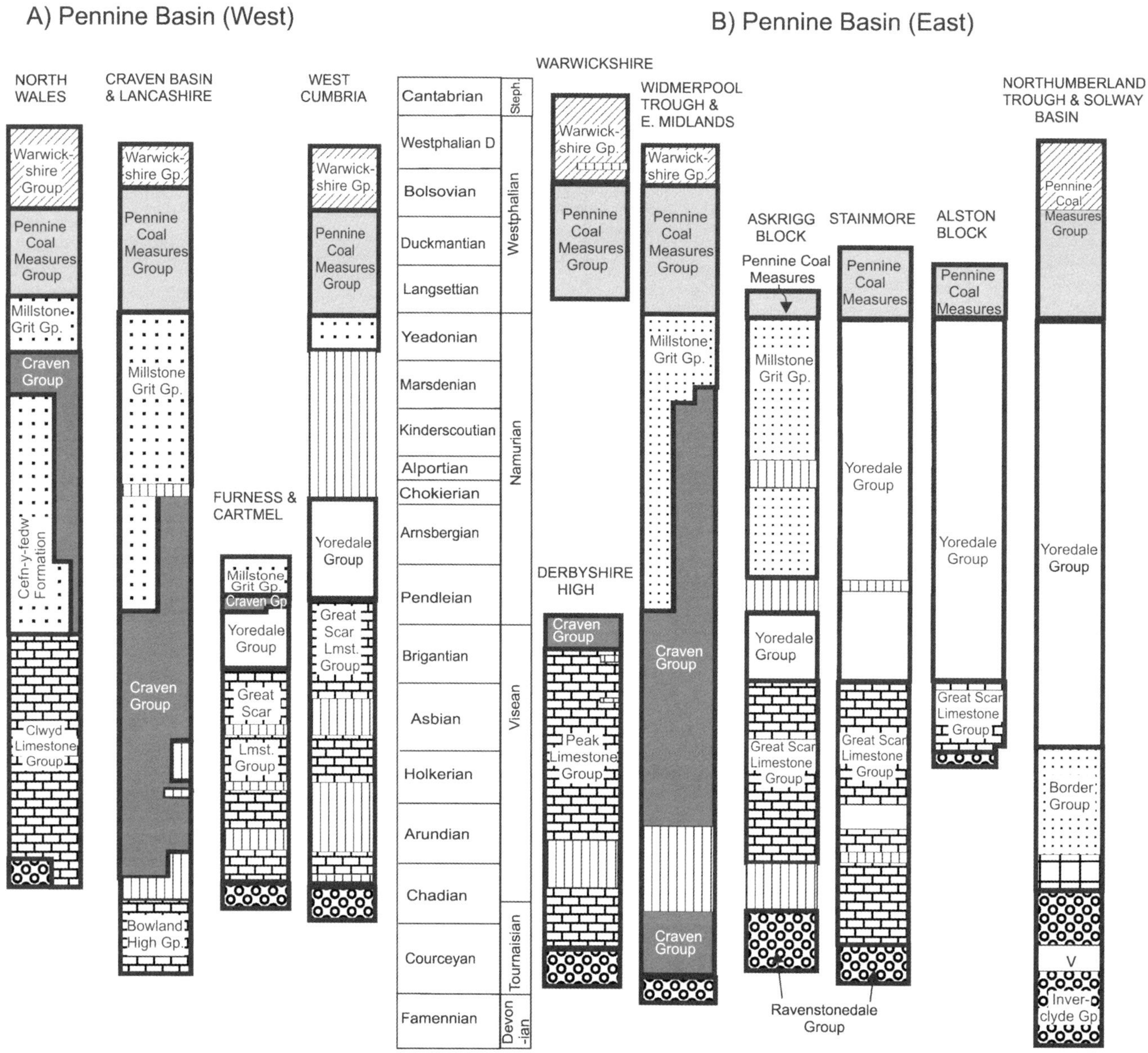

Fig. 9.8. Group lithostratigraphical framework for the Carboniferous of: (**A**) Western part of the Pennine Basin, from west Cumbria to North Wales; (**B**) Eastern part of the Pennine Basin, from Northumberland to the East Midlands; (**C**) South Wales and southern England, with relationships to lithofacies. Adapted from Waters *et al.* (2006*a*).

which has its thickest development in the East Midlands, notably in the Gainsborough Trough (Fig. 9.4). Closer to the Pennine Basin depocentre, the deep burial of source rocks and maturation prior to Variscan structural trap development severely limits oil potential in Lancashire, Yorkshire and North Staffordshire (Fraser *et al.* 1990).

Igneous rocks have been worked locally for roadstone and high specification aggregate, mainly within Dinantian lavas and sills of the Peak District and the Westphalian sills of the West Midlands.

Water resources

Carboniferous rocks are classified as minor aquifers, comprising a heterogeneous multilayered aquifer system (Jones *et al.* 2000). Sandstones and limestones typically form aquifers, whereas argillaceous strata mainly form aquitards or aquicludes. Carboniferous strata typically lack significant intergranular permeability. Groundwater movement occurs mainly through joints and fractures, which tend to be prevalent within sandstone and limestone beds. Dissolution within the thick platform carbonates of the Carboniferous Limestone Supergroup can result in open joints and cave systems along which groundwater flow rates are high. The prevalence of faulting and the lateral impersistence of sandstones, notably in the Coal Measures, affect the interconnectivity of aquifers and can significantly reduce yields, but fracturing associated with mining subsidence has enhanced groundwater flows in some areas of Coal Measures and overlying Warwickshire Group. Groundwater extraction from the Carboniferous aquifers is largely by industrial users, although this is in decline, resulting in the phenomenon of rising water tables in many of the industrial coalfield areas. Coal Measures aquifers tend to have non-potable groundwater, with poor water quality due to acid mine drainage and other anthropogenic contamination.

Stratigraphical framework

The economic importance of strata of Carboniferous age has resulted in over 200 years of research, and in numerous

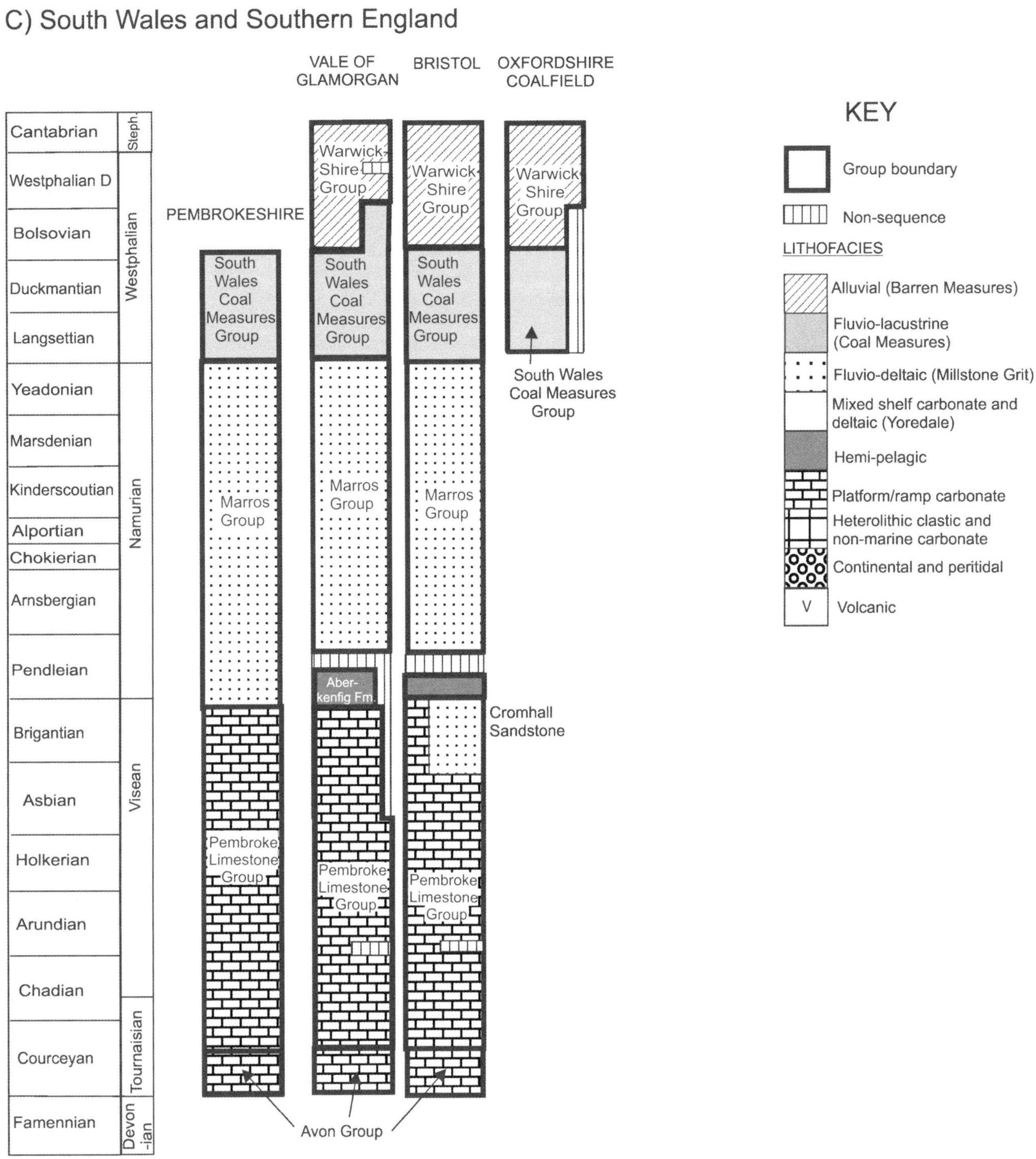

Fig. 9.8. Continued.

attempts to classify them. Much of this work was done long before procedures for naming rock units had been codified, and a haphazard approach to the establishment of the hierarchy of units prevailed. The earliest framework was relatively simple and essentially lithostratigraphical, but advances in the understanding of Carboniferous biostratigraphy, especially in the first half of the 20th century, led to a reappraisal and redefinition of lithostratigraphical terms as if they were chronostratigraphical units. Also, the localized nature of research produced numerous local names for essentially the same unit. The complexity in nomenclature has, to an extent, hindered the regional understanding of the Carboniferous successions of the UK.

The Geological Society Special Reports for the Dinantian and Silesian (George *et al.* 1976; Ramsbottom *et al.* 1978) provide useful stratigraphical correlations between key sections across the United Kingdom and Ireland. The reports aimed to integrate litho-, chrono- and biostratigraphy in a unifying scheme. However, the reports simply indicated how existing lithostratigraphical nomenclature related to a newly defined chronostratigraphical framework.

Lithostratigraphical framework for the Carboniferous

The British Geological Survey recently rationalized the nomenclature of Carboniferous lithostratigraphical units in the UK

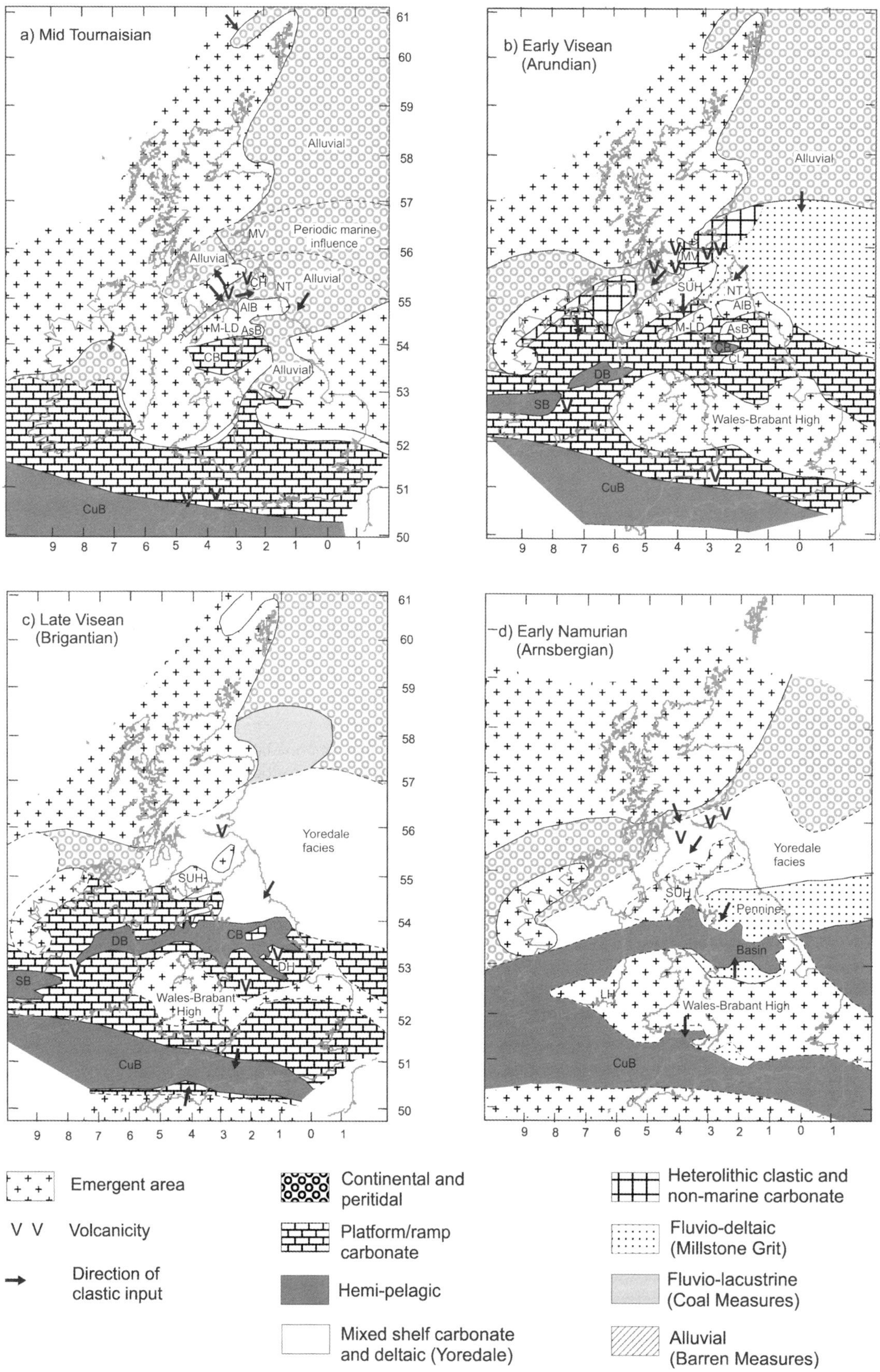

Fig. 9.9. Palaeogeographical reconstructions for England and Wales and adjacent offshore areas, showing disposition of the main lithofacies shown in Figure 9.8. (**a**) Mid Tournaisian; (**b**) Arundian; (**c**) Brigantian; (**d**) Arnsbergian; (**e**) Langsettian; (**f**) Westphalian D. Adapted from Cope *et al.* (1992). AlB, Alston Block; AsB, Askrigg Block; CB, Craven Basin; CH, Cheviot High; CL, Central Lancashire High; CuB, Culm Basin; DB, Dublin Basin; LH, Leinster High; M–LD, Manx–Lake District High; MV, Midland Valley; NT, Northumberland Trough; SB, Shannon Basin; SUH, Southern Uplands High.

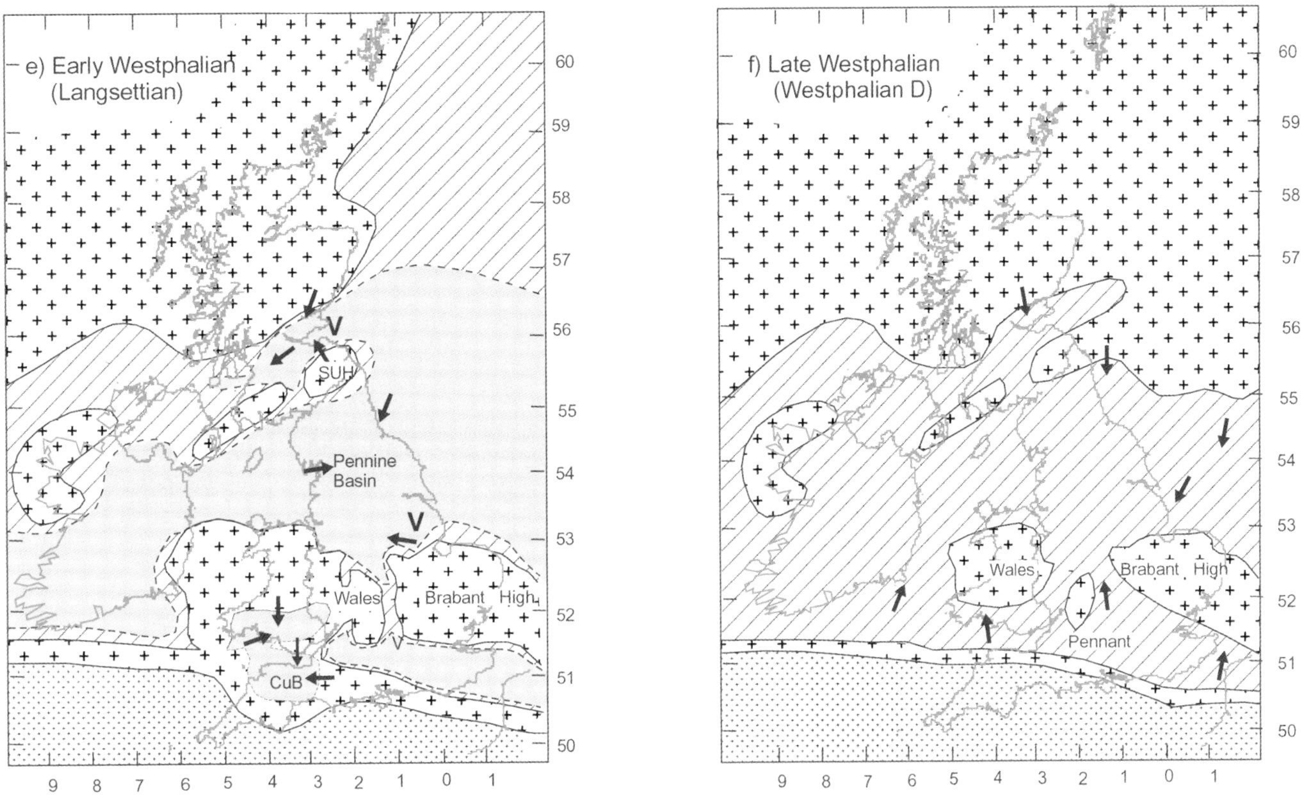

Fig. 9.9. Continued

(Waters *et al.* 2006*a*). The new scheme, presented here, equates lithostratigraphical groups with eight broad types of lithofacies associations. Distinct group names have been identified for each lithofacies grouping, with several names defined where the lithofacies developed in isolated tectonic regions during the Carboniferous (Fig. 9.8 and 9.9). The eight principal lithofacies associations are as follows.

Continental and peritidal lithofacies association

The lithofacies association was impersistently developed from Late Devonian to Viséan times across northern England and North Wales and comprises two facies that are commonly interdigitated in the Northumberland–Solway Basin.

a) *Continental fluvial clastic ('Cornstone') facies* comprises combinations of purple-red conglomerate, sandstone and red mudstone. The depositional environments range from alluvial fan, fluvial channel and floodplain overbank, respectively. The calcretes ('Cornstones') that characterize the facies are nodules and thin beds of concretionary carbonate. These carbonates represent immature–mature soil profiles developed on floodplains under the influence of a fluctuating water table in a semi-arid climate.

b) *Peritidal marine and evaporite ('Cementstone') facies* comprises grey mudstone, siltstone and sandstone, characterized by the presence of beds, and locally nodules, of ferroan dolostone ('cementstones') and evaporites (mainly gypsum and anhydrite). These were deposited in alluvial plains and marginal marine flats subject to periodic desiccation and fluctuating salinity, also indicating a semi-arid climate.

The continental fluvial clastic ('Cornstone') facies commonly forms the first element of basin infill and may extend onto horst and tilt-block highs, whereas the peritidal marine and evaporite ('cementstone') facies is generally limited to basins. The lithofacies association as a whole is represented by the Inverclyde Group within the Northumberland-Solway Basin, the Ravenstonedale Group in the Stainmore Trough and numerous formations representing isolated alluvial fan deposits, notably along the northern margin of the Wales–Brabant High (Figs 9.8 and 9.9a).

Heterolithic clastic and non-marine carbonate lithofacies association

The lithofacies association consists of interbedded grey sandstone, siltstone, mudstone and subordinate thin beds of lacustrine limestone and dolostone, seatearth, coal and sideritic ironstone. The range of depositional environments includes alternating fluviatile, deltaic, lacustrine and relatively rare marine mudstone deposits, often in thin cycles. This facies, more extensively developed in the Midland Valley of Scotland (Fig. 9.9b), is also present in Viséan strata of the Border Group, restricted to the Northumberland Trough (Fig. 9.8b).

Platform carbonate lithofacies association

Platform carbonates were mainly deposited on footwall highs during the Dinantian. There are two facies, resolved by the basic morphology of the platform (Kendall & Schlager 1981):

a) *Carbonate ramp facies* comprises calcareous mudstone with common dark bituminous and bioclastic limestone units, and locally the presence of carbonate breccias. Large carbonate mud-mounds, also referred to as 'knoll reefs' or 'Waulsortian reefs', also occur in this facies during the early Dinantian (Bridges *et al.* 1995). The facies represents deposition on gently inclined marine slopes, which may range from shallow-water to deeper-water, low-energy environments.

b) *Carbonate shelf facies* includes blanket bioclastic carbonates with crinoid banks, small bioherms, shelly or coral biostromes and algal (*Girvanella*) bands. This facies was deposited during the Viséan in a tropical shallow-marine environment on shelves and footwall highs, with a steep slope separating platform and basin environments. The facies often includes potholed bedding surfaces overlain by thin bentonite clays ('clay wayboards'), interpreted, respectively, as emergent surfaces and palaeosols in volcanic ash (Walkden 1972, 1974). Marginal shelf 'Cracoean' buildups developed during the late Dinantian (Mundy 1994).

Dinantian carbonate platforms commonly display a predictable sequence of evolutionary changes (Kendall & Schlager 1981). Initially, a ramp-like structure developed with relatively steep gradients and rapid lateral facies transitions. This evolved into a steep-sided, broad flat shelf. The lithofacies association is represented across England and Wales by the Carboniferous Limestone Supergroup, which comprises several geographically separated groups. The Great Scar Limestone Group occurs across the Alston and Askrigg blocks and fringing the Manx–Lake District High (Figure 9.8). The Peak Limestone Group of the East Midlands and Clwyd Limestone Group of North Wales occupy parts of the northern margin of the Wales–Brabant High. These groups are not laterally continuous and, hence have separate group names. Isolated platform carbonates also developed on the Holme High and Central Lancashire High within an otherwise basinal environment. The Pembroke Limestone Group occupies the southern flank of the Wales–Brabant High, from South Wales to Bristol, overlying a mud-dominated Carbonate ramp facies of the Avon Group (Figure 9.8c).

Hemi-pelagic lithofacies association
This lithofacies association comprises dark grey–black mudstones, in part calcareous, with calcareous nodules ('bullions'). Thin sandstone and limestone beds are locally common and breccias may sometimes be present. The facies was deposited in quiet and relatively deep basinal environments and was associated with minor sand-rich turbidites within pro-delta environments, or carbonate-rich turbidites on carbonate slopes. Breccias may represent proximal turbidites, debris flows or slump features. The facies is found in the Viséan and Namurian gulf areas of central England and North Wales as the Craven Group (Figs 9.8 and 9.9) and in most of the Carboniferous of the Culm Basin of SW England.

Mixed shelf carbonate and deltaic ('Yoredale') lithofacies association
This lithofacies association comprises typically upward-coarsening successions of limestone, shale, sandstone, seatearth, ganister and coal. The limestone, shale and some sandstones are marine. The clastic component was derived from the progradation of high-constructive lobate deltas into the marine environment. The delta tops were often colonized by vegetation, resulting in the generation of seatearth palaeosols and coals. This facies, represented by the Yoredale Group, was widespread in the Viséan and early Namurian across northern England, as far south as the Craven Fault System (Figure 9.8 and 9.9c).

Fluvio-deltaic ('Millstone Grit') lithofacies association
The Millstone Grit lithofacies association comprises typically upward-coarsening successions of black shale, grey mudstone, siltstone, fine- to very-coarse-grained sandstone, with seatearths and relatively thin coal seams. Initial black shale deposition as 'marine bands' reflects marine transgression and delta abandonment; subsequent coarsening upwards into sandstones indicates delta progradation.

There are two facies:

The deep-water turbidite-fronted lobate delta facies is more typical of the lower part of the lithofacies association, representing initial basin infill. Coals and seatearths are typically rare.
The shallow-water sheet-like delta facies is more typical of the upper part of the lithofacies association, representing reduced accommodation within the basin. Coals and seatearths are common, but typically thin.
This is represented by the Millstone Grit Group, which extended across northern and central England during the Namurian–early Westphalian (Figs 9.8 and 9.9d). This group includes predominantly northerly-sourced, quartz-feldspathic sandstones. A similar facies, but with predominantly quartzitic sandstones derived from erosion of the Wales–Brabant High, forms the Marros Group south of the high (Figs 9.8c & 9.9d), from South Wales to Bristol. On the northern flank of the High similar protoquartzitic sandstones, identified as the Cefn-y-fedw Sandstone Formation of North Wales and Morridge Formation of the East Midlands, are present (Fig. 9.8).

Fluvio-lacustrine ('Coal Measures') lithofacies association
The lithofacies association comprises both upward-fining and upward-coarsening successions typically of grey–black mudstone, grey siltstone, fine- to medium-grained sandstone, seatearths and relatively thick coal seams. Cyclic successions are thinner and more numerous than in the underlying Fluvio-deltaic ('Millstone Grit') lithofacies association. Environments include wetland forest and soils (coal and seatearth), floodplains (plant-rich or rooted siltstone and mudstone), river and delta distributary channels (thick sandstones), prograding lacustrine deltas (upward-coarsening sequences) and shallow lakes (mudstones with non-marine faunas). Marine bands are generally much less common than in the Fluvio-deltaic ('Millstone Grit') lithofacies association. The Fluvio-lacustrine ('Coal Measures') lithofacies association was mainly deposited during the Westphalian. It occurs as the Pennine Coal Measures Group of northern and central England and North Wales (Figs 9.8a, b & 9.9e) and as the South Wales Coal Measures Group to the south of the Wales–Brabant High (Figs 9.8c and 9.9e), from South Wales to Kent. It recurs as the Grovesend Formation in late Westphalian and Stephanian times, south of the Wales–Brabant High.

Alluvial ('Barren Measures') lithofacies association
This facies association ranges from the late Westphalian to early Permian and occurs as two facies:

a) *The 'Red-bed' facies* consists of predominantly red, brown or purple-grey mudstone, siltstone and sandstone. Pebbly sandstone, conglomerate and breccia are locally developed. Minor components comprise grey mudstone, thin coals, lacustrine limestone ('*Spirorbis* limestone') and pedogenic limestone (calcrete). Environments of deposition include alluvial fans, fluvial channels, overbank areas and lakes. The 'Red-bed' facies has undergone oxidation at, or close to, the time of deposition, indicating a semi-arid climate.
b) *'Pennant' facies* consists of grey sandstone, pebbly sandstone–granulestone, characteristically an immature lithic arenite, commonly thick-, massive- or cross-bedded. Subordinate components comprise grey mudstone and thin coal. The facies represents fluvial channel and fluvial overbank deposits derived from the south.

This lithofacies, as the Warwickshire Group, is present on the northern and southern fringes of the Wales–Brabant High (Fig. 9.8). The 'Red-bed' facies is dominant along the southern margins of the Pennine Basin, but is also present on the southern margin of the Southern Uplands High as the Whitehaven Sandstone Formation. The 'Pennant' facies is mainly found south of the Wales–Brabant High in the South Wales, Bristol and Oxfordshire Coalfield, deposited during Bolsovian–Westphalian D as the Pennant Sandstone Formation. The facies was also deposited locally in the southern parts of the Pennine Basin, during Westphalian D times, as the Halesowen Formation.

Chronostratigraphical Framework

The term Carboniferous was created as a formal stratigraphical term by Conybeare & Phillips (1822) for strata present in England and Wales and was first referred to as a System by Phillips (1835). Broad similarities within the successions of England and Wales with the rest of Western Europe have allowed development of a regionally applicable chronostratigraphy.

In Western Europe the Carboniferous System traditionally comprises two subsystems, an older Dinantian and younger Silesian, corresponding to Lower Carboniferous and Upper Carboniferous, respectively (Fig. 9.10). The Dinantian–Silesian boundary was chosen to represent a regional facies transition from dominantly carbonate to terrigenous clastic strata and does not reflect a global change in flora or fauna. The lower boundary of the Silesian is defined as the base of the ammonoid *Cravenoceras leion* Zone.

The Dinantian is subdivided into the Tournaisian and Viséan Series, whereas the Silesian is subdivided into three series, Namurian, Westphalian and Stephanian. These series were originally chosen to represent prominent facies variations in Western Europe.

The stage nomenclature used in Western Europe is based on basal stratotypes defined by George *et al.* (1976) for the Viséan, Ramsbottom (1981) for the Namurian and Owens *et al.* (1985) for the Westphalian, largely from localities in northern England. The Stephanian Series is poorly represented in England and Wales, with only strata from the oldest Cantabrian Stage recognized. The distribution at outcrop of the main chronostratigraphical units and the location of the stage stratotypes is shown in Fig. 9.11.

The Mississippian and Pennsylvanian of the USA have become recognized internationally as subsystems and strictly represent Lower and Upper Carboniferous, respectively, in international usage. The mid-Carboniferous boundary separating the two subsystems occurs within the Chokierian Stage of the Namurian Series in Western Europe. International stage names have been created (Fig 9.10), though not all have been defined (Heckel 2004). Difficulties in direct comparisons between international and Western European stages have resulted in the UK largely maintaining usage of the regional Western European chronostratigraphical nomenclature (but see Cossey *et al.* 2004).

Biostratigraphical features

Ammonoids (goniatites) are an essential component of Carboniferous biostratigraphy within the late Dinantian, Namurian and early Westphalian Series, where they provide the greatest biostratigraphical resolution. Some ammonoid biozones can be recognized across Western Europe and some biozones are applicable globally. Thick-shelled ammonoids occur within thin hemi-pelagic marine beds ('marine bands') that were deposited during marine transgressions. Marine bands typically contain distinctive ammonoid faunas. Ammonoid biozones are intervals defined by the successive first appearance of ammonoid taxa, with the base of the biozones coinciding with the bases of specific marine bands.

Corals have been of historical importance in the biozonation of Dinantian platform carbonates (Vaughan 1905; Garwood 1913), although they are now considered strongly facies-controlled. Foraminifera are also of biostratigraphical importance in Dinantian carbonates. They are particularly abundant in mid-ramp and platform settings, and also present in basinal deposits in limestone turbidites (Riley 1993). The formal foraminifera zonation for Belgium, established as the standard for NW Europe, has been applied to British and Irish sequences by Conil *et al.* (1980).

Conodonts are present in marine deposits, notably carbonate turbidites and hemi-pelagic shales. Conodont zones are particularly important for Dinantian correlation (Varker & Sevastopulo 1985).

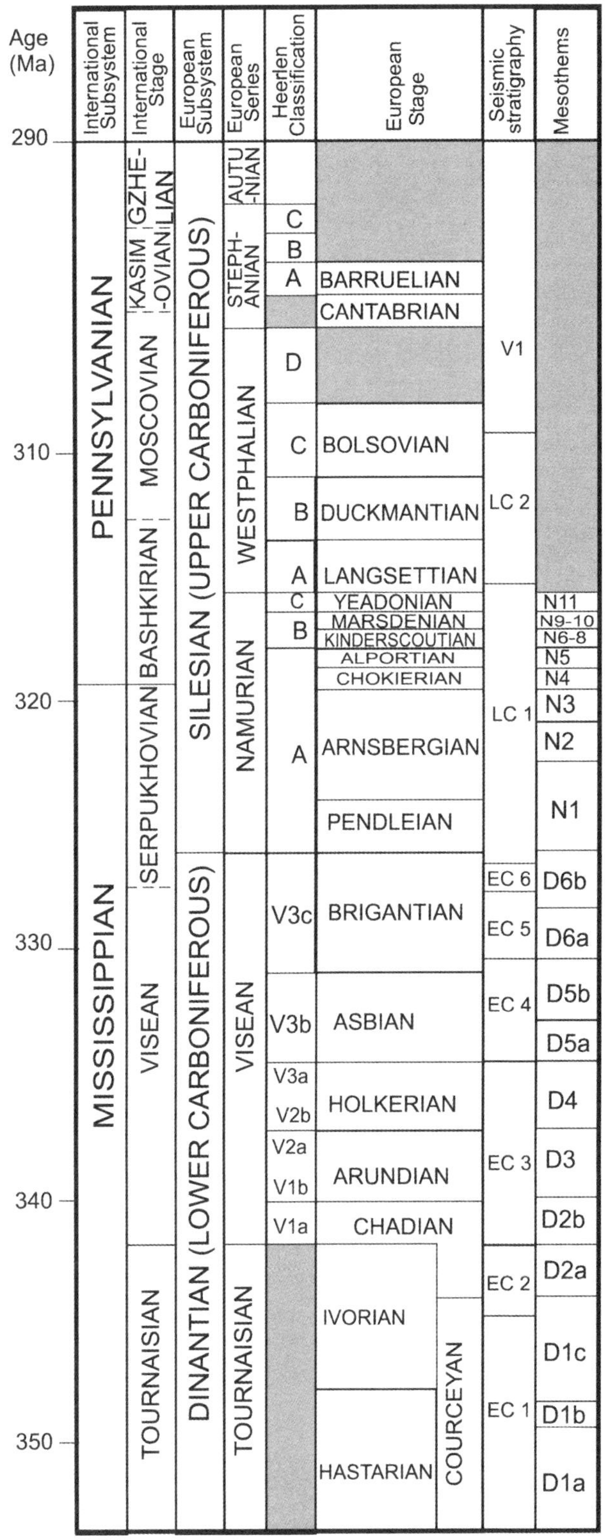

Fig. 9.10. Chronostratigraphical framework for the Carboniferous of England and Wales. Ages derived from Menning *et al.* (2000). Seismic sequences from Fraser *et al.* (1990). Mesothems from Ramsbottom (1973, 1977). International nomenclature that of Heckel (2004).

Palynomorphs (miospores) are present in both marine and terrestrial environments and have been used for biozonation up to and including the Westphalian D. Palynomorphs have proved important zonal indicators in the Northumberland Trough, where ammonoid faunal bands are rare. Palynology has been particularly useful for establishing the stratigraphy

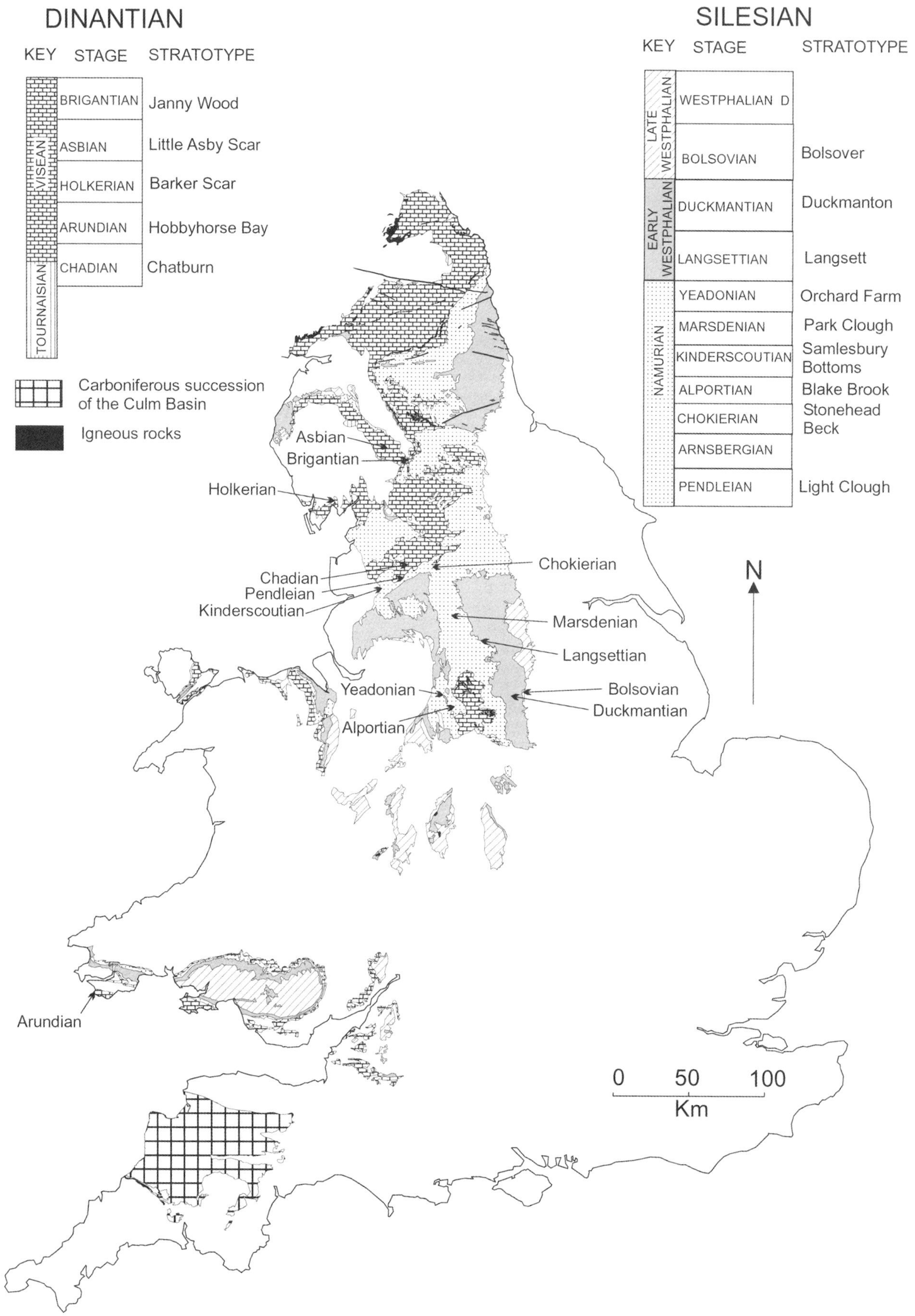

Fig. 9.11. Geological map showing the distribution of Carboniferous strata, subdivided into the main chronostratigraphical units. The map also shows the location of the stage stratotypes.

for the offshore Carboniferous where macrofossils are more difficult to obtain. Macrofloras are becoming increasingly important for the biozonation of Westphalian D and Stephanian strata (Cleal 1991).

In the Dinantian and Namurian, marine bivalves present within hemi-pelagic shales in association with ammonoids are of some stratigraphical importance (Riley 1993). In the Westphalian, the non-marine bivalves assume great importance (Trueman & Weir 1946). They tend to occur in association with fish and ostracodes. Estheriids ('*Estheria*'), small crustaceans that occupied brackish waters, can occur as prominent marker bands.

Sequence stratigraphy

The Carboniferous successions of England and Wales have been a fertile testing ground for the application of sequence stratigraphic methodologies in outcrop and subsurface studies largely for Silesian successions (Read 1991; Maynard 1992; Martinsen 1993; Church & Gawthorpe 1994; Hampson 1995; Hampson *et al.* 1997; Brettle 2001). Ramsbottom (1973, 1977) anticipated the advent of sequence stratigraphy in the 1990s (Van Wagoner *et al.* 1990). He recognized larger scale cycles in Carboniferous sequences ('mesothems') and defined the boundaries of each of these major cycles as widespread disconformities in shelf areas (Fig. 9.10). The mesothem model was criticized (Holdsworth & Collinson 1988; Martinsen 1990), and Namurian stage boundaries were never adjusted to coincide with mesothem boundaries, as proposed by Ramsbottom (1977) and Ramsbottom *et al.* (1978). Marine bands remain the most reliable means of local and regional correlation for Carboniferous successions and many individual marine bands can be traced within and between basins, providing a high-resolution biostratigraphical framework (Ramsbottom *et al.* 1978; Riley *et al.* 1995). However, sequence stratigraphic methodology enhances these established correlation methods and enables the important re-interpretation of critical elements of the sedimentary strata.

Sequence stratigraphic methodology uses a range of key surfaces, not just the marine bands, to examine facies relationships. These key surfaces include regionally widespread surfaces of erosion and emergence, referred to as sequence boundaries (Posmentier & Vail 1988; Van Wagoner *et al.* 1988) and a hierarchy of transgressive surfaces that are considered to reflect changes in relative sea level.

The presence of surfaces of erosion within the strata between marine bands has implications for timing and linkage of depositional environments. The strata between marine bands are referred to as cycles or cyclothems and have been considered to represent genetically-related facies. The facies-pattern variability that dominates the Yoredale, Millstone Grit and Coal Measures groups (see the section on 'Lithostratigraphical framework for the Carboniferous') has been interpreted in terms of differing positions within the depositional systems. In this traditional model, 5–30 m thick, late Viséan–early Namurian cyclothems comprising marine carbonates and non-marine clastic sediments were attributed to autocyclic channel avulsion and lobe switching (Leeder & Strudwick 1987). Erosively based sandstone units within cyclothems have traditionally been interpreted as the deposits of fluvial systems that were an integral part of the depositional system, e.g. distributary channels on delta plains (Elliott 1975). An alternative interpretation is that some thicker sandstones with substantial basal erosional relief could be interpreted as valleys incised during periods of sea-level fall and filled during the early rise.

Thick, sandstone-dominated multistorey fluvial systems are a common feature of Silesian strata and a number of characteristics can be used distinguish those that are valley fills from others of distributary channel origin. The four main characteristics of valley fill sandstones are defined by Hampson *et al.* (1997). (1) The basal surface to the valley fill is a regionally extensive high relief erosion surface that can be correlated laterally to an interfluve surface; the erosion surface should be more extensive than those surfaces associated with individual channels. (2) The facies associations above the erosional unconformity will differ greatly from the underlying associations. (3) The erosional unconformity removes underlying strata, sometimes at a scale that produces an identifiable time gap recognized by the removal of marine bands. (4) Valley fills have a distinctive internal architecture, they are commonly multistorey and record a trend of increasing accommodation evident from, for example, increasing preservation of channel members, that reflects rising sea level during the valley fill phase and which culminates in the initial flooding surface.

The recognition of interfluves is a critical test for the interpretation of erosional unconformities as sequence boundaries. Interfluves generally have well-developed, mature profiles, reflecting a longevity of exposure, that is equivalent to the period of valley incision and fill, and the palaeosol is overlain directly by a flooding surface. Relatively few good interfluve examples have been reported so far (Davies & Elliott 1996; Hampson 1998), possibly because many palaeosols may have been overprinted during the rising relative sea level associated with the transgressive surface.

The identification of erosional unconformities and interfluves within Carboniferous cyclothems indicates that, within a cyclothem, it cannot be assumed that sediments were genetically related and, in fact, the unconformity-bound sequences should be identified as the units of linked strata. The erosion and palaeosol surfaces (sequence boundaries) representing periods of sea-level fall are equally as important as the ammonoid bearing marine bands (maximum flooding surfaces) representing rising sea level. Major palaeogeographical re-organizations generally occurred during the development of the sequence boundaries and the horizons are also associated with potentially significant stratal omission.

The main implication of using sequence stratigraphy alongside more established correlation techniques, is that predictive models can be developed for the timing of depositional facies. Specific stratigraphical intervals containing major sandstone reservoir facies can be predicted by extending a high-resolution sequence stratigraphic framework developed in one area into basins containing time-equivalent strata.

Geochronology

Radiometric ages for the Carboniferous have been principally derived from thin tuffs, tonsteins and bentonites, which can provide ages of eruption in sedimentary successions with good biostratigraphical control. The main sources of internationally recognized ages are U–Pb zircon and Ar/Ar sanidine, although inconsistencies are recognized between sources and laboratory techniques (Menning *et al.* 2000). Much of the published data comes from Central and Western Europe, although as chronostratigraphical correlations with England and Wales are well established, these ages are applicable in the UK. A significant SHRIMP ion microprobe ^{206}Pb–^{238}U zircon date of 314.5 ± 4.6 Ma has been determined for biostratigraphically well-constrained Arnsbergian bentonites from the Pennine Basin (Riley *et al.* 1995). This date is some 5 Ma younger than an Ar/Ar plateau date for Arnsbergian strata in central Europe (Menning *et al.* 2000). Because of questions raised on the accuracy of SHRIMP dating techniques, the age scale presented on Fig. 9.10 is based upon Ar/Ar dates (Menning *et al.* 2000).

In the UK, much reliance has been placed on K–Ar dates determined from intrusive dolerites (Fitch *et al.* 1970). However, many of these dates have been affected by subsequent argon mobility and consequently are of little significance in calibration of the Carboniferous Period and its component series and stages. For example, the Kelso Lavas and

Birrenswark Volcanic formations, which are undoubtedly of Tournaisian age, have been dated by this method as late Devonian, at 360 Ma (de Souza 1975) and 361 ± 7 Ma (de Souza 1982), respectively. However, in SE Wales, a biotite K–Ar date of 336 ± 7 Ma (Fitch *et al.* 1969) from the Golden Hill diatreme pipe is in agreement with Viséan age fossils found in a block of limestone within the pipe (Bevins 2003). In Derbyshire, the Lower Miller's Dale Lava, of Asbian age, has a much younger whole-rock K–Ar age of 315 ± 12 Ma (Fitch *et al.* 1970), interpreted by these authors as a close approximation to the age of hydrothermal alteration, rather than an extrusion date. Derbyshire intrusions have typically yielded whole-rock K–Ar ages of 287 ± 13 Ma (Fitch *et al.* 1970) to an average of 311 ± 6 Ma (Stevenson *et al.* 1970), suggesting an age range of Bolsovian–Autunian. However, in Derbyshire no dolerite intrusions are found in strata of Namurian or Westphalian age (Fig. 9.7), and geochemical similarities between intrusions and the Dinantian lavas suggest the two are genetically linked. This suggests that the dates have been reset to a younger age by argon loss.

A whole-rock K–Ar date of 308 ± 10 Ma has been determined on a dolerite from Barrow Hill, near Dudley (Fitch *et al.* 1970). Extrusive volcaniclastic deposits associated with the vent are interbedded with sedimentary strata of Bolsovian age. This suggests that the radiometric age probably represents a close minimum for the true intrusive age. Fitch *et al.* (1970) also provided K–Ar radiometric dates for the sills intruded nearby within Bolsovian strata in the West Midlands, giving apparent minimum ages for intrusion of 295 ± 5 (Stephanian) to 265 ± 5 Ma (Permian). However, intrusions with this geochemistry are not found in strata younger than Bolsovian age. K–Ar dates for the Great Whin Sill (Plate 19) and the Little Whin Sill, both of which have undergone post-crystallization metasomatism, have been determined as 295 ± 19 Ma (Fitch & Miller 1964).

Dinantian: platform carbonates and clastic/carbonate filled basins

Introduction

The Pennine Basin includes the depositional area between the Southern Uplands and Wales–Brabant High (Fig. 9.2). During the Dinantian, the Pennine Basin was dominated by fault-controlled troughs and highs ('block and basin' topography). Significant troughs include the Northumberland Trough-Solway Basin, Stainmore Trough and Craven Basin. The Craven Basin is used as a generic name for a series of linked embayments, including the Bowland Trough, Harrogate Basin, Edale Gulf, Gainsborough Trough and Widmerpool Gulf and intervening structural highs (Fig. 9.2). In central and northern England these basins contain relatively continuous Dinantian successions ranging from 1.5 to 3 km in thickness (Fraser *et al.* 1990; Leeder 1992). To the south of the Wales–Brabant High, shelf/ramp carbonates accumulated in South Wales and in the Bristol and Mendips areas, in turn passing southwards into the deep-water marine shales and turbidites of the foreland Culm Basin of Cornwall and Devon.

The arid continental environment established during the Devonian continued into the early Tournaisian. Sedimentation was dominated by alluvial fan accumulations along basin margins. Also, there was localized development of fluvial and playa-lake deposits within small internally drained basins, with thick evaporites locally formed (e.g. flanks of the Derbyshire High).

A late Tournaisian–early Chadian marine transgression resulted in the deposition of peritidal marine sabkha deposits within those basins open to marine influence. The reduced topography of the block areas resulted in a diminished influx of finer-grained alluvial sediments. Isolated marine carbonate platforms became established as highs were submerged. Platform carbonates also accumulated on both the northern flank (North Wales and East Midlands) and southern flank (South Wales–Bristol and the Mendips areas) of the Wales–Brabant High.

The early Dinantian basins were gulf-like, initially hypersaline seas, apparently tideless, but with variable influx from freshwater river systems. Thick, open marine platform carbonates thinned onto the footwalls and up the hanging-wall dip slopes (Figs 9.2a & 9.3). Evidence of intermittent subaerial exposure and minor cyclicity occurs within these carbonate successions.

During the Tournaisian, sedimentation in the Northumberland Trough evolved from predominantly alluvial deposition to finer-grained deltaic deposition with subordinate marine influence. An early Tournaisian axial drainage pattern developed, with lateral alluvial fans on the southern margin of the Southern Uplands and in the Vale of Eden (Fig. 9.9a). By late Tournaisian–early Chadian times, mudstone-dominated ferroan dolostone beds ('cementstones') were deposited on alluvial plains and marginal marine flats subject to periodic desiccation and fluctuating salinity. These deposits may also have extended northwards into the Midland Valley of Scotland. In the central Northumberland Trough a deltaic system prograded west at this time, whilst periodic marine incursions flooded eastwards. During the late Tournaisian the diminishing relief of the Southern Uplands High discharged smaller volumes of sediment to the alluvial fans and fluvial systems on the northern margin of the Northumberland Trough/Solway Basin. Further to the south, in the Stainmore Trough, peritidal carbonates, comparable to the cementstones, were deposited and were interbedded with alluvial fan conglomerates.

To the north of the Wales–Brabant High, alluvial deposits accumulated in small, possibly internally drained basins. Local development of anhydrites probably reflects deposition in a sabkha environment. Late Tournaisian–early Chadian marine transgression is evident in North Wales, where red clastic beds are overlain by dolomitized carbonates deposited in a near-shore high-energy environment. Possible Waulsortian mudmounds were present offshore to the north (Somerville *et al.* 1989). In South Wales the Tournaisian is characterized by a fully marine shelf succession.

In England and North Wales, a period of sustained marine transgression marked the Arundian (Fig. 9.9b). The Askrigg block was submerged and about 150 m thickness of Arundian–Holkerian platform carbonates were deposited. Water depths in the Stainmore Trough increased (Higgins & Varker 1982). In the Craven Basin Waulsortian mud-mounds, which had developed during the Late Chadian, were replaced in Arundian times by limestone turbidite deposition. Low relief carbonate ramps dominated late Devonian–Holkerian deposition in South Wales and Bristol regions (Wright 1987). Evaporitic deposition ceased on the flanks of the Derbyshire High and marine conditions were established. In North Wales Arundian limestones overlapped and overstepped earlier Carboniferous strata.

From the early Asbian, carbonate platforms throughout England and Wales show a consistent evolution from gently sloping ramps into steep-margined, flat-topped shelves (Walkden 1987). This is interpreted by Wright & Vanstone (2001) as reflecting a global climate change from relatively stable sea levels to a period of high frequency sea-level oscillation driven by the development of a major ice-sheet in the southern hemisphere (Fig. 9.6). Major Brigantian-aged deltas prograded southwards but were periodically drowned by marine incursions, producing the cyclic *Mixed shelf carbonate and deltaic ('Yoredale') lithofacies association.* In northern England this lithofacies association extends across the region and into the North Sea (Fig. 9.9c).

Basinal sedimentation, characterized by hemi-pelagic shales and carbonate turbidites, extended from the Dublin Basin, across the Bowland Trough to the Gainsborough and Widmerpool gulfs. To the south of this basinal area, extensive platform carbonate deposition continued over the East Midlands Shelf, Derbyshire High and North Wales. The Wales–Brabant High was a subdued feature at this time.

The main control on Dinantian volcanicity throughout England and Wales was the N–S lithospheric stretching and thinning associated with the formation of faulted blocks and basins (Leeder 1982). Much of the activity occurred along lines of pre-existing basement lineaments which commonly bound the main blocks and basins (Francis 1970). The main centre of igneous activity at this time was in Derbyshire, with minor volcanism in the Bristol/Gloucester and Wenlock areas and in the East Midlands (Fig. 9.7).

Chronostratigraphy

The Dinantian, introduced by Munier-Chalmas & de Lapparent (1893), has long been subdivided into the Tournaisian and Viséan series in Europe. However, George *et al.* (1976) were not prepared to use these terms in their review of British chronostratigraphy, indicating inconsistencies in the definition of these series. Instead, they defined six regional stages for the Dinantian (Fig. 9.10), based upon stratotype stage boundaries. The base of each stage is taken at a lithological change below the entry of a diagnostic faunal group.

The Courceyan Stage, defined by George *et al.* (1976), broadly corresponds biostratigraphically with the Tournaisian Series (Ramsbottom & Mitchell 1980). The Hastarian and Ivorian stages of Conil *et al.* (1977), defined in Belgium, have succeeded the Courceyan Stage on the continent and have been applied increasingly in Britain. However, definition of these stages remain problematical and the zonal boundaries used to define the base of the Ivorian Stage are poorly represented in Britain, so the Courceyan has been retained here. The remaining five stages, all with stratotypes defined in England and Wales, are components of the Viséan Series. They are in ascending order: Chadian, Arundian, Holkerian, Asbian and Brigantian stages (George *et al.* 1976). They have no formal international recognition and several need to be redefined (Riley 1993). Each of the stratotype sections is described in detail by Cossey *et al.* (2004).

The basal Chadian Stratotype at Chatburn, near Clitheroe, Lancashire was incorrectly located by George *et al.* (1976) within the Chatburn Limestone Formation, of Tournaisian age. Riley (1993) has indicated that the true entry of the index fossil is some 300 m higher in the succession, within the Hodder Mudstone Formation. He has proposed the terms early and late Chadian, to correspond to the Tournaisian and Viséan components of the stage, respectively. This distinction between a chronostratigraphical and biostratigraphical base for the Chadian Stage has resulted in considerable confusion, which could be avoided by redefining the Chadian stratotype to concur with the base of the Viséan Series.

The Arundian stratotype is at Hobbyhorse Bay, Dyfed, South Wales (Fig. 9.11), defined at the base of the Pen-y-Holt Limestone above the dolomitized Hobbyhorse Bay Limestone. The Holkerian stratotype is at Barker Scar, south Cumbria, defined close to the base of the Park Limestone Formation. The Asbian stratotype, at Little Asby Scar, Ravenstonedale, Cumbria, is taken at the base of the Potts Beck Limestone Formation. The Brigantian stratotype is at Janny Wood, near Dent, Cumbria, the base of the stage defined by George *et al.* (1976) as coinciding with the base of the Peghorn Limestone. However, it has been proposed that the boundary be moved to a lower level at the base of the Birkdale Limestone (Cózar & Somerville 2004).

Biostratigraphy

The ammonoid faunas of the early Dinantian are best known from the deep marine Culm Basin of Devon and Cornwall, and are poorly developed in northern Britain and Ireland. The ranges of ammonoid zones in Britain are shown in Figure 9.12, for the Dinantian after Riley (1993). The base of the *Gattendorfia subinvoluta* Ammonoid Zone equates with the base of the Tournaisian Series.

Fewtrell *et al.* (1981) reported the stratigraphical distribution of British Dinantian foramanifera and recognized problems with the Chadian and Asbian stratotypes of George *et al.* (1976) and indicated discrepancies with the Belgium zonation scheme (Conil *et al.* 1977). A British zonation scheme has yet to emerge. However, Figure 9.12 provides the ranges of Dinantian foraminifera using the zonal scheme for Belgium and applied to the British Isles by Conil *et al.* (1980). The boundary between Hastarian and Ivorian stages of the Tournaisian corresponds with the boundary between *Chernyshinella* (Cf1) and *Paraendothyra* Biozones (Cf2), the position of which remains uncertain in England and Wales. The base of the Chadian Stage was defined by the first change in lithology below the entry of *Eoparastaffella* (George *et al.* 1976), also used to define the base of the international Viséan Series. The base of the Arundian Stage is defined by the first change in lithology below the entry of the family *Archaediscidae* (George *et al.* 1976).

The Dinantian conodont biozones for the Britsh Isles are shown in Figure 9.12, modified from Varker & Sevastopulo (1985). The base of the Tournaisian Series, and hence the Carboniferous System, was redefined by Conil *et al.* (1977) to occur at the base of the *Siphonodella sulcata* Conodont Zone. The boundary between Hastarian and Ivorian stages of the Tournaisian is recognized as corresponding with the boundary between the *Siphonodella* and *Dollymae hassi* Biozones. Important regional studies of conodonts include Armstrong & Purnell (1987) for the Northumberland Trough, Metcalfe (1981) for Dinantian–early Namurian strata of the Craven Basin and Riley in Chisholm *et al.* (1988) for the Tournaisian–Viséan boundary in the Peak District.

A complete miospore zonation for the Dinantian of Britain was proposed by Neves *et al.* (1972, 1973) and developed by Clayton *et al.* (1978). This zonal scheme is summarized in Figure 9.12. The base of the *Vallatisporites vallatus–Retusotriletes incohatus* (VI) Miospore Zone equates with the base of the Tournaisian.

The pioneering work of Vaughan (1905) defined coral assemblage zones for the Bristol area. Garwood (1913) initiated a comparable zonation for northern Britain, defined by faunal marker bands. The index names used for these zones, shown in Figure 9.12, are still widely referred to in modern literature. Courceyan zones were revised as assemblage biozones by Ramsbottom & Mitchell (1980), Mitchell (1981) and Sevastopulo & Nudds (1987). The distribution of Viséan rugose corals and heterocorals are summarized by Mitchell (1989) and Sutherland & Mitchell (1980), respectively. The base of the Holkerian Stage is defined as the lithological change that occurs below the incoming of diagnostic corals *Carcinophyllum vaughani* and *Lithostrotion minus* and brachiopods *Davidsonina carbonaria*, *Composita ficoides* and *Linoprotonia corrugatohemispherica* (George *et al.* 1976) and broadly corresponds with the S2 Zone. The base of the Asbian Stage was defined by George *et al.* (1976) as the lithological change which occurs below the incoming of diagnostic corals *Dibunophyllum bourtonense*, *Siphonodendron (Lithostrotion) pauciradiale*, *S. junceum*, *Palaeosmilia murchisoni* and brachiopods *Linoprotonia hemisphaerica*, *Daviesiella llangollensis*. The base of the Brigantian Stage occurs at the lithological change below the incoming of diagnostic coral fauna *Diphyphyllum lateseptatum*, *Actinocyathus (Lonsdaleia) floriformis*,

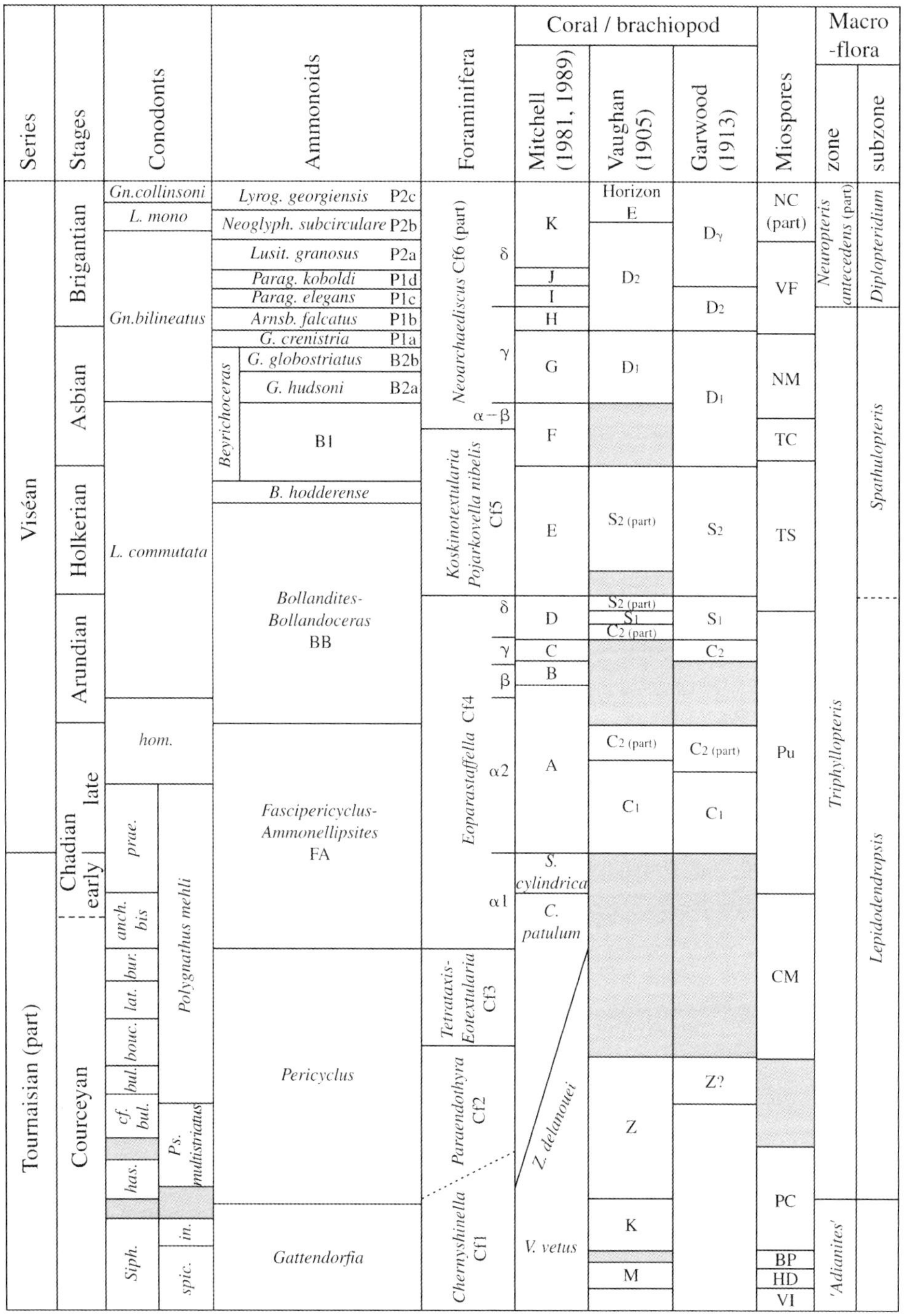

Fig. 9.12. Dinantian biostratigraphal zonations, (derived from Riley 1993, with permission from the Geological Society, London 2004). Stipple ornament shows interzones (conodonts and miospores) or non-sequences (corals and brachiopods). anch. bis., *Scaliognathus anchoralis–Polygnathus bischoffi*; Arnsb., *Arnsbergites*; B., *Bollandoceras*; bouc., *Dollymae bouckaerti*; bul., *Eotaphrus bultyncki*; bur., *Eotaphrus burlingtonensis*; C., *Caninophyllum*; coll., *Gnathodus girtyi collinsoni*; has., *Dollymae hassi*; hom., *Gnathodus homopunctatus*; in., *Polygnathus inornatus/Siphonodella*; L., *Lochriea*; mono., *mononodosa*; lat., *Doliognathus latus*; Lusit., *Lusitanoceras*; Lyrog., *Lyrogoniatites*; Neoglyph., *Neoglyphioceras*; Parag., *Paraglyphioceras*; prae., *Mestognathus praebeckmanni*; S., *Siphonophyllia*; Siph., *Siphonodella*; spic., *Polygnathus spicatus*; V., *Vaughania*; Z., *Zaphrentis*.

Nemistium edmondsi, *Orionastraea* spp., *Palastraea regia* and brachiopods *Productus hispidus*, *Productus productus* and *Pugilis pugilis*.

Sequence stratigraphy

The Dinantian strata of England and Wales display common patterns that suggest an allocyclic control, for example global sea-level change, on deposition (Fig. 9.6). A Tournaisian marine transgression resulted in widespread flooding across the 'Old Red Sandstone' terrestrial environment and subsequent development of ramp carbonates in northern and central England and North Wales. Many Asbian–Brigantian shelf carbonates display repeated shallowing upward cycles, which commonly culminate in palaeokarst surfaces formed during emergence (Walkden 1987; Walkden & Walkden 1990). There is a progressive increase in cycle thickness from 1–5 m in the early Asbian to greater than 10 m in the early Brigantian, and a concomitant trend towards deeper water sedimentation at the cycle bases (Walkden 1987).

The types and distribution of cycles present in British Dinantian successions have been reviewed by Walkden (1987) and computer-modelled by Walkden & Walkden (1990). These authors conclude that small-scale, glacio-eustatic sea-level oscillations were the principal cause of late Dinantian cyclicity but that the incremental accumulation of the cyclic sequences was accommodated by regional subsidence. Late Asbian successions reflect relatively small, but relatively frequent sea-level fluctuations in which rapid transgressions and regressions precluded the formation or preservation of peritidal facies (Walkden & Walkden 1990).

Changes in cycle style at the Asbian–Brigantian boundary (Somerville & Strank 1984*a*; Walkden 1987) also suggest that regional controls were operating. The strongly asymmetric form of Brigantian cycles, dominated by thick sequences of deeper water platform facies, indicates that transgressive inundations were larger, more rapid and protracted than those of Asbian age (Walkden 1987). This suggests that eustatic sea-level rises during the Brigantian were enhanced by regional downwarping (Ramsbottom 1981; Walkden 1987; Walkden & Walkden 1990) possibly marking the onset of thermal subsidence, which later dominated Silesian basin development (Leeder 1982).

The subdivision of British Carboniferous successions using event stratigraphy was initiated by Ramsbottom (1973, 1977, 1979). He identified cyclical facies changes (mesothems), which he argued resulted from eustatic transgressive–regressive pulses (Fig. 9.10). The pronounced thickness and facies variations of Dinantian strata, between individual blocks and basins, implies that the local development of accommodation also influenced sediment distribution (George 1958, George 1978). Leeder (1988), Fraser & Gawthorpe (1990) and Fraser *et al.* (1990) have suggested that many of the cyclical and widespread facies changes reflect discrete periods of rifting and subsidence alternating with episodes of tectonic quiescence, although these cannot explain the small scale but reasonably widespread cycles. Davies *et al.* (2004) suggest that facies changes correlated within separate Dinantian block successions may, in part at least, reflect parallel evolutionary trends operating largely independently of small-scale eustatic and tectonic events.

A seismostratigraphical classification for the Dinantian of the East Midlands (Ebdon *et al.* 1990) and entire Carboniferous of northern England (Fraser *et al.* 1990) recognized six Dinantian (broadly Late Devonian–late Brigantian) sequences, separated by seismic sequence boundaries (Fig. 9.10). These sequences were interpreted as synrift deposits, with eustatic sea-level changes considered insignificant. The resolution is coarse, in comparison with the existing chronostratigraphical subdivision based upon biostratigraphy, as argued by Riley (1993), but is valid for seismic interpretation and demonstrates the importance of a structural control.

Regional developments

Northumberland–Solway Basin

Inverclyde Group

Carboniferous deposition commenced during the Tournaisian with alluvial, lacustrine and evaporite sediments (continental and peritidal lithofacies association) of the Inverclyde Group, deposited along parts of the northern margin of the Northumberland Trough–Solway Basin. The Inverclyde Group was defined in the Midland Valley of Scotland, but is known to extend into the northern part of the Northumberland Trough and Solway Basin (Fig. 9.8) and the term is used to replace former names, such as the Berwick Cementstone Group (Waters *et al.* 2006*a*). The group is up to 900 m thick in the Tweed Valley and up to 640 m in the Solway Firth area, and typically comprises a lower Kinnesswood Formation and an upper Ballagan Formation, locally separated by volcanic rocks (Figs 9.13 & 9.14).

Initially, during the early Tournaisian, alluvial fan deposits of the Kinnesswood Formation were deposited in a series of small, linked basins with internal drainage that formed during the early stages of crustal extension (Chadwick *et al.* 1995). The formation comprises red sandstones, siltstones and conglomerates with calcretes, resting unconformably upon Silurian strata.

A phase of volcanic activity occurred during the initial phase of extensional faulting in Tournaisian times. The Birrenswark and Kelso Volcanic formations locally overlie the Kinnesswood Formation in the northern part of the Solway Basin (Lumsden *et al.* 1967) and Tweed Basin (Greig 1988), respectively (Fig. 9.14). The Cottonshope basalts, located on the SW flank of the Cheviot Block, and Cockermouth Lavas, present along the southern margin of the Solway Basin also belong to this episode. These volcanic formations comprise typically tholeiitic olivine-basalts and subordinate tuffs and sedimentary strata.

Subsequent deposition was characterized by an influx of alluvial fans, fluvial and fluvio-deltaic sediments derived from the Southern Uplands, intercalated with lacustrine and arid coastal plain deposits (Deegan 1973; Leeder 1974). This is the Tournaisian–Chadian Ballagan Formation, comprising interbedded sandstone, shale, limestone and anhydrite. In the central Northumberland Trough a lateral facies change sees the Ballagan Formation pass into the increasingly marine-influenced Lyne Formation of the Border Group (Fig. 9.14).

Border Group

The early Viséan Border Group is limited to the Northumberland Trough–Solway Basin and comprises components of the Heterolithic clastic and non-marine carbonate lithofacies and Fluvio-deltaic ('Millstone Grit') lithofacies associations. This group is a redefinition of the Lower, Middle and Upper Border groups of Day *et al.* (1970), from the Bewcastle area of the Northumberland Trough (Waters *et al.* 2006*a*). The component formations are the Lyne Formation, overlain diachronously by the Fell Sandstone Formation (Fig. 9.13).

Deposition of the Border Group commenced within the axial part of the Northumberland Trough with peritidal, deltaic and fluvial deposits of the Lyne Formation. The late Tournaisian–Chadian formation comprises cyclical sequences of fine-grained subarkosic sandstone, siltstone, mudstone and thin limestone. The peritidal limestones contain common thin oolitic pellet beds, stromatolites and vermetid gastropod bioherms and biostromes, and in the Solway Basin quasi-marine shelly faunas contain abundant brachiopods *Antiquatonia (Dictyoclostus) teres*. The first marine limestones developed later towards the NE of the basin. The Lyne Formation sandstones were deposited from lobate deltas that periodically migrated along the basin axis from NE to SW (Leeder 1974). A second clastic source, represented by the Whita Sandstone, is present in the northern part of the basin (Fig. 9.14) derived from a local northerly source in the Southern Uplands High. The Lyne Formation is at least 890 m thick, the base not proved, in the central part of the trough.

The Fell Sandstone Formation comprises fluvial deposits (Monro 1986, Turner *et al.* 1993), which accumulated initially in the east of the basin (Chadian–Holkerian) and which pass westward diachronously into a succession of fluvio-deltaic and shallow-marine deposits (Arundian–Holkerian). In the NE of the Northumberland Trough, the formation comprises between 130 and 300 m of fine- to medium-grained, mica-poor, subarkosic sandstone with sparse interbeds of red mudstone. Seatearths are common, but only associated with thin coals in the upper part of the formation. There is an unconformity at the base of the formation, with the Lyne Formation absent. Here, the formation is interpreted as deposits of several braided rivers,

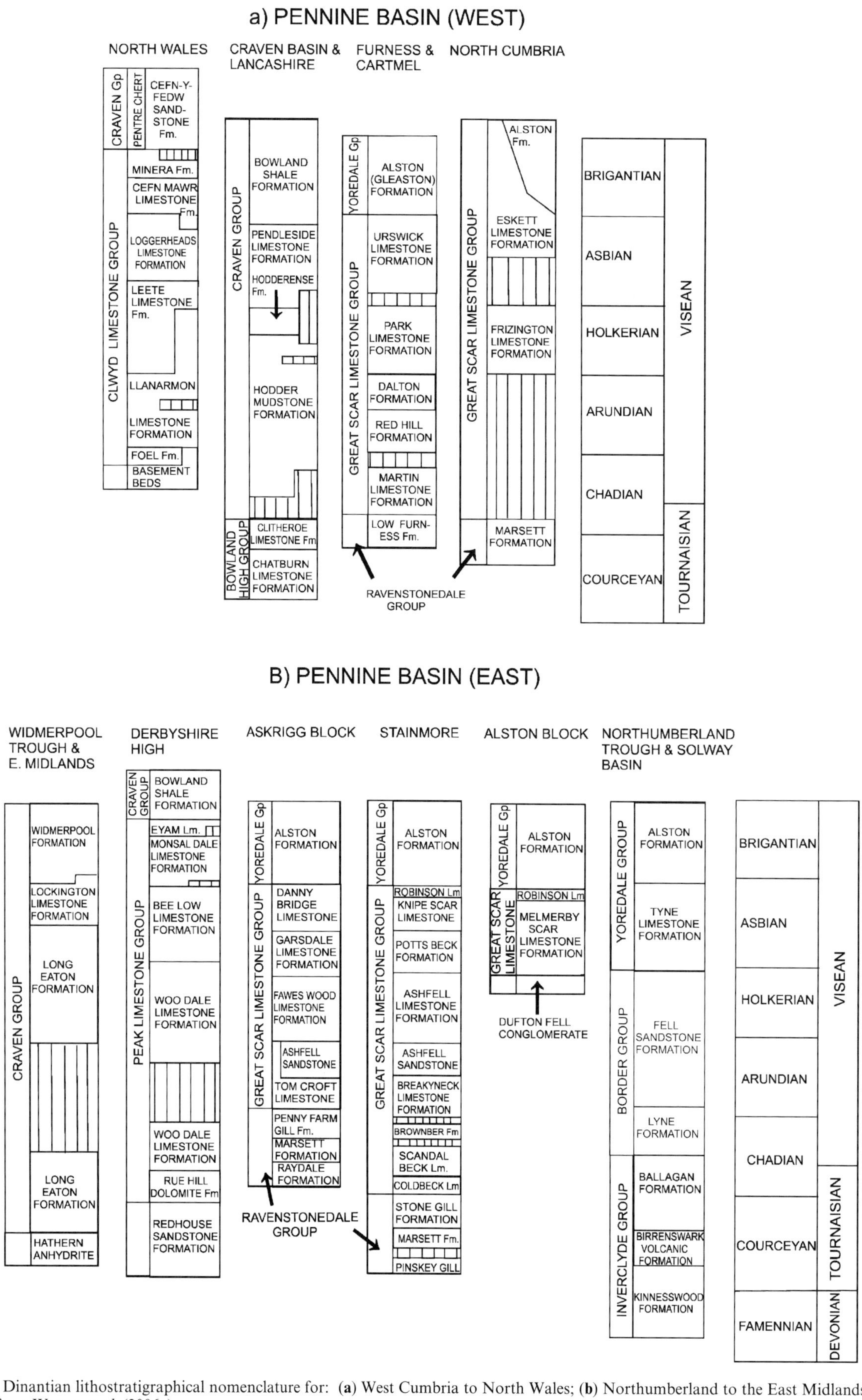

Fig. 9.13. Dinantian lithostratigraphical nomenclature for: (**a**) West Cumbria to North Wales; (**b**) Northumberland to the East Midlands. Adapted from Waters *et al.* (2006*a*).

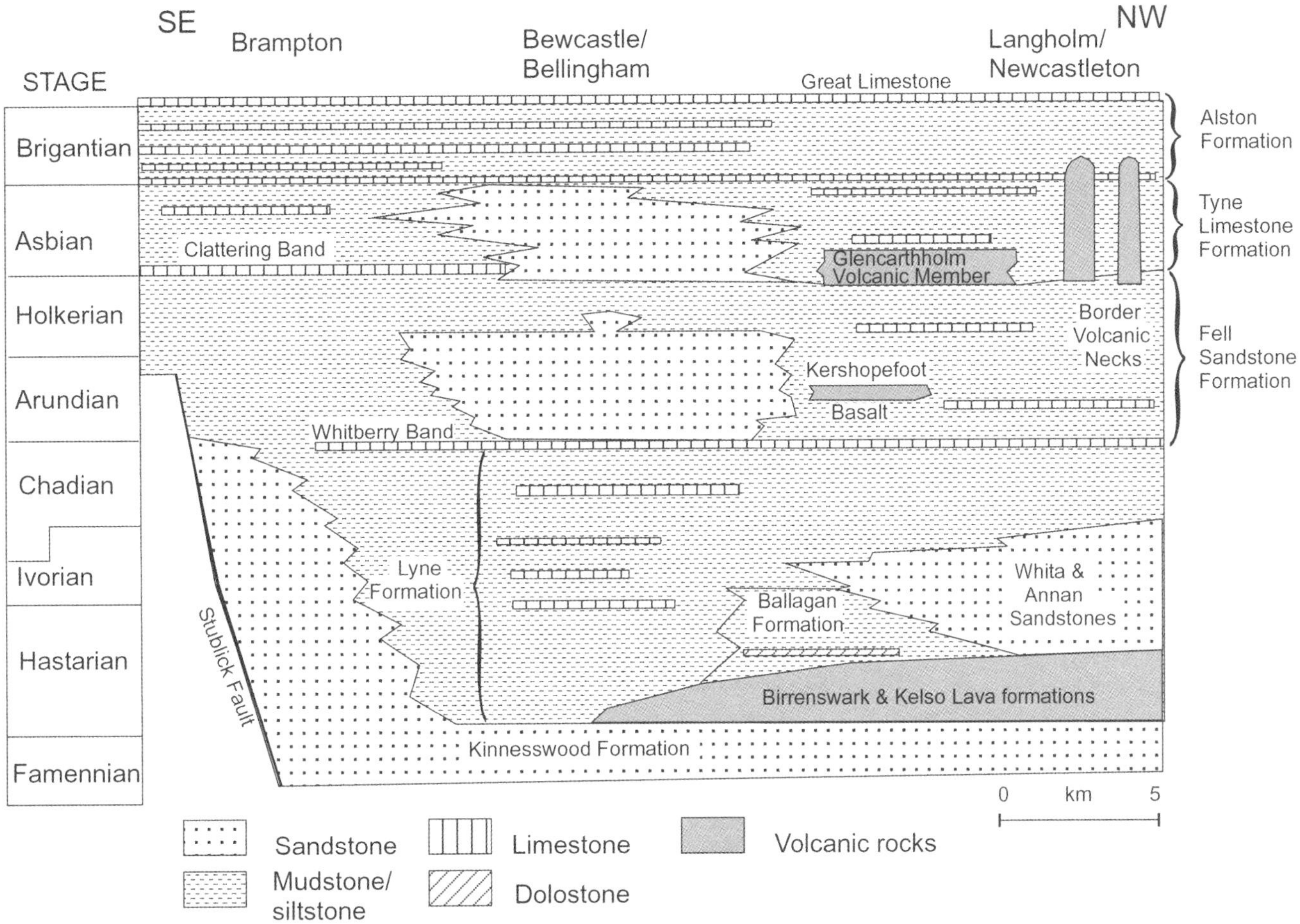

Fig. 9.14. Correlation of Dinantian strata within the Northumberland Trough–Solway Basin. A schematic NW–SE section, adapted from Gawthorpe *et al.* (1989).

occupying belts several kilometres wide, locally confined by intrabasinal syndepositional normal faults along the axial part of (Turner *et al.* 1993). Toward the central part of the trough the formation passes into a 300–450 m-thick succession of cyclical mixed fluvio-deltaic and shallow-marine sediments. This succession is similar to the underlying Lyne Formation, although the sandstones are thicker and limestones thinner in the Fell Sandstone Formation (Day *et al.* 1970). Within the central part of the trough, the base of the Fell Sandstone Formation is conformable and defined at the base of the Whitberry Band (Fig. 9.14).

A phase of volcanic activity within the Fell Sandstone Formation, known only from the northern margin of the Solway Basin, is evident as the Kershopefoot Basalt (Fig. 9.14).

Yoredale Group

The Yoredale Group is dominated by Mixed shelf carbonate and deltaic ('Yoredale') lithofacies association deposits and comprises three formations, namely the Tyne Limestone, Alston and Stainmore formations (the latter is Namurian in age). The group is the first lithostratigraphical unit within the Northumberland Trough–Solway Basin to extend beyond the confines of the graben. The group covered northern England during late Viséan–late Namurian times (Fig. 9.3). This distribution reflects a diminution in the topographical effects of the block and basin topography. However, tectonic controls on sedimentation persisted, with marked thickness variations across the Stublick–Ninety Fathom Fault, separating the Northumberland Trough from the Alston Block (Chadwick *et al.* 1995).

During the Asbian a heterogeneous succession, referred to as the Tyne Limestone Formation (Waters *et al.* 2006*a*), includes elements of a *Mixed shelf carbonate and deltaic 'Yoredale' lithofacies association*, which became established for the first time within the central basin. Localized lacustrine and deltaic deposits were deposited in the eastern part of the trough. The formation comprises typically upward-coarsening cycles of basal thin, laterally extensive marine limestone, marine shale often bioturbated, thin sandstone occasionally topped with seatearth and ganister, and an overlying coal. The base of the formation is the Clattering Band (Fig. 9.14), or its correlative Kingbridge Limestone, with distinctive fauna of *Siphonodendron (Lithostrotion) martini, Lithostrotion portlocki* and *Semiplanus*, marking the basal Asbian transgression (Day *et al.* 1970). In the NE Northumberland Trough there is a distinctive facies variation in the lower part of the formation, comprising up to 300 m of lacustrine shales and limestones, deltaic sandstones and coals.

A phase of volcanic activity restricted to the northern margin of the Solway Basin comprises the Glencartholm Volcanic Member (Fig. 9.14). Present at the base of the Tyne Limestone Formation, the succession of interbedded basaltic and trachytic pyroclastic rocks occur along with other volcaniclastic and sedimentary units and are up to 180 m thick (Lumsden *et al.* 1967). Numerous intrusive bodies, commonly referred to as 'volcanic necks', link the outcrops of the Kelso and Birrenswark Volcanic formations, and are probably associated with both the Holkerian–Asbian and earlier Tournaisian volcanic episodes.

The Alston Formation is described in detail for the area to the south of the Northumberland Trough, which includes its

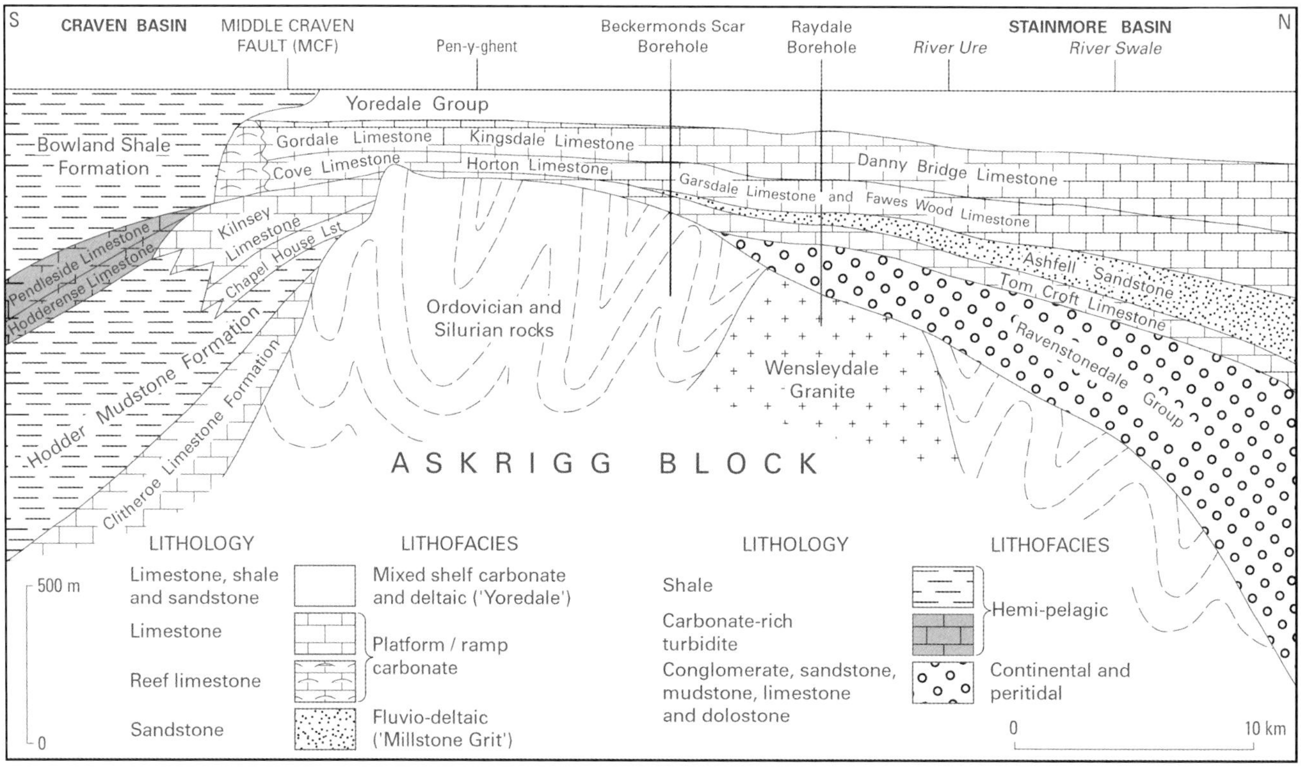

Fig. 9.15. Correlation of Dinantian strata within the northern part of the Craven Basin, Askrigg Block and Stainmore Trough. A schematic N–S section, adapted from Aitkenhead *et al.* (2002).

type area on the Askrigg Block. The base of the formation is taken at marker limestones used to define the base of the Brigantian Stage. These include the Callant Limestone of the Solway Basin, the Low Tipalt Limestone of the central part of the Northumberland Trough and a Spirifer Band a few metres below the Watchlaw Limestone of the NE part of the Northumberland Trough (Cossey *et al.* 2004).

Stainmore Trough, Askrigg–Alston blocks and Lake District High

Ravenstonedale Group

Carboniferous deposition commenced in this region during the Tournaisian with the *Continental and peritidal lithofacies* deposits of the Ravenstonedale Group (Fig. 9.8). The group is thickest, approximately 380 m, within the Stainmore Trough, thinning as the group onlaps onto the Askrigg Block (Fig. 9.15). The base of the group is everywhere marked by an unconformity upon Ordovician–Devonian strata. The onset of sedimentation within the Stainmore Trough is locally represented by a peritidal marine succession of grey, dolomitic limestone and dolostone, interbedded with calcareous mudstone and silty sandstone of the Pinskey Gill Formation. The formation lithologies yield a miospore assemblage broadly characteristic of the CM Zone (Holliday *et al.* 1979; Johnson & Marshall 1971) and a conodont assemblage also suggesting a Tournaisian age (Varker & Higgins 1979). However, the alluvial fan deposits of the Marsett Formation form the basal unit over much of the trough. The overlying Stone Gill Formation comprises grey, porcellanous limestone and thin dolostone, with beds of argillaceous limestone, sandstone and mudstone, representing a return to peritidal deposition.

Elsewhere, the Continental fluvial clastic ('Cornstone') facies is preserved in local basins, whose distribution adjacent to the Pennine/Dent and Lake District High boundary faults suggests that they were preserved in half-grabens formed during the initial, Late Devonian–Early Carboniferous, stages of extension. The Low Furness Formation of south Cumbria (Rose & Dunham 1977) comprises a succession of conglomerate, sandstone and shale with subordinate limestone and gypsum, up to 240 m thick, and contains miospores characteristic of the Tournaisian CM Zone (Holliday *et al.* 1979), comparable in age to the alluvial fan deposits of the Stainmore Trough. However, the Mell Fell Conglomerate Formation (and equivalent Sedbergh Conglomerates and Shap Wells Conglomerates) were deposited on the margins of the Lake District High prior to Tournaisian volcanism and were tilted and eroded before the phase of Tournaisian marine transgression (Waters *et al.* 2006). These conglomerates are attributed to the Devonian Old Red Sandstone Group and are older than the alluvial fan deposits assigned to the Ravenstonedale Group. Tournaisian volcanism included the Cockermouth Lavas Formation of north Cumbria.

Deposition of the Ravenstonedale Group on the Askrigg Block followed an early Viséan sea-level rise. The Chadian–early Holkerian succession (Burgess 1986) was deposited in marginal marine conditions in comparison to the more open marine setting of the Stainmore Trough. Similar depositional environments of peritidal marine and fluvial environment with intervening periods of alluvial fan deposition are present on both the Askrigg Block and Stainmore Trough, although the succession is younger on the block. Initial deposition of hypersaline deposits of the Raydale Dolomite Formation comprises thinly bedded and nodular dolostones interbedded with siltstone and sandstone. The overlying Marsett Formation comprises reddish brown and greenish grey sandstone and conglomerate with rare dolostone beds, deposited within an alluvial environment. The uppermost Penny Farm Gill Formation consists of interbedded limestone, dolostone, sandstone and siltstone, often rhythmically bedded, representing a return to peritidal deposition. Locally, a nodular dolostone bed with rhizoliths, indicative of emergence, marks the top of the Ravenstonedale Group on the Askrigg Block (Burgess 1986).

Great Scar Limestone Group
During the Viséan a combination of marine inundation of block areas and reduced influx of siliciclastic material permitted establishment of a series of thick limestones, forming the Platform carbonate lithofacies association strata of the Great Scar Limestone Group. The group has a widespread distribution extending from south Cumbria, Askrigg Block (the type area), Stainmore Trough, Alston Block and Lake District High (Figs 9.8 & 9.13). The group name was formalized on the Askrigg Block by George *et al.* (1976) and is now used to define the facies development across northern England (Waters *et al.* 2006*a*). It replaces the Orton Group and part of the Alston Group in Stainmore and the Chief Limestone Group in west Cumbria.

Deposition of the Great Scar Limestone Group in the Stainmore Trough commenced during the Tournaisian with the Coldbeck Limestone Formation (Dunham & Wilson 1985). Here, the base of the group, taken at the top of the Algal Nodular Beds, is conformable upon the Ravenstonedale Group. The Great Scar Limestone Group is thickest in the Stainmore Trough, at about 800 m and also has the longest age-range for the group, from Tournaisian to Asbian. The Martin Limestone Formation comprises the earliest deposits in south Cumbria (Rose & Dunham 1977; Johnson *et al.* 2001). Both the Coldbeck and Martin Limestone formations are carbonate-dominated, deposited within a nearshore–peritidal, restricted marine environment with common stromatolites and oncolites.

A basal Arundian sea-level lowstand produced an unconformity at the base of the Breakyneck Limestone Formation, within the Stainmore Trough, the base of the Red Hill Formation in south Cumbria and base of the Frizington Limestone Formation in north and west Cumbria (Fig. 9.13). The subsequent sea-level rise established an open, shallow-marine environment and thick bioclastic limestones were deposited. During the Arundian–Holkerian these carbonates extended across the Stainmore Trough (Breakyneck Limestone and Ashfell Limestone formations), onto the flanks of the Lake District High (Red Hill, Dalton and Park Limestone formations in south Cumbria and Frizington Limestone Formation in north and west Cumbria) and onlapped the Askrigg Block (Tom Croft Limestone and Fawes Wood Limestone formations). The succession comprises typically dark grey bioclastic and highly bioturbated limestone, with crinoid banks, shelly or coral biostromes and algal (*Girvanella*) bands. Within the Stainmore Trough, there are brief incursions of siliciclastic deposits. The Brownber Formation represents alluvial deposits and the Ashfell Sandstone Formation comprises fluvio-deltaic strata,which encroached into the Stainmore Trough from the north, possibly as the distal extension of the Fell Sandstone Formation.

The Great Scar Limestone Group shows an evolution from thick, dark grey Arundian–Holkerian carbonates to pale grey Asbian–Brigantian cyclic limestones with eight major potholed bedding surfaces overlain by thin mudstones (Waltham 1971). The palaeokarst surfaces indicate periodic emergence. This succession is represented on the Askrigg Block by the Garsdale Limestone and Danny Bridge Limestone formations, and in south Cumbria and north and west Cumbria by the Urswick Limestone and Eskett Limestone formations, respectively. From late Holkerian to early Brigantian times, 'Cracoean' buildups of knoll reefs and shelf-edge tracts formed a marginal facies along the southern margin of the Askrigg Block carbonate shelf (Arthurton *et al.* 1988, Mundy 1994). The Alston Block was the final horst to be inundated, and deposition of the Melmerby Scar Limestone Formation both commenced and terminated during the Asbian Stage. Consequently, the group only reaches a maximum thickness of 107 m across the Alston Block (Dunham 1990).

Yoredale Group
Deposition of platform carbonates ceased over most of northern England during the late Viséan in response to the southward encroachment of the Mixed shelf carbonate and deltaic ('Yoredale') lithofacies association of the Yoredale Group. In west Cumbria, platform carbonate deposition persisted into early Namurian (Pendleian) times.

The Yoredale Group extends across northern England and ranges from late Viséan to late Namurian age (Fig. 9.8). The group is named following the long-established usage of the term Yoredale facies, based upon the description of Yoredale cycles as early as Phillips (1836). This term replaces the essentially chronostratigraphical Wensleydale Group of the Askrigg Block, Alston Group of the Alston Block and Lower and Upper Liddesdale groups of the Northumberland Trough (Waters *et al.* 2006*a*).

The Yoredale Group comprises typically upward-coarsening cycles of basal thin, laterally extensive marine limestone, marine shale, thin sandstone frequently topped with seatearth and ganister, and an overlying coal. The limestones are typically dark blue-grey, thin bedded and biomicritic, with a benthonic fauna and rare ammonoids. The sandstones are typically pale grey, fine–medium grained and quartzitic–subarkosic. The cycles, which range from 15 to 90 m thick, are named after the limestone present at the base of the cyclothem. Details of limestone nomenclature and correlations, and cyclothem thicknesses, are provided by Dunham & Wilson (1985) for the Askrigg Block and Stainmore Trough, and by Dunham (1990) for the Alston Block. The progradation of high-constructive lobate deltas deposited the clastic sediments, with extensive shallow-marine reworking following delta abandonment (Elliott 1975). The marine limestones were deposited as a consequence of sea-level rise and switching or abandonment of the delta lobes.

The conformable base of the Yoredale Group is typically drawn at the base of marine limestone marker bands. On the Alston Block the base of the group is taken at the base of the Peghorn (Lower Smiddy) Limestone, defined as the Brigantian Stage Basal Stratotype (George *et al.* 1976). The base of the group on the Askrigg Block and Stainmore Trough occurs at the base of the Lower Hawes Limestone. The greatest thickness for the Dinantian component of the group is 1219 m, and was proved in the eastern part of the Stainmore Trough (Dunham & Wilson 1985). The group, formations and individual sandstones tend to thicken into troughs and half-grabens.

The Brigantian Alston Formation is distinguished from the Namurian components of the group (see Namurian: advancing deltas) by the common presence of thick, bioclastic limestones. In the absence of the *Cravenoceras leion* Marine Band, the base of the Namurian is taken near to the base of the Great Limestone of the Alston Block–Stainmore Trough and equivalent Main Limestone of the Askrigg Block.

Craven Basin

Within the overall basinal environment of the Craven Basin, isolated areas of platform carbonates developed on small footwall highs. Some of these platforms were short-lived, becoming submerged and buried beneath Dinantian basinal shales of the Craven Group in response to both active rifting (Gawthorpe 1987) and periodic eustatic sea-level rises.

Bowland High Group
At least 2300 m of Tournaisian carbonates, including the Chatburn Limestone and Clitheroe Limestone formations, accumulated on the Bowland High. The Chatburn Limestone Formation comprises mostly medium-dark grey, bioclastic and peloidal packstone and wackestone limestone, locally with abundant algal oncolites, with thin subordinate beds of calcareous mudstone (Arthurton *et al.* 1988, Facies associations 1 and 2 of Gawthorpe 1986). The formation was deposited on a

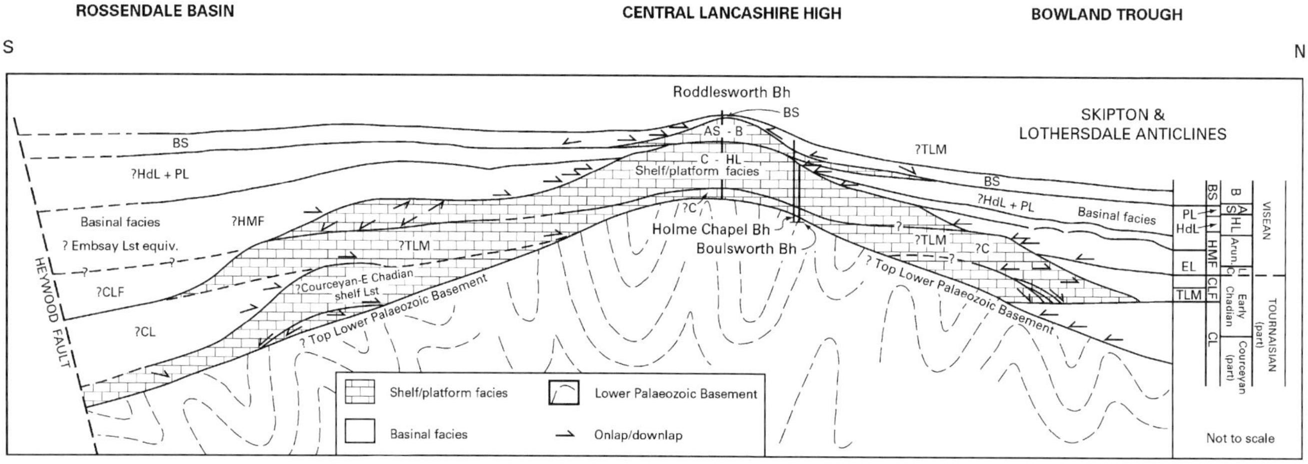

Fig. 9.16. Correlation of Dinantian strata within the western part of the Craven Basin. A N–S composite seismic reflection profile, adapted from Kirby *et al.* (2000). Lithostratigraphy: BS, Bowland Shale Formation; CL, Chatburn Limestone Formation; CLF, Clitheroe Limestone Formation; EL, Embsay Limestone Formation; HdL, Hodderense Limestone Formation; HMF, Hodder Mudstone Formation; PL, Pendleside Limestone Formation; TLM, Thornton Limestone Member. Chronostratigraphy: Arun., Arundian; AS, Asbian; B, Brigantian; C, Courceyan; HL, Holkerian; LC, Late Chadian.

shallow carbonate shelf (Riley 1990) or gently sloping seafloor ranging from 75–100 m depth, with a significant contribution of extra-basinal, fine terrigenous material at greater depths (Gawthorpe 1986). The overlying Clitheroe Limestone Formation comprises pale grey packstone, wackestone and floatstone with subordinate mudstone, with rapid lateral changes in thickness and facies (Facies association 7 of Gawthorpe 1986). Waulsortian limestones (knoll reefs or mud mounds) are a distinctive component of the formation. The Clitheroe Limestone is associated with the development of a carbonate ramp and northward retreat of shelf carbonates to be replaced by Waulsortian limestones (Riley 1990). The buildup of Waulsortian limestones was briefly interrupted by a phase of shallowing and development of storm-generated carbonates.

Seismic and geophysical logs have been used to identify further small accumulations of platform carbonates in the subsurface (Fig. 9.16). Successions of Chadian–Asbian age for the Central Lancashire High and Holme High have been recognized (Evans & Kirby 1999).

Craven Group

During the Viséan, Hemi-pelagic lithofacies association deposits of the Craven Group dominated within the Craven Basin. The Craven Group replaces numerous geographically localized group names for a Dinantian and Namurian succession of this facies, located between the platform carbonates of northern England and the northern flank of the Wales–Brabant High (Waters *et al.* 2006*a*). Sediments accumulated predominantly from suspension in moderately deep water, largely below the storm wave-base. Thin sandstone and limestone beds were introduced, possibly by storms and/or as turbidites.

The lower part of the Craven Group comprises calcareous mudstone and siltstone, interbedded with wackestone limestone and subordinate limestone breccias, conglomerates and sandstones (Gawthorpe 1986; Arthurton *et al.* 1988; Riley 1990). The formations are broadly defined by the relative abundance of limestone and mudstone. The limestone beds (e.g. within the Hodder Mudstone and Pendleside Limestone formations) represent slope carbonate turbidites developed as aprons basinward to the 'Cracoean reef' facies of the Askrigg Block (Fig. 9.15) or adjacent to the Central Lancashire High (Fig. 9.16). The presence of slumps, debris flows and gravity slides in the Craven Group (Gawthorpe & Clemmey 1985; Riley 1990) are evidence for relatively steep slopes and have been considered the product of slope instability induced by late Chadian–early Arundian and late Asbian–early Brigantian tectonic activity (Gawthorpe 1987). The upper part of the Craven Group is the Bowland Shale Formation (Fig. 9.13), of late Brigantian and Namurian age. The formation comprises dark grey and black organic-rich mudstone, with subordinate beds of siltstone, sandstone and dolomitic limestone. The formation shows an increase in siliciclastic sandstone turbidites derived from deltas accumulating on the margins of the basin.

The Craven Group typically rests unconformably upon submerged platform carbonates, including the Clitheroe Limestone Formation of the Craven Basin (Fig. 9.15) and Courceyan–Asbian carbonates upon the Central Lancashire High (Fig. 9.16). Unconformities occur internally within the group at the base of the Pendleside Limestone Formation in Lancashire and Pentre Chert Formation in North Wales. The absence of proven late Chadian and Arundian strata within the Long Eaton Formation of the Widmerpool half-graben (Fig. 9.17) may also indicate the presence of an unconformity. This unconformity may relate to a late Holkerian phase of basin inversion, also seen on the margin of the Wales–Brabant High (Fraser & Gawthorpe 1990).

Derbyshire

During the Tournaisian, strata of the Continental and peritidal lithofacies association were deposited adjacent to the Derbyshire High (Figs 9.9a, 9.13 and 9.17). Alluvial fan gravels, the Redhouse Sandstone Formation, have been proved in the subsurface (Aitkenhead & Chisholm 1982) and comprise at least 170 m of reddish brown pebbly sandstone with beds siltstone and mudstone and nodular dolomite recording calcrete development. Evaporites proved in the subsurface at Eyam (the Middleton Dale Anhydrite Formation) also belong to this phase.

Peak Limestone Group

A late Tournaisian marine transgression established the platform carbonates of the Peak Limestone Group, which generally rest unconformably upon deformed Lower Palaeozoic strata, or conformably upon strata of the Continental and peritidal lithofacies association. Internally, the group includes several local disconformities (Fig. 9.17). Subaerial palaeokarstic dissolution surfaces are also common, as in North Wales, in the Asbian and Brigantian limestones (Walkden 1974). The lithostratigraphical nomenclature for the Peak District

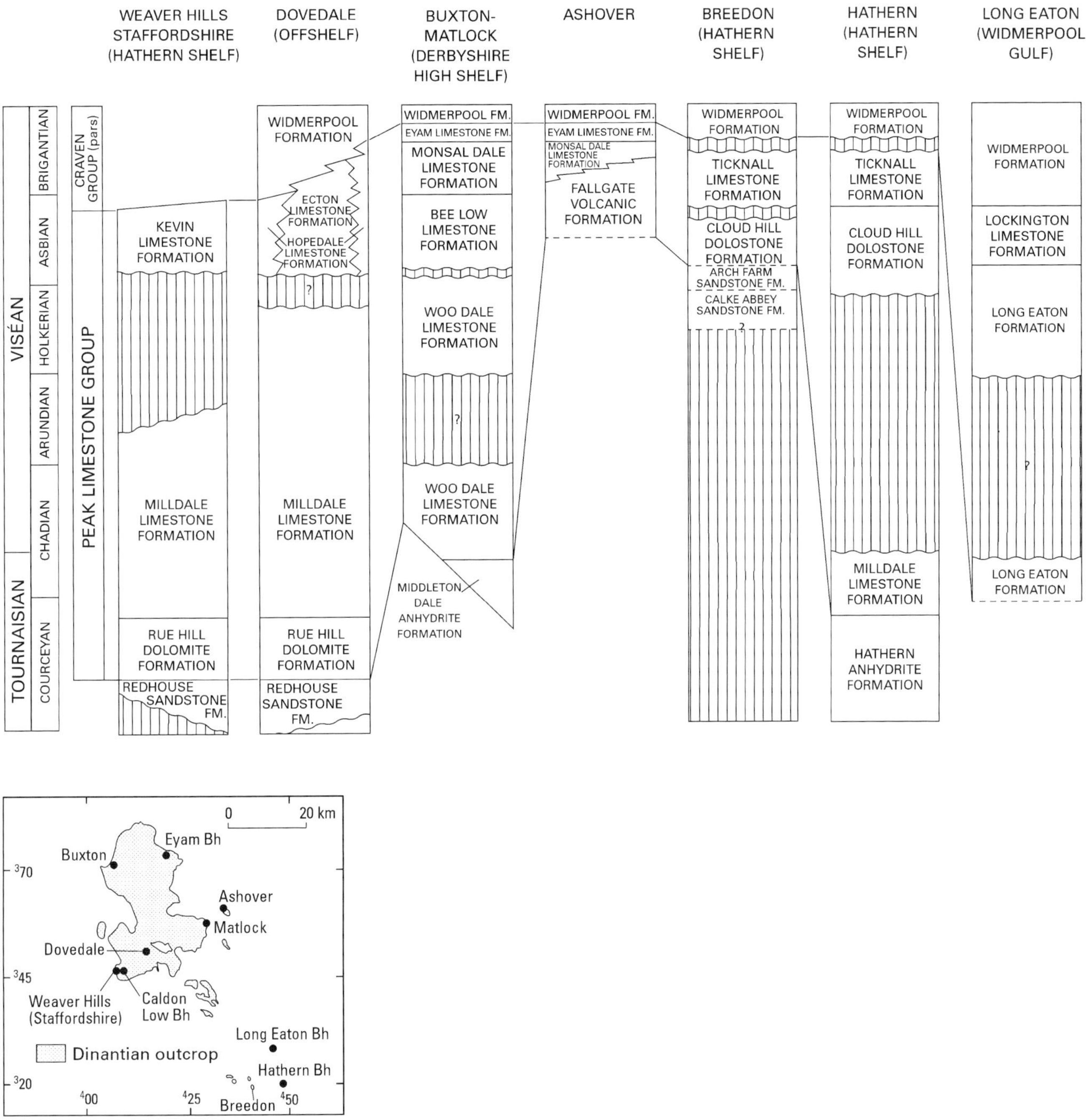

Fig. 9.17. Correlation of Dinantian strata in the Peak District of Derbyshire and Staffordshire. Adapted from Aitkenhead & Chisholm (1982).

(Fig. 9.17) is adapted by Waters *et al.* (2006*a*) from Aitkenhead & Chisholm (1982).

The Milldale Limestone Formation comprises limestone of both Waulsortian reef and inter-reef facies. The single or compound mud-mounds are composed of fossiliferous, massive micrite with common spar-filled cavities, whereas the inter-reef facies consists of well-bedded crinoidal biosparite and subordinate dark grey, cherty, micritic limestone. This deeper water succession, deposited at water depths of 220–280 m (Bridges & Chapman 1988), passes laterally and gradationally into the shallower water Woo Dale Limestone Formation (Fig. 9.17). The lower Woo Dale Limestone Formation, which comprises a grey-brown–dark grey dolomitic limestone, passes up into medium and pale grey, thick-bedded biosparites. The presence of 'birdseye' structures and rare coals within the top part of the formation suggests a regressive event associated with development of peritidal and restricted lagoonal environments during Holkerian times (Schofield & Adams 1985).

By Asbian times there was a clear differentiation between a shelf area, fringed by apron-reefs, and an off-shelf area where ramp carbonates are overlain by basinal deposits of the Craven Group. The shelf area includes the Asbian Bee Low Limestone Formation and the Brigantian Monsal Dale Limestone and Eyam Limestone formations (Fig. 9.17). The Bee Low Limestone Formation consists of very thick beds of pale grey biosparite and biopelsparite calcarenites, with pedogenic crusts and palaeokarstic surfaces, commonly overlain by thin, red-brown and grey-green bentonites. The Monsal Dale Limestone Formation comprises a heterogeneous succession of pale–medium grey, thin- to thick-bedded limestones with sporadic cherts, and dark grey, thin-bedded, cherty limestones with argillaceous partings. The basal limestone often includes abundant *Girvanella* or *Saccaminopsis*. The Eyam Limestone Formation typically comprises dark grey, thin-bedded, cherty bioclastic limestone with dark grey mudstone intercalations. A reef facies comprises pale grey limestone in small knolls, typically of massive biomicrite, with a distinctive brachiopod fauna. These reefs were deposited in comparatively shallow water with evidence of periodic emergence, in contrast to the deeper water Waulsortian mud-mounds of the Milldale Limestone Formation. In the off-shelf area, there is an Asbian–Brigantian succession of carbonate turbidites, which show an upward transition from limestone-dominated (Ecton Limestone Formation) passing up into deeper water mudstones with limestone and sandstone turbidites (Widmerpool Formation). The Hopedale Limestone Formation comprises Cracoean reef and inter-reef facies developed between the shelf and off-shelf areas.

In Derbyshire basaltic lavas, sills and pyroclastic rocks are common. There are several centres of igneous activity, including Matlock and Miller's Dale. The eruptions from the different centres are not contemporaneous and flows cannot be correlated from one centre to another (Aitkenhead *et al.* 1985). Walters & Ineson (1981) recognized 30 distinct lavas and beds of pyroclastic rock and Francis (1970) identified at least 14 agglomerate vents. In the subsurface to the east (Fig. 9.17), the proportion of volcanic rocks to limestone increases, and the name Fallgate Volcanic Formation has been given to the succession there (Aitkenhead & Chisholm 1982). The thickest section (293 m, not bottomed) was proved in a borehole near Ashover, where tuffs and basalts were interbedded with thin limestones. Much of the lava near Ashover was erupted subaqueously (Ramsbottom *et al.* 1962). Locally, the upper part of the Widmerpool Formation in the adjacent Widmerpool Gulf includes Brigantian subaqueous basaltic lavas, tuffs, agglomerates and hyaloclastites.

The majority of igneous activity is of tholeiitic affinity (Macdonald *et al.* 1984) and occurred during early Brigantian times (Fig. 9.7), although extrusive igneous rocks are also associated with carbonates of late Holkerian–late Brigantian age (Walters & Ineson 1981). The lavas are thin, typically a few tens of metres, but with some discrete flows up to 42 m thick (Francis 1970) and multiple flows in places (Walters & Ineson 1981; Macdonald *et al.* 1984). They are usually highly altered, fine-grained, vesicular, olivine-phyric and aphyric basalt. Many lavas are thought to have been subaerial flows, although some lavas may have flowed across wet sediments and terminated locally in shallow water (Walkden 1977; Macdonald *et al.* 1984). Other lavas are subaqueously erupted hyaloclastites, as suggested by Elliott (in Ramsbottom *et al.* 1962) and illustrated by Aitkenhead *et al.* (1985, plate 9). The Calton Hill Complex, located about 6 km east of Buxton, comprises Brigantian tuffs, agglomerates and lavas of the Upper Miller's Dale Lava, intruded by alkaline basanite sills, which host spinel lherzolite and harzburgite nodules. These nodules, brought to the surface during the volcanic activity, represent the only English examples of mantle-derived material (Donaldson 1978).

Tuffs are typically subordinate to lavas and sills. K-bentonites are widespread, most notably in the Bee Low Limestones of Asbian age. Between 30 and 40 beds, generally less than 3 cm thick, although locally reaching up to 1.25 m are present within this formation (Aitkenhead *et al.* 1985; Walkden 1972, 1977). These are thought to represent fine volcanic ash of distant origin, possibly not related to the Derbyshire volcanism.

The sills are dominantly medium- to coarse-grained olivine–dolerites, distinguished by the presence of altered olivine phenocrysts, ophitic intergrowths of clinopyroxene and plagioclase, and the absence or rarity of vesicles and amygdales (Macdonald *et al.* 1984). The majority of these dolerite intrusions were emplaced along planes of weakness between lavas and limestones. The sills appear to be genetically related to the extrusive rocks and are probably also of Dinantian age, although whole-rock K–Ar dates on the sills provides dates that are considerably younger than the lavas (Fig. 9.7). The discrepancy is probably a function of argon loss during hydrothermal alteration.

North Wales

Along the northern flank of the Wales–Brabant High 'Basement Beds' of North Wales (Fig. 9.18) represent a number

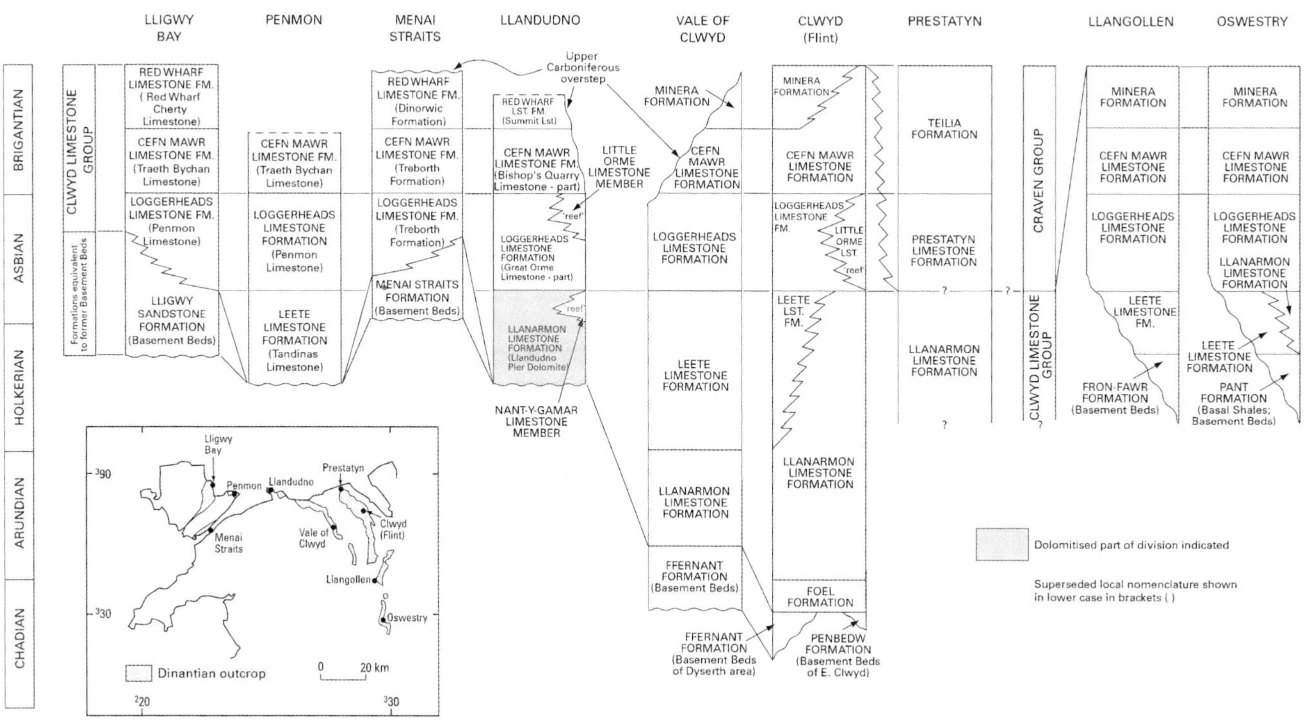

Fig. 9.18. Correlation of Dinantian strata in North Wales (Waters *et al.* 2006).

of isolated alluvial deposits infilling an incised, possibly fault-influenced, topography at different times (Davies *et al.* 2004). Clasts were derived from local upland areas and composition varies between these isolated deposits. Locally, principally in the Vale of Clwyd, the succession comprises up to 75 m of reddened alluvial breccias, conglomerates, sandstones, siltstones and mudstones. Calcareous nodules and thin beds of variegated, red and green, argillaceous, nodular limestone or dolomite record varying stages in calcrete development. The 'Basement Beds' of North Wales are unfossiliferous, as old as Tournaisian in age.

Clwyd Limestone Group

The Clwyd Limestone Group comprises up to 900 m of dominantly shallow-marine ramp and platform carbonates, which rest locally upon 'Basement Beds', or elsewhere with marked unconformity on Silurian rocks. Strata now included within the Clwyd Limestone Group were thought to comprise only Asbian and Brigantian strata (George *et al.* 1976). However, earlier Chadian, Arundian and Holkerian strata have subsequently been discovered (Davies *et al.* 1989; Somerville & Strank 1984*a*, *b*). The lithostratigraphical nomenclature for North Wales (Fig. 9.18) described below is that of Davies *et al.* (2004).

In North Wales, carbonate sedimentation commenced during late Chadian times with deposition of the Foel Formation in a lagoon or intertidal and supratidal carbonate mud flat. These deposits alternate with facies interpreted as high-energy and open marine deposits that record invasion of sediment derived from sheltering shoals or barriers (Davies *et al.* 1989). The formation comprises porcellaneous calcite mudstone and wackestone, argillaceous packstone, peloidal and locally ooidal grainstone, cryptalgal laminites and oncolitic floatstone, with intercalated red and green variegated mudstone, siltstone and calcareous sandstone.

The major early Arundian marine transgression deposited the Llanarmon Limestone Formation lithologies under relatively high-energy, shallow-marine conditions. The lower part of the formation developed as an eastward facing carbonate ramp (Somerville *et al.* 1989), which by the late Arundian had evolved into a carbonate platform. The formation comprises pale, thick-bedded, peloidal and skeletal grainstones, with subordinate dark, thinner bedded packstones. The upper part of the formation intertongues with, and is progressively replaced by, the Leete Limestone towards the south.

During the Holkerian–early Asbian, the Leete Limestone Formation records a return to deposition in a restricted, probably hypersaline, lagoonal and peritidal setting along the margin of the Wales–Brabant High (Somerville & Gray 1984). The formation comprises rhythmic units of dark, argillaceous skeletal packstone and paler grainstone, overlain by porcellaneous limestone (Davies *et al.* 1989; Somerville & Strank 1984*b*).

During the late Asbian–early Brigantian, the cyclic sequences of the Loggerheads Limestone and overlying Cefn Mawr Limestone represent shoaling-upwards rhythms developed as the product of consecutive marine inundations. The Loggerheads Limestone Formation consists mainly of pale, thick-bedded, commonly pseudo-brecciated, skeletal and peloidal packstone, present in cycles typically topped by palaeokarstic surfaces, and overlain by mudstone palaeosols. Coral and brachiopod assemblages are consistent with the late Asbian D_1 Subzone (George *et al.* 1976). The overlying Cefn Mawr Limestone Formation consists of wackestone, packstone and grainstone arranged in cyclic sequences, the boundaries of which are defined by correlatable calcrete and palaeokarstic features locally overlain by bentonitic clay palaeosols. Black, replacive chert nodules are common. The formation includes a rich coral assemblage, detailed by Somerville (1979), typical of the D_2 Subzone of the *Dibunophyllum* Biozone and the Brigantian Stage. The presence of karstic surfaces, bentonitic soils and well-developed calcrete profiles demonstrate long emergent intervals in response to marine regressions (Walkden 1974, 1977).

The cyclic shoaling upwards successions continue into late Brigantian times, with deposition of Minera Formation limestones and sandstones. The limestone beds are arranged in wackestone, packstone and grainstone cycles, which range from storm wave-base to fair-weather wave-base deposits. Thick, calcareous sandstones commonly developed at the top of the cycles, were deposited within upper shoreface and beach environments (Davies *et al.* 2004).

Late Dinantian igneous activity, equivalent to that seen in Derbyshire, is largely absent in North Wales. The only exception is the Little Wenlock Lava of the Welsh Borders, a Brigantian vesicular microporphyritic olivine-basalt, up to 30 m thick (Francis 1970).

South Wales and Bristol

Avon Group

A major phase of northward-directed marine transgression during the Tournaisian resulted in deposition within a shelf setting of the mudstone-dominated Avon Group, with strata onlapping the southern flank of the Wales–Brabant High. The group mainly comprises dark grey or black mudstone with interbedded crinoidal bioclastic or oolitic limestone. The Avon Group (former Lower Limestone Shale and Cefn Bryn Shales) is named from the type area of the Avon Gorge, south of Bristol, where it is up to 150 m thick, but it also extends across South Wales (Waters *et al.* 2006*a*). The Avon Group is entirely Tournaisian in age (Davies *et al.* 1991).

In the Vale of Glamorgan, South Wales, the Avon Group has been subdivided into three formations (Fig. 9.19a). Initial marine transgression is associated with deposition of interbedded thin skeletal packstone limestone and grey mudstone, with subordinate calcareous sandstone, oolite and hematitic skeletal grainstones (Gayer *et al.* 1973; Waters & Lawrence 1987) of the Tongwynlais Formation. The formation is absent in the North Crop of South Wales, where the middle unit, the Castell Coch Limestone, rests disconformably upon Devonian Upper Old Red Sandstone. The Castell Coch Limestone Formation represents a regional shallowing event, with progradation of high-energy oolite shoals across the Dinantian shelf. The formation comprises thick to well-bedded, commonly cross-bedded, skeletal and oolitic grainstones, rich in crinoidal debris and shows secondary reddening, commonly associated with dolomitization (Waters & Lawrence 1987; Wilson *et al.* 1990). The highest unit, the Cwmyniscoy Mudstone Formation, represents a further marine transgressive deposit (Waters & Lawrence 1987; Barclay *et al.* 1989; Wilson *et al.* 1990). The formation comprises dark grey silty mudstone, sparsely fossiliferous, with subordinate thin bioclastic limestone and calcareous siltstone beds. The mudstone beds are thinner toward the gradational boundary with the overlying Pembroke Limestone Group.

In the Bristol area the Avon Group includes a lower Shirehampton Formation (Fig. 9.19b), which comprises coarse bioclastic and ooidal limestone with subordinate calcareous sandstone and mudstone that has a brackish water fauna with *Lingula* (Kellaway & Welch 1993), mainly deposited in a marine influenced lagoonal environment. Some interbeds of sandstone and red and green mudstone occur locally. A reddened crinoidal limestone, the 'Bryozoa Bed' occurs at the top of the formation. A non-sequence is overlain by greenish grey mudstone with interbedded black crinoidal limestone with a rich shelly fauna of the undivided Avon Group (Kellaway & Welch 1993).

Pembroke Limestone Group

During the Tournaisian the mudstone-dominated shelf deposits of the Avon Group pass upwards into the shallow-marine carbonate-dominated deposits of the Pembroke Limestone Group (Waters *et al.* 2006*a*). This group is characterized by

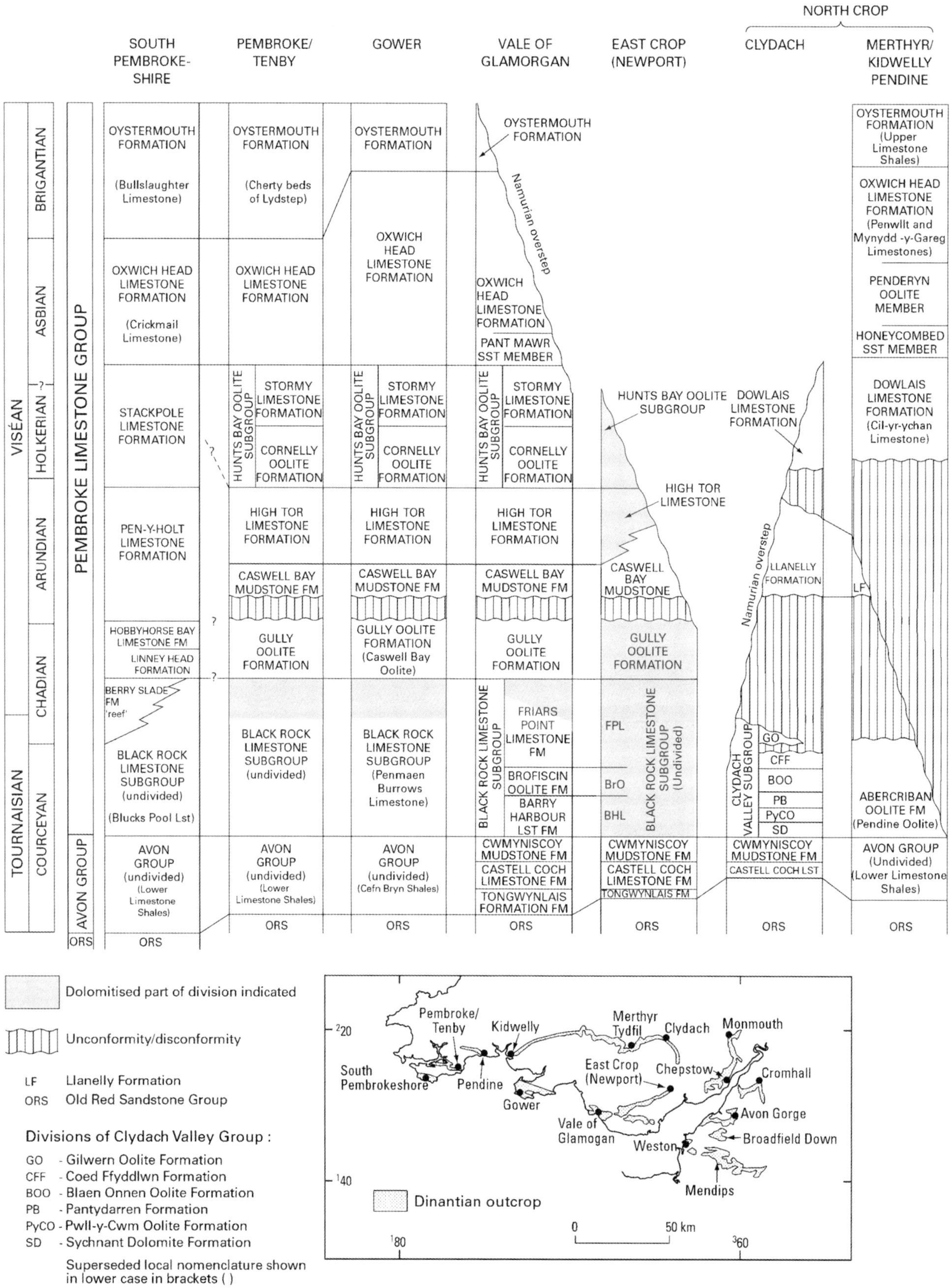

Fig. 9.19. Dinantian lithostratigraphical nomenclature for (**a**) South Wales and (**b**) southern England (Waters *et al.* in press); (**c**) N–S sections showing Dinantian facies variations for South Wales (after Wright 1987).

the Platform carbonate lithofacies association and shows an evolution from Carbonate ramp facies during the Courceyan–Arundian to Carbonate shelf facies during the Holkerian and Asbian (Fig. 9.19c). The lowermost part of the group is identified as the Black Rock Limestone Subgroup. First named in the Bristol region, the subgroup comprises predominantly

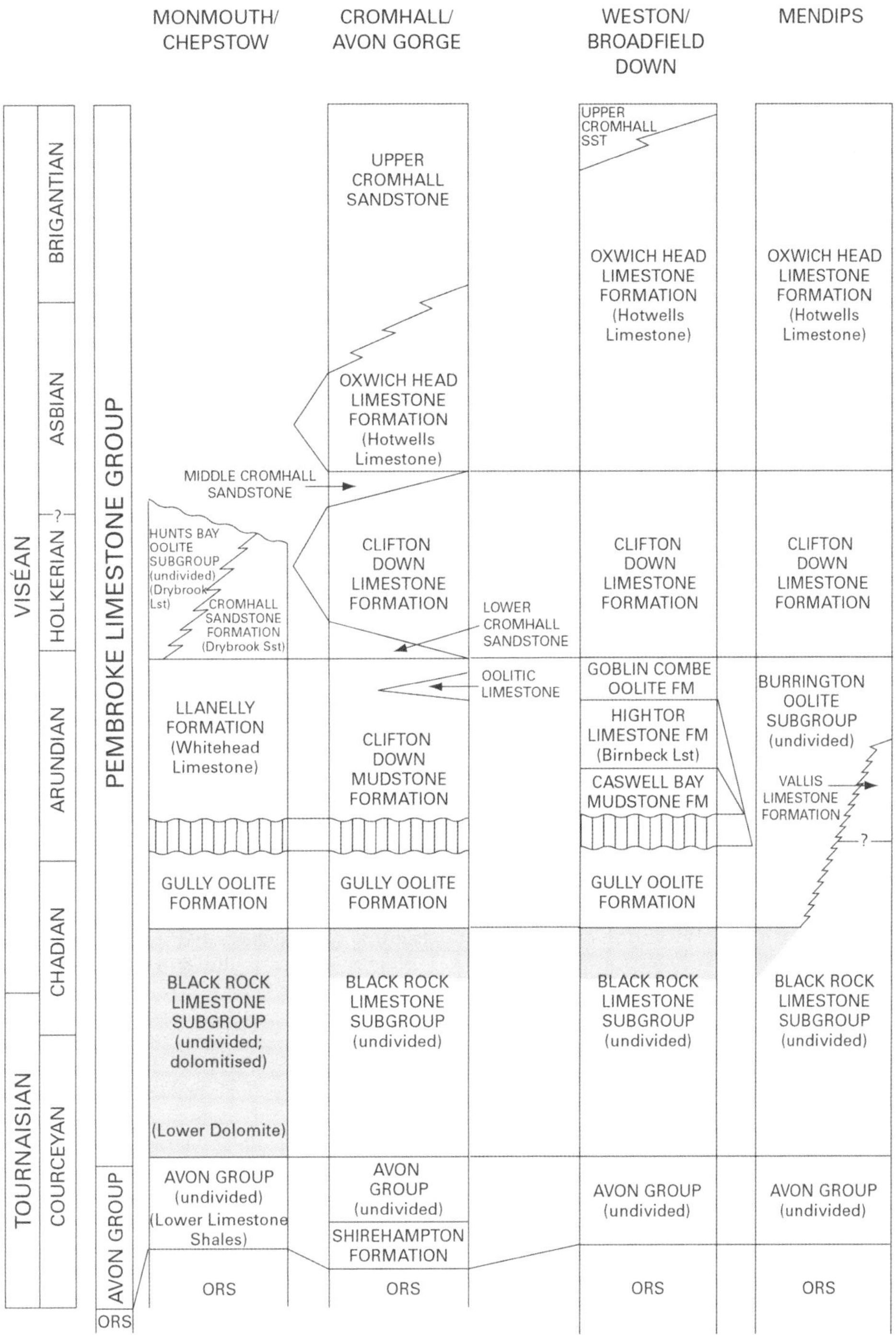

Fig. 9.19b. Continued.

bioclastic limestones, deposited in a ramp setting, that also extend across the South Crop of South Wales (Fig. 9.9a, b). This subgroup passes northward into the Clydach Valley Subgroup of the North Crop of the South Wales Coalfield, dominated by oolitic shoals and peritidal deposits, with abundant palaeosols (Barclay *et al.* 1989). Towards the east of Abergavenny the succession becomes entirely dolomitic.

The Black Rock Limestone Subgroup contains a rich fauna of conodonts, corals and foraminifera and ranges from Tournaisian to Chadian age. The subgroup shows a prominent southwards thickening, to a maximum of 500 m in the Vale of Glamorgan (Waters & Lawrence 1987; Wilson *et al.* 1990) (Fig. 9.19c), and more than 250 m in the Mendips. This contrasts with the condensed, approximately 45 m thick succession of the Clydach Valley Subgroup (Barclay *et al.* 1989). The Black Rock Limestone Subgroup comprises dark grey–black, poorly sorted crinoidal limestone, locally with chert bands and nodules. The limestone is commonly dolomitized, with dolostone generally present in the upper part of the subgroup over the Bristol–Mendips areas and South Crop of South Wales and throughout the succession in the East Crop of South Wales and the Forest of Dean. In the Vale of Glamorgan, where dolomitization is less pervasive, component formations are recognized (Waters & Lawrence 1987; Wilson *et al.* 1990). The lowest, the Barry Harbour Limestone Formation, is a high-energy storm deposit comprising thin-bedded, dark grey,

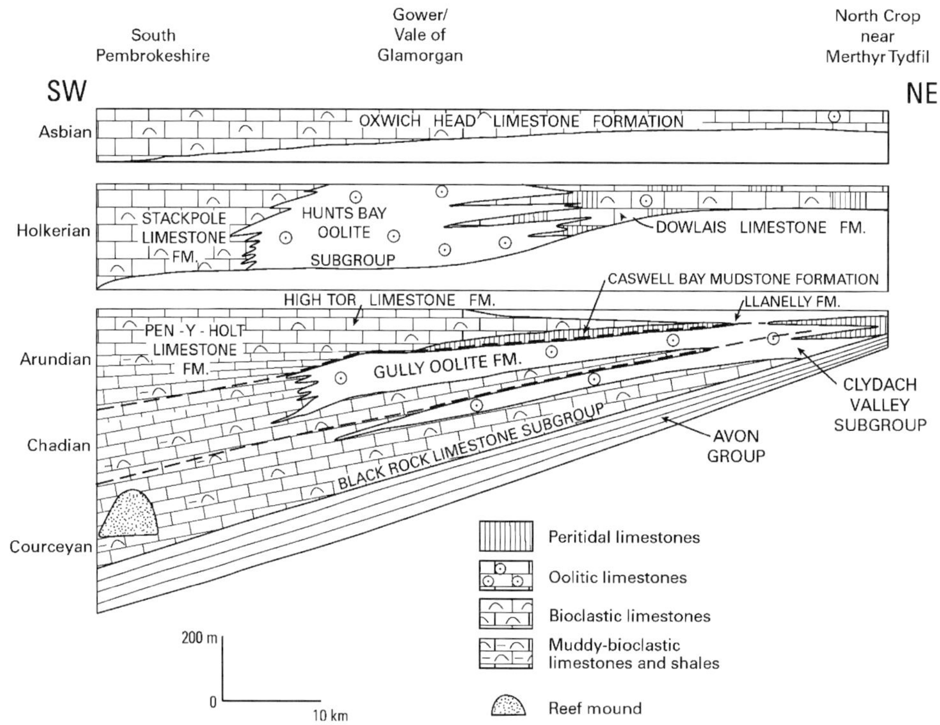

Fig. 9.19c. Continued.

fine- to coarse-grained, crinoidal, skeletal packstone with thin beds of shaly, calcareous mudstone and common replacive chert lenses. The overlying oolite shoals of Brofiscin Oolite Formation were deposited in relatively shallow-marine waters. The formation comprises pale–dark grey, massive to thick-bedded, well-sorted, predominantly ooidal and in part skeletal grainstone. Renewed transgression and re-establishment of deeper-water shelf conditions is associated with deposition of richly fossiliferous, mainly crinoidal, and strongly bioturbated limestones of the Friars Point Limestone Formation. The formation comprises thickly bedded, dark grey–black, fine-grained, skeletal packstones with subordinate thin interbeds of shaly, argillaceous, skeletal packstone.

A late Chadian marine regression is associated with deposition of high-energy carbonates of the Gully Oolite Formation. The regression culminated in a prominent non-sequence at the top of the formation, evident as an irregular and pitted palaeokarst surface sometimes referred to as the mid-Avonian unconformity. The formation, first defined in the Bristol area, is also found across South Wales and the Forest of Dean (Fig. 9.19a). It consists of a distinctive pale grey, massive, medium- to fine-grained, cross-bedded ooid, peloid and skeletal grainstone, locally in coarsening-upward cycles (Waters & Lawrence 1987). The formation shows a southward thickening (Fig. 9.19c), with a maximum thickness of 83 m recorded in the Cardiff area, passing southward into bioclastic limestone of the Linney Head Limestone and Hobbyhorse Bay Limestone formations in south Pembrokeshire.

A basal Arundian marine transgression deposited a mudstone-dominated succession within a lagoonal/peritidal setting. In South Wales and the Weston-super-Mare district this succession is seen as the Caswell Bay Mudstone Formation, which appears to be laterally equivalent to the Clifton Down Mudstone of Bristol, though the latter formation is thicker and has a greater chronostratigraphical range (late Chadian–Arundian). These formations are pale grey–green, porcellaneous limestone, interbedded with dark grey and brown shale and argillaceous skeletal wackestone and packstone. Cryptalgal lamination and birdseye structures are common. The Caswell Bay Mudstone Formation, up to 15 m thick in the Vale of Glamorgan, thins southward and is absent from the Mendips and the southern parts of the Cardiff and Bridgend areas (Fig. 9.19a, b & c).

During the Arundian marine transgression a barrier complex developed, which limited the peritidal deposits of the Caswell Bay Mudstone to the north (Fig. 9.19c). During continued transgression the barrier complex migrated northwards over the peritidal facies (Wilson *et al.* 1990). The barrier complex is defined by the lower part of the High Tor Limestone Formation in South Wales. The High Tor Limestone in South Wales consists of thin- to thick-bedded, bioclastic limestone with a few thin beds of shaly dolomite mudstone and siltstone, whereas in the Weston-super-Mare district the formation (formerly the Birnbeck Limestone) comprises grey–dark grey, thick- to very-thick-bedded bioclastic limestone where the lower part is commonly oolitic (Whittaker & Green 1983).

A late Arundian marine regression resulted in the return of high-energy shoreface deposits of the upper part of the High Tor Limestone Formation in South Wales and Goblin Combe Oolite Formation in the Weston-super-Mare area. The Goblin Combe Oolite comprises pale grey–black, massive, medium- to coarse-grained, ooidal and crinoidal limestone. The High Tor Limestone and Goblin Combe Oolite formations are absent north of Bristol, where they pass laterally into the Clifton Down Mudstone (Fig. 9.19b).

The Llanelly Formation of the North Crop of South Wales shows a comparable history of Arundian marine transgression, with basal fluvial deposits overlain by a shallow-marine peritidal complex, with subsequent regression indicated by re-establishment of fluvial floodplain deposition. The formation rests upon a prominent palaeokarstic surface, above a pedified horizon at the top of the Clydach Valley Subgroup (Wright 1981).

Late Arundian–Holkerian oolitic limestone, with subordinate skeletal, peloidal, oncolitic and micritic limestone present in South Wales are recognized as the Hunts Bay Oolite Subgroup (Fig. 9.19a), comprising two formations in the Vale of Glamorgan; the Cornelly Oolite and Stormy Limestone formations (Wilson *et al.* 1990), with the latter deposited in a lagoonal/peritidal back-barrier environment, which developed behind the Cornelly Oolite shoal. In the Bristol region the strata of equivalent age are identified as the Clifton Down Limestone Formation (Fig. 9.19b). The predominantly bioclastic Stackpole Limestone Formation of south Pembrokeshire, deposited in a front barrier, barrier and back barrier setting and the Dowlais Limestone Formation of the North Crop, a succession of foetid and bituminous peloidal grainstone, packstone and wackestone with dark grey shale interbeds representing lagoonal and peritidal deposits (Barclay *et al.* 1988, 1989), are also the same age as the Hunts Bay Oolite Subgroup. The thickest development of the subgroup is 245 m in the Bridgend area, thinning to the east and removed in the East Crop by the Namurian overstep (Fig. 9.19a). The base of the subgroup is typically conformable. However, the Dowlais Limestone Formation of the North Crop rests upon an unconformity, overstepping older strata toward the west, ranging from Llanelly Formation to the Avon Group.

Continued late Arundian shoaling is associated with deposition of the lower part of the Cornelly Oolite Formation, which comprises a thickly bedded, cross-bedded, ooid grainstone. The marine regression locally culminated in emergence and palaeokarst development. The succeeding strata record a basal Holkerian marine transgression and re-establishment of oolite shoals. The Stormy Limestone and Dowlais Limestone formations are interpreted as deposits within a lagoonal/peritidal back-barrier environment, which developed behind the Cornelly Oolite shoal (Barclay *et al.* 1988, 1989; Wilson *et al.* 1990). The Stormy Limestone Formation comprises a heterolithic succession of interbedded porcellaneous limestone with common stromatolitic lamination and birdseye structures, bioclastic limestone, and packstone-grainstone with skeletal grains, ooids, peloids and intraclasts. Locally, thrombolitic algal bioherms up to 2 m high are present. The Dowlais Limestone is dominated by dark grey, bituminous, bioclastic packstone-grainstone.

The lower part of the Clifton Down Limestone Formation is a relatively unfossiliferous, lagoonal, porcellaneous limestone succession. These limestones are overlain by shelly limestone, algal beds and cross-bedded oolites ('*Seminula* Oolite'), interpreted as tidal channel or bar deposits. In the Bristol region the base of the Holkerian Stage is taken at the sharp base of the '*Seminula* Oolite' (George *et al.* 1976). The succession represents a progressive change to a more open sea environment during the basal Holkerian marine transgression. The uppermost part of the formation comprises a thick calcareous algal limestone sequence.

The Asbian succession is dominated by high-energy marine limestones of the Oxwich Head Limestone Formation in South Wales and Bristol district (former Hotwells Limestone). The limestones accumulated on a mature carbonate platform during a period of overall marine transgression, in which periodic sea-level falls resulted in emergence. The Oxwich Head Limestone Formation in South Wales thickens southward (Fig. 9.19c) and comprises mainly thickly bedded, fine- to coarse-grained, recrystallized skeletal packstone, with distinctive pale–dark grey pseudobrecciation. The limestone contains conspicuous hummocky and pitted palaeokarstic surfaces overlain by thin, red and grey mottled clay beds, interpreted as palaeosols. In the Bristol region, the formation comprises hard, massive, grey–dark grey, ooidal and crinoidal limestone with an abundant coral and brachiopod fauna. The formation is about 225 m thick on Broadfield Down, but thins northward and is absent north of Bristol, where the formation passes laterally northward into the Upper Cromhall Sandstone (Fig. 9.19b). A non-sequence at the base of the formation in the Bristol region may be indicated by the absence of diagnostic early Asbian fauna (Kellaway & Welch 1993).

During the Brigantian there was a significant increase in the introduction of terrigenous material onto the carbonate shelf. In South Wales the mud-dominated succession of the Oystermouth Formation (formerly Upper Limestone Shales) comprises interbedded thin limestone and mudstone beds, silicified limestone and chert. The sand-dominated succession of the Cromhall Sandstone Formation consists of up to three leaves south of Bristol, ranging from Holkerian to Brigantian age. The leaves merge together to the north of Bristol, towards the southern flank of the Wales–Brabant High. This formation comprises grey, red weathering, coarse-grained quartzitic and calcareous sandstone, sandy crinoidal and ooidal limestone, dark grey mudstone and grey, rootlet-bearing fireclay. The sequence is markedly cyclic, a typical cycle comprising limestone passing up into mudstone, siltstone and sandstone and overlain by a seatearth (Kellaway & Welch 1993). The formation represents deposition in marginal marine and estuarine deltaic conditions. These sandstones probably represent terrigenous sands introduced onto an emergent carbonate shelf during marine regression at the end of the Holkerian, and subsequently marine-reworked during base Asbian transgression (Wilson *et al.* 1990).

Olivine-basalt lavas and tuffs are associated with the limestones of the Pembroke Limestone Group and some represent submarine eruptions. These late Tournaisian–early Viséan igneous rocks are largely restricted to the Bristol and Gloucester districts (Fig. 9.7). The Middle Hope Volcanic Member, within the Tournaisian Black Rock Limestone Subgroup, is a succession of up to 37 m of ash-tuffs, lapilli-tuffs and pillow basalts present in the Weston-super-Mare area (Whittaker & Green 1983; Wright & Cossey 2003). Basal tuffs record the onset of volcanic activity on the outer part of a sloping, shallow-marine shelf below storm wave-base (Faulkner 1989). Overlying pillow basalts also indicate subaqueous igneous activity. The volcanic centre developed in progressively shallower water as a result of volcanic updoming and/or development of the volcanic cone, approaching the sea surface before finally subsiding. Lavas and tuffs of Arundian age from the Bristol district at Spring Cove, Goblin Combe, Broadfield Down, Cadbury Camp and Tickenham indicate that volcanic activity occurred over a relatively wide area and included several small vents (Whittaker & Green 1983).

There is relatively little evidence for Carboniferous volcanism in South Wales. The Mathry quartz–dolerite dyke (Cave *et al.* 1989), of probable Carboniferous age, can be traced for some 40 km across SW Wales. Three minor dykes in the Welsh Borders cut strata of Devonian age and are considered to have been intruded during the Carboniferous (Francis 1970). These are the analcime gabbro Brockhill Dyke, the olivine–dolerite Bartestree Dyke and the monchiquitic Llanllywel Dyke (Fig. 9.7). The latter dyke is associated lithologically and geographically with the monchiquite intrusion and volcanic pipe at Golden Hill Quarry, Monmouthshire. The supposed Viséan volcanic diatreme pipe at Golden Hill contains an agglomeratic facies and a NW-orientated dyke-like facies, cutting through Devonian sandstones (Bevins 2003). The pipe and a nearby associated dyke include ultramafic mantle and lower crustal xenoliths (Upton *et al.* 1983).

Namurian: advancing deltas

Introduction

The onset and evolution of broad thermally subsiding basins dominated Namurian sedimentation in England and Wales. The 'block and basin' topography developed during the Dinantian, which resulted in isolated deposition within small sub-basins or as distinct carbonate platforms, was gradually infilled by predominantly fluvio-deltaic deposits. Depositional systems became established over broad areas. Three distinct provinces can be recognized: northern England, the Central

Pennine Basin of central England and North Wales, and the South Wales Basin of South Wales and southern England.

Namurian sedimentation in northern England was dominated by comparatively shallow-water, predominantly non-marine, stacked fluvio-deltaic successions with near-coastal marine intervals. A particularly condensed succession was deposited during Chokierian–Marsdenian times, probably reflecting a period of low subsidence rates and low sediment influx, possibly due to bypass into the Central Pennine Basin to the south. Basinal marine shales are rare and, more typically, marine transgressions are represented by brachiopod-crinoid-mollusc communities developed close to the coast.

The Central Pennine Basin was a broad depositional area extending from the Craven Fault System to the northern flank of the Wales–Brabant High. Eustatically controlled sea-level rises resulted in the deposition of marine shales with acme ammonoid faunas. Fluvio-deltaic sediments, sourced mainly from the north, infilled the remnant basin topography and new accommodation generated during sea-level highstands. In these relatively deep basins (up to a few hundred metres) of the early Namurian, the deltaic systems prograded only a short distance southward, and argillaceous sediments were deposited widely (Fig. 9.9d). The initial phase of coarse clastic basin sedimentation was dominated by accumulation of thick deposits (up to 550 m) from turbidity flows, a notable contrast to deposition within northern England. By late Kinderscoutian times sedimentation had reduced basinal topography and deltas prograded rapidly across the Central Pennine Basin, and commonly formed a sheet-like geometry. During phases of regression, rivers incised into their deltas and palaeosols developed on interfluves. Rivers probably directly fed pro-delta turbidite lobes in the basins; a model particularly invoked for deposition of the Kinderscoutian part of the Millstone Grit Group (Hampson 1997). By Marsdenian times the deltas had reached the northern margin of the Wales–Brabant High and by the Yeadonian, the first influx of westerly sourced fluviodeltaic sediments became apparent in the western part of the basin.

In addition to the northern-sourced sediments, the emergent Wales–Brabant High shed quartzitic sands, deposited as small fluviodeltaic systems, northwards into the Central Pennine Basin (Fig. 9.9d). Much of the sediment may have been derived from about 200 m of uplift associated with a Variscan foreland basin forebulge (Burgess & Gayer 2000). The small deltas fringing the southern part of the Pennine Basin were progressively onlapped by larger delta systems that spread across the basin from the north (Collinson 1988).

The South Wales Basin was a small, partly enclosed basin that was probably linked with the evolving Variscan Foreland Basin, and the Culm Basin in Devon and Cornwall (see Chapter 10). Quartz-rich sediments were shed southwards from the Wales–Brabant High into the basin. Namurian deposition in South Wales was influenced by tectonically-driven differential subsidence, induced by northward propagation of the Variscan Front and reactivation of Caledonian (NE–SW) lineaments, such as the Carreg Cennen, Swansea Valley and Neath disturbances (Fig 9.5). The Pembrokeshire sub-basin was isolated from the main South Wales area of deposition by the Penlan Axis, a growth anticline, which developed in response to movement on the Carreg Cennen Disturbance. Similarly, deposition in South Wales and Bristol was separated by the development of the Usk Anticline. The growth anticlines appear to have acted as local sources of sand, derived from uplifted and eroded Devonian Old Red Sandstone (George 2000). Braid deltas transported sands into a high-energy coastal zone, subject to wave- and storm-generated currents, which passed southward into the deep marine environment of the Culm Basin. In contrast with the Pennine Basin, the South Wales Basin was located at the margins of an open marine connection (Fig. 9.9d) and consequently wave and storm influences were enhanced.

Chronostratigraphy

The Namurian Series comprises seven stages coincident with groups of ammonoid zones: Pendleian (E_1), Arnsbergian (E_2), Chokierian (H_1), Alportian (H_2), Kinderscoutian (R_1), Marsdenian (R_2) and Yeadonian (G_1). Boundary stratotypes for all but the Arnsbergian have been defined by Ramsbottom (1981) in central and northern England (Fig. 9.11). The bases of the stages are taken at the bases of marine bands that display the earliest occurrence of certain diagnostic ammonoid faunas (Fig. 9.20). These are the *Cravenoceras leion* (Pendleian), *Cravenoceras cowlingense* (Arnsbergian), *Isohomoceras subglobosum* (formerly *Homoceras subglobosum*) (Chokierian), *Hudsonoceras proteum* (formerly *Hudsonoceras proteus*) (Alportian), *Hodsonites magistrorum* (Kinderscoutian), *Bilinguites gracilis* (formerly *Reticuloceras gracile*) (Marsdenian) and *Cancelloceras cancellatum* (formerly *Gastrioceras cancellatum*) (Yeadonian).

The earliest occurrence of *Cravenoceras leion* is taken as the base of the Namurian Series. The global Mid-Carboniferous Boundary between the Mississippian and Pennsylvanian subsystems, marked by the first appearance of the conodont *Declinognathodus noduliferus*, occurs within the Chokierian Stage.

Biostratigraphy

Thick-shelled ammonoids occur within thin hemi-pelagic shale beds, commonly referred to as 'marine bands'. In the Central Pennine Basin there are about 60 marine bands (Holdsworth & Collinson 1988). Ammonoid evolution rates were so great during the Namurian that typically each marine band comprises a distinct ammonoid fauna (Fig. 9.20). The nekto-pelagic habit of ammonoids allows the biozones to be recognized across Western Europe and some are applicable globally. The extensive distribution of the marine bands makes them of primary stratigraphical importance. Namurian ammonoid biostratigraphy was developed by Bisat (1924, 1928) and Bisat & Hudson (1943), and was later refined by Ramsbottom (1969, 1971). In northern England, by contrast, ammonoid-bearing marine shales are comparatively rare.

Owens *et al.* (1977) produced a palynostratigraphy for the Namurian of northern England and Scotland, with five miospore biozones encompassing the series (Fig. 9.20). Within the Central Pennine Basin the palynomorph zones have been closely related to ammonoid zones. Palynomorphs have proved particularly important zonal indicators in the Northumberland Trough, where ammonoid-bearing marine bands are rare. The base of the NC Biozone is located in the late Brigantian P_2 ammonoid Biozone, and extends throughout the Pendleian. The base of the TK Biozone occurs near to the base of the Arnsbergian Stage. The base of the overlying SO Biozone has been recognized at the High Wood Marine Band of the Stainmore area, and is considered to be located within the Arnsbergian E_{2b} ammonoid subzone. The base of the KV Biozone occurs at the base of the Kinderscoutian Stage, extending into early Marsdenian strata (Clayton *et al.* 1977). The base of the FR Biozone, in the absence of constraining marine bands, has been taken to approximate to the mid-Marsdenian and extends to the base of the Westphalian Series (base of the *Gastrioceras subcrenatum* Marine Band).

Five conodont biozones are recognized for the Namurian Series (Fig. 9.20). Conodonts have much longer stratigraphical ranges than ammonoids during the Namurian and therefore are of less stratigraphical importance. However, the exact position of the mid-Carboniferous boundary has been defined as the appearance of the conodont *Declinognathodus noduliferus* (Lane *et al.* 1985). This places the mid-Carboniferous boundary close to the base of the Chokierian Stage.

STAGE	ZONE		WESTERN EUROPEAN MARINE BANDS		MIOSPORES	MACROFLORA		CONODONTS
	Index	Ammonoid	Ammonoid	Index		Zone	Subzone	
YEADONIAN	G_{1b}	*Cancelloceras cumbriense*	*Ca. cumbriense*	$G_{1b}1$	*Raistrickia fulva* – *Reticulatisporites reticulatus* (FR)	*Pecopteris aspera*	*Neuralethopteris larischii*	*Idiognathoides sinuatus* – *Idiognathoides primulus*
	G_{1a}	*Cancelloceras cancellatum*	*Ca. cancellatum*	$G_{1a}1$				
MARSDENIAN	R_{2c}	*Bilinguites superbilinguis*	*Verneulites sigma*	$R_{2c}2$				
			B. superbilinguis	$R_{2c}1$				
	R_{2b}	*Bilinguites bilinguis*	*B. metabilinguis*	$R_{2b}5$	*Crassispora kosankei* – *Grumosisporites varioreticulatus* (KV)		*Sigillaria elegans*	
			B. eometabilinguis	$R_{2b}4$				
			B. bilinguis	$R_{2b}3$				
			B. bilinguis	$R_{2b}2$				
			B. bilinguis	$R_{2b}1$				
	R_{2a}	*Bilinguites gracilis*	*B. gracilis*	$R_{2a}1$				
KINDERSCOUTIAN	R_{1c}	*Reticuloceras reticulatum*	*R. coreticulatum*	$R_{1c}4$				*Idiognathoides corrugatus* – *Idiognathoides sulcatus*
			R. reticulatum	$R_{1c}3$				
			R. reticulatum	$R_{1c}2$				
			R. reticulatum	$R_{1c}1$				
	R_{1b}	*Reticuloceras eoreticulatum*	*R. stubblefieldi*	$R_{1b}3$				
			R. nodosum	$R_{1b}2$				
			R. eoreticulatum	$R_{1b}1$				
	R_{1a}	*Hodsonites magistrorum*	*R. dubium*	$R_{1a}5$				
			R. todmordenense	$R_{1a}4$				
			R. subreticulatum	$R_{1a}3$				
			R. circumplicatile	$R_{1a}2$				
			Ho. magistrorum	$R_{1a}1$				
ALPORTIAN	H_{2c}	*Vallites eostriolatus*	*Homoceratoides prereticulatus*	$H_{2c}2$	*Lycospora subtriquetra* – *Kraeuselisporites ornatus* (SO)			*Declinognathus noduliferus*
			V. eostriolatus	$H_{2c}1$				
	H_{2b}	*Homoceras undulatum*	*H. undulatum*	$H_{2b}1$				
	H_{2a}	*Hudsonoceras proteum*	*Hd. proteum*	$H_{2a}1$				
CHOKIERIAN	H_{1b}	*Homoceras beyrichianum*	*I. sp. nov.*	$H_{1b}2$				
			H. beyrichianum	$H_{1b}1$				
	H_{1a}	*Isohomoceras subglobosum*	*I. subglobosum*	$H_{1a}3$				
			I. subglobosum	$H_{1a}2$				
			I. subglobosum	$H_{1a}1$				
ARNSBERGIAN	E_{2c}	*Nuculoceras stellarum*	*N. nuculum*	$E_{2c}4$		*Lyginopteris larischii*		*Gnathodus bilineatus bollandensis*
			N. nuculum	$E_{2c}3$				
			N. nuculum	$E_{2c}2$				
			N. stellarum	$E_{2c}1$				
	E_{2b}	*Cravenocera-toides edalensis*	*Ct. nititoides*	$E_{2b}3$	*Stenozonotriletes triangulus* – *Rotaspora knoxi* (TK)	*Lyginopteris stangeri*		
			Ct. nitidus	$E_{2b}2$				
			Ct. edalensis	$E_{2b}1$				
	E_{2a}	*Cravenoceras cowlingense*	*Eumorphoceras yatesae*	$E_{2a}3$				
			C. gressinghamense	$E_{2a}2a$				
			Eumorphoceras ferrimontanum	$E_{2a}2$				
			C. cowlingense	$E_{2a}1$				
PENDLEIAN	E_{1c}	*Cravenoceras malhamense*	*C. malhamense*	$E_{1c}1$	*Bellispores nitidus* – *Reticulatisporites carnosus* (NC) part	*Neuropteris antecedens* (part)	*Lyginopteris fragilis*	*Kladognathus* – *Gnathodus girtyi simplex*
	E_{1b}	*Cravenoceras brandoni*	*Tumulites pseudobilinguis*	$E_{1b}2$				
			C. brandoni	$E_{1b}1$				
	E_{1a}	*Cravenoceras leion*	*C. leion*	$E_{1a}1$				

Fig. 9.20. Summary of the chronostratigraphical units of the Namurian and the main biozones for the most important fossil groups, derived from authors referred to within text. The macrofloral zonations are derived from Cleal (1991).

Sequence stratigraphy

A number of Namurian sandstone bodies that satisfy two or more of the key characteristics for sediments deposited in valleys generated by incision during sea-level fall are recognized. Most Namurian-aged multistorey sandstone bodies are 10–30 km wide and the preserved thickness is 25–35 m (Davies *et al.* 1999).

In northern England, north of the Craven Fault zone, major fluvial sandstone bodies occurred in the uppermost Pendleian; up to 30–40 m of erosional relief over several kilometres can be documented where correlative interfluves are identified. Sandstones in the Kinderscoutian, Marsdenian and Yeadonian of the Central Pennine Basin have been subject to the most scrutiny. Hampson (1997) documents two potential valley sandstones in the Lower Kinderscout Grit. The recognition of major erosional unconformities (sequence boundaries) produced by fluvial incision suggests that, during falling sea-level, sediment bypassed the shelf and was transported directly into the deep basin (Hampson 1997). Therefore the Kinderscoutian succession does not represent a simple southward progradation of a turbidite-fronted delta (Walker 1966*a*).

Those valley fills associated with turbidite-fronted deltas, for example in the Lower Kinderscout Grit (Hampson 1997) and in the Marsdenian Roaches Grit (Jones & Chisholm 1997) thicken to 50–80 m around the area 2–5 km adjacent to their mouths (Hampson *et al.* 1999). Other Marsdenian sandstones, for example the Midgley Grit, are interpreted as fluvial systems deposited following a major sea-level fall but deposited in a platform location and are much more laterally extensive, reaching widths of up to 90 km (Brettle 2001).

Maynard (1992) and Hampson *et al.* (1996) have interpreted the erosional base of the late Yeadonian Rough Rock as a sequence boundary that can be correlated to an interfluve palaeosol. However, Bristow (1988, 1993) does not regard the sheet-like braided river deposit, represented by the Rough Rock, as the product of a fall in sea level. The Farewell Rock of South Wales, which is the same age as the Rough Rock of the Pennine Basin, has also been interpreted in terms of deposition within a valley (Hampson 1998). The base of the Farewell Rock has major erosional relief, associated with an identifiable time gap and an angular discordance in bedding (Hampson 1998). The Farewell Rock can be correlated with the Cumbriense Quartzite of the North Crop, east of Pembrokeshire, which contains several mature palaeosols, interpreted as a well-developed interfluve (Hampson 1998).

Ramsbottom (1977) proposed eleven mesothemic cycles for the British Namurian linked to the appearance of new ammonoid genera, and controlled by eustatic sea-level fluctuations (Fig. 9.10). The boundaries of the mesothems are marked by widespread disconformities in shelf areas. Although the mesothem model for the British Namurian has been criticized (Holdsworth & Collinson 1988; Martinsen 1990), the underlying concept of eustasy as an influence on sedimentation, particularly with respect to widespread marine transgressions, is now widely accepted.

Regional Developments

Northern England

Yoredale Group

In northern England deposition of the Mixed shelf carbonate and deltaic ('Yoredale') lithofacies association of the Yoredale Group continued into the Namurian and extended across the entire region. The Namurian part of the Yoredale Group is referred to as the Stainmore Formation and is distinguished from the underlying Dinantian Alston Formation by a decrease in both the number and thickness of limestones (Waters *et al.* 2006*a*). Limestones in the Stainmore Formation also tend to be darker grey than the Alston Formation limestones. The base of the Namurian Series is taken at the base of the prominent and laterally extensive Great Limestone (Main Limestone).

On the Askrigg Block there are up to four main levels of chert of Pendleian age, typically developed above the limestone component of the cyclothem. Details of limestone nomenclature and correlations, and cyclothem thicknesses, are provided by Dunham & Wilson (1985) for the Askrigg Block and Stainmore Trough, and by Dunham (1990) for the Alston Block (Fig. 9.21).

Millstone Grit Group

The Millstone Grit Group is characterized by cyclic successions of quartz-feldspathic sandstone, grey mudstone, thin coals and prominent seatearths of Fluvio-deltaic ('Millstone Grit') lithofacies association. The Millstone Grit Group is Pendleian–Yeadonian in age on the Askrigg Block. Fluvial sandstones, with palaeocurrents from channelized sandstones suggest a north or NE source. Some seatearths may result from prolonged periods of emergence, e.g. the Mirk Fell Ganister, a lateral correlative of the Pendleian Lower Howgate Edge Grit, and a potential interfluve sequence boundary. The group is about 500 m thick and significantly thinner than the equivalent strata present within the Central Pennine Basin to the south (Figs 9.21 & 9.22). The base of the Millstone Grit Group is marked by a slightly angular intra-Pendleian (E_{1c}) unconformity above the Yoredale Group (Brandon *et al.* 1995). The presence of a Mid-Carboniferous unconformity is indicated by the absence of Alportian strata across the Askrigg Block (Ramsbottom 1977). In west Cumbria there is a large non-sequence in which strata of Chokierian–Marsdenian age are absent (Akhurst *et al.* 1997). The equivalent of the Millstone Grit Group is represented by less than 10 m thickness of dark grey carbonaceous mudstone grading up into siltstone and seatearth.

To the north of the Askrigg Block, the Millstone Grit Group is less well constrained, typically comprising sandy incised valley fills within the Mixed shelf carbonate and deltaic ('Yoredale') lithofacies association. The probable equivalent of the Millstone Grit Group is recognized in the upper part of the Stainmore Formation (Waters *et al.* 2006*a*). Limestones are either very thin or absent and thick, coarse-grained; often pebbly sandstones with ribbon-like geometries are characteristic. These fluvial sandstones have historically been mapped as First and Second Grit (Ramsbottom *et al.* 1978), although the use of such a simple nomenclature for a channel sandstone complex is now questioned.

Central Pennine Basin

Craven Group

Hemi-pelagic lithofacies-association-sedimentation of the Craven Group continued from the Dinantian into the Namurian in the Central Pennine Basin. The Namurian component of this association has historically been named Upper Bowland Shale in the Craven Basin, Edale Shale in Derbyshire (Plate 16) and Holywell Shales in North Wales. These mudstones are laterally contiguous in the subsurface and the use of a single formation name, Bowland Shale Formation, is now preferred (Waters *et al.* 2006*a*). These deposits accumulated predominantly from suspension in moderately deep water, possibly a few hundred metres deep and certainly below storm wave-base. The formation comprises dark grey and black, organic-rich mudstone, with subordinate beds of siltstone, sandstone, ironstone and dolomitic limestone. Thin sandstone and limestone beds were possibly introduced by storms and/or as turbidites. During the late Brigantian–early Pendleian the Bowland Shale Formation shows a decrease in carbonate turbidites and concomitant increase in siliciclastic sandstone turbidites derived from encroaching Millstone Grit Group deltas accumulating on the margins of the Central Pennine Basin. In North Wales impersistent coal seams and seatearths

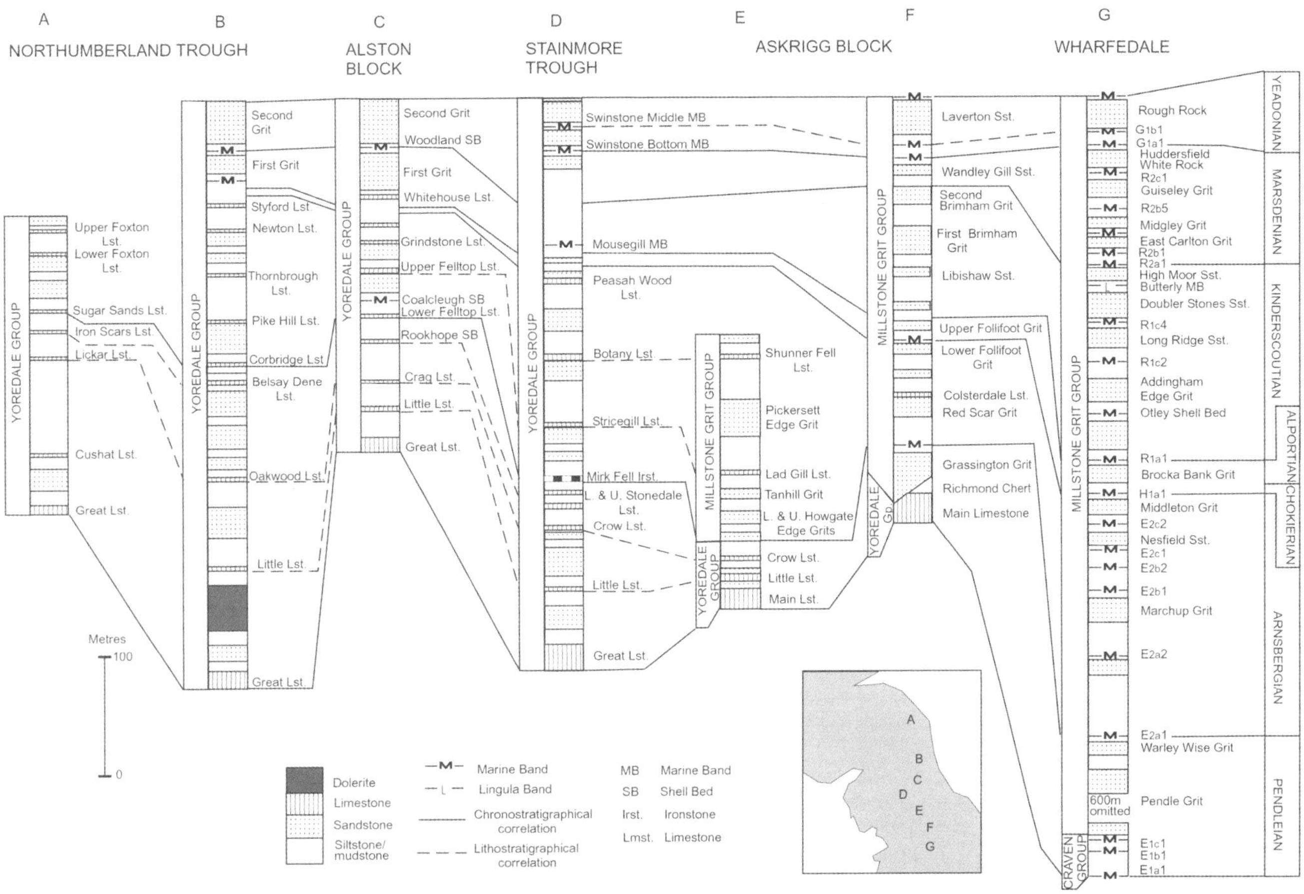

Fig. 9.21. Correlation of Namurian strata across northern England, adapted from Ramsbottom *et al.* (1978).

are present, which suggests a nearshore and/or lacustrine setting for the fine-grained sediments in this region. The newly defined Bowland Shale Formation ranges from late Brigantian to Yeadonian age, with the youngest strata present in the southern part of the Central Pennine Basin.

Millstone Grit Group

The Craven Group is overlain conformably and diachronously by the Millstone Grit Group of the Fluvio-deltaic ('Millstone Grit') lithofacies association. The Millstone Grit Series was until recently defined as a chronostratigraphical unit, equating with strata of Namurian age. The name Millstone Grit Group is now used for a lithostratigraphical unit (Waters *et al.* 2006*a*). The group has been traditionally described as locally named sandstone 'beds', with the intervening argillaceous succession commonly unnamed. Attempts to subdivide the thick cyclic succession have concentrated on the recognition of essentially chronostratigraphical divisions bounded by widespread marine marker bands. This has led to a proposal for component formations, in ascending order, of Pendleton (Pendleian), Silsden (Arnsbergian), Samlesbury (Chokierian and Alportian), Hebden (Kinderscoutian), Marsden (Marsdenian) and Rossendale (Yeadonian) formations (Fig. 9.22).

The Millstone Grit Group extends across most of the Central Pennine Basin. The base of the group is taken at the base of the first thick quartz-feldspathic sandstone, of Namurian age, typically present above the Bowland Shale Formation. The base of the group is markedly diachronous, ranging from Pendleian along the northern margin of the Central Pennine Basin to Yeadonian in North Wales (Fig. 9.22). The thickest development, 1225 m, is recorded in Wharfedale (Ramsbottom *et al.* 1978), in the northern part of the Central Pennine Basin (Fig. 9.21), where deposition commenced first and continued throughout the Namurian.

The fluvial channels recognized on the Askrigg Block supplied sediment to the northern margin of the Central Pennine Basin during Pendleian times. Thick fluvio-deltaic sequences of the Millstone Grit Group were deposited during repeated phases of delta progradation. Deep-water turbidite-fronted lobate delta facies, commonly several hundred metres thick, e.g. Pendle Grit (550 m), Mam Tor Sandstone (Plate 16) to Grindslow Shales (300 m), Roaches and Ashover grits (360 m), characterize early basin sedimentation (Fig. 9.22). These represent systems fed by large distributaries transporting coarse sands directly to submarine feeder channels, by-passing the delta slope to be deposited in a delta-front apron of coalescing turbidite lobes (Walker 1966*b*; Collinson 1969; McCabe 1978; Jones 1980). Following initial basin infill, sedimentation rates began to broadly match subsidence rates and shallow-water sheet-like delta facies were deposited, commonly in cycles tens of metres thick. Significant transport of sediment by turbidity currents was largely absent. Mouth-bar deposits dominate the lower part of each cycle and are overlain by generally sheet-like and laterally extensive sandstones deposited by braided fluvial systems, e.g. Guiseley Grit, Chatsworth Grit, Rough Rock. However, elongate deltas are locally developed, e.g. East Carlton Grit, Haslingden Flags (Collinson 1988). Along the northern margin of the Central Pennine Basin there was complex interdigitation of the deep-water and delta platform deposits during the early Namurian (Fig. 9.22), probably reflecting high-magnitude sea-level rises and falls.

The Namurian deltas were river-dominated with subordinate wave influence. The apparent absence of tidal influence has

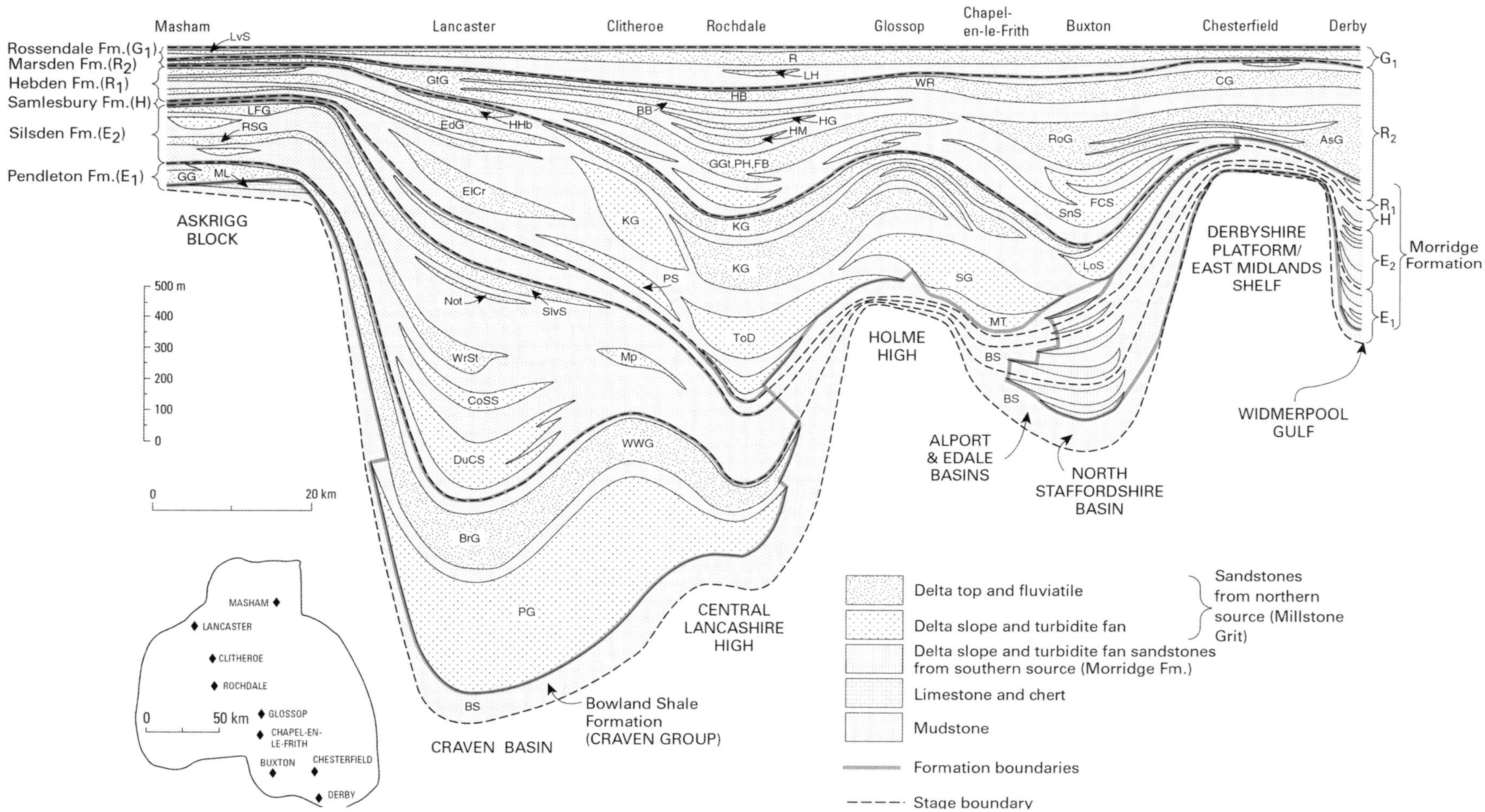

Fig. 9.22. Distribution and lithofacies of the main sandstones of the Millstone Grit Group of the Central Pennine Basin (from Aitkenhead *et al.* 2002). AsG, Ashover Grit; BB, Brooksbottoms Grit; BrG, Brennand Grit; BS, Bowland Shale Formation; CG, Chatsworth Grit; CoSS, Cocklett Scar Sandstones; DuCS- Dure Clough Sandstones; EdG, Eldroth Grit; ElCr, Ellel Crag Sandstone; FB, Fletcher Bank Grit; FCS, Five Clouds Sandstones; GG, Grassington Grit; GGt, Gorpley Grit; GtG, Greta Grits; HB, Holcombe Brook Grit; HG, Hazel Greave Grit; HHb, Heysham Harbour Sandstone; HM, Helmshore Grit; KG, Kinderscout Grit; LFG, Lower Follifoot Grit; LH, Lower Haslingden Flags; LoS, Longnor Sandstones; LvS, Laverton Sandstone; ML, Main Limestone; Mp, Marchup Grit; MT, Mam Tor Sandstones; Not, Nottage Crag Grit; PG, Pendle Grit; PS, Parsonage Sandstone; PH, Pule Hill Grit; R, Rough Rock; RoG, Roaches Grit; RSG, Red Scar Grit; SG, Shale Grit; SlvS, Silver Hills Sandstone; SnS, Sheen Sandstones; ToD, Todmorden Grit; WR, Huddersfield White Rock; WrSt, Ward's Stone Sandstone; WWG, Warley Wise Grit.

been attributed to the location of the Central Pennine Basin within a basin largely isolated from the oceans (Collinson 1988). However, possible tidal features have been recognized within mouth bar sediments deposited by buoyant effluent flumes within Marsdenian and Kinderscoutian deltas in Yorkshire (Aitkenhead & Riley 1996; Brettle *et al.* 2002).

The Millstone Grit Group fluvio-deltaic successions typically show broadly southward flow and a northern provenance. Heavy mineral studies (Drewery 1987; Hallsworth & Chisholm 2000), whole-rock Sm–Nd isotopic ages (Glover *et al.* 1996; Leng *et al.* 1999) and U–Pb radiometric age determinations from zircons and monazites (Drewery 1987; Cliff *et al.* 1991; Hallsworth *et al.* 2000; Evans *et al.* 2001) have recognized sediment contributions from a range of sources. The main sources recognized are Archaean granulite-facies basement, mid–late Proterozoic metasediments and granites of Silurian age associated with the Scandian orogen. Sediment was most probably derived from the margins of the Laurentia and Baltica plates in Greenland and Norway. Locally, deltas appear to infill the Widmerpool Gulf, flowing along the axis of the trough, giving palaeocurrents towards the NW, but have typical northerly-sourced petrography (Jones & Chisholm 1997). The Upper and Lower Haslingden Flags are interpreted as deposits within a birdsfoot delta (Collinson & Banks 1975), with a westerly provenance, shown by McLean & Chisholm (1996).

The earliest Millstone Grit Group succession represents a sudden influx of coarse siliciclastic sediment during the late Pendleian (E_{1c} ammonoid zone), and deposition was restricted to the northern margin of the Central Pennine Basin (Figs 9.21 & 9.22). The initial deposition is represented by the Pendle Grit, which comprises thinly interbedded silty mudstone, siltstone and fine-grained sandstone, cut by massive, laterally impersistent, coarse-grained, pebbly sandstones. These deposits are interpreted as turbidite aprons cut by submarine turbidite channels. The northern part of the basin was rapidly infilled by sediment and the overlying Warley Wise Grit is a shallow-water coarse-grained, cross-bedded sandstone, with periodic emergence indicated by the development of thin coals. Hemi-pelagic mudstones of the Bowland Shale Formation were deposited to the south of the deltas at this time.

During the Arnsbergian there was a return to deposition predominantly from turbidity currents. The succession is dominated by a great thickness of comparatively deep-water siltstone and thin sandstone with some development of laterally impersistent shallow-water cross-bedded sandstones present along the northern margin of the Central Pennine Basin. Thick hemi-pelagic mudstones, including the Caton Shales and Sabden Shales of Lancashire, developed to the south of the turbidite aprons.

During the Chokierian and Alportian comparatively small volumes of sand were introduced into the basin from the north. In Lancashire and Derbyshire a complete succession of hemi-pelagic, dark grey shaly mudstone with numerous marine bands was deposited. In Yorkshire the Brocka Bank Grit, interpreted as a fluvial system, is present in the northern part of the Harrogate Basin, and part of the Alportian succession appears to be absent, possibly representing the mid-Carboniferous unconformity.

During the Kinderscoutian a broadly upwards-coarsening succession reflects a return to a high influx of coarse sands, which extend across most of the Central Pennine Basin (Fig. 9.22). In Derbyshire the lowermost distal turbidite apron is evident as the thin-bedded siltstone and sandstone of the Mam Tor Sandstone (Plate 16). The overlying proximal tur bidite lobes of the Shale Grit comprise thick-bedded, coarse-grained sandstone. Overlying thin-bedded siltstones of the Grindslow Shales represent the delta slope, with feeder channels. In turn, the overlying very thick-bedded, cross-bedded and very coarse-grained sandstones of the Kinderscout Grit were deposited within major fluvial distributaries. The strata show a marked southward thickening up to about 600 m in north Derbyshire, infilling available accommodation space south of the Pendleian and Arnsbergian deltas. The Kinderscoutian succession has been envisaged to represent a simple southward progradation of a turbidite-fronted delta (Walker 1966*a*). However, recent studies suggest that major erosional unconformities (sequence boundaries) were produced by fluvial incision during falling relative sea levels and that sediments bypassed the delta through channels directly into the deep basin (Hampson 1997).

The Marsdenian succession includes laterally extensive sandstones and typically thinner cycles than within the underlying strata (Fig. 9.22). The Marsdenian strata thicken up to 600 m to the west of the Pennines, infilling the accommodation to the south and west of the thick Kinderscoutian succession (Collinson *et al.* 1977). Mouth bar-dominated deltas predominate during the early Marsdenian. There is evidence for incised valley formation and development of forced regressive mouth bars during lowstands (Brettle *et al.* 2002). During rising relative sea level, flooding of the incised valleys results in the formation of a shallow coastline embayment infilled with tidally influenced mouth bar deposits.

The Yeadonian succession typically comprises a lower thick succession of dark shale including the *Cancelloceras cancellatum* and *Cancelloceras cumbriense* marine bands. Small contributions of sediment from the west are recognized in the birdsfoot, fluvially dominated deltas of the Lower and Upper Haslingden Flags of Lancashire. The very coarse-grained sheet sandstone of the Rough Rock dominates the upper part of the formation and is present almost basin-wide. The Rough Rock has been interpreted as the fluvial infill of a lowstand incision surface (Maynard 1992) or lateral migration of a widespread sheet sandstone over an alluvial plain, with low gradients and little incision (Bristow 1988). In North Wales the typical northerly-sourced Millstone Grit Group is represented by the lower part of the Gwespyr Sandstone, which comprises fine-grained sandstone beds with subordinate thinly interbedded sequences of sandstone and mudstone and locally thick units of grey mudstone. Thin coals and seatearths are intermittently developed.

Along the southern margin of the Central Pennine Basin, quartzitic sandstones of the Cefn-y-fedw Sandstone Formation of North Wales and Morridge Formation of Staffordshire, were derived from the Wales–Brabant High to the south. Similarities in age and sedimentary architecture resulted in them being traditionally referred to as Millstone Grit. However, these strata are petrographically and provincially quite distinct.

These formations were deposited coevally and record the northward progradation of fluvio-deltaic sediments during Pendleian–late Marsdenian times. The formations are thickest toward the southern margins of the basin where shallow-water sheet-like delta facies predominate, with a thinner succession of predominantly deep-water turbidite-fronted lobate delta facies deposited to the north (Fig. 9.22). In the Widmerpool Gulf intermixing obscures the distinction between sandstones of southern and northern provenance. The formations comprise up to 600 m thickness of quartzose sandstone, locally pebbly, and thin beds of quartz conglomerate, interbedded with mudstone, siltstone and subordinate chert, commonly arranged in upwards-coarsening cycles up to 20 m thick. The quartzose sands appear to have been derived by reworking Old Red Sandstone deposits present on the Wales–Brabant High to the south.

The Cefn-y-fedw Sandstone is thickest north of the Bala Lineament, where it occurs as three separate sandstone units. The Cefn-y-fedw Sandstone intertongues with and onlaps the Pentre Chert Formation, and both intertdigitate with, and are conformably succeeded by, the Bowland Shale Formation. The three sandstone leaves merge into a single, thick sandstone succession to the south. The Morridge Formation conformably overlies, and intertongues with, the Bowland Shale Formation.

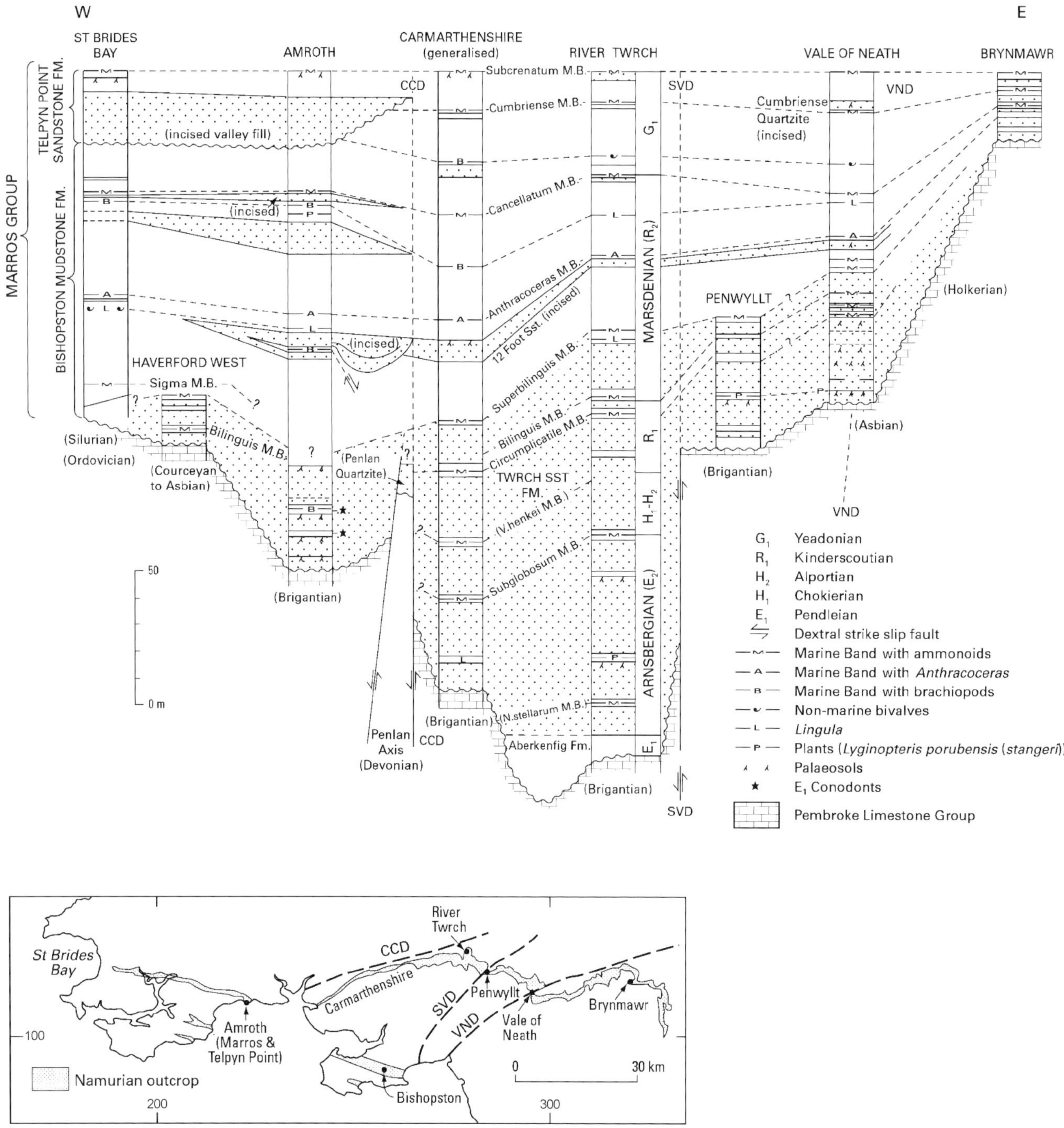

Fig. 9.23. Stratigraphy and correlation of Namurian strata along the North Crop of South Wales, adapted from George (2000). CCD, Carreg Cennan Disturbance; SVD, Swansea Valley Disturbance; VND, Vale of Neath Disturbance.

The youngest sandstone of the Morridge Formation, the Brockholes Sandstones, is of mid-Marsdenian (R_{2b}) age and interdigitates with the Roaches Grit, the local basal sandstone of the Marsden Formation.

South Wales–Southern England

Marros Group

The Namurian succession of South Wales and the Bristol area has traditionally been named Millstone Grit and has been treated as a chronostratigraphical unit equating with the entire Namurian Series. The newly defined Marros Group (Waters *et al.* 2006*a*) represents a lithostratigraphical unit comprising a cyclic lithofacies, similar to the Millstone Grit Group of the Pennines, but dominated by quartzose sandstones derived from the Wales–Brabant High to the north. The source for the Marros Group sediments is likely to be the same as for the Cefn-y-fedw and Morridge formations of the Central Pennine Basin.

Namurian deposition in South Wales commenced with the deposition of up to 35 m of dark grey shales with thin cherts and coarse-grained sandstones (Wilson *et al.* 1990) of the Aberkenfig Formation (the Plastic Clays of Ramsbottom *et al.* 1978) (Fig. 9.23). The formation is probably late Brigantian–early Pendleian in age and is equivalent in facies to the Craven Group developed on the northern flank of the Wales–Brabant High in North Wales. Similar deposits have been recognized in the Bristol area (Kellaway & Welch 1993).

The Marros Group is thickest, about 700 m, in the Gower of South Wales. The group thins and shows several depositional breaks both northwards, approaching the Wales–Brabant High, and eastward, approaching the Usk Anticline (George 1970, Figure 25).

Initial deposition of the Marros Group is characterized by fluvio-deltaic distributary channels, delta front and shoreface–foreshore deposits of the Twrch Sandstone Formation (former Basal Grits). In the North Crop the Twrch Sandstone overlies a prominent unconformity, which has removed the underlying Aberkenfig Formation in the East Crop (Fig. 9.23). In the type area near Ammanford, North Crop, the Twrch Sandstone Formation is dominated by lenticular–tabular quartzose sandstones with thin dark grey mudstone interbeds, some containing ammonoids and *Lingula*, plus minor seatearths and rare thin coals. The sandstones are medium- to coarse-grained, commonly pebbly and conglomeratic, trough cross-bedded and parallel laminated and may occur in upward-coarsening or upward-fining cycles. The fine-grained, well-sorted quartz arenites display hummocky and swaley cross-stratification and parallel to low-angle lamination, with complex palaeocurrent vectors. These deposits are interpreted as storm-dominated delta front and shoreface-foreshore deposits (George 2000). The coarse-grained, poorly sorted sandstones display channelized bases and sharp tops with mature palaeosols, and unimodal southerly-directed palaeocurrents suggesting deposition within braided fluvio-deltaic distributary channels (George 2000). In the type area, the formation ranges from Pendleian to Marsdenian in age, occurring below the *Bilinguites superbilinguis* Marine Band and has a maximum thickness of 190 m. Further to the east, successive overstep in the formation toward the Usk Anticline results in thinning, and progressive eastward younging of the base of the formation from Arnsbergian to Yeadonian age. The formation thins westwards over the Penlan Axis, into the Pembrokeshire sub-basin (Fig. 9.23). The Twrch Sandstone passes laterally westward and southward into the Bishopston Mudstone Formation and is absent in the basin depocentre, around the Gower.

The Bishopston Mudstone Formation (former Middle Shales) comprises three major coarsening-upward deltaic sequences, with interbedded basinal mudstones. In the North Crop, where this formation is comparatively thin and resting conformably upon the Twrch Sandstone, the eroded tops of cycles are overlain by brackish water interdistributary bay muds or marine bands containing benthic brachiopods, crinoids and bivalves indicative of a shallow, near-shore marine environment (Ramsbottom 1969). To the south and SW, where the formation is in excess of 700 m in Pembrokeshire (Ramsbottom *et al.* 1978), the marine bands are characterized by an open marine assemblage of ammonoids and pectinoids. In the East Crop and South Crop, quartzitic sands were entering the basin from the east and south, respectively, continuing through the Marsdenian and Yeadonian (Oguike 1969). The 9 m-thick transgressive systems tract developed beneath the *Cancelloceras cancellatum* Marine Band in Pembrokeshire records the northward migration of lagoonal and accompanying beach deposits and outer barrier bar sands associated with development of a lag-pavement (George 1970). This sea-level rise resulted in widespread deposition of muds, with thin beds of silt and sand attributed to density flows or turbidity flows at the base of pro-delta slopes. In the North Crop input of sands was almost entirely confined to distributary channels, which fed southward and southwestward advancing deltas. Locally, sand-filled channels incised into sheet-like bodies of littoral sand and silt.

During the late Namurian there is a localized return to fluvial sand-dominated deposition within the SW part of the basin (Fig. 9.23). The Telpyn Point Sandstone Formation (formerly one of several 'Farewell Rocks') comprises massive, cross-bedded, upward-fining sandstones, interpreted as fluvial infills of incised valleys (Hampson 1998). Palaeoflows within the incised valleys show a downflow change from southward to eastward, possibly reflecting a diversion around a growth fold evolving within the hanging-wall of the Ritec Fault (George 2001). The channel sandstone is overlain abruptly by argillaceous floodplain and/or estuarine bay sediments. The formation, of Yeadonian age, is limited to the Pembrokeshire sub-basin, where it is up to 30 m thick (Kelling 1974). To the east of Pembrokeshire, the Cumbriense Quartzite of the North Crop is of comparable age to the Telpyn Point Sandstone (Fig. 9.23). The Cumbriense Quartzite is interpreted as a storm-influenced, forced regressive barrier developed on the northern margin of a shallow bay (George 2001), or as a well-developed interfluve (Hampson 1998).

In the Bristol Coalfield, the Quartzitic Sandstone Formation forms the entire thickness of the Marros Group and is similar in appearance to the Twrch Sandstone Formation of South Wales. It consists of hard, pale grey quartzitic sandstone, sometimes conglomeratic with pebbles of white quartz, chert and ironstone, particularly above erosive surfaces (Kellaway & Welch 1993). Grey mudstone, including some *Lingula* bands, and seatearths with thin carbonaceous or coaly beds, occur as interbeds. The formation is up to 300 m thick in the northern part of the Bristol Coalfield, thinning southward.

Westphalian and Stephanian: coal swamps and redbeds

Introduction

Westphalian and Stephanian strata in England and Wales were deposited in three distinct depositional areas; the Pennine Basin extending from the Southern Uplands of Scotland to the Wales–Brabant High, and the Variscan foreland basins located to the south of the Wales–Brabant High, including the South Wales and Culm basins.

Although the Pennine Basin and South Wales Basin were linked at times, there was relatively little flux of sediment between them, as the Wales–Brabant High was an upland area for most of the time. These basins show a common evolution from grey Fluvio-lacustrine ('Coal Measures') lithofacies association during early Westphalian times to Alluvial (Barren Measures) lithofacies in late Westphalian and Stephanian times (Figs 9.9e, f).

The Pennine Basin continued to subside during early Westphalian times as a consequence of thermal crustal sagging caused by cooling of the asthenosphere beneath the thinned lithosphere. Deposition occurred in a fluvio-lacustrine environment, with marine transgressions that became less common through time (Calver 1968, 1969). The humid tropical climate resulted in a largely waterlogged depositional plain. The northerly sediment source persisted from Namurian times into the Westphalian. In the northern part of the Pennine Basin (Northumberland and Durham) this northerly source continued to Bolsovian times. However, in the southern part of the Pennine Basin a western source became predominant during the Langsettian and Duckmantian. This westerly source persisted until Bolsovian times when a source from the SE supplied additional sediment supply. Towards the southern and northern margins of the basin the Coal Measures successions are condensed and rest unconformably upon older rocks. Primary red bed facies associated with periodic well-drained soils commenced early in the Westphalian at the extreme southern margin of the Pennine Basin and became more extensive by Bolsovian times. This marked the demise of the predominantly fluvio-lacustrine environments and the onset of predominantly alluvial sedimentation. The reasons for the change are complex, involving factors that include tectonic inversion associated with the Variscan Orogeny, and the introduction of an increasingly arid climate to the north of the developing

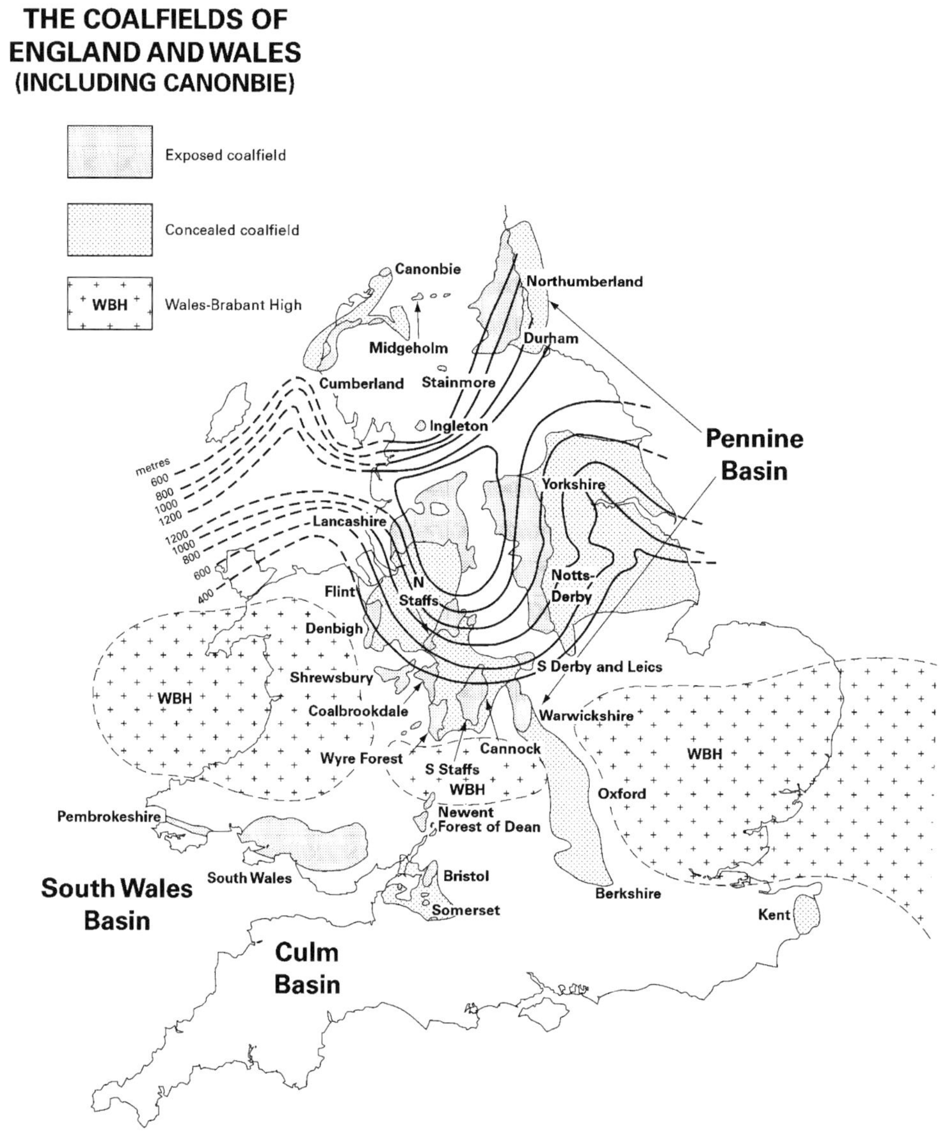

Fig. 9.24. Distribution of Westphalian coalfields at outcrop and within the subsurface of England and Wales (after Ramsbottom *et al.* 1978 and Powell *et al.* 2000*a*, with additions). Restored and generalized isopachs of the Lower and Middle Coal Measures within the Pennine Basin, modified from Calver (1968) and Kirby *et al.* (2000).

Variscan mountain belt (Besly 1988*a*). The Variscan deformation isolated previously contiguous Coal Measures strata into a number of separate coalfields (Fig. 9.24).

The South Wales and Culm basins evolved during the Westphalian and Stephanian as flexural foreland basins, developed in front of a northward propagating Variscan Orogen. The fluvial 'Pennant' facies, derived from the evolving Variscan thrust-fold belt to the south, prograded northwards across southern England and South Wales during late Westphalian and Stephanian times, eventually spilling across the Wales–Brabant High barrier into the southern margin of the Pennine Basin.

There was a resurgence in volcanic activity, notably along the southern margin of the Pennine Basin, after a period of quiescence during the Namurian. Alkaline igneous activity produced dolerite sills and lavas in the East Midlands, and sills in the West Midlands. In addition, explosive volcanicity produced tuffs and thin ash-fall clays, generally a few millimetres–centimetres thick, which are referred to as bentonites or tonsteins, and are typical Westphalian volcanic products. Acidic ash-fall deposits generally cover very large areas and have been associated with Variscan volcanic activity at a destructive plate margin to the south of Britain (Spears & Lyons 1995). Basic bentonites are more locally developed. Thermal subsidence is not normally associated with igneous activity (Leeder 1982), leading to suggestions that Variscan deformation south of the Wales–Brabant High resulted in reactivation of deep-seated basement lineaments, which acted as conduits for upward migration of magma (Waters *et al.* 1994).

Chronostratigraphy

The base of the Westphalian Series is taken at the base of the *Gastrioceras subcrenatum* ammonoid Zone, which broadly equates with the first incoming of thick coal seams in England and Wales. The series is subdivided into four stages (Fig. 9.25). The informal nomenclature of A, B, C and D has been superseded by the formal definition of the Langsettian (Westphalian A), Duckmantian (B) and Bolsovian (C) stages with boundary stratotypes (Fig. 9.11) defined in the East Pennine Coalfield of England (Owens *et al.* 1985). Westphalian D has, as yet, no formally defined stratotype and remains unnamed. The bases of the three formally defined stages are taken at the base of marine strata (marine bands), which display the earliest occurrence

STAGE	AMMONOIDS	CONODONTS	PALYNOMORPHS Index	PALYNOMORPHS Zone	MACROFLORA	MACROFLORA	NON-MARINE BIVALVES Zone	NON-MARINE BIVALVES Subzone
Cantabrian					*Odontopteris cantabrica*		*Anthraconauta tenuis*	
					Lobatopteris vestita	*Dicksonites plueckenetii*		
Westphalian D			XI	*Thymospora obscura*		*Lobatopteris micromiltoni*		
					Linopteris bunburii			
Bolsovian			X	*Torispora securis*	*Paripteris linguaefolia*		*Anthraconauta phillipsi*	
							'Upper *similis-pulchra*'	
			IX	*Vestispora magna*				*adamsi-hindi*
Duckmantian							'Lower *similis-pulchra*'	*atra*
			VIII	*Dictyotriletes bireticulatus*	*Lonchopteris rugosa*	*Sphenophyllum majus*		
								caledonica
						Neuropteris hollandica		*phrygiana*
					Lyginopteris hoeninghausii	*Laveineopteris loshii*	*Anthraconaia modiolaris*	*ovum*
			VII	*Schulzospora rara*				*regularis*
Langsettian								*cristagalli*
			VI	*Radiizonates aligerens*			*Carbonicola communis*	*pseudorobosta*
						Neuralethopteris jongmansii		*bipennis*
								torus
			SS	*Triquitrites sinan – Cirratriradites saturni*			*Carbonicola lenisulcata*	*proxima*
	Gastrioceras listeri	*Idiognathoides sulcatus parvus*						
								extenuata
	Gastrioceras subcrenatum							*fallax-protea*
		Idiognathoides sinuatus – Idiognathoides primulus (part)						

Fig. 9.25. Westphalian chronostratigraphy and biostratigraphical zonations. Biozonation is derived from authors referred to in the text.

of certain diagnostic ammonoid faunas that can be recognized extensively across Western Europe. The key faunas are those of *Gastrioceras subcrenatum* (base Langsettian), *Anthracoceratites vanderbeckei* (base Duckmantian) and *Donetzoceras aegiranum* (base Bolsovian) marine bands.

The Stephanian Series was originally defined in the Central Massif of France with three stages, referred as Stephanian A, B and C. The Stephanian A has been formally renamed the Barruelian Stage. The recognition of a non-sequence in the Central Massif and identification of an additional Stephanian succession in Cantabria, northern Spain, led to the recognition of a Cantabrian stage, which is older than the Barruelian. Only strata of Cantabrian age have been recognized in England and Wales.

Biostratigraphy

Much reliance is placed on the thick-shelled ammonoid biostratigraphy in the definition of the Langsettian, Duckmantian and Bolsovian stages (see above). Ammonoid-bearing marine bands have also been used as marker beds to allow correlation of successions within and between coalfields. These key marine bands, identified on Fig. 9.25 after Ramsbottom *et al.* (1978), are named after diagnostic ammonoid species. In contrast to the Namurian, the Westphalian Series includes relatively few marine bands with thick-shelled ammonoids. Marine bands broadly decrease in abundance upwards, with none recorded in Britain above the Bolsovian Cambriense Marine Band. Correlation in the Westphalian is aided by the presence of an additional framework of marine and near-marine bands that contain less diagnostic faunas, such as thin-shelled ammonoids (*Anthracoceras*), marine bivalves (e.g. *Dunbarella*), brachiopods (*Lingula*) and crustacea (ostracodes, estheriids), which are invaluable for detailed correlation. These marine bands typically have geographical names (see Fig. 9.26).

Non-marine bivalve biozones for the Westphalian were defined by Trueman & Weir (1946), see Figure 9.25, though with time the zonal boundaries have been fixed at prominent coals or marine bands and recognized as chronozones. The zonation is of particular importance in the correlation of late Westphalian strata, which lack marine bands. Notably, the base of Westphalian D is taken to correspond with that of the *Anthraconauta tenuis* Zone (Calver 1969).

Miospore zonations have been developed for Westphalian coal seams by Smith & Butterworth (1967), but realize greatest importance within late Westphalian strata lacking alternative biostratigraphical control (Fig. 9.25). The base of the *Thymospora obscura* miospore Zone equates with that of the non-marine bivalve *Anthraconauta tenuis* Zone (Calver 1969), at the base of Westphalian D. Advances in microfloral zonation have provided an improvement to the early attempts at macrofloral biostratigraphical zonation, initially developed by Kidston (1905) and refined by Dix (1934) and Crookall (1955), which can be used for coarse correlation. Recent advances in macrofloral zonation show the importance of plant fossil biostratigraphy, particularly for Westphalian D and the Cantabrian (Cleal 1991).

Sequence Stratigraphy

The application of sequence stratigraphic principles to the Coal Measures of the Pennine Basin has proved less successful than within the underlying Millstone Grit Group. Marine bands, representing maximum flooding surfaces, are present within the Coal Measures succession. However, the decrease upward in number of marine bands indicates a reduction in marine influence within the basin with time. Lowstands of sea level produce incision of rivers and the development of well-drained palaeosols on interfluves. However, regionally developed well-drained palaeosols are not common and major fluvial sandbodies show only limited basal incision of up to 5 m, exceptionally up to 8 m (Rippon 1996). This may reflect distance from the sea, with incision of river channels having insufficient time to work upstream from the coast before the next flooding

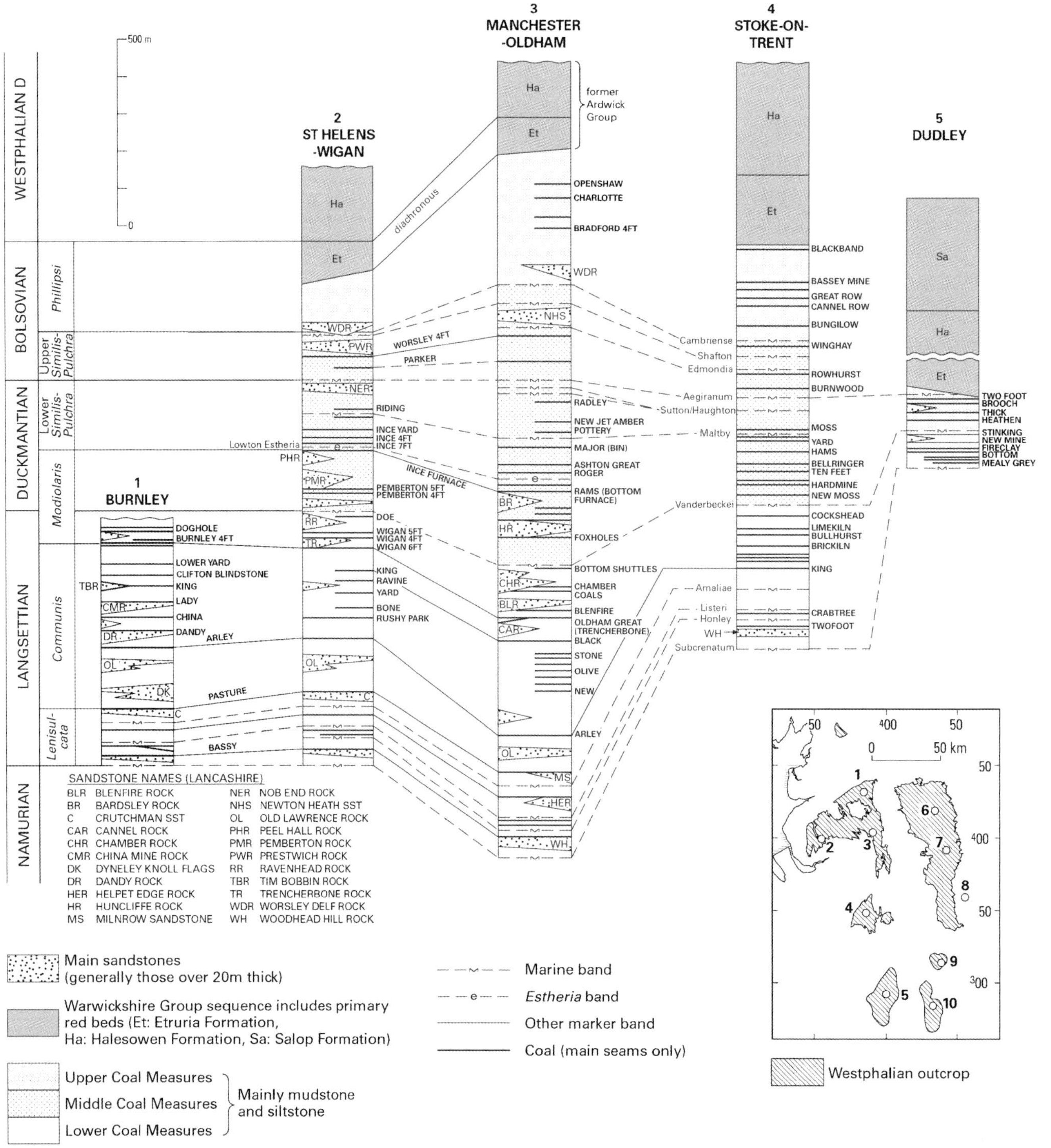

Fig. 9.26. Correlation of Pennine Coal Measures Group sandstones, coals and marine bands within: (**a**) the western part of the Pennine Basin; (**b**) eastern part of the Pennine Basin (amended after Aitkenhead *et al.* 2002).

event (Aitkenhead *et al.* 2002). Alternatively, it may relate to the enclosed nature of the basin. Only the highest global sea level rises would result in a rapid base-level rise within the basin, with the subsequent fall in sea level leaving an isolated basin for which base levels may fall comparatively slowly as it continues to be fed water by rivers.

The Langsettian Crawshaw Sandstone of the East Midlands Shelf is less than 70 km wide and has well defined margins with identifiable interfluves (Church & Gawthorpe 1994; Hampson *et al.* 1999). This is the last representative of this style of fluvial deposition, at least from the onshore UK. Subsequent fluvial systems are characterized by less laterally extensive sandstone bodies associated with less basal erosion, that were developed on a low-lying alluvial (?coastal) plain (Davies *et al.* 1999).

Duckmantian and Bolsovian sandstones examined by Aitken & Chisholm (1999, 2000) and Aitken *et al.* (1999) can also have local basal erosional relief and are complex, multistorey sandstone bodies. The basal Duckmantian Thornhill Rock cuts through the Vanderbeckei Marine Band (Lake 1999) and reaches thicknesses of 37–45 m in the Huddersfield and Wakefield areas. The Late Duckmantian Woolley Edge Rock has a locally erosive base and averages 30 m in thickness.

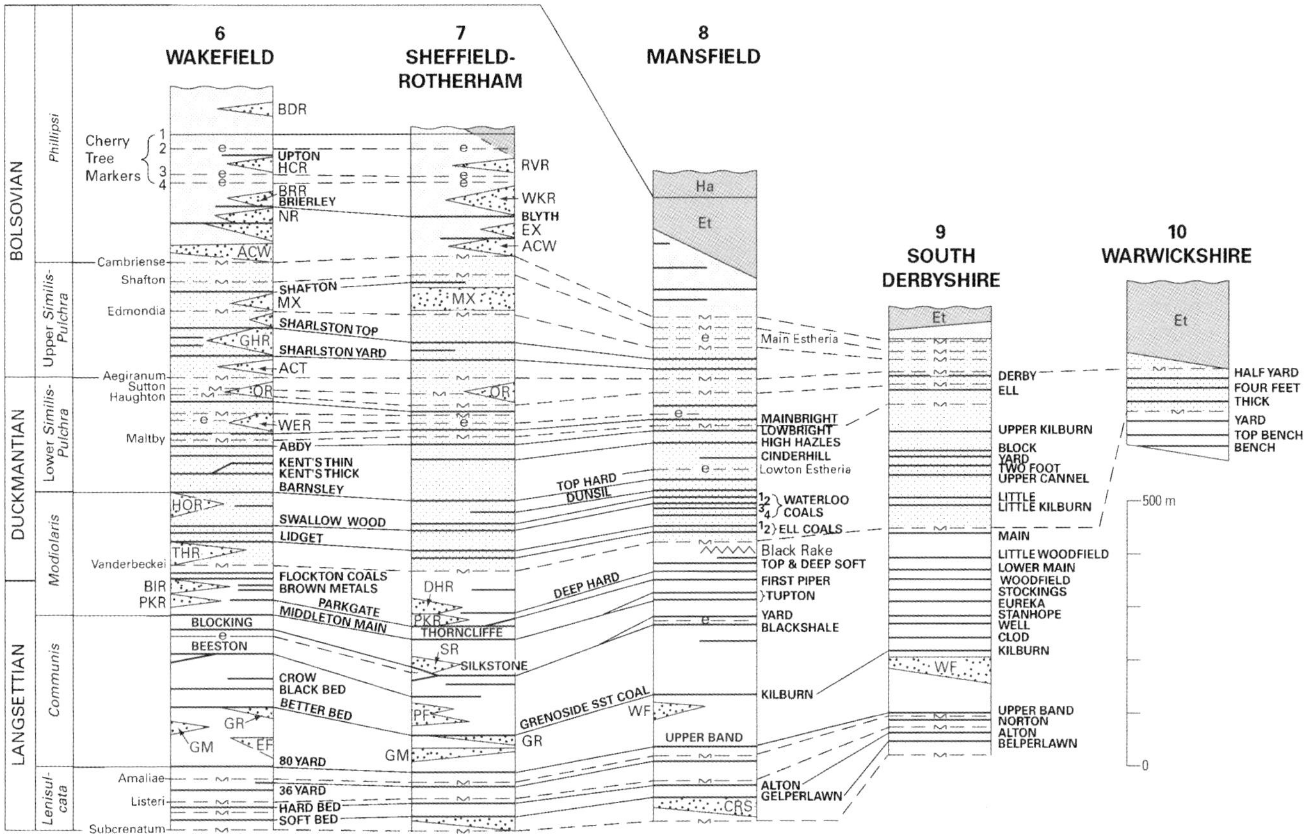

SANDSTONE NAMES (EAST PENNINES)

ACT	ACKTON ROCK	HCR	HOUGHTON COMMON ROCK
ACW	ACKWORTH ROCK	HOR	HORBURY ROCK
BDR	BADSWORTH ROCK	MX	MEXBOROUGH ROCK
BIR	BIRSTALL ROCK	NR	NEWSTEAD (PONTEFRACT) ROCK
BRR	BRIERLEY ROCK	OR	OAKS ROCK
CRS	CRAWSHAW SANDSTONE	PF	PENISTONE FLAGS
DHR	DEEP HARD ROCK	PKR	PARKGATE ROCK
EF	ELLAND FLAGS	RVR	RAVENFIELD ROCK
EX	'EXTRA' ROCK	SR	SILKSTONE ROCK
GHR	GLASS HOUGHTON ROCK	THR	THORNHILL ROCK
GM	GREENMOOR ROCK	WER	WOOLLEY EDGE ROCK
GR	GRENOSIDE SANDSTONE	WF	WINGFIELD FLAGS
		WKR	WICKERSLEY ROCK

Volcanic bed

Fig. 9.26. Continued.

Outcrop and subsurface mapping suggest an approximate width for this sandstone of 23 km. The Bolsovian Glass Houghton Rock and Mexborough Rock, also average 30 m in thickness but reach maximum thicknesses of 60–80 m locally. The width of the Glass Houghton Rock cannot be determined, but is at least 15 km. The Mexborough Rock is 15–30 km wide and both bodies can be traced for some 80–90 km in a depositional dip sense. Research on these sandstones suggest a relationship between the channel sandstones and their adjacent strata that includes high ash contents in coals, coal splitting towards channels and an increase in interbedded sandstone layers in proximity to the channel bodies (Bedrock 1984; Guion *et al.* 1995; Aitken *et al.* 1999). These observations imply that overbank flooding events from the channels occurred during peat accumulation and leads these authors to believe that the channel systems were aggradational, as opposed to having filled previously incised valley systems.

Westphalian cyclicity is characterized by the Coal Measures coarsening-upward cyclothem (Plate 15), analysed by Duff & Walton (1962) for the East Pennines Coalfield. In excess of 80% of the approximately 100 cycles recognized within the East Pennines Coalfield contain non-marine shales at their base. It is probable that deposition in these cyclothems occurred in areas distant from marine influences, with the cyclicity reflecting autocyclic delta lobe switching, augmented by local tectonic influences. Duff & Walton (1962) determined that non-marine cycles have a mean thickness of 7.3 m, as opposed to 11.6 m for the much rarer marine cycles. This suggests cyclothems generated in response to sea-level rise provided greater accommodation space for the deltaic sediments. The Subcrenatum, Vanderbeckei and Aegiranum marine bands occur within significantly thicker cyclothems, on average 18.6 m thick, suggesting that these reflect extreme sea-level rises.

A regular rhythmic repetition of arenaceous and argillaceous beds within a small interval of the Lower Coal Measures (Langsettian) of Lancashire has been interpreted as annual monsoonal deposition, of sand during the wet seasons and silt during dry seasons (Broadhurst *et al.* 1980). Groupings of thicker and thinner sandstone beds suggest systematic climatic variations over longer periods.

Regional Developments

Pennine Basin

Pennine Coal Measures Group

The Westphalian Pennine Coal Measures Group of the Pennine Basin extends at outcrop or at subcrop across northern England, to the west and east of the Pennine Anticline, central

England and North Wales (Fig. 9.24). The Coal Measures have historically had a chronostratigraphical connotation synonymous with Westphalian plus Stephanian strata, including both coal-bearing and barren 'red-bed' successions. However, the name has now been defined lithostratigraphically, to describe the main body of coal-bearing strata of Westphalian age (Powell *et al.* 2000*a*).

The Coal Measures have traditionally been divided into three formations, as defined by Stubblefield & Trotter (1957): Lower Coal Measures (Langsettian age); Middle Coal Measures (Duckmantian–Bolsovian age); Upper Coal Measures (Bolsovian–Westphalian D in age). The bases of the two lower formations are taken at the bases of key marine bands; the Subcrenatum Marine Band for the Lower Coal Measures, Vanderbeckei Marine Band for the Middle Coal Measures. The Upper Coal Measures are defined as lacking marine shales, the base of the formation being drawn at the top the Cumbriense Marine Band. The formation is limited to strata of the Fluvio-lacustrine ('Coal Measures') lithofacies association, so excludes barren or red measures in the Pennine Basin and 'Pennant' measures in the South Wales Basin.

By the end of the Namurian, the inherited Dinantian basinal topography had been largely infilled by fluvio-deltaic sediments of the Millstone Grit Group. Early Westphalian strata are marked by a gradual change to deposition in a delta-plain environment with lakes and distributary channels (Guion & Fielding 1988; Guion *et al.* 1995). This Pennine Coal Measures Group comprises cyclic sequences that are thinner, tens of metres, and more numerous than in the underlying Millstone Grit Group, the sandstones are typically finer, and widespread marine transgressions are less common.

The main distributary channels of the Coal Measures, up to 20 m thick and 10 km wide, were filled by relatively thick, sharp-based sands. Between the channels were freshwater lakes where mudstones were deposited. Upward-coarsening siltstones–sandstones, deposited as small deltas and crevasse splays, filled the lakes. Near-emergent surfaces were colonized by plants and became swamps or raised peat bogs, which formed coals following burial. Subsidence rates were lower along the southern margin of the Pennine Basin resulting in a relatively few thick seams (Fig. 9.26a, b). Northwards, towards the basin depocentre, subsidence rates were greater and seams split. Some of the cycles, particularly in the early Langsettian, commence with laterally widespread marine bands. Within the main basin depocentre, the Subcrenatum, Listeri, Vanderbeckei, Aegiranum and Cambriense Marine Bands represent notably widespread marine flooding events that are useful for correlation between separate coalfields (Fig. 9.26a, b). In basin-margin areas, marine influence was less strong and marine bands are dominated by foraminifera, *Lingula*, fish remains and shallow-marine, benthonic productoid faunas; correlation is less certain here.

The Coal Measures have been subdivided into deposits of 'lower and upper delta-plain' environments (Guion & Fielding 1988). Within the lower delta plain the number and duration of marine flooding events were greater and the time available for peat accumulation, principally as low-lying swamps, was shorter, resulting in coals that are thin and of relatively poor quality. Within the upper delta plain, marine influences were less frequent, and prolonged periods of peat accumulation, mainly as raised mires, resulted in the formation of thick coals with comparatively low ash and sulphur contents.

The lithofacies comprise cyclothems of alternating sandstone, siltstone and mudstone, with frequent coal seams, ironstone nodules or beds, and seatearth (palaeosol) horizons (Plate 15). The base of each cycle is marked by dark grey mudstones, commonly with non-marine faunas, less commonly with marine faunas. Both are of importance in the correlation of strata. The mudstones are typically grey and well-bedded, commonly carbonaceous and sometimes fissile. They include ironstones comprising siderite with clay and silt (clayband ironstone) or carbonaceous laminae (blackband ironstone). Palaeosols present within argillaceous rocks may occur as unleached gley seatearths with abundant roots and polished (listric) surfaces, which formed where water tables were at or above ground level for prolonged periods. Pale grey and ochreous mottled seatearths (fireclays) also occur where palaeosols were better drained. Siltstones are usually grey, laminated and sometimes burrowed, and grade into mudstone or sandstone. Sandstones are typically pale grey, buff-weathered, very-fine- to fine-grained and constitute less than 30% of the total cycle thickness. Sandstones in the Durham–Northumberland coalfields are more commonly coarser and may form up to 50% of the cycle thickness (Rippon 1996). Sandstone palaeosols range from rooty sandstones to leached ganisters. Coal seams typically rest on seatearths and can have a wide range in thickness. Many of the worked seams had a typical thickness of 1–3 m, but the Thick Coal of South Staffordshire represents an amalgamation of coal seams to form a single coal up to 17 m thick (Fig. 9.26a). Most of the coals in England and Wales are bituminous, although higher rank anthracite coals are present in South Wales.

The thickest development of the group is up to 1900 m near Manchester, located in the basin depocentre (Fig. 9.24). A condensed succession is developed at the basin margins, with as little as 150 m present in the South Staffordshire Coalfield (Fig. 9.26). The base of the group generally rests conformably upon the Millstone Grit Group. Along the southern margin of the basin, the Pennine Coal Measures Group oversteps the Millstone Grit Group and rests unconformably upon pre-Carboniferous strata of the Wales–Brabant High. This topographical feature, persistent throughout most of the Carbonifereous, appears to have been subdued, and provided little clastic material into the Pennine Basin during the Westphalian. In many areas the top of the group is defined by the sub-Permian unconformity, which developed in response to end-Carboniferous basin inversion. A variable thickness of secondary reddening is evident beneath this unconformity. In other places, especially at the basin margins the Pennine Coal Measures Group is overlain conformably, or locally unconformably, by primary Westphalian strata of 'Red-bed' facies, including the Warwickshire Group of the English Midlands (Plate 15) and the Whitehaven Sandstone Formation in Cumbria.

The lower part of the Lower Coal Measures, broadly equating with the non-marine bivalve *Lenisulcata* Zone, shows comparable features across the Pennine Basin and maintains aspects of deposition established during the late Namurian. Within the basin depocentre, this interval, between the Subcrenatum Marine Band and the 80 Yard (Pasture) Coal (Fig. 9.26a, b), includes numerous marine bands, the sandstones are typically micaceous, ganisters are common and coals thin (Aitkenhead *et al.* 2002). The feldspathic sandstones are considered to have the same source from the north or NE as the underlying Millstone Grit Group (Chisholm *et al.* 1996; Hallsworth & Chisholm 2000; Hallsworth *et al.* 2000). The Crawshaw Sandstone of Derbyshire is a coarse-grained sandstone, which has the appearance of a Millstone Grit sandstone. It shows palaeocurrents from the east, which appear to be tectonically constrained within the Edale Gulf and Widmerpool Gulf (Guion & Fielding 1988), but was also derived from the same northern source. In the Durham Coalfield this interval is recognized as lower delta-plain deposits by Fielding (1984). These deposits comprise major distributary channels, including a set of coarse-grained sandstones known as the 'Third Grit', with palaeocurrents predominantly towards the south. The interdistributary bay fill sediments show common wave reworking and the presence of brackish water faunas. Quartz arenitic sandstone bodies represent shoreline and shallow-marine wave-reworked deltaic sands. Marine bands are more common within this interval, though not as abundant as within the equivalent interval near the basin depocentre. In comparison with the southern sections, coal seams are relatively uncommon, and where present are generally thin (up to 0.8 m) and laterally discontinuous within this interval.

Within the basin depocentre the middle part of the Lower Coal Measures, between the 80 Yard Coal and Kilburn (Better Bed, Arley) Coal, comprises three basin-wide cyclothems characterized by thin and rare coal seams and very restricted marine band faunas (Fig. 9.26a, b). The sandstones are either micaceous and sourced from the north, or greenish grey, weakly micaceous and clay-rich, sourced from the west (Chisholm 1990; Chisholm *et al.* 1996; Hallsworth & Chisholm 2000). The latter contain grains of chrome spinel indicative of an origin from ophiolitic rocks, possibly as distant as the Appalachian–Labrador–Newfoundland area. The upper part of the Lower Coal Measures, from the Kilburn Coal to the base of the Vanderbeckei Marine Band, is a succession of laterally impersistent cyclothems, which lack true marine bands and have thick coals with many seam splits. The sandstones in this interval continue to be sourced from the west (Chisholm *et al.* 1996; Hallsworth & Chisholm 2000; Hallsworth *et al.* 2000). This source direction agrees with the predominance of W–E-oriented distributary channels across the region (Rippon 1996). Within the Durham Coalfield the middle and upper part of the formation are interpreted as upper delta-plain deposits (Fielding 1984) dominated by distributary sandbodies with variable palaeocurrents. These distributaries display both diagonal and vertical stacking patterns, indicative of compaction-assisted subsidence and tectonic subsidence, respectively. The inter-distributary lacustrine deposits lack marine influence. Coal seams are common, laterally extensive and may be up to 4 m thick within this interval.

The lower part of the Middle Coal Measures is similar across much of the Pennine Basin and represents a continuation of the features evident within the upper part of the Lower Coal Measures. Within the basin depocentre of Lancashire and Yorkshire this succession, between the Vanderbeckei Marine Band and Maltby Marine Band (Fig. 9.26a, b), includes sandstones sourced from the west (Hallsworth & Chisholm 2000). In the succession above the Maltby Marine Band, marine and estheriid faunal bands are common and coals are comparatively thin, interpreted as a return to lower delta-plain deposition in the Durham Coalfield (Fielding 1984). In Yorkshire, sandstones are generally thicker and coarser than is typical of the underlying Coal Measures (Aitkenhead *et al.* 2002). Heavy minerals suggest the start of a new influx, this time from the east and SE (Hallsworth & Chisholm 2000). This new source appears to be in central Europe, probably associated with the Variscan fold-belt, with isotopic signatures indicative of erosion of rocks affected by Variscan, Cadomian, Acadian and Sveco-Norwegian orogenies (Evans *et al.* 2001). The river systems rising from the Variscan Highlands appear to have passed to the east of the Wales–Brabant High into the eastern part of the Pennine Basin.

The Upper Coal Measures, here, as elsewhere throughout England and Wales, lack marine bands, although beds with estheriids are common, and coal seams are thin. In Yorkshire there are locally common medium-grained sandstones with continued derivation from the east or SE (Fig. 9.26b). Over much of the west Cumbria coalfield strata from the Upper *Similis-pulchra* Zone appear to be cut out below the unconformable base of the Whitehaven Sandstone.

Following a period of practically no igneous activity within the Pennine Basin during the Namurian, the Westphalian saw resurgence in activity along the southern margins of the basin. In the East Midlands evidence of igneous activity during the Westphalian is limited to the subsurface, revealed by coal and oil exploration (Burgess 1982). Olivine–basalt lavas occur within the Langsettian Lower Coal Measures. Extrusive igneous activity appears to have terminated abruptly at the end of the Langsettian. Olivine–dolerite sills, up to 40 m thick also occur, intruded into strata up to early Duckmantian in age. Both sills and lavas vary in composition from tholeiitic to alkaline basanite and basalt, basaltic hawaiite and hawaiite (Kirton 1984).

Warwickshire Group

The predominantly primary 'Red-bed' facies strata that overlie the Pennine Coal Measures Group have, in the past, been referred to collectively by informal terms such as the 'Barren (Coal) Measures' and 'Red Measures', and a new formal name, the Warwickshire Group, has been introduced to replace these (Powell *et al.* 2000*a*). The group is best developed along the southern margin of the Pennine Basin, notably in Warwickshire (Plate 15), the type area, in Staffordshire and North Wales (Fig. 9.27). The group is also proved, mainly in the subsurface, in Lancashire, South Yorkshire, Nottinghamshire and Lincolnshire. The group includes red-bed formations and interbedded coal-poor, grey formations, and ranges in age from Duckmantian to Autunian. Primary red measures are also found in the northern margin of the Pennine Basin, with the Bolsovian–Westphalian D Whitehaven Sandstone Formation present in west Cumbria. These Warwickshire Group 'red-beds', have undergone oxidation at, or close to, the time of deposition. The sediments were deposited within alluvial and lacustrine environments, with no marine influences. The Warwickshire Group consists of interbedded mudstone, siltstone, sandstone and conglomerate, typically red, brown, purple or grey, often mottled. The succession is divided into formations, some of which can be correlated between coalfields (Fig. 9.27). The formations typically include siliciclastic material derived from erosion of the Wales–Brabant High; the exception is the Halesowen Formation, which includes lithic arenites similar to the Pennant Sandstone Formation and indicative of a more distant southerly source.

The Warwickshire Group has its thickest development, 1225 m, in the Warwickshire Coalfield (Fig. 9.27). The lower boundary is taken generally at the base of the lowest red-bed formation (generally Etruria Formation) of late Carboniferous age (Plate 15). There is commonly a passage by alternation from grey strata of the Pennine Coal Measures Group into the Etruria Formation. Locally, the base is erosional (the Symon Unconformity) in Warwickshire (Plate 15), Coalbrookdale and Lincolnshire. A higher unconformity, below the Westphalian D Halesowen Formation, is traceable throughout most of the southern part the Pennine Basin, but has not been recognized in the northern basinal successions of South Yorkshire and Lancashire (Waters *et al.* 1994). A further unconformity (or disconformity) is present at the base of the Clent Formation in the southern part of the South Staffordshire Coalfield. The boundary becomes gradational in the north of the coalfield. The upper boundary of the group is taken at the base of unconformable (post-Variscan) Permian, or younger strata.

The well-drained alluvial floodplain deposits of the Etruria Formation mark the onset of red-bed' deposition. The formation comprises predominantly mudstone, coloured red, purple, brown, ochreous, green and grey mottled, with pedogenic horizons common (Glover *et al.* 1993). Subordinate lenticular sandstones and conglomerates (Plate 15) commonly include locally derived volcanic and lithic clasts. Thin coal seams are present in places. The Etruria Formation is generally Bolsovian in age, but ranges from late Duckmantian to early Westphalian D. The lower boundary is highly diachronous (Fig. 9.27), with primary red-beds forming earlier in the Pennine Basin margins (Besly 1988*b*; Waters *et al.* 1994). In Lancashire, the base of the formation contains non-marine bivalves of the *A. phillipsii* and *A. tenuis* zones (Trotter 1954), suggesting a Bolsovian–Westphalian D age for the onset of primary 'Red-bed' facies deposition (Fig. 9.26a).

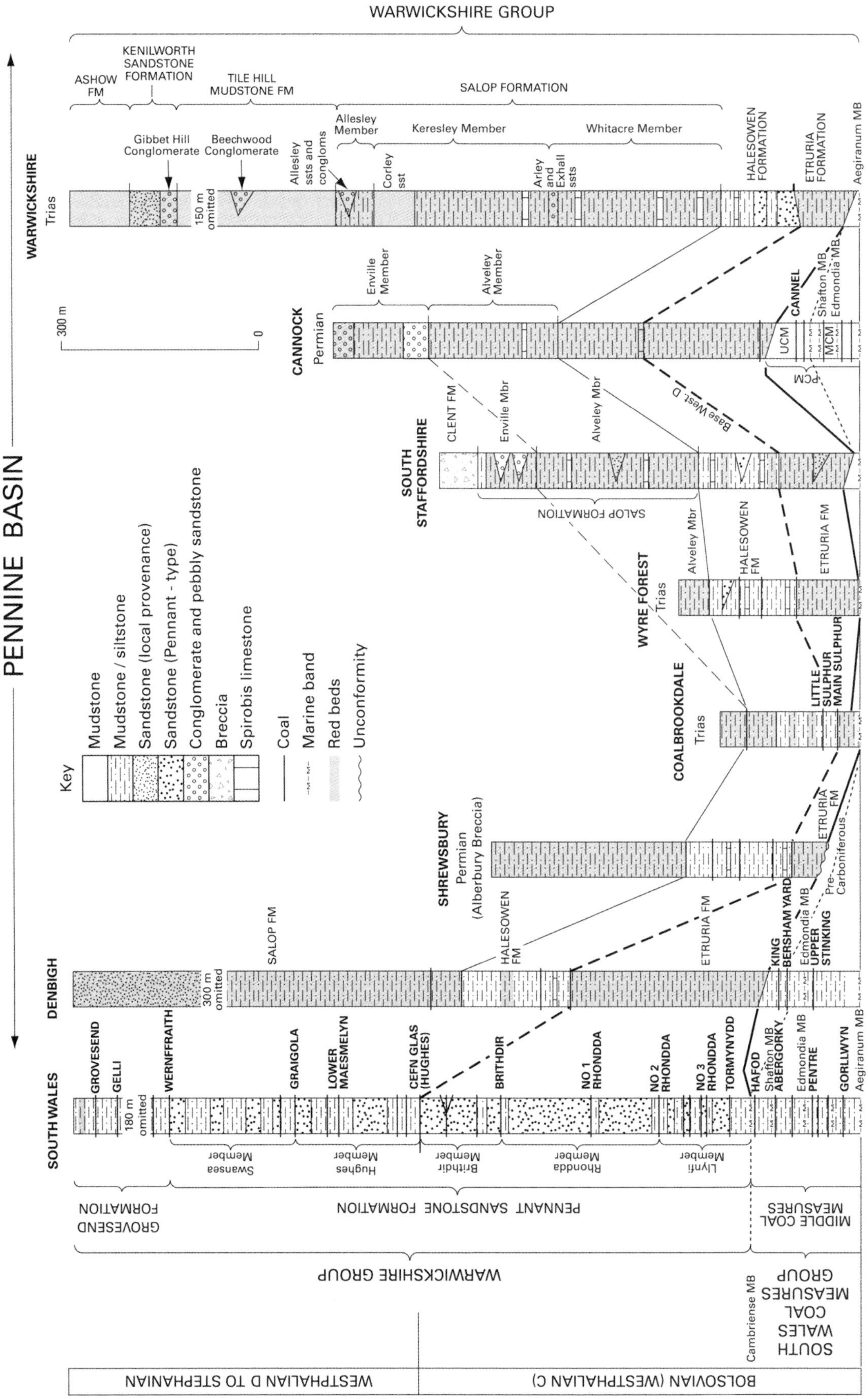

Fig. 9.27. Correlation of late Westphalian to early Permian Warwickshire Group from the southern part of the Pennine Basin and South Wales (from Powell *et al.* 2000*a*).

The succeeding Halesowen Formation represents a widespread return to fluvial deposition in association with relatively high water tables. The formation comprises grey-green, micaceous sandstone (litharenite) and grey-green mudstone, with thin coals, beds of *Spirorbis* limestone, local intraformational conglomerate, and pedogenic carbonate (caliche). The Halesowen Formation is interpreted as the northward equivalent of the 'Pennant' facies sandstone of the South Wales Basin (Besly 1988b). The Halesowen Formation has yielded miospores (Butterworth & Smith 1976; Smith & Butterworth 1967), plant macrofossils (Besly & Cleal 1997; Clayton *et al.* 1977; Cleal 1984) and non-marine bivalve faunas (Edwards 1951) with *Anthraconauta tenuis*, that indicate a Westphalian D age. Zircon grain ages indicate a source terrain of Variscan and Cadomian age, possibly from the Armorican Massif to the south (Hallsworth *et al.* 2000). A strong similarity in zircon isotopic dates suggest the Halesowen Formation has a similar source to the late Duckmantian and Bolsovian sandstones of Yorkshire (Hallsworth *et al.* 2000). However, northerly palaeocurrents within the Halesowen Formation suggest that these river systems followed a different path, crossing breaches within the Wales–Brabant High.

The Salop Formation and Tile Hill Mudstone formations represent a return to well-drained, alluvial plain settings with localized shallow lakes, and semi-arid conditions. The Salop Formation comprises red and red-brown interbedded mudstone and sandstone, with beds of pebbly sandstone and conglomerate. Thin *Spirorbis* limestone beds, calcrete and rare thin coals are present in the lower part. The sandstones are mostly sublitharenite, and conglomerate clasts include Carboniferous limestone and chert. The Tile Hill Mudstone Formation is limited to the southern part of the Warwickshire Coalfield where it rests conformably upon the Salop Formation (Fig. 9.27). It comprises red-brown mudstone with subordinate thin red-brown and green flaggy sandstones and sparse thin conglomeratic lenses. The Salop and Tile Hill formations are Westphalian D–Stephanian or Early Permian in age (Waters *et al.* 1995; Besly & Cleal 1997). Poorly preserved plant material from the Wyre Coalfield suggests the inclusion of early Stephanian (Cantabrian) beds in the Salop Formation (Besly & Cleal 1997).

In South Staffordshire the Salop Formation is overlain by the Clent Formation, which shows a rapid northward transition from breccia to red mudstone (Fig. 9.27). The Clent Formation is interpreted as proximal alluvial fan deposits with pebble clasts derived locally from uplifted hinterland blocks associated with the Wales–Brabant High to the south. The Clent Formation is probably Early Permian, or possibly late Stephanian in age (Waters *et al.* 1995).

A limited development of primary or early diagenetic red measures comparable to the Warwickshire Group occurs along the northern margin of the Pennine Basin, in west Cumbria. This is the Whitehaven Sandstone Formation, which is at least 300 m thick, and unconformably overlies the Pennine Coal Measures Group. The lower 100 m consists of red–deep purple or purplish brown, cross-bedded, micaceous, medium- to coarse-grained sandstone, known as the Whitehaven Sandstone (Akhurst *et al.* 1997). There are interbeds of pink–red or grey mudstone and siltstone, and thin palaeosols are present locally. The sandstones are overlain by a heterogeneous, dominantly red succession of mudstone, sandstone and marl with thin coals and limestones with *Spirorbis*. The lower part of the formation is interpreted as deposits from a major braided river system that flowed from the NE. The upper part of the formation represents deposition in interdistributary bay or lacustrine environments with minor river channels (Akhurst *et al.* 1997). Plant remains and the presence of the nonmarine bivalve *Anthraconauta phillipsii* (Eastwood *et al.* 1931) indicate a late Bolsovian–early Westphalian D age for the formation. To the NE similar strata have a Westphalian D *Tenuis* Zone fauna (Eastwood *et al.* 1968).

The Carboniferous igneous rocks of the West Midlands are predominantly alkaline olivine–dolerite sills (Kirton 1984). The sills are believed in many cases to have been emplaced into still-wet sediment of Bolsovian age. The sills are more limited in composition than those in the East Midlands, ranging from basaltic hawaiites to hawaiites.

The Barrow Hill Complex, located west of Dudley, is notably distinct from the other West Midlands intrusions in comprising vent agglomerate and dolerite with associated volcaniclastic rocks (Glover *et al.* 1993; Waters 2003*b*). The country rock of Etruria Formation, of Bolsovian age, appears to have been relatively wet and unlithified at the time of intrusion, suggesting emplacement at or near the contemporaneous ground surface. The interaction of hot, gassy magma with groundwater resulted in explosive activity in which dykes of sedimentary and igneous material were forced underground for some distance marginal to the vent. The surface expression of the explosive activity is seen as ash-fall deposits, hot, gaseous surge deposits and a thick breccia, similar in composition to the agglomerate present in the vent. The site is also of importance for the presence of anatomically well-preserved, conifer-like stems, which were buried by ash falls from this volcanic centre (Galtier *et al.* 1992).

In northern England a short-lived period of tholeiitic magmatism during the Stephanian resulted in intrusion of the quartz–dolerite Whin Sill-complex (Plate 19), together with associated ESE- and ENE-trending dyke-swarms (Fig. 9.11). There are no associated extrusive rocks. The Great Whin Sill is considered to be saucer-shaped (Francis 1982) and intruded into strata ranging in age from Dinantian to Westphalian. The intrusion changes stratigraphical level in a series of commonly fault-controlled transgressive steps.

Southern England–South Wales

South Wales Coal Measures Group

During early Westphalian times, the deposition of the South Wales Coal Measures Group in South Wales and southern England includes lithologies and environments of deposition that are essentially the same as those described for the Pennine Basin. Despite the broad similarity of Fluvio-lacustrine ('Coal Measures') lithofacies association to the north and south of the Wales–Brabant High, the evolution of the two successions in largely isolated basins has warranted the definition of two separate groups (Waters *et al.* 2006*a*). The South Wales Coal Measures Group of the Variscan Foreland Basin extends across South Wales, the Bristol region and the subsurface Berkshire and Kent coalfields.

Transitions between lower and upper delta plain appear to be synchronous between the two basins, suggesting a long-term eustatic control (Hartley 1993*a*). The group is predominantly argillaceous, with widespread sandstones rare. The succession is dominated by grey or black lacustrine mudstone, coarsening-upwards into lacustrine deltaic siltstones and sandstones. The sandstones are typically dark grey with variable grain-size, common lithic fragments and a dominant palaeocurrent direction towards the NE (Rippon 1996). The group is thickest in the west and SW of the South Wales Coalfield (about 900 m) and attenuated in the East Crop (about 240 m), adjacent to the Usk Anticline (Fig. 9.28). Coal seams thicken and amalgamate towards this anticline. The Pembrokeshire Coalfield is an attenuated westward extension of the main South Wales Coalfield. In England the group ranges from about 550 m thick in the Bristol area, 141 m in the Berkshire Coalfield to 285 m thick in the Kent Coalfield (Fig. 9.29a & b). In South Wales and the Bristol region the group rests conformably upon Namurian strata of the Marros Group, whereas in the Berkshire and Kent coalfields the Lower Coal Measures rest unconformably upon Dinantian or Devonian strata.

During late Langsettian–late Bolsovian times, E–W growth folds such as the Pontypridd Anticline developed in response to

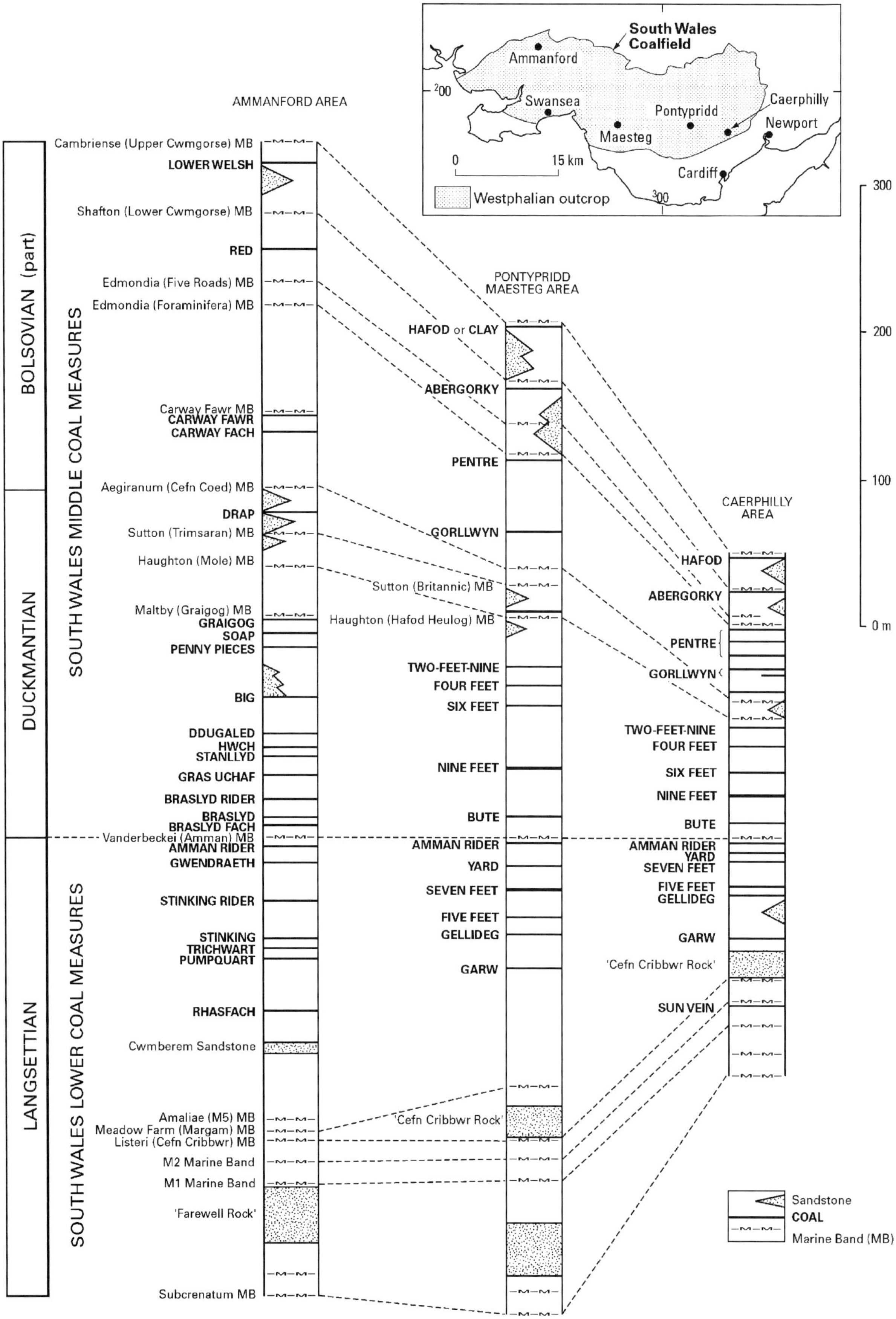

Fig. 9.28. Correlation of Westphalian strata of the South Wales Coal Measures Group within the South Wales Coalfield, adapted from Thomas (1974).

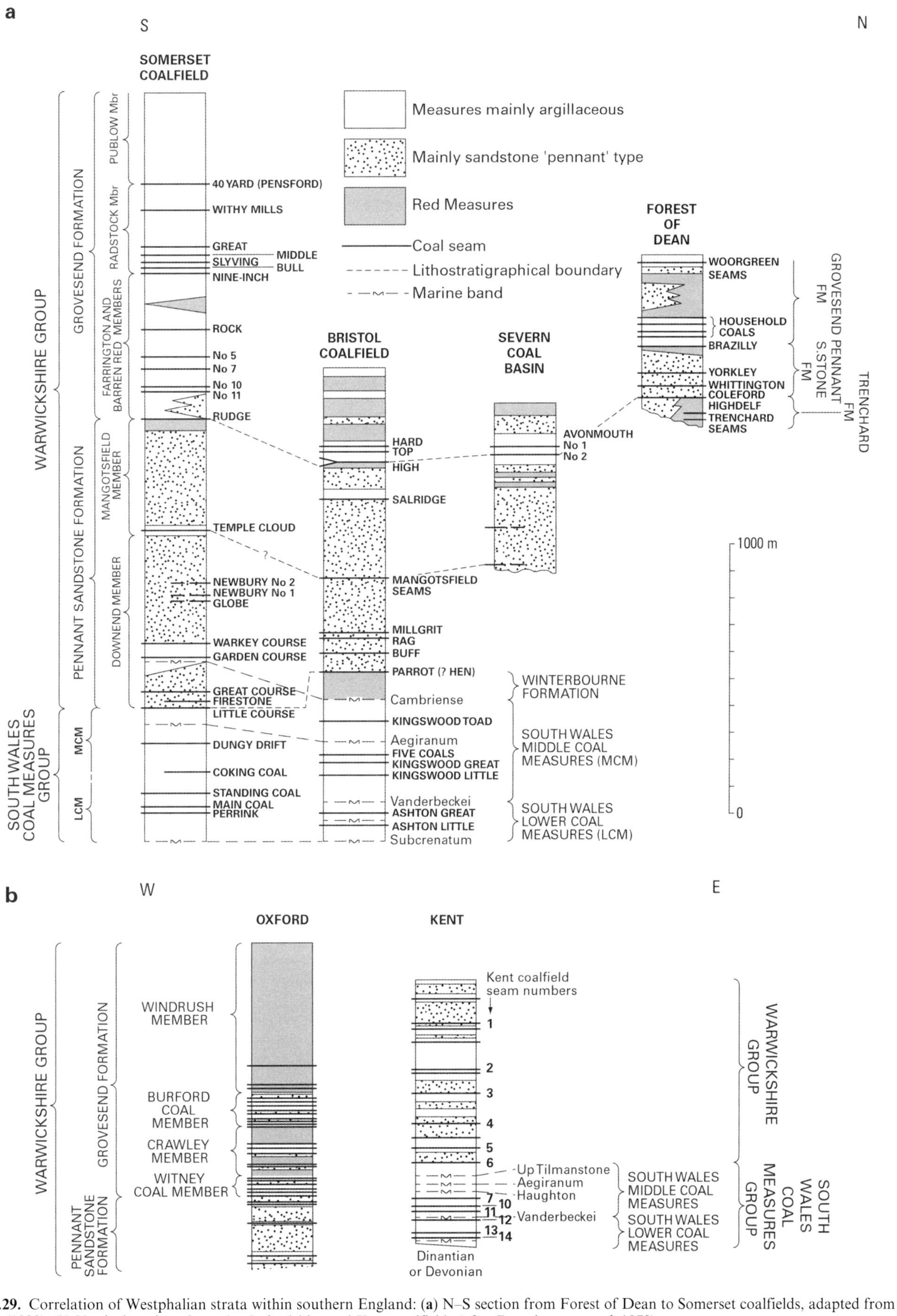

Fig. 9.29. Correlation of Westphalian strata within southern England: (**a**) N–S section from Forest of Dean to Somerset coalfields, adapted from Green (1992); (**b**) Vertical sections from the Oxfordshire and Kent coalfields (after Ramsbottom *et al.* 1978).

the early stages of Variscan compression. These folds actively controlled drainage patterns, with coal seam splits occurring within synclines (Jones 1989*a*). Seam splits and vertically stacked channel washouts also appear to be influenced by episodic movements on NNW–SSE oriented faults, inferred to be synsedimentary transtensional faults (Hartley 1993*a*).

A notable feature of the Coal Measures of South Wales is the increase in coal rank from bituminous, within the South and East crops, to anthracite along the North Crop, especially within the Pembrokeshire sub-basin. Illite crystallinity and vitrinite reflectance values for coals near the base of the group indicate maximum maturity temperatures for anthracites of 200–250 °C, whereas fluid inclusion studies suggest temperatures for bituminous coals of *c.*150 °C (Bevins *et al.* 1996). This does not appear to solely reflect depths of burial, as the depocentre occurs to the south of the area of highest rank. The heating appears to have been rapid, occurring soon after deposition and with high geothermal gradients, atypical of geothermal gradients within a foreland basin setting. The relationship of the thermal event to Variscan deformation has not been fully ascertained. Bevins *et al.* (1996) contend that the metamorphism occurred prior to deformation, but migration of hot fluids along thrusts has also been invoked.

The South Wales Lower Coal Measures show two distinct lithological variations in South Wales. As with the Pennine Basin, there is a lower succession, which equates with the *Lenisulcata* Biozone, that includes thin and impersistent coal seams and relatively common *Lingula*-bearing marine bands (Fig. 9.28). Sandstones within this interval, collectively known as the Farewell Rock, represent mainly southward-prograding delta lobes, or fluvial sands with channelled bases. Some thick, lenticular, coarse and pebbly, quartzitic sandstones on the East and South Crops formed as littoral or sublittoral sandbodies (Kelling & Collinson 1992). The sequence thins eastward toward the Usk Anticline, marine bands become impersistent and coals thin. The upper part of the formation is distinguished by thick coals, seatearths and sideritic ironstones, and relatively few marine bands. Locally, in the Berkshire Coalfield, Langsettian basic volcanic and intrusive rocks are dominant (Foster *et al.* 1989), including the 100 m-thick Aston Tirrold Volcanic Formation.

The South Wales Middle Coal Measures maintain broad lithological similarities with the upper part of the Lower Coal Measures. Thick, extensive coal seams, seatearths and numerous ironstone bands are common and a minor component of sandstone is present. A return to numerous marine bands and thin coals is recorded within the upper part of the formation. Southward-prograding delta lobes with high sinuosity fluvial channels are localized in the NE of the South Wales Coalfield, sourced from a subdued Wales–Brabant High (Hartley 1993*a*). However, in the west and notably in the Pembrokeshire Coalfield, fluvio-deltaic sandbodies appear to be northward prograding and include lithic arenites, precursors to the fluvial deposits of the Pennant Sandstone Formation.

The South Wales Upper Coal Measures is now restricted to the grey measures above the Cambriense (Upper Cwmgorse) Marine Band within a small area of the eastern part of the South Wales Coalfield. The formation is up to 14 m thick in the Kent Coalfield (Bisson *et al.* 1967). In the concealed Berkshire Coalfield the equivalent strata are absent beneath the Warwickshire Group (Foster *et al.* 1989).

The Warwickshire Group barren red-measures succession is present between the Cambriense Marine Band and the base of the Pennant Sandstone Formation in both the East Crop of the South Wales Coalfield (Deri Formation), Bristol Coalfield (Winterbourne Formation) and Forest of Dean (Trenchard Formation). These 'red-beds' are comparable to those present within the Warwickshire Group of the Pennine Basin and were deposited on a well-drained alluvial plain in the vicinity of the modern Severn Estuary. The raised topography associated with the Usk Anticline/Severn Axis enhanced drainage, resulting in a primary reddened succession, and providing a local source for conglomerate. The Deri Formation of the East Crop of the South Wales Coalfield consists of interbedded red, purple and green mottled mudstones and siltstones and channelized quartz arenites that are commonly pebbly and conglomeratic. The formation thickens eastward towards the Usk Anticline, but thin coals progressively fail toward this active structure. The conformable and diachronous base of the Deri Formation rises westwards above the Upper Coal Measures, the western limits of the formation being in the Taff Valley area (Downing & Squirrell 1965). The Winterbourne Formation of the Bristol Coalfield is up to 180 m thick (Fig. 9.29a) and comprises grey and red mudstone with common thin and lenticular beds of quartz-conglomerate and pebbly sandstone. The Deri and Winterbourne formations are of late Bolsovian to possibly early Westphalian D age. The Trenchard Formation of the Forest of Dean Coalfield consists of grey or pinkish grey quartzose sandstone, with quartzose conglomerate beds common at the base of the formation, increasing in thickness and clast size toward the north (Jones 1972). This formation, of possible late Westphalian D age, rests unconformably upon Dinantian and older strata. The formation is up to 120 m thick in the north of the coalfield, thinning southwards as it passes into the Pennant Sandstone Formation (Fig. 9.29a).

The Pennant Sandstone Formation comprises strata dominated by 'Pennant' facies sandstones, located south of the Wales–Brabant High. The formation extends across South Wales, Bristol, Somerset and the Forest of Dean, and the concealed Oxfordshire, Berkshire and Kent coalfields. A mudstone-dominated late Westphalian D–Stephanian succession is identified in South Wales as the Grovesend Formation. This name is now extended to include similar facies found in the Somerset, Bristol, Forest of Dean and Oxfordshire coalfields, formerly referred to as Supra-Pennant Measures. The Grovesend Formation of Bristol, Somerset and Oxfordshire includes some red-bed members.

The late Westphalian sedimentation of the area south of the Wales–Brabant High was dominated by a broad alluvial braidplain, with low- to moderate-sinuousity rivers flowing northward from the rising Hercynian mountain belt to the south (Kelling & Collinson 1992). This component of the Warwickshire Group is dominated by thick litharenite sandstone beds, with subordinate mudstone and thick coal seams in its lower part (Pennant Sandstone Formation), and a mudstone-dominated succession with subordinate sandstone in its upper part (Grovesend Formation). The Pennant Sandstone Formation is subdivided into members using marker coal seams, whereas the Grovesend Formation is subdivided into members using lithological variations between coal-bearing and barren measures (Fig. 9.27 & 9.29). In the Bristol region, marker coal seams can rarely be correlated with any confidence between coalfields. The Oxfordshire succession shows strong similarities with that in the Bristol region (Poole 1969). In the Kent Coalfield the Warwickshire Group cannot be subdivided into the component formations seen elsewhere. The group rests above a non-sequence upon Upper Coal Measures in the Kent Coalfield.

The Pennant Sandstone Formation is characterized by the presence of bluish grey, weathering green, coarse-grained, locally conglomeratic, feldspathic and micaceous lithic arenites of southern provenance. These sandstones are distinct from the quartz arenites of the South Wales Coal Measures Group, which are of northern provenance. There is a slight upward increase in grain size within the sandstones, representing northward progradation of alluvial facies with time. Typically, the formation thins northward towards the Wales–Brabant High (Fig. 9.29), but also over the Usk Anticline. The base of the formation is diachronous, with a broadly northward younging of the first incoming of 'Pennant'-type sandstones. The base of the formation in the SW of the South Wales Coalfield occurs at, or just above, the top of the Cambriense Marine Band. The base ranges from up to 180 m below the Cambriense Marine Band in the Somerset Coalfield, to 180 m above the same marine band in the Bristol area. In the Oxfordshire Coalfield the formation rests unconformably upon strata of Devonian or older age. The formation is thickest in the Swansea area of the South Wales Coalfield, up to 1150 m, decreasing in thickness towards the East Crop. The formation is about 1100 m thick in the Somerset Coalfield, whereas further towards the north in the Forest of Dean it is only 180–300 m, thinning towards the north as it onlaps the Wales–Brabant High (Fig. 9.29a). Within the concealed Oxfordshire Coalfield, the thickness is 335 m (Fig. 9.29b). The formation ranges from Bolsovian to late Westphalian D.

During the late Westphalian D there was a change to dominantly floodplain sedimentation of the Grovesend Formation, evident as mainly shales and sandy shales, with subordinate 'Pennant' type sandstone beds and thin coals. In the Somerset, Bristol and Oxfordshire coalfields, red mottled mudstone, seatearth, 'Pennant' type sandstone, but no coal seams typify the barren members of the formation. The formation is about 1400 m thick in the Somerset Coalfield, thinning northwards to 340 m in the Forest of Dean (Fig. 9.29a). In the Oxfordshire Coalfield, there is a comparable northwards thinning, with a maximum of about 1000 m recorded (Fig. 9.29b). The thickness is about 450 m in South Wales (Fig. 9.27). The Grovesend Formation is late Westphalian D in age, possibly in part Cantabrian (early Stephanian) in age in South Wales and the Forest of Dean (Ramsbottom *et al.* 1978) and Oxfordshire (Besly & Cleal, 2006). The top of the Warwickshire Group is truncated by the Permo-Triassic unconformity in England.

10 The Variscan Orogeny: the development and deformation of Devonian/Carboniferous basins in SW England and South Wales

B. E. LEVERIDGE & A. J. HARTLEY

The upper Palaeozoic Orogenic Province of SW England is a part of a belt of Devonian and Carboniferous basins that extended from Devon and Cornwall through to Germany, some 800 km to the east. Their complex sequence of basin development and phases of deformation, described in this chapter cumulatively comprise the Variscan Orogeny in this region.

Synchronously with the Devonian events within the Variscan Orogen, the mainly fluvial facies of the Old Red Sandstone filled basins in the Avalonian continent north of the Variscan front (Chapter 6). During the succeeding Carboniferous, basins within the continent were mainly extensional in origin, until a period in the late Carboniferous when many basement faults were inverted (Chapter 7) resulting in uplift of the basin fill, that initiated a new palaeogeography at the start of the Permian.

The South Wales Basin represents a transitional zone between the mobile Variscan belt and the continent to the north. This transitional position is reflected in the Devonian by the interdigitationof the Old Red Sandstone facies and marine sediments at the northern margins of the Variscan basins (Chapter 6). Throughout the Dinantian and Namurian the succession within the South Wales basin had much in common with successions in basins within the continent to the north (Chapter 7). It was not until the Silesian that Variscan deformation affected basin development and caused its deformation (see this chapter).

The Variscan of SW England

The Variscan orogen in Britain and mainland Europe

The Upper Palaeozoic massif of SW England (Fig. 10.1), is situated on the northern margin and forms an integral part of the Variscan orogen of central and western Europe. The orogen is some 1000 km wide, and represents tectonic activity from the Early Palaeozoic to the Early Permian, and was in large part contemporaneous, and interactive with the Caledonides (e.g. Soper 1986*b*). Correlation of the province with the Rheinisches Shiefergebirge of Germany, some 800 km to the east, is established (e.g. Franke & Engel 1982; Holder & Leveridge 1986*a*) and together they make up a major part of the Rhenohercynian Zone. That is the northernmost of the palaeogeographical and tectonic zones proposed by Kossmat (1927), which, with the Saxothuringian and Moldanubian zones, constitute the orogenic belt in Germany and central Europe (Fig. 10.2).

Bard *et al.* (1980) and Matte (1986) proposed interpretation of Kossmat's zones in terms of plate tectonics. At that time postulated settings for SW England were diverse; an ensialic back arc basin (Floyd 1982), part of a trans-Europe oblique shear zone (Badham 1982), and an intracontinental fold belt (Matthews 1984). The Gramscatho flysch (Hendriks 1949) and associated Lizard ophiolite (Strong *et al.* 1975; Styles & Kirby 1980) in southern Cornwall were seen by Barnes & Andrews (1986) as representing a small intracontinental basin generated in an E–W strike–slip zone across northern Europe. In contrast Holder & Leveridge (1986*b*) interpreted those as products of the closure of a more extensive oceanized basin. Correlation and continuity of stratigraphies and structures between SW England and central Europe (Holder & Leveridge 1986*a*) appeared to enhance the concept of an E–W linear Europe-wide Rhenohercynian oceanized zone and collision belt (Franke 1989). By extrapolation of Kossmat's zones through western Europe into Iberia (Fig. 10.2), Franke (1989, 1992) and Matte *et al.* (1990) defined sublinear microplates and separating collision belts, with southerly subduction zones in the Rhenohercynian and Saxothuringian and northerly subduction beneath the Moldanubian. Dextral interplate movements and small oblique basins were reinvoked, however, for SW England by Holdsworth (1989). He speculated that the E–W 'Start–Perranporth Line' might represent a terrane boundary, between the 'Old Red Sandstone' continent to the north and an Armorican microplate to the south, with transtension actively controlling intervening Devonian basin evolution.

The Rhenohercynian Zone in continental Europe is now considered to lie at or near the southern margin of the Eastern Avalonia plate (Franke 2000), the southerly of the three plates,

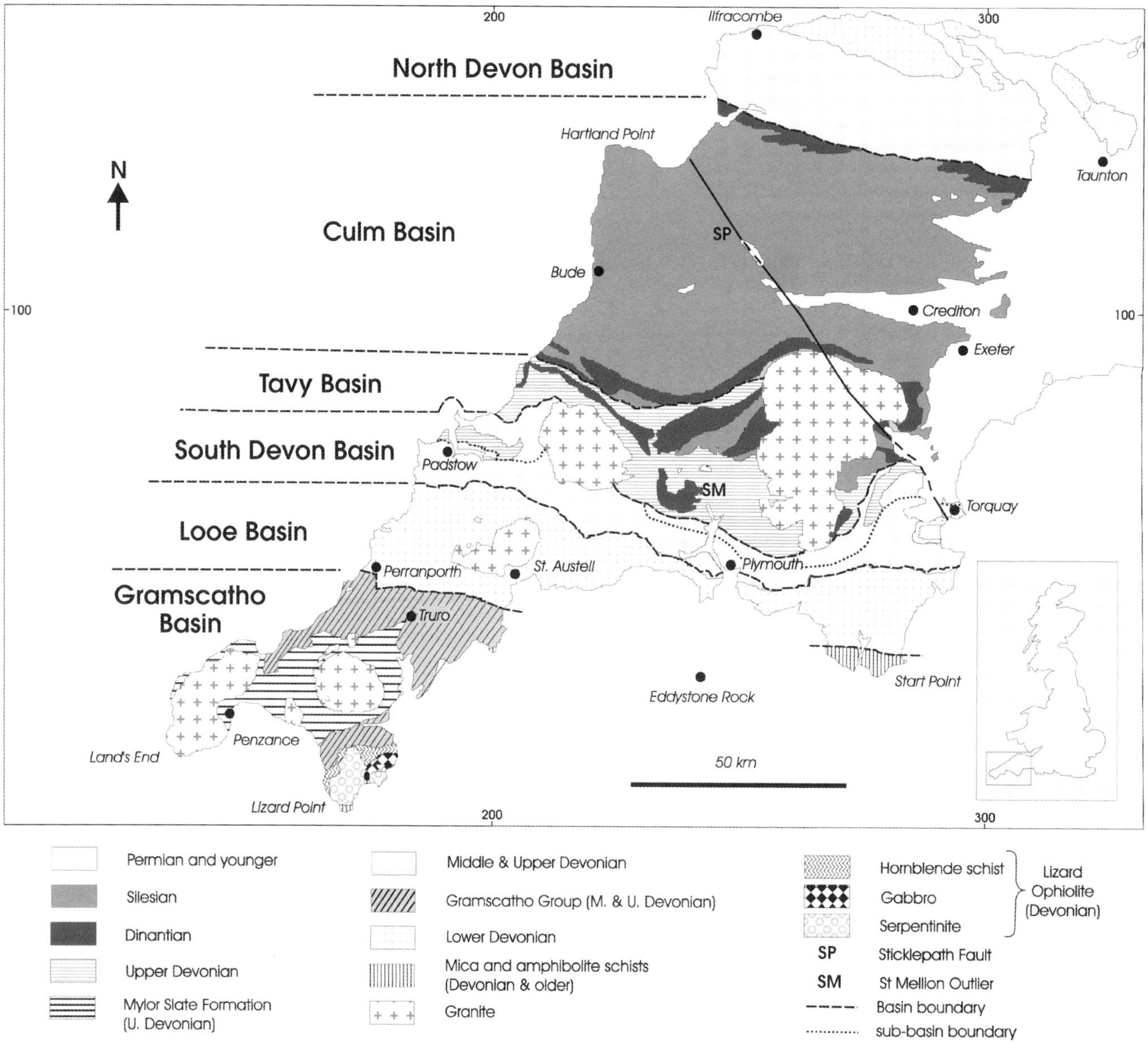

Fig. 10.1. Geological sketch map, based on BGS maps, showing the basins of SW England.

with Laurentia and Baltica, which collided to form the Caledonides (e.g. Soper 1986; McKerrow *et al.* 1991). Avalonia and Armorica were derived from Gondwana when in a southern peri-polar position in the Lower Palaeozoic, drifted north at differing rates (e.g. Trench & Torsvik 1991) and reassembled later in the Palaeozoic in equatorial regions. Armorica comprises an assemblage of continental microplates, with intervening marginal, basinal and ophiolite remnants, of the Saxothuringian and Moldanubian zones and their correlatives, which amalgamated during that process (Tait *et al.* 1997). The major Variscan oceanic separation, the Rheic Ocean, was proposed by Scotese *et al.* (1979), Cocks & Fortey (1982), and Van Der Voo (1983), to lie between the 'ORS' (Eastern Avalonia) continent and Armorica on the basis of palaeomagnetic and biostratigraphical evidence. More recent palaeomagnetic data have supported that view (see Tait *et al.* 1997). Recognition of an oceanic suture at the southern margin of the Rhenohercynian (Holder & Leveridge 1986*a*; Franke 1989) led some to propose that it was that of the Rheic Ocean (e.g. Oczlon 1993). The lack of a paired metamorphic belt, and apparently restricted related arc magmatism, indicated to Franke (1989) that it was a narrow ocean, up to 500 km being proposed by Franke *et al.* (1995). In Germany the ENE–WSE-trending Rhenohercynian suture lies between the Northern Phyllite Zone and the Mid German Crystalline High (of the Saxothuringian Zone) and within those tectonic units there are Ordovician and Silurian arc volcanics trending NE–SW. Those have been attributed to the closure of the Rheic Ocean (Franke & Onken 1995; Franke *et al.* 1995; Oncken 1997). Closure is thought to have been at the end of the Lower Palaeozoic (Tait 1999) immediately preceding Rhenohercynian rifting in the early Devonian along or close to the Rheic suture (Franke 2000).

McKerrow *et al.* (2000*a*) have accepted a close coincidence of the Rheic suture and the Rhenohercynian suture in SW England. Geological evidence confirming such coincidence, or the presence of a pre-Devonian Rheic suture immediately south of or within the province (beneath the Start–Perranporth zone with its locally strong secondary Variscan structures?) is awaited.

Devonian and Carboniferous basins of SW England

The Devonian and Carboniferous rocks of SW England are disposed within six juxtaposed E–W-trending basins (Fig. 10.1).

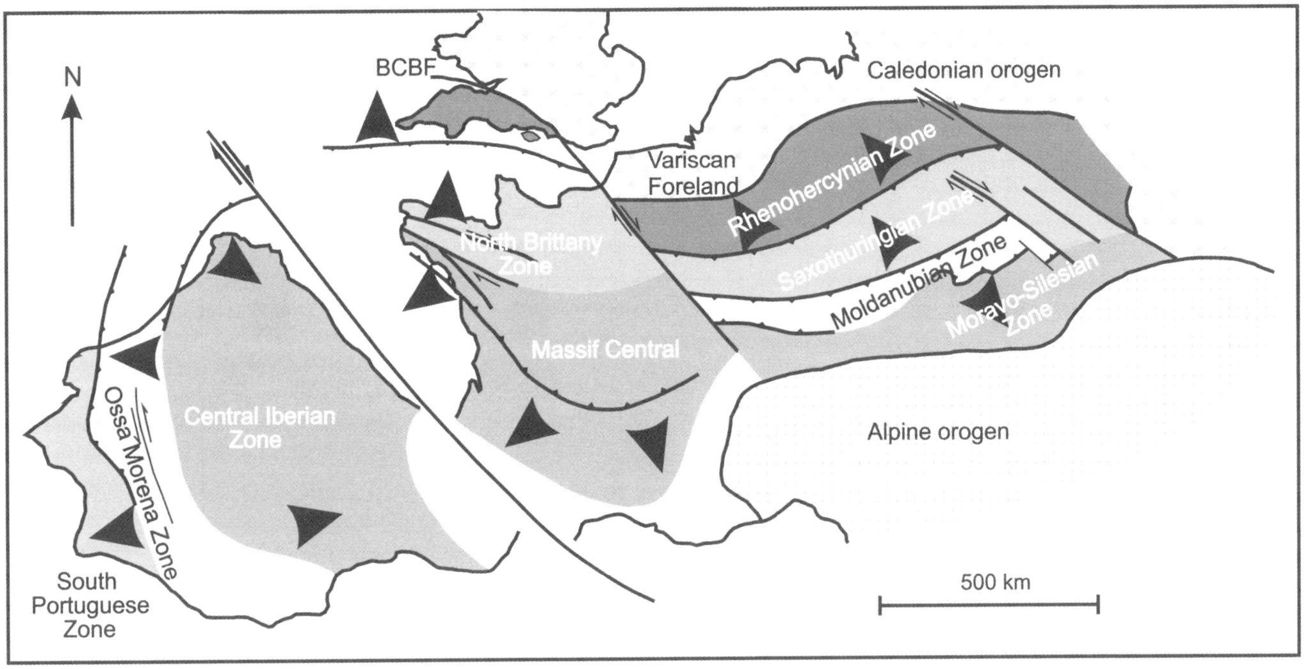

Fig. 10.2. European Variscan tectonic zonation, after Franke (1989), based on Kossmat (1927), as amended by Holder and Leveridge (1994). BCBF, Bristol Channel/Bray Fault Zone.

The evolution of those basins broadly reflects: (i) episodic Early Devonian–Dinantian continental rifting that developed a broad, complex, passive margin with oceanic lithosphere in the south; and (ii) Late Devonian continental collision and Carboniferous deformation of the passive margin.

In the southern part of the area now represented by the Gramscatho Basin and Lizard and Start complexes (Fig. 10.1), there was an intermediate stage. Here early rifting and the generation of oceanic lithosphere were succeeded by the formation of an active continental margin by the early Mid-Devonian. As a consequence, lithostratigraphical units here provide evidence of convergence-related deformation during Mid–Late Devonian times prior to continental collision.

Although the Gramscatho Basin formed during Early Devonian rifting, its infill (Figs 10.1 & 10.6) was largely derived during Mid–Late Devonian convergence, from its active southern continental margin. However, remnants of the earlier rift history, including oceanic lithosphere, are preserved. The Devonian–Dinantian infill of the other basins to the north (Figs 10.1, 10.8 & 10.10) reflect sedimentation and magmatic activity during the Early Devonian-Dinantian rifting alone. The Culm Basin initiated during this rift episode, but, following collision, and progressive basin closures and inversions to the south, it became a foreland basin in the Silesian.

This review emphasizes the role of extensional and convergent tectonics in determining the various stratigraphical successions of the province. The role of global sea-level changes due to eustatic, climatic and extraterrestrial events, and their possible bearing on Devonian successions of the province, have been discussed at length by House (e.g. 1983, 1992, 1996). The precise interrelationship or interplay between these factors within the province has yet to be determined.

Regional development

Devonian rifting and Mid–Late Devonian convergence

It has been recognized for some time that the Devonian–Dinantian lithostratigraphical divisions of SW England are compatible with the formation and development of a series of sedimentary basins during continental rifting (e.g. Matthews 1977). The Lizard ophiolite, in association with Gramscatho Basin tectonostratigraphy, has been interpreted as evidence that such rifting led to the oceanization of that basin (e.g. Barnes & Andrews 1986; Holder & Leveridge 1986*b*). The 'Basin and Rise' concept, imported to SW England from the Rheinisches Schiefergebirge of Germany (Goldring 1962), explained the common Mid–Late Devonian association of contemporaneous shallow-marine platform and basinal deposits between the Gramscatho and Culm basins (e.g. House & Selwood 1966; House 1975; Selwood *et al.* 1984). Most of these sequences were formerly assigned to the Trevone Basin (Matthews 1977) in the west or the South Devon Basin (Selwood & Durrance 1982) in the east (e.g. Selwood 1990). However, three major E–W Devonian basins, including previously unassigned Lower Devonian successions north of the Gramscatho Basin, are now identified (Leveridge *et al.* 2002); the Looe, South Devon and Tavy basins (Fig. 10.1).

The Gramscatho, Looe, South Devon and Tavy basins developed sequentially northwards from the Early Devonian, and rifting culminated with formation of the Dinantian Culm Basin. Pre-rift basement is not exposed, but is probably present at shallow depth locally (Leake *et al.* 1988). The timing of basin development is indicated and constrained by factors such as sediment supply blocking, and the availability of sediments derived from developing or isolated highs. The basins and associated highs reflect the formation of one or more half graben and/or graben and have their own characteristic stratigraphy, i.e. the province cannot be represented by a single 'layer-cake' model (Leveridge *et al.* 2003*b*).

The basins were variously interconnected, the Looe Basin being largely marine from the late Pragian and others essentially marine from the late Emsian (Fig. 10.3). Lithostratigraphical divisions of hemipelagic sedimentary rocks are recognized to form parts of more than one basinal succession. There was also along-strike variation of successions resulting from an overall westward deepening of the Looe and South Devon basins and greater submergence of intervening highs. All successions and part successions are fault bounded, dismembered or interdigitated by between one and three major episodes

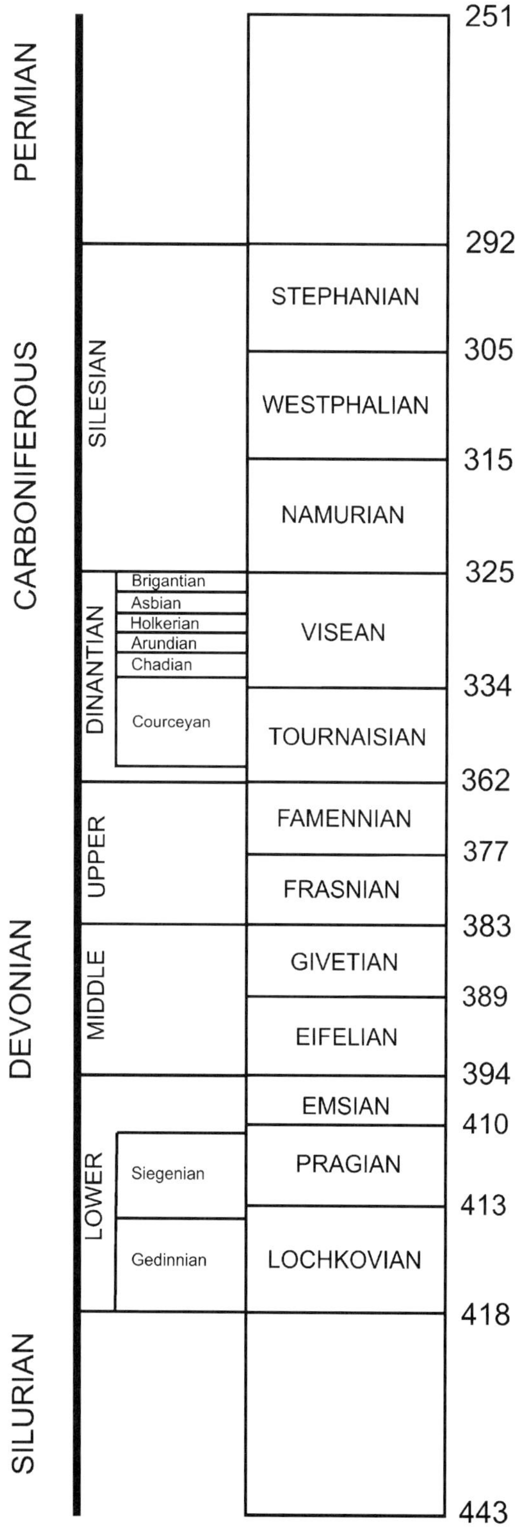

Fig. 10.3. Chronostratigraphic ages (Ma) of stages and series of Devonian and Carboniferous rocks, after McKerrow & Van Staal (2000), and of the Permo-Carboniferous boundary, after Menning *et al.* (2000).

of regional thrusting verging northwards and/or southwards. The relationship of principal lithostratigraphical divisions to this basin and high framework is indicated in Figure 10.8.

The connection of the apparently continuous North Devon 'Basin' succession to those of the contemporaneous rift basins to the south crop is not constrained, because of the crop of the intervening Culm Basin deposits, that link them only in the Dinantian. There are significant indicators however suggesting that they were indeed part of the same province. The major transgressions of north Devon, in the Late Emsian (?), Early Givetian and Late Famennian, show correspondence with the times of formation of the South Devon, Tavy and Culm basins respectively (see below, Fig. 10.11). This suggests that the passive margin subsided as a whole following major episodes of rifting and new basin formation to the south. There is also a direct link between basins in the Frasnian and Famennian. Purple and green sedimentary rocks are common to the neritic and continental facies of the Morte Slates and Pickwell Down Sandstones of the North Devon Basin and the Upper Devonian basinal facies of the South Devon and Tavy basins. Indeed, they extend into the deep-water Portscatho Formation sequence (Falmouth Series: Hill & MacAlister 1906) of the Gramscatho Basin.

Gramscatho Basin

Recognition that the association of peridotite, gabbro and mafic sheeted-dykes within the Lizard Complex in south Cornwall represented an obducted ophiolite (Strong *et al.* 1975; Bromley 1979; Kirby 1979; Styles & Kirby 1980) provided an important pointer to the setting of the province. In association with the dynamic stratigraphy of the Gramscatho Basin (Hendriks 1971; Holder & Leveridge 1986*b*) of south Cornwall, and its correlative lithostratigraphical development in Germany (e.g. Floyd *et al.* 1991), the significance of the ophiolite is enhanced in relation to the evolution of the Rhenohercynian.

Lizard Complex. The complex comprises the Lizard Ophiolite and representatives of possibly older basement (Fig. 10.4). Main components, (after Floyd *et al.* 1993*a*) are: (i) partially serpentinized peridotite; (ii) gabbros; (iii) sheeted basaltic dykes; (iv) meta-cumulates (Traboe Hornblende Schist); (v) mixed magma intrusions (Kennack Gneiss); (vi) metabasalt (Landewednack Hornblende Schist); (vii) metasediments (Old Lizard Head Series, OLHS); and (viii) orthogneiss (Man of War Gneiss, MOWG). The attribution of lithological units to tectonic units constituting the complex, the ophiolite, subjacent metamorphics or crystalline basement, remains a matter of debate, and indeed the credentials of the mafic rocks as an ophiolite have been questioned (Cook *et al.* 2002). The ophiolite, as generally understood, consists of (i)–(v) above, assigned to one (e.g. Leake & Styles 1984) or two (Bromley 1979) tectonic units, within which there are probably several thrust sheets (Jones 1997), with a structural thickness of a few hundred metres only (Styles & Kirby 1980; Rollin 1986). It rests upon a lower unit that has been considered to be (vi) and (vii) above, but may include (iv) (Power *et al.* 1996) or exclude (vi) as part of the ophiolite (e.g. Cook *et al.* 2002). Recent workers have agreed that locally the units structurally overlie the MOWG, part of a crystalline basement, forming inshore skerries off Lizard Point. The secondary nature of this faulted contact may, however, be indicated by associated semi-ductile/brittle fabrics (Jones 1997), and its obliquity to dated early Variscan fabric in the MOWG (Sandeman *et al.* 1997).

The peridotite comprises coarse lherzolitic, tremolite and dunite serpentinites. Separable stages of high-temperature and -pressure crystallization and recrystallization (Green 1964) and associated steeply inclined fabrics, mylonitic in part, orientated NNW–SSE, are attributed to progressive exhumation of mantle prior to obduction (Jones 1997; Cook *et al.* 2000). The Traboe Hornblende Schists are essentially metamorphosed cumulates, layered gabbros, dunites, pyroxenites and anorthosites, interpreted by Leake & Styles (1984) to be the first products of the ocean crust-forming process accompanying continental rifting. At Porthkerris these amphibolitic rocks are mylonitic, attributed to high-temperature ductile extension of the oceanic crust (Gibbons & Thompson 1991). A temporally associated plagiogranite vein has provided a constrained age of crust formation for the ophiolite (Clark *et al.* 1998*a*), which

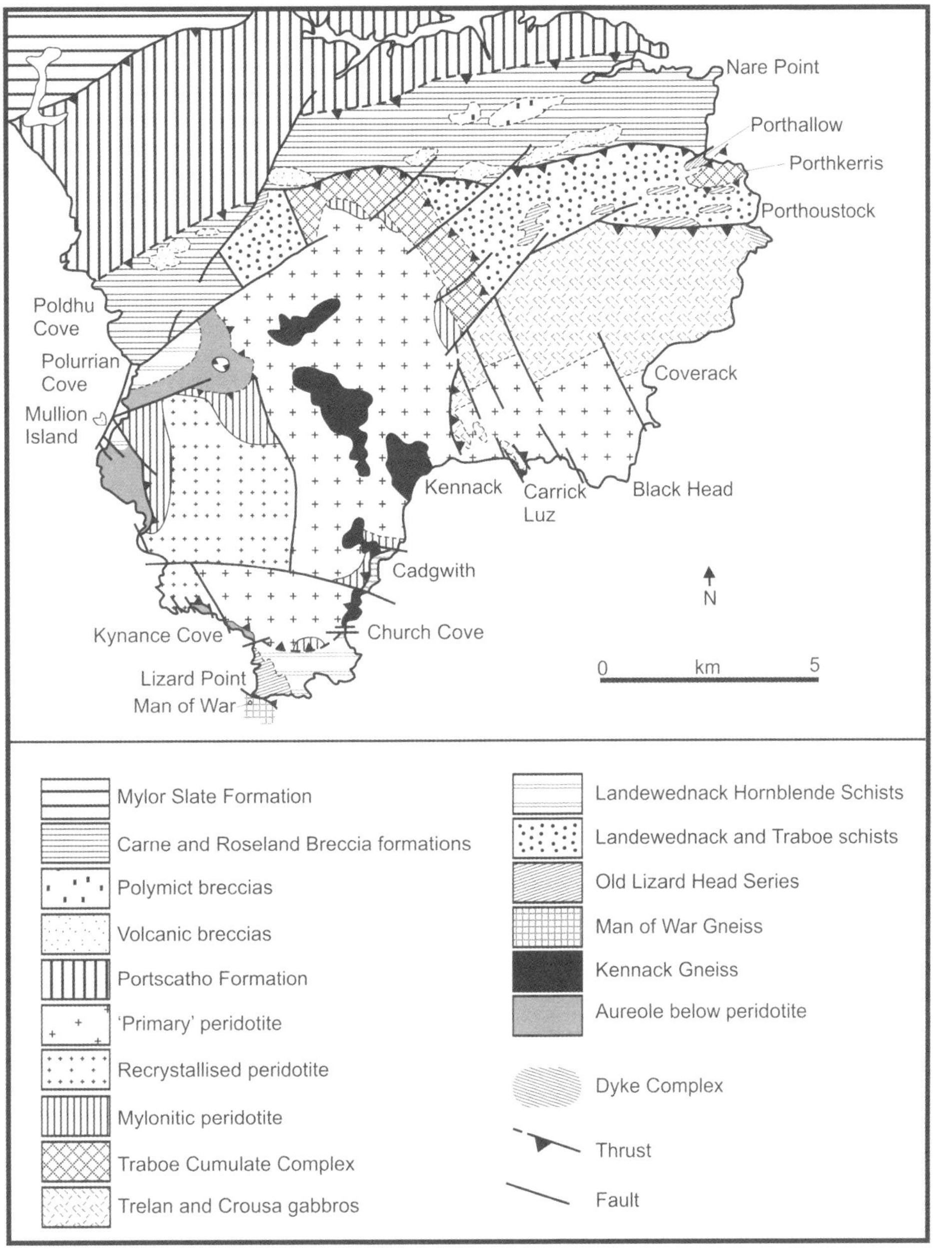

Fig. 10.4. Geological sketch map of the Lizard Complex, based on Floyd *et al.* (1993*a*) and Cooke *et al.* (2000).

is 397 ± 2 Ma (U–Pb single zircon), i.e. Emsian. The Crousa Gabbro, rather than superseding the mantle peridotite, intrudes it, and is cut by phases of basaltic dykes, trending NW–SE. The dykes, of N-MORB (MORB, mid-oceanic ridge basalt) composition, locally forming 80% of crop at Porthoustock, have been interpreted as a sheeted dyke swarm, as associated with a spreading centre (Kirby 1979). Amphibolite-grade metamorphism within these rocks, dated at approximately 390 Ma (U–Pb), has been attributed to their exhumation (Nutman *et al.* 2001; Cook *et al.* 2002).

The structurally lower unit of the complex comprises the OLHS and Landewednack Hornblende Schists to the south where they are in faulted contact (Jones 1997), and a similar association in the NE part of the complex (Floyd *et al.* 1993*a*; Power *et al.* 1996). The OLHS is largely metasedimetary quartz, mica and hornblende schists, locally with staurolite, kyanite and sillimanite (see Flett 1946), and the Landewednack Hornblende Schists, comprise meta-mafic volcanic and intrusive rocks of MORB affinities (Floyd 1976; Sandeman 1988). Designated by Floyd *et al.* (1993*a*) as sea-floor sediments and associated ocean floor basalts, or by Kirby (1979) as part of the ophiolite, an earlier history has been proposed for these rocks on the basis of early steep fabrics (e.g. Jones 1997; Cook *et al.* 2002). The OLHS is intruded by the granitoid Lizard Head Sill which has been dated at approximately 500 Ma (U–Pb) based on zircons interpreted by Nutman *et al.* (2001) to be primary magmatic, although zircons of similar age in the Landewednack and Traboe schists are interpreted as inherited. Later gently inclined mylonitic amphibolite facies metamorphic fabrics, with northwards transport indicators, within these rocks have yielded $^{40}Ar/^{39}Ar$ ages of *c.* 380 ± (av.5) Ma (Clark *et al.* 1998*b*) and slightly older (*c.* 390 Ma, with larger error margins) U–Pb ages (Nutman *et al.* 2001).

The Kennack Gneiss is a commingled magma intrusion at the thrust junction of the units, composed of E-MORB and crustally derived granitic melts. Intrusive and metamorphic age dates of 376.4 ± 1.4 Ma (U–Pb; Sandeman *et al.* 2000) and *c.* 370 Ma (Rb–Sr Styles & Rundle 1984; $^{40}Ar/^{39}Ar$ Sandeman *et al.* 1995), respectively, reflect its amphibolite facies metamorphism syn- and immediately post-emplacement (Sandeman *et al.* 2000).

The MOWG, composed of igneous rocks of gabbroic–tonalitic compositions, forms the undisputed base of the complex off Lizard Point. Geochronology by Sandeman *et al.* (1997) established that the gneiss, with an arc-like geochemistry, has a probable intrusive age (U–Pb, zircon) of about 500 Ma (Late Cambrian–earliest Ordovician), and a metamorphic cooling age (Ar^{40}/Ar^{39}, amphibole) of *c.* 374 Ma (Famennian) for amphibolite-grade recrystallization. This basement segment has Avalonian geochemical affinities (Sandeman *et al.* 1997) but its local status is uncertain. Possibilities are that it is part of the extensive nappe of basement mapped offshore (e.g. BGS 2000), a continental crustal block remnant of the Gramscatho Basin floor, or even a large olistolith within the underlying flysch nappe stack (see below).

The younger metamorphic dates within the lower tectonic unit, the Kennack Gneiss and the MOWG are generally attributed to Late Devonian thrust emplacement of the complex. The Kennack Gneiss represents magma comingling during that process, the felsic component derived by partial melting of basement or Gramscatho sediments, and mafic component derived from a mantle source similar to the ophiolitic rocks (Sandeman *et al.* 2000). Earlier dates within the ophiolite have been linked to mantle exhumation and uplift after Early Devonian oceanic crust formation and prior to obduction (Nutman *et al.* 2001).

Interpretations of the setting and generation of the oceanic lithosphere of the Lizard have focussed either on its internal fabrics and dyke orientations or the broader provincial and Rhenohercynian geology. Based on the former, a NNW–ESE spreading ridge axis (e.g. Gibbons & Thompson 1991), or small pull-apart basin (Badham 1982) linked to an E–W dextral intracontinental transform system (Barnes & Andrews 1986), or shear zone (Cook *et al.* 2002), have been proposed. In contrast basins to the north of the oceanized basin are E–W parallel, rather than oblique, within the province (see below). Available structural and stratigraphical data point to contemporaneous and ongoing pure shear extension, with formation within the province of E–W parallel, rather than oblique basins, in the Devonian and Dinantian (e.g. Leveridge *et al.* 2002). Early Devonian (Pragian) marine sediments (e.g. Meadfoot Group) straddle the 'Start–Perranporth Line' (Holdsworth 1989) and evidence for penecontempoaraneous independent Devonian basin evolution across it remains only speculative. The Lizard ophiolite was a small part of the Rhenohercynian ocean floor represented by MOR basalts associated with flysch nappes of Mid and Late Devonian deposits derived from its floor over 1000 km along strike (Floyd 1984, 1995; Holder & Leveridge 1986*a*; Franke 1989). A corollary of ophiolite obduction, as recognized by Coleman (1977), is the limited part of ocean floor evolution that it might represent, in terms of typicality, time and location of formation, and exhumation history. The interpreted strike-slip basin origin of the Lizard ophiolite and the orthogonal extensional rift origin of basins to the north, have yet to be satisfactorily reconciled. The possibility of ophiolite rotation during obduction is not fully constrained (e.g. Hailwood *et al.* 1984), and the consequences of its formation on the southern side of the basin at the time of transition from extensional to contractional regimes (see below) have yet to be addressed.

Gramscatho Basin, the rift phase. The main deposits of the Gramscatho Basin (Figs 10.1, 10.5 & 10.6), were interpreted as products of its active southern margin during basin closure and ophiolite obduction by Barnes & Andrews (1986), Holder & Leveridge (1986*b*) and Leveridge *et al.* (1990). MORB with deep-water sediments within the thrust stack also point to the oceanization of the basin, and Shail (1992) proposed that elements of both allochthonous and parautochonous sequences of the basin are relateable to the earlier rift phase of basin evolution.

The oldest biostratigraphically dated allochthonous rocks of the Gramscatho Basin are the early Mid–Late Eifelian (Sadler 1973; Leveridge 1974) deposits of the Pendower Formation (Figs 10.5 and 10.6) at the base of the Veryan Nappe. The formation is a hemipelagic shielded basin sequence of grey, green and brown slaty metalliferous (Mn, Cu, Ni, V, Zn) mudstones and interbedded radiolarian chert, with turbiditic limestone of pelagic platform provenance and subordinate thin beds of coarse lithic greywacke sandstone (Holder & Leveridge 1986*b*; Shail 1992). Near its lower thrust boundary pillowed basalt, LIL-enriched MORB at Tubbs Mill (Floyd 1984), interdigitates with the sedimentary rocks, and in sequence above the formation is the southerly sourced flysch of the Carne Formation (see below). Whilst attributing the MORB, chert and metalliferous mudstone association to the proximality of a spreading centre, the Pendower calciclastics and siliciclastics have been described by Shail (1992) as synrift deposits, sourced by tectonically controlled erosion of a within-basin remnant block of continental basement, capped by pelagic carbonate sediments. Deposition was interpreted to be within a northerly sub-basin, with its incipient Eifelian oceanic crust formation, of an asymmetrical Gramscatho Basin, the earlier (Emsian) Lizard ophiolite flooring part of a dominant southern sub-basin.

The parautochthonous succession (Holder & Leveridge 1986*b*) was probably deposited on the thinned northern continental passive margin of the basin. This is reflected in the presence in its uppermost formation, the Mylor Slate Formation (Figs 10.5 & 10.6), of contemporaneous intraplate tholeiitic basaltic rocks (Floyd & Al-Samman 1980). Locally, included tectonized granitoid xenoliths with model ages of 625 Ma and 851 Ma (T^{Nd}) have been interpreted as representatives of continental crystalline basement (Goode & Merriman 1987; Goode *et al.* 1987). Shail (1992) also proposed that some deposits of the underlying succession, the Porthtowan and Grampound formations (Fig. 10.5), were synrift sediments derived from the northern margin of the basin. Those included some of the thicker and coarser sandstone beds towards the top of the Porthtowan Formation and the Treworgans Sandstone Member of the Grampound Formation (Fig. 10.6). Constraints on the interpretation of northerly sourcing are the age (older than Frasnian but not known in detail), provenance (lithic greywackes compared to the few quartzo-feldspathic sandstones in passive margin sequences), and pathways of the sediments across the rifting margin to the north (see below).

Gramscatho Basin, the convergent phase. The infill of the basin was dominated by sedimentation generated along its active southern margin. Obduction of the Lizard ophiolite from the oceanized floor of the Gramscatho Basin during its closure has been proposed or modelled by several authors (e.g. Sanderson 1984; Barnes & Andrews 1986; Holder & Leveridge 1986*b*). Currently considered to have a late Early Devonian generation age (Clark *et al.* 1998*a*) it is located within a family of northward-verging thrust nappes (Fig. 10.5; Leveridge *et al.* 1984, 1990). These are major gently to moderately inclined structures not only mapped onshore but also traceable offshore in the South West Approaches Traverse (BIRPS & ECORS 1986) and commercial seismic profiles (e.g. Hillis & Chapman 1992) across some 300 km of the continental shelf.

Within the sedimentary rocks of the nappe pile, dip and younging are to the SE. The Carrick Nappe (Fig. 10.6), at the base of the allochthon (Holder & Leveridge 1986*b*: Leveridge *et al.* 1990), has a structural thickness up to 5.4 km onshore. It consists of the Portscatho Formation that is essentially Frasnian in age, but is possibly older in part, and probably extends up into the Famennian (Wilkinson & Knight 1989). This flysch facies sequence, of interbedded greywacke sandstone turbidites and dark grey mudstone, demonstrates progradation from outer fan to mid-fan depositional regimes. The overlying Veryan Nappe (Fig. 10.6), comprises the

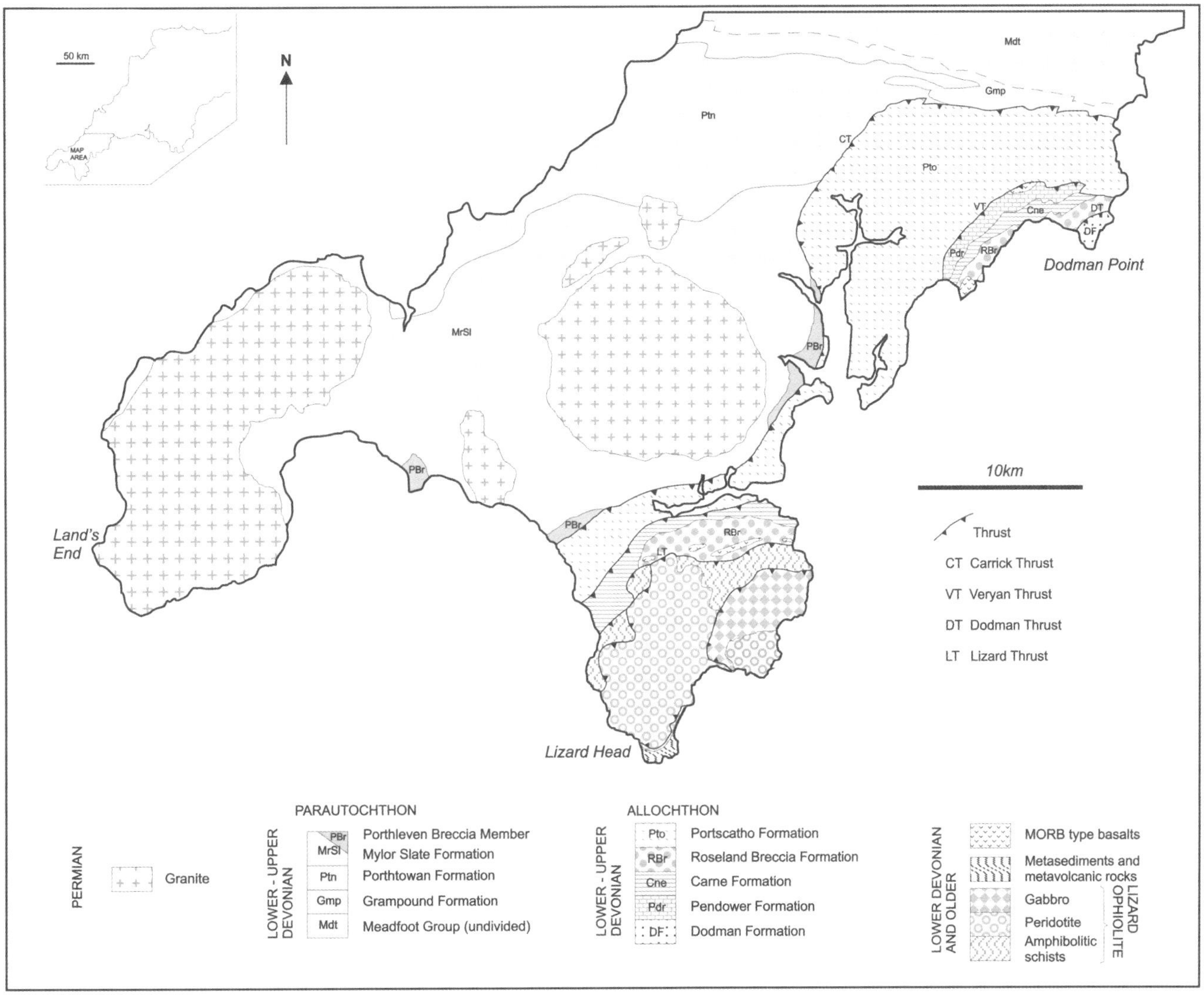

Fig. 10.5. Sketch map of the tectonostratigraphical divisions of south Cornwall, based on BGS maps, after Leveridge *et al.* (1990).

Pendower, Carne and Roseland Breccia formations, a succession some 2.6 km thick, ranging from the early Mid Eifelian–Givetian (Sadler 1973; Leveridge 1974), and probably early Frasnian (Hendriks *et al.* 1971). The latter attribution was unaffected by the subsequent revision upwards of the Givetian–Frasnian boundary by the International Subcommission on Devonian Stratigraphy (see Ziegler & Klapper 1985), as it was based on the conodont *Ancyrodella buckeyensis* (Stauffer) which is still assigned to the Frasnian (Over & Rhodes 2000). The Pendower Formation (see above) at the base of the nappe is succeeeded by the Carne Formation, a sequence of greywacke sandstone turbidites, olistostrome, channel-fill sandstone, slumped sandstone, and interlaminated mudstone, siltstone and sandstone. It is interpreted as an association of middle and upper fan distributary and interchannel deposits, slope deposits and proximal turbidites. The Roseland (and Meneage) Breccia Formation is a major olistostrome (e.g. Barnes 1983), composed of grey silty mudstone with dispersed clasts of sedimentary, metamorphic, volcanic and magmatic rocks, with interbedded sedimentary rocks, including breccias and conglomerates, and acid and basic volcanic rocks. Extrabasinal clasts, introduced by slumping, slides and sediment gravity flows, include shelf sediments of Lower Devonian, Silurian and Ordovician age, and a variety of schists and granitoids. Contemporaneous basic and acidic lavas, volcanic breccias and minor intrusions have MOR and within plate, calc alkaline, affinities (Barreiro 1996), and some larger clasts are ocean floor basalts (e.g. Nare Head, Roseland: Barnes 1983; Floyd 1984). Within this nappe sequence there is dramatic progradation from shielded basin sediments to climactic slope deposition. The overlying Dodman Nappe (Fig. 10.6), comprises an undated succession (Barnes 1983) of interbedded greywacke sandstone and mudstone. A reverse metamorphic gradient, from epizone to higher greenschist-grade mica schist, is present onshore in the Dodman Formation. Offshore data suggests that the Start Schists are part of this nappe. Hornblendic schists associated with mica schists at Start, which are also of flyschoid appearance, have N-MORB signatures (Floyd *et al.* 1993*b*; Merriman *et al.* 2000). Comparability with some schists of the lower tectonic unit beneath the Lizard ophiolite cannot yet be excluded (see above). A further nappe of crystalline rocks, the Normannian Nappe (Holder & Leveridge 1986*b*), is mapped offshore (SWAT 9: BIRPS & ECORS 1986; Edwards *et al.* 1989). It has surface expression in the garnetiferous gneiss of Eddystone Rock in Plymouth Bay, probably the Man of War Gneiss off Old Lizard Head and possibly the Old Lizard Head Series (Nutman *et al.* 2001). Locally interposed between the Dodman and Normannian nappes are the tectonic units of the Lizard Complex.

The flysch source, defined by sandstone-framework modal grains and their geochemistry, was a dissected continental margin magmatic arc (Floyd & Leveridge 1987), essentially of

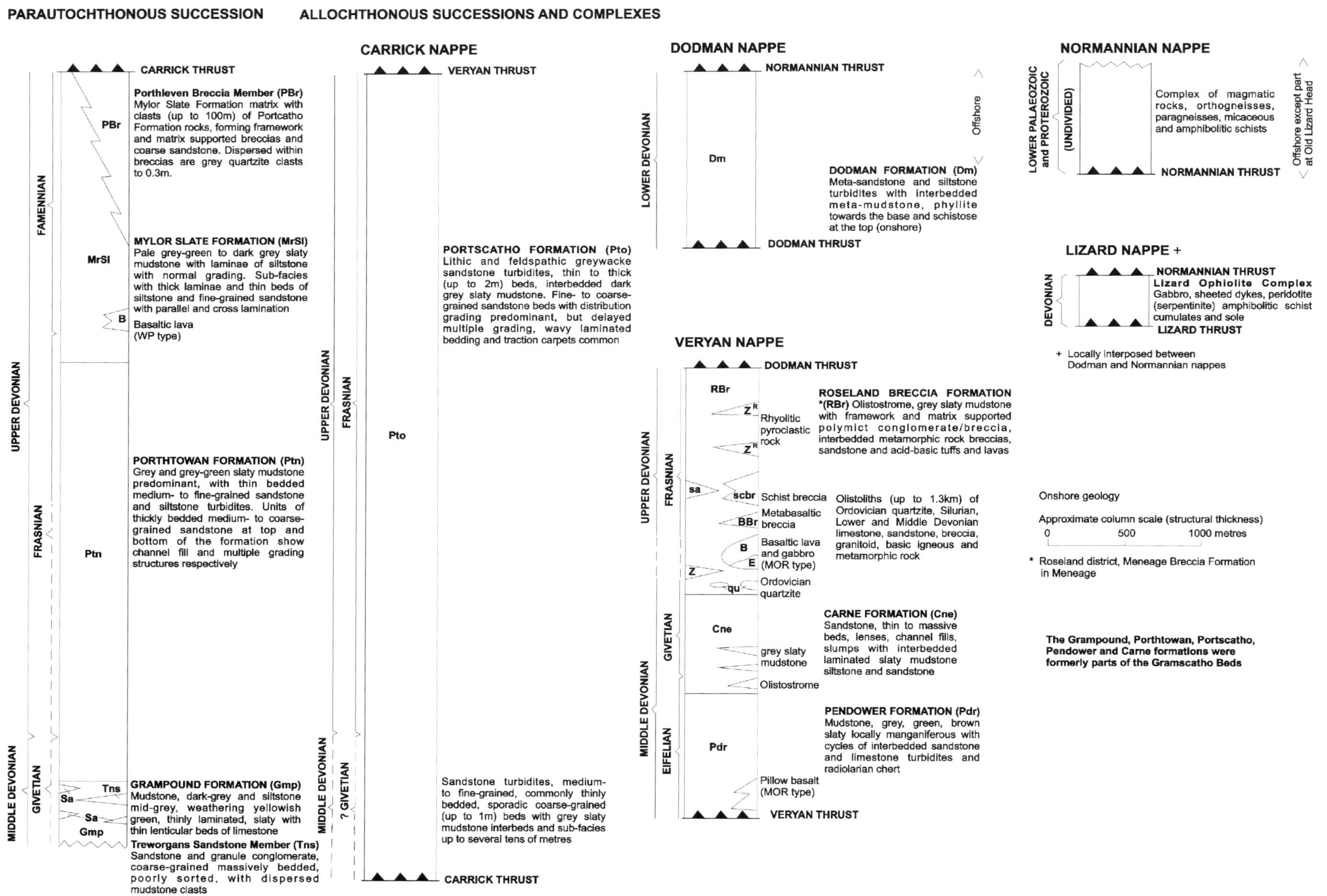

Fig. 10.6. Nappe and parautochthon successions in south Cornwall, based on BGS (1990) and BGS (2000).

pre-Devonian origin (Floyd *et al.* 1991). The extrabasinal clasts, including Lower Palaeozoic and Lower Devonian shelf sediments, schists and granitoids, are comparable to basement and Palaeozoic sequences in Armorica, and olistolith faunas also have southern affinities (Bohemian, Hendriks 1937; Armorican, Sadler 1973). A potential source between Cornwall and Brittany was identified by Ziegler (1982) as the Normannian High, an extension of the Mid German Crystalline High.

The thrust nappe stack (Fig. 10.5) has a transgressive ramped contact with the parautochthon. To the north of the Lizard it rests in a pseudoconformable manner on thick (up to 1.3 km) sedimentary breccias (Porthleven Breccia Member, Mylor Slate Formation), of Famennian age, at the top of the parautochthonous sequence (Leveridge & Holder 1985). The breccia includes clasts of the allochthonous Portscatho Formation showing pre-incorporation early (D_1) structures, in a matrix also deformed by D_1. The Mylor Slate Formation comprises mainly interlaminated and thinly interbedded dark grey mudstone, graded and locally cross-laminated siltstone and fine-grained sandstone. It is interpreted as representing, in its lower part (Shail 1989), an association of rise-slope proximal channel deposits and low-density turbidites and hemipelagites, and in its main part as distal turbidites (Leveridge *et al.* 1990). It overlies the Porthtowan Formation, an estimated 2.8 km-thick sequence of grey and grey-green mudstone with subordinate greywacke sandstone, part of the Gramscatho Group. It correlates in large part with the Portscatho Formation, but generally finer grain sizes and turbidite sedimentary characteristics are consistent with more distal deposition. The lower boundary is conformable with the Grampound Formation (Fig. 10.5), comprising coarse-grained sandstone grain-flow beds (up to 6 m) interdigitating with mudstone and wispy laminated siltstone and sandstone. The age and relationship of this formation and the Meadfoot Group (see below) immediately to the north has yet to be established. Significant however is that Meadfoot Group marine shelf rocks (Pragian–Emsian) extend from the north to the south of the 'Start Perranporth Line', suggesting that it did not have a significant influence on development of the basin, in terms of its constraint (see Holdsworth 1989) or sediment sourcing (Shail 1992).

Early, D_1, deformation in the flysch nappes is characterized by slaty cleavage and tight to isoclinal sheath folds that record NNW overriding translation (Leveridge *et al.* 1984). Dodson & Rex (1971) dated pervasive low-grade greenschist metamorphism (Warr *et al.* 1991) as a late Devonian–early Carboniferous event (K–Ar 370–350 Ma), but modern data (Clark *et al.* 1998*b*) extends that event back to the mid-Devonian ($^{40}Ar/^{39}Ar$ 385 ± 2 Ma) in the Dodman Formation. Radiometric determinations show a broad pattern of younging uplift metamorphic ages, late mid-Devonian–early Carboniferous (*c.* 385–355 Ma), from SSE to NNW through the nappe pile.

The close temporal relationship of climactic sedimentation and deformation in south Cornwall indicates causal connection. The Devonian–earliest Carboniferous evolution of the dynamic stratigraphy of south Cornwall was modelled by Leveridge *et al.* (1990) by integration of nappe stratigraphies and the timing of deformation (Fig. 10.7). Although supported by Clarke *et al.* (1998*b*) and Sandeman *et al.* (2000) the orthogonal extent of the ocean and also the internal basin organization implied by that sequence diagram have been questioned (see above). However, there is concurrence (see Shail 1992) on the link between flysch sedimentation, northward tectonic migration, obduction of the Lizard ophiolite, and closure of the Gramscatho Basin. Those processes began with flysch deposition at the southern margin of the basin, sourced by the Normannian High, probably late in the Early Devonian, a timing suggested by the in-sequence nature of the onshore nappes. Northerly overthrusting of the Normannian Nappe, accompanied by sedimentation at its leading edge, initiated southerly subduction of the ocean floor. Detachment of the Dodman and Lizard nappes, inundation of the Pendower hemipelagic environment, and probable filling of the oceanized basin with flysch extending onto the northern passive margin of the basin, took place in the Mid Devonian. During the Frasnian forward propagation transferred motion to the new Veryan Thrust, carrying the Veryan Nappe onto more distal flysch. Further forward propagation took place in the late Famennian–early Carboniferous when the major Carrick Nappe was thrust out of the basin. Erosion produced a thick olistostrome at the thrust front and this was overridden by continuing nappe movement. Sandstone clasts within the olistostrome showing pre-incorporation deformation illustrate the penecontemporaneity of sedimentation and deformation within the migrating nappe pile. The emergence of the nappe onto the marginal sequence effectively marked the closure of the oceanized Gramscatho Basin, a process that apparently took some 35 Ma (Fig. 10.7), and the onset of continental collision.

Looe Basin

The southernmost of the major basins on the passive margin of the Gramscatho Basin is the Looe Basin (Figs 10.1 & 10.9), whose succession illustrates clearly the basin forming and filling processes. Within the basin there is in excess of 5 km of Lower Devonian and early Middle Devonian sedimentary rocks. These include the major lithostratigraphical divisions of the Lower Devonian, identified and mapped across the province by Ussher (e.g. 1890, 1903, 1907), the Dartmouth Slates, Meadfoot Beds and Staddon Grits. These have since become known as the Dartmouth Group, and the Bovisand and Staddon formations of the succeeding Meadfoot Group, respectively. Completing the basin succession is dark grey mudstone, of latest Emsian and Eifelian age (e.g. Jennycliff 'Slates' division of the Saltash Formation at Plymouth), attributed to the variously named formations that persist through the Devonian (Fig. 10.8). Commonly this upper part of the Looe Basin succession is occluded by overthrust Staddon Formation.

The Dartmouth Group, locally subdivided into formations (Dineley 1966; Seago & Chapman 1988), comprises terrestrial deposits in which there is a restricted non-marine fauna and fish remains. The latter include such as *Rhinopteraspis dunensis* and *Althaspis leachi* (White 1956; Forey in Ivimey-Cook 1992) that are indicative of the early Mid Siegenian (Pragian). As they are found at higher levels in the group, it may extend down into the Lochkovian (cf. House & Selwood 1966). Considered by Dineley (1966) to represent stable coastal mudflat regime, with lagoonal and fluviatile deposits, rocks of the group are now interpreted as the products of a more dynamic setting. To the east, the deposits of perennial lakes are interspersed with fluvial deposits (Smith & Humphreys 1989, 1991), whereas to the west the lacustrine regime was semi-permanent (Jones 1992; Leveridge *et al.* 2002). There in a sequence up to 3.3 km thick, with no base exposed, red and green mudstone and siltstone, deposited from suspension, constitute the lacustrine facies. Interbedded siltstone and sandstone, as laminae, lenses, sheet-beds, and amalgamated units, represent distal and more proximal deposition from fluvial-fed traction currents and underflows. Regularly developed quartzite interbeds indicate deposition of fluvial sediment from confined channel flow. Through the sequence pebbly mudstone units, with clast dispersion and composition indicative of mass-flow emplacement and intrabasinal derivation, appear to be the product of slope collapse. These features, together with growth faults and associated penecontemporaneous microgabbroic intrusions, point to rapid subsidence and instability of the basin. Cyclic sequences in the upper parts of the group include grey marine mudstone within the lacustrine and fluvial sediments, indicating hydrological continuity between the lacustrine regime and open marine conditions during periodic downfaulting. Contemporaneous volcanic rocks, predominantly basic lavas and tuffs, are alkaline and calc-alkaline basaltic rocks typical of a rifting regime with crustal contamination at an early stage of the process (Merriman *et al.* 2000).

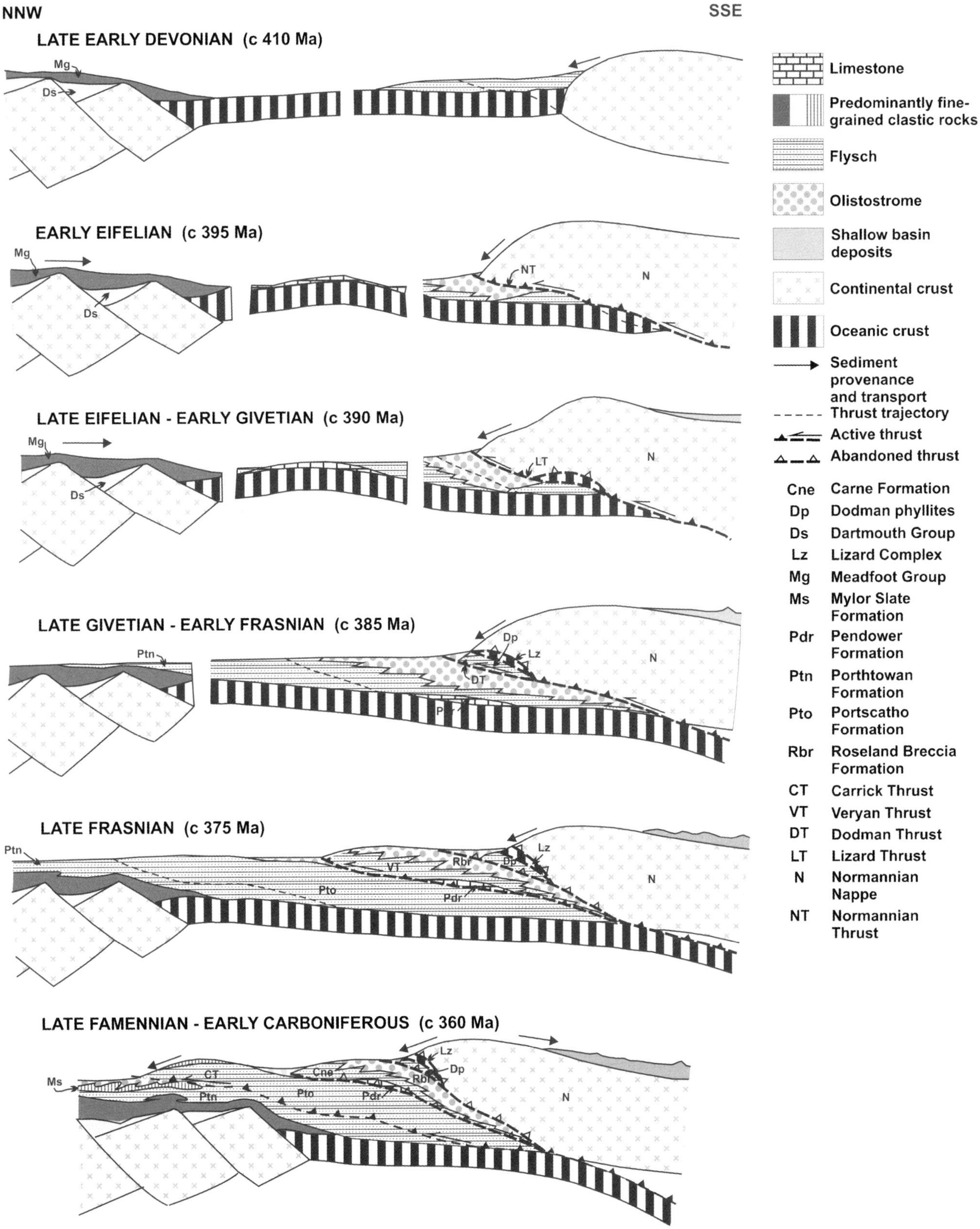

Fig. 10.7. Time sequence diagram indicating development of tectonostratigraphical divisions of south Cornwall, accommodating closure of the Gramscatho Basin, after Leveridge *et al.* (1990).

The Meadfoot Group represents establishment of the marine environment, a conformable base reflecting rapid inundation. Although extending locally onto the northerly marginal platform (Leveridge *et al.* 2003*a*) the group is largely confined within the Looe Basin where it comprises the Bovisand Formation and Staddon Formation (Harwood 1976; Seago & Chapman 1988).

The Bovisand Formation is a Pragian–mid-Emsian sequence representing a variety of offshore environments with sporadic lower shoreface reworking. Sandstone dominant members

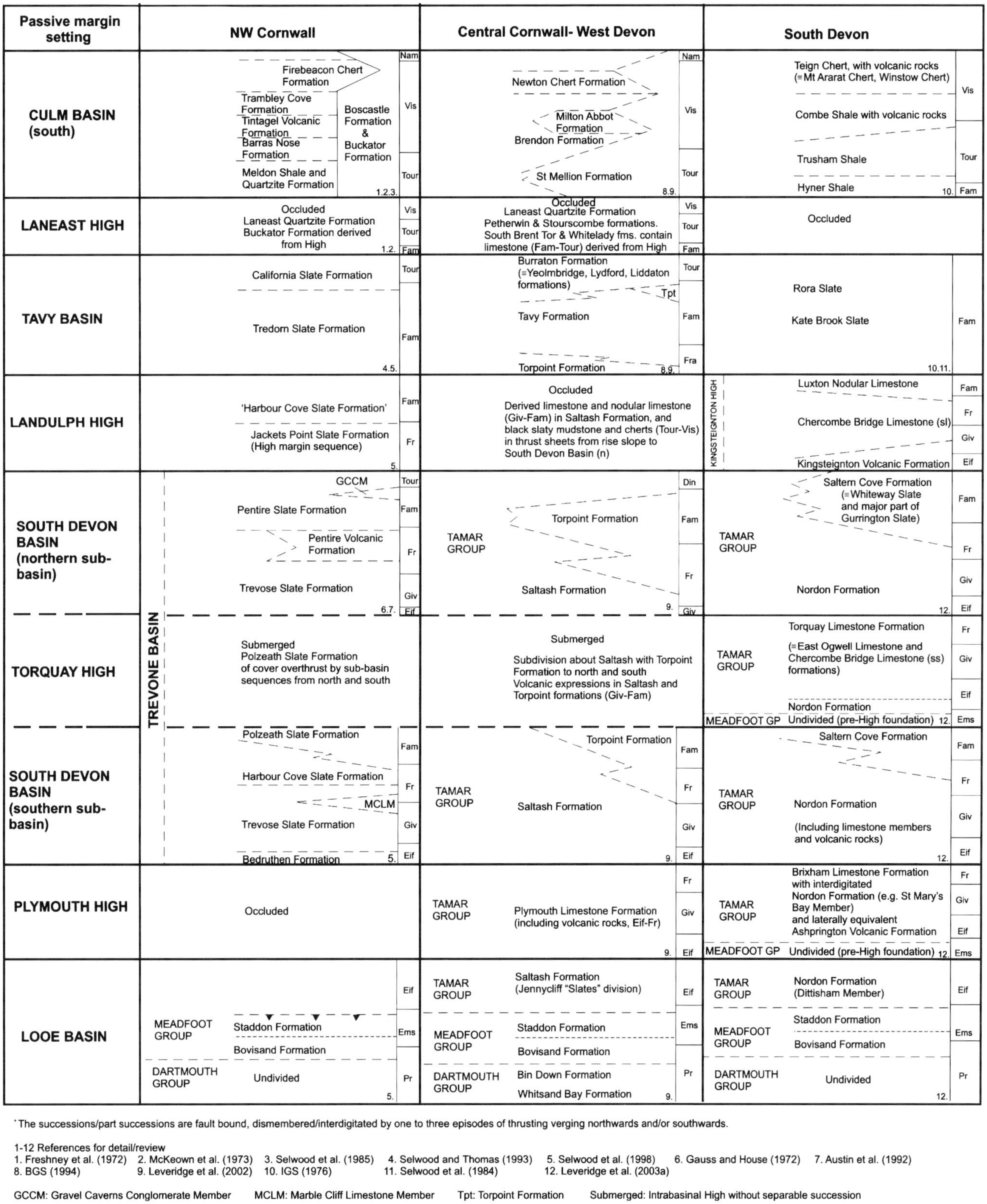

Fig. 10.8. Devonian–Dinantian successions of the passive margin, in relation to its basin and high framework.

(up to *c.* 350 m), comprising cyclic sequences ranging from offshore-shelf facies to wave dominated and storm-dominated shoreline facies associations, reflect numerous phases of gradual progradation and rapid transgression.

The late Emsian Staddon Formation, with a thickness of 400 m (Selwood *et al.* 1998) or more, represents a major phase of such southward progradation. Its component fine- to coarse-grained sandstone beds record the transition from shallow-marine setting to high-energy regime. Deposition in shallow-marine embayments, with storm-wave-generated bars (Pound 1983), is succeeded by fluvial channel, levee and overbank deposition in a non-marine setting (Humphreys & Smith 1988), and the establishment of a substantial fluvial coastal sand plain.

Completing the Looe Basin succession is the lowest part of a predominantly grey slaty mudstone sequence that forms the marine background sedimentation of the passive margin southern crop throughout the Devonian. Variously termed the Trevose Slate Formation (see Selwood *et al.* 1998), the Nordon Formation (Selwood *et al.* 1984) and Saltash Formation (Leveridge *et al.* 2002), the main expression is within the South Devon Basin (Fig. 10.3). About Plymouth Sound the mudstone sequence, is the Jennycliff ('Slates') division of the Saltash Formation (Leveridge *et al.* 2002), which on the basis of conodonts (Orchard 1977, 1978; Dean 1994) and palynomorphs (Molyneux 1990; Dean 1992) ranges from late Emsian to late Eifelian. Abundant *Spirophyton*, with gutter casting and hummocky cross-stratification of sandstones low in the division are compatible with encroachment of the sea over the fluvial Staddon Formation and the recycling of sand from the drowned coastal plain by storm activity. In the upper mud-rich part of the sequence are interbeds of limestone and units of hyaloclastitic basalt lava and bedded tuff. Features of the limestone are indicative of deposition from high-energy currents with minimal transport of crinoid, brachiopod, coral, stromatoporoid and bryozoan debris from a shallow source.

The Looe Basin is interpreted as having developed on the extending and subsiding continental margin of the oceanized Gramscatho Basin. Initially terrestrial (Fig. 10.9a), the onset of marine conditions in the late Pragian indicates that the rate of subsidence of the basin exceeded sediment supply. The polycyclic shoreline developments of the Bovisand Formation represent episodic (sub)basin infilling and subsidence. Delta-top and fluvial sandstone of the Staddon Formation show that, in late Emsian times, deposition at the northern edge of the Looe Basin had outstripped subsidence. Rivers were feeding siliciclastic sediment into the basin across a piedmont from a source area to the north (Leveridge *et al.* 2002). With further regional extension the South Devon Basin developed to the north, isolating the Plymouth High between the basins, and shielding the Looe Basin from coarse clastic sediments. Related subsidence in the Looe Basin produced the starved basin facies at the top of its succession, interrupted by early Eifelian carbonate storm deposits derived from the shallows developing above the nascent Plymouth High.

South Devon Basin

The South Devon Basin (Figs 10.1 & 10.9) is composite, with northern and southern sub-basins. These are separated to the east about Torbay (Leveridge *et al.* 2003*a*) by a subdividing high, which is recognized by lithostratigraphical segregation near Plymouth, and in the west where it is also reflected in structural facing confrontation near Polzeath. The basin shows an overall deepening from east to west. Basin formation by half-graben fault-block rotation initiated during the late Emsian and continued into the Mid Devonian producing the complementary E–W linear Plymouth High (Fig. 10.9b) separating the South Devon Basin from the Looe Basin to the south. This southerly bounding Plymouth High, and the Torquay High that formed a sub-basinal divide to the east, have coeval successions different to those in the basin. However, basin and high successions are linked by clastic sediments derived from the highs, aprons to the volcanic rock edifices associated with the highs extending out into the basins, and sporadic incursions of basinal muds into the successions on the highs. These are detached from their foundations, each being present within two or more northward transported thrust sheets, but generally remain in original relative positions with regard to adjacent basinal successions. The Plymouth High succession is occluded westwards from Plymouth by overthrust deposits of the Looe Basin (Leveridge *et al.* 2002).

Plymouth High succession. The sequence of rocks developed on the high comprises essentially reefal deposits with peripheral carbonate sediments (Plymouth Limestone Formation; Leveridge *et al.* 2002), or, to the east, biogenic bank complex (Brixham Limestone Formation: Drummond 1982; Leveridge *et al.* 2003*b*). The limestones do, however, interdigitate with, or are laterally replaced by, basaltic volcanic rocks (e.g. Ashprington Volcanic Formation).

The Plymouth Limestone Formation is present within two thin thrust sheets, extending east–west through Plymouth. Originally assigned to the Middle Devonian (Ussher 1907) on the basis of a rich coral, brachiopod and gastropod fauna, its rugose corals indicated a Middle–Upper Devonian age to Taylor (1951). Conodonts reveal that it ranges from mid Eifelian to the late Frasnian and possibly the Famennian (Orchard 1977, 1978). A progressive carbonate build-up in the Eifelian produced a blanketed, shallow carbonate platform by the early Givetian, on which coral/stromatoporoid reefs (Orchard 1977) developed with an extensive peripheral facies of reworked crinoids, stromatoporoids and corals, in the late Givetian (Leveridge *et al.* 2002). Succeeding Frasnian deposits, with a distinctive fauna of *Amphipora*, *Stringocephalus* brachiopods and turreted gastropods, desiccation fenestrae and cryptalgal lamination, reflect a restricted and emergent back reef setting (Garland *et al.* 1996) and southward progradation. Fissures, yielding Famennian conodonts (Orchard 1978), indicate foundering of the limestone build-up from mid Frasnian times, termination of carbonate deposition and inundation by red mud deposits.

The Brixham Limestone Formation, present in three thrust sheets (Leveridge *et al.* 2003*a*), is of early/mid Eifelian–Frasnian age. It similarly represents progradation of a biogenic bank complex, but rather than being formed of massive reefs, passes up to massive beds, of *in situ* worked crinoidal debris bound by laminar stromatoporoids, of a reefal flat (Mayall 1979). A Givetian back-reef lagoonal facies, and Frasnian high-energy platform deposits with features of emergence are also present (Drummond 1982).

The Ashprington Volcanic Formation forms an extensive thrust sheet to the south of Totnes in south Devon. Interdigitating with and of similar age to the Brixham Limestone Formation it is a thin slice of a thick sequence of alkaline basalt lavas and tuffs thrust over the South Devon Basin deposits. It represents a major volcanic expression, associated with the Plymouth High, which was emergent, and at times extended into the basin sequences to the north and south (Leveridge *et al.* 2003*a*).

Torquay High succession. The deposits of the high that separated the southern and northern sub-basins of the South Devon Basin, on the eastern side of the province, are present within three or more northerly transported thrust sheets (Leveridge *et al.* 2003*a*). East of the Sticklepath Fault it is represented essentially by the Torquay Limestone Formation, and to the west by the correlative Chercombe Bridge and East Ogwell limestone formations.

The Torquay Limestone Formation, subdivided in part into members, based on lithofacies changes (Scrutton 1977), ranges from early Eifelian (*Polygnathus costatus partitus* Zone: Castle 1982) to early Frasnian (Lower *Polygnathus asymmetricus* Zone: Castle 1982). A similar carbonate build-up to that on the Plymouth High is recorded, with carbonate bank and patch reefs in the Eifelian, giving way to major stromatoporoid reef development in the Givetian, and 'high-energy' conditions in the Frasnian with bioherms within worked crinoidal sand (Scrutton 1977). Locally in the upper part of the formation red silty mudstone, interbedded with and surrounding masses of limestone, yields Mid Frasnian (*Manticoceras cordatum* Zone) ammonoids (House 1963).

The equivalent succession west of the Sticklepath Fault is represented in thrust sheets to the SW of Newton Abbot by the Denbury Crinoidal Limestone, Chercombe Bridge Limestone Formation (ss) and East Ogwell Limestone Formation, and locally developed Kingsteignton Volcanic 'Group' (Selwood *et al.* 1984). Local reef development is characterized by massive corals (e.g. *Disphyllum caespitosum*) rather than stromotoporoids (Scrutton 1977).

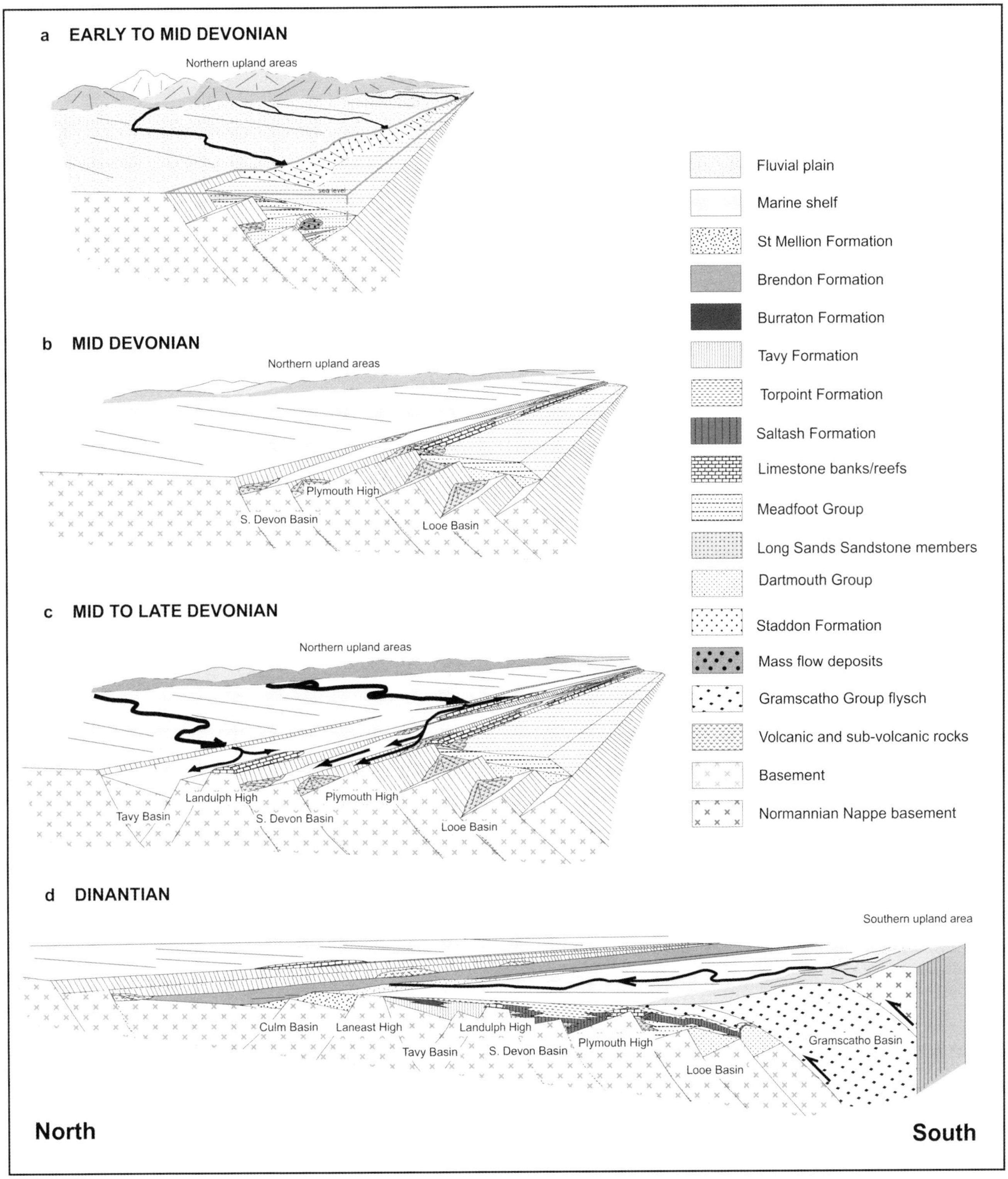

Fig. 10.9. Sequence diagram showing the progressive northward rifting of the passive margin. Sections show basins, sub-basins, highs, derived terrestrial sediment transport and the main lithostratigraphical units of the central part of the sub-province. Based on Leveridge *et al.* (2002).

To the west the Torquay High deposits are occluded by overthrusting, and towards the River Tamar the mid-basin rise was too deep to support biogenic carbonate production (Fig. 10.9b), but related extrusive volcanic rocks and basinal mudstone facies distribution indicates its presence.

Basin succession. The basinal deposits of the South Devon Basin are largely grey, reddish purple and green silty mudstones. Predominant are the grey facies hemipelagic rocks variously termed, from west to east across the province, the Trevose Slate Formation, Saltash Formation and Nordon Formation. Within the South Devon Basin they range from the late Emsian to the Tournaisian in the Saltash Formation of the Plymouth district (Leveridge *et al.* 2002). Upper Devonian occurrences have locally been separately named (e.g. Pentire Slates: Gauss & House 1972) or included with the purple and green mudstones (e.g. Gurrington Slate: Selwood *et al.* 1984). The Trevose Slate Formation (see Selwood *et al.* 1998), some 3.9 km thick, comprises largely submillimetric laminae couplets of fine-grained grey siltstone and dark grey silty mudstone, which is bioturbated and destructured in parts. Subordinate are siltstone and fine-grained sandstone turbidites, and sporadic slump beds. Knox (in Goode & Leveridge 1991) compared the sediments to those of the Santa Barbara Basin in offshore California (Hulsemann & Emery 1961) where laminae couplets are hemipelagic annual deposits in anoxic conditions in a rapidly subsiding barred basin some 500 m deep. Eastwards along the basin laminated sequences become subordinate to homogeneous mudstone. On the southern side of the South Devon Basin, low in the sequence (Eifelian), there are beds of richly fossiliferous siltstone, sandstone and limestone. These are interbedded turbidites in the Plymouth district (Leveridge *et al.* 2002), and storm-generated deposits constituting the separate Bedruthan Formation in North Cornwall (Selwood *et al.* 1998). Crinoidal limestone turbidites form the Marble Cliff Limestone Member (Kirchgasser 1970; Mouravieff 1977), on the western coast near Trevone, an inverted 80 m sequence of Givetian and Early Frasnian age (House *et al.* 1978: Austin *et al.* 1985). Tucker (1969) suggested derivation from a southerly sub-littoral environment, possibly the occluded Plymouth High. In the southern sub-basin between Dartington and Torquay (BGS 2003) limestone members of the Nordon Formation are of Eifelian, early Givetian, late Givetian and Famennian age (Drummond 1982). Of similar age to sequence breaks in the Torquay Limestone Formation, they are interpreted as storm current, turbid and debris flow deposits largely derived from the mid-basin Torquay High (Leveridge *et al.* 2003*a*).

Purplish red mudstone and fine-grained siltstone with subordinate green coarse-grained siltstone and fine-grained sandstone constitute the other major sedimentary facies of the basin. Principal occurrences across the basin are the Polzeath Slates to the west, the Torpoint Formation about the Tamar, and the Saltern Cove Formation (and Rora Slate) to the east (Fig. 10.8). Predominantly of Frasnian and Famennian age, there is some evidence of red mudstone being interbedded with late Givetian limestones of the highs, and of it extending into the Tournaisian (Gurrington Slate: Selwood *et al.* 1984). Sedimentary features are indicative of deposition from distal turbid flows and larger suspended turbid clouds (Leveridge *et al.* 2002). Thin mass-flow breccia deposits with Frasnian and Famennian limestone clasts are also locally present at Saltern Cove, Torbay (Van Straaten & Tucker 1972). A sparse benthos but rich planktonic fauna, including entomozoacean ostracodes (e.g. *Nehdentomis serratostriata*) and ammonoids (e.g. *Manticoceras cordatum*) indicate disaerobic bottom conditions in an open marine environment. Water depths of 500–1000 m have been estimated for this same facies, the *Cypridienschiefer*, in southern Rhineland (Bandel & Becker 1975).

The grey hemipelagic facies and the red facies sedimentary rocks are coeval through the Upper Devonian. The two interdigitate on millimetre–kilometre scales, but the red facies peters-out along the basin and sub-basins from areas of maximum development (Leveridge *et al.* 2002). Rich in ferric iron, and a geochemistry indicative of an evolved crustal source, the distribution of red sediments indicate introduction at various points along the northern margin of the basin (Fig. 10.9c). Thickest developments are about the centres of the depositional basin and sub-basins. That they do not form part of a layer-cake succession explains some earlier stratigraphical problems and anomalies of this belt (e.g. Gauss & House 1972; Selwood *et al.* 1993, 1998).

Associated with the basinal sediments there are significant expressions of alkali basaltic lavas, pillowed in parts, hyaloclastites, and tuffs. Major developments are about the mid-basin high (Leveridge *et al.* 2002) and the northern margin of the basin (e.g. Pentire Volcanic Formation: Frasnian; Selwood *et al.* 1998).

Completing the basinal succession about Landulph in the Tamar Valley is a sequence of grey basinal Upper Devonian mudstone passing to dark grey and black mudstone of Tournaisian age, and cherts of Tournaisian and probable Visean age. It is within thin slices derived from the rise slope to the high forming the northern boundary of the South Devon Basin.

Landulph High

Slumped limestone blocks, limestone turbidites, and *in situ* nodular limestones within the sequence at Landulph, indicate the presence of the Landulph High carbonate platform source from the Givetian to the Famennian, subsequently occluded by overthrusting (Leveridge *et al.* 2002). That this was a horst rather than a high generated by block rotation (Fig. 10.9c), is suggested by the associated D_1 structure-confrontation (Leveridge *et al.* 2002). To the east a more complete Late Eifelian–Famennian succession, including the Kingsteignton Volcanic Group, Chercombe Bridge Limestone (s.l.) and the Luxton Nodular Limestone of the Newton Abbot district can be attributed to this high. Part of the Ugbrooke–East Ogwell succession of Selwood *et al.* (1984), it is bound to the SE by overthrust South Devon Basin deposits, and itself is thrust northwestwards over the Kate Brook Slate of the Tavy Basin. Tuffs and lavas of the volcanic group pass up in the early Givetian to limestones that are locally thickly bedded with bioherms up to 6 m thick, and massive stromatoporoidal limestone. The overlying Luxton Nodular Limestone is a condensed mid Frasian and Famennian sequence with a rich pelagic fauna (House & Butcher 1973) representing a marked change of water depth. To the west of the province the high is represented by the Jackets Point Slate Formation (Selwood & Thomas 1986). Comprising mudstone, sandstone, and conglomerate, tuffs and agglomerates, and containing locally abundant shelly faunas, the formation has been interpreted as an association high on the basin shelf margin (Selwood *et al.* 1998).

Tavy Basin

The Tavy Basin (Fig. 10.9c) formed in the Givetian, when the Landulph High became isolated from clastic sediments and developed as a carbonate platform. The principal basinal deposit is green mudstone, the main occurrences across the peninsula being termed the Tredorn Slate Formation (see Selwood *et al.* 1998) in north Cornwall, the Tavy Formation (Leveridge *et al.* 2002) north of Plymouth, and the Kate Brook Slate (Waters 1974) in the Newton Abbot area. The formation comprises pale green–grey-green chloritic silty mudstone, faintly banded in part, with subordinate sandstone, as thin laminated turbidites, sporadic lenses, burrow fills, locally constituting a mappable member. Pyrite is ubiquitous, cubes to 50 mm being common. A restricted benthic macrofauna, including rhynchonelloids and spiriferoids (e.g. *Cyrtospirifer verneuili*) is not definitive, but a late Frasnian–late Famennian age is indicated by its microfossils, conodonts (Stewart 1981*a*), ostracodes (Whiteley 1983) and palynomorphs (Dean 1992). Selwood *et al.* (1998) have proposed an outer shelf setting for the formation, and lithological uniformity of thick sequences to the south led Leveridge *et al.* (2002) to propose rapid sedimentation from turbid flow rather than quiet water settlement from suspension. The chlorite and pyrite content of the rocks indicate the role of reducing conditions during or after deposition, and suggest differing original organic content, burial rates and/or sourcing from the contemporaneous purple facies rocks of the South Devon Basin.

In the northernmost crops of the above dominant facies, there are beds of bioclastic limestone (Selwood *et al.* 1998) and limestone rich sequences, such as the Petherwin and Stourscombe beds (Selwood 1960). These are richly fossilliferous condensed shell concentrate and cephalopod limestones typifying deposition upon or on the flanks of a high, with goniatite, cephalopod and conodont faunas indicating Famennian and early Tournaisian ages (Stewart 1981).

Sporadically intercalated with the predominant facies along the basin are purple and green mudstones (e.g. Rora Slate of the Newton Abbot district) equivalent to those of the South Devon Basin. Planktonic ostracodes show a similar Frasnian (Gooday 1973) to late Famennian (Wilkinson 1990) age range to the major facies. To the west bluish grey mudstone interdigitated with the green constitutes the Delabole Member of the Tredorn Slate Formation (Selwood *et al.* 1998).

Across the basin the green facies rocks pass up to grey facies mudstones such as the Burraton Formation and Yeolmbridge Formation (Selwood 1971). Within these grey–dark grey mudstones, the uppermost deposits of the basin, the transition from the Famennian to the Tournaisian is recorded in goniatite and trilobite faunas (Selwood 1960).

Laneast High

The presence of a high, bounding the Tavy Basin to the north, is intimated, if not fixed, by the limestone sequences of Petherwin and Stourscombe (Stewart 1981*a*, *b*), and possibly by the Buckator Formation (Freshney *et al.* 1972; Selwood *et al.* 1985). Southward-transporting nappes of the Devonian and Carboniferous rocks of the Culm Basin and late extensional faulting in this zone have obscured and occluded much of the body of the high and its deposits. The Laneast Quartzite Formation, of uncertain but probable late Devonian/Dinantian age (Selwood *et al.* 1985), is in thrust contact with Upper Devonian Tavy Basin rocks and basinal Dinantian rocks to the NE of the Bodmin Moor Granite (BGS 1994). Thickly bedded quartzite, coarse-grained, with a high degree of rounding and sorting, is interbedded with subordinate dark grey mudstone with abundant plant debris (Selwood *et al.* 1985). Those authors suggest a high-energy environment of deposition with active reworking, a near shore setting being considered probable. It is suggested here that the formation is a remnant of the shelf bounding the Tavy Basin prior to further rifting and the formation of the Culm Basin that isolated the intervening Laneast High in the Upper Devonian (Fig. 10.9d).

The Petherwin Formation, mid Fammenian-late Tournaisian, Stewart 1981b) and the Stourcombe Formation (late Fammenian, Selwood 1960) are penecontemporaneous deposits on and about the high where it was more deeply submerged. The former formation comprises a variety of lithofacies, nodular cephalopod limestone, calcareous sandstone, fine congomerate, brachiopod-rich, mudstone and the the latter comprises silicified cephalopod (e.g *Gonioclymenia (kalloclymenia)frechi lange, Wochlumeria sphaeroides* (Richter)) limestone, silicified mudstone and trilobite (e.g. *Phacops (phacops)granulatus*(Munster)) rich mudstone. They are characterized by both stratigraphic condensation and reworking in parts. Thus, the earlier shelf sands and contemporaneous deposits of the platform, being variably subject to reworking, sourced sediment into the adjacent basins, and such is manifest in the Buckator Formation (see below: Fig. 10.8)

Culm Basin

The Culm Basin (Figs 10.1 & 10.9) formed as an extensional basin (see below) in the Upper Devonian in sequence with the basins to the south (Leveridge *et al.* 2002). The transition to the Dinantian is recorded within the deposits of both the Tavy and South Devon basins. The main occurrences of Dinantian rocks are present in a polyphase series of E–W thrust slices near the southern margin of the main Carboniferous tract and segregation there of those not attributable to the Culm Basin is presently unfeasible. They are also in thrust nappes, with both Culm Basin and Tavy Basin rocks, that were transported to the south over the Tavy Basin (Fig. 10.1). Sequences originally near the southern margin of the basin are now up to 22 km further south in the St Mellion Outlier (Fig. 10.1) (Whiteley 1983).

Sequences in the thrust sheets extending eastwards from north Cornwall (Freshney *et al.* 1972) north of the Dartmoor Granite (Edmonds *et al.* 1968) to its eastern flank (Selwood *et al.* 1984) are largely similar. Grey mudstone passes up to black mudstone, succeeded by lava and tuff, black mudstone and chert (Fig. 10.8), a sequence generally comparable to the North Devon Basin succession in the paucity of coarse clastic sediments, a result of the 'Bathyal Lull' of Goldring (1962). To the east the greenish grey mudstone of the Trusham Shale yields Tournaisian ostracodes (e.g. *Maternella circumcostata, Richterina (R.) latior*) but the succeeding bluish black mudstone of the Combe Shale has proved barren, attributed to prevailing euxinic conditions by Selwood *et al.* (1984). Equivalent black slaty mudstone at Lydford in the central part of the province contains olistostromes, conglomerates and volcanic debris flows (Isaac *et al.* 1982) indicative of late Tournaisian and Visean instability. On the western coast the black mudstone Barras Nose and Trambley Cove formations has yielded conodonts from limestone lenses indicative of the late Tournaisian–early Visean (Austin & Matthews 1967). Between those two formations is the Tintagel Volcanic Formation, comprising variable proportions of basaltic tuffs and pillow lavas, and represented in the east of the province by lava and high-level microgabbroic sills. There, volcanicity, represented by rhyolitic tuffs and basaltic vesicular lava, extends into the succeeding Teign Chert, and local correlatives. The Teign Chert, consisting of laminae and thin beds of chert, variably coloured, white, black, green and red, and locally rich in radiolaria, is interbedded with black siliceous mudstone. Its Visean age is constrained by a pelagic fauna, in beds of mudstone, chert and limestone at its top, which includes the Zone fossil *Posidonia becheri* (Selwood *et al.* 1984). Similar *Posidonia* beds with the Fire Beacon Chert to the west (McKeown *et al.* 1973) have yielded late Visean bivalves (e.g. *Posidonia corrugata*) and goniatites (e.g. *Mesoglyphioceras granosum*).

Whilst that succession is recognized in occurrences across the peninsula, to the north of Dartmoor and also westwards, the Meldon Shale and Quartzite Formation (Edmonds *et al.* 1968), of probable Tournaisian age (Isaac 1985), consists of grey mudstone with thin beds and lenses of quartzitic sandstone. Such siliciclastic input in the Dinantian is more significant in the Boscastle Formation forming some 4 km of the highly faulted, multithrust, west coastal section between Boscastle and the Rusey Fault Zone (Freshney *et al.* 1972). The formation comprises pyritous black mudstone, with laminated siltstone and packets of greywacke sandstone turbidites. Attributed to the Silesian Crackington Formation by those authors it was reassigned to the Dinantian by Selwood *et al.* (1985), but the early Namurian goniatites (e.g. *Nuculoceras nuculum*) retrieved by Freshney *et al.* (1972) suggest a sequence transitional between the Dinantian and Namurian south of the Rusey Fault.

Interdigitating and stratigraphically conformable with the Boscastle Formation is the Buckator Formation (Freshney *et al.* 1972; Selwood *et al.* 1985), composed of grey-green mudstones with sporadic sandstone beds and limestone lenses with facies-related conodont species indicating a shallow-water sub-tidal depositional setting. The formation ranges from the Fammenian *Scaphignathus velifer* Zone (Selwood *et al.* 1985) to the *Gnathodus bilineatus* Zone (Austin & Matthews 1967).

In the St Mellion Outlier, well to the south of the main crop, three fault-bound Carboniferous formations, the St Mellion Formation, Brendon Formation and Newton Chert Formation are disposed in a series of thrust slices, in an imbricate stack with Torpoint and Tavy formation rocks (Fig. 10.17). In flat lying sheets, transported from the north, bedding is steeply overturned southwards and thick sequences are exposed (Leveridge *et al.* 2002). The major component of the outlier is the St Mellion Formation, composed of sandstone turbidites, slumped plant-rich sandstone, and interbedded dark grey mudstone. Retrieved conodonts indicated a Tournaisian age

(*Siphonella sandbergi* Zone), and a spore flora, including *Lycospora pusilla*, a late Visean age to Whiteley (1983). N. Riley (in Leveridge *et al.* 2002), revising ammonoid identifications by Matthews (1970), determined the presence of the *Nuculoceras nuculum* marine band of the early Namurian. The Brendon Formation, dark-grey siliceous mudstone with sporadic packets of blue-grey coarse-grained greywacke sandstone turbidites, is of largely equivalent age (Tournaisian and Visean: Whiteley 1983) and the Newton Chert Formation, with 'Posidonia Beds' and a rich conodont fauna (Whiteley 1983), compares and correlates with cherts to the north. On sedimentological evidence Jones (1993) described the St Mellion Formation as a pro-delta succession linked to coastal deltaic systems, sourced from the south. The Brendon Formation, juxtaposed by southward thrusting, represents a more distal basinal sequence with sporadic turbidite incursions reaching out into the basin from the delta complex. Abundant plant debris demonstrates a continental source, and grains of low-grade metamorphic rocks and potassium feldspar from exposed granitic rocks, indicate the probable source to have been the up-thrust Normannian and Gramscatho Basin nappes to the south (Fig. 10.9d). The rocks of the St Mellion Outlier are the original marginal facies of the Culm Basin, whereas the Dinantian successions further north represent more distal deposits of the basin (Leveridge *et al.* 2002).

The succession between Boscastle and the Rusey Fault on the west coast, which was also deformed by the main early deformation of the province (see below), is interpreted as the deposits of the southern sub-basin of the Culm Basin (Fig 10.14).

The Dinantian volcanic rocks, Tintagel Volcanic Formation and equivalents, are representative of a wide expression of volcanism in the Rhenohercynian Zone. It has been linked to high silica content in basinal waters promoting formation of associated radiolarian chert (Ziegler 1982). As with most of the Devonian volcanic rocks they are alkaline basalts (Floyd 1984) with Ocean Island Basalt characteristics and a mantle source (Merriman *et al.* 2000), typical of an extending rift regime with β (stretch factor) >2 (McKenzie & Bickle 1988). The Culm Basin thus originated as one of the passive margin rift basins. The volcanics are also the last of the province, and of the extensional regime that characterized most of the Devonian and Dinantian prior to regional shortening, basin and conduit closures and deformation.

North Devon 'Basin'

The Devonian–Dinantian rocks of north Devon and west Somerset (Fig. 10.10) constitute the succession of the North Devon Basin (Fig. 10.1). The limits and nature of this depositional basin are very poorly constrained. The area has been considered (e.g. Edmonds *et al.* 1975; Tunbridge 1978; Selwood & Durrance 1982) to have lain immediately to the south of Wales with the transition from the continental Old Red Sandstone facies to the Devonian marine facies being accommodated in the Bristol Channel area. Certainly, through the Devonian, its rocks do reflect the interplay of those facies. However, it has been proposed that such interplay took place up to 400 km to the SE (Holder & Leveridge 1986a, 1987). Sequences would therefore equate with the 'Old Red Sandstone' (Eastern Avalonia) derived rocks of the southern Ardennes (Behr *et al.* 1984), before Carboniferous displacement along the Bristol Channel–Bray Fault (Kellaway & Hancock 1983; Higgs 1986, fig. 2). The succession, although deformed by major–minor E–W northward-verging and upright gently plunging folds (Sanderson & Dearman 1973) with an axial plane slaty cleavage, youngs from north to south in a simple manner (Fig. 10.10). Edmonds *et al.* (1985) disputed the contention of others (e.g. Reading 1965; Holwill *et al.* 1969) that thrusting repeated sequences significantly, and suggested a minimum total thickness for the succession of approximately 6700 m (Fig. 10.11). Limited evidence from the Devonian of south Devon points to thin shelf expressions to the north of lithostratigraphical divisions formed mainly in the developing

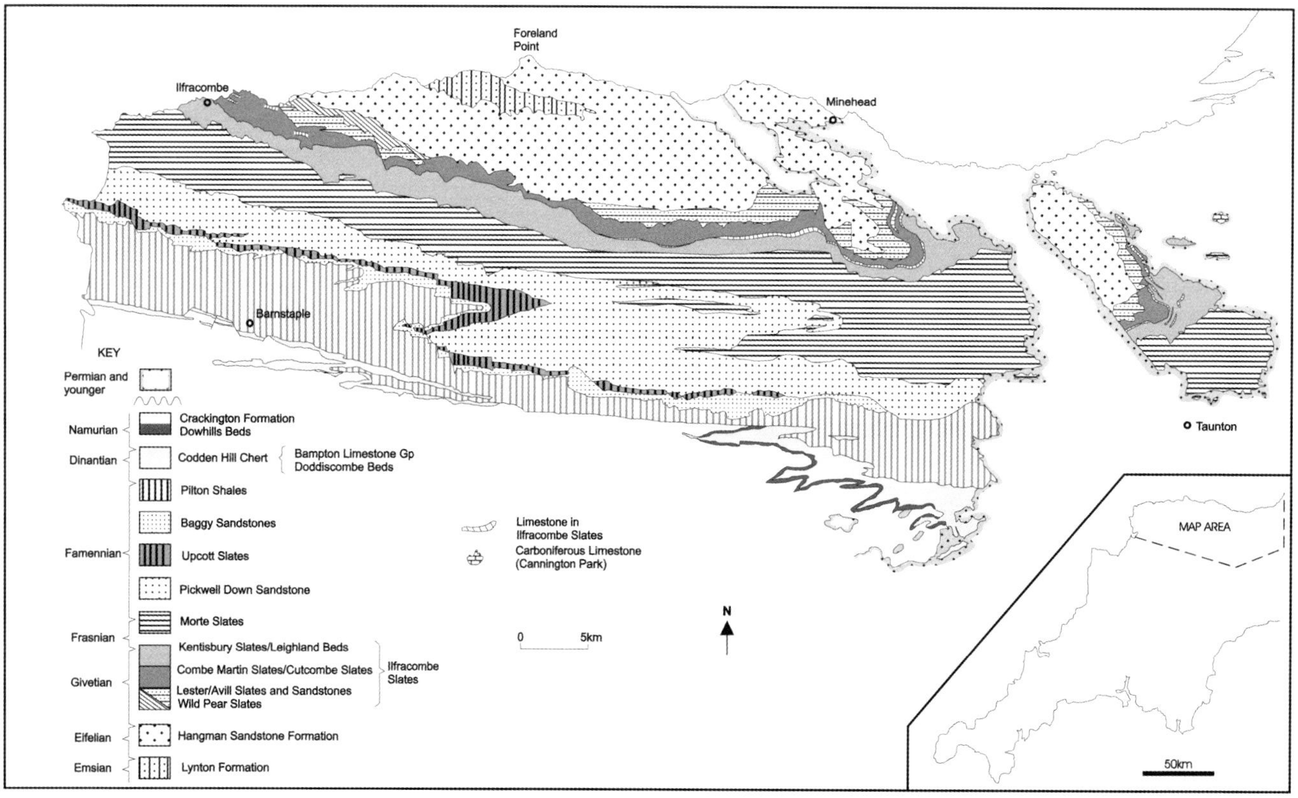

Fig. 10.10. Geological sketch map of the north crop Devonian and Dinantian succession, based on BGS maps.

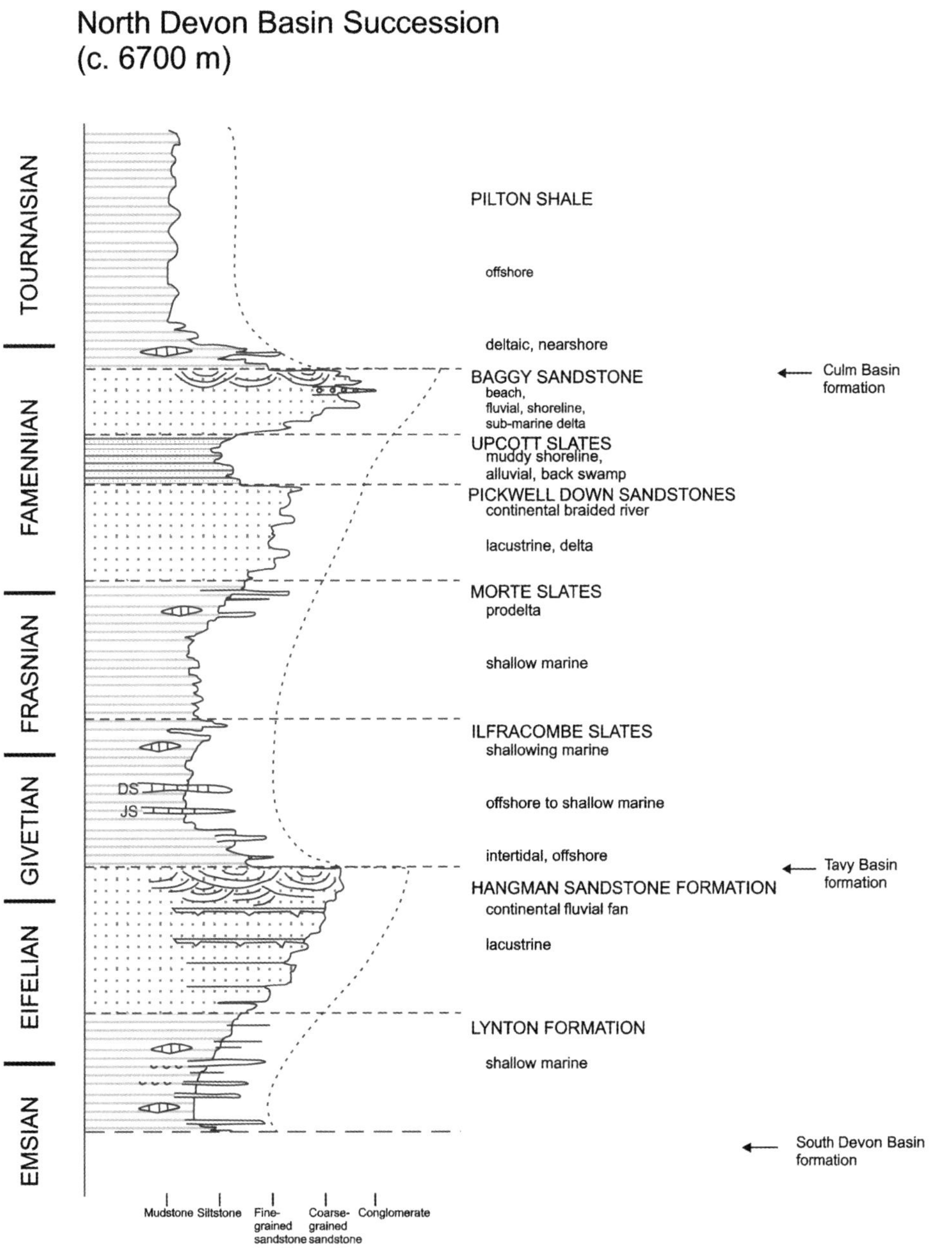

Fig. 10.11. Graphic log of the North Devon Basin succession.

basins (Leveridge *et al.* 2003a). This suggests that the main depositional axis of this basin was located in the north where sediment debouched before an elevated northerly hinterland.

The oldest rocks of the north crop form the Lynton Formation (Edwards 1999), lying within the faulted core of the Lynton Anticline, a major fold plunging gently east-southeastwards from Lynmouth Bay. The formation comprises largely interlaminated grey mudstone and sandstone, with the latter predominating in the lower and uppermost parts, also included are beds of quartzite and thin beds and lenses of bioclastic limestone. Bioturbation and burrows, including *Chondrites* (Simpson 1964) are common, and bivalves, brachiopods, bryozoans and crinoid debris constitute the limestone. Brachiopods including *Platyorthis longisulcata* and *Chonetes sarcinulatus* indicated a late Emsian–early Eifelian age to Evans (1983), a range confirmed by the shallow-water conodont faunas retrieved by Knight (1990). The shallow-marine setting of the deposits (Simpson 1964; Edmonds *et al.* 1985) was considered by Selwood & Durrance (1982) to represent 'almost certainly' the initial marine transgression of the Old Red Sandstone continent.

Re-establishment of continental conditions is represented in the succeeding Hangman Sandstone Formation (Edwards 1999). Variously estimated to be between approximately 1650 (Tunbridge 1978) and 2500 m (Edmonds *et al.* 1985) thick, the formation comprises predominantly purple, green and grey fine- to coarse-grained sandstone, with subordinate reddish brown mudstone. The mudstone, variably finely interlaminated with sandstone or homogeneous with extrabasinal clasts or pedogenic carbonate nodules (Jones 1995), shows desiccation cracking and bioturbation. Plant stem fragments, trace fossils, *Arenicolites* and the non-marine *Beaconites*, and gastropods *Naticopsis* and *Bellerophon* occur sporadically through the formation, and the bivalve *Myalina* is common towards the top, but none are definitive of age. However, palynomorph assemblages from the upper part of the formation have indicated a late Eifelian/early Givetian age to Knight (1990). Sedimentological analyses by Tunbridge (1981*a*, 1984) and Jones (1995) have defined a coarser facies of channel and sheet sandstones, with palaeocurrents indicating transport to the south, and a finer facies of mudstones deposited in lakes, or as terrestrial mass flows. Deposition was largely within a distal fluvial fan environment (Jones 1995) into which there were brief marine incursions and on which lakes and mudflats developed at intervals (Edwards 1999). Heterolithic beds and cross-bedded sandstone, with bimodal current directions, and the presence of locally abundant *Myalina* and bryozoans in the uppermost part of the formation, herald the next major marine incursion of the area (Tunbridge 1978).

The establishment of fully marine conditions is recorded in the Ilfracombe Slates. Mappable subdivisions (Edmonds *et al.* 1985) of the Slates (Fig. 10.10), similar to those described by Holwill (1963), reflect the deepening and shallowing of that marine environment. The silver-grey mudstone, with thin beds of sandstone siltstone and limestone of the Wild Pear Slates indicate shallow-water deposition (Edmonds *et al.* 1985), with a W–E along-strike variation from delta front to intertidal suggested by Webby (1965*a*). The Lester Slates-and-Sandstones contain a greater proportion of cross-bedded sandstone and thin bioclastic limestone beds within the mudstone, in which *Chondrites* is abundant. The sandstones and shelly limestones, with coral, brachiopod, bryozoan and crinoid debris, are interpreted as storm generated deposits in an offshore setting. The Combe Martin Slates comprise the silver-grey mudstone with sporadic thin sandstone beds and limestone beds, the latter forming two units up to 10 m thick mapped across the province (Fig. 10.10), namely the lower Jenny Start Limestone and the higher David's Stone Limestone. Corals and stromatoporoids, commonly in growth positions, characterize both. Rugose corals, such as *Disphyllum aequiseptatum* and *Thamnophyllum caespitosum*, are preponderant over tabulate corals in the lower unit, and the reverse relationship applies in the upper unit, with tabulate corals such as *Thamnopera cervicornis*, exceeding single corals. This has been regarded as an indicator of shallowing in the sequence (Webby 1966; Edmonds *et al.* 1985). The Givetian age indicated by the corals is confirmed by conodonts assigned to the Lower/Middle *Polygnathus varcus* Subzone (Knight 1990). The superceding Kentisbury Slates, comprising silver-grey mudstone and subordinate sandstone rich units, the former characterized by *Chondrites* burrows, are interpreted as representing regression in the early Frasnian (Edmonds *et al.* 1985). Within the division there are argentiferous lead deposits extending across the crop from Combe Martin, regarded by Scrivener & Bennett (1980) as sedimentary exhalative (Sedex) deposits linked to the Cockercombe Tuff (Webby 1965*b*) in the Quantock Hills.

How the succeeding Morte Slates relate to the cycles of regression and transgression is uncertain. Comprising some 1500 m of silver, grey-green and purple mudstone, sporadic thin cross-laminated sandstone beds in lowermost and central parts, and sporadic limestone nodules, they carry a sparse impoverished fauna, which includes *Lingula* and *Cyrtospirifer verneuili*. They are regarded as shallow-water marine deposits, with a continued regression (Selwood & Durrance 1982), or further transgression (Edmonds *et al.* 1985), represented by the prodelta and delta platform setting proposed by Webby (1965*c*, 1966). The depleted macrofauna, suggested by Selwood & Durrance (1982) to be linked to rapid sedimentation and reduced salinity at a delta mouth, is reflected in the depletion of the miospore and acritarch assemblages, that typify the Frasnian and earliest Famennian (Knight 1990).

The overlying Pickwell Down Sandstones consist of purple, red, brown and greenish grey sandstone, with subordinate red and grey mudstone near the top and base of the formation. Also, at the base is the Bittadon 'Felsite', a tuff up to 8 m thick that extends across the north crop, that has yielded remains of Upper Devonian armoured fish such as *Bothriolepis, Holoptychius and Coccosteus* (Rogers 1919). The main body of the 1200 m-thick formation comprises fine- to medium-grained thick to massive sandstone beds with wavy and cross-lamination, trough cross-bedding, erosive bases, and ripple marking. It is considered (Selwood & Durrance 1982; Edmonds *et al.* 1985) to represent predominantly continental braided river deposits, with subordinate lacustrine and deltaic facies. The formation contains few macrofossils apart from sporadic fish and plant remains, but it is assigned to the Famennian as it is in sequence with dated mudstones below and higher in the succession. There is upward passage to the Upcott Slates, a sequence of mudstones and silty mudstones, variegated cream, buff, green, grey and purple, with sporadic thin fine-grained cross-laminated sandstone beds, that is up to 250 m thick. The slates were thought by Goldring (1971) to have been deposited in alluvial, back-swamp, lacustrine, and muddy shoreline environments.

The Baggy Sandstones, overlying the Upcott Slates, comprise up to 450 m of sandstone, thick cross-bedded cosets to thinly bedded, siltstone and mudstone, with minor developments of intraformational conglomerate and slumped sediments, and sporadic thin beds of crinoidal and gastropodal limestone. Relatively rich in macrofossils and trace fossils, Goldring (1971) considered it an exemplar of shallow-water sedimentation. Within the framework of preponderant fine-grained facies associations towards top and base, and sandstone facies dominating in the main part of the formation, he defined a variety of subfacies and linked depositional environments. Prevailing submarine delta platform sedimentation is interrupted by repeated shoreline advances, represented by a diversity of deposits in fluvial distributary, estuarine channel, lagoonal and beach settings. A close relationship between environments and fossils was defined: plant debris (*Sphenopteridium rigidum* and *Knorria*) with distributaries and lakes, trace fossils (*Arenicolites curvata*, *Diplocraterion yoyo*) with a sub-beach setting, gastropods and bivalves (e.g. *Dolabra 'Cucullaea' unilateralis*) with near shore intertidal channels, and productellids and bivalves (e.g. *Ptychopteria damnoniensis*) with open marine conditions. The upper boundary of the Baggy Sandstones with the Pilton Shales is transitional and poorly defined (Edmonds *et al.* 1985). However, Goldring (1971) reported palynomorph and conodont evidence that the formation extended to the topmost Devonian '*Wocklumeria*' Stufe. Conodont analysis by Austin *et al.* (1970) indicated that a younger late Devonian biozone (Lower *Bispathodus costatus* Zone: Austin *et al.* 1985) extended across the boundary of the formations.

The lithostratigraphical base of the Pilton Shales is arbitrarily taken to be immediately above the uppermost sandstone unit attributable to the Baggy Sandstones (Edmonds *et al.* 1985). The formation extends up to the dark grey and black shaly mudstone of the succeeding Codden Hill Chert and correlatives within the Dinantian. Some 500 m thick, the formation comprises grey shaly mudstone, with thin beds of siltstone and sandstone, thin beds and lenses of limestone, and sporadic calcareous sandstone beds and composite units in its lower part. The sequence is rich in fossils (see Goldring 1970; Edmonds *et al.* 1985), particularly brachiopods, bivalves, trilobites and goniatites. Goldring (1970) linked lithological and faunal changes through the formation to major transgression, with residual near shore and deltaic sediments and associated brachiopods and bivalves, giving way to an offshore finer facies with trilobites and goniatites. The lower third of the formation Goldring (1957, 1970) attributed to the Famennian *Wocklumeria* Stufe on the basis of its productellid and trilobite (e.g. *Phacops accipitrinus accipitrinus*) fauna. The overlying sequence represented the Dinantian *Gattendorfia* and possiby the *Ammonellipsites* ammonoid zones (Tournaisian–early Visean).

The Devonian succession (Fig. 10.11) constitutes two major sedimentary cycles, with the timing of marine transgressions coinciding with basin formation to the south. This suggests general subsidence of the passive margin accompanying the release of extensional stress.

The Dinantian succession of the north crop is completed by a sequence of dark grey and black shaly mudstone, chert and limestone. These are designated the Codden Hill Chert (to the west and north of the E–W Brushford Fault), and the Dowhills Beds and Bampton Limestone 'Group' (to the east and south of the Brushford Fault). The Codden Hill Chert comprises some 150 m of dark grey–black shaly mudstone, locally siliceous, and chert, each of which is locally predominant (Prentice 1960*a*), with sporadic lenses of limestone. The chert, in part rich in radiolaria, is generally pale weathering. *Posidonia becheri* is common through the sequence, and both trilobites and goniatites (e.g. *Goniatites spiralis*) confirm a Mid–Late Visean attribution (P_1–P_2: Prentice 1960*a*). Correlatives to the east are the Doddiscombe Beds and Bampton Limestone Group (including the Westleigh Limestones). The Doddiscombe Beds, comprise pale and dark grey laminated mudstone, which is unfossiliferous and variably cleaved. The underlying Pilton Shales yield Tournaisian *Siphonodella* Zone conodonts (Matthews & Thomas 1968) and the superceding Bampton Limestone Group is of Visean age (Thomas 1963*a*). The Bampton sequence consists of interbedded chert, limestone and shale. Further to the east at Westleigh turbidite limestone beds, up to 6 m thick and locally conglomeratic, are interbedded with the shaly laminated mudstone. Pelagic (orthocones and *Posidonia*) and subordinate benthonic faunas in the shales, indicative of accumulation in quiet water conditions, contrast with the oolitic material, rolled coral colonies and abraded brachiopods in the limestone beds derived from a shallow higher energy shelf regime (Thomas 1963*b*).

Continental collision and deformation of the passive margin

The successions of the southern Devonian–Dinantian sub-province are complexly deformed in a fold and thrust terrain (e.g. Isaac *et al.* 1982; Coward & McClay 1983). Early fold facing and thrust transport are northwards in a southern domain and southwards in a northern domain (e.g. Sanderson & Dearman 1973; Seago & Chapman 1988). This overall structure of the belt was interpreted in terms of forward propagating thrusts originating in the south, with backfolding and backthrusting of an essentially layer-cake stratigraphy by Coward & Smallwood (1984) and Seago & Chapman (1988). The significant role of basin architecture in influencing structure development during basin inversion was proposed by Selwood (1990), who attributed southward-verging structures in the Trevone Basin (South Devon Basin) to backfolding associated with basin inversion and northward out-thrusting on its southerly inclined northern margin fault. Hartley & Warr (1990) and Warr (1993) also attributed southerly backthrusting, and consequently the facing confrontation of the Trevone Basin (e.g. Roberts & Sanderson 1971; Selwood *et al.* 1993), to the butressing effect of a rise bounding the rift basin to the north. The controlling influence of basin architecture on the development of major structures during inversion, with basin margin faults, not only becoming thrusts bounding basinal successions, but also controlling regional structure facing and structure confrontations, was proposed by Leveridge *et al.* (2002).

Closure of the Gramscatho Basin was accommodated by northward tectonic migration (see above; Fig. 10.7). The concept that such migration operated through the whole of the southern crop of Devonian and Dinantian rocks (e.g. Dearman 1971; Shackleton *et al.* 1982) was founded upon a perceived structural continuity, and a general progressive northward younging of metamorphic radiometric ages. Structural continuity is now established through this sub-province, structural sequences in both active and passive margin successions being similar, with three major deformation episodes (e.g. Alexander & Shail 1996; Leveridge *et al.* 2002) common to both, and subordinate phases locally developed (Leveridge *et al.* 1990; Selwood *et al.* 1998). The whole-rock K–Ar dating of Dodson & Rex (1971), adjusted by Warr *et al.* (1991) using revised decay constants, remains an important constraint on the setting-time of metamorphism. The low-greenschist grade of metamorphism within the belt is considered to be a function of both burial and thrust nappe loading (Warr *et al.* 1991), with slaty cleavage development of the first deformation, D_1, promoting strain enhanced crystal growth for high anchizonal and epizonal grades (cf. Merriman *et al.* 1995). The age range of slates within the belt is 345–325 Ma, spanning the late Tournaisian–early Namurian, and this is when the passive margin was first deformed. Younger narrow E–W zones along the northern margin of the belt (295–280 Ma) and between the Start peninsula and Perranporth area (315–295 Ma) are attributable to secondary (D_2) transpositions of earlier fabrics (e.g. Holdsworth 1989; Pamplin 1990).

Collision to the south in late Famennian/early Dinantian times was thus not a terminal event, and deformation migrated northwards. Structures of D_1 in the passive margin basin terrain are an ubiquitous penetrative cleavage, close to isoclinal folds, thrusts and strike-slip faults on mesoscopic and macroscopic scales. The prevailing orientation of folds and thrusts is E–W (e.g. Alexander & Shail 1996; Selwood *et al.* 1998), but there is incipient sheath folding, particularly in the vicinity of major thrusts and in higher strain zones to north and south (Dearman 1969; Leveridge *et al.* 1990). Related extension lineations are orientated NNW–SSE. Major D_1 thrusts bound basin and sub-basin successions (Leveridge *et al.* 2002). They have major antiformal folds in hanging walls at thrust fronts, a notable example being the Bovisand/Man Sands Antiform at the northern margin of the Looe Basin succession. This is a product of the inversion of half-graben and full-graben and the translation of each bounding rift fault into a basal thrust (Fig. 10.12). The process progressed with each extensional basin closing, inverting, and 'locking-up' before stress transmission to its neighbour

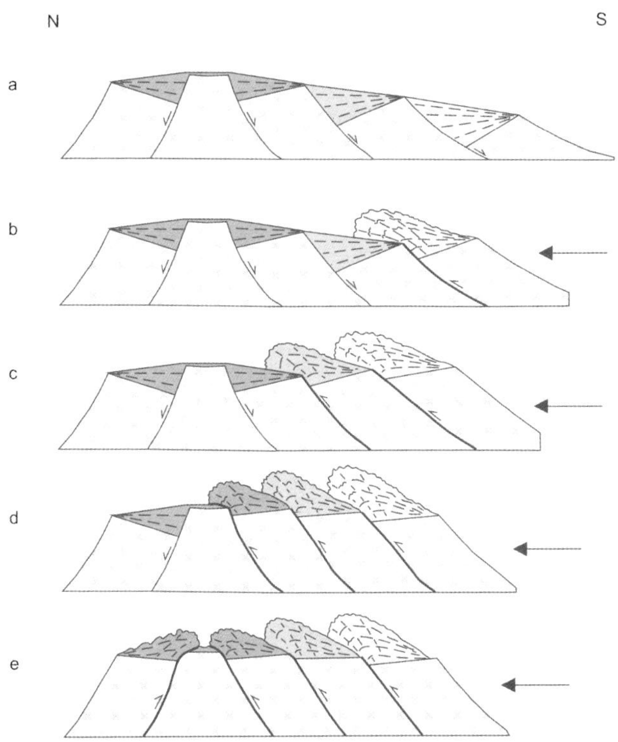

Fig. 10.12. Cartoon illustrating sequential half-graben inversion during D_1, with consequent thrusting, bounding successions, and major inversion antiforms. Facing confrontation is linked to larger horst blocks.

(Leveridge *et al.* 2002). Basin contents were not expelled over the major highs in south Devon, but the successions of the highs (Plymouth and Torquay) were detached, apparently by footwall short-cut thrusts, and translated over adjacent deposits to the north in the South Devon Basin (Fig. 10.15) (Leveridge *et al.* 2003*a*). In a similar manner the reversal of facing at or near the southern margin of the Tavy Basin (Selwood *et al.* 1984; Leveridge *et al.* 2002) within the northern domain is attributable to a northerly inclined bounding fault at the Landulph High. D_1 structures face southwards over much of the crop of the Tavy Basin deposits (Seago & Chapman 1988; Selwood *et al.* 1998). A further reversal of facing in the north (Isaac *et al.* 1982), where largely obscured by overthrust Culm Basin deposits, probably reflects a full-graben geometry for the basin. The facing confrontation between South Devon Basin and Tavy Basin successions across much of the peninsula is accommodated without interference, D_1 cleavage being common across the zone (Leveridge *et al.* 2002).

The confrontation of early structure facing on the north Cornwall coast, variously known as the Polzeath or Padstow Facing Confrontation, is within the South Devon Basin (Trevone Basin; see Selwood *et al.* 1998). Subsequent to its description by Gauss (1967) as two separate tectonic domains juxtaposed by later thrusting, and as a zone of overlap of northward and southward facing structures (Roberts & Sanderson 1971), debate centred on the relative timing of those early structures (e.g. Pamplin & Andrews 1988; Durning 1989). The confrontation has been attributed to backfolding (Andrews *et al.* 1988) and to backfolding and thrusting associated with basin inversion (Selwood 1990) or backthrusting caused by buttressing against earlier extensional faults to the north of the basin (Warr 1993), all mechanisms being ill constrained. Overlap fabrics of the zone can now be largely attributed to regional D_2 deformation associated with the northward transporting Trebetherick Thrust, that juxtaposes reversed facing, but beneath which the reversal-transition of primary facing is recorded (Selwood *et al.* 1998). The coeval, but locally distinct facies to north and south in the basin (Gauss & House 1972; Selwood *et al.* 1993) reflect division of the basin apparent to the east. The confrontation is compatible with inversion about a horst-block sub-basinal divide.

During deformation of the Devonian rocks there was deposition within the Culm Basin, with coarse clastics at the southern margin of its southern half-graben, passing northwards to finer grained condensed basinal deposits. At least the southern part of the basinal succession, up to and including early Namurian sediments, was then deformed in sequence by D_1, while the spread of thrust sheets of this succession across the basins to the south is in large part a result of D_2 and D_3 (see below). However, the southward facing and overturning of all these rocks (Seago & Chapman 1988; Leveridge *et al.* 2002; BGS unpublished data), rather than northward (Isaac *et al.* 1982), is a D_1 feature, resulting from the formation of major southward-verging antiformal folding at the southern basin margin during inversion. D_1 structures, overturned and southward facing, are developed (with D_2) within the Boscastle Formation (Selwood *et al.* 1985) up to the Rusey Fault zone (Freshney *et al.* 1972). The opposed vergence of structures in Culm Basin rocks, southwards on the southern margin, and northwards to the north (Dearman 1971), although including Silesian rocks deformed after the D_1 inversion to the south, is an indicator that the extensional basin was a full graben.

The D_1 transport direction, between north and NNW, is approximately normal to the main basinal architectural elements of the province, and thus transpressive fabrics are not typical of the deformation. At a late stage of D_1 the prominent NW–SE dextral strike-slip faults of the province developed (Leveridge *et al.* 2002), possibly as the extensional basins became largely locked. They are part of a wide family of stuctures across the Variscan belt, which includes the Bristol Channel–Bray Fault (Fig. 10.2), and along which associated ductile tectonic reworking has been dated at approximately 320 Ma (Matte *et al.* 1986), mid Namurian.

The Silesian Culm Basin

The major proportion of the Namurian and all of the Westphalian rocks of the Culm Basin had not been deposited when the Dinantian and lowermost Namurian succession near its southern margin was inverted by D_1. Continuing tectonic migration, consequent uplift and sedimentation were therefore concurrent (Warr 1993). With progressive closure of the extensional Culm Basin and the presence of inverted basins to the south, the Culm depositional area became a foreland basin during the Silesian (Hartley & Warr 1990; Hartley 1993*b*).

The principal formations of the foreland Culm Basin are the Crackington, Bude and Bideford formations (Edmonds 1974; Thomas 1988). Early Namurian mudstone sequences, such as the Dowhills Beds (Webby & Thomas 1965) to the north, Ashton Formation (Chesher 1968) to the south, are essentially conformable with Dinantian sequences and products of the bathyal lull, prior to the incursion of coarser synorogenic clastics of the main crop (Fig. 10.13). They have been regarded as sandstone-free, or sandstone-sparse, parts of the Crackington Formation (Edmonds 1974). The Bealsmill Formation between Dartmoor and Bodmin Moor (BGS 1994), constituting the Blackdown Nappe (Isaac 1985), derived from the north (Seago & Chapman 1988; BGS unpublished data), is a proximal facies (Selwood & Thomas 1988) correlative of the Crackington Formation. The affinity of the Early–Mid Namurian (House & Butcher 1973) proximal submarine-fan deposits of the Ugbrooke Sandstone in the Newton Abbot district, to either the Crackington Formation or the St Mellion Formation, is uncertain.

The Crackington Formation, lowermost of the formations of the Culm Basin, is present at the southern side and northern side of, and in antiformal inliers between, the main crop of the Silesian (Fig. 10.13). It comprises interbedded sandstone and dark grey–black pyritous shaly mudstone, in locally variable proportions, with sandstone becoming dominant higher in the sequence. The sandstone is a fine-grained quartz-rich greywacke forming laminae to thick beds, predominantly medium beds, of increasing thickness upwards (McKeown *et al.* 1973). Interpreted as distal turbidites, most beds are massive with delayed grading, others show distribution grading, and commonly beds have sharply defined bases with load casting and cross-laminated tops. Sole marks, groove casts and flute molds, prod marks, largely indicate easterly or westerly flow, with subordinate transport directions to the NE and south (Mackintosh 1964; Edmonds *et al.* 1968). The formation is characterized by a restricted marine benthos, ammonoids, fish and bivalves, and dispersed plant debris. Turbidite deposition commenced in the late Early Namurian (*Eumorphoceras* (E_2) subzone: Edmonds *et al.* 1968) in the south and generally in the early Late Namurian (*Reticuloceras* (R_2) subzone: Webby & Thomas 1965) in the north. Exceptionally, at Venn in the north (Edmonds *et al.* 1985), one sandstone occurrence yields *Homoceras* (Mid Namurian). It continued into the Early Westphalian ('A'; *Gastrioceras amaliae* subzone: Edmonds *et al.* 1979). Sedimentation is interpreted to have been in 'relatively' deep water, largely anoxic, with turbidite flow predominantly parallel to the basin axis. Sourcing has been attributed in small part to collapse from the inverted southern basin margin (see Thomas 1988), and largely to an emergent continental terrain to the north (Melvin 1986) and east (Freshney & Taylor 1980), possibly linked to the contemporaneous activity of the Bristol Channel/Bray Fault.

The Bideford Formation has a relatively restricted east–west linear distribution extending from the coast just to the south of Westward Ho! It is composed of coarsening-up sedimentary cycles (De Raaf *et al.* 1965), from black mudstone with sporadic

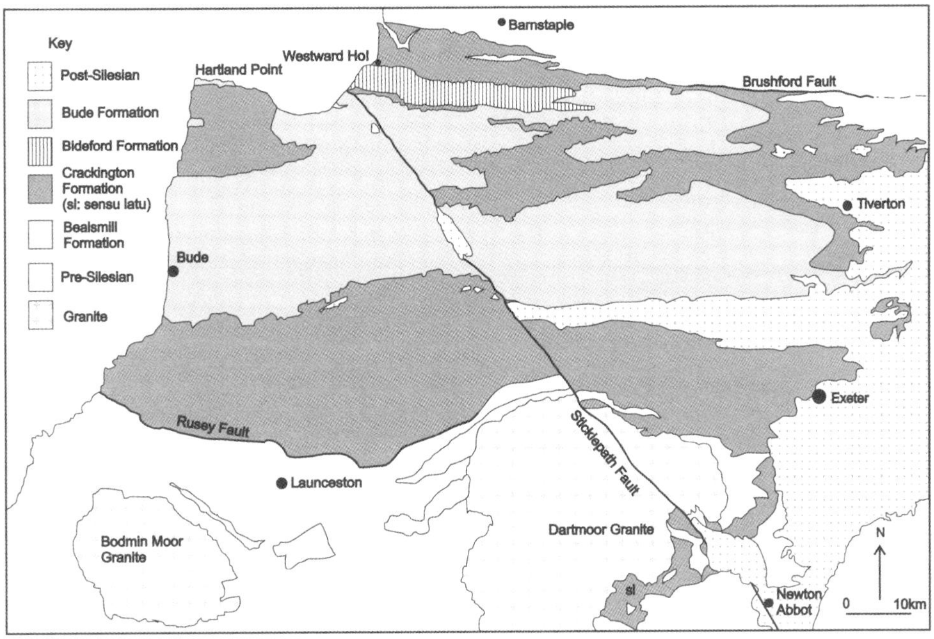

Fig. 10.13. Sketch geological map showing the distribution of the lithostratigraphical divisions of the Silesian, and the main mapped faults. Based on BGS maps, and Thomas (1988).

distal sandstone turbidites, through silty mudstone, siltstone, siltstone with sandstone, to medium-grained feldspathic sandstone. The upper sandstone units, sourced from the north (Prentice 1962) are commonly cross-bedded with channel features or wave-rippling, and rarely capped by rootlet beds, seat earths and thin seams of coal. A non-marine fauna (e.g. *Carbonicola*) is present at intervals through the sequence, but scarce marine fossils are present. *Gastrioceras amaliae* is recorded high in the formation (Edmonds *et al.* 1979) indicating its correlation in part with the Crackington Formation and the Bude Formation. The sequence represents sedimentation in a deltaic setting (Prentice 1960b), with associated nearshore and shoreline beach facies (Elliott 1976), formed in a basin sheltered from vigorous wave action.

The Bude Formation succeeds the Crackington Formation conformably over much of the basin. It is characterized by thick beds (to 5 m) and amalgamated units (to 20 m) of poorly sorted feldspathic sandstone with extrabasinal clasts (Freshney *et al.* 1979*a*, *b*), dark laminated mudstone and siltstone with interbeds of graded sandstone, black shaly mudstone, and 'slumped beds' (Freshney *et al.* 1972). Amalgamated units were considered by Melvin (1986) to lack organization, but Higgs (1991) identified coarsening up/fining up sequences as typical. Sandstone beds are locally erosive, bases exhibiting grove casts and flute molds, indicative of palaeocurrent transport from the north. Sporadically developed are sandstone beds with abundant plant remains and coal laminae (Edmonds *et al.* 1968), and rain spotting is recorded on some surfaces (Lloyd & Chinnery 2002). King (1966) interpreted locally abundant trace fossils as xiphosurid (king crab) trails, an indicator of a shallow-water setting. Fauna is generally sparse, apart from within a few extensive black shale 'marine bands' (Freshney & Taylor 1972) rich in goniatites. These indicate that the formation extends above the *Anthracoceras aegiranum* horizon and the Westphalian 'B/C' junction (Thomas 1988). With background sediments, including distal turbidites, indicating some continuity of the Crackington conditions, interpretations of the sedimentary environment of the formation based on sandstone sedimentology have varied between fluviodeltaic (e.g. Prentice 1962) and a deep-water (e.g. Lovell 1965) setting. A more recent consensus, that sandstone deposition was from turbidity currents, was not extended to the setting. A shallow basin pro-delta slope environment was favoured by Melvin (1986), supplied by a delta to the north, with levéed channels, and marine bands representing sedimentation after lobe abandonment. Higgs (1991) proposed a major freshwater lake setting, in which fluvial-fed sandstone turbidites were deposited above storm wave-base, and marine incursions occurring only at times of high eustatic sea level. However, the coaly deposits recorded by Edmonds *et al.* (1968) in the southern synformal crop of the Bude Formation, east of the Sticklepath Fault, may also indicate deltaic progradation across part of the basin late in the development of the formation.

The configuration of the extensional Culm Basin as a broadly symmetrical full-graben, with a deeper inner basin, flanked by shallower half-graben sub-basins (Fig. 10.14), is constrained by its inversion structures, and the facies distribution of its infilling sediments (see below). The early–mid Namurian inversion of the southern sub-basin left the inner-basin, bound to the south by the Rusey Fault, and northern sub-basin as the main areas of Silesian deposition. Continuing D_1 contraction resulted in progressive inversion of these basins during the Namurian and Westphalian (Fig. 10.14), with sedimentation in increasingly shallow water. Northerly-directed synsedimentary thrusting, with some décollments in units previously termed 'slump beds' in the Bude Formation, has been detailed (Enfield *et al.* 1985; Whalley & Lloyd 1986; Lloyd & Chinnery 2002). A very localized incipient cleavage appears to relate this deformation to ongoing D_1 in the regional structure sequence. The southerly inverted area effectively shielded the main basinal area from southerly sources, the Bealsmill Formation probably representing trapped sediments that were subsequently out-thrust to the south. Emergence of a source to the north supplied sediment to the basin in the Namurian (Hartley & Warr 1990). This is compatible with northerly overthrusting along the E–W Bristol Channel section (Brooks *et al.* 1993; Edwards 1999) of the Bristol Channel/Bray Fault, accommodating some of the major NW–SE dextral strike-slip displacement along that fault (Holder & Leveridge 1986*a*). The relationship of the three Silesian formations has not been closely defined. It is evident

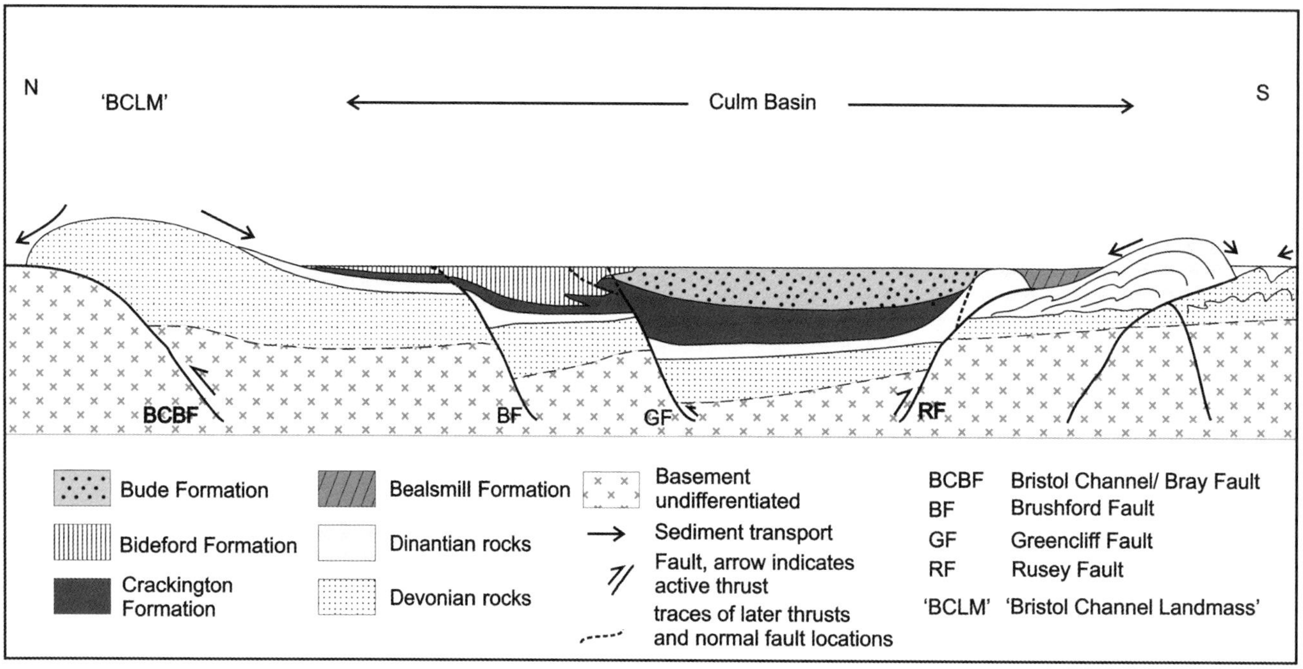

Fig. 10.14. Diagrammatic section of the Culm Basin rift graben during the Early Westphalian, showing the inverted Dinantian marginal sequence to the south, overthrusting on the Bristol Channel/Bray Fault, and progressive inversion on the basin-controlling faults.

that Crackington sedimentation commenced in the inner basin, only overflowing later in the Namurian into the northern sub-basin, where earlier sandstone, as at Venn, may represent a feeder system. The Bideford Formation was largely confined to the restricted northern sub-basin, deltaic sediments accumulating there, whilst Crackington turbidite sedimentation continued in the main basin. The Bude Formation represents the overspilling of the northerly deltaic sediments into the main shallowing central basin. Apart from differing interpretations of depositional setting for the same coastal section, there are a variety of deposits and thus environments recorded across the crop of the Bude Formation. This variation, as the basin became shallow and paralic, would result from such as fluvial input or penecontemporaneous movement on the crossing NW–SE Sticklepath Fault constraining depositional domains.

Major thrust-nappe loading affected only the southern part of the passive margin. Elsewhere D_1 inversion produced significant thrust stacking only in the Looe Basin (Fig. 10.16). Originating as an extensional basin, loading was not a major factor in producing the Culm Basin (see Gayer & Jones 1989), but possibly played a role in maintaining open marine conditions in the South Devon and Tavy basins as they inverted in the late Dinantian. Despite such inversions much of the Culm Basin Silesian sediment has a northerly derivation, thus whilst certainly a foreland basin, because of the province translation, it is not a classical example (Allen *et al.* 1986). Nevertheless, as indicated by Lloyd & Chinnery (2002), basin sedimentation and synsedimentary structures do represent an 'orogenic front propagating into its own foreland'.

Late Variscan deformation and related events in the basins of SW England

Contractional deformation, its continuation. The late Variscan orogenic episode of SW England took place in the Carboniferous when the extensional sub-basins of the Culm Basin began to lock-up during their D_1 inversion. Continuing shortening generated deformation throughout the province. For southern crop Devonian and Dinantian rocks, this was a regional second (D_2) deformation, but, for Silesian sequences, it was the first deformation of consolidated rocks (cf. Lloyd & Chinnery 2002), as it was for the north crop Devonian where deformation phases merged. As a single 'closed' system the province deformed within a short time interval. That episode is constrained by the latest deformed and earliest succeeding undeformed sediments, dating of the metamorphism associated with structure development, and intrusion of the early granite of the province.

The youngest deformed rocks of the Culm Basin are those of the Bude Formation of early Westphalian C age (see above), and the oldest unconformable overlying rocks currently recognized are possibly Late Stephanian (Edwards & Scrivener 1999). The latter is on the basis that a considerable red-bed sequence is mapped beneath a biotite–lamprophyre lava flow dated at 290.8 ± 0.8 Ma (Chesley in Edwards & Scrivener 1999) in the Crediton Trough. The low grade of metamorphism of the deformed Silesian rocks, late diagenetic and low anchizone (Warr *et al.* 1991), has precluded its radiometric dating. A few age-determinations from slates in the north crop Devonian (Dodson & Rex 1971) group around 305 Ma (Warr *et al.* 1991). This compares with dates from a 'Start–Perranporth' zone (315–295 Ma) where secondary transposition fabrics are strongly developed (Holdsworth 1989; Steele 1994). Between the Rusey Fault and Tintagel, some 8 km to the SW, the 'Tintagel High Strain Zone' (Sanderson 1979) has been attributed to tectonic reworking during this deformation by underthrusting of the Culm Basin (Andrews *et al.* 1988; Warr 1989). K–Ar whole-rock dates from the slates of the zone are younger (279–289 Ma: Warr 1991) but the high closing temperatures recorded (Primmer 1985) and extended phase of metamorphism (Warr *et al.* 1991) suggest the overprinting effect of epizonal metamorphism associated with later (Bodmin Moor) granite emplacement and cooling. The oldest granite cupola, Carnmenellis, with an emplacement age of 294 ± 1 Ma (Chesley *et al.* 1993), cuts structures assigned to the regional D_3 deformation (Alexander & Shail 1996). The Bodmin Moor Granite (287 ± 2 Ma: Darbyshire & Shepherd 1985) sourced the extrusive rhyolite lavas of the Rame Peninsula (Leveridge *et al.* 2002) that rest unconformably on deformed Devonian rocks. The deformation thus appears to have been at *c.* 305 ± 5 Ma, the Late Westphalian–Early Stephanian.

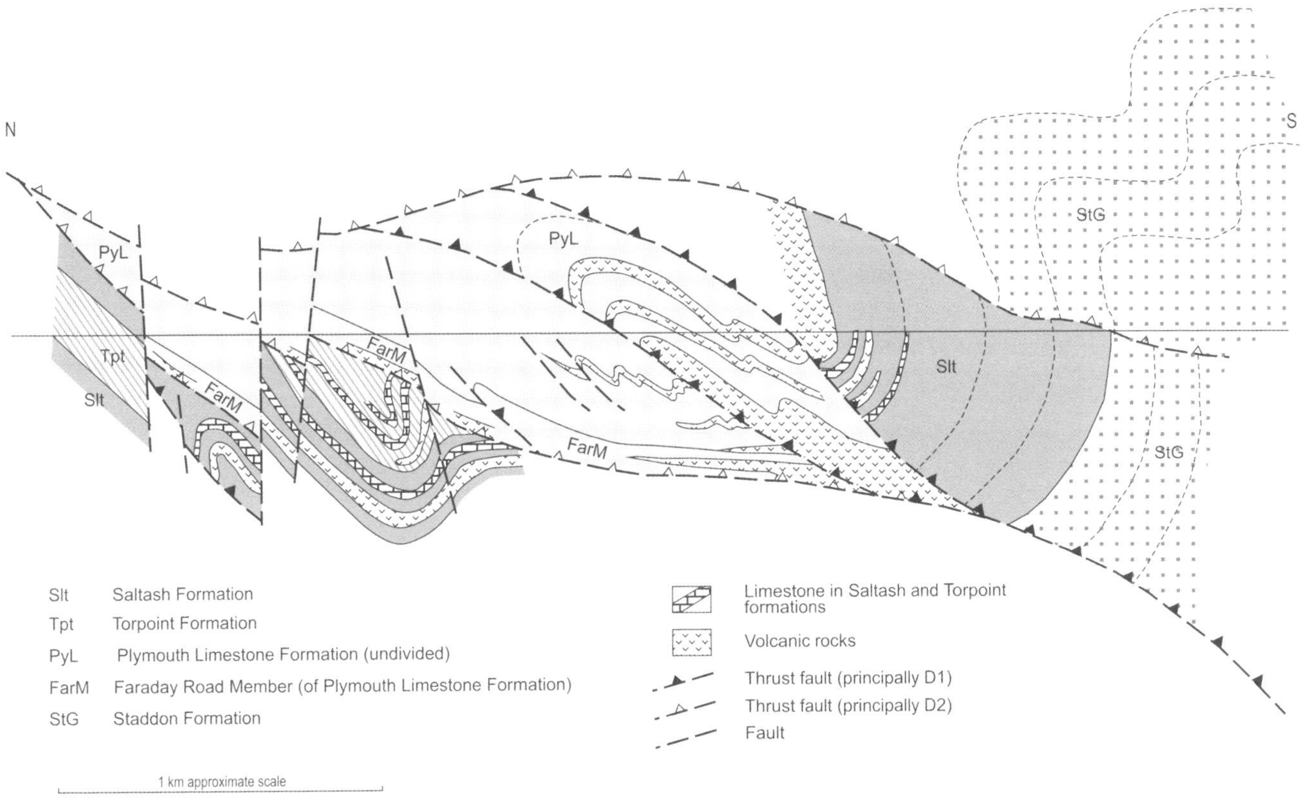

Fig. 10.15. Stuctural section through Plymouth and Plymouth Sound, showing the Looe Basin inversion antiform to the south, and the role of D_1 and D_2 out of sequence thrusting in accommodating the overthrusting of Looe Basin and Plymouth High deposits on to the South Devon Basin sequences to the north. After Leveridge *et al.* (2002).

In the south crop Devonian and Dinantian rocks D_2 structures are not ubiquitous, comprising sporadic zones of mesoscopic folds with associated cleavage, and thrust faults. Minor folds, generally trending between E and NE, verge northwestwards (e.g. Freshney *et al.* 1972; Alexander & Shail 1996). The gently to steeply southerly inclined axial plane cleavage is a crenulation cleavage that generally transposes D_1 fabrics only very locally. However, it can be the predominant fabric more extensively, as on the west Cornwall coast between Perran Sands and Holywell Bay, where an E–W zone some 1.5 km wide has strong transposition and dextral transpressive fabrics (Holdsworth 1989; Steele 1994: BGS unpublished data). Thrust faults are commonly associated with the fold zones, and over much of the subprovince it is the dominant expression of D_2. Thrusts are generally subhorizontal to gently inclined transporting northwards, cutting across early structures (Fig. 10.15), with displacements up to a few kilometres occluding major parts of successions (Leveridge *et al.* 2002, 2003*a*). They are significant structures in the sub-province. Such an out-of-sequence thrust (subsequently rotated to verticality by D_3) juxtaposes the Start Schists with lower grade rocks to the north. On the north Cornwall coast, the Trebetherick Thrust accommodates closer juxtaposition of north- facing (hanging wall) and south-facing (footwall) structures at the 'Padstow Confrontation' (Selwood *et al.* 1998).

D_2 structures are in parts of the province coaxial with D_1 but elsewhere have an anticlockwise divergence by up to 30°. Where there is obliquity with earlier basinal and deformational structures transpressive fabrics are developed (e.g. Andrews 1993; Steele 1994).

The Silesian deposits of the Culm Basin, north of the Rusey Fault, were deformed for the first time during this episode. Although D_2 structures just south of the fault verge northwards (Freshney *et al.* 1972), its northerly dip dictated the southerly vergence of stuctures of its hanging wall along the southern margin of the Silesian fill of the basin during inversion. There is evidence of southerly out-thrusting of Silesian rocks during this process (Warr 1989; Andrews *et al.* 1988) but remnants of the inverted limb of the major inversion antiform are present over an 8 km section north of the Rusey Fault (Freshney *et al.* 1972). South-facing recumbent folding of this zone (Plate 17), becomes generally upright chevron folding in the centre of the basin between Bude and Welcome Mouth, and further north overturning northwards predominates in Silesian (Freshney *et al.* 1979) and North Devon Basin rocks (Edmonds *et al.* 1975). Sanderson (1984) attributed this fan of structural facing to the deformation and closure of a filled Culm Basin graben. A strong spaced cleavage to the south decreases in intensity to the basin centre and then increases northwards (Freshney *et al.* 1979*b*). The slaty cleavage and epizone/high anchizone metamorphism of the pre-Silesian rocks to the north is largely a function of burial (Warr *et al.* 1991) and tectonic loading. This loading, by inverted Silesian rocks, is poorly constrained by lack of modern analysis, but is intimated in the NW area of the Silesian crop where thrust nappes probably represent inversion on earlier basin faults. There the Bideford Formation is bound to the north by a sequence of northerly transporting thrusts (Prentice 1960*a*; Cornford *et al.* 1987), along strike from the Brushford Fault (Fig. 10.13), and to the south by major steep E–W faulting (Burne & Moore 1971), referred to as the Greencliff Fault. Cornford *et al.* (1987) showed, on the basis of vitrinite reflectance, that the Bideford Formation had been buried less deeply than correlative rocks just to the south (4.5/5.5 km rather than 5.7/7.0 km) and proposed juxtaposition by major overthrusting from the south.

At the end of the Westphalian all basins and deposits had been inverted and deformed (Fig. 10.16). It is probable that the inverted Culm Basin formed a substantial edifice, and collision to the south had produced significant thrust nappe loading

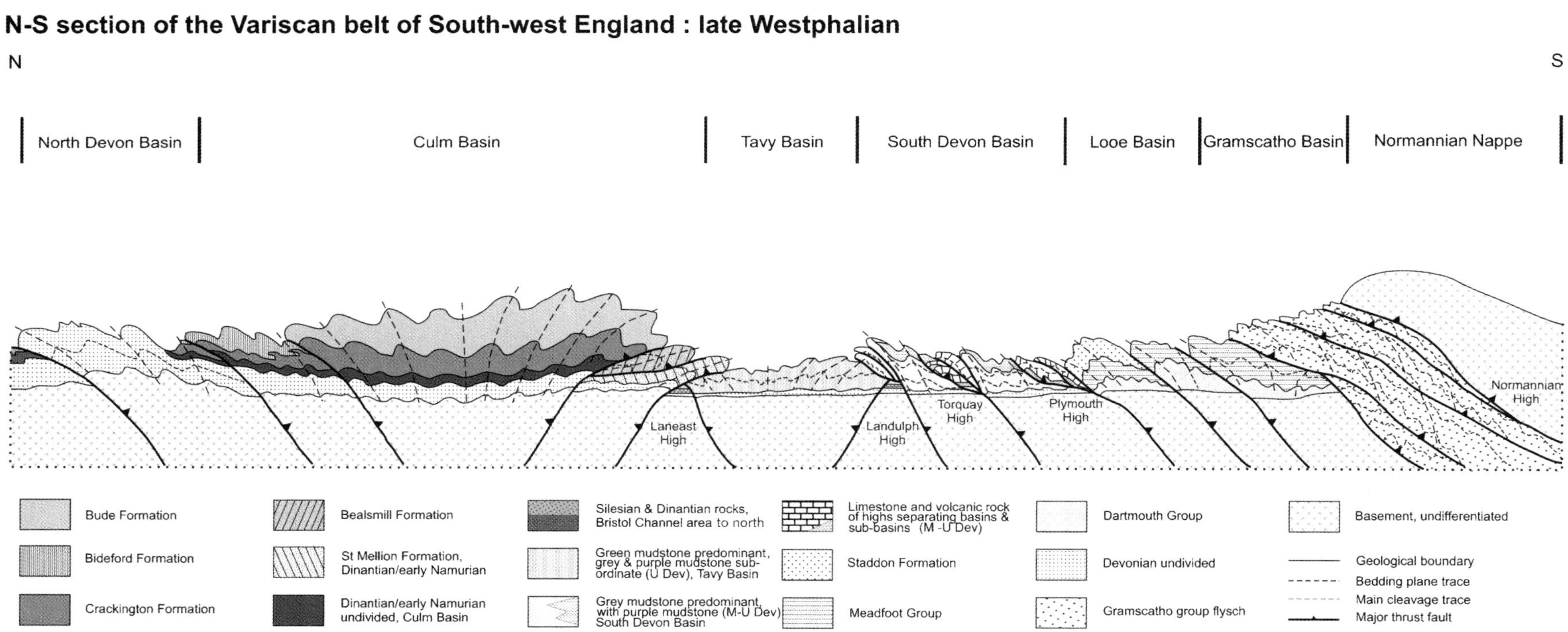

Fig. 10.16. Sketch cross-section of the Variscan belt of SW England, late Westphalian.

of the passive margin, but there does not appear to have been major crustal thickening during the protracted contractional deformation of the province.

Late Variscan orogenic collapse. The final phase of Variscan deformation was active in the brief interval between the inversion of the Silesian Culm basin and the intrusion of the earliest granite cupolas at the time of the Carboniferous–Permian transition. It is the D_3 regional deformation that, in the southern part of the province, represents a reversal of structure vergence (e.g. Alexander & Shail 1996). In south-crop pre-Silesian rocks this deformation is commonly represented by narrow zones of tight–open folds, with associated northerly inclined crenulation cleavage, or minor thrust duplex systems (e.g. Richter 1969; Goode & Leveridge 1991). It is also expressed in rotations of both D_1 and D_2 structures into major southward-verging monoformal/antiformal folds. Attributed to 'backfolding' by Coward & McClay (1983) this is the most notable of these structures with a steeply inclined limb, 1.5 km or more across, extending from north of the Start Peninsula westwards across the peninsula (Leveridge *et al.* 2002) to north of Perranporth (Hobson 1976).

Within the Silesian rocks the fan of chevron folds across the basin is affected by secondary structures. Described by Freshney *et al.* (1979b) as anomalous folds, namely 'box folds, crumples and recumbent folds', those authors ascribed them to either static load under low confining pressure, or for the latter, gravity sliding. Lloyd & Whalley (1986) attributed these structures to the south of Bude to modifications of the earlier chevron folds by a later phase of simple southerly shear. This was designated the D_3 deformation of Silesian rocks by Lloyd & Chinnery (2002), who counted synsedimentary thrusting as D_1 structures. Freshney *et al.* (1979b) noted that these late structures in the northern part of the Silesian crop indicated northerly translation.

Alexander & Shail (1995, 1996) described D_3 in terms of zones of distributed shear, detachments and brittle listric faults, all displaying top-to-the-SE sense of shear in the south of the province. The emplacement of the St Mellion Outlier some 22 km south of the Culm Basin main crop, on flat-lying thrusts (Fig. 10.17), has been attributed to this deformation by Leveridge *et al.* (2002) who invoked post-D_2 gravitational collapse of inverted Culm Basin deposits. This was the early stage of the deformation process that became progressively

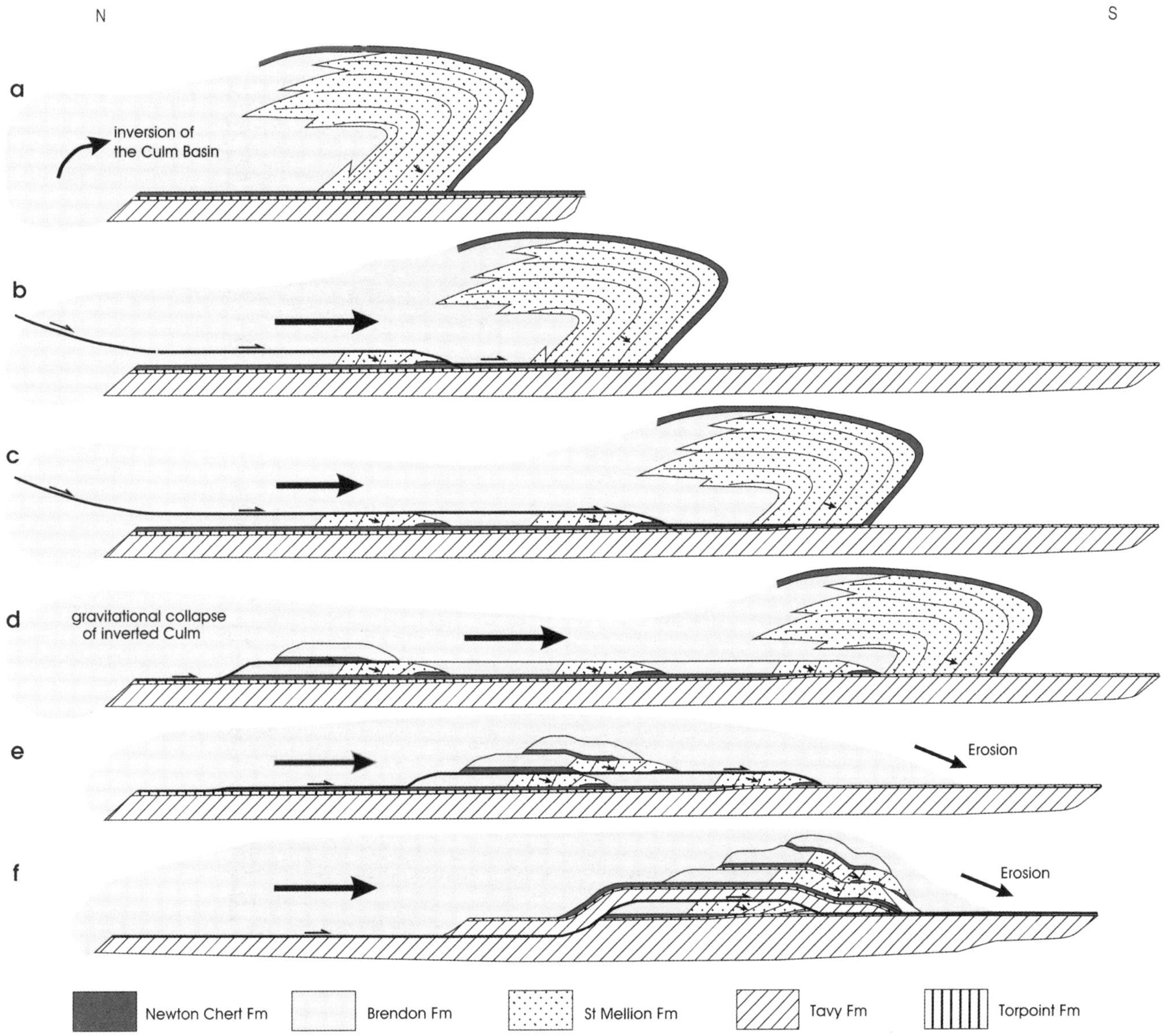

Fig. 10.17. Sequence diagram of the inversion of the southern marginal Culm Basin Dinantian deposits, and subsequent gravitational collapse, which produced the family of thrust slices of inverted rocks of the St Mellion Outlier some 22 km to the south. After Leveridge *et al.* (2002).

more brittle (Alexander & Shail 1996), with extension on earlier thrusts and then development of moderate to steep extensional faults (e.g. Freshney *et al.* 1972; Leveridge *et al.* 2002), some rooting into original basin controlling faults (Fig. 10.14). Continuing into the Permian, with coeval basin formation and granite intrusion at the front of the extending major southerly nappes (Holder & Leveridge 1994; Shail & Wilkinson 1994), it represented orogenic collapse.

The Variscan basin of South Wales

During the Silesian, South Wales comprised one of a series of foreland basins developed in front of the northwardly propagating Variscan orogenic belt. These basins essentially trend E–W and extended from southern England across northern Europe to the Urals. Silesian sediments within South Wales are exposed in a structurally complex WNW–ESE-trending and inwardly plunging synformal structure which extends westwards into SW Dyfed (Fig. 10.18). The structure is asymmetric, the southern limb (the south crop), dips steeply north, and the northern limb (north crop) dips gently south. The succession represents an erosional basin-remnant preserved within the Variscan fold belt. The impact of the Variscan Orogeny on basin development in South Wales is recorded by thickness and lithological changes in the Silesian succession (e.g. Kelling 1988).

Silesian sedimentation in South Wales

The preserved Silesian succession in the South Wales Coalfield and SW Dyfed is up to 3.5 km thick, ranges in age from basal Namurian to early Stephanian (Figs 9.22, 9.26 & 9.27) and was deposited in just over 21 Ma (time scale of Lippolt *et al.* 1984). The stratigraphy and sedimentology of the basin are described in Chapter 9. A brief summary of the sedimentology relevant to foreland basin development is presented here. The succession can be divided into three informal units.

Pendleian–mid-Yeadonian (Cumbriense Marine Band)
Pendleian sediments only conformably overlie Dinantian strata in the basin centre (see Fig. 10.19, 9.22 and Jones 1974 for further details;). Sedimentation was restricted initially to a narrow NE–SW-trending trough located between the Careg Cennan Distrubance (CCD) and the Tawe Disturbance (TD) for the Pendleian and much of the Arnsbergian (Fig. 10.19). The progressive expansion of the basin took place in Chokerian–early Marsdenian times with basinwide deposition by Superbilinguis Marine Band times. By mid-Marsdenian times sedimentation probably extended north of the present day Silesian outcrop (Fig. 10.19).

Pendleian–mid-Marsdenian sediments comprise interbedded quartzitic sandstones, conglomerates and mudstones that record deposition as a series of braid deltas that prograded into a storm-dominated, shallow-marine environment (George 2000).

Lateral facies variations reflect the palaeotopography of the basin. Sandstone deposition occurred intially throughout the narrow NE–SW trending Pendleian trough. However, as the basin continued to subside and/or sea-level rose, deeper water covered the Gower depocentre and shelfal mudstone sedimentation prevailed with sandstone deposition restricted to the basin flanks. Deeper water sedimentation persisted up to the mid-Yeadonian with deposition of inner shelf–offshore transition

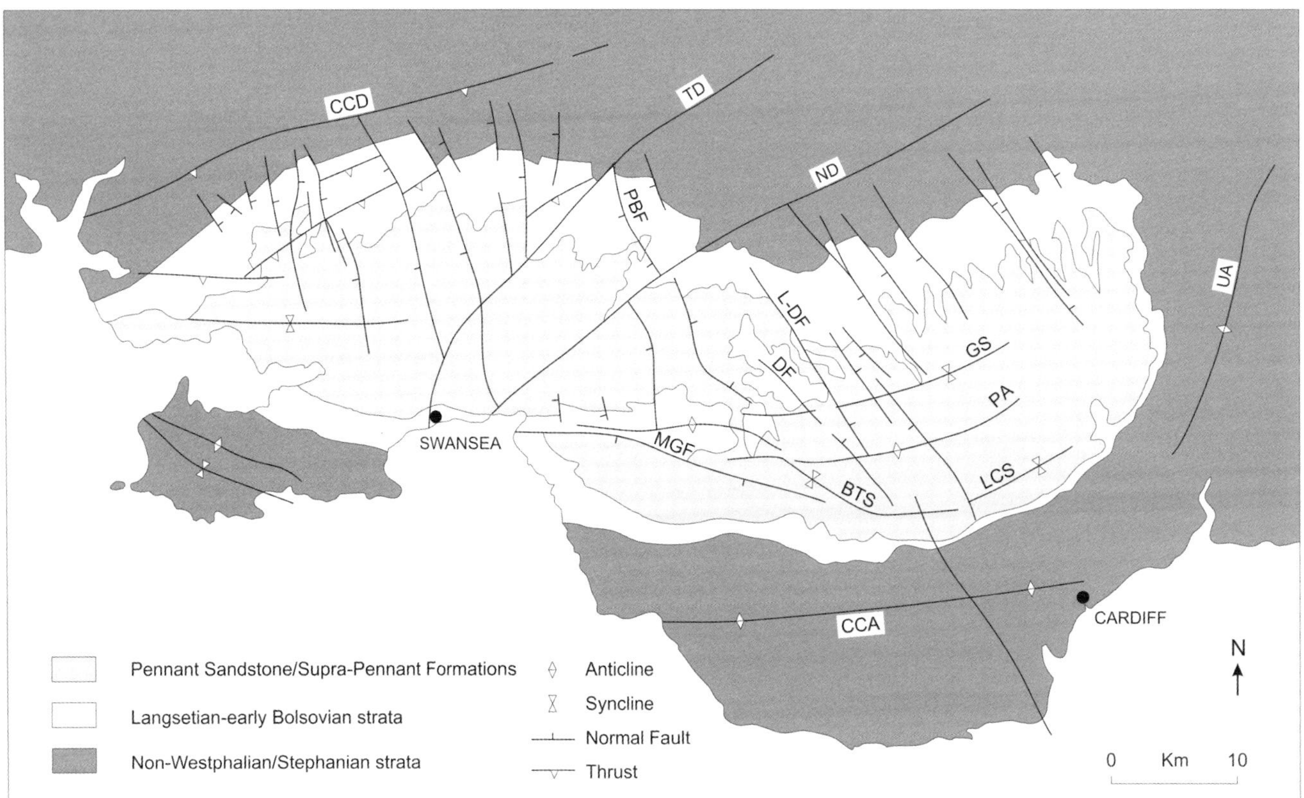

Fig. 10.18. Geological map of the South Wales Coalfield showing the main structures that are thought to have influenced sedimentation during deposition of the Langsettian–early Bolsovian interval. CCD, Carreg Cennan Disturbance; TD, Tawe Disturbance; PBF, Pwllau Bach Fault; L–DF, Llanwonno–Daren-Ddu Fault System; DF, Dinas Fault; MGF, Moel Gilau Fault; PA, Pontypridd Anticline; GS, Gelligaer Syncline; CCA, Cardiff–Cowbridge Anticline; UA, Usk Axis; BTS, Bettws–Tonyrefail Syncline; LCS, Llantwit–Caerphilly Syncline (after Hartley 1993*a*).

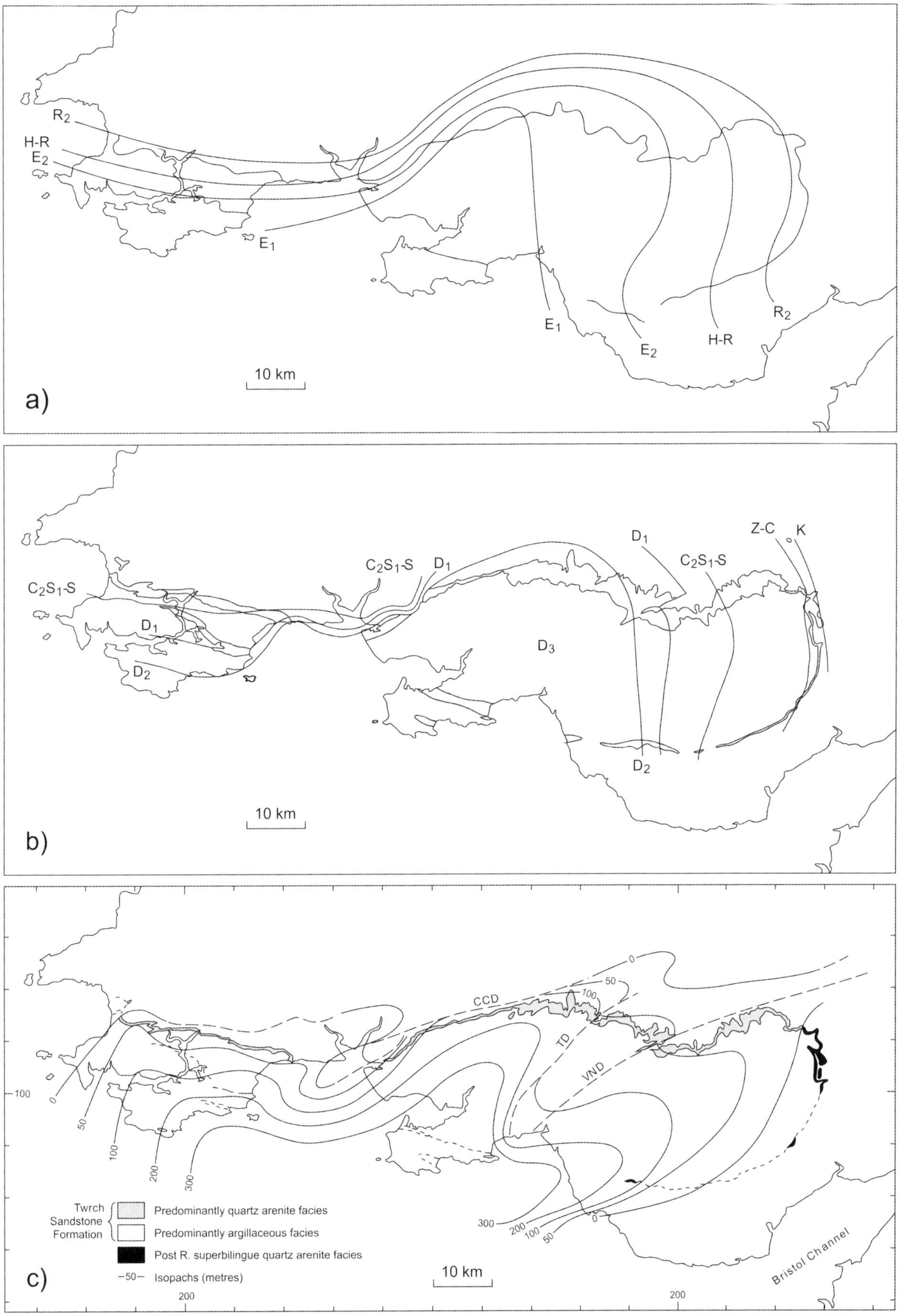

Fig. 10.19. (**a**) Progressive onlap of Namurian strata from the central part of the basin from Jones (1974). (**b**) Subcrop age of the Dinantian strata beneath the Namurian unconformity from George (1970). (**c**) outcrop and isopach map of the Pendleian–mid Marsdenian (*R. superbilinguis*) interval for South Wales, modified from George (2000).

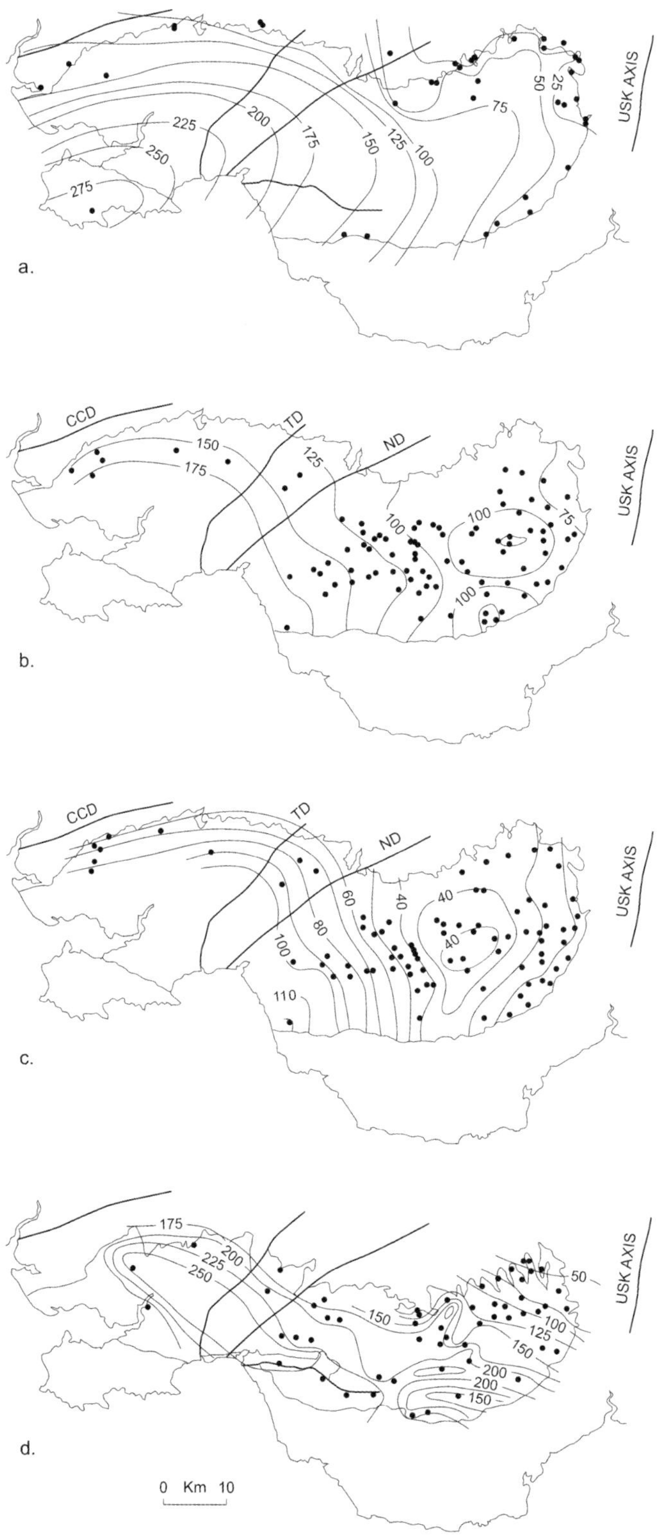

Fig. 10.20. Isopach maps (in metres) for the intervals: (**a**) Mid-Marsdenian–Mid Yeadonian; (**b**) Five Feet/Gellideg to Amman (Vanderbeckei) marine band; (**c**) Amman marine band to Two Feet Nine Coal; (**d**) Brithdir Member of the Pennant Sandstone Formation. Dots refer to control points. Thicknesses corrected for tectonic dip and repetition due to thrust faulting, data from Squirrel & Downing (1964, 1969), Woodland & Evans (1964), Parry (1966), Adams (1967), Archer (1968), Barclay (1989) and unpublished British Coal borehole data.

zone mudstones including marine bands (Fig. 10.20). Occasional fine- to medium-grained sharp-based/incised quartzite bodies are interbedded with the mudstones and represent lowstand shoreline/incised valley-fill deposits (Hampson 1998; George 2000).

Mid-Yeadonian–early Bolsovian

Following deposition of the Cumbriense Marine Band a significant relative sea-level fall resulted in a change from a mainly shelfal environment to a lower coastal plain environment. Coastal plain deposits then dominated the basin-fill succession until the early Bolsovian (Cambriense or Upper Cwmgors Marine Band; Figs 9.22 & 9.27). Both upper and lower coastal plain deposits can be recognized. Lower coastal plain deposits occur from mid-Yeadonian to mid-Langsettian and late/ Duckmantian–early Bolsovian times. They are characterized by horizontally laminated mudstones with marine bands, erosive-based, medium-coarse grained, cross-stratified quartzitic sandstone bodies, rippled siltstones, fine-grained sheet sandstone beds and thin coal seams (Jones 1989*a*, *b*; Hartley 1993*a*, *b*). Quartzite-bodies are interpreted as low-sinuosity fluvial deposits some of which form incised valley fills (Hampson 1998; George 2001). Upper coastal plain deposition prevailed in mid-Langsettian–late Duckmantian times and is characterized by well developed 5–30 m thick, mudstone-dominated coarsening upwards cycles containing only brackish water fauna. Mudstones grade through rippled siltstones to medium-grained sheet sandstones. Sandstones have rootleted tops and are overlain by thick coal seams and mudstones with a brackish fauna. Occasional siltstone or coarse-grained sandstone channel-fill units are present. Basin-wide coal seams overlain by brackish water fauna indicate the termination of peat accumulation by basin-wide flooding that resulted in lake development. Re-establishment of the clastic supply infilled lakes through overbank flood events (sheet sandstones) supplied by low-sinuosity fluvial systems (Hartley 1993*a*).

Early Bolsovian–early Stephanian

Following deposition of the Upper Cwmgors/Cambriense Marine Band (Fig. 9.26) a marked change in sedimentation took place. The preceding lower coastal plain succession was replaced by lithic sandstones of the Pennant and Supra-Pennant Sandstone formations. The Pennant Sandstone Formation is dominated by thick, medium- to coarse-grained, locally pebbly, channelized sandstone bodies interbedded with occasional horizontally laminated mudstones, cross-laminated siltstones and coals. They are interpreted to represent large-scale, bedload-dominated, braided fluvial systems deposited on an alluvial braidplain (for further details see Kelling 1968, 1974; Jones 1989*a*, *b*; Jones & Hartley 1993; Hartley 1993*b*). The mudstone-dominated Supra-Pennant Formation indicates a change to a more floodplain-dominated high-sinuosity fluvial succession. Sedimentation continued into the Stephanian although the top of the basin-fill sequence is truncated.

Subsidence patterns and rates

Prior to the onset of Silesian sedimentation in South Wales, regional deformation resulted in uplift and exposure of the underlying Dinantian carbonate succession. This is revealed by the erosion and karstification of the Dinanatian carbonates and the subcrop map of Dinantian strata beneath the Namurian (Fig. 10.19). This map shows the progressive onlap of Namurian strata northwards, eastwards and westwards from a basin depocentre located close to the Gower (Ware 1939; George 1956). Isopach maps for the Dinantian succession show regional thickening to the south of Pembrokeshire into the present day Bristol Channel area (George 1974), such that the basin depocentre narrowed and shifted to the NE from late Dinantian to early Namurian times. The subcrop map also illustrates the extent and location of major uplift axes which coincide with major lineaments notably the Carreg Cennan Disturbance, the Tawe Disturbance and Usk Axis, suggesting that movement on these structures was responsible for uplift and basin reorganization during late Dinantian–early Namurian times.

Further evidence for continued tectonic activity during sedimentation is provided by isopach and subcrop maps for selected stratigraphic intervals (Figs 10.19 & 10.20). These show a number of common characteristics, including:

(1) The NE–SW oriented trough between the Swansea/Gower area and Ammanford established in the early Namurian formed the basin centre throughout the Silesian.
(2) Thicknesses decreased markedly away from the depocentre particularly to the east and more gradually to the north and west.
(3) Isopach maps for all time intervals up to the Brithdir Member can be seen to parallel the Usk Axis to the east and the Carreg Cennan Disturbance to the NW of the Coalfield. It is likely therefore that these features were active during sedimentation particularly as rapid thickness changes occur adjacent to these structures. The Neath Disturbance may have also acted as a positive area as Westphalian sediments thin across it.
(4) The isopach maps suggest that there is a northwards shift of the basin depocentre (to just north of Swansea) as indicated by the closure of isopachs around the southern basin margin by the end of deposition of the Brithdir Member of the Pennant Sandstone Formation. This supports the observations of Kelling (1988).

Localized variations in thickness are common and are considered to be related to synsedimentary tectonism (Jones 1989*a*; Hartley & Gillespie 1990; Hartley 1993*a*). In particular, patterns of seam-splits, thickness variations and channel orientation in Langsettian and Duckmantian upper coastal plain deposits indicate tectonic control on sedimentation by a number of active structures. Thinning occurs over E–W-trending structures such as the Pontypridd Anticline with thickening into synclines such as the Bettws-Tonyrefail, Gelligaer and Caerphilly synclines; these growth folds have been ascribed to the reactivation of basement structures during early Variscan compression (Jones 1989*a*; Hartley 1993*a*). A number of NW–SE-trending faults are thought to have influenced sedimentation by controlling the position of seam-splits (Hartley 1993*a*). These faults show normal displacement at the present day, but are considered to represent originally reverse-fault bounded blocks that have been subsequently reactivated as extensional faults (Jones 1991; Hartley 1993*a*). The presence of soft-sediment deformation and gravity slides with channel-fill deposits preferentially developed on footwall blocks also indirectly suggests that tectonic activity was active during sedimentation (Hartley & Gillespie 1990).

Subsidence curves for the Silesian section in South Wales have been presented by Kelling (1988) and Burgess & Gayer (2000). They indicate relatively slow rates of subsidence through the Namurian, with a rapid increase at or shortly after the Namurian–Langsettian boundary (Fig. 10.21). The relatively rapid subsidence rates persisted throughout the Silesian, with no significant change associated with onset of Pennant Sandstone Formation deposition. Both sets of authors also indicate a northward migration of a depocentre through the Silesian.

Drainage systems and provenance

Palaeocurrent data for the Silesian succession has been published by a number of authors (Bluck 1961; Bluck & Kelling 1963; Thomas 1967; Kelling 1968, 1974; Williams 1968; George 1970, 1982, 2000, 2001; George & Kelling 1982; Jones 1989*a*, *b*; Hartley 1993*b*; Hampson 1998). Namurian sandstones were derived from the north, east, west and SE basin margins with an open marine connection south of the Gower area. A similar scenario prevailed throughout the Langsettian–late Duckmantian, with coarse-grained, lithic sandstones derived from the south whilst finer-grained channel-fill units were sourced from the east, north and west. Studies of late Duckmantian–early Bolsovian fluvial systems indicate that coarse-grained lithic sandstones were largely derived from the south and to a lesser extent from the east.

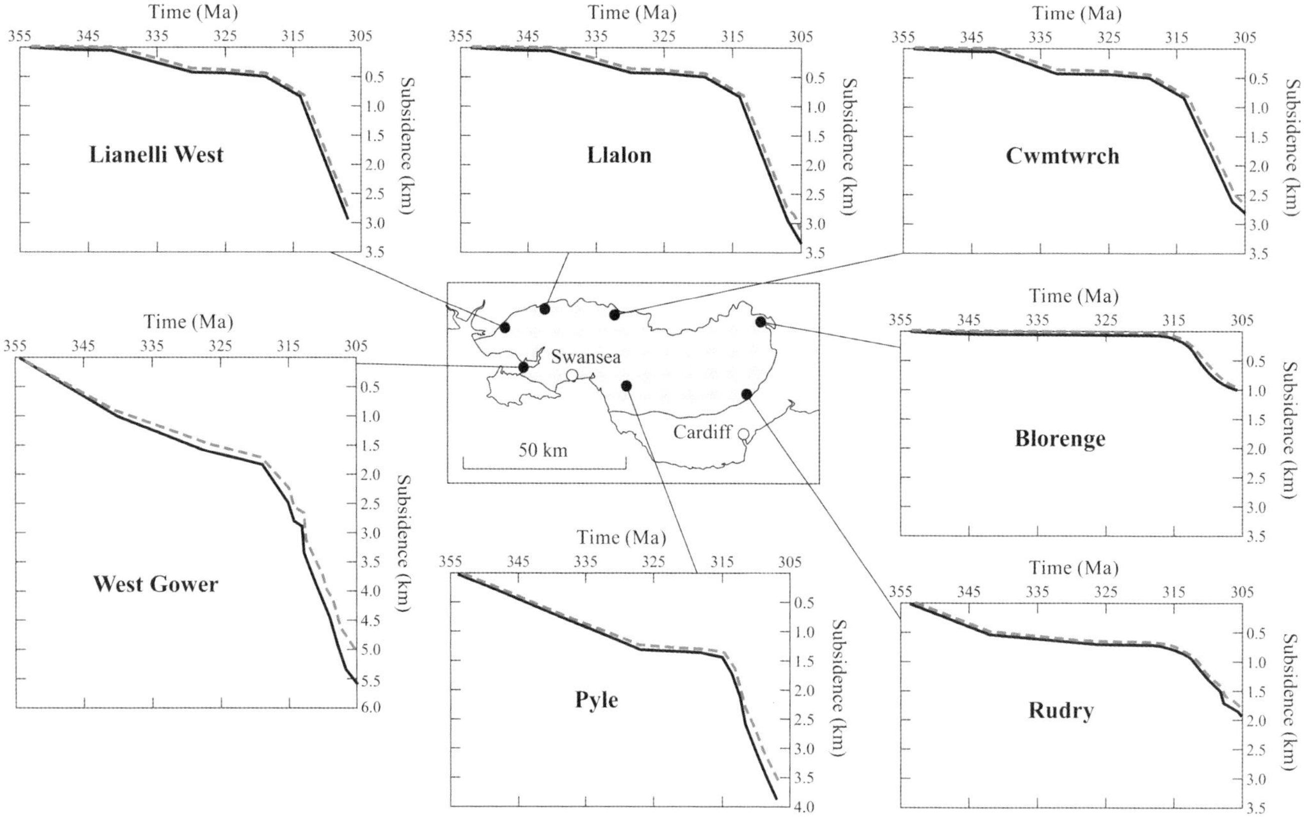

Fig. 10.21. Backstripped subsidence curves, solid line is total subsidence and dashed line is tectonic subsidence, from Burgess & Gayer (2000).

These palaeocurrent data derived from direct outcrop measurements suggest a predominantly centripetal drainage system throughout the Namurian, Langsettian, Duckmantian and into the early Bolsovian. Sediment derived from the basin flanks was transported towards the south/SW towards the marine connection and basin depocentre. This interpretation is in direct contrast to the NW–SE drainage direction proposed by Rippon (1996) from coal seam washout orientations in late Langsettian and Duckmantian strata in the east of the South Wales Coalfield. Rippon based his interpretation on washout trends and presented no palaeocurrent data. It seems unlikely that fluvial channels would effectively drain up-dip, away from the basin depocentre towards the basin margin. Unfortunately, the drainage pattern proposed by Rippon has been adopted by other authors (e.g. Guion *et al.* 2000).

Palaeocurrent and provenance studies of the Pennant Measures indicate derivation from a predominantly southerly source area composed of immature lithic detritus of possible Devonian–Upper Carboniferous age (Heard 1922). Limited quartzitic detritus was supplied to the basin from the east during deposition of the Llynfi and Rhondda Members (Kelling 1974). Abundant coal clasts in channel lag deposits of the Rhondda Member of the Pennant Sandstone Formation were derived from Langsettian coal seams located south of the present-day coalfield where they may have been uplifted by thrusting (Gayer & Pesek 1992). The influx of coarse-grained detritus from the south indicates rapid uplift of a southerly source area.

Synthesis of the South Wales Silesian basin-fill

The Silesian succession in South Wales was deposited during the northward propagation of the Variscan orogenic belt. The foreland basin interpretation is supported by:

(1) The progressive northward migration of the basin depocentre (e.g. Kelling 1988).
(2) A progressive increase in subsidence rates (particularly from the late Namurian early Westphalian onwards), suggestive of increased lithospheric flexure (e.g. Kelling 1988; Burgess & Gayer 2000).
(3) The general east–west orientation of the basin parallel to the inferred Variscan thrust front.
(4) The stratigraphic evolution of the basin-fill from Dinantian carbonates, Namurian shallow-marine and muddy shelfal deposits, through to coal-bearing coastal plain deposits and coarse-grained immature, high-energy braided fluvial systems is typical of the evolution of a foreland basins in general (e.g. the Alps and Pyrenees) and particularly for Variscan foreland basins (Hartley & Otava 2001).
(5) Coal clasts in channel lag deposits within the Rhondda Member of the Pennant Sandstone Formation were derived from erosion of Langsettian coal seams to the south of the present day coalfield indicating rapid burial and uplift (within 5 Ma) probably by thrusting (Gayer & Pesek 1992).
(6) Evidence from the inferred orogenic load in SW England for mid-Namurian age thrusting (Dodson & Rex 1971; Burgess & Gayer 2000).

The influence of the orogenic load on foreland basin subsidence and sedimentological development is interesting. Following uplift and reorganization of the basin in the late Dinantian, relatively low subsidence rates persisted throughout much of the Namurian and coincided with a general deepening of the basin suggesting that subsidence outpaced sediment supply. The main increase in subsidence rate in the basin occurred in late Namurian–early Westphlian times and coincided with a relative shallowing of the basin-fill succession suggesting a possible relationship between enhanced subsidence and increased sediment supply. Rapid subsidence during the remainder of the

Westphalian coincided initially with low-lying coastal plain sedimentation, suggesting that sediment supply kept pace with the increased rate of subsidence at least until the early Bolsovian. The influx of lithic sandstones from a southerly source in the early Bolsovian (Pennant Sandstone Formation), does not coincide with an increase in subsidence, however, the basin-fill changes rapidly from coastal to alluvial plain deposition suggesting that sediment supply outpaced subsidence from Bolsovian to Stephanian times.

Subsidence patterns were also strongly influenced by a number of major long-lived lineaments the most important of which appear to have been the Carreg Cennan Disturbance and the Usk Axis, which controlled the location of the NW and eastern basin margins, respectively. The Tawe and Neath Disturbances appear to have had some influence on sedimentation although not as marked as the other major structures. On a smaller-scale other structures (e.g. E–W-trending anticlines and synclines, NW–SE-trending faults, e.g. Jones 1989*a*, *b*; Hartley 1993*a*) were also active during sedimentation locally controlling channel courses, seam splits and thicknesses.

The marked cyclicity present within the Pendleian–early Bolsovian part of the basin-fill has been attributed to sea-level change driven by glacio-eustatic fluctuations (e.g. Ramsbottom 1978; Hartley 1993*a*; Hampson 1998; George 2000, 2001). This is supported by the basin-wide nature of many of the cycles, although, cycle development was locally affected by synsedimentary faulting and folding in places.

Post-Depositonal Thermal and Structural Development

Crustal thickening across SW Britain led to the progressive lowering of geothermal gradients. Estimates of 40–50 °C km^{-1} have been suggested for the South Wales Coalfield, which may reflect the high degree of crustal thinning that occurred prior to collision (Warr 2000). Although the degree of coalification generally increases with burial depth the NW part of the basin does show an anomalous increase in temperature as indicated by the presence of anthracite grade coal. It is possible that this thermal anomaly reflects a focal point for the upward migration of hotter fluids, driven in fort of the advancing thrust wedge. These fluids infiltrated the coal seams along brittle fractures, causing hydrothermal mineralization and the precipitation of gold before the onset of seam-parallel thrusting (Gayer *et al.* 1998).

The South Wales area forms part of the foreland to the main Variscan deformation belt (represented by the Rhenohercynian zone in SW England). The Variscan deformation front is considered to lie just south of the South Wales Basin (Fig. 9.4). The main phase of deformation within South Wales took place following deposition of the Supra Pennant Formation in the late Stephanian–early Permian. Evidence suggests that deformation of the Silesian succession varied both stratigraphically and geographically (Jones 1989*a*). The amount of deformation and coal rank increase to the west. The relatively incompetent Langsettian and Duckmantian (Productive Measures) strata are the most intensely deformed. The more competent Pennant Sandstone Formation is much less deformed (approximately 6% shortening, with just broad, gentle folds), and acted as a passive roof duplex to an underthrust wedge. Deformation in the coal-bearing succession is dominated by WNW–ESE striking thrusts and folds which produced approximately 30% shortening in the east of the coalfield (Jones 1991) locally rising to 50–67% in the west (Frodsham & Gayer 1997). Structures verge to the north in the north crop with an important south-verging back-thrust zone forming the south crop. Regionally the thrusts propagate to the north in a piggy-back sequence but locally a variety of complex structures evolved, controlled by the position and geometry of coal seams. Thrusts initially moved along the floors of the coal horizons until they locked up and produced hangingwall folds. Further displacements eventually caused the faults to break through the fold hinges, and to initiate new break-back thrusts in either the hangingwall or footwall segments leading to numerous repetitions of coal seams (Frodsham *et al.* 1993; Frodsham & Gayer 1997). The structures are compartmentalized by NW–SE cross-faults.

11 Cornubian granites and mineralization of SW England

R. C. SCRIVENER

The essentially E–W-trending structural grain of the Devonian and Carboniferous Variscan orogen of mainland SW England (see Chapter 10) is punctuated by five large granite plutons (Fig. 11.1), which are now known to have been emplaced during very latest Carboniferous–Early Permian times. This chronological frame is after the final compressional stage of the Variscan orogeny, and it is now considered that the Cornubian granites were associated with crustal extension and orogenic collapse – this is discussed more fully later. From east to west the plutons are: Dartmoor (650 km^2), Bodmin Moor (220 km^2), St Austell (85 km^2), Carnmenellis (135 km^2) and Land's End (190 km^2). Offshore, to the west of Land's End, the Isles of Scilly pluton has a seabed crop of about 120 km^2. Onshore, there are a number of smaller granite bodies, including Hemerdon Ball (south of Dartmoor), Hingston Down, Kit Hill and Gunnislake (between Dartmoor and Bodmin Moor), Belowda and Castle-an-Dinas (to the north of St Austell), Carn Brea, Carn Marth, St Agnes and Cligga Head (to the north of Carnmenellis), St Michael's Mount and Tregonning–Godolphin (between Carnmenellis and Land's End).

The higher parts of the granite outcrops are characterized by moorland, with thin peaty soils from which arise sporadic tors and their associated boulder screes, 'clitter' in the local dialect. The coastal outcrops of the Land's End and Tregonning–Godolphin granites show dramatic cliffs and rocky inlets known as 'zawns'. Evidence of former metal mining, made manifest by ruined engine houses, spoil dumps, disused shafts and open workings, is present within and around the granite outcrops in many districts. There are also working and disused china clay pits, particularly in the western part of the St Austell Granite and the southernmost area of Dartmoor, where extensive areas of granite have been argillized by hydrothermal and supergene processes. The present day china clay industry produces some 2.3 million tonnes (mt) of refined kaolin annually.

The Cornubian Batholith

In his account of the granite bodies of Devon and Cornwall, De la Beche (1839) speculated that the various outcrops were conjoined at depth beneath the surrounding slate. This was confirmed by the gravity survey work of Bott *et al.* (1958), with later gravity and seismic work and re-interpretation summarized in Brooks *et al.* (1983), Edwards (1984) and Evans (1990). The Cornubian Batholith is now considered to have an E–W extent of over 250 km from south Devon, where it terminates abruptly just to the east of the Dartmoor Granite, to some distance west of the Isles of Scilly. The batholith varies in width between 40 and 60 km, whilst the thickness is considered to reduce from 20 km in the east to about 10 km in the west: over this interval, the base of the crust is between 25 and 30 km deep. The nature of the base of the batholith has been the subject of some discussion: the view of Brooks *et al.* (1984) that a prominent seismic reflector at between 10 and 15 km deep represents a basal thrust has been disputed (Floyd *et al.* 1993). The exposed parts of the batholith include the plutons and cupolas listed above, together with local swarms of quartz–porphyry or felsite dykes, which cut the granites and their envelope of Carboniferous and Devonian rocks. The strike of the dyke swarms is generally roughly parallel with the axis of the batholith.

The essentially prismatic form of the batholith, as suggested by geophysics, is known to be cut by a series of NW–SE-trending faults (Fig. 11.2), some with considerable lateral displacements, such as the Sticklepath–Lustleigh Fault, which cuts the eastern part of the Dartmoor Granite, with a right-lateral displacement of some 3 km.

A number of features suggest that the granites were emplaced in the crust at high level: these include the narrow width of the metamorphic aureoles in areas with steep granite contacts, and the low thermal metamorphic grade indicated by their mineral assemblages (Phillips 1964). Broader tracts of country rock affected by thermal metamorphism, as for example to the west of the St Austell Granite, are taken to indicate the presence of buried granite at shallow depth. The contacts of the granite against the country rocks are generally sharp and without significant evidence of mobilization or migmatization, although Floyd *et al.* (1993) describe minor migmatization at Leusdon Common, southern Dartmoor. More typical of the region is the aureole of metasedimentary rocks of northern Dartmoor, in Late Carboniferous Crackington Formation shale and thin turbidite sandstones, where a proximal zone of cordierite–biotite hornfels up to 200 m wide gives way to spotted chiastolite hornfels up to 300 m wide and, the outermost zone, 500–850 m wide, where splintery unspotted hornfels with local silicification and sericitization have been developed (Edmonds *et al.* 1968; Edwards & Scrivener 1999). Although stoping has been suggested as an important mechanism of emplacement (e.g. Brammall & Harwood 1932), identifiable xenoliths of country rock are relatively rarely seen in exposures of granite throughout the peninsula and it is probable that the final stages of emplacement, at high crustal levels, was essentially by displacement. Shail & Wilkinson (1994) envisage emplacement of the granites in an extensional regime involving a regional pattern of ENE–WSW dip-slip faults interacting with NW–SE strike-slip faults to provide pull-aparts and space for the rising magma.

In terms of composition, the plutons of SW England are all granites *sensu stricto*, without the diorite or granodiorite intrusions commonly present in cordilleran batholiths elsewhere

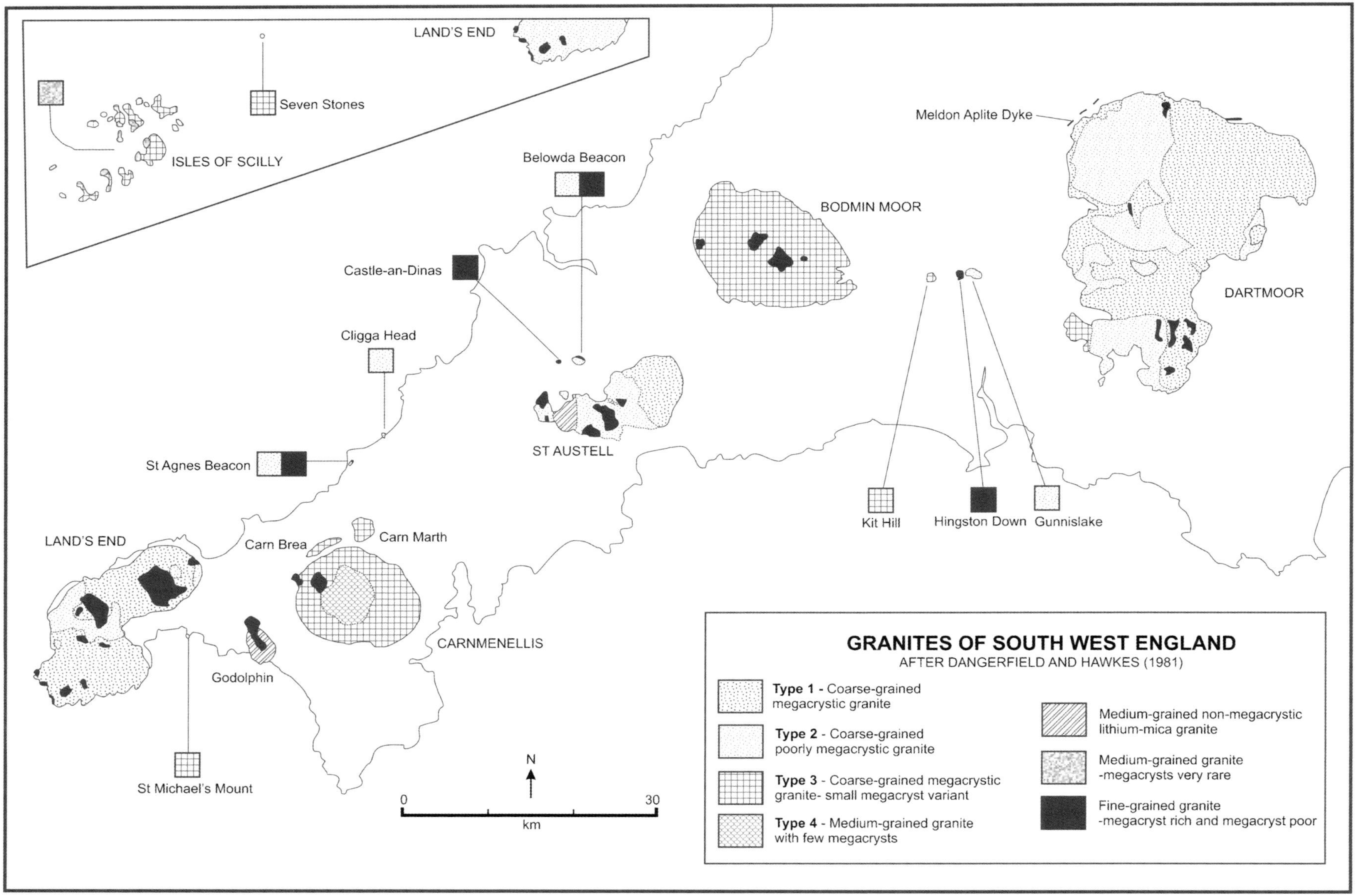

Fig. 11.1. Granite outcrops of the Cornubian Batholith, and their textural variation (after Dangerfield & Hawkes 1981).

Fig. 11.2. Major NW–SE-trending fault zones of the Cornubian region (derived from BGS mapping).

in the world. In terms of their chemistry the granites are peraluminous (Table 11.1), have high K_2O/Na_2O and low Fe_2O_3/FeO ratios, and are enriched in elements such as B, F, Li, U, Th, Sn, Rb and Pb (Jefferies 1984; Floyd *et al.*, 1993). The unusually high U content (up to 35 ppm, with a mean value for all plutons of 12 ppm) is largely contained as inclusions of uraninite, monazite and other mineral species within biotite crystals (Jefferies 1984). In most of the granites, biotite is the most abundant mica, commonly occurring together with muscovite: in certain cases lithium-micas predominate. Granite varieties with abundant large feldspar 'megacrysts' are common, particularly in the Dartmoor, St Austell and Land's End plutons. In all granites, hornblende, sphene and magnetite are rare or absent, and tourmaline is abundant (Floyd *et al.* 1993). These features are all indicative of 'S-type' granites of Chappell & White (1974), and the presence of varieties including tourmaline, lithium-micas, fluorite and topaz demonstrate that they are highly evolved (e.g. Stone & Exley 1986).

The granites and their history of intrusion

The work of Dangerfield & Hawkes (1981) provides, with some minor amendments in Fig. 11.1, a general scheme for the range of granite types in the exposed parts of the Cornubian Batholith. Their classification is essentially based on grain size and the abundance of 'megacrysts' of K-feldspar. Another criterion is the predominant variety of mica, either biotite or the less abundant Li-mica variants, which are essentially restricted to the St Austell and Godolphin plutons. The scheme of granite varieties as adapted for this review includes:

coarse-grained biotite granite – megacrysts abundant (Type 1);
coarse-grained biotite granite – megacrysts sparse (Type 2);
coarse-grained biotite granite – small megacrysts (Type 3);
medium-grained biotite granite – megacrysts sparse (Type 4);
medium-grained lithium-mica granite;
granite porphyry;
microgranite and aplite.

In the following discussion, the main features of the larger plutons are described in terms of these variants.

Dartmoor

Much of northern and central Dartmoor is underlain by Type 1 and Type 2 granites, which were considered by Brammall & Harwood (1932) to represent successive intrusive events, with Type 1 preceding Type 2. More recent mapping in northern Dartmoor, reported in Edmonds *et al.* (1968), has demonstrated that the megacryst-rich rocks grade downwards into megacryst-sparse varieties with no evidence of intrusive contacts. A particular feature of northern and central Dartmoor is the local abundance of sheet-like bodies of aplite, commonly rich in tourmaline and lacking biotite (Scrivener 1982), which clearly post-date the coarse biotite granite host. Within the proximal metamorphic aureole of NW Dartmoor is the 'Meldon Aplite', a dyke, up to 20 m wide, of equigranular aplite composed of albite–quartz–lepidolite–orthoclase, with accessory tourmaline, apatite and fluorite (Edmonds *et al.* 1968). The tourmaline occurrence, of pink and green lithian varieties is unique in SW England, in contrast to the ubiquitous schorl. In

Table 11.1 *Typical geochemical analyses of Cornubian granites. Data sources: (1) to (7) inclusive and (9) – Darbyshire and Shepherd, 1985; (8), (10) and (11) – Stone, 1987. Localities: (1) Blackingstone Quarry, (2) Luxulyan Quarry, (3) Geevor Tin Mine, (4) Merrivale Quarry, (5) Porth Nanven, (6) De Lank Quarry, (7) Carnsew Quarry, (8) average of 11 samples, (9) Castle-an-Dinas Quarry, (10) average of 14 samples (11) average of 12 samples.*

	TYPE 1			TYPE 2		TYPE 3		TYPE 4	FINE GRAINED GRANITE	MICROGRANITE AND APLITE	QUARTZ–PORPHYRY DYKES
	DARTMOOR	ST AUSTELL	LANDS END	DARTMOOR	LANDS END	BODMIN MOOR	CARNMENELLIS	CARNMENELLIS	LANDS END	CARNMENELLIS	CARNMENELLIS
	(1)	(2)	(3)	(4)	(5)	(6)	(7)	(8)	(9)	(10)	(11)
SiO_2	71.16	71.5	71.15	74.67	74.61	72.03	72.25	72.84	72.75	75.41	73.91
TiO_2	0.41	0.32	0.31	0.18	0.12	0.20	0.25	0.17	0.14	0.07	0.14
Al_2O_3	14.31	14.44	14.61	13.08	13.53	14.89	14.83	14.57	14.6	14.45	14.37
Fe_2O_3	3.05	2.45	2.23	2.09	1.49	1.68	1.76	1.36	1.77	1.04	1.39
MnO	0.07	0.06	0.05	0.09	0.03	0.05	0.08	0.03	0.06	0.03	0.04
MgO	0.49	0.45	0.41	0.14	0.14	0.35	0.45	0.38	0.23	0.13	0.25
CaO	1.33	0.98	0.84	0.43	0.35	0.78	0.93	0.68	0.58	0.57	0.45
Na_2O	3.24	3.44	2.88	3.20	3.27	3.59	2.93	3.18	3.42	3.76	0.71
K_2O	5.03	5.09	5.83	4.82	5.00	5.26	5.06	4.91	5.09	4.29	7.33
P_2O_5	0.19	0.25	0.30	0.19	0.23	0.24	0.23	0.22	0.24	0.20	0.23
Ba	229	246	238	78	8	191	212	223	140	111	219
Ce	72	73	103	27	16	52	63	59	45	19	42
Cs	40	34	34	82	47	32	38	58	51	46	59
Ga	21	22	24	22	28	27	27	21	24	–	–
Li	–	–	–	–	–	–	–	412	–	295	129
La	32	31	44	22	8	25	32	27	12	1	19
Pb	29	27	16	25	14	33	32	37	23	15	21
Rb	367	446	473	600	664	410	425	469	476	604	767
Sn	15	7	8	19	8	18	10	17	21	15	44
Sr	96	79	78	21	10	92	92	78	52	27	60
Th	21	19	27	15	6	12	14	8	10	3	19
U	9	12	9	18	13	11	12	10	13	12	13
Zn	57	40	25	35	24	47	44	57	48	48	55
Zr	138	148	158	104	54	100	115	74	70	25	73
Y	33	29	28	32	24	18	17	14	22	5	10
K_2O/Na_2O	1.55	1.48	2.02	1.51	1.53	1.46	1.73	1.54	1.49	1.14	10.32
Fe_2O_3/FeO	0.24	–	–	0.28	–	–	–	0.58	–	1.16	1.56
A/CNK	1.08	1.12	1.16	1.16	1.19	1.14	1.23	1.17	1.20	1.21	1.19

southern Dartmoor, there is evidence of successive intrusive events, which have been mapped in some detail by Knox & Jackson (1990). Distinct phases following the early Type 1 granite, are the Wigford Down granite (Type 3), granite porphyry and microgranite. The intrusive complex of SW Dartmoor is associated with locally intense kaolinization, which is worked at Lee Moor and Cornwood pits. Tourmaline is a very abundant mineral in the Dartmoor pluton, occurring as aggregates in all the granite varieties, especially the late-stage aplites and microgranites. Tourmalinite (quartz–tourmaline) veins and replacement bodies are also common, and there are recorded examples of tourmaline-rich hydrothermal breccias (Brammall & Harwood 1925; Lister 1978; Goode & Taylor 1980). Cassiterite-bearing veins are common in the Dartmoor pluton and have resulted in a long history of tin mining, both from placer gravels and from later vein mining. The greatest development of Sn-bearing veins is the Birch Tor and Vitifer complex of central Dartmoor, where narrow cassiterite–quartz–tourmaline veins also carry specular haematite and chlorite (Beer & Scrivener 1982). Farther to the east, around Lustleigh and Hennock, the Sn-bearing orebodies give way to steeply dipping veins carrying soft, fine-grained micaceous haematite, formerly used in the manufacture of paint (Selwood *et al.* 1984).

Bodmin Moor

The predominant lithology of the Bodmin Moor pluton is Type 3, with three areas of microgranite (Selwood *et al.* 1998). As in the case of the coarse-grained granites of Dartmoor, the early view of Ghosh (1927) that the Type 3 granite resulted from two intrusive events, is not supported by the most recent mapping (Selwood *et al.* 1998). An interesting feature of the Bodmin Moor pluton is the presence of a marginal zone of foliated granite, described in detail in Selwood *et al.* (1998). Late-stage veins and sills of aplite and microgranite, also quartz–porphyry dykes cut the main granite varieties of the Bodmin Moor pluton in places, and there are local occurrences of tourmaline veins and replacement bodies, and of Sn- and W-bearing veins associated with tourmalinization and greisening. These late-stage features are, however, less common than in the neighbouring Dartmoor and St Austell plutons.

St Austell

Dangerfield & Hawkes (1981) represented the St Austell pluton (Fig. 11.1) as Type 1 granite in the east and extreme west, with a central zone of Type 2 intruded by bodies of microgranite, and a western area of lithium-mica granite. The pluton has been the subject of intense study, due to the considerable production of china clay from kaolinized zones in its western part, and subsequent recent work has demonstrated a more complex scenario. Manning & Exley (1984) suggested a model in which two stages of magmatic activity were followed by pneumatolytic and hydrothermal alteration. The first magmatic stage involved the emplacement of several pulses of biotite granite magma, followed by the injection of non-megacrystic lithium-mica granite, as a volatile-rich differentiate of the biotite granite magma. Subsequent pervasive pneumatolytic/hydrothermal activity the produced patched of fluorite granite within the original lithium-mica granite, and also produced a halo of megacrystic lithium-mica granite around the non-megacrystic second intrusion. The later work of Manning *et al.* (1996) further identified varieties of tourmaline granite and topaz granite as part of the magmatic/hydrothermal sequence in the western part of the St Austell pluton in the following sequence from oldest to youngest:

1) biotite granite (Type 1);
2) lithium-mica granite;
3) tourmaline granite – globular quartz (Type 2);
4) tourmaline granite – equigranular (Type 2);
5) fine-grained tourmaline granite;
6) topaz granite.

The field relationships of these rocks are shown in Figure. 11.3. Tourmaline is extremely abundant in the St Austell Granite, occurring not only as intergranular aggregates and disseminations, but as tourmalinite (quartz–tourmaline) veins, replacement bodies and breccia pipes. A notable tourmalinite replacement is Roche Rock, on the southern margin of the village of Roche described in detail by Floyd *et al.* (1993), while a spectacular tourmaline breccia pipe is exposed within the Wheal Remfry china clay pit (Allman-Ward *et al.*, 1982). The St Austell granite is also host to numerous examples of metalliferous mineralization, most notably the sheeted tin-bearing greisen-bordered veins of the Goonbarrow china clay pit.

Carnmenellis

Much of the essentially circular outcrop of the Carnmenellis Granite, also the satellite bodies of Carn Brea and Carn Marth to the north, is composed of Type 3 granite, with a central boss of Type 4 granite. There are two areas of late-stage, leucocratic fine grained granite in the west of the pluton near Boswyn and Praze-an-Beeble. In addition, Leveridge *et al.* (1990), reporting on the most recent geological survey of the district (Fig. 11.4), distinguish a sparsely megacrystic variant of Type 3 around the eastern margin of the pluton: they also record a body of coarse-grained biotite granite, with abundant very small feldspar megacrysts extending from north to south within the eastern part of the pluton. In his classical account, Ghosh (1934) describes this last variety as being intrusive into the Type 3 granite, as also is the Type 4 granite: he further considered the fine-grained granite of Boswyn and Praze to be a differentiate of the Type 3 granite, which predates the Type 4 intrusion. Leveridge *et al.* (1990) agree in essence with this interpretation, but suggest that the Boswyn and Praze outcrops are foundered xenoliths within the Type 3 granite. The Carnmenellis granite hosts a small number of aplite dykes, which are sparsely distributed mostly in the Type 3 granites. There is a considerable suite of quartz–porphyry dykes, and these are particularly numerous in the northern part of the pluton, and in the envelope of Mylor Slate. Most of these dykes trend roughly ENE–WSW and they have a close spatial association with the polymetallic Sn–W–Cu–As–Zn–Pb vein systems of the intensely mineralized Redruth–Camborne district.

Tregonning–Godolphin

This small pluton includes two components, the Tregonning lithium-mica granite, which forms a V-shaped outcrop, and the fine-grained Godolphin biotite granite enclosed within its southern part. This body has been described in detail by Stone (1975) who has drawn attention to the complex sheeted nature of the roof zone of the Tregonning granite, with alternating bands of aplite and pegmatite that commonly include pelitic xenoliths of the enclosing Mylor Slate. Stone (1975), in consideration of the field evidence, concluded that the Tregonning lithium mica granite is later than the Godophin biotite granite, into which it is intrusive. The Tregonning–Godolphin Granite is cut by numerous Sn–Cu veins and also quartz–porphyry dykes. A notable feature in the Mylor Slate metasediments to the south of the pluton, is the complex of pegmatite–aplite–granite sheets at Megilliggar Rocks; these contain spectacular tourmaline crystals and unusual phosphate minerals such as triplite and amblygonite (Stone 1969; Floyd *et al.* 1993).

Land's End

Figure 11.1 shows that coarse-grained megacrystic biotite granite is the predominant component of the Land's End pluton, enclosing a number of outcrops of fine-grained granites of variable megacryst content. The most recent mapping by the British

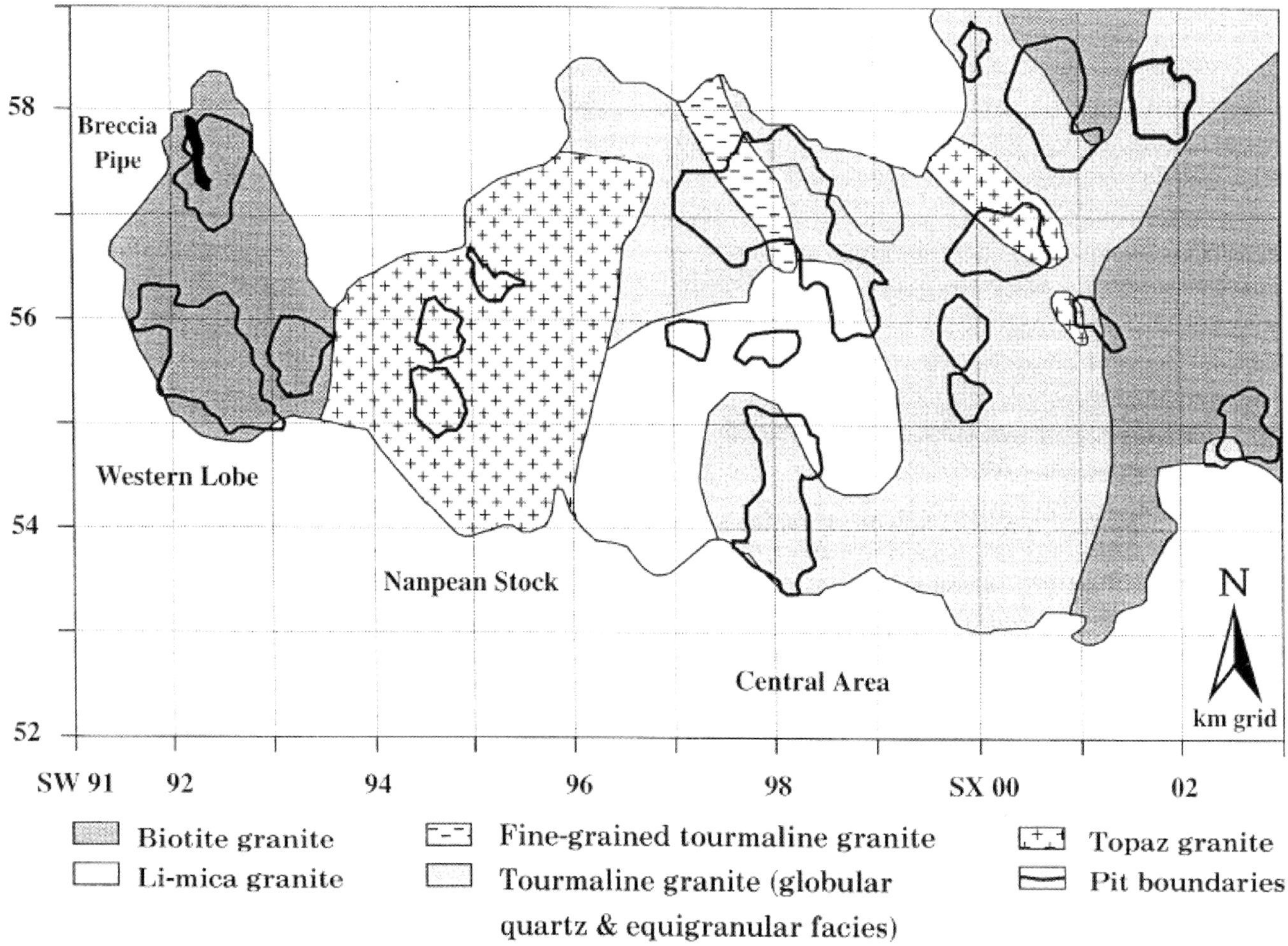

Fig. 11.3. Granite types of the western part of the St Austell Granite (after Manning *et al.* 1996).

Geological Survey, records a central E–W zone in the coarse biotite granite, in which megacrysts are sparse (Goode & Taylor 1988). Relationships between the coarse- and fine-grained granite varieties are complex: the Geological Survey Sheet 351/358 (England and Wales) shows that the boundaries are partly faulted along NW–SE-trending fractures. Elsewhere, Goode & Taylor (1988) describe sheets of fine-grained granite within the coarser varieties as enclaves with, in places, gradational contacts. The complex nature of the Land's End pluton has been illustrated by a number of workers in recent years. Salmon & Powell (1998) demonstrated that the fine-grained granites differ one from another in textural features and composition and were emplaced as separate events within the time span of formation of the pluton. They suggested that the variations in texture and composition may, in part, be due to magma mixing before or during emplacement of the various fine-grained bodies. In a sequel to this work, Powell *et al.* (1999) provided evidence that the Land's End granite can be considered as three units: the 'Zennor lobe' and 'St Buryan lobe' in the north and south, respectively, separated by a body of aphyric granite, the 'St Just wedge'. Age relations from isotopic evidence suggest that the Zennor lobe is 2–3 Ma older than the St Buryan lobe, and that the St Just wedge is younger than both. An important conclusion of this work, based on the recognition of a large number of discrete intrusions within the Land's End pluton, is the negation of emplacement of a single large diapiric magma body.

Isles of Scilly

Dangerfield & Hawkes (1981) record that most of the outcrops or the Isles of Scilly are of Type 3 granite, with some patches of medium- to fine-grained granite, and note similarities between the general lithologies of the Scilly granites and those of Bodmin Moor and Carnmenellis. As at Bodmin Moor, they recorded the presence of a zone of foliated granite, some 500–800 m broad and extending from White Island to North Bryher. Within this foliated zone are intermittent shear zones, in which quartz and tourmaline are developed. More recent work by Mullis *et al.* (2001), based on detailed mapping, has demonstrated considerable complexity in the Isles of Scilly pluton, which is considered to be built up from many small intrusions, albeit of the same general composition. The presence of multiple bodies of magma is deduced from the nature of contact relationships. There are minor occurrences of tourmaline and mineralized veins, and quartz–porphyry dykes are noted cutting the Type 3 granites of St Mary's, Annet and St Agnes.

Quartz–porphyry dykes, 'elvans'

The granite plutons of the Cornubian Batholith and the surrounding and intervening envelope rocks are cut by late quartz–porphyry or quartz–feldspar–porphyry dykes, locally called 'elvans'. These rocks are generally very fine-grained, show chilled margins with evidence of flow-banding and commonly pass from a non-porphyritic marginal zone to a core more or less rich in quartz and/or K-feldspar phenocrysts (Goode 1973). Some dykes show evidence of multiple intrusion. Individual dykes range in thickness from a few centimetres to 40 m or more. In some cases dykes show tourmalinization and, more rarely, impregnation by cassiterite (Dines 1956; Hosking 1964), while many are affected by chloritic or argillic alteration. This

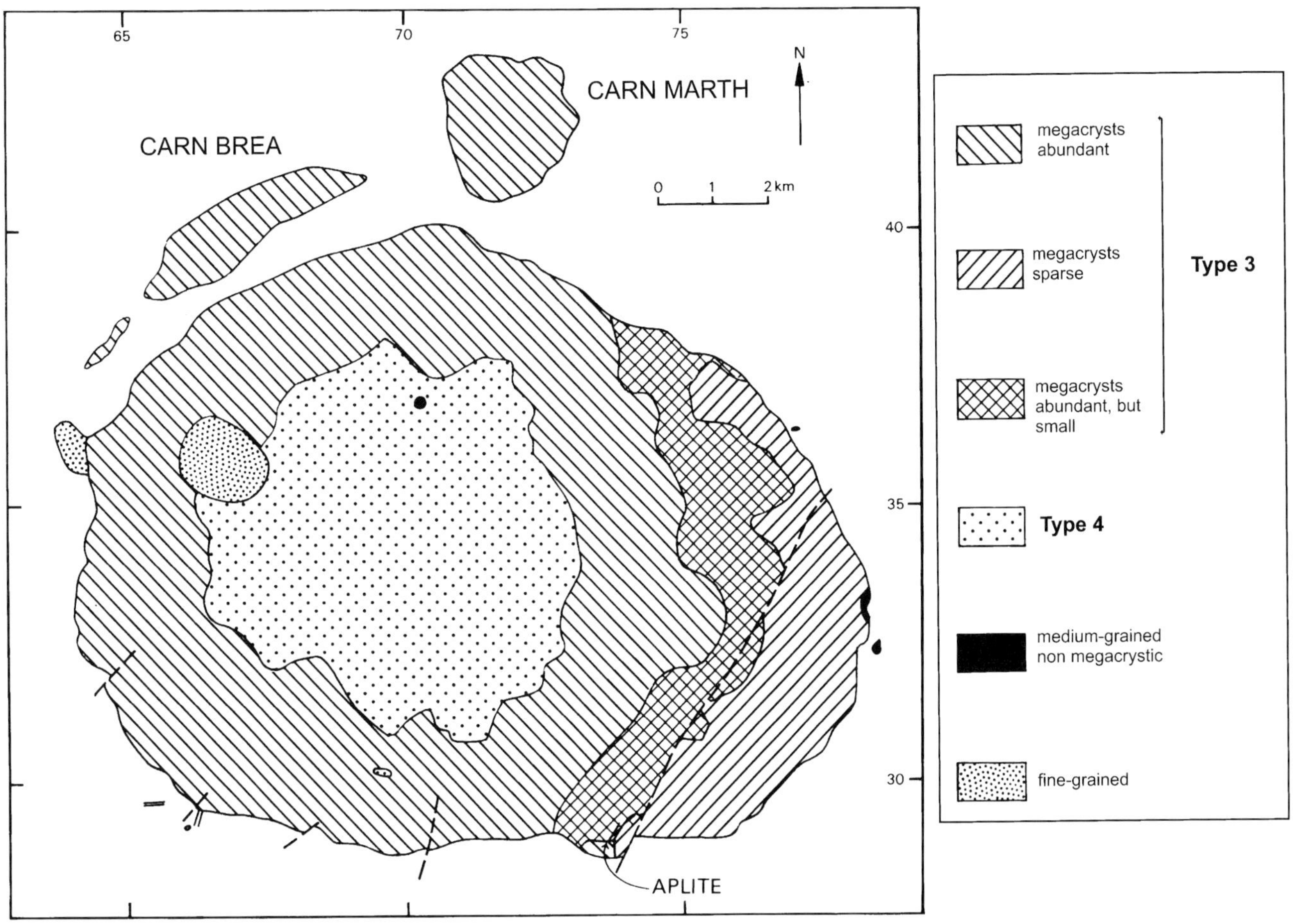

Fig. 11.4. Granite types of the Carnmenellis Pluton (after Leveridge *et al.* 1990).

variation in degree of alteration provides examples which are fresh and hard, through those which have been worked for carving, to those which are entirely altered to clay minerals.

Examination of the Geological Survey maps for the region show that elvan dykes within the granites are more common in the older Bodmin Moor and Carnmenellis granite plutons, and are sparse or absent from large areas of the Dartmoor, Land's End and St Austell masses. Most of the elvan dykes trend between E–W and ENE–WSW, though there are local variations. Strike lengths are also very variable, with some examples shown on the various Geological Survey maps to exceed 5 km. In many areas, there is a close similarity in trend between the elvans and the main-stage polymetallic mineralization, though it can usually be demonstrated that the mineralization post-dates the emplacement of the elvan dykes. The field relationships of elvans, and the close association with mineralization suggest that they are late in the scheme of igneous events in the province, and this is borne out by the isotopic evidence (Darbyshire & Shepherd 1985; Leveridge *et al.* 1990). Such geochemical data as are available suggest that the elvans are mostly directly derived from a biotite granite parent magma, although commonly with some K enrichment. Henley (1974), in discussion of examples in the Perranporth area of south Cornwall, suggests that this is the result of interaction with exsolving fluids from the cooling magma: he further supports Goode (1973), who describes elvan emplacement in terms of fluidized events linked to hydrothermal brecciation and the exsolution of high-temperature mineralizing fluids. The tectonic environment in which elvan dykes were emplaced has been discussed by Shail & Wilkinson (1994) and Alexander & Shail (1995), who concluded that they were synchronous with, or post-dated faulting in a NNW–SSE extensional regime associated with orogenic collapse.

Mineralization in SW England

Older models of metallic mineralization in SW England ascribe almost all the deposits to the migration of fluids of direct magmatic departure from 'emanative centres' within the granites (eg. Dewey 1925; Davison 1926; Dines 1934). More modern work has indicated a scenario with a much longer-lived time frame and more complex evolution, involving the deposition of metal ores from early hydrothermal systems associated with Devonian and Carboniferous volcanism (Leake *et al.* 1985; Alderton 1993; Leveridge *et al.* 2002) followed by much later 'cross-course' vein generation resulting from low-temperature, high-salinity brines which originated in the Permo-Triassic red-bed basins flanking and overlying the Variscan Orogen (Scrivener *et al.* 1994; Leveridge *et al.* 2002). The full sequence of events responsible for metallization in the SW England region can now be extended from the Early Devonian to at least the mid-Triassic (Table 11.2).

Of greatest economic importance are the tin–tungsten–arsenic–base metal deposits, associated genetically and spatially with the emplacement and cooling of the granite plutons and their satellites. Modern accounts of metalliferous mineralization in SW England are by Jackson *et al.* (1989) and Alderton (1993). Most occurrences of granite-related mineralization are accommodated within the classification which follows:

Skarn deposits – metallized exoskarns have locally been formed by the thermal alteration of carbonate-rich strata to assemblages of calc-silicate minerals that have subsequently been penetrated by early, high-temperature, generally boron-rich fluids. Such deposits are commonly of limited size (reflecting the dearth of carbonate-rich rocks in the province) and low economic importance. Examples

Table 11.2 *Generalized sequence of metalliferous mineralization.*

Pre-granite	Syngenetic and epigenetic metal concentrations	Pb, Zn, Ba, Mn, Fe
	Pre-granite veins	Sb, Pb, Fe, Au
Granite-related	Skarn deposits	Fe, Sn, Cu, As
	Greisen-bordered veins	Sn, W, Cu
	Tourmaline–quartz orebodies	Sn, Fe
	Main-stage polymetallic veins	Sn, Cu, As, Zn, Pb
	Major argillic alteration	Kaolin
Post-granite	Cross-course veins	Pb, Ag, Zn, Cu, Ba, F, Fe
	Low-temperature replacements	Mn, Fe
	Gold-carbonate veins	Au

include the cassiterite–magnetite–tourmaline 'floors' of Grylls Bunny, near St Just-in-Penwith (Jackson 1974), the cassiterite–magnetite ore beds of Magdalen Mine, Ponsanooth (Dines 1956), the magnetite–hornblende deposits of Haytor Vale Iron Mine, east Dartmoor (Scrivener *et al.* 1987) and the chalcopyrite–arsenopyrite–pyrrhotite skarns of the Sticklepath–Belstone–Meldon districts of north Dartmoor (Edmonds *et al.* 1968).

Greisen-bordered vein deposits – examples of quartz–cassiterite and/or wolframite veins with characteristic selvedges of white mica, quartz ± topaz are common within the granites and in the proximal envelope of country rocks. Field relationships suggest that this style of mineralization is early, and commonly accompanies hydraulic fracturing of the host rock. Sheeted vein complexes have been described from St Michael's Mount (Floyd *et al.* 1993), Cligga Head, near Perranporth (Moore & Jackson 1977), Goonbarrow China Clay Pit, near Bugle and Hemerdon Mine in the southernmost part of Dartmoor (Beer & Scrivener 1982). Veins in which greisen material is included within later tourmaline-dominated assemblages are common, particularly in and around the Bodmin Moor Granite.

Tourmalinite–tin deposits (tourmaline–quartz ± cassiterite ± Fe oxides) – these are very widely distributed in the province as fissure veins, replacement bodies and breccia pipes, both within the granite, and in the more proximal aureole rocks. Notable examples include the earlier suite of Sn veins (the 'numbered' lodes) of South Crofty Mine (Bromley 1989) and the Birch Tor and Vitifer mines of Central Dartmoor in Devon (Scrivener 1982). These veins show highly brecciated textures within tourmaline-impregnated wallrocks and, where they occur within the granite plutons, are generally deficient in sulphide minerals.

The 'Main stage' polymetallic lodes – these are commonly fissure veins bearing combinations of Sn–Cu–As–Zn–Pb minerals, and have been the main sources of ore throughout the historic mining activity of the past two centuries. These are commonly complex orebodies with an extended history of hydrothermal activity; they may include components of early greisen and tourmalinite mineralization, as described above, in addition to later stages in which base metal sulphide assemblages predominate, with chlorite and fluorite as important gangue minerals. The main stage orebodies may locally show vertical and lateral element zoning, though this is not as common as suggested by some earlier workers (Dines 1956), or may be 'telescoped' with early minerals such as cassiterite, arsenopyrite and wolframite on the vein walls, and successive generations base metal sulphide mineralization giving place to spar minerals in the core. Main stage veins of the Camborne–Redruth district commonly show vertical zoning, with sulphides tending to give way to cassiterite at depth, while the veins of the Hingston Down–Gunnislake area tend to show telescoping (Beer & Scrivener 1982).

The origin of the ore-forming fluids has been the subject of considerable research, with early models (e.g. Dines 1956) suggesting fluids of direct magmatic departure migrating outwards from 'emanative centres' within the granite plutons, forming essentially concentric zones of mineral deposition and accounting for all the vein deposits of the region. More recent workers (e.g. Simpson *et al.* 1979) have described systems in which most of the mineralization is ascribed to the large scale induction and circulation of meteoric waters, driven by convection in and around the granite plutons by their radiogenic heat output. A substantial body of work on fluid inclusions and fluid chemistry in recent years has greatly improved knowledge of the complex interaction of magmatic fluids, meteoric waters and evolved basinal brines that resulted in the formation of the Cornubian orefield (e.g. Shepherd *et al.* 1985; Gleeson *et al.* 2000). In essence, the early greisen- and tourmaline-related mineralization resulted from high-temperature magmatic fluids, which on exsolution from the parent granites evolved by unmixing into two components: one relatively low-density, CO_2-rich and the other of greater salinity. Of these, the former is responsible for the formation of greisen mineralization and the latter for tourmalinite bodies (Fig. 11.5). These early magmatic fluids then underwent mixing with meteoric water and further circulation to form the main-stage deposits. Shepherd & Scrivener (1987) have drawn attention to the early involvement of basinal brines in the main-stage mineralization of the Hingston Down–Gunnislake district of east Cornwall and have speculated that local variations in style of the main-stage deposits may be ascribed, in part, to the contribution of meteoric/basinal components. Scrivener *et al.* (1994) demonstrated that the N–S-trending vein systems, which typically carry Pb–Zn–Ag–Ba–F, termed 'cross-courses', were entirely due to the movement of low-temperature, high-salinity, $CaCl_2$-rich brines, which originated in nearby Permo-Triassic red-bed basins, and mark the end of granite-related vein deposits in the region.

Kaolinization

Large-scale alteration of the feldspars to clay minerals, dominated by kaolinite, is a feature of parts of the Cornubian Batholith. The phenomenon is most marked in those granites showing evidence of multiple intrusion, notably the western part of the St Austell Granite (Fig. 11.3), southern Dartmoor and the fringes of the St Just mining district of the Land's End Granite. There is also patchy kaolinization in the Bodmin Moor Granite. As kaolinite deposits are worked for china clay on a very large scale in the St Austell district and in the southern part of Dartmoor, there has been considerable study of the kaolinized districts, and much speculation as to their age and origin. The general form of the kaolinized zones in the St Austell district has been defined by mapping and exploration drilling. Bristow & Exley (1994) report that individual kaolinite deposits have an essentially conical form, narrowing with increasing depth. These zones may exceed 300 m in width at the top, and extend to depths of more than 200 m. Within the zones the argillic alteration is variable, and there are usually large inclusions of less altered granite, of no economic worth. The kaolinized zones are enclosed by granite in which much of the

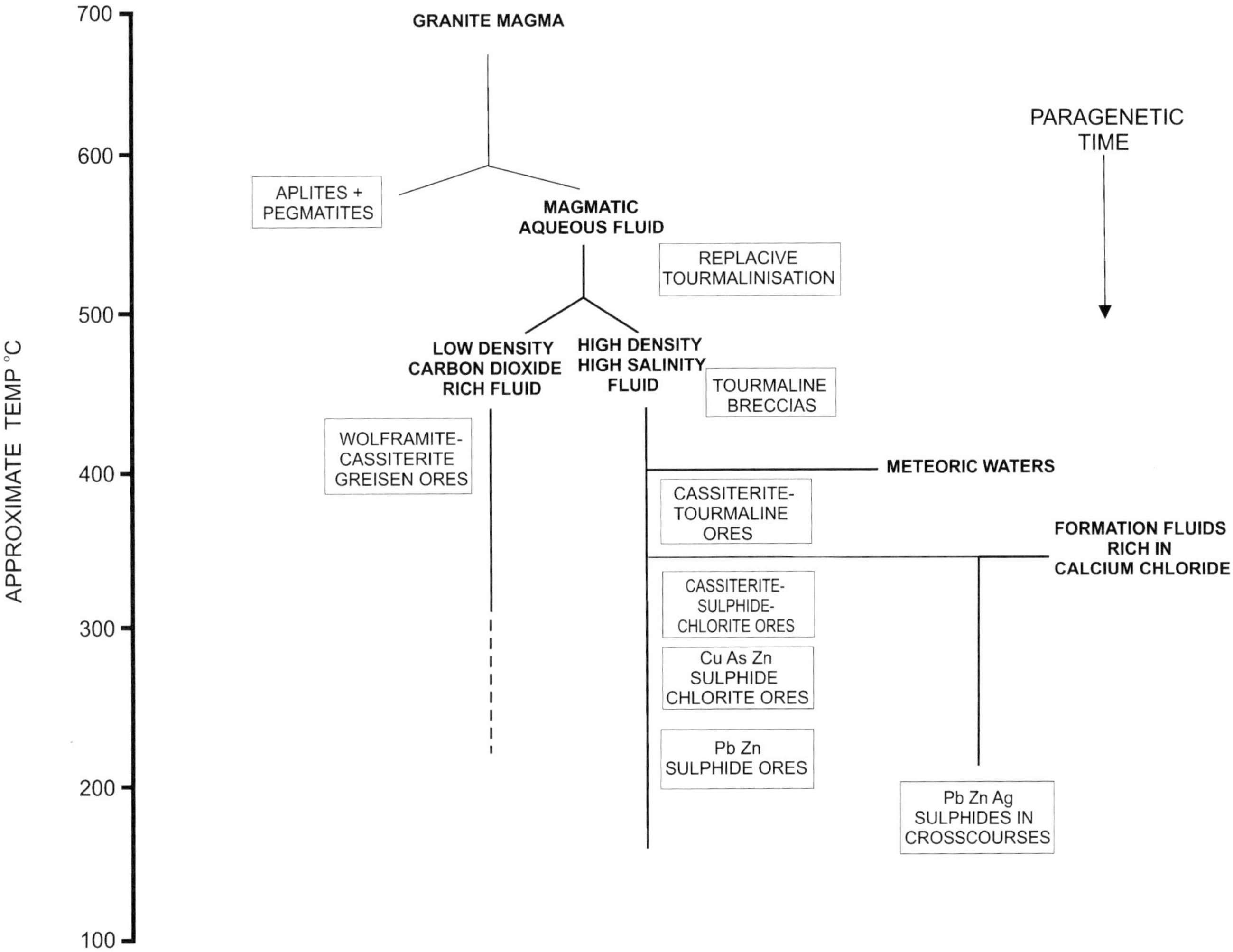

Fig. 11.5. Synoptic evolution of ore fluids in SW England (after Leveridge *et al.* 1990).

feldspar is altered to secondary mica variously described as 'sericite', 'gilbertite' or 'hydromuscovite'. Early workers, such as Hickling (1908), favoured genesis of the kaolinite deposits through deep weathering of the host granite, a view supported by the stable isotope work of Sheppard (1977). A second school suggests an entirely hydrothermal origin from low-temperature fluids circulating in the aftermath of the main-stage metalliferous mineralization of the region (e.g. Bray & Spooner 1983). Bristow & Exley (1994) describe a complex history for the formation of the kaolinite deposits, with successive early stages of low-temperature hydrothermal alteration, producing first smectite-illite clays, with later alteration to kaolinite-dominated assemblages. Further kaolinite is considered to have been produced, and earlier kaolinite b-axis disordered, during deep Tertiary weathering.

Granite and mineralization chronology.

Modern work on the radiomentric dating of the Cornubian batholith has elucidated the timing of pluton emplacement, subsequent cooling history and timing of the protracted history of mineralization associated with the plutons. Major contributions on Rb–Sr isotope studies are by Darbyshire & Shepherd (1985, 1987) and on U–Pb and Ar/Ar isotope studies by Chen *et al.* (1993) and Chesley *et al.* (1993). In spite of some differences in interpretation, this considerable body of work demonstrates a complex history of emplacement extending over the interval between 300 and 270 Ma (Table 11.3), with coeval and subsequent granite-related mineralization extending to around 250 Ma. There is no pattern of progressive pluton emplacement from one end of the batholith to the other, but there is clear

Table 11.3 *Synoptic geochronology of the major granite plutons.*

	Darbyshire & Shepherd (1985) Rb/Sr preferred age	Chesley *et al.* (1993) U/Pb preferred age (range)	Chen *et al.* (1993) U/Pb preferred age (range)
Dartmoor	280	278 to 280	280 to 285
Bodmin Moor	287	291	281
St Austell	285	282	278 to 281
Carnmenellis	290	294	291 to 294
Land's End	272	275 to 279	274 to 277
Isles of Scilly	No data	No data	290

evidence that the Type 3-dominated plutons – Bodmin Moor, Carnmenellis and Isles of Scilly – were emplaced earlier than the Dartmoor, St Austell and Land's End plutons.

A significant feature of the modern work on dating is the improvement in precision, which has permitted estimation of the cooling rates of the plutons, and the detailed investigation of the chronology of mineralization. For the major plutons of the Cornubian Batholith, Chen *et al.* (1993) estimated a cooling rate of 80–100 °C Ma^{-1} between the U–Pb closure temperature at 725 °C and the Ar/Ar closure temperature at 320 °C.

In respect of the age of mineralization, the northern and central parts of the Dartmoor Granite yield consistent emplacement ages of 280 Ma, using Rb–Sr, U–Pb and Ar/Ar analyses (Darbyshire & Shepherd 1985; Chesley *et al.* 1993). The age of Sn mineralization in the Birch Tor and Vitifer has been obtained from Ar/Ar analyses (Chesley *et al.* 1993) and from U–Pb measurements on inclusion fluids (Miller 1994). Both techniques provide an age of 277 Ma, and there is no evidence for later thermal activation. This indication of a 3–4 Ma (including errors) history of emplacement at approximately 800 °C to mineralization at 300–400 °C is in contrast to the very protracted history of emplacement and subsequent mineralization at South Crofty Mine, which extends over 40 Ma from granite emplacement at 295 Ma to latest main-stage mineralization at 255 Ma (Chesley *et al.* 1993). This extended history reflects not only successive magmatic events but also the prolonged contribution of radiogenic heat to a system that remained at depth beneath a cover of envelope rocks.

Petrogenesis

Most of the modern work on the origins of the Cornubian Batholith has focused on two areas; the source of the magma and the tectonic framework in which the magma was formed. Watson *et al.* (1984), in a geochemical comparison of the Grampian and Cornubian granites, came to the conclusion that the most probable source for the magma was mantle material previously enriched in volatile elements by metasomatism. Problems with this model are the fate of the basic residuum and the source of the volatile enrichment. More recent studies have concluded that the granite magma was essentially derived from partial melting of the lower crust; Stone & Exley (1986) proposed a source from garnet granulite at 7–8 kbar pressure and at 800 °C, while Charoy (1986) suggested cordierite–sillimanite gneiss at 5 kbar and 800 °C. It is unlikely that the considerable volume of magma required was derived from a single source, and there remains the problem of the source of the volatile and incompatible elements. Darbyshire & Shepherd (1994) confirm the essentially lower crustal origin of the granites in a study of Nd–Sm and Rb–Sr systematics. They suggest that a heterogeneous Lower Proterozoic source predominated, and that the rocks of the high-level Devonian and Carboniferous envelope did not contribute significantly to the melt through stoping. This study also indicated a mantle component in all the granites, particularly the younger Dartmoor, St Austell and Land's End plutons. Mantle input to the granites was confirmed by Stimac *et al.* (1995) in a study of the geochemistry of enclaves in the batholith. Two groups of enclaves were identified: the most abundant 'non-igneous enclaves' include affinities with high-grade metasedimentary source rocks, while the second group 'microgranular enclaves' were considered to originate from mantle-derived basalt, injected into the melting crust and possible related to the more basaltic components of the Permian Exeter Volcanic Rocks, which are contemporaneous with granite emplacement (Edwards & Scrivener 1999). Floyd *et al.* (1993) in considering the geochemistry of biotites in the granite and in enclaves suggest an upper mantle source for the U, Sn and W enrichments in the batholith, but are equivocal as to the source of the volatiles B and F. The combination of active mantle magma systems and partial melting of the lower crust as a precursor to granite magma migration provides clues as to the tectonic setting of the batholith, which radiomentric dating has shown to be post-collisional. The suggestion of Shail & Wilkinson (1994) that mantle convection beneath tectonically thickened crust combines the possibility of contemporaneous partial melting of the lower crust, as well as permitting mantle-derived magma to ascend as lamprophyric dykes, or the extrusive basalts of the Exeter Volcanic Rocks (Leat *et al.* 1987). Within the framework of orogenic collapse (*sensu* Dewey 1988), Holder & Leveridge (1994) suggested that the tectonic unloading of the Rhenohercynian passive margin, during extension, created the *P–T* (Pressure–Temperature) instability that generated crustal melts beneath the northern margin of the major southerly flysch nappes.

Suprabatholithic volcanism and unroofing history

In situ evidence for suprabatholithic volcanism is to be found at three sites in the Rame Peninsula of east Cornwall: Kingsand, Bridgemoor and Withnoe (Cosgrove & Elliot 1976). At Kingsand, undeformed banded rhyolite rests on the Sandway Cellar Conglomerate (of probable Early Permian age), while at Bridgemoor and Withnoe smaller outcrops of rhyolite rest on Early Devonian slate and sandstone of the Whitsand Bay Formation (Leveridge *et al.* 2002). At Withnoe, the rhyolite is cut by a complex of quartz–porphyry dykes: geochemical evidence presented by Leveridge *et al.* (2002) demonstrates a close genetic affinity between all three rhyolite outcrops and the elvan dykes. K–Ar ages for biotites from the Kingsand rhyolite recorded by Rundle (1981) give a mean age of 289 ± 4 Ma, in the Early Permian, and close to the age suggested above for the emplacement of the Bodmin Moor Granite. Leveridge *et al.* (2002) in consideration of Sm–Nd isotope studies, confirm that $\sum_{Nd}(T)$ for the Kingsand and Withnoe elvans and rhyolites fall in the range -6.5 to -7.1, close to that for the Bodmin Granite, with a value of -7.0, rather than that for the Dartmoor Granite at -4.7 (Darbyshire & Shepherd 1994).

In contrast, Late Permian breccias of the Exeter Group in the Crediton Trough to the north of Dartmoor (Edwards *et al.* 1997), have clast populations which include abundant acid igneous debris (Edwards & Scrivener 1999). The Newton St Cyres Breccia contains abundant cleavage fragments of K-feldspar, called 'murchisonite' in the early literature (Levy 1827), as well as fragments of granite, which Dangerfield & Hawkes (1969) matched petrographically with the Dartmoor Granite. The Crediton Breccia, beneath the Newton St Cyres Breccia, yields abundant fragments of rhyolite, acid tuff, vent agglomerate and felsite. Results of Sm–Nd isotope studies (Edwards & Scrivener 1999) demonstrate that $\sum_{Nd}(T)$ for examples of these clasts fall in the range -5.2 to -4.4, which bracket the mean value for the Dartmoor Granite of -4.7 (Darbyshire and Shepherd 1994). K–Ar and Ar/Ar ages for biotite from two clasts from the Kennford Breccia (Exeter district) and Newton St Cyres Breccia, respectively, are closely in accord with the mean age of the Dartmoor Granite of 280 Ma (Edwards & Scrivener 1999). It therefore appears that the Dartmoor Granite was exposed by erosion in the Late Permian, after the removal of a substantial volume of superincumbent acid igneous debris. The red-bed sediments of the Crediton trough record this unroofing in an inverted stratigraphy.

The lowest strata of the Exeter Group in the Crediton Trough have been demonstrated to be of Early Permian or possible latest Carboniferous age: both the Bow and Cadbury breccias lie stratigraphically beneath the level of the Killerton lamprophyric (minette) lava of the Exeter Volcanic Rocks (Edwards & Scrivener 1999), for which an Ar/Ar plateau age for biotite of 290.8 +/− 0.8 Ma, has been determined. Furthermore, the Bow

Breccia includes fragments of rhyolite and felsite, which are isotopically similar to the Kingsand and Withnoe rhyolite and quartz–porphyry occurrences, in that $\sum_{Nd}$ (T) for examples analysed, fall in the range -6.7 to -7.0 (Edwards & Scrivener 1999). These data suggest that suprabatholithic equivalents of the Bodmin Moor pluton were being eroded in the Early Permian, and that red bed sedimentation was already well-established prior to the emplacement of the Dartmoor granite at 280 Ma. It further highlights the rapid uplift and erosion of the components of the Cornubian Batholith during the N–S extensional regime of the orogenic collapse (Shail & Wilkinson 1994).

12 Permian: arid basins and hypersaline seas

A. H. RUFFELL, D. W. HOLLIDAY & D. B. SMITH

The distribution of Permian rocks in England and Wales (Fig. 12.1) is more complex than that of the overlying Mesozoic formations, which in the onshore area form a broad swathe displaying, as William Smith noted, an overall NE–SW-trending strike. Permian sediments, by contrast, are more patchily developed but rest in many locations on Carboniferous rocks, with a palaeotopography generated by Variscan mountain building and later erosion. Assessment of how the Permian landscape might have appeared is best achieved through consideration of the sedimentary evidence from both the Permian and the preceding Carboniferous strata. There are strong indications in the rock record of a changing tectonic and palaeoclimatic regime in NW Europe during this time, which reflected broader, even global, events. The general tectonic scene was one in which the southern supercontinent Gondwana moved north through Carboniferous time to collide with its northern counterpart Laurasia in the latest Carboniferous and earliest Permian. This continental collision was achieved as the Devonian–Carboniferous Rheic Ocean closed and Pangaea formed (Fig. 12.2). Simultaneously, the Ural Mountains were forming as the Kazakstan microplate collided with Fennoscandia, the final coalescence of Pangaea. The sedimentary fill of the Rheic Ocean is now preserved as deformed and locally metamorphosed pre-Permian successions in Cornwall, Devon, northern France, Belgium and Germany.

An angular unconformity between folded and faulted Devonian–Carboniferous (and older) strata and a horizontal or shallow-dipping Permian succession is common through England and Wales. Conversely, there are many locations where Stephanian (latest Carboniferous) red-beds pass into Permian strata of similar lithology with no obvious or significant break, begging the question "where is the Carboniferous–Permian boundary?" when no biostratigraphically-useful fossils or radiometrically dated materials are present. For instance, the Enville Member of the north Midlands (Powell *et al.* 2000*b*) has historically been placed in either system, although it is currently regarded as Carboniferous (Chapter 11). The lithological similarity and imprecise dating of the Carboniferous–Permian succession is not unique to the Midlands; the offshore Plymouth Bay Basin (see below) contains a 10 km-thick succession of Carboniferous, Permian and Triassic red-beds in which the erathem and system boundaries have not been defined accurately.

The reader should now have a picture of a poorly-dated, sometimes unfossiliferous dominantly red-bed succession deposited during a time of significant tectonic change. Preserved early Permian sedimentary rocks are most likely the syntectonic product of block-faulting, activated during and after orogenesis. Thus the original land surface is now largely eroded away. The basin and block tectonics of the early and mid Carboniferous in many areas gave way to the low-relief landscape of the later Carboniferous Coal Measures. In SW England, as the Rheic Ocean closed, foreland basins developed through the Westphalian until Stephanian uplift out-paced foreland basin subsidence in most areas. A landscape of fault-scarps, horsts, graben and larger foreland basins developed in the Stephanian and continued into the early Permian when, as Gondwanan–Laurasian collision ended, extensional tectonics and thermal relaxation began, allowing the preservation of Permian sediments in isolated basins throughout northern Europe, including England and Wales (Fig. 12.3).

The work of the British Geological Survey, most especially in the Exeter and Birmingham areas (Edwards & Scrivener 1999; Powell *et al.* 2000*b*), has confirmed the importance of latest Carboniferous–earliest Permian volcanic activity and reinforces Coward's (1995) view of a contemporaneous hotspot, centred on the North Sea. This formed part of a widespread chain of igneous events across NW Europe. These began with the Great Whin Sill (Plate 19) and associated intrusives (e.g. the Lindisfarne Dyke) around 300 Ma–290 Ma ago, with a mean age of 295 Ma for the Whin Sill. The Exeter Volcanics began to form at about 290–292 Ma and were short-lived, with few ages younger than about 280 Ma: their appearance was roughly synchronous with intrusion of the Bodmin Granite (dated at about 288 Ma) but generally preceded the main Dartmoor Granite intrusive phase at 280 Ma. Rhyolites of the Danish Horn Graben (Southern Permian Basin) began to form at about the same time as the Exeter Volcanics but continued to accumulate for much longer. The Danish rhyolites occur in two regions of the graben, denoted C1, where vulcanicity continued to about the early–late Permian boundary (roughly 269 Ma) and R-1, where it continued well into the late Permian (some 265 Ma). In Scotland and the rest of the southern North Sea, Permian igneous activity commenced later than in the Danish Horn Graben, with the Glasgow/Paisley and Mauchline basalts dated at 280–276 and 276–274 Ma, respectively. Basaltic igneous rocks from Altmark and the German Horn Graben (Southern Permian Basin of Fig. 12.3) erupted later still, at 261–265 Ma, well into the late Permian. A second phase of igneous activity occured in the Danish Horn Graben at about the same time, but this time was basaltic (not rhyolitic) in nature, like the German material.

Latest Carboniferous–early Permian tectonic changes were accompanied by a changing climatic regime. As Gondwana drifted north, closing the Rheic Ocean, maritime influence diminished and the largely amalgamated Gondwanan–Laurasian landmass moved north, possibly into the rain-shadow of the growing Variscan mountains. This northward drift took the present-day England and Wales from the palaeoequatorial (warm, humid) subtropical belt, into the warm, arid subequatorial circumtropical belt. Namurian–Westphalian (Carboniferous) humid climate deposits (coals, deeply weathered soils) were succeeded by Stephanian–early Permian arid-style sediments (red-beds, dune sands, evaporites). It is hard to decide whether declining maritime influence (ocean closure, sea-level fall), the rain-shadow effect or the northwards plate movement had the dominant effect on this mid Carboniferous–early Permian sedimentary change. In the southern hemisphere at this time, the southern parts of Gondwana were undergoing a prolonged

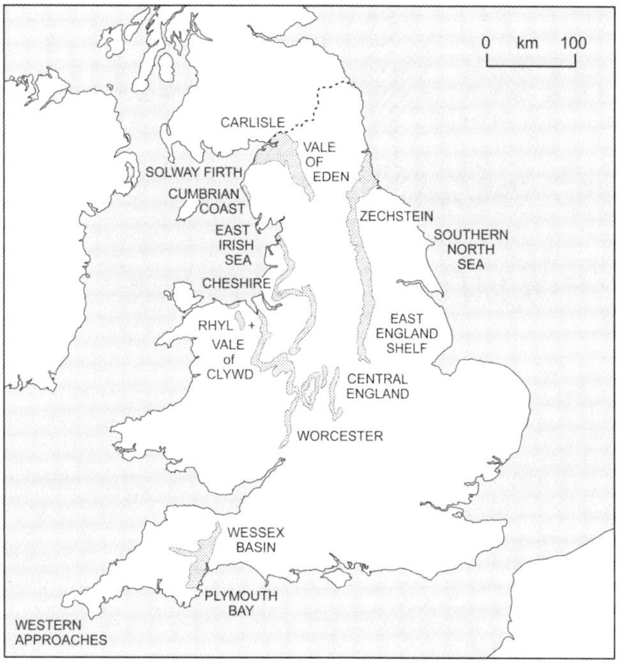

Fig. 12.1. Outcrop of Permian sediments in the onshore area.

and widespread continental glaciation. This presumably caused a significant fall in sea level and reduction of a maritime climate in northern Europe (and elsewhere). Desert erosion and sedimentation continued intermittently for the whole of the Permian in much of England and Wales and surrounding areas, and over time unfossiliferous desert sediments accumulated to considerable thickness in a number of low-lying and subsiding areas. In the NW and NE, however, desert conditions were abruptly ended by marine transgression and the formation of extensive epicontinental seas in NW and eastern England, wherein up to 1200 m of carbonate and evaporite rocks (Plate 20) were formed in the closing phases of the period.

In the West Midlands and SW England, the close of the Permian was in stark contrast to events at the end of the Carboniferous. While the Carboniferous–Permian was marked by Variscan tectonics, igneous activity and the resultant major unconformity in many basins, the Permian–Triassic saw limited tectonic or palaeoclimatic change. Some changes occurred in rates and styles of basin subsidence, and an increase in fluvial activity may be alluded to. Conversely, there is little tangible evidence in the rock record of England and Wales of the end Permian extinction, an event which caused the demise of up to 90% of marine species, 60% of reptile and amphibian families and most of the known terrestrial land plants (Benton & Twitchett 2003; Heydari & Hassanzadeh 2003). This, the largest mass extinction of the Phanerozoic has had regular debates about its origins (Erwin *et al.* 2002). The idea of the extinction being linked to the coalescence of Pangaea and the associated reduction in continental shelf coverage and biotic diversity has largely been discarded, due to the apparent rapidity of the extinctions. Instead, extraterrestrial impact (Becker *et al.* 2001), massive CO_2 release from the basaltic eruption of the Siberian Traps (Wignall 2001*a*), or mechanisms of release of methane from methane hydrate are now more favoured (Hallam & Wignall 1999; Heydari & Hassanzadeh 2003).

Extensional tectonics and major rifts

This section introduces and describes the basins and groups of basins (Fig. 12.3) whose successions are described later in this chapter. These basins are known partly from outcrop (Fig. 12.1) and also from seismic, gravity and borehole data, especially offshore. In NW England and the adjacent offshore area (e.g. East Irish Sea), Permian strata rest unconformably on Carboniferous and older rocks. This unconformity marks a period of erosion that began in this area in late Carboniferous times during the Variscan Orogeny and continued into the Early and perhaps Mid Permian. Small-scale irregular relief (a few metres) has been observed locally on this unconformity, but regionally, as indicated by seismic reflection profiles, the surface is relatively smooth suggesting widespread peneplanation: as shown below, there is evidence of greater relief in the sediment sourcelands beyond the depositional basins. Below the unconformity, the Carboniferous and older rocks have been extensively secondarily oxidized and reddened in a zone ranging between a few centimetres to over 500 m in thickness. As a result, the position of the unconformity can be difficult to detect precisely in open-hole deep boreholes and, until recently, reddened Carboniferous strata had been erroneously included with the Permian rocks in several areas (Jackson *et al.* 1995).

The Cheshire and East Irish Sea basins are within the Clyde Belt of tectonic fabrics and deposition, a NW–SE-trending series of sub-basins that extend from the Highland Boundary

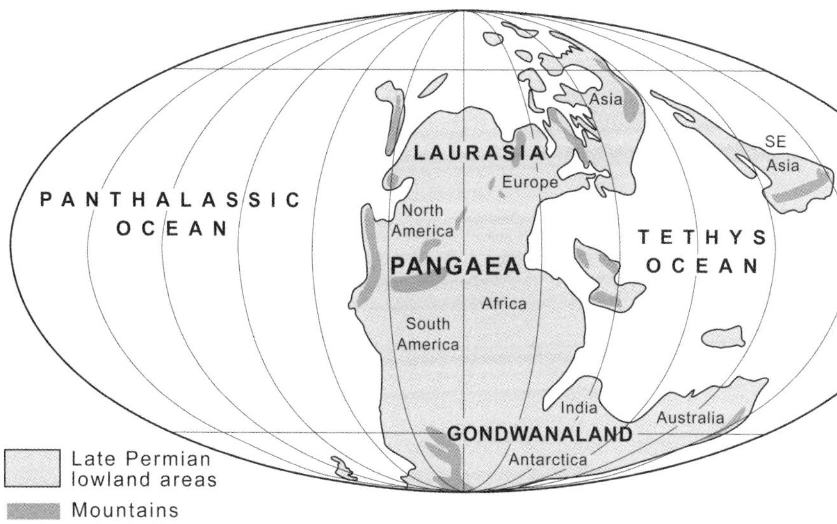

Fig. 12.2. Palaeogeography and palaeocoastlines of the Permian showing Pangaea and its subsequent rifting. (Redrawn after Benton *et al.* 2002.)

Fig. 12.3. Permian sedimentary basins of England and Wales in their regional context.

Fault to the southern edge of the Cheshire Basin. These basins originated during early Atlantic rifting when inheritance of underlying structures strongly influenced the orientation and structural style (Anderson *et al.* 1995). The Cheshire Basin is a classic example, with the Wem–Red Rock Fault system bounding its eastern edge (Fig. 12.4) and probably originating as a Carboniferous structure. The western edge of the basin shows gentle onlap, thus describing a classic half-graben (Fig. 12.4). The northern part of the East Irish Sea is transected by a zone of uplift, the Anglesey–Isle of Man to Rhinns of Galloway structure, that attenuates the Permian succession; basins to the north contain thick (2–3 km) Triassic with only thin Permian (where identified).

The Vale of Eden is a small, broadly N–S-trending fault-bounded basin flanked by the Alston and Askrigg blocks, with the complex Dent-Pennine Fault System on the eastern side of the basin showing evidence of late Carboniferous (Variscan) strike-slip, followed by Permian–Triassic extension; a tectonic history recorded in other basins. Permian sediments rest unconformably upon Carboniferous along the western margin of the basin, implying a 'trap–door' type geometry, probably with fault-scarp parallel river systems that show S–N palaeocurrent directions (Coward 1995).

The southern North Sea and eastern England collectively form part of the Southern Permian Basin and adjacent East England Shelf. This region of subsidence extended from the Pennines eastward to Germany and Poland, and was bounded to the north by the Mid North Sea High and to the south by the London Brabant Massif. Late Carboniferous–early Permian igneous activity in this area was linked to a NW European hotspot by Coward (1995). The basin is a broad, thick depression suggestive of thermal subsidence. However, the existence of fault topography and synfault deposition also indicates some mechanical stretching in common with the western Permian basins.

The Worcester Basin is a N–S-trending graben with Precambrian and Lower Palaeozoic rocks to the west brought into juxtaposition with the 3.1 km of Permian and Triassic red-beds (e.g. Kidderminster, Wildmoor sandstones) by the East Malvern Fault; concealed Palaeozoic rocks of the London Platform (London–Brabant Massif) are also fault-bounded at the eastern margin of the graben.

In the Wessex Basin, Permian strata are preserved in palaeovalleys and troughs, some of which have structural control, while others are apparently erosional with no underlying weakness. Permian strata are preserved in the west of the Wessex Basin (and associated graben, erosional valleys, etc.) and offshore in the Plymouth Bay Basin; thin Permian successions occur in other adjacent basins. In the east of the Wessex Basin, below Bournemouth Bay, the Permian is overstepped by the basal calcareous clastics of the Triassic Sherwood

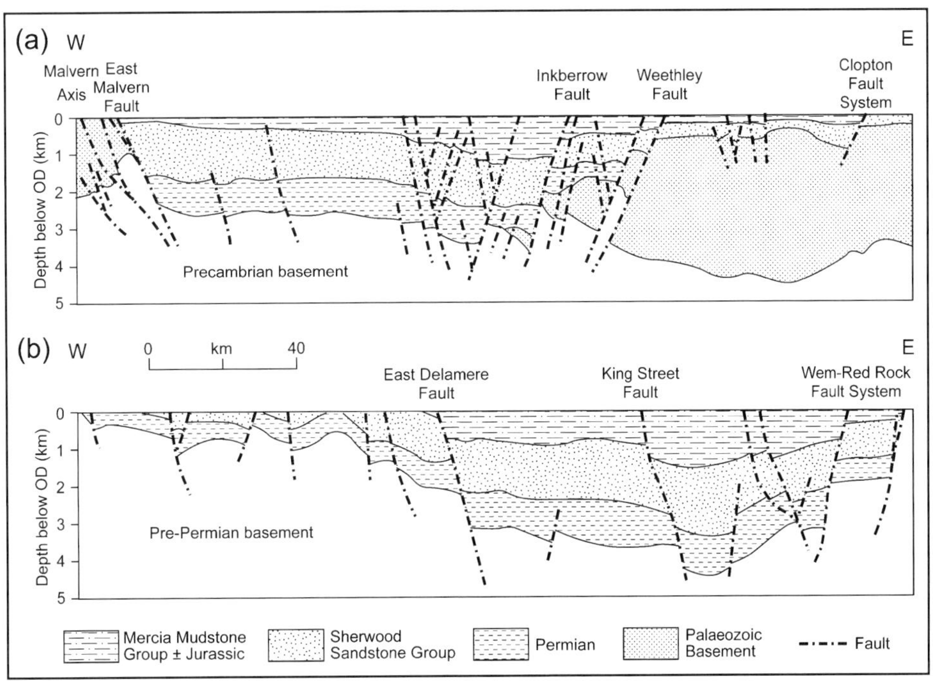

Fig. 12.4. Cross-sections through (**a**) the Worcester Basin and (**b**) the Cheshire Basin. (After Chadwick & Evans 1995.)

Sandstone Group. The basins of the Western Approaches have a complex history of subsidence and inversion that has resulted in thin Permian sequences being preserved in some areas (e.g. parts of the Brittany and South Celtic Sea basins) and none in others, or in very thick Carboniferous–Jurassic red beds accumulating that presumably include a Permian succession (e.g. Plymouth Bay Basin).

Permian stratigraphy

The Permian strata of our region have long been grouped into lower and upper 'series', conventionally divided at the incoming of marine strata in the north (Smith *et al.* 1974) and at supposedly equivalent levels elsewhere. Those divisions are retained informally here, although it should be noted that the International Commission on Stratigraphy has accepted a three-fold division of the Permian into Cisuralian, Guadalupian and Lopingian series, divided into a total of nine stages (Fig. 12.5). Throughout this chapter, these international stage names are referred to as far as possible. But it should be stressed that in view of the uncertain correlation between British Permian strata and international stratotypes, the identification of Permian system and stage boundaries in England and Wales cannot be achieved with certainty. Thus, for example, the base of the Permian is placed in the English Midlands on the evidence of scanty plant and reptilian remains, and in SW England by reference to the radiometric ages of about 285 Ma from volcanic rocks above the base of this red-bed succession. Here, numerical dating of volcaniclastic rocks may be used very effectively in conjunction with plant microfossils in defining the basal Permian (Warrington & Scrivener 1990; Edwards & Scrivener 1999). Elsewhere, only the evidence of unconformity and lithology is available. Controversy also surrounds the placing of the top of the Permian in rocks in England and Wales; this succession is largely barren of dateable materials, such that some authors (e.g. Smith *et al.* 1974) have arbitrarily taken the junction at the base of the Pebble Beds of the Sherwood Sandstone Group and their equivalents. An alternative view is that it should be taken at the top of the youngest Zechstein evaporites although correlation with the succession of the Dutch North Sea shows the Permian–Triassic boundary there lies within the lowest beds of the Dutch 'Bunter Sandstone'. Both views are difficult to apply in areas where neither lithological division is present.

The great lateral variability in facies and absence of fossils in most of the continental Permian rocks of England and Wales make correlation between strata of the various basins particularly difficult, although broadly similar lithological successions in separate basins probably evolved in parallel as landscapes and palaeoclimate changed. Enough macro-and microfossils have been found in lower Permian rocks of the English Midlands and SW England to allow those strata to be correlated tentatively with regional stages recognized in the marine facies of the USA and Russia (Fig. 12.5). Correlation between the upper Permian marine successions in the NW (Bakevillia Sea: Fig. 12.6) and NE (Zechstein Sea: Fig. 12.6) is firmer than between the various continental rocks, there being a measure of palaeontological control, and is reasonably firm between English Zechstein strata and those of The Netherlands, Germany, and Poland. Correlation with Permian rocks in the type area of Russia remains uncertain in the absence of fusulinids and useful ammonoids in Zechstein strata.

Early Permian environments and correlation

For perhaps 45 Ma, from late Carboniferous time until the late Permian marine incursions, England and Wales formed part of one of the great deserts of world history. Comparable in size and variety with the Sahara, and occupying a similar position relative to the contemporary equator and climatic belts, this desert was the scene for much early Permian mass downwasting of an initially immature landscape of which little has survived and, in later early Permian time, of the patchy accumulation of thick successions of continental sediments. There are indications that differential uplift and movement along some major faults may have continued from Carboniferous into earliest Permian time, and locally affected the course of erosion and sedimentation, but undoubtedly the most significant early Permian earth movement was extension and attenuation associated with the formation and persistent subsidence of a number of troughs (several of the trap-door type, faulted on the east) and inland drainage basins in the areas of the Irish Sea, western England and the North Sea; these became the main repositories for later lower Permian sedimentary rocks, including the principal reservoir rocks of the southern North Sea gas fields, and their formation may have been connected with the onset of rifting and the break-up of Laurasia (Smith & Taylor 1992).

Igneous activity (especially volcanicity), which contributed to the late Carboniferous and early Permian scene in SW Scotland and Northern Ireland, was restricted in England and Wales to

INTERNATIONAL SERIES		INTERNATIONAL STAGES	RUSSIAN STAGES	NORTH AMERICAN STAGES	GERMANIC DIVISIONS	UK INFORMAL DIVISIONS
Lopingian (Upper)	251 254	Changhsingian	Tatarian	Ochoan	Zechstein	upper Permian
	260	Wuchiapingian				
Guadalupian (Middle)	266	Capitanean		Capitanean		
	268	Wordian	Kazanian	Wordian	Upper Rotliegendes (Weissliegendes)	lower Permian
	271	Roadian	Ufimian	Roadian		
Cisuralian (Lower)	276	Kungurian	Kungurian	Leonardian	*unconformity*	
	284	Artinskian	Artinskian			
	295	Sakmarian	Sakmarian	Wolfcampian	Lower Rotliegendes	
	299	Asselian	Asselian			

Fig. 12.5. Permian chronostratigraphy of the main reference locations of the world. (After Benton *et al.* 2002.)

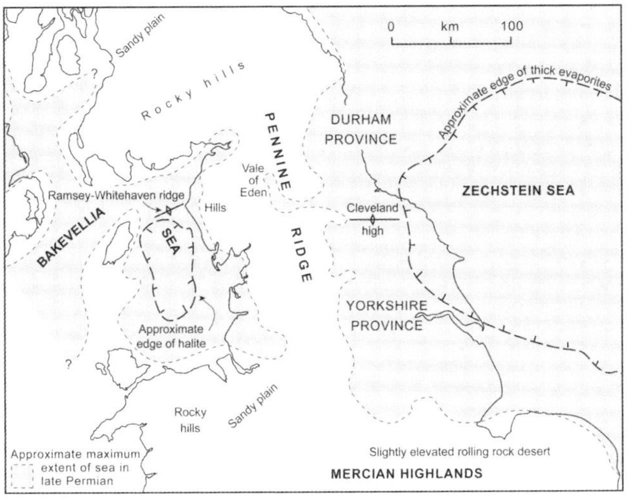

Fig. 12.6. Permian palaeogeography of northern England. The map is relevant for the time of the Zechstein transgression (in general outline) and for the time of deposition of the Ford Formation, Hartlepool Anhydrites, Fordon Evaporites and Boulby Halites.

the SW, especially the area around Exeter, and locally in the eastern Irish Sea. Although the record is scant, the effects of this igneous activity are widespread. New geochronologies (radiometric, plant microfossil), compared to the sedimentary record (Edwards & Scrivener 1999; Glennie *et al.* 2003) now allow the most precise comparison of igneous and sedimentary events yet achieved. The 'lower' Permian strata of SW England (Knowle, Thorverton sandstones) are broadly equivalent to the Lower Rotliegendes of the North Sea succession. In addition, deposition of these early sediments occurred when the broadly contemporaneous emplacement of Dartmoor granites and Danish rhyolites took place. Finally, it seems likely that no Upper Rotliegendes equivalent has been identified in SW England (Glennie *et al.* 2003).

The earliest Permian, judging from the scant surviving evidence, appears to have been a time of stark contrasts, in which areas of Lower Palaeozoic rocks that had not been covered by Carboniferous strata (or from which such strata had subsequently been removed) remained as wild uplands, whereas desert weathering of Carboniferous rocks in remaining areas resulted in landscapes dominated by immature landforms. Rainfall, attracted by relief that may locally have reached 3000 m, was sufficient to maintain a sparse population of plants and animals, but, like that of many modern deserts, was probably episodic and with a high rate of runoff. Dewfall was probably appreciable and the diurnal temperature range extreme. In this harsh setting, physical erosion may have been dominant, with detritus riven from soil-free and almost unvegetated rocks accumulating as screes before being swept into adjacent valleys by runoff, sheetfloods and debris flows; streams at this time would have been almost all short and ephemeral, and first cycle sediments accumulating on innumerable alluvial fans and valley bases were all locally derived. Most of this sediment was coarse and angular, but with a higher proportion of finer material in more distal fan and valley-floor environments with varying amounts of infiltrated fines in the more proximal parts. Reddening of the sediments, as in comparable modern environments probably began soon after deposition (Glennie *et al.* 1978). Aeolian processes at this stage probably caused erosion and transported much material: common ventifacts indicate widespread abrasion. As a depositional agent, the wind was important in some areas (Cumbria), mixed with fluvial reworking in others (Devon). Pediplains were probably not extensive at this time.

This partial reconstruction of the earliest Permian in England and Wales in the light of modern desert environments is based on the limited evidence of palaeotopography and overlying sediments in the Vale of Eden, West Cumbria, parts of the English Midlands, and in SW England, and accords with similar evidence from SW Scotland, the North Sea and much of Western Europe. It is suggested that the pattern revealed by these four areas was typical of much of England and Wales, but that vistas gradually changed as landscapes lowered and matured below the level of earlier valley fill; much of the evidence having thus been removed, only the sketchiest outlines of earliest Permian palaeogeography can be inferred.

Later early Permian palaeogeography may be reconstructed with rather more confidence for, despite contemporaneous erosion and reworking of sediments as landscapes continued to evolve, more deposits of this age and larger areas of the buried land surface have survived. This evidence shows that patterns of sedimentation at this time differed considerably between west and east. In Wales (where the only surviving direct evidence is from the Vale of Clwyd in the north) and much of western England, continental sedimentation continued within the framework established earlier and thick successions of lower Permian sediments accumulated in valleys and in a series of subsiding inland drainage basins. Initially these valleys and basins were separate, and the course of sedimentation in each was probably governed by a unique combination of local factors; deposits of this age in the various western basins were therefore extremely variable and are difficult to correlate in the almost total absence of useful fossils. Later, however, some of the sediment catchments merged as intervening barriers wasted or were buried and pediplains extended and matured. Many late early Permian sediments are continuous from basin to basin, include a proportion of far-travelled debris, and are more readily correlated with each other.

The evolution of the later early Permian landscape and climate of Wales and western England as mountainous areas were reduced and lowered, is reflected in a common succession of sediment types in each of the larger adjacent depositional areas. Earliest sediments were almost everywhere of coarse angular local detritus deposited on alluvial fans and plains, and in the wider basins these pass laterally into sands, silts, and clays. Evaporitic muds accumulated in saline playas on the floor of the large Irish Sea Basin and perhaps elsewhere. Water-laid, possibly partly vegetated sediment completely filled the smaller valleys and troughs, but was subordinate to, and commonly buried by, onlapping wind-blown sand in higher parts of the succession in all the larger basins. Shotton's (1956) pioneering work suggested that such sands accumulated on transverse dunes under the influence of prevailing easterly and northeasterly winds but work by Sneh (1988) has suggested that many of the early Permian dunes were of the oblique type and were formed under the influence of mainly northerly winds. Whether or not the whole of the western part of England became one great sand sea, as suggested by Shotton (1956), cannot now be determined, but there can be little doubt that the desert dunes of the late early Permian were much more extensive than the present outcrops of dune sandstone.

In the Southern Permian Basin (Fig. 12.3: southern North Sea) the picture is clearer, thanks to extensive drilling for hydrocarbons. In the Upper Rotliegendes, geophysical log correlation, combined with micropalaeontology, allows us to recognize the likely chronostratigraphic equivalence of red sandstones in the southern areas to halite-bearing desert lake sediments, resting on Lower Rotliegendes volcanics in the north. The two facies belts are separated by anhydrites, interpreted to be a sabkha. The uppermost beds of the red sandstone facies contain well-preserved duneforms ('dunes' on Fig. 12.7) termed the Weissliegend in the offshore successions. It is the orientation of aeolian cross-stratification in these beds that provides the most convincing evidence of E–W directed wind directions in the

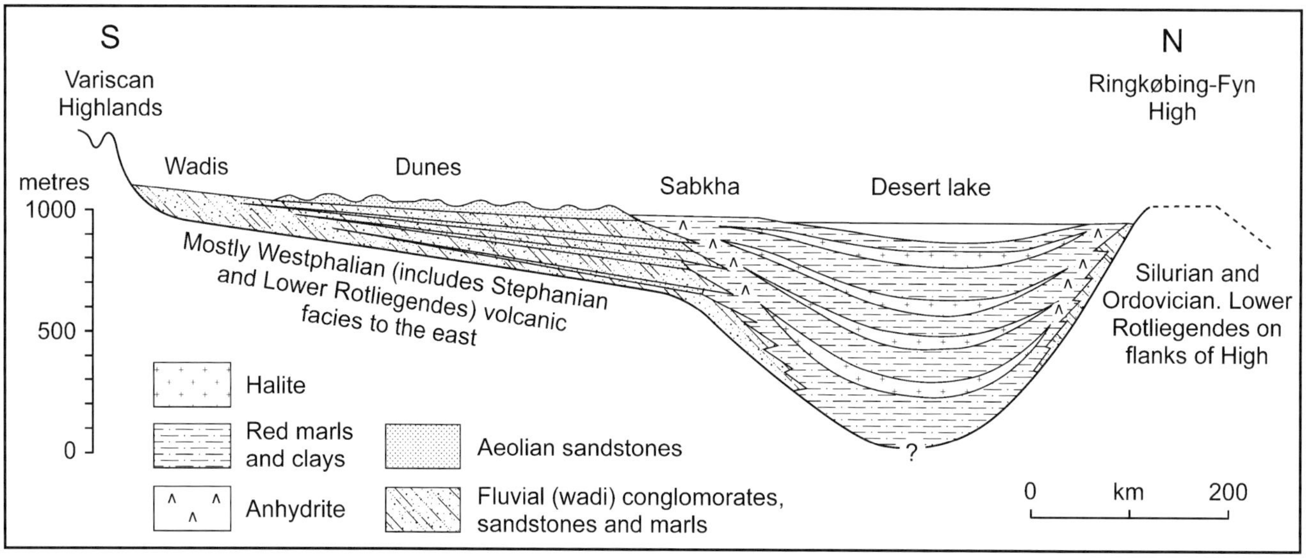

Fig. 12.7. Lithofacies relationships south to north through the Southern Permian Basin, showing how Glennie (1972) recognized the likely time-equivalence of the two facies of the Rotliegendes.

Permian. Plate tectonic reconstructions from palaeomagnetic measurements show that eastern England lay at a palaeolatitude of about 20°N in early Permian times, making the reconstructed wind directions consistent with palaeo-trade winds. Fortuitously, when the Zechstein transgression encroached upon these dunes, their form was often preserved: in eastern England synsedimentary slumps and up to 1 m of marine-reworked sand occurs at the Rotliengendes (Yellow Sands in this area)–Zechstein transition. In western Britain, earlier Permian breccia wedges in marginal parts of several of the dune fields show that substantial elevated sediment source areas persisted beyond the basin margins, despite the long continued levelling effects of erosion, pediplanation and filling of depressions.

Landscapes in the eastern part of England and the southern North Sea (as suggested by the morphology of the Permian land surface beneath Permian and Triassic deposits), evolved to far greater maturity than in the west before being overwhelmed in places by marine sediments. The evidence suggests that virtually the entire area was reduced to a vast rolling pediplain by the coalescence of a complex of earlier pediments, and that this surface was subsequently gently warped by subsidence in the southern North Sea Basin so as to form a major area of internal drainage. Here, probably somewhat later than in the west of Britain, continental sediments typical of a wide range of desert environments (Glennie *et al.* 2003) accumulated to thicknesses locally exceeding 300 m. Evaporitic playa deposits formed extensive spreads on the floor of the basin, which may have lain well below contemporary sea level (Smith & Taylor 1992), whilst bare rock desert or thin piedmont gravels occupied large parts of the marginal area. Geophysical, borehole and outcrop evidence suggest that the western sites of Permian deposition were dominated by fault-controlled subsidence, whereas the eastern areas (including much of the North Sea) experienced more regional subsidence. Exceptions, of course, occur with basin margins in the North Sea dominated by 'trap-door' style fault-controlled sedimentation, and broad areas of subsidence with minor normal fault control occurring in the Western Approaches and Irish Sea.

Early Permian palaeontology

In common with most modern deserts, it seems likely that the early Permian desert in England and Wales supported a relatively diverse specialized flora and fauna. However, the inferred slow rate of deposition, the abrasion and repeated reworking of sediment, and the prevailing strongly oxidising environment all conspired against the preservation of plant and animal remains and only trace fossils and the most durable organic debris were preserved.

With this unpromising background it is unlikely that the few fossils found in the lower Permian rocks of England and Wales are representative of the full contemporary biota. Fossils have been recorded in the Vale of Eden, the Warwickshire area of the English Midlands, and in SW England. Only the Midlands have yielded what might be termed a diverse assemblage of species, some represented by a single specimen, and the full stratigraphic range of these is poorly known. A summary of the fossil types is given in Smith (1992). Body fossils from this area comprise the pelycosaur reptiles *Haptodus grandis* Paton, *Sphenacodon britannicus* von Huene, and *Ophiacodon* sp., together with the labyrinthodont *Dasycleps bucklandi*, all from the lowest Permian Kenilworth Breccia Group, and the conifers *Lebachia* or *Waichiapiniformis* (Schlotheim) pars Florin, and *L. frondosa* (Renault) var *zeilleri* from the overlying Ashow Group. Footprints from the Kenilworth Breccias have been assigned to the 'Autunian' (low Permian) by Haubold & Katzung (1975) who also ascribed an 'Autunian' age to rocks of the underlying Gibbet Hill Group which are more commonly regarded as late Carboniferous (Smith *et al.* 1974). Footprints are the only fossils found in the lower Permian rocks of the Vale of Eden, whereas reptilian or amphibian burrows (Edwards & Scrivener 1999) constitute the only reported evidence of animal life in the higher Carboniferous or lower Permian rocks of SW England. The unrepresentativeness of the exceptionally short faunal list in the Midlands was noted by Paton (1974), who pointed out that the few forms found suggest a fairly varied fauna; 'there must', she wrote, 'have been plenty of food to support two species of carnivorous pelycosaurs'. Presumably some of the lower members of this food chain were herbivorous, but nevertheless direct evidence of root structures, for example, has yet to be reported from lower Permian rocks of England and Wales.

Late Permian environments and correlation

In much of England and Wales the continental conditions that characterized the early Permian persisted throughout the late Permian, but they ended abruptly in the NW and NE of England and much of the North Sea area when the Bakevellia and Zechstein seas (Fig. 12.6) were formed. These two seas are thought to have originated when the Boreal Ocean broke in from the north and flooded the inferred sub sea-level inland drainage basins there (Smith 1992), but for most of their existence they were separated by a persistent Pennine land barrier. Whether the Bakevellia Sea was fed directly from the Boreal Ocean, or was wholly or in part fed from the Zechstein Sea, is a matter of debate (Jackson 1994), but no unequivocal evidence for a cross-Pennine marine link has been established.

The deposits of both the Bakevellia and Zechstein seas are strongly cyclic in response to alternating phases of marine expansion and contraction. Such cycles were previously denoted EZ1–EZ4 but are now more commonly interpreted in terms of sequence stratigraphy. In the old concept of cycles, each comprised carbonate beds succeeded gradationally (commonly with diachronism) by beds of sulphates (gypsum or anhydrite), chlorides (generally halite), and finally highly soluble salts of potassium and magnesium. Where such a succession incomplete, it is generally the halites and potash that are absent: in the cycles concept, this was ascribed to incomplete evaporation or by later re-solution. In the sequence stratigraphic interpretation (Tucker 1991), this incomplete succession may be due to exposure and dissolution removal at a sequence boundary, or to the halites, potash, etc., forming part of a lowstand wedge, deposited shoreward of the rest of the succession and thus not always observed. The Bakevellia Sea (NW England) deposits were interpreted (Smith 1992) as comprising three incomplete cycles and those of the much larger and deeper Zechstein Sea (NE England and North Sea) comprise four relatively complete cycles and a number of minor cycles that are now often regarded as parasequences.

The succession of the Bakevellia Sea probably predates the Seaham Formation of the Zechstein and the time of deposition of the latter may not be represented by marine strata in most of the Bakevellia Sea Basin. The early marine deposits of the Bakevellia and Zechstein seas are generally similar to those being formed today in shallow epicontinental tropical and suptropical seas, and contemporary conditions in northern England must have been very like those in and around the modern Arabian Gulf. Evaporites make up a large proportion of the rocks of some cycles in each basin, however, and it is apparent that both seas on several occasions became highly saline and probably partly or wholly isolated from world oceans. Isolation of the Zechstein Sea may have resulted from re-emergence of the original rim of the inferred chain of lower Permian North Sea basins, and it has been argued that Zechstein sea levels periodically fluctuated by more than 100 m (Smith 1992). In both basins the main cause of such changes was probably slight oscillations of world ocean level relative to the original basin rims: glacioeustasy may still be inferred some 40 Ma after the peak of the Carboniferous–Permian southern hemisphere glaciations. At times of low ocean level, complete or almost complete emergence of the rims would have led to a reduction in the rate of sea-water influx below the rate of evaporation and a decline in basinal sea level ('evaporative drawdown') would have ensued; a subsequent rise in world ocean level relative to that of the basin rims would have resulted in reflooding and the beginning of the next major cycle.

As mentioned above, such repetitive stratigraphy can be interpreted in a different order to the former concept of cycles by applying the principles of sequence stratigraphy, wherein the halite–potash deposits are removed from the end of the cycle's sea-level highstand to the beginning of the overlying lowstand (Tucker 1991). This interpretation fits some of the outcrop data. Previously, exposure and erosion surfaces were recognized but only minor importance placed on them. In the sequence stratigraphic model, some of the more widespread (and obvious) surfaces are interpreted as sequence boundaries. Likewise, the sequence stratigraphic model alleviates the problem of accommodation space for the chloride–potash succession: previously these salts were placed at the top of the cycle, when waters were at their shallowest, requiring repeated replenishment by successive episodes of non-erosional marine inundation, or perhaps subsidence. Now such salts are accommodated in a shelf-margin wedge, where they do not have to be 'squeezed in' to the top of each cycle. During phases of drawdown the two seas probably dwindled to a series of shallow hypersaline playas and salinas surrounded by enormous salt flats; only the most adaptable and specialized organisms could have survived in these unremittingly harsh environments. Phases of freshening of the Zechstein sea water have also been inferred (Smith & Taylor 1992).

The formation of the late Permian seas led to a marked rise in the regional ground water level and to a climatic amelioration that profoundly changed the pattern of sedimentation and consequently the landscape in surrounding areas; these changes were augmented by the cutting off of the supply of wind-blown sand from the newly inundated areas and by the establishment of a sediment-stabilizing plant cover. In the rock sequences the changes are indicated by the incoming of extensive deposits of water-laid continental sediments, but oxidizing conditions apparently continued to prevail, for these rocks, like those of the lower Permian, are almost devoid of fossils and are ubiquitously reddened.

Reconstruction of the early late Permian scene from the evidence of the distribution and character of the continental sediments suggests that much of England at this time had a gently rolling landscape, with a progressively diminishing relief as sedimentary plains extended, merged, and gradually buried many of the surviving eminences. Upland areas such as most of Wales and the Welsh Borders, the Lake District, the Pennine Ridge, (the upland areas on Fig. 12.6) nevertheless remained uncovered except locally, and doubtless continued to supply sediment to the surrounding plains throughout the late Permian. Locally, the early Permian sand seas survived into the late Permian and there are indications that sandy desert conditions were temporarily re-established in several areas when, through periodic shrinkage and desiccation of the northern seas, marine influences waned. Thus extensive deposits of wind-blown and water-lain (Smith 1992) silt and fine sand were formed in NW England after the sea and lakes there finally dried up: similar deposits interdigitate with younger parts of the Zechstein succession in NE England.

With the ending of marine influences as the Zechstein Sea too dwindled and dried up, and with a continuing decline in the diversifying influence of uplands, the whole of England and Wales once again became dominated by desert conditions in the latest Permian. Under the influence of a generally high water table resulting from the low relief, the desolate and virtually lifeless sedimentary plains of the earlier late Permian extended into the former marine basins and vast spreads of uniform mainly water-laid clays, silts, and fine-grained sands were formed; ephemeral saline lakes in which gypsum and perhaps halite were precipitated lay only in the most rapidly subsiding depocentres, and these too eventually disappeared as fluvial and sometimes aeolian influxes of sand (destined to become the Sherwood Sandstone Group of probable Triassic age) were swept diachronously into the basins.

Late Permian palaeontology

Marine invertebrate fossils and the remains of land plants are present in the upper Permian carbonates of both NW (Bakevellia) and NE (Zechstein) England. Fossil fish and reptiles are recorded in the lower parts of the Zechstein succession. Most of these fossils were described by King (1850) in a monograph that remains the standard work for several fossil groups; discoveries by later workers, especially the work of Hollingworth, have extended the list of species identified. This summary is based on that provided by Smith (1992). Lack of space precludes full discussion of these biotas but it is important to note the close relationship between the facies and faunas of the northern seas and that a full and varied marine macrofauna is found only in marginal carbonate rocks of the lower beds of the English Zechstein; here brachiopods (especially productids such as *Horridonia*), gastropods, and bivalves are the most abundant macrofossils, but bryozoans (especially pinnate forms such as *Fenestella* and ramose forms such as *Acanthocladia*) are abundant in shelf and shelf-edge reefs where crinoids are also present (Hollingworth & Pettigrew 1988). Equivalent inshore Zechstein strata, and rocks formed in the more restricted parts of the Bakevellia Sea, have a limited molluscan fauna of which the bivalve *Bakevellia binneyi* (Brown) is the most widespread form and in places occurs alone (Pattison 1970, p. 160). Rocks of the Roker Dolomite–Concretionary Limestone and above contain only a limited range of indigenous gastropods and bivalves and one species of brachiopod, whilst those of the Seaham Formation and above contain only a few species of gastropods and bivalves.

The limited biostratigraphical evidence of the age of the Appleby Group (NW England, especially Cumbria) has been reviewed by Warrington *in* Holliday *et al.* (2004). Vertebrate tracks from near Penrith and in SW Scotland (Hickling 1909; Delair & Sarjeant 1985) have been assigned to *Chelichnus*, an ichnogenus known also from the Cornberger Sandstone at the top of the Rotliegendes succession in Hessen, Germany (McKeever & Haubold 1996). This formation is placed above the Illawara magnetic reversal (near the top of the Wordian Stage of the Guadalupian (Middle Permian) (Menning 1995), suggesting that the main part of the Appleby Group may, like the succeeding Cumbrian Coast Group (Holliday *et al.* 2001), be late Permian in age. Those parts of the Appleby Group that are lateral equivalents of the Cumbrian Coast Group, can be directly inferred to be late Permian in age. This view is supported by miospore assemblages recovered by Warrington (2002) from Appleby Group strata in the Vale of Eden interfingering with the Eden Shales Formation (Cumbrian Coast Group). Locally, as in west Cumbria, where beds of the Appleby Group are lateral equivalents of the lower part of the Sherwood Sandstone Group, they are of Early Triassic age. The limited marine faunas of the St Bees Evaporites in the lower part of the Cumbrian Coast Group are of little chronstratigraphical significance, other than to indicate broad correlation within the region and with the Permian successions east of the Pennines. The presence of such bivalves as *Bakevellia binneyi*, *Permophorus costatus* and *Schizodus obscurus* in the St Bees Evaporites suggests correlation with the Z1 and/or Z2 cycles of the southern North Sea Basin (e.g. Pattison 1970; Pattison *et al.* 1973; Arthurton *et al.* 1978; Jackson & Johnson 1996). A different fauna, including *Liebea squammosa* and *Calcinema permiana* in the Belah Dolomite (Eden Shales) of the Vale of Eden Basin, is taken to indicate correlation of this bed with the middle formations of the North Sea Zechstein carbonates. From this evidence, the lower part of the Cumbrian Coast Group is, therefore, likely to be of the same age as the Zechstein rocks, i.e. Wuchiapingian (e.g. Henderson & Mei 2000). This is supported by the palynological assemblages that have been obtained in the lower part of the group at several localities. These microfloras, in which the bisaccate pollen *Lueckisporites virkkiae* is prominent, have been ascribed a Late Permian, Wuchiapingian or possibly Changhsingian, age by Warrington (in Holliday *et al.* 2001 and in Holliday *et al.* 2004). They indicate that a significant vegetation cover existed locally: a supposition supported by the plant fossils of the calcareous mudstones ('Hampbole Beds') of Yorkshire.

The higher parts of the Cumbrian Coast Group, above the St Bees Evaporite Formation and its equivalents, and the equivalents of the Belah Dolomite, have proved barren. However, comparison of geophysical log traces suggests that correlatives of the Bröckelschiefer Member of the southern North Sea Basin, the base of which has been commonly regarded as marking the base of the Triassic (Geiger & Hopping 1968; Rhys 1974; Warrington *et al.* 1980), can be recognized in the Vale of Eden (*c.* 35 m above the Belah Dolomite), Carlisle and East Irish Sea basins (Holliday 1993; Jackson & Johnson 1996). This implies that the Bröckelschiefer equivalents, and higher parts of the Cumbrian Coast Group, are Early Triassic in age, an interpretation supported by stratigraphic studies from Germany.

NW England and the eastern Irish Sea

Sedimentation in the Irish Sea–NW England region was renewed in the Permian as a result of broadly E–W-directed extension which reactivated basement fractures and initiated a series of rift-basins, locally with associated volcanic activity (Chadwick & Evans 1995; Jackson *et al.* 1995, 1997). The syndepositional normal faults controlling these basins have a broadly N–S orientation, ranging between NE and NW, probably inherited from the Caledonide transfer direction in underlying Lower Palaeozoic rocks. The ENE-trending Permo-Triassic Peel, Solway Firth and Carlisle basins contrast with the above basins in being aligned broadly with the main Caledonide structural features. These basins overly, and are on the same trend as, the Lower Palaeozoic–Early Devonian Iapetus Convergence Zone and the Carboniferous Northumberland–Solway Basin. The influence of the Caledonide trend is also evident in the ENE orientation of the Ramsey–Whitehaven Ridge that separates the East Irish Sea and Solway Firth basins (Fig. 12.3). Permian rocks were probably more widespread than their present-day limits, as suggested by the presence of scattered small outliers, e.g. around Clitheroe and Ingleton, and local reddening of Carboniferous rocks.

In most parts of NW England and adjacent areas, the Permian strata have been divided into two principal groups, the Appleby and Cumbrian Coast groups, which comprise strata formerly classified as 'lower Permian' and 'upper Permian', respectively (Jackson & Johnson 1996) (Fig. 12.8). The boundary between these groups is commonly sharp, perhaps indicating a non-sequence (Jackson & Johnson 1996). However, at some basin margins, where the Cumbrian Coast Group passes into sandstones and breccias of Appleby Group facies, the boundary is diachronous and less distinct (e.g. Jackson *et al.* 1995; Akhurst *et al.* 1997). Locally (e.g. Vale of Clwyd and the southern Cheshire Basin), correlatives of both groups are included, as the Kinnerton Sandstone Formation, in the Sherwood Sandstone Group. Over much of the region the Appleby Group is thin or absent. It is overlapped by the Cumbrian Coast Group, which was deposited over a much wider area. Prominent structural highs, such as North Wales, the Pennines, the Lake District, the Isle of Man and the Southern Uplands of Scotland, many of which are cored by Caledonian granitic intrusions, were emergent features during deposition of the Appleby Group, but by the end of the Permian they had been reduced to areas of very low relief or had become sites of deposition (Smith & Taylor 1992; Jackson *et al.* 1995).

The Appleby Group

The principal component of the Appleby Group is sandstone. Conglomerate and breccia are locally common additional lithologies at the base of the group and, in some areas at basin margins, comprise much or all of the group. Argillaceous rocks are rare to absent. Basaltic volcanic rocks and/or contemporaneously reworked volcanic clasts are of local importance. The sandstones of the group are generally red in colour, and fine–coarse grained. Typically, they are well-sorted or bimodal,

<table>
<tr><th colspan="2" rowspan="2">WEST LANCASHIRE AND CHESHIRE BASIN</th><th colspan="4">EAST IRISH SEA AND SOLWAY FIRTH BASINS</th><th colspan="2" rowspan="2">CARLISLE and VALE OF EDEN BASINS</th><th rowspan="2">GROUP</th></tr>
<tr><th colspan="2">OFFSHORE</th><th colspan="2">ONSHORE West Cumbria</th></tr>
<tr><td rowspan="3">Kinnerton Sst</td><td rowspan="2">Manchester Marls Formation</td><td>Barrowmouth Mudstone Formation</td><td>BM3 BM2 BM1</td><td>St Bees Shale Formation</td><td rowspan="2">Brockram</td><td rowspan="2">Eden Shales Formation</td><td rowspan="2">ES3 ES2 ES1</td><td rowspan="2">CUMBRIAN COAST GROUP</td></tr>
<tr><td colspan="2">St Bees Evaporites Formation</td><td>St Bees Evaporite Formation</td></tr>
<tr><td>Collyhurst Sandstone Formation</td><td></td><td>Collyhurst Sandstone Formation</td><td colspan="2">Brockram</td><td>Penrith Sandstone Formation and Brockram</td><td></td><td>APPLEBY GROUP</td></tr>
</table>

Fig. 12.8. Lithostratigraphical subdivision of Permian rocks in NW England and adjacent offshore areas: data sources include: Arthurton *et al.* (1978); Jackson *et al.* (1995); Jackson & Johnson (1996); Akhurst *et al.* (1997); Holliday *et al.* (2001).

devoid of mica and contain well-rounded grains in large (metre scale) dune sets thought to be of aeolian origin. Foreset orientation in such deposits suggests that palaeowind directions were generally directed from the east (Waugh 1970; Smith 1972; Brookfield 1978, 1980; Arthurton & Wadge 1981; Jackson & Johnson 1996). Locally, these rocks contain irregularly bedded sandstones with fine-grained angular debris mixed, or interlayed with, coarser rounded 'millet seed' grains. These are thought to be damp interdune deposits. Thin beds of parallel-bedded, well-sorted sandstone are probably dry interdune deposits (Jones *et al.* 1999; Brookfield 2000). Sandstones, evenly-bedded or in moderate sized cosets, mainly with poorly sorted angular grains, but commonly also containing medium- to coarse-grained 'millet seed' grains, are inferred to be of fluvial (wadi) origin, the rounded grains indicative of the reworking of nearby aeolian deposits (Burgess & Holliday 1979). Rocks of fluvial origin are commonly associated with breccia and conglomerate deposits, particularly near the base of the group, but also locally near the top (Burgess & Holliday 1979; Jackson *et al.* 1995). In some areas, fluvial strata are more extensively developed throughout the group. The rudaceous rocks are mainly matrix-, but locally clast-supported and contain only locally-derived hard, resistant lithologies, e.g. Carboniferous Limestone, granite, Lower Palaeozoic greywacke and volcanic rocks (see below). For the most part, they are interpreted as alluvial fan deposits (Jackson *et al.* 1995; Akhurst *et al.* 1997), which formed at basin margins and adjacent to localized areas of higher relief and active fault-scarps. In some areas, where the Appleby Group comprises thin (<1 m thick) breccias, an origin as *remanié* deposits on a bare desert surface seems likely.

The rocks of the Appleby Group appear to have formed in a series of isolated intermontane basins, some of which may have been below sea level (Smith 1972; Brookfield 1980; Smith & Taylor 1992; Jackson *et al.* 1995; Jackson & Johnson 1996). Some thin localized occurrences of the Appleby Group may have resulted from the erosion of local areas of relief and infill of low hollows on the sub-Permian floor. Where Carboniferous Limestone subcrops the Appleby Group, these depressions may have been of karstic origin (Chadwick *et al.* 2001). The Appleby Group is characterized by abrupt thickness changes; the thickness is commonly less than 10 m, but is locally up to 1150 m (Jackson *et al.* 1995; Jackson & Johnson 1996; Chadwick *et al.* 2001). Overall, the pattern of thickness changes and distribution of fluvial and rudaceous rocks, suggest that the deposition of the Appleby Group was initiated and controlled by syndepositional normal faulting (e.g. Figs 12.9 & 12.10). The idealized model of the facies relationships within the Appleby Group of

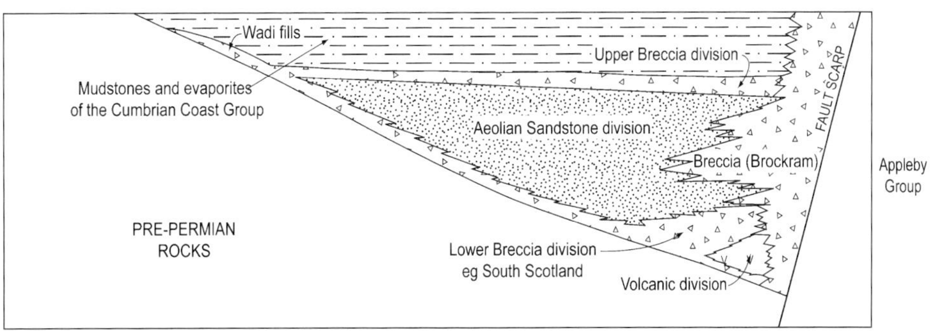

Fig. 12.9. Schematic cross-section of a typical half-graben in the East Irish Sea Basin, showing facies relationships in the Appleby Group. (Modified from Jackson *et al.* 1995, fig. 39.)

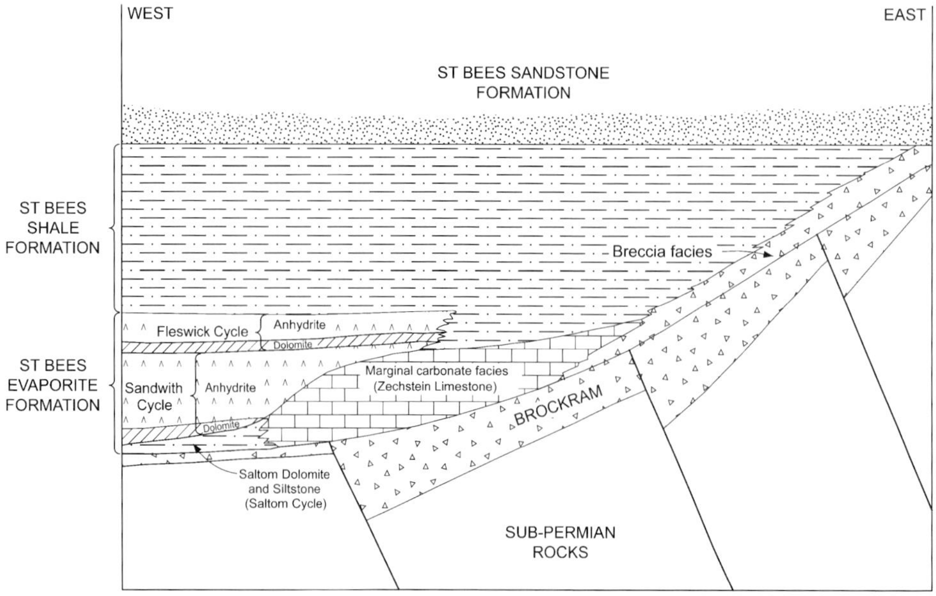

Fig. 12.10. Stratigraphical relationships in the Appleby and Cumbrian Coast groups of west Cumbria (not to scale). (Modified from Akhurst *et al.* 1997, fig. 31.)

the eastern Irish Sea and adjacent areas proposed by Jackson *et al.* (1995; see also Jackson & Johnson 1996) is widely, although not everywhere applicable (Fig. 12.9).

Cheshire Basin
The Appleby Group is represented in the north of the Cheshire Basin by the Collyhurst Sandstone Formation (Warrington *et al.* 1999), only locally being overlapped by the Manchester Marls Formation (Cumbrian Coast Group). However, towards the south of the basin, the latter group passes into sandstone, and the Collyhurst Sandstone, the Manchester Marls equivalent and overlying sandstones form a single sandstone unit, the Kinnerton Sandstone Formation (Fig. 12.8). Detailed information on the Appleby Group is restricted in much of the basin, because, away from outcrop, it has been proved by only a few deep boreholes in central and southern areas.

The fine- to coarse-grained sandstones of the Collyhurst Sandstone, and equivalent parts of the Kinnerton Sandstone, have been interpreted as being dominantly of easterly-derived aeolian facies. The presence of transverse dunes with barchanoid crests or, locally, linear seif dunes has been suggested by Thompson (1985). Interdune deposits have also been recognized (Jones *et al.* 1999). Some local occurrences of breccia, containing nearby-derived detritus, have been recorded towards the base of the sandstone at outcrop, but more generally rudaceous rocks appear to be rare in the basin (Tonks *et al.* 1931; Taylor *et al.* 1963; Warrington *et al.* 1999). The Collyhurst Sandstone thickens southwards to a maximum of approximately 550–600 m in the hanging wall of the NE-trending Wem–Bridgemere–Red Rock Fault System. This fault system forms the eastern and southern margins of the Cheshire Basin and was the dominating influence during Permian deposition (Evans *et al.* 1993; Chadwick *et al.* 1999; Warrington *et al.* 1999). The Collyhurst Sandstone is absent from the footwall of the fault system. Elsewhere in the basin, borehole and field evidence of thickness changes in the formation have been taken as indicative of intra-basin syndepositional faulting (Chadwick *et al.* 1999; Warrington *et al.* 1999). The lack of recorded rudaceous rocks may result from the limited amount of subsurface data in many areas, but may also indicate that the relief of the surrounding hinterland was generally low and that contemporary faulting did not generate major fault-scarps.

The East Irish Sea Basin
The Appleby Group is locally well developed in the various graben, half-graben and intra-basin highs that together make up the East Irish Sea Basin, including the adjacent onshore in North Wales, west Lancashire, south and west Cumbria (Jackson *et al.* 1995). Much of the information on the group in this basin has been inferred from seismic reflection data, supplemented by the results of hydrocarbon exploration boreholes and from scattered onshore outcrops. Most of the rocks encountered have been referred to the Collyhurst Sandstone Formation (Jackson & Johnson 1996) and are dominantly of aeolian origin. The first offshore exploration wells were thought to have proved a thick argillaceous playa-lake facies at this stratigraphical level (Colter & Barr 1975; Colter 1978), but these strata are now known to be reddened Carboniferous rocks (Jackson *et al.* 1995). Breccia and conglomeratic deposits occur in several areas, particularly in Cumbria where they are informally classified as Brockram (Fig. 12.8). Basic volcanic rocks, e.g. Tormentil Volcanics, are locally present or occur as probable penecontemporaneous clasts in breccias near the base of the group (Rose & Dunham 1977; Penn *et al.* 1983; Jackson & Johnson 1996). In some southerly areas, as in the Cheshire Basin, notably in North Wales around Rhyl and in the Vale of Clwyd, the Appleby and Cumbrian Coast groups pass into a sandstone facies and cannot be separated from the Kinnerton Sandstone Formation of the Sherwood Sandstone Group (Jackson *et al.* 1995).

Over much of the central part of the basin, thin aeolian sandstone (0–10 m thick), locally accompanied by breccia and water-lain sandstone, accumulated in small depressions. These hollows were separated by broad rocky surfaces covered by a thin layer of *remanié* stones coated in desert varnish (Chadwick *et al.* 2001). Thicker deposits of aeolian sandstone, commonly with an underlying breccia and fluvial sandstone facies, occur in the south of the basin, and also in the north. The southerly occurrences are strongly influenced by N–S syndepositional faulting cutting across a large ENE-trending palaeodepression, coincident with a Variscan syncline in underlying Carboniferous rocks, extending from east of Anglesey to Formby in west Lancashire (Jackson *et al.* 1995, fig. 36). The thickness of the Collyhurst Sandstone in this area is locally highly variable, due to the influence of syndepositional faulting, and is said to range up to 1150 m (Jackson & Johnson 1996). In the northern part of the East Irish Sea Basin, probably in a half-graben adjacent to the Ramsey–Whitehaven Ridge, there are similar local fault-controlled changes of thickness, but overall the succession is thinner than in the south (locally less than *c.* 700 m) (Chadwick *et al.* 2001).

Rather different, rudaceous strata of the Appleby Group are encountered at outcrop, at the NE margin of the basin and the more southerly parts of west Cumbria. The review by Akhurst *et al.* (1997, table 6) of the breccias and conglomerates making up the Brockram of west Cumbria is probably applicable to the other areas. In many basin margin areas, the Brockram facies does not exceed a few metres in thickness. Locally, the thickness of Brockram is greater, less than 26 m in south Cumbria (Rose & Dunham 1977) and approximately 150 m in west Cumbria (Akhurst *et al.* 1997). From the detailed study of outcrops, and of cored and geophysically logged boreholes in west Cumbria, it has been possible to show that the Brockram not only underlies the Cumbrian Coast Group but also towards the east is its lateral equivalent (Trotter *et al.* 1937; Akhurst *et al.* 1997, figs 31 & 32). This was probably also the case in other marginal areas (Jackson *et al.* 1995, fig. 39). These rudaceous rocks were derived from the erosion of nearby areas of higher relief (e.g. Isle of Man, Lake District) surrounding the basin, but sharp, localized thickness changes suggest that their distribution was also influenced by syndepositional faulting.

Ramsey–Whitehaven Ridge, Solway Firth, Carlisle and adjacent basins in southern Scotland

In the offshore area to the north of the East Irish Sea Basin, there is only limited information on the Appleby Group. This dearth of data is partially offset by information from onshore outcrops and boreholes. From the scattered well data, it appears that the group is thin or was not deposited over much of the offshore area, though, because of faulting in several of the boreholes, its absence cannot always be proved (Chadwick *et al.* 2001). The Ramsey–Whitehaven Ridge and Solway Firth and Carlisle basins have a marked ENE (Caledonian) orientation, and, although similarly orientated faults may have been locally of influence, onshore evidence points to the greater importance of broadly N–S-trending (NNE–NNW) syndepositional faults (Holliday *et al.* 2004). This latter trend is particularly marked in the adjacent basins of SW Scotland (e.g. Stranraer, Dumfries, Lochmaben).

The group has not been proved in offshore boreholes on the Ramsey–Whitehaven Ridge, although Brockram facies is present on the Isle of Man, less than 10 m thick (Chadwick *et al.* 2001), and in northernmost west Cumbria, less than 3 m thick (Arthurton & Hemingway 1972; Akhurst *et al.* 1997). Some of the breccias on the Isle of Man may be lateral equivalents of the Cumbrian Coast Group. It is probable that the ridge was mainly emergent during the deposition of the Appleby Group and, locally, all or part of the Cumbrian Coast Group, with breccia deposition restricted to local hollows and the ridge-flanks.

Jackson *et al.* (1995, figs 36 & 38) suggested, largely by comparison with adjacent onshore basins, that the Appleby Group (Collyhurst Sandstone) in the Solway Firth Basin contained 'thick sandstone', from 50 to more than 200 m thick, probably mainly of aeolian origin. This has not been confirmed by drilling, and the relevant interval is cut out by faulting in the main deep well (112/15-1) of the basin (Chadwick *et al.* 2001, fig. 21).

Along strike, the Appleby Group does not crop out at the margins of the Carlisle Basin, the onshore continuation of the Solway Firth Basin, except for a few local pockets of Brockram and aeolian sandstones, which are less than about 10 m thick. However, seismic reflection data, confirmed by the Silloth 1-A Borehole, show that the Penrith Sandstone Formation, up to about 500 m thick, is locally present in the western part of the basin, probably extending into the western Solway Firth Basin (Holliday *et al.* 2004). Thickness changes over NNW-trending faults are indicative of contemporary faulting. The formation was not contiguous with Appleby Group strata in the Vale of Eden and Lochmaben (SW Scotland) basins, although it may formerly have been contiguous with sediments in the Dumfries Basin. Much of the Penrith Sandstone appears to be of aeolian origin but there is a substantial thickness (*c.* 130 m) of breccia and water-laid sandstone facies at the base (Holliday *et al.* 2004). Indications of bedded anhydrite in the Penrith Sandstone of the Silloth 1-A Borehole, if confirmed, would suggest that continental inter-dune sabkha or saline lake conditions had been locally established.

Although outside the scope of this volume, the adjacent basins of SW Scotland are of some interest and relevance. The location of breccia facies in these basins provides good evidence of NW-trending syndepositional faulting, and the distribution of water-laid sandstones is indicative of a significant sediment source in the Southern Uplands (Brookfield 1980, 2000; Jackson *et al.* 1995; Chadwick *et al.* 2001).

Vale of Eden Basin

The Vale of Eden Basin, between the Lake District and Alston Blocks, occurs in the hangingwall of the Pennine Fault and shares its marked NNW-trend. The Appleby Group, here in its eponymous area, ranges up to at least 300 m in thickness, and comprises the Brockram and the Penrith Sandstone Formation (Fig. 12.11; Plate 21). The latter formation, particularly in the north of the basin, is made up of red aeolian sandstone, fine–coarse grained, with prominent rounded 'millet-seed' quartz grains (Burgess & Holliday 1979; Arthurton & Wadge 1981). At surface, the sandstones are generally soft and poorly cemented, although some boreholes suggest that, at depth, anhydrite is a common and pervasive cement (Turner *et al.* 1995). Locally, the Penrith Sandstone is characterized by the formation of secondary silica overgrowths on the quartz grains, infilling much of the original porosity (Waugh 1970). Where this alteration has taken place, the sandstones, being harder than the more sparsely silicified beds, form areas of relatively higher relief and have been extensively quarried for building stone, for which abundant evidence can be found in the towns of Carlisle (sometimes also Dumfries Stone), Penrith, Appleby and Kirkby Lonsdale. Sandstones, of inferred fluvial origin, commonly associated with alluvial fan breccio-conglomerates (brockram), occur towards the base of the formation and at basin margins. More particularly, towards the southern end of the basin, there is a more general lateral facies change from aeolian strata into water-laid, evenly-bedded sandstones with associated brockram. The higher part of the Appleby Group, in the south, is inferred to be the lateral equivalent of the lower part of the Cumbrian Coast Group (Fig. 12.8) (Burgess & Holliday 1979, figs 42 & 43).

Most palaeogeographical reconstructions of the basin assume that the Lake District and Pennine blocks were emergent during at least part of Appleby Group deposition (Burgess & Holliday 1979; Arthurton & Wadge 1981; Smith & Taylor 1992). This is supported by field and provenance evidence, but, as noted above, the most actively eroding sediment source area was to the south of the basin. The basin also terminated towards the northern end of the present-day outcrop, where the Appleby Group is overlapped by the Cumbrian Coast Group (Trotter & Hollingworth 1932; Holliday *et al.* 2004). The role of the Pennine Fault at this time is unclear. Burgess & Holliday (1979) and Arthurton & Wadge (1981) found little or no definite field evidence to indicate major syndepositional fault movements. More recently, comparison with the contemporary basins in adjacent areas described above, has suggested that the Vale of Eden Basin was probably also fault-controlled, with the Pennine Fault being a major influence on basin origin and evolution at this time. It is unfortunate that there are no seismic reflection data to confirm or reject this view.

(A)

(B)

Fig. 12.11. The Permian in Cumbria. A = the Brockram exposed 2 km south of Appleby (this is a dedicated climbing wall, leased by the Climbing Club of Great Britain). B = the Penrith Sandstone at Appleby.

The Cumbrian Coast Group

The rocks of the Cumbrian Coast Group are lithologically more varied, were laid down in a greater range of palaeoenvironments and are aerially more widespread than those of the underlying Appleby Group. The group includes a variety of wind-and water-lain siliciclastic lithologies, ranging from mudstone to breccio-conglomerate. There are three main facies types; an irregularly laminated facies, a regularly laminated facies, and a claystone–siltstone facies (Burgess & Holliday 1974; Akhurst *et al.* 1997; Holliday *et al.* 2001). Interbedded with them, towards still emergent areas such as the Lake District Block and the Isle of Man, are beds of matrix- and clast-supported breccia (described above). Evaporite rocks, locally associated with dolomite, are also widely present, particularly in the lower half of the group. Late diagenetic, displacive fibrous gypsum veins are common near outcrop and in shallow boreholes. As indicated by the names of some of its component formations, e.g. Eden Shales, St Bees Shales, Manchester Marls and Barrowmouth Mudstone, the Cumbrian Coast Group is finer grained overall than the Appleby Group, although sandstones and sandy beds are important components of all these formations. The Cumbrian Coast Group shows significant lateral facies changes but, unlike the Appleby Group, is less subject to abrupt thickness changes. It is commonly 100–150 m thick, but ranges up to about 550 m offshore in areas of marked halokinesis (Jackson *et al.* 1995; Jackson & Johnson 1996; Chadwick *et al.* 2001). As noted above, the boundary with the Appleby Group is commonly sharp, but it is diachronous and less well-defined in proximal areas. From its geophysical log character, Jackson & Johnson (1996) and Chadwick *et al.* (2001) suggested that the boundary with the overlying Triassic Sherwood Sandstone Group is sharp and erosional in offshore areas. However, numerous natural exposures and cored boreholes show that onshore, at least, this boundary is gradational (Holliday 1993; Akhurst *et al.* 1997; Holliday *et al.* 2001).

The relative thickness of evaporite deposits and the nature and proportion of associated lithologies is one of the principal characters defining the lithostratigraphical subdivision of the group. In areas proximal to siliciclastic sediment supply, the evaporite beds (gypsum–anhydrite) are relatively thin and separated by significant thicknesses of sandstone, siltstone and mudstone. In such situations, only one formation, e.g. Eden Shales or Manchester Marls, is recognized, with, in some areas, each of the gypsum–anhydrite strata being given named formal bed status. In distal areas, where the lower part of the formation largely comprises dolomite and anhydrite or anhydrite and halite, with minor mudstone, the group is divided into two formations. The lower, dominantly carbonate–evaporite unit is everywhere called the St Bees Evaporite (or Evaporites) Formation, and the upper siliciclastic unit either the St Bees Shale Formation or Barrowmouth Mudstone Formation (Fig. 12.13) (Arthurton & Hemingway 1972; Jackson & Johnson 1996; Akhurst *et al.* 1997).

The irregularly laminated siliciclastic facies, characterized by the presence of adhesion ripples, consists of beds of siltstone and very fine-grained sandstone 0.5–5 m thick. These beds have a characteristic irregular wavy lamination due to the presence of contorted, convoluted lenses of fine-to medium-grained (locally coarse-grained) sandstone, 1–2 cm thick and 3–6 cm in length, that typically contain granule-sized lithoclasts. The surrounding sediment is more fine-grained; an admixture of laminated siltstone and very fine-grained sandstone. The facies is interpreted as the product of sediment deposition from a combination of waterlain and windblown processes on an irregular surface that was probably covered by an irregular efflorescent salt crust formed by evaporation of saline groundwater (Burgess & Holliday 1974; Akhurst *et al.* 1997). During periods of sheetflooding, lenses of rippled sand infilled the hollows within the crusts; subsequent dissolution of the salt crusts led to the deformation and convolution of the sand lenses. Wind-blown silt adhered to the wet sediment surface to form the thin laminae that coated the salt crust and sand lenses.

The regularly laminated siliciclastic facies comprises of reddish brown, laminated siltstone and very fine-grained sandstone with subordinate claystone. These lithologies commonly form upward-fining units, some of which have granules or coarse-grained sandstone at the base. Cross-lamination and climbing ripple cross-lamination occur within sandstone and siltstone; load casts, convolute lamination, desiccation cracks and dewatering structures may also be present. Claystone, locally with desiccation cracks, occurs as thin–thick laminae. The laminated nature of this facies indicates deposition from unconfined sheetfloods. Ripple cross-stratification demonstrates that tractional processes were dominant; high rates of fallout from suspension accompanying the ripple migration led to the formation of climbing ripples. The sheetfloods were either deposited on a subaerial mudflat or in a lake, with some ponding of water subsequent to flooding allowing deposition of mud from suspension. The presence of desiccation cracks indicates that these water bodies were ephemeral features.

The claystone–siltstone facies comprises reddish brown, structureless, interbedded claystone and siltstone. Sandstone lenses may be locally present. Characteristically, sedimentary structures are uncommon but desiccation cracks have been noted. Carbonate and gypsum/anhydrite nodules are common in places, and siliceous nodules (probably replaced evaporites) occur at some levels. Accumulation was probably by accretion of fine-grained sediment supplied as windblown dust and by settling from suspension after periodic sheetfloods. The massive nature of the facies is probably a function of complete disruption of the sediment by the combined effects of desiccation and the interstitial growth and dissolution of evaporite minerals.

The evaporites are best known from mine workings and cored boreholes in Cumbria, where beds of anhydrite, and their secondary gypsum replacements, show a wide range of sedimentary features, e.g. nodular, laminated, 'varved', and provide evidence of complex diagenetic histories (Arthurton 1971; Arthurton & Hemingway 1972; Burgess & Holliday 1974). Evaporite lithologies in offshore areas are less well known and have been inferred principally from cuttings and geophysical log traces. Anhydrite rocks appear similar to those onshore. Beds of halite, locally in excess of 100 m thick, are characteristic of much of the offshore area. These typically are coarse-grained and pure, although minor amounts of clay, carbonate and anhydrite are present interstitially and as thin interbeds. Where such impurities are more abundant, they commonly enclose euhedral crystals of displacive halite (haselgebirge). Polyhalite and other potash minerals occur rarely with the halite (Jackson & Johnson 1996). The evaporites are thought to have been deposited in shallow seas and saline lakes, and in surrounding coastal and continental sabkhas. Near to outcrop, and in other areas where they have been in contact with groundwater during their diagenetic history, evaporite strata have been extensively removed from the section, commonly leading to the brecciation of overlying strata. In offshore areas, they have also been subjected locally to extensive halokinetic disruption.

Limestones and dolomites form a major part of the succession in south and west Cumbria, but elsewhere they are of much less significance. The complex diagenetic history of these commonly masks the details of original depositional fabrics and mineralogy. They contain a limited, but locally abundant fauna of bivalves, suggestive of marine deposition. The local presence of stromatolites, birdseye structure, faecal pellets and ooids suggests that deposition was dominantly in shallow water.

The lower part of the Cumbrian Coast Group (St Bees Evaporite Formation) contains the sediments deposited as a result of three major marine transgressions and subsequent retreats. At times of relatively high sea level, marine deposits were laid down principally in the Solway Firth and East Irish Sea basins. Carbonate platforms developed in many marginal marine areas, e.g. west and south Cumbria, Ramsey–Whitehaven Ridge, passing into siltstones in the more rapidly subsiding deeper basinal areas (Arthurton *et al.* 1978). Extensive evaporitic sabkhas (gypsum–anhydrite) surrounded the marine areas with siliciclastic sabkhas and alluvial fans at basin margins adjacent to, or over, the main structural highs. At times of marine retreat, evaporitic sabkhas (halite and anhydrite) extended over much of the basin centres, passing into siliciclastic sabkhas and alluvial fans at basin margins (Jackson *et al.* 1995; Jackson & Johnson 1996).

In the upper part of the group, there was little or only local direct marine influence on sedimentation. Siliciclastic sabkhas, with both water-and wind-laid strata, spread across much of the region, and alluvial fan deposition continued near the basin margins. The only clear evidence of marine transgression into the area is provided by the Belah Dolomite and the D-Bed Gypsum–anhydrite, which are found only in the Vale of Eden (Figs 12.12 & 12.13), and are probably indicative of a relatively short-lived cross-Pennine marine connection through the area of the present-day Stainmore Gap. In the highest beds of the group, wind-laid deposits become increasing rare; sandstones become more abundant and, in onshore areas at least, there is a gradual passage into the overlying fluvial deposits of the Sherwood Sandstone Group (Burgess & Holliday 1974; Arthurton *et al.* 1978; Akhurst *et al.* 1997; Holliday *et al.* 2001).

The palaeogeography of the more southerly parts of the region during the deposition of the group differs significantly from the above pattern. In the south of the Cheshire Basin, aeolian dune sands (Kinnerton Sandstone) were deposited (Warrington *et al.* 1999). In the upper part of the group, this facies spread northwards into the remainder of the basin and, as the Freshfield Sandstone Member, into the southern and SE parts of the East Irish Sea Basin (Jackson & Johnson 1996).

Cheshire Basin

The group is represented in the northern part of the Cheshire Basin by the Manchester Marls Formation, which locally overlaps the underlying Appleby Group. This formation comprises red, locally green, mudstone, siltstone and sandstone. In the lower part there are lenses and beds of calcareous mudstone, dolomitic and iron-stained limestones. These contain a microflora, shelly macrofauna and calcareous microfauna (Pattison 1970; Pattison *et al.* 1973; Warrington *et al.* 1999), and are the correlatives of the St Bees Evaporites of the East Irish Sea Basin (Fig. 12.9). Also present in these lower beds are plant remains transported from the adjacent hinterland. Alhough the lower beds are commonly gypsiferous, beds of gypsum–anhydrite do not seem to be present. The formation becomes increasingly arenaceous upwards and passes into the Kinnerton Sandstone of the Sherwood Sandstone Group. The maximum thickness of the formation is approximately 150 m in the Knutsford Borehole, but it is thinner (*c.* 40–75 m) towards the basin margin (coincident with present-day outcrop) around Manchester.

Largely from borehole evidence, the Manchester Marls Formation is believed to become more arenaceous towards the south of the basin, where its correlatives are distinguished with difficulty from the underlying Collyhurst and overlying Kinnerton sandstones. Probable arenaceous and silty equivalents of the Manchester marls, up to 70 m thick, have been identified in boreholes in the south of the basin, and named the Bold Formation (Evans *et al.* 1993; Warrington *et al.* 1999). These strata contain both aeolian and fluvial sheetflood deposits.

There is insufficient evidence to indicate whether the deposition of the Manchester Marls and its equivalents was influenced by syndepositional faulting, although their correlatives appear to be absent south and SE of the Wem–Bridgemere–Red Rock Fault System. The lack of recorded rudaceous rocks, apart from a thin (>1 m) breccia at or near the base of the Manchester Marls in the north of the basin, said to be derived from the Midlands (Taylor *et al.* 1963), may indicate that the relief of the surrounding hinterland was generally low and that any contemporary faulting did not generate major fault-scarps.

The East Irish Sea Basin

The Cumbrian Coast Group is concealed almost everywhere in the East Irish Sea Basin, except in a few marginal areas. However, principally because of the search for oil and gas, both onshore and offshore, the group has been penetrated by numerous boreholes and its distribution can, at least in part, be

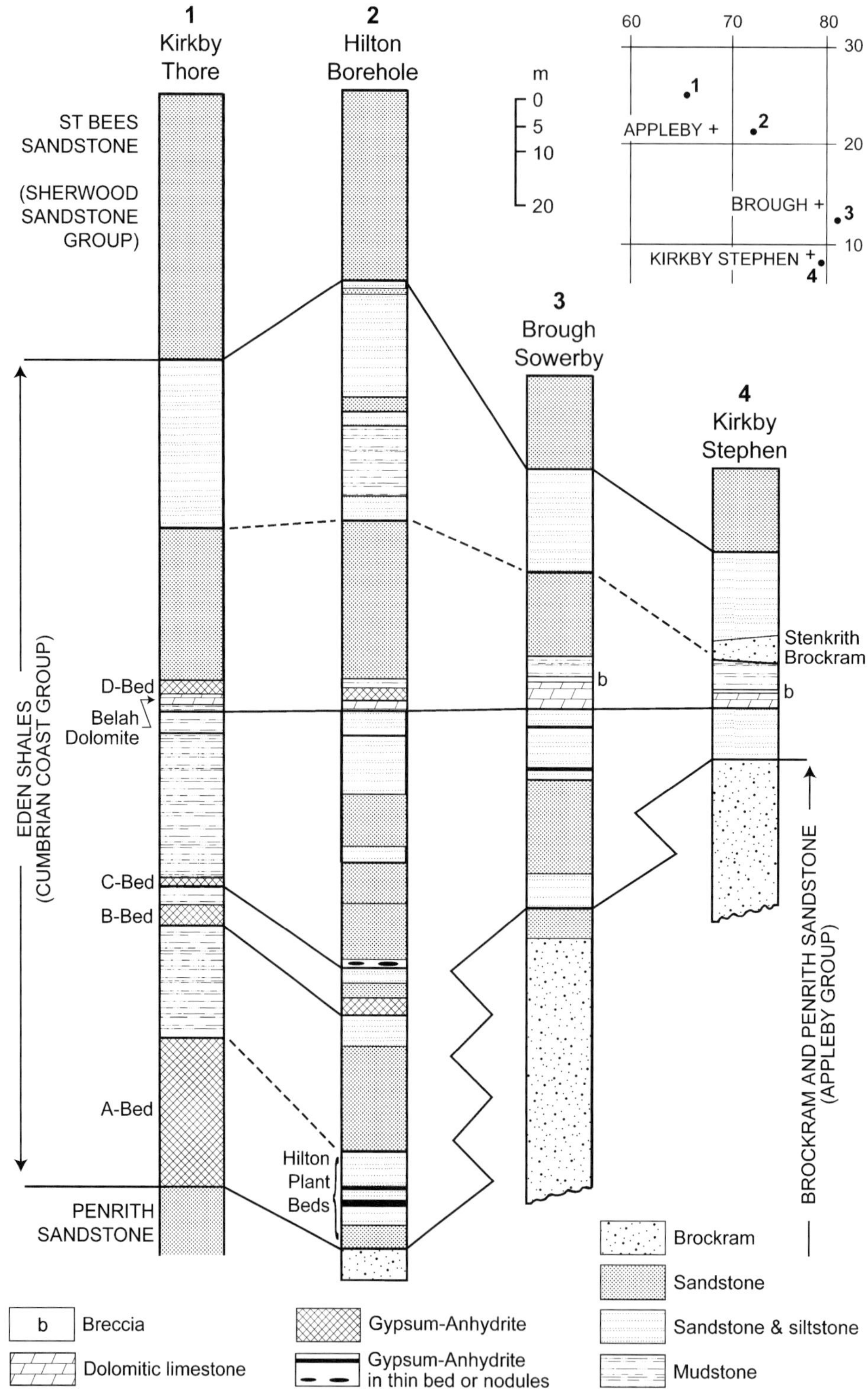

Fig. 12.12. Comparative vertical sections showing the lateral passage between the Eden Shales (Cumbrian Coast Group) and the Appleby Group in the southern part of the Vale of Eden Basin; modified from Burgess & Holliday (1979, fig. 42).

inferred from an extensive grid of seismic reflection data. The development of the group within the basin has been reviewed in some detail above (see also Jackson *et al.* 1995; Jackson & Johnson 1996; Akhurst *et al.* 1997; Chadwick *et al.* 2001) and is not considered again here.

The Manchester Marls Formation occurs over much of the southern part of the East Irish Sea Basin, and also at its eastern and western margins. The formation is lithologically similar to its development in the Cheshire Basin, with evaporite beds becoming more important northwards in the lower part, towards the lateral transition to the St Bees Evaporite Formation. As in the Cheshire Basin, the formation passes southwards into aeolian sandstones (Jackson *et al.* 1995). Offshore, these in-part correlatives of the Kinnerton Formation have been named the Freshfield Sandstone Member (Jackson & Johnson 1996). In the NE (SW and west Cumbria) there is a passage into Brockram at the basin margins (Rose & Dunham 1977; Akhurst *et al.* 1997). In the basin centre, and in some

VALE OF EDEN		CARLISLE BASIN (Chapelcross)		WEST CUMBRIA (St Bees)			SOUTH CUMBRIA		EAST IRISH SEA BASIN OFFSHORE				CHESHIRE BASIN	
			ES3									BM3		
			ES2		St Bees Shales Formation		St Bees Shale Formation				Barrow-mouth Mudstone Formation	BM2		
Eden Shales Fm	D-Bed/Belah Dol	Eden Shales Fm.		Brockram					Freshfield Sandstone Member	Manchester Marls Formation		BM1	Kinnerton Sandstone Formation/ Bold Formation	Manchester Marls Formation
	C-Bed					Fleswick Anhyd. & Dolomite		Roosecote Anhyd. & Dolomite				Upper Halite Unit		
	B-Bed		ES1		St Bees Evap-orites Fm.	Sandwith Anhyd. & Dolomite	St Bees Evaporite Fm	Haverigg Haws Anhyd.			St Bees Evaporites Fm	Lower Halite Unit		
	A-Bed & Hilton Plant Beds					Saltom Dolomite & Siltst		Gleaston Dolomite				Basal Siltst/ Carbonate Unit		
								Grey Beds						

Fig. 12.13. Classification and correlation of the Cumbrian Coast Group in NW England and the East Irish Sea Basin: data sources include: Arthurton *et al.* (1978); Jackson & Johnson (1996); Akhurst *et al.* (1997); Holliday *et al.* (2001).

marginal areas (Cumbria), the group is divided into a lower carbonate–evaporite formation, the St Bees Evaporite Formation, and an upper Barrowmouth Mudstone Formation offshore and the St Bees Shale Formation onshore (Fig. 12.13) (Jackson & Johnson 1996; Akhurst *et al.* 1997). The St Bees Evaporite Formation is characterized by thick beds of halite in the more rapidly subsiding areas, and thinner dolomite and gypsum–anhydrite in more marginal areas. The Barrowmouth Mudstone and St Bees Shale formations are lithologically similar to the upper part of the Manchester Marls Formation. An informal subdivision of the Barrowmouth Mudstone, in ascending order BM1, BM2 and BM3, based on gamma-ray log traces, has proved useful in offshore, subsurface mapping (Jackson & Johnson 1996).

Ramsey–Whitehaven Ridge, Solway Firth and Carlisle basins
Little is certainly known of the Cumbrian Coast Group offshore on the Ramsey–Whitehaven Ridge and in the Solway Firth Basin, because of limited hydrocarbon exploration. Knowledge to date has been reviewed by Newman (1999), Chadwick *et al.* (2001) and Floodpage *et al.* (2001), who suggested that the rocks are similar to their correlatives in the East Irish Sea Basin. The group is better known onshore in the area around Whitehaven (Akhurst *et al.* 1997) and in the Carlisle Basin (Holliday *et al.* 2001, 2004).

The succession around Whitehaven (Arthurton & Hemingway 1972; Arthurton *et al.* 1978) is probably typical of the Ramsey–Whitehaven Ridge. Over much of the ridge, the twofold subdivision into the St Bees Evaporite Formation (20–30 m thick) below and Barrowmouth Mudstone/St Bees Shale Formation (50–100 m thick) above can be recognized. The former formation comprises dolomite, dolomitic siltstone and mudstone and gypsum–anhydrite. Towards the Isle of Man, a single formation, the Manchester Marls, is recognized where the carbonate-evaporite rocks die out; towards the Lake District both the St Bees Evaporite and St Bees Shale formations pass into Brockram.

Only one offshore borehole (112/15-1) has cut the Cumbrian Coast Group in the Solway Firth Basin, in an incomplete section with the lower boundary faulted (Chadwick *et al.* 2001, fig. 21). In this borehole, the Barrowmouth Mudstone Formation is about 110 m thick overlying approximately 60 m, mainly halite, of the St Bees Evaporite Formation. This indicates that thick halite deposits are probably also characteristic of the more rapidly subsiding parts of this basin.

It is presumed that the St Bees Evaporite Formation of the Solway Firth Basin passes into a more gypsum–anhydrite-rich facies and then into a Manchester Marl equivalent, but this is unproved. Whether the St Bees Evaporite Formation continues along strike into the onshore Carlisle Basin is also uncertain as this interval appears to be faulted out in the Silloth 1-A Borehole at the coast (Holliday *et al.* 2001). From outcrops and nearby boreholes, only one formation has been recognized so far in the Cumbrian Coast Group in the Carlisle Basin, the Eden Shales Formation, which is about 100 m thick. An informal tripartite subdivision, ES1-3, of the formation has been made by Holliday *et al.* (2001) (Fig. 12.13). ES1 is an approximate equivalent of the St Bees Evaporite Formation. Beds of gypsum–anhydrite occur in the lower part of this formation (ES1), particularly on the northern, Scottish, margin of the basin, but no certain associated marine strata have been proved. In other parts of the outcrop, the evaporite beds have been removed by dissolution. Notably in the SE part of the basin, towards the Lake District Massif, the potentially evaporite-bearing strata of ES1 are overlapped by younger parts of the Eden Shales (Dixon *et al.* 1926; Holliday *et al.* 2004). Several stream sections and boreholes prove a gradational junction with the overlying Early Triassic St Bees Sandstone Formation (Sherwood Sandstone Group).

Vale of Eden Basin
The Cumbrian Coast Group (Eden Shales Formation) is not well exposed in the Vale of Eden. The succession (Fig. 12.13), 150–200 m thick, is best known from boreholes and, to a lesser extent, gypsum mine workings. There are a few notable stream and river sections but these are generally discontinuous and, in part, dislocated due to evaporite dissolution (Burgess & Holliday 1974, 1979; Arthurton & Wadge 1981).

The relationship between the Eden Shales and the Penrith Sandstone and Brockram of the Appleby Group is complex, and the base of the Eden Shales commonly appears to be diachronous (Fig. 12.12). In northern and central parts of the basin (NW of Appleby: Fig. 12.12), the junction is relatively sharp, with the Eden Shales overlapping and infilling hollows on the uneven surface of the silicified older dune sandstones. In the south of the basin, the junction is gradational and the Eden Shales interfinger with and pass laterally into the fluvial sandstone (Fig. 12.14) and Brockram facies of the Appleby Group. The lower part (*c.* 10 m thick) of the Eden Shales in the north of the basin comprises red, green and grey laminated,

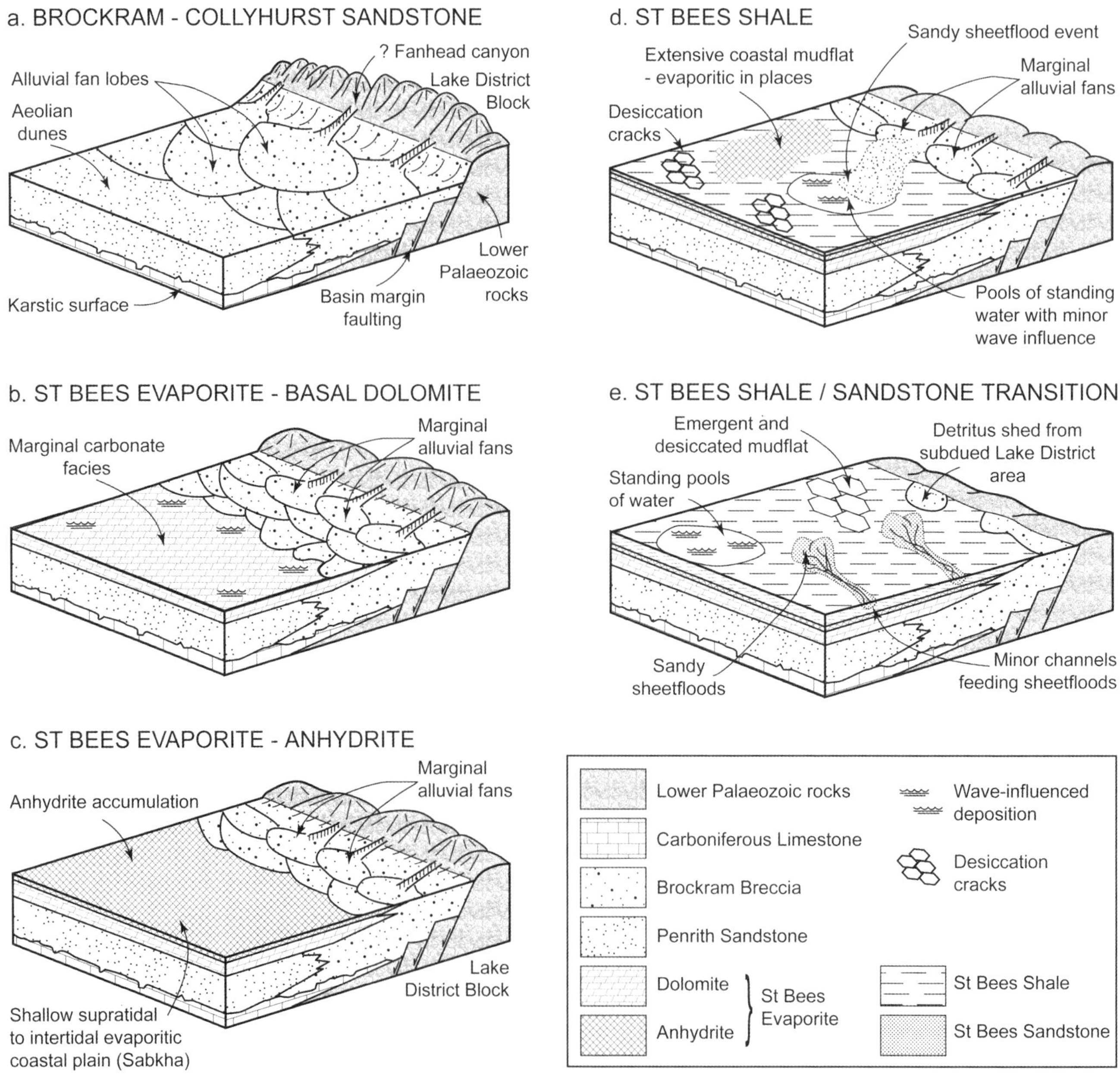

Fig. 12.14. Depositional model for Permian rocks of NW England. (Modified from Akhurst *et al.* 1997, fig. 30.)

mudstones, siltstones and sandstones, locally with scattered gypsum–anhydrite nodules, stromatolitic layers, scattered plant debris and rare *Estheria*. These are interpreted as playa and lacustrine deposits. To the south, around the Kirby Thore Gypsum Mine, the lower beds, in which gypsum with stromatolitic layers predominates, are thicker (30–40 m) and form the A-Bed gypsum–anhydrite which locally contains some halite (Worley 2005) (Fig. 12.15). They appear to be largely saline lake deposits infilling a major hollow in the pre-Cumbrian Coast Group surface. Still farther south, A-Bed passes into laminated sandstones, with subordinate calcareous or dolomitic siltstones and mudstones, and sporadic gypsum-anhydrite nodules and stromatolitic layers. These beds, in which transported plant debris is commonly abundant, were previously known as the 'Hilton Plant Beds', and have been interpreted as sheet flood deposits laid down on the margins of the A-Bed Lake. At the southern end of the basin, the Hilton Plant Beds pass into alluvial fan Brockram of the Appleby Group. No certain marine indicators have been found in A-Bed and the Hilton Plant Beds. The Vale of Eden was probably not flooded by the Bakevellia Sea and there seems to have been no direct connection to the Zechstein Sea at this time. Nevertheless, the preponderance of plant-bearing water-laid sediments towards the base of the Eden Shales is indicative of a significant climate change and rise in water table in the Vale of Eden and adjacent areas, probably contemporary with similar changes (Glennie & Hurst 2006) and the transgressions of the Zechstein and Bakevellia seas into the southern North Sea and east Irish Sea basins, respectively.

There is a similar, although less extreme, facies pattern in the remainder of the Eden Shales (Burgess & Holliday 1974, 1979, fig. 42). The formation is generally sandier towards the south and passes into Brockram (Figs 12.12 & 12.13). There is some borehole evidence that a similar interfingering with Brockram occurs adjacent to the Pennine Fault, which may have been active at this time. Much of the formation comprises alternations of water-laid and wind-laid siliciclastic strata, interpreted as continental sabkha/playa deposits, in which fibrous gypsum veins are common but gypsum–anhydrite nodules relatively

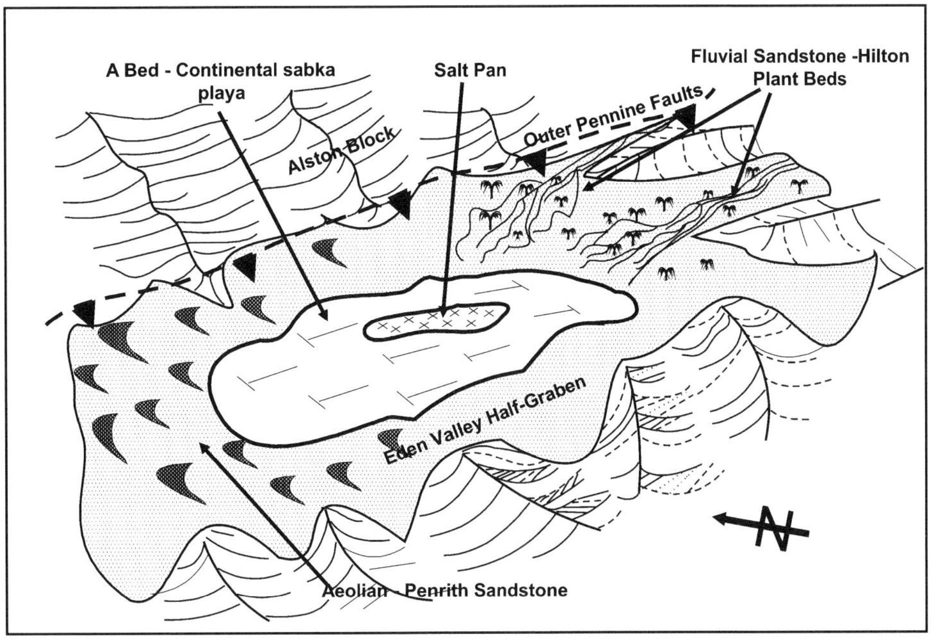

Fig. 12.15. Inferred palaeogeography of the Vale of Eden at the time of deposition of the A-Bed gypsum-anhydrite and the Hilton Plant Beds. Reproduced from Worley (2005, fig. 3) by permission of the Council of the Yorkshire Geological Society.

rare. However, there are three beds of gypsum–anhydrite, B-, C-and D-beds, sabkha and saline lake deposits, which are found basin-wide and are indicative of higher regional water tables. B-Bed is the most important of these, ranging up to 7 m in thickness. There are also several more local beds of gypsum–anhydrite (Arthurton & Wadge 1981, fig. 29). There are no certain indications of marine influx during the deposition of B-, C-and the localized evaporite beds. However, the former were probably continuous with gypsum–anhydrite beds in the St Bees Evaporites of the Bakevellia Sea, and contemporaneous with marine incursions into the southern North Sea Basin. By way of contrast, D-Bed overlies the marine Belah Dolomite. Together, these beds, which die out northwards, indicate a significant cross-Pennine marine transgression. In the higher parts of the formation, above D-Bed, there is a gradual-diminution in the role of wind in sediment deposition and entrapment, and a gradational upwards passage into the fluvial strata of the Early Triassic St Bees Sandstone Formation (Sherwood Sandstone Group).

Eastern England and the southern North Sea Basin

Lower Permian

Although similar in its general evolution to the smaller basins of western Britain, that of the southern North Sea (Southern Permian Basin and Eastern England Shelf of Fig. 12.3) differs in that almost all traces of the immature phase of desert erosion were removed and a mature pediplain was established before the lowest lower Permian sediments now preserved were deposited. This scenario holds true for the main area of extensive outcrop (the coast north from Hartlepool) as well as the subsurface of east Yorkshire, into the North Sea and north Nottinghamshire. Judging from the relief displayed at outcrop in NE England, this pediplain was gently rolling, with elevated tracts generally coinciding with resistant Lower Carboniferous rocks and with scattered hills up to perhaps 200 m high; offshore boreholes farther east show that the basin extended far across the present North Sea (Fig. 12.6) and was bounded to the north and south by topographic and structural highs which continued to supply sediment throughout the late early Permian. No indigenous fossils have been reported in any of the lower Permian rocks of this region.

The earliest widespread lower Permian deposit in NE England is a multicoloured breccia or breccia–conglomerate, locally 10 m thick but generally less than 1 m, which has also been reported in many North Sea boreholes. It contains abundant far-travelled (probably recycled) pebbles, some of them wind-abraded and coated with desert varnish, in a sandy calcareous matrix locally rich in frosted quartz grains. Like the earliest Permian conglomerates (see NW England descriptions) and similar breccias ('brockram') elsewhere, this thin breccia may once have been stabilized lag gravel, similar to those commonly found covering modern desert pediplains.

Higher lower Permian sediments, proved by the intensive search for hydrocarbons, comprise a full suite of desert sediments but, because of the generally lower relief, have less coarse detritus than in the western basins. Glennie *et al.* (2003) suggested that the sediments that accumulated on the SW flank of the basin comprise a complex of interdigitating sheet flood and fluviatile (wadi and wadi-fan) sandstones and fine breccias with some aeolian sand, and that these mainly water-laid rocks are peripheral to a belt of thick (locally 200 + m) and generally younger wind-blown sandstones that are the main gas reservoir of the southern North Sea Basin and extend on land as the Basal Permian Sands of Yorkshire (0–30 m); the equivalent Yellow Sands of County Durham (0–60 m) were deposited in a marginal embayment of the southern North Sea Basin (Glennie *et al.* 2003). In the latter, these water-laid and aeolian sandy deposits together comprise the Leman Sandstone Formation (Rotliegendes Group) (Rhys 1974). These and equivalent sandstones commonly overlie Carboniferous coals, forming a highly efficient gas source and reservoir succession.

Glennie *et al.* (2003) showed that the most rapidly subsiding interior of the southern North Sea Basin was occupied by a great inland sabkha plain composed of evaporite-bearing clays, silts, and fine sands (the Silverpit Formation of the Rotliegendes Group: Rhys 1974), with an extensive but perhaps ephemeral central playa. From published descriptions it seems probable that some of the halite and gypsum was precipitated directly in the playa but that much of it crystallized and grew displacively from saline groundwater within the sabkha and playa sediments. Smaller sabkhas may, according to Glennie *et al.* (2003), also have been present in deflation hollows between barchans of the dune belt, as in many modern deserts.

The petrology of the aeolian Yellow Sands in Durham (equivalent to grey and white strata at the top of the red beds of the southern North Sea Basin) is summarized by Price (1996) who showed them to be composed mainly of subrounded monocrystalline and polycrystalline quartz grains with less than 10% of feldspars and a varied suite of heavy minerals; ferromagnesian minerals are uncommon, possibly through post-depositional leaching, and detrital micas occur in trace quantities only. Laminae and isolated grains of coarse well-rounded quartz are conspicuous minor components in Durham, and up to 20% of small rock fragments are locally present in Yorkshire where the sands were redistributed during the Zechstein transgression and may be considered late Permian in age. Only the uppermost part of the Durham Yellow Sands is generally thought to have been thus redistributed, although in some places in the North Sea a considerable thickness (up to 80 m) of Permian continental rocks are believed to have been so affected. The characteristic colour of the Yellow Sands at outcrop results from the surface oxidation of pyrite and ferrous oxides that pervade the formation in the subsurface where it is grey. There are two possible explanations for the distribution of these colours. First, it has been suggested that much or all of the formation may at one time have been red like its counterparts in the other early Permian basins (Smith 1992) and that the complex diagenetic history of the rocks has resulted in the various facies ('Weissliegend') responding (and being reduced) differently to burial, probably because of their different primary and secondary mineralogy. A second possibility is that reddening occurred below the water table, to be halted at the time of the Zechstein transgression and the reducing seal of Kupferschiefer (Marl Slate). In the UK North Sea, Rotliegendes deposition lasted for maybe only one quarter of the time of that in Germany (terminated here also by the Zechstein transgression at 257.3 Ma). The Yellow Sands are extracted in open-cast quarries for building sand.

Upper Permian

The Zechstein Sea Basin occupied most of the present day North Sea and extended eastwards into parts of Holland, Germany, Denmark and Poland. It was divided into several sub-basins by ridges ('highs', 'swells') where subsidence was relatively slower than in the centres of the sub-basins, and upper Permian rocks of eastern England and the southern North Sea were formed in the English Basin. This initially measured some 250 × 350 km and was generally coincident with the pre-existing lower Permian Basin. The early Zechstein Sea could, at times, have been 200–250 m deep, increasing perhaps to 300 m as subsidence outstripped sedimentation. It appears to have become sediment-starved soon after initiation and at times anoxic in its deeper parts (Smith 1992). Subsequently, the basin filled when thick evaporites were formed in the interior at a rate exceeding that of subsidence and, with the original topographic hollow thus virtually eliminated, space for sediment accumulation in the later sequences was created by continuing subsidence. These sequences of sedimentation spanned perhaps the last 6 Ma years of Permian time (Smith 1992). The broad ridges between the various sub-basins were probably also high points on the early Permian deserts, and became shoal areas where the thin Zechstein rocks are mainly shallow-water and peritidal carbonates and evaporites. As such they probably emerged first when relative sea levels fell, causing these crestal successions to be stratigraphically incomplete. Because the sub-basins had marine connection during the early phases of marine inundation, the deposition of sediments at these times was influenced by basin-wide as well as local factors: their faunas and lithological succession may be similar in places as far apart as Greenland and Poland. Increasing isolation and perhaps eventual separation of the basins during sea-level lowstands resulted in reduced continuity of the more saline rocks, which were influenced to a greater degree by local conditions during deposition and are less readily correlated between sub-basins.

Zechstein strata crop out almost continuously in a narrow belt near the western margin of the English Zechstein Basin (Fig. 12.6) and extend across the North Sea beneath later rocks. At outcrop they comprise a series of progradational belts, mainly of limestone and dolomite (Plate 20), which are separated by emersion and erosion surfaces or by red beds containing evaporites or evaporite solution residues. These rocks record the evidence of repeated expansion and contraction of the Zechstein Sea, and its eventual demise. A minor barrier (the Cleveland High, Fig. 12.6) lying across North Yorkshire partly separated a Durham Province to the north from a Yorkshire Province, Zechstein rocks exposed in Durham having been formed, in general, farther offshore than those at outcrop in Yorkshire. The nomenclature used here for the English Zechstein succession (Figs 12.16 & 12.17) is based on Tucker (1991). The extent of these deposits in the offshore of much of the North Sea was reviewed by Glennie *et al.* (2003).

The lithology and biota of the initial marine deposit, the Marl Slate Formation, indicate that anoxic conditions were rapidly established in deeper parts of the newly formed Zechstein Sea. The Marl Slate forms the base of Tucker's (1991) sequence ZS1. It onlaps the Yellow Sands (onshore in Yorkshire and Durham) as well as the Rotliegendes (offshore). It is a thin sapropelic silty laminated limestone or dolomite with a fauna of *Lingula*, reptiles, and palaeoniscid fish; its content of metallic elements is much lower than that of the equivalent Kupferschiefer of the Zechstein succession in Germany (Smith 1992). Perhaps the best exposure of this succession is at Frenchman's Bay (South Shields: NZ 3886 6629: Fig. 12.18).

The Marl Slate is absent from the more freely aerated, warm, shallow waters at the basin margin and over some of the shoals. These were the areas that were ideal for the formation of a wide range of oolitic carbonate sediments, which rapidly built a flat-topped shelf up to 50 km across. At outcrop in Yorkshire, these shelf rocks form the Cadeby Formation, which is divided into two members (Wetherby and Sprotborough) by an erosion surface that is thought to have been cut when sea level fell by at least 5 m (Smith 1992). Tucker (1991) interpreted this surface as a sequence boundary between ZS1 and ZS2. Shelf oolites of the Wetherby Member have a predominantly bivalve fauna and contain bryozoan-algal patch reefs. Laterally equivalent strata deposited in deeper water on outer parts of the shelf and on the basin slope are exposed in Durham and form the Raisby Formation. This contains a relatively varied brachiopod–bryozoan–bivalve fauna, is commonly coarsely mottled and displays evidence of extensive and locally massive downslope sliding.

Sequence ZS2 is represented in the Yorkshire Province by the Sprotborough Member, a fine oolitic dolomite, probably deposited in sand waves on a broad shelf under at least 12 m of water (Smith 1992). The equivalent Ford Formation of the Durham Province comprises mainly oolitic shelf dolomites protected on its seaward flank by a massive bryozoan–algal shelf-edge reef up to 100 m thick and more than 30 km long (Tucker 1991; Smith & Taylor 1992); biostromal algal–stromatolitic dolomite covers the flat top of this reef.

Sequence ZS3 has at its base the massive Hartlepool (=Hayton) Anhydrite Formation, which rings the Zechstein basin and has a displacive fabric that was at first thought to be consistent with formation in a sabkha environment (see references in Smith 1992). Smith (1989) suspected that it may have been a primary gypsum sea-floor precipitate. The seaward face of the anhydrite plunges steeply to the basin floor where the formation passes into about 20 m of anhydrite–carbonate laminates in which Smith & Taylor (1992) described four widespread minor cycles, or parasequences. Halite is generally absent from the rocks of the first two depositional sequences

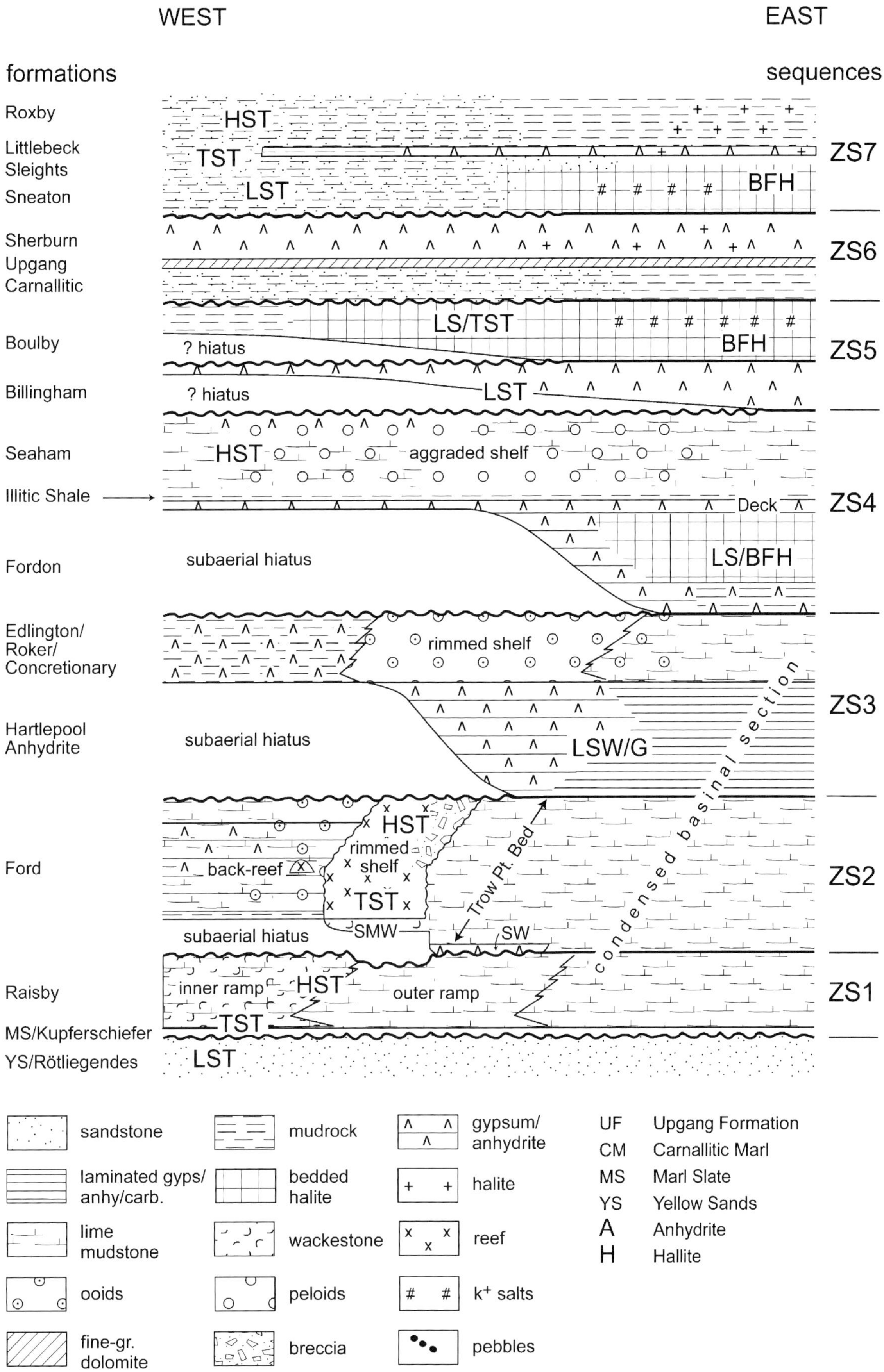

Fig. 12.16. Late Permian (Zechstein) lithostratigraphy and sequence stratigraphical interpretation in NE England. (After Tucker 1991.) Stratigraphical abbreviations: MS, Marl Slate; Kup., Kupferschiefer; YS, Yellow Sands; Rotl, Rotliegendes. Sequence stratigraphy: LS-BFH, lowstand systems tract–basin-fill halite; TST, transgressive systems tract; SMW, shelf-margin wedge; HST, highstand systems tract.

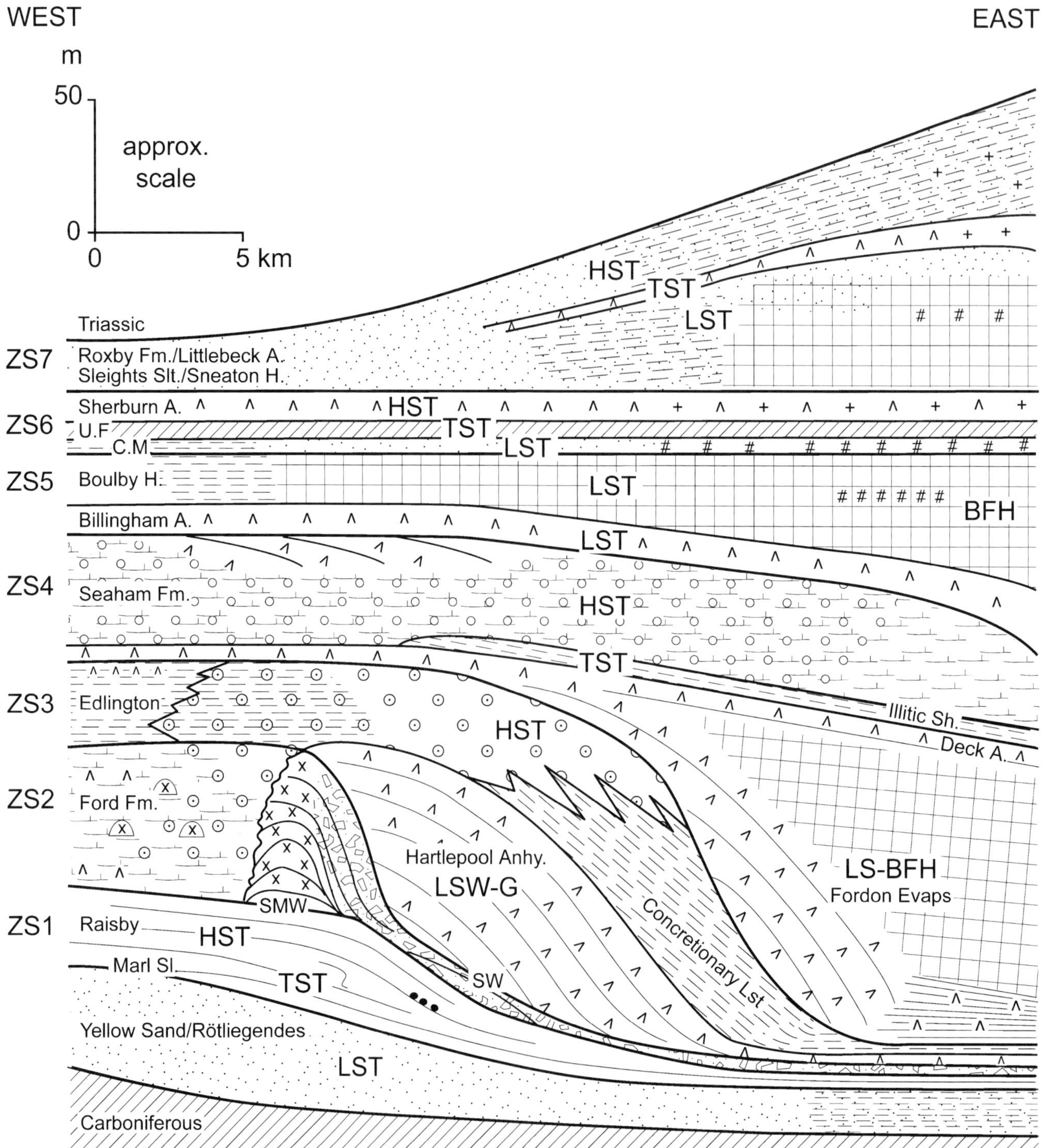

Fig. 12.17. Sequence stratigraphical interpretation of the geometry and correlation of Zechstein Group lithostratigraphy in NE England. (After Tucker 1991.) Stratigraphical abbreviations as for Fig.12.16, plus: SW, slope wedge; A. or Anhy., anhydrite; H, halite; Slt., siltstone; Sh., shale; Sl., slate; Lst., limestone; Dol., dolomite; Fm., formation; evaps, evaporates; U.F., Upgang Formation; C.M., Carnallitic Marl. Sequence stratigraphical abbreviations also follow Fig.12.16, with these additions: sb, sequence boundary; ts, transgressive surface.

(ZS1 and ZS2) but it forms lenses within the Hartlepool Anhydrite Formation (ZS3) in the southern North Sea (Smith 1995). Above the Hartlepool Anhydrite, predominantly shelf carbonates form the remainder of Tucker's (1991) sequence ZS3 and include the Concretionary Limestone Formation, which is thought to be a chronostratigraphic equivalent to the (up-dip) Roker Dolomite and Edlington formations, and the Kirkham Abbey Formation of Yorkshire (Tucker 1991). The Roker Dolomite Formation comprises mainly shallow-water oolitic dolomite with a scanty bivalve–gastropod–ostracod fauna, and was probably formed near sea level in shelf-edge shoals or barrier bars. It passes landwards into an almost barren succession of red and grey coastal plain and lagoonal mudstones and siltstones (the Edlington Formation or Middle Marls) that contain both primary and secondary anhydrite and some (mainly secondary) halite. The equivalent basin-slope rocks (Concretionary Limestone Formation) are mainly of fine-grained partly laminated dolomite and limestone and feature large and complex calcite concretions and evidence of repeated submarine slumping. They pass on the basin floor into a thin succession of

Fig. 12.18. The topmost Yellow Sands (Rotliegend), Marl Slate (Kupferscheifer) and basal Zechstein Group (Raisby Formation) exposed at Frenchman's Bay (South Shields) (NZ3886 6629). Regionally, the Marl Slate onlaps the Yellow Sands/Rotliegendes: here it bears abundant scales of fossil fish. The Raisby Formation is also exposed on the adjacent foreshore where tepee-like structures can be seen. Horizontal scale is 3.5 m.

brown fetid bituminous calcite laminates that may be a hydrocarbon source rock (seen in the Fulwell Hills Quarries, e.g. Southwick, NZ3859). Widespread dissolution of the underlying Hartlepool Anhydrite has led to brecciation, dedolomitization, and foundering of the ZS3 carbonate rocks and to the spectacular collapse-breccias of the Durham coast and beyond (e.g. in the Middle and Upper Concretionary Limestone at Souter Point).

Sequence ZS4 commences with the thick (>300 m), progradational, Fordon Evaporites Formation, which embraces several subcycles or parasequences. It is composed mainly of halite with basin-margin successions rich in anhydrite and polyhalite. Some of the salts in this cycle may be of deep-water origin, but others have fabrics consistent with formation in shallow water or on salt flats and thus may indicate major falls in sea level and phases of desiccation. The thicker salts of this cycle are the main component and driving force of the many salt domes of the southern North Sea.

A renewed marine transgression, generally beyond the shorelines of the earlier formations, led to the deposition of a fine-grained limestone (now partly dolomitic), the Seaham Formation (Fig. 12.16, sequence ZS4) (=Brotherton Formation in Yorkshire). This formation has a distinctive biota of two species of bivalves and the supposed alga *Calcinema*. It was deposited on a mainly shallow-water carbonate shelf, whose extraordinary width (up to 100 km) is probably a result of the low depositional slope following infilling of the original basin by the Hartlepool Anhydrite, and indicates unusually uniform conditions around the margins of the basin. In the central part of the basin, a thin clay, the Illitic Shale, forms part of the transgressive systems tract above the lowstand Fordon Evaporites (Fig. 12.17). The shelf edge was well seaward of that of the two preceding sequences, but the shelf rocks thin upwards and pass into a few metres of argillaceous limestone and calcareous mudstone on the basin floor.

Subsequent emergence of much of the shelf at the ZS4–ZS5 sequence boundary resulted in the establishment there of a widespread sabkha on which the succeeding nodular Billingham Anhydrite Formation was formed and this in turn was followed by the extensive Boulby Halite Formation and the more variable (in thickness) Boulby Potash Formation. An origin close to sea level, in and on vast salt flats and in shallow brine pools, is envisaged for both these chloride deposits in basin-margin areas, but their present mineralogy owes much to a long and complex series of diagenetic changes. Little is known of the mineralogy or depositional environment of these two deposits in the broad interior of the English Zechstein basin but their combined thickness there (more than 150 m) implies access to a large source of salts and at least limited connection with the main part of the Zechstein Sea and world oceans. The Boulby evaporites and overlying calcareous mudrocks are observed in boreholes and the potash mine at Boulby itself. The value of potash as a fertilizer has led to the Boulby mine being extended some kilometres east under the North Sea.

Above the Boulby Halite is the red blocky siltstone of the Carnallitic Marl Formation (base of ZS6), a salt-rich unit comparable with the broadly time-equivalent blocky aeolian siltstone and sandstones of the East Irish Sea Basin. It records a phase when marginal flats extended far into the basin and the Zechstein Sea dwindled to a series of shallow playas; the evaporites of these playas form a large part of the Carnallitic Marl in the basin centre (Tucker 1991). The overlying Upgang Formation has yielded no fossils and appears to be the product of a highly saline but extensive shallow inland sea. It is succeeded by a flooding surface across much of the basin and then by the halite-rich Sherburn Anhydrite Formation (the highstand systems tract equivalent of the Pegmatitanhydrit of the Zechstein in sequence in Germany).

At the end of Sherburn Anhydrite deposition, Zechstein salt sedimentation was again interrupted by a phase when silts and muds (now red) spread across much of the basin to form the Sleights Siltstone Formation, which marks the beginning of sequence ZS7. Depositional conditions were probably similar to those prevailing during deposition of the Carnallitic Marl, with a similarly high content of wind-blown siliciclastic or salt dust and secondary evaporite minerals. The thick Sneaton (Upper) Halite and the interbedded Sneaton (Upper) Potash are more uniform in thickness and lithology than the Boulby Halite and Boulby Potash, but both bear features consistent with deposition on extensive marginal salt flats (Smith 1992) and evidence of a complex diagenetic history. It may be that the Sneaton Potash is part of an incomplete depositional sequence or possibly parasequence, for the succeeding halite has a high content of red siltstone suggesting that it may have been formed mainly displacively on a subaerially exposed brine-soaked plain. The silty halite contains abundant potash salts in the interior of the basin.

The siltstones, mudstones and thin evaporates of the Littlebeck and Roxby formations in the upper part of ZS7 are generally taken to mark the end of the depositional history of the Zechstein Sea, although extensive thin lenses of anhydrite in succeeding red beds of the Bacton Group beneath the North Sea clearly point to continuing episodic subsidence and flooding of former Zechstein depocentres. The Permian–Triassic boundary is believed to lie low in the Bacton Group (Chapter 13).

The West Midlands

Continental, mainly red, sedimentary rocks believed to be of early Permian age and forming the upper part of the Warwickshire Group occur in several interconnected basins across the west Midlands, from the Staffordshire Basin to the Worcester Basin (see Chapter 13, Fig. 13.6 for their location). At least some of these basins originated in late Carboniferous times, and intermittent movement along boundary faults and synsedimentary subsidence and tilting are suggested by numerous non-sequences and unconformities. Regional correlations (Smith *et al.* 1974) show that perhaps the upper third of the lower Permian and almost all of the upper Permian are absent in the West Midlands. Hence the regional descriptions herein are necessarily brief. There is only a limited outcrop, but the subcrop is known from boreholes, particularly in the coalfield areas.

The base of the Permian is currently placed at the base of the Kenilworth Sandstone and Clent formations (Powell *et al.* 2000*a*), but there is only scanty fossil evidence from the Kenilworth Formation (Smith *et al.* 1974) and none from the Clent Formation (Powell *et al.* 2000*b*). The succession across the Carboniferous–Permian boundary appears to be most complete in the Warwickshire coalfield area, in the eastern part of the Knowle Basin (Fig. 12.19). Here the sandstones,

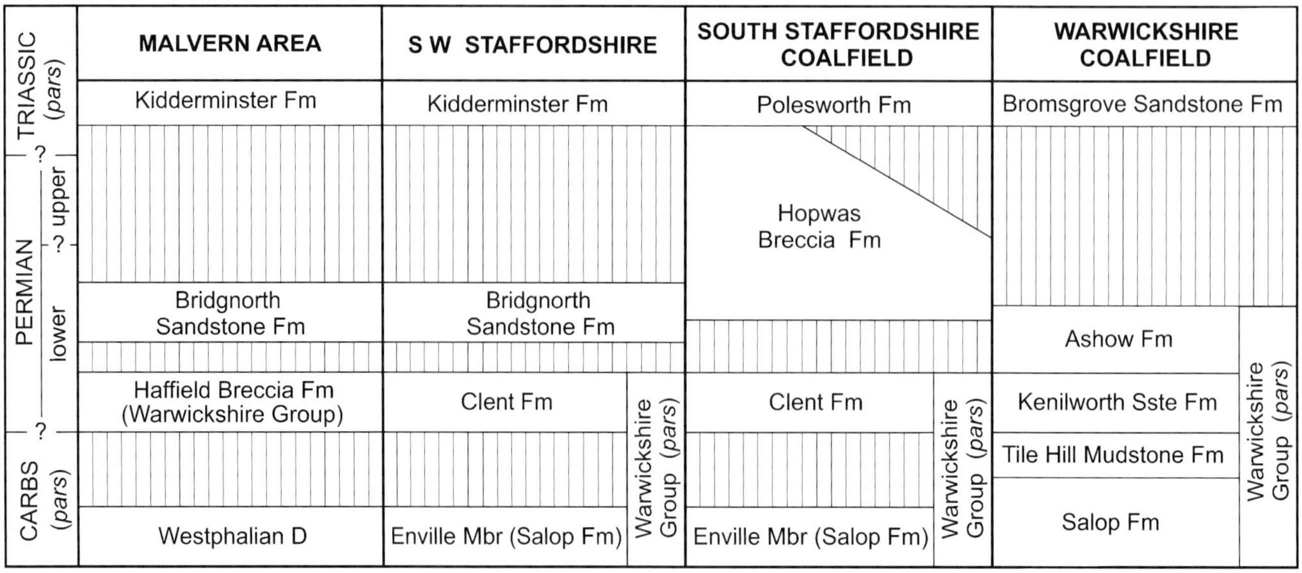

Fig. 12.19. Lithostratigraphical division of the Permian of the West Midlands.

local conglomerates and breccias of the Kenilworth Sandstone Formation (*c*. 100 m) rest on the finer-grained Tile Hill Mudstone Formation. Elsewhere the lowest Permian is represented mainly by breccias which, under several local names (e.g. Clent Formation, Haffield Breccia Formation) overlie a major unconformity at basin margins. These represent accumulations of of up to about 240 m of detritus derived from surrounding highlands and deposited on coalescing fans and in wadis at the edge of the various depositional basins. The clasts indicate a variety of sources; the Clent Formation breccias, for example, are composed mainly of angular fragments of Precambrian volcanic rocks while the Haffield breccias are more varied, with igneous and Lower Palaeozoic clasts.

The breccias pass laterally into sandstones and finer-grained sediments towards the basin centres. In common with those of many early Permian intermontane-type basins, the Midlands successions also generally fine-up into sandstones or mudstones. In the Warwickshire coalfield these are represented by the Ashow Formation, a sequence of up to 250 m of red-brown mudstones with subordinate sandstones. Over the Worcester and Stafford basins the equally thick Bridgnorth Sandstone rests unconformably on the underlying breccias. The aeolian origin of this sandstone formation was proved convincingly by the pioneer work of Shotton (1937).

The Hopwas Breccias Formation of the South Staffordshire coalfield and adjacent areas consists of up to 30 m of mainly locally-derived coarse breccias, with interbeds of sandstone and mudstone. It rests unconformably on the Clent Formation or older beds and is overlain by the Lower Triassic Polesworth Formation (Figure 12.19) but its exact age remains uncertain; it may be anything from Early Permian to early Triassic.

Southwest England and the Western Approaches Basins

The outcrop of upper Carboniferous and Permian continental red beds in SW England stretches across Somerset and Devonshire and beneath the English Channel. Their eastward extension is concealed by younger strata, but they probably extend east to Purbeck and beyond and underlie much of the adjoining English Channel. At outcrop they are seen, perhaps more clearly than anywhere else in England and Wales, to have accumulated in a series of valleys and troughs (many fault-bounded) and to have been banked against and progressively buried a reddened hilly land surface. Recent work by the BGS (Edwards & Scrivener 1999) has now clearly established a second link: that between the two main phases of deposition in the Exeter Group (broadly, early Permian and late Permian), the intrusion–extrusion of the Exeter Volcanic suite simultaneously to deposition of earlier Permian strata (Fig. 12.20), and the intervening intrusion of the Dartmoor Granites (at around 280 Ma). In addition, the overstep of later over earlier Permian strata across the Crediton Trough shows that N–S- and NE–SW-oriented 'Wessex Basin style' extension is related to the later phases of Permian extension, rather than to the earlier episodes (Miliorizos & Ruffell 1998). As in other areas, the earliest sediments (e.g. Exeter Group of Edwards & Scrivener 1999) are locally derived fan and wadi breccias, including sheetflood deposits. The merging of basins as they filled up and interfluves declined, permitted influxes of more distant materials, which can be seen as finer-grained beds (generally a few tens of centimetres to a few metres thick) in the upper beds of the breccia-dominated succession. Aeolian sandstones were initially subordinate but came to dominate higher parts of the succession in the large basins as gradients lowered and hilly source areas dwindled. The fault-bounded Permian palaeovalleys of the Crediton and Tiverton troughs, as well as the Western Permian–Triassic Basin (between the Quantock and Brendon hills) show many similarities, originating as half-graben filled initially by coarse debris that fines-up (Figs 12.21 & 12.22). Landscape rejuvenation caused by sporadic tectonism and associated volcanism is recorded by numerous non-sequences and disconformities.

The large quantity of volcanic debris in the continental breccias near the base of the succession in the Exeter area suggests that volcanism was a major feature of the contemporary local scene and its products were once much more widespread than now. Permian volcanic rocks in a number of other areas in SW England are probably slightly younger than those around Exeter. It was formerly thought that the radiometric age (291 ± 6 Ma) of early lavas near Exeter (Miller & Mohr 1964) helped to establish the position of the base of the Permian here, but this evidence is now thought to be unreliable (Warrington & Scrivener 1990). More recently, further palynological evidence (in Edwards & Scrivener 1999) has shown the Exeter Volcanics to be truly Permian in age, with the underlying Cadbury Breccia (near Thorverton, SS 910 050) to be very close to, if not astride, the Carboniferous–Permian boundary (Fig. 12.20). As in the Midlands, the early Permian breccias and sandstones have many local names and in each area contain a suite of Devonian

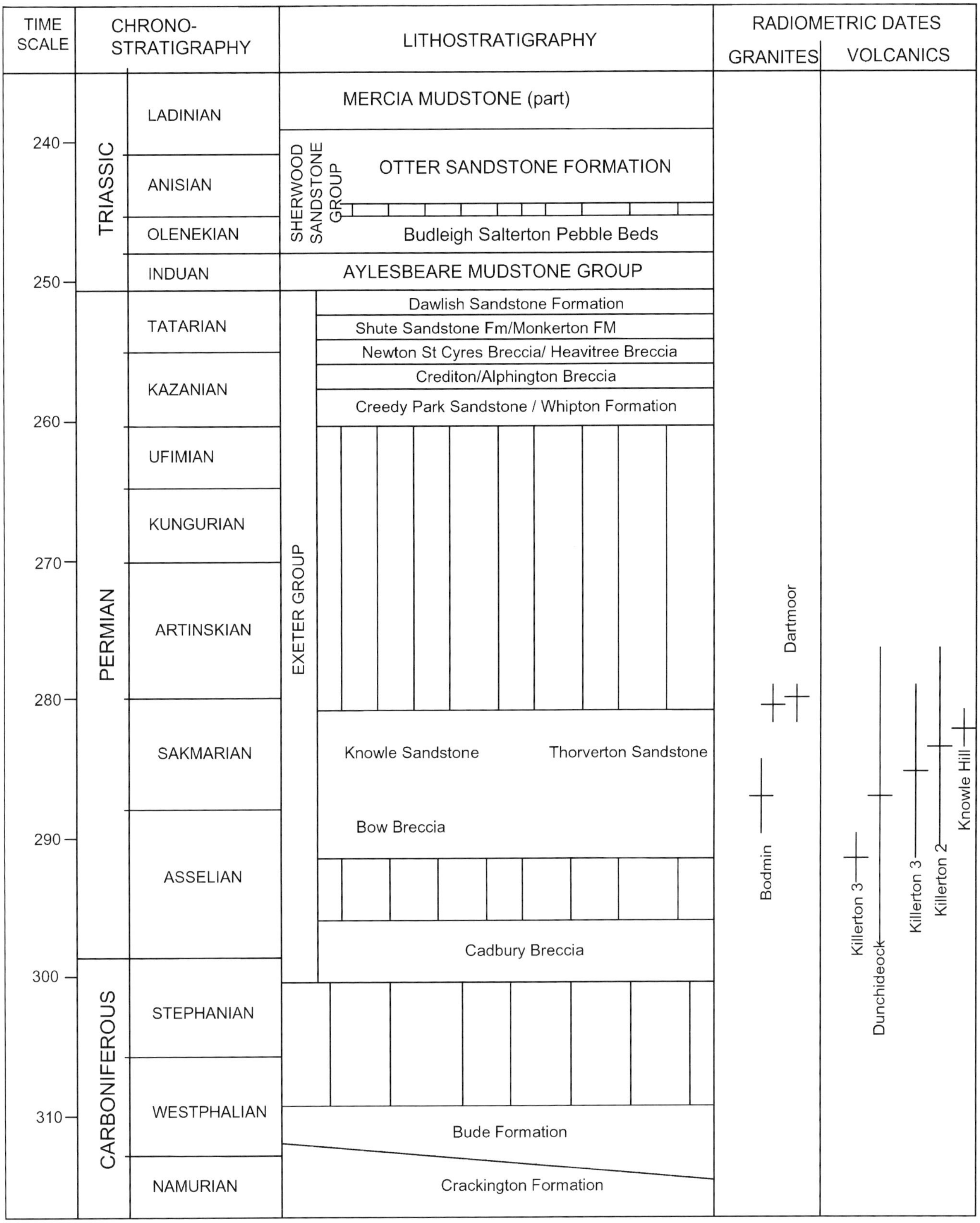

Fig. 12.20. Stratigraphy of the Permian succession in the Exeter area. (After Edwards & Scrivener 1999.)

and/or Carboniferous rock fragments, with or without contemporary volcanic debris; in places the mixture is so distinctive that individual fans may be recognized (Laming 1966). Clasts are smaller and more rounded in distal parts of fans than in proximal parts, although poor grading is prevalent, and sandstones are relatively more abundant distally. Despite active downwasting, hilly areas to the west continued to supply coarse detritus that accumulated as fan breccias at the margins of all the basins: breccia was, however, quantitatively overwhelmed in time by huge amounts of mainly aeolian sand, which accumulated to thicknesses of up to 500 m the western part of the main basin and there forms the youngest part of the presumed lower

Fig. 12.21. Typical 'southwestern basins' (Crediton, Tiverton) Permian half-graben and graben lithofacies and stratigraphy (north Somerset). (**a**) Shows the railway cutting east of Wiveliscombe at Hillacre Farm Crossroads (GR10302768) with the lowest formation of the Permian, the Wiveliscombe Sandstone Formation (rests unconformably on Devonian/Carboniferous) with the Vexford Breccia Formation above. Cutting is approximately 6 m high. (**b**) Shows Vexford Breccia Formation at Grant's Lane, near Wiveliscombe Castle, Somerset. GR 0830 2889. Lens cap is 2.4 cm in diameter.

Permian sequence. The top of the Permian in this area is based (lithostratigraphically) at the top of the Exeter Group (Dawlish Sandstone Formation). Breccias and sandstones interdigitate at basin margins, and isolated dunes on breccia fans are preserved locally.

Offshore in the complex of basins in the Western Approaches of the English Channel (Evans 1990) thick Permian–Triassic successions are known but proven (dated) Permian records are rare. Permian volcanics are known from the Melville Basin, but otherwise Permian strata are implied as opposed to proven. Probable Permian successions include the Aylesbeare Group, recorded in the Melville Basin (see Evans 1990 for details and location), and part of the thick red-bed sequence recorded in the St. Mary's and Plymouth Bay basins (Evans 1990; Harvey *et al.* 1994). The thickest succession (9 km in total) of Permian–Triassic rocks is preserved within the Plymouth Bay Basin. A combination of unfossiliferous sediments and few borehole penetrations means that we know very little of the internal lithostratigraphy or chronostratigraphy. Estimates of Permian–Triassic erosion from the surrounding landmasses (e.g. Cornubia) come up with similar figures to that of the sediment-fill (5.8–7 km, up to 10 km), suggesting that a simple mass-balancing of sediment removal and deposition was underway. This scenario may be true also for areas other than Plymouth Bay, where later Jurassic–Cretaceous and Cenozoic uplift and erosion may have removed significant thicknesses of strata. The borehole records we do have show many similarities between the Plymouth Bay and onshore successions, with earliest (presumed Permian) volcanics and volcaniclastic breccias or conglomerates, overlain by anhydritic mudstones and siltstones (often termed the Aylesbeare Group). Evans *et al.* (1990) demonstrated how such coarse-grained sediments may occur also in adjacent basins, in successions thought to be of earliest Triassic age. The lack of biostratigraphic or magnetostratigraphic dating is apparent. The situation is even more dire in the offshore Bristol Channel–Welsh basins where no dated Permian rocks are known, but again strongly implied (Tappin *et al.* 1994) in one well (93/6-1) of the South Celtic Sea and one well (103/2-1) of the St George's Channel Basin.

Economic geology

The Permian rocks of England and Wales, and adjacent offshore areas, have been of major significance to the economic development of Britain. Not only are many of the principal Permian lithologies of intrinsic value in themselves, but also they locally act as hosts to valuable fluids or deposits originating from these fluids. The principal uses of these rocks are summarized briefly here.

The most notable economic interest in the Permian succession is without doubt the hydrocarbon geology of the southern North Sea. The lower Permian sandstones of the Leman Sandstone Formation, as the principal reservoir rocks of the

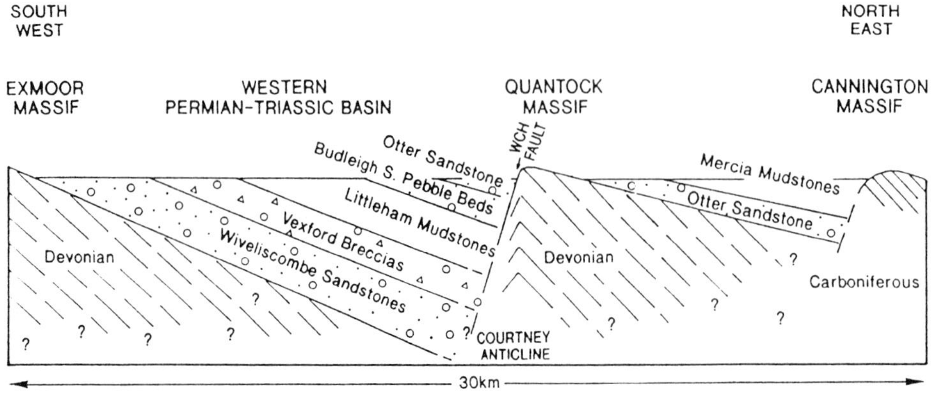

Fig. 12.22. Section across a typical Permian half-graben of SW England: the Western Permian–Triassic Basin of Miliorizos & Ruffell (1999).

southern North Sea gas fields (Cameron *et al.* 1992), make a major contribution to the UK economy, and some small accumulations also occur in fractured Zechstein carbonate rocks. The excellent reservoir characteristics of the Permian sandstones also make them particularly useful for gas storage schemes such as the Rough Field off eastern Yorkshire (Stuart 1993). Gas storage caverns have also been created by the dissolution of the EZ2 halite at Hornsea in eastern Yorkshire (British Geological Survey 2004*b*).

The Zechstein evaporites provide the seal for the gas fields. But evaporite minerals have also been extensively extracted from Permian rocks and their presence in the subsurface was the prime reason for the establishment of the chemical industry around Teesside. Permian gypsum has been mined at many localities in Cumbria and Yorkshire in the past but presently extraction is restricted to the mines in the A-Bed of the Kirby Thore area in the Vale of Eden. Gypsum has been put to many uses, but it is principally required by the building industry as plaster and plaster board (Highley *et al.* 2004*a*). There is currently no commercial extraction of anhydrite in Britain but there were significant workings in the middle decades of the 20th century at Billingham (Billingham Anhydrite) and Whitehaven (Sandwith Anhydrite), principally for sulphuric acid and NH_4SO_4 fertilizer manufacture. Salt and brine have been extracted from the Permian halite beds of Teesside since the late 19th century. Potash (principally sylvite, KCl) and more recently salt have been extracted from Permian (EZ_3) rocks deep under the North Yorkshire Moors at the Boulby Mine since the early 1970s (Highley *et al.* 2004*b*).

Permian sandstones, particularly in NW and SW England, are locally important aquifers (Allen *et al.* 1997), in many areas having relatively high intergranular and fracture porosities and permeabilities. In NE England, fractured Zechstein carbonate formations also locally make a significant contribution to water supply. At depth, Permian aquifers have some potential as a source of geothermal energy (Smith 1986).

Where Permian sandstones and breccias are sufficiently well-cemented and have suitable physical properties, they have been extensively worked at outcrop for building stone and aggregate in SW and NW England (including adjacent parts of SW Scotland), and many large quarries are presently in still in operation. Uncemented dune sands are worked at several localities in Durham and Yorkshire. Numerous quarries, many still working, mark the outcrop of the upper Permian Zechstein dolomitic carbonate rocks, or 'Magnesian Limestone' of former usage. The rocks extracted have been and still are used for a very wide range of purposes, including refractories, fluxes, roadstone, hardcore, concrete aggregate, rock fill and building stone (Highley *et al.* 2004*c*, 2006). Many notable buildings near to the outcrop were constructed using Magnesian Limestone, for example York, Beverley and Southwell minsters (Lott 2001), but it has also been transported and employed much farther afield, for example the Houses of Parliament (Lott & Richardson 1997) and Eton College.

Metalliferous mineralization in the form of veins and replacements is found locally in Permian rocks and has been worked in the past, although no such occurrence is now active.

13 Triassic: seasonal rivers, dusty deserts and saline lakes

M. W. HOUNSLOW & A. H. RUFFELL

The collision of Siberia and the Kazakstan microplate with the eastern side of the Fennoscandia continent in the Permian amalgamated the last major continental fragments to produce the supercontinent Pangaea, which persisted into the Jurassic (Fig. 13.1). During the last phases of this collision, during the latest Permian–Early Triassic, extrusion of massive amounts of flood basalts occurred in Siberia, to the east of the Urals (Otto & Bailey 1995). Some have proposed this event as one of the key processes controlling the largest extinction in Earth's history at the Permian–Triassic boundary (Wignall 2001*a*; Benton & Twitchett 2003).

During the Triassic, England and Wales lay beyond the western termination of the Tethys Ocean, which was divided into a northern part, the Palaeotethys, and a southern part, the Neotethys (Fig. 13.1). Between these oceans occurred the Cimmerian terrains; several now widely separated continental fragments which had rifted from the northern fringe of Gondwana in the Permian (Stampfli & Borel 2002). The Triassic witnessed the northward drift of these Cimmerian terrains, and the northward subduction of the Palaeotethys, which was mostly completed by the Late Triassic. However, in the western Tethys, some Triassic ocean basin fragments (e.g. Pindos, Meliata) persisted into the Jurassic (Fig. 13.1) (Stampfli & Borel 2002). The southern margin of the Neotethys Ocean stretched southeastwards through North Africa, Arabia, northern India and NW Australia and shows clastic and carbonate Triassic successions typical of a passive margin (Weidlich & Bernecker 2003), contrasting strongly with the destructive northern margin of Palaeotethys.

Extensive rift basins formed in Europe during the Triassic, stretching from the Boreal Ocean in the north, through the Greenland–Norwegian gateway into many of the bordering regions around Britain and Ireland, and extending south and SE into France, Iberia and central Europe (Fig. 13.2). This rifting extension was a continuation of Permian rifting and segmented both Caledonide and Variscan basement.

In the same way that the start of the Triassic was heralded by the flood basalts of west Siberia, the end of the Triassic and start of the Jurassic was marked by the extrusion of extensive flood basalts of the central Atlantic magmatic province, which extended along the boundary of the Africa–Americas continental join, from Iberia in the north to Brazil in the south (Marzoli *et al.* 1999). This extensive flood basalt province, of which only fragments are now seen, has been suggested as one of the decisive elements responsible for the floral and faunal turnover at the end of the Triassic (Wignall 2001*a*; Cohen & Coe 2002; Hesselbo *et al.* 2002; Tanner *et al.* 2004).

European palaeoenvironmental signatures

The Permian witnessed the fragmentation of the core of the Variscan Mountains into a number of separate areas, such as the Armorican, Bohemian and Iberian massifs (Fig. 13.2). These supplied sediment northwards into adjacent basins during much of Early and Mid Triassic time. It was only marginal to the Fennoscandia uplands, and during parts of the Carnian (Late Triassic) in Central Europe, that sediment sources switched to the north or NE (Ziegler 1990).

The south polar glaciations, which influenced the Permian, were not a factor in the Triassic, which has no sedimentary evidence of polar ice caps. The Pangaea supercontinent during the Triassic severely perturbed the latitudinal changes in palaeoclimate (Parrish 1993). The interior northern regions of the supercontinent would have experienced low-pressure conditions in the summer, drawing humid air from the Tethys Ocean. Much of this moisture may have precipitated on the remnant Variscan mountains in southern-central Europe and Iberia, leading to northwards seasonal transport of sediment by major river systems. In the winter, the northern interiors would have been sources of dry, externally directed winds. This strongly seasonal monsoonal climate was at its peak during the Triassic and was enhanced by the equatorial position of Tethys, flanked to the north and south by large continental masses (Parrish 1993). Cross-stratification measurements on dune systems in the upper parts of the Sherwood Sandstone Group (Middle Triassic) mostly demonstrate that they were deposited by winds blowing from the Triassic ENE to NNE (Thompson 1970*b*; Meadows & Beach 1993; Jones & Ambrose 1994; Mountney & Thompson 2002). These may have been the monsoonal dry winter winds emanating from the interior of northern Pangaea. This monsoonal climate would give a strong seasonality, possibly even at the equator (Parrish 1993). Consequently, even though England and Wales drifted northwards during the Triassic from a palaeolatitude of about 16°N–34°N, this drift may not have radically altered the climate, which was dominated by strong monsoonal circulation. Interior parts of Pangaea may have experienced arid and semi-arid conditions, stretching from near the equator to around 50°N, and perhaps extending over 80% of Pangaea in the Early Triassic (Chumakov & Zharkov 2004).

During the Triassic much of southern Europe (south of the Alps and Caucasus Mountains) comprised a series of fault-bounded blocks at the SW termination of the Tethys Ocean (Figs 13.1 & 13.2). The Mid and Late Triassic saw these blocks fragment, and subside, producing fringing and ramp carbonate reefs with intervening deep marine basins.

Within a zone across central Europe (Poland–Germany–Denmark–France–England–Wales–Spain), Triassic basinal

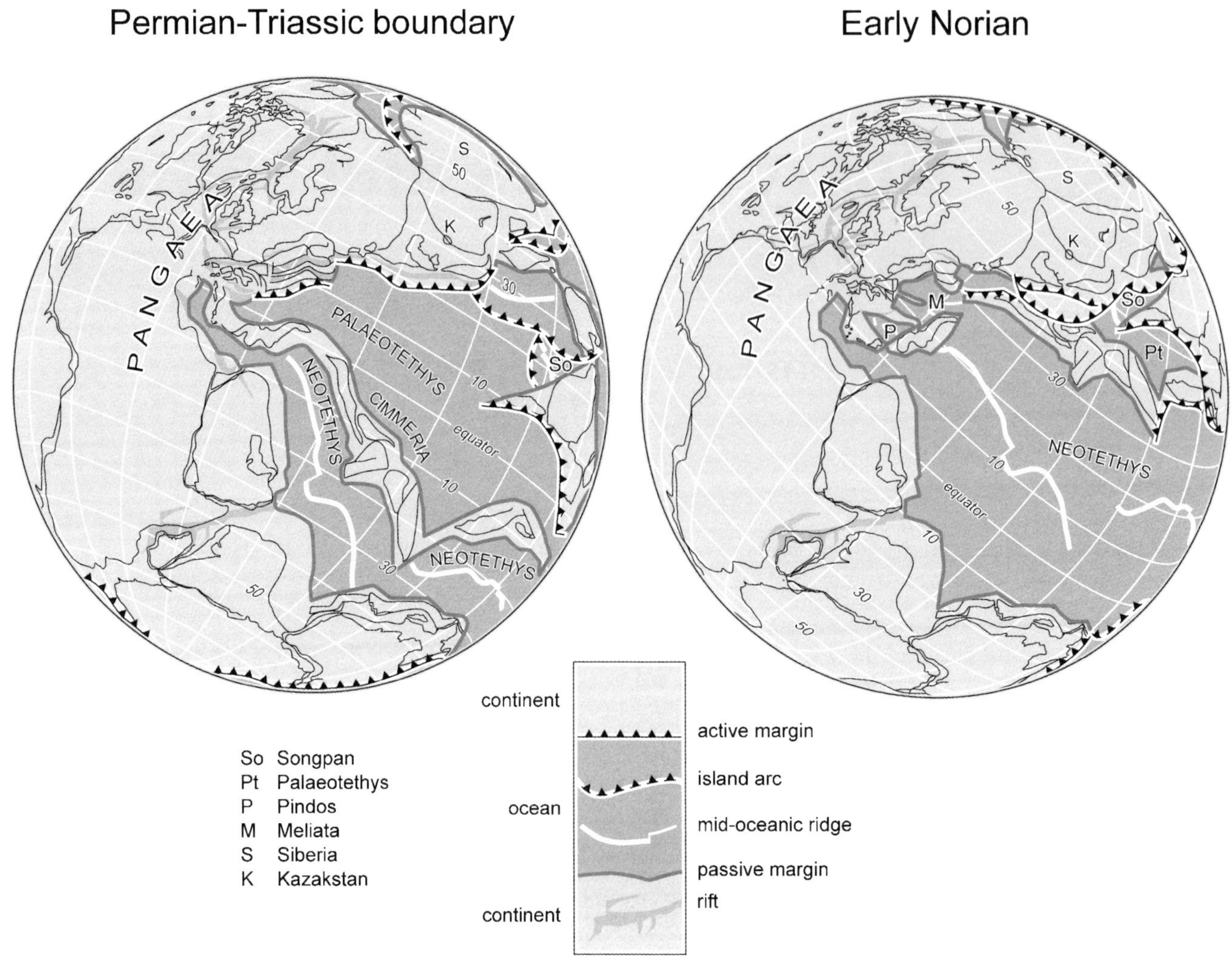

Fig. 13.1. Plate tectonic reconstruction for the Tethys region for the Permian–Triassic boundary and the early Norian. Modified from Stampfli & Borel (2002).

successions can be broadly divided into a Lower Triassic fluvial sandstone, red-bed dominated succession, overlain by a Middle and Upper Triassic succession that is dominated by mudstones, carbonates and evaporites. The extent to which this reflects palaeoclimatic or eustatic sea-level changes, end-Permian biotic recovery or regional changes in the tectonic stress regime is debatable (Ziegler 1990; Jacquin & de Graciansky 1998; Benton & Twitchett 2003). The timing of this switch, at the transition from the Lower to Middle Triassic in the red-bed succession of central Europe, also coincides with changes in the depositional style in many Tethyan successions on the southern European Margin (Ruffer & Zuhlke 1995; Gianolla & Jacquin 1998). Central European areas that were adjacent to local clastic sources, such as the fringes of Fennoscandia, or the dissected fragments of the Variscan Mountains, may not show the same pattern and timing of changes.

In the Boreal Ocean to the north, widespread marine clastic sedimentation persisted on the extensive Arctic and Barents Sea shelves throughout the Permian into the Early and Mid Triassic (Fig. 13.2) (Doré 1991). Early and Mid Triassic marine sedimentation may have episodically extended southwards through the Norway–Greenland rift system (Doré 1991). In contrast to much of northern Europe, the Boreal Ocean experienced its major environmental change during the Late Triassic (Carnian), when extensive deltaic systems prograded over many marine basins, which had existed for much of the preceding Permian and Triassic (Doré 1991; Skjold *et al.* 1998).

The red-bed Triassic successions in Europe were characteristically deposited in graben and extensional basins, with a variety of orientations, which may reflect both underlying basement structures and regional stress fields (Fig. 13.2) (Peacock 2004). The relative timing of basin formation and filling was quite variable. Locally an Early Triassic formation and filling-phase occurred in areas such as the Irish and Scottish Atlantic margins and northern North Sea. Major extension does not appear to have taken place in many European areas until Mid or Late Triassic times (Zeigler 1990), as in many graben systems in North Africa and the eastern seaboard of North America.

The last stage of the Triassic (the Rhaetian) witnessed marine flooding of many of the continental basins of Europe, with the exception of some in Eastern Europe and those fringing Fennoscandia. This pre-empted the marine connection of the Tethys and Boreal Oceans in the Jurassic through the Greenland–Norway rift system.

Triassic stratigraphy

The Triassic, as originally proposed by Friedrich August von Alberti in 1834, comprised three lithostratigraphic subdivisions based on the German succession, the Bunter, Muschelkalk and Keuper. The Triassic Period is now subdivided into seven stages (Induan, Olenekian, Anisian, Ladinian, Carnian, Norian and

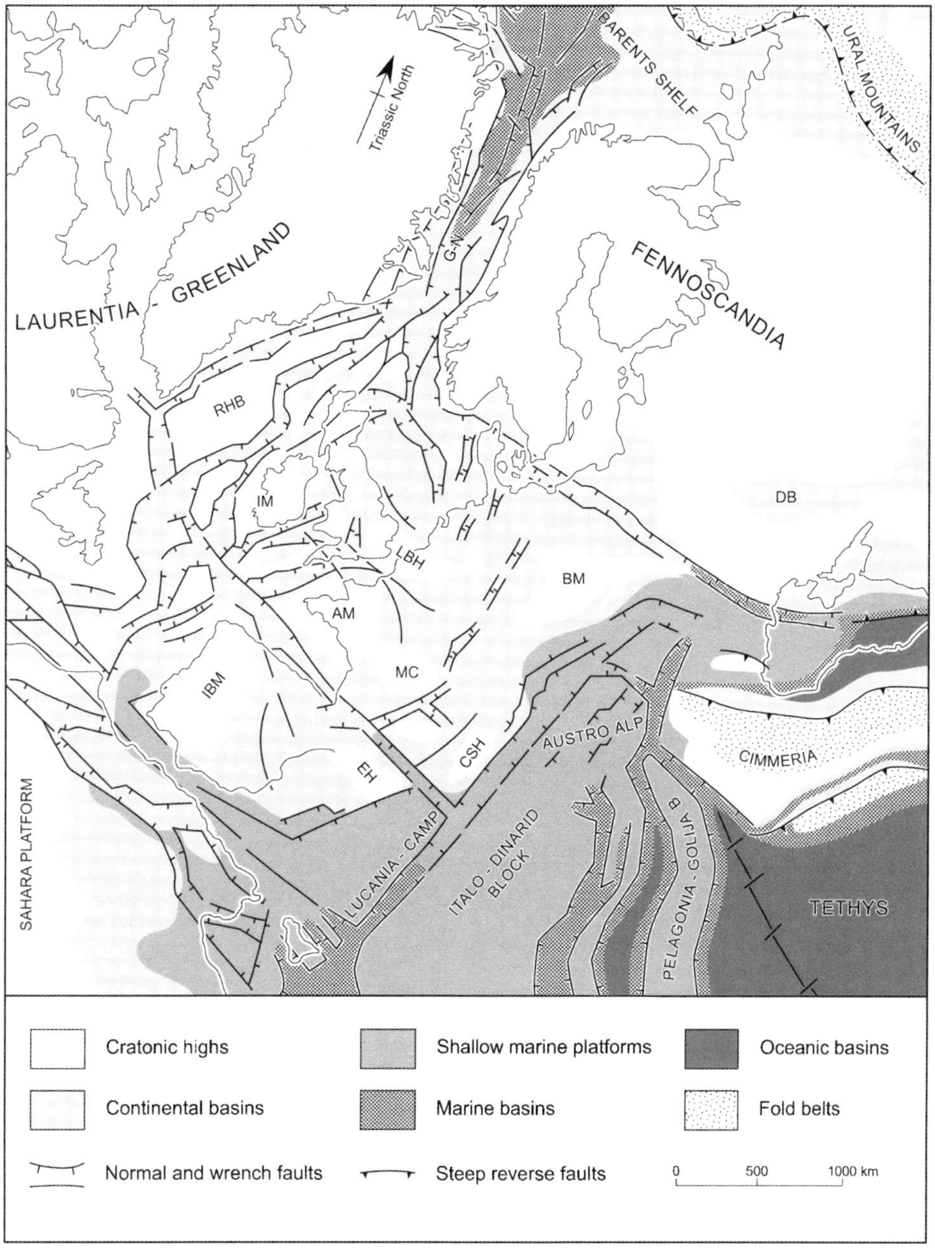

Fig. 13.2. Triassic basins and palaeogeography of Europe and the North Atlantic region in the Lower Triassic. Major rift basins are also indicated with fault ornament. Major 'upland' areas are AM, Armorican Massif; BM, Bohemian Massif; CSH, Corsica–Sardinia High; EH, Ebro High; IBM, Iberian Mesata; IM, Irish Massif; LBH, London–Brabant High; MC, Massif Central High; RHB, Rockall–Hatton Bank. Basins areas are: DB, Donetz Basin; G–N, Greenland–Norwegian rift. Modified from Ziegler (1990).

Rhaetian). Chronostratigraphic division of the Triassic succession in England and Wales is made problematic by the rarity of fossils, hence the British onshore Triassic is broadly divided into three lithostratigraphic units, the Sherwood Sandstone Group, the Mercia Mudstone Group and the Penarth Group (Figs 13.3 & 13.4). In this work we have used a subdivision of these groups based on the schemes of Warrington *et al.* (1980), Ritchie *et al.* (1996) and Howard *et al.* (2006), which attempt to define a UK wide system of formation subdivision, which is still in the process of refinement (Figs 13.3, 13.4).

Significant advances have begun to resolve some of the correlation and dating problems in the Triassic. Palynologists using fossil spore and pollen assemblages have begun to give sometimes broad and occasionally precise chronostratigraphic definitions, through links to section-based studies, outside the UK, that have ammonite or conodont-dated successions. (Warrington 1996). The use of borehole sonic and gamma-ray log signatures in the subsurface has also substantially improved lithostratigraphic schemes and intra-basin correlation (Lott *et al.* 1982; Howard *et al.* 2006). The use of magnetostratigraphy in the next decades holds the promise of providing a more precise chronostratigraphy, at the substage level (or better) for much of the unfossiliferous and poorly dated British Triassic (Hounslow & McIntosh 2003; Hounslow *et al.* 2004).

Defining the base of the Triassic in England and Wales has proved problematic in many areas, in that at best a succession of unfossiliferous red-beds marks the transition from the fossiliferous Zechstein into the Triassic. Magnetostratigraphic and biostratigraphic work in Germany accurately places the Permian–Triassic boundary there within the lowest part of the Calvörde Formation (Lower Buntsandstein; Szurlies *et al.*

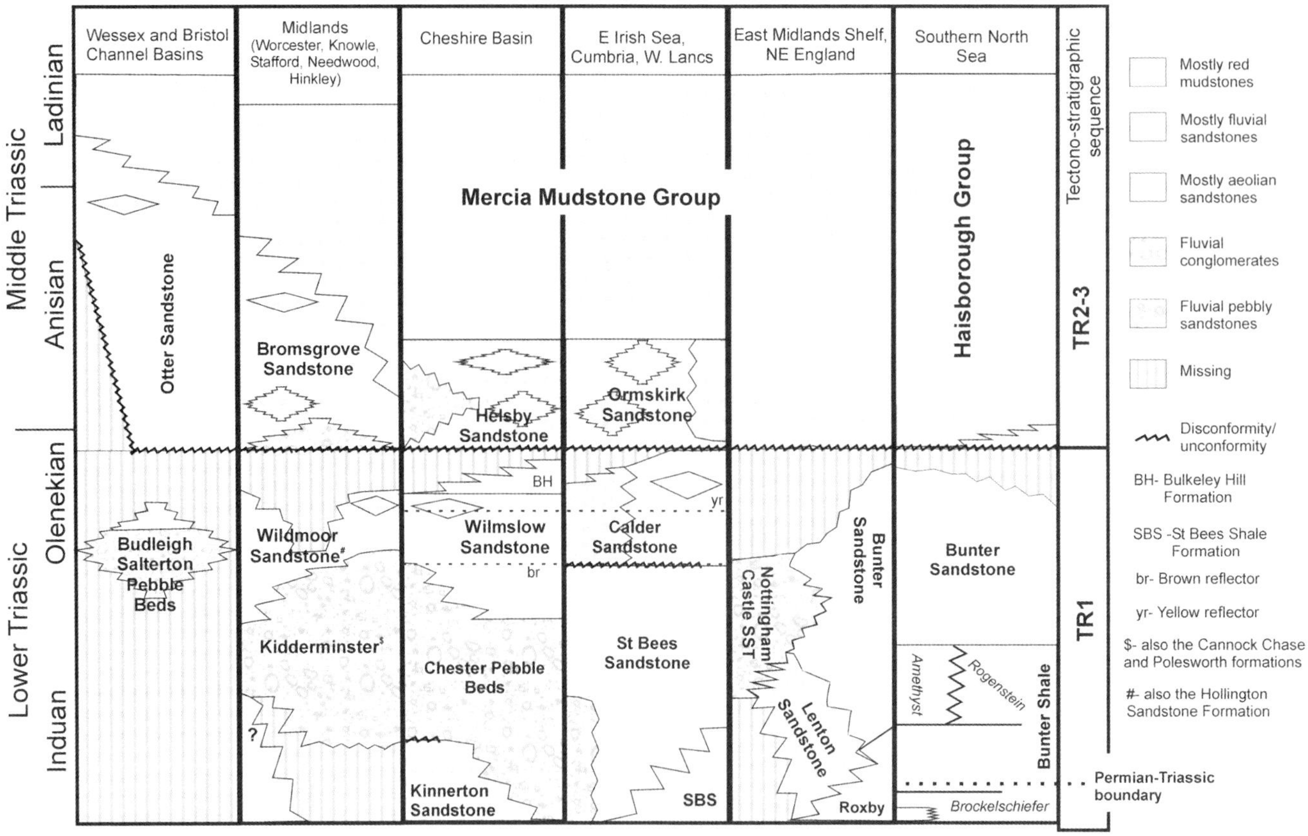

Fig. 13.3. Stratigraphy of the Sherwood Sandstone Group in England and Wales. In the southern North Sea the Bunter Sandstone and Shale comprise the Bacton Group. Based on Warrington *et al.* (1980, 1999), Cameron *et al.* (1992), Barnes *et al.* (1994), Jackson *et al.* (1995), Ritchie *et al.* (1996) and Rees & Wilson (1998).

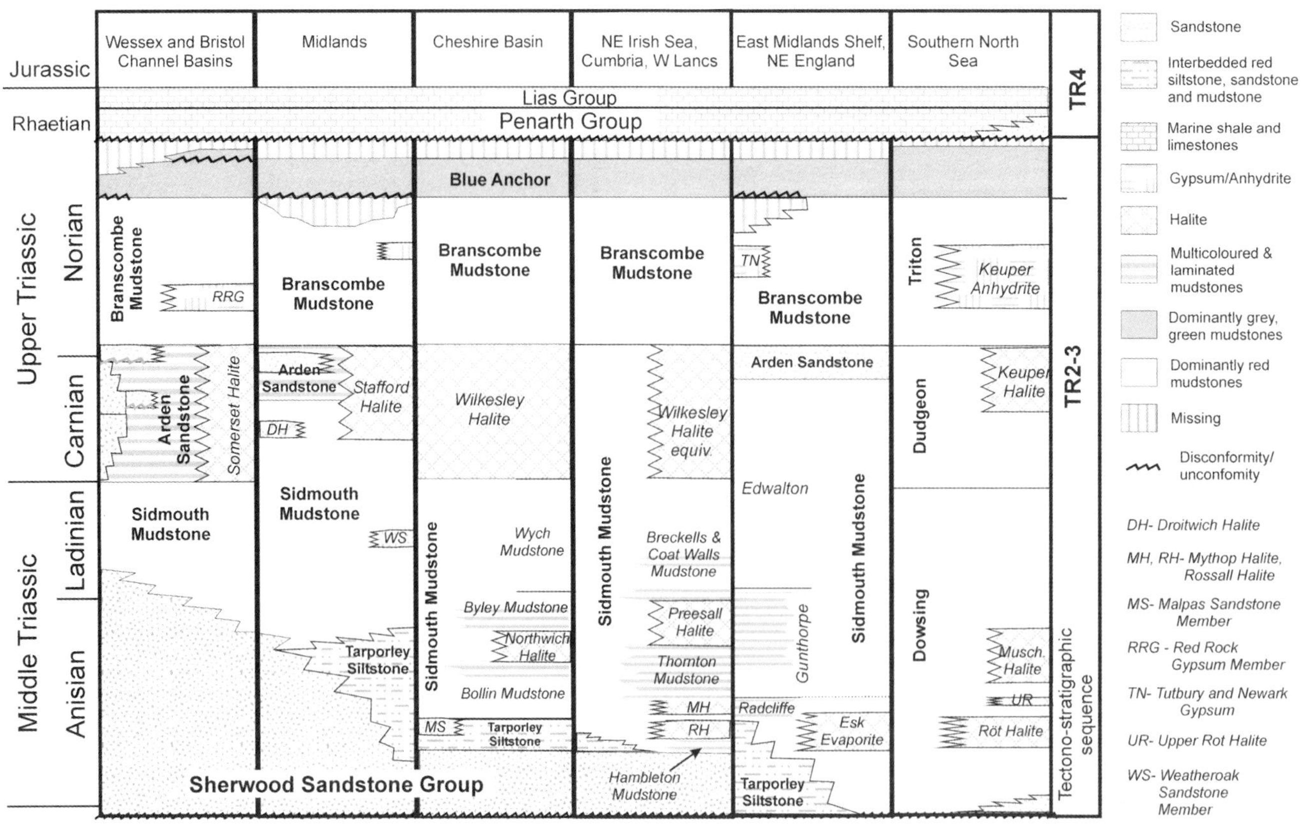

Fig. 13.4. Stratigraphy of the Middle and Upper Triassic in England and Wales. Based on Old *et al.* (1991), Cameron *et al.* (1992), Richie *et al.* (1996), Barclay *et al.* (1997) and Howard *et al.* (2006).

2003). Based on geophysical log correlation, this equates in the Dutch southern North Sea to cycle 5 of the Lower Buntsandstein – a level near the base of the UK offshore Bunter Shale Formation (part of the Bacton Group; Fig. 13.3), a short distance above the Bröckelschiefer Member (Cameron *et al.* 1992). This is a level that can also be recognized locally in the East Irish Sea Basin (Jackson *et al.* 1995). In other areas major changes in lithology or well-log response, which mark the base of the Sherwood Sandstone Group, have been chosen arbitrarily (Fig. 13.3).

Triassic palaeontology

The Triassic witnessed the lowest diversity of invertebrate families since the Cambrian, a reflection of the slow recovery rate of biota after the massive extinctions of the Late Permian (Chapter 12). In the UK the first major evidence of the new Mesozoic flora and faunas occurred during the Anisian (Bromsgrove Sandstone Formation, etc.). Punctuated extinctions appear to have characterized the Late Triassic, even though there is scant evidence of this in the British Triassic (Benton *et al.* 2002; Benton & Twitchett 2003). These Late Triassic extinctions, which include the extinctions at the Triassic–Jurassic boundary, globally show a larger loss of invertebrate families than that produced at the end of the Mesozoic (Hallam 2002; Tanner *et al.* 2004).

Early–Mid Triassic palaeontology

The Lower Triassic units of the Sherwood Sandstone Group are largely devoid of significant fossils, except for rare insects and invertebrates trace fossils, vertebrate footprints, conchostracans, (crustaceans), and a single fish (King & Benton 1996; Powell *et al.* 2000*b*) (Fig. 13.5a). Some of these records of vertebrate tracks have been discounted by King & Benton (1996) as inorganic structures. It is not until the Anisian age units of the Sherwood Sandstone Group that significant numbers of fossils are found in the UK (Fig. 13.5b). The reasons for this may be twofold. First, the biotic recovery from the end-Permian extinctions appears to have been slow, and diversity probably did not recover to Late Permian levels until the Mid Triassic (Erwin *et al.* 2002; Benton & Twitchett 2003). Secondly, fluvial systems which dominate the Early Triassic across many European basinal areas was not a good environment for fossil preservation.

The Otter Sandstone Formation (Wessex Basin), Bromsgrove Sandstone Formation (Midlands) and the Helsby Sandstone Formation (Cheshire Basin) all contain tetrapod remains and associated trace fossils (Benton 1997; Warrington *et al.* 1999). These faunas are dominated by reptilian (rhynchosaur) remains and include two or three types of temnospondyl amphibians as well as fish species (Milner *et al.* 1990; Benton 1997; Spenser & Storrs 2002). Footprints left by tetrapods occur in Cheshire (e.g. the Helsby Sandstone) and in other units of Anisian age (Benton *et al.* 1994; Benton & Spenser

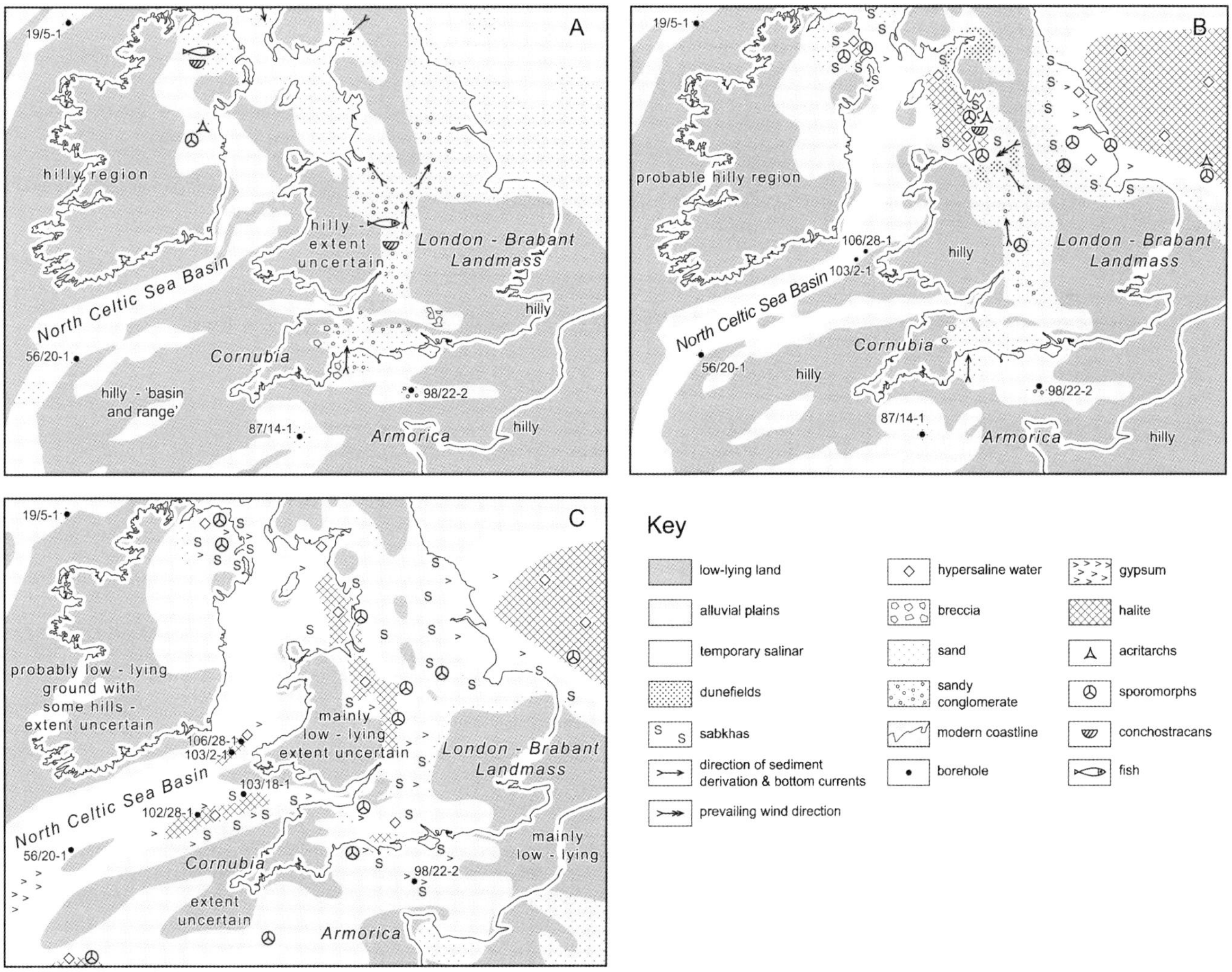

Fig. 13.5. Palaeogeography and palaeoenvironments during: (**a**) Early Triassic (Kidderminster Formation and equivalents); (**b**) early Anisian (Röt Halite Member and equivalents); (**c**) Carnian (Keuper Halite Member and equivalents). Modified from Warrington & Ivimey-Cook (1992).

1995). Micro- and macrofloral remains are common to rare in interpreted flood-plain deposits of meandering river systems. These include equisetalean plants of damp environments and conifers of dryer environments. The microfloral remains suggest dominance by gymnosperms, principally conifers, along with cycadalean and ginkgoalean components (Warrington & Ivimey-Cook 1992). Other fossils known from these and adjacent units are conchostracans (Fig. 13.7b), a scorpion, a single *Mytilus* (Wills 1970) and ostracods (Old *et al.* 1991; Warrington & Ivimey-Cook 1992). Equivalent basinal facies are largely devoid of fossils except for rare intervals of land-derived miospores. The Tarporley Siltstone Formation (Fig. 13.4) in Cheshire contains plant remains, inarticulate brachiopods (*Lingula*) and, rarely, *Rhynchosaurus* sp. tetrapods (Warrington *et al.* 1980).

Late Triassic palaeontology

Within the Mercia Mudstone Group two formations stand out as being particularly important for containing fossils, the Arden Sandstone Formation and the Blue Anchor Formation (Fig. 13.4). Other units are generally barren except for rare *Botryococcus* and *Plaesiodictyon* (green algae) fragments found in some units (Talbot *et al.* 1994; Barclay *et al.* 1997), and some land-derived spore and pollen-rich intervals (Warrington & Ivimey-Cook 1992; Barclay *et al.* 1997). Land vertebrates were present, evident by tracks left behind in environments marginal to the red mudstones in South Wales (Tucker & Burchette 1977).

The Carnian Arden Sandstone Formation and equivalent intervals have yielded plant megafossils (stems, fructifications), and miospores indicating bryophytes, sphenopsids and pteropsids (Old *et al.* 1991). The fauna includes possibly marine bivalves (*Nucula*?, *Pholadomya*?), conchostracans (*Euestheria*) and shark (teeth) remains (Old *et al.* 1991). Land vertebrates are represented by labyrinthodont amphibians and rhynchosauroid remains, along with vertebrate tracks. Soft-bodied organisms were also widespread, evident by a rich diversity of trace fossils (Fig. 13.7d) (Old *et al.* 1991; Barclay *et al.* 1997). Reworked fossil material may occurs at the base of the sandstones in the Arden Sandstone Formation, as abraded and concentrated vertebrate material (Ruffell 1991).

In the British Triassic, the Penarth Group contains the most diverse macro- and microfossils. Consequently it has attracted a long history of study, much of it described in detail in Swift & Martill (1999) and Benton *et al.* (2002).

The palynological (terrestrial spore and pollen) records from the upper part of the Branscombe Mudstone Formation, through the Blue Anchor Formation to the latest Triassic in the basal parts of the Lias Group have been intensively studied at many sections and boreholes in England and Wales (Warrington & Whittaker 1984; Waters & Lawrence 1987; Warrington & Ivimey-Cook 1995; Edwards 1999). Where these records appear to be complete, such as in north Somerset and South Wales, the spore and pollen diversity increases from the Branscombe Mudstone Formation, through the Blue Anchor Formation into the upper part of the Penarth Group. Many of the taxa that contribute to this diversification are species of trilete spore genera. Most of the common taxa range upwards through the Blue Anchor Formation, into the Penarth Group and in some cases, into the Lias Group. An abrupt reduction in microfloral diversity occurs in the uppermost parts of the Penarth Group and into the lowermost Lias Group. Above this level, miospore diversity is much lower, a feature which persists well into the Lower Jurassic. This decrease in diversity is also reflected in the bivalves, since the Penarth Group bivalve assemblages are distinct from the lower diversity assemblages that characterize the pre-planorbis parts of the Lias Group (Whittaker 1978). One of the key, stratigraphically useful, microfossils of this interval is the dinoflagellate cyst *Rhaetogonyaulax rhaetica*, which ranges from the higher part of the Blue Anchor Formation into the lowest part of the Lias Group, and which allows correlation to unequivocal Rhaetian strata elsewhere in Europe (Warrington & Ivimey-Cook 1995; Swift 1999). Conodonts, which became extinct prior to the Jurassic, are typically recovered from the upper parts of the Penarth Group and in some areas from the Pre-Planorbis Beds of the Lias Group (Swift 1995*a*).

Tectono-stratigraphic sequences

Within England and Wales the Triassic can be divided into three first-order tectono-stratigraphic sequences, each bounded at the base by a major regional unconformity, overlain by a transgressive succession which may overstep onto older rocks (Figs 13.3 & 13.4). The sequence naming utilizes that used in the European-wide regional synthesis by Gianolla & Jacquin (1998). In the case of the British Triassic the unconformities bounding the first-order sequences also separate sedimentary successions that display major differences in environmental conditions.

Sequence TR1. The lower boundary in basinal areas separates the Permian Zechstein from the overlying units of the Sherwood Sandstone Group. Within the southern North Sea the base of this sequence is the base of the Bacton Group, which may show a conformable contact with the underlying Zechstein Group (Fig. 13.3). In the East Irish Sea Basin away from basin margins, the base of TR1 may be within the St Bees Shale. Within many non-basinal areas the base of sequence TR1 corresponds with the locally unconformable base of the Sherwood Sandstone Group which may overstep onto older rocks. Disconformities within this sequence are known in the East Irish Sea Basin at the base of the Calder Sandstone Formation (Barnes *et al.* 1994), and in the succession on the eastern margin of the UK southern North Sea, where unconformities in the German and Dutch Lower Triassic can be traced into UK waters using wire-line logs (Geluk 2005*a*, *b*).

Sequence TR2–3. The Sherwood Sandstone Group in western and SW Britain is divided by the unconformable base of sequence TR2–3 (Fig. 13.3). In eastern England (east of Charnwood) and the southern North Sea this unconformity lies either at or slightly below the base of the Mercia Mudstone Group (or Haisborough Group offshore; Cameron *et al.* 1992; Geluk & Röhling 1999). This unconformity is equivalent to the Hardegsen unconformity (H-unconformity) of the Dutch and central European basins (Wills 1970; Holloway *et al.* 1989; Cameron *et al.* 1992; Geluk 2005*b*). The age of the H-unconformity in the German basin is thought to be late Olenekian (latest Early Triassic; Gianolla & Jacquin 1998; Kozur 1999). In some UK areas such as the East Irish Sea Basin, the base of TR2-3 is rather more disconformable than unconformable (Jackson *et al.* 1995). In basins adjacent to the East Irish Sea, this unconformity has been difficult to identify, leading some workers to deny its existence in these areas (Meadows pers comm. 2000). However, onshore basins, such as those in the Midlands and the Cheshire Basin, indicate the unconformable nature of the base of TR2–3. This unconformity is here referred to as the H-unconformity. One or more other disconformities or unconformities within TR2–3 may have regional significance in England and Wales, to subdivide the sequence. Workers in other European basins subdivide this sequence into separate sequences TR2 and TR3 with a boundary either in the late Ladinian or early Carnian (Gianolla & Jacquin 1998). The significance of a major unconformity or disconformity at this level in the UK is unproven, but may correspond to a level within the Arden Sandstone Formation at the base of one or more of the sandstone units (Fig. 13.4) (Ruffell & Shelton 1999). A better case can be made for a regional disconformity at the base of the Blue Anchor Formation (see below).

In the Dutch sector of the southern North Sea the early Kimmerian I unconformity occurs above the Red Keuper Evaporite Member (Geluk 2005*b*; equivalent to the Keuper Anhydrite Formation of the Triton Formation), which is currently unrecognized in the UK onshore. Future work will probably allow a subdivision of TR2–3 to be established in the UK.

Sequence TR4. The base of the Penarth Group locally forms an unconformity, or locally disconformable boundary, with a substantial part of the underlying beds removed by erosion (Fig. 13.4) (Mayall 1981; Hounslow *et al.* 2004). The equivalent of the Penarth Group in the southern North Sea and Germany, the Sleen and Exter formations of the Dutch and German sectors respectively, are separated from underlying units by the Kimmerian II unconformity of Ziegler (1990) and Geluk (2005*b*) and the equivalent early Cimmerian unconformity of Jacquin & de Graciansky (1998). Sequence TR4 extends into the Middle Jurassic as the 'Ligurian cycle' of Jacquin & de Graciansky (1998). In the England and Wales the Triassic part may be subdivided by a disconformity within the upper part of the Penarth Group (Hesselbo *et al.* 2004).

Tectono-stratigraphic sequence TR1 shows the largest regional differences in lithostratigraphic terminology and lithology (Fig. 13.3), reflecting the fact that separate basins had to a large extent separate sedimentation histories. This regional variation in geology is progressively lost in younger strata upwards through sequence TR2–3 (Fig. 13.4), such that by the time of deposition of the Blue Anchor Formation major regional differences are limited. For this reason the regional differences in the Triassic below the Blue Anchor Formation are discussed in separate regional sections below, while the Blue Anchor Formation and Penarth Group are discussed in a single section each.

Triassic environments in the Sherwood Sandstone Group

Within sequence TR1

Where the base of the Triassic can be located confidently (such as in the southern North Sea and East Irish Sea basins), fluvial sandstone-dominated units overlie a mudstone dominated unit in the lower parts of the British Lower Triassic (Fig. 13.3). These sandstones show northwards directed palaeocurrents (Fig. 13.5a) (Cameron *et al.* 1992; Jones & Ambrose 1994). Within the onshore basins (where the Permian–Triassic boundary can be less confidently located) the fluvial sandstones below the H-unconformity are dominated by evidence of a river system, which Wills (1956) termed the 'Budleighensis' river system (Fig. 13.5a). Within the western onshore basins this river system seems to have been one which was sourced from the Variscan uplands of Brittany and the southern English Channel (possibly also Iberia), and then flowed northwards through the Wessex Basin, through the Worcester Basin, the Midland basins (Knowle, Needwood, Stafford, Cheshire) and on into the East Irish Sea Basin (Figs 13.5a & 13.6). A distributary branch perhaps flowed periodically eastwards onto the East Midlands Shelf through the Charnwood Gap. Evidence for the generally northwards directed palaeocurrents in the Midland basins (Steel & Thompson 1983; Chisholm *et al.* 1988; Rees & Wilson 1998) is reinforced by the general southerly source of pebbles and heavy minerals (Fig. 13.5a) (Thomas 1909; Cocks 1993).

Within the Midlands these deposits are dominated by pebbly sandstones and conglomerates (Fig. 13.3), which have been interpreted as the deposits of a braided river, flowing within well-confined channels (Steel & Thompson 1983). This river system probably had large seasonal changes in discharge, evident by cross-stratified sandstones deposited at stages of lower flow, although it is perhaps doubtful if this system was ephemeral in nature (Smith 1990) for there is little evidence of interbedded aeolian sandstones in basinal settings (Rees & Wilson 1998). However, at basin margins in Cheshire and the East Midlands the oldest units of the Triassic are predominantly aeolian sandstones, whose deposition perhaps extended uninterrupted from the Permian. These are the Kinnerton Sandstone Formation (and Permian Collyhurst Sandstone Formation) in Cheshire and the Lenton Sandstone Formation (in part Permian) in the East Midlands (Fig. 13.3), which interdigitate laterally with fluvial sandstones and red playa mudstones.

Within each of the basins it is clear that locally sourced material was also an important component of the fluvial systems (Chisholm *et al.* 1988; Rees & Wilson 1998; Smith & Edwards 1991), indicating drainage relief at the basin margins. However, Triassic palaeovalleys have been recognized at only a few locations (Taylor 1968; Chisholm *et al.* 1988; Worssam & Old 1988), although these must have been relatively common features in the Early Triassic landscape before being erased by later events. These pebbly-sandstone deposits pass upwards into finer-grained sandstones with mudstone intervals (e.g. Wildmoor and Wilmslow Sandstone formations), indicating the upwards transition from braided to meandering river systems.

In the more northerly basins (East Irish Sea, Solway Firth and Vale of Eden basins; Fig. 13.6) Early Triassic river systems were sand dominated, evidence of the lower flow regime of the river system (Fig. 13.5) (Brookfield 2004). Stacked channel sandstones formed predominantly in a low sinuosity sandy-braided river system (Jones & Ambrose 1994) with local evidence of aeolian, sand-sheet and playa-type environments (Meadows & Beach 1993; Brookfield 2004). The aeolian-derived component may increase upwards until it forms the major component in some basin-margin situations, such as in West Cumbria (Jones & Ambrose 1994). The Budleighensis river system may have terminated in the East Irish Sea or Solway Firth basins as a series of continental playa basins, or perhaps more likely flowed westwards or northwards towards the present-day proto Atlantic rift.

Within sequence TR2-3

The 'Budleighensis' river system continued to flow northwards through the Wessex and Midlands basins, evident by palaeocurrent trends (Henson 1972; Ireland *et al.* 1978; Meadows & Beach 1993; Powell *et al.* 2000*b*; Svendson & Hartley 2001), but the depositional style radically changed, to fluvial sandstone dominated in the southern UK and Midlands basins, to aeolian-sandstone dominated in the NW England basins (Fig. 13.5b). This shift was perhaps a consequence of a major reduction in the gradient of the Budleighensis river system. The Otter Sandstone and Bromsgrove Sandstone formations of the southern basins are mostly stacked channel sandstones, formed in sinuous or straight channelled river systems, which produced copious amounts of intraclastic debris, both from pedogenic carbonates and erosion of overbank mudstones (Old *et al.* 1991). The abundance of pedogenic carbonates in these Anisian sandstones indicates that stable landscape surfaces (i.e. soils) were able to develop. Also for the first time in the British Triassic, significant faunal and floral elements are found including plants, invertebrates, arthropods, fishes and reptiles, all evidence of a fluvial landscape punctuated by lakes and plant growth more extensive than in the Early Triassic (Purvis & Wright 1991; Benton & Spenser 1995; Spenser & Storrs 2002). River systems from locally derived sources remained important at basin margins, as indicated by locally sourced breccias and conglomerates forming the major part of some fluvial successions (Fig. 13.5b) (Barclay *et al.* 1997; Bridge *et al.* 1998).

In the northern basins (Fig. 13.6; Cheshire, East Irish Sea, Vale of Eden basins) an interfingering succession of fluvial channel sandstones and aeolian sandstones dominate the succession. Fluvial systems in the East Irish Sea Basin were

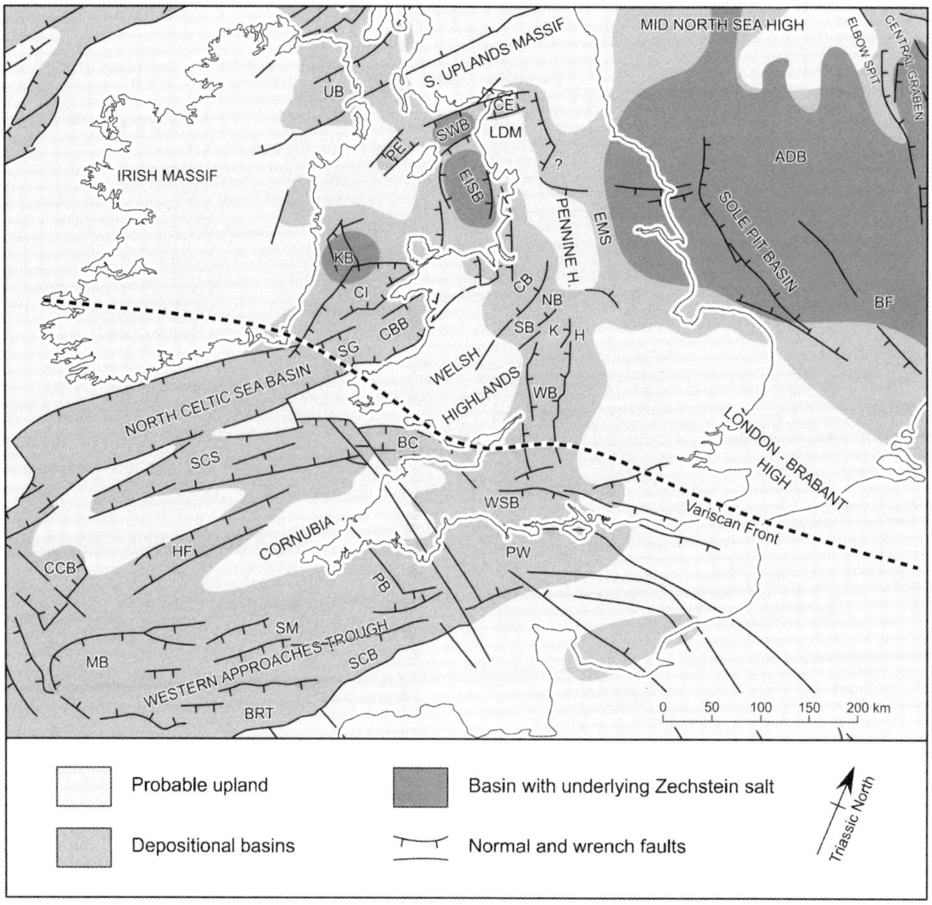

Fig. 13.6. Important Triassic basins and highs of England and Wales. Basins: ADB, Anglo Dutch; BC, Bristol Channel; BRT, Britanny Basin; CB, Cheshire; CBB, Cardigan Bay; CCB, Cockburn CE, Carlisle/Vale of Eden; CI, Central Irish Sea; EISB, East Irish Sea; EMS, East Midlands Shelf; H, Hinkley; HF, Haig Frais; K, Knowle; NB, Needwood; KB, Kish Bank Basin; MB, Melville; PB, Plymouth Bay; PE, Peel; PW, Portland–Wight; SB, Stafford; SCB, South West English Channel; SCS, South Celtic Sea; SG, St Georges'; SM, St Mary's; SWB, Solway Firth; UB, Ulster; WB, Worcester; WSB, Wessex. Faults and highs: BF, British Forties High; LDM, Lake District Massif.

located adjacent to bounding faults, indicating contemporaneous fault activity (Meadows & Beach 1993; Holliday *et al.* 2004). At the basin margin in west Cumbria, the river channels flowed into the East Irish Sea Basin in a WSW direction either due to control by wind-aligned linear dune systems or due to the local palaeoslope (Jones & Ambrose 1994). These channels formed part of large low gradient sand-flat systems, which flooded seasonally and strongly controlled the geometry of the aeolian bedforms both within the East Irish Sea Basin and farther south in the Cheshire Basin (Goodall *et al.* 2000; Mountney & Thompson 2002). These aeolian sand seas seem to have been most prevalent in basin margin locations such as those fringing the Lake District (Arthurton *et al.* 1978; Brookfield 2004).

Triassic environments in the Mercia Mudstone Group

The transitional interval between the sandstone-dominated Sherwood Sandstone Group and the mudstone-dominated Mercia Mudstone Group has always posed problems for subdivision and interpretation. In the older literature this transitional interval is occupied by the 'waterstones' in the Midlands and NW England (Wills 1970) but has been more recently assigned to the Tarporley Siltstone Formation and grouped with the Mercia Mudstone Group (Howard *et al.* 2006). This interval is characterized by interbedded red mudstones/siltstones and fine-grained sandstones, in approximately equal proportions, locally with high abundances of mica on bedding planes. It typically has numerous well preserved sedimentary structures, including laminations, ripples, mudcracks, slumping, gypsum nodules, salt psuedomorphs and invertebrate trace fossils (Elliot 1961; Wills 1970; Ireland *et al.* 1978). Ireland *et al.* (1978) interpreted the Tarporley Siltstone Formation of the Cheshire Basin as 'intertidal', although it probably represents the deposits of a lake-margin with interdigitation of fluvial sheet-floods and a playa–mudflat system. The transitional lithologies are diachronous (Fig. 13.4), and have been locally assigned to either the Mercia Mudstone or the Sherwood Sandstone (Old *et al.* 1987; Barclay *et al.* 1997; Powell *et al.* 2000*b*; Hounslow & McIntosh 2003; Gallois 2004).

The 'red' mudstones

During the Mid–Late Triassic extensive successions of predominantly red mudstones developed in England and other parts of NW Europe. There are broadly three types of mudstone facies present.

Blocky mudstones (Fig. 13.7a). These are calcareous and dolomitic silty mudstones, which commonly weather at outcrop to small blocks up to several cm in size (Arthurton 1980). They are predominantly dark–pale brownish red in colour, although pale grey–dark greenish grey mudstones are also not uncommon. The grey and grey-green coloured blocky mudstones are dominant in the Arden Sandstone Formation and the Blue Anchor Formation. The mudstones are a mixture of quartz silt, clay,

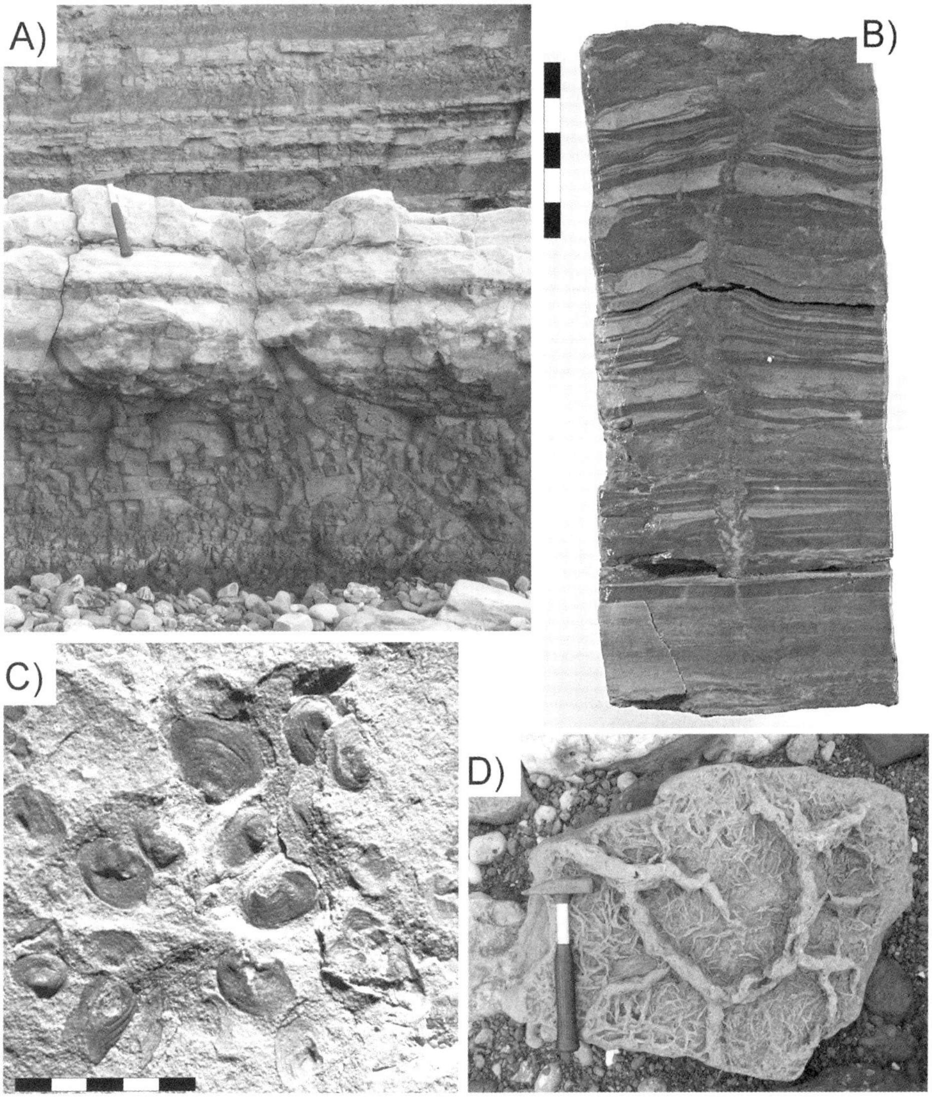

Fig. 13.7. (**a**) Pale carbonate-rich beds (bed A11 of Whittaker & Green 1983) interbedded with blocky-fracturing red and green mudstones. Branscombe Mudstone Formation, St Audries Bay, North Somerset. (**b**) Laminated siltstone-mudstones couplets, showing evidence of poorly preserved ripples, bisected by a large vertical desiccation crack (Thornton Mudstone Member, Churchtown borehole, Lancashire, 108 m). (**c**) Conchostracans (crustaceans – *Euestheria minuta*), from Cycle F of the Thornton Mudstone Member, Churchtown Borehole, depth 122.8 m. (**d**) Large reticular mudcracks with smaller intervening burrows of trace fossil *Fuersichnus* (?) on the base of a fallen sandstone bed from the Arden Sandstone Formation, Hook Ebb, South Devon. Photographs (b) and (c) reproduced by permission of the British Geological Survey, ©NERC, IPR/53-43c, all rights reserved. Photographs (a) and (d) courtesy of Richard Porter. Hammer 25 cm in length for scale in (a) and (d). Photograph (b) scale in cm, photograph (c) scale in mm.

dolomite and locally calcite, often with a few well-rounded scattered aeolian sand grains. Rarely do the blocky mudstones contain macrostructures, which can be easily observed at outcrop, other than cyclical variations in the size of the blocks the mudstones break into. The blocks commonly develop in weathered outcrops, and may in part be due to dissolution of variable amounts of evaporite or carbonate minerals. However, on sectioned surfaces, the mudstones may contain millimetre- or submillimetre-scale structures which may be evident by small changes in clay/silt or carbonate content. These sedimentary structures can be broadly classified into two end-member groups. Firstly there is a more common group of structures of branching clay films with wispy or ghost-like vague laminae. Microstructures include blocky structure (peds), clay coatings on coarser grains and former voids (Wilson 1990, 1993; Talbot *et al.* 1994; Barclay *et al.* 1997). Secondly, some beds may show a variety of breccia-like fabrics. Elliot (1961) recognized two types of intraformational breccias, which he termed 'vein breccias' and 'flow-breccias'. His 'vein breccias' were mudstone-filled, near-vertical cracks or veins separating paler and larger angular mudstone blocks. These features probably represent extensive brecciation due to penetrating, multiple generation shrinkage (Barclay *et al.* 1997), although the vein-like features are somewhat reminiscent of clay cutans in palaeosols. Elliot's 'flow-breccias' display a range of structures, from relatively undeformed beds displaying slickensided listric surfaces, and microscale faulting to true breccias displaying horizontal arrangement of angular siltstone and mudstone fragments. These fabrics may represent palaeosols (vertisols) or the results of penecontemporaneous halite or gypsum dissolution (Barclay *et al.* 1997; Wilson 1990, 1993). In unweathered outcrops or cores the mudstone beds may contain gypsum veins, nodules or dispersed gypsum grains (Wilson & Evans 1990; Talbot *et al.* 1994; Barclay *et al.* 1997). From the SW margin of the

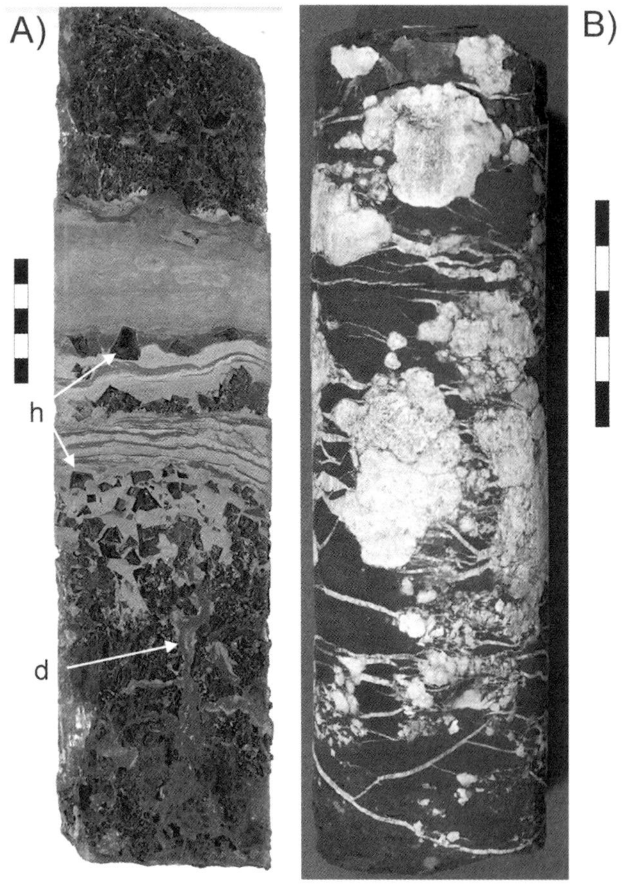

Fig. 13.8. (**a**) Grey mudstone bed sandwiched between upper and lower halite beds (darker grey). The mudstone shows large halite crystals (h) due to *in situ* production (Haselgebirge halite). Note the vertical, red-mud-filled structures (desiccation crack (d) ?) within the lower halite bed. Mythop Halite Member, 301.3 m in Thornton Cleveleys borehole, Lancashire. (**b**) Nodular and vein gypsum within red mudstones (Breckells Mudstone Member. 147.2 m in Hacensall Hall borehole). Reproduced by permission of the British Geological Survey, ©NERC, IPR/53-43c, all rights reserved.

Worcester Basin, Milroy (1998) has described a variety of features such as curvilinear slickensided planes, pseudo-anticlines, root traces, colour mottling and pedo-tubules, thought to be related to soil formation.

Laminated mudstones. These form a heterogeneous mudstone and siltstone facies characterized by siltstone (or sandstone) laminae, sometimes graded and up to 20 cm thick (Fig. 13.7b). The siltstones may be cross-laminated, and display current or wave-ripples, load structures, clay-breccias, small channel-like structures or bioturbation (Klein 1962; Talbot *et al.* 1994; Wilson & Evans 1990). The lamination may be distorted and broken by polygonal mud-crack networks, or slumping (Fig. 13.13). Psuedomorphs after anhydrite and halite were recorded by Arthurton (1980) and Wilson (1990). The colour varies from dark red to pale grey to dark grey and dark grey laminae may be pyrite bearing. This facies appears to be more common in basinal successions such as in the East Irish Sea, or Cheshire basins, where it may be interbedded with the blocky-mudstones (Fig. 13.13), but is also common in units approximately at the level of the Arden Sandstone Formation, or some areas flanking the southern North Sea (Plate 22).

Carbonate-rich mudstones. These are typically very pale grey–white beds, rich in calcite or dolomite (Fig. 13.7a). In many respects these are carbonate-rich variants of the laminated or blocky mudstone facies, displaying a similar range of macroscale sedimentary structures, including graded laminae, wave and current ripples, intraformational breccias, channel structures and mud cracks. In addition they may contain primary detrital carbonate grains such as pedogenic concretional carbonate and ooids (Talbot *et al.* 1994). This facies is not common, but is more typical of the upper part of the Mercia Mudstone Group, particularly the Blue Anchor Formation and parts of the Arden Sandstone Formation.

Both the laminated facies and the carbonate-rich facies that show abundant sedimentary structures, were probably formed in a shallow lacustrine environment. The waters of these lakes periodically dried up, shown by the common desiccation cracks (Arthurton 1980; Wilson & Evans 1990; Talbot *et al.* 1994). The nearest modern analogue may be the Quaternary saline lakes of Australia (Rosen 1991; Talbot *et al.* 1994; Magee *et al.* 1995). The Mercia Mudstone carbonate facies possibly represents wetter climatic intervals than the clastic facies with the resulting lakes being either runoff-fed, or possibly groundwater fed (Talbot *et al.* 1994), producing brines of varying salinity (Taylor 1983).

Environmental significance of blocky mudstones

Opinions differ as to the depositional environment of the blocky mudstone facies; they have been interpreted as the deposits of distal alluvial playas (Tucker 1978; Talbot *et al.* 1994), permanent lakes or hypersaline epeiric seas (Wills 1970; Jeans 1978; Warrington *et al.* 1980) and soils (Milroy 1998). Talbot *et al.* (1994) addressed in detail the sedimentology of the red mudstones of the Branscombe Mudstone Formation in north Somerset, indicating that they lack the variety of sedimentary structures and lithofacies typical of Quaternary lacustrine deposits. Rather, a variety of microstructures including breccia-like structures (peds?), clay coatings on coarser grains and former voids, branching clay films and wispy laminae, were interpreted as products of the soil environment. They envisaged that the red mudstones were formed on low-angle alluvial fans with the clay and silt transported in ephemeral rivers as mud aggregates, an environment comparable to that found in the Late Quaternary of east-central Australia (Magee *et al.* 1995). However, the alluvial input was probably supplemented and modified by major aeolian reworking, as is the case today in the Australian analogue (Dare-Edwards 1984; Jefferson *et al.* 2000). If this analogue were applied literally it would suggest that much of the clay and silt was transported into the basins as mud pellets via alluvial or aeolian transport (cf. Dare-Edwards 1984). Indeed, the aeolian origin of the blocky mudstones is a persistent interpretation from earlier studies (Lomas 1907; Wills 1970) that have suggested this facies is produced by dry deposition from dust storms, much like the Quaternary Australian parna deposits (Jefferson *et al.* 2000). The lack of preservation of macroscale structures, Talbot *et al.* (1994) suggested, was due to activity of plants and burrowing insects assisted by the development of efflorescent salt crusts (Smoot & Castens-Seidell 1994; Goodall *et al.* 2000) formed by contemporaneous precipitation of groundwater salts within the mudstones (Arthurton 1980). If these deposits were transported by water and wind as mud aggregates, the chances of preservation of sedimentary structures are indeed slim, and the size of the larger particles might provide a better guide to depositional particle size (Dare-Edwards 1984).

Macroscale sedimentary structures intercalated within the blocky mudstones, often on silty or sandy horizons, are dominated by wave-produced microripples, associated with thin intercalations of the laminated or carbonate facies, often heavily mudcracked (Klein 1962; Arthurton 1980; Wilson 1990, 1993; Wilson & Evans 1990). In the marginal successions in South Wales and the Bristol region, the wave-dominated environment of some upper Mercia Mudstone deposits is indicated by wave cut-notches and lake-shoreline deposits (Tucker 1978; Milroy & Wright 2000). These contrast with the mudstones described and interpreted by Milroy (1998) as displaying a strong pedogenic overprint. These and other successions such as those in NW England, suggest small-scale cyclicity in environmental conditions evident by interbedding of blocky and laminated mudstones (Fig. 13.13) (Arthurton 1980; Wilson 1990, 1993).

A single environmental interpretation for the blocky red mudstone units of the Mercia Mudstone Group is too simplistic, the environment probably fluctuated between periods of extensive saline–brackish lake systems to other times when the lakes were restricted to basin centres, and pedogenic processes and dust transport were widespread on the basin margins. The constant influences were probably long-distance aeolian transport of silt and clay (perhaps as aggregates, Dare-Edwards 1984; Jefferson *et al.* 2000), along with ephemeral alluvial redistribution of mud and silt by flash floods in a low gradient desert environment. The primary depositional process responsible is commonly concealed because of pervasive fabric disruption caused by efflorescent crusts, the burrowing activity of plants and animals, soil development and the original sediment transportation as mud aggregates.

The evaporites

These occur as beds of either gypsum-anhydrite or halite. Halite is present both as relatively pure layered halite (< 5% sulphate) or interbeds of mudstone and halite, with large halite crystals (Fig. 13.8a). This latter Haselgebirge texture indicates that the halite was produced within the mudstones, from saline groundwater (Arthurton 1973). Tucker (1981) described giant (*c.* 10 m width) polygons within halite beds, indicative of long-term shrinkage by desiccation or de-watering. The halite is usually interbedded with blocky mudstones, but interbedding with the laminated facies is not uncommon (Arthurton 1980; Wilson 1990, 1993; Jackson *et al.* 1995). Laminated halite may show evidence of penecontemporaneous dissolution, erosion and collapse (Arthurton 1973; Wilson 1990, 1993). Halite-bearing mudstones are most common at or below the level of the Arden Sandstone Formation.

Gypsum (and celestite) occurs either as nodules or as more substantial beds, sometimes interbedded with mudstones (Curtis 1982). Secondary forms of either gypsum veins or dispersed gypsum crystals also occur in the other facies. It may display chicken-wire texture, or enterolithic texture (Tucker 1978), sometimes replaced by dolomite (Fig. 13.8b). These textures have been interpreted as a result of precipitation from saline groundwater, due to pedogenic or early diagenetic activity (Wright & Sandler 1994). The cliff sections east of Blue Anchor, on the north Somerset coast (Bristol Channel) are some of the best locations to observe these textures.

Environmental significance of evaporites

The occurrence of thick halite units within basinal successions of the Mercia Mudstone Group, has prompted some to suggest a marine water source for such units, based on the large quantity of salt required to form such deposits (Wills 1970; Arthurton 1980), and the Br, Sr, K and Mg content of some of the halites (Thompson 1990). However, the stratigraphic separation of halite and sulphate deposits has always posed problems for this hypothesis (Wills 1970). Taylor (1983), in an isotopic study of sulphates in the Branscombe Mudstone Formation and Blue Anchor Formation of the East Midlands, concluded that some of the sulphates were sourced from continental brines and others from marine brines. An input from distant marine-sourced brines is not unusual in present-day interior rift basins, due to down-wind derivation of marine aerosols or halite dust (Clark & Smith 1988; Rosen 1991). The presence of Mg-enriched clays has also been used by Jeans (1978) and Leslie *et al.* (1993) to suggest a marine-source of the Mg. Wright & Sandler (1994) have challenged this view and implicated a model using evolution in continental groundwater that explained both sulphates and Mg-rich clays.

Marine microplankton are known from some mudstone members in the Cheshire Basin and the Midlands (Earp & Taylor 1986; Wilson & Evans 1990; Warrington & Ivimey-Cook 1992; Barclay *et al.* 1997). In addition, the Tarporley Siltstone Formation possesses organic walled acritarchs and tasminid algae (Warrington 1974; Rees & Wilson 1998), a *Lingula* fauna (Poole 1978) and a 'marine-like' trace fossil assemblage (Ireland *et al.* 1978). Hence, the evidence, other than presence of halite, would suggest the most likely interval for 'marine incursions' would be in the Anisian, rather than within the Carnian where halites are more widespread. Yet the abundant mudstone layers within Anisian (and Carnian) halites would require many seawater inundations, without leaving substantive sedimentary evidence of major floods or the seawater-derived sulphates. It seems more probable that the halite may have been aeolian-sourced along with the red desert dust (cf. Dare-Edwards 1984; Clark & Smith 1988), perhaps from deflation of down-wind, marine-derived saline lakes in the North Sea Basin, or by monsoonal winds from the Tethys. The Southern North Sea Basin contains major marine halites starting in the Anisian. The marine microplankton may also have arrived in the dust, but found better conditions for preservation in the non-red mudstones (Glennie & Evans 1976). The aeolian salts may also have come from uplifted and eroded Zechstein salts (Wills 1970).

The sandstones

Sandstone intervals within the Mercia Mudstone Group are most common at about the level of the Arden Sandstone Formation (Ruffell 1991; Barclay *et al.* 1997). These are usually white to pale grey to green, very-fine-grained, well-sorted sandstones, commonly laminated or cross-laminated, and occasionally containing mudstone rip-up clasts, erosional bases and mudcracks (Fig. 13.7d, Plate 22) (Klein 1962; Milroy 1998) The sandstones are commonly intimately interbedded with laminated to structureless grey, green and purple mudstones and siltstones. The thicker sandstones may display a wide variety of trace fossils, often of the *Skolithos*, *Planolites* or *Phycodes* associations (Ruffel 1991; Barclay *et al.* 1997; Porter pers. comm. 2005). Some intervals may display soft-sediment deformation, or the development of rootletted soils (Barclay *et al.* 1997; Porter pers. comm.). The sandstones within the Arden Sandstone Formation are frequently channelized deposits, which display a limited lateral extent of at best a few kilometres (Simms & Ruffell 1990; Ruffell 1991; Barclay *et al.* 1997). The sandstones probably represent a variety of channel and splay deposits of a fresh–brackish water river system, representing a wetter period in the Carnian, when eastwards-directed fluvial systems at the Welsh and Cornubian massif margins disgorged into lake systems which had persistently higher levels than at other times in the Triassic (Ruffell 1991). This interlude may correlate with the Schilfsandstein in the Central European Basin, when extensive fluvial systems extended southwestwards from Fennoscandia (Beutler & Häussaer 1981; Geluk 2005*b*). Thin pale grey–green sandstone and siltstone intervals (mostly less than 0.3 m but occasionally up to 1–3 m thick) are also known from the Sidmouth Mudstone and Branscombe Mudstone formations particularly in the Midlands where they are referred to as 'skerries' (Elliot 1961; Worssam & Old 1988; Milroy 1998; see below). The lateral continuity of these is limited, indicating the locally sourced nature of these coarser particles, perhaps from riverine input into the Mercia Mudstone playas.

Triassic extensional basins

Many of the Triassic rifts and basins in England and Wales had a common origin from the Permian (or even the Late Carboniferous; Chadwick & Evans 1995); regional differences are examined here.

The southern North Sea and eastern England in the Triassic comprised an area of subsidence that extended from the eastern Pennines into Poland, which in the UK is bounded to the north by the Mid North Sea High and to the SW by the London Brabant High (Fig. 13.6). This basin is a broad, thick depression suggestive of thermal subsidence. However, the existence of fault topography and synfault deposition indicates that mechanical stretching (in common with the western Permian basins) was a feature of Triassic tectonics. This aspect has only recently become clear (Griffiths *et al.* 1995), since salt tectonics (halokinesis) from movement of the underlying Zechstein has obscured the Triassic thickness distributions, observed from borehole correlation and on seismic data. This halokinesis problem has obscured the synsedimentary faults which were particularly active in the Mid–Late Triassic (Allen *et al.* 1994).

As noted in Chapter 12, the Cheshire and East Irish Sea basins (Fig. 13.6) are within the tectonic- faulting province known as the Clyde Belt (Anderson *et al.* 1995), and originated during Permian rifting. The effects of this rifting became very pronounced throughout the Triassic, with dramatic synsedimentary thickness changes occurring across major fault systems. The Wem–Red Rock Fault system (Cheshire Basin) is an example of such a fault, with over 1000 m of throw in some locations preserving a similar thickness of Triassic sediment on its western (hanging wall) side.

The Vale of Eden is a NNW–SSE-trending fault-bounded basin flanked by the Alston and Askrigg blocks (Holliday *et al.* 2004), and shows evidence of Late Carboniferous (Variscan) strike-slip, followed by Permian–Triassic extension, a history seen in other basins. In contrast the Solway Firth and Carlisle basins (Fig. 13.6) show a very strong alignment to the underlying NE or ENE trending Lower Palaeozoic structure.

In central England, the thickest Triassic sequences are in the N–S-trending Worcester Basin. Here, Precambrian and Lower Palaeozoic rocks were juxtaposed with 1.5 km of Triassic redbeds by the East Malvern Fault (Fig. 13.6). The basin is also fault-bounded to the east, juxtaposing the Triassic succession against concealed Palaeozoic rocks of the London–Brabant Massif (Sumbler *et al.* 2000).

In the Wessex Basin (Fig. 13.6), the thickest Triassic succession is preserved in the western part, thinning towards the east and the south beneath the Isle of Wight and towards the French side of the English Channel (Butler 1998). The basins of the Western Approaches, Bristol Channel, Celtic Sea and Cardigan Bay have a complex history of subsidence and inversion that has resulted in variable preservation of the Triassic, except within the deeper offshore depocentres such as the Plymouth Bay and Melville basins of SW England. In the Plymouth Bay Basin a Permian and Triassic red bed succession of over 9 km in thickness occurs, making this the deepest 'red-bed' basin of the post-Devonian in NW Europe (Harvey *et al.* 1994; Tappin *et al.* 1994). Extensional faults that were active during Triassic deposition in basins offshore of SW England and South Wales were often inherited from underlying ESE-oriented Variscan thrusts and extensional faults. During the Lower Triassic these produced presumably subdued foreland basins receiving sediment from the surrounding Variscan mountains. It is probably not until the deposition of the Mercia Mudstone Group that some of these southern basins were linked.

Eastern England and the Southern North Sea Basin

The eastern England Triassic succession is contiguous with that of the Southern North Sea Basin. Onshore the thickest Triassic is present in Yorkshire and Lincolnshire, and offshore in the Sole Pit Trough, where up to 1.6 km occurs. As a result of North Sea hydrocarbon exploration the regional variations in the Triassic succession are well understood, although there is a separate lithostratigraphy for onshore and offshore areas (Figs 13.3 & 13.4). The offshore Triassic below the H-unconformity forms the Bacton Group, and that above the Haisborough Group (Cameron *et al.* 1992). Much of the information for both offshore and onshore successions is derived from borehole records because of the extensive cover of later sediments (Warrington 1974; Warrington *et al.* 1980). The best-described outcrop successions occur on the Triassic basin margin in Nottinghamshire (Benton *et al.* 2002).

The base of the Triassic in the southern North Sea can be located accurately within cycle 5 of the Lower Buntsandstein a short distance (*c.* 30 m) above the log-marker represented by the Bröckelscheifer (Geluk 2005*b*). The Bröckelscheifer is a siltstone unit whose equivalent in the German succession is assigned to the uppermost Zechstein (Szurlies *et al.* 2003). The silty nature of this unit can be recognized on borehole logs in the lower part of the Roxby Formation in NE England (Fig. 13.3) (Berridge & Pattison 1994). Hence, the base of the Triassic equates to a horizon within the upper part of the Roxby Formation (Eskdale Group) of NE England, and within the lower part of the Bunter Shale Formation (part of the Bacton Group) in the UK offshore succession (Fig. 13.9). Both the Roxby Formation and the Bunter Shale Formation are dominated by red–red-brown silty mudstones, with rare dolomite and anhydrite-rich layers.

Southwards and westwards from the coastal regions of NE England the Roxby Formation progressively interdigitates with the Sherwood Sandstone Group, so that by the Retford area, the Roxby Formation is absent. Southwest of this area the Sherwood Sandstone oversteps older units of the Upper Permian (Warrington *et al.* 1980; Berridge *et al.* 1999). Hence towards the SW parts of the East Midlands Shelf the Permian–Triassic boundary is either within the lower part of the Sherwood Sandstone Group, or represented by a disconformity at the base.

Sherwood Sandstone and Bacton Groups

In Nottinghamshire the Sherwood Sandstone Group is represented by the Lenton Sandstone Formation (which is in part Permian) and the overlying Nottingham Castle Sandstone Formation (Fig. 13.3). The Lenton Sandstone is predominantly a cross-stratified reddish-brown, fine- to medium-grained sandstone, with layers of mud-cracked mudstone and siltstone. The sandstones are dominantly aeolian in origin (Howard per comm. 2005) although the formation contains intervals of stacked fluvial fining-up successions, with basal lags of mudstone clasts, and rare quartzite clasts (Warrington 1974). It may also include thin intervals of breccia, and rare ventifact layers.

The Nottingham Castle Sandstone Formation is a medium–coarse pebbly sandstone with beds of conglomerate; pebbles are well-rounded and consist of quartzite and limestone (Warrington 1974). The sandstones are arranged in 2–4 m-thick cross-stratified sets, with bounding and foreset surfaces commonly containing pebbles and mudstone clasts. Rare palaeosols and ventifacts have been reported (Taylor 1968; Benton *et al.* 2002). The distribution and palaeocurrent measurements of the Nottingham Castle Sandstone Formation shows it formed a NE- to E-directed wide braided alluvial fan, which extended into Humberside and to the Lincolnshire coast (Fig. 13.5a) (Warrington 1974). North of Doncaster and beyond the fan it passes laterally into the onshore undivided Sherwood Sandstone Group, and to the east it passes offshore into the Bunter Sandstone Formation (Fig. 13.3) (Warrington & Ivimey-Cook 1992). This is achieved by a progressive reduction in pebble size, and abundance, and an increase in the proportion of finer-grained sandstones and mudstone-rich intervals (Warrington 1974; Warrington *et al.* 1980).

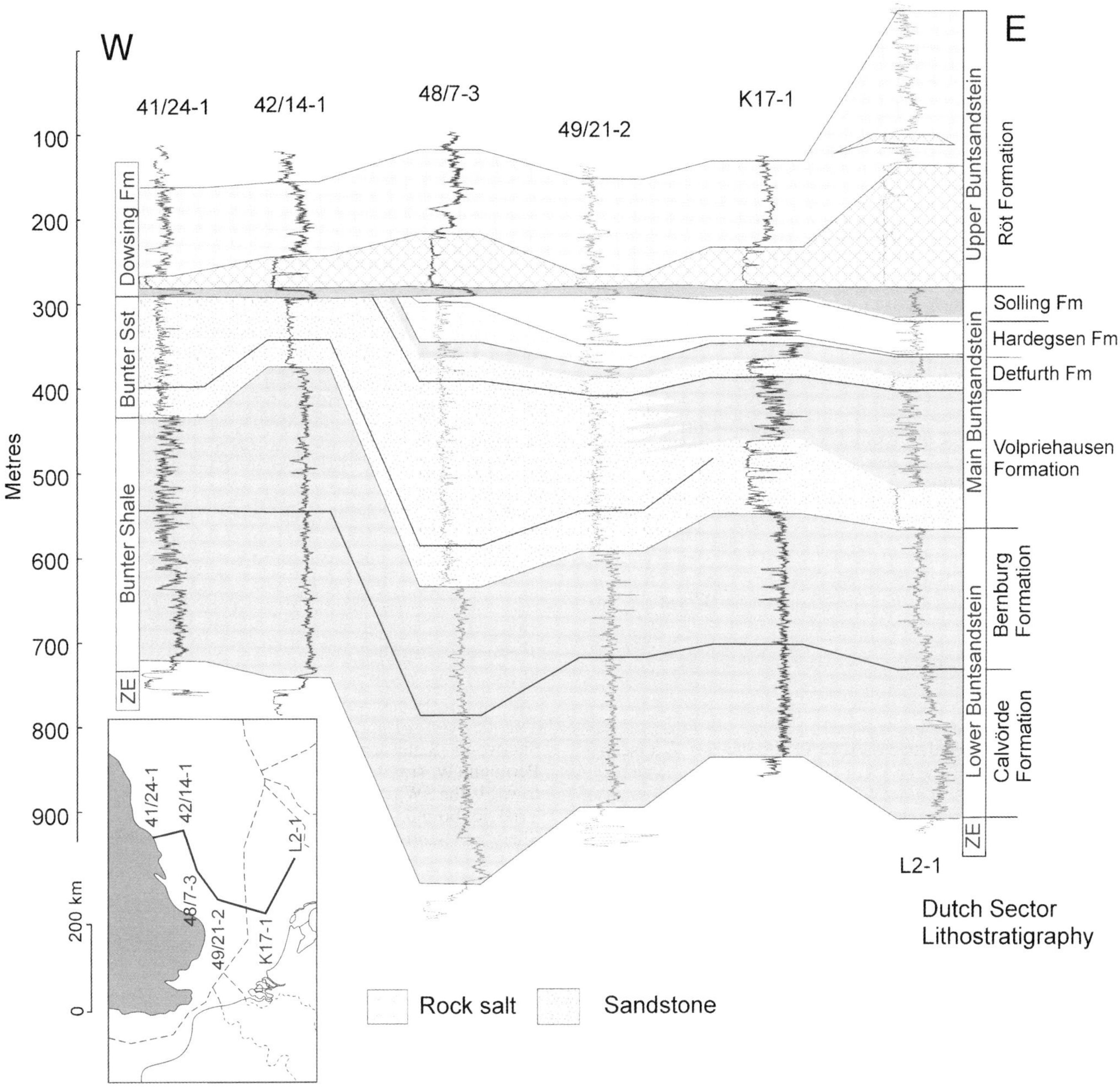

Fig. 13.9. Log correlation of the Lower Triassic succession from the English part of the Southern North Sea into the Dutch sector, showing the transition from the Bunter Sandstone Formation in the English sector into the classic Buntsandstein log-stratigraphy in the Dutch sector (modified from Geluk 2005*a*). Mudstone units in various shades of grey. ZE, Zechstein.

Palaeocurrent measurements from the Sherwood Sandstone Group throughout onshore eastern England indicate that fluvial transport systems were directed in a broad swath to the NE–SE (Taylor 1968; Warrington 1974; Gaunt 1994; Benton *et al.* 2002). This suggests much of this sand was derived from the region of the Pennine high to the west as well as from the distal Budleighensis river system draining into the East Midlands Shelf from the West Midlands. The rarity of pebbles in the Pennine-derived Sherwood Sandstone Group north of Doncaster probably attests either to the low relief of the Pennine High or the limited catchment of the river systems supplying sediment eastwards.

The Bunter Shale Formation of the southern North Sea is between 200 and 500 m thick with dramatic thickness changes indicative of synsedimentary activity, commonly obscured in basinal areas by extensive salt movement from the underlying Permian and overlying later Triassic successions. The offshore Bunter Shale Formation is the equivalent of the upper part of the onshore Roxby Formation and the lower part of the Sherwood Sandstone Group (Fig. 13.3). The younger part of this interdigitation is represented by the Amethyst Member, a succession of interbedded fluvial channel and sheet-flood sandstones and lacustrine mudstones (Fisher & Mudge 1998). The Amethyst Member is up to 340 m thick immediately offshore from Lincolnshire, and is laterally equivalent to the Rogenstein Member of the Bacton Group (Cameron *et al.* 1992), in turn the lateral equivalent of the Bernburg Formation of the German Basin, which is Dienerian (late Induan) in age (Fig. 13.9) (Szurlies 2004).

The Bunter Sandstone Formation is a predominantly fine-grained, or occasionally medium-grained, cross-stratified sandstone with common mudstone interbeds. Close to the English coast it is a general coarsening-up unit, but lacks recognisable units on wire-line logs that could be used for subdivision

(Fig. 13.9). However, at its eastern and NE borders with the Dutch sectors of the southern North Sea, it shows three major fining-up cycles, which can be linked to the Germanic and Dutch basin lithostratigraphy using wireline logs (Fisher & Mudge 1998; Geluk & Röhling 1999; Michelsen & Clausen 2002; Geluk 2005*b*). These three cycles represent the Volpriehausen, Detfurth and Hardegsen formations of the German and Dutch basins (Fig. 13.9) (Geluk & Röhling 1999; Szurlies *et al.* 2003; Szurlies 2004). Each of these cycles is a northwards-prograding alluvial-fan or braided river complex, separated by flood-plain or lacustrine mudstones, sourced from the basement highs to the south (Fisher & Mudge 1998). Tectonism seems to have been important in producing these cycles, for each is initiated with an erosional unconformity, preceded by slight regional folding (Geluk & Röhling 1999; Geluk 2005*b*).

Mercia Mudstone Group/Haisborough Group

The events responsible for the H-unconformity in the southern North Sea must have produced considerable relief, for the predominantly fine-grained unit that followed it, the Dowsing Formation (Upper Buntsandstein, in Dutch sector) locally cuts down through substantial parts of the Lower Triassic succession throughout the southern North Sea (Fig. 13.9) (Geluk & Röhling 1999; Geluk 2005*b*). Onshore the extent of this erosion is uncertain, but in parts of the Midlands, for example, it seems probable that an equivalent of the Wildmoor Sandstone Formation may have been entirely removed, due to erosion down into the pebble-bearing beds lower in the Sherwood Sandstone Group. In the NE part of the UK sector of the southern North Sea, the base of the Haisborough Group is diachronous, for a sandstone equivalent to the Solling Formation of the Dutch and German basins is present at the base (Cameron *et al.* 1992; Geluk & Röhling 1999). Westwards this passes into the Tarporley Siltstone Formation of the Nottingham area, which is in part the time equivalent of the Bromsgrove Sandstone Formation of the West Midlands (Figs 13.4 & 13.10).

A period of erosion at the base of the Mercia Mudstone Group is indicated in the Nottingham area by dreikanter-bearing gravels at the junction with the underlying sandstones (Swinnerton 1918; Warrington 1974). In the southern North Sea the Haisborough Group approaches 1000 m in maximum thickness in the Sole Pit Trough area of the southern North Sea, and is dominated by red silty mudstones (Fig. 13.11). The offshore Dowsing and Dudgeon formations are punctuated by four halite units, between which the succession is divided into a large number of mudstone and dolomitic units. The lowest halite, the Main Röt Halite Member, is the most widespread of the halites, extending into NE England as the Esk Evaporite Member (Fig. 13.4). This unit is composed of five cycles of marine flooding and evaporation, with contemporary anhydritic and dolomitic mudstone sabkha-facies on basin margins (Figs 13.5b & 13.11). The overlying Upper Röt Halite does not extend into onshore England, being restricted to the northern half of the Anglo-Dutch Basin (Figs. 13.6 & 13.11). The Muschelkalk Halite Member is divided by numerous thin red mudstone beds which become less common upwards through the member, and as with lower units the halite laterally

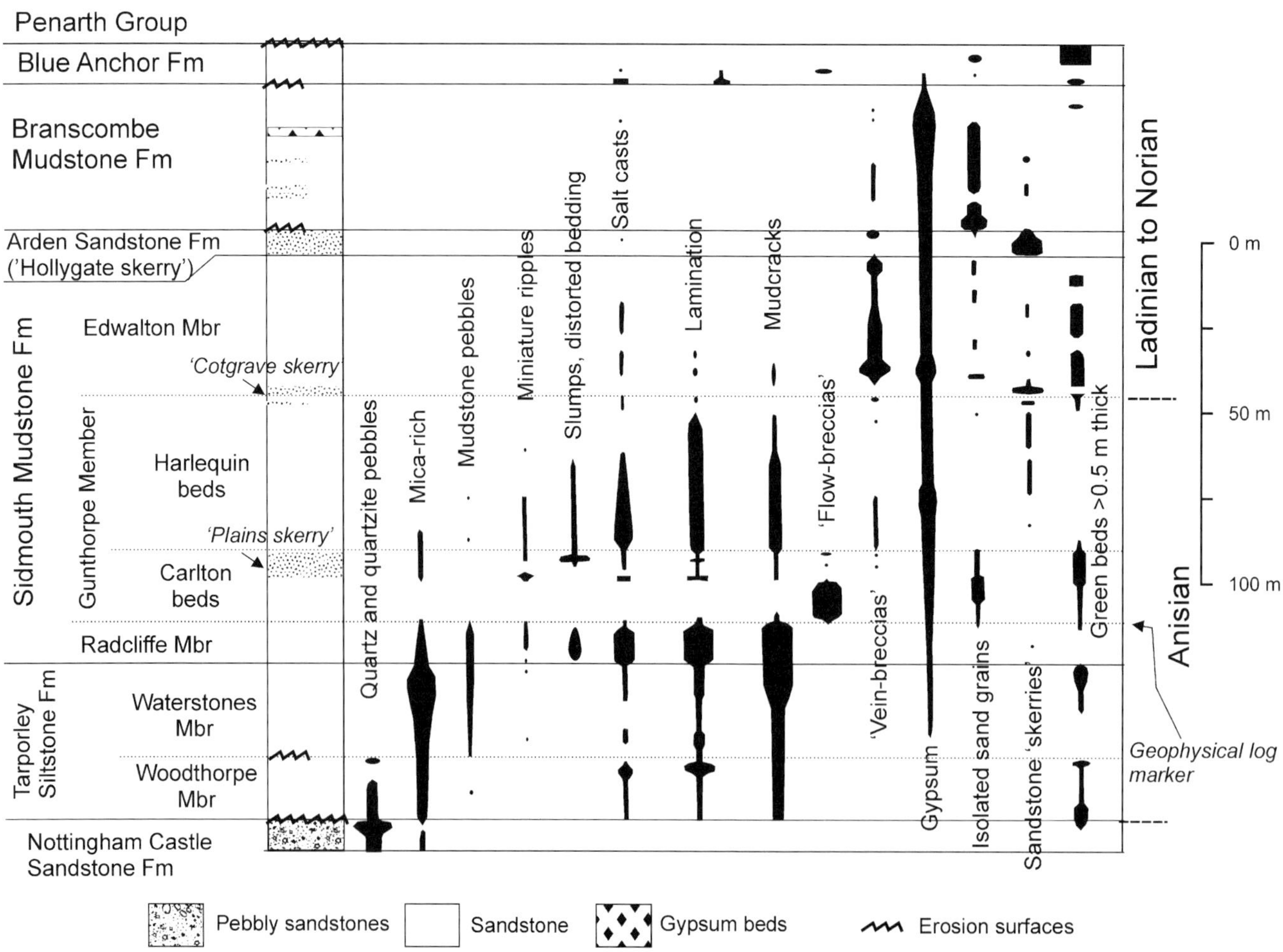

Fig. 13.10. Detail of the sedimentological structures in the Mercia Mudstone Group in borehole cores from the Nottingham area. Modified from Elliot (1961), using the lithostratigraphy of Howard *et al.* (2006).

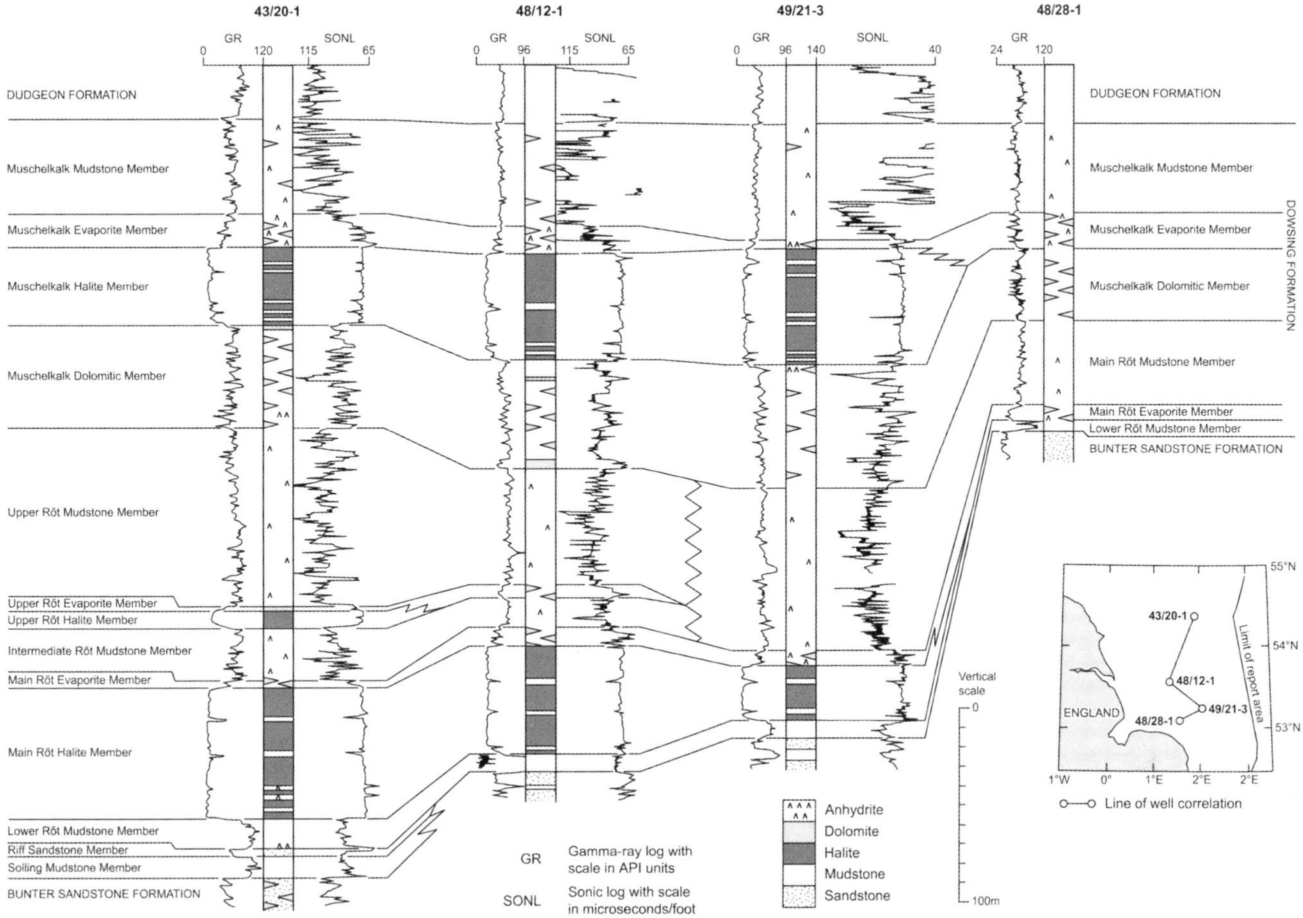

Fig. 13.11. Wireline log correlation of the Dowsing Formation, and its associated sub-divisions in the southern North Sea (modified from Cameron *et al.* 1992).

interfingers with sabkha dolomitic and anhydritic mudstones towards the basin flanks. The Keuper Halite Member (up to 300 m thick) is largely restricted to the Sole Pit Trough in the UK southern North Sea (Fig. 13.5c). Owing to large thickness changes, and lateral impersistance of the halite beds, the Keuper Halite appears to have been deposited in a large number of small shallow saline lakes (reflecting the palaeotopography), rather than laterally extensive salinas like in the older halites of the Dowsing Formation (Fisher & Mudge 1998). The Keuper Halite also appears to have been fed by riverine flow or ground waters, since lateral marine connection to the east was disrupted during the Carnian (Zeigler 1990; Geluk 2005*a*).

The thickness and depositional limits of units of the Haisborough Group are strongly affected by the Dowsing Fault Zone, which delimits the western boundary of the Sole Pit Trough, indicating that contemporaneous fault movements dictated deposition rates. The Triton Formation is dominated by red anhydrite-bearing mudstones, with anhydrite layers numerous enough in the middle parts of the formation to define the Keuper Anhydrite Member, which has a wide distribution throughout the offshore region. In onshore regions in eastern England the Keuper Anhydrite is probably represented in Nottinghamshire, Leicestershire and Staffordshire by the Tutbury and Newark Gypsum, that occurs in the upper 40 m of the Branscombe Mudstone Formation (Fig. 13.4) (Taylor 1968).

The lithostratigraphy of the Mercia Mudstone Group in onshore eastern England has largely evolved around that first erected by Elliot (1961). He divided the succession below the Arden Sandstone Formation (his Hollygate skerry) into four units, two of which are dominated by laminated mudstones (now the Radcliffe Member and Harlequin beds; Berridge *et al.* 1999; Howard *et al.* 2006) and two that are dominated by non-laminated mudstones (Carlton beds and Edwalton Member; Fig. 13.10). He also proposed units for the mudstones (now Branscombe Mudstone Formation) above the Hollygate skerry and two units which are now assigned to the Tarporley Siltstone Formation (Fig. 13.10). The vein breccias of Elliot (1961) are common in the Edwalton Member (Fig. 13.10). Elliot's 'flow-breccias' are in beds up to 4 m thick, interbedded with less disturbed mudstone beds, and are common in the lower part of the Carlton beds. From Nottingham westwards, Elliot's laminated and non-laminated mudstone divisions can be traced perhaps as far west as the Coventry area (Old *et al.* 1987), and north as far as the Grimsby area (Berridge & Pattison 1994).

The boundary between the Gunthorpe and Radcliffe members provides a gamma-ray peak in wire-line logs, which has provided an alternative means to correlate this lithostratigraphy into other onshore areas in eastern England (Berridge & Pattison 1994). Traced northwards this regional gamma marker appears to equate with an interval above the Esk Evaporite Member (Howard pers comm.). This later correlation is compatible with dating of these successions by miospores which indicates the succession near the base of the Edwalton Formation probably corresponds to the Anisian–Ladinian boundary (Worssam & Old 1988).

Economic Geology

Triassic hydrocarbon reservoirs in the southern North Sea are minor in comparison to the large number of Permian (Rotliegende) gas reservoirs (Fig. 13.12), these all being based on Late Carboniferous sourced gas migrating into the Permian

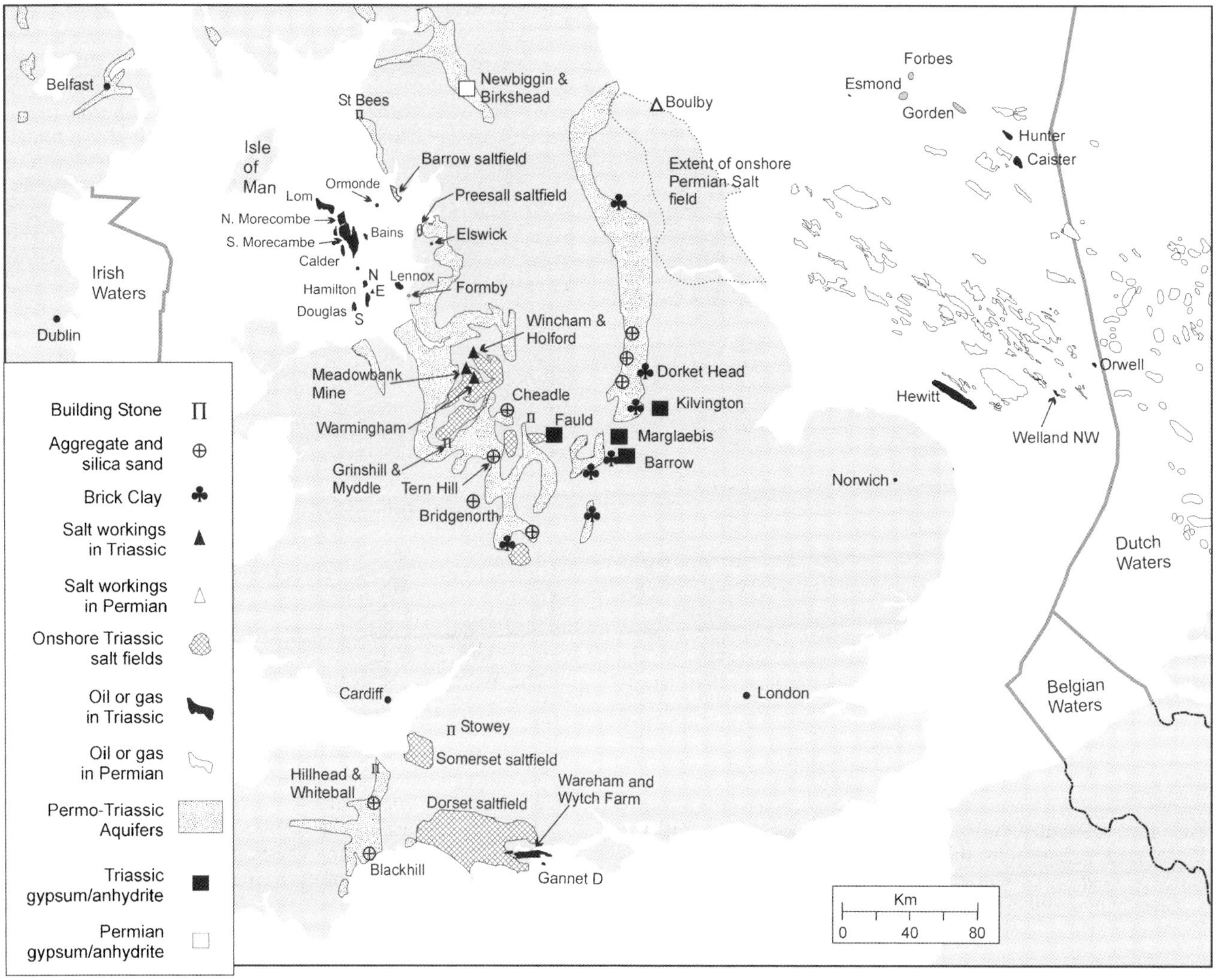

Fig. 13.12. Location of main worked economic resources in the Triassic, with some Permian resources also indicated. Building stone, aggregates, silica sand and brick clay only indicated for the Triassic. Triassic oil and gas fields decommissioned in grey.

and Triassic reservoirs (Fisher & Mudge 1998). The Triassic Bunter Sandstone gas reservoirs are relatively poor quality fluvial sandstones, with overlying mudstones of the Haisborough Group as the topseal (Ritchie & Pratsides 1993). The earliest of these discoveries was the Hewitt field (latest Permian in this account) in 1966, followed by the Caister, Gorden and Forbes fields in the late 1960s–early 1970s (Bifani 1986), with the last Triassic discovery being the Hunter gas field in 1992.

Just as the Bunter Sandstone reservoir and Haisborough Group mudstones form an effective topseal for the hydrocarbon play in the southern North Sea, the same two units, onshore and at shallower depths, form aquifer and aquiclude for water abstraction. In the UK as a whole, groundwater resources in Permo-Triassic sandstone aquifers are the second most important after the Chalk, providing some 25% of the total licensed abstractions. This resource in northern and central England, around the flanks of the Pennines, is particularly important in providing water supplies for industrial and potable uses in cities such as Manchester, Liverpool, Nottingham and Birmingham, located over the aquifer.

In Nottinghamshire the Sherwood Sandstone aquifer (which includes the Lenton Sandstone) is a heavily used supply of potable water (Trowsdale & Lerner 2003); further NE in Yorkshire, the finer-grained nature of the Sherwood Sandstone along with the lower water quality (higher salinities, harder water) limit abstractions to industrial uses. Generally, fractures in the Sherwood Sandstone contribute significantly to transmissivity, lesser so in the south around Nottingham where intergranular flow is important (Allen *et al.* 1997).

In Nottinghamshire poorly consolidated parts of the Sherwood Sandstone Group are an important source of sand and gravel accounting for some 30% of the county's output. Most of the quarries are located between Nottingham and Mansfield and mostly utilize the Nottingham Castle Sandstone Formation. A single site (Ratcher Hill quarry) also produces specialist (non-aggregate) silica sand (high silica content sand) for a variety of purposes (Highley *et al.* 2004*c*). In the past, both Scrooby Top and Styrrup quarries in Nottinghamshire were important sources of red loose sand and some sandstone; both now have preserved faces under the Geological Conservation Review of sites of special scientific importance.

Onshore the Branscombe Mudstone Formation of the Mercia Mudstone Group locally hosts beds of gypsum–$CaSO_4{\cdot}H_2O$ (anhydrite–$CaSO_4$ below 40–50 m), which are used commercially for the production of plaster, plasterboard and cement (Highley *et al.* 2004*a*). The most important resources are those associated with the Tutbury Gypsum in Leicestershire, Nottinghamshire (and Staffordshire) and the Newark Gypsum in east Nottinghamshire. The Tutbury Gypsum is a bed up to 3.5 m thick and is worked underground using the pillar and stall method at the Barrow Mine in Leicestershire and the Marglaebis Mine in Nottinghamshire (also the Fauld

Mine in Staffordshire; Fig. 13.12). The Newark Gypsum in Nottinghamshire comprises multiple beds and nodular bands of gypsum of variable thickness and purity, spread over some 15–18 m of the Branscombe Mudstone Formation, and is worked at the Kilvington open cast quarry. The individual worked beds range from about 0.3 to 2 m in thickness with some of the beds producing the highest quality UK gypsum (Highley *et al.* 2004*a*). The amount of gypsum extracted has declined appreciably over the last 25 years due to importation of cheaper gypsum and availability of gypsum as a by-product from flue gas desulphurization of coal-fired power stations.

One of the last remaining brick clay producing sites from the Mercia Mudstone Group in eastern England is the Dorket Head quarry, to the north of Nottingham, which works the lower part of the Sidmouth Mudstone Formation, producing facing bricks.

Northwest England, North Wales and the East Irish Sea Basin

The basins and sub-basins in NW England and the East Irish Sea contain some of the thickest proven Triassic successions in onshore or offshore UK basins, with up to 2 km of Sherwood Sandstone Group and 3.2 km of Mercia Mudstone Group sediments (Jackson *et al.* 1995). Much of this succession is very poorly exposed, other than fragments of the Sherwood Sandstone Group within the Cheshire Basin and on the Lancashire and west Cumbria coasts, so the succession is largely known from boreholes. The Penarth Group and upper parts of the Mercia Mudstone Group were probably deposited across the current Lower and Middle Triassic outcrop, but are now only known from isolated outliers in the Cheshire Basin, the Carlisle Basin and offshore in the Solway-Firth and Keys basins (Jackson *et al.* 1995; Warrington *et al.* 1999).

There is no fossil evidence to firmly place the position of the Permian–Triassic boundary in the successions in NW England. However, based on wireline-log similarities to the southern North Sea, the base of the Triassic is probably within the St Bees Shale, but it is commonly placed for convenience at the base of the St Bees Sandstone Formation (Jackson *et al.* 1987), even though this boundary is diachronous (Holliday *et al.* 2001). Within the onshore basins, the aeolian sandstones of the Kinnerton Sandstone Formation (partly Permian in age) rest conformably upon the Permian Manchester Marl and Bold Formation in the Cheshire Basin and west Lancashire (Fig. 13.3; Plate 23). The Sherwood Sandstone Group in NW England below the H-unconformity (Fig. 13.3) has failed to provide any biostratigraphic evidence of age, although magnetostratigraphy indicates its Early Triassic credentials (Mange *et al.* 1999; Hounslow & McIntosh 2003).

Sherwood Sandstone Group in sequence TR1

Within the Cheshire and East Irish Sea basins (EISB) Jackson *et al.* (1987, 1995) divided the succession into two units, above and below a seismic reflector termed the 'brown reflector', which appears to be related to the top of a widespread interval of silicified sandstone (Colter & Barr 1975). The lower unit, below the brown reflector, is the St Bees Sandstone Formation (Barnes *et al.* 1994), which is composed of stacked fluvial sandstones formed in low sinuosity braided rivers (Plate 24; Jones & Ambrose 1994). The sandstones are commonly distinctively micaceous and are rather similar to those farther north in the Carlisle Basin (Brookfield 2004; Holliday *et al.* 2004). On the Cumbrian coast, within the St Bees Sandstone, detritus derived from the Lake District High diminishes in importance upwards; this parallels a change from sheetflood sandstones in the lower parts to more common stacked channel sandstones in the upper parts (Jones & Ambrose 1994) (Plate 24).

In the EISB the unit above the brown reflector is the Calder Sandstone Formation (Barnes *et al.* 1994; Akhurst *et al.* 1997) and within the northern part of the Cheshire Basin this correlates with an interval within the Wilmslow Sandstone Formation (Fig. 13.3; Evans *et al.* 1993). Within the EISB the base of the Calder Sandstone appears to be a regional disconformity marked by an upwards decrease in gamma-ray values and sharp increase in sonic velocity (Barnes *et al.* 1994). In the southern parts of the Cheshire Basin, the brown reflector may correspond with the top of the Chester Pebble Beds Formation (Jackson *et al.* 1987). An overlying 'yellow' seismic reflector in the Calder Sandstone Formation of the EISB marks the top of a widespread well-cemented fluvial sandstone within the formation (Fig. 13.3) (Akhurst *et al.* 1997).

The Chester Pebble Beds Formation, some 90–305 m in thickness (Fig. 13.3), is dominated by red-brown, cross-stratified, pebbly, medium- to coarse-grained sandstones with conglomerate layers, and may locally rest on Carboniferous rocks at the margin of the Cheshire Basin (Plate 23). Steel & Thompson (1983) and Rees & Wilson (1998) interpreted the deposits of the formation as migrating sand and gravel bars within a braided river system. Relatively uncommon parallel laminated sandstone and mudstone attest to periods of low river discharge, or to deposition in abandoned channels. Rare interbedded aeolian cross-stratified sandstones (Rees & Wilson 1998) and the presence of aeolian-rounded sand grains indicate some reworking of the fluvial deposits by wind winnowing.

The Wilmslow Sandstone Formation comprises red, non-pebbly cross-stratified fine- to medium-grained sandstone. Rees & Wilson (1998) and Thompson (1970*a*, *b*) recognized a fluvial facies, composed of poorly sorted sandstone with mud clasts, and a well-sorted, paler-coloured aeolian sandstone facies. The fluvial sandstones dominate the interval below the brown reflector, whereas the aeolian sandstones are more prevalent above this and at the margin of the Cheshire Basin (Steel & Thompson 1983). The Wilmslow Sandstone Formation varies greatly in thickness from 900 m in the centre of the Cheshire Basin to absent at some basin margin locations. This appears to be due to both truncation below the Helsby Sandstone Formation and primary depositional thinning (Evans *et al.* 1993; Rees & Wilson 1998).

In the northern parts of the Cheshire Basin and southern part of the EISB there is a lateral transition from the Chester Pebble Beds and Wilmslow Sandstone of the Cheshire Basin into the St Bees Sandstone Formation of the EISB. This is indicated by the progressive northwards loss of quartzite pebbles within the St Bees Sandstone Formation in the southern parts of the EISB (Jackson *et al.* 1995). In the southern parts of the EISB heavy mineral data show the sediment source of the St Bees Sandstone to be the Welsh Massif. Evidently, the Armorican sourced detritus in the Midlands and Cheshire Basins was considerably diluted by material derived from local highs by the time it reached the central EISB.

To the west of the Cheshire Basin, in the Vale of Clwyd, Permo-Triassic sandstones of uncertain age, mainly of aeolian origin, occur beneath thick drift deposits. These may be equivalent in age to the Kinnerton Sandstone Formation of the Cheshire Basin, since they contain no known pebble-bearing beds equivalent with the Chester Pebble Beds (Warrington *et al.* 1980).

Near the Cumbrian coast the Calder Sandstone Formation is dominated by aeolian dune sandstones and various wet interdune facies, whereas within the western and central EISB, it contains much larger proportions of fluvial sandstones, which were deposited in channel systems directing flow to the WSW. This unit is also recognized to the north in the western parts of the Carlisle Basin, and appears to be the equivalent of part of the aeolian Kirklington Sandstone Formation of northern Cumbria (Holliday *et al.* 2004). The upland areas to the east of the EISB seem to have been a major source for the wind-derived grains in the Calder Sandstone that were transported by north-easterly winds (Jones & Ambrose 1994).

The transition from the Calder Sandstone Formation into the Ormskirk Sandstone Formation of the EISB is not marked by a major unconformity (Akhurst *et al.* 1997), whereas the equivalent boundary in the Cheshire Basin, shows removal of up to about 900 m of strata. In the EISB the Calder Sandstone–Ormskirk Sandstone boundary is a sharp disconformable boundary, indicated on wire-line logs by an upwards increase in sonic velocity and gamma-ray values, which are concomitant with a major environmental change, and a shift in the palaeocurrent directions. In the Cheshire Basin the Bulkeley Hill Formation locally underlies the H-unconformity (Warrington *et al.* 1980), may be up to 220 m thick (Evans *et al.* 1993) and is equivalent, in an uncertain way, to the upper part of the Calder Sandstone Formation.

Sherwood Sandstone Group in sequence TR2-3

In the Cheshire Basin the topmost unit of the Sherwood Sandstone Group, the Helsby Sandstone Formation (up to *c.* 250 m thick), contains Anisian tetrapods and palynomorphs, as well as crustaceans and ichnofossils (Benton & Spenser 1995; Warrington *et al.* 1999; King & Thompson 2000). In the centre and NW of the Cheshire Basin the sandstone units adjacent to the H-unconformity are divided into three members, a lower aeolian and fluvial unit (the Thurstaston Sandstone Member), a middle fluvial pebbly sandstone unit (the Delamere Member) and an upper aeolian sandstone (the Frodsham Member; Thompson 1970*b*; Warrington *et al.* 1980). The lowest member has been assigned by the British Geological Survey to the Wilmslow Sandstone Formation and the higher two members to the Helsby Sandstone Formation. The precise placement of the H-unconformity seen on seismic reflection profiles in the Cheshire Basin is uncertain. Thompson (1970*b*) and Mountney & Thompson (2002) place it as the base of the Thurstaston Sandstone Member, whereas the Geological Survey (Poole & Whiteman 1966; Evans *et al.* 1993) placed it more naturally at the erosive base of the fluvial pebbly sandstones of the Delamere Member (base of Helsby Sandstone Formation). The aeolian sandstones of these units were formed by large dome-shaped dunes that migrated from east to west (Brookfield 1977). The interbedded fluvial sandstone units were formed by NW-migrating sandbars in a sinuous river system (Thompson 1970*a*, *b*). The Helsby Sandstone Formation away from the north and NW of the Cheshire Basin is dominated by pebbly fluvial sandstones, of which the quartzite pebbles form a distinctive part and were probably in part derived from reworking of underlying parts of the Sherwood Sandstone (Rees & Wilson 1998).

The Ormskirk Sandstone Formation of the EISB, west Lancashire and Cumbrian coasts has been extensively investigated by boreholes for it forms the major reservoir of the Morecambe Bay gas fields (Cowan 1993; Meadows & Beach 1993; Stuart 1993). It is distinguished from the underlying Calder Sandstone Formation by being generally finer-grained and better sorted, and containing common frosted 1–2 mm diameter quartz grains (Akhurst *et al.* 1997). It is mostly a fine- to medium-grained pale grey or brown sandstone, with thin (< 1 m) interbeds of green or red-brown mudstone. Within the central EISB the sandstone units appear to be dominated by fluvial channel deposits, produced by SW- to west-flowing river systems (Cowan 1993; Meadows & Beach 1993; Stuart 1993). This is in marked contrast to the Cumbrian coast, Isle of Man and the Carlisle Basin, where the unit is dominated by aeolian sandstones, produced by SW migrating dunes (Jones & Ambrose 1994; Jackson *et al.* 1995; Brookfield 2004). This difference may in part be due to localized control of the distributary fluvial channels by synsedimentary fault networks (Meadows & Beach 1993), or perhaps due to uncertainty in the correct environmental interpretation of 'sheetflood' type sandstone intervals common in this formation (Goodall *et al.* 2000).

Mercia Mudstone Group

It is probable that the Mercia Mudstone Group was originally deposited across the entire subcrop and outcrop region of the Triassic, with the thickest deposits in the East Irish Sea, Solway Firth and Cheshire basins (Jackson *et al.* 1995; Holliday *et al.* 2004). The Mercia Mudstone Group is dominated by red mudstones, but can include substantial thicknesses of halite, and in most areas intervals of non-red mudstones.

The boundary between the sandstones of the Sherwood Sandstone Group and the red mudstones of the Mercia Mudstone Group is at the base of the Tarporley Siltstone Formation (Howard *et al.* 2006), although it is a transitional lithological boundary that is diachronous. The Tarporley Siltstone Formation comprises reddish-brown mudstones and siltstones with interbedded paler, fine–very fine sandstones, characteristically with a high content of mica on bedding surfaces (Fig. 13.14) (Ireland *et al.* 1978). Intraformational conglomerates are common along with wave and current-ripple laminated sandstones and dessication cracks and a rich variety of ichnofossils. The unit is at most 150 m thick within the Cheshire Basin, but is not regionally differentiated within the EISB (Jackson *et al.* 1995; Howard *et al.* 2006). In west Lancashire and Merseyside the Tarporley Siltstone Formation passes into laminated mudstones (Wilson & Evans 1990); further north and west in the EISB it passes into dolomite and anhydrite-rich mudstones (Jackson *et al.* 1995). The laminated mudstones of the Lancashire coast comprise interbedded and flaser-laminated grey mudstones and siltstones, with halite psuedomorphs, along with mudcracks, current ripples and clay-flake conglomerates (Figs 13.13 & 13.15) (Wilson & Evans 1990).

The successions in NW England are among the few places in the UK where the Anisian–Ladinian boundary can be defined using biostratigraphy. Within the Cheshire Basin, palynological evidence suggests this stage boundary occurs at the top of the Byley Mudstones Member or in the bottom part of the Wych Mudstones Member (Warrington *et al.* 1999). Farther north on the Lancashire coast, the Anisian–Ladinian boundary is located within the Coat Walls Mudstone Member, which overlies the Preesall Halite (Fig. 13.4). The overlying Wilkesley Halite may be late Ladinian, or entirely Carnian in age.

The succession below the Wilkesley Halite includes a widespread halite marker interval the Northwich Halite of the Cheshire Basin and the equivalent Preesall Halite of the EISB and west Lancashire coast (Fig. 13.4) (Wilson 1990, 1993; Jackson *et al.* 1995). These halites contain approximately 25% interbedded red mudstones, which progressively thicken, with the loss of the interbedded halite, towards basin margin locations. Within several hundred metres of the surface the halites may be replaced by solution collapse breccias due to post-Triassic ingress of water (Wilson 1993).

Below the Preesall Halite in the EISB and west Lancashire coast is a succession of interbedded halites and mudstones. The halite units comprise upwards of 50% halite, whereas the mudstone units comprise lesser amounts. Within the deeper parts of these basins up to three halite units, the Mythop Halite, Rossall Halite-Lower and Rossall Halite-Upper, can be recognized using wireline logs (Wilson 1990; Jackson *et al.* 1995). Sedimentological details about these units are known from cored boreholes in west Lancashire and SW Cumbria (Wilson 1990). Like the Preesall Halite, the halite units thin towards basin margins, with some of this loss due to penecontemporaneous dissolution of the halite, evident from collapse breccias (Wilson 1990; Wilson & Evans 1990). The halite units below the Preesall Halite are dominantly in the Haselgebirge lithology. These halite units extend offshore to the Isle of Man, SW Scotland and Northern Ireland, and are locally mined in

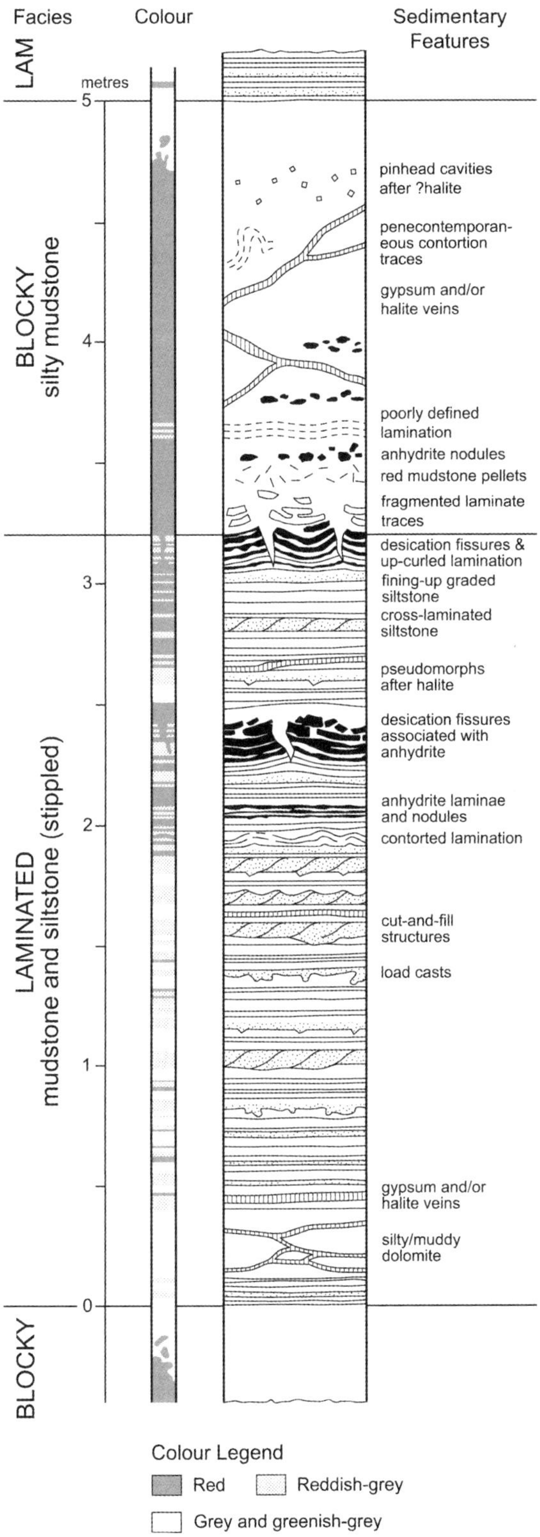

Fig. 13.13. Idealized representation of a sedimentary cycle between laminated mudstones facies and blocky mudstone facies, based on borehole records from the mudstones in the vicinity of the Northwich Halite (from Arthurton 1980).

Northern Ireland (at Kilroot), suggesting these deposits were once widespread.

Most of the intervening mudstone units are massive red mudstones, with subordinate interbeds of laminated siltstones. Exceptions are the mudstone units that overlie and underlie the Preesall Halite in west Lancashire and the Northwich Halite in Cheshire (Fig. 13.4) (Wilson 1990, 1993). These show vivid interbedding of reddish-brown and greenish grey mudstones interlaminated with thin (< 1 cm thick) siltstones (Fig. 13.15). In west Lancashire the total thickness of the siltstones decreases towards the centre of the EISB, and the most commonly displayed sedimentary structure are desiccation cracks (Wilson & Evans 1990). In the Thornton Mudstones (below the Preesall Halite; Fig. 13.4), which display the laminated facies to their best, the red and green-grey mudstones are arranged in a series of 12 cycles, which can be correlated from Walney island in SW Cumbria to the Blackpool area (Fig. 13.15) (Wilson 1990; Wilson & Evans 1990). The upper part of the Bollin Mudstones (Fig. 13.4) of the Cheshire Basin shows a similar prominence of laminated mudstones as in the EISB, and a number of correlated red-green cycles (Arthurton 1980; Wilson 1993, 2003). The Byley Mudstone Member of the Cheshire Basin and the Coat Walls Mudstone Member, both of which overlie the Northwich and Preesall halites, have a lesser development of the laminated facies and are mostly blocky red mudstones (Arthurton 1980; Rees & Wilson 1998). The Wych Mudstone Member of the Cheshire Basin, and its equivalent in the EISB and west Lancashire (Breckells Mudstone Member; Fig. 13.4), marks a complete return to deposition of the blocky red mudstone facies, with a few halite and anhydrite layers (Wilson 1990, 1993).

The Wilkesley Halite is known mostly from its type area in the Cheshire Basin where it is up to 404 m thick. It consists of beds of halite up to 26 m thick with numerous inter-beds of blocky red mudstones up to 12 m thick (Wilson 1990). In contrast to the Northwich Halite, it also contains sandstones up to 1 m thick, presumably representing short influxes from marginal successions, which may correlate with sandstone intervals in the Arden Sandstone Formation of the Midlands.

The overlying Branscombe Mudstone Formation is dominated by blocky red mudstones. It also locally contains thin greenish-grey mudstone intervals, thin sandstones, anhydrite nodules and veins, and halite veins (Wilson 1993).

Economic Geology

Offshore from Wales and NW England there are three groups of hydrocarbon discoveries hosted by Triassic rocks. The first comprises the various gas fields in the central part of the eastern Irish Sea region (North and South Morecambe, Dalton, Calder, Lom, Ormonde, Bains and the onshore Elswick field; Fig. 13.12) (Stuart 1993; Jackson *et al.* 1995) with the first discovery being the South Morecambe field in 1974. The second comprises the oil and gas producing fields further south in Liverpool Bay (Hamilton, Lennox, Douglas and onshore the now decommissioned Formby Oilfield); most of these offshore fields were discovered in the 1990s (Pearson 1995). The third comprise smaller, mostly oil discoveries and indications in the Cardigan Bay–Celtic Sea areas, many of which have yet to be fully developed. The Irish Sea gas discoveries comprise mainly Westphalian coal and shale source rocks underlying Triassic Sherwood Sandstone Group reservoirs, in turn overlain by Mercia Mudstone Group topseal, with local salt seals at faulted contacts. The Morecambe North and South discoveries were based on a large shallow domal anticline observed on seismic data, with the chief reservoir being the Ormskirk Sandstone. Acidic waters were crucial in dissolving carbonate and feldspar in the reservoir sandstone, prior to an early fill with liquid hydrocarbons during the Early Jurassic. Uplift and breaching occurred toward the end of the Jurassic and the present gas filling did not take place until Cretaceous and early Tertiary burial (Jackson *et al.* 1995). The Formby Oilfield (now decommissioned) is located over natural oil seepages, the products of which have been used since AD 1637. Production boreholes in the 1940s located oil in the Tarporley Siltstone Formation and

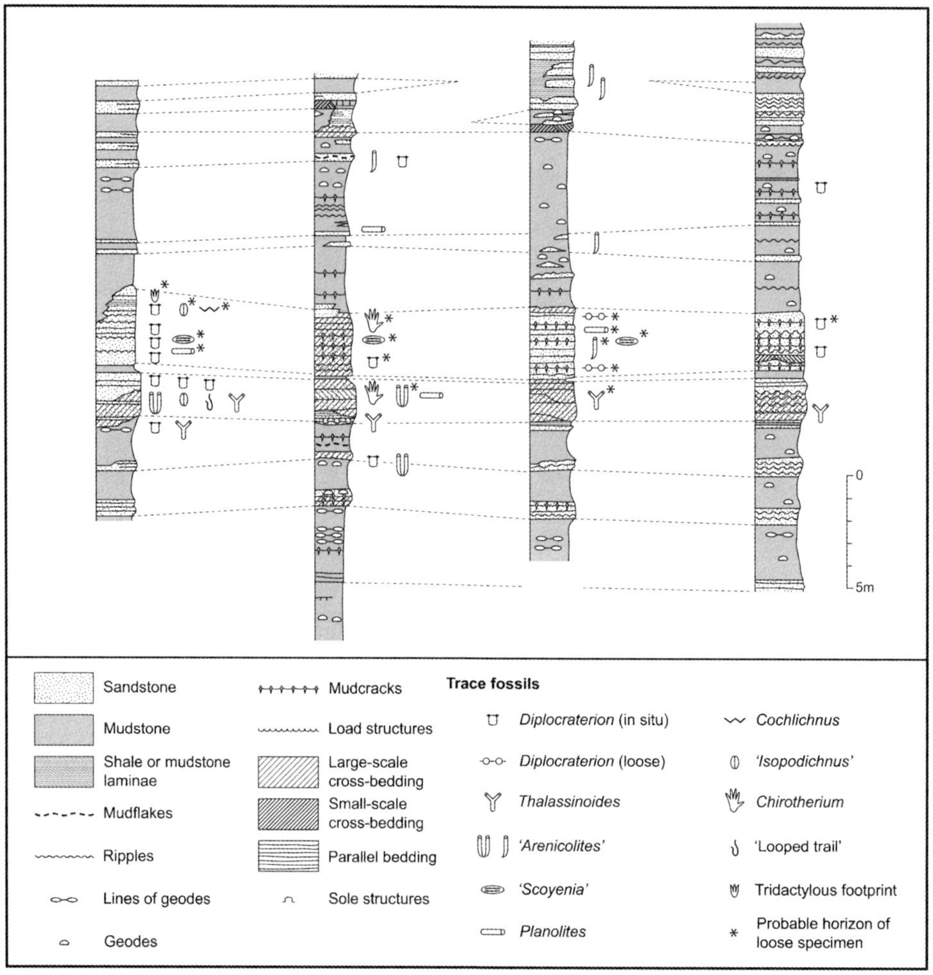

Fig. 13.14. Detailed sedimentary structures and correlation in the Tarporley Siltstone Formation at the Red Hill cutting, Cheshire. Modified from Ireland *et al.* (1978).

the uppermost Ormskirk Sandstone, in a faulted monocline. The Formby Oilfield contained a light crude oil, similar in composition to the East Midlands Carboniferous-hosted oilfields, which similarly are thought be derived from Namurian marine black shales (Jackson *et al.* 1995).

The Permo-Triassic sandstone aquifers are a heavily utilized groundwater resource in NW England, accounting for some 78% of licensed abstractions for public water supply from groundwater (Thewsey pers comm. 2005). The groundwater resources are segmented into a number of separate regions corresponding to the outcrop (Fig. 13.12). In Cheshire and around the Liverpool- Manchester area the Sherwood Sandstone is the major aquifer, although it is not a uniform unit for groundwater flow, most of the flow being restricted to depths down to 200 m below the surface, where the flow is primarily through fractures (Allen *et al.* 1997). Further north in west Lancashire and west Cumbria the intergranular hydraulic conductivity decreases, due to better cementation and the finer grain size of the Sherwood Sandstone, and fracture flow is the most significant effect on transmissivity (Allen *et al.* 1997). In northern Cumbria the absence of mudstone interbeds, the coarser-grained and less-cemented nature of the Kirklington Sandstone make it a better groundwater reservoir than the St Bees Sandstone. The Sherwood Sandstone in the Vale of Clwyd is the major aquifer in North Wales and is confined by a significant thickness of glacial till, giving rise to artesian conditions in the centre of the basin (Allen *et al.* 1997).

Salt (NaCl) in the UK is currently extracted either using brine pumping (70% of production) or through mining methods which extract rock salt (Highley *et al.* 2004*b*). The only working mine in the Triassic in England is the Meadowbank Mine (near Winsford, Cheshire; Fig. 13.12), which utilizes several beds in the Northwich Halite Formation and since 1928 has been the UK major source of rock salt for road de-icing. Natural (or 'wild') brine in the subsurface is generated by ground water dissolution of salt beds. In the past these natural brines were utilized, but produced uncontrolled collapse of the resulting cavities, which produced hazardous surface subsidence. Thus in areas of old workings in Cheshire (around Northwich and Nantwich), some areas have been prone to sudden and catastrophic collapses of the ground, sometimes with buildings and roads collapsing into the hole. All solution mining is now by controlled pumping, in which a single borehole, with a triple tube system, combined with a compressed air blanket to limit upward dissolution, allows complete control of the brine extraction process and the size of the resulting cavity. The cavities produced are designed to withstand the weight of the overlying strata, so collapse does not occur (Highley *et al.* 2004*b*). Controlled brine pumping takes place at the Holford and Warmingham brine fields from the Northwich Halite (Fig. 13.12). In the past, controlled brine pumping also occurred at Preesall in Lancashire and near Barrow-in-Furness in Cumbria (Fig. 13.12). The stable cavities produced by rock salt mining and controlled brine pumping are themselves important economic assets for storage of natural gas, hydrogen, various organic chemicals, documents, and various types of waste.

In the past, sandstones from the Sherwood Sandstone Group in Cheshire, Shropshire and Cumbria were extensively used as

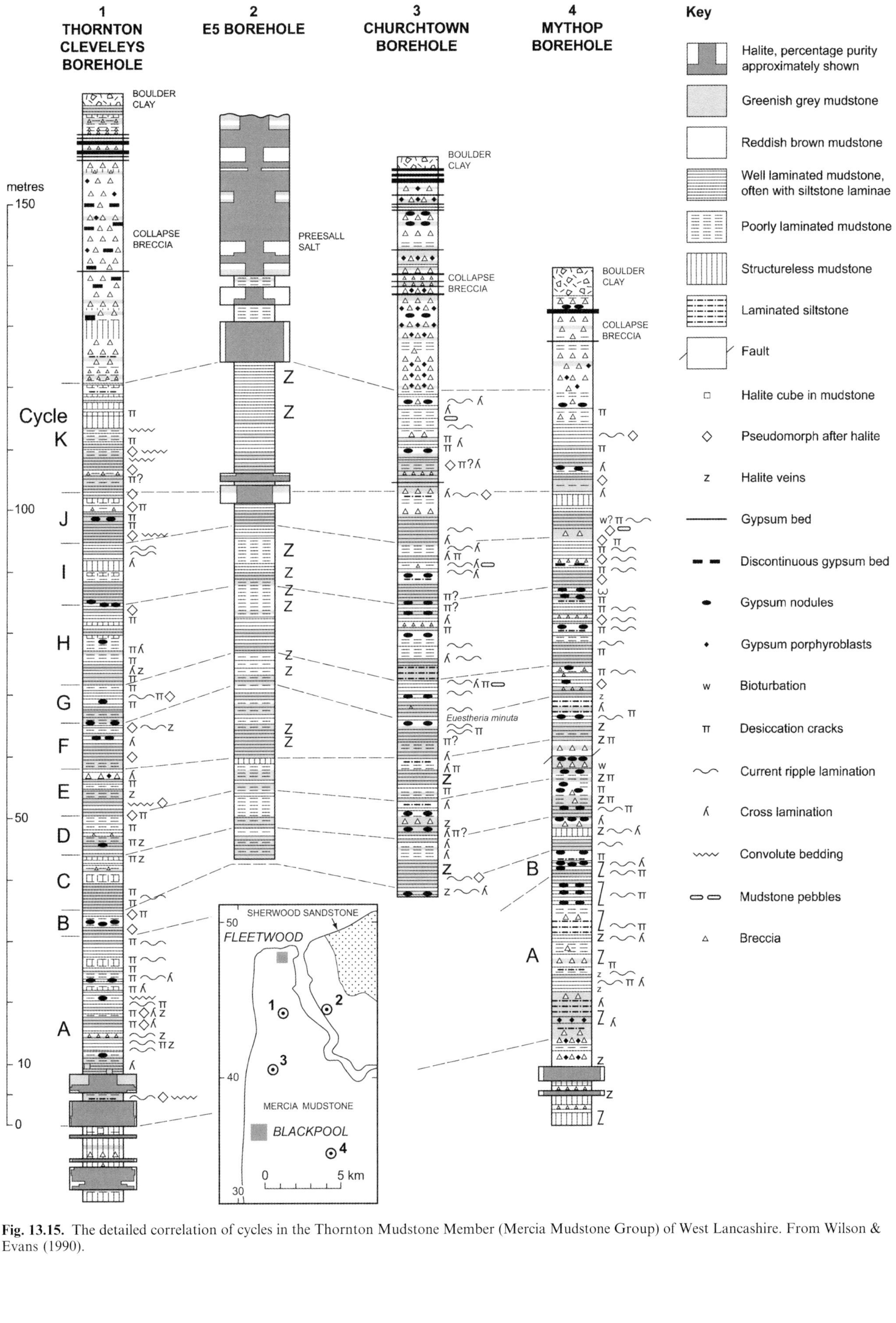

Fig. 13.15. The detailed correlation of cycles in the Thornton Mudstone Member (Mercia Mudstone Group) of West Lancashire. From Wilson & Evans (1990).

building stone, with over 20 sites in the Cheshire–Shropshire area alone (British Geological Survey 2001*a*). Now the only working sources of building stone are small quarries at St Bees in Cumbria (working the St Bees Sandstone), and quarries at Grinshill and nearby Myddle in Shropshire (Fig. 13.12), which utilize sawn white and red coloured sandstone from the Helsby Sandstone Formation, mostly for building restoration.

Central England

The most complete Triassic successions of central England are located in the basin depocentres of the Worcester Basin, Knowle Basin, Needwood Basin and Hinckley Basin (Fig. 13.6). The thickest succession preserved is that in the Worcester Basin where the Triassic is up to 1.5 km thick, but it thins considerably by onlap on to the adjacent Palaeozoic and Precambrian highs. This onlap appears to have been punctuated but progressive, so that the youngest units of the Triassic have the largest areal extent onto the basin margins (Sumbler *et al.* 2000).

Evidence of synsedimentary fault activity is abundant, with seismic lines demonstrating thickness changes of many hundreds of metres, with the thickest successions adjacent to major faults. These basins are generally aligned N–S, with fault-bounded Palaeozoic horsts. These predominantly N–S structures are terminated to the north by an array of approximately E–W aligned fault structures forming the north termination of the Needwood Basin against the Pennine High (Fig. 13.6) (Chisholm *et al.* 1988).

Sherwood Sandstone Group in sequence TR1

It is only within the deepest parts of the Worcester Basin that the boundary between the Permian Bridgnorth Sandstone and the Triassic Sherwood Sandstone Group may be conformable, through a mudstone succession that may be comparable to the Permian–Triassic transition interval in NW England and the southern North Sea (Sumbler *et al.* 2000).

Outside the centre of the Worcester Basin, the lowest beds of the Triassic are commonly poorly sorted breccias or conglomeratic sandstones, with a high content of locally derived, angular clasts. These basal breccias have been given a variety of local regional names (such as Moira and Hopwas breccias). The thinner breccia successions may represent regolith or pediment gravels, whereas the thicker deposits represent local alluvial fans shedding material from the highs into the basins. Their age is uncertain, and may be in part Permian probably extending into the Mid or Late Triassic, since locally they pass laterally into the Mercia Mudstone Group (Chisholm *et al.* 1988; Worssam & Old 1988; Powell *et al.* 2000*b*).

The lowest beds of the Sherwood Sandstone Group are typically conglomerates, overlain by pebbly sandstones (Kidderminster, Polesworth and Cannock Chase formations; Fig. 13.3), which pass upwards to largely pebble free, cross-stratified fluvial sandstones (Wildmoor Sandstone and Hollington Sandstone formations; Fig. 13.3). The Kidderminster Formation and its equivalents possess well-rounded pebbles dominated by vein quartz and quartzite, with lesser amounts of locally derived clasts. Petrography, heavy minerals, isotopic data from micas and derived fossils in clasts attest to the southerly-derived nature of these sandstones and their pebbles (Wills 1956; Mitchell & Taka 1984; Jeans *et al.* 1993; Powell *et al.* 2000*b*). The regional spread of these pebbly sandstones indicates the establishment of a predominantly N-directed braided fluvial system (Budleighensis river of Wills 1956), with far-travelled and well-rounded clasts. The thickness of the pebble-bearing units varies widely from 70 m in the centre of the Worcester Basin to 150 and 200 m in the Needwood and Hinckley basins. The Kidderminster Formation consists of a stacked succession of fluvial fining-up cycles, each with conglomeratic lower parts and either sandstone or mudstone tops. Wills (1970), in his seminal work on the English Midlands Triassic, proposed that regional correlations could be achieved in more basinal successions by grouping these cycles into bundles (which presumably reflect a climatic imprint), although his study has not been repeated using modern techniques. To the NE, in the Hinckley Basin, the abundance of mudstone intervals increases (Worssam & Old 1988).

The pebble-bearing sandstones pass upwards through a gradational boundary into the overlying largely pebble-free Wildmoor Sandstone Formation (Fig. 13.3). This is an orange-red coloured, fine-grained, commonly micacous sandstone, with a few red and green mudstone beds. The unit has a maximum thickness of some 120–250 m in the Worcester and Knowle Basins, and may be diachronous northeastwards into the conglomerate-bearing sandstones of the Hinckley and Needwood Basins, which probably indicates the local provenance of substantial proportions of the clastics from the London–Brabant High. However, the heavy mineral content suggests a similar southerly source like the underlying Kidderminster Formation (Powell *et al.* 2000*b*).

The thickness and preservation of the successions below the H-unconformity was strongly controlled by both faulting and erosion prior to the deposition of the overlying Bromsgrove Sandstone. This appears to have restricted the older units of the Sherwood Sandstone Group to basin depocentres, and led to the removal of the Wildmoor Sandstone Formation in some areas (Wills 1970; Bridge *et al.* 1998). These younger units were perhaps deposited more widely on the basin margins in the Triassic, but have been removed during the uplift phase that is represented by the H-unconformity. This situation is reminiscent of the sometimes large erosional truncation of the Dutch and German Middle Buntsandstein that occurs in the southern North sea below the Hardegsen unconformity (Geluk & Röhling 1999).

Sherwood Sandstone Group in sequence TR2-3

The unconformity at the base of the Bromsgrove Sandstone Formation is particularly marked in the West Midlands, where it oversteps onto Lower Palaeozoic strata. In Warwickshire the basal beds locally contain wind-faceted dreikanter pebbles (Wills 1970). The formation exhibits fluvial fining-up cycles, 2–10 m thick, ranging from conglomerate (or breccia), with mudstone clasts at the erosive base, to mudstone at the top (Worssam & Old 1988). The abundance of mudstone beds increases upwards, and in parts of the Hinckley Basin and south of Warwick, may comprise up to 30–40% of the formation (Old *et al.* 1987; Worssam & Old 1988). The distinctive feature of the sandstones is their feldspar content, which can range up to 50%, and coarser intervals may contain feldspathic pebbles (Wills 1970; Old *et al.* 1987; Powell *et al.* 2000*b*). The formation is up to 500 m thick in the Worcester Graben, and 200–300 m thick in the vicinity of Birmingham, with rapid thinning to 20–50 m onto basement highs. In its type area south of Birmingham, and also in boreholes in the Hinckley Basin, three subdivisions are recognized (Worssam & Old 1988; Old *et al.* 1991; Powell *et al.* 2000*b*). The lower division (Burcot Member) is dominated by sandstones with common intervals of conglomerate containing mud-chip breccia, reworked caliche and vertebrate fragments. The middle division (Finstall Member) is dominated by sandstone–mudstone fining-up cycles, with common plant remains. The upper division (Sugarbrook Member–'waterstones' facies) is composed of argillacous mica-rich sandstones and siltstones, which may overstep onto Palaeozoic units. Both the middle and upper divisions locally contain sulphate nodules (Old *et al.* 1991), and mudstone beds may show evidence of brecciation due to pedogenesis and desiccation. The Finstall Member contains the most diverse fossil assemblages including, annelids, bivalves (*Mytilus*), arthropods (Crustacea and Scorpionida), amphibians, fish (*Acrodus* and *Ceradotus*) and reptiles, along with evidence of a diverse flora including lycopsids, sphenopsids, pteropsids, gymnosperms and conifers (Old *et al.* 1991). South of Birmingham, cross-stratification indicates sediment transport to the north and NE (Old *et al.* 1991), although on the flanks of the Worcester Basin the sandstone is interbedded with locally-sourced breccia showing that sediment transport was also locally derived from basement highs (Barclay *et al.* 1997).

Mercia Mudstone Group

The Mercia Mudstone Group in the Midlands is mostly known in detail from boreholes, and is up to about 650 m thick in the Worcester Basin with a similar thickness in the Hinckley Basin. It oversteps the older units of the Triassic to rest on the basement at basin margins, and probably totally covered some basement highs like Charnwood in the east. The subdivisions recognized by Elliot (1961) around Nottingham, can be identified in the Sidmouth Mudstone Formation of the Hinckley Basin and in adjacent parts of the West Midlands (Worssam & Old 1988). These divisions progressively become indistinguishable westwards, so that the succession below the Arden Sandstone Formation is mostly undivided blocky red mudstones. Within southern and western parts of the Midlands the Arden Sandstone Formation (probably equivalent to Hollygate/Dane Hills sandstones of the East Midlands) can be identified, but this disappears northwards, and may be partly equivalent to the Stafford Halite in the Stafford Basin (Fig. 13.4).

A distinctive feature of the Mercia Mudstone Group in the Midlands is the presence of a number of sandstone or siltstone beds ('skerries'), which have provided locally useful markers for mapping by the Geological Survey and others. The thickest of these are up to about 10 m, but many are less than 0.3 m thick. These intervals are pale brown–pale green in colour and range from siltstones to fine-grained sandstones. These are well-sorted, rich in dolomite, calcite or ankerite, with less common anhydrite and rare baryte (Old *et al.* 1987; Worssam & Old 1988; Barclay *et al.* 1997), commonly with alternating carbonate-rich and quartz-rich laminae. Dolomite may form 40–95% of the matrix, and dolomitized ooids are rarely present (Worssam & Old 1988). These coarser units are characteristically thin-bedded intervals, sometimes with wave/current ripples, planar and cross-lamination, with intervening thin red/green mudstone intervals. Mudcracks and halite pseudomorphs are also present.

The Arden Sandstone Formation is a unit of multicoloured mudstones with abundant discontinuous sandstones, which has

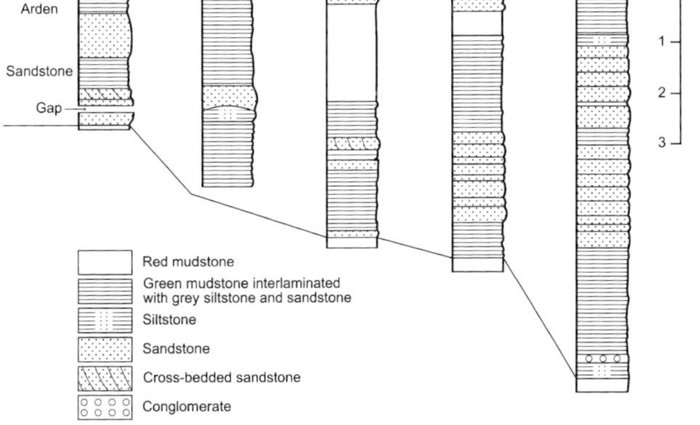

Fig. 13.16. Comparative sections in the Arden Sandstone Formation in its type area in Warwickshire (from Old *et al.* 1991).

a type area in the South Midlands (Fig. 13.16). The thicker sandstone intervals in the Arden Sandstone Formation are cross-stratified and have erosional bases (Old *et al.* 1987; Barclay *et al.* 1997). The sandstone intervals are interbedded with dark green–grey laminated mudstones and siltstones, with subordinate purple and grey mudstones (Fig. 13.16). In some areas of the Midlands (i.e. deeper basins), the sandstone intervals may be reduced to little more than a few thin sandstone beds within a predominantly non-red mudstone succession. Cross-stratification measurements in the west Midlands suggests the Arden Sandstone detritus was transported by easterly- or NE-directed currents (Old *et al.* 1991). This and the distribution of the sandstone suggests that it may have been sourced from clastics shed from the Welsh High and the London–Brabant High.

Economic Geology

The Sherwood Sandstone Group to the west of the Birmingham Fault is a major aquifer, which largely acts as a single hydrological unit, although impermeable faults, finer-grained units and cemented layers within the sandstone may locally segment the aquifer both vertically and horizontally (Allen *et al.* 1997). Until the end of the 19th century, the city of Birmingham used the Sherwood Sandstone aquifer extensively for public supply; since then water reservoirs in Wales or the River Severn have been used. However, during years of drought on the River Severn (such as 1976, 1989 and 1995) the Triassic sandstone aquifer of north Shropshire is utilized to top-up the river for domestic and industrial supply in the West Midlands. Now in Birmingham almost all groundwater abstraction is for industrial use, although this use has been reducing since the 1970s. With the reduction of underground water abstraction there has been a steady rise in the height of the water table to the pre-industrial levels (Ford & Tellam 1994). This has caused flooding of cellars in domestic housing and created the potential for the groundwater to come into contact with contaminated ground, previously above the water table. Groundwater contamination around Coventry and Birmingham is particularly acute, with some of the Sherwood Sandstone aquifer compromised by leakage of light hydrocarbons and solvents into the subsurface (Ford & Tellam 1994; Tait *et al.* 2004). Limited, agricultural water abstraction using wind-pumps does occur in the Midlands and SW England from the Arden Sandstone Formation, which is enclosed by Mercia Mudstone aquicludes and is thus less liable to pollution.

Historically, building stone from the Bromsgrove Sandstone Formation was locally important, with Norbury Church and Calke Abbey (both in Derbyshire) as some of the historically important buildings of this material (British Geological Survey 2001*a*). The loosely consolidated conglomerate-bearing lower intervals of the Sherwood Sandstone Group are still important sources of aggregate in the Midlands, with various locations being quarried (Fig. 13.12), such as at Tern Hill (near Market Drayton), Barnsley Lane east of Bridgenorth and several quarries SE of Cheadle in Staffordshire (Huntley Wood, Croxden and Freehay Quarries).

Southwest England, South Wales and the offshore basins

The outcrop of the Triassic in SW England stretches across Somerset and Devon and beneath the English Channel. Eastward extension is concealed by younger strata, but Triassic rocks in the subsurface extend eastwards to Purbeck and beyond (Butler 1998). Triassic is known from South Wales and its extension westwards in various offshore basins in the Bristol Channel, Western Approaches, southern Celtic Sea and Cardigan Bay (Evans 1990; Hamblin *et al.* 1992; Tappin *et al.* 1994).

The onshore Triassic is divided into a number of predominantly E–W trending sub-basins, which include the Portland–Wight Basin and the Bristol Channel Basin (Fig. 13.6) (Hamblin *et al.* 1992). In the western approaches the sub-basins of the Western Approaches Trough (the Melville, St Marys, SW Channel and Plymouth basins; Fig. 13.6) are elongated in a more SW–NW trend, reflecting the underlying Variscan structure. These contain Triassic successions which are relatively poorly known, but are similar in some ways to the Midlands and NW England Triassic, in that they contain relatively thick equivalents of the Sherwood Sandstone Group (Tappin *et al.* 1994). Minor syn-Triassic grabens and half-grabens throughout the greater SW English–Welsh area show variable orientations, probably as a result of underlying basement structural control (Miliorizos & Ruffell 1998).

Sherwood Sandstone Group

In the Wessex Basin, the Sherwood Sandstone Group is divided into a lower conglomeratic unit, the Budleigh Salterton Pebble Beds and an overlying sandstone-dominated unit, the Otter Sandstone Formation (Fig. 13.3). This division also appears to be present in the Bristol Channel Basin, and the offshore St Georges Channel and South Celtic Sea basins (Fig. 13.6) (Tappin *et al.* 1994). The boundary between the Budleigh Salterton Pebble Beds and the underlying Aylesbeare Mudstone Group within the Wessex Basin is an unconformity (Holloway *et al.* 1989), which is most clearly apparent from seismic surveys and well data further east in the basin (Butler 1998). Within the deeper part of the basin in west Dorset an additional sandstone or siltstone interval (up to 70 m thick) rich in anhydrite rests conformably on the Budleigh Salterton Pebble Beds, which is unconformably overlain by the Otter Sandstone Formation (Fig. 13.3) (Butler 1998). Holloway *et al.* (1989) and Butler (1998) have suggested that the base Otter Sandstone unconformity seen on seismics is equivalent to the Germanic Basin H-unconformity, although magnetostratigraphy suggests that the base of the Otter Sandstone Formation in the outcrops in S. Devon corresponds to a slightly younger level (Hounslow & McIntosh 2003). A prominent unconformity within the Permo-Triassic successions in the Melville and Plymouth Bay basins (Western Approaches Trough) has been broadly equated with the Aylebeare Mudstone Group–Sherwood Sandstone Group boundary based on a similar seismic signature to that seen in the Wessex Basin (Evans 1990; Harvey *et al.* 1994).

The Sherwood Sandstone Group thins from up to 250 m in the western part of the Portland–Wight Basin to 170 m in the Wytch Farm area, and is absent east of the Isle of Wight and in the SE part of the English Channel (Hamblin *et al.* 1992). In the western Approaches Trough, the Sherwood Sandstone Group ranges from over 900 m in the Melville Basin to a more typically thickness of 100–300 m in other sub-basins in the Trough (Evans 1990). Depositional continuity from the Wessex Basin to the offshore St Georges Channel and South Celtic Sea Basins was probably not maintained through the Bristol Channel Basin, which acted as a high until deposition of the Mercia Mudstone Group (Tappin *et al.* 1994).

Budleigh Salterton Pebble Beds – sequence TR1

The Budleigh Salterton Pebble Beds (Fig. 13.17a) are composed of ortho-conglomerates and subordinate sandstones and are the deposits of a braidplain or wet alluvial fan (Smith 1990; Smith & Edwards 1991). Magnetostratigraphy suggests the Budleigh Salterton Pebble Beds represent only a short interval of the Early Triassic (Hounslow & McIntosh 2003). The quartzite clasts within the Budleigh Salterton Pebble Beds are famous for containing an exotic Ordovician and Devonian brachiopod fauna from Brittany and Normandy (Cocks 1993). The beds also contain tourmalinized hornfels and quartz–porphyry clasts, and an Armorican-derived heavy mineral suite (Thomas 1909;

Fig. 13.17. (**a**) H-disconformity (arrowed) between the Budleigh Salterton Pebble Beds and the Otter Sandstone Formation. At this location the first beds in the Otter Sandstone are dune-bedded sandstones that occupy the upper part of the photo. There is a thick impersistant sandstone interval in this section of the upper most part of the Budleigh Salterton Pebble Beds; Budleigh Salterton, Devon. (**b**) Differential weathering of vertical rhizoconcretions from Unit B of the Otter Sandstone Formation, that are particularly marked in one bed (arrowed), Coal Beach, near Budleigh Salterton. People used for scale.

Wills 1956; Campbell-Smith 1963; Jeans *et al.* 1993). The Budleigh Salterton Pebble Beds represent the deposits of a river system which drained the Armorican land-mass located in western France and the southern English Channel, and transported exotic clastics northwards into the Wessex Basin (Holloway *et al.* 1989; Smith & Edwards 1991). Within the northerly part of the Budleigh Salterton Pebble Beds outcrop in Devon, locally derived Carboniferous and Devonian sandstone and limestone clasts together with palaeocurrent measurements show that tributaries from upland areas to the west in Cornubia also contributed locally (Smith & Edwards 1991).

A horizon of wind-faceted pebbles (ventifacts) is regionally important at the top of the Budleigh Salterton Pebble Beds (Leonard *et al.* 1982; Smith & Edwards 1991), and is associated in the coastal outcrops with a desert palaeosol (Fig. 13.17a) (Wright *et al.* 1991), which appears to indicate a major hiatus. Seismic evidence in the central and western parts of the Wessex Basin suggests that this major unconformity at the base of the Otter Sandstone Formation locally truncates faults in the underlying succession (Butler 1998).

Conglomerate-bearing sandstones equivalent with the lower part of the Sherwood Sandstone Group (of the Wessex Basin) are known from the Melville Basin of the Western Approaches Trough, the Wessex Basin, the South Celtic Sea and the St Georges Channel basins (up to 75 m thick). It may be that this equivalent of the Budleigh Salterton Pebble Beds is only present in more strongly subsiding sub-basins, perhaps indicating the original restriction of deposition to palaeovalleys in the Early Triassic landscape, or major truncation at the H-unconformity.

Otter Sandstone Formation–sequence TR2–3

The Otter Sandstone Formation forms most of the Sherwood Sandstone Group within the Wessex Basin, and largely consists of stacked fluvial channel sandstones, with minor mudstones and some aeolian sandstones. At outcrop in south Devon the age of the Otter Sandstone Formation ranges from the latest Early Triassic to near the Anisian–Ladinian boundary (Hounslow & McIntosh 2003). Farther east in the Wessex Basin around the Wytch Farm Oilfield (Fig. 13.12), seismic evidence suggests the upper part of the Sherwood Sandstone Group passes westwards into the Mercia Mudstone Group, indicating that the upper boundary of the Sherwood Sandstone Group is probably diachronous (Butler 1998).

Within the coastal outcrop in Devon the Otter Sandstone Formation has been divided into four units, A, B, C and the Pennington Point Member (Hounslow & McIntosh 2003; Gallois 2004). These loosely define an overall fining-up succession, which is also typical of the Otter Sandstone in the Wytch Farm area (Lott & Strong 1982; Holloway *et al.* 1989; McKie *et al.* 1998; Svendsen & Hartley 2001, 2002). In the coastal outcrops, the oldest parts of the Otter Sandstone Formation are a heterogeneous unit (*c.* 20 m thick), consisting of a basal aeolian sandstone overlain by non-calcrete bearing fluvial channel sandstones (Fig. 13.17a). At outcrop these pass upwards into Unit B of the Otter Sandstone Formation, characterized by multistorey stacked channel sandstones, with a paucity of lacustrine or floodplain mudstones. Unit B contains abundant reworked calcrete (from rhizoconcretions; Purvis & Wright 1991) and mudstone clasts in the basal lags of channel sandstones. *In situ* calcretes common in unit B are seen to be both vertical calcrete rhizoconcretions formed around the taproots of phreatophytic plants (Fig. 13.17b), and horizontal sheets of calcrete formed at contemporary water table surfaces (Purvis & Wright 1991). Reptilian remains are mostly found within the channel lags, which include rhynchosaurs, temnospondyl amphibians and fish, most commonly comprising disarticulated skull elements (Milner *et al.* 1990; Benton 1997; Spenser & Storrs 2002).

The upper part of the Otter Sandstone Formation (Unit C) has more abundant mudstone and sandstone filled abandoned channels, with lesser amounts of *in situ* and reworked calcrete, perhaps representing a higher sinuosity river system (Svendsen & Hartley 2001). Minor aeolian sandstones also occur in the upper part of the Otter Sandstone at Wytch Farm or are known from marginal areas (McKie *et al.* 1998; Edwards 1999). Gallois (2004) has referred the uppermost unit in the Otter Sandstone at outcrop to the Pennington Point Member. This is a succession of fluvial channel and sheetflood sandstones interbedded with numerous red mudstones, whose sandstones lack reworked calcrete, and may be heavily channelized. This is similar in character to the uppermost units in the Otter Sandstone Formation at Wytch Farm that are interpreted as strongly lacustrine in character (McKie *et al.* 1998).

In the Burton Row borehole in Somerset Whittaker *et al.* (1985) divided the equivalent of the Otter Sandstone Formation into two units on the basis geophysical logs, a lower sand-dominated interval and an upper mudstone-rich interval. This division is also apparent in wells in the St Georges Channel and South Celtic Sea Basins (Tappin *et al.* 1994). This two fold division may correspond to the Unit A+B and Unit C+Pennington Point Member division known from the Devon outcrop sections, perhaps implying a regional response to a climatic event, or a consistent regional relaxation phase in basin tectonics. Miospore dating of the Sherwood Sandstone Group in the offshore basins suggests the age of the upper boundary of the Sherwood Sandstone Group may range from Anisian to Ladinian (Tappin *et al.* 1994).

Mercia Mudstone Group

The Mercia Mudstone Group (below the Blue Anchor Formation) within the Wessex and Bristol Channel basins is divided into three units, comprising upper and lower red-mudstone dominated units, separated by a multicoloured mudstone unit containing sandstone intervals (Arden Sandstone Formation; Fig. 13.4), which in basin depocentres contains halite (Gallois 2003). The Sidmouth and Branscombe Mudstone formations, which have a type area in south Devon, are dominated by blocky red mudstone and represent most of the Mercia Mudstone Group at outcrop in SW England. It is only within the multicoloured mudstone unit (some 30 m thick) that substantially different environmental conditions developed.

In the cliffs east of Sidmouth the Sidmouth Mudstone Formation comprises blocky red mudstones, with a few interbeds of green mudstones and siltstones, and some thin sandy siltstones. Gypsum nodules and veins are particularly common in the upper part of the succession (Gallois 2001). The Sidmouth Mudstone Formation may correspond to lithological units A–C defined by Lott *et al.* (1982) from gamma ray and sonic velocity measurements in boreholes in the Wessex Basin. Unit C includes substantial quantities of halite in the deeper parts of the basin, which Gallois (2003) has instead correlated to his Dunscombe Mudstone Formation (=Arden Sandstone Formation). Inter-borehole correlation is made difficult by listric faults that become bedding-parallel in the halite units in the Mercia Mudstone Group (Butler 1998).

The Arden Sandstone Formation at outcrop on the south Devon coast near Weston Mouth is a distinctive unit of interbedded greenish-grey, purple grey, red-brown and dark grey mudstones and siltstones with several sandstone beds. It contains a rich diversity of macroscale sedimentary structures including lamination, cross-lamination, invertebrate trace fossils, mudcracks, mudstone rip-up clasts, brecciated beds and immature palaeosols (Fig. 13.7c). Some of the brecciated beds are penecontemporaneous halite collapse breccias (Gallois 2003). Strata equivalent to part or all of the Arden Sandstone Formation in west Somerset display strongly incised fluvial channels, indicating easterly directed-palaeoflow away from the margin of Cornubia (Ruffell 1991). Simms & Ruffell (1990) interpreted the increased fluvial inputs in the Carnian as representing a period of increased rainfall. This would certainly accord with the widespread development of the Arden Sandstone Formation in the Midlands.

The predominantly red-mudstones of the Branscombe Mudstone Formation are interrupted by minor thin (<1 m thick) sandstones, and a major interval of gypsum, the Red Rock Gypsum Member (Gallois 2001). This same unit is also present in the subsurface farther east (Gallois 2001), and may be the equivalent of the Yate sulphate bed in the Bristol region, the Tutbury Gypsum of the East Midlands and parts of the Keuper Anhydrite in the southern North Sea. At St Audrie's Bay in north Somerset, the uppermost 99 m of the Branscombe Mudstone Formation comprise red-brown dolomitic mudstones and siltstones with irregularly interbedded, green or greenish-grey mudstones and pale dolomites, with rare desiccation cracks, wave ripples and burrows (Talbot *et al.* 1994; Hounslow *et al.* 2004). The former presence of evaporates (sulphates) is indicated by dissolution voids and small collapse structures. The upper parts of the Branscombe Mudstone Formation can also be seen in South Wales, west of Penarth, and at Aust Cliff, near Bristol (Curtis 1982; Waters & Lawrence 1987). In north Somerset, the upper part of the Branscombe Mudstone has been correlated, using magnetostratigraphy, with the mid–late Norian conodont-bearing successions of the Tethys (Hounslow *et al.* 2004).

Although the presence of gypsum is a persistent feature of the Branscombe Mudstone Formation in borehole cores, the preservation of gypsum at outcrop is quite variable over distances of several hundred metres. This can be seen readily at Aust Cliff (near Bristol) and Seven Sisters Bay (near Penarth), where substantial beds of gypsum pass laterally within a few metres into green mudstones.

The tripartite subdivision of the Mercia Mudstone Group in the Wessex and Central Somerset basins can be applied to the Mercia Mudstone Group in the St Georges Channel, South Celtic Sea and Bristol Channel basins (Tappin *et al.* 1994). The middle division (equivalent to the Arden Sandstone Formation) contains substantial quantities of interbedded evaporates, and may comprise up to 70% of the thickness of the Mercia Mudstone Group in some wells in the deeper parts of the basins. The age control that is available in these offshore basins implies that the majority of the halite-bearing middle unit is Carnian in age, although in some boreholes it may extend into the Norian, and possibly down into the Early Triassic (Tappin *et al.* 1994).

Economic Geology

The Wytch Farm and Wareham oilfields (discovered in 1973) extends beneath Poole Harbour (and surrounding lands), Studland in Dorset and offshore below eastern Bournemouth Bay. Along with the Gannet D satellite field discovered in 1987 (Fig. 13.12), these fields are estimated to contain some 65 million tonnes of oil and are expected to remain productive until about 2035. The oilfield contains two reservoirs, the principal one (90% of reserves) being the Sherwood Sandstone Group, with production also from the Jurassic Bridport Sands. The Sherwood Sandstone Group reservoir is segmented by extensional faults which are oriented E–W and are influenced by underlying Variscan thrusts (Bowman *et al.* 1993). Oil was generated in the downfaulted Lias to the south of Wytch Farm in the Late Cretaceous and migrated upwards and northwards to its present position. The best quality reservoir is found in the lower half of the Sherwood Sandstone reservoir, interpreted by McKie *et al.* (1998) as a floodplain–sheetflood facies that was successively incised by stacked, channel-fill sandstones.

The main development of groundwater resources in the Permo-Triassic aquifers in SW England are in the Sherwood Sandstone, mostly in the Otter Valley north of Budleigh Salterton. Here the aquifer is about 150 m thick, and may be locally separated into a number of flow units by mudstones horizons near the boundary of the Otter Sandstone–Budleigh Salterton Pebble Beds and within the Otter Sandstone. Potable water in this aquifer is limited to within 5 km east of the basal outcrop of the Mercia Mudstone by high dissolved solids (Allen *et al.* 1997).

The Budleigh Salterton Pebble Beds are utilized as a source of sand and gravel in Devon with quarries at Blackhill, Hillhead and Whiteball exploiting the full stratigraphic thickness of the deposit (Fig. 13.12). Small supplies of building and walling stone are derived from the Budleigh Salterton Pebble Beds, worked at Capton near Williton (Somerset), and Langport Member limestones ('white Lias') from the Stowey Quarry near Bishop Sutton (Avon).

In the past, salt from natural brines was extracted from the Somerset salt field, hosted in the Mercia Mudstone Group, and a similarly large deposit of salt occurs in Dorset, buried at greater depth (Fig. 13.12). In the upper parts of the Branscombe Mudstone Formation, celestite (strontium sulphate) was worked for over 100 years, extracted from shallow pits mostly to the west of Yate (north of Bristol), from the Yate sulphate bed. Production ceased in the mid 1990s, but this area was once the world's leading producer of celestite.

Blue Anchor Formation

During deposition of the Mercia Mudstone Group there was a progressive loss, in younger strata, of regional dissimilarities which first makes its mark during deposition of the Branscombe Mudstone Formation, which everywhere in the UK is dominated by red mudstones with gypsum and anhydrite. However, it is unclear if the onset of Branscombe Mudstone deposition is the same throughout England and Wales, since this formation lacks age control apart from magnetostratigraphy in north Somerset and some sporadic miospore dating (Warrington 1981, 1984; Hounslow *et al.* 2004).

At the type area of the Blue Anchor Formation in north Somerset, the base is placed at the top of the highest major red mudstone bed (Whittaker & Green 1983; Warrington & Whittaker 1984) (Plate 25). In north Somerset it is divided into the Rydon and overlying Williton members (Mayall 1981), where the Rydon Member (29 m thick) comprises the 'tea green marls' and most of the overlying 'grey marls' of older terminology (Richardson 1911; Warrington *et al.* 1980; Warrington & Ivimey-Cook 1995) (Fig. 13.18). Here the 'tea green marls' include a succession of interbedded, green, dark grey and black dolomitic and calcitic mudstones, whereas the overlying 'grey marls' are dominated by interbedded pale grey–black dolomitic mudstones and dolostones (Plate 25). A prominent dolostone with solution cavities marks the base of the 'grey-marls' (Whittaker & Green 1983). Laminated silty mudstone beds, brecciated beds and mudcracked intervals are also present. In the subsurface and at Blue Anchor the Rydon Member contains gypsum nodules and beds, but at St Audrie's Bay, and in the Blue Anchor Formation at Penarth (South Wales; Waters & Lawrence 1987) solution features and collapse structures reflect the former presence of evaporite minerals.

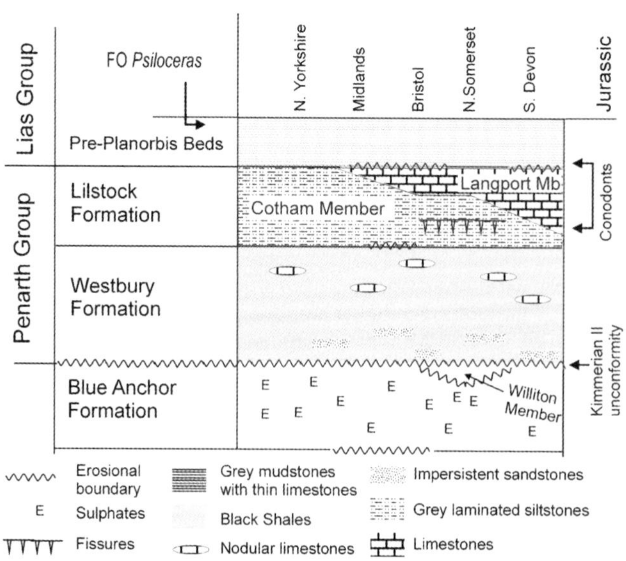

Fig. 13.18. Lithostratigraphy and regional variations in the Blue Anchor Formation and Penarth Group. FO, first occurrence.

In other areas, such as South Wales, south Devon, Avon and Gloucestershire, the Blue Anchor Formation includes substantially less of the 'tea green marls' type of lithology. In these areas the green mudstones, with interbedded red mudstones are included within the Branscombe Mudstone Formation (Old *et al.* 1987, 1991; Gallois 2001). In South Wales the Blue Anchor Formation also includes thin (<0.3 m thick) intervals of red mudstone. Locally, in places such as Warwickshire and at basin margin locations such as in South Wales, the Blue Anchor Formation has an erosive relationship to the underlying strata (Old *et al.* 1987; Water & Lawrence 1987; Barclay *et al.* 1997), or may overstep onto pre-Triassic units. Hence, the Blue Anchor Formation probably represents a separate tectonostratigraphic unit of sequence TR2–3, whose base may equate with one or more of the sequence boundaries at about this level in the southern North Sea.

In north Somerset the Williton Member (Mayall 1981; Edwards 1999) comprises the highest beds of the former 'Grey Marls' which thickens westwards into the Bristol Channel Basin (Fig. 13.18). It represents the first truly marine deposit of the UK Upper Triassic, in that it consistently contains marine microplankton and macrofauna (Warrington 1981; Warrington & Whittaker 1984). The Williton Member comprises grey silty shales with fine, flaser-bedded silts and sands and is overlain disconformably by the Westbury Formation (Penarth Group). However, it cannot be identified outside the north Somerset region, probably due to its erosion in other areas prior to the deposition of sequence TR4 (Penarth Group). Chronostratigraphic dating of the Blue Anchor Formation is problematic, mainly because the criteria that should be used to define the base of the Rhaetian have not been decided, even though magnetostratigraphy from this unit allows correlation with successions in other parts of the world that have been loosely referred to as Rhaetian or late Norian (Hounslow *et al.* 2004).

Triassic basin margin deposits

The nature of the margins of the Mid and Late Triassic basins are generally poorly understood, because for the most part the marginal deposits are poorly preserved. Locally-derived breccias, commonly matrix-supported (both sandstone and mudstone matrix) are characteristic of some regions (Chisholm *et al.* 1988; Worssam & Old 1988). These are thin deposits, laterally very variably, that have been interpreted as local alluvial fans, or fossil screes. Brookfield (2004) considered similar deposits as regolith and pediment channel deposits on an exposed desert surface, much like in the present day Egyptian desert. Around Charnwood in the East Midlands, the Mercia Mudstone abuts directly against basement, with a few metres of laterally intervening brecciated basement. The basement outcrops show features indicating *in situ* weathering, and the mudstones may be cross-stratified (Worssam & Old 1988), perhaps suggesting that aeolian deposition from mud aggregates, rather than alluvial processes, was most important in covering these basement outcrops.

In South Wales and in parts of the Bristol region, marginal facies of the Mercia Mudstone Group lacustrine basins are preserved, and provide the opportunity of understanding the interface zone between Triassic 'upland areas' and the lacustrine systems (Fig. 13.19). In both these regions the shore-zone deposits pass within a short distance into blocky red dolomitic mudstones of the Branscombe Formation, or into the grey-green mudstones of the Blue Anchor Formation (Waters & Lawrence 1987). In South Wales, there are broadly two groups of marginal facies. Around Barry Island, the Wenvoe Valley, and around Bridgend, lacustrine shore-zone sediments dominate, which are composed of breccias, conglomerates, calcareous sandstones and siltstones, with subordinate fenestral and cryptalgal carbonates. The carbonates generally occur

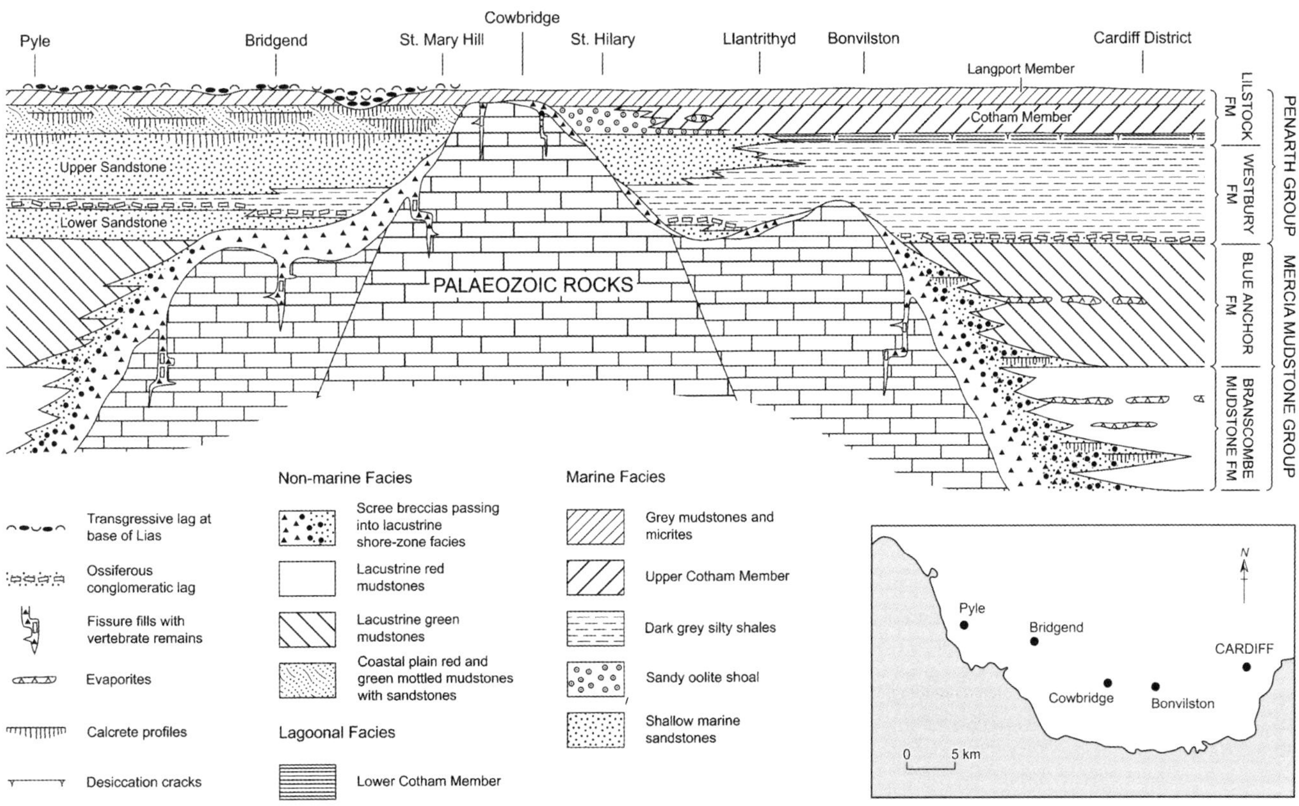

Fig. 13.19. The distribution of marginal facies in the Late Triassic around the 'Cowbridge Island' in South Wales. In the Mercia Mudstone Group, alluvial fan breccias and shore-zone lacustrine facies are limited to a thin zone at the basin margin. Modified from Waters & Lawrence (1987).

basinward of the clastics but in depositional continuity with the Branscombe Mudstone or Blue Anchor formations (Fig. 13.19). Nodular calcretes and replaced evaporates also occur (Tucker 1978; Waters & Lawrence 1987; Wilson *et al.* 1990). The carbonates are interpreted to have formed in a low-salinity lacustrine system (Leslie *et al.* 1992). These deposits formed around small contemporaneous islands composed of Carboniferous limestones and Devonian sandstones (Tucker 1978; Waters & Lawrence 1987). These islands had up to five wave-cut terraces and notches, and the deposits can within 10 m pass laterally into red and green mudstones with evaporites (Leslie *et al.* 1992). The marginal facies of the Blue Anchor Formation display a distinctively different lithology, being dominated by pale grey limestone-chip breccias and conglomerates, interbedded with pale brown and green calcitic and dolomitic mudstones with fenestral carbonates (Wilson *et al.* 1990).

South of Cardiff the lacustrine shore-zone deposits in turn pass northwards into a SSE–NNW-oriented palaeovalley filled with a series of breccias, fluvial sandstones and conglomerates, with local development of calcrete and dinosaur footprints (Tucker & Burchette 1977; Tucker 1978; Waters & Lawrence 1987). These deposits represent deposition in alluvial fans, augmented by sheet-floods during periods of intense rainfall (Tucker 1978), which transported sediment eastwards into the basin (Fig. 13.19). The interbedding of the lacustrine shore-zone carbonates with alluvial fan deposits, together with the wave-cut platforms and structures indicating desiccation, attest to the fluctuating level of the Late Triassic lake system.

In the Bristol region there is a similar succession of alluvial fan-like deposits (dolomitic conglomerate) composed of a variety of breccias, conglomerates and sandstones (Kellaway & Welch 1993). In the Clevedon area these occur with shore-zone clastics, including ooid-bearing deposits, which in some cases represent cementation in wave-dominated beach systems (Milroy & Wright 2000). The age of the ooid-bearing carbonates described by Milroy & Wright (2000) is less than certain and these may be coeval with the Blue Anchor Formation. The alluvial-fan deposits vary greatly in thickness due to both filling of probable palaeotopography and faulting. The nature and disposition of these marginal deposits in both South Wales and the Bristol region are perhaps the clearest evidence that during deposition of the upper part of the Mercia Mudstone Group, substantial water bodies existed, which have left little apparent imprint on the blocky red mudstones which dominate the Mercia Mudstone Group.

The Penarth Group

There were major environmental changes in England and Wales, and the rest of the world, during the transition from the Late Triassic to the Jurassic. The extinctions in Triassic biota and flora that precede the start of the Jurassic are one of the five major extinctions in Earth's history (see below: Hallam 2002; Tanner *et al.* 2004). The Rhaetian seems to have been a time of globally low sea level (Hallam & Wignall 1999), since there are few marine sections in the world that record an uninterrupted transition from the Rhaetian into the Jurassic. England and Wales is fortunate in this respect, in that a relatively complete record of the changes from the latest Triassic into the Jurassic is preserved in the sediments. Latest Triassic environmental changes are broadly reflected in England and Wales by the transition from the red 'desert' mudstones of the Branscombe Mudstone Formation to the lacustrine-dominated sediments of the Blue Anchor Formation (Plate 25), then to marine conditions, which generally began with the Penarth Group (Fig. 13.18).

The erosional unconformity between the Penarth Group and the underlying Blue Anchor Formation is universally recognized through England and Wales, and is the Kimmerian II unconformity of Zeigler (1990), widely recognized through northern Europe (Gianolla & Jacquin 1998; Goggin & Jacquin 1998). In the UK this unconformity represents a substantial time gap and has probably removed a variable proportion of the Blue Anchor Formation (Hounslow *et al.* 2004). The Kimmerian II unconformity heralds a major change to deposition of black marine shales of the Westbury Formation of the Penarth Group throughout England and Wales, similar to that seen throughout many basinal areas of northern and central Europe (Geluk 2005*b*).

The Westbury Formation consists of dark grey and black shaly mudstones and subordinate calcareous silty mudstones and grey, concretionary limestones. Also present are thin, wave-rippled sandstones, some of which are pebbly and contain remains of fish and other vertebrates, forming tempestite 'bone-beds' (Warrington & Ivimey-Cook 1995; Swift & Martill 1999) probably as a result of monsoon storms (MacQuaker 1994*b*).

The Cotham Member of the Lilstock Formation comprises pale grey-green calcareous mudstones with subordinate limestones, siltstones and sandstones commonly with mudcracks (Fig. 13.18). Stromatolites occur in these deposits around Bristol (Mayall 1983; Swift 1999). A prominent bed of disturbed sediment commonly occurs in the middle of the member, below a bedding surface penetrated by deep shrinkage cracks (Mayall 1983; Hesselbo *et al.* 2002; Simms 2003). This surface shows local evidence of erosion, although the deep shrinkage cracks are limited to the Bristol Channel region, and do not occur in the Cotham Member on the south Devon coast (Mayall 1983). The Cotham Member formed in shallower water than the underlying Westbury Formation and contains a marine microflora and fauna and was deposited largely in marine environments (Warrington 1981; Swift 1999). The origin of the disturbed bed in the Cotham Member has been variously interpreted as due to local seismic disturbance or shock waves from a bolide impact (Mayall 1983; Simms 2003). Hesselbo *et al.* (2004) have proposed that this shrinkage-cracked bed represents a sea level lowstand, with the overlying units (into the Jurassic) representing a transgressive system tract (Fig. 13.20).

The overlying Langport Member varies considerably on a regional scale. In southern England it is represented by microspar and peloidal limestones, that are commonly laminated and formed in shallow lagoons in warm, carbonate-rich marine water. These are well exposed in Pinhay Bay at Lyme Regis, where they display breccias, slumped units and micritized cross-stratified grainstone beds (Wignall 2001*b*; Hesselbo *et al.* 2004). Farther north this limestone facies is progressively replaced by siltstones and mudstones, until the Nottinghamshire area northwards, beyond which the Lilstock Formation is entirely represented by siltstones and mudstones (Fig. 13.18) (Swift 1995*b*).

Marginal facies of the Penarth Group are known at relatively few locations, which include boreholes on the northern margin of the London–Brabant High (Warrington *et al.* 1980; Cameron *et al.* 1992), in North Wales (Mochras borehole) and at Vallis Vale near Frome (Swift 1999). The most extensive marginal deposits are those around Bridgend in South Wales. Here, up to 14 m of sandstones overlain by 6 m of mostly green, red and buff-coloured mudstones form the lateral equivalent of the Westbury and Lilstock Formations respectively (Fig. 13.19) (Wilson *et al.* 1990). The sandstones equivalent to the Westbury Formation consist of an upper and lower unit separated by black mudstone, with the upper sandstone locally overstepping onto older rocks. The sandstone contains low-angle and hummocky cross-stratification with common conglomeratic lags, indicating deposition in a storm-dominated shoreface environment. The marginal facies of the Cotham Member comprises green, red and buff calcareous mudstones, with siltstone and sandstone lenses. Both the top of the upper sandstone and the Cotham Member equivalent units display dolomitic calcretes

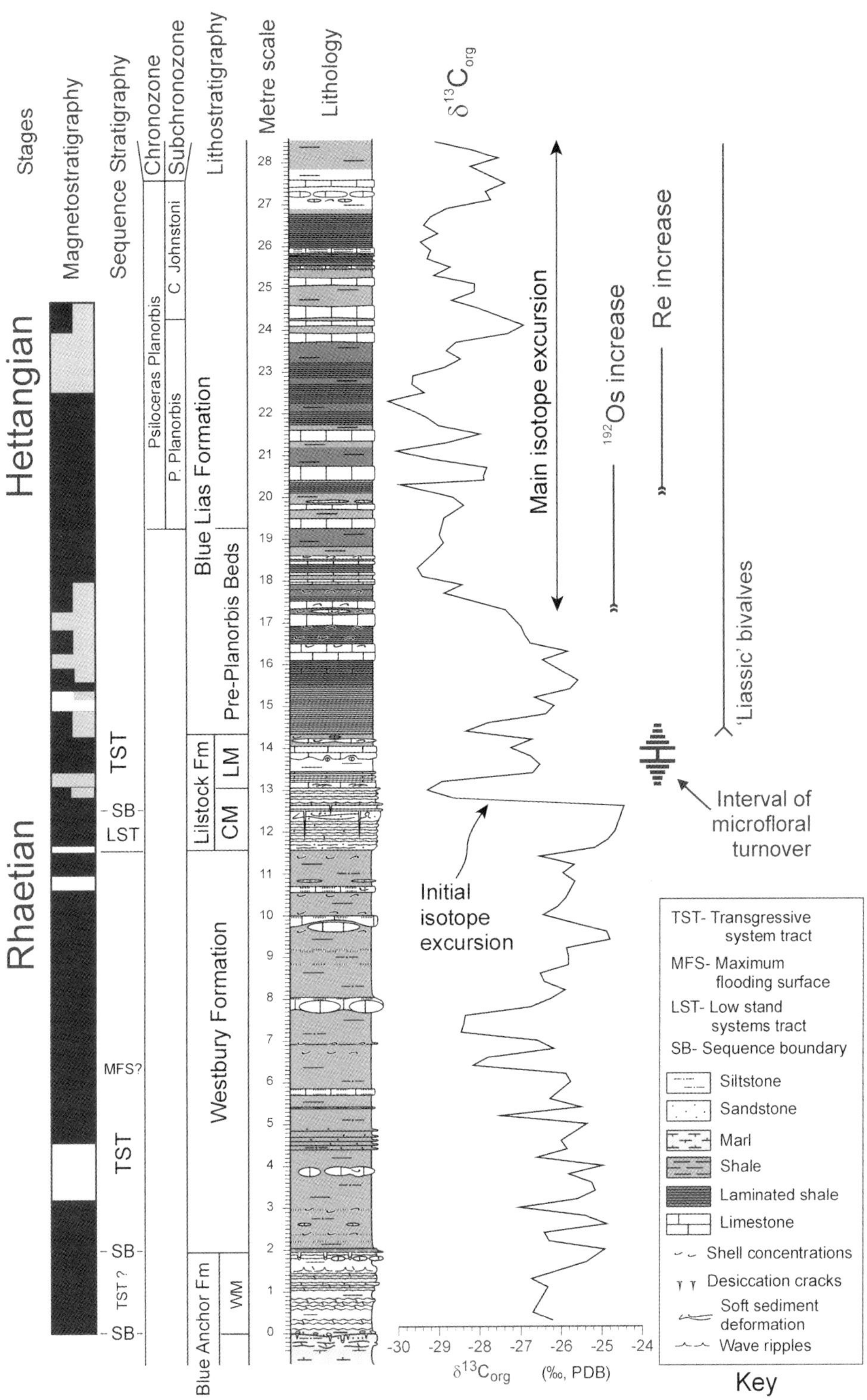

Fig. 13.20. The stratigraphy and indicators of environmental change (organic carbon isotopes – $\delta^{13}C_{org}$, flora changes, etc.) in the Penarth Group and basal Blue Lias Formation (St Audries Bay, North Somerset). CM, Cotham Member; LM, Langport Member; WM, Williton Member. Compiled from Cohen & Coe (2002), Hesselbo *et al.* (2002, 2004) and Hounslow *et al.* (2004).

profiles, attesting to periods of emergence and soil development, prior to deposition of the shelly grey mudstones of the Langport Member (Fig. 13.19) (Wilson *et al.* 1990). In NE England the Cotham Member contains intervals of reddish-brown mudstone, which suggests a return here to playa-type depositional conditions similar to those of the Mercia Mudstone Group (Swift & Martill 1999).

The Triassic–Jurassic boundary succession and global environmental change

The demise of ceratitine ammonoids, bivalves, conodonts, coral and crinoid groups from the Mid Triassic to the end-Triassic contributed greatly to the apparent extinctions of Triassic fauna before the Jurassic commenced. Raup & Sepkowski (1984) estimated that 76% of all species became extinct from the Mid to the latest Triassic. Tanner *et al.* (2004) reviewed the nature of these extinctions by considering both the global statistics on genera and species abundance, in addition to fossil records from sections. This conjunctive approach led them to reiterate that the Late Triassic saw a demise in both terrestrial and marine organisms during, possibly in an out of phase, stepwise manner. In British strata this stepwise loss in species cannot be evaluated, since much of the Upper Triassic succession is devoid of fossils until the Rhaetian Penarth Group. Hence, the rapid changes in both flora and fauna seen in Britain are restricted to the last phase of a punctuated or gradual demise. The causes of this demise are uncertain.

Firm evidence has been found of bolide-impact ejecta spherules and shocked quartz, probably from the Manicougan impact site in northern Canada, in a Norian red-bed succession from the Branscombe Mudstone Formation in the Bristol region (Walkden *et al.* 2002). This event predates the base of the Jurassic by about 14 Ma, and hence could in part contribute to the stepped-extinction hypothesis. One of the major losses in diversity in the marine environment took place in the transition from the Norian into the Rhaetian (Tanner *et al.* 2004). The postulated global sea-level drop during the Rhaetian may be responsible for this loss through reduction of marine habitats, although this cannot explain the changes in the terrestrial realm (Tanner *et al.* 2004).

The causes of the extinctions across the Rhaetian–Jurassic boundary are uncertain, but processes which have been implicated are: (a) bolide impact and a consequent increase in atmospheric opacity; (b) major changes in CO_2 content of the atmosphere (McElwain *et al.* 1999; Beerling & Berner 2002); (c) expulsion and out-gassing of large quantities of flood basalts from the central Atlantic magmatic province (Marzoli *et al.* 2004); (d) sea-level change; and (e) release of methane from sea-floor methane hydrates (Wignall 2001*a*).

Speculative evidence of a bolide impact occurs in the form of deformed and slumped beds in the Cotham Member in England and Northern Ireland (Simms 2003). This appears to occur at approximately the same level as a modest Ir-anomaly in some eastern North America sections, thought to be associated with the impactor material (Olsen *et al.* 2002).

During the latest Rhaetian and earliest Jurassic, there were two major perturbations in the global carbon cycle, evident by two negative excursions in the organic carbon isotopic signal seen in British sections (Fig. 13.20). The first of these, of apparently shorter duration, occurs within the Cotham Member, above the shrinkage-cracked bed. The second, of perhaps longer duration, starts within the Pre-Planorbis Beds of the Lias Group and appears to be coincident with Os and Re abundance changes in the mudstones. Cohen & Coe (2002) have interpreted the Os and Re abundance changes as indicating the initial weathering of the flood basalts from the Central Atlantic magmatic province (CAMP), suggesting that massive extrusion of basalts from CAMP and their out-gassing were responsible for the younger isotopic excursion. The origin of the older isotopic excursion is uncertain, but may also be linked with outgassing of flood basalts in North Africa (Marzoli *et al.* 2004).

The spore and pollen data record a major turnover of Triassic microflora within the middle–upper part of the Lilstock Formation (Edwards 1999; Hounslow *et al.* 2004), prior to the younger isotopic excursion in the Pre-Planorbis Beds. The major spore and pollen changes are slightly younger than the initial perturbation in the organic carbon isotope record in the Cotham Member (Hesselbo *et al.* 2002). Hence, in England and Wales there appears to be a sequence of significant 'environmental events' (Fig. 13.20). The earlier carbon isotopic event pre-dates fundamental changes in the spore and pollen flora (and perhaps the marine microfauna), and closely post-dates the lowstand in sea level in the Cotham Member. The later carbon isotopic event follows a transition interval into a typical 'Jurassic-type' bivalve fauna within the Pre-Planorbis Beds. Establishment of ammonoids that are considered to be typically Jurassic in character follows later when *Psiloceras* appears. The changes in atmospheric CO_2 content inferred by McElwain *et al.* (1999), from changes in fossil-leaf stomata size, in strata spanning the Triassic–Jurassic boundary in Greenland, cannot easily be related to the sequence of events seen in the UK. The increase in inferred atmospheric CO_2 in Greenland is probably across a minor hiatus but is also linked to a major turnover from Triassic to Jurassic-type flora (Hesselbo *et al.* 2002), implying there were also major climatic changes in the transition from the Rhaetian into the Jurassic.

14 Jurassic: the returning seas

J. C. W. COPE

The Jurassic Period in NW Europe began with an interval of pulsed subsidence that may be linked to local extensional fault movements. This parallels movements farther south where the western end of the Tethys opened in the Early Jurassic with a period of discontinuous rifting that lasted some 40 Ma and that is referred to as the Ligurian cycle; this phase of extensional tectonism resulted in block-faulting and tilting and was followed by rapid subsidence and the development of regional erosional surfaces (De Graciansky *et al.* 1998). The Tethys Ocean had long been established as a large oceanic area contiguous with the great Panthalassa Ocean to the east, but in the Early Jurassic it was bounded by continental areas on all other sides, with little marine connection across these areas.

Britain was situated in a seaway between the Laurentian margin and Scandinavia that allowed limited interchange of water northwards from the Tethys into the Arctic Sea; the latter extended from about 60°N to the pole. This seaway was interrupted by a series of islands of different sizes, including in the western European area the Spanish Meseta, the French Massif Central and Brittany, the Anglo-Brabant Massif, Cornubia, much of Ireland, the Scottish Highlands and Rockall. Interchange of waters between the Tethys and the Arctic sea was of fundamental importance in governing the faunas of the area; this movement was forced by differences in water-densities across the area, water flowing towards the areas of greater density. The Arctic sea received a considerable input of freshwater from high precipitation and continental run-off. This resulted in a predominant cold light N–S lower salinity surface current system countered by a deeper more saline S–N warm current from the Tethys, but with a net southerly flow (Miller 1990); this counter-current was reinforced during the high sea levels of the latest Jurassic. Occasionally, however, Arctic waters were denser than Tethyan and this resulted in a net northward flow (Bjerrum *et al.* 2001). In the Mid Jurassic uplift preceding volcanicity blocked the North Sea rift as a seaway, thus restricting water circulation to the western side of Britain.

Within the Jurassic Period there were several large-scale flood-basalt extrusions. These include the Central Atlantic Province, now split between South America and West Africa, that was initiated in the late Rhaetian, but continued through Hettangian time. This event is implicated in the end-Triassic extinctions (Chapter 13; Courtillot 1999) that rank as the fifth most important Phanerozoic extinction (Hallam 1996). At higher southern latitudes, the Karoo and Ferrar Provinces of continental flood basalts were a globally recognized short-lived series of eruptions dated around the Pliensbachian–Toarcian boundary whose onset was correlated with a rapid rise of $^{87}Sr/^{86}Sr$ in sea water (Palfy & Smith 2000) and linked later with a Global Oceanic Anoxic Event in the Toarcian (Jenkyns 1988; Jenkyns & Clayton 1997). These events are contemporary with a second-order extinction event in which about 5% of marine families were lost, but up to 12.8% of continental families in the Toarcian (Benton 1995). In the British area the North Sea volcanicity in the Mid Jurassic had a significant regional effect (see below) and there are other much less significant unidentified centres of volcanic activity that existed for short intervals of time.

Climate in the Jurassic was warm and for the most part humid; CO_2 levels may have been as much as four times higher than those of today (Chandler *et al.* 1992). Following the long period of aridity in the Permian and Triassic periods, the latest Triassic Rhaetian Age marked the transition to a wetter climate. Clay mineral assemblages reflect this by their increasingly significant amounts of kaolinite, but proximity to land areas can also be an important contributory feature to kaolinite content, irrespective of climate. The combination of subtropical temperatures and high humidity reached its peak in the Late Callovian–Early Oxfordian and was followed by a progressive warming and increasing aridity over extra-polar latitudes; these changes appear related to the progressive modification of the global palaeogeographical picture ensuing from the progressing break-up of Pangaea (Abbink *et al.* 2001) and led to the northward spread of the low-latitude arid belt. These changes have been recorded in the clay mineral assemblages from the Central Atlantic, where high levels of kaolinite in the Callovian declined throughout the Oxfordian and Kimmeridgian (Chamley *et al.* 1983). In continental areas, the spread of aridity can be traced first to central Asia where evaporite deposition was initiated in the Oxfordian (Hallam 1984); farther west, conditions were semi-arid rather than arid and red-bed fluvial sequences are recorded from Portugal (Wilson 1971; Wright & Wilson 1987). Evaporite deposition began in the Late Kimmeridgian in Poland (Dembowska 1976), Germany (Jordan 1971) and around the same time in the Celtic Sea; in the last area evaporites are associated with red beds, although their occurrence there was ascribed to the influence of proto-Atlantic rifting by Millson (1987*b*).

In marine successions the onset of the aridity may also be recognized by reduction and ultimate loss of kaolinite from clay mineral assemblages; whereas loss of kaolinite occurs at the base of the Tithonian in southern France (Deconnink 1987) it was probably some 3 Ma later that the same change happened in southern England (Wignall & Ruffell 1990). Aridity and maximum temperatures occurred during the Tithonian (= Bolonian and Portlandian Secondary Standard Ages; see below) which was a time of high sea levels, and may be due to the arrival of cooler high-latitude waters in the North Sea region following the re-opening of the seaway connecting the Tethys to the Boreal Ocean. At the close of the Jurassic Period minor evaporite deposition was widespread over southern England and northern France.

It has for long been assumed that the Jurassic climatic zonation was much broader than that of the present with less marked contrasts between equatorial and polar regions and with no polar ice (Hallam 1975, 1984, 1985; Frakes 1979). However, evidence is now emerging that suggests that there may have been at times regimes cold enough to produce seasonal ice and perhaps even ice-caps in the polar regions. In a review of the evidence for polar ice during the Mesozoic, Price (1999) concluded that the records of tillites and dropstones provided firm evidence of polar ice during the Bajocian–Bathonian and the Tithonian, with the Pliensbachian another (although less certain) candidate. The evidence for both the Pliensbachian and Bajocian–Bathonian ice-caps was also supported by the abundance of glendonites (star shaped concretions after ikaite), which were notably absent during the Tithonian.

Latitudinal temperature gradients were considered by Hallam (1984) to be of lesser importance in controlling faunal and floral differences than at the present day, although he recognized a clear tropical zone comprising the area between the 30°N and 30°S within which hermatypic corals, thick-shelled bivalves and larger foraminifera abounded. As part of this tropical zone, the Tethys was characterized by abundant hydrozoans and dasycladacean algae together with lituoloidean foraminiferans zone (Hallam 1977). To the north of the Tethys a distinct Boreal realm is distinguishable from Mid Jurassic times onwards, although in part this may be due to the protracted geographical near isolation of the Boreal Sea (Callomon 1985). To the south of Tethys a southern high-latitude Austral realm has also been recognized (Stevens 1973; Crame 1993). It has also been noted that some bivalve taxa show bipolarity in their distribution in both the Pliensbachian and Tithonian (Crame 1993); the latter author suggested that increasing latitudinal temperature contrasts could lead to loss of low-latitude populations in a hitherto cosmopolitan species, leading to a bipolar distribution. Some ammonite taxa also show clear evidence of provincialism, but it should be noted that the phylloceratids and lytoceratids, often quoted as typically Tethyan elements of the Jurassic ammonite fauna, were controlled by water depth and not latitude. They are both characteristic of open waters and are present at the highest northern and southern latitudes when open-water conditions obtained.

Palaeogeography and depositional patterns

Towards the end of the Triassic Period England and Wales constituted a desert plain of generally low relief that had been flooded by the shallow latest Triassic Rhaetian transgression. The Rhaetian transgression marked the end of a long period of red-bed sedimentation in England and Wales (Chapter 13) and transgression continued into the Early Jurassic with overlap of the margins of the Rhaetian seas and progressive overstep onto the Upper Palaeozoic basement in some marginal areas (e.g. Donovan *et al.* 1979). This marine transgression reflects the continued rifting of the Pangaean continent in the Early Jurassic. Widespread extensional faulting reflects the persistence of rifting in the Jurassic Period, but lithospheric thinning caused thermally induced subsidence to continue initially in the Jurassic in areas such as the southern North Sea.

Throughout the Jurassic Period sediments show the proximity of land by the presence of drifted plant material and occasional preservation of insects and terrestrial vertebrates in marine sediments. Despite this, only two persistent land areas can be identified during the Jurassic in the Anglo-Welsh area; these are a large part of Devon and Cornwall (Cornubia) (Cope 2004) and most of East Anglia and the London area (Sumbler 1996) that formed the western end of a significant Anglo-Brabant landmass, the limits of which varied with time throughout the Jurassic Period (Fig. 14.1). Both of these land areas contributed sediment to the Jurassic rocks near their margins at various times during the period. But they were not the source of the majority of the Jurassic clay, which, from considerations of volume alone, is the product of a major river or river systems larger than any in the immediate vicinity. The clay is thus far-travelled. Its source is not clear, but Scandinavia and adjacent areas appear to be the most probable candidate source areas to the east, while a westerly source from Laurentia is also possible.

Wales had traditionally been thought of as a Jurassic land area (e.g. Wills 1951), but the sinking of the Mochras Borehole (Woodland 1971), close to the edge of the Harlech Dome, proved an extremely thick Lower Jurassic sequence, developed in typical offshore facies, showing that much of Wales must have had a Jurassic sedimentary cover, although its extent and thickness remain conjectural. The suggestion that North Wales had basinal thicknesses of Lower Jurassic rocks (Dobson & Whittington 1979) has not been supported by other authors. Cope (1984, 1998*a*) suggested that as North Wales lay between the basinal areas of Cardigan Bay and the Cheshire Basin, it was likely to have acted as a positive area during the Jurassic and received a thinner and perhaps a more attenuated Jurassic succession than the neighbouring basinal areas. Areas of Mid- and South Wales may have been land for parts at least of the Jurassic Period but were almost certainly submerged at times of high sea level such as the mid Toarcian. From the Vale of Glamorgan across to the Mendip Hills an archipelago of islands existed in earliest Jurassic times and the Jurassic shoreline can be seen around them. At other times in the Jurassic, however, parts of Wales supplied sediment into the Jurassic basins of the Bristol Channel and St George's Channel.

The Pennine Hills have previously been considered as a likely Jurassic land area (e.g. Arkell 1933) but recent palaeogeographical reconstructions suggest (Bradshaw *et al.* 1992) that they too were only land for some intervals of Jurassic time, as a southern extension of a Scottish landmass.

Viewed as a whole the Jurassic palaeogeography of England and Wales was affected by a series of transgressive/regressive events:

(i) the initial Hettangian transgression that continued through to the Early Pliensbachian;
(ii) a Late Pliensbachian regression followed by further transgression in the Toarcian;
(iii) a major Late Toarcian–Aalenian regression occasioned by the updoming of the Rattray Head volcanic centre, at the southern end of the Northern North Sea Basin, which caused uplift over an area some 800 km across (Underhill & Partington 1993). This is known as the Mid-Cimmerian uplift;
(iv) renewed transgression from Callovian through to the Kimmeridgian ages, so that by Early Tithonian (Bolonian) times the transgression was as far-reaching as that of the Toarcian;
(v) major regression in the Late Tithonian (Portlandian) so that marine sedimentation was eventually confined to a small area of East Anglia and south Lincolnshire. This regression coincided with the first phase of the Late Cimmerian movements and was accompanied by block-faulting.

This sequence of events may be linked to the global picture of Jurassic eustatic sea-level changes (Fig.14.2).

The patterns of Jurassic deposition were inherited from those of the Triassic Period with rapidly subsiding basins, often with marginal growth faults separating areas of relative crustal stability where subsidence was on a much smaller scale. This 'basin and swell' type of deposition can be traced back as far as Permian times in some areas; in other areas the basins became established later. Thus we find, in general, that where Jurassic deposits are thick, so are those of the underlying Triassic and where Triassic deposits are thin or absent, the Jurassic is also thinly developed. Such depositional differences have been ascribed to posthumous movement along pre-existing 'axes of uplift' (see for example Arkell 1933) and it was Hallam (1958) who first showed that differential subsidence rather than

Fig. 14.1. Palaeogeographical reconstructions for selected Jurassic intervals. Note persistence of land areas over London area (western end of the Anglo-Brabant landmass) and Cornubia. Modified after Cope *et al.* (1992).

positive movement was the critical factor. The most important 'swells' in the Anglo-Welsh area are the Market Weighton 'Swell' in the East Riding of Yorkshire, the Mendip 'Swell' in Somerset and the Vale of Moreton 'Swell' on the Gloucestershire–Oxfordshire border. The first owes its positive nature to its position above a basement granite (Bott *et al.* 1978) and persisted as a 'swell' area right through the Jurassic Period and into the Early Cretaceous; the second existed as an influence on sedimentation from earliest Jurassic through to Bathonian times, whilst the last, which consists of a series of en echelon N–S faults marking the eastern margin of the Worcester Basin (Sumbler 1996, fig. 8) affected sedimentation over a similar interval of time. In the cases of the first two of these 'swells', their effects can be seen even on the small scale of the 1:625 000 Geological Survey map sheet 2. These so-called 'swells' separate areas of thicker basinal deposition and so they are, in effect, essentially the basin margins.

In southern England the Weald and Wessex basins are effectively northerly sub-basins of the larger English Channel Basin to the south. This latter basin is floored by a Variscan basement, structures of which are aligned predominantly E–W, in line with the Variscan deformation over the whole of southern Britain (Lake & Karner 1987). The largely offshore Hampshire–Dieppe High, that extends into eastern Dorset, is terminated southwestwards by the Wight–Bray Fault; it separates the Wessex Basin from the Weald Basin. It has become clear that both the E–W faults and the predominantly NW–SE-trending sinistral strike-slip faults in the Jurassic of the

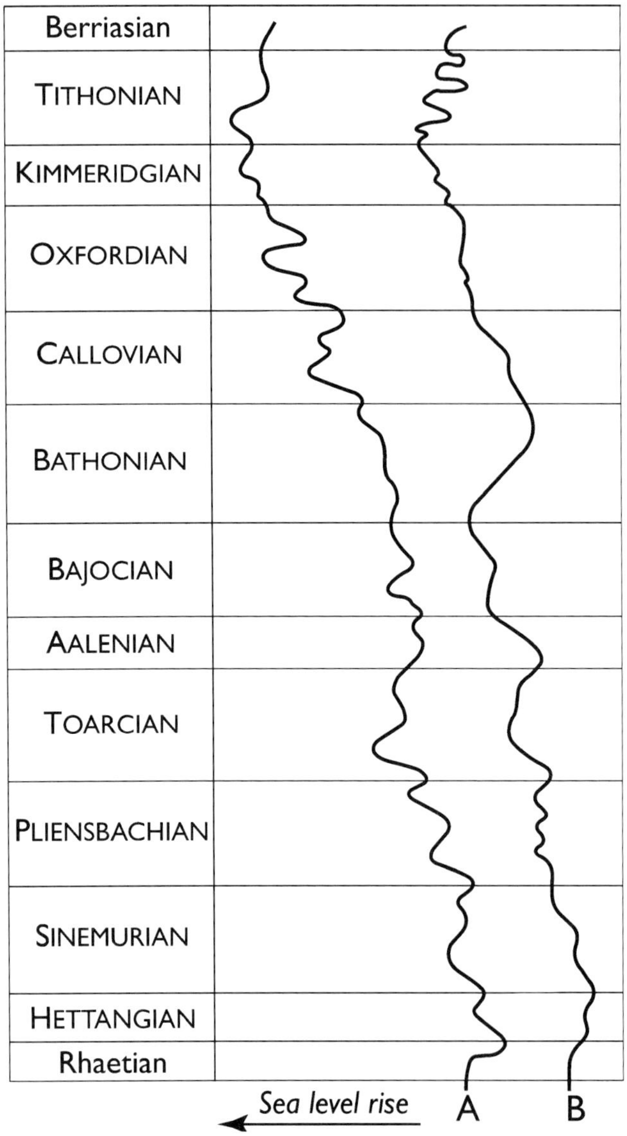

Fig. 14.2. Sea level curves for the Jurassic. Based on: (**A**). Hallam (1988) and (**B**). Haq *et al.* (1987). Diagram modified after Hallam (1992*a*).

whole English Channel Basin are inherited from this Variscan lineage (Chadwick *et al.* 1983; Whittaker 1985; Lake & Karner 1987). The Weald Basin abuts northwards onto the edge of the Anglo-Brabant landmass and part of the boundary was probably fault-controlled. The Wessex Basin was limited northwards by the Mendip 'swell', also probably fault-controlled, and westwards by the Cornubian landmass. The English Channel Basin connected westwards to the Western Approaches Basin that was separated by Cornubia from the Celtic Sea basins. The South Celtic Sea Basin extends eastwards into the Bristol Channel where basement Variscan faulting again played a significant role during the Jurassic Period. The North Celtic Sea Basin extends north-eastwards into the Cardigan Bay Basin where caledonoid basement structures provided the principal structural control (Tappin *et al.* 1994).

In the Eastern Irish Sea Basin little Jurassic rock is now preserved, but apatite fission track analysis has shown that a considerable inversion of that basin took place early in the Palaeogene and indicates the loss of up to 3 km of cover (Green *et al.* 1998). Ware & Turner (2002) have suggested that on the basis of sonic velocity analysis this may be an over-estimate and that generally the loss of cover was less than 1.5 km, although they calculated a maximum figure of 2132 m lost. Cope (1998*a*) has suggested that a considerable thickness of Jurassic rocks must have originally been deposited in the Eastern Irish Sea Basin; southeastwards an outlier of Jurassic rocks is preserved in the Cheshire Basin (Poole & Whiteman 1966). Outliers of Jurassic are also preserved in the Solway Firth Basin and its landward extension in the Carlisle Outlier (Ivimey-Cook *et al.* 1995; Holliday *et al.* 2004).

The Southern North Sea Basin extends from the NE side of the Anglo-Brabant landmass northwards to the Mid North Sea High; contemporaneous movement on its fault-defined western margin at times provided a facies boundary westwards with the East Midlands Shelf and northwestwards with the Cleveland Basin. During the Jurassic Period, the basin was far more extensive, and it extended across the southern North Sea to The Netherlands and Germany. Erosion following the Late Jurassic–Early Cretaceous Late Cimmerian movements fragmented the Jurassic outcrop and the only preserved Jurassic rocks within the UK sector of the southern North Sea lie directly offshore from the East Midlands and Cleveland Basin outcrops of which they are a natural eastward extension. The Jurassic rocks are relatively poorly known over this area because they have nowhere in the region provided a target in hydrocarbon exploration, and there is little published detail.

Unlike the Wessex Basin, where basin development was initiated in immediate post-Variscan times in the Permian Period, the development of the Cleveland Basin was essentially from the late Triassic (Kent 1980). Its margins to the west are the Pennine High and to the north the Mid-North Sea High, whilst it is bounded to the south by a series of E–W faults lying a short distance to the north of the Market Weighton 'Swell'. It is contiguous southeastwards with the Sole Pit Trough of the Southern North Sea Basin. Within the Cleveland Basin the NNW–SSE-trending Peak Fault system periodically influenced Jurassic sedimentation, especially in the Mid Jurassic, whilst the complex of E–W faults north of the Market Weighton Swell were principally active in the Late Jurassic and Early Cretaceous.

Jurassic lithostratigraphy

Jurassic rocks crop out extensively in England, with the principal outcrop running from the Dorset coast up to the Yorkshire coast, with major outliers in the Carlisle area, around Prees in Shropshire and in the Vale of Glamorgan in South Wales (Fig. 14.3); from the Midlands northwards, however, there is extensive drift cover. The Jurassic subcrop is extensive in southern and eastern England. Offshore, Jurassic rocks crop out on the sea-bed in the Straits of Dover, the western part of the English Channel, the Bristol Channel, St George's Channel and Cardigan Bay Basins and the NE portion of the Southern North Sea Basin; in the latter area again much of the outcrop is obscured by drift and modern sea-bed sediments. The offshore subcrop is more extensive (Fig. 14.3).

The main lithostratigraphical divisions are summarized in Figure 14.4.

The Jurassic rocks of the Anglo-Welsh area have long been famed for the wealth of beautifully preserved fossils that they contain; because of this these rocks have attracted attention for some 200 years, and much of the development of geology and of stratigraphy in particular is due to the fossiliferous nature, and in particular the abundance of ammonites, of the Jurassic sediments. It was on the Jurassic rocks of the Bath area that William Smith formulated his ideas on the superposition of rocks and the fact that formations could be identified by the fossils that they contained. Many of the names that appear in William Smith's table of strata (MS 1799, published 1815 – see Smith 1799) such as Fuller's Earth, Forest Marble or Cornbrash are still in use at the present day, and to Thomas Webster (1816) we owe the introduction of such terms as Kimmeridge, Portland and Purbeck into the geological lexicon.

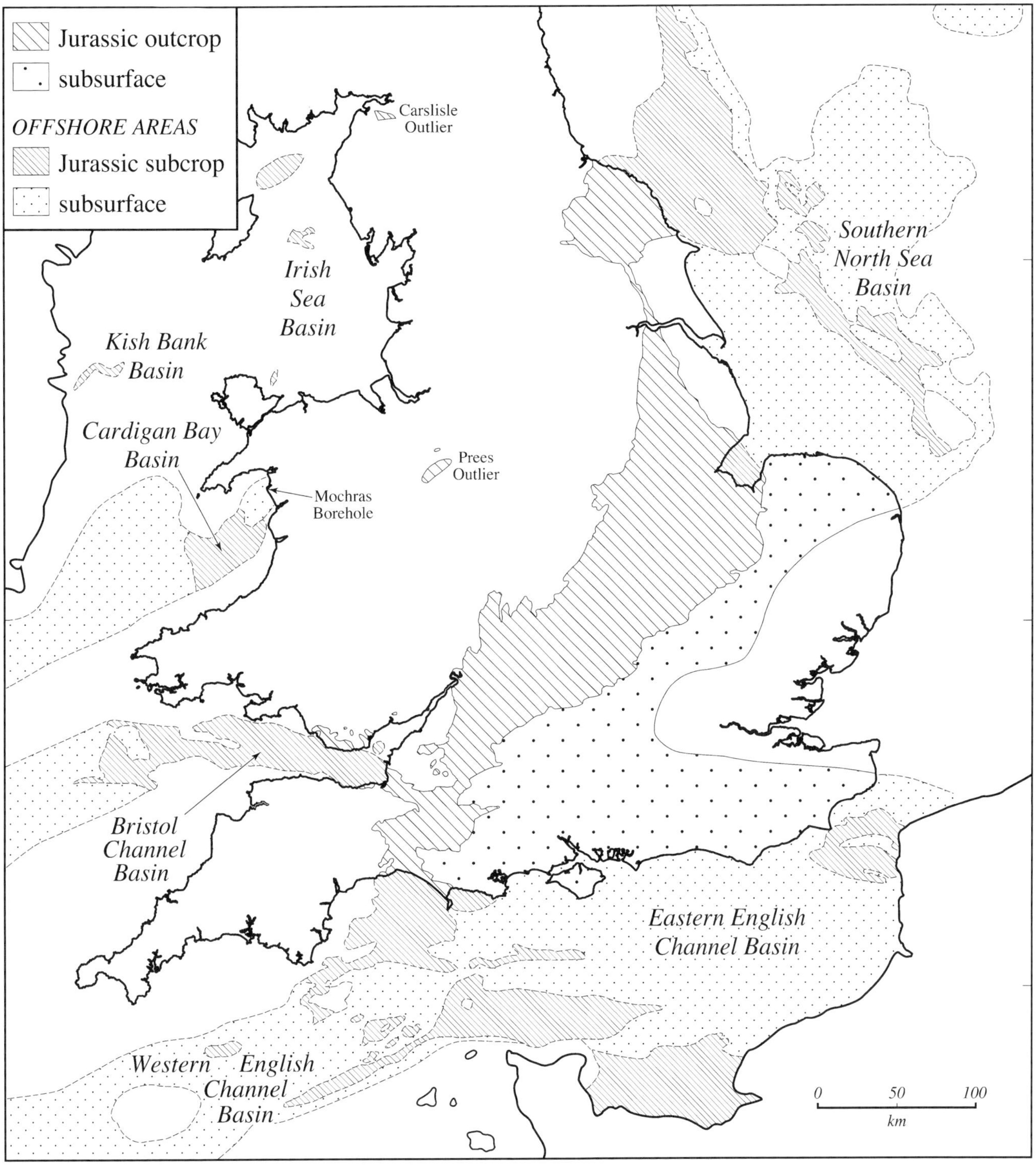

Fig. 14.3. Map showing the distribution of Jurassic rocks at outcrop and subcrop of England and Wales and offshore areas.

The variety and preservation of the Jurassic fossils is due to a combination of factors. The latitude of England and Wales at about 35°N was about 20° further south than now. The combination of latitude, climate and proximity to land has resulted in sedimentary facies that are dominated by siliciclastic rocks, particularly mudstones, with subordinate sandstones. At times, however, carbonate sedimentation predominated with extensive oolite shoals and occasional periods of reef coral growth, but at no time was the whole of the Anglo-Welsh area under the influence of carbonate deposition. Each of these facies has its own faunas.

In the mudrocks, bivalves are frequently the commonest fossils, but there are also abundant ammonites, belemnites and fish; many of the clays are dark and pyritic and fossils are frequently pyritized. The presence of calcareous nodules in many clays has often provided a source of further well-preserved fossils and such nodules often have a long and complex diagenetic history (Curtis *et al.* 2000). Most pyritic mudrocks have both abundant benthos and bioturbation suggesting that the reducing conditions that produced the pyrite only obtained below the sediment–water interface, but within a minority of clays, the presence of finely divided pyrite and the absence of both

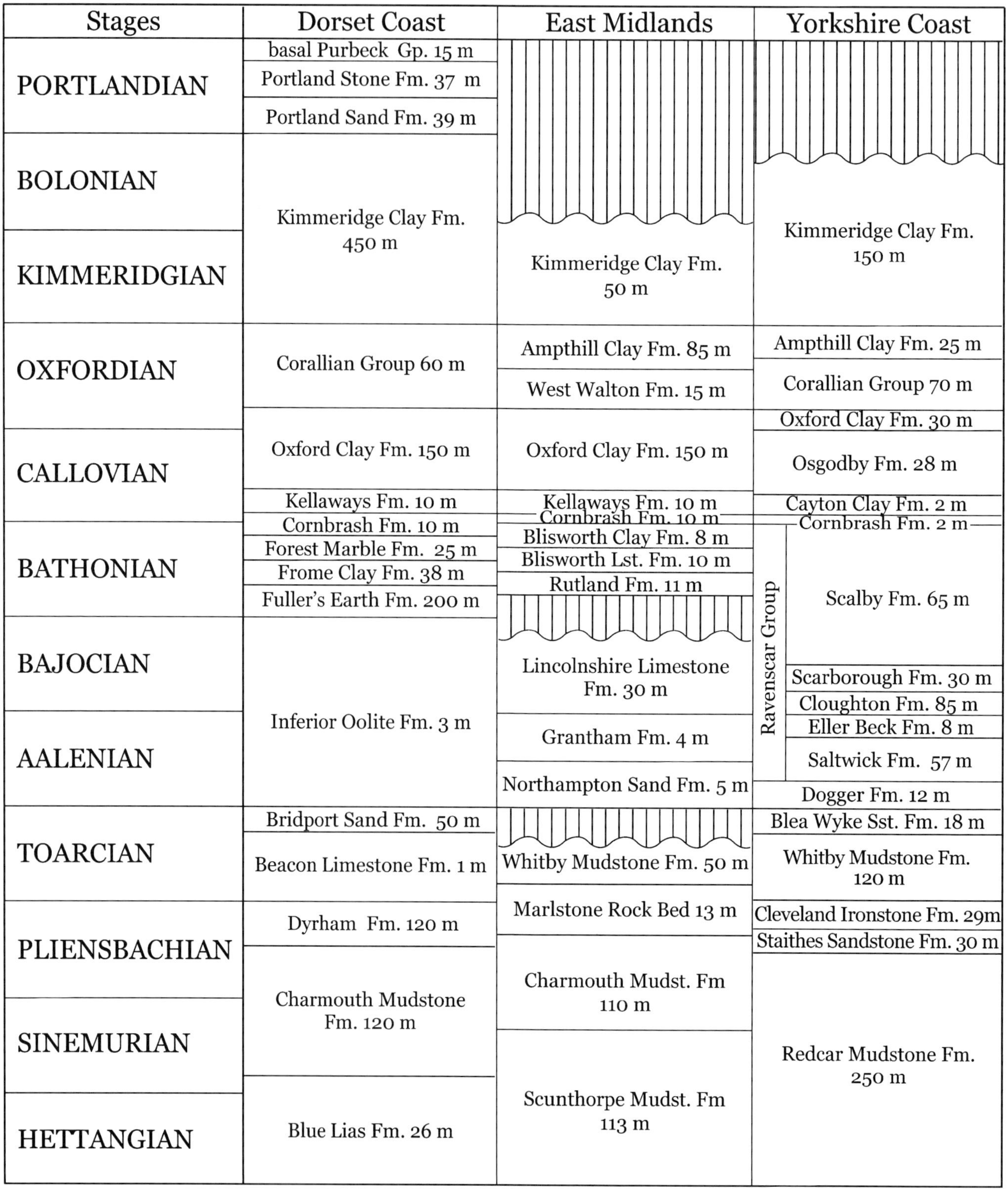

Fig. 14.4. Representative successions of Jurassic rocks in England.

benthos and bioturbation imply anaerobic bottom conditions. Some mudrocks have a high organic content, including parts of the Kimmeridge Clay which is the principal oil source rock in the northern North Sea, but which, over the area included in this volume appears to be always immature. However older Jurassic oil-shales have reached maturity in some areas of onshore England as in the Wytch Farm Oilfield of Dorset.

Jurassic sands are typically fine-grained and often of a surprisingly bright yellow appearance at outcrop, although generally a shade of grey when unweathered; the colour difference reflects the oxidation states of iron minerals. They often show sedimentary features typical of high-energy environments (occasionally of storm origin) and the fossils they yield are those that are likely to survive under such conditions, such as thick-shelled bivalves, brachiopods and belemnites. The sands are commonly cemented, generally by calcium carbonate, to form hard concretions ('doggers') that weather out differentially.

Limestones are of two principal types: firstly, ones that are thin and with frequent non-sequences; such condensed deposits are often ferruginous and frequently replete with a wide range of molluscs, brachiopods, echinoderms, bryozoans and sponges. The iron was originally deposited as ooids of berthierine (a hydrated alumino-silicate of iron and magnesium, belonging to the serpentine group of phyllosilicates), but ooids of goethite (a hydrous iron oxide) may also have been generated primarily (Young 1993). Secondary iron minerals in ooids include siderite (formed during early diagenesis) and limonite which is a widespread weathering product. Berthierine oolites have provided locally significant iron ores that were extensively worked up to the last quarter of the 20th century. When the ferruginous ooids are more scattered, the rock frequently has a speckled appearance, referred to as 'ironshot' in the literature. The second type of limestone tends to be thicker and frequently has evidence to show that it accumulated more rapidly; such limestones are frequently ooidal and are generally poorly fossiliferous, although they may locally contain rich faunas of bivalves, gastropods, brachiopods and echinoids; corals occasionally flourished in shallow-water areas, particularly those fringing the land areas.

Clays, sands and limestones are often arranged in cycles; the clays becoming progressively more silty upwards and the sands more calcareous towards the top. This cyclicity reflects the effects of both eustatic changes in sea level and local controls on sedimentation, such as synsedimentary faulting, that can introduce very local facies such as ironstones or ooidal limestones into mudrock-dominated successions. The importance of cyclicity in correlating Jurassic sediments is discussed below.

Jurassic chronostratigraphy and biostratigraphy

Series and Stages

The Series of the Jurassic System are Lower, Middle and Upper. These do not correspond to the continental terms Lias, Dogger and Malm, which are essentially lithostratigraphical terms, despite their continued use by some European workers as chronostratigraphical divisions. The stages of the Jurassic System are based on d'Orbigny's (1850) stages, although with some later modifications (Fig. 14.5) (see Orbigny 1842–1851).

There is a special case with the terminal Jurassic. d'Orbigny's Purbeckian Stage was abandoned as it was realized that the Purbeck facies is a widespread largely non-marine diachronous latest Jurassic–earliest Cretaceous facies. This left d'Orbigny's Portlandian Stage as the terminal Jurassic stage, but within years of its proposal, it presented problems. On the one hand, d'Orbigny's (1850) definition stated its equivalence to the English Portland Sand and Portland Stone of Fitton (1836); this presented an unequivocal interpretation of the term to the British, but d'Orbigny's (1850) citation of *Ammonites gigas, irius* and *gravesiana* as typical ammonite species of the stage, presented an equally unequivocal interpretation to stratigraphers in northern France and Germany. These ammonite species occur in the middle of the English Kimmeridge Clay as first demonstrated by Salfeld (1913), meaning that by d'Orbigny's second criterion these rocks were already Portlandian. This ambiguity was first perceived by Blake (1880) and has led to a long and confusing dual interpretation of the Kimmeridgian Stage; the longer *sensu anglico* meaning, or the shorter *sensu gallico* meaning. Meanwhile, for most of the rest of the world, the Tithonian Stage (Oppel 1865) had long been used for the equivalent Tethyan facies predominantly of white limestones.

SERIES	STANDARD STAGES	SECONDARY STANDARD STAGES
UPPER	Tithonian	Portlandian Bolonian
	Kimmeridgian	
	Oxfordian	
MIDDLE	Callovian	
	Bathonian	
	Bajocian	
	Aalenian	
LOWER	Toarcian	
	Pliensbachian	
	Sinemurian	
	Hettangian	

Fig. 14.5. Series and stages of the Jurassic System. The Secondary Standard Stages used for the terminal Jurassic in Britain are necessary because of pronounced ammonite provincialism. The base of the Volgian Secondary Standard Stage is almost certainly older than the base of the Tithonian, and its top lies well within the Berriasian Stage, the Primary Standard Stage for the base of the Cretaceous. Faunas of Upper Volgian aspect occur in Eastern England and the Southern North Sea.

Extreme Late Jurassic ammonite provincialism prohibits the correlation of Tithonian faunas with those areas of the Boreal Realm for which the Portlandian Stage is used (northern France, Britain and Greenland), and with those of Russian and Poland for which a Volgian Stage (Nikitin 1881) is used. The Tithonian Stage has recently been accepted internationally as the terminal Jurassic Stage; thus the underlying Kimmeridgian Stage can only be used in its shortened, *sensu gallico*, interpretation. Cope (1993*b*, 1995, 1996) suggested that as it was still not possible to correlate the Tithonian ammonite faunas with those to the north of the Tethys, a series of Secondary Standard Stages was needed for those areas, at least as a short-term measure. In addition, as the Kimmeridgian Stage could no longer be interpreted in its longer, *sensu anglico*, way, Blake's proposal (1880, 1881) for a Bolonian Stage, equivalent to the Upper Kimmeridgian *sensu anglico* and the Lower Portlandian *sensu gallico*, was the way forward. Thus for most of England and Wales the Bolonian and Portlandian Secondary Standard Stages are used as local equivalent of the Tithonian Stage, which is the international standard for the terminal Jurassic stage. In Norfolk and Lincolnshire the topmost Jurassic (equivalent to the basal Purbeck facies to the S) yields ammonites of Upper Volgian aspect, thus that Secondary Standard Stage is used. Volgian is also used routinely in the southern North Sea, although the ammonites are entirely of Bolonian and Portlandian type up to the Spilsby Sandstone (see p. 363).

Ammonite faunas and chronozones

Ammonites provide the basic tool for Jurassic biostratigraphy and chronostratigraphy; there are a few marine facies in which they do not occur (see below), but they are frequently abundant. The increasing resolution of ammonites in biostratigraphy can be judged from the fact that Oppel's (1858) table of Jurassic zones listed 33 zones (not all of which had an ammonite as the index fossil). By the time of Arkell's monumental *The Jurassic System in Great Britain* (1933) there were 57 Jurassic ammonite zones; today the total is 76. Most of these ammonite zones are divided into subzones, allowing even greater discrimination. It is important to point out that Jurassic ammonite zones are chronostratigraphically significant Standard Zones (chronozones) and not biozones (Callomon 1995); they are frequently divided into Standard Subzones (subchronozones). The finest scale units used in some parts of the Jurassic System are ammonite horizons (or biohorizons), which are discrete and widely correlatable units each characterized by a distinct ammonite fauna (e.g. Callomon & Chandler 1990; Page 1992, 1995) that allow very high degrees of temporal resolution; in some cases this may approach periods as short as 40 Ka (Cope 1993*a*). Such ammonite horizons are a modern interpretation of divisions first introduced by Buckman (1893) as *hemerae*, for the smallest unit of geological time represented by the acme of abundance of a single ammonite species. Callomon & Chandler (1990) proposed abandonment of the hemera with its connotation of acme of a species, that is in practice difficult to determine, and its stratal equivalent the *epibole* (Trueman 1923). They proposed that the divisions be named *ammonite horizons* that together represented an incomplete record of time, since the time intervals between successive ammonite horizons were of unknown duration. These intervals may be non-sequences or beds without ammonites. It is important to note that horizons differ in this respect from the chronostratigraphical Standard Zones and Subzones which encompass an interval of time without breaks; contiguous chronostratigraphical divisions represent a continuous period of time. It is also important to point out that calculations of the average duration of, for example, ammonite-based standard zones, have assumed that all such units were of approximately equal duration. In fact many of the radiometric dates quoted for parts of the Jurassic Period are extrapolations from known radiometric dates to other boundaries based on the number of contained ammonite Standard Zones or Subzones, assuming equal duration for each. McArthur *et al.* (2000) recently cautioned the use of such assumptions and suggested, on the basis of determinations based on isotopes of strontium, that ammonite subzones in the Toarcian could differ in duration by a factor of 30. However, Palfy & Smith. (2000), using radiometric dating based on U–Pb and $^{40}Ar/^{39}Ar$, estimated zonal duration to vary from 0.4 to 1.6 Ma, a factor of 4.

The latitude of the Anglo-Welsh area and its connection with the Tethys Ocean to the south, via the Paris Basin and with the Boreal Sea to the north via the Faeroe Rift and (at times) the Viking Graben, ensured that ammonites could migrate freely across the area and, as a result, the faunas are often mixed. The Anglo-Welsh area is often a critical area allowing correlation between Boreal and Tethyan faunas. Tethyan ammonite faunas are for the most part uncommon, but at certain times dominant; Tethyan immigrants are well known from the Sinemurian–Aalenian interval of areas from Dorset (e.g. Page 1992) as far north as Yorkshire (Howarth 1976). A Toarcian immigration coincided with a major transgressive phase that allowed the typically open water genus *Phylloceras* into the Yorkshire area (*Phylloceras* in fact has a Yorkshire type species – *Phylloceras heterophyllum* (Sowerby, 1820)). The first Boreal influences appear in the Hettangian with the southward migration of the bivalve *Gryphaea* (see below), whilst the first truly Boreal ammonite is the Toarcian *Pseudolioceras* which migrated southwards via the Viking Strait and is known from Yorkshire and spread as far south as the southern Paris Basin (Bjerrum *et al.* 2001); it is often confined to a single bed, emphasizing the transitory nature of this Boreal incursion. At other times Boreal elements were dominant, notably in the Callovian when cardioceratid and kosmoceratid ammonites spread southwards (in the latter case as far south as the Algarve), but the latest Callovian shows a notable Tethyan invasion. Boreal domination became evident again in the Oxfordian when the cardioceratids continued to spread southwards from the Arctic (Callomon 1985). The Oxfordian is a particularly interesting case as the ammonite faunas of England are similar to those of northern France and Swabia; these differ from the faunas of Sub-Mediterranean Europe on the one hand and from those of Scotland and the Boreal region on the other. However, enough cardioceratids have been found in southern England to allow the Boreal zonal scheme to be applied there. In turn the southern English faunas allow correlation with the Sub-Mediterranean province (Cope 1980*b*, fig. 10). Even greater provincialism existed in the latest Jurassic, with distinct ammonite faunal provinces separating the faunas of Britain, northern France and Greenland from those of the Tethyan area on the one hand, and from Russia and Poland on the other. Correlation between these areas is still not possible by ammonites as the faunas are so different, and this has led to the retention of separate secondary standard stages in each area (see above, p. 331).

At certain times, or in certain places, in the British Jurassic rocks ammonites are not available for biostratigraphical purposes. This may be because of non-marine facies (e.g. most of the Midlands and Yorkshire Bathonian, or the latest Jurassic–earliest Cretaceous Purbeck Group) or because of marine facies inimical to ammonites such as oolite shoals or coral reefs. In these cases other fossils are sometimes used biostratigraphically. Microfaunas and microfloras are routinely used, especially in offshore operations by oil companies, but the degree of resolution they offer is not as great as that available from ammonites (see below). Belemnites have good potential and are used in areas where ammonites are rare or absent; only at certain times, however, was evolution rapid enough to provide time constraints similar to that of ammonite zones (Doyle & Bennett 1995; Combémorel 1997). Brachiopods can provide good resolution, at least locally, as for the Cornbrash Formation (see below) and also have limited biostratigraphical use in the Inferior Oolite Group of the Cotswold Hills where ammonites are often rare. In a review of the use of brachiopods in Jurassic biostratigraphy, Ager (1978) concluded that because of their greater diversity and more rapid evolution, rhynchonellids provided the best biostratigraphical indicators, although their resolution did not compare with that provided by ammonites. Nerineid gastropods are useful in the Great Oolite Group of Central England, as first demonstrated by Arkell (1931). In the Purbeck Group of southern England ostracods have been widely used. Several bivalves are of limited local biostratigraphical use, especially the oyster group. Elsewhere, individual fossil species have local biostratigraphical significance and give their name to individual fossil beds, such as the Middle Jurassic Boueti Bed that yields a distinctive faunal assemblage including the eponymous brachiopod *Goniorhynchia boueti* and which can be traced northwards for some 70 km from the Dorset coast.

Microfossils

Microfossils have one major advantage over ammonite faunas in biostratigraphy: that is that their small size means that they may be readily obtained in large numbers from cuttings obtained from boreholes, so that they are normally readily available in petroleum exploration and provide fundamental tools in correlating and dating offshore successions. Their disadvantage is that the degree of resolution currently obtainable from their use is not as great as that obtainable from ammonites. Woollam & Riding (1983) looked at the use of dinoflagellate cysts for rocks from latest Triassic to basal Cretaceous; for the Jurassic Period and into the basal Berriasian they recognized 16 assemblage zones, with 29 subzones over the same time interval. They calibrated these against the Standard Ammonite Zones; the duration of one of their dinoflagellate subzones is roughly equivalent to three Standard Ammonite Zones. Ainsworth *et al.* (1998), in a study applicable to the English Channel and adjoining areas, proposed subdivision of the Jurassic into 52 dinoflagellate zones and, over the same interval, 30 ostracod zones and 25 foraminiferid zones.

Calcareous Nannofossils

Calcareous nannofossils (coccolithophores and nannoliths) are extremely small (15–100 μm) fossils that appeared in the Late

Triassic. They are extremely abundant in marine sediments and this, together with their wide distribution through their planktic habit and rapid evolution, gives them a good biostratigraphical potential. However, in the Jurassic they were not as diverse or rapidly evolving as in Cretaceous or later times. Nevertheless, Bown *et al.* (1988) and Bown & Cooper (1998) produced a biostratigraphical scheme for the Jurassic of the Boreal region (including the Anglo-Welsh area) that recognized 18 zones, eight of which were divided into two subzones. Zonal resolution is thus about a quarter of that achieved from ammonites and approximates to substage level.

Non-biostratigraphic correlation

Chemostratigraphy

The increasing number of geochemical analyses of sediments and/or their contained fossils has resulted in the production of a series of plots of concentration of selected chemical elements (or their isotopes) against standard chronostratigraphical time-scales. The results show that these concentrations can vary in a time-significant way and thus have application to chronostratigraphy in dating and time-correlating rocks on a basinal to global scale. Conversely, the interpretation of the changes can give insights into the evolution of coupled environmental variables such as climate, cratonic erosion and ocean-floor creation, oceanic circulation, and the biogenic carbon cycle. The progress so far in this field has been admirably summarized by Jenkyns *et al.* (2002) who demonstrated that in some cases chemostratigraphy is able to provide time-resolution approaching that currently achievable biostratigraphically with ammonites.

High concentrations of organic carbon in marine Jurassic sediments have been found to be widespread and often occur synchronously across wide areas. One of the best examples of this occurs in the Toarcian Stage where the anomalously high Total Organic Carbon (TOC) of the Jet Rock, developed in the lower part of the Mulgrave Shale Member of the Whitby Mudstone Formation (Jenkyns & Clayton 1997; Jenkyns *et al.* 2001) (see below, p. 339), correlates precisely with similar high TOC values in France, northern and southern Germany and Switzerland. It is now recognized that these high TOC values, associated with characteristic excursions of their carbon isotope-ratios (see below), are even more widely distributed and reflect what appears to have been a global Toarcian Anoxic Event (Jenkyns & Clayton 1997). Such correlations based on physical evidence can make important contributions to the problems of biostratigraphical correlation between different ammonite biogeographical provinces and their standard zonations.

Another Jurassic formation whose organic carbon has been of major interest is the Kimmeridgian–Bolonian Kimmeridge Clay, within which are many horizons with anomalously high TOC values. As the Kimmeridge Clay is the source rock for most of the North Sea oil, this has generated a great deal of interest in its organic carbon. The high TOC levels are traceable across Britain, into the North Sea and across the whole of NW Europe (Gallois 1979; Herbin & Geyssant 1993; Herbin *et al.* 1993, 1995). Unlike the Toarcian example, however, the Kimmeridge Clay high TOC excursions appear to be only basin-wide and cannot be applied to the geographically contiguous Volga Basin with its similar facies (Hantzpergue *et al.* 1998); thus, although high TOC values appear widespread in this part of the Upper Jurassic, they are not manifestations of any global anoxic event.

Carbon isotope-ratios offer another means of correlation. They vary in response to mass-dependent dynamic isotope-effects in the chemical reactions that transport the element round the world's carbon cycle. These effects lead to the preferential incorporation of ^{12}C into the organic matter of the ocean-atmosphere system via photosynthesis affecting the $\delta^{13}C$ value of sea-water and thereby reflecting the partitioning of the global carbon pool between oxidized or reduced reservoirs. The former include carbonates, bicarbonates and carbon dioxide, whilst the latter is principally locked in organic carbon. An increase in $\delta^{13}C$ values results from increase of the global biomass or from burial of biogenic carbon, whereas decrease in biomass, oxidation of previously buried carbon or introduction of other reduced carbon into the system will decrease the $\delta^{13}C$ values. Thus the Toarcian Global Anoxic Event is accompanied by a positive carbon isotope excursion. Jenkyns *et al.* (2002, fig. 7) show from a carbon isotope plot for the Jurassic, that broad trends are traceable throughout the Jurassic Period. The Sinemurian–Pliensbachian boundary is characterized by a negative $\delta^{13}C$ excursion, whilst the early Toarcian is marked by a positive $\delta^{13}C$ excursion. Significantly higher values for $\delta^{13}C$ have been obtained from oysters than from belemnites, which may be due to diagenetic alteration. Results so far show that carbon isotope-ratios clearly offer significant potential in Jurassic stratigraphy, but more work is needed to refine and increase the data, particularly from the Callovian.

Strontium isotope stratigraphy has provided one of the best-constrained results for the Jurassic. It is based on the marine $^{87}Sr/^{86}Sr$ isotopic abundance-ratio and assumes that even on the shortest geological time-scales it was homogeneous then as it is in modern seas, that it has varied continuously through time and that the values determined from calcium minerals precipitated from sea-water are truly representative of that ratio. The first high-resolution strontium isotope curve for the Jurassic appeared a decade ago (Jones *et al.* 1994*a*, *b*) and many more data have since been added (Fig. 14.6). Because the data are well constrained, sudden 'jumps' in the curve, when taken from contiguous beds in a section, can identify depositional breaks that may not be biostratigraphically recognizable and sudden changes in sedimentary accumulation rates (see examples in Jenkyns *et al.* 2002, figs. 12 & 13). At present the strontium isotope curve is still short of data for the Bathonian, Callovian and upper part of the Kimmeridgian and within the Upper Jurassic it is clear that, as in the case of carbon isotopes, data from oysters in general plot higher than those from belemnites. The reason for these latter discrepancies may be due to diagenetic changes or possibly to ecological differentiation between nekton and benthos.

Other chemostratigraphical methods include the use of isotopes of nitrogen, oxygen, sulphur, neodymium and osmium. The last shows much promise and may be obtainable from metalliferous minerals. Details of the osmium composition of Jurassic sea-water are discussed by Cohen *et al.* (1999) and Cohen (2004). The rhenium–osmium isotope system can also be used for dating organic-rich sedimentary rocks. Refinements in analytical te chniques and the use of thermally immature samples have produced results that are indistinguishable from those obtained from high precision U–Pb and Ar/Ar ages (Cohen 2004). The lowest value of the $^{87}Sr/^{86}Sr$ ratio of the whole Phanerozoic occurred during the Callovian–Kimmeridgian (Jones *et al.* 1994*a*; Podlaha *et al.* 1998) and Oxford Clay mudrocks from the Midlands yielded results suggesting that the osmium-isotope composition of Callovian sea-water was exceptionally unradiogenic (Cohen 2004). This author concluded that the exceptionally low levels of radiogenic Sr and Os in the oceans were the result of low levels of continental weathering worldwide.

Sedimentary cycles

The earliest report that the Jurassic rocks of England and Wales display cyclic repetition of facies was made as far back as 1822 by Conybeare & Phillips, who recognized that clay–sand–limestone cycles occur in many parts of the Jurassic, and by 1871 John Phillips reported that in the Oxford area he was able to recognize five such cycles from the Upper Lias to the Portland Stone. By the time that Arkell (1933) examined

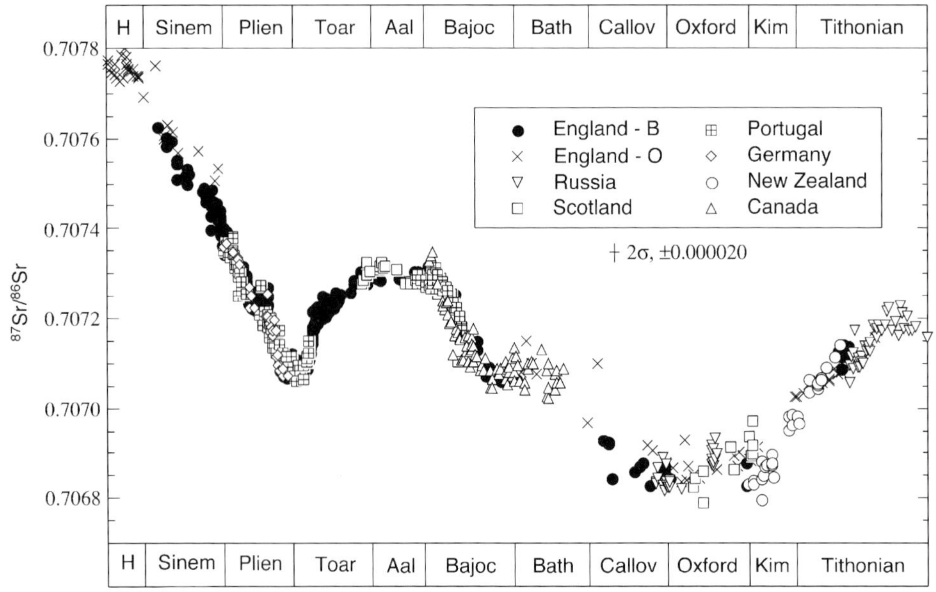

Fig. 14.6. Strontium isotope curve for the Jurassic System. Reproduced after Jenkyns *et al.* (2002), with permission.

the problem it was clear that sea-level change was one of the fundamental factors in such cyclicity and Hallam (1978) demonstrated that major sedimentary cycles in the Jurassic could be recognized on different plates and were also synchronous. Such fundamental first-order cycles can be identified on sea-level charts (see Fig. 14.2).

Second-order cycles within these may often be recognized by local–regional disconformities within sedimentary packages that may also be related to sea-level rise, whilst third-order cycles may reflect either sea-level change or areally restricted effects such as local tectonism. Such cycles provide a valuable tool for correlation in, for example, the Cleveland Ironstone Formation (Howard 1985) and the Kimmeridge Clay Formation (Gallois & Cox 1976), in the latter case from the Dorset coast to the Wash.

Metric scale cycles, such as those of the Blue Lias may well be due to changes in solar radiation induced by variations in Earth's orbit (Milankovitch cycles). These have been identified in various parts of the Jurassic, especially in mud/carbonate successions (House 1985; Weedon & Jenkyns 1990; Weedon *et al.* 1999, 2004). They provide not only a fine-scale correlation on a regional or local scale but also a 'hanging' numerical timescale for parts of the Jurassic.

Geophysical logs

Many boreholes have been drilled in recent years through the Jurassic rocks of England and Wales and it is usual to employ downhole logging techniques, especially when there may be some hydrocarbon interest in the formations through which the borehole passes; such borehole logs have considerable importance in stratigraphical correlation and in recognising lithologies and sedimentary cycles. Many types of log are used, but perhaps the most significant ones for much of the onshore Jurassic are the gamma ray and sonic logs. Gamma ray emissions provide lithological indicators as they shows variation in the amount of clay (the principal source of radioactive K, U and Th) within the sediment. The sensitivity allows limestone beds only 20 cm thick to be identified within the Blue Lias of southern England. Sonic logs measure the speed of sound through a formation, which is dependent upon lithology and porosity; where the lithology is known, porosity may be determined. Modern sonic logs are compensated for irregularities in the borehole wall (Whittaker *et al.* 1985 – see also for different types of geophysical logs and their applications).

Characteristic log signatures are now recognized throughout the onshore Jurassic of England and Wales and provide excellent tools for correlation. Thus the five divisions of the lower part of the Lias Group may be followed across country, although frequently the four divisions of the Charmouth Mudstone Formation cannot be mapped due to lack of exposure. These divisions may even be recognized within with the Redcar Mudstone Formation Cleveland Basin, despite the succession being more silty there. Suggested correlations have been confirmed by comparison with the ammonite chronostratigraphy (Whittaker *et al.* 1985). Figure 14.7 demonstrates the potential of geophysical logs in correlation.

Economic geology

Throughout the Jurassic System in England and Wales are rocks and minerals that have been and remain of economic importance; these are dealt with individually in the descriptive part of this chapter, but are summarized here.

Clays from all parts of the Jurassic System have been used for brick making in the past, but with the increasing scale of the industry, smaller workings have largely disappeared. At present the Oxford Clay of the East Midlands is worked on a large scale and provides the principal source of bricks in the UK, whilst smaller scale working survives in brickworks in the lower part of the Lias Group. Clays are also used in cement making; either mixed with co-occurring limestones (as in the Blue Lias Formation), or used as an addition to limestones such as the Lincolnshire Limestone, or Chalk, as in the case of the Kimmeridge Clay in Wiltshire. Some clays, notably parts of the Whitby Mudstone Formation of Yorkshire, provided a source of alum, used in tanning leather and as a mordant for dyes in the past. Other clays have also been worked on smaller scales for alum, such as the Kimmeridge Clay in Dorset. The Fuller's Earth Formation of the Bath area formerly provided a commercial fuller's earth whose property of adsorbing oils was used in the woollen industry.

Many Jurassic clays are rich in organic matter and in some cases are potential oil source-rocks. The most notable is the Kimmeridge Clay that is the principal oil source-rock in the northern North Sea, but over the southern North Sea, and the whole of England and offshore, has not been sufficiently

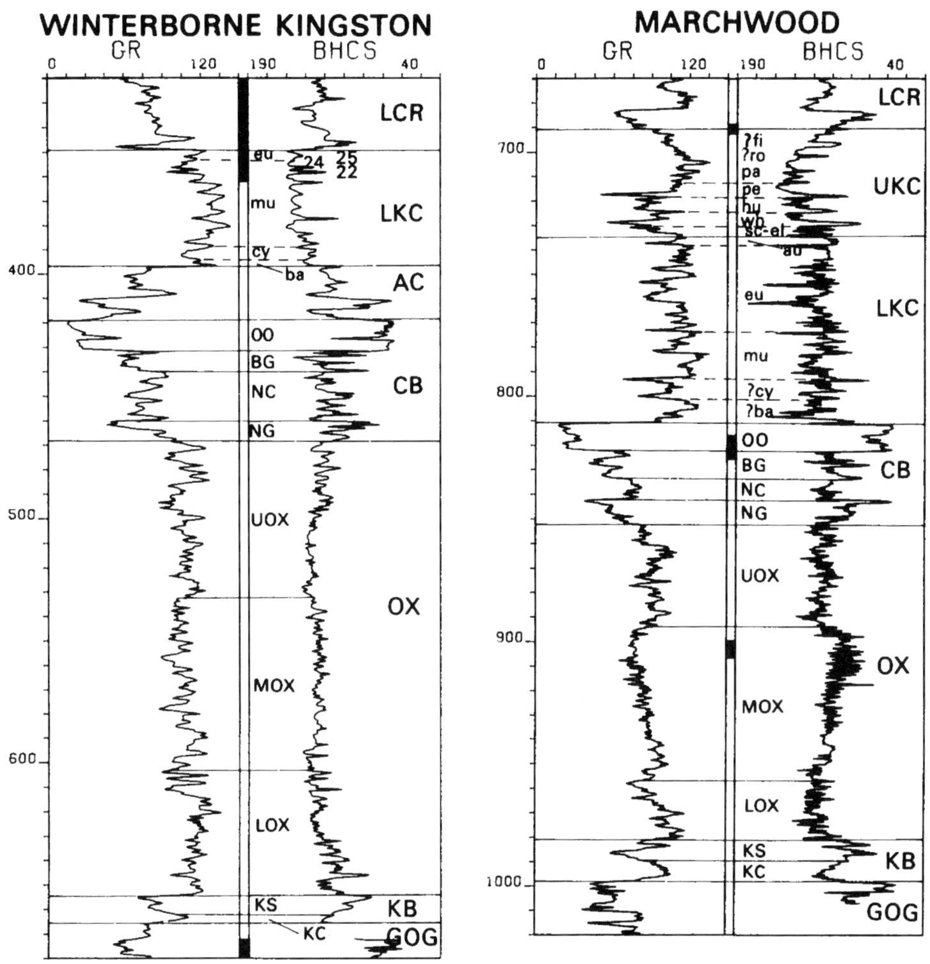

Fig. 14.7. Upper Jurassic log signatures of the Winterborne Kingston (Dorset) and Marchwood (Hampshire) boreholes showing Gamma Ray logs (GR) and Borehole Compensated Sonic logs (BHCS). GOG, Great Oolite Group; KB Kellaways Beds; KC and KS, Kellaways Clay and Sand; OX, Oxford Clay; LOX, MOX and UOX, Lower , Middle and Upper Oxford Clay; CB, Corallian Beds: NG, Nothe Grit: BG, Bencliff Grit; OO, Osmington Oolite; AC, Ampthill Clay: LKC, Lower Kimmeridge Clay (initial letters of zones as in Fig. 14.18); LCR, Lower Cretaceous. After Whittaker *et al.* (1985) with permission.

deeply buried to become mature. However, clays of the lower part of the Lias Group are mature source rocks in eastern Dorset and provide the source of oil in the Wytch Farm and Kimmeridge oilfields; Jurassic rocks provide oil reservoirs in both cases. At Wytch Farm the higher reservoir horizon is the Bridport Sands at the top of the Lias Group, whilst in the Kimmeridge oilfield the Cornbrash Limestones are the reservoir. The Cornbrash is also a good small-scale aquifer across England whilst the Lincolnshire Limestone is a more important aquifer.

Possibly every Jurassic limestone has been used in the past as a building stone, mainly on a domestic scale. Such small-scale industries still survive with, for example, limestones in the Blue Lias, Lincolnshire Limestone and Purbeck Group. Fissile limestones have provided roofing materials, such as within the Forest Marble and Purbeck Beds. But the Jurassic is particularly noted for dimension stone, such as the Bath Stone, used extensively in Georgian Bath, and more particularly the Portland Stone of Portland that has been used in civic buildings throughout the UK for almost 400 years, and remains important today.

Jurassic ironstones were worked extensively until, within the third quarter of the 20th century, the advent of higher quality ores from abroad made them uneconomic. Ironstones from the Lias Group were worked extensively across the South and East Midlands and ironstones from horizons in the Middle and Upper Jurassic were worked locally.

The latest Jurassic rocks are also small-scale sources of gypsum and workings in the Purbeck Group for this mineral persist as mines in the Weald.

Lower Jurassic Series

Lower Jurassic rocks crop out extensively in England and South Wales and are also preserved in outliers indicating their former extent beyond the main outcrop that runs from the Devon–Dorset coast to the Yorkshire coast, with the principal outliers in the Vale of Glamorgan, in the southern part of the Cheshire Basin around Prees (Shropshire) and near Carlisle. Offshore Lower Jurassic outcrops occur in the Solway Basin, the Eastern Irish Sea, Cardigan Bay and St George's Channel basins, the Bristol Channel, the English Channel and the Southern North Sea Basin. The Lower Jurassic subcrop extends eastwards from its outcrop to the margins of the Anglo-Brabant Massif and under the Southern North Sea Basin (Fig. 14.3). Chronostratigraphical subdivision of the Lower Jurassic is shown in Figure 14.8.

Hesselbo & Jenkyns (1998) divided the Lower Jurassic into four large-scale (second-order) lithological cycles that they recognized in all British onshore basins; these had durations in the order of 3–10 Ma. They identified sequence boundaries by unconformities or by successions of facies that indicated minimum accommodation space within the mid Sinemurian,

STAGE	STANDARD ZONES
TOARCIAN	*Pleydellia* AALENSIS
	Dumortieria PSEUDORADIOSA
	Phlyseogrammoceras DISPANSUM
	Grammoceras THOUARSENSE
	Haugia VARIABILIS
	Hildoceras BIFRONS
	Harpoceras SERPENTINUM
	Dactylioceras TENUICOSTATUM
PLIENSBACHIAN	*Pleuroceras* SPINATUM
	Amaltheus MARGARITATUS
	Prodactylioceras DAVOEI
	Tragophylloceras IBEX
	Uptonia JAMESONI
SINEMURIAN	*Echioceras* RARICOSTATUM
	Oxynoticeras OXYNOTUM
	Asteroceras OBTUSUM
	Caenisites TURNERI
	Arnioceras SEMICOSTATUM
	Arietites BUCKLANDI
HETTANGIAN	*Schlotheimia* ANGULATA
	Alsatites LIASICUS
	Psiloceras PLANORBIS

Fig. 14.8. Stages and Standard Ammonite Zonation of the Lower Jurassic Series. Substage boundaries for the Sinemurian, Pliensbachian and Toarcian stages are indicated by short lines at the right hand side of the stage column.

Upper Sinemurian, mid Pliensbachian and Upper Toarcian. Within the British area the Wessex Basin exhibited the most distal patterns where rise in sea level, or highstand, was represented by condensed sections in limestone or mudrock and sea-level fall, or lowstand, by thickened successions of mudrocks or sandstones. In contrast, the Hebridean Basin of western Scotland shows the most proximal setting in which expanded succession of mudstones and sandstones characterized lowstands and condensed sandstones the highstands. The Cleveland Basin is intermediate in character between these to up to the beginning of the Toarcian, at which point the greater sediment input into the Cleveland Basin shows it to be within the most proximal setting. Within the larger sedimentary cycles Hesselbo & Jenkyns (1998) recognized smaller (third-order) cycles with durations of 300–500 Ka – about the same duration as ammonite zones or subzones; these they found to become less pronounced upwards through the Lower Jurassic successions and in general could be less readily related to sea-level change.

The Lower Jurassic Series is often referred to as the Lias, a lithostratigraphical term of Group status that may be a quarrymen's corruption of the word 'layers', but the Lias only approximates to the Lower Jurassic. In particular, the Triassic–Jurassic boundary, as at present recognized, falls within the basal part of the Lias Group. George *et al.* (1969) selected a paper-shale marking the base of the Lias Group at Watchet on the Somerset coast of the Bristol Channel as the basal marker of the Jurassic System. As pointed out by Torrens & Getty (in Cope 1980*a*) this is a departure from normal stratigraphical practice and there appeared to be good reason to define this boundary by the base of the Planorbis Subzone. At the time that was written, *Psiloceras planorbis* was believed to be the first ammonite species to appear in the British succession after the Rhaetian transgression. However, Page & Bloos (1998) reported the presence of three separate ammonite horizons within the Lias Group, below that of *Psiloceras planorbis*; these are the *Erugatum*, *Imitans* and *Antecedens* horizons. All three occur a few metres above the base of the Lias, so that whichever of these ammonite faunas may ultimately be selected as providing the basal Jurassic ammonite zonal marker, the basal beds of the Lias Group, in most Anglo-Welsh sections, are of latest Triassic age and, indeed, at one locality the basal bed of the Lias Group has yielded the latest Triassic conodont zonal index *Misikella posthernsteini* (Swift 1989).

The basal formation of the Lias Group, everywhere south of the Market Weighton Swell, is the Blue Lias, an alternation of limestones and shales, up to 130 m thick in the expanded successions of the Bristol Channel area, but thinner on the Devon–Dorset coast. Controversy over the origin of the alternations from shale or marl to calcilutite, centred on whether the alternation was primary or whether the carbonate segregated in early diagenesis, resulting in a secondary origin. In many cases trace-fossil burrows pass from shale to limestone or vice versa and different coloured infills provide evidence of primary differences, but many nodular limestones are clearly of secondary origin. Many shales and some limestones are finely laminated and have a high organic content; these same shales are largely devoid of benthos, suggesting that there were periods of restricted circulation of bottom waters producing periodic anaerobic conditions (Hallam 1958, 1964). House (1985) suggested that much of the primary alternation could be due to Milankovitch cyclicity and Weedon *et al.* (1999) suggested that the 38 Ka obliquity cycle was the dominant factor, but they recognized smaller cycles related to orbital eccentricity (95 Ka) and orbital precession (20 Ka). In a study of the clay minerals in the Blue Lias of Dorset and Somerset, Deconnink *et al.* (2003) found that the predominant species were chlorite, illite, illite-smectite and kaolinite. Kaolinite increased in proportion upwards and this was correlated with progressive inundation of Variscan massifs, as the sea levels rose, and with increase in humidity.

The Blue Lias is a remarkably persistent facies, recognizable lithologically not only across much of the Anglo-Welsh area, but in other areas of NW Europe. Its fossils are predominantly calcitic bivalves, particularly oysters, especially the curved *Gryphaea* together with *Plagiostoma* and *Pinna*. Trueman's (1922) hypothesis that *Gryphaea* evolved from the flat *Liostrea* was shown to be poorly founded by Hallam (1959), but many years previously (and overlooked in the West) Kisparova (1936) had already shown that Siberian Triassic *Gryphaea* were remarkably similar to *G. arcuata*, the Jurassic type species of the genus; thus its sudden appearance in the Blue Lias reflects nothing more than a southerly migration in the Early Jurassic. Other calcitic fossils are rhynchonellid brachiopods and crinoid stem ossicles. Ammonites are most frequently preserved as moulds, as are nautilids. Locally, as in South Wales, silicification has preserved a window on the otherwise largely unpreserved aragonitic faunas; these include a wide variety of bivalves (Hodges 1996, 2000) and gastropods.

Economically the Blue Lias has been used as a building stone (and is still used for this purpose on a small scale) and for cement making. This quarrying produced an abundance of reptilian remains in the 19th century, particularly from the Dorset coast, the Street area of Somerset, and from Barrow upon Soar, Leicestershire. The use of the Blue Lias for cement manufacture persists in places such as the Vale of Glamorgan and around Rugby.

The Hettangian–Sinemurian boundary lies within the upper part of the Blue Lias and the GSSP (Global Stratotype Section and Point) for the base of the Sinemurian has now been defined on the north Somerset coast at East Quantoxhead (Bloos & Page 2002).

South of the Market Weighton Swell the clay-dominated succession above the Blue Lias, is now assigned to the Charmouth Mudstone Formation (Cox *et al.* 1999). Within the Wessex Basin many of the shales are locally laminated and bituminous, again reflecting poor bottom-water circulation and such shales provide the source rock for the Wytch Farm Oilfield of eastern Dorset (Colter & Havard 1981).

The Charmouth Mudstone Formation is well exposed on the Dorset coast and the faunas, first collected systematically by Lang and described by him with various co-authors (1914–1936) are now well established. The Shales-with-'Beef' Member (Lang *et al.* 1923) is a succession of pyritic shales with limestone bands and nodules, and with 'beef' – diagenetically produced bands of calcite whose fibrous crystals are aligned vertically. The clays of the overlying Black Ven Marl Member (Lang & Spath 1926) are frequently pyritic and in many places have yielded beautifully preserved ammonite faunas; those from the coastal sections are best known and include faunas from nodules that have undergone a complex diagenetic history (Curtis *et al.* 2000). Sometimes the nodules are largely composed of ammonites, such as in the case of the Ammonite Marble of Marston Magna, Somerset. Drifted insects, plants and terrestrial reptiles (*Scoledosaurus*) show the proximity of the Dorset coast sections to the Cornubian landmass. On the Dorset coast a major non-sequence involves the top of the Obtusum Zone and whole of the Oxynotum Zone and a further non-sequence affects the upper part of the Raricostatum Zone, but these non-sequences are less pronounced only a few kilometres to the north. The succeeding Belemnite Marl Member (Lang *et al.* 1928) consists of alternations of calcareous and non-calcareous pale mudstones, with a fauna dominated by belemnites. The alternation of facies has been ascribed to Milankovitch cyclicity (Weedon & Jenkyns 1990) with bedding couplets recording the precession cycles. The youngest member of the Charmouth Mudstones is the Green Ammonite Beds Member (Lang 1936), so called because some of the nodules bear ammonites with a greenish tinge to the calcite filling their phragmocones.

Inland sections in the Blue Lias and Charmouth Mudstone formations are few; stream sections have limited exposure in various places, but in general little is seen of them between the south coast and the Mendip Hills.

Around the Mendip Hills and in South Wales, an archipelago of islands, principally of Carboniferous Limestone, existed in the Early Jurassic. Marginal facies around them can be best seen on the western Vale of Glamorgan coast. The importance of these sections is that they provide the only exposures of Jurassic shorelines in the Anglo-Welsh area.

The relationships between the facies were first explained by Trueman (1922) who showed that a shoreline facies of white conglomeratic limestone (Sutton Stone) was everywhere in basal contact with Carboniferous Limestone (Fig. 14.9). This shoreline facies contains a wide variety of fossils, including many genera of corals that are otherwise rare in the Anglo-Welsh Lower Jurassic. The nature of the rocky shoreline and its fauna have been described by Johnson & McKerrow (1995). On the coast many of the corals are preserved as large fragments clearly torn from a reef in the vicinity; they are frequently replaced by barite and this secondary recrystallization has destroyed much of their original structure, leaving largely hollow cylindrical tubes, so that for many years they were identified as serpulid colonies. The unequivocally coralline nature of these fossils has been demonstrated by Simms *et al.* (2002). The Sutton Stone facies passes upwards and laterally into a more offshore facies that Tawney (1866) called the Southerndown Beds. These are finely conglomeratic and contain clasts of Carboniferous chert and have thin shaly interbeds; they pass laterally and vertically into the offshore Blue Lias facies. Most of the conglomerates are clast-supported, but the presence of localized matrix-supported beds suggests that mass-flow deposits were periodic features. These marginal facies, now referred to the Sutton Limestone Formation, have yielded enough ammonites to prove Trueman's original (1922) thesis, that the islands were submerged progressively with ongoing transgression, during the Hettangian and early Sinemurian, although in the case of one of the largest islands, in the eastern Mendips, submergence was not complete until the Late Bajocian (*q.v.*). Trueman (1922) believed that the South Wales islands fringed a Welsh mainland that extended over most of the Principality, a view that was generally accepted, so that palaeogeographical reconstructions (e.g. Wills 1951) always showed Wales as a persistent landmass throughout the Jurassic Period. This concept had to be abandoned, however, when the Mochras Borehole, on the shore of Cardigan Bay, proved an extremely thick succession of the Lias Group, thicker than any

Fig. 14.9. Sutton Formation (littoral basal Jurassic) resting unconformably on Carboniferous Limestone, Southerndown, Vale of Glamorgan.

known onshore, all in offshore facies (Woodland 1971), thus precluding the possibility of there being any substantial Jurassic landmass over Wales.

In the eastern part of the Vale of Glamorgan the Blue Lias was subdivided into three formations, now best treated as members of the Blue Lias Formation, by Waters & Lawrence (1987); similar divisions were recognized to the west, away from the marginal facies, by Wilson *et al.* (1990). The proportion of limestone increases upwards through the Blue Lias, apparently due to local uplift that is also reflected in a progradation of the marginal facies into the offshore areas at this time (Lawrence & Waters 1978).

Similar littoral facies are developed around the Mendip Hills where the Downside Stone consists of massive limestones and has yielded ammonites varying in age from Hettangian through to Pliensbachian, although not all zones are represented in this facies. Around Radstock, however, immediately north of the Mendip Hills, is a peculiar local condensed facies. Up to the Lower Pliensbachian Davoei Zone the succession is predominantly in limestone facies and is highly condensed with several non-sequences and horizons of derived phosphatized fossils. The area is famous for the abundance of well-preserved *Spiriferina walcotti*, although this brachiopod species does occur at other places in the British Lower Jurassic. Tutcher & Trueman (1925) showed that the ammonite successions and bed thicknesses varied from quarry to quarry (see also Edmunds *et al.* 2003). The Davoei Zone is in clay facies and its thickness of 37 m is several times greater than the whole of the rest of the succession.

Northwards along the main outcrop the few natural sections, such as those on the banks of the Severn, are supplemented by occasional brickpits (as at Blockley, Gloucestershire). Long-vanished brickpits have provided local details and local fossil abundances, such as those of the thick-shelled bivalve *Hippopodium* at Battledown, Cheltenham. Occasional fossil occurrences also provide local marker beds at particular horizons, such as the latest Sinemurian Coral Bed of the Severn Basin (Wright 1860) with abundant *Stylophyllopsis*, or the Banbury Marble, a local shelly Lower Pliensbachian limestone with corals, formerly cut and polished. Details of thickness are hard to establish over much of the outcrop except through various boreholes that have established zonal thicknesses across the outcrop. The thickest succession on the main outcrop is in the Vale of Moreton, where the Stowell Park Borehole proved a combined thickness of 341.78 m for the Blue Lias and Charmouth Mudstone formations (Sumbler *et al.* 2000).

This total is well exceeded by that of the Mochras Borehole, Gwynedd, which proved no less than 896 m for the equivalent strata. Here the Blue Lias Formation is represented by mudstones with thin interbedded limestones; the remainder of the succession is mudstones with an increase in silt at the top. The remarkable thickness penetrated here must be exceeded by some margin within a total Lias Group thickness of some 2500 m recorded in the middle of the Cardigan Bay Basin; such an extraordinary thickness is explained by synsedimentary movement producing a bathymetric deep on the hanging-wall of the Bala Fault (Turner 1996). In the Bristol Channel Basin the corresponding thickness is 375 m, including over 200 m of Blue Lias equivalents with a thinner development of the Charmouth Mudstone Formation.

Progressive onlap of the lower part of the Lias Group onto the northern and western sides of the London landmass has been shown by boreholes (Donovan *et al.* 1979), and similar types of relationships appear to occur to the south. At the western end of the landmass, a borehole at Chalgrove, near Oxford, showed 7 m of the Lower Pliensbachian Jamesoni Zone overlying probable Old Red Sandstone and overlain directly by Inferior Oolite Group (Sumbler 1996, fig. 10). The Warlingham Borehole, near Croydon, showed an uncertain horizon in the Sinemurian resting on probable Rhaetian (Worssam & Ivimey-Cook 1971) and similar successions are known from boreholes in Kent, but in the Portsdown Borehole (Taitt & Kent 1958), typical and complete Blue Lias Formation was overlain by Charmouth Mudstone Formation showing no influence of the London landmass. In the Bolney Borehole, in the centre of the Weald Basin, some 160 m of Blue Lias was overlain by 115 m of Charmouth Mudstone (Gallois & Worssam 1993).

To the north of the area under the influence of the London landmass the succession is seemingly complete and on the East Midlands Shelf a thickness of about 223 m was recorded for the rocks from the base of the Lias Group up to the latest Pliensbachian Marlstone Rock Bed in a borehole north of Grantham (Brandon *et al.* 1990). Two formations are recognized, a lower Scunthorpe Mudstone Formation (113 m) beneath the Charmouth Mudstone Formation (110 m). The former is divided into five members, of which the lowest, the Barnstone Member, is in typical Blue Lias facies (Brandon *et al.* 1990). Northwards again the Blue Lias and Charmouth Mudstone formations thin to a maximum thickness of less than 100 m in north Lincolnshire and are very thin or absent across the Market Weighton Swell. The area south of the Humber, on the southern flank of the Market Weighton Swell, is notable for the Frodingham Ironstone that is developed as a unit up to 12 m thick and appears to be the lateral equivalent of the upper part of the Foston Member of the Scunthorpe Mudstone (Cox *et al.* 1999, p. 5).

Within the Hettangian–Pliensbachian succession in the Cleveland Basin Van Buchem & Knox (1998) recognized three large-scale second-order depositional cycles, bounded by erosional planes, that they interpreted as recording three major sea-level rises (in the Planorbis-Liasicus, Semicostatum and Jamesoni zones). Within these cycles they identified third-order cycles that they ascribed to local tectonic control and eustasy.

The fine coastal sections of Robin Hood's Bay in North Yorkshire expose approximately 160 m of Sinemurian and Lower Pliensbachian shales with rich ammonite faunas. These are assigned to the Redcar Mudstone Formation which was divided by Buckman (1915) into four units, now treated as members. The lowest member, the Calcareous Shales is predominantly of mudstones with thin bioclastic limestones; only the higher parts are seen in Robin Hood's Bay, the lower parts (but not the lower Hettangian part) crop out at Redcar. The Hettangian has been penetrated in boreholes and ammonites from the lowest horizons, including *Psiloceras erugatum*, are known from the local drift. The Calcareous Shale Member is of Hettangian and Sinemurian age, its top lying just within the Upper Sinemurian. The succeeding Siliceous Shales are silty shales with siltstone and fine sandstone interbeds that become predominant upwards and show evidence of storm generation (Rawson & Wright 1995); their environment was more proximal than that of the top of the Calcareous Shales (Hesselbo & Jenkyns 1995). The GSSP for the base of the Pliensbachian has been proposed at Wine Haven, Robin Hood's Bay (Meister *et al.* 2003). Deeper water conditions in the Pliensbachian saw the deposition of the Pyritous Shales with their small pyritized fauna and ichnofauna, much of it under conditions of very rapid deposition (Hesselbo & Jenkyns 1995). At the top of the Redcar Mudstone the Ironstone Shales are shales with silts and bands of ironstone concretions; the silts thicken upwards where there is a transitional boundary with the Staithes Sandstone Formation. Systematic collection from the Redcar Mudstone was the life's work of Leslie Bairstow; this work remained largely unpublished until Howarth (2002), using Bairstow's data and material, described the sections and figured some of the principal ammonite species.

The lower part of the Lias Group in the Prees Outlier in Shropshire has a basinal thickness totalling some 597 m; exposures are limited to a few stream sections and most information comes from boreholes, one of which, at Wilkesley proved the ammonite *Psiloceras erugatum* at the base of the

Lias (Poole & Whiteman 1966; Bloos & Page 2002). Even less is known of the Carlisle Outlier where the outcrop has a thick drift mantle. Recent boreholes have established the extent of the outcrop and proved up to 110 m of the lowest part of the Lias (Ivimey-Cook *et al.* 1995); this is largely in mudstone facies and there is no Blue Lias recognized. Offshore work has shown that the Carlisle Outlier is within an onshore extension of the Solway Firth Basin and there is a further Lias outlier offshore (Holliday *et al.* 2004). Similar small outliers occur in the eastern Irish Sea in the Keys Basin and possible also the Berw Basin (Warrington 1998).

Above the mudstone formations in the lower part of the Lias Group, successions right across England show evidence of depositional shallowing. They comprise silty and sandy mudstones, silts and sands, and include high-energy facies with a fauna of thick-shelled bivalves, together with brachiopods and belemnites. Storm-generated sands have been identified both in Dorset, where they smothered a sea-bed with frequent brittlestars and in Yorkshire where the Staithes Sandstone Formation displays hummocky cross-bedding. Thicknesses are generally of the order of 30 m, but the Dyrham Formation on the Dorset coast succession is over 120 m thick. In the Mochras Borehole on the shore of Cardigan Bay, some 148 m predominantly of interbedded siltstones and mudstones is ascribed to the Upper Pliensbachian, but even this thickness must be exceeded within the Lias Group total of some 2500 m in the Cardigan Bay Basin in the hanging-wall of the Bala Fault (Turner 1996).

In much of southern and central England, the Dyrham Formation is succeeded by the Marlstone Rock Formation, a frequently ferruginous limestone, up to about 8 m thick, with ammonites of the Upper Pliensbachian Spinatum Zone, but much thinner in Dorset (see below). In the Radstock area these Upper Pliensbachian arenaceous and ferruginous beds are missing, as they are around the Mendip Hills generally due to the effects of the Mendip Swell. Over much of the southern and eastern part of the Midlands the Marlstone is a berthierine oolite with a siderite cement. This ferruginous facies yields abundant brachiopods, particularly *Tetrarhynchia tetrahedra* and *Lobothyris punctata* and was formerly worked as ironstone in north Oxfordshire, Warwickshire, Northamptonshire, Leicestershire and south Lincolnshire. Over parts of the East Midlands the topmost part of the Marlstone is of earliest Toarcian age. The Marlstone Rock Formation is the highest horizon preserved in the Prees Outlier in Shropshire.

The Marlstone Rock Formation, along with the Dyrham Formation, is absent over the Market Weighton Swell (Howard 1985; Copestake & Johnson 1989). Farther north, in North Yorkshire, the Staithes Sandstone Formation (up to 30 m) succeeds the Redcar Mudstone Formation; its lithology of bioturbated fine sandstones and argillaceous silts recalls the Dyrham Formation to the south of the Market Weighton swell. The sandstones show hummocky cross stratification together with wave-ripple laminations and gutter casts (Rawson *et al.* 1983) and have been interpreted as indicating a storm-influenced lower shoreface setting (Brenchley & Gowland 1985). The Staithes Formation is succeeded by a ferruginous facies, the Cleveland Ironstone Formation, which represents the Spinatum Zone as well as most of the underlying Margaritatus Zone. The ironstone was worked for well over a century, until the 1950s, and provided the basis for the iron industry of Teesside. The ironstone occurs in several distinct seams interbedded with shales in a cyclic succession (Rawson *et al.* 1983; Howard 1985); ironstones form up to a third of the thickness which reaches a maximum of 28 m at Staithes, but is less than 2 m at the southern basin margin. The bivalve fauna indicates normal marine conditions and sedimentary structures suggest a shallow storm-prone shelf. The iron was sourced from a tropically weathered hinterland.

The upper part of the Lias Group, mainly of Toarcian age, is most fully developed in the Cleveland Basin. Most of the succession comprises the approximately 120 m-thick Whitby Mudstone Formation; many horizons within it yield calcareous nodules and these are the principal source of ammonites found in museums worldwide. The basal Grey Shale Member consists of shales and siltstones with calcareous and sideritic nodules. Its fauna includes abundant belemnites including *Passaloteuthis bisulcata* and the bivalve *Bositra buchi* that is a widespread Jurassic species characteristic of fine muddy facies, both argillaceous and micritic. Jefferies & Minton (1965) suggested that its habit was nektonic or pseudo-planktonic, but it is possible that it was adapted to living within the upper part of 'soupy' sediments, where the sediment/water interface was indistinct and where other species could not survive.

Succeeding the Grey Shale Member is the Mulgrave Shale Member (Plate 26), the lower part of which is a laminated mudstone with a high organic content whose laminae of 20 to each millimetre can be probably ascribed to annual sedimentary rhythms (Cope 1998*b*; *contra* Hallam 1997). This represents a highly carbonaceous level that accumulated during a global Toarcian Oceanic Anoxic Event (Jenkyns 1988; Jenkyns & Clayton 1997) that developed during the maximum Early Toarcian sea-level rise in the Exaratum Subzone of the Serpentinum Zone. From analysis of fossil wood from these sediments, Hesselbo *et al.* (2000) found that isotopically light carbon dominated all the oceanic, biospheric and atmospheric carbon reservoirs even though this interval was characterized by enhanced burial of organic carbon; their explanation for the anomaly was that there had been a massive and rapid discharge of methane from gas hydrates from sediments on the continental margins. Jenkyns *et al.* (2001) found that this anoxic event correlated with a positive $\delta^{15}N_{tot}$ excursion that they explained by the upwelling of a deoxygenated water-mass that had undergone partial denitrification. This lower part of the Mulgrave Shale is known as the Jet Rock, after the occurrence within it of a black compact form of lignite, derived from araucarian wood, which can take a polish and was popular in Victorian jewellery. The Jet Rock contains several layers of calcareous concretions that mark probable non-sequences (Cope 1998*b*). The concretions themselves are often pyrite-coated and appear to have formed under sulphate reduction conditions, but the lowest concretions lack the pyrite coating and seem more likely to result from carbonate redistribution following methane generation (Raiswell 1988). The concretions yield abundant ammonites that are the principal fossils of this part of the succession, together with the bivalve *Pseudomytiloides dubius*. The upper part of the Mulgrave Shale is less bituminous and lamination less pronounced, marking the lessening of the anoxia.

The succession continues from the Mulgrave Shale into the overlying Alum Shale Member (almost 50 m thick), formerly worked (Plate 26) to produce alum (potassium aluminium sulphate); this industry lasted from around 1600 to 1870 (Rawson & Wright 1995). Its upper part contains limestone bands that were worked in the past for cement. The upward facies change in the Alum Shale reflects increasing oxygenation of the seabed (Pye & Krinsley 1986). The Alum Shale is normally the highest part of the Lias Group to be preserved beneath the unconformity at the base of the Middle Jurassic Dogger Formation. The remainder of the Lias Group occurs only in the Peak Trough, to the SE of the active Peak Fault (Milsom & Rawson 1989). Here the Alum Shale passes up into the Peak Mudstone Member, comprising micaceous silty mudstones with calcareous and sideritic nodules and a fauna dominated by bivalves, marking a shallowing of the water. This shallowing continued upwards into the top member of the Whitby Mudstone Formation, the approximately 10 m-thick Fox Cliff Siltstone. The overlying beds are assigned to the Blea Wyke Sandstone Formation. This consists of coarsening-upwards siltstones, paler in colour and *c.* 18 m thick; it is divisible into two members by their weathered colour, a lower Grey Sandstone Member and an upper Yellow Sandstone Member. Their fauna is dominated by brachiopods and bivalves and there is extensive bioturbation. These appear to be the earliest sediments preserved in the present-day onshore area that were derived from the rising Rattray Volcanic dome to the NE (see below).

Southwards from North Yorkshire these formations thin dramatically or disappear towards the Market Weighton Swell where about 0.5 m of Lower Toarcian clays are overlain directly by Upper Albian sediments (Mitchell 1996); the presence of the Lower Toarcian over this swell reflects the high sea levels of that time, but in the light of the succeeding regression, the absence of younger Toarcian sediments over the swell is not surprising. Indeed, southwards from the Whitby area to Oxfordshire the Upper Toarcian is totally missing due to latest Toarcian erosion following uplift centred on the Rattray Volcanic Centre in the North Sea. The doming preceding the eruptions of the Rattray Volcanic Centre and its subsequent erosion have been documented by Underhill & Partington (1993) who showed that in some offshore areas of the North Sea, the erosion had removed the whole of the Lower Jurassic, but on land the greatest erosion is seen in the East Riding of Yorkshire where it locally reached down to the Upper Pliensbachian. The uplift, which Underhill & Partington (1993) ascribed to a mantle plume, initially caused erosion over a wide area and the onset of clastic sedimentation as the dome was eroded. The effects were widespread and, most significantly, closed the previously open seaway connecting the southern North Sea to the Arctic Ocean. Migration of faunas both northwards and southwards was thus then limited to migration through presumed shallower areas to the west of Britain (Bradshaw *et al.* 1992, maps J4–J7).

Southwards from the Market Weighton area the Whitby Mudstone Formation thickens to *c.* 54 m in the Grantham area (Berridge *et al.* 1999), but southwards from here there is further thinning, to around 10 m in Oxfordshire, not because of any deeper late Toarcian erosion, but presumably reflecting the influence of the London landmass.

In contrast to these areas, along the Jurassic outcrop from Worcestershire southwards to the Dorset coast the Toarcian succession is essentially complete. At Bredon Hill, Worcestershire, the Whitby Mudstone Formation is over 110 m thick (Whittaker 1972). The exception to this picture is over the Mendip Swell where the succession is very condensed.

In Dorset most of the Toarcian time interval is represented by the Beacon Limestone Formation, of which the Upper Pliensbachian Marlstone Rock, here a maximum of only some 0.6 m thick, is the basal member. It is succeeded by 0.6–1.5 m of condensed limestones (Eype Mouth Limestone Member) with an ammonite-dominated fauna that clearly accumulated in deep water that was starved of clastic input. This formation was formerly known as the Junction Bed across its outcrop in Dorset and Somerset as its limestones encompass the boundary between the middle and upper parts of the Lias Group; revision of the nomenclature follows Cox *et al.* (1999). Beds of the Beacon Limestone show signs of early lithification and erosion as ammonites were frequently planed off before deposition of the succeeding bed. Stromatolitic crusts and iron and pink manganese staining characterize much of this succession (Hallam 1967) which was deposited in oxygenated conditions and which, although very condensed, contains representatives of most Toarcian ammonite zones. The thickest development of the formation is close to the Watton Cliff Fault, which was clearly active at the time, and resulted in fissuring of the already deposited limestone and deposition of younger ammonite faunas apparently within older limestones. The complex history of the opening and re-opening of the fissures gave rise to difficulties in interpreting the ammonite succession (e.g. Buckman 1910; Jackson 1922, 1926). Jenkyns & Senior (1977) reported on these relationships. Inland the formation forms the top of a prominent flat-topped escarpment that is traceable across its outcrop in Dorset and into Somerset.

Inland in NE Dorset the Beacon Limestone Formation is represented by some 85 m of mudstones with silts and limestone beds and nodules in the Winterborne Kingston Borehole (Ivimey-Cook 1982). Within this thickness the Toarcian Anoxic Event can be recognized by a horizon with TOC levels up to 4% (Jenkyns *et al.* 2001).

The Beacon Limestone is succeeded on the coast by the Down Cliff Clay (21 m), marking a return to more normal depositional rates. This is the basal member of the Bridport Sand Formation, the overlying beds of which comprise up to 30 m of sands with regular hard bands that are well displayed in striking vertical yellow cliffs between West Bay and Burton Bradstock. The hard bands of the Bridport Sands are more calcareous than the intervening units and show abundant evidence of large-scale bioturbation. Fossils are infrequently preserved, although ammonites of the upper subzone of the Pseudoradiosa Zone and the Aalensis Zone have been recorded, together with belemnites. The top 2 m of the Bridport Sand are of Aalenian age as they yield abundant and well-preserved specimens of *Leioceras opalinum* signifying the basal subzone of the Opalinum Zone.

The Bridport Sand, at its type locality on the Dorset coast, marks the southern end of an outcrop of similar sands that can be traced continuously from the Worcester Basin where they were formerly known as the Cotteswold Sands. Their subcrop extends well to the east, as the formation provides one of the principal reservoirs in the Wytch Farm Oilfield of eastern Dorset (Colter & Havard 1981). As early as 1889, Buckman showed, by means of the contained ammonites, that these sands became progressively younger southwards, and that sand deposition had already stopped in the Worcester Basin before it had commenced in south Dorset. The sedimentology of the sands was studied by Davies (1967) who suggested that they represented the deposition of a southward-migrating sand bar. Large-scale undulations in the sands, well exposed at the base of the cliffs on the coast, were described as scour structures by Davies (1967) but were re-interpreted as erosional bedforms produced below standing waves by Pickering (1995). In the area west of Yeovil the Ham Hill Stone is a local calcareous development within the formation (there previously known as the Yeovil Sands). This bioclastic limestone has been used as a building stone since Roman times and has been used extensively for building far beyond the Yeovil region. The topmost part of the Bridport Sand Formation in the Cotswold Hills is the Cephalopod Bed, a condensed limestone yielding ammonites of the Upper Toarcian Thouarsense and Levesquei zones, that is overlain directly by the Inferior Oolite Group.

The Toarcian is also essentially complete in the Cardigan Bay Basin; here it was too far west to be affected by the Rattray updoming (Underhill & Partington 1993). The succession is very thick in the Mochras Borehole (272 m) and only the youngest preserved Jurassic in the borehole, the latest Toarcian Aalensis Zone, shows any sign of shallowing; its top is missing. Farther out in the Cardigan Bay Basin the Toarcian must be even thicker considering that the total Lias thickness is almost twice that of the Mochras Borehole (Turner 1996); here it is succeeded by the Aalenian. In the Bristol Channel Basin Toarcian shales may be up to 180 m in thickness, but this figure may include some Upper Pliensbachian sediments.

In the Southern North Sea Basin the Lias Group is well developed. 744 m of Lias Group was proved in well 49/16-5 (Cameron *et al.* 1992), but generally thicknesses are less than 300 m. Lott & Knox (1994) recognized four formations within the group. The Penda and Offa formations are equivalent to the Redcar Mudstone Formation; the succeeding Ida Formation is ferruginous silts and limestones equivalent to the Cleveland Ironstone and Staithes formations. The Cerdic Formation is mudstone equivalent to the Whitby Mudstone Formation (Lott & Knox 1994). Over most of this area the Upper Toarcian is missing, occasioned by erosion following the doming centred on the Rattray volcanic centre to the north, but the sandy Phillips Member is locally preserved at the top the Cerdic Formation and provides an offshore equivalent of the Blea Wyke Sandstone Formation.

Middle Jurassic Series

Chronostratigraphical subdivisions of the Middle Jurassic Series are shown in Figure 14.10.

STAGE	STANDARD ZONES
CALLOVIAN	*Quenstedtoceras* LAMBERTI
	Peltoceras ATHLETA
	Erymnoceras CORONATUM
	Kosmoceras JASON
	Sigaloceras CALLOVIENSE
	Proplanulites KOENIGI
	Macrocephalites HERVEYI
BATHONIAN	*Clydoniceras* DISCUS
	Prohecticoceras RETROCOSTATUM
	Cadomites BREMERI
	Morrisiceras MORRISI
	Tulites SUBCONTRACTUS
	Procerites PROGRACILIS
	Asphinctites TENUIPLICATUS
	Zigzagiceras ZIGZAG
BAJOCIAN	*Parkinsonia* PARKINSONI
	Strenoceras GARANTIANA
	Strenoceras SUBFURCATUM
	Stephanoceras HUMPHRIESIANUM
	Emileia SAUZEI
	Witchellia LAEVIUSCULA
	Sonninia OVALIS
	Hyperlioceras DISCITES
AALENIAN	*Graphoceras* CONCAVUM
	Brasilia BRADFORDENSIS
	Ludwigia MURCHISONAE
	Tmetoceras SCISSUM
	Leioceras OPALINUM

Fig. 14.10. Stages and Standard Ammonite Zonation of the Middle Jurassic Series. Substage boundaries for the Bajocian, Bathonian and Callovian stages are indicated by short lines at the right hand side of the stage column.

The Inferior Oolite Group

Over the whole of the Anglo-Welsh area the sediments of the Aalenian–Bajocian interval reflect the alternation of transgression and regression (Hallam 1978). Overprinting the eustatic sea-level curve locally were the effects of the North Sea doming centred on the Rattray Volcanic centre. The Aalenian began with a transgression after the Late Toarcian regression resulting from the Rattray doming and ended with a regression resulting in little or no deposition on highs, such as in south Dorset; in the east there was a major sediment influx from the North Sea. A further transgression in the Early Bajocian brought renewed deposition over the highs, although the water probably remained very shallow over the East Midlands. A further period of regression, that restricted deposition to a few areas, was followed by a Late Bajocian transgression (the Vesulian transgression) that proved far more extensive than those of the Aalenian and Early Bajocian, and resulted in marginal overlap and in places overstep onto the Palaeozoic basement. This latter transgression did not occur until late in the Late Bajocian and appears to be invariably of youngest Garantiana Zone Acris Subzone age.

The Weald–Wessex Basin

The Aalenian and Bajocian rocks of the Weald–Wessex Basin are best known from the western margin of the basin in Dorset and Somerset where they form the greater part of the Inferior Oolite Group (or Formation: see below), traceable northwards as a discontinuous outcrop from the coast south of Bridport to the Mendip Hills. The 'inferior' refers to the quality of the stone in the Bath area where Townsend (1813) stated that it was unfit for use as a building stone; the term supplants William Smith's earlier proposed 'Under Oolite'. The Aalenian Stage is largely equivalent to the Lower Inferior Oolite, the Lower Bajocian Substage to the Middle Inferior Oolite and the Upper Bajocian Substage to the Upper Inferior Oolite. Over the whole of its outcrop area, south of Bristol, the Inferior Oolite is so thin that it is customary to refer to it as a Formation. It consists of a series of condensed limestone deposits often ooidal and with ferruginous ooids ('ironshot' of the older literature) with a rich and often spectacularly well-preserved fauna of all types of molluscs, especially ammonites, together with brachiopods, echinoids, bryozoans and sponges. During the 19th century a plethora of small quarries existed along this outcrop and it was from those, in the area around Bradford Abbas in north Dorset, that Buckman started his meticulous bed-by-bed collecting of the ammonite faunas; that led to his monumental paper (Buckman 1893) detailing the succession of ammonite faunas in the area. Buckman recognized that each bed of limestone had its own individual ammonite fauna and was separated from contiguous beds by non-sequences; comparison of neighbouring quarries showed that no two successions were exactly the same, but by recording the succession of ammonite faunas from each bed throughout the area, a more complete succession of ammonite faunas could be established. Buckman proposed the term *hemera* for the smallest unit of geological time represented by the acme of an ammonite species and drew up hemeral tables to encompass the ammonite faunas of the Inferior Oolite. The significance of this work was not appreciated at the time and it fell into disrepute, largely through Buckman's own later work in which he proposed hemeral successions for the whole of the Jurassic Period in Britain as in the final volumes of *Type Ammonites* (1909–1930), many of which were not based on primary field observations and were clearly flawed. More recently Buckman's work has been revisited, initially by Parsons (e.g. 1974, 1976, 1979; results summarized in Cope

1980*b*); since then Callomon & Chandler (1990) have documented many of Buckman's old localities (often involving excavation) and have recorded many new, often temporary, sections. The result has been to show that Buckman's original work could not only be duplicated (and his conclusions confirmed) but that the succession of ammonite faunas was capable of even finer subdivision. Callomon & Chandler (1990) noted that each bed of the Inferior Oolite of Dorset and Somerset is usually homogenized through bioturbation and its ammonite fauna always constitutes a single horizon that cannot be further subdivided. A total of 53 horizons is now recognized for the Aalenian and Bajocian stages (Callomon & Cope 1995) (see Fig. 14.11).

On the coast in the Bridport area, the Inferior Oolite attains a maximum thickness of 3 m and here the uppermost limestones of the Upper Inferior Oolite are of earliest Bathonian age. Over southern Dorset thicknesses vary rapidly locally (Fig. 14.12) and are not infrequently less than a metre in total, reflecting the sediment starvation of this area throughout the Aalenian and Bajocian ages. Boreholes in Lyme Bay, in the Weymouth Peninsula and in the Jurassic subcrop to the east have shown that the succession in general thickens eastwards and southwards and

LOWER BATHONIAN

Bt-3	*Oxycerites yeovilensis*	Zigzag	Yeovilensis
Bt-2	*Morphoceras macrescens*		Macrescens
Bt-1	*Parkinsonia convergens*		Convergens

UPPER BAJOCIAN

Bj-28	*Parkinsonia bomfordi*	Parkinsoni	Bomfordi
Bj-27c	*Parkinsonia pseudoferruginea*		Truellei
Bj-27b	*Strigoceras truellei*		
Bj-27a	*Parkinsonia parkinsoni* α		
Bj-26b	*Parkinsonia rarecostata*	Garantiana	Acris
Bj-25	*Garantiana tetragona*		Tetragona
Bj-24	*Garantiana dichotoma*		Dichotoma
Bj-23	*Leptosphinctes davidsoni*	Subfurcatum	Baculata
Bj-22	*Caumontisphinctes polygyralis*		Polygyralis
Bj-21	*Caumontisphinctes aplous*		Banksi
Bj-20	*Teloceras banksi*		

LOWER BAJOCIAN

Bj-19	*Teloceras coronatum*	Humphriesianum	Blagdeni
Bj-18	*Teloceras blagdeni*		
Bj-17	*Stephanoceras blagdeniforme*		Humphriesianum
Bj-16	*Stephanoceras gibbosum*		
Bj-15	*Stephanoceras humphriesianum*		
Bj-14b	*Chondroceras wrighti*		Romani
Bj-14a	*Chondroceras delphinum*		
Bj-13	*Stephanoceras umbilicum*		
Bj-12	*Stephanoceras rhytum*	Sauzei	
Bj-11b	*Nannina evoluta*		
Bj-11a	*Otoites sauzei*		
Bj-10	*Witchellia laeviuscula*	Laeviuscula	Laeviuscula
Bj-9	*Witchellia ruber*		
Bj-8b	*Shirbuirnia trigonalis*		Trigonalis
Bj-8a	*Witchellia nodatipinguis*		
Bj-7b	*Witchellia connata*		Sayni
Bj-7a	*Witchellia gelasina*		

to be continued

LOWER BAJOCIAN *(continued)*

Bj-6c	*Witchellia "pseudoromani"* MS	Ovalis	
Bj-6b	*Fissilobiceras gingense*		
Bj-6a	*Euhoploceras zugophorum*		
Bj-5	*Witchellia romanoides*		
Bj-4	*Bradfordia inclusa*		
Bj-3	*Hyperlioceras subsectum*	Discites	
Bj-2b	*Hyperlioceras rudidiscites*		
Bj-2a	*Hyperlioceras walkeri*		
Bj-1	*Hyperlioceras politum*		

AALENIAN

Aa-16	*Euhoploceras acanthodes*	Concavum	Formosum
Aa-15	*Graphoceras formosum*		
Aa-14	*Graphoceras concavum*		Concavum
Aa-13	*Graphoceras cavatum*		
Aa-12	*Brasilia decipiens*	Bradfordensis	Gigantea
Aa-11	*Brasilia gigantea*		
Aa-10	*Brasilia bradfordensis, similis*		Bradfordensis
Aa-9	*Brasilia bradfordensis, baylii*		
Aa-8	*Brasilia bradfordensis, subcornuta*		
Aa-7	*Ludwigia murchisonae*	Murchisonae	Murchisonae
Aa-6	*Ludwigia patellaria*		
Aa-5	*Ludwigia obtusiformis*		Obtusiforme
Aa-4	*Ancolioceras opalinoides*		Haugi
Aa-3	*Leioceras bifidatum*	Scissum	
Aa-2	*Leioceras lineatum*		
Aa-1	*Leioceras opalinum*	Opalinum	

Fig. 14.11. Ammonite faunal horizons of the Inferior Oolite (Aalenian, Bajocian and basal Bathonian stages) and basal part of Great Oolite (upper part of lowest Bathonian Stage). After Callomon & Cope (1995).

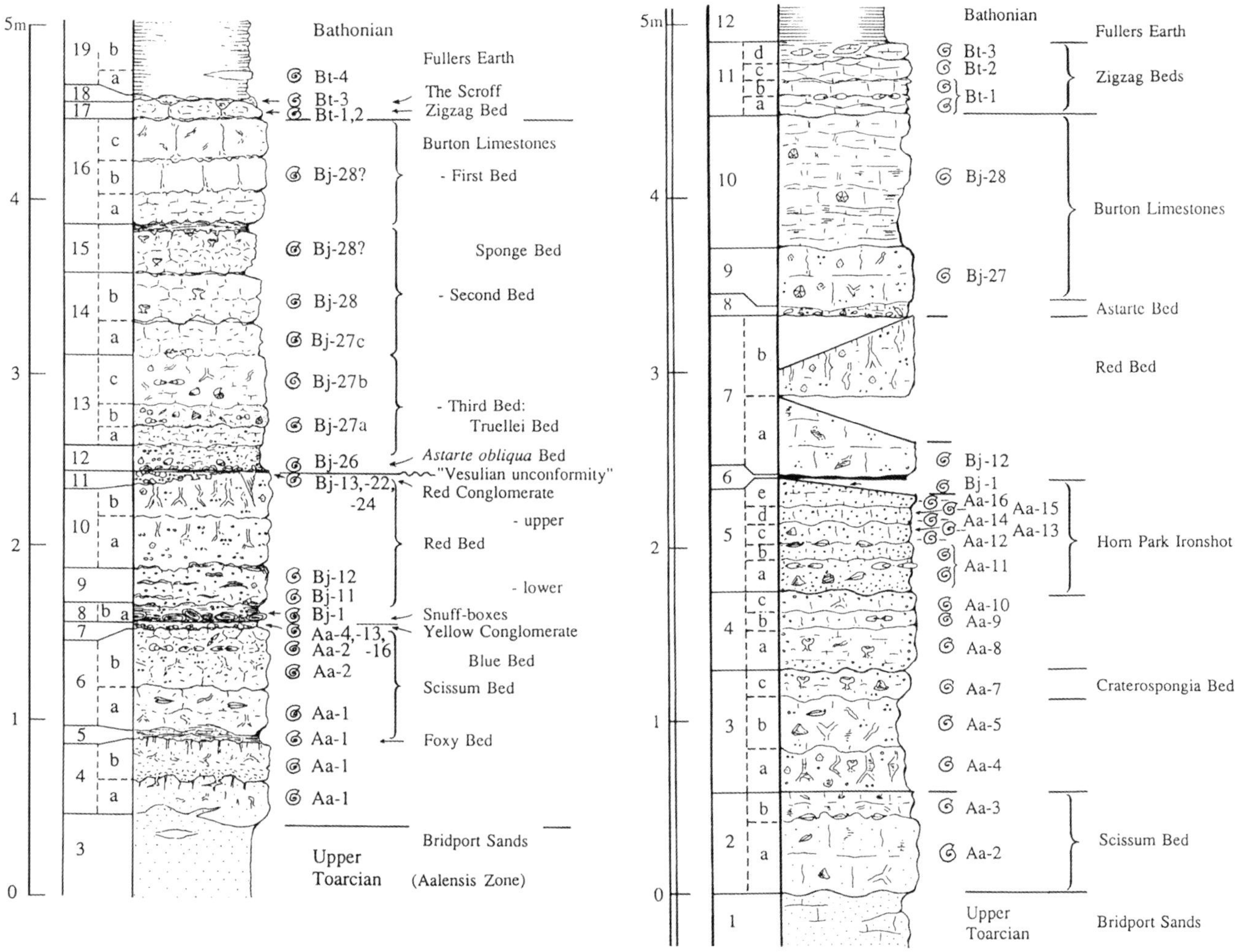

Fig. 14.12. Comparison between the development of the Inferior Oolite in two localities in South Dorset. Left-hand column: coastal section at Burton Bradstock; right-hand column: Horn Park Quarry, Beaminster. Note much more complete development of Aalenian horizons at Horn Park; more complete development of Upper Bajocian horizons at Burton Bradstock; severe attenuation of Lower Bajocian in both sections. After Callomon & Cope (1995).

that the thin successions of South Dorset reflect the control of the South Dorset Swell (Penn 1982) (Fig. 14.13). The Lyme Bay Borehole shows that, unlike in the Inferior Oolite of the rest of the Wessex Basin, kaolinite here is a major clay mineral; this was interpreted as showing greater proximity to a land area (in this case Cornubia) by Jones & Sellwood (1989). At outcrop most successions are highly incomplete (see Cope 1980*b*, fig. 3a) and the most complete succession exposed is that around Oborne in north Dorset (Fig. 14.13) where the Inferior Oolite reaches almost 40 m in thickness, but even here there are stratigraphical gaps. In the Purse Caundle Borehole (6 km east of Sherborne) 44.2 m was proved (Bristow *et al.* 1995, table 3) and was more complete. This thin ammonitiferous sediment-starved Dorset-Somerset facies continues northwards as far as the outlier of Dundry Hill, south of Bristol, described by Buckman & Wilson (1896) and revised by Parsons (1974, 1979) (see also Cox & Sumbler 2002).

Over the Mendip Hills the Lower and Middle Inferior Oolite are cut out by the effects of the Mendip Swell and the Upper Inferior Oolite rests with strong unconformity on the Carboniferous Limestone in the area of Vallis Vale, west of Nunney (Cox & Sumbler 2002, 101–103). This transgression at the base of the Acris Subzone (the Vesulian transgression) is marked by borings into the peneplaned Carboniferous Limestone surface and by oyster encrustation. In the eastern Mendips the Upper Inferior Oolite totals up to 11 m in thickness.

In the eastern part of the Weald–Wessex Basin, the Inferior Oolite has been encountered in many boreholes to the south of the London landmass. These typically show a facies similar to that developed in the Cotswold Hills (see below) with oolites and sands of Aalenian Lower Inferior Oolite overlain with a break by Lower Bajocian Middle Inferior Oolite and these in turn disconformably overlain by thick Upper Inferior Oolite limestones of Late Bajocian age, presumed to belong to the Acris Subzone, at their base. The thickest development seen was in the Collendean Farm Borehole where 169.2 m was recorded, with all divisions of the formation well-developed (Gallois & Worrsam 1993); these authors suggested the formation could be approaching 200 m in thickness in some parts of the surrounding area.

The Cotswold Hills

Northeast of Bristol the Inferior Oolite takes on a different facies; instead of the thin ammonite-rich limestones, a much thicker limestone succession is developed that totals up to about 110 m in the most complete sections. It is responsible for the steep escarpment of the Cotswold Hills, that attains a maximum height of 330 m. The Inferior Oolite Group here includes thicknesses of cross-bedded oolites and beds containing corals – both facies largely inimical to ammonites. Several of these oolites have been worked in the past for building stones and the quarries have provided a great deal of the information on the

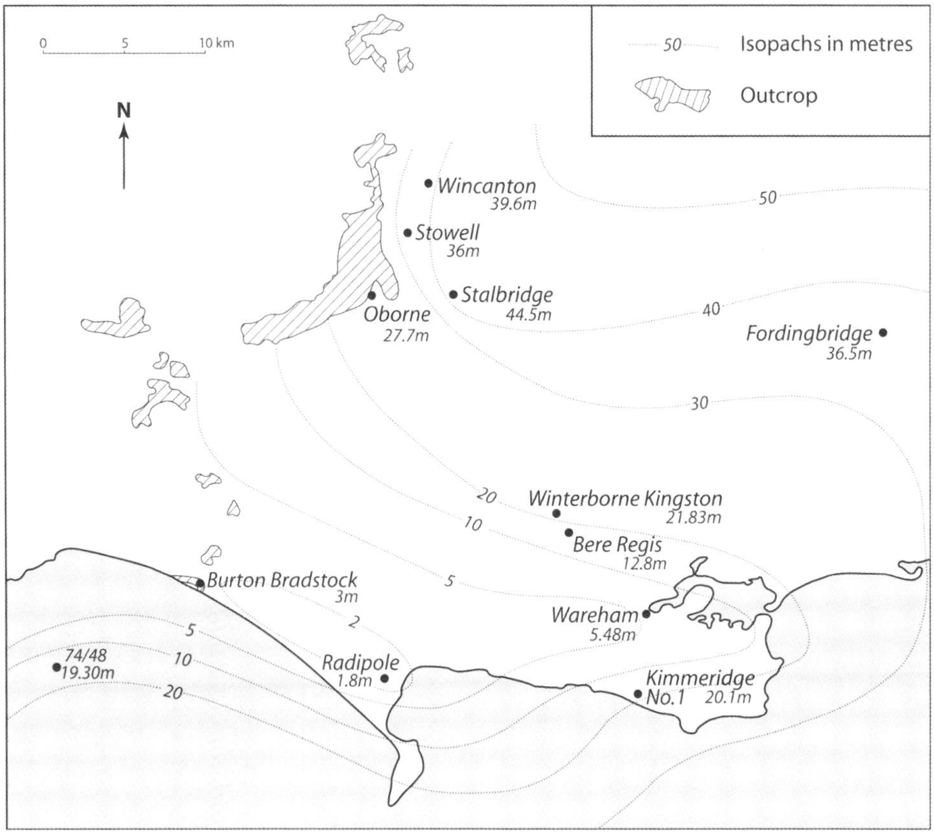

Fig. 14.13. The development of the Inferior Oolite in the sediment-starved region of the western Wessex Basin. Isopachs in metres. Based on Penn (1982, fig. 9) by permission of the British Geological Survey.

complex of facies changes within the area. Some horizons yield autochthonous ammonites fairly frequently, and in others ammonites are very rare and the few that have been found may well be drifted shells; sufficient numbers of these have been found to date most horizons into the zonal (and in some cases subzonal) scheme. Over much of the area the Middle Inferior Oolite is missing and the Upper Inferior Oolite rests on various thicknesses of Lower Inferior Oolite. The exception to this picture is around Cleeve Hill in Gloucestershire, where within a pre-Vesulian syncline some 22 m of Middle Inferior Oolite are preserved (Fig. 14.14); even here there is a major disconformity at the base of the Upper Inferior Oolite, with the top of the Lower Bajocian and much of the Upper Bajocian missing.

The lithostratigraphy of the Cotswolds' Inferior Oolite has recently been revised by Barron *et al.* (1997), who have rationalized the often-conflicting previous lithostratigraphical schemes. They divided the group into three formations separated by non-sequences: the Birdlip Limestone Formation, equivalent to the Lower Inferior Oolite is of Aalenian age; it was formerly well exposed in quarries throughout the area and consequently is the best known part of the succession. The succeeding Aston Limestone Formation is equivalent to the Middle Inferior Oolite and is of Early Bajocian age; it is only preserved within a shallow downfold in the north Cotswolds around and to the east of Cheltenham. It is overlapped by the Salperton Limestone Formation that is equivalent to the Upper Inferior Oolite and is of Late Bajocian–earliest Bathonian age. The Salperton Limestone oversteps the Aston Limestone to rest on either the Birdlip Limestone Formation or the Lias Group (Fig. 14.14). These breaks between the formations are marked by hardground formation, with borings and an encrusting epifauna, and although regionally there is major overstep, eventually onto the Carboniferous Limestone, little or no angular discordance between the formations is obvious in the field. Details of the revised lithostratigraphy are shown in Figure 14.15.

Eastwards the Inferior Oolite thins towards the western end of the London landmass and at Hook Norton (Oxfordshire) is reduced to some 6 m, half of which is of Early Aalenian age represented by the Northampton Sand Formation and the remainder, belonging to the Clypeus Grit, is of latest Bajocian age.

The East Midlands

Across the East Midlands the early Aalenian Northampton Sand Formation, including the facies known as the Northampton Ironstone, is overlain by the Grantham Formation, that must also be of Aalenian age. The Grantham Formation (Lower Estuarine Series of Kent 1975) consists of shoreface silts or sands overlain by coastal plain silts and muds with rootlets; this intercalation, that extends from north Lincolnshire as far south as the Peterborough area, reflects the persisting influence of the uplift centred on the Rattray Volcanic Centre in the southern part of the northern North Sea. The Harford Member of the Birdlip Limestone Formation of the Cotswolds includes facies that are similar to those of the Grantham Formation; the two are more or less coeval. These Grantham Formation sands are difficult to date directly due to the absence of ammonites in both the sands and the basal parts of the overlying Lincolnshire Limestone. However, as the Lincolnshire Limestone Formation has yielded ammonites of the basal Bajocian Discites Zone, at some distance above its base, the Grantham Formation must be of Aalenian age.

The Lincolnshire Limestone Formation has been extensively quarried, both as a building stone, but more recently for cement; it is also important as an aquifer. In places its junction with the underlying Grantham Formation is non-sequential. The formation is readily divided into two; the Lower Lincolnshire Limestone consists of carbonate mudstones, wackestones and packstones, typical of low-energy lagoonal

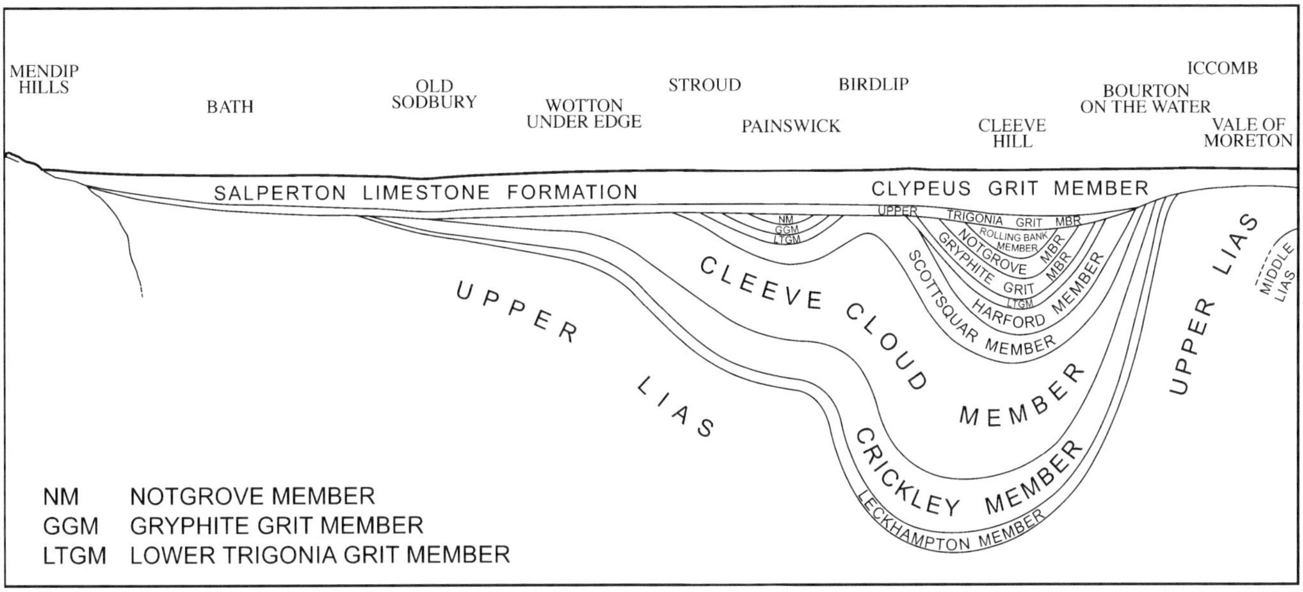

Fig. 14.14. Section from the Mendip Hills to the Vale of Moreton showing the development of the Inferior Oolite Group in the Cotswold Hills. Modified after Arkell (1933).

complexes with a maximum thickness of some 21 m. The Upper Lincolnshire Limestone (maximum 22 m) succeeds, frequently with an undulating erosional contact; it consists of higher energy cross-bedded ooidal grainstones, of marine barrier bar facies, sometimes with a particularly coarse 'rag' facies at the top with local names, such as the Ancaster Stone. At the southern end of its outcrop the Lincolnshire Limestone averages about 10 m in thickness, but thickens northwards to attain a thickness of over 30 m between Grantham and Lincoln (Sylvester-Bradley 1968; Berridge *et al.* 1999). Only a handful of ammonites have been obtained from the formation and these have been crucial in assigning it principally to the Lower Bajocian (Ashton 1980; Cox & Sumbler 2002). The fauna consists principally of bivalves, together with corals, bryozoans, brachiopods, gastropods and echinoderms. Around Collyweston (northern Northamptonshire) a local facies, the Collyweston Slate, is developed within the basal beds of the Lincolnshire Limestone. The 'Slate' is worked underground and the massive fine-grained sandy limestone is then exposed to weathering, when it becomes readily fissile and splits into thin roofing 'slates'. This is the same technique formerly used to work the Bathonian Stonesfield Slate (see below). In the Humber area and northwards to Market Weighton the Lincolnshire Limestone is developed in an ooidal limestone facies, the Cave Oolite, up to some 7 m thick.

Over most of the East Midlands, the uppermost part of the Lower Bajocian and the whole of the Upper Bajocian are missing and there is a major non-sequence that possibly also embraces the lowest Bathonian.

The Bristol Channel, Cardigan Bay and St George's Channel basins

Overlying the Toarcian sediments of the eastern Bristol Channel is a thick clay succession that has yielded foraminifera of Aalenian and Bajocian age (Evans & Thompson 1979). This succession is totally unlike that of the Cotswold Hills to the east and represents an entirely local facies. In borehole 103/18-1, south of the Pembroke Ridge 54 m of siltstones were recorded. In the St George's Channel Basin the Lower and Middle Inferior Oolite are represented by a thick (124 m) sequence of sandstone with some mudstone, with 133 m of calcareous siltstones above, representing the Upper Inferior Oolite. Within Cardigan Bay Aalenian and Bajocian rocks were encountered in several boreholes, including 107/16-1 where sandstones and siltstones attain 230 m in thickness. Further north, off St Tudwal's Arch, Penn & Evans (1976) recorded Bajocian or Bathonian rocks resting directly on Sinemurian strata. In the context of the regional geology of southern Britain, this is almost certainly the Upper Bajocian Acris Subzone, representing the Vesulian transgression, that is recognized in virtually every Bajocian section, as suggested by Cope (1998*a*).

Great Oolite Group

The Great Oolite Group succeeds the Inferior Oolite Group in England south of the Market Weighton Swell. The term 'Great Oolite' was originally applied by Warner (1811) to the Bath Freestone in the Bath area. Later, the Geological Survey, following Judd (1875), used the term 'Great Oolite Series' much in the sense of the present group and its modern usage follows Cave (1977). Over much of the outcrop exposures are poor and the facies frequently change rapidly laterally. Many of these facies were inimical to ammonites, which are generally rare in the Great Oolite Group so that the scheme of Standard Zones for the English Bathonian has evolved slowly.

The time interval during which the Great Oolite Group accumulated essentially embraces the Bathonian and earliest Callovian. The stage boundary lies within the Cornbrash Formation at the top of the Great Oolite Group. The Cornbrash was named by William Smith as a limestone that provides good ground for growing cereal crops and which gives rise to a soil with a great deal of stony debris (brash). The formation is traceable right across England from south Dorset to the Yorkshire coast and although it is generally less than 10 m thick it gives rise to a small escarpment as it is more resistant than the clays that are normally found both above and below. It is locally a good aquifer on a small scale and this has given rise to a large number of small villages along its outcrop. William Smith himself noted that there was a faunal distinction between the lower and upper parts; the lower part is Bathonian and the upper part of Callovian age.

Dorset and Somerset to the Mendips

The lower part of the Great Oolite Group comprises a thick clay formation, the Fuller's Earth, which was originally named by William Smith in the Bath area, where these beds formerly

Formation	Member	Description
SALPERTON LIMESTONE FORMATION	Clypeus Grit Member	Pale grey to brown rubbly ooidal, peloidal and shelly packstones to grainstones with *Clypeus ploti*. 10-16 m
	Upper Trigonia Grit Member	Very hard ooidal shelly grainstone or packstone with trigoniid bivalves in mouldic preservation. 1-5 m
	major erosion surface	
ASTON LIMESTONE FORMATION	Rolling Bank Member	Grey rubbly limestones, yellow and ironshot in lower part, shelly in middle and upper parts. < 8·4 m
	Notgrove Member	Pale cross-bedded peloidal and ooidal grainstones, few fossils. 4-10 m
	Gryphite Grit Member	Grey and brown peloidal and shelly grainstones, packstones and wackestones with muddy or sandy interbeds. Abundant *Gryphaea bilobata*. 3-7 m
	Lower Trigonia Grit Member	Shelly, sandy peloidal wackestone, packstone or grainstone, marly or sandy interbeds and frequent trigoniids. 1-3 m
	erosion surface	
BIRDLIP LIMESTONE FORMATION	Harford Member	Marine to ?non-marine fine sands, brown mudstones and sandy ooidal limestones. 0-5 m
	Scottsquar Member	Peloid, ooid and shelly packstone and grainstone with marl or carbonate mudstone. 2·4-30 m
	Cleeve Cloud Member	Pale well-sorted and well-bedded ooidal grainstones, sandy in E & N. 25-51 m
	Crickley Member	Grey to yellow brown limestones, ooidal grainstones below and pisoidal grainstones and packstones above. 7-15·7 m
	Leckhampton Member	Grey, yellow-weathering limestones with thin marly interbeds. 2-10 m

Fig. 14.15. Lithostratigraphical nomenclature of the Inferior Oolite Group in the Cotswold Hills.

yielded a commercial fuller's earth (i.e. a clay with the properties of adsorbing oil). Nowhere on the coast is the middle part of this formation exposed and only the upper and lower parts are known; boreholes have proved a maximum thickness of 215 m. Resting on the topmost limestones of the Inferior Oolite Formation, that are of earliest Bathonian age, is a succession of poorly exposed and monotonous pale grey conchoidally fracturing clays comprising the Lower Fuller's Earth Member. In its uppermost part is a horizon with the oyster *Liostrea acuminata*, the Acuminata Beds, that can be traced from north Dorset into the southern Cotswolds. These beds have yielded ammonites typical of the Progracilis Zone at several localities and are a good marker horizon. Above the Lower Fuller's Earth Member is a rubbly non-ooidal limestone that can be traced from the Mendips southwards to a point some 5 km to the south of Yeovil; this is the Fuller's Earth Rock Member and it reaches a maximum of 11 m in thickness. Southwards from where the Fuller's Earth Rock wedges out, as far as the Dorset coast, the Fuller's Earth Formation is in unbroken clay facies up to some thin limestones intercalated with clays and yielding a brachiopod-dominated fauna including the zeilleriid *Wattonithyris*. These Wattonensis Beds crop out continuously from the Dorset coast west of Bridport, into north Dorset, and form the basal member of the succeeding Frome Clay Formation. On the coast they are some 7–8 m thick, but to the east, in subcrop, they were 15 m thick in the Lulworth Banks Borehole (Martin 1967). In north Dorset the Wattonensis Beds can be seen to lie above the Fuller's Earth Rock and the clays between the two (up to 21 m) comprise the Upper Fuller's Earth Member. Formerly these were referred to the Middle Fuller's Earth, but the classification of these beds has been revised following the introduction of the term Frome Clay Formation by Penn & Wyatt (1979). The latter is referred to in older literature as the Upper Fuller's Earth, but this latter term is now used in a more restricted sense and is equivalent to the Middle Fuller's Earth of earlier authors. The Frome Clay maintains a thickness of

30–45 m across its outcrop in the area, but exceeded 80 m in thickness in the Lulworth Banks and Kimmeridge No 3 boreholes of eastern Dorset (Martin 1967). It is succeeded by the Boueti Bed, a thin (0.3 m) bed largely composed of brachiopods including the eponymous *Goniorhynchia boueti*; the fauna is frequently epifaunally encrusted and suggests a relatively long period of slow deposition. The distinctive Boueti Bed fauna has been traced from the Dorset coast as far north as the north side of the Mendip Hills and was recorded in the subcrop to the east, in the Kimmeridge No 3 Borehole, where it was almost a metre thick (Martin 1967).

The Boueti Bed is the basal bed of the Forest Marble Formation, another of William Smith's original stratal terms, taking its name from the Forest of Wychwood (Oxfordshire) where bioclastic limestones were formerly cut and polished as ornamental stone. The Forest Marble falls readily into three parts on the Dorset coast west of West Bay. The lower part is dominantly of clays with occasional calcareous siltstone bands; these yield an abundant shallow-water trace-fossil assemblage. The middle part of the formation consists of cross-bedded bioclastic limestones, consisting principally of oyster debris; they yield abundant wood and occasional vertebrate teeth, including those of theraspids and mammals. The upper part of the succession, seen on the coast west of West Bay, consists of further clays with siltstones, but these are replaced east of Bridport by limestones that were formerly extensively quarried. These limestones form a conspicuous feature along the outcrop northwards to the Mendips although locally much of the Forest Marble succession consists of mudstones that, because of poor exposure, are poorly known (Bristow *et al.* 1995). The formation is generally 30–45 m thick, but in the Weymouth anticline exceeds 70 m and is there principally in clay facies.

The Lower Cornbrash (*c.* 3 m) is typically pale bioclastic micritic and marly nodular limestones that belong to the Discus Subzone of the Discus Zone. Two brachiopod biozones have been recognized within the member, a lower one characterized by *Cererithyris intermedia* and an upper one of *Obovothyris obovata*. The Upper Cornbrash is characteristically calcareous sands and calcarenites that embrace most of the Herveyi Zone. Again, two brachiopod biozones are recognized, a lower one of *Microthyridina siddingtonensis* and an upper one of *M. lagenalis*. The Upper Cornbrash is up to 6 m thick on the Dorset coast.

Bath area

In the Bath area, the Fuller's Earth Rock, here some 5 m thick, again divides the Fuller's Earth into Lower and Upper members. Within the latter, the commercial fuller's earth was shown by Jeans *et al.* (1977) to be a smectite and to contain devitrified glass shards and must have been derived from a volcanic source of trachytic nature (although none was known in the British area at the time that paper was written). Its source is unlikely to be from the distant Rattray Volcanic centre, but is more likely to have lain in the Celtic Sea or Western Approaches. The Bath area is notable for the development of the limestones, known historically as the Great Oolite, which crop out across the western Cotswold Hills. They were renamed the Chalfield Oolite Formation by Wyatt & Cave (2002), partly to avoid confusion with the Great Oolite Group. In the Bath area this formation includes the Bath Freestone from which Georgian Bath was constructed. The Freestone is a creamy yellow oolite that was quarried in underground galleries to the east of the city. When freshly mined and before it has lost its contained water, the rock is soft and readily cut with a hand-saw and carved. On weathering the stone becomes hardened and resistant to further weathering. The strength of the Bath Freestone is due to its sparry cement; the ooliths are soft and tend to weather out (Palmer 2004). Although there is little biostratigraphical control, it seems likely that these limestones are lateral equivalents of the Frome Clay (Penn & Wyatt 1979). The latter authors showed that the top member of the Great Oolite Limestones of Green & Donovan (1969), the Ragstones, was better regarded as the basal member of the Forest Marble Formation. This correlation is supported by changes in thickness, as the Frome Clay is reduced to about 12 m and the overlying Forest Marble is reduced from about 45 to 23 m where the Chalfield Oolite limestones appear. There are a number of impersistent clay beds within the basal part of the Forest Marble of this area; one of these, the Bradford Clay, is well known for its abundance of beautifully preserved thecae of the millericrinid *Apiocrinites*.

The Lower Cornbrash has thinned to some 1.5 m at Corsham, east of Bath; north of here it thickens again, with massive limestones, the Corston Beds, making up most of the succession. The Upper Cornbrash thins out completely in south Wiltshire; but then re-appears in north Wiltshire and north Gloucestershire.

The Cotswold Hills

North of Bath and along the southern part of the Cotswold escarpment to Oxfordshire, limestones occur both above and below the Chalfield Oolite, but correlation is difficult owing to the near-absence of ammonites and it is only by detailed mapping, complicated by extensive cambering of the ground, that the complex facies relationships can start to be understood (see for example: Arkell & Donovan 1952; Cave 1977; Cox & Sumbler 2002). Hardgrounds have been recognized as providing regionally persistent depositional breaks of correlative significance.

In contrast to the deeper water of the Wessex Basin, the Great Oolite Group of the southern Cotswolds accumulated in a shallow-water subtropical sea on a shallow stable shelf that extended westwards from the London landmass. Ooids generated in the shallow waters were redistributed by the combined action of tides, wind and waves and storm activity resulted in large-scale movement of oolite shoals. The result is a complex of interdigitating facies that has taken detailed mapping combined with the limited biostratigraphy much time to resolve.

Over this area, the Fuller's Earth Formation is equivalent to only the Lower Fuller's Earth Member south of Bath and is succeeded by a complex of several, predominantly limestone facies, totalling about 25 m thick, that progressively replace the rest of the Fuller's Earth Formation. In south Gloucestershire these facies comprise, in ascending order, the Taynton Limestone, Tresham Rock and Athelstan Oolite formations, each of which youngs diachronously southwestwards (Fig. 14.16). The Tresham Rock is a pale grey siltstone or calcite mudstone with a sparse molluscan fauna (Cave 1977). The Athelstan Oolite is a fine white even-grained oolite in the south; northwards it includes beds of coarse bioclastic cross-bedded oolites. It was deposited as a barrier–bar complex of mobile carbonate sands flanking a lagoon to the east (Wyatt & Sumbler in Cox & Sumbler 2002). Periods of sea-bed stability allowed the establishment of a rich and varied fauna of gastropods and bivalves; these were collected from a complex of former quarries on Minchinhampton Common, SE of Stroud, and monographed by Morris & Lycett (1851–1855). The Chalfield Oolite gradually replaces the Athelstan Oolite upwards when progressing northeastwards and its uppermost (Corsham) member succeeds it. The Forest Marble Formation rests on an eroded surface of the Chalfield Oolite (Fig. 14.16).

In the north Gloucestershire–Oxfordshire region the Fuller's Earth Formation wedges out rapidly northeastwards. At the base of the succession is the Chipping Norton Limestone whose base may be of latest Bajocian age; it rests on Leckhampton Member or Northampton Sand. The Chipping Norton Limestone is an ooidal limestone deposited in high-energy conditions recalling the mobile carbonate sands of the Bahamas (Sellwood & McKerrow 1974); northeastwards it passes laterally into the Horsehay Sand Formation, formerly known as the White Sands, that consists of up to 11 m of high-energy sands with

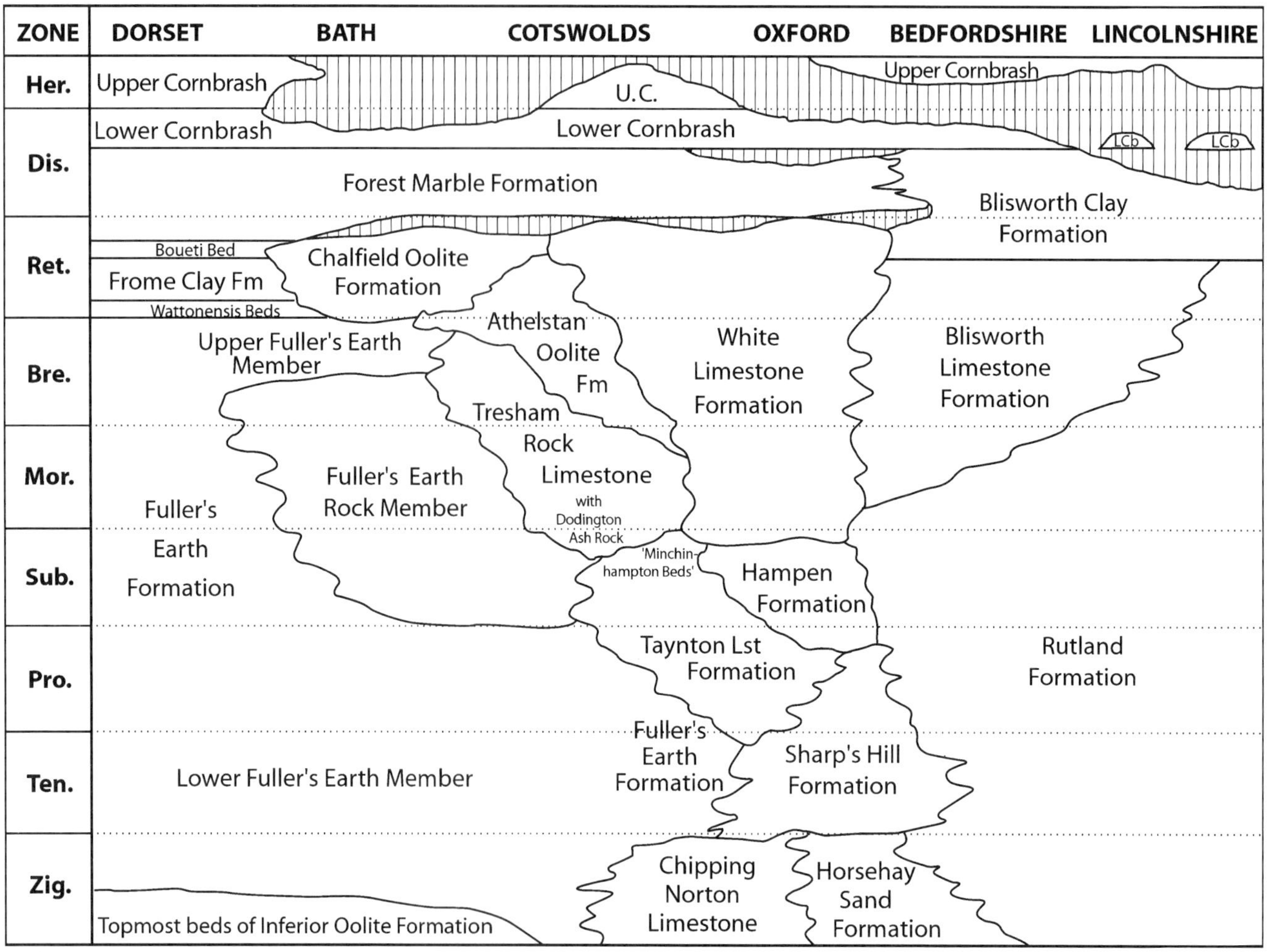

Fig. 14.16. Diagram to show lateral facies variation in the Great Oolite Group from Dorset to Lincolnshire; some simplifications have been made. Based on information supplied by B. M. Cox and M. G. Sumbler.

a low-diversity bivalve fauna indicating a marginal marine environment (Cox & Sumbler 2002). Dinoflagellate cysts in the more northerly outcrops indicate marine conditions and a late Bajocian–Bathonian age, suggesting more or less exact equivalence to the Chipping Norton Limestone, but in the south the microflora is dominated by terrestrial palynomorphs and a brackish coastal marsh environment is envisaged (Fenton *et al.* 1994).

Both Chipping Norton Limestone and Horsehay Sand are overlain by the thin (<5 m) Sharp's Hill Formation that includes a variety of lithologies from dark clays through to shelly limestones; sand is present in several parts of the succession. The fauna is predominantly marine and includes corals, but there are beds containing freshwater gastropods, ostracods and charophytes and also abundant plant fragments and stegosaur bones that have been drifted in from land nearby (Cox & Sumbler 2002). Seat-earths and rootlet beds are reminiscent of the very shallow-water conditions of the Rutland Formation to the NE (see below). Above is the Taynton Limestone which is a shallow-water high-energy bioclastic carbonate sand that has been shown (Boneham & Wyatt 1993) to include three horizons from which the Stonesfield Slate had been formerly won. The 'slates' are calcareous sandstones and siltstones with ooids that may be scattered or concentrated into laminae or thin beds of oolite; sometimes they occur as concretions within less cemented sands. The beds were mined and left above ground to be exposed to frost before splitting as a roofing material. These 'slates' forms the roofs of many of the typical Cotswolds' buildings. The beds yield an extensive molluscan fauna, including ammonites of the Progracilis Zone, but the Stonesfield Slate is internationally important for its vertebrate fauna that includes three species of mammals, together with mammal-like reptiles, dinosaurs, pterosaurs, marine reptiles and fish. Additionally it yields insects and a varied flora indicating proximity to the London landmass.

The Taynton Limestone is succeeded by the Hampen Formation (<9 m) consisting of ooidal limestones with marl intercalations; both limestones and marls contain quartz sand (Sumbler & Barron 1996). Above is the White Limestone Formation, up to 25 m thick, which has yielded ammonites allowing limited chronostratigraphical control. Three members are recognized and correlation is aided by the extensive faunas of nerineid gastropods that give their name to hardground marker beds (Barker 1976).

The overlying Forest Marble in its type area is channelled into the top of the White Limestone; in places this channelling is deep enough locally to penetrate the youngest (Bladon) member of the White Limestone (Sumbler *et al.* 2000, fig. 4.4). At its base, north and NW of Fairford, Gloucestershire, the Fairford Coral Bed is developed, and has yielded over 60 species of corals (Negus & Beauvais 1975). This basal horizon has small patch reefs developed from east of Bath northwards (Wyatt & Cave 2002). The Forest Marble is mudstone-dominated over much of this area and its thickness varies from 10–30 m; the mudstones when unweathered are typically greenish in colour. Limestones occur commonly, but they are laterally impersistent; they are typically cross-bedded bioclastic ooidal grainstones with mudflakes and lignite, and have commonly been used for walling, paving stones, and locally as roofing tiles.

Across this region, the Lower Cornbrash is developed as limestones up to 5 m thick. The Upper Cornbrash is absent or very thin. A maximum of 1.2 m was recorded in the Oxford area. As pointed out by Page (1989) *Microthyridina siddingtonensis*, the lower of the two brachiopod biozonal indices for the Upper Cornbrash, sometimes occurs in the Lower Cornbrash, as at Long Hanborough, Oxfordshire, causing confusion to earlier workers.

London landmass

Boreholes around the London landmass have shown that Great Oolite Group transgresses farther onto the landmass on its western and southern margins, than had the beds of the Lias Group, although its margin fell short of that of the Lias on the landmass' northern side (Hallam 1992*b*, fig 11.10). One of the most famous boreholes, as it was the first to reach the Palaeozoic basement, was that in Tottenham Court Road in 1878, that proved a thin development of 'Great Oolite Limestone' (initially identified as Lower Greensand) directly overlying Devonian Old Red Sandstone. This transgressive event appears to be a continuation of the Acris Subzone 'Vesulian Transgression' discussed above and is notable as this is one of the few areas where Middle Jurassic marginal facies have been preserved in the Anglo-Welsh area. To the south, the Bolney Borehole proved 95.3 m of Great Oolite Group (Gallois & Worssam 1993).

East Midlands

Northwards along the outcrop from Oxfordshire, the lower part of the Great Oolite Group comprises the Rutland Formation, formerly known as the Upper Estuarine Series and totalling up to 11 m thick. The basal Stamford Member is principally composed of non-marine silts with rootlet horizons and interpreted as a freshwater lake infill (Bradshaw 1978). Above is a succession of brightly coloured clays that are predominantly non-marine asymmetric sedimentary cycles of brackish and freshwater sediments. Bradshaw (1978) showed that each of the (at least) seven sedimentary cycles was initiated by a short-lived marine incursion; the remainder of the cycle could be either land-derived clastic sediment, or limestone, but the cycle often ended with a rootlet-bearing clay. Individual cycles are distinguishable for some 150 km along the outcrop from Oxfordshire to Lincolnshire. Between each cycle is a non-sequence of unknown duration. Bradshaw postulated a swamp environment in a very shallow embayment with a variable but lowered salinity and a humid climate. The more marine of the beds include the oyster *Praeexogyra hebridica*, and rhynchonellid and lingulide brachiopods; there are bivalve-dominated shell plasters.

At the top of the Rutland Formation the beds become more calcareous and there is a transition into the succeeding Blisworth Limestone Formation; this is the time-equivalent of the White Limestone Formation farther south and up to 10 m thick. This limestone has a more fully marine aspect and yields a fauna including bivalves, terebratulid brachiopods, echinoids and serpulids. The succeeding Blisworth Clay Formation (up to 8 m thick) is broadly equivalent to the Forest Marble. Near its base are ironstone nodules, once worked economically; palaeontologically it is poorly known, but yields *Praeexogyra hebridica*.

The Lower Cornbrash is only preserved in isolated parts of the East Midlands, whilst the Upper Cornbrash, although always present, is generally thinner than 3 m.

Ravenscar Group

Most of the Aalenian and Bajocian rocks to the north of the Market Weighton Swell are assigned to the Ravenscar Group, which extends upwards to the base of the Callovian. Over this area the effects of the end-Toarcian uplift of the Rattray Volcanic centre become most apparent. The Cleveland Basin received fluviatile sediments draining from surrounding land areas. Here a succession of sands accumulated in which a wide variety of fluviatile facies have been recognized (Leeder & Nami 1979). With these are associated a spectacularly varied flora (Harris 1961; van Konijnenburg-van Cittert & Morgans 1999) and footprints and trackways of terrestrial animals including sauropods (Fig. 14.17) (Whyte & Romano 1993; Romano & Whyte 2003). Three or four marine horizons are intercalated within the fluviatile facies.

The flora gives good indications of the climate. Up to the mid Bajocian the plants indicate a seasonal climate with wet winters, but with increasing aridity evident from the late Bajocian into the Bathonian (Morgans *et al.* 1999). Periodic abundance of water is indicated by freshwater bivalves, authigenic siderite and the large size of logs drifted into the succession, whilst dryer periods are represented by desiccation cracks and charcoal. In the Early Bajocian coal was dominant over charcoal, but as the climate became dryer charcoal became dominant (Hesselbo *et al.* 2003). Cope (1993*c*) suggested that the conifer–ginkgophyte–cycadophyte flora was a fire-controlled climax floral assemblage.

Beneath the Ravenscar Group the Dogger Formation has yielded Lower Aalenian ammonites and rests with minor unconformity on gently warped Toarcian sediments, the junction with the Lias Group being marked by an erosional surface and derived Toarcian ammonites in a basal pebble bed. The break is more pronounced farther to the west, where subaerial erosion appears likely (Hemingway 1974). The Dogger is a sideritic sandstone some 1–2 m thick but on the coast, in the Peak Trough (to the east of the Peak Fault), it thickens to some 12 m. The succession is marine and dominated by molluscs and brachiopods; ammonites indicate the Opalinum and (in the NW) the Murchisonae zones. Although the Dogger is thin there is considerable facies variation and it clearly represents a considerable period of time. There was immediate post-depositional uplift and subaerial erosion with locally a thin coal resting on the top of the Dogger (Alexander 1989).

The succeeding Ravenscar Group represents a predominantly non-marine succession that ranges in age from Aalenian through the Bajocian and Bathonian. There has been considerable controversy over the past century over the environments of deposition. Initially named the Estuarine 'Series,' this group was then named the Deltaic 'Series'. The beds total some 200 m

Fig. 14.17. Tridactyl footprint in fallen block from the Long Nab Member of the Scalby Formation, Burniston Bay, near Scarborough. P. F. Rawson collection.

in total thickness; the non-marine sediments are interspersed with several marine incursions, two of which are widespread and form the basis of the subdivision of the group into five formations.

The basal Saltwick Formation (up to 57 m thick on the coast) consists of cyclothems of argillaceous sandstone–siltstone–underclay–carbonaceous clay/coal. Particularly to the west of the Whitby Fault and to the east of the Peak Fault channels sometimes penetrate both the Saltwick Formation and the underlying Dogger; their infills are typically sandstones. The succeeding marine Eller Beck Formation (*c.* 8 m) consists of shales and ironstones with a bivalve and gastropod fauna; it has yielded no ammonites. The marine incursion appears to have come from an easterly or SE direction.

The Cloughton Formation marks a return to non-marine facies similar to those of the Saltwick Formation and although it is cut by channels these do not penetrate it deeply. In the southern part of the basin there is a marine intercalation that separates the Cloughton Formation into lower Sycarham (*c.* 50 m) and upper Gristhorpe (*c.* 30 m) members. The intervening marine intercalation is referred to the Lebberston Member; as originally proposed (Hemingway & Knox 1973) it consisted of two parts. The lower Millepore Bed is a calcareous sandstone that thins and becomes more calcareous northwards; but the upper part, the Yons Nab Beds, are hardly marine in places and are now regarded as forming the basal part of the Gristhorpe Member (Knox in Rawson & Wright 1992). The Millepore Bed is named after a bryozoan (now the genus: *Collapora*) and is up to 9 m thick on the coast. To the SW, in the Howardian Hills, the Lebberston Member is represented by the Whitwell Oolite (8 m), consisting of calcareous sandstones and oolites; this seems to correlate well with the Cave Oolite to the south of the Market Weighton Swell. The Gristhorpe Member is interpreted as a fluvio-deltaic succession prograding from marginal marine sands (Yons Nab Beds) through crevasse-splay sheet sands into lacustrine clays that have yielded a prolific flora (Rawson & Wright 1992).

Above the Cloughton Formation, the succeeding Scarborough Formation (3–30 m) is more thoroughly marine and has yielded rare ammonites indicating subzones of the Humphriesianum Zone. The fauna is dominated by bivalves and the belemnite *Megateuthis*. It is best developed on the coast at Hundale Point where seven members are recognized (Gowland & Riding 1991); some of these are recognizable in inland sections, such as the Bogmire Gill Member, but otherwise an inland succession is distinguishable and has its own nomenclature (e.g. Crinoid Grit and Brandsby Roadstone).

Above the Scarborough Formation is the Scalby Formation, a thick (>60 m) fluviatile succession divided into two members: a lower Moor Grit Member, up to 8 m of cross-bedded sandstones which in some parts pass into more massive orthoquartzites that were formerly quarried as building stones; the Moor Grit Member appears to have been deposited within broad channels of low sinuosity (Leeder & Nami 1979). The upper Long Nab Member consists of clays and silts with planar-bedded sheet sandstones, with a well-documented meander belt in the lower part. Some 5 m above this is the horizon that has yielded saurian footprints (Whyte & Romano 1993; Romano & Whyte 2003) and, above that, a sandstone has yielded trace fossils that may be of marine origin and this, together with the reported presence of marine palynomorphs in the overbank clays (Fisher & Hancock 1985) suggests intermittently saline swamp conditions rather than the entirely alluvial model of Nami & Leeder (1978). The higher horizons in the Long Nab Member are, however, undoubtedly alluvial in origin. Overall, the evidence suggests small prograding deltas that ultimately coalesced into a large coastal alluvial plain, rather than a large river system feeding a large delta (Rawson & Wright 1992, 1995).

The age of the Scalby Formation is contentious. The traditional view is that the formation represents the whole time interval from the Late Bajocian, through the entire Bathonian and into the earliest Callovian. Leeder & Nami (1979) argued, however, that the basal Moor Grit Member rested disconformably on the Scarborough Formation and that its relatively high depositional rates suggested that the whole Scalby Formation was deposited in less than 1 Ma and thus that most of Bathonian time was unrepresented in the Yorkshire succession. Inland, the upper part of the Long Nab Member has yielded Bathonian dinoflagellate cysts (Riding & Wright 1989) and these authors noted that the top of the member was lithified and eroded prior to the deposition of the Cornbrash; they thus concluded that the Scalby Formation ranged from Late Bajocian to Early Bathonian. In contradiction of this, however, is the work of Hogg (1993) who concluded on palynological evidence that the whole of the formation was of Late Bathonian age. Cox & Sumbler (*in* Cope in press) concluded that 'the Scalby Formation...probably represents a long period of considerably interrupted sedimentation spanning the latest Bajocian and much of the Bathonian'.

Over the Market Weighton swell both Lower and Upper Cornbrash are absent and in the Cleveland Basin the Lower Cornbrash is still absent (unless represented within the Scalby Formation). The Upper Cornbrash here consists of limestones that reach a maximum of 3.6 m at Newtondale (Wright 1978); they are overlain by up to 3 m of shales that are assigned to the Cayton Clay Formation, whose ammonite fauna is that of the basal part of the Kellaways Formation outside of the Cleveland Basin.

West Sole Group

In the Southern North Sea Basin the strata between the Aalenian and Lower Callovian are grouped together within the West Sole Group. In the north and east of the basin the succession is dominated by fluvio-deltaic sandstones and mudstones but these pass, in the south and west, into shallow-marine carbonates and estuarine clastics (Lott & Knox 1994). These authors subdivided the Group into four formations. The basal Wroot Formation is dominantly mudstones with some siltstones and is equivalent to the Dogger and Saltwick formations of the Cleveland Basin; it is succeeded by the sands and silts of the Strangways Formation, equivalent of the Cloughton and Scarborough formations onshore. The overlying Hudleston Formation is again dominantly paralic mudstones that pass laterally into the Scalby Formation. The youngest member of the succession is the Leckenby Formation, 15 m of shallow-marine sandstones, equivalent of the Redcliff Rock Member of the Osgodby Formation (see below) that is readily identifiable in well logs since it is a prominent geophysical marker throughout the area.

The Kellaways Formation

Named after the village of Kellaways (Roman Callovium), Wiltshire, by William Smith the stratigraphy of this thin formation has been revised by Page (1989) who showed that the ammonite faunas allowed the facies relationships to be understood. The lower part of the formation, in particular, is generally poorly exposed and much detail has been obtained from temporary exposures and the occasional borehole, such as that at Tytherton, Wiltshire (Cave & Cox 1975). A succession of silty or sandy pyritic clays up to a maximum of some 37 m thick, the Kellaways Clay Member, with a diverse benthic fauna, is developed between the Dorset coast and Gloucestershire; from North Dorset northwards to the Yorkshire coast the basal 3 m or so of the Kellaways Clay can be distinguished from the remainder of the Kellaways Formation by its phosphatic nodules and a low-diversity benthic fauna. From Wiltshire northwards the Kellaways Sand Member of sands and silts with occasional clay bands or sometimes, as in the type area, calcareous sandstones (= Kellaways Rock *auctt.*) at first succeeds the Kellaways Clay Member, but then becomes its lateral equivalent, and, from north Gloucester northwards, rests directly upon the phosphatic lower part of the Kellaways Clay. South of the Humber, a local Cave Rock Member of sands, locally with calcareous cement and a rich fauna belongs entirely to the Enodatum Subzone; it is therefore younger than the 'Kellaways Rock' farther south. The Kellaways Formation is absent over the Market Weighton swell and on the Yorkshire coast is represented by the massive sandstones of the Redcliff Rock Member of the Osgodby Formation (see below). These sandstones rest on the Cayton Clay Formation, which is equivalent to the lowest phosphatic part of the Kellaways Clay Member.

The Oxford Clay Formation

One of the thicker clay formations of the English Jurassic, the Oxford Clay crops out from the Dorset coast northwards to the Yorkshire coast. In the past much of the Oxford Clay outcrop was dotted with brick-pits, but with the increased mechanization of brick-making these have become restricted to large-scale pits in the East Midlands where brick-making still continues. The advantages of the Oxford Clay for brick making is that its lowest member has a high Total Organic Content (TOC) and a high percentage of fine silt and clay-sized particles of quartz (22.4–23.7%, Norry *et al.* 1994). The latter are important in bricks, as firing produces a glass – the primary source of bricks' strength, whilst the former is economically important, as the organic content burns with an exothermic reaction, sufficient to reach the 1050 °C firing temperature; fuel is only necessary to maintain this temperature for the 24 h spent in the kiln, reducing costs considerably. Pyrite in the clay is oxidized to haematite during firing, producing red bricks, but liberating sulphur dioxide in the process.

The Oxford Clay has yielded some remarkably preserved fossils (Martill & Hudson 1991). Although some are collected from the brick pits today, and also from temporary exposures of the clays revealed by extraction of overlying Pleistocene gravels in the South Midlands, some of the most remarkably preserved came from borrow-pits excavated along the Great Western Railway main line from London to Bristol in the 19th century, around Christian Malford in Wiltshire, and have found their way into collections all around the world. The rich molluscan fauna is often crushed, but otherwise perfectly preserved in largely original aragonite, and includes belemnites and teuthids with their arms and hooks preserved. Other important fossils include the fish and marine reptilian faunas that were collected by the Leeds brothers before the mechanization of clay extraction from the brickpits of the East Midlands (see Martill & Hudson 1991).

The lithostratigraphy of the Oxford Clay has been revised by Cox *et al.* (1993); it had been traditionally divided into three parts: Lower, Middle and Upper Oxford Clay. The Oxford Clay has been regarded as a formation in this revision (as it is a distinct mappable unit) with three subdivisions treated as members (Peterborough, Stewartby and Weymouth members, respectively). The Peterborough and Stewartby members are of Callovian age, whilst the Weymouth Member is of early Oxfordian age.

The Peterborough Member (= Lower Oxford Clay) is up to about 65 m of predominantly fissile organic-rich mudstones, but there is a wide variation in lithology and silty mudstones and muddy sandstones also occur (MacQuaker 1994). The lithologies are laterally persistent and beds can be traced from one brickpit to another (Hudson & Martill 1994). MacQuaker (1994) linked changes in sediment type to changes in the length of sediment transport paths, surface productivity and sediment accumulation rates; factors influencing these were accommodation space and relative sea level. He concluded that the Peterborough Member accumulated in a shallow epeiric sea with high productivity and oxic bottom waters.

The fauna consists of crushed aragonitic ammonites and bivalves often in shell beds; the basal part is commonly silty and has *Gryphaea*-rich horizons. The preservation of aragonite suggests a maximum burial depth of less than 500 m for the Oxford Clay of the East Midlands (in the Late Cretaceous) as first suggested by Hudson (1978). This figure is seriously at variance with the 1450 m obtained from Apatite Fission Track Analysis for the area (Green *et al.* 2001), but is supported by stratigraphical considerations (Cope 1994).

The Peterborough Member is the best known part of the Oxford Clay, largely because its organic-rich beds have been extensively worked for bricks, and it is at present the principal brick clay in Britain; its total organic carbon (TOC) ranges up to 16.6% (Kenig *et al.* 1994). Large clay pits have major advantages for many types of geological investigation because the continuous working ensures a continuous supply of pristine material, although mechanical methods of working the clay render recovery of larger vertebrates, for example, more difficult. Important palaeontological work has been based on collections from the Peterborough Member. The first of these was Brinkmann's (1929) work on kosmoceratid ammonites, collected centimetre by centimetre, through some 13 m of the Peterborough Member; he obtained some 3000 mature ammonites that allowed recognition of some 27 separate ammonite biohorizons. Statistical work on this fauna showed systematic changes in end-diameters and parallel breaks in the graphs indicated not only likely non-sequences, but were one of the features that led to the now near universal recognition of sexual dimorphism in Jurassic ammonites following Callomon (1955, 1963). Duff's (1975) work on the palaeoecology of the Peterborough Member led to recognition of ten separate biofacies dominated by infaunal deposit feeders and high-level epifaunal suspension feeders. His study of benthic communities was the first such work on the Oxford Clay. This work was succeeded by a paper on the trophic structure of the biota (Martill *et al.* 1994) that showed that phytoplankton formed the basis of the food-chain, with vertebrate predators at the top.

Oxygen isotope derived palaeotemperatures indicate that the Peterborough Member seawater was thermally stratified, with *Kosmoceras* aragonite suggesting a near-surface temperature of 16–28°C, whilst calcite from belemnites and aragonitic benthic bivalves gave ranges of 12–19°C, suggesting that the former were nektobenthic (Anderson *et al.* 1994). Vertebrate phosphate results were similar to those of *Kosmoceras* suggesting that the fish and reptiles also lived in near-surface waters, but presumed deeper-water fish yielded palaeotemperature results in the same range.

The overlying Stewartby Member (= Middle Oxford Clay) is approximately 50 m of greyish sometimes silty, calcareous mudstones. Its ammonite fauna is preserved largely as pyritic internal moulds; these are associated with *Gryphaea* and silty mudstones with abundant *Bositra buchii*. At the top of the Stewartby Member in the South Midlands is developed a thin limestone band, the Lamberti Limestone. This has yielded a prolific ammonite fauna. As well as indicating the southward spread of cardioceratid ammonites from the Arctic, this bed also yields the last of the kosmoceratids and a rich mix of Tethyan ammonites, including perisphinctids, oppeliids, aspidoceratids, reineckiids and pachyceratids. With these are phylloceratids and lytoceratids (Callomon 1968).

The succeeding Weymouth Member (= Upper Oxford Clay) is approximately 70 m of mainly pale grey calcareous mudstones with thin dark carbonaceous beds. There may be thin calcareous siltstones but otherwise the silt content is low. There are often abundant large *Gryphaea* and the ammonite fauna is often pyritized or may be preserved in sideritic mudstone nodules.

From Lincolnshire northwards to the Market Weighton swell the Peterborough and Stewartby members become much reduced in thickness and are altogether lacking in places, whilst the Weymouth Member is always present, although locally thinned. North of the Vale of Pickering and onto the Yorkshire coast, the Peterborough and Stewartby members correlate with the sandstones with berthierine oolites of the Langdale and Hackness Rock members of the Osgodby Formation. The Oxford Clay of this Yorkshire area is thus equivalent to only the Weymouth Member of areas farther south. The Weymouth Member here is also thinner than its counterpart to the south as arenaceous sedimentation of the Corallian Group began earlier in the Cleveland Basin. Its clays are silty and in Cayton Bay display light/dark alternations that may be related to Milankovitch cyclicity (Rawson & Wright 1995). Thickness differences in the Weymouth Member over the Cleveland Basin (Wright 1983) were ascribed by Coe (1995) as probably due to erosion at a disconformable junction between the Weymouth Member and the succeeding Lower Calcareous Grit, but Rawson & Wright (1995) describe the contact between the two as transitional.

Oxford Clay is well developed in the Bristol Channel Basin, with the formation totalling some 88 m thick in the western part of the basin, south of Pembrokeshire. It thickens to the east, however, to some 370 m with facies similar to those of southern England in the eastern Bristol Channel Basin. In borehole 107/21-1 in the St George's Channel Basin the Oxford Clay is 88 m thick and the succession starts with calcareous mudstones passing up into fine sandstones that are in turn succeeded by silts and mudstones (Tappin *et al.* 1994).

Upper Jurassic Series

For convenience the Weymouth Member of the Oxford Clay Formation is dealt with above, as part of the description of that formation. Everywhere it is, however, of Oxfordian age and thus belongs to the Upper Jurassic Series. Chronostratigraphical subdivisions of the Upper Jurassic are shown in Figure 14.18.

STAGE	STANDARD ZONES
PORTLANDIAN*	*Titanites* ANGUIFORMIS
	Galbanites KERBERUS
	Galbanites OKUSENSIS
	Glaucolithites GLAUCOLITHUS
	Progalbanites ALBANI
BOLONIAN*	*Virgatopavlovia* FITTONI
	Pavlovia ROTUNDA
	Pavlovia PALLASIOIDES
	Pectinatites PECTINATUS
	Pectinatites HUDLESTONI
	Pectinatites WHEATLEYENSIS
	Pectinatites SCITULUS
	Pectinatites ELEGANS
KIMMERIDGIAN	*Aulacostephanus* AUTISSIODORENSIS
	Aulacostephanus EUDOXUS
	Aulacostephanoides MUTABILIS
	Rasenia CYMODOCE
	Pictonia BAYLEI
OXFORDIAN	*Amoeboceras* ROSENKRANTZI
	Amoeboceras REGULARE
	Amoeboceras SERRATUM
	Amoeboceras GLOSENSE
	Cardioceras TENUISERRATUM
	Cardioceras DENSIPLICATUM
	Cardioceras CORDATUM
	Quenstedtoceras MARIAE

Fig. 14.18. Stages and Standard Ammonite Zonation of the Upper Jurassic Series. Stages marked * are Secondary Standard stages approximating to the Tithonian Standard Stage. Substage boundaries for the Oxfordian Stage are indicated by short lines at the right hand side of the stage column. Note that for the Middle and Upper Oxfordian the Boreal zonal scheme is shown on this figure; correlation with the Northwest European Scheme is shown on Fig. 14.19. The Middle-Upper Oxfordian boundary is fixed by the base of the Cautisnigrae Zone in the latter scheme.

The Corallian Group

Over much of the area north of the Tethys coralline facies became very widespread in the later parts of the Oxfordian Age. England was no exception to this and coralline facies were widespread in the shallows to the west of the London landmass and again in North Yorkshire. In general the coral reefs do not fringe the coastlines, but are found in positive areas that are often at a high angle to the coastlines (Wilson 1968*a*). The Corallian Group shows a variety of facies from muds through to micritic limestone and oolites, to bioclastic limestones, with interbeds of silts and sands, some possibly of lagoonal origin. The presence of coralline facies and cross-bedded oolites, both facies inimical to ammonites, means that these fossils are frequently rare. In addition, faunal provincialism of ammonite faunas became extreme at this time and separate zonal schemes are used for Boreal, Northwest European and Submediterranean areas. England lies across the boundaries of the Boreal and North-west European provinces and whilst Boreal ammonites are rare in Dorset, they are the norm in the Oxfordian rocks of Skye. The correlation of these different zonal schemes is shown in Figure 14.19.

Everywhere in England the varied Oxfordian succession of carbonate and siliciclastic sediments is interrupted by disconformities that may be correlated biostratigraphically; these have been recorded by several previous workers (e.g. Talbot 1973; Sun 1989). These disconformities define packets of strata that record sea-level change, with the period of maximum regression at the depositional breaks. Thus these breaks form sequence boundaries in terms of a sequence stratigraphical model. Coe (1992) recognized six such disconformities in her study of these rocks, but Newell (2000) recognized only four.

The classical division of the Corallian Group in the older literature was into Lower Calcareous Grit, Coral Rag and Upper Calcareous Grit. The so-called grits are usually sandy limestones with clays, and the Coral Rag usually coralliferous limestone with an abundant fauna of bivalves and the hemicidaroid *Plegiocidaris florigemma* (Phillips).

	Subzone	Zone
Upper Oxfordian	Evoluta	Pseudocordata
	Pseudocordata	
	Pseudoyo	
	Caledonica	
	Variocostatus	Cautisnigrae
	Cautisnigrae	
Middle Oxfordian	Nunningtonense	Pumilus
	Parandieri	
	Antecedens	Plicatilis
	Vertebrale	

Zone	Subzone	
Rosenkrantzi		Upper Oxfordian
Regulare		
Serratum	Serratum	
	Koldeweyense	
Glosense	Glosense	
	Ilovaiskii	
Tenuiserratum	Blakei	Middle Oxfordian
	Tenuiserratum	
Densiplicatum	Maltonense	
	Vertebrale	
Cordatum	Cordatum	Lower Oxfordian
	Costicardia	
	Bukowskii	
Mariae	Praecordatum	
	Scarburgense	

Fig. 14.19. Standard Oxfordian Ammonite Zonal schemes for the Boreal and Sub-Boreal Realm (right-handside) and for the Northwest European Realm (left-handside). For simplicity the Boreal scheme has been adopted across England, but, in general, ammonites of the Boreal Realm become less common (and correlation with this standard less easy) farther south. Modified after Wright & Cox (2001) with permission of JNCC.

On the Dorset coast there are magnificent sections through the whole of the Corallian Group, here some 60 m thick, but corals are virtually absent. The best sections are to the north-east of Weymouth, between Bowleaze Cove and Ringstead Bay. The succession has received a great deal of attention since the classic account of Arkell (1947*a*) and was divided into six formations by Wright (1986) in a major revision; these have since been rationalized to four formations (Wright & Cox 2001), which largely correspond with the four sequences of the succession proposed by Newell (2000; but see also Wright 2001) (see Fig. 14.20). The basal Nothe Grit Member of the Redcliff Formation succeeds the Weymouth Member of the Oxford Clay with a sharp erosive junction. It is a shallow-water bioturbated sandstone, up to 8 m thick, with spherical concretions and with common *Gryphaea dilatata*, interpreted as a lowstand systems tract by Newell (2000). The junction with the overlying Preston Grit Member (or Trigonia hudlestoni Bed) is sharp and coincides with an ammonite zonal boundary and marks a major transgressive flooding surface following erosion. The upper part of the Member displays spectacular trace fossils such as *Rhizocorallium*, *Diplocraterion* and *Thalassinoides*. Body fossils include the eponymous *Myophorella hudlestoni* (Lycett) and a wide variety of other burrowing bivalves together with *Gryphaea*. There is a varied ammonite fauna that includes both cardioceratids and perisphinctids. The succeeding Nothe Clay Member (12 m) succeeds transitionally; it has oolites developed in its lower part in the more westerly coastal sections that have yielded many ammonites, and represents quieter offshore waters of a highstand. The transition to the overlying 4 m of sands and large doggers of the Bencliff Grit Member appears gradual, although it marks a major change of facies from offshore mud to high-energy shallow sands of a shallower regime induced by sea-level fall. The sands yield only trace fossils together with ostracods and foraminiferans that could tolerate fresh water together with abundant spores and pollen, although there is a varied dinoflagellate flora. Sedimentary structures within the Bencliff Grit include swaley cross-bedding, indicating storm activity (Sun 1989; Allen & Underhill 1999); combination of all this evidence suggests that the beds may be a lagoonal deposit, although Coe (1995) suggested an estuarine environment.

The base of the Osmington Oolite Formation is sharp, with burrows into the top of the Bencliff Grit. There is major erosion at this interval that is most plausibly explained by a sea-level fall producing emersion (Hallam 1999). The Osmington Oolite is initiated by transgressive sandy bioclastic limestones overlain by clays representing deeper water; these pass up into a highstand unit including often spectacularly cross-bedded white oolites. The formation totals about 20 m in thickness and embraces three ammonite subzones; it yields a fauna of bivalves, gastropods and echinoids with rarer ammonites. Wright (2001) suggested that top member of the Osmington Oolite (the Nodular Rubble Member) represented the highstand sytems tract of a further sequence not recognized by Newell (2000).

The major non-sequence between the Osmington Oolite and the overlying Clavellata Formation is traceable across the Wessex Basin and marks a transgression at the base of the third sequence of Newell (2000) with a change to predominantly siliciclastic sedimentation. The lower part of the formation, the Clavellata Member, is up to about 7 m thick, consisting of transgressive sandy and ooidal limestones with marl partings and includes shell banks largely composed of disarticulated valves of the bivalve *Myophorella clavellata* together with *Trigonia* (*Trigonia*), numerous other bivalves and occasional ammonites including large perisphinctids. The upper part of the formation, comprising the Sandsfoot Clay Member, is most thickly developed in the more poorly exposed section on the southern side of Weymouth where it may be up to 12 m thick. It has a transitional boundary with the top of the bioclastic limestones beneath and consists of clays and mudstones with an abundant bivalve fauna, especially in the lower part, and is intensely bioturbated. It is interpreted as the highstand systems tract of the third sequence, but both Talbot (1973) and Brookfield (1978) suggested that it represented a low-salinity bay or lagoonal deposit because of the low diversity of the bivalve fauna, but bivalve diversity is believed to have declined more rapidly with depth during the Mesozoic than at present (Hallam 1975, 1997).

The Corallian succession terminates with the Sandsfoot Formation comprising two members. A sharp transgressive contact at the base of the Sandsfoot Grit Member marks a major depositional break during which the Sandsfoot Clay underwent erosion, particularly at the Black Head section, NE of Weymouth, where it has been reduced to some 3.5 m. The Sandsfoot Grit Member comprises up to 7.5 m of yellow and

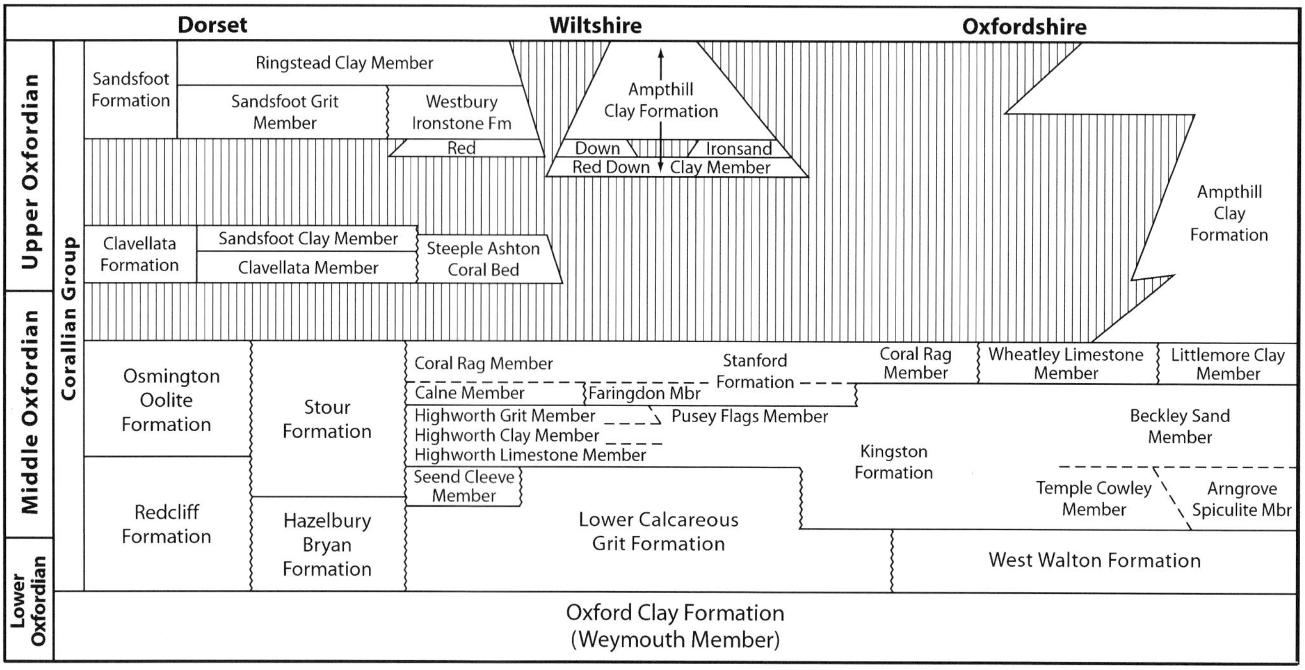

Fig. 14.20. Development of the Corallian Group across north Dorset. Note how thicknesses vary across the syndepositional faults. Based on Bristow *et al.* (1995, fig. 38) by permission of the British Geological Survey.

brown bioturbated sands with limonite ooids and yielding ammonites including cardioceratids of the Rosenkrantzi Zone. This member is best developed at its type locality in southern Weymouth and is thinner and poorly exposed NE of Weymouth. The succession is intensely bioturbated throughout and its upper unit is an ooidal sand with phosphate, representing a relatively condensed phase. This was interpreted as a beach barrier bar complex by Brookfield (1973). The Ringstead Clay Member (up to 3.5 m) consists of mudstones deposited under highstand conditions; in the upper part these have abundant discoidal orange-weathering sideritic concretions. At the top is the Osmington Mills Ironstone Member with the Ringstead Coral Bed – up to 0.3 m of limestone with bivalves and fragmented corals, particularly species of *Thamnasteria* and *Protoseris*, frequently bored. Wright (2001) pointed out that Newell (2000) had erroneously included beds of the basal Kimmeridgian in his fourth cycle, overlooking the fact that a widespread and well-developed basal Kimmeridgian erosion surface is traceable as far north as Lincolnshire (Wignall 1990*a*).

Northwards, the outcrop of the Corallian Group disappears below Cretaceous rocks, but re-appears in North Dorset where the succession can be correlated well with that to the south although there are differences. The stratigraphy of this region has been revised by Wright (1981) and Bristow *et al.* (1995, 1999). The total thickness of the group varies considerably, in part due to syndepositional faulting; 74.4 m of Corallian Group was recorded in the Wyke Brewery Borehole (Bristow *et al.* 1999), but to the east, in the Shaftesbury district the maximum thickness is 90 m (Bristow *et al.* 1995). The basal Hazelbury Bryan Formation (formerly Lower Calcareous Grit) comprises clays, clayey sands and sands, the latter locally calcareous, arranged in up to four coarsening-upwards cycles that overlie the top of the Oxford Clay with sideritic nodules as at Weymouth. The number of sands and the thickness of the formation are greatest on the hanging walls of major faults, indicating syndepositional movement (Fig. 14.20). Above this is the Stour Formation that consists of alternations of calcareous and sandy clays with oolites and pisolites. Six members are recognized, but nowhere are all six developed. The limestones, such as the Cucklington Oolite, up to 11 m of oolite, pisolite and calcareous sandstone, or the Todber Freestone (<5 m of cross-bedded oosparite) have been quarried locally. Above the Stour Formation there is a non-sequence and in some sections a thin (<3 m) Coral Rag occurs beneath the sharp base of the Clavellata Formation, here shelly oolites and *Rhaxella* biomicrites. The Clavellata Member is up to 7 m thick and the formation locally contains the Eccliffe Member, up to 9 m of fine-grained oolite. It is succeeded by the locally developed Sandsfoot Clay Member that reaches a maximum thickness of 3 m. The Sandsfoot Grit Member of the Sandsfoot Formation consists of sands and clays with local ferruginous oolites; thicknesses vary from 2–12 m. The Ringstead Clay Member is well developed in some localities and has sideritic concretions in its upper part as to the south; it marks the local top of the Corallian Group.

Across Wiltshire the Lower Calcareous Grit Formation is usually succeeded by the Kingston Formation, up to 10 m of clays and limestones, but these are locally absent. They are overlain disconformably by the Coral Rag Member of the Stanford Formation, usually developed as a coralliferous micrite, but locally, the Calne Member, up to 7.5 m of oolites and pisolite directly underlies the Coral Rag. The Calne and the Coral Rag members both correlate with the upper part of the Osmington Oolite Formation and appear to represent a transgressive event that lies between sequences 2 and 3 of Newell (2000). The Steeple Ashton Coral Bed, resting disconformably on the Stanford Formation, has yielded a uniquely rich fauna of corals for the British Oxfordian, with 46 species belonging to 23 genera (Negus & Beauvais 1979); it passes southwards into a limestone with *Myophorella clavellata*, showing its equivalence to the Clavellata Formation (Fig. 14.21). It is succeeded by the Sandsfoot Formation, that includes in its upper part, in the area around Westbury, the Westbury Ironstone Member that yields up to 45% iron. It was worked commercially in the 19th century, during which time an estimated 2 million tonnes (mt) of ore were extracted, and is correlated, on the basis of its ammonite fauna, with the Sandsfoot Grit Member to the south. It is succeeded by the Ringstead Clay Member.

Across southern Oxfordshire pre-Kimmeridgian erosion has removed virtually all of the Upper Oxfordian, but small patches of Upper Oxfordian mudstones are locally preserved beneath

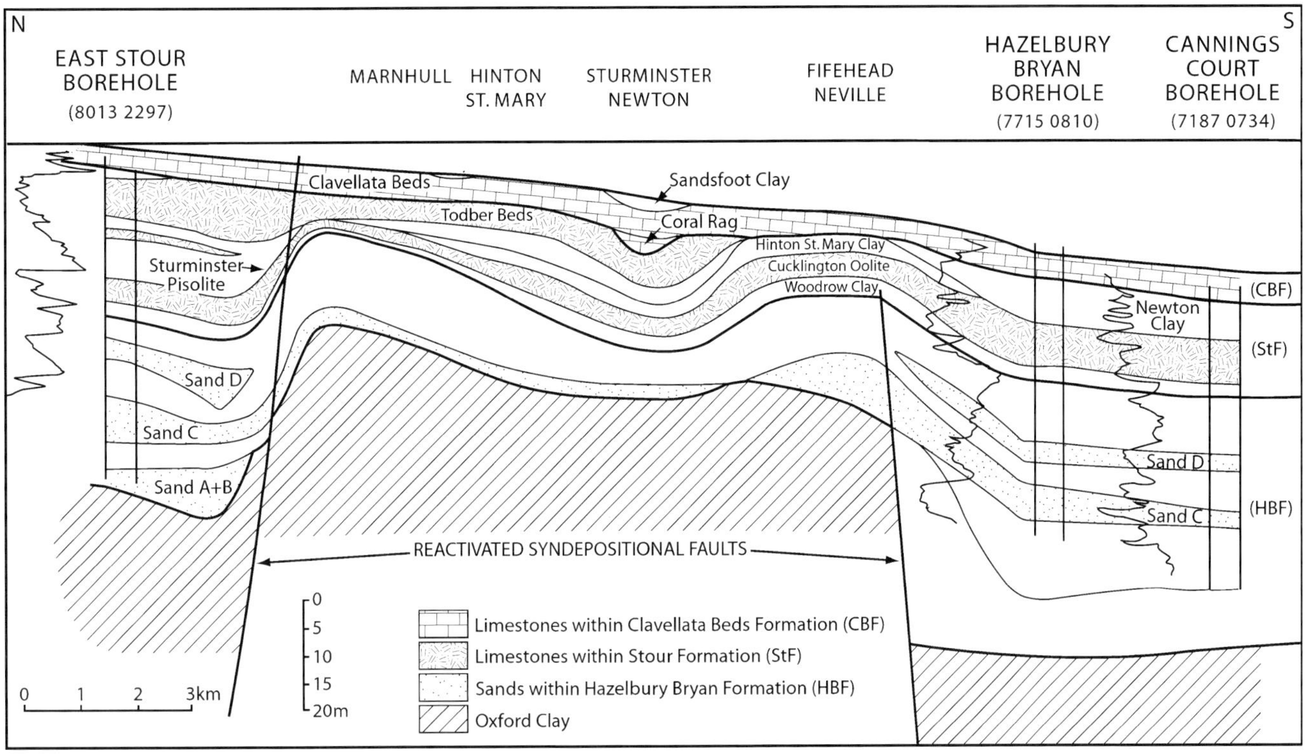

Fig. 14.21. Correlation of the Corallian Group along the outcrop from north Dorset to Oxfordshire. Modified after Wright & Cox (2001) with permission of JNCC.

the Kimmeridge Clay, as at Abingdon and Cumnor. Over this area the Lower Calcareous Grit Formation is well developed, locally reaching 24 m. It is overlain by the Kingston Formation, up to 15 m of oolites, clays and sands; in turn, these are overlain by the Stanford Formation comprising at first the Coral Rag Member.

Proceeding northeastwards along the outcrop major lateral facies changes occur throughout the Corallian Group, and a complex of other facies appears. These changes coincide with the Wheatley Fault Zone, developed above a basement fault, as first surmised by Arkell (1947*b*). Above a thin development of the West Walton Formation is the Temple Cowley Member of the Kingston Formation, consisting of fine-grained silty sand and clayey silts, passing laterally into the Arngrove Spiculite Member composed largely of tetractinellid sponge spicules and quartz sand. This resistant rock forms an escarpment in the area north of Oxford to Beckley and Brill. The overlying Beckley Sand Member comprises up to 7 m of sands and sandstones with limestone at the top. This is succeeded by the Wheatley Limestone Member of the Stanford Formation that consists of platformal limestones with coral colonies (reaching 20 m in thickness in the Cumnor Borehole). This is the lateral equivalent of the Coral Rag and passes laterally northeastwards into the Littlemore Clay Member, comprising clays of possible lagoonal origin (Horton *et al.* 1995). This change coincides with the abrupt disappearance of the Corallian facies northeastwards (Fig. 14.20).

Over the East Midlands the Oxford Clay is succeeded with minor disconformity by the West Walton Formation, a succession of alternations of calcareous mudstones and silty mudstones with doggers of cementstone or siltstone; it is predominantly of Mid Oxfordian age. The beds contain both plant and shell debris and yield cardioceratid ammonites and *Gryphaea dilatata*. The formation continues northeastward to occupy a position between the Oxford and Ampthill Clay formations over the Fenland district (Gallois & Cox 1977). On the basis of borehole cores these authors recognized 16 beds within the formation that could be traced across the Fenland and recorded that its thickness varied from about 10 to 15 m.

The succeeding Ampthill Clay is entirely of Late Oxfordian age. Its lowest part consists of dark silty clays and mudstones, the middle of smooth-textured clays, whilst its upper portion consists of paler calcareous silty clays and mudstones containing several horizons of phosphatic nodules indicating minor non-sequences. The formation was divided by Gallois & Cox (1977) into 42 laterally persistent units; it is up to about 18 m thick in Bedfordshire (Shephard-Thorn *et al.* 1994), but up to 85 m thick in Lincolnshire. The Geological Survey found that the West Walton Formation had to be mapped with the Ampthill Clay, as there was no field evidence for the boundary.

Thus, over eastern England, an essentially clay succession persists from the early Callovian to the late Bolonian and this, together with the shallow dip, explains the lack of any marked topography over the Fenland country. As much of this area is drift-covered, parts of the succession have never been seen and it is only through recent boreholes that the stratigraphy has been elucidated in detail. Gaunt *et al.* (1992) grouped these clay formations from the Kellaways Beds to the Kimmeridge Clay together as the Ancholme Group, and this name has been widely applied in areas of eastern England where the Corallian Group is absent. In the area around Thame, Horton *et al.* (1995) recognized the Ancholme Group as an entirely clay facies in the east of the area, but in the west recognized the Corallian Group as dividing the Ancholme Group into two parts.

The one exception in eastern England to this clay-dominated succession is around Upware, Cambridgeshire, where for some 6 km along the strike there is a return to coralline facies. Here, the West Walton Formation begins with 2 m of dark clay, succeeded by some 4 m of grey marls with limestones and lenses of coralliferous limestone. In turn this is overlain by 9 m of limestones with a prolific fauna including corals, ammonites (including both Boreal and Tethyan elements), bivalves, gastropods, echinoids and sponges (Wright *et al.* 2000). The corals here are not the typical domical forms of Corallian reefs, but thin platey colonies characteristic of deeper water; thus the Upware reef developed in an area of deeper water with little clastic input. North of Upware the clay facies returns and is maintained as far north as the Market Weighton swell.

North of the Market Weighton swell Corallian facies return above the Oxford Clay. Much of the rock is highly resistant and forms the Howardian Hills, to the south of the Vale of Pickering, and the Hambleton and Tabular Hills to the north. The Yorkshire Corallian has been studied extensively (see Wright 1972, 1996*a*, *b* and references therein). The main lithological subdivisions and their inter-relationships are shown on Figure 14.22.

The Lower Calcareous Grit Formation is the same age as the top of the Oxford Clay farther south; it consists for the most part of fine- to medium-grained sandstones and is developed most thickly in the Howardian and Tabular Hills, where it is up to 42 m thick. It is succeeded by the Coralline Oolite Formation that consists of thick cross-bedded oolites with coralliferous limestones of latest early and mid Oxfordian ages; these ooidal facies are largely devoid of ammonites.

At the base of the Coralline Oolite, the Hambleton Oolite (up to 30 m) is developed from the Vale of Pickering northwards, but south of the vale, the Birdsall Calcareous Grit (27 m max.) largely takes its place; this is a fine sand derived from uplift of the Market Weighton Swell. The local thin basal development of the Hambleton Oolite beneath this is referred to as the Lower Leaf of the Hambleton Oolite as distinct from the Upper Leaf above the Birdsall Calcareous Grit. The sharp base of the Lower Leaf correlates with the transgression at the base of the Nothe Grit (Coe 1995). On the coast at Filey the junction appears to be a minor non-sequence, but in the Hambleton Hills the Costicardia Subzone is missing beneath the Hambleton Oolite, showing a significant time-break. The Birdsall Calcareous Grit is a sandstone with ooids; some parts show cross-stratification whilst others are strongly bioturbated. It attains a maximum thickness of 27 m in the Howardian Hills.

The overlying Malton Oolite is up to 40 m thick and occurs across much of the region, as does the overlying Coral Rag (up to 9 m), but 5 km south of Malton and southwards erosion has removed both the Coral Rag and a large part of the Malton Oolite. The lower part of the Upper Oxfordian is represented by cementstones in the Howardian Hills, but in the Vale of Pickering and the Hambleton and Tabular Hills, the Upper Calcareous Grit is in sandstone facies. Across the region the upper part of the Upper Oxfordian is represented by Ampthill Clay, that in some parts of the region, at least, probably persists to the base of the Kimmeridgian.

South of the London landmass was the now separate Weald Basin. By plotting of isopachs for the Weald and Wessex areas, Wilson (1968*a*) was able to demonstrate that the Portsdown Swell effectively separated the two basins. The Warlingham Borehole, near Croydon, showed Upper Oxfordian mudstones resting disconformably on Middle Oxfordian coralliferous micrites (Worssam & Ivimey-Cook 1971). The whole of the Lower Oxfordian is in Oxford Clay facies. East of there, in the Kent coalfield, coralliferous limestones are again developed in the Middle Oxfordian; these are overlain by clays and limestones in the Upper Oxfordian with the development of an ironstone reminiscent of the Westbury Ironstone of Wiltshire (Lamplugh & Kitchin 1911; Lamplugh *et al.* 1923). In contrast to these successions, the Ashdown boreholes of Sussex showed an almost entirely argillaceous succession throughout the Oxfordian (Bristow & Bazley 1972); see also Figure 14.23.

In the Southern North Sea Basin the Oxfordian sediments are part of the Humber Group, a major group including the Oxford Clay, Corallian Beds and Kimmeridge Clay and some 200–300 m thick. In the northern part of the basin Corallian facies of ooidal grainstones, sandstones and clays are developed as in the Cleveland Basin. Southwards these are gradually transitional into lower energy mudstones and siltstones of the Seeley and Woodward formations, the offshore equivalents of the West Walton and Ampthill Clay formations respectively (Lott & Knox 1994); there is no offshore equivalent of the

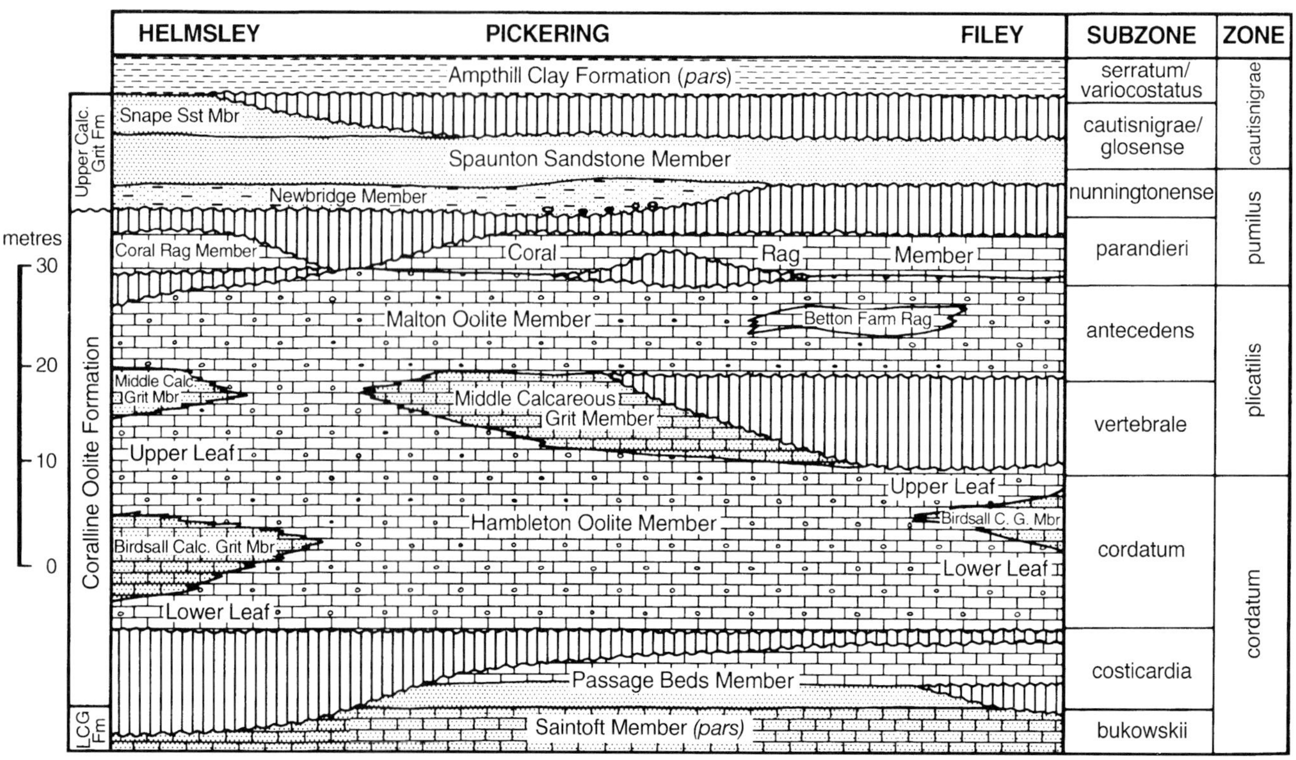

Fig. 14.22. W–E section through the Corallian Group on the north side of the Vale of Pickering, Yorkshire. Length of section is 50 km. LCG = Lower Calcareous Grit Formation. Modified after Rawson & Wright (1995).

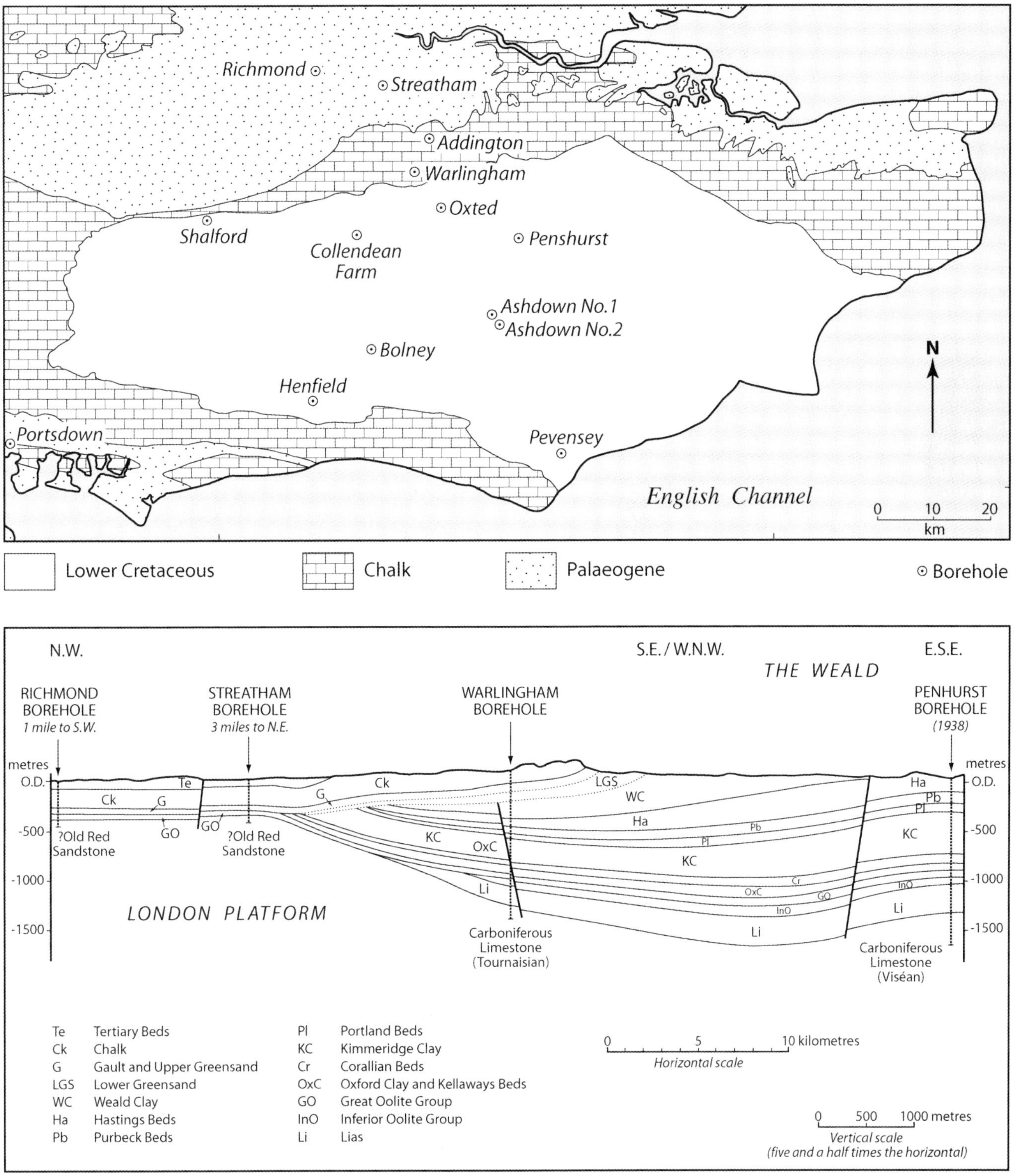

Fig. 14.23. Section showing thinning of the Jurassic rocks towards the London landmass and their development in the northern Weald. Map shows boreholes in the region referred to in the text. Based on Worssam & Ivimey-Cook (1971, figs 2 & 3) by permission of the British Geological Survey.

Market Weighton swell to provide a distinct facies boundary. The Yorkshire facies re-appears locally in the West Sole Gas Field where borehole 48/6-5 proved some 27 m of limestones including ooidal and bioclastic grainstones (Cameron *et al.* 1992).

In the Bristol Channel the Oxfordian Stage is represented by a thick succession of fine-grained sandstones with glauconite and lignite, totalling 520 m (Evans & Thompson 1979); there is a restricted molluscan fauna and the beds are interpreted as a non-marine to transitional marine environment, possibly estuarine or deltaic. This thick facies suggests that there was uplift on the flanks of the Bristol Channel Basin and the material may be derived from an emergent land area over Mid-or South Wales. Cope (1984) suggested that this material could have been largely derived from reworking of earlier Jurassic sediments deposited over Wales. In the St George's Channel Basin, soft grey calcareous mudstones of the West Walton Formation overlie Oxford Clay (Weymouth Member). In borehole 107/21-1 the Weymouth Member was 45 m thick and overlain by 66 m of the West Walton Formation, in turn this was overlain by some 214 m of calcareous mudstones and well-cemented sandstones of fluvio-deltaic aspect that were assigned to the Ampthill Clay (Tappin *et al.* 1994).

The Kimmeridge Clay Formation

The Kimmeridge Clay is one of the thickest and most widely distributed clay formations of the British Jurassic. It takes its name from the village of Kimmeridge in the Isle of Purbeck, Dorset. Its first use in a geological context was by Webster (1816) who referred to Kimmeridge Strata; the first detailed descriptions were by Fitton (1836). When d'Orbigny proposed his Jurassic stages (1850), he stated that the Kimmeridgian Stage was the equivalent of Fitton's (1836) Weymouth Beds and Kimmeridge Clay. However, currently the Kimmeridgian Stage corresponds closely to only the Lower Kimmeridge Clay and the Upper Kimmeridge Clay is assigned to the Bolonian (Secondary Standard) Stage (see discussion above, p. 331). When Salfeld (1913) defined the zones of the Kimmeridgian Stage he proposed the Baylei Zone as its basal zone; as stages are defined by their bases, the base of the Baylei Zone thus defines the base of the Kimmeridgian Stage. This fact is important in view of the ongoing debate about the selection of a GSSP for the base of the Kimmeridgian Stage; historical precedent dictates that it should be selected where the base of the Baylei Zone can be defined.

The Kimmeridge Clay has been subject to particular scrutiny since it was identified as the principal oil source rock in the northern North Sea, although over onshore England and Wales and in the offshore areas considered in this account it is nowhere mature. It is still highly bituminous in part and at Kimmeridge itself the richest oil-shale horizon, the Blackstone or Kimmeridge Coal, was worked until the early years of the 20th century, both as a local 'coal' and as a source of fuel for various industrial enterprises including an alum works at Kimmeridge, where the local shale was used as a raw material. The oil-shale was also exported to France as a source of gas for lighting the streets of Paris. The distribution of organic material in the Kimmeridge Clay has been the subject of important studies by a French team (Herbin *et al.* 1991, 1993, 1995; Herbin & Geyssant 1993).

Lithologically the Kimmeridge Clay Formation can be divided informally into four major lithological units, with gradational boundaries. The base of the formation upwards to about one third of the way up the Eudoxus Zone consists predominantly of bioturbated shelly clays; intercalated with these are various horizons of oil and bituminous shale. The upper two thirds of the Eudoxus Zone up to the upper part of the Pectinatus Zone consists of cyclical arrangements of clay, bituminous clay, oil-shale and coccolith limestone; the last-named lithology is often lacking from the cycle. From the upper part of the Pectinatus Zone up to the Rotunda Nodule Bed the clays become more shelly and have fewer oil-shales; Wignall & Ruffell (1990) correlated this latter change with the increasing aridity that was reflected in loss of kaolinite from the clay mineral assemblages. Above this level up to the base of the overlying Portland Sand, the beds become progressively more silty and there are few bituminous horizons. Throughout the formation are stone bands, many of which are named, that are frequently dolomitized; most of them appear to be secondary carbonate replacements for shales or mudstones and their origin is attributed to diagenesis in the methanogenic zone, under deep burial (Irwin *et al.* 1977; Scotchman 1989).

The lithologies of the Kimmeridge Clay (apart from the secondary stone bands) are laterally persistent across considerable distances and Gallois & Cox (1976), Cox & Gallois (1979, 1981) and Gallois (2000) divided the whole of the Kimmeridge Clay into 63 packages of laterally persistent strata varying in thickness from less than one metre up to more than 12 m. To these they assigned bed numbers (nos KC1 at the base to KC63 at the top).

The origin of the sedimentary cycles was investigated by Weedon *et al.* (2004) who ascribed the larger cycles, of the range 1.87–4.05 m wavelengths, to orbital obliquity and the smaller cycles, of about half this wavelength, to precession. These authors recognized 95 longer wavelength cycles for the Kimmeridgian and 103 for the Bolonian, from which they extrapolated a duration of 3.6 Ma for the Kimmeridgian and 3.9 Ma for the Bolonian.

There have been several sequence stratigraphical studies of the formation; that of Wignall (1991) recognized nine sequences separated by hiatuses, but these did not always correspond to the eustatic sea-level chart of Haq *et al.* (1989). Wignall suggested that the sequence boundaries up to the top of the Mutabilis Zone (his 1991 K5 boundary) could be traced from basin margin to centre by a marked upward replacement of mudstone by siltstone, with all the associated sequences being mudstone-dominated. Sequence boundaries K6–K10 (Elegans–Fittoni zones), in contrast, were well developed and often combined within hiatuses at the basin margins, but were only represented in the basins by minor facies changes. He concluded (Wignall 1991, p. 438) that the basins became more distinct as a result of a more pronounced topography from the Eudoxus Zone upwards. A later examination of the sequences within the Kimmeridge Clay by Taylor *et al.* (2001) recognized 11 depositional sequences and 12 sequence boundaries. Their conclusions on changes to basin geometry, leading to unconformities no longer being of basin-wide extent, was basically the same as that of Wignall's (1991). Taylor *et al.* (2001) concluded that as the Kimmeridge Clay was deposited in a partly enclosed sea, its basin dynamics were not typical of those of a continental shelf margin, but were more like those of Recent meromictic lakes. Williams *et al.* (2001) found that quartz silt-rich horizons in the basinal successions correlated with shallow-water sands on the basin margins; thus silty levels in the basins correlated with lowstands. Rather than a single deep-water phase lasting from the Eudoxus to the end of the Pectinatus Zone, recognized by earlier workers, Williams *et al.* (2001) identified a period of significant shallowing that lasted through most of the Scitulus and Wheatleyensis chrons.

During the later part of the Bolonian Stage increasing tectonic activity became evident through the more widespread occurrence of marginal facies and non-sequences in the vicinity of the principal landmasses, particularly the London landmass. The principal hiatus occurs above the level of the Eastlecottensis Subzone of the Pectinatus Zone and represents a period of regional sea-level fall correlated with the Late Cimmerian movements (Rawson & Riley 1982). In the deeper basinal areas, such as in the Wessex Basin, the initial effect is a change from a predominantly infaunal to an epifaunal benthos; this appears to correlate with a reduction in sedimentation rates to produce a firmer sea-bed (Wignall 1990*b*; Wignall & Ruffell 1990). Then, from the Rotunda Zone upwards, silt becomes an increasing component of the succession; as shallowing continued, so deposition became more restricted, so that the top Bolonian Fittoni Zone has so far only been identified in south Dorset.

The fauna of the Kimmeridge Clay is dominated by bivalves. These were the subject of a study by Wignall (1990*b*) who recognized a variety of both dysaerobic and aerobic fossil associations, in many of which one or two species predominated. In addition to the bivalves, ammonites are common fossils and, apart from at a few nodular horizons, are mainly crushed. Vertebrate remains are also common and include a wide variety of fish and reptiles, whilst some of the most spectacular remains are those of teuthids, preserved in exquisite detail and currently undergoing description.

There has been a great deal of discussion of the depositional conditions of the Kimmeridge Clay; most are based on a model of a stratified water column first proposed by Tyson *et al.* (1979), with vertical changes in lithology produced by migration of the oxic–anoxic boundary through both the sediment and the sediment-water interface (Irwin 1979; Tyson *et al.* 1979; Myers & Wignall 1987). Periodically storms were able to disrupt this as rip-up clasts, sea-bed erosion and graded mud horizons demonstrate that deepest parts of the basin were in the range of storm wave-base during the largest storms (Wignall 1989); such disturbances introduced oxygen to the sea-bed. Shallower areas were more susceptible to storm disruption of the water column, so that they had a substantially smaller proportion of organic-rich shales within them (Wignall 1989). Although the Tyson *et al.* (1979) model has gained a general acceptance, debate involving ideas of water depth and of water column productivity continues. Such discussions centre on the abundance of oil-shales and bituminous shales, necessarily implying bottom-water anoxia, but such anoxia could equally occur in shallow or deep water as long as the bottom-waters were stagnant. Thus, Hallam (1975) and Aigner (1980) proposed very shallow depths of less than 10 m up to a few tens of metres; at the other end of the scale, Tyson *et al.* (1979) suggested water depths of several hundred metres. Palaeogeographical changes from the preceding Oxfordian Stage through to the Bolonian Stage appear small (see Cope & Rawson 1992, maps J9 and J10); some transgression of marginal land areas is seen, but certainly not on the scale that would be necessary if the water depth much exceeded, say, 100 m. The implication is that changes in water depths during this time interval may have been minimal, suggesting that the underlying cause of the periodic anoxia in the Kimmeridge Clay may simply have been restricted water-column circulation.

Sedimentation of the Kimmeridge Clay is closely related to subsidence; in the tectonically stable areas the mudstones are generally carbonate-rich, implying that the sedimentation rate equalled the subsidence rate, whereas in the basins the organic-rich horizons developed when subsidence rates exceeded depositional rates, producing water deepening. Wignall (1989) showed that exactly the same proportion of organic-rich sediments accumulated in two coeval successions, some 15 km apart, although one was only a quarter of the thickness of the other, suggesting that both must have accumulated in water of about the same depth, and that the differences in thickness were governed principally by differential subsidence. In swell areas, where subsidence did not keep pace with sedimentation, sediment types were local in lateral extent and emergence produced localized erosion. These differences are also reflected in the diagenetic history of the sediments; dolostone bands, formed in the methanogenic zone, are restricted to the basins, whilst swell areas are characterized by carbonate concretions formed in the sulphate-reduction zone (Wignall 1989).

The richness of the organic levels in the Kimmeridge Clay was related to high productivity by Gallois (1976), but Weedon *et al.* (2004) calculated the organic productivity as averaging some 220 g m^{-2} $year^{-1}$, which is less than the average productivity of modern continental shelves. Thus they concluded that the high organic content of the Kimmeridge Clay (average 3.8% TOC) is not due to high productivity, but to low dilution by siliciclastic minerals probably accompanied by enhanced organic preservation in dysaerobic or anaerobic bottom waters (Weedon *et al.* 2004). Tyson (2004) came to similar conclusions.

A recent development of the Tyson *et al.* (1979) model of the Kimmeridge Clay cyclicity by Sælen *et al.* (2000) suggested that when stratification of the water column occurred, nutrients became enriched in the bottom-waters, whereas they were relatively depleted in the upper waters. This depletion favoured the proliferation of a planktonic microflora dominated by dinoflagellates, cyanobacteria and bacterioplankton, all of which can regulate their position in the water column. Under influence of strong winds, mixing of the waters allowed nutrients to escape into the surface waters, producing widespread dinoflagellate blooms and, if the conditions persisted, coccolith blooms. This large-scale organic productivity, in turn, led to loss of oxygen through decay, causing anoxia to rise from the sediments up to shallow depths, perhaps as little as 10 m. These latter conditions favoured one group of phytoplankton, the Chlorobiaceae, causing them to bloom. The Chlorobiaceae are phototrophic oxidizing bacteria and their biomass can exceed that of other phytoplankton; they appear to be intimately involved in the production of oil-shales, as they develop during photic zone anoxia; they are traceable from aryl isoprenoid biomarkers present in the Kimmeridge oil-shales. Further wind-induced mixing of the water column eliminated the Chlorobiaceaea and the abundance of nutrients produced a normal food-chain, followed by algal blooms and then back to conditions favouring the Chlorobiaceae.

The Weald–Wessex Basin

At the type-section at Kimmeridge, in the Isle of Purbeck (Plate 27), the Kimmeridge Clay reaches a thickness of 500 m, but the base is not exposed here; it has, however been reached in boreholes including the Metherhills Borehole (Gallois 2000). The section stretches for 8 km and is disposed on either side of an anticline; the beds to the east are more thickly developed than those to the west (Arkell 1947*a*). The lowest beds exposed are within the Kimmeridgian Eudoxus Zone. Blake's Bed 42, marking the base of the Bolonian (Secondary Standard) Stage and the base of the Upper Kimmeridge Clay also marks the base of the succession of ammonite zones based on the ammonite genus *Pectinatites* (Cope 1967). The thick Blackstone oil-shale (see above) occurs at the top of the Wheatleyensis Zone. Above this level coccolith limestone bands are more common and the distinctive White Stone Band, at the base of the Pectinatus Zone, is a notable example of that lithology. Above the Pectinatus Zone are the Pallasioides and Rotunda zones (based on species of *Pavlovia*) and then the Fittoni Zone that has not been recorded outside of Dorset (Cope 1978). At this point in the succession, the beds become dominated by silts and the base of the Portlandian (Secondary Standard) Stage is marked by the prominent Massive Bed that yields *Progalbanites* indicative of the Albani Zone (Arkell 1935; Wimbledon & Cope 1978).

Westwards from the type section the Kimmeridge Clay thins rapidly so that in the area northeast of Weymouth it is about half of the Isle of Purbeck thickness, the thinning being due to local differential subsidence (Wignall 1989). The basal beds are well exposed around Ringstead Bay and these sections are being considered as candidate basal boundary stratotype sections (GSSP) for the base of the Kimmeridgian Stage. At the base is the Inconstans Bed, a bed yielding the asymmetric rhynchonellid *Torquirhynchia inconstans*, resting on the Osmington Mills Ironstone, and then a succession of clays and shales, that includes several more marker beds. The upper beds of the Kimmeridge Clay are exposed on the slopes below the Cretaceous unconformity at the eastern end of Ringstead Bay on Burning Cliff, so-called after a spontaneous combustion of the oil-shales in 1826. The principal stone bands may be identified here, but differences from Purbeck include two phosphatic nodule beds developed in the *Pavlovia* zones, the higher of these probably correlates with the Rotunda Nodule Bed of Purbeck, whilst the lower may correlate with Gallois' (2000) Chapman's Pool Pebble Bed, a horizon with phosphatized body-chambers of *Pavlovia*.

South of Weymouth the basal beds of the Kimmeridge Clay are exposed intermittently around Wyke Regis and Chesil Beach, the shingle bar that connects Portland to the mainland, here rests on the Kimmeridge Clay. Estimates of thickness, based on outcrop width and dip at the base and top of the formation, suggest that the formation may approach 400 m. On the northern part of Portland there are sporadic outcrops of the

Kimmeridge Clay, but most of the outcrop is hidden under landslips. Occasionally slips and temporary exposures reveal local details, such as that around the Pallasioides–Rotunda zonal boundary reported by Wignall (1990*b*) that showed thicknesses slightly less than those of Purbeck. An oil-shale horizon, possibly the Blackstone, has been worked intermittently as a fuel around Castletown since at least the 16th century and on the northwest coast landslipped sections have provided specimens of *Virgatopavlovia* indicating the Fittoni Zone (Cope 1978).

Northwest of Weymouth the thickness of the Kimmeridge Clay is close to 350 m; it is poorly exposed, but the Blackstone is well developed and was sporadically worked locally. The principal interest in the succession here lies in the facies developed locally, against the Abbotsbury and Litton Cheney faults. There the clay facies passes laterally into ooidal ironstones that provide evidence of syndepositional movement of these faults (Jenkyns & Senior 1991). The ironstones were worked around Abbotsbury in the late 19th century, but although containing up to 40% of iron, were too siliceous to be worked profitably. The ironstone has a distinctive fauna of brachiopods including species of *Ornithella, Rugitela* and *Rhynchonella*, together with a varied bivalve and gastropod fauna and ammonites proving the Cymodoce Zone. In the small outcrop around Litton Cheney the rocks are less ferruginous and the molluscan fauna more diverse (Cope 1971).

Northwards the Kimmeridge Clay disappears beneath the Cretaceous rocks and re-appears in Blackmore Vale of north Dorset. It is ill exposed here, but the basal clays were worked in a brickpit at Gillingham until the latter half of the last century. Exposures in the area have proved many of the marker beds in the Lower Kimmeridge Clay, including the basal Inconstans Bed (Bristow *et al.* 1995). The highest horizons exposed in the area, to the south of Shaftesbury, are in the Pectinatus Zone of the Upper Kimmeridge Clay.

In the Vale of Wardour, higher horizons crop out, but are very poorly exposed. A borehole at Chicksgrove, proved silts below the presumed lowest Portlandian; their correlation with the Fittoni Zone by Gallois & Etches (2001) is suppositional, as no ammonites were recovered from that part of the section.

The next exposures northwards are around Westbury, Wiltshire, where Lower Kimmeridge clay is used for mixing with Chalk for cement manufacture; the succession proved in a borehole showed some 56 m of Lower Kimmeridge Clay up to a position within the Eudoxus Zone (Birkelund *et al.* 1983), although the pit exposes mainly the underlying Mutabilis Zone. Farther north, around Swindon, there were several brickpits working up to the early years of the 20th century. These have all disappeared, but excavations for landfill sites at present provide temporary exposures in the area. The succession was clarified by a borehole (Gallois & Cox 1994) that showed the formation to be just over 100 m thick in the area; clearly subsidence was much less than in Dorset and the area to the west of the London landmass seems to have acted in the manner of a swell. Here the facies becomes silty and then sandy above the Wheatleyensis Zone (the Lower Cemetery Beds) denoting the western limit of a sandy input from the London landmass; these sands probably belong to the upper Wheatleyensis Zone and Hudlestoni Zone. The Upper Cemetery Beds succeed disconformably with a minor break; they are assigned to the lower Pectinatus Zone (Eastlecottensis Subzone). The Lower Lydite Bed may represent the remainder of the Pectinatus Zone and the lower part of the Pallasioides Zone, as the succeeding Swindon Clay belongs to the Pallasioides Zone. The Portland Group follows disconformably above.

Northeastwards from here the Kimmeridge Clay emerges intermittently from beneath a Cretaceous cover, but is poorly known. Former brickpits in the area around Oxford showed that the formation has thinned to about 25 m, indicating proximity to the London landmass, and breaks are more significant, with sandy horizons in the Pectinatus Zone and nothing representative of the Hudlestoni Zone. The Oxford area yielded ammonites from the Wheatleyensis and Pectinatus zones figured by Buckman (1922–1930) and Neaverson (1925); the latter's unsatisfactory zonal scheme was based largely on these incomplete successions.

In the Aylesbury region Oates (1991) introduced the name Hartwell Silt as a member of the Kimmeridge Clay; it was the source of Neaverson's holotype of *Pavlovia pallasioides* and more recent temporary sections have yielded a pyritized Pallasioides Zone fauna overlying a thin and condensed succession containing another lower silty member (Oates 1991) with the Upper Kimmeridge Clay totalling 31 m in thickness. In the nearby Hartwell Borehole Cox *et al.* (1994) recorded approximately 48 m of Kimmeridge Clay Formation resting disconformably on the Ampthill Clay.

In the SE of England Kimmeridge Clay has been met in several boreholes and south of the London landmass it has a wide subcrop (Fig. 14.23). In the Warlingham Borehole close to the southern side of the landmass 217 m of the formation was proved, but a fault cut out most of the Mutabilis Zone and the remainder of the formation beneath. There were two lydite beds, suggesting similarities with the Oxford area, but an absence of any sandy facies. Zones from the Pectinatus Zone down to the Mutabilis Zone were all proven (Callomon & Cope 1971). In Kent the formation was considerably thicker at Ashdown with 560.5 m proved in the Ashdown No 1 Borehole (Bristow & Bazley 1972) and 443.5 m in the Penshurst Borehole (Dines *et al.* 1969). In the Bolney Borehole, some 12 km south of Crawley, 521 m of Kimmeridge Clay was recorded (Gallois & Worssam 1993).

Eastern England and the Southern North Sea

The Kimmeridge Clay extends across the Fenland of East Anglia, but is poorly exposed. A single pit in the Lower Kimmeridge Clay in the Ely area is one of the few exposures in the area; it has long been used for building embankments along local drainage systems (Wright & Cox 2001). It has yielded a varied fauna of marine and terrestrial reptiles (Benton & Spencer 1995) and, although present exposures are in the Mutabilis and Eudoxus Zones, the presence of *Torquirhynchia inconstans* in old collections suggests that the basal clays were formerly exposed. A series of boreholes in the Wash area proved the formation was up to 110 m thick and was terminated upwards at the lower part of the Pectinatus Zone (Cope 1974*a*). On the north side of the Wash the Kimmeridge Clay is again poorly known, with the overlying Spilsby Sandstone containing a pebble-bed at its base that yields derived pavloviid ammonites from the top of the Bolonian and the lowest Portlandian stages. Lower Kimmeridge Clay is again exposed at the cement-works at South Ferriby just south of the Humber, where some 7.5 m of mudstones, of the Baylei and Cymodoce zones, rest on Ampthill Clay; they are overlain directly by Cretaceous Carstone (Gaunt *et al.* 1992).

On the Yorkshire coast poor exposures are seen intermittently cropping out from beneath the Cretaceous Speeton Clay, some 12 km north of Flamborough Head. Ammonites indicative of zones from the Autissiodorensis Zone to the lower part of the Pectinatus Zone have been recorded (Cope 1974*b*), but the succession may not be complete. The thickness of 385 m given for the Kimmeridge Clay in the Fordon No. 1 Borehole (Falcon & Kent 1960) is excessive, in that the thickness is now known to include the underlying Ampthill Clay; nevertheless, the thickness here is greater than any from north Dorset northwards. Former brickpits in the Vale of Pickering exposed various horizons in the Lower and Upper Kimmeridge Clay up to the level of the lower part of the Pectinatus Zone (Cope 1974*b*). Wignall's (1993) doubts of the validity of the subzonal scheme for the Bolonian are unfounded and are based on his misidentification of largely homeomorphic species of *Pectinatites*. Herbin *et al.* (1991) recorded an indeterminate *Pavlovia* that was taken to be of Pallasioides Zone age from the

Flixton 87 Borehole on the north side of the Vale of Pickering; this was correlated with beds from the nearby Ebberston boreholes which did not yield ammonites. The Pallasioides Zone is the youngest Bolonian yet recorded in the region. Geyssant (1994) recorded successive invasions of southern ammonite genera into Yorkshire, from the Mutabilis Zone up to the Elegans Zone; she concluded that these forms were likely to have migrated to the area through the Hesse Strait at the eastern end of the London–Brabant Massif and *via* its northeastern flank into the North Sea (Geyssant 1994, fig. 1).

In the Southern North Sea Basin the Kimmeridge Clay consists of its usual mix of clay lithologies, including kerogen-rich shales with calcareous mudstones and cementstones. It is likely that the higher parts of the formation only occur within the Wash (Cope 1974*a*).

Bristol Channel and environs

In the Bristol Channel syncline, Kimmeridge Clay is known from the core of the fold in the area south of Swansea. Typical offshore Kimmeridge Clay facies is developed, together with oil-shales, suggesting that the Welsh Massif was under water at this time and did not materially contribute sediment into the area; the thickness totals some 330 m (Evans & Thompson 1979). Northwestwards, however, records of the Kimmeridge Clay from the St George's Channel Basin show that the formation there includes land-derived sands and lignite; the beds are often orange or red in colour and anhydrite occurs locally (Tappin *et al.* 1994). There are no oil-shales here and the fauna is predominantly freshwater (Millson 1987*a*). In borehole 106/24-1, off the north Pembrokeshire coast, a thickness of 958 m was recorded (Penn 1987). Although local sediment input from the west Wales coast is possible, the terrestrial material seems more likely to have been derived from eastern Ireland, as the Bristol Channel shows normal Kimmeridge Clay facies, suggesting Wales was not emergent. Further support for an Irish source comes from the North Celtic Sea Basin, where the Kimmeridge Clay has anhydrite throughout with red beds, suggesting a sabkha facies fringe to the Irish Massif (Millson 1987*b*). The whole of the Celtic Sea area was enclosed by land on three sides and normal marine claystone facies appear confined to the axial region of the North Celtic Sea Basin (Cope & Rawson 1992, map J10). Over all these areas the Kimmeridge Clay, as onshore, is always insufficiently deeply buried to be mature.

The Portland Group

Continuing Late Cimmerian movement had effectively brought marine sedimentation to a halt over northern and western areas of Britain by the end of Bolonian times. In areas from which Jurassic sediments have since been removed by erosion, such as the Eastern Irish Sea Basin, figures obtained from apatite fission track analysis suggest a Late Cimmerian inversion may have been less than 300 m (Hardman *et al.* 1993) or may have approached 1000 m (Bushell 1986). Cope (1998*a*) suggested that these figures could be reconciled. This inversion left much of England and Wales as land (maps J11a–d *in* Cope *et al.* 1992).

Above the Kimmeridge Clay in the Weald–Wessex Basin is a succession of silts and carbonates that accumulated in a geographically restricted basin (maps J11a–d *in* Cope *et al.* 1992). The Portland Group is only known from this basin, the Bristol Channel Basin and the St George's Channel Basin; much of the rest of England and Wales was undergoing erosion linked with the initial phase of the Late Cimmerian movements. As noted above, the top of the Kimmeridge Clay became silty and the succeeding Portland Sand Formation is a continuation of this silty deposition with carbonates, chiefly dolomites, in the upper part. The Portland Sand Formation reaches a maximum of some 39 m in thickness in the Isle of Purbeck (Plate 27), the upper dolomitic member being named as the Black Sandstones by Arkell (1935, 1947*a*). The faunas are generally poorly preserved and are generally found as moulds. The succession is thinner on Portland Isle, although with similar dolomites (the West Weare Sandstones) in the upper part. Coe (1996) suggested that the St Alban's Head Marls of Purbeck were missing in a non-sequence on Portland; but undescribed ammonites from the Cast Beds, beneath the West Weare Sandstones, appear to be the same as those of the St Alban's Head Marls, but the succession is certainly much thinner on Portland. Farther north, in the Vale of Wardour and at Swindon the Portland Sand equivalents are limestones, frequently glauconitic, and with a better-preserved molluscan fauna, although the beds are here considerably thinner.

The Portland Stone is well known as one of the prime dimension stones used for public buildings throughout Britain, but the commercially valuable stone makes up only a small part of the succession. The Portland Stone can be conveniently divided into two formations in Dorset, a lower Portland Cherty Formation and an upper Portland Freestone Formation; both formations are dominated by limestones. The Portland Cherty Formation has abundant nodules or in some cases beds of black chert. These cherts have been derived from siliceous sponge spicules and the silica has been redistributed during diagenesis. Frequently burrow systems provide the foci for the cherts, which often preserved the burrows' shape, but in other cases nodules are large or tabular. The overlying Portland Freestone Formation is predominantly ooidal grainstones, although there are still occasional bands of chert. In Dorset there is a considerable difference in thickness between the successions of the Isle of Portland in the west and the Isle of Purbeck in the east. What is most noticeable is that the Portland Cherty Formation is significantly thicker in Purbeck, whilst on Portland (and also at Ringstead) there is a clay, the Portland Clay, developed beneath the Portland Stone. The Portland Clay is well-developed throughout the island (contra Arkell 1947*a*, p. 121). At the base of the Portland Cherty Formation on Portland is the Basal Shell Bed, a composite three-bedded condensed unit with a distinctive and varied ammonite fauna, including species of *Titanites* (Fig. 14.24) and *Galbanites*; the bed also yields a varied fauna of bivalves and gastropods (Cox 1925). The same ammonite fauna is found in Arkell's (1935) beds J and J′ in Purbeck, which are also remarkably similar in lithology to the Basal Shell Bed of Portland and which lie in the middle of the Portland Cherty Formation of Purbeck (Cope & Wimbledon 1973). Thus the lower part of the latter formation of the Isle of Purbeck seems to equate in large measure with the Portland Clay of Portland, but Coe (1996) argued that there was a break in the Portland Isle succession below the Basal Shell Bed and that the Portland Clay was equivalent to the top of the Portland Sand of Purbeck.

The Portland Freestone Member is best developed as a dimension stone on the Isle of Portland where it has been quarried for more than a millennium. Its use outside of Dorset was promoted by the rebuilding of London after the Great Fire of 1666 and was specified by Christopher Wren for St Paul's cathedral, although earlier (1619–1622) Inigo Jones had used it for the Banqueting Hall in Whitehall. Since then its use has become extensive throughout Britain where it forms the principal dimension stone.

A significant part of the Isle of Portland is scarred by old quarries, whilst quarrying still continues apace at further sites. The principal commercial horizons are the Whit Bed, a fine white oolite some 3 m thick, and the Base Bed, up to 2.5 m thick. To reach these horizons, the overburden of beds of the Purbeck Group that crop out over most of Portland must be removed. This material, which is of no commercial use, is stacked up in the old quarries, and was formerly dumped over the cliffs, forming the 'weares' around both east and west sides of the island. The stone is worked along the master joints that transect the island; these are oriented approximately NNE–SSW and are frequently over 0.5 m wide (Coombe 1981). A secondary set is developed at approximate right angles to the master joints

Fig. 14.24. *Titanites anguiformis* Wimbledon. Holotype, diameter 665 mm. Roach, Portland Stone, Portland, Dorset.

and this allows the stone to be quarried in cuboids up to about 3 m maximum dimension. Like any good dimension stone the Portland Stone is relatively free of fossils, the occasional specimen of the ammonite *Titanites* being the principal exception.

Between the Whit Bed and the Purbeck Group is the Portland Roach, best developed at the northern part of the island. This is a cream oolite, up to a metre thick, largely composed of moulds of bivalves (especially trigoniids) and the nerineid gastropod *Aptyxiella portlandica*, known by quarrymen as the Portland Screw. The shell material of these molluscs was leached out by fresh water from the former overlying Purbeck lagoon. Beneath the Whit Bed are up to 2.5 m of micritic and ooidal limestone with chert, of no commercial value, that in turn overlie the Base Bed.

Unlike dimension stones like the Bath Stone, Portland Stone has little cement; its strength lies within the ooliths that have been extensively bored by microendoliths whose microborings formed the focus of subsequent crystalline calcite growth (Palmer 2004).

The Portland Stone has been extensively quarried too on the Isle of Purbeck, mainly along the coastal cliffs, where galleries follow the principal freestones for some distance underground. Here the freestone is more extensively jointed than on Portland and has never provided dimension stone of the same quality as that of Portland. The succession was well described by Arkell (1935).

In the Vale of Wardour the succession is well established through quarries that have existed in the area for several centuries; the quarries at Chilmark provided the principal stone used in construction of Salisbury cathedral. Most of the building stones are glauconitic limestones that are equivalent to the upper part of the Portland Sand farther south. The overlying beds are the Ragstones, which at their base yield fish and reptile bones; these are overlain by a micrite with freshwater gastropods. This horizon suggests that Purbeck Group facies were introduced here much earlier. However, they are succeeded by further micritic limestones that yield a Kerberus Zone and, right at the top, an Anguiformis Zone ammonite fauna. Above this are typical Purbeck Group lithologies.

Further to the NE the Portland Group is known at Swindon, where it was formerly quarried; here the total thickness is approximately 13 m and probably embraces several non-sequences. At the base of the thin Portland Sand Formation is the phosphatic Upper Lydite Bed with derived ammonites from the *Pavlovia* zones of the underlying Kimmeridge Clay and probably also from the basal Portlandian Albani Zone. The Glauconitic Member comprises some 1.5 m of glauconitic sands and sandstone. The Portland Stone Formation comprises three members. The basal Cockly Member is a condensed limestone succession, some 1.5 m thick, with a profuse fauna of the Okusensis and the basal part of the Kerberus Zones. It is succeeded by the Swindon Sand and Stone Member of the Kerberus Zone (8 m), and this in turn by the Town Gardens Member (2 m) of uncertain age, but which contains horizons with freshwater gastropods. The overlying Purbeck Limestone facies, totalling up to about 10 m, is considered to be equivalent in age to the Portland freestones of Dorset.

Through Oxfordshire and into Buckinghamshire the Portland Group appears again from beneath the Lower Greensand and Gault and the outcrop continues northwards as a series of small outliers across the region (Horton *et al.* 1995; Shephard-Thorn *et al.* 1994). Exposures were formerly seen in a series of quarries that yielded many ammonites figured in *Type Ammonites* by Buckman in the 1920s. The succession is rarely more than 12 m thick and often much less; for this reason the Portland Group has sometimes been treated in this region as a formation, with Portland Sand and Stone as members of it (Horton *et al.* 1995). The succession was interpreted by the latter authors as representing two regressive cycles and accurate thicknesses, in an area where there are now few exposures, were obtained from the cores of the BGS Hartwell Borehole (Cox *et al.* 1994). The Portland Sand Formation succeeds the Hartwell Silt Member of the Kimmeridge Clay disconformably, the break encompassing the Rotunda, Fittoni and, usually, the Albani zones. At its base is the Upper Lydite Bed, here consisting of chert and quartz pebbles, together with black phosphatic fragments in a matrix of glauconitic silt or sandy clay; locally, above it there are thin limestone lenses with mouldic preservation of *Aptyxiella portlandica*, reminiscent of the (younger) Portland Isle Roach. Faunas of the Upper Lydite Bed generally indicate the Glaucolithus Zone, but locally the Albani Zone may be present. These form the basal bed of the Glauconitic Member, almost 6 m of glauconitic sands and sandstone (Horton *et al.* 1995). The succeeding Portland Stone Formation begins with the Aylesbury (or Rubbly) Limestone Member, some 2 m of pale grey sandy limestones. The overlying Crendon Sand Member is some 2.3 m of grey sands and silts, weathering yellow-brown, that introduces the second cycle; in the area SW of Haddenham they directly succeed the Glauconitic Member as the Aylesbury Limestone wedges out. The Crendon Sand passes up into the Creamy Limestone Member, which is up to *c.* 3 m thick. The ammonite faunas of the latter belong entirely to the Kerberus zone. As at Swindon, the succeeding basal Purbeck Limestone Formation may well be equivalent to the Dorset Portland freestones.

In the Weald, various boreholes have penetrated the Portland Group; most show a thicker development of Portland Sand with a thinner Portland Stone above, but biostratigraphical control is weak. At Warlingham (Worssam & Ivimey-Cook 1971) the thickness of 125 feet (38 m) includes about 6 m of what could be called Portland Stone; this includes the Kerberus Zone at the top and this is succeeded by Purbeck Beds. Beneath are mudstones, dolomitic sandstones and siltstones of the Portland Sand Formation.

In the Bristol Channel Basin a Portland Group succession was recorded by Evans & Thompson (1979) and there have been several subsequent boreholes into the Group. Most of the 45 m succession comprises calcareous mudstones and siltstones, but there is a limestone at the top. How far these correlate with the Portland Sands and Portland Stone formations is unknown. In borehole 103/18-1 these beds rest directly on Oxford Clay, but it seems clear that this is as a result of a faulted contact and not from overstep of the Corallian and Kimmeridge Clay by the Portland Group as suggested by Kamerling (1979).

Equivalents of the Portland Group also occur in East Anglia. As a result of the East Coast floods of 1953 a major drainage scheme was instituted for the Wash area. During this, and the later construction of the King's Lynn By-Pass road, ammonites were discovered in the lower parts of the Sandringham Sand of Norfolk, showing that the basal part of that formation was of latest Jurassic age. Re-interpretation of the largely coeval Spilsby Sandstone ammonites from Lincolnshire showed that the genus *Paracraspedites* was not related to the later *Subcraspedites* as had been supposed, but was instead a Jurassic pavloviid (Casey 1973), so that now, latest Jurassic ammonites were recognized both north and south of the Wash. Casey's (1973) species *Paracraspedites oppressus* was claimed by him to occur both in East Anglia and at the top of the Portland Stone of the Isle of Purbeck and he proposed an Oppressus Zone to accommodate it. Extensive collections in Dorset have now revealed that *P. oppressus* is in fact the inner whorls (or perhaps the microconch) of the titanitids of the Anguiformis Zone and extends considerably lower than the Shrimp Bed at the top of the Portland Stone of Purbeck which was the source of Casey's holotype and believed by him to be the sole horizon for it in Dorset; it thus does not merit separate zonal status. Nevertheless, this ammonite provides evidence of a marine connection from Purbeck through to East Anglia and invaluable correlation at this level.

It is likely that part of the Spilsby Sandstone Formation in the Southern North Sea Basin is also of latest Jurassic age. However, this part of the succession has not been cored and so the practice has been to take the top of the Humber Group as forming the top of the Jurassic.

The Purbeck Group

Overlying the Portland Group is a succession of predominantly lagoonal carbonates and muds that were clearly deposited under conditions of varying salinity, from hypersaline through to fresh-water, as well as terrestrial horizons. Apart from one or two horizons, they are devoid of marine fossils and the few marine species that are present do not include any biostratigraphically useful forms. As these beds overlie Upper Jurassic marine carbonates and are themselves overlain by non-marine sands and clays (the Wealden Group) there is a long interval of essentially non-marine deposition from Portlandian through to Aptian, and the position of the Jurassic–Cretaceous boundary within this succession has long been debated (Chapter 15). It is now believed to lie low in the Purbeck Group, which is therefore discussed in the following chapter.

The continuing marine succession in East Anglia, albeit condensed, provides a marine record through to the earliest Cretaceous within the Spilsby Sandstone and Sandringham Sand formations. These sands lie with a major erosive break on the Kimmeridge Clay Formation and mark the beginning of a new cycle of sedimentation that continued through most of Early Cretaceous times. They too are discussed in the next chapter.

15 Cretaceous: sea levels peak as the North Atlantic opens

P. F. RAWSON

The Cretaceous was a particularly eventful period of geological time. It saw the continuing break-up of the former supercontinents of Laurasia and Gondwana, accompanied by major volcanic eruptions, experienced arguably the highest sea levels of Phanerozoic time, and endured an extraterrestrial bombardment. These aspects of Cretaceous history are reviewed briefly below, but for a comprehensive account the reader is referred to Skelton's (2003) review of the Cretaceous world.

The Cretaceous climate is often described as an equable, 'greenhouse' one, although this view is being challenged increasingly. Certainly the Cretaceous world was noticeably warmer than today's, with temperatures apparently rising overall through Early Cretaceous times to reach a peak in the Late Cretaceous. At mid to low latitudes highest temperatures were reached during the Cenomanian–Turonian interval, while at high latitudes they persisted to the earliest Campanian (Gale 2000) (see Figs 15.5–15.7 for Cretaceous stages/ages). At low to mid latitudes, Early Cretaceous climate fluctuated between arid and humid phases, but aridity dominated through most of Late Cretaceous times, although there was a humid belt further north in which coniferous forests thrived, at times extending up to at least 85° N (e.g. Spicer & Corfield 1992; Spicer 2000).

There is no firm evidence of polar ice, although the possibility has been suggested (e.g. Kemper 1987; Frakes & Francis 1988), especially for the Early Cretaceous when well-marked, latitudinally orientated climatic belts were developed with considerable N–S temperature gradients (Mutterlose & Kessels 2000). The waxing and waning of ice-sheets is the most likely driving force for the frequent changes in sea level, which may have fluctuated by about 50 m over intervals of 200–300 Ka (Stoll & Schrag 1996). Recent oxygen, carbon and strontium isotope studies also point to the possible existence of Late Jurassic and Early Cretaceous polar ice- or snow-sheets (e.g. Ditchfield 1997; Gröcke *et al.* 2003; Mutterlose *et al.* 2003) especially in Antarctica. Climate-modelling lends support to this, at least for the Late Jurassic (Sellwood *et al.* 2000). Stoll & Schrag (1996) suggested that an Antarctic ice sheet may have persisted through the Late Cretaceous, 'despite overall climatic warmth'. The occurrence of Late Cretaceous very high latitude forests in the northern hemisphere suggest that there was no Arctic ice sheet at that time.

In many parts of the world there was a significant phase of rifting across the Jurassic–Cretaceous boundary, known as the Late Cimmerian movements. This tensional phase was probably related to the early development of the Atlantic Ocean. Following its onset in the central Atlantic in Mid Jurassic times, sea-floor spreading extended further north and south during the Cretaceous until by Late Cretaceous times there were continuous marine connections from North to South polar regions through the developing North and South Atlantic oceans. Probably linked to these events, sea levels fell markedly across the Jurassic–Cretaceous boundary before rising again through much of the Cretaceous to reach a probable Phanerozoic peak during the Late Campanian. It is very difficult to estimate how much higher that was compared with present sea level, and widely fluctuating figures have been suggested (Hancock 1989), but it may have been between 150 and 300 m. At times anoxia developed in some of the main ocean basins, and these oceanic anoxic events (OAEs) sometimes spread onto adjacent shelves so that their effects are seen, for example, in England and Wales.

Volcanicity occurred through most of the Cretaceous, and produced three of the world's most extensive areas of flood basalts. The Paraná–Etendeka traps are continental flood basalts; a huge lava pile that accumulated over parts of Brazil, Paraguay, NE Argentina and Namibia during the Valanginian. The eruptions were associated with initial opening of the South Atlantic, which eventually divided the volcanic province. Subsequently, during the Early Aptian, the Ontong Java basalts were emplaced under water to form a huge oceanic plateau in the Pacific (Shipboard Scientific Party, 2001), apparently over a mantle plume. Finally, the continental flood basalts of the Deccan traps erupted at the end of the Cretaceous, to form a series of lava flows covering an area the size of France and totalling over 2000 m in thickness (Courtillot 1999).

All these events may have influenced the evolution of Cretaceous life, and the end-Cretaceous Deccan vulcanicity has been linked with sea-level fall and climate change to help explain the increasing level of extinction in many fossil groups in very late Cretaceous times (Macleod *et al.* 1999). But extra-terrestrial events may also have played a crucial role in the final extinction. The evidence that at least one bolide impact, at Chicxulub on the Yucatan Peninsula in Mexico, occurred at about the end of the Cretaceous appears incontrovertible, although how much of Earth was affected is more debatable. The link between the impact and the end Cretaceous extinctions also appears convincing, although this too is currently disputed. While most researchers accept such a link, Keller *et al.* (2003*a*, *b*) claim that radiometric dating of the impact-created microtektites (glass spherules) shows that the Chicxulub impact predated the extinctions by some 300 000 years. These authors suggest therefore that the world-wide iridium layer that marks the K/T (Cretaceous– Tertiary) boundary must have been caused by a separate impact. But Smit (2004) has refuted much of Keller *et al.*'s (2003*a*, *b*) geological evidence and concluded that production of the microtektite- and iridium-bearing layers was simultaneous.

Other Cretaceous impact sites of varying ages are now known including two at about the Jurassic–Cretaceous boundary, Morokweng in South Africa (Koeberl *et al.* 1997) and Mjølnir in the Barents Sea (Smelror *et al.* 2001).

The regional picture

Many of these global events are reflected in the Cretaceous history of England and Wales. At the beginning of the Cretaceous the area lay at a latitude some 10° further south than at the present day, and experienced a generally warm climate that fluctuated noticeably through time. A phase of aridity across the Jurassic–Cretaceous boundary was marked at times by the development of evaporites in the Purbeck lagoons of southern England. There followed a shift to a wetter, more humid climate by the Valanginian (Batten 2002), before more arid conditions prevailed during the Barremian and Early Aptian (Ruffell & Rawson 1994; Allen 1998). Humid conditions then returned for a while, but aridity probably prevailed through much of Late Cretaceous times, when most of the area became submerged beneath shallow–moderately deep, warm, clear waters in which biogenic oozes accumulated to form the Chalk.

On the NW European shelf, bottom waters were generally well oxygenated, but several OAEs affected the area, resulting in the deposition of organic-rich black shales. The first occurred during the mid Barremian, when the 'Blätterton' facies spread across the North Sea and North German basins (Rawson & Mutterlose 1983). There was a second in Early Aptian times, marked in the same areas by the Fischschiefer facies. Then the most widespread OAE of all occurred across the Cenomanian/Turonian boundary, marked by deposition of the Plenus Marls sequence within the Chalk Group across most of the onshore and offshore area.

There was considerable tectonic activity in the region through the Early Cretaceous, so that sediments accumulated in several different basins (Fig. 15.1). The Late Cimmerian movements led to renewed rifting in most of the adjacent offshore basins, peaking across the Jurassic–Cretaceous boundary but continuing with diminishing effect until mid Cretaceous times. By the latest Jurassic rifting had started on the continental margin to the west of Britain, and during the Early Cretaceous the main focus of extensional tectonics moved here as rifting gradually died down in the North Sea region. Uplift that accompanied the Atlantic rifting particularly affected the northern basins of the Western Approaches Trough (Evans 1990; McMahon & Turner 1998).

In Late Cretaceous times extensional activity continued to the west of Britain as the North Atlantic continued to widen. Elsewhere, rising sea levels gradually submerged many of the former structural highs and thick chalk sequences accumulated to smooth out the underlying topography, although the newly buried highs continued to influence sediment thickness. Thus for most of our area the Late Cretaceous has often been described as a tectonically quiet phase. However, there is increasing evidence of significant compressional movements – reflecting events much further south, as the African Plate began to converge with the European Plate. It is now clear that three important compressional phases occurred across the Anglo-Paris basin, through the North Sea and into North Germany (Mortimore & Pomerol 1997; Mortimore *et al.* 1998). The earliest occurred in latest Turonian–Coniacian times (Ilsede Phase), the next was Late Santonian–Early Campanian (Wernigerode Phase) and the last was latest Campanian (Peine Phase). All affected chalk sedimentation, mainly through the rejuvenation of Variscan lineaments, to cause lateral changes in thickness and lithology, slumping and local lacunae (e.g. Gale 1980; Mortimore & Pomerol 1997; Evans *et al.* 2003). On a broader scale, basin inversion began in some areas in latest Cretaceous times, including the St George's Channel Basin (Williams *et al.* 2005), the Sole Pit Basin and possibly the Cleveland Basin (Kent 1980).

Salt tectonics also played a role in Cretaceous times. In the southern North Sea salt movement started by the mid Triassic and has continued episodically to produce pillows and walls. Movement during the Cretaceous is indicated by rapid changes in sediment thickness adjacent to some of these structures. The highest Cretaceous beds are sometimes missing over the top of a structure, although that could reflect post-Cretaceous erosion (Cameron *et al.* 1992, p. 80).

It is only over the last 30 years that Cretaceous volcanic activity has been recognized in the UK area. Intrusive and extrusive volcanic rocks are now known from both the Western Approaches Trough (Evans 1990) and the North Sea Basin (see Jeans *et al.* 2000, fig. 1). There are no volcanic rocks in the onshore area, but it has been increasingly recognized that some, if not all, of the bentonitic clays known from so many levels represent the waterlain alteration products of air-born volcanic ash, sometimes still preserving shards of volcanic glass in siliceous nodules (see Jeans *et al.* 1977). Some of the clays occur as thick but localized deposits, such as some of the Aptian fuller's earths. Others form very thin but laterally extensive layers that provide important marker horizons, especially some of the Chalk marl bands (e.g. Wray & Wood 1998; Wray 1999). Goldring (1999) favoured a westerly origin for the Aptian deposits, probably from a source in the Western Approaches Trough. On the other hand, Jeans *et al.* (2000) preferred an easterly, possibly North Sea Basin, origin. Wray & Wood (1998) suggested that the Turonian–Coniacian bentonites may have had a subaerial source in the North Atlantic area.

The palaeogeographical setting

The very late Jurassic eustatic fall in sea level together with the Late Cimmerian movements led to a dramatic shrinkage of the seas across England and Wales, and by mid Volgian (latest Jurassic) times much of the area was emergent (maps J10 and J11 in Bradshaw *et al.* 1992). The palaeogeographical pattern thus established (Fig. 15.2a) continued through Early Cretaceous times until it began to break down during the Aptian. During this time the Anglo-Brabant and adjacent Welsh and Pennine highs divided England and Wales into two separate areas with completely different depositional histories. To the NE lay the North Sea Basin, which remained marine throughout, while to the south and west lay a series of interconnected basins which experienced a long period of non-marine deposition before the seas began to return in very early Aptian times. The 'Bedfordshire Straits' provided an intermittent link between the two.

Landmasses marginal to the northern area were mainly low-lying, and there is little evidence of a major river system. In contrast, to the south rivers transported sediment to the non-marine basins off adjacent highlands, including the Welsh, Cornubian and Armorican highlands and the intermittently uplifted 'London uplands' along the southern margin of the Anglo-Brabant Landmass (Rawson 1992*b*, maps K1 and K2).

The pattern began to change early in the Aptian, when a marine transgression flooded southern England (Fig. 15.2b). The initial marine influx may have come from the north, via the 'Bedfordshire Straits', but there was soon a marine spillover from the Paris Basin too. Subsequent sea-level rises from then until late in the Cretaceous resulted in northeastward transgression from the continental margin to the SW of Britain into the Celtic Sea basins and Western Approaches Trough to link with the seas further east over the Wessex Basin. This was

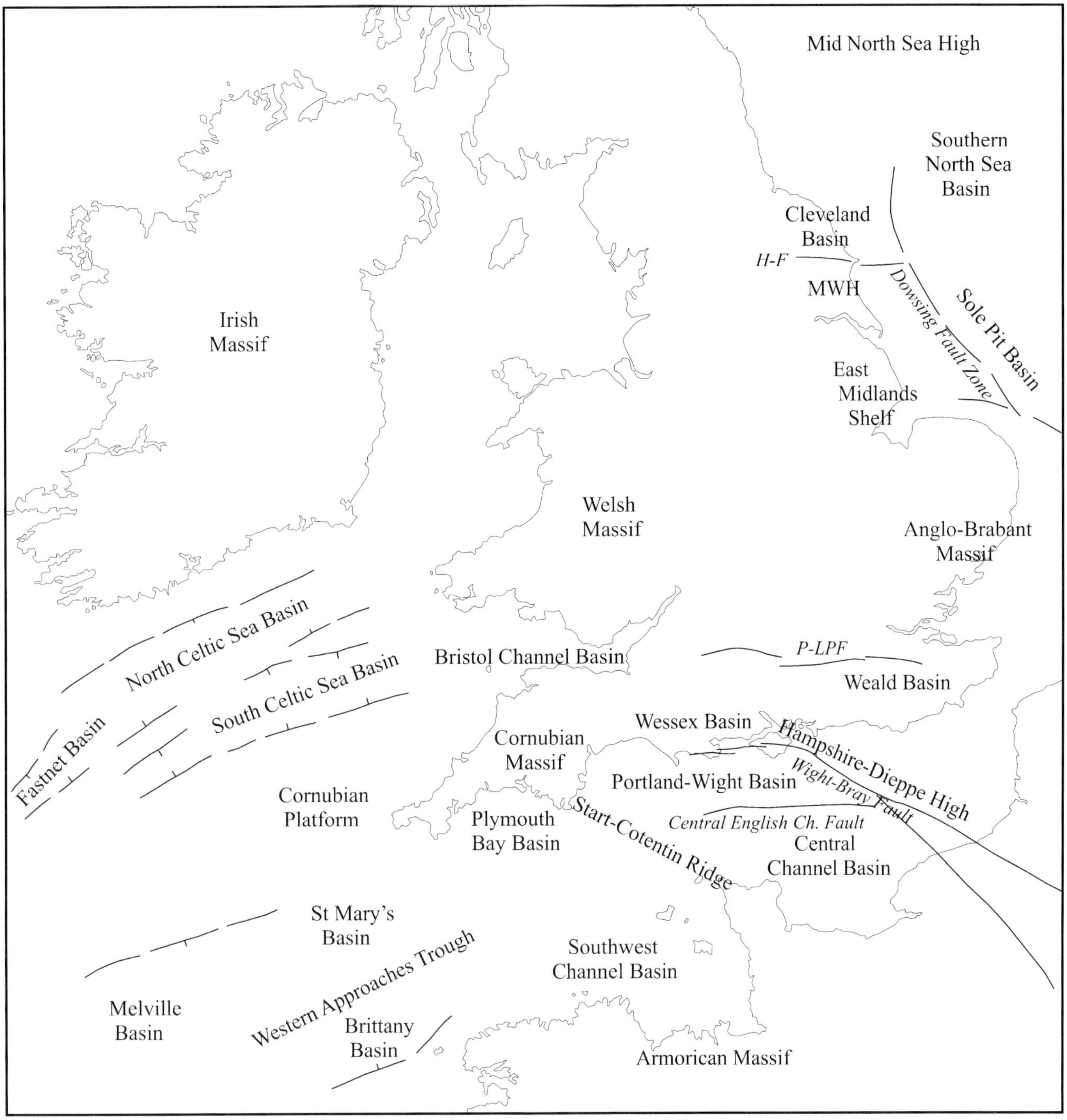

Fig. 15.1. Cretaceous structural framework. H-F, Howardian–Flamborough Fault Zone; MWH, Market Weighton High; P-LPF, Pewsey–London Platform Faults.

followed by submergence of the Market Weighton High in Mid Albian and the East Anglian Massif in Late Albian times, by which time the western shoreline to the North Sea Basin was also retreating. This Late Albian transgression established a continuous marine regime from the North Sea across eastern England to the Channel (Fig. 15.2c) (Rawson 1992*b*, map K3). Marine transgression continued through much of the Late Cretaceous until by the time of maximum transgression in the Late Campanian probably only the highest parts of the Welsh Massif may have remained above sea level. But the extent of any residual Late Campanian landmass may have been less than that indicated on Figure 15.2d (based on Hancock 1992, maps K4a and b) if North Wales had become an inversion basin by latest Cretaceous times (Cope 1994).

Biogeography

The occurence of taxa in the English Cretaceous reflects global biogeography, the pattern of which varies from one taxonomic group to another through both time and space. After the pioneering work of Neumayr (1883) the study of Mesozoic marine biogeography focused for a long time principally on ammonite distributions, and ammonite workers normally recognize only Boreal and Tethyan Realms. The Boreal Realm embraced the more northerly parts of the northern hemisphere, the Tethyan Realm the rest of the world. The distribution of other fossil groups was often 'fitted' to this framework. But an increasing amount of research on many other fossil groups, from belemnites to nannofossils, has shown that a higher-latitude southern realm, the Austral, can be recognized also. In such a model,

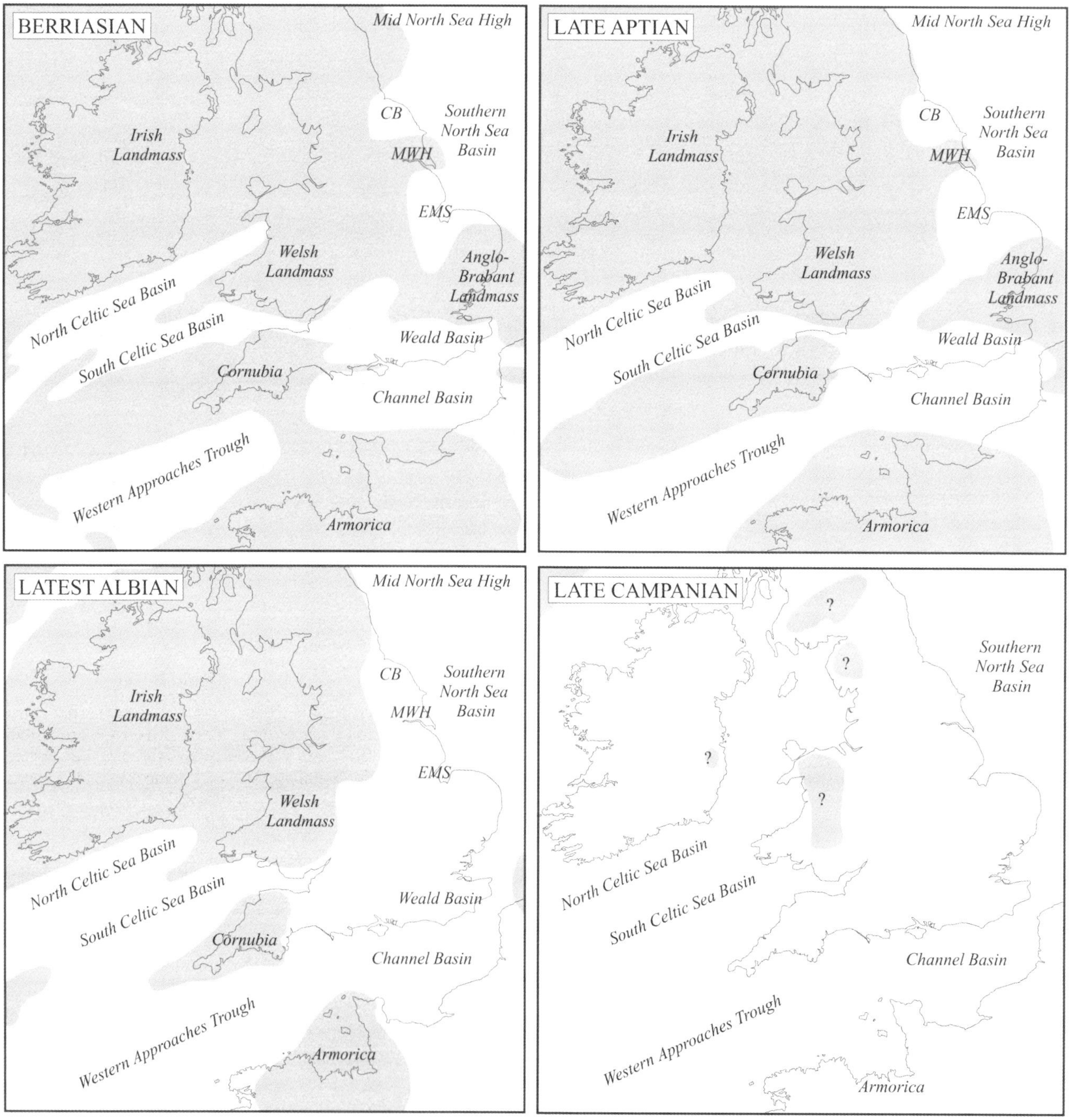

Fig. 15.2. The evolving palaeogeography. The stippled areas on the Berriasian map indicate areas of non-marine deposition. (Simplified from Rawson 1992*b* and Hancock 1992.) CB, Cleveland Basin; MWH, Market Weighton High; EMS, East Midlands Shelf.

the Tethyan Realm is seen as a low-latitude belt whose limits fluctuated considerably through time.

In the northern hemisphere, the difference between Boreal and Tethyan realms was most marked across the Jurassic–Cretaceous boundary and remained strong through most of pre-Aptian time, although some faunal and floral mixing occurred. With rising sea levels from the Late Barremian onward the distinctions began to blur, although there was still some significant biogeographic differentation through the remainder of the Cretaceous.

In pre-Aptian times NW Europe, including the North Sea Basin, formed a peripheral area of the Boreal Realm, the West European Province of Rawson (1981). In this area faunas and microfloras were dominantly boreal, although endemic forms developed and there was also a clear Tethyan influence among many fossil groups. In particular, the occurrence of Tethyan ammonites at some levels facilitates correlation with the 'standard' sections in southern France (Kemper *et al.* 1981; Rawson 1995). Southern England at that time was non-marine, but rising sea-levels in the Aptian flooded the area.

During the Aptian and Albian transgressive pulses over Europe continued to flood increasingly larger areas, so the biogeographical patterns changed. From England and north Germany to northern Iran similar ammonites flourished in the interconnected shelf seas, while planktonic microrganisms were able to migrate more freely from Tethys. A similar pattern continued during the Late Cretaceous, when global sea levels were even higher. The whole of Europe then belonged to the

North European Province of the North Temperate (= Boreal) Realm (*sensu* Kauffman 1973; Christensen 1997). But while many taxa spread over the whole of the area, there was some latitudinal differentiation. Based on belemnite distributions, three subprovinces have been recognized (Christensen 1976, 1997), the Central European, Central Russian and Balto-scandian subprovinces. The Central European Subprovince embraces the area of the British Isles eastwards to the Caucasus. But even within this region, some differences become apparent. Thus in the English outcrops two faunal (and sedimentological) 'provinces' are distinguished, a northern one embracing Yorkshire to north Norfolk (and extending through north Germany to eastern Europe) and a southern one including southern England and the Paris Basin. There is an intermediate area from Berkshire to Norfolk (Mortimore *et al.* 2001, fig. 1.6). The 'provinces' are distinguished by differences in constituent taxa or relative abundance of various invertebrate groups, not just belemnites.

The Stratigraphical Framework

The Cretaceous Stages

The current usage of the standard Cretaceous stage terms is discussed fully in Rawson *et al.* (1996), since when the bases of the Cenomanian, Turonian and Maastrichtian stages have been defined formally by a GSSP or 'golden spike' (Fig. 15.3). None of the stages are based on British sections and there are problems in defining their boundaries accurately within the UK sequences because of correlation problems. These are particularly acute across the Jurassic–Cretaceous boundary, as a result of which differing stage nomenclatures are used for Tethyan and Boreal realms. The North Sea Basin faunal and floral sequences are primarily boreal in origin, so there it is common practice to recognize the Ryazanian as the basal stage of the Cretaceous. In southern England the problem is even more difficult because the boundary sequences are non-marine, but there the approximate base of the 'standard', Berriasian, stage can be recognized (see below). The application of the Upper Cretaceous stage and substage boundaries within the British sequence is discussed in detail by Mortimore *et al.* (2001, pp. 479–485). Current UK usage is summarized in Figs 15.4–15.6.

Biostratigraphy

Ammonites provide a standard zonal scheme for the English Lower Cretaceous successions (Figs 15.4 & 15.5) and in practice are common enough in most marine strata for subdivision and correlation of onshore successions to be made with tolerable accuracy. Belemnites are of some value too (Mutterlose 1990), especially for the Barremian where ammonites are rare (Rawson & Mutterlose 1983). Among microfossils, nannofossils (Bown *et al.* 1998) and dinocysts (e.g. Duxbury 1977) are the primary biostratigraphic tools, and are especially important for studying the offshore marine sequences. Foraminifera (mainly benthic) provide a broader zonation for both onshore sequences (Hart *et al.* 1989) and the southern North Sea (King *et al.* 1989). Cypridean ostracods provide a zonation for the non-marine Purbeck/Wealden facies (e.g. Anderson 1985; Horne 1995). Correlation with non-marine facies in other parts of Europe has been attempted using the cyprideans but the main biostratigraphical evidence for correlating these non-marine beds with marine strata comes from plant microfossils (e.g. Worssam 1978; Feist *et al.* 1995).

Many of the Upper Cretaceous macrofossil zones in current use (Fig. 15.6) stem from a scheme first worked out in France and then introduced to England 130 years ago by Hébert (1874). Index taxa include ammonites, belemnites, a bivalve, echinoids, crinoids and a brachiopod: this strange mixture primarily represents the vagaries of preservation and occurrence in the Chalk. For example, in the Cenomanian of southern England ammonites are sufficiently common to provide a detailed modern zonation there (Fig. 15.6), whereas they are very rare at equivalent levels in the Northern Province. In the past 40 years, work in Germany and North America in particular has demonstrated the value of inoceramid bivalves in Upper Cretaceous biostratigraphy, and these are being used increasingly in England, where they are 'perhaps the most consistently abundant fossils' (Mortimore *et al.* 2001, p. 45). Like the ammonites, belemnites are common at some levels and rare or missing at others. But a NW European zonation has been built up based mainly on species of *Gonioteuthis*, *Belemnitella* and *Belemnella* (Christensen 1997), and this can be used in parts of the English succession (Fig. 15.6) (Mortimore *et al.* 2001, pp. 43–45).

Like the Lower Cretaceous sequences, the Upper Cretaceous offshore sequences are correlated using microfossils, primarily nannofossils (Burnett 1998). Planktonic foraminifers have provided an international scheme (e.g. Robaszynski & Caron 1995) but are of limited use in our area, although some key species are known; benthic foraminifers provide a more useful means of inter-regional correlation (Hart *et al.* 1989).

Non-biostratigraphical correlation

Although correlation of the onshore Cretaceous sequences traditionally relied primarily on biostratigraphy, some of the most extensively exposed and thickest sequences, the Purbeck, Wealden and Chalk, are not readily correlated using fossils alone. Alternative methods of correlation are increasingly being applied, and this is revolutionizing our understanding of the Chalk Group in particular. As a result, the initially very different approaches to correlation onshore and offshore are converging. The most important non-biostratigraphic methods have been described by Rawson *et al.* (2002). Among these, sequence and seismic stratigraphy and the application of a 'global' sea-level curve provide both a useful and a contentious method.

While it is widely accepted that global sea level rose through much of the Cretaceous, there is much debate on the validity

STAGE	BOUNDARY MARKER	HORIZON AND LOCALITY
MAASTRICHTIAN	Mean of 12 biostratigraphical criteria; including just below the first appearance of the ammonite *Pachydiscus neubergicus*	115.2 m level, Garnd Carri
TURONIAN	Lowest occurrence of the ammonite *Watinoceras devonense*	Base of bed 120, Rock Canyon anticline, east of Pueblo, Colorado, USA
CENOMANIAN	Lowest occurrence of the planktonic foraminifera *Rotalipora globotruncanoides*	36 m below Marnes Bleues Formation, Mont Risou, Rosans, Haute-Alpes, SE France

Fig. 15.3. Approved Cretaceous GSSPs.

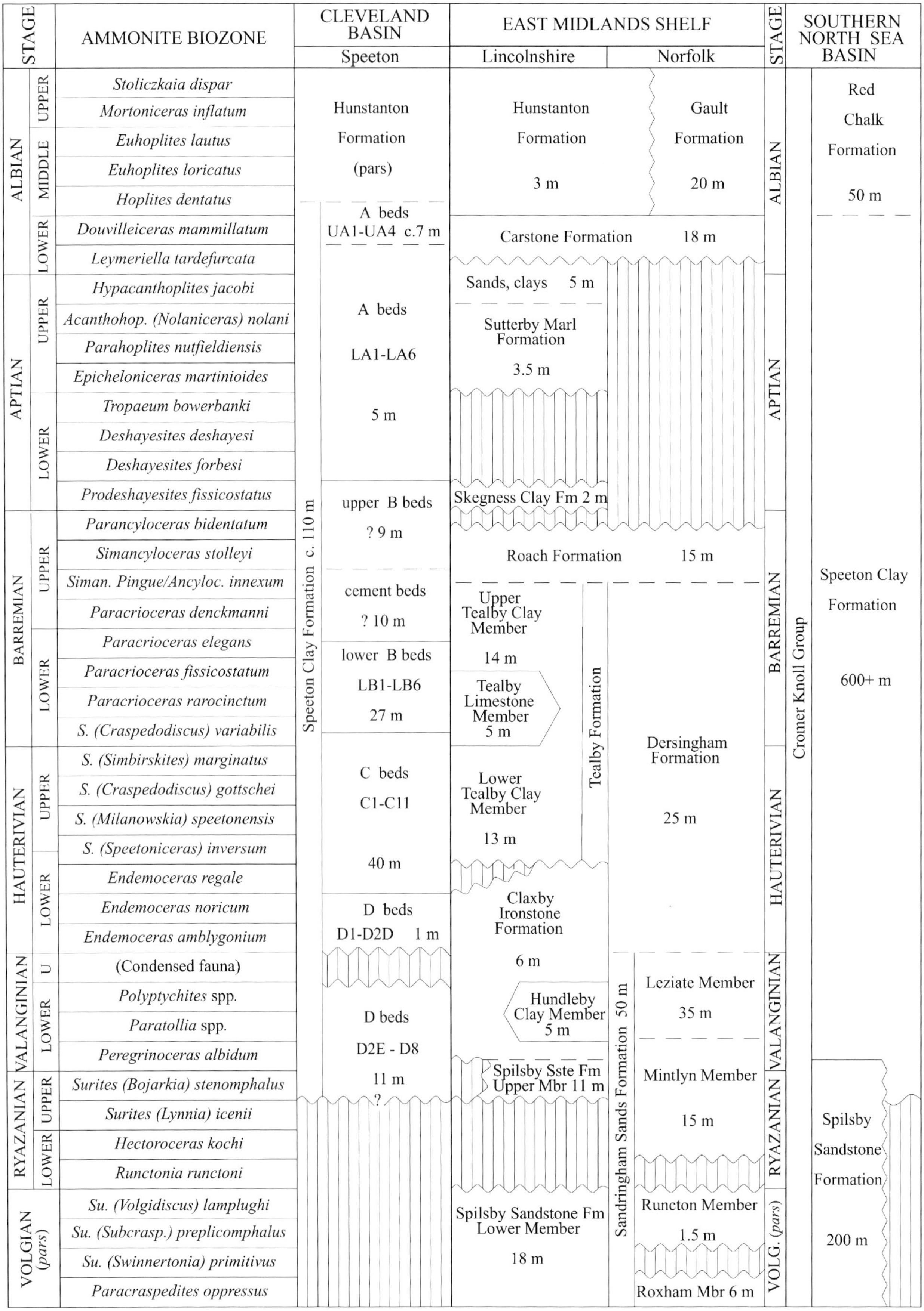

Fig. 15.4. Correlation of Lower Cretaceous strata: northern area. All thicknesses are maximum thicknesses.

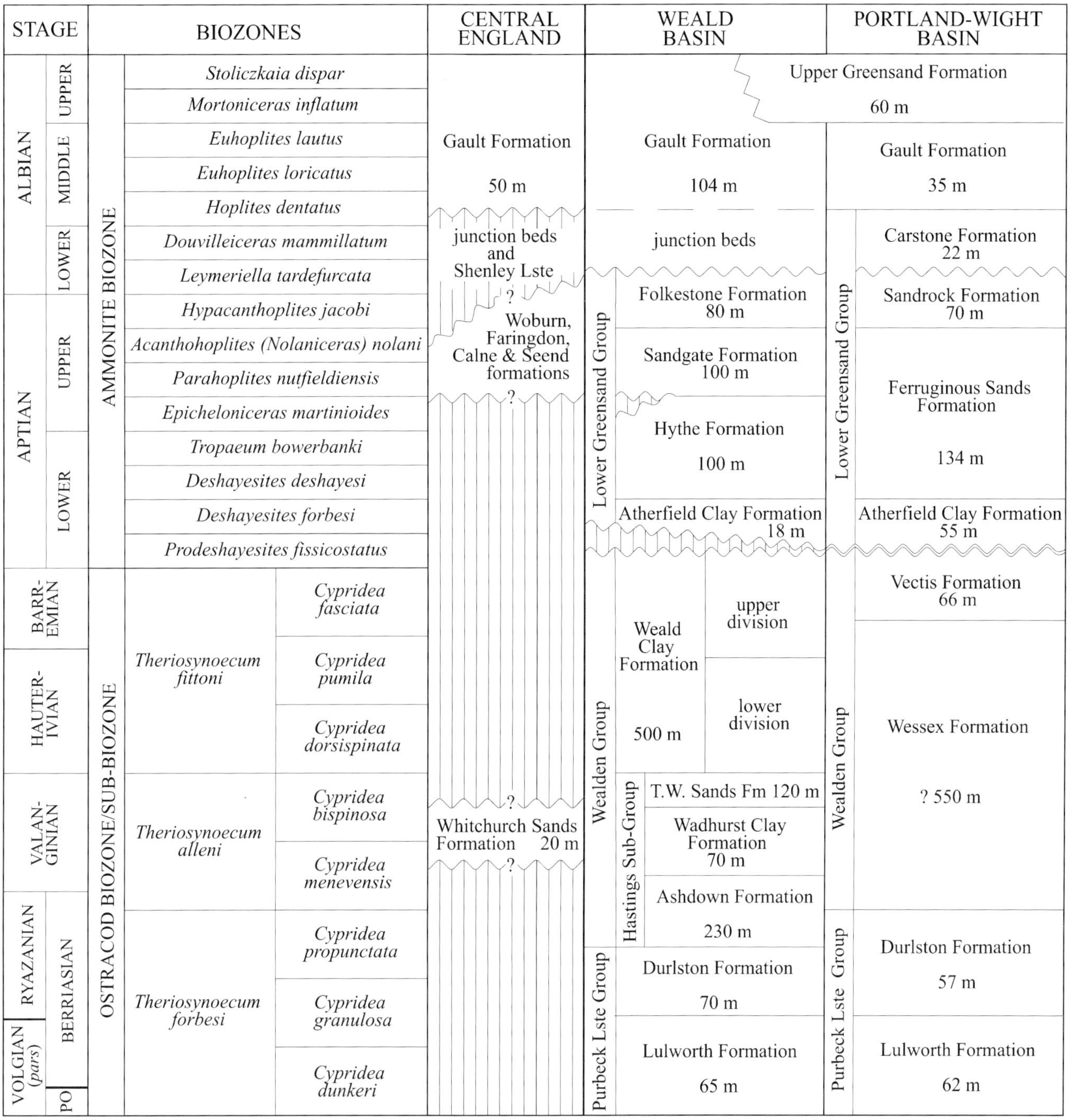

Fig. 15.5. Correlation of Lower Cretaceous strata: central and southern areas. All thicknesses are maximum thicknesses.

and applicability to correlation of a 'global' sea-level curve (Haq *et al.* 1988), based on the belief that individual falls and rises were synchronous in different areas and of probable eustatic origin. A product of Cretaceous sea-level change was the development at basin margins of unconformities and transgressive/regressive sequences. However, the same events are often detectable away from the margins by sedimentary breaks, condensed horizons, sharp facies changes, rapid faunal turnovers, etc. Using such criteria a large number of sea-level events have been recorded in the Cretaceous of NW Europe, some of them recognized long before the development of the sequence stratigraphic concepts that are now being applied to onshore and offshore successions. What has become increasingly apparent as a result is that many successions can be divided into 'packages' of sediment separated by significant stratigraphic (time) breaks, and that the commencement of sedimentation after such breaks was essentially isochronous across basins and often from basin to basin. How one interprets these observations in terms of the alternative sequence stratigraphic models of Galloway (1989) or the Exxon Group (e.g. Vail *et al.* 1977) is again a matter of debate.

For the North Sea area, Rawson & Riley (1982) produced a sea-level curve and documented 12 early Cretaceous 'events' – horizons of apparently rapid sea-level rise leading to significant transgression at basin margins. These represent flooding surfaces or maximum flooding surfaces and are thus the bounding surfaces of genetic stratigraphic sequences *sensu* Galloway (1989). Some additional surfaces have since been recognized. Such surfaces are readily identified offshore and are now widely applied in subsurface correlation (Copestake *et al.* 2003).

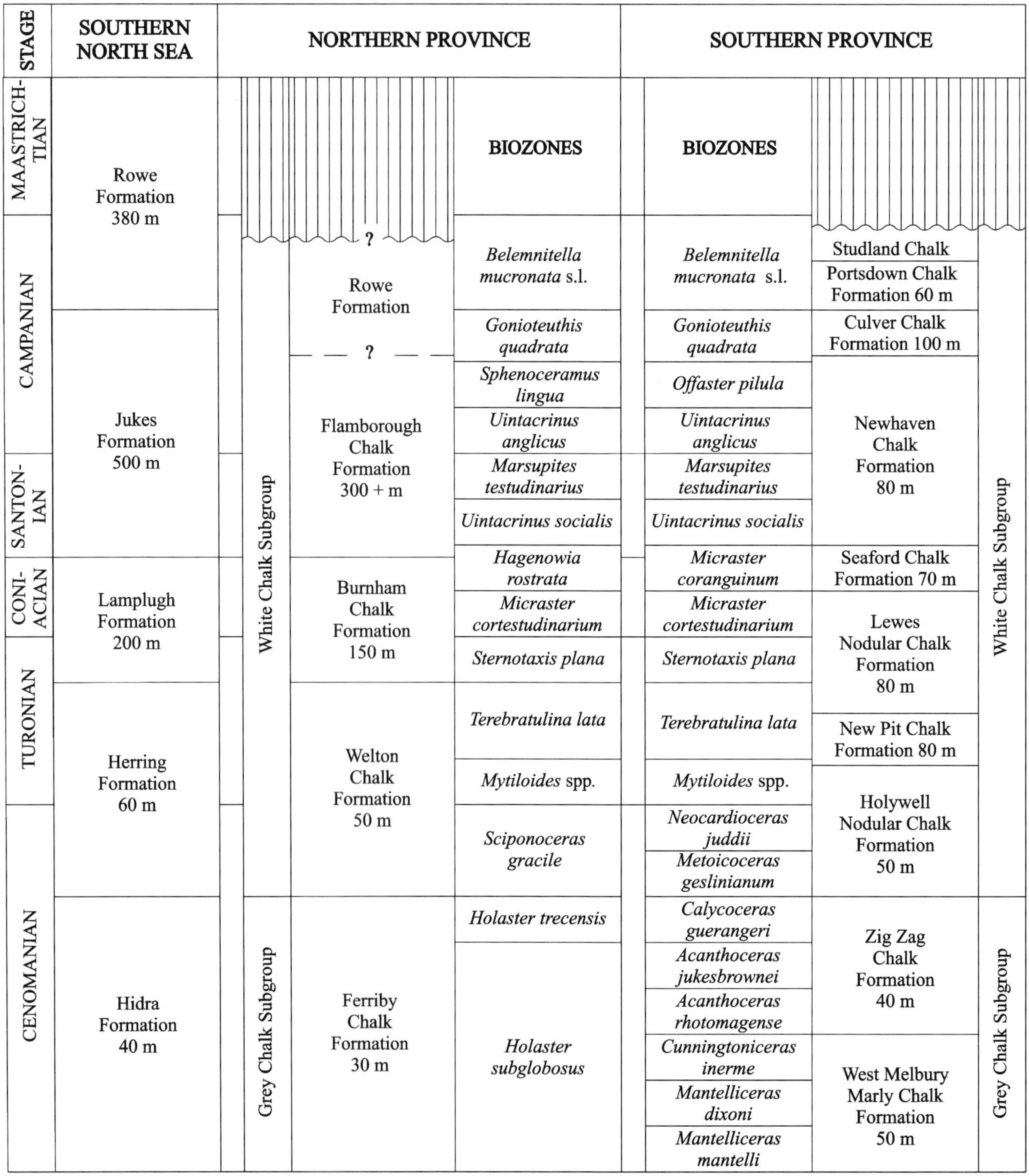

Fig. 15.6. Correlation of Upper Cretaceous strata. All thicknesses are maximum thicknesses.

Most can be recognized farther afield and the late Ryazanian, mid-Valanginian, mid-Hauterivian and some of the Aptian and Albian ones appear to be of global significance.

Rawson & Riley (1982) drew some of their evidence from onshore sequences. The Exxon sequence stratigraphic model, where sequence boundaries correspond with lowstands supposedly reflecting a rapid fall in sea level, has been tested on some of these (e.g. Hesselbo *et al.* 1990; Ruffell & Wach 1991). This has brought out the difficulty of applying some aspects of the Exxon model to the relatively shallow-water Lower Greensand sequences of the Wessex Basin, where the sequence boundaries commonly fall within a condensed level while some boundaries appear to be related to transgression rather than lowstand. But again the re-commencement of sedimentation appears to have been isochronous (Hesselbo *et al.* 1990, fig. 3; Ruffell & Wach 1991, fig. 4).

For the late Cretaceous, Hancock & Kauffman (1979) produced an initial north European sea-level curve with several significant peaks. In a more detailed study in NW Germany, Ernst *et al.* (1983) distinguished at least 11 'eustato-events' in the Cenomanian and Turonian alone. Since then the importance of stratigraphical breaks in correlating the English Chalk has been increasingly recognized. They are often marked by distinctive hardgrounds in shallow-water areas that can be traced into nodular chalks in the basins (Bromley & Gale 1982).

At a smaller scale, the emphasis on parasequences in the Exxon model has helped to renew interest in the small-scale (third-order) cyclicity seen in so many Cretaceous sediments and expressed in a variety of ways. For example, in the Lower Greensand it is often represented by vertically stacked coarsening-upward sequences, many terminating in a firmground or hardground (Ruffell & Wach 1998). In the Gault Clay fining-upward cycles occur (Gallois & Morter 1982). Cyclicity in the Coniacian chalks of the Anglo-Paris Basin is more subtle but eight third-order cycles have been distinguished (Grant *et al.* 1999) while Cenomanian chalks are characterized by smaller-scale marl–limestone couplets (Gale *et al.* 1999). Many, probably all, such cycles are on a Milankovitch scale and in every case they can be used in detailed correlation.

Geophysical techniques widely utilized by the oil industry offshore are increasingly applied to onshore successions. This has been particularly important in the Chalk, where Gray's (1958) pioneering work demonstrated the value of wireline logs in correlating boreholes in the London area. These are of particular value in demonstrating the lateral continuity of many of the marl bands but can also pick out lithological differences in the chalks (Whittaker *et al.* 1985; Mortimore & Pomerol 1987). Isotope stratigraphy (Fig. 15.7) is becoming increasingly important, based on recognizing both overall trends and distinctive excursions in carbon, oxygen and strontium isotope ratios through time (e.g. Weissert *et al.* 1998; McArthur *et al.* 2004). Finally, both magnetostratigraphy and tephrostratigraphy has provided some firm isochronous lines of correlation in both Lower and Upper Cretaceous sequences.

The results of applying these various methods of correlation to the sequences in England and Wales are discussed in appropriate sections below.

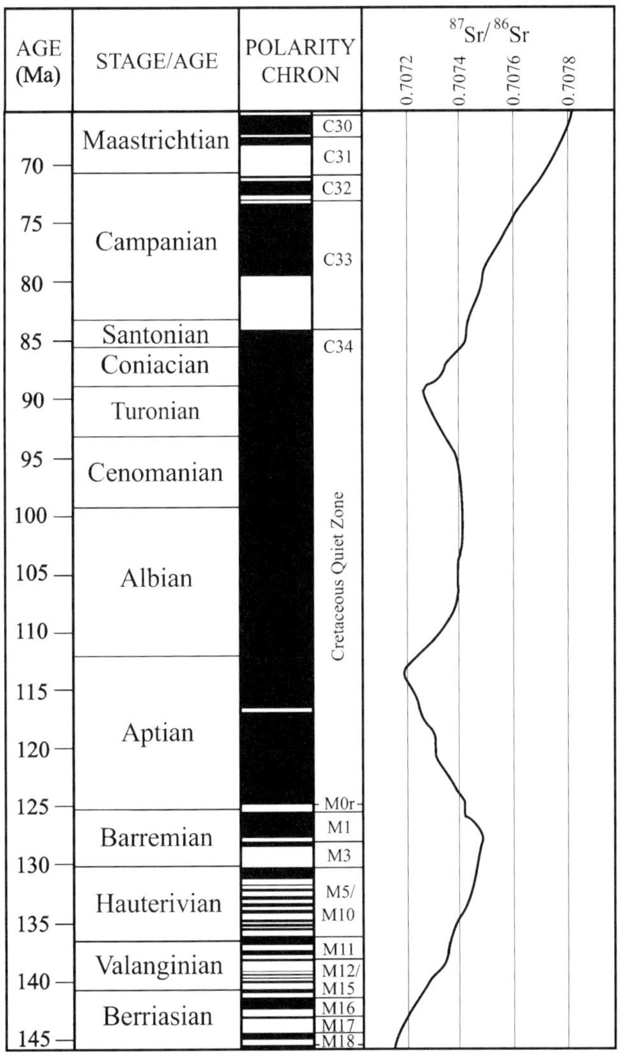

Fig. 15.7. Cretaceous numerical timescale, geomagnetic polarity timescale and strontium isotope curve. (Modified from figs 19.1 & 19.3 of Gradstein *et al.* 2004.)

The lithostratigraphical framework

The onshore successions

For the present-day onshore area, most of the long-established names for Lower Cretaceous lithostratigraphical units remain in use, although gradually 'formalized' by the addition of the appropriate term in the lithostratigraphical hierarchy. The divisions shown in Figs 15.4 and 15.5 are based mainly on Rawson's (1992*a*) table 12.2, with modifications based on the BGS Lexicon and references therein; the use of some names is discussed in more detail in the appropriate section below.

Historically, the Upper Cretaceous Chalk Group proved much more difficult to subdivided lithologically, because of its superficially homogenous nature. The long established division into Lower, Middle and Upper Chalk was based effectively on a mixture of biostratigraphical and lithological criteria and was used inconsistently. However, over the last 25 years there has been a revolution in Chalk studies, with the recognition of a series of mappable lithological units in both the Northern and Southern provinces. Coupled with this, the meticulous recording of numerous sections has shown that there are many lithological markers (e.g. marl and flint bands, bioclastic horizons and small-scale sedimentary cycles) that show a remarkable lateral continuity and thus facilitate a detailed 'event' correlation across the country. The more important flint and marl bands have been formally named.

Wood and Smith (1978) divided the Northern Province chalks into four formations, the Ferriby Chalk, Welton Chalk, Burnham Chalk and Flamborough Chalk Formations. Subsequently, a fifth unit, the Rowe Formation, has been recognized (Fig. 15.6), although it is known only from boreholes beneath the Pleistocene deposits.

Formational division of the southern province chalks stems from Mortimore's (1983, 1986*a*, *b*) studies in which six members, some divided into 'beds', were proposed for his 'Sussex White Chalk Formation' (now the White Chalk Subgroup). Subsequent mapping by members of the British Geological Survey proved that most of Mortimore's divisions were mappable over wide areas but raised queries about how they would best fit into a lithostratigraphical hierarchy. Hence in 1999 the Geological Society's Stratigraphy Commission called a meeting of all interested specialists at which agreement was reached that the Chalk Group throughout the onshore area falls naturally into two major divisions, the Grey and White Chalk Subgroups, and that the old 'Lower', 'Middle' and 'Upper' chalk divisions should be abandoned (Rawson *et al.* 2001). Five members and two of the 'beds' in Mortimore's (1983, 1986*a*) original scheme for the White Chalk Subgroup were raised to formational status (Fig. 15.6), while Bristow's (1989; Bristow *et al.* 1995) division of the Grey Chalk Subgroup into two formations was retained. The main modification was that the Holywell Nodular Chalk Formation was expanded downward to include the Plenus Marls Member at its base; Mortimore (1983, 1986*a*) had left this unit in the 'Lower Chalk' beneath, so that Bristow *et al.* (1995) originally included it in the top of their Zig Zag Chalk Formation.

The offshore successions

Lithostratigraphical nomenclature for most of the offshore areas of England and Wales is related to that of adjacent onshore schemes. For the UK sector of the southern North Sea, Rhys (1974) proposed the Cromer Knoll Group for the whole Lower Cretaceous succession, but applied the nomenclature of the adjacent onshore areas for individual formations. Thus the mudrocks that form the bulk of the succession were placed in the Speeton Clay Formation. This was followed by Cameron *et al.* (1992) in their account of the sequence, but not by Lott & Knox (1994) who adopted the central North Sea terminology of Johnson & Lott (1993). As a result the Speeton Clay Formation changes its name at the present coastal boundary! The earlier position is followed in this chapter.

For the Lower Cretaceous of the Channel Basin the British Geological Survey retained the old established Purbeck, Wealden, Lower Greensand and Gault divisions without formalising them as formations, etc. (Hamblin *et al.* 1992). A similar procedure has been followed in the other basins offshore from SW England and Wales.

The southern North Sea Upper Cretaceous Chalk sequence has not been studied in detail and until recently the succession had not been subdivided. However, the Black Band equivalent is indicated in many wells by a distinctive gamma-ray and sonic response (Cameron *et al.* 1992, fig. 81). This has led some companies to extend part of the central North Sea nomenclature to the area, referring the chalks beneath the Black Band to the Hidra Formation and those immediately above to the Herring Formation (Lott & Knox 1994). Subsequently, Lott & Knox (1994) have placed the scattered Danian (Paleocene) chalks in the Ekofisk Formation and introduced three new formational names for the Cretaceous chalks between the Ekofisk and Herring formations as changes in log characteristics do not allow the intervening central North Sea formations to be distinguished. As a result, both the sub-Danian southern North Sea and the adjacent onshore sequences are divided into five formations which correlate closely or exactly (Fig. 15.6). The highest (Rowe) formation is recognized both offshore and onshore; the names of the other four again change at the present coastline.

In the south and SW basins the Chalk sequence is either undivided or the former 'Lower, Middle and Upper' units and their subdivisions are recognized.

The Lower Cretaceous Series

The Southern North Sea Basin (UK sector)

Although Lower Cretaceous marine sediments extend over most of the UK sector of the Southern North Sea Basin and some of the adjacent onshore areas, there were several different depocentres where there was quite rapid, often fault-controlled subsidence, separated by shelf areas or more local highs characterized by slow subsidence and attenuated sequences. The Cromer Knoll Group and its onshore equivalents accumulated over most of the UK sector, either resting with slight unconformity on the Kimmeridge Clay Formation (Upper Jurassic) or transgressing over older rocks (Cameron *et al.* 1992, fig. 78). The succession consists predominantly of mudrocks, with shallower-water sands, silts and clay ironstones in more marginal areas. The sediments are generally less than 100 m thick (Cameron *et al.* 1992, fig. 79) but up to 900 m have been drilled adjacent to the Dowsing Fault Line (Cameron *et al.* 1992), while there may be up to 1000 m locally in the Flamborough Fault Belt (Kirby & Swallow 1987).

There were three main depositional areas in the UK sector; the East Midlands Shelf and the more rapidly subsiding Cleveland Basin, both of which extend from onshore to offshore areas, and the offshore Sole Pit Basin and adjacent shelves (Fig. 15.8). Because of our detailed knowledge of the stratigraphy and palaeontology of the onshore sections from Yorkshire to Norfolk, these areas have provided the key for understanding much of the Early Cretaceous history of the North Sea area.

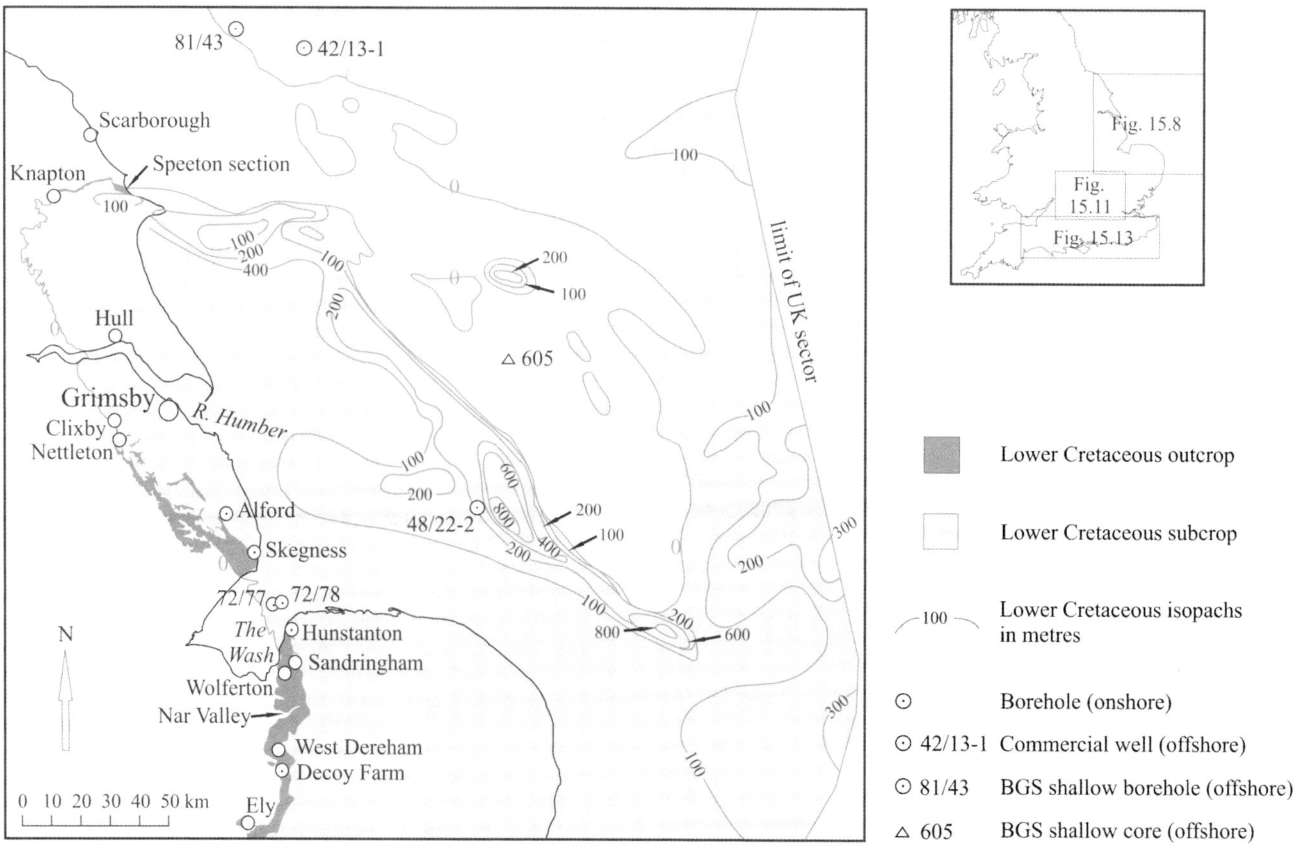

Fig. 15.8. Lower Cretaceous rocks of the Southern North Sea Basin and adjacent areas: Distribution, thickness and localities. (Based partly on Cameron *et al.* 1992, fig. 79.)

Not only are the faunal and microfloral sequences well established but many of the stratigraphic breaks and major facies changes can be dated with great precision. Offshore sequences are known is less detail as although numerous wells have penetrated the succession few have been cored and most have been drilled on structural highs where the sequences are thin and incomplete (Cameron *et al.* 1992).

The Cleveland Basin

The Cleveland Basin lies at the western margin of the Southern North Sea Basin. It fringes the northern margin of the East Midlands Shelf and in the onshore region it is bounded to the south by the downfaulted northern margin of the Market Weighton High, while offshore it is bounded to the east by the Central Fault Zone and passes southeastwards into the Sole Pit Basin.

Within the Cleveland Basin almost the whole Lower Cretaceous sequence consists of a typical southern North Sea mudrock, the Speeton Clay Formation. In the only permanent exposure, the type section along the SE part of Filey Bay, the formation is only about 100 m thick. Here it rests nonconformably on the Kimmeridge Clay Formation, the boundary being marked by a 10 cm thick phosphatic nodule band. This stratigraphical break represents an interval of nondeposition or erosion spanning some 10 Ma.

All six Lower Cretaceous stages are represented in the Speeton Clay, but there are several breaks in the succession and the higher beds (mid-Barremian–mid-Albian) are mainly hidden beneath landslips and poorly exposed. Despite these limitations, Speeton is a key section for the understanding of many aspects of the Early Cretaceous history of NW Europe and provides important data for correlation with sequences elsewhere. Fossils are common at many levels and the Speeton section is the type locality for many of the English Valanginian–early Barremian ammonite zones.

Lamplugh (1889) divided the sequence into four main units, the D (base)–A (top) beds, based on their belemnite faunas. He also recognized finer, lithological, divisions, especially in the D and C beds, which were divided into D8–D1 and C11–C1, respectively. Lamplugh's bed-by-bed annotation is still used (Fig. 15.4) with later, more detailed refinements summarized by Rawson *et al.* (1978), Rawson & Mutterlose (1983) and Mitchell (1995*a*). Phosphatic nodule beds and glauconite concentrates indicate pauses in sedimentation, and these, together with mottled beds (*Chondrites*), silty layers and rapid vertical colour changes, facilitate the lithological subdivision (Fig. 15.9; Plate 29). At some levels thin (average *c.* 0.5 m) rhythmic alternations of pale/dark (more calcareous/less calcareous) clays occur; such couplets are best seen in the top C and lowest B beds, i.e. across the Hauterivian–Barremian boundary, where each couplet probably represents the 22 300 year Milankovitch precession cycle (Mutterlose & Ruffell 1999). Coarsening-upward cycles higher the lower B beds correlate with the north German Blatterton facies (Rawson & Mutterlose 1983) and probably reflect the same precession cycle.

Fig. 15.9. Bed D6, Ryazanian, Speeton. These pale clays form a distinctive marker in the lower part of the D beds. The folding reflects Early Paleocene compression.

The Speeton Clay Formation may have been deposited in water depths of less than 100 m, the clay accumulating close to a low-relief landmass (the Market Weighton High) that usually supplied only clay-grade detritus to the adjacent basin. In the latest Ryazanian times there was a brackish interlude, and in early Barremian times the bottom waters may have been oxygen-deficient during deposition of the 'Blatterton' equivalent.

Inland, the formation forms a long narrow outcrop on the Wolds scarp, eventually thinning westward to disappear near Knapton, where the overlying Hunstanton Formation comes to rest on the Kimmeridge Formation. Southwards from the outcrop a seismic profile indicates that within the Flamborough Fault Zone the formation thickens to as much as 1 km along the southern margin of a listric fault (Kirby & Swallow 1987) before disappearing southward over the Market Weighton High.

Offshore from Speeton, in block 42/12, BGS borehole 81/43 entered the Speeton Clay Formation beneath 4.5 m of seabed sand (Lott *et al.* 1986). It proved a sequence of about 90 m of clays of Late Ryazanian–Early Barremian age (equivalent to the 54 m of beds D8–LB6 at Speeton). Lithological units correlate very closely with the Speeton type sequence but are somewhat thicker. Both here and at Speeton several thin volcanogenic (bentonitic) clays occur in the Ryazanian–Hauterivian sequence.

BP well 42/13-1 was also re-interpreted by Lott *et al.* (1986). Here the Speeton Clay Formation is 250 m thick, much of the increase apparently because of the preservation of Barremian–Aptian sediments; log correlation between the two boreholes suggests these are about 137 m thick.

The Speeton Clay Formation grades up into the Hunstanton Formation (formerly 'Red Chalk'), which is well exposed in the vicinity of Speeton cliffs. It consists of red marly and nodular chalks, with discrete marly clay bands at some levels. Nodular chalks set in marl drapes are common (griotte facies) and flaser-bedded chalks occur. The Hunstanton Formation here is usually dated as Late Albian alone (Rawson *et al.* 1978), but Mitchell (1995*a*) has shown that the base is of Mid Albian age, correlating closely with the base of the same formation on the East Midlands Shelf. Unfortunately the transitional beds are rarely exposed, but Mitchell (1995*a*) drew the boundary at a level that few previous workers had seen, where there is a change from clays to marls with chalk nodules (base of his Queens Rock Member). Two metres higher in the succession, still within the lower part of the member, the sediments become more nodular (bed QR3) and an erosion horizon at the base of this bed appears to mark the widespread basal Upper Albian (*cristatum* Subzone) erosion surface. Hence most of the Middle Albian is cut out. Mitchell (1995*a*) also included in the upper part of the formation beds that some previous workers had included in the overlying Chalk (Ferriby Formation). He recognized a total thickness of 24 m, and divided the formation into five members, distinguished primarily by the relative development of marl–chalk rhythms. Rhythmic bedding occurs on at least two scales, the larger (0.5–7 m) marked by upward decrease in clay content reflecting decreased local terrigenous input as sea level rose (Mitchell 1995*a*). The Albian–Cenomanian boundary lies high in the fourth (Weather Castle) member.

Correlative red marls with subordinate chalks are widely distributed over the North Sea region (Red Chalk and Rødby formations), and in the UK sector are normally 20–30 m thick, exceptionally more than 50 m (Cameron *et al.* 1992).

In the offshore part of the Cleveland Basin, well 42/13-1 (as re-interpreted by Lott *et al.* 1986, fig. 2) includes about 19 m of red chalk.

The East Midlands Shelf

Immediately north of the Anglo-Brabant Massif lay the East Midlands Shelf, bounded to the north and east by active fault systems – the Howardian–Flamborough and Dowsing fault belts (Fig. 15.1). Over the southern part of the shelf shallow-marine sediments accumulated: this depositional area is sometimes referred to as the Spilsby Basin. The northern part of the shelf was emergent and formed the Market Weighton High, a hinge with a fault-bounded northern margin and a gentle southern slope merging into the shelf. Over the onshore part of the shelf a thin (maximum about 100 m) but varied series of shallow-marine sediments accumulated (Figs 15.4, 15.8 & 15.10). Facies and faunal distributions indicate nearshore environments in north Norfolk and north Lincolnshire with a broad intervening area of offshore, inner–mid-shelf sediments. The coastline may have been only a few kilometres west of the present outcrop, which forms a narrow band through Lincolnshire before passing beneath the Fen deposits to re-emerge as a short N–S strip in north Norfolk (Fig. 15.8).

The pre-Albian succession begins with the widely-distributed sands of the latest Jurassic–earliest Cretaceous Spilsby Sandstone Formation and its Norfolk correlatives, the Roxham, Runcton and Mintlyn members of the Sandringham Sands Formation. At the base a phosphatic nodule bed everywhere rests on the Kimmeridge Clay Formation and similar nodule beds higher in the sands mark widespread pauses in deposition. They are particularly frequent in the attenuated North Norfolk sequence, the lower part of which (Roxham Member) extends as far south as the Ely district (Gallois 1988). At outcrop the sands are generally medium-grained, glauconitic, sometimes argillaceous, incoherent sediments. Because of extensive bioturbation, few sedimentary structures are visible, though, in the northern part of the outcrop traces of bedding suggest possible deposition in offshore sand bars. At some levels calcareous concretions occur and are the main source of the diverse bivalve faunas (Kelly 1983). The bivalve assemblages generally suggest an unstable substrate. Ammonites occur through most of the sequence and indicate a Volgian (latest Jurassic)–Ryazanian age. Unfortunately, some of the most important faunas were obtained from temporary exposures in north Norfolk, from horizons not exposed elsewhere (Casey 1973; Gallois 1994).

To the east of the outcrop the sands extend to the margin of the shelf in the offshore area, where they are generally less than 50 m thick but reach a maximum of about 200 m in the vicinity of the Dowsing Fault line. They also extend eastwards along the northern margin of the Anglo-Brabant Massif (Glennie & Boegner 1981, fig. 3c; Cameron *et al.* 1992, fig. 79; Lott & Knox 1994).

In the Skegness–Wash area the overlying strata are predominantly sandy clays with scattered iron ooliths and some sandy or calcareous horizons. This was probably the area of maximum subsidence, as it was during deposition of the preceding Upper Jurassic Humber Group (Cameron *et al.* 1992, fig. 72). To the NW there is a gradual thinning with accompanying facies changes. Ferruginous ooliths become more abundant and even form a local ironstone (the Claxby Ironstone), impure limestones appear (the Tealby Limestone Member) and non-sequences become more significant as individual units lap out against the Market Weighton High. Much of this more marginal succession was exposed in the 1950s and 1960s during extensive quarrying for the Claxby Ironstone around Nettleton (Penny & Rawson 1969). A subsequent (now filled) exposure at Nettleton Top was described by Kelly & Rawson (1983) and Gaunt *et al.* (1992). The ironstone yielded rich brachiopod and molluscan faunas, including numerous ammonites that allowed accurate dating of the non-sequences exposed there. Erosion horizons associated with the non-sequences may extend over much of the basin; those of mid-Hauterivian and Early and Late Aptian age are particularly prominent. Above the mid-Hauterivian erosion surface the Lower Tealby Clay Member represents the maximum spread of clays over the shelf following a major sea-level rise.

Southwards into Norfolk most horizons become more sandy until the (incomplete) southernmost sequence is almost wholly arenaceous. The main facies changes take place over a short distance beneath the Wash, and this together with the local thinning of the sequence at Hunstanton suggests that there may have been some tectonic control on deposition. The position of the lateral passage from Claxby to Leziate beds indicated by Gallois (1994, fig. 27) closely parallels the SW–NE-trending 'Wash Line' which Jeans (1980) postulated to account for thickness changes in individual units in the younger Hunstanton and Ferriby Formations. Such a line may be linked to a pre-Mesozoic rise beneath north Norfolk which brings probable Precambrian rocks close to the surface in the North Creake borehole (Kent 1947). Gallois (1994, fig. 7) indicated possible faults in the same area but with NW–SE trends.

The lower part of the Norfolk sequence is formed by the Leziate Member, the highest and thickest (to *c.* 30 m) division of the Sandringham Sands Formation (Fig. 15.4). This differs

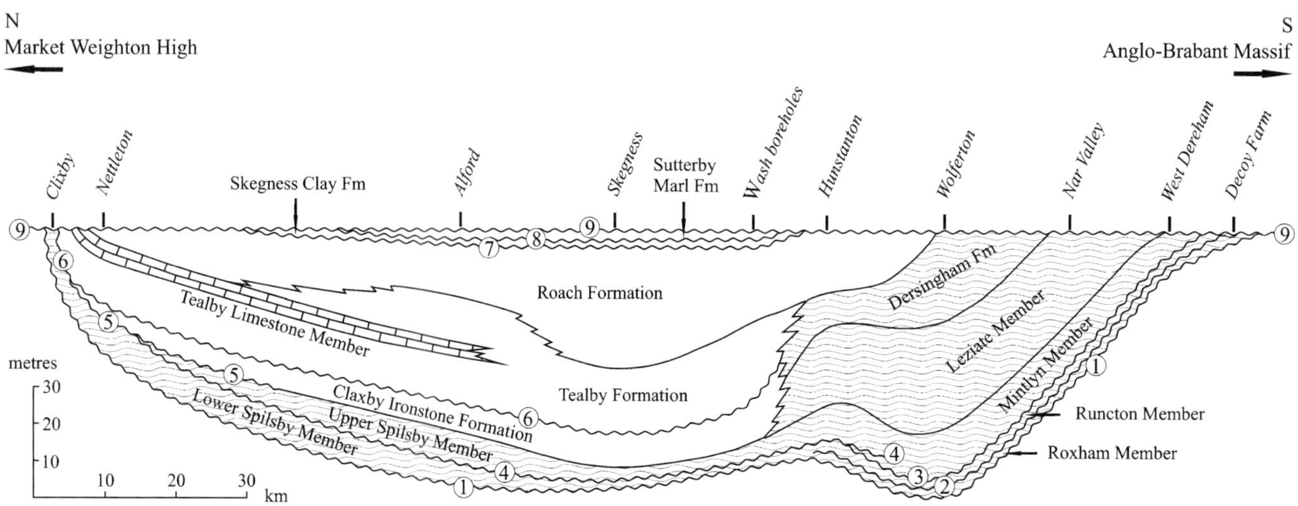

Fig. 15.10. Cross-section of the Lower Cretaceous strata of the East Midlands Shelf. Unconformities: 1 mid Volgian; 2 late Volgian; 3 early Ryazanian; 4 late Ryazanian; 5 latest Ryazanian; 6 mid Hauterivian; 7 earliest Aptian; 8 late Aptian; 9 early Albian (base of Carstone Fm).

from the underlying sandy members in consisting of clean-washed, predominantly fine-grained, grey or white sands. These are beautifully exposed in extensive quarries in the Sandringham area, described in detail by Gallois (1994). There they are locally glauconitic and exhibit planar, often herringbone, cross-stratification and occasional rippled horizons. Unfortunately they are almost unfossiliferous and their exact age is therefore uncertain, although a single? *Polyptychites* apparently from this member would support the presumed Valanginian age (Gallois 1994, p. 75). The overlying Dersingham Formation is more variable, with fine sands, silts and clays; it becomes more sandy towards the southern limit of the outcrop. The highest beds are argillaceous, forming the Snettisham Clay, which in the 19th century yielded an interesting fauna including Barremian heteromorph ammonites. At Hunstanton, the Dersingham Formation is overlain by the argillaceous Roach Formation, but this is cut out southwards by an unconformity at the base of the overlying Carstone Formation.

Eastwards from outcrop into the offshore area these varied sediments overlying the Spilsby Sandstone Formation pass laterally into argillaceous sediments assigned to the Speeton Clay Formation (Rhys 1974; Cameron *et al.* 1992). In 'type' well 48/22-2, near the edge of the shelf, the formation is 236 m thick, with thin bands of silty sandstone in the lower part (Rhys 1974, fig. 7), but towards the Dowsing Fault Line thickens to over 600 m.

The source of sediment supply to the East Midlands Shelf has been much debated. Schwarzacher (1953) studied cross-bedding in the Sandringham Sands Formation (mainly the Leziate Member) and recorded a predominantly easterly dip and postulated a N–S shoreline in the vicinity of The Wash. But heavy mineral suites in the Spilsby Formation include abundant fresh kyanite and staurolite (Versey & Carter 1926; Ingham 1929), for which Allen (1967) suggested a Scottish–Scandinavian origin. However, the palaeogeographical setting of the sands makes this unlikely. The northward diminution of these metamorphic minerals indicates that they may have been derived from the London Brabant Massif. This is supported by the more recent discovery of (buried) Carboniferous and Triassic rocks fringing the NW margin of the massif that contain 'clasts of Precambrian and Palaeozoic rocks, together with heavy minerals derived from them' (Gallois 1994, p. 78).

An important early Albian transgression marked the beginning of a new and more widespread phase of sedimentation. A basal pebbly sand, the Carstone Formation (maximum thickness 12 m), oversteps earlier Cretaceous beds to rest on the Jurassic at both ends of the outcrop: it extends from the southern part of the Market Weighton High to Cambridgeshire, and scattered patches occur over the crest of the high. It has not been proved in offshore wells but 'similar sediments may occur near the coast between Flamborough Head and the Humber' (Cameron *et al.* 1992, p. 86). Further offshore it probably passes laterally into clays forming the highest part of the Speeton Clay Formation.

The Carstone Formation is generally a medium- to coarse-grained, pebbly, glauconitic, quartz sand with a pale green ferruginous cement that weathers orange-brown. Distinct pebble bands occur and towards the margins of the outcrop the base is conglomeratic. Cross-bedding is fauntly visible at Nettleton, north Lincolnshire, where there is a southerly dip, and at Hunstanton, where dips are apparently to the SW (Versey & Carter 1926). Much of the sediment may have been reworked from the underlying Spilsby sands, for they have similar heavy mineral suites, and derived Lower Spilsby ammonites occur in the basal Carstone conglomerate in north Lincolnshire. Indigenous fossils are known only from the Humber area, where the formation is finer-grained and slightly argillaceous. They include brachiopods, bivalves, ostracods and foraminifera (Owen *et al.* 1968; Gaunt *et al.* 1992).

Over most of the shelf the Carstone Formation grades up into the Hunstanton Formation (Plate 35), but at its southern extent it is overlain by the Gault Clay Formation. The Hunstanton Formation extends from the Cleveland Basin to north Norfolk and over the offshore part of the shelf: its continuity over the Market Weighton High marks the final submergence of that feature by the Mid Albian transgression (Fig. 15.2c).

Over the East Midlands Shelf the Hunstanton Formation forms a thin (generally 1–3 m) sequence of impure, nodular and massive chalks and marls, usually of a bright red colour, although locally it is greenish-grey. Non-sequences occur, while penecontemporaneous reworking is indicated by included sand grains and small pebbles, which diminish in quantity upward, and by epizoan-encrusted chalk pebbles. Algal mats indicate deposition in the photic zone (Jeans 1973; Gallois & Morter 1982). Ammonites are generally very rare, but together with belemnites and some bivalves they show that the formation is highly condensed sequence of Mid–Late Albian age (see Gaunt *et al.* 1992; Gallois 1994).

In north Norfolk the Hunstanton Formation passes laterally southwards into the Gault Clay Formation in the vicinity of Sandringham (Gallois 1994). There the Gault is only 2 m thick, but it thickens steadily southwards (see below).

The Sole Pit Trough and adjacent areas

Offshore, the Speeton Clay Formation extends from the Cleveland Basin and East Midlands Shelf towards the southern margin of the mid North Sea High and southeastwards towards the Sole Pit Trough but like the overlying Red Chalk Formation has been removed from the main part of the trough following later inversion.

In Late Jurassic times the Sole Pit Trough was a rapidly subsiding area whose SW margin was delimited by the Dowsing Fault Line. The Swarte Bank Hinge Line (a series of en-echelon basement faults) formed the NE depositional margin of the basin (Walker & Cooper 1987). The trough was partially isolated by the Late Cimmerian movements when the area to the NW was uplifted (Bulat & Stoker 1987) and extensively eroded, locally as deep as the Bunter Shale Formation. The Early Cretaceous history of the trough remains uncertain as Lower Cretaceous sediments were almost completely stripped off as the trough was inverted in later Cretaceous and Early Tertiary times. Although it has often been assumed that the Cromer Knoll Group thickened to the centre of the trough, Cameron *et al.* (1992) suggested that Late Jurassic uplift of the trough may have had a major influence on Lower Cretaceous sedimentation as individual Lower Cretaceous stages appear to thin abruptly at the flanks of the trough. While there was probably almost uninterrupted sedimentation from late Jurassic to mid Cretaceous times along the axial region, younger units (Hauterivian–Aptian) onlap onto the NE flank, as in the Indefatigable Field (France 1975). Lower–Middle Albian sediments apparently overstepped the older units eastwards.

Because of the uplift and erosion only a few small outliers of Lower Cretaceous sediments are preserved within the trough (Fig. 15.8). In one of these BGS core 605 penetrated 0.3 m of mudstone and siltstone of late Ryazanian/Valanginian age, 'preserved from erosion in the hanging wall of a later Cretaceous fault' (Cameron *et al.* 1992, p. 84).

Crittenden (1982) summarized several wells located over the shelf area to the NE of the Trough, in UK block 49. He suggested that the Dutch nomenclature could usefully be applied here and recognized both the Holland and Vlieland Formations. He noted that sedimentation was strongly influenced by pre-Cretaceous highs which the Lower Cretaceous sediments onlap, until by Late Aptian times the highs were covered. From foraminiferal evidence Crittenden (1982) dated the Lower Holland Marl Member as Early Aptian and the Middle Holland Shale Member as Early Albian: hence the whole of the Late Aptian interval is apparently missing. The base of the Middle Holland Member is often sandy or conglomeratic. Crittenden (1982) also observed that the Upper Holland Marl Member is equivalent to the Red Chalk Formation, describing it as a calcareous, red-brown mudstone in which the carbonate content gradually increases upward. It is dated as Mid–Late Albian.

Central and eastern England

Between the NW margin of the Anglo-Brabant Massif and the Welsh Landmass was a lower-lying region that opened out northeastward to the seas over the East Midlands Shelf and southwestward into the Wessex Basin (Fig. 14.2a, b) (Rawson 1992*b*, maps K1 and K2). It extends from Cambridgeshire southwestwards to Wiltshire, but can be divided into two main areas that appear to have had quite distinct histories until Mid Albian times. From the Ely district to Leighton Buzzard is the 'Cambridge–Bedford province' of Casey (1961) or 'Bedfordshire Straits' *sensu* Rawson (1992*a*, *b*). This shows affinities with the East Midlands Shelf to the north. The more southerly part (the 'western outliers' of Casey 1961) effectively formed a marginal embayment off the Wessex Basin. The boundary between the two areas lies in the vicinity of 'Rastall's axes', a pair of NW–SE-orientated structural highs in the vicinity of Leighton Buzzard and Sandy (Rastall 1927) which may have formed an intermittent barrier.

Over the whole of this intermediate region the pre-Upper Aptian sedimentary record is very patchy (Fig. 15.11) and is limited to the more southerly area. Here, non-marine sedimentation started earlier than in the Wessex Basin, for Purbeck Group sediments preserved in several scattered outcrops from Buckinghamshire to Swindon (Arkell 1933; Wimbledon in Cope *et al.* 1980) are everywhere dated as Jurassic, essentially correlating with the highest part of the Portland Limestone Formation of Dorset (Wimbledon, fig. 15 in Cope *et al.* 1980). The most northerly exposure consists of less than 2 m of thinly-bedded clays, marls and limestones exposed at Stewkley in Buckinghamshire (Shephard-Thorn *et al.* 1994, p. 32). Above occur several isolated patches of fine- to medium-grained, ferruginous quartz sands with subordinate clays, scattered along the western margin of the Cretaceous outcrop, from Stewkley (Shephard-Thorn *et al.* 1994) to Shotover Hill near Oxford (Horton *et al.* 1995). Most were placed by Casey & Bristow (1964) into the Whitchurch Sands, which at Whitchurch yielded a diverse, quasi-marine fauna that indicated a correlation with the Cinder Bed in the middle of the Purbeck Group. But at Shotover Hill the 'Shotover Ironsands' yield a sparse non-marine fauna of bivalves and gastropods that are usually assigned to the Wealden. However, Horton *et al.* (1995) noted that these various sands shared a similar petrology and lithology and suggested that they represent the erosional relics of a formerly continuous sheet of predominantly fluvial sands up to 20 m thick which they assigned to the Whitchurch Sands Formation. The heavy mineral assemblage is similar to those of Mesozoic sands fringing the northern margin of the London Platform, such as the Spilsby Sandstone Formation, rather than with those of the Wealden beds. The age of the sands remains problematic. They are generally separated from both the Upper Jurassic beds beneath and the Upper Aptian or Albian beds above by unconformities, and Horton *et al.* (1995) have suggested that they are of Early Cretaceous, probably Valanginian, age. Two other relics of what appears to be the same formation occurs much further to the SE, at Swindon and near Devizes (Casey & Bristow 1964; BGS Lexicon).

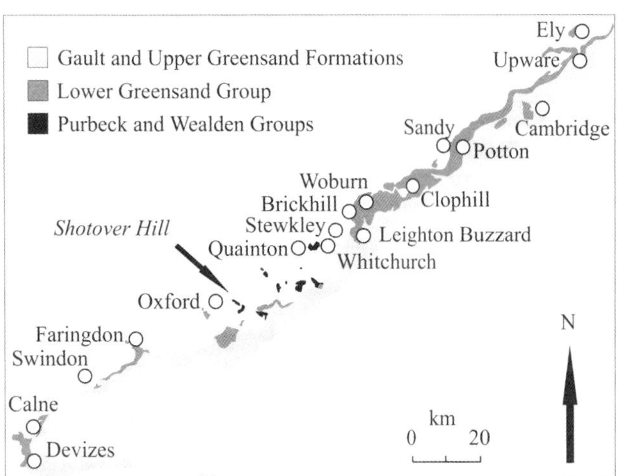

Fig. 15.11. Lower Cretaceous rocks of Central England: distribution and localities.

There are no other sediments of pre-Late Aptian age in the area, but circumstantial evidence indicates that periodic marine spillovers crossed the region during Berriasian–Barremian times. Remanié Spilsby Sandstone fossils are preserved at the base of the Woburn Sands Formation as far south as Brickhill (Keeping 1883; Middlemiss 1962), indicating a significant marine transgression from the north that may be represented in the Wessex Basin by the quasi-marine Cinder Bed in the Purbeck Group (Casey 1963). Subsequent north to south transgressions are indicated by marine or near-marine horizons in the Purbeck and Wealden groups of the Wessex Basin showing an apparent northward increase in salinity, as in the Warlingham borehole (Worrsam & Ivimey-Cook 1971).

At the beginning of the Aptian seawater apparently surged southward again to flood the Wessex Basin and bring non-marine sedimentation to an end there. Within the Bedfordshire Straits remanié Early Aptian faunas indicate that sedimentation occurred at least as far south as Potton (Middlemiss 1962). Then they apparently closed again briefly in mid Aptian times before reopening at the beginning of the Late Aptian.

The Late Aptian history of the region is complex and there appears to have been significant fault control on the accumulation and preservation of sediments. An initial transgression (*nutfieldiensis* Zone) removed any earlier Aptian sediments and incorporated their reworked faunas with older remanié ones in nodule beds at the base of the Woburn Sands; rich derived faunas preserved at Little Brickhill and Upware were described by Keeping (1883). The Woburn Sands Formation forms an almost continuous outcrop from the vicinity of Ely (Gallois 1988) to the Leighton Buzzard area (Fig. 15.11); the SW termination is bounded by a significant structural line, the Woburn–Aspley fault (Ruffell & Wignall 1990, fig. 4), which coincides with one of Rastall's 'axes'; however, it remains debatable whether this controlled sedimentation (Shephard-Thorn *et al.* 1994, p. 83). The formation accumulated on an irregular surface and shows considerable variation in thickness, generally reaching a maximum of about 70 m, but thickening to about 120 m in the Woburn area. It consists of yellow–reddish, locally silver, fine- to coarse-grained, quartz sands, sometimes glauconitic and with seams of iron-cemented 'carstone'. Large-scale planar and trough cross-bedding is common (Fig. 15.12, and Shephard-Thorn *et al.* 1994, plates 4–6) and indicates that currents came predominantly from the north. The style of bedding indicates deposition under a strong tidal influence, but interpretation of the overall environment has fluctuated from fully marine to estuarine, especially for the well-exposed sections in the Leighton Buzzard area (Shephard-Thorn *et al.* 1994; Wonham & Elliott 1996). The interpretation has relied mainly on sedimentological evidence as, except at the base, the sands have yielded very few indigenous macrofossils. But rare *nutfieldiensis* Zone ammonites are recorded at Clophill (Cox 1990), which (unless drifted in) would indicate a marine influx, and other rare marine fossils occur further north (Gallois 1998). Furthermore, at Quainton, only some 15 km west of the southern margin of the Woburn Sands outcrop, Casey & Bristow (1964) recorded a 1.2 m thick ferruginous sandstone with marine fossils, including the ammonite *Parahoplites* and the rudist *Toucasia lonsdalii*, resting with an erosive base on the Whitchurch Sands Formation. They compared this isolated outcrop with the Seend Ironsands (see below) and dated it to the *nutfieldiensis* Zone. Thus it could represent a marginal facies of the Woburn Sands.

Fig. 15.12. Planar and trough cross-bedding in the Woburn Sands, Leighton Buzzard, Bedfordshire. Photograph by Charlie Bristow.

Fuller's earth seams occur in the Woburn Sands around Woburn, some thick enough to have been extracted commercially. Both here and in the Weald Basin the fuller's earths are concentrated in lenticular seams. It is generally believed that the earths represent windborne ashes that were deposited originally on adjacent landmasses, then washed into the shallow-marine waters nearby (e.g. Jeans *et al.* 1977; Young & Morgan 1981). But Goldring (1999) suggested that although the bulk of the earths were of reworked origin, the primary ash apparently accumulated on a shallow sea-floor where little other sediment was accumulating, and was then redeposited in local depressions. Probably both processes operated.

Sediments of similar age to the Woburn Sands occur further to the SW, in Berkshire and Wiltshire, close to the NW margin of the Wessex Basin. Here there were several local marine embayments, each representing a small sub-basin partially bounded by a listric fault (Ruffell & Wignall 1990). The best known is at Faringdon, Berkshire, where sediments are preserved in a shallow, NNW–SSE-trending asymmetric trough cut into Upper Jurassic rocks (Poole *et al.* 1971) and fault-bounded to the NW. At the base occurs the Faringdon Sponge Gravels Formation, a unit of cross-bedded sands and gravels composed largely of sponge fragments. These are overlain by, or pass laterally into, finer-grained sands and sandy clays with some seams of fuller's earth, which are overlain in turn by ferruginous sands and sandstones. Ammonites indicate the *nutfieldiensis* Zone (Casey 1961, p. 562).

In the area around Calne (Wiltshire), two closely adjacent NW–SE troughs mark palaeovalleys cut into the Kimmeridge Clay and later infilled by Upper Aptian sediments. In one, the ferruginous sands of the Seend Ironsands Formation accumulated; they have yielded a rich molluscan and brachiopod fauna again indicating the *nutfieldiensis* Zone (Casey 1961, p. 563). In the other trough there are some quite clean-washed, cross-bedded sands that Hesselbo *et al.* (1990) named the Calne Sands Formation. These accumulated under tidal influence with, unusually, currents flowing towards the NW. Dinocyst assemblages suggest the *jacobi* Zone; hence the sands probably correlate with the lower part of the Folkestone Sands Formation of the southern region.

The Bedforshire Straits remained open during the Early Albian. The most interesting deposits of that age occur in the vicinity of Leighton Buzzard, where a thin (*c.* 5 m maximum) but complex series of sediments occurs at the top of the Woburn Sands Formation. They show rapid lateral variation and include ferruginous sands and clays, phosphatic nodule beds and lenticles of fossiliferous Shenley Limestone. Most workers place them all in the 'Junction Beds' at the base of the Gault (Owen 1972: Shephard-Thorn *et al.* 1994; Eyers 1995; Smart 1997). But the position of the Shenley Limestone lenticles and associated ironstone is problematic. They rest on isolated 'mounds' of the 'Silty Beds' of the Woburn Sands Formation *sensu* Shephard-Thorn *et al.* (1994), and Wonham & Elliott (1996, fig. 4) have suggested that they are a lateral equivalent of part of the overlying 'Red Sands' at the top of the formation. That would support their view that the erosion surface (sequence boundary) at the base of the 'Red Sands' is of Early Albian age and possibly equivalent to the major surface at the base of the Carstone Formation of the East Midlands Shelf. It should be noted that Eyers (1995) regarded the 'Silty Beds' as part of the 'Junction Beds' sequence.

Further to the SW, at Calne, the Calne Sands Formation is overlain by a poorly-sorted, iron-rich sandstone which Hesselbo *et al.* (1990) assigned to the Carstone Formation. No age-diagnostic fossils occur. Glauconitic sands and ferruginous silty clays occur immediately beneath the Gault at several localities from Buckinghamshire to Wiltshire, some with faunas providing evidence of a *mammillatum* Zone transgression (Casey 1961). Elsewhere, any patches of Aptian/Albian sediments that may occur have been overstepped by the Gault Formation.

With the basal Gault and subsequent transgressions the separate identity of the central England region was lost as the Anglo-Brabant Massif to the east was flooded. The Gault succession is therefore discussed below.

The Weald and Wessex basins

From the Pewsey–London Platform fault system southwards, renewed extensional movement in latest Jurassic–Early Cretaceous times led to the fault-controlled development of several rapidly subsiding basins in both onshore and adjacent offshore areas (Fig. 15.13). By Aptian times the pattern was changing to one of broader regional subsidence with almost no faulting, in response to the development of active sea-floor spreading in the North Atlantic (Hamblin *et al.* 1992). The basin terminology for the onshore area used here follows that of Underhill & Stoneley (1998) and Underhill (2002), who distinguish two principal basins, the Weald and Wessex basins. In Early Cretaceous times the Portland–Wight Basin was the main depocentre within the Wessex Basin, although the Pewsey Basin was still subsiding differentially in Purbeck–Wealden times (Whittaker 1985, map 20).

Significant variations in thickness and facies in all the main Lower Cretaceous units reflect this complex tectonic picture. However, although each basin evolved differently, the succession everywhere begins with a thick sequence of non-marine beds (Purbeck and Wealden groups), the lowest part of which is of latest Jurassic age. This is overlain by a marine succession (Lower Greensand Group and Gault and Upper Greensand Formations: Fig. 15.5), deposition of which usually started in Early Aptian times but sometimes commenced as late as the Albian.

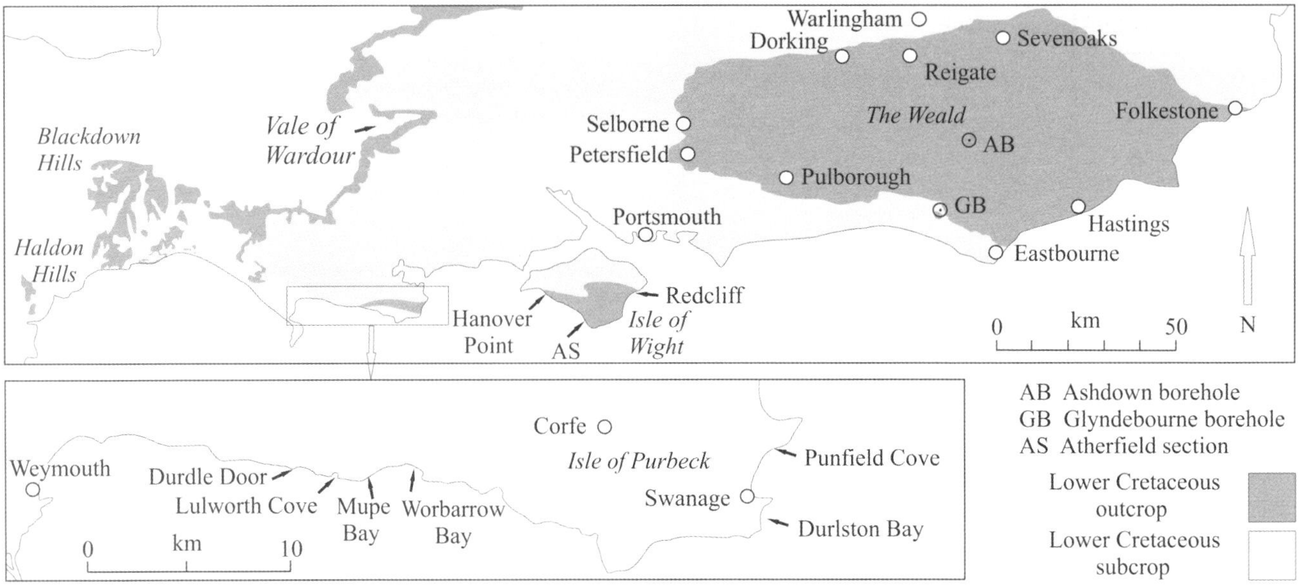

Fig. 15.13. Lower Cretaceous rocks of the onshore part of the Weald and Wessex basins: distribution and localities.

Sediments are exposed in both the Weald (see Ruffell *et al.* 1996) and Wessex basins; in the latter they form scattered outcrops in the western outliers and a narrow strip running from the southern part of the Isle of Wight (see Insole *et al.* 1998) to Dorset (see House 1989) and south Devon, forming the onshore part of the Portland–Wight Basin. A few deep boreholes through the Chalk give glimpses of the covered area. Offshore, there are extensive areas of sea-floor outcrop: numerous BGS cores and a few BGS and commercial boreholes provide a good general picture of the succession there (Hamblin *et al.* 1992).

Within the Weald Basin maximum sedimentation occurred along what is now the axis of the Wealden anticline, especially in the more westerly part, so that the greatest thicknesses of preserved sediments are in the western Weald (Fig. 15.14). In the Wessex Basin, maximum sedimentation occurred in the Portland–Wight Basin, probably from east of Swanage to south of the Isle of Wight. Between the Weald and Portland–Wight Basins lay the Hampshire–Dieppe High (Fig. 15.1) (Hamblin *et al.* 1992, fig. 14a), which may have formed a land area during part of Lower Greensand times and on other occasions hindered the movement of sediment from one basin to the other. The Central English Channel High separates the Portland–Wight Basin from a third, poorly documented, depocentre, the Central English Channel Basin.

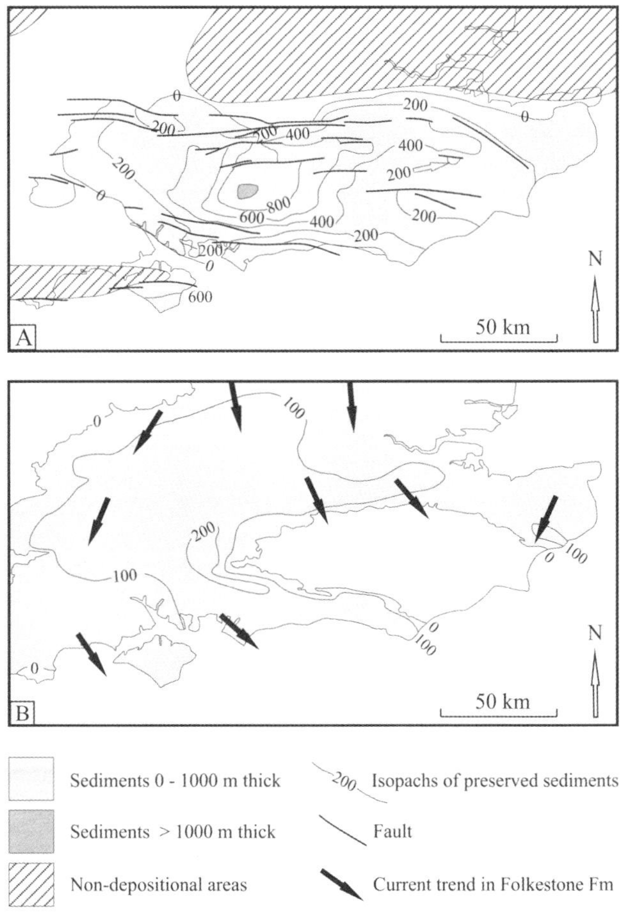

Fig. 15.14. (**a**) Isopachs of Purbeck and Wealden Groups (modified from Whittaker 1985, map 20). (**b**) Isopachs of Lower Greensand Group, Gault and Upper Greensand formations (modified from Whittaker 1985, map 23), with current trends for the Folkestone Formation based on Allen (1982).

The non-marine succession

The non-marine sediments of the Weald and Wessex basins fall into two divisions, the lower essentially calcareous and argillaceous and the upper arenaceous and argillaceous. To these the names 'Purbeck' and 'Wealden' have long been applied, either as lithostratigraphic names or as facies terms. The latter usage remains valuable, although both are used now also as group names. The boundary between the Purbeck and Wealden groups is transitional and therefore hard to define, while interpretations are further complicated by the significant lateral variation in facies between one basin and another, and also within the individual basins. Lithostratigraphic subdivisions currently used are summarized in Fig. 15.5.

Correlation within the Purbeck and Wealden groups has long been based primarily on ostracods. In a series of detailed studies, Anderson (e.g. 1971, 1985) proposed both zones (based on *Cypridea*) and a finer-scale division based on faunicycles. Each (named) faunicycle consists of one or more C- and S-phase couplets. Most of the faunicycles were recorded in both the Weald and Wessex basins, hence providing an apparently refined means of correlation. However, Horne (1995) has pointed

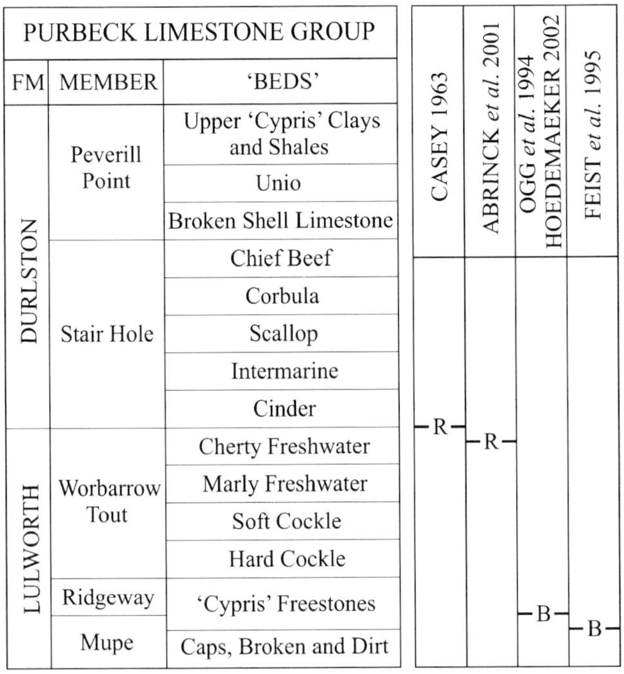

Fig. 15.15. Subdivision of the Purbeck Limestone Group of the Portland–Wight Basin (Westhead & Mather 1996) and position of base Ryazanian [R] and base Berriasian [B] according to various authors.

out that because of inadequate and sometimes conflicting definitions of individual faunicycles the correlations that have been based upon them are probably unrepeatable. He also noted that Anderson's ostracod zones were also inadequately defined, and thus used species ranges determined from Anderson's (1985) final account to propose a revised zonation with three zones based on species of an S-phase genus, *Theriosynoecum*, and eight subzones based on species of *Cypridea* (Fig. 15.5).

Palynomorphs and charophytes also provide some biostratigraphical control (e.g. Hunt 1985; Feist *et al.* 1995). Other means of correlation use such features as distinctive quasi-marine bands and other lithostratigraphical markers, some of which are mentioned below.

Applying the standard chronostratigraphical scale, based on Mediterranean (Tethyan) marine sequences, to these non-marine strata is even more difficult. Biostratigraphical correlations have been attempted using near-marine ostracods in the S-phases, palynomorphs and charophytes. But there is only very limited evidence to work from. There is probably not an ostracod in common with nearby marine areas (Horne 1995) and pollen and dinoflagellate records have so far given only a few slender clues. Charophyte floras (Feist *et al.* 1995) have proved more useful and provide evidence consistent with that obtained from a magnetostratigraphical study of the Purbeck–Wealden sequence (Ogg *et al.* 1994). Carbon-isotope data is also proving valuable (Robinson & Hesselbo 2004). Attempts have also been made to link some of the quasi-marine bands with regional or global sea-level rises, but their use has to be treated with great caution as evidenced by the varying correlations suggested for some of the Purbeck horizons, including the well-marked Cinder Bed.

An additional point to emphasize is that some correlations have been made with 'boreal' areas to the north, which are not easy to correlate with the 'standard' Cretaceous successions of the Mediterranean area. This is particularly important to bear in mind when one considers the placing of the Jurassic–Cretaceous boundary within the Purbeck–Wealden sequence. For well over a hundred years it was drawn at the base or even the top of the Wealden beds. Arkell (1933, pp. 544–556) gave a detailed historical summary of the varying views and brought together the evidence for placing the boundary between the Purbeck and Wealden groups. However, 30 years later, Casey (1963) put forward a radically different interpretation, suggesting that the Cinder Bed, a distinctive oyster-rich horizon in the middle of the Purbeck sequence, represents a brief marine incursion from the north which probably marks the basal 'boreal Cretaceous' transgression, and that the upper half of the Purbeck Group, the Durlston Formation, is therefore wholly Cretaceous in age. That is now generally accepted, although the exact age of the Cinder Bed event remains debatable (Casey 1973; Rawson & Riley 1982; Wimbledon & Hunt 1983). But, as Casey (1963, 1973) demonstrated, circumstantial evidence indicates that the base of the 'boreal Cretaceous' (i.e. base Ryazanian) is younger than the base of the 'standard' Cretaceous (base Berriasian), implying that the latter lies even lower within the Purbeck succession, somewhere within the Lulworth Formation. More recent work, based on palynomorphs (Hunt 1987), charophytes (Feist *et al.* 1995), sea-level events (Hoedemaeker 2002) and palaeomagnetism (Ogg *et al.* 1994), shows that the 'standard' boundary (base of the *jacobi* Zone) correlates with a level at or just above the base of the Cypris Freestones (Fig. 15.15). The base of the Cypris Freestones has also been correlated with a level low in the Upper Volgian *Subcraspedites preplicomphalus* Zone of the southern North Sea area (Abbink *et al.* 2001).

Thus it now seems that the great bulk of the Purbeck Group is of Cretaceous (Berriasian) age. Only the few metres of beds beneath the Cypris Freestones can be assigned to the Jurassic, and both magnetostratigraphy and charophytes indicate that these correlate with the *Durangites* Zone at the top of the Upper Tithonian Substage (Ogg *et al.* 1994; Feist *et al.* 1995). Hoedemaeker (1991) and Feist *et al.* (1995) suggested that the Cinder Bed correlates approximately with the base of the *occitanica* Zone (i.e. base of *subalpina* Subzone) of the Mediterranean Berriasian, but Ogg *et al.* (1994) put it slightly above, at the base of the *privasensis* Subzone.

The Hastings Sub-Group of the Weald Basin is approximately late Ryazanian and Valanginian in age and the Weald Clay Formation Hauterivian–Barremian (Rawson *et al.* 1978). The Hauterivian–Barremian boundary may lie at about the base of the Lower/Upper Weald Clay subdivisions (Worssam 1978; Allen & Wimbledon 1991; Feist *et al.* 1995). In the Portland–Wight Basin, the transition from Purbeck to Wealden conditions apparently started later than in the Weald Basin; the Wessex Formation is of Valanginian–Barremian age (Allen & Wimbledon 1991), although the Valanginian part may be condensed or incomplete (Robinson & Hesselbo 2004). The Vectis Formation was formerly dated as Barremian too, the Barremian–Aptian boundary being placed within the hiatus between the Vectis and Lower Greensand formations (e.g. Casey 1961; Rawson *et al.* 1978). But palaeomagnetic evidence indicates that the base of the Aptian lies high in the Vectis Formation of the Isle of Wight (Kerth & Hailwood 1988) and this is supported from carbon isotope data (Gröcke *et al.* 1999; Robinson & Hesselbo 2004), which implies that the highest Weald Clay is also earliest Aptian (Allen & Wimbledon 1991, fig. 2).

The Purbeck Limestone Group. This consists of a varied series of non-marine limestones, shales and evaporates, with some well-developed sandstone horizons in the Weald Basin. The sediments rest comformably but normally with sharp contact on varying horizons in the Portland Stone Formation. They are missing on some of the main structural highs due to non-deposition or subsequent erosion (Underhill 2002, fig. 7B).

The Group extends over most of the Weald Basin and has been penetrated by numerous boreholes (Anderson & Bazley 1971), in one of which (Ashdown) it reaches its maximum 'onshore' thickness of about 191 m. But exposure is limited to three small inliers in Sussex and the lowest beds are not exposed (Howitt 1964). The top of the underlying Portland Limestone Formation is a calcareous sandstone, which contrasts markedly with the overlying lowest Purbeck beds. The first 15 m of the latter consist of silty limestones and calcareous shales with seams of anhydrite and gypsum, some thick enough to be mined. The remainder of the Lulworth Formation consists of some 40 m of shales and interbedded limestones, the latter becoming more common upward, overlain by a mudstone-dominated sequence about 12 m thick.

A thin (*c.* 1.5 m) unit of calcareous shales and limestones with scattered *Praeexogyra distorta* (J. de C. Sowerby) is regarded as the correlative of the Cinder Bed and placed with the overlying beds into the Durlston Formation (*c.* 70 m). The lower part of the formation (*c.* 30 m) includes cross-bedded sandstones capping small-scale cycles in the 'Arenaceous Beds' and 'Beef Beds' (Anderson & Bazley 1971, p. 13). Above, bands of shelly, argillaceous limestone appear, becoming increasingly numerous upwards; these form the 'Greys Limestones' (*c.* 20 m). These are overlain by a mudstone-dominated succession up to 18 m thick, the top of which is formed by a thick bed of clay that is overlain by sands and silts at the base of the Ashdown Sands Formation.

There are excellent coastal exposures in the Portland–Wight Basin, especially in the type area of the Isle of Purbeck. Here the sediments accumulated during a phase of extensional tectonics, and the Purbeck Fault in particular had a significant influence on thicknesses and lithofacies developments (Underhill 2002). The thickest onshore development (*c.* 120 m) is at Durlston Bay, Swanage, where numerous lithological subdivisions can be distinguished (see Clements 1993); despite lateral facies changes and thinning, some of these can be traced westward at least as far as Lulworth Cove. Eastward from Swanage into the offshore area there are over 200 m of Purbeck sediments in the centre of the Portland–Wight Basin (Underhill 2002, fig. 7B).

The Lulworth Formation (normally 34–63 m thick) consists of white-weathering, generally micritic, limestones interbedded with dark mudstones (Westhead & Mather 1996). It embraces eight of the 'beds' that have been recognized since the mid 19th century, grouped by Westhead & Mather (1996) into the Mupe, Ridgeway and Worbarrow Tout members (Fig. 15.15). The lowest part of the sequence, like its counterpart in the Weald Basin, includes evaporite levels. The gypsum has often been dissolved away, to be replaced by calcite and other secondary minerals (West 1975). In the Great Dirt Bed there are are some spectacular occurrences of coniferous tree stumps, some silicified, others represented by hollows (moulds)–preserved because algal stromatolites grew around the original stumps as the forest was flooded (Francis 1983). Longer sections of tree trunk are also found. There is considerable lithological variation higher in the Lulworth Formation sequence, with ostracod-rich limestones (Cypris Freestones), chert-rich limestones (Cherty Freshwater beds) and calcareous mudstones and marls. Pseudomorphs after rock salt occur at many levels (West 1975; Radley 1992). Two beds in the Cherty Freshwater beds in Durlston Bay (DB91 and DB 98 of Clements 1993) are believed to be of volcanogenic origin (Jeans *et al.* 2000, 2005).

The overlying Durlston Formation consists mainly of thin alternations of limestones and shales with several distinctive fossil horizons. The basal Cinder Bed is an oyster (*Praeexogyra distorta*) lumachelle which also includes apparently marine taxa such as the bivalve *Laevitrigonia cineris* Kelly and, in Durlston Bay, the regular echinoid *Hemicidaris purbeckensis* Forbes. Some higher horizons, especially the Scallop Bed, also yield marine or quasi-marine bivalves, but the bulk of the Durlston Formation is characterized by non-marine ostracods and gastropods. Near the top of the sequence, beds DB224 and DB 239 in the Upper 'Cypris' Clays and Shales are also of volcanogenic origin (Jeans *et al.* 2000), and may equate with bentonitic beds D7D and D6J at Speeton and their equivalents in BGS borehole 81/43 in the southern North Sea (Jeans *et al.* 2005).

The Wealden Group. The passage from Purbeck to Wealden groups marks a return to completely clastic sedimentation, including the deposition of major sand bodies. The latter are particularly well developed in the Weald Basin, and because of the facies differences between this and the Portland–Wight Basin, different lithostratigraphic names are used in each basin. In the Weald Basin the lower part of the succession is a sand-dominated sequence of ferruginous sandstones, siltstones and shales assigned to the Hastings Sub-Group. This embraces the thick, arenaceous Ashdown and Tunbridge Wells formations, separated by an argillaceous unit, the Wadhurst Clay Formation. This clay extends over the whole Weald. Higher in the succession another argillaceous unit, the Grinstead Clay Formation, occurs in the western Weald, where it splits the Tunbridge Wells sands before petering out eastwards. Hence in the eastern Weald the upper and lower members of the Tunbridge Wells Formation cannot be separated. Although these major units can be placed in vertical succession their boundaries are diachronous: the vertical lithological changes represent two major fining-upward cycles (Ashdown and Wadhurst formations; lower Tunbridge Wells sands and Wadhurst Clay Formation) and part of a third (Allen 1959; Bristow & Bazley 1972). Superimposed on this broad cyclicity is a smaller-scale cyclicity in some units which, coupled with lateral facies changes, results in the recognition of numerous minor, local members and beds within the Hastings Sub-Group.

The Ashdown Sands Formation is a thick (maximum *c.* 230 m) sequence of fine-grained sandstones, siltstones and subordinate shales and mudstones. The lower part tends to be the finer grained (siltstones and shales), especially on the coast near Hastings, where these argillaceous beds, sometimes referred to as the 'Fairlight Clays', are the principal source of Wealden plants (Watson & Sincock 1992, p. 2). The upper part shows a series of coarsening-upward units, each terminating in a massive sandstone (Bristow & Bazley 1972). At the top of the formation is a widespread pebble bed, known as the 'Top Ashdown Pebble Bed' but taken to mark the base of the Wadhurst Clay Formation. The latter consists mainly of up to about 70 m of grey mudstones and siltstones with thin shelly limestone, ironstone and sandstone beds and is generally poorly exposed. It is overlain by the Tunbridge Wells Sand Formation which consists mainly of fine-grained sandstones, with subordinate siltstones, silty clays and mudstones. Where it is split by the Grinstead Clay Formation (maximum *c.* 18 m) in the western Weald, the upper part of the lower unit is a massive sandstone, the Ardingly Sandstone Member. Where the Grinstead Clay disappears eastwards the Ardingley Sandstone is no longer mapped, although some massive sandstones do occur about 18 m above the base of the Tunbridge Wells Sandstone Formation.

Above the Hastings Sub-Group is a thick argillaceous sequence, the Weald Clay Formation. This reaches its maximum thickness of about 500 m in the western Weald. It consists mainly of clays and silty clays with subordinate thin bands of sandstone, shelly limestone and clay ironstone (Worssam 1978). Cyclicity occurs on three scales, the smallest being an alternation of laminated silty clay layers (0.3–1.5 m) and thinner, less silty clay beds with fewer laminae. The intermediate scale consists of an essentially fining-upward sequence culminating in a thin 'Paludina' limestone, while on the largest scale the whole formation can be divided into two informal units, the lower characterized by small-'Paludina' limestones and the higher by large-'Paludina' limestones (Worssam 1978). Red clay bands are developed in more marginal areas and are useful for local correlation. Sand bodies still occur but are much less prominent than in the Hastings Group. The Weald Clay has yielded a varied flora and fauna, including well-preserved insect remains (e.g. Jarzembowski 1991).

The Wealden Group of the Portland–Wight Basin is divided into two units, a lower of marls and sands called the Wessex Formation, and an upper of shales, the Vectis Formation (Daley & Stewart 1979). The Isle of Wight sequence is the best exposed and is described in detail by Insole *et al.* (1998). The main depocentre lay to the south of the Wight Fault, which together with the contiguous Purbeck and Bray faults formed the southern margin of the Hampshire–Dieppe High. Here the Wessex Formation may be as much as 550 m thick, but only the top 180 m are exposed. It is composed of predominantly reddish-coloured marls and marly shales with subordinate sands. The mudstones were deposited on an alluvial plain crossed by occasional meandering channels. The channels are sometimes filled by fining upward arenaceous sequences; these include the Sudmoor Point Member, which consists mainly of point bar sands deposited in a meander belt at least 1–8 km wide (Stewart 1981). At Hanover Point the 'Pine Raft' is a plant bed yielding logs, up to 3 m long and 1 m in diameter, of gymnosperms (Francis 1987; Insole *et al.* 1998, p. 92).

Traced westward into Dorset the Wessex Formation becomes more varied – the 'variegated marls and sands' (see Arkell 1947). Sandstones, grits and conglomerates appear and these coarsen and increase in proportion westward towards Cornubia as the overall thickness decreases from some 700 m at Swanage to 65 m at Durdle Door. The formation disappears west of Durdle Door, through erosion and Gault overstep. Most beds cannot be traced far laterally, but the Coarse Quartz Grit Member, a band of ferruginous coarse quartz grit with interbedded sand and clay seams in the middle of the formation, extends for some 23 km along the whole outcrop (Arkell 1947); it is 8.5 m thick at Durdle Door, close to the source area, but thins eastwards into the basin. This distinctive unit indicates a significantly increased flow of water, probably reflecting a sudden tectonic or climatic event in the source area (Allen & Wimbledon 1991). The 'Pine Raft' of the Isle of Wight has been suggested to be a more distal correlative (Hughes & McDougall 1990), but carbon isotope evidence suggests that while the Coarse Quartz Grit is probably mid to late Hauterivian, the 'Pine Raft' is younger, lying at about the Hauterivian–Barremian boundary (Robinson & Hesselbo 2004). Fossils are not common in the Wessex Formation, but ostracods, *Unio* and gastropods occur. The remains of dinosaurs (especially *Iguanodon*) and their footprints, crocodiles and turtles are recorded from the Isle of Wight (Swinton 1936; Radley 1994).

The Vectis Formation contrasts markedly in colour with the Wessex Formation: the boundary is marked by a sharp change from reddish beds beneath to dark grey beds above. Ruffell & Wach (1991) interpret this boundary as a transgressive surface of Late Barremian age. The Vectis Formation is fairly uniform over most of the basin, consisting of mudstones and siltstones with occasional thin shelly limestones and minor sands. But in the vicinity of Corfe the shales pass laterally into sandy clays and shales difficult to separate from the variegated beds below. There is a gradual westerly thinning from a maximum of about 66 m at Atherfield to only 3–5 m at Corfe: the beds are not known farther west.

On the Isle of Wight much of the formation consists of thin (70–90 cm) fining-upward cycles (sandy mud–mud) assigned to the Cowleaze Chine and Shepherd's Chine members. The intervening Barnes High Sandstone Member, towards the base of the formation, consists of one or more thick, coarsening-upward sandstone beds: bedding structures change upwards from lenticular bedding through wavy bedding to large-scale cross-bedding (Stewart 1981; Insole *et al.* 1998).

Ostracods are abundant in the Vectis Formation, while the molluscan faunas are dominated by *Viviparis* and *Filosina*. Brackish beds with *Liostrea* and other mollusca occur in the top of the Shepherds Chine Member.

In the offshore part of the Portland–Wight Basin the Wessex and Vectis formations have not been distinguished separately. Here the Wealden reaches a thickness of over 500 m but in the limited number of well sections, sandstones are scattered throughout, especially towards the Hampshire–Dieppe High (Hamblin *et al.* 1992).

The sediments that poured into the Weald and Wessex basins were derived from several different sources, as attested by the nature of the detritus, facies distributions (particularly the lateral changes in distribution and form of sand bodies) and the many associated sedimentary structures. The main source areas were the London Uplands, Cornubia and Armorica (Allen 1976; Garden 1991), but there were more local sources too, such as the Hampshire–Dieppe High, which may have provided sand to the adjacent part of the Portland–Wight Basin (Hamblin *et al.* 1992). The input from individual sources varied through time in response to both local tectonics and regional climate change. For much of the time the London Platform was the major contributor of sediment to the Weald Basin, while Cornubia fulfilled a similar role for the Portland–Wight Basin. But through time the Cornubian influence extended eastward so that towards the end of the Valanginian its detritus began to spill into the western Weald to become the dominant component there in Hauterivian–Barremian (Weald Clay Formation) times (Allen 1976, 1998). There was also a 'boreal' input which diminishes southward from the NW Weald, complementing the facies evidence of transgressions from the same direction and the faunal evidence of diminishing salinity southwards.

The environment in which the Purbeck–Wealden sediments accumulated has been much debated. They certainly represent a long period, some 30 Ma, of non-marine sedimentation, and there is considerable evidence that during that time climate changed from arid to humid, and possibly back to semi-arid (Ruffell & Rawson 1994; Allen 1998). The facies change from the Portland Limestone Formation to the Purbeck Limestone Group represents a change from shallow-marine carbonate deposition to non-marine deposition in very shallow lagoons and lakes of highly variable salinity, from freshwater to hypersaline. But the degree of salinity and the durability of water bodies is open to question. Anderson (1971, 1985) thought that each of his C- and S-phase couplets reflected an alternation of less and more saline phases, though some authors have queried the salinity tolerances of some of the S-phase genera. More recent work on the ecology and especially the reproductive strategies of different ostracod groups has led Horne (1995) to suggest that salinity fluctuations were less important than the permanence of water bodies; cyprideans could survive periodic drying out and therefore inhabit ephemeral waters, while S-phase forms required a 'permanent' water body to survive in. On the other hand, most cypridean taxa were probably freshwater (Horne 2002) while some of the other Purbeck/Wealden ostracods could probably tolerate saline waters, or fluctuating salinities. Whether there was salinity fluctuation on a cyclic basis or not, there is no doubt that there were both lateral and temporal changes from fresh to saline water and from ephemeral to more 'permanent' water bodies in response to phases of marine or freshwater flooding and intervals of evaporation during more arid phases.

Purbeck water depths were probably no more than 1 m (Batten 2002) and Wealden depths were probably similar (Allen 1975). But in the predominantly semi-arid, warm climate that lasted for much of Purbeck time the basins dried out periodically to produce evaporites and soil horizons. Coniferous forests grew in the latter, and Underhill (2002) has demonstrated how the rapid submergence of such forests in the Lulworth area was caused by contemporaneous movement along the Purbeck Fault.

In mid Berriasian times the climate became more humid (Batten 2002), leading to the deposition of Wealden clastic sediments. Wright *et al.*'s (2000) study of the Isle of Wight Wealden suggests that rainfall was seasonal, with arid periods intervening, and that the overall environment represented seasonal wetlands analogous to those occurring in tropical and subtropical areas today. But climate modelling indicates high precipitation throughout the year, the apparent summer aridity representing times when evaporation rates were also very high (Haywood *et al.* 2004).

Whichever interpretation is preferred, both proxy and modelling evidence indicates that Wealden sediments accumulated under a much wetter regime than the Purbeck ones did. Rainfall increased further at times in response to periodic uplift of the highs surrounding the Wessex Basin. This increased rainfall was often storm-induced, so that rivers discharged sediment over the basin at very fluctuating rates. The major studies by Allen (e.g. 1976, 1981, 1998) envisaged a fluvial/lacustrine environment with fringing lagoons where the basin opened out towards the 'Boreal sea' (Fig. 15.2a). Within the Weald Basin, marginal areas were crossed by major streams that deposited braided floodplain-sands traversed by channels. Towards the centre of the basin silts and clays accumulated in a meander plain region, crossed by sinuous streams feeding into numerous lakes. During the wetter phases, the braidplain sands from the north (London Uplands) and south (Armorica) transgressed over the meander plain, sometimes meeting and even overlapping along the axis of the basin. Periods of particularly severe flooding are marked by geographically widespread pebble beds that provide regional markers at the base of major cycles (e.g. top Ashdown pebble bed). At other times the more argillaceous, meanderplain, facies transgressed rapidly over the sandy facies to cover much of the Weald Basin (Wadhurst, Grinstead and Wealden Formations). These transgressions may have resulted from a reduced input of sand rather than from rises in sea level in adjacent marine basins, although the beginning of Weald Clay sedimentation may reflect the basal Hauterivian transgression.

The more detailed interpretation of sequences in, for example, the Isle of Wight supports Allen's overall model.

The marine successions

Lower Greensand Group. The 30 Ma interval of non-marine deposition was brought to an end by a series of eustatic sea-level rises that commenced in very early Aptian times. The incoming seas gently planed the top of the Wealden so that there is a disconformable relationship between that and the overlying Lower Greensand Group. Throughout southern England and adjacent offshore areas the initial marine sediments were argillaceous, but sands soon prograded over the area to form a thick (maximum 300 m) pile of shallow-marine sediments from which the Lower Greensand derives its name. However, the glauconite which imparts the green colour has usually been oxidized so that the bulk of the sands are yellow or brown.

The Hampshire–Dieppe High continued to separate the Weald and Portland–Wight basins to affect facies and thicknesses and may have been emergent at times. It forms a likely source of land-fall ash redeposited as pockets of fuller's earth in the southern part of the Weald Basin (Young & Morgan 1981) and the Isle of Wight (Ruffell *et al.* 2002).

The Lower Greensand Group crops out from Kent to Dorset, and is exposed in both coastal cliffs and inland quarries. Casey's (1961) masterly synthesis of its stratigraphy and palaeontology, with accounts of all the main exposures then visible, forms the starting point for all subsequent work. The key section is the 5.5 km-long cliff exposure from Atherfield Point to Rocken End in the Isle of Wight, known as the Atherfield (Chale Bay) section (Fig. 15.16; Plate 28). This exposes the thickest Lower Greensand sequence in the Portland–Wight Basin. It not only serves as a standard of reference for the basin but is the type locality for many of the standard ammonite zones of the English Aptian. Since Casey's (1961) review, many new ammonites have been collected and necessitated some alterations to the zonal scheme (Casey *et al.* 1998); the modified scheme is incorporated in Figure 15.5.

At the base of the Lower Greensand Group, the Atherfield Clay Formation extends over both the Weald and Portland–Wight basins. It reaches its maximum onshore thickness of approximately 55 m at the type locality, in the Isle of Wight, but 66.35 m were proved in BGS borehole 55/35 just offshore to the east of the island. The base of the formation is diachronous; it generally lies in the *obsoletus* Subzone (*fissicostatum* Zone), but in east Kent sedimentation started a little later (?*kiliani* Subzone of the *forbesi* Zone). The formation consists predominantly of silty clays, with local seams of fuller's earth in west Kent. But at Atherfield it is lithologically more varied, consisting of silty clays, silts and fine sands, some forming coarsening-upward sequences (Insole *et al.* 1998). Here the formation has long been divided into five units that are now treated as local members (Simpson 1985: Insole *et al.* 1998). Most cannot be traced with confidence in other sections, where the formation tends to be more uniform lithologically. But the thin, richly fossiliferous Perna Beds Member which marks the initial marine transgression at the base of the formation can be traced more widely in the Isle of Wight and parts of the Weald Basin (Casey 1961; Simpson 1985). Fossils are common at some levels, especially in the Perna Beds and Crackers members: the faunas include ammonites, bivalves and crustaceans. Both sedimentological and faunal evidence indicates deposition in a shallow-marine environment affected by intermittent storms.

At Punfield Cove in Dorset there is an interesting horizon at the top of the Atherfield Clay, the 'Punfield Marine Band', or Punfield Bed, consisting of 0.3 m of argillaceous and sandy limestones with lignite streaks. It contains a diverse shallow-marine fauna (Simpson 1983) including decapod crustaceans, gastropods, bivalves and the ammonites *Deshayesites* and *Roloboceras*. Further west at Corfe the same bed is represented by a thin clay ironstone with a similar fauna. But at Worbarrow Bay and Lulworth Cove the possible correlative yields bivalves of brackish-water aspect (*Ceratostreon*, *Eomiodon* and *Filosina*). Casey (1961) suggested that the faunal change may represent a passage from Wealden to Lower Greensand conditions along the estuary of an eastward-flowing river which briefly discharged sand (the Crackers Member) as far as the Isle of Wight.

The sandy beds above the Atherfield Clay Formation show considerable variation, so that different lithostratigraphical schemes are used for the Weald and Portland–Wight basins (Fig. 15.5). In the Weald Basin, three formations are distinguished over most of the area. The lowest is the Hythe Formation, which rests disconformably on the beds beneath. It reaches a maximum thickness of about 100 m in the western Weald. The formation occurs in two distinct facies, both of which exhibit coarsening-upwards cycles (Ruffell 1992). The easterly facies (Kent and East Sussex: Ruffell 1992, fig. 4) is the 'rag and hassock', which consists of alternating layers (15–60 cm thick) of hard, sandy blue-hearted limestones (the 'Rag') and poorly cemented, grey, glauconitic sandy mudstones and argillaceous sandstones (the 'Hassock'). This facies is very fossiliferous, with rich bivalve, ammonite, belemnite and brachiopod faunas. Further west, in Surrey and West Sussex, the equivalent beds become much less calcareous, consisting of greenish-grey, sparsely glauconitic, fine- to medium-grained sandstones with lenticular beds of chert. Clay interebeds occur, some of fuller's earth. Fossils are much less common in this facies.

The overlying Sandgate Formation is transgressive; in marginal areas of the basin the basal nodule beds (*nutfieldiensis* Zone) incorporate faunas reworked from the Hythe Formation beneath (Casey 1961). The Sandgate Formation reaches a maximum thickness of about 100 m in the SW Weald and is very variable in both lithology and thickness. It consists predominantly of fine sands, silts and silty clays, often glauconitic and poorly fossiliferous. This facies persists through Kent, but then around Reigate (Surrey) bands of fuller's earth and cherty and glauconitic sandstones and limestones appear, overlain by glauconitic argillaceous sands. The fuller's earths have long been exploited in the Reigate area, where they form thin, lenticular beds in the *nutfieldiensis* Zone (Dines & Edmunds 1933). The main seam is over 4 m thick and extends laterally for several kilometres.

West of Dorking (Surrey) the Sandgate facies change again. Here, bands and concretions of calcareous stone appear, the Bargate Member, overlain by the ferruginous argillaceous sands of the Puttenham Member (Kirkaldy 1933). Fossils are common in the Bargate Member, which is famed for its brachiopod faunas (Middlemiss 1959). In the SW part of the outcrop, between Peterfield and Pulborough, there are further lithological changes and several local units have been recognized (Kirkaldy 1933; BGS Lexicon), all regarded now as local members (BGS Lexicon). East of Pulborough the Sandgate and Hythe Formations cannot be separated with confidence (Kirkaldy 1933; Young & Lake 1988).

The Folkestone Formation (maximum 80 m) typically consists of generally unfossiliferous, medium- to coarse-grained, sands with occasional pebbly and silty seams. Irregular veins of secondary iron form 'boxstones'. Over much of the NE and western Weald the sands show spectacularly preserved, mud-draped cross-stratified units, which Allen (1982) interpreted to represent tidally-deposited sandwaves probably showing neap-spring cycles, the general transport direction being from the north. The facies changes in east Kent, where the sediments accumulated closer to the shoreline. Here, at the type locality, the Folkestone Formation consists of 20 m of coarse, yellowish greensands with bands of calcareous and glauconitic sandstone, sometimes strongly bioturbated. This sequence is richly fossiliferous (Casey 1961) with numerous species of ammonites and bivalves. Within the formation a significant erosion surface marks the 'mid-*tardefurcata* break' (Casey 1961, p. 501).

The Lower Greensand formations become increasingly difficult to distinguish as they are traced eastward along the southern margin of the Weald Basin, where the sediments thin and become more fine-grained (Lake *et al.* 1987). This probably reflects proximity to the northern margin of the Hampshire–Dieppe High.

In the Portland–Wight Basin three sandy formations are again distinguished above the Atherfield Clay Formation. The lowest, the Ferruginous Sands Formation, varies from medium to coarse glauconitic sands to sandy mudrocks, often extensively bioturbated. Eleven local members are currently recognized in the Isle of Wight, where the formation consists essentially of a series of coarsening-upwards units (parasequences), from sandy muds or argillaceous sands to glauconitic sands, often capped by a firmground (Insole *et al.* 1998; Ruffell & Wach 1998). Here it reaches a maximum thickness of about 134 m at Atherfield but thins rapidly westward into east Dorset where it is only about 45 m thick at Punfield Cove, Swanage. Further west the apparent correlative thins to 25 m at Worbarrow Bay: there and at Mupe Bay the development of brilliant colours and thin ironstone seams gives a distinct Wealden aspect (Arkell 1947, p. 176). One thick (maximum 6 m) pebbly sandstone near the top is apparently traceable from Punfield (bed 13) to Worbarrow (bed 14), and Arkell (1947, p. 174) suggested that it may be represented in the Isle of Wight by 'Groups' XIII and XIV (= New Walpen Chine Member and Member XIV in Insole *et al.* 1998).

Body fossils occur mainly in calcareous concretions in the upper part of coarsening upward cycles, but the large oyster *Aetostreon* is common on the top surface of some firmgrounds (Insole *et al.* 1998). Extensive bioturbation has destroyed most primary sedimentary structures.

There is a significant lithological change at the top of the Ferruginous Sands Formation, with the appearance of white and yellow quartz sands, often compacted into 'sandrock' – the Sandrock Formation. This occurs mainly in the Isle of Wight, where it reaches a maximum thickness of about 70 m at Atherfield. It is composed of three or four large-scale (15–30 m) coarsening-upward units, each with a pebbly erosion surface at the base. The lower part of each shows massive, dark-coloured glauconitic mudstones alternating with thinly laminated units of fine sand and dark mudrock, grading up through silty sands into the well-sorted, cross-stratified quartz sands characteristic of the formation, which was probably deposited under a strong tidal influence. At Punfield Cove the apparent equivalent is some 3 m of poorly exposed yellow sand, apparently overlain by the Gault Formation, which oversteps this horizon further west. The Sandrock Formation contains few body macrofossils but a varied trace-fossil assemblage and a microfauna indicating that it was apparently deposited in an estuarine environment, the thicker cross-bedded sands accumulating in subtidal channels (Wach & Ruffell 1991; Insole *et al.* 1998).

The overlying Carstone Formation (maximum 22 m) consists of gritty, red-brown, medium- to coarse-grained, ferruginous sands and sandstones with pebbles and phosphatic nodules. It shows an overall fining-upward trend. Body fossils are rare but bioturbation is common. It rests disconformably on the Sandrock Formation and is the only Lower Greensand formation to transgress northward across the Wight Fault (Ruffell 1991). This is probably another reflection of the Early Albian mid-*tardefurcata* break. As Insole *et al.* (1998, p. 10) have pointed out, although the formation is included within the Lower Greensand Group it really represents a trangressive unit genetically linked with the overlying Gault Formation, into which it grades. Although the Carstone Formation of eastern England is not continuous with the Isle of Wight unit it is of comparable facies and age and shows a similar unconformable relationship with underlying units and a gradational one with the overlying Gault/Hunstanton formations. Relics of a similar facies occur also in the intervening area, around Leighton Buzzard in Bedfordshire.

The Lower Greensand Group was deposited in shallow-marine and, sometimes, estuarine waters subject to strong tidal influence which appears to have increased in strength through time, from microtidal at the beginning of the Aptian times to macrotidal across the Aptian–Albian boundary (Ruffell & Wach 1991). The initiation of strong tides in the Aptian may be linked with graben collapse in the Western Approaches which would have allowed the penetration of currents from the opening Atlantic Ocean (Anderton *et al.* 1979, p. 237) through the Western Approaches Trough (Rawson 1992*b*, map K2b). The sands were derived from a variety of sources. The predominant one was probably from the west (Cornubia), following on the pattern established in later Wealden times. But the slightly rejuvenated 'London Uplands' provided some sediment to the Weald Basin (Middlemiss 1976), although Ruffell (1992) suggested a westerly source too. The Hampshire-Dieppe High probably provided sediment to adjacent parts of both the Weald and Portland–Wight basins.

The Gault Formation. Across most of the outcrop and subcrop there is a rapid transition from the sands of the Lower Greensand to the clays of the Gault Formation, normally marked by 1–2 m of rather condensed silty clays and fine sands with bands of phosphatic nodules – the 'junction beds' (Casey 1961; Owen 1992). But in East Anglia the Gault oversteps older Mesozoic sediments to rest on Palaeozoic rocks at the core of the buried Anglo-Brabant Massif, while in Dorset and Devon the formation thins westward as it oversteps successively older beds. This expansion of the depositional area marks the beginning of a major transgression that spread progressively further west and north as sea levels rose through the Albian to flood many of the former landmasses (Rawson 1992*b*, map K3).

The typical Gault lithology is a pale–dark calcareous clay, with phosphatic, pyritic and calcareous nodules. Two informal divisions are recognized, the Lower and Upper Gault. The Lower Gault and lower part of the Upper Gault are lithologically more varied than the paler, more calcareous highest beds and usually contain numerous phosphatic nodule bands marking erosion levels or pauses in deposition.

From Norfolk to Bedfordshire, the Gault can be divided into 18 beds, with a 19th developed locally (Gallois & Morter 1982). At outcrop, it thickens from 30 m at Cambridge to 50 m in Buckinghamshire. Eastwards from outcrop it thins rapidly onto the Anglo-Brabant Massif, due at least in part to subsequent (base Chalk) erosion. In this region the Lower Gault and lower Upper Gault are composed of a series of 13 small-scale (1–2 m) fining-upward cycles, each coinciding with one or more of the individual beds (Gallois & Morter 1982). There is an erosion horizon at the base of each cycle, normally marked by an overlying band of phosphatic nodules. The base of bed 11 is the most significant one, forming the Lower–Upper Gault boundary and traceable across the whole of East Anglia and the Wessex Basin. This level marks an important phase of erosion and transgression in early *cristatum* Subzone times throughout the Anglo-Paris Basin. The Upper Gault oversteps the Lower Gault to rest on Palaeozoic rocks at the core of the Anglo-Brabant Massif; it marks the Late Albian (base of *cristatus* Subzone and *inflatum* Zone) transgression which flooded the whole of eastern England (Fig. 15.2c) to submerge the last remnants of the massif, where there is also a lateral passage from the Upper Gault to the Hunstanton Formation in north Norfolk (Rawson 1992*b*, map K3).

Within the Weald Basin, E–W-trending faults ('axes') continued to affect Lower Gault (Middle Albian) sedimentation (Owen 1971). Here the Gault outcrop forms a narrow strip beneath the Chalk escarpments of the Downs. The clays are exposed in a few inland quarries but are best known from the type section at Copt Point, Folkestone, where the formation has long been divided into 13 beds (I–XIII: see Smart *et al.* 1966; Owen 1971), some of which can be recognized inland. In addition, Owen (1975) divided the Upper Gault into four informal units recognizable over a broad area of the Weald. The Gault is about 45 m thick at Folkestone, but is twice this thickness in the western Weald, where it reaches a maximum of 104 m in the Glyndebourne (Sussex) borehole (Lake *et al.* 1987).

The lowest beds (to at least the *spathi* Subzone of the *dentatus* Zone) thin over the Hampshire–Dieppe High, then the sequence thickens again into the Portland–Wight Basin. Here the Gault is normally poorly exposed, but at Redcliff (Isle of Wight) landslipping in the winter of 1994–95 provided an excellent exposure of most of the sequence (Gale *et al.* 1996). This and data from other sections and boreholes shows that the Gault formation of the Isle of Wight has a significant sand component, which increases in proportion towards the probable source area to the west and SW. At Redcliff, there is a gradational contact with the underlying Carstone. The Lower Gault is thicker here than at Folkestone and has few phosphatic nodules, while the base of the Upper Gault (*cristatum* Subzone) is probably represented by a 20 cm thick green glauconitic sand resting on an omission surface (Gale *et al.* 1996).

The base of the Gault becomes younger as it thins westward through Dorset and Devon, to disappear completely in the vicinity of Beer, where it is overstepped by the Upper Greensand Formation. Exposures are limited, but the clays are normally quite sandy. At the best exposure, in Worbarrow Bay, the combined Gault Clay–Upper Greensand sequence is 57 m thick (Wright in Arkell 1947), of which probably 34.5 m belongs to the Gault.

The 'typical' Gault clay is a mid or outer shelf mud. Estimates of water depth have fluctuated considerably, usually in the range of 90–350 m, but Knight (1997) suggested that the common occurrence of algal micro-borings in shell material indicates that the sea floor normally lay in the photic zone, with a depth of less than 80 m. He also concluded that the floor was swept by cold, nutrient-rich currents that fluctuated in strength. Waters shallowed towards the western margins of the Wessex Basin and over the London Platform.

The Upper Greensand Formation. The westward increase in amount of silt and sand within the Gault clays in both the Weald (Owen 1971, p. 147) and Portland–Wight basins foreshadows a major influx of sands from the west in Late Albian times, to form the Upper Greensand Formation. The sands prograded progressively eastward so that the diachronous base of the formation youngs in that direction. Eventually they extended approximately to a line drawn through Dunstable, Sevenoaks and Eastbourne (Fig. 15.2c) (Rawson 1992*b*, map K3): east of this the Upper Albian sediments are wholly in Gault Clay facies. The boundary between the two formations is a gradational one.

The Upper Greensand is the least-studied part of the Lower Cretaceous sequence. In the Weald Basin, it normally consists of poorly consolidated siltstones in the lower part, overlain by fine-grained sands and sandstones, sometimes glauconitic, and irregularly cemented by calcareous or siliceous material – the latter probably derived from the siliceous sponge spicules that are so common in this formation. This is the 'Malmstone' facies. Exposures are poor, but the formation is thickest in the western Weald, where it reaches about 60 m in the Selborne district (Gallois 1965).

The formation thins over the Hampshire–Dieppe High then thickens again in the Portland–Wight Basin. Here, parts of the sequence are visible in the Isle of Wight (Insole *et al.* 1998), where it is up to 45 m thick. Again the lower beds are finer-grained, varying from silts to sandy mudrocks and argillaceous sands. Above are fine-grained, glauconitic, silts, sands and sandstones with bands of calcareous concretions ('Malm Rock'); then a similar sequence but with siliceous bands too ('Chert Beds'). Much of the succession is built up of thin rhythmic units; each rhythm is laminated below, and intensely bioturbated above (Insole *et al.* 1998).

There are scattered coastal and inland sections of the Upper Greensand Formation in Dorset (Arkell 1947; House 1989), where House (1989, table V) indicated a maximum thickness of 60 m. In Worbarrow Bay, the formation appears to be only about 22.5 m thick. But in Devon it again thickens to about 60 m: here it oversteps the Lower Gault clays, and in the Haldon Hills outlier rests on the Permian Teignmouth Breccias (Hamblin & Wood 1976). In eastern Devon the lower part of the sequence consists of glauconitic sands and beds of calcareous sandstone (the 'Foxmould'). Above the sediments are more varied and sometimes coarser-grained, with layers and nodules of chert ('Chert Beds'). The facies changes to the NW, particularly in the Blackdown and Haldon Hills, where the sands are non-calcareous and coarser than usual, even pebbly (the 'Blackdown facies' of Tresise 1960). In the Haldon Hills four local members are recognized, the highest of which could be of Early Cenomanian age (Hamblin & Wood 1976).

In both the Gault and Upper Greensand Formations molluscs generally dominate the macrofauna.

Ammonites are common in the former formation but sparse in the latter, while bivalves occur everywhere. However, Gault bivalves are predominantly infaunal forms while epifaunal, often strongly ornamented, species are common in the Upper Greensand. The Blackdown sands are famed for their beautifully preserved silicified fossils, among which bivalves and gastropods are predominant. In the Haldon Hills, littoral faunas, including corals, indicate closer proximity to the shore.

The offshore basins

Offshore from SW England and Wales Lower Cretaceous sediments are known from some of the basins in the Western Approaches Trough, and in the Celtic Sea and Bristol Channel basins, as well as in the Fastnet Basin (Irish Sector). No sediments are known in the Cardigan Bay and Irish Sea Basins; the former is inferred to have been a site of non-marine deposition and the latter a land area in Early Cretaceous times (Rawson 1992*b*, maps K1–K3).

The Western Approaches Trough opened towards the Bay of Biscay and was separated from the Wessex/Central Channel basins by a high, the Start–Cotentin Ridge, at least in earliest Cretaceous times. Latest Jurassic–earliest Cretaceous uplift centred on the Cornubian Platform (McMahon & Turner 1998) stripped most of the Jurassic strata from the northern part of the Trough, so that there the earliest Cretaceous sediments are missing and there is a marked unconformity (Evans 1990). Along the southern margin there is a more continuous Upper Jurassic and Lower Cretaceous sequence in the Southwest Channel and Brittany basins, where subsidence outpaced erosion (Evans 1990; Ruffell 1995). There the Upper Jurassic consists of shallow-marine limestones, anhydrites and non-marine clastics. Then argillaceous and sandy beds appear (Wealden facies). These reach at least 1200 m in the Southwest Channel Basin, and include at the base a Berriasian silty clay in 'a freshwater facies similar to the middle and upper Purbeck in Dorset' (Evans 1990, p. 53). Gradually sedimentation spread northwards during the Hauterivian and Barremian, represented by thin, local accumulations in the Melville, St Mary's and Plymouth Bay basins. Here the sediments may have been more extensive, only to be removed by intra-Early Aptian erosion (Ziegler 1987; Evans 1990). Wells in the Brittany Trough show an increasing marine influence to the SW, in the direction of the open sea (Evans 1990).

A major marine transgression started during the Aptian, and spread northeastwards to link the trough with the Wessex and Central Channel basins (Rawson 1992*b*, map K2b). A thin (maximum 100 m) veneer of marine sandstones, some glauconitic, was deposited over much of the trough (Evans 1990), although non-marine sediments continued to accumulate in the Southwest Channel Basin until mid Albian times. Exact dating of the marine sandstones is difficult, but they appear to be of essentially Late Aptian–Mid Albian age. As sea-level rise continued the southern and western margins of the Cornubian Platform were onlapped during the Albian (Lott *et al.* 1980). Here, Middle Albian sandy, carbonaceous clays (15 m) occur and show some marine influence. They are overlain by a massive, partly bioclastic, open-marine limestone (21 m) of Late Albian–Early Cenomanian age (Lott *et al.* 1980). In the centre of the Trough, the Upper Albian sequence is argillaceous (Rawson 1992*b*, map K3). Adjacent to the Cornubian landmass sandstones, sometimes glauconitic, are widespread, representing an offshore development of the Upper Greensand Formation.

Separated from the Western Approaches Trough by the Cornubian Platform, the Celtic Sea basins probably drained eastward through the Bristol Channel Basin to the NW margin of the Wessex Basin. The Lower Cretaceous sequence is most completely developed in the North Celtic Sea Basin, in the Irish sector, and is described in detail by Tappin *et al.* (1994). Here Purbeck sediments occur in the centre of the basin, overlain by up to 2000 m of non-marine sands and mudrocks of Ryazanian–Early Albian age, representing the Wealden Group. There is evidence of periodic marine influence from the SW starting in Late Hauterivian times (McMahon & Turner 1998), and of shallow-marine sedimentation in the southern part of the basin during the Aptian, but in other areas Wealden conditions only terminated early in the Albian when marine mudrocks and sands were deposited across the basin. The latter include the main reservoir for the Kinsale Head Gasfield.

In the South Celtic Sea Basin (UK sector) the sequence is less complete and there is a major hiatus between the Jurassic and Cretaceous beds (Tappin *et al.* 1994). No Purbeck sediments occur, but the Wealden is represented by a predominantly sandy sequence (to 500 m) derived from the Welsh and Cornubian landmasses and dated as Aptian. It is interpreted as a deltaic sequence, with both channel sands and interdistributory facies preserved. It is overlain by some 50 m of ?Mid Albian–Early Cenomanian clastics (siltstones, glauconitic sandstones and mudstones) and limestones.

In the Bristol Channel Basin the Purbeck Group was penetrated in well 103/18-1 and showed a succession comparable to that in Dorset (Penn 1987). The overlying Wealden is almost 200 m thick in the same well and has been proved in several BGS boreholes (Tappin *et al.* 1994). It consists of interbedded claystones, siltstones, and sandstones deposited on a floodplain crossed by distributory channels. The sources were again the Welsh and Cornubian landmasses.

It is uncertain when the highs surrounding the various basins began to be buried, but Tappin *et al.* (1994) suggested from provenance studies that the non-marine depositional area onlapping Cornubia increased from Ryazanian to Aptian times, while much of the Cornubian Platform was flooded by the sea by Late Albian times (Lott *et al.* 1980).

There is evidence of Early Cretaceous vulcanicity on both the continental margin and close to Cornubia. There are at least 35 m of porphyritic olivine basalts interbedded with Aptian–Albian sediments closed to the shelf edge (Bennet *et al.* 1985; Evans 1990), and these may be linked to sills of Berriasian–Barremian age in the Brittany Basin (Ziegler 1987; Evans 1990). Phonolites of Berriasian–Aptian age, apparently representing volcanic necks, are recorded from Wolf Rock and the adjacent Epson Shoal (Harrison *et al.* 1977; Evans 1990).

The Upper Cretaceous Series

By latest Albian times five main lithofacies developments covered England and the adjacent offshore basins – reddish-coloured calcareous marls and shallow-water limestones from north Norfolk to Yorkshire and the Southern North Sea Basin (Hunstanton and Rødby formations): mudrocks, often slightly calcareous, from Norfolk to Kent (Gault Formation): glauconitic sands from Surrey to Devon (Upper Greensand Formation): bioclastic limestones to the west of Cornubia; and mixed clastics in the Western Approaches, Celtic Sea and Bristol Channel basins (Rawson 1992*b*, map K3). Yet this heterogeneity virtually vanished over a very short interval of time across the Albian–Cenomanian boundary, when everywhere except in the vicinity of Cornubia deposition of Chalk commenced (Hancock 1992, map K4a).

This dramatic change reflects environmental changes on a very extensive scale, for chalk seas became established from Ireland eastward as far as Kazakhstan, over a distance of some 7000 km. Causative factors include the rapid eustatic sea-level rises that commenced in the Late Albian, when shorelines began to retreat and many highs were partially or wholly submerged – including much of the Anglo-Brabant and Cornubian massifs. Just as the sources of clastic sediment supply began to diminish climate also began to change towards a more arid one that lasted through most of Late Cretaceous times. In the Cenomanian there was still a minor mud input over much of our area, apparently derived from the east or SE (Jeans 1968), and Cornubia continued to supply sand to adjacent shorelines. But by the Turonian an almost pure white carbonate mud (>99% $CaCO_3$) was being deposited everywhere, to form the White Chalk Subgroup.

With continuing sea-level rise almost the whole of England and Wales may have been submerged by the Late Campanian (Fig. 15.2d) (Hancock 1992, fig. K4b) though sediments are preserved only over part of the area (Fig. 15.16). Chalk sedimentation continued into the Maastrichtian in Norfolk and some of the offshore areas, but elsewhere the higher levels have been removed by erosion so that the former extent of the Maastrichtian remains speculative. On a global scale there was a Maastrichtian regression, so that within the British Isles the depositional area may have shrunk after the Late Campanian.

Chalk

Chalk is typically a pure white, micritic limestone of biogenic origin. It is composed essentially of detritus from calcareous

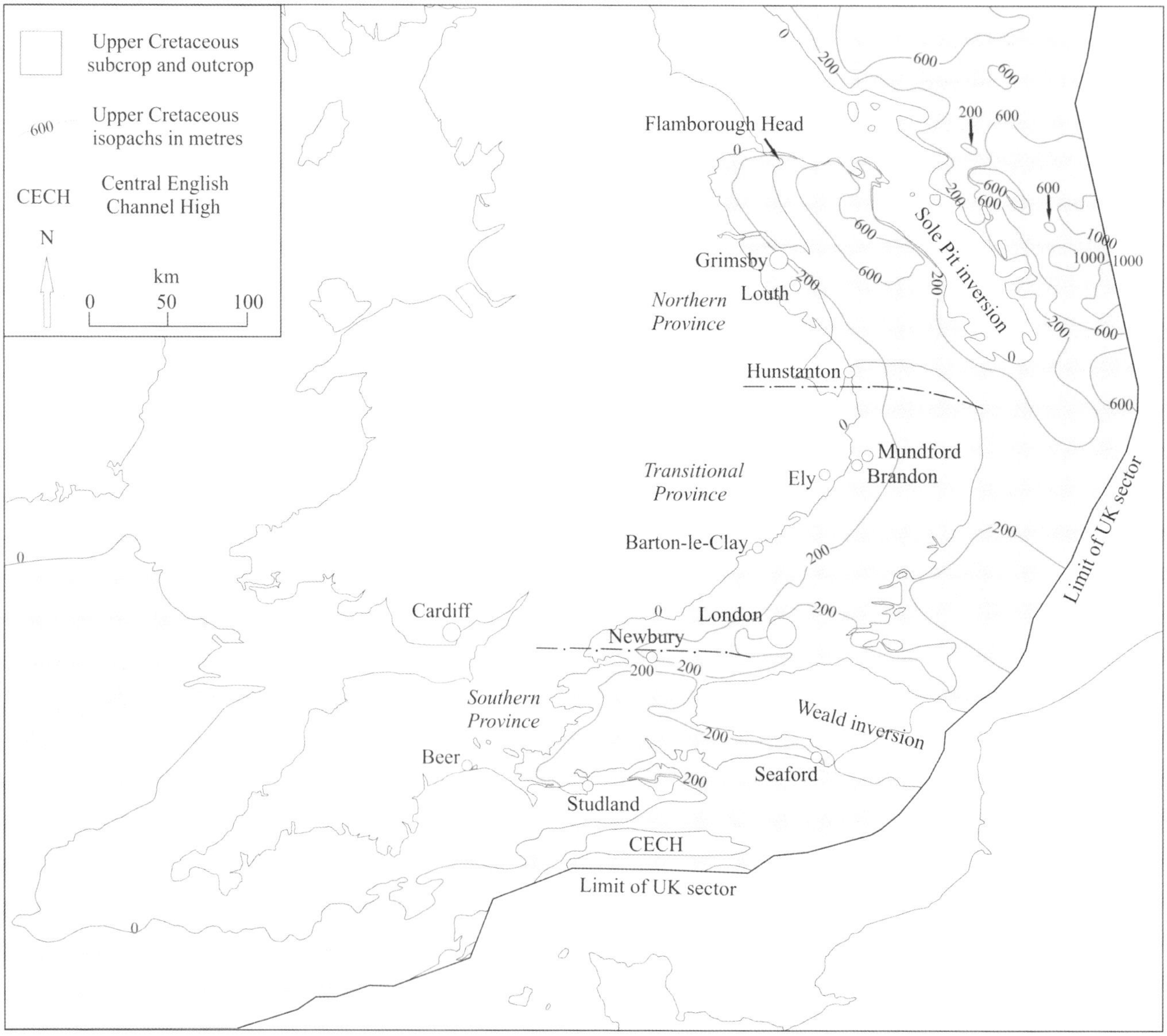

Fig. 15.16. The Chalk Group: distribution, thickness and localities. (Based on Whittaker 1985, map 25; Evans 1990; Hamblin *et al.* 1992; Tappin *et al.* 1994.)

algae (coccolithophores), mainly in the form of simple, plate-like crystals known as coccoliths. The algae lived in the surface waters (photic zone) of the chalk seas where they formed a food supply for many organisms, including copepod crustaceans which deposited coccoliths on the sea floor via their faecal pellets. The coarser fraction of chalk (10–100 μm) includes foraminifers, ostracods, calcispheres (calcareous dinoflagellates), bryozoans, echinoid plates and bivalve fragments (especially inoceramid prisms).

The main inorganic 'impurities' are clay minerals and silt or sand grains, which normally occur in very minor amounts. Clay minerals are scattered through the Grey Chalk Subgroup, but are particularly concentrated in thin interbeds of marl which ocur throughout the Chalk Group. The marl seams normally vary from a few millimetres to several centimetres in thickness and some of the more prominent ones appear to be laterally continuous for several hundred kilometres. Some appear to be of detrital origin, but others are now believed to represent altered, air-born volcanic ash (e.g. Wray 1999). They thus provide excellent isochronous marker bands ('tephro-events') for long-distance correlation, from the Northern and Southern provinces (Fig. 15.17) to Germany and the Paris Basin (Wray 1999; Mortimore *et al.* 2001). Their correlation value is further enhanced by the distinct 'spikes' that they generate in wireline logs, representing high gamma-ray and low sonic-velocity values (Gray 1965; Whittaker *et al.* 1985; Lott & Knox 1994). This facilitates correlation not only from one province to another in the onshore area but also between onshore and offshore sequences.

Flint is also common at many levels. Flint is a hard, brittle, siliceous mineral that occurs predominantly in layers parallel to the bedding, where it may form continuous sheets at one extreme and small, scattered, discrete nodules at the other. It sometimes forms vertical cylinders (paramoudras) up to 1 m or more in height, with a core of hard chalk which may contain a very narrow (less than 1 cm) burrow. The silica is of biogenic origin and apparently replaced chalk and infilled pore spaces quite early in the diagenetic history of the Chalk. The replacement occurred particularly at burrowed horizons, the flint nodules nucleating along the burrow walls and growing beyond them (Bromley 1967; Bromley & Ekdale 1986). Some of these horizons extend for considerable distances and hence, like the marl layers, flint bands are of considerable value in correlation.

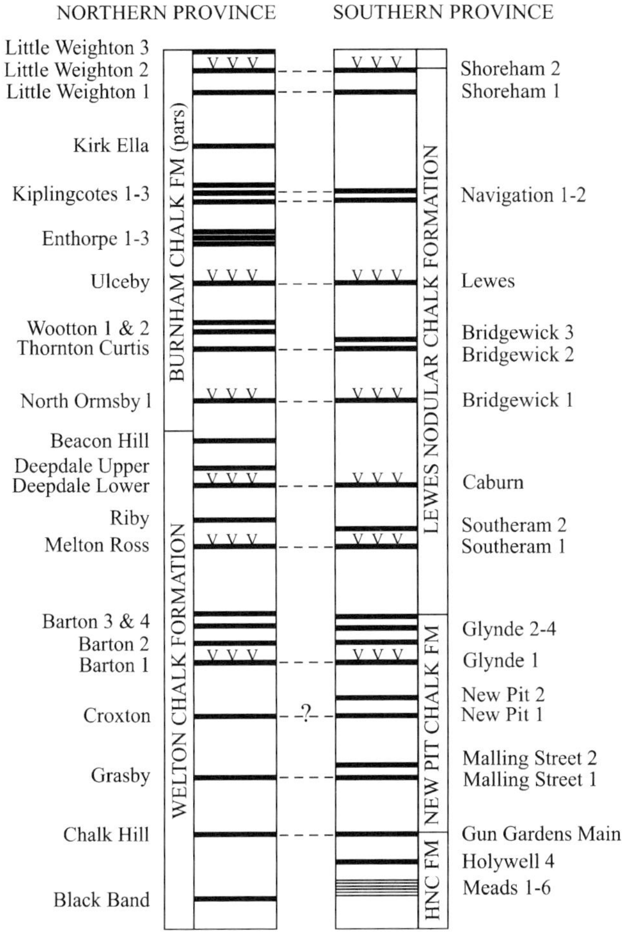

Fig. 15.17. North–South correlation of marl bands in part of the Chalk Group. V = Volcanogenic marl. Data from Wray & Wood (1998), Wood (1999) and Mortimore *et al.* (2001).

Chalk was originally regarded as a very homogeneous rock deposited more or less continuously in a stable environment – which is why Chalk faunas played such an important role in early palaeoevolutionary studies. However, several different chalk lithologies/textures are now recognized, reflecting a variety of depositional environments. The most important ones are:

Bioclastic chalk: composed mainly of shell fragments (larger than sand grade), frequently of inoceramid bivalves.

Calcarenitic chalk: gritty chalk composed of sand-grade calcareous (bio)clasts.

Griotte chalk: nodules of chalk surrounded by marl flasers, often splitting into 'horsetails'.

Nodular chalk: chalks composed of harder nodules set in a softer, burrow-fill chalk. The boundaries between nodules and softer chalk are often indistinct, and there is every gradation from nodular to fully cemented, hardground chalk.

Laminate chalk: chalks with fine laminae.

It is also clear that far from chalk sedimentation being continuous, there were frequent pauses, some lasting hundreds of thousands of years (Gale 1996). These are marked by omission surfaces at the top of nodular chalks and hardgrounds. Nodular chalks were formed by selective early diagenetic cementation of the chalk ooze just beneath the sea floor. If lithification continued long enough the sea floor would become a completely

Fig. 15.18. Milankovitch-scale cyclicity in the Grey Chalk Subgroup, Compton Bay, Isle of Wight. Photograph by Rory Mortimore.

cemented hardground, which was often subject to phosphatization and glauconitization. Although they could be formed at almost any depth, nodular chalks and hardgrounds are particularly common in the shallower-water areas over submerged highs. Their formation can be linked with both sea-level fall and transgression – i.e. they may represent either Exxon-type sequence boundaries or form part of a transgressive systems tract (Gale 1996).

At many levels in the Chalk Group there is also evidence of rhythmic sedimentation. In particular, much of the lower part of the Chalk is composed of decimetre-scale rhythmic marl/limestone couplets, normally bounded by omission surfaces. Couplets in the Grey Chalk Subgroup (Fig. 15.18) reflect the 20 ka Milankovitch precession cycle, while bundles of couplets are bounded by marl bands that represent the short eccentricity (100 ka) cycles (Gale 1989, 1996; Gale *et al.* 1999). Other marls, higher in the succession, may be episodic rather than cyclic.

The typical white chalk is neither a deep oceanic ooze nor a deposit of very shallow waters. In general, it was probably deposited under normal marine salinity at depths of 100–500 m. The sea floor was usually very soft initially, and some bivalves were specially adapted to 'float' on the sediment surface or to lie half buried within it. Other organisms burrowed into the sediment. At times, low oxygen conditions developed on the sea floor and some of the inoceramid bivalves were adapted to survive these conditions, for example in the Holywell Nodular Chalk (Mortimore *et al.* 2001). Conversely, as nodular chalks and hardgrounds developed, faunas adapted: burrowing crustaceans modified their feeding habits to keep their burrows open as the surrounding sediment was lithified (Bromley 1967) and eventually boring organisms (especially sponges) and encrusters such as oysters took advantage of the hardground surfaces.

The Chalk Group was deposited in two lithological/faunal provinces, northern and southern, with a transitional region between, where shallow-water chalks accumulated over the submerged Anglo-Brabant Massif (Fig. 15.16). The Northern Province chalks are generally harder than their southern correlatives and may have accumulated in deeper water: they show strong faunal links with north Germany. The Southern Province chalks are faunally and lithologically linked with the Paris Basin.

The Northern Province

The Northern Province chalks extend from Yorkshire, through Lincolnshire to north Norfolk and eastwards across the Southern North Sea Basin.

The onshore sequence: Yorkshire–north Norfolk
In the onshore area the Chalk Group is beautifully exposed in extensive cliff sections either side of Flamborough Head, Yorkshire (itineraries 11–13 in Rawson & Wright 2000) and the lowest beds are well seen at Hunstanton in north Norfolk (Gallois 1994). In the intervening areas there are several important quarry sections, especially either side of the Humber. The succession over much of east Yorkshire has been described in detail by Whitham (1991, 1993) while much of the area from the Humber to the Wash has been described in recent British Geological Survey memoirs (Wood in Gaunt *et al.* 1992; Berridge & Pattison 1994; Gallois 1994).

In general, the northern chalks are not highly fossiliferous and faunal diversity is low, with inoceramid bivalves, brachiopods and echinoids predominating. But some thin, highly fossiliferous beds, plus many marl and flint bands extend over much of the region and even into the Southern Province, thus providing valuable correlation levels. The more prominent markers have been formally named as beds.

The Grey Chalk Subgroup. This is represented by a single formation, the Ferriby Chalk Formation. This is a slightly marly, greyish-white, flintless chalk with finely disseminated iron, and hence is a little darker than overlying chalks. It is 33 m thick at Speeton on the southern margin of the Cleveland Basin, thins over the Market Weighton High, then thickens to some 21–24 m over most of the East Midlands Shelf. Hunstanton cliff in north Norfolk marks the most southerly exposure of the formation. Here, only the lower part of the sequence is preserved, and it is quite condensed. Further east, in the Trunch borehole, it reaches its minimum thickness of only 11 m. At both localities the chalk accumulated over a high buoyed up by granite beneath (Gallois 1994; Mortimore *et al.* 2001, fig. 1.15).

Wood (in Kent 1980; in Gaunt *et al.* 1992) has described in detail the main lithological characteristics of the Ferriby Chalk. Some of the more distinctive marker beds, such as the Paradoxica Bed at the base, have long been recognized and traced over a wide area. There are well-marked shell bands and bioclastic horizons (inoceramid grit), while two distinctive, hard bands of sand-grade chalk occur, the lower of which is identified as a thin representative of the Totternhoe Stone (Whitham 1991; Wood in Gaunt *et al.* 1992) while the upper is named the Nettleton Stone (Wood in Gaunt *et al.* 1992). Other markers are of more limited distribution; in particular, two distinctive pink bands are well developed in the Louth district, one between the Totternhoe and Nettleton stones, and the other at the top of the formation (Bower & Farmery 1910; Wood in Gaunt *et al.* 1992, fig. 30). Several erosion surfaces occur, the most prominent being at the base of the Paradoxica Bed and the base of the Totternhoe Stone (Wood in Gaunt *et al.* 1992, fig. 30; Gallois 1994, fig. 38). Fossils are common at some horizons and inoceramids and brachiopods occur through most of the sequence.

The White Chalk Subgroup. This is divided into four formations, of which the lowest is the Welton Chalk Formation (Plate 30). This reaches a maximum thickness of about 53 m and consists principally of thick-bedded to massive chalks, but rare laminate chalks appear. The lowest 4–5 m are flintless, but the rest of the formation is characterized by flint nodules of thalassinidean burrow-fill type. Flints are rare in some parts of the succession but there are several well-marked nodular bands which form excellent (named) markers. The base of the formation is defined by a glauconitized, pebble-strewn convolute erosion surface forming the top surface of the underlying Ferriby Chalk. The lowest beds of the Welton Chalk consist of about 0.5 m of variegated, occasionally laminated, marls that are now assigned to the Plenus Marls Member (Rawson *et al.* 2001). Only the lower part of this unit appears to be of the same age as the Plenus Marls Member of the Southern Province; the upper part, including the thin, very dark grey carbonaceous marl known as the 'Black Band', is slightly younger, correlating with a level just above the Plenus Marls to about the lowest Meads Marls (Wood & Mortimore 1995). Several other named marl bands occur higher in the Welton Chalk (Fig. 15.17). There is a low-diversity fauna of inoceramids, small brachiopods and thin-shelled echinoids, often concentrated in thin horizons with few fossils between (Wood in Gaunt *et al.* 1992).

The Burnham Chalk Formation is some 130–150 m thick and is predominantly thin-bedded with numerous layers of laminate chalk. In contrast with the nodular flints of the Welton Chalk, the characteristic flints of the Burnham Chalk occur as continuous bands of tabular and lenticular flints. A distinctive series of such bands forms a well-marked mappable feature in the lower part of the formation and represents a late Turonian flint maximum (Mortimore & Wood 1986). There are also several well-defined named marls, some bentonitic (Fig. 15.17). The fauna is generally more abundant and diverse than in formations above and below, but unevenly distributed. The echinoid genus *Infulaster* is particularly characteristic.

The Flamborough Chalk Formation marks a return to flintless chalks. It crops out only along the southern part of the Wolds Syncline, in east Yorkshire. Boreholes near the Holderness coast suggest that it is up to 280 m thick, but only the lowest 165 m are exposed in the cliff sections south of Flamborough Head and another 40 m may be exposed in inland sections (Whitham 1993). The formation consists of well-bedded, pure white flintless chalks with numerous thin marl bands and partings, and numerous stylolitic surfaces. Whitham (1993) described the type section in detail, naming three local members and several marl bands that provide useful markers. Minor modifications to his sequence were made by Mitchell (1994, 1995*b*). Whitham (1993) noted that the marl bands averaged one per 1.01 m, compared with one per 2.59 m for the lower three formations.

Faunas are varied, and include the unusual echinoid *Hagenowia* which is characteristic of the Northern Province. The well-known Flamborough sponge beds (about 10 m thick) appear about 120 m above the base of the formation and higher still there are horizons rich in scaphitid ammonites.

The upper part of the formation is concealed beneath the drift, but log data from uncored boreholes show that it is overlain by up to 70 m of flinty chalks (Mortimore & Wood 1986; Mortimore *et al.* 2001, p. 427). These are now placed in the Rowe Formation (Sumbler 1996; Rawson *et al.* 2001), which offshore marks a return to chalks with abundant flint bands throughout.

The offshore sequence: Southern North Sea Basin
Offshore the Chalk Group extends over most of the Southern North Sea Basin, although it has been removed from much of the Sole Pit Trough and Cleveland Basin following Tertiary inversion. It attains a maximum thickness of over 1200 m to the east of the Sole Pit Trough (Cameron *et al.* 1992, fig. 82) where the thickness is enhanced by both the preservation of Maastrichtian chalks and localized thickening into the rim synclines of rising salt diapirs. There are also localized patches of Danian (earliest Tertiary) chalks in the Cleaver Bank Basin (Lott & Knox 1994, fig. 2).

The lowest unit, the Hidra Formation (Fig. 15.6), is the equivalent of the Ferriby Chalk Formation and like that formation is characterised by chalks and argillaceous chalks, occasionally pink or green, with marl bands but almost no flints. It attains a maximum thickness of about 40 m. The base of the overlying Herring Formation is defined by the base of the Black Band, which is readily recognized on gamma and sonic logs. It thus lies just a few centimetres higher than the base of the equivalent Welton Formation onshore. The Herring Formation is up to about 60 m thick and, like the Welton Formation, is characterized by chalks with abundant flints. The overlying Lamplugh Formation consists of up to 200 m of a softer chalk with flint bands throughout – a characteristic also of the approximately equivalent Burnham Chalk Formation onshore. The Jukes Formation is a hard, almost flint-free chalk like its approximate equivalent, the Flamborough Chalk Formation. It has a maximum thickness of 500 m. Finally, the Rowe Formation marks a return to chalks with abundant flint bands throughout. It reaches a maximum thickness of about 380 m. This unit is also recognized in onshore boreholes SE of the Chalk outcrop (see above).

The intermediate region

The chalk strata cropping out in a wide NE–SW-orientated belt from East Anglian through the Chilterns to the Berkshire Downs were deposited over the buried NW margin of the Anglo-Brabant Massif and are thinner than chalks to the north and south. Both lithologically and faunally the sequence is transitional between the Northern and Southern provinces (Mortimore & Wood 1986; Mortimore *et al.* 2001). There is lateral and vertical variability across the region which makes it very difficult to trace some of the distinctive marker beds across the whole area or correlate them with their basinal equivalents further south. There was also some tectonic control on sedimentation. In particular, the southern margin of the region is marked by the Vale of Pewsey fault system, across which it is difficult to correlate successions (Mortimore *et al.* 2001, p. 87). The Lilley Bottom structure in the Chilterns also appears to have played a significant role at times (Shephard-Thorn *et al.* 1994; Hopson *et al.* 1996).

Details of local successions across the region, and their possible correlations, are given in Mortimore *et al.* (2001). These authors applied the Southern Province lithostratigraphy (Fig. 15.6) over most of the area, although noting that some individual formations differ noticeably from equivalent units deposited in the deeper waters further south. Furthermore, post-Cretaceous erosion has removed a significant part of the succession over most of the region, so that normally the highest beds preserved belong to the Lewes or Seaford Chalk formations (Coniacian–lowest Santonian). However, higher beds occur in East Anglia, especially in Norfolk.

The Grey Chalk Subgroup. In the Hunstanton area the subgroup is in a rather condensed northern (Ferriby Chalk) facies, but southwards the lithology changes and the sequence thickens to reach up to 70–80 m in the Chilterns. Over the whole of the intermediate region the sequence has long been divided into Chalk Marl, Totternhoe Stone and Grey Chalk. Mortimore *et al.* (2001, p. 299) correlate the Totternhoe Stone with the basal part of the Zig Zag Formation of the Southern Province.

The lowest Upper Cretaceous beds are glauconitic marls. From Ely into Bedfordshire these are placed in the Cambridge Greensand Bed. This very thin (0.1–1 m) unit contains abundant phosphatic nodules, some representing reworked Albian ammonites, including taxa not known elsewhere in England. Indigenous fossils suggest that the matrix sediment may range in age from latest Albian to earliest Cenomanian (Morter & Wood 1983). The greensand is well known for both its vertebrate fauna and its extraordinary variety of erratics – both angular rock fragments and rounded pebbles (Jukes-Browne & Hill 1903; Hawkes 1943) – which apparently drifted into the sea from the nearby landmasses entangled in tree roots.

The Cambridge Greensand disappears near Barton-le-Clay (Bedfordshire), in the vicinity of the Lilley Bottom structure, which may have partly controlled its distribution. To the SW of the structure the 'Chalk Marl' (now the West Melbury Marly Chalk Formation) rests directly on Gault or Upper Greesand, locally with a very thin representative of the Glauconitic Marl Member at the base (Shephard-Thorn *et al.* 1994): the latter is more extensively developed in the Southern Province.

The West Melbury Marly Chalk Formation is thickest from Cambridgeshire to the Chilterns (Mortimore *et al.* 2001, p. 298). It consists of numerous thin (to 1 m) cycles of calcareous, bioturbated marl grading up into marly chalk, and there is an overall decrease in marl content up through the formation. Marker beds include two highly fossiliferous, variably cemented harder beds, the Doolittle Limestone and the Dixoni Limestone. The overlying Zig Zag Formation is a more uniformly marly chalk. The Totternhoe Stone, a calcarenitic chalk at the base of the formation, can be traced over the whole region, and into the Northern Province. It forms an intermittently mappable feature in Cambridgeshire and the Chilterns (e.g. Hopson *et al.* 1996). By the time this unit was deposited, the area from the Chilterns to the Berkshire Downs had become a structural high where more condensed sequences developed – a situation that continued during deposition of the overlying White Chalk Subgroup (Mortimore *et al.* 2001, p. 298).

The White Chalk Subgroup. The Southern Province formations (Fig. 15.6) can be recognized over most of the area, although lithological changes towards north Norfolk make it difficult to recognize them there (Mortimore *et al.* 2001, p. 299). The Holywell Nodular Chalk Formation begins with the Plenus Marls Member (named after the occurrence in the upper part of the belemnite *Praeactinocamax plenus*). In the southern part of the area, as in the Southern Province, eight beds can be recognized in the Plenus Marls (Jefferies 1963), but north of Hitchin lithologies start to change and some individual beds are eventually cut out as the member thins and becomes increasingly condensed northwards from Cambridge towards Norfolk (Mortimore *et al.* 2001, p. 304).

The nodular, bioclastic chalks above the Plenus Marls are more condensed than their equivalents in the basin to the south. They include at their base the Melbourn Rock, a well-cemented chalk that can be mapped from Norfolk to the Chilterns and was used to mark the base of the former 'Middle Chalk'. In places the upper part of the Holywell Formation contains flints, and their development apparently reflects local tectonics as they occur in the thicker, less condensed sections that accumulated between minor anticlines.

The base of the overlying New Pit Chalk Formation is taken where shell-detrital chalks disappear. Unlike its basinal counterpart, the formation here is much more flinty. Generally it is poorly exposed so that for both this formation and the immediately underlying and overlying beds the names for distinctive marl, flint, and fossil bands are based on borehole sequences in south Norfolk (Ward *et al.* 1968; Mortimore *et al.* 2001, p. 305).

The Lewes Nodular Chalk Formation is again characterized by nodular and gritty chalks. It includes in the lower part the Chalk Rock Member (Bromley & Gale 1982; Gale 1996), which marks the base of the former 'Upper Chalk'. The Chalk Rock is a condensed nodular chalk which at its most complete development, in Berkshire and Wiltshire, embraces several hardgrounds. Some of these are traceable for up to 300 km, from Hertfordshire to Dorset. Southeastward, the Chalk Rock facies passes into typical Lewes Formation nodular chalks in the Southern Province. In the Brandon area in Norfolk, just beneath the local Chalk Rock horizon, the Turonian flint maximum is represented by the Brandon flint complex, mined at Grimes Graves in Neolithic times (Mortimore & Wood 1986).

Another marker higher in the Lewes Nodular Chalk Formation is the Top Rock hardground complex. This is well developed in the Chilterns and extends as far northwards as Mundford in south Norfolk. Above are some flinty chalks which are not as nodular as their correlatives further south.

Over most of the intermediate region outcrop higher chalks have been eroded away. But in places they are preserved, such as around Newbury on the Berkshire Downs, where the Newhaven Chalk Formation is represented by small patches of thin, highly condensed, flintless, brown, phosphate-rich, highly fossiliferous chalks which normally rest on erosion or hardground surfaces. There are more extensive developments in East Anglia, where probably 'at least the Seaford Chalk Formation will be mapped through this region in the future' (Mortimore *et al.* 2001, p. 309). Yet higher chalks from boreholes, former workings and isolated exposures in Norfolk cannot be correlated readily with those further south, and includes beds younger than any preserved in the Southern Province. To these, the informal name 'Norwich Chalk' (Formation) has been applied, with several local members (Peake & Hancock 1961; Wood 1988; Mortimore *et al.* 2001, pp. 20 and 310). These chalks are characterized by their prolific and beautifully preserved Late Campanian faunas, including echinoids, asteroids, belemnites, ammonites and brachiopods. Even younger chalks, of Maastrichtian age, are known from the Trunch borehole (Wood *et al.* 1994) and are visible as erratic masses, some very fossiliferous, on the Norfolk coast from Overstrand to Trimingham. This section is described in detail by Mortimore *et al.* (2001, pp. 359–373).

The Southern Province

The onshore area

The chalks of the Southern Province are generally softer than those of the Northern Province, hardgrounds are much commoner and continuous tabular flints are uncommon. However, both provinces, together with the intermediate region, share some general trends; discrete marl seams are best developed in the Turonian and lower Campanian chalks and there is a high Turonian level of maximum flint development (Mortimore & Wood 1986). In addition, some distinctive marl seams extend from Yorkshire to the Dorset or Sussex coasts and provide a valuable check on correlation between the two provinces (Fig. 15.17).

Chalk is the dominant lithofacies, but marginal sands and sandy limestones occur close to the Cornubian massif. The most rapidly subsiding area lay over SE England where almost 400 m of soft chalk accumulated in Sussex. Overall, the successions thin northwards to Kent and southwestwards towards Devon, but some individual units do not follow this general trend (Mortimore *et al.* 2001, p. 92). Hardgrounds and condensed, hard nodular chalks, generally characteristic of shallower waters, are exposed along the western outcrops, from the Berkshire–Chilterns shelf of the intermediate region to Dorset (Bromley & Gale 1982; Mortimore 1983).

Although the lowest unit, the West Melbury Marly Chalk Formation, occurs over much of the province, younger chalks lap onto the flanks of the mid-Dorset Swell and eventually onto the adjoining shelf to cover the more clastic facies beneath. Thus the base of the chalk facies is diachronous (Kennedy 1970), becoming progressively younger westwards until at Beer it is Turonian, with the New Pit Chalk overstepping Holywell Chalk onto Triassic sediments.

The Grey Chalk Subgroup. This is virtually flintless. Recently it has been divided into two formations over most of the area, the West Melbury Marly Chalk and the Zig Zag Chalk (Bristow 1989; Bristow *et al.* 1997; Rawson *et al.* 2001), based originally on mapping in Dorset but now applied throughout the Southern Province (Mortimore *et al.* 2001). Almost everywhere the West Melbury Marly Chalk (20–50 m) succession commences with 1–5 m of marly, glauconitic chalk with phosphatic nodules, the Glauconitic Marl Member. This unit generally has an erosive base. Above these basal beds the bulk of the West Melbury Marly Chalk Formation embraces most of the old 'Chalk Marl' division, consisting of soft, marly chalks that include only a few thin beds of harder chalk. As in the intermediate area to the north, marl-chalk rhythms occur throughout. The sequence is most complete in Sussex and east Kent.

The overlying Zig Zag Chalk Formation (15–75 m) here embraces the top part of the former 'Chalk Marl' and the overlying 'Grey Chalk'. The Zig Zag Chalk is composed mainly of grey and off-white, blocky chalks, but with some more marly interbeds in the lowest part. The Totternhoe Stone, which marks the base of the formation further north, disappears in the deeper-water facies of the Southern Province. Here the base of the formation is taken at the base of the Cast Bed, a widespread horizon with an abundant and diverse brachiopod fauna which lies directly on the locally developed Tenuis Limestone (Bristow *et al.* 1997). The base of the Cast Bed correlates with the base of the Totternhoe Stone (Mortimore *et al.* 2001, fig. 4.3).

The Grey Chalk Subgroup in west Dorset and Devon differs considerably from that elsewhere, reflecting both the proximity of land and local tectonic influences (Drummond 1970). Over the mid-Dorset swell, the West Melbury Marly Chalk is cut out and the Zig Zag Chalk comes to rest on the Upper Greensand Formation (Bristow *et al.* 1997). A more sandy sequence is seen in a series of scattered outliers west of the main outcrop, where there is a highly condensed Cenomanian sequence consisting of about 5 m of sandy limestones, assigned to the Beer Head Limestone Formation (Jarvis & Woodruff 1984), and correlative sands (the 'Wilmington facies'). At Hooken Cliff, near Beer (Devon), there is a lateral passage from one to the other across the fault bounded 'Hooken–Wilmington trough' (Mortimore *et al.* 2001, pp. 116–123). The sands are also exposed near Honiton (Mortimore *et al.* 2001, pp. 123–133), and have yielded particularly rich and diverse Cenomanian faunas, particularly echinoids (Smith *et al.* 1988) and crabs (Wright & Collins 1972). Some 30 km west of Beer, the Haldon Hills are group of small outliers exposing the 'Haldon Sands Formation' of Hamblin & Wood (1976), now included in the Upper Greensand Formation. The lower part of the sequence is of Late Albian age, but the highest sands are Cenomanian.

The White Chalk Subgroup. (Plate 28). This shows a broadly similar character over most of the province. The lowest part is generally flintless, the remainder characterized by numerous flint bands. But the level at which the first bands appear varies, being lowest in the shallower water areas of SW England.

The lowest division is the Holywell Nodular Chalk Formation (15–50 m), with the Plenus Marls Member at the base. The Plenus Marls consists of a thin (maximum 11 m) sequence of alternating marls and chalks with a major erosion surface (sequence boundary) at the base. They form a very distinctive lithological and geophysical marker. The sequence was described in detail by Jefferies (1963), whose beds 1–8 are recognized over wide areas of the Southern Province.

The remainder of the Holywell Nodular Chalk Formation consists of gritty, nodular, shell-detrital chalk with several named marl bands. It shows small-scale (to 1 m) rhythmic couplets, each composed of a thin marl band and chalk (Gale 1996). The formation is thickest in east Sussex, in the deepest-water area. Here it is flintless, but to the west, on the Dorset coast, flints appear at the base and near the top (Mortimore & Pomerol 1987; Bristow *et al.* 1997).

The Holywell Chalk is overlain by smoother, white, flaggy chalks with numerous marl seams, assigned to the New Pit Chalk Formation (20–50 m). This again shows rhythmic couplets in the lower part (Gale 1996). Flint bands occur through much of the formation in Dorset and Devon, where the proportion of flint increases westward (Bristow *et al.* 1997). But in the basinal facies they appear only at the base of the formation (Glyndebourne Flints), which is otherwise flintless.

Above the New Pit Chalk Formation the remainder of the White Chalk Subgroup is markedly flinty. The lowest unit, the Lewes Nodular Chalk Formation (maximum 80 m in East Sussex) is readily distinguished from the formations below and above by its characteristic nodularity. It consists mainly of rhythmic units composed of soft chalk passing up into nodular chalk and terminating in an omission surface or hardground, sometimes with an overlying marl band. The base of the formation is marked in Dorset and the Isle of Wight by the 'Spurious Chalk Rock', which correlates with the Ogbourne hardground at the base of the Chalk Rock further north (Mortimore *et al.* 2001, p. 94). A little higher in the sequence, particularly in Kent, several closely-spaced flint courses (the 'Basal Complex') mark the high Turonian flint maximum, and contrast markedly with the chalk below with only small, inconspicuous flints. In Sussex the contrast is much less because flints are also common in the beds below (Mortimore & Wood 1986). It is this complex that provided a mappable base for the former 'Upper Chalk'.

The remaining formations are very different from the nodular chalks just beneath, consisting predominantly of soft white chalks with numerous flint bands. The flints vary in character through the sequence, but the formations are distinguished primarily on the presence or absence of well-marked marl bands. The Seaford Chalk Formation (70 m) has some prominent marls in the basal part, the lowest of which is of volcanic origin, but they effectively disappear upward and the bulk of the formation is characterized by conspicuous bands of large flint nodules. The base of the overlying Newhaven Chalk Formation (40–80 m) marks the reappearance of well-marked marl seams, and the base of the lowest one, the Buckle Marl 1, is taken as the base of the formation. Both marl seams and flint bands may disappear over local highs, where it can then be impossible to separate the Seaford and Newhaven chalks (Bristow *et al.* 1997). At Whitcliff, Isle of Wight, the Newhaven Chalk is condensed, with phosphatic chalks and hardgrounds (Mortimore *et al.* 2001, p. 99).

In Kent the remainder of the Chalk sequence is truncated by the Tertiary erosion surface, but elsewhere higher beds are exposed. In the Culver Chalk Formation (80–100 m), marls occur only in the lowest few metres then disappear through the bulk of the formation, which consists of soft white chalks containing numerous nodular flint bands, some with paramoudra flints.

Marls reappear as a series of conspicuous seams in the Portsdown Chalk Formation (60 m). This is a white chalk with common marl and nodular flint seams which is of fairly constant thickness across much of southern England (Bristow *et al.* 1997). This is normally the highest chalk exposed, but at Studland (Dorset) and in the Isle of Wight it is overlain by a few metres of chalk without marl seams (Mortimore 1983; Mortimore *et al.* 2001, p. 102), the 'Studland Chalk Member' of Gale *et al.* (1988).

No Maastrichtian chalks are known in the onshore area, but derived Maastrichtian faunas occur in post-Cretaceous flint gravels in Hampshire (Rawson *et al.* 1978, p. 49) and the stage is represented offshore.

The offshore sequences

Chalk was deposited over most of the areas offshore from southern England and Wales, extending over a considerably greater area than the earlier Cretaceous sediments as the rising chalk seas submerged most of the former land areas surrounding the Early Cretaceous basins. Hence the basal beds smooth out the underlying topography and onlap localstructural/topographic highs. Nevertheless the thickest sequences tend to lie over the buried Early Cretaceous basins. As in other parts of the UK, early Paleogene erosion has often removed the highest beds, but Maastrichtian chalks are preserved in some areas and there are relics of Danian (early Paleocene) chalks in parts of the Melville and Plymouth Bay basins (Evans 1990).

In the Hampshire–Dieppe Basin the maximum thickness of chalk is over 400 m, while in the Portland–Wight Basin well 98/11-1 proved 477 m (Hamblin *et al.* 1992). The sequence thins over the Start–Cotentin ridge but thickens again into the Western Approaches Trough, where the succession is between 350 and 550 m thick when fully developed (Evans 1990). Here the thickest sequences are preserved in the northern basins, which were less affected by Tertiary uplift and erosion than the southern ones. Although the distinctive log signature of the Plenus Marl can be recognized in some wells in the Western Approaches Trough, the Chalk is not formally subdivided into lithostratigraphical units here, though individual stages have been distinguished. Along the trough, the lower stages thin and eventually disappear southwestward towards the open ocean, while the higher stages thicken in the same direction. Although the bulk of the Western Approaches sequence is chalk, at the northern margin of the trough, around the Cornubian Platform, deposition of sandy limestones and sandstones continued from the Albian into the Cenomanian (Lott *et al.* 1980). A Cenomanian glauconitic sand up to 67 m thick also occurs locally further offshore, in the eastern part of the Trough, while sands and shales are interbedded with Cenomanian–Santonian chalks in well 73/13-1 in the outer part of the trough (Evans 1990, fig. 39). Some 45 km south of the Scilly Isles, seismic profiles indicate an area where some of the younger chalks include reefal banks and channels.

In the South Celtic Sea and Bristol Channel basins the Chalk sequence is up to 1000 m thick (Tappin *et al.* 1994). In the former basin local variations in thickness along the southern margin reflect penecontemporaneous salt movement. Here and in the Bristol Channel Basin the Chalk Group is divided into the old Lower, Middle and Upper divisions by comparison with the sequence in the onshore Winterborne Kingstone borehole (Tappin *et al.* 1994). The lowest beds are shallow-water calcareous greensands contiguous with the Albian strata immediately below. These are overlain by a 3–4 m sequence of glauconitic limestone or marl representing the Glauconitic Marl of onshore areas. Above are 30–50 m of calcareous silts and clays equivalent to the 'Chalk Marl'. Then a mid-Cenomanian transgression led to the commencement of chalk deposition. Within the Chalk Group several of the informal divisions used in southern England can be recognized, including the Melbourn Rock and Spurious Chalk Rock (Tappin *et al.* 1994). The older chalks (Cenomanian–Coniacian) are harder than the younger ones.

Economic significance of the Cretaceous rocks

Lower Cretaceous sediments provide aggregates, glass sands and, in minor quantities, building materials and fuller's earth. Lower Cretaceous sands are currently quarried in Lincolnshire (Spilsby Sandstone Formation), Norfolk (Sandringham Sands Formation) the South Midlands (Woburn Sands Formation) and the Weald (various levels in the Lower Greensand Group). The Leziate Member of the Sandringham Sands provides good quality sand for glass-making. The beds of fuller's earth within the Woburn Sands and Lower Greensand were formerly quarried at several localities from Bedfordshire to Surrey, but the only remaining works are at Baulking in Oxfordshire.

The Lower Cretaceous sands also provide local aquifers, although their importance is overshadowed by that of the Chalk, which is the largest aquifer in Britain (Grey *et al.* 1995). It currently provides over 50% of the water for England and Wales. The Chalk Group is also extensively quarried along much of the outcrop, to provide materials for cement-making, agricultural applications and constructional use. In 2003 the total chalk production was just over 8 million tonnes, of which some 66% was used for cement making, while about 7% went to the construction industry (BGS 2005).

16 Paleogene and Neogene: uplift and a cooling climate

C. KING

The K/T (Cretaceous–Tertiary) boundary (more correctly the Cretaceous–Paleogene boundary) is defined by a mass extinction event, whose cause and characteristics is one of the most thoroughly investigated research topics of recent years. The extinction of almost all calcareous plankton produced lithological changes in open marine sediments which are also recognizable world-wide. In England and Wales and adjacent areas, a widespread sub-Paleogene unconformity has, however, removed any evidence of this event.

In terms of geological processes the Early Paleogene is essentially a continuation of the Late Cretaceous, with similar high sea levels and warm climates to the Late Cretaceous 'greenhouse world'. These are generally believed to be due to the absence of permanent polar ice sheets, although the existence of an Antarctic ice sheet in the Late Cretaceous (and presumably the early Paleogene) has been suggested recently (see Chapter 15). A very brief but intense warming episode at the Paleocene–Eocene boundary is believed to be due to the rapid release of large volumes of methane from sea-floor sediments.

From the Mid Eocene a slow cooling set in, with a marked temperature drop at the Eocene–Oligocene boundary. These events mark the change to an 'icehouse' world, and are believed to reflect the first appearance and growth of a major permanent Antarctic icesheet. Oligocene and Neogene climates are characterized by fluctuating but overall lower temperatures, culminating in the Quatenary with the inception of extensive Pleistocene glaciation. In NW Europe there is a marked difference between water temperatures on the North Atlantic margin, influenced by warm currents, and the semi-enclosed North Sea Basin, open mainly to more northern oceans, and these are reflected in the marine biota.

The opening of the North Atlantic, beginning in the Jurassic, continued into the Paleocene, followed in the Early Eocene by crustal separation and sea-floor spreading in the Norwegian–Greenland Sea. These led to a predominantly tensional regional tectonic regime in NW Europe, with the sedimentary basins which had begun in the Mesozoic continuing to subside. The early phases of Alpine tectonics, due to African–Eurasian continental collision, led to compressional episodes with localized basin inversion. From the Mid Eocene onwards, basin subsidence took place mainly by thermal relaxation. Through the later Paleogene and Neogene compressional Alpine tectonics became the dominant regime, with intermittent inversion and uplift, although in larger basins sedimentation continued intermittently. Details are controversial, due to the predominantly erosional regime, and to the stripping off of much of the sedimentary evidence during inversion and tilting in the mid to late Cenozoic.

Paleogene and Neogene palaeogeography continues to be controversial for similar reasons, as it relies greatly on interpretation of the timing and intensity of uplift and erosion, but it is clear that the major Caledonian and Variscan massifs, already evident in the Mesozoic, continued to be areas of erosion or limited sedimentation, periodically rejuvenated by uplift, and supplying sediments to the surrounding basins. Debate still continues on the nature and age of erosional surfaces in these areas.

North Atlantic spreading was associated with the development of a 'hotspot' beneath Greenland, resulting in the development of a major igneous province in the Late Paleocene and Early Eocene in northern Britain and adjacent areas. In the North Sea Basin and adjacent areas it is manifested mainly by repeated ash-falls. Some dykes extended from igneous centres into North Wales and northern England, and the Lundy Granite was intruded at this time.

The Paleogene 'hotspot' is believed to have led to tilting of northern Britain, influencing sedimentation as far away as southern England. The relative importance of regional tectonics and eustatic sea-level fluctuations in the development of the well-documented depositional sequences in this areas is still unresolved.

A supposed Paleogene impact structure recently discovered in southern North Sea aroused much initial interest, but further study suggests it may have a purely terrestrial origin.

Recent progress in Paleogene and Neogene Geology

During the past twenty-five years, since the publication of Curry *et al*' s (1978) review of the correlation of Tertiary rocks in the British Isles, the Paleogene and Neogene geology and geological history of the British Isles and surrounding areas has been revolutionized by a flood of new data and interpretations. Some of these were incorporated in Curry's (1992) chapter in the previous edition of this book, but active research has continued, and major aspects of Cenozoic research have reached relative maturity only recently. Many aspects of Cenozoic tectonic and sedimentary history in England and Wales and their surroundings are still controversial or inadequately documented, but a much more balanced picture of onshore and offshore areas can now be given.

The recently acquired data have been derived from newly available sections (both offshore and onshore), re-study of supposedly well-known outcrops, seismic surveys, new sedimentological interpretations, and improved techniques of dating and correlation. These techniques include microfaunal and microfloral biostratigraphy, magnetostratigraphy, isotopic analysis, radiometric dating and fission-track analysis. Onshore and offshore mapping by the British Geological Survey has involved extensive sampling and many cored boreholes. An increasing volume of seismic and well data from onshore and offshore petroleum exploration has become available, covering almost every major Cenozoic basin. Much of the data are from outside the area covered in this text, but they help to place the Cenozoic of England and Wales and their immediate surroundings in an increasingly well-defined tectonic and palaeogeographic context.

Recent reviews or summaries of Cenozoic geology in parts of southern Britain and surrounding areas are given by Evans (1990), Cameron *et al.* (1992), Hamblin *et al.* (1992), Knox & Holloway (1992), Gatliff (1994), Lott & Knox (1994), Tappin (1994), Knox (1996*a*), Daley (1999) and Daley & Balson (1999). Wider areas are covered by Vinken (1988) and Ziegler (1990). Vandenberghe *et al.* (1998) is relevant to the Cenozoic of the southern North Sea and SE England. The journal *Tertiary Research* (1976–2003) and its successor *Cainozoic Research* (2003–) contain papers on all aspects of Cenozoic geology, mainly in Britain and adjacent areas of western Europe. These publications provide profuse references to earlier literature.

Chronostratigraphy

The Cenozoic Era was formerly subdivided into the Tertiary and Quaternary Sub-Eras (see Curry *et al.* 1978). It is currently divided into the Paleogene, Neogene and Quaternary Systems (the previously preferred British spelling of Pal*a*eogene is no longer in use). The term 'Tertiary' is thus obsolete, but is still used informally. The Paleogene includes the Paleocene, Eocene and Oligocene Series, the Neogene the Miocene and Pliocene Series. The nomenclature of Cenozoic stages was formerly 'in a state of confusion' (Curry 1992, p. 389), with the use of many 'regional stages' and with different limits assigned to the same stage in different areas. It has now been stabilized through successive decisions of the International Commission on Stratigraphy (ICS), a constituent body of the IUGS (International Union of Geological Sciences).

The current chronostratigraphic scale is summarized in Figures 16.1 & 16.2. The Paleogene stages were almost all defined originally in NW Europe, but mostly in areas with relatively shallow marine or marginal-marine facies. These often lack adequate records of planktonic microfossils. Their boundaries were frequently erosional surfaces, marking significant stratigraphic gaps. The boundaries of these and the Neogene stages (originally defined in southern France and Italy) have been subsequently modified for international use and redefined mainly by nannofossil or planktonic foraminiferid events in open oceanic low-latitude areas (Berggren *et al.* 1995). Stratotypes have been ratified by the ICS for the boundaries of some series and stages; others are still under study. Some earlier ICS decisions on stage boundaries may need reconsideration in the light of more recent data, particularly from isotope stratigraphy.

The time-scale and chronostratigraphic terminology used here follows Berggren *et al.* (1995), the most recent comprehensive synthesis, except for the following boundaries:

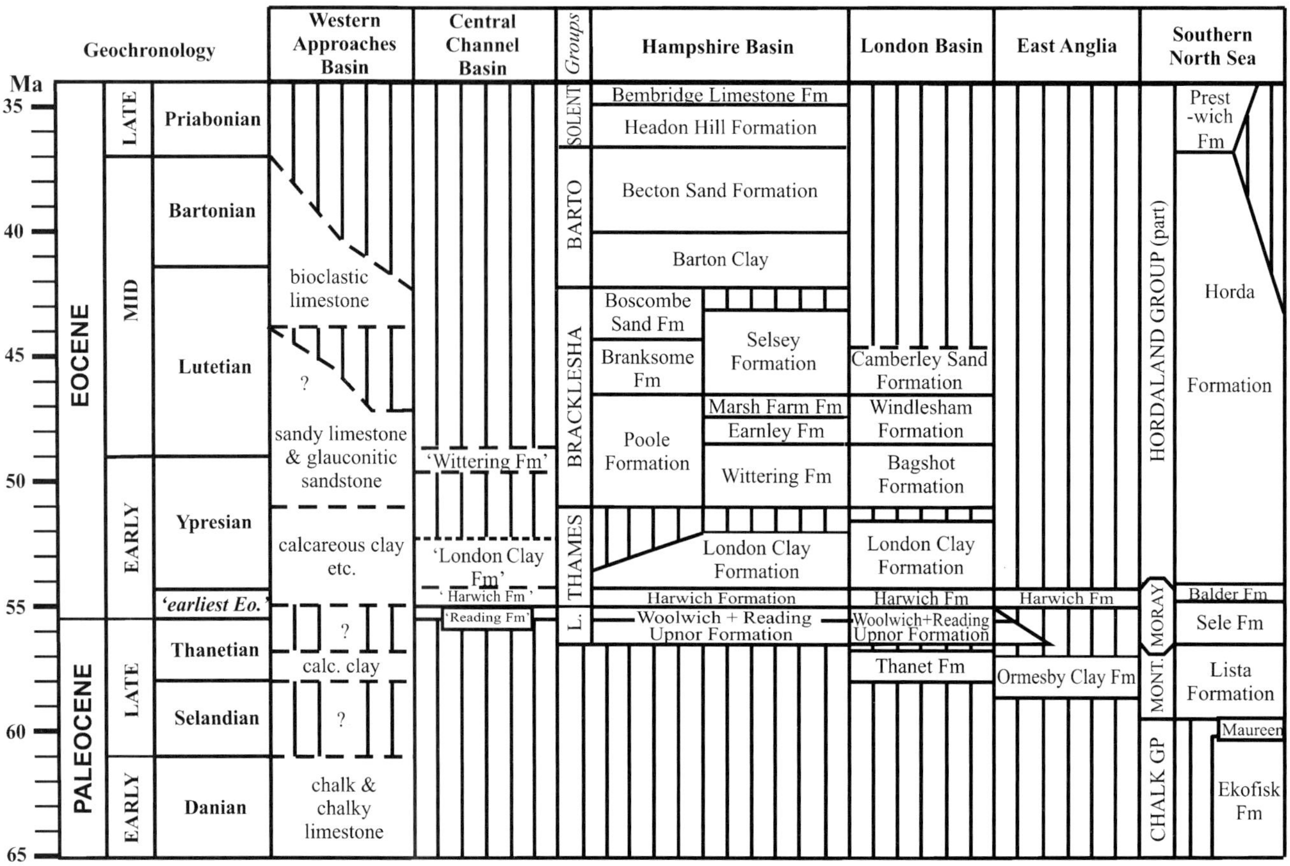

Fig. 16.1. Paleocene and Eocene chronostratigraphy and lithostratigraphy in southern Britain and adjacent areas. Geochronology based mainly on Berggren *et al.* (1995) (see comments in the text). L., Lambeth Group. MONT., Montrose Group. 'Groups' applies to Hampshire Basin, London Basin and East Anglia. Terminology for the Central Channel Basin is informal. NB. Correlation with the standard biostratigraphic/chronostratigraphic scale, both here and in Figure 16.2, varies from relatively accurate to very imprecise.

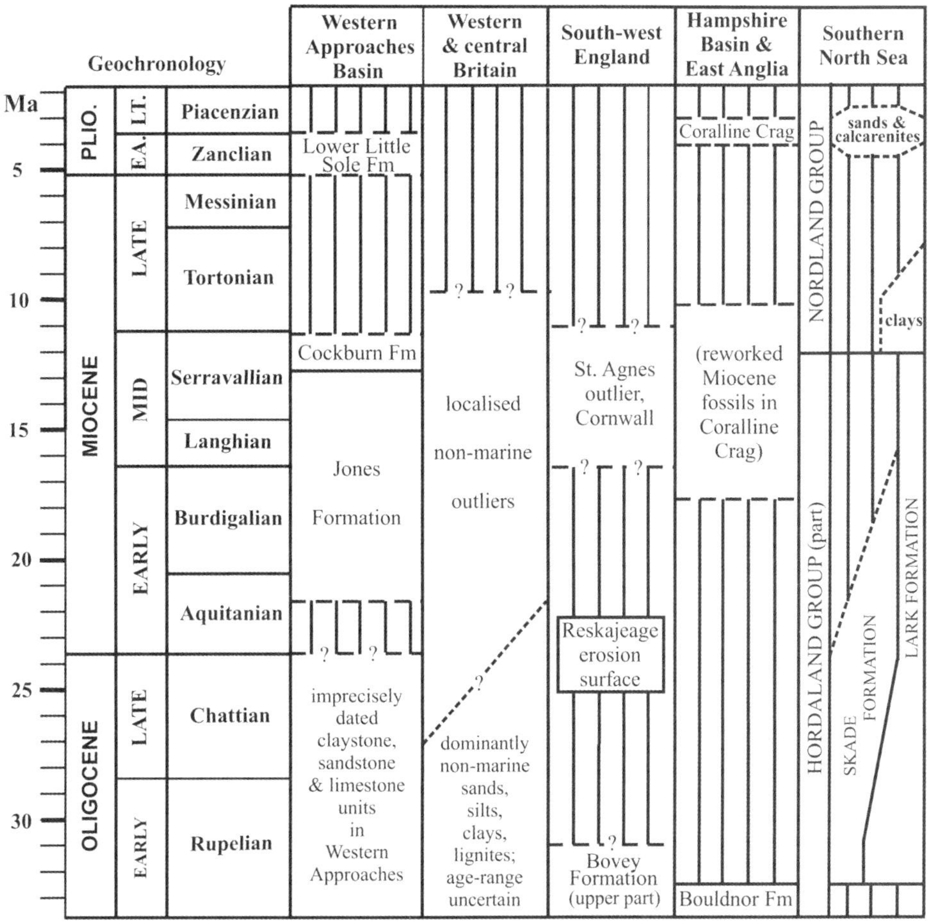

Fig. 16.2. Oligocene and Neogene chronostratigraphy and lithostratigraphy in southern Britain and adjacent areas. Geochronology based mainly on Berggren *et al.* (1995) (see comments in the text). The age of the non-marine Miocene outliers is imprecisely known.

1. *Danian–Selandian boundary*. This was placed by Berggren *et al.* (1995) at the base of planktonic foraminiferid zone P3. This event is now known to lie within the 'Danian' as originally defined in Denmark. A recent proposal by the Paleocene Working Group of the International Subcommission on Paleogene Stratigraphy (ISPS) to place the boundary at a slightly higher level, which would correspond approximately to the original (lithologically based) boundary, is followed here.
2. *Paleocene–Eocene boundary*. A stratotype for this boundary has recently been ratified by the IUGS, at the Dababiya quarry (near Luxor, Egypt), at the base of a prominent negative carbon isotope excursion which is recognizable worldwide (see Aubry *et al.* 2003). In southern England, this boundary lies within the Reading Formation. This is lower than the currently defined base of the Ypresian, the earliest Eocene stage, which lies approximately at the base of the London Clay Formation (see Fig. 16.1). The classification of the currently un-named interval below the Ypresian (here termed 'earliest Eocene'), formerly generally included in the Thanetian, is still under discussion by the ISPS (Aubry *et al.* 2003).
3. *Pliocene–Pleistocene boundary*. This was ratified by the ICS in 1985, but at a level (at *c.* 1.64 Ma) much younger than that used previously by northern European workers. The International Union for Quaternary Research (INQUA) is currently (2003) seeking to reinstate the 'traditional' boundary, now known to correspond approximately to the Gauss/Matuyama (G/M) magnetic reversal (*c.* 2.6 Ma). This is characterized by well-defined environmental and biotic events, particularly the earliest significant climatic deterioration in the northern hemisphere (see full discussion in Chapter 17). Here and in the following Chapter, the G/M boundary (currently the base of the Gelasian stage) is taken as the Pliocene–Pleistocene boundary. In East Anglia, this corresponds approximately to the base of the Red Crag Formation (see Fig. 16.2).

A detailed summary of Paleogene and Neogene chronostratigraphy, biostratigraphy and lithostratigraphy in NW Europe will be given in the forthcoming Geological Society Special Report 'Correlation of Paleogene and Neogene rocks of the British Isles and adjacent areas'.

Biostratigraphy

Most outcropping Cenozoic sediments in southern England and adjacent areas were deposited in marginal marine or shallow marine environments. Molluscs are generally the commonest and most prominent macrofossils in these facies, associated at some levels in the Paleogene with large foraminiferids (mainly *Nummulites*). These groups were used as the primary means of biostratigraphic correlation until the 1960s (Curry 1966). Microfossils (mainly benthic foraminiferids) were, however, already used for subsurface correlation and dating in the 1930s and 1940s, in petroleum exploration in Germany and The Netherlands.

Offshore and onshore petroleum exploration and oilfield development has vastly increased micropalaeontological research since the late 1960s. Initially this concentrated mainly on foraminiferids, accompanied later by dinoflagellate cysts and calcareous nannofossils. This has led to dramatic improvements in the biostratigraphic subdivision and correlation of marine Cenozoic sediments over the last 30 years. In non-marine and marginal-marine sediments, enhanced techniques of palynological analysis and sampling for mammalian fossils have also greatly increased biostratigraphic resolution.

Marine biostratigraphy

Biogeographic factors

Two distinct marine provinces can be differentiated in southern Britain and adjacent areas. An '*Atlantic province*', including the Celtic Sea basins and the Western Approaches Basin, comprises basins and shelves marginal to the North Atlantic Ocean, influenced by open oceanic water circulation patterns. Here relatively diverse planktonic foraminiferid and calcareous nannofossil assemblages, particularly in the Paleogene, are of 'low-latitude' aspect, similar to those in the Aquitaine Basin and in circum-Mediterranean areas. They can generally be directly correlated with the low-latitude zonal schemes which form the international biostratigraphic standard.

A '*North Sea province*' comprises the North Sea Basin and adjacent areas, including the Hampshire and London 'basins'. The biostratigraphy of marine sediments in this area is constrained by its partial isolation from the world ocean, with limited and intermittent watermass exchange. This led to lowered diversity in planktonic microfossil groups, with vertical ranges of individual species which are often shorter than in open oceanic areas. The low diversity is accentuated by the high palaeolatitude of its main gateway to oceanic enironments (King 1989). As a result, the standard low-latitude international zonations based on calcareous nannofossils (NP zones) and planktonic foraminiferids (P and N zones) (see Berggren *et al.* 1995) are only imperfectly recognizable this area. The following summary refers mainly to the 'North Sea province', as published data from the 'Atlantic province' are very limited.

Calcareous nannofossils

Calcareous nannofossils are present in most calcareous marine Paleogene and Neogene sediments, but are generally relatively sparse and low in diversity in inner neritic environments. Reworking is also frequent, particularly at and above sequence boundaries and in transgressive units. They are also subject to dissolution or overgrowth during diagenesis. The nannofossil record in the North Sea Basin and adjacent areas is thus imperfect. Nevertheless, in the Paleogene and early Neogene, nannofossils provide almost the only first-order correlations with the standard biostratigraphic scale. Most of the standard nannofossil zones (see Perch-Nielsen 1985) can be identified (Aubry 1983), although some index species are not present due to the high latitude and environmental controls. More refined subdivision is possible in parts of the Paleocene and Eocene (e.g. Steurbaut 1998). From the Mid Miocene onwards, only sparse assemblages generally occur in the North Sea Basin, with few short-ranging taxa.

Diatoms

Diatoms were originally widespread in most marine and marginal-marine sediments in the North Sea Basin, but diagenetic dissolution of their siliceous skeletons is characteristic of most facies. This has led to their absence in many stratigraphic units, or to preservation only as pyrite moulds. However, a number of morphologically distinctive and short-ranging taxa of large benthic diatoms preserved in this manner have been utilized as biostratigraphic markers, particularly in the Late Paleocene and Early Eocene of the central and northern North Sea (e.g. Mitlehner 1996). Some of these also occur in the southern North Sea and in the London and Hampshire basins.

Bolboformids

Bolboforma is a calcareous marine microfossil of uncertain affinities, probably a planktonic algal cyst. It ranges from Mid Eocene to Late Pliocene, and is most abundant in high latitudes. Species are mostly short-ranging, with wide geographic distributions. A *Bolboforma* biozonation has been developed (e.g. Spiegler 1999), which can be directly calibrated in open oceanic sections (mainly North Atlantic ODP sites) with the standard calcareous nannofossil zones. Most *Bolboforma* zones have been identified in the North Sea Basin, and they provide currently the most accurate second-order correlation for much of the Neogene in this area. *Bolboforma* is abundant at some levels in the Miocene of the southern and central North Sea Basin (King 1983), and occurs in the Pliocene of East Anglia.

Dinoflagellate cysts

Dinoflagellates are the most widespread and diverse planktonic group in the Cenozoic of NW Europe, and their cysts are abundant in most marine environments. However, the temporal and geographic distribution of many taxa is clearly influenced by temperature, salinity and other water-mass characteristics. These factors limit their biostratigraphic use at a worldwide level, but zonations have been proposed for the whole or part of the Paleogene and Neogene in NW Europe (e.g. Bujak *et al.* 1980; Costa & Manum 1988; Powell 1992). These can be recognized in part elsewhere. Dinoflagellates are particularly important for biostratigraphy in sediments in which calcareous plankton is rare or absent, due to ecological factors or post-depositional dissolution (as in the Paleogene of the central North Sea).

Planktonic foraminiferids

Planktonic foraminiferids are common to abundant in many deeper-water marine sediments of the North Sea Basin, but their diversity is often low. The standard planktonic foraminiferid zones (see Berggren *et al.* 1995) are based on predominantly low-latitude species. Few of these occur in the North Sea Basin, although correlation based on subsidiary species has been attempted. Several regional planktonic foraminiferid zonations for this area have been proposed (e.g. King 1983, 1989; Spiegler *et al.* 1988). Provincial North Atlantic Pliocene planktonic foraminiferid datums, calibrated by nannofossils and magnetostratigraphy, can be partially identified in East Anglia and the southern North Sea Basin. More diverse assemblages, with low-latitude index species, occur at some levels in the Western Approaches Basin.

Larger rotaliid foraminiferids

Large rotaliid foraminiferids (*Nummulites*, etc.) are characteristic of shallow, warm, generally carbonate-dominated environments. They evolved rapidly and are used primarily for subdivision and correlation of low-latitude shallow-marine Paleogene and Neogene environments, in which planktonic microfossils are generally rare or absent. In the 'North Sea province', larger rotaliids are mainly confined to the Eocene in the extreme south and SE of the Basin and its dependencies (including the Paris Basin). They probably migrated from the Atlantic margins through an English Channel 'gateway' during periods of high sea level. In the Atlantic margin basins, larger rotaliids are more widespread both geographically and temporally (from the Paleocene–Eocene boundary to the Late Eocene). They are important biostratigraphically, particularly in shallow carbonate environments (e.g. the Middle–Late Eocene of the western Channel) where other age-diagnostic fossils are rare.

Benthic foraminiferids

Benthic foraminiferid taxa are often long-ranging, and their distribution is strongly influenced by environmental factors. However, in the North Sea Basin many taxa have relatively short vertical ranges, and many first and last appearances appear to be approximately synchronous. This is partly a reflection of the semi-isolated nature of the basin: most taxa probably evolved elsewhere and migrated into the area during periods of high sea level.

A benthic foraminiferid biozonation for the Cenozoic of the North Sea and adjacent areas (King 1983, 1989) is applicable mainly to deeper-water environments, but can be applied also to the Neogene (and Quaternary) of East Anglia. Gramann & von Daniels (1988) proposed an alternative zonal scheme for the North Sea Basin, largely based on assemblages in shallow marine facies. This can be partially identified in southern England and the southern North Sea.

Molluscs
Molluscs are common in many marginal-marine to mid-neritic facies in the North Sea Basin. Many short-ranging species were recognized already in the early nineteenth century and used as biostratigraphic index-fossils. The original division of the Tertiary into Eocene, Miocene and Pliocene by Lyell (1830–1833) was based on the proportion of living species of molluscs in successive faunas, partly from the North Sea Basin. Molluscs played a leading role in establishing many local and regional correlations, before the study of planktonic microfossils and nannofossils had begun. Regional correlation based on mollusc assemblages reached its most sophisticated phase in the use of correlation coefficients for comparison of Mid Eocene faunas of southern England, Belgium and the Paris Basin (Curry 1966). A benthic mollusc biozonation for neritic facies of the North Sea Basin was proposed by Hinsch (1988).

Correlations based on benthic molluscs are generally imprecise and often misleading, due to the strong environmental controls on their distribution. They are most reliable within individual stratigraphic units (e.g. the Lower Eocene London Clay facies: King 1981). In this context they can supplement planktonic fossil groups, which are often rare or poorly diversified in the nearshore environments in which molluscs are most common. Mollusc moulds can survive in outcrops leached of carbonate and oxidized by surface weathering, and can provide the only biostratigraphic information in these contexts (e.g. in the Miocene–Lower Pliocene Lenham Beds of SE England).

Planktonic molluscs (pteropods) occur in most marine units, although often poorly preserved due to their small size (usually <5 mm) and thin fragile shells. The wide geographic distribution and short range of many taxa gives them great value in inter-regional correlation; 17 pteropod biozones have been differentiated in the Cenozoic of the North Sea Basin (Janssen & King 1988).

Terrestrial biostratigraphy

Terrestrial palynomorphs (pollen and spores)
Pollen and spores potentially provide a means of correlating marine and non-marine sediments, as they are distributed widely in marine sediments by wind and currents. Taxonomic problems, and the relatively long ranges of most taxa, have impeded their use in the past. Nevertheless regional zonal schemes based on pollen assemblages have been in use for many years, and spore/pollen zones for NW Europe were established by IGCP Project 124 (Meyer 1988). These are mainly assemblage zones, with diffuse boundaries, and often imperfectly calibrated to zonal schemes used in marine environments. More detailed correlations can be made by detailed quantitative analysis of pollen assemblages. These are particularly applicable to intervals in which climate change is rapid or extreme, due to the sensitivity of plant communities to climatic fluctuations. These include the PETM (Paleocene–Eocene Thermal Maximum) (Beerling & Jolley 1998) and the 'Terminal Eocene (cooling) Event(s)' (Prothero & Berggren 1992), but similar techniques can also be effective in other contexts (Jolley 1996). Pollen and spores are the main tool for dating the non-marine basins of western Britain.

Charophytes
The oogonia (calcified fruiting bodies) of charophytes (freshwater green algae), have high preservation potential. They are most common in freshwater limestones, but also occur in other marginal-marine and non-marine facies. Individual species have relatively short ranges, and a Paleogene charophyte biozonation has been established for NW Europe (Riveline 1984). It is applicable mainly to the Paris Basin, where freshwater limestones and marls occur at many levels. Charophytes also occur commonly at some levels in the Upper Eocene and lowest Oligocene successions of the Hampshire Basin.

Land mammals
Paleogene land mammal biozones and 'reference levels' have been established in parts of western Europe. Mammals have been recorded from a number of levels in the Paleogene of the London and Hampshire basins (e.g. Hooker 1986, 1992, 1998). The sequence of faunas in the Upper Eocene and lowest Oligocene of the Hampshire Basin is the best documented in western Europe (e.g. Hooker *et al.* 2004). Although some faunas are endemic, most can be correlated with continental European faunas. Neogene land mammals are represented only by rare reworked Miocene and Pliocene bones and teeth in the Lower Pleistocene of East Anglia.

Other Stratigraphic Techniques

Magnetostratigraphy

Magnetostratigraphy has proved an important technique for correlating the Upper Paleocene–low Middle Eocene successions, both regionally (within the Hampshire and London basins and East Anglia) and with the standard geomagnetic scale (Ali *et al.* 1993; Ali & Jolley 1996). During this interval, reversal are mostly spaced relatively widely, enabling confident calibration with the geomagnetic scale. In the later Paleogene and Neogene, reversals tend to be more closely spaced, and thus more difficult to identify.

Radiometric dating

Radiometric dates of glauconies from the Paleogene of southern England were used in the past in attempts to construct a geochronologic scale (e.g. Curry & Odin 1982). These suffered from problems due to the 'leakage' of radio-isotopes, probable reworking of some glaucony grains, and inconsistencies in techniques. More sophisticated techniques of isotopic analysis on igneous rocks, and improved methods of international correlation, have superseded these methods.

Isotope stratigraphy

The use of isotope stratigraphy is exemplified by the identification of the CIE at the Paleocene–Eocene boundary in southern England and adjacent areas, and its use in correlation between the London and Paris basins (Thiry *et al.* 1998). There is little published research on other aspects of Cenozoic isotope stratigraphy in this area, such as strontium isotope stratigraphy (see Rawson *et al.* 2002).

Tephrostratigraphy

Correlation of packages of volcanic ash beds, and of individual beds, through lithological and geochemical comparisons, has played an important part in the dating and correlation of the highest Paleocene and lowest Eocene of the North Sea Basin (e.g. Knox & Morton 1988).

The distribution of Paleogene and Neogene Sediments

Onshore areas

Paleogene and Neogene sediments occupy rather limited areas within England and Wales (Fig 16.3). The Paleogene is largely confined to the synclinal Hampshire and London 'basins' of south and SE England, with a few small outliers further west and north. The Neogene is represented almost entirely by a narrow strip along the coast of East Anglia. These sediments are generally unconsolidated, and mostly form areas of low relief. They are often covered by spreads of Pleistocene fluvial gravels, and where permeable are usually deeply oxidized and decalcified. Exposures of unweathered sediments are limited mainly to coastal cliffs and foreshores. These are often partly obscured by slumping and talus, and due to their relatively rapid erosion, are often in danger from coastal protection schemes. Man-made exposures (mainly in sand and clay quarries) were once common, but are now rare. Temporary excavations for road and rail construction, landfill sites or other civil engineering works have however provided many excellent temporary exposures, and major rail construction schemes in the London area have recently provided good quality borehole core through the Upper Paleocene and lowest Eocene. Cored boreholes drilled by the British Geological Survey in the London and Hampshire basins over the past 35 years have greatly improved stratigraphic resolution of the Upper Paleocene and Eocene.

Offshore areas

Seismic surveys, sea-floor sampling, boreholes and petroleum exploration wells have shown that England and Wales are almost surrounded by Cenozoic basins of varying sizes and depths, filled by marine or non-marine sediments (Figs 16.3 & 16.4). The main onshore Paleogene 'basins' (the Hampshire and London 'basins') now appear in this wider context to be parts of a formerly continuous depositional area, fragmented by mid-Cenozoic tectonics, which is part of the marginal fringe of the North Sea Basin. The centre of the North Sea Basin lay between Scotland and Norway. It was periodically connected through the central English Channel to the North Atlantic Ocean, via the Western Approaches Basin on the North Atlantic margin.

The small onshore Cenozoic basins of western Britain (e.g. the Bovey Basin) are now known to be part of a series of non-marine basins extending from Cornwall through the Celtic and Irish Seas to the Sea of the Hebrides. It is probable that the 'upland' areas of western Britain (e.g. the Welsh massif) have been above sea level for much of the Cenozoic.

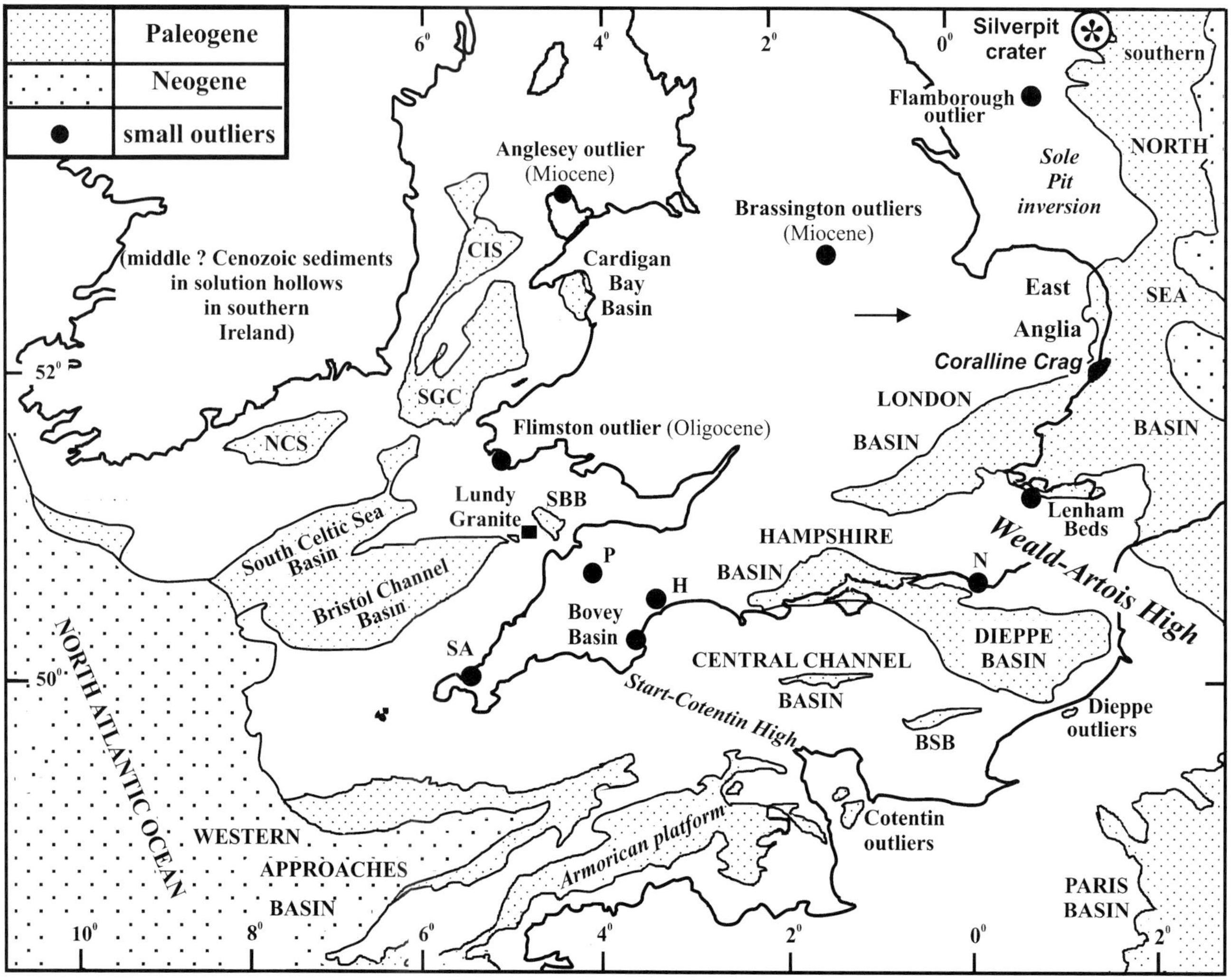

Fig. 16.3. Paleogene and Neogene outcrops in southern Britain and adjacent areas. Overlying Pleistocene and Holocene sediments are omitted. CIS, Central Irish Sea Basin. BSB, Baie de la Seine Basin. H, Haldon Gravels. N, Newhaven Outlier. NCS, North Celtic Sea Basin. P, Petrockstow Basin. SA, St. Agnes Outlier. SBB, Stanley Bank Basin. SGC, St Georges Channel Basin. Some minor outliers are omitted, particularly in northern France. The Hampshire and London Basins are shown in greater detail in Figures 16.11 & 16.13. The location of offshore borehole 81/46A is shown on Figure 16.10.

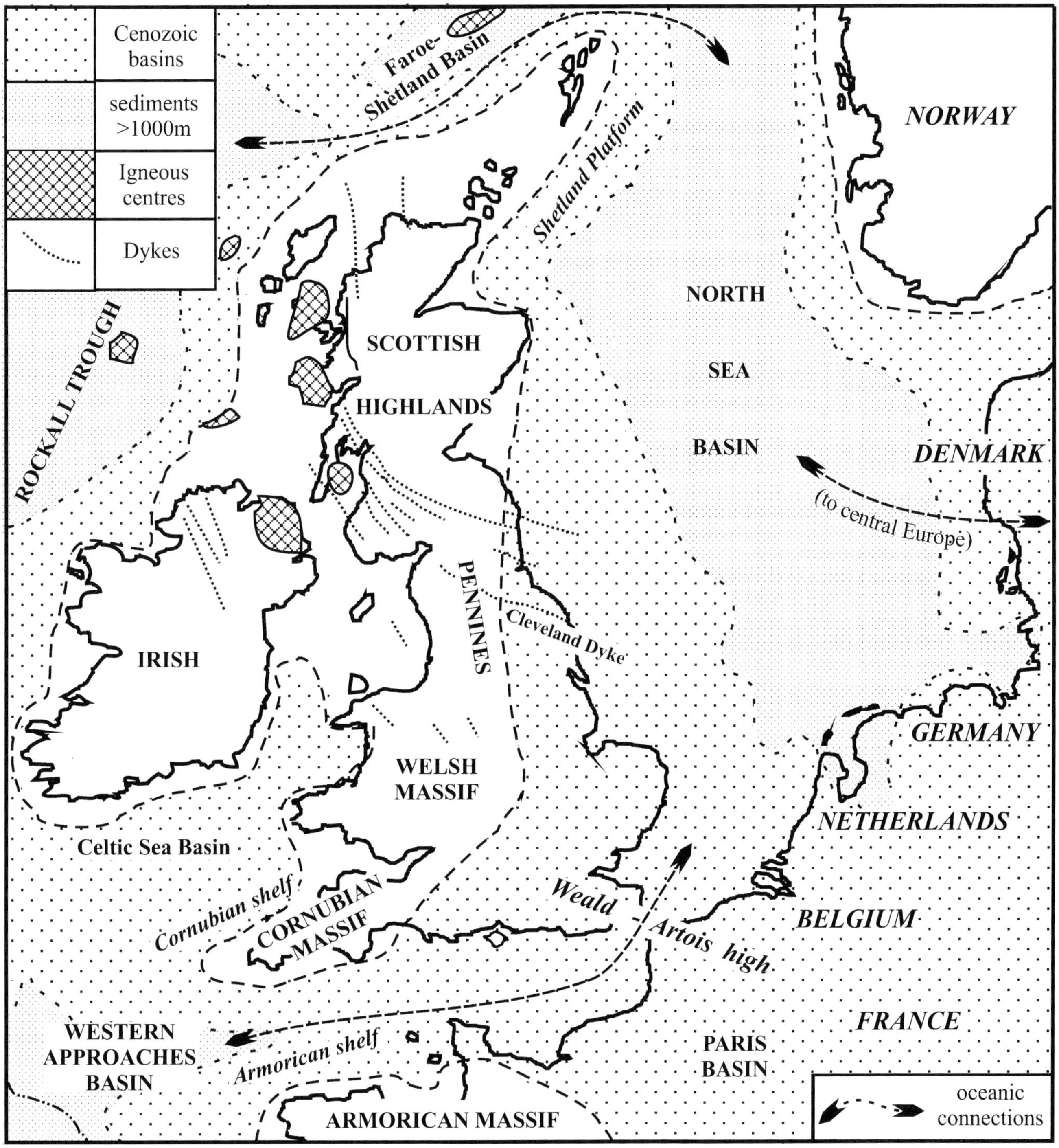

Fig. 16.4. Major Cenozoic basins and intervening 'highs' in the British Isles and surrounding areas (modified after Ziegler 1990). The basin margins shown here represent the maximum probable extent of Cenozoic marine and marginal-marine sedimentation, and in many areas should be regarded only as rough approximations. The main igneous centres and dyke swarms are also shown.

Cenozoic basins

The North Sea Basin and its margins

North Sea Basin

This is the largest Cenozoic basin in NW Europe. It is centred on a series of underlying Mesozoic rift systems, which extend from the Netherlands through the central North Sea to the Norwegian Sea. Following the cessation of active rifting in the mid-Cretaceous, thermal relaxation associated with relatively high sea levels led to regional subsidence. Thus a single marine depositional area extended from the northern North Sea to Poland, including parts of Denmark, Germany, the Netherlands and Belgium, eastern and southern England and northern France. Renewed rifting during the Late Paleocene in the northern North Sea led to very thick clastic accumulations, but these are to the north of our area. In the centre of the basin an almost continuous section (>3000 m) of mainly deepwater Cenozoic (including Pleistocene) sediments was deposited. Away from the axis of the basin, in areas adjacent to eastern England, much thinner sediments are present. These were deposited mainly in mainly neritic environments, with frequent condensation and stratigraphic breaks, due to a combination of eustatic sea-level fluctuations and intermittent tectonic activity. Only the southern part of the North Sea Basin is included in this chapter.

In the northern and central part of the North Sea Basin, Paleocene and Eocene stratigraphy is documented by many petroleum exploration wells and some short cored boreholes, but in the southern North Sea Basin less information is available, and post-Eocene stratigraphy is documented only in outline. The western margin of the Cenozoic outcrop runs southwards through the southern North Sea, converging on the English coast in East Anglia, where a thin strip of Paleogene and Neogene sediments is preserved, before running westward to form the 'embayment' of the synclinal London 'Basin', centred on the Thames Valley.

London 'Basin'

This has in the past been considered as a distinct sedimentary basin, but it is more probable that it is an integral part of the North Sea Basin, with its present configuration resulting mainly from mid-Cenozoic folding, as discussed below. It is bounded on the south by the 'Weald–Artois High [axis]', an inversion structure overlying the Mesozoic Weald Basin. The southern margin of the Cenozoic outcrop of the London Basin follows the northern edge of the Weald–Artois High into SW Belgium.

Only lower Paleogene (Upper Paleocene to Middle Eocene) sediments are preserved in the London Basin. These were deposited mainly in inner neritic to coastal plain environments, with a deeper-water unit (London Clay Formation) deposited during the high sea levels of the Early Eocene (early to mid Ypresian). Many exposures and boreholes have enabled a relatively complete picture of the stratigraphy.

The timing of the inversion of the Mesozoic Weald Basin is uncertain. It has often regarded as early Cenozoic, with the Weald–Artois High forming an isthmus or island intermittently separating the London and Hampshire–Dieppe 'basins' (and implicitly forming the southern margin of the North Sea Basin). Evidence discussed here suggests it was probably later (see below).

Hampshire–Dieppe 'Basin'

This is a large synclinal outlier of Paleogene sediments. It comprises two 'sub-basins'. The *Hampshire 'Basin'*, on the coast of central southern England, is relatively narrow, with a steep southern border formed by the Purbeck–Isle of Wight 'Alpine' folds, which overlie a major Mesozoic fault zone. Approximately 700 m of Paleogene sediments are preserved along the basin axis, which is close to the southern border of the basin. An almost continuous succession from Upper Paleocene (upper Thanetian) to Lower Oligocene (lower Rupelian) is preserved. Lower Paleogene sediments were deposited predominantly in inner neritic to coastal plain environments, deepening to mid-neritic during the mid- presian (as in the London Basin). Upper Eocene and Lower Oligocene sediments are largely marginal-marine and non-marine. Excellent coastal exposures are supplemented by a number of cored boreholes and petroleum exploration wells.

The Hampshire Basin narrows eastwards, expanding again southeastwards to form the wider and shallower *Dieppe 'Basin'*. This lies almost entirely beneath the eastern English Channel, with an estimated 500 m of Paleogene (Upper Paleocene to high Middle Eocene) sediments in the centre. Its SW border is the 'Bembridge–St Valery line', a continuation of the Purbeck–Isle of Wight structures. Small Paleogene outcrops on the English Channel coast (e.g. at Newhaven), and on the Normandy coast (near Dieppe) are outliers of this basin. The offshore stratigraphy is very imperfectly documented, as information has been obtained largely from shallow seismic surveys and sea-bed sampling.

The Hampshire–Dieppe Basin has been generally interpreted (as in the case of the London Basin) as a true sedimentary basin, becoming progressively isolated by the uplift of the Weald–Artois High (e.g. Murray 1992). More recent evidence, summarized below, suggests that it was at most times palaeogeographically and tectonically part of the North Sea Basin, with open marine connections to Belgium to the east and the Paris Basin to the south. This indicates that it is essentially an outlier isolated by post-depositional uplift and erosion during the mid–late Cenozoic.

Central Channel 'Basin'

This synclinal lower Paleogene outlier lies between the Isle of Wight and Normandy. It is in a similar structural context to the Hampshire 'Basin. An Upper Paleocene (upper Thanetian) to Lower Eocene (upper Ypresian) succession is preserved, very similar in facies to the corresponding interval in the Hampshire Basin. Its stratigraphy is known only from sea-bed sampling and a single cored borehole. Its present isolation is again due to mid-Cenozoic inversion and erosion. A similar but smaller synclinal outlier, the Baie de la Seine 'Basin', lies to the SE, in the French offshore sector.

North Atlantic margin basins

The North Sea Basin was characterized by dominantly clastic sedimentation from the mid-Paleocene onwards, including significant sand accumulations near the basin margins. This contrasts strongly with the 'Atlantic margin' basins to the west and SW of southern Britain, in which fine-grained, often carbonate-rich sediments predominate.

Western Approaches Basin

This long, broadly synclinal basin lies between the Cornubian and Armorican massifs, overlying a rifted Mesozoic basin complex. It deepens and widens to the SW, where it terminates at the North Atlantic shelf-edge. It is almost entirely offshore, in the western English Channel and Western Approaches, except for small outliers on the Cotentin peninsula of Normandy. Lower Paleocene to Upper Pliocene sediments are represented, mainly mid to outer neritic marls and argillaceous limestones, with stratigraphic breaks at a number of levels, reflecting episodic inversion. A major bioclastic limestone unit is developed high in the Middle Eocene, which transgressed widely onto the Armorican Platform (the northern margin of the Armorican massif). A maximum thickness of approximately 2000 m of Cenozoic is present in the Western Approaches, thinning to c. 650 m in the western English Channel, where only Paleogene sediments are preserved. In the Western Approaches, a number of petroleum exploration wells have provided stratigraphic information, but in the western English Channel data are derived mainly from extensive sea-bed sampling and some short cored boreholes.

South Celtic Sea Basin and Bristol Channel Basin

These basins lie to the north of the Cornubian High, overlying a Mesozoic rifted basin. Here only Paleogene marginal-marine sediments are preserved, overlapping southwards onto the Cornubian shelf. They are generally thin (<100 m), but thicken locally to about 300 m, due to synsedimentary halokinesis of thick underlying Triassic halite.

Non-marine Paleogene basins

A series of mainly small (but often deep) fault-bounded basins in western Britain are filled by dominantly clastic Eocene and Oligocene non-marine sediments. These include the Bovey Basin in Devon and the Cardigan Bay Basin in the Irish Sea. Similar basins occur further north, in Northern Ireland and off the western Scottish mainland. These were initiated and maintained mainly by strike-slip fault movements, reactivating pre-existing structural lineaments.

Tectonics

Cenozoic plate tectonics in western Europe

The distribution of Paleogene and Neogene sediments in southern Britain and adjacent areas is largely conditioned by the position of the British Isles on the NW margin of the Eurasian Plate. During the early Cenozoic the tectonic environment was characterized by the interplay between two forces: the tensional regime due to North Atlantic–Arctic crustal separation to the west and north, and the compressional regime deriving from periodic phases of Alpine tectonics to the south and SE, as the African and Eurasian plates converged and collided.

Plate reorganization throughout the Paleogene and Neogene resulted in transpressional features and local inversion events, but crustal separation in the North Atlantic area was completed by the Early Eocene. From the Eocene onwards, the influence of Alpine tectonics was probably the dominant control on stress patterns in this area. The subsequent stress regime was predominantly due to increasingly intense Alpine tectonic phases, which culminated in the mid-Cenozoic, but was also influenced by changing rates of sea-floor spreading in the North Atlantic, and crustal separation and sea-floor spreading in the Norwegian–Greenland Sea (Ziegler 1990).

The tectonic framework of Cenozoic sedimentation

The long-continued marine carbonate sedimentation (chalk facies) of the Late Cretaceous continued in the main basins around the British Isles through the Early Paleocene. In the mid Paleocene, major uplift of northern Britain led to intense erosion, and shedding of large volumes of sediment into the North Sea and other basins. This is interpreted as due to the development of a mantle plume beneath Greenland. It coincided with a phase of Alpine tectonism, causing inversion of basins throughout western Europe. These factors together led to the replacement of carbonate by clastic sedimentation, which persisted throughout the Cenozoic, except on the North Atlantic margins.

Most Cenozoic basins around Britain were inherited from the Mesozoic rifted basins developed as a result of North Atlantic sea-floor spreading. Their Cenozoic history is predominantly one of thermal subsidence, with episodes of inversion and renewal of movement along inherited fault lines, due to successive phases of compressive Alpine tectonics and North Atlantic–Arctic sea-floor spreading. These tectonic controls, together with successive eustatic sea-level fluctuations, led to the development of well-defined depositional sequences. These are recognizable both in deepwater areas of the North Sea Basin, but also in more marginal areas.

Cenozoic tectonic history

Paleocene

Subsidence due to thermal relaxation continued in the major basins from the Late Cretaceous into and through the Early Paleocene. It was interrupted in the Late Paleocene by major phases of uplift and erosion of their bounding Variscan and Caledonian massifs, probably including the Welsh Massif and Armorican Massif (Ziegler 1990), associated with the inversion of some basinal areas. These were the consequence of thermal updoming centred on Greenland (attributed to a mantle plume, see below) and possibly also in the the Irish Sea (Cope 1994), together with the peripheral effects of the mid Paleocene (Laramide) Alpine orogenic phase. The relative importance of Alpine tectonics and thermal updoming in sedimentation and tectonics in southern England and adjacent areas is still uncertain. This updoming is now believed to have occurred in several discrete phases, rather than a single continuous episode (Knox 1996*a*), each phase associated with extrusive and intrusive igneous activity. The initial episode occurred in the early Selandian (*c.* 59 Ma), and another in the late Thanetian (*c.* 56 Ma). They involved the inversion of Mesozoic basins, leading to erosion with amplitudes of hundreds of metres in some areas.

The 'sub-Paleogene unconformity' in southern England and adjacent areas, discussed below, has been suggested to date from the latest Cretaceous (Jones 1999*b*), but is much more likely to reflect the mid Paleocene (early Selandian) tectonic episode.

Eocene to Early Oligocene

Thermal updoming ended in the earliest Eocene, at the same time as crustal separation between Norway and Greenland initiated sea-floor spreading in this area. The following phase of thermal relaxation led to regional subsidence, with major transgression at the basin margins. The shifting of sea-floor spreading axes in the North Atlantic–Arctic region during the Mid Eocene to Oligocene (Ziegler 1990) changed the regional stress regime. This induced reactivation of previous Variscan fracture systems in western Britain as wrench (strike-slip) faults, with the development of pull-apart and other fault-bounded non-marine basins in this area. Cyclic alluvial-fan sedimentation in some of these basins indicates repeated movement on their bounding faults.

An early compressional episode in the Pyrenean orogenic phase (Mid Eocene to Early Oligocene) (Ziegler 1990) may be responsible for localized reactivation of faults in southern England (Plint 1982; Gale *et al.* 1999). A later episode is believed to be responsible for the several phases of localized inversion leading to regional regression, uplift and erosion during the Late Eocene and Early Oligocene. These are identified in Belgium, the Netherlands, parts of the central and southern North Sea (Ziegler 1990, fig. 86), the Western Approaches Basin and the Celtic Sea Basin. In the southern North Sea area and the Celtic Sea Basin, inversion was probably enhanced by halokinesis.

Late Oligocene to Pliocene

During this period, transmitted stresses from repeated episodes of Alpine tectonism were the dominant control on tectonics in southern Britain. This is reflected by episodic inversion and uplift in this and surrounding areas, including the Weald, central Channel, Western Approaches Basin, and the Sole Pit Basin and other areas of the southern North Sea (Cameron *et al.* 1992). Timing of these events is difficult, due to the subsequent erosion of most of the sediments which may have been deposited during this time. They are generally regarded as Mid or Late Miocene, related to African–Eurasian continental collision.

Intermittent differential uplift continued during the Pliocene, as indicated by non-marine Miocene sediments formerly near sea level, approximately 450 m above sea level in Derbyshire (Boulter *et al.* 1971) and Lower Pliocene shallow marine sediments well above present sea level in SE England.

Cenozoic uplift and denudation history

Southwest England and Wales

A series of almost planar erosion surfaces between 50 and 150 m above present sea level, identified by early studies in these areas, were originally believed to be mainly marine and probably Cenozoic in age. Later studies favoured a subaerial origin for most, but the latest (lowest) surfaces were still believed to be marine, formed during phases of high sea level in the Pliocene or Early Pleistocene. More recent interpretations have thrown doubt on the identity of many of the surfaces, suggesting that some may have been formed by marine processes as early as the Jurassic, and exhumed by early Cenozoic uplift and erosion (Campbell *et al.* 1998; Walsh *et al.* 1999). The correlation of surfaces over wide areas also now seems doubtful, as evidence has accumulated for regional warping and for differing tectonic histories in adjacent areas. Walsh *et al.* (1999) now recognize

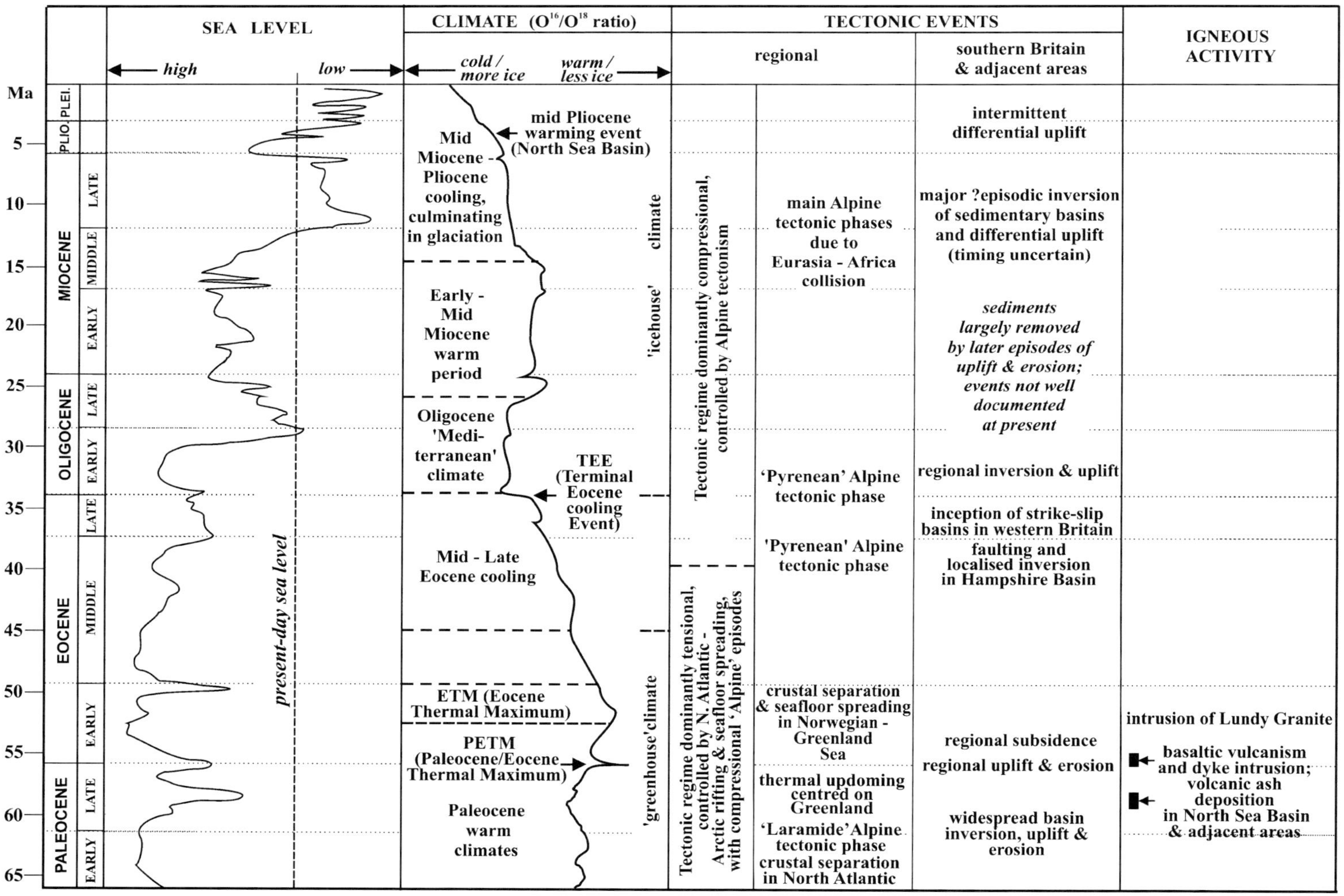

Fig. 16.5. Paleogene and Neogene sea levels, temperatures, and tectonic and igneous events relevant to southern Britain and adjacent areas. The relative sea-level curve is modified from Haq *et al.* (1988), adjusted as far as possible to the timescale of Berggren *et al.* (1995). The deep-sea oxygen isotope scale is after Zachos *et al.* (2001), and does not necessarily apply fully in the North Sea Basin and adjacent areas, in which water circulation patterns were influenced by tectonic controls.

only one main erosional surface, the Reskajeage Surface, in SW England.

It has frequently been suggested (e.g. Campbell *et al.* 1998) that the uplift and erosion of the Cretaceous cover in SW England began in the Early Paleocene or even the Late Cretaceous. Evidence from areas around the British Isles indicates, however, that marine carbonate-dominated sedimentation continued in basinal areas until the end of the Early Paleocene. In the Western Approaches Basin, mid-neritic Danian carbonates outcrop less than 50 km from the present coast of SW England. The absence of marginal facies of the Danian in the UK offshore areas, the very low clastic input, and the presence of reworked Danian microfossils in areas now far from any Danian outcrop (in the London Basin), all suggest that Danian marine limestones were deposited over much of southern and eastern England (see Fig. 16.6). This strongly indicates that uplift and erosion did not begin until later.

Soil profiles and Chalk residual deposits in east Devon, post-dating the main phase of uplift and erosion, were developed during a phase of tropical weathering, indicating a pre-Oligocene date (Edwards & Freshney 1982). They have been interpreted as Danian in age, by comparison with the interbasaltic laterites in the Antrim Volcanic Group (Northern Ireland), but these are now dated as Late Paleocene (late Selandian or early Thanetian). A better analogy may be with the lowest Eocene 'red-beds' of SE England (Reading Formation), deposited during the earliest Eocene thermal maximum (see later). This suggests that this first phase of uplift was synchronous with the mid-Paleocene (early Selandian) uplift and erosion seen elsewhere in England and further afield.

The residual deposits in east Devon apparently pre-date the initiation of the Bovey Basin, which was mid Eocene or older. Subsidence in this and associated fault-controlled basins continued into the Early and perhaps Late Oligocene. It has been suggested that the main morphological features of SW England were developed during this phase (Campbell *et al.* 1998) rather than later, in the Neogene or Quaternary, as previously proposed. This is supported by the presence of remnants of unconsolidated non-marine Oligocene and Miocene sediments, originally deposited close to sea level, at c. 75m above present-day sea level in west Cornwall. The Reskajeage Surface underlies the Miocene sediments, indicating that there has been little erosional modification of SW England since their deposition (Walsh *et al.* 1999), but apparently post-dates nearby Oligocene sediments (see the comments below on the dating of the Oligocene non-marine sediments of western Britain)

In Wales, Oligocene and Miocene non-marine sediments are known only from small pockets preserved in solution hollows. These cannot be directly related to erosional surfaces, but again the main erosional surface appears to post-date the Early Oligocene and pre-date the Miocene sediments (Walsh *et al.* 1999).

These data indicate that two main episodes of subaerial erosion are recognizable in western Britain, in the mid Paleocene and the Late Oligocene or Miocene, respectively. These probably correlate with the main episodes of uplift and erosion in SE England and adjacent areas. Earlier (Late Cretaceous–Danian) and later (Pliocene–Pleistocene) major erosional episodes previously proposed now seem improbable.

Southeastern England and the eastern English Channel

Cenozoic deformation is most clearly recognizable in the south and SE of England. Here it is associated with the compressional reactivation of Mesozoic (Triassic–Early Cretaceous) extensional faults and the inversion of Mesozoic basins. The Mesozoic faults were themselves controlled by major Variscan thrusts in the Palaeozoic basement. Reactivation of these structural lines is expressed in the Upper Cretaceous and Paleogene cover mainly by asymmetrical E–W or NW–SE-trending folds. These folds include the Purbeck–Isle of Wight structures bordering the Hampshire 'Basin', the Hogs Back structure bordering part of the London 'Basin', and the structures bordering the Central Channel 'Basin'. Both the Hampshire and London 'basins' are partly superimposed on underlying Palaeozoic horst blocks (the Hampshire–Dieppe High and the London Platform). The intervening areas, superimposed on Mesozoic sedimentary basins (the Weald Basin and Channel Basin), suffered major Cenozoic inversion (Chadwick 1993). In the central Channel, this inversion was associated with uplift of about 1500 m (Lake & Karner 1987).

Wooldridge & Linton (1955), in a classic synthesis of Cenozoic landscape development, identified a series of erosional surfaces on the Upper Cretaceous Chalk outcrops in SE England. The earliest, the ('sub-Eocene') [sub-Paleogene] surface, was interpreted as warped by mid Cenozoic Alpine tectonics, centred on the Weald, followed by a prolonged period of subaerial denudation, resulting in the erosional surface of the 'Miocene–Pliocene peneplain'. This surface was partially inundated and eroded by a widespread 'Calabrian' (Late Pliocene–Early Pleistocene) transgression from the North Sea. The marine Lenham Beds on the Chalk of north Kent and supposedly synchronous deposits elsewhere were regarded as a remnant of this transgression. This model remained in favour for many years, but the age and identity of the erosional surfaces have since been widely disputed (see Jones 1999*a*). It is also clear that tectonic activity and warping, which Wooldridge & Linton (1955) believed to be due to the 'outer ripples of the Alpine Storm', was largely an effect of vertical movements due to basin inversion, and was both much more pulsed, and much less uniform regionally, than previously believed.

The Lenham Beds, at 180 m above sea level, are now regarded as probably latest Miocene or Early Pliocene, while the other supposedly synchronous marine remnants are younger (Late Pliocene–Pleistocene), and in East Anglia range from near sea level to over 100 m higher. This indicates that their location is the result of differential Pleistocene warping, rather than high sea levels as Wooldridge & Linton (1955) supposed.

Dating of the main phase(s) of inversion and warping is still controversial (see Jones 1999*a*, *b*). This is due to the two major intervals without dated sediments in this area, between the Campanian and Late Paleocene (late Thanetian), and between the earliest Oligocene and the ?latest Miocene (Lenham Beds). Wooldridge & Linton (1955) identified two main episodes of uplift and erosion of the Weald: 'pre-Eocene' (pre-Late Paleocene), and mid-Cenozoic (Oligocene–Miocene), bur regarded the latter as the more important. Inversion of the Weald–Artois High and the southern margin of the Hampshire Basin is still generally interpreted as initiated in the Paleocene (e.g. Curry 1992, fig. 13.6). It has been regarded as continuing episodically through the Eocene (Plint 1982), creating the Hampshire and London basins as distinct depocentres during the early Paleogene, and culminating in the Miocene.

A modified version of the Wooldridge & Linton (1955) model has recently been strongly advocated by Jones (1999*a*, *b*), who pointed out that there is increasing evidence for differential uplift already in the Late Cretaceous. His model proposes intermittent pulsed tectonic activity throughout the Cenozoic, with major episodes of uplift and inversion in the latest Cretaceous and mid Cenozoic. The alternative view, that only one major mid Cenozoic inversion episode is involved, has had less support. King (1981) concluded from stratigraphic data that there was no evidence for differential elevation of the Weald area during the Early Eocene, proposing that inversion dated mainly from the mid Cenozoic (presumably Miocene). Chadwick (1993) also concluded that a Miocene date for the inversion was most probable.

Recent fission track analysis of Mesozoic sediments from deep boreholes in and adjacent to the Hampshire Basin and the Central Channel Basin (Bray *et al.* 1998), has provided no evidence for early Paleogene inversion. It indicates a regional cooling episode, related to uplift and erosion, in the mid Cenozoic (*c.* 40 Ma or younger, probably Oligocene). A Miocene episode of cooling (*c.* 20 Ma) is documented from only one site. This suggests an earlier mid Cenozoic inversion than previously suspected, and supports the alternative model, although may not be directly relevant to the Weald itself.

These divergent models result in widely differing palaeogeographic reconstructions for the Paleogene. They are evaluated in more detail below (see 'Palaeogeographic models for southern England and adjacent areas'), with the 'alternative' model preferred. The preferred geological history, utilized in the interpretation of depositional environments and palaeogeography in this chapter (Figs 16.6–16.8), comprises:

1. Marine carbonate sedimentation continuing regionally into the Danian.
2. Early Selandian regional warping and uplift.
3. Thanetian to Rupelian marine and marginal-marine sedimentation in southern England, in a wide embayment of the North Sea Basin. Episodic localized inversion along individual structures, but not forming the dominant control on depositional patterns.
4. Major inversion, uplift and erosion in the mid-Cenozoic.

Central and eastern England and the southern North Sea
Fission-track, vitrinite reflectance and sediment compaction studies have suggested up to 3 km of Cenozoic uplift and exhumation erosion in some areas of eastern England and adjacent areas of the southern North Sea, beginning in the mid-Paleocene (see Japsen 1997). This does not agree with estimates based on palaeogeographic reconstructions, which suggest a much thinner former sedimentary cover, and a later date for the major exhumation episode, and this has limited their acceptance.

Re-evaluation of the onshore data and integration with offshore well data (Japsen 1997) indicates that the main exhumation probably occurred during the Neogene, superimposed on minor and more localized Paleogene uplift. This is supported by the recent discovery of the Paleogene Flamborough Outlier west of the Sole Pit Inversion, off the Yorkshire coast (Stewart & Bailey 1996) (Fig. 16.3). Recent studies based on a wider geographic range of fission-track analysis and vitrinite reflectance (e.g. Green *et al.* 2001) indicate significant differential uplift. Areas of the Midlands and northern England underlain by Carboniferous basins underwent major Cenozoic inversion, while intervening Late Palaeozoic highs such as the London Platform were more stable.

In the inverted areas, an early Cenozoic cooling episode beginning at about 65–55 Ma (Paleocene), and a later one at <25 Ma (Miocene or younger) have been identified. The earlier of these seems to calibrate with the relaxation following the main plume-related updoming of northern Britain. The latter may be related to the Miocene Alpine compression seen in southern England, but could be later. Upper Miocene non-marine sediments in Derbyshire (the Brassington outliers, Fig. 16.3), probably deposited near sea level, were subsequently uplifted by at least 450 m, probably as a result of Pliocene–Pleistocene neotectonics (Walsh *et al.* 1999).

These studies also confirm the former existence of thick Cenozoic sediments in eastern England; objections previously raised on palaeogeographic grounds are considered invalid, as discussed below ('Paleogene shorelines in SE and E England').

Igneous activity and related uplift

The North Atlantic Igneous Province

A major 'hot spot' (the Iceland mantle plume), initiated in latest Cretaceous time, is believed to have impinged on the base of the lithosphere at approximately 62 Ma (Danian) beneath East Greenland (Knott *et al.* 1993; Ritchie *et al.* 1999). It is interpreted as the cause of the extensive volcanism and associated intrusive activity in the NW British Isles and adjacent areas, extending as far as Greenland. This comprises the NAIP (North Atlantic Igneous Province), which includes the British Tertiary Igneous Province (BTIP). A vast outpouring of predominantly subaerial basaltic lavas, associated with sill and dyke intrusion and pyroclastic ash falls, took place from a number of igneous centres (see Fig.16.4), prior to and accompanying North Atlantic–Arctic crustal separation. Intermittent igneous activity began in mid Danian (62–63 Ma), and was most intense during two phases, at approximately 58–59 Ma (Selandian–Thanetian boundary) and at c. 55–56 Ma (Paleocene–Eocene boundary) (Knox 1996*a*; Bell & Jolley 1997). These were probably associated with changes in the locus of sea-floor spreading. The latest phase may reflect the final phase of crustal readjustment before North Atlantic continental break-up.

Igneous activity in southern Britain

Igneous rocks related to the BTIP are represented in southern Britain and adjacent areas mainly by the distal members of dyke swarms (Fig. 16.3), and by ash-falls in marine sediments. Dykes originating in the Mull Igneous Centre, particularly the Cleveland Dyke and the Blyth and Accrington dyke groups, have been traced up to 800 km from their source (Knox 1996*a*). Dykes probably sourced from the Antrim Igneous Centre have been identified in North Wales and the northern Irish Sea (Francis 1992). The Lundy Igneous Complex, in the SW Bristol Channel, comprises the Lundy Granite intrusion and an associated dyke swarm, dated as approximately 52 Ma, and may be associated with the second intense phase of igneous activity.

Volcanic ash falls, which were probably derived from igneous centres north of Scotland, covered large areas of NW Europe, and have been identified as far away as the Bay of Biscay and Austria (Egger *et al.* 2000). In the London Basin and East Anglia, they are represented mainly by very thin bentonitic clay beds, derived from chemically altered ash. Isolated ash beds have been recognized in the late Selandian and early Thanetian, representing pyroclastic phase 1 of Knox & Morton (1988) and Knox (1996*a*). Phase 2.1 is represented by occasional ash beds in the late Thanetian and earliest Eocene. The maximum, Early Eocene, pyroclastic phase (phase 2.2) is identified throughout the North Sea within the Balder Formation, including BGS borehole 81/46A, off the Yorkshire coast (Unit 2 of Lott *et al.* 1983) (Fig. 16.10). In East Anglia it is represented by dispersed ash in the lower Harwich Formation, followed by up to 40 ash beds in the upper Harwich Formation. A final, much weaker phase of explosive volcanic activity is represented by a few ash beds in the Lower Eocene (middle Ypresian) London Clay Formation of the eastern London Basin and in the basal Horda Formation of the southern North Sea.

Uplift related to the Iceland mantle plume

During the Late Paleocene an area centred on Greenland was uplifted over a radius of up to 2000 km. This has been interpreted as due to the Iceland mantle plume, with magmatic underplating of the crust combined with the thermal buoyancy of the plume (Ritchie & Hitchen 1996). The magnitude of the uplift is estimated at 1–2 km in the Scottish Highlands, resulting in the tilting of Scotland and Northern Ireland to the SE, accompanied by deep erosion and the shedding of huge volumes of clastic sediment into the northern North Sea and other areas. Knox (1996*a*) has proposed that the uplift was pulsed, in relation to successive phases of magmatism.

Southern Britain would have been on the margins of this uplift, but it is probable that it was a significant factor in the development of successive Late Paleocene depositional sequences in the North Sea Basin, including the Hampshire and London 'basins'. Much of the clastic sediment deposited in these areas was derived originally from the Scottish Highlands (Morton 1982*a*, *b*). Cope (1994) has proposed that a second hotspot of a similar date located beneath the Irish Sea was responsible for approximately 2 km of uplift in this area, and the eastward tilting of England and Wales. This is not yet generally accepted, but was considered by Knox (1996*a*) as possibly a contributory factor to the pattern of Paleogene uplift and erosion seen in SE England and adjacent areas.

A Paleocene meteoritic impact?

A possible buried impact structure, the Silverpit Crater, 3 km in diameter and 300 m deep, has recently been discovered by seismic surveys in the southern North Sea (Fig.16. 3) (Stewart & Allen 2002). It post-dates the Chalk Group (here probably intra-Maastrichtian at the top), and pre-dates the Moray Group (earliest Eocene). More detailed study has, however, suggested that this circumferentially faulted circular depression resulted from a combination of Paleogene strike-slip faulting and halokinesis (Smith 2004).

Sea-level changes and sequence stratigraphy

Cenozoic sea-level fluctuations

Worldwide eustatic sea-level models (Haq *et al.* 1988) (Fig. 16.5) indicate relatively high sea levels through most of the Paleogene (only slightly lower than during the Late Cretaceous, when the highest sea levels in the Phanerozoic were reached). Maximum sea levels tended to decrease from the late Danian, with a major sea-level fall in the early Thanetian, recovering in the earliest Eocene. Early Eocene sea levels were the highest in the Cenozoic since the Danian, and were followed by somewhat lower sea levels through the Mid Eocene–Early Oligocene. A very marked sea-level fall in the mid Oligocene was followed by a recovery into the Early and Mid Miocene, but from the Late Miocene increasing continental glaciation resulted in generally lower sea levels. These culminated in the very low levels during glacial phases of the Pleistocene.

Some of these features are recognizable in the relative coastal onlap of marine Cenozoic sediments in southern Britain and adjacent areas. For instance, relatively high sea levels in the Early Eocene are reflected by deep-water clays through most of the southern North Sea Basin and the deeper parts of the Western Approaches Basin. However, other features, such as the widespread development of marginal-marine and non-marine facies in latest Thanetian and earliest Eocene times, are out of phase with the oceanic record. In these cases an over-riding tectonic control is probable, as noted above.

Controls on depositional sequence development

The North Atlantic–Arctic and Alpine plate-tectonic events affecting Britain during the Paleogene imply that sea-level changes during this period are likely to reflect a combination of eustatic and tectonic control. Major episodes of uplift have been documented in Britain and adjacent areas. Most recent studies interpret the larger-scale Paleocene and lowest Eocene sequences in this area (third-order sequences in the terminology of Haq *et al.* 1988) as 'the direct result of tectonic activity, rather than tectonic events superimposed on major eustatic events' (Knox 1996*a*, p. 227). This implies that direct correlation with eustatic sea-level fluctuations is unlikely. The origin of the smaller-scale (fourth-order) sequences is uncertain, and considered more likely to reflect eustatic sea-level change. In the 'icehouse' world of the Neogene these small-scale sequences are attributed to fluctuations in the size of continental ice sheets, but the controlling mechanism in the 'greenhouse' world of the early Paleogene (with apparently no significant continental glaciation) is not yet clear.

Sequence stratigraphic models

There are repeated lateral and vertical facies changes in the shallow marine to non-marine Paleogene sediments of the southern North Sea Basin and adjacent areas (Hampshire Basin, London Basin, Belgium and Paris Basin). These were recognized already in the early nineteenth century to comprise depositional sequences, with boundaries usually defined by disconformities. They were classified into transgressive-regressive depositional sequences by Leriche (1906) and Stamp (1921). The boundaries between these sequences, together with their contained macrofaunas, were used to define regional 'stages'. These have, in many cases, become transmuted into the standard chronostratigraphic stages in use today. Even at this time, the relative importance of tectonic and eustatic controls on depositional patterns were debated. These early but far-sighted models were largely neglected until the 1970s, with the 'rediscovery' of sequence stratigraphy. Sedimentological analysis of the Bracklesham Group in the Hampshire Basin (Plint 1983, 1988*a*) identified depositional sequences, transgressive surfaces and regressive (erosional) surfaces marking sequence boundaries. These events were calibrated to the world-wide eustatic sea-level scale (Haq *et al.* 1988). This study has been followed by many others.

Recent studies have been to a significant degree generated by the petroleum industry, particularly in order to understand the stratigraphic context of important Paleocene and Eocene hydrocarbon accumulations in the northern North Sea Basin. These have resulted in a variety of sequence stratigraphic interpretations, both for specific areas and also as broad overviews, for part or all of the Paleocene and Eocene of the North Sea and adjacent areas (e.g. Gramann & Kockel 1988; Knox 1996*a*; Neal 1996; Vandenberghe *et al.* 1998). They have relied on an increasingly detailed biostratigraphic framework for regional correlation.

Some studies have correlated their sequences with those of Haq *et al.*, but as the latter was partly based on data from the North Sea, an element of circular reasoning may be involved! Others have erected independent regional schemes. Most use 'Exxon-type' sequences, delimited by sequence boundaries (sea-level fall), but others have preferred to identify 'Galloway-type' sequences, with boundaries taken at transgressive events. The most ambitious North Sea Basin-wide scheme, that of Neal (1996), recognizes major 'R/TF' (regressive/transgressive facies) cycles, with their bases defined at regressive events, each divided into smaller-scale 'sequence cycles'.

Development of sequence stratigraphic models in the North Sea Basin is impeded by a number of factors. It is difficult to interpret facies changes in the relatively continuous deep-water sediments of the central North Sea Basin in terms of sea-level change, while many stratigraphic breaks are identified by biostratigraphy or sedimentology in more proximal environments. Correlation of smaller-scale sequences is often below the resolution of existing biostratigraphic data, and there is inadequate documentation of lithostratigraphy and biostratigraphy in many key areas. Most important is the uncertainty, discussed earlier, over the relative importance of tectonic versus eustatic controls in effecting changes in relative sea level. This implies that sequences may be only locally developed, or diachronous on a regional scale.

As pointed out by Knox (1996*b*, p. 5) 'an interregional sequence stratigraphic scheme will inevitably be hampered by geographic variation in the amount of uplift, and by the interplay between uplift and eustatic sea-level change'. The effect of these factors can be seen in the differing interpretations of the same areas (compare Knox 1996*a* and Powell *et al.* 1996). In spite of these problems, there is general agreement on the major Paleocene and Eocene sequence boundaries and transgressive surfaces, and on their calibration with the geological timescale, but there are major discrepancies in the identification of smaller-scale sequences. The depositional sequences identified in the London and Hampshire 'basins' are summarized below (see Fig. 16.20).

Oligocene and younger sediments are poorly represented in the southern North Sea and adjacent onshore areas of England, and no sequence stratigraphic framework has been proposed, although detailed studies have been carried out in adjacent areas of the North Sea and continental Europe (e.g. Vandenberghe *et al.* 1998). In the Western Approaches Basin, major episodes of uplift and erosion have been documented, but the data currently available do not permit any satisfactory sequence stratigraphic analysis.

Palaeogeography

Regional context

Cenozoic palaeogeography is controlled by the position of the British Isles between the North Atlantic Ocean with its marginal basins to the west, and the intracontinental North Sea Basin to the east (Fig. 16.4). The North Sea Basin was connected to oceanic areas through the English Channel, the Faroe–Shetland Basin and (from the Eocene onwards) the Norwegian Sea, and indirectly to Tethys and its successors eastwards through central Europe. Water circulation through these portals was intermittent and frequently interrupted, due to sea-level fluctuations and tectonic uplift. This is reflected in microfaunal assemblage fluctuations (King 1989), and in episodes of reduced salinity and central basin stagnation, particularly during periods of lowered sea level.

In the centre of the North Sea Basin, bathyal water depths persisted through much of the Cenozoic, while to the west of the British Isles such depths were reached only on the Atlantic continental slope and beyond. The southern North Sea Basin and Western Approaches Basin subsided intermittently, sedimentation generally keeping pace with subsidence, but with several episodes of inversion and uplift. The main depocentres were in the central North Sea and the western part of the Western Approaches Basin, in which marine environments probably existed through the entire Cenozoic. In more marginal areas of the basins, the interplay of tectonics and sea-level fluctuations resulted in continual environmental shifts, and it is here that depositional sequences are best developed.

The present 'upland' areas of the British Isles (the Caledonian and Variscan blocks) had been uplifted already in Mesozoic time, and were subject to several phases of differential uplift and erosion through the Cenozoic. The Armorican, Cornubian and Welsh Massifs and the Scottish Highlands supplied sediment intermittently to the surrounding basins, as indicated by heavy-mineral studies (Morton 1982*a*, *b*) and were probably above sea level at almost all times. Much more problematic is the extent to which the shallow seas of the North Sea Basin margins transgressed onto the present land areas of southern and eastern England. Present-day marine Paleogene outcrops are confined to relatively small areas of SE England; the clearest evidence of a coastline is seen in the Middle and Upper Eocene marginal-marine and coastal plain/fluvial sediments of the western and southern Hampshire Basin. Interpretation of the location of Paleogene shorelines is heavily dependent on interpretation of the timing and nature of the structural evolution of England and Wales, which is itself still debated, as discussed earlier.

Mid Cenozoic inversion and uplift significantly restricted the area of marine sedimentation, but little evidence remains for the position of Miocene and Pliocene shorelines. A moderately deep (outer neritic) marine environment persisted in the Western Approaches Basin and parts of the southern North Sea Basin until the end of the Pliocene.

Palaeogeographic models for southern England and adjacent areas

The tectonically-dominated 'semi-closed' model

The 'classic' interpretation of the Hampshire and London 'basins' (e.g. Wooldridge & Linton 1955; see the discussion earlier in this chapter) regarded them as initiated by pre-Cenozoic and early Cenozoic tectonics, and functioning as distinct depocentres throughout the Paleogene. Their sediments were regarded as derived mainly through 'proto-Thames' and 'proto-Solent' river systems from the emergent uplands of Cornubia and Wales to the west. They were envisaged as separated by the Weald–Artois High, with intermittent connection between the basins around the western margin of a 'Wealden island'. The Purbeck–Isle of Wight folds, delimiting the southern margin of the Hampshire Basin, were interpreted as the margin of a land area which functioned as a sediment source.

This 'semi-closed' model was partly based on reconstructed depositional patterns of Paleogene stratigraphic units, (e.g. Hester 1965). These appeared to demonstrate thinning of individual stratigraphic units from the centre of the London Basin towards its northern and southern margins. Plint (1982) similarly regarded the Hampshire Basin in the Mid Eocene as a narrow elongated inlet of the Dieppe Basin, readily isolated by intermittent tectonic uplift or sea-level fall, with sediment supplied by rivers from the west. Similar reconstructions were presented by Edwards & Freshney (1987*a*) for the middle Lutetian Selsey Formation and by Keen (1977) for the Upper Eocene Colwell Bay Member [Middle Headon Beds]. These interpretations were followed by Murray (1992), with maps showing progressive regression, beginning in the Mid Eocene, with the London Basin becoming a land area by the Bartonian or later. The Weald–Artois Axis (High) was shown as an emergent barrier between the Hampshire Basin and the North Sea Basin, perhaps from the Mid Eocene onwards.

A very similar 'semi-closed' model, accompanied by palaeogeographic maps resembling those of Murray (1992) has again been proposed recently (Gibbard & Lewin 2003), suggesting that the main drainage systems of southern England were already established in the Paleogene.

These 'semi-closed' models rely on tectonic and stratigraphic interpretations which are believed to be erroneous, as discussed below.

A somewhat divergent, partially 'semi-closed' interpretation, was recently proposed for the Upper Paleocene and Lower Eocene successions of the Hampshire Basin (Newell 2001), envisaging open marine connection to the north, but deltaic progradation from the south and SW. It is based on lithostratigraphic and biostratigraphic correlations radically different from those in all other recent studies and from those accepted here.

The 'open' (North Sea Basin margin) model

The 'semi-closed' reconstructions were criticized by King (1981), whose study of the Early Eocene failed to substantiate these previous interpretations. Curry (1992) also preferred a more 'open' interpretation, with a much smaller 'Wealden island' forming an extension of the Artois High at the eastern end of the Weald, and a single depositional area extending at most times from the southern North Sea across SE England and adjacent areas. This was based on faunal evidence of marine connections between southern England and the North Sea area at many levels in the Paleogene. Knox (1996*a*) also preferred a much more 'open' interpretation of Thanetian palaeogeography.

The 'open' palaeogeography preferred here is based on many lines of evidence. These include:

1. The close similarity between the Upper Paleocene to Middle Eocene stratigraphy of the London and Hampshire basins.
   Facies boundaries in these areas, where identifiable, trend N–S, at right angles to the dominant structural trends. This similarity applies both to mid-neritic units and those deposited in marginal-marine environments. A striking example is the lowest Eocene marginal-marine Lambeth Group. The western limit of the upper and lower 'tongues' of the Woolwich Formation within the Reading Formation (see 'Regional Stratigraphy', below) in the Hampshire Basin and the Paleogene outliers further east, are almost exactly due south of their limits in the London Basin. This is precisely the same trend as seen in the later deep-water London Clay Formation (King 1981), and must reflect the regional basinal architecture. Other examples are noted in the 'Regional Stratigraphy' section below; even the youngest Paleogene stratigraphic unit preserved in the London Basin is almost identical to its correlative in the Hampshire Basin. These observations would appear to preclude any significant differential uplift of the Weald during this period, except possibly at its eastern end. It is surprising to see palaeogeographic reconstructions still being published (Gibbard & Lewin 2003, fig. 6) which (for instance) show the London and Hampshire basins entirely separated by a 'Weald–Artois High' during the Lutetian and Bartonian. There is no evidence whatsoever for such a barrier, as discussed below.
2. The periodic influxes of clastic material derived from Scottish sources into the London and Hampshire basins (Morton 1982*a*, *b*).
   These tend to support an 'open' interpretation.
3. The similarity of microplankton, molluscs and other taxa in Upper Eocene and lowest Oligocene marginal-marine facies between the Hampshire Basin and Paris Basin (e.g. Chateauneuf 1980), and the Priabonian 'Brockenhurst Bed' transgression from the NE (see details below).
   These imply that even at this relatively late stage the Hampshire Basin was by no means an isolated or semi-isolated depositional area.
4. Re-evaluation of the validity of the main lines of support for the 'semi-closed' model: the early inversion of the Weald–Artois High and a Paleogene 'Wealden island', and Paleogene tectonics in the Hampshire Basin.

These are discussed below.

The Weald–Artois High and the 'Wealden Island'

The concept of early updoming of the Weald–Artois High , with the development of a Paleogene 'Wealden Island', originated with Prestwich (1852). It was reinforced by Stamp (1921) and Wooldridge & Linton (1955). As noted earlier, Jones (1999*b*) strongly favoured early inversion and uplift of the Weald, and, along with Curry (1992, fig. 13.6) and Murray (1992), retained a 'Wealden island' or 'shoal', although in a more restricted geographic and temporal sense than some previous reconstructions. The 'Artois axis', the eastern section of the Weald–Artois High, has also been generally regarded as playing a major role in Paleogene sedimentation and tectonics (e.g. Pomerol 1982).

The 'sub-Paleogene unconformity surface' is interpreted as indicating Early Paleocene or even latest Cretaceous inversion of the Weald High (Jones 1999*b*), with development of an emergent 'Wealden Island'. Evidence of major early differential SW–NE uplift and erosion across areas of the London Basin and the Weald–Artois High is clearly demonstrated by mapping the age of the Upper Cretaceous Chalk underlying the sub-Paleogene surface (e.g. Curry 1992, fig. 13.2a). However, the apparent axis of maximum uplift appears to be unrelated to known Mesozoic and Cenozoic structural trends, or to the orientation of the later (mid-Cenozoic) inversion of the Weald–Artois High. It seems improbable that this unconformity marks early inversion of the Weald. The validity of the estimates of Chalk thicknesses eroded in this area, and hence relative regional uplift, may require modification when more recent data on Chalk depositional patterns are taken into account.

The other main lines of evidence used to support early uplift of the Weald are:

1. The southward overstep of the lowest Eocene Blackheath Beds pebble gravels onto the Chalk south of the London Basin (see Curry 1965).
   The outliers in north Kent where this has been noted are, however, all disturbed by post-depositional solution of the underlying chalk. The apparent contacts of Blackheath Beds on Chalk may be due to washing out of formerly underlying Paleocene sands, or to redistribution of Blackheath Beds pebbles by Pleistocene solifluction. Alternatively, they may fill deep channels incised into the Chalk, as suggested by Curry (1965, p.158); the possibility that these sediments are actually incised Thanetian (Upnor Formation) pebble-gravels has also been suggested (Ellison & Skipper pers. comm.). The assumption of a progressive southward overstep of Paleogene sediments onto a Wealden 'island' based on these outliers (Jones 1999*b*) is entirely speculative.
2. The occurrence of Paleogene sediments in the London Basin apparently derived from erosion of the Weald uplift. These include the flint pebble gravels of the Blackheath Beds (Early Eocene) noted above, which appear to be restricted to localized areas near the southern margin of the London Basin. These certainly indicate erosion of the Chalk, but they have been highly abraded in a high-energy littoral environment, and may have been transported by longshore drift for long distances before deposition in deeply incised channels. Localized pebble accumulations of similar age and in comparable depositional contexts occur in East Anglia (Suffolk Pebble Beds) and in Normandy. This suggests that a more regional explanation for their origin must be sought.
   Chert pebbles, apparently derived from the Hythe Formation (Aptian) of the Weald, are reported in the basal pebble bed of the Camberley Sand Formation (Middle Eocene) in the London Basin. This has been regarded as conclusive evidence of the uplift and deep erosion of the Wealden anticline, but does not prove more than localized and perhaps short-lived uplift, as documented in the Hampshire Basin at this time. Curry (1992, fig. 13.6) suggested that the pebbles were derived from localized uplift in the SE corner of the Weald. As quartz pebbles are also reported from this bed, derivation of its components may be more complex, and requires re-investigation.

Over the eastern part of the Weald–Artois High, in northern France, the wide distribution of small outliers of 'Thanetien' (mainly upper Thanetian) marine sands (Mégnien 1980, plate T2), indicates that there was no differential uplift at this time. The evidence for uplift in the Early Eocene in this area was the barrier formerly supposed to separate the non-marine 'Sparnacian' of the Paris Basin from the marine Ieper Clay of northernmost France and Belgium (e.g. Pomerol 1982, fig. 4.3). These stratigraphic units are now known to be of different ages. More recent study (Steurbaut & King 1994) indicates that there was certainly continuity of Ypresian marine sedimentation across the Artois High, although little trace of Ypresian sediments remains. This suggests an episode of post-Ypresian, pre-Lutetian erosion. This could be related to the eustatic sea level fall at this time, but was probably tectonically enhanced, as suggested by an angular unconformity between the middle Ypresian and lower Lutetian in the Paris Basin (unpublished data). A small outlier of upper Lutetian carbonates in the eastern Channel near Boulogne, apparently resting unconformably on the Lower Cretaceous (Bignot 1972), also implies pre-Mid Eocene uplift in this area.

Early Lutetian depositional continuity is evident from the occurrence of *Nummulites laevigatus* both north and south of the Weald–Artois High, and remnants of lower Lutetian marine sediments occur widely over the Weald–Artois High in northern France. By the mid Lutetian the generally accepted interpretation is that 'the Paris and Belgian basins had become separated by the Artois anticline' (Pomerol 1982, p. 69). However, the differences in facies through the Middle Eocene–Oligocene between the Paris Basin and Belgium can all be explained by lateral facies change, and do not require a physical barrier.

Paleogene tectonics in the Hampshire Basin
Evidence for intra-Paleogene uplift and erosion along Alpine structures is now well documented in the Middle Eocene of the Hampshire Basin (Plint 1982; Gale *et al.* 1999), but it must be emphasized that this proves only localized uplift (e.g. on the Sandown Anticline, Gale *et al.* 1999, fig. 9). There is no stratigraphic evidence for the Middle Eocene reconstructions of Plint (1982) and Murray (1992, map Pg2c), showing the southern margin of the Hampshire Basin along the Purbeck–Isle of Wight structures. At Whitecliff Bay, there is evidence of localized uplift and erosion at several levels in the Bartonian. Here (and at Alum Bay, in a similar structural context) there is, however, little indication from the depositional environments that they are only hundreds of metres from the largest Cenozoic structural lineament in southern England. The fission-track data cited earlier also provide no evidence for major early inversion.

Paleogene shorelines in SE and E England
Many previous reconstructions of Late Paleocene to Mid Eocene palaeogeography (e.g. Curry 1992; Murray 1992) have shown shorelines or basin margins running northeastwards through East Anglia into the North Sea. Thus much of eastern England north of East Anglia would have been above sea level through most or all of the Cenozoic.

Estimates of the former depth of burial of outcropping Middle and Upper Jurassic sediments in eastern England and the East Midlands have been based on their thermal maturity. These imply that up to 1.5 km of sediments have been eroded from this area (Green *et al.* 2001). Subtracting the estimated former thickness of Cretaceous sediments in eastern England suggests that at least 0.5 km of this must be Cenozoic; indicating the former presence of a thick and probably marine Paleogene section, as a thick Neogene section is regionally improbable. As this contradicts the generally accepted palaeogeographic reconstructions, there has been much debate about the results; 'the discrepancy between stratigraphic and palaeo-thermal reconstruction of former burial depths remains unresolved' (Green *et al.* 2001, p. 69).

This discrepancy can be resolved by re-evaluation of the palaeogeographic models, which were based largely on the interpretation of the London Basin and Hampshire Basin as synsedimentary Paleogene basins ('semi-closed model'). If this is discounted in favour of the 'open model', the stratigraphic data favours a more linear basin margin, perhaps running northeastwards from Dorset through the Midlands, and then northwards along the eastern side of the Pennines, as modelled by Knox (1996*a*, fig. 7) for the Thanetian. This reconstruction is consistent with the N–S facies trends discussed above, and with the dominance of heavy minerals derived from Scottish sources at various levels in the Hampshire Basin and London Basin (Morton 1982*a*, *b*). It is also supported by the discovery of the offshore Flamborough Paleogene outlier near the Yorkshire coast (Stewart & Bailey 1996). It implies that a Paleogene section of comparable thickness and facies to that in the Hampshire and London basins (maximum thickness of *c.* 500 m) could well have extended far northwards along the western margin of the North Sea Basin (Figs 16.6 & 16.7).

Marine connection between the North Sea Basin and the North Atlantic
Above the Lower Paleocene, there are major differences between the stratigraphic successions in the Western Approaches Basin and other marginal Atlantic areas, and the southern North Sea Basin. No Cenozoic sediments are preserved in the central Channel, over the Start–Cotentin High, which runs NW–SE across the English Channel (Evans 1990), and forms the boundary between the North Sea Basin and the 'Atlantic margin' basins. These are characterized, respectively, by almost wholly clastic marine sediments with 'boreal' biotic assemblages, and intermittently carbonate-dominated sediments with 'Tethyan' or 'Atlantic' biotas with low-latitude elements. The high appears to have formed a largely 'passive' feature, which acted as a barrier at times of low relative sea level. During high eustatic sea levels (e.g. during the mid Ypresian and Lutetian) watermass exchange across the high took place, allowing 'leakage' of tethyan faunal elements and carbonate-dominated facies into the southern North Sea Basin (e.g. Middle Eocene carbonates of the Paris Basin). The timing and duration of these episodes of marine connection are, however, still disputed. Identifying them is important for their bearing on water circulation patterns and faunal migrations, both marine and terrestrial.

Paleogene palaeogeography

Early Paleocene (Danian)

Danian sediments are represented in the Western Approaches Basin, and in small areas of the UK southern North Sea; the latter are outliers of the extensive Danian subcrop in the central North Sea. In these areas there was apparently no major stratigraphic break at the Cretaceous–Paleogene boundary. The former extent of marine sedimentation is very uncertain, as all preserved sediments in this area are mid to outer neritic chalks. More proximal inner neritic to marginal marine carbonate facies of the Danian occur in small outliers in the Mons area (southern Belgium) and in the Paris Basin.

Most reconstructions (e.g. Murray 1992, map Pg2a) show early Paleogene marine deposition only in a very narrow strait connecting the Paris Basin to the Atlantic, either isolated from the North Sea Basin, or with a narrow connection to it (Curry 1992, fig. 13.2). These are based on the assumption that the early episodes of Alpine tectonics affecting this area began in the latest Cretaceous or Early Paleocene. However, as noted earlier, there is very little clastic input, except at the extreme margins of the outcrops in the eastern Paris Basin. The similarity of facies in the Western Approaches Basin to those in the southern North Sea indicates that the marine depositional area was probably extensive, and the relief of land areas low, as in the Late Cretaceous. The major Alpine (Laramide) and north Atlantic tectonic events appear to have been in the mid Paleocene, as noted earlier. This suggests a pattern of land and sea similar to the Late Cretaceous, modified by regression in the early Danian (Haq *et al.* 1988).

Marine Danian carbonates were probably deposited throughout the southern North Sea Basin, including SE and, perhaps, eastern England, across northern France and through the English Channel to the Atlantic (Fig. 16.6a). This interpretation is supported by reworked Danian planktonic foraminiferids at several levels in the Thanetian of the London Basin (Curry 1992, fig. 13.2). These are more than 300 km from the nearest present-day Danian outcrops.

Late Paleocene (Selandian and early Thanetian)

Mid Paleocene uplift and erosion in England and northern France probably resulted in the closure of the Atlantic connection. The 'sub-Paleogene surface' of southern England and adjacent areas probably originated at this time, with uplift and erosion of approximately 200 m of chalk in some areas (Jones 1981). Selandian marine sediments in the southern North

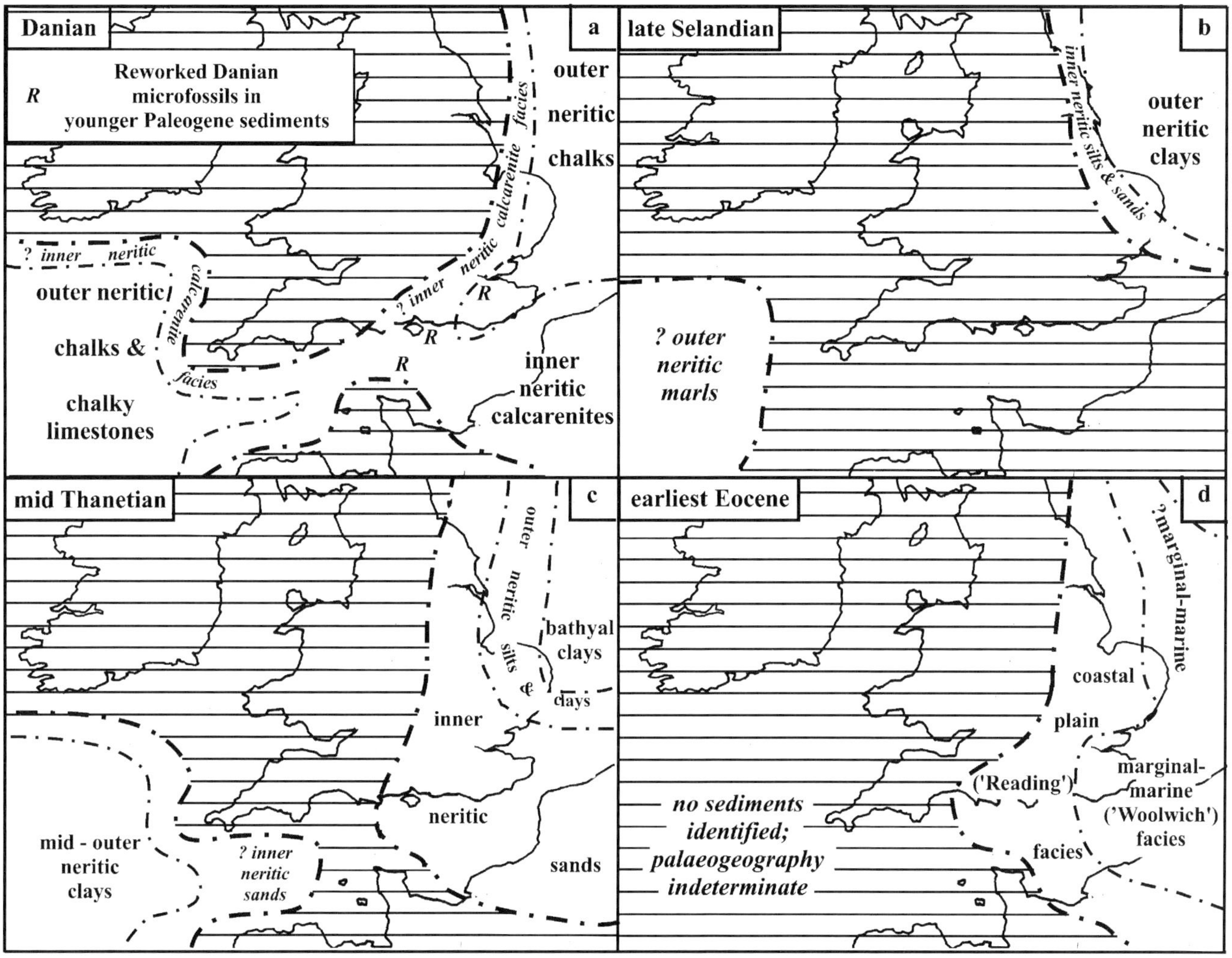

Fig. 16.6. Paleocene and earliest Eocene palaeogeography and lithofacies in southern Britain and adjacent areas. Land areas are shown by horizontal lines. Facies boundaries and shorelines are largely speculative, as discussed in the text, and significantly different interpretations have been proposed (e.g. Murray 1992).

Sea Basin and (possible) Selandian sediments in the Western Approaches Basin are at least 600 km apart, and erosion probably continued through much of this period throughout southern England and adjacent areas (Fig. 16.6b). Pulsed marine transgressions in the late Selandian and early Thanetian resulted in the much wider distribution of marine Thanetian sediments (Fig. 16.6c). Early Thanetian faunas in the North Sea Basin (including the Paris Basin) are of 'boreal' aspect, but 'warm-water' calcareous algae and low-latitude planktonic foraminiferids (*Morozovella*) occur in the southern North Sea Basin in the later Thanetian. This indicates that open marine circulation through the Channel area was probably re-established by this time.

Latest Paleocene–early Eocene

Regression in the terminal Paleocene, probably tectonically controlled, associated with localized inversion, resulted in the retreat of the southern shoreline of the North Sea Basin several hundred kilometres northeastwards (Knox 1996*a*, fig. 7). The North Sea Basin became almost cut off from oceanic areas, and restricted marine environments developed even in the basin centre. Marginal-marine and coastal plain sediments probably covered most of southern and eastern England and adjacent areas (Fig. 16.6d). The synchronous development and very wide extent of these facies (including the Woolwich and Reading Formations, the 'Argile Plastique' of the Paris Basin and the Tienen Formation of Belgium) suggests very low relief, as a result of late Thanetian peneplanation. There was probably a similar regression in the Western Approaches Basin, as sediments of this age have not yet been identified there.

The subsequent early Ypresian eustatic rise in sea level, perhaps enhanced by thermal relaxation following the ending of mantle plume uplift, led to successive marine transgressions. Initially, restricted marine environments persisted, but eventually marine connection to the North Atlantic was re-established.

The highest sea levels in the Paleogene were reached in the mid Ypresian. At this time, deep-water marine clay sediments, previously confined to the central North Sea Basin, probably extended throughout SE England (London Clay Formation) and further south to Normandy (Formation de Varengeville). They grade into shallow marine glauconitic sands in the Paris Basin and eastern Belgium. Similar facies occur in the Western Approaches Basin (Fig. 16.7a). Open marine circulation to the North Atlantic through the Channel is indicated by the appearance of low-latitude rotaliid foraminiferids (*Nummulites*, etc.) and planktonic foraminiferids (*Morozovella*) in the southern North Sea Basin.

Sea-level fall in the late Ypresian led to regional regression, with progressive retreat of deepwater clays towards the centre of the North Sea Basin. Sediments of this age in the southern North Sea Basin and the Western Approaches Basin are generally shallow or marginal marine. Coastal plain and fluvial sediments in the western Hampshire Basin indicate the position of

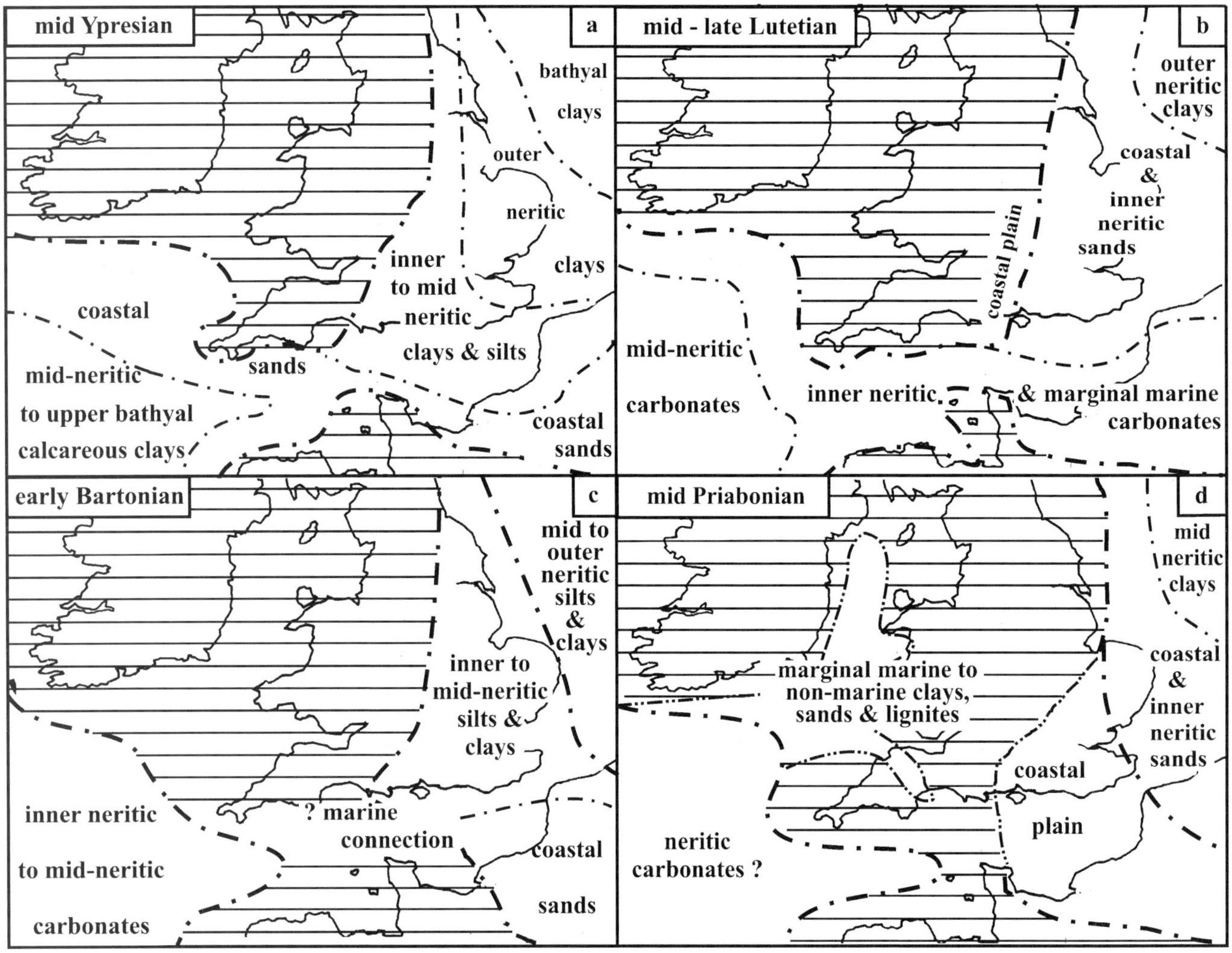

Fig. 16.7. Eocene palaeogeography and lithofacies in southern Britain and adjacent areas. Comments as for Figure 16.6.

the shoreline in this area. Inversion during this period resulted in differential uplift and erosion, with the development of unconformities, as in the Paris Basin and parts of the Dieppe Basin, but marine connections through the Channel are indicated by the reappearance of *Nummulites* and *Morozovella* in the Hampshire Basin and elsewhere.

Mid Eocene

Renewed transgression in the early Lutetian re-established shallow marine environments throughout the southern fringes of the North Sea Basin, the Paris Basin, and the Western Approaches Basin. Water depths were sufficient for the re-entry of *Nummulites* and other low-latitude foraminiferids through the Channel. Carbonate-dominated facies were developed in the Western Approaches Basin and the Paris Basin, but clastic-dominated facies further north. Marginal-marine facies in the western Hampshire Basin again indicate a coastline in this area. Evidence of localized uplift and erosion is seen in this area, but as noted earlier, the close similarity of facies in the London and Hampshire basins indicates that the Weald–Artois High was not a positive feature, as often interpreted (e.g. Murray 1992, map Pg2c). Intermittent localized uplift on pre-existing structural lines is evidenced in the Middle Eocene in the Hampshire Basin (Plint 1982; Gale *et al.* 1999), but there is no positive evidence for uplift on the Portsdown Anticline, which was interpreted by Plint (1982) as forming the northern margin of the Hampshire Basin at this time.

Further sea-level rise in the late Lutetian led to transgression onto the margins of the Armorican Massif, with bioclastic carbonates spreading again from the Western Approaches Basin through the Channel and into the Paris Basin (Fig. 16.7b).

Latest Lutetian uplift and erosion in some areas was followed by renewed transgression. Marine Bartonian sediments are in entirely clastic facies in the northern part of the Hampshire–Dieppe Basin. In the Western Approaches Basin, deposition of shallow-water carbonates continued until at least the end of the Middle Eocene, and perhaps into the Late Eocene. Identical species of *Nummulites* in the lower Bartonian in the Hampshire–Dieppe Basin and the Western Approaches Basin indicate that these areas maintained a marine connection. Sedimentological and palaeontological links between the lower Bartonian ('Lower Barton Beds') of the Hampshire Basin and Belgium suggest that this area was still connected to the main North Sea Basin (Fig. 16.7c). The mammals of the Creechbarrow Limestone in the Hampshire Basin (see Regional Stratigraphy, below) show, however, relatively little endemism with respect to continental Europe, and indicate that a land bridge to France was in existence at some time within the Bartonian (Hooker 1986). It seems likely that this post-dates the early Bartonian.

The non-marine basins of western Britain may have been initiated during the Mid Eocene. The location of shorelines at this time is even more uncertain than for earlier periods; any reconstruction is largely speculative, except in the western Hampshire Basin and the Paris Basin.

Late Eocene and earliest Oligocene

Sediments of this age are preserved only in limited areas. In northern parts of the southern North Sea Basin, they are probably represented by marine clays, with shallow-marine sands prograding from the west. Shallow-marine Upper Eocene and lowest Oligocene (lower Rupelian) sediments occur in the southernmost North Sea and NW Belgium.

In the Hampshire Basin and the Paris Basin, there is a shift at about the Middle–Upper Eocene boundary from dominantly marine to dominantly fine-grained coastal plain, marginal-marine and non-marine sediments (Solent Group and 'Ludian' respectively). These include evaporites in the Paris Basin. The development of these facies has generally been attributed to the isolation of these areas from the North Sea Basin by uplift along the Weald–Artois High (e.g. Pomerol 1982), but their inception corresponds to a significant eustatic sea-level fall (Fig. 16.5). Similar facies of Early Oligocene age are present in eastern Belgium, north of the 'High'. This indicates that the 'semi-closed' palaeogeographic reconstructions of the Hampshire Basin at this time (e.g. Murray & Wright 1974; Keen 1977) are probably unfounded, as discussed earlier. The generally 'layer-cake' stratigraphy of the Solent Group, with no evidence for coarser marginal facies except in localized channelled units, indicates that it extended well beyond the limits of the present outcrop (Fig. 16.7d).

The most prominent lateral facies change within the Solent Group, in the Colwell Bay Member, is marked by a marine incursion which indicates transgression from the NE. Its faunal links with marine sediments of similar age in Belgium, and the occurrence of planktonic foraminiferids, indicate open marine connection to the southern North Sea Basin at this time, as pointed out by Curry (1992). It is considered here that the Upper Eocene and lowest Oligocene marginal-marine and non-marine sediments of southern England and adjacent areas of mainland Europe were deposited in a wide coastal-plain environment fringing the shoreline of the southern North Sea Basin. In the Western Approaches Basin, marine sedimentation probably continued, but evidence is fragmentary.

Oligocene

The highest Paleogene sediments preserved onshore (except in the fault-bounded basins of western Britain) are Lower Oligocene (lower Rupelian) dominantly marginal-marine to coastal plain sediments in the Hampshire Basin, in depositional continuity with the underlying Priabonian. The highest levels are shallow marine, with faunas indicating marine connections to the Paris Basin and Belgium. Sediments of this age have been removed by Early Oligocene (probably mid Rupelian) uplift and erosion in most areas of the southern North Sea Basin, but in northern Belgium they are represented by shallow marine glauconitic sands.

Following the mid Rupelian (late Pyrenean) tectonic phase, a major eustatic sea-level rise led to widespread transgression around the basin margins, with deposition of marine Rupelian clays over a wide area. As this transgression is represented by shallow marine sands in the Paris Basin, it is quite probable that marine sediments of this age were deposited in southern England (Fig. 16.8a).

Upper Oligocene sediments are confined to parts of the southern North Sea and adjacent areas of continental Europe.

Non-marine basins of western Britain

These basins were probably initiated during the Eocene; their early fill is normally of locally derived immature conglomerates and breccias, but their later (Late Eocene and Oligocene) fill is dominated by low-energy alluvial plain sediments. This progressive fining-upwards, culminating in episodic clastic starvation, with widespread deposition of lignites, suggests that individual basins may have widened and eventually merged. This may have led eventually to a very extensive alluvial plain covering much of the Irish Sea and northern Celtic Sea, and adjacent lowland areas of Wales and Ireland during the Early Oligocene, with 'fingers' extending south through the Cornubian massif (Fig. 16.8a). This may reflect the steadily rising eustatic sea levels in the Late Oligocene (Chattian) (Fig. 16.5), and is supported by indications of a marine environment in the highest levels of some basins. This model is compatible with the interpretation of Walsh *et al.* (1999, p.60) that at

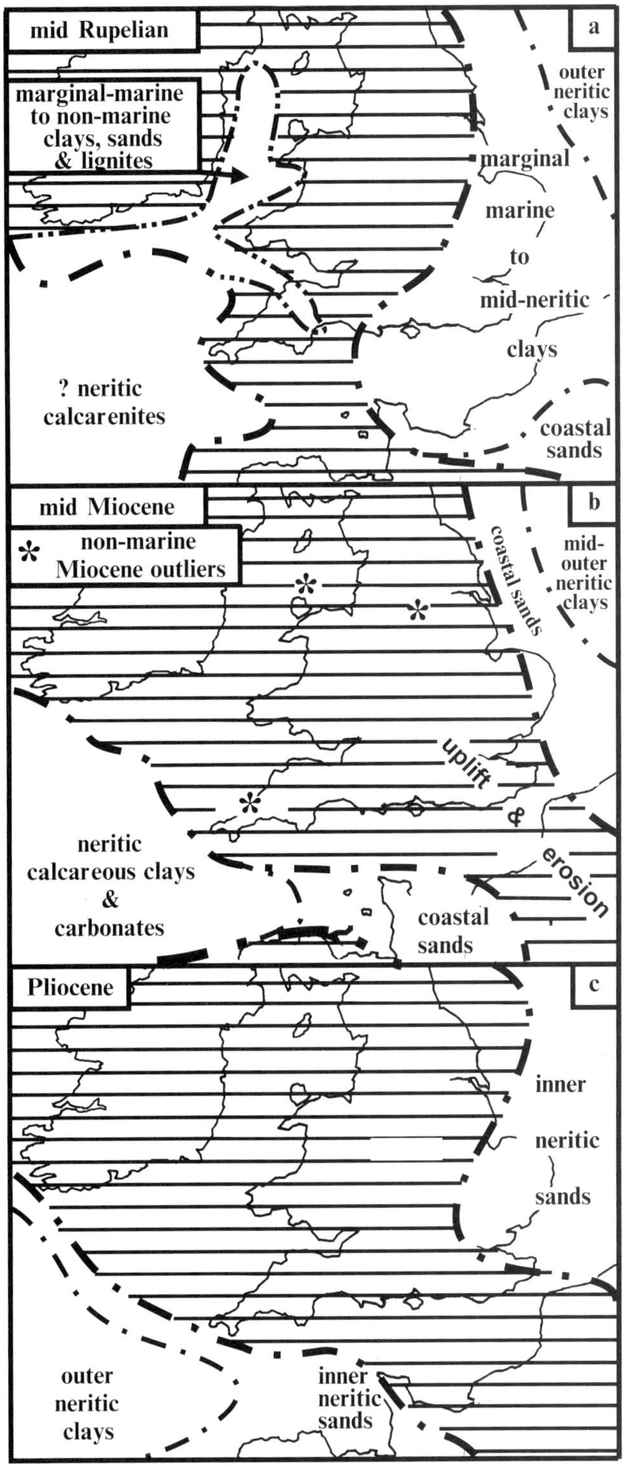

Fig. 16.8. Oligocene, Miocene and Pliocene palaeogeography and lithofacies in southern Britain and adjacent areas. Comments as for Figure 16.6.

this time western Britain and Ireland formed 'a broad forested plain. . .on [the] continental margin'. The occurrence of glauconite in the Stanley Bank Basin (Fig. 16.3) suggests that the coastline was in this area; shallow marine Oligocene sediments are documented in the areas to the south and west.

Neogene palaeogeography

Miocene

A marine, neritic environment existed in axial areas of the southern North Sea, but further south and west the limits of marine deposition are uncertain due to removal by later uplift and erosion. The most southerly marine Miocene sediments preserved, in northern Belgium, are inner neritic sands. Similar shallow marine environments probably extended to East Anglia, evidenced by Miocene fossils reworked into the Pliocene. West of Britain, marine neritic Miocene sediments extended over the North Atlantic margins into the Western Approaches Basin and western Celtic Sea (Fig. 16.8b).

Miocene shorelines are speculative, but the reconstruction of Murray (1992) in which they approximate the present-day coastline of the British Isles, is a generally accepted model. Non-marine Miocene sediments, deposited originally near sea level, are known from SW England, the Pennines and North Wales. As these lie in present-day 'upland' areas, they suggest that the land surface had been significantly planed down by previous episodes of erosion, with extensive swampy coastal plains bordering the sea.

Pliocene

Pliocene shallow marine sediments, mainly sands and bioclastic calcarenites, are patchily preserved through the southern UK North Sea and adjacent areas of northern Belgium and East Anglia. These latter areas are generally interpreted as lying within an embayment of the southern North Sea Basin (e.g. Murray 1992, map Ng10). Previous interpretations of Pliocene shorelines in southern England (e.g. Wooldridge & Linton 1955) were based on identification of a supposed Pliocene erosional platform, and on the presence of isolated patches of probable Pliocene marine sediments in the area around the eastern London Basin, envisaged as lying close to the shoreline. However, the 'Pliocene peneplain' is now less confidently identified (Jones 1981), and the 'Pliocene outliers' are now known to be of differing ages. The 'Atlantic' affinities of the Lenham Beds (Daley & Balson 1999), if confirmed by further analysis, are a severe blow to the 'embayment' model, and suggest that there may have been (intermittent) connection between the southern North Sea Basin and the Atlantic at this time. Marine mid-neritic Pliocene sediments occur in the Western Approaches (Fig. 16.8c).

Palaeoclimate

General context

In the Paleocene Britain was at approximately 40°N, about the present-day latitude of the central Mediterranean. The higher temperatures this implies were enhanced by a lower temperature gradient from equator to pole, due to the absence of large continental ice sheets. This 'hothouse' or 'greenhouse' world is confirmed by the high atmospheric CO_2 levels identified by isotopic analysis. The highest Cenozoic temperatures are recorded during the Early Eocene. In the late Mid Eocene progressive but stepwise cooling led to the 'icehouse' world, characterized by increasingly extensive polar ice sheets and greater latitudinal temperature gradients. This stepwise cooling reflects episodes of ice-sheet growth and shrinkage.

Oceanic temperature changes during the Cenozoic are becoming increasingly well documented, through carbon and oxygen isotopic analysis of deep-water benthic and planktonic foraminiferids (Zachos *et al.* 2001) (Fig. 16.5). Some isotopic data from molluscs has given indications of temperature ranges in shallower seas of the North Sea Basin (Buchardt 1978; Andreasson & Schmitz 2000). The relationship of the isotopic ratios in the shells of marine organisms to the ambient water temperature is subject to some uncertainties, but the results agree well with data from other sources. These include comparisons of temperature-sensitive groups of terrestrial vertebrates, molluscs and plants with their modern equivalents, and identification of climatically-controlled soil types and sediments such as evaporites.

Water temperatures in the North Sea Basin were modified by its partial isolation from the world ocean; water circulation through its southern gateway to the North Atlantic through the English Channel was periodically interrupted, and its main (northern) gateway was to the high-latitude Norwegian Sea. Thus its marine fauna often has a 'cooler' aspect than at the same latitude in the North Atlantic. 'Low-latitude' forms such as *Nummulites* occur on the North Atlantic margin as far north as the Faroes–Shetland Basin (present-day latitude 60°N) (unpublished data), while in the North Sea Basin they only reach 53°N.

Paleocene climate and the Paleocene–Eocene boundary thermal maximum

Paleocene temperature estimates in the North Sea Basin have been affected by the discrepancies between the 'cool' mollusc assemblages of Denmark (Selandian) and southern England (early Thanetian) and the 'warm' assemblages of Belgium (late Danian) and the Paris Basin (late Thanetian). These can be resolved in terms of differences in age and water depths, with water circulation patterns in the North Sea Basin varying with tectonic and eustatic controls.

A brief but very pronounced carbon isotope excursion (the CIE) has been identified world-wide at the Paleocene–Eocene boundary, associated with a prominent oxygen isotope excursion (Fig. 16.5). This is interpreted as reflecting a short-term (*c.* 200 Ka) but extreme warming episode, designated the PETM (Paleocene–Eocene boundary Thermal Maximum); formerly the LPTM (Late Paleocene Thermal Maximum), but renamed due to relocation of the Paleocene–Eocene boundary. It was most probably related to the release of methane by rapid dissociation of gas hydrates in sea-floor sediments (Schmitz *et al.* 2000). The associated climatically controlled biotic changes include a major extinction of benthic foraminiferids and a massive bloom of the thermophilic dinoflagellate *Apectodinium*, both extending to high latitudes. A long-recognized turnover in land mammal assemblages related to extracontinental dispersal, the MDE (mammalian dispersal event) is now also known to be synchronous with the CIE (Hooker 1998). Increased kaolinite influx into marine environments at this time, as seen in the central North Sea and elsewhere, has also been attributed to increased humidity and continental runoff (Knox 1996*b*). In marine and marginal-marine environments in the North Sea Basin, the PETM is marked by organic-rich facies. On the surrounding coastal plains, intense tropical pedogenesis took place (Skipper 1999).

Warm Eocene climates and the transition to an icehouse world

After a brief fall in the early Ypresian, relatively high temperatures returned during the Early and Mid Eocene (the ETM: Eocene Thermal Maximum). A tropical rainforest environment has long been proposed for the Early Eocene London Clay Formation, on the basis of its seed and fruit assemblages, and the most recent analysis confirms the existence of coastal mangrove swamps with probably 'dense megathermal rain forest' behind

(Collinson 2000). The large *Nypa* (mangrove) fruits characteristic of this assemblage occur throughout the Early and early Mid Eocene in western Europe. High temperatures are also indicated by the large rotaliid foraminiferids. However, isotopic studies of shallow marine gastropods from the Mid Eocene of southern England and the Paris Basin indicate a seasonal temperature difference of at least 10 °C, and a maximum temperature of approximately 27 °C (Andreasson & Schmitz 2000).

Worldwide oceanic cooling, which began intermittently in the late Mid Eocene, is believed to be due to the first establishment of Antarctic ice sheets. After the first brief cooling episode, the Late Eocene Hampshire Basin terrestrial flora and vertebrate fauna indicate extratropical forest (Hooker 1992). Sedimentological and faunal evidence from the Bembridge Limestone (Priabonian) indicates alternating wetter and drier phases; the development of evaporites in the Paris Basin at this time also indicates periodic aridity. A further rapid cooling event in the latest Eocene and earliest Oligocene, the Terminal Eocene Event (TEE) is due to circulation of cold oceanic bottom waters linked to an extension of Antarctic glaciation (Prothero & Berggren 1992). This is marked in NW Europe by microfloral, macrofloral and mammalian assemblage changes (Collinson 1992; Hooker *et al.* 2004) which also indicate increasing aridity. It coincides with the 'Grande Coupure', a major break in European mammalian faunas due to immigration from Asia, suggested to be related to the climatic change, with forest dwellers replaced by savannah-dwellers (Hooker 1992).

Mid and Late Cenozoic icehouse; cooling climates

Oligocene and Miocene climatic data are mainly from areas outside Britain, particularly from palynological and macrofloral studies in Germany and the Netherlands (e.g. Utescher *et al.* 2000). Oligocene climates were characterized by 'Mediterranean' temperatures, with minor fluctuations. With the closing of the marine connection between the southern North Sea Basin and the North Atlantic in the Oligocene, tropical and subtropical faunal elements among the marine fauna disappeared. Renewed cooling in the Late Oligocene to a temperate to warm-temperate climate was followed by warming again in the Early and Mid Miocene to warm temperate/subtropical temperatures. This was the warmest period since the Early Eocene.

A pronounced cooling episode at the Mid–Late Miocene boundary is related to an increase in Antarctic ice volume. It continued into the Early Pliocene. A brief mid-Pliocene warming (at *c.* 3 Ma) is related to tectonically controlled water circulation changes in the North Atlantic. This is reflected in the warm ('Mediterranean') Coralline Crag mollusc assemblage, associated with warm dinoflagellate and pollen assemblages, indicating temperatures at least 5 °C warmer than today (Head 1997).

At the Pliocene–Pleistocene boundary episodic but progressive climatic deterioration began, due to the establishment and growth of Arctic ice sheets, culminating in the glacial–interglacial climatic fluctuations of the Pleistocene.

Regional stratigraphy

Lithostratigraphic terminology

The lithostratigraphic terminology of onshore areas of southern England was established mainly in the nineteenth century, but was formalized only relatively recently (King 1981; Insole & Daley 1985; Hooker 1986; Edwards & Freshney 1987*b*; Ellison *et al.* 1994; Daley & Balson 1999). This has involved many minor changes in the definition of previously named units, and the introduction of a number of new names. Current and previous classifications are summarized in Figure 16.9. The Paleogene lithostratigraphic terminology of the southern North Sea Basin is summarized by Lott & Knox (1994). In other offshore areas (Dieppe Basin, Central Channel Basin and Western Approaches Basin) no formal terminology has yet been established. Paleogene and Neogene lithostratigraphic units are summarized in Figures 16.1 & 16.2.

The southern North Sea Basin and adjacent areas (including the London 'Basin', Hampshire–Dieppe 'Basin', and Central Channel 'Basin')

Lower Paleocene (Danian)

Lower Paleocene sediments are preserved only in offshore areas. In the central North Sea Basin, Danian chalks (Ekofisk Formation) deposited in mid to outer neritic environments apparently overlie upper Maastrichtian chalks without significant disconformity. The Danian chalks are characteristically more argillaceous than the upper Maastrichtian chalks. In the southern North Sea, similar Danian chalks occur as isolated outliers. Their present limits are clearly due to subsequent (probably mid Paleocene) uplift and erosion. Reworked Danian planktonic foraminiferids are recorded at several levels in the Upper Paleocene of the London Basin and the Lower Eocene of the Central Channel outlier (Curry 1992), providing evidence for the former extension of similar facies over a much wider area.

The sub-Paleogene unconformity surface

The Danian is overstepped by progressively younger Upper Paleocene sediments southwards and westwards from the central North Sea. Together with this, the degree of erosion of the Upper Cretaceous succession increases. Thus in Norfolk, highest Selandian sediments overlie Early Maastrichtian chalks, while in the western London Basin and Hampshire Basin the 'Cretaceous–Tertiary unconformity' ('sub-Paleogene surface') is marked by a major stratigraphic break between the Santonian or Campanian and the highest Thanetian (Fig. 16.11, Plate 31). Mapping the age of the Chalk at this 'sub-Paleogene surface' indicates that it was differentially uplifted and deeply eroded along a wide belt from NW of London to northern France (Curry 1992, fig. 13.2). As discussed earlier, uplift has been suggested to begin in the Late Cretaceous, but a mid Paleocene (Selandian) date is more probable.

In East Anglia and the eastern London Basin, upper Selandian or lower Thanetian sediments overlie the sub-Paleogene surface. These were probably originally more extensive. Elsewhere, they were removed during a further phase of uplift and erosion during the mid Thanetian, followed by renewed transgression in the late Thanetian. The surface is thus regionally diachronous. The chalk surface is generally planar, but locally deep solution hollows are preserved, indicative of paleokarst.

Upper Paleocene (lower Selandian)

Lower Selandian marls and calcareous claystones deposited in outer neritic to upper bathyal environments (Maureen Formation) are widespread in the central North Sea, conformably overlying the Ekofisk Formation. In the southern North Sea these occur mainly near the median line (UK offshore boundary). Their present limit is due to mid Paleocene uplift and erosion.

Upper Paleocene (upper Selandian and lower Thanetian)

The Lista Formation comprises non-calcareous upper Selandian and lower Thanetian claystones deposited in bathyal environments, and extends through much of the North Sea. In the southern North Sea it oversteps the Maureen Formation to rest unconformably on Maastrichtian chalks. In the Flamborough outlier (Stewart & Bailey 1996) it is represented by rather more coarse-grained sediments, suggesting a more proximal facies. In East Anglia its equivalent is the Ormesby Clay Formation, glauconitic, mainly non-calcareous clays deposited in outer neritic to upper bathyal environments, unconformably overlying

central & eastern HAMPSHIRE BASIN		
HAMSTEAD BEDS	*Upper Hamstead Beds*	Bouldnor Formation
	Lower Hamstead Beds	
BEMBRIDGE BEDS	*Bembridge Marls*	
	Bembridge Limestone	Bembridge Lst. Fm
	Osborne Beds	Headon Hill Formation
HEADON BEDS	*Upper Headon Beds*	
	Middle Headon Beds	
	Lower Headon Beds	
BARTON SAND	*Upper Barton Beds*	Becton Sand Formation
BARTON CLAY	*Middle Barton Beds*	Barton Clay Formation
	Lower Barton Beds	
BRACKLESHAM BEDS	*Huntingbridge Div.*	
	Selsey Division	Selsey Formation
	Earnley Division	Marsh Farm Fm
		Earnley Formation
	Wittering Division	Wittering Formation
LONDON CLAY	*London Clay & 'Bagshot Sands'*	London Clay Formation
	L. C. Basement Bed	Harwich Formation
READING BEDS	*Reading Beds (s.s.)*	Reading Formation
	Reading Bottom Bed	Upnor Formation

LONDON BASIN & EAST ANGLIA	
Upper Bagshot / Barton Beds	Camberley Sand Formation
Middle Bagshot / Bracklesham Beds	Windlesham Formation
(Lower) Bagshot Beds / Sands	Bagshot / Virginia Water Formation
London Clay	London Clay Fm
L.C. Basement Bed / Oldhaven Beds	Harwich Formation
Reading Beds	Reading Formation
Woolwich Beds	Woolwich Formation
Woolwich / Reading Bottom Bed	Upnor Formation
Thanet Sands / Beds	Thanet Sand Fm
Ormesby Clay	Ormesby Clay Fm

Fig. 16.9. Paleogene lithostratigraphic terminology of the Hampshire and London 'Basins'. Relationship between the current classification and previously used names (shown in italics) which are now obsolete. The layout of this table is not intended to indicate correlations between these areas. Terminology of other lithostratigraphic units of the western Hampshire Basin is discussed in the text (see also Dayley & Balson 1999).

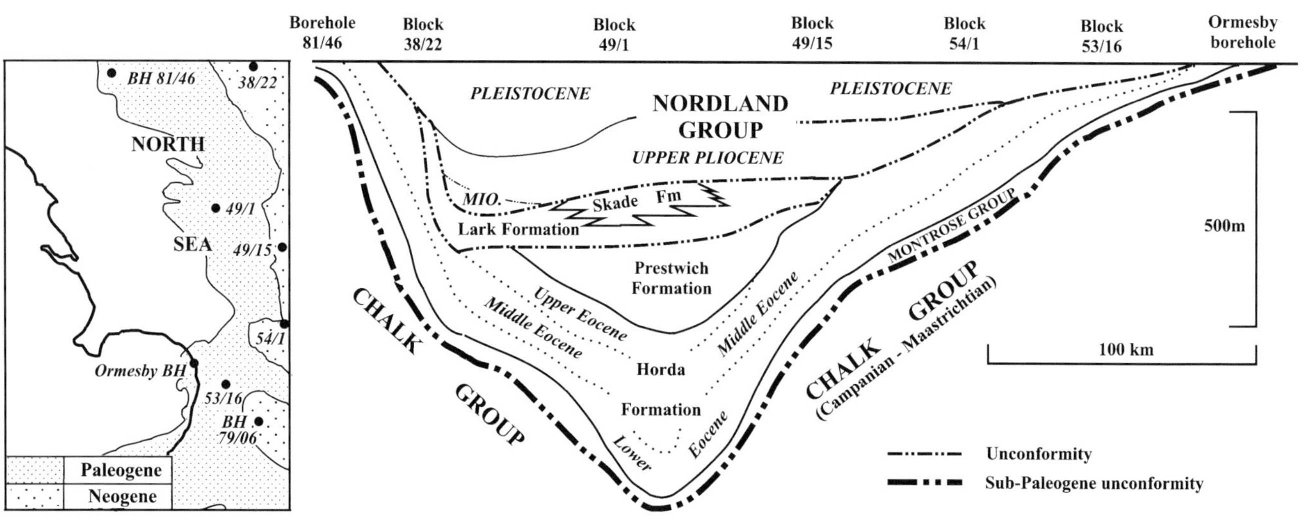

Fig. 16.10. Cenozoic stratigraphy of the southern North Sea. 38/22, etc, are UK North Sea quadrangle and block numbers (designated for petroleum exploration). NB. The vertical scale is highly exaggerated.

Campanian and Maastrichtian chalk. This is the oldest Cenozoic sedimentary unit represented onshore, and is known almost entirely from boreholes. Its lowest part is latest Selandian in age, as determined by magnetostratigraphy and palynology, somewhat younger than the base of the Lista Formation. Argillized volcanic ash layers in the lower part represent the

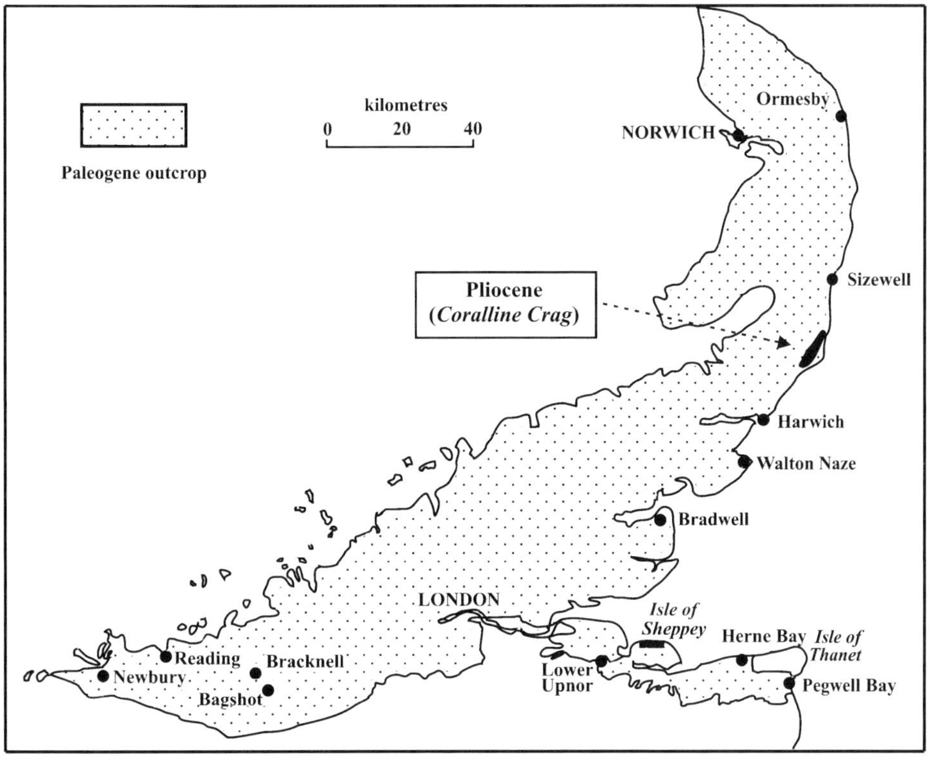

Fig. 16.11. Paleogene and Neogene outcrops and key localities of the London 'Basin' and East Anglia. Other localities include Tilehurst (a suburb of Reading), Windlesham and Camberley (close to Bagshot), Woolwich and Lambeth (districts of London) and Oldhaven Gap (close to Herne Bay).

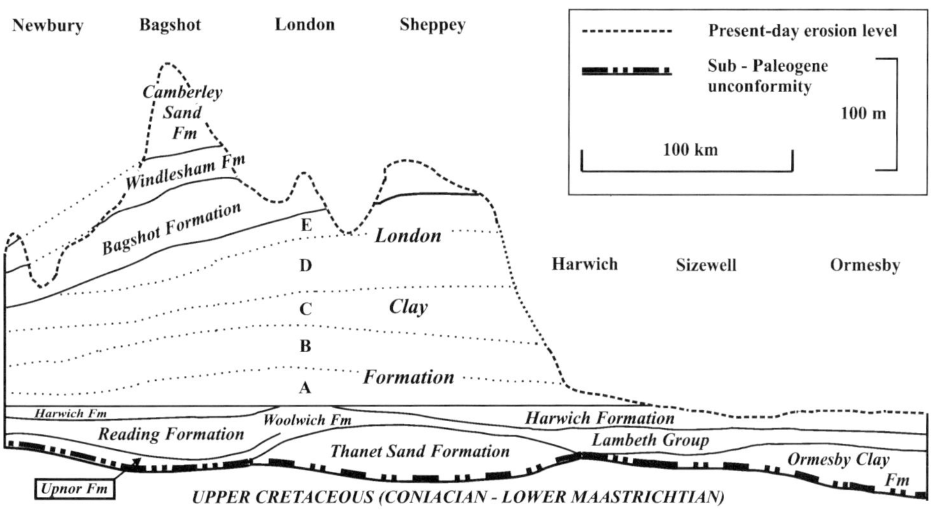

Fig. 16.12. Paleogene stratigraphy of the London 'Basin' and East Anglia. Localities cited are shown in Figure 16.11. The horizontal datum is the base of the London Clay Formation. The stratigraphy of the Sizewell and Ormesby boreholes is based on Ali & Jolley (1996), Jolley (1996) and Knox (1996*a*). NB. The vertical scale is highly exaggerated.

earliest pyroclastic phase in the BTIP (British Tertiary Igneous Province). A distinctive red clay unit in the lower part is also recognizable in parts of the central North Sea.

There is a rapid southward transition in the south of East Anglia from the deep water Ormesby Clay to silts and glauconitic sands deposited in shallow marine environments (Thanet (Sand) Formation). Their base is early Thanetian in age. A basal bed of large unabraded flint nodules ('Bullhead Bed') probably represents a winnowed solution residue of the topmost Chalk. The Thanet Sand Formation is restricted to the eastern London Basin and adjacent parts of East Anglia. The main outcrop sections are at Pegwell Bay and Herne Bay in north Kent.

The final part of the first BITP pyroclastic phase is represented by degraded volcanic ash in the base of the Thanet Formation. Higher levels are locally silica-cemented, due to dissolution and reprecipitation of biogenic silica from sponge spicules, a feature widespread in this interval over extensive areas of the southern North Sea Basin. Three depositional sequences are differentiated by Knox (1996*a*). The third sequence is only locally preserved, but faunal evidence indicates that it represents the highest sea level in the Thanetian.

The Thanet Sand Formation becomes coarser-grained westwards towards the London area, where it is abruptly truncated beneath the Upnor Formation, probably reflecting mid

Thanetian uplift and erosion. It is absent in the Hampshire Basin and probably in the Dieppe Basin, but a small erosional remnant on the Normandy coast near Dieppe (Dieppe Sands), and the presence of coeval sands in the Paris Basin, indicate that marine sands of this age were formerly much more widespread. Heavy minerals in the London Basin indicate that these sands were derived from Scottish sources (Morton 1982*a*).

Upper Thanetian unconformity

In the central North Sea there is an abrupt biotic and sedimentary change at the boundary between the Lista Formation and the overlying Sele Formation. This is marked by an influx of terrestrial palynomorphs, a major decrease in diversity of dinoflagellates and foraminiferids, and an increase in silt and organic content, followed by the incoming of significant proportions of kaolinite. These features probably reflect a major sea-level fall (Knox 1996*a*). The Sele Formation was deposited in restricted marine environments of uncertain depth.

In East Anglia and the eastern London Basin, there is a stratigraphic break between the Thanet Sand Formation/Ormesby Clay Formation and the Lambeth Group. It is marked by the differential uplift and erosion of the underlying sediments. Evidence of subaerial exposure and leaching by meteoric groundwaters (Morton 1982*a*) implies that uplift was perhaps in excess of 100 m. This event can probably be correlated with the sea-level fall in the central part of the basin, and indicate that it was a tectonically controlled event, which has been related to the postulated Greenland mantle plume (Knox 1996*a*). Similar uplift and erosion probably occurred through much of southern England and adjacent areas, removing most traces of any lower Thanetian sediments previously deposited. Differential uplift in the southern North Sea (the Sole Pit inversion) and adjacent areas of eastern England took place at this time, approximately coincident with intrusion of the Cleveland Dyke and the parallel Blyth–Acklington Dykes (Cameron *et al.* 1992).

Late Paleocene (late Thanetian)

In the central North Sea, the Sele Formation comprises non-calcareous silty claystones with some thin ash layers. It is divided into successive units S1a, S1b, S2 and S3 (Knox & Holloway 1992). Unit S1a is a claystone deposited in a restricted marine environment of uncertain depth. When traced into the southern North Sea, a hiatus develops between the Lista Formation and the upper part of Unit S1a. The upper part of S1a and S1b appear to pass into shallow-marine and non-marine sediments assigned to the Lambeth 'Formation' (Group) (Lott & Knox 1994).

In southern England the Upnor Formation is the probable correlative of Unit S1a. Its base is a major transgressive surface. In parts of East Anglia and in the eastern London Basin, it overlies the Ormesby Clay Formation or Thanet Sand Formation, overstepping in the western London Basin and the Hampshire Basin onto the Chalk. It comprises glauconitic sands ('greensands'), generally coarser and less well-sorted than the Thanet Sand Formation, with rounded-pebble layers and oyster shell-beds, deposited in an inner neritic environment. It is well-exposed at Herne Bay and in the type section in the Lower Upnor quarry, north Kent. It has not been identified in the offshore outcrops of the Dieppe Basin, but this may reflect limited sampling. Similar facies of the same age occur in outliers on the Normandy coast (Sables de Creil) and in the Paris Basin (Sables de Bracheux), indicating that it was originally very widespread. In the London Basin heavy minerals (Morton 1982*a*) indicate a major shift in probable source, from the Scottish Highlands to the Armorican or German (Rhenish) massifs. This probably reflects the preceding Late Thanetian uplift.

Earliest Eocene

In the central North Sea, the base of Sele Formation Unit S1b is marked by a change from bioturbated to laminated claystone and an increase in organic content, reflected in a high gamma-ray response. This is associated with an influx of the thermophilic dinoflagellate cyst *Apectodinium*, comprising over 90% of the dinoflagellate assemblage at some levels. This reflects the rapid warming marking the PETM. The carbon isotope excursion (CIE) which marks this event has recently been identified in the central North Sea (Beerling & Jolley 1998). In the southern North Sea, Unit S1b passes gradationally southwards into marginal-marine and non-marine sediments (upper Lambeth Group), which apparently extend widely into this area as a result of the earlier Sole Pit Inversion.

In the London Basin, the upper part of the Upnor Formation is subaerially weathered and pedogenized, indicating a fall in sea level. The overlying Reading Formation comprises mainly colour-mottled clays and silts deposited in an aggrading coastal plain environment subjected to periodic tropical pedogenesis. In the London area it interfingers eastwards with the Woolwich Formation, dark organic-rich shelly clays and laminated heterolithic sediments, with subordinate fine sand beds and coarser sand-filled channels, deposited predominantly in low-energy marginal marine environments. Pedogenic calcretes are locally developed in the lower part of the Reading Formation, immediately below the base of the lowest Woolwich Formation 'tongue'. The CIE has been identified at this level in London and in a coeval lignite in north Kent (Thiry *et al.* 1998; Collinson *et al.* 2003). The base of the Woolwich Formation is an omission surface, with *Thalassinoides* burrow systems extending deeply into the underlying Reading Formation. The *Apectodinium* influx occurs at this level. Much recent data has been obtained from major tunnelling projects in the London area (Skipper 1999), but there are few exposures; the type-section at Charlton, SE London, is still exposed. The clay component of the 'Clay-with-Flints', which covers extensive areas of the Upper Chalk, particularly beyond the northern fringe of the London Basin, is probably a relic of a former extension of the Reading Formation outcrop (Bateman 1988).

The Woolwich Formation and Reading Formation are locally preserved in East Anglia, although largely eroded by pre-Harwich Formation uplift. Only the Reading Formation is present throughout most of the Hampshire Basin, where it is well exposed in the coastal sections at Alum Bay and Whitecliff Bay (Isle of Wight). Upper and lower 'tongues' of the Woolwich Formation have been identified in the most eastern parts of the basin, and in outliers still further east (e.g. Newhaven) these are amalgamated. The Reading Formation occurs in the northern Dieppe Basin and in the Central Channel Basin. Very similar facies are present in the Paris Basin (Argile Plastique). The Woolwich Formation is identifiable in the southern Dieppe Basin, and similar facies occur in Normandy (Sables et Argiles d'Ostracodes et Mollusques) (Dupuis *et al.* 1998) and the northern Paris Basin (Fausses Glaises) (Mégnien 1980).

The 'Reading' and 'Woolwich' facies are here interpreted as deposited in a wide and shallow embayment of the southern North Sea Basin, with the proximal 'Reading' facies interfingering basinwards with the marginal-marine 'Woolwich' facies.

'Base Ypresian' discontinuity

In the central North Sea, there is no evidence for significant stratigraphic discontinuity within the Sele Formation, but towards the southern margins of the North Sea Basin there is a stratigraphic break within the lowest Eocene, marking a tectonically enhanced sequence boundary. In the Hampshire and London basins this is represented by the erosional contact between the Lambeth Group and the overlying Harwich Formation. A sea-level fall is evidenced by localized but deep channels, particularly on the southern margin of the London Basin in SE London and north Kent. These were subsequently filled by the Blackheath 'Beds', sands and rounded-pebble gravels forming a series of cross-cutting channel fills, deposited in an estuarine environment. The earliest substantial Paleogene mammal fauna in England has been recovered from this unit. Localized incised channel-fills in a similar context, but not necessarily exactly synchronous, occur in East Anglia ('Suffolk Pebble Bed') and in the Hampshire Basin (unpublished data).

Significant differential uplift and erosion is marked by the progressive truncation of the Lambeth Group east and northeastwards from London; in north Kent and Norfolk the Harwich Formation overlies the Upnor Formation or its equivalent. This may reflect renewed uplift on the Sole Pit structure.

Lower Eocene (lower and middle Ypresian)

Unit S2 of the Sele Formation is a claystone, present in the central North Sea and northwards. It is apparently not represented in the southern North Sea or onshore areas; it is interpreted by Knox (1996*a*) as deposited during the hiatus seen in more proximal areas ('base Eocene discontinuity', see above). In the central North Sea and the southern North Sea, the upper Sele Formation (Unit S3) comprises claystones with disseminated volcanic ash and several discrete ash layers (pyroclastic phase 2.2a). It conformably overlies Sele Unit S2 in the central North Sea, and the Lambeth Group in the southern North Sea, but locally oversteps onto the Chalk offshore from Yorkshire (e.g. in borehole 81/46A, Lott *et al.* 1983).

Sele Unit S3 is correlated on the basis of tephrostratigraphy and palynology with part of the 'Hales Clay' of northern East Anglia and the lower Harwich Formation (Orwell Member) of southern East Anglia (Jolley 1996; Knox 1996*a*), whose base is slightly younger. These are similar in lithology to Unit S3, but siltier and less organic-rich. The overstep of Unit S3 and, in more marginal areas, the change from the marginal-marine facies of the Lambeth Group to an inner to mid-neritic environment, mark a transgressive surface indicating the first phase of the progressive Ypresian sea-level rise. The base of the Harwich Formation and its lateral equivalents is everywhere marked by an erosional surface, with a basal glauconitic and pebbly bed often including reworked clasts and fossils from the Lambeth Group.

The overlying lower Balder Formation (Unit B1) occurs throughout the central and southern North Sea, comprising silty claystones with numerous thin basaltic ash layers (volcanic phase 2.2b), deposited in similar environments to the upper Sele Formation. It is represented onshore in East Anglia by the upper Harwich Formation (Wrabness Member), deposited in a mid-neritic environment. Here its base is a minor transgressive surface. It is well exposed in cliffs at Wrabness and Walton Naze, north Essex. The 'Harwich Stone Band', a tabular argillaceous limestone bed with a central ash band, is near the base of the Wrabness Member. Rare terrestrial mammals and marine turtles have been obtained from this bed. The ash beds of this phase represent a long-continued episode of massive explosive volcanism in volcanic centres to the north and NW of Scotland. Ashes from these sources have been identified as far as the Bay of Biscay and Austria, but in proximal marine facies the ash beds are often rendered unrecognizable by bioturbation and chemical alteration.

The Harwich Formation becomes coarser grained and the clays pass laterally southwards in the London Basin into thin shallow sublittoral glauconitic sands and silts with shell beds (Fig. 16.13), including the Oldhaven Member (well exposed at Herne Bay, Kent), Swanscombe Member and Tilehurst Member (former 'London Clay Basement Bed'). The Tilehurst Member is also represented in the Hampshire Basin, well exposed at Whitecliff Bay. Sediments similar in facies to the Swanscombe Member and Oldhaven Member, and probably their correlatives, are also represented in outliers near Dieppe (Normandy) (Formation de Varengeville inferieur) and in the Central Channel Outlier and the northern Paris Basin (unpublished data). This indicates that marine sediments of this age were formerly widespread on the southern margin of the North Sea Basin.

Fig. 16.13. Earliest Eocene sediments at Lower Upnor quarry, Kent (London 'Basin'). Marginal-marine sands of the upper Woolwich Formation are overlain by pebbly and shelly marine glauconitic sands of the Oldhaven Member (Harwich Formation). Their contact marks the 'base Ypresian' tectonically enhanced sequence boundary, recognisable throughout marginal areas of the southern North Sea Basin and adjacent areas. Here the base of the Oldhaven Member contains abundant molluscs reworked from the Woolwich Formation and flint pebbles reworked from the Blackheath Member, a localized estuarine unit at the base of the Harwich Formation, mixed with indigenous marine molluscs. The base of the overlying glauconitic silts of the Swanscombe Member marks a transgressive surface (parasequence boundary) within the Harwich Formation. Photograph of a sedimentary peel, reproduced by kind permission of David J. Ward. Scale bar is 10 cm.

In the central and southern North Sea, the upper Balder Formation (Unit B2) is a thin claystone, deposited in a less restricted marine environment than Unit B1, and post-dating volcanic phase 2.2b. It is represented in southern England by the basal unit of the London Clay Formation (Walton Member, King 1981), characterized by abundant large benthic diatoms.

The Balder Formation is overlain by the Horda Formation. The base of the Horda Formation is marked by an abrupt facies change to bioturbated calcareous claystones with a low organic content, and an influx of planktonic foraminiferids. This reflects a further major sea-level rise, ending the restricted marine environments of the earliest Eocene. It indicates renewed open marine water circulation between the North Sea Basin and oceanic areas.

The lowest part of the Horda Formation is a highly condensed hemipelagic unit, deposited in upper bathyal environments, with abundant planktonic and benthic foraminiferids and nannofossils. Its red or brown coloration is due to high iron content, reflecting clastic starvation ('Unit 3' of Lott *et al.* 1983 in borehole 81/46A). It thickens southwards into the outer neritic grey and grey-brown clays of the London Clay Formation (above the basal Walton Member).

The London Clay Formation is the thickest and most extensive Paleogene marine unit in marginal areas of the southern North Sea Basin. It extends from the southern North Sea into the London Basin, the Hampshire–Dieppe Basin, the Central Channel Basin, and (under different names) into Belgium, The Netherlands and Germany. It passes into siltier and sandier mid to inner neritic sediments in the western London Basin and Hampshire Basin, and into inner neritic glauconitic sands in the eastern Dieppe Basin, northern France and eastern Belgium ('Cuisian' sands and Mons-en-Pevele Formation). These all formed part of a single depositional area. In the London Basin the London Clay Formation crops out in the extensive Isle of Sheppey coastal sections, long famous for their extensive fauna and terrestrial flora of seeds and fruits (see Daley & Balson 1999). In the Hampshire Basin, it is well exposed at Alum Bay and Whitecliff Bay (Isle of Wight).

In southern England the expanded succession and the location near the basin margin enables the details of Ypresian depositional sequences to be unravelled (King 1981) (Fig.16.14). The London Clay Formation comprises several stacked lower-order sequences (divisions A–E of King 1981) delimited by transgressive surfaces with thin glauconitic/pebble beds. Inner neritic silts and sands prograded eastwards episodically from the margin of the basin. Sedimentation was initially retrogradational, with successive sea-level rises, until the greatest water depths were reached in the mid Ypresian. This was followed by a progradational phase, with increasing volumes of coarse estuarine sands deposited in channels incised during episodes of sea-level fall. Prograding silts and sands eventually reached as far as the eastern London Basin in the upper London Clay Formation.

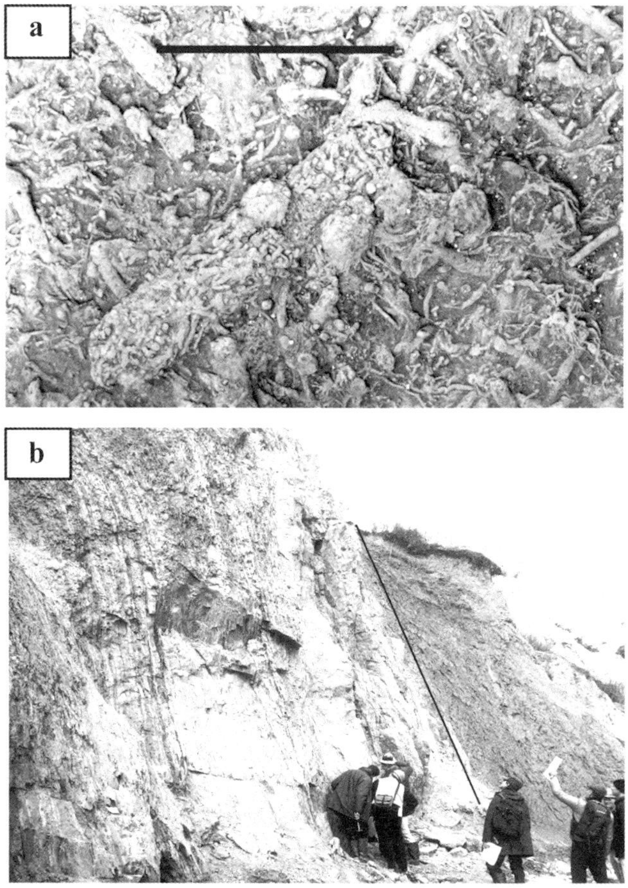

Fig. 16.14. Early Eocene (mid-Ypresian) sediments of the southern North Sea Basin (London Clay Formation). (**a**) Deep-water (outer neritic) clay facies. Intensely bioturbated clays deposited on a well-oxygenated sea floor during the highest sea levels in the Paleogene. Bioturbation revealed by differential cementation on the surface of a calcareous concretion. Isle of Sheppey, Kent (London 'Basin'). This facies is characteristic of most of the London Clay Formation in the eastern London 'Basin'. (**b**) Depositional sequences in the London Clay Formation, closer to the margin of the North Sea Basin. On the left of the picture, the upper part of the highstand tract of sequence B, coarsening-upwards mid-neritic silts and clays prograding from the basin margin, with thin storm-generated sands becoming thicker and more numerous upwards. The base of sequence C (thick line) is a combined sequence boundary and transgressive surface, with a thin transgressive unit of glauconitic pebbly silt, followed to the right by dark grey homogenous clays of the highstand tract. Whitecliff Bay, Isle of Wight.

Lower Eocene (upper Ypresian)

In the central and southern North Sea, the calcareous claystones of the basal Horda Formation are overlain by non-calcareous claystones with agglutinating foraminiferids, indicating restricted circulation in bathyal environments (Lott *et al.* 1983, Unit 4; King 1989). This probably reflects the late Ypresian eustatic sea-level fall.

In the London and Hampshire basins, a major sequence boundary within the upper Ypresian is represented by the erosional and locally deeply channelled surface between the London Clay Formation and the overlying Bracklesham Group. This event does not appear to have a tectonic component, and probably corresponds to the prominent but brief eustatic sea-level fall in the late Early Eocene (Fig. 16.5). The channels were filled and overtopped by relatively coarse cross-stratified sands, deposited in estuarine and other coastal environments. These are the basal units of the Wittering Formation of the Hampshire Basin and the Bagshot/Virginia Water Formation of the London Basin. They gradually pass up to finer grained coastal sediments. In the eastern Hampshire Basin these are terminated by a transgressive surface, overlain by glauconitic clays and sands ('Fisher Bed IV') deposited in shallow marine environments, with planktonic foraminiferids and *Nummulites*, indicating a marine connection to the North Atlantic. At its maximum extent, this marine transgression extended into the western Hampshire Basin. The upper part of the Wittering Formation marks a return to marginal-marine clays and laminated clays (Fig 16.17a); organic-rich clays and a thick autochthonous lignite (the Whitecliff Bay Bed) are quite widespread (Plint 1983, 1988*a*). Plint (1983) regarded these facies (and the similar Marsh Farm Formation at a higher level) as restricted bay sediments indicating a tectonically virtually isolated Hampshire Basin, but they are present also in the London Basin (Swinley Clay Member, Ellison & Williamson 1999). Depositional continuity between these areas is very probable.

In the Hampshire Basin the Wittering Formation interfingers westwards with the lower part of the Poole Formation (Fig. 16.16). This comprises several units of clays, variable in lithology, partly organic and laminated, partly pale and kaolinitic, and partly reddened and pedogenized, interbedded with thick medium to coarse grained cross-stratified sands (Plate 32). The Poole Formation (former 'Bagshot Sands/Pipeclay Series) has generally been regarded as 'non-marine', largely fluvial and alluvial/lacustrine. This interpretation was reinforced by Plint (1983), but the occurrence of bioturbation and marine dinoflagellates in the supposedly 'overbank' clay units (Bristow *et al.* 1991) and teredinid (marine shipworm)-bored logs in the supposedly fluvial sands (Plint & Pickerill 1985) requires re-assessment of this model. The Poole Formation can be re-interpreted as comprising several stacked depositional

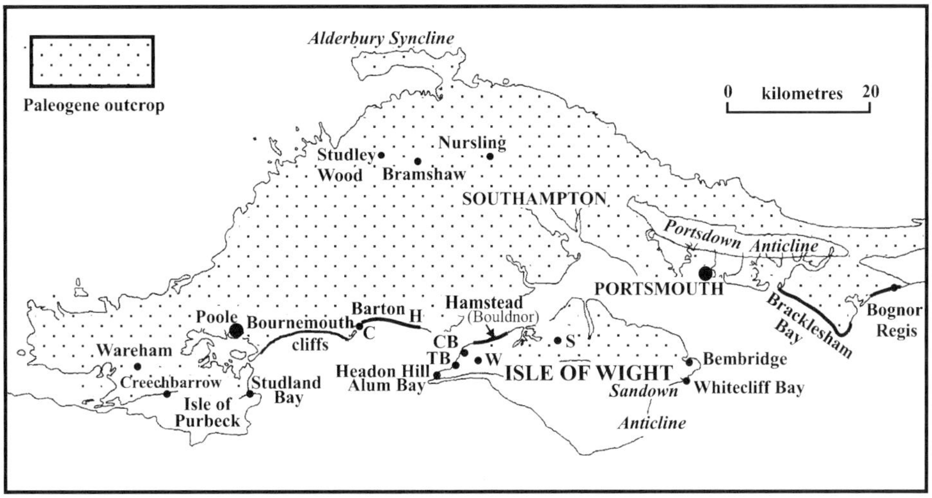

Fig. 16.15. Paleogene outcrop and key Paleogene localities in the Hampshire 'Basin'. C, Christchurch Borehole. CB, Colwell Bay. H, Hordle cliffs. S, Sandhills Borehole. TB, Totland Bay. W, Wilmingham Borehole.

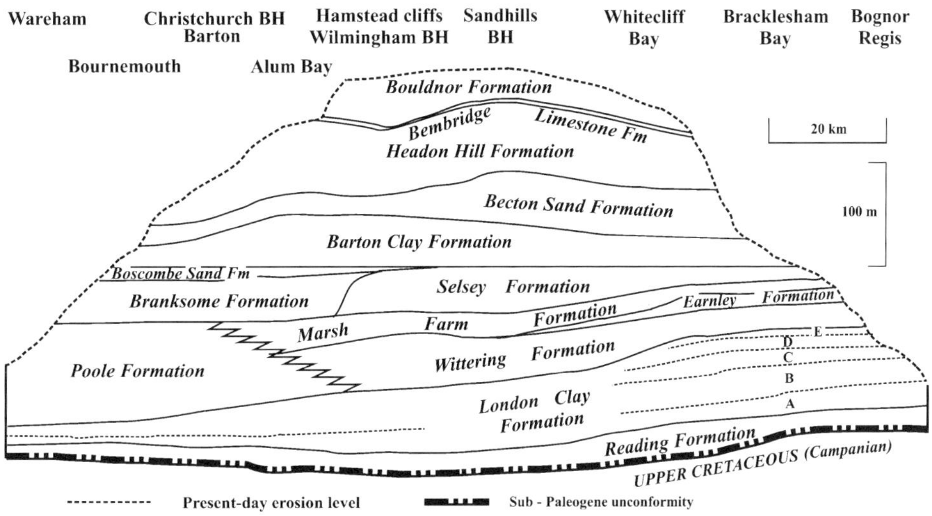

Fig. 16.16. Paleogene stratigraphy of the Hampshire 'Basin'. A W–E cross-section. Localities cited are shown on Figure 16.15. The horizontal datum is the base of the Barton Clay Formation. Christchurch Borehole and Sandhills Borehole are based on Edwards & Freshney (1987*a*, *b*) and Bristow *et al.* (1991). Wilmingham Borehole data is from composite well log and sample examination. NB. The vertical scale is highly exaggerated.

sequences. Each sequence has a deeply channelled base, followed by sands deposited mainly in estuarine environments, giving way with rising sea levels to low-energy coastal-plain and marginal-marine sediments. These can be correlated with depositional sequences in the more marine facies of the Bracklesham Group further east.

A major hiatus has been suggested to exist in the Hampshire and London basins at the level of the Whitecliff Bay Bed (Aubry *et al.* 1986), based largely on magnetostratigraphic data (the apparent absence of Chron C22n). The magnetostratigraphic data are however ambiguous, and any hiatus is more likely to be at the base of the overlying Earnley Formation (see Ali *et al.* 1993).

The Wittering Formation and higher units in the Bracklesham Group are well exposed at Whitecliff Bay and Alum Bay, and much of the Bracklesham Group was formerly well exposed in coastal sections further west in the Bournemouth–Poole area.

In the Dieppe Basin this interval is documented only by sea-bed samples of glauconitic sandy clay, similar lithologically and palaeontologically to 'Fisher Bed IV'. In the Central Channel Outlier, sea-bed cores of silty shelly clay overlying the London Clay Formation contain a diverse marine microfauna, including planktonic foraminiferids and *Nummulites aquitanicus*. These were assigned to the Wittering Formation by Curry *et al.* (1978), and again appear to correlate with the 'Fisher Bed IV' transgression. This is the youngest unit preserved in this outlier.

Middle Eocene (Lutetian)

In the central North Sea Basin, the non-calcareous clays of the Horda Formation persist through the Middle and Upper Eocene. In the northern part of the southern North Sea they are replaced in the middle Lutetian by calcareous clays with benthic and planktonic foraminiferids, indicating better oxygenation at the basin floor and more open circulation. This may reflect relatively high sea level at this time. Further south, Lutetian sediments are poorly documented, but several cored boreholes (Cameron *et al.* 1992) have proved silty and sandy clays presumably deposited in mid-neritic environments.

In the eastern Hampshire Basin, where the best-documented onshore record exists, an early Lutetian transgressive surface marks the base of the Earnley Formation. This comprises highly glauconitic shelly sands (greensands) deposited in inner neritic environments (Fig. 16.17b). With progressive transgression, the immigration of abundant larger foraminiferids (*Nummulites laevigatus*) indicates re-opening of the connection to the North Atlantic. An almost identical unit is present in the London Basin, where it is classified as part of the Windlesham Formation (former 'Middle Bagshot Beds'), confirming the existence of a single depositional area in SE England. In the eastern Hampshire Basin the Earnley Formation is overlain by the marginal-marine sediments (laminated clays and channelled estuarine sands) of the Marsh Farm Formation (middle Lutetian). These were deposited in similar depositional environments to the Wittering Formation (Plint 1983, 1988*a*) and similarly include a wedge of glauconitic shelly clays from the east, with abundant *Nummulites*. The Marsh Farm Formation passes westwards into the upper part of the Poole Formation, coastal plain and marginal-marine clays and sands. Again an almost identical unit is present in the London Basin (part of the Windlesham Formation).

A transgressive surface overlies the Marsh Farm Formation and coeval units in the London Basin. In the central and eastern Hampshire Basin it forms the base of the marine Selsey Formation (middle–upper Lutetian). Its lower transgressive part begins with shelly greensands, similar to the Earnley Formation, followed by stacked parasequences of sandy and shelly glauconitic clays grading up to silty clays. The top of this interval probably marks the maximum flooding surface, and is followed by prograding parasequences of silts and sands, often with bioclastic debris (including the '*Nummulites variolarius* Bed'). In the lower part of this prograding interval there is an influx of 'warm-water' benthic foraminiferids, associated with other 'warm' taxa including brachiopods and calcareous algae, as seen at this level in the Paris Basin, Normandy and the Western Approaches Basin. This indicates the opening of a marine connection with the North Atlantic. Progradation culminates in a wedge of fine shelly inner neritic sands with *Nummulites* aff. *prestwichianus* (Studley Wood Nummulite Bed).

Fig. 16.17. Middle Eocene sediments in the Hampshire Basin. (**a**) Marginal marine heterolithic tidally influenced sands and clays of the Wittering Formation. Alum Bay, Isle of Wight. Much of the late Early Eocene and Middle Eocene in the western Hampshire Basin was deposited in similar environments. Length of ruler: 1 m. (**b**) Glauconitic, shelly, poorly sorted, inner neritic sands of the Earnley Formation. Whiteliff Bay, Isle of Wight. Abundant lensoid cross-sections of the large rotaliid foraminiferid *Nummulites laevigatus*, a warm-water immigrant. Similar facies are typical of the marine late Early Eocene and Middle Eocene of the eastern Hampshire Basin. Scale bar is 5 cm.

In the western Hampshire Basin the Selsey Formation is replaced by the Branksome Formation (former 'Bournemouth Freshwater' and 'Bournemouth Marine' Beds), a complex of laterally variable and predominantly heterolithic sediments, deposited in fluvial to marginal-marine environments (Plint 1988*b*; Bristow *et al.* 1991).

The Camberley Sand Formation (former 'Upper Bagshot Sands') of the western London Basin comprises fine and silty glauconitic sands with a basal pebble bed. It is almost everywhere decalcified, with poorly preserved molluscs of uncertain age (Curry 1965). A calcareous macrofauna and microfauna, indicating correlation with the lower and middle Selsey Formation of the Hampshire Basin (Kemp & King unpublished data), has recently been recovered from the lower part of the Camberley Sand Formation. This clearly indicates depositional continuity between these areas. The Camberley Sand Formation is the youngest Paleogene unit preserved in the London Basin.

The limited data from the UK sector of the Dieppe Basin indicate the presence of the Earnley and Selsey Formations in areas near the English coast (unpublished data). In the French sector, sea-bed samples of lower and middle Lutetian glauconitic sandstones and sandy limestones with *Nummulites laevigatus* and *N. variolarius* suggest a progressive change to the more calcareous Lutetian facies of the Paris Basin. Reworked Ypresian *Nummulites* in these samples indicate an erosional contact with the underlying Ypresian, as in the Paris Basin.

Samples of glauconitic clayey sand and clay from higher levels in the French sector are lithologically and faunally very similar to the Selsey Formation. At the eastern end of the English Channel, near the French coast, sea-bed samples of sandy bioclastic Lutetian limestones have been collected (Bignot 1972). This isolated outlier is important for its location near the axis of the Weald-Artois High, as noted earlier; the Lutetian appears to overlie the Lower Cretaceous unconformably.

Middle Eocene (upper Lutetian–Bartonian)

In the northern part of the southern North Sea, the Bartonian is represented by calcareous clays deposited in outer sublittoral environments, similar in facies to the Lutetian. Further south it is mostly poorly documented, but a cored borehole close to the UK–Netherlands median line offshore from East Anglia (Hughes 1981) penetrated 70 m of alternating silty/sandy clays and fine glauconitic sands with Bartonian dinoflagellates. These lithologies are very similar to coeval Bartonian sediments in northern Belgium, deposited in mid-neritic environments.

In the Hampshire Basin, the Bracklesham Group is overlain by the Barton Clay Formation. In the eastern and central Hampshire Basin this contact is a combined sequence boundary and transgressive surface, with differential uplift and erosion of the underlying Selsey Formation (Todd 1990), perhaps related to movement on the Sandown anticline and nearby structures (Gale *et al.* 1999). In the western Hampshire Basin, a thick sheet of estuarine sands and pebble gravels, the Boscombe Sand, rests with an erosional contact on the Branksome Sand (Bristow *et al.* 1991, fig. 16) and is overlain by the Barton Clay Formation. This erosional contact was interpreted by Plint (1988*a*, *b*) as a transgressive surface, but is here regarded as a sequence boundary, with the Boscombe Sand as an incised-valley fill marking the early phase of sea-level rise. (The Boscombe Sand was classified as a Member of the Barton Clay Formation by Edwards & Freshney 1987*b*, but is here regarded as a Formation within the Bracklesham Group)

The Barton Clay Formation comprises two minor depositional sequences (Hooker 1986). The first sequence comprises the former highest Bracklesham Beds ('Huntingbridge Division') and 'Lower Barton Beds'. Its basal unit is a transgressive pebbly and glauconitic shelly sandy clay (the Huntingbridge Bed). Above this is an interval of stacked and probably retrogradational parasequences of glauconitic and locally pebbly sandy clays passing up to silty clays, as in the middle Selsey Formation. These are overlain by a thin highly glauconitic unit with abundant *Nummulites* (N. prestwichianus Bed) (formerly the base of the 'Barton Clay'/'Lower Barton Beds'). This unit probably marks the maximum flooding surface. It is followed by a progradational sequence of silty clays passing up to sandy clays with storm-generated laminated sand and shell beds.

The second sequence corresponds approximately to the former 'Middle Barton Beds'. Its base (base of Bed B at Barton, see Daley & Balson 1999) is a complex *Thalassinoides*-burrowed omission surface. It is overlain by a package of retrogradational parasequences similar to those in the first sequence, culminating in a probably condensed shelly glauconitic sandy clay (Bed E) representing the maximum flooding surface. This is followed by a progradational unit of silty clay with thin storm-generated sands and winnowed shell layers in the upper part, culminating in a thin coquina ('Tellina Bed').

The Barton Clay Formation is now rather poorly exposed in the type-section (Mudeford to Barton-on-Sea cliff section, shown as 'Barton' in Fig. 16.15) due to coastal protection works, but is mostly well exposed at Alum Bay. At Whitecliff Bay, an intra-Bartonian episode of uplift and erosion, probably related to uplift of the adjacent Sandown anticline, is inferred from the absence of the 'Middle Barton Beds', with a thin sand containing reworked Lower and Middle Eocene clasts and fossils at the base of the 'Upper Barton Beds' (Gale *et al.* 1999).

The Creechbarrow Formation, a thin freshwater limestone, is preserved only as a small hilltop outlier in the western Hampshire Basin in east Dorset. It overlies non-marine Eocene sediments of uncertain age. It was originally regarded as an outlier of the (Priabonian) Bembridge Limestone, but the diverse mammalian fauna indicates a Bartonian age (Hooker 1986), probably coeval with part of the Barton Clay Formation.

At Alum Bay and Barton, the Becton Sand Formation (former 'Barton Sand/Upper Barton Beds', including the Chama Sand Formation of Edwards & Freshney 1987*b*) overlies the Barton Clay Formation. It marks an abrupt, but conformable, shift to inner neritic and coastal sand-dominated sediments, interpreted here as probably a sequence boundary. It is a complex unit including sandy clays, very fine to coarse sands, and an intercalated shelly clay unit (the Becton Bunny Bed). Faunas in the lower part indicate a fully marine environment, but in the upper part they are dominantly brackish.

Bartonian sediments in the Dieppe Basin are proved only from sea-bed samples with diagnostic species of *Nummulites* in the French sector, and from a cored borehole near the centre of the Basin in the UK sector (Warrington & Owens 1977). This penetrated approximately 56 m of glauconitic silty sand and sandy silt, with dinoflagellates and *Nummulites* indicating an age corresponding to that of the middle Barton Clay Formation, but in shallower marine facies. This is the youngest Paleogene unit identified in the Dieppe Basin; younger units may be preserved in the basin centre, but no sampling has been possible due to the cover of recent sediments.

Upper Eocene (Priabonian)–lowest Oligocene (lower Rupelian)

In the southern North Sea, Upper Eocene sediments have been largely removed by Early Oligocene uplift and erosion, but in some areas significant thicknesses are still preserved. Although little biostratigraphic data are available, the sediments apparently comprise marine clays, with thick wedges of inner neritic sands (Prestwich Formation) prograding from the western basin margin (Lott & Knox 1994). Sands with *Nummulites*, immediately beneath the mid Rupelian unconformity in BGS offshore borehole 79/06 east of East Anglia (Hughes 1981) (Fig 16.9) probably correlate with the Upper Eocene Bassevelde Sand of northern Belgium.

In the Hampshire Basin, there is an abrupt change in depositional pattern at about the Middle–Upper Eocene boundary, with the predominantly marine environments of the Bartonian giving way to dominantly fine-grained coastal plain sediments of the Solent Group. These continue into the lowest Oligocene. The Solent Group comprises the Headon Hill Formation, Bembridge Limestone Formation and Bouldnor Formation. It is excellently exposed in coastal sections in the Isle of Wight and on the Hampshire coast (Hordle Cliff). It comprises several depositional sequences. Ideally each sequence boundary is marked by localized incision of sand-filled fluvial or estuarine channels, followed by an episode of clastic starvation marked by freshwater limestone, followed in turn by a transgressive surface with a basal lag. The overlying marginal-marine sediments grade rapidly up into freshwater or low-salinity lacustrine and lagoonal clays and silts, and finally into pedogenized colour-mottled clays.

The stratigraphy of the Solent Group is markedly 'layer-cake', with little lateral facies change, and lateral persistence of many thin units. This, and the predominately fine grained facies, suggests that the relatively limited area of the Solent Group preserved (only *c.* 45 km wide) is a small part of a formerly extensive coastal plain of low relief, as discussed earlier. Former 'autocyclic' models for the alternation on medium and small scales of depositional environments have more recently been replaced by 'allocyclic' models, suggesting that eustatic sea-level change was the main controlling factor (e.g. Daley 1999). There is also evidence for localized uplift and erosion, including unworn flint pebbles derived from erosion of the Chalk at some levels. The Solent Group has yielded an important series of terrestrial mammal faunas and macrofloras, which apart from their biostratigraphic significance are critical indicators of climatically controlled environmental change.

The Totland Bay Member (former 'Lower Headon Beds'), the lowest unit of the Headon Hill Formation, comprises predominantly clays and silts, with some localized channelled sands, and a thick freshwater limestone (How Ledge Limestone) at the top. Extensive mammal faunas have been obtained from several levels. The base of the overlying Colwell Bay Member ('Middle Headon Beds') is a *Thalassinoides*-burrowed omission surface marking a transgressive surface. An overlying glauconitic and shelly sandy clay (Brockenhurst Bed) has a diverse marine fauna, including planktonic foraminiferids, and nannofossils indicating a mid–late Priabonian age. This is the only unit within the Solent Group deposited in an open marine environment; elements in the fauna indicate open marine connection to the North Sea Basin. The Brockenhurst Bed and the overlying marine Roydon Bed are restricted to the NE part of the outcrop; they are overlain and laterally replaced by sandy clays deposited in coastal hyposaline environments. During a subsequent transgressive phase the sandy clays of the brackish/marine '*Venus* Bed' were deposited. The upper part of the Colwell Bay Member comprises mainly brackish and non-marine clays and silts.

A wide sand-filled channel (Linstone Chine Member) is incised into the Colwell Bay Member, overlain by a thick but localized freshwater limestone (Hatherwood Limestone Member, former 'Headon Hill Limestone') with a diverse mammalian fauna. The overlying Cliff End Member, Fishbourne Member and Osborne Member comprise alternations of shelly clays with freshwater and brackish mollusc and ostracod assemblages and colour-mottled pedogenized clays, with localized freshwater limestones. A localized sand-filled channel (Seagrove Bay Member) is incised into the Osborne Member.

The Bembridge Limestone Formation is a thick freshwater limestone, with well-developed cyclicity indicating repeated change from shallow-water to emergent, with periodic development of calcretes (Daley 1999). The occurrence of foraminiferids and brackish molluscs at some levels emphasizes its close proximity to a marine environment (Plate 33).

The Bouldnor Formation (former 'Bembridge Marls' and 'Hamstead Beds') comprises mostly cyclic alternations of clays and silts with brackish or freshwater molluscs and ostracods and plant debris, with colour-mottled, leached and pedogenized clays. The brackish faunas occur mainly in the lower part; near the base is a shelly calcareous sand (Bembridge Oyster Bed) with brackish and marine molluscs and ostracods. A marked cooling event has been identified at the 'Black Band' (former base of the 'Hamstead Beds') (Machin 1971), and mammalian faunas indicate that the 'Grande Coupure' lies somewhere near this level (Hooker 1992). These events correlate approximately with the Eocene–Oligocene boundary.

In the highest part of the Bouldnor Formation (Cranmore Member) there is a return to clays with brackish and finally shallow marine faunas. Molluscs and dinoflagellates indicate an early Rupelian age, and a marine connection to the Paris Basin. This is the youngest Paleogene unit preserved in the Hampshire Basin.

Lower Oligocene unconformity
In the central and southern North Sea, there is a hiatus and a slight angular unconformity between the Horda Formation and the overlying Lark Formation, apparently reflecting regional inversion. In the central North Sea, where the hiatus is shortest, it can be dated as intra-Rupelian. Over most of the southern North Sea, the underlying lower Rupelian and Upper Eocene sediments have been eroded. In the southernmost North Sea, middleRupelian clays unconformably overlie probable Bartonian–?Priabonian sediments (Hughes 1981). This unconformity can be correlated with the stratigraphic break at the base of the Rupelian Boom (Rupel) Clay in the Netherlands and adjacent areas (Vinken 1988).

Oligocene–?Early Miocene
In the central North Sea, the Lark Formation comprises lower Oligocene (middle Rupelian) to lower Miocene claystones deposited in bathyal environments. In the southern North Sea it is siltier, often glauconitic, and deposited in shallower mid- to outer neritic environments; Oligocene sediments in the UK sector are restricted to small outliers and to areas adjacent to the median line. Offshore East Anglia, BGS Borehole 79/06 (Hughes 1981) penetrated Lower Oligocene (Rupelian) clays very similar in lithology and microfauna to the Boom Clay Formation of Belgium. These are part of a very widespread Lower Oligocene marine neritic clay unit extending through the Netherlands and Germany as far as Poland.

Clay deposition continued in axial areas of the southern North Sea Basin into Late Oligocene (Chattian) times, but in more proximal areas of the southern North Sea inner neritic glauconitic sands (Skade Formation) prograded from the basin margins. Progradation may have been initiated by the major mid Oligocene sea-level fall (Haq *et al.* 1988). Similar sands occur in northern Belgium (Voort Formation). Biostratigraphic data in the southern North Sea is limited, and it is possible that sand deposition continued into the early Miocene.

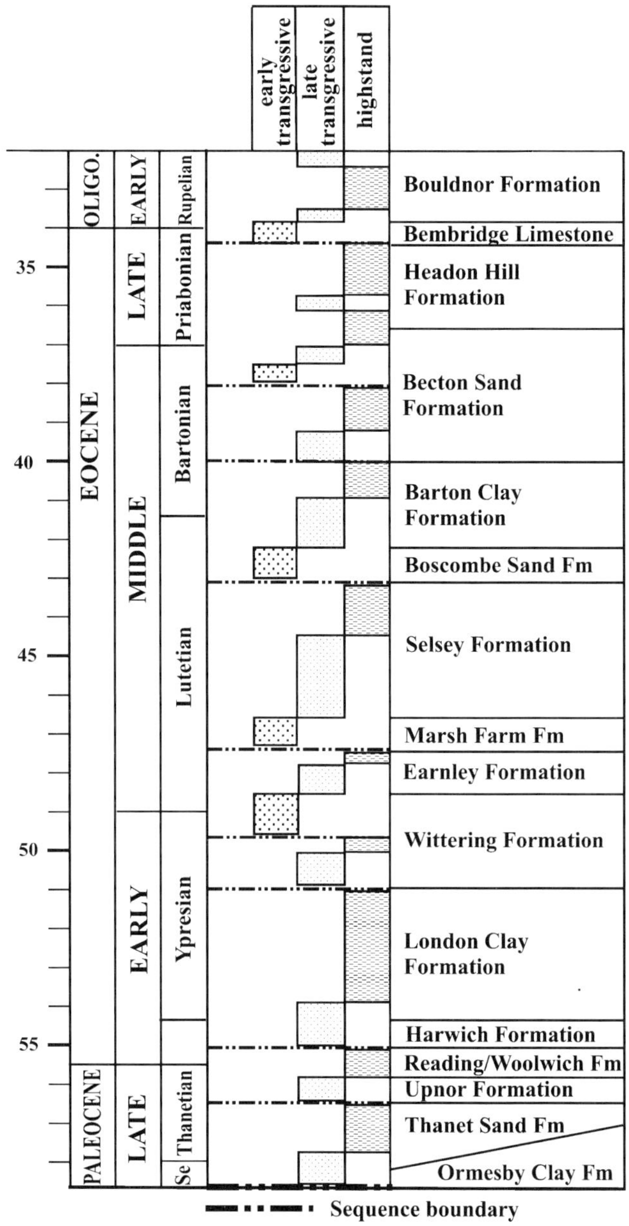

Fig. 16.18. Paleogene depositional sequences of the Hampshire and London 'Basins'. A generalized and tentative interpretation based on published sources and unpublished data. Only major sequences are differentiated, and stratigraphic breaks at sequence boundaries are not shown. 'Early transgressive' sediments include estuarine/tidal sediments forming incised valley fills. Base of 'late transgressive' sediments is taken at the transgressive surface. Note the apparent absence of lowstand sediments.

Mid Cenozoic unconformity
A major episode of inversion in the mid Cenozoic, associated with halokinetic movements, led to uplift and erosion of large areas in the southern North Sea, including the Sole Pit Trough (Cameron *et al.* 1992). Dating of this episode (or series of episodes) is difficult due to the limited extent of mid Cenozoic sediments preserved in this area. Evidence from adjoining onshore areas of continental Europe indicates that it is probably intra-Early Miocene.

Miocene
Miocene sediments have an extensive subcrop in adjacent sectors of the southern North Sea, but in the UK sector are

restricted to the most northerly part, near the UK median line. Very little stratigraphic information is available, but in adjacent areas to the north they are clays deposited in outer neritic environments. Probable reworked marine Miocene fossils in East Anglia, and extensive marine Miocene sands in northern Belgium and The Netherlands, indicate that most of the southern North Sea was probably covered by shallow marine Miocene sediments.

The Lenham Beds of north Kent are sands and limonite-cemented glauconitic sandstones, preserved only as displaced blocks within solution pipes in the surface of the Chalk escarpment of the North Downs. Their age, based on a diverse but poorly preserved marine mollusc assemblage, has been disputed, but is currently regarded as latest Miocene or Early Pliocene. The molluscs are apparently more closely related to contemporaneous faunas of Brittany than to those of East Anglia (Daley & Balson 1999). This suggests that the 'Wealden uplift' may have been at least partially breached by this time.

Pliocene

Pliocene marine sediments are widespread in the adjacent Netherlands sector of the southern North Sea, but in the UK sector they are restricted to isolated outliers (Cameron *et al.* 1992, fig. 85). One of these outliers intersects the coast of East Anglia, where it outcrops in a narrow strip (Coralline Crag Formation). This comprises bioclastic calcarenites and silty sands, with abundant bryozoan and shell debris, and well-preserved and diverse molluscs in some facies, deposited in shallow marine, mainly high-energy, environments. It has an erosional contact with the underlying Lower Eocene London Clay Formation, with a basal polygenetic pebble lag. Faunal and microfloral evidence indicates that it was deposited during the brief Early Pliocene 'warm' episode, when a widespread marine transgression resulted from partial melting of the polar ice sheets (Head 1997).

The offshore outliers, where sampled, are similar in age to the Coralline Crag, but with somewhat more distal facies, including glauconitic shelly sands and clayey sands. Many similarly occupy channels eroded into the Eocene. The widespread occurrence of Lower Pliocene fossils and glauconitic sands reworked into present-day sea-floor sediments, and of Lower Pliocene sediments of similar lithology in northern Belgium, suggests that a sheet of Lower Pliocene shallow marine glauconitic sands formerly covered this area (Cameron *et al.* 1992).

Non-marine Cenozoic of western Britain

Paleogene outliers of East Devon

There are several small outliers of presumed Paleogene non-marine sediments in south and SE Devon, including the Haldon Hill Gravels (Fig. 16.3). They rest on a high-level horizontal surface incised into Cretaceous sediments (Daley & Balson 1999). They comprise residual gravels of unabraded flint and chert, originating from *in situ* weathering of the Cretaceous (Chalk and Upper Greensand), and associated paleosols, including clays and silcretes, representing several episodes of pedogenesis. They are overlain by gravels largely derived by fluvial reworking from the residual deposits, but including some pebbles of Palaeozoic rocks. No fossils have been recorded, as these sediments are oxidized and decalcified.

Evidence of a tropical weathering regime indicates a pre-Oligocene age, and dates from Late Cretaceous to Mid Eocene have been suggested in the past, but (as discussed earlier) a Late Paleocene age is most probable. The only Paleogene silcretes in a firm stratigraphic context in southern England (in the London Basin) are of this age. As the Haldon Hill Gravels are correlated lithologically with the lowest exposed unit in the Bovey Basin (see below), they must be Mid Eocene or older (see the discussion in Daley & Balson 1999).

Paleogene non-marine and marginal-marine basins

The non-marine fault-controlled Paleogene basins of western Britain, mainly half-grabens, lie in a narrow N–S zone from south Devon to the Hebrides. Their general depositional characteristics have been discussed above. The great thickness of sediments in some basins (up to 1500 m) indicates initial rapid fault-controlled subsidence and deposition; the later stages suggest more extensive but no longer fault-controlled aggradation and eventually clastic starvation. Due to lack of fossils, the coarser basal facies have not been dated biostratigraphically in any basin, but palynomorphs in higher levels consistently indicate a probable Late Eocene or Oligocene age. Many have been previously regarded as Late Oligocene (Chattian) (e.g. Walsh *et al.* 1999) but the widespread record of the pollen *Boehlensipollis* indicates an Early Oligocene (Rupelian) age in most cases; it is the index taxon of the NW European spore/pollen Zone SP7 (Meyer 1988). In the highest parts of the fill of several basins, marine microplankton indicate marginal-marine environments.

The Bovey Basin, Petrockstow Basin (SW England), Stanley Bank Basin (Bristol Channel) and probably the Flimston Basin in SW Wales, all lie along or adjacent to the Sticklepath–Lustleigh (strike-slip) Fault. The best-documented of these, the Bovey Basin, is a deep and narrow half-graben, with a maximum sediment thickness of about 1100 m, estimated from gravity survey, overlying Devonian. Only the upper part of the basin fill is exposed or proved by boreholes; the lithology of the lowest *c.* 800 m is unknown (Edwards & Freshney 1982). The fluvial Aller Gravel (cross-stratified sand and gravel), on the margins of the basin, is probably a marginal representative of the first phase of the basin fill; similar facies may underlie the whole basin. The (probably) overlying Bovey Formation comprises dominantly, organic and kaolinitic clays, with lignite beds (dominantly allochthonous, but occasionally in situ), and subordinate sand and gravel units, mainly towards the margins of the basin. Deposition took place in floodplain and lacustrine environments. Palynological data indicate an Oligocene age for the exposed Bovey Formation, originally interpreted as Chattian, but more probably Rupelian. The deepest levels penetrated by boreholes yielded probable Mid Eocene palynomorphs. The age of the underlying undrilled section is unknown. The clays are derived mainly from erosion of Carboniferous sediments. In the Stanley Bank Basin, dinoflagellate cysts in the upper part of the fill indicate marine influence.

The small Beacon Cottage Farm Outlier in west Cornwall (Walsh *et al.* 1999) (close to the Miocene St Agnes Outlier, Fig. 16.3) is formed by approximately 10 m of silty, locally lignitic, clay, with a thin basal pebbly sand. The terrestrial palynomorphs indicate an early Oligocene (Rupelian) age. Several other very small outliers are recorded in this area, mostly undated, but one has yielded a Miocene palynoflora (see below).

Other offshore non-marine basins in the Bristol Channel, the Irish Sea and adjacent areas include the Bristol Channel Marginal Basin, small basins along the West Lundy Fault zone, the South Celtic Sea Basin, and the Central Irish Sea Basin (Tappin *et al.* 1994). The large St George's Channel Basin is filled by up to 1500 m of fluvial and alluvial plain sediments, comprising interbedded units of sandstone, siltstone, claystone and lignite, generally fining upwards. In the small Teifi Basin near the Welsh coast, the youngest sediments are lignitic clay and laminated clay with plant debris and glauconite. The Cardigan Bay Basin is mainly offshore, but a small area is beneath coastal area of North Wales. Here the Mochras Borehole penetrated over 500 m of cyclically deposited but overall fining-upwards sediments, based by alluvial-fan cobble conglomerates, sands and silts through floodplain silt and clay to 'swamp-plain' sand, clay and lignite beds. The middle and upper parts of the basin fill are Early and ?Late Oligocene in age.

Neogene non-marine outliers
In the small St Agnes outlier (west Cornwall) (Fig. 16.3), 10 m of sand and silt, interpreted as aeolian and ?lacustrine sediments, overlie the Palaeozoic (Walsh *et al.* 1987). A probable Miocene palynoflora is recorded from one sample. A Miocene palynoflora has also been obtained from a small pocket of sediments in a solution hollow in Anglesey (Walsh *et al.* 1999).

In south Derbyshire, 'pocket deposits' of non-marine sediments (Brassington Formation) are preserved in deep solution hollows in Carboniferous limestone (Boulter *et al.* 1971). They comprise sand and gravel, overlain by colour-mottled (pedogenized?) clay, with plant-bearing clay at the top containing a Late Miocene macroflora and palynoflora.

Western Approaches Basin

The geology of this area is summarized by Evans (1990). Published information is relatively generalized; although unpublished data are incorporated here, stratigraphic resolution is limited at present.

Lower Paleocene
Danian chalks and chalky limestones, deposited in mid to outer neritic environments, very similar to the Danian in the North Sea Basin, overlie Upper Cretaceous chalks. The youngest dated Cretaceous chalks are Maastrichtian in age at most sites, suggesting at best only a minor hiatus at the Cretaceous–Paleogene boundary. The Danian is preserved only in the basin axis, and is overstepped by the Upper Paleocene towards the northern margin of the basin.

Upper Paleocene
Calcareous clays and marls, deposited in mid to outer neritic environments, have been penetrated by several wells in the Western Approaches. Planktonic foraminiferids indicate a Thanetian age. It is not clear if the lower part is Selandian in age, and so the nature of the contact with the Danian is unknown. Marine Thanetian clays were formerly much more extensive through the basin, as there is an isolated record in sea-bed samples north of Brittany, 300 km further east (Andreieff & Lefort 1972).

Base Eocene unconformity
Paleocene sediments are overstepped by the Eocene away from the basin axis. Seismic profiles show the basal Eocene surface to be smooth and nearly parallel with the reflectors in the underlying Chalk, but cut by isolated valleys. This suggests a sequence boundary with localized incised valley fills. This event may correspond to the 'Late-Thanetian unconformity' of the North Sea Basin and surrounding areas.

Lower Eocene
A thin calcareous glauconitic sandstone with abundant *Nummulites* cf. *rockallensis* occurs in the Western Approaches in the basin axis. It overlies Thanetian clays, overstepping onto Maastrichtian chalk (unpublished data). This unit is probably of earliest Eocene age, and probably overlies the seismically identified 'base Eocene' reflector. It is interpreted as a transgressive shallow marine unit. It is overlain and overlapped by Ypresian calcareous clays, with thin argillaceous limestone beds in some wells, deposited in mid to outer neritic environments; the upper part becomes more silty and less calcareous, reflecting progressive shallowing. These relatively deep-water sediments reflect, as in the North Sea Basin, the high early and mid Ypresian sea levels. In the western English Channel, Ypresian bioclastic limestones and clays are recorded in the basin axis, but few details are available. They are mostly overlapped by Middle Eocene sediments towards the basin margins. There is an outcrop of sandstones with *Nummulites planulatus* (probably middle Ypresian) on the northern side of the basin in the area south of the Scilly Isles. Reworked specimens of *N. planulatus* are reported in the Lower Lutetian at a site on the Armorican Platform (north of Brittany), indicating that sediments of this age were formerly widespread.

Upper lower Eocene–lower Middle Eocene
In the Western Approaches, the Ypresian clay unit is overlain by an interval of sandy glauconitic limestone and fine glauconitic sandstone. This is poorly dated, but is probably late Ypresian or early Lutetian in age. Similar sediments are recorded from boreholes in the western English Channel, but again without well-defined dates. These sediments include coarser clastic components than those above and beneath, and were probably deposited during the relatively low sea level of this interval. Disconformities may be present, but cannot be resolved in the present state of information. A small outlier of lower Lutetian sandy limestone with abundant *Nummulites laevigatus* is recorded on the Armorican Platform north of Brittany.

Middle Eocene
In the Western Approaches, this interval is represented by a thick unit of chalky bioclastic limestone, often weakly cemented, with abundant bryozoan and echinoderm debris, deposited in mid-neritic environments, shallowing upwards to inner neritic. Planktic and benthic foraminiferids indicate an age range of mid or late Lutetian to Bartonian.

Similar bioclastic limestones, but coarser-grained and predominantly inner neritic, extend eastwards through the western English Channel. They overstep older Eocene sediments southwards widely onto the Armorican Platform, incorporating quartz sand and locally pebbles at their margins. They crop out over very wide areas of the western Channel, and (apart from small outliers of younger Paleogene sediments) are the youngest Cenozoic sediments in this area. They are predominantly late Lutetian in age; the benthic foraminiferids have a markedly 'warm-water' aspect. There are outliers of these limestones on the Cotentin Peninsula (Normandy), unconformably overlying Jurassic and Palaeozoic. This strongly transgressive unit, with its low clastic content, indicates a major rise in relative sea level during the Lutetian.

Samples of finer-grained bioclastic limestones and calcareous sandstones with Bartonian species of *Nummulites* are recorded from sea-bed samples in the western Channel on the northern side of the Basin.

Upper Eocene
Upper Eocene sediments have not been identified in the Western Approaches, probably reflecting pre-Oligocene uplift and erosion. In the western English Channel the Late Eocene *Nummulites fabianii* has been recorded from sea-floor samples on the northern edge of the basin.

Oligocene unconformity
There is a major regional pre–(mid?) Oligocene erosional surface throughout the Western Approaches, marked by a prominent seismic reflector (Evans 1990). It is due to inversion, marked by faulting and uplift, both in this area and further east, with up to 2 km of movement on the Alderney–Ouessant fault line and its western continuation. This was followed by deep erosion, cutting down to the Mesozoic in some areas. Calculations suggesting that up to 800 m of sediments were eroded in the Western Approaches (see Evans 1990, p. 40) are however not supported by the stratigraphic evidence.

Oligocene
In the Western Approaches, Oligocene sediments are poorly documented. Rupelian marine claystones have been identified in one well. Overlying marine sandstones and argillaceous limestones have been assigned a Chattian age, but published biostratigraphic evidence is inconclusive. In the western English

Channel, a number of sea-bed samples from the Armorican Platform, overlying the Lutetian carbonates, are probably of Priabonian or early Rupelian age. They include freshwater limestones, clays with marginal-marine microfaunas, and bioclastic limestone with abundant shallow-water foraminiferids similar to assemblages from the Rupelian of the Aquitaine Basin. A small outlier of shelly clays in the Cotentin peninsula, deposited in a brackish–marine environment, is of either Priabonian or early Rupelian age; the warm-water foraminiferid *Halkyardia* indicates a marine connection through the Western Approaches Basin. These poorly dated records are significant palaeogeographically, as they suggest that marginal-marine and non-marine facies, similar in part to those of the Late Eocene and Early Oligocene in the Hampshire and Paris basins, were also deposited in this area.

Miocene

Miocene sediments are confined to the Western Approaches, comprising two seismically defined 'formations'. In the western part of the Western Approaches, Miocene sediments apparently overly Oligocene ones conformably, but further east they overstep progressively onto the Eocene, the Upper Cretaceous (over the SW Cornubian platform) and onto Variscan basement (in the Haig Fras area). The basal contact is generally planar, although locally a karstic surface is developed where it overlies Eocene limestones (Evans & Hughes 1984). This transgressive surface is probably due to earliest Miocene eustatic sea-level rise.

The Lower–lower Middle Miocene (Aquitanian–Langhian) Jones 'Formation' (former 'Globigerina silts') forms a westward thickening wedge in the axis of the basin. It comprises silty marls and marly limestones with some thin sandstone beds, deposited in mid- to outer neritic environments.

In the marginal St Mary's Bay Basin there is an intra-Middle Miocene erosion surface incised deeply into the Jones Formation, but elsewhere this is not clearly recognizable. The overlying upper Middle Miocene (Serravallian) Cockburn 'Formation' comprises silty bioclastic limestones, deposited in mid-neritic environments.

Pliocene

There is a regional erosional surface between the Miocene and Pliocene, reflected by the absence of Upper Miocene sediments. The overlying (seismically defined) Little Sole 'Formation' is restricted to the western end of the Basin, where it extends along the outer shelf onto the Atlantic continental slope. It is divided seismically into two units. The lower lensoid, 'wedge-like' unit (Evans & Hughes 1984), the lower Little Sole 'Formation', is confined to the Atlantic shelf edge. It has not been sampled, but from its relative stratigraphic position it is inferred to be Pliocene. It is overlain disconformably by the 'sheet-like' upper Little Sole Formation. This is much more widespread, but only two sea-bed samples have been recovered. These are clayey sands dated as Early (earliest) Pleistocene (Evans & Hughes 1984), similar in age to the St Erth Beds of Cornwall (see Chapter 17).

There are several outliers of shallow-marine, probably Upper Pliocene, clays and sands in the Cotentin peninsula. These outliers indicate that shallow marine Pliocene sediments were probably deposited throughout the western English Channel.

South Celtic Sea Basin

Here, Upper Eocene–Oligocene sediments rest unconformably on Upper Cretaceous chalk or Variscan basement to the NW of Cornwall (Evans 1990). Several shallow cored boreholes and petroleum exploration wells have penetrated interbedded clays, glauconitic sandstones and lignites. The facies, and sparse dinoflagellate assemblages, indicate a marginal-marine to probably inner neritic environment. This probably marks the boundary between the non-marine facies in basins to the east, and the marine Paleogene seen to the west, in the Fastnet Basin (Irish offshore). In the adjacent Irish offshore, the Paleogene sediments are overstepped by marine Miocene sediments similar to those in the Western Approaches Basin.

Economic Geology

Paleogene and Neogene sediments have been extensively exploited in the past for raw materials, including clay for bricks, tiles, ceramic products and cement production, and sand for the construction industry. The brick and tile industry in the Hampshire and London basins, concentrated mainly on the Reading Formation and London Clay Formation, has almost entirely disappeared during the last 30 years, except for a few small specialist works. The London Clay is still quarried at South Ockendon (Essex) for use in cement production, but quarrying at High Ongar (Essex) for production of light-weight clay aggregate (LECA) ended in the 1990s. Kaolinitic clays ('ball clay' and 'pipeclay') are currently still exploited in the Bovey Basin (Devon) (Bovey Formation), the Petrockstow Basin (Devon) and in the Wareham area of Dorset (Poole Formation), mainly for the ceramics industry.

Many (mostly small) sand quarries formerly existed, but almost all are now abandoned or used for landfill. Sand is still extracted at a few sites, mainly from the Poole Formation in the Bournemouth–Poole area, the Whitecliff Member (London Clay Formation) in the Southampton area, and the Bagshot Formation in the western London Basin.

London Clay is quarried at Emsworth (Hampshire) as a liner for landfill sites. There are a number of landfill sites excavated in Paleogene sediments in the Hampshire and London basins.

The offshore Paleogene and Neogene sediments, in the southern North Sea and elsewhere, have no hydrocarbons sourcing potential, due to their generally low organic content and shallow burial, and no hydrocarbons reservoirs have been identified.

17 Quaternary: ice sheets and their legacy

J. A. CATT, P. L. GIBBARD, J. J. LOWE, D. McCARROLL, J. D. SCOURSE, M. J. C. WALKER & J. J. WYMER‡

The Quaternary, or final period of geological time, has been popularly equated for the last 150 years with the 'ice age', when glaciers invaded many high latitude and high altitude parts of the Earth's surface not previously glaciated since at least the Permo-Carboniferous. Early in the 19th century a distinction was drawn between the 'solid' rock formations, which often show a regular stratigraphic order and uniform thickness because of deposition in extensive marine basins, and the thinner, unconsolidated and much more variable 'drift' or superficial formations now known to result from more recent deposition mainly in glacial and other non-marine environments. The terms solid and drift are still preserved in the legends of quite recently published British Geological Survey (BGS) maps, although since 2004 they have been replaced by 'Bedrock' and 'Superficial Deposits', respectively. Over most of England and Wales they correspond to pre-Quaternary and Quaternary deposits.

The term drift (or diluvium) originally implied deposition by waters of the Biblical Flood, but with increasing exploration of polar regions in the 19th century it became popular to invoke floating ice as a depositional agent, accounting especially for the large blocks of identifiable rock types (erratics) displaced long distances from their nearest known outcrops. However, both flood and floating ice implied an unlikely submergence of great depth in order to deposit erratics and other drifts on mountains well above present sea level, for example on Moel Tryfan in North Wales, where Quaternary marine molluscs occur at 430 m OD. This problem was resolved by Louis Agassiz and other Swiss geologists of the 19th century, who proposed that land-based glaciers, such as those presently occupying many Alpine valleys, had previously been much more extensive and thus able to transport the erratic blocks long distances. With the assistance of William Buckland and Charles Lyell, Agassiz introduced the land ice hypothesis to Britain in 1840 (Boylan 1981), although it did not fully displace the diluvial (floating ice) theory among British geologists until 1870 or even later.

Early recognition that drift deposits can be separated into the older 'diluvium', and younger 'alluvium' led to subdivision of Quaternary time into the Pleistocene and Holocene. This is still widely recognized, although some prefer to use only Pleistocene, which then equates with the Quaternary, partly because the Holocene is much shorter (only 10 000 ^{14}C years by current definition) and seems to be a natural continuation of previous Quaternary events.

Stratigraphic studies of Quaternary deposits began in the mid-19th century and soon provided a wide range of evidence for two or more glaciations in many areas beyond the present ice margins. Fossil remains obtained from peat, lake and river sediments lying between glacial deposits suggested that in the intervening ('interglacial') periods the climate was approximately the same as at the present day. Consequently, the Holocene is now regarded as merely the latest in a succession of interglacials, although one which has been extensively influenced by human activities. During glacial periods expansion of the intensely cold polar regions led to equator-ward compression of the major climatic zones, and during interglacials the zonal boundaries returned to approximately their present positions. With its mid-latitude position, Britain therefore experienced strong climatic oscillations throughout the Quaternary, which are reflected in major biological, sedimentological and geomorphological fluctuations.

Despite its promising geographical position, consistent glacial-interglacial sequences of deposits have proved especially difficult to establish in Britain. The glacial and other terrestrial deposits of cold periods contain few indigenous fossils and are often laterally variable in lithology, and the interglacial deposits of rivers, lakes, mires, caves and even shallow marine (beach) environments have very limited lateral continuity. Also, standard biostratigraphic methods for correlation and relative dating are difficult to apply even to richly fossiliferous interglacial deposits, because there have been few true evolutionary changes in the relatively short timespan of the Quaternary. Almost all the biostratigraphic variation in the Quaternary deposits of England and Wales results in fact from climatic change over time and local differences in depositional environments. The main exception are the mammals, which do show some evolutionary changes useful for dating. They also show stronger changes in assemblage composition between successive interglacials than other animal and plant groups, and even some changes within interglacials (Schreve 2001*a*).

There are also problems with most of the isotopic and other numerical dating methods developed for use with Quaternary deposits. Some (e.g. radiocarbon dating) can only be used over limited parts of the late Quaternary and others, such as the thermoluminescence (TL) and amino-acid methods, are reliable only when applied to specific materials. The sequence of well-dated reversals of the Earth's magnetic field provides a satisfactory means of subdividing Quaternary time as a whole, but is only suitable for dating individual beds where there is a fairly continuous succession of deposits, a situation that is rare in Britain. Because of these problems, anomalous dating results are common in Quaternary sequences.

Radiocarbon dates do not equate with calendar years, because the concentration of ^{14}C in the atmosphere varied in the past as a result of variations in solar activity, changes in the geomagnetic field strength and different rates of $^{14}CO_2$ exchange between the atmosphere and oceans. Dates are expressed as ^{14}C years before present (BP), where 'present' is the standard year 1950 AD. In recent decades considerable advances have been made in calibrating the ^{14}C timescale to calendar years (van der Plicht, 2002), using annual tree-rings counted back either from

‡ Deceased

the present or from other known dates (dendrochronology), corals dated by both ^{14}C and uranium decay series (U-series), and annually laminated lake sediments. The most accurate part of the calibration curve (0 to approximately 11 ka BP) is that based on dendrochronology. The ^{14}C dates given in later parts of this chapter are shown as both the original measurements and as calendar years (cal.) calibrated by the currently recommended calibration model (IntCal04, version 5.0.1) of Reimer *et al.* (2004).

British Quaternary Stages

Despite the various difficulties of correlation and dating, work on subdividing British Quaternary deposits was sufficiently advanced by the mid-20th century for West (1963) and Mitchell *et al.* (1973) to propose 14 separate Quaternary climatic stages, seven cold and seven warm, based on major glacial-interglacial fluctuations (Fig. 17.1). With some possible additions, the same sequence of stages is still widely used today and forms the basis of this chapter. However, it has major weaknesses (Bowen 1999*a*, chapter 1). First, it hides large gaps in the succession. For example, as Zagwijn (1975) and Gibbard *et al.* (1991) have pointed out, comparison with the much more complete Dutch sequence suggests that a major hiatus about 1 Ma long and possibly others exist in earlier parts of the British Quaternary sequence.

A second weakness results from the stages having been defined mainly from sequences of pollen assemblage biozones, which reflect changes in essentially local vegetation over time. Because of various climatic and edaphic factors, the assemblages varied laterally so that the biozone sequences are difficult to use for long distance correlation. Also in some regions there is considerable similarity in assemblage biozone sequences between interglacial stages, leading to the possibility of erroneously conflating the deposits of two or more different interglacials. Where non-palynological evidence suggests that this has happened, but the climatic sequences are still incompletely resolved, we use the original stage names followed by the word 'Complex'.

However, within areas of minimal climatic and edaphic variability, such as East Anglia and the Midlands, detailed differences in plant succession and the presence or absence of certain taxa have been used fairly successfully to distinguish different interglacials and thus assist correlation and relative dating (West 1980a; Thomas 2001). Some of these differences probably resulted from changes in sea level leading to the isolation of Britain from the European mainland at various times in different interglacials. In the Early and Mid Pleistocene different warm stages have been distinguished partly on the basis of the reappearance of certain thermophilous Tertiary relict tree genera, such as hemlock (*Tsuga*) and Caucasian wingnut (*Pterocarya*). In contrast, the so-called 'full-glacial' vegetation was strikingly similar in most cold stages, often including a mixture of herbaceous species typical of various modern phytogeographical areas, including arctic, alpine, steppe, maritime, marsh-loving and even temperate genera. Consequently, cold stages are even more difficult to distinguish palynologically than the interglacials.

A third problem is that the climostratigraphic stages have diachronous boundaries and are not strictly suitable as chronostratigraphic units. Climatic amelioration at the beginning of an interglacial was recorded significantly earlier in the south than the north of the country, because migration rates of the warm-loving indicator plants were slow. Also, although the cold stages are often referred to as glaciations or glacials, glaciers actually arrived in lowland areas many thousands of years after the initial climatic deterioration, because they advanced very slowly from the mountainous ice accumulation areas in the north of Britain. True glacial deposits consequently represent only small and usually late parts of the cold stages

	British Terrestrial Stages	Marine Isotope Stages	Middle Thames Terrace deposits	Lower Thames Terrace deposits
HOLOCENE	^{w}Holocene	1		
UPPER PLEISTOCENE	^{c}Devensian	5d-2	Shepperton Kempton Park	Shepperton East Tilbury
	^{w}Ipswichian	5e		Trafalgar Square
MIDDLE PLEISTOCENE	^{c}Wolstonian	10-6	Taplow Lynch Hill Boyn Hill	Mucking Corbetts Tey Wigborough/ Orsett Heath
	^{w}Hoxnian	11		Swanscombe
	^{c}Anglian	12	Black Park Winter Hill/ Westmill	Asheldham/ Dartford Heath St Osyth/Holland
	^{w}Cromerian	20-13	Gerrards Cross Beaconsfield Satwell Westland Green	Wivenhoe Ardleigh Waldringfield
LOWER PLEISTOCENE	^{c}Beestonian	62-21 (including 1 Ma hiatus)		Moreton Bures Stebbing Westland Green
	^{w}Pastonian	67-63	Stoke Row	Stoke Row
	^{c}Baventian	70-68	Pebble Gravels	Westleton Beds
	^{w}Antian/ Bramertonian	73-71		Chillesford Clay
	^{c}Thurnian	76-74		
	^{w}Ludhamian	80-75		
	^{c}Pre-Ludhamian	100-81	Nettlebed	Red Crag

Fig. 17.1. British Quaternary Stages (Mitchell *et al.* 1973), approximate Marine Isotope Stage (MIS) equivalents and terrace sequences of the Middle and Lower Thames (based on Westaway *et al.* 2002). ^{c} = cold, ^{w} = warm.

after which they are named, although they are often thicker, more extensive and better preserved than the non-glacial (fluvial and periglacial) deposits of cold stages.

The onshore distribution of the numerous types of Quaternary deposits in England and Wales is shown in the current series of 1:50 000 BGS maps. A national synopsis is provided by the southern Quaternary sheet at 1:625 000 scale. Over the past 25 years BGS has also mapped and sampled the offshore Quaternary deposits of the UK sector of the NW European continental shelf. The results, published at 1:250 000 scale, represent one of the most detailed shelf mapping programmes yet completed. This has extended the limits of glaciations previously known only from onshore deposits and clarified the depositional and erosional processes resulting from the essentially eustatic changes of sea level associated with the glacial-interglacial cycles.

During glacial lowstands, when sea level fell by up to 135 m (Clark & Mix 2002), the coastline migrated basin-wards to expose much of the continental shelf. Many major NW European rivers then extended to the current shelf-break (Bourillet *et al.* 2003) and deposited their loads on the outer shelf or continental slope, rather than within estuaries as at times of interglacial highstands. Over glaciated parts of the shelf, the British and Fennoscandian ice sheets depressed the land surface, giving regionally high relative sea levels despite eustatic lowering worldwide (Lambeck 1995; Lambeck & Purcell 2001). Consequently depocentres were created in the Irish and North Sea basins, in which thick glaciomarine sediments accumulated.

There were therefore two main sources of sediment reaching the continental slope during cold stages of the Quaternary – fluvial in the southern part of the UK sector and glacial in the northern (Scottish) part. Most of the Quaternary stratigraphic units in the UK sector of the continental shelf are consequently terrestrial in origin, glacial to the north and fluvial to the south. As such they have more in common with onshore Quaternary sequences than with the marine Quaternary of the North Atlantic, and even the marine influence on the shelf during interglacial highstands is often difficult to decipher because of glacial and fluvial erosion in subsequent cold stages. The only good record of highstand events on the shelf is that of the Holocene (Scourse & Austin 2002).

Many of the continental slope accumulations are unstable and subject to redistribution by slides, debris flows and turbidity currents triggered by seismic events, sea-level changes and the escape of gas-hydrates (Bunz *et al.* 2003). During the later Quaternary some slope failures around NW Europe created extensive tsunamis, such as that resulting from the mid-Holocene Storegga Slide, which deposited sand along the east coast of Scotland and Northumberland (Smith *et al.* 2004).

The marine isotope stages

Compared with the British terrestrial Quaternary record, a much more complete succession of glacial-interglacial cycles has emerged over the last 50 years from micropalaeontological and oxygen and other isotopic studies of deep oceanic sediment cores. This evidence now suggests that there have been about 50 separate major cold stages over the last 2.6 Ma, and that these were driven by variations in solar radiation input resulting from perturbations in the Earth's orbit (Milankovitch cycles). The marine isotope record is divided into isotope stages (MISs) numbered backwards from the Holocene (MIS 1), so that major warm stages (interglacials) have odd numbers and the cold stages even numbers (Fig. 17.1), with the exception of MIS 3, which equates with the middle part of the terrestrial Devensian (last cold) Stage of Mitchell *et al.* (1973). The MIS boundaries are by convention placed at mid-points between temperature maxima and minima. As defined in oceanic sequences they are assumed to be globally isochronous, but may not be equivalent to stage boundaries defined in terrestrial sequences, which are based on a range of different criteria.

The most complete onshore Quaternary sequence in Britain is probably that of the Thames terraces (Fig. 17.1), which have been tentatively correlated with the MIS sequence by Westaway *et al.* (2002). Most of the other onshore deposits in England and Wales provide only snapshots of a limited number of often imprecisely dated Quaternary events. However, using a range of dating results, Bowen (1999*a*, tables 2–22) attempted to relate many of the younger onshore Quaternary lithostratigraphic units to MISs. The thickest Quaternary sediment accumulation in NW Europe (>1 km) occurs offshore in the southern North Sea Basin (Cameron *et al.* 1992), but precise correlation of the seismic-stratigraphical units defined here with the more detailed lithostratigraphic successions defined onshore is difficult at present.

The Plio-Pleistocene (Neogene-Quaternary) boundary

Another major problem of the Quaternary is where to draw the base (i.e. the Plio-Pleistocene boundary). As currently accepted by the International Union of Geological Sciences, it is defined in the coastal section at Vrica, southern Italy, at a level just below the top of the Olduvai palaeomagnetic Subchron, where a climatic deterioration, probably at the boundary between MISs 63 and 62, is indicated by a change in microfauna (Aguirre & Pasini 1985). The boundary was originally dated to 1.64 Ma, but revision of this part of the palaeomagnetic timescale, based on tuning the magnetic events recognized in oceanic sediments to the astronomical (Milankovitch) cycles, suggests it should be somewhat older, *c.* 1.81 Ma (Hilgen 1991) or 1.905 Ma (Shackleton *et al.* 1990).

However, most Quaternary scientists prefer to place the Plio-Pleistocene boundary at an earlier climatic deterioration close to the Gauss–Matuyama (G/M) reversal (Gibbard *et al.* 2005). The G/M boundary is dated to *c.* 2.6 Ma on the astronomically-tuned timescale (Shackleton *et al.* 1990) and lies within MIS 103. There are various reasons for preferring this earlier boundary. First, it equates with a stronger cooling episode that led to a greater microfaunal change in oceanic cores from the North Atlantic and North Pacific and to a major increase in ice-rafted debris into the Rockall Basin and other parts of the North Atlantic. Second, it coincides approximately with the beginning of the Pretiglian Stage of the Dutch sequence, when there was the first appearance of a cold climate vegetation in NW Europe (De Jong 1988) and a large change in shallow marine and terrestrial faunas (Gibbard *et al.* 1991). Third, at this time widespread loess deposition began in East Asia as a result of a change to drier, colder conditions. Fourth, in long lake sequences from Colombia, Japan and Israel, there was a greater vegetational change at the G/M boundary than towards the end of the Olduvai Event. Fifth, the G/M boundary coincides with the first appearance of the important early Quaternary hominid *Homo habilis* in Africa. Sixth, at the 1948 International Geological Congress in London, the base of the Quaternary in Britain was placed at the base of the Red Crag in East Anglia (Boswell 1952), which is now thought to date from just before 2.6 Ma, and this boundary was later used in both of the Geological Society's Special Reports on correlation of Quaternary deposits in Britain (Mitchell *et al.* 1973; Bowen 1999*a*) and in the Quaternary chapter of the first edition of *Geology of England and Wales* (Boulton 1992). It has also been followed by almost all authors describing the British Quaternary. We therefore draw the base of the Quaternary in England at the base of the Red Crag Formation, even though this conflicts with current international practice. Consequently it includes deposits assigned to the Gelasian Stage, and equivalent in age to Calcareous Nannoplankton Biozones NN17 and

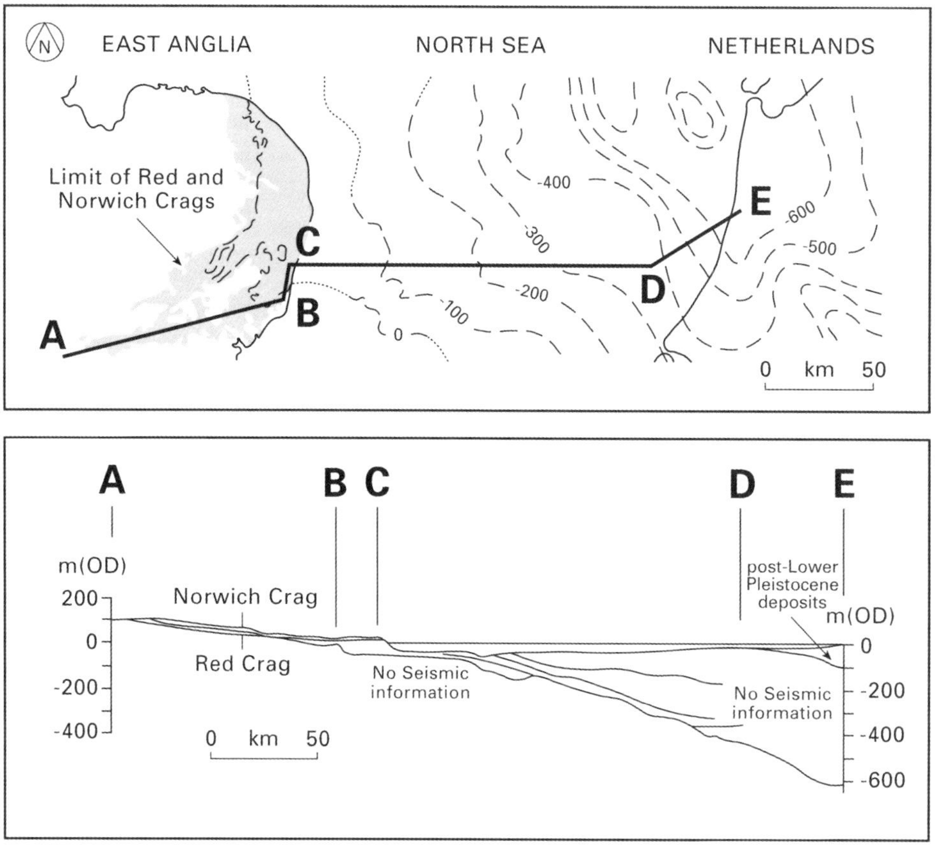

Fig. 17.2. Contours on the base of the Red Crag and Norwich Crag Formations in East Anglia and their equivalents in the southern North Sea and The Netherlands (from Mathers & Zalasiewicz 1988). The section A–E shows the extent of seismic information.

NN18 and Planktonic Foraminiferal Biozone PL6 plus the upper part of PL5 in marine sequences.

This chapter describes the Quaternary deposits of England and Wales and what is currently known of their equivalents in the UK offshore sector. It expands and brings up to date the last five sections (pp. 425–439) of Boulton (1992) in the first edition of *Geology of England and Wales*.

The Lower Pleistocene

Using the date of the Plio-Pleistocene boundary defined above, the Early Pleistocene includes the period between the G/M palaeomagnetic reversal (*c.* 2.6 Ma) and the Matuyama/Brunhes (M/B) reversal (*c.* 0.78 Ma), which is accepted by The International Union for Quaternary Research (INQUA) as the boundary between the Early and Mid Pleistocene (Richmond 1996). Onshore the Lower Pleistocene deposits occur mainly in East Anglia, where essentially coarse-grained marine sediments (the Crag Group) were laid down near the western margin of the southern North Sea, and fluvial gravels (the Kesgrave Formation) were deposited by the proto-Thames before it was diverted southwards into its present course by the Middle Pleistocene Anglian glaciation. The Crag Group rests on an eroded surface of Cretaceous and Early Palaeogene strata. In the later Palaeogene and early Neogene much of eastern England was probably emergent and eroded subaerially, although a brief late Miocene or early Pliocene transgression into SE Suffolk is suggested by reworked clasts with derived fossils, which are the only known remnants of a former deposit termed the Trimley Sands (Balson 1990).

The Crag Group

The Crag Group is much thicker and better developed offshore in the southern North Sea than in East Anglia (Fig. 17.2), but correlation between the two areas is hampered by the different information available in each situation. Offshore the sequence accumulated in a multistage Palaeozoic–Holocene rift basin trending NNW–SSE approximately parallel to the present English coastline. It includes prominent seismically-defined bounding surfaces, which probably represent regional disconformities (Cameron *et al.* 1984), but there is little lithological or palaeontological information available. Some of the seismic-stratigraphic units are linked by widely-spaced boreholes, but few have been studied lithologically or biostratigraphically (Ekman & Scourse 1993). In contrast, onshore much is known about the sediments and fossils, but major non-sequences are difficult to distinguish from lower-order erosional discontinuities. The offshore disconformities tend to coalesce westwards towards East Anglia, suggesting that the onshore sequence is incomplete, with parts of the Pliocene and Early Pleistocene absent.

Onshore the Crag Group is divided into five formations, the Pliocene Coralline Crag Formation (see Chapter 16), and the Early Pleistocene Red Crag, Norwich Crag, Wroxham Crag and Cromer Forest-bed formations. Offshore these are approximately equivalent to the Westkapelle Ground, Smith's Knoll, Ijmuiden Ground, Winterton Shoal, Markham's Hole, Outer Silver Pit and Aurora formations (Fig. 17.3), which consist of thick marine units deposited by two northward prograding delta complexes, a large eastern delta fed by sediment from the European mainland and a smaller western delta composed of sediment from eastern Britain (Balson & Cameron 1985). The 190 m-thick Ijmuiden Ground Formation oversteps

Formations	Depositional Environments	Stages
Bligh Bank	Marine	HOLOCENE
Indefatigable Grounds	Marine	
Nieuw Zeeland Gronden	Marine	
Well Hole	Marine	
Elbow	Freshwater & intertidal	
Sunderland Ground	Glaciolacustrine	UPPER DEVENSIAN
Botney Cut	Glaciomarine & glaciolacustrine	
Kreftenheye	Fluvial	
Twente	Periglacial aeolian	
Well Ground	Glaciofluvial	
Dogger Bank	Glaciomarine & glaciolacustrine	
Bolders Bank	Subglacial	
Brown Bank	Marine & lacustrine	LOWER DEVENSIAN
Eem	Marine	IPSWICHIAN
Tea Kettle Hole	Periglacial aeolian	WOLSTONIAN
Cleaver Bank	Glaciomarine	
Egmond Ground	Marine	HOXNIAN
Sand Hole		
Swarte Bank	Subglacial	ANGLIAN
Yarmouth Roads	Fluvial	LOWER PLEISTOCENE TO LOW MIDDLE PLEISTOCENE
Aurora	Marine	
Outer Silver Pit	Marine	
Markham's Hole	Marine	
Winterton Shoal	Marine	
Ijmuiden Ground	Marine	
Smith's Knoll	Marine	
Westkapelle Ground	Marine	
Red Crag	Marine	

Fig. 17.3. Quaternary formations of the southern North Sea (based on Cameron *et al.* 1992).

the Smith's Knoll Formation and consists entirely of European mainland sediment. Sands and clays in its upper part suggest a late Tiglian age (1.8–2.0 Ma).

The Red Crag, Norwich Crag and Wroxham Crag formations form the bulk of the onshore marine Early Pleistocene succession. They comprise sands, often shelly, with some clays and gravels, which thicken eastwards to more than 70 m at the coast. A northward decline in the number of extinct and southern (warm water) mollusc species was originally used to suggest that the deposits became younger northwards (Harmer 1902), but studies of pollen (West 1961), foraminifera (Funnell 1961) and molluscs (Norton & Beck 1972) indicated alternations of warm and cool episodes, and a system of formal stages (Fig. 17.1) was erected on this basis (Mitchell *et al.* 1973). Mathers & Zalasiewicz (1988) established a reliable lithostratigraphy in southern East Anglia, but this is difficult to extend northwards because of facies changes and limited exposure.

The Red Crag Formation

In Essex and SE Suffolk the Red Crag Formation consists of a basal lag gravel rich in phosphatic nodules overlain by shelly, glauconitic and sometimes gravelly sands. The lower part of the sands, as at Orford Lodge (TM 425512) near Chillesford (Fig. 17.4), shows tabular cross-sets up to 4 m thick with almost no mud drapes (Fig. 17.5), and was probably deposited by southwestward migration of large submarine sand waves in an open marine embayment under the influence of powerful rotary tidal currents (Balson *et al.* 1991). The upper part has a greater range of sedimentary structures, including cross-sets, ripples and planar bedding, with numerous mud drapes locally arranged in packets, probably reflecting spring–neap tidal cycles. Sedimentation of this part was by essentially rectilinear tidal currents in a shallower, more constricted estuarine environment.

The oldest (warmest water) Red Crag at Walton-on-the-Naze (TM 2622) has yielded a dinoflagellate flora and single pollen spectrum suggesting correlation with the Reuverian (latest Pliocene) Stage of The Netherlands (Gibbard *et al.* 1991). An upper part of the formation in Suffolk and Essex, the Sizewell Member of Zalasiewicz *et al.* (1988), is correlated with the Westkapelle Ground Formation (Fig. 17.3), and is correlated on foraminiferal evidence (Beck *et al.* 1972) with the British Pre-Ludhamian Stage of the Ludham borehole (TG 385199), Norfolk (West 1961). The Pre-Ludhamian deposits in this borehole contain pollen of pine (*Pinus*), birch (*Betula*), spruce (*Picea*), heathers (Ericales) and meadow grasses (Poaceae), indicating boreal forest on adjacent land areas. On this basis the Pre-Ludhamian is thought to be equivalent to the Dutch Pretiglian Stage (Funnell 1995). In the same borehole, the Pre-Ludhamian is overlain by the Ludham Crag Member of early Ludhamian age, which is probably equivalent to the temperate Tiglian A Substage of The Netherlands (Funnell 1987). Where it occurs at the surface in north Essex and Suffolk, the Red Crag is strongly iron-stained, because of either deep weathering during later Pleistocene interglacials or diagenetic oxidation by groundwater.

Inland from the East Anglian coast, the Red Crag forms an ascending sheet of sands beneath later estuarine, fluvial and glacial deposits. Its base rises from close to sea level on the East Anglian coast to an altitude of 90 m OD at Stansted Mountfitchet, Essex, and the thickness decreases in the same direction from *c.* 40 m to 10 m. Even further west, isolated blocks of ferruginous sandstone with a mollusc fauna similar to that of the Red Crag were found at 131 m OD on the Chiltern Hills at Rothamsted (TL 118143), Hertfordshire (Dines & Chatwin 1930). The generally increasing elevation westwards results mainly from post-depositional tilting of

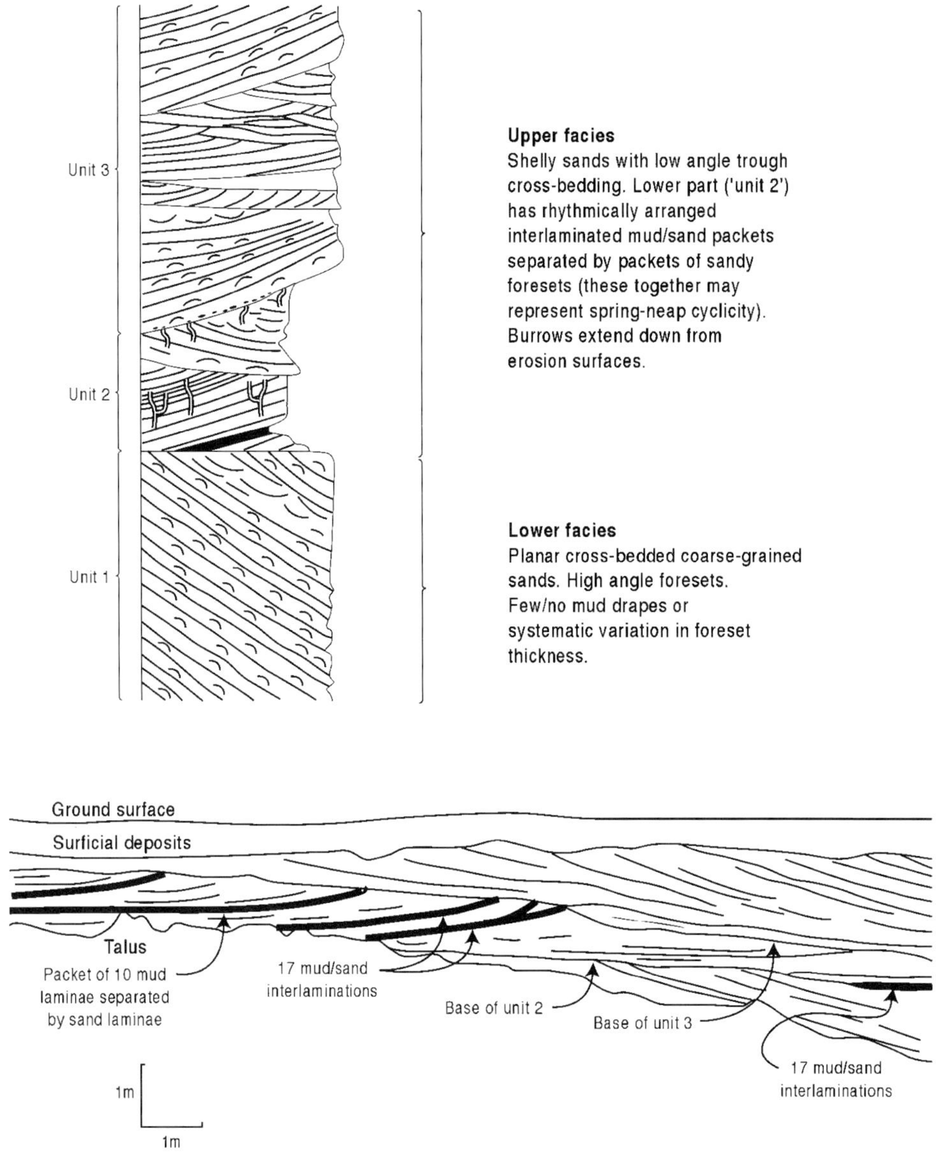

Fig. 17.4. Facies of the Red Crag Formation at Orford Lodge Pit, Chillesford (from Balson *et al.* 1991).

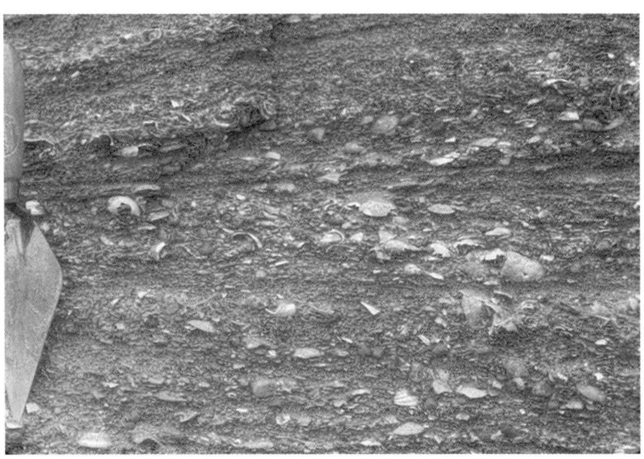

Fig. 17.5. Red Crag Formation exposed in 1978 at Butley (TL 371510), Suffolk, showing large scale cross-stratification interpreted as sandwave deposition in a shallow marine environment. The laminae dip towards the SW indicating a dominant current direction similar to that of the present-day southern North Sea. Photograph by Philip Gibbard.

southern England caused by uplift in the west, although offshore this was emphasized by subsidence of the southern North Sea Basin. Between Stowmarket and SE Norfolk the base of the Crag Group is a complex of ridges and enclosed basins suggesting tectonic deformation, probably at various times during and since deposition (Hamblin *et al.* 1997; Rose *et al.* 2002).

The Norwich Crag Formation

In southern parts of East Anglia, the Red Crag is overlain by the Chillesford Sand Member of the Norwich Crag Formation (Fig. 17.6), which consists of sparsely fossiliferous fine sands showing horizontal bedding, ripples, flaser bedding, water escape structures, trough cross-sets up to 0.3 m thick, vertical burrows and isolated clay drapes, together suggesting rapid deposition in a tidal flat environment (Mathers & Zalasiewicz 1988). Northwards in the coastal regions of north Suffolk, the Chillesford Sand passes laterally into coarser, more shelly, glauconitic, deeper water sands (Fig. 17.7). Most surface exposures of the Norwich Crag have, where fossiliferous, been assigned to either the Antian or Bramertonian Stages (Fig. 17.1), which are now thought to represent two parts of the same warm climatic episode, equivalent to part of the Tiglian C3 Substage of The

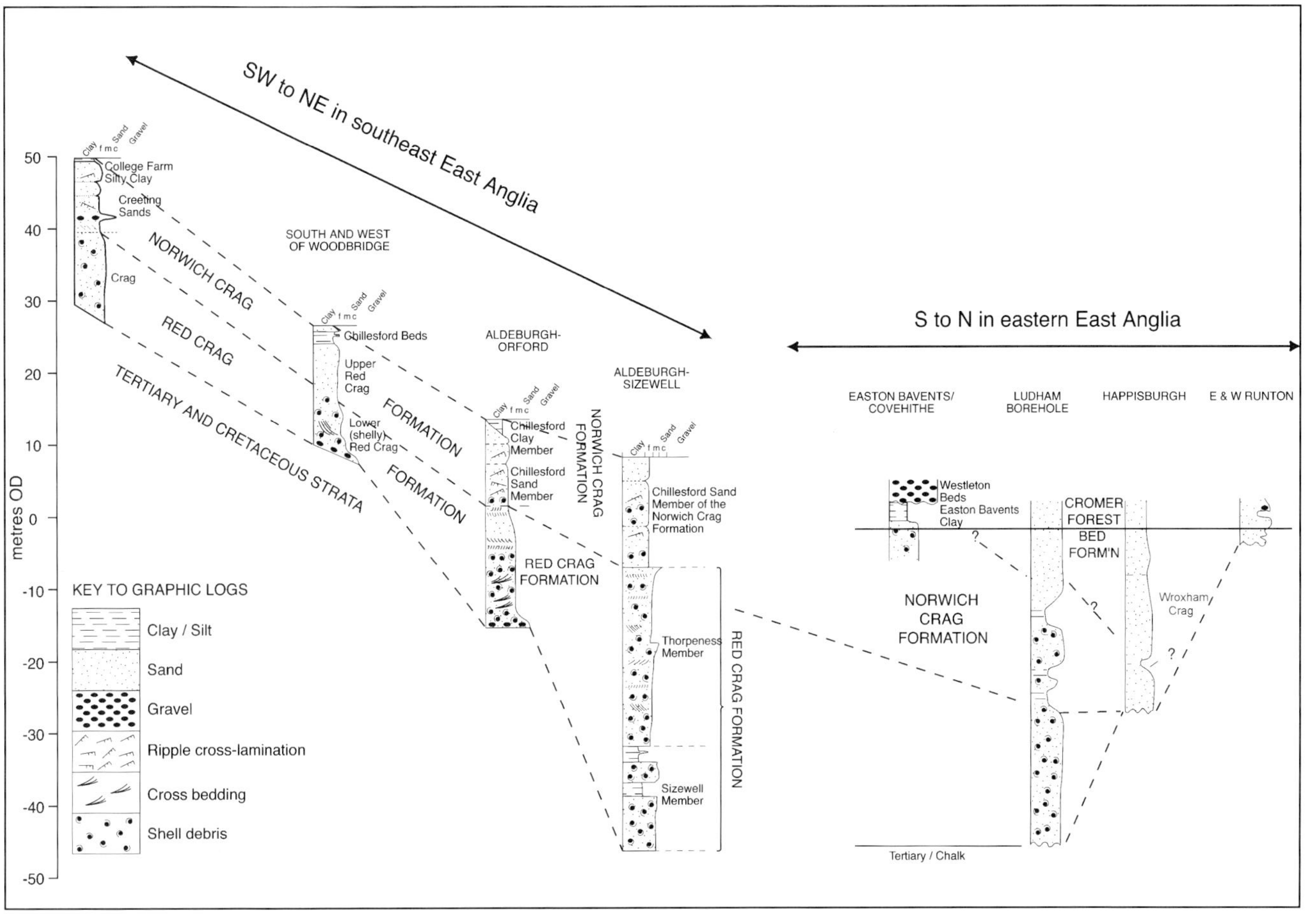

Fig. 17.6. Stratigraphical relationships of the Red Crag, Norwich Crag and Cromer Forest-bed Formations in various parts of East Anglia (from Mathers & Zalasiewicz (1988) and Zalasiewicz *et al.* (1991)).

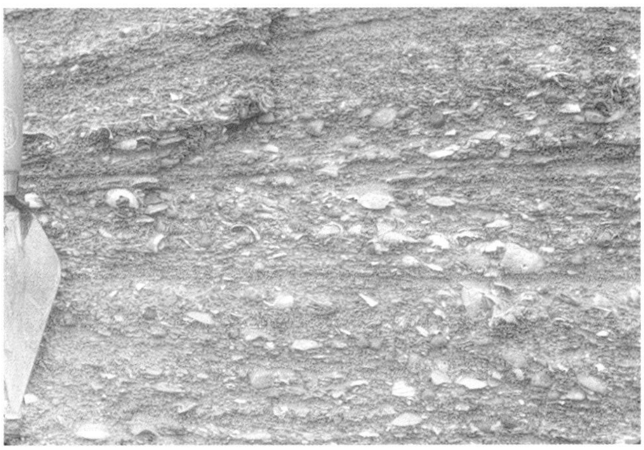

Fig. 17.7. Exposure of the Chillesford Sand facies of the Norwich Crag Formation at Wangford (TM 463778), north Suffolk, showing shell-rich sand interpreted as deposited in a shallow, current-swept marine environment. Photograph by Philip Gibbard.

Netherlands (Zalasiewicz *et al.* 1991), which is dated to *c.* 2.0–2.1 Ma. In the Norwich Crag of the Ludham borehole, West (1961) also recognized a cold stage (the Thurnian) between the Ludhamian and Antian deposits. Pollen evidence suggests that the offshore delta-front Smith's Knoll Formation (Fig. 17.3) is Antian in age.

Several isolated clay beds occur locally in Suffolk at or near the top of the Norwich Crag. These include the Chillesford Clay Member in the Aldeburgh–Orford area, with a type section at TM 388526 (Zalasiewicz *et al.* 1991; Hamblin *et al.* 1997), the Easton Bavents Clay of the north Suffolk coast near Easton Bavents (TM 518787) and the College Farm Member in the Great Blakenham–Creeting St Mary area (TM 111503) near Ipswich (Fig. 17.8). The first two of these are marine and contain the same cold-climate assemblage of pollen, foraminifera and molluscs. They are tentatively assigned to the Baventian cold stage (Fig. 17.1), and probably correlate with deposits in The Netherlands of a marine regression during the Tiglian C4c Substage (Fig. 17.9), thought to be equivalent to MIS 68 (Funnell 1995) and dated to *c.* 1.86 Ma. At Easton Bavents ice wedge casts provide further evidence for cold conditions. In contrast, the College Farm Member is a thin (<3 m) temperate-climate freshwater lagoonal clay capping the Chillesford Sand of the Ipswich area, and is somewhat older than the other two, as it contains abundant remains of the water fern *Azolla tegeliensis* and pollen of hemlock (*Tsuga*), a Tertiary relic now extinct in Europe. These suggest correlation with the Tiglian C3 Substage of The Netherlands (De Jong 1988), and Gibbard *et al.* (1996*b*) proposed that the College Farm Member was deposited late in the Antian/Bramertonian Stage *c.* 2.0 Ma. A similar age was supported by Maher & Hallam (2005) using palaeomagnetic evidence.

The Westleton Member, which unconformably overlies the Easton Bavents Member, has an extensive outcrop along the north Suffolk coast, and also extends inland, where it overlies shelly sands of the Norwich Crag. Probably of late Baventian age, it consists almost entirely of flint pebbles, locally with small amounts of white quartz and *Rhaxella* chert from NE Yorkshire or the North Sea floor. These components were

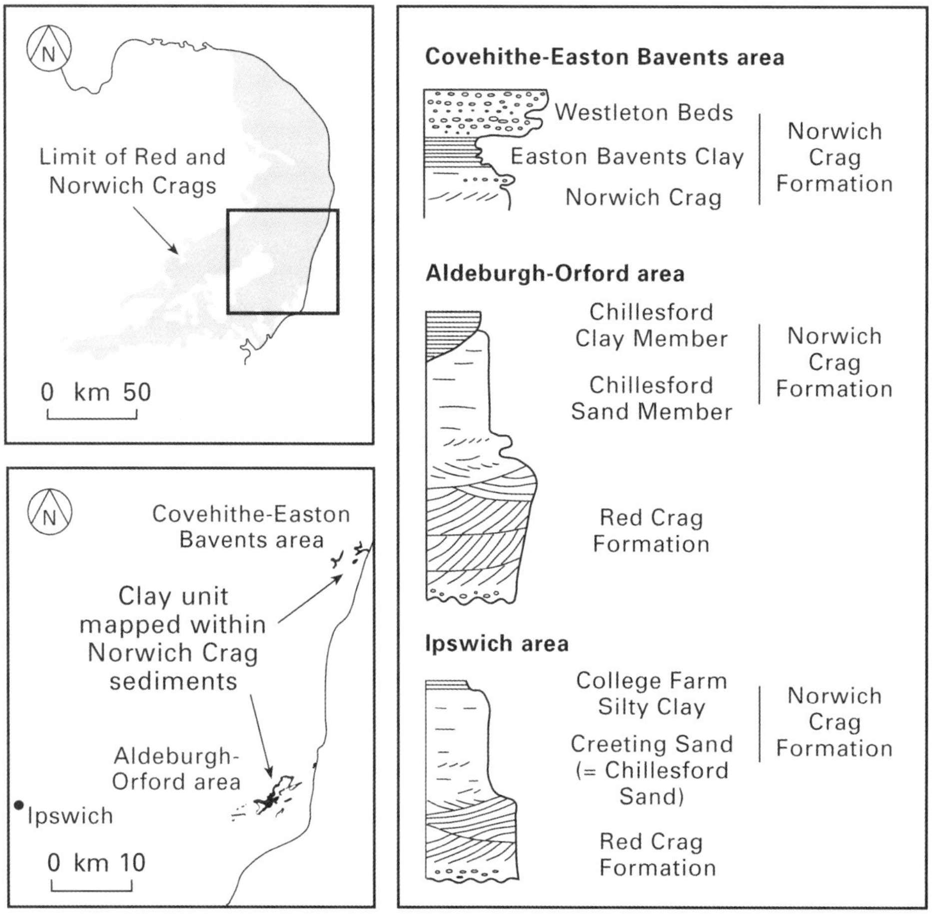

Fig. 17.8. Occurrence of clay members at the top of the Norwich Crag Formation (from Zalasiewicz *et al.* 1991).

probably derived fluvially from the north and then carried southwards by longshore drift (Hey 1976), eventually being deposited as eastward-prograding beachface gravels or tidal deltas. To the east (seawards) the gravels often occur in shallow, rhythmically-spaced channels, which probably originated as high-energy rip channels. The Westleton Member is usually unfossiliferous, although marine molluscs and bones of whale and other mammals have been recorded, and at Thorington (TM 420725), near Halesworth, a lens of laminated clay within the gravels contains pollen and macroscopic plant remains suggesting accumulation in a shallow lagoon under a cold climate. As originally suggested by Prestwich (1890), the Westleton Beds are probably equivalent in age to the earliest fluvial gravels of the proto-Thames, the Nettlebed and Pebble Gravel Formations (Whiteman & Rose 1992).

The Wroxham Crag Formation

This was originally recognized in parts of Norfolk by Rose *et al.* (2001, 2002). It overlies the Norwich Crag at sites such as Bramerton (TG 295060) near Norwich (Fig. 17.10), and consists of littoral shelly marine and estuarine sands and gravels ranging in height from OD near the Norfolk–Suffolk border to 25 m OD near Norwich. It probably accumulated during multiple marine transgressions resulting from local tectonic depression post-dating the Norwich Crag but predating and coeval with the transgressions of the Cromer Forest-bed Formation. The Wroxham Crag is distinguished from the Norwich Crag by its pebble types, which consist mainly of flint with up to 40% quartz + quartzite and small amounts of Spilsby Sandstone and Carboniferous, Greensand and *Rhaxella* cherts. It has also yielded a few igneous and metamorphic clasts from Wales. The far-travelled components probably resulted from high-energy river influxes from the north and west into the littoral environment. The fauna indicates that the climate during deposition was cooler than in Norwich Crag times. At Dobb's Plantation Pit near Wroxham (TG 273158) it includes the bivalve *Macoma balthica*, which is absent from the Red and Norwich Crags, and is thought to have first appeared about half way through the Late Tiglian C4c Substage of The Netherlands (Gibbard *et al.* 1991). This suggests deposition during the Baventian cold stage *c.* 1.86 Ma (Fig. 17.1).

A marginal crag facies, known as the Weybourne Crag or Sidestrand Member, occurs in northern parts of East Anglia, where it rests mainly on an eroded surface of Chalk, with a remanié deposit (the Cromer Stone Bed) at its base. It also contains *Macoma balthica*, suggesting a Baventian age. West (1980*b*) assigned the Sidestrand and Chillesford Members to a separate post-Baventian stage, the Pre-Pastonian *a*, but it is likely that the Pre-Pastonian *a* is merely a late part of the Baventian (Zalasiewicz *et al.* 1991).

The Cromer Forest-bed Formation

This consists of marine, brackish and freshwater sediments deposited in a wide embayment approximately coincident with the coastal region of northern East Anglia. They are usually exposed in cliff sections at various sites between Weybourne (TG 1143), Norfolk, and Kessingland (TM 536870), Suffolk, where they rest on Chalk and are overlain by glacial deposits of the Anglian Stage. Inland there is little information on their distribution. Although fairly thin (<8 m), they represent a complex sequence of environmental changes (West 1980*b*, 1996), which is broadly contemporaneous with the marine Wroxham Crag and later parts of the fluvial proto-Thames Kesgrave Formation in Essex.

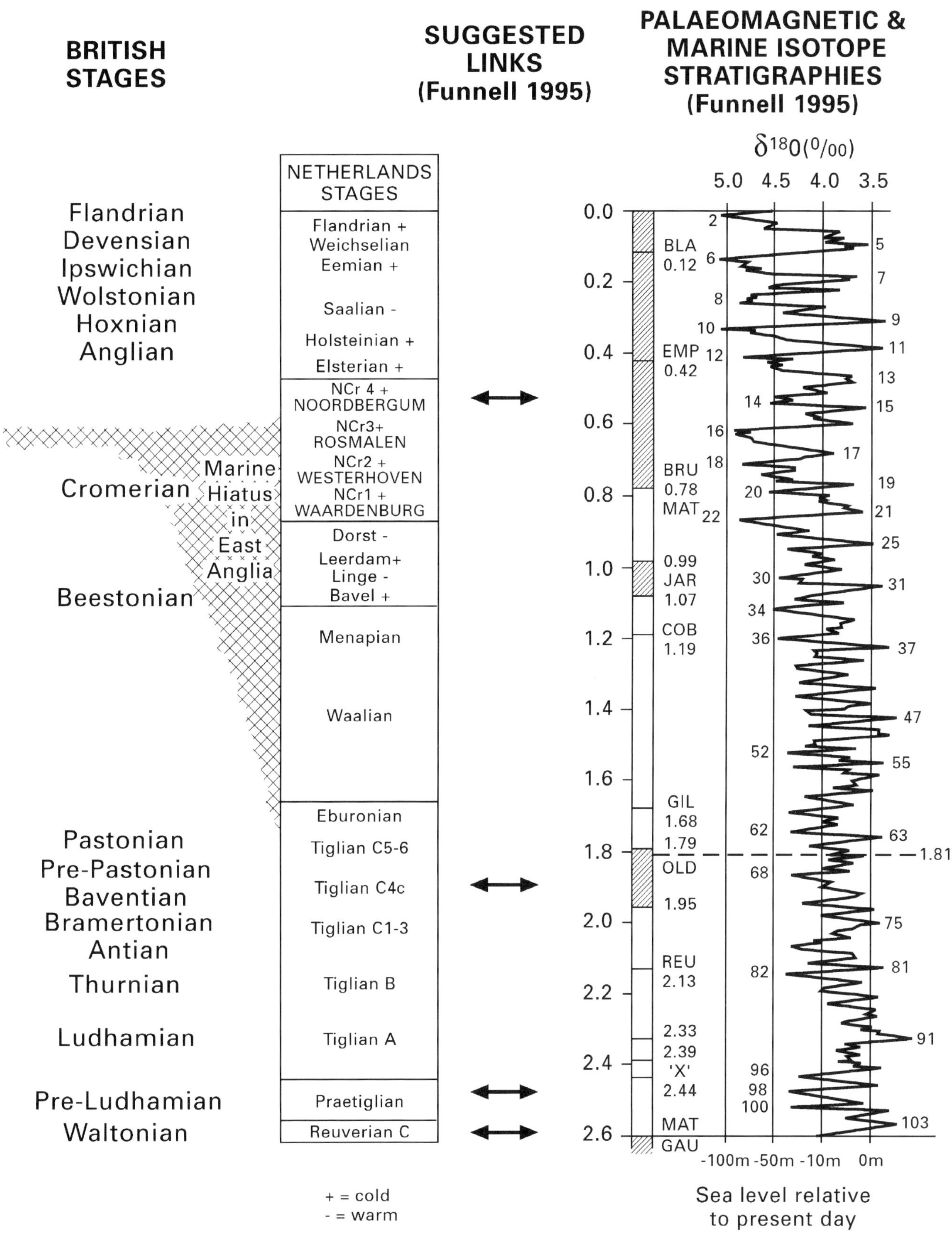

Fig. 17.9. Correlation of the British and Dutch Quaternary chronostratigraphies and their marine isotope stage (MIS) equivalents (from Funnell 1995).

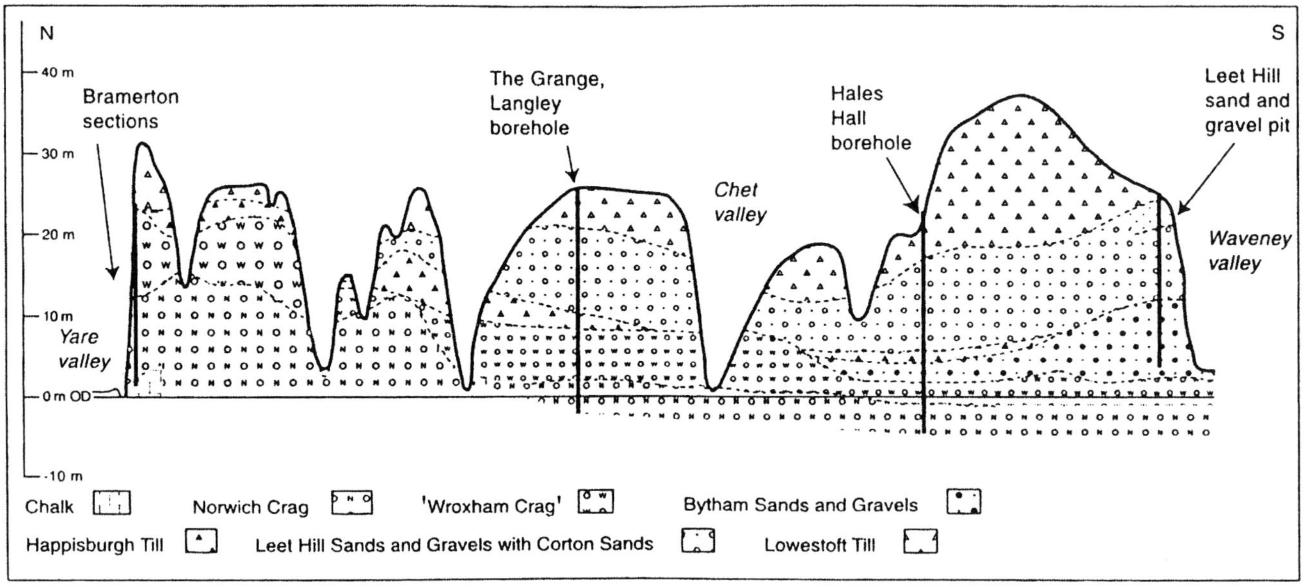

Fig. 17.10. Section from Bramerton, near Norwich, to the Norfolk–Suffolk border in the Waveney Valley to show the sequence from the Norwich and Wroxham Crag Formations to the Lowestoft Formation (from Rose *et al.* 2002).

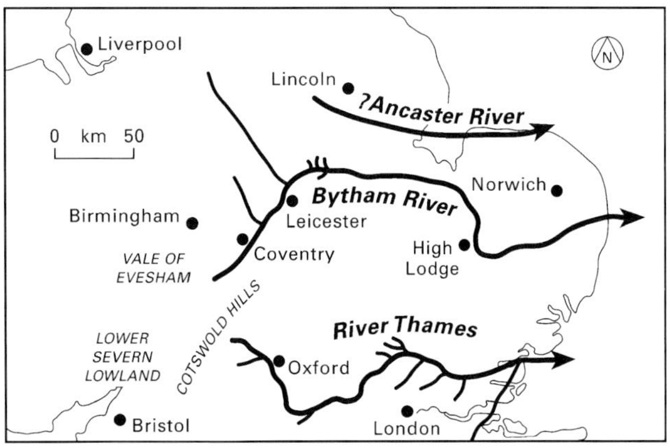

Fig. 17.11. Approximate courses of the pre-Anglian proto-Thames and Bytham Rivers (from Rose 1994).

The formation has been divided into the Sheringham, Paston, Runton, West Runton, Mundesley and Bacton members. Of these, the Paston, Runton and Mundesley members are marine. Their exact relationship to the coastal Wroxham Crag is unclear at present, but further work may show that the Wroxham Crag Formation is superfluous as a lithostratigraphic unit. The various members contain pollen, mollusc and mammal assemblages indicating deposition over at least three cold periods separated by two or more warm periods, so the formation may be equivalent to at least five oceanic MISs. However, it probably spans many more, as the large hiatus of *c.* 1 Ma mentioned previously seems to occur within the Runton Member (Gibbard *et al.* 1991).

The earliest (Pre-Pastonian *b–d*) deposits of the Cromer Forest-bed Formation, forming the Sheringham Member, consist of up to 2.5 m of freshwater sands and clays resulting from initiation of cold-climate fluvial sedimentation in the Crag Basin following an erosional break after the Pre-Pastonian *a* Stage (Baventian). The break is indicated by the irregular topography of the underlying surface, by the facies variability of the deposits above it and by a marked biostratigraphic change (West 1980*b*). Locally sedimentation was continuous from the Sheringham Member into the succeeding Paston Member, which consists of 3 m of tidally laminated silts and cross-bedded shelly sands indicating a marine transgression and development of a large tidal estuary in a temperate climate. Sea level at the climatic optimum of the Pastonian Stage reached *c.* 5 m OD, and a climax temperate mixed oak woodland developed on surrounding land areas. From botanical and faunal evidence the Pastonian has been correlated with the Tiglian C5 Substage of The Netherlands (Gibbard *et al.* 1991), dated to approximately 1.78–1.80 Ma.

Towards the end of the Pastonian there was a marine regression, and a climatic deterioration leading to deposition of the Runton Member in the cold Beestonian Stage is indicated by pollen of dominantly coniferous vegetation similar to that now found in parts of Finland. The succession within the Runton Member is complex and includes laminated freshwater muds, shelly marine sands and beach gravels. Ice wedge casts at several levels within the sequence indicate repeated development of permafrost, but the lack of fossil evidence makes it difficult to reconstruct a clear environmental history. Where best developed, between Beeston Regis (TG 174433) and West Runton (TG 189431), the sequence shows two episodes of fluvial erosion and deposition interrupted by a marine incursion. Marl deposited in the later fluvial episode contains a herb-dominated cold-climate pollen assemblage. The exact position of the 1 Ma hiatus within this sequence is uncertain (Gibbard *et al.* 1991), but it is likely that the Early–Middle Pleistocene boundary (*c.* 0.78 Ma) occurs within the hiatus. The remaining parts of the Cromer Forest-bed Formation (the West Runton, Mundesley and Bacton Members) are therefore Middle Pleistocene, and are discussed below under that subheading.

The Kesgrave–Bytham Group

The Kesgrave Formation

This formation, defined by Rose & Allen (1977), covers >3000 km^2 in southeastern parts of East Anglia and reaches a thickness of about 30 m beneath a cover of Anglian glacial sediments. It is the terrestrial (entirely fluvial) equivalent of the later Early Pleistocene parts of the Crag Group further north, notably the Wroxham Crag (Rose *et al.* 2001). It consists of flint-, quartz- and quartzite-rich gravels deposited by the proto-Thames before the river was diverted southwards into its present course by the Anglian glaciation (Fig. 17.11). There is

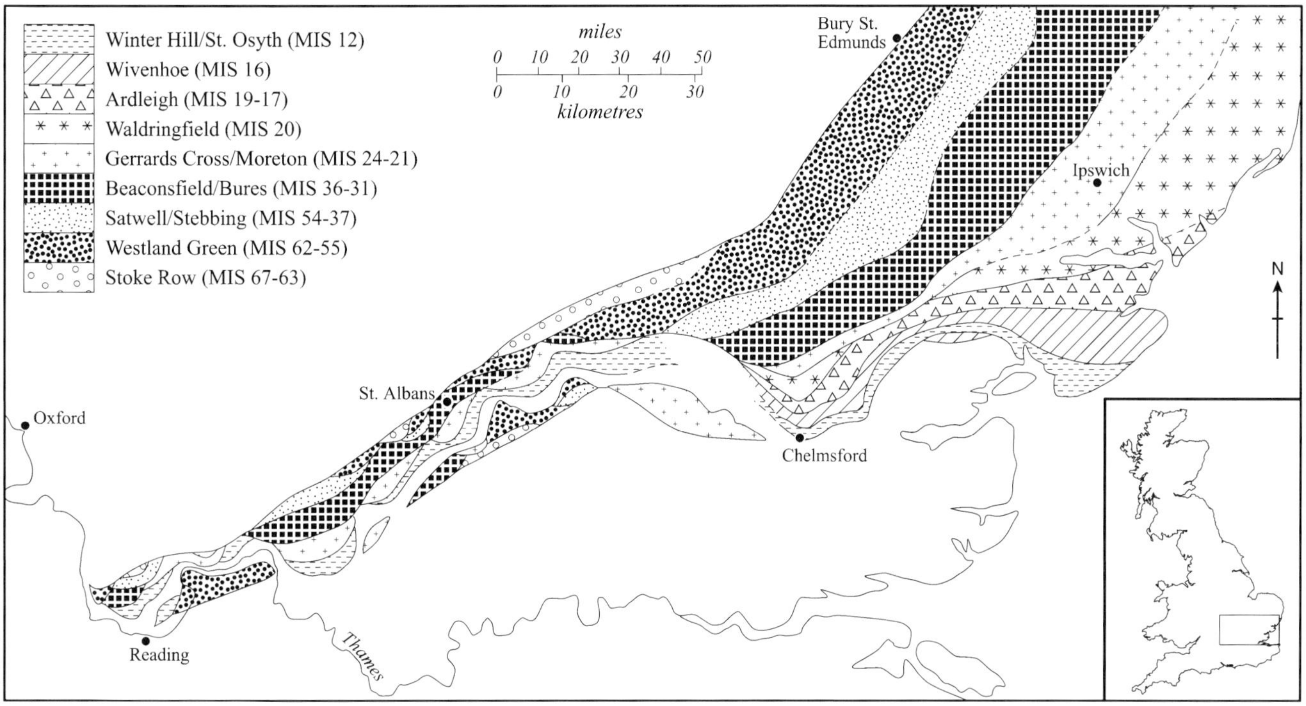

Fig. 17.12. Distribution of members of the Kesgrave Formation deposited in the middle and lower Thames valley before the Anglian glaciation (from Whiteman & Rose 1992) and tentative correlations with marine isotope stages (from Westaway *et al.* 2002).

evidence of deposition in periglacial conditions when the Thames adopted a braided mode. The formation can be divided into ten lithologically and altitudinally separate members (Whiteman & Rose 1992), of which nine are extensive in the middle and lower proto-Thames valley (Fig. 17.12). They represent separate terrace aggradations formed in successive cold stages of the Early and Middle Pleistocene as the river progressively incised its bed in response to the uplift of SE England and the subsidence of the southern North Sea.

Thames terrace aggradations predating the Kesgrave sediments are included within the Pebble Gravel Formation on the southern bank in south Hertfordshire and the Nettlebed Formation on the northern bank near Reading (Gibbard 1989). They are dominantly composed of flint pebbles, suggesting a catchment confined to the London Basin. In contrast, the abundant quartz and quartzite pebbles in the Kesgrave Formation indicate exploitation of Triassic, Carboniferous and Devonian rocks in the West Midlands, either by headwater extension and capture of Midlands rivers or by Midlands glaciation (Whiteman & Rose 1992). Many of the gravel sequences also contain clasts of acid volcanic rocks from North Wales (Hey & Brenchley 1977; Green *et al.* 1980), suggesting multiple early glaciations of the Welsh mountains. One such glaciation may have reached the Cotswolds during formation of the Ardleigh Terrace (MIS 18?) of east Essex (Fig. 17.1), depositing till in the Evenlode Valley north of Oxford (Hey 1986).

The first influx of quartz-rich gravel into the Wroxham Crag in northern East Anglia is in sediments of Pre-Pastonian *a* (Baventian) age (Hey 1976; Gibbard *et al.* 1991), suggesting that the Kesgrave Formation and parts of the Wroxham Formation accumulated during much of the later Early Pleistocene, including probably the 1 Ma hiatus in the sequence of northern East Anglia. The Kesgrave Formation also extends into the early Middle Pleistocene, because its younger members in Essex contain intercalated fossiliferous sediments of Cromerian age (Gibbard *et al.* 1996*a*).

The Bytham Formation

This consists of terrace sands and quartzose gravels similar to those of the Kesgrave Formation but deposited by the Bytham River (Fig. 17.11), which drained an extensive area of the North Midlands until its course was completely overridden by Anglian glacial deposits probably in MIS 12 (Rose 1994). The deposits have been recognized at many localities in the Midlands and northern parts of East Anglia, and a sequence of six terraces has been distinguished on height. At Knettishall (TL 951798), Suffolk (Lewis *et al.* 1999) and Leet Hill (TM 384926), SE Norfolk (Rose *et al.* 1999*a*), gravels of the third youngest (Timworth) terrace (Lee *et al.* 2004) are overlain by Anglian tills and glaciofluvial deposits. A proto-Soar River flowing northeastwards through the Wolston (SP 4175) and Leicester (SK 5804) areas joined the Bytham River west of Melton Mowbray (near SK 6519) as a south bank tributary. Rivers of approximately similar age flowing southeastwards across Jurassic strata in upper parts of the present Welland and Nene basins near Brigstock (SP 9585) and Milton Malsor (SP 7355) deposited the Milton Formation, which consists of sands with pebbles of local Jurassic rocks (Belshaw *et al.* 2005).

Sediments of the Kesgrave–Bytham Group extend eastwards and become finer beneath the North Sea off East Anglia, forming a substantial part of the seismically varied but extensive and largely non-marine delta-top sequence of the Yarmouth Roads Formation (Fig. 17.3) (Cameron *et al.* 1992). This reaches a total thickness of *c.* 160 m, and is strongly diachronous, younging northwards as the delta prograded in response to inputs of sediment from large NW European rivers, such as the Rhine, Meuse and Thames (Gibbard 1988). The highstand coastline lay just north of the current Thames estuary at 2.3 Ma but, after numerous fluctuations related mainly to eustatic changes in sea level, had reached almost 55°N by about 1.0 Ma (Fig. 17.13). Over this period the southern North Sea was an extensive wetland complex of delta-top sediments, which has been called Ur-Frisia (Jeffery & Long 1989) or the Eridanos Delta (Overeem *et al.* 2001). In the Southern Bight, the lower

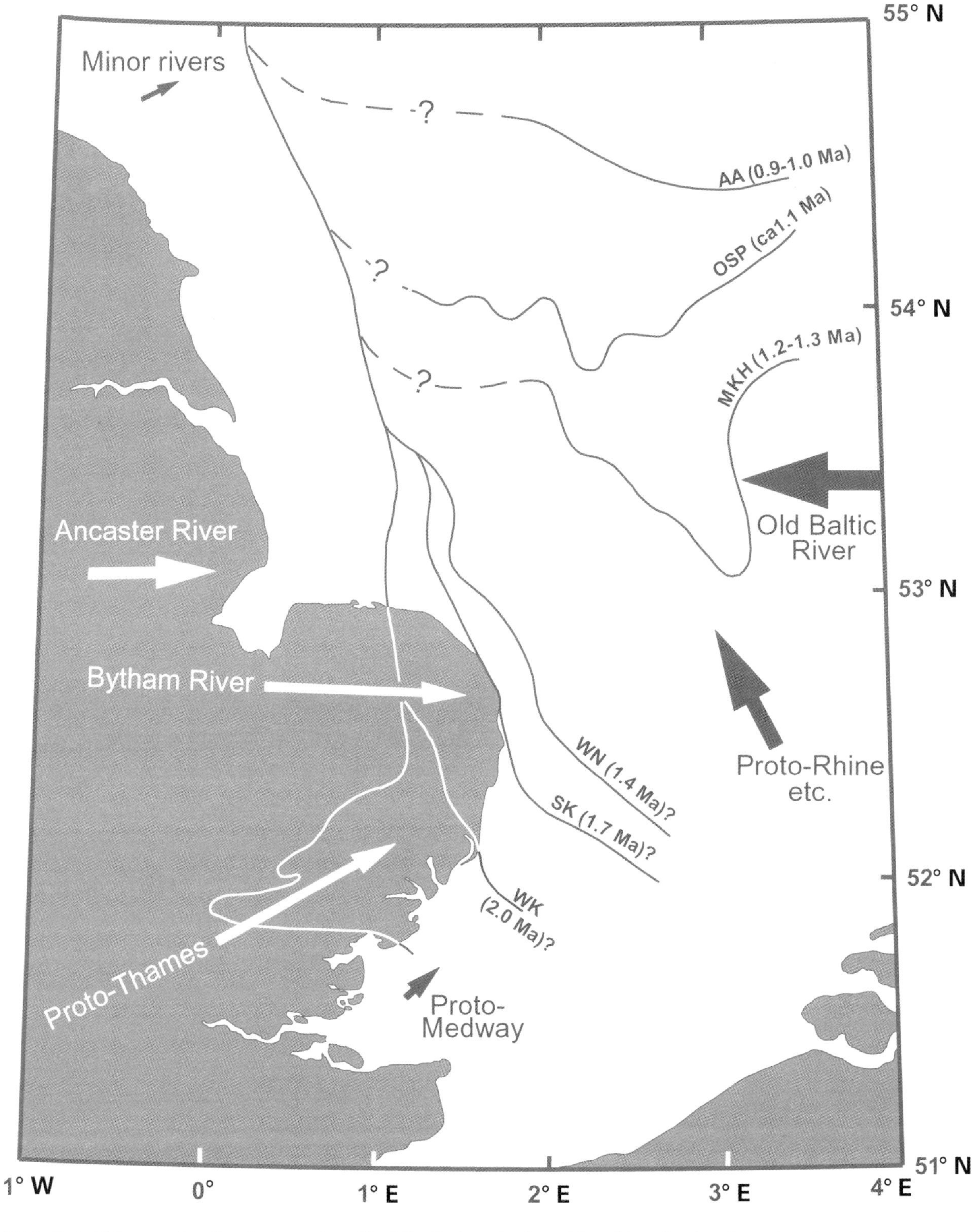

Fig. 17.13. Early Pleistocene shorelines in the southern North Sea Basin at the time of deposition of various formations (based on Cameron *et al.* 1992). Dating based partly on Cameron *et al.* (1984), and shorelines on land from Mathers & Zalasiewicz (1985) and Harland *et al.* (1991). WK, Westkapelle Ground Formation; SK, Smith's Knoll Formation; WN, Winterton Shoal Formation; MKH, Markham's Hole Formation; OSP, Outer Silver Pit Formation; AA, Aurora Formation.

part of the Yarmouth Roads Formation is coeval with the Markham's Hole Formation, but at 54°N it is equivalent to the Aurora Formation (Fig. 17.3). The sediments are intertidal, supratidal and terrestrial, suggesting that the rate of sediment supply exceeded any eustatic rise of sea level as well as the tectonic subsidence.

The St Erth Beds, west Cornwall

This small outlier (SW 5535) on the northern side of the Cornubian massif comprises 7 m of sands and clays resting on a surface eroded across the Variscan basement at *c.* 30 m OD. The clay was deposited in a marine, mid-neritic environment, and contains diverse mollusc and ostracod assemblages. Planktonic foraminifera, calcareous nannofossils and palynomorphs (Jenkins & Houghton 1987; Head 1999) suggest a Gelasian age.

The Middle Pleistocene

Worldwide the base of the Middle Pleistocene is drawn by the INQUA Stratigraphic Commission at the B/M magnetic reversal (*c.* 780 ka, within MIS 19). At about this time there was a change in the Earth's climatic cyclicity, with a decline in emphasis of the rather low amplitude 41 ka Milankovitch obliquity cycle and an increase in the strength of the 100 ka eccentricity cycle. The latter had a strong effect on climate and led to more extensive glaciations in the northern hemisphere. In Britain the B/M reversal probably lies between the Waldringfield and Ardleigh Members of the Kesgrave Formation of east Essex (Fig. 17.1), and within the Runton Member (Beestonian) of the Cromer Forest-bed Formation in Norfolk, but its precise position has not been recognized anywhere in the country. The top of the Middle Pleistocene is placed at the beginning of the Ipswichian temperate Stage (i.e. close to the MIS 6–5e boundary at *c.* 128 ka). In the deep oceanic sequence, the 652 000 years of the Middle Pleistocene therefore includes seven major cold stages (MISs 6, 8, 10, 12, 14, 16 and 18) and seven interglacials (MISs 7, 9, 11, 13, 15, 17 and the later part of 19). However, the sequence is not fully represented by any terrestrial deposits known in England and Wales.

In the central North Sea, close to the axis of maximum Quaternary subsidence, the B/M reversal has been identified in a core penetrating shallow marine sediments deposited in *c.* 15 m of water but now lying at −270 m OD (Stoker *et al.* 1985). This suggests that *c.* 255 m of subsidence has occurred in 0.78 Ma, a mean rate of approximately 0.3 mm year^{-1} (adapted from Cameron *et al.* 1992).

Pre-Anglian (Cromerian Complex) Stages

On the NE Norfolk and north Suffolk coast the West Runton, Mundesley and Bacton members of the Cromer Forest-bed-Formation exposed near West Runton (TG 189431), Sidestrand (TG 265400) and Ostend (TG 365325) were originally interpreted by West (1980*b*) as representing a single interglacial climatic sequence resulting from a marine transgression–regression cycle immediately preceding the Anglian cold stage. However, it is clear that they represent fragments of a Cromerian Complex spanning several stages, as discussed by Gibbard *et al.* (1991), Turner (1996), Preece (2001) and Westaway (2001). The West Runton Member is separated from the underlying Beestonian sediments by a non-sequence, and consists of 2 m of freshwater organic muds, including the West Runton Freshwater Bed and the Corton Rootlet Bed, deposited in isolated alluvial channels. The pollen indicates that the vegetation at this time was dominated by temperate woodland, with pine (*Pinus*), oak (*Quercus*), elm (*Ulmus*) and later also lime (*Tilia*), and the rich vertebrate and invertebrate fauna from the West Runton Freshwater Bed indicates a temperate climate similar to that of today. The Mundesley Member comprises 2.5 m of marine beach sands and laminated marine clays, including the *Yoldia (Leda) myalis* Bed at West Runton. The sea level at this time reached 7.5 m OD. Mixed oak woodland persisted on surrounding land areas but, in addition to the above genera, contained hornbeam (*Carpinus*) and later silver fir (*Abies*), which were slower to migrate into the area than other trees. The Bacton Member was deposited after a marine regression; it includes the Arctic Freshwater Bed, which consists of 1.5 m of clay containing pollen indicating a progressive climatic deterioration.

In The Netherlands, deposits of similar early Mid Pleistocene age between the B/M reversal and the Elsterian glacial (probably MIS 12) contain evidence for several climatic oscillations, including four warm climate events (Interglacials I–IV), separated by three cold stages (Glacials A–C). Interglacial I is older than the B/M reversal (within MIS 19) and therefore Lower Pleistocene (Zagwijn 1996), so Glacials A–C could be equivalent to MISs 18, 16 and 14, respectively, and Interglacials II–IV to MISs 17, 15 and 13, respectively. However, Westaway (2001) pointed out that MIS 18 was a very weak cold stage, so that A, B and C may correlate with MISs 20, 16 and 14, respectively, and Interglacial II may span MISs 19–17. The Middle Pleistocene part of the Cromer Forest-bed Formation is probably equivalent to the same MIS sequence, but exact correlation with the Dutch succession remains uncertain. The marine isotope curve for the early Mid Pleistocene (*c.* 780 ka–450 ka) is especially complex and, from molluscan and mammalian evidence, Preece (2001) has suggested that the British Cromerian Complex includes as many as five interglacials all within the Brunhes Chron.

Other deposits in England and Wales attributed to parts of the Cromerian Complex include: (a) gravels and silts filling a channel cut into marine deposits between Pakefield (TM 537895) and Kessingland (TM 536870) on the Suffolk coast; (b) a similar sequence at Sidestrand; (c) a similar sequence beneath Anglian glacial deposits at Corton (TM 543977), Suffolk; (d) sands and silts filling a channel of the proto-Thames close to its confluence with the extended River Medway at Little Oakley (TM 223294), Essex; (e) organic fluvial sediments filling a channel cut in bedrock near Sugworth Farm (SP 513007) south of Oxford; (f) glacially disturbed clays and silts at High Lodge (TL 739754) near Mildenhall, Suffolk; (g) the fluviatile Baginton Formation between Stratford-upon-Avon and Leicester, consisting of gravels and sands of the Thurmaston Member underlain at Waverley Wood (SP 365715) by fossiliferous silts of the Waverley Wood Member; (h) raised beach gravels, sands and silts (the Slindon Member) resting on a chalk platform at 42 m OD at Boxgrove (SU 920085) near Chichester, West Sussex; (i) fine-grained river sediments at Mathon (SO 729441), Herefordshire; (j) cave deposits with numerous mammal bones at Westbury-sub-Mendip, Somerset (ST 507504); (k) the Basal Breccia (a series of debris flows) in Kent's Cavern (SX 934642), south Devon; (l) the Kenn Formation at Kenn (ST 4269) and other sites in Avon and north Somerset; and (m) the Penfro Formation at Llandre (SN 093203) and other sites in south Pembrokeshire and west Carmarthenshire. Later members of the Kesgrave–Bytham Group, such as the Waldringfield, Ardleigh and Wivenhoe Members of the Kesgrave Formation in east Essex, must also date from this part of the early Mid Pleistocene.

As at West Runton, the precise age of all these deposits is uncertain. However, in Germany there was a change in interglacial water vole populations during the early Mid Pleistocene from an earlier group with *Mimomys savini* (possessing rooted teeth) to a later group containing *Arvicola terrestris cantiana* (with rootless, permanently growing teeth), and Stuart & Lister (2001) have used the occurrence of these and other mammal and mollusc species to divide the English sites of the Cromerian Complex into at least six groups. The earlier sites with *Mimomys* include West Runton, Sugworth, Pakefield/Kessingland, Corton and Little Oakley, and the later ones with *Arvicola* include Sidestrand, Ostend, Westbury-sub-Mendip and Boxgrove.

The Kenn Formation includes glaciofluvial gravels (the Bleadon and Nightingale members at Bleadon Hill (ST 350573) and Kenn, respectively), a till *c.* 2 m thick with large erratics (the Kennpier Member) overlying and interdigitating with the Nightingale Member, and shelly glaciomarine clays and sands containing cold, brackish water foraminifera (the Kenn Court Member) at Kenn Court (ST 414688) (Gilbertson & Hawkins 1978). At Yew Tree Farm (ST 423693) the Kenn Formation is overlain by deposits (the Yew Tree Formation), which contain pollen, molluscs and dinoflagellate cysts indicating a temperate interglacial climate. Amino-acid racemization studies of the mollusc shells suggest that the Yew Tree Formation is equivalent to part of the Cromerian Complex (Andrews *et al.* 1984), possibly MIS 15. This correlation has been questioned by Scourse (1997) but, if correct, it implies that the underlying Kenn Formation is MIS 16 or older, and represents the oldest known Pleistocene Irish Sea glaciation in Britain. The Penfro Formation of SW Wales could also have been deposited in MIS 16 by an Irish Sea glacier, as it consists of 5 m of gravels containing erratics from north Pembrokeshire (Bowen 1994).

Pre-Anglian archaeology

The deposits of the Cromerian Complex at Pakefield/Kessingland, High Lodge, Boxgrove, Waverley Wood, Westbury-sub-Mendip, Kent's Cavern and some exposures of the Bytham Formation in East Anglia all contain humanly worked flint tools or waste flakes. These sites provide the earliest evidence for a human presence (probably *Homo heidelbergensis*) in Britain. Boxgrove has also yielded human teeth and a tibia (Roberts & Parfitt 1999). The Pakefield/Kessingland site is unique in NW Europe in yielding human artefacts associated with *Mimomys*.

The most characteristic (Acheulian) artefacts found at these and other sites in England dating from later parts of the Mid Pleistocene are hand-axes. These tools vary considerably in shape (pointed, cordate, ovate) and size, and at some localities there is a strong predominance of one type. Most were made from fresh flint obtained directly from the Chalk, but a few are of chert, quartzite or silcrete. They are often associated with numerous waste flakes, and many have been found in river gravels, where they are usually abraded and could be derived from earlier deposits. Studies of edge and surface microwear patterns on flint hand-axes from lake sites, such as Hoxne (Suffolk), where they are completely unabraded, suggest they had numerous uses, including butchering, hide-cutting and scraping, bone and wood boring, wood chopping, splitting and whittling and plant gathering (Keeley & Newcomer 1977).

Cromerian palaeosols

In the proto-Thames valley, further evidence for the Cromerian is provided by the reddened and clay-enriched interglacial Valley Farm Soil (stratotype at the now-infilled Valley Farm Pit, TM 116433) developed on the upper surface of members of the Kesgrave Formation (Rose & Allen 1977). However, this is not a simple (monogenetic) soil formed in a single period of uniform climate. Even on the youngest pre-Anglian terrace surface, it contains evidence for early Anglian periglacial frost disturbance in the form of cryoturbations and ice-wedge casts filled with aeolian sand. As at Barham (TM 133515), Suffolk, these features are superimposed on the earlier interglacial soil, and constitute the Barham Arctic Structure Soil (Rose *et al.* 1985). In addition, on older terrace surfaces the soil contains micromorphological evidence for repeated cycles of temperate- and cold-climate evolution (Kemp 1985, 1987). In lower parts of the proto-Thames valley, the soil was buried by Anglian glacial deposits, but upstream of the Anglian ice margin near St Albans, it remains at the surface as part of a thick relict palaeosol developed on pre-Anglian terraces. Similar thick relict palaeosols developed in a remnant veneer of basal Palaeogene deposits (mainly Upnor and Reading Formations) on parts of the Chilterns dipslope above the highest proto-Thames terrace, on the North Downs and other parts of the Chalk outcrop in southern England, are equivalent to the material shown on BGS maps as Clay-with-flints (Catt 1986).

The Anglian Stage

The Quaternary climatic history indicated by the MIS sequence suggests that parts of the British Isles were sufficiently cold to have been glaciated several times in the Mid Pleistocene. However, in eastern and central England and central and southern parts of the North Sea Basin, the only firmly dated, widespread Mid Pleistocene glaciation was the Anglian, which is generally equated with MIS 12 (commencing *c.* 450 ka). This is the earliest definite glaciation of eastern England, although beneath the northern part of the North Sea there is evidence for a pre-Anglian (Early Pleistocene) glaciation at about 1 Ma (Sejrup *et al.* 1995).

Onshore, the Anglian glaciation resulted in deposition of the North Sea Drift and Lowestoft formations in Norfolk and of probable equivalents of the latter in Lincolnshire, the east Midlands and parts of East Anglia as far south as Essex and Hertfordshire, making the Anglian the most extensive glaciation in eastern England (Fig. 17.14). The Wolston Formation of Leicestershire, Warwickshire and adjacent areas has also been correlated with the Anglian (Maddy 1999), although it was previously the stratotype of a later cold stage, the Wolstonian (=Saalian of north Germany, Denmark, The Netherlands and Poland), now known to include the three cold MISs 10, 8 and 6. The Risbury Formation of the Risbury area (SO 5453–SO 5471) in the SW Midlands, the Seisdon Formation near Trysull (SJ 8494) in the Staffordshire lowlands, the Nurseries Formation at Quinton (SO 992847) near Birmingham and the Llanddewi Formation of the Hills Farm borehole (SS 452861) in the SW Gower Peninsula (South Wales) have also been attributed to the Anglian.

No Anglian glacial deposits are known at present in areas further north, even though they would have been deposited there because glaciers must have crossed northern England to reach East Anglia and the Midlands. It is therefore likely that any Anglian deposits were removed from northern areas by repeated glacial and periglacial erosion in subsequent cold stages. In mountainous regions, the classic major features of glacial erosion, such as the U-shaped troughs of the Llanberis Pass (SH 6455–SH 5562) and Nant Ffrancon (SH 6460–SH 6265) cutting through parts of Snowdonia, also probably resulted from repeated glacial scouring in several Mid and Late Pleistocene cold stages.

The Lowestoft Formation

The main component of the Lowestoft Formation is the Lowestoft Till Member, which is a moderately homogeneous grey chalky till extensive in East Anglia and parts of the East Midlands, including west Lincolnshire. At many localities from Kirmington (TA 102116) in north Lincolnshire to Hatfield (TL 212075) in Hertfordshire and Marks Tey (TL 912242) in Essex, it is overlain by organic lake (kettle-hole) deposits providing palynological evidence for a characteristic vegetational sequence attributed to the Hoxnian temperate Stage (probably = MIS 11). Minor lateral variations in the almost entirely British erratic suite, the preferred orientation of stones and the mineralogical and granulometric composition of the till matrix (Perrin *et al.* 1979) suggest that the Lowestoft Till was deposited by a British ice sheet that flowed down the eastern side of England and western part of the North Sea Basin, then created and passed through the Fenland gap in the Chalk escarpment, before fanning out to invade the East Midlands and East Anglia. The greatly subdued aspect of the Chalk escarpment in East Anglia and the way it is setback to the east are probably the result of erosion by this ice sheet. Indeed, Clayton (2000) calculated that the total volume of Lowestoft Till in eastern England (363 km^3) is approximately equal to that of the Chalk and Mesozoic mudrocks removed from the Wash embayment and Fenland area, assuming that relief typical of the Lincolnshire Wolds and Chilterns continued across East Anglia before the Anglian Stage.

On the Chalk outcrop of the Lincolnshire Wolds and in some parts of East Anglia, the Lowestoft Till is white and intensely chalky, a facies known as the Marly Drift in East Anglia and Calcethorpe Till in Lincolnshire. In other parts of East Anglia,

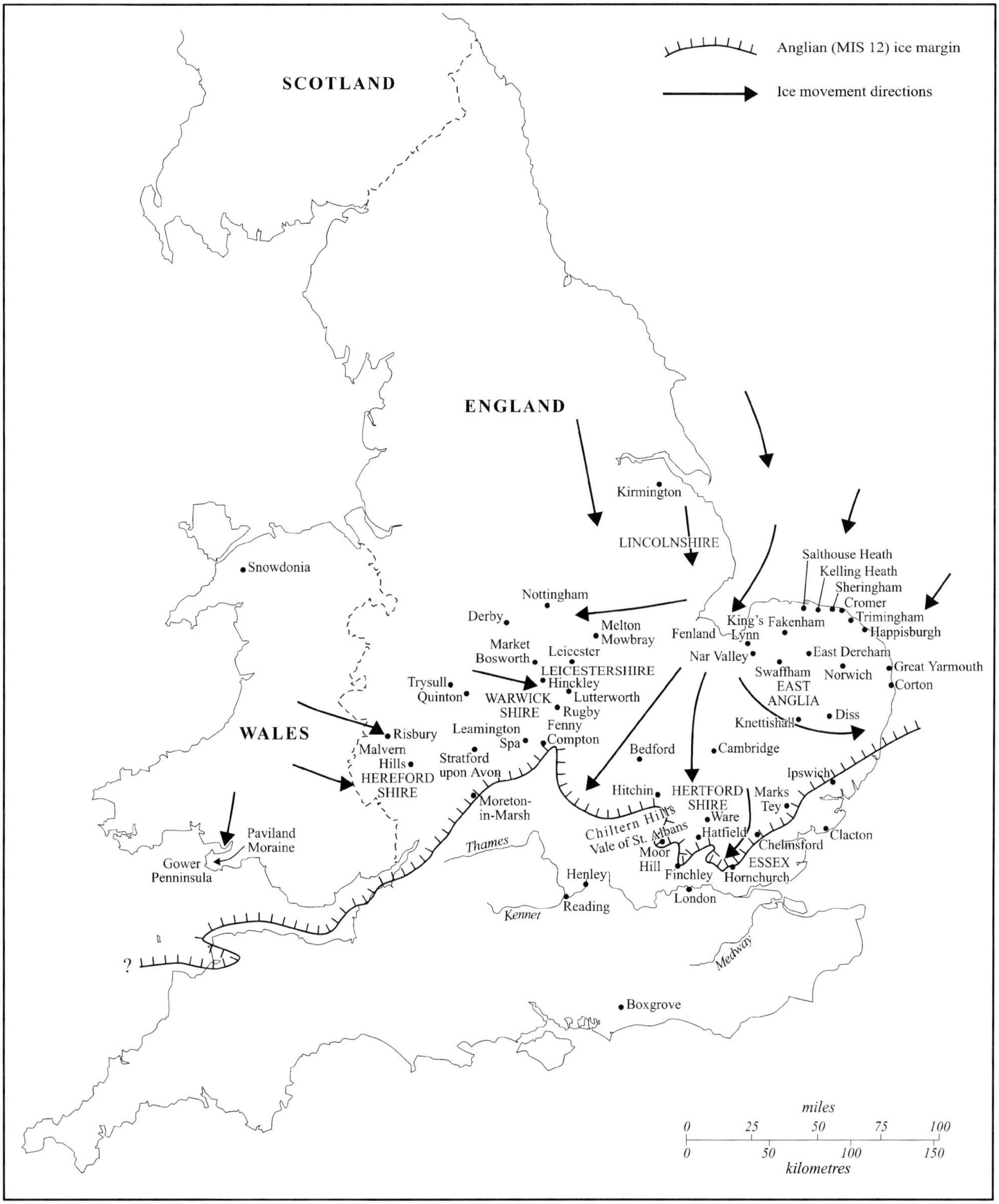

Fig. 17.14. Important Anglian sites, margin of the Anglian (MIS 12) ice sheet in England and Wales, based on Bowen *et al.* (1986), and approximate directions of ice movement indicated by erratics and directions of preferred stone orientation (macrofabric) in tills.

different facies (lodgement, meltout and flow till units) have been distinguished according to mode of deposition and incorporation of local materials. Minor fluctuations of the ice sheet margin in Hertfordshire led to deposition of four petrographically distinct lodgement tills, each separated by outwash gravels (Allen *et al.* 1991). One of these, the Stortford Till, was deposited by marginal ice lobes reaching Finchley (TQ 253907), North London, and Hornchurch (TQ 547874), Essex. These form the most southern ice margin in eastern England at any time in the Quaternary (Fig. 17.14).

Near Kings Lynn (TF 6220), Norfolk and in the Gipping Valley upstream of Ipswich (TM 1644), Suffolk, the Marly Drift overlies typical grey Lowestoft Till. Ehlers *et al.* (1987) consequently suggested that it represents a late readvance phase, which did not reach as far south as the original Lowestoft advance. During retreat it was responsible for depositing extensive outwash sands and gravels south of the Cromer Ridge (TG 051415–TG292385), for example in the East Dereham (TF 9913), Swaffham (TF 8209) and Fakenham (TF 9129) areas.

Most ice marginal features left by the Anglian glacier have been removed by subsequent erosion, and the glacial limit shown in Fig. 17.14 is based on the known distribution of the Lowestoft Till, although the ice could have extended several kilometres beyond this. Perhaps the most convincing marginal features are an end-moraine-like ridge in the Hendon–Finchley area of North London (TQ 2389–2692) and the marginal meltwater–fan complex of the Danbury–Tiptree Ridge (TL 7704–TL 8817), east of Chelmsford. The latter consists of >30 m of sands and gravels, which accumulated on the eastern side of the ice margin where its southward progress was prevented by a range of London Clay hills on the southern bank of the Anglian proto-Thames valley.

Beneath the thick British ice sheet of the Anglian Stage in East Anglia, meltwater streams under hydrostatic pressure cut a radiating pattern of deep 'tunnel valleys' in the Chalk bedrock (Woodland 1970). These buried valleys approximately coincide with the regional dip of the Chalk, and are filled mainly with sands and gravels deposited by meltwater. In Denmark, northern Germany and Poland, tunnel valleys of similar dimensions and infill characteristics are common beneath the Elsterian glacial deposits, which also probably date from MIS 12 (Ehlers *et al.* 1984). This suggests a widespread uniformity of glaciological behaviour in the continental ice sheet of the Elsterian/Anglian Stage.

The British ice which invaded the proto-Thames valley in Essex and Hertfordshire (the Vale of St Albans) deposited Lowestoft Till on the terrace aggradations of the Kesgrave Formation, blocked the river's exit to the North Sea and impounded lakes at Ware (TL 341149) and Moor Mill (TL 143025) south of St Albans, in which laminated clays (the Watton Road and Moor Mill Members, respectively) accumulated (Gibbard 1977). Other lakes were also created in valleys to the east, which had carried rivers such as the Mole and Wey northwards into the proto-Thames. These lakes progressively overspilled eastwards through temporary spillways until the water reached the first valley not blocked by ice (the Medway Valley). This led to the establishment of a new course for the Thames approximately equivalent to the present one through London. A well-preserved sequence of terrace deposits in east Essex indicates that for much of the Early and early Mid Pleistocene the Medway had a pre-Anglian course approximately parallel to the proto-Thames (Bridgland 2003). However, by late in the Cromerian Complex the two had joined south of Ipswich (TM 1644), and in the early Anglian the confluence was near Clacton (TM 1714). After the Anglian diversion the joint course across east Essex progressively migrated southwards, forming the present Thames estuary in the late Wolstonian, possibly in MIS 6.

During melting of the British Anglian ice sheet in East Anglia, meltwater was concentrated into large channels, which formed a broadly radiating pattern again parallel to the dip of the underlying Chalk bedrock. Much of the present drainage pattern was inherited from these channels, the alignment of many modern rivers (e.g. the Waveney, Wensum, Gipping and Stort) following the courses of earlier subglacial tunnel valleys. However, in north Norfolk extensive proglacial outwash sandur plains developed, particularly at Kelling Heath (TG 0942) and Salthouse Heath (TG 0742); these may have accumulated in front of stillstand positions of the retreating ice front. In the Nar Valley (TF 61–TF 71), NW Norfolk, a proglacial lake was formed where a tunnel valley deepens towards the west in the direction of ice retreat. The thinning ice initially floated on the lake, depositing subaquatic till, then later retreated further so that laminated clays accumulated; complete deglaciation eventually drained the lake and fluvial sedimentation was then established (Ventris 1996).

The North Sea Drift Formation

The other important Anglian deposit in East Anglia, the North Sea Drift Formation, is confined to NE Norfolk, where it is exposed in the coastal cliffs between Great Yarmouth (TG 5307) and Sheringham (TG 1643) (Lunkka 1994), and can be traced inland to Diss (TM 1179), Norwich (TG 2308) and Fakenham. Weathered equivalents at the surface in areas inland were previously termed the Norwich Brickearth (Rose *et al.* 1999*b*). Shell fragments, black flints and occasional Norwegian igneous rocks, such as larvikite, rhomb porphyry and Fredriksvärn porphyry (Harmer 1909; Slater 1929), suggest that it was deposited by a North Sea ice sheet originating partly in SW Scandinavia, which arrived in Norfolk slightly earlier than the British (Lowestoft Till) ice. However, dolerite from the Whin Sill and Cheviot igneous rocks are more common as erratics. So, despite the earlier records, Lee *et al.* (2002) suggested that the ice originated entirely in Scotland and northern England, because they found no Scandinavian erratics in parts of the formation exposed on the coast at Happisburgh, Trimingham and Corton.

The North Sea Drift Formation consists of three brown sandy subglacial tills, the Happisburgh, Walcott and Cromer members, a waterlain (glaciolacustrine) till (the Mundesley–Trimingham Member) and various intercalated waterlain gravels, sands, silts and clays. This complex sequence suggests at least three cycles of ice advance and retreat in NE Norfolk but, apart from the final stillstand that resulted in meltwater aggradation and the thrust moraine of the Cromer Ridge, no precise margins are known for the individual advances. The westernmost occurrence of till is at Knettishall, Suffolk (TL 951798), where ice invaded the Bytham Valley to deposit the Starston Till, which probably equates with the coastal Walcott Member (Lewis *et al.* 1999). Norwegian erratics have been found as far south as Cambridge (Rastall & Romanes 1909), Bedford, Hitchin and Ipswich (Ehlers & Gibbard 1991), but they could have been carried on the last part of their journey by the slightly later Lowestoft ice sheet.

Parts of the North Sea Drift Formation between Sheringham (TG 1643) and Cromer (TG 2242) are known as the Cortorted Drift, as they are strongly disturbed by irregular compaction and thrusts, folds and other glaciotectonic structures (Banham 1988), which resulted from gravitational loading by a heavy overburden when the sediments were still soft and saturated and from oscillations of the ice sheet margin. The Cromer Member also contains large rafts of chalk up to 100 m long, which were thrust up from fractured bedrock on the floor of the North Sea. Following final retreat of the North Sea ice sheet, NE Norfolk was invaded from the west by the British ice sheet, here depositing Marly Drift.

As the North Sea ice margin retreated eastwards and the British ice sheet advanced from the west, a lake was impounded between the two. Meltwater streams from the North Sea ice carried sand into the lake to deposit deltaic sands of the Corton Member (Ehlers & Gibbard 1991). At the Corton type site (TM 548965) in NE Suffolk, the sands contain thin silty bands with arctic plant remains. Interaction of the Scandinavian and British ice sheets along the western margin of the North Sea Basin could have led to the unusual pattern of movement of the British ice, invading the Fenland area and then spreading in a fanlike lobe into the Midlands and East Anglia.

The Anglian of eastern offshore areas

Beneath the southern North Sea the Anglian glacial sediments form part of the Swarte Bank Formation (Fig. 17.3), which unconformably overlies the Lower–Middle Pleistocene deltaic formations. The unconformity forms a fan-like array of infilled 'scaphiform' (boat-shaped) channels up to 12 km wide and 450 m deep with irregular long profiles (Wingfield 1990), which are similar in shape and dimensions to the tunnel valleys beneath the Anglian deposits in East Anglia and the Elsterian in northern Germany. The Swarte Bank Formation filling the channels is a deglacial sequence consisting of three members: a basal stiff dark grey diamicton correlated with the Lowestoft Till and North Sea Drift Formations of East Anglia (Scourse *et al.* 1999), overlain by strongly laminated grey glaciolacustrine muds and finally shallow marine clays attributed to the Hoxnian Stage (Gibbard *et al.* 1991). The continuity of the lowest member across the North Sea north of 53°N provides further evidence that the British ice sheet was confluent with the Fennoscandian ice sheet, which deposited the Elsterian sequence of northern Germany.

The origin of the Dover Strait and English Channel

Before the Anglian Stage the Chalk of the Weald–Artois anticline formed an unbroken barrier across what is now the Dover Strait, and the rivers of SE England, such as the Thames and Medway, were confluent with the Rhine, Meuse and other north European rivers flowing into the southern North Sea basin to create the deltaic Yarmouth Roads Formation. However, the southward advance of the confluent Fennoscandian and British ice into the North Sea Basin blocked the northward escape of river water, so that a large lake was impounded between the ice and Chalk uplands of the Weald–Artois anticline (Fig. 17.15). The Corton Sands and similar glacio-lacustrine and associated deltaic deposits in the correlative Peelo Formation of The Netherlands (Ter Wee 1983) were probably deposited in parts of the glacial lake. The current altitudinal distribution of these sediments suggest that the lake water rose to a maximum of *c.* 25 m above present sea level. Overflow from this lake eventually breached the anticline to initiate cutting of the Dover Strait (Gibbard 1995).

During global glacioeustatic sea-level lowstands the English Channel floor was exposed subaerially and drained by a 'Channel River' or 'Fleuve Manche', which had existed in cold periods since the Early Pleistocene. The floor of the central and eastern Channel is incised by an anastomosing network of deep

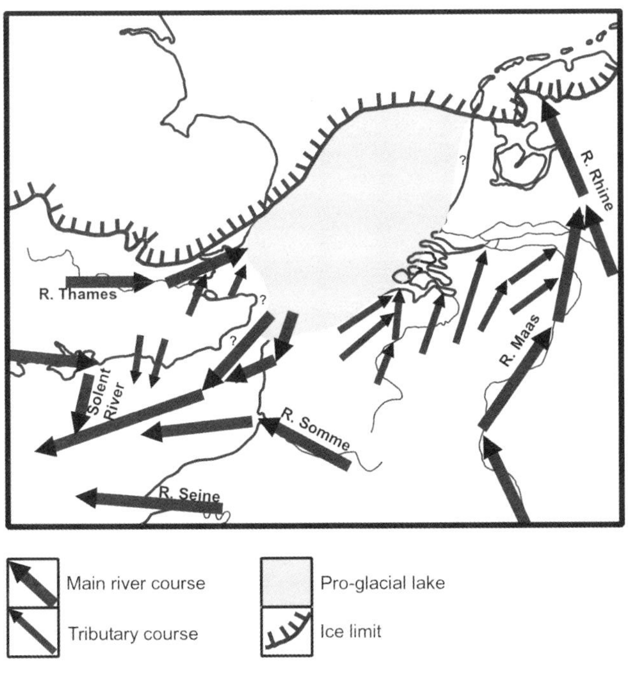

Fig. 17.15. Palaeogeography of the southern North Sea and English Channel during the Anglian glacial maximum (based on Gibbard 1995 and Cohen *et al.* 2006.)

narrow troughs (fosses), the largest of which is the Hurd Deep. The troughs are extensions of the channels of the Solent and other rivers draining southern England (Smith 1985; Bellamy 1995) and northern France (Antoine *et al.* 2003*a*; Lericolais *et al.* 2003). During the Anglian, the overflowing lake water from the southern North Sea augmented this channel network, so that the main late- and post-Anglian drainage of NW Europe became routed through the Channel River (Gibbard 1988). In the SW Approaches this combined flow cut distinctive canyon networks and built large deep-sea fans on the continental slope and adjacent abyssal plain (Bourillet *et al.* 2003).

Other onshore Anglian deposits

The Wolston Formation of the East Midlands, originally described as the Wolston Series by Shotton (1953), consists of two tills, the Thrussington and Oadby members, with sands and gravels below, between and above (Rice & Douglas 1991). The Thrussington Till is reddish brown and mainly Trias-derived, whereas the Oadby Till contains chalk, flint and Jurassic limestone clasts in a grey matrix derived mainly from Liassic clays. These components suggest deposition by ice from western and eastern sources, respectively. The Thrussington Till was deposited as far east as Melton Mowbray (SK 7519), Rugby (SP 5075) and Fenny Compton (SP 4152), and its meltwater formed a large ice-dammed lake between Leicester (SK 5804) and Market Bosworth (SK 4003), in which up to 25 m of laminated clays and silts (the Bosworth Member) accumulated. The lake was then overrun by outwash streams from the same ice sheet, depositing a similar thickness of sand and gravel (the Wigston Member). Another lake (Lake Harrison) was impounded between the ice margin and the Cotswold scarp near Fenny Compton (SP 4352) and the Moreton Gap (SP 23); lacustrine clays and shoreline benches suggest that Lake Harrison reached a height of 125 m OD (Bishop 1957).

The Oadby Till was then deposited over much of the East Midlands as far north as Nottingham (SK 5740) and Derby (SK 3536) and as far west as Stratford-upon-Avon (SP 2055). It also extends southwards to Moreton-in-Marsh (SP 2032), where flint presumed to originate from its associated outwash was incorporated into the Wolvercote Terrace of the Upper Thames. As this has been correlated with terraces of the Middle Thames as young as the Lynch Hill (?MIS 8) Terrace (Gibbard 1985), Sumbler (2001) has argued that the Oadby Till at Moreton-in-Marsh is younger than Anglian, possibly MIS 10. Between Rugby and Lutterworth (SP 5484), the Oadby Till passes laterally into flint-containing gravels (the Shawell Member), which were deposited as proximal outwash of the eastern ice sheet. Flinty outwash was also incorporated into the Dunsmore Member, which is the oldest deposit of the Warwickshire Avon in the area between Hinckley (SP 4294), Leamington Spa (SP 3165) and Rugby. The lithological similarity of the Oadby Till to the Lowestoft Till of East Anglia suggests that it was deposited by a westward extension of the same Anglian ice sheet (Perrin *et al.* 1979), which crossed the East Midlands after deposition of the Thrussington Till by western ice.

Outwash from an Anglian ice sheet in NE Herefordshire deposited the Risbury Formation, which consists of up to 30 m of glaciofluvial and ice-contact glaciodeltaic gravels (Richards 1999). The gravels contain clasts of Devonian and Lower Palaeozoic rocks derived from the west and southwest, suggesting that the glacier originated in central Wales. However, Welsh ice seems to have extended no further east than the Malvern Hills (SO 7534–SO 7647). In the Mathon Valley immediately west of the Malverns, the formation locally overlies fluvial silts containing a pre-Anglian interglacial fauna and flora (Coope *et al.* 2002), and at the head of the Cradley Brook (SO 7439) it is overlain by silts containing pollen and molluscs attributed to the late Anglian and early Hoxnian (Barclay *et al.* 1992).

The Seisdon Formation consists of 27 m of proximal outwash gravels filling the Seisdon–Stourbridge Channel at Trysull, Staffordshire (Worsley 1991*a*). The gravels are overlain unconformably by a thin cover of Late Devensian till (the Stockport Formation), and between the two Morgan (1973) recorded silts (the Trysull Member) containing pollen indicating a Hoxnian age. At Quinton, the Nurseries Formation, a brown till containing many local erratics mainly from the Coal Measures, is also thought to be Anglian in age, because it is overlain by pollen-bearing silts and clays (the Quinton Formation) attributed to the Hoxnian Stage.

In Wales, the only possible Anglian deposit is the Llanddewi Formation of SW Gower, which consists mainly of glacial outwash gravels containing erratics of Coal Measure sandstone, coal, flint, basalt and rhyolite. However, this dating is very tentative because the deposit is not related to any interglacial sediments. In boreholes through the Paviland Moraine, which is possibly the Anglian end-moraine in South Wales, the gravels are 21.5 m thick and overlie 1.5 m of red clay (till?) resting on Coal Measure shales (Bowen 1999*b*). The ice sheet depositing these probably came from the north, suggesting that the Welsh mountains had an extensive ice cover in the Anglian, but elsewhere the depositional evidence for it has probably been removed by later glacial and periglacial erosion.

The Anglian of western offshore areas

Although Anglian ice sheets must have extended onto the floors of the Irish Sea and Bristol Channel, glacial deposits here are limited in extent, probably because of subsequent erosion. However, Bowen *et al.* (1986) suggested that ice in the Bristol Channel reached as far as the northern coast of SW England (Fig. 17.14). Beneath central parts of the Irish Sea the glacial deposits occur mainly in three buried channels, the Western, St George's and Celtic Deep Troughs. These are flanked by the low-relief shelves of the Irish, Eastern and Welsh platforms, which are covered by thin, discontinuous Quaternary deposits. The troughs contain a total Quaternary infill up to 300 m thick (Jackson *et al.* 1995), of which the earliest unit is the Bardsley Loom Formation (Fig. 17.16). In the Celtic Deep Trough this consists of variable cold stage facies, including fluvial deposits, of possible pre-Anglian age (Tappin *et al.* 1994). The overlying units are mainly glacial in origin and, as in upper parts of the North Sea sequence, are subdivided by reference to unconformities resulting from erosion by subglacial meltwater streams (Boulton & Hindmarsh 1987). Three generations of incisions provide a crude event stratigraphy (Jackson *et al.* 1995). The Caernarfon Bay Formation consists of two glacial members separated by the oldest incision event, all of which are attributed to the Anglian (Wingfield 1989).

Anglian fluvial deposits

During the Anglian Stage, the rivers of SE England had multithread braided courses, and deposited coarse, often flinty gravels under periglacial conditions. These form the Dartford Heath and Asheldham members in the Lower Thames Valley, the Winter Hill and Black Park members in the Middle Thames (Fig. 17.1), the Wallingford Member in the Upper Thames, the St George's Hill Member in the Mole and Wey Valleys and the Silchester Member in the Kennett Valley. The Westmill Member in the Vale of St Albans is the downstream preglacial proto-Thames equivalent of the Winter Hill Member. Some of these terrace gravels, such as the Black Park Member in the abandoned Caversham Channel of the Thames between Reading and Henley (SU 67–SU 78), have yielded abundant Acheulian artefacts.

Anglian periglacial deposits

At Boxgrove (SU 920085), West Sussex, the Slindon Member (Cromerian Complex) is overlain unconformably by up to 12 m of coarse gravel (the Eartham Member), which is composed of angular fragments of frost-shattered chalk and flint with intercalated layers of yellowish silt. This typical 'coombe deposit' probably accumulated during the Anglian and perhaps later cold stages by gelifluction from the almost bare chalk slopes of the adjacent South Downs. The silty layers resemble periglacial loess, which was probably deposited by wind on the chalk slopes and then incorporated into the gelifluction gravels. Loess of probable Anglian age also occurs locally in Cambridgeshire, Essex and Suffolk, and is preserved in dolines penetrating the Clay-with-flints of the Chiltern Hills (Antoine *et al.* 2003*b*).

The Hoxnian Stage

The stratotype for this stage in Britain is at Oakley Park Pit, Hoxne (TM 176769), Suffolk, where the Hoxne Formation, comprising up to 15 m of lacustrine clay and peat and fluvial sands and gravels, overlies the Lowestoft Formation, and is overlain by gelifluction gravels (West 1956). Flint artefacts

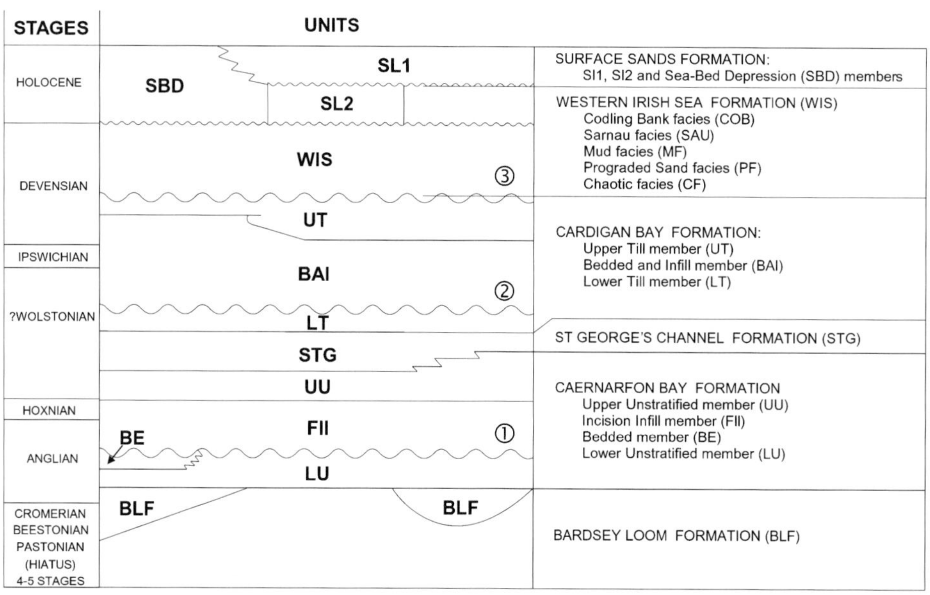

Fig. 17.16. Quaternary stratigraphy of the Irish Sea Basin showing relationships of the main stratigraphical units (based on Jackson *et al.* 1995).

occur at various horizons on or above the top of the lacustrine clays (Singer *et al.* 1993). A thicker (35 m) and more complete interglacial lacustrine sequence occurs at the parastratotype of Mark's Tey (TL 912242), Essex (Turner 1970), where laminations resulting partly from lush seasonal growth of the diatom *Stephanodiscus astraea* var. *minutula* suggest accumulation over *c.* 20 000 years. At both sites U-series dates indicate deposition in MIS 11 at *c.* 400 ka BP (Rowe *et al.* 1999; Grün & Schwarcz 2000).

At Barnfield Pit (TQ 595745), Swanscombe, Kent, in the Boyn Hill Terrace of the Lower Thames Valley, a sequence of fluvial and other deposits forming the Swanscombe Member of the Lower Thames Formation, has also been attributed to MIS 11 (Gibbard 1994). The upper parts of the sequence have yielded numerous pointed handaxes and an occipital and two parietal bones of a female human skull (Ovey 1964). The lowest deposits contain abundant mammal remains (Fig. 17.17), including a large form of fallow deer (*Dama dama clactoniana*), *Arvicola terrestris cantiana* and the earliest known cave bear in western Europe (*Ursus spelaeus*). The human artefacts at this level consist of flakes and cores but no handaxes, an assemblage resembling that found in Hoxnian fluvial gravels at Clacton-on-Sea (TM 1413–TM 1714), Essex, and attributed to an industry (the Clactonian) distinct from the Acheulian (Wymer 1985). As Acheulian handaxes are most common in western and southern Europe, whereas flake-core assemblages resembling the Clactonian occur mainly in eastern Europe and Asia, it seems that in the Hoxnian SE England was at the frontier between two different Early Palaeolithic human provinces. The youngest deposits at Swanscombe contain characteristic twisted ovate handaxes, which seem to be restricted in Britain to late MIS 11 or early MIS 10 (White 1998).

Lacustrine deposits, often infilling kettle-holes on the surface of the Lowestoft Formation and with pollen sequences similar to those at Hoxne and Mark's Tey, are also known from Stansted Airport (TL 538238) and Rivenhall End (TL 839165) in Essex, and from Gosmore, Hitchin (TL 119277), Hatfield (TL 212075), Fishers Green, Stevenage (TL 223260) and Bell Lane, St Albans (TL 1802) in Hertfordshire. At many of these sites near the southern limit of the Lowestoft Formation, the lower part of the Hoxnian sequence contains white calcareous silts composed almost entirely of charophyte oogonia. At Oughtonhead Lane, Hitchin (TL 172299), a tufa overlying Anglian glaciofluvial gravels has also been attributed to the Hoxnian. Interglacial lacustrine sites correlated with the Hoxnian on pollen evidence in other parts of East Anglia include Athelington (TM 222710), Elveden (TL 809804), Barnham (TL 875787) and St Cross South Elmham (TM 303840) in Suffolk and Barford (TG 111069), Trimingham (TG 268397) and Tottenhill (TF 635115) in Norfolk. Outside East Anglia Hoxnian lacustrine deposits have been recognized at Kirmington, north Lincolnshire, Quinton and Nechells (SP 087881), Birmingham, and in the proto-Thames valley near Denham (TQ 018897).

As the sea reinvaded the North Sea Basin after the Anglian glaciation, it deposited first the uppermost member of the Swarte Bank Formation, then up to 20 m of laminated clays, the Sand Hole Formation, in a bowl-shaped depression around the Silver Pit 25–50 km east of Spurn Head, and finally the more widespread Egmond Ground Formation (Fig. 17.3). The last has a conspicuous basal seismic reflector truncating the upper part of the Swarte Bank Formation, and increases in thickness eastwards from about 8 to 20 m, though it is locally thicker in some of the incompletely filled Anglian tunnel valleys.

In the middle part of the Swanscombe sequence and also in the Hoxnian deposits at Clacton and Tillingham (TL 982041), Essex, the mollusc assemblages include *Corbicula fluminalis*, *Viviparus diluvianus*, *Belgrandia marginata*, *Valvata naticina*, *Unio crassus* and *Theodoxus serratiliniformis*, which are characteristic of the Mid Pleistocene Rhenish (Rhine) fauna. This suggests a connection between the Thames and mainland European rivers during one or more parts of the Hoxnian (Roe 1999), probably through the complex interplay of sea-level change, isostatic rebound after melting of the Elsterian/Anglian glacier and fluvio-estuarine progradation through the interglacial.

In southern England the approximate position of the Hoxnian coastline is indicated by shallow marine, estuarine and channel-fill deposits of the Thames in east Essex (Roe 1999) and by beach gravels (the Aldingbourne Member) resting on a marine platform cut in Chalk at 17–25 m OD on the West Sussex coastal plain (Bates *et al.* 1997). Elsewhere little is known about its position. At Aldingbourne Park (SU 931071) the Aldingbourne Member contains Acheulian flint artefacts and boulders of far-travelled rocks, such as granite, basalt and quartzite. The erratics were perhaps deposited from icebergs, suggesting that the beach accumulated partly in periglacial conditions early in MIS 10 before the eustatic lowering of sea level associated with worldwide growth of extensive glaciers had occurred.

The Wolstonian Complex

The Wolstonian was originally defined as a cold stage between the Hoxnian and Ipswichian Interglacials (Mitchell *et al.* 1973), but dating of these two interglacials and comparison with more complete sequences in northern Europe has since shown that it covers a long period of complex climatic change lasting about 230 000 years. In the oceanic sequence this covers three cold stages (MISs 10, 8 and 6) and two interglacials (MISs 9 and 7). Various deposits in England and Wales have been attributed to all five MISs, but none of the stages within the complex has yet been formally named. Continued use of the name Wolstonian is deemed inappropriate by some, because it was originally used for glacial deposits in the Midlands (the Wolston Formation), which are now thought to be partly or entirely Anglian (Perrin *et al.* 1979; Rose 1987) and therefore older than any part of the Wolstonian Complex. However, in the absence of formally defined British equivalents of MISs 10–6, the name is retained here for the period of complex climatic change, part of which it was originally thought to represent (Gibbard & Turner 1990).

Wolstonian fluvial sequences

In Britain the fullest sequence of the Wolstonian Complex is probably represented by the succession of terrace gravels in the Lower and Middle Thames: the Boyn Hill and Orsett Heath members (probably MIS 10), the Lynch Hill and Corbets Tey members (probably MIS 8) and the Taplow and Mucking members (probably MIS 6), with interglacial deposits probably equivalent to MIS 9 at Purfleet (TQ 568789) and Little Thurrock (TQ 625783), and to MIS 7 at Aveley (TQ 552807) and West Thurrock (TQ 597779). The interglacial deposits contain two different mammalian assemblages (Fig. 17.17), which have been used to distinguish them from interglacials at Swanscombe (MIS 11) and Trafalgar Square (TQ 300804) in central London (MIS 5e) (Schreve 2001*b*). However, Gibbard (1994) considered the Purfleet, Little Thurrock, Aveley and West Thurrock deposits as all Ipswichian.

Less well dated fluvial sequences of this period were also deposited by the River Severn (Maddy *et al.* 1995), the Warwickshire Avon (Maddy *et al.* 1991) and a Mid Pleistocene proto-Solent river in the New Forest area (Allen & Gibbard 1994). In the Severn Valley, Anglian glaciofluvial sediments of the Wolston Formation form the highest recognizable (Woolridge) terrace in the area south of Tewskesbury (SO 8932). Six younger terraces at progressively decreasing heights above the present river level are composed of gravels, which together form the Severn Valley Formation, and of these the three highest and oldest (Spring Hill, Bushley Green and Kidderminster Station members) are attributed to the Wolstonian Complex. The youngest (the Kidderminster Station

Species	Common name	MIS 11	MIS 9	MIS 7	MIS 5e
Palaeoloxodon antiquus	straight-tusked elephant	x	x	x	x
Mammuthus primigenius	woolly mammoth	-	-	x	-
Mammuthus trogontherii	small-toothed mammoth	-	-	x	-
Stephanorhinus kirchbergensis	Merck's rhinoceros	x	x	x	-
Stephanorhinus hemitoechus	narrow-nosed rhinoceros	x	-	x	x
Coelodonta antiquitatus	woolly rhinoceros	-	-	x	-
Hippopotamus amphibius	hippopotamus	-	-	-	x
Panthera leo	lion	x	-	x	x
Felis sylvestris	wild cat	x	x	-	x
Felis chaus	jungle cat	-	-	x	-
Panthera pardus	leopard	-	-	x	-
Bos primigenius	aurochs	x	-	x	x
Ovibos moschatus	musk ox	-	-	x	-
Bison priscus	bison	x	-	x	x
Equus hydruntinus	stenonid horse	x	-	-	-
Equus ferus	horse	x	x	x	-
Premegaceros verticornis	verticornis deer	-	-	x	-
Cervus elaphus	red deer	x	x	x	x
Dama dama clactoniana	large fallow deer	x	-	-	-
Dama dama	fallow deer	x	x	-	x
Megaloceros giganteus	giant deer	x	-	x	x
Capreolus capreolus	roe deer	x	x	-	-
Ursus spelaeus	cave bear	x	-	-	-
Ursus arctos	brown bear	-	x	x	x
Canis lupus	wolf	x	-	x	x
Vulpes vulpes	red fox	-	-	x	x
Crocuta crocuta	spotted hyaena	-	x	x	x
Cyrnaonyx antiqua	clawless otter	-	-	x	-
Macaca sylvanus	macaque monkey	x	x	-	-
Sus scrofa	wild boar	x	-	-	x
Lepus timidus	hare	-	-	-	x
Oryctolagus cuniculus	rabbit	x	-	-	-
Meles meles	badger	-	-	-	x
Citellus citellus	ground squirrel	-	-	x	-
Castor fiber	European beaver	x	x	x	x
Trogontherium cuvieri	giant beaver	x	-	-	-
Talpa minor	extinct small mole	x	-	-	-
Talpa cf. europaea	common mole	-	-	x	-
Arvicola terrestris cantiana	water vole	x	x	x	x
Microtus agrestis	field vole	x	x	x	x
Microtus (Terricola) subterraneus	European pine vole	x	-	-	-
Microtus oeconomus	northern vole	x	-	x	-
Microtus arvalis	common vole	x	x	x	-
Clethrionomys glareolus	bank vole	x	x	x	x
Apodemus sylvaticus	wood mouse	x	x	x	x
Neomys browni	water shrew	-	x	-	-
Sorex araneus	Eurasian common shrew	-	x	x	x
Sorex minutus	pygmy shrew	-	x	x	-
Crocidura russula	white-toothed shrew	-	-	x	-
Dicrostonyx torquatus	collared lemming	-	-	x	-
Lemmus lemmus	Norway lemming	x	-	x	-
Homo spp.	humans (mainly artefacts)	x	x	x	-

Fig. 17.17. Mammal assemblages of interglacials correlated with MIS 11-5e in SE England (x, present; -, as yet unrecorded) (based on Stuart 1976; Schreve 2001*a*, *b*, 2004 *a–c*, Currant & Jacobi 2001).

Member) contains Uriconian volcanic clasts thought to have been glacially eroded from the Permian Clent Breccia in the late Wolstonian (MIS 6). At Marsworth (SP 933143), Buckinghamshire, the fluvial deposits of two interglacials, attributed to MISs 7 and 5e (Murton *et al.* 2001), are separated by a chalky gelifluction deposit of MIS 6.

Some of the later fluvial deposits of Wolstonian age in SE England contain Lower Palaeolithic artefacts produced by the Levallois technique. This involved preshaping an oval flint core by flaking a nodule to remove all of its entire cortex, preparing a striking platform at one end and then detaching a large flake by a carefully directed blow on the platform. Evidence for this technique first appeared in the Thames and Medway sequences in MIS 9 and is quite widespread in deposits thought to date from MIS 7 and later.

In the Stour Valley, Suffolk, fossiliferous interglacial deposits at Brundon (TL 863417) and Stutton (TM 150330) probably date from MIS 7 (Wymer 1985). They both contain Levallois-type flint artefacts, remains of mammoth (*Mammuthus primigenius*), a small horse (*Equus* cf. *ferus*) and the fluvial Rhenish bivalve *Corbicula fluminalis*, which is now restricted to the Nile Estuary. All of these are common in Thames deposits of MIS 7, and U-series dates on mammoth bones from Brundon (Szabo & Collins 1975) support correlation with MIS 7. Loamy deposits with remains of horse and mammoth exposed at the southern end of Stoke railway tunnel (TM 162434), Ipswich, also probably date from MIS 7.

Wolstonian raised beaches

On the West Sussex coastal plain between Brighton (TQ 3303) and Ports Down (SU 6406), two marine beach deposits resting on platforms cut in the Chalk at levels lower than the Aldingbourne Member were probably deposited in MISs 9 and 7 (Bates *et al.* 1997). The MIS 9 beach (the Cams Bridge Member) occurs at 15–18 m OD and that dating from MIS 7 (the Norton Member) is at 5.5–9.5 m OD. The Norton Member and an associated chalk cliff are exposed at Black Rock (TQ 334033), Brighton. All three raised beaches are covered by chalky gelifluction ('coombe') deposits formed during succeeding cold stages (MISs 10, 8 and 6 or even younger).

Further west along the Channel coast, raised beaches tentatively correlated with MIS 7 occur at Portland Bill (SY 675686) (14 m OD), Hope's Nose, Torquay (SX 949637) (9–12 m OD), Godrevy (SW 581430–SW 582423) (4–10 m OD), Porth Nanven (SW 356308) (8.5 m OD), Fistral Bay (SW 799625–SW 801623) (6 m OD) and Gerrans Bay (SW 902382) (4 m OD). On the Bristol Channel coasts further possible MIS 7 raised beaches occur at Saunton Sands (SS 437379), north Devon, and Minchin Hole (SS 562868), South Wales. The beach deposits at Minchin Hole (the Minchin Hole Member) occur at about 10 m OD, and are dated to MIS 7 by the amino acid and TL methods (Bowen 1999*b*). They are overlain by a cave earth containing a large form of the northern vole *Microtus oeconomus* indicative of MIS 6. Those at Saunton Sands are 3–4 m thick and rest on a wave-cut platform at 5 m OD. They contain large iceberg-transported erratics of gneiss, and are overlain by approximately 10 m of aeolian sand locally cemented into calcareous 'sandrock'. At all these sites the raised beach deposits are overlain by gelifluction deposits ('head') composed of local frost-shattered rock material derived from adjacent coastal slopes.

The Pontnewydd Cave Site

A cave sequence at Pontnewydd (SH 014710), beside a tributary of the River Clwyd, NE Wales, has yielded evidence of human occupation in the form of teeth and jaw fragments and numerous artefacts fabricated mainly from glacial erratics. Mammalian remains, U-series and TL dates suggest a depositional history spanning the later Mid Pleistocene and part of the Late Pleistocene, but the human settlement probably dates from 250–225 ka (Stephen Green 1984).

Possible Wolstonian glacial deposits

Although the Drenthe or late Saalian (probably MIS 6) glaciation was very extensive in northern Germany and The Netherlands, few glacial deposits in England and Wales can be definitely related to either this cold stage or to MIS 8 or 10. Shotton (1953) originally attributed the Wolston Formation of the East Midlands to a cold stage immediately before the Ipswichian, and Mitchell *et al.* (1973) made the Wolston Pit in Warwickshire the stratotype of their Wolstonian Stage, but this formation is now widely regarded as Anglian (MIS 12) in age. Saalian till (the Borkumriff Formation) occurs beneath the eastern (Dutch) sector of the North Sea (Joon *et al.* 1990), but cannot be traced into the British sector. It extends less than 100 km NW of the present Dutch coastline, where an ice margin is indicated by lateral passage into thin proglacial and glaciomarine sediments of the Cleaver Bank Formation (Fig. 17.3).

In many parts of England poorly-dated glacial deposits of restricted lateral extent have also been attributed to a glaciation between the Ipswichian and Hoxnian Interglacial Stages, but only because they seem to be either older than Ipswichian or younger than Hoxnian (never both), or are lithologically distinct from extensive Anglian and Devensian glacial deposits nearby. Some also lie outside the accepted Devensian glacial limit. They include: (a) the Bridlington Member or Basement Till of East Yorkshire and lithologically similar deposits at Welton-le-Wold (TF 282884), Lincolnshire (the Welton Member) and Warren House Gill (NZ 447423) on the coast of Co. Durham (the Warren House Formation); (b) the Oakwood Formation at Chelford (SJ 824717), Cheshire; (c) the Ridgacre Formation overlying the Hoxnian Quinton Formation at Quinton, Birmingham; (d) the reddish brown till-like Fremington Member of Barnstaple Bay (SS 4931-SS 5231), north Devon; (e) the Pilkenzane Formation at Pilkenzane (SE 112007), Lancashire; (f) the Thornsgill Formation exposed in the valley of the Thornsgill Beck (NY 381243), Cumbria; (g) the Balby Formation at Balby (SE 562004) and Brayton Barff (SE 585305) in the Vale of York; and (h) the Bakewell Formation at Shining Bank Quarry (SK 230652), north Derbyshire. Maddy *et al.* (1995) suggested that the Ridgacre Formation is equivalent to the Kidderminster Station Member of the Severn Valley Formation, which was probably deposited in MIS 6 by a glacier that occupied the Stour Valley as far south as its present confluence with the River Severn near Kidderminster.

Perhaps the most likely candidate for a Wolstonian till in eastern England is the Bridlington Member. Catt & Penny (1966) recorded temporary shore exposures showing that it underlies the Ipswichian (MIS 5e) raised beach at Sewerby (TA 199686) near Bridlington, and at Speeton (TA 147758) in Filey Bay it overlies the estuarine Raincliff Formation (Speeton Shell Bed), which contains estuarine shells attrubuted to MIS 7 by the amino-acid method (Wilson 1991). However, there are doubts over its age because: (a) the till and the Raincliff Formation at Speeton could have been ice-rafted to their present position in the Late Devensian; and (b) at Dimlington (TA 376237) in SE Holderness the Bridlington Member contains marine shells attributed to MIS 2 by the amino-acid method (Eyles *et al.* 1994), although these could have been intruded into the till when it was disturbed by Late Devensian ice. Offshore, part of the extensive Bolders Bank Formation (Fig. 17.3) may be equivalent to the Bridlington Member (Scourse *et al.* 1999), although Cameron *et al.* (1992) attributed it to the Late Devensian.

Several glacial outwash gravels containing erratics have also been attributed to the Wolstonian. For example, at Tottenhill (TF 635115), NW Norfolk, approximately 10 m of coarse deltaic gravels (the Tottenhill Member) overlie the Hoxnian Nar Valley Member and were probably deposited in an ice marginal lake (Gibbard *et al.* 1992). The Welton (= Bridlington)

Member occurs slightly further north in Lincolnshire, at Welton-le-Wold near Louth, but is unknown further south, so the Tottenhill Member could be outwash from the ice sheet that deposited the Bridlington and Welton Members.

Other offshore Wolstonian glacial deposits
Beneath the Irish Sea both the St George's Channel Formation and parts of the overlying Cardigan Bay Formation have been attributed to the Wolstonian (Fig. 17.16), the latter consisting of three glacial members including a strongly channeled unconformity (Jackson *et al.* 1995). The St George's Channel Formation, which overlies the Anglian Caernarfon Bay Formation, was originally attributed to an interglacial marine event, but re-evaluation of the microfauna indicated overwhelmingly cold water conditions during deposition, so an early (preglacial) Wolstonian age has been suggested (Jackson *et al.* 1995).

Wolstonian periglacial deposits
Small patches of Wolstonian (probably MIS 6) loess occur beneath Ipswichian fluvial deposits at Northfleet (TQ 612740), NW Kent and on Mid Pleistocene gravel terraces of the New Forest area (Antoine *et al.* 2003*b*). The pre-Ipswichian age of the latter is indicated by development in its upper layers of a typical interglacial soil profile, which is buried locally beneath a younger (Devensian) loess showing Holocene soil features. Wolstonian periglacial activity also resulted in some coastal head (gelifluction) deposits, such as the High Tor Member, which overlies the Minchin Hole Member (MIS 7) at Minchin Hole (SS 562868) and is overlain unconformably by the Ipswichian Hunts Bay Member.

Beneath the North Sea the Tea Kettle Hole Formation (Fig. 17.3), which is approximately equivalent in age to the proglacial and glaciomarine Cleaver Bank Formation, consists of fine periglacial aeolian sands. It is up to 6 m thick in the Dutch sector, but thin or absent in the British sector because of later erosion.

An alternative history of Mid Pleistocene glaciations

Recent BGS fieldwork in the East Midlands and East Anglia (Hamblin *et al.* 2000) suggests that the Mid Pleistocene history of one main (Anglian) glaciation and a possible later (Wolstonian) glaciation of some eastern areas, as summarized above, may be over-simplified. As presented by Clark *et al.* (2004), the new work suggests four Mid Pleistocene glaciations in these areas, in MISs 16, 12, 10 and 6. They propose that the MIS 16 (*c.* 630 ka) glaciation resulted in deposition of the lowest (Happisburgh) member of the North Sea Drift Formation, because: (a) this till underlies the Anglian Leet Hill Sands and Gravels in the Hales Hall borehole (Fig. 17.10); and (b) at Leet Hill (TM 384926) clasts of it occur in the 3rd periglacial (Timworth) terrace of the Bytham River (Lee *et al.* 2004), and this terrace is thought to have formed two cold stages before burial of the Bytham valley in the Anglian. Clark *et al.* (2004) also suggested that the MIS 12 glaciation deposited the whole of the Lowestoft Till Member in East Anglia (but only its lower part in areas further west), the second (Walcott) member of the North Sea Drift Formation and the Thrussington Till of the West Midlands. The MIS 10 glaciation then deposited the third (Bacton Green) member of the North Sea Drift Formation, the upper part of the Lowestoft Till west of the Chalk escarpment, the Oadby Till of the Midlands and the Marly Drift in NE Norfolk (Pawley *et al.* 2004). Finally, they suggest that the MIS 6 glaciation deposited the Bridlington, Welton and Tottenhill members in east Yorkshire, east Lincolnshire and west Norfolk, and also the Britons Lane Sands and Gravels forming the Cromer Ridge on the north Norfolk coast, all of which have a partly Scandinavian origin.

The detailed evidence on which this revised model of Mid Pleistocene glaciations is based has been published only in part. Many aspects of it have been challenged by Banham *et al.* (2001), especially on biostratigraphic grounds, and Preece (2001) has pointed out that the Happisburgh Member cannot be as old as MIS 16 because it overlies the Sidestrand *Unio*-Bed, a late component of the Cromer Forest-bed Formation, which contains freshwater molluscs and mammal remains indicating an age late in the Cromerian Complex. Consequently, we feel it is premature at present to rewrite the conventional Mid Pleistocene model outlined previously.

The Upper Pleistocene

The lower limit of the Upper Pleistocene is placed either at the MIS 6–5e boundary (*c.* 128 ka BP) in the oceanic record or at the base of the European Eemian (=Ipswichian) Stage, which has been dated to about 126 ka BP in sediments off the coast of Portugal (Gibbard 2003; Shackleton *et al.* 2003). The upper limit, at the beginning of the Holocene, is placed at 10 000 ^{14}C years BP (*c.* 11 530 cal.), although this post-dates the MIS 2/1 boundary by about 1000 years. This period of *c.* 115 000 years includes two of the terrestrial Pleistocene stages originally defined by Mitchell *et al.* (1973), the Ipswichian (warm) and Devensian (cold), and four of the MISs (5, 4, 3 and 2).

The Ipswichian Stage

The British stratotype of the Ipswichian is at Bobbitshole (TM 148414) near Ipswich, Suffolk, where the Bobbitshole Formation consists of 3.5 m of lacustrine silts and clays. Amino acid racemization studies of freshwater molluscs from these deposits confirm correlation with MIS 5e (Bowen *et al.* 1989), but the pollen analyses reported by West (1957) suggest that the formation represents only the early part of a temperate episode. Coleoptera suggest that mean summer temperatures were approximately 3 °C higher than in Britain at the present day (Coope 1974), and the youngest deposits, probably representing the climatic optimum of the interglacial, contain water plants, such as *Najas minor* and *Salvinia natans*, which today occur only as far north as The Netherlands.

Interglacial fluvial or lacustrine sequences with pollen assemblages broadly similar to those at Bobbitshole are widespread in England and Wales, but at no site is there a complete sequence through the interglacial. The full vegetational sequence for the Ipswichian has therefore been compiled from fragments at numerous sites in East Anglia (West 1980*a*). However, this has led to difficulties in distinguishing Ipswichian (MIS 5e) deposits from those of the previous interglacial (MIS 7), which may have had a broadly similar vegetational history. Some deposits originally attributed to the Ipswichian may therefore date from the earlier interglacial.

The Ipswichian fauna
However, these two interglacials have been distinguished more successfully on the basis of mammalian faunas, especially in the Thames valley (Schreve 2001*b*), where deposits with similar pollen sequences occur below gravels forming terraces at two distinct levels. Below the lower terrace, known as the Kempton Park, Upper Floodplain or Isleworth Terrace (MIS 3), interglacial deposits, such as the Trafalgar Square Member previously exposed at Rwanda/Uganda House (TQ 301804), central London, have been dated to MIS 5e (Bowen *et al.* 1989; Gibbard 1994) and contain mammals, such as hippopotamus (*Hippopotamus amphibius*), wild boar (*Sus scrofa*), badger (*Meles meles*) and hare (*Lepus timidus*), which are rare or absent in the preceding later Mid Pleistocene interglacials (Fig. 17.17). The higher or Mucking (=Taplow) Terrace was probably aggraded mainly in MIS 6, but at Aveley (TQ 551808) and West Thurrock (TQ 598781), Essex, has been interpreted as incorporating MIS 7 sediments containing mammoth (*Mammuthus*

primigenius and *Mammuthus trogontherii*), Merck's rhinoceros (*Stephanorhinus kirchbergensis*), common mole (*Talpa* cf. *europaea*), verticornis deer (*Premagaceros verticornis*), horse (*Equus ferus*), white-toothed shrew (*Crocidura russula*) and northern vole (*Microtus oeconomus*). All of these are absent from the Trafalgar Square Member and suggest somewhat cooler summers than during the Ipswichian.

In the Ipswichian, hippopotamus was surprisingly widespread in England and Wales, even occurring in the Pennine Hills at Victoria Cave (SD 837651), north Yorkshire. It also reached England sometime in the Cromerian Complex, but it has never been found at sites dated to MIS 7, 9, or 11, and in later parts of the Pleistocene is therefore regarded as diagnostic of Ipswichian deposits, especially those without well-preserved pollen sequences. For example, it occurs in the raised beach (Sewerby Member) exposed beneath Devensian till at Sewerby (TA 199686), east Yorkshire (Catt & Penny 1966), where aeolian sand between the beach deposit and the till has been dated by TL methods to 121 ka (Bateman & Catt 1996). Schreve (2001*b*, p. 1700) listed other sites where U-series dates, often on cave stalagmites associated with hippopotamus, indicate that the Ipswichian is equivalent in age to MIS 5e.

In Britain interglacial deposits thought to date from MIS 7 and 5e also have distinctly different coleopteran faunas. That of MIS 5e includes many typical mediterranean species, such as the scarab beetles *Oniticellus fulvus, Caccobius schreberi, Onthophagus furcatus, O. massai* and *Aphodius carpetanus*, which are absent from deposits of MIS 7 and are also unknown in Britain today. Other warm indicators found in English Ipswichian deposits, but absent from those of MIS 7, include the European pond tortoise (*Emys orbicularis*), Cory's shearwater (*Calonectris diomeda*) and the lesser water parsnip (*Berula erecta*).

The more precise allocation of Ipswichian deposits resulting from studies of mammal remains has shown that, despite the unusually warm conditions, there is no definite evidence for *Homo* in Britain during MIS 5e. This is surprising because the often abundant artefacts from deposits of earlier warm and even cold stages suggest an almost continuous human presence (originally *Homo heidelbergensis* and later *H. neanderthalensis*) in Britain between retreat of the Anglian glaciers in MIS 12 until the Mousterian industry some time in MIS 6 (Ashton & Lewis 2002). Also, Mid Palaeolithic humans (probably *H. neanderthalensis*) later returned to Britain in the Mid Devensian (MIS 3) between approximately 50 and 38 ka (Jacobi *et al.* 1998). It has been suggested that late in the Wolstonian (MIS 6) *Homo* was driven southwards across the English Channel by increasingly cold conditions, as indicated by the possible glacier in Lincolnshire, deposition of loess in NW Kent and the New Forest and coastal gelifluction in southern England and South Wales. Because of the eustatic fall of sea level, the English Channel was then dry, but an abrupt transgression at the beginning of the interglacial rapidly separated Britain from the European continent and prevented *Homo* from returning when the climate improved. However, this explanation is unconvincing in that the sea-level rise was not so rapid as to prevent hippo, rhino, elephant and other thermophilous land animals from reaching Britain in the Ipswichian.

Ipswichian raised beaches

The sea-level rise in the Ipswichian led to formation of cliffs and raised beaches often quite close to the present coastline of England and Wales. On the east coast the raised beach (Sewerby Member) can be followed from Sewerby on the southern side of Flamborough Head southwards around the western margins of Holderness (Catt & Penny 1966) and the Lincolnshire Marsh to the Fenland (West *et al.* 1995) and coastal areas of east Essex and the Thames Estuary (Jardine 1979). It then reappears on the West Sussex coastal plain as the Pagham Member (Bates *et al.* 1997), in Torbay (SX 95–SX 96), South Devon, as the Thatcher Stone Member (Mottershead *et al.* 1987) and at various sites along the South Wales coast as the Hunts Bay Member. Amino-acid ratios of shells from sites in the last two areas confirm correlation with the Ipswichian (Bowen *et al.* 1985). At Bacon Hole (SS 561868), Gower, the fauna of the Hunts Bay Member lacks *Hippopotamus amphibius* and contains *Mammuthus primigenius*, roe deer (*Capreolus capreolus*) and *Microtus oeconomus* as well as *Stephanorhinus hemitoechus* and *Palaeoloxodon antiquus*, so it probably dates from a cool phase of MIS 5 after MIS 5e (Currant & Jacobi 2001). Because of irregular crustal movements since the Ipswichian, fragments of the marine platform beneath the various Ipswichian beach deposits occur at heights ranging from below present sea level to higher than +10 m OD.

Offshore Ipswichian

Beneath the northern North Sea, the Eem and Brown Bank Formations (Fig. 17.3) are tentatively dated as Ipswichian (Cameron *et al.* 1992). The Eem Formation unconformably overlies the Egmond Ground Formation and consists of up to 20 m of shelly shallow-water sands and intertidal muds. The Brown Bank Formation consists of approximately 5 m of laminated brackish water silty clays, which were probably deposited as sea-level fell late in the Ipswichian and Early Devensian.

No Ipswichian deposits are known from other parts of the British offshore sector. This is probably because over much of the shelf bottom currents driven by tidal dynamics or by storm and wave resuspension processes entrained and remobilized bottom sediments to create either erosion surfaces or local thin condensed sequences of coarse shelly gravel (Curry 1989).

Ipswichian palaeosols

Onshore, buried palaeosols formed during the Ipswichian are preserved locally, for example at Northfleet in NW Kent (Kemp 1991) and on the surface of the Thornsgill Formation (till) in Cumbria (Boardman 1985). Ipswichian pedogenesis may also have contributed to development of the relict (unburied) paleo-argillic soils mapped by the National Soil Resources Institute (previously Soil Survey of England and Wales) on plateau surfaces in many parts of southern England (Avery 1985). However, the parent materials of most of the paleo-argillic soils are Anglian or older, so they could have begun development in MIS 11, 9 or 7. Some relict soils on mountain tops in Wales and northern England contain gibbsite, suggesting weathering in warm humid conditions. These could have survived on nunataks standing above the limits of Devensian glaciers, but may also have formed over several interglacials (McCarroll & Ballantyne 2000).

The Devensian Stage

The Devensian Stage of Mitchell *et al.* (1973) spans three MISs (4–2) and four of the substages (the later part of 5d and 5c–5a) of MIS 5. The most extensive deposits resulted from the main event of the Devensian, namely glaciation early in MIS 2 during the Dimlington Stadial, which is often known worldwide as the Last Glacial Maximum (LGM). The periglacial, fluvial, lacustrine and cave deposits of earlier parts of the Devensian have much more limited local distributions and inter-site correlation is difficult.

The Early Devensian

During MIS 5d–4 (*c.* 115–60 ka BP) a sequence of climatic oscillations is indicated principally by palynological studies of fluvial deposits occurring in low terraces of several English rivers. For example, the deposits of the River Wissey near Wretton (TL 6899), Norfolk, indicate an initially cold episode with herbaceous tundra vegetation after the Ipswichian. This was followed by two periods with birch-pine woodland separated by a second tundra grassland episode (West *et al.* 1974). The two woodland periods were tentatively equated with the Amersfoort and Brørup Interstadials of The Netherlands, which are probably equivalent to the temperate MI Substages 5c and 5a, respectively. During the tundra grassland episodes, which probably correspond to the colder MI Substages 5d and 5b, the river was braided and deposited gravels that were disturbed by periglacial structures such as ground ice mounds (pingos). A similar sequence of stadials and interstadials underlying dated Middle Devensian deposits occurs at Brimpton (SU 568651), Berkshire (Bryant *et al.* 1983).

The pollen sequence of the later interstadial at Wretton is similar to that of the Farm Wood Member (peat with stumps of Scots pine in growth position) of the Chelford Formation at Chelford (SJ 822720), Cheshire, originally described by Simpson & West (1958). A similar interstadial was also reported in a depression resulting from collapse of a ground ice mound at Beetley (TF 986180), Norfolk, with grass- and herb-dominated pollen assemblages occurring both before and after the interstadial (Phillips 1976).

Lower Devensian deposits are also known from several Pennine caves, such as Stump Cross Cavern (SE 089635), north Yorkshire, where bones of subarctic mammals including reindeer (*Rangifer tarandus*) and wolverine (*Gulo gulo*) are enclosed in flowstone dated by the U-series method to about 83 ka BP, i.e. within MI Substage 5b (Sutcliffe *et al.* 1985). In South Wales the sedimentology and mammal faunas of the Bacon Hole Member, which overlies the Ipswichian shelly sand in the sea cave of Bacon Hole, Gower, also suggest complex climatic changes in the Early Devensian (Stringer *et al.* 1986), and a stalagmite horizon here was dated to *c.* 81 ka BP. At Pen-y-Bryn brickworks (SH 490615), Gwynedd (North Wales), a sequence of peats, muds and sands underlying Late Devensian glacial deposits has been provisionally dated to MI Substages 5c and 5a (Chambers *et al.* 1995).

The Mid Devensian

This is equated with MIS 3 (60–26 ka BP). In England it is mainly represented by fluvial and cave deposits. Incomplete sequences are known from various sites in the Midlands, Lincolnshire, Yorkshire, the Thames Valley, East Anglia, the Mendips and South Wales. The Mid Devensian age of these deposits has often been established by ^{14}C dating, but most are only just within the range of this method, so that the dates are imprecise, with large standard errors, and are at present impossible to calibrate to calendar years.

In the Midlands the Mid Devensian is represented at several sites by organic deposits of a relatively mild interval, known as the Upton Warren Interstadial after the site in Worcestershire (SO 935673) where it was first described (Coope *et al.* 1961). Cryoturbated lenses of organic silt, which accumulated in shallow ponds in the valley of the River Salwarpe at Upton Warren were ^{14}C-dated to 41 900 $\pm$ 800 and 41 500 $\pm$ 1200 ^{14}C BP, and contain a mixed assemblage of temperate and arctic molluscs, beetles and herbaceous plant remains, with several large mammals typical of cool conditions, such as *Mammuthus primigenius, Bison priscus, Rangifer tarandus* and woolly rhinoceros (*Coelodonta antiquitatus*). A similar assemblage dated to 38 000 $\pm$ 700 ^{14}C BP occurs at Fladbury (SO 995462), Worcestershire (Coope 1962), and at Four Ashes (SJ 925085), south Staffordshire, there is a more complete Middle Devensian succession suggesting that the interstadial lasted from 42 500 to 38 500 ^{14}C BP and was preceded and followed by much colder conditions (Morgan 1973). Organic horizons in gravels of the Lower Bain Valley at Tattershall (TF 210570), Lincolnshire, also indicate cold conditions with a sparse insect fauna before 42 100 ^{14}C BP, followed by a short, imprecisely dated interval of much milder interstadial conditions with beetles that now inhabit southern Europe (Girling 1974). The Upton Warren Interstadial is also represented by organic silts within the Middle Devensian Kempton Park Gravel (Upper Floodplain Terrace) of the River Thames at the Ismaili Centre (TQ 269789), South Kensington (Coope *et al.* 1997). Here the beetle assemblage indicates a mean mid-winter temperature of −4 °C and a mean July temperature of 17 °C at 38 000 $\pm$ 2000 ^{14}C BP, compared with −22 and 9 °C, respectively, before the interstadial. Despite the mild climate, there is no evidence for woodland at any of the Upton Warren Interstadial sites, probably because the climatic amelioration was sudden and too short for trees to migrate into Britain from refuges well to the south (Coope 2000).

Cave sequences in NW and SW England also indicate slightly warmer conditions in the later part of the Mid Devensian from *c.* 40 to 30 ka ^{14}C BP. At this time there was sufficient production of soil CO_2 to dissolve limestones and allow intermittent growth of speleothems, whereas almost no speleothems were formed in earlier parts of the Devensian or from *c.* 30 to 13 ka ^{14}C BP (Atkinson *et al.* 1986), when permafrost inhibited the microbiological oxidation of soil organic matter. Caves such as Pin Hole (SK 533742) and Robin Hood Cave (SK 534742) in the Permian dolomitic limestone at Creswell Crags near the Nottinghamshire–Derbyshire border, Badger Hole (ST 532479) in the Carboniferous Limestone of the Mendip Hills and Kent's Cavern, south Devon, in Devonian limestone, also contain late Mid Palaeolithic flint and quartzite artifacts often associated with mammal bones dated to the late Mid Devensian (Jacobi & Grün 2003). A cave in the Carboniferous Limestone at Coygan (SN 284091), Carmarthenshire, has yielded evidence of Mousterian human occupation close to a Mid Devensian hyaena den. U-series, electron spin resonance and radiocarbon dates suggest that the human occupation of the cave was early in the period between 64 and 38 ka ^{14}C BP (Aldhouse-Green *et al.* 1995). The Upton Warren Interstadial was therefore mild enough to allow *Homo* to migrate into England and Wales, even if for much of the time it was necessary to seek the warmth and protection of caves.

The Late Devensian

In Britain this part of the Devensian, equivalent to MIS 2, has been divided into two cold periods, the Dimlington Stadial or Last Glacial Maximum (LGM) (26 ka–13 500 ^{14}C BP) and the Loch Lomond Stadial (11 000–10 000 ^{14}C years BP = 12 970–11 530 cal.), separated by the warmer Windermere Interstadial (16 810–12 970 cal.). Glaciers existed in England and Wales during both stadials, but those of the Loch Lomond Stadial were much more restricted in extent than the Dimlington Stadial ice sheets, which covered northern England, almost the whole of Wales and large areas of the North and Irish Seas. The Windermere Interstadial and Loch Lomond Stadial are usually grouped together as the Devensian Late-glacial.

Glacial deposits of the Last Glacial Maximum. The Dimlington Stadial is named after a village on the Holderness coast (Fig. 17.18), now lost to coast erosion, where the cliffs (TA 399205–TA 386224) expose two Late Devensian tills. These overlie lacustrine silts and sands containing an arctic insect fauna and remains of a moss typical of cold wet conditions, such as glacial meltwater streams. The moss was ^{14}C-dated to 18 500 $\pm$ 400 (20 757–22 953 cal.) and 18 240 $\pm$ 250 BP (20 931–22 281 cal.) (Penny *et al.*, 1969). The two tills are the Skipsea and Withernsea Members of the Holderness Formation. In Holderness, the Withernsea Member is restricted to an arcuate area close to the present coast between Cowden (TA 242419) and Easington (TA 403200). However, the underlying Skipsea Member extends westwards to the Devensian till margin on the dipslope of the Yorkshire and Lincolnshire Wolds and southwards to the coastal fringe of north Norfolk (Madgett & Catt 1978). Greyish brown till resembling the Skipsea Member can be traced northwards along the east coast into Durham and Northumberland, and contains essentially

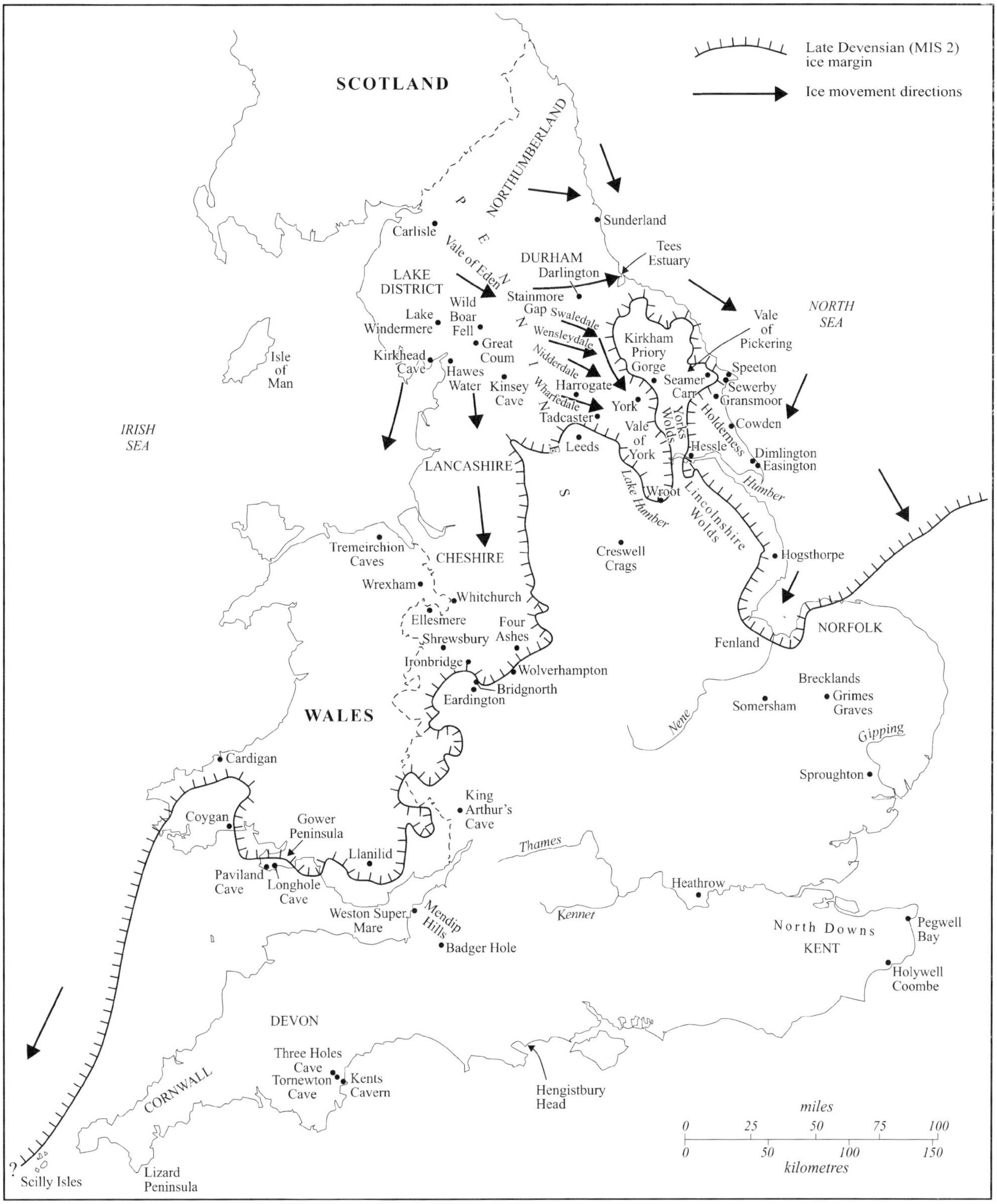

Fig. 17.18. Important Devensian sites, margin of the Late Devensian Dimlington Stadial (MIS 2) glacier in England (based on Bowen *et al.* 1986), and directions of ice movement inferred from erratics and directions of preferred stone orientation (macrofabric) in tills. Figure 17.20 shows further details for Wales.

east coast and North Sea erratics, so it was probably deposited by east coast ice originating in southern Scotland and the Cheviot Hills (NT 70–NT 92). However, the reddish brown Withernsea Member extends no further north than the Tees Estuary (NZ 52), and is thought to have been deposited by ice that originated in NW England and crossed the northern Pennines via the Stainmore Gap (NY 91), picking up reddish material from the Triassic rocks of the Darlington area (NZ 21).

The uppermost parts of whichever Dimlington Stadial till occurs at the surface were weathered in the Holocene, often to a depth of 5 m or more (Madgett & Catt 1978). Because both tills contain easily weathered iron minerals, such as pyrite, siderite and chamosite, which were oxidized to reddish-brown iron oxides, the weathered material is slightly redder than either till. Because of this and other features resulting from Holocene pedogenesis, the uppermost 5 m were previously incorrectly identified as a third lithostratigraphic unit of the Dimlington Stadial, known as the Hessle Till in east Yorkshire and Lincolnshire. Eyles & Sladen (1981) suggested that a similar Holocene weathering profile accounts for what had previously been recognized as a separate reddish brown upper till in Northumberland and Durham. As in Holderness, the thick weathered layer is often separated from the greyish brown, unweathered Dimlington Stadial till in these areas by sands and gravels, which provide a lower limit to the weathering processes. These and other sand and gravel bodies within and between the Skipsea and Withernsea Tills were probably deposited in subglacial or englacial meltwater channels during melting of the Dimlington Stadial ice sheet.

During the Dimlington Stadial, ice from NW England also entered the Vale of York (SE 56–SE 62), where it was joined on its western flank by Pennine ice moving southeastwards down Swaledale (SD 99–SE 19), Wensleydale (SD 89–SE 18) (Plate 34), Nidderdale (SE 07–SE 26) and Wharfedale (SD 97–SE 44) (Fig. 17.18). As the coastal (Skipsea) ice of the Dimlington Stadial blocked the major drainage outflow through the Humber Gap at Hessle (TA 02), an extensive lake (Lake Humber) was created in the southern part of the Vale of York. Gaunt (1976) suggested that the Vale of York ice surged into this lake as far south as Wroot (SE 714030), near Doncaster, although till with erratics of Carboniferous limestones and sandstones and occasional pieces of Shap Granite and Borrowdale Volcanic rocks (the Escrick Member) extends no further south than the Escrick Moraine (SE 4158–SE 6352–SE 6960) a few km south of York (SE 6051). Up to 20 m of laminated clays and silts (the Hemingbrough Formation) were deposited in Lake Humber. They and the Escrick Member are overlain by glaciofluvial sands and gravels deposited by meltwater issuing from the retreating ice front.

The east coast ice also blocked the eastern end of the Vale of Pickering (TA 1070), creating Lake Pickering, which over-flowed westwards through the Kirkham Priory Gorge (SE 7365–SE 7667) into Lake Humber. In the Fenland embayment, the LGM ice depositing the Skipsea Till similarly impounded Lake Sparks for a period of about 70 years soon after the ^{14}C date of 18 310 + 1980 − 1440 BP (17 846–25 980 cal.), which West (1993) obtained on willow (*Salix*) leaves from a channel sand beneath the varved lake sediments at Somersham (TL 371800).

Although eastward movement of the Dimlington Stadial ice on the lower eastern slopes of the Pennines was topographically controlled by major valleys such as Wensleydale, at greater heights in the west Pennines and Lake District there was much less topographic control, especially in the early stages of the glaciation. The long axes of the classic drumlin landforms in the west Pennines indicate a complex flow pattern, which changed slightly during the course of the glaciation (Mitchell 1994). The drumlins occur on high interfluves as well as in valleys, but are absent from narrow linear zones that mark the divides between ice flowing westwards towards the Irish Sea, northwards to the Stainmore Gap and eastwards into Wensleydale. The ice divides continue across interfluves and valleys, and the main divide traced northwards from Great Coum (SD 7083) to Wild Boar Fell (SD 7598) is contiguous with the main ice divide of the Lake District, suggesting that the flow of the ice sheet that developed initially in these upland areas was not topographically controlled. This may account for local uphill movement of erratics, such as the large masses of Silurian Austwick Grit from Crummack Dale deposited on Norber Brow (SD 7669), north Yorkshire (Fig. 17.19).

Beneath the North Sea, the Bolders Bank, Dogger Bank and Botney Cut formations (Fig. 17.3) were probably all deposited during the LGM (Cameron *et al.* 1992). The Bolders Bank Formation is generally 5–15 m thick, and resembles the Skipsea and Withernsea Tills of east Yorkshire in containing clasts of chalk and other rocks from northern England in a reddish to greyish brown matrix. However, off the coast of Holderness its seismically determined base is at 32–55 m below present sea level, which suggests it also includes the Bridlington Member, because at Dimlington the Bridlington Member extends from 2–3 m above sea level almost down to the Chalk surface at −30 to −35 m OD (Catt & Digby 1988). The Dogger Bank Formation only occurs east of 1°E and north of 54°30'N. It is a laminated glaciomarine clay up to 42 m thick, which interdigitates with the Bolders Bank Formation, and was probably deposited in an ice-marginal remnant of the North Sea.

Fig. 17.19. Erratics of Silurian Austwick Gritstone derived from Crummack Dale (low ground in distance) and deposited 60–90 m higher on Carboniferous Limestone, Norber Brow (SD 7669), north Yorkshire. Photograph by John Catt.

The Botney Cut Formation occupies numerous scaphiform channels up to 100 m deep and 8 km wide cut into the Bolders Bank and Dogger Bank Formations. The channels probably originated subglacially, as they are filled partly with stiff reddish brown till like that of the Bolders Bank Formation and partly with similarly coloured, laminated glaciolacustrine or glaciomarine clays. Also, in size and shape they resemble Weichselian (Devensian) subglacial channels in northern Germany and Poland (Ehlers & Wingfield 1991). East of Sunderland a trough approximately 75 km long, 15 km wide and 80 m deep cut into evaporites of the Zechstein Group is partly filled with Botney Cut Formation overlain by about 25 m of glaciolacustrine clays of the Sunderland Ground Formation (Fig. 17.3). Erosion by the LGM glacier removed the Egmond Ground, Cleaver Bank, Eem and Brown Bank formations from most of the North Sea floor north of the Dogger Bank. From evidence in the northern North Sea, Sejrup *et al.* (2000) suggested that the British LGM glacier was confluent with the much larger Fennoscandian LGM ice sheet between about 22 and 28 ka ^{14}C BP, but that by the time the British ice had reached Holderness after 18 ka ^{14}C BP the two had disengaged. South of the glacial limit, the Rhine and other northern European rivers flowed into the Atlantic through the English Channel, and in the Southern Bight deposited fluvial sands up to 8 m thick, which form the Kreftenheye Formation (Fig. 17.3).

On the western side of the Pennines, LGM ice moving southwards from SW Scotland and the Lake District into the Irish Sea Basin (Fig. 17.18) deposited tills and outwash gravels of the Irthing Formation in the Carlisle area (NY 35–NY 45), the Selker Formation in the western Lake District (NY 03–SD 18), the Penrith Formation in the Vale of Eden (NY 35–NY 71), all described by Huddart (1991), the Kirkham Formation in central and southern Lancashire (Longworth 1985) and the Stockport Formation in Cheshire (Worsley 1991*a*). The Kirkham and Stockport formations often consist of two reddish-brown (Permo-Triassic-derived) tills separated by 'Middle Sands', a sequence previously used to postulate two extensive ice advances in the LGM. However, as in other parts of northern England where two or more LGM tills can be distinguished, such as east Yorkshire (Madgett & Catt 1978), these are now attributed to a single Dimlington Stadial ice advance (Earp & Taylor 1986).

Till of the Stockport Formation covers much of mid-Cheshire and forms a prominent arcuate moraine stretching from Wrexham (SJ 3350) through Ellesmere (SJ 3934) to Whitchurch (SJ 5441). Beyond this feature the till extends as far south as Bridgnorth (SO 7193) and Wolverhampton (SO 9198), and at Four Ashes, 8 km north of Wolverhampton, it overlies deposits of the Middle Devensian Upton Warren Interstadial (Morgan 1973). At the 'Wolverhampton line' the glacier forced the upper Severn, which earlier had probably flowed north to join the River Dee, to cut southwards through the former watershed, thus creating the Ironbridge Gorge (SJ 6603). At Eardington (SO 728893), till of the Stockport Formation interdigitates with, and contributed Scottish and Lake District erratics to, the Late Devensian Holt Heath Member (Main Terrace) of the Severn Valley Formation (Maddy *et al.* 1995).

In NW Wales the Irish Sea ice crossed Anglesey (SH 36–SH 49) and the Lleyn Peninsula (SH 12–SH 45) (Fig. 17.20), depositing reddish brown till and glaciofluvial deposits of the St Asaph Formation. This includes marine shells, clasts of local rocks and also erratics from Eskdale (Lake District) and Criffel–Dalbeattie (SW Scotland). The formation occurs mainly below 60 m OD along the coast, but rises to more than 180 m OD east of the Vale of Clwyd (SJ 17–SJ 16). The Irish Sea ice was probably responsible for transporting shells to 430 m OD on Moel Tryfan (GR 520560) and other north-facing mountains in NW Wales. It also sealed the Tremeirchion Caves (SJ 085725) in the Vale of Clywd and Pontnewydd Cave (SH 014710) in the Elwy Valley near Rhyl. A ^{14}C date of 18 000 + 1400 – 1200 BP (18 599–25 008 cal.) on a mammoth carpal from one of the Tremeirchion caves (Rowlands 1971) suggests the ice reached this area at approximately the same time as the east coast glacier reached Dimlington, east Yorkshire.

The contact between the Welsh and Irish Sea ice sheets is well expressed in the Lleyn Peninsula of NW Wales, where Irish Sea ice crossed the peninsula from the northeast and Welsh ice simultaneously extended from the Vale of Ffestiniog (SH 6239–SH 6842) as far west as St Tudwall's Peninsula (SH 22–SH 32). Large meltwater channels were formed beneath both glaciers, and a series of sand and gravel terraces mark stages in the fall in water level of a lake trapped between them (Young *et al.* 2002).

On the southern side of Cardigan Bay near Cardigan (SN 1746), up to 75 m of glaciolacustrine deposits accumulated in a lake occupying the lower Teifi Valley, which was impounded by the Irish Sea glacier (Hambrey *et al.* 2001). Further south the Irish Sea ice also impinged on the north Pembrokeshire coast, but deposits of this ice stream do not extend further inland, because during the LGM the Welsh uplands nourished one or more independent ice accumulation areas, from which glaciers spread in various directions down major valleys onto lower ground (Fig. 17.20). The thickest Welsh ice may have been SE of Snowdonia (SH 65), whence it spread westwards across the Rhinog Hills (SH 62) to St Tudwall's Peninsula and eastwards across the summits of the Aran Mountains (SH 82) and Arenig Hills (SH 8236–SH 8241) to the Welsh Borderland. The altitude of the surface of the Welsh ice in Snowdonia was mapped by McCarroll & Ballantyne (2000) based on the contrast between ice-scoured bedrock below and intensely frost-shattered periglacial terrain above the glacial 'trimline'.

With the possible exception of Plynlimon (SN 7885–SN 8087), the mountains of mid-Wales were completely ice-covered, with major outlet glaciers flowing westwards into Cardigan Bay, south-westwards into Carmarthen Bay and eastwards into the Welsh Marches, utilizing the Severn and Wye Valleys (Fig. 17.20). On the eastern margin of Cardigan Bay (SN 35), ridges of boulders ('sarns') stretching west for at least 8 km into the Irish Sea are thought to be medial moraines formed where ice tongues coalesced after flowing down the major valleys from mid-Wales. The Brecon Beacons (SO 01–SO 12), Fforest Fawr (SN 81–SN 82) and Black Mountains (SO 22–SO 23) were another major Welsh ice centre, which blocked ice from the north and fed the outlet glaciers of the South Wales valleys (Fig. 17.20).

In the higher areas of mid-Wales, the local ice deposited variable tills grouped as the Elenid Formation, and in NW Wales ice originating mainly in Snowdonia (SH 65) and Rhobell Fawr (SH 7825) deposited tills and glaciofluvial gravels of the Eryri and Meirion formations. Locally these units include deposits of small cirque glaciers dating from the Loch Lomond Stadial. They have also been partly rearranged by subsequent periglacial mass movement on slopes (Potts 1971) to form paraglacial gelifluction deposits. Along the North Wales coastal plain and in boreholes near Llandudno (SH 7882), the Meirion Formation is overlain by the St Asaph Formation of the Irish Sea glacier, but in the Vale of Clwyd (SJ 07–SJ 15) interbedding of the Welsh and Irish Sea tills suggests local mixing and interfingering of the two ice sheets (Warren *et al.* 1984).

In the Wrexham–Shrewbury area (SJ 35–SJ 41) to the NE, the Welsh ice impinged on Scottish ice moving southwards from Cheshire. Here it deposited the Shrewsbury Formation, which overlies and locally interdigitates with the Stockport Formation (Worsley 1991*a*). Ice originating on the Brecon Beacons and Fforest Fawr (Fig. 17.20) deposited till and outwash gravels of the Brecknockshire Formation in many of the valleys of SE Wales and as far south as a complex moraine belt between Margam (SS 7887) and Pyle (SS 8282) on the South Wales coastal plain. These deposits contain abundant Upper Palaeozoic (especially Devonian) erratics, but no Lower Palaeozoic rocks (Bowen 1970). On the Gower Peninsula (SS 48–SS 59) west of Swansea, the position of the ice limit remains uncertain, although patches of till and glacial gravel of the Brecknockshire Formation occur at Rotherslade (SS 613871) in Langland Bay and at Broughton Bay (SS 4193).

Beneath the Irish Sea, the LGM Western Irish Sea Formation includes two glacial members, each underlain by strongly incised unconformities (Fig. 17.16), suggesting two separate ice advances. Lithologically they both resemble the Irish Sea till found onshore. Thin discontinuous patches of these tills, known as the Melville Till, occur as far south as the south-central Celtic Sea (Scourse *et al.* 1990), and have been correlated with the Late Devensian Scilly Member, which occurs on northern parts of the Scilly Isles (SV 91) (Scourse 1991). To the SW it grades into glaciomarine sediment, the Melville Laminated Clay, the distribution of which indicates the lobate marine terminus of the LGM Irish Sea ice stream (Scourse & Furze 2001). This outlet contributed sediment to the Celtic Deep Sea Fan (Bourillet *et al.* 2003) both before and during Heinrich Event 2 (a major depositional episode of iceberg-rafted debris in the North Atlantic) at *c.* 22 ka BP (Scourse *et al.* 2000). Also it could have been the source of the sand forming large linear ridges associated with the terminal zone of the ice sheet on the central and outer parts of the Celtic shelf (Scourse *et al.* 1990), although the ridges were last reshaped by tidal processes (Reynaud *et al.* 2003).

Eyles & McCabe (1989) proposed that the LGM tills around the Irish Sea Basin are glaciomarine in origin, implying a very high relative sea level resulting from strong glacioisostatic depression. This in turn implies a very thick LGM ice sheet in the Irish Sea basin. However, various aspects of this interpretation have been disputed (Lambeck & Purcell 2001; McCarroll 2001; Scourse & Furze 2001), and the tills and associated deposits are now thought to be mainly terrestrial. Nevertheless, an

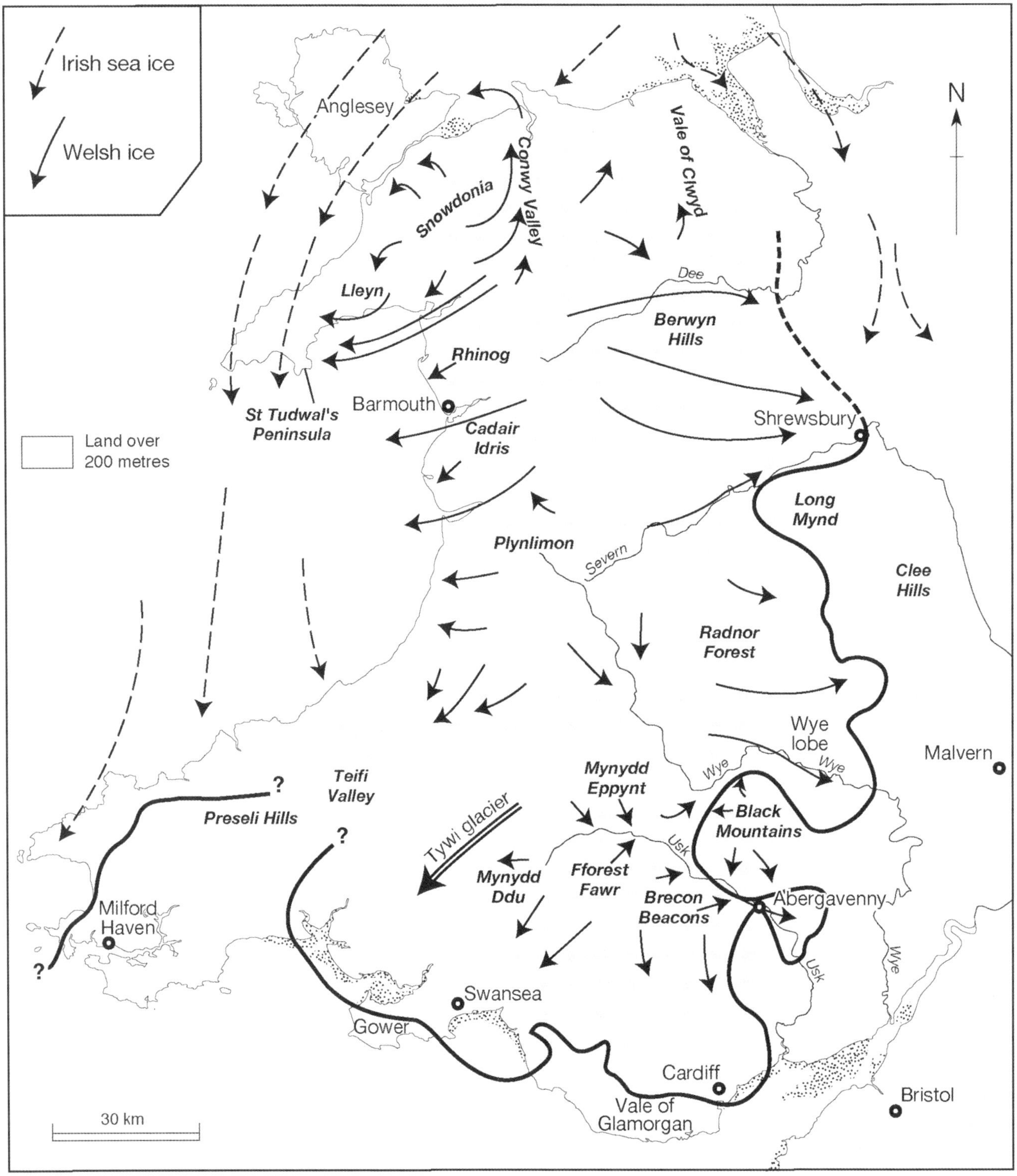

Fig. 17.20. Irish Sea and Welsh ice flow directions in Wales during the Dimlington Stadial.

actively calving tidewater glacier probably existed early in the deglaciation below present sea level close to the terminal zone of the ice sheet in the south-central Celtic Sea (Scourse & Furze 2001), and proximal glaciomarine sequences occur in the northern Irish Sea where glacioisostatic depression was indeed significant (McCabe 1997). Readvances during deglaciation in the Irish Sea Basin resulted in the formation of proglacial push moraines, such as the Bride Moraine of the Isle of Man (NX 4201–SC 4599) (McCabe *et al.* 1998). Thomas *et al.* (2004) recognized three LGM advance episodes in the northern part of the Isle of Man, the first depositing the Shellag Formation (probably also responsible for the Bride Moraine), the second the Orrisdale Formation, and the third the Jurby Formation, which in turn is overlain by organic muds with ^{14}C dates of 15 150 ± 350 BP (17 624–18 578 cal.) and 12 890 ± 360 BP (14 370–15 992 cal.).

The form and flow pattern of the Dimlington Stadial glacier in Britain have been modelled by Boulton *et al.* (1977, 1985,

1991) and Lambeck (1993). The models are based on: (a) ice margins and flow lines indicated by geomorphological features, erratics and till margins; (b) estimates of glacioisostatic depression and recovery based on contemporary shorelines and crustal and mantle rheology; and (c) the assumption that longitudinal ice profiles and glacier dynamics were similar to those of modern ice sheets. However, an ice lobe depositing the Skipsea and Withernsea Tills in the south-western North Sea Basin from the Tees Estuary as far south as north Norfolk is difficult to explain in terms of modern glacier dynamics, and is generally attributed to a surge over soft, saturated marine sediments.

Possible Devensian glaciations preceding the Dimlington Stadial. Early or Mid Devensian glaciations have been proposed in parts of south Lincolnshire (TA 57–TA 35) and north Norfolk (TF 74–TG 04), south Yorkshire (the Leeds–Tadcaster–Harrogate area, SE 34–SE 44), the West Midlands and North Wales. The original evidence given for these was mainly geomorphological, namely the presence of a prominent, undegraded hummocky ridge (assumed to be the LGM end moraine) some distance north of the feather-edge of the Devensian till. Near the feather-edge the relief is much more subdued on account of the presumed longer period that the till has been subject to subaerial erosion. Examples of these ridges are the Hogsthorpe Moraine (TA 4078–TA 5372), south Lincolnshire (*c.* 25 km north of the till margin, which here lies a short distance south of the north Norfolk coast), and the Wrexham–Ellesmere–Whitchurch Moraine, Cheshire (north of the till margin at the 'Wolverhampton line'). However, the proposed age difference between moraine and till margin is nowhere supported by lithostratigraphic evidence, in that the till forming the moraine or occurring on the land surface north of it is lithologically similar to that at the feather-edge. Also the assumption that fresh hummocky moraines are good indicators of the maximal Late Devensian ice limit is questionable, as they can form elsewhere. For example, Thomas (1989) suggested that the Ellesmere Moraine was probably formed by supraglacial sedimentation close to a temporarily stable ice margin during retreat from the LGM limit, and a similar origin has been proposed for the prominent transverse hummocky moraines occurring in Wensleydale (Plate 1) and other Pennine valleys some distance up-glacier from the known Late Devensian ice limit.

Worsley (1991*b*) reviewed the stratigraphic and dating evidence for pre-LGM Devensian glaciations in various onshore parts of the British Isles, and concluded that the concept remains unproved. On parts of the continental shelf with fairly complete Late Quaternary successions, such as the North Sea (Fig. 17.3) and Irish Sea (Fig. 17.16), the evidence for an Early or Mid Devensian glaciation is also uncertain at present. However, the idea was revived by Bowen *et al.* (2002), who suggested a widespread Mid Devensian glaciation based on clusters of Ipswichian amino-acid dates obtained on derived marine shells found in Devensian tills south of the hummocky moraines in Cheshire, Norfolk and other parts of Britain. Fundamental to this interpretation is the concept that the youngest cluster of amino-acid dates for an assemblage of derived shells is only slightly older than the ice advance that transported the shells. However, McCarroll (2002) has questioned the statistical reliability of the clusters, and even if they do exist they could predate the ice advance by any length of time, so that they provide weak evidence for a Mid Devensian glaciation. In addition, as far as S Lincolnshire and Norfolk is concerned, the idea is disproved by the ^{14}C date indicating that Lake Sparks was created after 18 310 ^{14}C BP (West 1993), because the lake could only have been impounded by ice completely sealing the Fenland embayment, and an LGM ice front at the Hogsthorpe Moraine would not have achieved this. Consequently, all the Devensian glacial deposits in England and Wales are most reasonably attributed to the Dimlington Stadial.

Periglacial features of the Last Glacial Maximum. Periglacial deposits and ground ice structures, such as ice wedge casts, cryoturbations and pingos, occurring in many parts of S England and S Wales outside the LGM ice limit (Boulton 1992, fig. 14.13) have been attributed mainly to the Dimlington Stadial, although some may date from earlier cold periods or from the Loch Lomond Stadial. The most extensive periglacial deposits are loess (aeolian silt) and gelifluction deposits ('head') on slopes. Organic deposits are much rarer, even in southern parts of the country. Examples from the Scilly Isles dating from before about 21 ka ^{14}C BP (Scourse 1991) contain 'full-glacial' pollen assemblages indicating completely open tundra grassland conditions.

At many sites loess of the Pegwell Formation (usually shown as brickearth on BGS maps) is dated to the LGM by TL methods (Wintle 1981; Parks & Rendell 1992), and its mineralogical composition suggests it was derived mainly from outwash of the Dimlington Stadial ice sheet in the North Sea area. It reaches a maximum thickness of 4 m at Pegwell Bay (TR 3564) in east Kent (Fig. 17.21). Thinner layers are widespread in eastern and southern England, especially on outcrops of the Chalk and other limestones (Catt 1984). As it usually forms the present land surface in these areas, it is an agriculturally important soil-forming material, and has also been extensively reworked and incorporated into Upper Devensian gelifluction deposits, Holocene river alluvium and footslope colluvial deposits ('hillwash') resulting from soil erosion at various times since the Neolithic period. Loess of probably the same age is also widespread on the Channel Islands and the Lizard Peninsula (SW 62–SW 71), S Cornwall (the St Saviour Beds and Lizard Formation, respectively).

The LGM gelifluction deposits are less well dated than the loess, and are very variable in lithology, being derived from whatever rock types occur upslope. Consequently, numerous lithostratigraphical units have been recognized. For example, slate-rich deposits in Cornwall are known as the Tregunna Member (Scourse 1996), and those derived from granite on the Scilly Isles constitute the Bread and Cheese Member (Scourse 1991). Similar accumulations of poorly-sorted, angular, frost-shattered stony materials exposed in the coastal sections of south and north Devon have been termed the Start Point and Croyde members, respectively. In SW Wales they are grouped as the Pendreudraeth Member, in NW Wales as the Lethyr Member and in SE Wales as the Pontypool Member. In the chalkland areas of south and east England, deposits of frost-shattered and geliflucted chalk and flint ('coombe deposits') often overlie interglacial coastal cliffs and shore platforms, and inland are widespread on lower scarp slopes and the floors of dry valleys.

Fig. 17.21. Late Devensian (Dimlington Stadial) loess overlying Palaeocene Thanet Sand Formation, Pegwell Bay (TR 351642), east Kent. Photograph by John Catt.

Periglacial aeolian sands (coversands) of Dimlington Stadial age are thinner and have a more patchy distribution in southern England than the silt-rich loess. The most extensive are in the Brecklands (TL 78–TF 88) in East Anglia, where they are cryoturbated into upper layers of chalky gelifluction deposits. Bateman (1995) obtained TL dates of 14 600–13 400 from these sands at Grimes Graves (TL 8189) near Thetford, Norfolk. On the floor of the North Sea, scattered patches of sand less than 1 m thick, interpreted as aeolian, occur south of the limit of the Bolders Bank Formation and have been correlated with the Twente Formation of The Netherlands (Cameron *et al.* 1989).

The Devensian Late-glacial

Subdivision of the Late-glacial. Devensian Late-glacial successions in Britain were originally divided into three main pollen assemblage biozones similar to those recognized in Scandinavia in the early 20th century. Zone I, equivalent to the later parts of the Dimlington Stadial, was characterized by tundra plant communities, Zone II (the Windermere Interstadial) by pollen of woodland, mainly birch (*Betula*), and scrub vegetation, and Zone III (the Loch Lomond Stadial) by a return to tundra vegetation. Sediments from the floor of Lake Windermere (Cumbria) were proposed as the type sequence of the Windermere Interstadial (Coope & Pennington 1977), and west central Scotland as the type area for the Loch Lomond Stadial (Rose 1989).

However, the boundaries between Zones I, II and III are time-transgressive, because they reflect slow vegetational responses to the climatic changes. Although the major temperature changes in the Late-glacial were very rapid, often occurring in only a few decades, *Betula* and other trees migrated very slowly in response to them. More mobile animals, such as the Coleoptera, could migrate much more rapidly. Indeed, coleopteran assemblages suggest that the rapid warming at the beginning of the Windermere Interstadial occurred approximately 1500 ^{14}C years earlier than the first appearance of *Betula* in England, and that the woodland had become strongly established only when the climate was beginning to deteriorate towards the Loch Lomond Stadial (Coope 1977).

Recently, a different approach to subdivision of the Devensian Late-glacial has been based on the δ^{18}O record of the GRIP ice core through the Greenland ice sheet (Björck *et al.* 1998). Stadial and interstadial (and substadial and subinterstadial) events recognized in this and other Greenland ice cores are numbered backwards in time, and can be dated very precisely by counting annual ice layers and correlating the δ^{18}O record in the ice with that in Atlantic marine sediments and lacustrine deposits in Switzerland. Greenland Stadial 1 (GS - 1, 12 650–11 500 ice core years ago) is broadly equivalent to the Loch Lomond Stadial; the preceding Greenland Interstadial 1 (GI - 1, 14 700–12 650 ice core years) is equivalent to the Windermere Interstadial, and is divided into five sub-interstadial episodes (GI - 1a–GI - 1e); the earlier GS - 2 is equivalent to the Dimlington Stadial (Fig. 17.22). Because the emphasis is on events rather than the boundaries between events, as is usual in subdividing sedimentary sequences, the problem of diachronism is less acute with this approach. Also it provides a more meaningful record of environmental change and surface processes. Consequently the GRIP ice core has been proposed as the formal stratotype for the Devensian Late-glacial in the whole North Atlantic region (Walker *et al.* 1999), including England and Wales.

Late-glacial lacustrine sequences. The main Late-glacial sequences in England and Wales are from lake basins, in which the deposits typically include a basal inorganic sedimentary unit followed by more organic sediments and then further inorganic deposits, which correspond with GS - 2, GI - 1 and GS - 1, respectively. Pre-GS - 2 lacustrine deposits are never present, because the basins originated either as kettle holes by melting and collapse of buried blocks of GS - 2 ice or by glacial or periglacial erosion during GS - 2. Examples include Gransmoor (TA 113597), east Yorkshire (Walker *et al.* 1993) and Llanilid (SS 944818), S Wales (Walker *et al.* 2003). Many of the lake basins, especially in the Lake District, Wales and East Anglia,

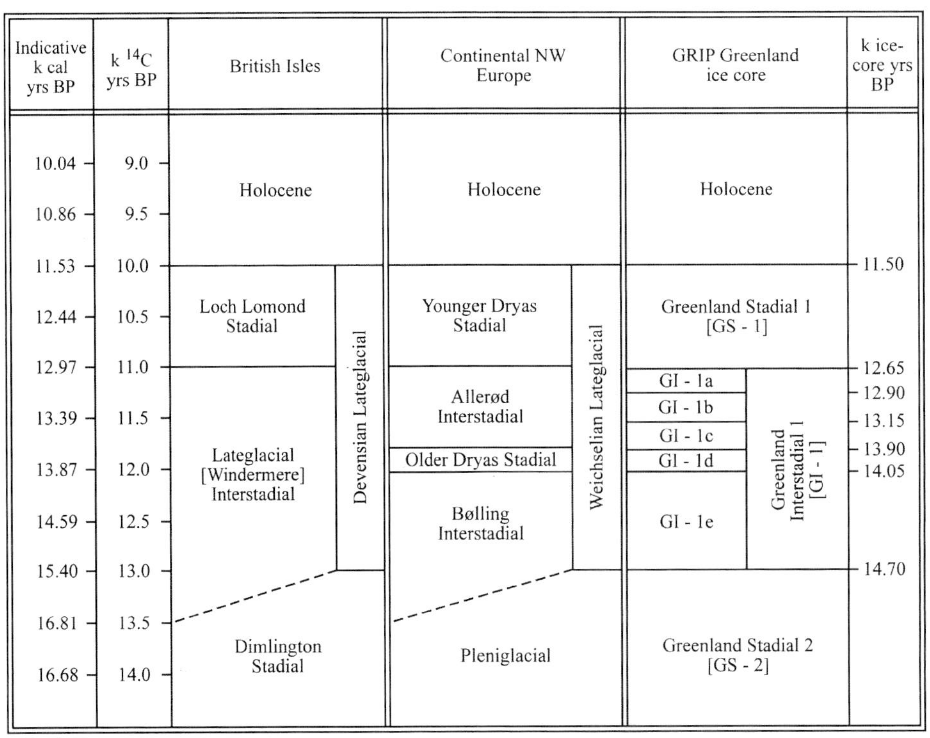

Fig. 17.22. Subdivision of the Late-glacial in Britain and adjacent areas of NW Europe, with dating based on ^{14}C (calibrated according to Reimer *et al.* 2004) and the Greenland ice cores (dating in ice core years).

persist today as areas of open water, but some have been infilled through hydroseral plant succession, with later Holocene peats and early Holocene limnic sediments overlying the Late-glacial sequence.

The GS - 2 deposits contain pollen assemblages dominated by herbaceous taxa, such as Poaceae, Caryophyllaceae, Cruciferae, dock (*Rumex*) and meadow rue (*Thalictrum*), indicating an open tree-less landscape with areas of bare, unstable ground. Erosion of these areas resulted in deposition of the basal inorganic lake sediments. During GI - 1 an initial episode of scrub dominated by juniper (*Juniperus communis*) was followed by expansion of patchy *Betula* or *Betula–Salix* woodland, especially in southern areas. This stabilized the land surface, increased lake inputs of organic matter and decreased inwash of inorganic sediment. The pollen of GS - 1 is dominated by Poaceae, Cyperaceae, herbs of disturbed ground such as mugwort (*Artemisia*) and *Rumex*, and others typical of tundra plant communities.

Further evidence of environmental change during the Late-glacial is provided at many sites by Coleoptera, molluscs, ostracods, oxygen and carbon isotopes and mineral magnetic data (Marshall *et al.* 2002). The Coleoptera indicate that at the onset of GI - 1 the temperature at sites in S England (Coope, 1998), east Yorkshire (Walker *et al.* 1993) and S Wales (Walker *et al.* 2003) rose as rapidly as 1 °C per century, to give an overall average increase of 7 °C and an increase of as much as 20 °C in minimum winter temperatures. The Coleoptera also indicate that the climatic deterioration from GI - 1 to GS - 1 occurred in two stages (Coope 1977), an initial decrease of about 3 °C in summer temperatures around 12 000 ^{14}C BP (*c.* 13 870 cal.) and a later decrease of about 4 °C at 11 000 ^{14}C BP (*c.* 12 970 cal.). During GS - 1 they suggest that July temperatures were 9–12 °C in Wales and central and southern England, and no higher than 10 °C in northern England.

Glaciers of the Devensian Late-glacial. Early in GI - 1 glaciers disappeared completely from England and Wales, even from typical upland ice-accumulation areas (Lowe & Lowe 1989). However, they reformed during GS - 1 (the Loch Lomond Stadial), mainly in cirques and valley heads. The area of most intensive glacial activity was the Lake District (NY 10–NY 42), where 80 individual glaciers have been identified (Sissons 1980; Wilson 2002). A 55 km^2 plateau icefield developed over the central fells, with smaller (<10 km^2) icefields to the west (McDougall 2001). Five small (mean area 0.29 km^2) cirque glaciers also formed in the west Pennines (SD 78-SD 79) (Mitchell 1996); Fig. 17.23 shows the basin left by one of these on the eastern edge of Swarth Fell (SD 7596), Cumbria. In Wales, Gray (1982) mapped 35 separate glaciers (mean area 0.5 km^2, total area 17.5 km^2) in Snowdonia (SH 65) (Fig. 17.24), and other small (<0.3 km^2) glaciers formed in the Brecon Beacons (SO 01–SO 12) (Shakesby 1992) and on the Aran (SH 82) and Arenig (SH 83–SH 84) mountains, North Wales (Hughes 2002).

Fig. 17.23. Cirque basin formed in Loch Lomond Stadial (GS - 1), Swarth Fell (SD 7596), Cumbria. Photograph by John Catt.

The margins of the GS - 1 glaciers are usually marked by smaller but geomorphologically sharper and less degraded end- and lateral-moraines than those of GS - 2. Often the deposits of the two glaciations can also be distinguished on the basis of shorter (Holocene only) lacustrine sequences inside the GS - 1 limits and longer (Late-glacial + Holocene) sequences outside them (Fig. 17.24). Palaeotemperatures inferred from the elevation of the reconstructed GS - 1 glaciers are similar to those indicated by Coleopteran assemblages, with mean summer maxima of *c.* 8 °C in the Lake District (Sissons 1980), 9 °C in Snowdonia (Gray 1982) and *c.* 8.5 °C in the Brecon Beacons, S Wales (Shakesby & Matthews 1993). The temperature difference of 7–8 °C between the thermal maximum of GI - 1 and the minimum of GS - 1 indicated by both coleopteran and glaciological evidence agrees with that indicated by an oxygen isotope record from Hawes Water (SD 478767) in north Lancashire (Marshall *et al.* 2002). The similarity of these temperature signals and that from the Greenland ice cores suggests that the Late-glacial climatic changes in England and Wales were part of a sequence of climatic events affecting the entire North Atlantic province (Lowe *et al.* 2001).

Periglacial features of the Devensian Late-glacial. The distribution of ground-ice features that can be dated to GS - 1 suggests that permafrost was widespread in England and Wales at this time. These include cryoturbated palaeosols formed in GI - 1 (Rose *et al.* 2000), and valley bulges, cambered strata (Hutchinson & Coope 2002), relict pingo scars or ground ice depressions (Sparks *et al.* 1972; Watson & Watson 1974) and extensive clay-rich gelifluction deposits overlying GI - 1 organic deposits (Skempton & Weeks 1976). It is also likely than many undated periglacial features of upland areas in the Lake District, Pennines, Wales and SW England, such as soil patterns (polygons, nets and stripes), blockfields, talus (scree) accumulations, rock glaciers and protalus (pronival) ramparts, either developed or were last active in GS - 1 (Ballantyne & Harris 1994). Other features that may date from this period include some deep-seated rotational rock failures and mudslides (Shakesby & Matthews 1996; Mitchell 1996).

Windblown coversands dated to GS - 1 by TL techniques or locally overlying organic sediments ^{14}C-dated to GI - 1, are widespread in north Lincolnshire (Bateman 1998), the Vale of York (Gaunt *et al.* 1971) and SW Lancashire (Wilson *et al.* 1981). In some areas they form dune-like mounds (Matthews 1970), but these could have resulted from localized aeolian reworking in the Holocene (Bateman *et al.* 2000), a process that continues today in the 'sandblows' of dry windy spells when the cultivated soil surface is eroded where it is not protected by a vegetation cover (Bateman & Godby 2004).

Late-glacial fluvial deposits. The major climatic fluctuations of the Late-glacial strongly influenced fluvial incision rates, terrace formation, channel migration and sediment loading. In upland regions, the fluvial sediments of GS - 1 and GS - 2 are dominantly gravels derived from glacial outwash or valley-side gelifluction deposits and deposited by braided (multithread) systems of anastomosing channels. Valley-side alluvial fans of GS - 1 age are common in the Lake District, Pennines and Welsh uplands (Ballantyne & Harris 1994). At the beginning of GI - 1, decreasing discharge and sediment loads led to a change to single-thread rivers with more sinuous channels and an essentially erosional regime (Macklin & Lewin 1986), with only localized preservation of fine sediments (Lewin & Macklin 2003).

In lowland areas the Late-glacial fluvial sequences consist of generally finer sediments than in the uplands, and these often contain organic muds. In many valleys a phase of incision

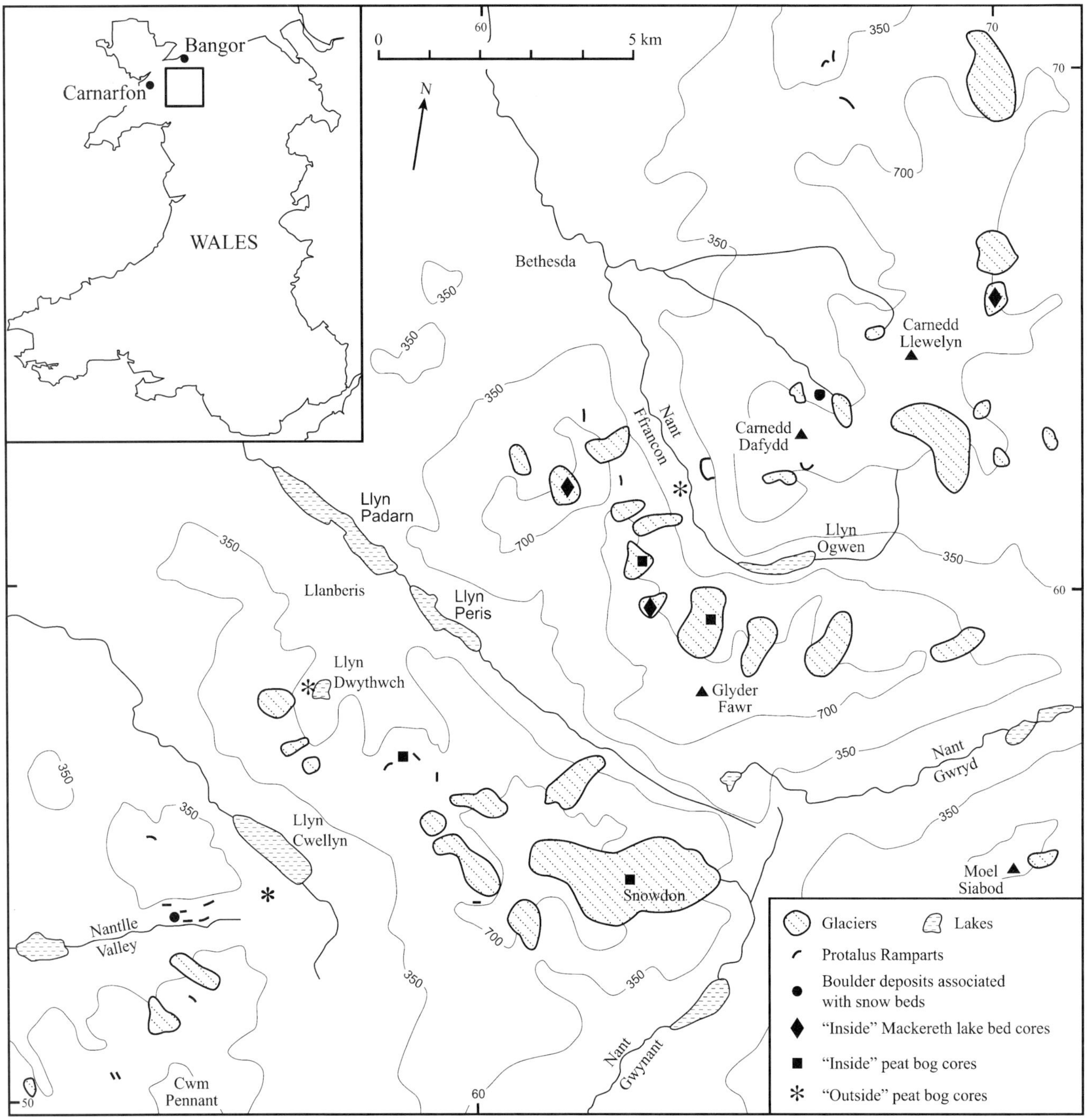

Fig. 17.24. The distribution of Loch Lomond Stadial (GS - 1) glaciers, protalus ramparts and lake and peat sequences inside and outside the GS - 1 glacial limits in Snowdonia, North Wales (based on Gray 1982).

during GS - 2 was followed by a shift to a braided aggradational river regime with widespread deposition of fluvial gravels (Rose 1995). During GI - 1 lowland floodplains were characterized by meandering rivers, with fine-grained lacustrine sediments accumulating in backswamp areas as, for example, in the Gipping Valley (Rose *et al.* 1980), or in abandoned channels, as in the Kennet (SU 16-SU 67) and Nene Valleys (Brown *et al.* 1994; Collins *et al.* 1996). As climatic conditions deteriorated during the later part of GI - 1 and early in GS - 1, there was a general episode of incision because of increasing snow-melt discharge without a proportionate increase in valley-side sediment supply (Rose *et al.* 1980), but greater sediment loading later resulted in further aggradation in most lowland valleys.

Late-glacial shorelines and marine deposits. The rapid eustatic rise in sea level, which began with wasting of the great LGM continental ice sheets and continued through the Late-glacial and early Holocene, strongly influenced the coastal configuration of England and Wales and the nearshore sedimentary record. At the LGM the worldwide sea level was about 130 m lower than at present, and at the end of GS - 1 (beginning of the Holocene) the rapid warming led to a rise from approximately 65 to 25 m below present within a few centuries at the most. The broad patterns of shoreline migration around Britain resulting from the eustatic rise, crustal rebound resulting from glacial unloading of northern Britain, meltwater loading of adjacent seas and remote glacioisostatic changes caused by melting of the

North American and Fennoscandian ice sheets were modelled by Lambeck (1995, 1996). Such models are valuable because of the insights they provide into the rheology of the crust and mantle and the evidence for Late Devensian ice thickness and extent as well as the palaeoceanography of the shelf seas.

The modelling indicates that the southern North Sea was dry throughout the Late Devensian, with maximum emergence at *c*. 15 ka ^{14}C BP shortly after the start of deglaciation. A narrow estuary persisted east of the British ice sheet to the east of Scotland from 22 to 14 ka ^{14}C BP (*c*. 25 000–16 680 cal.), then gradually broadened and extended southwards between the Dogger Bank and the mainland. However, the southern North Sea area remained dry with no connection to the English Channel developing until the mid Holocene.

Between Wales and S Ireland a wide but low land bridge persisted from 20 to 14 ka ^{14}C BP (*c*. 23 300–16 680 cal.), separating the Celtic Sea to the south from a large freshwater lake in the central Irish Sea area, which was impounded by the Irish Sea glacier to the north. By 13 000 ^{14}C BP (*c*. 15 400 cal.) the ice had retreated to SW Scotland and the mountains of central Wales, and the land bridge had been breached, so that a narrow seaway completely separated Ireland from the remainder of Britain. However, the Isle of Man (SC 61-NX 40) was still linked to NW England until 11 000 ^{14}C BP (*c*. 12 970 cal.).

Late-glacial palaeosols. Under glacial and periglacial conditions, almost the only soil process occurring during GS - 2 and GS - 1 was frost disturbance. However, warmer conditions and plant growth led to widespread weakly developed soils in GI - 1. The most obvious of these are the dark grey rendzinas developed on Chalk or chalky gelifluction deposits of GS - 2 and buried by further slope deposits of GS - 1. The main process involved in formation of rendzinas is incorporation of partially decomposed organic matter by strong soil faunal activity. ^{14}C dates for charcoal, fruits of sedge (*Carex*) or rush (*Scirpus*) and terrestrial molluscs such as *Arianta arbustorum* from the GI - 1 rendzina occurring on slopes at Holywell Coombe (TR 220379), Folkestone (Kent), fall within the range 11 810 ± 120 (13 393–13 917 cal.) to 11 370 ± 150 BP (12 946–13 543 cal.) (Switsur & Housley 1998). Dates within this range have also been obtained from buried rendzinas at three other sites on the North Downs in Kent (Preece 1994).

On lower valley-side slopes at Holywell Coombe, the GI - 1 palaeosol developed in thin coombe deposits over Gault Clay is a stagnogleyic argillic brown earth (Catt & Staines 1998), which indicates other processes of soil development, including decalcification, clay illuviation and weak gleying (mobilization of iron in seasonally waterlogged and reducing conditions). Clay illuviation (downward movement of clay particles in percolating water) is indicated by the presence of oriented (birefringent) clay accumulations seen in thin section. Rose *et al.* (2000) reported similar microscopic evidence for clay illuviation in the GI - 1 soil developed on Late Devensian loess at Prospect Park (TQ 059783) near Heathrow Airport. This pedogenic process occurs mainly under woodland, which in GI - 1 was probably the *Betula* woodland suggested by pollen analyses of lake deposits in S England.

Homo sapiens *in the Late Devensian: the Late Palaeolithic.* The stratigraphic distribution of human artefacts of Late Devensian age in England and Wales suggests two main periods of occupation, the earlier Late Palaeolithic predating the Dimlington Stadial glaciation, and the later Late Palaeolithic of GI - 1 and GS - 1 (Campbell 1977).

The earlier Late Palaeolithic is known mainly from cave sites, many of which had also been occupied in the late Mid Palaeolithic (Mid Devensian), such as Kent's Cavern, Badger Hole, Robin Hood Cave and Pin Hole. Other important earlier Late Palaeolithic cave sites include Wookey Hole (ST 532479) in the Mendips, Uphill Cave (ST 316548) near Weston-super-Mare, Long Hole (SS 451851) and Paviland Cave (SS 437859), S Wales and Tremeirchion Caves, North Wales. Some typological aspects of the artefacts, notably the numerous stout scrapers and occasional examples of bifacial leaf-points and 'burins busqués', suggest similarities to the Aurignacian culture of continental Europe. They were made mainly from flint, but some are of Carboniferous chert or Greensand chert. Excavations in the 19th century at Paviland Cave exposed the skeleton of a young adult male preserved in an ochreous deposit and originally interpreted as a female, the so-called 'red lady' of Paviland (Aldhouse-Green 2000). It was subsequently ^{14}C-dated to *c*. 26 ka BP (Green & Walker 1991).

The later Late Palaeolithic is also known mainly from cave sites, although there are some important open sites as well. The cave sites include Kent's Cavern, Three Holes Cave (SX 814675) and Tornewton Cave (SX 816674) in S Devon, Aveline's Hole (ST 476587), Sun Hole (ST 467541), Soldier's Hole (ST 469540) and Gough's Cave (ST 467539) in the Mendips, Cathole (SS 538900), Paviland Cave and Hoyle's Mouth (SN 111003) in S Wales, King Arthur's Cave (SO 547157) in Herefordshire, Kirkhead Cave (SD 391756) in S Cumbria, Kinsey Cave (SD 804657) in the Pennines, and Langwith Cave (SK 517695), Mother Grundy's Parlour (SK 536743), Pin Hole and Robin Hood Cave in the Creswell Crags area.

Because of their association principally with the Cheddar (Mendip) and Creswell groups of caves, the later Late Palaeolithic artefacts were originally attributed to two different industries known as Cheddarian and Creswellian. However, there are few significant differences between them and both are now referred to as Creswellian. In terms of dating and many aspects of artefact typology, the assemblage has strong affinities with the Magdalenian of France, Belgium and The Netherlands. As in the earlier Late Palaeolithic, migration from the continent would have been easy because the eastern part of the English Channel was dry land. The artefact assemblage consists mainly of flint tools, such as the trapezoidal backed blades with either double truncation (Cheddar points) or single truncation (Creswell points), burins, awls and steeply angled scrapers, with rare harpoons or barbed points made from bone or antler. Fewer of the tools are made from chert than in the earlier Late Palaeolithic. On the walls of Church Hole (SK 533741) and other caves at Creswell Crags, Britain's oldest examples of rock art, including engravings of bison and birds, probably also date from the later Late Palaeolithic.

The main open sites of the later Late Palaeolithic are at Sproughton (TM 133443), Suffolk, Hengistbury Head (SZ 179906), Hampshire and Seamer Carr (TA 0281) in the Vale of Pickering, north Yorkshire. Like the cave sites, these probably date mainly from early GI - 1 between *c*. 12 600 and 12 000 BP (*c*. 14 700–13 870 cal.) (Barton *et al.* 2003). However, the 700+ artefacts from Seamer Carr were from around a stone-lined hearth within a thin peat, which gave ^{14}C dates as young as 10 200 ^{14}C BP (*c*. 11 900 cal.) and was covered by a frost-disturbed sand, suggesting brief summer occupation in GS - 1.

The Holocene Stage

Subdivision of the Holocene

Stratigraphic subdivision of the Holocene (= Flandrian) Stage in England and Wales has been based almost entirely on pollen assemblage biozones, which reflect a combination of climatic change and human influence on the vegetation. Five zones (numbered IV–VIII, continuing those of the Late-glacial) have also been equated with the Blytt–Sernander scheme of climatic phases derived from palaeoecological analyses of Scandinavian peat sequences, namely (in order of decreasing age) the Pre-Boreal, Boreal, Atlantic, Sub-Boreal and Sub-Atlantic phases. With the advent of radiocarbon dating, the Holocene biostratigraphic record was divided into chronozones following the division of earlier interglacial temperate stages: a pre-temperate substage (Flandrian F - 1), an early temperate substage (F - 2) and a late-temperate substage (F - 3). Mangerud *et al.* (1974) proposed a three-fold subdivision of the Holocene in NW Europe: a Lower Holocene Substage (10 000–8000 ^{14}C BP, *c.* 11 530–8850 cal., the Preboreal and Boreal), a Mid Holocene Substage (8000–2500 ^{14}C BP, *c.* 8850–2587 cal., the Atlantic and Sub-Boreal) and an Upper Holocene Substage (after 2500 ^{14}C BP, the Sub-Atlantic).

However, as in the Late-glacial, problems of time-transgression serve to undermine these pollen-based stratigraphic schemes. Indeed well-dated pollen records show that, because of human influences and slow migration of trees, forest vegetation has varied almost continually in time and space throughout the Holocene (Bennett 1988). No type section has been defined, and there is widespread doubt that a type site would serve any useful purpose (West 1979). Hence, in practical terms, there is little to be gained by attempting to erect a formal chronostratigraphy for the Holocene (Birks 1982).

Holocene lacustrine deposits

As in the Late-glacial, lake sediments provide the main evidence for environmental change in England and Wales during the Holocene. Pollen analyses of these sequences show that, following the rapid warming after GS - 1, there was a sudden expansion of *Juniperus* scrub, quickly followed by the spread of *Betula, Salix* and hazel (*Corylus*). Other broad-leaved trees appeared later, elm (*Ulmus*) by 9500 ^{14}C BP (*c.* 10 860 cal.), oak (*Quercus*) by 9100 ^{14}C BP (*c.* 10 300 cal.) and lime (*Tilia*) in southern areas by 7800 ^{14}C BP (*c.* 8575 cal.) (Birks 1989). Alder (*Alnus*) became increasingly widespread on poorly drained soils after 7800 ^{14}C BP, and after 5800 ^{14}C BP (*c.* 6585 cal.) there was a rapid decline in *Ulmus*, probably attributable to disease transmitted by the bark beetle *Scolytus scolytus* (Girling 1988), although woodland clearance for agriculture, selective exploitation of elm leaves for cattle fodder or climatic change could also have been contributory factors (Parker *et al.* 2002). Minor deforestation, possibly to assist hunting and food-gathering activities, occurred in the later early Holocene and early mid Holocene (the Mesolithic), but following the *Ulmus* decline there is increasing evidence for woodland clearance for pasture and cereal cultivation in the Neolithic. Much of S England had been stripped of woodland by the late Neolithic, and there was even more widespread clearance after 4500 BP (*c.* 5140 cal.) during the early Bronze Age (Parker-Pearson 1999) and again in the Iron Age after 2600 ^{14}C BP (*c.* 2760 cal.). Consequently, by the beginning of the historic period all but the more remote upland areas of England and Wales bore the imprint of human activity (Dark 2000).

Coleoptera from lake deposits indicate a rapid rise of more than 10 °C in mean annual temperature at the onset of the Holocene (Atkinson *et al.* 1987), which is similar to the increase of *c.* 7 °C in less than 50 years indicated by oxygen isotope studies of Greenland ice cores (Alley 2000). Several proxy-climatic indicators (pollen, molluscs, Coleoptera) suggest that mean temperatures during the early and early mid Holocene were 1–3 °C above present-day levels (Bell & Walker 2005). However, from 5000 ^{14}C BP (*c.* 5750 cal.) onwards the temperature decreased progressively, culminating in the Little Ice Age of the 14th–18th centuries AD, when lowland England and Wales was 1–2 °C cooler than present and upland areas even colder (Fagan 2000).

Holocene peats

Peats were much more widespread in the Holocene than in the Late-glacial. They formed in a range of permanently wet situations that encouraged plant growth but prevented the rapid decomposition of organic matter. This occurred: (a) in valleys or enclosed basins (topogenous mires); (b) where infiltration was impeded by clay-rich or other slowly permeable deposits (soligenous mires); or (c) where rapid plant growth was sustained by high rainfall and was not dependent on a supply of ground water (ombrogenous mires). The most extensive are the ombrogenous blanket bogs of upland plateaux in the wetter western regions. However, many of these began development only after a mid Holocene episode of upland deforestation, which led to modification of the local hydrology (Moore 1993).

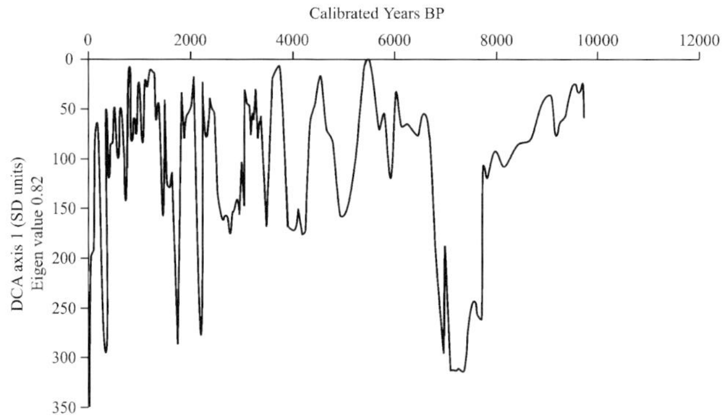

Fig. 17.25. Changes in mire surface wetness and implied rainfall at Walton Moss, Cumbria during the Holocene (from Hughes *et al.* 2000). Greater values of Axis 1 of detrended correspondence analysis (DCA) shown on the *y*-axis indicate greater wetness. DCA is a standard multivariate technique used to analyse the variation in plant macrofossil components through the peat profile. The position of plant taxa along DCA Axis 1 suggest that this represents water level, so that the sample scores can be used as an index of bog surface wetness and hence precipitation.

Recent work on peat stratigraphy has involved a wide range of palaeoenvironmental evidence provided by pollen, plant macrofossils, Coleoptera, testate Amoebae (Protozoa), isotopic and biochemical data and differences in the extent of peat humification. These have confirmed many of the local and regional vegetational changes resulting from climatic change (Blackford 2000) and human impacts (Dumayne-Peaty 2000) identified in Holocene lacustrine sequences. Differences in the surface wetness of ombrogenous bogs, as indicated by well-dated palaeobotanical changes, can be attributed to variations in rainfall. At sites such as Walton Moss (NY 504667) (Hughes *et al.* 2000) (Fig. 17.25) and Bolton Fell Moss (NY 495695) (Barber *et al.* 2003), Cumbria, these indicate that, after a fairly dry early Holocene, there were periods of increased rainfall, often lasting several centuries, starting at approximately 7800 ^{14}C BP (*c.* 8575 cal.), 6000 ^{14}C BP (*c.* 6840 cal.), 5300 ^{14}C BP (*c.* 6070 cal.), 4500 ^{14}C BP (*c.* 5140 cal.), 3500 ^{14}C BP (*c.* 3775 cal.), 2850 ^{14}C BP (*c.* 3000 cal.), 2250 ^{14}C BP (*c.* 2245 cal.), 1750 ^{14}C BP (*c.* 1680 cal.), 1450 ^{14}C BP (*c.* 1350 cal.), 1000 ^{14}C BP (*c.* 896 cal.) and 700 ^{14}C BP (*c.* 675 cal.). Since 4000 ^{14}C BP (*c.* 4460 cal.) these show some correspondence with periods of increased flooding indicated by dated alluvial sediment units (Macklin & Lewin 2003), and the most recent wet period coincides with the beginning of the Little Ice Age. Spectral analysis of these results has suggested rainfall cycles of approximately 600, 800 and 1100 ^{14}C years, which resemble cycles detected in Holocene sequences of North Atlantic Ocean sediments (Chapman & Shackleton 2000) and Greenland ice cores (Mayewski *et al.* 1997). As in the Late-glacial, the close parallels in these cyclical changes from very different depositional contexts suggest a climatic forcing mechanism that affected not just Britain but the whole North Atlantic region.

Well-preserved human remains or 'bog bodies' (Turner & Scaife 1995), such as the ritually murdered Late Iron Age or Roman man found at Lindow Moss (SJ 8281), Cheshire (Stead *et al.* 1986), provide a further cultural dimension to studies of peat stratigraphy. In Hartlepool Bay (NZ 5231) a human skeleton found in peat and dated to 4680 ± 60 BP (5489–5307 cal.) was probably an *in situ* Neolithic wetland inhumation (Tooley 1978).

Icelandic tephras, such as the Glen Garry (2050 ^{14}C BP, 2020 cal.) and Hekla-4 (4260 ^{14}C BP, *c.* 4880 cal.) tephras, have been detected in Holocene upland blanket peats in northern England (Pilcher & Hall 1996; Wells *et al.* 1997), and others are known from Wales and Dartmoor (Buckley & Walker 2002; Hall & Pilcher 2002). However, they are less widespread in England and Wales than in Ireland and Scotland.

Holocene fluvial deposits

During the Holocene, there was a marked contrast in fluvial activity between upland and lowland areas. In upland drainage basins, there was rapid incision in the early Holocene in response to declining sediment supply and glacioisostatic recovery, leaving terraces composed of Upper Devensian deposits, which now lie a few metres above the present channel (Howard *et al.* 1999). The alternating reaches of narrow and wide valley floors characteristic of many upland rivers resulted from differential glacial erosion, sequences of cross-valley Late Devensian moraines or constriction by tributary alluvial fans. However, after about 6000 ^{14}C BP (*c.* 6840 cal.) episodic aggradation resulted from the increased sediment delivery from deforested hillslopes and from flood events reflecting climatic fluctuations (Lewin & Macklin 2003).

In lowland areas, by contrast, the early Holocene rivers remote from the influence of low relative sea level were often multichannel anastomosing systems inherited from Late Devensian braided valley-floor landforms. These achieved very little downcutting. In the Soar (SP 59) and Nene (SP 66–TL 19) valleys of the East Midlands, for example, there was episodic aggradation and infilling of floodplain channels (Brown *et al.* 1994), and further north in the lowlands of the Yorkshire Ouse (SE 4760–SE 8623) fine alluvium was deposited in overbank environments (Taylor *et al.* 2000). In some valleys of south and eastern England, the early Holocene sediment load was so small that the floodplain was dominated by organic accumulations (Rose 1995). Topogenous (valley-floor) peat formation in these situations was often dependent on the rising soil water table determined by rising sea level, so that the basal peats decrease in age with increasing distance from the present coastline (Shennan *et al.* 2000*c*). Early Holocene downcutting was significant mainly where large rivers, such as the Humber (SE 8623–TA 31), were flowing over soft Pleistocene sediments under the influence of a low relative sea level (Long *et al.* 1998). However, during the later Holocene, especially after *c.* 4500 BP (*c.* 5140 cal.), the increase in runoff and sediment load resulting from conversion of forest to grassland or arable fields led to widespread deposition of fine mineral alluvium, and in catchments such as the Tyne (Rumsby & Macklin 1994) made river sedimentation more sensitive to small changes in rainfall.

Holocene sea-level change and marine deposits

The rate of eustatic sea-level rise during the early Holocene was 50–100 mm year^{-1}, but it had decreased to less than 5 mm year^{-1} between 7500 and 4000 ^{14}C BP (*c.* 8300–4460 cal.), and has averaged approximately 1 mm year^{-1} since then. At 7500 ^{14}C BP (*c.* 8300 cal.) worldwide sea level was about 20 m below present, and by 4000 ^{14}C BP (*c.* 4460 cal.) it was within 3 m of the present level. Around the coast of S England and Wales from the Humber and Mersey (SJ 39) southwards, the eustatic rise was the main factor determining Holocene relative sea-level trends, as determined from the heights of ^{14}C-dated horizons marking a transition from terrestrial to marine deposits or vice versa (Shennan & Horton 2002). However, at sites in northern England the trends are complicated by glacioisostatic recovery since the LGM. In the early and mid Holocene, this was much slower than the eustatic rise, and made little difference to the trend compared with that for southern areas. But when the rate of eustatic rise decreased after *c.* 7500 ^{14}C BP (*c.* 8300 cal.), it was overtaken by continuing glacioisostatic recovery, with the result that relative sea level was slightly higher than present for much of the later Holocene. For example, in north Northumberland (NU 04) relative sea level was 1.5–2.5 m above present between about 7200 and 3500 ^{14}C BP (*c.* 8050 and *c.* 3775 cal.) (Shennan *et al.* 2000*a*).

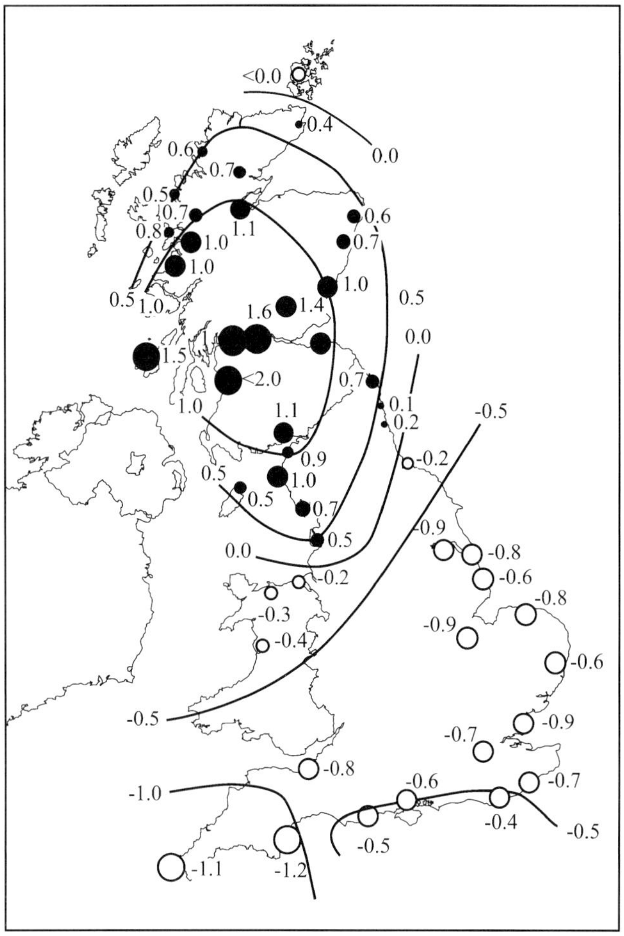

Fig. 17.26. Late Holocene relative sea-level changes (mm year^{-1}) in Britain, from Shennan & Horton (2002). Positive values indicate relative sea-level fall or land uplift; negative values indicate relative sea-level rise or land subsidence. Contours drawn by eye.

Shennan & Horton (2002) estimated rates of current relative sea-level change at various coastal sites based on: (a) ^{14}C dates younger than 4000 cal. years for basal peat layers over sand or bedrock (i.e. avoiding the problem of sample levels modified by soft sediment compaction); and (b) calculations using the RSL4 model of Shennan *et al.* (2002). The estimates omit the effect of global warming on accelerated sea-level rise in the 20th century. They show (Fig. 17.26) maximum sea-level decrease rates (land uplift rates) of 1–2 mm year^{-1} in western Scotland centred over the area occupied by the Loch Lomond Stadial ice cap. In northern England uplift rates decrease southwards to zero values in the Tees (NZ 52) and Mersey Estuaries, but further south sea-level rise (land subsidence) rates increase southwards to maxima of 1–1.2 mm year^{-1} in SW England and 0.7–0.9 mm year^{-1} in the Humber Estuary, Fenland (TF 41–TF 61) and Thames Estuary.

The geological evidence indicating late Holocene sea-level change reflects storm magnitude and frequency, variations in terrestrial sediment delivery to the coastal zone and rates of peat accumulation as well as sea level trends (Allen 2000). It includes evidence for past sea levels obtained from: (a) dated organic and mineral sediment sequences in coastal embayments, such as the Somerset Levels (ST 34–ST 44) (Haslett *et al.* 1998) and Fenland (Waller 1994); (b) estuarine sediment sequences, such as those of the Humber (Long *et al.* 1998) and Severn (ST 58-SO

70) (Allen 2001); and (c) gravel storm beach barriers, as at Dungeness (TR 0917), Kent (Waller & Long 2003) and Porlock (SS 8648–SS 8848), north Somerset (Jennings *et al.* 1998). Within the last millennium, human activities have also affected sedimentation and shoreline positions through reclamation of tidal flats and saltmarshes by artificial drainage schemes, trapping of floodtide sediment (known as 'warping' in the Humber basin) and coastal defence projects. The mid and late Holocene high-intertidal estuarine sediments of both anthropogenic and natural origin are commonly well laminated. In the anthropogenic (warp) sediments, upward fining laminae resulted from different sedimentation rates of clay and coarser particles following each tidal inundation, but the natural layering in the estuarine deposits of the Severn probably reflects the influence of seasonal changes in windiness and water temperature on grain settling rates (Dark & Allen 2005).

For the earlier Holocene, very few peat or other sediments from offshore areas have been dated (Kirby & Oele 1975), so the early Holocene shorelines are not well defined. The pattern modelled for the North Sea basin by Lambeck (1995) suggests that the shallow estuary west of the Dogger Bank (1°20'E; 54–55°N) inherited from the Late-glacial gradually extended southwards and then eastwards, so that by *c.* 9100 ^{14}C BP (*c.* 10 300 cal.) a low-lying land area over the Dogger Bank was isolated from the European mainland by the 'Strait of Dogger', and by *c.* 7000 ^{14}C BP (*c.* 7800 cal.) it had become completely submerged (Shennan *et al.* 2000*b*). The North Sea was probably connected to the English Channel by a narrow channel originating in the Preboreal period between 9700 and 9400 ^{14}C BP (*c.* 11 150 and 10 630 cal.), but the full separation of Britain from continental Europe did not occur until *c.* 6000 ^{14}C BP (*c.* 6840 cal.), when sea levels around England and Wales were within 8 m of contemporary levels and the shoreline lay only a short distance seaward of the present coast.

In the southern North Sea area, the basal peats, brackish water clays and muddy intertidal sands deposited before and during the Early Holocene transgression are grouped as the Elbow Formation (Fig. 17.3), which is up to 6 m thick in the eastern part of the UK sector. The overlying fully marine deposits include the Well Hole Formation (fine sands and sandy muds infilling depressions in Devensian glacial deposits), the Nieuw Zeeland Gronden Formation (marine sands and silts derived from Devensian glacial and periglacial deposits), the Indefatigable Grounds Formation (gravels and coarse sands forming a thin veneer over Pleistocene tills) and the Bligh Bank Formation or Young Sea Sands (extensive fine sands forming the slowly mobile sediment over much of the present sea floor of the North Sea).

Because the North Sea is very shallow, tidal currents even near the sea floor can reach 1 m s^{-1}, and the surface of the Bligh Bank Formation shows various bedforms, including sand ribbons and sand banks parallel to the dominant tidal flow, and sand waves elongated transverse to the flow (Cameron *et al.* 1992). The sand ribbons are up to several km long and tens of metres wide but less than 1 m high, and occur off Flamborough Head and northern East Anglia and in the Dover Strait. The sand banks are much larger, up to 65 km long, 2 km wide and 40 m high (Caston 1972), and occur off NE Norfolk (the Norfolk Banks), west of the Dogger Bank (the Sand Hills) and in the outer Thames Estuary. Comparison of hydrographic charts for 1851 and 1967 suggested that some of the Norfolk Banks have extended to the NW, and some of the nearshore banks, such as Haisborough Sand and Smith's Knoll, have migrated hundreds of metres to the NE (Caston 1972). The sand waves are intermediate in size, up to 500 m wide and 16 m high. They are widespread, covering large areas of the North Sea floor. Many occur on the flanks of sand banks, and where the banks are close to the shore, such as the Goodwin Sands in the Dover Strait, the sand waves are often exposed at low tide.

Large areas of gravelly sea-floor sediments occur close to the Yorkshire, Lincolnshire and Norfolk coasts, off the S Suffolk coast and in the Dover Strait (Cameron *et al.* 1992). In the first area, much of the gravel consists of rocks from northern England and S Scotland, and probably results from marine reworking of Late Devensian glacial deposits during the Holocene transgression. Off the Suffolk coast, rounded flint, quartz and quartzite pebbles are the dominant components and are probably derived from the Wroxham Crag, Kesgrave and Yarmouth Roads formations. In the Dover Strait and other parts of the eastern English Channel, the gravels are mainly composed of flint and chalk, probably derived from periglacial gravels deposited by rivers that traversed the Chalk outcrop.

In the Irish Sea basin, the N–S seaway inherited from the Late-glacial gradually widened through the Holocene, and by 6000 ^{14}C BP (*c.* 6840 cal.) the shorelines were very similar to those of the present. As in the North Sea, the distribution of Holocene sediment types and bedforms on the floor of the Irish Sea is determined mainly by tidal bottom currents of the present day; those of earlier periods formed features that are now moribund and immobile (Scourse *et al.* 2002). Tidal dynamics also control the location of seasonal tidal mixing fronts. These highly productive zones supply organic particles to the seabed and thereby influence sediment character. For instance, most of the major mudbanks in the Irish Sea are now known to be located beneath seasonally stratified water masses. Holocene records through these deposits indicate a record of tidal mixing front migration driven by sea-level change, a pattern predicted earlier by palaeotidal models (Austin 1991).

Homo *in the Early and Mid Holocene: the Mesolithic, Neolithic and Bronze Ages*

The rapid climatic amelioration at the beginning of the Holocene and an accompanying increase in diversity of plant and animal life probably led to a major increase in the human population of England, Wales and adjacent land areas from a few thousand to over 100 000 (Evison 1999). However, there seem to have been no major changes in activities or tool typology, which were still principally oriented towards a nomadic life of hunting and gathering food.

The earliest Holocene tools found in England and Wales are those of the Mesolithic Maglemosian culture found at open lakeside or riverside sites, such as Star Carr (TA 027810) (Clark 1954; Day & Mellars 1994) and Seamer Carr (TA 0381) (Schadla-Hall 1989) in the Vale of Pickering, and Thatcham (SU 500671) in the Kennet Valley, Berkshire (Healy *et al.* 1992). They consist mainly of small flint blades (microliths), some of which were mounted in wood as arrowheads, cutting or engraving tools, but also include barbed bone and antler spears and fishing harpoons and flint axe-heads used for felling and shaping timber. Sites such as Star Carr were originally thought to be hunting camps occupied only seasonally, but subsequent studies of the associated faunal remains and the discovery of stone-lined hearths suggest that they were occupied throughout the year. Isotopic evidence suggests that seafood was a major component of the diet of domesticated dogs (and presumably of their owners) at Star Carr and Seamer Carr (Clutton-Brock & Noe-Nygaard 1990), and marine resources were also exploited in Cumbria (Bonsall 1980) as well as many coastal sites in Scotland and Scandinavia. At inland sites, such as Thatcham, the diet seems to have been derived mainly from forest plants including hazel nuts.

A range of Maglemosian artefacts also occurs extensively in many other lowland southern and eastern areas, including Holderness, east Yorkshire (Davis-King 1980), tributary valleys of the Thames in Hertfordshire and Buckinghamshire and numerous sites on the Lower Greensand in the western Weald, such as Iping Common (SU 8525), Sussex (Keef *et al.* 1965). The main type of Mesolithic artefact found in the Thames deposits in the tranchet axe (i.e. with a transverse cutting edge). Other axe-heads have been found more widely, for example in west Yorkshire, Pembrokeshire, Wiltshire, Dorset, Devon and Cornwall, all at the western extremity of a Maglemosian province that extended across the then dry southern North Sea to Denmark, northern Germany and Estonia. Most are of flint, but in SW England many artefacts of the later Mesolithic were fashioned from Portland Chert (Palmer 1970).

On the higher ground of the Pennines, North York Moors, Dartmoor and parts of Wales, a slightly different Mesolithic industry (the Sauveterrian) lacks the heavy axe-heads and contains instead a profusion of microliths. This developed somewhat later in the Holocene, from *c.* 8000 ^{14}C BP (*c.* 8850 cal.) to the mid Holocene climatic optimum of the Atlantic period, and was associated with local deforestation mainly by natural or deliberate fires, as the horizons with microliths often contain charcoal. Also, although there was a decrease in pollen of *Quercus*, *Alnus* and some other trees, there was a relative increase in genera that can regrow rapidly after fire, such as *Betula* and *Corylus* (Walker 1956). Removal of the rapidly transpiring trees in these wet upland regions probably led to increased soil waterlogging, decreased nutrient cycling in the soil profile and greater soil acidity, which together encouraged the development of the ombrogenous blanket peat.

The start of the Neolithic Period in England and Wales at *c.* 5500 ^{14}C BP (*c.* 6300 cal.) resulted from the arrival of immigrants from the European continent and ultimately from the Middle East, who had agricultural skills unknown to the Mesolithic peoples. The Neolithic migration may still be detected in a SE–NW gradient of human DNA characteristics across Europe (Evison 1999), as the Mesolithic–Neolithic transition is thought to be the only time when a sufficiently large and distinctive population from the Middle East could have mixed with a sufficiently small indigenous population to create this type of pattern. In England and Wales the initiation of farming resulted in more widespread deforestation (Brown 1997). The earliest farming was mainly pastoral, because the decreases in tree pollen are accompanied in peat and lacustrine deposits by the first appearance of the pollen of pasture weeds such as dock (*Rumex*) and ribwort plantain (*Plantago lanceolata*). Slightly later soil cultivation is often indicated by the subsequent appearance of small amounts of cereal pollen. However, at a few sites in NW England and on the Isle of Man, cereal pollen occurs immediately after a phase of forest disturbance with charcoal (Innes *et al.* 2003).

Neolithic woodland clearance and cereal cultivation are detected in pollen sequences from northern England rather later than in the south. For example, in the lower Tyne Valley (NY 96–NZ 16), where it resulted in a major episode of alluvial deposition, clearance commenced at 4800±80 ^{14}C BP (5434–5662 cal.) (Passmore *et al.* 1992). However, in northern upland areas, such as the Cheviots (Tipping 1996) and North York Moors (Simmons *et al.* 1993), there was much less Neolithic pressure on woodland. Often there is palynological evidence for only sporadic episodes of slash-and-burn ('landnam'), in which small clearings were abandoned after the soil fertility conferred by incorporation of wood-ash had been depleted by cropping for a few years, so that the woodland regenerated naturally.

Neolithic artefacts include axe-heads made from flint and a wide range of other hard rocks, including fine-grained igneous and metamorphic rocks (Houlder 1961), and various stone, antler, bone and wooden tools for soil cultivation and crop-gathering (e.g. primitive ploughs and sickles). At many sites, such as Grimes Graves (TL 8189), Norfolk, and Harrow Hill (TQ 081100), Sussex, fresh flint was mined from the Chalk (Holgate 1991) at depths below where it had been fractured by Pleistocene frost action. However, although Neolithic sites are common on the Yorkshire and Lincolnshire Wolds, there are no flint mines, possibly because the chalk here is harder and the flint more brittle than in S England.

In Chalk and other limestone areas, where pollen is poorly preserved because of the high soil pH, a history of deforestation, pasture and subsequent arable cultivation has been reconstructed from terrestrial mollusc assemblages (Evans 1972). The soils buried beneath Neolithic barrows in chalkland areas such as Salisbury Plain often contain snails, initially typical of woodland and later of open grassland, suggesting that many of the burial mounds were built in open grassy areas after deforestation. In S England, palaeo-environmental data from numerous sites suggest a patchwork landscape with small dispersed communities practising regionally diverse activities (Whittle 1996).

Many Neolithic sites include above-ground artificial structures, such as elongated burial mounds (long barrows), and ditches and embankments for enclosures, which are unknown from the Mesolithic. These suggest a sedentary way of life and some concern for ancestry and the sacred (Edmonds 1999). Another characteristic feature of the Neolithic are henge monuments, such as Stonehenge (SU 123422) and the great circle and avenue at Avebury (SU 1069) on the Chalk landscape of Wiltshire. The main rock type used in Wiltshire was sarsen, a silicified sandstone probably formed in a veneer of basal Palaeogene sediments covering the Chalk of the Marlborough Downs. At Stonehenge large blocks of 'bluestone' (dolerite) and rhyolite from the Preseli Hills (SN 13–SN 23), SW Wales were also used (Green 1997), indicating that considerable cost and effort were expended on constructing these monuments. Some were built piecemeal or rebuilt over periods of several centuries. For example, the original henge ditch at Stonehenge was constructed close to 5000 ^{14}C BP (*c.* 5750 cal.) (Cleal *et al.* 1995), and dating of antlers from Y hole 30 suggest the final activity on the site was in the early Bronze Age *c.* 3500 ^{14}C BP (*c.* 3775 cal.).

Although the Bronze Age is marked archaeologically by the introduction to Britain of bronze tools and weapons *c.* 3800 ^{14}C BP (*c.* 4220 cal.), its main effect on the landscape was of increased woodland clearance, leading in particular to the disappearance of *Tilia* from many areas. Increased amounts of cereal pollen occur in sequences from upland areas such as the North York Moors (Jones 1978), although pollen of pastoral weeds often rises to such abundance that pasture must have been the main upland land-use (Simmons *et al.* 1993). In some upland areas of SW England and NE Yorkshire, where the soils of the areas newly cleared for agriculture were bouldery, the stones were often moved to create cairns and field walls (Fleming 1971), which still persist today. Repeated cultivation of the Chalk downlands in the later Bronze Age is indicated by the presence of 'celtic fields', often associated with enclosed settlements. Frequent ploughing of these fields created marginal lynchets and resulted in widespread deposition of colluvium in dry valleys (Bell 1983) and alluvium in lower valley sections.

During the Bronze Age there was also a considerable increase in monument building, notably the construction of round barrows, and in the later Bronze Age large amounts of timber were used to create palisaded enclosures, primitive hillforts and wetland trackways, such as the Sweet Trackway across the Somerset Levels (Coles *et al.* 1984).

Holocene soils

Extensive evidence for Holocene pedogenesis is provided by the unburied surface soils widely developed in Late Devensian glacial and periglacial sediments throughout England and Wales. Typically these show evidence for decalcification to depths of 0.7–1.0 m, clay illuviation increasing the clay content of subsoil (Bt) horizons by up to 8%, chemical weathering of pyroxenes, amphiboles, plagioclase feldspars and some layer silicates (where these are present in the parent material), gleying where affected by seasonal waterlogging, and podzolization (downward leaching of Fe and Al in sandy soils under acidic heathland or pine forest).

Early and mid Holocene buried soils are also quite common in England and Wales. Most were buried by human activities, such as construction of burial mounds or deposition of colluvium (hillwash) resulting from Neolithic or later soil erosion. In Kent soils formed in Late Devensian loess and buried by colluvium contain evidence for clay illuviation (Weir *et al.* 1971; Catt *et al.* 1987), which probably occurred under the deciduous woodland that developed in the early Holocene. Subsequent pedogenesis in the overlying colluvium has resulted in further clay illuviation, although usually at a much slower rate than in the early Holocene. In many areas of non-calcareous sandy parent materials, with surface podzols under heathland, the mid Holocene barrows and other monuments, particularly those constructed in the Bronze Age, cover less acidic and more nutrient-rich early Holocene soils, such as brown earths with subsurface horizons enriched in illuvial clay (Dimbleby 1962). This suggests that the soils have become progressively more acidic through the Holocene, although whether this resulted from climatic deterioration or the decrease in soil fertility by anthropogenic activities such as deforestation and primitive agriculture is unclear.

The characteristics of unburied surface soils important for ecology, agriculture, water movement and other environmental factors in England and Wales are largely determined by the particle size distribution and mineralogy of the soil parent materials. The properties of Quaternary sediments and pre-Quaternary strata occurring at the land surface are in fact utilized by the National Soil Resources Institute for differentiating soils at the lowest level of classification (soil series), which is used for depicting soil variation on their 1:25 000 and 1:50 000 maps (Clayden & Hollis 1984). This is possible because the Holocene was too short and its climate too mild for the particle size distribution and bulk mineralogy of most soil parent materials to be altered significantly. The main soils on the present land surface inherited from pre-Holocene periods are the paleo-argillic brown earths and paleo-argillic podzols (Avery 1985) occurring on upland plateaux (e.g. the Chalk dipslope) and pre-Devensian river terraces south of the Anglian (MIS 12) glacial limit. These older soils have subsurface horizons that have been reddened and enriched in illuvial clay more strongly than any Holocene soils, and indicate areas that escaped significant glacial and periglacial erosion, particularly in the Devensian Stage.

Applications of Quaternary studies

Quaternary deposits are important sources of several natural materials, especially sand and gravel for aggregate (Smith & Collis 1993). Because of depleting resources and planning permission problems, increasing quantities of aggregate are now being obtained from bedrock sources or offshore Holocene gravels rather than Pleistocene deposits on land. However, in many parts of England and Wales, coarse Pleistocene fluvial, glaciofluvial and marine deposits still provide convenient local sources of sand and gravel. Studies indicating ice movement and river flow directions are useful for predicting clast sources and thus the content of components which affect the quality and suitability of deposits as aggregate. Dating of deposits can also be useful, in that prolonged weathering of older deposits in one or more interglacials has often removed unwanted components, such as soft chalk or pyrite-containing clasts. In northern England, Late Devensian coversands have also been exploited for building, moulding or glass-making, and in SE England small brickmaking enterprises use Devensian and older loess deposits.

Local knowledge of Quaternary history is important in civil engineering projects, such as road-building, housing construction, siting of waste disposal facilities and dam construction for reservoirs (Hutchinson 2001). For example, in S England and the Midlands, numerous site stability problems have arisen from reactivation of shallow periglacial (usually GS - 1) translational landslips and mudflows by cuts and fills on gentle clay slopes. Other landslips in upland glaciated valleys have resulted from steepening and smoothing of valley sides by glacial erosion (Waltham & Dixon 2000) or by release of lateral pressure on the bedrock after melting of the ice (sheeting). Also, prediction of the likely lateral and vertical variation in deposits from knowledge of a site's Quaternary history (e.g. occurrence of glaciofluvial gravel bodies within an over-consolidated subglacial till) is of special value in geotechnical site evaluation.

Because it often determines the nature and variation of soil types in England and Wales, the Quaternary history of deposition, erosion and pedogenesis is central also to understanding many present-day environmental problems. Soils are both sources and sinks of numerous possible atmospheric and hydrospheric pollutants, such as CO_2, CH_4, N_2O, Rn, P, NO_3^- and pesticides, and their behaviour with respect to these is influenced by factors such as clay, carbonate and organic matter contents, which depend strongly on Quaternary history. The considerable variation in soil properties resulting from Quaternary processes also presents problems in intensive agricultural production, because the different soil types, often forming intricate patterns within a single field, require different management techniques for optimum crop production. Methods for achieving this based upon detailed maps of soil properties and the use of computer-controlled farm machinery, known by the general term 'precision agriculture', are currently being developed. The intricate patterns of soil inhomogeneity leading to the necessity for precision agriculture results mainly from short-range variation in processes of glacial deposition, periglacial soil disturbances (e.g. cryoturbations, ice-wedge polygons), gelifluction deposits derived from different sources upslope and archaeological features such as pits and buried foundations (Catt 1992).

Establishing a robust and complete stratigraphy for the Quaternary also provides an essential template for studying many important ecological and environmental processes, especially those that are sensitive to climate change. These include species migrations, groundwater depletion and recharge, sea-level variations, responses of fluvial regimes (e.g. flood frequency) and the rate and causes of abrupt climatic events. Given the growing concern about the possibility of rapid environmental change in the near future, it is important that we understand the mechanisms and thresholds that determine these processes and the interactions between them. Study of the Quaternary record is especially valuable in this context, because it includes the likely boundary conditions under which future changes will occur, although there are no Quaternary analogues for some changes, such as the current and expected future levels of atmospheric CO_2 and CH_4.

Conclusions

Because of the numerous major climatic oscillations and associated changes in terrestrial depositional environments, most Quaternary formations of England and Wales are much more variable, both vertically and laterally, than those of any previous period. The main exceptions are the lithologically homogeneous periglacial aeolian deposits (loess and coversands) and the tills of some lowland areas, such as the Anglian Lowestoft Till and the Late Devensian Skipsea Till in eastern England. Almost all other Quaternary deposits are either very restricted in distribution or show considerable short-range variability, which creates problems in geotechnical interpretation, assessing environmental impacts and agricultural exploitation. It also means correlation by lithology is often difficult and unreliable.

Despite these problems, it is important to reconstruct Quaternary environmental history in detail, because it links previous geological periods to the present and explains the origin of modern environments. The latter is especially significant because it helps predict future changes in climate and sea level by distinguishing the anthropogenic effects from the moving background of natural changes.

These and other applications of Quaternary geology have driven a considerable recent expansion in research on the causes and worldwide effects of Quaternary climatic change and also on the detailed Quaternary history of England and Wales. As a result, Quaternary science has become increasingly multidisciplinary with inputs from, and benefits to, anthropologists, archaeologists, civil engineers, climate modellers, ecologists, geomorphologists, hydrologists, meteorologists, phenologists, oceanographers and soil scientists, as well as the usual range of geological sub-disciplines. This trend is clearly set to continue and lead to expansion and intensification of links between geology and other sciences.

Consequently the recent proposal by the International Commission on Stratigraphy to abandon the term Quaternary as a formal time unit is seen as inappropriate by almost everyone studying recent Earth history (Gibbard *et al.* 2005). The wide use of the term among numerous groups of professional scientists, the ubiquitous influence of characteristic Quaternary processes on terrestrial and marine environments, and the immense socio-economic importance of changes in these environments together indicate that this valuable term must be retained. Any contrary promulgation will cause confusion and bring derision on the geological profession.

18 The future: climate and sea-level change, glaciation and a northward drift

P. L. GIBBARD, J. E. RAWSON & A. G. SMITH

Although geologists are used to looking backwards in time, to reconstruct past events and processes from the preserved rock record, one of the roles of the scientist is to use their evidence to predict the consequences of their observations. One of the most important lessons from the geological history of the British landmass is that the dynamic interaction of processes that operate at the Earth's surface can produce highly differing but also predictable patterns of sedimentary sequences under varied environmental conditions. Moreover, the present, if it is indeed the key to the past in the true Huttonian sense, may also be seen as a key to predicting our geological future, albeit with the modifying effect of the interference of humans in the operation of natural processes.

In this chapter an attempt is made to offer some insight into potential future geological and environmental developments. These ideas are based on a forward projection of processes operating today, or in the recent geological past, and give rise to what is hoped are interesting insights. For the sake of simplicity, three timescales are examined. First, the climate changes predicted to occur over the next two centuries are discussed, mainly focusing on the role of greenhouse-gas emissions on our modern climate and its implications. This is followed by a short review of the changes that might be expected over the next 130 ka, based on climate simulation modelling. Finally, a speculative review of long-timescale developments is presented.

Climate change over the next two centuries

Over the last century the mean surface air temperature of the Earth has increased by 0.58 °C for Europe (IPCC 1998). This could represent natural warming since the end of the Little Ice Age (1590–1850), but investigation of ice-core samples from the Vostok site in Antarctica has shown that at the same time, because of increasing industrial emissions, the atmospheric content of CO_2 is also increasing and has now reached a level greater than at any time over approximately the last 400 ka (Raynaud *et al.* 2003). The key property of CO_2 (and other 'greenhouse gases') in the atmosphere is that it absorbs low-wavelength radiation (heat) and thereby prevents heat loss from the Earth's atmosphere. A given change in CO_2 concentration causes a direct change in the global average surface temperature, although details of the precise mechanism remain uncertain. However, the complexity of the relationship is illustrated by comparison of mean annual temperature during the last interglacial in England and Wales and beyond, which was 3–4 °C higher than today, with today's higher concentrations that have yet to produce unprecedented warmth. Nevertheless, future projections, published by the IPCC (IPCC 1995, 2000, 2001) in the Special Report on Emissions Scenarios (SRES), show that mean surface warming is expected to increase by as much as 5 °C by 2100 as a consequence of continuing emissions of CO_2 and other greenhouse gases. The spatial patterns generated by these modelling exercises suggest that surface warming is likely over most of the Earth's surface, with the greatest increases of up to 10 °C occurring in high northern latitudes. The consequent melting of sea ice is expected to reduce the albedo of polar regions thereby increasing heat absorption. Differential heating of the land in comparison to the sea results in the land warming 80% faster than the oceans. Exposure of land surfaces previously covered by ice, especially in the polar regions, is therefore likely to lead to rapid warming in those areas.

The increase in global temperatures is also likely to be accompanied by increasing global mean precipitation. Although it is difficult to obtain predictions as as reliable as those for temperature, SRES predictions indicate the largest regional changes in precipitation occur in tropical regions. In the 21st century, substantial increases are likely in eastern Asia, Central Africa, eastern South America and high northern latitudes, whilst substantial decreases will be seen in southern Europe, southern Africa, Australia, Central America and northern South America. Precipitation increase in western Europe could potentially change river flow regimes and flood frequencies, which would in turn result in increased soil erosion and therefore sediment transfer (Bell & Walker 2005). Changes in mean annual temperature, seasonal variability, precipitation and related phenomena are expected to be accompanied by changes in plant and animal distribution, increased rates of chemical weathering and soil development.

However, the most important consequence from the geoscience perspective is sea-level rise. Increases in global eustatic sea level are a consequence of the interplay of thermal expansion of the oceans, the melting of sea ice and mountain glaciers, but above all melting of the west Antarctic and Greenland ice sheets; the latter likely to be completed by 5 ka from now (Berger & Loutre 2002). Simulations indicate that sea-ice cover, particularly at the highest latitudes in the Arctic Ocean, is likely to be drastically reduced (Fig. 18.1). Although sea level is predicted to rise virtually throughout the world, there is considerable local variation, with some regions experiencing little or no increase, while others may see rises of as much as twice the global mean value.

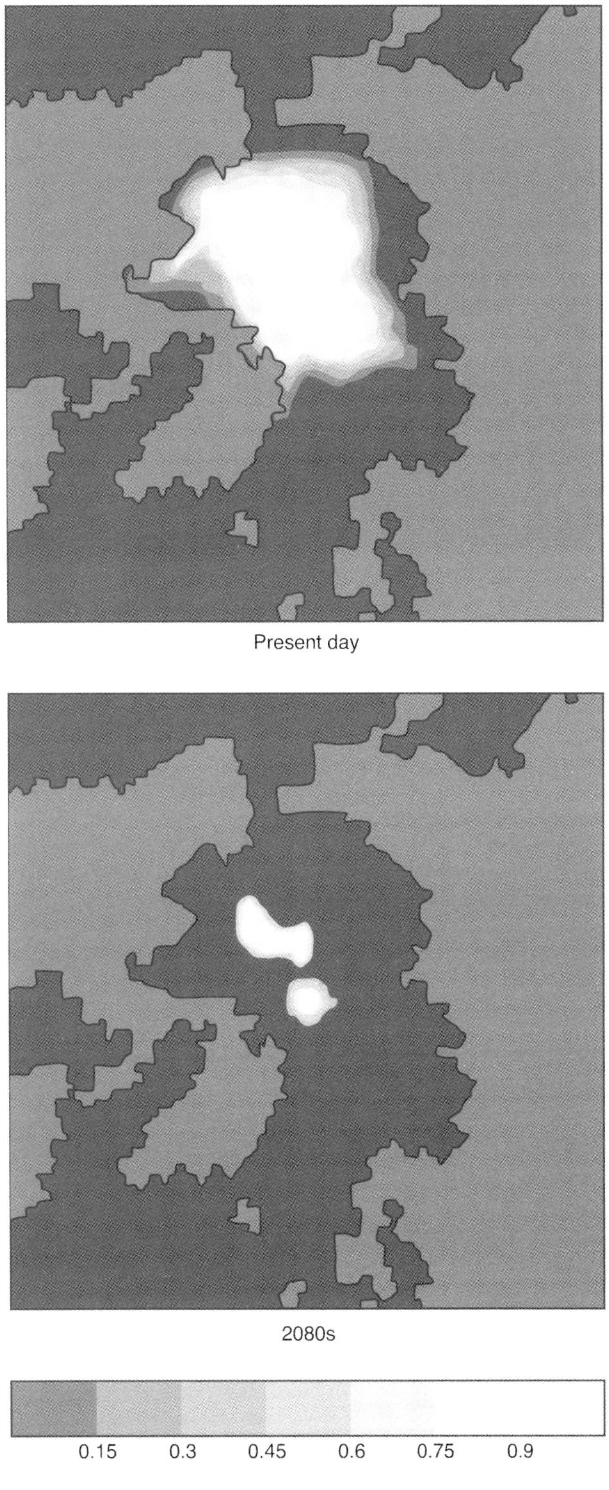

Fig. 18.1. Climate model estimates of September Northern Hemisphere sea-ice cover (fractional) at the present day and at and after 2080 for the A1F1 emissions simulation. The clustering of ice around the North Pole in the future plot may be a modelling artefact (source Meteorological Office: 2003).

One change that is likely to have profoundly important consequences, particularly in the North Atlantic Ocean and surrounding areas, is in ocean current patterns. This is because thermohaline circulation is the main engine for heat transfer on the planet's surface. All the SRES scenarios predict that the magnitude of the thermohaline circulation is likely to decrease by about 20% during the 21st century. In the upper ocean, this corresponds to a reduction in the Gulf Stream, which transports heat from the Gulf of Mexico to northern Europe. Whilst this reduction will act to cool northern Europe, with some workers predicting a deterioration of conditions similar to those in present-day Labrador, any cooling will be more than offset by greenhouse warming, so that both the ocean and the continent warm substantially.

All these climate-change predictions assume that developments are likely to be slow. However, there are strong reasons for thinking that the unprecedented atmospheric conditions could be of sufficient magnitude that an abrupt change could be triggered. Current lack of knowledge on biosphere responses to such events could mean that, among other things, an abrupt climatic shift could trigger an ecological disaster large enough to be truly a mass extinction (Overpeck *et al.* 2003).

Implicit in these short-term predictions of climate change, because of the substantial inertia of the ocean-atmosphere system, are effects which will lead to marked modification of future longer-term natural climatic changes. Even if present levels of greenhouse-gas concentrations are stabilised at today's levels, the world is committed to 1 °C of additional global warming, and about 1 m of sea-level rise, from emissions that have already occurred over the last century. Moreover, global sea level is likely to continue rising for many centuries after greenhouse-gas concentrations have been stabilized.

Sea-level change in England and Wales

Present situation and trends

More recent anthropogenic influences and longer-term effects of natural global climate change are changing the physical pressures on our coastline. The average rate of relative sea-level rise during the last century around the UK coastline was approximately 1 mm year^{-1} (Hulme *et al.* 2002); this average takes account of significant regional variation, partly resulting from isostatic rebound prompted by glacial retreat following the last ice-age (Chapter 17). Depending on the region, this effect either exacerbates the effect of rising sea level, or works to negate it – maximum relative land uplift occurs in central and western Scotland, at around +1.6 mm year^{-1}, and maximum subsidence is in SW England, at −1.2 mm year^{-1} (Shennan & Horton 2002). In addition, consolidation of Holocene sediments, resulting from compaction and land drainage, increases subsidence in areas with thick sequences of these sediments, with an average effect equivalent to, at least, approximately an extra 0.2 mm year^{-1} subsidence, but more in parts of SE England, ranging from 0.5 to 1.1 mm year^{-1} (Shennan & Horton 2002).

The powerful action of waves during storms is a key factor in eroding the coast and increasing flood risk, and average significant wave height has also increased. Over the northern North Sea, for the relatively active January–March period, there has been an upward trend for the last 30 years (Hulme *et al.* 2002). Evidence from the meteorological office shows the average number of winter storms has also increased significantly, with the largest increases in the south (Meteorological Office 2003).

When viewed over geological timescales, human occupation of England and Wales has taken place very recently, and any attempts on our part to influence the long-term future evolution of our coasts and seas may appear insignificant. However, over the next 200 years or so, in addition to directly affecting global climate change, the way in which humans manipulate natural coastal processes will continue to influence the shape and nature of our coastline.

England and Wales have a coastline approximately 3700 km long, and coastal defence has a long history. During the 20th century a significant increase in investment was promoted after the World War II to decrease loss of land and promote agricultural production. After the east coast storm surge of 1953, when 300 people lost their lives as defences were breached in 1200 places, and more than 25 000 houses were flooded, investment in sea defences was again increased. There was further drainage of low-lying land, building of embankments, reclamation of intertidal areas and strengthening of sea walls. Much of this land remains at risk of flooding and a legacy of building houses behind defences means approximately 10% of the population is now at risk of river or coastal flooding. One third of the population live within 10 km of the coast, much of which is also susceptible to erosion (Environment Agency pers comm).

Natural ongoing patterns of erosion and accretion, coupled with the increasing rate of sea-level rise and storminess, continue to shape our coastline. At present approximately 28% of the coastline experiences erosion greater than 10 cm year^{-1}; but as a significant proportion is being protected by defences, a more realistic estimate of potential erosion is that 67% is under threat (Halcrow 2002).

Gradual erosion of hard and soft rock cliffs has affected many parts of the coast, with some areas often losing towns and villages over the last few hundred years – particular examples are areas of the Holderness coast and those in Norfolk and Suffolk, where erosion rates often exceed 1 m year^{-1}.

The loss of beaches fronting low lying areas is also widespread; in many coastal and estuarine areas the process of coastal 'squeeze' accounts for the loss of intertidal deposits. This results from the dual effects of sea-level rise and human interference in altering the balance of processes at the coast. In low-lying areas such as the Essex estuaries, over 950 ha of saltmarsh has been lost since 1973 (University of Newcastle 1998). Marsh and mudflat accretion rates may keep pace with sea-level rise if sufficient sediment is available. However, if the rate of rise increases, or if sediment sources are insufficient to keep pace, these areas can be 'drowned'. Even if vertical accretion does keep pace with rising sea levels, coastal defences may prevent the lateral migration of these marshes, which are then 'squeezed' by rising sea levels, leading to erosion and intertidal lowering.

Another widespread contemporary effect is coastal steepening, demonstrated by the rate of recession of Mean Low (MLW) Water level exceeding that of High Water (HW) level. This has recently been shown to affect 61% of the coastline of England and Wales. Rates of MHW retreat, typically highest on the east coast, average –0.3 m year^{-1}, with a maximum of –0.5 m year^{-1}. The average MLW recession rate of –0.9 m year^{-1} shows 'steepening', with figures calculated for each area using an average of 113 years' data. The widespread nature of steepening as a phenomenon shows that it can not be explained solely by underlying geology or tidal range, and may only in part be explained by 'squeeze' against defences. Steepening remains the subject of ongoing research, and similar results have also been demonstrated elsewhere in Europe (Taylor *et al.* 2004).

Predictions for the future

Relative sea level (including the effect of land movements) will continue to rise around most of the UK shoreline, the rate depending on region and actual emissions of greenhouse gases. By the 2080s, sea level may be between 26 and 86 cm above the current level in SE England (Hulme *et al.* 2002).

Extreme sea levels, occurring through combinations of high tides, sea-level rise and changes in winds, will be experienced more frequently in many coastal locations. For example, for some east coast locations a water level that at present has a 2% probability of occurring in any given year, will have an annual occurrence probability of 33% by the 2080s for the medium-high emissions scenario (Hulme *et al.* 2002). The most damaging events, comprising surge tides and high waves, will occur more frequently in the coming decades so increasing the risk of flooding and erosion.

Over the next 30–100 years the threat posed by accelerated rates of erosion which will be greatest along the east coast and in the major estuaries, and the natural response of the coastline to sea-level rise will be to migrate landward (Evans *et al.* 2004).

Impacts on geomorphology and geology

A period of accelerated sea level rise could result in widespread destabilisation of coastal defences and trigger far more active shoreline evolution, exacerbated by a general reduction in onshore transport of sediments, and a decrease in alongshore transport because of defence and other engineering works (Hosking & McInnes 2002). More frequent storms may mean less 'recovery time' for beach systems between events, with long term or permanent losses of material offshore rather than the cyclical on-offshore movements which are common today. The major effects of surge events on sediment transport in the southern North Sea have recently been modelled (HRWallingford 2002), suggesting that any increase in frequency could significantly alter the nature and location of offshore banks, with knock-on impacts for coastal evolution.

Clearly the sensitivity of each type of coastal feature will determine its response to these changes. Hard rock cliffs are subject to infrequent failures and their strength tends to make them less susceptible to increased wave action or inundation. Soft rock cliffs are more vulnerable to these forces, and also to other erosive forces such as groundwater flow and aeolian erosion. Soft cliffs can be relatively simple systems which recede on a gradual and ongoing basis through falls, slumps or slides of material; where soft strata are interspersed with harder bands, failure can be more sporadic. Soft rock coastal cliffs are more likely to retreat as they are directly eroded by increased wave action, and as increased rainfall alters groundwater patterns and flow. Retreat rates for unprotected cliffs are likely to double in future, depending on local conditions (Bray & Hooke 1997), although there might be a lag before the cliff top responds to conditions at the toe. Erosion processes will also increase potential for reactivation of coastal slopes presently covered by relict slides. Landslides such as those at Lyme Regis and on the north coast of the Isle of Wight following the wet winter of 2000–2001 may occur more frequently in the future (Hosking & McInnes 2002).

The continually evolving nature of beaches fronting lowlands will make them very sensitive to the changing pressures of the next 200 years. Rates of saltmarsh erosion in estuaries are likely to accelerate in future with increasing coastal 'squeeze' and drowning of these systems. Based on continuation of present rates of loss of saltmarsh deposits, a recent study concluded that all such areas in the Stour and Orwell estuaries, Hamford Water, the Crouch and Roach and Thames Estuary could be lost within the next 50 years (Gibberd & Brew 2004).

Recent forecasts of future coastal erosion rates have taken into account likely changes in sea-level rise, surge activity, wave heights, littoral drift and movement of the shoreline as part of the 'Foresight' study of future flooding (Evans *et al.* 2004). This study has recognized that human activity over the next decades will both influence climate change and determine how much we are affected by it. Four potential scenarios of the future are represented by different societal values and types of governance. 'World Markets' and 'National Enterprise' represent 'consumerist' values, while 'Global Sustainability' and 'Local Stewardship' represent 'community-oriented' values (Evans *et al.* 2004).

Results show that, depending on the climate scenario adopted, over the next 100 years England and Wales could see an average of 99 m of coastal erosion, up to an average of 175 m. This compares to a predicted average of between 20 and 67 m if contemporary rates of erosion are not affected by additional climate change (Fig. 18.2) (Evans *et al.* 2004). As many areas will experience little or no erosion, these figures mean that others will experience erosion of several hundred metres. Figure 18.3 demonstrates both large variation in rates of erosion around our coastline for any one model, and variation in risk for any one area depending on the model adopted.

Management response

The predicted impacts of climate change and ongoing long-term coastal evolution will significantly add to the risk of the destabilisation of the existing system of sea defences. It is

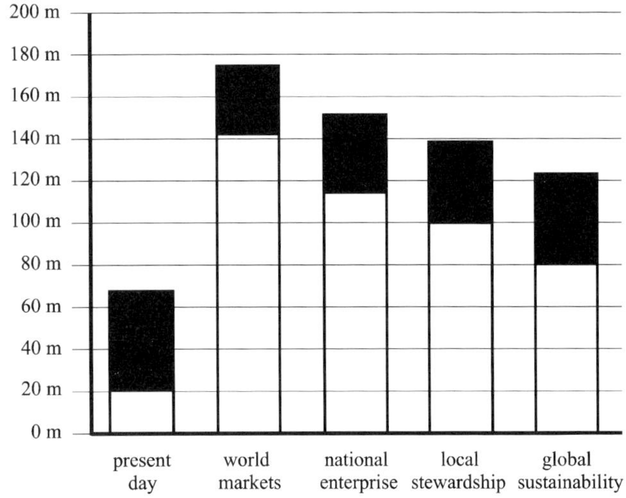

Fig. 18.2. Average coastal erosion of England and Wales over the next 100 years (metres) – the black bars show the range of possible values for each scenario (after Evans *et al.* 2004).

thought that 92 000 houses may be at direct risk from erosion by 2100 if we do nothing (Defra 2004), and many more from tidal flooding. Where it is necessary to hold the present coastline in place, it will become increasingly difficult and expensive to do so. Obvious concerns are that defences will simply be overtopped, but losses of beaches and marshes, increased direct wave action in deeper water close to the shore, changed patterns of drift and changed wave foci along the coast will all challenge our ability to manage the coast. They will also challenge our decisions to do so – it is clear from the climate change scenarios examined in the Foresight Study that it is going to be neither economically feasible nor technically sensible to attempt to hold the entire coastline in its present position. To do so would result in massive engineering structures standing in deep water having a significant effect on coastal processes and risking major losses through catastrophic failure. If our river and coastal flood and erosion management policies remain unchanged from the present, and present levels of investment continue, estimated annual damage will rise from under £1000m at present to around £25 000m in 2080 under the most extreme consumerist scenario. This means the number of people at high risk of river and coastal flooding could increase from 1.6 million today, to between 2.3 and 3.6 million by the 2080s (Evans *et al.* 2004).

Change over the next 130 000 years

The recent geological record of climatic changes in the latest Pleistocene suggests that the Holocene is in fact an interglacial with all the characteristics of previous events typified by the last or Ipswichian Stage interglacial (126–115 ka: approximately equivalent to Marine Isotope Substage see Chapter 17, p. 450),

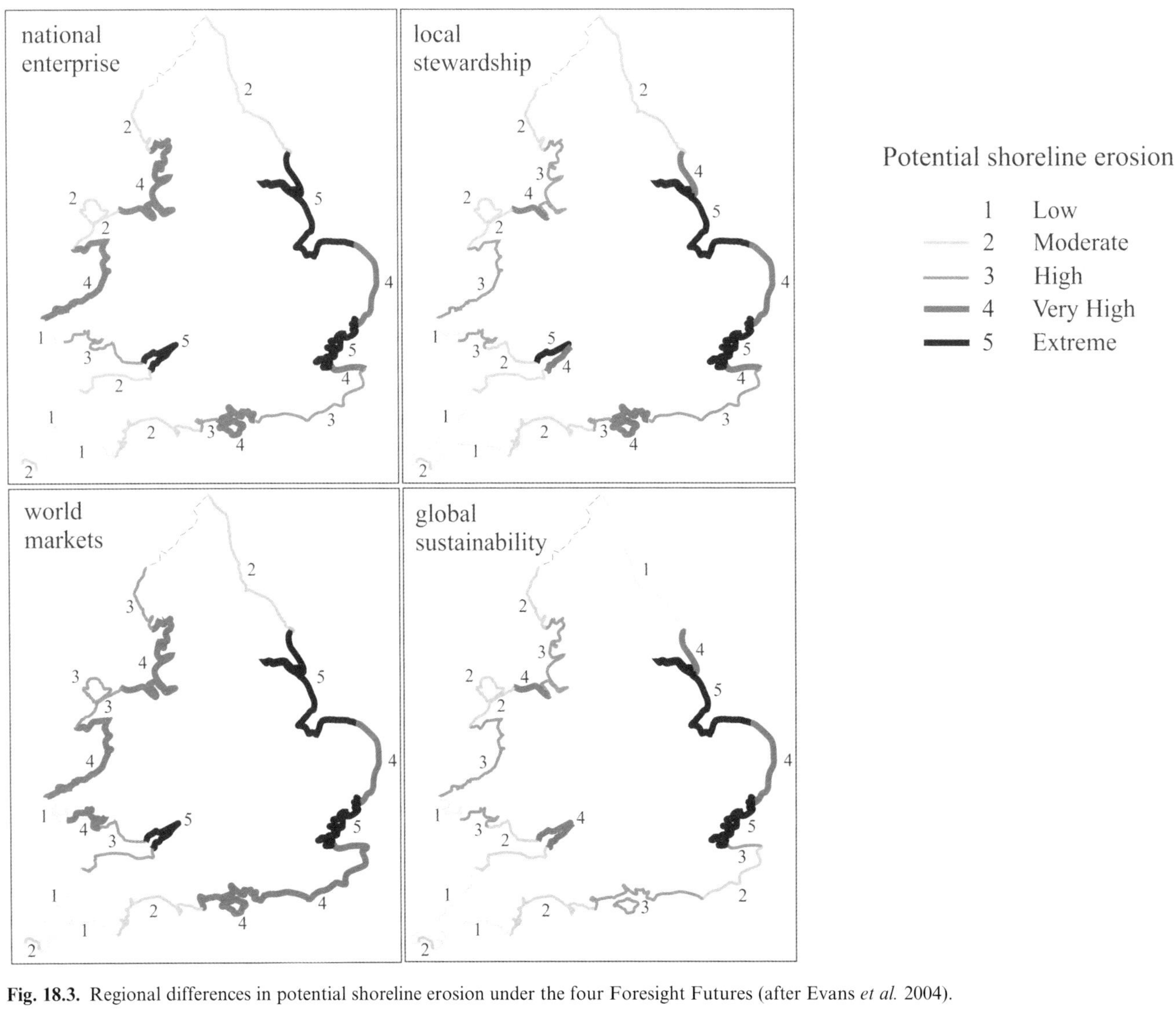

Fig. 18.3. Regional differences in potential shoreline erosion under the four Foresight Futures (after Evans *et al.* 2004).

although it more closely resembles the late Middle Pleistocene Hoxnian Stage (*c.* 350 ka: equivalent to part of Marine Isotope Stage 11). The duration of the Ipswichian was approximately 11.0 ka and therefore it is reasonable to suppose that in the next few thousand years the earth should experience climatic cooling leading to the arrival of the next glacial or cold period.

During the last two decades several attempts have been made to predict the possible future course of large-scale climate changes based on simple statistical models that did not include CO_2-concentration forcing. These models have suggested that a significant cold period will not arrive before 20 ka and a major glacation will not occur before 60 ka (Joussaume 1999). However, as the variation in insolation (solar radiation received at the top of the atmosphere) caused by the Milankovitch orbital cycles drives the large-scale glacial–interglacial climatic alternations, it is possible to predict the course of future climate changes on a geologically significant timescale for the next 130 ka by projecting forward these variations. Recent simulations for this period by Berger & Loutre (2002) have shown that the amplitude of insolation will be relatively small, compared to that during the Ipswichian interglacial (Fig. 18.4). Moreover they suggest that the present interglacial (Holocene) will be extended to an exceptionally long 50 ka, i.e. almost 5 times longer than the Ipswichian. This is because the climate system may take 50 ka to assimilate the impacts of anthropogenic activities. These simulations also predict that the next glacial maximum will occur in approximately 100 ka, with a short, about 5 ka interstadial warm event 60 ka from now.

It is probable that an initial cooling, following this 'long-Holocene' interglacial but before the interstadial, is likely to be similar to the Little Ice Age (1550–1850 AD: Grove 1988, 2004) that occurred during recent history. This period was one of a series of Neoglacial events during cooler periods within the Holocene recorded from peat bog, lacustrine sediment sequences and glacial ice-front oscillations (cf. Chapter 17). The Little Ice Age saw a cooling of mean global temperature by 1 °C, with more severe winters, advances of glaciers and withdrawal of the treeline in mountain and northern latitude regions (Grove 1988, 2004; Roberts 1998).

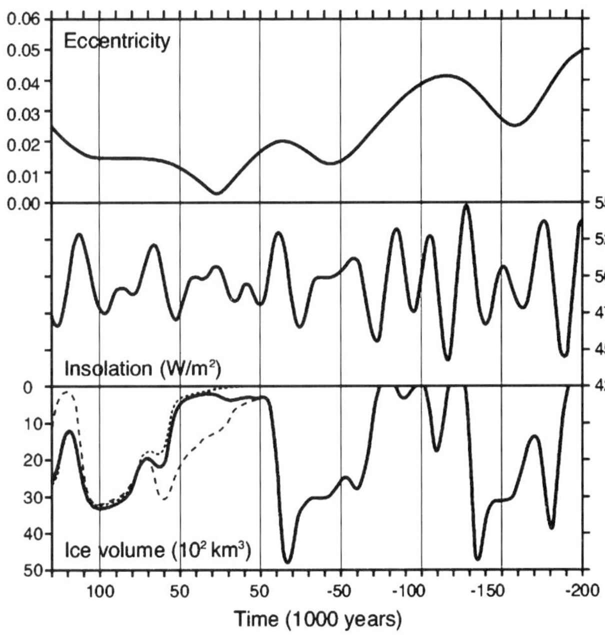

Fig. 18.4. Long-term variations of orbital eccentricity (periodicity 100 ka) (top), June insolation at 65°N (middle) and simulated northern hemisphere ice volume (increasing downwards) (bottom) for 200 ka before to 130 ka after present. Time is negative in the past and positive in the future. For the latter, three CO_2 scenarios were used: last glacial-interglacial values (solid line), human-induced concentration of 750 ppmv (dashed line) and a constant concentration of 210 ppmv (dotted line) (after Berger & Loutre 2002).

Continued cooling will see the southward retreat of forests in England and Wales and their replacement by shrub-dominated steppe grassland. After the short respite of the interstadial, once the climate is sufficiently severe, the grassland will give way to patchy tundra-like vegetation. Animals typical of present-day Britain will also retreat southwards to southern refugia, leaving the region populated only by those currently typically found in polar regions, like reindeer, arctic fox and lemming. The reduction in mean annual temperature of over 12 °C will result in re-establishment of physical processes typical of polar regions, with ground-water becoming permanently frozen. The whole region will take on an appearance comparable to that during much of the last glacial period (Devensian). The fall of sea level, that inevitably follows from the build-up of glacial ice initially on land areas beyond Britain, will result in exposure of the shelves surrounding Britain, reuniting it with the continent once more. The rivers, like the Thames, will extend their courses across the shelves to follow the retreating sea. In the process they will reunite with those like the Rhine and the Seine from the near continent to form substantial water courses such as the Channel River not seen since the Late Devensian *c.* 20 ka earlier. Rivers swollen by spring meltwaters will carry coarse detrital materials and adopt a braided form, after first having incised their valleys and in the process having destroyed much of the pre-existing interglacial floodplain sediments. Ultimately the ice sheets may arrive once more, initially in the mountains of Wales, Scotland and Scandinavia, then expanding and coalescing to occupy lowland England and surrounding areas.

Change over millions of years

A continuing ice age

Predicting long-term geological developments over the coming millions of years is highly speculative. Nevertheless, certain patterns can be suggested from the geological record. For example, the operation of plate tectonics can, like Milankovitch cyclicity, be projected into the future, on the basis of recent tendencies and geological histories. The present distribution of continents and oceans is fundamentally important for determining the future evolution of Earth's environments, in particular climatic evolution in the 'geological-scale' future. This is because the efficiency of heat transfer from tropical to polar regions is determined by the thermohaline oceanic circulation and to a lesser extent by winds. The patterns of both are controlled by the form and orientation of the ocean basins and mountain chains, respectively. The latter profoundly influences the monsoonal oscillation, particularly in eastern Asia.

Seen in the context of long-term development from the 'hot-house' climate of the Mesozoic, the progressive climatic deterioration through the Tertiary that culminated in the 'ice-house' climate of the Quaternary can be seen as a response to the changing distribution of land and oceans on the planet's surface. This change began in the late Cretaceous with the closing of the mighty E–W-orientated Tethys Ocean and its progressive replacement by the Alpine–Himalayan mountain chain in the late Cenozoic. At much the same time Antarctica passed over the south pole and the Atlantic Ocean began to open. This critical alignment of N–S-orientated ocean basins and continental land masses resulted in the thermal isolation of the polar regions. The result was initially seen in Antarctica where the East Antarctic ice sheet began to form in the Oligocene, although small glaciers had existed on the continent earlier. Substantial glacio-eustatic sea-level fall followed as an automatic consequence of storage of water from the oceans on the land, as it does today.

The continued closing of corridors by which heat could be exchanged by the ocean–atmosphere system within the tropical regions was completed by closure of the Isthmus of Panama and the raising of the Himalayan mountains in the late Cenozoic.

It is this distribution of land and sea that gives rise to the strongly zoned climate system that characterised the Quaternary and allows the Milankovitch cyclicity to drive the climatic oscillations that are expressed as the glacial–interglacial oscillations. There is no reason to suppose that the large-scale climatic oscillations that have typified the last 1 Ma of later Quaternary time will end in the foreseeable future. This is because the circumstances outlined above that initiated the decline from the Mesozoic 'hot-house' to the Quaternary 'ice-house' conditions have not changed. Indeed it is likely that once the Earth has negated the effects of human interference with the climate system, Quaternary-type climates will continue for millions of years into the future. This is supported by comparison with the long duration of earlier 'ice ages' such as that during the Permo-Carboniferous. A return to a more equitable, less strongly-zoned climate, accompanied by higher eustatic sea levels like those of the Palaeogene, can only be expected when the poles are less isolated from ocean–atmosphere system heat transfer. That will occur when Antarctica moves away from the South Pole, the Arctic Ocean becomes less land-locked or an E–W-orientated corridor opens to link the major ocean basins.

Possible worlds 100 and 200 Ma in the future

Predicting the future positions of the continents is an untestable theoretical activity and therefore, some would say, unscientific. However, there are some principles, mostly qualitative, that suggest what the future configuration of the continents may be. We can assume that plate tectonics will be applicable over this period, i.e. the Earth will still be hot enough for plates to exist and move. Also plates will have divergent, convergent and conservative margins, and ocean floor will be created at the same rate as old surface areas are removed by subduction.

Short-term plate motions are well known (DeMets *et al.* 1990) and are being progressively refined. They can be extrapolated forward a few million years, but not hundreds of million years into the future. Yet, if we are to see significant changes in continental distributions we need to consider intervals of this duration. Continental motions relative to most reference frames are generally much less than 50 mm year^{-1}, so changes of say, 3300 km, or about 30° of latitude, in the positions of the continents will take over 60 Ma to accomplish. Thus we need to travel 100 Ma or more into the future to begin to see obvious differences in continental positions.

The Atlantic–Indian Ocean hot-spots frame (Muller *et al.* 1993) provides a longer-term past reference frame, but it is unclear what it represents, particularly for future periods. Consequently we adopt the palaeomagnetic reference frame because this is directly related to the Earth's spin axis. The fundamental assumption of palaeomagnetism is that, on average, the Earth's magnetic field is a geocentric axisymmetric dipole field, or GADF (e.g. McElhinny & McFadden 2000). This field model can be used to turn a reassembly without latitudes into a palaeocontinental map with latitudes. Palaeolongitudes cannot be found from the palaeomagnetic record, but are irrelevant if all continental fragments are repositioned relative to a reference continent.

About 100 Ma ago the global mean palaeomagnetic pole relative to Eurasia was at about 77°N, 203°E. Linear extrapolation into the future with an average rate of change in pole position of about 15 mm year^{-1}gives a pole position at +200 Ma of 63°N, 43°E. Eurasia can then be repositioned relative to the geographic poles at any time in the next 200 Ma, so that any reassembly of the continents relative to Eurasia can be mapped.

The positions of the other continents and associated fragments on the +100 Ma (Fig. 18.5) and +200 Ma (Fig. 18.6) 'maps' are guided by the following very general, non-quantitative principles, whose justification requires a longer discussion than is possible here. All passive continental margins appear to become active margins within 250 Ma of their creation; subduction zones become inactive only as a result

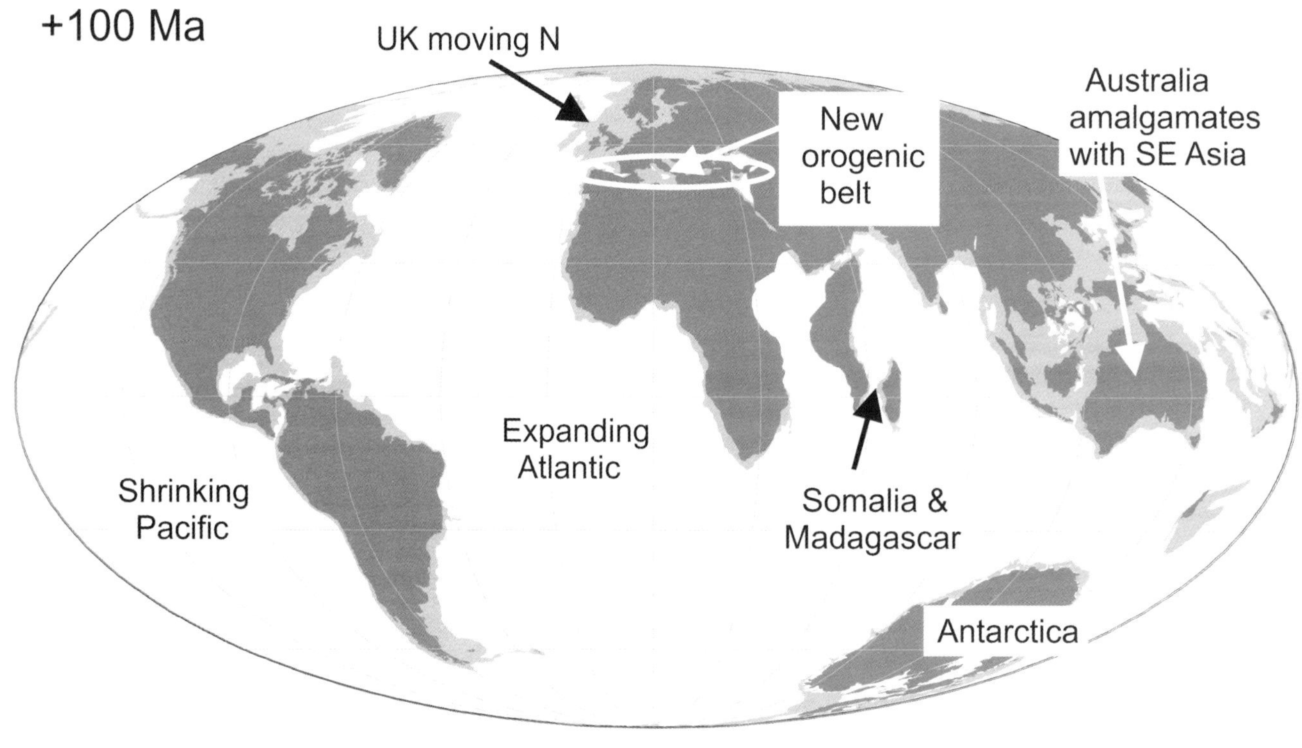

Fig. 18.5. Possible continental configuration at +100 Ma, i.e. 100 Ma into the future (showing present-day coastlines and present-day 2000 m bathymetric contour). See text for details. Mollweide projection (equal-area, global in extent).

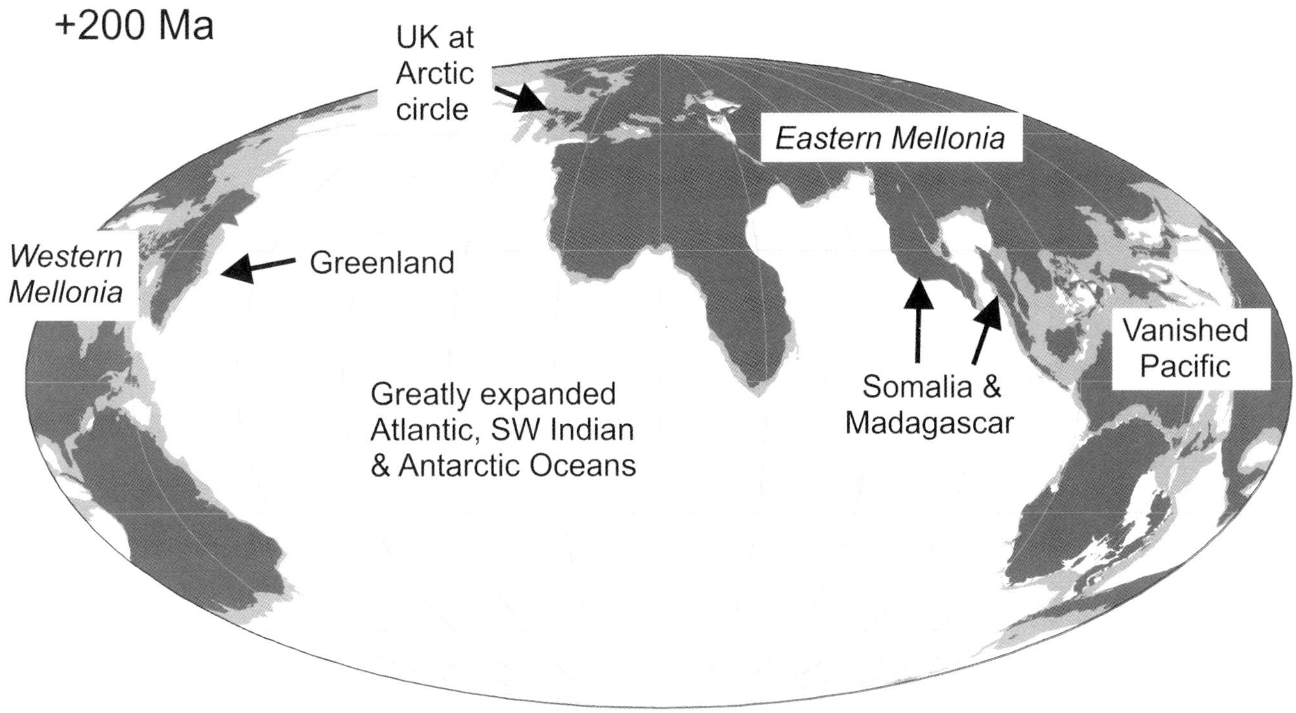

Fig. 18.6. Possible continental configuration at +200 Ma, i.e. 100 Ma into the future (showing present-day coastlines and present-day 2000 m bathymetric contour). See text for details. Mollweide projection (equal-area, global in extent).

of some sort of collision; and any ocean floor trapped between two colliding continents is eventually removed by breaking up the continents around it, as in the present-day Mediterranean, Black and Caspian seas. Such ocean basins behave as self-sealing black holes that swallow up all their surrounding ocean floor.

Even with these limiting principles, there are many possible future continental configurations. The model selected here shows how a new Pangea could form by closing the Pacific and part of the Atlantic Ocean. Since this Pangea is in the future, it is named Mellonia, drawing on Moores' (1986) suggested name for the geological future. In the model, the Australasian region completes its collision with eastern Asia (highly probable) by about +100 Ma (Fig. 18.5); North America moves closer to eastern Australasia (probable) and South America rotates and collides with eastern North America (possible). Eventually they coalesce to form a new supercontinent, Mellonia, by +200 Ma (possible).

West from Australasia, the East African Rift finally turns into an ocean that spreads as Somalia moves away and collides with western India (possible, but note that the North Sea rift has not yet developed into an ocean). Further west still, the Mediterranean, Black and Caspian seas all close (highly probable), and Iberia reverses the motion that created the Bay of Biscay (probable). This tectonic mayhem will give rise to a high, long and wide mountain chain that, were it there today, would be a much sought after skiing and holiday region (highly probable).

Britain, on the fringes of all this activity, does not escape entirely (highly probable). Given that the ocean floor next to the UK margin is about 55 Ma old, it follows that subduction will start at the continental margin (inevitable), no later than +150–+200 Ma (highly probable), with all that that entails in terms of explosive eruptions and earthquakes. At +200 Ma, the Arctic Circle may run along the length of Britain (possible). Thus, given its latitude, the UK at +200 Ma may have a climate resembling that of present-day Kamchatka (probable), which, for many, is not a particularly appealing prospect. This conclusion appears to be relatively 'robust' so long as the UK remains part of continental Europe, irrespective of what happens to the other continents in the interval.

References

ABBINK, O. A., CALLOMON, J. H., RIDING, J. B., WILLIAMS, P. D. B. & WOLFARD, A. 2000. Biostratigraphy of Jurassic–Cretaceous boundary strata in the Terschelling Basin, The Netherlands. *Proceedings of the Yorkshire Geological Society*, **53**, 275–302.

ABBINK, O. A., TARGARONA, J., BRINKHUIS, H. & VISSCHER, H. 2001. Late Jurassic to earliest Cretaceous palaeoclimate evolution of the southern North Sea. *Global and Planetary Change*, **30**, 231–256.

ADAMS, H. F. 1967. The seams of the South Wales Coalfield. Monograph of the Institute of Mining Engineers.

AGER, D. V. 1978. The stratigraphical distribution of Jurassic Brachiopoda. *Recent Researches in Geology*, **4**, 6–21.

AGUIRRE, E. & PASINI, G. 1985. The Plio-Pleistocene boundary. *Episodes*, **8**, 116–120.

AIGNER, T. 1980. Biofacies and stratinomy of the Lower Kimmeridge Clay (Upper Jurassic, Dorset, England) *Neues Jahrbuch für Geologie und Paläontologie, Abhandlungen*, **159**, 324–388.

AINSWORTH, N. R., BRAHAM, W., GREGORY, J., JOHNSON, B. & KING, C. 1998. A proposed latest Triassic to earliest Cretaceous microfossil biozonation for the English Channel and its adjacent areas. *In*: UNDERHILL, J. R. (ed.) *Development, Evolution and Petroleum Geology of the Wessex Basin*. Geological Society, London, Special Publications, **133**, 87–102.

AITKEN, J. F. & CHISHOLM, J. I. 1999. Aspects of the Middle Coal Measures sedimentation in the Sheffield/Barnsley district, South Yorkshire. Geological Society Coal Geology Group Field Meeting. October 1999. Unpublished Field Trip Guide.

AITKEN, J. F. & CHISHOLM, J. I. 2000. Lower Westphalian C (Bolsovian) major fluvial sandbodies in the Sheffield–Barnsley district, South Yorkshire. BSRG 39th Annual Field Meeting Field Trip F3 December 2000. Unpublished Field Trip Guide.

AITKEN, J. F., QUIRK, D. G. & GUION, P. D. 1999. Regional Correlation of Westphalian sandbodies onshore UK: implications for reservoirs in the Southern North Sea. *In*: FLEET, A. J. & BOLDY, S. A. R. (eds) *Petroleum Geology of Northwest Europe*. Proceedings of the 5th Conference. Geological Society, London, 747–756.

AITKENHEAD, N. & CHISHOLM, J. I. 1982. *A standard nomenclature for the Dinantian formations of the Peak District of Derbyshire and Staffordshire*. Report of the Institute of Geological Sciences, **82/8**, 17.

AITKENHEAD, N. & RILEY, N. J. 1996. Kinderscoutian and Marsdenian successions in the Bradup and Hag Farm boreholes, near Ilkley, West Yorkshire. *Proceedings of the Yorkshire Geological Society*, **51**, 115–125.

AITKENHEAD, N., CHISHOLM, J. I. & STEPHENSON, I. P. 1985. *Geology of the Country Around Buxton, Leek and Bakewell*. Memoir of the British Geological Survey, Sheet 111 (E&W), HMSO, London.

AITKENHEAD, N., BARCLAY, W. J., BRANDON, A., CHADWICK, R. A., CHISHOLM, J. I., COOPER, A. H. & JOHNSON, E. W. (eds). 2002. *British Regional Geology: The Pennines and Adjacent Areas*, 4th edn. HMSO, London.

AKHURST, M. C., CHADWICK, R. A., HOLLIDAY, D. W., MCCORMACK, M., MCMILLAN, A. A., MILLWARD, D. & YOUNG, B. 1997. *Geology of the West Cumbria District*. Memoir of the British Geological Survey, sheets 28 (Whitehaven), 37 (Gosforth) and 47 (Bootle). HMSO, London.

AL JAWADI, A. F. 1987. *Minor igneous intrusions of the Lake District: geochronology, geochemistry and petrology*. PhD thesis, University of Newcastle.

ALDER, K. E. 1976. The geology of the area around Bettws Gwerfil Goch and Melin y Wig, Clwyd. PhD thesis, University of Cambridge, UK.

ALDERTON, D. H. M. 1993. Mineralization associated with the Cornubian granite batholith. *In*: PATTRICK, R. A. D. & POLYA, D. A. (eds) *Mineralization in the British Isles*. Chapman & Hall, London, 270–354.

ALDHOUSE-GREEN, S. 2000. *Paviland Cave and the 'Red Lady': A Definitive Report*. The Western Academic and Specialist Press, Bristol.

ALDHOUSE-GREEN, S., SCOTT, K. *ET AL.* 1995. Coygan Cave, Lougharne, south Wales, a Mousterian site and hyaena den: a report on the University of Cambridge excavations. *Proceedings of the Prehistoric Society*, **61**, 37–79.

ALDRIDGE, R. J. 2000. The Llandovery Series. *In*: ALDRIDGE, R. J., SIVETER, D. J., SIVETER, D. J., LANE, P. D., PALMER, D. & WOODCOCK, N. H. (eds) *British Silurian Stratigraphy*. Joint Nature Conservation Committee. Geological Conservation Review Series, **19**, 69–180.

ALDRIDGE, R. J., JEPPSON, L. & DORNING, K. J. 1993. Early Silurian oceanic episodes and events. *Journal of the Geological Society, London*, **150**, 501–513.

ALDRIDGE, R. J., BASSETT, M. G. *ET AL.* 2002. Telychian Rocks in the British Isles. *In*: HOLLAND, C. H. & BASSETT, M. G. (eds) *Telychian Rocks of the British Isles and China*. Silurian, Llandovery Series, National Museums & Galleries of Wales, Geological Series, **21**, 43–72.

ALDRIDGE, R. J., SIVETER, D. J., SIVETER, D. J., LANE, P. D., PALMER, D. & WOODCOCK, N. H. (eds). 2000. *British Silurian Stratigraphy*. Joint Nature Conservation Committee. Geological Conservation Review Series, **19**.

ALEXANDER, A. C. & SHAIL, R. K. 1995. Late Variscan structures on the coast between Perranporth and St. Ives, Cornwall. *Proceedings of the Ussher Society*, **8**, 398–404.

ALEXANDER, A. C. & SHAIL, R. K. 1996. Late- to post-Variscan structures on the coast between Penzance and Pentewan, south Cornwall. *Proceedings of the Ussher Society*, **9**, 72–78.

ALEXANDER, J. 1989. Delta or coastal plain? With an example of the controversy from the Middle Jurassic of Yorkshire. *In*: WHATELEY, M. K. G. & PICKERING, K. T. (eds) *Deltas: Sites and Traps For Fossils Fuels*. Geological Society, London, Special Publications, **41**, 1119.

ALI, J. R. & JOLLEY, D. W. 1996. Chronostratigraphic framework for the Thanetian and lower Ypresian deposits of southern England. *In*: KNOX. R. W. O'B., CORFIELD, R. M. & DUNAY, R. E. (eds) *Correlation of the Early Paleogene in Northwest Europe*. Geological Society, London, Special Publications, **101**, 129–144.

ALI, J. R., KING, C. & HAILWOOD, E. A. 1993. Magnetostratigraphic calibration of early Eocene depositional sequences in the southern North Sea Basin. *In*: HAILWOOD, E. A. & KIDD, R. B. (eds) *High Resolution Stratigraphy*. Geological Society, London, Special Publications, **70**, 99–125.

ALLEN, D. J., BREWERTON, L. J., COLEBY, L. M., GIBBS, B. R., LEWIS, M. A., MACDONALD, A. M., WAGSTAFF, S. J. & WILLIAMS, A. T. 1997. *The physical properties of major aquifers in England and Wales*. British Geological Survey Technical Report, **WD/97/34**. Environment Agency Research and Development Publication, **8**.

Allen, J. R. L. 1965*a*. The sedimentation and palaeogeography of the Old Red Sandstone of Anglesey, North Wales. *Proceedings of the Yorkshire Geological Society*, **35**, 139–185.

Allen, J. R. L. 1965*b*. Upper Old Red Sandstone (Farlovian) palaeogeography in South Wales and the Welsh Borderland. *Journal of Sedimentary Petrology*, **35**, 167–195.

Allen, J. R. L. 1970. Studies in fluviatile sedimentation: a comparison of fining-upwards cyclothems with reference to coarse member composition and interpretation. *Journal of Sedimentary Petrology*, **40**, 298–323.

Allen, J. R. L. 1974*a*. Studies in fluviatile sedimentation: implications of pedogenic carbonate units, Lower Old Red Sandstone, Anglo-Welsh outcrop. *Geological Journal*, **9**, 181–208.

Allen, J. R. L. 1974*b*. Sedimentology of the Old Red Sandstone (Siluro-Devonian) in the Clee Hills area, Shropshire, England. *Sedimentary Geology*, **12**, 73–167.

Allen, J. R. L. 1975. Source rocks of the Lower Old Red Sandstone: the Llanishen Conglomerate of the Cardiff area, South Wales. *Proceedings of the Geologists' Association*, **83**, 63–76.

Allen, J. R. L. 1979. Old Red Sandstone Facies in External Basins, with particular reference to southern Britain. *Special Papers in Palaeontology*, **23**, 65–80.

Allen, J. R. L. 1982. Mud drapes in sand-wave deposits: a physical model with application to the Folkestone Beds (Early Cretaceous, southeast England). *Philosophical Transactions of the Royal Society, London*, **A306**, 291–345.

Allen, J. R. L. 1983. Studies in fluvial sedimentation: bars, bar-complexes and sandstone sheets (low-sinuosity braided streams) in the Brownstones (L. Devonian), Welsh Borders. *Sedimentary Geology*, **33**, 237–293.

Allen, J. R. L. 1985. Marine to fresh water: the sedimentology of the interrupted environmental transition (Ludlow–Siegenian) in the Anglo-Welsh region. *Philosophical Transactions of the Royal Society, London*, **B309**, 85–104.

Allen, J. R. L. 2000. Morphodynamics of Holocene salt marshes; a review sketch from the Atlantic and southern North Sea coasts of Europe. *Quaternary Science Reviews*, **19**, 1155–1231.

Allen, J. R. L. 2001. Late Quaternary stratigraphy in the Gwent Levels (southeast Wales): the subsurface evidence. *Proceedings of the Geologists' Association*, **112**, 289–316.

Allen, J. R. L. & Crowley, S. F. 1983. Lower Old Red Sandstone fluvial dispersal systems in the British Isles. *Transactions of the Royal Society of Edinburgh*, **74**, 61–68.

Allen, J. R. L. & Dineley, D. L. 1976. The succession of the Lower Old Red Sandstone (Siluro-Devonian) along the Ross-Tewkesbury Spur Motorway (M50), Hereford and Worcester. *Geological Journal*, **11**, 1–14.

Allen, J. R. L. & Gibbard, P. L. 1994. Pleistocene evolution of the Solent River of southern England. *Quaternary Science Reviews*, **12**, 503–528.

Allen, J. R. L. & Williams, B. P. J. 1978. The sequence of the earlier Lower Old Red Sandstone (Siluro-Devonian), north of Milford Haven, southwest Dyfed (Wales). *Geological Journal*, **13**, 113–136.

Allen, J. R. L. & Williams, B. P. J. 1979. Interfluvial drainage on Siluro-Devonian alluvial plains in Wales and the Welsh Borders. *Journal of the Geological Society, London*, **136**, 361–366.

Allen, J. R. L. & Williams, B. P. J. 1981. Sedimentology and stratigraphy of the Townsend Tuff Bed (Lower Old Red Sandstone) in South Wales and the Welsh Borders. *Journal of the Geological Society, London*, **138**, 15–29.

Allen, J. R. L. & Williams, B. P. J. 1982. The architecture of an alluvial suite: rocks between the Townsend Tuff and Pickard Bay Tuff Beds (early Devonian), southwest Wales. *Philosophical Transactions of the Royal Society, London*, **B287**, 51–89.

Allen, J. R. L., Halstead, L. B. & Turner, S. 1968. Dittonian ostracoderm fauna from the Brownstones of Wilderness Quarry, Mitcheldean, Gloucestershire. *Proceedings of the Geological Society, London*, **1649**, 141–53.

Allen, J. R. L., Elliot, T. & Williams, B. P. J. 1981. Old Red Sandstone and Carboniferous fluvial sediments in South Wales. *In*: Elliot, T. (ed.) *Field Guides to Modern and Ancient Fluvial Systems in Britain and Spain*. Proceedings of the 3rd International Symposium on Fluvial Sedimentology, Keele, 1–39.

Allen, J. R. L., Thomas, R. G. & Williams, B. P. J. 1982. The Old Red Sandstone north of Milford Haven. *In*: Bassett, M. G. (ed.) *Geological Excursions in Dyfed, Southwest Wales*. National Museum of Wales, Cardiff, 123–149.

Allen, M. R., Griffiths, P. A. Craig, J., Fitches, W. R. & Whittington, R. J. 1994. Halokinetic initiation of Mesozoic tectonics in the Southern North Sea: a regional model. *Geological Magazine*, **131**, 559–561.

Allen, P. 1959. The Wealden environment: Anglo-Paris Basin. *Philosophical Transactions of the Royal Society, London*, **B242**, 283–346.

Allen, P. 1967. Origin of the Hastings facies in north-west Europe. *Proceedings of the Geologists' Association*, **78**, 27–105.

Allen, P. 1976. Wealden of the Weald: a new model. *Proceedings of the Geologists' Association*, **86**, 389–437.

Allen, P. 1981. Pursuit of Wealden models. *Journal of the Geological Society, London*, **138**, 375–405.

Allen, P. 1998. Purbeck–Wealden (early Cretaceous) climates. *Proceedings of the Geologists' Association*, **109**, 197–236.

Allen, P. A. & Underhill, J. R. 1999. Swaley cross-stratification produced by unidirectional flows, Bencliff Grit, (Upper Jurassic), Dorset, UK. *Journal of the Geological Society, London*, **156**, 241–252.

Allen, P. & Wimbledon, W. A. 1991. Correlation of NW European Purbeck–Wealden (nonmarine Lower Cretaceous) as seen from the English type-areas. *Cretaceous Research*, **12**, 511–526.

Allen, P., Cheshire, D. A. & Whiteman, C. A. 1991. The tills of southern East Anglia. *In*: Ehlers, J., Gibbard, P. L. & Rose, J. (eds) *Glacial deposits in Great Britain and Ireland*. Balkema, Rotterdam, 255–278.

Allen, P. A., Homewood, P. & Williams, G. D. 1986. Foreland basins: an introduction. *In*: Allen, P. A. & Homewood, P. (eds) *Foreland Basins*. International Association of Sedimentologists, Special Publications, **8**, 3–12.

Allen, P. M. & Jackson, A. A. 1985. *Geology of the country around Harlech*. Memoir of the British Geological Survey, Sheet 135 with part of 149 (England & Wales), HMSO, London.

Allen, P. M. & Cooper, D. C. 1986. The stratigraphy and composition of the Latterbarrow and Redmain Sandstones, Lake District, England. *Geological Journal*, **21**, 59–76.

Allen, P. M. & Jackson, A. A. 1978. Bryn-Teg Borehole, North Wales. *Bulletin of the Geological Survey of Great Britain*, **61**, 1–52.

Allen, P. M., Jackson. A. A. & Rushton, A. W. A. 1981. The stratigraphy of the Mawdach Group in the Cambrian succession of North Wales. *Proceedings of the Yorkshire Geological Society*, **43**, 295–329.

Allen, P. P. Cooper, D. C., Fortey N. J. 1987. Composite lava flows of Ordovician age in the English Lake District. *Journal of the Geological Society, London*, **21**, 59–76.

Alley, R. B. 2000. The Younger Dryas as viewed from central Greenland. *Quaternary Science Reviews*, **19**, 213–235.

Allman-Ward, P., Halls, C. & Rankin, A. H. 1982. An intrusive hydrothermal breccia at Wheal Remfry in the western part of the St. Austell pluton, Cornwall, England. *In*: Evans, A. M. (ed.) *Metallization Associated with Acid Magmatism*. Wiley, Chichester.

Allsop, J. M. 1987. Patterns of late Caledonian intrusive activity in eastern and northern England from geophysics, radiometric dating and basement geology. *Proceedings of the Yorkshire Geological Society*, **46**, 335–353.

Almond, J., Williams, B. P. J. & Woodcock, N. H. 1993. The Old Red Sandstone of the Brecon Beacons to Black Mountains area. *In*: Woodcock, N. H. & Bassett, M. G. (eds) *Geological Excursions in Powys*. University of Wales Press, National Museum of Wales, Cardiff, 311–330.

Anderson, F. W. 1971. Appendix B. The sequence of ostracod faunas in the Wealden and Purbeck of the Warlingham borehole. *Bulletin of the Geological Survey of Great Britain*, **36**, 122–138.

ANDERSON, F. W. 1985. Ostracod faunas in the Purbeck and Wealden of England. *Journal of Micropalaeontology*, **4**, 1–68.

ANDERSON, F. W. & BAZLEY, R. A. B. 1971. The Purbeck Beds of the Weald (England). *Bulletin of the Geological Survey of Great Britain*, **34**, 1–173.

ANDERSON, T. B., PARNELL, J. T. & RUFFELL, A. H. 1995. Influence of Basement on the Geometry of Permo–Triassic Basins in the North-West British Isles. *In*: BOLDY, S. A. R. (ed.) *Permian–Triassic Rifting*. Geological Society, London, Special Publications, **91**, 103–122.

ANDERSON, T. F., POPP, B. N., WILLIAMS, A. C., HO, L.-Z. & HUDSON, J. D. 1994. The stable isotope records of fossils from the Peterborough Member, Oxford Clay Formation (Jurassic), UK: palaeoenvironmental implications. *Journal of the Geological Society, London*, **151**, 125–138.

ANDERTON, R., BRIDGES, P. H., LEEDER, M. R. & SELLWOOD, B. W. 1979. *A Dynamic Stratigraphy of the British Isles*. Allen & Unwin, London.

ANDREASSON, F. P. & SCHMITZ, B. 2000. Temperature seasonality in the early Middle Eocene from $\delta^{18}O$ profiles of marine gastropod shells. *GFF*, **122**, 9–10.

ANDREIEFF, P. & LEFORT, J. P. 1972. Contribution à l'étude stratigraphique des terraines secondaires et tertiares affleurant en Manche orientale. *Mémoire du Bureau de Recherche Géologiques et Minières*, **79**, 49–56.

ANDREWS, J. R. 1993. Evidence for Variscan dextral transpression in the Pilton Shales, Croyde Bay, north Devon. *Proceedings of the Ussher Society*, **8**, 198–199.

ANDREWS, J. R., BARKER, A. J. & PAMPLIN, C. F. 1988. A reappraisal of the facing confrontation in north Cornwall: fold or thrust dominated tectonics? *Journal of the Geological Society, London*, **145**, 777–778.

ANDREWS, J. T., GILBERTSON, D. D. & HAWKINS, A. B. 1984. The Pleistocene succession of the Severn Estuary; a revised model based on amino-acid racemization studies. *Journal of the Geological Society, London*, **141**, 967–974.

ANKETELL, J. M. 1987. On the geological succession and structure of South-Central Wales. *Geological Journal*, **22**, 155–165.

ANTIA, D. D. J. 1981. The Temeside Bone-Bed and associated sediments from Wales and the Welsh Borderland. *Mercian Geologist*, **8**, 163–215.

ANTOINE, P., COUTARD, J. P., GIBBARD, P. L., HALLÉGOUËT, B., LAUTRIDOU, J. P. & OZOUF, J. C. 2003*a*. The Pleistocene rivers of the English Channel region. *Journal of Quaternary Science*, **18**, 227–243.

ANTOINE, P., CATT, J., LAUTRIDOU, J.-P. & SOMMÉ, J. 2003*b*. The loess and coversands of northern France and southern England. *Journal of Quaternary Science*, **18**, 309–318.

ARCHER, A. A. 1968. *Geology of the South Wales Coalfield (Special Memoir): The Gwendraeth Valley and Adjoining Areas*. Memoir of the Geological Survey of Great Britain, HMSO, London.

ARKELL, W. J. 1931. The Upper Great Oolite, Bradford Beds and Forest Marble of South Oxfordshire, and the succession of gastropod faunas in the Great Oolite. *Quarterly Journal of the Geological Society of London*, **87**, 563–629.

ARKELL, W. J. 1933. *The Jurassic System in Great Britain*. Clarendon Press, Oxford.

ARKELL, W. J. 1935. The Portland Beds of the Dorset Mainland. *Proceedings of the Geologists' Association*, **46**, 301–347, plates 19–26.

ARKELL, W. J. 1947*a*. *The Geology of the Country Around Weymouth, Swanage, Corfe and Lulworth*. Memoir of the Geological Survey of Great Britain, Sheets 341, 342 and 343, HMSO, London.

ARKELL, W. J. 1947*b*. *The Geology of Oxford*. Clarendon Press, Oxford.

ARKELL, W. J. & DONOVAN, D. T. 1952. The Fuller's Earth of the Cotswolds and its relation to the Great Oolite. *Quarterly Journal of the Geological Society of London*, **107**, 227–253, plates 13–16.

ARMSTRONG, H. A. & PURNELL, M. A. 1987. Dinantian conodont biostratigraphy of the Northumberland Trough. *Journal of Micropalaeontology*, **6**, 97–112.

ARTHURTON, R. S. 1971. *The Permian evaporites of the Langwathby Borehole, Cumberland*. Institute of Geological Sciences Report 71/17.

ARTHURTON, R. S. 1973. Experimentally produced halite compared with Triassic layered halite rock from Cheshire, England. *Sedimentology*, **20**, 145–160.

ARTHURTON, R. S. 1980. Rhythmic sedimentary sequences in the Triassic Keuper marl (Mercia Mudstone Group) of Cheshire, northwest England. *Geological Journal*, **15**, 43–58.

ARTHURTON, R. S. & HEMINGWAY, J. E. 1972. The St Bees Evaporites – a carbonate-evaporite formation of Upper Permian age in West Cumberland. *Proceedings of the Yorkshire Geological Society*, **38**, 565–592.

ARTHURTON, R. S. & WADGE, A. J. 1981. *Geology of the Country Around Penrith*. Memoir of the Geological Survey of Great Britain, England and Wales, Sheet 24, HMSO, London.

ARTHURTON, R. S., BURGESS, I. C. & HOLLIDAY, D. W. 1978. Permian and Triassic. *In*: MOSELEY, F. (ed.) *Geology of the Lake District*, Yorkshire Geological Society Occasional Publication, **3**, 189–206.

ARTHURTON, R. S., JOHNSON, E. W. & MUNDY, D. J. 1988. *Geology of the Country Around Settle*. Memoir of the British Geological Survey, Sheet 60 (England and Wales), HMSO, London.

ASHTON, M. 1980. The stratigraphy of the Lincolnshire Limestone Formation (Bajocian) in Lincolnshire and Rutland (Leicestershire). *Proceedings of the Geologists' Association*, **91**, 203–223.

ASHTON, N. M. & LEWIS, S. G. 2002. Deserted Britain: declining populations in the British Late Middle Pleistocene. *Antiquity*, **76**, 388–396.

ATKINSON, T. C., LAWSON, T. J., SMART, P. L., HARMON, R. S. & HESS, J. W. 1986. New data on speleothem deposition and palaeoclimate in Britain over the last forty thousand years. *Journal of Quaternary Science*, **1**, 67–72.

ATKINSON, T. C., BRIFFA, K. R. & COOPE, G. R. 1987. Seasonal temperatures in Britain during the last 22,000 years, reconstructed using beetle remains. *Nature*, **325**, 587–592.

AUBRY, M.-P. 1983. Biostratigraphie du Paléogene epicontinental de l'Europe du Nord-ouest. Etude fondée sur les nannofossiles calcaires. *Documents des Laboratories de Géologie de Lyon*, **89**, 1–317.

AUBRY, M.-P., BERGGREN, W. A. *ET AL*. 2003. Chronostratigraphic terminology at the Paleocene/Eocene boundary. *In*: WING, S. L., GINGERICH, P. D., SCHMITZ, B. & THOMAS, E. (eds) *Causes and Consequences of Globally Warm Climates in the Early Paleogene*. Geological Society of America, Special Paper, **369**, 551–566.

AUBRY, M.-P., HAILWOOD, E. A. & TOWNSEND, H. A. 1986. Magnetic and calcareous-nannofossil stratigraphy of the lower Palaeogene formations of the Hampshire and London Basins. *Journal of the Geological Society, London*, **143**, 729–735.

AUSTIN, R. L. & MATTHEWS, S. C. 1967. *In*: *Annual Report of the Institute of Geological Sciences for 1966*, 88.

AUSTIN, R. L., DREESDEN, R., SELWOOD, E. B. & THOMAS, J. M. 1992. New conodont information relating to the Devonian stratigraphy of the Trevone Basin, north Cornwall, south-west England. *Proceedings of the Ussher Society*, **8**, 23–28.

AUSTIN, R. L., DRUCE, E. C., RHODES, F. H. C. & WILLIAMS, J. A. 1970. The value of conodonts in the recognition of the Devonian – Carboniferous boundary, with particular reference to Great Britain. *Comptes Rendus 6'eme Congres Internationale de Stratigraphie et de Géologique du Carbonifère*, Sheffield, for 1967, **2**, 431–444.

AUSTIN, R. L., ORCHARD, M. J. & STEWART, I. J. 1985. Conodonts of the Devonian system from Great Britain. *In*: HIGGINS, A. C. & AUSTIN, R. L. (eds) *Stratigraphical Index of Conodonts*. British Micropalaeontological Society, Ellis Horwood, 93–166.

AUSTIN, R. M. 1991. Modelling Holocene tides on the NW European continental shelf. *Terra Nova*, **3**, 276–288.

AUVREY, B., CHARLOT, R. & VIDAL, P. 1980. Données nouvelles sur le Poterozique inférieur du domaine Nord-Amoricain (France): âge et significance. *Canadian Journal of Earth Sciences*, **17**, 532–538.

AVERY, B. W. 1985. Argillic horizons and their significance in England and Wales. *In*: BOARDMAN, J. (ed.) *Soils and Quaternary Landscape Evolution*. Wiley, Chichester, 69–86.

AWAN, M. A. & WOODCOCK, N. H. 1993. Structural arcuation in the Berwyn Hills, North Wales. *Geological Journal*, **28**, 179–189.

BADHAM, J. P. N. 1982. Strike-slip orogens – an explanation for the Hercynides. *Journal of the Geological Society, London*, **139**, 493–504.

BAILEY, R. J. 1969. Ludlovian sedimentation in south central Wales. *In*: WOOD, A. (ed.). *The Pre-Cambrian and Lower Palaeozoic rocks of Wales*. University of Wales Press, Cardiff, 283–304.

BAIRD, A. W. & MCCAFFREY, K. J. W. 1999. Polyphase deformation and metamorphism in the Llyn Ogwen area of Snowdonia, North Wales. *Journal of the Geological Society, London*, **156**, 11–14.

BAKER, J. W. 1969. Correlation problems of unmetamorphosed Pre-Cambrian rocks in Wales and southeast Ireland. *Geological Magazine*, **106**, 246–259.

BALL, T. K, DAVIES, J. R, WATERS, R. A. & ZALASIEWICZ, J. A. 1992. Geochemical discrimination of Silurian mudstones according to depositional process and provenance within the Southern Welsh Basin. *Geological Magazine*, **129**, 567–572.

BALLANTYNE, C. K. & HARRIS, C. 1994. *The Periglaciation of Great Britain*. Cambridge University Press, Cambridge.

BALSON, P. S. 1990. The 'Trimley Sands': a former marine Neogene deposit from eastern England. *Tertiary Research*, **11**, 148–158.

BALSON, P. S. & CAMERON, T. D. J. 1985. Quaternary mapping offshore East Anglia. *Modern Geology*, **9**, 221–239.

BALSON, P. S., HUMPHREYS, B. & ZALASIEWICZ, J. A. 1991. Coralline and Red Crags in East Anglia. *In*: *Field Guide No. 3*, 13th International Sedimentological Congress, Nottingham 1990. British Sedimentological Research Group.

BANCROFT, B. B. 1928. On the unconformity at the base of the Ashgillian in the Bala district. *Geological magazine*, **65**, 484–93.

BANCROFT, B. B. 1929. Some new species of *Cryptolithus* (s.l..), from the Upper Ordovician. *Transactions of the Manchester Literary and Philosophical Society*, **73**, 67–98.

BANCROFT, B. B. 1933. *Correlation tables of the stages Costonian to Onnian in England and Wales*. Blakeney, Gloucestershire (privately printed).

BANCROFT, B. B. 1945. Zonal indices of the stages Costonian to Onnian in Britain. *Journal of Palaeontology*, **19**, 181–252.

BANDEL, K. & BECKER, G. 1975. Ostracoden aus Palaozoischen-pelagischen der karnischen Alpen (Silurium bis Unterkarbon), *Senkenbergiana lethaea*, **156**, 1–283. [In German.]

BANHAM, P. H. 1988. Polyphase glaciotectonic deformation in the Contorted Drift of Norfolk. *In*: CROOT, D. (ed.) *Glaciotectonics; Forms and Processes*. Balkema, Rotterdam, 27–32.

BANHAM, P. H., GIBBARD, P. L., LUNKKA, J. P., PARFITT, S. A., PREECE, R. C. & TURNER, C. 2001. A critical assessment of 'A New Glacial Stratigraphy for Eastern England'. *Quaternary Newsletter*, **93**, 5–14.

BARBER, A. J. & MAX, M. D. 1979. A new look at the Mona Complex (Anglesey, North Wales). *Journal of the Geological Society, London*, **136**, 407–432.

BARBER, K. E., CHAMBERS, F. M. & MADDY, D. 2003. Holocene palaeoclimates from peat stratigraphy: macrofossil proxy climate records from three oceanic raised bogs in England and Ireland. *Quaternary Science Reviews*, **22**, 521–539.

BARCLAY, W. J. 1989. *Geology of the South Wales Coalfield, Part II, the country around Abergavenny*, 3rd edn, Sheet 232, British Geological Survey.

BARCLAY, W. J. & SMITH, N. J. P. 2002. *Geology of the country between Hereford and Ross-on-Wye*. Sheet Explanation of the British Geological Survey. Sheet 215 (England and Wales), HMSO, London.

BARCLAY, W. J., AMBROSE, K., CHADWICK, R. A. & PHARAOH. T. C. 1997. *Geology of the country around Worcester*. Memoir of the British Geological Survey, Sheet 199 (England and Wales). HMSO, London.

BARCLAY, W. J., BRANDON, A., ELLISON, R. A. & MOORLOCK, B. S. P. 1992. Pleistocene palaeovalley-fill west of the Malvern Hills. *Journal of the Geological Society, London*, **149**, 75–92.

BARCLAY, W. J., JACKSON, D. I. *ET AL*. 1989. *Geology of the South Wales Coalfield: Part II the Country Around Abergavenny*. Memoir of the British Geological Survey, Sheet 232 (England and Wales), HMSO, London.

BARCLAY, W. J., RATHBONE, P. A.,WHITE D. E. & RICHARDSON, J. B. 1994. Brackish water faunas from the St. Maughans Formation: the Old Red Sandstone section at Ammons Hill, Hereford and Worcester, UK re-examined. *Geological Journal*, **29**, 369–379.

BARCLAY, W. J., TAYLOR, K. & THOMAS, L. P. 1988. *Geology of the South Wales Coalfield, Part V, the Country around Merthyr Tydfil*. Memoir of the British Geological Survey, 1:50 000 Sheet 231 (England and Wales), HMSO, London.

BARD, J. P., BURG, J. P., MATTE, P. & RIBIERO, A. 1980. La Chaine hercynienne d'Europe occidentale en termes de tectoniques des plaques. *In*: COGNE, J. & SLANZKY, M. (eds) *Geologie de l'Europe*. 26th International Geological Congress, Paris, 90–111. [In French.]

BARKER, M. J. 1976. *A stratigraphical, palaeoecological and biometrical study of some English Bathonian Gastropoda (especially Nerineacea)*. PhD thesis, University of Keele.

BARNES, R. P. 1983. The stratigraphy of the sedimentary mélange and associated deposits in South Cornwall, England. *Proceedings of the Geologists' Association*, **94**, 217–229.

BARNES, R. P. & ANDREWS, J. R. 1986. Upper Palaeozoic ophiolite generation and obduction in south Cornwall. *Journal of the Geological Society, London*, **143**, 117–124.

BARNES, R. P., AMBROSE, K., HOLLIDAY, D. W. & JONES, N. S. 1994. Lithostratigraphical subdivision of the Triassic Sherwood Sandstone group in west Cumbria. *Proceedings Yorkshire Geological Society*, **50**, 51–60.

BARNES, R. P., LINTERN, B. C. & STONE, P. 1989. Timing and regional implications of deformation in the Southern Uplands of Scotland. *Journal of the Geological Society, London*, **146**, 905–908.

BARREIRO, B. 1996. Sr and Nd isotopic compositions of igneous rock fragments from the Roseland Breccia and other related units, Cornwall. *NERC Isotope Geosciences Laboratory Report Series*, **84**.

BARRON, A. J. M., SUMBLER, M. G. & MORIGI, A. N. 1997. A revised lithostratigraphy for the Inferior Oolite Group (Middle Jurassic) of the Cotswolds, England. *Proceedings of the Geologists' Association*, **108**, 269–285.

BARTON, R. N. E., JACOBI, R. M., STAPERT, D. & STREET, M. J. 2003. The Late-glacial reoccupation of the British Isles and the Creswellian. *Journal of Quaternary Science*, **18**, 631–643.

BASSETT, D. A. 1969. Some of the major structures of early Palaeozoic age in Wales and the Welsh Borderland: An historical essay. *In*: WOOD, A. (ed.) *The Precambrian and Lower Palaeozoic Rocks of Wales*. University of Wales Press, Cardiff, 67–116.

BASSETT, D. A. 1991. Roderick Murchison's Silurian System: a sesquicentennial tribute. *In*: BASSETT, M. G., LANE, P. D & EDWARDS, D. (eds) *The Murchison Symposium: Proceedings of an International Conference on the Silurian System*, 1989, Keele University. *Special Papers in Palaeontology*, **44**, 7–90.

BASSETT, D. A., WHITTINGTON, H. B. & WILLIAMS, A. 1966. The stratigraphy of the Bala district, Merionethshire. *Quarterly Journal of the Geological Society of London*, **122**, 219–21.

BASSETT, M. G. 1969. The age of the oldest Silurian beds of the Rhumney (Cardiff) Silurian inlier. *Geological Magazine*, **106**, 90–92.

BASSETT, M. G. 1974. Review of the stratigraphy of the Wenlock Series in the Welsh Borderland. *Palaeontology*, **17**, 745–777.

BASSETT, M. G. 1979. 100 years of Ordovician geology. *Episodes*, **8** no.2, 18–21.

BASSETT, M. G. 1985. Towards a 'common language' in stratigraphy. *Episodes*, **8**, 87–92.

BASSETT, M. G. 1989*a*. Brachiopods. *In*: HOLLAND C. H. & BASSETT, M. G. (eds) *A Global Standard for the Silurian System*. National Museum of Wales, Cardiff, Geological Series, **9**, 232–242.

BASSETT, M. G. 1989*b*. The Wenlock Series in the Wenlock area. *In*: HOLLAND, C. H. & BASSETT, M. G. (eds) *A Global Standard for the*

Silurian System. National Museum of Wales, Cardiff, Geological Series, **9**, 51–73.

Bassett, M. G., Cocks, L. R. M., Holland, C. H., Rickards, R. B. & Warren, P. T. 1975. *The type Wenlock Series*. Institute of Geological Sciences Report, **75/13**.

Bassett, M. G., Owens, R. M. & Rushton, A. W. A. 1976. Lower Cambrian fossils from the Hell's Mouth Grits, St Tudwal's Peninsula, North Wales. *Journal of the Geological Society, London*, **136**, 623–44.

Bassett, M. G., Lawson, J. D. & White, D. E. 1982. The Downton Series as the fourth series of the Silurian System. *Lethaia*, **15**, 1–24.

Bassett, M. G., Bluck, B. J., Cave, R., Holland, C. H. & Lawson, J. D. 1992*a*. Silurian. *In*: Cope, J. C. W., Ingham, J. K. & Rawson, P. F. (eds) *Atlas of Palaeogeography and Lithofacies*. Geological Society, London, Memoir, **13**, 37–56.

Bassett, M. G., Wright, A. D. & Duff, P. McL. D. 1992*b*. Ordovician. *In*: Duff, P. McL. D. & Smith, A. J. (eds) *Geology of England and Wales*. 1st edn, Geological Society, London, 63–125.

Bateman, M. D. 1995. Thermoluminescence dating of the British coversand deposits. *Quaternary Science Reviews*, **14**, 791–798.

Bateman, M. D. 1998. The origin and age of coversand in north Lincolnshire, UK. *Permafrost and Periglacial Processes*, **9**, 313–325.

Bateman, M. D. & Catt, J. A. 1996. An absolute chronology for the raised beach and associated deposits at Sewerby, East Yorkshire, England. *Journal of Quaternary Science*, **11**, 389–395.

Bateman, M. D. & Godby, S. P. 2004. Late-Holocene inland dune activity in the UK: a case study from Breckland, East Anglia. *The Holocene*, **14**, 579–588.

Bateman, M. D., Murton, J. B. & Crowe, W. 2000. Late Devensian and Holocene depositional environments associated with the coversand around Caistor, north Lincolnshire, UK. *Boreas*, **16**, 1–16.

Bateman, R. N. 1988. Relationship of the Woolwich and Reading Formation (Late Palaeocene) to the Upper Chalk (Late Cretaceous) and Clay-with-flints sensu lato (Quaternary) in the Chiltern Hills, southern England. *Tertiary Research*, **10**, 53–63.

Bates, D. E. B. 1972. The stratigraphy of the Ordovician rocks of Anglesey. *Geological Journal*, **8**, 29–58.

Bates, D. E. B. 1974. The structure of the Lower Palaeozoic rocks of Anglesey, with special reference to faulting. *Geological Journal*, **9**, 39–60.

Bates, M. R., Parfitt, S. A. & Roberts, M. B. 1997. The chronology, palaeogeography and archaeological significance of the marine Quaternary record of the West Sussex coastal plain, Southern England, UK. *Quaternary Science Reviews*, **16**, 1227–1252.

Batten, D. J. 2002. Palaeoenvironmental setting of the Purbeck Limestone Group of Dorset, southern England. *Special Papers in Palaeontology*, **68**, 13–20.

Beck, R. B., Funnell, B. M. & Lord, A. R. 1972. Correlation of the Lower Pleistocene Crag at depth in Suffolk. *Geological Magazine*, **109**, 137–139.

Becker, L., Poreda, R. J., Hunt, A. G., Bunch, T. E. & Rampino, M. 2001. Impact event at the Permian-Triassic boundary: Evidence from extraterrestrial noble gases in fullerenes. *Science*, **291**, 1530–1533.

Beckinsale, R. D. & Thorpe, R. S. 1979. Rubidium-strontium whole-rock isochron evidence for the age of metamorphism and magmatism in the Mona Complex of Anglesey. *Journal of the Geological Society, London*, **136**, 433–439.

Beckly, A. J. 1987. Basin development in North Wales during the Arenig. *Geological Journal*, **22**, 19–30.

Beddoe-Stephens, B. & Mason, I. 1991. The volcanogenetic significance of garnet-bearing minor intrusions within the Borrowdale Volcanic Group, Eskdale area, Cumbria. *Geological Magazine*, **128**, 505–516.

Beddoe-Stephens, B., Petterson, M. G., Millward, D. & Marriner, G. F. 1995. Geochemical variation and magmatic cyclicity within an Ordovician continental-arc volcanic field: the lower Borrowdale Volcanic Group, English Lake District. *Journal of Volcanology and Geothermal Research*, **65**, 81–110.

Bedrock, M. 1984. *Sedimentology of some Westphalian sequences in the Yorkshire Coalfield*. Unpublished thesis, University of Imperial College, London.

Beer, K. E. & Scrivener, R. C. 1982. Metalliferous mineralisation. *In*: Durrance, E. M. & Laming, D. J. C. (eds) *The Geology of Devon*. University of Exeter, Exeter,117–147.

Beerling, D. J. & Berner, R. A. 2002. Biogeochemical constraints on the Triassic–Jurassic boundary carbon cycle event. *Global Biogeochemical Cycles*, **16**, 10–13.

Beerling, D. J. & Jolley, D. W. 1998. Fossil plants record an atmospheric $^{12}CO_2$ and temperature spike across the Palaeocene-Eocene transition in NW Europe. *Journal of the Geological Society, London*, **155**, 591–594.

Behr, H. J., Engel, W., Giese, P. & Weber, K. 1984. The Variscan Belt in central Europe: main structures, geodynamic implications, open questions. *Tectonophysics*, **109**, 15–40.

Bell, B. R. & Jolley, D. W. 1997. Application of palynological data to the chronology of the Paleogene lava fields of the British Province: implications for magmatic stratigraphy. *Journal of the Geological Society, London*, **154**, 701–708.

Bell, M. 1983. Valley sediments as evidence of prehistoric land-use on the South Downs. *Proceedings of the Prehistoric Society*, **49**, 119–150.

Bell, M. & Walker, M. J. C. 1992. *Late Quaternary Environmental Change: Physical and Human Perspectives*. Longman, London.

Bell, M. & Walker, M. J. C. 2005. *Late Quaternary Environmental Change: Physical And Human Perspectives*. Pearson Education, Harlow.

Bellamy, A. G. 1995. Extension of the British landmass: evidence from shelf sediment bodies in the English Channel. *In*: Preece, R. C. (ed.) *Island Britain: A Quaternary Perspective*. Geological Society, London, Special Publications, **96**, 47–62.

Belshaw, R. K., Hackney, G. D. & Smith, K. A. 2005. The evolution of the drainage pattern of the English Midlands from the Late Tertiary to the Early Pleistocene: the significance of the Milton Formation. *Quaternary Newsletter*, **105**, 16–31.

Bennet, G., Copestake, P. & Hooker, N. P. 1985. Stratigraphy of the Britoil 72/10-1A well, Western Approaches. *Proceedings of the Geologists' Association*, **96**, 255–261.

Bennett, K. D. 1988. Holocene pollen stratigraphy of central East Anglia, England, and comparison of pollen zones across the British Isles. *New Phytologist*, **109**, 237–253.

Bennett, M. A. 1987. Genesis and diagenesis of the Cambrian manganese deposits, Harlech, North Wales. *Geological Journal*, **22**, Spring Thematic Issue, 7–18.

Benton, M. J. 1995. *The Fossil Record 2*. Chapman & Hall, London.

Benton, M. J. 1997. The Triassic reptiles from Devon. *Proceedings of the Ussher Society*, **9**, 141–152.

Benton, M. J & Hiscock, C. 1996. Lower Silurian trace fossils and the Eocoelia community in the Tortworth Inlier, SW England. *Proceedings of the Geologists' Association*, **107**, 199–208.

Benton, M. J. & Spenser, P. S. 1995. *Fossil Reptiles of Great Britain*. Geological Conservation Review Series, **10**.

Benton, M. J. & Thackeray, E. 1995. Shelf storm beds on the southern margin of the Welsh Basin: the Wenlock of the Tortworth Inlier, England. *Proceedings of the Geologists' Association*, **106**, 81–92.

Benton, M. J. & Twitchett, R. J. 2003. How to kill (almost) all life: the end Permian extinction event. *Trends in Ecology and Evolution*, **18**, 358–365.

Benton, M. J., Warrington, G., Newell, A. J. & Spenser, P. S. 1994. A review of the British Triassic tetrapod assemblages. *In*: Fraser, N. C. & Sues, H. D. (eds) *In the Shadow of the Dinosaurs*. Cambridge University Press, Cambridge, 131–160.

Benton, M. J., Cook, E. & Turner, P. 2002. *Permian and Triassic Red Beds and the Penarth Group of Great Britain*. Joint Nature Conservation Committee. Geological Conservation Review Series, **24**.

Berger, A. & Loutre, M. F. 2002. An exceptionally long interglacial ahead? *Science*, **297**, 1287–1288.

Berggren, W. A., Kent, D. V., Swisher, C. C. & Aubry, M.-P. 1995. A revised Cenozoic geochronology and chronostratigraphy.

In: Berggen, W. A., Kent, D. V., Aubry, M. P. & Hardenbol, J. (eds) *Geochronology, Time Scales and Global Stratigraphic Correlation*. SEPM, Special Publications, **54**, 129–212.

Bergström, S. M. 1964. Remarks on some Ordovician conodont faunas from Wales. *Acta Universitatis Lundensis*, Section II **3**, 1–66.

Berridge, N. G. & Pattison, J. 1994. *Geology of the Country Around Grimsby and Patrington*. Memoir for 1:50 000 geological sheets 90 and 91 and 81 and 82 (England and Wales). HMSO, London.

Berridge, N. G., Pattison, J., Samuel, M. D. A., Brandon, A., Howard, A. S., Pharaoh, T. C. & Riley, N. J. 1999. *Geology of the Grantham District*. Memoir of the British Geological Survey, Sheet 127 (England & Wales), HMSO, London.

Besly, B. M. 1988*a*. Late Carboniferous sedimentation in northwest Europe: an introduction. *In*: Besly, B. M. & Kelling, G. (eds) *Sedimentation in a Synorogenic Basin Complex: The Upper Carboniferous of Northwest Europe*. Blackie, Glasgow, 1–7.

Besly, B. M. 1988*b*. Palaeogeographic implications of late Westphalian to early Permian red-beds, central England. *In*: Besly, B. M. & Kelling, G. (eds) *Sedimentation in a Synorogenic Basin Complex: The Upper Carboniferous of Northwest Europe*. Blackie, Glasgow, 200–221.

Besly, B. M. & Cleal, C. J. 1997. Upper Carboniferous stratigraphy of the West Midlands (UK) revised in the light of borehole geophysical logs and detrital compositional suites. *Geological Journal*, **32**, 85–118.

Besly, B. M. & Cleal, C. J. in press. *Lithostratigraphy and macrofloral biostratigraphy of the Late Carboniferous of the Midlands and Oxfordshire Coalfields*. National Museums and Galleries of Wales, Cardiff.

Bettley, R. M., Fortey, R. A., Siveter, D. J. 2001. High resolution correlation of Anglo-Welsh Middle to Upper Ordovician sequences and its relevance to international chronostratigraphy. *Journal of the Geological Society, London*, **158**, 937–952.

Beutler, G. & Häusser, I. 1981. Über den Schilsandstein der DDR. *Zeitschrift für Geologie Geschichtswissenschaft*, **10**, 511–525.

Bevins, R. E. 2003. Golden Hill Quarry, Monmouthshire. *In*: Stephenson, D., Loughlin, S. C., Millward, D., Waters, C. N. & Williamson, I. T. (eds) *Carboniferous and Permian Igneous Rocks of Great Britain North of the Variscan Front*. Joint Nature Conservation Committee. Geological Conservation Review Series, **27**, 312–316.

Bevins, R. E. & Robinson, D. 1988. Low grade metamorphism of the Welsh Basin Lower Palaeozoic succession: an example of diastathermal metamorphism? *Journal of the Geological Society, London*, **145**, 363–366.

Bevins, R. E. & Robinson, D. 1989. Discussion on low grade metamorphism of the Welsh Basin Lower Palaeozoic succession: an example of diastathermal metamorphism? *Journal of the Geological Society, London*, **146**, 888–890.

Bevins, R. E. & Rowbotham, G. 1983. Low-grade metamorphism within the Welsh sector of the paratectonic Caledonides. *Geological Journal*, **18**, 141–167.

Bevins, R. E., Pharaoh, T. C., Cope, J. C. W. & Brewer, T. S. 1995*a*. Geochemical character of Neoproterozoic rocks in southwest Wales. *Geological Magazine*, **132**, 330–349.

Bevins, R. E., Pharaoh, T. C., Cope, J. C. W. & Brewer, T. S. 1995*b*. Geochemical character of Neoproterozoic volcanic rocks in southwest Wales. *Geological Magazine*, **132**, 339–349.

Bevins, R. E., White, S. C. & Robinson, D. 1996. The South Wales Coalfield: low grade metamorphism in a foreland basin setting. *Geological Magazine*, **133**, Pt 6, 739–749.

Bifani, R. 1986. Esmond Gas Complex. *In*: Brooks J. (ed.) *Habitat of Palaeozoic gas in NW Europe*. Geological Society, London, Special Publications, **23**, 209–221.

Bignot, G. 1972. Etude biostratigraphique des prélèvements Eocènes. *Mémoire du Bureau de Recherches Géologiques et Minières*, **79**, 153–156.

Birkelund, T., Callomon, J. H., Clausen, C. K., Nøhr Hansen, H. & Salinas, I. 1983. The Lower Kimmeridge Clay at Westbury, Wiltshire, England. *Proceedings of the Geologists' Association*, **94**, 289–309.

Birks, H. J. B. 1982. Holocene (Flandrian) chronostratigraphy of the British Isles: a review. *Striae*, **16**, 95–105.

Birks, H. J. B. 1989. Holocene isochrone maps and patterns of tree-spreading in the British Isles. *Journal of Biogeography*, **16**, 503–540.

BIRPS & ECORS. 1986. Deep seismic reflection profiling between England, France and Ireland. *Journal of the Geological Society, London*, **143**, 45–52.

Bisat, W. S. 1924. The Carboniferous goniatites of the north of England and their zones. *Proceedings of the Yorkshire Geological Society*, **20**, 40–124.

Bisat, W. S. 1928. The Carboniferous goniatite zones of England and their continental equivalents. *Congrès pour l'Avancement des Etudes de Stratigraphie Carbonifère, Compte Rendu, Heerlen 1927*, 117–133.

Bisat, W. S. & Hudson, R. G. S. 1943. The Lower Reticuloceras (R1) Goniatite succession in the Namurian of the north of England. *Proceedings of the Yorkshire Geological Society*, **24**, 383–440.

Bishop, W. W. 1957. The Pleistocene geology and geomorphology of three gaps in the Midland Jurassic escarpment. *Philosophical Transactions of the Royal Society of London*, **B241**, 255–306.

Bisson, G., Lamb, R. K. & Calver, M. A. 1967. Boreholes in the concealed Kent coalfield between 1948 and 1959. *Bulletin of the Geological Survey of Great Britain*, **26**, 99–166.

Bjerrum, C. J., Surlyk, F., Callomon, J. H. & Slingerland, R. L. 2001. Numerical palaeoceanographic study of the Early Jurassic transcontinental Laurasian Seaway. *Palaeoceanography*, **16**, 390–404.

Björck, S., Walker, M. J. C., Cwynar, L. C., Johnsen, S., Knudsen, K.-L., Lowe, J. J., Wohlfarth, B. & INTIMATE Members. 1998. An event stratigraphy for the Last Termination in the North Atlantic region based on the Greenland ice-core record: a proposal by the INTIMATE group. *Journal of Quaternary Science*, **13**, 283–292.

Blackford, J. 2000. Palaeoclimatic records from peat bogs. *Trends in Ecology and Evolution*, **15**, 193–194.

Blake, J. F. 1880. On the Portland rocks of England. *Quarterly Journal of the Geological Society of London*, **36**, 189–236, plates 8–10.

Blake, J. F. 1881. On correlation of the Kimmeridge and Portland rocks of England with those of the continent. Part 1: The Paris Basin. *Quarterly Journal of the Geological Society of London*, **37**, 497–587.

Blake, J. F. 1890. On the Monian and basal Cambrian rocks of Shropshire. *Quarterly Journal of the Geological Society of London*, **46**, 386–420.

Bland, B. H. 1984. Arumberia Glaessner and Walter, a review of its potential for correlation in the region of the Precambrian-Cambrian boundary. *Geological Magazine*, **121**, 625–633.

Bland, B. H. & Goldring, R. 1995. Teichichnus rectus Seilacher from the Charnian of Leicestershire. *Neues Jahrbuch fur Geologie und Palaontologie Abhandlungen*, **195**, 5–23.

Bloos, G. & Page, K. N. 2000. The basal Jurassic ammonite succession in the North-West European Province – review and new results. *GeoResearch Forum*, **6**, 27–39.

Bloos, G. & Page, K. N. 2002. Global Stratotype Section and Point for base of the Sinemurian Stage (Lower Jurassic). *Episodes*, **25**, 22–28.

Bloxham, T. W. & Dirk, M. H. J. 1988. The petrology and geochemistry of the St. David's Granophyre and the Cwm Bach Rhyolite, Pembrokeshire, Dyfed. *Minerological Magazine*, **52**, 563–575.

Bluck, B. J. 1961. *The sedimentary history of the rocks between the horizon G. subcrenatum and the Garw Coal in the South Wales Coalfield*. PhD Thesis, University of Wales (Swansea).

Bluck, B. J. 2000. Old Red Sandstone basins and alluvial systems of the Midland Valley. *In*: Friend, P. F. & Williams, B. P. J. (eds) *New Perspectives on the Old Red Sandstone*. Geological Society, London, Special Publications, **180**, 417–437.

Bluck, B. J. & Kelling, G. 1963. Channels from the Upper Carboniferous Coal Measures of South Wales. *Sedimentology*, **2**, 29–53.

Bluck, B. J., Cope, J. C. W. & Scrutton, C. T. 1992. Devonian. *In*: Cope, J. C. W., Ingham, J. K. & Rawson, P. F. (eds) *Atlas of*

Palaeogeography and Lithofacies. Geological Society, London, Memoir, **13**, 57–66.

BOARDMAN, J. 1985. The Troutbeck Paleosol, Cumbria, England. *In*: BOARDMAN, J. (ed.) *Soils and Quaternary Landscape Evolution*. Wiley, Chichester, 231–260.

BONEHAM, B. F. W. & WYATT, R. J. 1993. The stratigraphical position of the Middle Jurassic (Bathonian) Stonesfield Slate of Stonesfield, Oxfordshire, UK. *Proceedings of the Geologists' Association*, **104**, 123–136.

BONNEY, T. G. 1879. Notes on the microscopic structure of some Shropshire rocks. *Quarterly Journal of the Geological Society of London*, **35**, 662–669.

BONNICHSEN, B. 1982. The Bruneau–Jarbidge eruptive centre, South-western Idaho. *In*: BONNICHSEN, B. & BRECKENRIDGE, R. M. (eds) *Cenozoic Geology of Idaho*. Idaho Bureau of Mines and Geology Bulletin, **26**, 237–254.

BONSALL, C. 1980. The coastal factor in the Mesolithic settlement of north-west England. *Veröffenlichungen des Museums für Ur- und Frühgeschichte Potsdam*, **14/15**, 451–472.

BOSE, P. K. 1983. A reappraisal of the conditions of deposition of the Maentwrog Beds (Upper Cambrian) at Port Ceiriad, North Wales. *Geological Magazine*, **120**, 73–9.

BOSWELL, P. G. H. 1952. The Plio–Pleistocene boundary in the east of England. *Proceedings of the Geologists' Association*, **63**, 301–312.

BOSWORTH, T. O. 1912. *The Keuper Marls Around Charnwood Forest*. Leicester Literary and Philosophical Society, Thornley, Leicestershire, 1–129.

BOTT, M. H. P. 1967. Geophysical investigations of the northern Pennine basement rocks. *Proceedings of the Yorkshire Geological Society*, **36**, 139–168.

BOTT, M. H. P. 1974. The geological interpretation of a gravity survey of the English Lake District and the Vale of Eden. *Journal of the Geological Society, London*, **130**, 309–331.

BOTT, M. H. P., DAY, A. A. & MASSON-SMITH, D. 1958. The geological interpretation of gravity and magnetic surveys in Devon and Cornwall. *Philosophical Transactions of the Royal Society of London*, **A251**, 161–91.

BOTT, M. H. P., ROBINSON, J. & KOHNSTAMM, M. A. 1978. Granite beneath Market Weighton. *Journal of the Geological Society, London*, **135**, 535–543.

BOUCOT, A. J. 1975. *Evolution and Extinction Rate Controls*. Elsevier, Amsterdam.

BOULTER, M. C., FORD, T. D., ITABA, M. & WALSH, E. T. 1971. Brassington Formation: a newly recognised Tertiary Formation in the Southern Pennines. *Nature (Physical Science)*, **231**, 134–136.

BOULTON, G. S. 1992. Quaternary. *In*: DUFF, P. MCL. D. & SMITH, A. J. (eds) *The Geology of England and Wales*. 1st edn, Geological Society, London, 413–444.

BOULTON, G. S. & HINDMARSH, R. C. A. 1987. Sediment deformation beneath glaciers: rheology and geological consequences. *Journal of Geophysical Research*, **92**, 9059–9082.

BOULTON, G. S., JONES, A. S., CLAYTON, K. M. & KENNING, M. J. 1977. A British ice sheet model and patterns of glacial erosion and deposition in Britain. *In*: SHOTTON, F. W. (ed.) *British Quaternary Studies*. Oxford University Press, Oxford, 231–246.

BOULTON, G. S., SMITH, G. D., JONES, A. S. & NEWSOME, J. 1985. Glacial geology and glaciology of the last mid-latitude ice sheets. *Journal of the Geological Society, London*, **142**, 447–474.

BOULTON, G. S., PEACOCK, J. D. & SUTHERLAND, D. G. 1991. Quaternary. *In*: CRAIG, G. Y. (ed.) *The Geology of Scotland*, 3rd edn. Geological Society, London, 503–542.

BOURILLET, J. F., REYNAUD, J. Y., BALTZER, A. & ZARAGOSI, S. 2003. The 'Fleuve Manche'; the submarine sedimentary features from the outer shelf to the deep sea fans. *Journal of Quaternary Science*, **18**, 261–282.

BOWDLER-HICKS, A., INGHAM. J. K. & OWEN, A. W. 2002. The taxonomy and stratigraphic significance of the Anglo-Welsh Cryptolithinae (Trinucleidae, Trilobita). *Palaeontology*, **45**, 1075–1105.

BOWEN, D. Q. 1970. South-east and Central Wales. *In*: LEWIS, C. A. (ed.) *The Glaciations of Wales and Adjoining regions*. Longman, London, 197–228.

BOWEN, D. Q. 1994. Late Cenozoic Wales and South-West England. *Proceedings of the Ussher Society*, **8**, 209–213.

BOWEN, D. Q. (ed.). 1999*a*. *A Revised Correlation of Quaternary Deposits in the British Isles*. Geological Society Special Report, **23**.

BOWEN, D. Q. 1999*b*. Wales. *In*: BOWEN, D. Q. (ed.) *A Revised Correlation of Quaternary Deposits in the British Isles*. Geological Society Special Report, **23**, 79–90.

BOWEN, D. Q., HUGHES, S., SYKES, G. A. & MILLER, G. H. 1989. Land-sea correlations in the Pleistocene based on isoleucine epimerization in non-marine molluscs. *Nature*, **340**, 49–51.

BOWEN, D. Q., PHILLIPS, F. M., MCCABE, A. M., KNUTZ, P. C. & SYKES, G. A. 2002. New data for the Last Glacial Maximum in Great Britain and Ireland. *Quaternary Science Reviews*, **21**, 89–101.

BOWEN, D. Q., ROSE, J. & MCCABE, A. M. 1986. Correlation of Quaternary glaciations in England, Ireland, Scotland and Wales. *Quaternary Science Reviews*, **5**, 299–340.

BOWEN, D. Q., SYKES, G. A., REEVES, A., MILLER, G. H., ANDREWS, J. T., BREW, J. S. & HARE, P. E. 1985. Amino acid geochronology of raised beaches in south-west Britain. *Quaternary Science Reviews*, **4**, 279–318.

BOWER, C. R. & FARMERY, J. R. 1910. The zones of the Lower Chalk of Lincolnshire. *Proceedings of the Geologists' Association*, **21**, 333–359.

BOWMAN, M. B. J., MCCLURE, N. M. & WILKINSON, D. W. 1993. Wytch Farm Oilfield: deterministic reservoir description of the Triassic Sherwood Sandstone. *In*: PARKER, J. R. (ed.) *Petroleum Geology of North-west Europe: Proceedings of the 4th Conference*. Geological Society, London, 1513–1517.

BOWN, P. R. & COOPER, M. K. E. 1998. Jurassic. *In*: BOWN, P. R. (ed.) *Calcareous Nannofossil Biostratigraphy*. British Micropalaeontological Society, Publications Series, 30–85.

BOWN, P. R., COOPER, M. K. E. & LORD, A. R. 1988. Calcareous nannofossil scheme for the early to mid Mesozoic. *Newsletters on Stratigraphy*, **20**, 91–114.

BOWN, P. R., RUTLEDGE, D. C., CRUX, J. & GALLAGHER, T. 1998. Lower Cretaceous. *In*: BOWN, P. R. (ed.) *Calcareous Nannofossil Biostratigraphy*. British Micropalaeontological Series, 86–131.

BOWN, T. M. & KRAUS, M. J. 1987. Integration of channel and flood-plain suites, I. Developmental sequence and lateral relations of alluvial paleosols. *Journal of Sedimentary Petrology*, **57**, 587–601.

BOYLAN, P. J. 1981. The role of William Buckland (1784–1856) in the recognition of glaciation in Great Britain. *In*: NEALE, J. & FLENLEY, J. (eds) *The Quaternary in Britain*. Pergamon, Oxford, 1–8.

BOYNTON, H. E. 1978. Fossils from the Precambrian of Charnwood Forest, Leicestershire, England. *Mercian Geologist*, **14**, 197–201.

BOYNTON, H. E. & FORD, T. D. 1995. Ediacaran fossils from the Charnian Supergroup of Charnwood Forest, Leicestershire. *Mercian Geologist*, **13**, 165–182.

BRADFIELD, K. E. S. 1999. Fossil communities of the late Ludlow and early Přídolí in Wales and the Welsh Borderland. *In*: BOUCOT, A. J. & LAWSON, J. D. (eds) *Paleocommunities – A Case Study From the Silurian and Early Devonian*. Cambridge University Press, Cambridge, 370–372.

BRADSHAW. M. J. 1978. *A facies analysis of the Bathonian of eastern England*. DPhil thesis, University of Oxford.

BRADSHAW, M. J., COPE, J. C. W. *ET AL*. 1992. Jurassic. *In*: COPE, J. C. W., INGHAM, J. K. & RAWSON, P. F. (eds) 1992. *Atlas of Palaeogeography and Lithofacies*. Geological Society, London, Memoirs, **13**, 107–129.

BRAMMALL, A. & HARWOOD, H. F. 1923. The Dartmoor Granite: its mineralogy, structure and petrology. *Mineralogical Magazine*, **20**, 39–53.

BRAMMALL, A. & HARWOOD, H. F. 1925. Tourmalinisation in the Dartmoor Granite. *Mineralogical Magazine*, **20**, 319–330.

BRAMMALL, A. & HARWOOD, H. F. 1932. The Dartmoor Granites: their genetic relationships. *Quarterly Journal of the Geological Society of London*, **88**, 171–237.

BRANDON, A. 1989. *Geology of the Country Between Hereford and Leominster*. Memoir of the British Geological Survey, Sheet 198 (England and Wales), HMSO, London.

BRANDON, A., RILEY, N. J., WILSON, A. A. & ELLISON, R. A. 1995. Three new early Namurian (E1c-E2a) marine bands in central and northern England, UK, and their bearing on correlations with the Askrigg Block. *Proceedings of the Yorkshire Geological Society*, **50**, 333–355.

BRANDON, A., SUMBLER, M. G. & IVIMEY-COOK, H. C. 1990. A revised lithostratigraphy for the Lower and Middle Lias (Lower Jurassic) east of Nottingham, England. *Proceedings of the Yorkshire Geological Society*, **48**, 121–141.

BRANNEY, M. J. 1988. The subaerial setting of the Borrowdale Volcanic Group, English Lake District. *Journal of the Geological Society, London*, **145**, 887–890.

BRANNEY, M. J. 1991. Eruption and depositional facies of the Whorneyside Tuff: an exceptionally large-magnitude phreatoplinian eruption. *Geological Society of America Bulletin*, **203**, 886–897.

BRANNEY, M. J. & KOKELAAR, P. 1994. Volcanotectonic faulting, soft-state deformation and rheomorphism of tuffs during development of a piecemeal caldera: English Lake District. *Geological Society of America Bulletin*, **106**, 507–530.

BRANNEY, M. J. & SOPER, N. J. 1988. Ordovician volcano-tectonics in the English Lake District. *Journal of the Geological Society, London*, **145**, 367–376.

BRANNEY, M. J. & SUTHREN, R. J. 1988. High-level peperitic sills in the English Lake District: distinction from block lavas and implications for Borrowdale Volcanic Group stratigraphy. *Geological Journal*, **23**, 171–187.

BRANNEY, M. J., KOKELAAR, B. P. & MCCONNELL, B. J. 1992. The Bad Step Tuff: a lava-like ignimbrite in a calc-alkaline piecemeal caldera, English Lake District. *Bulletin of Volcanology*, **54**, 187–199.

BRASIER, M. D. 1984. Microfossils and small shelly fossils from the Lower Cambrian *Hyolithes* Limestone at Nuneaton, English Midlands. *Geological Magazine*, **121**, 229–53.

BRASIER, M. D. 1989. Sections in England and their correlation. *In*: COWIE, J. W. & BRASIER, M. D. (eds) *The Precambrian-Cambrian Boundary*, Oxford University Press, 82–104.

BRASIER, M. D., INGHAM, J. K. & RUSHTON, A. W. A. 1992. Cambrian. *In*: COPE, J. C. W., INGHAM, J. K. & RAWSON P. F. (eds) *Atlas of Palaeogeography and Lithofacies*, Geological Society of London Memoir, **13**, 3–18.

BRASIER, M. D., COWIE, J. W. & TAYLOR, M. E. 1994. Decision on the Precambrian–Cambrian boundary. *Episodes*, **17**, 3–8.

BRAY, C. J. & SPOONER, E. T. C. 1983. Sheeted vein Sn-W mineralisation and greisenisation associated with economic kaolinisation, Goonbarrow China Clay Pit, St. Austell, England. *Economic Geology*, **78**, 1064–1089.

BRAY, M. J. & HOOKE, J. M. (1997) Prediction of soft cliff retreat with accelerating sea-level rise. *Journal of Coastal Research*, **13**, 453–467.

BRAY, R. J., DUDDY, I. R. & GREEN, P. F. 1998. Multiple heating episodes in the Wessex Basin: implications for geological evolution and hydrocarbon generation. *In*: UNDERHILL, J. R. (ed.) *Development, Evolution and Petroleum Geology of the Wessex Basin*. Geological Society, London, Special Publications, **133**, 199–213.

BRENCHLEY, P. J. 1972. The Cwm Clwyd Tuff, North Wales: a palaeogeographical interpretation of some Ordovician ash-shower deposits. *Proceedings of the Yorkshire Geological Society*, **39**, 199–224.

BRENCHLEY, P. J. 1978. The Caradocian rocks of the north and west Berwyn Hills, North Wales. *Geological Journal*, **13**, 137–164.

BRENCHLEY, P. J. 1988. Environmental changes close to the Ordovician–Silurian boundary. *Bulletin of the British Museum (Natural History) London, Geology Series*, **43**, 377–385.

BRENCHLEY, P. J. 1999. Pandy. *In*: STEPHENSON, D., BEVINS, R. E., MILLWARD, D., HIGHTON, A. J., PARSONS, I., STONE, P. & WADSWORTH., W. J. (eds) *Caledonian Igneous Rocks of Great Britain*. Joint Nature Conservation Committee. Geological Conservation Review Series, **17**, 321–325.

BRENCHLEY, P. J. & CULLEN, B. 1984. The environmental distribution of associations belonging to the *Hirnantia fauna*: evidence from North Wales and Norway. *In*: BRUTON, D.L. (ed.) *Aspects of the Ordovician System*, Palaeontological Contributions from the University of Oslo, **295**, 113–25.

BRENCHLEY, P. J. & GOWLAND, S. 1985. The sedimentology of wave and storm influenced shelf facies: Carboniferous of Northumberland and Jurassic of Yorkshire. Poroperm Excursion Guide, 8, Poroperm, Chester.

BRENCHLEY, P. J. & NEWALL G. 1980. A facies analysis of Upper Ordovician regressive sequences in the Oslo region, Norway – a record of glacio-eustatic changes. *Palaeogeography, Palaeoclimatology, Palaeoecology*, **31**, 1–38.

BRENCHLEY, P. J. & NEWALL, G. 1982. Storm influenced inner-shelf sand lobes in the Caradoc (Ordovician) of Shropshire, England. *Journal of Sedimentary Geology*, **52**, 1257–1269.

BRENCHLEY, P. J. & NEWALL, G. 1984. Late Ordovician environmental changes and their effect on faunas. *In*: BRUTON, D. L. (ed.) *Aspects of the Ordovician System*. Universitetsforlaget, Oslo, 65–79.

BRENCHLEY, P. J., MARSHALL, A. D., HARPER, D. A. T., BUTLER, C. J. & UNDERWOOD, C. J. 2006. A late Ordovician (Hirnantian) karstic surface in a submarine channel, recording glacio-eustatic sea-level changes: Meifod, central Wales. *Geological Journal*, **41**, 1–22.

BRETT, C. E., BOUCOT, A. J. & JONES, B. 1993. Absolute depths of Silurian benthic assemblages. *Lethaia*, **26**, 25–40.

BRETTLE, M. J. 2001. *Sedimentology and high-resolution sequence stratigraphy of shallow water delta systems in the early Marsdenian (Namurian) Pennine basin, Northern England*. Unpublished PhD thesis, University of Liverpool, UK.

BRETTLE, M. J., MCILROY, D., ELLIOTT, T., DAVIES, S. J. & WATERS, C. N. 2002. Identifying cryptic tidal influences within deltaic successions: an example from the Marsdenian (Namurian) interval of the Pennine Basin, UK. *Journal of the Geological Society, London*, **159**, 379–391.

BRIDGE, D. MCC., CARNEY, J. N., LAWLEY, R. S. & RUSHTON, A. W. A. 1998. *The geology of the Country Around Coventry and Nuneaton*. Memoir of the British Geological Survey, Sheet 169 (England & Wales), 1–185, HMSO, London.

BRIDGES, P. H. 1975. The transgression of a hard substrate shelf: the Llandovery (Lower Silurian) of the Welsh Borderland. *Journal of Sedimentary Petrology*, **45**, 79–84.

BRIDGES, P. H. 1976. Lower Silurian transgressive barrier islands, southwest Wales. *Sedimentology*, **23**, 347–362.

BRIDGES, P. H. & CHAPMAN, A. J. 1988. The anatomy of a deep water mud-mound complex on the SW margin of the Dinantian platform in Derbyshire. *Sedimentology*, **35**, Pt 1, 139–162.

BRIDGES, P. H., GUTTERIDGE, P. & PICKARD, N. A. H. 1995. The environmental setting of Early Carboniferous mud-mounds. *In*: MONTY, C. V. L., BOSENCE, D. W. J., BRIDGES, P. H. & PRATT, B. R. (eds) *Carbonate mudmounds; their origin and evolution*. International Association of Sedimentologists, Special Publications, **23**, 171–190.

BRIDGLAND, D. R. 2003. The evolution of the River Medway, SE England, in the context of Quaternary palaeoclimate and the Palaeolithic occupation of NW Europe. *Proceedings of the Geologists' Association*, **114**, 23–48.

BRINKMANN, R. 1929. Statistisch-biostratigraphische Untersuchungen an mitteljurassischen Ammoniten über Artebegriff und Stammesentwicklung. Abhandlungen der Gesellschaft der Wissenschaft, Göttingen, mathematische und physische Klasse. *Neue Folge*, **13**, 1–249.

BRISTOW, C. M. & EXLEY, C. S. 1994. Historical and geological aspects of the china-clay industry of Southwest England. *Transactions of the Royal Geological Society of Cornwall*, **21**, 247–314.

BRISTOW, C. R. 1989. *Geology of the East Stow–Shaftesbury district (Dorset)*. British Geological Survey Technical Report, **WA/89/58**.

BRISTOW, C. R. & BAZLEY, R. A. 1972. *Geology of the country around Royal Tunbridge Wells*. Memoir of the Geological Survey of Great Britain, HMSO, London.

Bristow, C. R., Freshney, E. C. & Penn, I. E. 1991. *Geology of the country around Bournemouth*. Memoir of the British Geological Survey, Sheet 329 (England and Wales), HMSO, London.

Bristow, C. R., Barton, C. M. *et al.* 1995. *Geology of the Country around Shaftesbury*, Sheet 313 (England & Wales). British Geological Survey, HMSO, London.

Bristow, C. R., Barton, C. M., Westhead, R. K., Freshney, E. C., Cox, B. M. & Woods, M. A. 1999. *The Wincanton District – A Concise Account of the Geology*. Memoir of the British Geological Survey, Sheet 297 (England & Wales), HMSO, London.

Bristow, C. R., Mortimore, R. & Wood, C. J. 1997. Lithostratigraphy for mapping the Chalk of southern England. *Proceedings of the Geologists' Association*, **108**, 293–315.

Bristow, C. S. 1988. Controls on the sedimentation of the Rough Rock Group (Namurian) from the Pennine Basin of northern England. *In*: Besly, B. M. & Kelling, G. (eds) *Sedimentation in a Synorogenic Basin Complex: The Upper Carboniferous of Northwest Europe*. Blackie, Glasgow, 114–131.

Bristow, C. S. 1993. Sedimentology of the Rough Rock: a Carboniferous braided river sheet sandstone in northern England. *In*: Best, J. L. & Bristow, C. S. (eds) *Braided Rivers*. Geological Society, London, Special Publications, 291–304.

British Geological Survey Lexicon. *The British Geological Survey Lexicon of Named Rock Units*. www.bgs.ac.uk/lexicon/lexicon_intro.html

British Geological Survey. 1988. Classical areas of British Geology: Llyn Padarn. 1:25,000 scale map.

British Geological Survey. 1990. *Falmouth*. England and Wales Sheet 352. Solid and Drift Geology. 1:50 000. British Geological Survey, Nottingham.

British Geological Survey. 1993. Corwen. England and Wales Sheet 120, Solid and drift geology, 1:50 000. (Southampton: Ordnance Survey for British Geological Survey)

British Geological Survey. 1994. *Tavistock*. England and Wales Sheet 337. Solid and Drift geology. 150 000. Provisional Series. British Geological Survey, Nottingham.

British Geological Survey. 1996. Tectonic map of Britain, Ireland and adjacent areas. British Geological Survey, Keyworth, Nottingham.

British Geological Survey. 2000. *Mevagissey*. England and Wales Sheet 353. Solid and Drift Geology. 1:50 000. British Geological Survey, Nottingham.

British Geological Survey. 2001*a*. Building stone resources map of the United Kingdom. 1:1,000,000 Geological Map. NERC.

British Geological Survey. 2001*b*. Isle of Man: solid and drift geology, 1:50,000 geological map. British Geological Survey, Keyworth, Nottingham.

British Geological Survey. 2003. *Torquay*. England and Wales Sheet 350. Solid and Drift Geology. 1:50 000. British Geological Survey, Nottingham.

British Geological Survey. 2005*a*. *United Kingdom Minerals Yearbook 2004*. British Geological Survey, Keyworth, Nottingham.

British Geological Survey. 2005*b*. Builth Wells. England and Wales Sheet 196, Solid geology, 1:50 000. British Geological Survey, Keyworth, Nottingham.

Broadhurst, F. W., Simpson, I. M. & Hardy, P. G. 1980. Seasonal sedimentation in the Upper Carboniferous of England. *Journal of Geology*, **88**, 639–651.

Bromley, A. V. 1979. Ophiolitic origin of the Lizard Complex. *Journal of Camborne School of Mines*, **79**, 35–38.

Bromley, A. V. 1989. *Field Guide to the Cornubian Orefield. 6th International Symposium On Water–Rock Interaction, Malvern, 1989*. International Association of Geochemistry and Cosmochemistry.

Bromley, R. G. 1967. Some observations on burrows of thalassinidean Crustacea in chalk hardgrounds. *Quarterly Journal of the Geological Society of London*, **123**, 157–182.

Bromley, R. G. & Ekdale, A. A. 1986. Flint and fabric in the European Chalk. *In*: Sieveking, G. de G. & Hart, M. B. (eds) *The scientific study of flint and chalk*. Cambridge University Press, Cambridge, 71–82.

Bromley, R. G. & Gale, A. A. 1982. The Lithostratigraphy of the English Chalk Rock. *Cretaceous Research*, **3**, 273–306.

Brookfield, M. E. 1973. Palaeogeography of the Upper Oxfordian and Lower Kimmeridgian (Jurassic) in Britain. *Palaeogeography, Palaeoclimatology, Palaeoecology*, **14**, 137–167.

Brookfield, M. E. 1977. The origin of bounding surfaces in ancient aeolian sandstones. *Sedimentology*, **24**, 303–332.

Brookfield, M. E. 1978. Revision of the stratigraphy of Permian and supposed Permian rocks of southern Scotland. *Geologische Rundschau*, **67**, 110–149.

Brookfield, M. E. 1980. Permian intermontane basin sedimentation in southern Scotland. *Sedimentary Geology*, **27**, 176–194.

Brookfield, M. E. 2000. Temporary desert lake deposits, Lower Permian (Rotliegendes), Southern Scotland, UK. *In*: Gierlowski-Kordesch, E. H. & Kelts, K. R. (eds) *Lake Basins Through Space and Time*. AAPG Studies in Geology, **46**, 67–74.

Brookfield, M. E. 2004. The enigma of fine-grained alluvial basin fills: the Permo-Triassic (Cumbrian Coastal and Sherwood Sandstone Groups) of the Solway Basin, NW England and SW Scotland. *International Journal of Earth Science*, **93**, 282–296.

Brooks, M., Doody, J. J. & Al-Rawi, F. J. R. 1984. Major crustal reflectors beneath SW England. *Journal of the Geological Society, London*, **141**, 97–103.

Brooks, M., Hillier, B. V. & Miliorizos, M. 1993. New seismic evidence for a major geological boundary at shallow depth, N. Devon. *Journal of the Geological Society, London*, **151**, 909–917.

Brooks, M., Mechie, J. & Llewellyn, D. J. 1983. Geophysical reflectors in the Variscan of south-west Britain. *In*: Hancock, P. L. (ed.) *The Variscan Fold Belt of the British Isles*. Adam Hilger, Bristol, 186–197.

Brown, A. G. 1997. Clearances and clearings: deforestation in Mesolithic–Neolithic Britain. *Oxford Journal of Archaeology*, **16**, 133–146.

Brown, A. G., Keough, M. K. & Rice, R. J. 1994. Floodplain evolution in the East Midlands, United Kingdom: the Lateglacial and Flandrian alluvial record from the Soar and Nene valleys. *Philosophical Transactions of the Royal Society of London*, **A348**, 261–293.

Brown, R. J. 2001. *Eruption history and depositional processes of the Poris ignimbrite of Tenerife and the Glaramara Tuff of the English Lake District*. Unpublished PhD thesis, University of Leicester.

Bruckschen, P. & Veizer, J. 1997. Oxygen and carbon isotopic composition of Dinantian brachiopods: paleoenvironmental implications for the Lower Carboniferous of Western Europe. *Palaeogeography, Palaeoclimatology, Palaeoecology*, **132**, 243–264.

Bryant, I. D., Holyoak, D. T. & Moseley, K. A. 1983. Late Pleistocene deposits at Brimpton, Berkshire, England. *Proceedings of the Geologists' Association*, **94**, 321–343.

Buchardt, B. 1978. Oxygen isotope palaeotemperatures from the Tertiary period in the North Sea area. *Nature*, **275**, 121–123.

Buckley, S. L. & Walker, M. J. C. 2002. A mid-Flandrian tephra horizon, Cambrian Mountains, west Wales. *Quaternary Newsletter*, **96**, 5–11.

Buckman, S. S. 1889. On the Cotteswold, Midford and Yeovil Sands, and the division between the Lias and the Oolite. *Quarterly Journal of the Geological Society of London*, **45**, 440–473.

Buckman, S. S. 1893. The Bajocian of the Sherborne district; its relation to subjacent and superjacent strata. *Quarterly Journal of the Geological Society of London*, **49**, 479–522.

Buckman, S. S. 1909–30. *Yorkshire Type Ammonites*, [continued as] Type Ammonites. 7 vols. Published by the author. London & Thame.

Buckman, S. S. 1910. Certain Jurassic (Lias-Oolite) strata of south Dorset, and their correlation. *Quarterly Journal of the Geological Society of London*, **66**, 52–89.

Buckman, S. S. 1915. A palaeontological classification of the Jurassic rocks of the Whitby district, with a zonal table of Liassic ammonites. *In*: Fox-Strangways, C. & Barrow, G. (eds) *The geology of the country between Whitby and Scarborough*. Memoir of the British Geological Survey, 2nd edn, HMSO, London, 59–102.

Buckman, S. S. & Wilson, E. 1896. Dundry Hill: its upper portion, or the beds marked as Inferior Oolite (g^5) in the maps of the Geological Survey. *Quarterly Journal of the Geological Society of London*, **52**, 669–720.

Bujak, J. P., Downie, C., Eaton, G. L. & Williams, G. L. 1980. *Dinoflagellate Cysts and Acritarchs from the Eocene of Southern England*. Palaeontology, Special Paper, **24**.

Bulat, J. & Stoker, S. J. 1987. Uplift determination from interval velocity studies, UK southern North Sea. *In*: Brooks, J. & Glennie, K. W. (eds) *Petroleum Geology of North West Europe*. Graham & Trotman, 293–305.

Bull, B. 1982. *Mineralisation of an area around Tavistock, South-west England*. Unpublished PhD thesis, University of Exeter.

Bulman, O. M. B. & Rushton, A. W. A. 1973. Tremadoc faunas from Boreholes in Central England. *Bulletin of the Geological Survey of Great Britain*, **43**, 1–40.

Bunz, S., Meinert, J. & Berndt, C. 2003. Geological controls on the Storegga gas-hydrate system of the mid-Norwegian continental margin. *Earth and Planetary Science Letters*, **209**, 291–307.

Burdon, E. S. 1999. *Contact metamorphism and fluid–rock interaction associated with the Skiddaw Granite intrusion, Cumbria, Northwest England*. PhD thesis, University of Derby.

Burgess, I. C. 1971. *Synopses of logs of Allenheads No 1 Borehole ([8604 4539]; OD 406. 76m) and No 2 Borehole ([8715 4505]; OD 536. 45m)*. Annual Report of the Institute of Geological Sciences for 1970, **33**.

Burgess, I. C. 1982. The stratigraphical distribution of Westphalian volcanic rocks to the east and south of Nottingham. *Proceedings of the Yorkshire Geological Society*, **44**, 29–44.

Burgess, I. C. 1986. Lower Carboniferous sections in the Sedbergh district, Cumbria. *Transactions of the Leeds Geological Association*, **11**, 1–23.

Burgess, I. C. & Holliday, D. W. 1974. The Permo-Triassic rocks of the Hilton Borehole, Westmorland. *Bulletin of the Geological Survey of Great Britain*, **46**, 1–34.

Burgess, I. C. & Holliday, D. W. 1979. *Geology of the Country Around Brough-under-Stainmore*. Memoir of the Geological Survey of Great Britain, England and Wales, Sheet 31, HMSO, London.

Burgess, P. M. & Gayer, R. A. 2000. Late Carboniferous tectonic subsidence in South Wales: implications for Variscan basin evolution and tectonic history in SW Britain. *Journal of the Geological Society, London*, **157**, 93–104.

Burne, R. V. & Moore, L. J. 1971. The Upper Carboniferous rocks of Devon and north Cornwall. *Proceedings of the Ussher Society*, **2**, 288–298.

Burnett, D. J. & Quirk, D. G. 2001. Turbidite provenance in the Lower Palaeozoic Manx Group, Isle of Man: implications for the tectonic setting of Eastern Avalonia. *Journal of the Geological Society, London*, **158**, 913–924.

Burnett, J. A. 1998. Upper Cretaceous. *In*: Bown, P. R. (ed.) *Calcareous Nannofossil Biostratigraphy*. British Micropalaeontological Series, Chapman & Hall, 132–199.

Busby, J. P., Kimbell, G. S. & Pharaoh, T. C. 1993. Integrated geophysical/geological modelling in southern Britain. *Geological Magazine*, **130**, 593–604.

Bushell, T. P. 1986. Reservoir geology of the Morecambe Field. *In*: Brooks, J., Goff, J. C. & Van Hoorn, B. (eds) *Habitat of Palaeozoic Gas in NW Europe*. Geological Society, London, Special Publications, **26**, 189–208.

Butler, A. J., Woodcock, N. H. & Stewart, D. M. 1997. The Woolhope and Usk Basins: Silurian rift basins revealed by subsurface mapping of the southern Welsh Borderland. *Journal of the Geological Society, London*, **154**, 209–223.

Butler, M. 1998. The geological history and the southern Wessex Basin – a review of new information from oil exploration. *In*: Underhill, J. R. (ed.) *Development, Evolution and Petroleum Geology of the Wessex Basin*. Geological Society, London, Special Publications, **133**, 67–86.

Butterworth, M. A. & Smith, A. H. V. 1976. The age of the British Upper Coal Measures with reference to their miospore content. *Review of Palaeobotany and Palynology*, **22**, 281–306.

Calef, C. E. & Hancock, N. J. 1974. Wenlock and Ludlow marine communities in Wales and the Welsh Borderland. *Palaeontology*, **17**, 779–810.

Callomon, J. H. 1955. The ammonite succession in the Lower Oxford Clay and Kellaways Beds at Kidlington, Oxfordshire and the zones of the Callovian Stage. *Philosophical Transactions of the Royal Society of London*, **B239**, 215–264.

Callomon, J. H. 1963. Sexual dimorphism in Jurassic ammonites. *Transactions of the Leicester Literary and Philosophical Society*, **57**, 22–56.

Callomon, J. H. 1968. The Kellaways Beds and the Oxford Clay. *In*: Sylvester-Bradley, P. C. & Ford, T. D. (eds) *The Geology of the East Midlands*. Leicester University Press, 264–290.

Callomon, J. H. 1985. The evolution of the Jurassic ammonite family Cardioceratidae. *In*: Cope, J. C. W. & Skelton, P. W. (eds) *Evolutionary Case Histories From the Fossil Record. Special Papers in Palaeontology*, **33**, 49–90.

Callomon, J. H. 1995. Time from fossils: S. S. Buckman and Jurassic high-resolution geochronology. *In*: Le Bas, M. J. (ed.) *Milestones in Geology*. Geological Society, London, Memoirs, **16**, 127–150.

Callomon, J. H. & Chandler, R. B. 1990. A review of the ammonite horizons of the Aalenian-Lower Bajocian Stages in the Middle Jurassic of southern England. *In*: Cresta, S. & Pavia, G. (eds) *Atti del meeting sulla stratigrafia del Baiociano. Memorie Descrittive della Carta Geologica d'Italia*, **40**, 85–111.

Callomon, J. H. & Cope, J. C. W. 1971. The stratigraphy and ammonite succession of the Oxford and Kimmeridge Clays in the Warlingham Borehole. *Bulletin of the Geological Survey of Great Britain*, **36**, 147–176, pls. 6–12.

Callomon, J. H. & Cope, J. C. W. 1995. The Jurassic geology of Dorset. *In*: Taylor, P. D. (ed.) *Field Geology of the British Jurassic*. Geological Society, London, 51–103.

Calner, M., Jeppson, L. & Munnecke, A. 2004. The Silurian of Gotland PtII, Guide to the IGCP field meeting, 2004, Erlangen Geologische Abhandlungen – Sondeband 5, Erlangen.

Calver, M. A. 1968. Distribution of Westphalian marine faunas in northern England and adjoining areas. *Proceedings of the Yorkshire Geological Society*, **37**, 1–72.

Calver, M. A. 1969. Westphalian of Britain. *Compte Rendu 6e Congrès Internationale Stratigraphie et Géologie Carbonifère, Sheffield, 1967*, **I**, 233–254.

Cameron, T. D. J., Bonny, A. P., Gregory, D. M. & Harland, R. 1984. Lower Pleistocene dinoflagellate cysts, foraminiferal and pollen assemblages in four boreholes in the southern North Sea. *Geological Magazine*, **121**, 85–97.

Cameron, T. D. J., Schüttenhelm, R. T. E. & Laban, C. 1989. Middle and Upper Pleistocene and Holocene stratigraphy in the southern North Sea between 52° and 54°N, 2° to 4°E. *In*: Henriet, J. P. & De Moor, G. (eds) *The Quaternary and Tertiary Geology of the Southern Bight, North Sea*. Belgian Geological Survey, Brussels, 119–135.

Cameron, T. D. J., Crosby, A., Balson, P. S., Jeffery, D. H., Lott, G. K., Bulat, J. & Harrison, D. J. 1992. *The geology of the southern North Sea*. United Kingdom Offshore Regional Report. British Geological Survey, HMSO, London.

Campbell, J. B. 1977. *The Upper Palaeolithic of Britain*. Clarendon Press, Oxford.

Campbell, S. D. G. 1983. Outline of geology. *In*: British Geological Survey 1:50 000 Series England and Wales Sheet 120 Corwen, Provisional Drift and Solid Edition. Southampton Ordinance Survey.

Campbell, S. D. G. 1984. Aspects of dynamic stratigraphy (Caradoc-Ashgill)in the northern part of the Welsh marginal basin. *Proceedings of the Geologists' Association*, **95**, 390–391.

Campbell, S. D. G. 1993. Outline of geology, *In: British Geological Survey 1:50 000 Series England and Wales, Sheet 120 (Corwen)*. Provisional solid and drift edition. Southampton, Ordnance Survey.

CAMPBELL, S. D. G., REEDMAN, A. J. & HOWELLS, M. F. 1985. Regional variations in cleavage and fold development in North Wales. *Geological Journal*, **20**, 43–52.

CAMPBELL, S. D. G, REEDMAN, A. J., HOWELLS, M. F. & MANN, A. C. 1987. The emplacement of geochemically distinct groups of rhyolites during the evolution of the Lower Rhyolite Tuff Formation caldera (Ordovician), North Wales, UK. *Geological Magazine*, **124**, 501–511.

CAMPBELL, S., HUNT, C. O., SCOURSE, J. D., KEEN, D. H. & STEPHENS, N. 1998. *Quaternary of South-west England*. Joint Nature Conservation Committee. Geological Conservation Review Series, **14**.

CAMPBELL-SMITH, W. 1963. Description of the Igneous Rocks Represented Among Pebbles From the Bunter Pebble Beds of the Midlands of England. *Bulletin of the British Museum of Natural History (Mineralogy)*, **2**, 1–17.

CANTRILL, T. C., DIXON, E. E. L., THOMAS, H. H. & JONES, O. T. 1916. *The Geology of the South Wales Coalfield. Part XII: The country around Milford*. Memoir of the Geological Survey of Great Britain, HMSO, London.

CAPEWELL, J. G. 1954. The post-Silurian pre-marine Carboniferous sedimentary rocks of the eastern side of the English Lake District. *Quarterly Journal of the Geological Society of London*, **111**, 23–46.

CAPUTO, M. V. 1998. Ordovician-Silurian glaciations and global sea-level changes. *In:* Silurian cycles: linkages of dynamic stratigraphy with atmospheric, oceanic, and tectonic changes. *Bulletin of the New York State Museum*, **491**, 15–25.

CAPUTO, M. V. & CROWELL, J. C. 1985. Migration of glacial centres across Gondwana during the Palaeozoic era. *Geological Society of America Bulletin*, **96**, 1020–1036.

CARNEY, J. N. 2002. Charnwood Forest. *In*: CARNEY, J. N., HORÁK, J., PHARAOH, T. C., GIBBONS, W., WILSON, D., BARCLAY, W. J. & BEVINS, R. E. (ed.) *Precambrian rocks of England and Wales*. Joint Nature Conservation Committee. Geological Conservation Review Series, **20**.

CARNEY, J. N. & PHARAOH, T. C. 1993. *Geology and structure of Precambrian rocks in quarries north-west of Nuneaton*. British Geological Survey Technical Report, WA/93/94.

CARNEY, J. N., HORÁK, J. M. *ET AL*. 2000. *Precambrian Rocks of England and Wales*. Joint Nature Conservation Committee. Geological Conservation Review Series, **20**.

CASEY, R. 1961. The stratigraphical palaeontology of the Lower Greensand. *Palaeontology*, **3**, 487–621.

CASEY, R. 1963. The dawn of the Cretaceous period in Britain. *Bulletin of the South-eastern Union of Scientific Societies*, **117**, 1–15.

CASEY, R. 1973. The ammonite succession at the Jurassic-Cretaceous boundary in eastern England. *In*: CASEY, R. & RAWSON, P. F. (eds) *The Boreal Lower Cretaceous*. *Geological Journal*, Special Issue **5**, 193–266, plates 1–10.

CASEY, R. & BRISTOW, C. R. 1964. Notes on some ferruginous strata in Buckinghamshire and Wiltshire. *Geological Magazine*, **101**, 116–128.

CASEY, R., BAYLISS, H. M. & SIMPSON, M. I. 1998. Observations on the lithostratigraphy and ammonite succession of the Aptian (Lower Cretaceous) Lower Greensand of Chale Bay, Isle of Wight, UK. *Cretaceous Research*, **19**, 511–535.

CASTLE, C. 1982. *Middle and Upper Devonian conodont biostratigraphy of the Torquay area, South Devon*. Unpublished PhD Thesis, University of Hull.

CASTON, V. N. D. 1972. Linear sand banks in the southern North Sea. *Sedimentology*, **18**, 63–78.

CATT, J. A. 1984. Soil particle size distribution and mineralogy as indicators of pedogenic and geomorphic history: examples from the loessial soils of England and Wales. *In*: RICHARDS, K. S., ARNETT, R. R. & ELLIS, S. (eds) *Geomorphology and Soils*. G. Allen & Unwin, London, 202–218.

CATT, J. A. 1986. The nature, origin and geomorphological significance of clay-with-flints. *In*: SIEVEKING, G. DE G. & HART, M. B. (eds) *The Scientific Study of Flint and Chert*. Cambridge University Press, Cambridge, 151–159.

CATT, J. A. 1992. Quaternary environments and their impact on British soils and agriculture. *Quaternary Proceedings*, **2**, 17–24.

CATT, J. A. & DIGBY, P. G. N. 1988. Boreholes in the Wolstonian Basement Till at Easington, Holderness, July 1985. *Proceedings of the Yorkshire Geological Society*, **47**, 21–27.

CATT, J. A. & PENNY, L. F. 1966. The Pleistocene deposits of Holderness, East Yorkshire. *Proceedings of the Yorkshire Geological Society*, **35**, 375–420.

CATT, J. A. & STAINES, S. J. 1998. Petrography of sediments and buried soils. *In*: PREECE, R. C. & BRIDGLAND, D. R. (eds) *Late Quaternary Environmental Change in North-west Europe: Excavations at Holywell Coombe, South-east England*. Chapman & Hall, London, 69–85.

CATT, J. A., BATEMAN, R. M., WINTLE, A. G. & MURPHY, C. P. 1987. The 'loess' section at Borden, Kent, SE England. *Journal of Quaternary Science*, **2**, 141–147.

CAVE, R. 1965. The Nod Glas sediments of Caradoc age in North Wales. *Geological Journal*, **4**, 279–98.

CAVE, R. 1977. *The Geology of the Malmesbury District*. Memoir of the Geological Survey of Great Britain, Sheet 251 (England and Wales), HMSO, London.

CAVE, R. 1979. Sedimentary environments of the basinal Llandovery of mid-Wales. *In*: HARRIS, A. L., HOLLAND, C. H. & LEAKE, B. E. (eds) *The Caledonides of the British Isles, Reviewed*. Geological Society, London, Special Publications, 8, 517–526.

CAVE, R. & COX, B. M. 1975. The Kellaways Beds of the area between Chippenham and Malmesbury. *Bulletin of the Geological Survey of Great Britain*, **54**, 41–66.

CAVE, R. & DIXON, R. J. 1993. The Ordovician and Silurian of the Welshpool area. *In*: WOODCOCK, N. H. & BASSETT, M. G. (eds) *Geological Excursions in Powys, Central Wales*. University of Wales Press, Cardiff, 51–84.

CAVE, R. & HAINS, B. A. 1986. *Geology of the Country Between Aberystwyth and Machynlleth*. Memoir of the British Geological Survey, Sheet 163 (England and Wales), HMSO, London.

CAVE, R. & HAINS, B. A. 2001. *Geology of the Country Around Montgomery and the Ordovician rocks of the Shelve area*. Memoir of the British Geological Survey sheet 165 and part of sheet 151 (Welshpool) (England and Wales) HMSO, London.

CAVE, R. & LOYDELL, D. K. 1998. Wenlock volcanism in the Welsh Basin. *Geological Journal*, **33**, 107–120.

CAVE, R. & RUSHTON, A. W. A. 1996. The Llandeilo (Ordovician) Series in the core of the Tywi Anticline, Llanwrtyd, Powys, UK. *Geological Journal*, **31**, 47–60.

CAVE, R. & WHITE, D. E. 1968. *The Brookend borehole*. Report of the Institute of Geological Sciences (GB) for 1967, 75–76.

CAVE, R. & WHITE, D. E. 1971. The exposure of Ludlow rocks and associated beds at Tites Point, and near Newham, Gloucestershire. *Geological Journal*, **7**, 239–254.

CAVE, R., CORNWELL, J. D. & EVANS, A. D. 1989. The Mathry Dyke, a quartz-dlerite intrusion of probable Carboniferous age in Southwest Wales. *Geological Magazine*, **126**, 715–721.

CAVE, R., EVANS, J. A. & CAMPBELL, S. D. G. 1992. Garn Prys: a mid-Silurian canyon, feeder to the Denbigh Grits of North Wales. *Geological Journal*, **27**, 301–315.

CHADWICK, R. A. 1985. End Jurassic-early Cretaceous sedimentation and subsidence (late Portlandian to Barremian), and the late-Cimmerian unconformity. *In*: WHITTAKER, A. (ed.) *Atlas of Onshore Sedimentary Basins in England and Wales*. Blackie, Glasgow, 52–56.

CHADWICK, R. A. 1993. Aspects of basin inversion in southern Britain. *Journal of the Geological Society, London*, **150**, 311–322.

CHADWICK, R. A. & EVANS, D. J. 1995. The timing and direction of Permo-Triassic extension in southern Britain. *In*: BOLDY, S. A. R. (ed.) *Permian and Triassic Rifting in Northwest Europe*. Geological Society, London, Special Publications, **91**, 161–192.

CHADWICK, R. A. & HOLLIDAY, D. W. 1991. Deep crustal structure and Carboniferous basin development within the Iapetus convergence zone, northern England. *Journal of the Geological Society, London*, **148**, 41–53.

Chadwick, R. A., Kenolty, N. & Whittaker, A. 1983. Crustal structure beneath southern England from deep seismic reflection profiles. *Journal of the Geological Society, London*, **140**, 893–911.

Chadwick, R. A., Holliday, D. W., Holloway, S. & Hulbert, A. G. 1995. *The Structure and Evolution of the Northumberland–Solway Basin and Adjacent Areas*. British Geological Survey Subsurface Memoir. HMSO, London.

Chadwick, R. A., Evans, D. J., Rowley, W. J., Smith, N. J. P., Walker, A. S. D., Birch, B. & Bulat, J. 1999. Structure and evolution of the basin. *In*: Plant, J. A., Jones, D. G. & Haslam, H. W. (eds) *The Cheshire Basin: Basin Evolution, Fluid Movement and Mineral Resources in a Permo-Triassic Rift Setting*. British Geological Survey, Keyworth, Nottingham, 41–89.

Chadwick, R. A., Jackson, D. I. *et al.* 2001. *Geology of the Isle of Man and its offshore area*. British Geological Survey Research Report, RR/01/06.

Challinor, J. 1927. Further note on the age of the rocks around the Teifi Estuary. *Geological Magazine*, **64**, 322–4.

Chambers, F. M., Addison, K., Blackford, J. J. & Edge, M. J. 1995. Palynology of organic beds below Devensian glacigenic sediments at Pen-y-Bryn, Gwynedd, North Wales. *Journal of Quaternary Science*, **10**, 157–173.

Chamley, H., Debrabant, P., Cadillier, A.-M. & Foulon, J. 1983. Clay mineralogy and inorganic geochemical stratigraphy of Blake-Bahama Basin since the Callovian, site 534, Deep Sea Drilling Project leg 76. *Initial Reports of the Deep Sea Drilling Project*, **76**, 437–451.

Chandler, M. A., Rind, D. & Ruedy, R. 1992. Pangean climate during the Early Jurassic: GCM simulations and the sedimentary record of paleoclimate. *Bulletin of the Geological Society of America*, **104**, 543–559.

Channell, J. E. T. & McCabe, C. 1992. Palaeomagnetic data form the Borrowdale Volcanic Group: volcano-tectonics and Late Ordovician palaeolatitudes. *Journal of the Geological Society, London*, **149**, 881–888.

Channel, J. E. T., McCabe, C. & Woodcock, N. H. 1992. An Early Devonian (pre-Acadian) magnetization component recorded in the Lower Old Red Sandstone of South Wales (UK). *Geophysical Journal International*, **108**, 883–894.

Chapman, M. R. & Shackleton, N. J. 2000. Evidence of 550-year and 1000-year cyclicities in North Atlantic circulation patterns during the Holocene. *The Holocene*, **10**, 287–292.

Chappell, B. W. & White, A. J. R. 1974. Two contrasting granite types. *Pacific Geology*, **8**, 173–174.

Charoy, B. 1986. The genesis of the Cornubian Batholith (south-west England): the example of the Carnmenellis Granite. *Journal of Petrology*, **27**, 571–604.

Chateauneuf, J. J. 1980. Palynostratigraphie et paléoclimatologie de l' Eocène supérieur et le l' Oligocène du Bassin de Paris (France). *Memoire du Bureau de Recherches Géologiques et Minières*, **116**.

Chen, Y., Clark, A. H., Farrar, E., Wasteneys, H. A. H. P., Hodgson, M. J. & Bromley, A. V. 1993. Diachronous and independent histories of plutonism and mineralization in the Cornubian Batholith, southwest England. *Journal of the Geological Society, London*, **150**, 1183–1191.

Cherns, L. 1980. Hardgrounds in the Lower Leintwardine Beds (Silurian) of the Welsh Borderland. *Geological Magazine*, **117**, 311–326.

Cherns, L. 1988. Faunal and facies dynamics in the Upper Silurian of the Anglo-Welsh basin. *Palaeontology*, **31**, 451–502.

Cherns, L. 1999. Faunal associations of the Lower Leintwardine Formation of the Anglo-Welsh Basin. *In*: Boucot, A. J. & Lawson, J. D. (eds) *Paleocommunities – A Case Study From the Silurian and Early Devonian*. Cambridge University Press, Cambridge, 373–379.

Cherns, L. & Bassett, M. G. 1999. Facies evolution and faunal dynamics in the Wenlock and Ludlow of the Anglo-Welsh shelf. *In*: Boucot, A. J. & Lawson, J. D. (eds) *Paleocommunities – A Case Study From the Silurian and Early Devonian*. Cambridge University Press, Cambridge, 20–31.

Chesher, J. A. 1968. The geology of the Middle Teign Valley. *Proceedings of the Ussher Society*, **2**, 15–17.

Chesley, J. T., Halliday, A. N., Snee, L. W., Mezger, K., Shepherd, T. J. & Scrivener, R. C. 1993. Thermochronology of the Cornubian batholith in southwest England: implications for pluton emplacement and protracted hydrothermal mineralisation. *Geochimica et Cosmochimica Acta*, **57**, 1817–1835.

Chisholm, J. I. 1990. The Upper Band-Better Bed sequence (Lower Coal Measures, Westphalian A) in the central and south Pennine area of England. *Geological Magazine*, **127**, 55–74.

Chisholm, J. I., Charsley, T. J. & Aitkenhead, N. 1988. *Geology of the Country Around Ashborne and Cheadle*, British Geological Survey Memoir for Sheet 124. HMSO, London.

Chisholm, J. I., Waters, C. N., Hallsworth, C. R., Turner, N., Strong, G. E. & Jones, N. S. 1996. Provenance of Lower Coal Measures around Bradford, West Yorkshire. *Proceedings of the Yorkshire Geological Society*, **51**, 153–166.

Christensen, W. K. 1976. Palaeobiogeography of Late Cretaceous belemnites of Europe. *Paläontologische Zeitschrift*, **50**, 113–129.

Christensen, W. K. 1997. Palaeobiogeography and migration in the Late Cretaceous familiy Belemnitellidae. *Acta Palaeontologica Polonica*, **42**, 457–495.

Chumakov, N. M. & Zharkov, M. A. 2004. Climate during the Permian–Triassic biosphere reorganizations. Article 2. Climate of the Late Permian and Early Triassic: General inferences. *Stratigraphy and Geological Correlation*, **11**, 361–375.

Church, K. D. & Gawthorpe, R. L. 1994. High resolution sequence stratigraphy of the late Namurian in the Widmerpool Gulf (East Midlands, UK). *Marine and Petroleum Geology*, **11**, 528–544.

Claoué-Long, J. C., Compston, W., Roberts, J. & Fanning, C. M. 1995. *Two Carboniferous ages: a comparison of SHRIMP zircon dating with conventional zircon ages and 40Ar/39Ar analysis*. SEPM, Special Publications, **54**, 3–21.

Clapham, M. E., Narbonne, G. M. & Gehling, J. G. 2003. Paleoecology of the oldest known animal communities: Ediacaran assemblages at Mistaken Point, Newfoundland. *Paleobiology*, **29**, 527–544.

Clark, A. H., Scott, D. J., Sandeman, H. A., Bromley, A. V. & Farrar, E. 1998*a*. Siegenian generation of the Lizard ophiolite: U-Pb zircon age data for plagiogranite, Porthkerris, Cornwall. *Journal of the Geological Society, London*, **155**, 595–598.

Clark, A. H., Sandeman, H. A., Liu, C., Scott, D. J., Farrar, E., Archibald, D. A., Bromley, A. V., Jones, K. A. & Warr, L. N. 1998*b*. An emerging geochronological record of the construction and emplacement of the Lizard ophiolite, SW Cornwall (Extended Abstract). *Proceedings of the Ussher Society*, **9**, 276–277.

Clark, C. D., Gibbard, P. L. & Rose, J. 2004. Pleistocene glacial limits in England, Scotland and Wales. *In*: Ehlers, J. & Gibbard, P. L. (eds) *Quaternary Glaciations: Extent and Chronology. Part 1 – Europe*. Elsevier Science, Amsterdam, 47–82.

Clark, J. G. D. 1954. *Excavations at Star Carr*. Cambridge University Press, Cambridge.

Clark, N. J. & Smith, R. B. 1988. Observations of salt particles in the atmosphere in the vicinity of an inland salt lake: Lake Eyre, South Australia. *Atmospheric Research*, **22**, 91–104.

Clark, P. U. & Mix, A. C. 2002. Ice sheets and sea level of the Last Glacial Maximum. *Quaternary Science Reviews*, **21**, 1–7.

Clayden, B. & Hollis, J. M. 1984. *Criteria for Differentiating Soil Series*. Soil Survey of England and Wales, Harpenden. Soil Survey Technical Monograph, **17**.

Clayton, C. 1994. Contrasting sediment gravity flow processes in the late Llandovery Rhuddnant Grits turbidite system, Welsh Basin. *Geological Journal*, **29**, 167–181.

Clayton, G., Coquel, R., Doubinger, J., Gueinn, K. J., Loboziak, S., Owens, B. & Streel, M. 1977. Carboniferous miospores of Western Europe: illustration and zonation. *Mededelingen Rijks Geologische Dienst*, **29**, 1–71.

Clayton, G., Higgs, K., Keegan, J. B. & Sevastopulo, G. D. 1978. Correlation of the palynological zonation of the Dinantian of the British Isles. *Palinologia, numero extraordinario*, **1**, 137–147.

CLAYTON, K. M. 2000. Glacial erosion of the Wash and Fen basin and the deposition of the chalky till of eastern England. *Quaternary Science Reviews*, **19**, 811–822.

CLEAL, C. J. 1984. The recognition of the base of the Westphalian D Stage in Britain. *Geological Magazine*, **121**, 125–129.

CLEAL, C. J. 1991. Carboniferous and Permian biostratigraphy. *In*: CLEAL, C. J. (ed.) *Plant Fossils in Geological Investigation: The Palaeozoic*. Ellis Horwood, Chichester, 182–215.

CLEAL, C. J. & THOMAS, B. A. 1996. *British Upper Carboniferous Stratigraphy*. Geological Conservation Review Series, **11**.

CLEAL, R. M. J., WALKER, K. E. & MONTAGUE, R. 1995. *Stonehenge in its Landscape. Twentieth-century Excavations*. English Heritage Archaeological Report, **10**. English Heritage, London.

CLEMENTS, R. G. 1993. Type-section of the Purbeck Limestone Group, Durlston Bay, Swanage, Dorset. *Proceedings of the Dorset Natural History and Archaeological Society*, **114**, 181–206.

CLIFF, R. A., DREWERY, S. E. & LEEDER, M. R. 1991. Sourcelands for the Carboniferous Pennine river system: constraints from sedimentary evidence and U–Pb geochronology using zircon and monazite. *In*: MORTON, A. C., TODD, S. P. & HAUGHTON, P. D. W. (eds) *Developments in Sedimentary Provenance Studies*. Geological Society, London, Special Publications, **57**, 137–159.

CLUTTON-BROCK, J. & NOE-NYGAARD, N. 1990. New osteological and C-isotope evidence on Mesolithic dogs; companions to hunters and fishers at Star Carr, Seamer Carr and Kongemose. *Journal of Archaeological Science*, **17**, 643–653.

COBBOLD, E. S. 1900. Geology. *In*: CAMPBELL-HYSLOP, C. W. (ed.) *Church Stretton*. L. Wilding, Shrewsbury, 1–115.

COBBOLD, E. S. 1921. The Cambrian horizons of Comley (Shropshire) and their Brachiopoda, Pteropoda, Gasteropoda, etc. *Quarterly Journal of the Geological Society of London*, **76**, 325–386.

COCKS, L. R. M. 1989. The Llandovery Series in the Llandovery area. *In*: HOLLAND, C. H. & BASSETT, M. G. (eds) *A Global Standard for the Silurian System*. National Museum of Wales, Cardiff, Geological Series, **10**.

COCKS, L. R. M. 1990. The geology of South Shropshire. *Proceedings of the Geologists' Association*, **100** (for 1989), 505–519.

COCKS, L. R. M. 1993. Triassic pebbles, derived fossils and the Ordovician to Devonian palaeogeography of Europe. *Journal of the Geological Society, London*, **150**, 219–226.

COCKS, L. R. M. 2000. The Early Palaeozoic Geography of Europe. *Journal of the Geological Society, London*, **157**, 1–10.

COCKS, L. R. M. & FORTEY, R. A. 1982. Faunal evidence for oceanic separations in the Palaeozoic of Britain. *Journal of the Geological Society, London*, **139**, 465–478.

COCKS, L. R. M. & MCKERROW, W. S. 1978. Silurian. *In*: W. S. MCKERROW (ed.) *The Ecology of Fossils*. MIT Press, Cambridge, MA, 93–124.

COCKS, L. R. M. & MCKERROW, W. S. 1984. Review of the distribution of the commoner animals in Lower Silurian marine benthic communities. *Palaeontology*, **27**, 663–670.

COCKS, L. R. M. & PRICE, D. 1975. The biostratigraphy of the Upper Ordovician and Lower Silurian of southwest Dyfed, with comments on the Hirnantia Fauna. *Palaeontology*, **18**, 703–724.

COCKS, L. R. M. & RICKARDS, R. B. 1969. Five boreholes in Shropshire and the relationships of shelly and graptolitic facies in the Lower Silurian. *Quarterly Journal of the Geological Society of London*, **124**, 213–238.

COCKS, L. R. M. & TORSVIK,T. H. 2002. Earth Geography from 500 to 400 million years ago: a faunal and palaeomagnetic review. *Journal of the Geological Society, London*, **159**, 631–644.

COCKS, L. R. M., HOLLAND, C. H., RICKARDS, R. B. & WARREN, P. T. 1975. *The type Wenlock Series*. Institute of Geological Sciences Report, **75/13**, 1–33.

COCKS, L. R. M., WOODCOCK, N. H., RICKARDS, R. B., TEMPLE, J. T. & LANE, P. D. 1984. The Llandovery series of the type area. Bulletin of the British Museum (Natural History), *Geology*, **38**, 131–182.

COCKS, L. R. M., HOLLAND, C. H. & RICKARDS, R. B. 1992. *A revised correlation of Silurian rocks in the British Isles*. Geological Society, London, Special Report, **21**, 1–32.

COE, A. L. 1992. *Unconformities Within the Upper Jurassic of the Wessex Basin, Southern England*. DPhil thesis, University of Oxford.

COE, A. L. 1995. A comparison of the Oxfordian successions of Dorset, Oxfordshire, and Dorset. *In*: TAYLOR, P. D. (ed.) *Field Geology of the British Jurassic*. Geological Society, London, 151–172.

COE, A. L. 1996. Unconformities within the Portlandian Stage of the Wessex Basin and their sequence-stratigraphical significance. *In*: HESSELBO, S. P. & PARKINSON, D. N. (eds) *Sequence Stratigraphy in British Geology*. Geological Society, London, Special Publications, **103**, 109–143.

COHEN, A. S. 2004. The rhenium–osmium isotope system: applications to geochronological and palaeoenvironmental problems. *Journal of the Geological Society, London*, **161**, 729–734.

COHEN, A. S. & COE, A. L. 2002. New geochemical evidence for the onset of volcanism in the central Atlantic magnetic province and environmental change at the Triassic–Jurassic boundary. *Geology*, **30**, 267–270.

COHEN, A. S., COE, A. L., BARTLETT, J. M. & HAWKESWORTH, C. J. 1999. Precise Re–Os ages of organic-rich mudrocks and the Os isotope composition of Jurassic seawater. *Earth and Planetary Science Letters*, **167**, 159–173.

COHEN, A. S., COE, A. L., HARDING, S. M. & SCHWARK, L. 2004. Osmium isotope evidence for the regulation of atmospheric CO_2 by continental weathering. *Geology*, **32**, 157–160.

COLEMAN, R. G. 1977. *Ophiolites*. Springer, Berlin.

COLES, J., ORME, B. J. & ROUILLARD, S. E. (eds) 1984. *Somerset Levels Papers 10*. Somerset Levels Project, Cambridge.

COLLINS, A. S. & BUCHAN, C. 2004. Provenance and age constaints of the South Stack Group, Anglesey, UK: U–Pb SIMS detrital zircon data. *Journal of the Geological Society, London*, **161**, 743–746.

COLLINS, P. E. F., FENWICK, I. M., KEITH-LUCAS, D. M. & WORSLEY, P. 1996. Late Devensian river and floodplain dynamics and related environmental change in northwest Europe, with particular reference to a site at Woolhampton, Berkshire. *Journal of Quaternary Science*, **11**, 357–376.

COLLINSON, J. D. 1969. The sedimentology of the Grindslow Shales and the Kinderscout Grit: A delta complex in the Namurian of Northern England. *Journal of Sedimentary Petrology*, **39**, 194–221.

COLLINSON, J. D. 1988. Controls on Namurian sedimentation in the Central Province basins of northern England. *In*: BESLY, B. M. & KELLING, G. (eds) *Sedimentation in a Synorogenic Basin Complex: the Upper Carboniferous of Northwest Europe*. Blackie, Glasgow, 85–101.

COLLINSON, J. D. & BANKS, N. L. 1975. The Haslingden Flags (Namurian G1) of South East Lancashire: Barfinger sands in the Pennine Basin. *Proceedings of the Yorkshire Geological Society*, **40**, 431–458.

COLLINSON, J. D., JONES, C. M. & WILSON, A. A. 1977. The Marsdenian (Namurian R2) succession west of Blackburn: implications for the evolution of Pennine Delta Systems. *Geological Journal*, **12**, 59–76.

COLLINSON, M. E. 1992. Vegetational and floristic changes around the Eocene–Oligocene boundary in western and central Europe. *In*: PROTHERO, D. R. &. BERGGREN, W. A. (eds) *Eocene–Oligocene Climatic and Biotic Evolution*. Princeton University Press, Princeton, NJ, 437–450.

COLLINSON, M. E. 2000. Fruit and seed floras from the Palaeocene/Eocene transition and subsequent Eocene in southern England: Comparison & palaeoenvironmental implications. *GFF*, **122**, 36–37.

COLLINSON, M. E., HOOKER, J. J. & GRÖCKE, D. R. 2003. Cobham Lignite Bed and penecontemporaneous macrofloras of southern England: A record of vegetation and fire across the Paleocene–Eocene Thermal Maximum. *In*: WING, S. L, GINGERICH, P. D., SCHMITZ, B. & THOMAS, E. (eds) *Causes and Consequences of Globally Warm Climates in the Early Paleogene*. Geological Society of America, Special Paper, **369**, 333–349.

COLTER, V. S. 1978. Exploration for gas in the Irish Sea. *Geologie en Mijnbouw*, **57**, 503–516.

Colter V. S. & Barr, K. W. 1975. Recent developments in the geology of the Irish Sea and Cheshire basins. *In*: Woodland, A. W. (ed.) *Petroleum Geology and the Continental Shelf of North-west Europe*. Applied Science, London, 61–75.

Colter, V. S. & Havard, D. J. 1981. The Wytch Farm oilfield, Dorset. *In*: Illing, L. V. & Hobson, G. D. (eds) *Petroleum Geology of the Continental Shelf of North-west Europe*. Heyden & Son, London, 494–503.

Combémorel, R. 1997. Bélemnites. *In*: Cariou, E. & Hantzpergue, P. (eds) *Biostratigraphie du Jurassique ouest-européen et méditerranéen*. Bulletin du Centre de Recherche, Elf Exploration Production, Mémoire, **17**, 157–167.

Compston, W., Wright, A. E. & Toghill, P. 2002. Dating the Late Precambrian volcanicity of England and Wales. *Journal of the Geological Society, London*, **159**, 323–339.

Conil, R., Groessens, E. & Pirlet, H. 1977. Nouvelle charte stratigraphique du Dinantien type de la Belgique. *Annales de la Societé géologique Nord*, **96**, 363–371.

Conil, R., Longerstaey, P. J. & Ramsbottom, W. H. C. 1980. Materiaux pour l'étude micropaléontologique du Dinantien de Grande-Bretagne. *Memoires de l'Institute Géologique de l'Université de Louvain*, **30**, 1–186.

Conybeare, W. D. & Phillips, W. 1922. *Outlines of the Geology of England and Wales*. Wm. Phillips, London.

Cook, C. A., Holdsworth, R. E. & Styles, M. T. 2002. The emplacement of peridotites and associated oceanic rocks from the lizard Complex, SW England. *Geological Magazine*, **139**, 27–45.

Cook, C. A., Holdsworth, R. E., Styles, M. T. & Pearce, J. A. 2000. Pre-emplacement structural history recorded by mantle peridotites: an example from the Lizard Complex, SW England. *Journal of the Geological Society, London*, **157**, 1049–1064.

Coombe, E. D. K. 1981. *Some aspects of coastal landslips and cliff falls at Portland*. DPhil Thesis, University of Oxford.

Coope, G. R. 1962. A Pleistocene coleopterous fauna with arctic affinities from Fladbury, Worcestershire. *Quarterly Journal of the Geological Society of London*, **118**, 103–123.

Coope, G. R. 1974. Interglacial coleoptera from Bobbitshole, Ipswich, Suffolk. *Journal of the Geological Society, London*, **130**, 333–340.

Coope, G. R. 1977. Fossil coleopteran assemblages as sensitive indicators of climatic changes during the Devensian (Last) cold stage. *Philosophical Transactions of the Royal Society of London*, **B280**, 313–337.

Coope, G. R. 1998. Insects. *In*: Preece, R. C. & Bridgland, D. R. (eds) *Late Quaternary Environmental Change in North-west Europe: Exacavations at Holywell Coombe, South-east England*. Chapman & Hall, London, 213–232.

Coope, G. R. 2000. Middle Devensian (Weichselian) coleopteran assemblages from Earith, Cambridgeshire (UK) and their bearing on the interpretation of 'Full glacial' floras and faunas. *Journal of Quaternary Science*, **15**, 779–788.

Coope, G. R. & Pennington, W. 1977. The Windermere Interstadial of the Late Devensian. *Philosophical Transactions of the Royal Society of London*, **B280**, 337–339.

Coope, G. R., Shotton, F. W. & Strachan, I. 1961. A Late Pleistocene fauna and flora from Upton Warren, Worcestershire. *Philosophical Transactions of the Royal Society of London*, **B244**, 379–421.

Coope, G. R., Gibbard, P. L., Hall, A. R., Preece, R. C., Robinson, J. E. & Sutcliffe, A. J. 1997. Climatic and environmental reconstructions based on fossil assemblages from Middle Devensian (Weichselian) deposits of the River Thames at South Kensington, Central London, UK. *Quaternary Science Reviews*, **16**, 1163–1195.

Coope, G. R., Field, M. H., Gibbard, P. L., Greenwood, M. & Richards, A. E. 2002. Palaeontology and biostratigraphy of Middle Pleistocene river sediment in the Mathon Member, at Mathon, Herefordshire, England. *Proceedings of the Geologists' Association*, **113**, 237–258.

Cooper, A. H. & Hughes, R. A. 1993. Discussion of Millward, D. & Molyneux, S. G. 1992. Filed and biostratigraphic evidence for an unconformity at the base of the Eycott Volcanic Group in the English Lake District. *Geological Magazine*, **130**, 271–272.

Cooper, A. H. & Molyneux, S. G. 1990. The age and correlation of the Skiddaw Group (Early Ordovician) sediments of the Cross Fell inlier (northern England). *Geological Magazine*, **127**, 147–157.

Cooper, A. H., Rushton, A. W. A., Molyneux, S. G., Hughes, R. A., Moore, R. & Webb, B. C. 1995. The stratigraphy and correlation of the Skiddaw Group (Ordovician) of the English Lake District. *Geological Magazine*, **132**, 185–211.

Cooper, A. H., Fortey, N. J., Hughes, R. A., Molyneux, S. G., Moore, R. M., Rushton, A. W. A. & Stone, P. 2004. *The Skiddaw Group of the English Lake District*. Memoir of the British Geological Survey (for parts of sheets 22, 23, 28, 29, 30, 31, and 48 England & Wales). HMSO, London.

Cooper, D. C., Lee, M. K., Fortey, N. J., Cooper, A. H., Rundle, C. C., Webb, B. C. & Allen, P. M. 1988. The Crummock Water aureole: a zone of metasomatism and source of ore metals in the English Lake District. *Journal of the Geological Society, London*, **145**, 523–540.

Cope, J. C. W. 1967. The palaeontology and stratigraphy of the lower part of the Upper Kimmeridge Clay of Dorset. *Bulletin of the British Museum (Natural History), Geology Series*, **15**, 1–79, plates 1–33.

Cope, J. C. W. 1971. Abbotsbury Iron Ore at Litton Cheney. *In*: Geology Report for 1970. *Proceedings of the Dorset Natural History and Archaeological Society*, **92**, 42.

Cope, J. C. W. 1974*a*. The Upper Kimmeridgian ammonite faunas of the Wash area and a subzonal scheme for the lower part of the Upper Kimmeridgian. *Bulletin of the Geological Survey of Great Britain*, **47**, 29–37, plates 1–3.

Cope, J. C. W. 1974*b*. New information on the Kimmeridge Clay of Yorkshire. *Proceedings of the Geologists' Association*, **85**, 211–221.

Cope, J. C. W. 1977. An Ediacara type fauna from South Wales. *Nature, London*, **264**, 624.

Cope, J. C. W. 1978. Ammonite faunas and stratigraphy of the upper part of the Upper Kimmeridge Clay of Dorset. *Palaeontology*, **21**, 469–534, plates 45–56.

Cope, J. C. W. 1979. Early history of the southern margin of the Tywi Anticline in the Carmarthen area, South Wales. *In*: Harris, A. L., Holland, C. H. & Leake, B. E. (ed.) *Caledonides of the British Isles: Reviewed*. Geological Society, London, Special Publications, **8**, 527–532.

Cope, J. C. W. (ed. and convener). 1980*a*. *A correlation of the Jurassic rocks in the British Isles. Part One. Introduction and Lower Jurassic*. Geological Society, London, Special Report, **14**.

Cope, J. C. W. (ed. and convener). 1980*b*. *A correlation of the Jurassic rocks in the British Isles. Part Two. Middle and Upper Jurassic*. Geological Society, London, Special Report, **15**.

Cope, J. C. W. 1983. Precambrian faunas from the Carmarthen district. *Nature in Wales*, **1**, 11–16.

Cope, J. C. W. 1984. The Mesozoic history of Wales. *Proceedings of the Geologists' Association*, **95**, 373–385.

Cope, J. C. W. 1993*a*. High resolution biostratigraphy. *In*: Hailwood, E. A. & Kidd, R. B. (eds) *High Resolution Stratigraphy*. Geological Society, London, Special Publications, **70**, 257–265.

Cope, J. C. W. 1993*b*. The Bolonian Stage: an old answer to an old problem. *Newsletters on Stratigraphy*, **28**, 151–156.

Cope, J. C. W. 1994. A latest Cretaceous hotspot and the south-easterly tilt of Britain. *Journal of the Geological Society, London*, **151**, 905–908.

Cope, J. C. W. 1995. Towards a unified Kimmeridgian Stage. *Petroleum Geoscience*, **1**, 351–354.

Cope, J. C. W. 1996*a*. Early Ordovician (Arenig) bivalves from the Llangynog inlier, South Wales. *Palaeontology*, **39**, 979–1029.

Cope, J. C. W. 1996*b*. The role of the Secondary Standard in stratigraphy. *Geological Magazine*, **133**, 107–110.

Cope, J. C. W. 1998*a*. The Mesozoic and Tertiary history of the Irish Sea. *In*: Meadows, N. S., Trueblood, S., Hardman, M. & Cowan, G. (eds) *The Petroleum Geology of the Irish Sea and Adjacent Areas*. Geological Society, London, Special Publications, **124**, 47–59. [Dated 1997.]

COPE, J. C. W. 1998*b*. Discussion on estimates of the amount and rate of sea-level change across the Rhaetian–Hettangian and Pliensbachian–Toarcian boundaries (latest Triassic to Early Jurassic). *Journal of the Geological Society, London*, **155**, 421–422.

COPE, J. C. W. 2004. A new look at the post-Variscan evolution of south-west England. *Geoscience in South-west England*, **11**, 1–9.

COPE, J. C. W. (ed.) in press. *A revised correlation of Jurassic rocks in the British Isles*. Geological Society Special Report.

COPE, J. C. W. & BASSETT, M. G. 1987. Sediment sources and Palaeozoic history of the Bristol Channel area. *Proceedings of the Geologists' Association*, **98**, 315–330.

COPE, J. C. W. & BEVINS, R. E. 1993. The stratigraphy and setting of the Precambrian rocks of the Llangynog Inlier, Dyfed, South Wales. *Geological Magazine*, **130**, 101–111.

COPE, J. C. W. & MCILROY, D. 1998. On the occurrence of foraminiferans in the Lower Cambrian of the Langynog inlier, south Wales. *Geological Magazine*, **135**, 227–9.

COPE, J. C. W. & RAWSON, P. F. 1992. Late Kimmeridgian. *In*: COPE, J. C. W., INGHAM, J. K. & RAWSON, P. F. (eds) *Atlas of Palaeogeography and Lithofacies*. Geological Society, London, Memoirs, **13**, 122, 124, map J10.

COPE, J. C. W. & RUSHTON, A. W. A. 1992. Cambrian and early Tremadoc rocks of Llangynog inlier, Dyfed, South Wales. *Geological Magazine*, **129**, 543–52.

COPE, J. C. W. & WIMBLEDON, W. A. 1973. The ammonite faunas of the uppermost Kimmeridge Clay, Portland Sand and Portland Stone of Dorset. *Proceedings of the Ussher Society*, **2**, 593–598.

COPE, J. C. W., DUFF, K. L., PARSONS, C. F., TORRENS, H. S., WIMBLEDON, W. A. & WRIGHT, J. K. 1980. *A correlation of Jurassic rocks in the British Isles. Part Two: Middle and Upper Jurassic*. Geological Society Special Report, **15**.

COPE, J. C. W., INGHAM, J. K. & RAWSON, P. F. (eds) 1992. *Atlas of Palaeogeography and Lithofacies*. Geological Society, London, Memoirs, **13**.

COPE, M. J. 1993*c*. A preliminary study of charcoalified plant fossils from the Middle Jurassic Scalby Formation of North Yorkshire. *In*: COLLINSON, M. E. & SCOTT, A. C. (eds) *Studies in Palaeobotany in Honour of Professor W. G. Chaloner, FRS. Special Papers in Palaeontology*, **49**, 101–111.

COPESTAKE, P. & JOHNSON, B. 1989. The Hettangian to Toarcian (Lower Jurassic) *In*: JENKINS, D. G. & MURRAY, J. W. (eds) *Stratigraphical Atlas of Fossil Foraminifera*. Ellis Horwood, Chichester, 129–188.

COPESTAKE, P., SIMS, A. P., CRITTENDEN, S., HAMAR, G. P., INESON, J. R., ROSE, P. T. & TRINGHAM, M. E. 2003. Lower Cretaceous. *In*: EVANS, D., GRAHAM, C., ARMOUR, A. & BATHURST, P. (eds & co-ordinators) *The Millenium Atlas: Petroleum Geology of the Central and Northern North Sea*. Geological Society, London, 191–211.

CORNFORD, C., YARNELL, L. & MURCHISON, D. G. 1987. Initial vitrinite reflectance results from the Carboniferous of north Devon and north Cornwall. *Proceedings of the Ussher Society*, **6**, 461–467.

CORNWELL, J. D. 1972. A gravity survey of the Isle of Man. *Proceedings of the Yorkshire Geological Society*, **39**, 93–106.

COSGROVE, J. 1980. The tectonic implications of some small scale structures on the Mona Complex, Holy Island, North Wales. *Journal of Structural Geology*, **2**, 383–396.

COSGROVE, M. E. & ELLIOT, M. H. 1976. Surabatholithic volcanism of the South west England granites. *Proceedings of the Ussher Society*, **3**, 391–399.

COSSEY, P. J., ADAMS, A. E., PURNELL, M. A., WHITELEY, M. J., WHYTE, M. A. & WRIGHT, V. P. 2004. *British Lower Carboniferous Stratigraphy*. Joint Nature Conservation Committee. Geological Conservation Review Series, **29**.

COSTA, L. I. & MANUM, S. B. 1988. The description of the interregional zonation of the Paleogene (D1 –D15) and the Miocene (D16–D20). *Geologisches Jahrbuch*, **A100**, 321–330.

COURTILLOT, V. 1999. *Evolutionary Catastrophes: The Science of Mass Extinctions*. Cambridge University Press, Cambridge.

COURTILLOT, V. 2004. *Evolutionary Catastrophes: The Science of Mass Extinctions*. Cambridge University Press, Cambridge.

COWAN, G. 1993. Identification and significance of aeolian deposits within the dominantly fluvial Sherwood Sandstone Group of the East Irish Sea Basin. *In*: NORTH, C. P. & PROSSER, D. J. (eds) *Characterization of Fluvial and Aeolian Reservoirs*. Geological Society, London, Special Publications, **73**, 231–245.

COWARD, M. P. 1995. Tectonic background to Permian and Triassic rifting in NW Europe. *In*: BOLDY, S. A. R. (ed.) *Permian–Triassic Rifting*. Geological Society, London, Special Publications, **91**, 1–22.

COWARD, M. P. & MCCLAY, K. R. 1983. Thrust tectonics of S Devon. *Journal of the Geological Society, London*, **140**, 214–228.

COWARD, M. P. & SIDDANS, A. W. B. 1979. The tectonic evolution of the Welsh Caledonides. *In*: HARRIS, A. L., HOLLAND, C. H. & LEAKE, B. E. (eds) *Caledonides of the British Isles: Reviewed*. Geological Society, London, Special Publications, **8**, 187–198.

COWARD, M. P. & SMALLWOOD, S. 1984. An interpretation of the Variscan tectonics of SW Britain. *In*: HUTTON, D. H. & SANDERSON, D. J. (eds) *Variscan Tectonics of the North Atlantic Region*. Geological Society, London, Special Publications, **14**, 89–102.

COWIE, J. W., RUSHTON, A. W. A. & STUBBLEFIELD, C. J. 1972. *A correlation of the Cambrian rocks of the British Isles*. Geological Society, London, Special Report, **2**.

COX, B. M. 1990. *Lower Greensand ammonoids from Clophill, Bedfordshire*. British Geological Survey, Technical Report, WH/90/276R.

COX, B. M. & GALLOIS, R. W. 1979. *Description of the standard stratigraphical sequences of the Upper Kimmeridge Clay, Ampthill Clay and West Walton Beds*. Report of the Institute of Geological Sciences, London, **78/19**, 68–72.

COX, B. M. & GALLOIS, R. W. 1981. *The stratigraphy of the Kimmeridge Clay of the Dorset type area and its correlation with some other Kimmeridgian sequences*. Report of the Institute of Geological Sciences, London, **80/4**, 1–44.

COX, B. M. & SUMBLER, M. G. 2002. *British Middle Jurassic Stratigraphy*. Joint Nature Conservation Committee. Geological Conservation Review Series, **26**.

COX, B. M., HUDSON, J. D. & MARTILL, D. M. 1993. Lithostratigraphic nomenclature of the Oxford Clay (Jurassic). *Proceedings of the Geologists' Association*, **103** [for 1992], 343–345.

COX, B. M., GALLOIS, R. W. & SUMBLER, M. G. 1994. The stratigraphy of the BGS Hartwell borehole, near Aylesbury, Buckinghamshire. *Proceedings of the Geologists' Association*, **105**, 209–224.

COX, B. M., SUMBLER, M. G. & IVIMEY-COOK, H. C. 1999. *A formational framework for the Lower Jurassic of England and Wales*. British Geological Survey Research Report, RR/99/01.

COX, L. R. 1925. The fauna of the basal Shell Bed of the Portland Stone, Isle of Portland. *Proceedings of the Dorset Natural History and Archaeological Society*, **46**, 113–172.

CÓZAR, P. & SOMERVILLE, I. D. 2004. New algal and foraminiferal assemblages and evidence for recognition of the Asbian–Brigantian boundary in northern England. *Proceedings of the Yorkshire Geological Society*, **55**, 43–65.

CRAIG, J. 1987. The structure of the Llangranog Lineament, West Wales: a Caledonian transpression zone. *Geological Journal*, **22**, 167–181.

CRAME, J. A. 1993. Bipolar molluscs and their evolutionary implications. *Journal of Biogeography*, **20**, 145–161.

CRAMPTON, S. L. & ALLEN, P. A. 1995. Recognition of forebulge unconformities associated with early stage foreland basin development: example from the North Alpine foreland basin. *AAPG Bulletin*, **79**, 1495–1514.

CRIMES, T. P. 1970*a*. A facies analysis of the Arenig of Western Lleyn, North Wales. *Proceedings of the Geologists' Association*, **81**, 221–240.

CRIMES, T. P. 1970*b*. A facies analysis of the Cambrian of Wales. *Paleaogeography, Palaeoclimatology, Palaeoecology*, **7**, 113–170.

CRIMES, T. P. & DHONAU, N. B. 1967. The Precambrian and Lower Palaeozoic rocks of south-east Co. Wexford, Eire. *Geological Magazine*, **104**, 213–221.

Critchley, M. F. 1984. Variscan tectonics of the Alston Block, northern England. *In*: Hutton, D. H. W. & Sanderson, D. J. (eds) *Variscan tectonics of the North Atlantic region*. Geological Society, London, Special Publications, **14**, 139–146.

Crittenden, S. 1982. Lower Cretaceous lithostratigraphy NE of the Sole Pit area in the UK southern North Sea. *Journal of Petroleum Geology*, **5**, 191–202.

Crookall, R. 1955. *Fossil Plants of the Carboniferous rocks of Great Britain*. Geological Survey of Great Britain Palaeontological Memoir. HMSO, London.

Crowley, T. J. & Baum, S. K. 1991. Estimating Carboniferous sea level fluctuations from Gondwanan ice extent. *Geology*, **19**, 975–977.

Cummins, W. A. 1959. The Lower Ludlow Grits in Wales. *Liverpool and Manchester Geological Journal*, **2**, 168–179.

Currant, A. & Jacobi, R. 2001. A formal mammalian biostratigraphy for the Late Pleistocene of Britain. *Quaternary Science Reviews*, **20**, 1707–1716.

Curry, D. 1965. The Palaeogene beds of southeast England. *Proceedings of the Geologists' Association*, **76**, 151–174.

Curry, D. 1966. Problems of correlation in the Anglo-Paris–Belgian Basin. *Proceedings of the Geologists' Association*, **77**, 437–467.

Curry, D. 1989. The rock floor of the English Channel and its significance for the interpretation of marine unconformities. *Proceedings of the Geologists' Association*, **100**, 339–352.

Curry, D. 1992. Tertiary. *In*: Duff, P. M. D. & Smith, A. J. (eds) *Geology of England and Wales*. 1st edn, Geological Society, London, 389–411.

Curry, D. & Odin, G. S. 1982. Dating of the Palaeogene. *In*: Odin, G. S. (ed.) *Numerical Dating in Stratigraphy*. Wiley, Chichester, 607–630.

Curry, D., Adams, C. G., Boulter, M. C., Dilley, F. C., Eames, F. E., Funnell, B. M. & Wells, M. K. 1978. *A Correlation of Tertiary Rocks in the British Isles*. Geological Society, London, Special Report, **12**.

Curtis, C. D., Cope, J. C. W., Plant, D. & MacQuaker, J. H. S. 2000. "Instantaneous" sedimentation, early microbial sediment strengthening and a lengthy record of chemical diagenesis preserved in Lower Jurassic ammonitiferous concretions from Dorset. *Journal of the Geological Society, London*, **157**, 165–172.

Curtis, M. L. K. 1972. The Silurian rocks of the Tortworth Inlier, Gloucestershire. *Proceedings of the Geologists' Association*, **83**, 1–35.

Curtis, M. T. 1982. Playa cycles in the Mercia Mudstone (Keuper Marl) of Aust Cliff, Avon. *Proceedings of the Bristol Naturalist Society*, **42**, 13–22.

Daley, B. 1999. Palaeogene sections in the Isle of Wight. A revision of their description and significance in the light of research undertaken over recent decades. *Tertiary Research*, **19**, 1–69.

Daley, B. & Balson, P. 1999. *British Tertiary Stratigraphy*. Joint Nature Conservancy Committee. Geological Conservation Review Series, **15**.

Daley, B. & Stewart, D. J. 1979. Weekend field meeting: The Wealden Group in the Isle of Wight 17–19 June 1977. *Proceedings of the Geologists' Association*, **90**, 51–54.

Dallmeyer, R. D. & Gibbons, W. 1987. The age of bluschist metamorphism in Anglesey, North Wales: evidence from $^{40}Ar/^{39}Ar$ mineral dates of the Penmynydd schists. *Journal of the Geological Society, London*, **144**, 843–850.

Dangerfield, J. & Hawkes, J. R. 1969. Unroofing of the Dartmoor Granite and possible consequences with regard to mineralization. *Proceedings of the Ussher Society*, **2**, 122–131.

Dangerfield, J. & Hawkes, J. R. 1981. The Variscan granites of south-west England: further information. *Proceedings of the Ussher Society*, **5**, 116–120.

Darbyshire, D. P. F. & Shepherd, T. J. 1985. Chronology of granite magmatism and associated mineralization, SW England. *Journal of the Geological Society, London*, **142**, 1159–1177.

Darbyshire, D. P. F. & Shepherd, T. J. 1987. Chronology of magmatism in south-west England: the minor intrusions. *Proceedings of the Ussher Society*, **6**, 431–438.

Darbyshire, D. P. F. & Shepherd, T. J. 1994. Nd and Sr isotope constraints on the origin of the Cornubian Batholith, SW England. *Journal of the Geological Society, London*, **151**, 795–802.

Dare-Edwards, A. J. 1984. Aeolian clay deposits of south-eastern Australia: parna or loessic clay. *Transactions of the Institute of British Geographers*, **9**, 337–344.

Dark, P. 2000. *The Environment of Britain in the First Millennium AD*. Duckworth, London.

Dark, P. & Allen, J. R. L. 2005. Seasonal deposition of Holocene banded sediments in the Severn Estuary Levels (southwest Britain); palynological and sedimentological evidence. *Quaternary Science Reviews*, **24**, 11–33.

Davidek, K., Landing. E., Bowring, S. A., Westrop, S. R., Rushton, A. W. A., Fortey, R. A. & Adrain, A. 1998. New uppermost Cambrian U-Pb date from Avalonian Wales and the age of the Cambrian-Ordovician boundary. *Geological magazine*, **135**, 227–9.

Davies, D. A. B. 1936. The Ordovician rocks of the Trevriw district (North Wales) *Quarterly Journal of the Geological Society of London*, **92**, 62–90.

Davies, D. K. 1967. Origin of friable sandstone–calcareous sandstone rhythms in the Upper Lias of England. *Journal of Sedimentary Petrology*, **37**, 1179–1188.

Davies, G. R., Gledhill, A. & Hawksworth, C. J. 1985. Upper crustal recycling in southern Britain: evidence from neodymium and strontium isotopes. *Earth and Planetary Science Letters*, **75**, 1–12.

Davies, J. H. 1981. *Structure and sedimentology of Lower Palaeozoic rocks in Brecknock, Wales*. Unpublished PhD thesis, University of London.

Davies, J. R. & Waters, R. A. 1995. The Caban Conglomerate and Ystrad Meurig Grits Formation – nested channels and lobe switching on a mud-dominated latest Ashgill to Llandovery slope apron, Welsh Basin, UK. *In*: Pickering, K. T., Hiscott, R. N., Kenyon, N. H., Ricci Lucci, F. & Smith, R. D. A. (eds) *Atlas of Deep Water Environments: Architectural Style in Turbidite Systems*. Chapman & Hall, London, 184–193.

Davies, J. R., Fletcher, C. J. N., Waters, R. A., Wilson, D., Woodhall, D. G. & Zalasiewicz, J. A. 1997. *Geology of the country around Llanilar and Rhayader*. Memoir of the British Geological Survey, Sheets 178 & 179 (England & Wales), HMSO, London.

Davies, J. R., McNestry, A. & Waters, R. A. 1991. Palaeoenvironments and palynofacies of a pulsed transgression: the late Devonian and early Dinantian (Lower Carboniferous) rocks of southeast Wales. *Geological Magazine*, **128**, Pt 4, 355–380.

Davies, J. R., Riley, N. J. & Wilson, D. 1989. The distribution of Chadian and earliest Arundian strata in North Wales: implications for Dinantian (Carboniferous) lithostratigraphy and palaeogeography. *Geological Journal*, **24**, 31–47.

Davies, J. R., Waters, R. A., Wilby, P. R., Williams, M. & Wilson, D. 2002. *The Cardigan and Dinas Island district – A Brief Explanation of the Geology*. Sheet Explanation of the British Geological Survey. 1:50 000 Sheet 193 (including part of Sheet 210) Cardigan and Dinas Island (England and Wales).

Davies, J. R, Wilson, D. & Williamson, I. T. 2004. *Geology of the country around Flint*. Memoir of the British Geological Survey, Sheet 108, England and Wales. HMSO, London.

Davies, S. J. & Elliott, T. 1996. Spectral gamma ray characterization of high resolution sequence stratigraphy: examples from Upper Carboniferous fluvio-deltaic systems, County Clare, Ireland. *In*: Howell, J. A. & Aitken, J. F. (eds) *High Resolution Sequence Stratigraphy: Innovations and Applications*. Geological Society, London, 25–35.

Davies, S. J., Hampson, G. J., Flint, S. S. & Elliott, T. 1999. Continental-scale sequence stratigraphy of the Namurian, Upper Carboniferous and its applications to reservoir prediction. *In*: Fleet, A. J. & Boldy, S. A. R. (eds) *Petroleum Geology of Northwest Europe: Proceedings of the 5th Conference*. Geological Society, London, 757–770.

DAVIS-KING, S. 1980. A note on new barbed points from Brandesburton, North Humberside. *Archaeological Journal*, **137**, 22–26.

DAVISION, E. H. 1926. A study of the Cornish granite, its variation and its relation with the occurrences of tin and other metallic ores. *Transactions of the Royal Geological Society of Cornwall*, **15**, 578–592.

DAY, J. B. W., LAND, D. H. & RAMSBOTTOM, W. H. C. 1970. *Geology of the country around Bewcastle*. Memoir of the Geological Survey of Great Britain, Sheet 12 (England & Wales), HMSO, London.

DAY, S. P. & MELLARS, P. 1994. 'Absolute' dating of Mesolithic human activities at Star Carr, Yorkshire: new palaeo-ecological studies and identification of the 9600 BP radiocarbon plateau. *Proceedings of the Prehistoric Society*, **60**, 417–422.

DE GRACIANSKY, P.-C., JACQUIN, T. & HESSELBO, S. P. 1998. The Ligurian cycle: an overview of Lower Jurassic 2nd order transgressive-regressive facies cycles in western Europe. *In*: DE GRACIANSKY, P.-C., HARDENBOL, J. JACQUIN, T. & VAIL, P. R. (eds) *Mesozoic–Cenozoic Sequence Stratigraphy of European Basins*. Society for Sedimentary Geology, Special Publications, **60**, 468–479.

DE JONG, J. 1988. Climatic variability during the past three million years, as indicated by vegetational evolution in northwest Europe and with emphasis on data from The Netherlands. *Philosophical Transactions of the Royal Society of London*, **B318**, 603–617.

DE LA BECHE, H. T. 1839. *Report on the Geology of Cornwall, Devon and West Somerset*. Longmans, London.

DE PERMENTIER, J. & VERNIERS, J. 2002. Chitinozoans from the global stratigraphical sections and points (GSSP) of the bases of the Aeronian and the Telychian (Llandovery) in Wales, UK. *In*: *Palaeozoic Palynology in the Third Millenium: New Directions in Acritarch, Chitinozoan and Miospore Research*. International Meeting and Workshops of the Commission International de Microflore du Paleozoique, 5–7 September 2002, Lille, France.

DE RAAF, J. F. M., READING, H. G. & WALKER, R. G. 1965. Cyclic sedimentation in the lower Westphalian of north Devon, England. *Sedimentology*, **4**, 1–52.

DE SOUZA, H. A. F. 1975. *K–Ar ages of Carboniferous igneous rocks from E. Lothian and the south of Scotland*. Unpublished MSc thesis, University of University of Leeds.

DE SOUZA, H. A. F. 1982. Age data from Scotland and the Carboniferous time scale. *In*: ODIN, G. S. (ed.) *Numerical Dating in Stratigraphy*. Wiley, Chichester, 456–465.

DEAN, A. 1992. *Report on samples processed for palynomorphs 1/10/87 – 1/7/91*. Report of the British Geological Survey, Biostratigraphy Series, **WH/93/197R**.

DEAN, M. T. 1994. *Interim report: conodont biostratigraphic control for the Devonian rocks outcropping at Plymouth Sound and south-east Cornwall*. Technical Report of the British Geological Survey, Stratigraphy Series, **WH/94/258R**.

DEAN, W. T. 1958. The faunal succession in the Caradoc Series of South Shropshire. *Bulletin of the British Museum of Natural History* (Geology), **3**, 191–231.

DEAN, W. T. 1964. The geology of the Ordovician and adjacent strata in the southern Caradoc district of Shropshire. *Bulletin of the British Museum of Natural History (Geology)*, **9**, 527–596.

DEARMAN, W. R. 1969. Tergiversate folds from south-west England. *Proceedings of the Ussher Society*, **1**, 79–81.

DEARMAN, W. R. 1971. A general view of the structure of Cornubia. *Proceedings of the Ussher Society*, **2**, 220–236.

DEARNLEY, R. 1990. *Malvern Hills Igneous Suite*. British Geological Survey Technical Report, **WG/90/11**.

DECONNINK, J.-F. 1987. Identification de l'origine détritique ou diagénétique des assemblages argileux: le cas des alternances marne-calcaire du Crétacé inférieur subalpin. *Bulletin de la Société Géologique de France* (8), **3**, 139–145.

DECONNINK, J.-F., HESSELBO, S. P., DEBUISSER, N., AVERBUCH, O., BAUDIN, F. & BESSA, J. Environmental constraints on clay mineralogy of an Early Jurassic mudrock (Blue Lias Formation, southern England). *International Journal of Earth Sciences*, **92**, 255–266.

DEEGAN, C. E. 1973. Tectonic control of sedimentation at the margin of a Carboniferous depositional basin in Kirkcudbrightshire. *Scottish Journal of Geology*, **9**, 1–28.

DEFRA 2004. *National Assessment of Defence Needs and Costs For Flood and Coastal Erosion Management (NADNAC)*. Summary Report. Flood Management Division, Department for Environment, Food and Rural Affairs, London.

DELAIR, J. B. & SARJEANT, W. A. S. 1985. History and bibliography of the study of fossil vertebrate footprints in the British Isles: supplement 1973–1983. *Palaeogeography, Palaeoclimatology, Palaeoecology*, **49**, 123–160.

DEMBOWSKA, J. 1976. The Jurassic. *In*: SOKOWSKI, S., CIESLINSKI, S. & CZERMINSKI, J. (eds) *Geology of Poland – Volume 1: Stratigraphy, Part 2: Mesozoic*. Wydawnictawa Geologiczne, Warsaw.

DEMETS, C., GORDON, R. G., ARGUS, D. F. & STEIN, S. 1990. Current plate motions. *Geophysical Journal International*, **101**, 425–478.

DEWEY, H. 1925. The mineral zones of Cornwall. *Proceedings of the Geologist's Association*, **36**, 107–135.

DEWEY, J. F. 1969*a*. Evolution of the Caledonian/Appalachian orogen. *Nature*, **222**, 124–129.

DEWEY, J. F. 1969*b*. *Structure and sequence in paratectonic British Caledonides*. AAPG, Memoir, **12**, 309–335.

DEWEY, J. F. 1982. Plate tectonics and the evolution of the British Isles. *Journal of the Geological Society, London*, **139**, 371–414.

DEWEY, J. F. 1988. Extensional collapse of orogens. *Tectonics*, **7**, 1123–1139.

DEWEY, J. F. & STRACHAN, R. A. 2003. Changing Silurian–Devonian relative plate motion in the Caledonides: sinistral transpression to sinistral transtension. *Journal of the Geological Society, London*, **160**, 219–229.

DIMBERLINE, A. J. & WOODCOCK, N. H. 1987. The south-east margin of the Wenlock turbidite system. *Geological Journal*, **22**, 61–71.

DIMBERLINE, A. J., BELL, A. & WOODCOCK, N. H. 1990. A laminated hemipelagic facies from the Wenlock and Ludlow of the Welsh Basin. *Journal of the Geological Society, London*, **147**, 693–701.

DIMBLEBY, G. W. 1962. *The Development of British Heathlands and Their Soils*. Oxford Forestry Memoir, **23**.

DINELEY, D. L. 1966. The Dartmouth Beds of Bigbury Bay, south Devon. *Quarterly Journal of the Geological Society of London*, **122**, 187–217.

DINELEY, D. L. 1999*a*. Late Silurian fossil fishes sites of the Welsh Borders. *In*: DINELEY, D. L. & METCALFE, S. J. (eds) *Fossil Fishes of Great Britain*. Nature Conservancy Council. Geological Conservation Review, **16**, 65–106.

DINELEY, D. L. 1999*b*. Early Devonian fossil fishes sites of the Welsh Borders. *In*: DINELEY, D. L. & METCALFE, S. J. (eds) *Fossil Fishes of Great Britain*. Nature Conservancy Council. Geological Conservation Review, **16**, 109–144.

DINELEY, D. L. & LOEFFLER, E. J. 1999. Upper Silurian and Lower Devonian vertebrate communities from the Anglo-Welsh Basin. *In*: BOUCOT, A. J. & LAWSON, J. D. (eds) *Palaeocommunities – A Case Study From the Silurian and Lower Devonian*. Cambridge University Press, Cambridge, 425–437.

DINES, H. G. 1934. The lateral extent of the ore-shoots in the primary depth zones of Cornwall. *Transactions of the Royal Geological Society of Cornwall*, **16**, 279–96.

DINES, H. G. 1956. *The Metalliferous Mining Region of South-west England*. 2 vols. HMSO, London.

DINES, H. G. & CHATWIN, C. P. 1930. Pliocene sandstone from Rothamsted (Hertfordshire). *Geological Survey Summary of Progress*, **1929**, 1–7.

DINES, H. G. & EDMUNDS, F. H. 1933. *Geology of the Country Around Reigate and Dorking*. Memoirs of the Geological Survey of England and Wales, explanation of Sheet 286.

DINES, H. G., BUCHAN, S., HOLMES, S. C. A. & BRISTOW, C. R. 1969. *Geology of the country around Sevenoaks and Tonbridge*. Memoir of the Geological Survey of Great Britain, HMSO, London.

DITCHFIELD, P. W. 1997. High northern palaeolatitude Jurassic-Cretaceous palaeotemperature variation: new data from Kong Karls Land, Svalbard. *Palaeogeography, Palaeoclimatology, Palaeoecology*, **130**, 163–175.

DIX, E. 1934. The sequence of Floras in the Upper Carboniferous, with special reference to South Wales. *Transactions of the Royal Society of Edinburgh*, **57**, 787–838.

DIXON, E. E. L., MADEN, J., TROTTER, F. M., HOLLINGWORTH, S. E. & TONKS, L. H. 1926. *The Geology of the Carlisle, Longtown and Silloth District*. Memoir of the Geological Survey of Great Britain, England and Wales, Sheets 11, 16 and 17, HMSO, London.

DIXON, R. J. 1988, The Ordovician (Caradoc) volcanic rocks of Montgomery, Powys, N. Wales. *Geological Journal*, **23**, 149–156.

DOBSON, M. R. & WHITTINGTON, R. 1979. The geology of the Kish Bank Basin. *Journal of the Geological Society, London*, **136**, 243–249.

DODD, C. 1986. *The sedimentology of fluvio-aeolian interactions with geological examples from the British Isles*. PhD Thesis, Plymouth Polytechnic.

DODSON, H. R. & REX, D. C. 1971 (for 1970). Potassium–argon ages of slates and phyllites from south-west England. *Quarterly Journal of the Geological Society of London*, **126**, 464–499.

DONALDSON, C. H. 1978. Petrology of the uppermost upper mantle deduced from Spinel–Lherzolite and Harzburgite nodules at Calton Hill, Derbyshire. *Contributions to Mineralogy and Petrology*, **65**, 363–377.

DONG, H., HALL, C. M., HALLIDAY, A. N., PEACOR, D. R., MERRIMAN, R. J. & ROBERTS, B. 1997. $^{40}Ar/^{39}Ar$ illite dating of Late Caledonian (Acadian) metamorphism and cooling of K-bentonites and slates from the Welsh Basin, UK. *Earth and Planetary Science Letters*, **150**, 337–351.

DONOVAN, D. T., HORTON, A. & IVIMEY-COOK, H. C. 1979. The transgression of the Lower Lias over the northern flank of the London Platform. *Journal of the Geological Society, London*, **136**, 165–73.

DONOVAN, S. K. & COPE J. C. W. 1989. A new camerate crinoid from the Ordovician of South Wales. *Palaeontology*, **32**, 101–110.

DORÉ, A. G. 1991. The structural foundation and evolution of the Mesozic seaways between Europe and the Arctic. *Palaeogeography, Palaeoclimatology, Palaeoecology*, **87**, 441–492.

DOWNIE, C. & SOPER, N. J. 1972. Age of the Eycott Volcanic Group and its conformable relationship to the Skiddaw Slates in the English Lake District. *Geological Magazine*, **109**, 259–268.

DOWNING, R. A. & SQUIRRELL, H. C. 1965. On the red and green beds in the Upper Coal Measures of the eastern part of the South Wales Coalfield. *Bulletin of the Geological Survey of Great Britain*, **23**, 45–56.

DOYLE, P. & BENNETT, M. R. 1995. Belemnites in biostratigraphy. *Palaeontology*, **38**, 815–829.

DREWERY, S. A. 1987. Provenance of Carboniferous sandstones from U–Pb of detrital zircons. *Nature*, **325**, 50–53.

DRUITT, T. H., EDWARDS, L. *ET AL*. 1999. *Santorini Volcano*. Geological Society, London, Memoirs, **19**, 1–165.

DRUMMOND, M. E. 1982. *The geology of Devonian limestones of the Brixham–Dartington area, South Devon*. Unpublished PhD thesis, University of Newcastle-upon-Tyne.

DRUMMOND, P. V. O. 1970. The Mid-Dorset Swell. Evidence of Albian–Cenomanian movements in Wessex. *Proceedings of the Geologists' Association*, **81**, 679–714.

DUFF, K. L. 1975. Palaeoecology of a bituminous shale – the Lower Oxford Clay of Central England. *Palaeontology*, **18**, 443–482.

DUFF, P. M. D. & WALTON, E. K. 1962. Statistical basis for cyclothems: A quantitative study of the sedimentary succession in the East Pennine Coalfield. *Sedimentology*, **1**, 235–255.

DUMAYNE-PEATY, L. 2000. Human impact on vegetation. *In*: BROTHWELL, D. R. & POLLARD, A. M. (eds) *Handbook of Archaeological Sciences*. Wiley, Chichester, 15–76.

DUNHAM, K. C. 1990. *Geology of the Northern Pennine Orefield; Volume 1, Tyne to Stainmore*. Economic Memoir of the British Geological Survey, England and Wales, Sheets 19 and 25, and parts of 13, 24, 26, 31, 32. Second Edition. HMSO, London.

DUNHAM, K. C. & WILSON, A. A. 1985. *Geology of the Northern Pennine Orefield: Volume 2, Stainmore to Craven*. Economic Memoir of the British Geological Survey, England and Wales, Sheets 40, 41, 50, and parts of 31, 32, 51, 60 and 61. HMSO, London.

DUNHAM, K. C., DUNHAM, A. C., HODGE, B. L. & JOHNSON, G. A. L. 1965. Granite beneath Visean sediments with mineralisation at Rookhope, north Pennines. *Quarterly Journal of the Geological Society of London*, **121**, 383–417.

DUNLOP, J. A. & SELDEN, P. A. 2004. A trigonotarbid arachnid from the Lower Devonian of Tredomen, Wales. *Palaeontology*, **47**, 1469–1476.

DUPUIS, C., STEURBAUT, E., DE CONINCK, J. & RIVELINE, J. 1998. The western Argiles à lignites facies–Cap d'Ailly sections. *In*: THIRY, M. & DUPUIS, C. (eds) *The Palaeocene–Eocene Boundary in the Paris Basin: The Sparnacian Deposits*. Field trip guide. Memoires des Sciences de la Terre, Ecole des Mines de Paris, **34**, 60–71.

DURNING, B. 1989. A new model for the development of the Variscan facing confrontation at Padstow, north Cornwall. *Proceedings of the Ussher Society*, **7**, 141–145.

DUXBURY, S. 1977. A palynostratigraphy of the Berriasian to Barremian of the Speeton Clay of Speeton, England. *Palaeontographica*, **160**, 17–67.

EARP, J. R. & TAYLOR, B. J. 1986. *Geology of the Country Around Chester and Winsford*. Memoir of British Geological Survey, Sheet 109 (England & Wales). HMSO, London.

EASTWOOD, T., DIXON, E. E. L. & HOLLINGSWORTH, S. 1931. *The Geology of the Whitehaven and Workington District*. Memoir of the Geological Survey, Sheet 28, England & Wales, HMSO, London.

EASTWOOD, T., HOLLINGSWORTH, S. E., ROSE, W. C. C. & TROTTER, F. M. 1968. *Geology of the Country Around Cockermouth and Caldbeck*. Memoir of the Geological Survey, Sheet 23 (England and Wales), HMSO, London.

EBDON, C. C., FRASER, A. J., HIGGINS, A. C., MITCHENER, B. C. & STRANK, A. R. E. 1990. The Dinantian stratigraphy of the East Midlands: A seismostratigraphic approach. *Journal of the Geological Society, London*, **147**, 519–536.

EDMONDS, E. A. 1974. *Classification of the Carboniferous rocks of south-west England*. Report of the Institute of Geological Sciences, **74/13**.

EDMONDS, E. A., MCKEOWN, M. C. & WILLIAMS, M. 1975. *British Regional Geology: South-West England*, 4th edn. British Geological Survey, HMSO, London.

EDMONDS, E. A., WHITTAKER, A. & WILLIAMS, B. J. 1985. *Geology of the Country Around Ilfracombe and Barnstaple*. Memoir of the British Geological Survey, Sheets 277 and 273, England and Wales, HMSO, London.

EDMONDS, E. A., WILLIAMS, B. J. & TAYLOR, R. T. 1979. *Geology of Bideford and Lundy Island*. Memoir of the Geological Survey of Great Britain, Sheets 292 with 275, 276, 291 and part of 308, England and Wales, HMSO, London.

EDMONDS, E. A., WRIGHT, J. E., BEER, K. E., HAWKES, J. R., WILLIAMS, M., FRESHNEY, E. C. & FENNING, P. J. 1968. *The Geology of the country around Okehampton*. Memoir of the Geological Survey of Great Britain, Sheet 324, England and Wales, HMSO, London.

EDMONDS, M. 1999. Inhabiting Neolithic landscapes. *Journal of Quaternary Science*, **14**, 485–492.

EDMUNDS, M., VARAH, M. & BENTLEY, A. 2003. The ammonite biostratigraphy of the Lower Lias 'Armatum Bed' (Upper Sinemurian–Lower Pliensbachian) at St Peter's Field, Radstock, Somerset. *Proceedings of the Geologists' Association*, **114**, 65–96.

EDWARDS, D. 1970. Fertile Rhyniophytina from the Lower Devonian of Britain. *Palaeontology*, **13**, 451–461.

EDWARDS, D. 1979. A Late Silurian flora from the Lower Old Red Sandstone of southwest Dyfed. *Palaeontology*, **22**, 23–52.

EDWARDS, D. & RICHARDSON, J. B. 1978. Brecon Beacons Quarry (Storey Arms), Powys. *In*: FRIEND, P. F. & WILLIAMS, B. P. J. (eds) *A Field Guide to Selected Outcrop Areas of the Devonian of Scotland, the Welsh Borderland and South Wales*. Palaeontological Association, London, 74–75.

EDWARDS, D. & RICHARDSON, J. B. 2004. Silurian and Lower Devonian plant assemblages from the Anglo-Welsh Basin: a palaeobotanical and palynological synthesis. *Geological Journal*, **39**, 375–402.

Edwards, D., Bassett, M. G. & Rogerson, E. C. W. 1979. The earliest vascular plants: continuing the search for proof. *Lethaia*, **12**, 313–324.

Edwards, D., Kenrick, P. & Carluccio, L. M. 1989. A reconsideration of cf. Psilophyton princeps (Croft and Lang, 1942), a zosterophyll widespread in the Lower Old Red sandstone of South Wales. *Botanical Journal of the Linnean Society*, **100**, 293–318.

Edwards, D., Selden, P. A., Richardson, J. B. & Axe, L. 1995. Coprolites as evidence for plant–animal interaction in Siluro-Devonian terrestrial ecosystems. *Nature*, **337**, 329–331.

Edwards, J. W. F. 1984 Interpretation of seismic and gravity surveys over the eastern part of the Cornubian platform. *In*: Hutton, D. H. W. & Sanderson, D. J. (eds) *Variscan Tectonics of the North Atlantic Region*. Geological Society, London, Special Publications, **14**, 119–124.

Edwards, J. W. F., Day, G. A. & Leveridge, B. E. 1989. Thrusts under Mount's Bay and Plymouth Bay. *Proceedings of the Ussher Society*, **7**, 131–135.

Edwards, R. A. 1999. *The Minehead District: A Concise Account of the Geology*. Memoir of the British Geological Survey 1:50,000 Geological Sheet 278 and part of Sheet 294 (England and Wales), HMSO, London.

Edwards, R. A. & Freshney, E. C. 1982. The Tertiary sedimentary rocks. *In*: Durrance, E. M. & Laming, D. J. C. (eds) *Geology of Devon*. University of Exeter, Exeter, 204–237.

Edwards, R. A. & Freshney, E. C. 1987*a*. *Geology of the country around Southampton*. Memoir of the British Geological Survey, Sheet 315 (England and Wales), HMSO, London.

Edwards, R. A. & Freshney, E. C. 1987*b*. Lithostratigraphical classification of the Hampshire Basin Palaeogene deposits (Reading Formation to Headon Formation). *Tertiary Research*, **8**, 43–73.

Edwards, R. A. & Scrivener, R. C. 1999. *The Geology of the country around Exeter*. Memoir of the British Geological Survey, Sheet 325 (England and Wales), HMSO, London.

Edwards, R. A., Warrington, G., Scrivener, R. C., Jones, N. S., Haslam, H. W. & Ault, L. 1997. The Exeter Group, south Devon, England: a contribution to the early post-Variscan stratigraphy of northwest Europe. *Geological Magazine*, **134**, 177–197.

Edwards, W. N. 1951. *The Concealed Coalfield of Yorkshire and Nottinghamshire*, 3rd edn. Memoir of the Geological Survey of Great Britain, HMSO, London.

Egger, H., Heilmann-Clausen, C. & Schmitz, B. 2000. The Paleocene–Eocene boundary interval of a tethyan deep-sea section and its correlation with the North Sea Basin. *Bulletin de la Société Géologique de France*, **171**, 207–216.

Ehlers, J. & Gibbard, P. 1991. Anglian glacial deposits in Britain and the adjoining offshore regions. *In*: Ehlers, J., Gibbard, P. L. & Rose, J. (eds) *Glacial Deposits in Great Britain and Ireland*. Balkema, Rotterdam, 17–24.

Ehlers, J. & Wingfield, R. 1991. The extension of the Late Weichselian/Late Devensian ice sheets in the North Sea Basin. *Journal of Quaternary Science*, **6**, 313–326.

Ehlers, J., Meyer, K. D. & Stephan, H. J. 1984. The pre-Weichselian glaciations of north-west Europe. *Quaternary Science Reviews*, **3**, 1–40.

Ehlers, J., Gibbard, P. L. & Whiteman, C. A. 1987. Recent investigations of the Marly Drift of northwest Norfolk, England. *In*: Van der Meer, J. J. M. (ed.) *Tills and Glaciotectonics*. Balkema, Rotterdam, 39–54.

Ekes, C. 1993. Bedload transported pedogenic mud aggregates in the Lower Old Red Sandstone in southwest Wales. *Journal of the Geological Society, London*, **150**, 469–472.

Ekman, S. R. & Scourse, J. D. 1993. Early and Middle Pleistocene pollen stratigraphy from British Geological Survey borehole 81/26, Fladen Ground, central North Sea. *Review of Palaeobotany and Palynology*, **79**, 285–295.

Elles, G. L. 1909. The relation of the Ordovicianand Silurian rocks of Conway (North Wales). *Quarterly Journal of the Geological Society of London*, **65**, 169–94.

Elliot, R. E. 1961. The stratigraphy of the Keuper series in southern Nottinghamshire. *Proceedings of the Yorkshire Geological Society*, **33**, 197–234.

Elliott, T. 1975. The sedimentary history of a delta lobe from a Yoredale (Carboniferous) cyclothem. *Proceedings of the Yorkshire Geological Society*, **40**, 505–536.

Elliott, T. 1976. Upper Carboniferous sedimentary cycles produced in river-dominated, elongate deltas. *Journal of the Geological Society, London*, **132**, 199–208.

Ellison, R. A. & Williamson, I. 1999. *Geology of the Windsor and Bracknell District – A Brief Explanation of the Geological Map*. British Geological Survey, Sheet 269 Windsor, HMSO, London.

Ellison, R. A., Knox, R. W., Jolley, D. W. &. King, C. 1994. A revision of the lithostratigraphic classification of the early Palaeogene strata in the London Basin and East Anglia. *Proceedings of the Geologists' Association*, **105**, 187–197.

Enfield, M. A., Gilchrist, J. R., Palmer, S. N. & Whalley, J. S. 1985. Structural and sedimentary evidence for the early tectonic history of the Bude and Crackington formations, north Cornwall and Devon. *Proceedings of the Ussher Society*, **6**, 165–172.

England, R. W. 1995. The seismic signature of the Iapetus Suture beneath the North Sea. *Studia Geophysica et Geodaetica*, **39**, 235–240.

Ensom, P. 2002. The Purbeck Limestone Group of Dorset, Southern England: a guide to lithostratigraphic terms. *Special Papers in Palaeontology*, **68**, 7–11.

Ernst, G., Schmid, F. & Seibertz, E. 1983. Event-Stratigraphie im Cenoman und Turon von NW-Deutschland. *Zitteliana*, **10**, 531–554.

Erwin, D. H., Bowring, S. A. & Yugan, J. 2002. End-Permian mass extinction: a review. *In*: Koeberl, C. & MacLoeod, K. G. (eds) *Catastrophic Events and Mass Extinction: Impacts and Beyond*. Geological Society of America, Special Paper, **356**, 363–383.

Evans, C. D. R. 1990. *The Geology of the Western English Channel and its Western Approaches*. United Kingdom Offshore Regional Report. HMSO, London.

Evans, C. D. R. & Hughes, M. J. 1984. The Neogene succession of the South Western Approaches, Great Britain. *Journal of the Geological Society, London*, **141**, 315–326.

Evans, D. C. 1906. The Ordovician rocks of western Caermarthenshire. *Quarterly Journal of the Geological Society of London*, **62**, 597–643.

Evans, D. J. & Kirby, G. A. 1999. The architecture of concealed Dinantian carbonate sequences over the Central Lancashire and Holme highs, northern England. *Proceedings of the Yorkshire Geological Society*, **52**, 297–312.

Evans, D. J. & Thompson, M. S. 1979. The geology of the central Bristol Channel and the Lundy area, South Western Approaches, British Isles. *Proceedings of the Geologists' Association*, **90**, 1–14.

Evans, D. J., Hopson, P. M., Kirby, G. A. & Bristow, C. R. 2003. The development and seismic expression of synsedimentary features within the Chalk of southern England. *Journal of the Geological Society, London*, **160**, 797–813.

Evans, D. J., Rees, J. G. & Holloway, S. 1993. The Permian to Jurassic stratigraphy and structural evolution of the central Cheshire Basin. *Journal of the Geological Society London*, **150**, 857–870.

Evans, D. J., Rowley, W. J., Chadwick, R. A., Kimbell G. S. & Millward, D. 1994. Seismic reflection data and the internal structure of the Lake District batholith, Cumbria, northern England. *Proceedings of the Yorkshire Geological Society*, **50**, 11–24.

Evans, E., Ashley, R., Hall, J., Penning-Rowsell, E., Saul, A., Sayers, P., Thorne, C. & Watkinson, A. 2004. *Foresight. Future Flooding. Scientific Summary: Volume 1 – Future Risks and Their Drivers*. Office of Science and Technology, London.

Evans, J. A. 1989*a*. A note on Rb–Sr whole-rock ages from cleaved mudrocks in the Welsh Basin. *Journal of the Geological Society, London*, **146**, 901–904.

Evans, J. A. 1989*b*. Resetting of the Rb–Sr whole-rock isotope system of an Ordovician microgranite during Devonian low-grade metamorphismm. *Geological Magazine*, **126**, 675–679.

EVANS, J. A. 1991. Resetting of Rb–Sr whole-rock ages during Acadian low-grade metamorphism in North Wales. *Journal of the Geological Society, London*, **148**, 703–710.

Evans, J. A. 1996. Dating the transition of smectite to illite in Palaeozoic mudrocks using the Rb–Sr whole-rock technique. *Journal of the Geological Society, London*, **153**, 101–108.

EVANS, J. A. & ZALASIEWICZ, J. 1996. U–Pb, Pb–Pb and Sm–Nd dating of authigenic monazite: implications for the diagenetic evolution of the Welsh Basin. *Earth and Planetary Science Letters*, **144**, 421–433.

EVANS, J. A., CHISHOLM, J. I. & LENG, M. J. 2001. How U–Pb detrital monazite ages contribute to the interpretation of the Pennine Basin infill. *Journal of the Geological Society, London*, **158**, 741–744.

EVANS, J. A., MILLAR, I. L. & NOBLE, S. R. 1995. Hydration during uplift is recorded by reset Rb–Sr whole-rock ages. *Journal of the Geological Society, London*, **152**, 209–212.

EVANS, J. G. 1972. *Land Snails in Archaeology*. Seminar Press, London.

EVANS, K. M. 1983. Note on the age and fauna of the Lynton Beds (lower Devonian) of north Devon. *Geological Journal*, **18**, 297–305.

EVANS, W. D. 1945. The geology of the Prescelly Hills, north Pembrokeshire. *Quarterly Journal of the Geological Society of London*, **101**, 89–110.

EVISON, M. P. 1999. Perspectives on the Holocene in Britain; Human DNA. *Journal of Quaternary Science*, **14**, 615–623.

EYERS, J. 1995. The Silty Beds: a tidal flat sequence at the junction of the Lower Greensand and Gault (Albian, Lower Cretaceous) of Bedfordshire, England. *Proceedings of the Geologists' Association*, **106**, 107–118.

EYLES, N. & MCCABE, A. M. 1989. The Late Devensian (<22,000 BP) Irish Sea Basin: the sedimentary record of a collapsed ice sheet margin. *Quaternary Science Reviews*, **8**, 307–351.

EYLES, N. & SLADEN, J. A. 1981. Stratigraphy and geotechnical properties of weathered lodgement till in Northumberland, England. *Quarterly Journal of Engineering Geology*, **14**, 129–141.

EYLES, N., MCCABE, A. M. & BOWEN, D. Q. 1994. The stratigraphic and sedimentological significance of Late Devensian ice sheet surging in Holderness, Yorkshire, UK. *Quaternary Science Reviews*, **13**, 727–759.

FAGAN, B. 2000. *The Little Ice Age*. Basic Books, New York.

FALCON, N. L. & KENT, P. E. 1960. *Geological Results of Petroleum Exploration in Britain*. Geological Society, London, Memoirs, **2**, 1–56, plates 1–5.

FALCON-LANG, H. J. 1999. The Early Carboniferous (Asbian–Brigantian) seasonal tropical climate of northern Britain. *Palaios*, **14**, 116–126.

FAULKNER, T. J. 1989. The early Carboniferous (Courceyan) Middle Hope volcanics of Weston-super-mare: development and demise of an offshore volcanic high. *Proceedings of the Geologists' Association*, **100**, 93–106.

FEARNSIDES, W. G. 1910. The Tremadoc slates and associated rocks of Southeast Carnarvonshire. *Quarterly Journal of the Geological Society of London*, **66**, 142–188.

FEDONKIN, M. A. 1981. *The Vendian White Sea biota*. Trudy Geol. Inst Akad. Nauk SSSR, **342**, 1–100.

FEIST, M., LAKE, R. D. & WOOD, C. J. 1995. Charophyte biostratigraphy of the Purbeck and Wealden of Southern England. *Palaeontology*, **38**, 407–442.

FENTON, J. P. G., RIDING, J. B. & WYATT, R. J. 1994. Palynostratigraphy of the Middle Jurassic 'White Sands' of central England. *Proceedings of the Geologists' Association*, **105**, 225–230.

FEWTRELL, M. D., RAMSBOTTOM, W. H. C. & STRANK, A. R. E. 1981. Carboniferous. *In*: JENKINS, D. G. & MURRAY, J. W. (eds) *Stratigraphical atlas of fossil Foraminifera*. British Micropalaeontological Society Series. Ellis Horwood, Chichester, 13–69.

FIELDING, C. R. 1984. A coal depositional model for the Durham Coal Measures of NE England. *Journal of the Geological Society, London*, **141**, 919–931.

FIRMAN, R. J. 1957. Fissure metasomatism in volcanic rocks adjacent to the Shap granite, Westmorland. *Quarterly Journal of the Geological Society of London*, **113**, 205–222.

FIRMAN, R. J. 1978. Epigenetic mineralization. *In*: MOSELEY, F. (ed.) *The Geology of the Lake District*. Yorkshire Geological Society, Occasional Publication, **3**, 226–241.

FISHER, M. J. & HANCOCK, N. J. 1985. The Scalby Formation (Middle Jurassic, Ravenscar Group) of Yorkshire: reassessment of age and depositional environment. *Proceedings of the Yorkshire Geological Society*, **45**, 293–298.

FISHER, M. J. & MUDGE, D. C. 1998. Triassic. *In*: GLENNIE, K. W. (ed.) *Petroleum Geology of the North Sea*, 3rd edn. Blackwell, Oxford, 212–244.

FITCH, F. J. & MILLER, J. A. 1964. The age of the paroxysmal Variscan Orogeny in England. *In*: HARLAND, W. B., GILBERT SMITH, A. & WILCOCK, B. (eds) *The Phanerozoic Time-scale*. Geological Society, London, Special Publications, **120**, 159–173.

FITCH, F. J., EVANS, A. L., GRASTY, R. L. & MENEISY, M. Y. 1969. Isotopic age determinations on rocks from Wales and the Welsh Borders. *In*: WOOD, A. (ed.) *The Precambrian and Lower Palaeozoic Rocks of Wales*. University of Wales Press, Cardiff, 23–45.

FITCH, F. J., MILLER, J. A. & WILLIAMS, S. C. 1970. Isotopic ages of British Carboniferous rocks. *Sixième Congrès International de Stratigraphie et de Géologie du Carbonifère*, **2**, 771–789.

FITCHES, W. R. & CAMPBELL, S. D. G. 1987. Tectonic evolution of the Bala Lineament in the Welsh Basin. *Geological Journal*, **22**, 131–153.

FITCHES, W. R. & JOHNSON, R. 1978. Cleavage-fold relationships in the Aberystwyth Grits – a preliminary report. *Journal of the Geological Society, London*, **135**, 250.

FITCHES, W. R., BARNES, R. P. & MORRIS, J. H. 1999. Geological structure and tectonic evolution of the Lower Palaeozoic rocks of the Isle of Man. *In*: WOODCOCK, N. H., QUIRK, D. G., FITCHES, W. R. & BARNES, R. P. (eds) *In Sight of the Suture: The Palaeozoic Geology of the Isle of Man in its Iapetus Ocean Context*. Geological Society, London, Special Publications, **160**, 259–287.

FITTON, J. G. 1971. *The petrogenesis of the calc-alkaline Borrowdale Volcanic Group, Northern England*. PhD thesis, University of Durham.

FITTON, J. G. & HUGHES, D. J. 1970. Volcanism and plate tectonics in the British Ordovician. *Earth and Planetary Science Letters*, **8**, 223–228.

FITTON, J. G., THIRLWALL, M. F. & HUGHES, D. J. 1982. Volcanism in the Caledonian orogenic belt of Britain. *In*: THORPE, R. S. (ed.) *Andesites*. Wiley, New York, 611–636.

FITTON, W. H. 1836. Observations on some of the strata between the Chalk and the Oxford Oolite, in the south-east of England. *Transactions of the Geological Society of London*, (2) **4**, 103–388.

FLEMING, A. 1971. Bronze Age agriculture on the marginal lands of north-east Yorkshire. *Agricultural History Review*, **19**, 1–24.

FLETT, J. S. 1946. *Geology of the Lizard and Meneage* (Sheet 359), 2nd edn. Memoir of the Geological Survey of Great Britain. HMSO, London.

FLOODPAGE J., NEWMAN, P. & WHITE, J. 2001. Hydrocarbon prospectivity of the Irish Sea area: insights from recent exploration of the Central Irish Sea, Peel and Solway basins. *In*: SHANNON, C. M., HAUGHTON, P. D. W. & CORCORAN, P. V. (eds) *The Petroleum Exploration of Ireland's Offshore Basins*. Geological Society, London, Special Publications, **188**, 107–134.

FLOYD, P. A. 1976. Geochemical variation in the greenstones of SW England. *Journal of Petrology*, **17**, 522–545.

FLOYD, P. A. 1982. Chemical variation in Hercynian basalt relative to plate tectonics. *Journal of the Geological Society, London*, **139**, 505–520.

FLOYD, P. A. 1984. Geochemical characteristics and comparison of the basic rocks of the Lizard Complex and the basaltic lavas within the Hercynian troughs of SW England. *Journal of the Geological Society, London*, **141**, 61–70.

FLOYD, P. A. 1995. Rhenohercynian foldbelt: autochthon and non-metamorphic nappe units – igneous activity. *In*: DALLMEYER, D., FRANKE, W. & WEBER, K. (eds) *Pre-Permian Geology of Central and Western Europe*. Springer, Berlin, 59–81.

FLOYD, P. A. & AL-SAMMAN, A. M. 1980. Primary and secondary chemical variation exhibited by some west Cornish volcanic rocks. *Proceedings of the Ussher Society*, **5**, 68–75.

FLOYD, P. A. & LEVERIDGE, B. E. 1987. Tectonic environment of the Devonian Gramscatho basin, S. Cornwall; framework mode and geochemical evidence from turbiditic sandstones. *Journal of the Geological Society, London*, **144**, 532–542.

FLOYD, P. A., EXLEY, C. S. & STYLES, M. T. 1993*a*. *Igneous Rocks of South-west England*. Geological Conservation Review Series. Chapman & Hall, London.

FLOYD, P. A., HOLDSWORTH, R. E. & STEELE, S. A. 1993*b*. Geochemistry of the Start Complex greenschists: Rhenohercynian MORB? *Geological Magazine*, **130**, 345–352.

FLOYD, P. A., SHAIL, R., LEVERIDGE, B. E. & FRANKE, W. 1991. Geochemistry and provenance of Rhenohercynian synorogenic sandstones: implications for tectonic environment discrimination. *In*: MORTON, A. C., TODD, S. P. & HAUGHTON, P. D. W. (eds) *Developments in Sedimentary Provenance Studies*. Geological Society, London, Special Publications, **57**, 173–188.

FORD, M. & TELLAM, J. H. 1994. Source, type and extent of inorganic contamination within the Birmingham urban aquifer system. *Journal of Hydrology*, **156**, 101–135.

FORD, T. D. 1958. Precambrian fossils from Charnwood Forest. *Proceedings of the Yorkshire Geological Society*, **31**, 211–217.

FORTEY, N. J., COOPER, A. H., HENNEY, P. J., COLMAN, T. & NANCARROW, P. H. A. 1994. Appinitic intrusions in the English Lake District. *Mineralogy and Petrology*, **51**, 355–375.

FORTEY, N. J. MERRIMAN, R. J. & HUFF, W. D. 1996. Silurian and late-Ordovician K-bentonites as a record of late-Caledonian volcanism in the British Isles. *Transactions of the Royal Society of Edinburgh*, **86**, 167–180.

FORTEY, N. J., ROBERTS, B. & HIRONS, S. R. 1993. Relationship between metamorphism and structure in the Skiddaw Group, English Lake District. *Geological Magazine*, **130**, 631–638.

FORTEY, R. A. 1984. Global earlier Ordovician transgressions and regressions and their biological implications. *In*: BRUTON, D. L. (ed.) *Aspects of the Ordovician System*. Palaeontological Contributions from the University of Oslo 295. 37–50.

FORTEY, R. A. & COCKS, L. R. M. 1986. Marginal faunal belts and their structural implications, with examples from the Lower Palaeozoic. *Journal of the Geological Society, London*, **143**, 151–160.

FORTEY, R. A. & OWENS, R. M. 1987. The Arenig Series in South Wales. *Bulletin of The British Museum of Natural History (Geology)*, **41**, 69–307.

FORTEY, R. A. & OWENS, R. M. 1991. A trilobite fauna from the highest Shineton Shales in Shropshire and the correlation of the latest Tremadoc. *Geological Magazine*, **128**, 437–64.

FORTEY, R. A. & OWENS, R. M. 1992. The Habberley Formation: youngest Tremadoc in the Welsh Borderlands. *Geological Magazine*, **129**, 553–66.

FORTEY, R. A. & RUSHTON, A. W. A. 1992. Southern British Isles Arenig. *In*: COPE, J. C. W., INGHAM, J. K. & RAWSON, P. F. (eds) *Atlas of Palaeogeography and Lithofacies*. Geological Society, London, Memoirs, **13**, 26–28.

FORTEY, R. A., BASSETT, M. G., HARPER, D. A. T., HUGHES, R. A., INGHAM, J. K., MOLYNEUX, S. G., OWEN, A. W., OWENS, R. M., RUSHTON, A. W. A. & SHELDON, P. 1991. Progress and problems in the selection of stratotypes for the bases of Ordovician series in the type area UK. *In*: BARNES C. R. & WILLIAMS S. H. (eds) *Advances in Ordovician Geology*. Geological Survey of Canada, Paper, **90–9**. 5–25.

FORTEY, R. A., HARPER, D. A, T., INGHAM. J. K., OWEN, M. A. & RUSHTON, A. W. A. 1995. A revision of Ordovician series and stages from the historical type area. *Geological Magazine*, **132**, 15–30.

FORTEY, R. A., HARPER, D. A. T., INGHAM, J. K., OWEN, A. W., PARKES, M. A., RUSHTON A. W. A. & WOODCOCK, N. H. 2000. *A Revised Correlation of Ordovician rocks in the British Isles*. Geological Society, London, Special Report, **24**.

FORTEY, R. A., HARPER, D. A. T., INGHAM, J. K., OWEN, A. W. & RUSHTON, A. W. A. 1995. A revision of the Ordovician series and stages from the historical type areas. *Geological Magazine*, **132**, 15–30.

FORTEY, R. A., OWENS, R. M. & RUSHTON, A. W. A. 1989. The palaeogeographic position of the Lake District in the early Ordovician. *Geological Magazine*, **126**, 9–17.

FOSTER, D., HOLLIDAY, D. W., JONES, C. M., OWENS, B. & WELSH, A. 1989. The concealed Upper Palaeozoic rocks of Berkshire and South Oxfordshire. *Proceedings of the Geologists' Association*, **100**, 395–407.

FRAKES, L. A. 1979. *Climates Through Geologic Time*. Elsevier, Amsterdam.

FRAKES, L. A. & FRANCIS, J. E. 1988. A guide to Phanerozoic cold polar climates from high-latitude ice-rafting in the Cretaceous. *Nature*, **333**, 547–549.

FRANCE, D. S. 1975. The geology of the Indefatigable Gas-Field. *In*: WOODLAND, A. W. (ed.) *Petroleum and the Continental Shelf of North-west Europe*. Applied Science Publishers, 233–239.

FRANCIS, E. H. 1970. Review of Carboniferous volcanism in England and Wales. *Journal of Earth Sciences, Leeds Geological Association*, **8**, 41–56.

FRANCIS, E. H. 1982. Magma and sediment – 1. Emplacement mechanism of late Carboniferous tholeiite sills in northern Britain. *Journal of the Geological Society, London*, **139**, 1–20.

FRANCIS, E. H. 1992. Igneous rocks. *In*: DUFF, P. M. D. & SMITH, A. J. (eds) *Geology of England and Wales*. Geological Society, London, 489–521.

FRANCIS, J. E. 1983. The dominant conifer of the Jurassic Purbeck Formation, England. *Palaeontology*, **26**, 277–294.

FRANCIS, J. E. 1987. Growth rings in upper Jurassic and Lower Cretaceous fossil wood from southern England and their palaeoclimatic significance. *In*: WARD, R. G. (ed.) *Applications of Tree-Ring Studies*. British Archaeological Report Series, **333**, 21–36.

FRANKE, W. 1989. *Tectonostratigraphic units in the Variscan belt of central Europe*. Geological Society of America, Special Paper, **230**, 67–90.

FRANKE, W. 1992. Phanerozoic structures and events in Central Europe. *In*: BLUNDELL, D., FREEMAN, R. & MUELLER, ST. (eds) *A Continent Revealed*. Cambridge University Press, Cambridge, 164–179.

FRANKE, W. 2000. The mid-European segment of the Variscides: tectonostratigraphic units, terrane boundaries and plate tectonic evolution. *In*: FRANKE, W., HAAK, V., ONCKEN, O. & TANNER, D. (eds) *Quantification and Modelling in the Variscan Belt*. Geological Society, London, Special Publications, **179**, 35–61.

FRANKE, W. & ENGEL, W. 1982. Variscan sedimentary basins on the Continent and relations with south-west England. *Proceedings of the Ussher Society*, **5**, 259–269.

FRANKE, W. & ONKEN, O. 1995. Zur prädevonischen Gesschichte des Rhenohercynischen Beckens. *Nova Acta Leopoldina*, Neue Folge, **71**, 53–72. [In German.]

FRANKE, W., DALLMEYER, R. D. & WEBER, K. 1995. Geodynamic Evolution. *In*: DALLMEYER, R. D., FRANKE, W. & WEBER, K. (eds) *Pre-Permian Geology of Central and Eastern Europe*. Springer, Berlin, 579–593.

FRASER, A. J. & GAWTHORPE, R. L. 1990. Tectono-stratigraphic development and hydrocarbon habitat of the Carboniferous in northern England. *In*: HARDMAN, R. F. P. & BROOKS, J. (eds) *Tectonic Events Responsible for Britain's Oil and Gas Reserves*. Geological Society, London, Special Publications, 49–86.

FRASER, A. J. & GAWTHORPE, R. L. 2003. *An Atlas of Carboniferous Basin Evolution in Northern England*. Geological Society, London, Memoirs, **28**.

FRASER, A. J., NASH, D. F., STEELE, R. P. & EBDON, C. C. 1990. A regional assessment of the intra-Carboniferous play of northern England. *In*: BROOKS, J. (ed.) *Classic Petroleum Provinces*. Geological Society, London, Special Publications, 417–440.

FRESHNEY, E. C. & TAYLOR, R. T. 1972. The Upper Carboniferous stratigraphy of north Cornwall and west Devon. *Proceedings of the Ussher Society*, **2**, 464–471.

FRESHNEY, E. C. & TAYLOR, R. T. 1980. The Variscides of southwest Britain. *In*: OWEN, T. R. (ed.) *United Kingdom: introduction to General Geology and Guides to Excursions 002, 055, 093 and 151*. 26th International Geological Congress, Paris. Institute of Geological Sciences, London, 49–57.

FRESHNEY, E. C., MCKEOWN, M. C. & WILLIAMS, M. 1972. *Geology of the coast between Tintagel and Bude*. Memoir of the Geological Survey of Great Britain, Sheet 322, England and Wales, HMSO, London.

FRESHNEY, E. C., BEER, K. E. & WRIGHT, J. E. 1979*a*. *Geology of the country around Chulmleigh*. Memoir of the Geological Survey of Great Britain, Sheet 309, England and Wales, HMSO, London.

FRESHNEY, E. C., EDMONDS, E. A., TAYLOR, R. T. & WILLIAMS, B. J. 1979*b*. *Geology of the country around Bude and Bradworthy*. Memoir of the Geological Survey of Great Britain, Sheets 307 and 308, England and Wales, HMSO, London.

FRIEND, P. F. 1969. Tectonic features of Old Red Sandstone sedimentation in North Atlantic Borders. *In*: KAY, M. (ed.) *North Atlantic Geology and Continental Drift*. AAPG, Memoir, **12**, 703–710.

FRIEND, P. F., WILLIAMS, B. P. J., FORD, M. & WILLIAMS, E. A. 2000. Kinematics and dynamics of Old Red Sandstone basins. *In*: FRIEND, P. F. & WILLIAMS, B. P. J. (eds) *New Perspectives on the Old Red Sandstone*. Geological Society, London, Special Publications, **180**, 29–60.

FRODSHAM, K. & GAYER, R. A. 1997. Variscan conpressional structures within the main productive coal-bearing strata of South Wales. *Journal of the Geological Society, London*, **154**, 195–208.

FRODSHAM, K., GAYER, R. A., JAMES, J. E. & PRYCE, R. 1993. Variscan thrust deformation in the South Wales Coalfield: a case study from Ffos Las opencast coal site. *In*: GAYER, R. A. & GREILING, R. O. (eds) *The Rhenohercynian and Subvariscan Fold Belts*. Earth Evolution Science Series, Vieweg, Braunschweig, 515–348.

FUNNELL, B. M. 1961. The Palaeogene and Early Pleistocene of Norfolk. *Transactions of the Norfolk and Norwich Naturalists' Society*, **19**, 340–364.

FUNNELL, B. M. 1987. Late Pliocene and Early Pleistocene Stages of East Anglia and the adjacent North Sea. *Quaternary Newsletter*, **52**, 1–11.

FUNNELL, B. M. 1995. Global sea-level and the (pen)insularity of late Cenozoic Britain. *In*: PREECE, R. C. (ed.) *Island Britain: a Quaternary Perspective*. Geological Society, London, Special Publications, **96**, 3–13.

FURNESS, R. R. 1965. The petrography and provenance of the Coniston Grits east of the Lune Valley, Westmorland. *Geological Magazine*, **102**, 253–260.

GALE, A. S. 1980. Penecontemporaneous folding, sedimentation and erosion in Campanian Chalk near Portsmouth, England. *Sedimentology*, **27**, 137–151.

GALE, A. S. 1989. A Milankovitch scale for Cenomanian time. *Terra Nova*, **1**, 400–425.

GALE, A. S. 1996. Turonian correlation and sequence stratigraphy of the Chalk in southern England. *In*: HESSELBO, S. P. & PARKINSON, D. N. (eds) *Sequence Stratigraphy in British Geology*. Geological Society, London, Special Publications, **103**, 177–195.

GALE, A. S. 2000. The Cretaceous World. *In*: CULVER, S. & RAWSON, P. F. (eds) *Biotic Response to Global Change: the Last 145 Million Years*. Cambridge University Press, Cambridge, 4–19.

GALE, A. S., HUGGETT, J. M. & GILL, M. 1996. The stratigraphy and petrography of the Gault Clay Formation (Albian, Cretaceous) at Redcliff, Isle of Wight. *Proceedings of the Geologists' Association*, **107**, 287–298.

GALE, A. S., JEFFERY, P. A., HUGGETT, J. M. & CONNOLLY, P. 1999. Eocene inversion history of the Sandown Pericline, Isle of Wight, southern England. *Journal of the Geological Society, London*, **156**, 327–339.

GALE, A. S., WOOD, C. J. & BROMLEY, R. G. 1988. The Lithostratigraphy and Marker Bed correlation of the White Chalk (Late Cenomanian–Campanian) in Southern England. *Mesozoic Research*, **1**, 107–118.

GALE, A. S., YOUNG, J. R., SHACKLETON, N. J., CROWHURST, S. J. & WRAY, D. S. 1999*b*. Orbital tuning of Cenomanian marly chalk successions: towards a Milankovitch time-scale for the Late Cretaceous. *Philosophical Transactions of the Royal Society of London*, **A 357**, 1815–1829.

GALE, N. H., BECKINSALE, R. D. & WADGE, A. J. 1979. A Rb–Sr whole-rock isochron for the Stockdale Rhyolite of the English Lake District and a revised Mid-Palaeozoic time-scale. *Journal of the Geological Society, London*, **136**, 235–242.

GALLOIS, R. W. 1965. *The Wealden District*, 4th edn. British Regional Geology, Institute of Geological Sciences.

GALLOIS, R. W. 1976. Coccolith blooms in the Kimmeridge Clay and the origin of North Sea oil. *Nature*, **259**, 473–475.

GALLOIS, R. W. 1979. *Oil shale resources in Great Britain*. 2 vols. Internal Report of the Institute of Geological Sciences.

GALLOIS, R. W. 1988. *Geology of the Country Around Ely*. British Geological Survey, Sheet 173 (England & Wales), HMSO, London.

GALLOIS, R. W. 1994. *Geology of the Country Around Kings Lynn and The Wash*. British Geological Survey, Sheets 145 and part of 129, HMSO, London.

GALLOIS, R. W. 2000. The stratigraphy of the Kimmeridge Clay Formation (Upper Jurassic) in the RGGE Project boreholes at Swanworth Quarry and Metherhills, south Dorset. *Proceedings of the Geologists' Association*, **111**, 265–280.

GALLOIS, R. W. 2001. The lithostratigraphy of the Mercia Mudstone Group (mid to late Triassic) of the South Devon Coast. *Proceedings of the Ussher Society*, **10**, 195–204.

GALLOIS, R. W. 2003. The distribution of halite (rock salt) in the Mercia Mudstone Group (mid to late Triassic) in SW England. *Geoscience in South-west England*, **10**, 383–389.

GALLOIS, R. W. 2004. The type section of the junction of the Otter Sandstone Formation and the Mercia Mudstone Group (mid Triassic) at Pennington Point, Sidmouth. *Geoscience in South-west England*, **11**, 51–58.

GALLOIS, R. W. & COX, B. M. 1976. The stratigraphy of the Lower Kimmeridge Clay of eastern England. *Proceedings of the Yorkshire Geological Society*, **41**, 13–26.

GALLOIS, R. W. & COX, B. M. 1977. The stratigraphy of the Middle and Upper Oxfordian sediments of Fenland. *Proceedings of the Geologists' Association*, **88**, 207–228.

GALLOIS, R. W. & COX, B. M. 1994. The Kimmeridge Clay and underlying strata (Upper Jurassic) at Swindon, Wiltshire. *Proceedings of the Geologists' Association*, **105**, 99–110.

GALLOIS, R. W. & ETCHES, S. 2001. The stratigraphy of the youngest part of the Kimmeridge Clay Formation (Upper Jurassic) of the Dorset type area. *Proceedings of the Geologists' Association*, **112**, 169–182.

GALLOIS, R. W. & MORTER, A. A. 1982. The stratigraphy of the Gault of East Anglia. *Proceedings of the Geologists' Association*, **93**, 351–368.

GALLOIS, R. W. & WORSSAM, B. C. 1993. *Geology of the country around Horsham*. Memoir of the British Geological Survey, Sheet 302 (England & Wales), HMSO, London.

GALLOWAY, W. E. 1989. Genetic stratigraphic sequences in basin analysis i: architecture and genesis of flooding – surface bounded depositional units. *AAPG Bulletin*, **73**, 125–142.

GALTIER, J., SCOTT, A. C., POWELL, J. H., GLOVER, B. W. & WATERS, C. N. 1992. Anatomically preserved conifer-like stems from the Upper Carboniferous. *Proceedings of the Royal Society of London*, **247**, 211–214.

GARDEN, I. R. 1991. Changes in the provenance of pebbly detritus in southern Britain and northern France associated with basin rifting. *In*: MORTON, A. C., TODD, S. P. & HAUGHTON, R. D. W. (eds) *Developments in Sedimentary Provenance Studies*. Geological Society, London, Special Publications, **57**, 273–289.

GARDINER, P. R. R. 1974 The Duncannon Group; an upper Ordovician unit in south-west County Wexford. *Bulletin of the Geological Society of Ireland*, **1**, 429–446.

GARLAND, J., TUCKER, M. E. & SCRUTTON, C. T. 1996. Microfacies analysis and metre scale cyclicity in the Givetian back-reef sediments of SE Devon. *Proceedings of the Ussher Society*, **9**, 31–36.

GARWOOD, E. J. 1913. The Lower Carboniferous succession in the north-west of England. *Quarterly Journal of the Geological Society of London*, **68**, 449–572.

GATLIFF, R. W. 1994. *The geology of the central North Sea*. United Kingdom Offshore Regional Report. British Geological Survey.

GAUNT, G. D. 1976. The Devensian maximum ice limit in the Vale of York. *Proceedings of the Yorkshire Geological Society*, **40**, 631–637.

GAUNT, G. D. 1994. *Geology of the Country Around Goole, Doncaster and the Isle of Axholme*. Memoir of British Geological Survey, Sheets 79 and 88 (England and Wales). HMSO, London.

GAUNT, G. D., JARVIS, R. A. & MATTHEWS, B. 1971. The Late-Weichselian sequence in the Vale of York. *Proceedings of the Yorkshire Geological Society*, **38**, 281–284.

GAUNT, G. D., FLETCHER, T. P. & WOOD, C. J. 1992. *Geology of the Country Around Kingston upon Hull and Brigg*. Memoir of the British Geological Survey Sheets 80 and 89 (England & Wales). HMSO, London.

GAUSS, G. A. 1967. Structural aspects of the Padstow area, North Cornwall. *Proceedings of the Ussher Society*, **1**, 284–285.

GAUSS, G. A. & HOUSE, M. R. 1972. The Devonian successions in the Padstow area, North Cornwall. *Journal of the Geological Society, London*, **128**, 151–172.

GAWTHORPE, R. L. 1986. Sedimentation during carbonate ramp-to-slope evolution in a tectonically active area: Bowland Basin (Dinantian), N. England. *Sedimentology*, **33**, 185–206.

GAWTHORPE, R. L. 1987. Tectono-sedimentary evolution of the Bowland Basin, northern England, during the Dinantian. *Journal of the Geological Society, London*, **144**, 59–71.

GAWTHORPE, R. L. & CLEMMEY, H. 1985. Geometry of submarine slides in the Bowland Basin (Dinantian) and their relation to debris flows. *Journal of the Geological Society, London*, **142**, 555–565.

GAWTHORPE, R. L., GUTTERIDGE, P. & LEEDER, M. R. 1989. Late Devonian and Dinantian basin evolution in northern England and North Wales. *In*: ARTHURTON, R. S., GUTTERIDGE, P. & NOLAN, S. C. (eds) *The Role of Tectonics in Devonian and Carboniferous Sedimentation in the British Isles*. Yorkshire Geological Society, Occasional Publication, **6**, 1–23.

GAYER, R. & JONES, J. 1989. The Variscan foreland in South Wales. *Proceedings of the Ussher Society*, **7**, 177–179.

GAYER, R. A. & PESEK, J. 1992. Cannibalisation of Coal Measures in the South Wales Coalfield – significance for foreland basin evolution. *Proceedings of the Ussher Society*, **8**, 44–49.

GAYER, R. A., ALLEN, K. C., BASSETT, M. G. & EDWARDS, D. 1973. The structure of the Taff Gorge area, Glamorgan, and the stratigraphy of the Old Red Sandstone–Carboniferous Limestone transition. *Geological Journal*, **8**, 345–374.

GAYER, R. A., GARVEN, G. & RICKARD, D. 1998. Fluid migration and coal-rank development in foreland basins. *Geology*, **26**, 679–682.

GEIGER, M. E. & HOPPING, C. A. 1968. Triassic stratigraphy of the southern North Sea Basin. *Philosophical Transactions of the Royal Society*, **B 254**, 1–36.

GELUK, M. G. 2005*a*. *Stratigraphy and tectonics of Permian and Triassic basins in the Netherlands and surrounding areas*. PhD thesis, University of Utrecht.

GELUK, M. G. 2005*b*. Triassic. *In*: WONG, TH. E., BATJES, D. A. J. & DE JADER, J. (eds) *Geology of the Netherlands*. Royal Dutch Academy of Science and Arts, Amsterdam.

GELUK, M. G. & RÖHLING, H.-G. 1999. High resolution sequence stratigraphy of the Lower Triassic Buntsandstein: a new tool for basin analysis. *In*: BACHMANN, G. H. & LERCHE, I. (eds) *Epicontinental Triassic, Zentralblatt für Geologie und Paläontologie T1*, Jb 1998, H7–8, 727–745.

GEORGE, G. T. 1982. Sedimentology of the Upper Sandstone Group (Namurian G1) in south-west Dyfed: a case study. *In*: BASSETT, M. G. (ed.) *Geological Excursions in Dyfed, South-west Wales*. National Museum of Wales, Cardiff, 203–214.

GEORGE, G. T. 2000. Characterisation and high resolution sequence stratigraphy of storm dominated and braid delta shoreface sequences from the Basal grit Group (Namurian) of the South Wales Variscan peripheral foreland basin. *Marine and Petroleum Geology*, **17**, 445–475.

GEORGE, G. T. 2001. Late Yeadonian (Upper Sandstone Group) Incised valley supply and depositional systems in the South Wales peripheral foreland basin: implications for the evolution of the Culm Basin and for the Silesian hydrocarbon plays of onshore and offshore UK. *Marine and Petroleum Geology*, **18**, 671–705.

GEORGE, G. T. & KELLING, G. 1982. Stratigraphy and sedimentology of Upper Carboniferous sequences in the coalfield of south-west Dyfed. *In*: BASSETT, M. G. (ed.) *Geological Excursions in Dyfed, South-west Wales*. National Museum of Wales, Cardiff, 175–202.

GEORGE, T. N. 1956. The Namurian Usk Anticline. *Proceedings of the Geologists' Association*, **66**, 297–316.

GEORGE, T. N. 1958. Lower Carboniferous palaeogeography of the British Isles. *Proceedings of the Yorkshire Geological Society*, **31**, 227–318.

GEORGE, T. N. 1970. *South Wales*, 3rd edn. British Regional Geology, British Geological Survey. HMSO, London.

GEORGE, T. N. 1974. The Lower Carboniferous rocks in Wales. *In*: OWEN, T. R. (ed.) *The Upper Palaeozoic and Post-Palaeozoic Rocks of Wales*. University of Wales Press, Cardiff, 85–116.

GEORGE, T. N. 1978. Eustasy and tectonics: sedimentary rhythms and stratigraphical units in British Dinantian correlation. *Proceedings of the Yorkshire Geological Society*, **42**, 229–253.

GEORGE, T. N., HARLAND, W. B. *ET AL*. 1969. Recommendations on stratigraphical usage. *Proceedings of the Geological Society of London*, **156**, 139–166.

GEORGE, T. N., JOHNSON, G. A. L., MITCHELL, M., PRENTICE, J. E., RAMSBOTTOM, W. H. C., SEVASTOPULO, G. D. & WILSON, R. B. 1976. *A Correlation of Dinantian rocks in the British Isles*. Geological Society, London, Special Reports, **7**.

GEYSSANT, J. R. 1994. Colonisation par des ammonites méridionales des mers subboréales Kimméridgiennes du Yorkshire (Angleterre). *Géobios Mémoire Spécial*, **17**, 245–254.

GHOSH, P. K. 1927. Petrology of the Bodmin Moor Granite, Cornwall. *Mineralogical Magazine*, **21**, 285–309.

GHOSH, P. K. 1934. The Carnmenellis granite: its petrology, metamorphism and tectonics. *Quarterly Journal of the Geological Society of London*, **90**, 240–276.

GIANOLLA, P. & JACQUIN, T. 1998. Triassic sequence stratigraphic framework of western European basins. *In*: DE GRACIANSKY, P.-C., HARDENBOL, J., JACQUIN, T. &VAIL, P. R. (eds) *Mesozoic and Cenozoic Sequence Stratigraphy of European Basins*. SEPM, Special Publications, **60**, 643–650.

GIBBARD, P. L. 1977. The Pleistocene history of the Vale of St Albans. *Philosophical Transactions of the Royal Society of London*, **B280**, 445–483.

GIBBARD, P. L. 1985. *The Pleistocene History of the Middle Thames Valley*. Cambridge University Press, Cambridge.

GIBBARD, P. L. 1988. The history of the great northwest European rivers during the past three million years. *Philosophical Transactions of the Royal Society of London*, **B318**, 559–602.

GIBBARD, P. L. 1989. The geomorphology of a part of the Middle Thames forty years on: a reappraisal of the work of F. Kenneth Hare. *Proceedings of the Geologists' Association*, **100**, 481–503.

GIBBARD, P. L. 1994. *Pleistocene history of the Lower Thames Valley*. Cambridge University Press, Cambridge.

GIBBARD, P. L. 1995. The formation of the Strait of Dover. *In*: PREECE, R. C. (ed.) *Island Britain: a Quaternary Perspective*. Geological Society, London, Special Publications, **96**, 15–26.

GIBBARD, P. L. 2003. Definition of the Middle–Upper Pleistocene boundary. *Global and Planetary Change*, **36**, 201–208.

GIBBARD, P. L. & LEWIN, J. 2003. The history of the major rivers of southern Britain during the Tertiary. *Journal of the Geological Society, London*, **160**, 829–845.

GIBBARD, P. L. & TURNER, C. 1990. Cold stage type sections: some thoughts on a difficult problem. *Quaternaire*, **1**, 33–40.

Gibbard, P. L., Aalto, M. M. *et al.* 1996*a*. Early Middle Pleistocene fossiliferous sediments in the Kesgrave Formation at Broomfield, Essex, England. *In*: Turner, C. (ed.) *The Early Middle Pleistocene in Europe*. Balkema, Rotterdam, 83–119.

Gibbard, P. L., Allen, P., Field, M. F. & Hallam, D. 1996*b*. Early Middle Pleistocene deposits at Great Blakenham, Suffolk, England. *Quaternary Science Reviews*, **15**, 413–424.

Gibbard, P. L., West, R. G. Andrew, R. & Pettit, M. 1992. The margin of Middle Pleistocene ice advance at Tottenhill, Norfolk, England. *Geological Magazine*, **129**, 59–76.

Gibbard, P. L., West, R. G. *et al.* 1991. Early and Middle Pleistocene correlations in the southern North Sea Basin. *Quaternary Science Reviews*, **10**, 23–52.

Gibbard, P. L., Smith, A. G. *et al.* 2005. What status for the Quaternary? *Boreas*, **34**, 1–6.

Gibberd, B. & Brew, D. 2004. *Coastal Squeeze, Saltmarsh Loss and Special Protection Areas*. Report to English Nature, Peterborough.

Gibbons, W. 1983*a*. The Monian 'Penmynydd Zone of Metamorphism' in Llyn, North Wales. *Geological Journal*, **18**, 21–41.

Gibbons, W. 1983*b*. Stratigraphy, subduction and strike-slip faulting in the Mona Complex of North Wales – a review. *Proceedings of the Geologists' Association*, **94**, 147–163.

Gibbons, W. 1987. Menai Strait fault system: An early Caledonian terrane boundary in North Wales. *Geology*, **15**, 744–747.

Gibbons, W. 1989. Basement-cover relationships around Aberdaron, Wales, UK: the fault-reactivated northwestern margin of the Welsh Basin. *Geological Magazine*, **126**, 363–372.

Gibbons, W. 1990. Pre-Arenig terranes of northwest Wales. *In*: Strachan, R. A. & Taylor, G. K. (eds) *Avalonian and Cadomian Geology of the North Atlantic*. Blackie, Glasgow, 28–48.

Gibbons, W. & Ball, M. J. 1991. A discussion of the Monian Supergroup stratigraphy in NW Wales. *Journal of the Geological Society, London*, **148**, 5–8.

Gibbons, W. & Gyopari, M. C. 1986. A greenschist protolith for blueschists in Anglesey, UK. *In*: Evans, B. W. & Brown, E. H. (eds) *Blueschists and Eclogites*. Geological Society of America, Memoir, **164**, 217–228.

Gibbons, W. & Harris, A. 1994. *A revised correlation of Precambrian rocks in the British Isles*. Geological Society, London, Special Report, **22**.

Gibbons, W. & Horák, J. M. 1990. Contrasting metamorphic terranes in north-west Wales. *In*: D'Lemos, R. S., Strachan, R. A. & Topley, C. G. (eds) *The Cadomian Orogeny*. Geological Society, London, Special Publications, **51**, 315–327.

Gibbons, W. & Horák, J. M. 1996. The evolution of the Neoproterozoic subduction system: Evidence from the British Isles. *In*: Nance, R. D. & Thompson, M. D. (eds) *Avalonian and Related Peri-Gondwanan Terranes of the Circum-North Atlantic*. Geological Society of America, Special Papers, **304**, 269–280.

Gibbons, W. & Mann, A. 1983. Pre-Mesozoic lawsonite in Anglesey, northern Wales: preservation of ancient blueschists. *Geology*, **11**, 3–6.

Gibbons, W. & McCarroll, D. 1993. *Geology of the Country Around Aberdaron, including Bardsey Island*. Memoir of the British Geological Survey, Sheet 133, England and Wales, HMSO, London.

Gibbons, W. & Thompson, L. 1991. Ophiolitic mylonites in the Lizard complex: Ductile extension in the lower oceanic crust. *Geology*, **19**, 1000–1012.

Gibbons, W., Tietzsch-Tyler, D., Horák, J. M. & Murphy, F. C. 1994. The Precambrian geology of Anglesey, Southwest Llyn & SE Ireland. *In*: Gibbons, W. & Harris A. L. (eds) *A Correlation of the Precambrian Rocks in the British Isles*, 2nd edn. Geological Society, Special Reports, **22**, 75–84.

Gilbertson, D. D. & Hawkins, A. B. 1978. The Pleistocene succession at Kenn, Somerset. *Bulletin of the Geological Survey of Great Britain*, **66**, 1–41.

Girling, M. A. 1974. Evidence from Lincolnshire of the age and intensity of the mid-Devensian temperate episode. *Nature*, **250**, 270.

Girling, M. A. 1988. The bark beetle Scolytus scolytus (Fabricius) and the possible role of elm disease in the early Neolithic. *In*: Jones, M. (ed.) *Archaeology and the Flora of the British Isles*. Oxford Committee for Archaeology Monograph 14, Oxford, 34–38.

Gleeson, S. A., Wilkinson, J. J., Shaw, H. F. & Herrington, R. J. 2000. Post-magmatic hydrothermal circulation and the origin of base metal mineralization, Cornwall, UK. *Journal of the Geological Society, London*, **157**, 589–601.

Glennie, K. W. 1972. Permian Rotliegendes of North-western Europe interpreted in the light of modern desert sedimentation studies. *AAPG Bulletin*, **56**, 1046–1071.

Glennie, K. W. & Boegner, P. L. E. 1981. Sole Pit Inversion tectonics. *In*: Illing, L. V. & Hobson, G. D. (eds) *Petroleum geology of the continental shelf of North-West Europe*, Heyden & Son, 110–120.

Glennie, K. W. & Evans, G. 1976. A reconnaissance of the Recent sediments of the Ranns of Kutch, India. *Sedimentology*, **23**, 625–647.

Glennie, K. W. & Hurst, A. 2006. Fluidisation and associated soft-sediment deformation in eolian sandstones: Hopeman sandstone (Permian), Scotland, and Rotliegende, North Sea. *In*: Hurst, A. & Cartwright, J. (eds) *Sand Injectites: Implications for Hydrocarbon Exploration and Production*, American Association of Petroleum Geologists, Memoir **87**, 1–8.

Glennie, K. W., Higham, J. & Stemmerik, L. 2003. Permian (Chapter 8). *In*: *The Millenium Atlas: Petroleum Geology of the Central and Northern North Sea*. Geological Society, London.

Glennie, K. W., Mudd, G. C. & Nagtegaal, P. J. C. 1978. Depositional environment and diagenesis of Permian Rotliegendes sandstones in Leman Bank and Sole Pit areas of the UK, southern North Sea. *Journal of the Geological Society, London*, **135**, 25–34.

Glover, B. W., Powell, J. H. & Waters, C. N. 1993. Etruria Formation (Westphalian C) palaeoenvironments and volcanicity on the southern margins of the Pennine Basin, South Staffordshire, England. *Journal of the Geological Society, London*, **150**, 737–750.

Glover, B. W., Leng, M. J. & Chisholm, J. I. 1996. A second major fluvial sourceland for the Silesian Pennine Basin of northern England. *Journal of the Geological Society, London*, **153**, 901–906.

Goggin, V. & Jacquin, T. 1998. A sequence stratigraphic framework of the marine and continental Triassic series in the Paris Basin, France. *In*: de Graciansky, P.-C., Hardenbol, J., Jacquin, T. & Vail, P. R. (eds) *Mesozoic and Cenozoic Sequence Stratigraphy of European Basins*. SEPM Special Publications, **60**, 667–690.

Goldring, R. 1957. The last toothed Productelliniae in Europe (Brachiopoda, Upper Devonian). *Paläontologische Zeitschrift*, **36**, 232–251.

Goldring, R. 1962. The bathyl lull: Upper Devonian and Lower Carboniferous sedimentation in the Variscan geosyncline. *In*: Coe, K. (ed.) *Some Aspects of the Variscan Fold Belt*. Manchester University Press, Manchester, 75–91.

Goldring. R. 1970. The stratigraphy about the Devonian-Carboniferous boundary in the Barnstaple area of north Devon, England. *Comptes Rendus 6'eme Congres Internationale de Stratigraphie et de Géologique du Carbonifère, Sheffield, for 1967*, **2**, 807–816.

Goldring, R. 1971. *Shallow-water Sedimentation; As Illustrated in the Upper Devonian Baggy Beds*. Geological Society, London, Memoirs, **5**.

Goldring, R. 1999. Sedimentological aspects and preservation of Lower Cretaceous (Aptian) bentonites (fuller's earth) in southern England. *Neues Jahrbuch für Geologie und Paläontologie*, **214**, 3–24.

Gonzalez-Bonorino, G. & Eyles, N. 1995. Inverse relation between ice extent and the late Paleozoic glacial record of Gondwana. *Geology*, **23**, 1015–1018.

Goodall, T. M., North, C. P. & Glennie, K. W. 2000. Surface and sub-surface sedimentary structures produced by salt crusts. *Sedimentology*, **47**, 99–118.

Gooday, A. J. 1973. *Taxonomic and stratigraphic studies on Upper Devonian and Lower Carboniferous Entomozoidae and Rhomboentomozoidae (Ostracoda,?Myodocopida) from south-western England*. Unpublished PhD thesis, University of Exeter.

Goode, A. J. J. 1973. *The mode of intrusion of Cornish elvans*. Report of the Institute of Geological Sciences, **73/7**.

Goode, A. J. J. & Leveridge, B. E. 1991. *Geological notes and details for 1:10 000 sheets SW 86 NW and NE, SW 87 NE, SW and SE, SW 96 NW and NE and SW 97 NW and SW*. Technical Report of the British Geological Survey, Onshore Series, **WA/92/11**.

Goode, A. J. J. & Merriman, R. J. 1987. Evidence of crystalline basement west of the Land's End granite. *Proceedings of the Geologists' Association*, **98**, 39–43.

Goode, A. J. J. & Taylor, R. T. 1988. *Geology of the country around Penzance*. Memoir of the Geological Survey, sheets 351 and 358 (England and Wales), HMSO, London.

Goode, A. J. J., Merriman, R. J. & Darbyshire, D. P. F. 1987. Evidence of crystalline basement west of the Land's End granite, Cornwall: a reply. *Proceedings of the Geologists' Association*, **98**, 272.

Gowland, S. & Riding, J. B. 1991. Stratigraphy, sedimentology and palaeontology of the Scarborough Formation (Middle Jurassic) at Hundale Point, North Yorkshire. *Proceedings of the Yorkshire Geological Society*, **48**, 375–392.

Gradstein, F. M. & Ogg, J. 1996. A Phanerozoic time-scale. *Episodes*, **19**, 3–5.

Gradstein, F. M., Ogg, J. G., Smith, A. G. *et al.* 2004. *A Geologic Time Scale 2004*. Cambridge University Press, Cambridge.

Gramann, F. & Kockel, F. 1988. Palaeogeographical,lithological, palaeoecological and palaeoclimatic development of the Northwest European Tertiary Basin. *In*: Vinken, R. (compiler) *The Northwest European Tertiary Basin*. Results of the IGCP Project No. 124. *Geologisches Jahrbuch*, **A100**, 428–441.

Gramann, F. & von Daniels, C. H. 1988. Foraminifera. *In*: Vinken, R. (compiler) *The Northwest European Tertiary Basin*. Results of the IGCP Project No. 124. *Geologisches Jahrbuch*, **A100**, 145–225.

Grant, S. F., Coe, A. L. & Armstrong, H. A. 1999. Sequence stratigraphy of the Coniacian succession of the Anglo-Paris Basin. *Geological Magazine*, **136**, 17–38.

Gray, D. A. 1958. Electrical resistivity marker bands in the Lower and Middle Chalk of the London Basin. *Bulletin of the Geological Survey of Great Britain*, **15**, 85–95.

Gray, D. A. 1965. The stratigraphical significance of electrical resistivity marker bands in the Cretaceous strata of the Leatherhead (Fetcham Mill) Bprehole, Surrey. *Bulletin of the Geological Survey of Great Britain*, **23**, 65–115.

Gray, J. M. 1982. The last glaciers (Loch Lomond Readvance) in Snowdonia, North Wales. *Geological Journal*, **17**, 111–133.

Green, C. P. 1997. Stonehenge: geology and prehistory. *Proceedings of the Geologists' Association*, **108**, 1–10.

Green, C. P., Hey, R. W. & McGregor, D. F. M. 1980. Volcanic pebbles in Pleistocene gravels of the Thames in Buckinghamshire and Hertfordshire. *Geological Magazine*, **117**, 59–64.

Green, D. H. 1964. The petrogenesis of the high-temperature peridotite intrusion in the Lizard area, Cornwall. *Journal of Petrology*, **5**, 134–188.

Green, G. W. 1992. *Bristol and Gloucester Region*, 3rd edn. British Regional Geology, British Geological Survey. HMSO, London.

Green, G. W. & Donovan, D. T. 1969. The Great Oolite of the Bath area. *Bulletin of the Geological Survey of Great Britain*, **30**, 1–63.

Green, J. F. N. 1908. The geological structure of the St. David's area, Pembrokeshire. *Quarterly Journal of the Geological Society of London*, **64**, 363–383.

Green, P. F., Duddy, I. R. & Bray, R. J. 1998. Variation in thermal history styles around the Irish Sea and adjacent areas; implications for hydrocarbon occurrence and tectonic evolution. *In*: Meadows, N. S., Trueblood, S., Hardman, M. & Cowan, (eds) *The Petroleum Geology of the Irish Sea and Adjacent Areas*. Geological Society, London, Special Publications, **124**, 73–93. [Dated 1997.]

Green, P. F., Thomson, K. & Hudson, J. D. 2001. Recognition of tectonic events in undeformed regions: contrasting results from the Midland Platform and East Midlands Shelf, Central England. *Journal of the Geological Society, London*, **158**, 59–73.

Green, S. & Walker, E. 1991. *Ice Age Hunters, Neanderthals and Early Modern Hunters in Wales*. National Museum of Wales, Cardiff.

Greenly, E. 1919. *The Geology of Anglesey*. Memoir of the Geological Survey of Great Britain, **78**, 1–980, HMSO, London.

Greig, D. C. 1988. *Geology of the Eyemouth District*. Memoir of the British Geological Survey, Sheet 34 (Scotland), HMSO, London.

Greig, D. C., Wright, J. E., Hains, B. A. & Mitchell, G. H. 1968. *Geology of the Country Around Church Stretton, Craven Arms, Wenlock Edge and Brown Clee*. Memoir of the Geological Survey of Great Britain. HMSO, London.

Grey, D. R. C., Kinniburgh, D. G., Barker, J. A. & Bloomfield, J. P. 1995. *Groundwater in the UK. A strategic study*. Issues and Research Needs. Groundwater Forum Report FR/GF 1.

Griffiths, J. C. 1987. *Sedimentology of the Lower and Middle Cambrian of North Wales*. PhD thesis, University of Liverpool, UK.

Griffiths, P. A., Whittington R. J., Allen M. R., Craig J. & Fitches W. R. 1995. Distinction between fault and salt control of Mesozoic sedimentation on the southern margin of the Mid-North Sea High. *In*: Boldy S. A. R. (ed.) *Permian and Triassic Rifting in Northwest Europe*. Geological Society, London, Special Publications, **91**, 145–159.

Gröcke, D. R., Hesselbo, S. P. & Jenkyns, H. C. 1999. Carbon-isotope composition of Lower Cretaceous fossil wood: Ocean-atmosphere chemistry and relation to sea-level change. *Geology*, **27**, 155–158.

Gröcke, D. R., Price, G. D., Ruffell, A. H., Mutterlose, J. & Baraboshkin, E. 2003. Isotopic Evidence for Late Jurassic–Early Cretaceous climate change. *Palaeogeography, Palaeoclimatology, Palaeoecology*, **202**, 97–118.

Grötzinger, J. P. Bowring, S. A,, Saylor, B. Z. & Kaufman, A. J. 1995. Biostratigraphic and geochronic constraits on early animal evolution. *Science*, **270**, 598–604.

Grove, J. M. 1988. *The Little Ice Age*. Methuen, London.

Grove, J. M. 2004. *Little Ice Ages – Ancient and Modern*, 2nd edition, Volume 1. Routledge, London.

Grün, R. & Schwarcz, H. P. 2000. Revised open-system U-series/ESR age calculations for teeth from Stratum C at the Hoxnian Interglacial type locality, England. *Quaternary Science Reviews*, **19**, 1151–1154.

Guion, P. D. & Fielding, C. R. 1988. Westphalian A and B sedimentation in the Pennine Basin, UK. *In*: Besly, B. M. & Kelling, G. (eds) *Sedimentation in a Synorogenic Basin Complex: The Upper Carboniferous of North West Europe*. Blackie, Glasgow, 153–177.

Guion, P. D., Fulton, I. M. & Jones, N. S. 1995. Sedimentary facies of the coal-bearing Westphalian A and B north of the Wales-Brabant High. *In*: Whateley, M. K. G. & Spears, D. A. (eds) *European Coal Geology*. Geological Society, London, Special Publications, **82**, 45–78.

Guion, P. D., Gutteridge, P. & Davies, S. J. 2000. Carboniferous sedimentation and volcanism along the Laurussian margin. *In*: Woodcock, N. H. & Strachan, R. A. (eds) *Geological History of Britain and Ireland*. Blackwell, Oxford, 227–270.

Hagadorn, J. W. & Bottjer, D. J. 1999. Restriction of a late Neoproterozoic biotope: Suspect-Microbial structures and trace fossils at the Vendian–Cambrian transition. *Palaios*, **14**, 73–86.

Hailwood, E. A., Gash, P. J. R., Andresen, P. C. & Badham, J. P. N. 1984. Palaeomagnetism of the Lizard Complex, SW England. *Journal of the Geological Society, London*, **141**, 27–35.

Halcrow 2002. Prediction of Future Coastal Evolution for SMP Review: Final Project Report FD2002. 'Futurecoast' CD format; Defra.

Hall, J., Brewer, J. A., Matthews, D. H. & Warner, M. R. 1984. Crustal structure across the Caledonides from the 'WINCH' seismic reflection profile: influences on the evolution of the Midland Valley of Scotland. *Transactions of the Royal Society of Edinburgh*, **75**, 97–109.

Hall, V. A. & Pilcher, J. R. 2002. Late-Quaternary icelandic tephras in Ireland and Great Britain: detection, characterisation and usefulness. *The Holocene*, **12**, 223–230.

Hallam, A. 1958. The concept of Jurassic axes of uplift. *Science Progress*, **46**, 441–449.

HALLAM, A. 1959. On the supposed evolution of Gryphaea in the Lias. *Geological Magazine*, **96**, 99–108.

HALLAM, A. 1964. Origin of the limestone-shale rhythms in the Blue Lias of England a composite theory. *Journal of Geology*, **72**, 157–169.

HALLAM, A. 1967. An environmental study of the Upper Domerian and Lower Toarcian in Great Britain. *Philosophical Transactions of the Royal Society of London*, **B252**, 393–445.

HALLAM, A. 1975. *Jurassic environments*. Cambridge University Press, Cambridge. ix + 269 pp.

HALLAM, A. 1977. Jurassic bivalve biogeography. *Paleobiology*, **3**, 58–73.

HALLAM, A. 1978. Eustatic cycles in the Jurassic. *Palaeogeography, Palaeclimatology, Palaeoecology*, **23**, 1–32.

HALLAM, A. 1984. Distribution of fossil marine invertebrates in relation to climate. *In*: BRENCHLEY, P. J. (ed.) *Fossils and Climate*. Wiley, Chichester, 107–125.

HALLAM, A. 1985. A review of Mesozoic climates. *Journal of the Geological Society, London*, **142**, 433–445.

HALLAM, A. 1986. Pliensbachian and Tithonian extinction events. *Nature*, **319**, 765–768.

HALLAM, A. 1988. A re-evaluation of Jurassic eustasy in the light of new data and the revised Exxon curve. *In*: WILGUS, C. K., HASTINGS, B. S., KENDALL, G. ST C., POSAMENTIER, H. W., ROSS, C. A. & VAN WAGONER, J. C. (eds) *Sea-level Changes: An Integrated Approach*. SEPM, Special Publications, **42**, 261–273.

HALLAM, A. 1992*a*. *Phanerozoic Sea-level Changes*. Columbia University Press, New York.

HALLAM, A. 1992*b*. Jurassic. *In*: DUFF, P. MCL. D. & SMITH, A. J. (eds) *Geology of England and Wales*. Geological Society, London, 325–354.

HALLAM, A. 1996. Recovery of the marine fauna in Europe after the end-Triassic and early Toarcian extinction events. *In*: HART, M. B. (ed.) *Biotic Recovery From Mass Extinctions*. Geological Society, London, Special Publications, **102**, 231–236.

HALLAM, A. 1997. Estimates of the amount of time and rate of sea-level change across the Rhaetian–Hettangian and Pliensbachian–Toarcian boundaries (latest Triassic to early Jurassic). *Journal of the Geological Society, London*, **154**, 773–780.

HALLAM, A. 1999. Evidence of sea-level fall in sequence stratigraphy: examples from the Jurassic. *Geology*, **27**, 343–346.

HALLAM, A. 2002. How catastrophic was the end-Triassic mass extinction. *Lethaia*, **35**, 147–157.

HALLAM, A. & WIGNALL, P. B. 1999. Mass extinctions and sea level changes. *Earth Science Reviews*, **48**, 217–250.

HALLSWORTH, C. R. & CHISHOLM, J. I. 2000. Stratigraphic evolution of provenance characteristics in Westphalian sandstones of the Yorkshire Coalfield. *Proceedings of the Yorkshire Geological Society*, **53**, 43–72.

HALLSWORTH, C. R., MORTON, A. C., CLAOUÉ-LONG, J. C. & FANNING, C. M. 2000. Carboniferous sand provenance in the Pennine Basin, UK: constraints from heavy mineral and detrital zircon age data. *Sedimentary Geology*, **137**, 147–185.

HAMBLIN, R. J. O. & WOOD, C. J. 1976. The Cretaceous (Albian–Cenomanian) stratigraphy of the Haldon Hills, South Devon, England. *Newsletters on Stratigraphy*, **4** (3), 135–149.

HAMBLIN, R. J. O., CROSBY, A., ALSON, R. S., JONES, S. M., CHADWICK, R. A., PENN, I. E. & ARTHUR, M. J. 1992. *The Geology of the English Channel*. United Kingdom Offshore Report, Geological Survey of Great Britain. HMSO, London.

HAMBLIN, R. J. O., MOORLOCK, B. S. P., BOOTH, S. J., JEFFERY, D. H. & MORIGI, A. N. 1997. The Red Crag and Norwich Crag formations in eastern Suffolk. *Proceedings of the Geologists' Association*, **108**, 11–23.

HAMBLIN, R. J. O., MOORLOCK, B. S. P. & ROSE, J. 2000. A new glacial stratigraphy for eastern England. *Quaternary Newsletter*, **92**, 35–44.

HAMBREY, M. J., DAVIES, J. R., GLASSER, N. F., WATERS, R. A., DOWDESWELL, J. A., WILBY, P. R., WILSON, D. & ETIENNE, J. L. 2001. Late Devensian glacigenic sedimentation and landscape evolution in the Cardigan area of southwest Wales. *Journal of Quaternary Science*, **16**, 455–482.

HAMPSON, G. J. 1995. Discrimination of regionally-extensive coals in the Upper Carboniferous of the Pennine Basin, UK. using high-resolution sequence stratigraphic concepts. *In*: WHATELEY, M. K. G. & SPEARS, D. A. (eds) *European Coal Geology*. Geological Society, London, Special Publications, **82**, 79–97.

HAMPSON, G. J. 1997. A sequence stratigraphic model for deposition of the Lower Kinderscout Delta, an Upper Carboniferous turbidite-fronted delta. *Proceedings of the Yorkshire Geological Society*, **51**, 273–296.

HAMPSON, G. J. 1998. Evidence for relative sea-level falls during deposition of the Upper Carboniferous Millstone Grit, South Wales. *Geological Journal*, **33**, 243–266.

HAMPSON, G. J., DAVIES, S. J., ELLIOTT, T., FLINT, S. S. & STOLLHOFEN, H. G. 1999. Incised valley fill sandstone bodies in Upper Carboniferous fluvio-deltaic strata: recognition and reservoir characterization of Southern North Sea analogues. *In*: FLEET, A. J. & BOLDY, S. A. R. (eds) *Petroleum Geology of Northwest Europe: Proceedings of the 5th Conference*. Geological Society, London, 771–788.

HAMPSON, G. J., ELLIOTT, T. & DAVIES, S. J. 1997. The application of sequence stratigraphy to upper Carboniferous fluvio-deltaic strata of the onshore UK and Ireland: Implications for the southern North Sea. *Journal of the Geological Society, London*, **154**, 719–733.

HAMPSON, G. J., ELLIOTT, T. & FLINT, S. S. 1996. Critical application of high resolution sequence stratigraphic concepts to the Rough Rock Group (Upper Carboniferous) of northern England. *In*: HOWELL, J. A. & AITKEN, J. F. (eds) *High Resolution Sequence Stratigraphy: Innovations and Applications*. Geological Society, London, Special Publications, 221–246.

HANCOCK, J. M. 1989. Sea-level changes in the British region during the Late Cretaceous. *Proceedings of the Geologists' Association*, **100**, 565–594.

HANCOCK, J. M. 1992. Late Cretaceous. *In*: COPE, J. C. W., INGHAM, J. K. & RAWSON, P. F. (eds) *Atlas of Palaeogeography and Lithofacies*. Geological Society, London, Memoir, **13**, 134–139.

HANCOCK, J. M. & KAUFFMAN, E. 1979. The great transgressions of the Late Cretaceous. *Journal of the Geological Society, London*, **136**, 175–186.

HANCOCK, N. J. 1982. Stratigraphy, palaeogeography and structure of the East Mendips Silurian inlier. *Proceedings of the Geologists' Association*, **93**, 247–261.

HANTZPERGUE, P., BAUDIN, F., MITTA, A., OLFERIEV, A. & ZAKHAROV, V. 1998. The Upper Jurassic of the Volga Basin: ammonite biostratigraphy and occurrence of organic-carbon rich facies. Correlations between boreal-subboreal and submediterranean provinces. *In*: CRASQUIN-SOLEAU, S. & BARRIER, É. (eds) *Péri-Tethys Memoir 4: Epicratonic Basins of Peri-tethyan Platforms*. Mémoires du Muséum National d'Histoire Naturelle, **179**, 9–33.

HAQ, B. U., HARDENBOL, J. & VAIL, P. R. 1987. Chronology of fluctuating sea levels since the Triassic. *Science*, **235**, 1156–1167.

HAQ, B. U., HARDENBOL, J. & VAIL, P. 1988. *Mesozoic and Cenozoic Chronostratigraphy and Eustatic Cycles*. SEPM, Special Publications, **42**, 71–108.

HAQ, B. U., HARDENBOL, J. & VAIL, P. R. 1989. Mesozoic and Cenozoic chronostratigraphy and cycles of relative sea-level change. *In*: WILGUS, C. K., HASTINGS, B. S., KENDALL, G. ST C., POSAMENTIER, H. W., ROSS, C. A. & VAN WAGONER, J. C. (eds) *Sea-level Changes: An Integrated Approach*. SEPM, Special Publications, **42**, 71–108.

HARDMAN, M., BUCHANAN, J., HERRINGTON, P. & CARR, A. 1993. Geochemical modelling of the East Irish Sea Basin: its influence on predicting hydrocarbon type and quality. *In*: PARKER, J. R. (ed.) *Petroleum Geology of North-west Europe: Proceedings of the 4th Conference*. Geological Society, London, 809–821.

HARLAND, W. B. & GAYER, R. A. 1972. The Arctic Caledonides and earlier oceans. *Geological Magazine*, **116**, 145–153.

Harland, R., Bonney, A. P., Hughes, M. J. & Morigi, A. N. 1991. The Lower Pleistocene stratigraphy of the Ormesby borehole, Norfolk, England. *Geological Magazine*, **128**, 647–660.

Harland, W. B., Armstrong, R. L., Cox, A. V., Craig, L. E., Smith, A. G. & Smith, D. G. 1989. *A geologic time scale 1989*. Cambridge University Press, Cambridge.

Harmer, F. W. 1902. A sketch of the later Tertiary history of East Anglia. *Proceedings of the Geologists' Association*, **17**, 416–479.

Harmer, F. W. 1909. The Pleistocene Period in the eastern counties of England. *In*: Monckton, H. W. & Herries, R. S. (eds) *Geology in the Field: The Jubilee Volume of the Geologists' Association (1858–1908)*. E. Stanford, London, 103–123.

Harper, J. C. 1956. The Ordovician succession near Llanystwmdwy, Caernarvonshire. *Liverpool and Manchester Geological Journal*, **1**, 385–393.

Harris, A. L. 1985. *The Nature and Timing of Orogenic Activity in The Caledonian Rocks of The British Isles*. Geological Society, London, Memoir, **9**, 1–53.

Harris, P. & Dagger, G. W. 1987. The intrusion of the Carrock Fell Gabbro Series (Cumbria) as a sub-horizontal tabular body. *Proceedings of the Yorkshire Geological Society*, **46**, 371–380.

Harris, T. M. 1961. *The Yorkshire Jurassic flora*. I. Thallophyta – Pteridophyta, ix + 212pp. II. Caytoniales, Cycadales and Pteridosperms, viii + 191 pp., 7 pls. III. Bennettitales, vi + 186 pp., 7 pls. IV. (with contributions from Miller, J. & Millington, W.) Ginkgoales and Czekanowskiales, viii + 150 pp., 8 pls. V. Coniferales, vi + 166 pp., 7 pls. British Museum (Natural History), London.

Harrison, R. K., Snelling, N. J., Merriman, R. J., Morgan, G. E. & Goode, A. J. J. 1997. The Wolf Rock, Cornwall: new chemical, isotopic age and palaeomagnetic data. *Geological Magazine*, **114**, 249–264.

Hart, M. B., Bailey, H. W., Crittenden, S., Fletcher, B. N., Price, R. J. & Swiecicki, A. 1989. Cretaceous. *In*: Jenkins, G. D. & Murray, J. W. (eds) *Stratigraphical Atlas of Fossil Foraminifera*, 2nd edn. Ellis Horwood, Chichester, 273–371.

Hartley, A. J. 1993*a*. A depositional model for the Mid-Westphalian A to Late Westphalian B Coal Measures of South Wales. *Journal of the Geological Society, London*, **150**, 1121–1136.

Hartley, A. J. 1993*b*. Silesian sedimentation in South-West Britain: sedimentary responses to the developing Variscan Orogeny. *In*: Gayer, R. A. & Greiling, R. (eds) *The Rhenohercynian and Subvariscan Fold Belts*. Earth Evolution Series. Springer, Berlin, 159–196.

Hartley, A. J. & Gillespie, P. 1990. Controls on alluvial architecture by synsedimentary faults in the Coal Measures of South Wales. *Geological Journal*, **25**, 189–197.

Hartley, A. J. & Otava, J.: 2001. Sediment provenance and dispersal in a deep marine foreland basin: the Lower Carboniferous Culm Basin, Czech Republic. *Journal of the Geological Society, London*, **158**, 137–150.

Hartley, A. J. & Warr, L. N. 1990. Upper Carboniferous foreland basin evolution in SW Britain. *Proceedings of the Ussher Society*, **7**, 212–216.

Hartz, E. H. & Torvik, T. H. 2002. Baltica upside down: A new plate tectonic model for Rodinia and the Iapetus Ocean. *Geology*, **30**, 255–258.

Harvey, M., Stewart, S., Wilkinson, J. & Ruffell, A. H. 1994. Variscan structural control on the Plymouth Bay collapse basin, English Channel. *Proceedings of the Ussher Society*, **8**, 271–278.

Harwood, G. M. 1976. The Staddon Grits – or Meadfoot Beds? *Proceedings of the Ussher Society*, **3**, 333–338.

Haslett, S. K., Davies, P., Curr, R. H. F., Davies, C. F. C., Kennington, K., King, C. P. & Margets, A. J. 1998. Evaluating late-Holocene relative-sea-level change in the Somerset Levels, southwest Britain. *The Holocene*, **8**, 197–208.

Haubold, H. & Katzung, G. 1975. Die position der Autun/Saxon-Grenze (Unteres Perm0 in Europa und Nordamerika. *Schriftenr geol Wiss. Berlin*, **3**, S. 87–138.

Hawkes, L. 1943. The erratics of the Cambridge Greensand – their nature, provenance and mode of transport. *Quarterly Journal of the Geological Society, London*, **99**, 93–104.

Hawley, D. & Owen, G. 2006. Old Red Sandstone of the Black Mountains. *In*: Bassett, M. G. & Owens, R. M. (eds) *Geological Excursions of South-east Wales*. National Museum of Wales, Cardiff, in press.

Haywood, A. M., Valdes, P. J. & Markwick, P. J. 2004. Cretaceous (Wealden) climates: a modelling perspective. *Cretaceous Research*, **25**, 303–311.

Head, M. J. 1997. Thermophilic dinoflagellate assemblages from the mid-Pliocene of eastern England. *Journal of Paleontology*, **71**, 165–193.

Head, M. J. 1999. The Late Pliocene St. Erth Beds of Cornwall: a review of the palynology and reappraisal of the dinoflagellates. *In*: Scourse, J. D. & Furze, M. F. A. (eds) *The Quaternary of West Cornwall*. Quaternary Research Association, London, 88–92.

Healy, F., Heaton, M. & Lobb, S. J. 1992. Excavation of a Mesolithic site at Thatcham, Berkshire. *Proceedings of the Prehistoric Society*, **58**, 41–76.

Heard, A. 1922. The petrology of the Pennant Measures. *Geological Magazine*, **59**, 83–92.

Hebért, E. 1874. Comparison de la craie des côte d'Angleterre avec celle de la France. *Bulletin de la Société géologique de France*, series 3, **2**, 416–428.

Heckel, P. H. 1977. Origin of phosphatic black shale facies in Pennsylvanian cyclothems of -Continent North America. *Bulletin of the American Association of Petroleum Geologists*, **61**, 1045–68.

Heckel, P. H. 2004. Chairman's Column. *Newsletter on Carboniferous Stratigraphy*, **22**, 1–3.

Helm, D. G. 1970. Stratigraphy and structure of the Black Combe inlier (English Lake District) on the northern margin of Eastern Avalonia. *Proceedings of the Yorkshire Geological Society*, **38**, 105–148.

Helm, D. G., Roberts, B. & Simpson, A. 1963. Polyphase folding in the Caledonides south of the Scottish Highlands. *Nature*, **200**, 1060–1062.

Hemingway, J. E. 1974. Jurassic. *In*: Rayner, D. H. & Hemingway, J. E. (eds) *The Geology and Mineral Resources of Yorkshire*. Yorkshire Geological Society, Leeds, 161–223.

Hemingway, J. E. & Knox, R. W. O'B. 1973. Lithostratigraphical nomenclature of the Middle Jurassic of the Yorkshire Basin of north-east England. *Proceedings of the Yorkshire Geological Society*, **39**, 527–535.

Henderson, C. M. & Mei, S. 2000. Preliminary cool water Permian conodont zonation in North Pangea: a review. *Permpohiles*, **36**, 16–23.

Hendriks, E. M. L. 1937. Rock succession and structure in south Cornwall, a revision. With notes on the central European facies and Variscan there present. *Quarterly Journal of the Geological Society of London*, **93**, 322–360.

Hendriks, E. M. L. 1939. The Start–Dodman–Lizard Boundary Zone in Relation to the Alpine Structure of Cornwall. *Geological Magazine*, **76**, 385–402.

Hendriks, E. M. L. 1949. The Gramscatho Series. *Transactions of the Royal Geological Society of Cornwall*, **18**, 50–61.

Hendriks, E. M. L. 1971. Facies variation in relation to tectonic evolution in Cornwall. *Transactions of the Royal Geological Society of Cornwall*, **20**, 115–152.

Hendriks, E. M. L., House, M. R. & Rhodes, F. H. T. 1971. Evidence bearing on the stratigraphical successions in South Cornwall. *Proceedings of the Ussher Society*, **2**, 270–275.

Henley, S. 1974. Geochemistry and petrogenesis of elvan dykes in the Perranporth area, Cornwall. *Proceedings of the Ussher Society*, **3**, 136–145.

Henson, M. R. 1972. The form of the Permo–Triassic basin in south-east Devon. *Proceedings of the Ussher Society*, **3**, 447–457.

Herbin, J.-P. & Geyssant, J. R. 1993. "Ceintures organiques" au Kimméridgien/Tithonien en Angleterre (Yorkshire, Dorset) et en France (Boulonnais). *Comptes rendus de l'Academie des Sciences, Paris*, (2) **317**, 1309–1316.

HERBIN, J.-P., FERNANDEZ-MARTINEZ, J. L., GEYSSANT, J. R., ALBANI, A. EL., DECONINCK, J. F., PROUST, J. N., COLBEAUX, J. P. & VIDIER, J. P. 1995. Sequence stratigraphy of source rocks applied to the study of the Kimmeridgian/Tithonian in the north-west European shelf (Dorset/UK, Yorkshire/UK and Boulonnais/France). *Marine and Petroleum Geology*, **12**, 177–194.

HERBIN, J.-P., MÜLLER, C., GEYSSANT, J. R., MÉLIÈRES, F., PENN, I. E. & YORKIM GROUP. 1993. Variation of the distribution of organic matter within a transgressive system tract: Kimmeridge Clay (Jurassic) England. *In*: KATZ, B. J. & PRATT, L. M. (eds) *Source Rocks in a Sequence Stratigraphic Framework*. AAPG, Studies in Geology, **37**, 67–100.

HERBIN, J.-P., MÜLLER, C., PENN, I. E., GEYSSANT, J. R., MÉLIÈRES, F. & YORKIM GROUP. 1991. Hétérogénéité quantitative et qualitative de la matière organique dans les argiles du Kimméridgien du Val de Pickering (Yorkshire, UK). *Revue de l'Institut Français du Pétrole*, **46**, 675–712.

HESSELBO, S. P. & JENKYNS, H. C. 1995. A comparison of the Hettangian to Bajocian successions of Dorset and Yorkshire. *In*: TAYLOR, P. D. (ed.) *Field Geology of the British Jurassic*. Geological Society, London, 105–150.

HESSELBO, S. P. & JENKYNS, H. C. 1998. British Lower Jurassic sequence stratigraphy. *In*: DE GRACIANSKY, P.-C., HARDENBOL, J., JACQUIN, T. & VAIL, P. (eds) *Mesozoic–Cenozoic Sequence Stratigraphy of European Basins*. SEPM, Special Publications, **60**, 561–581.

HESSELBO, S. P., COE., A. L., BATTEN, D. J. & WACH, G. D. 1990. Stratigraphic relations of the Lower Greensand (Lower Cretaceous) of the Calne area, Wiltshire. *Proceedings of the Geologists' Association*, **101**, 265–278.

HESSELBO, S. P., GRÖCKE, D. R., JENKYNS, H. C., BJERRUM, C. J., FARRIMOND, P., MORGANS-BELL, H. S. & GREEN, O. R. 2000. Massive dissociation of gas hydrate during a Jurassic anoxic event. *Nature*, **406**, 392–395.

HESSELBO, S. P., MORGANS-BELL, H. S., MCELWAIN, J. J., REES, P. MCA., ROBINSON, S. A. & ROSS, C. E. 2003. Carbon-cycle perturbation in the Middle Jurassic and accompanying changes in the terrestrial environment. *Journal of Geology*, **111**, 259–276.

HESSELBO, S. P., ROBINSON, S. A. & SURLYK, F. 2004. Sea-level change and facies development across potential Triassic–Jurassic boundary horizons, S. W. Britain. *Journal of the Geological Society, London*, **161**, 365–379.

HESSELBO, S. P., ROBINSON, S. A., SURLYK, F. & PIASECKI, S. 2002. Terrestrial and marine extinction at the Triassic–Jurassic boundary synchronized with major carbon-cycle perturbation: a link to initiation of massive volcanism? *Geology*, **30**, 251–254.

HESTER, S. W. 1965. Stratigraphy and palaeogeography of the Woolwich and Reading Beds. *Bulletin of the Geological Survey of Great Britain*, **23**, 117–137.

HEY, R. W. 1976. Provenance of far-travelled pebbles in the Pre-Anglian Pleistocene of East Anglia. *Proceedings of the Geologists' Association*, **87**, 69–82.

HEY, R. W. 1986. A re-examination of the Northern Drift of Oxfordshire. *Proceedings of the Geologists' Association*, **97**, 291–301.

HEYDARI, E. & HASSANZADEH, J. 2003. Deev Jahi model of the Permian – Triassic boundary mass extinction: a case for gas hydrates as the main cause of biological crisis on Earth. *Sedimentary Geology*, **163**, 147–163.

HEY, R. W. & BRENCHLEY, P. J. 1977. Volcanic pebbles from Pleistocene gravels in Norfolk and Essex. *Geological Magazine*, **114**, 219–225.

HICKLING, G. 1908. China Clay: its nature and origin. *Transactions of the Institute of Mining Engineering*, **36**, 10–33.

HICKLING, G. 1909. British Permian footprints. *Memoirs of the Manchester Literary and Philosophical Society*, **53**, 1–30.

HICKS, H. 1877. On the Precambrian (Dimentian and Pebidian) rocks of St. David's. *Quarterly Journal of the Geological Society of London*, **33**, 299–241.

HICKS, H. 1878. On the Precambrian rocks of Pembrokeshire with especial reference to the St. David's district. *Quarterly Journal of the Geological Society of London*, **40**, 507–560.

HICKS, H. 1892. The fauna of the Olenellus Zone in Wales. *Geological Magazine*, decade 3, **9**, 21–24.

HIGGINS, A. C. & VARKER, W. J. 1982. Lower Carboniferous conodont faunas from Ravenstonedale, Cumbria. *Palaeontology*, **25**, 145–166.

HIGGS, K. T. 2004. An early Devonian (Lochkovian) microflora from the Freshwater West Formation, Lower Old Red Sandstone, southwest Wales. *Geological Journal*, **39**, 359–374.

HIGGS, R. 1986. The 'South British Fault': 200 km of Stephanian-early Permian dextral wrenching on an E-W fault through the Bristol Channel. *In*: BESLEY, B. M. *ET AL*. (eds) *Controls of Upper Carboniferous Sedimentation in North West Europe*. University of Keele, 19–22.

HIGGS, R. 1991. The Bude Formation (Lower Westphalian), SW England – siliciclastic shelf sedimentation in a large equatorial lake. *Sedimentology*, **38**, 445–452.

HIGHLEY, D., BLOODWORTH, A., BATE, R. 2004*a*. *Gypsum: Mineral Planning Factsheet*. British Geological Survey for the Office of the Deputy Prime Minister.

HIGHLEY, D., BLOODWORTH, A., BATE, R. 2004*b*. *Salt: Mineral Planning Factsheet*. British Geological Survey for the Office of the Deputy Prime Minister.

HIGHLEY, D., BLOODWORTH, A., BATE, R. 2004*c*. *Silica sand: mineral planning factsheet*. British Geological Survey for the Office of the Deputy Prime Minister.

HIGHLEY, D., BLOODWORTH, A. & BATE, R. 2006. Potash: Mineral Planning Factsheet. British Geological Survey & Office of the Deputy Prime Minister HILLER, N. 1981. The Ashgillrocks of the Glyn Ceiriog district, North Wales. *Geological Journal*, **16**, 181–200.

HILGEN, F. J. 1991. Astronomical calibration of Gauss to Matuyama sapropels in the Mediterranean and implications for the Geomagnetic Polarity Time Scale. *Earth and Planetary Science Letters*, **104**, 226–244.

HILL, J. B. & MACALISTER. D. A. 1906. *The Geology of Falmouth and Camborne and of the Mining District of Camborne and Redruth*. Memoir of the Geological Survey of Great Britain, Sheet 352 (England and Wales), HMSO, London.

HILLIER, R. D. 2000. Silurian marginal marine sedimentation and the anatomy of the marine–Old Red Sandstone transition in Pembrokeshire, SW Wales. *In*: FRIEND, P. F. & WILLIAMS, B. P. J. (eds) *New Perspectives on the Old Red Sandstone*. Geological Society, London, Special Publications, **180**, 343–354.

HILLIER, R. D. 2002. Depositional environment and sequence architecture of the Silurian Coralliferous Group, Southern Pembrokeshire, UK. *Geological Journal*, **37**, 247–268.

HILLIER, R. D. & WILLIAMS, B. P. J. 2004. Sedimentation and tectonics: the marine Silurian–basal Lower Old Red Sandstone transition in southwest Wales. *Geological Journal*, **39**, 237–256.

HILLIS, R. R. & CHAPMAN, T. J. 1992. Variscan structure and its influence on post-Carboniferous basin developmant, Western Approaches Basin, SW UK Continental Shelf. *Journal of the Geological Society, London*, **149**, 413–417.

HINSCH, W. 1988. Benthic molluscs (Pelecypods, Gastropods) The description of the interregional zonation (BM zones) and its correlation with the regional lithostratigraphy. *In*: VINKEN, R. (compiler) *The Northwest European Tertiary Basin*. Results of the IGCP Project No. 124. *Geologisches Jahrbuch*, **A100**, 344–356.

HOBSON, D. M. 1976. The structure of the Dartmouth antiform. *Proceedings of the Ussher Society*, **3**, 320–323.

HODGES, P. 1996. *Bivalves from the Lower Jurassic of South Wales*. PhD thesis, University of Wales.

HODGES, P. 2000. *The Early Jurassic Bivalvia From the Hettangian and Lower Sinemurian of South-west Britain*. Part 1. Monograph of the Palaeontographical Society, London, **154**, no. 614.

HOEDEMAEKER, PH. J. 1991. Tethyan-Boreal correlations and the Jurassic–Cretaceous boundary. *Newsletters on Stratigraphy*, **25**, 37–60.

HOEDEMAEKER, PH. J. 2002. Correlating the uncorrelatables: a Tethyan-Boreal correlation of pre-Aptian strata. *In*: MICHALÍK, J.

(ed.) *Tethyan/Boreal Cretaceous Correlation*. VEDA, Bratislava, 235–284.

Hogg, N. M. 1993. *A palynological investigation of the Scalby Formation and adjacent strata (Ravenscar Group, Middle Jurassic) from the Cleveland Basin, North Yorkshire*. PhD thesis, University of Sheffield.

Holder, M. T. & Leveridge, B. E. 1986*a*. Correlation of the Rhenohercynian Variscides. *Journal of the Geological Society, London*, **143**, 141–147.

Holder, M. T. & Leveridge, B. E. 1986*b*. A model for the tectonic evolution of south Cornwall. *Journal of the Geological Society, London*, **143**, 125–134.

Holder, M. T. & Leveridge, B. E. 1987. Discussion on the correlation of Rhenohercynian Variscides. *Journal of the Geological Society, London*, **144**, 353–354.

Holder, M. T. & Leveridge, B. E. 1994. *A framework for the European Variscides*. Technical Report of the British Geological Survey of Great Britain, **WA/94/24**.

Holdsworth, B. K. & Collinson, J. D. 1988. Millstone Grit cyclicity revisited. *In*: Besly, B. M. & Kelling, G. (eds) *Sedimentation in a Synorogenic Basin Complex: The Upper Carboniferous of Northwest Europe*. Blackie, Glasgow, 132–152.

Holdsworth, R. E. 1989. The Start–Perranporth line: a Devonian terrane boundary in the Variscan orogen of SW England. *Journal of the Geological Society, London*, **146**, 419–421.

Holgate, N. & Hallowes, K. A. K. 1941. The igneous rocks of the Stanner–Hanter District, Radnorshire. *Geological Magazine*, **78**, 241–267.

Holgate, R. 1991. *Prehistoric Flint Mines*. Shire Books, Aylesbury.

Holland, C. H. 1959. The Ludlovian and Downtonian of the Knighton district, Radnorshire. *Quarterly Journal of the Geological Society of London*, **114**, 449–482.

Holland, C. H. 1985. Series and stages of the Silurian System. *Episodes*, **8**,101–103.

Holland, C. H. & Bassett, M. G. (eds). 1989. *A Global Standard for the Silurian System*. National Museum of Wales, Cardiff, Geological Series, **9**.

Holland, C. H. & Lawson, J. D. 1963. Facies patterns in the Ludlovian of Wales and the Welsh Borderland. *Geological Journal*, **3**, 269–288.

Holland, C. H. & Williams, E. M. 1985. The Ludlow–Downton transition at Kington, Herefordshire. *Geological Journal*, **20**, 31–41.

Holland, J. G. & Lambert, R. S. J. 1970. Weardale Granite. *In*: Johnson, G. A. L. (ed.) *Geology of Durham County*. Transactions of the Natural History Society of Northumberland, Durham and Newcastle upon Tyne, **41**, 103–118.

Holliday, D. W. 1993. Geophysical log signatures in the Eden Shales (Permo-Triassic) of Cumbria and their regional significance. *Proceedings of the Yorkshire Geological Society*, **49**, 345–354.

Holliday, D. W., Neves, R. & Owens, B. 1979. Stratigraphy and palynology of early Dinantian (Carboniferous) strata in shallow boreholes near Ravenstonedale, Cumbria. *Proceedings of the Yorkshire Geological Society*, **42**, 343–356.

Holliday, D. W., Warrington, G., Brookfield, M. E., McMillan, A. A. & Holloway, S. 2001. Permo-Triassic rocks in boreholes in the Annan–Canonbie area, Dumfries and Galloway, southern Scotland. *Scottish Journal Geology*, **37**, 97–113.

Holliday, D. W., Holloway, S., McMillan, A. A., Jones, N. S., Warrington, G. & Akhurst, M. C. 2004. The evolution of the Carlisle Basin, NW England and SW Scotland. *Proceedings of the Yorkshire Geological Society*, **55**, 1–19.

Hollingworth, N. & Pettigrew, T. 1988. *Field Guides to Fossils: Zechstein Reef Fossils and Their Palaeoecology*. Palaeontological Association Field Guides to Fossils Series.

Holloway, S., Milodowski, A. E., Strong, G. E. & Warrington, G. 1989. The Sherwood Sandstone Group of the Wessex Basin, southern England. *Proceedings of the Geologists' Association*, **100**, 383–394.

Holwill, F. J. W. 1963. The succession of limestones within the Ilfracombe Beds (Devonian) of north Devon. *Proceedings of the Geologists' Association*, **73**, 281–293.

Holwill, F. J. W., House, M. R., Lane, R., Gauss, G. A., Hendriks, E. M. L. & Dearman, W. R. 1969. Summer (1966) field meeting in Devon and Cornwall. *Proceedings of the Geologists' Association*, **80**, 44–63.

Hooker, J. J. 1986. Mammals from the Bartonian (middle/late Eocene) of the Hampshire Basin, southern England. *Bulletin of the British Museum (Natural History)*, **39**, 191–478.

Hooker, J. J. 1992. British mammalian paleocommunities across the Eocene–Oligocene transition and their environmental implications. *In*: Prothero, D. R. & Berggren, W. A. (eds) *Eocene–Oligocene Climatic and Biotic Evolution*. Princeton University Press, Princeton, NJ, 494–511.

Hooker, J. J. 1998. Mammalian faunal change across the Paleocene–Eocene transition in Europe. *In*: Aubry, M.-P., Lucas, S. G. & Berggren, W. A. (eds) *Late Paleocene–Early Eocene Climatic And Biotic Events in the Marine and Terrestrial Records*. Columbia University Press, New York, 91–102.

Hooker, J. J., Collinson, M. E. & Sille, N. P. 2004. Eocene–Oligocene mammalian faunal turnover in the Hampshire Basin, UK: calibration to the global time scale and the major cooling event. *Journal of the Geological Society, London*, **161**, 161–172.

Hopson, P. M., Aldiss, D. T. & Smith, A. 1996. *Geology of the Country around Hitchin*. Memoir of the British Geological Survey, Sheet 221 (England & Wales), HMSO, London.

Horák, J. M. 1993. *The late Precambrian Coedana and Sarn Complexes, NW Wales- a geochemical and petrological study*. Unpublished Ph.D. thesis, University of Wales.

Horák, J. M. 2003. Discussion on dating the late Precambrian volcanicity of England and Wales. *Journal of the Geological Society, London*, **160**, 329–330.

Horák, J. M., Doig, R. Evans, J. A. & Gibbons, W. 1996. Avalonian magmatism and terrane linkage: new isotopic data from the Precambrian of North Wales. *Journal of the Geological Society, London*, **153**, 91–99.

Horne, D. J. 1995. A revised ostracod biostratigraphy for the Purbeck–Wealden of England. *Cretaceous Research*, **16**, 639–663.

Horne, D. J. 2002. Ostracod biostratigraphy and palaeoecology of the Purbeck Limestone Group in southern England. *Special Papers in Palaeontology*, **68**, 13–20.

Horton, A., Sumbler, M. G., Cox, B. M. & Ambrose, K. 1995. *Geology of the Country Around Thame*. Memoir of the British Geological Survey, Sheet 237 (England & Wales), HMSO, London.

Hosking, A. & McInnes, R. 2002. Preparing for the impacts of climate change on the central south coast of England: a framework for future risk management. *Journal of Coastal Research*, Special Issue, **36**, 381–389.

Hosking, K. F. G. 1964. Permo-Carboniferous and later mineralisation of Cornwall and south-west Devon. *In*: Hosking, K. F. G. & Shrimpton, G. H. (eds) *Present Views on some Aspects of the Geology of Cornwall and Devon*. Royal Geological Society of Cornwall, Truro, 201–245.

Houghton, B. F., Wilson, C. J. N., McWilliams, M. O. Lanphere, M. A., Weaver, S. D., Briggs, R. M. & Pringle, M. S. 1995. Chronology and dynamics of a large silicic magmatic system: Central Taupo Volcanic Zone, New Zealand. *Geology*, **23**, 13–16.

Houlder, C. H. 1961. The excavation of a Neolithic stone implement factory on Mynydd Rhiw in Caernarvonshire. *Proceedings of the Prehistoric Society*, **27**, 108–143.

Hounslow, M. W. & McIntosh, G. 2003. Magnetostratigraphy of the Sherwood Sandstone Group (Lower and Middle Triassic): South Devon, UK. Detailed correlation of the marine and non-marine Anisian. *Palaeogeography, Palaeoclimatology, Palaeoecology*, **193**, 325–348.

Hounslow, M. W., Posen, P. E. & Warrington, G. 2004. The magnetostratigraphy and biostratigraaphy of the Norian to lowest Jurassic succession of the St Audries Bay, North Somerset

(UK). *Palaeogeography, Palaeoclimatology, Palaeoecology*, **213**, 331–358.

House, C. H. 2000. Chronostratigraphic framework for the Devonian and Old Red Sandstone. *In*: Friend, P. F. & Williams, B. P. J. (eds) *New Perspectives on the Old Red Sandstone*. Geological Society, London, Special Publications, **180**, 23–27.

House, C. H., Richardson, J. B., Chaloner, W. G., Allen, J. R. L. Holland, C. H. & Westoll, T. S. 1977. *A correlation of Devonian Rocks in the British Isles*. Geological Society, London, Special Reports, **8**.

House, M. R. 1963. Devonian ammonoid successions and facies in Devon and Cornwall. *Quarterly Journal of the Geological Society of London*, **119**, 1–27.

House, M. R. 1975. Facies and time in the marine Devonian. *Proceedings of the Yorkshire Geological Society*, **40**, 233–288.

House, M. R. 1983. Devonian eustatic events. *Proceedings of the Ussher Society*, **5**, 396–405.

House, M. R. 1985. A new approach to an absolute timescale from measurement of orbital cycles and sedimentary microrhythms. *Nature*, **315**, 721–725.

House, M. R. 1989. *Geology of the Dorset Coast*. Geologists' Association Guide.

House, M. R. 1992. Devonian sedimenary microrhythms and a Givetian time scale. *Proceedings of the Ussher Society*, **7**, 392–395.

House, M. R. 1996. The middle Devonian Kačák event. *Proceedings of the Ussher Society*, **9**, 79–84.

House, M. R. & Butcher, N. E. 1973. Excavations in the Upper Devonian and Carboniferous rocks near Chudleigh, south Devon. *Transactions of the Royal Geological Society of Cornwall*, **20**, 199–220.

House, M. R. & Selwood. E. B. 1966. Palaeozoic palaeontology in Devon and Cornwall. *In*: Hosking, K. F. G. & Shrimpton, G. J. (eds) *Present Views on Some Aspects of The Geology Of Cornwall and Devon*. 150th Anniversary Volume. Royal Geological Society of Cornwall, Penzance, 48–86.

House, M. R., Mouravieff, N. & Beese, A. P. 1978. North Cornwall. *In*: Scrutton, C. T. (ed.) *A Field Guide to Selected Areas of the Devonian of South-west England*, International Symposium on the Devonian System. Palaeontological Association, London, 57–68.

Howard, A. J., Bateman, M. D., Garton, D., Green, F. M. L., Wagner, P. & Priest, V. 1999. Evidence of Late Devensian and early Flandrian processes and environments in the Idle Valley at Tiln, North Nottinghamshire. *Proceedings of the Yorkshire Geological Society*, **52**, 383–393.

Howard, A. S. 1985. Lithostratigraphy of the Staithes Sandstone and Cleveland Ironstone formations (Lower Jurassic) of north-east Yorkshire. *Proceedings of the Yorkshire Geological Society*, **45**, 261–275.

Howard, A. S., Warrington, G., Ambrose, K. & Rees, J. G. 2006. *A Formational Framework for the Mercia Mudstone Group of England and Wales*. British Geological Survey Research Report, HMSO, London.

Howarth, M. K. 1976. An occurrence of the Tethyan ammonite Meneghiniceras in the Upper Lias of the Yorkshire coast. *Palaeontology*, **19**, 773–777.

Howarth, M. K. 2002. The Lower Lias of Robin Hood's Bay, Yorkshire and the work of Leslie Bairstow. *Bulletin of the British Museum (Natural History), Geology Series*, **58**, 81–152.

Howe, M. P. A. 1999. The Silurian fauna (graptolite and nautiloid) of the Niarbyl Formation, Isle of Man. *In*: Woodcock, N. H., Quirk, D. G., Fitches, W. R. & Barnes, R. P. (eds) *In Sight of the Suture: the Palaeozoic Geology of the Isle of Man in its Iapetus Ocean Context*. Geological Society, London, Special Publications, **160**, 177–187.

Howells, M. F. & Smith, M. 1997. *The Geology of The country around Snowdon*. Memoir of the British Geological Survey. HMSO, London.

Howells, M. F., Reedman, A. J. & Campbell, S. D. G. 1986. The submarine eruption and emplacement of the Lower Rhyolitic Tuff Formation (Ordovician) N.Wales. *Journal of the Geological Society of London*, **143**, 411–424.

Howells, M. F., Reedman, A. J. & Campbell, S. D. G. 1991. *Ordovician (Caradoc) Marginal Basin Volcanism in Snowdonia (North-west Wales)*. British Geological Survey of Great Britain. HMSO, London.

Howells, M. F., Reedman, A. J. & Leveridge, B. E. 1985. *Geology of the country around Bangor*. Explanation of Sheet 106 (England and Wales). HMSO, London.

Howitt, F. 1964. Stratigraphy and structure of the Purbeck inliers of Sussex, England. *Quarterly Journal of the Geological Society of London*, **120**, 77–113.

Huddart, D. 1991. The glacial history and glacial deposits of the North and West Cumbrian lowlands. *In*: Ehlers, J., Gibbard, P. L. & Rose, J. (eds) *Glacial Deposits in Great Britain and Ireland*. Balkema, Rotterdam, 151–168.

Hudson, J. D. 1978. Concretions, isotopes, and the diagenetic history of the Oxford Clay (Jurassic) of central England. *Sedimentology*, **24**, 339–370.

Hudson, J. D. & Martill, D. M. 1994. The Peterborough Member (Callovian, Middle Jurassic) of the Oxford Clay Formation at Peterborough, UK. *Journal of the Geological Society, London*, **151**, 113–124.

Hughes, M. J. 1981. Contribution to the Oligocene and Eocene microfaunas of the southern North Sea. *In*: Neale, J. W. & Brasier, M. D. (eds) *Microfossils From Recent and Fossil Shelf Seas*. Ellis Horwood, Chichester, 186–195.

Hughes, N. F. & McDougall, A. B. 1990. New Wealden correlation for the Wessex Basin. *Proceedings of the Geologists' Association*, **101**, 85–90.

Hughes, P. D. 2002. Loch Lomond Stadial glaciers in the Aran and Arenig Mountains, North Wales, Great Britain. *Geological Journal*, **37**, 9–15.

Hughes, P. D., Mauquoy, D., Barber, K. E. & Langdon, P. G. 2000. Mire-development pathways and palaeoclimatic records from a full Holocene peat archive at Walton Moss, Cumbria, England. *The Holocene*, **10**, 465–479.

Hughes, R. A. & Kokelaar, B. P. 1993. The timing of Ordovician magmatism in the English Lake District and Cross Fell inliers. *Geological Magazine*, **130**, 369–377.

Hughes, R. A., Cooper, A. H. & Stone, P. 1993. Structural evolution of the Skiddaw Group (English Lake District) on the northern margin of eastern Avalonia. *Geological Magazine*, **130**, 621–629.

Hughes, R. A., Evans, J. A., Noble, S. R. & Rundle, C. C. 1996. U–Pb chronology of the Ennerdale and Eskdale intrusions supports sub-volcanic relationships with the Borrowdale Volcanic Group (Ordovician, English Lake District). *Journal of the Geological Society, London*, **153**, 33–38.

Hulme, M., Jenkins, G. J., Lu, X., Turnpenny, J. R., Mitchell, T. D., Jones, R. G., Lowe, J., Murphy, J. M., Hassell, D., Boorman, P., McDonald, R. & Hill, S. 2002. *Climate Change Scenarios for the United Kingdom: The UKCIP02 Scientific Report*. Tyndall Centre for Climate Change Research, School of Environmental Sciences, University of East Anglia, Norwich, UK.

Hulseman, J. & Emery K. O. 1961. Stratification in Recent sediments of the Santa Barbara basin as controlled by organisms and water character. *Journal of Geology*, **69**, 279–290.

Humphreys, B. & Smith, S. A. 1988. *The sedimentology of a regressive-transgressive sequence: the Middle Devonian Meadfoot Beds, Staddon Grits and Jennycliff Slates, Plymouth Sound*. Technical Report of the British Geological Survey of Great Britain, Stratigraphy Series, **WH88/256C**.

Hunt, C. O. 1985. Miospores from the Portland Stone Formation and the lower part of the Purbeck Formation (Upper Jurassic/Lower Cretaceous) from Dorset, England. *Pollen et Spores*, **27**, 419–451.

Hunt, C. O. 1987. Dinoflagellate cysts and acritarch assemblages in shallow-marine and marginal-marine carbonates: the Portland Sand, Portland Stone and Purbeck formations (Upper Jurassic/ Lower Cretaceous) of southern England and northern France. *In*: Hart, M. B. (ed.) *Micropalaeontology of Carbonate Environments*. Elllis Horwood, Chichester, 208–225.

Hunter, R. H. 1980. *The petrology and geochemistry of the Carrock Fell gabbro–granophyre Complex, Cumbria*. PhD thesis, University of Durham.

Hurst, J. M. 1975. Wenlock carbonate, level bottom brachiopod-dominated communities from Wales and the Welsh Borderland. *Palaeogeography, Palaeoclimatology, Palaeoecology*, **17**, 227–255.

Hurst, J. M. 1979*a*. Evolution, succession and replacement in the type upper Caradoc (Ordovician) benthic faunas of England. *Palaeogeography, Palaeoclimatology, Palaeoecology*, **27**, 189–246.

Hurst, J. M. 1979*b*. The stratigraphy and brachiopods of the upper part of the type Caradoc of south Salop. *Bulletin of the British Museum of Natural History (Geology)*, **32**, 183–304.

Hurst, J. M. 1979*c*. The environment of deposition of the Caradoc Alternata Limestone and contiguous deposits of Salop. *Geological Journal*, **14**, 15–40.

Hurst, J. M., Hancock, N. & McKerrow, W. S. 1978. Wenlock stratigraphy and palaeogeography of Wales and the Welsh Borderland. *Proceedings of the Geologists' Association*, **89**, 197–226.

Hutchinson, J. N. 2001. Reading the ground: morphology and geology in site appraisal. *Quarterly Journal of Engineering Geology and Hydrogeology*, **33**, 325–334.

Hutchinson, J. N. & Coope, G. R. 2002. Cambering and valley bulging, periglacial solifluction and Lateglacial Coleoptera at Dowdeswell, near Cheltenham. *Proceedings of the Geologists' Association*, **113**, 291–300.

Hutton, D. H. W. 1987. Strike-slip terranes and a model for the evolution of the British and Irish Caledonides. *Geological Magazine*, **124**, 405–425.

IGS. 1976. *Newton Abbot*. England and Wales Sheet 339. Ordnance Survey for Institute of Geological Sciences, Southampton.

Ingham, F. T. 1929. The Petrography of the Spilsby Sandstone. *Proceedings of the Geologists' Association*, **40**, 1–17.

Ingham, J. K. 1966. The Ordovician rocks of the Cautley and Dent districts of Westmorland and Yorkshire. *Proceedings of the Yorkshire Geological Society*, **35**, 455–505.

Ingham, J. K. & McNamara, K. J. 1978. The Coniston Limestone Group. *In*: Moseley, F. (ed.) *The Geology of the Lake District*. Yorkshire Geological Society, Leeds, 121–139.

Ingham, J. K. & Wright, A. D. 1970. A revised classification of the Ashgill Series. *Lethaia*, **3**, 233–242.

Ingham, J. K. & Rickards, R. B. 1974. Lower Palaeozoic rocks. *In*: Rayner, D. H. & Hemingway, J. E. (ed.) *The Geology and Mineral Resources of Yorkshire*. Yorkshire Geological Society, Leeds, 29–44.

Innes, J. B., Blackford, J. J. & Davey, P. J. 2003. Dating the introduction of cereal cultivation to the British Isles: early palaeoecological evidence from the Isle of Man. *Journal of Quaternary Science*, **18**, 603–613.

Insole, A. N. & Daley, B. 1985. A revision of the lithostratigraphical nomenclature of the late Eocene and early Oligocene strata of the Hampshire Basin, southern England. *Tertiary Research*, **7**, 67–100.

Insole, A. N., Daley, B. & Gale, A. 1998. *The Isle of Wight*. Geologists' Association Guide No. 60.

Institute of Geological Sciences. 1967. Geological Survey of Great Britain (England and Wales) 1:63360, Sheet **229** Carmarthen. Southampton Ordnance Survey.

Institute of Geological Sciences. 1973. St David's 1:25000 special sheet.

Institute of Geological Sciences. 1981. Geological Survey of Great Britain (England and Wales). 1:25000 Sheet **76**, Dolgarrog. Southampton Ordnance Survey.

IPCC 1995. *The Science of Climate Change – Contribution of Working Group I to The Second Assessment Report*. Intergovernmental Panel on Climate Change Special Report. Cambridge University Press, Cambridge.

IPCC 1998. *The Regional Impacts of Climate Change: An Assessment of Vulnerability*. Intergovernmental Panel on Climate Change Special Report. Cambridge University Press, Cambridge.

IPCC 2000. *Special Report on Emissions Scenarios*. Intergovernmental Panel on Climate Change Special Report. Cambridge University Press, Cambridge.

IPCC 2001. *Climate Change 2001 – The Scientific Basis*. Intergovernmental Panel on Climate Change Special Report. Cambridge University Press, Cambridge.

Ireland, R. J., Pollard, J. E. Steel, R. J. & Thompson, D. B. 1978. Intertidal sediments and trace fossils from the Waterstones (Scythian–Anisian) at Daresbury, Cheshire. *Proceedings Yorkshire Geological Society*, **41**, 399–444.

Irwin, H. 1979. On an environmental model for the type Kimmeridge Clay – a discussion. *Nature*, **279**, 819.

Irwin, H., Curtis, C. D. & Coleman, M. 1977. Isotopic evidence for source of diagenetic carbonates formed during burial of organic-rich sediments. *Nature*, **269**, 209–213.

Isaac, K. P. 1985. Thrust and nappe tectonics of west Devon. *Proceedings of the Geologists' Association*, **96**, 109–127.

Isaac, K. P., Turner, P. J. & Stewart, I. J. 1982. The evolution of the Hercynides of central SW England. *Journal of the Geological Society, London*, **138**, 521–531.

Ivimey-Cook, H. C. 1982. *Biostratigraphy of the Lower Jurassic and Upper Triassic (Rhaetian) of the Winterborne Kingston Borehole, Dorset*. Report of the Institute of Geological Sciences, **81/3**, 97–106.

Ivimey-Cook, H. C. 1992. *Devonian fish remains from 'The Long Stone', Cornwall*. Technical Report of the British Geological Geological Survey, Stratigraphy Series, WH/92/59R.

Ivimey-Cook, H. C., Warrington, G., Worley, N. E., Holloway, S. & Young, B. 1995. Rocks of Late Triassic and Early Jurassic age in the Carlisle Basin, Cumbria (north-west England). *Proceedings of the Yorkshire Geological Society*, **50**, 305–316.

Jackson, D. I. & Johnson, H. 1996. *Lithostratigraphic nomenclature of the Triassic, Permian and Carboniferous of the UK offshore East Irish Sea Basin*. British Geological Survey, Nottingham.

Jackson, D. I. 1994. Discussion of 'Geophysical log signatures in the Eden Shales (Permo-Triassic) of Cumbria and their regional significance', Proceedings, Vol. 49, pp. 345–354, 1993. *Proceedings of the Yorkshire Geological Society*, **50**, 173–184.

Jackson, D. I., Jackson, A. A., Evans, D., Wingfield, R. T. R., Barnes, R. P. & Arthur, M. J. 1995. *The Geology of The Irish Sea*. United Kingdom Offshore Regional Reports, British Geological Survey. HMSO, London.

Jackson, D. I., Jones, S. M. & Warrington, G. 1987. The geological framework of the east Irish Sea Basin. *In*: Brooks, J. & Glennie, K. W. (eds) *Petroleum Geology of North-west Europe*. Graham and Trotman, London, 191–203.

Jackson, J. F. 1922. Sections of the Junction Bed and contiguous deposits. *Quarterly Journal of the Geological Society of London*, **78**, 436–448.

Jackson, J. F. 1926. The Junction Bed of the Middle and Upper Lias of the Dorset coast. *Quarterly Journal of the Geological Society of London*, **82**, 490–525.

Jackson, N. J. 1974. Grylls Bunny a 'tin floor' at Botallack. *Proceedings of the Ussher Society*, **3**, 186–188.

Jackson, N. J., Willis-Richards, J., Manning, D. A. C. & Sams, M. 1989. Evolution of the Cornubian orefield, south-west England: mineral deposits and ore-forming processes. *Economic Geology*, **84**, 1101–1133.

Jacobi, R. M. & Grün, R. 2003. ESR dates from Robin Hood Cave, Creswell Crags, Derbyshire, UK and the age of its early human occupation. *Quaternary Newsletter*, **100**, 1–12.

Jacobi, R. M., Rowe, P. J., Gilmour, M. A., Grün, R. & Atkinson, T. C. 1998. Radiometric dating of the Middle Palaeolithic tool industry and associated fauna of Pin Hole Cave, Creswell Crags, England. *Journal of Quaternary Science*, **13**, 29–42.

Jacquin, T. & de Graciansky, P.-C. 1998. Major transgressive /regressive cycles: the stratigraphic signature of European basin development. *In*: de Graciansky, P.-C., Hardenbol, J., Jacquin, T. & Vail, P. R. (eds) *Mesozoic and Cenozoic Sequence Stratigraphy of European Basins*. SEPM, Special Publications, **60**, 15–29.

James, D. M. D. 1983. Observations and speculations on the northeast Towy 'axis', mid-Wales. *Geological Journal*, **18**, 283–296.

James, D. M. D. 1986. The Rhiwnant Inlier, Powys, Mid-Wales. *Geological Magazine*, **123**, 585–587.

James, D. M. D. 1987. Tectonics and sedimentation in the Lower Palaeozoic back-arc basin of S. Wales, UK: some quantitative aspects of basin development. *Norsk Geologisk Tidsskrift*, **67**, 419–428.

James, D. M. D. 2003. *Geological studies in the late Ordovician–early Silurian of south and central Wales, with particular reference to basinal topography*. DPhil thesis, University of Wales, Cardiff.

James, D. M. D. & James, J. 1969. The influence of deep fractures on some areas of Ashgillian–Llandoverian sedimentation in Wales. *Geological Magazine*, **106**, 562–582.

James, J. H. 1952. *The structure and stratigraphy of the Pre-Cambrian of the Longmynd area, Shropshire*. PhD thesis, University of Bristol.

James, J. M. D. 1975. Caradoc turbidites at Poppit Sands (Pembrokeshire). *Geological Magazine*, **112**, 295–304.

James, D. M. D. 1990. Late Ordovician stratigraphy northwest of Sugarloaf, mid-Wales—a discussion. *Geological Journal*, **25**, 199–204.

Janssen, A. W. & King, C. 1988. Planktonic molluscs (Pteropoda). *In*: Vinken, R. (compiler) *The Northwest European Tertiary Basin*. Results of the IGCP Project No. 124. *Geologisches Jahrbuch*, **A100**, 356–368.

Japsen, P. 1997. Regional Neogene exhumation of Britain and the western North Sea. *Journal of the Geological Society, London*, **154**, 239–247.

Jardine, G. 1979. The western (United Kingdom) shore of the North Sea in Late Pleistocene and Holocene times. *In*: Oele, E., Schüttenhelm, R. T. E. & Wiggers, A. J. (eds) *The Quaternary History of the North Sea*. University of Uppsala, Uppsala, 159–174.

Jarvis, I. & Woodruff, P. B. 1984. Stratigraphy of the Cenomanian and basla Turonian (Upper Cretaceous) between Branscombe and Seaton, SE Devon, England. *Proceedings of the Geologists' Association*, **95**, 193–215.

Jarzembowski, E. A. 1991. New insects from the Weald Clay of the Weald. *Proceedings of the Geologists' Association*, **102**, 93–108.

Jeans, C. V. 1973. The Market Weighton Structure: tectonics, sedimentation and diagenesis during the Cretaceous. *Proceedings of the Yorkshire Geological Society*, **39**, 409–444.

Jeans, C. V. 1978. The origin of the Triassic clay assemblages of Europe with special reference to the Keuper Marl and Rhaetic of parts of England. *Philosophical Transactions of the Royal Society of London*, **A289**, 549–639.

Jeans, C. V. 1980. Early submarine lithification in the Red Chalk and Lower Chalk of Eastern England: a bacterial control model and its implications. *Proceedings of the Yorkshire Geological Society*, **43**, 81–157.

Jeans, C. V., Merriman, R. J. & Mitchell, J. G. 1977. Origin of Middle Jurassic and Lower Cretaceous Fuller's Earths in England. *Clay Minerals*, **12**, 11–44.

Jeans, C. V. Reed, S. J. B. & Xing, M. 1993. Heavy mineral stratigraphy in the UK Trias: Western approaches, onshore England and the Central North Sea. *In*: Parker, J. R. (ed.) *Petroleum Geology of Northwest Europe: Proceedings of the 4th Conference, Volume 1*. Geological Society, London, 609–624.

Jeans, C. V., Wray, D. S., Merriman, R. J. & Fisher, M. J. 2000. Volcanogenic clays in Jurassic and Cretaceous strata of England and the North Sea Basin. *Clay Minerals*, **35**, 25–55.

Jeans, C. V., Wray, D. S., Mitchell, J. G. & Ditchfield, P. 2005. Correlation between the basal part of the Lower Cretaceous Speeton Clay Formation, Yorkshire, and the Purbeck Limestone Group, Dorset: a bentonite tie line. *Proceedings of the Yorkshire Geological Society*, **55**, 183–197.

Jefferies, N. L. 1984. The radioactive accessory mineral assemblage of the Carnmenellis Granite, Cornwall. *Proceedings of the Ussher Society*, **6**, 35–41.

Jefferies, R. P. S. 1963. The stratigraphy of the Actinocamax plenus Subzone (Turonian) in the Anglo-Paris Basin. *Proceedings of the Geologists' Association*, **74**, 1–33.

Jefferies, R. P. S. & Minton, P. 1965. The mode of life of two Jurassic species of '*Posidonia*' (Bivalvia). *Palaeontology*, **8**, 156–185.

Jefferson, I., Rosenbaum, M. & Smalley, I. 2000. Mercia Mudstone as a Triassic aeolian desert sediment. *Mercian Geologist*, **15**, 157–162.

Jeffery, D. H. & Long, D. 1989. Early Pleistocene sedimentation and geographic change in the UK sector of the North Sea. *Terra Abstracts*, **1**, 425.

Jehu, R. M. 1926. The geology of the district around Tywyn and Abergynolwyn (Merioneth). *Quarterly Journal of the Geological Society of London*, **82**, 465–89.

Jenkins, D. G. & Houghton, S. D. 1987. Age, correlation and paleoecology of the St. Erth Beds and the Coralline Crag of England. *Mededelingen van de Werkgroep voor Tertiare en Kwartaire Geologie*, **24**, 147–156.

Jenkins, G. 1998. *An investigation of marine influence during deposition of the Lower Old Red Sandstone, Anglo-Welsh Basin, UK*. PhD thesis, University of Wales, Cardiff.

Jenkins, R. J. F. 1992. Functional and ecological aspects of Ediacaran assemblages. *In*: Lipps, J. H. & Signor, P. W. (eds) *Origin and Early Evolution of the Metazoa*. Plenum Press, New York, 131–176.

Jenkyns, H. C. 1988. The Early Toarcian (Jurassic) anoxic event: stratigraphic, sedimentary and geochemical evidence. *American Journal of Science*, **288**, 101–151.

Jenkyns, H. C. & Clayton, C. J. 1997. Lower Jurassic epicontinental carbonates and mudstones from England and Wales: chemostratigraphic signals and the early Toarcian anoxic event. *Sedimentology*, **144**, 687–706.

Jenkyns, H. C. & Senior, J. R. 1977. A Jurassic palaeofault from Dorset. *Geological Magazine*, **114**, 47–52.

Jenkyns, H. C. & Senior, J. R. 1991. Geological evidence for intra-Jurassic faulting in the Wessex Basin and its margins. *Journal of the Geological Society, London*, **148**, 245–260.

Jenkyns, H. C., Jones, C. E., Gröcke, D. R. & Hesselbo, S. P. 2001. Nitrogen isotope evidence for water mass denitrification during the early Toarcian (Jurassic) anoxic event. *Paleoceanography*, **16**, 593–603.

Jenkyns, H. C., Jones, C. E., Gröcke, D. R., Hesselbo, S. P. & Parkinson, D. N. 2002. Chemostratigraphy of the Jurassic System: applications, limitations and implications for palaeoceanography. *Journal of the Geological Society, London*, **159**, 351–378.

Jennings, S., Orford, J. D., Canti, M., Devoy, R. J. N. & Straker, V. 1998. The role of relative sea-level rise and changing sediment supply on Holocene gravel barrier development: the example of Porlock, Somerset, UK. *The Holocene*, **8**, 165–182.

Jeppson, L. 1990. An oceanic model for lithological and faunal changes tested on the Silurian record. *Journal of the Geological Society, London*, **147**, 663–674.

Jeppson, L., Aldridge, R. J. & Dorning, R. K. 1995. Wenlock (Silurian) ocean episodes and events. *Journal of the Geological Society, London*, **152**, 487–498

Jeram, A. J., Selden, P. A. & Edwards, D. 1990. Land animals in the Silurian: arachnids and myriapods from Shropshire, England. *Science*, **250**, 658–661.

Jhingran, A. G. 1942. The Cheviot Granite. *Quarterly Journal of the Geological Society of London*, **98**, 241–254.

Johnson, E. W. 1992. *Geology of Stoupdale area, Black Combe, S.W. Cumbria*. British Geological Survey Technical Report, **WA/92/71**.

Johnson, E. W., Soper, N. J., Burgess, I. C. et al. 2001. *Geology of the country around Ulverston*. Memoir of the British Geological Survey, Sheet 48 (England and Wales), HMSO, London.

Johnson, G. A. L. 1961. Skiddaw Slates proved in the Teesdale inlier. *Nature*, **190**, 996.

Johnson, G. A. L. & Marshall, A. E. 1971. Tournaisian beds in Ravenstonedale, Westmorland. *Proceedings of the Yorkshire Geological Society*, **38**, 261–280.

Johnson, H. J. & Lott, G. K. 1993. Cretaceous of the Central and Northern North Sea. *In*: Knox, R. W. O'B. & Cordey, W. G. (eds) *Lithostratigraphic Nomenclature of the UK North Sea*. British Geological Survey on behalf of the UK Offshore Operators Association.

Johnson, J. G., Klapper, G. & Sandberg, C. A. 1985. Devonian eustatic fluctuations in Euamerica. *Bulletin of the Geological Society of America*, **96**, 567–587.

Johnson, M. E. 1996. Stable cratonic sequences and a standard for Silurian eustasy. *In*: Witzke, B. J., Ludvigson, G. A. & Day, J. (eds) *Paleozoic Sequence Stratigraphy; Views From the North American Craton*. Geological Society of America, Special Paper, **306**, 203–211.

Johnson, M. E. & McKerrow, W. S. 1995. The Sutton Stone: an early Jurassic rocky shore deposit in South Wales. *Palaeontology*, **38**, 529–541.

Johnson, M. E., Jia-Yu, R. & Kershaw, S. 1998. Calibrating Silurian eustasy against the erosion and burial of coastal paleotopography. *In*: Landing, E. & Johnson, M. E. (eds) *Silurian Cycles*. University of the State of New York Bulletin, **491**, 3–13.

Johnson, M. E., Kaljo, D. & Rong, J.-Y. 1991. Silurian eustasy. *In*: Bassett, M. G., Lane, P. D. & Edwards, D. (eds) *The Murchison Symposium: Proceedings of an international conference on the Silurian System. Special Papers in Palaeontology*, **44**, 145–163.

Johnson, T. E. 1992. *Cleavage-transected folds in Wales*. Unpublished PhD thesis, University of Cambridge.

Jolley, D. W. 1996. The earliest Eocene sediments of eastern England: an ultra-high resolution palynological correlation. *In*: Knox. R. W. O'B., Corfield, R. M. & Dunay, R. E. (eds) *Correlation of the Early Paleogene in Northwest Europe*. Geological Society, London, Special Publications, **101**, 219–254.

Jones, C. E., Jenkyns, H. C., Coe, A. L. & Hesselbo, S. P. 1994*a*. Sr-isotopic variations in Jurassic and Cretaceous seawater. *Geochimica et Cosmochimica Acta*, **58**, 3061–3074.

Jones, C. E., Jenkyns, H. C. & Hesselbo, S. P. 1994*b*. Strontium isotopes in Early Jurassic seawater. *Geochimica et Cosmochimica Acta*, **58**, 1285–1301.

Jones, C. M. 1980. Deltaic sedimentation in the Roaches Grit and associated sediments (Namurian R2b) in the South-west Pennines. *Proceedings of the Yorkshire Geological Society*, **43**, 39–67.

Jones, C. M. & Chisholm, J. I. 1997. The Roaches and Ashover Grits: sequence stratigraphic interpretation of a 'turbidite fronted delta' system. *Geological Journal*, **32**, 45–68.

Jones, D. G. 1974. The Namurian Series in South Wales. *In*: Owen, T. R. (ed.) *The Upper Paleozoic and Post-Palaeozoic Rocks of Wales*. University of Wales Press, Cardiff, 117–132.

Jones, D. G., Morton, A. C., Leng, M. J., Haslam, H. W., Milodowski, A. E., Strong, G. E. & Kemp, S. J. 1999. Provenance of the basin fill. *In*: Plant, J. A., Jones, D. G. & Haslam, H. W. (eds) *The Cheshire Basin: Basin Evolution, Fluid Movement and Mineral Resources in a Permo-Triassic Rift Setting*. British Geological Survey, Keyworth, Nottingham, 90–124.

Jones, D. K. C. 1981. *Southeast and Southern England*. The Geomorphology of the British Isles. Methuen, London.

Jones, D. K. C. 1999*a*. Evolving models of the Tertiary evolutionary geomorphology of southern England, with special reference to the Chalklands. *In*: Smith, B. J., Whalley, W. B. & Warke, P. A. (eds) *Uplift, Erosion and Stability: Perspectives on Long-term Landscape Development*. Geological Society, London, Special Publications, **162**, 1–23.

Jones, D. K. C. 1999*b*. On the uplift and denudation of the Weald. *In*: Smith, B. J., Whalley, W. B. & Warke, P. A. (eds) *Uplift, Erosion and Stability: Perspectives on Long-term Landscape Development*. Geological Society, London, Special Publications, **162**, 25–43.

Jones, H. K., Morris, B. L. *et al.* 2000. *The physical properties of minor aquifers in England and Wales*. British Geological Survey Technical Report, **WD/00/4**. Environment Agency R&D Publication, **68**.

Jones, J. A. 1989*a*. The influence of contemporaneous tectonic activity on Westphalian sedimentation in the South Wales Coalfield. *In*: Athurton, R. S., Gutteridge, P. & Nolan, S. (eds) *Devonian and Carboniferous Tectonics and Sedimentation*. Occasional Publication of the Yorkshire Geological Society, **6**, 243–253.

Jones, J. A. 1989*b*. *Sedimentation and tectonics in the eastern part of the South Wales Coalfield*. PhD thesis, University of Wales, Cardiff.

Jones, J. A. 1991. A mountain front model for the Variscan deformation of the South Wales Coalfield. *Journal of the Geological Society, London*, **148**, 881–891.

Jones, J. A. & Hartley, A. J. 1993. Reservoir characteristics of a braidplain depositional system: the Upper Carboniferous Pennant Sandstone of South Wales. *In*: North, C. P. & Prosser, D. J. (eds) *Characterisation of Fluvial and Aeolian Reservoirs*. Geological Society, London, Special Publications, **73**, 145–158.

Jones, K. A. 1997. Deformation and emplacement of the Lizard Ophiolite Complex, SW England, based on evidence from the Basal Unit. *Journal of the Geological Society, London*, **154**, 871–885.

Jones, L. E. & Sellwood, B. W. 1989. Palaeogeographic significance of clay mineral distributions in the Inferior Oolite (Mid Jurassic) of southern England. *Clay Minerals*, **24**, 91–105.

Jones, N. S. 1992. *Lacustrine and distal fan sediments and processes, lower Devonian dartmouth Group, SE Cornwall*. Technical Report of the British Geological Survey, Stratigraphy Series, **WH/92/111R**.

Jones, N. S. 1993. *Sedimentology of Devonian and Carboniferous strata from the Plymouth area, south-west Englanda:*. Technical Report of the British Geological Survey, Stratigraphy Series, **WH/93/77R**.

Jones, N. S. 1995. *Shallow marine sedimentation in the Lower Devonian Meadfoot Group (Bovisand Formation) from the Whitsand Bay area, with sedimentological notes on other Devonian strata examined within the area*. Technical Report of the British Geological Survey, Stratigrphy Series, **WH/95/210R**.

Jones, N. S. & Ambrose, K. 1994. Triassic sandy braidplain and Aeolian sedimentation in the Sherwood Sandstone Group of the Sellafield area, west Cumbria. *Proceedings Yorkshire Geological Society*, **50**, 61–76.

Jones, O. T. 1925. The geology of the Llandovery district: Part I. The southern area. *Quarterly Journal of the Geological Society of London*, **81**, 344–388.

Jones, O. T. 1949. The geology of the Llandovery district: Part II. The southern area. *Quarterly Journal of the Geological Society of London*, **105**, 43–64.

Jones, O. T. & Pugh, W. J. 1942. An early Ordovician shoreline in Radnorshire, near Builth Wells. *Quarterly Journal of the Geological Society of London*, **105**, 65–99.

Jones, P. C. 1972. Quartzarenite and litharenite facies in the fluvial foreland deposits of the Trenchard Group (Westphalian), Forest of Dean, England. *Sedimentary Geology*, **8**, 177–198.

Jones, R. L. 1978. Late Quaternary vegetational history of the North York Moors VI. The Cleveland Moors. *Journal of Biogeography*, **5**, 81–92.

Joon, B., Laban, C. & Van der Meer, J. J. M. 1990. The Saalian glaciation in the Dutch part of the North Sea. *Geologie en Mijnbouw*, **69**, 151–158.

Jordan, R. 1971. Zur Salinität des Meeres im höheren Oberen Jura Nordwest-Deutschlands. *Zeitschrifte der Deutschlands Geologisches Gesellschaft*, **122**, 231–241.

Joussaume, S. 1999. *Climat d'hier et à demain*. CNRS editions, Paris.

Judd, J. W. 1875. *The Geology of Rutland and the Parts Of Lincoln, Leicester, Northampton, Huntingdon and Cambridge Included in Sheet 64 of the One-inch Map of the Geological Survey*. Memoir of the Geological Survey of Great Britain, HMSO, London.

Jukes-Browne, A. J. & Hill, W. 1903. *The Cretaceous Rocks of Britain. Volume II: The Lower and Middle Chalk of England*. Memoirs of the Geological Survey of the United Kingdom.

KALJO, D., MARTMA, T MÄNNIK, P. & VIIRA, V. 2003. *Bulletin Societé Géologique Français*, **174**, 59–66.

KAMERLING, P. 1979. The geology and hydrocarbon habitat of the Bristol Channel. *Journal of Petroleum Geology*, **2**, 75–93.

KANARIS-SOTIRIOU, R., MILLWARD, D. & RUNDLE C. C. 1991. The Great Whinscale Dacite – an enigmatic lava flow from the Borrowdale Volcanic Group, English Lake District. *Proceedings of the Yorkshire Geological Society*, **48**, 393–408.

KANARIS-SOTIRIOU, R., MATHIESON, N. A., JOHNSON, E. W. & MILLWARD, D. 2002. The geochemistry and significance of sills within the Ordovician Borrowdale Volcanic Group around Black Combe, SW English Lake District. *Proceedings of the Yorkshire Geological Society*, **54**, 95–110.

KAUFFMAN, E. G. 1973. Cretaceous Bivalvia. *In*: HALLAM, A. (ed.) *Atlas of Palaeobiogeography*. Elsevier, Amsterdam, 353–383.

KEEF, P. A. M., WYMER, J. J. & DIMBLEBY, G. W. 1965. A Mesolithic site on Iping Common, Sussex, England. *Proceedings of the Prehistoric Society*, **31**, 85–92.

KEELEY, L. H. & NEWCOMER, M. H. J. 1977. Microwear analysis of experimental flint tools: a test case. *Archaeological Science*, **4**, 29–62.

KEEN, M. C. 1977. Ostracod assemblages and the depositional environments of the Headon, Osborne and Bembridge Beds (upper Eocene) of the Hampshire Basin. *Palaeontology*, **20**, 405–445.

KEEPING, W. 1883. *The fossils and palaeontological affinities of the Neocomian deposits of Upware and Brickhill*. Sedgwick Prize Essay for 1879, Cambridge.

KELLAWAY, G. A. & HANCOCK, P. L. 1983. Structure of the Bristol District, the Forest of Dean and the Malvern Fault Zone. *In*: HANCOCK, P. L. (ed.) *The Variscan Fold Belt in the British Isles*. Adam Hilger, Bristol, 88–107.

KELLAWAY, G. A. & WELCH, F. B. A. 1993. *Geology of the Bristol District*. Memoir of the British Geological Survey, HMSO, London.

KELLER, G., ADATTE, T. & STINNESBECK, W. 2003*a*. The non-smoking gun. *Geoscientist*, **13** (11), 9–11.

KELLER, G., STINNESBECK, W. ET AL. 2003*b*. Spherule deposits in Cretaceous–Tertiary boundary sediments in Belize and Guatemala. *Journal of the Geological Society, London*, **160**, 783–795.

KELLING, G. 1968. Patterns of sedimentation in the Rhondda Beds of South Wales. *American Association of Petroleum Geologists Bulletin*, **52**, 2369–2386.

KELLING, G. 1974. Upper Carboniferous sedimentation in South Wales. *In*: OWEN, T. R. (ed.) *The Upper Palaeozoic and Post-Palaeozoic Rocks of Wales*. University of Wales, Cardiff, 185–224.

KELLING, G. 1988. Silesian sedimentation and tectonics in the South Wales Basin: a brief review. *In*: BESLY, B. & KELLING, G. (eds) *Sedimentation in a Synorogenic Basin Complex: The Upper Carboniferous of Northwest Europe*. Blackie, Glasgow, 38–42.

KELLING, G. & COLLINSON, J. D. 1992. Silesian. *In*: DUFF, P. M. D. & SMITH, A. J. (eds) *Geology of England and Wales*. Geological Society, London, 239–273.

KELLING, G. & WOOLLANDS, M. A. 1969. The stratigraphy and sedimentation of the Llandoverian rocks of the Rhayader district. *In*: WOOD, A. (ed.) *The Pre-Cambrian and Lower Palaeozoic rocks of Wales*. University of Wales, Cardiff, 255–282.

KELLY, S. R. A. 1983. Boreal influences on English Ryazanian bivalves. *Zitteliana*, **10**, 285–292.

KELLY, S. R. A. & RAWSON, P. F. 1983. Some late Jurassic–mid Cretaceous sections on the East Midlands Shelf, England, as demonstrated on a Field Meeting, 18–20 May 1979. *Proceedings of the Geologists' Association*, **94**, 65–73.

KEMP, A. E. S. 1987. Evolution of Silurian depositional systems in the Southern Uplands, Scotland. *In*: LEGGETT, J. K. & ZUFFA, G. G. (ed.) *Marine Clastic Sedimentology*. Graham & Trotman, London, 124–155.

KEMP, A. E. S. 1991. Mid Silurian pelagic and hemipelagic sedimentation and paleoceanography. *Special Papers in Palaeontology*, **44**, 261–299.

KEMP, R. A. 1985. The Valley Farm Soil in southern East Anglia. *In*: BOARDMAN, J. (ed.) *Soils and Quaternary Landscape Evolution*. Wiley, Chichester, 179–196.

KEMP, R. A. 1987. Genesis and environmental significance of a buried Middle Pleistocene soil in eastern England. *Geoderma*, **41**, 49–77.

KEMP, R. A. 1991. Micromorphology of the buried Quaternary soil within Burchell's 'Ebbsfleet Channel', Kent. *Proceedings of the Geologists' Association*, **102**, 275–287.

KEMPER, E. 1987. Das Klima der Kreide-Zeit. *Geologisches Jahrbuch*, **A96**, 5–185.

KEMPER, E., RAWSON, P. F. & THIEULOY, J.-P. 1981. Ammonites of Tethyan ancestry in the early Lower Cretaceous of north-west Europe. *Palaeontology*, **24**, 251–311.

KENDALL, C. G. & SCHLAGER, W. 1981. Carbonates and relative changes in sea-level. *Marine Geology*, **44**, 181–212.

KENIG, F., HAYES, J. M., POPP, B. N. & SUMMONS, R. E. 1994. Isotopic biogeochemistry of the Oxford Clay Formation (Jurassic), UK. *Journal of the Geological Society, London*, **151**, 139–152.

KENNAN, P. S. & MORRIS, J. H. 1999. Manganiferous ironstones in the early Ordovician Manx Group, Isle of Man: a protolith of coticule? *In*: WOODCOCK, N. H., QUIRK, D. G., FITCHES, W. R. & BARNES, R. P. (eds) *In Sight of the Suture: The Palaeozoic Geology of the Isle of Man in its Iapetus Ocean Context*. Geological Society, London, Special Publications, **160**, 109–119.

KENNEDY, W. J. 1970. A correlation of the uppermost Albian and the Cenomanian of South-West England. *Proceedings of the Geologists' Association*,

KENRICK, P. & EDWARDS, D. 1988. A new zosterophyll from a recently discovered exposure of the Lower Devonian Senni Beds in Dyfed, Wales. *Botanical Journal of the Linnean Society*, **98**, 97–115.

KENT, P. E. 1947. A deep boring at North Creake, Norfolk. *Geological Magazine*, **84**, 2–18.

KENT, P. E. 1975. The Grantham Formation in the east Midlands: revision of the Middle Jurassic Lower Estuarine Beds. *Mercian Geologist*, **5**, 305–327.

KENT, P. E. 1980*a*. Subsidence and uplift in East Yorkshire and Lincolnshire: a double inversion. *Proceedings of the Yorkshire Geological Society*, **42**, 505–524.

KENT, P. E. 1980*b*. *Eastern England from the Tees to the Wash*. British Regional Geology. HMSO, London.

KERTH, M. & HAILWOOD, E. A. 1988. Magnetostratigraphy of the Lower Cretaceous Vectis Formation (Wealden Group) on the Isle of Wight, Southern England. *Journal of the Geological Society, London*, **145**, 351–360.

KIDSTON, R. 1905. Divisions and correlation of the upper portion of the Coal Measures. *Quarterly Journal of the Geological Society of London*, **61**, 319–323.

KIMBELL, G. S. & QUIRK, D. G. 1999. Crustal magnetic structure of the Irish Sea region: evidence for a major basement boundary beneath the Isle of Man. *In*: WOODCOCK, N. H., QUIRK, D. G., FITCHES, W. R. & BARNES, R. P. (ed.) *In Sight of the Suture: The Palaeozoic Geology of the Isle of Man in its Iapetus Ocean Context*. Geological Society, London, Special Publications, **160**, 227–238.

KING, A. F. 1966. Structure and stratigraphy of the Upper Carboniferous Bude sandstones, North Cornwall. *Proceedings of the Ussher Society*, **1**, 229–232.

KING, C. 1981. *The Stratigraphy of the London Clay and associated deposits*. Tertiary Research Special Paper, **6**. Backhuys, Rotterdam.

KING, C. 1983. *Cainozoic micropaleontological biostratigraphy of the North Sea*. Institute of Geological Sciences Report, **82/7**.

KING, C. 1989. Cenozoic of the North Sea. *In*: JENKINS, D. & MURRAY, J. (eds) *Stratigraphical Atlas of Fossil Foraminifera*, 2nd edn. Ellis Horwood, Chichester, 418–489.

KING, C., BAILEY, H. W., BURTON, C. A. & KING, A. D. 1989. Cretaceous of the North Sea. *In*: JENKINS, G. D. & MURRAY, J. W. (eds) *Stratigraphical Atlas of Fossil Foraminifera*, 2nd edn. Ellis Horwood, Chichester, 372–417.

KING, L. M. 1992. *A basin study of the early Palaeozoic Windermere Group, NW England*. PhD thesis, University of Cambridge.

KING, L. M. 1994*a*. Subsidence analysis of Eastern Avalonian sequences: implications for Iapetus closure. *Journal of the Geological Society, London*, **151**, 647–657.

KING, L. M. 1994*b*. Turbidite to storm transition in a migrating foreland basin: the Kendal Group (Upper Silurian), northwest England. *Geological Magazine*, **131**, 255–267.

KING, M. J. & BENTON, M. J. 1996. Dinosaurs in the Early and Mid Triassic? – the footprint evidence from Britain. *Palaeogeography, Palaeoclimatology, Palaeecology*, **122**, 213–225.

KING, M. J. & THOMPSON, D. B. 2000. Triassic vertebrate footprints from the Sherwood Sandstone group, Hilbre, Wirral, northwest England. *Proceedings of the Geologists' Association*, **111**, 111–132.

KING, W. 1850. *A Monograph of the Permian Fossils of England*. Palaeontographical Society Monograph.

KING, W. B. R. 1928. The geology of the district around Meifod (Montgomeryshire). *Quarterly Journal of the Geological Society of London*, **84**, 671–702.

KIRBY, G. 1979. The Lizard Complex as an ophiolite. *Nature*, **282**, 58–60.

KIRBY, G. A. & SWALLOW, P. 1987. Tectonism and sedimentation in the Flamborough Head region of north-east England. *Proceedings of the Yorkshire Geological Society*, **46**, 301–309.

KIRBY, G. A., BAILY, H. E. *ET AL*. 2000. *The Structure and Evolution of the Craven Basin and Adjacent Areas*. British Geological Survey Subsurface Memoir. HMSO, London.

KIRBY, R. & OELE, E. 1975. The geological history of the Sandettie–Fairy Bank area, southern North Sea. *Philosophical Transactions of the Royal Society of London*, **A279**, 257–267.

KIRCHGASSER, W. T. 1970. Conodonts from near the Middle/Upper Devonian boundary in North Cornwall. *Palaeontology*, **13**, 335–354.

KIRKALDY, J. F. 1933. The Sandgate Beds of the Western Weald. *Proceedings of the Geologists' Association*, **34**, 270–311.

KIRTON, S. R. 1984. Carboniferous volcanicity in England, with special reference to the Westphalian of the East Midlands. *Journal of the Geological Society, London*, **141**, 161–170.

KISPAROVA, L. D. 1936. Upper Triassic pelecypods from the Kolyma-Indigirka land. *Transactions of the Arctic Institute of the USSR*, Geology Series, **30**, 71–136. [In Russian, English summary pp. 115–128.]

KLEIN, G. DEV. 1962. Sedimentary structures in the Keuper marl (Upper Triassic). *Geological Magazine*, **9**, 137–144.

KNELLER, B. C. 1990. *The Ludlow rocks of Sheet 38 (Ambleside)*. British Geological Survey, Technical Report, **WA/90/62**.

KNELLER, B. C. 1991. A foreland basin on the southern margin of Iapetus. *Journal of the Geological Society, London*, **148**, 207–210.

KNELLER, B. C. & BELL, A. M. 1993. An Acadian mountain front in the English Lake District: the Westmorland Monocline. *Geological Magazine*, **130**, 203–213.

KNELLER, B. C., EDWARDS, D., MCCAFFREY, W. & MOORE, R. 1991. Oblique reflection of turbidity currents. *Geology*, **14**, 250–252.

KNELLER, B. C., KING, L. M. & BELL, A. M. 1993. Foreland basin development and tectonics on the northwest margin of eastern Avalonia. *Geological Magazine*, **130**, 691–697.

KNELLER, B. C., SCOTT, R. W., SOPER, N. J., JOHNSON, E. W. & ALLEN, P. M. 1994. Lithostratigraphy of the Windermere Supergroup, Northern England. *Geological Journal*, **29**, 219–240.

KNIGHT, R. I. 1997. Benthic palaeoecology of the Gault Clay Formation (Mid- and basal Upper Albian) of the western Anglo-Paris Basin. *Proceedings of the Geologists' Association*, **108**, 81–103.

KNIGHT, R. R. W. 1990. *Aspects of the micropalaeontology of Devonian strata of North Devon*. PhD thesis, University of Southampton.

KNOTT, S. D., BURCHELL, M. T, JOLLEY, E. J. & FRASER, A. J. 1993. Mesozoic to Cenozoic plate reconstructions of the North Atlantic and hydrocarbon plays of the Atlantic margins. *In*: PARKER, J. R. (ed.) *Petroleum Geology of Northwest Europe: Proceedings of the 4th Conference*. Geological Society, London, 953–974.

KNOX, D. A. & JACKSON, N. J. 1990. Composite granite intrusions of SW Dartmoor, Devon. *Proceedings of the Ussher Society*, **7**, 246–251.

KNOX, R. W. O'B. 1996*a*. Tectonic controls on sequence development in the Palaeocene and earliest Eocene of southeast England: implications for North Sea stratigraphy. *In*: HESSELBO, S. P. & PARKINSON, D. N. (eds) *Sequence Stratigraphy in British Geology*. Geological Society, London, Special Publications, **103**, 209–230.

KNOX, R. W. O'B. 1996*b*. Correlation of the early Paleogene in northwest Europe: an overview. *In*: KNOX, R. W. O'B., CORFIELD, R. M. & DUNAY, R. E. (eds) *Correlation of the Early Palaeogene in Northwest Europe*. Geological Society, London, Special Publications, **101**, 1–11.

KNOX, R. W. O'B. & HOLLOWAY, S. 1992. *Lithostratigraphic Nomenclature of the UK. North Sea. 1. Paleogene of the Central and Northern North Sea*. British Geological Survey.

KNOX, R. W. O'B. & MORTON, A. C. 1988. The record of early Tertiary N. Atlantic volcanism in sediments of the North Sea Basin. *In*: MORTON, A. C. & PARSON, L. M. (eds) *Early Tertiary volcanism and the opening of the N E Atlantic*. Geological Society, London, Special Publications, **39**, 407–419.

KOEBERL, C., ARMSTRONG, R. A. & REIMOLD, W. U. 1997. Morokweng, South Africa; a large impact structure of Jurassic–Cretaceous boundary age. *Geology*, **25**, 731–734.

KOKELAAR, B. P. 1982. Fluidization of wet sediment during the emplacement and cooling of various igneous bodies. *Journal of the Geological Society, London*, **139**, 21–33.

KOKELAAR, B. P. 1988. Tectonic controls of Ordovician arc and marginal basin volcanism in Wales. *Journal of the Geological Society, London*, **145**, 759–776.

KOKELAAR, P. 1992. Ordovician marine volcanic record of rifting and volcanotectonism: Snowdon, Wales, United Kingdom. *Geological Society of America Bulletin*, 104, 1433–1455.

KOKELAAR, P. & BRANNEY, M. J. (1999) *Inside Silicic Calderas (Snowdon, Scafell and Glencoe, UK): interactions of caldera development, tectonism and hydrovolcanism*. Field guidebook of the IUGG/CEV Field Workshop, 7–18 July 1999.

KOKELAAR, P. & KÖNIGER, S. 2000. Marine emplacement of welded ignimbrite: the Ordovician Pitts Head Tuff, North Wales. *Journal of the Geological Society, London*, **157**, 517–536.

KOKELAAR, B. P., BEVINS, R. E.& ROACH, R. A. 1985. Submarine silicic volcanism and associated sedimentary and tectonic processes, Ramsey Island, SW Wales. *Journal of the Geological Society, London*, **142**, 591–613.

KOKELAAR, B. P., HOWELLS, M. F., BEVINS, R. E., ROACH, R.A. & DUNKLEY, P. N. 1984. The Ordovician marginal basin in Wales. *In*: KOKELAAR, B. P. & HOWELLS, M. F. (eds) *Marginal Basin Geology*. Geological Society, London, Special Publications, **16**, 254–269.

KOKELAAR, P., RAINE, P. & BRANNEY, M. J. 2006. Incursion of a large-volume, spatter-bearing pyroclastic density current into a caldera lake: the Pavey Arc ignimbrite of Scafell caldera, England. *Bulletin of Volcanology*, in press.

KOSSMAT, F. 1927. Gliederung des varistischen Gebirgsbaues. *Abhandlungen des Sächsischen Geologischen landesamtes*, **1**, 1–39. [In German.]

KOZUR, H. W. 1999. The correlation of the Germanic Buntsandstein and the Muscelkalk with the Tethyan scale. *In*: BACHMANN, G. H. & LERCHE, I. (eds) *Epicontinental Triassic*. Zentralblatt fur Geologie und Palaontologie T1, Jb 1998, H7–8, 701–726.

LAKE, R. D. 1999. *The Wakefield district – a concise account of the geology*. Memoir of the British Geological Survey, Sheet 78 (England and Wales), HMSO, London.

LAKE, R. D., YOUNG, B., WOOD, C. J. & MORTIMORE, R. N. 1987. *Geology of the country around Lewes*. (Sheet 319) Memoir of the British Geological Survey, HMSO, London.

LAKE, S. D. & KARNER, G. 1987. The structure and evolution of the Wessex Basin, southern England: an example of inversion tectonics. *Tectonophysics*, **137**, 347–378.

LAMBECK, K. 1993. Glacial rebound of the British Isles. II. A high-resolution, high-precision model. *Geophysical Journal International*, **115**, 960–990.

LAMBECK, K. 1995. Late Devensian and Holocene shorelines of the British Isles and North Sea from models of glacio-hydro-isostatic rebound. *Journal of the Geological Society, London*, **152**, 437–448.

LAMBECK, K. 1996. Glaciation and sea-level change for Ireland and the Irish Sea since late Devensian/Midlandian time. *Journal of the Geological Society, London*, **153**, 853–872.

LAMBECK, K. & PURCELL, A. P. 2001. Sea-level change in the Irish Sea since the Last Glacial Maximum: constraints from isostatic modelling. *Journal of Quaternary Science*, **16**, 497–507.

LAMING, D. J. C. 1966. Imbrications, palaeocurrents and other sedimentary features in the Lower New Red Sandstone, Devonshire, England. *Journal of Sedimentary Petrology*, **36**, 940–959.

LAMPLUGH, G. W. 1889. On the subdivisions of the Speeton Clay. *Quarterly Journal of the Geological Society, London*, **45**, 575–618.

LAMPLUGH, G. W. 1903. *The Geology of the Isle of Man*. Memoir of the Geological Survey of Great Britain, HMSO, London.

LAMPLUGH, G. W. & KITCHIN, M. A. 1911. *On the Mesozoic Rocks in Some of the Coal Explorations in Kent*. Memoir of the Geological Survey of Great Britain, HMSO, London.

LAMPLUGH, G. W., KITCHIN, M. A. & PRINGLE, J. 1923. *The Concealed Mesozoic rocks in Kent*. Memoir of the Geological Survey of Great Britain, HMSO, London.

LANDING, E. 1992. The Lower Cambrian of southeastern New foundland. Epeirogeny and Lazarus faunas, lithofacies-biofacies linkages and the myth of a global chronostratigraphy. *In:* LIPPS, J. H. & SIGNOR P. W. (Eds) *Origin and Early Evolution of the Metazoa*, Plenum Press, New York, 283–309.

LANDING, E. 1993. Cambrian-Ordovician boundary in the Taconic allochthon, eastern New York, and its correlation. *Journal of Palaeontology*, **67**, 1–19.

LANDING, E. 1996. Avalon: insular continent by the latest Precambrian. *In*: NANCE, R. D. & THOMPSON, M. D. (eds) *Avalonian and Related Peri-Gondwanan Terranes of the Circum-North Atlantic*. Geological Society of America, Special Papers, **304**, 29–64.

LANDING, E., MYROW, P., BENUS, A. P. & NARBONNE, G. M. 1989. The Placentian Series: appearance of the oldest skeletalized faunas in southeastern Newfoundland. *Journal of Palaeontology*, **63**, 739–769.

LANDING, E., BOWRING, S. A., DAVIDEK, K. L., WESTROP, S. R., GEYER, G. & HELDMAIER, W. 1998. Duration of the Early Cambrian: U-Pb ages of the volcanic ashes from Avalon and Gondwana. *Canadian Journal of Earth Sciences*, **35**, 329–338.

LANDING, E., BOWRING, S. A., DAVIDEK, K. L., RUSHTON, A. W. A., FORTEY, R. A. & WIMBLEDON, W. A. P. 2000. Cambrian-Ordovician boundary age and duration of the lowest Ordovician Tremadoc Series based on U-Pb zircon dates from AvalonianWales. *Geological Magazine*, **137**, 485–494.

LANDING, E., BOWRING, S. A., DAVIDEK, K. L., RUSHTON, A. W. A., FORTEY, R. A. & WIMBLEDON, W. A. P. 2000. Cambrian-Ordovician boundary age and duration of the lowest Ordovician Tremadoc Series based on U-Pb zircon dates from AvalonianWales. *Geological Magazine*, **137**, 485–494.

LANE, H. R., BOUCKAERT, J. *ET AL*. 1985. Proposal for an international Mid-Carboniferous boundary. *Compte Rendu 10me Congrès International de Stratigraphie et de Gèologie du Carbonifère, Madrid*, 1983, **4**, 323–339.

LANE, P. D. 2000. The Přídolí Series. *In*: ALDRIDGE, R. J., SIVETER, D. J., SIVETER, D. J., LANE, P. D., PALMER, D. & WOODCOCK, N. H. (eds) *British Silurian Stratigraphy*. Joint Nature Conservation Committee, Peterborough, 427–458.

LANG, W. D. 1914. The geology of the Charmouth cliffs, beach and foreshore. *Proceedings of the Geologists' Association*, **25**, 293–360.

LANG, W. D. 1924. The Blue Lias of the Devon and Dorset coasts. *Proceedings of the Geologists' Association*, **35**, 169–185.

LANG, W. D. 1936. The Green Ammonite Beds of the Dorset Lias. *Quarterly Journal of the Geological Society of London*, **92**, 423–437.

LANG, W. D. & SPATH, L. F. 1926. The Black Marl of Black Ven and Stonebarrow, in the Lias of the Dorset coast. With notes on the Lamellibranchia by L. R. Cox; on the Brachiopoda by H. M. Muir-Wood; on certain Echioceratidae by A. E. Trueman and D. M. Williams. *Quarterly Journal of the Geological Society of London*, **82**, 144–187, plates 8–11.

LANG, W. D., SPATH, L. F. & RICHARDSON, W. A. 1923. Shales-with-'Beef', a sequence in the Lower Lias of the Dorset coast. *Quarterly Journal of the Geological Society of London*, **79**, 47–99, pls 3–4.

LANG, W. D., SPATH, L. F., COX, L. R. & MUIR-WOOD, H. M. 1928. The Belemnite Marls of Charmouth, a series in the Lias of the Dorset coast. *Quarterly Journal of the Geological Society of London*, **84**, 179–257, plates 13–18.

LAPWORTH, C. 1879. On the Tripartite division of the Lower Palaeozoic Rocks. *Geological Magazine*, **6**, 1–15.

LATTER, M. P. 1925. Note on the age of the rocks around the Teifi Estuary. *Geological Magazine*, **62**, 187–88.

LAWRENCE, D. J. D. & WATERS, R. A. 1978. St Fagan's Borehole. *In*: *IGS Boreholes, 1977*. Report of the Institute of Geological Sciences, London, **78/21**, 10–11.

LAWRENCE, D. J. D., WEBB, B. C., YOUNG, B. & WHITE, D. E. 1986. *The geology of the late Ordovician and Silurian rocks (Windermere Group) in the area around Kentmere and Crook*. Report, British Geological Survey, **18**(5), 1–32.

LAWSON, J. D. 1954. The Silurian succession at Gorsley (Herefordshire). *Geological Magazine*, **91**, 227–237.

LAWSON, J. D. 1955. The geology of the May Hill Inlier. *Quarterly Journal of the Geological Society, London*, **111**, 85–116.

LAWSON, J. D. 1973. Faunal and facies changes in the Ludlovian rocks of Aymestrey, Herefordshire. *Geological Journal*, **8**, 247–278.

LAWSON, J. D. 1999. Atkins' Lower Bringewood associations of Wales and the Welsh Borderland. *In*: BOUCOT, A. J. & LAWSON, J. D. (eds) *Paleocommunities – A Case Study From the Silurian and Early Devonian*. Cambridge University Press, Cambridge, 380–383.

LAWSON, J. D. & WHITE, D. E. 1989. The Ludlow Series in the Ludlow area. *In*: HOLLAND, C. H. & BASSETT, M. G. (eds) *A Global Standard for the Silurian System*. National Museum of Wales, Geological Series, Cardiff, **9**, 73–90.

LE BAS, M. J. 1972. Caledonian igneous rocks beneath Central and Eastern England. *Proceedings of the Yorkshire Geological Society*, **39**, 71–86.

LEAKE, R. C. & STYLES, M. T. 1984. Borehole sections through the Traboe hornblende schists, a cumulate complex overlying the Lizard peridotite. *Journal of the Geological Society, London*, **141**, 41–52.

LEAKE, R. L., BROWN, M. J. & SMITH, K. 1985. *Volcanogenic and exhalative mineralisation within Devonian rocks of the South Hams district of Devon*. Mineral Reconnaissance Programme Report, **78**, British Geological Survey, Keyworth.

LEAT, P. T., THOMPSON, R. N, MORRISON, M. A., HENDRY, G. L. & TRAYHORN, S. C. 1987. Geodynamic significance of post-Variscan intrusive and extrusive potassic magmatism in SW England. *Transactions of the Royal Society of Edinburgh: Earth Sciences*, **77**, 349–360.

LEE, J. R., ROSE, J., HAMBLIN, R. J. O. & MOORLOCK, B. S. P. 2004. Dating the earliest lowland glaciation of eastern England: a pre-MIS 12 early Middle Pleistocene Happisburgh glaciation. *Quaternary Science Reviews*, **23**, 1551–1566.

LEE, J. R., ROSE, J., RIDING, J. B., MOORLOCK, B. S. P. & HAMBLIN, R. J. O. 2002. Testing the case for a Middle Pleistocene Scandinavian glaciation in Eastern England: evidence for a Scottish ice source for tills within the Corton Formation of East Anglia, UK. *Boreas*, **31**, 345–355.

LEE, M. K. 1986. A new gravity survey of the Lake District and three-dimensional model of the granite batholith. *Journal of the Geological Society, London*, **143**, 425–435.

LEE, M. K. 1989. *Upper crustal structure of the Lake District from modelling and image processing of potential field data*. British Geological Survey, Technical Report, **WK/89/1**.

LEE, M. K., PHARAOH, T. C. & GREEN, C. A. 1991. Structural trends in the concealed basement of eastern England from images of regional potential field data. *Annales de la Societe Geologique de Belgique*, **114**, 45–62.

LEE, M. K., PHARAOH, T. C. & SOPER, N. J. 1990. Structural trends in central Britain from images of gravity and aeromagnetic fields. *Journal of the Geological Society, London*, **147**, 241–258.

LEEDAL, G. P. & WALKER, G. P. L. 1950. A restudy of the Ingletonian Series of Yorkshire. *Geological Magazine*, **87**, 57–66.

LEEDER, M. R. 1974. Lower Border Group (Tournaisian) fluviodeltaic sedimentation and the palaeogeography of the Northumberland Basin. *Proceedings of the Yorkshire Geological Society*, **40**, 129–180.

LEEDER, M. R. 1982. Upper Palaeozoic basins of the British Isles: Caledonide inheritance versus Hercynian plate margin processes. *Journal of the Geological Society, London*, **139**, 481–494.

LEEDER, M. R. 1988. Recent developments in Carboniferous geology: a critical review with implications for the British Isles and NW Europe. *Proceedings of the Geologists' Association*, **99**, Pt 2, 73–100.

LEEDER, M. R. 1992. Dinantian. *In*: DUFF, P. M. D. & SMITH, A. J. (eds) *Geology of England and Wales*. 1st edn, Geological Society, London, 207–238.

LEEDER, M. R. & NAMI, M. 1979. Sedimentary models for the non-marine Scalby formation (Middle Jurassic) and evidence for late Bajocian/Bathonian uplift of the Yorkshire Basin. *Proceedings of the Yorkshire Geological Society*, **42**, 461–482.

LEEDER, M. R. & STRUDWICK, A. E. 1987. Delta-marine interactions: a discussion of sedimentary models for Yoredale-type cyclicity in the Dinantian of northern England. *In*: MILLER, J., ADAMS, A. E. & WRIGHT, V. P. (eds) *European Dinantian Environments*. Wiley, Chichester, 115–130.

LEGGETT, J. K. 1978. Eustacy and pelagic regimes in the Iapetuc Ocean during the Ordovician and Silurian. *Earth and Planetary Science Letters*, **41**, 163–9.

LEGGETT, J. K. 1980. British Lower Palaeozoic black shales and their palaeo-oceanographic significance. *Journal of the Geological Society, London*, **137**, 139–156.

LEGGETT, J. K., MCKERROW, W. S., COCKS, L. R. M. & RICKARDS, R. B. 1981. Periodicityin the early Palaeozoic marine realm. *Journal of the Geological Society of London*, **138**, 167–76.

LENG, M. J., GLOVER, B. W. & CHISHOLM, J. I. 1999. Nd and Sr isotopes as clastic provenance indicators in the Upper Carboniferous of Britain. *Petroleum Geoscience*, **5**, 293–301.

LEONARD, A. J. MOORE, A. G. & SELWOOD, B. E. 1982. Ventifacts from a deflation surface marking the top of the Budleigh Salterton Pebble Beds, east Devon. *Proceedings of the Ussher Society*, **5**, 333–339.

LERICHE, M. 1906. Observations sur la classification des assises paléocenes et éocenes du Bassin de Paris. *Annales de la Sociète Géologique du Nord*, **34**, 383–392.

LERICOLAIS, G., AUFFRET, J.-P. & BOURILLET, J.-F. 2003. The Quaternary Channel River : seismic stratigraphy of its palaeo-valleys and deeps. *Journal of Quaternary Science*, **18**, 245–260.

LESLIE, A. B., TUCKER, M. E. & SPIRO, B. 1992, Sedimentological and stable isotope study of travertines and associated sediments within Upper Triassic lacustrine sediments, S. Wales, UK. *Sedimentology*, **39**, 613–629.

LESLIE, A. B., SPIRO, B. & TUCKER, M. E. 1993. Geochemical and mineralogical variations in the upper Mercia Mudstone Group (late Triassic), Southwest Britain: correlation of outcrop sequences with borehole geophysical logs. *Journal of the Geological Society, London*, **150**, 67–75.

LEUY, A. 1827. On a new mineral substance called Murchisonite. *Philosophical Magazine*, **1**, 448–452.

LEVERIDGE, B. E. 1974. *The tectonics of the Roseland coastal section, south Cornwall*. PhD thesis, University of Newcastle-upon-Tyne.

LEVERIDGE, B. E. & HOLDER, M. T. 1985. Olistostromic breccias at the Mylor/Gramscatho boundary, south Cornwall. *Proceedings of the Ussher Society*, **6**, 147–154.

LEVERIDGE, B. E., HOLDER, M. T. & DAY, G. A. 1984. Thrust nappe tectonics in the Devonian of south Cornwall and the western English Channel. *In*: HUTTON, D. H. W. & SANDERSON, D. J. (eds) *Variscan Tectonics of the North Atlantic Region*. Geological Society, London, Special Publications, **14**, 103–112.

LEVERIDGE, B. E., HOLDER, M. T. & GOODE, A. J. J. 1990. *Geology of the country around Falmouth*. Memoir of the British Geological Survey, Sheet 352, England and Wales, HMSO, London.

LEVERIDGE, B. E., HOLDER, M. T., GOODE, A. J. J., SCRIVENER, R. C. JONES, N. S. & MERRIMAN, R. J. 2002. *The Plymouth and South-east Cornwall Area – A Concise Account of Geology*. Memoir of the British Geological Survey, Sheet 348, England and Wales, HMSO, London.

LEVERIDGE, B. E., SCRIVENER, R. C., GOODE, A. J. J. & MERRIMAN, R. J. 2003*a*. *Geology of the Torquay District*. Sheet Description of the British Geological Survey. 1:50 000 Series, Sheet 350 Torquay (England and Wales), HMSO, London.

LEVERIDGE, B. E., SCRIVENER, R. C., GOODE, A. J. J. & MERRIMAN, R. J. 2003*b*. *Geology of the Torquay District – A Brief Explanation of The Geological Map*. Sheet Explanation of the British Geological Survey. 1:50 000 Series, Sheet 350 Torquay (England and Wales), HMSO, London.

LEWIN, J. & MACKLIN, M. G. 2003. Preservation potential for Late Quaternary river alluvium. *Journal of Quaternary Science*, **18**, 107–120.

LEWIS, S. G., ROSE, J. & DAVIES, H. 1999. Pre-Anglian fluvial and Anglian glaciogenic sediments, Knettishall, Suffolk, England. *Proceedings of the Geologists' Association*, **110**, 17–32.

LIPMAN, P. W. 1984. Roots of ash-flow tuffs calderas in western North America; windows into the tops of granitic batholiths. *Journal of Geophysical Research*, **89**, 8801–8841.

LIPPOLT, H. J., HESS, J. C. & BURGER, K. 1984. Isotopische Alter von pyroklastischen Sanidinen aus Kaolin-Kohlentonsteinene als Korrelationsmarken fur das mitteleuropaische Oberkarbon. *Fortschritte in der Geologie von Rheinlandund Westfalen*, **32**, 119–150.

LISTER, C. J. 1978. Some tourmalinised rocks from Devon and Cornwall. *Proceedings of the Ussher Society*, **4**, 211–214.

LLOYD, G. E. & CHINNERY, N. 2002. The Bude Formation, SW England – A three-dimensional, intra-formational Variscan inbricate stack? *Journal of Structural Geology*, **24**, 1259–1280.

LLOYD, G. E. & WHALLEY, J. S. 1986. The modification of chevron folds by simple shear: examples from north Cornwall and Devon. *Journal of the Geological Society, London*, **143**, 89–94.

LOCKLEY, M. G. 1980. The geology of the Llanwchllyn to Llanymawddwy area, south Gwynedd, North Wales. *Geological Journal*, **15**, 21–41.

LOEFFLER, E. J. & THOMAS, R. G. 1980. A new pteraspidid ostracoderm from the Devonian Senni Beds Formation of South Wales and its stratigraphic significance. *Palaeontology*, **23**, 287–296.

LOMAS, J. 1907. Desert conditions and the origin of the British Trias. *Proceedings of the Liverpool Geological Society*, **10**, 172.

LONG, A. J., INNES, J. B., KIRBY, J. R., LLOYD, J. M., RUTHERFORD, M. M., SHENNAN, I. & TOOLEY, M. J. 1998. Holocene sea-level change and coastal evolution in the Humber Estuary, eastern England: an assessment of rapid coastal change. *The Holocene*, **8**, 229–247.

LONGWORTH, D. 1985. The Quaternary history of the Lancashire Plain. *In*: ROBINSON, R. H. (ed.) *The Geomorphology of North-west England*. Manchester University Press, Manchester, 178–200.

LOTT, G. K. 2001. *Building Stone Resources Map of Britain*. Geological Map, NERC.

LOTT, G. K. & KNOX, R. W. O'B. 1994. 7. Post-Triassic of the Southern North Sea. *In*: KNOX, R. W. O'B. & CORDEY, W. G. (eds) *Lithostratigraphic Nomenclature of the UK North Sea*. British Geological Survey.

LOTT, G. K. & RICHARDSON, C. 1997. Yorkshire stone for building the Houses of Parliamemnt (1939–c. 1952). *Proceedings of the Yorkshire Geological Society*, **51**, 265–272.

LOTT, G. K. & STRONG, G. E. 1982. *The petrology and petrography of the Sherwood Sandstone (?Middle Triassic) of the Winterborne Kingston Borehole, Dorset*. Institute of Geological Sciences Report, **81**, No. 3, 135–142.

LOTT, G. K., FLETCHER, B. N. & WILKINSON, I. P. 1986. The stratigraphy of the Lower Cretaceous Speeton Clay Formation in a cored borehole off the coast of north-east England. *Proceedings of the Yorkshire Geological Society*, **46**, 39–56.

LOTT, G. K., KNOX, R. W. O'B., BIGG, P. J., DAVEY, R. J. & MORTON, A. C. 1980. *Aptian–Cenomanian stratigraphy in boreholes from offshore south-west England*. Institute of Geological Sciences Report **80/8**, 1–12.

Lott, G. K., Knox, R. W. O'B., Harland, R. & Hughes, M. J. 1983. *The stratigraphy of Palaeogene sediments in a cored borehole off the coast of north-east Yorkshire*. Institute of Geological Sciences Report, **83/9**.

Lott, G. K. Sobey, R. A. Warrington, G. & Whittaker, A. 1982. The Mercia Mudstone Group (Triassic) in the western Wessex Basin (UK). *Proceedings of the Ussher Society*, **5**, 340–346.

Love, S. E. & Williams, B. P. J. 2000. Sedimentology, cyclicity and floodplain architecture in the Lower Old Red Sandstone of SW Wales. *In*: Friend, P. F. & Williams, B. P. J. (eds) *New Perspectives on the Old Red Sandstone*. Geological Society, London, Special Publications, **180**, 371–388.

Lovell, J. P. B. 1965. The Bude Sandstones from Bude to Widemouth, north Cornwall. *Proceedings of the Ussher Society*, **1**, 172–174.

Lovell, R. W. W. 1978*a*. *The sedimentology and stratigraphy of the Upper Old Red Sandstone and Lower Limestone Shales of the South Wales Coalfield*. PhD thesis, University of Bristol.

Lovell, R. W. W. 1978*b*. Duffryn Crawnon, Powys. *In*: Friend, P. F. & Williams, B. P. J. (eds) *A Field Guide to Selected Outcrop Areas of the Devonian of Scotland, the Welsh Borderland and South Wales*. Palaeontological Association, London, 70–72.

Lowe, J. J. & Lowe, S. 1989. Interpretation of the pollen stratigraphy of Late Devensian lateglacial and Flandrian sediments at Llyn Gwernan, near Cader Idris, North Wales. *New Phytologist*, **113**, 391–408.

Lowe, J. J., Hoek, W. & INTIMATE Group 2001. Inter-regional correlation of palaeoclimatic records for the Last Glacial-Interglacial Transition: a protocol for improved precision recommended by the INTIMATE project group. *Quaternary Science Reviews*, **20**, 1175–1188.

Loydell, D. K. 1991. The biostratigraphy and formational relationships of the upper Aeronian and lower Telychian (Llandovery, Silurian) formations of western mid-Wales. *Geological Journal*, **l. 26**, 209–244.

Loydell, D. K. 1992. Upper Aeronian and Lower Telychian (Llandovery) graptolites from western mid-Wales. Part 1. *Monograph of the Palaeontographical Society, London*, **146**, 1–55.

Loydell, D. K. 1998. Early Silurian sea-level changes. *Geological Magazine*, **135**, 447–471.

Lumsden, G. I., Tulloch, W., Howells, M. F. & Davies, A. 1967. *The Geology of the Neighbourhood of Langholm*. Memoir of the Geological Survey of Great Britain, HMSO, London.

Lunkka, J. P. 1994. Sedimentation and lithostratigraphy of the North Sea Drift and Lowestoft Till Formations in the coastal cliffs of northeast Norfolk, England. *Journal of Quaternary Science*, **9**, 209–233.

Lyell, C. 1830–1833. *Principles of Geology*, Volumes 1–3. Murray, London.

Macdonald, R., Gass, K. N., Thorpe, R. S. & Gass, I. G. 1984. Geochemistry and petrogenesis of the Derbyshire Carboniferous basalts. *Journal of the Geological Society, London*, **141**, 147–159.

Macdonald, R., Millward, D., Beddoe-Stephens, B., Laybourn-Parry, J. 1988. The role of tholeiitic magmatism in the English Lake District: evidence from dykes in Eskdale. *Mineralogical Magazine*, **52**, 459–472.

Macdonald, R., Thorpe, R. S., Gasgarth, J. W. & Grindrod, A. R. 1985. Multi-component origin of Caledonian lamprophyres of northern England. *Mineralogical Magazine*, **49**, 485–494.

Mackie, A. H. & Smallwood, S. D. 1987. A revised stratigraphy of the Abergwesyn-Pumpsaint area, mid Wales. *Geological Journal*, **15**, 21–41.

Machin, J. 1971. Plant microfossils from the Tertiary deposits of the Isle of Wight. *New Phytologist*, **70**, 851–872.

Mackintosh, D. M. 1964. The sedimentation of the Crackington Measures. *Proceedings of the Ussher Society*, **1**, 88–89.

Macklin, M. G. & Lewin, J. 1986. Terrace fills of Pleistocene and Holocene age in the Rheidol Valley, Wales. *Journal of Quaternary Science*, **1**, 21–34.

Macklin, M. G. & Lewin, J. 2003. River sediments, great floods and centennial-scale Holocene climate change. *Journal of Quaternary Science*, **18**, 101–105.

Macleod, N., Rawson, P. F. *et al*. 1997. The Cretaceous–Tertiary biotic transition. *Journal of the Geological Society, London*, **154**, 265–292.

MacQuaker, J. H. S. 1994*a*. A lithofacies study of the Peterborough Member, Oxford Clay Formation (Jurassic), UK: an example of sediment bypass in a mudstone succession. *Journal of the Geological Society, London*, **151**, 161–172.

MacQuaker, J. H. S. 1994*b*. Palaeoenvironmental significance of 'bone-beds' in organic rich mudstone successions: an example from the upper Triassic of south-west Britain. *Zoological Journal of the Linnean Society*, **112**, 285–308.

Maddy, D. 1999. English Midlands. *In*: Bowen, D. Q. (ed.) *A Revised Correlation of Quaternary Deposits in the British Isles*. Geological Society, London, Special Report, **23**, 28–44.

Maddy, D., Green, C. P., Lewis, S. G. & Bowen, D. Q. 1995. Pleistocene geology of the Lower Severn Valley, UK. *Quaternary Science Reviews*, **14**, 209–222.

Maddy, D., Keen, D. H., Bridgland, D. R. & Green, C. P. 1991. A revised model for the Pleistocene development of the River Avon, Warwickshire. *Journal of the Geological Society, London*, **148**, 473–484.

Madgett, P. A. & Catt, J. A. 1978. Petrography, stratigraphy and weathering of Late Pleistocene tills in East Yorkshire, Lincolnshire and north Norfolk. *Proceedings of the Yorkshire Geological Society*, **42**, 55–108.

Magee, J. W., Bowler, J. M., Miller, G. H. & Williams, D. L. G. 1995. Stratigraphy, sedimentology, chronology and palaeohydrology of Quaternary lacustrine deposits at Madigan Gulf, Lake Eyre, South Australia. *Palaeogeography, Palaeoclimatology, Palaeoecology*, **113**, 42.

Maher, B. A. & Hallam, D. F. 2005. Palaeomagnetic correlation and dating of Plio–Pleistocene sediments at the southern margins of the North Sea Basin. *Journal of Quaternary Science*, **20**, 67–77.

Mange, M., Turner, P. Ince, D. Pugh, J. & Wright, D. 1999. A new perspective on the zonation and correlation of barren strata: an integrated heavy mineral and palaeomagnetic study of the Sherwood Sandstone Group, East Irish Sea Basin and surrounding areas. *Journal of Petroleum Geology*, **22**, 325–348.

Mangerud, J., Andersen, S. T., Berglund, B. E. & Donner, J. J. 1974. Quaternary stratigraphy of Norden: a proposal for terminology and classification. *Boreas*, **4**, 109–128.

Mann, A. 1986. *Geological Studies within the Mona Complex of Central Anglesey, North Wales*. PhD thesis, University of Wales.

Manning, D. A. C. & Exley, C. S. 1984. The origins of late-stage rocks in the St Austell granite – a re-interpretation. *Journal of the Geological Society, London*, **141**, 581–591.

Manning, D. A. C., Hill, P. A. & Howe, J. H. 1996. Primary lithological variation in the kaolinized St Austell Granite, Cornwall, England. *Journal of the Geological Society, London*, **153**, 827–838.

Marriott, S. B. & Wright, V. P. 1993. Palaeosols as indicators of geomorphic stability in two Old Red Sandstone alluvial suites, South Wales. *Journal of the Geological Society, London*, **150**, 1109–1120.

Marriott, S. B. & Wright, V. P. 1996. Sediment recycling on Siluro-Devonian floodplains. *Journal of the Geological Society, London*, **153**, 661–664.

Marriott, S. B. & Wright, V. P. 2004. Mudrock deposition in an ancient dryland system: Moor Cliffs Formation, Lower Old Red Sandstone, southwest Wales, UK. *Geological Journal*, **39**, 277–298.

Marriott, S. B., Wright, V. P. & Williams, B. P. J. 2005. A new evaluation of fining-upward sequences in a mud-rock dominated succession of the Lower Old Red Sandstone of South Wales, UK. *In*: Blum, M. D. & Marriott, S. B. (eds) *Fluvial Sedimentology VII*. International Association of Sedimentologists, Special Publication, **35**, 517–529.

MARSHALL, J. D. 2000*a*. Fault-bounded basin fill: fluvial response to tectonic controls in the Skrinkle Sandstones of SW Pembrokeshire, Wales. *In*: FRIEND, P. F. & WILLIAMS, B. P. J. (eds) *New Perspectives on the Old Red Sandstone*. Geological Society, London, Special Publications, **180**, 401–416.

MARSHALL, J. D. 2000*b*. Sedimentology of a Devonian fault-bounded braidplain and lacustrine fill in the lower part of the Skrinkle Sandstones, Dyfed, Wales. *Sedimentology*, **47**, 325–342.

MARSHALL, J. D., JONES, R. T., CROWLEY, S. F., OLDFIELD, F., NASH, S. & BEDFORD, A. 2002. A high resolution, Late-Glacial isotopic record from Hawes Water, Northwest England: climatic oscillations, calibration and comparison of palaeotemperature proxies. *Palaeogeography, Palaeoclimatology, Palaeoecology*, **185**, 25–40.

MARTILL, D. M. & HUDSON, J. D. 1991. *Fossils of the Oxford Clay*. Palaeontological Association Field Guide to Fossils, **4**.

MARTILL, D. M., TAYLOR, M. A. & DUFF, K. L. 1994. The trophic structure of the biota of the Peterborough Member, Oxford Clay Formation (Jurassic), UK. *Journal of the Geological Society, London*, **151**, 173–194.

MARTIN, A. J. 1967. Bathonian sedimentation in southern England. *Proceedings of the Geologists' Association*, **78**, 473–488.

MARTINSEN, O. J. 1990. *Interaction between eustacy, tectonics and sedimentation with particular reference to the Namurian E1c-H2c of the Craven-Askrigg area, northern England*. Unpublished DrSc thesis, University of Bergen, Norway.

MARTINSEN, O. J. 1993. Namurian (late Carboniferous) depositional systems of the Craven–Askrigg area, northern England: implications for sequence stratigraphic models. *In*: POSAMENTIER, H. W., SUMMERHAYES, C. P., HAQ, B. U. & ALLEN, G. P. (eds) *Stratigraphy and Facies Associations in a Sequence Stratigraphic Framework*, 247–281.

MARZOLI, A., RENNE, P. R., PICCIRILLO, E. M., ERNESTO, M., BELLIENI, G. & DE MIN, A. 1999. Extensive 200 million year old continental flood basalts of the Central Atlantic Magmatic Province. *Science*, **284**, 616–618.

MARZOLI, A., BERTRAND, H. ET AL. 2004. Synchrony of the Central Atlantic magmatic province and the Triassic–Jurassic boundary climatic and biotic crisis. *Geology*, **32**, 973–976.

MATHERS, S. J. & ZALASIEWICZ, J. A. 1985. Producing a comprehensive geological map: a case study – the Aldeburgh–Orford area of East Anglia. *Modern Geology*, **9**, 207–220.

MATHERS, S. J. & ZALASIEWICZ, J. A. 1988. The Red Crag and Norwich Crag Formations of southern East Anglia. *Proceedings of the Geologists' Association*, **99**, 261–278.

MATHIESON, N. A. 1986. *Geology, structure and geochemistry of the Ordovician volcanic succession in SW Cumbria*. PhD thesis, University of Sheffield.

MATLEY, C. A. 1899. The geology of Northern Anglesey, part 1. *Quarterly Journal of the Geological Society of London*, **55**, 635–380.

MATLEY, C. A. 1900. The geology of Northern Anglesey, part 2. *Quarterly Journal of the Geological Society of London*, **56**, 233–256.

MATLEY, C. A. 1938. The geology around Pwllheli, Llanbedrog and Madryn, south-west Carnarvonshire. *Quarterly Journal of the Geological Society of London*, **94**, 555–606.

MATTE, P. 1986. Tectonics and plate tectonics for the variscan belt of Europe. *Tectonophysics*, **126**, 329–374.

MATTE, P., RESPAULT, J.-P, MALUSKI, H., LANCELOT, J. R. & BRUNEL, M. 1986. La faille NW-SE du Pays de Bray, un décrochement ductile dextre hercynien: deformation a 320 Ma d'un granite a 570 Ma dans le sondage Pays de Bray 201. *Bulletin Societé géologique de la France*, **8**, 69–77. [In French.]

MATTE, P., MALUSKI, H., RAJLICH, P. & FRANKE, W. 1990. Terrane boundaries in the Bohemian massif: Result of large-scale Variscan shearing. *Tectonophysics*, **177**, 151–170.

MATTHEWS, B. 1970. Age and origin of aeolian sands in the Vale of York. *Nature*, **227**, 1234–1236.

MATTHEWS, S. C. 1970. A new cephalopod fauna from the Lower Carboniferous of east Cornwall. *Palaeontology*, **13**, 262–275.

MATTHEWS, S. C. 1977. The Variscan fold belt in southwest England. *Neues Jahrbuch fur Geologie und Palaeontologie*, **154**, 94–127.

MATTHEWS, S. C. 1984. Northern margins of the Variscides in the North Atlantic region: comments on the tectonic context of the problem. *In*: HUTTON, D. H. W. & SANDERSON, D. J. (eds) *Variscan Tectonics of the North Atlantic Region*. Geological Society, London, Special Publications, **14**.

MATTHEWS, S. C. & THOMAS, J. M. 1968. Conodonts from the Dinantian of N. E. Devon. *Proceedings of the Ussher Society*, **2**, 27–28.

MAX, M. D., BARBER, A. J. & MARTINEZ, J. 1990. Terrane assemblage of the Leinster Massif, SE Ireland, during the lower Palaeozoic. *Journal of the Geological Society, London*, **147**, 1035–1050.

MAYALL, M. J. 1979. Facies and sedimentology of part of the Middle Devonian limestones of Brixham, South Devon, England. *Proceedings of the Geologists' Association*, **90**, 171–179.

MAYALL, M. J. 1981. The late Triassic Blue Anchor Formation and the initial Rhaetian marine transgression in south-west Britain. *Geological Magazine*, **118**, 377–384.

MAYALL, M. J. 1983. An earthquake origin for syn-sedimentary deformation in a late Triassic (Rhaetian) lagoonal sequence, southwest Britain. *Geological Magazine*, **120**, 613–622.

MAYEWSKI, P., MEEKER, L. D., TWICKLER, M. S., WHITLOW, S., YANG, Q., LYONS, W. B. & PRENTICE, M. 1997. Major features and forcing of high latitude northern hemisphere circulation using a 110,000-year-long glaciological series. *Journal of Geophysical Research*, **102**, 345–366.

MAYNARD, J. R. 1992. Sequence stratigraphy of the Upper Yeadonian of northern England. *Marine and Petroleum Geology*, **9**, 197–207.

MAYNARD, J. R. & LEEDER, M. 1992. On the periodicity and magnitude of Late Carboniferous glacio-eustatic sea-level changes. *Journal of the Geological Society, London*, **149**, 303–311.

MCARTHUR, J. M., DONOVAN, D. T., THIRLWALL, M. F., FOUKE, B. W. & MATTEY, D. 2000. Strontium isotope profile of the early Toarcian (Jurassic) oceanic anoxic event, the duration of ammonite biozones, and belemnite palaeotemperatures. *Earth and Planetary Science Letters*, **179**, 269–285.

MCARTHUR, J. M., MUTTERLOSE, J., PRICE, G. D., RAWSON, P. F., RUFFELL, A. & THIRLWALL, M. F. 2004. Belemnites of Valanginian, Hauterivian and Barremian age: Sr-isotope stratigraphy, composition ($^{87}Sr/^{86}Sr$, $\delta^{13}C$, $\delta^{18}O$, Na, Sr, Mg), and palaeo-oceanography. *Palaeogeography, Palaeoclimatology, Palaeoecology*, **202**, 253–272.

MCCABE, A. M. 1997. Geological constraints on geophysical models of relative sea-level change during deglaciation of the western Irish Sea basin. *Journal of the Geological Society, London*, **154**, 601–604.

MCCABE, A. M., KNIGHT, J. & MCCARRON, S. 1998. Evidence for Heinrich Event 1 in the British Isles. *Journal of Quaternary Science*, **13**, 549–568.

MCCABE, P. J. 1972. The Wenlock and Lower Ludlow strata of the Austwick and Horton-in-Ribblesdale Inlier of north-west Yorkshire. *Proceedings of the Yorkshire Geological Society*, **39**, 167–174.

MCCABE, P. J. 1978. The Kinderscoutian delta (Carboniferous) of northern England; A slope influenced by density currents. *In*: STANLEY, D. J. & KELLING, G. (eds) *Sedimentation in Submarine Canyons, Fans and Trenches*. Dowden, Hutchinson and Ross, Stroudsburg, 116–126.

MCCAFFREY, W. D. 1994. Sm–Nd isotopic characteristics of sedimentary provenance: the Windermere Supergroup of NW England. *Journal of the Geological Society, London*, **151**, 1017–1021.

MCCAFFREY, W. D. & KNELLER, B. C. 1996. Silurian turbidite provenance on the Northern Avalonian margin. *Journal of the Geological Society, London*, **153**, 437–450.

MCCAFFREY, W. D., BARRON, H. F., MOLYNEUX, S. G. & KNELLER, B. C. 1992. Recycled acritarchs as provenance indicators: implications for Caledonian terrane reconstruction. *Geological Magazine*, **129**, 457–464.

McCann, T. 1992. The stratigraphy of the Ordovician rocks around Cardigan, Wales. *Proceedings of the Yorkshire Geological Society*, **49**, 37–65.

McCarroll, D. 2001. Deglaciation of the Irish Sea Basin: a critique of the glaciomarine hypothesis. *Journal of Quaternary Science*, **16**, 393–404.

McCarroll, D. 2002. Amino-acid geochronology and the British Pleistocene: secure stratigraphical framework or a case of circular reasoning? *Journal of Quaternary Science*, **17**, 647–651.

McCarroll, D. & Ballantyne, C. K. 2000. The last ice sheet in Snowdonia. *Journal of Quaternary Science*, **15**, 765–778.

McConnell, B. & Morris, J. 1997. Initiation of Iapetus subduction under Irish Avalonia. *Geological Magazine*, **134**, 213–218.

McConnell B. J., Menuge J. F. & Hertogen, J. 2002. Andesite petrogenesis in the Ordovician Borrowdale Volcanic Group of the Lake District by fractionation, assimilation and mixing. *Journal of the Geological Society, London*, **159**, 417–424.

McCormac, M. 2001. *The Upper Palaeozoic rocks of the Shap and Penrith District, Edenside, Cumbria*. British Geological Survey Research Report, RR/01/10.

McDougall, D. A. 2001. The geomorphological impact of Loch Lomond (Younger Dryas) Stadial plateau icefields in the central Lake District, northwest England. *Journal of Quaternary Science*, **16**, 531–543.

McElhinny, M. W. & McFadden, P. L. 2000. *Paleomagnetism: Continents and Oceans*. Academic Press, San Diego.

McElwain, J. C., Beerling, D. J. & Woodward, F. I. 1999. Fossil plants and global warming at the Triassic–Jurassic boundary. *Science*, **285**, 386–390.

McIlroy, D. & Logan, G. A. 1999. The impact of bioturbation on infaunal ecology and evolution during the Proterozoic–Cambrian transition. *Palaios*, **14**, 58–72.

McIlroy, D. & Walter, M. R. 1997. A reconsideration of the biogenicity of *Arumberia banksi* Glaessner & Walter. *Alcheringa*, **21**, 79–80.

McIlroy, D., Brasier, M. D. & Moseley, J. B. 1998. The Proterozoic–Cambrian transition within the 'Charnian Supergroup' of central England and the antiquity of the Ediacara fauna. *Journal of the Geological Society, London*, **155**, 401–411.

McIlroy, D., Green, O. R. & Brasier, M. D. 2001. Palaeobiology and evolution of the earliest agglutinated Foraminifera: *Platysolenites, Spirosolenites* and related forms. *Lethaia*, **34**, 13–29.

McIlroy, D., Crimes, T. P. & Pauley, J. C. 2005. Fossils and matgrounds from the Neoproterozoic Longmyndian Supergroup, Shropshire, UK. *Geological Magazine*, **42**, 441–455.

McKeever, P. J. & Haubold, H. 1996. Reclassification of vertebrate trackways from the Permian of Scotland and related froms from Arizona and Germany. *Journal of Paleontology*, **70**, 1011–1022.

McKensie, D. & Bickle, M. J. 1988. The volume and composition of melt generated by extension of the Lithosphere. *Journal of Petrology*, **29**, 625–679.

McKeown, M. C., Edmonds, E. A., Williams, M., Freshney, E. C. & Masson Smith, B. A. 1973. *Geology of the country around Boscastle and Holsworthy*. Memoir of the Geological Survey of Great Britain, Sheets 322 and 323, England and Wales, HMSO, London.

McKerrow, W. S. 1982. *The northwest margin of the Iapetus Ocean during the early Palaeozoic*. AAPG, Memoir, **34**, 521–533.

McKerrow, W. S. 1988. Wenlock to Givetian deformation in the British Isles and the Canadian Appalachians. *In*: Harris, A. L. & Fettes, D. J. (eds) *The Caledonian–Appalachian Orogen*. Geological Society, London, Special Publications, **38**, 437–448.

McKerrow, W. S. & Van Staal, C. R. 2000*b*. The Palaeozoic time scale reviewed. *In*: Franke, W., Haak, V., Oncken, O. & Tanner, D. (eds) *Orogenic Processes: Quantification and Modelling in the Variscan Belt*. Geological Society, London, Special Publications, **179**, 5–8.

McKerrow, W. S., Dewey, J. F. & Scotese, C. F. 1991. The Ordovician and Silurian development of the Iapetus Ocean. *In*: Bassett, M. G., Lane, P. & Edwards, D. (eds) *The Murchison Symposium*. Special paper in Palaeontology, **44**, 165–178.

McKerrow, W. S., MacNiocaill, C. & Dewey, J. T. 2000*a*. The Caledonian Orogeny redefined. *Journal of the Geological Society, London*, **157**, 1149–1154.

McKerrow, W. S., MacNiocaill, C., Ahlberg, P. E., Clayton, G., Cleal, C. J. & Eagar, R. M. C. 2000*b*. The Late Palaeozoic relations between Gondwana and Laurussia. *In*: Franke, W., Haak, V., Oncken, O. & Tanner, D. (eds) *Quantification and Modelling in the Variscan Belt*. Geological Society, London, Special Publications, **179**, 9–20.

McKie, T. Aggett, J. & Hogg, J. 1998. Reservoir architecture of the upper Sherwood Sandstone, Wytch Farm, southern England. *In*: Underhill, J. R. (ed.) *Development, evolution and Petroleum Geology of the Wessex Basin*. Geological Society, London, Special Publications, **133**, 133–406.

McLean, D. & Chisholm, J. I. 1996. Reworked palynomorphs as provenance indicators in the Yeadonian of the Pennine Basin. *Proceedings of the Yorkshire Geological Society*, **51**, 141–151.

McMahon, N. A. & Turner, J. 1998. The documentation of a latest Jurassic-earliest Cretaceous uplift throughout southern England and adjacent offshore areas. *In*: Underhill, J. R. (ed.) *Development, Evolution and Petroleum Geology of the Wessex Basin*. Geological Society, London, Special Publications, **133**, 215–240.

McNamara, K. J. 1979. The age, stratigraphy and genesis of the Coniston Limestone group in the southern Lake District. *Geological Journal*, **14**, 41–69.

Meadows, N. S. & Beach, A. 1993. Structural and climatic controls on facies distribution in a mixed fluvial and aeolian reservoir: the Triassic Sherwood Sandstone in the Irish Sea. *In*: North, C. P. & Prosser, D. J. (eds) *Characterization of fluvial and aeolian reservoirs*. Geological Society, London, Special Publications, **73**, 247–264.

Mégnien, C. 1980. *Synthese Géologique du Bassin de Paris. Vol. III, Atlas*. Mémoire du Bureau de Recherche Géologiques et Minières, **102**.

Meister, C., Blau, J., Dommergues, J.-L., Feist-Burkhardt, S., Hart, M., Hesselbo, S. P., Hylton, M., Page, K. N. & Price, G. 2003. A proposal for the Global Stratotype Section and Point (GSSP) for the base of the Pliensbachian (Lower Jurassic). *Eclogae Geologica Helvetica*, **96**, 275–298.

Mellors, R. A. & Sparks, R. S. J. 1991. Spatter-rich pyroclastic flow deposits on Santorini, Greece. *Bulletin of Volcanology*, **53**, 327–342.

Melvin, J. 1986. Upper Carboniferous fine-grained turbiditic sandstones from southwest England: a model for growth in an ancient, delta-fed subsea fan. *Journal of Sedimentary Petrology*, **56**, 19–34.

Menning, M. 1995. A numerical time scale for the Permian and Triassic periods: an integrated time analysis. *In*: Scholle, P. A., Peryt, T. M. & Ulmer-Scholle, D. S. (eds) *The Permian of Northern Pangea, 1*. Springer, Berlin, 77–97.

Menning, M., Weyer, D., Drozdzewski, G., Van Ameron, H. W. J. & Wendt, I. 2000. A Carboniferous Time Scale 2000: Discussion and use of geological parameters as time indicators from central and western Europe. *Geologisches Jahrbuch*, Hannover, **156**, 3–44.

Merriman, R. J. 2002. Contrasting clay mineral assemblages in British Lower Palaeozoic slate belts: the influence of geotectonic setting. *Clay Minerals*, **37**, 207–219.

Merriman, R. J. & Frey, M. 1999. Patterns of very low grade metamorphism in metapelitic rocks. *In*: Frey, M. & Robinson, D. (eds) *Low-grade Metamorphism*. Blackwell, Oxford, 61–107.

Merriman, R. J. & Peacor, D. R. 1999. Very low-grade metapelites: mineralogy, microfabrics and measuring reaction progress. *In*: Frey, M. & Robinson, D. (eds) *Low-grade Metamorphism*. Blackwell, Oxford, 10–60.

Merriman, R. J. & Roberts, B. 1985. A survey of white-mica crystallinity and polytypes in pelitic rocks of Snowdonia and Llyn, North Wales. *Mineralogical Magazine*, **49**, 321–333.

Merriman, R. J., Pharaoh, T. C., Woodcock, N. H. & Daly, P. 1993. Metamorphic history of the concealed Caledonides of eastern England and their foreland. *Geological Magazine*, **130**, 613–620.

MERRIMAN, R. J., REX, D. C., SOPER, N. J. & PEACOR, D. R. 1995*a*. The age of Acadian cleavage in northern England, UK: K–Ar and TEM analysis of a Silurian metabentonite. *Proceedings of the Yorkshire Geological Society*, **50**, 255–265.

MERRIMAN, R. J., ROBERTS, B., PEACOR, D. R. & HIRONS, S. R. 1995*b*. Strain-related differences in the crystal growth of white mica and chlorite: a TEM and XRD study of the development of metapelite microfabrics in the Southern Uplands thrust terrane, Scotland. *Journal of Metamorphic Geology*, **13**, 449–576.

MERRIMAN, R. J., EVANS, J. A. & LEVERIDGE, B. E. 2000. Devonian and Carboniferous volcanic rocks associated with the passive margin sequences of SW England; some geochemical perspectives. *Proceedings of the Ussher Society*, **10**, 77–85.

METCALFE, I. 1981. *Conodont zonation and correlation of the Dinantian and early Namurian strata of the Craven lowlands of Northern England*. Report of the Institute of Geological Sciences, **80/10**, 1–70.

METEOROLOGICAL OFFICE. 2003. *Climate Change Observations and Predictions: Recent Research on Climate Change Science from the Hadley Centre, December 2003*. Meteorological Office, Bracknell.

MEYER, K.-J. 1988. Pollen and spores. *In*: VINKEN, R. (compiler) *The Northwest European Tertiary Basin*. Results of the IGCP Project No. 124. *Geologisches Jahrbuch*, **A100**, 288–294.

MICHELSEN, O. & CLAUSEN, O. R. 2002. Detailed stratigraphic subdivision and regional correlations of the Southern Danish Succession. *Marine and Petroleum Geology*, **19**, 563–587.

MIDDLEMISS, F. A. 1959. English Aptian Terebratulidae. *Palaeontology*, **2**, 94–142.

MIDDLEMISS, F. A. 1962. Brachiopods and shorelines in the English Lower Cretaceous. *Annals and Magazine of Natural History, series 13*, **4**, 613–626.

MIDDLEMISS, F. A. 1976. Studies in the sedimentation of the Lower Greensand of the Weald, 1875–1975: a review and commentary. *Proceedings of the Geologists' Association*, **86**, 457–473.

MILIORIZOS, M. & RUFFELL, A. 1998. Kinematics and geometry of the Watchet–Cothelstone–Hatch Fault System from outcrop and seismic data in the Bristol Channel and Wessex Basins. *In*: UNDERHILL, J. R. (ed.) *Development and Evolution of the Wessex Basin*. Geological Society, Special Publications, **133**, 311–330.

MILLER, C. G. 1995. Ostracode and conodont distribution across the Ludlow–Přídolí boundary of Wales and the Welsh Borderland. *Palaeontology*, **38**, 341–384.

MILLER, C. G., SUTHERLAND S. J. E. & DORNING, K. J. 1997. Late Silurian (Ludlow–PYídolí) microfossils and sedimentation in the Welsh Basin near Clun, Shropshire. *Geological Journal*, **32**, 69–83.

MILLER, J. A. & GREEN, D. H. 1961. Age determinations of rocks in the Lizard (Cornwall) area. *Nature*, **192**, 1175–1176.

MILLER, J. A. & MOHR, P. A. 1964. Potassium–argon measurements on the granites and some associated rocks from south-west England. *Geological Journal*, **4**, 105–126.

MILLER, M. F. 1994. *Geochemical and isotopic characteristics of palaeohydrothermal fluids related to granite magmatism, SW England*. PhD thesis, The Open University.

MILLER, R. G. & O'NIONS, R. K. 1984. The provenance and crustal residence ages of British sediments in relation to palaeogeographic reconstructions. *Earth and Planetary Science Letters*, **68**, 459–470.

MILLER, R. G. A. 1990. A paleoceanographic approach to the Kimmeridge Clay Formation. *In*: HUC, A. Y. (ed.) *Deposition of Organic Facies*. AAPG, Studies in Geology, **30**, 13–26.

MILLSON, J. A. 1987*a*. *The sedimentology of the Celtic Sea Jurassic*. PhD thesis, University of Wales.

MILLSON, J. A. 1987*b*. The Jurassic evolution of the Celtic Sea basins. *In*: BROOKS, J. & GLENNIE, K. W. (eds) *Petroleum geology of North West Europe*. Graham & Trotman, London, 599–610.

MILLWARD, D. 2002. Early Palaeozoic magmatism in the English Lake District. *Proceedings of the Yorkshire Geological Society*, **54**, 65–93.

MILLWARD D. 2004. The Caradoc volcanoes of the English Lake District. *Proceedings of the Yorkshire Geological Society*, **55**, 145–177.

MILLWARD, D. & EVANS, J. A. 2003. U–Pb chronology and duration of late-Ordovician magmatism in the English Lake District. *Journal of the Geological Society, London*, **160**, 773–781.

MILLWARD, D. & LAWRENCE, D. J. D. 1985. The Stockdale (Yarlside) Rhyolite – a rheomorphic ignimbrite? *Proceedings of the Yorkshire Geological Society*, **45**, 299–306.

MILLWARD, D. & MOLYNEUX, S. G. 1992. Field and biostratigraphic evidence for an unconformity at the base of the Eycott Volcanic Group in the English Lake District. *Geological Magazine*, **129**, 77–92.

MILLWARD, D., BEDDOE-STEPHENS, B. & YOUNG B. 1999. Pre-Acadian copper mineralisation in the English Lake District. *Geological Magazine*, **136**, 159–176.

MILLWARD, D., JOHNSON, E. W. BEDDOE-STEPHENS, B. *ET AL*. 2000*a*. *Geology of the Ambleside District*. Memoir of the British Geological Survey, Sheet 38, England and Wales, HMSO, London.

MILLWARD, D., MARRINER, G.F. & BEDDOE-STEPHENS, B. 2000*b*. The Eycott Volcanic Group, an Ordovician continental margin andesite suite in the English Lake District. *Proceedings of the Yorkshire Geological Society*, **53**, 81–96.

MILNER, A. R., GARDINER, B. G., FRASER, N. C. & TAYLOR, M. A. 1990. Vertebrates from the Middle Triassic Otter Sandstone Formation of Devon. *Palaeontology*, **33**, 873–892.

MILROY, P. G. & WRIGHT, V. P. 2000. A highstand oolitic sequence and associated facies from a Late Triassic lake Basin, south-west England. *Sedimentology*, **47**, 187–209.

MILROY, P. G. 1998. *Palaeoenvironmental analysis of the Mercia Mudstone Group (Upper Triassic) of the Severn Basin and Bristol Channel Region, Southwest Britain*. PhD thesis, Bristol University.

MILSOM, J. & RAWSON, P. F. 1989. The Peak Trough – a major control on the geology of the North Yorkshire coast. *Geological Magazine*, **126**, 699–705.

MITCHELL, G. F., PENNY, L. F., SHOTTON, F. W. & WEST, R. G. 1973. *A Correlation of Quaternary Deposits in the British Isles*. Geological Society, London, Special Report, **4**.

MITCHELL, J. G. & TAKA, A. S. 1984. Potassium and argon loss patterns in weathered micas: implications for detrital mineral studies, with particular reference to the Triassic palaeogeography of the British Isles. *Sedimentary Geology*, **39**, 27–52.

MITCHELL, M. 1981. The distribution of Tournaisian and early Visean (Carbonifeous) coral faunas from the Bristol and South Wales areas of Britain. *Acta Palaeontologica Polonica*, **25**, Pt 3–4, 577–585.

MITCHELL, M. 1989. Biostratigraphy of Visean (Dinantian) rugose coral faunas from Britain. *Proceedings of the Yorkshire Geological Society*, **47**, 233–247.

MITCHELL, M., TAYLOR, B. J. & RAMSBOTTOM, W. H. C. 1978. Carboniferous. *In*: MOSELEY, F. (ed.) *The Geology of the Lake District*. Occasional Publication of the Yorkshire Geological Society, **3**, 168–177.

MITCHELL, S. F. 1994. New data on the biostratigraphy of the Flamborough Chalk Formation (Santonian, Upper Cretaceous) between South Landing and Danes Dyke, North Yorkshire. *Proceedings of the Yorkshire Geological Society*, **50**, 113–118.

MITCHELL, S. F. 1995*a*. Lithostratigraphy and biostratigraphy of the Hunstanton Formation (Red Chalk, Cretaceous) succession at Speeton, North Yorkshire, England. *Proceedings of the Yorkshire Geological Society*, **50**, 285–303.

MITCHELL, S. F. 1995*b*. Uintacrinus anglicus Rasmussen from the Upper Cretaceous Flamborough Chalk Formation of Yorkshire: implications for the position of the Santonian–Campanian boundary. *Cretaceous Research*, **16**, 745–756.

MITCHELL, S. F. 1996. The Jurassic–Cretaceous succession at Rifle Butts Quarry (SSSI) and its implications for sedimentation on the Market Weighton structure. *Proceedings of the Geologists' Association*, **107**, 161–166.

MITCHELL, W. A. 1994. Drumlins in ice sheet reconstructions, with reference to the western Pennines, northern England. *Sedimentary Geology*, **91**, 313–331.

Mitchell, W. A. 1996. Significance of snowblow in the generation of Loch Lomond Stadial (Younger Dryas) glaciers in the western Pennines, northern England. *Journal of Quaternary Science*, **11**, 233–248.

Mitlehner, A. G. 1996. Palaeoenvironments in the North Sea Basin around the Paleocene–Eocene boundary: evidence from diatoms and other siliceous microfossils. *In*: Knox, R. W. O'B., Corfield, R. M. & Dunay, R. E. (eds) *Correlation of the Early Paleogene in Northwest Europe*. Geological Society, London, Special Publications, **101**, 255–273.

Molyneux, S. G. 1979. New evidence for the age of the Manx Group, Isle of Man. *In*: Harris, A. L., Holland, C. H. & Leake, B. E. (eds) *The Caledonides of the British Isles Reviewed*. Geological Society, London, Special Publications, **8**, 415–421.

Molyneux, S. G. 1988. Micropalaeontological evidence for the age of the Borrowdale Volcanic Group. *Geological Magazine*, **125**, 541–542.

Molyneux, S. G. 1990. *Palynological analysis of Devonian samples from 1:10 000 sheet SX 45NW*. Technical Report of the British Geological Survey, Stratigraphy Series, **WH/90/74**.

Molyneux, S. G. 1991. The contribution of palaeontological data to an understanding of the early Palaeozoic framework of eastern England. *Annales de la Société Géologique be Belgique*, **114**, 93–105.

Molyneux, S. G. 1999. A reassessment of Manx Group acritarchs, Isle of Man. *In*: Woodcock, N. H., Quirk, D. G., Fitches, W. R. & Barnes, R. P. (eds) *In Sight of the Suture: the Palaeozoic geology of the Isle of Man in its Iapetus Ocean context*. Geological Society, London, Special Publications, **160**, 23–32.

Molyneux, S. G. 2001. *Palynology of the Manx Group, Isle of Man*. British Geological Survey Report, **CR/01/21**.

Monro, M. 1986. *The sedimentology of the Fell Sandstone Group, Northumberland*. PhD thesis, University of Newcastle-upon-Tyne.

Moorbath, S. & Shackleton, R. M. 1966. Isotopic ages from the Precambrian Mona Complex of Anglesey, North Wales (Great Britain). *Earth and Planetary Science Letters*, **1**, 113–117.

Moore, J. M. & Jackson, N. J. 1997. Structure and mineralization in the Cligga Granite stock. *Journal of the Geological Society, London*, **133**, 467–480.

Moore, P. D. 1993. The origin of blanket mire revisited. *In*: Chambers, F. M. (ed.) *Climate Change and Human Impact on the Landscape*. Chapman & Hall, London, 217–224.

Moore, R. M. 1992. *The Skiddaw Group of Cumbria: Early Ordovician turbidite sedimentation and provenance on an evolving microcontinental margin*. PhD thesis, University of Leeds.

Moores, E. M. 1986. Geofuture defined. *Geology*, **14**, 368.

Morgan, A. V. 1973. The Pleistocene geology of the area north and west of Wolverhampton, Staffordshire, England. *Philosophical Transactions of the Royal Society of London*, **B265**, 233–297.

Morgans, H. S., Hesselbo, S. P. & Spicer, R. S. 1999. The seasonal climate of the early-Middle Jurassic, Cleveland Basin, England. *Palaios*, **14**, 261–272.

Morris, J. & Lycett, J. 1851–1855. *A Monograph of the Mollusca from the Great Oolite, Chiefly From Minchinhampton and the Coast of Yorkshire*. Monograph of the Palaeontographical Society, London.

Morris, J. H., Woodcock, N. H. & Howe, M. P. A. 1999. The Silurian succession of the Isle of Man: the late Wenlock Niarbyl Formation, Dalby Group. *In*: Woodcock, N. H., Quirk, D. G., Fitches, W. R. & Barnes, R. P. (eds) *In Sight of the Suture: The Palaeozoic Geology of the Isle of Man in its Iapetus Ocean Context*. Geological Society, London, Special Publications, **160**, 189–211.

Morris, T. O. & Fearnsides, W. G. 1926. The stratigraphy and structure of the Cambrian Slate Belt of Nantlle (Caernarvonshire). *Quarterly Journal of the Geological Society of London*, **82**, 250–303.

Morrissey, L. B. & Braddy S. J. 2004. Terrestrial trace fossils from the Lower Old Red Sandstone of southwest Wales. *Geological Journal*, **39**, 315–336.

Morrissey, L. B., Braddy, S. J., Bennett, J. P., Marriott, S. B. & Tarrant, P. R. 2004. Fish trails from the Lower Old Red Sandstone of Tredomen Quarry, Powys, southeast Wales. *Geological Journal*, **39**, 337–358.

Morter, A. A. & Wood, C. J. 1983. The biostratigraphy of Upper Albian–Lower Cenomanian Aucellina in Europe. *Zitteliana*, **10**, 515–529.

Mortimore, R., Wood, C., Pomerol, B. & Ernst, G. 1998. Dating the phases of the Subhercynian epoch: Late Cretaceous tectonics and eustatics in the Cretaceous basins of northern Germany compared with the Anglo-Paris Basin. *Zentralblatt für Geologie und Paläontologie*, Teil 1, 1996, 1349–1401.

Mortimore, R. N. 1983. The stratigraphy and sedimentation of the Turonian–Campanian in the Southern Province of England. *Zitteliana*, **10**, 27–41.

Mortimore, R. N. 1986*a*. Stratigraphy of the Upper Cretaceous White Chalk of Sussex. *Proceedings of the Geologists' Association*, **97**, 97–139.

Mortimore, R. N. 1986*b*. Controls on Upper Cretaceous sedimentation in the South Downs, with particular reference to flint distribution. *In*: Sieveking, G. de G. & Hart, M. B. (eds) *The Scientific Study of Flint and Chalk*. Cambridge University Press, Cambridge, 21–42.

Mortimore, R. N. & Pomerol, B. 1987. Correlation of the Upper Cretaceous White Chalk (Turonian to Campanian) in the Anglo-Paris Basin. *Proceedings of the Geologists' Association*, **98**, 97–143.

Mortimore, R. N. & Pomerol, B. 1997. Upper Cretaceous tectonic phases and end Cretaceous inversion in the Chalk of the Anglo-Paris Basin. *Proceedings of the Geologists' Association*, **108**, 231–255.

Mortimore, R. N. & Wood, C. J. 1986. The distributionof flint in the English Chalk, with particular reference to the 'Brandon Flint Series' and the high Turonian flint maximum. *In*: Sieveking, G. de G. & Hart, M. B. (eds) *The scientific study of flint and chalk*. Cambridge University Press, Cambridge, 7–20.

Mortimore, R. N., Wood, C. J. & Gallois, R. W. 2001. *British Upper Cretaceous Stratigraphy*. Joint Nature Conservation Committee. Geological Conservation Review Series, **23**.

Morton, A. C. 1982*a*. The provenance and diagenesis of Palaeogene sandstones of southeast England as indicated by heavy mineral analysis. *Proceedings of the Geologists' Association*, **93**, 263–274.

Morton, A. C. 1982*b*. Heavy minerals of Hampshire Basin Palaeogene strata. *Geological Magazine*, **119**, 463–476.

Morton, A. C., Davies J. R. & Waters, R. A. 1992. Heavy minerals as a guide to turbidite provenance in the Lower Palaeozoic Southern Welsh Basin: a pilot study. *Geological Magazine*, **129**, 573–580.

Morton, R. A. & Suter, J. R. 1996. Sequence stratigraphy and composition of Late Quaternary shelf-margin deltas, Northern Gulf of Mexico. *AAPG Bulletin*, **80**, 505–530.

Mosar, J., Lewis, G. & Torsvik, T. H. 2002. North Atlantic sea-floor spreading rates: implications for the Tertiary development of the Norwegian–Greenland Sea. *Journal of the Geological Society, London*, **159**, 503–515.

Moseley, F. 1964. The succession and structure of the Borrowdale volcanic rocks northwest of Ullswater. *Geological Journal*, **4**, 127–142.

Moseley, F. 1968. Joints and other structures in the Silurian rocks of the southern Shap Fells, Westmorland. *Geological Journal*, **6**, 79–96.

Mottershead, D. N., Gilbertson, D. D. & Keen, D. H. 1987. The raised beaches and shore platforms of Torbay: a re-evaluation. *Proceedings of the Geologists' Association*, **98**, 241–257.

Mountney, N. P. & Thompson, D. B. 2002. Stratigraphic evolution and preservation of Aeolian dune and damp/wet interdune strata: an example from the Triassic Helsby Sandstone Formation, Cheshire Basin, UK. *Sedimentology*, **49**, 805–833.

Mouravieff, N. A. 1977. Additional conodonts from near the Middle/Upper Devonian boundary in North Cornwall. *Palaeontology*, **13**, 335–354.

MUIR, M. D., BLISS, G. M., GRANT, P. R. & FISHER, M. J. 1979. Palaeontological evidence for the age of some supposedly Precambrian rocks in Anglesey, North Wales. *Journal of the Geological Society, London*, **136**, 61–64.

MULLER, R. D., ROYER, J. Y. & JAWVER, L. A. 1993. Revised plate motions relative to the hotspots from combined Atlantic and Indian Ocean hotspot tracks. *Geology*, **21**, 275–278.

MULLINS, G. L. & ALDRIDGE, R. J. 2004. Chitozoan biostratigraphy of the basal Wenlock Series (Silurian) Global Stratotype Section and Point. *Palaeontology*, **47**, 745–773.

MULLINS, G. L. & LOYDELL, D. K. 2001. Integrated Silurian chitinozoan and graptolite biostratigraphy of the Banwy River section, Wales. *Palaeontology*, **44**, 731–781.

MULLINS, G. L. & LOYDELL, D. K. 2002. Integrated Lower Silurian chitinozoan and graptolite biostratigraphy of Buttington Brick Pit, Wales. *Geological Magazine*, **139**, 89–96.

MULLIS, S. J. L., SALMON, S. & POWELL, T. 2001. Insights into the formation of the Isles of Scilly Pluton. *Geoscience in South-west England*, **10**, 230–238.

MUNDY, D. J. C. 1994. Microbialite-sponge-bryozoan-coral framestones in Lower Carboniferous (late Viséan) buildups of northern England (UK) *In*: BEAUCHAMP, B., EMBRY, A. F. & GLASS, D. J. (eds) *Pangea: Global Environments and Resources*. Canadian Society of Petroleum Geologists, Memoir, **17**, 713–729.

MUNIER-CHALMAS, E. & DE LAPPARENT, H. 1893. Note sur la nomenclature des terrains sedimentaire. *Bulletin de la Société géologique de France* (3), **21**, 40.

MURCHISON, R. I. 1833. On the sedimentary deposits which occupy the western parts of Shropshire and Herefordshire, and are prolonged from NE to SW, through Radnor, Brecknock and Caermarthenshire, with descriptions of the accompanying rocks of intrusive or igneous character. *Proceedings of the Geological Society of London*, **1**, 474–477.

MURCHISON, R. I. 1835. On the Silurian system of rocks. *London and Edinburgh Philosophical Magazine*, **7**, 46–52.

MURCHISON, R. I. 1839. *The Silurian System*. John Murray, London.

MURPHY, F. C. 1990. Basement–cover relationships of a reactivated Cadomian mylonite zone: Rosslare Complex, SE Ireland. *In*: D'LEMOS, R. S., STRACHAN, R. A. & TOPLEY, C. G. (eds) *The Cadomian Orogeny*. Geological Society, London, Special Publications, **51**, 329–339.

MURPHY, F. C. & HUTTON, D. H. W. 1986. Is the Southern Uplands of Scotland really an accretionary prism? *Geology*, **14**, 354–357.

MURPHY, F. C., ANDERSON, T. B. *ET AL*. J. P. 1991. An appraisal of Caledonian suspect terranes in Ireland. *Irish Journal of Earth Sciences*, **11**, 11–41.

MURPHY, J. B. & KEPPIE, J. D. 2005. The Acadian Orogeny in the northern Appalachians. *International Geology Review*, **47**, 663–687.

MURPHY, J. B. & NANCE, R. D. 1989. Model of the Avalonian-Cadomian belt. *Geology*, **17**, 375–378.

MURPHY, J. B., VAN STAAL, C. R. & KEPPIE, J. D. 1999. Middle to late Paleozoic Acadian orogeny in the northern Appalachians: A Laramide-style plume-modified orogeny? *Geology*, **27**, 653–656.

MURPHY, J. B., STRACHAN, R. A., NANCE, R. D., PARKER, D. K. & FOWLER, M. B. 2000. Proto-Avalonia: A ca. 1. 2-1. 0 Ga tectonothermal event and constrains for the evolution of Rodinia? *Geology*, **28**, 1071–1074.

MURRAY, J. W. 1992. Palaeogene and Neogene. *In*: COPE, J. C. W., INGHAM, J. K. &. RAWSON, P. F. (eds) *Atlas of Palaeogeography and Lithofacies*. Geological Society, London, Memoir, **13**, 141–147.

MURRAY, J. W & WRIGHT, C. A. 1974. Palaeogene Foraminiferida and palaeoecology, Hampshire and Paris Basins and the English Channel. *Special Papers in Palaeontology* 14.

MURTON, J. B., BAKER, A. *ET AL*. 2001. A late Middle Pleistocene temperate-periglacial-temperate sequence (Oxygen Isotope Stages 7-5e) near Marsworth, Buckinghamshire, UK. *Quaternary Science Reviews*, **20**, 1787–1925.

MUSGROVE F. W., MURDOCH, L. M. & LENEHAN, T. 1995. The Variscan fold-thrust belt of southeast Ireland and its control on early Mesozoic extension and deposition: a method to predict the Sherwood Sandstone. *In*: CROKER P. F. (ed.) *The Petroleum Geology of Ireland's Offshore Basins*. Geological Society, London, Special Publications, **93**, 81–100.

MUTTERLOSE, J. 1990. A belemnite scale for the Lower Cretaceous. *Cretaceous Research*, **11**, 1–15. MUTTERLOSE, J. & KESSELS, K. 2000. Early Cretaceous calcareous nannofossils from high latitudes: implications for palaeobiogeography and palaeoclimate. *Palaeogeography, Palaeoclimatology, Palaeoecology*, **160**, 347–372.

MUTTERLOSE, J. & RUFFELL, A. H. 1999. Milankovitch-scale palaeoclimate changes in pale-dark bedding rhythms from the Early Cretaceous (Hauterivian & Barremian) of eastern England and northern Germany. *Palaeogeography, Palaeoclimatology, Palaeoecology*, **154**, 133–160.

MUTTERLOSE, J., BRUMSACK, H. *ET AL*. 2003. The Greenland–Norway Seaway: A key area for understanding Late Jurassic to Early Cretaceous paleoenvironments. *Paleoceanography*, **18**, 10-1–10-26.

MYERS, K. & WIGNALL, P. B. 1987. Understanding Jurassic organic-rich mudrocks: new concepts using gamma-ray spectrometry and palaeoecology: examples from the Kimmeridge Clay of Dorset and the Jet Rock of Yorkshire. *In*: LEGGETT, J. K. & ZUFFA, G. G. (eds) *Marine Clastic Sedimentology: Concepts and Case Studies*. Graham and Trotman, London, 172–189.

NAKAMURA, N. & BORRADAILE, G. 2001. Do reduction spots predate finite strain? A magnetic diagnosis of Cambrian slates in North Wales. *Tectonophysics*, **340**, 133–139.

NAMI, M. & LEEDER, M. R. 1978. Changing channel morphology and magnitude in the Scalby Formation (M. Jurassic) of Yorkshire, England. *In*: MIALL, A. D. (ed.) *Fluvial Sedimentology*. Canadian Society of Petroleum Geologists, Memoir, **5**, 431–440.

NANCE, R. D. & MURPHY, J. B. 1996. Basement isotopic signatures and Neoproterozoic palaeogeography of Avalonian–Cadomian and related terranes in the circum-North Atlantic. *In*: NANCE, R. D. & THOMPSON, M. D. (eds) *Avalonian and Related Peri-Gondwanan Terranes of the Circum-North Atlantic*. Geological Society of America, Special Papers, **304**, 333–346.

NANCE, R. D., MURPHY, J. D., STRACHAN, R. A., D'LEMOS, R. S. & TAYLOR, G. K. 1991. Late Proterozoic tectonostratigraphic evolution of the Avalonian and Cadomian terranes. *Precambrian Research*, **53**, 41–78.

NEAL, J. E. 1996. A summary of Paleogene sequence stratigraphy in northwest Europe and the North Sea. *In*: KNOX, R. W. O'B., CORFIELD, R. M & DUNAY, R. E. (eds) *Correlation of the Early Paleogene in Northwest Europe*. Geological Society, London, Special Publications, **101**, 15–42.

NEAVERSON, E. 1925. Ammonites from the Upper Kimmeridge Clay. *Papers from the Geology Department of the University of Liverpool*, **1**, 1–52.

NEGUS, P. E. & BEAUVAIS, L. 1975. The Fairford Coral Bed (English Bathonian), Gloucestershire. *Proceedings of the Geologists' Association*, **86**, 183–204.

NEGUS, P. E. & BEAUVAIS, L. 1979. The corals of Steeple Ashton (English Upper Oxfordian). *Proceedings of the Geologists' Association*, **90**, 213–227.

NEUMAN, R. B. & BATES, D. E. B. 1978. Reassessment of Arenig and Llanvirn age of early Ordovician brachiopods from Anglesey, north-west Wales. *Palaeontology*, **21**, 571–613.

NEUMAYR, M. 1883. Über klimatische Zonen während der Jura – und Kreidezeit. Denkschriften. Akademie der Wissenschaften in Wien. *Mathematisch-Naturwissenschaftliche Klasse*, **47**, 277–310.

NEVES, R., GUEINN, K. J., CLAYTON, G., IOANNIDES, N. S. & NEVILLE, R. S. W. 1972. A scheme of miospore zones for the British Dinantian. *Septième Congrès International de Stratigraphie et de Géologie du Carbonifère, Krefeld, 1971, Compte Rendu*, **1**, 347–353.

NEVES, R., GUEINN, K. J., CLAYTON, G., IOANNIDES, N. S., NEVILLE, R. S. W. & KRUSZEWSKA, K. 1973. Palynological correlations within the Lower Carboniferous of Scotland and northern England. *Transactions of the Royal Society of Edinburgh*, **69**, 23–70.

NEWELL, A. J. 2000. Discussion on fault activity and sedimentation in a marine rift basin (Upper Jurassic, Wessex basin, UK). *Journal of the Geological Society, London*, **157**, 83–92.

NEWELL, A. J. 2001. Construction of a Palaeogene tide-dominated shelf: influence of Top Chalk topography and sediment supply (Wessex Basin, UK). *Journal of the Geological Society, London*, **158**, 379–390.

NEWHALL, C. G. & SELF, S. 1982. The volcanic explosivity index (VEI): an estimate of explosive magnitude for historical volcanism. *Journal of Geophysical Research*, **87**, 1231–1238.

NEWMAN, P. J. 1999. The geology and hydrocarbon potential of the Peel and Solway Basins, East Irish Sea. *Journal of Petroleum Geology*, **22**, 305–324.

NICHOLAS, T. C. 1915. The geology of the St Tudwal's Peninsula (Carnarvonshire). *Quarterly Journal of the Geological Society of London*, **71**, 83–143.

NIELSEN, A. T. 1996. Iltsvind, sort slam og trilobiter. *Varv*, **1996**, 3–39.

NIKITIN, S. 1881. Die Jura-Ablagerungen zwischen Rybinsk, Mologa und Myschkin. *Mémoires de l'Academie impériale des Sciences de St. Pétersburg*, (7) **18**, no. 5.

NIXON, P. H., REX, D. C. & CONDLIFFE, E. 1984. A note on the age and petrogenesis of lamprophyre dykes of the Cautley area, Yorkshire Dales National Park. *Transactions of the Leeds Geological Association*, **10**, (4), 40–45.

NOBLE, S. R., TUCKER, R. D. & PHARAOH, T. C. 1993. Lower Palaeozoic and Precambrian igneous rocks from eastern England, and their bearing on late Ordovician closure of the Tornquist Sea: constraints from U–Pb and Nd isotopes. *Geological Magazine*, **130**, 835–846.

NORFORD, B. S. 1988. Introduction to papers on the Cambrian–Ordovician boundary. *Geological Magazine*, **125**, 323–326.

NORRY, M. J., DUNHAM, A. C. & HUDSON, J. D. 1994. Mineralogy and geochemistry of the Peterborough Member, Oxford Clay Formation, Jurassic, UK: element fractionation during mudrock sedimentation. *Journal of the Geological Society, London*, **151**, 195–207.

NORTON, P. E. P. & BECK, R. B. 1972. Lower Pleistocene molluscan assemblages and pollen from the Crag of Aldeby, and Easton Bavents, Suffolk. *Bulletin of the Geological Society of Norfolk*, **22**, 11–31.

NOTHOLT, J. G. 1980. Economic phosphatic sediments : mode of occurrence and stratigraphic distribution. *Journal of the Geological Society of London*, **94**, 555–606.

NUTMAN, A. P., GREEN, D. H., COOK, C. A., STYLES, M. T. & HOLDSWORTH, R. E. 2001. SHRIMP U–Pb zircon dating of the exhumation of the Lizard Peridotite and its emplacement over crustal rocks: constraints for tectonic models. *Journal of the Geological Society, London*, **158**, 809–820.

O'BRIEN, C., PLANT, J. A., SIMPSON, P. R. & TARNEY, J. 1985. The geochemistry, metasomatism and petrogenesis of the granites of the English Lake District. *Journal of the Geological Society, London*, **142**, 1139–1157.

OATES, M. J. 1991. Upper Kimmeridgian stratigraphy of Aylesbury, Buckinghamshire. *Proceedings of the Geologists' Association*, **102**, 185–199.

OCZLON, M. S. 1993. Palaeogeographic and metamorphic eveolution of the Ligerian Belt of Europe. *In*: GAYER, M. S., GREILING, R. O. & VOGEL. A. (eds) *The Rhenohercynian and Sub-Variscan Fold belts*. Vieweg, Wiesbaden, 83–130.

OGG, J. G., HASENYAGER, II, R. W. & WIMBLEDON, W. A. 1994. Jurassic–Cretaceous boundary: Portland–Purbeck magnetostratigraphy and possible correlation to the Tethyan faunal realm. *Geobios*, Mémoire Spécial, **17**, 519–527.

OGUIKE, R. O. 1969. *Sedimentation of the Middle Shales (Upper Namurian) of the South Wales Coalfield*. PhD thesis, University of Wales, Swansea.

OLD, R. A. HAMBLIN, R. J. O., AMBROSE, K. & WARRINGTON, G. 1991. *Geology of the country around Redditch*. Memoir of the British Geological Survey, Sheet 183 (England and Wales). HMSO, London.

OLD, R. A., SUMBLER, M. G. & AMBROSE, K. 1987. *The Geology of the country around Warwick*. Memoir of the British Geological Survey, Sheet 184 (England and Wales). HMSO, London.

OLIVER, R. 1961 The Borrowdale Volcanic and associated rocks of the Scafell area, English Lake District. *Quarterly Journal of the Geological Society of London*, **117**, 377–417.

OLSEN, P. E., KENT, D. V. *ET AL*. 2002. Ascent of Dinosaurs linked to an Iridium anomaly at the Triassic–Jurassic boundary. *Science*, **296**, 1305–1307.

ONCKEN, O. 1997. Transformation of a magmatic arc and orogenic root during the oblique collision and its consequences for the evolution of the European Variscides (Mid-German Crystalline Rise). *Geologische Rundschau*, **86**, 2–21.

OPPEL, A. 1858. *Die Juraformation Englands, Frankreichs und des sudwestlichen Deutschlands*. Stuttgart.

OPPEL, A. 1865. Die Tithonische Etage. *Zeitschrift der Deutschen geologischen Gesellschaft*, **17**, 535–588.

ORBIGNY, A. D'. 1842–1851. *Paléontologie Française: Terrains jurassiques, I. Céphalopodes*. Paris.

ORCHARD, M. J. 1977. Plymouth–Tamar. *In*: HOUSE, M. R., RICHARDSON, J. B., CHALONER, W. G., ALLEN, J. R. L., HOLLAND, C. H. & WESTOLL, T. S. (eds) *A Correlation of Devonian rocks of the British Isles*. Geological Society, London, Special Report, **8**.

ORCHARD, M. J. 1978. The conodont biostratigraphy of the Devonian Plymouth Limestone, south Devon. *Palaeontology*, **21**, 907–955.

ORCHARD, M. J. 1980. Upper Ordovician conodonts from England and Wales. *Geologica et Palaeontologica*, **14**, 9–44.

ORTON, G. 1988. A spectrum of mid-Ordovician fan and braidplain deltaic sequences, North Wales: a consequence of varying fluvial input. *In*: NEMEX, E. & STEEL, R. J. (eds) *Fan Deltas: Tectonic Setting, Sedimentology and Recogition*. Blackie, Glasgow, 23–49.

ORTON, G. J. & READING, H. G. 1993. Variability of deltaic processes in terms of sediment supply, with particular emphasis on grain size. *Sedimentology*, **40**, 475–512.

OTTO, S. C & BAILEY, R. J. 1995. Tectonic evolution of the North Urals Orogen. *Journal of the Geological Society, London*, **152**, 903–906.

OUCHI, S. 1985. Response of alluvial rivers to slow active tectonic movement. *Geological Society of America Bulletin*, **96**, 504–515.

OVER, D. J. & RHODES, M. K. 2000. Conodonts from the Upper Olentangy Shale (Upper Devonian, Central Ohio) and stratigraphy across the Frasnian–Famennian Boundary. *Journal of Palaeontology*, **74**, 101–112.

OVEREEM, I., WELTJE, G. J., BISHOP-KAY, C. & KROONENBERG, S. B. 2001. The Eridanos delta system in the southern North Sea Basin: a climate signal in sediment supply? *Basin Research*, **13**, 293–312.

OVERPECK, J., WHITLOCK, C. & HUNTLEY, B. 2003. Terrestrial biosphere dynamics in the climate system: past and future. *In*: ALVERSON, K. D., BRADLEY, R. S. & PEDERSON, T. F. (eds) *Paleoclimate, Global Change and the Future*. Springer, Berlin, 82–103.

OVEY, C. D. (ed.) 1964. *The Swanscombe Skull: A Survey of Research on a Pleistocene Site*. Royal Anthropological Institute, Occasional Paper, **21**, 1–215.

OWEN, E. F., RAWSON, P. F. & WHITHAM, F. 1968. The Carstone (Lower Cretacous) of Melton, East Yorkshire, and its Brachiopod Fauna. *Proceedings of the Yorkshire Geological Society*, **36**, 513–524.

OWEN, G. 1995. Senni Beds of the Devonian Old Red Sandstone, Dyfed, Wales: anatomy of a semi-arid floodplain. *Sedimentary Geology*, **95**, 221–235.

OWEN, G. & HAWLEY, D. 2000. Depositional setting of the Lower Old Red Sandstone at Pantymaes Quarry, central South Wales: new perspectives on the significance and occurrence of 'Senni Beds' facies. *In*: FRIEND, P. F. & WILLIAMS, B. P. J. (eds) *New Perspectives on the Old Red Sandstone*. Geological Society, London, Special Publications, **180**, 389–400.

OWEN, H. G. 1971. Middle Albian stratigraphy in the Anglo-Paris Basin. *Bulletin of the British Museum (Natural History) Geology*, Supplement 8.

OWEN, H. G. 1972. The Gault and its junction with the Woburn Sands in the Leighton Buxzzard area, Bedfordshire and

Bucklinghamshire. *Proceedings of the Geologists' Association*, **83**, 287–312.

Owen, H. G. 1975. The stratigraphy of the Gault and Upper Greensand of the Weald. *Proceedings of the Geologists' Association*, **86**, 475–498.

Owen, H. G. 1992. The Gault–Lower Greensand Junction Beds in the northern Weald (England) and Wissant (France), and their depositional environment. *Proceedings of the Geologists' Association*, **103**, 93–110.

Owens, B., Neves, R., Gueinn, K. J., Mishell, D. R. F., Sabry, H. S. M. Z. & Williams, J. E. 1977. Palynological division of the Namurian of northern England and Scotland. *Proceedings of the Yorkshire Geological Society*, **41**, 381–398.

Owens, B., Riley, N. J. & Calver, M. A. 1985. Boundary stratotypes and new stage names for the Lower and Middle Westphalian sequences in Britain. *Compte Rendu 10e Congrès Strat. Géol. Carbonifère, Madrid, 1983*, 461–472.

Owens, R. M. 2002. Cyclopygid trilobites from the Ordovician Builth–Llandrindod inlier, central Wales. *Palaeontology*, **45**, 469–485.

Owens, R. M., Fortey, R. A., Cope, J. C. W., Rushton, A. W. A. & Bassett, M. G. 1982. Tremadoc faunas from the Carmarthen district, South Wales. *Geological Magazine*, **119**, 1–38.

Page, K. N. 1989. A stratigraphical revision for the English Lower Callovian. *Proceedings of the Geologists' Association*, **100**, 363–382.

Page, K. N. 1992. The sequence of ammonite correlated horizons in the British Sinemurian (Lower Jurassic). *Newsletters on Stratigraphy*, **27**, 129–156.

Page, K. N. 1995. Biohorizons and zonules: intra-subzonal units in Jurassic ammonite stratigraphy. *Palaeontology*, **38**, 801–814.

Page, K. N. & Bloos, G. 1998. The base of the Jurassic System in west Somerset, south-west England – new observations on the succession of ammonite faunas of the lowest Hettangian Stage. *Geoscience in South-west England*, **9**, 231–235.

Palfy, J. & Smith, P. L. 2000. Synchrony between Early Jurassic extinction, oceanic anoxic event, and the Karoo–Ferrar flood basalt volcanism. *Geology*, **28**, 747–750.

Palmer, S. 1970. The stone age industries of the Isle of Portland, Dorset and the utilisation of Portland Chert as artificial material in southern England. *Proceedings of the Prehistoric Society*, **36**, 82–115.

Palmer, T. J. 2004. What's so good about Portland Stone? *Geoscientist*, **14**, (3), 4–7.

Pamplin, C. F. 1990. A model for the tectono-thermal evolution of north Cornwall. *Proceedings of the Ussher Society*, **7**, 206–211.

Pamplin, C. F. & Andrews, J. R. 1988. Timing and sense of shear in the Padstow Confrontation Zone, north Cornwall. *Proceedings of the Ussher Society*, **7**, 73–76.

Parker, A. G., Goudie, A. S., Andersen, D. E., Robinson, M. A. & Bonsall, C. 2002. A review of the mid-Holocene elm decline in the British Isles. *Progress in Physical Geography*, **26**, 1–45.

Parker-Pearson, M. 1999. The Earlier Bronze Age. *In*: Hunter, J. & Ralston, I. (eds) *The Archaeology of Britain*. Routledge, London, 77–94.

Parks, D. A. & Rendell, H. M. 1992. Thermoluminescence dating and geochemistry of loessic deposits in southeast England. *Journal of Quaternary Science*, **7**, 99–107.

Parrish, J. T. 1993. Climate of the supercontinent Pangaea. *Journal of Geology*, **101**, 215–233.

Parry, C. 1966. *The effects of differential subsidence on the principal coal seams of Ammanian age in the eastern part of the South Wales Coalfield and the factors controlling sedimentation*. MSc thesis, University of Wales, Swansea.

Parsons, C. F. 1974. The sauzei and 'so-called' sowerbyi Zones of the Lower Bajocian. *Newsletters on Stratigraphy*, **3**, 152–179.

Parsons, C. F. 1976. A stratigraphical revision of the humphriesianum-subfurcatum Zone Rocks (Bajocian Stage, Middle Jurassic) of southern England. *Newsletters on Stratigraphy*, **5**, 114–142.

Parsons, C. F. 1979. A stratigraphic revision of the Inferior Oolite of Dundry Hill, Bristol. *Proceedings of the Geologists' Association*, **90**, 133–151, pls 13–14.

Passmore, D. G., Macklin, M. G., Stevenson, A. C., O'Brien, C. F. & Davies, B. A. S. 1992. A Holocene alluvial sequence in the Lower Tyne Valley, northern Britain: a record of river response to environmental change. *The Holocene*, **2**, 138–147.

Patchett, P. J. & Jocelyn, J. 1979. U–Pb zircon ages for late Precambrian igneous rocks in South Wales. *Journal of the Geological Society, London*, **136**, 13–19.

Patchett, P. J., Gale, N. H., Goodwin, R. & Humm, M. J. 1980. Rb–Sr whole rock isochron ages of late Precambrian to Cambrian igneous rocks from southern Britain. *Journal of the Geological Society, London*, **137**, 469–656.

Paton, R. L. 1974. Lower Permian pelycosaurs from the English Midlands. *Palaeontology*, **17**, 541–552.

Pattison, J. 1970. A review of the marine fossils from the Upper Permian rocks of Northern Ireland and north-west England. *Bulletin of the Geological Survey of Great Britain*, **32**, 123–165.

Pattison, J., Smith, D. B. & Warrington, G. 1973. A review of late Permian and early Triassic biostratigraphy in the British Isles. *In*: Logan A. V. & Mills, L. V. (eds) *The Permian and Triassic Systems and Their Mutual Boundary*. Canadian Society of Petroleum Geologists, Memoir, **2**, 220–260.

Paul, C. R. C. & Cope, J. C. W. 1982. A new parablastoid from the Ordovician of South Wales. *Palaeontology*, **25**, 499–507, pl. 51.

Pauley, J. C. 1990. The Longmyndian Supergroup and related Precambrian sediments of England and Wales. *In*: Strachan, R. A. & Taylor, G. K. (eds) *Avalonian and Cadomian Geology of the North Atlantic*. Blackie, Glasgow, 5–27.

Pauley, J. C. 1991. A revision of the stratigraphy of the Longmyndian Supergroup, Welsh Borderland, and its relationship to the Uriconian Volcanic Complex. *Geological Journal*, **26**, 167–183.

Pawley, S. M., Rose, J., Lee, J. R., Moorlock, B. S. P. & Hamblin, R. J. O. 2004. Middle Pleistocene sedimentology and lithostratigraphy of Weybourne, northeast Norfolk, England. *Proceedings of the Geologists' Association*, **115**, 25–42.

Peacock, D. C. P. 2004. The post Variscan development of the British Isles within a regional transfer zone influenced by orogensis. *Journal of Structural Geology*, **26**, 2225–2231.

Peake, N. B. & Hancock, J. M. 1961. The Upper Cretaceous of Norfolk. *Transactions of the Norfolk and Norwich Naturalists' Society*, **19**, 293–339.

Pearson, P. N. & Nicholas, C. J. 1992. Defining the base of the Cambrian: the Hicks–Geikie confrontation of April 1883. *Earth Sciences History*, **11**, 70–80.

Pearson, R. 1995. Liverpool Bay development takes shape in Irish Sea. *Oil and Gas Journal*, **93**, (35), 60–66.

Peat, C. J. 1984. Precambrian microfossils from the Longmyndian of Shropshire. *Proceedings of the Geologists' Association*, **95**, 17–22.

Penn, I. E. 1982. Middle Jurassic stratigraphy and correlation of the Winterborne Kingston borehole. *In*: Rhys, G. H., Lott, G. K. & Calver, M. A. (eds) *The Winterborne Kingston borehole, Dorset, England*. Report of the Institute of Geological Sciences, London, **81/3**, 53–76.

Penn, I. E. 1987. Geophysical logs in the stratigraphy of Wales and adjacent areas. *Proceedings of the Geologists' Association*, **98**, 275–314.

Penn, I. E. & Evans, C. D. R. 1976. *The Middle Jurassic (mainly Bathonian) of Cardigan Bay and its palaeogeographical significance*. Report of the Institute of Geological Sciences, London, **76/6**.

Penn, I. E. & Wyatt, R. J. 1979. *Stratigraphy and correlation of the Bathonian strata in the Bath–Frome area*. Report of the Institute of Geological Sciences, London, **78/22**, 88 pp.

Penn, I. E., Holliday, D. W. *et al.* 1983. The Larne No. 2 Borehole: discovery of a new Permian volcanic centre. *Scottish Journal of Geology*, **19**, 333–346.

Penny, L. F. & Rawson, P. F. 1969. Field meeting in East Yorkshire and North Lincolnshire. *Proceedings of the Geologists' Association*, **80**, 193–218.

Penny, L. F., Coope, G. R. & Catt, J. A. 1969. Age and insect fauna of the Dimlington Silts, East Yorkshire. *Nature*, **224**, 65–67.

PERRIN, R. M. S., ROSE, J. & DAVIES, H. 1979. The distribution, variation and origins of pre-Devensian tills in eastern England. *Philosophical Transactions of the Royal Society of London*, **B287**, 535–570.

PETTERSON, M. G., BEDDOE-STEPHENS, B., MILLWARD, D., JOHNSON, E. W. 1992. The pre-caldera plateau–andesite field in the Borrowdale Volcanic Group of the English Lake District. *Journal of the Geological Society, London*, **149**, 889–906.

PHARAOH, T. C. & CARNEY, J. N. 2000. Introduction to the Precambrian rocks of England and Wales. *In*: CARNEY, J. N., HORÁK, J. M., PHARAOH, T. C., GIBBONS, W., WILSON, D. BARCLAY, W. J. & BEVINS, R. E. (eds) *The Precambrian Rocks of England and Wales*. Joint Nature Conservation Review Committee. Geological Conservation Review Series, **20**, 1–17.

PHARAOH, T. C. & EVANS, C. J. 1987. Morley Quarry No. 1 Borehole: Geological well completion report. *Investigation of the Geothermal Potential of the UK*. British Geological Survey Report, 1–43.

PHARAOH, T. C. & GIBBONS, W. 1994. Precambrian rocks in England and Wales south of the Menai Strait Fault System. *In*: GIBBONS, W. & HARRIS, A. L. (eds) *A Revised Correlation of the Precambrian Rocks in the British Isles*. Geological Society, London, Special Reports, **22**, 85–97.

PHARAOH, T. C., ALLSOP, J. M. *ET AL*. 1997. The Moorby Microgranite: a deformed high level intrusion of Ordovician age in the concealed Caledonian basement of Lincolnshire. *Proceedings of the Yorkshire Geological Society*, **51**, 329–342.

PHARAOH, T. C., BREWER, T. S. & WEBB, P. C. 1993. Subduction-related magmatism of late Ordovician age in eastern England. *Geological Magazine*, **130**, 647–656.

PHAROAH, T. C., ENGLAND, R. & LEE, M. 1995. The concealed Caledonide basement of eastern England and the southern North Sea – a review. *Studia Geophysica et Geodaetica*, **39**, 330–346.

PHARAOH, T. C., MERRIMAN, R. J., EVANS, J. A., BREWER, T. S., WEBB, P. C. & SMITH, N. J. P. 1991. Early Palaeozoic arc-related volcanism in the concealed Caledonides of Southern Britain. *Annales de la Société Géologique de Belgique*, **114**, 63–91.

PHARAOH, T. C., MERRIMAN, R. J., WEBB, P. C & BECKINSDALE, R. D. 1987*a*. The concealed Caledonides of eastern England; preliminary results of a multidisciplinary study. *Proceedings of the Yorkshire Geological Society*, **46**, 355–369.

PHARAOH, T. C., WEBB, P. C., THORPE, R. S. & BECKINSDALE, R. D. 1987*b*. Geochemical evidence for the tectonic setting of late Proterozoic volcanic suites in central England. *In*: PHARAOH, T. C., BECKINSDALE, R. D. & RICKARD, D. (eds) *Geochemistry and Mineralization of Proterozoic Volcanic Suites*. Geological Society, London, Special Publications, **33**, 541–552.

PHILLIPS, E. 1991*a*. Progressive deformation of the South Stack and New Harbour Groups, Holy Island, western Anglesey, North Wales. *Journal of the Geological Society, London*, **148**, 1091–1100.

PHILLIPS, E. 1991*b*. The lithostratigraphy, sedimentology and tectonic setting of the Monian Supergroup, western Anglesey, North Wales. *Journal of the Geological Society, London*, **148**, 1079–1090.

PHILLIPS, F. C. 1964. Metamorphism in South-west England. *In*: HOSKING K. F. G. & SHRIMPTON G. H. (eds) *Present Views on some Aspects of the Geology of Cornwall and Devon*. Royal Geological Society of Cornwall, Truro, 185–200.

PHILLIPS, E. 1989. *The geology of the Monian Supergroup, Western Anglesey, North Wales*. PhD thesis, University of Wales.

PHILLIPS, J. 1835. *Guide to Geology*, 2nd edn. London.

PHILLIPS, J. 1836. *Illustrations of the Geology of Yorkshire. Part 2 – The Mountain Limestone District*. Murray, London.

PHILLIPS, J. 1871. *Geology of Oxford and the Valley of the Thames*.

PHILLIPS, L. 1976. Pleistocene vegetational history and geology in Norfolk. *Philosophical Transactions of the Royal Society of London*, **B275**, 215–286.

PHILLIPS, W. E. A., STILLMAN, C. J. & MURPHY, T. 1976. A Caledonian plate tectonic model. *Journal of the Geological Society, London*, **132**, 579–609.

PHIPPS, C. B. & REEVE, F. A. E. 1967. Stratigraphy and geological history of the Malvern, Abberley and Ledbury Hills. *Geological Journal*, **5**, 339–368.

PICK, M. C. 1964. The stratigraphy and sedimentary features of the Old Red Sandstone, Portishead coastal section, northeast Somerset. *Proceedings of the Geologists' Association*, **75**, 199–221.

PICKERILL, R. K. & BRENCHLEY, P. J. 1979. Caradoc marine benthic communities of the south Berwyn Hills, North Wales. *Palaeontology*, **22**, 229–264.

PICKERING, K. T. 1995. Are enigmatic erosional sandy wave-like bedforms in Jurassic Bridport Sands, Dorset, due to standing waves? *Journal of the Geological Society, London*, **152**, 481–485.

PICKERING, K. T. & SMITH, A. G. 1995. Arcs and backarc basins in the Early Paleozoic Iapetus Ocean. *The Island Arc*, **4**, 1–67.

PIDGEON, R. G. & AFTALION, M. 1978. Cogenetic and inherited zircon U–Pb systems in granites of Scotland and England. *In*: BOWES, D. R. & LEAKE, B. E. (eds) *Crustal Evolution in Northwestern Britain and Adjacent Regions*. *Geological Journal*, Special Issue, **10**, 183–220.

PILCHER, J. R. & HALL, V. A. 1996. Tephrochronological studies in northern England. *The Holocene*, **6**, 100–105.

PIPER, J. D. A. 1997. Tectonic rotation within the British paratectonic Caledonides and Early Palaeozoic location of the orogen. *Journal of the Geological Society, London*, **154**, 9–13.

PIPER, J. D. A. & CROWLEY, S. F. 1999. Palaeomagnetism of (Palaeozoic) Peel Sandstones and Langness Conglomerate Formation, Isle of Man: implications for the age and regional diagenesis of Manx red beds. *In*: WOODCOCK, N. H., QUIRK, D. G., FITCHES, W. R. & BARNES, R. P. (eds) *In Sight of the Suture: The Palaeozoic Geology of the Isle of Man in its Iapetus Ocean Context*. Geological Society, London, Special Publications, **160**, 213–226.

PIPER, J. D. A., STEPHEN J. C. & BRANNEY, M. J. 1997. Palaeomagnetism of the Borrowdale and Eycott Volcanic Groups, English Lake District: primary and secondary magnetisation during a single Upper Ordovician polarity chron. *Geological Magazine*, **134**, 481–506.

PIQUÉ, A. 1982. Northwestern Africa and the Avalon plate: Relations during late Precambrian and Late Palaeozoic time. *Geology*, **9**, 319–322.

PLINT, A. G. 1982. Eocene sedimentation and tectonics in the Hampshire Basin. *Journal of the Geological Society, London*, **139**, 239–254.

PLINT, A. G. 1983. Facies, environments and sedimentary cycles in the Middle Eocene, Bracklesham Formation of the Hampshire Basin: evidence for global sea-level changes? *Sedimentology*, **30**, 625–653.

PLINT, A. G. 1988*a*. Global eustacy and the Eocene sequence in the Hampshire Basin, England. *Basin Research*, **1**, 11–22.

PLINT, A. G. 1988*b*. Sedimentology of the Eocene Strata exposed between Poole Harbour and High Cliff, Dorset, UK. *Tertiary Research*, **10**, 107–145.

PLINT, A. G. & PICKERILL, R. K. 1985. Non-marine *Teredolites* from the Middle Eocene of southern England. *Lethaia*, **18**, 341–347.

PODLAHA, O. G., MUTTERLOSE, J. & VEIZER, J. 1998. Preservation of $\delta^{18}O$ and $\delta^{13}C$ in belemnite rostra from the Jurassic/early Cretaceous successions. *American Journal of Science*, **298**, 324–347.

POMEROL, C. 1982. *The Cenozoic Era: Tertiary and Quaternary*. Ellis Horwood, Chichester.

POOLE, E. G. 1969. The stratigraphy of the Geological Survey Apley Barn Borehole, Witney, Oxfordshire. *Bulletin of the Geological Survey of Great Britain*, **29**, 1–66.

POOLE, E. G. 1978. Stratigraphy of the Withycombe Farm borehole, near Banbury, Oxfordshire. *Bulletin of the Geological Survey of Great Britain*, 68.

POOLE, E. G. & WHITEMAN, A. J. 1966. *Geology of the country around Nantwich and Whitchurch*. Memoir of the British Geological Survey, Sheet 122 (England and Wales). HMSO, London.

POOLE, E. G., KELK, B., BAIN, J. A. & MORGAN, D. J. 1971. *Calcium montmorillonite (fullers' earth) in the Lower Greensand of the Farnham area, Berkshire*. Report of the Institute of Geological Sciences, **77/12**.

POSAMENTIER, H. W. & VAIL, P. R. 1988. Eustatic controls on clastic deposition II-sequence and tract models. *In*: WILGNUS, C. K., HASTINGS, C. G. S. C., KENDALL, H. W., POSAMENTIER, H. W.,

Ross, C. A. & Van Wagoner, J. C. (eds) *Sea-level Changes: An Integreated Approach*. SEPM, Special Publications, **42**, 124–154.

Potter, J. F. & Price, J. H. 1965. Comparative sections through rocks of Ludlovian–Downtonian age in the Llandovery and Llandeilo districts. *Proceedings of the Geologists' Association*, **76**, 379–402.

Potts, A. S. 1971. Fossil cryonival features in central Wales. *Geografiska Annaler*, **53A**, 39–51.

Pound. C. J. 1983. The sedimentology of the Lower–Middle Devonian Staddon Grits and Jennycliff Slates on the eastern side of Plymouth Sound, Devon. *Proceedings of the Ussher Society*, **5**, 465–472.

Powell, A. J. 1992. Dinoflagellate cysts of the Tertiary System. *In*: Powell, A. J. (ed.) *A Stratigraphic Index of Dinoflagellate Cysts*. Chapman & Hall, London, 155–251.

Powell, A. J., Brinkhuis, H. & Bujak, J. P. 1996. Upper Paleocene–Lower Eocene dinoflagellate sequence biostratigraphy of south-east England. *In*: Knox, R. W. O'B., Corfield, R. M. & Dunay, R. E. (eds) *Correlation of the Early Paleogene in Northwest Europe*. Geological Society, London, Special Publications, **101**, 145–183.

Powell, C. M. 1987. Inversion tectonics in southwest Dyfed. *Proceedings of the Geologists' Association*, **98**, 193–203.

Powell, C. M. 1989. Structural controls on Palaeozoic basin evolution and inversion in southwest Wales. *Journal of the Geological Society, London*, **146**, 439–446.

Powell, J. H., Chisholm, J. I., Bridge, D. M., Rees, J G, Glover, B. W. & Besly, B. M. 2000*a*. *Stratigraphical framework for Westphalian to Early Permian red-bed successions of the Pennine Basin*. British Geological Survey Technical Report, **WA/99/101**.

Powell, J. H., Glover, B. W. & Waters, C. N. 2000*b*. *Geology of the Birmingham Area*. Memoir of the British Geological Survey, Sheet 168, England and Wales. HMSO, London.

Powell, T., Salmon, S., Clark, A. H. & Shail, R. K. 1999. Emplacement styles within the Land's End Granite, West Cornwall. *Geoscience in South-west England*, **9**, 333–339.

Power, G. M. & Barnes, R. P. 1999. Relationships between metamorphism and structure on the northern edge of Eastern Avalonia in the Manx Group, Isle of Man. *In*: Woodcock, N. H., Quirk, D. G., Fitches, W. R. & Barnes, R. P. (eds) *In Sight of the Suture: The Palaeozoic Geology of the Isle of Man in its Iapetus Ocean Context*. Geological Society, London, Special Publications, **160**, 289–305.

Power, M. R., Alexander, A. C., Shail, R. K. & Scott, P. W. 1996. A re-interpretation of the internal structure of the Lizard complex ophiolite, south Cornwall. *Proceedings of the Ussher Society*, **9**, 63–67.

Pratt, W. T. 1992. The use of kink bands to constrain fault displacements: an example from the Bala Lineament, Wales. *Geological Magazine*, **129**, 625–632.

Pratt, W. T. & Fitches, W. R. 1993. Transected folds from the western part of the Bala Lineament, Wales. *Journal of Structural Geology*, **15**, 55–68.

Pratt, W. T., Woodall, D. G. & Howells, M. F. 1995. Geology of the country around Cadair Idris. *Memoir of The British Geological Survey*, Sheet **149** (England and Wales). HMSO, London.

Preece, R. C. 1994. Radiocarbon dates from the 'Allerød soil' in Kent. *Proceedings of the Geologists' Association*, **105**, 111–123.

Preece, R. C. 2001. Molluscan evidence for differentiation of interglacials within the 'Cromerian Complex'. *Quaternary Science Reviews*, **20**, 1643–1656.

Prentice, J. E. 1960*a* (for 1959). Dinantian, Namurian and Westphalian rocks of the district south-west of Barnstaple, north Devon. *Quarterly Journal of the Geological Society of London*, **115**, 261–289.

Prentice, J. E. 1960*b*. The stratigraphy of the Upper Carboniferous rocks of the Bideford region, north Devon. *Quarterly Journal of the Geological Society of London*, **116**, 397–408.

Prentice, J. E. 1962. The sedimentation history of the Carboniferous in Devon. *In*: Coe, K. (ed.) *Some Aspects of the Variscan Fold Belt*. Exeter University, Manchester University Press, 93–108.

Prestwich, J. 1852. On the structure of the strata between the London Clay and the Chalk in the London and Hampshire Tertiary Systems. *Quarterly Journal of the Geological Society of London*, **8**, 235–264.

Prestwich, J. 1890. On the relation of the Westleton Beds, or pebbly sands of Suffolk, to those of Norfolk, and their extension inland. *Quarterly Journal of the Geological Society of London*, **46**, 84–154.

Price, D. 1973. The age and stratigraphy of the Sholeshook Limestone of southwest Wales. *Geological Journal*, **8**, 225–46.

Price, D. 1977. Species of *Tretaspis* (Trilobita) from the Ashgill Series in Wales. *Palaeontology*, **20**, 763–92.

Price, D. 1980*a*. The Ordovician trilobite fauna of the Sholeshook Limestone Formation, South Wales. *Palaeontology*, **23**, 839–87.

Price, D. 1980*b*. A revised age and correlation for the topmost Sholeshook Limestone Formation (Ashgill) of South Wales. *Geological Magazine*, **117**, 485–9.

Price, D. 1984. The Pusgillian Stage in Wales. *Geological Magazine*, **121**, 99–105.

Price, D. & Magor, P. M. 1984. The ecological significance of variation in generic composition of Rawtheyan (late Ordovician) trilobite faunas from North Wales, UK. *Geological Journal*, **19**, 167–200.

Price, G. D. 1996. Significance of infiltrated clays within the Lower Permian Yellow Sands of north-east England. *Geological Journal*, **31**, 189–195.

Price, G. D. 1999. The evidence and implications of polar ice during the Mesozoic. *Earth Science Reviews*, **48**, 183–210.

Prigmore, J. K., Butler, A. J. & Woodcock, N. H. 1997. Rifting during separation of Eastern Avalonia from Gondwana: Evidence from subsidence analysis. *Geology*, **25**, 203–206.

Primmer, T. J. 1985. The pressure–temperature history of the Tintagel district, Cornwall: metamorphic evidence on the evolution of the area. *Proceedings of the Ussher Society*, **6**, 218–223.

Prothero, D. R. &. Berggren, W. A. (eds). 1992. *Eocene–Oligocene Climatic and Biotic Evolution*. Princeton University Press, Princton, NJ.

Pugh, W. J. 1923. The geology of the district around Corris and Aberllefenni (Merionethshire). *Quarterly Journal of the Geological Society of London*, **79**, 508–41.

Pugh, W. J. 1928. The geology of the district around Dinas Mawddwy (Merioneth). *Quarterly Journal of the Geological Society of London*, **84**, 345–81.

Pugh, W. J. 1929. The geology of the district between Llanymawddwy and Llanwchllyn (Merioneth). *Quarterly Journal of the Geological Society of London*, **85**, 242–306.

Purvis, K. & Wright, V. P. 1991. Calcretes related to phreatophytic vegetation from the Middle Triassic Otter Sandstone of south-western England. *Sedimentology*, **38**, 539–551.

Pye, K. & Krinsley, D. 1986. Microfabric, mineralogy and early diagenetic history of the Whitby Mudstone Formation (Toarcian), Cleveland Basin, UK. *Geological Magazine*, **123**, 191–203.

Pyle, D. 1989. The thickness, volume and grainsize of tephra-fall deposits. *Bulletin of Volcanology*, **51**, 1–15.

Quirk, D. G., Burnett, D. J., Kimbell, G. S., Murphy, C. A. & Varley, J. S. 1999. Shallow geophysical and geological evidence for a regional-scale fault duplex in the Lower Palaeozoic of the Isle of Man. *In*: Woodcock, N. H., Quirk, D. G., Fitches, W. R. & Barnes, R. P. (eds) *In Sight of the Suture: The Palaeozoic Geology of the Isle of Man in its Iapetus Ocean Context*. Geological Society, London, Special Publications, **160**, 239–257.

Radley, J. D. 1992. Halite pseudomorphs from the Cherty Freshwater Member (Purbeck Limestone Formation) of Lulworth Cove, Dorset. *Proceedings of the Dorset Natural History and Archaeological Society*, **113**, 195–196.

Radley, J. D. 1994. Stratigraphy, palaeontology and palaeoenvironment of the Wessex Formation (Wealden Group, Lower Cretaceous) at Yaverland, Isle of Wight, southern England. *Proceedings of the Geologists' Association*, **105**, 199–208.

Raiswell, R. 1988. Chemical model for the origin of minor limestone-shale cycles by anaerobic methane oxidation. *Geology*, **16**, 641–644.

RAMSAY, J. G. 1962. The geometry and mechanics of 'similar' type folds. *Journal of Geology*, **70**, 309–327.

RAMSBOTTOM, W. H. C. 1969. The Namurian of Britain. *Compte Rendue 6 ème Congrès International de Stratigraphie et de Geologie du Carbonifère, Sheffield, 1967*.

RAMSBOTTOM, W. H. C. 1971. Palaeogeography and goniatite palaeogeography in the Namurian and early Westphalian. *Compte Rendue, 6 ème Congrès International de Stratigraphie et de Geologie du Carbonifère, Sheffield, 1967*, 1396–1399.

RAMSBOTTOM, W. H. C. 1973. Transgressions and regression in the Dinantian: a new synthesis of British Dinantian stratigraphy. *Proceedings of the Yorkshire Geological Society*, **39**, 567–607.

RAMSBOTTOM, W. H. C. 1977. Major cycles of transgression and regression (mesothems) in the Namurian. *Proceedings of the Yorkshire Geological Society*, **41**, 261–291.

RAMSBOTTOM, W. H. C. 1978. Namurian mesothems in South Wales and northern France. *Journal of the Geological Society, London*, **135**, 307–312.

RAMSBOTTOM, W. H. C. 1979. Rates of transgression and regression in the Carboniferous of NW Europe. *Quarterly Journal of the Geological Society of London*, **136**, 147–154.

RAMSBOTTOM, W. H. C. 1981. Fieldguide to the boundary stratotypes of the Carboniferous stages in Britain. Biennial Meeting of the Subcommisson of Carboniferous Stratigraphy, Leeds.

RAMSBOTTOM, W. H. C. & MITCHELL, M. 1980. The Recognition of the Tournaisian Series in Britain. *Journal of the Geological Society, London*, **137**, 61–63.

RAMSBOTTOM, W. H. C., RHYS, G. H. & SMITH, E. G. 1962. Boreholes in the Carboniferous rocks of the Ashover district, Derbyshire. *Bulletin of the Geological Survey of Great Britain*, **19**, 75–168.

RAMSBOTTOM, W. H. C., CALVER, M. A., EAGAR, R. M. C., HODSON, F., HOLLIDAY, D. W., STUBBLEFIELD, C. J. & WILSON, R. B. 1978. *A correlation of Silesian rocks in the British Isles*. Geological Society, London, Special Reports **10**.

RASTALL, R. H. 1927. The Underground Structure of Eastern England. *Geological Magazine*, **64**, 10–26.

RASTALL, R. H. & ROMANES, J. 1909. On the boulders of the Cambridge drift: their distribution and origin. *Quarterly Journal of the Geological Society of London*, **65**, 246–264.

RAUP, D. M. & SEPKOWSKI JR, J. J. 1984. Periodicity in extinctions in the geologic past. *Proceedings National Academy Science, USA*, **81**, 801–805.

RAWSON, P. F. 1981. Early Cretaceous ammonite biostratigraphy and biogeography. *In*: HOUSE, M. R. & SENIOR, J. R. (eds) *The Ammonoidea. Systematics Association*, Special Volume, **18**, 499–529

RAWSON, P. F. 1992*a*. The Cretaceous. *In*: DUFF, P. MCL. D. & SMITH, A. J. (eds) *Geology of England and Wales*. 1st edn. Geological Society, London, 355–388.

RAWSON, P. F. 1992*b*. Early Cretaceous. *In*: COPE, J. C. W., INGHAM, J. K. & RAWSON, P. F. (eds) *Atlas of Palaeogeography and Lithofacies*. Geological Society, London, Memoirs, **13**, 131–137.

RAWSON, P. F. 1995. The "Boreal" Early Cretaceous (Pre-Aptian) ammonite sequences of NW Europe and their correlation with the Western Mediterranean faunas. *Memorie Descrittive della Carta Geologica d'Italia*, **51**, 121–130.

RAWSON, P. F. & MUTTERLOSE, J. 1983. Stratigraphy of the Lower B and basal Cement Beds (Barremian) of the Speeton Clay, Yorkshire, England. *Proceedings of the Geologists' Association*, **94**, 133–146.

RAWSON, P. F. & RILEY, L. A. 1982. Latest Jurassic–Early Cretaceous events and the 'Late Cimmerian Unconformity' in the North Sea area. *AAPG Bulletin*, **66**, 2628–2648.

RAWSON, P. F. & WRIGHT, J. K. 1992. *The Yorkshire Coast*. Geologists' Association, Field Guide, **34**.

RAWSON, P. F. & WRIGHT, J. K. 1995. Jurassic of the Cleveland Basin, North Yorkshire. *In*: TAYLOR, P. D. (ed.) *Field Geology of the British Jurassic*. Geological Society, London, 173–208.

RAWSON, P. F., ALLEN, P. & GALE, A. 2001. The Chalk Group: a revised lithostratigraphy. *Geoscientist*, **11**, (1), 21.

RAWSON, P. F., ALLEN, P. M. *ET AL*. 2002. *Stratigraphical Procedure*. Geological Society, London, Professional Handbook.

RAWSON, P. F., CURRY, D. *ET AL*. 1978. *A correlation of Cretaceous rocks in the British Isles*. Geological Society, London, Special Reports, *9*.

RAWSON, P. F., DHONDT, A. V., HANCOCK, J. M. & KENNEDY, W. J. (eds) 1996. Proceedings 'Second International Symposium on Cretaceous Stage Boundaries' Brussels 8–16 September 1995. *Bulletin de l'Institut Royal Des Sciences Naturelles de Belgique*, **66** (Supplement).

RAWSON, P. F., GREENSMITH, J. T. & SHALABY, S. E. 1983. Coarsening upward cycles in the uppermost Staithes and Cleveland Ironstone formations (Lower Jurassic) of the Yorkshire coast, England. *Proceedings of the Geologists' Association*, **94**, 91–93.

RAYMOND, A., KELLEY, P. H. & LUTKEN, C. B. 1989. Polar glaciers and life at the Equator – the history of Dinantian and Namurian (Carboniferous) climate. *Geology*, **17**, 408–411.

RAYNAUD, D., BLUNIER, T., ONO. Y. & DELMAS, R. J. 2003. The late Quaternary history of atmospheric trace gases and aerosols: interactions between climate and biogeochemical cycles. *In*: ALVERSON, K. D., BRADLEY, R. S. & PEDERSON, T. F. (eds) *Paleoclimate, Global Change and the Future*. Springer, Berlin, 13–31.

READ, W. A. 1991. The Millstone Grit (Namurian) of the southern Pennines viewed in the light of eustatically controlled sequence stratigraphy. *Geological Journal*, **26**, 157–166.

READING. H. G. 1965. Recent finds in the Upper Carboniferous of south-west England and their significance. *Nature*, **208**, 745–748.

REEDMAN, A. J., LEVERIDGE, B. E. & EVANS, R. B. 1984. The Arfon Group ('Arvonian') of North Wales. *Proceedings of the Geologists' Association*, **95**, 313–321.

REES, J. G. & WILSON, A. A. 1998. *Geology of the Country Around Stoke on Trent*. Memoir of the British Geological Survey Sheet 123 (England & Wales). HMSO, London.

REIMER, P. J., BAILLIE, M. G. L. *ET AL*. 2004. IntCal04: terrestrial radiocarbon calibration 26-0ka BP. *Radiocarbon*, **46**, 1029–1058.

REYNAUD, J. Y., TESSIER, B., AUFFRET, J. P., BERNÉ, D., DE BATIST, M., MARSSET, T. & WALKER, P. 2003. The offshore Quaternary sediment bodies of the English Channel and its Western Approaches. *Journal of Quaternary Science*, **18**, 361–371.

RHYS, G. H. (ed.) 1974. *A proposed standard lithostratigraphic nomenclature for the southern North Sea and an outline structural nomenclature for the whole of the (UK) North Sea*. Report of the Institute of Geological Sciences, **74/8**.

RICE, R. J. & DOUGLAS, T. 1991. Wolstonian glacial deposits and glaciation in Britain. *In*: EHLERS, J., GIBBARD, P. L. & ROSE, J. (eds) *Glacial Deposits in Great Britain and Ireland*. Balkema, Rotterdam, 25–35.

RICHARDS, A. E. 1999. Middle Pleistocene glaciation in Herefordshire: sedimentological and structural evidence from the Risbury Formation. *Proceedings of the Geologists' Association*, **110**, 173–192.

RICHARDSON, J. B. & EDWARDS, D. 1989. Sporomorphs and plant megafossils. *In*: HOLLAND C. H. & BASSETT, M. G. (eds) *A Global Standard for the Silurian System*. National Museum of Wales, Cardiff, Geological Series, **9**, 216–226.

RICHARDSON, J. B. & MCGREGOR, D. C. 1986. Silurian and Devonian spore zones of the Old Red Sandstone continent and adjacent regions. *Bulletin of the Geological Survey of Canada*, **364**, 1–79.

RICHARDSON, J. B. & RASUL, S. M. 1990. Palynofacies in a Late Silurian regressive sequence in the Welsh Borderland and Wales. *Journal of the Geological Society, London*, **147**, 675–686.

RICHARDSON, J. B., RODRIGUEZ, R. M. & SUTHERLAND, S. J. E. 2000. Palynology and recognition of the Silurian/Devonian boundary in some British terrestrial sediments by correlation with Cantabrian and other European marine sequences – a progress report. *Courier Forschungsistitut Senckenberg*, **220**, 1–7.

RICHARDSON, L. 1911. The Rhaetic and continuous deposits of west, mid and part of east Somerset. *Quarterly Journal of the Geological Society of London*, **67**, 1–74.

RICHMOND, G. M. 1996. The INQUA-approved provisional Lower–Middle Pleistocene boundary. *In*: TURNER, C. (ed.) *The Early Middle Pleistocene in Europe*. Balkema, Rotterdam, 319–327.

Richmond, L. K. & Williams, B. P. J. 2000. A new terrane in the Old Red Sandstone of the Dingle Peninsula, SW Ireland. *In*: Friend, P. F. & Williams, B. P. J. (eds) *New Perspectives on the Old Red Sandstone*. Geological Society, London, Special Publications, **180**, 147–183.

Richter, D. 1969. Structure and metamorphism of the Devonian rocks south of Torquay, south-east Devon (England). *Geologische Mittelungen*, **9**, 109–178.

Rickards, R. B. 1964. The graptolitic mudstone and associated facies in the Silurian strata of the Howgill Fells. *Geological Magazine*, **101**, 435–451.

Rickards, R. B. 1973. On some highest Llandovery red beds and graptolite assemblages in Britain and Eire. *Geological Magazine*, **110**, 70–72.

Rickards, R. B. 1976. The sequence of Silurian graptolite zones in the British Isles. *Geological Journal*, **11**, 153–187.

Rickards, R. B. 1978. Silurian. *In*: Moseley, F. (ed.) *The Geology of the Lake District*. Yorkshire Geological Society, 130–145.

Rickards, R. B. 1989. Northern England. *In*: Holland, C. H. & Bassett, M. G. (eds) *A Global Standard for the Silurian System*. National Museum of Wales, Cardiff, 267–274.

Rickards, R. B. 2002. The graptolitic age of the type Ashgill Series (Ordovician), Cumbria, UK. *Proceedings of the Yorkshire Geological Society*, **54**, 1–16.

Rickards, R. B. & Woodcock, N. H. 2005. Stratigraphic revision of the Windermere Supergroup (Late Ordovician–Silurian) in the southern Howgill Fells NW England. *Proceedings of the Yorkshire Geological Society*, **55**, 263–285.

Ridgeway, J. 1975. The stratigraphy of Ordovician volcanic rocks around south and east of Harlech Dome in Merionethshire, North Wales. *Geological Journal*, **10**, 87–106.

Riding, J. B. & Wright, J. K. 1989. Palynostratigraphy of the Scalby Formation (Middle Jurassic) of the Cleveland Basin, north-east Yorkshire. *Proceedings of the Yorkshire Geological Society*, **47**, 31–38.

Riley, N. J. 1990. Stratigraphy of the Worston Shale Group (Dinantian) Craven Basin, north-west England. *Proceedings of the Yorkshire Geological Society*, **48**, 163–187.

Riley, N. J. 1993. Dinantian (Lower Carboniferous) biostratigraphy and chronostratigraphy in the British Isles. *Journal of the Geological Society, London*, **150**, 427–446.

Riley, N. J., Claoué-Long, J. C., Higgins, A. C., Owens, B., Spears, A., Taylor, L. & Varker, W. J. 1995. Geochronometry and geochemistry of the European mid-Carboniferous boundary global stratotype proposal, Stonehead Beck, North Yorkshire, UK. *Annales de la Société géologique de Belgique*, **116**, 275–289.

Rippon, J. H. 1996. Sand body orientation, palaeoslope analysis and basin-fill implications in the Westphalian A–C of Great Britain. *Journal of the Geological Society, London*, **153**, 881–900.

Ritchie, J. D. & Hitchen, K. 1996. Early Paleogene offshore igneous activity to the northwest of the UK margin and its relationship to the North Atlantic Igneous Province. *In*: Knox, R. W. O'B., Corfield, R. M. & Dunay, R. E. (eds) *Correlation of the Early Paleogene in Northwest Europe*. Geological Society, London, Special Publications, **101**, 63–78.

Ritchie, J. D., Gatcliff, R. W. & Riding, J. B. 1996. Pre-Tertiary lithostratigraphy. *In: Stratigraphic Nomenclature of the UK North West Margin*. British Geological Survey. HMSO, London.

Ritchie, J. S. & Pratsides, P. 1993. The Caister Fields, Block 44/23a, UK North Sea. *In*: Parker, J. R. (ed.) *Petroleum Geology of Northwest Europe: Proceedings of the 4th Conference, Volume 1*. Geological Society, London, 759–769.

Ritchie, J. D., Gatliff, R. W. & Richards, P. C. 1999. Early Tertiary magmatism in the offshore NW UK margin and surrounds. *In*: Fleet, A. J. & Boldy, S. A. R. (eds) *Petroleum Geology of Northwest Europe*. Proceedings of the 5th Conference, Geological Society, London, 573–584.

Riveline, J. 1984. Les gisements à charophytes du Cenozoique (Danien à Burdigalien) d' Europe occidentale: Lithostratigraphie, Biostratigraphie, Chronostratigraphie. *Bulletin d'information des Géologues du Bassin de Paris*, Memoire hors serie 4.

Robaszynski, F. & Caron, M. 1995. Foraminifères planctoniques du Crétacé: commentaire de la zonation Europe-Méditerranée. *Bulletin de la Société géologique de la France*, **166**, 681–692.

Roberts, B. 1981. Low grade and very low grade regional metabasic Ordovician rocks of Llyn and Snowdonia, Gwynedd, North Wales. *Geological Magazine*, **118**, 189–200.

Roberts, B. & Merriman, R. J. 1985. The distinction between Caledonian burial and regional metamorphism in metapelites from North Wales: an analysis of isocryst patterns. *Journal of the Geological Society, London*, **142**, 615–624.

Roberts, B. & Merriman, R. J. 1990. Cambrian and Ordovician metabentonites and their relevance to the origins of associated mudrocks in the northern sector of the Lower Palaeozoic Welsh marginal basin. *Geological Magazine*, **127**, 31–43.

Roberts, B., Evans, J. A., Merriman, R. J. & Smith, M. 1989. Discussion on low grade metamorphism of the Welsh Basin Lower Palaeozoic succession: an example of diastathermal metamorphism. *Journal of the Geological Society, London*, **146**, 885–890.

Roberts, B., Merriman, R. J., Hirons, S. R., Fletcher, C. J. N. & Wilson, D. 1996. Synchronous very low-grade metamorphism, contraction and inversion in the central part of the Welsh Lower Palaeozoic Basin. *Journal of the Geological Society, London*, **153**, 277–285.

Roberts, B., Merriman, R. J. & Pratt, W. 1991. The influence of strain, lithology and stratigraphical depth in white mica (illite) crystallinity in mudrocks from the vicinity of the Corris Slate Belt, Wales: the implications for the timing of metamorphism in the Welsh Basin. *Geological Magazine*, **128**, 633–645.

Roberts, B., Morrison, C. & Hirons, S. 1990. Low grade metamorphism of the Manx Group, Isle of Man: a comparative study of white mica 'crystallinity' techniques. *Journal of the Geological Society, London*, **147**, 271–277.

Roberts, D. E. 1992. River Caldew and Raven Crags, Mungrisdale. *In*: Treagus, J. (ed.) *Caledonian Structures In Britain, South of the Midland Valley*. Geological Conservation Review Series, 3, Chapman & Hall, London, for Nature Conservancy Council, 67–73.

Roberts, J. L. & Sanderson. 1971. Polyphase development of slaty cleavage and the confrontation of facing directions in the Devonian rocks of North Cornwall. *Nature*, **230**, 87–89.

Roberts, M. B. & Parfitt, S. A. 1999. *Boxgrove. A Middle Pleistocene hominid site at Eartham Quarry, Boxgrove, West Sussex*. English Heritage Archaeological Report, **17**. English Heritage, London.

Roberts, N. 1998. *The Holocene*, 2nd edn. Blackwell, Oxford.

Robinson, D. & Bevins, R. E. 1986. Incipient metamorphism in the Lower Palaeozoic marginal basin of Wales. *Journal of Metamorphic Geology*, **4**, 101–113.

Robinson, D., Reverdatto, V. V., Bevins, R. E., Polyansky, O. P. & Sheplev, V. S. 1999. Thermal modeling of convergent and extensional tectonic settings for the development of low-grade metamorphism in the Welsh Basin. *Journal of Geophysical Research*, **B104**, 23069–23079.

Robinson, S. A. & Hesselbo, S. P. 2004. Fossil-wood carbon-isotope stratigraphy of the non-marine Wealden Group (Lower Cretaceous, southern England). *Journal of the Geological Society, London*, **161**, 133–145.

Rock, N. M. S., Gaskarth, J. W. & Rundle, C. C. 1986. Late Caledonian dyke-swarms in Southern Scotland: a regional zone of primitive K-rich lamprophyres and associated vents. *Journal of Geology*, **94**, 505–522.

Roe, H. M. 1999. Late Middle Pleistocene sea-level change in the southern North Sea: the record from eastern Essex, UK. *Quaternary International*, **55**, 115–128.

Rogers, I. 1919. Fossil fishes in the Devonian of north Devon. *Geological Magazine*, **56**, 100–101.

Rollin, K. E. 1986. Geophysical surveys on the Lizard Complex, Cornwall. *Journal of the Geological Society, London*, **143**, 437–446.

Romano, M. & Whyte, M. A. 2003. Jurassic dinosaur tracks and trackways of the Cleveland Basin, Yorkshire: preservation, diversity and distribution. *Proceedings of the Yorkshire Geological Society*, **54**, 185–215.

Rose, J. 1987. Status of the Wolstonian glaciation in the British Quaternary. *Quaternary Newsletter*, **53**, 1–9.

Rose, J. 1989. Stadial type sections in the British Quaternary. *In*: Rose, J. & Schlüchter, C. (eds) *Quaternary Type Sections; Imagination or Reality*. Balkema, Rotterdam, 45–67.

Rose, J. 1994. Major river systems of central and southern Britain during the Early and Middle Pleistocene. *Terra Nova*, **6**, 435–443.

Rose, J. 1995. Lateglacial and early Holocene river activity in lowland Britain. *Paläoclimatoforschung*, **14**, 51–74.

Rose, J. & Allen, P. 1977. Middle Pleistocene stratigraphy in southeast Suffolk. *Journal of the Geological Society, London*, **133**, 83–102.

Rose, J., Allen, P., Kemp, R. A., Whiteman, C. A. & Owen, N. 1985. The early Anglian Barham Soil of eastern England. *In*: Boardman, J. (ed.) *Soils and Quaternary Landscape Evolution*. Wiley, Chichester, 197–229.

Rose, J., Candy, I., Moorlock, B. S. P., Wilkins, H., Lee, J. A., Hamblin, R. J. O., Lee, J. R., Riding, J. B. & Morigi, A. N. 2002. Early and Middle Pleistocene river, coastal and neotectonic processes, southeast Norfolk, England. *Proceedings of the Geologists' Association*, **113**, 47–67.

Rose, J., Lee, J. A., Candy, I. & Lewis, S. G. 1999*a*. Early and Middle Pleistocene river systems in eastern England: evidence from Leet Hill, southern Norfolk, England. *Journal of Quaternary Science*, **14**, 347–360.

Rose, J., Lee, J. A., Kemp, R. A. & Harding, P. A. 2000. Palaeoclimate, sedimentation and soil development during the Last Glacial Stage (Devensian), Heathrow Airport, London, UK. *Quaternary Science Reviews*, **19**, 827–847.

Rose, J., Lee, J. A., Moorlock, B. S. P. & Hamblin, R. J. O. 1999*b*. The origin of the Norwich Brickearth: micromorphological evidence for pedological alteration of sandy Anglian Till in northeast Norfolk. *Proceedings of the Geologists' Association*, **110**, 1–8.

Rose, J., Moorlock, B. S. P. & Hamblin, R. J. O. 2001. Pre-Anglian fluvial and coastal deposits in Eastern England: lithostratigraphy and palaeoenvironments. *Quaternary International*, **79**, 5–22.

Rose, J., Turner, C., Coope, G. R. & Bryant, M. D. 1980. Channel changes in a lowland river catchment over the last 13,000 years. *In*: Cullingford, R. A., Davidson, D. A. & Lewin, J. (eds) *Timescales in Geomorphology*. Wiley, Chichester, 159–175.

Rose, W. C. C. & Dunham, K. C. 1977. *Geology and Hematite Deposits of South Cumbria*. Economic Memoir of the Geological Survey of Great Britain, Sheet 58 and part of 48.

Rosen M. R. 1991. Sedimentologic and geochemical constraints on the evolution of Bristol Dry Lake basin, California, USA. *Palaeogeography, Palaeoclimatology, Palaeoecology*, **84**, 229–257.

Rosi, M., Vezzoli, L., Aleotti, P. & De Censi, M. 1996. Interaction between caldera collapse and eruptive dynamics during the Campanian ignimbrite eruptions, Phlegrean Fields, Italy. *Bulletin of Volcanology*, **57**, 541–554.

Ross, J. R. P. & Ross, C. A. 1992. Ordovician sea-level fluctuations. *In*: Webby, B. D. & Laurie, J. R. (eds) *Global Perspectives on Ordovician Geology*. Balkema, Rotterdam/ Brookfield, 327–326.

Rowe, P. J., Atkinson, T. C. & Turner, C. 1999. U-series dating of Hoxnian interglacial deposits at Marks Tey, Essex, England. *Journal of Quaternary Science*, **14**, 693–702.

Rowlands, B. M. 1971. Radiocarbon evidence of the age of an Irish Sea glaciation of the Vale of Clwyd. *Nature*, **230**, 9–11.

Ruffell, A. H. 1991. Palaeoenvironmental analysis of the late Triassic succession in the Wessex Basin and correlation with surrounding areas. *Proceedings of the Ussher Society*, **7**, 402–407.

Ruffell, A. H. 1992. Correlation of the Hythe Beds Formation (Lower Greensand Group: early-mid-Aptian), southern England. *Proceedings of the Geologists' Association*, **103**, 273–291.

Ruffell, A. H. 1995. Evolution and hydrocarbon prospectivity of the Brittany Basin (Western Approaches Trough), off-shore northwest France. *Marine and Petroleum Geology*, **12**, 387–407.

Ruffell, A. H. & Rawson, P. F. 1994. Palaeoclimate control on sequence stratigraphic patterns in the late Jurassic to mid-Cretaceous, with a case study from Eastern England. *Palaeogeography, Palaeoclimatology, Palaeoecology*, **110**, 43–54.

Ruffell, A. H. & Shelton, R. 1999. The control of sedimentary facies by climate during phases of crustal extension: examples from the Triassic of onshore and offshore England and Northern Ireland. *Journal of the Geological Society, London*, **156**, 779–789.

Ruffell, A. H. & Shelton, R. G. 2000. Permian to late Triassic post-orogenic collapse, early Atlantic rifting, deserts, evaporating seas and mass extinctions. *In*: Woodcock, N. & Strahan, R. (eds) *Geological History of Britain and Ireland*. Blackwell Science.

Ruffell, A. H. & Wach, G. 1991. Sequence stratigraphic analysis of the Aptian–Albian Lower Greensand in southern England. *Marine and Petroleum Geology*, **8**, 341–353.

Ruffell, A. H. & Wach, G. 1998. Firmgrounds – key surfaces in the recognition of parasequences in the Aptian Lower Greensand Group, Isle of Wight (southern England). *Sedimentology*, **45**, 91–107.

Ruffell, A. H. & Wignall, P. B. 1990. Depositional trends in the Upper Jurassic–Lower Cretaceous of the northern margin of the Wessex Basin. *Proceedings of the Geologists' Association*, **101**, 279–288.

Ruffell, A. H., Ross, A. & Taylor, K. 1996. *Early Cretaceous Environments of the Weald*. Geologists' Association Guide no. 55.

Ruffell, A. H., Hesselbo, S. P., Wach, G. D., Simpson, M. I. & Wray, D. S. 2002. Fuller's earth (bentonite) in the Lower Cretaceous (Upper Aptian) of Shanklin (Isle of Wight, southern England). *Proceedings of the Geologists' Association*, **113**, 281–290.

Ruffer T. & Zuhlke R. 1995. Sequence stratigraphy and sea-level changes in the early to middle Triassic of the Alps: a global comparison. *In*: Haq, B. U. (ed.) *Sequence Stratigraphy and Depositional Response to Eustatic, Tectonic and Climatic Forcing*. Kluwer, Amsterdam, 161–207.

Rumsby, B. T. & Macklin, M. G. 1994. Channel and floodplain response to recent abrupt climate change; the Tyne basin, northern England. *Earth Surface Processes and Landforms*, **19**, 499–515.

Rundle, C. C. 1979. Ordovician intrusions in the English Lake District. *Journal of the Geological Society, London*, **136**, 29–38.

Rundle, C. C. 1981*a*. The significance of isotopic dates from the English Lake District for the Ordovician–Silurian time-scale. *Journal of the Geological Society, London*, **138**, 569–572.

Rundle, C. C. 1981*b*. *K:Ar ages for micas from S. W. England*. Institute of Geological Sciences, Isotope Geology Unit, Report No. 81/10.

Rundle, C. C. 1987. *Isotopic analyses of rocks from the English Lake District. Preliminary report and summary of results from the 1986 rock collection*. Report of the Isotope Geology Unit, British Geological Survey, **87/6**.

Rundle, C. C. 1992. *Review and assessment of isotopic ages from the English Lake District*. Technical Report, British Geological Survey, **WA/92/38**.

Rushton, A. W. A. 1966. *Cambrian trilobites from the Purley Shales of Warwickshire*. Monograph of the Palaeontographical Society, London, Publ. no.**511**, 55pp, 6pls.

Rushton, A. W. A. 1974. The Cambrian of Wales and England. *In*: Holland, C. H. (ed.) *Cambrian of the British Isles, Norden and Spitzbergen*. Wiley and Sons, London, 43–121.

Rushton, A. W. A. 1978. Fossils from the Middle-upper transition in the Nuneaton district. *Palaeontology*, **21**, 245–83.

Rushton, A. W. A. 1982. The biostratigraphy and correlation of the Merioneth-Tremadoc Series boundary in North Wales. *In*: Bassett, M. G. & Dean, W. T. (eds) *The Cambrian-Ordovician boundary: sections, fossil distributions and correlations*. National Museum of Wales, Geological Series No. 3 Cardiff, pp 41–59.

Rushton, A. W. A. 1983. Trilobites from the Upper Cambrian Olenus Zone in central England. *Special Papers in Palaeontology*, **30**, 107–139, pls 14–19.

Rushton, A. W. A. 1988. Tremadoc trilobites from the Skiddaw Group in the English Lake District. *Palaeontology*, **31**, 677–698.

Rushton, A. W. A. 1993. Graptolites from the Manx Group. *Proceedings of the Yorkshire Geological Society*, **49**, 259–262.

Rushton A. W. A. & Howells, M. F. 1999. *Stratigraphic framework for the Ordovician of Snowdonia and the Lleyn Peninsula. A discussion of Tremadoc to Caradoc rocks lying between the Menai Straits and the Llanderfel Syncline and including an appendix on Cambrian rocks*. Version 2. British Geological Survey, Research Report RR / 99/08.

Rushton, A. W. A. & Molyneux, S. G. 1989. The biostratigraphic age of the Ordovician Skiddaw Group in the Black Combe inlier, English Lake District. *Proceedings of the Yorkshire Geological Society*, **47**, 267–276.

Rushton, A. W. A., Hamblin, R. J. O. & Strong, G. E. 1988. *The Croft borehole in the Lilleshall inlier of north Shropshire*. Report of the British Geological Survey, **19**.

Rushton, A. W. A., Owen, A. W., Owens, R.M. & Prigmore, J. K. 2000. *British Cambrian to Ordovician Stratigraphy*. Geological Conservation Review Series, No. **18**, Joint Nature Conservancy Committee, Peterborough [dated 1999].

Sadler, P. M. 1973. An interpretation of new stratigraphic evidence from south Cornwall. *Proceeedings of the Ussher Society*, **3**, 535–550.

Sælen, G., Tyson, R. V., Telnæs, N. & Talbot, M. R. 2000. Contrasting watermass conditions during deposition of the Whitby Mudstone (Lower Jurassic) and Kimmeridge Clay (Upper Jurassic) formations, UK. *Palaeogeography, Palaeoclimatology, Palaeoecology*, **163**, 163–196.

Salfeld, H. 1913. Certain Upper Jurassic strata of England. *Quarterly Journal of the Geological Society of London*, **69**, 423–432, pls 41–2.

Salmon, S. & Powell, T. 1998. Variation in the fine-grained granites of the Land's End Pluton. *Geoscience in south-west England*, **9**, 157–164.

Salter, J. W. 1856. On fossil remains in the Cambrian (sic) rocks of the Longmynd and North Wales. *Quarterly Journal of the Geological Society of London*, **12**, 246–251.

Salter, J. W. 1857. On annelide burrows and surface markings from the Cambrian rocks of the Longmynd. *Quarterly Journal of the Geological Society of London*, **13**, 199–206.

Samson, D. D. & D'Lemos, R. S. 1998. U–Pb chronology and Sm–Nd isotopic composition of Proterozoic gneisses, Channel Islands. *Journal of the Geological Society, London*, **155**, 609–618.

Sandeman, H. A. I. 1988. A field, petrographical and geochemical investigation of the Kennack Gneiss, Lizard peninsula, south-west England. MSc Thesis, Memorial University, Newfoundland.

Sandeman, H. A. I., Chen, Y., Clark, A. H. & Farrar, E. 1995. Constraints on the P–T conditions and age of emplacement of the Lizard ophiolite, Cornwall: amphibole–plagioclase thermobarometry and $^{40}Ar/^{39}Ar$ geochronology of basal amphibolites. *Canadian Journal of Earth Sciences*, **32**, 261–272.

Sandeman, H. A. I., Clark, A. H., Scott, D. J. & Malpas, J. G. 2000. The Kennack Gneiss of the Lizard peninsula, Cornwall, SW England: commingling and mixing of mafic and felsic magmas accompanying Givetian continental incorporation of the Lizard ophiolite. *Journal of the Geological Society, London*, **157**, 1227–1242.

Sandeman, H. A., Clark, A. H., Styles, M. T., Scott, D. J., Malpas, J. G. & Farrar, E. 1997. Geochemistry and U–Pb and $^{40}Ar–^{39}Ar$ geochronology of the Man of War Gneiss, Lizard Complex, SW England: pre-Hercynian arc-type crust with a Sudetan–Iberian connection. *Journal of the Geological Society, London*, **154**, 403–417.

Sanderson, D. J. 1979. The transition from upright to recumbent folding in the Variscan fold belt of southwest England; a model based on the kinematics of simple shear. *Journal of Structural Geology*, **1**, 171–180.

Sanderson, D. J. 1984. Structural variation across the northern margin of the Variscides of NW Europe. *In*: Hutton, D. H. W. & Sanderson, D. J. (eds) *Variscan Tectonics of the North Atlantic Region*. Geological Society, London, Special Publications, **14**, 149–165.

Sanderson, D. J. & Dearman. W. R. 1973. Structural zones of the Variscan fold belt in SW England: their location and development. *Journal of the Geological Society, London*, **129**, 527–533.

Sanderson, R. W. & Cave, R. 1980. Silurian volcanism in the central Welsh Borderland. *Proceedings of the Geologists' Asoociation*, **117**, 455–462.

Sanzen-Baker, I. 1972. Stratigraphic relationships and sedimentary environments of the Silurian–early Old Red Sandstone of Pembrokeshire. *Proceedings of the Geologists' Association*, **83**, 139–164.

Savage, N. M. & Bassett, M. G. 1985. Caradoc- Ashgill conodont faunas from wales and the Welsh Borderland . *Palaeontology*, **28** (4), 679–713. Siveter, D. J. 2000*a*. The Wenlock Series. *In*: Aldridge, R.J., Siveter, D. J., Lane, P. D., Palmer, D. & Woodcock, N. H. (eds) *British Silurian Statigraphy*. Joint Nature Conservation Committee, Peterborough, 181–323.

Schadla-Hall, R. T. 1989. The Vale of Pickering in the early Mesolithic in context. *In*: Bonsall, C. (ed.) *The Mesolithic in Europe*. John Donald, Edinburgh, 218–224.

Schmitz, B., Sundquist, B. & Andreasson, F. P. 2000. Early Paleogene warm climates and biosphere dynamics. *GFF*, **122**, 1–192.

Schofield, K. & Adams, A. E. 1985. Stratigraphy and depositional environments of the Woo Dale Limestone Formation (Dinantian), Derbyshire. *Proceedings of the Yorkshire Geological Society*, **45**, 225–233.

Schreve, D. C. 2001*a*. Mammalian evidence from Middle Pleistocene fluvial sequences for complex environmental change at the oxygen isotope substage level. *Quaternary International*, **79**, 65–74.

Schreve, D. C. 2001*b*. Differentiation of the British late Middle Pleistocene interglacials; the evidence from mammalian biostratigraphy. *Quaternary Science Reviews*, **20**, 1693–1705.

Schreve, D. C. 2004*a*. The mammalian fauna of Barnfield Pit, Swanscombe, Kent. *In*: Schreve, D. C. (ed.) *The Quaternary Mammals of Southern and Eastern England*. Quaternary Research Association, London, 29–48.

Schreve, D. C. 2004*b*. Late Middle Pleistocene (MIS 9) River Thames terrace deposits at Purfleet, Essex. *In*: Schreve, D. C. (ed.) *The Quaternary Mammals of Southern and Eastern England*. Quaternary Research Association, London, 49–67.

Schwarzacher, W. 1953. Cross-bedding and grain size in the Lower Cretaceous sands of East Anglia. *Geological Magazine*, **90**, 322–330.

Schreve, D. C. 2004*c*. The mammalian fauna of the penultimate (MIS 7) interglacial in thje Lower Thames Valley. *In*: Schreve, D. C. (ed.) *The Quaternary Mammals of Southern and Eastern England*. Quaternary Research Association, London, 69–79.

Scotchman, I. C. 1989. Diagenesis of the Kimmeridge Clay Formation, Onshore UK. *Journal of the Geological Society, London*, **146**, 285–303.

Scotese, C. R. & McKerrow, W. S. 1990. Revised World maps and introduction. *In*: McKerrow, W. S. & Scotese, C. R. (eds) *Palaeozoic Palaeogeography and Biogeography*. Geological Society, London, Memoir, **12**, 1–21.

Scotese, C. R. & McKerrow, W. S. 1991. Ordovician plate tectonic reconstructions. *In*: Barnes, C. R. & Williams, S. H. (eds) *Advances in Ordovician Geology*. Geological Survey of Canada, Paper, **90–9**, 271–282.

Scotese, C. R., Bambach, R. K., Barton, C., Van Der Voo, R. & Ziegler, R. M. 1979. Palaeozoic base maps. *Journal of Geology*, **87**, 217–277.

Scourse, J. D. 1991. Late Pleistocene stratigraphy and palaeobotany of the Isles of Scilly. *Philosophical Transactions of the Royal Society of London*, **B334**, 405–448.

Scourse, J. D. 1996. Late Pleistocene stratigraphy and palaeobotany of north and west Cornwall. *Transactions of the Royal Geological Society of Cornwall*, **22**, 2–56.

Scourse, J. D. 1997. Transport of the Stonehenge bluestones: testing the glacial hypothesis. *Proceedings of the British Academy*, **92**, 271–314.

Scourse, J. D. & Austin, W. E. N. 2002. Quaternary shelf sea palaeoceanography: recent developments in Europe. *Marine Geology*, **191**, 87–94.

Scourse, J. D. & Furze, M. F. A. 2001. A critical review of the glaciomarine model for Irish Sea deglaciation: evidence from southern Britain, the Celtic shelf and adjacent continental slope. *Journal of Quaternary Science*, **16**, 419–434.

Scourse, J. D., Austin, W. E. N., Bateman, R. M., Catt, J. A., Evans, C. D. R., Robinson, J. E. & Young, J. R. 1990. Sedimentology and micropalaeontology of glacimarine sediments from the Central and Southwestern Celtic Sea. *In*: Dowdeswell, J. A. & Scourse, J. D. (eds) *Glacimarine environments: processes and sediments*. Geological Society, London, Special Publications, **53**, 329–347.

Scourse, J. D., Austin, W. E. N., Sejrup, H. P. & Ansari, M. H. 1999. Foraminiferal isoleucine epimerization determinations from the Nar Valley Clay, Norfolk, UK: implications for Quaternary correlations in the southern North Sea basin. *Geological Magazine*, **136**, 543–560.

Scourse, J. D., Austin, W. E. N., Long, B. T., Assinder, D. J. & Huws, D. 2002. Holocene evolution of seasonal stratification in the Celtic Sea: refined age model, mixing depths and foraminiferal stratigraphy. *Marine Geology*, **191**, 119–145.

Scourse, J. D., Hall, I. R., McCave, I. N., Young, J. R. & Sugdon, C. 2000. The origin of Heinrich layers; evidence from H2 for European precursor events. *Earth and Planetary Science Letters*, **182**, 187–195.

Scrivener, R. C. 1982. *Tin and related mineralisation of the Dartmoor granite*. Unpublished PhD thesis, University of Exeter.

Scrivener, R. C. & Bennett, M. J. 1980. Ore genesis and controls on mineralisation in the Upper Palaeozoic rocks of north Devon. *Proceedings of the Ussher Society*, **5**, 54–58.

Scrivener, R. C., Cooper, B. V. & George, M. C. 1987. Mineralogy and paragenesis of the Haytor iron ore deposit. *Proceedings of the Ussher Society*, **6**, 558.

Scrivener, R. C., Darbyshire, D. P. F. & Shepherd, T. J. 1994. Timing and significance of crosscourse mineralization in SW England. *Journal of the Geological Society, London*, **151**, 587–590.

Scrivener, R. C., Leake, R. C., Leveridge, B. E. & Shepherd, T. J. 1989. Volcanic-exhalative mineralisation in the variscan province of SW England. *Terra Abstracts*, **1**, 125.

Scrutton, C. T. 1977. Facies variations in the Devonian limestones of eastern South Devon. *Geological Magazine*, **114**, 165–193.

Scrutton, C. T. 1996. Ecophenotypic variation in the early Silurian rugose coral *Palaeocyclus porpita*. *Proceedings of the Yorkshire Geological Society*, **51**, 1–8.

Seago, R. D. & Chapman, T. J. 1988. The confrontation of structural styles and the evolution of a foreland basin in central south-west England. *Journal of the Geological Society, London*, **145**, 789–800.

Sedgwick, A. 1852. On the classification of the fossiliferous slates of North Wales, Cumberland, Westmoreland and Lancashire. *Quarterly Journal of the Geological Society of London*, **3**, 133–164.

Sedgwick, A. & Murchison, R. I. 1835. On the Silurian and Cambrian Systems exhibiting the order in which the older sedimentary strata succeeded each other in England and Wales. *London and Edinburgh Philosophical Magazine*, **7**, 483–485.

Sejrup, H. P., Aarseth, I. *et al.* 1995. Quaternary history of the Norwegian Channel; glaciation history and palaeoceanography. *Norsk Geologisk Tidsskrift*, **75**, 65–87.

Sejrup, H. P., Larsen, E., Landvik, J., King, E. L., Haflidason, H. & Nesje, A. 2000. Quaternary glaciations in southern Fennoscandia: evidence from southwestern Norway and the northern North Sea region. *Quaternary Science Reviews*, **19**, 667–685.

Sellwood, B. W. & McKerrow, W. S. 1974. Depositional environments in the lower part of the Great Oolite Group of Oxfordshire and north Gloucestershire. *Proceedings of the Geologists' Association*, **85**, 189–210.

Sellwood, B. W., Valdes, P. J. & Price, G. D. 2000. Geological evaluation of multiple general circulation model simulations of Late Jurassic palaeoclimate. *Palaeogeography, Palaeoclimatology, Palaeoecology*, **156**, 147–160.

Selwood, E. B. 1960. Ammonoids and trilobites from the Upper Devonian and Lower Carboniferous of the Launceston area, Cornwall. *Palaeontology*, **3**, 153–185.

Selwood, E. B. 1971. Successions at the Devonian–Carboniferous boundary between Boscastle and Dartmoor. *Proceedings of the Ussher Society*, **2**, 275–285.

Selwood, E. B. 1990. A review of basin development in Central SW England. *Proceedings of the Ussher Society*, **7**, 199–205.

Selwood, E. B. & Durrance, E. M. 1982. The Devonian Rocks. *In*: Durrance, E. M. & Laming, D. J. C. (eds) *The Geology of Devon*. University of Exeter, Exeter, 15–41.

Selwood, E. B. & Thomas, J. M. 1986. Upper Palaeozoic successions and nappe structures in north Cornwall. *Journal of the Geological Society, London*, **143**, 75–82.

Selwood, E. B. & Thomas, J. M. 1988. A tectonic framework of Upper Carboniferous sedimentation in central southwest England. *In*: Besley, B. M. & Kelling, G. (eds) *Sedimentation in a synorogenic basin complex: the Upper Carboniferous of northwest Europe*. Blackie, Glasgow, 18–23.

Selwood, E. B. & Thomas, J. M. 1993. The Tredorn Nappe, north Cornwall: a review. *Proceedings of the Ussher Society*, **8**, 89–93.

Selwood, E. B., Edwards, R. A. *et al.* 1984. *Geology of the country around Newton Abbot*. Memoir of the British Geological Survey, Sheet 339 (England and Wales), HMSO, London.

Selwood, E. B., Stewart, I. J. & Thomas, J. M. 1985. Upper Palaeozoic sediments and structure in north Cornwall – a reinterpretation. *Proceedings of the Geologists' Association*, **96**, 129–141.

Selwood, E. B., Thomas, J. M., Borley, G. D. & Dean, A. 1993. A revision of the Upper Palaeozoic stratigraphy of the Trevone Basin, north Cornwall, and its regional significance. *Proceedings of the Geologists' Association*, **104**, 137–148.

Selwood, E. B., Thomas, J. M., Williams, B. J., Clayton, R. E., Durning, B., Smith, O. & Warr, L. N. 1998. *Geology of the country around Trevose Head and Camelford*. Memoir of the British Geological Survey, Sheets 335 and 336 (England and Wales), HMSO, London.

Sevastopulo, G. D. & Nudds, J. R. 1987. Courceyan (early Dinantian) biostratigraphy of Britain and Ireland: coral and conodont zones compared. *Courier Forschungsinstitut Senckenberg*, **98**, 39–46.

Shackleton, N. J., Berger, A. & Peltier, W. R. 1990. An alternative astronomical calibration of the Lower Pleistocene timescale based on ODP Site 677. *Transactions of the Royal Society of Edinburgh: Earth Sciences*, **81**, 251–261.

Shackleton, N. J., Sánchez-Goñi, M. F., Pailler, D. & Lancelot, Y. 2003. Marine Isotope Substage 5e and the Eemian Interglacial. *Global and Planetary Change*, **36**, 151–155.

Shackleton, R. M. 1954. The structural evolution of North Wales. *Liverpool and Manchester Geological Journal*, **1**, 261–297.

Shackleton, R. M. 1954. The structure and succession of Anglesey and the Lleyn peninsula. *Journal for the Advancement of Science, London*, **11**, 106–108.

Shackleton, R. M. 1969. The Precambrian of North Wales. *In*: Wood, A. (ed.) *The Precambrian and Lower Palaeozoic Rocks of Wales*. University of Wales Press, Cardiff, 1–22.

Shackleton, R. M., Ries, A. C. & Coward, M. P. 1982. An interpretation of the Variscan structures in SW England. *Journal of the Geological Society, London*, **139**, 533–541.

Shail, R. 1989. Gramscatho–Mylor facies relationships; Hayle, south Cornwall. *Proceedings of the Ussher Society*, **7**, 125–130.

Shail, R. K. 1992. *Provenance of the Gramscatho Group, south Cornwall*. Unpublished PhD thesis, University of Keele.

Shail, R. & Wilkinson, J. J. 1994. Late- to post-Variscan extensional tectonics in south Cornwall. *Proceedings of the Ussher Society*, **8**, 262–270.

Shakesby, R. A. 1992. *Classic Landforms of the Brecon Beacons*. Classic Landform Guides, **13**. Geographical Association, Sheffield.

Shakesby, R. A. & Matthews, J. A. 1993. Loch Lomond Stadial glacier at Fan Hir, Mynydd Du (Brecon Beacons), South Wales: critical evidence and palaeoclimatic implications. *Geological Journal*, **28**, 69–79.

SHAKESBY, R. A. & MATTHEWS, J. A. 1996. Glacial activity and paraglacial landsliding in the Devensian Late Glacial: evidence from Craig Cerrig-gleisiad and Fan Dringarth, Fforest Fawr (Brecon Beacons), South Wales. *Geological Journal*, **31**, 143–157.

SHAW, R. W. L. 1971. The faunal stratigraphy of the Kirkby Moor Flags of the type area near Kendal, Westmorland. *Geological Journal*, 7, 359–380.

SHEAR, W. A. & SELDEN, P. A. 2001. Rustling in the undergrowth: animals in early terrestrial environments. *In*: GENSEL, P. G. & EDWARDS, D. (eds) *Plants Invade the Land: Evolutionary and Environmental Perspectives*. Columbia University Press, New York, 29–51.

SHENNAN, I. & HORTON, B. 2002. Holocene land- and sea-level changes in Great Britain. *Journal of Quaternary Science*, **17**, 511–526.

SHENNAN, I., HORTON, B., INNES, J., GEHRELS, R., LLOYD, J., MCARTHUR, J. & RUTHERFORD, M. 2000*a*. Late Quaternary sea-level changes, crustal movements and coastal evolution in Northumberland, UK. *Journal of Quaternary Science*, **15**, 215–237.

SHENNAN, I., LAMBECK, K. ET AL. 2000*b*. Modelling western North Sea palaeogeographies and tidal changes during the Holocene. *In*: SHENNAN, I. S. & ANDREWS, J. E. (eds) *Holocene Land–Ocean Interaction and Environmental Change Around the North Sea*. Geological Society, London, Special Publications, **166**, 299–319.

SHENNAN, I., LAMBECK, K., HORTON, B. P., INNES, J. B., LLOYD, J. M., MCARTHUR, J. J. & RUTHERFORD, M. M. 2000*c*. Holocene isostasy and relative sea-level on the east coast of England. *In*: SHENNAN, I. S. & ANDREWS, J. E. (eds) *Holocene Land–Ocean Interaction and Environmental Change Around the North Sea*. Geological Society, London, Special Publications, **166**, 275–298.

SHENNAN, I., PELTIER, W. R., DRUMMOND, R. & HORTON, B. 2002. Global to local scale parameters determining relative sea-level changes and the post-glacial isostatic adjustment of Great Britain. *Quaternary Science Reviews*, **21**, 397–408.

SHEPHARD-THORN, E. R., MOORLOCK, B. S. P., COX, B. M., ALLSOP, J. M. & WOOD, C. J. 1994. *Geology of the country around Leighton Buzzard*. Memoir of the British Geological Survey, Sheet 220 (England & Wales), HMSO, London.

SHEPHERD, T. J. & SCRIVENER, R. C. 1987. Role of basinal brines in the genesis of polymetallic vein deposits, Kit Hill–Gunnislake area, SW England. *Proceedings of the Ussher Society*, **6**, 491–497.

SHEPHERD, T. J., BECKINSALE, R. D., RUNDLE, C. C. & DURHAM, J. 1976. Genesis of Carrock Fell tungsten deposits Cumbria: fluid inclusion and isotopic study. *Transactions of the Institution of Mining and Metallurgy*, **85**, B63–73.

SHEPHERD, T. J., MILLER, M. F., SCRIVENER, R. C. & DARBYSHIRE, D. P. F. 1985. Hydrothermal fluid evolution in relation to mineralisation in southwest England, with special reference to the Dartmoor–Bodmin area. *In*: *High Heat Production Granites, Hydrothermal Circulation and Ore Genesis*. Institution of Mine and Metallurgy, London, 345–364.

SHEPPARD, S. M. F. 1977. The Cornubian Batholith, SW England: D/H and $^{18}O/^{16}O$ studies of kaolinite and other alteration minerals. *Journal of the Geological Society, London*, **133**, 573–591.

SHERGOLD, J. H. & SHIRLEY, J. 1968. The faunal-stratigraphy of the Ludlovian rocks between Craven Arms and Bourton, near Much Wenlock, Shropshire. *Geological Journal*, **6**, 119–138.

SHERLOCK, S. C., JONES, K. A. & KELLEY, S. P. 2002. Fingerprinting polyorogenic detritus using the $^{40}Ar/^{39}Ar$ ultraviolet laser microprobe. *Geology*, **30**, 515–518.

SHERLOCK, S. C., KELLEY, S. P., ZALASIEWICZ, J. A., SCHOFIELD, D. I., EVANS, J. A., MERRIMAN, R. J. & KEMP, S. J. 2003. Precise dating of low-temperature deformation: Strain-fringe analysis by ^{40}Ar–^{39}Ar laser microprobe. *Geology*, **31**, 219–222.

SHIPBOARD SCIENTIFIC PARTY. 2001. Leg 192 Summary. *In*: MAHONEY, J. J., FITTON, ET AL. *Proceeedings ODP, Initial Reports*, **192**, 1–75.

SHOTTON, F. W. 1937. The Lower Bunter Sandstone of north Worcestershire and east Shropshire. *Geological Magazine*, **74**, 534–553.

SHOTTON, F. W. 1953. The Pleistocene deposits of the area between Coventry, Rugby and Leamington, and their bearing upon the topographic development of the Midlands. *Philosophical Transactions of the Royal Society of London*, **B237**, 209–260.

SHOTTON, F. W. 1956. Some aspects of the New Red Desert in Britain. *Liverpool and Manchester Geological Journal*, **1**, 450–465.

SIMMONS, I. G., ATHERDEN, M. A., CLOUTMAN, E. W., CUNDILL, P. R., INNES, J. B. & JONES, R. L. 1993. Prehistoric environments. *In*: SPRATT, D. A. (ed.) *Prehistoric and Roman Archaeology of North-east Yorkshire*. Council for British Archaeology, London, British Archaeology Research Report, **87**, 15–50.

SIMMS, M. J. 2003. Uniquely extensive seismite from the latest Triassic of the United Kingdom: Evidence for bolide impact. *Geology*, **31**, 557–560.

SIMMS, M. J. & RUFFELL, A. 1990. Climatic and biotic change in the late Triassic. *Journal of the Geological Society, London*, **147**, 321–327.

SIMMS, M. J., LITTLE, C. T. S. & ROSEN, B. R. 2002. Corals not serpulids: mineralised colonial fossils in the Lower Jurassic marginal facies of South Wales. *Proceedings of the Geologists' Association*, **113**, 31–36.

SIMON, J. B. & BLUCK, B. J. 1982. Palaeodrainage of the southern margin of the Caledonide mountain chain in the northern British Isles. *Transactions of the Royal Society of Edinburgh: Earth Sciences*, **73**, 11–15.

SIMPSON, A. 1963. The stratigraphy and tectonics of the Manx Slate Series. *Quarterly Journal of the Geological Society of London*, **119**, 367–400.

SIMPSON, A. 1965. The syn-tectonic Foxdale–Archallagan granite and its metamorphic aureole. *Geological Journal*, **4**, 189–206.

SIMPSON, A. 1967. The stratigraphy and tectonics of the Skiddaw Slates and the relationship of the overlying Borrowdale Volcanic Series in part of the Lake District. *Geological Journal*, **5**, 391–418.

SIMPSON, I. M. & WEST, R. G. 1958. On the stratigraphy and palaeobotany of a late-Pleistocene organic deposit at Chelford, Cheshire. *New Phytologist*, **57**, 239–250.

SIMPSON, M. I. 1983. Decapod Crustacea and Associated Fauna from the Punfield Marine Band (Lower Cretaceous; Lower Aptian), Punfield, Dorset. *Proceedings of the Dorset Natural History and Archaeological Society*, **104**, 143–146.

SIMPSON, M. I. 1985. The stratigraphy of the Atherfield Clay Formation (Lower Aptian, Lower Cretaceous) at the type and other localities in southern England. *Proceedings of the Geologists' Association*, **96**, 23–45.

SIMPSON, P. R., BROWN, G. C., PLANT J. & OSTLE, D. 1979. Uranium mineralization and granite in the British Isles. *Philosophical Transactions of the Royal Society*, **A291**, 385–412.

SIMPSON, S. 1964. The Lynton Beds of north Devon. *Proceedings of the Ussher Society*, **2**, 121–122.

SINGER, R., GLADFELTER, B. G. & WYMER, J. J. 1993. *The Lower Palaeolithic Site at Hoxne, England*. University of Chicago Press, Chicago, IL.

SISSONS, J. B. 1980. The Loch Lomond Advance in the Lake District, northern England. *Transactions of the Royal Society of Edinburgh: Earth Sciences*, **71**, 13–27.

SIVETER, D. J. 2000*b*. The Ludlow Series. *In*: ALDRIDGE, R. J., SIVETER, D. J., SIVETER, D. J., LANE, P. D., PALMER, D. & WOODCOCK, N. H. (eds) *British Silurian Stratigraphy*. Joint Nature Conservation Committee, Peterborough, 325–426.

SIVETER, D. J., OWENS, R. M. & THOMAS, A. T. 1989. *Silurian Field Excursions: A Geotraverse Across Wales and the Welsh Borderland*. National Museum of Wales, Cardiff, Geological Series, **10**.

SIVETER, D. J. & WILLIAMS, M. 1995. An early Cambrian assignment for the Caerfai Group of South Wales. *Journal of the Geological Society, London*, **152**, 221–224.

SIVETER, D. J., LANE, P. D., PALMER, D. & WOODCOCK, N. H. (eds) 2000. *British Silurian Stratigraphy*. Joint Nature Conservation Committee. Geological Conservation Review Series, **19**, 183–323.

SKEMPTON, A. W. & WEEKS, A. G. 1976. The Quaternary history of the Lower Greensand escarpment and Weald Clay vale near Sevenoaks, Kent. *Philosophical Transactions of the Royal Society of London*, **A283**, 493–526.

SKEVINGTON, D. 1974. Controls influencing the composition and distribution of Ordovician graptolite faunal provinces. *Special Papers in Palaeontology*, **13**, 59–73.

SKELTON, P. (ed.) 2003. *The Cretaceous World*. Cambridge University Press, Cambridge.

SKIPPER, J. A. E. 1999. *The stratigraphy of the Lambeth Group (Palaeocene) of southern England*. PhD thesis, University of Greenwich.

SKJOLD, L. J., VAN VEEN, P. M., KRISTENSEN, S.-E. & RASMUSSEN, A. R. 1998. Triassic sequence stratigraphy of the southwestern Barents Sea. *In*: DE GRACIANSKY, P.-C., HARDENBOL, J., JACQUIN, T. & VAIL P. R. (eds) *Mesozoic and Cenozoic Sequence Stratigraphy of European Basins*. SEPM, Special Publications, **60**, 652–667.

SLATER, G. 1929. Quaternary Period. *In*: EVANS, J. W. & STUBBLEFIELD, C. J. (eds) *Handbook of the Geology of Great Britain*. T. Murby, London, 457–510.

SMALLWOOD, S. D. 1986. *Lower Paleozoic history of the Tywi Lineament, Mid Wales*. PhD thesis, University of Cambridge.

SMART, J. G. O., BISSON, G. & WORSSAM, B. C. 1966. *Geology of the Country Around Canterbury and Folkestone*. Memoirs of the Geological Survey of Great Britain, Sheets 289, 305 and 306.

SMART, P. J. 1997. The basal Gault and Gault–Woburn Sands junction beds in Chamberlain's Barn quarry, Leighton Buzzard, Bedfordshire, England. *Proceedings of the Geologists' Association*, **108**, 287–292.

SMELROR, M., KELLY, S. R. A., DYPVIK, H., MØRK, A., NAGY, J. & TSIKALAS, F. 2001. Mjølnir (Barents Sea) meteorite impact ejecta offers a Volgian–Ryazanian boundary marker. *Newsletters on Stratigraphy*, **38**, 129–140.

SMIT, J. 2004. And yet it smokes! *Geoscientist*, **14** (1), 8–9, 13, 17.

SMITH, A., BRADDY, S. J., MARRIOTT, S. B. & BRIGGS, D. E. G. 2003. Arthropod trackways from the Early Devonian of South Wales: a functional analysis of producers and their behaviour. *Geological Magazine*, **140**, 63–72.

SMITH, A. B., PAUL, C. R. C., GALE, A. S. & DONOVAN, S. K. 1988. Cenomanian and Lower Turonian echinoderms from Wilmington, south-east Devon, England. *Bulletin of the British Museum (Natural History), Geology*, **42**, 1–245.

SMITH, A. H. V. & BUTTERWORTH, M. A. 1967. *Miospores in Coal Seams of the Carboniferous of Great Britain*. Palaeontological Association, London, *Special Papers in Palaeontology*, **1**, 1–324.

SMITH, A. J. 1985. Catastrophic origin for the palaeovalley system of the eastern English Channel. *Marine Geology*, **64**, 65–75.

SMITH, D. B. 1972. The British Isles. *In*: FALKE, H. (ed.) *Rotliegend; essays on European Lower Permian*. Brill, Leiden, 1–33.

SMITH, D. B. 1989. The late Permian palaeogeography of north-east England. *Proceedings of the Yorkshire Geological Society*, **47**, 285–312.

SMITH, D. B. 1992. Permian. *In*: DUFF, P. MCL. D & SMITH, A. J. (eds) *The Geology of England and Wales*. 1st edn, Geological Society, London, 275–306.

SMITH, D. B. 1995. *Marine Permian of England*. Geological Conservation Review Series. No. 8. pp. 205.

SMITH, D. B. & TAYLOR, J. C. M. 1992. Permian. *In*: COPE, J. C. W., INGHAM, J. K. & RAWSON, P. F. (eds) *Atlas of Palaeogeography and Lithofacies*. Memoir of the Geological Society, London, 13, 87–96.

SMITH, D. B., BRUNSTROM, R. G. W., MANNING, P. I., SIMPSON, S. & SHOTTON, F. W. 1974. A correlation of Permian rocks in the British Isles. *Journal of the Geological Society, London*, **130**, 1–45; & Geological Society Special Reports, **5**, 45pp.

SMITH, D. E., SHI, S., CULLINGFORD, R. A., DAWSON, A. G., DAWSON, S., FIRTH, C. R., FOSTER, I. D. L., FRETWELL, P. T., HAGGART, B. A., HOLLOWAY, L. K. & LONG, D. 2004. The Holocene Storegga Slide tsunami in the United Kingdom. *Quaternary Science Reviews*, **23**, 2291–2321.

SMITH, I. F. 1986. Mesozoic Basins. *In*: DOWNING, R. A. & GRAY, D. A. (eds) *Geothermal energy – the potential in the United Kingdom*. HMSO, London, 42–83.

SMITH, K. 2004. The North Sea Silverpit Crater: impact structure or pull-apart basin? *Journal of the Geological Society, London*, **161**, 593–602.

SMITH, M. 1987. The Tremadoc 'Thrust' Zone in southern central Snowdonia. *Geological Journal*, **22**, 119–129.

SMITH, M. R. & COLLIS, L. (eds) 1993. *Aggregates. Sand, gravel and crushed rock aggregates for construction purposes* (2nd edition). Geological Society Engineering Geology Special Publication 9, Geological Society, London.

SMITH, N. J. P. & RUSHTON, A. W. A. 1993. Cambrian and Ordovician stratigraphy related to structure and seismic profiles in the western part of the English Midlands. *Geological Magazine*, **130**, 665–671.

SMITH, R. A. 1995. The Siluro–Devonian evolution of the southern Midland Valley of Scotland. *Geological Magazine*, **132**, 503–513.

SMITH, R. D. A. 1987*a*. The griestoniensis Zone Turbidite System, Welsh Basin. *In*: LEGGETT, J. K. & ZUFFA, G. G. (eds) *Marine Clastic Sedimentation*. Graham and Trotman, London, 89–107.

SMITH, R. D. A. 1987*b*. Early diagenetic phosphate cements in a turbidite basin. *In*: MARHSALL, J. D. (ed.) *Diagenesis of Sedimentary Sequences*. Geological Society, London, Special Publications, 36. 141–156.

SMITH, R. D. A. 1987*c*. Structure and deformation history of the Central Wales synclinorium, northeast Dyfed: implications for a long-lived basement structure. *Geological Journal*, **22**, 183–198.

SMITH, R. D. A. 2004. Turbidite systems influenced by structurally induced topography in the multi-sourced Wesh Basin. *In*: LOMAS, S. A. & JOSEPH, P. (eds) *Confined Turbidite Systems*. Geological Society, London, Special Publications, **222**, 209–228.

SMITH, R. D. A. & AINSWORTH, R. B. 1989. Hummocky cross-stratification in the Downton of the Welsh Borderland. *Journal of the Geological Society, London*, **146**, 897–900.

SMITH, S. A. 1990. The sedimentology and accretionary styles of an ancient gravel-bed stream: the Budleigh Salterton Pebble Beds (Lower Triassic), southwest England. *Sedimentary Geology*, **67**, 199–219.

SMITH, S. A. & EDWARDS, R. A. 1991. Regional sedimentological variations in Lower Triassic fluvial conglomerates (Budleigh Salterton Pebble Beds), southwest England: some implications for palaeogeography and basin evolution. *Geological Journal*, **26**, 65–83.

SMITH, S. A. & HUMPHREYS, B. 1989. Lakes and alluvial sandflat-playas in the Dartmouth Group, south-west England. *Proceedings of the Ussher Society*, **7**, 118–124.

SMITH, S. A. & HUMPHREYS, B. 1991. Sedimentology and depositional setting of the Dartmouth Group, Bigbury Bay, south Devon. *Journal of the Geological Society, London*, **148**, 235–244.

SMITH, W. 1799. MS first published by Smith in 1815 as Table 1 in *A Memoir to the Map and Delineation of the Strata of England and Wales, with part of Scotland*. John Cary, London.

SMOOT, J. P. & CASTENS-SEIDELL, B. 1994. Sedimentary features produced by efflorescent salt crusts, Saline Valley and Death Valley, California. *In*: RENAULT, R. W. (ed.) *Sedimentology and Geochemistry of Modern and Ancient Saline Lakes*, SEPM, Special Publications, **50**, 73–90.

SNEH, A. 1988. Permian dune patterns in northwestern Europe challenged. *Journal of Sedimentary Petrology*, **58**, 44–51.

SOMERVILLE, I. D. 1979. A cyclicity in the early Brigantian (D2) limestones east of the Clwydian Range, North Wales and its use in correlation. *Geological Journal*, **14**, Pt 1, 69–86.

SOMERVILLE, I. D. & GRAY, D. I. 1984. *Peritidal Cyclic Sedimentation in Some Late Dinantian Platform Limestones of Northern Britain*. European Dinantian Environments Meeting. Department of Earth Sciences, Open University, 137–142.

SOMERVILLE, I. D. & STRANK, A. R. E. 1984*a*. The recognition of the Asbian/Brigantian boundary fauna and marker horizons in the Dinantian of North Wales. *Geological Journal*, **19**, 227–237.

SOMERVILLE, I. D. & STRANK, A. R. E. 1984*b*. Faunal discoveries from Visean limestones of North Wales. *Proceedings of the Geologists' Association*, **95**, 394–395.

Somerville, I. D., Strank, A. R. E. & Welsh, A. 1989. Chadian faunas and flora from Dyserth: Depositional environments and palaeogeographic setting of Visean strata in northeast Wales. *Geological Journal*, **24**, 49–66.

Soper, N. J. 1970. Three critical localities on the junction of the Borrowdale Volcanic rocks with the Skiddaw Slates in the Lake District. *Proceedings of the Yorkshire Geological Society*, **37**, 461–493.

Soper, N. J. 1986*a*. Geometry of transecting, anastomosing solution cleavage in transpression zones. *Journal of Structural Geology*, **8**, 937–940.

Soper, N. J. 1986*b*. The Newer Granite problem: a geotectonic view. *Geological Magazine*, **123**, 227–236.

Soper, N. J. 1999. *The Windermere Supergroup of 1:25,000 sheets NY50 and NY60, Southern Shap Fells and Northern Howgill Fells, Cumbria*. British Geological Survey Technical Report, **WA/99/35**. British Geological Survey, Edinburgh.

Soper, N. J. & Hutton, D. H. W. 1984. Late Caledonian sinistral displacements in Britain: implications for a three-plate collision model. *Tectonics*, **3**, 781–794.

Soper, N. J. & Kneller, B. C. 1990. Cleaved microgranite dykes of the Shap swarm in the Silurian of NW England. *Geological Journal*, **25**, 161–170.

Soper, N. J. & Roberts, D. E. 1971. Age of cleavage in the Skiddaw Slates in relation to the Skiddaw aureole. *Geological Magazine*, **108**, 293–302.

Soper, N. J. & Woodcock, N. H. 1990. Silurian collision and sediment dispersal patterns in southern Britain. *Geological Magazine*, **127**, 527–542.

Soper, N. J. & Woodcock, N. H. 2003. The lost Lower Old Red Sandstone of England and Wales: a record of post Iapetan flexure or Early Devonian transtension? *Geological Magazine*, **140**, 627–647.

Soper N. J., Branney M. J., Mathieson, N. A. & Davis, N. C. 1987*a*. The Ordovician batholith of the English Lake District. *Geological Magazine*, **124**, 483–484.

Soper, N. J., England, R. W., Snyder, D. B. & Ryan, P. D. 1992*a*. The Iapetus suture zone in England, Scotland and eastern Ireland: a reconciliation of geological and deep seismic data. *Journal of the Geological Society, London*, **149**, 697–700.

Soper, N. J., Strachan, R. A., Holdsworth, R. E., Gayer, R. A. & Greiling, R. O. 1992*b*. Sinistral transpression and the Silurian closure of Iapetus. *Journal of the Geological Society, London*, **149**, 871–880.

Soper, N. J., Webb, B. C. & Woodcock, N. H. 1987*b*. Late Caledonian (Acadian) transpression in North West England: timing, geometry and geotectonic significance. *Proceedings of the Yorkshire Geological Society*, **46**, 175–192.

Sparks, B. W., Williams, R. B. G. & Bell, F. G. 1972. Presumed ground ice depressions in East Anglia. *Proceedings of the Royal Society of London*, **A327**, 329–343.

Spears, D. A. & Lyons, P. C. 1995. An update on British Tonsteins. *In*: Whateley, M. K. G. & Spears, D. A. (eds) *European Coal Geology*. Geological Society, London, Special Publications, **82**, 137–146.

Spears, D. A., Kanaris-Sotiriou, R., Riley, N. & Krause, P. 1999. Namurian bentonites in the Pennine Basin, UK – origin and magmatic affinities. *Sedimentology*, **46**, 385–401.

Spenser, P. S. & Storrs, G. W. 2002. A re-evaluation of small tetrapods from the Middle Triassic Otter Sandstone Formation of Devon, England. *Palaeontology*, **45**, 447–467.

Spicer, R. A. 2000. Leaf physiognomy and climate change. *In*: Culver, S. & Rawson, P. F. (eds) *Biotic Response to Global Change. The Last 145 Million Years*. Cambridge University Press, Cambridge, 244–264.

Spicer, R. A. & Corfield, R. M. 1992. Review of terrestrial and marine climates in the Cretaceous with implications for modelling the 'Greenhouse Earth'. *Geological Magazine*, **129**, 169–180.

Spiegler, D. 1999. *Bolboforma* biostratigraphy from the Hatton–Rockall Basin (North Atlantic). *In*: Jansen, E., Raymo, M. E. & Blum, P. (eds) *Proceedings of the ODP, Scientific Results*, **162**, 35–49.

Spiegler, D., Gramann, F. & von Daniels, C. H. 1988. Planktonic foraminifera: the description of the interregional zonation. *In*: Vinken, R. (compiler) *The Northwest European Tertiary Basin*. Results of the IGCP Project No. 124. *Geologisches Jahrbuch*, **A100**, 152–160.

Squirrel, H. C. & Downing, R. A. 1964. The attenuation of the Coal Measures in the south-east part of the South Wales Coalfield. *Geological Survey of Great Britain Bulletin*, **21**, 119–132.

Squirrel, H. C. & Downing, R. A. 1969. *Geology of the South Wales Coalfield, Part I, the country around Newport* (Mon.), 3rd edn. Memoir of the Geological Survey of Great Britain Sheet 249, HMSO, London.

Squirrel, H. C. & Tucker, E. V. 1960. The geology of the Woolhope Inlier, Herefordshire. *Quarterly Journal of the Geological Society of London*, **116**, 139–185.

Stamp, L. D. 1921. On Cycles of Sedimentation in the Eocene strata of the Anglo-Franco-Belgian Basin. *Geological Magazine*, **58**, 108–114, 146–157, 194–200.

Stampfli, G. M. & Borel, G. D. 2002. A plate tectonic,model for the Palaeozoic and Mesozoic constrained by dynamic plate boundaries and restored synthetic oceanic magnetic isochrones. *Earth and Planetary Science Letters*, **196**, 17–33.

Stanley, C. J. & Vaughan, D. J. 1982. Copper, lead, zinc and cobalt mineralization in the English Lake District: classification, conditions of formation and genesis. *Journal of the Geological Society, London*, **139**, 569–579.

Stead, I. M., Bourke, J. B. & Brothwell, D. R. 1986. *Lindow Man: The Body in the Bog*. British Museum, London.

Steel, R. J. & Thompson, D. B. 1983. Structures and textures in Triassic braided stream conglomerates ('Bunter Pebble Beds') in the Sherwood Sandstone Group, north Staffordshire, England. *Sedimentology*, **30**, 341–367.

Steele, S. A. 1994. *The Start–Perranporth Zone: transpression reactivation across a major basement fault in the Variscan orogen of SW England*. PhD thesis, University of Durham.

Stephen Green, H. 1984. *Pontnewydd Cave. A Lower Palaeolithic Hominid Site in Wales*. National Museum of Wales, Cardiff.

Stephenson, D., Bevins, R. E., Millward, D., Highton, A. J., Parsons, I., Stone, P., Wadsworth, W. J. 1999. *Caledonian Igneous Rocks in Britain*. Joint Nature Conservation Committee. Geological Conservation Review, **17**.

Stephenson, D., Loughlin, S. C., Millward, D., Waters, C. N. & Williamson, I. T. 2003. *Carboniferous and Permian Igneous Rocks of Great Britain North of the Variscan Front*. Joint Nature Conservation Committee. Geological Conservation Review Series, **27**.

Steurbaut, E. 1998. High-resolution biostratigraphy of Middle Paleocene to Early Eocene strata in Belgium and adjacent areas. *Palaeontographica*, **A 247**.

Steurbaut, E. & King, C. 1994. Integrated stratigraphy of the Mont–Panisel borehole section (151E340), Ypresian (Early Eocene) of the Mons Basin, SW Belgium. *Bulletin de la Societé belge de Géologie*, **102**, 175–202.

Stevens, G. R. 1973. Cretaceous belemnites. *In*: Hallam, A. (ed.) *Atlas of Palaeobiogeography*. Elsevier, Amsterdam, 385–401.

Stevenson, I. P. 1971. The Ordovician rocks of the country between Dwygyfylchi and Dolgarrog, Carnarvonshire. *Proceedings of the Yorkshire Geological Society*, **38**, 517–548.

Stevenson, I. P., Harrison, R. K. & Snelling, N. J. 1970. Potassium–argon age determination of the Waterswallows Sill, Buxton, Derbyshire. *Proceedings of the Yorkshire Geological Society*, **37**, 445–447.

Stewart, D. J. 1981. A meander-belt sandstone of the Lower Cretaceous of Southern England. *Sedimentology*, **28**, 1–20.

Stewart, I. J. 1981*a*. The structure, stratigraphy and conodont biostratigraphy of the north-eastern margin of Bodmin Moor and adjacent areas. PhD thesis, University of Exeter.

Stewart, I. J. 1981*b*. Late Devonian and Lower Carboniferous conodonts from north Cornwall and their stratigraphical significance. *Proceedings of the Ussher Society*, **5**, 178–185.

Stewart, S. A. & Allen, P. J. 2002. A 20-km-diameter multi-ringed impact structure in the North Sea. *Nature*, **418**, 520–523.

Stewart, S. A. & Bailey, H. W. 1996. The Flamborough Tertiary outlier, UK southern North Sea. *Journal of the Geological Society, London*, **153**, 163–175.

Stimac, J. A., Clark, A. H., Chen, Y. & Garcia, S. 1995. Enclaves and their bearing on the origin of the Cornubian batholith, southwest England. *Mineralogical Magazine*, **59**, 273–296.

Stoker, M. S., Long, D. & Fyfe, J. A. 1985. *A revised Quaternary stratigraphy for the central North Sea*. Report of the British Geological Survey, **17/2**.

Stoll, H. M. & Schrag, D. P. 1996. Evidence for glacial control of rapid sea level changes in the Early Cretaceous. *Science*, **272**, 1771–1774.

Stone, M. 1969. Nature and origin of banding in the granitic sheets, Tremearne, porthleven, Cornwall. *Geological Magazine*, **103**, 440–460.

Stone, M. 1975. Structure and petrology of the Tregonning-Godolphin granite, Cornwall. *Proceedings of the Geologists Association*, **86**, 155–170.

Stone, M. 1987. Geochemistry and origin of the Carnmenellis pluton, Cornwall: further considerations. *Proceedings of the Ussher Society*, **6**, 454–460.

Stone, M. & Exley, C. S. 1986. High heat production granites of south-west England and their mineralization: a review. *Transactions of the Institute of Mining and Metallurgy*, Section B, **95**, B25–B36.

Stone, P. & Evans, J. A. 1995. Nd-isotope study of provenance patterns across the British sector of the Iapetus Suture. *Geological Magazine*, **132**, 571–580.

Stone, P. & Evans, J. A. 1997. A comparison of the Skiddaw and Manx Groups (English Lake District and Isle of Man) using neodymium isotopes. *Proceedings of the Yorkshire Geological Society*, **51**, 343–347.

Stone, P. & Evans, J. A. 2002. Neodymium isotope characteristics of Ordovician provenance on the Avalonian margin of the Iapetus Ocean. *Scottish Journal of Geology*, **38**, 143–153.

Stone, P., Cooper, A. H. & Evans, J. A. 1999. The Skiddaw Group (English Lake District) reviewed: early Palaeozoic sedimentation and tectonism at the northern margin of Avalonia. *In*: Woodcock, N. H., Quirk, D. G., Fitches, W. R. & Barnes, R. P. (eds) *In sight of the suture: the Palaeozoic geology of the Isle of Man in its Iapetus Ocean context*. Geological Society, London, Special Publications, **160**, 325–336.

Stone, P., Floyd, J. D., Barnes, R. P. & Lintern, B. C. 1987. A sequential back-arc and foreland basin thrust duplex model for the Southern Uplands of Scotland. *Journal of the Geological Society, London*, **144**, 753–764.

Strachan, R. A., Nance, R. D., Dallmeyer, R. D., D'Lemos, R. S., Murphy, J. B. & Watt, G. R. 1996. Late Precambrian tectonothermal evolution of the Malverns Complex. *Journal of the Geological Society, London*, **153**, 589–600.

Strachan, R. A., Smith, M., Harris, A. L. & Fettes, D. J. 2002. The Northern Highland and Grampian terranes. *In*: Trewin, N. (ed.) *Geology of Scotland*. Geological Society, London, 87–147.

Strahan, A., Cantrill, T. C., Dixon, E. E. L.& Thomas, H. H. 1909. *The geology of the South Wales Coalfield, Part 10. The country around Carmarthen*. Memoir of the British Geological Survey (England and Wales), HMSO, London.

Strahan, A., Cantrill,T. C., Dixon, E. E. L., Thomas, H. H. & Jones, O. T. 1914. *The geology of the South Wales Coalfield, Part 11, The country around Haverfordwest*, Memoir of the British Geological Survey (England and Wales), HMSO, London.

Strachan, I., Temple, J. & Williams, A. 1948. The age of the neptunean dyke at Hazler Hill, South Shropshire. *Geological Magazine*, **85**, 276–287.

Straw, S. H. 1933. The fauna of the Palaeozoic rocks of the Little Missenden boring, with a report on the fish remains, by Arthur Smith Woodward. *Geological Survey of Great Britain*, Summary of Progress, 1932 (2), 112–142.

Streel, M., Higgs, K., Loboziak, S., Riegel, W. & Steemans, P. 1987. Spore stratigraphy and correlation with faunas and floras in the type marine Devonian of the Ardenne–Rhenish regions. *Review of Palaeobotany and Palynology*, **50**, 211–229.

Stringer, C. B., Currant, A. P., Schwarz, H. P. & Collcutt, S. N. 1986. Age of Pleistocene faunas from Bacon Hole, Wales. *Nature*, **320**, 59–62.

Strong, D. F., Stevens, R. K., Malpas, J. G. & Badham, J. P. N. 1975. A new tale for the Lizard (abstract). *Proceedings of the Ussher Society*, **3**, 265.

Stuart, A. J. 1976. The history of the mammal fauna during the Ipswichian/Last Interglacial in England. *Philosophical Transactions of the Royal Society of London*, **B276**, 221–250.

Stuart, A. J. & Lister, A. M. 2001. The mammalian faunas of Pakefield/Kessingland and Corton, Suffolk, UK: evidence for a new temperate episode in the British early Middle Pleistocene. *Quaternary Science Reviews*, **20**, 1677–1692.

Stuart, I. A. 1993. The geology of the Morecambe Bay gas field, east Irish Sea basin. *In*: Parker, J. R. (ed.) *Petroleum Geology of North-west Europe: Proceedings of the 4th Conference*. Geological Society, London, 883–895.

Stubblefield, C. J. 1930. A new Upper Cambrian section in south Shropshire, *Summary of progress of the Geological Survey for* 1929, Part 11, 54–62.

Stubblefield, C. J. & Bulman, O. M. B. 1927. The Shineton Shales of the Wrekin district: with notes on their development in other parts of Shropshire and Herefordshire. *Quarterly Journal of the Geological Society, London*, **83**, 96–146.

Stubblefield, C. J. & Trotter, F. M. 1957. Divisions of the Coal Measures on Geological Survey maps of England and Wales. *Bulletin of the Geological Survey of Great Britain*, **13**, 1–5.

Styles, M. T. & Kirby, G. A. 1980. New investigations of the Lizard Complex, Cornwall, England, and a discussion of an ophiolite model. *In*: Panayiotou, A. (ed.) *Ophiolites. Proceedings of the International Symposium, Cyprus 1979*. Cyprus Geological Survey Department, 517–526.

Styles, M. T. & Rundle, C. C. 1984. The Rb–Sr isochron age of the Kennack Gneiss and its bearing on the age of the Lizard Complex, Cornwall. *Journal of the Geological Society, London*, **141**, 15–19.

Sumbler, M. G. 1996*a*. *London and the Thames Valley*. *British Regional Geology*, 4th edn. HMSO, London.

Sumbler, M. G. 1996*b*. *The stratigraphy of the Chalk Group in Yorkshire, Humberside and Lincolnshire*. British Geological Survey, Technical Report, **WA/96/26C**.

Sumbler, M. G. 2001. The Moreton Drift: a further clue to glacial chronology in central England. *Proceedings of the Geologists' Association*, **112**, 13–27.

Sumbler, M. G. & Barron, A. J. M. 1996. The type section of the Hampen Formation (Middle Jurassic, Great Oolite Group) at Hampen Cutting, Gloucestershire. *Proceedings of the Cotteswold Naturalists' Field Club*, **40**, 118–128.

Sumbler, M. G., Barron, A. J. M. & Morigi, A. N. 2000. *Geology of the Cirencester District*. Memoir of the British Geological Survey, HMSO, London.

Sun, S. Q. 1989. A new interpretation of the Corallian (Upper Jurassic) cycles of the Dorset coast, southern England. *Geological Journal*, **24**, 139–158.

Surlyk, F. 1978. Submarine fan sedimentation along fault scarps on tilted fault blocks at the Jurassic/Cretaceous boundary, East Greenland. *Bulletin Grønlands Geologiske Undersøgelse*, **128**, 1–108.

Sutcliffe, A. J., Lord, T. C., Harmon, R. S., Ivanovich, M., Rae, A. & Hess, J. W. 1985. Wolverine in northern England at about 83,000 yrs BP: faunal evidence for climatic change during Isotope Stage 5. *Quaternary Research*, **24**, 73–86.

Sutherland, P. K. & Mitchell, M. 1980. *Distribution of the coelenterate order Heterocorallia in the Carboniferous of the British Isles*. Report of the British Geological Survey, **80/3**.

Sutherland, S. J. E. 1994. *Ludlow chitinozoans from the type area and adjacent regions*. Monograph of the Palaeontographical Society, London, **148**, 1–104.

SUTHREN, R. J. 1977. *Volcanic and sedimentary facies of part of the Borrowdale Volcanic Group, Cumbria*. PhD thesis, University of Keele, UK.

SVENDSEN, J. B. & HARTLEY, N. R. 2001. Comparison between outcrop-spectral gamma ray logging and whole rock geochemistry: implications for quantitative reservoir characterization in continental sequences. *Marine and Petroleum Geology*, **18**, 657–670.

SVENDSEN, J. B. & HARTLEY, N. R. 2002. Synthetic heavy mineral stratigraphy: applications and limitations. *Marine and Petroleum Geology*, **19**, 389–405.

SWIFT, A. 1989. First records of conodonts from the late Triassic of Britain. *Palaeontology*, **32**, 325–333.

SWIFT, A. 1995*a*. *Conodonts From the Late Permian and Late Triassic of Britain*. Monograph of the Palaeontological Society London, **598** (part of Volume 149 for 1995).

SWIFT, A. 1995*b*. A review of the nature and outcrop of the 'White Lias' facies of the Langport Member (Penarth group: Upper Triassic) in Britain. *Proceedings of the Geologists' Association*, **106**, 247–258.

SWIFT, A. 1999. Stratigraphy (including biostratigraphy) *In*: SWIFT, A. & MARTILL, D. M. (eds) *Fossils of the Rhaetian Penarth Group*. Palaeontological Association, London, 15–30.

SWIFT, A. & MARTILL, D. M. 1999. *Fossils of the Rhaetian Penarth Group*. Palaeontological Association, London.

SWINNERTON, H. H. 1918. The Keuper basement beds near Nottingham. *Proceedings of the Geologists' Association*, **29**, 16–28.

SWINTON, W. E. 1936. The Dinosaurs of the Isle of Wight. *Proceedings of the Geologists' Association*, **47**, 204–220.

SWITSUR, V. R. & HOUSLEY, R. A. 1998. Radiocarbon dating. *In*: PREECE, R. C. & BRIDGLAND, D. R. (eds) *Late Quaternary Environmental Change in North-west Europe: Excavations at Holywell Coombe, South-east England*. Chapman & Hall, London, 107–119.

SYLVESTER-BRADLEY, P. C. 1968. The Inferior Oolite Series. *In*: SYLVESTER-BRADLEY, P. C. & FORD, T. D. (eds) *Geology of the East Midlands*. Leicester University Press, 211–226.

SZABO, B. J. & COLLINS, D. 1975. Ages of fossil bones from British interglacial sites. *Nature*, **254**, 680–682.

SZURLIES, M. 2004. Magnetostratigraphy: the key to a global correlation of the classic Germanic Trias-case study Volprehausen Formation (Middle Buntsandstein), Central Germany. *Earth and Planetary Science Letters*, **227**, 395–410.

SZURLIES, M., BACHMANN, G. H, MENNING, M., NOWACZYK, N. R. & KÄDING, K.-C. 2003. Magnetostratigraphy and high-resolution lithostratigraphy of the Permian–Triassic boundary interval in Central Germany. *Earth and Planetary Science Letters*, **212**, 263–278.

TAIT, J. 1999. New Early Devonian palaeomagnetic data from NW France: Palaeogeography and implications for the Armorican microplate hypothesis. *Journal of Geophysical Research*, **104**, 2831–2839.

TAIT, J. A., BACHTADSE, W., FRANKE, W. & SOFFEL, H. C. 1997. Geodynamic evolution of the European Variscan fold belt: palaeomagnetic and geological constraints. *Geologische Rundschau*, **86**, 585–598.

TAIT, N. G., LERNER, D. N. SMITH, J. W. N. & LEHARNE, S. A. 2004. Prioritisation of abstraction in boreholes at risk from chlorinated solvent contamination on the UK Permo–Triassic Sandstone aquifer using a GIS. *Science of the Total Environment*, **319**, 77–98.

TAITT, A. H. & KENT, P. E. 1958. *Deep Boreholes at Portsdown (Hants.) and Henfield (Sussex)*. British Petroleum Co. Ltd, London.

TALBOT, M. R. 1973. Major sedimentary cycles in the Corallian Beds (Upper Oxfordian) of southern England. *Journal of Sedimentary Petrology*, **41**, 261–273.

TALBOT, M. R., HOLM, K. & WILLIAMS, M. A. J. 1994. Sedimentation in low gradient desert margins systems: a comparison of the late Triassic of northwest Somerset (England) and the late Quaternary of east central Australia. *In*: ROSEN, M. R. (ed.) *Palaeoclimate and Basin Evolution of Playa Systems*. Geological Society America, Special Paper, **289**, 87–119.

TANNER, L. H., LUCAS, S. & CHAPMAN, M. G. 2004. Assessing the record and causes of late Triassic extinctions. *Earth Science Reviews*, **65**, 103–139.

TAPPIN, D. I. 1994. *The Geology of Cardigan Bay and the Bristol Channel*. United Kingdom Offshore Regional Report. British Geological Survey.

TAPPIN, D. R., CHADWICK, R. A., JACKSON, A. A., WINGFIELD, R. T. R. & SMITH, N. J. P. 1994. *The Geology of Cardigan Bay and the Bristol Channel*. UK Offshore Regional Report. British Geological Survey, HMSO, London.

TAWNEY, E. B. 1866. On the western limit of the Rhaetic Beds in South Wales, and on the position of the Sutton Stone. *Quarterly Journal of the Geological Society of London*, **22**, 69–93.

TAYLOR, B. J., PRICE, R. H. & TROTTER, F. M. 1963. *Geology of the Country Around Stockport and Knutsford*. Memoir of the Geological Survey of Great Britain, England and Wales, Sheet 98, HMSO, London.

TAYLOR, F. M. 1968. Permian and Triassic Formations, *In*: SYLVESTER-BRADLEY, P. C. & FORD, T. D. (eds) *Geology of the East Midlands*. Leicester University Press, 149–173.

TAYLOR, J. A., MURDOCK, A. P. & PONTEE, N. I. 2004. A macroscale analysis of coastal steepening around the coast of England and Wales. *The Geographical Journal*, **170**, No. 3.

TAYLOR, K. & RUSHTON, A. W. A. 1972. The pre-Westphalian geology of the Warwickshire Coalfield. *Bulletin of The Geological Survey of Great Britain*, **35**. [dated 1971].

TAYLOR, K. & THOMAS, L. P. 1974. Field meeting in the Upper Old Red Sandstone of southern Breconshire. *Proceedings of the Geologists' Association*, **85**, 423–432.

TAYLOR, K. & THOMAS, L. P. 1975. Geological Survey boreholes in south Brecknockshire and their bearings on the stratigraphy of the Upper Old Red Sandstone and the Carboniferous–Old Red Sandstone boundary east of the Afon Hepste. *Bulletin of the Geological Survey of Great Britain*, **54**, 1–39.

TAYLOR, M. P., MACKLIN, M. G. & HUDSON-EDWARDS, K. 2000. River sedimentation and fluvial response to Holocene environmental change in the Yorkshire Ouse Basin, northern England. *The Holocene*, **10**, 201–213.

TAYLOR, P. W. 1951. The Plymouth Limestone: and the Devonian tetracorals of the Plymouth Limestone. *Transactions of the Royal Geological Society of Cornwall*, **18**, 146–214.

TAYLOR, S. P., SELLWOOD, B. W., GALLOIS, R. W. & CHAMBERS, M. H. 2001. Sequence stratigraphy of the Kimmeridgian and Bolonian stages (late Jurassic): Wessex–Weald Basin, southern England. *Journal of the Geological Society, London*, **158**, 179–192.

TAYLOR, S. R. 1983. A stable isotope study of the Mercia Mudstones (Keuper marl) and associated sulphate horizons in the English Midlands. *Sedimentology*, **30**, 11–31.

TEALE, C. Y. & SPEARS, D. A. 1986. The mineralogy and origin of some Silurian bentonites, Welsh Borderland, UK. *Sedimentology*, **36**, 757–765.

TER WEE, W. M. 1983. The Elsterian glaciation in The Netherlands. *In*: EHLERS, J. (ed.) *Glacial Deposits in North-west Europe*. Balkema, Rotterdam, 413–415.

THIRLWALL, M. F. 1988. Geochronology of Late Caledonian magmatism in northern Britain. *Journal of the Geological Society, London*, **145**, 951–967.

THIRLWALL, M. F. 1989. Movement on proposed terrane boundaries in northern Britain: constraints from Ordovician–Devonian igneous rocks. *Journal of the Geological Society, London*, **146**, 373–376.

THIRLWALL, M. F. & FITTON, J. G. 1983. Sm–Nd garnet age for the Ordovician Borrowdale Volcanic Group in Northern Britain. *Journal of the Geological Society, London*, **140**, 511–518.

THIRY, M., DUPUIS, C. *ET AL*. 1998. Tentative correlations between continental deposits of the argiles plastiques (Paris Basin) and Reading Beds (London Basin), based on chemostratigraphy. *Strata*, **9**, 125–129.

THOMAS, A. T., OWENS, R. M. & RUSHTON, A. W. A. 1984. *Trilobites in British Stratigraphy*. Geological Society, London, Special Report, **16**.

THOMAS, G. N. 2001. Late Middle Pleistocene pollen biostratigraphy in Britain: pitfalls and possibilities in the separation of interglacial sequences. *Quaternary Science Reviews*, **20**, 1621–1630.

THOMAS, G. S. P. 1989. The Late Devensian glaciation along the western margin of the Cheshire–Shropshire lowland. *Journal of Quaternary Science*, **4**, 167–181.

THOMAS, G. S. P., CHIVERRELL, R. C. & HUDDART, D. 2004. Ice-marginal depositional responses to readvance episodes in the Late Devensian deglaciation of the Isle of Man. *Quaternary Science Reviews*, **23**, 85–106.

THOMAS, H. H. 1909. A contribution to the petrography of the New Red Sandstone in the west of England. *Quarterly Journal of the Geological Society of London*, **65**, 230–245.

THOMAS, J. M. 1963*a*. The Culm Measures succession in Noth-East Devon and North-West Somerset. *Proceedings of the Ussher Society*, **1**, 63–64.

THOMAS, J. M. 1963*b*. Sedimentation of the Lower Culm Measures around Westleigh, Northeast Devon. *Proceedings of the Ussher Society*, **1**, 71–72.

THOMAS, J. M. 1988. Basin history of the Culm Trough of southwest England. *In*: BESLEY, B. M. & KELLING, G. (eds) *Sedimentation in a Synorogenic Basin Complex: the Upper Carboniferous of Northwest Europe*. Blackie, Glasgow, 24–37.

THOMAS, L. P. 1967. *A sedimentary study of the sandstones between the horizons of the Four-Foot Coal and the Gorllwyn Coal of the Middle Coal Measures of the South Wales Coalfield*. PhD thesis, University of Wales, Swansea.

THOMAS, L. P. 1974. The Westphalian (Coal Measures) in South Wales. *In*: OWEN, T. R. (ed.) *The Upper Palaeozoic and Post-Palaeozoic Rocks of Wales*. University of Wales Press, 131–160.

THOMAS, R. G. 1978. *The stratigraphy, palynology and sedimentology of the lower Old Red Sandstone Cosheston Group, S. W. Dyfed, Wales*. PhD thesis, University of Bristol.

THOMPSON, D. B. 1970*a*. Sedimentation of the Triassic (Scythian) red pebbly sandstones in the Cheshire Basin and its margins. *Geological Journal*, **7**, 183–216.

THOMPSON, D. B. 1970*b*. The stratigraphy of the so-called Keuper Sandstone Formation (Scythian to Anisian) in the Permo Triassic of Cheshire basin. *Journal of the Geological Society London*, **126**, 151–181.

THOMPSON, D. B. 1990. The geology of the neighborhood of Chester – an essay review. *Amateur Geologist*, **13**, 45–54.

THOMPSON, D. B. 1985. *Field excursion to the Permo-Triassic of the Cheshire, East Irish Sea, Needwood and Stafford basins*. Poroperm Excursion Guide No. 4. Poroperm-Geochem Ltd, Chester.

THOROGOOD, E. J. 1990. Provenance of pre-Devonian sediments of England and Wales: Sm–Nd isotopic evidence. *Journal of the Geological Society, London*, **147**, 591–594.

THORPE, R. S. 1972. Ocean floor basalt affinity of Precambrian glaucophane schist from Anglesey. *Nature*, Physical Sciences, **240**, 164–166.

THORPE, R. S. 1974. Aspects of magmatism and plate tectonics in the Precambrian of England and Wales. *Geological Journal*, **9**, 115–135.

THORPE, R. S. 1987. Pseudotachylite from a Precambrian shear zone in the Malvern Hills. *Proceedings of the Geologists' Association*, **98**, 205–210.

THORPE, R. S., BECKINSALE, R. D., PATCHETT, P. J., PIPER, J. A., DAVIS, G. R. & EVANS, J. A. 1984. Crustal growth and late Precambrian – early Palaeozoic plate tectonic evolution of England and Wales. *Journal of the Geological Society, London*, **141**, 521–536.

THORPE, R. S., LEAT, P. T., BEVINS, R. E. & HUGHES, D. J. 1989. Late-orogenic alkaline/subalkaline Silurian volcanism of the Skomer Volcanic Group in the Caledonides of south Wales. *Journal of the Geological Society, London*, **146**, 125–132.

THORPE, R. S., GASKARTH, J. W. & HENNEY, P. J. 1993. Composite Ordovician lamprophyre (spessartite) intrusions around the Midlands Microcraton in central Britain. *Geological Magazine*, **130**, 657–663.

TIETZSCH-TYLER, D. & PHILLIPS, E. R. 1989. Correlation of the Monian Supergroup in NW Anglesey with the Cahore Group in SE Ireland. *Journal of the Geological Society, London*, **141**, 417–418.

TIPPING, R. M. 1996. The Neolithic landscapes of the Cheviot Hills and hinterland: palaeoenvironmental evidence. *In*: FRODSHAM, P. (ed.) *Neolithic Studies in No-Man's Land. Northern Archaeology*, **13/14**, 17–33.

TODD, J. A. 1990. The stratigraphy and correlation of the Selsey Formation and Barton Clay Formation (M. Eocene) of Studley Wood, Hampshire. *Tertiary Research*, **12**, 37–50.

TODD, S. P. 2000. Taking the roof off a suture zone: basin setting and provenance of conglomerates in the ORS Dingle Basin of SW Ireland. *In*: FRIEND, P. F. & WILLIAMS, B. P. J. (ed.) *New Perspectives on the Old Red Sandstone*. Geological Society, London, Special Publications, **180**, 185–222.

TOGHILL, P. 1992. The Shelveian event, a late Ordovician tectonic episode in southern Britain (Eastern Avalonia). *Proceedings of the Geologists' Association*, **103**, 31–35.

TONKS, L. H., JONES, R. C. B., LLOYD, W. & SHERLOCK, R. L. 1931. *The Geology of Manchester and the South-east Lancashire Coalfield*. Memoir of the Geological Survey of Great Britain, Sheet 85, England and Wales, HMSO, London.

TOOLEY, M. J. 1978. The history of Hartlepool Bay. *International Journal of Nautical Archaeology and Underwater Exploration*, **7**, 71–87.

TORSVIK, T. H. & COCKS, L. R. M. 2004. Earth geography from 400 to 250 Ma: a palaeomagnetic, faunal and facies review. *Journal of the Geological Society, London*, **161**, 555–572.

TORSVIK, T. H., TRENCH, A. & MCKERROW, W. S. 1994. Implications of palaeomagnetic data from the Tortworth Silurian Inlier (southern Britain) to palaeogeography and Variscan tectonism. *Geophysics Journal International*, **119**, 91–100.

TOWNSEND, J. 1813. *The Character of Moses Established for Veracity as an Historian Recording Events From the Creation to the Deluge*. 1, M. Gye, Bath, Longman, London.

TRAYNOR, J.-J. 1988. The Arenig of South Wales: sedimentary and volcanic processes during the initiation of a marginal basin. *Geological Journal*, **23**, 275–292.

TRAYNOR, J.-J. 1990. Arenig sedimentation and basin tectonics in the Harlech Dome area (Dolgellau Basin), Nort Wales. *Geological Magazine*, **127**, 13–30.

TREAGUS, S. H., TREAGUS, J. E. & DROOP, G. T. R. 2003. Superimposed deformations and their hybrid effects: the Rhoscolyn Anticline unravelled. *Journal of the Geological Society, London*, **160**, 137–150.

TRENCH, A. & TORSVIK, T. H. 1991. A revised Paleozoic apparent polar wander path for south Britain (eastern Avalonia). *Geophysical Journal International*, **104**, 227–233.

TRESISE, G. R. 1960. Aspects of the lithology of the Wessex Upper Greensand. *Proceedings of the Geologists' Association*, **71**, 316–339.

TRENCH, A. & TORSVIK, T. H. 1992. The closure of the Iapetus Ocean and Tornquist Sea: new palaeomagnetic constraints. *Journal of the Geological Society, London*, **149**, 867–870.

TROTTER, F. M. 1954. Reddened beds in the Coal Measures of south Lancashire. *Bulletin of the Geological Survey of Great Britain*, **5**, 61–80.

TROTTER, F. M. & HOLLINGWORTH, S. E. 1932. *The Geology of the Brampton District*. Memoir of the Geological Survey of Great Britain, Sheet 18, England and Wales, HMSO, London.

TROTTER, F. M., HOLLINGWORTH, S. E., EASTWOOD, T. & ROSE, W. C. C. 1937. *The geology of the Gosforth district*. Memoir of the Geological Survey of Great Britain, Sheet 37, England and Wales, HMSO, London.

TROWSDALE, S. A. & LERNER, D. N. 2003. Implications of flow in the sandstone aquifer beneath the mature conurbation of Nottingham for source protection. *Quarterly Journal Engineering Geology and Hydrogeology*, **36**, 197–206.

TRUEMAN, A. E. 1922. The Liassic rocks of Glamorgan. *Proceedings of the Geologists' Association*, **33**, 245–284, pls 8–10.

Trueman, A. E. 1923. Some theoretical aspects of correlation. *Proceedings of the Geologists' Association*, **34**, 193–206.

Trueman, A. E. & Weir, J. 1946. *A Monograph of British Carboniferous Non-marine Lamellibranchia*. Palaeontographical Society Monograph, **32**, Part 1.

Tucker, M. E. 1969. Crinoidal turbidites from the Devonian of Cornwall and their palaeogeographic significance. *Sedimentology*, **13**, 281–290.

Tucker, M. E. 1978. The marginal Triassic deposits of south Wales: shore zone clastics, evaporates and carbonates. *In*: Matter, A. & Tucker, M. E. (eds) *Modern and Ancient Lake Sediments*. International Association of Sedimentologists, Special Publication, **2**, 205–224.

Tucker, M. E. 1991. Sequence stratigraphy of carbonate – evaporite basins: models and applications to the Upper Permian (Zechstein) of northeast England and adjoining North Sea. *Journal of the Geological Society, London*, **148**, 1019–1036.

Tucker, M. E & Burchette, T. P. 1977. Triassic dinosaur footprints from South Wales: their context and preservation. *Palaeogeography, Palaeoclimatology, Palaeoecology*, **22**, 195–208.

Tucker, R. D. & McKerrow, W. S. 1995. Early Palaeozoic chronology: a review in the light of new U–Pb zircon ages from Newfoundland and Britain. *Canadian Journal of Earth Sciences*, **32**, 368–379.

Tucker, R. D. & Pharaoh, T. C. 1991. U–Pb zircon ages for Late Precambrian rocks in southern Britain. *Journal of the Geological Society, London*, **148**, 435–443.

Tucker, R. D., Bradley, D. C., Ver Straeten, C. A., Harris, A. G., Ebert, J. R. & McCutcheon, S. R. 1998. New U–Pb zircon ages and the duration and division of Devonian time. *Earth and Planetary Science Letters*, **158**, 175–186.

Tucker, R. M. 1981. Giant polygons in the Triassic salt of Cheshire, England: a thermal contraction model for their origin. *Journal of Sedimentary Petrology*, **51**, 779–786.

Tunbridge, I. P. 1978. North Devon: Lower and Middle Devonian. Lynton Beds and Hangman Sandstone Group. *In*: Scrutton, C. T. (ed.) *International Symposium on the Devonian System. A Field Guide to Selected Areas of South-West England*. Palaeontological Association, 8–13.

Tunbridge, I. P. 1981*a*. Sandy high-energy flood sedimentation – some criteria for recognition, with an example from the Devonian of S. W. England. *Sedimentary Geology*, **28**, 79–95.

Tunbridge, I. P. 1981*b*. Old Red Sandstone sedimentation – an example from the Brownstones (highest Lower Old Red Sandstone) of south central Wales. *Geological Journal*, **16**, 111–124.

Tunbridge, I. P. 1984. Facies model for a sandy ephemeral stream and clay playa complex; the Middle Devonian Trentishoe Formation of north Devon, UK. *Sedimentology*, **31**, 697–715.

Tunbridge, I. P. 1986. Mid-Devonian tectonics and sedimentation in the Bristol Channel area. *Journal of the Geological Society, London*, **143**, 107–115.

Turner, B. R., Younger, P. L. & Fordham, C. E. 1993. Fell Sandstone Group lithostratigraphy south-west of Berwick-upon-Tweed: implications for the regional development of the Fell Sandstone. *Proceedings of the Yorkshire Geological Society*, **49**, 269–281.

Turner, C. 1970. The Middle Pleistocene deposits at Mark's Tey, Essex. *Philosophical Transactions of the Royal Society of London*, **B257**, 373–440.

Turner, C. 1996. A brief survey of the early Middle Pleistocene in Europe. *In*: Turner, C. (ed.) *The Early Middle Pleistocene in Europe*. Balkema, Rotterdam, 295–317.

Turner, J. P. 1996. Gravity-driven nappes and their relation to palaeobathymetry: examples from West Africa and Cardigan Bay, UK. *In*: Buchanan, P. G. & Nieuwland, D. A. (eds) *Modern Developments in Structural Interpretation, Validation and Modelling*. Geological Society, London, Special Publications, **99**, 345–362.

Turner, P., Burley, S. D., Rey, D. & Prosser, J. 1995. Burial History of the Penrith Sandstone (Lower Permian) Deduced From the Combined Study of Fluid Inclusion and Palaeomagnetic Data. *In*: Turner, P. & Turner, A. (eds) *Palaeomagnetic Applications in Hydrocarbon Exploration and Production*. Geological Society, London, Special Publications, **98**, 43–78.

Turner, R. C. & Scaife, R. G. (eds) 1995. *Bog Bodies: New Discoveries and New Perspectives*. British Museum, London.

Turner, S. 1949. The deeper structure of central and northern England. *Proceedings of the Yorkshire Geological Society*, **27**, 280–297.

Turner, S. 1973. Siluro-Devonian thelodonts from the Welsh Borderland. *Journal of the Geological Society, London*, **129**, 557–584.

Turner, S. 1984. *Studies on Palaeozoic Thelodonti* (Craniata: Agnatha). PhD thesis, University of Newcastle-upon-Tyne.

Turner, S. 2000. New Llandovery to early Přídolí microvertebrates including Lower Silurian zone fossil, Loganellia avonia nov. sp., from Britain. *Courier Forschungs-Institut Senckenberg*, **223**, 91–127.

Tutcher, J. W. & Trueman, A. E. 1925. The Liassic rocks of the Radstock district, Somerset. *Quarterly Journal of the Geological Society of London*, **81**, 595–666, pls 38–41.

Tyler, J. E. & Woodcock, N. H. 1987. Bailey Hill Formation: Ludlow Series turbidites in the Welsh Borderland, reinterpreted as distal storm deposits. *Geological Journal*, **22**, 73–86.

Tyson, R. V. 2004. Variation in marine total organic carbon through the type Kimmeridge Clay Formation (Late Jurassic), Dorset, UK. *Journal of the Geological Society, London*, **161**, 667–673.

Tyson, R. V., Wilson, R. C. L. & Downie, C. 1979. A stratified water column environmental model for the type Kimmeridge Clay. *Nature*, **277**, 377–380.

Underhill, J. R. 2002. *Evidence for structural controls on the deposition of the Late Jurassic–Early Cretaceous Purbeck Limestone Group, Dorset, Southern England. Special Papers in Palaeontology*, **68**, 21–40.

Underhill, J. R. & Partington, M. A. 1993. Jurassic thermal doming and deflation in the North Sea: implications of the sequence stratigraphic evidence. *In*: Parker, J. R. (ed.) *Petroleum Geology of Northwest Europe; Proceedings of the 4th Conference*. Geological Society, London, 337–345.

Underhill, J. R. & Stoneley, R. 1998. Introduction to the development, evolution and petroleum geology of the Wessex Basin. *In*: Underhill, J. R. (ed.) *Development, Evolution and Petroleum Geology of the Wessex Basin*. Geological Society, London, Special Publications, **133**, 1–18.

University of Newcastle. 1998. *Erosion of the saltmarshes of Essex between 1988 and 1998*. Internal report to the Environment Agency, Anglian Region.

Upton, B. G. J., Aspen, P. & Chapman, N. A. 1983. The upper mantle–deep crust beneath the British Isles: Evidence from inclusion suites in volcanic rocks. *Journal of the Geological Society, London*, **140**, 105–122.

Upton, B. G. J. 1982. Carboniferous to Permian volcanism in the stable foreland. *In*: Sutherland, D. S. (ed.) *Igneous rocks of the British Isles*. Wiley, Chichester, 255–275.

Ussher, W. A. E. 1890. The Devonian rocks of south Devon. *Quarterly Journal of the Geological Society of London*, **46**, 487–517.

Ussher, W. A. E. 1903. *The Geology of the country around Torquay*. Memoir of the Geological Survey of Great Britain, Sheet 350, England and Wales, HMSO, London.

Ussher, W. A. E. 1907. *The Geology of the country around Plymouth and Liskeard*. Memoir of the Geological Survey of Great Britain, Sheet 348, England and Wales, HMSO, London.

Utescher, T., Mosbrugger, V. & Ashraf, A. R. 2000. Terrestrial climate evolution in Northwest Germany over the last 25 milion years. *Palaios*, **15**, 430–449.

Vail, P. R., Mitchum, R. M. *et al.* 1977. *Seismic Stratigraphy and Global Changes of Sea Level*. AAPG, Memoir, **26**, 49–212.

Van Buchem, F. S. P & Knox, R. W. O'B. 1998. Lower and Middle Jurassic depositional sequences of Yorkshire (UK) *In*: de Graciansky, P.-C., Hardenbol, J., Jacquin, T. & Vail, P. R. (eds) *Mesozoic – Cenozoic Sequence Stratigraphy of European Basins*. SEPM, Special Publications, **60**, 545–559.

Van der Plicht, H. 2002. Calibrating the 14C time scale: towards the complete dating range. *Netherlands Journal of Geosciences*, **81**, 85–96.

Van der Voo, R. 1983. Palaeomagnetic constraints on the assembly of the Old Red Continent. *Tectonophysics*, **9**, 271–283.

Van Konijnenburg-van Cittert, J. H. A. & Morgans, H. S. 1999. *The Jurassic Flora of Yorkshire*. Palaeontological Association Field Guide to Fossils, **8**.

van Staal, C. R., Dewey, J. F., MacNiocaill, C. & McKerrow, W. S. 1998. The Cambrian–Silurian tectonic evolution of the Northern Appalachians and Caledonides: history of a complex, west and southwest Pacific-type segment of Iapetus. *In*: Blundell, D. J. & Scott, A. C. (eds) *Lyell: The Past is the Key to the Present*. Geological Society, London, Special Publications, **143**, 199–242.

Van Straaten, P. & Tucker, M. E. 1972. The Upper Devonian Saltern Cove goniatite bed as an intraformational slump. *Palaeontology*, **15**, 430–438.

Van Wagoner, J., Mitchum, V., Campion, K. & Rahmanian, V. 1990. *Siliciclastic Sequence Stratigraphy in Well Logs and Outcrops*. AAPG, Methods in Exploration Series, **7**.

Van Wagoner, J., Posamentier, H., Mitchum, R., Vail, P., Sarg, J., Loutit, T. & Hardenbol, J. 1988. An overview of the fundamentals of sequence stratigraphy and key definitions. *In*: Wilgnus, C. K., Hastings, C. G. S. C., Kendall, H. W., Posamentier, H. W., Ross, C. A. & Van Wagoner, J. C. (eds) *Sea-level Changes: An Integrated approach*. SEPM, Special Publications, **42**, 39–45.

Vandenberghe, N., Laga, P., Steurbaut, E., Hardenbol, J. & Vail, P. R. 1998. Tertiary sequence stratigraphy at the southern border of the North Sea Basin in Belgium. *In*: De Graciansky, P. C., Hardenbol, J., Vail, P. R. & Jacquin, T. (eds) *Sequence Stratigraphy of Mesozoic and Cenozoic European Basins*. SEPM Special Publications, **60**, 119–154.

Varker, W. J. & Higgins, A. C. 1979. Conodont evidence for the age of the Pinskey Gill Beds of Ravenstonedale, north-west England. *Proceedings of the Yorkshire Geological Society*, **42**, 357–369.

Varker, W. J. & Sevastopulo, G. D. 1985. The Carboniferous System: part 1 – conodonts of the Dinantian Subsystem from Great Britain and Ireland. *In*: Higgins, A. C. & Austin, R. L. (eds) *A Stratigraphical Index of Conodonts*. Ellis Horwood, Chichester, 167–209.

Vaughan, A. 1905. Palaeontological sequence in the Carboniferous Limestone of the Bristol area. *Quarterly Journal of the Geological Society of London*, **61**, 181–307.

Veevers, J. J. & Powell, C. M. 1987. Late Paleozoic glacial episodes in gondwanaland reflected in transgressive–regressive depositional sequences in Euramerica. *Geological Society of America Bulletin*, **98**, 475–487.

Ventris, P. A. 1996. Hoxnian Interglacial freshwater and marine deposits in Northwest Norfolk, England, and their implications for sea-level reconstruction. *Quaternary Science Reviews*, **15**, 437–451.

Vergoossen, J. M. J. 1999. Siluro-Devonian microfossils of acanthodii and chondrichthyes (pisces) from the Welsh Borderlan/ South Wales. *Modern Geology*, **24**, 23–90.

Vergoossen, J. M. J. 2000. Acanthodian and chondrichthyan microremains in the Siluro-Devonian of the Welsh Borderland, Great Britain, and their biostratigraphical potential. *Courier Forschungsinstitut Senckenberg*, **223**, 175–199.

Verniers, J. 1999. Calibration of Chitinozoa versus graptolite biozonation in the Wenlock of the Builth Wells district (Wales, UK.), compared with other areas in Avalonia and Baltica. *Bollettino della Societá Paleontologica Italiana*, **38**, 359–380.

Versey, H. C. & Carter, C. 1926. The Petrography of the Carstone and associated beds in Yorkshire and Lincolnshire. *Proceedings of the Yorkshire Geological Society*, **20**, 349–365.

Vinken, R. (compiler) 1988. *The Northwest European Tertiary Basin*. Results of the IGCP Project No. 124. *Geologisches Jahrbuch*, **A100**.

Wach, G. D. & Ruffell, A. H. 1991. *Sedimentology and sequence stratigraphy of a Lower Cretaceous tide and storm-dominated clastic succession, Isle of Wight and S.E. England*. Field Guide 4, 13th Congress, International Association of Sedimentologists, Nottingham.

Wade, A. 1911. The Landovery and associated rocks of northeastern, Montgomeryshire. *Quarterly Journal of the Geological Society*, **67**, 415–59.

Wadge, A. J. 1972. Sections through the Skiddaw–Borrowdale unconformity in eastern Lakeland. *Proceedings of the Yorkshire Geological Society*, **39**, 179–198.

Wadge, A. J., Nutt, M. J. C. & Skevington, D. 1972. Geology of the Tarn Moor Tunnel in the Lake District. *Bulletin of the Geological Survey of Great Britain*, **41**, 55–73.

Wadge, A. J., Nutt, M. J. C., Lister, T. R. & Skevington, D. 1969. A probable Didymograptus murchisoni Zone fauna from the Lake District. *Geological Magazine*, **106**, 595–598.

Walkden, G. M. 1972. The mineralogy and origin of interbedded clay wayboards in the Lower Carboniferous of the Derbyshire Dome. *Geological Journal*, **8**, 143–160.

Walkden, G. M. 1974. Palaeokarstic surfaces in upper Visean (Carboniferous) Limestones of the Derbyshire Block, England. *Journal of Sedimentary Petrology*, **44**, 1232–1247.

Walkden, G. M. 1977. Volcanic and erosive events on an Upper Viséan carbonate platform, North Derbyshire. *Proceedings of the Yorkshire Geological Society*, **41**, 347–367.

Walkden, G. M. 1987. Sedimentary and diagenetic styles in late Dinantian carbonates of Britain. *In*: Miller, J., Adams, A. E. & Wright, V. P. (eds) *European Dinantian Environments*. Wiley, Chichester, 131–155.

Walkden, G. M., Parker, J. & Kelley, S. 2002. A late Triassic impact ejecta layer in southwestern Britain. *Science*, **298**, 2185–2188.

Walkden, G. M. & Walkden, G. D. 1990. Cyclic sedimentation in carbonate and mixed carbonate clastic environments: four simulation programs for a desk-top computer. *In*: Tucker, M. E., Wilson, J. L., Crevello, P. D., Sarj, J. R. & Read, J. F. (eds) *Carbonate Platforms: Facies, Sequences and Evolution*. Association of Sedimentologists, Special Publications, **9**, 55–78.

Walker, D. 1956. A site at Stump Cross, near Grassington, Yorkshire, and the age of the Pennine microlithic industry. *Proceedings of the Prehistoric Society*, **22**, 23–28.

Walker, E. E. 1904. Notes on the garnet-bearing and associated rocks of the Borrowdale Volcanic Series. *Quarterly Journal of the Geological Society of London*, **60**, 70–104.

Walker, I. M. & Cooper, W. G. 1987. The structural and stratigraphic evolution of the northeast margin of the Sole Pit Basin. *In*: Brooks, J. & Glennie, K. W. (eds) *Petroleum Geology of North West Europe*. Graham and Trotman, London, 263–275.

Walker, M. J. C., Coope, G. R. & Lowe, J. J. 1993. The Devensian (Weichselian) Lateglacial palaeoenvironmental record from Gransmoor, East Yorkshire. *Quaternary Science Reviews*, **12**, 659–680.

Walker, M. J. C., Björck, S. *et al.*1999. Isotopic 'events' in the GRIP ice core: a stratotype for the Late Pleistocene. *Quaternary Science Reviews*, **18**, 1143–1150.

Walker, M. J. C., Coope, G. R., Sheldrick, C., Turney, C. S. M., Lowe, J. J., Blockley, S. P. E. & Harkness, D. D. 2003. Devensian Lateglacial environmental changes in Britain: a multi-proxy record from Llanilid, South Wales, UK. *Quaternary Science Reviews*, **22**, 475–520.

Walker, R. G. 1966*a*. Shale Grit and Grindsow Shales: Transition from turbidite to shallow water sediments in the Upper Carboniferous of Northern England. *Journal of Sedimentary Petrology*, **36**, 90–114.

Walker, R. G. 1966*b*. Deep channels in turbidite-bearing formations. *AAPG Bulletin*, **50**, 1899–1917.

Waller, M. 1994. *The Fenland Project, Number 9: Flandrian Environmental Change in Fenland, East Anglia*. Archaeology Report 70, Fenland Project Committee. Cambridgeshire County Council, Cambridge.

Waller, M. P. & Long, A. J. 2003. Holocene coastal evolution and sea-level change on the southern coast of England: a review. *Journal of Quaternary Science*, **18**, 351–360.

Walmsley, V. G. 1959. The geology of the Usk inlier. *Quarterly Journal of the Geological Society of London*, **114**, 483–521.

WALMSLEY, V. G. & BASSETT, M. G. 1976. Biostratigraphy and correlation of the Coralliferous Group and Gray Sandstone Group (Silurian) of Pembrokeshire, Wales. *Proceedings of the Geologists' Association*, **87**, 191–220.

WALSH, P. T., ATKINSON, K., BOULTER, M. C. & SHAKESBY, R. A. 1987. The Oligocene and Miocene outliers of west Cornwall and their bearing on the geomorphological evolution of Oldland Britain. *Philosophical Transactions of the Royal Society of London, Series A*, **323**, 211–245.

WALSH, P., BOULTER, M. & MORAWIECKA, I. 1999. Evolving models of the Tertiary evolutionary geomorphology of southern England, with special reference to the Chalklands. *In*: SMITH, B. J., WHALLEY, W. B. & WARKE, P. A. (eds) *Uplift, Erosion and Stability: Perspectives on Long-term Landscape Development*. Geological Society, London, Special Publications, **162**, 45–63.

WALTERS, S. G. & INESON, P. R. 1981. A review of the distribution of and the correlation of igneous rocks in Derbyshire, England. *Mercian Geologist*, **8**, 81–132.

WALTHAM, A. C. 1971. Shale units in the Great Scar Limestone of the southern Askrigg Block. *Proceedings of the Yorkshire Geological Society*, **38**, 285–292.

WALTHAM, A. C. & DIXON, N. 2000. Movement of the Mam Tor landslide, Derbyshire, UK. *Quarterly Journal of Engineering Geology and Hydrogeology*, **33**, 105–123.

WARD, W. H., BURLAND, J. B. & GALLOIS, R. W. 1968. Geotechnical assessment of a site at Mundford, Norfolk, for a large proton accelerator. *Géotechnique*, **18**, 399–431.

WARE, P. D. & TURNER, J. P. 2002. Sonic velocity analysis of the Tertiary denudation of the Irish Sea Basin. *In*: DORÉ, A. G., CARTWRIGHT, J. A., STOKER, M. S., TURNER, J. P. & WHITE, N. (eds) *Exhumation of the Northern Atlantic Margin: Timing, Mechanisms and Implications for Petroleum Exploration*. Geological Society, London, Special Publications, **196**, 355–370.

WARE, W. D. 1939. The Millstone grit of Carmarthenshire. *Proceedings of the Geologists' Association*, **50**, 168–204.

WARNER, R. 1811. *A New Guide Through Bath and its Environs*. Bath.

WARR, L. N. 1989. The structural evolution of the Davidstow Anticline, and its relationship to the Southern Culm Overfold, north Cornwall. *Proceedings of the Ussher Society*, **7**, 136–140.

WARR, L. N. 1993. Basin inversion and foreland basin development in the Rhenohercynian of south-west England. *In*: GAYER, R. A., GREILING, R. O. & VOGEL, A. K. (eds) *The Rhenohercynian and Sub-Variscan Fold belts*. Vieweg,Wiesbaden, 197–224.

WARR, L. N. 2000. The Variscan Orogeny: the welding of Pangaea. *In*: WOODCOCK, N. & STRAHAN, R. (eds) *Geological History of Britain and Ireland*. Blackwell, Oxford, 271–294.

WARR, L. N., PRIMMER, T. J. & ROBINSON, D. 1991. Variscan very low-grade metamorphism in southwest England: a diastathermal and thrust related origin. *Journal of Metamorphic Geology*, **9**, 751–764.

WARREN, P. T., PRICE, D., NUTT, M. J. C. & SMITH, E. G. 1984. *Geology of the Country Around Rhyl and Denbigh*. Memoir of the British Geological Survey, Sheets 95 and 107 and parts of sheets 94 and 106, England and Wales, HMSO, London.

WARRINGTON, G. 1974. Trias. *In*: RAYNER, D. H. & HEMINGWAY, J. E. (eds) *Geology and Mineral Resources of Yorkshire*. Yorkshire Geological Society, 145–160.

WARRINGTON, G. 1981. The indigenous micropalaeontology of British Triassic shelf sea deposits. *In*: NEALE, J. W. & BRASIER, M. D. (eds) *Microfossils from Recent and Fossil Shelf Seas*. Ellis Horwood for the British Micropalaeontological Society, Chichester, 61–70.

WARRINGTON, G. 1984. Late Triassic palynomorph records from Somerset. *Proceedings of the Ussher Society*, **6**, 29–34.

WARRINGTON, G. 1996. Triassic spores and pollen, *In*: JANONIUS, J. & MCGREGOR, D. C. (ed.) *Palynology Principles and Applications*. America Association of Stratigraphic Palynologists Foundation, **2**, 755–766.

WARRINGTON, G. 1998. The Penarth Group – Lias Group succession (Late Triassic – Early Jurassic) in the East Irish Sea Basin and neighbouring areas: a stratigraphical review. *In*: MEADOWS, N. S., TRUEBLOOD, S., HARDMAN, M. & COWAN, G. (eds) *The Petroleum Geology of the Irish Sea and Adjacent Areas*. Geological Society, London, Special Publications, **124**, 33–46. [Dated 1997.]

WARRINGTON, G. 2002. *Palynology report: the Hilton Borehole, Cumbria*. British Geological Survey Technical Report, IR/02/060.

WARRINGTON, G. & IVIMEY-COOK, H. C. 1992. Triassic, *In*: COPE, J. C. W., INGHAM, J. K. & RAWSON, P. F. (eds) *Atlas of palaeogeography and Lithofacies*. Geological Society, London, Memoirs, **13**, 97–106.

WARRINGTON, G. & IVIMEY-COOK, H. C. 1995. The Late Triassic and Early Jurassic of coastal sections in west Somerset and South and Mid-Glamorgan. *In*: TAYLOR, P. D. (ed.) *Field Geology of the British Jurassic*. Geological Society, London, 9–30.

WARRINGTON, G. & OWENS, B. (compilers). 1977. *Micropalaeontological Biostratigraphy of offshore samples from south-west Britain*. Institute of Geological Sciences Report, **77/7**.

WARRINGTON, G. & SCRIVENER, R. C. 1990. The Permian of Devon, England. *Review of Palaeobotany and Palynology*, **66**, 263–272.

WARRINGTON, G. & WHITTAKER, A. 1984. The Blue Anchor Formation (Late Triassic) in Somerset. *Proceedings of the Ussher Society*, **6**, 100–107.

WARRINGTON, G., AUDLEY-CHARLES, M. G. *ET AL*. 1980. *A Correlation of Triassic Rocks in the British Isles*. Geological Society, London, Special Reports, **13**.

WARRINGTON, G., WILSON, A. A., JONES, N. S., YOUNG, S. R. & HASLAM, H. W. 1999. Stratigraphy and sedimentology in the Cheshire Basin. *In*: PLANT, J. A., JONES, D. G. & HASLAM, H. W. (eds) *Basin Evolution, Fluid Movement and Mineral Resources in a Permo-Triassic Rift Setting*. British Geological Survey. HMSO, London, 23–46.

WATERS, C. N. 2003*a*. Introduction. *In*: STEPHENSON, D., LOUGHLIN, S. C., MILLWARD, D., WATERS, C. N. & WILLIAMSON, I. T. (eds) *Carboniferous and Permian Igneous Rocks of Great Britain North of the Variscan Front*. Joint Nature Conservation Committee. Geological Conservation review Series, **27**, 281–285.

WATERS, C. N. 2003*b*. Barrow Hill, Dudley. *In*: STEPHENSON, D., LOUGHLIN, S. C., MILLWARD, D., WATERS, C. N. & WILLIAMSON, I. T. (eds) *Carboniferous and Permian Igneous Rocks of Great Britain North of the Variscan Front*. Joint Nature Conservation Committee. Geological Conservation Review Series, **27**, 301–305.

WATERS, C. N., GLOVER, B. W. & POWELL, J. H. 1994. Structural synthesis of S Staffordshire, UK: implications for the Variscan evolution of the Pennine Basin. *Journal of the Geological Society London*, **151**, 697–713.

WATERS, C. N., GLOVER, B. W. & POWELL, J. H. 1995. Discussion on structural synthesis of south Staffordshire, UK: implications for the Variscan evolution of the Pennine Basin. *Journal of the Geological Society, London*, **152**, 197–200.

WATERS, C. N., BROWNE, M. A. E., DEAN, M. T. & POWELL, J. H. 2006. *BGS Lithostratigraphical framework for Carboniferous successions of Great Britain (Onshore)*. British Geological Survey Research Report, **RR/05/06**.

WATERS, C. N., BARCLAY, W. J., DAVIES, J. R. & WATERS, R. A. In press. *Stratigraphical framework for Carboniferous successions of Southern Great Britain (Onshore)*. British Geological Survey Research Report.

WATERS, R. A. 1974. *Palaeozoic successions and structures between Little Haldon and Dartmoor*. Unpublished PhD thesis, University of Exeter.

WATERS, R. A. & LAWRENCE, D. J. D. 1987. *Geology of the South Wales Coalfield, Part III, the Country Around Cardiff*, 3rd edn. Memoir of the British Geological Survey, Sheet 263 (England & Wales). HMSO, London.

WATERS, R. A., DAVIES, J. R., FLETCHER, C. J. N., WILSON, D., ZALASIEWICZ, J. A. & CAVE, R. 1993. Llandovery basinal and slope sequences of the Rhayader district. *In*: WOODCOCK, N. H. & BASSETT, M. G. (eds) *Geological Excursions in Powys*. University of Wales Press, National Museum of Wales, Cardiff, 155–182.

WATKINS, R. 1979. Benthic community organization in the Ludlow Series of the Welsh Borderland. *Bulletin of the British Museum (Natural History), Geology*, **31**, 3.

WATKINS, R. & AITHIE, C. J. 1980. Carbonate shelf environments and faunal communities in the Upper Bringewood Beds of the British Silurian. *Palaeogeography, Palaeoclimatology, Palaeoecology*, **39**, 342–368.

WATSON, E. & WATSON, S. 1974. Remains of pingos in the Cletwr Basin, southwest Wales. *Geografiska Annaler*, **56A**, 213–225.

WATSON, J. & SINCOCK, C. A. 1992. *Bennettitales of the English Wealden*. Monograph of the Palaeontographical Society, 1–228.

WATSON, J. V. 1984. The ending of the Caledonian Orogeny in Scotland. *Journal of the Geological Society, London*, **141**, 193–214.

WATSON, J. V., FOWLER, M. B. & PLANT, J. A. 1984. Variscan–Caledonian comparisons: late-orogenic granites. *Proceedings of the Ussher Society*, **6**, 2–12.

WATTS, W. W. 1903. Charnwood Forest: A buried Triassic lanscape. *Geographical Journal*, **21**, 623–636.

WATTS, W. W. 1925. The geology of south Shropshire. *Proceedings of the Geologists' Association*, **36**, 322–393.

WAUGH, B. 1970. Petrology, provenance and silica diagenesis of the Penrith Sandstone (Lower Permian) of north-west England. *Journal of Sedimentary Petrology*, **40**, 1226–1240.

WEBB, B. C. & COOPER, A. H. 1988. Slump folds and gravity slide structures in a Lower Palaeozoic marginal basin sequence (the Skiddaw Group) N. W. England. *Journal of Structural Geology*, **10**, 463–472.

WEBBY, B. D. 1965*a*. The Middle Devonian marine transgression in north Devon and west Somerset. *Geological Magazine*, **102**, 478–488.

WEBBY, B. D. 1965*b*. The stratigraphy and structure of the Devonian rocks in the Quantock Hills, west Somerset. *Proceedings of the Geologists' Association*, **76**, 321–343.

WEBBY, B. D. 1965*c*. The stratigraphy and structure of the Devonian rocks in the Brendon Hills, west Somerset. *Proceedings of the Geologists' Association*, **76**, 39–60.

WEBBY, B. D. 1966. Middle–Upper Devonian palaeogeography of north Devon and west Somerset, England. *Palaeogeography, Palaeoclimatology, Palaeoecology*, **2**, 27–46.

WEBBY, B. D. & THOMAS, J. M. 1965. Whitsun Field meeting: Devonian of West Somerset and Carboniferous of North-West Devon, 15–18 may 1964. *Proceedings of the Geologists' Association*, **76**, 179–194.

WEBSTER, T. 1816. *In*: ENGLEFIELD, H. C. *A description of the principal picturesque beauties, antiquities, and geological phenomena of the Isle of Wight*. With additional observations on the strata of the island, and their continuation in the adjacent parts of Dorsetshire.

WEDD, C. B., SMITH, B. & WILLS, L. J. 1927. *The country around Wrexham.*, Part 1. Memoir of the British Geological Survey, England and Wales (Explanation of Sheet **121**), HMSO, London.

WEEDON, G. P. 1986. Hemipelagic shelf sedimentation and climatic cycles: the basal Jurassic (Blue Lias) of south Britain. *Earth and Planetary Science Letters*, **76**, 321–335.

WEEDON, G. P. & JENKYNS, H. C. 1990. Regular and irregular climatic cycles and the Belemnite Marls (Pliensbachian, Wessex Basin, UK). *Journal of the Geological Society, London*, **147**, 915–918.

WEEDON, G. P., JENKYNS, H. C., COE, A. L. & HESSELBO, S. P. 1999. Astronomical calibration of the Jurassic time scale from cyclostratigraphy in British mudrock formations. *Philosophical Transactions of the Royal Society of London*, **A357**, 1787–1813.

WEEDON, G. P., COE, A. L. & GALLOIS, R. W. 2004. Cyclostratigraphy, orbital tuning and inferred productivity for the type Kimmeridge Clay (Late Jurassic), Southern England. *Journal of the Geological Society, London*, **161**, 655–666.

WEIDLICH, O. & BERNECKER, M. 2003. Supersequence and composite sequence carbonate platform growth: Permian and Triassic outcrop data of the Arabian platform and Neo-Tethys. *Sedimentary Geology*, **158**, 87–116.

WEIR, A. H., CATT, J. A. & MADGETT, P. A. 1971. Postglacial soil formation in the loess of Pegwell Bay, Kent (England). *Geoderma*, **5**, 131–149.

WEISSERT, H., LINI, A., FÖLLMI, K. B. & KUHN, O. 1998. Correlation of Early Cretaceous carbon-isotope stratigraphy and platform drowning events: a possible link? *Palaeogeography, Palaeoclimatology, Palaeoecology*, **137**, 189–203.

WELLMAN, C. H., HABGOOD, K., JENKINS, G. & RICHARDSON, J. B. 2000. A new plant assemblage (microfossil and megafossil) from the Lower Old Red Sandstone of the Anglo-Welsh Basin: its implications for the palaeoecology of early terrestrial ecosystems. *Review of Palaeobotany and Palynology*, **109**, 161–196.

WELLMAN, C. H., THOMAS, R. G., EDWARDS, D. & KENRICK, P. 1998. The Cosheston Group (Lower Old Red Sandstone) in southwest Wales: age, correlation and palaeobotanical significance. *Geological Magazine*, **135**, 397–412.

WELLS, C., HUCKERBY, E. & HALL, V. A. 1997. Mid- and Late-Holocene vegetation history and tephra studies at Fenton Cottage, Lancashire, UK. *Vegetation History and Archaeobotany*, **6**, 153–166.

WEST, I. M. 1975. Evaporites and associated sediments of the basal Purbeck Formation (Upper Jurassic) of Dorset. *Proceedings of the Geologists' Association*, **86**, 205–225.

WEST, R. G. 1956. The Quaternary deposits at Hoxne, Suffolk. *Philosophical Transactions of the Royal Society of London*, **B239**, 265–356.

WEST, R. G. 1957. Interglacial deposits at Bobbitshole, Ipswich. *Philosophical Transactions of the Royal Society of London*, **B241**, 33–44.

WEST, R. G. 1961. Vegetational history of the Early Pleistocene of the Royal Society borehole at Ludham, Norfolk. *Proceedings of the Royal Society of London*, **B155**, 437–453.

WEST, R. G. 1963. Problems of the British Quaternary. *Proceedings of the Geologists' Association*, **74**, 147–186.

WEST, R. G. 1979. Further on the Flandrian. *Boreas*, **8**, 426.

WEST, R. G. 1980*a*. Pleistocene forest history in East Anglia. *New Phytologist*, **87**, 127–137.

WEST, R. G. 1980*b*. *The Pre-glacial Pleistocene of the Norfolk and Suffolk Coasts*. Cambridge University Press, Cambridge.

WEST, R. G. 1993. On the history of the Late Devensian Lake Sparks in southern Fenland, Cambridgeshire, England. *Journal of Quaternary Science*, **8**, 217–234.

WEST, R. G. 1996. Outline of the stratigraphy and vegetational history of the Cromer Forest-bed Formation. *In*: TURNER, C. (ed.) *The Early Middle Pleistocene in Europe*. Balkema, Rotterdam, 1–8.

WEST, R. G., ANDREWS, R., KNUDSEN, K.-L., PEGLAR, S. M. & PETTIT, M. E. 1995. Late Pleistocene deposits at Chatteris, March and Wimblington, Cambridgeshire, UK. *Proceedings of the Geologists' Association*, **106**, 195–210.

WEST, R. G., DICKSON, C. A., CATT, J. A., WEIR, A. H. & SPARKS, B. W. 1974. Late Pleistocene deposits at Wretton, Norfolk. II. Devensian deposits. *Philosophical Transactions of the Royal Society of London*, **B267**, 337–420.

WESTAWAY, R. 2001. Flow in the lower continental crust as a mechanism for the Quaternary uplift of the Rhenish Massif, north-west Europe. *In*: MADDY, D., MACKLIN, M. & WOODWARD, J. (eds) *River Basin Sediment Systems: Archives of Environmental Change*. Balkema, Abingdon, 87–167.

WESTAWAY, R., MADDY, D. & BRIDGLAND, D. 2002. Flow in the lower continental crust as a mechanism for the Quaternary uplift of south-east England: constraints from the Thames terrace record. *Quaternary Science Reviews*, **21**, 559–603.

WESTHEAD, R. K. & MATHER, A. E. 1996. An updated lithostratigraphy for the Purbeck Limestone Group in the Dorset type-area. *Proceedings of the Geologists' Association*, **107**, 117–128.

WHALLEY, J. S. & LLOYD, G. E. 1986. Tectonics of the Bude Formation, N. Cornwall; the recognition of northerly directed decollement. *Journal of the Geological Society, London*, **143**, 82–88.

WHITAKER, J. H. McD. 1962. The geology of the area around Leintwardine, Herefordshire. *Quarterly Journal of the Geological Society of London*, **118**, 319–351.

WHITAKER, J. H. McD. 1994. Silurian basin slope sedimentation and mass movement in the Wigmore Rolls area, central Welsh Borderland. *Journal of the Geological Society, London*, **151**, 27–40.

White, D. E. & Lawson, J. D. 1989. The Přídolí Series in the Welsh Borderland and south-central Wales. *In*: Holland, C. H. & Bassett, M. G. (eds) *A Global Standard for the Silurian System*. National Museum of Wales, Cardiff, Geological Series, **9**, 131–141.

White, E. I. 1956. Preliminary note on the range of Pteraspids in Western Europe. *Bulletin d'Institut royal des Sciences naturelles de Belgique*, **32**, 1–10.

White, M. J. 1998. Twisted ovate bifaces in the British Lower Palaeolithic; some observations and implications. *In*: Ashton, N., Healy, F. & Pettitt, P. (eds) *Stone Age Archaeology*. Essays in Honour of J. Wymer. Lithic Studies Society Occasional Paper, 6. Oxbow Books, Oxford, 98–104.

Whiteley, M. J. 1983. *The geology of the St Mellion outlier, Cornwall, and its regional setting*. Unpublished PhD thesis, University of Exeter.

Whiteman, C. A. & Rose, J. 1992. Thames river sediments of the British Early and Middle Pleistocene. *Quaternary Science Reviews*, **11**, 363–375.

Whitham, F. 1991. The stratigraphy of the Upper Cretaceous Ferriby, Welton and Burnham Formations north of the Humber, north-east England. *Proceedings of the Yorkshire Geological Society*, **48**, 227–254.

Whitham, F. 1993. The stratigraphy of the Upper Cretaceous Flamborough Chalk Formation north of the Humber, north-east England. *Proceedings of the Yorkshire Geological Society*, **49**, 235–258.

Whittaker, A. 1972. Geology of Bredon Hill. *Bulletin of the Geological Survey of Great Britain*, **42**, 1–49.

Whittaker, A. 1978. The lithostratigraphical correlation of the uppermost Rhaetic and lowermost Liassic strata of the W Somerset and Glamorgan coasts. *Geological Magazine*, **115**, 63–67.

Whittaker, A. (ed.). 1985. *Atlas of Onshore Sedimentary Basins in England and Wales*. Blackie, Glasgow.

Whittaker, A. & Green, G. W. 1983. *Geology of the country around Weston-super-Mare*. Memoir of the Geological Survey of Great Britain, Sheet 279 with parts of Sheets 263 and 295 (England and Wales). HMSO, London.

Whittaker, A., Holliday, D. W. & Penn, I. E. 1985. *Geophysical Logs in British Stratigraphy*. Geological Society Special Report, **18**.

Whittard, W. F. 1955–1967. *The Ordovician trilobites of the Shelve inlier, west Shropshire*. Parts 1–9 (1955, 1958, 1960, 1961 (Parts 5 and 6) 1964, 1966, 1967). Monograph of the Palaeontographical Society, London.

Whittard, W. F. 1979. An account of the Ordovician rocks of the Shelve Inlier in west Salop and parts of north Powys (by the late W. F. Whittard, compiled by W. T. Dean). *Bulletin of the British Museum of Natural History* (Geology), **33**, 1–69.

Whittington, H. B. et al. 1997. *Trilobita. Introduction, Order Agnostida, Order Redlichiida*. Treatise on Invertebrate Paleontology, Part O, revised, Vol.1. University of Kansas and Geological Society of America.

Whittington, H. B. 1938. The geology of the district around Llansantfraid ym Mechain, Montgomeryshire. *Quarterly Journal of the Geological Society of London*, **94**, 423–57.

Whittington, H. B., Chatterton, B. D. E., Speyer, S. E. et al. 1997. *Trilobita, Introduction Order Agnostina, Order Redlichiida*. Treatise on Invertebrate Palaeontology; Part O, revised, University of Kansas and Geological Society of America.

Whittle, A. W. R. 1996. *Europe in the Neolithic: the Making of New Worlds*. Cambridge University Press, Cambridge.

Whyte, M. A. & Romano, M. 1993. Footprints of a sauropods dinosaur from the Middle Jurassic of Yorkshire. *Proceedings of the Geologists' Association*, **104**, 195–199.

Wignall, P. B. 1989. Sedimentary dynamics of the Kimmeridge Clay. *Journal of the Geological Society, London*, **146**, 273–284.

Wignall, P. B. 1990*a*. Depositional history and palaeoecology of the Oxfordian/Kimmeridgian boundary beds at South Ferriby, South Humberside. *Proceedings of the Yorkshire Geological Society*, **48**, 197–208.

Wignall, P. B. 1990*b*. *Benthic Palaeoecology of the Late Jurassic Kimmeridge Clay of England. Special Papers in Palaeontology*, **43**.

Wignall, P. B. 1991. Test of the concepts of sequence stratigraphy in the Kimmeridgian (Late Jurassic) of England and northern France. *Marine and Petroleum Geology*, **8**, 430–441.

Wignall, P. B. 1993. The stratigraphy of the Upper Kimmeridge Clay (late Jurassic) of Golden Hill, Vale of Pickering, North Yorkshire. *Proceedings of the Yorkshire Geological Society*, **49**, 207–214.

Wignall, P. B. 2001*a*. Large igneous provinces and mass extinctions. *Earth Science Reviews*, **53**, 1–33.

Wignall, P. B. 2001*b*. Sedimentology of the Triassic–Jurassic boundary beds in Pinhay Bay (Devon, SW England). *Proceedings of the Geologists' Association*, **112**, 349–360.

Wignall, P. B. & Ruffell, A. H. 1990. The influence of a sudden climatic change on marine deposition in the Kimmeridgian of northwest Europe. *Journal of the Geological Society, London*, **147**, 365–371.

Wilkinson, I. 1987. A finite strain study of the Ordovician volcanic rocks of Snowdonia, North Wales, and its implications for a regional strain model. *Geological Journal*, **22**, 95–105.

Wilkinson, I. & Smith, M. 1988. Basement fractures in North Wales: recognition and control on Caledonian deformation. *Geological Magazine*, **125**, 301–306.

Wilkinson, I. P. 1990. *Entomozoacean Ostracoda and Cricoconarida from the Upper Devonian purple and green slates of the Tamar estuary*. Report of the British Geological Survey, Stratigraphy Series, **WH/90/155R**.

Wilkinson, J. J. & Knight, R. R. W. 1989. Palynological evidence from the Porthleven area, south Cornwall: implications for Devonian stratigraphy and Hercynian structural evolution. *Journal of the Geological Society, London*, **146**, 739–742.

Williams, A. 1951. Llandovery brachiopods from Wales with special reference to the Llandovery district. *Quarterly Journal of the Geological Society of London*, **107**, 85–136.

Williams, A. 1953. The geology of the Llandeilo district, Carmarthenshire. *Quarterly Journal of the Geological Society of London*, **108**, 177–208.

Williams, A. & Bulman, O. M. B. 1931. The geology of the Dolwyddelan syncline (North Wales) *Quarterly Journal of the Geological Society of London*, **87**, 425–58.

Williams, A. & Wright, A. D. 1981. The Ordovician-Silurian boundary in the Garth area of southwest Powys, Wales. *Geological Journal*, **16**, 1–39.

Williams, A., Lockley, M. G. & Hurst, J. M. 1981. Benthic palaeocommunities represented in the Ffairfach group and coeval Ordovician successions of Wales. *Palaeontology*, **24**, 661–694.

Williams, B. P. J. 1971. Sedimentary features of the Old Red Sandstone and Lower Limestone Shales of south Pembrokeshire, south of the Ritec Fault. *In*: Bassett, D. A. & Bassett, M. G. (eds) *Geological Excursions in South Wales and the Forest of Dean*. Geologists' Association, South Wales Group, Cardiff, 222–239.

Williams, B. P. J. & Hancock P. L. 1977. The sedimentology and structure of the Upper Palaeozoic rocks at Portishead. *In*: Savage, R. J. G. (ed.) *Geological Excursions in the Bristol District*. University of Bristol, Bristol, 29–46.

Williams, B. P. J. & Hillier, R. D. 2004. Variable alluvial sandstone architecture within the Lower Old Red Sandstone, southwest Wales. *Geological Journal*, **39**, 257–275.

Williams, B. P. J., Allen, J. R. L. & Marshall, J. D. 1982. Old Red Sandstone facies of the Pembroke Peninsula, south of the Ritec Fault. *In*: Bassett, M. G. (ed.) *Geological Excursions in Dyfed, Southwest Wales*. National Museum of Wales, Cardiff, 151–174.

Williams, C. J., Hesselbo, S. P., Jenkyns, H. C. & Morgans-Bell, H. S. 2001. Quartz silt in mudrocks as a key to sequence stratigraphy (Kimmeridge Clay Formation, Late Jurassic, Wessex Basin, UK). *Terra Nova*, **13**, 449–455.

Williams, E. A. 2000. Flexural cantilever models of extensional subsidence in the Munster Basin (SW Ireland) and the Old Red Sandstone fluvial dispersal systems. *In*: Friend, P. F. & Williams, B. P. J. (eds) *New Perspectives on the Old Red Sandstone*. Geological Society, London, Special Publications, **180**, 239–268.

WILLIAMS, E. A., FRIEND, P. F. & WILLIAMS, B. P. J. 2000. A review of Devonian time scales: databases, construction and new data. *In*: FRIEND, P. F. & WILLIAMS, B. P. J. (ed.) *New perspectives on the Old Red Sandstone*. Geological Society, London, Special Publications, **180**, 1–21.

WILLIAMS, G. A., TURNER, J. P. & HOLFORD, S. P. 2005. Inversion and exhumation of the St George's Channel basin, offshore Wales, UK. *Journal of the Geological Society, London*, **162**, 97–110.

WILLIAMS, H. & BULMAN, O. M. B. 1931. The geology of the Dolwyddelan Syncline (North Wales). *Quarterly Journal of the Geological Society, London*, **87**, 425–428.

WILLIAMS, H. & HATCHER, R. D. 1982. Suspect terranes and accretionary history of the Appalachian Orogen. *Geology*, **10**, 530–536.

WILLIAMS, M., DAVIES, J. R., WATERS, R. A., RUSHTON, A. W. A. & WILBY, P. R. 2003. Stratigraphical and palaeoecological importance of Caradoc (Upper Ordovician) graptolites from the Cardigan area, southwest Wale. *Geological Magazine*, **140**, 549–571.

WILLIAMS, M. & SIVETER, D. J. 1998. *British Cambrian and Tremadoc bradoriid and phosphatocopid arthropods*. Monograph of the Palaeontographical Society, London, **607**.

WILLIAMS, P. F. 1968. The sedimentation of Wesphalian (Ammanian) Measures in the Little Haven–Amroth Coalfield, Pembrokeshire. *Journal of Sedimentary Petrology*, **38**, 332–362.

WILLIAMS, S. H. & INGHAM, J. K. 1989. The Ordovician–Silurian boundary stratotype at Dob's Linn, southern Scotland. *In*: HOLLAND, C. H. & BASSETT, M. G. (eds) *A Global Standard for the Silurian System*. National Museum of Wales, Cardiff, Geological Series, **9**, 27–35.

WILLS, L. J. 1951. *A Palaeogeographical Atlas of the British Isles and Adjacent Parts of Europe*. Blackie, Glasgow.

WILLS, L. J. 1956. *Concealed Coalfields*. Blackie, Glasgow.

WILLS, L. J. 1970. The Triassic succession in the central Midlands in its regional setting. *Journal of the Geological Society, London*, **126**, 225–283.

WILLS, L. J. & SMITH, B. 1922. The Lower Palaeozoic rocks of the Llangollen area, with special reference to the tectonics. *Quarterly Journal of the Geological Society of London*, **78**, 176–226.

WILSON, A. A. 1990. The Mercia Mudstone Group (Triassic) of the East Irish Sea Basin. *Proceedings of the Yorkshire Geological Society*, **48**, 1–22.

WILSON, A. A. 1993. The Mercia Mudstone Group (Triassic) of the Cheshire basin. *Proceedings of the Yorkshire Geological Society*, **49**, 171–188.

WILSON, A. A. 2003. The Mercia Mudstone Group (Triassic) of the Manchester Airport, Second Runway. *Proceedings of the Yorkshire Geological Society*, **54**, 129–145.

WILSON, A. A. & CORNWELL, J. D. 1982. The IGS borehole at Beckermonds Scar, North Yorkshire. *Proceedings of the Yorkshire Geological Society*, **44**, 59–88.

WILSON, A. A. & EVANS, W. B. 1990. *Geology of the Country Around Blackpool*. Memoir of the British Geological Survey, Sheet 66. HMSO, London.

WILSON, D., DAVIES, J. R., SMITH, M. & WATERS, R. A. 1988. Structural controls on Upper Palaeozoic sedimentation in southeast Wales. *Journal of the Geological Society, London*, **145**, 901–914.

WILSON, D., DAVIES, J. R., FLETCHER, C. J. N. & SMITH, M. 1990. *Geology of the South Wales Coalfield – Part VI: the Country Around Bridgend*. Memoir of British Geological Survey, Sheets 261 and 262 (England & Wales). HMSO, London.

WILSON, D., DAVIES, J. R., WATERS, R. A. & ZALASIEWICZ, J. A. 1992. A fault-controlled depositional model for the Aberystwyth Grits turbidite system. *Geological Magazine*, **129**, 595–607.

WILSON, J. T. 1966. Did the Atlantic close and then re-open? *Nature*, **211**, 676–681.

WILSON, P. 2002. Morphology and significance of some Loch Lomond Stadial moraines in the south-central Lake District, England. *Proceedings of the Geologists' Association*, **113**, 9–22.

WILSON, P., BATEMAN, R. M. & CATT, J. A. 1981. Petrography, origin and environment of deposition of the Shirdley Hill Sands of southwest Lancashire, England. *Proceedings of the Geologists' Association*, **92**, 211–229.

WILSON, R. C. L. 1968*a*. Upper Oxfordian palaeogeography of southern England. *Palaeogeography, Palaeoclimatology, Palaeoecology*, **4**, 5–28.

WILSON, R. C. L. 1968*b*. Carbonate facies variation within the Osmington Oolite series in southern England. *Palaeogeography, Palaeoclimatology, Palaeoecology*, **4**, 89–123.

WILSON, R. C. L. 1971. Upper Jurassic red beds and carbonates of the Asturias and Lusitanian basins, Iberian peninsula. *Abstracts of the 7th International Sedimentological Congress, Heidelberg*, 110.

WILSON, S. J. 1991. The correlation of the Speeton Shell Bed, Filey Bay, Yorkshire, to an oxygen isotope stage. *Proceedings of the Yorkshire Geological Society*, **48**, 223–226.

WIMBLEDON, W. A. & COPE, J. C. W. 1978. The ammonite faunas of the English Portland Beds and the zones of the Portlandian Stage. *Journal of the Geological Society, London*, **135**, 183–190.

WIMBLEDON, W. A. & HUNT, C. O. 1983. The Portland–Purbeck junction (Portlandian–Berriasian) in the Weald, and correlation of latest Jurassic-early Cretaceous rocks in southern England. *Geological Magazine*, **120**, 267–280.

WINGFIELD, R. T. R. 1989. Glacial incisions indicating Middle and Upper Pleistocene ice limits off Britain. *Terra Nova*, **1**, 538–548.

WINGFIELD, R. T. R. 1990. The origin of major incisions within the Pleistocene deposits of the North Sea. *Marine Geology*, **91**, 31–52.

WINTLE, A. G. 1981. Thermoluminescence dating of Late Devensian loesses in southern England. *Nature*, **289**, 479–480.

WITZKE, B. J. 1990. Palaeoclimate constraints for palaeolatiudes of Laurentia and Euramerica. *In*: MCKERROW, W. S. & SCOTESE, C. R. (eds) *Palaeozoic Palaeogeography and Biogeography*. Memoir of the Geological Society, **12**, 57–73.

WONHAM, J. P. & ELLIOT, T. 1996. High-resolution sequence stratigraphy of a mid-Cretaceous estuarine complex: the Woburn Sands of the Leighton Buzzard area, southern England. *In*: HESSELBO, S. P. & PARKINSON, D. N. (eds) *Sequence Stratigraphy in British Geology*. Geological Society, London, Special Publications, **103**, 41–62.

WOOD, A. & SMITH, A. J. 1959. The sedimentation and sedimentary history of the Aberystwyth Grits (Upper Llandoverian). *Quarterly Journal of the Geological Society of London*, **114**, 163–195.

WOOD, C. J. 1988. The stratigraphy of the Chalk of Norwich. *Bulletin of the Geological Society of Norfolk*, **38**, 3–120.

WOOD, C. J. & MORTIMORE, R. N. 1995. An anomalous Black Band succession (Cenomanian–Turonian boundary interval) at Melton Ross, Lincolnshire, eastern England and its international significance. *Berliner geowissenschaftliche Abhandlungen*, Reihe E, **16** (1), 277–287.

WOOD, C. J. & SMITH, E. G. 1978. Lithostratigraphical classification of the Chalk in North Yorkshire, Humberside and Lincolnshire. *Proceedings of the Yorkshire Geological Society*, **42**, 263–287.

WOOD, C. J., MORTER, A. A. & GALLOIS, R. W. 1994. Appendix 1. Upper Cretaceous stratigraphy of the Trunch borehole(TG 23 SE 8) *In*: *Geology of the Country Around Great Yarmouth*. British Geological Survey, Sheet 162.

WOOD, D. S. 1971. *Studies of strain and slaty cleavage in the Caledonides of Northwest Europe and the Eastern United States*. Unpublished PhD thesis, University of Leeds.

WOOD, D. S. 1974. Ophiolites, mélanges, blueschists and ignimbrites: early Caledonian subduction in Wales? *In*: DOTT, R. H. & SHAVER, R. H. (eds) *Modern and Ancient Geosynclinal Sedimentation*. SEPM, Special Publications, **19**, 187–194.

WOOD, D. S. & HARPER, J. C. 1962. Notes on a temporary section in the Ordovician at Conway, North Wales. *Liverpool and Manchester Geological Journal*, **3**, 177–86.

WOOD, D. S. & NICHOLLS, G. D. 1973. Precambrian stromatolitic limestones from Northern Anglesey. *Nature, London*, **241**, 65.

WOODCOCK, N. H. 1976. Ludlow Series slumps and turbidites and the form of the Montgomery Trough, Powys, Wales. *Proceedings of the Geologists' Association of London*, **87**, 169–182.

Woodcock, N. H. 1984*a*. Early Palaeozoic sedimentation and tectonics in Wales. *Proceedings of the Geologists' Association*, **95**, 323–335.

Woodcock, N. H. 1984*b*. The Pontesford Lineament, Welsh Borderland. *Journal of the Geological Society, London*, **141**, 1001–1014.

Woodcock, N. H. 1987. Structural geology of the Llandovery Series in the type area, Dyfed, Wales. *Geological Journal*, **22**, 199–209.

Woodcock, N. H. 1988. Strike-slip faulting along the Church Stretton Lineament, Old Radnor, Welsh Borderland. *Journal of the Geological Society, London*, **145**, 925–933.

Woodcock, N. H. 1990*a*. Sequence stratigraphy of the Welsh Basin. *Journal of the Geological Society, London*, **147**, 537–547.

Woodcock, N. H. 1990*b*. Transpressive Acadian deformation across the Central Wales Lineament. *Journal of Structural Geology*, **12**, 329–337.

Woodcock, N. H. 1991. The Welsh, Anglian and Belgian Caledonides compared. *Annales de la Societe Geologique de Belgique*, **114**, 5–17.

Woodcock, N. H. 2000*a*. Terranes in the British and Irish Ordovician. *In: A revised correlation of Ordovician rocks in the British Isles*, Geological Society, London, Special Report, **24**, 8–12.

Woodcock, N. H. 2000*b*. Ordovician volcanism and sedimentation in Eastern Avalonia. *In*: Woodcock, N. H. & Strachan, R. 2000. *Geological History of Britain and Ireland*. Blackwell, Oxford, 153–167.

Woodcock, N. H. 2000*c*. Introduction to the Silurian. *In*: Aldridge, R. J., Siveter, D. J., Siveter, D. J., Lane, P. D., Palmer, D. & Woodcock, N. H. (eds) *British Silurian Stratigraphy*. Joint Nature Conservation Committee. Geological Conservation Review Series, **19**, 3–22.

Woodcock, N. H. 2000*d*. Devonian sedimentation and volcanism of the Old Red Sandstone continent. *In*: Strachan, R. A. (ed.) *Geological History of Britain and Ireland*. Blackwell, Oxford, 207–223.

Woodcock, N. H. & Barnes, R. P. 1999. An early Ordovician turbidite system on the Gondwana margin: the southeastern Manx Group, Isle of Man. *In*: Woodcock, N. H., Quirk, D. G., Fitches, W. R. & Barnes, R. P. (eds) *In Sight of the Suture: The Palaeozoic geology of the Isle of Man in its Iapetus Ocean Context*. Geological Society, London, Special Publications, **160**, 89–107.

Woodcock, N. H. & Gibbons, W. 1988. Is the Welsh Borderland Fault System a terrane boundary? *Journal of the Geological Society, London*, **145**, 915–923.

Woodcock, N. H. & Morris, J. H. 1999. Debris flows on the Ordovician margin of Avalonia: Lady Port Formation, Manx Group, Isle of Man. *In*: Woodcock, N. H., Quirk, D. G., Fitches, W. R. & Barnes, R. P. (eds) *In Sight of the Suture: The Palaeozoic Geology of the Isle of Man in its Iapetus Ocean Context*. Geological Society, London, Special Publications, **160**, 121–138.

Woodcock, N. H. & Pharaoh, T. C. 1993. Silurian facies beneath East Anglia. *Geological Magazine*, **130**, 681–690.

Woodcock, N. H. & Rickards, R. B. 1999. Geological notes and local details for 1:10 000 sheets: Sheet SD 69 NE (Westerdale) and parts of Sheets SD 69 NW (Howgill), SD 69 SW (Firbank) and SD 69 SE (Sedbergh). British Geological Survey Technical Report, Edinburgh, **44**.

Woodcock, N. H. & Strachan, R. 2000. *Geological History of Britain and Ireland*. Blackwell, Oxford.

Woodcock, N. H. & Smallwood, S. 1987. Late Ordovician shallow marine environments due to glacio-eustatic regression: Scratch Formation, Mid-Wales. *Journal of the Geological Society, London*, **144**, 393–400.

Woodcock, N. H., Awan, M. A., Johnson, T. E., Mackie, A. H. & Smith, R. D. A. 1988. Acadian tectonics in Wales during Avalonia/Laurentia convergence. *Tectonics*, **7**, 483–495.

Woodcock, N. H., Butler, A. J., Davies, J. R. & Waters, R. A. 1996. *Sequence Stratigraphical Analysis of Late Ordovician and Early Silurian Depositional Systems in the Welsh Basin: A Critical Assessment*. Geological Society, London, Special Publications, **103**, 197–208.

Woodcock, N. H., Morris, J. H. et al. 1999*b*. Revised lithostratigraphy of the Manx Group, Isle of Man. *In*: Woodcock, N. H., Quirk, D. G., Fitches, W. R. & Barnes, R. P. (eds) *In Sight of the Suture: The Palaeozoic Geology of the Isle of Man in its Iapetus Ocean Context*. Geological Society, London, Special Publications, **160**, 45–68.

Woodcock, N. H., Quirk, D. G., Fitches, W. R. & Barnes, R. P. (eds) 1999*a*. *In Sight of the Suture: The Palaeozoic Geology of the Isle of Man in its Iapetus Ocean Context*. Geological Society, London, Special Publications, **160**.

Woodland, A. W. 1970. The buried tunnel valleys of East Anglia. *Proceedings of the Yorkshire Geological Society*, **37**, 521–578.

Woodland, A. W. (ed.) 1971. *The Llanbedr (Mochras Farm) Borehole*. Report of the Institute of Geological Sciences, London, **71/18**.

Woodland, A. W. & Evans, W. B. 1964. *The Geology of the South Wales Coalfield – Part IV: the Country Around Pontypridd and Maesteg*, 3rd edn. Memoir of the Geological Survey of Great Britain, HMSO, London.

Woolacot, D. 1923. On a boring at Roddymoor Colliery, near Crook, Co. Durham. *Geological Magazine*, **60**, 50–62.

Wooldridge, S. W. & Linton, D. L. 1955. *Structure, Surface and Drainage in Southeast England*. George Philip, London.

Woollam, R. & Riding, J. B. 1983. *Dinoflagellate cyst zonation of the English Jurassic*. Report of the Institute of Geological Sciences, **83/2**.

Worley, N. E. 2005. The occurrence of halite in the Permian 'A' bed evaporite, Kirkby Thore, Cumbria. *Proceedings of the Yorkshire Geological Society*, **55**, 199–203.

Worsley, P. 1991*a*. Glacial deposits of the lowlands between the Mersey and Severn rivers. *In*: Ehlers, J., Gibbard, P. L. & Rose, J. (eds) *Glacial Deposits in Great Britain and Ireland*. Balkema, Rotterdam, 203–211.

Worsley, P. 1991*b*. Possible early Devensian glacial deposits in the British Isles. *In*: Ehlers, J., Gibbard, P. L. & Rose, J. (eds) *Glacial Deposits in Great Britain and Ireland*. Balkema, Rotterdam, 47–51.

Worssam, B. 1978. *The Stratigraphy of the Weald Clayal:*. Report of the Institute of Geological Sciences, **78/11**.

Worssam, B. C. & Old, R. A. 1988. *Geology of the Country Around Coalville*. Memoir of British Geological Survey, Sheet 155. HMSO, London.

Worssam, B. C. & Ivimey-Cook, H. C. 1971. The stratigraphy of the Geological Survey borehole at Warlingham, Surrey. *Bulletin of the Geological Survey of Great Britain*, **36**, 1–146.

Worssam, B. C., Ellison, R. A. & Moorlock, B. S. P. 1989. *Geology of the country around Tewkesury*, Memoir of the British Geological Survey, sheet 216 (England and Wales). HMSO, London.

Wray, D. S. 1999. Identification and long-range correlation of bentonites in Turonian–Coniacian (Upper Cretaceous) chalks of northwest Europe. *Geological Magazine*, **136**, 361–371.

Wray, D. S. & Wood, C. J. 1998. Distinction between detrital and volcanogenic clay-rich beds in Turonian–Coniacian chalks of eastern England. *Proceedings of the Yorkshire Geological Society*, **52**, 95–105.

Wright, A. E. 1969. *Precambrian Rocks of England and Wales and South-east Ireland*. AAPG, Memoir, **12**, 93–109.

Wright, C. W. & Collins, J. S. H. 1972. *British Cretaceous Crabs*. Monograph of the Palaeontographical Society, London, 1–114.

Wright, J. K. 1972. The stratigraphy of the Yorkshire Corallian. *Proceedings of the Yorkshire Geological Society*, **39**, 225–266.

Wright, J. K. 1978. The Cornbrash Formation (Callovian) in North Yorkshire and Cleveland. *Proceedings of the Yorkshire Geological Society*, **41**, 325–346.

Wright, J. K. 1981. The Corallian rocks of north Dorset. *Proceedings of the Geologists' Association*, **92**, 17–32.

Wright, J. K. 1983. The Lower Oxfordian (Upper Jurassic) of North Yorkshire. *Proceedings of the Yorkshire Geological Society*, **44**, 249–281.

Wright, J. K. 1986. A new look at the stratigraphy, sedimentology and ammonite fauna of the Corallian Group (Oxfordian) of south Dorset. *Proceedings of the Geologists' Association*, **97**, 1–21.

Wright, J. K. 1996*a*. Perisphinctid ammonites of the Upper Calcareous Grit (Upper Oxfordian) of North Yorkshire. *Palaeontology*, **39**, 433–469.

Wright, J. K. 1996*b*. The *Amoeboceras* faunas of the Upper Calcareous Grit Formation (Jurassic, Upper Oxfordian) of Yorkshire. *Proceedings of the Yorkshire Geological Society*, **51**, 33–43.

Wright, J. K. 2001. Discussion on fault activity and sedimentation in a marine rift basin (Upper Jurassic, Wessex basin, UK). *Journal of the Geological Society, London*, **158**, 391–392.

Wright, J. K. & Cox, B. M. 2001. *British Upper Jurassic stratigraphy (Oxfordian to Kimmeridgian)*. Joint Nature Conservation Committee. Geological Conservation Review Series, **21**.

Wright, J. K., Kelly, S. R. A. & Page, K. N. 2000. The stratigraphy of the 'Corallian' facies Middle Oxfordian (Upper Jurassic) deposits at Upware, Cambridgeshire, England. *Proceedings of the Geologists' Association*, **111**, 97–110.

Wright, T. 1860. On the zone of Avicula contorta, and the Lower Lias of the south of England. *Quarterly Journal of the Geological Society of London*, **16**, 374–411.

Wright, V. P. 1981. *The stratigraphy and sedimentology of the Llanelly Formation between Penderyn and Blorenge, South Wales*. PhD thesis, University of Wales.

Wright, V. P. 1987. The evolution of the early Carboniferous limestone province in SW Britain. *Geological Magazine*, **124**, 477–480.

Wright, V. P. 1990. Equatorial aridity and climatic oscillations during the early Carboniferous, southern Britain. *Journal of the Geological Society, London*, **147**, 359–363.

Wright, V. P. & Cossey, P. J. 2003. Middle Hope, North Somerset. *In*: Stephenson, D., Loughlin, S. C., Millward, D., Waters, C. N. & Williamson, I. T. (eds) *Carboniferous and Permian Igneous Rocks of Great Britain North of the Variscan Front*. Joint Nature Conservation Committee. Geological Conservation Review Series, **27**, 305–309.

Wright, V. P. & Marriott, S. B. 1996*a*. A quantitative approach to soil occurrence in alluvial deposits and its application to the Old Red Sandstone of Britain. *Journal of the Geological Society, London*, **153**, 907–913.

Wright, V. P. & Marriott, S. B. 1996*b*. Discussion of 'Brackish water faunas from the St. Maughans Formation: the Old Red Sandstone section at Ammons Hill, Hereford and Worcester, UK' by W. J. Barclay, P. A. Rathbone, D. E. White and J. B. Richardson. *Geological Journal*, **31**, 89–94.

Wright, V. P. & Sandler, A. 1994. A hydrogeological model for the early diagenesis of late Triassic alluvial sediments. *Journal of the Geological Society, London*, **151**, 897–900.

Wright, V. P. & Vanstone, S. D. 2001. Onset of Late Palaeozoic glacio-eustasy and the evolving climates of low latitude areas: a synthesis of current understanding. *Journal of the Geological Society, London*, **158**, 579–582.

Wright, V. P. & Wilson, R. C. L. 1987. *Paralic Carbonate Facies and Source Rocks, Upper Jurassic, Portugal*. SEPM, Mid-year meeting, Abstract volume, Austin, TX, 93.

Wright, V. P., Marriot, S. B. & Vanstone, S. D. 1991. A reg palaeosol from the Lower Triassic of south Devon: stratigraphic and palaeoclimatic implications. *Geological Magazine*, **128**, 517–523.

Wright, V. P., Taylor, K. G. & Beck, V. H. 2000. The palaeohydrology of Lower Cretaceous seasonal wetlands, Isle of Wight, southern England. *Journal of Sedimentary Research*, **70**, 619–632.

Wyatt, R. J. & Cave, R. 2002. The Chalfield Oolite Formation (Bathonian, Middle Jurassic) and the Forest Marble overstep in the South Cotswolds and the stratigraphical position of the Fairford Coral Bed. *Proceedings of the Geologists' Association*, **113**, 139–152.

Wymer, J. J. 1985. *Palaeolithic Sites of East Anglia*. Geobooks, Norwich.

Young, B. & Lake, R. D. 1988. *Geology of the Country around Brighton and Worthing*. British Geological Survey, Sheets 318 and 333.

Young, B. & Morgan, D. J. 1981. The Aptian Lower Greensand fuller's earths of Bognor Common, West Sussex. *Proceedings of the Geologists' Association*, **92**, 33–37.

Young, B., Ansari, S. M. & Firman, R. J. 1988. Field relationships, mineralogy and chemistry of the greisens and related rocks associated with the Eskdale Granite, Cumbria. *Proceedings of the Yorkshire Geological Society*, **47**, 109–123.

Young, T. P. 1993. Sedimentary iron ores. *In*: Pattrick, R. A. D. & Polya, D. (eds) *Mineralization in the British Isles*. Chapman & Hall, London, 446–489.

Young, T., Martin, F., Dean, W. T. & Rushton, A. W. A. 1994. Cambrian stratigraphy of St Tudwal's Peninsula, Gwynedd, northwest Wales. *Geological Magazine*, **131**, 335–360.

Young, T. P., Gibbons, W. & McCarroll, D. 2002. *Geology of the Country Around Pwllheli*. Memoir of the British Geological Survey, Sheet 134 (England & Wales), HMSO, London.

Zachos, J. C., Pagani, M., Sloan, L., Thomas, E. & Billups, K. 2001. Trends, rhythms and aberrations in global climate 65 Ma to present. *Science*, **292**, 686–693.

Zagwijn, W. H. 1975. Variations in climate as shown by pollen analysis, especially in the lower Pleistocene of Europe. *In*: Wright, A. E. & Moseley, F. (eds) *Ice Ages: Ancient and Modern*. *Geological Journal*, Special Issue, **6**, 137–152.

Zagwijn, W. H. 1996. The Cromerian Complex stage of the Netherlands and correlation with other areas in Europe. *In*: Turner, C. (ed.) *The Early Middle Pleistocene in Europe*. Balkema, Rotterdam, 145–172.

Zalasiewicz, J. A. 1984. A re-examination of the type Arenig Series. *Geological Journal*, **19**, 105–124.

Zalaseiwicz, J. A. 1989. *Distribution of graptolites through Caradoc strata west of Whitland, Dyfed*. British Geological Survey, Technical Report, Stratigraphy Series, No.W.H./89/475R, 1–9.

Zalasiewicz, J. A. 1990. Silurian graptolite biostratigraphy in the Welsh Basin. *Journal of the Geological Society, London*, **147**, 619–622.

Zalasiewicz, J. A. & Williams, M. 1999. Graptolite biozonation of the Wenlock Series (Silurian) rocks of the Builth Wells district, central Wales. *Geological Magazine*, **136**, 263–283.

Zalasiewicz, J. A., Mathers, S. J. *et al.* 1988. Stratigraphy and palaeoenvironments of the Red Crag and Norwich Crag Formations between Aldeburgh and Sizewell, Suffolk, England. *Philosophical Transactions of the Royal Society of London*, **B322**, 221–272.

Zalasiewicz, J. A., Mathers, S. J. *et al.* 1991. Age and relationships of the Chillesford Clay (early Pleistocene: Suffolk, England). *Philosophical Transactions of the Royal Society of London*, **B333**, 81–100.

Zalasiewicz, J., Rushton, A. W. A. & Owen, A. W. 1995. Late Caradoc graptolite faunal gradients across the Iapetus Ocean. *Geological Magazine*, **132**, 611–7.

Ziegler, A. M. 1966. The Silurian brachiopod Eocoelia hemisphaerica (J. de C. Sowerby) and related species. *Palaeontology*, **9**, 523–543.

Ziegler, A. M. & McKerrow, W. S. 1975. Silurian marine red beds. *American Journal of Science*, **275**, 31–56.

Ziegler, A. M., Cocks, L. R. M. & Bambach, R. K. 1968*a*. The composition and structure of Lower Silurian marine communities. *Lethaia*, **1**, 1–27.

Ziegler, A. M., Cocks, L. R. M. & McKerrow, W. S. 1968*b*. The Llandovery transgression of the Welsh Borderland. *Palaeontology*, **11**, 736–782.

Ziegler, A. M., McKerrow, W. S., Burne, R. V. & Baker, P. E. 1969. Correlation and environmental setting of the Skomer Volcanic Group, Pembrokeshire. *Proceedings of the Geologists' Association*, **80**, 409–439.

Ziegler, P. A. 1982. *Geological atlas of Western and central Europe*. Elsevier for Shell International Petroleum Maatschapij BV, Amsterdam.

Ziegler, P. A. 1987. Evolution of the Western Approaches Trough. *Tectonophysics*, **137**, 341–346.

Ziegler, P. A. 1990. *Geological Atlas of Western and Central Europe*, 2nd edn. Shell International Petroleum Maatschappij B. V, Amsterdam.

Ziegler, W. & Klapper, G. 1985. Stages of the Devonian System. *Episodes*, **8**, 104–109.

Index

Note: page numbers in *italic* type refer to illustrations.